DIRECTORY OF GEOSCIENCE DEPARTMENTS

54th Edition

Carolyn Wilson, Editor
Christopher M. Keane, Editor

American Geosciences Institute
Alexandria, Virginia

Directory of Geoscience Departments 2019, 54th Edition

Edited by Carolyn Wilson

ISBN-13: 978-0-922152-30-8
ISSN: 0364-7811

Published and printed in the United States of America. All rights reserved. No part of this work may be reproduced or transmitted in any form or by any means, electronic or mechanical, recording, or any information storage and retrieval system without the expressed written consent of the publisher.

Typeset in Times New Roman and Arial using Adobe InDesign CC.
Layout and programming: Christopher M. Keane
Advertising: John P. Rasanen

For more information on the American Geosciences Institute and its publications check us out at https://www.americangeosciences.org

© 2019 American Geosciences Institute.

Cover
Ellen Reat, University of Utah Ph.D. student, operates "Winnie the Drone" to create photogrammetry models of Katjiesberg deltaic outcrops, Tanqua Karoo, South Africa, May 2018. Credit: Ellen Jayne Reat, submitted to AGI's 2019 Life as a Geoscientist contest.

Introduction

Thank you for purchasing the 54th edition of the Directory of Geoscience Departments. We, at the American Geosciences Institute (AGI), are grateful to all the departments who updated their information and provided additional historical data of enrollments and awarded degrees for their department. We would also like to thank our various interns within the AGI Workforce Program during 2018 for their contributions to updating the database.

We have continued the enhancements introduced last year, including the expanded list of research specialties to better represent the various areas of research currently conducted in the geosciences, along with information about specific programs and certificates offered by individual programs, and indications of whether programs offered online or blended courses and degree or certificate programs.

This edition includes a listing of U.S. and Canadian student theses and dissertations from 2016 that have been reported to GeoRef Information Services. We would like to thank GeoRef for their diligent work compiling this list for use in this edition.

The data used to compile the directory is provided by the individuals and departments listed. AGI only edits the information for consistency and format. Any errors or omissions most likely reflect the entries of the individuals or the primary department contact. If you identify any issues, please email dgd@americangoesciences.org with information about a current responsible contact for the department in question, so that we may further encourage updates to the next edition.

The American Geosciences Institute does not warrant the accuracy of any of the self-reported information, but we do encourage individuals and departments to update the material they believe is out-of-date.

Some basic statistics for the 54th edition of the Directory of Geoscience Departments: 1947 academic departments and programs globally, with 984 of those departments in the United States. Three hundred thirty-three departments in the United States are two-year institutions. There at 18,008 individuals globally identified in the Directory, with 13,384 in the United States and 1384 in Canada.

Christopher M. Keane
Co-Editor

DIRECTORY OF GEOSCIENCE DEPARTMENTS
54th Edition 2019

Table of Contents

Directory of Geoscience Departments
    U.S. Section 1
    Non-U.S. Section 173
Theses and Dissertations...271
U.S. State Geological Surveys 286
U.S. Federal Agencies and National Laboratories 301
List of Faculty Specialty Codes 311
Faculty by Specialty Index..312
Faculty Index 417

Usage Key

Degrees Offered
A - Associate's or 2-year degree
B - Bachelor's or equivalent undergraduate degree
M - Master's Degree
D - Doctorate

Basic enrollment and degrees-granted data for the most recent year reported are shown below department information: degree level, enrollment, and degrees granted (in parentheses).

📖 indicates thesis and dissertations were provided.
● indicates department offers a field camp only for their majors.
○ indicates department offers a field camp with open enrollment.
☒ indicates department does not offer any online courses
⊘ indicates department offers blended learning courses
🖥 indicates department offers fully online courses
🖥 indicates department offers fully online programs or certificates

The letters following individual faculty listings indicates their research specialties. Capital letters indicate general area, lowercase letters indicate subspecialty. Two or more consecutive lowercase letters refer to multiple subspecialties in the same major focus area referred to by the preceding capital letter.

DEPARTMENTS AND FACULTIES

This section contains a global listing of academic geoscience departments. Universities and colleges in the United States are arranged alphabetically by state. Non-U.S. institutions are listed alphabetically by country following the U.S. section.

All data is as reported by the departments and faculty themselves and current as of February 12, 2019. If you are aware of updates, corrections, or other means to ensure continued improvement of this listing, please email dgd@americangeosciences.org with the appropriate information so that it can be addressed during the next revision.

Alabama

Alabama A&M University
Dept of Biological and Environmental Sciences (A,B,M,D) (2015)
P.O. Box 1208
Normal, AL 35762
  p. (256) 372-4214
  wubishet.tadesse@aamu.edu
  http://www.aamu.edu
Chair:
  Anthony Overton
Professor:
  Tommy L. Coleman, (D), Iowa State, 1980, So
  Florence A. Okafor, (D), Nigeria, 1995, ZnEn
  Govind Sharma, (D), Kansas State, 1970, Zn
  James W. Shuford, (D), Penn State, 1975, So
Associate Professor:
  Monday O. Mbila, (D), Iowa State, 2000, Sd
Secretary:
  Martha Palmer

Auburn University
Dept of Crop, Soil & Environmental Sciences (B,M,D) (2019)
201 Funchess Hall
Auburn University, AL 36849
  p. (334) 844-4100
  jpb0035@auburn.edu
  Programs: Crop and Soil Sciences (B.S.)
  Crop, Soil & Environmental Sciences (M.S., M.Ag., Ph.D.)
  Enrollment (2006): M: 8 (2) D: 6 (3)
Professor:
  Yucheng Feng, (D), Penn State, 1995, SbCb
  Elizabeth A. Guertal, (D), Oklahoma State, 1993, Sc
  Joey N. Shaw, (D), Georgia, 1997, Sd
Associate Professor:
  Gobena E. Huluka, (D), Auburn, 1990, Sc
Assistant Professor:
  Audrey V. Gamble, (D), Delaware, 2017, Sc
  Thorsten J. Knappenberger, (D), Hohenheim, 2009, Sp
  Rishi Prasad, (D), Florida, 2014, Sb
  Di Tian, (D), Florida, 2014, At
  Matthew N. Waters, (D), N Carolina, 2007, Gn

Dauphin Island Sea Lab
Marine Science Program (M,D) (2015)
P.O. Box 369
101 Bienville Boulevard
Dauphin Island, AL 36528
  p. (334) 861-7528
  langelo@disl.org
  http://www.disl.org/aboutus.html
  Department Secretary: Carolyn F. Wood
Director:
  William W. Schroeder, (D), Texas A&M, 1971, Og
Professor:
  Thomas S. Hopkins, (D), California (San Diego), 1967, Ob
Associate Professor:
  Jonathan R. Pennock, (D), Delaware, 1983, Oc
Librarian:
  Dennis Patronas, Zn

University of Alabama
Dept of Geological Sciences (B,M,D) O☒ (2018)
Box 870338
201 7th Avenue
Room 2003 Bevill Bldg.
Tuscaloosa, AL 35487-0338
  p. (205) 348-5095
  geology@geo.ua.edu
  http://www.geo.ua.edu
  t: @UA_GEOLOGY
  Programs: Geology; Marine Science Geology
  Enrollment (2018): B: 65 (22) M: 13 (13) D: 1 (1)
Associate Dean of the Graduate School:
  Andrew Goodliffe, (D), Hawaii, 1998, YgeGt
Chair:
  C. Fred T. Andrus, (D), Georgia, 2000, PcGaCs
Professor:
  Ibrahim Çemen, (D), Penn State, 1983, Gct
  Rona J. Donahoe, (D), Stanford, 1984, Cl
  Delores Robinson, (D), Arizona, 2001, Gc
  Harold H. Stowell, (D), Princeton, 1987, GpCcGt
  Nick Tew, (D), Alabama, 1999, EoGro
  Chunmiao Zheng, (D), Wisconsin, 1988, Hw
Associate Professor:
  Samantha E. Hansen, (D), California (Santa Cruz), 2007, YsGt
  Yuehan Lu, (D), Michigan, 2008, GgOuGe
  Alberto Perez-Huerta, (D), Oregon, 2004, PgClGz
  Geoffrey Tick, (D), Arizona, 2003, Hw
  Yong Zhang, (D), Nanjing, 1998, Hqw
Assistant Professor:
  Julia Cartwright, (D), Manchester, 2010, XmgCc
  Natasha T Dimova, (D), Florida State, 2010, HwCcm
  Kimberly Genareau, (D), Arizona State, 2009, Gvi
  Deborah A. Keene, (D), Georgia, 2002, Gga
  Rezene Mahatsente, (D), Clausthal Tech, 1998, YgmYv
  Marcello Minzoni, Kansas, 2007, GosEo
  Rebecca T. Minzoni, (D), Rice, 2015, GsPme
  Grey S. Nearing, (D), Arizona, 2013, HsqZr
  Alain Plattner, (D), ETH Zurich, 2011, YmuYg
  Tom S. Tobin, (D), Washington, 2014, PsCsPe
  Matthew Wielicki, (D), California (Los Angeles), CcgGt
  Michelle Wielicki, (D), Colorado, 2014, CcXgGi
  Bo Zhang, (D), Oklahoma, 2014, YeGoEo
Stable Isotope Research Scientist:
  W. Joseph Lambert, (D), Alabama, 2010, CsPeCg
Emeritus:
  Paul Aharon, (D), Australian Nat, 1980, ClmCq
  Donald J. Benson, (D), Cincinnati, 1976, Gd
  Richard H. Groshong, Jr., (D), Brown, 1971, Gc
  W. Gary Hooks, (D), North Carolina, 1961, Gm
  Ernest A. Mancini, (D), Texas A&M, 1974, Gro
  Carl W. Stock, (D), North Carolina, 1977, Pi
Geochemical Research Lab Manager:
  Sidhartha Bhattacharyya, (D), Alabama, 2010, CatGe

University of South Alabama
Dept of Earth Sciences (B) O (2018)
5871 USA Dr. N.
Room 136
Mobile, AL 36688-0002
  p. (251) 460-6381
  dbeebe@southalabama.edu
  http://www.usouthal.edu/earthsci

Programs: Geology (B); Geography (B); Meteorology (B); GIS (minor)
Certificates: GIS
Enrollment (2017): B: 20 (20)
Chair:
   Sytske K. Kimball, (D), Penn State, 2000, Am
Professor:
   Roy Ryder, (D), Florida, 1989, SdZry
Associate Professor:
   David T. Allison, (D), Florida State, 1992, Gc
   Murlene W. Clark, (D), Florida State, 1983, PiGrg
   Carol F. Sawyer, (D), Texas State, 2007, GmZye
Assistant Professor:
   Alex Beebe, (D), Clemson, 2013, Ge
   William T. Jackson, (D), Alabama, 2017, GsCc
   Steven R. Schultze, (D), Michigan State, 2015, At
   Wes Terwey, (D), Colorado State, 2007, Am
Instructor:
   Karen J. Jordan, (M), Alabama, 2005, Zyr
   Andrew Murray, (M), Florida State, 2010, Am
   Diana Sturm, (D), Alabama, 2000, GeZge
Emeritus:
   Keith G. Blackwell, (D), Texas A&M, 1990, Am
   Douglas W. Haywick, (D), James Cook, 1990, Gs
   Aaron Williams, (D), Oklahoma, 1971, Am

Dept of Marine Sciences (M,D) (2019)
307 University Drive
Mobile, AL 36688-0002
   p. (334) 460-7136
   keyser@southalabama.edu
   Administrative Assistant: Jan Keyser
Chair:
   Sean Powers, (D), Ob
Professor:
   Ronald P. Kiene, (D), SUNY (Stony Brook), 1986, Oc
Adjunct Professor:
   Douglas W. Haywick, (D), James Cook, 1990, Gs
   Erich M. Mueller, (D), Miami, 1983, Ob

University of West Alabama
Dept of Biological & Environmental Sciences (B) (2019)
UWA Station 7
Livingston, AL 35470
   p. (205) 652-3416
   arindsberg@uwa.edu
   http://www.uwa.edu/Biological_and_Environmental_Sciences.aspx
   Programs: Environmental Science
Director, Black Belt Museum:
   James P. Lamb, (B), Alabama, PvePg
Professor:
   Andrew K. Rindsberg, (D), Colorado Mines, 1986, PeGePi

Alaska

Prince William Sound Community College
Dept of Natural Sciences (2015)
P.O. Box 97
Valdez, AK 99686
   p. (907) 822-3673
   rsbenda@pwscc.edu
   http://www.pwscc.edu

University of Alaska, Anchorage
Dept of Geological Sciences (B) ● (2016)
3211 Providence Drive
CPISB 101
Anchorage, AK 99508-4670
   p. (907) 786-1298
   uaa_mns@uaa.alaska.edu
   http://www.uaa.alaska.edu/geology/
   f: www.facebook.com/UAAGeologicalSciences
   t: @UAAGeology
   Enrollment (2015): B: 60 (8)
Chair:
   Kristine J. Crossen, (D), Washington, 1997, Gl
Professor:
   LeeAnn Munk, (D), Ohio State, 2001, Cle
Associate Professor:
   Jennifer Aschoff, (D), Colorado Mines, 2008, GrdGo
   Donald Matt Reeves, (D), Nevada, 2006, GeHgZn
Term Assistant Professor:
   Terry R. Naumann, (D), Idaho, 1998, Gi
Assistant Professor:
   Erin Shea, (D), MIT, 2014, CcGcCg
Term Instructor:
   Peter J. Oswald, (M), Idaho, Gg
   Mark Rivera, (M), New Mexico State, 2000, Gg

University of Alaska, Fairbanks
Alaska Quaternary Center (2015)
907 Yukon Drive
Fairbanks, AK 99775-1200
   p. (907) 474-7758
   rmtopp@alaska.edu
   Administrative Assistant: Leicha Welton
Professor:
   W. Scott Ambruster, (D), California (Davis), 1981, Zn
   Bruce P. Finney, (D), Oregon State, 1987, Ou
   David M. Hopkins, (D), Harvard, 1955, Gg
   Roger W. Powers, (D), Wisconsin, 1973, Zn
   Dan L. Wetzel, (B), Alaska (Fairbanks), 1975, Zn
Research Associate:
   Wendy H. Arundale, (D), Michigan State, 1976, Zn
   Peter Bowers, (M), Washington State, 1980, Ga
   Owen Mason, (D), Alaska, 1990, Ga
Adjunct Professor:
   Daniel H. Mann, (D), Washington, 1983, Pe
Geology Librarian:
   Judy Triplehorn, Zn

Dept of Civil & Environmental Engineering (2015)
Fairbanks, AK 99775-5900
   jlhulsey@alaska.edu

Dept of Geosciences (B,M,D) O (2019)
1930 Yukon Drive
P.O. Box 755780
Fairbanks, AK 99775-5780
   p. (907) 474-7565
   pjmccarthy@alaska.edu
   http://www.uaf.edu/geology/
   Programs: Geology; Geophysics; Earth Science; Geography
   Enrollment (2017): B: 83 (9) M: 35 (11) D: 42 (10)
Chair:
   Paul McCarthy, (D), Guelph, 1995, GsSaPc
Professor:
   Bernard Coakley, (D), Columbia, 1991, Yg
   Sarah J. Fowell, (D), Columbia, 1994, Pl
   Regine M. Hock, (D), Swiss Fed Inst Tech, 1997, Gl
   Jessica F. Larsen, (D), California (Santa Cruz), 1996, Gv
   Rainer J. Newberry, (D), Stanford, 1980, Em
   Vladimir Romanovsky, (D), Alaska (Fairbanks), 1996, Yg
   Michael T. Whalen, (D), Syracuse, 1993, Gs
Associate Professor:
   Cary de Wit, (D), Kansas, 1997, Zg
   Patrick S. Druckenmiller, (D), Calgary, 2006, Pv
   Dan Mann, (D), Washington, 1983, Zy
   Franz J. Meyer, (D), Tech (Munich), 2004, Zr
   Elisabeth S. Nadin, (D), Caltech, 2007, Gtc
   Carl Tape, (D), Caltech, 2009, Ys
Assistant Professor:
   Chris Maio, (D), Massachusetts (Boston), 2014, Zy
Instructor:
   Jochen E. Mezger, (D), Alberta, 1997, GcpCc
Emeritus:
   James E. Beget, (D), Washington, 1981, Gm
   Douglas Christensen, (D), Michigan, 1987, Ys
   Catherine L. Hanks, (D), Alaska (Fairbanks), 1991, Go
   Mary J. Keskinen, (D), Stanford, 1979, Gxz
   Kenneth P. Severin, (D), California (Davis), 1987

Instructor:
   Joann Deakin, (M), Mast, Ggg

Dine' College
Science Dept (A,B) (2017)
#1 Circle Drive
Tsaile, Navajo Nation, AZ 86556
   p. (928) 724-6721
   cacate@dinecollege.edu
   http://www.dinecollege.edu
Chair:
   Don Robinson, (D), Maharishi Management
Instructor:
   Margaret Mayer, (M), Rhode Island, GeZn

Mesa Community College
Dept of Physical Science (A) (2015)
1833 W. Southern Avenue
Mesa, AZ 85202
   p. (480) 461-7015
   kelli.santistevan@mesacc.edu
   http://www.mesacc.edu/dept/d43/glg/
Professor:
   Donna M. Benson, (M), Arizona State, 1996, Gg
   Kaatje Kraft, (M), Arizona State, 1999, Gg
   Robert S. Leighty, (D), Arizona State, 1997, Zg
Adjunct Professor:
   Zack Bowles, (M), Arizona State, Gg
   Chloe Branciforte, (M), SD Mines, Gg
   Mike Grivois, (M), Texas Christian
   Steve Guggino
   Jack Kepper
   Jill Lockard
   Tony Occhiuzzi, (M), Arizona State, Zn
   Donna Pollard, (M)
   Joanna Scheffler, (M), Washington State, Gg
   Melinda Shimizu, (M), Arizona State, 2008, Gg
   Carolyn Taylor, (M), Georgia State, Gg
   Kelli Wakefield, (M), Arizona State, Gg
Emeritus:
   Armand J. Lombard, (M), Arizona, 1977, Gg

Northern Arizona University
Dept of Geography, Planning, and Recreation (B,M) (2015)
P.O. Box 15016
Flagstaff, AZ 86011
   p. (928) 523-2650
   geog@nau.edu
   http://nau.edu/sbs/gpr/
   Enrollment (2013): B: 39 (15) M: 44 (17)
Professor and Chair:
   Pamela Foti, (D), Wisconsin, 1988, Zn
Park Ranger Training Program Director:
   Mark J. Maciha, (M), N Arizona, 2009, ZnnZn
Coordinator, Parks and Recreation Management Program:
   Charles Hammersley, (D), New Mexico, 1988, Zn
Professor:
   Alan A. Lew, (D), Oregon, 1986, Zn
Associate Professor:
   Rebecca D. Hawley, (D), Arizona State, 1994, Zn
   Ruihong Huang, (D), Wisconsin (Milwaukee), 2003
Associate Scientist:
   Erik Schiefer, (D), British Columbia, 2004, GmZyi
Senior Lecturer:
   Judith Montoya, (M), New Mexico, 1985, Zy
Lecturer - Outdoor Leadership and Education:
   Aaron Divine, (M), N Arizona, 2004, ZnnZn
Distance Learning Lecturer:
   R. Marieke Taney, (M), N Arizona, 2002, ZnnZn
Emeritus:
   Graydon L. Berlin, (D), Tennessee, 1970, Zr
   Robert O. Clark, (D), Denver, 1970, Zn
   Carolyn M. Daugherty, (D), Arizona State, 1987, Zn
   Leland R. Dexter, (D), Colorado, 1986, Zy
   Christina B. Kennedy, (D), Arizona, 1989, Zn
   Stanley W. Swarts, (D), California (Los Angeles), 1975, Zy
   George A. Van Otten, (D), Oregon State, 1977, Zn

Division of Geosciences, School of Earth and Sustainability (B,M,D) (2019)
P O Box 4099
Building 12, Room 100
625 S Knoles Dr
Flagstaff, AZ 86011-4099
   p. 928-523-9333
   SES.Admin@nau.edu
   https://nau.edu/ses/division-of-geosciences/
   Enrollment (2018): B: 120 (32) M: 26 (9)
Professor:
   David K. Elliott, (D), Bristol (UK), 1979, Pv
   Thomas D. Hoisch, (D), S California, 1985, Gp
   Darrell S. Kaufman, (D), Colorado, 1991, Zc
   Michael H. Ort, (D), California (Santa Barbara), 1991, Gv
   Roderic A. Parnell, (D), Dartmouth, 1982, Cl
   Mary R. Reid, (D), MIT, 1987, CucGi
   Nancy Riggs, (D), California (Santa Barbara), 1991, Gv
   James C. Sample, (D), California (Santa Cruz), 1986, GtClOu
   Abraham E. Springer, (D), Ohio State, 1994, Hw
   Paul J. Umhoefer, (D), Washington, 1989, Gt
Associate Professor:
   Laura Wasylenki, (D), Caltech, 1999, ClOoCs
Assistant Professor:
   Nicholas McKay, (D), Arizona, 2012, PcZo
   Ryan Porter, (D), Arizona, 2011, YgsGt
   Michael E. Smith, (D), Wisconsin, 2007, Gsr
Research Associate:
   Joseph E. Hazel, Jr., (M), N Arizona, 1991, Gs
   Matthew A. Kaplinski, (M), N Arizona, 1990, Yg
Lecturer:
   Lisa Skinner, (M), ZeGvZi
   Kelsey Winsor, (D), Wisconsin, 2014, PcGm
Adjunct Professor:
   Wendell A. Duffield, (D), Stanford, 1967, Gv
   David D. Gillette, (D), S Methodist, 1974, Pv
   James Wittke, (D), Texas, 1984, Gi
Emeritus:
   Charles W. Barnes, (D), Wisconsin, 1965, GcXg
   Ronald C. Blakey, (D), Iowa, 1973, Gr
   David S. Brumbaugh, (D), Indiana, 1972, Ys
   Ernest M. Duebendorfer, (D), Wyoming, 1986, Gct
   Richard F. Holm, (D), Washington, 1969, Gv
   James I. Mead, (D), Arizona, 1983, Pg
   Larry T. Middleton, (D), Wyoming, 1979, Gs
   Paul Morgan, (D), London, 1973, Yh
Business Manager:
   Sarah Colby, Zn

Phoenix College
Dept of Physical Sciences (A) (2016)
1202 West Thomas Road
Phoenix, AZ 85013
   p. (602) 285-7244
   abeer.hamdan@phoenixcollege.edu
Chair:
   James J. White, (D), Arizona, 1986
Professor:
   Richard Cups, (M), Arizona State, 1983, Gg
   Don Speed, (M), Arizona, 1980, Gg

Pima Community College - Community Campus
Geology Dept (2015)
401 North Bonita Ave.
Tucson, AZ 85709-5000
   p. (520) 206-4500
   wrcavanaugh@pima.edu
   http://www.pima.edu

Pima Community College - West Campus
Geology Dept (2015)
2202 West Anklam Rd.
Tucson, AZ 85709-0001
   p. (520) 206-4500

David B. Stone, (D), Newcastle, 1963, Ym
Donald M. Triplehorn, (D), Illinois, 1961, Gs

Dept of Mining & Geological Engineering (B,M,D) (2015)
P.O. Box 755800
Fairbanks, AK 99775-5800
 p. (907) 474-7388
 mmdarrow@alaska.edu
 http://www.alaska.edu/uaf/cem/ge/
 Administrative Assistant: Judy L. Johnson
 Enrollment (2011): B: 10 (10) M: 2 (2) D: 2 (2)
Professor:
 Sukumar Bandopadhyay, (D), Penn State, 1982, Nm
 Gang Chen, (D), Virginia Tech, 1989, Nm
 Scott L. Huang, (D), Missouri (Rolla), 1981
 Hsing K. Lin, (D), Utah, 1985, Zn
 Paul A. Metz, (D), Imperial Coll (UK), 1991
 Debsmita Misra, (D), Minnesota, 1994, HwZrNg
 Daniel E. Walsh, (D), Alaska (Fairbanks), 1984, Zn
Assistant Professor:
 Margaret M. Darrow, (D), Znn
 Sabry Sabour Hafez, (D), Nm
Emeritus:
 Rajive Ganguli, (D), Kentucky, 1999, Nm

Dept of Petroleum Engineering (2015)
Fairbanks, AK 99775-5880
 adandekar@alaska.edu

University of Alaska Museum Earth Science (B,M,D) O☒ (2018)
Box 756960
Fairbanks, AK 99775-6960
 p. (907) 474-7505
 psdruckenmiller@alaska.edu
 http://www.uaf.edu/museum/collections/earth/
 Programs: Geology
 Enrollment (2017): M: 2 (0) D: 1 (0)
Curator of Earth Science:
 Patrick S. Druckenmiller, (D), Calgary, 2006, Pv

University of Alaska, Southeast
Dept of Natural Sciences (2015)
11120 Glacier Hwy
Juneau, AK 99801
 p. (907) 796-6163
 cathy.connor@uas.alaska.edu
 http://www.uas.alaska.edu

Arizona

Arizona State University
School of Earth and Space Exploration (B,M,D) ●⚐ (2019)
Box 871404
Tempe, AZ 85287-1404
 p. (480) 965-5081
 sese@asu.edu
 https://sese.asu.edu
 f: https://www.facebook.com/SESE.at.ASU
 t: @SESEASU
 Programs: Earth and Environmental Studies, Earth and Space Exploration, Astrobiology and Biogeosciences, Astrophysics, Exploration Systems Design, Geological Sciences
 Enrollment (2018): B: 255 (58) M: 14 (1) D: 84 (19)
Director:
 Linda Elkins-Tanton, (D), MIT, 2002
Professor:
 Ariel D. Anbar, (D), Caltech, 1996, CbsXb
 Ramon Arrowsmith, (D), Stanford, 1995, GcmGt
 James Bell, (D), Hawaii, 1992
 Donald M. Burt, (D), Harvard, 1972, EgCgGz
 Peter R. Buseck, (D), Columbia, 1962, Cg
 Philip R. Christensen, (D), California (Los Angeles), 1981, Xg
 Steven Desch, (D), Illinois, 1998, Xa
 Jack D. Farmer, (D), California (Davis), 1978, Pg
 Edward Garnero, (D), Caltech, 1994, Ys
 Arjun Heimsath, (D), California (Berkeley), 1999, Gc
 Richard Hervig, (D), Chicago, 1979, Cg
 Kip V. Hodges, (D), MIT, 1982, CcGtXg
 Philip Mauskopf, (D), California (Berkeley), 1997
 Stephen J. Reynolds, (D), Arizona, 1982, GctZe
 Mark Robinson, (D), Hawaii (Honolulu), 1993, Zn
 Steven Semken, (D), MIT, 1989, ZeGgZg
 Thomas G. Sharp, (D), Arizona State, 1990, Gz
 Everett Shock, (D), California (Berkeley), 1987, Zn
 Sumner Starrfield, (D), California (Los Angeles), 1969, YhsYv
 Francis Timmes, (D), California (Santa Cruz), 1992
 James A. Tyburczy, (D), Oregon, 1983, GyzYx
 Enrique Vivoni, (D), MIT, 2003, Hg
 Meenakshi Wadhwa, (D), Washington Univ (St. Louis), 1994, Zn
 Kelin Whipple, (D), Washington, 1994, Gt
 Stanley N. Williams, (D), Dartmouth, 1983, Gv
 Rogier Windhorst, (D), Leiden, 1984
Associate Professor:
 Judd Bowman, (D), MIT, 2007
 Nathaniel Butler, (D), MIT, 2003
 Amanda Clarke, (D), Penn State, 2002, Gv
 Christopher Groppi, (D), Arizona, 2003, XaZrn
 Hilairy E. Hartnett, (D), Washington, 1998, ComOc
 Jennifer Patience, (D), California (Los Angeles), 2000
 Evan Scannapieco, (D), California (Berkeley), 2001, Xa
 Sang-Heon Shim, (D), Princeton, 2011, Gz
 Heather Throop, (D), Stony Brook, 2002
 Patrick Young, (D), Arizona, 2004, Zn
Assistant Professor:
 Melanie Barboni, (D), Lusanne, 2011, Gi
 Sanchayeeta Borthakur, (D), Massachusetts, 2010
 Maitrayee Bose, (D), Washington Univ (St. Louis), 2011
 Craig Hardgrove, (D), Tennessee, 2011
 Daniel Jacobs, (D), Pennsylvania, 2011
 Mingming Li, (D), Arizona State, 2015
 Michael Line, (D), Caltech, 2013
 Manoochehr Shirzaei, (D), Potsdam, 2010
 Evgenya Shkolnik, (D), British Columbia, 2004
 Christy Till, (D), MIT, 2011
 Sara Walker, (D), Dartmouth, 2010
Emeritus:
 L. Paul Knauth, (D), Caltech, 1979, Cs
 Robert F. Lundin, (D), Illinois, 1962, Pm
 Edmund Stump, (D), Ohio State, 1976

Arizona Western College
Dept of Geosciences (A) ⚐ (2017)
2020 S. Avenue 8E
Yuma, AZ 85365
 p. (928) 317-6000
 fred.croxen@azwestern.edu
 Programs: Geology (A); Geography (A); Applied Geology (A)
 Certificates: GIS Technology
 Enrollment (2017): A: 1 (0)
Professor:
 Fred W. Croxen III, (M), N Arizona, 1977, HwPvZi
 Catherine R. Hill, (M), GgOgHw
Adjunct Professor:
 Kelly L. Esslinger, (M), AsOpGg
 Maureen Garrett, (M), GgeOg
 Todd Pinnt, (M), Zyi

Central Arizona College
Dvision of Science (2015)
Coolidge, AZ 85228
 p. (602) 426-4444
 diane.beecroft@centralaz.edu
Instructor:
 Allan E. Morton, (M), Brigham Young, 1975, Ze

Cochise College
Geology Dept (A) (2016)
901 North Colombo Avenue
Sierra Vista, AZ 85635
 p. (520) 515-5425
 priddise@cochise.edu
 http://skywalker.cochise.edu/wellerr/aawellerweb.htm

Chair, Global Change IDP:
  Katherine K. Hirschboeck, (D), Arizona, 1985, AtHsRn
Emeritus:
  Robert A. Clark, (D), Texas A&M, 1964, Hs
  Lucien Duckstein, (D), Colorado State, 1962, Hq
  Simon Ince, (D), Iowa, 1953, Hs
  Austin Long, (D), Arizona State, 1966, Cl
  Ernest T. Smerdon, (D), Missouri State, 1959, Hs
  Arthur W. Warrick, (D), Iowa State, 1967, Sp
  Peter J. Wierenga, (D), California (Davis), 1968, Sp
  Lorne G. Wilson, (D), California (Davis), 1962, Hw
Cooperating Faculty:
  Mark L. Brusseau, (D), Florida, 1989, Hw
  Jonathan D. Chorover, (D), California (Berkeley), 1993, Cg
  Bonnie C. Colby, (D), Wisconsin, 1983, Zn
  R. B. Hawkins, (D), Colorado State, 1968, Hs
  Kevin E. Lansey, (D), Texas, 1987, Hs
  Sharon B. Megdal, (D), Princeton, 1981, Hg
  Steven L. Mullen, (D), Washington, 1985, As
  Robert G. Varady, (D), Arizona, 1981, Zn
  Marvin Waterstone, (D), Rutgers, 1983, Zn

Dept of Mining & Geological Engineering (B,M,D) (2018)
1235 E. James E. Rogers Way
Mines Building, Room 229
PO Box 210012
Tucson, AZ 85721-0012
  p. (520) 621-6063
  ENGR-Mining@email.arizona.edu
  http://www.mge.engineering.arizona.edu
  f: https://www.facebook.com/UAMining/
  Programs: Mining Engineering, Mining & Geological Engineering
  Certificates: Mineral Processing and Extractive Metallurgy, Geomechanics and Rock Mechanics, Mine Production & Information Technology, Mining Occupational Health & Safety
  Enrollment (2017): B: 71 (26) M: 24 (8) D: 11 (4)
Head:
  John M. Kemeny, (D), California (Berkeley), 1986, Nr
Professor of Practice:
  Victor O. Tenorio, (D), Arizona, 2012, Nmx
Director, Lab for Advanced Subsurface Imaging:
  Ben K. Sternberg, (D), Wisconsin, 1977, Ye
Professor:
  Pinnaduwa H. S. W. Kulatilake, (D), Ohio State, 1981, Nr
Associate Professor:
  Jaeheon Lee, (D), Arizona, 2004, Nx
  Moe Momayez, (D), McGill, NmYxNr
  Jinhong Zhang, (D), Virginia Tech, 2006, Nm
Assistant Professor:
  Isabel Barton, (D), Arizona
  Kwangmin Kim, (D), Arizona, 2012, NmrNx
Emeritus:
  J. Brent Hiskey, (D), Utah, 1973, Nx
  Mary M. Poulton, (D), Arizona, 1990, Ng

Dept of Soil, Water & Environmental Science (B,M,D) ☒ (2019)
PO Box 210038
Shantz 429
Tucson, AZ 85721-0038
  p. (520) 621-1646
  wrhaley@email.arizona.edu
  https://swes.cals.arizona.edu
  Programs: Environmental Science (undergrad and grad), Sustainable Plant Systems (undergrad), Subsurface science (grad), Soil, plant, atmosphere systems (grad)
Head:
  Jonathan Chorover, (D), California (Berkeley), 1993, ScClo
Associate Director, SAHRA Center:
  Katharine L. Jacobs, (M), California (Berkeley), 1981, Hg
Professor:
  Mark L. Brusseau, (D), Florida, 1989, Hw
  Kevin Fitzsimmons, (D), Arizona, 1999, Zn
  Charles P. Gerba, (D), Miami, 1973, Sb
  Martha C. Hawes, (D)
  Raina M. Maier, (D), Rutgers, 1988, Zn
  Ian L. Pepper, (D), Ohio State, 1975, Zn
  Craig Rasmussen, (D), GmCl
  Charles Sanchez, (D), Iowa State, 1986, Zn
  Jeffrey C. Silvertooth, (D), Oklahoma State, 1986, So
  Markus Tuller, (D), Zn
  James Walworth, (D), Georgia, 1985, Soc
  Arthur W. Warrick, (D), Iowa State, 1967, Sp
Associate Professor:
  Joan E. Curry, (D), California (Davis), 1992, Sc
  Marcel Schaap, (D), So
Assistant Professor:
  Karletta Chief, (D), Arizona, 2007, Zn
Associate Research Scientist:
  Janick F. Artiola, (D), Arizona, 1980, ScHsCa
Associate Professor of Practice:
  Thomas B. Wilson, (D), Zn
Adjunct Professor:
  Floyd Adamsen, (D), Colorado State, 1983, So
  Paul (Ty) Ferre, (D), So
  Jim Yeh, (D), Sp
Emeritus:
  Ed Glenn, (D), Hawaii, 1978, Zn
  Donald F. Post, (D), Purdue, 1967, Sd
  James Riley, (D), Arizona, 1968, Zn
  Peter J. Wierenga, (D), California, 1968, Sp
Extension Specialist:
  Paul W. Brown, (D), Wisconsin, 1981, Am

Laboratory of Tree Ring Research ○ (2018)
Bryant Bannister Tree-Ring Bldg
1215 E. Lowell St.
Tucson, AZ 85721
  p. (520) 621-1608
  webmaster@ltrr.arizona.edu
  http://ltrr.arizona.edu/
  f: https://www.facebook.com/pages/Laboratory-of-Tree-Ring-Research/112628272090047
  t: @TreeRingLabUA
  Certificates: Graduate Certificate in Dendrochronology
Professor:
  David C. Frank, (D), Bern, 2005
Professor:
  Malcolm K. Hughes, (D), Durham (UK), 1970, Pe
  Steven W. Leavitt, (D), Arizona, 1982, Csc
  Russ Monson, (D), Washington State, 1982
Associate Professor:
  Bryan A. Black, (D), 2003
  Paul Sheppard, (D), 1995
  Ronald Towner, (D), Arizona, 1997, Gae
  Valerie Trouet, (D), KU Leuven (Belgium), 2004, PeZyAs
Assistant Professor:
  Charlotte Pearson, (D), Reading, 2003
Research Associate:
  Matthew W. Salzer, (D), Arizona, 2000, PeGav
  Christopher H. Guiterman, (D), Arizona, 2016
  Kiyomi A. Morino, (D), Arizona, 2008
Emeritus:
  Jeffrey S. Dean, (D), Arizona, 1967, Zn
  Harold C. Fritts, (D), Ohio State, 1956, Pe
  Katherine K. Hirschboeck, (D), Arizona, 1985, AsHsZc
  Charles W. Stockton, (D), Arizona, 1971, Hq
  Thomas W. Swetnam, (D), Arizona, 1987, Pe
Research Professor:
  David M. Meko, (D), Arizona, 1981, Hq
  Ramzi Touchan, (D), Arizona, 1991, PcHsAs
Curator, Collections:
  Peter Brewer, (D), Reading, 2004
Associate Professor:
  Pearce Paul Creasman, (D), Texas A&M, 2008
Assoc. Research Professor:
  Irina Panyushkina, (D), Sukachev Inst Forest, 1997
  Tomasz Wazny, (D), Hamburg, 1990
Assistant Research Professor:
  Margaret Evans, (D), Arizona, 2003

Lunar and Planetary Laboratory (M,D) (2015)
1629 E. University Boulevard
Tucson, AZ 85721-0092

jrasmussen@pima.edu
http://www.pima.edu

Prescott College
Dept of Environmental Studies (B) (2015)
220 Grove Avenue
Prescott, AZ 86301
p. (928) 350-2256
kurt.refsnider@prescott.edu
http://www.prescott.edu
Professor:
 Kurt Refsnider, (D), Colorado, 2012, GmlPe

Scottsdale Community College
Dept of Mathematics and Sciences (2015)
9000 East Chaparral Road
Scottsdale, AZ 85256-2626
p. (480) 423-6111
roberto.ribas@scottsdalecc.edu
http://www.scottsdalecc.edu/academics/departments/math-sciences

South Mountain Community College
Geology Program (2015)
7050 South 24th Street
Phoenix, AZ 85042-5806
p. (602) 243-8290
sian.proctor@southmountaincc.edu
http://www.southmountaincc.edu/math-science-engineering/program/geology/

University of Arizona
Dept of Geosciences (B,M,D) O (2015)
Gould-Simpson Building
1040 E. Fourth Street
Tucson, AZ 85721-0077
p. (520) 621-6000
bcarrapa@email.arizona.edu
http://www.geo.arizona.edu
Administrative Assistant: Anne Chase
Administrative Assistant: Sylvia Quintero
Enrollment (2006): B: 70 (0) M: 23 (0) D: 52 (0)
Head:
 Peter Reiners, (D), Washington, Cg
Professor:
 Victor R. Baker, (D), Colorado, 1971, GmXgRh
 Mark D. Barton, (D), Chicago, 1981, Eg
 Susan L. Beck, (D), Michigan, 1987, YsGt
 Richard Bennett, (D), MIT, 1995, YdGt
 Jon Chorover, California (Berkeley), 1993, Sc
 Andrew S. Cohen, (D), California (Davis), 1982, Ps
 Julia E. Cole, (D), Columbia, 1992, PeCsAs
 Owen K. Davis, (D), Minnesota, 1981, Pl
 Peter G. DeCelles, (D), Indiana, 1984, GtsGc
 Robert T. Downs, (D), Virginia Tech, 1992, Gi
 Mihai N. Ducea, (D), Caltech, 1998, Gt
 Karl W. Flessa, (D), Brown, 1973, PgeGe
 Jiba Ganguly, (D), Chicago, 1967, Gx
 George E. Gehrels, (D), Caltech, 1986, Gt
 Vance T. Holliday, (D), Colorado, 1982, GamSa
 Roy A. Johnson, (D), Wyoming, 1984, Ys
 A. J. Timothy Jull, (D), Bristol, 1976, CclCg
 Paul A. Kapp, (D), California (Los Angeles), 2001, GtcGm
 Alfred S. McEwen, (D), Arizona State, 1988, XgGmv
 Jonathan T. Overpeck, (D), Brown, 1985, As
 Jon D. Pelletier, (D), Cornell, 1997, Gm
 Mary Poulton, (D), Arizona, 1990, Ng
 Jay Quade, (D), Utah, 1990, Sc
 Randall M. Richardson, (D), MIT, 1978, Yg
 Joaquin Ruiz, (D), Michigan, 1983, Cgc
 Timothy Swindle, (D), Washington, 1986, Xc
 Connie Woodhouse, (D), Arizona, 1996, Pe
Associate Specialist:
 Michael A. Crimmins, (D), Arizona, 2004, Zc
Associate Professor:
 Barbara Carrapa, (D), Vrije (Amsterdam), 2002, Gs
 Joellen Russell, (D), Scripps, 1999, Oc
 Eric Seedorff, (D), Stanford, 1987, Eg
 Jessica Tierney, (D), Brown, 2010, Co
 Marek Zreda, (D), New Mexico Tech, 1994, Cc
Assistant Professor:
 Jennifer McIntosh, (D), Michigan, 2004, Hy
 Jianjun Yin, (D), Illinois, Og
Lecturer:
 Paul Goodman, Washington, 2000, As
 Jessica Kapp, (D), California (Los Angeles), Gg
Adjunct Professor:
 Robert J. Kamilli, (D), Harvard, 1976, EgmCg
 Charles Prewitt, (D), MIT, 1969, Zn
 Marc Sbar, (D), Columbia, 1972, Yes
Emeritus:
 William B. Bull, (D), Stanford, 1960, Gm
 Clement G. Chase, (D), California (San Diego), 1970, Yg
 George H. Davis, (D), Michigan, 1971, Gc
 William R. Dickinson, (D), Stanford, 1958, Gt
 John M. Guilbert, (D), Wisconsin, 1962, Em
 DeVerle P. Harris, (D), Penn State, 1965, Eg
 C. Vance Haynes, Jr., (D), Arizona, 1965, Ga
 Everett H. Lindsay, (D), California (Berkeley), 1967, Pv
 Edgar J. McCullough, Jr., (D), Arizona, 1963, Ge
 Jonathan Patchett, (D), Edinburgh (UK), 1976
 Joseph Schreiber, (D)
 Spencer R. Titley, (D), Arizona, 1958, Eg
 George Zandt, (D), MIT, 1978, Ys
Researcher:
 David Dettman, (D), Michigan, 1994, Cs
 Chris J. Eastoe, (D), Tasmania, 1979, CsGg

Dept of Hydrology & Atmospheric Sciences (B,M,D) ● (2015)
226A Harshbarger Building #11
PO Box 210011
Tucson, AZ 85721-0011
p. (520) 621-5082
better@email.arizona.edu
http://www.hwr.arizona.edu
Administrative Assistant: Erma Santander
Enrollment (2014): B: 0 (2) M: 23 (2) D: 27 (1)
Head:
 Thomas Maddock, III, (D), Harvard, 1972, Hq
Regents Professor:
 Victor R. Baker, (D), Colorado, 1971, GmXgRh
 Shlomo P. Neuman, (D), California (Berkeley), 1968, HwqGq
 W. James Shuttleworth, (D), Manchester (UK), 1971, Hs
Director, SAHRA Center:
 Juan B. Valdes, (D), MIT, 1976, Hs
Professor:
 Paul D. Brooks, (D), Colorado, 1995, Cg
 Paul A. Ferre, (D), Waterloo, 1997, Hw
 Hoshin V. Gupta, (D), Case Western, 1984, HqZgAs
 Thomas Meixner, (D), Arizona, 1999, CgRw
 Peter A. Troch, (D), Ghent, 1993, Hs
 Tian-Chyi Jim Yeh, (D), New Mexico Tech, 1983, Hq
 Marek G. Zreda, (D), New Mexico Tech, 1994, Hw
Associate Professor:
 Michael D. Bradley, (D), Michigan, 1971, Zn
 Jennifer C. McIntosh, (D), Michigan, 2004, Cg
Associate Director, SAHRA Center:
 Gary C. Woodard, (D), Michigan, 1981, Zn
Adjunct Associate Professor:
 David C. Goodrich, (D), Arizona, 1990, Hs
 Robert H. Webb, (D), Arizona, 1985, Gm
Adjunct Assistant Professor:
 James E. Smith, (D), Waterloo, 1995, Hw
Adjunct Professor:
 Roger C. Bales, (D), Caltech, 1984, Cg
 David Hargis, (D), Arizona, 1979, Hwq
 Leo S. Leonhart, (D), Arizona, 1978, Hw
 Robert MacNish, (D), Michigan, 1966, Hw
 Soroosh Sorooshian, (D), California (Los Angeles), 1978, Hg
 Donald W. Young, (D), Arizona, 1994, Zn

p. (520) 621-6963
acad_info@lpl.arizona.edu
http://www.lpl.arizona.edu
Head:
   Michael J. Drake, (D), Oregon, 1972, Xc
Research Professor:
   Martin G. Tomasko, (D), Princeton, 1969, As
Professor:
   Erik I. Asphaug, (D), Arizona, 1993, XygXa
   Victor R. Baker, (D), Colorado, 1971, XgGmRh
   William V. Boynton, (D), Carnegie Mellon, 1971, XcZr
   Robert H. Brown, (D), Hawaii, 1982, Xc
   Uwe Fink, (D), Penn State, 1965, As
   Tom Gehrels, (D), Chicago, 1956, Zn
   Richard J. Greenberg, (D), MIT, 1972, Xy
   Donald M. Hunten, (D), McGill, 1950, As
   Jack R. Jokipii, (D), Caltech, 1965, Xy
   Harold P. Larson, (D), Purdue, 1967, Zn
   Dante Lauretta, (D), Washington Univ (St. Louis), 1997, XcmXg
   Jonathan I. Lunine, (D), Caltech, 1985, Xc
   Renu Malhotra, (D), Cornell, 1988
   Alfred McEwen, (D), Arizona State, 1988, Xg
   Henry J. Melosh, (D), Caltech, 1972, Gt
   George H. Rieke, (D), Harvard, 1969, Zn
   Timothy D. Swindle, (D), Washington (St. Louis), 1986, XmCc
   Roger Yelle, (D), Wisconsin, 1984, Zn
Senior Research Scientist:
   Lonnie L. Hood, (D), California (Los Angeles), 1979, XyAs
   Larry A. Lebofsky, (D), MIT, 1974, ZngXm
Senior Scientist:
   Jozsef Kota, (D), Roland Eotvos, 1980, As
   David Kring, (D), Harvard, 1989, XcGiXg
Associate Professor:
   Caitlin Griffith, (D), SUNY (Stony Brook), 1991, Zn
   Robert Kursinski, (D), Caltech, 1997, As
Associate Research Scientist:
   Robert S. McMillan, (D), Texas, 1977, ZnnZn
Assistant Professor:
   Joe Giacalone, (D), Kansas, 1991, Xy
   Adam Showman, (D), Caltech, 1998, Zn
Senior Research Scientist:
   Lyle A. Broadfoot, (D), Saskatchewan, 1963, Zr
   Jay B. Holberg, (D), California (Berkeley), 1974, As
   Bill R. Sandel, (D), Rice, 1972, As
Research Associate:
   Alexander Dessler, (D), Duke, 1956, Zn
   Ralph Lorenz, (D), Kent (UK), 1994, Xg
   Elizabeth Turtle, (D), Arizona, 1998, Xg
Associate Research Scientist:
   Peter Smith, (M), Arizona, 1977, Zr
Emeritus:
   William B. Hubbard, (D), California (Berkeley), 1967, XyYvAs
   John S. Lewis, (D), California (San Diego), 1968, Xcm
   Elizabeth Roemer, (D), California (Berkeley), 1955, Zn
   Charles P. Sonett, (D), California (Los Angeles), 1954, Yg
   Robert G. Strom, (M), Stanford, 1957, Xg
Graduate Secretary:
   Mary Guerrieri, Zn

School of Geography and Development (B,M,D) (2015)
Harvill Building
PO Box 210076
Tucson, AZ 85721
   p. (520) 621-1652
   lstaeheli@email.arizona.edu
   http://geography.arizona.edu/
   Enrollment (2010): B: 350 (0) M: 20 (0) D: 54 (0)
Dean:
   John Paul Jones, III, (D), Ohio State, 1984, Zn
Director:
   Paul Robbins, (D), Clark, 1996
Co-Director of the University of Arizona Institute of the Environment :
   Diana Liverman, (D), California (Los Angeles), 1984
Professor:
   Michael E. Bonine, (D), Texas, 1975, Zu
   Andrew Comrie, (D), Pennsylvania, 1992, Zy
   Sallie A. Marston, (D), Colorado, 1986, Zu
   Beth A. Mitchneck, (D), Columbia, 1990, Zn
   David A. Plane, (D), Pennsylvania, 1981, Zn
Associate Professor:
   Keiron D. Bailey, (D), Kentucky, 2002, Zi
   Carl J. Bauer, (D), California (Berkeley), 1995, Hs
   Sandy Dall'Erba, (D), Pau, 2004
   Elizabeth Oglesby, (D), California (Berkeley), 2002, Zn
   Christopher A. Scott, (D), Cornell, 1998, Hw
   Willem van Leeuwen, (D), Arizona, 1995, Zy
   Marvin Waterstone, (D), Rutgers, 1983, Zn
   Margaret Wilder, (D), Arizona, 2002, Zn
Director of GIST:
   Christopher Lukinbeal, (D), San Diego State, 2000, Zi
Assistant Professor:
   Jeffrey Banister, (D), Arizona, 2010
   Gary Christopherson, (D), Arizona, 2000, Zi
   Sarah Moore, (D), Kentucky, 2006
   Tracey Osborne, (D), California (Berkeley), 2010
   Daoqin Tong, (D), Ohio State, 2007, Zi
Lecturer:
   Dereka Rushbrook, (D), Arizona, 2005
Adjunct Professor:
   Julio Betancourt, (D), Arizona, 1989, Zn
   Vance Holliday, (D), Colorado, 1982, Ga
   Laura Huntoon, (D), Pennsylvania, 1991, Zn
   Charles F. Hutchinson, (D), California (Riverside), 1978, Zr
   Miranda Joseph, (D), Stanford, 1995, Zn
   Barbara Morehouse, (D), Arizona, 1993, Zn
   Thomas W. Swetnam, (D), Arizona, 1987, Zn
Emeritus:
   Stephen R. Yool, (D), California (Santa Barbara), 1985, Zri
Director, International Studies-External Affairs:
   Wayne Decker, (D), Johns Hopkins, 1979

Arkansas

Arkansas Tech University
Dept of Physical Sciences-Geology (B) (2015)
1701 N. Boulder Ave
McEver Science Building
Russellville, AR 72801
   p. (479) 968-0293
   jmusser@atu.edu
   http://www.atu.edu/physci
   Enrollment (2014): B: 44 (10)
Professor:
   Cathy Baker, (D), Iowa, 1986, Gg
Associate Professor:
   Jason A. Patton, (D), Arkansas, 2008, GgeGc
Assistant Professor:
   Genet Duke, (D), SD Mines

Northwest Arkansas Community College
Northwest Arkansas Community College (2015)
One College Drive
Bentonville, AR 72712
   p. (479) 631-8661
   dandroes@nwacc.edu
   http://www.nwacc.edu

Ozarka College
Math, Science, and Education (2015)
218 College Drive
Melbourne, AR 72556
   hayers@ozarka.edu
   http://www.ozarka.edu/

University of Arkansas at Little Rock
Dept of Earth Sciences (B) O (2018)
2801 South University Avenue
Little Rock, AR 72204-1099
   p. (501) 569-3546
   memcmillan@ualr.edu
   http://www.ualr.edu/earthsciences/
   Programs: Geology (B); Environmental Geology (B)

Enrollment (2018): B: 52 (10)
Chair:
    Margaret E. McMillan, (D), Wyoming, 2003, Gm
Professor:
    Jeffrey B. Connelly, (D), Tennessee, 1993, GcNg
Associate Professor:
    Michael T. DeAngelis, (D), Tennessee, 2011, GpCp
    Laura S. Ruhl, (D), Duke, 2012, GeClGb
Assistant Professor:
    Rene A. Shroat-Lewis, (D), Tennessee, 2011, PeZe
Instructor:
    Joshua C. Spinler, (D), Arizona, 2014, YdGt
Visiting Professor:
    Thomas A. Colby, (D), Boise State, 2016, GcNg
Emeritus:
    Michael T. Ledbetter, (D), Rhode Island, 1978, Og

University of Arkansas, Fayetteville
Arkansas Water Resources Center (2015)
113 Ozark Hall
Fayetteville, AR 72701
    p. (501) 575-4403
    haggard@uark.edu
    http://www.uark.edu/depts/awrc
    Administrative Assistant: Theresa J. Russell
Director:
    Kenneth F. Steele, (D), North Carolina, 1971, Cg
Research Associate:
    Terry E. Nichols, (D), Chicago, 1973, Zn
Laboratory Director:
    Marc A. Nelson, (D), Arkansas, 1992, Cl

Dept of Geosciences (B,M,D) ○⊘ (2018)
340 N. Campus Drive
216 Gearhart Hall
Fayetteville, AR 72701
    p. (479) 575-3355
    lmilliga@uark.edu
    http://fulbright.uark.edu/departments/geosciences/
    f: https://www.facebook.com/uageosciences/0396/
    Programs: Earth Science (B); Geology (B,M); Geography (B,M); Geosciences (D)
    Certificates: Geospatial Technologies (UG,G)
    Enrollment (2018): B: 160 (32) M: 69 (16) D: 19 (1)
Maurice Storm Endowed Chair:
    Christopher Liner, (D), Colorado Mines, 1989, YseGg
University Professor:
    William (. Limp, (D), Indiana, 1983, Zfi
    Thomas R. Paradise, (D), Arizona State, 1993, GmZyRn
Distinguished Professor:
    David W. Stahle, (D), Arizona State, 1990, GeZon
    Daniel Sui, (D), Georgia, 1993, Zf
Director of the Center for Advanced Spatial Technologies:
    Jackson D. Cothren, (D), Ohio State, 2004, Zir
Assistant Professor:
    Edward C. Holland, (D), Colorado (Boulder), unkn, Zy
Professor:
    Stephen K. Boss, (D), North Carolina, 1994, RcYgGg
    Jason A. Tullis, (D), South Carolina, 2003, ZriZf
Associate Professor:
    Matthew Covington, (D), California, 2008, HqGmg
    Fiona M. Davidson, (D), Nebraska, 1991, Zy
    Ralph K. Davis, (D), Nebraska, 1992, HwGgg
    Gregory Dumond, (D), Massachusetts, 2008, GtcGg
    Song Feng, (D), Chinese Acad of Sci, 1999, AsZyAm
    John B. Shaw, (D), Texas, 2013, GsrGg
    Celina Suarez, (D), Kansas, 2010, ClPqGg
Assistant Professor:
    Linyin Cheng, (D), California (Irvine), Zo
    Andrew P. Lamb, (D), Boise State, 2013, YumYv
    Jill A. Marshall, (D), Oregon, 2015, Gm
    Adriana Potra, (D), Florida Intl, 2011, EgCcGg
    Glenn R. Sharman, (D), Stanford, 2014, Go
Associate Research Professor:
    Phillip D. Hays, (D), Texas A&M, 1996, HwCsGg

Research Associate:
    Barry J. Shaulis, (D), Houston, 2013
Instructor:
    Paula Anderson, (M), Arkansas, 2012, Gg
    Rashauna Hintz, (M), Arkansas, Zu
    Henry Turner, III, (D), Arkansas, 2010, GgtGc
Adjunct Professor:
    Fred Paillet, (D), Rochester, 1974, YgHw
Emeritus:
    John Van Brahana, (D), Missouri, 1973, HwGe
    Malcolm K. Cleaveland, (D), Arizona, 1983, PcZc
    John C. Dixon, (D), Colorado, 1983, GmXgSo
    Thomas O. Graff, (D), Kansas, 1973, Zu
    Margaret J. Guccione, (D), Colorado, 1982, GmaGg
    John G. Hehr, (D), Michigan State, 1971, AmZy
    Ron Konig, (D), Cornell, 1959, Gc
    Walter Manger, (D), Iowa, Gr
    Richard Smith, N Colorado, Gm
    Ken Steele, (D), North Carolina, HgCgGg
    Doy L. Zachry, (D), Texas, 1969, Gr

California

American River College
Dept of Earth Science (A) (2015)
4700 College Oak Drive
Sacramento, CA 95841
    p. (916) 484-8107
    aubertj@arc.losrios.edu
    http://arc.losrios.edu/~earthsci/
Chair:
    Melissa H. Levy, (M), E Tennessee State, 1994, GgZgy
Professor:
    John E. Aubert, (M), California (Davis), 1994, Zy
    Charles E. Thomsen, (M), Cal State (Chico), 1994, Zy
GIS Coordinator:
    Hugh H. Howard, (D), Kansas, 2003, Zi
Assistant Professor:
    Glenn Jaecks, (D), California (Davis), 2002, GgOgPg
Adjunct:
    Paul M. Veisze, (M), California (Berkeley), 1985, Zi
Adjunct Professor:
    Terry J. Boroughs, (M), Ohio State, 1992, GgXgCg
    Beth Dushman, (M), California (Davis), 2007, Gg
    Robert Earle, (M), Zi
    Nathan Jennings, (M), Zi
    Tom Lupo, (M), Zi
    Richard L. Oldham, (M), Nevada, 1972, Gg
    Kimberly Olson, (M), Cal State (Chico), 2008, Zye
    Steven C. Smith, (M), Cal State (Chico), 1994, ZynZn

Antelope Valley College
Div-Geosciences Prog (2018)
3041 West Avenue K
Lancaster, CA 93536
    mpesses@avc.edu
    https://www.avc.edu/sites/default/files/catalog/geosciences.pdf
Associate Professor:
    Michael Pesses, Zyi
Assistant Professor:
    Aurora Burd, (D), Washington, 2013, GgZgYm

Bakersfield College
Physical Science Dept (2015)
1801 Panorama Drive
Bakersfield, CA 93305
    p. (661) 395-4391
    yadira.guerrero@bakersfieldcollege.edu
Acting Chair:
    Robert A. Schiffman, (M), California (Santa Barbara), 1971, Ga
Professor:
    John C. Lyman, (B), Cal State (San Diego), 1962, Gg
Instructor:
    Robert Lewy, (M), Cal State (Bakersfield), 1992, Gx
    Michael Oldershaw, (M), California (Davis), 1987, Eo

Cabrillo College
Dept of Geology, Oceanography and Environmental Science
(A) ⊘ (2017)
6500 Soquel Drive
Aptos, CA 95003
   p. (408) 479-6495
   daschwar@cabrillo.edu
   http://www.cabrillo.edu/academics/earthscience/
   Programs: Geology (A); Environmental Science (A)
   Enrollment (2015): A: 35 (0)
Chair:
   David Schwartz, (M), San Jose State, 1983, Gu

California Institute of Technology
Div of Geological & Planetary Sciences (B,M,D) ●☒ (2019)
1200 East California Boulevard
MC 170-25
Pasadena, CA 91125
   p. (626) 395-6111
   shechet@gps.caltech.edu
   http://www.gps.caltech.edu
   Enrollment (2018): B: 17 (0) D: 112 (0)
Chair:
   John P. Grotzinger, (D), Virginia Tech, 1985, GsPeXg
Professor:
   Jess F. Adkins, (D), MIT, 1997, Cm
   Jean-Paul Ampuero, (D), Paris VII, 2002, Ys
   Paul D. Asimow, (D), Caltech, 1997, CpGzy
   Jean-Philippe Avouac, (D), Inst Phys du Globe de Paris, 1991, GtYs
   Geoffrey A. Blake, (D), Caltech, 1985, Xc
   Simona Bordoni, (D), California (Los Angeles), 2007, Op
   Michael E. Brown, (D), California (Berkeley), 1994, Xy
   Robert W. Clayton, (D), Stanford, 1981, Ys
   Bethany L. Ehlmann, (D), Brown, 2010, XgZrGz
   John M. Eiler, (D), Wisconsin, 1994, Cg
   Kenneth A. Farley, (D), California (San Diego), 1991, Cg
   Woodward W. Fischer, (D), Harvard, 2007, Pg
   Christian Frankenberg, (D), Ruprecht-Karls Univ, 2005, Zr
   Michael C. Gurnis, (D), Australian Nat, 1987, Ys
   Thomas H. Heaton, (D), Caltech, 1979, Ne
   Andrew P. Ingersoll, (D), Harvard, 1966, As
   Jennifer M. Jackson, (D), Illinois, 2005, Gy
   Joseph L. Kirschvink, (D), Princeton, 1979, Pg
   Heather Knutson, (D), Harvard, 2009, Xg
   Shrinivas R. Kulkarni, (D), California (Berkeley), 1983, Xa
   Michael P. Lamb, (D), California (Berkeley), 2008, Gm
   Nadia Lapusta, (D), Harvard, 2001, GcYs
   Jared R. Leadbetter, (D), Michigan State, 1997, Pm
   Dianne K. Newman, (D), MIT, 1997, Pg
   Victoria Orphan, (D), California (Santa Barbara), 2001, Pg
   George R. Rossman, (D), Caltech, 1971, GzCa
   Tapio Schneider, (D), Princeton, 2001, AsmZg
   Alex L. Sessions, (D), Indiana, 2001, CosCb
   Mark Simons, (D), MIT, 1995, Ys
   David J. Stevenson, (D), Cornell, 1976, Xy
   Joann M. Stock, (D), MIT, 1988, GtcYr
   Edward M. Stolper, (D), Harvard, 1979, Cp
   Andrew F. Thompson, (D), Scripps, 2006, OpgZc
   Victor Tsai, (D), Harvard, 2009, Yg
   Paul O. Wennberg, (D), Harvard, 1994, As
   Brian P. Wernicke, (D), MIT, 1982, Gtc
   Yuk L. Yung, (D), Harvard, 1974, As
Assistant Professor:
   Konstantin Batygin, (D), Caltech, 2012, Xg
   Zhongwen Zhan, (D), Caltech, 2013, Ys
Research Professor:
   Egill Hauksson, (D), Columbia, 1981, YsGtYg
Emeritus:
   Arden L. Albee, (D), Harvard, 1957, GizXg
   Clarence R. Allen, (D), Caltech, 1954, Ys
   Donald S. Burnett, (D), California (Berkeley), 1963, Xc
   Charles Elachi, (D), Caltech, 1971, Xg
   Peter M. Goldreich, (D), Cornell, 1963, Xy
   Donald V. Helmberger, (D), California (San Diego), 1967, Ys
   Hiroo Kanamori, (D), Tokyo, 1964, Ysg
   Duane O. Muhleman, (D), Harvard, 1963, Xy
   Jason B. Saleeby, (D), California (Santa Barbara), 1975, Gc
   Leon T. Silver, (D), Caltech, 1955, Cc
   Hugh P. Taylor, (D), Caltech, 1959, Cs
   Peter J. Wyllie, (D), St. Andrews, 1958, Cp

California Lutheran University
Dept of Geology (B) ☒ (2018)
60 West Olsen Road #3700
Thousand Oaks, CA 91360-2787
   p. (805) 493-3264
   bilodeau@callutheran.edu
   http://www.callutheran.edu/admission/undergraduate/majors/geology/
   Programs: Geology (B)
   Enrollment (2018): B: 4 (5)
Chair:
   William L. Bilodeau, (D), Stanford, 1979, GcsGg
Emeritus:
   Linda A. Ritterbush, (D), California (Santa Barbara), 1990, Pi

California Polytechnic State University
Department of Natural Resources Management and Environmental Sciences (B,M) (2015)
One Grand Avenue
San Luis Obispo, CA 93407
   p. (805) 756-2261
   jstevens@calpoly.edu
   http://nres.calpoly.edu
Professor:
   Delmar D. Dingus, (D), Oregon State, 1975, So
   Brent G. Hallock, (D), California (Davis), 1979, Sf
   William L. Preston, (D), Oregon, 1979, Ze
   Thomas A. Ruehr, (D), Colorado State, 1975, Sb
   Terry L. Smith, (D), Nebraska, 1980, So
Assistant Professor:
   Christopher (Chip) S. Appel, (D), Florida, 2001, Sc
Emeritus:
   Ronald D. Taskey, (D), Oregon State, 1977, Sf
Administrative Coordinator:
   Joan M. Stevens, Zn
Instructional Tech:
   Craig Stubler, (B), California Polytech, 1996, So

Department of Physics (B) (2017)
1 Grand Ave
180 204
San Luis Obispo, CA 93407
   p. (805) 756-2448
   physics@calpoly.edu
   http://physics.calpoly.edu
   Administrative Assistant: Kathy Simon
Professor:
   Antonio F. Garcia, (D), California (Santa Barbara), 2001, Gmg
Associate Professor:
   John J. Jasbinsek, (D), Wyoming, 2008, Ysg
   Scott Johnston, (D), California (Santa Barbara), 2006, GcpCc
Emeritus:
   David H. Chipping, (D), Stanford, 1970, Grt

California State Polytechnic University, Pomona
Dept of Geological Sciences (B,M) (2015)
3801 West Temple Avenue
Pomona, CA 91768
   p. (909) 869-3454
   janourse@cpp.edu
   http://geology.csupomona.edu
   Administrative Assistant: Monica Baez
   Enrollment (2014): B: 130 (18) M: 28 (5)
Chair:
   Jonathan A. Nourse, (D), Caltech, 1989, GctEm
Professor:
   Jeffrey S. Marshall, (D), Penn State, 2000, GmtHs
   Jascha Polet, (D), Caltech, 1999, Ysg
Associate Professor:
   Stephen G. Osborn, (D), Arizona, 2010, HwClSp

Assistant Professor:
    Nicholas J. Van Buer, (D), Stanford, 2012, GxtGz
Emeritus:
    David R. Berry, (D), California (Los Angeles), 1983, PgGo
    David R. Jessey, (D), Missouri (Rolla), 1981, Eg
    John A. Klasik, (D), Louisiana State, 1976, Gu
Equipment Technician:
    Mike McAtee, (B), Cal Poly Pomona, 1978, Zn

California State University, Bakersfield
Dept of Geological Sciences (B,M) ☒ (2017)
9001 Stockdale Hwy, 66 SCI
Bakersfield, CA 93311
    p. (661) 654-3027
    arathburn@csub.edu
    http://www.csub.edu/Geology/
    f: https://www.facebook.com/groups/103146842589/
    Programs: Geology; Natural Science
    Administrative Assistant: Sue Holt
    Enrollment (2017): B: 76 (29) M: 35 (6)
Professor:
    Dirk Baron, (D), Oregon Grad Inst, 1996, HgGeZe
    Janice Gillespie, (D), Wyoming, 1991, Go
    Anthony E. Rathburn, (D), Duke, PmOoCb
Assistant Professor:
    Chandranath Basak, (D), Florida, 2011, CmOoCs
    Junhua Guo, (D), Missouri, Gs
    W Chris Krugh, Swiss Fed Inst Tech, 2008, GcmCc
Lecturer:
    David Miller, (D), Stanford, GgtGc
    Katie O'Sullivan, (D), Notre Dame, 2013, GviXg
Emeritus:
    Robert A. Horton, Jr., (D), Colorado Mines, 1985, GdsGo
    Robert M. Negrini, (D), California (Davis), 1986, YmgGn
Department IT Support:
    Elizabeth Powers, (B), N Arizona, 1998, Gg

California State University, Chico
Dept of Geological and Environmental Sciences (B,M) ●☒ (2019)
400 W First Street
Chico, CA 95929-0205
    p. (530) 898-5262
    geos@csuchico.edu
    http://www.csuchico.edu/geos/
    f: https://www.facebook.com/CSUChicoGEOS/
    Programs: Geology, Environmental Science
    Enrollment (2014): B: 200 (50) M: 30 (6)
Chair:
    Russell S. Shapiro, (D), California (Santa Barbara), 1998, PgGdXb
Professor:
    David L. Brown, (D), California (Berkeley), 1995, Hg
    Ann Bykerk-Kauffman, (D), Arizona, 1990, GcZe
    Julie Monet, (D), Rutgers, 2006, GgNg
Associate Professor:
    Todd Greene, (D), Stanford, 2000, Gsr
    Karin A. Hoover, (D), Johns Hopkins, 1998, Hw
    Shane D. Mayor, (D), Wisconsin, 2001, AsZrAp
    Rachel Teasdale, (D), Idaho, 2001, Gv
Assistant Professor:
    Hannah M. Aird, (D), Duke, 2014, GxEg
Lecturer:
    Carrie Monohan, (D), Washington, 2004, Hg
    Jochen Nuester, (D), Max Planck, 2005, Cm
Emeritus:
    Terence T. Kato, (D), California (Los Angeles), 1976, GtcGp

California State University, Dominguez Hills
Earth Sciences (B) ☒ (2018)
1000 E. Victoria Street
NSM B202
Carson, CA 90747
    p. (310) 243-3377
    cmtrujillo@csudh.edu
    Programs: Bachelor of Science in Earth Science, Minor in Earth Sciences, Bachelor of Arts in Geography, Minor in Geography
    Certificates: Certificate Program in Geotechniques
    Enrollment (2014): B: 55 (9)
Chair, Earth Sciences and Geography:
    Brendan A. McNulty, (D), California (Santa Cruz), 1996, Gc
Professor:
    Rodrick Hay, (D), Arizona State, 1996, Zr
    Ashish Sinha, (D), S California, 1997, PeCs
Associate Professor:
    John Keyantash, (D), California (Los Angeles), 2001, HgAsZe
    Ralph Saunders, (D), Arizona, 1996, Zg
Assistant Professor:
    Parveen Chhetri, (D), Texas A&M, 2017, Geg
Lecturer:
    Michael H. Ferris, (M), Cal State (Long Beach), 2012, Zr
Emeritus:
    David Sigurdson, (D), California (Riverside), 1974, GizEg

California State University, East Bay
Dept of Anthropology, Geography, and Environmental Studies (2015)
25800 Carlos Bee Blvd
Hayward, CA 94542
    p. (510) 885-3193
    david.larson@csueastbay.edu

    http://class.csueastbay.edu/geography
Chair:
    David J. Larson, (D), California (Berkeley), 1994, Zu
Professor:
    Scott W. Stine, (D), California (Berkeley), 1987, Gm
Associate Professor:
    Karina Garbesi, (D), California (Berkeley), 1994, Cg
    Michael D. Lee, (D), London Schl of Econ, 1990, Hg
    Gary Li, (D), SUNY (Buffalo), 1997, Zi
    David Woo, (D), California (Santa Barbara), 1991, Zr

Dept of Earth & Environmental Sciences (B,M) ●☒ (2018)
25800 Carlos Bee Blvd.
Hayward, CA 94542
    p. (510) 885-3486
    geology@csueastbay.edu
    http://www.sci.csueastbay.edu/earth
    Programs: Geology (B); Geoscience Education (B); Environmental Science (B); Environmental Geoscience (M); Environmental Health (B)
    Certificates: Geology minor
    Enrollment (2017): B: 81 (24) M: 20 (5)
Associate Professor and Chair:
    Michael S. Massey, (D), Stanford, 2013, ScGeCg
Professor:
    Mitchell S. Craig, (D), Georgia Tech, 1990, Yg
    Jean E. Moran, (D), Rochester, 1994, HwCl
    Jeffery C. Seitz, (D), Virginia Tech, 1994, CgZeGx
Associate Professor:
    Luther M. Strayer, (D), Minnesota, 1998, Gc
Assistant Professor:
    Patricia Oikawa, (D), Virginia, 2011, SbCoSf
    Jose Rosario, (D), Cornell, 2015, PsSaGr

California State University, Fresno
Dept of Earth & Environmental Sciences (B,M) (2015)
2345 E. San Ramon Avenue
MH24
Fresno, CA 93740-8031
    p. (559) 278-3086
    slewis@csufresno.edu
    http://www.csufresno.edu/ees/
    Enrollment (2009): B: 51 (13) M: 30 (5)
Chair:
    Stephen D. Lewis, (D), Columbia, 1982, Yg
Professor:
    Keith D. Putirka, (D), Columbia, 1997, GviGp
    Zhi (Luke) Wang, (D), Leuven (Belgium), 1997, SpHwZi
Associate Professor:
    Christopher J. Pluhar, (D), California (Santa Cruz), 2003, GtYmNg

John Wakabayashi, (D), California (Berkeley), GmcGt
Assistant Professor:
  Robert G. Dundas, (D), California (Berkeley), 1994, PvGr
  Mathieu Richaud, (D), N Illinois, 2006, GusCs
  Peter Van de Water, (D), Arizona, GrPbGf
Lecturer:
  Jeff Anglen, (M), Texas Tech, 2001, GsdPg
  Susan Bratcher, (M), Cal State (Fresno), 2008, Hw
  Kerry Workman-Ford, (M), Cal State (Fresno), 2003, Gc
Adjunct Professor:
  Jerry DeGraff, (M), Utah State, 1976, Zn
  Dong Wang, (D), HwSpZr
Emeritus:
  Jon C. Avent
  Bruce A. Blackerby, (D), California (Los Angeles), Giz
  Roland H. Brady, (D), California (Davis), 1986, NgZuGa
  Seymour Mack, YrHgOn
  Robert D. Merrill, (D), Texas, 1974, GsmPe
  C. John Suen, (D), MIT, 1978, HwCsGe

California State University, Fullerton
Dept of Geological Sciences (B,M) ●◌ (2018)
800 N. State College Boulevard
Geological Sciences, MH 254
Fullerton, CA 92831
  p. (657) 278-3882
  geology@fullerton.edu
  http://geology.fullerton.edu
  f: https://www.facebook.com/CalStateFullerton.Geology
  Programs: B.S. Geology
  B.A. Earth Science
  Enrollment (2014): B: 145 (9) M: 16 (0)
Chair:
  Jeffrey R. Knott, (D), California (Riverside), 1998, Gmg
Professor:
  Phillip A. Armstrong, (D), Utah, 1996, GcYg
  Diane Clemens-Knott, (D), Caltech, 1992, Gig
  Matthew E. Kirby, (D), Syracuse, 2001, GnPeGg
  Adam D. Woods, (D), S California, 1998, GsPe
Associate Professor:
  Nicole Bonuso, (D), S California, 2005, Pie
  W. Richard Laton, (D), W Michigan, 1997, HwGeHy
  James Parham, (D), California (Berkeley), 2003, PvgGg
Assistant Professor:
  Sinan Akciz, (D), MIT, GcRn
  Joe Carlin, (D), Texas A&M, 2013, GuOuGs
  Sean Loyd, (D), S California, 2010, CgGe
  Valbone Memeti, (D), S California, 2009, GivCa
Lecturer:
  Angela Aranda, (M), Cal State (Fullerton), 2016, Rn
  Freddi Jo Bruschke, (D), Gg
  Patricia M. Butcher, (M), Utah, 1993, Zeg
  Emily A. Hamecher, (D), Caltech, 2013, GgzGi
  Wayne G. Henderson, (M), George Washington, 1997, PgGg
  Joanna Hynes, (M), Cal State (Fullerton), Gg
  Scott Mata, (D), USC, Gg
  Carolyn Rath, (M), Cal State (Fullerton), Gg
  Kelly R. Ruppert, (M), California (Riverside), 2001, Gg
Emeritus:
  Galen R. Carlson, (D), Zg
  John H. Foster, (D)
  Prem K. Saint, (D), Minnesota, 1973, Hw

California State University, Long Beach
Dept of Geological Sciences (B,M) (2015)
1250 Bellflower Boulevard
Long Beach, CA 90840-3902
  p. (562) 985-4809
  rfrancis@csulb.edu
  http://www.csulb.edu/colleges/cnsm/departments/geology/
  Administrative Assistant: Margaret Costello
  Enrollment (2009): B: 67 (11) M: 16 (3)
Chair:
  Richard J. Behl, (D), California (Santa Cruz), 1992, GsPeGd
Conrey Endowed Chair:
  Matthew Becker, (D), Texas, 1996, Hwg
Professor:
  Stanley C. Finney, (D), Ohio State, 1977, Psi
  Robert D. Francis, (D), California (San Diego), 1980, Co
  Roswitha B. Grannell, (D), California (Riverside), 1969, Yv
  Jack Green, (D), Columbia, 1953, Gv
  Gregory J. Holk, (D), Caltech, 1997, CsGxEm
  Nate Onderdonk, (D), California (Santa Barbara), 2003, Gtm
Associate Professor:
  Thomas Kelty, (D), California (Los Angeles), 1998, Gc
Assistant Professor:
  Lora R. Stevens (Landon), (D), Minnesota, 1997, Gn
Lecturer:
  Bruce Perry, (M), Cal State (Long Beach), 1993, Gc
Emeritus:
  Kwan M. Chan, (D), Liverpool, 1966, Oc
  Paul J. Fritts, (D), Colorado, 1969, Pm
  Charles T. Walker, (D), Leeds, 1952, Cl
  Robert E. Winchell, (D), Ohio State, 1963, Gz

California State University, Los Angeles
Department of Geosciences and Environment (B,M) O (2016)
5151 State University Drive
Los Angeles, CA 90032
  p. (323) 343-2400
  mmurill@calstatela.edu
  http://www.calstatela.edu/academic/geos
  Enrollment (2016): B: 65 (0) M: 13 (0)
Department Chair:
  Hengchun Ye, (D), Delaware, 1995, ZyHs
Associate Dept Chair:
  Steven Mulherin, (D), Ohio State, 1999
Professor:
  Qiu Hong-lie, (D), Louisiana State, 1994
  Steve LaDochy, (D), Manitoba, 1985, Zn
  Pedro C. Ramirez, (D), California (Santa Cruz), 1990, Gs
Associate Dept Chair:
  Jennifer M. Garrison, (D), California (Los Angeles), 2004, GiCcGv
Associate Professor:
  Andre Ellis, (D), Illinois, 2003, HwCg
  Barry Hibbs, (D), Texas, 1993, Hw
Assistant Professor:
  Kris Bezdecny, (D), S Florida, 2011, ZniZn
  Jingjing Li, (D), California (Irvine), 2013
Instructor:
  Yoshie Hagiwara, (M)
  Angel Hamane, (D), Pepperdine, 2015, Ze
Adjunct Professor:
  Mohammad Hassan Rezaie-Boroon, (D), Erlangen (Germany), 1997, GeCg
Emeritus:
  Kim Bishop, (D), S California, 1994, Gc
  Ivan P. Colburn, (D), Stanford, 1961, Gs
  Robert J. Stull, (D), Washington, 1969, Gi

California State University, Northridge
Dept of Geography (B,M) ◌ (2018)
18111 Nordhoff Street
Northridge, CA 91330-8249
  p. (818) 677-3532
  geography@csun.edu
  http://www.csun.edu/social-behavioral-sciences/geography
  Programs: Undergraduate: Geography: Standard and GIS (BA/MA); Graduate (MA) GIS (MSc)
  Enrollment (2016): B: 105 (0) M: 29 (0)
Professor:
  Julie E. Laity, (D), California (Los Angeles), 1982, Zy
  Amalie Orme, (D), California (Los Angeles), 1983, Gm
  Yifei Sun, (D), SUNY (Buffalo), 2000, Zi
Associate Professor:
  Soheil Boroushaki, (D), W Ontario, 2010, Zi
  Mario A. Giraldo, (D), Georgia, 2007, ZriZg
  Regan Maas, (D), California (Los Angeles), 2010, Zi
Assistant Professor:
  Sanchayeeta Adhikari, (D), Florida, 2011, Zi
  Erin Bray, (D), California (Santa Barbara), 2013, HwsGm
  Luke Drake, (D), Rutgers, 2015, Zi

California

Dept of Geological Sciences (B,M) ●☒ (2019)
18111 Nordhoff Street
MD: 8266
Northridge, CA 91330-8266
   p. (818) 677-3541
   geology@csun.edu
   http://www.csun.edu/geology/
   Programs: Geology, Geophysics
   Administrative Assistant: Mari C. Flores-Garcia
   Enrollment (2012): B: 72 (0) M: 23 (0)
Retired:
   J. Douglas Yule, (D), Caltech, 1996, Gt
Professor:
   Kathleen M. Marsaglia, (D), California (Los Angeles), 1989, Gs
   Elena A. Miranda, (D), Wyoming, 2006, Gc
   Dayanthie Weeraratne, (D), Carnegie Inst, 2005, Yg
Associate Professor:
   M. Robinson Cecil, (D), Arizona, 2009, GiCcGz
   Matthew d'Alessio, (D), California (Berkeley), 2004, ZeGt
   Richard V. Heermance, (D), California (Santa Barbara), 2007, GrCc
   Joshua J. Schwartz, (D), Wyoming, 2007, GiCc
Assistant Professor:
   Jennifer Cotton, (D), Michigan, 2013, Pc
   Eileen L. Evans, (D), Harvard, 2014, YdGtYg
   Priya Ganguli, (D), California (Santa Cruz), 2013
   Scott Hauswirth, (D), N Carolina, 2014, HgCg
   Julian Lozos, (D), California (Riverside), 2013, YgGt
Emeritus:
   Herbert G. Adams, (D), California (Los Angeles), 1971, Ng
   Lorence G. Collins, (D), Illinois, 1959, Gzx
   George C. Dunne, (D), Rice, 1972, Gc
   Vicki A. Pedone, (D), SUNY (Stony Brook), 1990, GdCs
   Gerald W. Simila, (D), California (Berkeley), 1979, Yg
   Jon R. Sloan, (D), California (Davis), 1980, Pm
   Richard L. Squires, (D), Caltech, 1973, Pi

California State University, Sacramento
Dept of Geology (B,M) O (2015)
6000 J Street
Placer Hall
Sacramento, CA 95819-6043
   p. (916) 278-6337
   geology@csus.edu
   http://www.csus.edu/geology/
   Enrollment (2010): B: 77 (16) M: 18 (3)
Chair:
   Tim C. Horner, (D), Ohio State, 1992, HwGs
Professor:
   Kevin J. Cornwell, (D), Nebraska, 1994, GmeHs
   David G. Evans, (D), Louisiana State, 1989, HqYg
   Lisa Hammersley, (D), California (Berkeley), 2003, Gig
   Brian Hausback, (D), California (Berkeley), 1984, Gv
   Judith E. Kusnick, (D), California (Davis), 1996, Ze
Lecturer:
   Barbara J. Munn, (D), Virginia Tech, 1997, Ggp
Adjunct Professor:
   Brian Bergamaschi, (D), Washington, 1995, HwOc
Emeritus:
   Diane H. Carlson, (D), Washington State, 1984, Gc
   Charles C. Plummer, (D), Washington, 1969, Gp
   Greg Wheeler, (D), Washington, Em
Secretary:
   Stacy Lindley, (B), California (Sacramento), 2010, Zn
Instructional Support Tech:
   Steven W. Rounds, (B), Cal State (Sacramento), 1990, Gg

California State University, San Bernardino
Dept of Geological Sciences (B,M) ☒ (2019)
5500 University Parkway
San Bernardino, CA 92407-2397
   p. (909) 537-5336
   CPalmer@csusb.edu
   https://cns.csusb.edu/geology
   Administrative Assistant: Christina Palmer
   Enrollment (2018): B: 47 (7) M: 5 (0)
Professor:
   Joan E. Fryxell, (D), North Carolina, 1984, Gct
   Sally F. McGill, (D), Caltech, 1992, GtYdGm
   Erik Melchiorre, (D), Washington Univ (St. Louis), 1997, EmHwCl
Associate Professor:
   W. Britt Leatham, (D), Ohio State, 1987, PsmOg
Assistant Professor:
   Kerry Cato, (D), Texas A&M, 1991, Ng
   Codi Lazar, (D), California (Los Angeles), 2010, Cp
Emeritus:
   Louis A. Fernandez, (D), Syracuse, 1969, Gi
   Alan L. Smith, (D), California (Berkeley), 1969, GviGz
Cooperating Faculty:
   David Maynard, (D), California (Riverside), 1992, Zn
   Stuart S. Sumida, (D), California (Los Angeles), 1987, Pv

California State University, Stanislaus
Dept of Physics and Geology (B) ●☒ (2019)
One University Circle
Turlock, CA 95382
   p. (209) 667-3466
   HFerriz@csustan.edu
   https://www.csustan.edu/geology
   Programs: Geology (B); Applied Geology (B); Geology (minor)
   Enrollment (2018): B: 18 (6)
Chair:
   Horacio Ferriz, (D), Stanford, 1984, NgHgYe
Professor:
   Mario J. Giaramita, (D), California (Davis), 1989, Gpz
   Robert D. Rogers, (D), Texas, 2003, GcmGt
   Julia Sankey, (D), Louisiana State, 1998, Pve
Lecturer:
   Garry F. Hayes, (M), Nevada (Reno), 1988, Gg
   Michael Whittier, (M), Cal State (Stanislaus), 2005, Ggz

Cerritos College
Earth Science Dept (A) (2015)
11110 E. Alondra Boulevard
Norwalk, CA 90650
   p. (562) 860-2451
   ddekraker@cerritos.edu
   http://www.cerritos.edu/earth-science/
   Enrollment (2012): A: 9 (0)
Earth Science Dept Chair:
   Crystal LoVetere, (D), S California, Zy
Assistant Professor:
   Dan DeKraker, (M), Cal State (Fullerton), 2004, ZeOpGg
   Aline Gregorio, (M), Cal State (Fullerton), 2011
   Tor Lacy, (M), Cal State (Long Beach), 2005, Gg
Instructor:
   Gary D. Johnpeer, (M), Arizona State, 1977, Ng

Chabot College
Div of Mathematics & Science (A) (2015)
25555 Hesperian Boulevard
Hayward, CA 94545
   p. (415) 786-6865
   kbononcini@chabotcollege.edu
Instructor:
   Adolph A. Oliver, (M), Stanford, 1974, Ys
   David J. Perry, (M), San Jose State, 1967, Op

Chaffey College
Earth Science/Geology Dept (2015)
5885 Haven Avenue
Rancho Cucamonga, CA 91737-3002
   p. (909) 652-6402
   henry.shannon@chaffey.edu
   http://www.chaffey.edu/mathandscience/
Professor:
   Jane Warger, (M), Columbia, 1990, ZgGg

City College of San Francisco
Earth Sciences Dept (A) (2018)
Box S62

50 Phelan Avenue
San Francisco, CA 94112
    p. (415) 452-7014
    cjlewis@ccsf.edu
    http://www.ccsf.edu/Earth
    Programs: Associate degrees in Geography, Geology and Oceanography
    Certificates: GIS
    Enrollment (2018): A: 10 (1)
Professor:
    Darrel E. Hess, (M), California (Los Angeles), 1990, Zy
    Chris Lewis, (M), California (Berkeley), 1993, GgEmSo
    Katryn Wiese, (M), Oregon State, 1992, GiOg
Adjunct Professor:
    Ian Duncan, (D), San Francisco State, Zy
    James Kuwabara, (D), Caltech, 1980, RwOgCg
    Joyce Lucas-Clark, (D), Stanford, PgGg
    Russell McArthur, (M), California (Berkeley), Gg
    Elizabeth Proctor, (M), San Francisco State, Zi
    Gordon Ye, (M), California (Berkeley), 1993, Zi

College of Marin
Geology Dept (A) (2015)
835 College Ave
SC 192, Phone Ext: 7523
Kentfield, CA 94904
    p. (415) 457-8811
    dfoss@marin.edu
    http://marin.edu/~geology
Professor:
    Donald J. Foss, (M), Boise State, 1980, Zg
Emeritus:
    James L. Locke, (M), San Jose State, 1971, Zg

College of San Mateo
Geological Sciences (A) ☒ (2019)
1700 W. Hillsdale Blvd.
San Mateo, CA 94402
    p. (650) 574-6633
    hand@smccd.edu
    http://www.collegeofsanmateo.edu/geologicalsciences/
    Programs: Geological Sciences, Geology
    Certificates: none
    Enrollment (2018): A: 9 (1)
Professor:
    Linda M. Hand, (M), Texas A&M, 1988, GgPgOg

College of the Canyons
Dept of Earth, Space, and Environmental Sciences (A) (2015)
26455 Rockwell Canyon Road
Santa Clarita, CA 91355
    p. (661) 362-3658
    Vincent.Devlahovich@canyons.edu
    http://www.canyons.edu/departments/ESES/
    Enrollment (2012): A: 4 (0)
Chair:
    Vincent A. Devlahovich, (D), Cal State (Northridge), 2012, GgZye
Professor:
    Mary Bates, (M), Cal State (Northridge), Zyg

College of the Desert
Dept of Science (A) (2016)
43-500 Monterey Avenue
Palm Desert, CA 92260
    p. (760) 776-7272
    nmoll@collegeofthedesert.edu
    http://www.collegeofthedesert.edu/students/ss/ap/physsci/Pages/Geology.aspx
    Enrollment (2013): A: 3 (1)
Professor:
    Nancy E. Moll, (D), Washington, 1981, Gge
Adjunct Professor:
    Brian Koenig, (M), Arizona, 1978, Gge
    Robert Pellenbarg, (D), Delaware, 1976, OcGu

College of the Sequoias
Science Div (A) (2015)
Visalia, CA 93277
    p. (209) 730-3812
    erich@cos.edu
Chair:
    Eric D. Hetherington, (D), Minnesota, 1991, Gc
Emeritus:
    John R. Crain, (M), Nevada, 1961, Gg

College of the Siskiyous
Dept of Biological & Physical Sciences (A) (2016)
800 College Avenue
Weed, CA 96094
    p. (530) 938-5255
    hirt@siskiyous.edu
    http://www.siskiyous.edu/class/ess/
    Enrollment (2015): A: 1 (0)
Instructor:
    William H. Hirt, (D), California (Santa Barbara), 1989, Gi

Contra Costa College
Astronomy/Engineering/Geology/Physics Dept (A) (2015)
2600 Misson Bell Drive
San Pablo, CA 94806
    p. (510) 235-7800
    jsmithson@contracosta.edu
    http://www.contracosta.edu/home/programs-departments/engineering/
Professor:
    Mary Lewis, Gg
    Jayne Smithson, Gg

Cosumnes River College
Dept of Science, Mathematics & Engineering (A) (2015)
8401 Center Parkway
Sacramento, CA 95823
    p. (916) 691-7210
    MuranaB@crc.losrios.edu
    Department Secretary: Sue McCoy
Chair:
    Debra Sharkey, (M), California (Davis), 1994, Zy
Assistant Professor:
    Hiram Jackson, (M), California (Davis), 1992, Gg
Adjunct Professor:
    Gerry Drobny, (M), Washington State, 1981, Gg

Crafton Hills College
Geology Dept (A) (2016)
11711 Sand Canyon Road
Yucaipa, CA 92399
    p. (909) 794-2161
    rihughes@craftonhills.edu
    http://www.craftonhills.edu/courses_&_programs/Physical_Science/Geology/
    Enrollment (2013): A: 16 (4)
Professor:
    Richard O. Hughes III, (M), Ohio, 1994, GlZe

Cuesta College
Physical Sciences Div (A) (2015)
P. O. Box 8106
San Luis Obispo, CA 93403
    p. (805) 546-3230
    jgrover@cuesta.edu
    http://academic.cuesta.edu/physci/Geology/index.html
    Enrollment (2014): A: 6 (0)
Professor:
    Jeffrey A. Grover, (M), Arizona, 1982, Gg
    Debra Stakes, (D), Gg

Cuyamaca College
Dept of Science and Engineering (A) (2015)
900 Rancho San Diego Parkway

El Cajon, CA 92019
p. (619) 660-4345
Glenn.Thurman@gcccd.edu
http://www.cuyamaca.edu/
Instructor:
  Lisa Chaddock, Zy
  Michael Farrell, Gg
  Bryan Miller-Hicks, Gg
  Agatha Wein, Gg
  Ray Wolcott, Og
Emeritus:
  Waverly Ray, Gg

Cypress College
Geology Dept (A) (2015)
9200 Valley View Street
Cypress, CA 90630
p. (714) 484-7153
RArmale@CypressCollege.edu
Professor:
  Russell L. Flynn, (M), San Diego State, 1971, Op
Adjunct Professor:
  Hank Wadleigh, (B), Cal State (Long Beach), 1957, Gg
Visiting Professor:
  Curtis J. Williams, (M), Cal State (Los Angeles), 1996, Gg
Emeritus:
  Keith E. Green, (M), S California, 1958, Pg
  Altus Simpson, (M), S California, 1958, Go

Diablo Valley College
Div. of Physical Sciences (A) (2015)
321 Golf Club Road
Pleasant Hill, CA 94523
p. (925) 685-1230 x46
JHetherington@dvc.edu
http://www.dvc.edu/org/departments/physics/
Chair:
  Jean Hetherington, (M), Washington, 1983, Gg

El Camino College
Dept of Earth Sciences (A) (2015)
16007 Crenshaw Blvd.
Torrance, CA 90506
p. (310) 660-3593
jholliday@elcamino.edu
http://www.elcamino.edu/academics/naturalsciences/earth/
Professor:
  Jerry Brothen, (M), California (Los Angeles), Zy
  Sara Di Fiori, (M), California (Los Angeles), GgOg
  Matt Ebiner, (M), California (Los Angeles), Zy
  Joseph W. Holliday, (M), Oregon State, 1982, Og
Associate Professor:
  Chuck Herzig, (D), California (Riverside), GgOg
Instructor:
  Gary Booher, Zg
  Robin Bouse, Gg
  Charles Dong, Og
  Lynn Fielding, Zg
  Patricia Neumann, Zg
  Douglas Neves, Zg
  Jim Noyes, (D), Scripps, Og
  Ebenezer Peprah, Zg

Feather River College
Feather River College (A) (2015)
570 Golden Eagle Ave.
Quincy, CA 95971
dlerch@frc.edu
http://www.frc.edu/
Instructor:
  Derek Lerch, Zg

Folsom Lake College
Geosciences (A) ● (2015)
10 College Parkway

Folsom, CA 95630
p. (916) 608-6668
pittmaj@flc.losrios.edu
http://www.flc.losrios.edu/Academics/Geology.htm
Enrollment (2015): A: 7 (1)
Professor:
  Jason Pittman, (M), Oregon State, 1999, ZgiZi

Fresno City College
Earth & Physical Science Dept (A) ◊ (2019)
Math, Science & Engineering Div
1101 E. University Avenue
Fresno, CA 93741
p. (559) 442-4600
alexandra.priewisch@fresnocitycollege.edu
Programs: Geology (A)
Enrollment (2018): A: 40 (1)
Instructor:
  T. Craig Poole, (M), Cal State (Fresno), 1987, Gg
  Alexandra Priewisch, (D), New Mexico, 2014, Gg

Fullerton College
Div of Natural Sciences (Geology) (A) ⚐ (2018)
321 E. Chapman Ave
Fullerton, CA 92832
p. (714) 992-7445
rlozinsky@fullcoll.edu
http://natsci.fullcoll.edu/
Programs: AS, AS-T Geology
AS Earth Science
Enrollment (2018): A: 4 (1)
Professor:
  William S. Chamberlin, (D), S California, 1989, OgbZe
  Richard P. Lozinsky, (D), New Mexico Tech, 1988, GreGm
Associate Professor:
  Roman DeJesus, (D), Og
Assistant Professor:
  Marc Willis, (M), Gg

Gavilan College
Dept of Physical Sciences (2015)
5055 Santa Theresa Boulevard
Gilroy, CA 95020
p. (408) 848-4701
dwillahan@gavilan.edu
http://www.gavilan.edu/natural_sciences/
Instructor:
  Duane Willahan, (M), Santa Fe State, 1992, Gg

Golden West College
Physical Science Program (A) (2015)
15744 Golden West Street
Huntington Beach, CA 92647
p. (714) 892-7711 x51116
msouto@gwc.cccd.edu
http://www.goldenwestcollege.edu/campus/physicalscience.html
Professor:
  Ronald C. Gibson, (M), California (Riverside), 1964, Gc
  Bernard J. Gilpin, (M), California (Riverside), 1976, Ys

Grossmont College
Dept of Earth Sciences (A) O⚐ (2017)
8800 Grossmont College Drive
El Cajon, CA 92020
p. (619) 644-7887
gary.jacobson@gcccd.edu
https://www.grossmont.edu/academics/programs-departments/earth-sciences/default.aspx
f: https://www.facebook.com/grossmontearthsciences/
Programs: Geology (A); Geography (A); Oceanography (A)
Professor:
  Chris Hill, (D), S California, 2000, GgOg
Instructor:
  Gary L. Jacobson, (M), San Diego State, 1982, GucEm

Hartnell College
Div of Math and Science (A) (2015)
411 Central Avenue
Salinas, CA 93901
aramirez@hartnell.edu
http://www.hartnell.edu/academics/math.html
Instructor:
Robert Barminski, (M), Moss Landing Marine Lab, GgOg

Humboldt State University
Dept of Geology (B,M) O☒ (2018)
1 Harpst Street
Arcata, CA 95521-8299
p. (707) 826-3931
geology@humboldt.edu
http://www.humboldt.edu/geology/
Programs: Geology (B); Geoscience (B)
Enrollment (2018): B: 84 (29) M: 4 (3)
Chair:
Mark Hemphill-Haley, (D), Oregon, 2000, Gt
Associate Professor:
Brandon L. Browne, (D), Alaska (Fairbanks), 2005, GivZe
Assistant Professor:
Laura B. Levy, (D), Dartmouth, 2014, GslPc
Melanie Michalak, (D), California (Santa Cruz), 2013
Jasper Oshun, (D), California (Berkeley), 2015, HyGmg
Research Associate:
Eileen Hemphill-Haley, (D), California (Santa Cruz), 1991, Pm
Harvey M. Kelsey, (D), California (Santa Cruz), 1977, Gmt
Robert C. McPherson, (M), Humboldt State, 1989, Gt
Dallas D. Rhodes, (D), Syracuse, 1973, GmZre
Lecturer:
Amanda R. Admire, (M), Humboldt State, 2013, Zg
Adjunct Professor:
David Bazard, (D), Arizona, 1991
Jason R. Patton, (D), Oregon State, 2014, GuOuGg
Jon R. Pedicino, (D), Arizona, 1996
Brandon E. Schwab, (D), Oregon, 2000, GizGv
Robert R. Ziemer, (D), Colorado State, 1978, HgGmZn
Emeritus:
Raymond M. Burke, (D), Colorado, 1976, Gm
Gary A. Carver, (D), Washington, 1972, Gmt
Susan M. Cashman, (D), Washington, 1977, Gc
Lorinda Dengler, (D), California (Berkeley), 1978, Yg
Andre K. Lehre, (D), California (Berkeley), 1982, GmHs
Alistair W. McCrone, (D), Kansas, 1961, Gs
William C. Miller, (D), Tulane, 1984, Pi

Imperial Valley College
Science, Math & Engineering Div (A) ⌁ (2019)
380 E. Aten Rd.
Imperial, CA 92251
p. (760) 355-6304
ofelia.duarte@imperial.edu
http://www.imperial.edu/index.php?pid=343
Professor:
Kevin G. Marty, (M), New Orleans, 1994, GcZg
Instructor:
Kevin Marty, Gg

Irvine Valley College
Dept of Geology (A) ☒ (2018)
5500 Irvine Center Drive
Irvine, CA 92618
p. (949) 451-5561
astinson@ivc.edu
http://www.ivc.edu
Programs: Geology (A)
Enrollment (2007): A: 20 (5)
Professor:
Amy L. Stinson, (M), San Diego State, 1990, Gc
Adjunct Professor:
Mark Bordelon, (M)
Scott Mata, (D), USC
Doug Neves, (D)
Jim Schneider, (M)

Laney College
Dept of Geography/Geology (A) (2015)
900 Fallon Street
Oakland, CA 94607
p. (510) 464-3233
dwoodrow@peralta.edu
http://www.laney.peralta.edu/apps/comm.asp?Q=30123

Las Positas College
Dept of Geology (A) ⌁ (2018)
3000 Campus Hill Drive
Livermore, CA 94551
p. (925) 424-1319
rhanna@laspositascollege.edu
http://www.laspositascollege.edu/geology/
Programs: Geology (AS and AS-T)
Certificates: Geology Certificate
Enrollment (2018): A: 10 (2)
Instructor:
Ruth L. Hanna, (M), California (Davis), 1988, Gg
Geology Lab Technician:
Carol Edson, Gg

Loma Linda University
Dept of Earth and Biological Sciences (B,M,D) ● (2017)
Loma Linda University
Loma Linda, CA 92350
p. (909) 824-4530
pbuchheim@llu.edu
https://medicine.llu.edu/research/department-earth-and-biological-sciences
f: https://www.facebook.com/lomalindageology
Enrollment (2016): B: 1 (0) M: 1 (4) D: 11 (2)
Professor:
Leonard R. Brand, (D), Cornell, 1970, Pv
Paul Buchheim, (D), Wyoming, 1978, GnsPe
Leroy Leggitt, (D), Loma Linda, 2005, Pi
Senior Scientist:
Benjamin L. Clausen, (D), Colorado, 1987, GiCtYg
Associate Professor:
Kevin Nick, (D), Oklahoma, 1990, GsbGr
Assistant Professor:
Ronald Nalin, (D), Padova, 2006, Gsr
Adjunct Professor:
Raul Esperante, (D), Loma Linda, 2002, Pv

Los Angeles County Museum of Natural History
Research and Collections Branch (2016)
900 Exposition Boulevard
Los Angeles, CA 90007
p. (213) 763-3360
acelesti@nhm.org
https://www.nhm.org/site/research-collections/mineral-sciences
Associate Curator:
Luis M. Chiappe, (D), Buenos Aires, 1992, Pv
Curator:
Kenneth E. Campbell, (D), Florida, 1973, PvGgr
Anthony R. Kampf, (D), Chicago, 1976, Gz
Associate Curator:
Aaron Celestian, (D)
Nathan Smith, (D)
Collections Manager:
Samuel A. McLeod, (D), California (Berkeley), 1981, Pv
Christopher A. Shaw, (M), Cal State (Long Beach), 1981, Pv

Los Angeles Harbor College
Dept of Earth Science (A) (2015)
1111 Figueroa Place
Wilmington, CA 90744
p. (310) 233-4000
munasit@lahc.edu

http://www.lahc.cc.ca.us/
Instructor:
  Patricia Kellner, (D), Zg
  John Mack, Zg
  Tissa Munasinghe, (D), Zg
  Melanie Renfrew, (D), Zg
  Susan White, Zg

Los Angeles Pierce College
Physics and Planetary Sciences (A) (2015)
6201 Winnetka Ave
Woodland Hills, CA 91371
  p. (818) 710-2218
  zayacjm@piercecollege.edu
  Enrollment (2012): A: 11 (0)
Professor:
  Jason P. Finley, (M), California (Los Angeles), Am
  Stephen C. Lee, (B), Illinois, 1971, Og
  W. Craig Meyer, (M), S California, 1973, GeuPm
  John M. Zayac, (M), California (Santa Barbara), 2006, Ggv
Adjunct Professor:
  Harry Filkorn, (D), Kent State, GgPi
  James P. Krohn, (M), S California, 1974, NgGg
  Donald Prothero, (D), Columbia, Gg
Emeritus:
  Ruth Y. Lebow, (M), Chicago, 1941, Op
  Mark L. Powell, (M), Cal State (Northridge), 1967, As
  William H. Russell, (M), Cal State (Northridge), 1970, Am
  James Y. Vernon, (M), California (Los Angeles), 1951, Am

Los Angeles Southwest College
Geology Dept (A) (2015)
1600 West Imperial Highway
Los Angeles, CA 90047
  p. (323) 241-5297
  doosepr@lasc.edu
  http://www.lasc.edu
Chair:
  Glenn Yoshida, Gg
Professor:
  Paul R. Doose, (D), California (Los Angeles), 1980

Los Angeles Valley College
Earth Science (A) (2018)
5800 Fulton Ave.
Valley Glen, CA 91401
  p. (818) 778-5566
  hamsje@lavc.edu
  https://lavc.edu/earthscience/index.aspx
  Programs: AS-Transfer Geology
  AS Earth Science
  AA-Transfer Geography
  Enrollment (2014): A: 3 (0)
Chair:
  Jacquelyn E. Hams, (M), Cal State (Los Angeles), 1987, GgOuZe
Professor:
  Meredith L. Leonard, (M), Cal State (Northridge), AmZei
Academic Senate President:
  Donald J. Gauthier, (M), California (Los Angeles), 1993, ZiyAm

Mendocino College
Earth Science (A) (2018)
1000 Hensley Creek Road
Ukiah, CA 95482
  p. (707) 468-3002
  scardimo@mendocino.edu
  Programs: Earth Science (A); Natural Resources (A); Geology (A); Geography (A)
  Enrollment (2018): A: 74 (3)
Professor:
  Steve Cardimona, (D), Texas, 1992, Ys

Merced College
Science, Mathematics & Engineering (A) (2017)
3600 M Street
Merced, CA 95348
  p. (209) 384-6293
  kain.d@mccd.edu
  http://www.mccd.edu/academics/sme/index.html
  Programs: Geology

MiraCosta College
Dept of Geography (A) (2018)
1 Barnard Drive
Oceanside, CA 92056
  lmiller@miracosta.edu
  https://www.miracosta.edu/Instruction/Geography/
Chair:
  Herschel I. Stern, (D), Oregon, 1988, Zy

Dept of Physical Sciences (A) (2015)
1 Barnard Drive
Oceanside, CA 92056
  p. (760) 944-4449 x7738
  lmiller@miracosta.edu
  https://www.miracosta.edu/Instruction/Geology/
Professor:
  Keith H. Meldahl, (D), Arizona, 1990, Gg
  Christopher V. Metzler, (D), California (San Diego), 1987, Gg
  John Turbeville, GgOg
Adjunct Professor:
  Phil Farquharson, Gg
Laboratory Director:
  Larry Hernandez, Gg

Modesto Junior College
Dept of Science, Mathematics & Engineering (A) (2015)
435 College Avenue
Modesto, CA 95350
  p. (209) 575-6172
  hayesg@mjc.edu
  https://www.mjc.edu/instruction/sme/earthscience.php
Instructor:
  Donald C. Ahrens, (D), N Colorado, Zg
  Garry F. Hayes, (M), Nevada (Reno), 1985, Gg

Monterey Peninsula College
Earth Sciences Dept (A) (2018)
980 Fremont St.
Monterey, CA 93940
  p. (831) 646-4149
  ahochstaedter@mpc.edu
  http://www.mpc.edu/academics/academic-divisions/physical-science/earth-science-eart
  Programs: Geology
  Enrollment (2018): A: 1 (1)
Chair:
  Alfred Hochstaedter, (D), California (Santa Cruz), 1991, GgOgZg

Moorpark College
Geology Program (A) (2016)
7075 Campus Road
Moorpark, CA 93021
  p. (805) 553-4161
  rharma@vcccd.edu
  http://www.moorparkcollege.edu/geology
  Enrollment (2016): A: 24 (2)
Instructor:
  Roberta L. Harma, (M), Hawaii, 1982, Gg

Moss Landing Marine Laboratories
CSU Consortium for Marine Sciences (M) (2018)
8272 Moss Landing Road
Moss Landing, CA 95039
  p. (831) 771-4400
  frontdesk@mlml.calstate.edu
  http://www.mlml.calstate.edu
  Enrollment (2014): M: 7 (25)
Director:
  James T. Harvey, (D), Oregon, 1987, Ob

Librarian:
  Joan Parker, (M), California (Los Angeles), 1986, Zn
Professor:
  Ivano Aiello, (D), Bologna, 1997, OuGmu
  Kenneth H. Coale, (D), California (Santa Cruz), 1988, CtcOc
  Jonathan Geller, (D), California (Berkeley), 1988, Zn
  Michael Graham, (D), Scripps, 2000, Ob
  Nicholas A. Welschmeyer, (D), Washington, 1982, Ob
Research Faculty:
  John S. Oliver, (D), Scripps, 1980, Ob
Assistant Professor:
  Scott Hamilton, (D), California (Santa Barbara), 2008
  Birgitte McDonald, (D)
Research Faculty:
  Simona Bartl, (D), California (San Diego), 1989, Zn
  Laurence Breaker, (D), Naval Postgrad Sch, 1983, Op
  David Ebert, (D), Rhodes, 1990, Zn
  Stacy Kim, (D), WHOI, 1996, Zn
  Valerie Loeb, (D), Scripps, 1979, Zn
  Richard Starr, (D), Autó Baja California Sur, 2002, Zn
Emeritus:
  Gregor M. Cailliet, (D), California (Santa Barbara), 1972, Zn
  Michael Foster, (D), California (Santa Barbara), 1971, Zn
  H. Gary Greene, (D), Stanford, 1977, Gu
Diving Safety Officer:
  Diana Steller, (D), California (Santa Cruz), 2003, Zn

Mt. San Antonio College
Dept of Earth Sciences & Astronomy (A) (2017)
1100 North Grand Avenue
Walnut, CA 91789
  p. (909) 594-5611
  cwebb@mtsac.edu
  Enrollment (2010): A: 16 (0)
Acting Chair:
  Julie Ali-Bray, (M), S California, 1999, Zn
Professor:
  Micol Christopher, (D), Caltech, 2007, Zn
  Craig A. Webb, (M), Duke, 1996, ZgAmOg
Instructor:
  Mark Boryta, (D), New Mexico Tech, 1997
  Barbara Grubb, (M), Cal State (Long Beach), 1991, Ps
  Larry Mendenhall, (D), Oregon State, 1961, Am
  Charles Roberts, (M), Ohio, 1972, Pm
Emeritus:
  Hallock J. Bender, (D), San Gabriel, 1959, Gg
  John W. Burns, (M), Pittsburgh, 1960, Zn
  Damon P. Day, (M), Michigan Tech, 1965, Gg
  Ron N. Hartman, (M), Cal State (Los Angeles), 1973, Xm
  Kazimierz M. Pohopien, (M), McGill, 1951, Ge
  Harold V. Thurman, (M), Cal State (Los Angeles), 1966, Og
Geotechnician:
  Mark Koestel, (B), Arizona, 1978, Em

Naval Postgraduate School
Dept of Meteorology (M,D) (2015)
589 Dyer Road
Root Hall, Room 254
Monterey, CA 93943-5114
  p. (831) 656-2516
  nuss@nps.edu
  http://www.nps.edu/Academics/Schools/GSEAS/Departments/Meteorology/
  Enrollment (2012): M: 45 (36) D: 9 (2)
Dean:
  Philip A. Durkee, (D), Colorado State, 1984, AmZr
Chair:
  Wendell A. Nuss, (D), Washington, 1986, Am
Distinguished Professor:
  Michael T. Montgomery, (D), Harvard, 1990, AmsZg
Professor:
  Patrick A. Harr, (D), Naval Postgrad Sch, 1993, Am
  Qing Wang, (D), Penn State, 1993, Am
Associate Professor:
  Joshua P. Hacker, (D), British Columbia, 2001, Am

Assistant Professor:
  Richard W. Moore, (D), Colorado State, 2004, Am
  Barbara V. Scarnato, (D), ETH, 2008, Am
Research Associate:
  Hway-Jen Chen, (M), California (Los Angeles), 1993, Am
  Paul A. Frederickson, (M), Maryland, 1989, Am
  Mary S. Jordan, (M), Naval Postgrad Sch, 1985, Am
  Kurt E. Nielsen, (M), Oklahoma, 1988, Am
  Andrew Penny, (M), Arizona, 2009, Am
Emeritus:
  Robert Haney, (D), California (Los Angeles), 1971, Am
  Robert J. Renard, (D), Am
  Carlyle H. Wash, (D), Wisconsin, 1978, Am
  Forrest Williams, (M), MIT, 1972, Am
  Roger T. Williams, (D), California (Los Angeles), 1963, Am
Research Professor:
  Kenneth L. Davidson, (D), Michigan, 1970, Am
  Peter S. Guest, (D), Naval Postgrad Sch, 1992, Am
Research Associate Professor:
  James Thomas Murphree, (D), California (Davis), 1989, Am
NRC Postdoctoral Fellow:
  Myung-Sook Park, (D), Am
Distinguished Research Professor:
  Russell L. Elsberry, (D), Colorado State, 1968, Am
Distinguished Professor:
  Chih-Pei Chang, (D), Washington, 1972, Am
Meteorologist:
  Robert L. Creasey, (M), Am

Dept of Oceanography (M,D) ☒ (2019)
833 Dyer Road, Room 328
Monterey, CA 93943-5122
  p. (831) 656-2673
  pcchu@nps.edu
  http://www.nps.edu/Academics/GSEAS/Oceanography/
  Enrollment (2014): M: 14 (14) D: 2 (2)
Chair:
  Peter C. Chu, (D), Chicago, 1985, OpZr
Professor:
  John A. Closi, California (Santa Cruz), 1993
  Jamie MacMahan, (D), Florida, 2003, Onp
  Jeffrey D. Paduan, (D), Oregon State, 1987, Op
  Timour Radko, Florida State, 1997, Op
Assistant Professor:
  Derek Olson, (D), Penn State, 2017, Zn
Emeritus:
  Robert H. Bourke, (D), Oregon State, 1972, Op
  Ching-Sang Chiu, (D), MIT/WHOI, 1985, Op
  Curtis A. Collins, (D), Oregon State, 1967, Op
  Roland W. Garwood, (D), Washington, 1976, Op
  Thomas H. Herbers, (D), California (San Diego), 1990, Op
  Timothy P. Stanton, (M), Auckland, 1978, Op
  Edward B. Thornton, (D), Florida, 1970, On
  Robin T. Tokmakian, (D), Naval Postgrad Sch, 1997, Op
  Joseph J. Von Schwind, (D), Texas A&M, 1968, Op
Research Professor:
  Wieslaw Maslowski, (D), Alaska (Fairbanks), 1994, Op

Occidental College
Dept of Geology (B) ☒ (2019)
1600 Campus Road
Los Angeles, CA 90041
  p. (323) 259-2823
  jangarcia@oxy.edu
  http://www.oxy.edu/Geology.xml
  Administrative Assistant: Elisa Ruiz
  Enrollment (2018): B: 33 (0)
Chair:
  Christopher Oze, (D), Stanford
Professor:
  Scott W. Bogue, (D), California (Santa Cruz), 1982, Ym
  Darren J. Larsen, (D), Colorado
  Margaret E. Rusmore, (D), Washington, 1985, Gc
  James L. Sadd, (D), South Carolina, 1987, ZiGeZr
Related Staff:
  Jan Garcia, Zn

Ohlone College
Dept of Geology (A) ◎ (2018)
43600 Mission Boulevard
Fremont, CA 94539
   p. (510) 979-7938
   pbelasky@ohlone.edu
   http://www.ohlone.edu/instr/geology/
   Programs: Associate in Geology
   Certificates: Geology, Paleobiology, GIS (Geography Dept)
Professor:
   Paul Belasky, (D), California (Los Angeles), 1994, PgqGs

Orange Coast College
Div of Mathematics & Science (A) ☒ (2018)
2701 Fairview Road
Box 5005
Costa Mesa, CA 92626-5005
   p. (714) 432-5647
   ebender@occ.cccd.edu
   http://www.orangecoastcollege.edu/academics/divisions/math_science/geology/Pages/default.aspx
   Programs: Geology; Earth Science
   Enrollment (2018): A: 4 (1)
Chair:
   E. Erik Bender, (D), S California, 1994, GitGz
Assistant Professor:
   Christopher A. Berg, (D), Texas, 2007, GpZeGt
Lecturer:
   Taylor Kennedy, (M), Cal State (Fullerton), Ge
   Michael Van Ry, (M), Cal State (Fullerton), 2010, Gv

Palomar College
Dept of Earth, Space, and Aviation Sciences (A) (2016)
1140 West Mission Road
San Marcos, CA 92069
   p. (760) 744-1150
   sfigg@palomar.edu
   http://www.palomar.edu/earthscience/
   Administrative Assistant: Brenda Morris
   Enrollment (2014): A: 35 (0)
Professor:
   Wing H. Cheung, (D), California (Irvine), 2017, ZiRnZr
   Doug Key, (M), San Diego State, Zy
   Alan P. Trujillo, (M), N Arizona, 1984, OguZe
   Lisa Yon, (D), Brown, 1994, OgZg
Associate Professor:
   Wing Cheung, (M), Indiana, 2007, ZirZy
   Patricia A. Deen, (M), San Diego State, 1984, OgZg
   Mark Lane, (M), San Diego State, 1996, Zn
Assistant Professor:
   Sean Figg, (M), N Colorado, 2012, Gg
Emeritus:
   Jim Pesavento, (D), GgZg

Pasadena City College
Dept of Geology ● (2017)
NAtural Sciences Division
1570 E. Colorado Boulevard
Pasadena, CA 91106
   p. (626) 585-7138
   naturalsciences@pasadena.edu
   http://pasadena.edu/academics/divisions/natural-sciences/areas-of-study/geology.php
Dean, Natural Sciences:
   David N. Douglass, (D), Dartmouth, 1987, Ze
Acting Chair:
   Martha House, (D), MIT, Gg
Professor:
   Elizabeth Nagy-Shadman, (D), Caltech, Gg
   Yuet-Ling O'Connor, (D), S California, Ge
   Bryan Wilbur, (D), California (Los Angeles), Gg
Instructor:
   MIchael Vendrasco, (D), California (Los Angeles)
Emeritus:
   Gerald L. Lewis, (M), Cal State (Long Beach), 1965, Ps
Geology Lab Tech:
   Debra A. Cantarero, (A), Pasadena City Coll, 2000, Zn

Pomona College
Geology Dept (B) ☒ (2019)
185 East Sixth Street
Claremont, CA 91711-6339
   p. (909) 621-8675
   LKeala@pomona.edu
   http://www.geology.pomona.edu
   Programs: Geology
   Administrative Assistant: Lori Keala
   Enrollment (2018): B: 31 (6)
Chair:
   Eric B. Grosfils, (D), Brown, 1996, XgGvq
Professor:
   Robert R. Gaines, (D), California (Riverside), 2003, GsClPi
Associate Professor:
   Jade Star Lackey, (D), Wisconsin, 2005, GipCs
   Linda A. Reinen, (D), Brown, 1993, Gct
Department Technician:
   Jonathan Harris

Riverside City College
Dept of Physical Science: Geology (A) (2017)
4800 Magnolia Avenue
Riverside, CA 92506-1299
   p. (951) 222-8350
   william.phelps@rcc.edu
   Enrollment (2016): A: 11 (0)
Assistant Professor:
   William Phelps, (D), California (Riverside), 2007, PeOpPg

Sacramento City College
Dept of Physics, Astronomy, & Geology (A) ☒ (2018)
3835 Freeport Blvd.
Sacramento, CA 95822
   p. (916) 558-2343
   stantok@scc.losrios.edu
   http://www.scc.losrios.edu/pag/
   Enrollment (2016): A: 10 (0)
Professor:
   Kathryn Stanton, (D), California (Davis), 2006, GgPg

Saddleback Community College
Dept of Earth and Ocean Sciences (A) ☒ (2018)
28000 Marguerite Parkway
Mission Viejo, CA 92692
   p. (949) 582-4820
   jrepka@saddleback.edu
   http://www.saddleback.edu/mse/geo/
   f: https://www.facebook.com/Saddleback-College-Geology-731491460197935/
   Enrollment (2015): A: 7 (1)
Chair:
   James Repka, (D), California (Santa Cruz), 1998, GgmZe

San Bernardino County Museum
Geological Sciences Div (2015)
2024 Orange Tree Lane
Redlands, CA 92374
   p. (909) 307-2669
   kspringer@sbcm.sbcounty.gov
   http://www.co.san-bernardino.ca.us/museum/
Senior Curator:
   Kathleen B. Springer, (M), California (Riverside), Zg
Curator:
   J. Chris Sagebiel, (M), Texas, 1998, Pv

San Bernardino Valley College
Earth and Spatial Sciences Program (A) (2015)
701 South Mount Vernon Ave.

San Bernardino, CA 92410
    p. (909) 384-8638
    theibel@valleycollege.edu
Head:
    Todd Heibel, Zyi
Professor:
    Stephen H. Sandlin, Zy
Instructor:
    Gary M. Croft, Zy
    Vanessa Engstrom, Zy
    Walter Grossman, Og
    Jeffrey Krizek, Zi
    William Muir, Og
    Solomon Nana Kwaku Nimako, Zi
    Edmund Jekwu Ogbuchiekwe, Zy
    Lisa Schmidt, Zy
Adjunct Professor:
    Donald G. Buchanan, (M), Naval Postgrad Sch, 1975, Og

San Diego State University
Dept of Geological Sciences (B,M,D) ●☒ (2019)
5500 Campanile Drive
San Diego, CA 92182-1020
    p. (619) 594-5586
    dkimbrough@mail.sdsu.edu
    http://www.geology.sdsu.edu
    f: https://www.facebook.com/SDSU.Geology
    t: @sdsugeology
    Administrative Assistant: Irene Occhiello
    Administrative Assistant: Pia Parrish
    Enrollment (2015): B: 72 (9) M: 25 (12) D: 9 (0)
Chair:
    David L. Kimbrough, (D), California (Santa Barbara), 1982, Cc
Associate Dean, Division of Undergraduate Studies:
    Stephen A. Schellenberg, (D), S California, 2000, Pe
Professor:
    Eric G. Frost, (D), S California, 1983, Zr
    Gary H. Girty, (D), Columbia, 1983, Gc
    Kim B. Olsen, (D), Utah, 1994, Ye
    Thomas K. Rockwell, (D), California (Santa Barbara), 1983, Gm
Senior Scientist:
    Barry B. Hanan, (D), Virginia Tech, 1980, Cc
Associate Professor:
    Shuo Ma, (D), California (Santa Barbara), 2006, Ys
    Kathryn W. Thorbjarnarson, (D), California (Los Angeles), 1990, Hq
Lecturer:
    Victor E. Camp, (D), Washington State, 1976, Gv
    Kevin Robinson, (M), San Diego State, 1996, Gc
    Isabelle Sacramentogrilo, (M), San Diego State, 1999, Gg
Adjunct Professor:
    Mario V. Caputo, (D), Cincinnati, 1988, GsdOn
Emeritus:
    Patrick L. Abbott, (D), Texas, 1973, Gs
    Kathe K. Bertine, (D), Yale, 1970, Cl
    Steven M. Day, (D), California (San Diego), 1977, Ys
    Clive E. Dorman, (D), Oregon State, 1974, OpAmt
    George R. Jiracek, (D), California (Berkeley), 1972, Ye
    J. Philip Kern, (D), California (Los Angeles), 1968, Pg
    Daniel Krummenacher, (D), Geneva, 1959, Cc
    Monte Marshall, (D), Stanford, 1971, YmGtc
    Richard H. Miller, (D), California (Los Angeles), 1975, Ps
    Gary L. Peterson, (D), Washington, 1963, Gr
    Anton D. Ptacek, (D), Washington (St. Louis), 1965, Gx

San Francisco State University
Dept of Earth & Climate Sciences (B,M) (2015)
1600 Holloway Avenue TH 509
San Francisco, CA 94132
    p. (415) 338-2061
    geosci@sfsu.edu
    http://tornado.sfsu.edu
    Enrollment (2013): B: 56 (0) M: 22 (0)
Chair:
    Karen Grove, (D), Stanford, 1989, Gs
Professor:
    David P. Dempsey, (D), Washington, 1985, As
    Oswaldo Garcia, (D), SUNY (Albany), 1976, As
    Mary L. Leech, (D), Stanford, 1999, GptGz
    John P. Monteverdi, (D), California (Berkeley), 1977, As
    David A. Mustart, (D), Stanford, 1972, GizGv
    Raymond Pestrong, (D), Stanford, 1965, Ng
Associate Professor:
    John Caskey, (D), Nevada (Reno), 1996, Gc
    Petra Dekens, (D), California (Santa Cruz), 2007, Pe
    Newell (Toby) Garfield, (D), Rhode Island, 1990, Op
    Jason Gurdak, (D), Colorado Mines, 2006, HwCqAt
    Leonard Sklar, (D), California (Berkeley), 2003, Gm
Assistant Professor:
    Alexander Stine, (D), California (Berkeley), At
Adjunct Professor:
    E. Jan Null, (D), California (Davis), 1974, Ge
Emeritus:
    Charles E. Bickel, (D), Harvard, 1971, Gx
    York T. Mandra, (D), Stanford, 1958, Pg
    Erwin Seibel, (D), Michigan, 1972, On
    Raymond Sullivan, (D), Glasgow, 1960, Go
    Lisa D. White, (D), California (Santa Cruz), 1989, Pg
Geosciences Tech:
    Russell McArthur, Zn

San Joaquin Delta College
Dept of Geology (A) (2015)
Stockton, CA 95204
    p. (209) 954-5354
    gfrost@deltacollege.edu
    http://www.deltacollege.edu/div/scimath/geology.html
Professor:
    Gina Marie Frost, (D), California (Santa Cruz), Gg

San Jose City College
Dept of Earth and Space Sciences ☒ (2018)
2100 Moorpark Ave
San Jose, CA 95128
    p. (408) 288-3716
    jessica.smay@sjcc.edu
    http://www.sjcc.edu/academics/departments-divisions/physical-sciences
Associate Professor:
    Jessica J. Smay, (M), California (Santa Barbara), 2002, ZeGm

San Jose State University
Dept of Geology (B,M) O☒ (2019)
One Washington Square
San Jose, CA 95192-0102
    p. (408) 924-5050
    leslie.blum@sjsu.edu
    http://www.sjsu.edu/geology/
    Enrollment (2009): B: 26 (7) M: 24 (6)
Chair:
    Jonathan S. Miller, (D), North Carolina, 1994, GiCc
Professor:
    Emmanuel Gabet, (D), California (Santa Barbara), 2002, Gm
    Paula Messina, (D), CUNY, 1998, ZeGm
    Ellen P. Metzger, (D), Syracuse, 1984, GpZe
    Robert B. Miller, (D), Washington, 1980, Gct
    June A. Oberdorfer, (D), Hawaii, 1983, HwGeHy
    Donald L. Reed, (D), California (San Diego), 1985, GuYg
Assistant Professor:
    Kimberly Blisniuk, (D), California (Davis), 2011, Gmt
    Carlie Pietsch, (D), S California, 2015, PiePc
    Ryan Portner, (D), Macquarie, 2010, Gdv
Emeritus Professor:
    David W. Andersen, (D), Utah, 1973, GsCl
    Calvin H. Stevens, (D), S California, 1963, PsGrs
    John W. Williams, (D), Stanford, 1970, Ng

Santa Barbara City College
Dept of Earth & Planetary Sciences (A) (2015)
721 Cliff Drive
Santa Barbara, CA 93109
    p. (805) 965-0581

ebkitao@sbcc.edu
Chair:
   Michael A. Robinson, (D), California (Santa Barbara), 2009, ZyiAm
Professor:
   Jeffrey W. Meyer, (D), California (Santa Barbara), 1992, GgzGx
   Jan Schultz, (M), California (Santa Barbara), 1991, GgePg
Assistant Professor:
   William Dinklage, (D), California (Santa Barbara), Gge
Emeritus:
   Robert S. Gray, (D), Arizona, 1965, PvGzx

Santa Monica College
Earth Science Dept (A) O (2018)
1900 Pico Blvd.
Santa Monica, CA 90405
   p. (310) 434-8652
   drake_vicki@smc.edu
   http://www.smc.edu/AcademicPrograms/EarthScience/Pages/default.aspx
   Programs: Geography, Geospatial Technologies
   Certificates: Geospatial Technologies
Professor:
   Pete Morris, (M), Wisconsin, 1999, Zyc
Associate Professor:
   Jing Liu, (D), Wisconsin, 2017, ZyiZr
Adjunct Professor:
   Alessandro Grippo, (D), Gs

Santa Rosa Junior College
Earth and Space Sciences (A) (2015)
1501 Mendocino Avenue
Santa Rosa, CA 95401-4395
   p. (707) 527-4365
   llupa@santarosa.edu
   http://online.santarosa.edu/presentation/?2992

Santiago Canyon College
Dept of Earth, Space, & Physical Sciences (2015)
8045 East Chapman Avenue
Orange, CA 92869-4512
   p. (714) 564-4788
   hannes_susan@sccollege.edu
Chair:
   Debra A. Brooks, (M), Texas A&M, 1989, Yg
Adjunct Professor:
   Gail F. Montwill, (M), Cal State (Los Angeles), 1988, Gg
   Lizanne V. Simmons, (B), Brigham Young, 1979, Gr
   Amy L. Stinson, (M), San Diego State, 1990, Gc
   Adam D. Woods, (D), S California, 1998, Gs

Solano Community College
School of Mathematics and Science (2015)
4000 Suisun Valley Road
Fairfield, CA 94534
   p. (707) 864-7211
   John.Yu@solano.edu
   http://www.solano.edu/

Sonoma State University
Dept of Geology (B) (2017)
1801 East Cotati Avenue
Rohnert Park, CA 94928
   p. (707) 664-2334
   james@sonoma.edu
   http://www.sonoma.edu/geology/
   Enrollment (2017): B: 80 (30)
Department Chair:
   Matthew J. James, (D), California (Berkeley), 1987, PiRhPg
Professor:
   Matty Mookerjee, (D), Rochester, 2005, GcYgZi
Assistant Professor:
   Owen Anfinson, (D), Calgary, 2012, GsdGg
   Laura E. Waters, (D), Michigan, 2013, GviGz
Emeritus:
   Thomas B. Anderson, (D), Colorado, 1969, GsrGd

   Rolfe C. Erickson, (D), Arizona, 1968, Gx
Retired:
   Daniel B. Karner, (D), California (Berkeley), 1997, Cc
Instructional Support Technician:
   Phillip Mooney, (M), California (Davis), 2010, ZnGc

Stanford University
Dept of Geological and Environmental Sciences (B,M,D) (2019)
Stanford, CA 94305-2115
   p. (650) 723-0847
   jlpayne@stanford.edu
   http://pangea.stanford.edu/GES/
   Enrollment (2010): B: 21 (9) M: 11 (5) D: 45 (8)
Chair:
   Jonathan F. Stebbins, (D), California (Berkeley), 1983, Cg
Research Professor:
   J. Michael Moldowan, (D), Michigan, 1972, Co
Consulting Professor:
   Francois Farges, Yr
Professor:
   C. Page Chamberlain, (D), Harvard, 1985, Yr
   Robert B. Dunbar, (D), California (San Diego), 1981, On
   Marco T. Einaudi, (D), Harvard, 1969, Em
   W. Gary Ernst, (D), Johns Hopkins, 1959, Cp
   Rodney C. Ewing, (D), Stanford, 1974, GzeCl
   Steven M. Gorelick, (D), Stanford, 1981, Hw
   Stephan A. Graham, (D), Stanford, 1976, Go
   James C. Ingle, Jr., (D), S California, 1966, Pm
   Andre G. Journel, (D), Nancy (France), 1977, Gq
   Juhn G. Liou, (D), California (Los Angeles), 1970, Gp
   Keith Loague, (D), British Columbia, 1986, Hq
   Donald R. Lowe, (D), Illinois, 1967, Gd
   Gail A. Mahood, (D), California (Berkeley), 1979, Gi
   Pamela A. Matson, (D), Oregon State, 1983, Sb
   Michael O. McWilliams, (D), Australian Nat, 1978, CcYme
   Elizabeth L. Miller, (D), Rice, 1977, Gc
   David D. Pollard, (D), Stanford, 1969, Gc
Consulting Associate Professor:
   Thomas Holzer, (D), Stanford, 1970, Ng
   Joseph W. Ruetz, (M), Stanford, 1977, Gg
   Joseph Wooden, (D), North Carolina, 1975, Cg
Assistant Professor:
   Scott E. Fendorf, (D), Delaware, 1992, Sc
   Adina Paytan, (D), California (San Diego), 1996, Oc
Courtesy Professor:
   Peter K. Kitanidis, (D), Yr
   Stephen G. Monismith, Yr
Courtesy Associate Professor:
   David L. Freyberg, Yr
   Anders Nilsson, Yr
   Alfred Spormann, Yr
   Debra S. Stakes, (D), Oregon State, 1978, Ob
Courtesy Professor:
   James O. Leckie, (D), Harvard, 1970, Cl
Emeritus:
   Atilla Aydin, (D), Stanford, 1978, Gc
   Dennis K. Bird, (D), California (Berkeley), 1978, Cg
   Gordon E. Brown, Jr., (D), Virginia Tech, 1970, Gz
   Robert G. Coleman, (D), Stanford, 1957, Gi
   Robert R. Compton, (D), Stanford, 1949, Gx
   John W. Harbaugh, (D), Wisconsin, 1955, Gq
   Ronald J. P. Lyon, (D), California (Berkeley), 1954, Zr
   Irwin Remson, (D), Columbia, 1954, Hw
   Tjeerd H. Van Andel, (D), Groningen (Neth), 1950, Gu

Dept of Geophysics (B,M,D) (2018)
397 Panama Mall
Stanford, CA 94305-2215
   p. (650) 497-3498
   coreyann@stanford.edu
   https://earth.stanford.edu/geophysics/
   Programs: Geophysics
   Enrollment (2018): B: 2 (2) M: 7 (5) D: 76 (19)
Chair:
   Howard A. Zebker, (D), Stanford, 1984, Yx

Research Professor:
   William L. Ellsworth, (D), MIT, 1978, Ys
Courtesy Associate Professor:
   Simon L. Klemperer, (D), Cornell, 1985, GtYes
Professor:
   Gregory C. Beroza, (D), MIT, 1989, Ys
   Biondo L. Biondi, (D), Stanford, 1990, Ye
   Jerry M. Harris, (D), Caltech, 1980, Ys
   Rosemary J. Knight, (D), Stanford, 1985, Hy
   Paul Segall, (D), Stanford, 1981, Yg
   Norman H. Sleep, (D), MIT, 1972, Yg
   Mark D. Zoback, (D), Stanford, 1975, Yg
Associate Professor:
   Eric M. Dunham, (D), California (Santa Barbara), 2005, Ys
Assistant Professor:
   Dustin M. Schroeder, (D), Texas, 2014, Gl
   Jenny Suckale, (D), Yg
   Tiziana Vanorio, (D), Yx
Adjunct Professor:
   Steven Gorelick, Yr
Emeritus:
   Jon F. Claerbout, (D), MIT, 1967, Ye
   Antony C. Fraser-Smith, (D), Auckland, 1966, Ym
   Robert L. Kovach, (D), Caltech, 1962, Ys
   Gerald M. Mavko, (D), Stanford, 1977, Yg
   Amos M. Nur, (D), MIT, 1969, Yg
   Joan Roughgarden, (D), Harvard, 1971, Ob
Department Manager:
   Jen Kidwell, Zn

Taft College
Math/Science Div (2015)
29 Emmons Park Drive
Taft, CA 93268
   p. (661) 763-7932
   ggolling@taftcollege.edu

University of California, Berkeley
Dept of Earth & Planetary Science (B,M,D) (2015)
307 McCone
Berkeley, CA 94720-4767
   p. (510) 642-3993
   bbuffett@berkeley.edu
   http://eps.berkeley.edu
Chair:
   Hans-Rudolf Wenk, (D), Zurich, 1965, Gz
Professor:
   Walter Alvarez, (D), Princeton, 1967, GrRh
   Jillian Banfield, (D), Johns Hopkins, 1990, CoGz
   James K. Bishop, (D), MIT/WHOI, 1977, Oc
   George H. Brimhall, (D), California (Berkeley), 1972, Eg
   Roland Burgmann, (D), Stanford, 1993, GtYd
   Donald J. DePaolo, (D), Caltech, 1978, Cc
   William E. Dietrich, (D), Washington, 1982, Gm
   Inez Fung, (D), MIT, 1977, As
   B. Lynn Ingram, (D), Stanford, 1992, Cs
   Raymond Jeanloz, (D), Caltech, 1979, Gy
   James W. Kirchner, (D), California (Berkeley), 1990, Ge
   Michael Manga, (D), Harvard, 1994, GvXyHw
   Mark A. Richards, (D), Caltech, 1986, Yg
   Barbara A. Romanowicz, (D), Paris, 1979, Ys
Associate Professor:
   Kristie Boering, (D), Stanford, 1991, As
   Douglas S. Dreger, (D), Caltech, 1992, Ys
Assistant Professor:
   Richard M. Allen, (D), Princeton, 2001, Ys
   Burkhard Militzer, (D), Illinois (Urbana), 2000, Gy
Adjunct Associate Professor:
   David L. Alumbaugh, (D), California (Berkeley), 1993, Ye
   Paul Renne, (D), California (Berkeley), 1987, Cc
Adjunct Professor:
   Steven Pride, (D), Texas A&M, 1991, Ys
Visiting Professor:
   William D. Collins, (D), Chicago, 1988, As
Emeritus:
   Mark S. Bukowinski, (D), California (Los Angeles), 1975, Gy
   Ian S. Carmichael, (D), London, 1958, Gi
   Garniss H. Curtis, (D), California (Berkeley), 1951, Cc
   Lane R. Johnson, (D), Caltech, 1966, Ys
   Doris Sloan, (D), California (Berkeley), 1981, Pme
   Chi-Yuen Wang, (D), Harvard, 1964, Yg

Dept of Environmental Science, Policy and Management
(B,M,D) O☒ (2018)
140 Mulford Hall
Berkeley, CA 94720-3110
   p. (510) 643-3788
   earthy@berkeley.edu
   http://espm.berkeley.edu/
Enrollment (2018): B: 200 (200) D: 30 (10)

Dept of Integrative Biology (B,M,D) (2017)
3040 Valley Life Sciences Building #3140
Berkeley, CA 94720-3140
   p. (510) 502-5887
   johnh@berkeley.edu
   http://ib.berkeley.edu/
Curator:
   Carole S. Hickman, (D), Stanford, 1975, PgePg
Professor:
   F. Stuart Chapin, (D), Stanford, 1973, Pe
   William A. Clemens, (D), California (Berkeley), 1960, Pv
   Robert J. Full, (D), SUNY (Buffalo), 1984, Pi
   Harry W. Green, (D), Tennessee, 1977, Pv
   Ned K. Johnson, (D), California (Berkeley), 1961, Pv
   Mimi A. R. Koehl, (D), Duke, 1976, Pg
   Brent Mishler, (D), Harvard, 1984, Pe
   Kevin Padian, (D), Yale, 1980, Pv
   Wayne P. Sousa, (D), California (Santa Barbara), 1977, Ob
Associate Professor:
   Mary E. Power, (D), Washington, 1981, Pe
Emeritus:
   Roy L. Caldwell, (D), Iowa, 1969, Pi
   Carole S. Hickman, (D), Stanford, 1975, Pg
   William Z. Lidicker, Jr., (D), Illinois, 1957, Pv
   David R. Lindberg, (D), California (Santa Cruz), Pi
   James L. Patton, (D), Arizona, 1968, Pv
   Thomas M. Powell, (D), California (Berkeley), 1970, Op
   Montgomery Slatkin, (D), Harvard, 1970, Pg
   Glennis Thompson, (D), Melbourne, 1974, Zn
   James W. Valentine, (D), California (Los Angeles), 1958, Pg
   David B. Wake, (D), S California, 1964, Pg
   Marvalee H. Wake, (D), S California, 1968, Pg

Dept of Materials Science and Engineering (B,M,D) (2015)
577 Evans Hall
Berkeley, CA 94720-1760
   p. (510) 642-3801
   hfmorrison@berkeley.edu
   http://www.mse.berkeley.edu/
Chair:
   Robert O. Ritchie, (D), Cambridge, 1973, Zm
Professor:
   Alex Becker, (D), McGill, 1963, Yg
   George A. Cooper, (D), Cambridge, 1967, Np
   Didier deFontaine, (D), Northwestern, 1967, Zm
   Lutgard DeJonghe, (D), California (Berkeley), 1970, Zm
   Thomas M. Devine, (D), MIT, 1974, Zm
   Fiona M. Doyle, (D), Imperial Coll (UK), 1983, Nx
   James W. Evans, (D), SUNY (Buffalo), 1970, Nx
   Douglas W. Fuerstenau, (D), MIT, 1953, Nx
   Andreas Glaeser, (D), MIT, 1981, Zm
   Ronald Gronsky, (D), California (Berkeley), 1977, Zm
   Eugene Haller, (D), Basel, 1970, Zm
   J. W. Morris, Jr., (D), MIT, 1969, Zm
   H. Frank Morrison, (D), California (Berkeley), 1967, Yg
   T. N. Narasimhan, (D), California (Berkeley), 1975, Hw
   Timothy Sands, (D), California (Berkeley), 1984, Zm
   Kalanadh V. S. Sastry, (D), California (Berkeley), 1970, Nx
   Eicke Weber, (D), Cologne, 1976, Zm
Associate Professor:
   Tad W. Patzek, (D), Silesian Tech, 1979, Np
   James W. Rector, III, (D), Stanford, 1990, Ys

Assistant Professor:
  Daryl Chrzan, (D), California (Berkeley), 1989, Zm
  Mauro Ferrari, (D), California (Berkeley), 1989, Zm

GeoEngineering Program (B,M,D) (2015)
Berkeley, CA 94720
  p. (510) 642-3157
  pestana@ce.berkeley.edu
  http://www.ce.berkeley.edu/geo/
Chair:
  Lisa Alvarez-Cohen
Professor:
  Alex Becker, (D), McGill, 1964, Ye
  Huntly Frank Morrison, (D), California (Berkeley), 1967, Ye
Senior Scientist:
  Ki Ha Lee, (D), California (Berkeley), 1978, Ye
Assistant Professor:
  James W. Rector, (D), Stanford, 1990, Ye

Museum of Paleontology (D) (2016)
1101 Valley Life Sciences Bldg #4780
Berkeley, CA 94720-4780
  p. (510) 642-1821
  crmarshall@berkeley.edu
  http://www.ucmp.berkeley.edu
  Enrollment (2016): D: 25 (3)
Professor:
  Charles R. Marshall, (D), Chicago, 1989, PqgPi
Curator:
  Kevin Padian, (D), Yale, 1980, Pv
Principal Museum Scientist:
  Mark B. Goodwin, (D), California (Davis), 2008, Pv
Museum Scientist:
  Diane Erwin, (D), Alberta, 1990, Pb
  Pat Holroyd, (D), Duke, 1994, Pv
Associate Professor:
  Cynthia Looy, (D), Utrecht, 1995, Pb
Assistant Professor:
  Seth Finnegan, (D), California (Riverside), 1994, Pig
Curatorial Associate:
  Walter Alvarez, (D), Princeton, 1967, Gr
Curator:
  Anthony D. Barnosky, (D), Washington, 1983, Pv
  William A. Clemens, (D), California (Berkeley), 1960, Pv
  Leslea Hlusko, (D), Penn State, 2000, Znn David R. Lindberg, (D), California (Santa Cruz), 1983, Pg
Emeritus:
  Roy Caldwell, (D), Iowa, 1969, Pg
  James W. Valentine, (D), California (Los Angeles), 1958, Pg
Assistant Director:
  Lisa D. White, California (Santa Cruz), 1989, ZePim

University of California, Davis
Dept of Earth and Planetary Sciences (B,M,D) ●◌ (2018)
2119 Earth & Physical Sciences Building
One Shields Ave.
Davis, CA 95616-5270
  p. (530) 752-0350
  geology@ucdavis.edu
  http://geology.ucdavis.edu
  f: https://www.facebook.com/pages/Geology-Department-at-UC-Davis/87869598212
  Programs: Geology (B,M,D); Marine and Coastal Science (B); Natural Sciences (B)
  Enrollment (2018): B: 84 (25) M: 15 (3) D: 34 (4)
Chair:
  Michael Oskin, (D), Caltech, 2002, GmtGc
Professor:
  Sandra J. Carlson, (D), Michigan, 1986, Pg
  Kari M. Cooper, (D), California (Los Angeles), 2001, Gi
  Eric S. Cowgill, (D), California (Los Angeles), 2001, Gtc
  Tessa M. Hill, (D), California (Santa Barbara), 2004, Pe
  Louise H. Kellogg, (D), Cornell, 1988, Yg
  Charles E. Lesher, (D), Harvard, 1985, Gi
  James S. McClain, (D), Washington, 1979, Yr
  Isabel P. Montañez, (D), Virginia Tech, 1990, Gd
  Ryosuke Motani, (D), Toronto, 1997, Pv

  Sujoy Mukhopadhyay, (D), Caltech, 2002, Cg
  John Rundle, (D), California (Los Angeles), 1976, Yd
  Sarah T. Stewart, (D), Caltech, 2002
  Dawn Y. Sumner, (D), MIT, 1995, Gs
  Geerat J. Vermeij, (D), Yale, 1971, Pe
  Kenneth L. Verosub, (D), Stanford, 1973, Ym
  Qing-zhu Yin, (D), Max Planck, 1995, Cc
Assistant Professor:
  David Gold, (D), California (Los Angeles), 2014
  Maxwell L. Rudolph, (D), California (Berkeley), 2012, Yg
Assistant Project Scientist:
  Dylan Spaulding, (D), California (Berkeley), 2010
Research Associate:
  Oliver Kreylos, (D), California (Davis), 2003, Ng
  Ann D. Russell, (D), Washington, 1994, Cm
  Peter Thy, (D), Aarhus (Denmark), 1982, Gx
  Burak Yikilmaz, California (Davis), 2010, Gc
Adjunct Professor:
  Gordon M. Moore, (D), California (Berkeley), 1995, Gx
Emeritus:
  Cathy J. Busby, (D), Princeton, 1983, Gt
  Richard Cowen, (D), Cambridge, 1966, Pg
  John Dewey, (D), London (UK), 1960, Gt
  Charles G. Higgins, (D), California (Berkeley), 1950, Gm
  Jeffrey F. Mount, (D), California (Santa Cruz), 1980, Gm
  Sarah M. Roeske, (D), California (Santa Cruz), 1988, GtcGp
  James R. Rustad, (D), Minnesota, 1992, Cl
  Peter Schiffman, (D), Stanford, 1978, Gp
  Robert J. Twiss, (D), Princeton, 1970, Gc
  Robert A. Zierenberg, (D), Wisconsin, 1983, Cs
Cooperating Faculty:
  William H. Casey, (D), Penn State, 1985, Cl
  Graham E. Fogg, (D), Texas, 1986, Hw
  Alexandra Navrotsky, (D), Chicago, 1967, Zm

Dept of Land, Air & Water Resources (B,M,D) ○☒ (2017)
1110 Plant and Environmental Sciences Building
One Shields Avenue
Davis, CA 95616
  p. (530) 752-1130
  rjsouthard@ucdavis.edu
  http://lawr.ucdavis.edu/
  Programs: Atmospheric Science; Hydrology/Hydrologic Science; Environmental Science and Management; Soils and Biogeochemistry
  Enrollment (2014): B: 5 (5) M: 54 (19) D: 55 (8)
Specialist in Cooperative Extension:
  Thomas Harter, (D), Arizona, 1994, Hg
  Toby O'Geen, (D), Idaho, 2002, So
Professor:
  Cort Anastasio, (D), Duke, 1994, As
  Shu-Hua Chen, (D), Purdue, 1999, ZrAsm
  Randy A. Dahlgren, (D), Washington, 1987, So
  Graham E. Fogg, (D), Texas, 1986, Hw
  Richard Grotjahn, (D), Florida State, 1979, As
  Peter J. Hernes, (D), Washington, 1999, Hg
  Jan W. Hopmans, (D), Auburn, So
  William R. Horwath, (D), Michigan State, 1993, Sob
  Louise E. Jackson, (D), Washington, 1982, So
  Terrence R. Nathan, (D), SUNY (Albany), 1985, Asm
  Kate Scow, (D), Cornell, 1988, So
  Randal J. Southard, (D), North Carolina State, 1983, So
  Susan L. Ustin, (D), California (Davis), 1983, Zr
Associate Professor:
  Ian C. Faloona, (D), Penn State, 2001, AscAp
  Ben Houlton, (D), Princeton, So
  Sanjai Parikh, (D), Arizona, So
20% Cooperative Extension:
  Samuel Sandoval Solis, (D), Texas, 2011, Hg
Assistant Professor:
  Yufang Jin, (D), Boston, 2002, Zr
Adjunct Professor:
  Travis A. O'Brien, (D), California (Santa Cruz), 2011, AstAp
  Minghua Zhang, (D), California (Davis), 1993, Hg
Emeritus:
  James H. Richards, (D), Alberta, 1981, So

Asst Proj Scientist:
  Michael L. Grieneisen, (D), North Carolina (Chapel Hill), 1992, Hg
Asst Professional Researcher:
  Lucas CR Silva, (D), Guelph, 2011, Sb
Cooperating Faculty:
  Daniel Geisseler, (D), California (Davis), 2009, So
  Stephen R. Grattan, (D), California (Riverside), 1984, Sp
  Richard L. Snyder, (D), Iowa State, 1980, As
  Daniele Zaccaria, (D), Utah State, 2011, Hg

Dept of Land, Air, & Water Resources - Hydrology Program
(B,M,D) ☒ (2019)
1110A Plant Environmental Sciences
One Shields Avenue
Davis, CA 95616-8628
  p. (530) 752-3060
  radahlgren@ucdavis.edu
  http://lawr.ucdavis.edu/
  Enrollment (2012): B: 25 (0) M: 16 (0) D: 21 (0)
Chair:
  Randy Dahlgran, (D), Washington, 1987, Sc
  Jan W. Hopmans, (D), Auburn, 1985, Sp
Professor:
  William H. Casey, (D), Penn State, 1985, Sc
  Graham E. Fogg, (D), Texas, 1986, Hwg
  Mark E. Grismer, (D), Colorado State, 1984, Hw
  Miguel A. Marino, (D), California (Los Angeles), 1972, Hw
  Gregory B. Pasternack, (D), Johns Hopkins, 1998, GmHg
  Carlos E. Puente, (D), MIT, 1984, Hq
  Susan L. Ustin, (D), California (Davis), 1983, Zr
  Wesley W. Wallender, (D), Utah State, 1982, Hg
Water Management Specialist:
  Terry L. Prichard, (D), California (Davis), 1975, Zn
Plant-Water Relations Specialist:
  Stephen R. Grattan, (D), California (Riverside), 1984, Zn
Irrigation Specialist:
  Larry J. Schwankl, (D), California (Davis), 1991, Zn
Irrigation & Soil Specialist:
  David A. Goldhamer, (D), California (Davis), 1980, Zn
Groundwater Hydrology Specialist:
  Thomas L. Harter, (D), Arizona, 1993, Hw
Associate Professor:
  Helen E. Dahlke, (D), Cornell, 2011, Hgs
  Peter Hernes, (D), Washington, Og
Assistant Professor:
  Samuel Sandoval, (D), Texas, 2011, Hg
Irrigation & Drainage Specialist:
  Blaine R. Hanson, (D), Colorado State, 1977, Hg
Emeritus:
  Robert M. Hagan, California (Davis), 1948, Hg
  Theodore C. Hsiao, (D), Illinois (Urbana), 1963, SpAmZn
  Allen W. Knight, (D), Utah, 1965, Hs
  Donald R. Nielsen, (D), Iowa State, 1958, Sp
  Verne H. Scott, Colorado State, 1959, Hs

Graduate Group in Hydrologic Sciences (M,D) (2015)
1152 PES
One Shields Ave
Davis, CA 95616
  p. (530) 752-1669
  lawrgradadvising@ucdavis.edu
  http://hydscigrad.ucdavis.edu/
  Administrative Assistant: Diane Swindall
  Enrollment (2010): M: 14 (1) D: 14 (0)
Chair:
  Graham E. Fogg, (D), Texas, 1986, Hy
Professor:
  William H. Casey, (D), Pennsylvania, 1985, Cl
  Jeannie Darby, (D), Texas, 1988, Hg
  Mark E. Grismer, (D), Colorado State, 1984, Hw
  Jan W. Hopmans, (D), Auburn, 1985, So
  Theodore C. Hsiao, (D), Illinois, 1964, Zn
  B. E. Larock, (D), Stanford, 1966, Zn
  Jay R. Lund, (D), Washington, 1986, Hs
  M. A. Marino, (D), California (Los Angeles), 1972, Hw
  James S. McClain, (D), Washington, 1979, Yr
  Eldridge M. Moores, (D), Princeton, 1963, Gt

  Jeffrey F. Mount, (D), California (Santa Cruz), 1980, Gs
  Dennis E. Rolston, (D), California (Davis), 1971, So
  K. K. Tanji, (M), California (Davis), 1961, Zn
  Wes W. Wallender, (D), Utah State, 1982, Hg
  B. C. Weare, (D), SUNY (Buffalo), 1974, As
Associate Professor:
  Bruce Kutter, (D), Cambridge, 1983, Ne
  Carlos E. Puente, (D), MIT, 1984, Hg
Emeritus:
  Robert A. Matthews, (B), California (Berkeley), 1953, Ge

University of California, Irvine
Dept of Earth System Science (B,D) (2015)
School of Physical Sciences
Irvine, CA 92697-3100
  p. (949) 824-8794
  essinfo@ess.uci.edu
  http://www.ess.uci.edu
  Enrollment (2013): B: 166 (64) D: 49 (5)
Chair:
  Michael L. Goulden, (D), Stanford, 1992, Zg
Advance Professor:
  Ellen R. M. Druffel, (D), California (San Diego), 1980, Oc
Professor:
  James Famiglietti, (D), Princeton, Hw
  Gudrun Magnusdottir, (D), Colorado State, 1989, As
  Michael J. Prather, (D), Yale, 1975, As
  Eric Rignot, (D), S California, 1991, GlOpZr
  Eric S. Saltzman, (D), Rosenstiel, 1986, As
  Soroosh Sorooshian, (D), California (Los Angeles), 1978, Hg
  Susan E. Trumbore, (D), Columbia, 1989, Sc
  Jin-Yi Yu, (D), Washington, AsOpZc
  Charles Zender, (D), Colorado, 1996, As
Associate Professor:
  Steven J. Davis, (D), Stanford, 2008, GeCsZu
  Jefferson Keith Moore, (D), Oregon State, 1999, Op
  Francois Primeau, (D), MIT/WHOI, 1998, Op
Assistant Professor:
  Claudia Czimczik, (D)
  Todd Dupont, (D), Penn State, 2004, Gl
  Julie Ferguson, (D), Oxford, 2008, Ze
  Kathleen Johnson, (D), California (Berkeley), 2004, Zg
  Saewung Kim, (D), Georgia Tech, 2007, As
  Adam Martiny, (D), Tech (Denmark), 2003, Ob
  Michael Pritchard, (D)
  Isabella Velicogna, (D), Trieste, 1999, Zr
Laboratory Director:
  John Southon, (D), Auckland, 1976, CcOcPe

University of California, Los Angeles
Dept of Atmospheric and Oceanic Sciences (B,M,D) (2016)
7127 Math Sciences
Box 951565
Los Angeles, CA 90095
  p. (310) 825-1217
  deptinfo@atmos.ucla.edu
  http://www.atmos.ucla.edu
  f: https://www.facebook.com/AOS.UCLA
Chair:
  Jochen P. Stutz, (D), Heidelberg, 1996, As
Professor:
  Rong Fu
  Kuo Nan Liou, (D), New York, 1970, As
  Lawrence Lyons, (D), California (Los Angeles), 1972, As
  James McWilliams, (D), Harvard, 1971, OgAs
  Carlos R. Mechoso, (D), Princeton, 1978, As
  J. David Neelin, (D), Princeton, 1987, As
  Suzanne Paulson, (D), Caltech, 1991, As
  Richard M. Thorne, (D), MIT, 1968, As
  Yongkang Xue, (D)
Associate Professor:
  Jacob Bortnik, (D)
  Marcelo Chamecki
  Gang Chen
  Alex Hall, (D)
  Qinbin Li

Tina Treude
Assistant Professor:
   Daniele Bianchi , (D)
   Rob Eagle
   Jasper Kok , (D)
   Ulli Seibt , (D)
   Andrew Stewart, (D)
   Arahdna Tripati, (D)
Lecturer:
   Jeffrey Lew, (D), California (Los Angeles), 1985, As
Adjunct Professor:
   Yi Chao
Emeritus:
   Akio Arakawa, (D), Tokyo, 1961, As
   Robert Fovell, (D), Illinois, 1988, As
   Michael Ghil, (D), Courant Inst, 1975, As
   Richard Turco, (D), Illinois, 1971, As

Dept. of Earth, Planetary, and Space Sciences (B,M,D) O⊠ (2018)
595 Charles E Young Drive East
3806 Geology Building, Box 951567
Los Angeles, CA 90095-1567
   p. 310.825.3880
   info@epss.ucla.edu
   http://epss.ucla.edu
   f: https://www.facebook.com/uclaepss
   t: @uclaepss
   Programs: Geology (BS); Engineering Geology (BS); Geophysics (BS); Earth & Environmental Science (BA)
   Enrollment (2017): B: 108 (26) M: 3 (7) D: 68 (7)
Chair:
   Jean-Luc Margot, (D), Cornell, 1999, XyYdv
Professor:
   Vassilis Angelopoulos, (D), California (Los Angeles), 1993, XyAsYm
   Jonathan M. Aurnou, (D), Johns Hopkins, 1999, YgmYx
   Paul M. Davis, (D), Queensland, 1974, YsmYg
   T. Mark Harrison, (D), Australian Nat, 1981, CcGt
   David Jewitt, (D), Caltech, 1983, XyZnn
   Abby Kavner, (D), California (Berkeley), 1997, GyYg
   Carolina Lithgow-Bertelloni, (D)
   Craig E. Manning, (D), Stanford, 1989, GxCp
   Kevin D. McKeegan, (D), Washington Univ (St. Louis), 1987, XcCsc
   William I. Newman, (D), Cornell, 1979, Xy
   David A. Paige, (D), Caltech, 1985, Xy
   Gilles Peltzer, (D), Paris VII, 1987, ZrGt
   Christopher T. Russell, (D), California (Los Angeles), 1968, Xy
   Edwin A. Schauble, (D), Caltech, 2002, Cgs
   J. William Schopf, (D), Harvard, 1968, Pg
   Laurence C. Smith, (D), Cornell, 1996, Hs
   Lars Stixrude, (D)
   Tina Treude, (D), Max Planck, 2004, ObPgCm
   Marco Velli, (D), Pisa, 1985, Yg
   An Yin, (D), S California, 1988, Gt
   Edward D. Young, (D), S California, 1990, CsXc
Associate Professor:
   Caroline Beghein, (D), Utrecht, 2003, Ysg
   Jonathan Mitchell, (D), Chicago, 2007, AsXg
   Hilke Schlichting, (D), Caltech, 2009, Xy
   Ulrike Seibt, (D), Hamburg, 2003, PgCg
   Aradhna Tripati, (D), California (Santa Cruz), 2002, CmPeCs
Assistant Professor:
   Mackenzie Day, (D)
   Lingsen Meng, (D), Caltech, 2012, Ys
   Seulgi Moon, (D), Stanford, 2013, Gm
Adjunct Professor:
   Robert C. Newton, (D), California (Los Angeles), 1963, CpGp
   Edward J. Rhodes, (D), Oxford, 1990, CcGma
Emeritus:
   G. Peter Bird, (D), MIT, 1976, GtYsRn
   Paul J. Coleman, (D), California (Los Angeles), 1966, Xy
   Wayne A. Dollase, (D), MIT, 1966, Gz
   Clarence A. Hall, (D), Stanford, 1956, GgtGe
   Raymond V. Ingersoll, (D), Stanford, 1976, Gst
   David D. Jackson, (D), MIT, 1969, YsdYg
   Isaac R. Kaplan, (D), S California, 1962, Csg
   Margaret G. Kivelson, (D), Radcliffe, 1957, Xy
   Robert L. McPherron, (D), California (Berkeley), 1968, Xy
   Paul M. Merifield, (D), Colorado, 1963, Ng
   John L. Rosenfeld, (D), Harvard, 1954, Gp
   Bruce Runnegar, (D), Queensland, 1967, Pg
   Gerald Schubert, (D), California (Berkeley), 1964, YgXy
   Raymond J. Walker, (D), California (Los Angeles), 1973, Xy
   John T. Wasson, (D), MIT, 1958, Xmc

Inst of Geophysics & Planetary Physics (2015)
3845 Slichter Hall
Los Angeles, CA 90095-1567
   p. (310) 825-1664
   jnakatsu@igpp.ucla.edu
   http://www.igpp.ucla.edu
Professor:
   Vassilis Angelopoulos, (D), California (Los Angeles), 1993, Yg
   Maha Ashour-Abdalla, (D), Imperial Coll (UK), 1971, Zn
   Friedrich H. Busse, (D), Munich, 1967, Zn
   Richard E. Dickerson, (D), Minnesota, 1957, Zn
   Michael Ghil, (D), New York, 1975, AsYdOp
   T. Mark Harrison, (D), Australian Nat, 1980, Cc
   Charles F. Kennel, (D), Princeton, 1964, Zn
   Margaret G. Kivelson, (D), Radcliffe, 1957, Xy
   Robert L. McPherron, (D), California (Berkeley), 1968, Xy
   James C. McWilliams, (D), Harvard, 1971, As
   Paul H. Roberts, (D), Cambridge, 1967, Ym
   Bruce Runnegar, (D), Queensland, 1967, Pg
   Christopher T. Russell, (D), California (Los Angeles), 1968, Yg
   J. William Schopf, (D), Harvard, 1968, Pg
   Gerald Schubert, (D), California (Berkeley), 1964, Yg
   Karl O. Stetter, (D), Tech (Munich), 1973, Pg
   Richard Turco, (D), Illinois, 1971, As
   Raymond J. Walker, (D), California (Los Angeles), 1973, Zn
   John T. Wasson, (D), MIT, 1958, Xm
   Edward Young
Senior Scientist:
   Stanislav I. Braginsky, (D), Moscow Inst, 1948, Ym
   Robert J. Strangeway, (D), London, 1978, Zn
   M I. Venkatesan, (D), Madras, 1973, Co
   Paul Warren, (D), California (Los Angeles), 1979, Xm
Associate Professor:
   Abby Kavner, (D), California (Berkeley), 1997, Yg
   J. David Neelin, (D), Princeton, 1987, As
Associate Research Scientist:
   Jean Berchem, (D), California (Los Angeles), 1986, Zn
   Gregory Kallemeyn, (D), California (Los Angeles), 1982, Cg
   Frank T. Kyte, (D), California (Los Angeles), 1983, Ct
   Alan E. Rubin, (D), New Mexico, 1982, Xm
   David Schriver, (D), California (Los Angeles), 1988, Yg
   Fred Schwab, (B), California (Los Angeles), 1960, Ys
Assistant Research Scientist:
   Kayo Ide, (D), Caltech, 1990, Zg
   Krishan Khurana, (D), Durham, 1984, Zn
   Robert Richard, (D), California (Los Angeles), 1988, Yg
   William Smythe, (D), California (Los Angeles), 1979, Ys
   Ferenc D. Varadi, (D), California (Los Angeles), 1989, Zn
Emeritus:
   Orson L. Anderson, (D), Utah, 1951, Yx
   Paul J. Coleman, Jr., (D), California (Los Angeles), 1966, Gc
   Isaac R. Kaplan, (D), S California, 1962, Cg
   Leon Knopoff, (D), Caltech, 1949, Ys
   Ronald L. Shreve, (D), Caltech, 1958, Gm
Department Manager:
   James Nakatsuka, Zn

University of California, Riverside
Department of Earth Sciences (B,M,D) ●⁻⌂ (2019)
Department of Earth Sciences
University of California, Riverside
Riverside, CA 92521
   p. (951) 827-3434
   david.oglesby@ucr.edu
   http://earthsciences.ucr.edu
   f: https://www.facebook.com/ucrearthsciences/
   Programs: Geology
   Earth Sciences
   Geophysics

Enrollment (2013): B: 51 (13) M: 15 (8) D: 38 (4)
Chair:
  David D. Oglesby, (D), California (Santa Barbara), 1999, Ys
Distinguished Professor:
  Timothy W. Lyons, (D), Yale, 1992, CgOcCl
Professor:
  Mary L. Droser, (D), S California, 1987, Pg
  Nigel C. Hughes, (D), Bristol (UK), 1990, Pg
  Gordon D. Love, (D), Strathclyde, 1995, Cog
  Richard A. Minnich, (D), California (Los Angeles), 1978, Zy
  Andy Ridgwell, (D), East Anglia, 2001, Zn
  Peter M. Sadler, (D), Bristol (UK), 1973, GrPs
Associate Professor:
  Robert J. Allen, (D), Yale, 2009, As
  Andrey Bekker, (D), Virginia Tech, 2001, GsrCg
  Gareth J. Funning, (D), Oxford, 2005, YdgYs
  Abhijit Ghosh, (D), Washington, 2011, YsGt
  Michael A. McKibben, (D), Penn State, 1984, Cg
Assistant Professor:
  Nicolas Barth, (D), Otago, 2013, Gt
  Maryjo Brounce, (D), Rhode Island, 2014, Cg
  Roby Douilly, (D), Purdue, 2016, YsGtYg
  Heather Ford, (D), Brown, 2013, Ys
  Sandra Kirtland Turner, (D), Scripps, Cs
Adjunct Assistant Professor:
  Katherine J. Kendrick, (D), California (Riverside), 1999, Gm
  Thomas A. Scott, (D), California (Berkeley), 1987, Zy
Adjunct Professor:
  Elizabeth Cochran, California (Los Angeles), 2005, Ys
  Larissa F. Dobrzhinetskaya, (D), Inst of Physics of Earth (Moscow), 1978, Gx
  Douglas M. Morton, (D), California (Los Angeles), 1966, Gp
Professor of the Graduate Division:
  James H. Dieterich, (D), Yale, 1968, Yg
Emeritus:
  Wilfred A. Elders, (D), Durham, 1961, Gi
  Michael A. Murphy, (D), California (Los Angeles), 1954, Ps
  Stephen K. Park, (D), MIT, 1984, Ym
  Michael O. Woodburne, (D), California (Berkeley), 1966, Pv

University of California, San Diego
Scripps Institution of Oceanography (B,M,D) (2018)
Graduate Office
9500 Gilman Dr.
Mail Code 0208
La Jolla, CA 92093-0208
  p. (858) 534-3206
  cconstable@ucsd.edu
  https://scripps.ucsd.edu/education/
  Enrollment (2012): M: 7 (0) D: 62 (13)
Vice Chancellor:
  Margaret Leinen, (D), Rhode Island, 1980, OgPe
Director:
  T. Guy Masters, (D), Cambridge (UK), 1979, YsGy
Vice-Chair:
  Catherine G. Constable, (D), California (San Diego), 1987, Ymg
Director of MPL:
  William A. Kuperman, (D), Maryland, 1972, Op
Professor:
  Duncan C. Agnew, (D), California (San Diego), 1979, Ys
  Lihini I. Aluwihare, (D), MIT/WHOI, 1999, Gu
  Laurence Armi, (D), California (Berkeley), 1975, Op
  Farooq Azam, (D), Czech Acad of Sci, 1968, Ob
  Katherine A. Barbeau, (D), MIT/WHOI, 1998, Oc
  Douglas H. Bartlett, (D), Illinois, 1985, Ob
  Kevin M. Brown, (D), Durham (UK), 1987, Gu
  Michael J. Buckingham, (D), Reading, 1971, Op
  Ronald S. Burton, (D), Stanford, 1981, Ob
  Paterno R. Castillo, (D), Washington Univ (St. Louis), 1987, Giu
  Paola Cessi, (D), MIT, 1987, Op
  Christopher D. Charles, (D), Columbia, 1991, Pe
  Steven C. Constable, (D), Australian Nat, 1983, Yr
  Andrew G. Dickson, (D), Liverpool (UK), 1978, Cm
  Leroy M. Dorman, (D), Wisconsin, 1970, Yg
  Neal W. Driscoll, (D), Columbia, 1992, Gu
  William H. Fenical, (D), California (Riverside), 1968, Oc
  Yuri A. Fialko, (D), Princeton, 1998, YgdGt
  Peter J. S. Franks, (D), MIT/WHOI, 1990, Ob
  Helen A. Fricker, (D), Tasmania, 1999, Yr
  Terry Gaasterland, (D), Maryland, 1992, Ob
  Jeffrey S. Gee, (D), California (San Diego), 1991, Gu
  William H. Gerwick, (D), California (San Diego), 1981, Cm
  Carl H. Gibson, (D), Stanford, 1962, No
  Sarah T. Gille, (D), MIT/WHOI, 1995, Op
  Vicki Grassian, (D), California (Berkeley), 1987, Cm
  Philip A. Hastings, (D), Arizona, 1987, Ob
  Myrl C. Hendershott, (D), Harvard, 1966, Op
  John A. Hildebrand, (D), Stanford, 1983, Yr
  David R. Hilton, (D), Cambridge, 1985, Cs
  William S. Hodgkiss, Jr., (D), Duke, 1975, No
  Nicholas D. Holland, (D), Stanford, 1965, Ob
  Miriam Kastner, (D), Harvard, 1970, CmlCm
  Ralph F. Keeling, (D), Harvard, 1988, As
  Michael R. Landry, (D), Washington, 1976, Ob
  Gabi Laske, (D), Karlsruhe, 1993, Ys
  James J. Leichter, (D), Stanford, 1997, Ob
  Lisa A. Levin, (D), California (San Diego), 1982, Ob
  Peter F. Lonsdale, (D), California (San Diego), 1974, Gu
  W. Kendall Melville, (D), Southampton (UK), 1974, Op
  J. Bernard H. Minster, (D), Caltech, 1974, Ys
  Mario J. Molina, (D), California (Berkeley), 1972, As
  Bradley S. Moore, (D), Zurich, 1995, Cm
  Joel R. Norris, (D), Washington, 1997, As
  Richard D. Norris, (D), Harvard, 1990, Gu
  Mark D. Ohman, (D), Washington, 1983, Ob
  John A. Orcutt, (D), California (San Diego), 1976, YrsYd
  Brian Palenik, (D), MIT/WHOI, 1989, Ob
  Kimberly A. Prather, (D), California (Davis), 1990, AsCmHg
  V. Ramanathan, (D), SUNY (Stony Brook), 1973, As
  Dean H. Roemmich, (D), MIT/WHOI, 1980, Op
  Gregory W. Rouse, (D), Sydney, 1991, Ob
  Daniel L. Rudnick, (D), California (San Diego), 1987, Op
  Lynn M. Russell, (D), Caltech, 1995, As
  David T. Sandwell, (D), California (Los Angeles), 1981, Yr
  John G. Sclater, (D), Cambridge, 1966, Yh
  Uwe Send, (D), California (San Diego), 1988, Op
  Jeffrey P. Severinghaus, (D), Columbia, 1995, Pe
  Peter M. Shearer, (D), California (San Diego), 1986, Ysg
  Len Srnka, (D), Newcastle upon Tyne, 1974, Ym
  Dariusz Stramski, (D), Gdansk (Poland), 1985, Op
  George Sugihara, (D), Princeton, 1983, Ob
  Lynne D. Talley, (D), MIT/WHOI, 1982, OpZc
  Lisa Tauxe, (D), Columbia, 1983, Ym
  Bradley T. Werner, (D), Caltech, 1987, Gm
  William R. Young, (D), MIT/WHOI, 1981, Op
Associate Professor:
  Eric E. Allen, (D), California (San Diego), 2002, Ob
  Andreas Andersson, (D), Hawaii, 2006, Oc
  James Day, (D), Durham (UK), Cg
  Kerry Key, (D), Scripps, 2003, Yx
  Todd Martz, (D), Montana, 2005, Gg
  Jennifer E. Smith, (D), Hawaii, 2003, Obn
Assistant Professor:
  Adrian Borsa, (D), Scripps, 2005, Hg
  Chambers Hughes, (D), California (Berkeley), 2004, Oc
  Anne Pommier, (D), Orleans (France), 2009, Cg
  Dave Stegman, (D), California (Berkeley), Yd
  Jennifer K. Vanos, (D), Guelph (Canada), 2012, As
Adjunct Professor:
  Jay P. Barlow, (D), California (San Diego), 1982, Ob
Emeritus:
  Gustaf Arrhenius, (D), Stockholm, 1952, ClGzCm
  George E. Backus, (D), Chicago, 1956, YmsOp
  Jeffrey L. Bada, (D), California (San Diego), 1968, Co
  Wolfgang H. Berger, (D), California (San Diego), 1968, Gs
  Steven C. Cande, (D), Columbia, 1976, Yr
  David M. Checkley, (D), California (San Diego), 1978, Ob
  Paul Crutzen, (D), Oc
  Joseph R. Curray, (D), Scripps, 1959, Gu
  Paul K. Dayton, (D), Washington, 1970, Ob
  Horst Felbeck, (D), Muenster, 1979, Ob
  Joris M. Gieskes, (D), Manitoba, 1965, Oc
  Robert T. Guza, (D), California (San Diego), 1974, On
  James W. Hawkins, (D), Washington, 1963, OuGip

Anthony DJ Haymet, (D), Chicago, 1981, OcZgn
Robert R. Hessler, (D), Chicago, 1960, Ob
Glenn R. Ierley, (D), MIT, 1982, Op
Jeremy B. C. Jackson, (D), Yale, 1971, PgePi
Charles F. Kennel, (D), Princeton, 1964, As
Gerald L. Kooyman, (D), Arizona, 1966, Ob
J. Douglas Macdougall, (D), California (San Diego), 1972
John A. McGowan, (D), California (San Diego), 1960, Ob
Walter H. Munk, (D), Scripps, 1947, Op
William A. Newman, (D), California (Berkeley), 1962, Ob
Robert L. Parker, (D), Cambridge, 1966, Ym
Robert Pinkel, (D), California (San Diego), 1974, Op
Richard L. Salmon, (D), California (San Diego), 1976, Op
Richard C. J. Somerville, (D), New York, 1966, As
Victor D. Vacquier, (D), California (Berkeley), 1968, Ob
Ray F. Weiss, (D), California (San Diego), 1970, Cm
Clinton D. Winant, (D), S California, 1972, On

University of California, Santa Barbara
Dept of Earth Science (B,M,D) ☒ (2019)
Room 1006
Webb Hall
Santa Barbara, CA 93106-9630
p. (805) 893-4688
wyss@geol.ucsb.edu
http://www.geol.ucsb.edu
Enrollment (2012): B: 81 (22) M: 12 (2) D: 36 (3)
Professor:
Stanley M. Awramik, (D), Harvard, 1973, Pg
Jordan F. Clark, (D), Columbia, 1995, ClHg
John Cottle, (D), Oxford, 2007, Gi
Phillip B. Gans, (D), Stanford, 1987, Gt
Bradley R. Hacker, (D), California (Los Angeles), 1998, Gp
Matt Jackson, (D), Cu
Chen Ji, (D), Caltech, 2002, Ys
Edward A. Keller, (D), Purdue, 1973, Gm
David W. Lea, (D), MIT, 1990, PeCm
Lorraine Lisiecki, (D), Brown, 2005, Gn
Francis Macdonald, (D), Harvard, Gg
Susannah Porter, (D), Harvard, 2002, Gn
Roberta Rudnick, (D)
Toshiro Tanimoto, (D), California (Berkeley), 1982, Ys
Bruce H. Tiffney, (D), Harvard, 1977, Pb
David Valentine, (D), Cm
Andre R. Wyss, (D), Columbia, 1989, Pv
Associate Professor:
Robin Matoza, (D), Gv
Alexander Simms, (D), Rice, 2006, GsuGm
Syee Weldeab, (D), Tubingen, 2002, PeCm
Assistant Professor:
Zach Eilon, (D), Colombia, Ys
Kristen Morell, (D)
Morgan Raven, (D), Caltech, Cl
Lecturer:
Matt Rioux, (P), California (Santa Barbara), Cc
Emeritus:
Ralph J. Archuleta, (D), California (San Diego), 1976, Ys
Doug Burbank, (D), Dartmouth, 1982, GmtGs
Frank J. Spera, (D), California (Berkeley), 1977, Gi

Earth Research Institute (2015)
Mail Code 1100
Santa Barbara, CA 93106-1100
p. (805) 893-8231
davey@eri.ucsb.edu
http://www.crustal.ucsb.edu
Administrative Assistant: Kathy J. Scheidemen
Director:
Douglas W. Burbank, (D), Dartmouth, 1982, Gm
Associate Director:
Bradley R. Hacker, (D), California (Los Angeles), 1988, Gx
Professor:
Ralph J. Archuleta, (D), California (San Diego), 1976, Ys
Cathy Busby, (D), Princeton, 1983, Gs
Edward A. Keller, (D), Purdue, 1973, Gt
Bruce P. Luyendyk, (D), California (San Diego), 1969, Yr
Frank J. Spera, (D), California (Berkeley), 1977, Cp
Toshiro Tanimoto, (D), California (Berkeley), 1982, Ys
Associate Professor:
Jordan Clark, (D), Columbia, 1995, Cg
Phillip B. Gans, (D), Stanford, 1987, Gt
Assistant Research Scientist:
Christopher C. Sorlien, (D), California (Santa Barbara), 1994, Yr
Assistant Professor:
Bodo Bookhagen, (D), Potsdam, 2005, Gm
Research Seismologist:
Jamison H. Steidl, (D), California (Santa Barbara), 1995, YsNeYg
Assistant Research Scientist:
Daniel Lavallee, (D), McGill, 1991, YsnAs

University of California, Santa Cruz
Center for the Dynamics & Evolution of the Land-Sea Interface (2015)
1156 High Street
Room A234, Earth & Marine Sciences Building
Santa Cruz, CA 95064
p. (831) 459-4089
acr@es.ucsc.edu
Professor:
Robert S. Anderson, (D), Washington, 1986, Gm
Kenneth W. Bruland, (D), California (San Diego), 1974, OcCtm
Margaret L. Delaney, (D), MIT/WHOI, 1983, Oc
Russell Flegal, (D), Oregon State, Cg
Laurel R. Fox, (D), California (Santa Barbara), Zn
Dianne Gifford-Gonzalez, (D), California (Berkeley), Zn
Mark S. Mangel, (D), British Columbia, 1978, Zn
Donald C. Potts, (D), California (Santa Barbara), Ob
Mary W. Silver, (D), California (San Diego), 1971, Ob
Jonathan P. Zehr, (D), California (Davis), 1985, Ob
Associate Professor:
Mark Carr, (D), California (Santa Barbara), Ob
Brent Haddad, (D), California (Berkeley), 1996, Zn
Karen D. Holl, (D), Virginia Tech, 1994, Zu
Christina Ravelo, (D), Columbia, 1991, Cs
Donald R. Smith, (D), California (Santa Cruz), Zn
Assistant Professor:
Don Croll, (D), California (Santa Cruz), Ob
Raphael M. Kudela, (D), S California, 1995, Zr
Margaret A. McManus, (D), Old Dominion, 1996, On

Center for the Origin, Dynamics, & Evolution of Planets (2015)
1156 High Street
Room A234, Earth & Marine Sciences Building
Santa Cruz, CA 95064
p. (831) 459-4089
dkorycan@ucsc.edu
Professor:
Robert S. Anderson, (D), Washington, 1986, Gm
Peter Bodenheimer, (D), California (Berkeley), 1965, Xc
Frank Bridges, (D), California (San Diego), 1968, Yx
Douglas Lin, (D), Cambridge, 1976, Xc
Steven Vogt, (D), Texas, 1978, Xc
Other:
Don Korycansky, (D), California (Santa Cruz), Xc

Earth & Planetary Sciences Dept (B,M,D) ●✓🗋 (2019)
1156 High Street
Earth & Marine Sciences Bldg, Rm A232
Santa Cruz, CA 95064
p. (831) 459-4089
jzachos.ucsc.edu
http://www.eps.ucsc.edu
f: https://www.facebook.com/UcscEPS
t: @EpsUcsc
Programs: Earth Sciences B.S., Earth Sciences B.S.with Environmental Geology, Earth Sciences B.S. with Ocean Sciences, Earth Sciences B.S. with Planetary Sciences, Earth Sciences B.S. with Science Education, Earth Sciences/Anthropology BA, Earth Sciences Minor
Enrollment (2018): B: 220 (55) M: 8 (4) D: 48 (6)
Distinguished Professor, Department Chair:
Quentin Williams, (D), California (Berkeley), 1988, Gy
Distinguished Professor:
Thorne Lay, (D), Caltech, 1983, Ys

Professor:
    Emily Brodsky, (D), Caltech, 2001, YsgHw
    Patrick Y. Chuang, (D), Caltech, 1999, As
    Matthew Clapham, (D), USC, 2006
    Andrew T. Fisher, (D), Miami, 1989, Hw
    Garry Griggs, (D), Oregon State, 1968, On
    Elise Knittle, (D), California (Berkeley), 1988, Gy
    Paul L. Koch, (D), Michigan, 1989, Pv
    Francis Nimmo, Cambridge, 1996, Xy
    Susan Schwartz, (D), Michigan, 1988, Ys
    Eli A. Silver, (D), California (San Diego), 1969, Yr
    Othmar T. Tobisch, (D), Imperial Coll (UK), 1963, Gct
    Slawek Tulaczyk, (D), Caltech, 1998, GlmNg
    James Zachos, Rhode island, 1988, Pc
Associate Professor:
    Noah J. Finnegan, (D), Washington, 2007, Gm
    Ian Garrick-Bethell, (D), MIT, 2009, Xy
    Jeremy Hourigan, (D), Stanford, 2002, Cc
Assistant Professor:
    Terrence Blackburn, (D), MIT, 2012, Cc
    Nicole Feldl, (D), Washington, 2013, At
    Myriam Telus, (D), Hawaii, 2015, Xc
    Xi Zhang, (D), Caltech, 2013, AsXy
    Margaret Zimmer, (D), Duke, 2017, Hw
Senior Lecturer:
    Hilde Schwartz, (D), California (Santa Cruz), 1983, Pv
Emeritus:
    Erik Asphaug, (D), Arizona, 1993, Xg
    Robert S. Coe, (D), California (Berkeley), 1966, Ym
    Robert E. Garrison, (D), Princeton, 1964, Gs
    James B. Gill, (D), Australian Nat, 1972, Gi
    Leo F. Laporte, (D), Columbia, 1960, Pg
    J. Casey Moore, (D), Princeton, 1971, Gc
    Lisa C. Sloan, (D), Penn State, 1990, Pe
Graduate Program Advisor:
    Jennifer M. Fish, (B), California (Santa Cruz), Zn

Inst of Geophysics & Planetary Physics (2015)
1156 High Street
Earth & Marine Sciences Building, Room A234
Santa Cruz, CA 95064
    p. (831) 459-4089
    shalevst@gmail.com
    http://igpp.ucsc.edu
Director:
    Ana Christina Ravelo, (D), OcGe

University of San Diego
Dept of Environmental and Ocean Sciences (B,M) ☒ (2019)
Alcala Park
San Diego, CA 92110
    p. (619) 260-4795
    eosc@sandiego.edu
    https://www.sandiego.edu/cas/environmental-ocean-sciences/
    Programs: B.A. Marine Ecology, B.A. Environmental Science and B.A. Environmental Studies, M.S. Environmental and Ocean Science
Chair:
    Sarah Gray, (D), California (Santa Cruz), Zn
Professor:
    Zhi-Yong Yin, (D), Georgia, Zir
Graduate Director:
    Ronald S. Kaufmann, (D), California (San Diego), 1992, Ob
Associate Professor:
    Michel A. Boudrias, (D), California (San Diego), 1992, Ob
    Bethany O'Shea, (D), New South Wales, 2006, CqGeCt
    Nathalie Reyns, (D), North Carolina State
    Drew Talley, (D), California (San Diego), 2000, ObZe
Assistant Professor:
    Jennifer Prairie, (D), Scripps, Ob
    Suzanne Walther, (D), Oregon, GmZi
Adjunct Assistant Professor :
    LeeAnna Y. Chapman, (D), North Carolina State, 2017, GeZe
Adjunct Assistant Professor and Field Trip Coordinator:
    Elizabeth Baker-Trelor, (M), Scripps, 1995, Geu

Adjunct Assistant Professor :
    Eric Cathcart, (M), San Diego, 1996, Geg
    Andrew Nosal, (D), Scripps, Ob
    Steven Searcy, (D), North Carolina State, Ob

University of Southern California
Dept of Earth Sciences (B,M,D) ● (2016)
3651 Trousdale Parkway ,ZHS117
Los Angeles, CA 90089-0740
    p. (213) 740-6106
    waite@usc.edu
    http://www.usc.edu/dept/earth
    Administrative Assistant: Vardui Ter-Simonian
    Administrative Assistant: Cynthia H.. Waite
    Enrollment (2016): B: 30 (12) M: 0 (1) D: 61 (9)
Chair:
    William M. Berelson, (D), S California, 1986, Cm
Research Professor:
    David A. Okaya, (D), Stanford, 1985, Ys
Professor:
    Jan Amend, (D), California (Berkeley), 1995, ClmCo
    Yehuda Ben-Zion, (D), S California, 1990, Ys
    David J. Bottjer, (D), Indiana, 1978, Pe
    Frank A. Corsetti, (D), California (Santa Barbara), 1998, Gs
    James F. Dolan, (D), California (Santa Cruz), 1988, Gt
    Douglas E. Hammond, (D), Columbia, 1975, Cm
    Heidi B. Houston, (D), Caltech, 1987, YsGt
    Thomas H. Jordan, (D), Caltech, 1972, Ys
    Steven P. Lund, (D), Minnesota, 1981, Ym
    James W. Moffett, (D), Miami, 1986, OcCba
    Kenneth Nealson, (D), Chicago, 1969, Pg
    Scott R. Paterson, (D), California (Santa Cruz), 1986, Gc
    John P. Platt, (D), California (Santa Barbara), 1973, GctGp
    Charles G. Sammis, (D), Caltech, 1971, Yg
    Sergio Sanudo, (D), California (Santa Cruz), 1993, Pg
    Lowell D. Stott, (D), Rhode Island, 1989, Pm
    Ta-liang Teng, (D), Caltech, 1966, YsGtYe
    John E. Vidale, (D), Caltech, 1986, YsGtYg
Associate Professor:
    Julien Emile-Geay, (D), Columbia, 2006, Oc
    Sarah Feakins, (D), Columbia, 2006, CoGuCs
    A. Joshua (Josh) West, (D), Cambridge (UK), 2007, ClGmHs
Research Associate Professor:
    Yong-Gang Li, (D), S California, 1988, Ys
    Ellen S. Platzman, (D), ETH, 1990
Emeritus:
    Gregory A. Davis, (D), California (Berkeley), 1961, GtcGe
    Alfred G. Fischer, (D), Columbia, 1950, Gs
    Thomas L. Henyey, (D), Caltech, 1968, Yg
    Teh-Lung Ku, (D), Columbia, 1966, CgcGn
    Bernard W. Pipkin, (D), Arizona, 1964, Ng

University of the Pacific
Department of Geological & Environmental Sciences (B) ☒ (2018)
3601 Pacific Avenue
Stockton, CA 95211
    p. (209) 946-2482
    lrademacher@pacific.edu
    http://pacific.edu/GESC
    f: https://www.facebook.com/pacificees/
    t: @PacificGESC
    Programs: BS in Geological and Environmental Sciences with a concentration in Geology or Environmental Science; BA in Geological and Environmental Sciences
    Enrollment (2018): B: 45 (0)
Chair:
    Laura K. Rademacher, (D), California (Santa Barbara), 2002, ClHySo
Professor:
    Eugene F. Pearson, (D), Wyoming, 1972, GsPgZg
Associate Professor:
    Kurtis C. Burmeister, (D), Illinois, 2005, GctEg
    Lydia K. Fox, (D), California (Santa Barbara), 1989, GizZe
Visiting Professor:
    Karrigan Bork, (D), California (Davis), 2011, Hg
Emeritus:
    Roger T. Barnett, (D), California (Berkeley), 1973, Zy

J. Curtis Kramer, (D), California (Davis), 1976, Gg

Ventura College
Dept of Geology (2017)
4667 Telegraph Road
Ventura, CA 93003
    spalladino@vcccd.edu
Chair:
    Luke D. Hall, (M), W Kentucky, 1975, Zy
Assistant Professor:
    Steve D. Palladino, (M), California (Santa Barbara), 1994, Zi
Instructor:
    William Budke, (M), California Polytech, 2001, So

Victor Valley College
Victor Valley College (2015)
18422 Bear Valley Road
Victorville, CA 92395
    p. (415) 506-0234
    carol.delong@vvc.edu
    http://www.vvc.edu/

West Valley College
Dept of Geology (A) (2015)
14000 Fruitvale Avenue
Saratoga, CA 95070-5698
    p. (408) 741-2437
    robert_lopez@westvalley.edu
Instructor:
    Harry Shade, (M), Miami (Ohio), 1959, Gg
Emeritus:
    Theodore C. Herman, (M), Michigan, 1961, Ze

Whittier College
Environmental Science Program (B) (2015)
Whittier, CA 90608
    p. (562) 907-4220
    cswift@whittier.edu
    http://www.whittier.edu/Academics/EnvironmentalSciences/
Chair:
    Cheryl Swift
Visiting Professor:
    Andrew H. Wulff, (D), Massachusetts, 1998, Gi
Emeritus:
    William B. Wadsworth, (D), Northwestern, 1966, Gi

Colorado

Adams State University
Biology and Earth Sciences (B) ☒ (2019)
208 Edgemont Blvd
Suite 3060
Alamosa, CO 81101
    p. (719) 587-7256
    babrink@adams.edu
    https://www.adams.edu/academics/undergraduate/earth-science/
    Programs: Geology, Physical Geography and Conservation, Science Education (Secondary Teacher Licensure)
    Enrollment (2014): B: 34 (7)
Chair:
    Benita A. Brink, (D), Marquette, 1989, Zn
Professor:
    Robert G. Benson, (D), Colorado Mines, 1997, Gg
Assistant Professor:
    Chayan Lahiri, (D), Mississippi, 2018, HwZr

Aims Community College
Dept of Sciences (A) ☒ (2017)
5401 West 20th Street
Greeley, CO 80634
    p. (970) 339-6637
    jim.stone@aims.edu
    http://www.aims.edu/academics/sciences/index.php
Instructor:
    Jim Stone, Zg

Arapahoe Community College
Geology Program (A) (2015)
5900 South Santa Fe Drive
Littleton, CO 80120
    p. (303) 797-5831
    henry.weigel@arapahoe.edu
    http://www.arapahoe.edu/departments-and-programs/a-z-offerings/geology
Program Chair:
    Henry Weigel

Colorado College
Geology Dept (B) ☒ (2019)
14 E Cache La Poudre
Colorado Springs, CO 80903
    p. (719) 389-6621
    geology@coloradocollege.edu
    http://www.coloradocollege.edu/academics/dept/geology
    f: https://www.facebook.com/geodept.coloradocollege
    Programs: Geology
    Enrollment (2018): B: 24 (17)
Professor:
    Eric M. Leonard, (D), Colorado, 1981, Gl
    Paul M. Myrow, (D), Memorial, 1987, Gsr
    Jeffrey B. Noblett, (D), Stanford, 1980, Gxv
    Christine S. Siddoway, (D), California (Santa Barbara), 1995, GctGg
Department Chair:
    Henry C. Fricke, (D), Michigan, 1997, Cs
Technical Director:
    Stephen G. Weaver, (D), Colorado Mines, 1988, Gx

Colorado Mesa University
Dept of Physical & Environmental Sciences (A,B) O (2017)
1100 North Avenue
Grand Junction, CO 81501-3122
    p. (970) 248-1993
    rwalker@coloradomesa.edu
    http://www.coloradomesa.edu/geosciences/
    Programs: Geosciences (B); Geology (B); Environmental Geology (B); Secondary Education (B); Geology (A); Geology (minor); Watershed Science (minor); GIST (minor)
    Certificates: GIS Technology
    Enrollment (2017): B: 75 (15)
Professor:
    Andres Aslan, (D), Colorado, 1994, Gm
    Rex D. Cole, (D), Utah, 1975, GsrGo
    Verner C. Johnson, (D), Tennessee, 1975, Yg
    Richard F. Livaccari, (D), New Mexico, 1994, Gc
    Gigi A. Richard, (D), Colorado State, 2001, Hsg
Instructor:
    Cassandra Fenton, (D), Utah, 2002, CcGmCl
Lecturer:
    Lawrence S. Jones, (D), Wyoming, 1996, GsmGr
Adjunct Professor:
    William C. Hood, (D), Montana, 1964, Gg
    Julia McHugh, (D), Iowa, 2012, Pgv
    Dave Wolny, (B), Mesa State, 1992, Ys
Emeritus:
    James B. Johnson, (D), Colorado, 1979, Gl
    Jack E. Roadifer, (D), Arizona, 1966, Gx

Colorado Mountain College
Geology Div (A) (2015)
3000 County Road 114
Glenwood Springs, CO 81601
    p. (303) 945-7481 x263
    kdee@coloradomtn.edu
Professor:
    Garrett E. Zabel, (M), Houston, 1977, Gg

Colorado School of Mines
Department of Chemistry (B,M,D) ☒ (2018)

1012 14th Street, CO 204
Golden, CO 80401
 p. (303) 273-3610
 chemistry@mines.edu
 http://chemistry.mines.edu/
 f: https://www.facebook.com/csmchemistry/
 t: @MinesChemistry
 Enrollment (2016): B: 78 (27) M: 14 (5) D: 52 (10)
Department Head:
 Thomas Gennett, (D), Vermont
Professor:
 Mark E. Eberhart, (D), MIT, 1983, Zm
 Mark Jensen, (D), Florida State, 1994
 Daniel M. Knauss, (D), Virginia Tech, 1994
 James Ranville, (D), Colorado Mines, Ca
 Ryan M. Richards, (D), Michigan State, 2000
 Bettina Voelker, (D), Swiss Fed Inst Tech, 1994, Ca
 Kim R. Williams, (D), Michigan State, 1986, Ca
 David T. Wu, (D), California (Berkeley), 1991, Zn
Associate Professor:
 Stephen G. Boyes, (D), New South Wales, 2000
 Renee L. Falconer, (D), South Carolina, 1994
 Matthew C. Posewitz, (D), Dartmouth, 1995
 Mark R. Seger, (D), Colorado State
 Alan Sellinger, (D), Michigan, 1997
 Angela C. Sower, (D), New Mexico
Assistant Professor:
 Allison Caster, (D), California (Berkeley), 2010
 Dylan Domaille, (D), California (Berkeley)
 Amanda Jameer, (M)
 Svitlana Pylypenko, (D), New Mexico, Zmn
 Jenifer C. Shafer, (D), Washington State, 2010
 Brian G. Trewyn, (D), Iowa State, 2006
 Shubham Vyas, (D), Ohio State
 Yongan Yang, (D), Chinese Acad of Sci, 1999
Emeritus:
 Dean W. Dickerhoof, (D), Illinois, 1961
 Donald L. Macalady, (D), Wisconsin, 1969, Cl
 Patrick MacCarthy, (D), Cincinnati, 1975, Co
 Craig Simmons, (D), SUNY (Stony Brook), 1976, Cp
 Kent J. Voorhees, (D), Utah State, 1970, Co
 Thomas R. Wildeman, (D), Wisconsin, 1967, Ct

Dept of Geology & Geological Engineering (B,M,D) ●☒ (2019)
1516 Illinois Street
Golden, CO 80401-1887
 p. (303) 273-3800
 cmedford@mines.edu
 http://geology.mines.edu/
 Enrollment (2018): B: 82 (30) M: 76 (27) D: 51 (11)
Professor:
 Merritt S. Enders, (D), Arizona, 2000, Gzg
Professor:
 David A. Benson, (D), Nevada (Reno), 1998, Hq
 Zhaoshan Chang, (D), Peking, 1997, Eg
 Wendy J. Harrison, (D), Manchester, 1979, Cl
 Reed M. Maxwell, (D), California (Berkeley), 1998, Hs
 Alexei Milkov, (D), Texas A&M, 2001, Cg
 Paul M. Santi, (D), Colorado Mines, 1995, Ng
 Kamini Singha, (D), Stanford, 2005, Hw
 Stephen A. Sonnenberg, (D), Colorado Mines, 1981, Go
 Richard F. Wendlandt, (D), Penn State, 1978, GizCp
 Lesli Wood, (D), Colorado State, 1992, GsoGm
Associate Professor:
 Yvette D. Kuiper, (D), New Brunswick, 2003, Gc
 Thomas Monecke, (D), Germany, 2003, Em
 Alexis Navarre-Sitchler, (D), Penn State, 2008, Ca
 Piret Plink-Bjorklund, (D), Goteborg, 1998, Gs
 Bruce D. Trudgill, (D), Imperial Coll (UK), 1989, Gc
 Wendy W. Zhou, (D), Missouri S&T, 2001, NgZir
Assistant Professor:
 Alex Gysi, (D), Iceland, 2011, Cg
 Richard Palin, (D), Oxford, 2013, Gp
 Danica Roth, (D), California, 2016, GmYs
 Gabriel Walton, (D), Queen's, 2014, NgrYg
Lecturer:
 Christian V. Shorey, (D), Iowa, 2002, Gg

Emeritus:
 L. Graham Closs, (D), Queen's, 1973, Ce
 John B. Curtis, (D), Ohio State, 1989, GoCo
 Jerry D. Higgins, (D), Missouri (Rolla), 1980, NxgNt
 Keenan Lee, (D), Stanford, 1969, Zr
 Eileen P. Poeter, (D), Washington State, 1980, Ng
 Samuel B. Romberger, (D), Penn State, 1968, Em
 A. Keith Turner, (D), Purdue, 1969, Ng
 John E. Warme, (D), California (Los Angeles), 1966, GsPeGu
 Robert J. Weimer, (D), Stanford, 1953, Gr

Dept of Geophysics (B,M,D) ●☒ (2017)
924 16th Street
Golden, CO 80401
 p. (303) 273-3451
 geophysics@mines.edu
 http://geophysics.mines.edu/
 f: https://www.facebook.com/MinesGeophysics/
 t: @MinesGeophysics
 Programs: Geophysical Engineering (B,M,D); Petroleum Reservoir Systems (PSM); Geophysics (M,D)
 Enrollment (2017): B: 115 (44) M: 28 (24) D: 44 (4)
Head:
 John Bradford, (D), Rice, 1999, Yue
Department Head:
 John Bradford, (D), Rice, Yu
Professor:
 Yaoguo Li, (D), British Columbia, 1992, Yev
 Gary R. Olhoeft, (D), Toronto, 1975, YxmYu
 Roel Snieder, (D), Utrecht, 1987, Ye
 Ilya D. Tsvankin, (D), Moscow State, 1982, Ye
 Ali Tura, (D), California (Berkeley), YsNr
Senior Scientist:
 Warren Hamilton, (D), California (Los Angeles), 1951, Ys
Associate Professor:
 Brandon Dugan, (D), Penn State, 2003, GuYrHw
 Paul Sava, (D), Stanford, 2005, Ye
 Jeffrey Shragge, (D), Stanford, Ys
Assistant Professor:
 Ebru Bozdag, (D), Utrecht, 2009, Ys
 Andrei Swidinsky, (D), Toronto, 2011, Ye
 Whitney Trainor-Guitton, (D), Stanford, 2010, RwYe

Dept of Mining Engineering (B,M,D) ● (2016)
1600 Illinois
Golden, CO 80401
 p. (303) 273-3700
 pnelson@mines.edu
 http://mining.mines.edu/
 Enrollment (2015): B: 100 (41) M: 18 (18) D: 14 (6)
Department Head:
 Priscilla P. Nelson, (D), Cornell, 1983, NrgNm
Professor:
 Kadri Dagdelen, (D), Colorado Mines, 1985, Nm
 M. Ugur Ozbay, (D), Witwatersrand, 1988, Nr
Associate Professor:
 Mark Kuchta, (D), Lulea Univ of Tech, 1990, Nm
 Hugh Miller, (D), Colorado Mines, 1996, Nmg
 Masami Nakagawa, (D), Cornell, 1988, Ze
Assistant Professor:
 Elizabeth Holley, (D), Colorado Mines, 2012, Eg
 Rennie Kaunda, (D), W Michigan, 2007, Nm
 Eunhye Kim, (D), Penn State, 2010, NrmNt
Research Professor:
 Karl Zipf, (D), Penn State, 1988, Nm
Research Assistant Professor:
 Vilem Petr, (D), Colorado Mines, 2000, Nm
Manager of Earth Mechanics Institute:
 Brian Asbury, Nr
Research Associate:
 Jürgen F. Brune, (D), Tech (Clausthal), 1994, Nm

Colorado State University
Dept of Atmospheric Science (M,D) ☒ (2018)
1371 Campus Delivery
Fort Collins, CO 80523-1371

p. (970) 491-8360
info@atmos.colostate.edu
http://www.atmos.colostate.edu
Programs: Master of Science in Atmospheric Science
Doctorate in Atmospheric Science
Enrollment (2018): M: 34 (12) D: 43 (6)
Head:
   Jeffrey L. Collett, Jr., (D), Caltech, 1989, As
Professor:
   A. Scott Denning, (D), Colorado State, 1994, AsZg
   James Hurrell, (D), Purdue, 1990, As
   Sonia M. Kreidenweis, (D), Caltech, 1989, As
   Christian D. Kummerow, (D), Minnesota, 1987, Yr
   Eric Maloney, (D), Washington, 2000, As
   David A. Randall, (D), California (Los Angeles), 1976, As
   A.R. Ravishankara, (D), Florida, 1975, AsZg
   Steven A. Rutledge, (D), Washington, 1983, As
   David W. J. Thompson, (D), Washington, 2000, Yr
   Sue van den Heever, (D), Colorado State, 2001, As
   Peter Jan van Leeuwen, (D), Delft (Neth), 1992, As
Associate Professor:
   Elizabeth A. Barnes, (D), Washington, 2012, As
   Michael M. Bell, (D), Naval Postgrad Sch, 2010, As
   Christine Chiu, (D), Purdue, 2003, As
   Jeff Pierce, (D), Carnegie Mellon, 2008, As
   Russ Schumacher, (D), Colorado State, 2008, As
Assistant Professor:
   Emily Fischer, (D), Washington, 2010, As
   Kristen L. Rasmussen, (D), Washington, 2014, As

Dept of Geosciences (B,M,D) ● (2018)
1482 Campus Delivery
Fort Collins, CO 80523-1482
   p. (970) 491-5661
   WCNR_GEO_Info@mail.colostate.edu
   https://warnercnr.colostate.edu/geosciences/
   Programs: Geology (B); Geology, Environmental Geology Concentration (B); Geology, Geophysics Concentration (B); Geology, Hydrogeology Concentration (B); Geosciences (M,D)
   Administrative Assistant: Sharon Gale
   Enrollment (2018): B: 131 (45) M: 34 (11) D: 24 (1)
Head:
   Richard C. Aster, (D), California (San Diego), 1991, YsGv
Professor:
   Sven O. Egenhoff, (D), Tech (Berlin), 2000, Gsu
   Judith L. Hannah, (D), California (Davis), 1980, GiCc
   Dennis L. Harry, (D), Texas (Dallas), 1989, Yg
   Ellen E. Wohl, (D), Arizona, 1988, GmHs
Associate Professor:
   Jerry F. Magloughlin, (D), Minnesota, 1993, Gpz
   Sara L. Rathburn, (D), Colorado State, 2001, Ggm
   John R. Ridley, (D), Edinburgh, 1982, Eg
   Michael J. Ronayne, (D), Stanford, 2008, Hwq
   William E. Sanford, (D), Cornell, 1992, Hw
   Derek L. Schutt, (D), Oregon, 2000, Ysg
   Sally J. Sutton, (D), Cincinnati, 1987, GdCl
Assistant Professor:
   Sean Gallen, (D), North Carolina State, 2013, GemGt
   Daniel McGrath, (D), Colorado, 2013, YuGl
   John Singleton, (D), Texas, 2011, Gct
   Lisa Stright, (D), Stanford, 2011, GoNpYe
Research Associate:
   James R. Chappell, (B), Colorado State, 2000, Ge
   Svetoslav Georgiev, (D), ETH (Switzerland), 2008, CcGo
   Ronald J. Karpilo, Jr, (M), Denver, 2004, Zy
   Stephanie A. O'Meara, (M), Colorado State, 1997, Gc
   Trista L. Thornberry-Ehrlich, (M), Colorado State, 2001, GgcGx
   Gang Yang, (D), Sci & Tech (China), 2005, CcaCg
   Aaron Zimmerman, (M), Colorado State, 2006, Cc
Emeritus:
   Eric A. Erslev, (D), Harvard, 1981, GctGg
   Frank G. Ethridge, (D), Texas A&M, 1970, Gs
Academic Success Coordinator:
   Jill Putman, (M), Georgia, 2009

Community College of Aurora
Dept of Science (2015)
16000 East Centretech Parkway
Aurora, CO 80011
   p. (303) 361-7398
   jim.weedin@ccaurora.edu
   http://www.ccaurora.edu

Denver Museum of Nature & Science
Dept of Earth Sciences ⊠ (2017)
2001 Colorado Boulevard
Denver, CO 80205-5798
   p. (303) 370-6000
   Taylor.Foreman@dmns.org
   http://www.dmns.org
Director of Earth and Space Sciences Branch, Department Chair of Earth Sciences, Curator of Paleobotany:
   Ian Miller, (D), Yale, 2007, PbGg
Tim and Kathryn Ryan Curator of Geology:
   James W. Hagadorn, (D), S California, 1998, GgsPg
Preparator:
   Natalie Toth, (M), SD Mines, 2010, Pg
Curator of Vertebrate Paleontology:
   Tyler Lyson, (D), Yale, 2012, Pv
   Joseph Sertich, (D), Stony Brook, 2011, Pv
Collections Manager:
   Kristen Mackenzie, (M), Oregon, 2013, Zg
Business Specialist:
   Taylor Foreman, (B), Texas, 2012, Zn

Fort Lewis College
Dept of Geosciences (B) (2015)
1000 Rim Drive
Durango, CO 81301
   p. (970) 247-7278
   gonzales_d@fortlewis.edu
   http://geo.fortlewis.edu
   Enrollment (2012): B: 120 (22)
Chair:
   David A. Gonzales, (D), Kansas, 1997, Gx
Professor:
   James D. Collier, (D), Colorado Mines, 1982, Cg
   Gary Gianniny, (D), Wisconsin, 1995, Gs
   Kimberly Hannula, (D), Stanford, 1993, GctGp
   Ray Kenny, (D), Arizona State, 1991, Gm
   Scott White, (D), Utah, 2001, ZirZy
Instructor:
   Lauren Heerschap, (M), Colorado, 2014, GgEgGt
Adjunct Professor:
   Charles Burnham, (D), MIT, 1961, Gz
   Mary L. Gillam, (D), Colorado, 1998, Gm
Laboratory Director:
   Andrea J. Kirkpatrick,

Front Range Community College - Larimer
Natural Sciences (A) ⊠ (2017)
4616 S. Shields Street
Fort Collins, CO 80526
   p. (970) 204-8607
   stephanie.irwin@frontrange.edu
   http://www.frontrange.edu/Academics/Academic-Departments/Larimer-Campus/Natural-Applied-Environment-Science/
   Programs: Geology (A)
Instructor:
   Andy Caldwell, (M), N Colorado, 1999, Gg
   Mike Smith, (M), N Colorado, 1998, GgZe

Front Range Community College - Westminster
Science and Technology (A) (2018)
3645 West 112th Avenue
Westminster, CO 80031
   p. (303) 404-5279
   Angela.Greengarcia@frontrange.edu
   https://www.frontrange.edu/programs-and-courses/a-z-program-list/geology

Enrollment (2018): A: 2 (2)

Metropolitan State College of Denver
Earth & Atmospheric Sciences Dept (B) (2015)
P.O. Box 173362, Campus Box 22
Denver, CO 80217-3362
p. (303) 556-3143
jharr115@msudenver.edu
Administrative Assistant: Diane Hollenbeck
Chair:
  James M. Cronoble, (D), Colorado Mines, 1977, Gg
Professor:
  John R. Kilcoyne, (D), Washington, 1973, Zy
  Anthony A. Rockwood, (M), Colorado State, 1976, Am
  Roberta A. Smilnak, (D), Clark, 1973, Zg
Associate Professor:
  Thomas J. Corona, (M), Colorado State, 1978, Am
Assistant Professor:
  Robert E. Leitz, (M), California (Berkeley), 1974, Gz
  Rafael Moreno, (D), Colorado State, 1992, Zg
Emeritus:
  James MacLachlan, (D), Princeton, Gg
  H. Dixon Smith, (D), Minnesota, 1960, Zg

Northeastern Junior College
Math, Science, and Health (A) ☒ (2019)
100 College Avenue
Sterling, CO 80751
p. (970) 521-6753
david.coles@njc.edu
http://njc.edu
Programs: Associate of Science with Geology Designation

Red Rocks Community College
Geology Program, Science Dept (A) ●☒ (2018)
13300 West Sixth Avenue
Campus Box 20
Lakewood, CO 80228
p. (303) 914-6290
eleanor.camann@rrcc.edu
http://rrcc.edu/geology/
Programs: Geology
Enrollment (2018): A: 5 (6)
Professor:
  Eleanor J. Camann, (D), N Carolina, 2005, OnZeGs

United States Air Force Academy
Dept of Economics & Geosciences (B) (2015)
HQ USAFA/DFEG
2354 Fairchild Drive, Suite 6K110
USAF Academy, CO 80840-5701
p. (719) 333-3080
jennifer.alexander@usafa.edu
http://www.usafa.edu/df/dfeg/?catname=dfeg
Enrollment (2013): B: 90 (27)
Professor:
  Terry W. Haverluk, (D), Minnesota, 1993, Zg
Associate Professor:
  Steven J. Gordon, (D), Arizona State, 1999, Gm
  Thomas Koehler, (D), Wisconsin, As
Lt Col :
  Matthew Tracy, (D), Arizona State, 2008, Zn
Assistant Professor:
  Glen Gibson, (D), Virginia Tech, 2012, Zri
  Brett Machovina, (D), Denver, 2010, Zir
  Evan Palmer, (D), Arizona State, 2014, Zig
  Sarah Robinson, (D), Arizona State, GgZi
Instructor:
  Scott Dubsky, (M), North Dakota, 2006, ZyGm
Related Staff:
  Danny Portillo, (B), Zi

University of Colorado
Dept of Geography (B,M,D) ☒ (2019)
Campus Box 260
Boulder, CO 80309-0260
p. (303) 492-8312
darla.shatto@colorado.edu
www.colorado.edu/geography
Programs: Geography
Certificates: Certificate in GIS and Computational Science (Undergraduate)
Enrollment (2018): B: 150 (74) M: 23 (8) D: 38 (12)
Professor:
  Waleed Abdalati, (D), Colorado (Boulder), 1996, Zr
  Suzanne P. Anderson, (D), California (Berkeley), 1995, GmCl
  Peter D. Blanken, (D), British Columbia, 1997, HsAsm
  Mark Serreze, (D), Colorado (Boulder), 1989, As
  Thomas T. Veblen, (D), California (Berkeley), 1975, Zg
Associate Professor:
  Holly R. Barnard, (D), Oregon State, 2009, Hg
  Noah Molotch, (D), Arizona, 2004, Hg
Assistant Professor:
  Jennifer K. Balch, (D), Yale, 2008, Zg
  Katherine Lininger, (D), Colorado State, 2018, Gm
Emeritus:
  John Pitlick, (D), Colorado State, 1988, Gm
  Mark W. Williams, (D), California (Santa Barbara), 1990, Zg

Dept of Geological Sciences (B,M,D) (2015)
Campus Box 399
Boulder, CO 80309-0399
p. (303) 492-8141
shemin.ge@Colorado.edu
https://www.colorado.edu/geologicalsciences/
Administrative Assistant: Carmen Juszczyk
Enrollment (2014): B: 248 (29) M: 13 (7) D: 45 (10)
Associate Dean:
  Mary J. Kraus, (D), Colorado, 1983, GsSa
Professor:
  Robert Anderson, (D), Washington, 1986, Gm
  David A. Budd, (D), Texas, 1984, GdsEo
  Jaelyn J. Eberle, (D), Wyoming, 1996, Pv
  G. Lang Farmer, (D), California (Los Angeles), 1983, Cg
  Shemin Ge, (D), Johns Hopkins, 1990, HwEoYg
  Bruce M. Jakosky, (D), Caltech, 1982, Xg
  Craig H. Jones, (D), MIT, 1987, YsGtYg
  Gifford H. Miller, (D), Colorado, 1975, GlCc
  Stephen J. Mojzsis, (D), California (San Diego), 1997, XcCcGp
  Peter Molnar, (D), Columbia, 1970, Gt
  Karl J. Mueller, (D), Wyoming, 1992, GtcGm
  Anne F. Sheehan, (D), MIT, 1991, Yse
  Joseph R. Smyth, (D), Chicago, 1970, Gz
  Charles R. Stern, (D), Chicago, 1973, Gi
  Eric E. Tilton, (D), California (Santa Cruz), 1998, Hw
  Gregory E. Tucker, (D), Penn State, 1996, GmtGs
  Paul Weimer, (D), Texas, 1989, Gor
  James W. C. White, (D), Columbia, 1983, CsAt
Associate Professor:
  Karen Chin, (D), California (Santa Barbara), 1996, Pg
  Rebecca M. Flowers, (D), MIT, 2005, CgcGt
  Brian M. Hynek, (D), Washington, 2003, Xg
  Thomas M. Marchitto, (D), MIT, 1999, Oc
  Dena M. Smith, (D), Arizona, 2000, Pg
  Alexis Templeton, (D), Stanford, 2002, ClZaCq
Assistant Professor:
  Kevin H. Mahan, (D), Massachusetts, 2005, Gpc
  Julio C. Sepúlveda, (D), Bremen, 2008, Co
Instructor:
  Lon Abbott, (D), California (Santa Cruz), 1993, GgtGm
Emeritus:
  John T. Andrews, (D), Nottingham (UK), 1965, GluGs
  William W. Atkinson, Jr., (D), Harvard, 1973, EmCg
  Peter W. Birkeland, (D), Stanford, 1961, Sd
  William C. Bradley, (D), Stanford, 1956, Gm
  Don L. Eicher, (D), Yale, 1958, Pm
  Alexander Goetz, (D), Caltech, 1967, Zr
  Edwin E. Larson, (D), Colorado, 1965, Ym
  James L. Munoz, (D), Johns Hopkins, 1966, Cp
  Peter Robinson, (D), Yale, 1960, Pv
  Donald D. Runnells, (D), Harvard, 1964, Cl
  Hartmut A. Spetzler, (D), Caltech, 1969, Ys

Theodore R. Walker, (D), Wisconsin, 1952, Gd

Univ of Colorado Museum (2015)
Campus Box 265
Boulder, CO 80309-265
    p. (303) 492-6165
    linda.cordell@colorado.edu
    http://cumuseum.colorado.edu/
Associate Professor:
    Jaelyn J. Eberle, (D), Wyoming, 1996, Pve
Assistant Professor:
    Karen Chin, (D), California (Santa Barbara), 1996, Pg
Research Associate:
    Dena M. Smith, (D), Arizona, 2000, PeiPb
Adjunct Curator:
    Kenneth Carpenter, Pv
    Mary Dawson, Pv
    Trihn Dzanh, Pv
    Emmett Evanoff, Pi
    Jeff Indeck, Pv
    Jonathan Marcot, Pv
    Greg McDonald, Pv
    Karen Sears, Pv
Emeritus:
    Judith A. Harris, (D), Cambridge, 1972, Pv
    Peter Robinson, (D), Yale, 1960, Pv
Museum Associate:
    Emily Bray, Pg
    Frank Fisher, Pg
    Pat Monaco, Pg
    Steve Wallace, Pg
Collection Manager:
    Tonia Superchi-Culver, (M), SD Mines, 2001, Pg
Associate Curator, Micropaleontology:
    Donald Eicher, (D), Yale, 1958, Pm
Associate Curator, Fossil Primates:
    Herbert Covert, Pv

University of Colorado, Denver
Dept of Geography, & Environmental Science (2015)
CB 172
P.O. Box 173364
Denver, CO 80217-3364
    p. (303) 556-2276
    sue.eddleman@ucdenver.edu
Associate Professor:
    John W. Wyckoff, (D), Utah, 1980, Zy
Emeritus:
    Wesley E. LeMasurier, (D), Stanford, 1965, Gv
    Martin G. Lockley, (D), Birmingham (UK), 1977, Pi

University of Denver
Dept of Geography and the Environment (B,M,D) (2019)
2050 E. Iliff Avenue
Boettcher Center West, Room 120
Denver, CO 80208
    p. (303) 871-2654
    Michael.Keables@du.edu
    http://www.du.edu/geography
    f: https://www.facebook.com/DUGeography/
    Programs: Environmental Science (B); Geography (B,M); Geographic Information Science (M); Geography (D)
    Certificates: Geographic Information Systems (G)
    Enrollment (2018): B: 198 (34) M: 46 (23) D: 10 (1)
Professor:
    Andrew R. Goetz, (D), Ohio State, 1987, Zgu
    Paul C. Sutton, (D), California (Santa Barbara), 1999, Zc
    Matthew Taylor, (D), Arizona State, 2003, Zc
Teaching Professor:
    Hillary Hamann, (D), Colorado, 2002, HgZy
Teaching Associate Professor:
    Helen Hazen, (D), Minnesota, 2006, Zy
    Erika Trigoso, (D), Oxford, 2010, Zc
Associate Professor:
    E. Eric Boschmann, (D), Ohio State, 2008, Zc
    J. Michael Daniels, (D), Wisconsin, 2002, GmSp
    Michael J. Keables, (D), Wisconsin, 1986, AstAm
    Michael W. Kerwin, (D), Colorado, 2000, GgeZc
    Jing Li, (D), George Mason, 2012, Zi
    Rebecca Powell, (D), California (Santa Barbara), 2006, Zr
    Donald G. Sullivan, (D), California (Berkeley), 1988, Pc
Teaching Assistant Professor:
    Kristopher Kuzera, (D), San Diego State, 2011, ZyHg
Assistant Professor:
    Hanson Nyantakyi-Frimpong, (D), W Ontario, 2014, Zc
    Guiming Zhang, (D), Wisconsin, 2018, Zi
Adjunct Professor:
    G. Thomas Lavanchy, (D), Denver, 2015, Hg
    Michelle Moran-Taylor, (D), Arizona State, 2003, Zn
    Martha A. Narey, (D), Denver, 1999, Zg
    Sean Tierney, (D), Denver, 2009, Zc
Emeritus:
    David Longbrake, (D), Iowa, 1972, Zi
    Terrence J. Toy, (D), Denver, 1973, Gm

University of Northern Colorado
Earth and Atmospheric Sciences (B,M) (2018)
Campus Box 100
Greeley, CO 80639
    p. (970) 351-2647
    timothy.grover@unco.edu
    http://esci.unco.edu
    Programs: Environmental Earth Sciences (B); Geology (B); Meteorology (B); Secondary Teaching (B); Earth Sciences (M); Environmental Geosciences (PSM)
    Certificates: Safety Science Certificate (12 credits), including Industrial Safety course and HAZWOPER certificate (OSHA Hazardous Waste Operations and Emergency Response)
    Administrative Assistant: Vicki L.. Ouellette
    Enrollment (2018): B: 89 (22) M: 13 (14)
Chair:
    Timothy W. Grover, (D), Oregon, 1988, GpzGt
Professor:
    Steven W. Anderson, (D), Arizona, 1990, GvZeXg
    Graham Baird, (D), Minnesota, 2006, GcgGz
    William H. Hoyt, (D), Delaware, 1982, OgGsZe
    Lucinda Shellito, (D), California (Santa Cruz), 2004, AsPeAm
Associate Professor:
    Joe T. Elkins, (D), Georgia, 2002, ZeGgCl
    Emmett Evanoff, (D), Colorado, 1990, GrPgGs
    Wendilyn Flynn, (D), Illinois, 2012, Ams
    David G. Lerach, (D), Colorado State, 2012, Ams
Assistant Professor:
    Sharon Bywater-Reyes, (D), Montana, 2015, HqsHt
Senior Lecturer:
    Byron Straw, (M), N Colorado, 2010, ZeGgl
Lecturer:
    Carolyn D. Lambert, (M), Arizona, 2003, Hw
Adjunct Professor:
    Todd A. Dallegge, (D), Alaska (Fairbanks), 2002, GosGr
Emeritus:
    Richard D. Dietz, (D), Colorado, 1965, Xg
    Kenneth D. Hopkins, (D), Washington, 1976, Gml
    William D. Nesse, (D), Colorado, 1977, Gxz
    K. Lee Shropshire, (D), Colorado, 1974, Pg
Other:
    Walter A. Lyons, (D), Chicago, 1970, Asm

Western Colorado University
Dept of Geology (B) O (2018)
Gunnison, CO 81231
    p. (303) 943-2015
    astork@western.edu
    http://www.western.edu/geology
    Programs: Geology; Petroleum Geology; Geoarchaeology; Environmental Geology, Earth Science Education
    Enrollment (2018): B: 46 (14)
Professor:
    Robert P. Fillmore, (D), Kansas, 1994, Gs
    David W. Marchetti, (D), Utah, 2006, GmClc
    Allen L. Stork, (D), California (Santa Cruz), 1984, Giv

Rady Chair:
   Bradford R. Burton, (D), Wyoming, 1997, GocGt
Moncrief Chair:
   Elizabeth S. Petrie, (D), Utah State, 2014, GcoYe
Lecturer:
   Holly Brunkal, (D), Colorado Mines, 2015, NgGm

Connecticut

Central Connecticut State University
Dept of Geography (2015)
New Britain, CT 06050-4010
   p. (860) 832-2785
   cannatadij@ccsu.edu
   http://www.ccsu.edu/geography/index.html

Dept of Physics & Earth Sciences (B,M) (2015)
1615 Stanley Street
New Britain, CT 06050-4010
   p. (860) 832-2930
   antar@CCSU.EDU
   http://www.physics.ccsu.edu/
   Department Secretary: Sandra O'Day
Chair:
   Ali A. Antar, (D), Connecticut, Zn
Professor:
   Sandra Burns, (D), Connecticut, 1972, Ze
   Steven B. Newman, (D), SUNY (Albany), As
   Michael Wizevich, (D), Virginia Tech, GsdGm
Associate Professor:
   Marsha Bednarski, (D), Connecticut, 1997, Ze
   Mark Evans, (D), Pittsburgh, 1989, GciGz
   Kristine Larsen, (D), Connecticut, 1988, Xy
   Jennifer L. Piatex, (D), Pittsburgh, XgZr
Emeritus:
   Charles W. Dimmick, (D), Tulane, 1969, Ge
Related Staff:
   R. Craig Robinson, (B), Millersville, 1971, Xy

Eastern Connecticut State University
Environmental Earth Science Dept (B) ☒ (2018)
83 Windham Street
Willimantic, CT 06226
   p. (860) 465-4317
   cunninghamw@easternct.edu
   http://www1.easternct.edu/environmentalearthscience/
   f: https://www.facebook.com/groups/181927638678890/
   Programs: Environmental Earth Science; General Earth Science; Sustainable Energy Science
   Enrollment (2018): B: 90 (22)
Chair:
   Dickson Cunningham, (D), Texas, 1993, GtcGx
Professor:
   Catherine A. Carlson, (D), Michigan State, 1994, Hwg
   Peter A. Drzewiecki, (D), Wisconsin, 1996, GrsGd
   James A. Hyatt, (D), Queens, 1993, Gml
   Paul Torcellini, (D), Purdue, 1993, Zn
Associate Professor:
   Meredith Metcalf, (D), Connecticut, 2013, ZirHw
   Stephen Nathan , (D), Massachusetts, 2005, PmEo
   Bryan Oakley, (D), Rhode Island, 2012, OnGml
Adjunct Professor:
   Susan Bruening, (M), E Connecticut State, 1988, Zg
   Heath Carlson, (M), E Connecticut State, 2013, Zg
   Lynn-Ann DeLima, (M), S Connecticut State, 2000, Zg
   Vishnu R. Khade, (D), Cincinnati, 1987, Zg
   Emile Levasseur, (M), Sacred Heart, 1992, Zg
   Bruce Morton, (D), Connecticut, 1983, Zg
   James Motyka, (M), E Connecticut State, 1987, Zg
   Julie Sandeen, (M), Connecticut, 1991
   Wesley Winterbottom, (M), Connecticut, 1988, Zg
Emeritus:
   Sherman M. Clebnik, (D), Massachusetts, 1975, Gl
   Fred Loxsom, (D), Dartmouth, 1969, Zn
   Henry I. Snider, (D), New Mexico, 1966, Ge
   Roy R. Wilson, (D), Oregon State, 1984, Zi

Middlesex Community College
Div of Science, Allied Health and Engineering (A) (2015)
100 Training Hill Road
Middletown, CT 06457
   p. (860) 343-5779
   MBusa@mxcc.commnet.edu
   http://www.mxcc.commnet.edu/Content/Environmental_Science_1.asp
Professor:
   Mark Busa, (D), Connecticut, ZgEgYg
   Christine Witkowski, (M), Connecticut, 2002, GeZg

Naugatuck Valley Community College
Science, Technology, Engineering, and Math (STEM) (2015)
750 Chase Parkway
Waterbury, CT 06708
   p. (203) 596-8690
   cdonaldson@nv.edu
   http://www.nvcc.commnet.edu

Norwalk Community College
Sciences Deptartment (2015)
188 Richards Avenue
Norwalk, CT 06854
   p. (203) 857-7275
   mbarber@ncc.commnet.edu
   http://www.ncc.commnet.edu/dept/science/default.asp

Quinebaug Valley Community College
Dept of Environmental Science (2015)
742 Upper Maple Street
Danielson, CT 06239
   p. (860) 412-7230
   mvesligaj@gvcc.commnet.edu
   http://www.qvcc.commnet.edu

Southern Connecticut State University
Department of Earth Science (B) O☒ (2018)
501 Crescent Street
New Haven, CT 06515
   p. (203) 392-5835
   flemingt1@southernct.edu
   http://www.southernct.edu/academics/schools/arts/departments/earthscience/
   Programs: Geology (BS); Environmental Earth Science (BS); General Earth Science (BS); Earth Science Education (BS); Earth Science (BA)
   Enrollment (2014): B: 46 (7)
Professor:
   Cynthia R. Coron, (D), Toronto, 1982, Eg
   Thomas H. Fleming, (D), Ohio State, 1995, GiCgc
Associate Professor:
   James W. Fullmer, (D), MIT, 1979, Am
Assistant Professor:
   Dushmantha Jayawickreme, (D), Michigan State, 2008, YgSpNg
   Michael J. Knell, (D), Montana State, 2012, PvGsPe
Instructor:
   Bryan Adinolfi, (M), SUNY (Environ), 2006, GgZg
   Christopher Balsley, (M), Wesleyan, 1972, Gg
   Daniel Coburn, (M), C Connecticut, 2003, Ze
   Jennifer Cooper, (M), Missouri, 2006, GgcGi
   Yolanda Lee-Gorishti, (M), Connecticut, 2006, GaZe
   Julie Rumrill, (M), Vermont, 2009, GgeGl
Emeritus:
   John W. Drobnyk, (D), Rutgers, 1962, RhGs
   Robert Radulski, (D), Rhode Island, Og
   William Tolley, (M), Syracuse, Gg

University of Connecticut
Center for Integrative Geosciences (B,M,D) (2015)
354 Mansfield Road
U-1045
Storrs, CT 06269-1045

Connecticut

p. (860) 486-4432
geology@uconn.edu
http://www.geosciences.uconn.edu
Enrollment (2012): B: 31 (0) M: 10 (0) D: 7 (0)

Program Director:
Pieter Visscher, (D), Groningen, 1991, Co
Professor:
Vernon F. Cormier, (D), Columbia, 1976, YsGt
William F. Fitzgerald, (D), MIT/WHOI, 1970, Oc
Gary A. Robbins, (D), Texas A&M, 1983, Hw
Robert M. Thorson, (D), Washington, 1970, Gm
Associate Professor:
Andrew M. Bush, (D), Harvard, 2005, PgqPi
Timothy Byrne, (D), California (Santa Cruz), 1981, Gc
Jean M. Crespi, (D), Colorado, 1985, Gc
Lanbo Liu, (D), Stanford, 1993, Yg
Assistant Professor:
Christophe Dupraz, (D), Fribourg, 1999, Pg
Michael Hren, (D), Stanford, 2007, CslCo
William Ouimet, (D), MIT, 2007, GmZy
Emeritus:
Larry Frankel, (D), Nebraska, 1956, Pm
Alfred J. Frueh, (D), MIT, 1949, Gz
Norman H. Gray, (D), McGill, 1971, Gx
Raymond Joesten, (D), Caltech, 1974, GzZmGp
Homer C. Liese, (D), Utah, 1962, Ct
Anthony R. Philpotts, (D), Cambridge, 1963, Gi

Dept of Geography (2015)
Storrs, CT 06269-4148
p. (860) 486-2610
cindy.zhang@uconn.edu
https://geography.uconn.edu/

Dept of Marine Sciences (B,M,D) ⊘ (2019)
1080 Shennecossett Road
Groton, CT 06340
p. (860) 405-9152
marinesciences@uconn.edu
http://www.marinesciences.uconn.edu/
Programs: BS and BA - Marine Sciences
MS and PhD - Oceanography
Enrollment (2018): B: 50 (10) M: 9 (0) D: 33 (0)

Professor:
Ann Bucklin, (D), California (Berkeley), 1980, Ob
Timothy Byrne, (D), California (Santa Cruz), 1981, Ou
Hans G. Dam, (D), SUNY (Stony Brook), 1989, Ob
Heidi Dierssen, (D), California, 2000, Op
Senjie Lin, (D), SUNY (Stony Brook), 1995, Ob
Robert P. Mason, (D), Connecticut, 1991, OcCtm
George B. McManus, (D), SUNY (Stony Brook), 1986, Ob
James O'Donnell, (D), Delaware, 1986, Op
Sandra Shumway, (D), Coll of North Wales, 1976, Ob
Craig Tobias, (D), William & Mary, 1999, Oc
Pieter T. Visscher, (D), Groningen, 1991, Co
J. Evan Ward, (D), Delaware, 1989, Ob
Associate Professor:
Julie Granger, (D), British Columbia, 2006, Oc
David Lund, (D), MIT/WHOI, 2006, Oc
Annelie Skoog, (D), Göteborg (Sweden), Oc
Penny Vlahos, (D), Massachusetts, 2001, OcGeu
Michael Whitney, (D), Delaware, 2003, Op
Huan Zhang, (D), Tokyo Fisheries, 1995, Ob
Assistant Professor:
Hannes Baumann, (D), Hamburg, 2006, Ob
Kelly Lombardo, (D), Stony Brook, 2011, Asm
Catherine Matassa, (D), Northeastern, 2014, Ob
Samantha Siedlecki, (D), Chicago, 2010, On
Jamie Vaudrey, (D), Connecticut, 2007, Ob
Research Associate:
Zofia Baumann, (D), SUNY (Stony Brook), 2011, ObcCm
Emeritus:
Peter Auster, (D), National (Ireland), 2000, Ob
Walter F. Bohlen, (D), MIT/WHOI, 1969, Op
James Edson, (D), Penn State, Op
William F. Fitzgerald, (D), MIT/WHOI, 1970, Oc
Edward C. Monahan, (D), MIT, 1966, AsOp

University of New Haven

Dept of Environmental Sciences (B,M) ○⊠ (2017)
300 Boston Post Rd
West Haven, CT 06516
p. (203) 932-7101
rldavis@newhaven.edu
http://www.newhaven.edu
Programs: Environmental Science (BS, MS), Marine Biology (BS), Marine Affairs (BS), MS in Environmental Science offers concentrations in Environmental Geoscience, Environmental Ecology, GIS, Environmental Health and Management
Certificates: GIS
Enrollment (2016): B: 28 (5) M: 29 (8)

Provost and Senior Vice President of Academic Affairs:
Daniel J. May, (D), California (Santa Barbara), 1986, GtxGe
Coordinator-Graduate Environmental Science:
Roman N. Zajac, (D), Connecticut, 1985, Zni
Professor:
Carmela Cuomo, (D), Yale, CmmGu
R. Laurence Davis, (D), Rochester, 1980, GemHg
Coordinator-Undergraduate Environmental Science:
Kristen Przyborski, (D)
Associate Professor:
Amy L. Carlile, (D), Washington, ObZn
John Kelly, (D), California (Davis), ObZn
Lecturer:
Jean-Paul Simjouw, (D), Old Dominion, 2004, OcCma
Practitioner-in-Residence:
Paul Bartholemew, (D), British Columbia, GeZiGz

Wesleyan University

Dept of Earth & Environmental Sciences (B,M) ⊠ (2018)
265 Church Street
Room 455
Middletown, CT 06459-0139
p. (860) 685-2244
vharris@wesleyan.edu
http://www.wesleyan.edu/ees
Administrative Assistant: Virginia M. Harris
Enrollment (2018): B: 27 (11) M: 5 (4)

Chair:
Dana Royer, (D), Yale, 2002, PegPb
Professor:
Barry Chernoff, (D), Michigan, 1983, Ge
Martha S. Gilmore, (D), Brown, 1997, XgGmZr
Suzanne B. OConnell, (D), Columbia, 1986, GsuGe
Johan C. Varekamp, (D), Utrecht, 1979, CgGzv
Associate Professor:
Timothy C.W. Ku, (D), Michigan, 2001, Cl
Phillip G. Resor, (D), Stanford, 2003, Gct
Assistant Professor:
James P. Greenwood, (D), Brown, 1997, Xc
Emeritus:
James T. Gutmann, (D), Stanford, 1972, GviGz
Peter C. Patton, (D), Texas, 1976, Gm
Facilities Manager:
Joel LaBella, (B), S Connecticut, 1987, Zn

Yale University

Dept of Geology & Geophysics (B,D) ⊠ (2019)
210 Whitney Avenue
P.O. Box 208109
New Haven, CT 06520-8109
p. (203) 432-3114
rebecca.pocock@yale.edu
http://earth.yale.edu
Enrollment (2009): B: 13 (7) D: 55 (5)

Curator:
Derek E. G Briggs, (D), Cambridge (UK), 1976, PigPg
Professor:
Jay J. Ague, (D), California (Berkeley), 1987, Gx
David Bercovici, (D), California (Los Angeles), 1989, Yg
Ruth E. Blake, (D), Michigan, 1997, Cg
Mark T. Brandon, (D), Washington, 1984, Gc

Derek E G Briggs, (D), Cambridge, 1976, Yg
David A D Evans, (D), Caltech, 1998, YmGt
Alexey V. Fedorov, (D), Scripps, 1997, Op
Jacques Gauthier, (D), California (Berkeley), 1984, Pv
Shun-ichiro Karato, (D), Tokyo, 1977, Gy
Jun Korenaga, (D), MIT, 2000, YgsCg
Maureen D. Long, (D), MIT, 2006, Ysg
Jeffrey J. Park, (D), California (San Diego), 1985, YsZgc
Danny M. Rye, (D), Minnesota, 1972, Cs
Ronald B. Smith, (D), Johns Hopkins, 1975, As
Mary-Louise Timmermans, (D), Cambridge, 2000, Oc
John S. Wettlaufer, (D), Washington, 1991, Yg

Senior Research Scientist:
Edward W. Bolton, (D), California (Los Angeles), 1985, GqHqCq
Ellen Thomas, (D), Utrecht, 1979, Pm

Associate Professor:
Kanani K.M. Lee, (D), California (Berkeley), 2003, Gy

Assistant Professor:
Bhart-Anjan Bhullar, (D), Harvard, 2014
Pincelli Hull, (D), California (San Diego), 2010
Juan Lora, (D), Arizona, 2009, As
Noah Planavsky, (D), California (Riverside), 2012
Alan Rooney, (D), Durham (UK), 2011

Emeritus:
Robert B. Gordon, (D), Yale, 1955, Nr
Brian J. Skinner, (D), Harvard, 1955, CgEmGz
George Veronis, (D), Brown, 1954, Op
Elisabeth S. Vrba, (D), Cape Town, 1974, Pv

Research Affiliate:
William C. Graustein, (D), Yale, 1981, Cl

Related Staff:
James O. Eckert, (D), Texas A&M, 1988, Gp

Peabody Museum of Natural History ☒ (2017)
PO Box 208118
170 Whitney Avenue
New Haven, CT 06520-8118
p. (203) 432-3752
peabody.director@yale.edu
http://www.peabody.yale.edu/

Emeritus Curator:
Leo W. Buss, (D), Johns Hopkins, 1979, Pi
Elisabeth S. Vrba, (D), Cape Town, 1974, Pvg

Curator:
Jay J. Ague, (D), California (Berkeley), 1987, Gzx
Michael J. Donoghue, (D), Harvard, 1982, Pbg
Jacques A. Gauthier, (D), California (Berkeley), 1984, Pv

Senior Collections Manager:
Susan H. Butts, (D), Idaho, 2003, PiGsPe
Christopher A. Norris, (D), Oxford, 1992, Pv

Collections Manager:
Shusheng Hu, (D), Florida, 2006, PblGs
Stefan Nicolescu, (D), Gothenburg, 1998, CtGpz

Delaware

University of Delaware

Dept of Geography ●☒ (2018)
125 Academy Street
216 Pearson Hall
Newark, DE 19716-2541
p. (302) 831-3218
dlevia@udel.edu
http://www.ceoe.udel.edu/schools-departments/department-of-geography

Dept of Geological Sciences (B,M,D) ●☒ (2019)
103 Penny Hall
Newark, DE 19716
p. (302) 831-2569
smcgeary@udel.edu
http://www.ceoe.udel.edu/schools-departments/department-of-geological-sciences
f: https://www.facebook.com/UDCEOE
t: @udceoe
Programs: BS Geological Sciences, BA Geological Sciences, BS Earth Science Education, MS Geological Sciences, PhD Geological Sciences
Administrative Assistant: Cheryl Doherty
Enrollment (2018): B: 55 (12) M: 19 (5) D: 9 (3)

Director, DGS, and State Geologist:
David R. Wunsch, (D), Kentucky, 1992, Ge

Chair:
Neil C. Sturchio, (D), Washington Univ (St. Louis), 1983, ClcCa

Professor:
Eliot A. Atekwana, (D), W Michigan, 1996, CgsCq
Ronald E. Martin, (D), California (Berkeley), 1981, PmePg
Holly A. Michael, (D), MIT, 2005, Hwq
Michael ONeal, (D), Washington, 2005, Gm
James E. Pizzuto, (D), Minnesota, 1982, Gm

Associate Professor:
Clara S. Chan, (D), California (Berkeley), 2006, PoObGz
John A. Madsen, (D), Rhode Island, 1987, YrOuZe
Susan McGeary, (D), Stanford, 1984, Ze
Jessica Warren, (D), MIT/WHOI, 2007, GxCpGt

Assistant Professor:
Adam F. Wallace, (D), Virginia Tech, 2008, ClGzZm

Visiting Professor:
Claire J. O'Neal, (D), Washington, 2005, Ze

Director Emeritus, DGS:
Robert R. Jordan, (D), Bryn Mawr, 1964, Gr
John H. Talley, (M), Franklin & Marshall, 1974, Eg

Emeritus:
Billy P. Glass, (D), Columbia, 1968, XmcGz
John C. Kraft, (D), Minnesota, 1955, Gs
Peter B. Leavens, (D), Harvard, 1967, Gz
John F. Wehmiller, (D), Columbia, 1971, Cl

Laboratory Director:
Bill Parnella, (B), Buffalo, 1984, Gg

Cooperating Faculty:
A. Scott Andres, (M), Lehigh, 1984, Hw
Katharina Billups, (D), California (Santa Cruz), 1998, Ou
Shreeram Inamdar, (D), Virginia Tech, 1996, Hg
Deb Jaisi, (D), Miami, 2007, Cb
Thomas E. McKenna, Jr., (D), Texas, 1997, Hy
Peter P. McLaughlin, Jr., (D), Louisiana State, 1989, Gr
Jack Puleo, (D), Florida, 2004, Hw
Kelvin W. Ramsey, (D), Delaware, 1988, Gs
William S. Schenck, (M), Delaware, 1997, Gg
Angelia L. Seyfferth, (D), California (Riverside), 2008, Sc
Christopher K. Sommerfield, (D), SUNY (Stony Brook), 1997, On
Art Trembanis, (D), Virginia Inst of Marine Sci, 2004, Ou
William J. Ullman, (D), Chicago, 1982, Oc

Oceanography Program (M,D) (2015)
700 Pilottown Road
Lewes, DE 19958
p. (302) 645-4279
kbillups@udel.edu
http://www.ocean.udel.edu

Professor:
Thomas M. Church, (D), California (San Diego), 1970, Cm
Victor Klemas, (D), Braunschweig (Germany), 1965, Zr
George W. Luther, III, (D), Pittsburgh, 1972, Cm
Jonathan H. Sharp, (D), Dalhousie, 1972, Oc
Christopher K. Sommerfield, (D), SUNY (Stony Brook), 1997, OuGsCc
William J. Ullman, (D), Chicago, 1982, Cm
Ferris Webster, (D), MIT, 1961, Op
Xiao-Hai Yan, (D), SUNY (Stony Brook), 1989, Zr

Associate Professor:
Katharina Billups, (D), California (Santa Cruz), 1998, Pe
Douglas C. Miller, (D), Washington, 1985, Ob

Associate Scientist:
Charles H. Culberson, (D), Oregon State, 1972, Oc
Richard T. Field, (D), Delaware, 1994, Og

Assistant Professor:
Matthew J. Oliver, (D), Rutgers, 2006, ObZrOg

Adjunct Professor:
Richard B. Coffin, (D), Delaware, 1986, Zn
James Crease, (D), Cambridge, 1951, Op
Marilyn L. Fogel, (D), Texas, 1977, Oc
Norden E. Huang, (D), Johns Hopkins, 1967, Zn
David E. Krantz, (D), South Carolina, Cs

Kamlesh Lulla, (D), Indiana State, 1963, Zr
Donald B. Nuzzio, (D), Rutgers, 1982, Oc
Manmohan Sarin, (D), Gujarat (India), 1984, Zn
Alain J. Veron, (D), Paris, 1988, Oc
John F. Wehmiller, (D), Columbia, Gu

Emeritus:
Jin Wu, (D), Iowa, 1964, Op

Physical Ocean Science and Engineering (2015)
Lewes, DE 19958
p. (302) 831-6640
carcher@udel.edu
http://www.ceoe.udel.edu/schools-departments/school-of-marine-science-and-policy/pose-program

District of Columbia

Carnegie Institution for Science
Department of Terrestrial Magnetism ☒ (2019)
5241 Broad Branch Road, N.W.
Washington, DC 20015-1305
p. (202) 478-8820
jdunlap@carnegiescience.edu
https://dtm.carnegiescience.edu

Director:
Richard W. Carlson, (D), California (San Diego), 1980, Ccg
Senior Scientist:
Conel M. O'D Alexander, (D), Essex (UK), 1987, Xc
Alan P. Boss, (D), California (Santa Barbara), 1979, Xa
R. Paul Butler, (D), Maryland, 1993, Xa
John E. Chambers, (D), Manchester, 1994, Xa
Peter E. Driscoll, (D), Johns Hopkins, 2010, GtYm
Helene Le Mevel, (D), Wisconsin, 2016, Gv
Larry R. Nittler, (D), Washington (St. Louis), 1996, Xc
Diana C. Roman, (D), Oregon, 2004, Gv
Scott S. Sheppard, (D), Hawaii, 2004, Xa
Steven B. Shirey, (D), SUNY (Stony Brook), 1984, CcGiz
Peter E. vanKeken, (D), Utrecht, 1993, GtYsCg
Lara S. Wagner, (D), Arizona, 2005, YsgGt
Alycia J. Weinberger, (D), Caltech, 1998, Xa
Emeritus:
Alan T. Linde, (D), Queensland, 1972, Ys
I. Selwyn Sacks, (D), Witwatersrand, 1961, Ys
Fouad Tera, (D), Vienna, 1962, CcgXm
Senior Research Scientist and SIMS Lab Manager:
Jianhua Wang, (D), Chicago, 1995, Cs
Mass Spectrometry Laboratory Manager:
Timothy D. Mock, (M), Vermont, 1989, Cc
Geochemistry Laboratory Manager:
Mary F. Horan, (M), SUNY (Stony Brook), 1984, Cc
Assistant Controller:
Wan Kim, (M), Yonsei, 1989, Zn
Librarian:
Shaun J. Hardy, (M), SUNY (Buffalo), 1987, Zn

Geophysical Laboratory (2016)
5251 Broad Branch Road, N.W.
Washington, DC 20015-1305
p. (202) 478-8900
gcody@carnegiescience.edu
http://www.gl.ciw.edu/

Senior Scientist:
Ronald E. Cohen, (D), Harvard, 1985, Gy
Yingwei Fei, (D), CUNY, 1989, Cp
Alexander F. Goncharov, (D), Russian Acad of Sci, 1983, GyYxh
Robert M. Hazen, (D), Harvard, 1975, Gz
Ho-kwang Mao, (D), Rochester, 1968, Yx
Bjorn O. Mysen, (D), Penn State, 1974, Cp
Douglas Rumble, III, (D), Harvard, 1969, Gp
Anat Shahar, (D), California (Los Angeles), 2008, CgYxCs
Andrew Steele, (D), Portsmouth, 1996, Xy
Timothy A. Strobel, (D), Colorado Mines, 2008, Zm
Viktor V. Struzhkin, (D), Moscow Inst of Physics & Tech, 1991, Gy
Senior Scientist:
T. Neil Irvine, (D), Caltech, 1959, Gi
Research Scientist:
Jinfu Shu, (M), Wuhan, 1981, Yx

Director:
Wesley T. Huntress, Jr., (D), Stanford, 1968, Xc
Research Scientist:
Muhetaer Aihaiti, (D), Electrocomm Tokyo, 1996, Yg
Reinhard Boehler, (D), Tubingen, 1974, Yg
Xiaojia Chen, (D), Zhejiang, 1997, Ym
Dionysis Foustoukos, (D), Minnesota, 2005, Cg
Stephen Gramsch, (D), Chicago, 1994, Yg
Maddury Somayazulu, (D), Bombay, 1992, Yx
Changsheng Zha, (D), Beijing Inst Tech, 1969, YgZm
Director, HPCAT:
Guoyin Shen, (D), Uppsala, 1994, GzCg
Beamline Scientist, NSLS:
Zhenxian Liu, (D), Jilian (China), 1990, Gy
Beamline Scientist:
Paul Chow, (D), Illinois, 1988, Yg
Yue Meng, (D), Yg
Changyong Park, (D), Tohoku, 1998, Yg
Yuming Xiao, (D), California (Davis), 2007, Yg
Associate Director, HPCAT:
Stanislav Sinogeikin, (D), Yg
Acting Director:
George D. Cody, (D), Penn State, 1992, Co
Librarian:
Shaun J. Hardy, (M), SUNY (Buffalo), 1987, Zn

George Washington University
Dept of Geography (B,M) ☒ (2019)
2036 H St NW
Washington, DC 20052
p. (202) 994-6185
geog@gwu.edu
https://geography.columbian.gwu.edu/
f: https://www.facebook.com/GWGeography/
Enrollment (2018): B: 60 (0) M: 21 (0)

Chair:
Lisa Benton-Short, (D), Syracuse
Professor:
Elizabeth Chacko, (D), California (Los Angeles), 1997, Zy
Marie D. Price, (D), Syracuse, 1990, Zy
Associate Professor:
Mona Atia, (D), Washington
Ryan Engstrom, (D), San Diego State, Zr
David Rain, (D), Penn State
Nikolay Shiklomanov, (D)
Dmitry Streletskiy, (D)
Assistant Professor:
Ginger Allington, (D), Saint Louis
Melissa Keeley, Tech (Berlin)
Michael Mann, (D)

National Academy of Sciences, Engineering, and Medicine
Board on Earth Sciences and Resources ☒ (2018)
The National Academies Keck Center-6th Floor
500 5th Street, NW
Washington, DC 20001
p. (202) 334-2744
besr@nas.edu
http://dels.nas.edu/besr
f: https://www.facebook.com/Board-on-Earth-Sciences-and-Resources-260128822801/
t: @NASEM_Earth

Director:
Elizabeth A. Eide, (D), Stanford, 1993, GogGt
Senior Program Officer:
Sammantha L. Magsino, (M), Florida Intl, 1993, GvNg
Scholar:
Anne M. Linn, (D), California (Los Angeles), 1991, Gs
Senior Program Officer:
Deborah Glickson, (D), Washington, 2007, GutGg

Smithsonian Inst / National Air & Space Museum
Center for Earth & Planetary Studies (2016)
MRC 315, P.O. Box 37012

6th and Independence Ave., SW
Washington, DC 20013-7012
    p. (202) 633-2470
    campbellb@si.edu
    http://airandspace.si.edu/research/earth-and-planetary/
Geophysicist:
    Bruce A. Campbell, (D), Hawaii, 1991, Zr
Planetary Geologist:
    James R. Zimbelman, (D), Arizona State, 1984, Xg
Geologist:
    Robert A. Craddock, (D), Virginia, 1999, Xg
    John A. Grant, (D), Brown, 1990, Xg
    Rossman P. Irwin, (D), Virginia, 2005
    Thomas R. Watters, (D), George Washington, 1985, Xg
Geologist:
    Ted A. Maxwell, (D), Utah, 1977, Xg
Program Manager:
    Priscilla L. Strain, (B), Smith, 1974, Zr
Photo Librarian:
    Rosemary Aiello

Smithsonian Inst / Nat Museum of Natural History
Dept of Mineral Sciences ☒ (2018)
NHB MRC 119
10th & Constitution Avenue, NW
PO Box 37012
Washington, DC 20013-7012
    p. (202) 633-1860
    http://mineralsciences.si.edu/
Research Geologist:
    Jeffrey E. Post, (D), Arizona State, 1981, Gz
Research Geologist:
    Benjamin Andrews, (D), Texas, 2009, Gv
    Catherine Corrigan, (D), Case Western, 2004, XmcXg
    Elizabeth Cottrell, (D), Columbia, 2004, Cp
    Glenn J. MacPherson, (D), Princeton, 1981, XcGi
Senior Scientist:
    Timothy J. McCoy, (D), Hawaii, 1994, Xmg
Museum Specialist (Labs):
    Timothy Gooding, (B), Hampshire Coll, 1990
Museum Specialist (GVP):
    Sally K. Sennert, (M), Pittsburgh, 2003, GvZr
    Edward Venzke, (M), Minnesota (Duluth), 1993, Gv
Collection Manager:
    Cathe Brown, (M), Maryland, 1996, Gzi
    Russell Feather, (B), George Mason, 1982, Gz
Analytical Laboratories - Manager:
    Timothy Rose, (M), Maryland, 1991, CaGv
Research Collaborator:
    Steve Lynton, (D), Maryland, 2003
Postdoctoral Fellow:
    Marion Le Voyer, (D), Blaise Pascal, 2009, Cu
Contract Geologist:
    B. Carter Hearn Jr, (D), Johns Hopkins, 1959, Gic
Research Geologist:
    Sorena S. Sorensen, (D), California (Los Angeles), 1984, Gp
Contractor:
    Christine R. Webb, (B), Penn State, GzgZn
Research Geologist:
    Michael A. Wise, (D), Manitoba, 1987, Gz
Museum Specialist (IT):
    Adam Mansur, (M), Maryland, 2008, Zf
Collection Manager:
    Leslie J. Hale, (B), Maryland, 1989, GgZnRh

Dept of Paleobiology (2015)
Dept. of Paleobiology, MRC121
NMNH, Smithsonian Institution
P.O. Box 37012
Washington, DC 20013-7012
    p. (202) 633-1320
    cloydd@si.edu
    http://www.nmnh.si.edu/paleo/
Curator:
    Scott L. Wing, (D), Yale, 1981, Pb

Curator:
    Anna K. Behrensmeyer, (D), Harvard, 1973, Pv
    Martin A. Buzas, (D), Yale, 1963, Pm
    Matthew T. Carrano, (D), Chicago, 1998, Pv
    Alan H. Cheetham, (D), Columbia, 1959, Pg
    William A. DiMichele, (D), Illinois, 1979, Pb
    Robert Emry, (D), Columbia, 1970, Pv
    Douglas H. Erwin, (D), California (Santa Barbara), 1985, Pi
    Brian T. Huber, (D), Ohio State, 1988, Pm
    Gene Hunt, (D), Chicago, 2003, Pm
    Conrad C. Labandeira, (D), Chicago, 1991, Pi
    Ian G. Macintyre, (D), McGill, 1967, Gs
    Clayton E. Ray, (D), Harvard, 1962, Pv
    Daniel J. Stanley, (D), Grenoble, 1964, Ou
    Thomas R. Waller, (D), Columbia, 1966, Pi
Geologist:
    Thomas Dutro, (D), Pi
    Bevan French, (D), Xm
    John Pojeta, (D), Cincinnati, 1963, Pi
Associate Chair & Collections Manager:
    Jann W. M. Thompson, (B), George Washington, 1970, Pg

Florida

Broward College
Dept of Physical Sciences (Oceanography & Geology) (A,B) (2016)
3501 SW Davie Road
Davie, FL 33314
    p. (954) 201-6771
    jmuza@broward.edu
    http://www.broward.edu/academics/programs/Pages/science-technology-math-engineering-STEM
    Administrative Assistant: Nicki Pickett
Chair:
    Valerio Bartolucci, (M), Bologna, Ge
Professor:
    Lewis Fox, (D), Delaware, 1981, Oc
    Jay P. Muza, (D), Florida State, 1996, OuPmOo
    Laura Precedo, (D), Emory, 1993, Oc
Lab Manager:
    Lynn A. Curtis, (B), Florida Atlantic, 1997, GgPg

Natural Science Dept (A) (2015)
111 East Las Olas Boulevard
Fort Lauderdale, FL 33301
    p. (954) 201-7650
    mdugan@broward.edu
Instructor:
    Xenia Conquy, (M), Florida Atlantic, Gg
    Henri L. Liauw, (M), S Florida, Gg
    William Opperman, (M), Florida, Gg

Chipola College
Dept of Natural Science (A) (2015)
3094 Indian Circle
Mariana, FL 32446
    p. (850) 526-2761
    hiltond@chipola.edu
    http://www.chipola.edu/instruct/science/index.htm
Professor:
    Allan Tidwell, (M), Troy State, Zg

Daytona State College
School of Biological and Physical Sciences (A) (2017)
1200 W. International Speedway Blvd.
Daytona Beach, FL 32114
    horikas@daytonastate.edu
    http://www.daytonastate.edu/CampusDirectory/deptInfo.jsp?dept=SCI
Professor:
    Debra W. Woodall, (D), Florida Inst of Tech, GgOg

Eckerd College
Dept of Geosciences (B) ☒ (2019)
Galbraith Marine Science Laboratory

4200 54th Avenue South
St. Petersburg, FL 33711
    p. (727) 864-8200
    wetzellr@eckerd.edu
    https://www.eckerd.edu/marinescience/
    Programs: Bachelors degrees in Geosciences and Marine Science
    Enrollment (2018): B: 5 (5)
Professor:
    Gregg R. Brooks, (D), S Florida, 1986, Gu
    Joel B. Thompson, (D), Syracuse, 1989, Pg
    Laura R. Wetzel, (D), Washington Univ (St. Louis), 1997, YrOuGg

Florida Atlantic University
Dept of Geosciences (B,M,D) O (2019)
777 Glades Road
Boca Raton, FL 33431
    p. (561) 297-3250
    warburto@fau.edu
    http://www.geosciences.fau.edu
    Programs: BS Geoscience, MS Geoscience, PhD Geoscience
    Certificates: GIS certificate, Advanced GIS certificate
    Enrollment (2018): B: 103 (25) M: 27 (8) D: 34 (1)
Chair:
    Zhixiao Xie, (D), SUNY (Buffalo), 2002, ZirZy
Director, Florida Center for Environmental Studies:
    Colin Polsky, (D), Penn State, 2002, Znc
Associate Provost for Programs and Assessment :
    Russell L. Ivy, (D), Florida, 1992, Zn
Professor:
    Leonard Berry, (D), Bristol, 1969, Zg
Assistant Chair:
    David L. Warburton, (D), Chicago, 1978, CgGe
Associate Professor:
    Xavier Comas, (D), Rutgers, 2005, Yg
    Maria Fadiman, (D), Texas, 2003, Zn
    Scott H. Markwith, (D), Georgia, 2007, Zy
    Anton Oleinik, (D), Purdue, 1998, Pi
    Tara L. Root, (D), Wisconsin, 2005, Hg
    Caiyun Zhang, (D), Texas (Dallas), 2010, Zr
Associate Scientist:
    Tobin Hindle, (D), Florida Atlantic, 2006, ZyiZc
Assistant Professor:
    Tiffany Roberts M. Briggs, (D), S Florida, 2012, On
    Erik Johanson, (D), Tennessee, 2016, RnZy
    Weibo Liu, (D), Kansas, 2016, Zi
Instructor:
    James Gammack-Clark, (M), Florida Atlantic, 2001, Zri
Emeritus:
    Howard Hanson, (D), Miami, 1979, AsOg
    Edward J. Petuch, (D), Miami, 1980, Pg
    Jorge I. Restrepo, (D), Colorado State, 1987, Hqw
    Charles E. Roberts, (D), Penn State, 1991, Zri

Florida Gateway College
Mathematics/Science Div (A,B) (2016)
149 SE College Place
Lake City, FL 32025
    p. (386) 752-1822
    mustapha.kane@fgc.edu
    https://www.fgc.edu/academics/liberal-arts--sciences/math-and-science.aspx
Professor:
    Mustapha Kane, (D), Zg
Instructor:
    Avo Oymayan, Zg

Florida Institute of Technology
Ocean Engineering and Marine Science (B,M,D) O (2019)
150 West University Boulevard
Melbourne, FL 32901
    p. (321) 674-8034
    oems@fit.edu
    https://www.fit.edu/engineering-and-science/academics-and-learning/ocean-engineering-and-marine-sciences/
    f: https://www.facebook.com/pg/FLTechOEMS
    Programs: Biological Sciences, Ph.D., Conservation Technology, M.S., Earth Remote Sensing, M.S., Ecology, M.S., Environmental Resource Management, M.S., Environmental Science, B.S., Environmental Science, M.S., Environmental Science, Ph.D., Fisheries and Aquaculture, B.S., General Biology, B.S., Marine Biology, B.S., Marine Biology, M.S., Marine Conservation, B.S., Meteorology, B.S., Meteorology, M.S., Ocean Engineering, B.S., Ocean Engineering, M.S., Ocean Engineering, Ph.D., Oceanography, B.S., Oceanography, M.S., Oceanography, Ph.D., Sustainability Studies, B.S.
    Enrollment (2014): B: 100 (25) M: 60 (20)
Head:
    George A. Maul, (D), Miami, 1974, Op
Professor:
    Kevin Johnson, (D), Oregon, 1998, Ob
    Steven M. Lazarus, (D), Oklahoma, 1996, As
    Geoffrey Swain, (D), Southhampton, 1981, Ob
    John G. Windsor, Jr., (D), William & Mary, Oc
    Gary Zarillo, (D), Georgia, OupOn
Associate Professor:
    Charles Bostater, (D), Delaware, 1990, Op
Assistant Professor:
    Austin Fox, (D), Florida Inst of Tech, 2015, Oc
    Kelli Hunsucker, (D), Florida Inst of Tech, 2013, Ob
    Pallav Ray, (D), Miami, 2008, AstZo

Florida International University
Center for the Study of Matter at Extreme Conditions (M,D) (2016)
11200 SW 8th ST
VH 150, CeSMEC
Miami, FL 33199
    p. (305) 348-3030
    chenj@fiu.edu
    http://cesmec.fiu.edu
    Enrollment (2012): D: 2 (2)
Director:
    Jiuhua Chen, (D), Japan Grad Univ Adv Studies, 1994, Gyz
Emeritus:
    Surendra K. Saxena, (D), Uppsala, 1967, Gy

Dept of Earth & Environment (B,M,D) (2015)
Miami, FL 33199
    p. (305) 348-2365
    geology@fiu.edu
    http://www.fiu.edu/orgs/geology
    Enrollment (2007): B: 30 (7) M: 11 (5) D: 20 (5)
Chair:
    Rene Price, (D), Miami, 2001, HwClRw
Professor:
    William T. Anderson, (D), ETH (Switzerland), 2000, CsPcCl
    Grenville Draper, (D), West Indies, 1979, GctRh
    Rosemary Hickey-Vargas, (D), MIT, 1983, Cg
    Jose F. Longoria, (D), Texas (Dallas), 1972, ZeGoe
    Florentin J-M.R Maurrasse, (D), Columbia, 1973, PsGsCm
    Gautam Sen, (D), Texas (Dallas), 1981, Gi
    Michael C. Sukop, (D), Kentucky, 2001, HwqSp
    Dean Whitman, (D), Cornell, 1993, YgGet
Associate Professor:
    Laurel S. Collins, (D), Yale, 1988, Pg
    Michael Gross, (D), Penn State, 1993, Gc
    Andrew W. Macfarlane, (D), Harvard, 1989, Em
    Shimon Wdowinski, (D), Harvard, 1990, YdGt
Assistant Professor:
    Ping Zhu, (D), Miami, 2002, As
Distinguished Research Professor:
    Stephen E. Haggerty, (D), London, 1968, GiEgGz
    Hugh E. Willoughby, (D), Miami, 1977, As
Lecturer:
    Neptune Srimal, (D), Rochester, 1986, Gt
Adjunct Professor:
    Jose Antonio Barros, (D), Miami, 1995, Zr
    Michael Wacker, (M), Florida Intl, Gg
Research Scientist:
    Edward Robinson, (D), London, 1969, Pm
Research Lab Manager:
    Diane H. Pirie, (B), Florida Intl, 1980, Gg

Cooperating Faculty:
    Gabriel Guitierrez-Alonso, (D), Oviedo, 1992, Gc

Florida State University
Dept of Earth, Ocean, and Atmospheric Science (B,M,D) (2015)
108 Carraway Building
Tallahassee, FL 32306-4100
    p. (850) 644-5861
    odom@gly.fsu.edu
    http://www.gly.fsu.edu
    Program Assistant: Tami S. Karl
Chair:
    LeRoy A. Odom, (D), North Carolina, 1971, Cg
    James F. Tull, (D), Rice, 1973, Gct
Professor:
    James B. Cowart, (D), Florida State, 1974, Cc
    Lynn M. Dudley, (D), Washington State, 1983, Sc
    Philip Froelich, (D), Rhode Island, 1979, Og
    Bill X. Hu, (D), Purdue, 1996, Hw
    Munir Humayun, (D), Chicago, 1994, XcCgGi
    Yang Wang, (D), Utah, 1992, CgsPc
Associate Professor:
    Anthony J. Arnold, (D), Harvard, 1983, Pm
    Stephen A. Kish, (D), North Carolina, 1983, Eg
    William C. Parker, (D), Chicago, 1983, PqeGq
    Vincent J. Salters, (D), MIT, 1989, Cg
Assistant Professor:
    Jennifer Georgen, (D), MIT/WHOI, 2001, Gu
    Ming Ye, (D), Arizona, 2002, Hg
Emeritus:
    George W. Devore, (D), Chicago, 1952, Gp
    John K. Osmond, (D), Wisconsin, 1954, Cc
    Paul C. Ragland, (D), Rice, 1962, Ca
    Sherwood W. Wise, Jr, (D), Illinois, 1970, PmGu

Hillsborough Community College
Earth Science (A) (2015)
2112 North 15th St.
Tampa, FL 33605
    p. (813) 253-7647
    jolney2@hccfl.edu
    http://www.hccfl.edu
Instructor:
    Marianne O. Caldwell, (D), S Florida, 2012, Ggs
    James W. Fatherree, Zg
    Thomas M. Klee, (M), S Illinois, 1986, Gg
    Jessica L. Olney, (D), N Illinois, 2006, ZgGg
    Matthew J. Werhner, (M), Adelphi, 1974, Gg
    James F. Wysong, Jr., (M), S Florida, 1989, Am
Adjunct Professor:
    Joseph D. Brod, Zg
    Kyle M. Champion, Zg
    James R. Douthat, Jr., Zg
    Van E. Hayes, Zg
    James E. MacNeil, Zg
    Brian W. Marlowe, Zg
    Matthew P. Olney, (D), N Illinois, 2006, ZgGg
    Leonard T. Roth, Zg
    Norman E. Soash, Zg
    Poetchanaporn Tongdee, Zg
    Kevan A. Van Cleave, Zg
    William V. Wills, Zg

Jacksonville University
Dept of Biology & Marine Science (B,M) ☒ (2018)
2800 University Boulevard North
Jacksonville, FL 32211
    p. (904) 256-7302
    ngoldbe@ju.edu
    Programs: Biology; Marine Science
    Enrollment (2018): B: 10 (10)
Chair:
    Nisse Goldberg, (D)

Miami Dade College (Kendall Campus)
Chemistry-Physics & Earth Science Dept (A) (2015)
11011 S.W. 104 St.
Room 3291
Miami, FL 33176
    p. (305) 237-2492
    sorbon@mdc.edu
    http://www.mdc.edu/kendall/chmphy/
    Enrollment (2010): A: 100 (0)
Assistant Professor:
    Michael G. McGauley, (M), Miami, 2003, AmOp
Emeritus:
    John M. Steger, (D), Naval Postgrad Sch, 1997, OgAmZg

Miami-Dade College (Wolfson Campus)
Dept of Natural Sciences, Health & Wellness (A) (2017)
300 NE 2nd Avenue
Miami, FL 33132
    p. (305) 237-3658
    mkraus@mdc.edu
    Administrative Assistant: Ileana Baldizon
Professor:
    Tony Barros, (D), Miami, 1995, Og
    Michael Kaldor, (M), SUNY (Buffalo), 1969, Gg

Palm Beach Community College
Environmental Science (2015)
MS #56 4200 Congress Avenue
Lake Worth, FL 33467
    p. (561) 868-3475
    milesj@palmbeachstate.edu
    http://www.pbcc.edu

Pensacola Junior College
Dept of Physical Sciences (2015)
1000 Collge Blvd
Pensacola, FL 32504-8998
    p. (850) 484-1189
    estout@pensacolastate.edu
    Administrative Assistant: Kimberly LaFlamme
District Academic Department Head:
    Edwin Stout
Professor:
    Lois Dixon
    Brooke L. Towery, (M), Ball State, 1969, Gg
    Wayne Wooten
    Joseph Zayas
Assistant Professor:
    Thor Garber
    Timothy Hathaway
Instructor:
    Bobby Roberson
    Michael Stumpe
Emeritus:
    Thomas Gee
Science Lab Specialist:
    Darrell Kelly

Saint Petersburg College, Clearwater
Dept of Natural Sciences (A) (2015)
2465 Drew Street
Clearwater, FL 33765
    p. (727) 791-2534
    williams.john@spcollege.edu
    http://www.spcollege.edu/clw/science/
Professor:
    Carl Opper, (M), Florida, 1982, Hw
Adjunct Professor:
    Neva Duncan Tabb, Am
    Hilary Flower, (M), California (Santa Barbara), Gg
    Joseph C. Gould, (D), Nova, Gg
    Heather L. Judkins, (D), S Florida, Og

Sante Fe Community College
Physical Science Dept (2015)

3000 NW 83rd Street
Gainesville, FL 32606
p. (505) 428-1307
david.johnson@sfcc.edu
http://www.sfcollege.edu

Florida

University of Florida
Dept of Geological Sciences (B,M,D) O (2018)
P.O. Box 112120
241 Williamson Hall
1843 Stadium Road
Gainesville, FL 32611-2120
p. (352) 392-2231
info@geology.ufl.edu
http://web.geology.ufl.edu/
Programs: Geology
Certificates: Geology
Enrollment (2018): M: 9 (5) D: 29 (7)

Chair:
 David A. Foster, (D), SUNY (Albany), 1989, GtCc
Affiliate Faculty:
 Joann Mossa, (D), Louisiana State, 1990, GmHsZy
Professor:
 Thomas S. Bianchi, (D), Maryland, 1987, Co
 Mark Brenner, (D), Florida, 1983, Gn
 James E T Channell, (D), Newcastle upon Tyne, 1975, YmGt
 Douglas S. Jones, (D), Princeton, 1980, PeiPg
 Ellen E. Martin, (D), California (San Diego), 1993, Cl
 Jonathan B. Martin, (D), California (San Diego), 1993, ClHg
 Joseph G. Meert, (D), 1993, YgGt
 Paul A. Mueller, (D), Rice, 1971, CcGit
 Michael R. Perfit, (D), Columbia, 1977, Giu
 Elizabeth J. Screaton, (D), Lehigh, 1995, Hg
Associate Professor:
 Peter N. Adams, (D), California (Santa Cruz), 2004, Gm
 Paul F. Ciesielski, (D), Florida State, 1978, Pm
 John M. Jaeger, (D), SUNY (Stony Brook), 1998, Gs
 Raymond Russo, (D), Northwestern, 1990, Yg
 Andrew Zimmerman, (D), William & Mary, 2000, ComSb
Associate Scientist:
 Ann L. Heatherington, (D), Washington (St. Louis), 1988, CcGti
 Kyoungwon Min, (D), California (Berkeley), 2002, Cc
Assistant Professor:
 Andrea Dutton, (D), Michigan, 2003, CcGsPc
 Mark Panning, (D), California (Berkeley), 2004, Ys
Research Associate:
 Ray G. Thomas, (B), Florida, 1990, Zg
Lecturer:
 Matthew C. Smith, (D), Gi
 Jim Vogl, (D), California (Santa Barbara), 2000, Gct
Emeritus:
 Frank N. Blanchard, (D), Michigan, 1960, Gz
 Guerry H. McClellan, (D), Illinois, 1964, En
 Neil D. Opdyke, (D), Durham (UK), 1958, Ym
 Anthony F. Randazzo, (D), North Carolina, 1968, Gd
 Douglas L. Smith, (D), Minnesota, 1972, Yh
 Daniel P. Spangler, (D), Arizona, 1969, Hw
Laboratory Director:
 Jason H. Curtis, (D), Florida, 1997, Pe
 George D. Kamenov, (D), Florida, 2004, CgaEm
Assistant Curator:
 Jonathan Bloch, (D), Michigan, 2001, Pv
Affiliate Faculty:
 Micheal W. Binford, (D), Indiana, 1980, Zr
Cooperating Faculty:
 David L. Dilcher, (D), Yale, 1964, Pb
 Bruce J. McFadden, (D), Columbia, 1976, Pv
 Claire L. Schelske, (D), Michigan, 1961, Pe

Florida Museum of Natural History (2015)
PO Box 117800
Gainesville, FL 32611-7800
p. (352) 846-2000
bmacfadd@flmnh.ufl.edu
http://www.flmnh.ufl.edu
Administrative Assistant: Pam Dennis

Director:
 Douglas S. Jones, (D), Princeton, 1980, Pg
Associate Director:
 Graig D. Shaak, (D), Pittsburgh, 1972, Pe
Associate Curator:
 Bruce J. MacFadden, (D), Columbia, 1976, Pv
 David W. Steadman, (D), Arizona, 1982, PvGaPs
Professor:
 Steven R. Manchester, (D), Indiana, 1981, PbGs
Senior Scientist:
 Roger W. Portell, (B), Florida, 1985, PisPv
Graduate Research Professor:
 David L. Dilcher, (D), Yale, 1964, Pb
Research Associate:
 David M. Jarzen, (D), Toronto, 1973, Pb
Emeritus:
 Sylvia J. Scudder, (M), Florida, 1993, Pg
Other:
 Ann S. Cordell, (M), Florida, 1983, Gxa

Soil & Water Science Dept (B,M,D) (2018)
PO Box 110290
2181 McCarty Hall A
Gainesville, FL 32611-0290
p. (352) 294-3151
krr@ufl.edu
http://soils.ifas.ufl.edu

Chair:
 K Ramesh Reddy, (D), Louisiana State, 1976, CbSbc
Professor and Center Director:
 Nicholas B. Comerford, (D), SUNY (Syracuse), 1980, Sf
 John E. Rechcigl, (D), Virginia Tech, 1986, So
 Charley Wesley Wood, (D), So
Professor:
 Teri Balser, (D), Soo
 Mary E. Collins, (D), Iowa State, 1980, Sd
 James H. Graham, Jr., (D), Oregon State, 1980, Sb
 Sabine -. Grunwald, (D), Giessen (Germany), 1996, ZiSoZr
 Willie G. Harris, Jr., (D), Virginia Tech, 1984, Scd
 George J. Hochmuth, (D), Wisconsin, 1980, So
 Yuncong Li, (D), Maryland, 1993, Sco
 Lena Q. Ma, (D), Colorado State, 1991, Sc
 Peter Nkedi-Kizza, (D), California, 1979, Sp
 George A. O'Connor, (D), Colorado State, 1970, Sc
 Thomas A. Obreza, (D), Florida, 1983, So
 Andrew V. Ogram, (D), Tennessee, 1988, Sb
 Jerry B. Sartain, (D), North Carolina, 1973, So
Associate Professor:
 Zhenli He, (D), Zhejang Ag, 1988, Sc
 James W. Jawitz, (D), Florida, 1999, SpHsw
 Marc Kramer, (D), So
 S Rao Mylavarapu, (D), Clemson, 1996, Sc
 Arnold W. Schumann, (D), Georgia, 1997, So
 P. Christopher Wilson, (D), Clemson, 1999, So
Assistant Professor:
 Mark W. Clark, (D), Florida, 2000, Sf
 Samira H. Daroub, (D), Michigan State, 1994, So
 Stefan Gerber, (D), So
 Patrick Inglett, (D), Florida, 2005, Sf
 Cheryl Mackowiak, (D), Utah State, 2001, So
 Kelly Morgan, (D), So
 Maria Silveira, (D), Sao Paulo, 2003, So
 Max Teplitski, (D), Ohio State, 2002, Zn
 Gurpal Toor, (D), Lincoln, 2002, Sc
 Alan Wright, (D), Sc
Research Associate Professor:
 Vimala D. Nair, (D), Gottingen, 1978, Sc
 Ann C. Wilkie, (D), Univ Coll (Ireland), 1984, Sb
Lecturer:
 James Bonczek, (D), SoHsg
 Susan Curry, (M), ZiSo
Research Assistant Professor:
 Todd Osborne, (D), Sf
 John Thomas, (D), Sp
Assistant In:
 Mengsheng Gao, (D), Zn

University of Miami
Dept of Geography and Regional Studies (2015)
Coral Gables, FL 33124
   p. (305) 284-6695
   dofuller@miami.edu
   http://www.as.miami.edu/geography/

Dept of Geological Sciences (B,M,D) (2017)
1301 Memorial Drive
43 Cox Science Building
Coral Gables, FL 33124
   p. (305) 284-4253
   ruthgoodin@miami.edu
   http://www.as.miami.edu/geology/
   Administrative Assistant: Ruth Goodin
   Enrollment (2016): B: 37 (7)
Professor:
   Larry C. Peterson, (D), Brown, 1984, Gr
   Peter Swart
   Harold R. Wanless, (D), Johns Hopkins, 1971, GseOn
Senior Lecturer:
   Teresa A. Hood, (D), Miami, 1991, Gg
Lecturer:
   Peter J. Leech, (D), Georgia Tech, 2013
   Ta-Shana A. Taylor, (M), Arizona, 2006, ZePvGe
Emeritus:
   David E. Fisher, (D), Florida, 1958, Cs
   John R. Southam, (D), Illinois, 1974, Op

Dept of Marine Geosciences (B,M,D) ☒ (2019)
RSMAS
4600 Rickenbacker Causeway
Miami, FL 33149-1098
   p. (305) 421-4662
   spurkis@rsmas.miami.edu
   https://marine-geosciences.rsmas.miami.edu/
   Programs: PhD and MSc in Marine Geosciences
   BS Dual major in Marine Science/Geological Sciences
   BS in Geological Sciences
   BS/MS Marine Geology
   BA in Geological Sciences
   Certificates: The Certificate Program in "Applied Carbonate Geology
   Enrollment (2018): B: 30 (6) M: 0 (1) D: 14 (13)
Chair:
   Sam J. Purkis, (D), Vrije (Amsterdam), 2004, Gu
Professor:
   Falk Amelung, (D), Lous Pasteur, 1996, Zn
   Keir Becker, (D), Scripps, 1981, YrGu
   Gregor P. Eberli, (D), ETH (Switzerland), 1985, GusYx
   Larry C. Peterson, (D), Brown, 1984, GuPcm
   Pam Reid, (D), Miami, 1979, Gs
   Peter K. Swart, (D), London, 1980, ClPsCs
Scientist:
   Donald F. McNeill, (D), Miami, 1989, GsrGu
Associate Professor:
   James S. Klaus, (D), Illinois, 2005, Gg
Associate Scientist:
   Guoqing Lin, (D), Scripps, 2005, Yg
   Ali Pourmand, (D), Tulane, 2005, Cga
Assistant Professor:
   Amanda M. Oehlert, (D), Miami, RSMAS, 2014, Gu
Emeritus:
   Christopher G.A. Harrison, (D), 1964, Yr
Cooperating Faculty:
   Patricia L. Blackwelder, (D), South Carolina, 1976, Pg

University of South Florida
College of Marine Science (M,D) (2017)
140 7th Avenue South
St. Petersburg, FL 33701
   p. (727) 553-1130
   Mitchum@usf.edu
   http://www.marine.usf.edu
   Enrollment (2009): M: 30 (13) D: 50 (6)

Distinguished Research Professor:
   Robert H. Byrne, (D), Rhode Island, 1974, Cm
   John H. Paul, (D), Miami, 1980, Ob
   John J. Walsh, (D), Miami, 1969, Ob
   Robert H. Weisberg, (D), Rhode Island, 1975, Op
Professor:
   Luis H. Garcia-Rubio, (D), McMaster, 1981, Oc
   Pamela Hallock-Muller, (D), Hawaii, 1977, OuGsPm
   Gary T. Mitchum, (D), Florida State, 1984, Op
   Frank E. Muller-Karger, (D), Maryland, 1988, Zr
   Joseph J. Torres, (D), California, 1980, Ob
Associate Professor:
   Paula G. Coble, (D), MIT/WHOI, 1992, Cm
   Kendra Lee Daly, (D), Tennessee, 1995, Ob
   Boris Galperin, (D), Israel Inst of Tech, 1982, Op
   David J. Hollander, (D), Swiss Fed Inst Tech, 1989, Oc
   Mark E. Luther, (D), North Carolina, 1982, Op
   David F. Naar, (D), Scripps, 1990, Ou
   Ernst B. Peebles, (D), S Florida, 1996, ObCs
Assistant Professor:
   Mya Breitbart, (D), California (San Diego), 2006, Ob
   Ashanti J. Pyrtle, (D), Texas A&M, 1999, Oc
Research Associate:
   Rick Cole, (B), Florida Inst of Tech, 1983, Op
   Dwight A. Dieterle, (M), San Jose, 1979, Ob
   Jeff C. Donovan, (B), Florida, 1985, Op
   Debra E. Huffman, (D), S Florida, 1994, Zn
   Stanley D. Locker, (D), Rhode Island, 1989, Ou
   Wensheng Yao, (D), Miami, 1995, Oc
Research Assistant:
   David C. English, (M), Washington, 1983, Op
Adjunct Professor:
   Serge Andrefouet, (D), Polynesie Francaise, 1998, OpZr
   Bruce Barber, (D), Ob
   Leonard Ciaccio, (D)
   Thomas Cuba, (D), S Florida, Ob
   Christopher D'Elia, (D), Georgia, 1974, Ob
   George Denton, (D), Og
   Cynthia Heil, (D), S Florida, Ob
   Brian Keller, (D)
   John Lisle, (D)
   Anne Meylan, (D), Florida, 1984, Ob
   Terrence Quinn, (D), Ou
   Harunur Rashid, (D)
   Eugene Shinn, (B), Miami, 1957, Gg
   Randy Wells, (D), California (Santa Cruz), 1986, Ob
Emeritus:
   Peter R. Betzer, (D), Rhode Island, 1970, Cm
   Norman J. Blake, (D), Rhode Island, 1972, Ob
   John C. Briggs, (D), Stanford, 1951, Ob
   Kendall L. Carder, (D), Oregon State, 1970, Op
   Kent A. Fanning, (D), Rhode Island, 1973, OcGu
   Albert C. Hine, (D), South Carolina, 1975, Gus
   Thomas L. Hopkins, (D), Florida State, 1964, Ob
   Harold J. Humm, (D), Duke, 1945, Ob
   Edward S. Van Vleet, (D), Rhode Island, 1978, Co
   Gabriel A. Vargo, (D), Rhode Island, 1976, Ob

University of South Florida, Tampa
Dept of Geology (B,M,D) (2015)
SCA 528
4202 East Fowler Avenue
Tampa, FL 33620-5201
   p. (813) 974-2236
   mrains@usf.edu
   Administrative Assistant: Mary Haney
Chair:
   Charles B. Connor, (D), Dartmouth, 1987, Gv
Professor:
   Mark T. Stewart, (D), Wisconsin, 1976, Hw
   H. Leonard Vacher, (D), Northwestern, 1971, Hw
Associate Professor:
   Peter J. Harries, (D), Colorado, 1993, Pg
   Jeffrey G. Ryan, (D), Columbia, 1989, Ct
Assistant Professor:
   Sarah E. Kruse, (D), MIT, 1989, Yg
   Thomas Pichler, (D), Ottawa, 1998, Cl

Mark C. Rains, (D), California (Davis), 2002, Hw
Ping Wang, (D), S Florida, 1995, Gs
Instructor:
Thomas C. Juster, (D), S Florida, 1995, Hw
Eleanour Snow, (D), Brown, 1987, Gz
Emeritus:
Richard A. Davis, Jr., (D), Illinois, 1964, Gs
Assistant Curator:
Raymond Denicourt, Gz
Cooperating Faculty:
Jeffrey Heikoop, Cl
Lisa Robbins, (D), Miami, 1987, Pg
James E. Sorauf, (D), Kansas, 1962, Pg
Barbara W. Leyden, (D), Indiana, 1982, Pl
Robert Morton, (D), West Virginia, 1972, Gs
Thomas M. Scott, (D), Florida State, Gr
Sam B. Upchurch, (D), Northwestern, 1970, Hw

University of West Florida
Dept of Earth and Environmental Sciences (B,M) ☒ (2018)
11000 University Parkway
Bldg 13
Pensacola, FL 32514
  p. (850) 474-3377
  environmental@uwf.edu
  http://uwf.edu/cse/departments/earth-and-environmental-sciences/
  f: https://www.facebook.com/pages/UWF-Department-of-Environmental-Studies/205988446096964
  t: @UWF_EES
  Programs: Environmental Science (B,M); GIS Administration (M)
  Certificates: GIS (UG,G)
  Enrollment (2016): B: 133 (39) M: 40 (6)
Chair:
  Matthew C. Schwartz, (D), Delaware, 2002, OcCl
Professor:
  Johan Liebens, (D), Michigan State, 1995, So
Associate Professor:
  Zhiyong Hu, (D), Georgia, 2003, Zr
Assistant Professor:
  John D. Morgan, (D), Florida State, 2010, Zi
  Jason Ortegren, (D), North Carolina (Greensboro), 2008, PeZu
  Phillip P. Schmutz, (D), Louisiana State, 2014, Gm
Instructor:
  Chasidy Hobbs, (M), W Florida, 2005, ZgeZy
  Taylor Kirschenfeld, (M), W Florida, 1988, Ob
Adjunct Professor:
  Wilbur G. Hugli, (D), W Florida, 2001, As
  Hilde Snoeckx, (D), Michigan, 1995, OuZg
Online GIS Program Coordinator:
  Amber Bloechle, (M), W Florida, 2007, Zi

Valencia Community College
Valencia Community College (2015)
P.O. Box 3028
Orlando, FL 32802
  p. (407) 299-5000
  abosley@valenciacollege.edu
  http://www.valenciacc.edu

Georgia

Columbus State University
Earth & Space Sciences (B,M) ☒ (2018)
4225 University Avenue
Columbus, GA 31907-5645
  p. (706) 507-8091
  barineau_clinton@columbusstate.edu
  http://ess.columbusstate.edu/
  f: https://www.facebook.com/Earth-and-Space-Sciences-at-Columbus-State-University-GA-347767372234/
  Programs: AS Engineering Studies, BS Earth and Space Sciences with 4 concentrations: Astrophysics and Planetary Geology, Environmental Science, Geology, Secondary Education, MS Natural Sciences with 2 tracks: Environmental Science, Geosciences
  Certificates: Robotics
  Enrollment (2018): B: 78 (13) M: 10 (0)
Chair:
  Clinton I. Barineau, (D), Florida State, 2009, GctGg
Professor:
  Warren Church, (D), Yale, 1996, ZnPcZc
  Shawn Cruzen, (D), Nevada, 1997, Xa
  Zdeslav Hrepic, (D), Kansas State, 2004, Zn
  Troy Keller, (D), Michigan, 1997, RwZc
  Shaw Kimberly, (D), Florida State, 1997, Zn
  David R. Schwimmer, (D), SUNY (Stony Brook), 1973, Pv
  Abiye Seifu, (D), Rensselaer, 1991, Zn
  Rosa Williams, (D), Illinois, 1999, Xa
Associate Professor:
  Andrew Puckett, (D), Chicago, 2007, Xa
Assistant Professor:
  Stacey Sloan Blersch, (D), SUNY (Buffalo), 2016, HsgRw
  William Scott Gunter, (D), Texas Tech, 2015, Ams
  Diana Ortega-Ariza, (D), Kansas, 2016, GdgGr
  Lavi Zamstein, (D), Florida, 2009, Zn
Emeritus:
  William J. Frazier, (D), North Carolina, 1973, GsrGd
  Thomas B. Hanley, (D), Indiana, 1975, Gc

Dalton State Community College
Dept of Natural Sciences (A,B) ☒ (2019)
650 College Drive
Dalton, GA 30720
  p. (706) 272-4440
  jmjohnson@daltonstate.edu
  http://www.daltonstate.edu/natural-sciences/index.html
Professor:
  Jean M. Johnson, (D), Michigan, 1995, Zg

East Georgia State College
Science & Mathematics (A,B) ☒ (2017)
131 College Circle
Swainsboro, GA 30401
  p. (478) 289-2073
  stracher@ega.edu
  http://faculty.ega.edu/facweb/stracher/stracher.html
  Enrollment (2015): A: 3 (1)
Professor Emeritus:
  Glenn B. Stracher, (D), Nebraska, 1989, GycEc

Emory University
Dept of Environmental Sciences (B) (2015)
Mathematics & Science Center
400 Dowman Drive
Atlanta, GA 30322
  p. (404) 727-4216
  jwegner@emory.edu
  http://www.envs.emory.edu/
Head:
  Joy Budensiek, (D)
Chair:
  Lance Gunderson, (D), Florida, 1992, Sf
Professor:
  William B. Size, (D), Illinois, 1971, GivGa
Associate Professor:
  Thomas Gillespie, (D), Florida, 2004, Zn
Assistant Professor:
  Tracy Yandle, (D), Indiana, 2001, Ob
Instructor:
  Anne M. Hall, (M), Georgia Tech, 1985, Gz
Campus Env Officer:
  John Wegner, (D), Carleton, 1995, Sf
Lecturer:
  Anthony J. Martin, (D), Georgia, 1991, Pe
Adjunct Professor:
  Pamela J. W. Gore, (D), George Washington, 1983, Gs
Senior Lecturer:
  Charles W. Hickcox, (D), Rice, 1971, Gg
Emeritus:
  Howard R. Cramer, (D), Northwestern, 1954, Pg

Willard H. Grant, (D), Johns Hopkins 1993, Cg
Lore Ruttan (D), California (Davis), 1999, Ob

Fort Valley State University
Cooperative Developmental Energy Program ☒ (2017)
Box 5800 FVSU
1005 State University Drive
Fort Valley, GA 31030
p. (912) 825-6454
crumblyi@fvsu.edu
http://www.fvsu.edu/academics/cdep
Director:
  Isaac J. Crumbly, (D), North Dakota State, 1970, Zn
Assistant Director:
  Jackie Hodges, (B), Georgia Southern, 1983, Zn
Associate Professor:
  Aditya Kar, (D), Oklahoma, 1997, Ct

Georgia Highlands College
Div of Science and Physical Education (A,B) (2015)
Main Campus 3175 Hwy 27 South
P.O. Box 1864
Rome, GA 30162
p. (706) 368-7528
bmorris@highlands.edu
http://www.highlands.edu/site/geology
Associate Professor:
  Billy Morris, Gg
Instructor:
  Tracy Hall, Gg

Georgia Institute of Technology
School of Earth & Atmospheric Sciences (B,M,D) (2015)
311 Ferst Drive
Atlanta, GA 30332-0340
p. (404) 894-3893
greg.huey@eas.gatech.edu
http://www.eas.gatech.edu
  Enrollment (2011): B: 63 (21) M: 34 (19) D: 76 (13)
Chair:
  Judith A. Curry, (D), Chicago, 1992, AsZrn
Professor:
  Annalisa Bracco, (D), Genova (Italy), 2000, OpZnn
  Kim M. Cobb, (D), Scripps, 2002, PeCsOg
  Felix J. Herrmann, (D), Delft (Neth), 1997, YesZn
  Gregory L. Huey, (D), Wisconsin, 1992, As
  Ellery D. Ingall, (D), Yale, 1991, CmlOc
  Jean Lynch-Stieglitz, (D), Columbia, 1995, OoPcCs
  Andrew V. Newman, (D), Northwestern, 2000, YdsYg
  Edward Michael Perdue, (D), Georgia Tech, 1973, Co
  Irina Sokolik, (D), Russian Acad of Sci, 1989, AsZr
  Martial Taillefert, (D), Northwestern, 1997, ClmCa
  Rodney J. Weber, (D), Minnesota, 1995, As
  Peter J. Webster, (D), MIT, 1972, AsOp
Senior Scientist:
  Hai-ru Chang, (D), As
  Robert Stickel, (D), Rice, 1979, As
  Viatcheslav Tatarskii, (D), As
  Hsiang-Jui Wang, (D), Georgia Tech, 1995, As
Associate Professor:
  Michael H. Bergin, (D), Carnegie Mellon, 1995, As
  Robert X. Black, (D), MIT, 1990, As
  Yi Deng, (D), Illinois, 2005, AsZoAm
  Emanuele Di Lorenzo, (D), Scripps, 2003, OpZnn
  Takamitsu Ito, (D), MIT, 2005
  Athanasios Nenes, (D), Caltech, 2002, As
  Marc Stieglitz, (D), Columbia, 1995, Hsg
  Yuhang Wang, (D), Harvard, 1997, As
  James Wray, (D), Cornell, 2010
Assistant Professor:
  Josef Dufek, (D), Washington, 2006, GvZnn
  Carol M. Paty, (D), Washington, 2006, YmgZn
  Zhigang Peng, (D), S California, 2004, YsgZn
  Andrew Stack, (D), Wyoming, 2002, Cg
Research Scientist:
  Carlos Hoyos, (D), Georgia Tech, 2008, As
  Hyemi Kim, (D), As
  Jiping Liu, (D), Columbia, As
  Chao Luo, (D), Peking, 1990, As
  James C. St. John, (D), Georgia Tech, 1997, As
  Henian Zhang, (D), As
Adjunct Associate Professor:
  Jay Brandes, (D), Washington, Og
  Carmen Nappo, (D), Georgia Tech, 1989, As
  Valerie Thomas, (D), Cornell, 1987, Zu
Adjunct Assistant Professor:
  Karim Sabra, (D), Michigan, 2003, Yg
Adjunct Professor:
  Clark R. Alexander, (D), North Carolina State, 1990, Cl
  Dominic Assimaki, (D)
  Jackson O. Blanton, (D), Oregon State, 1968, Cl
  Thomas D. Christina, (D), Caltech, 1989, Cl
  James Crawford, (D), Georgia Tech, As
  Heidi Cullen, (D), Columbia, Am
  Rong Fu, (D)
  Leonid Germanovich, (D), Moscow State Mining (Russia), 1982, NrYg
  Gary G. Gimmestad, (D), Colorado, 1978, ZrAs
  Richard Jahnke, (D), Washington, 1976, Cl
  Yongqiang Liu, (D), Am
  Joseph Montoya, (D), Harvard, 1990, Cm
  Armistead G. Russell, (D), Caltech, 1985, As
  Stuart G. Wakeham, (D), Washington, 1976, Co
  Herbert L. Windom, (D), California (San Diego), 1968, Cm
Emeritus:
  Paul H. Wine, (D), Florida State, 1974, As

Georgia Southern University
Applied Coastal Research Laboratory (2015)
10 Ocean Science Circle
Savannah, GA 31411
p. (912) 598-2329
clarkalexander@georgiasouthern.edu
http://cosm.georgiasouthern.edu/icps/acrl/
Director:
  Clark R. Alexander, (D), North Carolina State, 1990, OuGsOn
Assistant Professor:
  Chester M. Jackson, (D), Georgia, 2010, OnGs
Research Associate:
  Michael Robinson, (B), Georgia Southern, 2002, ZiOn

Dept of Geology and Geography (B) (2019)
68 Georgia Avenue
Building 201
PO Box 8149
Statesboro, GA 30460-8149
p. (912) 478-5361
jreich@georgiasouthern.edu
http://cost.georgiasouthern.edu/geo/
  Enrollment (2016): B: 77 (11)
Chair:
  Stephen J. Underwood, (D), Georgia, 1999, Am
Professor:
  James S. Reichard, (D), Purdue, 1995, Hw
  Fredrick J. Rich, (D), Penn State, 1979, Pl
  Wei Tu, (D), Texas A&M, 2004, Zi
  Robert Kelly Vance, (D), New Mexico Tech, 1989, GiEgGg
  Mark R. Welford, (D), Illinois, 1993, Zy
Associate Professor:
  Robert A. Yarbrough, (D), Georgia, 2006, Zn
Assistant Professor:
  Christine Hladik, (D), Georgia, 2012, SbZr
  Chester W. Jackson, (D), Georgia, 2010, GsmGu
  Jacque L. Kelly, (D), Hawaii, 2012, Cl
  Kathlyn M. Smith, (D), Michigan, 2010, Pv
  John Van Stan, (D), Delaware, 2012, HgClSb
  Xiaolu Zhou, (D), Illinois, 2014, Zr
Emeritus:
  Gale A. Bishop, (D), Texas, 1971, Pi
  James H. Darrell, (D), Louisiana State, 1973, Pl
  Daniel B. Good, (D), Tennessee, 1973, Zn
  Dallas D. Rhodes, (D), Syracuse, 1973, Gm

Georgia Southwestern State University
Dept of Geology & Physics (B) (2015)
800 GSW State University Drive
Americus, GA 31709
  p. (229) 931-2353
  Deborah.Standridge@gsw.edu
  http://www.gsw.edu/%7Egeology/
  Administrative Assistant: Debbie Standridge
Professor:
  Burchard D. Carter, (D), West Virginia, 1981, Pg
  Thomas J. Weiland, (D), North Carolina, 1988, Gi
Associate Professor:
  Samuel T. Peavy, (D), Virginia Tech, 1997, Yg
Assistant Professor:
  Svilen Kostov, (D), CUNY, 1992, Zn
Emeritus:
  Daniel D. Arden, (D), California (Berkeley), 1961, Ps
  Harland E. Cofer, (D), Illinois, 1957, Gz
  John P. Manker, (D), Rice, 1975, Gs

Georgia State University
Dept of Geosciences (B,M,D) ○⊘ (2018)
730 Langdale Hall
PO Box 3965
38 Peachtree Center Avenue
Atlanta, GA 30302-3965
  p. (404) 413-5750
  khankins@gsu.edu
  http://geosciences.gsu.edu
  f: https://www.facebook.com/gsugeosciences
  t: @gsugeos
  Programs: B.S. in Geosciences with concentrations in Geography, Urban Studies, Environmental Geosciences, and Geology, M.S. in Geosciences with concentrations in Geography, Geology, or Water Science
  Certificates: undergraduate certificates: Water Science, GIS, Sustainability, graduate certificates: GIS
  Administrative Assistant: Basirat Lawal
  Enrollment (2018): B: 112 (33) M: 47 (27) D: 1 (1)
VPAA-Provost:
  Risa I. Palm, (D), Minnesota, 1972, Zn
Associate Dean:
  Daniel M. Deocampo, (D), Rutgers, 2001, Gsn
Chair:
  Katherine B. Hankins, (D), Georgia, 2004, Zn
Professor:
  Jeremy E. Diem, (D), Arizona, 2000, At
Associate Professor:
  Hassan A. Babaie, (D), Northwestern, 1984, GcZf
  Dajun Dai, (D), S Illinois, 2007, Zi
  W. Crawford Elliott, (D), Case Western, 1988, Clc
  Nadine Kabengi, (D), Florida, ScClGe
  Lawrence M. Kiage, (D), Louisiana State, 2007, ZyPlZr
Assistant Professor:
  Sarah Ledford, (D), Syracuse, 2016, HsRw
  Richard Milligan, (D), Georgia, 2016, Zn
  Luke Pangle, (D), Oregon State, 2013, HqCs
Lecturer:
  Paulo J. Hidalgo, (D), Michigan State, 2011, GivGz
  Brian Meyer, (D), Georgia State, 2013, GesHw
  Ricardo Nogueira, (D), Louisiana State, 2009, Asm
  Christy Visaggi, (D), North Carolina (Wilmington), 2011, PiZe
Adjunct Professor:
  J. Marion Wampler, (D), Columbia, 1963, Cc
Emeritus:
  Sanford H. Bederman, (D), Minnesota, 1973, Zn
  Timothy E. La Tour, (D), W Ontario, 1979, Gp
  W. Robert Power, (D), Johns Hopkins, 1959, En
  Seth E. Rose, (D), Arizona, 1987, Hw
Related Staff:
  Atieh Tajik, (M), Georgia State, 2005, Gg

Georgia State University, Perimeter College, Alpharetta Campus
Dept of Life & Earth Science (A) ⊠ (2019)
3705 Brookside Parkway
Alpharetta, GA 30022
  p. (678) 240-6227
  dstewart29@gsu.edu
  Programs: Geology
Professor:
  Dion C. Stewart, (D), Penn State, 1980, Gzi

Georgia State University, Perimeter College, Clarkston Campus
Dept of Life & Earth Science (A) ⊘ (2019)
555 North Indian Creek Drive
Clarkston, GA 30021
  p. (678) 891-3754
  pgore@gsu.edu
  Programs: Geology
  Enrollment (2007): A: 3 (0)
Professor:
  Pamela J. W. Gore, (D), George Washington, 1983, GsZeGn
Associate Professor:
  E. Lynn Zeigler, (M), Emory, 1989, Gd
Lecturer:
  Stephan Fitzpatrick, (M), Georgia, 2011, HwCgSo
  Nils W. Thompson, (M), Fort Hays State, 1988, GgZgGe

Georgia State University, Perimeter College, Decatur Campus
Dept of Life & Earth Science (A) ⊠ (2019)
3251 Panthersville Road
Decatur, GA 30034-3897
  p. (678) 891-2600
  dbulger1@gsu.edu
  Programs: Geology
Assistant Professor:
  Daniel E. Bulger, (D), Georgia, 2014, GgZn

Georgia State University, Perimeter College, Dunwoody Campus
Dept of Life & Earth Science (A) ⊘ (2019)
2101 Womack Road
Dunwoody, GA 30338-4497
  p. (770) 274-5050
  Programs: Geology
  Enrollment (2011): A: 4 (0)
Lecturer:
  Robin J. McDowell, (D), Kentucky, 1992, Gge
  Alexander D. Ullrich, (M), Florida, 2010, GglGu
Adjunct Professor:
  Virginia L. Mauldin-Kinney, (M), Georgia State, 2012, GgXg

Georgia State University, Perimeter College, Newton Campus
Dept of Geology (A) (2017)
239 Cedar Lane
Covington, GA 30014
  p. (770) 278-1263
  Polly.bouker@gpc.edu
  http://www.gpc.edu/~newsci/
Associate Professor:
  Polly A. Bouker, (M), Georgia, 2006, Gg

Georgia State University, Perimeter College, Online
Dept of Life & Earth Science (A) ⊘ (2019)
555 North Indian Creek Drive
Clarkston, GA 30021
  p. (678) 212-7577
  dballero@gsu.edu
  Programs: Geology
Associate Chair :
  Deniz Z. A. Ballero, (D), Georgia, 2013, GgPm
Instructor:
  Nils W. Thompson, (M), Fort Hays State, 1988, Gge

Adjunct Professor:
  Edward Albin, (D), Georgia, Xm
  Roy (Rick) A. Nixon III, (D), Emory, 1981, Gg
  Kimberly D. Schulte, (M), Gg
  Christine D. Valenti, Gg

Mercer University
Dept of Environmental Engineering (B,M) (2015)
1400 Coleman Avenue
Macon, GA 31207
    p. (912) 744-2597
    lackey_l@mercer.edu
    Department Secretary: Brenda Walraven
Professor:
  Bruce D. Dod, (D), S Mississippi, 1973, Zg
  Dan R. Quisenberry, (D), World Open, 1980, Zr
Assistant Professor:
  Geoffrey W. Hayden, (D), Pennsylvania, 1987, Zm
  Robert L. Huffman, (D), Massachusetts, 1985, Zn
Research Associate:
  Paul P. Sipiera, (D), Otago (NZ), 1985, Xm
Instructor:
  Barbara D. Henley, (M), Mercer, 1978, Gg

Middle Georgia College
Geology Dept (A) (2015)
1100 Second Street
Cochran, GA 31014
    tmahaffee@mgc.edu
    http://www.mgc.edu
Associate Professor:
  Tina Mahaffee, (M), Georgia State, Gg
Assistant Professor:
  Daniel Snyder, (D), Iowa, Gg

University of Georgia
Dept of Crop & Soil Science (B,M,D) (2015)
311 Plant Sciences Building
Athens, GA 30602-7272
    p. (706) 542-2461
    dgs@uga.edu
    http://www.cropsoil.uga.edu/
Chair:
  Don Shilling
Professor:
  Domy C. Adriano, (D), Kansas State, 1970, Sc
  Paul M. Bertsch, (D), Kentucky, 1983, Sc
  Gary J. Gascho, (D), Michigan State, 1968, Sc
  James E. Hook, (D), Penn State, 1975, Sp
  Edward T. Kanemasu, (D), Wisconsin, 1969, Sp
  David E. Kissel, (D), Kentucky, 1969, Sc
  William P. Miller, (D), Virginia Tech, 1981, Sc
  David E. Radcliffe, (D), Kentucky, 1984, Sp
  William P. Segars, (D), Clemson, 1972, Sc
  Larry M. Shuman, (D), Penn State, 1970, Sc
Associate Professor:
  Miguel L. Cabrera, (D), Kansas State, 1986, So
  Peter G. Hartel, (D), Oregon State, 1984, Sb
  Owen C. Plank, (D), Virginia Tech, 1973, Sc
  Larry T. West, (D), Texas A&M, 1986, Sd
Assistant Professor:
  Glendon H. Harris, (D), Michigan State, 1993, Sb

Dept of Geology (B,M,D) ●✓📖 (2018)
210 Field Street
Room 308
Athens, GA 30602
    p. (706) 542-2652
    geology@uga.edu
    http://geology.uga.edu/
    Enrollment (2011): B: 69 (9) M: 32 (11) D: 6 (0)
Head:
  Paul A. Schroeder, (D), Yale, 1992, GzEnCl
Grad Coordinator :
  Alberto E. Patino-Douce, (D), Oregon, 1990, ZnnZn
Professor:
  Douglas E. Crowe, (D), Wisconsin, 1990, Em
  Ervan G. Garrison, (D), Missouri, 1979, GamGn
  Susan T. Goldstein, (D), California (Berkeley), 1984, Pm
  Robert B. Hawman, (D), Princeton, 1988, YseYu
  Steven M. Holland, (D), Chicago, 1990, Gr
  John E. Noakes, (D), Texas A&M, 1962, Cc
  Valentine A. Nzengung, (D), Georgia Tech, 1993, ClGeCq
  L. Bruce Railsback, (D), Illinois, 1989, PcClGd
  Michael F. Roden, (D), MIT, 1982, Gi
  Sally E. Walker, (D), California (Berkeley), 1988, Pm
  James (Jim) E. Wright, (D), California (Santa Barbara), 1981, Gt
Associate Head:
  Adam Milewski, (D), W Michigan, 2008, Hw
Assistant Professor:
  Geoffrey Howarth, (D), Rhodes, 2013, Gi
  Christian Klimczak, (D), Nevada, 2011, GcmGt
Senior Lecturer:
  Marta Patino-Douce, (D), Buenos Aires, 1990, Gi
Adjunct Assistant Professor:
  Roger A. Burke, (D), S Florida, 1985, Cl
Adjunct Professor:
  Elizabeth J. Reitz, (D), Florida, 1979, Ga
Emeritus:
  Gilles O. Allard, (D), Johns Hopkins, 1956, Eg
  R. David Dallmeyer, (D), SUNY (Stony Brook), 1972, Gt
  John F. Dowd, (D), Yale, 1984, Hy
  Sam Swanson , (D)
  David B. Wenner, (D), Caltech, 1971, Cs
  James A. Whitney, (D), Stanford, 1972, GivCg

University of West Georgia
Dept of Geosciences (B) (2016)
1601 Maple Street
Callaway Building, Room 148
Carrollton, GA 30118
    p. (678) 839-6479
    jmayer@westga.edu
    http://www.westga.edu/geosci/
    Administrative Assistant: Anita M. Bryant
    Enrollment (2016): B: 122 (22)
Chair:
  James R. Mayer, (D), Texas, 1995, Hw
Professor:
  David M. Bush, (D), Duke, 1991, Ou
  Curtis L. Hollabaugh, (D), Washington State, 1980, GzCg
  Randal L. Kath, (D), SD Mines, 1990, GcNgHg
  Jeong C. Seong, (D), Georgia, 1999, Zir
Associate Professor:
  Brad Deline, (D), Cincinnati, 2009, Pi
  Georgina G. DeWeese, (D), Tennessee, 2007, Zy
  Hannes Gerhardt, (D), Arizona, 2007, Zn
  Leanna Shea Rose, (D), Florida State, 2008, ZyAsm
  Karen S. Tefend, (D), Michigan State, 2005, CgGe
  Nathan A. Walter, (D), Florida State, 2005, Zg
Assistant Professor:
  Christopher A. Berg, (D), Texas, 2007, Gpi
  Jessie Hong, (D), Colorado, 2012, ZieZy
Research Associate:
  Randa R. Harris, (M), Tennessee, 2001, Hs
Emeritus:
  Timothy M. Chowns, (D), Newcastle, 1968, GsrGg
  Thomas J. Crawford, (M), Emory, 1957, Eg
Lab Coordinator:
  John D. Congleton, (M), S Methodist, 1990, GeZiHs

Valdosta State University
Dept of Physics, Astronomy & Geosciences (B) (2015)
1500 North Patterson Street
Valdosta, GA 31698-0055
    p. (229) 333-5752
    echatela@valdosta.edu
    http://www.valdosta.edu/phy/
    Enrollment (2010): B: 77 (10)
Head:
  Edward E. Chatelain, (D), Iowa, 1984, PiGlPv

Professor:
　Cecilia S. Barnbaum, (D), California (Los Angeles), 1992, Zn
　Frank A. Flaherty, (D), Fordham, 1983, Zn
　Martha A. Leake, (D), Arizona, 1982, Xg
　Kenneth S. Rumstay, (D), Ohio State, 1984, Znn
Associate Professor:
　Can Denizman, (D), Florida, 1998, Hw
　Mary A. Fares, (D), Tennessee Tech, 1992, Zn
　Judy Grable, (D), Tennessee, 2001, Hs
　Mark S. Groszos, (D), Florida State, 1996, GctEg
　Michael G. Noll, (D), Kansas, 2000, Zn
　Paul C. Vincent, (D), Texas A&M, 2004, Zig
Pre-Engineering Director:
　Barry Hojjatie, (D), Florida, 1990, Zm
Assistant Professor:
　Jason Allard, (D), Penn State, 2006, AsZg
Instructor:
　Perry A. Baskin, (M), Valdosta State, 1972, Zn
　Donald Thieme, (D), Georgia, 2003, GaSdGm
Emeritus:
　Dennis W. Marks, (D), Michigan, 1970, Zn
　Arnold E. Somers, Jr., (D), Virginia Tech, 1975, Zn

Guam

University of Guam
College of Natural and Applied Sciences (B,M) (2015)
Mohammad H. Golabi, PhD
Mangilao, GU 96923
　p. (671) 734-9305
　mgolabi@uguam.uog.edu
Professor:
　Mohammad H. Golabi, (D), Georgia, 1991, Spc

Water & Environmental Research Institute of the Western Pacific (M) ☒ (2017)
UOG Station
Mangilao, GU 96923
　p. (671) 735-2685
　jjenson@triton.uog.edu
　http://www.weriguam.org/
　Programs: Environmental Science (M)
　Certificates: Groundwater Hydrology
　Enrollment (2017): M: 7 (2)
Director:
　John W. Jenson, (D), Oregon State, 1993, HwGel
Professor:
　Gary R. W. Denton, (D), London, 1974, Ob
　Joseph D. Rouse, (D), Hg
Associate Professor:
　Yuming Wen, (D), Rhode Island, 2004, Zir
Assistant Professor:
　Nathan C. Habana, (D), Guam, 2014, Hw
　Mark A. Lander, (D), Hawaii, 1990, Am
Research Associate:
　Yongsang (Barry) Kim, (D), Purdue, Hw
Professor of Hydrology:
　Leroy F. Heitz, (D), Idaho, Hg
Professor of Geology:
　Henry Galt Siegrist, jr, (D), Penn State, 1961, GdzHy

Hawaii

Honolulu Community College
Natural Sciences (A) (2017)
874 Dillingham Blvd.
Honolulu, HI 96817
　p. (808) 845-9488
　brill@hawaii.edu
　http://libart.honolulu.hawaii.edu/natsci/geology.php
Professor:
　Richard C. Brill, Jr., (M), Hawaii, Gg

University of Hawai'i, Hilo
Dept of Geology (B) (2015)
200 W. Kawili Street
Hilo, HI 96720-4091
　p. (808) 933-3383
　kenhon@hawaii.edu
Chair:
　Ken Hon, (D), Colorado, 1987, Gvi
Professor:
　Steven P. Lundblad, (D), North Carolina, 1994, Gra
　Jene D. Michaud, (D), Arizona, 1992, HqwGm
Associate Professor:
　James L. Anderson, (D), S California, 1987, GctYd
Affiliate Faculty:
　Cheryl A. Gansecki, (D), Stanford, 1998, GviZe
Affiliate Faculty:
　John P. Lockwood, (D), Princeton, 1966, Gvg
Emeritus:
　Joseph B. Halbig, (D), Penn State, 1969, Ca
Cooperating Faculty:
　Christina C. Heliker, (M), W Washington, 1984, Gv

University of Hawai'i, Manoa
Dept of Atmospheric Sciences (B,M,D) ☒ (2019)
2525 Correa Road, HIG 350
Honolulu, HI 96822
　p. (808) 956-8775
　metdept@hawaii.edu
　http://www.soest.hawaii.edu/MET
　Programs: Atmospheric Sciences BS - with NWS Forecasting or US Government Track available - ATMO Minor possible with other Geo-science, Engineering, or Math or Computer Programming degrees, Atmospheric Sciences Masters Plan A & B, Plan A is Thesis track, Plan B is non-thesis track. Requires a general capstone research project. Atmospheric Sciences Ph.D.
　Enrollment (2018): B: 13 (7) M: 17 (7) D: 14 (4)
Chair:
　Steven Businger, (D), Washington, 1986, Am
Professor:
　Yi-Leng Chen, (D), Illinois, 1980, Asm
　Pao-Shin Chu, (D), Wisconsin, 1981, As
　Fei Fei Jin, (D), Acad Sinica (Beijing), 1985, AmsZo
　Tim Li, (D), Hawaii, 1993, Asm
　Bin Wang, (D), Florida State, 1984, Am
　Yuqing Wang, (D), Monash, 1995, As
Associate Professor:
　Jennifer Small Griswold, (D), California (Santa Cruz), AmZr
Assistant Professor:
　Christina Karamperidou, (D), Columbia, 2012, AsZoc
　Alison D. Nugent, (D), Yale, 2014, AsZnn
　Giuseppe Torri , (D), Imperial Coll (UK), 2012, ApsAm
　Jingxia Zhao, (D), California (Los Angeles), 1993, AmmAc

Dept of Earth Sciences (B,M,D) ☒ (2019)
1680 East-West Road, POST 701
Honolulu, HI 96822
　p. (808) 956-7640
　earth-chair@hawaii.edu
　http://www.soest.hawaii.edu/gg
　Programs: Earth Sciences
　Earth and Planetary Sciences
　Enrollment (2018): B: 38 (9) M: 28 (7) D: 35 (5)
Chair:
　Paul Wessel, (D), Columbia, 1989, Yr
Dean:
　Brian Taylor, (D), Columbia, 1982, Gtu
Assoc. Dean:
　Charles H. Fletcher, (D), Delaware, 1986, GsOnZu
Professor:
　Robert A. Dunn, (D), Oregon, 1999, Yrs
　Aly I. El-Kadi, (D), Cornell, 1983, Hw
　L. Neil Frazer, (D), Princeton, 1978, Ys
　Eric J. Gaidos, (D), MIT, 1996, Pg
　Michael O. Garcia, (D), California (Los Angeles), 1976, Gi
　Craig R. Glenn, (D), Rhode Island Grad Sch Ocean, 1987, GeCmGs
　Julia E. Hammer, (D), Oregon, 1998, GiCpGv
　Bruce F. Houghton, (D), Otago (NZ), 1977, Gv
　Garrett T. Ito, (D), MIT/WHOI, 1996, YgrGt

Kevin T. M. Johnson, (D), MIT, 1990, Gi
Stephen J. Martel, (D), Stanford, 1987, Ng
Gregory F. Moore, (D), Cornell, 1977, Gt
Brian N. Popp, (D), Illinois, 1986, Cs
Scott K. Rowland, (D), Hawaii, 1987, Zn
Kenneth H. Rubin, (D), California (San Diego), 1991, CcGvCg
Associate Professor:
Henrietta Dulai, (D), Florida State, 2005, CcHw
Jasper G. Konter, (D), California (San Diego), 2007, CcGiCt
Gregory Ravizza, (D), Yale, 1991, Cm
Kathleen Ruttenberg, (D), Yale, 1990, Cg
Bridget R. Smith-Konter, (D), California (San Diego), 2005, YdXy
Assistant Professor:
Thomas Shea, (D), Hawaii, 2010, Gv
Emeritus:
Ralph Moberly, (D), Princeton, 1956, GutGs
John M. Sinton, (D), Otago (NZ), 1976, Gi
Director of Student Services:
Leona M. Anthony, (B), 1986, Zn
Cooperating Faculty:
David A. Clague, (D), California (San Diego), 1974, Gi
Eric H. De Carlo, (D), Hawaii, 1982, Ca
Margo H. Edwards, (D), Columbia, 1992, Gu
Sarah Fagents, (D), Lancaster (UK), 1994, Gv
Luke P. Flynn, (D), Hawaii, 1992, Zr
Patricia B. Fryer, (D), Hawaii, 1981, Gu
Milton A. Garces, (D), California (San Diego), 1995, Ys
Alexander N. Krot, (D), Moscow State (Russia), 1989, Xm
Paul G. Lucey, (D), Hawaii, 1986, Xy
Murli H. Manghnani, (D), Montana State, 1962, Yx
Fernando Martinez, (D), Columbia, 1988, Yr
Floyd W. McCoy, Jr., (D), Harvard, 1974, Gu
Peter J. Mouginis-Mark, (D), Lancaster (UK), 1977, Xg
Jane E. Schoonmaker, (D), Northwestern, 1981, Cl
Shiv K. Sharma, (D), Indian Inst of Tech, 1973, Gx
G. Jeffrey Taylor, (D), Rice, 1970, Xm
Donald M. Thomas, (D), Hawaii, 1977, Ca
Affiliate Graduate Faculty:
Donald A. Swanson, (D), Johns Hopkins, 1964, Gv

Dept of Oceanography (B,M,D) (2015)
1000 Pope Road
Honolulu, HI 96822
p. (808) 956-7633
ocean@soest.hawaii.edu
http://www.soest.hawaii.edu/oceanography/
Administrative Assistant: Kristin Momohara
Professor:
Barbara Bruno, (D), Hawaii, 1994, Ob
Eric H. De Carlo, (D), Hawaii, 1982, Cm
Jeffrey C. Drazen, (D), San Diego, 2000, Ob
Eric Firing, (D), MIT, 1978, Op
Pierre J. Flament, (D), California (San Diego), 1986, Op
David Ho, (D), Columbia, 2001, OcCb
David M. Karl, (D), California (San Diego), 1978, Ob
Paul Kemp, (D), Oregon State, 1985, Ob
Rudolf C. Kloosterziel, (D), Utrecht, 1990, Op
Douglas S. Luther, (D), MIT, 1980, Op
Julian P. McCreary, (D), California (San Diego), 1977, Op
Margaret Anne McManus, (D), Old Dominion, 1996, Op
Christopher Measures, (D), Southampton, 1978, Oc
Mark A. Merrifield, (D), California (San Diego), 1989, Op
Michael J. Mottl, (D), Harvard, 1976, OuCgm
Bo Qiu, (D), Kyoto, 1990, Op
Kelvin J. Richards, (D), Southampton, 1978, Op
Kathleen C. Ruttenberg, (D), Yale, Cg
Francis J. Sansone, (D), North Carolina, 1980, Cm
Niklas Schneider, (D), Hawaii, 1992, Og
Craig R. Smith, (D), California (San Diego), 1983, Ob
Grieg F. Steward, (D), California (San Diego), 1996, Ob
Axel Timmermann, (D), Hamburg (Germany), 1999, Op
Richard E. Zeebe, (D), Bremen, 1998, Ou
Associate Professor:
Glenn Carter, (D), Washington, 2005, Op
Matthew Church, (D), William & Mary, 2003, Ob
Brian Glazer, (D), Cm
Erica Goetze, (D), California (San Diego), 2004, Ob

Christopher Kelley, (D), Hawaii, 1995, Ob
Gary M. McMurtry, (D), Hawaii, 1979, CaGvu
Brian Powell, (D), Colorado, 2005, Op
Karen E. Selph, (D), Hawaii, 1999, Ob
Affiliate Graduate Faculty:
Allen H. Andrews, (D), Rhodes, 2009, Oc
Emeritus:
Paul K. Bienfang, (D), Hawaii, 1977, Ob
Antony D. Clarke, (D), Washington, 1983, As
Richard W. Grigg, (D), California (San Diego), 1969, Ob
Barry J. Huebert, (D), Northwestern, 1970, OcCg
Yuan-Hui Li, (D), Columbia, 1967, Cm
Roger Lukas, (D), Hawaii, 1981, Op
Fred T. Mackenzie, (D), Lehigh, 1962, CmGes
Lorenz Magaard, (D), Kiel, 1963, Op
Alexander Malahoff, (D), Hawaii, 1965, Cm
Peter Muller, (D), Hamburg, 1974, Op
Jane E. Schoonmaker, (D), Northwestern, 1981, Cm
Stephen V. Smith, (D), Hawaii, 1970, Ou
Richard E. Young, (D), Miami, 1968, Ob
Affiliate Graduate Faculty:
Russell E. Brainard, (D), Naval Postgrad Sch, 1994
Paul G. Falkowski, (D), British Columbia, 1975, Og
Carolyn Jones, (D), Georgia Tech, 2004
Dennis Moore, (D), Harvard, 1968, Og
Jaromir Ruzicka, (D), Tech (Czech), 1963, Zn
John R. Sibert, (D), Columbia, 1968, Ob
Kevin Weng, (D), Stanford, 2007, Ob
Cooperating Faculty:
Marlin J. Atkinson, (D), Hawaii, 1981, Ob
Whitlow W L. Au, (D), Washington State, 1970, Zn
Janet M. Becker, (D), California (San Diego), 1989, Og
Michael Cooney, (D), California (Davis), 1992, Og
Walter Dudley, (D), Hawaii, 1976, Ou
Eric J. Gaidos, (D), MIT, 1996, Yg
Ruth D. Gates, (D), Newcastle (UK), 1990, Ob
Petra H. Lenz, (D), California (Santa Barbara), 1983, Og
Jeffrey J. Polovina, (D), California (Berkeley), 1974, Zn
Brian N. Popp, (D), Illinois, 1986, Cm
Michael S. Rappe, (D), Oregon State, 1997, Og
Florence Thomas, (D), California (Berkeley), 1992, Ob
Robert Toonen, (D), California (Davis), 2001, Ob
John C. Wiltshire, (D), Hawaii, 1983, Og

Windward Community College
Natural Sciences (A) O-🔒 (2018)
45-720 Kea'ahala Rd.
Kane'ohe, HI 96744
p. (808) 236-9115
fmccoy@hawaii.edu
http://windward.hawaii.edu
Programs: AA in Natural Sciences
Certificates: certificate programs in Natural Science; Sustainability; Marine Science [Marine Option Program])
Enrollment (2018): A: 6 (4)
Professor:
Floyd W. McCoy, (D), Harvard, 1974, GgaGu

Idaho

Boise State University
Dept of Geosciences (B,M,D) ●-🔒 (2018)
1910 University Drive
Mail Stop 1535
Boise, ID 83725
p. (208) 426-1631
geosciences@boisestate.edu
http://earth.boisestate.edu/
Programs: Geology, Geophysics, Hydrology, Earth Science Education
Certificates: Geographic Information Systems
Administrative Assistant: Liz Johansen
Enrollment (2014): B: 143 (11) M: 29 (6) D: 20 (7)
Dean Graduate College:
John R. Pelton, (D), Utah, 1979, YgeYs

Idaho

Associate Dean/College of Arts & Sciences:
  Clyde J. Northrup, (D), MIT, 1996, Gc
Chair:
  James P. McNamara, (D), Alaska (Fairbanks), 1997, Hg
University Distinguished Professor:
  Matthew Kohn, (D), Rensselaer, 1991, CsGpPv
Professor:
  Shawn Benner, (D), Waterloo, 2000, Clt
  Nancy Glenn, (D), Nevada (Reno), 2000, ZrNgZi
  Mark D. Schmitz, (D), MIT, 2001, CcaCg
Senior Scientist:
  Jim Crowley, (D), Carleton, 1997, Ccg
Associate Professor:
  Brittany Brand, (D), Arizona State, 2008, Gsv
  Alejandro N. Flores, (D), MIT, 2009, HqAs
  Jeffrey Johnson, (D), Washington, 2000, YsGv
  Hans-Peter Marshall, (D), Colorado, YxGl
  Jennifer L. Pierce, (D), New Mexico, 2004, GmRcZc
  David E. Wilkins, (D), Utah, 1997, ZyGm
Assistant Professor:
  Ellyn Enderlin, (D), Ohio State, 2013, GlYg
  Dylan Mikesell, (D), Boise State, 2012, YsuYx
  Qifei iu Niu, (D), Hong Kong, 2014, Ygu
  Karen Viskupic, (D), MIT, 2002, ZeGc
  Dorsey Wanless, (D), Florida, 2010, Gi
Research Professor:
  Lee M. Liberty, (M), Wyoming, 1992, Ys
Adjunct Professor:
  Virginia S. Gillerman, (D), California (Berkeley), 1982, Eg
Emeritus Research Professor:
  Warren Barrash, (D), Idaho, 1986, HwYxGe
Emeritus:
  Paul R. Donaldson, (D), Colorado Mines, 1974, Ye
  Kenneth M. Hollenbaugh, (D), Idaho, 1968, Eg
  Paul Michaels, (D), Utah, 1993, Ne
  Walter S. Snyder, (D), Stanford, 1977, GrcGt
  Claude Spinosa, (D), Iowa, 1968, Pi
  Charles J. Waag, (D), Arizona, 1968, Gc
  Craig M. White, (D), Oregon, 1980, Giv
  Monte D. Wilson, (D), Idaho, 1969, Gm
  Spencer H. Wood, (D), Caltech, 1975, Gmm

Brigham Young University - Idaho
Dept of Geology (B) ●⌂ (2019)
525 South Center Street
Rexburg, ID 83460-0510
  p. (208) 496-7670
  roselleg@byui.edu
  http://www.byui.edu/Geology
  Programs: Geology; Earth Science Education; Environmental Geoscience; Geospatial Computing; GeoBusiness and Data Analytics
  Certificates: Geospatial Technology
  Enrollment (2018): B: 158 (20)
Chair:
  Gregory T. Roselle, (D), Wisconsin, 1997, GpCaHg
Professor:
  Daniel K. Moore, (D), Rensselaer, 1997, GxiCp
  Robert W. Clayton, (D), S California, 1993, GcYgNg
  Forest J. Gahn, (D), Michigan, 2004, PggGs
  William W. Little, (D), Colorado, 1995, GsdGm
  Mark D. Lovell, (M), Idaho, 1999, HyZiGo
  Megan Sjoblom, (D), Penn State, 2015, GiZeGe
  Julie B. Willis, (D), Utah, 2009, GtcZi

College of Southern Idaho
Dept of Physical Science (A) (2015)
Box 1238
315 Falls Avenue
Twin Falls, ID 83301
  p. (208) 732-6400
  swillsey@csi.edu
  http://physsci.csi.edu/geology/
  Enrollment (2013): A: 8 (0)
Professor:
  Shawn P. Willsey, (M), N Arizona, 2000, GgcGv

College of Western Idaho
Physical & Agricultural Sciences (A) ⌂ (2018)
5500 E Opportunity Drive
Nampa, ID 83687
  andreaschumaker@cwidaho.cc
  http://cwidaho.cc
  Programs: Geology; Geography; Biology-Natural Resources
  Certificates: GIS
  Enrollment (2018): A: 38 (3)
Assistant Professor:
  Ander Sundell, (M), Boise State, Ggc

Idaho State University
Dept of Geosciences (B,M,D) O (2016)
921 S. 8th Ave STOP 8072
Pocatello, ID 83209-8072
  p. (208) 282-3365
  geology@isu.edu
  http://geology.isu.edu/
  f: https://www.facebook.com/idahostategeosciences/
  t: @ISUGeoscience
  Enrollment (2016): B: 74 (25) M: 33 (12) D: 3 (1)
Professor:
  Michael O. McCurry, (D), California (Los Angeles), 1985, Gz
Professor:
  Paul K. Link, (D), California (Santa Barbara), 1982, Gs
  David W. Rodgers, (D), Stanford, 1987, Gct
  Glenn D. Thackray, (D), Oregon State, 1989, Gl
Associate Professor:
  Benjamin T. Crosby, (D), MIT, 2006, GmHsZy
  Leif Tapanila, (D), Utah, 2005, Pi
Assistant Professor:
  Donna M. Delparte, (D), Calgary, 2008, Zi
  Sarah E. Godsey, (D), California (Berkeley), 2009, HwGm
  Shannon E. Kobs-Nawotniak, (D), Buffalo, 2007, Gv
  David M. Pearson, (D), Arizona, 2012, Gc
Research Associate:
  Diana L. Boyack, (M), Idaho State, 2012, ZiGg
Lecturer:
  H. Carrie Bottenberg, (D), Missouri S&T, 2012, ZiGtc
  Lori Tapanila, (M), Utah, 2006, Gg
Emeritus:
  Scott S. Hughes, (D), Oregon State, 1983, GvXgGi
GIS Director:
  Keith Weber, (M), Montana, Zi
Financial Tech:
  Melissa Neiers
Related Staff:
  Kate Anthony-Zajanc, (M), Idaho State, 2013, Zi
Cooperating Faculty:
  John A. Welhan, (D), California, 1981, Hw

Lewis-Clark State College
Earth Sciences (B) ☒ (2019)
Div of Natural Science & Mathematics
500 8th Ave
Lewiston, ID 83501
  p. (208) 792-2283
  klschmidt@lcsc.edu
  http://www.lcsc.edu/science/degree-programs/earth-science/
  Programs: Earth Science (B), Geochemistry emphasis in Chemistry (B), Earth Science Secondary Education
  Certificates: GIS, Geology
  Enrollment (2018): B: 13 (3)
Professor:
  Keegan L. Schmidt, (D), S California, 2000, Gc

North Idaho College
Dept of Geosciences (A) ⌂ (2018)
1000 West Garden Avenue
Coeur d'Alene, ID 83814
  p. (208) 769-3477
  Bill_Richards@nic.edu

http://www.nic.edu/
Programs: Geology
Geography
Enrollment (2018): A: 3 (6)
Associate Professor:
Bill Richards, (M), Kansas State, 1984, GgZyGz

University of Idaho
College of Natural Resources (2015)
Moscow, ID 83844-3019
drbecker@uidaho.edu

Dept of Civil & Environmental Engineering (2015)
Moscow, ID 83844-1022
colberg@uidaho.edu

Dept of Geography (2018)
875 Perimeter Drive MS 3021
Moscow, ID 83844-3021
p. (208) 885-6216
geography@uidaho.edu
https://www.uidaho.edu/sci/geography
f: https://www.facebook.com/groups/502335576777518
t: @GeogUidaho
Programs: B.S. Geography, M.S. Geography, Ph.D. Geography
Certificates: GIS, Climate Change
Professor:
Raymond Dezzani, (D), California, 1996, RhZnn
Jeffrey Hicke, (D), Colorado, 2000, ZcyZn
Karen Humes, (D), Arizona, 1992, ZriZn
Associate Professor:
John Abatzoglou, (D), California, 2006, ZoyZc
Assistant Professor:
Chao Fan, Arizona State, 2016, Zir
Grant Harley, (D), Tennessee, 2012, ZynZn
Haifeng (Felix) Liao, (D), Utah, 2014, ZuiZn
Tom Ptak, (D), Oregon, 2016, Zu
Steven Radil, (D), Illinois, 2011, Zn
Research Scientist:
Katherine Hegewisch, (D), Washington State, 2010, Zo
Related Staff:
Patrick Olsen, (M), Idaho, 2014, Zi

Geological Sciences (B,M,D) O☒ (2019)
875 Perimeter Drive
MS 3022
Moscow, ID 83844-3022
p. (208) 885-6192
geology@uidaho.edu
http://www.uidaho.edu/sci/geology
Enrollment (2018): B: 42 (9) M: 8 (2) D: 8 (2)
Chair:
Leslie L. Baker, (D), Brown, 1996, ClXgGe
Distinguished Professor:
Robert Smith, (D), New Mexico Tech, 1984, Cb
Professor:
Jerry P. Fairley, (D), California (Berkeley), 2000, HwGv
Kenneth F. Sprenke, (D), Alberta, 1983, Ye
Clinical Associate Professor:
Thomas Williams, (D), Maryland, 1994, Gf
Assistant Professor:
Timothy Bartholomaus, (D), Alaska, 2013, Gl
Elizabeth J. Cassel, (D), Stanford, 2010, Gs
Jeffrey Langman, (D), Texas, 2008, ClHws
Eric L. Mittelstaedt, (D), Hawaii, 2008, Yr
Jessica Stanley, (D), Colorado, 2015, Gt
Instructor:
Renee Love, (D), Idaho, 2011, Pg
Malinda Ritts, (D), Stanford, 2009, GsCg
Emeritus University Distinguished Professor:
Mickey E. Gunter, (D), Virginia Tech, 1987, GzEnGb
Emeritus:
Dennis J. Geist, (D), Oregon, 1985, Giv
Peter E. Isaacson, (D), Oregon State, 1974, Psi

Soil Science Div (B,M,D) (2015)
Dept of Plant, Soil, & Entomological Sciences
Moscow, ID 83844-2339
p. (208) 885-7012
pses@uidaho.edu
Professor:
Guy Knudsen, (D), Cornell, 1984, Sb
Robert L. Mahler, (D), North Carolina State, 1980, Zn
Daniel G. Strawn, (D), Delaware, 1999, Sc
Associate Professor:
Bradford D. Brown, (D), Utah State, 1985, Zn
Matthew J. Morra, (D), Ohio State, 1986, Sb
Assistant Professor:
Jodi Johnson-Maynard, (D), California (Riverside), 1999, Sd
Paul A. McDaniel, (D), North Carolina State, 1989, Sd

Illinois

Augustana College
Department of Geology (B) ☒ (2018)
639 38th St.
Rock Island, IL 61201
p. (309) 794-7318
jeffreystrasser@augustana.edu
https://augustana.edu/academics/areas-of-study/geology
f: https://www.facebook.com/AugustanaGeology/
Programs: B.A. major in geology
Administrative Assistant: Jennifer Milner
Enrollment (2018): B: 24 (9)
Professor:
Jeffrey C. Strasser, (D), Lehigh, 1996, GmlGe
Michael B. Wolf, (D), Caltech, 1992, Gi
Assistant Professor:
Kelsey M. Arkle, (D), Cincinnati, 2015, PeGsg
Lab technician and part-time instructor:
Jenny C. Arkle, (M), California State (Fullerton), 2010, GtmCc
Emeritus:
William R. Hammer, (D), Wayne State, 1979, Pv
Assistant Curator:
John Oostenryk
Museum Educational Programs Coordinator:
Susan Kornreich Wolf, (B), Hamilton, 1987, Ze

Black Hawk College
Dept of Natural Science & Engineering (A) (2015)
6600 34th Avenue
Moline, IL 61265
p. (309) 796-5000
harwoodr@bhc.edu
http://www.bhc.edu/academics/departments/natural-science-and-engineering/
Administrative Assistant: Tara Carey
Chair:
Brian Glaser, (D), N Iowa, 1996, Zn
Professor:
Richard D. Harwood, (M), N Arizona, 1989, Gv

College of Lake County
Earth Science Dept (A) (2017)
19351 W. Washington Steet
Grayslake, IL 60030-1198
p. (847) 543-2504
eng499@clcillinois.edu
http://www.clcillinois.edu/programs/esc/
Professor:
Eric Priest, (M), Creighton, 1986, AmZe
Xiaoming Zhai, (D), California (Davis), 1997, Gp

Columbia College Chicago
Dept of Science & Mathematics (2015)
600 South Michigan Avenue
Chicago, IL 60605
p. (312) 369-7396
crasinariu@colum.edu
http://www.colum.edu

Professor:
  Gerald E. Adams, (D), Northwestern, 1985, Gi
Associate Professor:
  Robin L. Whatley, (D), California (Santa Barbara), 2005, PvZr

Eastern Illinois University
Geology and Geography Department (B,M) (2018)
600 Lincoln Avenue
Charleston, IL 61920-3099
  p. (217) 581-2626
  geoscience@eiu.edu
  http://www.eiu.edu/~geoscience/
  Programs: Geology (B.S); Geography (B.S.); Science Teacher Certification (B.S); Geographic Information Systems (MBA, PSM)
  Certificates: GISci Graduate Certificate, co-offer Certificate in Public Planning (with Political Science)
  Administrative Assistant: Susan Kile
  Enrollment (2014): B: 74 (23)
Professor:
  Michael W. Cornebise, (D), Tennessee, 2003, Zg
  Betty E. Smith, (D), SUNY (Buffalo), 1994, Zi
Associate Professor:
  Diane M. Burns, (D), Wyoming, 2006, Gsr
  James A. Davis, (D), Kansas State, 2001, ZgAm
  Barry J. Kronenfeld, (D), SUNY (Buffalo), 2004, Ziy
  Christopher R. Laingen, (D), Kansas State, 2009, SfZn
  Katherine Lewandowski, (D), Ohio State, 2008, Pmc
  Dave Viertel, (D), Texas State, 2008, Zru
Emeritus:
  Craig A. Chesner, (D), Michigan Tech, 1988, GivGz
  Belayet H. Khan, (D), Pittsburgh, 1985, ZyAm
  James F. Stratton, (D), Indiana, 1975, PgGu

Elgin Community College
Dept of Geology (2015)
1700 Spartan Drive
Elgin, IL 60123
  p. (847) 214-7359
  agoyal@elgin.edu
  http://www.elgin.edu/employeelisting/faculty/default.aspx
Adjunct Professor:
  Mark R. Kuntz, Gg
  Timothy M. Millen, (M), N Illinois, 1982, Gg
  Joseph E. Peterson, Gg

Field Museum of Natural History
Dept of Geology (2015)
1400 S. Lake Shore Drive
Chicago, IL 60605-2496
  p. (312) 665-7621
  klawson@fieldmuseum.org
  http://www.fieldmuseum.org
  Administrative Assistant: Karsten L. Lawson
  Administrative Assistant: Elaine Zeiger
Chair:
  Olivier C. Rieppel, (D), Basel, 1978, Pv
Curator, Meteoritics:
  Meenakshi Wadhwa, (D), Washington (St. Louis), 1994, Xm
Curator, Fossil Fishes:
  Lance Grande, (D), CUNY, 1983, Pv
Associate Curator, Fossil Invertebrates:
  Scott H. Lidgard, (D), Johns Hopkins, 1984, Pi
Curator, Fossil Amphibians & Reptiles:
  John R. Bolt, (D), Chicago, 1968, Pv
Emeritus:
  Matthew H. Nitecki, (D), Chicago, 1968, Pi
  Bertram G. Woodland, (D), Chicago, 1962, Gp
Associate Curator, Fossil Invertebrates:
  Peter J. Wagner, (D), Chicago, 1995, Pi

Harold Washington College
Physcial Science Dept (A) (2015)
30 E. Lake Street
Room 903
Chicago, IL 60601
  p. (312) 553-5791
  aescuadro@ccc.edu
  http://www.ccc.edu/colleges/washington/departments/Pages/Physical-Sciences.aspx
  Enrollment (2012): A: 17 (0)

Heartland Community College
Math And Science Div (A) (2015)
2400 Instructional Commons Building
1500 W Raab Road
Normal, IL 61761
  p. (309) 268-8640
  mark.finley@heartland.edu
  http://www.heartland.edu/ms/easc/
Professor:
  Robert Dennison, (M), S Mississippi, Zgn
  Mark Finley, (M), Iowa State, Gg
Adjunct Professor:
  Janet Beach Davis, (B), Illinois (Springfield), Zg
  Paul Ritter, (M), E Illinois, Zg
  Steven Travers, (M), Illinois State, Zg
  Mark Yacucci, (B), Youngstown State, Zg

Illinois Central College
Math, Science, and Engineering Dept (A) (2015)
One College Drive
East Peoria, IL 61611
  p. (309) 694-5364
  sdunham@icc.edu
  Administrative Assistant: Diane Weber
Professor:
  Martin A. Petit, (M), Illinois State, 1967, Ze
Associate Professor:
  Cheryl R. Emerson, (M), N Arizona, 1989, Ze
Assistant Professor:
  Ed Stermer, (M), Iowa, Zg
Adjunct Professor:
  Linda Aylward, (B), Bradley, 1980, Ze
  Sean Mulkey, (M), Iowa, 1994, Ze

Illinois State University
Dept of Geography, Geology, and the Environment (B,M) (2019)
101 South School Street
Campus Box 4400
206 Felmley Hall of Science
Normal, IL 61790-4400
  p. (309) 438-7649
  geo@illinoisstate.edu
  http://www.geo.ilstu.edu/contactus/
  f: https://www.facebook.com/groups/21051078663/
  Programs: Geography (BA or BS); Geography Education (BS); Earth and Space Science Education (BS); Environmental Systems and Sustainability (BS) Geology (BS); Hydrogeology (M), Minors in Geography, Geology, and Environmental Studies
  Certificates: Geographic Information Systems, Hydrogeology Geographic Information Systems
  Administrative Assistant: Karen K. Dunton
  Enrollment (2018): B: 65 (20) M: 18 (3)
Chair:
  Dagmar Budikova, (D), Calgary, 2001, AsZyi
University Professor:
  Eric W. Peterson, (D), Missouri, 2002, Hws
Distinguished University Professor:
  David H. Malone, (D), Wisconsin, 1994, GcsEg
Professor:
  James E. Day, (D), Iowa, 1988, PisPm
  John Kostelnick, (D), Kansas, 2006, Zi
Associate Professor:
  Catherine M. O'Reilly, (D), Arizona, 2001, HgCgGg
  Rex J. Rowley, (D), Kansas, 2009, Zi
  Jonathan B. Thayn, (D), Kansas, 2009, Zir
Assistant Professor:
  Tenley Banik, (D), Vanderbilt, 2015, GviCc

Alec Foster , (D), Temple, 2016, Zn
Wondwosen M. Seyoum, (D), Georgia, 2016, HwqZr
Lisa M. Tranel, (D), Virginia Tech, 2010, Gmt
Adjunct Professor:
Toby J. Dogwiler, (D), Missouri (Colombia), 2002, Gm
Walton R. Kelly, (D), Virginia, 1993, HgCg
Edward Mehnert, (D), Illinois, 1998, Hg
Andrew Stumpf, (D), New Brunswick, 2001, GgsCg
Steve Van Der Hoven, (D), Utah, 2000, HwSpYg
Emeritus:
Robert S. Nelson, (D), Iowa, 1970, Gm
Coordinator of Academic Services :
Paul A. Meister, (M), Illinois State, 2016, GgHgZn

Illinois Valley Community College
Dept of Geology (A) ☒ (2018)
Natural Sciences, Math, & Business Division
815 North Orlando Smith Road
Oglesby, IL 61348-9692
  p. (815) 224-0394
  Mike_Phillips@ivcc.edu
  http://www.ivcc.edu/geology.aspx?id=534
  Programs: Geology
  Enrollment (2018): A: 2 (2)
Professor:
  Michael Phillips, (M), S Illinois, 1990, GgeGm

Kaskaskia College
Life and Physical Sciences Dept (A) (2015)
27210 College Road
Centralia, IL 62801
  pvig@kaskaskia.edu
  http://www.kaskaskia.edu/LPDept/Default.aspx
Instructor:
  Pradeep K. Vig, Gg

Lincoln Land Community College
Math and Sciences (A) (2015)
Springfield, IL 62794-9256
  p. (217) 786-4923
  dean.butzow@llcc.edu
  http://www.llcc.edu/mtsc/Sciences/tabid/3280/Default.aspx
  Enrollment (2006): A: 7 (0)
Professor:
  Dean G. Butzow, (M), W Michigan, Zy
Instructor:
  Samantha Reif, (M), Michigan Tech, Gg

McHenry County College
Geography, Geology and Earth Science (A) (2015)
8900 US Hwy 14
Crystal Lake, IL 60012
  p. (815) 455-3700
  pstahmann@mchenry.edu
  http://www.mchenry.edu/EarthScience/index.asp
Instructor:
  Theodore Erski, (M), Akron, Zg
  Paul Hamill, (M), Wisconsin, AmZg
  Kate Kramer, (M), Indiana-Purdue, 2008, GgZg
  Paul Stahmann, (M), Brigham Young, 2000, Zyg

Moraine Valley Community College
Physical Sciences Dept (A) (2015)
9000 W. College Pkwy
Palos Hills, IL 60465-2478
  p. 708-974-5615
  syrup@morainevalley.edu
Professor:
  Krista A. Syrup, (M), W Michigan, 2002, Cs

Northeastern Illinois University
Dept of Earth Science (B) ●◯ (2018)
5500 N. St. Louis Avenue
Chicago, IL 60625
  p. (773) 442-6050
  E-Head@neiu.edu
  https://www.neiu.edu/academics/college-of-arts-and-sciences/departments/earth-science
  f: https://www.facebook.com/neiuearthscience/
  Programs: Earth Science (B)
  Administrative Assistant: Lidia Costea
  Enrollment (2018): B: 28 (6)
Professor:
  Laura L. Sanders, (D), Kent State, 1986, HwGe
Associate Professor:
  Elisabet M. Head, (D), GvZrCg
  Kenneth M. Voglesonger, (D), Arizona State, 2004, ClGe
Assistant Professor:
  Nadja Insel, (D), Michigan, 2011, Gt
Instructor:
  Mohammad Fariduddin, (D), N Illinois, 1999, Pe
  Rebekah Fitchett, (M), Illinois (Chicago), 2006, GgCg
  Jean M. Hemzacek Laukant, (M), N Illinois, 1986, Sc

Northern Illinois University
Dept of Geology and Environmental Geosciences (B,M,D)
◯ ⌂ (2018)
1425 W. Lincoln Hwy
De Kalb, IL 60115-2854
  p. (815) 753-1943
  askgeology@niu.edu
  http://www.niu.edu/geology/
  Programs: B.S. Geology and Environmental Geosciences
  High School Teacher Licensure in Earth and Space Science Education, M.S. Geology and Environmental Geosciences, Ph.D. Geology and Environmental Geosciences
  Certificates: Certificate of Professional Development in Environmental Geosciences
  Enrollment (2018): B: 57 (18) M: 20 (7) D: 5 (1)
Chair:
  Mark P. Fischer, (D), Penn State, 1994, GcoHy
Professor:
  Philip J. Carpenter, (D), New Mexico Tech, 1984, YgHwNg
  Ross D. Powell, (D), Ohio State, 1980, GslZc
  Reed P. Scherer, (D), Ohio State, 1991, Pm
Graduate Program Director:
  Mark R. Frank, (D), Maryland, 2001, Cp
Associate Professor:
  Justin P. Dodd, (D), New Mexico, 2011, Cs
  Melissa E. Lenczewski, (D), Tennessee, 2001, HwGeCo
Assistant Professor:
  Nicole D. LaDue, (D), Michigan State, 2013, Ze
  Nathan D. Stansell, (D), Pittsburg, 2009, Gl
Research Associate:
  Anna Buczynska, (D), Antwerp (Belgium), 2014, Csa
Adjunct Professor:
  B. Brandon Curry, (D), Illinois, 1995, GelPs
  Virginia Naples, (D), Massachusetts, 1980, Pv
  Karen Samonds, (D), Stony Brook, 2006, Pv
Office Manager:
  Nina B. Slack, (B)

Northwestern University
Earth and Planetary Sciences (B,D) ●☒ (2018)
2145 Sheridan Rd - TECH-F374
Evanston, IL 60208-3130
  p. (847) 491-3238
  earth@northwestern.edu
  http://www.earth.northwestern.edu/
  t: @NU_EARTHSCI
  Programs: Earth and Planetary Sciences (B)
  Administrative Assistant: Lisa J. Collins
  Administrative Assistant: Benjamin Rice
  Administrative Assistant: Robin Stark
  Enrollment (2018): B: 35 (11) D: 33 (3)
Director of the Environmental Sciences Program:
  Andrew D. Jacobson, (D), Michigan, 2001, ClPeCa
Director of Graduate Studies:
  Suzan van der Lee, (D), Princeton, 1996, YsxYg

Director of Graduate Admissions:
  Steven D. Jacobsen, (D), Colorado, 2001, GyzZm
Professor:
  Craig R. Bina, (D), Northwestern, 1987, YgZmGy
  Neal E. Blair, (D), Stanford, 1980, CoSbOc
  Matthew T. Hurtgen, (D), Penn State, 2003, GsPgGr
  Donna M. Jurdy, (D), Michigan, 1974, YgXyg
  Bradley B. Sageman, (D), Colorado, 1991, PsCbPc
  Seth Stein, (D), Caltech, 1978, YsdYg
Associate Professor:
  Yarrow L. Axford, (D), Colorado, 2007, GnPmi
Assistant Professor:
  Daniel E. Horton, (D), Michigan, 2011, AsPeZn
  Seth A. Jacobson, (D), Colorado, 2012, XacXg
  Magdalena Osburn, (D), Caltech, 2013, PoCbs
Radiogenic Lab Manager:
  Meagan Ankney, (D), CcqCa
IRMS Lab Manager:
  Andrew Masterson, (D), CslCa
Research Associate:
  Mitchell Barklage, (D), YgeYs
Director of Undergraduate Studies, Assistant Chair, Assistant Prof. of Instruction:
  Patricia A. Beddows, (D), Bristol (UK), 2004, HyGma
Adjunct Professor:
  Gilbert Klapper, (D), Iowa, 1962, Pm
Emeritus:
  Abraham Lerman, (D), Harvard, 1964, CgGnCm
  Emile A. Okal, (D), Caltech, 1978, YsgYr

Oakton Community College
Earth Science (A) ☒ (2019)
1600 E. Golf Road
Des Plaines, IL 60016
  p. (847) 376-7042
  clandrie@oakton.edu
  http://www.oakton.edu/academics/academic_departments/earth_science/
Chair:
  Chad Landrie, (D), Illinois (Chicago)
Lecturer:
  Richard DiMaio, (M), N Illinois
  David LoBue, (M), Kansas
  William Tong, (M), NE Illinois
  Raymond Wiggers, (B), Perdue
  Julie Wulff, (D), Ohio State

Olivet Nazarene University
Dept of Chemistry and Geosciences (B) ☒ (2019)
One University Avenue
Bourbonnais, IL 60914
  p. (815) 939-5394
  ccarriga@olivet.edu
  http://geology.olivet.edu/
  f: https://www.facebook.com/pages/Olivet-Nazarene-University-Geosciences/50458842809
  Programs: Geology (B.S.), Earth and Space Science (B.A.), Science Education (B.S.), Geography (B.A.), Environmental Science (B.S.)
  Enrollment (2018): B: 14 (4)
Acting Director:
  Charles W. Carrigan, (D), Michigan, 2005, GzCgGc
Director, Strickler Planetarium:
  Stephen Case, (D), Notre Dame, 2014, RhXg
Emeritus:
  Max W. Reams, (D), Washington (St. Louis), 1968, GsmPg

Prarie State College
Dept of Physical Science, Earth Science, and Geology (2015)
202 S. Halsted St
Chicago Heights, IL 60411
  p. (708) 709-3674
  lburrough@prairiestate.edu
  http://www.prairie.cc.il.us

Principia College
Biology and Natural Resources Department (B) ☒ (2018)
1 Maybeck Place
Elsah, IL 62028
  scott.eckert@principia.edu
  http://www.principiacollege.edu/biology-env-studies
Chair:
  Scott Eckert, (D), Georgia
Professor:
  Greg Bruland, (D), Duke
  Christine McAllister, (D), St. Louis
Instructor:
  John Lovseth, (D), S Illinois, 2018, Zni

Richard J. Daley Community College
Physical Sciences (2015)
7500 South Pulaski Rd
Chciago, IL 60652
  p. (773) 838-7636
  gguerra@ccc.edu
  http://daley.ccc.edu/

Richland Community College
Richland Community College (2015)
One College Park
Decatur, IL 62521
  p. (217) 875-7211
  jsmith@richland.edu
  http://www.richland.edu/

Rock Valley College
Rock Valley College (2015)
3301 North Mulford Road
Rockford, IL 61114
  J.Hankins@RockValleyCollege.edu
  http://ednet.rockvalleycollege.edu/

Sauk Valley Community College
Earth Science (2015)
173 IL Rt. 2
Dixon, IL 61021
  michael.j.bates@svcc.edu
  http://www.svcc.edu

South Suburban College
Dept of Physical Science (2015)
15800 South State Street
South Holland, IL 60473
  p. (708) 596-2000 x2417
  ahelwig@ssc.edu
  http://www.ssc.edu

Southern Illinois University Carbondale
Dept of Geology (B,M,D) ○☒ (2018)
1259 Lincoln Drive, Mailcode 4324
Parkinson 102
Carbondale, IL 62901
  p. (618) 453-3351
  geology@geo.siu.edu
  http://www.geology.siu.edu/
  f: https://www.facebook.com/Geology-at-Southern-Illinois-University-191268227586004/
  Programs: BA Geology, BS Geology, MA Geology, MS Geology, PhD Geosciences
  Administrative Assistant: Dana Wise
  Enrollment (2018): B: 30 (14) M: 22 (9) D: 8 (1)
Chair:
  Steven P. Esling, (D), Iowa, 1983, HwGl
Professor:
  Ken B. Anderson, (D), Melbourne, 1989, Co
  Scott E. Ishman, (D), Ohio State, 1990, Pmc
  Susan M. Rimmer, (D), Penn State, 1985, EcCgGo
  John L. Sexton, (D), Indiana, 1974, Ye

Associate Professor:
    James A. Conder, (D), Brown, 2001, Ysr
    Liliana Lefticariu, (D), N Illinois, 2004, CsbGe
Assistant Professor:
    Harvey Henson, (M), S Illinois, 1989, Yg
    Daniel R. Hummer, (D), Penn State, 2010, GzyCu
    Sally Potter-McIntyre, (D), Utah, 2013, Gs
Research Associate:
    William Huggett, (M), S Illinois, 1981, Ec
Emeritus:
    Richard Fifarek, (D), Oregon State, 1985, Em
    Charles O. Frank, (D), Syracuse, 1973, Gx
    Stanley E. Harris, Jr., (D), Iowa, 1947, Ge
    John E. Marzolf, (D), California (Los Angeles), 1970, Gs
    Paul D. Robinson, (M), S Illinois, 1963, Gz
    Jay Zimmerman, Jr., (D), Princeton, 1968, Gc

Southwestern Illinois College - Belleville Campus
Dept of Earth Science (2015)
2500 Carlyle Avenue
Belleville, IL 62221
    p. (618) 235-2700
    http://www.swic.edu

Southwestern Illinois College - Sam Wolf Granite City Campus
Physical Science (A) (2015)
4950 Maryville Road
Granite City, IL 62040
    p. (618) 235-2700
    http://www.swic.edu
Professor:
    Joy Branlund, (D), Washington Univ (St. Louis), 2008, Zg

Triton College
Triton College (2015)
2000 Fifth Ave.
River Grove, IL 60171
    p. (708) 456-0300
    austinweinstock@triton.edu
    http://www.triton.edu

University of Chicago
Dept of the Geophysical Sciences (B,M,D) (2015)
5734 S. Ellis Avenue
Chicago, IL 60637
    p. (773) 702-8101
    info@geosci.uchicago.edu
    http://geosci.uchicago.edu
    Administrative Assistant: David J. Taylor
    Enrollment (2009): B: 20 (9) D: 31 (4)
Professor:
    David Archer, (D), Washington, 1990, Yr
    Peter R. Crane, (D), Reading, 1981, Pg
    Andrew Davis, (D), Yale, 1977, Xc
    Michael J. Foote, (D), Chicago, 1989, Pg
    John E. Frederick, (D), Colorado, 1975, As
    Lawrence Grossman, (D), Yale, 1972, Xc
    David Jablonski, (D), Yale, 1979, Pi
    Susan M. Kidwell, (D), Yale, 1982, GsPe
    Douglas R. MacAyeal, (D), Princeton, 1983, Yr
    Raymond T. Pierrehumbert, (D), MIT, 1980, As
    Frank M. Richter, (D), Chicago, 1972, Gt
    David B. Rowley, (D), SUNY (Albany), 1983, GtgGc
Associate Professor:
    Dion L. Heinz, (D), California (Berkeley), 1986, GyzYx
    Noboru Nakamura, (D), Princeton, 1989, As
Assistant Professor:
    Charles Kevin Boyce, (D), Harvard, 2001, Pb
    Fred Ciesla, (D), Arizona, 2003, Xcg
    Albert Colman, (D), Yale, 2002, PgCs
    Nicolas Dauphas, (D), Ctr Pétro et Géochimiques, 2002, Xc
    Pamela Martin, (D), California (Santa Barbara), 2000, Pe
    Elisabeth Moyer, (D), Caltech, 2001, As
    Mark Webster, (D), California (Riverside), 2003, Pi

Visiting Professor:
    Ho-kwang (David) Mao, (D), Rochester, 1968, Yg
Emeritus:
    Alfred T. Anderson, Jr., (D), Princeton, 1963, Gv
    Robert N. Clayton, (D), Caltech, 1955, Cs
    Dave Fultz, (D), Chicago, 1947, As
    Michael C. LaBarbera, (D), Duke, 1976, Pi
    David M. Raup, (D), Harvard, 1957, Pg
    Ramesh C. Srivastava, (D), McGill, 1964, AsmAm

University of Illinois at Chicago
Dept of Earth and Environmental Sciences (B,M,D) ☒ (2018)
845 W. Taylor Street
2440 SES
MC 186
Chicago, IL 60607-7059
    p. (312) 996-3155
    klnagy@uic.edu
    http://eaes.uic.edu/
    Programs: Earth and Environmental Sciences
    Enrollment (2018): B: 79 (19) M: 3 (2) D: 14 (1)
Head:
    Kathryn L. Nagy, (D), Texas A&M, 1988, Cg
Professor:
    Fabien Kenig, (D), Orleans (France), 1991, CobOo
    Roy E. Plotnick, (D), Chicago, 1983, PgiGq
    Carol A. Stein, (D), Columbia, 1984, Yg
Associate Professor:
    Andrew J. Dombard, (D), Washington (St. Louis), 2000, Xy
    DArcy Meyer Dombard, (D), Washington (St. Louis), 2004, Pg
Assistant Professor:
    Max Berkelhammer, (D), S California, 2010, Cg
    Kimberly Van Meter, (D), Waterloo, 2017, Hs
Adjunct Assistant Professor:
    Andrew King, (D), California (San Diego), 2008, Og
    Ethan Theuerkauf, (D), North Carolina, 2016, On
Adjunct Professor:
    Peter T. Doran, (D), Nevada (Reno), 1996, HswAm
    Paul Fenter, (D), Pennsylvania, 1990, Gy
    Barry Lesht, (D), Chicago, 1977, Op
    Eugene Yan, (D), Ohio State, 1998, Hg
Visiting Assistant Professor:
    Andrew Malone, (D), Chicago, 2017, GlAt
Emeritus:
    Jean E. Bogner, (D), N Illinois, 1996, ClSbHy
    Martin F.J Flower, (D), Manchester, 1971, GiCuGt
    Stephen J. Guggenheim, (D), Wisconsin, 1976, Gz
    August F. Koster Van Groos, (D), Leiden (Neth), 1966, Cp
    Kelvin S. Rodolfo, (D), S California, 1967, GueGs
Clinical Assistant Professor:
    Stefany Sit, (D), Miami, 2013, Ygs
Office Support Specialist:
    Minnie O. Jones, Zn
Coordinator of Undergrad Labs:
    Lee Falkena, (M), Iowa, 2012, ZeGig
Asst to the Head/Business Manager:
    Edna L. Fuentes, (B), Illinois (Chicago), 2004, Zn

University of Illinois, Urbana-Champaign
Dept of Atmospheric Sciences (B,M,D) ●◯ (2018)
Nat Hist Bldg, Rm 3081, MC-104
1301 W Green Street
Urbana, IL 61801
    p. (217) 333-2046
    atmos-sci@illinois.edu
    http://www.atmos.illinois.edu
    f: https://www.facebook.com/dasuiuc/
    Administrative Assistant: Tammy R. Warf
    Enrollment (2018): B: 71 (17) M: 12 (10) D: 34 (8)
Director:
    Robert M. Rauber, (D), Colorado State, 1985, As
Head:
    Robert J. Trapp, (D), Oklahoma, 1994, As
Professor:
    Larry Di Girolamo, (D), McGill, 1996, Zr
    Atul K. Jain, (D), India, 1988, As

Sonia Lasher-Trapp, (D), Oklahoma, 1998, As
Greg M. McFarquhar, (D), Toronto, 1993, Asm
Donald J. Wuebbles, (D), California (Davis), 1983, As
Associate Professor:
Francina Dominguez, (D), Illinois, 2007, AsHg
Stephen W. Nesbitt, (D), Utah, 2003, As
Nicole Riemer, (D), Karlsruhe, 2002, As
Zhuo Wang, (D), Hawaii, 2004, As
Assistant Professor:
Deanna A. Hence, (D), Washington, 2011, AsmAp
Ryan Sriver, (D), Purdue, 2008, As
Research Scientist:
Brian Jewett, (D), Illinois, 1996, As
Jun shik Um, (D), Illinois, 2009
Research Associate:
Guangyu Zhao, (D), Illinois, 2006
Lecturer:
Donna J. Charlevoix, (M), California (Davis), 1996, As
Emeritus:
Kenneth V. Beard, (D), California (Los Angeles), 1970, As
Mankin Mak, (D), MIT, 1968, As
Walter A. Robinson, (D), Columbia, 1985, As
Robert B. Wilhelmson, (D), Illinois, 1972, As
Office Manager:
Joe Jeffries, (B)
Clinical Assistant Professor:
Jeffrey Frame, (D), Penn State, 2008, Asm

Dept of Geology (B,M,D) ● (2018)
1301 W Green Street
3081 Natural History Building
Urbana, IL 61801
   p. (217) 333-3540
   geology@illinois.edu
   http://www.geology.illinois.edu
   Programs: Geology (B, M, D); Geology and Geophysics (B); Earth Science Teaching (B)
   Enrollment (2016): B: 80 (32) M: 14 (7) D: 26 (0)
Head:
   Thomas M. Johnson, (D), California (Berkeley), 1995, HwCls
Threet Professor:
   James L. Best, (D), London, 1985, Gs
Johnson Professor:
   Gary Parker, (D), Minnesota, 1974, Gm
Grim Professor:
   Jay D. Bass, (D), SUNY (Stony Brook), 1982, Gy
   Feng-Sheng Hu, (D), Washington, 1994, Pe
Affiliate Faculty:
   Stanley Ambrose, (D), California (Santa Barbara)
   Kenneth T. Christensen, (D), Illinois
   Marcelo H. Garcia, (D), Minnesota, 1989, Gmu
   Bruce Rhoads, (D), Arizona, Gm
   Charles J. Werth, (D), Stanford
Professor:
   Bruce W. Fouke, (D), SUNY (Stony Brook), 1993, GsdGb
   Craig C. Lundstrom, (D), California (Santa Cruz), 1996, Cpg
   William R. Roy, (D), Illinois, 1985, ScGeCl
   Xiaodong Song, (D), Caltech, 1994, Ys
Research Associate Professor:
   Robert A. Sanford, (D), Michigan State, 1996, CbSbPo
Associate Head:
   Stephen P. Altaner, (D), Illinois, 1985, Gz
Affiliate Faculty:
   Scott Olson, (D), Illinois
Associate Professor:
   Alison M. Anders, (D), Washington, 2005, Gm
   Lijun Liu, (D), Caltech, 2010, YgGtm
Affiliate Faculty:
   Surangi W. Punyasena, (D), Chicago, 2007, PblPe
Assistant Professor:
   Jessica Conroy, (D), Arizona, 2011, Cs
   Jennifer Druhan, (D), California (Berkeley), 2012, HwCs
   Patricia Gregg, (D), MIT/WHOI, 2008, YgCg
   Wendy Yang, (D), California (Berkeley), 2010
Lecturer:
   Max L. Christie, (D), Penn State, 2017, PgGr
Adjunct Professor:
   Ercan Alp, (D), S Illinois, 1984
   Kurtis Burmeister, (D), Illinois, 2005
   Brandon Curry, Illinois, 1995
   Przemyslaw Dera, (D), Mickiewicz (Poland), 2000
   Robert J. Finley, (D), South Carolina, 1975, Go
   Leon R. Follmer, (D), Illinois, 1970, SaGlm
   Hannes E. Leetaru, (D), Illinois, 1997, GoYe
   Morris W. Leighton, (D), Chicago, 1951, Eo
   George Roadcap, (D), Illinois, 2004
   M. Scott Wilkerson, (D), Illinois, 1991, Gc
Walgreen Chair Emeritus Professor:
   Susan W. Kieffer, (D), Caltech, 1971, Gv
Grim Emeritus Professor:
   Craig M. Bethke, (D), Illinois, 1984, Hw
Emeritus:
   Thomas F. Anderson, (D), Columbia, 1967, CslCm
   Daniel B. Blake, (D), California (Berkeley), 1966, Pi
   Albert V. Carozzi, (D), Geneva, 1948, Gd
   Chu-Yung Chen, (D), MIT, 1983, Cg
   Wang-Ping Chen, (D), MIT, 1979, YsGt
   Donald L. Graf, (D), Columbia, 1950, Cl
   Ralph L. Langenheim, Jr., (D), Minnesota, 1951, Gr
   Stephen Marshak, (D), Columbia, 1983, Gct
   Alberto S. Nieto, (D), Illinois, 1974, Ng
   Philip A. Sandberg, (D), Stockholm, 1965, Gd
Teaching Specialist:
   Ann D. Long, (M), Leeds, 1981, Gm
Teaching Assistant Professor:
   J. Cory Pettijohn, (D), Boston, 2008, Hg
Research Scientist:
   Jonathan H. Tomkin, (D), Australian Nat, 2002, Gm

Dept of Natural Resources & Environmental Sciences
(B,M,D) (2015)
W-503 Turner Hall
1102 South Goodwin Avenue
Urbana, IL 61801-4798
   p. (217) 333-2770
   g-rolfe@uiuc.edu
   http://nres.illinois.edu/
Chair:
   Jeffrey Brawn
Professor:
   Charles W. Boast, (D), Iowa State, 1970, Sp
   Mark B. David, (D), New York, 1983, Sf
   John J. Hassett, (D), Utah State, 1970, Sc
   Robert L. Jones, (D), Illinois, 1962, Sc
   Richard L. Mulvaney, (D), Illinois, 1983, Sc
   Theodore R. Peck, (D), Wisconsin, 1962, Sc
   Joseph W. Stucki, (D), Purdue, 1975, Sc
Associate Professor:
   Robert G. Darmody, (D), Maryland, 1980, Sd
   Timothy R. Ellsworth, (D), California, 1989, Sp
   Kenneth R. Olson, (D), Cornell, 1983, Sc
   F. William Simmons, (D), North Carolina State, 1987, Sp
Assistant Professor:
   Robert J. M. Hudson, (D), MIT, 1989, Cg
   Gregory F. McIsaac, (D), Illinois, 1994, Hs

Waubonsee Community College
Earth Sciences (A) (2018)
Rt. 47 at Waubonsee Dr.
Sugar Grove, IL 60554
   p. (630) 466-2783
   dvoorhees@waubonsee.edu
   https://www.waubonsee.edu/programs-courses/programs-subject/science-technology-engineering-and-mathematics/earth-science
   Programs: Geology (A)
   Certificates: GIS
   Enrollment (2018): A: 6 (0)
Professor:
   David H. Voorhees, (M), Rensselaer, 1982, ZgGg
Associate Professor:
   Karl Schulze, (M), Texas A&M, 2003, ZgAms

Assistant Professor:
    Alfred W. Weiss, (M), S Illinois, 2000, Ziy

Western Illinois University
Department of Geography, Geographic Information Science and Meteorology (2015)
Macomb, IL 61455
    p. (309) 298-1648
    s-thompson@wiu.edu
    http://www.wiu.edu/cas/geography/

Dept of Geology (B) O-🏛 (2018)
1 University Circle
Macomb, IL 61455
    p. (309) 298-1151
    geology@wiu.edu
    http://www.wiu.edu/cas/geology/
    Programs: Geology; Paleontology
    Administrative Assistant: Cerese Wright
    Enrollment (2017): B: 16 (0)
Professor:
    Kyle R. Mayborn, (D), California (Davis), 2000, Gic
    Leslie A. Melim, (D), S Methodist, 1991, Gd
Associate Professor:
    Steven W. Bennett, (D), Indiana, 1994, Hw
Assistant Professor:
    Thomas A. Hegna, (D), Yale, 2012, PigPg
Instructor:
    Sara Bennett, (M), Indiana, Gg
Emeritus:
    Jack B. Bailey, (D), Illinois, 1975, Pg
    Peter L. Calengas, (D), Indiana, 1977, En

Wheaton College
Dept of Geology & Environmental Science (B) ☒ (2019)
501 E. College Ave.
Wheaton, IL 60187
    p. (630) 752-5063
    geology@wheaton.edu
    http://www.wheaton.edu/geology/
    Enrollment (2018): B: 20 (6)
Chair:
    Stephen O. Moshier, (D), Louisiana State, 1986, Gd
Professor:
    Chris Keil, (D), Illinois (Chicago), 1994, Rm
Assistant Professor:
    Andrew J. Luhmann, (D), Minnesota (Twin Cities), 2011, HwGmCl
    Kathryn A. Maneiro, (D), Boston, 2016, CcGic
Instructor:
    Lisa Heidlauf, (M), Illinois (Urbana), 1986, Gg
Visiting Professor:
    Gilles V. Tagne, (M), Université de Rennes I, 2013, HwZiGe
Emeritus:
    James A. Clark, (D), Colorado, 1977, GmYgZi
    Jeffrey K. Greenberg, (D), N Carolina, 1978, GicGe

Indiana

Ball State University
Dept of Geography (2015)
Muncie, IN 47306
    p. 765-285-1776
    turk@bsu.edu
    https://www.bsu.edu/academics/collegesanddepartments/geography

Ball State University
Dept of Geology (B,M,D) O (2016)
2000 University Avenue
AR117
Muncie, IN 47306
    p. (765) 285-8270
    geology@bsu.edu
    http://www.bsu.edu/geology
    Administrative Assistant: Brenda J. Rathel
    Enrollment (2016): B: 47 (8) M: 14 (4) D: 4 (0)
Chair:
    Richard H. Fluegeman, (D), Cincinnati, 1987, PmGro
Associate Dean:
    Jeffry D. Grigsby, (D), Cincinnati, 1989, GdClGo
Professor:
    Kirsten N. Nicholson, (D), Joseph Fourier, 1999, Gi
    R. Scott Rice-Snow, (D), Penn State, 1983, Gm
Associate Professor:
    Carolyn B. Dowling, (D), Rochester, 2002, HgGeCg
    Klaus Neumann, (D), Alabama, 1999, Cl
Assistant Professor:
    Shawn J. Malone, (D), Iowa, 2012, GtxGm
Emeritus:
    Alan C. Samuelson, (D), Penn State, 1972, Hw
Technician, Environmental Science:
    Eric Lange, (M), Ball State, 2014, Cg
Geological Tech:
    Michael Kutis, (B), Wisconsin (Platteville), 1989, Gg

College of the Holy Cross
Dept of Geosciences (A,B) ☒ (2018)
P.O. Box 308
Notre Dame, IN 46556
    p. (219) 239-8417
    smitchel@holycross.edu

DePauw University
Dept of Geosciences (B) ☒ (2018)
602 South College Avenue
Julian Science Center
Greencastle, IN 46135
    p. (765) 658-4654
    fsoster@depauw.edu
    http://www.depauw.edu/acad/geosciences/
    f: https://www.facebook.com/DePauwGeosciences/
    Programs: Geology, Environmental Geoscience, Earth Science
    Administrative Assistant: Rachel Curtis
    Enrollment (2018): B: 28 (14)
Professor:
    Tim D. Cope, (D), Stanford, 2003, GsZiGr
    James G. Mills, (D), Michigan State, 1991, Gi
    Frederick M. Soster, (D), Case Western, 1984, Gs
    M. Scott Wilkerson, (D), Illinois, 1991, Gct
Cooperating Faculty:
    Jeanette K. Pope, (D), Virginia Tech, 2002, Ge

Earlham College
Geology Dept (B) ☒ (2017)
801 National Road West
Richmond, IN 47374-4095
    p. (765) 983-1429
    streeme@earlham.edu
    http://www.earlham.edu/geology/
    Programs: Geology
    Enrollment (2017): B: 14 (4)
Associate Professor:
    Cynthia M. Fadem, (D), Washington Univ (St. Louis), 2009, GaSoCs
    Andrew Moore, (D), Washington, 1999, Gm
    Meg Streepey Smith, (D), Michigan, 2001, Gt
Emeritus:
    Jon W. Branstrator, (D), Cincinnati, 1975, Pg

Hanover College
Dept of Geology (B) (2015)
PO Box 890
Hanover, IN 47243-0890
    p. (812) 866-7306
    worcestr@hanover.edu
    Enrollment (2007): B: 12 (5)
Chair:
    Peter A. Worcester, (D), Miami (Ohio), 1976, Gi

Professor:
  Kenneth A. Bevis, (D), Oregon State, 1995, GlmGe
  Heyo Van Iten, (D), Michigan, 1989, PgHgRn
Emeritus:
  Stanley M. Totten, (D), Illinois, 1962, Gl

Indiana State University
Dept of Earth and Environmental Systems (B,M,D) (2015)
159 Science Building
Terre Haute, IN 47809
  p. (812) 237-2444
  isu-ees@mail.indstate.edu
  http://www.indstate.edu/ess/
  f: www.facebook.com/isu.ees
  Enrollment (2009): B: 87 (13) M: 16 (7) D: 6 (2)
Professor:
  Gregory Bierly, (D), Michigan State, Asm
  Sandra S. Brake, (D), Colorado Mines, 1989, GeEm
  Anthony Rathburn, (D), Duke, 1992, PgOu
  James Speer, (D), Tennessee, 2001, PeZyGe
  C. Russell Stafford, (D), Arizona State, 1981, Gam
  Qihao Weng, (D), Georgia, Zru
Associate Professor:
  Susan Berta, (D), Oklahoma, 1986, GmZr
  Kathleen M. Heath, (D), Utah, 2001, Pb
  Jennifer C. Latimer, (D), Indiana, 2004, CmGeb
  Nancy J. Obermeyer, (D), Chicago, 1987, Zi
  Shawn Phillips, (D), New York, 2001
Emeritus:
  William Dando, (D)
  Prodip K. Dutta, (D), Indiana, 1983, Gs

Indiana University / Purdue University, Indianapolis
Dept of Earth Sciences (B,M) ☒ (2019)
723 West Michigan Street
Indianapolis, IN 46202-5132
  p. (317) 274-7484
  geology@iupui.edu
  http://www.geology.iupui.edu
Professor:
  Andrew P. Barth, (D), S California, 1989, Gi
  Gabriel M. Filippelli, (D), California (Santa Cruz), 1994, CmGb
  Lin Li, (D), Brown, 2002, XgZrg
Associate Professor:
  Kathy J. Licht, (D), Colorado, 1999, Gl
Assistant Professor:
  Pierre-Andre Jacinthe, (D), Ohio State, 1995, Cl

Indiana University Northwest
Dept of Geosciences (A,B) (2017)
3400 Broadway
Gary, IN 46408
  p. (219) 980-6738
  zkilibar@iun.edu
  http://www.iun.edu/~geos/
  Programs: Geology (B); Environmental Science (B)
  Enrollment (2017): B: 21 (3)
Chair:
  Zoran Kilibarda, (D), Nebraska, 1994, GsmGd
Associate Professor:
  Erin Argyilan, (D), Illinois (Chicago), 2004, GeHwAm
  Kristin Huysken, (D), Michigan State, 1996, GizGc
Emeritus:
  Robert Votaw, (D), PiGrs

Indiana University, Bloomington
Dept of Earth and Atmospheric Sciences (B,M,D) O (2015)
1001 E. Tenth Street
Bloomington, IN 47405
  p. (812) 855-5582
  brophy@indiana.edu
  http://geology.indiana.edu/
  Enrollment (2014): B: 67 (20) M: 33 (0) D: 39 (0)
Chair:
  Lisa M. Pratt, (D), Princeton, 1981, Co
Professor:
  Abhijit Basu, (D), Indiana, 1975, Gd
  David L. Bish, (D), Penn State, 1977, GzySc
  Simon C. Brassell, (D), Bristol, 1980, Cob
  James G. Brophy, (D), Johns Hopkins, 1984, Gi
  Michael W. Hamburger, (D), Cornell, 1986, YsGtv
  Gary L. Pavlis, (D), Washington, 1982, YsGtYe
  P David Polly, (D), California (Berkeley), 1993, PvgPq
  Edward M. Ripley, (D), Penn State, 1976, Em
  Juergen Schieber, (D), Oregon, 1985, Gs
  Robert P. Wintsch, (D), Brown, 1975, GpcGt
  Chen Zhu, (D), Johns Hopkins, 1992
Academic Director Judson Mead Geologic Field Station:
  Bruce Douglas, (D), Princeton, 1983, Gc
Senior Scientist:
  Chusi Li, (D), Toronto, 1993, GiEm
  Arndt Schimmelmann, (D), California (Los Angeles), 1985, CsGsPe
Associate Professor:
  Claudia C. Johnson, (D), Colorado, 1993, Pe
  Kaj Johnson, (D), Stanford, 2004, Yg
  Laura E. Wasylenki, (D), Caltech, 1999, ClOoCs
Associate Scientist:
  Erika R. Elswick, (D), Cincinnati, 1998, Cl
  Edward W. Herrmann, (D), Indiana, 2013, Sa
  Peter Sauer, (D), Colorado, 1997, Pe
Assistant Professor:
  Douglas A. Edmonds, (D), Penn State, 2009, Gsm
  Julie C. Fosdick, (D), Stanford, 2011, Gt
  Chanh Q. Kieu, (D), Maryland, 2008, As
  Jackson K. Njau, (D), Rutgers, 2006, Pv
  Paul W. Staten, (D), Utah, 2013, As
Lab Manager - SIRF:
  Ben Underwood, (B), Indiana, 2009, Ca
Lab Manager - SESAME:
  Alice Hui, (M), Indiana, 2013, Cl
Lecturer:
  Cody Kirkpatrick, (D), Alabama (Huntsville), 2010, As
Adjunct Professor:
  Maria D. Mastalerz, (D), Poland, 1988, Ec
  Jeffrey White, (D), Syracuse, 1984, Hs
Emeritus:
  Robert F. Blakely, (D), Indiana, 1973, Ys
  J. Robert Dodd, (D), Caltech, 1961, Pg
  John B. Droste, (D), Illinois, 1956, Gs
  Jeremy D. Dunning, (D), North Carolina, 1978, Gc
  Donald E. Hattin, (D), Kansas, 1954, Gr
  John M. Hayes, (D), MIT, 1966, Co
  Erle G. Kauffman, (D), Michigan, 1961, Pg
  Enrique Merino, (D), California (Berkeley), 1973, Cl
  Greg A. Olyphant, (D), Iowa, 1979, Hw
  Lee J. Suttner, (D), Wisconsin, 1966, Gd
Senior Lecturer:
  Bruce Douglas, (D), Princeton, 1983, Gc
Facilities Administrator:
  John L. Hettle Jr.

School of Public & Environmental Affairs (B,M,D) (2015)
SPEA Building
Bloomington, IN 47408
  p. (812) 855-7485
  spea@indiana.edu
  http://www.spea.indiana.edu/home/
Geology Librarian:
  Christina Sheley, Zn
Chair:
  Philip S. Stevens, (D), Harvard, 1990, As
Professor:
  Ronald A. Hites, (D), MIT, 1968, Co
  Jeffrey R. White, (D), Syracuse, 1984, Hg
Associate Professor:
  Diane S. Henshel, (D), Washington, 1987, Zn
  Todd Royer, (D), Idaho State, 1999, HsZg
Assistant Professor:
  Vicky J. Meretsky, (D), Arizona, 1995, Sf
  Flynn W. Picardal, (D), Arizona, 1992, So
Lecturer:
  Melissa Clark, (M), Indiana, 1999, SfZg

Indiana University, Indianapolis
Dept of Geography (A,B) (2015)
213 Cavanaugh Hall
Indianapolis, IN 46202
   p. (317) 274-8877
   tbrother@iupui.edu
   http://www.iupui.edu/~geogdept/
Chair:
   Jeffey S. Wilson, (D), Indiana State, 1998, Zy
Professor:
   F. L. Bein, (D), Florida, 1974, Zy
Associate Professor:
   Timothy S. Brothers, (D), California (Los Angeles), 1985, Zy
   Catherine J. Souch, (D), British Columbia, 1990, Gm

Purdue University
Dept of Earth, Atmospheric, and Planetary Sciences (B,M,D) (2017)
550 Stadium Mall Drive
West Lafayette, IN 47907-2051
   p. (765) 494-3258
   eas-info@purdue.edu
   http://www.eaps.purdue.edu
   Enrollment (2014): M: 13 (13) D: 58 (6)
Department Head:
   Indrajeet Chaubey, (D), Oklahoma State, 1997, Hg
Associate Department Head:
   Darryl E. Granger, (D), California (Berkeley), 1996, As
   Harshvardhan, (D), SUNY (Stony Brook), 1976, As
University Distinguished Professor:
   H. Jay Melosh, (D), Caltech, 1972, XmgXy
Distinguished Professor:
   John Cushman, (D), Iowa State, 1978, Sp
   Paul B. Shepson, (D), Penn State, 1982, As
Director:
   Douglas R. Schmitt, (D), Caltech, 1987, YxNrYe
Professor:
   Ernest M. Agee, (D), Missouri, 1968, As
   Lawrence W. Braile, (D), Utah, 1973, Ys
   Maarten de Hoop, (D), Delft (Neth), 1992, Gq
   Timothy R. Filley, (D), Penn State, 1997, Co
   Andrew M. Freed, (D), Arizona, 1998, Gg
   Andrei Gabrielov, (D), Moscow State (Russia), 1973, Ys
   Alexander Gluhovsky, (D), USSR Acad of Sci, 1973, As
   Jon M. Harbor, (D), Washington, 1990, GmlGe
   Robert L. Nowack, (D), MIT, 1985, Ys
   James G. Ogg, (D), Scripps, 1981, PsGsYm
   Kenneth D. Ridgway, (D), Rochester, 1992, Gs
   Daniel P. Shepardson, (D), Iowa, 1990, Ze
   Yuch-Ning Shieh, (D), Caltech, 1968, CsGpv
   Wen-Yih Sun, (D), Chicago, 1975, AsZoAm
   Terry R. West, (D), Purdue, 1966, Ng
   Qianlai Zhuang, (D), Alaska (Fairbanks), As
   William J. Zinsmeister, (D), California (Riverside), 1974, Ps
Indiana State Climatologist:
   Dev Niyogi, (D), North Carolina State, 2000, As
Associate Professor:
   Chris Andronicos, (D), Princeton, 1999, Gg
   Michael Baldwin, (D), Oklahoma, 2003, As
   Lucy M. Flesch, (D), Stony Brook, 2002, Yg
   Hersh Gilbert, (D), Colorado, 2001, Ys
   Nathaniel A. Lifton, (D), Arizona, 1997, GmCc
   Greg Michalski, (D), California (San Diego), 2003, CsGeAs
   Wen-wen Tung, (D), California (Los Angeles), 2002, As
Assistant Professor:
   Julie Elliott, (D), Alaska (Fairbanks), 2011, Yd
   Marty Frisbee, (D), New Mexico Tech, 2010, Hws
   Saad Haq, (D), SUNY (Stony Brook), 2004, Gt
   Briony Horgan, (D), Cornell, 2010, XgZr
   David Minton, (D), Arizona, 2009, Xg
   Lisa Welp, (D), Caltech, 2006, Cs
   Yutian Wu, (D), Columbia, 2011, As
Emeritus:
   William J. Hinze, (D), Wisconsin, 1957, YevGt
   Arvid M. Johnson, (D), Penn State, 1965, Gc
   Gerald H. Krockover, (D), Iowa, 1970, Ze
   Darrell I. Leap, (D), Penn State, 1974, Hy
   Phillip J. Smith, (D), Wisconsin, 1967, As
   Thomas M. Tharp, (D), Wisconsin, 1978, Nm
   Dayton G. Vincent, (D), MIT, 1969, As

University of Notre Dame
Dept of Civil & Environmental Engineering & Earth Sciences (B,M,D) ☒ (2018)
156 Fitzpatrick Hall
Notre Dame, IN 46556
   p. (219) 631-5380
   ceees@nd.edu
   http://www.nd.edu/~ceees
   Programs: Environmental Earth Sciences; Environmental Science with a Concentration in Earth Sciences
   Certificates: Earth Sciences (minor)
Chair:
   Joannes J. Westerink, (D), MIT, 1984, On
Professor:
   Peter C. Burns, (D), Manitoba, 1994, Gz
   Jeremy B. Fein, (D), Northwestern, 1989, Cl
   Joe Fernando, (D), Johns Hopkins, 1983, As
   Clive R. Neal, (D), Leeds (UK), 1985, Gi
Associate Professor:
   Diogo Bolster, (D), California (San Diego), 2007, Hw
   Andrew Kennedy, (D), Monash, 1998, On
   Tony Simonetti, (D), Carleton, 1994, Ct
Assistant Professor:
   Melissa Berke, (D), Minnesota, 2011, Cs
   Alan Hamlet, (D), Washington, 2006, Hs
   Amy E. Hixon, (D), Clemson, 2013, ClqCc
   Marc Muller, (D), Rw
   David Richter, (D), Stanford, 2011, As
Instructor:
   Stephanie Simonetti, (D), McGill, 2002, Zn

University of Southern Indiana
Dept of Geology & Physics (B) ⚲ (2018)
8600 University Boulevard
Evansville, IN 47712
   p. (812) 464-1701
   wselliott@usi.edu
   http://www.usi.edu/science/geology-and-physics/
   Programs: Environmental Science; Geology
   Administrative Assistant: Kim E.. Schauss
   Enrollment (2018): B: 26 (5)
Chair:
   William S. Elliott, Jr., (D), Indiana, 2002, GsrCl
Professor:
   Joseph A. DiPietro, (D), Oregon State, 1990, GcpGm
   Paul K. Doss, (D), N Illinois, 1991, HwGeZu
Associate Professor:
   James Durbin, (D), Nebraska, 1999, Gm
   Tony Maria, (D), Rhode Island, 2000, Gv
Instructor:
   Carrie L. Wright, (M), Wright State, 2006, Ze
Emeritus:
   Norman R. King, (D), Indiana, 1973, Gr

Iowa

Cornell College
Dept of Geology (B) ●☒ (2017)
600 First Street SW
Mount Vernon, IA 52314
   p. (319) 895-4306
   rdenniston@cornellcollege.edu
   http://www.cornellcollege.edu/geology
   Programs: Geology
   Enrollment (2017): B: 18 (8)
Professor:
   Rhawn F. Denniston, (D), Iowa, 2000, GeCs
Associate Professor:
   Emily O. Walsh, (D), California (Santa Barbara), 2003, GptCc

Drake University
Dept of Environmental Science and Policy (B) (2015)
Des Moines, IA 50311
  p. (515) 271-2803
  thomas.rosburg@drake.edu

Iowa State University of Science & Technology
Dept of Agronomy (B,M,D) (2015)
2101 Agronomy Hall
Ames, IA 50011-1010
  p. (515) 294-1360
  agron@iastate.edu
  http://www.agron.iastate.edu/
  Department Secretary: Pam Hinderaker
Chair:
  Kendall Lamkey, (D), Iowa State, 1985
Director:
  Dennis R. Keeney, (D), Iowa, 1965, So
Professor:
  Raymond W. Arritt, (D), Colorado State, 1985, As
  Richard E. Carlson, (D), Iowa State, 1971, Am
  Richard M. Cruse, (D), Minnesota, 1978, So
  V. P. (Bill) Evangelou, (D), California (Davis), 1981, Sc
  William J. Gutowski, (D), MIT, 1984, Am
  Robert Horton, (D), New Mexico State, 1981, Sp
  Douglas L. Karlen, (D), Kansas State, 1978, So
  Randy J. Killorn, (D), Idaho, 1983, So
  Thomas B. Moorman, (D), Washington State, 1983, So
  Jonathan A. Sandor, (D), California (Berkeley), 1983, Sd
  Eugene S. Takle, (D), Iowa State, 1971, Asm
  Elwynn Taylor, (D), Washington (St. Louis), 1970, As
  Michael L. Thompson, (D), Ohio State, 1980, Sc
  Regis D. Voss, (D), Iowa State, 1962, So
  Douglas N. Yarger, (D), Arizona, 1967, As
Associate Professor:
  Cynthia Cambardella, (D), Colorado, 1991, So
  Thomas A. Kaspar, (D), Iowa State, 1982, So
  David A. Laird, (D), Iowa State, 1987, Sc
  Antonio W. Mallarino, (D), Iowa State, 1991, So
Assistant Professor:
  Lee Burras, (D), Ohio State, 1992, Sd
  Larry Halverson, (D), Wisconsin, 1991, So
  Stanley J. Henning, (D), Oregon State, 1975, So
  Thomas A. Polito, (D), Iowa State, 1987, So
Associate Dean, College of Agriculture:
  Gerald A. Miller, (D), Iowa State, 1974, SdoGm
Emeritus:
  Thomas E. Loynachan, (D), North Carolina State, 1975, So
Geology Librarian:
  Peter A. Peterson, (D), Illinois, 1953, Zn

Dept of Geological & Atmospheric Sciences (B,M,D) ○◐ (2019)
253 Science I
2237 Osborn Drive
Ames, IA 50011-3212
  p. (515) 294-4477
  geology@iastate.edu
  http://www.ge-at.iastate.edu/
  f: https://www.facebook.com/ISUgeology
  Programs: Geology; Earth Science, Meteorology
  Administrative Assistant: DeAnn M. Frisk
  Enrollment (2018): B: 136 (43) M: 26 (8) D: 22 (4)
Chair:
  Sven S. Morgan, (D), Virginia Tech, 1998, Gc
Professor:
  Igor A. Beresnev, (D), USSR Acad of Sci, 1986, YgsYe
  Cinzia C. Cervato, (D), ETH Zurich, 1990, GgZe
  William A. Gallus, (D), Colorado State, 1993, As
  William J. Gutowski, (D), MIT, 1984, ZoAtZc
  Neal R. Iverson, (D), Minnesota, 1989, Gl
  William W. Simpkins, (D), Wisconsin, 1989, HwGlCs
  Paul G. Spry, (D), Toronto, 1984, Em
  Xiaoqing Wu, (D), California (Los Angeles), 1992, As
Associate Professor:
  Kristie Franz, (D), California (Irvine), 2006, Hs
  Chris Harding, (D), Houston, 2001, Gq
  Alan D. Wanamaker, (D), Maine, 2007, CsPeOg
Assistant Professor:
  Beth E. Caissie, (D), Massachusetts, 2012, Pe
  Alex Gonzalez, (D), Colorado State, 2015, As
  Jacqueline Reber, (D), Oslo, 2012, Gct
  Yuyu Zhou, (D), Rhode Island, 2008, Zyi
Senior Lecturer:
  James V. Aanstoos, (D), Purdue, 1996, ZrAs
  Jane P. Dawson, (D), New Mexico, 1995, Gp
  David M. Flory, (M), Iowa State, 2003, AmsZn
Lecturer:
  Rachindra Mawalagedara, (D), Nebraska (Lincoln), 2013, As
  Aaron R. Wood, (D), Michigan, 2009, PvGrs
Adjunct Professor:
  Michael R. Burkart, (D), Iowa, 1976, Hy
Emeritus:
  Tsing-Chang Chen, (D), Michigan, 1975, As
  Robert Cody, (D), Colorado, 1968, Gz
  Carl E. Jacobson, (D), California (Los Angeles), 1980, GtcGp
  Karl E. Seifert, (D), Wisconsin, 1963, Ct
  Eugene S. Takle, (D), Iowa State, 1971, As
  Carl F. Vondra, (D), Nebraska, 1963, Gr
  Kenneth E. Windom, (D), Penn State, 1976, Cp
  Douglas N. Yarger, (D), Arizona, 1967, As
Lab Equipment Specialist:
  Alison Whale, (M), Iowa State, 2018

Scott Community College
Environmental Science (2015)
500 Belmont Road
Bettendorf, IA 52722
  p. 563-441-4001
  eiccinfo@eicc.edu
  http://www.eicc.edu

University of Iowa
Earth & Environmental Sciences (B,M,D) ●◐ (2018)
115 Trowbridge Hall
123 N. Capitol Street
Iowa City, IA 52242
  p. (319) 335-1818
  geology@uiowa.edu
  http://clas.uiowa.edu/ees/
  f: https://www.facebook.com/UIowaGeoscience
  Programs: Geoscience
  Administrative Assistant: Angela Bellew
  Enrollment (2018): B: 53 (11) M: 15 (8) D: 14 (1)
Chair:
  Charles T. Foster Jr., (D), Johns Hopkins, 1975, GptGg
Professor:
  Jonathan M. Adrain, (D), Alberta, 1993, Pi
  Christopher A. Brochu, (D), Texas, 1997, Pv
  Jane A. Gilotti, (D), Johns Hopkins, 1987, GctGp
  William C. McClelland, (D), Arizona, 1990, GtCcGc
  David W. Peate, (D), Open Univ (UK), 1989, CgGi
  Mark K. Reagan, (D), California (Santa Cruz), 1987, Gi
Associate Professor:
  Bradley D. Cramer, (D), Ohio State, 2009, GrPsGs
  Jeffrey A. Dorale, (D), Minnesota, 2001, Cs
  Emily Finzel, (D), Purdue, 2010, GstEo
  Ingrid Ukstins Peate, (D), Royal Holloway, 2003, GviXg
  Frank H. Weirich, (D), Toronto, 1982, Gm
Assistant Professor:
  William Barnhart, (D), Cornell, 2013, Gt
  Jessica R. Meyer, (D), Guelph, 2013, Gm
Lecturer:
  Mary Kosloski, (D), Cornell, 2012, Pig
  Kate Tierney, (D), Ohio State, 2010, GrCsGg
Adjunct Associate Professor:
  Brian J. Witzke, (D), Iowa, 1981, GrPsg
Adjunct Assistant Professor:
  Raymond R. Anderson, (D), Iowa, 1992, Gr
  Rhawn F. Denniston, (D), Iowa, 2000, Cs
  Keith E. Schilling, (D), Iowa, 2009, Hgs
  Emily O. Walsh, (D), California (Santa Barbara), 2003, GptCc

Adjunct Professor:
   David L. Campbell, (D), California (Berkeley), 1969, Yg
Emeritus:
   Richard G. Baker, (D), Colorado, 1969, PebPc
   E. Arthur Bettis III, (D), Iowa, 1995, SdGmSa
   Ann F. Budd, (D), Johns Hopkins, 1978, Pgi
   Robert S. Carmichael, (D), Pittsburgh, 1967, YexGg
   Lon D. Drake, (D), Ohio State, 1968, Hy
   Philip H. Heckel, (D), Rice, 1966, GrPsGd
   George R. McCormick, (D), Ohio State, 1964, GzEmCa
   Holmes A. Semken, Jr., (D), Michigan, 1965, Pv
   Keene Swett, (D), Edinburgh, 1965, Gs
   You-Kuan Zhang, (D), Arizona, 1990, Hw
Rock Lab Manager:
   Matthew J. Wortel, (M), Iowa, 2007, Gx
Collections Manager, Paleontology:
   Tiffany S. Adrain, (M), Iowa, 2003, Pg
Department Secretary:
   Christine K. Harms, (N)

University of Northern Iowa
Dept of Earth and Environmental Sciences (B) ⊘ (2018)
121 Latham Hall
Cedar Falls, IA 50614-0335
   p. (319) 273-2759
   siobahn.morgan@uni.edu
   http://www.earth.uni.edu/
   f: https://www.facebook.com/pages/University-of-Northern-Iowa-Earth-and-Environmental-Science/56860203682
   t: @uni_earthsci
   Programs: Earth Science; Earth Science Teaching; Environmental Sciences, Environmental Resource Management
   Administrative Assistant: Noel Graff
   Enrollment (2018): B: 80 (26)
Head:
   Siobahn M. Morgan, (D), Washington, 1991, Xa
Professor:
   Alan C. Czarnetzki, (D), Wisconsin, 1992, Am
   Thomas A. Hockey, (D), New Mexico State, 1988, Zn
   Mohammad Z. Iqbal, (D), Indiana, 1994, Hw
Associate Professor:
   Kyle R. Gray, (D), Akron, 2009, ZeGg
   Chad E. Heinzel, (D), N Illinois, 2005, GmaSa
Assistant Professor:
   Alexa Sedlacek, (D), Ohio State, 2013, CsGr
   Xinhua Shen, (D), Colorado State, 2011, Asm
Instructor:
   Lee S. Potter, (D), Texas, 1996, Gz
   Aaron Spurr, (M), N Iowa, 1997, Zeg
Emeritus:
   Wayne I. Anderson, (D), Iowa, 1964, Pg
   Lynn A. Brant, (D), Penn State, 1980, Gn
   Timothy M. Cooney, (D), N Colorado, 1976, Ze
   Walter E. De Kock, (D), Ohio State, 1972, Ze
   Kenneth J. De Nault, (D), Stanford, 1974, Gz
   James C. Walters, (D), Rutgers, 1975, Gm
Lab Tech:
   Steven J. Smith, (M), N Iowa, 2000, Zg

Kansas

Emporia State University
Earth Science Program (B,M) O⌂ (2019)
1 Kellogg Circle
Emporia, KS 66801-5087
   p. (620) 341-5330
   mmorales@emporia.edu
   http://www.emporia.edu/earthsci/
   Programs: Master's degree in Earth Science and Undergraduate degrees in Earth Science
   Certificates: GIS
   Enrollment (2018): B: 19 (9) M: 15 (5)
Professor:
   Marcia K. Schulmeister, (D), Kansas, 2000, HwClGe
   Richard O. Sleezer, (D), Kansas, 2001, ZiSdHs
Associate Professor:
   Michael A. Morales, (D), California (Berkeley), 1987, PvGrg
Assistant Professor:
   Alivia J. Allison, (D), Missouri (Kansas City), 2013, GgaGm
   Paul Zunkel, (D), Texas State, 2017, AmZig
Emeritus:
   James S. Aber, (D), Kansas, 1978, Gl
   Paul L. Johnston, (M), Kansas, 1959, Gg
   Kenneth W. Thompson, (D), Iowa State, 1991, Ze

Fort Hays State University
Department of Geosciences (B,M) O⌂ (2019)
600 Park Street
Tomanek Hall
Hays, KS 67601-4099
   p. (785) 628-5389
   geosciences@fhsu.edu
   http://www.fhsu.edu/geo
   f: https://www.facebook.com/GeoFHSU/
   t: @GeoFHSU
   Programs: Geology, Geography, Geographic Information Systems
   Certificates: Geographic Information Systems, Museum Studies
   Administrative Assistant: Patricia Duffey
   Enrollment (2018): B: 175 (28) M: 30 (9)
Chair:
   P. Grady Dixon, (D), Arizona State, 2005, AmZyAs
Emeritus:
   Kenneth R. Neuhauser, (D), South Carolina, 1973, GceGs
   Paul Phillips, (D), Kansas, 1977, Zen
   Richard J. Zakrzewski, (D), Michigan, 1968, PvsPe
Professor:
   Richard Lisichenko, (D), Kansas State, 2000, Zie
Associate Professor:
   Tom Schafer, (D), Kansas State, 2000, ZyiRn
   Laura E. Wilson, (D), Colorado (Boulder), 2012, PvePg
Assistant Professor:
   Henry Agbogun, (D), New Brunswick, 2012, GgYg
   Hendratta N. Ali, (D), Oklahoma State, 2010, GogYs
   Keith A. Bremer, (D), Texas State, 2011, ZuiZn
   Jonathan B. Sumrall, (D), Mississippi State, 2013, GgxZn
Instructor:
   Eamonn Coveney, (M), Fort Hays State, 2006, Zyn
   William H. Heimann, (M), Fort Hays State, 1987, HgwGm
   Kara Kuntz, (M), Kansas State, 1990, Zn
   Jami J. Seirer, (M), Montana, 2015, ZriZn
   Jeanne L. Sumrall, (D), Mississippi State, 2015, Zgn
Cooperating Faculty:
   Joseph R. Thomasson, (D), Iowa State, 1976, Pb

Johnson County Community College
Science Division (A) ⌂ (2019)
12345 College Blvd
Overland Park, KS 66210
   p. (913) 469-3826
   mwisgird@jccc.edu
   http://www.jccc.edu/academics/math-science/index.html
   Enrollment (2015): A: 7 (2)
Professor:
   Lynne Beatty, (M), S Illinois, 1985, GgZy
Assistant Professor:
   John P. Harty, (D), Zy
Adjunct Professor:
   John Maher, (D), Maryland, 1999, Zy

Kansas State University
Dept of Agronomy (B,M,D) (2016)
2004 Throckmorton
Manhattan, KS 66506
   p. (785) 532-6101
   agronomy@ksu.edu
   http://www.agronomy.k-state.edu/
   f: https://www.facebook.com/kstate.agronomy
   t: @KStateAgron
   Enrollment (2014): B: 70 (33) M: 41 (9) D: 43 (10)

Deparment Head:
 Gary M. Pierzynski, (D), Ohio State, 1989, Sc
Professor:
 Stewart Duncan, (D), Kansas State, 1991, So
 Walter H. Fick, (D), Texas Tech, 1978, Sf
 Dale Fjell, (D), Kansas State, 1982, So
 Allan Fritz, (D), Kansas State, 1994, So
 Mary Beth Kirkham, (D), Wisconsin, Sp
 Gerard J. Kluitenberg, (D), Iowa State, 1989, Sp
 David Mengel, (D), Purdue, 1975, So
 Clenton E. Owensby, (D), Kansas State, 1969, Sf
 Dallas Peterson, (D), North Dakota State, 1987, So
 Vara Prasad, (D), Reading, 1999, So
 Michel D. Ransom, (D), Ohio State, 1984, SdcGm
 Chuck W. Rice, (D), Kentucky, 1983, Sb
 Bill T. Schapaugh, (D), Purdue, 1979, So
 Alan Schlegel, (D), Purdue, 1985, So
 Phillip Stahlman, (D), Wyoming, 1989, So
 Curtis Thompson, (D), Idaho, 1993, So
 Steve M. Welch, (D), Michigan State, 1977, Sp
Associate Professor:
 Ignacio Ciampitti, (D), Purdue, 2012, So
 Gary Cramer, (D), Nebraska, 1998, So
 Ganga Hettiarachchi, (D), Kansas State, 2000, Sc
 John Holman, (D), Idaho, 2005, So
 Doo-Hong Min, (D), Maryland, 1998, So
 Nathan Nelson, (D), North Carolina State, 2004, So
 DeAnn Presley, (D), Kansas State, 2007, So
 Kraig Roozeboom, (D), Kansas State, 2006, So
 Dorivar Ruiz-Diaz, (D), Iowa State, 2007, So
 Gretchen Sassenrath, (D), Illinois, 1988, So
 Tesfaye Tesso, (D), Kansas State, 2002, So
Assistant Professor:
 Eric Adee, (D), Wisconsin, 1993, So
 Lucas Haag, (D), Kansas State, 2013, So
 Mithila Jugulam, (D), Guelph, 2004, So
 Xiaomao Lin, (D), Nebraska, 1999, Zn
 Colby Moorberg, (D), North Carolina, 2014, So
 Geoffrey Morris, (D), Chicago, 2007, So
 Augustine Obour, (D), Florida, 2010, So
 Ram Perumal, (D), Tamil Nadu Ag, 1993, So
 Eduardo Santos, (D), Guelph, 2011, So
 Peter Tomlinson, (D), Arkansas, 2006, So
 Guorong Zhang, (D), North Dakota State, 2007, So
Emeritus:
 Mark Claassen, (D), Iowa State, 1971, So
 Bill Eberle, (D), Illinois, 1973, SoZu
 Stan Ehler, (D), Missouri, 1975, So
 Barney Gordon, (D), South Dakota, 1990, So
 Keith Janssen, (D), Michigan State, 1973, So
 George Liang, (D), Wisconsin, 1965, So
 Gerry L. Posler, (D), Iowa State, 1969, Sf
 Kevin Price, (D), Utah, 1987, Zr
 David Regehr, (D), Illinois, 1975, So
 Jim Shroyer, (D), Iowa State, 1980, So
 Loyd Stone, (D), South Dakota, 1973, Sp
 Steve J. Thien, (D), Purdue, 1971, Sc
 Richard Vanderlip, (D), So
 D.A. Whitney, (D), Iowa State, 1966, So
Agronomist:
 Doug Shoup, (D), Kansas State, 2006, So
Librarian:
 Nancy William, (N), Zn

Dept of Geology (B,M) (2015)
108 Thompson Hall
Manhattan, KS 66506-3201
 p. (785) 532-6724
 rocknrat@ksu.edu
 http://www.k-state.edu/geology/
 Administrative Assistant: Lori Page-Willyard
 Enrollment (2013): B: 69 (15) M: 27 (5)
Department Head:
 Pamela D. Kempton, (D), S Methodist, 1984, GiCtc
Professor:
 Sambhudas Chaudhuri, (D), Ohio State, 1966, GeoCc

Associate Professor:
 Allen W. Archer, (D), Indiana, 1983, Gr
 Matthew E. Brueseke, (D), Miami (Ohio), 2006, GitGv
 Abdelmoneam E. Raef, (D), AGH (Poland), 2001, YesZr
 Matthew W. Totten, (D), Oklahoma, 1992, Gso
Assistant Professor:
 Saugata Datta, (D), W Ontario, 2001, ClHwGe
 Keith B. Miller, (D), Rochester, 1988, Pe
 Joel Q.G. Spencer, (D), Glasgow (UK), 1996, CcGs
Emeritus:
 Robert L. Cullers, (D), Wisconsin, 1971, Ct
 Charles G. Oviatt, (D), Utah, 1984, Gm
 Ronald R. West, (D), Oklahoma, 1970, PgeGs

University of Kansas
Dept of Geology (B,M,D) (2018)
1414 Naismith Dr
Room 254
Ritchie Hall
Lawrence, KS 66045-7613
 p. (785) 864-4974
 geology@ku.edu
 http://geo.ku.edu/
 f: https://www.facebook.com/KUGeology
 Programs: BA, BS, MS, PhD in Geology
 Professional Science MS in Environmental Geology
 Certificates: Environmental Geology
 Enrollment (2013): B: 72 (14) M: 50 (17) D: 34 (9)
Professor and Chair:
 Jennifer A. Roberts, (D), Texas, 2000, Cbl
Union Pacific Distinguished Professor:
 J. Douglas Walker, (D), MIT, 1985, Gc
Ritchie Distinguished Professor:
 Michael D. Blum, (D), Texas, 1997, GsrGm
Gulf-Hedberg Distinguished Professor:
 Paul A. Selden, (D), Cambridge (UK), 1979, Pig
Distinguished Professor:
 Robert H. Goldstein, (D), Wisconsin, 1986, Gs
Courtesy Professor:
 Rolfe Mandel, (D), Kansas, 1991, Gam
Professor:
 J. F. Devlin, (D), Waterloo, 1994, Hw
 David A. Fowle, (D), Notre Dame, 2000, Cb
 Evan K. Franseen, (D), Wisconsin, 1989, Gd
 Luis A. Gonzalez, (D), Michigan, 1989, CsGd
 Stephen T. Hasiotis, (D), Colorado, 1997, Pe
 Mary C. Hill, (D), Princeton, 1985, HgqHw
 Gwendolyn L. Macpherson, (D), Texas, 1989, Hw
 Gene Rankey, (D), Kansas, 1996, GsrGu
 Michael H. Taylor, (D), California (Los Angeles), 2004, GtmZr
 George P. Tsoflias, (D), Texas, 1999, YgeGe
Associate Professor:
 Craig Marshall, (D), Technology (Australia), 2001, GzZmXb
 Andreas Möller, (D), Christian-Albrechts (Germany), 1996, Gt
 Alison Olcott Marshall, (D), S California, 2006, PgCoPo
 Leigh Stearns, (D), Maine, 2008, GlZrf
 Anthony W. Walton, (D), Texas, 1972, GsEoGv
Associate Scientist:
 Diane Kamola, (D), Georgia, 1989, Gs
 Randy Stotler, (D), Waterloo, 2008, Hw
Assistant Professor:
 Noah McLean, (D), MIT, 2012, CcGq
 Chi Zhang, (D), Rutgers, 2012, YgxYu
Courtesy Professor:
 John H. Doveton, (D), Edinburgh, 1969, Gq
 William C. Johnson, (D), Wisconsin, 1976, Grm
 Leonard Krishtalka, (D), Kansas, 1976, Pv
 Bruce S. Lieberman, (D), Columbia, 1994, Pgi
 Gregory A. Ludvigson, (D), Iowa, 1988, Gs
 Richard D. Miller, (M), Kansas, 1983, Ye
 Edith Taylor, (D), Ohio State, 1983, Pb
 W. Lynn Watney, (D), Kansas, 1985, Gr
 Donald O. Whittemore, (D), Penn State, 1973, Cl
Courtesy Associate Professor:
 Geoff Bohling, (D), Kansas, 1999, Hw
 Gaisheng Liu, (D), Alabama, 2004, Hg

Courtesy Assistant Professor:
   Andrea Brookfield, (D), Waterloo, 2009, Hw
   Jon Smith, (D), Kansas, 2008, GsPe
Adjunct Professor:
   Timothy R. Carr, (D), Wisconsin, 1981, Go
   John Gosse, (D), Lehigh, 1994, Gm
Emeritus:
   Ernest E. Angino, (D), Kansas, 1961, Cl
   Ross A. Black, (D), Wyoming, 1989, Ye
   Wakefield Dort, Jr., (D), Stanford, 1955, Gm
   Gisela Dreschoff, (D), Tech (Braunschweig), 1972, Ce
   Paul Enos, (D), Yale, 1965, GsdGu
   Lee C. Gerhard, (D), Kansas, 1964, GsoGd
   Carl D. McElwee, (D), Kansas, 1970, HgYg
   Richard A. Robison, (D), Texas, 1962, PiGr
   Albert J. Rowell, (D), Leeds, 1953, Pi
   Don W. Steeples, (D), Stanford, 1975, YseGe
   W. Randall Van Schmus, (D), California (Los Angeles), 1964, Cc

Wichita State University
Dept of Geology (B,M) (2015)
1845 Fairmount
Wichita, KS 67260-0027
   p. (316) 978-3140
   john.gries@wichita.edu
   Administrative Assistant: K. L. Smith
Chair:
   William C. Parcell, (D), Alabama, 2000, Gr
Professor:
   William D. Bischoff, (D), Northwestern, 1985, Cl
   John C. Gries, (D), Texas, 1970, Gt
   Salvatore J. Mazzullo, (D), Rensselaer, 1974, Go
Associate Professor:
   Collette D. Burke, (D), Wisconsin (Milwaukee), 1983, Pm
Assistant Professor:
   Hongsheng Cao, (D), Florida State, 2001, Cl
   Wan Yang, (D), Texas, 1999, Gs
Lecturer:
   Toni K. Jackman, (M), Wichita State, 1984, Ge
   David L. Schaffer, (B), Utah, 1982, Am
Emeritus:
   James N. Gundersen, (D), Minnesota, 1958, Ga
   Daniel F. Merriam, (D), Kansas, 1961, Gr
   Peter G. Sutterlin, (D), Northwestern, 1958, Gd

Kentucky

Alice Lloyd College
Div of Natural Sciences & Mathematics (2015)
Pippa Passes, KY 41844-9701
   p. (606) 368-2101 x5405
   paulyeary@alc.edu

Bluegrass Community and Technical College
Environmental Science Technology Program (2015)
470 Cooper Dr
Lexington, KY 40506
   p. (859) 246-6448
   jean.watts@kctcs.edu
   http://www.bluegrass.kctcs.edu/Natural_Sciences/Environ-mental_Science_Technology.aspx

Eastern Kentucky University
Dept of Geosciences (B) (2019)
521 Lancaster Avenue
Science 2234
Richmond, KY 40475-3102
   p. (859) 622-1273
   melissa.dieckmann@eku.edu
   http://www.geosciences.eku.edu
   f: https://www.facebook.com/EKU-Geoscienc-es-447266678713162
   Programs: B.S. Geology (until Fall 2019)
   B.S. Environmental and Applied Geology (after Fall 2019)
   B.S. Geographic Information Science
   Certificates: GIS Certificate
   Enrollment (2018): B: 61 (20)
Chair:
   Melissa S. Dieckmann, (D), Notre Dame, 1995, ZeCl
Professor:
   Walter S. Borowski, (D), North Carolina, 1998, GsrOu
   Alice Jones, (D), Ohio State, 1998, Zyn
   John C. White, (D), Baylor, 2002, GiCgt
   Donald M. Yow, (D), South Carolina, 2003, AtZye
Associate Professor:
   F. Tyler Huffman, (D), Connecticut, 2006, Zi
   Kelly Watson, (D), Florida State, 2012, ZriZn
Assistant Professor:
   Jonathan Malzone, (D), Buffalo, 2015, HwgGm
Instructor:
   Cory BlackEagle, (M), E Kentucky, 1989, HwZg
   Ann Harris, (D), Kentucky, 2018, GgZy
   Sung Bae Jeon, (D), Zy
   Craig Webb, (M), W Kentucky, 1994, Zyn
   Peter Worcester, (D), Miami, 1977, GgOuZe
Senior Lecturer:
   Glenn A. Campbell, (M), Marshall, 1995, Zy
Emeritus:
   Gary L. Kuhnhenn, (D), Illinois, 1976, GgZe
   David Zurick, (D), Hawaii, 1986, Zn

Morehead State University
Dept of Earth and Space Sciences (B) (2015)
235 Martindale Drive
Morehead, KY 40351
   p. (606) 783-2381
   c.mason@moreheadstate.edu
   http://www.moreheadstate.edu/physsci/
   Administrative Assistant: Amanda Holbrook
   Enrollment (2014): B: 40 (4)
Department Chair:
   Benjamin K. Malphrus, (D), West Virginia, 1990
Professor:
   Charles E. Mason, (M), George Washington, 1981, PiGrs
Associate Professor:
   Marshall Chapman, (D), Massachusetts, 1996, Gv
   Eric Jerde, (D), California (Los Angeles), 1991, Cg
   Jennifer O'Keefe, (D), Kentucky, 2008, PlEcZe
   Steven K. Reid, (D), Texas A&M, 1991, GsoGe

Murray State University
Dept of Earth and Environmental Sciences (B,M) O (2019)
334 Blackburn Science Building
Murray, KY 42071
   p. (270) 809-2591
   qzhang@murraystate.edu
   murraystate.edu/ees
   f: https://www.facebook.com/murraystateees/
   t: @MurrayStateEES
   Programs: BS Tracks: Geology, Earth Science Education, Environmental Science, Geography & GIS, Archaeology
   MS Concentrations: Environmental Geology, Geoinformatics, Watershed Science, Archaeology
   Certificates: GIS (undergraduate), Geospatial Data Science (graduate)
   Enrollment (2012): B: 44 (9) M: 8 (4)
Chair:
   Qiaofeng (Robin) Zhang, (D), W Ontario, 2002, ZirZu
Professor:
   Haluk Cetin, (D), Purdue, 1993, ZrGe
Associate Professor:
   Anthony L. Ortmann, (D), Tulane, 2007, GaYg
Assistant Professor:
   Bassil El Masri, (D), Indiana, 2011, Zc
   Sung-ho Hong, (D), New Mexico Tech, 2008, HyGe
   Gary E. Stinchcomb, (D), Baylor, 2012, Ge
   Marcie Venter, (D), Kentucky, Ga
Lecturer:
   Jane L. Benson, (M), Murray State, 1986, Zi
   Michael R. Busby, (M), Murray State, 1996, Zi

Northern Kentucky University
Dept of Geology (B) O-☐ (2017)
204H Natural Sciences Center
Highland Heights, KY 41099
    p. (859) 572-5309
    rockawayj@nku.edu
    http://nku.edu/
    Programs: Geology; Earth Science Education
    Certificates: GIS
    Enrollment (2017): B: 30 (60)
Associate Professor:
    Janet Bertog, (D), Cincinnati, 2002, Gd
    John D. Rockaway, (D), Purdue, 1968, Ng
Assistant Professor:
    Trent Garrison, (D), Kentucky, 2015, GeHwEc
Lecturer:
    Sarah E. Johnson, (M), Purdue, 1997, GmNgZi

Owensboro Community and Technical College
Owensboro Community and Technical College (2015)
4800 New Hartford Road
Owensboro, KY 42303
    http://www.owensboro.kctcs.edu/

University of Kentucky
Dept of Earth and Environmental Sciences (B,M,D) O-☐ (2018)
101 Slone Research Building
121 Washington St.
Lexington, KY 40506-0053
    p. (859) 257-3758
    moker@uky.edu
    http://ees.as.uky.edu
    Programs: Geological Sciences
    Enrollment (2018): B: 47 (12) M: 24 (5) D: 14 (2)
Chair:
    David P. Moecher, (D), Michigan, 1988, GxtCg
Professor:
    Frank R. Ettensohn, (D), Illinois, 1975, PsGdPi
    William C. Haneberg, (D), Cincinnati, 1989, NgZi
    James C. Hower, (D), Penn State, 1978, Ec
    Dhananjay Ravat, (D), Purdue, 1989, YgmYv
    Edward W. Woolery, (D), Kentucky, 1998, Ys
Associate Professor:
    Alan E. Fryar, (D), Alberta, 1992, HwGeCl
    Kevin M. Yeager, (D), Texas A&M, 2002, GsCmc
Assistant Professor:
    John R. Bowersox, (D), South Florida, 2006, God
    Andrea M. Erhardt, (D), Stanford, 2013, CgsCm
    Rebecca Freeman, (D), Tulane, 2011, GrPi
    Michael M. McGlue, (D), Arizona, 2011, GnrGm
    Keely A. O'Farrell, (D), Toronto, 2013, Ygd
    Ryan Thigpen, (D), Virginia Tech, 2009, Gtc
Research Associate:
    Jordan S. Munizzi, (D), W Ontario, 2018, Cs
Instructor:
    Summer Brown, (M), Virginia Tech, 2010, GgZei
Lecturer:
    Kent Ratajeski, (D), North Carolina, 1999, Gi
Adjunct Professor:
    Cortland F. Eble, (D), West Virginia, 1988, Pl
    Stephen F. Greb, (D), Kentucky, 1992, GsEcGr
    Hickman B. John, (D), Kentucky, 2011, Gto
    Thomas M. Parris, (D), California (Santa Barbara), 1998, Cg
    Zhenming Wang, (D), Kentucky, 1998, YsNeRn
    Junfeng Zhu, (D), Arizona, 2005
Emeritus:
    William H. Blackburn, (D), MIT, 1967, GxCg
    Bruce R. Moore, (D), Melbourne, 1967, Gd
    Kieran D. O'Hara, (D), Brown, 1984, GcCg
    Lyle V. A. Sendlein, (D), Iowa State, 1964, Hw
    Ronald L. Street, (D), St. Louis, 1975, Ys
Laboratory Director:
    Peter J. Idstein, (M), E Kentucky, 1992, Gg

Dept of Mining Engineering (B,M,D) (2015)
230 Mining & Mineral Resources Building
Lexington, KY 40506-0107
    p. (859) 257-8026
    rick.honaker@uky.edu
    http://www.engr.uky.edu/mng/
    t: @UK_Mining
    Enrollment (2015): B: 125 (36) M: 14 (7) D: 17 (3)
Chair:
    Rick Honkaer, (D), Virginia Tech, 1992, Nm
Professor:
    Zach Agioutantis, (D), Virginia Tech, 1987, NrmEc
    Braden Lusk, (D), Missouri (Rolla), 2006, Nm
    Thomas Novak, (D), Penn State, 1984, Nm
    Joseph Sottile, Jr., (D), Penn State, 1991, Nm
Assistant Professor:
    Kyle Perry , (D), Kentucky, 2010, Nr
    Jhon Silva-Castro, (D), Kentucky, 2012, Nm
    William Chad Wedding, (D), Kentucky, 2014, Nm
Emeritus:
    Kot F. Unrug, (D), Acad Mining-Metallurgy, 1966, Nr
    Andrew M. Wala, (D), Acad Mining-Metallurgy, 1972, Nm

Western Kentucky University
Dept of Geography & Geology (B,M) O-☐ (2019)
1906 College Heights Blvd
#31066
Bowling Green, KY 42101-1066
    p. (270) 745-4555
    fred.siewers@wku.edu
    http://www.wku.edu/geoweb
    Programs: Geology, Geography, Meteorology, GIS
    Certificates: GIS, grad and undergrad
    Enrollment (2016): B: 110 (33) M: 25 (9)
Chair:
    Fredrick D. Siewers, (D), Illinois, 1995, GsPiZc
Professor:
    Catherine Algeo, (D), Louisiana State, 1997, Zi
    Stuart A. Foster, (D), Ohio State, 1988, At
    Margaret Gripshover, (D), Tennessee, Zy
    Christopher Groves, (D), Virginia, 1993, Hg
    David J. Keeling, (D), Oregon, 1992, Zg
    Michael May, (D), Indiana, 1993, Ge
    Jun Yan, (D), Buffalo, 2004, Zi
Associate Professor:
    Josh Durkee, (D), Georgia, 2008, AsZy
    Xingang Fan, (D), Lanzhou, As
    M. Royhan Gani, (D), Texas (Dallas), 2005, GsrGg
    Nahid Gani, (D), Texas (Dallas), Gc
    Greg Goodrich, (D), Arizona State, 2005, Ast
    Jason Polk, (D), S Florida, Gm
    Andrew Wulff, (D), Massachusetts, 1993, Gv
Associate Scientist:
    Leslie North, (D), S Florida, Zg
Instructor:
    William Blackburn, (M), W Kentucky, 2002, Zy
    Kevin Cary, (M), W Kentucky, 2000, Zi
    Margaret Crowder, (D), W Kentucky, 2012, Ze
    Scott Dobler, (M), Bowling Green, 1996, As
    Patricia Kambesis, (D), Mississippi State, 2014, ZigHw
Instructor:
    Amy Nemon, (M), W Kentucky, 2005, Zig
Emeritus:
    Nicholas Crawford, (D), Clark, 1977, Hg

Louisiana

Centenary College of Louisiana
Dept of Geology (B) ☒ (2018)
2911 Centenary Boulevard
Shreveport, LA 71104
    p. (318) 869-5234
    dbieler@centenary.edu
    Programs: Geology (B)
    Enrollment (2018): B: 20 (4)
Professor:
    Scott K. Vetter, (D), South Carolina, 1989, Gi

Associate Professor:
  David B. Bieler, (D), Illinois, 1983, GtrYe

Delgado Community College
Science & Math Div (A) (2015)
615 City Park Avenue
New Orleans, LA 70119
  p. (504) 671-6480
  jwood@dcc.edu
  http://www.dcc.edu/divisions/sciencemath/
Professor:
  Jacqueline Wood, (M), New Orleans, Gg

Louisiana State University
Dept of Geography & Anthropology (B,M,D) (2015)
227 Howe-Russell Geoscience Complex
Baton Rouge, LA 70803
  p. (225) 578-5942
  gachair@lsu.edu
  http://www.ga.lsu.edu
  Enrollment (2006): B: 54 (23) M: 12 (6) D: 22 (7)
Chair:
  Kevin Robbins
Professor:
  Patrick A. Hesp, (D), Sydney, 1981, Gm
  Richard H. Kesel, (D), Maryland, 1971, Hg
Associate Professor:
  Steven Namikas, (D), S California, 1999, Gms

Dept of Geology & Geophysics (B,M,D) O☒ (2019)
E235 Howe Russell Kniffen Geoscience Complex
Baton Rouge, LA 70803-4101
  p. (225) 578-3353
  geology@lsu.edu
  www.lsu.edu/science/geology/
  f: http://www.facebook.com/LSUGeology
  t: @LSUGeology
  Enrollment (2017): B: 86 (31) M: 28 (16) D: 28 (7)
Chair:
  Carol M. Wicks, (D), Virginia, 1992, HwClHy
Professor:
  Huiming Bao, (D), Princeton, 1998, CsAcCu
  Samuel J. Bentley, (D), SUNY (Stony Brook), 1998, Gu
  Peter Clift, (D), Edinburgh, 1993, GsuGo
  Peter Doran, (D), Nevada (Reno), 1996, Hws
  Barbara L. Dutrow, (D), S Methodist, 1985, Gz
  Brooks B. Ellwood, (D), Rhode Island, 1977, Ym
  Darrell J. Henry, (D), Wisconsin, 1981, Gp
Associate Professor:
  Philip J. Bart, (D), Rice, 1998, Gr
  Juan M. Lorenzo, (D), Columbia, 1991, Ys
  Sophie Warny, (D), Catholic (Belgium), 1999, Pl
Assistant Professor of Research:
  Xiaobin Cao, (D), Chinese Acad of Sci, 2012, Cs
  Yongbo Peng, (D), Louisiana State, Cg
Assistant Professor - Professional Practice:
  Amy Luther, (D), New Mexico Tech, GcZe
Assistant Professor:
  Adam Forte, (D), California (Davis), 2012, Gcm
  Achim Herrmann, (D), Penn State, 2005, PeGd
  Suniti Karuntillake, (D), Cornell, 2008, Xg
  Karen Luttrell, (D), Scripps, Yg
  Patricia Persaud, (D), Caltech, Ys
  Jianwei Wang, (D), Illinois (Urbana), 2004, ClZmGy
  Carol A. Wilson, (D), Boston, 2013, GsOnCc
  Guangsheng Zhuang, (D), California (Santa Cruz), 2011, CocGs
Instructor:
  Yanxia Ma, (D), Og
Associate Curator, LSU Museum of Natural Science:
  Judith A. Schiebout, (D), Texas, 1973, Pv
Emeritus:
  Ajoy K. Baksi, (D), Toronto, 1970, Cc
  Gary R. Byerly, (D), Michigan State, 1974, Gi
  Ray E. Ferrell, Jr., (D), Illinois, 1966, Cl
  Jeffrey S. Hanor, (D), Harvard, 1967, Cl
  George Hart, (D), Sheffield, 1961, Gg
  Clyde Moore, Zn
  Jeffrey A. Nunn, (D), Northwestern, 1981, Yg
  James E. Roche, (D), Illinois, 1969, Gg
  Barun K. Sen Gupta, (D), Indian Inst of Tech, 1963, Pm
Accountant Technician:
  Jeanne L. Johnson, Zn

Dept of Oceanography & Coastal Sciences (M,D) (2015)
1002 Energy Coast and Environment Building
Baton Rouge, LA 70803
  p. (225) 578-6308
  ocean@lsu.edu
  http://www.ocean.lsu.edu
  Administrative Assistant: Gaynell Gibbs
  Enrollment (2010): M: 63 (11) D: 92 (11)
Professor:
  Robert P. Gambrell, (D), North Carolina State, 1974, Cg
  Paul A. LaRock, (D), Rensselaer, 1968, Ob
  Irving A. Mendelssohn, (D), North Carolina State, 1974, Ob
  Richard F. Shaw, (D), Maine, 1981, Ob
  Robert E. Turner, (D), Georgia, 1974, Ob
Associate Professor:
  Donald M. Baltz, (D), California (Davis), 1980, Ob
  Robert S. Carney, (D), Oregon State, 1977, Ob
  Lawrence J. Rouse, Jr., (D), Louisiana State, 1969, Op
Assistant Professor:
  Mark C. Benfield, (D), Texas A&M, 1991, Ob
Adjunct Professor:
  Dubravko Justic, (D), Zagreb, 1989, Op
  Nancy N. Rabalais, (D), Texas, 1983, Ob
  Harry H. Roberts, (D), Louisiana State, 1969, Gs
  Paul W. Sammarco, (D), SUNY, 1978, Zn
  Nan D. Walker, (D), Cape Town, 1989, Og

School of Plant, Environmental and Soil Sciences (B,M,D) ☒ (2017)
104 M. B. Sturgis Hall
Baton Rouge, LA 70803
  p. (225) 578-2110
  spess@lsu.edu
  http://www.spess.lsu.edu
  f: https://www.facebook.com/LsuSchoolOfPlantEnvironmentalSoilSciences/
  Programs: Plant, Environmental Management and Soil Sciences (M,D)
  Enrollment (2017): B: 3 (0) M: 3 (3) D: 13 (0)
Professor:
  Lewis A. Gaston, (D), Florida, 1987, Sc
  Hussein M. Selim, (D), Iowa State, 1971, Sp
  Maud M. Walsh, (D), Louisiana State, 1989, GeRcZe
  Jim Wang, (D), Iowa State, 1990, Sc
Associate Professor:
  Brenda S. Tubana, (D), Oklahoma State, 2007, So
Assistant Professor:
  Lisa M. Fultz, Texas Tech, 2012, Sb

Louisiana Tech University
Geosciences Program (B,M) (2015)
600 W. Arizona Street
Ruston, LA 71272
  p. (318) 257-3972
  mm@engr.latech.edu
  http://www.coes.latech.edu/geo/
  Department Secretary: Connie McKenzie
Chair:
  Gary S. Zumwalt, (D), California (Davis), 1976, Pi
Associate Professor:
  Maureen McCurdy, (D), Wisconsin, 1990, Hw
Emeritus:
  Leo A. Herrmann, (D), Johns Hopkins, 1951, Go

Nicholls State University
Dept of Physical Sciences (2016)
P.O. Box 2022
Thibodaux, LA 70310
  p. (985) 448-4502

marguerite.moloney@nicholls.edu
http://www.nicholls.edu/phsc
Assistant Professor:
Marguerite M. Moloney, (M), S Illinois, 2004, GeZnn
Instructor:
Adam Beyer, (M), S Illinois, Gg

Northwestern State University
Dept of Chemistry and Physics (B) (2015)
Natchitoches, LA 71497
p. (318) 357-5501
chinc@nsula.edu
Director:
Paul Withey
Professor:
Carol S. Chin, (D)
Kelly Knowlton, (D), Texas A&M, 1991, Gg

South Louisiana Community College
South Louisiana Community College (2015)
320 Devalcourt
Lafayette, LA 70506
p. (337) 521-8983
http://www.slcc.cc.la.us

Tulane University
Dept of Earth and Environmental Sciences (B,M,D) ☒ (2019)
6823 St. Charles Ave.
101 Blessey Hall
New Orleans, LA 70118
p. (504) 865-5198
ees@tulane.edu
http://www.tulane.edu/sse/eens
f: https://www.facebook.com/EENS6823/
Programs: Geology BS, Environmental Earth Science BS, Earth & Env Sciences MS, Earth & Env Sciences PhD
Certificates: GIS Certificates (undergraduate and graduate certificates available)
Enrollment (2015): B: 6 (0) M: 0 (2) D: 0 (2)
Chair:
Torbjörn E. Törnqvist, (D), Utrecht, 1993, Gs
Professor:
Mead A. Allison, (D), SUNY (Stony Brook), 1993, Gs
Karen Haley Johannesson, (D), Nevada, 1993, HgCg
Associate Professor:
Nancye H. Dawers, (D), Columbia, 1997, GcoGm
George C. Flowers, (D), California (Berkeley), 1979, Hw
Associate Scientist:
Nicole Gasparini, (D), MIT, 2003, GmHs
Assistant Professor:
Brent M. Goehring, (D), Columbia, 2010, CcGlm
Kyle Martin Straub, (D), MIT, 2007, GgCg
Lecturer:
Jeffrey G. Agnew, (D), Louisiana State, 2008, Pge
Visiting Professor:
Reda Amer, (D), St. Louis, 2011, ZirGe
Emeritus:
Stephen A. Nelson, (D), California (Berkeley), 1979, GivRn

University of Louisiana at Lafayette
School of Geosciences (B,M) O☒ (2019)
BOX 43705
Lafayette, LA 70504-4530
p. (337) 482-6468
geology@louisiana.edu
http://geology.louisiana.edu
Programs: Bachelors of Science in Environmental Geology and Petroleum Geology, Master's in Geology, Ph.D. in Earth and Energy Sciences, Bachelors of Science in Environmental Sciences, Master's in Environmental Resource Science
Administrative Assistant: Nadean S.. Bienvenu
Administrative Assistant: Pauline R. Greene
Enrollment (2018): B: 209 (19) M: 113 (36)
Director:
Eric C. Ferre', (D), Toulouse, 1989, YmGc
Assistant Director:
Durga Poudel, (D), Georgia, 1998, Ge
Professor:
Gary L. Kinsland, (D), Rochester, 1974, YgeGt
Carl Richter, (D), Tubingen, 1990, Ym
Jenneke M. Visser, (D), Louisiana State, 1989, Ge
Senior Scientist:
James E. Martin, (D), Washington, 1979, PvGg
Resource Facilitator:
Jim Foret, (M), Iowa State, 1971, Ge
Associate Professor:
Timothy W. Duex, (D), Texas, 1983, HgGex
Geology Program Director:
Brian Schubert, (D), Binghamton, 2008, Cg
Field Camp Director:
Raphael Gottardi, (D), Minnesota, 2012, Gcp
Assistant Professor:
Katie H. Costigan, (D)
Aubrey Hillman, (D), Pittsburg, 2015, GneGa
Davide Oppo, (D), Bologna, 2012, Gs
Rui Zhang, (D), Houston, 2010, Ye
Instructor:
Kristie Cornell, (M), Louisiana (Lafayette), 2003, Gg
Jennifer E. Hargrave, (D), Oklahoma, 2009, PvGrZe
Emeritus:
Brian E. Lock, (D), Cambridge, 1969, Gs
Laboratory Director:
Yingfeng XU, (D), Tulane, 2004, ZgGe
Adjunct Instructor:
William R. Finley, (M), SW Louisiana, 1975, YsZi
Bernardo Teixeira, (M), Lisbon, 2012, GdrGn

University of Louisiana, Monroe
School of Science, Atmospheric Science Program (B) (2017)
700 University Avenue
Monroe, LA 71209-0550
p. (318) 342-1822
casehanks@ulm.edu
http://www.ulm.edu/atmos/
Enrollment (2012): B: 43 (9)
Associate Dean of Arts and Sciences:
Michael A. Camille, (D), Texas A&M, 1991, Zy
Professor:
Eric A. Pani, (D), Texas Tech, 1987, As
Associate Professor:
M. Sean Chenoweth, (D), Wisconsin (Milwaukee), 2003, GmZir
Department Head:
Anne T. Case Hanks, (D), Georgia Tech, 2008, AsZe
Assistant Professor:
Ken Leppert, (D), Alabama (Huntsville), As
Todd Murphy, (D), Alabama (Huntsville), As
Station Archeologist/Poverty Point:
Diana M. Greenlee, (D), Washington, 2002, GaCo

University of New Orleans
Dept of Earth and Environmental Sciences (B,M,D) (2015)
2000 Lakeshore Drive
New Orleans, LA 70148
p. (504) 280-6325
djreed@uno.edu
http://www.uno.edu/geology/
Professor:
William H. Busch, (D), Oregon State, 1981, Ou
Terry L. Pavlis, (D), Utah, 1982, Gt
Denise J. Reed, (D), Cambridge, 1986, GmOn
A. K. Mostofa Sarwar, (D), Indiana, 1983, Ye
Laura F. Serpa, (D), Cornell, 1986, Ys
William B. Simmons, (D), Michigan, 1973, Gz
Ronald K. Stoessell, (D), California (Berkeley), 1977, Cl
Associate Professor:
Kraig L. Derstler, (D), California (Davis), 1985, PgiPv
Frank R. Hall, (D), Rhode Island, 1991, Ym
Mark A. Kulp, (D), Kentucky, 2000, GrmGs
Assistant Professor:
Christopher D. Parkinson, (D), London (UK), 1991, Gp

Adjunct Professor:
    Miles O. Hayes, (D), Texas, 1965, Gs
    Karen L. Webber, (D), Rice, 1988, Gv
    Michael A. Wise, (D), Manitoba, 1987, Gz
Emeritus:
    Gary C. Allen, (D), North Carolina, 1968, Gp
    Jacqueline Michel, (D), South Carolina, 1980, Cg

Maine

Bates College
Dept of Geology (B) ☒ (2018)
Carnegie Science Center
44 Campus Avenue
Lewiston, ME 04240-6084
    p. (207) 786-6490
    http://www.bates.edu/geology/
    f: https://www.facebook.com/groups/243958415633596/
    Programs: Geology
    Administrative Assistant: Sylvia Deschaine
    Enrollment (2018): B: 15 (12)
Chair:
    J. Dykstra Eusden, (D), Dartmouth, 1988, Gc
Professor:
    Beverly J. Johnson, (D), Colorado, 1995, CsbCl
    Michael J. Retelle, (D), Massachusetts, 1985, Gl
Assistant Professor:
    Genevieve Robert, (D), Missouri (Columbia), 2014, CpGiz
Assistant Instructor:
    Marita Bryant, (M), Free (Berlin), 1984, Gg
Lecturer:
    Raj Saha, (D), N Carolina, 2011
Visiting Professor:
    Alice Doughty, Victoria (NZ), 2013, Gl
Emeritus:
    Gene A. Clough, (D), Caltech, 1978, Ym

Bowdoin College
Dept of Earth and Oceanographic Science (B) (2015)
6800 College Station
Brunswick, ME 04011
    p. (207) 725-3628
    mparker@bowdoin.edu
    http://www.bowdoin.edu/earth-oceanographic-science/
    Administrative Assistant: Marjorie Parker
    Enrollment (2014): B: 60 (27)
Chair:
    Collin Roesler, (D), Washington, 1992, OgpZr
Rusack Professor of Environmental Studies:
    Philip Camill III, (D), Duke, 1999, PebSb
Professor:
    Rachel J. Beane, (D), Stanford, 1997, GzxZe
Associate Professor:
    Peter D. Lea, (D), Colorado, 1989, Gl
Assistant Professor:
    Michéle LaVigne, (D), Rutgers, 2010, OcPe
    Emily M. Peterman, (D), California (Santa Barbara), 2009, Gtp
Service Learning Coord/Lab Instr:
    Cathryn K. Field, (M), Smith, 2000, Ge
Lab Instructor:
    Joanne Urquhart, (M), Dartmouth, 1987, Gge
Associate Professor:
    Edward P. Laine, (D), MIT, 1977, Gu

Colby College
Department of Geology (B) ☒ (2018)
5800 Mayflower Hill
Waterville, ME 04901-8858
    p. (207) 859-5800
    amridky@colby.edu
    http://www.colby.edu/geologydept/
    Administrative Assistant: Alice M. Ridky
    Enrollment (2017): B: 24 (6)
Chair:
    Walter Sullivan, (D), Wyoming, 2007, Gct

Professor:
    Robert A. Gastaldo, (D), S Illinois, 1978, PbGs
Associate Professor:
    Bess Koffman, (D), Maine, 2013, PcClZc
Assistant Professor:
    Tasha L. Dunn, (D), Tennessee, 2008, XmGiz
Visiting Professor:
    Bruce F. Rueger, (D), Colorado, 2002, PlGg

University of Maine
School of Earth and Climate Sciences (B,M,D) (2017)
5790 Bryand Global Sciences Center
Orono, ME 04469-5790
    p. (207) 581-2152
    johnsons@maine.edu
    http://www.umaine.edu/earthclimate/
    f: https://www.facebook.com/pages/UMaine-School-of-Earth-and-Climate-Sciences/238244500701
    Enrollment (2015): B: 42 (6) M: 13 (4) D: 14 (1)
Director:
    Scott E. Johnson, (D), James Cook, 1989, Gc
Research Professor:
    Edward S. Grew, (D), Harvard, 1971, Gpz
    Roger L. Hooke, (D), Caltech, 1965, Gm
Professor:
    George H. Denton, (D), Yale, 1965, Gl
    Brenda L. Hall, (D), Maine, 1997, Gl
    Gordon S. Hamilton, (D), Cambridge, 1992, Gl
    Joseph T. Kelley, (D), Lehigh, 1980, Gua
    Peter O. Koons, (D), ETH (Switzerland), 1982, Gt
    Karl J. Kreutz, (D), New Hampshire, 1998, Cs
    Daniel R. Lux, (D), Ohio State, 1981, Gi
    Kirk A. Maasch, (D), Yale, 1989, As
    Paul A. Mayewski, (D), Ohio State, 1973, Pe
    Aaron E. Putnam, (D), Maine, 2011, Gll
    Andrew S. Reeve, (D), Syracuse, 1996, Hw
Associate Professor:
    Christopher C. Gerbi, (D), Maine, 2005, Gt
Research Professor:
    Ellyn M. Enderlin, (D), Ohio State, 2013, GlZrg
Assistant Professor:
    Katherine A. Allen, (D), Columbia, 2013, Gu
    Alicia M. Cruz-Uribe, (D), Penn State, 2014, GpCpt
    Amanda A. Olsen, (D), Virginia Tech, 2007, Cl
    Sean Smith, (D), Johns Hopkins, 2010, Hs
Research Assistant Professor:
    Seth W. Campbell, (D), Maine, 2010, Gl
Researach Assistant Professor:
    Gordon R. M. Bromley, (D), Maine, 2010, Gl
Instructor:
    Alice R. Kelley, (D), Maine, 2006, GalGm
    Martin G. Yates, (D), Indiana, 1988, EgGzx
Emeritus:
    Daniel F. Belknap, (D), Delaware, 1979, GusGa
    Harold W. Borns, Jr., (D), Boston, 1959, Gl
    Joseph V. Chernosky, Jr., (D), MIT, 1973, Gg
    Terence J. Hughes, (D), Northwestern, 1968, Gl
    Stephen A. Norton, (D), Harvard, 1967, Cl
Research Assistant Professor:
    Sean Birkel, (D), Maine, 2010, Gl

School of Marine Sciences (B,M,D) O (2015)
5706 Aubert Hall, Rm 360
Orono, ME 04469-5706
    p. (207) 581-4381
    fchai@maine.edu
    http://www.umaine.edu/marine/
    Enrollment (2010): B: 145 (38) M: 56 (0) D: 9 (0)

University of Maine - Farmington
Dept of Geology (B) ☒ (2017)
173 High Street
Farmington, ME 04938
    p. (207) 778-7402
    dgibson@maine.edu
    http://sciences.umf.maine.edu

Programs: Geology (B); Earth and Environmental Science (B)
Enrollment (2017): B: 26 (8)
Professor:
David Gibson, (D), Queen's (Ireland), 1984, GigGz
Douglas N. Reusch, (D), Maine, 1998, GtCgOu
Associate Professor:
Julia F. Daly, (D), Maine, 2002, GmHsGu
Emeritus:
Thomas E. Eastler, (D), Columbia, 1970, GeZrGg

University of Maine, Presque Isle
Div of Mathematics & Science (B) (2015)
181 Main Street
Presque Isle, ME 04769
p. (207) 768-9482
kevin.mccartney@umpi.edu
http://www.umpi.edu/
Department Secretary: Connie Leveque
Chair:
Michael Knopp
Professor:
Kevin McCartney, (D), Florida State, 1988, Pm

Maryland

College of Southern Maryland
Biological and Physical Sciences (A) (2015)
8730 Mitchell Rd
La Plata, MD 20646
p. (301) 934-7841
jenniferh@csmd.edu
http://www.csmd.edu/bio/index.html
Professor:
Jean M. Russ, (M), Kutztown, Zey

Community College of Baltimore County, Catonsville
School of Mathematics & Science (A) ☒ (2017)
800 S. Rolling Road
MASH 015B
Catonsville, MD 21228
p. (443) 840-5935
DLudwikoski@ccbcmd.edu
http://www.ccbcmd.edu/math_science/geology.html
Programs: Earth Science; Geology
Certificates: GIS
Administrative Assistant: Annjeannette Black
Enrollment (2015): A: 3 (0)
Associate Professor:
David J. Ludwikoski, (M), Toledo, 1993, ZegZn

Frederick Community College
Science Dept (A) (2018)
7932 Opossumtown Pike
Frederick, MD 21702
p. (301) 846-2510
shsmith@frederick.edu
http://www.frederick.edu/courses_and_programs/dept_science.aspx
Professor:
Natasha Cleveland, (M), Utah, 2001, Zge

Frostburg State University
Dept of Geography (B) (2015)
101 Braddock Rd.
Frostburg, MD 21532
p. (301) 687-4369
jsaku@frostburg.edu
http://www.frostburg.edu/dept/geog/
Administrative Assistant: Gale Yutzy
Enrollment (2010): B: 65 (16)
Chair:
Craig L. Caupp, (D), Utah State, 1986, ZnHs

Professor:
Henry W. Bullamore, (D), Iowa, 1978, Zn
Francis L. Precht, (D), Georgia, 1989, Zy
James C. Saku, (D), Saskatchewan, 1995, Zn
Associate Professor:
Phillip Allen, (D), Coventry (UK), 2005, GmePe
Fritz Kessler, (D), Kansas, 1999, Zr
George W. White, (D), Oregon, 1994, Zn
Assistant Professor:
David L. Arnold, (D), Indiana, 1994, As
Matthew E. Ramspott, (D), Kansas, 2006, Zr

Howard Community College
Science, Engineering, and Technology Division (2015)
10901 Little Patuxent Pkwy
Columbia, MD 21044
p. (443) 518-1000
pturner@howardcc.edu
http://www.howardcc.edu/programs-courses/academics/academic-divisions/science-engineering-technology/sciences/

Johns Hopkins University
Dept of Environmental Health & Engineering (B,M,D) (2015)
313 Ames Hall
34th & Charles Streets
Baltimore, MD 21218-2681
p. (410) 516-7092
mwkarp@jhu.edu
http://www.jhu.edu/dogee
Professor:
William P. Ball, (D), Stanford, 1989, Zn
Edward Bouwer, (D), Stanford, 1982, RwHws
Grace S. Brush, (D), Harvard, 1956, PleGn
Hugh Ellis, (D), Waterloo, 1984, Zn
Steve H. Hanke, (D), Colorado, 1969, Zn
Benjamin F. Hobbs, (D), Cornell, 1983, Zn
A. Lynn Roberts, (D), MIT, 1991, Zn
Erica Schoenberger, (D), California (Berkeley), 1984, Zn
Alan T. Stone, (D), Caltech, 1983, ClSc
Peter W. Wilcock, (D), MIT, 1987, Gm
Associate Professor:
Markus Hilpert, (D), Zn
Assistant Professor:
Kai Loon Chen, (D), Yale, 2008, Zn
Seth Guikema, (D), Stanford, 2003, Zn
Catherine Norman, (D), California (Santa Barbara), 2005, Zn
Lecturer:
Hedy Alavi, (D), Ohio State, 1983, Zn
Emeritus:
John J. Boland, (D), Johns Hopkins, 1973, Zn
Charles R. O'Melia, (D), Michigan, 1963, Zn
Eugene D. Shchukin, (D), Moscow State, 1958, Zn
Senior Academic Program Coordinator:
Adena Rojas, (M), Zn

The Morton K. Blaustein Dept of Earth & Planetary Sciences (B,M,D) (2015)
3400 N Charles Street
301 Olin Hall
Baltimore, MD 21218
p. (410) 516-7135
jseat@jhu.edu
http://eps.jhu.edu/
Enrollment (2013): D: 33 (5)
Chair:
Thomas W.N. Haine, (D), Southampton, 1993, OpgAs
Professor:
Anand Gnanadesikan, (D), MIT/WHOI, 1994, OpcAs
Peter L. Olson, (D), California (Berkeley), 1977, Yg
Darrell F. Strobel, (D), Harvard, 1969, AsZnn
Dimitri A. Sverjensky, (D), Yale, 1980, Cl
Darryn W. Waugh, (D), Cambridge, 1991, As
Assistant Professor:
Sarah Horst, (D), Arizona, 2011, XaAcXb
Naomi Levin, (D), Utah, 2008, Gs
Kevin Lewis, (D), Caltech, 2009

Benjamin Zaitchik, (D), Yale, 2006, As
Research Professor:
Katalin Szlavecz, (D), Eotvos (Hungary), 1981, Pi

Montgomery College
Dept of Physics, Engineering & Geosciences (A) (2015)
51 Mannakee Street
Rockville, MD 20850
p. (301) 279-5230
Muhammad.Kehnemouyi@montgomerycollege.edu
http://www.montgomerycollege.edu/Departments/phengrv/
Administrative Assistant: Mary (Deep) McGregor
Professor:
Alan Cutler, (D), Geology, PgRh
Instructional Laboratory Coordinator for Geosciences:
Kimberly Kelly, (B), Pittsburg, Zn

St. Charles Community College
4601 Mid Rivers Mall Drive
Cottleville, MO 63376
p. (636) 922-8000
nfo_desk@stchas.edu
http://www.stchas.edu/

Towson University
Dept of Physics, Astronomy & Geosciences (B) ☒ (2018)
8000 York Road
Towson, MD 21252-0001
p. (410) 704-3020
dschaefer@towson.edu
http://wwwnew.towson.edu/physics/geosciences/
Programs: Geology, Earth Space Science (for secondary education majors as well as for science majors), Environmental Science, Physics
Enrollment (2018): B: 49 (10)
Professor:
Rachel J. Burks, (D), Texas, 1985, Gc
David A. Vanko, (D), Northwestern, 1982, Gxz
Assistant Professor:
Joel Moore, (D), Penn State, 2008, ClsSc
Wendy Nelson, (D), Penn State, 2009, GiCc
Lecturer:
Andrew Hawkins, (D), Virginia Tech, 2017, Pg

United States Naval Academy
Dept of Oceanography (B) (2015)
572C Holloway Road
Annapolis, MD 21402-5026
p. (410) 293-6550
natunewi@usna.edu
http://www.nadn.navy.mil/Oceanography/
Administrative Assistant: Cynthia A.. Ervin
Enrollment (2014): B: 216 (117)
Chair:
Cecily N. Steppe, (D), Delaware, 2001, Ob
Professor:
Peter L. Guth, (D), MIT, 1980, GcZiy
David R. Smith, (D), Texas A&M, 1979, Am
Associate Professor:
Andrew C. Muller, (D), Old Dominion, 1999, OpnOu
Assistant Professor:
Bradford S. Barrett, (D), Oklahoma, 2007, As
Gina R. Henderson, (D), Delaware, 2010, As
Emil T. Petruncio, (D), Naval Postgrad Sch, 1996, OpZr
Elizabeth R. Sanabia, (D), Naval Postgrad Sch, 2010, Am
William J. Schulz, (D), Old Dominion, 1999, OpAm
Joseph P. Smith, (D), Massachusetts (Boston), 2007, CmOg
Instructor:
Dwight E. Smith, (M), Naval Postgrad Sch, 2009, AmOg
Megan D. Thomas, (M), Naval Postgrad Sch, 2005, Og
Oceanography Reference Librarian:
Barbara Yoakum, (M), South Carolina, 1985, Zn

University of Maryland
Dept of Geographical Studies (2015)
College Park, MD 20742
p. (301) 405-1600
cjustice@umd.edu
https://geog.umd.edu/

Dept of Geology (B,M,D) ☒ (2019)
Geology Building (#237)
8000 Regents Drive
College Park, MD 20742
p. (301) 405-4082
geology@umd.edu
http://www.geol.umd.edu
f: https://www.facebook.com/UMDGeology
Administrative Assistant: Dorothy Brown
Enrollment (2018): B: 54 (12) M: 0 (3) D: 33 (2)
Chair:
Richard J. Walker, (D), SUNY (Stony Brook), 1984, CcXcCg
Affiliate Professor:
Fernando Mirales-Wilhelm, (D)
Raghuram G. Murtugudde, (D), Columbia, 1994, ObZr
Jessica Sunshine, (D), Brown, 1993
Ning Zeng, (D), Arizona, 1994, AsCgGl
Professor:
Michael Brown, (D), Keele, 1975, GxpGt
James Farquhar, (D), Alberta, 1995, Cs
Alan J. Kaufman, (D), Indiana, 1990, Csg
Daniel Lathrop, (D), Texas, 1991, Yg
William F. McDonough, (D), Australian Nat, 1988, Cg
Laurent G.J. Montesi, (D), MIT, 2002, YgXyGt
Wenlu Zhu, (D), SUNY (Stony Brook), 1996, YrNrHg
Senior Research Scientist:
Philip M. Piccoli, (D), Maryland, 1992, Cg
Igor Puchtel, (D), Russian Acad of Sci, 1992, CcgGi
Associate Professor:
Ricardo Arevalo, Jr, (D), Maryland, 2010
Michael N. Evans, (D), Columbia, 1999, PeAsCs
Sujay Kaushal, (D), Colorado, 2003, ZuGeHs
Sarah Penniston-Dorland, (D), Johns Hopkins, 2005, Gp
Karen L. Prestegaard, (D), California (Berkeley), 1982, Hgw
Associate Research Scientist:
Richard Ash, (D), Open, 1990, Ca
Affiliate Assistant Professor:
Derrick Lampkin, (D), GlAsZr
Assistant Professor:
Mong-Han Huang, (D), California (Berkeley), 2017
Vedran Lekic, (D), California (Berkeley), 2009, Ysg
Megan E. Newcombe, (D), Caltech, 2017
Nicholas C. Schmerr, (D), Arizona State, 2008, XyYs
Research Associate:
Jabrane Labidi, (D), Inst Physique du Globe de Paris, 2012, Cg
Research Assistant Scientist:
Katherine R. Bermingham, (D), Muenster, 2011, XcCcg
Shuiwang Duan, (D), Tulane, 2005, HsZu
Principal Lecturer:
Thomas R. Holtz, Jr., (D), Yale, 1992, PvgPg
John W. Merck, Jr., (D), Texas, 1997, Pv
Lecturer:
Tracey Centorbi, (B), Maryland, 2003, Cg
Adjunct Professor:
Anat Shahar, (D), California (Los Angeles), 2008, CspCa
Adjunct Associate Professor:
Elizabeth Cottrell, (D), Columbia, 2004, CpgGx
Adjunct Professor:
Yingwei Fei, (D), CUNY, 1989, CpGxCg
Jeffrey Plesia, (D)
Deborah Smith, (D), California (San Diego), 1985
Professor Emeritus:
Philip A. Candela, (D), Harvard, 1982, CpgEg
Galt Siegrist, (D)
Emeritus Professor:
Ann G. Wylie, (D), Columbia, 1972, Gz
Emeritus Affiliate Research Professor:
George Helz, (D), Penn State, 1970, HsGeCg
Associate Professor Emeritus:
Peter B. Stifel, (D), Utah, 1964, Pg

Adjunct Professor Emeritus:
 Roberta L. Rudnick, (D), Australian Nat, 1988, CgsCt
Faculty Research Assistant:
 Todd Karwoski, (B), Maryland, 2005, Gg
 Valentina Puchtel, (M), Moscow Geol Prospect Acad, 1983

Dept of Plant Science & Landscape Architecture (B,M,D) (2015)
2102 Plant Sciences Building
College Park, MD 20742-4432
 p. (301) 405-4356
 asmurphy@umd.edu
 http://www.psla.umd.edu/
Professor:
 Christopher Walsh, (D), Cornell, 1980, Zn
Associate Professor:
 Gary D. Coleman, (D), Nebraska, 1989, Zn
 Jack B. Sullivan, (M), Virginia, 1980, Zu
Program Management Specialist:
 Kathy Hunt, Zn
Coordinator:
 Sue Burk, Zn

Marine-Estuarine-Environmental Sciences Graduate Program (M,D) ☒ (2018)
1213 HJ Patterson Hall
University of Maryland
College Park, MD 20742
 p. (301) 405-6938
 mees@umd.edu
 http://www.mees.umd.edu
 Enrollment (2013): M: 6 (0) D: 12 (2)
Professor:
 Shenn-Yu Chao, (D), North Carolina State, 1979, Op
 Keith N. Eshleman, (D), MIT, 1985, Hg
 Thomas R. Fisher, Jr., (D), Duke, 1975, ObZiCl
 Patricia M. Gilbert, (D), Harvard, 1982, Ob
 Lawrence P. Sanford, (D), MIT, 1984, Op
 Diane Stoecker, (D), SUNY (Stony Brook), 1979, Ob
Associate Professor:
 William Boicourt, (D), Johns Hopkins, 1973, Op
 James Carton, (D), Princeton, 1983, Op
 Micheal S. Kearney, (D), Ontario, 1981, Gu
 Karen L. Prestegaard, (D), California (Berkeley), 1982, Hw
Research Associate Professor:
 Jeffery C. Cornwell, (D), Alaska, 1983, Oc
Assistant Professor:
 Mark S. Castro, (D), Virginia, 1991, As
 Raleigh Hood, (D), California (San Diego), 1990, Ob
 Alba Torrents, (D), Johns Hopkins, 1992, Sb
Research Associate Professor:
 Todd M. Kana, (D), Harvard, 1982, Ob

Massachusetts

Amherst College
Dept of Geology (B) ☒ (2019)
P.O. Box 2238
Amherst, MA 01002-5000
 p. (413) 542-2233
 taharms@amherst.edu
 https://www.amherst.edu/academiclife/departments/geology
 Enrollment (2018): B: 20 (0)
Professor:
 John T. Cheney, (D), Wisconsin, 1975, Gi
 Peter D. Crowley, (D), MIT, 1985, Gc
 Tekla A. Harms, (D), Arizona, 1986, Gt
 Anna M. Martini, (D), Michigan, 1997, Cl
Associate Scientist:
 David S. Jones, (D), Harvard, 2009, GsClGr
Assistant Professor:
 Victor E. Guevara, (D), Virginia Tech, 2017
Postdoctoral Fellow:
 Rachael E. Bernard, (D), Texas, 2018
Emeritus:
 Edward S. Belt, (D), Yale, 1963, Gs
 Margery C. Coombs, (D), Columbia, 1971, Pv

Bard College at Simon's Rock
Bard College at Simon's Rock (2015)
84 Alford Rd.
Great Barrington, MA 01230
 p. (413) 644-4400
 admin@simons-rock.edu
 http://www.simons-rock.edu/

Bentley University
Dept of Natural and Applied Sciences (B) ☒ (2018)
175 Forest Street
Waltham, MA 02452-4705
 p. (781) 891-2980
 pdavis@bentley.edu
 http://www.bentley.edu/academics/departments/natural-and-applied-sciences
 Programs: BA in environmental science and sustainability
 Administrative Assistant: Martha E. Keating
 Enrollment (2018): B: 1 (0)
Dean of Arts and Sciences:
 Rick Oches, (D), Massachusetts, 1994, ZcePe
Professor:
 P. Thompson Davis, (D), Colorado, 1980, GlmZc
Associate Professor:
 David Szymanski, (D), Michigan State, 2007, GifCg
Lecturer:
 George Fishman, (M), Boston, 1993, XgbXc
 Betsy Stoner, (D), Florida Intl, 2014, ObPe
Adjunct Professor:
 Janette Gartner, (M), Massachusetts, 2000, Hy
 Witt Taylor, (D), Arizona State, 2010, GgPbe

Berkshire Community College
Environmental and Life Science Dept (A) (2015)
Pittsfield, MA 01201
 p. (413) 236-4601
 saleksa@berkshirecc.edu
Chair:
 Clifford D. Myers, (D), Maine, Zn
Professor:
 Timothy Flanagan, (M), Antioch, 1983, Zei
 Thomas F. Tyning, (M), Zn
 Charles E. Weinstein, (M), Wisconsin, Zn
Emeritus:
 Richard L. Ferren, (M), Louisiana State, Zn
 George Hamilton, (M), North Adams State Coll, Zn
 Mary R. Mercuri, (M), Catholic

Boston College
Dept of Earth & Environmental Sciences (B,M) ☒ (2018)
140 Commonwealth Avenue
213 Devlin Hall
Chestnut Hill, MA 02467-3809
 p. (617) 552-3640
 baxteret@bc.edu
 http://www.bc.edu/geology
 Programs: Geological sciences, Environmental geoscience
 Environmental studies
 Enrollment (2018): B: 112 (0) M: 16 (0)
Chair:
 Ethan Baxter, (D), California (Berkeley), 2000, Ccg
Professor:
 John E. Ebel, (D), Caltech, 1981, Ys
 Gail C. Kineke, (D), Washington, 1993, On
Associate Professor:
 Mark Behn, (D), MIT/WHOI, 2002, YgrGt
 Rudolph Hon, (D), MIT, 1976, Cq
 Alan L. Kafka, (D), SUNY (Stony Brook), 1980, Yg
 Seth C. Kruckenberg, (D), Minnesota (Twin Cities), 2009, Gct
 Noah Snyder, (D), MIT, 2001, GgHg
Assistant Professor:
 Carling Hay, (D), Toronto, 2012, YgZoOn
 Tara Pisani-Gareau, (D), California (Santa Cruz), Ge
 Jeremy D. Shakun, (D), Oregon State, 2010, Gg

Lab Coordinator:
  Kenneth G. Galli, (D), Massachusetts, 2003, GsdGg
Adjunct Professor:
  Paul K. Strother, (D), Harvard, 1980, PlbPg
Emeritus:
  J. Christopher Hepburn, (D), Harvard, 1972, Gpt
  James W. Skehan, S.J., (D), Harvard, 1953, Gc

Weston Observatory (2015)
381 Concord Road
Weston, MA 02493
  p. (617) 552-8300
  weston.observatory@bc.edu
  http://www.bc.edu/westonobservatory
Director:
  Alan L. Kafka, (D), Stony Brook, 1980, Ys
Science Education:
  Michael Barnett, (D), Indiana, 2003, Ze
Senior Scientist:
  John E. Ebel, (D), Caltech, 1981, Yg
Seismology, Seismic Network Development:
  Michael Hagerty, (D), California (Santa Cruz), 1998, Ys
Seismic Analyst and Educational Seismologist:
  Anastasia Moulis, (M), Boston, 2003, Ys
Research Scientist:
  Seth Kruckenberg, (D), Minnesota (Twin Cities), 2009, Gc
Research Scientist:
  John J. Cipar, (D), Caltech, 1981, Ys
Research Associate:
  John H. Beck, (D), Boston, 1998, Pb
Adjunct Professor:
  Paul L. Strother, (D), Harvard, 1980, PlbPm
  Alfredo Urzua, (D), MIT, 1981, Ng
Visiting Professor:
  Vincent Murphy, (M), Boston, 1957, Yg
Emeritus:
  J. Christopher Hepburn, (D), Harvard, 1972, Gg
  James W. Skehan, (D), Harvard, 1953, Gc

Boston University

Center for Remote Sensing (2015)
685 Commonwealth Avenue
Room 433
Boston, MA 02215
  p. (617) 353-9709
  crsadmin@bu.edu
  Administrative Assistant: Emily P. Johnson
Director:
  Farouk El-Baz, (D), Missouri (Rolla), 1964, Zr
Professor:
  Sucharita Gopal, (D), California (Santa Barbara), 1988, Zi
  Alan Strahler, (D), Johns Hopkins, 1969, Zr
  Curtis E. Woodcock, (D), California (Santa Barbara), 1986, Zr
Research Associate Professor:
  Magaly Koch, (D), Boston, 1993, HyZri
Associate Professor:
  Mark A. Friedl, (D), California (Santa Barbara), 1994, Zr
  Kenneth L. Kvamme, (D), California (Santa Barbara), 1983, Ga
  Guido D. Salvucci, (D), MIT, 1994, Hq
Research Associate Professor:
  Cordula Robinson, (D), Univ Coll (UK), 1991, Gm
  Crystal Schaaf, (D), Boston, 1994, Zr
Research Associate:
  Eman Ghoneim, (D), Southampton, 2002, Gm
Geology Librarian:
  Nasim Momen, Zn

Dept of Earth & Environment (B,M,D) (2015)
675 Commonwealth Avenue
Boston, MA 02215
  p. (617) 353-2525
  earth@bu.edu
  http://www.bu.edu/earth/
  Enrollment (2015): B: 70 (15) M: 15 (10) D: 50 (8)
Professor:
  Bruce Anderson, (D), Scripps, 1998, AsOp
  James Lawford Anderson, (D), Gi
  Duncan M. FitzGerald, (D), South Carolina, 1977, OnGsu
  Mark Friedl, (D), California (Santa Barbara), 1993, Zr
  Sucharita Gopal, California (Santa Barbara), 1988, Zi
  Tony Janetos, (D), 1980, Zn
  Richard Murray, (D), California (Berkeley), 1991, Cm
  Ranga Myneni, (D), Antwerp, 1985, Zr
  Nathan Phillips, (D), Duke, 1997, Zy
  Guido D. Salvucci, (D), MIT, 1994, Hq
  Curtis Woodcock, (D), California (Santa Barbara), 1986, Zr
Associate Professor:
  Rachel Abercrombie, (D), Reading, 1991, Gt
  Michael Dietze, (D), Duke, 2006, SfZr
  Sergio Fagherazzi, (D), Padua, 1999, GmOnGs
  Robinson Fulweiler, (D), Rhode Island, 2007, Cm
  Lucy Hutyra, (D), Harvard, 2007, AsZr
  Andrew Kurtz, (D), Cornell, 2000, Clg
Assistant Professor:
  Dan Li, (D), Princeton, 2013, HqAs
  Christine Regalla, (D), Penn State, 2013, GtcGm
  Diane Thompson, Arizona, 2013, ObPe
Research Associate:
  Farouk El-Baz, (D), Missouri, 1964, Zr

Bridgewater State University

Dept of Geological Sciences (B) (2018)
Conant Science Building
Bridgewater, MA 02325
  p. (508) 531-1390
  Brenda.Flint@bridgew.edu
  http://www.bridgew.edu/academics/colleges-departments/department-geological-sciences
  Programs: Geological Sciences; Environmental Geosciences; Earth Science
  Administrative Assistant: Brenda Flint
  Enrollment (2017): B: 60 (9)
Chairperson:
  Robert D. Cicerone, (D), MIT, 1991, YsgXy
Professor:
  Richard L. Enright, (D), Rutgers, 1969, HgZrEg
  Michael A. Krol, (D), Lehigh, 1996, GzxGt
  Peter J. Saccocia, (D), Minnesota, 1991, CqsCm
Assistant Professor:
  Christine M. Brandon, (D), Massachusetts, 2015, Gs
Lecturer:
  Joseph Doyle, (M), New Hampshire, Gg
  Suzanne R. O'Brien, (M), New Hampshire, 1995, Zg
  Michael A. Penzo, (M), SUNY (Binghamton), 1981, Ge

Bristol Community College

Div of Mathematics, Science and Engineering (A) (2015)
777 Elsbree Street
Fall River, MA 02720
  p. (508) 678-2811
  John.Ahola@bristolcc.edu
Instructor:
  John Ahola, Gg

Cape Cod Community College

Environmental Technology Program (2015)
2240 Iyannough Rd
West Barnstable, MA 02668
  p. (508) 362-2131 x4468
  junderwood@capecod.edu
  http://www.capecod.edu/web/natsci/env

Fitchburg State University

Earth and Geographic Sciences (B) (2019)
160 Pearl Street
Fitchburg, MA 01420-2697
  p. (978) 665-4636
  egordon3@fitchburgstate.edu
  http://www.fitchburgstate.edu/academics/academic-departments/earth-and-geographic-sciences-dept/
  f: https://www.facebook.com/FitchburgStateUniversityEGS
  Programs: Environmental and Earth Science; Geographic

Science and Technology
Enrollment (2018): B: 43 (5)
Professor:
  Jane Huang, (D), Zi
Associate Professor:
  Elizabeth S. Gordon, (D), OgAsZg
Assistant Professor:
  Elyse Clark, (D), Cq
  Reid A. Parsons, (D), XgHs

Hampshire College
School of Natural Science (B) ☒ (2017)
Amherst, MA 01002
  p. (413) 582-5373
  sroof@hampshire.edu
Professor:
  Steven Roof, (D), Massachusetts, 1995, ZiPe
Associate Professor:
  Christina Cianfrani, (D), Vermont, Hg

Harvard University
Dept of Earth and Planetary Sciences (B,D) ●☒ (2019)
Hoffman Laboratory
20 Oxford Street
Cambridge, MA 02138-2902
  p. (617) 495-2351
  shaw@eps.harvard.edu
  http://www.eps.harvard.edu
  f: https://www.facebook.com/Harvard-University-Earth-and-Planetary-Sciences-6322073679/
  Programs: Earth and Planetary Sciences
  Enrollment (2018): B: 34 (6) D: 58 (10)
Chair:
  John Shaw, (D), Princeton, 1993, Gt
Affiliated Faculty with EPS:
  Naomi Oreskes, (D), Stanford, 1990, Rh
Professor:
  James G. Anderson, (D), Colorado, 1970, As
  Jeremy Bloxham, (D), Cambridge, 1985, Yg
  Brian F. Farrell, (D), Harvard, 1981, Am
  John P. Holdren, (D), Stanford, 1970, Zn
  Peter Huybers, (D), MIT, 2004, Pe
  Miaki Ishii, (D), Harvard, 2003, Yg
  Daniel J. Jacob, (D), Caltech, 1985, As
  Stein B. Jacobsen, (D), Caltech, 1980, Cc
  David T. Johnston, (D), Maryland, 2007, CgPg
  Andrew H. Knoll, (D), Harvard, 1977, Pb
  Zhiming Kuang, (D), Caltech, 2003, As
  Charles H. Langmuir, (D), SUNY (Stony Brook), 1980, Cg
  Scot T. Martin, (D), Caltech, 1995, Cg
  James T. McCarthy, (D), Scripps, 1971, Ob
  Michael B. McElroy, (D), Queen's, 1962, As
  Brendan Meade, (D), MIT, 2004, Yg
  Jerry X. Mitrovica, (D), Toronto, 1991, Yg
  Ann Pearson, (D), MIT/WHOI, 2000, Cb
  James R. Rice, (D), Lehigh, 1964, Ygu
  Daniel P. Schrag, (D), California (Berkeley), 1993, Cg
  Eli Tziperman, (D), MIT/WHOI, 1987, Op
  Steven C. Wofsy, (D), Harvard, 1971, As
Affiliated Faculty with EPS:
  Robin Wordsworth, (D), Oxford, 2008, Xg
Assistant Professor:
  Marine Denolle, (D), Stanford, 2014, Ys
  Rebecca A. Fischer, (D), Chicago, 2015, Gy
  Roger Fu, (D), MIT, 2015, YmXym
  Kaighin McColl, (D), MIT, 2016, Hg
Visiting Professor:
  Carl Wunsch, (D), MIT, 1966, Op
Emeritus:
  Charles W. Burnham, (D), MIT, 1961, GzyGe
  Paul F. Hoffman, (D), Johns Hopkins, 1970, Gc

Massachusetts Institute of Technology
Dept of Earth, Atmospheric, & Planetary Sciences (B,M,D) ●☒ (2018)
77 Massachusetts Avenue, 54-918
Cambridge, MA 02139
  p. (617) 253-2127
  eapsinfo@mit.edu
  http://eapsweb.mit.edu/
  f: https://www.facebook.com/EAPS.MIT
  t: @eapsmit
  Programs: Earth, Atmospheric and Planetary Sciences (B, M, D); Atmospheric Science (M,D); Climate Science (M,D); Geology (M, D); Geochemistry (M, D); Geobiology (M, D); Geophysics (M, D); Planetary Sciences (M, D)
  Certificates: none
  Enrollment (2018): B: 19 (10) M: 8 (7) D: 136 (20)
Vice-President for Research:
  Maria T. Zuber, (D), Brown, 1986, Xy
Department Head:
  Robert van der Hilst, (D), Utrecht, 1990, Ys
Professor:
  Richard P. Binzel, (D), Texas, 1986, Xm
  Samuel A. Bowring, (D), Kansas, 1985, Gg
  Edward A. Boyle, (D), MIT, 1976, Oc
  Kerry A. Emanuel, (D), MIT, 1978, AmRn
  Dara Entekhabi, (D), MIT, 1990, Hg
  Raffaele Ferrari, (D), Scripps, 2000, OpAsOb
  Glenn R. Flierl, (D), Harvard, 1975, Op
  Michael Follows, (D), E Angola (UK), 1991, Op
  Timothy L. Grove, (D), Harvard, 1976, Gi
  Bradford H. Hager, (D), Harvard, 1978, Ys
  Thomas A. Herring, (D), MIT, 1983, Yd
  Paola M. Malanotte-Rizzoli, (D), California (San Diego), 1978, Op
  John C. Marshall, (D), Imperial Coll (UK), 1980, Op
  F Dale Morgan, (D), MIT, 1981, Yg
  Ronald G. Prinn, (D), MIT, 1971, As
  Daniel H. Rothman, (D), Stanford, 1986, Yg
  Leigh H. Royden, (D), MIT, 1982, Gt
  Sara Seager, (D), Harvard, 1999, Xy
  Susan Solomon, (D), California (Berkeley), 1981, As
  Roger Summons, (D), New South Wales, 1972, CobCs
  Benjamin Weiss, (D), Caltech, 2003, Ym
  Jack Wisdom, (D), Caltech, 1981, Zn
Senior Research Scientist:
  William Durham, (D), MIT, 1975, Cg
  Michael Fehler, (D), MIT, 1979, Yg
  C. Adam Schlosser, (D), Maryland, 1995, As
Principal Research Scientist:
  Nilanjan Chatterjee, (D), CUNY, 1989, Cg
  Stephanie Dutkiewicz, (D), Rhode Island, 1997, Cm
  Robert W. King, (D), MIT, 1975, Yd
  Eduardo Andrade Lima, (D), Catholic (Brazil), Ym
  Sai Ravela, (D), Massachusetts, 2002, Zn
Associate Professor:
  Tanja Bosak, (D), Caltech, 2004, Pg
  Kerri Cahoy, (D), Stanford, 2008, Zn
  Dan Cziczo, (D), Chicago, 1999, As
  Laurent Demanet, (D), Caltech, 2006, Zn
  Colette Heald, (D), Harvard, 2005, As
  Oliver Jagoutz, (D), ETH (Switzerland), 2004, GitGc
  Ruben Juanes, (D), California (Berkeley), 2003, Zn
  David McGee, (D), Columbia, 2009, Cl
  Paul O'Gorman, (D), Caltech, 2004, As
  Shuhei Ono, (D), Penn State, 2001, PgCs
  Taylor Perron, (D), California (Berkeley), 2006, Gm
  Noelle Selin, (D), Harvard, 2007, As
Assistant Professor:
  Andrew Babbin, (D), Princeton, 2014, Cb
  Kristin Bergmann, (D), Caltech, 2013, GsCl
  Timothy W. Cronin, (D), MIT, 2014, As
  Julien de Wit, (D), MIT, 2014, Zn
  Gregory Fournier, (D), Connecticut, 2009, Pg
  Brent Minchew, (D), Caltech, 2016, Yg
  Matej Pec, (D), Basel, 2012, Nr
Senior Lecturer:
  Amanda Bosh, (D), MIT, 1994, Zn
  Lodovica Illari, (D), Imperial Coll (UK), 1982, Am
Emeritus:
  B. Clark Burchfiel, (D), Yale, 1961, Gc
  Charles C. Counselman, III, (D), 1969

J. Brian Evans, (D), MIT, 1978, Yx
Frederick A. Frey, (D), Wisconsin, 1967, CtGvi
Richard S. Lindzen, (D), Harvard, 1964, Am
Gordon H. Pettengill, (D), California (Berkeley), 1955, Zg
Raymond A. Plumb, (D), Manchester, 1972, As
M. Gene Simmons, (D), Harvard, 1962, Yg
John B. Southard, (D), Harvard, 1966, Gs
Peter H. Stone, (D), Harvard, 1964, As
M Nafi Toksoz, (D), Caltech, 1963, Ys
Carl I. Wunsch, (D), MIT, 1967, Opo
Principal Research Engineer:
Christopher Hill, (D), Zn

Mount Holyoke College
Dept of Geology (B) (2016)
50 College Street
Clapp Laboratory #304
South Hadley, MA 01075-6419
p. (413) 538-2278
rforjwuo@mtholyoke.edu
http://www.mtholyoke.edu/acad/geology
f: https://www.facebook.com/groups/mhcgeoalums/
Administrative Assistant: Rhodaline Forjwuor
Enrollment (2016): B: 22 (14)

Professor:
Steven R. Dunn, (D), Wisconsin, 1989, Gp
Girma Kebbede, (D), Syracuse, 1981, Zy
Mark McMenamin, (D), California (Santa Barbara), 1984, PgGst
Thomas L. Millette, (D), Clark, 1989, ZriGm
Alan Werner, (D), Colorado, 1988, GleGm
Associate Professor:
Michelle J. Markley, (D), Minnesota, 1998, GcgGt
Assistant Professor:
Serin D. Houston, (D), Syracuse, 2012, Zyn
Geoprocessing Lab Manager:
Eugenio J. Marcano, (D), Cornell, 1994, ZiSoZn
Visiting Assistant Professor:
Samuel Tuttle, (D), Boston, 2015, Gq
Emeritus:
Martha M. Godchaux, (D), Oregon, 1969, Gv
Laboratory Director:
Penny M. Taylor, (M), SUNY (Oneonta), 2000, Gg
Geology Technician:
Gerard Marchand, (B), Westfield State, 1990, Gga

Northeastern University
Dept of Marine and Environmental Sciences (B,M,D) O☒ (2018)
14 Holmes Hall
360 Huntington Ave
Boston, MA 02115
p. (617) 373-3176
environment@neu.edu
http://www.northeastern.edu/mes
f: https://www.facebook.com/northeastern.mes
Programs: Environmental Science (B); Environmental Studies (B); Marine Biology (B,M,D); Ecology (D); Evolution (D)
Administrative Assistant: Danielle Walquist. Lynch, M.Ed.
Enrollment (2018): B: 591 (154) M: 18 (1) D: 65 (13)

Chair:
Geoffrey Trussell, (D), William & Mary, 1998, Zn
Professor:
Joseph Ayers, (D), California (Santa Cruz), 1975, Zn
Richard H. Bailey, (D), North Carolina, 1973, Pi
William Detrich, (D), Yale, Zn
Brian Helmuth, (D), Washington, 1997, OnZcRc
Mark Patterson, (D), Harvard, 1995, Zn
Hanumant Singh, (D), MIT, 1995, Zn
Associate Chair:
Rebeca Rosengaus, (D), Boston, 1993, Zn
Associate Professor:
Jonathan Grabowski, (D), North Carolina (Chapel Hill), 2012, GuEg
Malcolm D. Hill, (D), California (Santa Cruz), 1979, ZiGze
Justin Ries, (D), Johns Hopkins, 2005, Cm
Martin E. Ross, (D), Idaho, 1978, Gie
Steven Scyphers, (D), S Alabama, 2012, Zn
Aron Stubbins, (D), Newcastle-upon-Tyne, 2001, CgbZc

Assistant Professor:
Jennifer Bowen, (D), Boston, Zn
Loretta Fernandez, (D), MIT, 2010, Co
Tarik Gouhier, (D), McGill, Zn
Randall Hughes, (D), California (Davis), 2006, Zn
David Kimbro, (D), California, Zn
Kathleen Lotterhos, (D), Florida State, 2011, Zn
Amy Mueller, (D), MIT, 2012, Zn
Samuel Munoz, (D), Wisconsin, 2015, GsPcRn
Samuel Scarpino, (D), Texas, Zn
Steve Vollmer, (D), Harvard, Zn
Lecturer:
Daniel C. Douglass, (D), Wisconsin, 2005, GlSdPc
Tara Duffy, (D), Stony Brook, Ob
Stephanie Eby, (D), Syracuse, 2010, Zn
Emeritus:
Donald Cheney, (D), S Florida, Zn
Gwilym Jones, (D), Indiana State, Zn
Peter S. Rosen, (D), William & Mary, 1976, OnZu
Business and Operations Manager:
Heather Sears, (D), MIT, 2004, Zn
Co-op Coordinator:
Sarah Klionsky, (M), Wisconsin, 2009

Salem State University
Geological Sciences Dept (B) O☒ (2019)
352 Lafayette Street
Salem, MA 01970
p. (978) 542-6282
bhubeny@salemstate.edu
Programs: BS Geology
BS Applied Geosciences: Sustainability
BS Applied Geosciences: Forensic
BS Earth Science Education
BA Geology
Enrollment (2018): B: 50 (17)

Chair:
J Bradford Hubeny, (D), Rhode Island, 2006, GeOnCs
Professor:
Douglas Allen, (D), Minnesota, 2003, CgHsCt
James L. Cullen, (D), Brown, 1984, GsPmOu
Assistant Professor:
Sara Mana, (D), Rutgers, 2013, GicGv
Emeritus:
Lindley S. Hanson, (D), Boston, 1988, GmgOn
Laboratory Director:
Renee Knudstrup, (B), Ca

Smith College
Dept of Geosciences (B) ☒ (2018)
Clark Science Center
44 College Lane
Northampton, MA 01063
p. (413) 585-3805
dkortes@smith.edu
https://www.smith.edu/academics/geosciences
Enrollment (2010): B: 27 (6)

Chair:
John B. Brady, (D), Harvard, 1975, GxzZe
Professor:
Bosiljka Glumac, (D), Tennessee, 1997, Gs
Robert M. Newton, (D), Massachusetts, 1978, Gm
Amy L. Rhodes, (D), Dartmouth, 1996, ClGe
Associate Professor:
Jack Loveless, (D), Cornell, 2008, Gtc
Sara B. Pruss, (D), S California, 2004, Pg
Lecturer:
Mark E. Brandriss, (D), Stanford, 1994, Gx
Emeritus:
H. Robert Burger, (D), Indiana, 1966, GcYeZi
H. Allen Curran, (D), North Carolina, 1968, Pg
Geoscience Technician:
Michael Vollinger, (B), Massachusetts, 1993

Massachusetts

Tufts University
Dept of Earth and Ocean Sciences (B) ☒ (2018)
Lane Hall
2 N Hill Road
Medford, MA 02155
    p. (617) 627-3494
    anne.gardulski@tufts.edu
    http://eos.tufts.edu/
    f: https://www.facebook.com/TuftsEOS
    Programs: Geological Sciences (B); Environmental Geology (B)
    Administrative Assistant: Lindsay Riordan
    Enrollment (2018): B: 13 (9)
Professor:
    G Garven, (D), British Columbia, 1982, HwqHy
    John C. Ridge, (D), Syracuse, 1985, GlnGg
Associate Professor:
    Anne F. Gardulski, (D), Syracuse, 1987, GrdGg
Assistant Professor:
    Andrew Kemp, (D), Pennsylvania, 2009, OnPmOg
Emeritus:
    Robert L. Reuss, (D), Michigan, 1970, Gzi

University of Massachusetts, Amherst
Dept of Geosciences (B,M,D) ●☒ (2019)
233 Morrill Science Center
627 North Pleasant St.
Amherst, MA 01003-9297
    p. (413) 545-2286
    juliebg@geo.umass.edu
    http://www.geo.umass.edu/
    f: https://www.facebook.com/UMass-Geosciences
    Enrollment (2018): B: 328 (83) M: 29 (13) D: 38 (3)
Department Head:
    Julie Brigham-Grette, (D), Colorado, 1985, GlPcGu
Distinguished University Professor:
    Raymond S. Bradley, (D), Colorado, 1974, As
Professor:
    Stephen J. Burns, (D), Duke, 1987, Cs
    Michele L. Cooke, (D), Stanford, 1996, Gc
    Robert DeConto, (D), Colorado, 1996, As
    Piper Gaubatz, (D), California (Berkeley), 1968, Zy
    R. Mark Leckie, (D), Colorado, 1984, PmGru
    J. Michael Rhodes, (D), Australian Nat, 1970, GviCa
    Sheila J. Seaman, (D), New Mexico, 1988, Gi
    Michael L. Williams, (D), New Mexico, 1987, Gc
Extension Associate Professor:
    Michael Rawlins, (D), New Hampshire, 2006, As
Extension Assistant Professor:
    Christine Hatch, (D), California (Santa Cruz), 2007, HwgSp
Associate Extension Professor:
    William P. Clement, (D), Wyoming, 1995, Yue
Associate Professor:
    David Boutt, (D), New Mexico Tech, 2004, HwqHy
    Steven Petsch, (D), Yale, 2000, Col
    Stan Stevens, (D), California (Berkeley), 1983, Zy
    Eve Vogel, (D), Oregon, 2007, Zy
    Jonathan D. Woodruff, (D), MIT, 2008, Gs
    Qian Yu, (D), California (Berkeley), 2005, Zri
Research Assistant Professor:
    Brian Yellen, (D), Univ, Gs
Assistant Professor:
    Isla Castaneda, (D), Minnesota, 2007, Ca
    Haiying Gao, (D), Rhode Island, 2012, Ys
    Isaac J. Larsen, (D), Washington, 2013, Gm
    Justin Richardson, (D), Dartmouth, 2015, CbSc
    Matthew Winnick, Stanford, 2015, Cbq
Research Associate:
    Tim Cook, (D), Massachusetts, 2009, Pc
Lab Manager:
    Jeff Salacup, (D), Brown, 2014, PcCb
Assistant Professor :
    Forrest J. Bowlick, (D), Texas A&M, 2016, ZieZg
Lecturer:
    Michael J. Jercinovic, (D), New Mexico, 1988, Gx
Adjunct Professor:
    Douglas R. Hardy, (D), Massachusetts, 1995, Hy
    Thomas C. Johnson, (D), California (San Diego), 1975, PeCoOu
    Douglas Kowaleski, (D), Boston, 2009, Gm
    Eileen McGowan, (D), Massachusetts, 2010, Xg
    Thomas L. Millette, (D), Clark, 1989, Zy
    Stephen Nathan, (D), Massachusetts, 2005, PmOg
    Liang Ning, (M), Nanjing, 2007, Am
    Peter T. Panish, (D), Massachusetts, 1989, Gp
    Nicholas Venti, (D), Delaware, 2012, Og
Emeritus:
    Laurie L. Brown, (D), Oregon State, 1974, Ymg
    Christopher D. Condit, (D), New Mexico, 1984, Gi
    James A. Hafner, (D), Michigan, 1970, Zn
    William D. McCoy, (D), Colorado, 1981, Zy
    George E. McGill, (D), Princeton, 1958, Gc
    Stearns A. Morse, (D), McGill, 1962, Gi
    Rutherford H. Platt, (D), Chicago, 1971, Zu
    Peter Robinson, (D), Harvard, 1963, Gc
    Richard W. Wilkie, (D), Washington, 1968, ZynZn
    Richard F. Yuretich, (D), Princeton, 1976, Cl
Massachusetts State Geologist:
    Stephen B. Mabee, (D), Massachusetts, 1992, GgHw

University of Massachusetts, Boston
School for the Environment (B,M,D) ☒ (2019)
100 Morrissey Boulevard
Boston, MA 02125
    p. (617) 287-7440
    sfe@umb.edu
    http://www.umb.edu/environment
    Enrollment (2010): B: 153 (27) M: 22 (12) D: 18 (4)
Dean:
    Robyn Hannigan, (D), Rochester, ClmCt
Professor:
    Bob Chen, (D), California (San Diego), 1992, CoOc
    Zhongping Lee, (D), S Florida, 1994, ZrOg
    William Robinson, (D), Northeastern, 1981, Zn
    Crystal Schaaf, (D), Boston, 1994, ZrAm
Director for Research, New England Aquarium:
    Michael Tlusty, (D), Syracuse, 1996, Ob
Director, Environmental Studies Program:
    Alan D. Christian, (D), Miami, 2002, HsGmCs
Associate Professor:
    Robert Bowen, (D), S California, 1981, ZnnZn
    Ellen Douglas, (D), Tufts, 2002, HwqHy
    John Duff, (D), Washington, 1995, ZnnZn
    Eugene Gallagher, (D), Washington, 1983, Ob
    Juanita Urban-Rich, (D), Memorial, Ob
Assistant Professor:
    Jennifer Bowen, (D), Boston, 2005, ObPgZn
    Helen Poynton, (D), California (Berkeley), Zn
Research Engineer:
    Francesco Peri, (M), Massachusetts (Boston), Zn
Manager, GIS Lab:
    Helenmary Hotz, (M), Massachusetts (Boston), Zi
Director, Green Harbors Project:
    Anamarija Frankic, (D), Virginia Inst of Marine Sci, Ze

University of Massachusetts, Lowell
Dept of Environmental, Earth, & Atmospheric Sciences
(B,M) ☒ (2019)
1 University Avenue
University of Massachusetts
Lowell, MA 01854
    p. (978) 934-3900
    Erica_Gavin@uml.edu
    https://www.uml.edu/sciences/eeas/
    f: https://www.facebook.com/Earth.Sciences.UMass.Lowell/
    Programs: Geology, Environmental Science, Atmospheric Science
    Enrollment (2017): B: 32 (0) M: 8 (0)
Chair:
    Daniel Obrist, (D), Nevada, 2002, As
Professor:
    Frank P. Colby, (D), MIT, 1983, Am

G. Nelson Eby, (D), Boston, 1971, CgGiCt
Associate Professor:
   Mathew Barlow, (D), Maryland, 1999, As
   Juliette Rooney-Varga, (D), ZcRc
Assistant Professor:
   Richard M. Gaschnig, (D), Washington State, 2010, CctCa
   James Heiss, (D), Delaware, 2017, Hw
   Christopher Skinner, (D), Stanford, 2014, At
   Kate Swanger, (D), Boston, 2009, Gl
Lecturer:
   Lori Weeden, (M), Boston Coll, 2002, GgeGs
Emeritus:
   Arnold L. O'Brien, (D), Boston, 1973, Hw

Wellesley College
Dept of Geosciences (B) ☒ (2019)
106 Central Street
Wellesley, MA 02481-8203
   p. (781) 283-3151
   dbraband@wellesley.edu
   http://www.wellesley.edu/Geosciences/
   Programs: Geosciences
   Administrative Assistant: Carol Gagosian
   Enrollment (2018): B: 24 (8)
Chair:
   Daniel J. Brabander, (D), Brown, 1997, Cl
Associate Scientist:
   Katrin Monecke, (D), ETH, Gsn
Instructor:
   Hilary Palevsky, (D), Washington, 2016, ObZo
Emeritus:
   James Besancon, (D), MIT, 1975, Hww
   David Hawkins, (D), MIT, 1996, Gzi
   Margaret D. Thompson, (D), Harvard, 1976, Gc
Instructor:
   Kathleen W. Gilbert, (M), Miami, 1995, Cs

Williams College
Dept of Geosciences (B) ☒ (2019)
947 Main Street
Williamstown, MA 01267
   p. (413) 597-2221
   patricia.e.acosta@williams.edu
   http://www.williams.edu/Geoscience
   f: https://www.facebook.com/groups/williamsgeosciences
   Programs: Geosciences
   Administrative Assistant: Patricia E. Acosta
   Enrollment (2018): B: 25 (12)
Chair:
   Mea S. Cook, (D), MIT/WHOI, 2006, OuGe
Professor:
   Ronadh Cox, (D), Stanford, 1993, GsXg
   Paul Karabinos, (D), Johns Hopkins, 1981, Gc
   Reinhard A. Wobus, (D), Stanford, 1966, GivGz
Associate Professor:
   Phoebe A. Cohen, (D), Harvard, 2010, Pgg
Assistant Professor:
   Alice Bradley, (D), Colorado, 2016, AtOn
   Jose A. Constantine, (D), California (Santa Barbara), 2008, GmZi
Research Associate:
   B. Gudveig Baarli, (D), Oslo, 1988, PssPi
   Mark E. Brandriss, (D), Stanford, 1993, Gi
Lecturer:
   Alex Apotsos, (D), MIT, 2007, Ge
Emeritus:
   David P. Dethier, (D), Washington, 1977, Gm
   William T. Fox, (D), Northwestern, 1962, Gs
   Markes E. Johnson, (D), Chicago, 1977, PseGm
Science Librarian:
   Helena F. Warburg, (M), Indiana Sch Lib Sci, 1987, Zn

Woods Hole Oceanographic Institution
Dept of Geology & Geophysics (D) (2018)
Woods Hole, MA 02543-1541
   p. (508) 289-2388
   mburke@whoi.edu
   http://www.whoi.edu/page.do?pid=7145
   Administrative Assistant: Maryanne F. Ferreira
   Enrollment (2014): D: 26 (0)
Chair:
   Robert L. Evans, (D), Cambridge (UK), 1991, Yrr
Senior Scientist:
   Mark D. Behn, (D), MIT/WHOI, 2002, Yr
   Joan M. Bernhard, (D), California (San Diego), 1990, Cg
   Henry J B. Dick, (D), Yale, 1975, GitGc
   Jeffrey Donnelly, (D), Brown, 2000, Gu
   Daniel J. Fornari, (D), Columbia, 1978, Gu
   Chris German, (D), Cambridge, 1988, Gt
   Susan E. Humphris, (D), MIT/WHOI, 1977, GuCm
   Lloyd D. Keigwin, (D), Rhode Island, 1979, CsGul
   Jian Lin, (D), Brown, 1988, Gt
   Olivier Marchal, (D), Paris, 1996, Pe
   Daniel C. McCorkle, (D), Washington, 1987, Cm
   Delia W. Oppo, (D), Columbia, 1989, Pm
   Ralph A. Stephen, (D), Cambridge, 1978, Ys
   Maurice A. Tivey, (D), Washington, 1988, Yrm
Tenured Associate Scientist:
   Andrew Ashton, (D), Duke, 2005, On
   Juan Pablo Canales Cisneros, (D), Barcelona, 1997, Yr
   Anne L. Cohen, (D), Cape Town, 1993, PecOo
   Sarah B. Das, (D), Penn State, 2003, GlAtZg
   Virginia Edgcomb, (D), Delaware, 1997, ObZc
   Glenn A. Gaetani, (D), MIT, 1996, Gi
   Liviu Giosan, (D), SUNY (Stony Brook), 2001, GusPc
   Daniel Lizarralde, (D), MIT/WHOI, 1997, Ys
   Jeffrey J. McGuire, (D), MIT, 2000, Ys
   Robert A. Sohn, (D), California (San Diego), 1996, Yr
   S. Adam Soule, (D), Oregon, 2003, Gvu
Associate Scientist:
   Weifu Guo, (D), Caltech, 2008, CslPe
   Sune G. Nielsen, (D), ETH (Switzerland), 2005, Cs
Assistant Scientist:
   Veronique Le Roux, (D), Macquarie, 2008, Gi
Senior Research Specialist:
   James E. Broda, (B), Penn State, 1970, Gs
   John A. Collins, (D), MIT/WHOI, 1989, Ys
   Ann P. McNichol, (D), MIT, 1986, Oc
   Mark L. Roberts, (D), Duke, 1988, Yg
Research Specialist:
   Jurek Blusztajn, (D), Polish Acad of Sci, 1985, CcGi
   Alan R. Gagnon, (B), New Hampshire, 1983, Cs
   Li Xu, (D), Xiamen, 1992, Yr
Engineer:
   Peter B. Landry, (B), NEIT, 1990, Ca
Research Associate:
   Kathryn L. Elder, (B), Massachusetts, 1985, Cs
   Kalina D. Gospodinova, (M), MIT/WHOI, 2012, Cm
   Joshua D. Hlavenka, (B), North Texas
   Peter C. Lemmond, (B), Lehigh, 1978, Zn
   Brett Longworth, (M), Massachusetts, 2005, Ou
   Brian D. Monteleone, (D), Syracuse, 2000, Cc
   Kathryn R. Pietro, (M), California (Davis), 2007, Gu
Adjunct Scientist:
   Peter D. Bromirski, (D), Hawaii, 1993
   Johnson R. Cann, (D), Cambridge, 1963, Ys
   Colin Devey, (D), Oxford, 1986
   Javier Escartin, (D), MIT, 1996, Gg
   Andrea D. Hawkes, (D), Pennsylvania, 2008
   Gregory Hirth, (D), Brown, 1991, Gc
   Kuo-Fang (Denner) Huang, (D), National Cheng Kung, 2007
   Peter B. Kelemen, (D), Washington, 1987, Gi
   Yajing Liu, (D), Harvard, 2007
   John Maclennan, (D), Cambridge, 2000
   Larry Mayer, (D), Scripps, 1979, Gg
   Andrew M. McCaig, (D), Cambridge, 1983
   Jerry F. McManus, (D), Columbia, 1997, Pe
   Uri S. ten Brink, (D), Columbia, 1986, Yg
   David Thornalley, (D), Churchill, 2008
   Masako Tominaga, (D), Texas A&M, 2009
Oceanographer Emeritus:
   Graham S. Giese, (D), Chicago, 1966, On
   Steven J. Manganini, (B), Nasson, 1974, Ou
   Robert J. Schneider, (D), Tufts, 1968, Cc

Stephen A. Swift, (D), MIT/WHOI, 1986, Yr
Karl F. Von Reden, (D), Hamburg, 1983, Ct
Emeritus:
   William A. Berggren, (D), Stockholm, 1962, Ps
   Carl O. Bowin, (D), Princeton, 1960, YrGtg
   William B. Curry, (D), Brown, 1980, Pm
   Stanley R. Hart, (D), MIT, 1960, CgGvCp
   Susumu Honjo, (D), Hokkaido, 1961, Ou
   George P. Lohmann, (D), Brown, 1972, Pe
   David A. Ross, (D), California (San Diego), 1965, GuOg
   Hans Schouten, (D), Utrecht, 1970, Yr
   Nobumichi Shimizu, (D), Tokyo, 1968, Ca
   William G. Thompson, (D), Columbia, 2005, PeCc
   Brian E. Tucholke, (D), MIT/WHOI, 1973, Gut
   Elazar Uchupi, (D), S California, 1962, Ou
   Frank B. Wooding, (B), Harvard, 1965, Yr

Dept of Marine Chemistry & Geochemistry (M,D) ☒ (2018)
360 Woods Hole Road
MS 25
Woods Hole, MA 02543-1541
   p. (508) 289-3696
   mcg@whoi.edu
   http://www.whoi.edu/page.do?pid=7146
   Programs: Chemical Oceanography
   Enrollment (2015): D: 28 (5)
Chair:
   Bernhard Peucker-Ehrenbrink, (D), Max Planck, 1994, CmlHs
Associate Dean:
   Margaret K. Tivey, (D), Washington, 1989, Cm
Senior Scientist:
   Ken O. Buesseler, (D), MIT/WHOI, 1986, Oc
   Matthew A. Charette, (D), Rhode Island, 1998, Cm
   Konrad A. Hughen, (D), Colorado, 1997, Cc
   Elizabeth B. Kujawinski, (D), MIT/WHOI, 2000, Oc
   Mark D. Kurz, (D), MIT/WHOI, 1982, Cc
   Christopher M. Reddy, (D), Rhode Island, 1997, Co
   Daniel J. Repeta, (D), MIT/WHOI, 1982, Co
   Mak A. Saito, (D), MIT/WHOI, 2001, Oc
   Jeffrey S. Seewald, (D), Minnesota, 1990, Cp
   Benjamin Van Mooy, (D), Washington, 2003, Oc
Associate Scientist w/Tenure:
   Valier Galy, (D), Inst Nat Polytechnique de Lorrain, 2007, Cm
   Colleen Hansel, (D), Stanford, 2004, Cm
   Julie Huber, (D), Washington, 2004
   Z. Aleck Wang, (D), Georgia, 2003, Cm
Associate Scientist:
   Amy Apprill, (D), Hawaii, 2009, Ob
   Frieder Klein, (D), Bremen, 2009, Cm
   David Nicholson, (D), Washington, 2009, Gu
   Scott Wankel, (D), Stanford, 2007, Cm
Assistant Scientist:
   Peter Barry, (D), California (San Diego), 2012, Cm
   Tristan Horner, (D), Oxford, 2012, Cm
   Matthew Long, (D), Virginia, 2013, Cm
   Collin Ward, (D), Michigan, 2015
Research Specialist:
   Heather Benway, (D), Oregon State, 2005, OcCmOo
   Helen Fredricks, (D), Plymouth (UK), 2000, Oc
   Ivan D. Lima, (D), Miami, 1999, Oc
   Krista Longnecker, (D), Oregon State, 2004, Oc
   Steven M. Pike, (M), Rhode Island, 1998, Ca
   Melissa Soule, Oc
Research Assistant:
   Kevin Cahill, Cm
   Jessica Drysdale, Oc
   Kelsey Gosselin, Oc
   Kate Morkeski
Research Associate:
   Maureen Auro, (M), San Francisco State, 2007
   Matthew Biddle, (M), Maryland, 2019
   Joshua M. Curtice, (B), Massachusetts, 1992, Cc
   Paul Henderson, Cm
   Mairead Mahegan
   Matt McIlvin, (D), Massachusetts, 2004, Oc
   Carolyn Miller
   Dawn Moran, Ob
   Jennie Rheuban, (M), Virginia, 2013, Oc
   Margrethe Serres
   Gretchen Swarr, Oc
   Sean Sylva, Oc
Adjunct Professor:
   Minhan Dai, Cm
   Thomas Trull, Cm
Scientist Emeritus:
   Michael P. Bacon, (D), MIT/WHOI, 1976, Oc
   Werner G. Deuser, (D), Penn State, 1963, CsGsu
   John W. Farrington, (D), Rhode Island, 1972, ComOc
   William R. Martin, (D), MIT/WHOI, 1985, Oc
   Frederick L. Sayles, (D), Manchester, 1968, Oc
   Geoffrey Thompson, (D), Manchester, 1965, Cm
   Oliver C. Zafiriou, (D), Johns Hopkins, 1966, Oc
Oceanographer Emeritus:
   Cyndy Chandler, (B), SUNY (Geneseo), 1975, Og
   Nelson M. Frew, (D), Washington, 1971, Ca
   David M. Glover, (D), Alaska, 1985, Oc
   Jean K. Whelan, (D), MIT, 1965, Co
Emeritus:
   Edward R. Sholkovitz, (D), California (San Diego), 1972, Oc
Research Specialist:
   Carl G. Johnson, (M), Maine, 1983, Ca
   Robert K. Nelson, (B), C Connecticut, 1980, Co
Research Assistant:
   Zoe Sandwith, Oc
Dept. Administrator:
   Mary Murphy
Senior Research Specialist (ret.):
   Dempsey E. Lott, (M), Florida State, 1973, Oc
Senior Research Assistant (ret.):
   Margaret Sulanowska, Cm

Worcester State University
Earth, Environment and Physics (B) ⚲ (2018)
486 Chandler Street
Worcester, MA 01602-2597
   p. (508) 929-8583
   whansen@worcester.edu
   http://www.worcester.edu/Earth-Environment-and-Physics/
   f: https://www.facebook.com/worcesterstatedeep
   Programs: Earth Science Education; Geography; Environmental Science
   Enrollment (2018): B: 120 (12)
Professor:
   Allison L. Dunn, (D), Harvard, 2006, AsZyHg
   William J. Hansen, (D), CUNY, 2002, ZiyZr
Associate Professor:
   Patricia A. Benjamin, (D), Clark, 2002, Zg
   Douglas E. Kowalewski, (D), Boston, 2009, GmlZy
Assistant Professor:
   Nabin Malakar, (D), SUNY (Albany), 2012, Ap
   Alexander Tarr, (D), California (Berkeley), 2014, Zg
Instructor:
   Mark O. Johnson, (D), Clark, 1993, Zy

Michigan

Adrian College
Geology Dept (A,B) ☒ (2019)
110 S. Madison St.
Adrian, MI 49221
   p. (517) 265-5161
   tmuntean@adrian.edu
   http://adrian.edu/academics/academic-departments/geology/
   f: https://www.facebook.com/groups/55153346895/
   Programs: Geology, Environmental Geology
   Enrollment (2018): B: 14 (4)
Chair:
   Thomas Muntean, (D), Nevada, 2012, Gde
Professor:
   Sarah L. Hanson, (D), Utah, 1995, Giz

Albion College
Dept of Geological Sciences (B) O☒ (2018)
611 E. Porter St.
Albion, MI 49224
    p. (517) 629-0759
    cmenold@albion.edu
    http://www.albion.edu/geology/
    Programs: Geology; Earth Science; Earth Science with Secondary Education Certification; Geology with Secondary Education Certification
    Certificates: Geology; GIS; Paleontology; Environmental Geology; Earth Science with Secondary Education Certification (minors)
    Enrollment (2018): B: 22 (3)
Chair:
    Carrie A. Menold, (D), California (Los Angeles), 2006, Gpz
Professor:
    William S. Bartels, (D), Michigan, 1986, Pv
    Beth Z. Lincoln, (D), California (Los Angeles), 1985, Gc
    Thomas I. Wilch, (D), New Mexico Tech, 1997, GlvGm
Assistant Professor:
    Michael McRivette, (D), California (Los Angeles), 2011, GtZir
Emeritus:
    Russell G. Clark, (D), Dartmouth, 1972, GiZi
    Lawrence D. Taylor, (D), Ohio State, 1962, Gl

Calvin College
Dept of Geology, Geography, & Environmental Studies (B) (2015)
3201 Burton SE
Grand Rapids, MI 49546
    p. (616) 526-8415
    jbascom@calvin.edu
    http://www.calvin.edu/academic/geology/
    Enrollment (2012): B: 60 (14)
Chair:
    Jason E. VanHorn, (D), Ohio State, 2007, ZirRh
Professor:
    Henry Aay, (D), Clark, 1978, Zyu
    Johnathan Bascom, (D), Iowa, 1989, Zy
    Janel M. Curry, (D), Minnesota, 1985, Zy
    Ralph F. Stearley, (D), Michigan, 1990, PvOgRh
Associate Professor:
    Deanna van Dijk, (D), Waterloo, 1998, GmOnZy
Assistant Professor:
    Kenneth A. Bergwerff, (M), Grand Valley State, 1988, Ze
    James Skillen, (D), Cornell, 2006, Zu
Laboratory Manager/Instructor:
    Margene Brewer, (M), W Michigan, 1991, ZnGgZn
Emeritus:
    Clarence Menninga, (D), Purdue, 1966, CcGg
    Gerald K. Van Kooten, (D), California (Santa Barbara), 1980, GoCePi
    Davis A. Young, (D), Brown, 1969, GiRhGz

Central Michigan University
Dept of Earth and Atmospheric Sciences (B) (2018)
314 Brooks Hall
Mount Pleasant, MI 48859
    p. (989) 774-3179
    l.d.lemke@cmich.edu
    https://www.cmich.edu/colleges/cst/earth_atmos/Pages/
    Programs: Geology; Meteorology; Environmental Science
    Enrollment (2018): B: 116 (12)
Professor:
    Martin Baxter, (D), St. Louis, 2006, Ams
    Lawrence D. Lemke, (D), Michigan, 2003, HwGer
    Mona Sirbescu, (D), Missouri, 2002, GiCup
Associate Professor:
    Anthony Chappaz, (D), Quebec, 2008, Cat
    Daria Kluver, (D), Delaware, 2011, Asm
Assistant Professor:
    John Allen, (D), Australia, 2013, Ams
    Jason Keeler, (D), Illinois, 2015, Am
    Wendy Robertson, (D), Texas, 2014, HwCg
    Megan Rohrssen, (D), California, 2013, Co
    Nicole West, (D), Penn State, 2014, Gm
    Natalia Zakharova, (D), Columbia, 2014, YgGe

Lecturer:
    Rachael Agardy, (M), Texas A&M, 2005, Ou
    Maria Mercedes Gonzalez, (D), Nacional del Sur (Argentina), 1997, Gd
Emeritus:
    Richard Mower, (D), Wisconsin, 1981, Am
Related Staff:
    James J. Student, (D), Virginia Tech, 2002, CaGiz

Charles Stewart Mott Community College
Divison of Science and Mathematics (A) (2018)
1401 East Court St.
Flint, MI 48503
    p. (810) 762-0279
    sheila.swyrtek@mcc.edu
    http://www.mcc.edu/science_math/index.shtml
Professor:
    Sheila Swyrtek, (M), Minnesota (Duluth), 1996, Ga

Concordia University
Div of Natural Sciences (A,B,M) (2015)
4090 Geddes Road
Ann Arbor, MI 48105-2750
    p. (734) 995-7300
    nskov@cuaa.edu
Assistant Professor:
    James L. Refenes, (M), E Michigan, 2009, Ze

Delta College
Dept of Geology (A) (2015)
1961 Delta Road
University Center, MI 48710-0002
    p. (989) 686-9252
    andreabair@delta.edu
    Department Secretary: Barb Jurmanovich
    Enrollment (2011): A: 8 (4)
Department Chair:
    Timothy L. Clarey, (D), W Michigan, 1996, HwGcPv
Lecturer:
    Mary C. Gorte, (M), Rensselaer, 1983, Gs
Emeritus:
    Barry A. Carlson, (D), Michigan State, 1974, Yg
    Paul A. Catacosinos, (D), Michigan State, 1972, Gr
Cooperating Faculty:
    Kevin T. Dehne, (M), E Michigan, 1992, Ze

Eastern Michigan University
Dept of Geography & Geology (B) (2015)
Ypsilanti, MI 48197
    p. (743) 487-8589
    rsambroo@emich.edu
Head:
    Michael Kasenow, (D), W Michigan, 1994, Hg
    Richard A. Sambrook, (D)
Professor:
    Eugene Jaworski, (D), Louisiana State, 1971, Zr
    Carl F. Ojala, (D), Georgia, 1972, As
    Constantine N. Raphael, (D), Louisiana State, 1967, Gu
Associate Professor:
    Steven T. LoDuca, (D), Rochester, 1990, Pi
Assistant Professor:
    Kevin Blake, (D), Wisconsin, 1998, Gl
    Michael Bradley, (D), Utah, 1988, Gc
    Maria-Serena Poli, (D), Padova, 1995, Pe

Ferris State University
Dept of Physical Sciences ☒ (2019)
ASC-3021
Big Rapids, MI 49307
    p. (231) 591-2588
    heckf@ferris.edu
    https://ferris.edu/arts-sciences/departments/physical-sciences/index.htm
Professor:
    Frederick R. Heck, (D), Northwestern, 1987, GgZg

Gogebic Community College
Math-Science Div (A) (2015)
E4946 Jackson Rd.
Ironwood, MI 49938
 p. (906) 932-4231
 admissions@gogebic.edu
 http://www.gogebic.edu/academics/Math_Science/
Professor:
 Bill Perkis, Gg

Grand Rapids Community College
Physical Sciences Dept (A) ☒ (2018)
143 Bostwick Avenue, NE
Grand Rapids, MI 49503
 p. (616) 234-4248
 jqualls@grcc.edu
 http://www.grcc.edu/physicalscience/geology
 f: https://www.facebook.com/GRCCphysicalsciences
 Programs: Geology
 Enrollment (2018): A: 9 (0)
Assistant Professor:
 Tari Mattox, (M), N Illinois, 1984, GgvCg

Grand Valley State University
Dept of Geology (B) ◎ (2019)
1 Campus Drive
118 Padnos Hall
Allendale, MI 49401
 p. (616) 331-3728
 geodept@gvsu.edu
 http://www.gvsu.edu/geology
 Programs: Geology; Geology with Environmental Emphasis; Geology-Chemistry
 Administrative Assistant: Janet H. Potgeter
 Enrollment (2018): B: 100 (23)
Head:
 Figen A. Mekik, (D), N Illinois, 1999, GuCm
Professor:
 Patrick M. Colgan, (D), Wisconsin, 1996, GmlPe
 Stephen R. Mattox, (D), N Illinois, 1992, ZeGv
 Virginia L. Peterson, (D), Massachusetts, 1992, GxcGt
 Peter J. Wampler, (D), Oregon State, 2004, HsZiGm
 John C. Weber, (D), Northwestern, 1995, GctYd
Associate Professor:
 Kevin C. Cole, (D), Arizona, 1990, Gz
 Peter E. Riemersma, (D), Wisconsin, 1997, Hw
Assistant Professor:
 Caitlin N. Callahan, (D), W Michigan, 2013, GgZg
 Tara A. Kneeshaw, (D), Texas A&M, 2008, ClGe
Instructor:
 Kelly L. Heid, (M), Mississippi State, 2011, Ze
 John M. VanRegenmorter, (M), Colorado, 2011, PvGsr
 Ian Z. Winkelstern, (D), Michigan, 2018, GsOo
Visiting Professor:
 Jeremy C. Gouldey, (D), Northwestern, 2015, CgPeCg
Emeritus:
 Thomas E. Hendrix, (D), Wisconsin, 1960, Gc
 William J. Neal, (D), Missouri, 1968, GsdGr
 Norman W. Ten Brink, (D), Washington, 1971, Gm
 Patricia E. Videtich, (D), Brown, 1982, Gd

Henry Ford Community College
Science Div (2015)
Dearborn, MI 48128
 p. (313) 845-9632
 cjacobs@hfcc.edu

Hope College
Dept of Geological & Environmental Sciences (B) ⚜ (2019)
P.O. Box 9000
Holland, MI 49422-9000
 p. (616) 395-7540
 bodenbender@hope.edu
 https://hope.edu/academics/geological-environmental-sciences/
 Programs: BA in Geology, BS in Geology, Minor in Geology, Minor in Environmental Science
 Enrollment (2018): B: 16 (7)
Chair:
 Brian E. Bodenbender, (D), Michigan, 1994, GsPiGe
Professor:
 Edward C. Hansen, (D), Chicago, 1983, GmpGc
 Jon W. Peterson, (D), Chicago, 1989, Ge
Adjunct Professor:
 Suzanne DeVries-Zimmerman, (M), Princeton, 1989, Gge

Jackson College
Dept of Geology & Geography (A,B) (2017)
2111 Emmons Road
Jackson, MI 49201
 p. ((51) 7) -787- (Ext. 00 x157)
 AlbeeScSteven@jccmi.edu
 http://www.jccmi.edu/academics/science/geo
 Enrollment (2012): A: 200 (20)
Wilbur L. Dungy Endowed Chair in the Sciences:
 Steven R. Albee-Scott, (D), Michigan, 2005, GePeSf

Lake Michigan College
Lake Michigan College (A) (2015)
Napier Avenue Campus
2755 E. Napier Avenue
Benton Harbor, MI 49022
 lovett@lakemichigancollege.edu
 http://www.lakemichigancollege.edu/index.php?option=com_content&task=view&id=509&Itemid=157
Instructor:
 Cole Lovett, (D), W Michigan, 1995, Gg

Lake Superior State University
Geology (B) ◯☒ (2019)
650 W. Easterday Avenue
Sault Ste. Marie, MI 49783
 p. (906) 635-2267
 pkelso@lssu.edu
 https://www.lssu.edu/geology/
 Programs: B.S. Geology
 B.S. Geology - Environmental option
 Certificates: GIS certificate
 GIS associates degree
 Administrative Assistant: Donna White
 Enrollment (2018): B: 38 (12)
Professor:
 Paul R. Kelso, (D), Minnesota, 1993, YmGct
Assistant Professor:
 William (Bill) Houston, (D), Michigan Tech, 2002, Gsg
 Hari Kandel, (D), Florida Intl, 2015, HgZi
 MaryKathryn (Kat) Rocheford, (D), Iowa, 2014, GeZiGa
 Matt Spencer, (D), Penn State, 2005, Gl
Emeritus:
 Lewis M. Brown, (D), New Mexico, 1973, PiZe
Science Lab Manager:
 Benjamin Southwell, (M), C Michigan, 2012, Zn

Macomb Community College, Center Campus
Dept of Science (Geology) (A) (2015)
44575 Garfield Road
Clinton Township, MI 48038-1139
 p. (586) 286-2154
 schaferc@macomb.edu
 Enrollment (2014): A: 9 (0)
Professor:
 Carl M. Schafer, (M), Montana, 1998, Gg

Michigan State University
Dept of Earth and Environmental Sciences (B,M,D) (2018)
288 Farm Lane
East Lansing, MI 48824-1115
 p. (517) 355-4626
 geosci@msu.edu
 https://glg.natsci.msu.edu/

Enrollment (2014): M: 4 (0) D: 20 (0)
Chair:
  Ralph E. Taggart, (D), Michigan State, 1971, Pb
Professor:
  jiquan Chen, (D), Washington, 1991, Zg
  Kazuya Fujita, (D), Northwestern, 1979, Ys
  Julie C. Libarkin, (D), Arizona, 1999, Zen
  David T. Long, (D), Kansas, 1977, ClGeb
  Michael A. Velbel, (D), Yale, 1984, ClGdSc
Associate Professor:
  Bruno Basso, (D), Michigan State, 2002, Zn
  Danita S. Brandt, (D), Yale, 1985, PiGr
  Michael D. Gottfried, (D), Kansas, 1991, Pv
  David W. Hyndman, (D), Stanford, 1995, Hy
Assistant Professor:
  Tyrone Rooney, (D), Penn State, 2006, Gi
  Matt Schrenk, (D), Washington, 2005, Pg
  Jay Zarnetske, (D), Oregon State, 2011, Hwg

Michigan Technological University
A. E. Seaman Mineral Museum ☒ (2018)
1404 E. Sharon Avenue
Houghton, MI 49931
  p. (906) 487-2572
  tjb@mtu.edu
  http://www.museum.mtu.edu
Director:
  Theodore J. Bornhorst, (D), New Mexico, 1980, EgCgGx
Associate Curator:
  Christopher J. Stefano, (D), Michigan, 2010, GzxCg

Dept of Geological & Mining Engineering & Sciences
(B,M,D) O☒ (2019)
1400 Townsend Drive
Dow Building, Room 630
Houghton, MI 49931-1295
  p. (906) 487-2531
  geo@mtu.edu
  http://www.mtu.edu/geo
  Enrollment (2018): B: 78 (30) M: 39 (5) D: 7 (5)
Provost and Acting Dean, Graduate School:
  Jacqueline E. Huntoon, (D), Penn State, 1990, GsZe
Chair & Professor:
  John S. Gierke, (D), Michigan Tech, 1990, Hy
Professor:
  Simon Carn, (D), Cambridge, 1999, ZrGv
  Alex S. Mayer, (D), North Carolina, 1992, Hy
  Aleksey V. Smirnov, (D), Rochester, 2002, YmgGt
  James R. Wood, (D), Johns Hopkins, 1973, Cl
Associate Professor:
  Thomas Oommen, (D), Tufts, 2009, NeZiNg
  Gregory P. Waite, (D), Utah, 2004, YsGv
  Shiliang Wu, (D), Harvard, 2007, AsCg
Assistant Professor:
  Roohollah Askari, (D), Calgary, 2013, YxxNp
  Snehamoy Chatterjee, (D), Indian Inst of Tech, 2007, NmGqNx
  Chad Deering, (D), Canterbury, 2009, Ggi
Department Facilities Manager:
  Robert Barron, (B), Michigan Tech, 1979, ZgGzi
Lecturer:
  Nathan Manser, (D), S Florida, 2015, GeNm
Adjunct Professor and Director, Great Lakes Research Ctr.:
  Guy A. Meadows, (D), Purdue, 1977, Gu
Senior Lecturer:
  Jeremy Shannon, (D), Michigan Tech, 2006, Gg
Postdoctoral Research Fellow and Temporary Faculty:
  Rudiger Escobar-Wolf, (D), Michigan Tech, 2013, Gv
Research Scientist:
  Carol Asiala, (B), Michigan Tech, 1985
Cooperating Faculty:
  William I. Rose, (D), Dartmouth, 1970, Gv
  Roger M. Turpening, (D), Michigan, 1966, Ye

Mott Community College
Science and Mathematics (A) (2016)
1401 East Court St.
Flint, MI 48503
  p. (810) 232-9312
  sheila.swyrtek@mcc.edu
  http://www.mcc.edu
  Enrollment (2015): A: 4 (0)
Professor:
  Sheila M. Swyrtek, (M), Minnesota (Duluth), 1997, GgZg

Muskegon Community College
Dept of Mathematics & Physical Sciences (A) (2016)
221 S. Quarterline Road
Muskegon, MI 49442
  p. (231) 777-0289
  amber.kumpf@muskegoncc.edu
  http://www.muskegoncc.edu/pages/652.asp
  Administrative Assistant: Tamera Owens
Instructor:
  Amber C. Kumpf, (M), Rhode Island, 2010, GgYrOu

Northwestern Michigan College
Northwestern Michigan College (2015)
1701 East Front Street
Traverse City, MI 49686
  p. (231) 995-1000
  information@nmc.edu
  http://www.nmc.edu

Oakland Community College
Natural Sciences Dept (A) ☒ (2017)
2900 Featherstone Road
Auburn Hills, MI 48326
  p. (248) 232-4538
  lgkodosk@oaklandcc.edu
  Programs: Science (A)

Schoolcraft College
Dept of Geology (A) (2015)
18600 Haggerty Road
Livonia, MI 48151
  p. (734) 462-4400
  jrexius@schoolcraft.edu
Instructor:
  James E. Rexius, (M), E Michigan, 1978, Gl

University of Michigan
Dept of Climate and Space Sciences and Engineering
(B,M,D) ☒ (2018)
2455 Hayward
Ann Arbor, MI 48109-2143
  p. (734) 615-3583
  clasp--UM@umich.edu
  http://clasp.engin.umich.edu/
  f: www.facebook.com/umclasp
  t: @umclasp
  Programs: Climate & Space Science & Engineering--Geoscience & Remote Sensing Concentration; Applied Climate MEng; Space Engineering MEng
  Enrollment (2018): B: 19 (12) M: 43 (33) D: 48 (14)
Chair:
  Tuija Pulkkinen, (D), Helsinki, 1992, Zn
Professor:
  Sushil Atreya, (D), Michigan, ApZy
  John Boyd, (D), Harvard, 1976, OpAs
  R. Paul Drake, (D), Johns Hopkins
  Lennard Fisk, (D), California (San Diego)
  Brian Gilchrist, (D), Stanford
  Tamas Gombosi, (D), Lóránd Eötvös
  Enrico Landi, (D), Florence
  Michael Liemohn, (D), Michigan
  Mark B. Moldwin, (D), Boston, 1993, Xp
  Joyce Penner, (D), Harvard
  Nilton Renno, (D), MIT
  Aaron Ridley, (D), Michigan
  Richard B. Rood, (D), Florida State, 1982, ZcAsm
  Christopher Ruf, (D), Massachusetts

Perry Samson, (D), Wisconsin, 1979, AsZeAm
James Slavin, (D), California (Los Angeles)
Associate Professor:
Jeremy Bassis, (D), Scripps, 2007, GlOgZg
Mark Flanner, (D), California (Irvine)
Xianglei Huang, (D), Caltech
Christiane Jablonowski, (D), Michigan, 2004, AsmZo
Xianzhe Jia, (D), California (Los Angeles)
Justin C. Kasper, (D), MIT, 2003
Susan Lepri, (D), Michigan
Allison Steiner, (D), Georgia Tech
Shasha Zou, (D), California (Los Angeles)
Assistant Professor:
Angel F. Adames-Corraliza, (D), Washington, Ast
Gretchen Keppel-Aleks, (D), Caltech
Eric A. Kort, (D), Harvard
Ashley E. Payne, California (Irvine), 2016, At

Dept of Earth and Environmental Sciences (B,M,D) (2015)
2534 C.C. Little Building
1100 North University Avenue
Ann Arbor, MI 48109-1005
  p. (734) 764-1435
  mukasa@umich.edu
  http://www.lsa.umich.edu/geo/
  Administrative Assistant: Robert J. Patterer
Professor:
  Tomasz K. Baumiller, (D), Chicago, 1990, Pg
  Joel D. Blum, (D), Caltech, 1990, CsaCl
  Maria Clara Castro, (D), Paris, 1995, Hy
  Daniel C. Fisher, (D), Harvard, 1975, Pi
  Stephen E. Kesler, (D), Stanford, 1966, Eg
  Rebecca A. Lange, (D), California (Berkeley), 1989, Gi
  Kyger C. Lohmann, (D), SUNY (Stony Brook), 1977, CsGsCl
  Samuel B. Mukasa, (D), California (Santa Barbara), 1984, Cc
  Henry N. Pollack, (D), Michigan, 1963, Yg
  Jeroen Ritsema, (D), California (Santa Cruz), 1995, Ysg
  Larry J. Ruff, (D), Caltech, 1981, Ys
  Gerald R. Smith, (D), Michigan, 1965, Ps
  Ben van der Pluijm, (D), New Brunswick, 1984, GcZce
  Lynn M. Walter, (D), Miami, 1983, Cl
  Youxue Zhang, (D), Columbia, 1989, Cu
Associate Professor:
  Udo Becker, (D), Virginia Tech, 1995, Zn
  Robyn J. Burnham, (D), Washington, 1987, Pb
  Todd A. Ehlers, (D), Utah, 2001, Yg
  Nathan A. Niemi, (D), Caltech, 2001, Gtc
  Benjamin Passey, (D), Utah, 2007, CsaCl
  Christopher J. Poulsen, (D), Penn State, 1999, Pe
  Peter J. van Keken, (D), Utrecht, 1989, Yg
Associate Research Scientist:
  Shaopeng Huang, (D), Acad Sinica, 1990, YhZcGt
Assistant Professor:
  Marin Clark, (D), MIT, 2003, Gm
  Ingrid Hendy, (D), California (Santa Barbara), 2000, Pg
  Jeffrey A. Wilson, (D), Chicago, 1999, Gg
Research Scientist:
  Jeffrey C. Alt, (D), Miami, 1984, Ou
Associate Research Scientist:
  Chris Hall, (D), Toronto, 1982, Cc
  Josep M. Pares, (D), Barcelona, 1988, Ym
Assistant Research Scientist:
  James D. Gleason, (D), Arizona, 1994, Cg
  Jie Lian, (D), Michigan, 2003, Gz
  Mirjam Schaller, (D), Bern, 2001, Gm
Adjunct Assistant Research Scientist:
  Roland C. Rouse, (D), Michigan, 1972, Gz
Adjunct Assistant Professor:
  Karen L. Webber, (D), Rice, 1988, Gv
Adjunct Professor:
  John W. Geissman, (D), Michigan, 1980, Ym
  William B. Simmons, (D), Michigan, 1973, Gz
Emeritus:
  Charles Beck, (D), Cornell, 1955, Pg
  William Farrand, (D), Michigan, 1960, Ga
  William C. Kelly, (D), Columbia, 1954, Eg
  Philip A. Meyers, (D), Rhode Island, 1972, CoGnu

Theodore C. Moore, (D), California (San Diego), 1968, OuPmc
James O'Neil, (D), Chicago, 1963, Cs
Robert M. Owen, (D), Wisconsin, 1975, CmGu
Donald R. Peacor, (D), MIT, 1962, Gz
David K. Rea, (D), Oregon State, 1974, Gu
Rob Van der Voo, (D), Utrecht, 1969, YmGtc
Bruce H. Wilkinson, (D), Texas, 1973, GseGt

University of Michigan, Dearborn
Dept of Natural Sciences (B,M) (2015)
4901 Evergreen Road
Dearborn, MI 48128
  p. (313) 593-5277
  kmurray@umd.umich.edu
  http://www.umd.umich.edu/?id=570101
  Enrollment (2013): B: 30 (4) M: 14 (3)
Acting Chair:
  Kent S. Murray, (D), California (Davis), 1981, Hw
Professor:
  Don Bord, (D), Dartmouth, 1976, Xy
  Jacob A. Napieralski, (D), Purdue, 2005, GmlZi
Associate Professor:
  John Riebesell, (D), Chicago, 1975, Zu
Lecturer:
  David Matzke, (M), Michigan, 1975, Xy
Adjunct Professor:
  Michael Favor, (B), Michigan, 1985, Zn

Washtenaw Community College
Dept of Geology (A) (2015)
4800 E. Huron River Drive
Ann Arbor, MI 48105
  p. (734) 677-5111
  salbach@wccnet.edu
Head:
  Suzanne M. Albach, (M), Mississippi State, 2005, ZegGe

Wayne State University
Geology Dept (B,M) ☒ (2019)
0224 Old Main Building
4831 Cass Avenue
Detroit, MI 48201
  p. (313) 577-2506
  baskaran@wayne.edu
  http://www.clas.wayne.edu/geology/
  t: @WSUgeology
  Programs: Geology; Environmental Science
  Certificates: Hoz
  Enrollment (2011): B: 101 (15) M: 9 (1)
Professor:
  Jeffrey L. Howard, (D), California (Santa Barbara), 1987, GsScGa
Associate Professor:
  Sarah J. BrownLee, (D), California (Berkeley), 2009, GxcGy
  Shirley Papuga, (D), Colorado, 2006, Hgs
Assistant Professor:
  Scott Burdick, (D), MIT, 2014, Ysg
Instructor:
  Walter Pociask, (M), Zn
Lecturer:
  Grazyna Sledzinski, (M), Wayne State, 1994, Gg
  Felice Sperone, (M), Calabria - Rende (Italy), 2003, GgZr
  John M. Zawiskie, (M), Wayne State, 1979, Gg
Adjunct Professor:
  Gi-Hoon Hong, (D), Alaska, 1986, OcCcm
  Bratton F. John, (D), California (Berkeley), 1997, Cg

Western Michigan University
Dept of Geosciences (B,M,D) O (2018)
1903 W Michigan Ave
Kalamazoo, MI 49008-5241
  p. (269) 387-5486
  kathryn.wright@wmich.edu
  http://www.wmich.edu/geology/
  f: https://www.facebook.com/wmugeosciences
  Programs: Earth Science (B); Earth Science Education (B);

Geochemistry (B); Geology (B); Geophysics (B); Hydrogeology (B); Accelerated MA in Earth Sciences (M), Earth Sciences (M), Geosciences (M, D)
Certificates: Applied Hydrogeology (UG, G)
Enrollment (2015): B: 108 (18) M: 51 (17) D: 12 (4)

Dean of the College of Arts and Sciences:
    Carla Koretsky, (D), Johns Hopkins, 1998, CglHs

Chair:
    Mohamed Sultan, (D), Washington, 1984, ZriGe

Professor:
    Alan E. Kehew, (D), Idaho, 1977, GmlHw
    Michelle A. Kominz, (D), Columbia, 1986, OuYgGu
    R. V. Krishnamurthy, (D), Phy Res Lab (India), 1984, Cs

Co-Director of the Hydrogeology Field Course:
    Thomas R. Howe III, (M), W Michigan, 2017, HwGem
    Donald Matthew M. Reeves, (D), Nevada (Reno), 2006, HwqNr

Associate Professor:
    Daniel P. Cassidy, (D), Notre Dame, 1995, Hw
    Johnson R. Haas, (D), Washington, 1993, Cl
    Duane R. Hampton, (D), Colorado State, 1989, HwSpHg
    William A. Sauck, (D), Arizona, 1972, YgeYv

Assistant Professor:
    Robb Gillespie, (D), SUNY, GsoGm
    Stephen E. Kaczmarek, (D), Michigan State, 2005, GdCgGz
    Joyashish Thakurta, (D), Indiana, 2008, GiEg

Senior Research Associate and Part-time Instructor:
    Andrew Caruthers, (D), British Columbia, 2013, GsCsGr

Director of Michigan Geological Repository for Research and Education and Professor Emeritus:
    William B. Harrison, III, (D), Cincinnati, 1974, GsrPg

Teaching Faculty Specialist and Director of CoreKids:
    Peter J. Voice, (D), Virginia Tech, 2010, GsZn

Minnesota

Bemidji State University
Center for Environmental, Economic, Earth, & Space Studies (A,B,M) (2017)
#27, 1500 Birchmont Drive NE
Bemidji, MN 56601
    p. (218) 755-2783
    jeffrey.ueland@bemidjistate.edu
    http://www.bemidjistate.edu/academics/departments/ceeess/
    Enrollment (2012): B: 109 (17) M: 23 (3)

Professor:
    Dragoljub D. Bilanovic, (D), Tech Israel Inst Tech, 1990, GePgCa

Assistant Professor:
    Carl Isaacson, (D), Oregon State, 2007, Ge
    Miriam Rios-Sanchez, (D), Michigan Tech, 2012, HwZrGg

Carleton College
Dept of Geology (B) ☒ (2019)
One North College Street
Northfield, MN 55057
    p. (507) 222-5769
    cdavidso@carleton.edu
    https://apps.carleton.edu/curricular/geol/
    Programs: Geology
    Administrative Assistant: Tamara Little
    Enrollment (2018): B: 38 (26)

Charles L. Denison Professor of Geology:
    Mary E. Savina, (D), California (Berkeley), 1982, Gm

Professor:
    Clinton A. Cowan, (D), Queen's, 1992, GsPe
    Cameron Davidson, (D), Princeton, 1991, GptCc
    Bereket Haileab, (D), Utah, 1994, GxzGi

Associate Professor:
    Sarah J. Titus, (D), Wisconsin, 2006, Gc

Director, Science Ed Resource Center:
    Cathryn A. Manduca, (D), Caltech, 1988, Gi

Emeritus:
    Caryl E. Buchwald, (D), Kansas, 1966, Gg

Technical Director:
    Jonathon L. Cooper, (B), W Washington, Gg

Century College
Earth Science (A) ◐ (2018)
3300 Century Avenue North
White Bear Lake, MN 55110
    p. (651) 779-3242
    joe.osborn@century.edu
    http://www.century.edu/futurestudents/programs/pnd.aspx?id=66

Head:
    Joe Osborn, (M), Zg

Instructor:
    Jill Bries-Korpik, (M), Zg

Fond du Lac Tribal and Community College
Fond du Lac Tribal and Community College (A) (2015)
2101 14th Street
Cloquet, MN 55720
    glang@fdltcc.edu
    http://www.fdltcc.edu/

Instructor:
    Glenn Langhorst, (M), Minnesota (Duluth), Gg

Gustavus Adolphus College
Dept of Geology (B) ☒ (2019)
800 West College Avenue
St Peter, MN 56082
    p. (507) 933-7333
    ltriplet@gustavus.edu
    http://www.gustavus.edu/geology
    Programs: Geology
    Administrative Assistant: Jennifer Kruse
    Enrollment (2018): B: 13 (7)

Chair:
    Laura Triplett, (D), Minnesota, 2008, GemCl

Associate Provost and Dean of Sciences and Education:
    Julie K. Bartley, (D), California (Los Angeles), 1994, GsPgo

Assistant Professor:
    Rory McFadden, (D), GxcGz

Emeritus:
    Keith J. Carlson, (D), Chicago, 1966, Pv
    James L. Welsh, (D), Wisconsin, 1982, Gxc

Cooperating Faculty:
    Daniel Mollner, Zn

Inver Hills Community College
Geology Dept (A) (2015)
2500 East 80th Street
Inver Grove Heights, MN 55076
    jkorpik@inverhills.mnscu.edu
    http://www.inverhills.edu/Departments/Geology/

Instructor:
    Jill Bries Korpik, Gg

Itasca Community College
Geography/Geographic Info Systems (GIS) (A) (2016)
1851 East Highway 169
Grand Rapids, MN 55744
    p. (218) 322-2364
    timothy.fox@itascacc.edu

Instructor:
    Mike LeClair
    Kenneth Tapp

Lake Superior College
Liberal Arts & Sciences Dept (A) (2015)
2101 Trinity Rd.
Duluth, MN 55811
    m.whitehill@lsc.edu
    http://www.lsc.edu

Instructor:
    Matthew Whitehill, Gg

Macalester College
Geology Dept (B) ☒ (2019)

1600 Grand Avenue
St Paul, MN 55105
    p. (651) 696-6000
    rogers@macalester.edu
    http://www.macalester.edu/geology/
    Enrollment (2018): B: 29 (3)
Chair:
    Raymond R. Rogers, (D), Chicago, 1995, Gs
Professor:
    Kristina A. Curry Rogers, (D), SUNY (Stony Brook), 2001, Pv
    Kelly MacGregor, (D), California (Santa Cruz), 2002, Gml
Associate Professor:
    Karl R. Wirth, (D), Cornell, 1991, GiZn
Assistant Professor:
    Alan Chapman, (D), Caltech, 2011, Gct
Geology Lab Supervisor:
    Jeffrey T. Thole, (M), Washington State, 1991, GgCaHw

Minneapolis Community and Technical College
Div of Arts and Sciences (2015)
1501 Hennepin Ave S
Minneapolis, MN 55403
    chuck.paulson@minneapolis.edu
    http://www.minneapolis.edu/academics/artsandsciences.cfm

Minnesota State University
Chemistry and Geology (B) (2018)
Department of Chemistry and Geology
241 Ford Hall
Mankato, MN 56001
    p. (507) 389-1963
    christine.cords@mnsu.edu
    http://cset.mnsu.edu/chemgeol/programs/geol/
    Programs: Geology (B); Earth Science Education (B); Earth Science (B)
    Certificates: Environmental Geology
Professor:
    Bryce W. Hoppie, (D), California (Santa Cruz), 1996, GeHw
    Steven Losh, (D), Yale, 1985, GzxGo
Associate Professor:
    Chad Wittkop, (D), Minnesota, GsCgGl

Normandale Community College
Dept of Geography and Geology (A) (2017)
9700 France Avenue South
Bloomington, MN 55431
    p. (952) 358-8668
    lindsay.iredale@normandale.edu
    http://www.normandale.edu/departments/stem-and-education/geography-and-geology
Instructor:
    David J. Berner, (M), Colorado
    Douglas J. Claycomb, (D), Texas A&M
    Richard P. Dunning, (D), Wisconsin
    Carolyn Dykoski, (M), Minnesota
    Annia Fayon, (D), Arizona State
    Lindsay Iredale, (M), Minnesota
    Paul D. Sabourin, (D), Minnesota
    Ronald D. Ward, (D), Georgia
Other:
    Cary Komoto, (D), Minnesota, 1994, ZnnZn

North Hennepin Community College
Earth & Environmental Science Dept (A) (2018)
7411 85th Avenue N.
Brooklyn Park, MN 55445
    p. (763) 424-0869
    megan.jones@nhcc.edu
    http://www.nhcc.edu/academic-programs/academic-departments/geology
    Programs: Environmental Science (AS)
    Enrollment (2017): A: 1 (0)

Northland Community & Technical College
Liberal Arts Program (2015)
2022 Central Avenue NE
East Grand Forks, MN 56721
    p. (218) 683-8694
    http://www.northlandcollege.edu

Rochester Community & Technical College
Department of Earth Sciences (2015)
851 30th Ave SE
Rochester, MN 55904
    p. (507) 285-7220
    Cory.Rubin@roch.edu
    http://www.rctc.edu
Instructor:
    John C. Tacinelli, (D), Minnesota (Twin Cities), 2000, Gig

Saint Cloud State University
Dept of Atmospheric and Hydrologic Sciences (B) (2015)
720 4th Avenue South
St Cloud, MN 56301-4498
    p. (320) 308-3260
    arhansen@stcloudstate.edu
    http://www.stcloudstate.edu/eas/
    Administrative Assistant: Debbie Schlumpberger
    Enrollment (2012): B: 119 (31)
Professor:
    Anthony R. Hansen, (D), Iowa State, 1981, Am
    Kate S. Pound, (D), Otago (NZ), 1993, GgZeGd
    Robert A. Weisman, (D), SUNY (Albany), 1988, Am
Associate Professor:
    Juan J. Fedele, (D), Illinois, 2003, Hg
    Jean L. Hoff, (D), North Dakota, 1989, HwZeGg
    Rodney Kubesh, (D), Illinois, 1991, Am
Assistant Professor:
    Brian J. Billings, (D), Nevada (Reno), 2009, Am

St. Cloud State University
Department of Geography & Planning (2018)
St. Cloud, MN 56301-4498
    p. (320) 308-3160
    geog@stcloudstate.edu
    https://www.stcloudstate.edu/geogplan/default.aspx

University of Minnesota, Duluth
Dept of Earth & Environmental Science (B,M,D) (2018)
229 Heller Hall
1114 Kirby Drive
Duluth, MN 55812
    p. (218) 726-8385
    dees@d.umn.edu
    http://www.d.umn.edu/dees/
    Programs: Geology; Environmental Sciences
    Administrative Assistant: Laura L. Chapin
    Administrative Assistant: Claudia J. Rock
    Enrollment (2018): B: 151 (23) M: 19 (10) D: 3 (0)
Professor:
    Erik T. Brown, (D), MIT, 1990, Og
    John W. Goodge, (D), California (Los Angeles), 1987, Gp
    Vicki L. Hansen, (D), California (Los Angeles), 1987, GtXgGc
    Howard Mooers, (D), Minnesota, 1988, GleHw
Associate Professor:
    Christina D. Gallup, (D), Minnesota, 1997, Cc
    Karen B. Gran, (D), Washington, 2005, Gms
    John B. Swenson, (D), Minnesota, 2000, Gr
    Nigel J. Wattrus, (D), Minnesota, 1984, Yr
Emeritus:
    Thomas Johnson, (D), Scripps, PcGn
    James Miller, (D), Minnesota (Twin Cities), GiEd
    Penelope Morton, (D), Carlton (Canada), EgCg
    Ronald Morton, (D), Carlton (Canada), GvEg
Assistant Professor:
    Latisha A. Brengman, (D), Tennessee, 2015, Gd
    Fred A. Davis, (D), Minnesota (Twin Cities), 2012, CpGi
    Salli F. Dymond, (D), Minnesota, 2014, Ge
    Christian Schardt, (D), Tasmania, EgCg
    Byron A. Steinman, (D), Pittsburg, 2011, Gn

Emeritus:
 John C. Green, (D), Harvard, 1960, GieGv
 Timothy Holst, (D), Minnesota (Twin Cities), GcYn

University of Minnesota, Morris
Div of Science & Mathematics (B) (2015)
Geology Discipline
Science Building
600 East 4th Street
Morris, MN 56267
 p. (320) 589-6300
 geol@mrs.umn.edu
 http://www.mrs.umn.edu
Coordinator:
 James F. Cotter, (D), Lehigh, 1984, Gl
Professor:
 Keith A. Brugger, (D), Minnesota, 1992, GllGm
 James B. Van Alstine, (D), North Dakota, 1980, Pg
Emeritus:
 Peter M. Whelan, (D), California (Santa Cruz), 1988, Gx

University of Minnesota, Twin Cities
Dept of Civil, Environmental and Geo-Engineering (B,M,D) (2015)
500 Pillsbury Drive SE
Minneapolis, MN 55455
 p. (612) 625-5522
 cege@umn.edu
 http://www.ce.umn.edu/
 Enrollment (2010): B: 26 (8) M: 11 (4) D: 49 (2)
Head:
 John S. Gulliver, (D), Minnesota, 1980, Hs
Professor:
 Emmanuel M. Detournay, (D), Minnesota, 1983, NrpNm
 Andrew Drescher, (D), Inst of Fund Tech Res (Poland), 1968, So
 Efi Foufoula-Georgiou, (D), Florida, 1985, Hg
 Joseph F. Labuz, (D), Northwestern, 1985, Nr
 Otto D. Strack, (D), Delft (Neth), 1973, Hw
 Vaughan R. Voller, (D), Sunderland, 1980, Zm
Associate Professor:
 Randal J. Barnes, (D), Colorado Mines, 1985, Zn
 Bojan B. Guzina, (D), Colorado, 1996, Nr
 Karl A. Smith, (D), Minnesota, 1980, Nx
Research Associate:
 Sonia Mogilevskaya, (D), Russian Acad of Sci, 1987, Ng
Adjunct Professor:
 Peter A. Cundall, (D), Nr

Dept of Earth Sciences (B,M,D) O☒ (2018)
116 Church Street SE
150 John T. Tate Hall
Minneapolis, MN 55455
 p. (612) 624-1333
 esci@umn.edu
 http://www.esci.umn.edu
 t: @UMNEarthScience
 Programs: Earth Sciences; Geology; Geophysics; Geobiology; Environmental Sciences; Hydrogeology
 Enrollment (2016): B: 48 (15) M: 15 (2) D: 41 (7)
Head:
 Donna L. Whitney, (D), Washington, 1991, Gpt
Teaching Professor:
 Kent C. Kirkby, (D), Wisconsin, 1994, Gg
Regents Professor:
 R. Lawrence Edwards, (D), Caltech, 1988, Cc
Director, Minnesota Geological Survey:
 Harvey Thorleifson, (D), Colorado, 1989, Gg
Director of IRM:
 Bruce M. Moskowitz, (D), Minnesota, 1980, Ym
Professor:
 Hai Cheng, (D), Nanjing (China), 1988, CcPe
 David L. Fox, (D), Michigan, 1999, Pg
 Marc M. Hirschmann, (D), Washington, 1992, Gi
 Peter J. Hudleston, (D), Imperial Coll (UK), 1969, Gc
 Emi Ito, (D), Chicago, 1979, ClGnCs
 David L. Kohlstedt, (D), Illinois, 1970, YxGzi
 Sally G. Kohlstedt, (D), Illinois, 1972, Zn
 Katsumi Matsumoto, (D), Columbia, 2000, Oc
 Samuel Mukasa, (D), Cg
 Christopher Paola, (D), MIT/WHOI, 1983, Gs
 Justin Revenaugh, (D), MIT, 1989, Ys
 William E. Seyfried, Jr., (D), S California, 1977, Cm
 Christian P. Teyssier, (D), Monash, 1986, GctGg
 David A. Yuen, (D), California (Los Angeles), 1978, Yg
Assoc Director of IRM:
 Joshua Feinberg, (D), California (Berkeley), 2005, YmGze
Associate Professor:
 Jake Bailey, (D), S California, 2008, Pg
Assistant Professor:
 Max Bezada, (D), Rice, 2010, Ys
 Jeff Havig, (D), GeCg
 Peter Kang, (D), Hg
 Crystal Ng, (D), MIT, 2008, Hy
 Cara M. Santelli, (D), MIT/WHOI, 2007, Pg
 Ikuko Wada, (D), Victoria, 2009, ZnYgGt
 Andrew D. Wickert, (D), Colorado (Boulder), 2014, Gml
XRCT Lab Manager:
 Brian Bagley, (D), Minnesota, 2011, YsZn
Research Fellow:
 Brad Herried, (M), Minnesota, 2010, Zi
Microprobe Manager:
 Anette von der Handt, (D), GiCt
Research Associate:
 Dario Bilardello, (D), Lehigh, 2009, Ym
 Randy Calcote, (D), Minnesota, 2000, Ple
 Kang Ding, (D), Acad Sinica, 1987, Cg
 Beverley Flood, (D), Pg
 Stefan Liess, (D), Hg
 Jed Mosenfelder, (D), Cp
 Amy Myrbo, (D), Minnesota, 2006, PeZe
 Mark Shapley, Minnesota, 2005, Gn
 Ivanka Stefanova, (D), Sofia, 1991, Pl
 Alexander Stone, (M), Gn
 Chunyang Tan, (D), Zhejiang, 2011, Cm
 Mark Zimmerman, (D), Minnesota, 1999, Yx
Adjunct Professor:
 James E. Almendinger, (D), Minnesota, 1988, Pe
 Val W. Chandler, (D), Purdue, 1977, Ye
 Mark B. Edlund, (D), Michigan, 1998, Pe
 Daniel R. Engstrom, (D), Minnesota, 1983, Pe
 Annia K. Fayon, (D), Arizona State, 1997, Gc
 Robert G. Johnson, (D), Iowa State, 1952, Pe
 Shenghua Mei, (D), Yx
 Kristina Curry Rogers, (D), Pg
 Raymond Rogers, (D), Pg
 Anthony Runkel, (D), Texas, 1988, Gs
Emeritus:
 E. Calvin Alexander, Jr., (D), Missouri (Rolla), 1970, HwCcl
 Subir K. Banerjee, (D), Cambridge, 1963, Ym
 Roger LeB Hooke, (D), Gl
 Karen L. Kleinspehn, (D), Princeton, 1982, Gt
 Hans Olaf Pfannkuch, (D), Paris, 1962, HwGeRh
 James H. Stout, (D), Harvard, 1970, GpzGt
 Paul W. Weiblen, (D), Minnesota, 1965, Gi
Geochemical Analyst:
 Elizabeth Lundstrom, (B), Carleton Coll, Ca
Student Services:
 Jennifer Petrie
Researcher:
 Michelle LaRue, (D), Zi
Remote Sensing Specialist:
 Cathleen Torres Parisian, (M), Zr
Postdoc:
 Andrea Biedermann, (D), Ym
 Joseph Byrnes, (D), Oregon, 2016, Ys
 Carla Rosenfeld, (D), Pg
 Michael Volk, (D), Ym
Director, CSDCO:
 Anders Noren, Gn
Dept Administrator:
 Sharon J. Kressler, (B), Zn
Department Safety Officer:
 Scott Alexander, (B), Hw

Curator, LacCore:
   Kristina Brady, (M), Minnesota, 2006, Gn
Agouron Institute Fellow:
   Dan Jones, (D), Pg
Research Fellow:
   Thomas Juntunen, (M), Minnesota, 2011, Zir
Geology Librarian:
   Carolyn Bishoff, (M), Zn
Cooperating Faculty:
   Martin O. Saar, (D), California (Berkeley), 2003, Hg

Dept of Soil, Water, & Climate (B,M,D) ●⊘ (2018)
1991 Upper Buford Circle
St. Paul, MN 55108-6028
   p. (612) 625-8114
   crosen@umn.edu
   http://www.swac.umn.edu
   Programs: Environmental Science, Policy and Management (BS); Land and Atmospheric Science (MS, PhD)
   Administrative Assistant: Marjorie J. Bonse
   Enrollment (2018): B: 334 (67) M: 12 (3) D: 20 (5)
Head:
   Carl J. Rosen, (D), California (Davis), 1983, Sc
Professor:
   James C. Bell, (D), Penn State, 1990, SdfZe
   Timothy J. Griffis, (D), McMaster, 2000, As
   Satish C. Gupta, (D), Utah State, 1972, Sp
   Dylan B. Millet, (D), California (Berkeley), 2003, As
   David J. Mulla, (D), Purdue, 1983, Sp
   Edward A. Nater, (D), California (Davis), 1987, Sd
   Michael J. Sadowsky, (D), Hawaii, 1983, Sb
   Michael A. Schmitt, (D), Illinois, 1985, Sc
   Jeffrey S. Strock, (D), North Carolina State, 1999, So
Associate Professor:
   Melinda L. Erickson, Minnesota, 2005, Hw
   Daniel E. Kaiser, (D), Iowa State, 2007, Sc
   Albert L. Sims, (D), North Carolina State, 1992, Sc
   Peter K. Snyder, (D), Wisconsin, 2003, As
   Brandy M. Toner, (D), California (Berkeley), 2004, ClmSc
   Tracy E. Twine, (D), Wisconsin, 2003, As
   Kyungsoo Yoo, (D), California (Berkeley), 2003, Sd
Assistant Professor:
   Fabian G. Fernandez, (D), Purdue, 2006, ScbSo
   Jessica L. M. Gutknecht, (D), Wisconsin, 2007, Sb
   Satoshi Ishii, (D), Minnesota, 2007, SbPg
   Nicolas Jelinski, (D), Minnesota, 2014, SdCs
   Yuxin Miao, (D), Minnesota, 2005, ScZf
   Paulo H. Pagliari, (D), Wisconsin, 2012, Sc
   Vasudha Sharma, (D), Nebraska, 2018, SpHg
   Melissa L. Wilson, (D), Minnesota, 2012, Sco
Adjunct Professor:
   John M. Baker, (D), Texas A&M, 1987, Sp
   Gary Feyereisen, (D), Minnesota, 2005, Sp
   Jane M F Johnson, (D), Minnesota, 1995, So
   Randall K. Kolka, (D), Minnesota, 1996, So
   Pamela J. Rice, (D), Iowa State, 1996, So
   Kurt Spokas, (D), Minnesota, 2005, Spo
   Rodney T. Venterea, (D), California (Davis), 2000, Sp
Emeritus:
   Deborah L. Allan, (D), California (Riverside), 1987, Sb
   James L. Anderson, (D), Wisconsin, 1976, Sd
   Paul R. Bloom, (D), Cornell, 1978, Sc
   H. H. Cheng, (D), Illinois, 1961, Sb
   Terence H. Cooper, (D), Michigan State, 1975, Sd
   Robert H. Dowdy, (D), Michigan State, 1966, Sc
   William C. Koskinen, (D), Washington State, 1980, Sc
   John A. Lamb, (D), Nebraska, 1984, Sc
   Gary L. Malzer, (D), Purdue, 1973, Sc
   Jean-Alex E. Molina, (D), Cornell, 1967, SbCgSo
   John F. Moncrief, (D), Wisconsin, 1981, Sp
   Gyles W. Randall, (D), Wisconsin, 1972, So
   George W. Rehm, (D), Minnesota, 1969, Sc
   Michael P. Russelle, (D), Nebraska, 1982, Sc
   Mark W. Seeley, (D), Nebraska, 1977, As
Executive Secretary:
   Kari A. Jarcho, (B), Zn

University of Saint Thomas
Dept of Geology (B) (2015)
OWS 153
2115 Summit Avenue
St Paul, MN 55105
   p. (651) 962-5241
   kmtheissen@stthomas.edu
   http://www.stthomas.edu/geology/
   Enrollment (2010): B: 16 (0)
Professor:
   Melissa A. Lamb, (D), Stanford, 1998, GtcGs
Associate Professor:
   Thomas A. Hickson, (D), Stanford, 1999, GsmGe
   Kevin Theissen, (D), Stanford, GnOg
Assistant Professor:
   Jennifer McGuire, (D), Michigan State, CgHsGb
Laboratory Director:
   Erik Smith, (B), Zn

Vermilion Community College
Vermilion Community College (2015)
1900 East Camp Street
Ely, MN 55731
   p. (218) 235-2173
   http://www.vcc.edu/

Winona State University
Dept of Geoscience (B) ⊠ (2018)
P.O. Box 5838
Winona, MN 55987
   p. (507) 457-5260
   geoscience@winona.edu
   https://www.winona.edu/geoscience/
   Programs: Geoscience (B); Geology (B); Environmental Science (B); Earth Science Teaching (B)
   Enrollment (2018): B: 68 (15)
Professor:
   Stephen T. Allard, (D), Wyoming, 2003, GctGx
   Jennifer LB Anderson, (D), Brown, 2004, XgYgZe
   Candace L. Kairies-Beatty, (D), Pittsburgh, 2003, GeClRw
Associate Professor:
   William L. Beatty, (D), Pittsburgh, 2003, PgiGs
Assistant Professor:
   Dylan Blumentritt, (D), Minnesota, 2013, HsGm
Emeritus:
   John F. Donovan, (D), Cornell, 1963, Eg
   Nancy O. Jannik, (D), New Mexico Tech, 1989, Hw
   Jamie Ann Meyers, (D), Indiana, 1971, Gsd
   Dennis N. Nielsen, (D), North Dakota, 1973, Gm
College Laboratory Services Specialist:
   Luke Zwiefelhofer, (B), Wyoming, 2004, Zn

Mississippi

Jackson State University
Chemistry, Physics, and Atmospheric Sciences (B,M,D) ⊠ (2017)
1400 J. R. Lynch St.
JSU Box 17660
Jackson, MS 39217
   p. (601) 979-7012
   mfadavi@jsums.edu
   http://www.jsums.edu/cset/phyat.htm
   Programs: Physics (B); Chemistry (B); Meteorology (B); Earth System Science (B); Chemistry (M,D)
   Enrollment (2014): B: 9 (4)

Millsaps College
Dept of Geosciences (B) O⊠ (2018)
Box 150648
1701 North State Street
Jackson, MS 39210
   p. (601) 974-1344
   musseza@millsaps.edu
   http://www.millsaps.edu/geology/
   Enrollment (2018): B: 8 (5)

Professor:
: Stan Galicki, (D), Mississippi, 2002, Ged
: James B. Harris, (D), Kentucky, 1992, Ye
Chair:
: Zachary A. Musselman, (D), Kentucky, 2006, Gm
Emeritus:
: Delbert E. Gann, (D), Missouri Sch of Mines, 1976, Gz

Mississippi State University
Dept of Geosciences (B,M,D) (2015)
P. O. Box 5448
108 Hilbun Hall
East Lee Blvd.
Mississippi State, MS 39762
: p. (662) 325-3915
: rclary@geosci.msstate.edu
: http://www.geosciences.msstate.edu/
: f: https://www.facebook.com/MSUGeosciences/
: *Enrollment (2014): B: 505 (86) M: 224 (103) D: 24 (4)*

Professor:
: Michael E. Brown, (D), North Carolina, 1999, Am
: Brenda L. Kirkland, (D), Louisiana State, 1992, GdEoPo
: Darrel W. Schmitz, (D), Texas A&M, 1991, Hw
Associate Professor:
: Shrinidhi Ambinakudige, (D), Florida State, 2006, ZiyZr
: Renee M. Clary, (D), Louisiana State, 2003, ZePgGe
: Jamie Dyer, (D), Georgia, 2005, Am
: Andrew Mercer, (D), Oklahoma, 2008, Ams
: John C. Rodgers, (D), Georgia, 1999, Zy
: Kathleen M. Sherman-Morris, (D), Florida State, 2006, As
Assistant Professor:
: Padmanava Dash, (D), Louisiana State, 2011, ZrHg
: Christopher Fuhrmann, (D), 2011, As
: Rinat Gabitov, (D), Rensselaer, 2005, Cas
: Qingmin Meng, (D), Georgia, 2006, Zig
: Adam Skarke, (D), Delaware, 2013, GuYrOn
: Kim Wood, (D), Arizona, 2012, AsmZr
Research Associate:
: Katarzynz Grala, (M), Iowa State, 2004, Zi
Instructor:
: Christa M. Haney, (M), Mississippi State, 1999, Am
: Amy P. Moe-Hoffman, (M), Colorado, 2002, Zn
: John A. Morris, (M), Mississippi State, 2007, ZirAm
: Lindsey Morschauser, (M), Mississippi State, Am
: Athena Nagel, (D), Mississippi State, 2014, ZgiZr
: Greg Nordstrom, (M), Mississippi State, 2007, Am
: Tim Wallace, (M), Mississippi State, 1994, Am
Adjunct Professor:
: Paul J. Croft, (D), Rutgers, 1991, Am
: Patrick J. Fitzpatrick, (D), Colorado State, 1995, Am
: James May, (D), Texas A&M, 1988, Hw
: Jack C. Pashin, (D), Kentucky, 1990, GoEcGs
: Janet E. Simms, (D), Texas A&M, 1991, Yg
: Jayaram Veeramony, (D), Delaware, 1999, Og
Emeritus:
: John M. Kaye, (D), Louisiana State, 1974, Gg
: John E. Mylroie, (D), Rensselaer, 1977, GmHyGs
: Charles L. Wax, (D), Louisiana State, 1977, As
Distance Academic Coordinator:
: Mary A. Dean
Collection Manager:
: Amy P. Moe-Hoffman, (M), Colorado, 2002, PgZe
Business Manager:
: Jerri Wright, (B), Mississippi Univ for Women, 1996
Academic Coordinator:
: Tina Davis, (M), Mississippi Univ for Women
Related Staff:
: Cynthia Bell

University of Mississippi
Dept of Geology & Geological Engineering (B,M,D) ● (2017)
School of Engineering
P.O. Box 1848
120A Carrier Hall
University, MS 38677-1848
: p. (662) 915-7498
: geology@olemiss.edu
: http://www.engineering.olemiss.edu/gge/
: *Enrollment (2016): B: 301 (45) M: 21 (6) D: 5 (0)*

Professor:
: Adnan Aydin, (D), Memorial, 1994, NgrYe
: Gregg R. Davidson, (D), Arizona, 1995, Hw
: Gregory L. Easson, (D), Missouri (Rolla), 1996, Zr
: Robert M. Holt, (D), New Mexico Tech, 2000, Hq
Associate Professor:
: Louis G. Zachos, (D), Texas, 2008, PiGeNx
Assistant Professor:
: Jennifer N. Gifford, (D), Florida, 2013, GtxCg
: Andrew M. O'Reilly, (D), Florida, 2012, Hwq
: Brian F. Platt, (D), Kansas, 2009, GsPe
: Lance D. Yarbrough, (D), Mississippi, 2006, Ng
Instructor:
: Inoka Widanagamage, (D), Kent State, 2015, ClGep
Lecturer:
: Cathy A. Grace, (M), Mississippi, 1996, Gg

University of Southern Mississippi
Dept of Geography and Geology (B,M,D) O (2018)
118 College Drive, Box 5051
Walker Science Building, Room 127
Hattiesburg, MS 39406
: p. (601) 266-4729
: david.cochran@usm.edu
: http://www.usm.edu/geography-geology
: f: https://www.facebook.com/groups/Geography.Geology-USM/
: t: @geousm
: Programs: Geology (B, M); Geography (B, M, D)
: Certificates: Geographic Information Technologies
: *Enrollment (2014): B: 114 (19) M: 30 (5) D: 4 (1)*

Chair:
: Andy Reese, (D), Louisiana State, 2003, Zy
Professor:
: Greg Carter, (D), Wyoming, 1985, Zyi
: Clifton Dixon, (D), Texas A&M, 1988, Zn
: Maurice A. Meylan, (D), Hawaii, 1978, Gu
: Mark Miller, (D), Arizona, 1988, Zn
: David M. Patrick, (D), Oklahoma, 1972, Ng
: Mark Puckett, (D), Alabama, PmiGg
Associate Professor:
: Jerry Bass, (D), Texas, 2003, Zn
: David Cochran, (D), Kansas, 2005, Zn
: Franklin Heitmuller, (D), Texas, 2009, GmsGe
: David Holt, (D), Arkansas, 2002, Zyi
: Bandana Kar, (D), South Carolina, 2008, Zi
: George Raber, (D), South Carolina, 2003, Zi
Assistant Professor:
: Jeremy Deans, (D), Texas Tech, 2016, YrGc
Instructor:
: Evan Bagley, (M), GgEo
: Alyson Brink, (D), Texas Tech, 2016, PgvGg

Gulf Coast Research Laboratory (2015)
Department of Coastal Sciences
Geology Section
Ocean Springs, MS 39566-7000
: p. (601) 872-4200
: joe.griffitt@usm.edu
: http://www.coms.usm.edu
: Administrative Assistant: Angelia Bone

Emeritus:
: Ervin G. Otvos, (D), Massachusetts, 1964, OnGsm

Marine Science (B,M,D) ⊠ (2018)
1020 Balch Boulevard
National Aeronautics and Space Administration
Stennis Space Center, MS 39529
: p. (228) 688-3177
: marine.science@usm.edu
: http://www.usm.edu/marine
: f: https://www.facebook.com/SouthernMissDivisionOfMarineScience/

t: @USMMarineSci
Programs: Marine Science; Hydrographic Science; Ocean Engineering
Certificates: UMS (Unmanned Maritime Systems Tier 1 Certification Program)
Enrollment (2018): B: 33 (4) M: 32 (12) D: 13 (2)
School of Ocean Science and Engineering:
  William (Monty) Graham, (D), California (Santa Cruz), 1994, Ob
Hydrographic Science Research Center:
  CAPT (Ret) Brian D. Connon, (M), S Mississippi, 2000, ZnnOg
Associate Director of the School of Ocean Science and Engineering:
  Jerry D. Wiggert, (D), S California, 1995, Op
Director of the Center for Gulf Studies:
  Denis A. Wiesenburg, (D), 1980, Og
Professor:
  Donald G. Redalje, (D), Hawaii, 1980, Ob
  Alan M. Shiller, (D), California (San Diego), 1982, OcCtHs
Associate Professor:
  Stephan Howden, (D), Rhode Island, 1996, Op
  Scott P. Milroy, (D), S Florida (St. Petersburg), 2007, Ob
  Dmitri Nechaev, (D), Shirshov Inst, 1986, Op
Assistant Professor:
  Maarten Buijsman, (D), Utrecht, 2007, Op
  Mustafa Kemal Cambazoglu, (D), Georgia Tech, 2009, No
  Christopher T. Hayes, (D), 2013, Oc
  Anand Devappa Hiroji, (D), 2016, Yd
  Gero Nootz, (D), C Florida, 2010, No
  Davin Wallace, (D), Rice, 2010, Ou
Research Professor:
  Robert (Bob) Arnone, (M), Georgia Tech, 1974, Op
Program Coordinator of the Hydrographic Science Degree Programs:
  Maxim F. van Norden, (M), Naval Postgrad Sch, 1979, Zn
Instructor:
  Danielle Greenhow, (D), S Florida (St. Petersburg), 2013, Ob

Missouri

Metropolitan Community College-Blue River
Geology and Geography Program (A) (2015)
20301 E. 78 Highway
Independence, MO 64057-2053
  p. (816) 220-6622
  benjamin.wolfe@mcckc.edu
  http://www.mcckc.edu/progs/geol/geology/overview.asp

Metropolitan Community College-Kansas City
Geology Dept (A) (2015)
3200 Broadway
Kansas City, MO 64111
  p. (816) 604-3335
  melissa.renfrow@mcckc.edu
  http://www.mcckc.edu/programs/geology/
Instructor:
  Alice Fuerst, Gg
  John Horn, (D), Nebraska, GgZy
  Carl Priesendorf, (M), Missouri, 1987, GgZyGe
  Laura Veverka, (M), Missouri (Kansas City), Zy
  Ben Wolfe, (M), Alaska (Fairbanks), 2001, GgZyg
Adjunct Professor:
  Janet Raymer, (M), Missouri S&T, GgEo

Mineral Area College
Science Dept (A) (2019)
5270 Flat River Road
P.O. Box 1000
Park Hills, MO 63601
  p. (573) 518-2314
  bscheidt@MineralArea.edu
  http://www.mineralarea.edu/faculty/academicDepartments/science.aspx
  Enrollment (2016): A: 2 (2)
Assistant Professor:
  Brian Scheidt, (M), S Illinois, Hw

Missouri State University
Dept of Geography, Geology & Planning (B,M) (2018)
901 S. National
Springfield, MO 65897
  p. (417) 836-5800
  ggp@missouristate.edu
  https://geosciences.missouristate.edu
  f: https://www.facebook.com/MSUggp/
  t: @MSUggp
  Programs: https://geosciences.missouristate.edu
  Certificates: https://geosciences.missouristate.edu
  Administrative Assistant: Tracy Carroll
  Administrative Assistant: Deana Gibson
  Enrollment (2016): B: 185 (56) M: 38 (8)
Department Head:
  Toby Dogwiler, (D), Missouri, 2002, ZiGmZy
Distinguished Professor:
  Kevin L. Mickus, (D), Texas (El Paso), 1989, GtYv
  Robert T. Pavlowsky, (D), Wisconsin, 1995, Gm
Professor:
  Kevin R. Evans, (D), Kansas, 1997, GrsGt
  Douglas R. Gouzie, (D), Kentucky, 1986, HwGeCl
  Melida Gutierrez, (D), Texas (El Paso), 1992, CgHy
  Rajinder S. Jutla, (D), Virginia Tech, 1995, Zn
  Jun Luo, (D), Wisconsin (Milwaukee), Zi
  Judith Meyer, (D), Wisconsin, 1994, Zn
  Xin Miao, (D), California (Berkeley), 2005, Zri
  Charles W. Rovey, (D), Wisconsin (Milwaukee), 1990, HyGle
Associate Professor:
  Alice (Jill) Black, (D), Missouri, 2003, ZeAsHw
Assistant Professor:
  Mario Daoust, (D), McGill, 1992, AtZy
  Krista M. Evans, (D), Clemson, 2017, Zn
  Ron Malega, (D), Georgia, Zn
  Matthew P. McKay, (D), West Virginia, 2015, Gct
  Gary Michelfelder, (D), Montana State, 2015, GviCg
  David R. Perkins, (D)
Senior Instructor:
  Damon Bassett, (M), Missouri, 2003, Gg
  Deborah Corcoran, (M), Michigan State, 1980, Zn
  Linnea Iantria, (M), George Washington, Zn
Instructor:
  Melanie E. Carden-Jessen, (M), Missouri State, 1998, Ze
Emeritus:
  David A. Castillon, (D), Michigan State, 1972, Gm
  John C. Catau, (D), Michigan State, 1973, Zn
  William H. Cheek, (D), Michigan State, 1976, Zn
  William Corcoran, (D), Michigan State, 1981, As
  Dimitri Ioannides, (D), Rutgers, 1994, Zn
  Elias Johnson, (D), Oklahoma, 1977, Zr
  Vincent E. Kurtz, (D), Oklahoma, 1960, Ps
  Erwin J. Mantei, (D), Missouri (Rolla), 1965, EdGez
  Diane M. May, (M), S Illinois (Edwardsville), 1974, Zn
  James F. Miller, (D), Wisconsin, 1970, Ps
  Thomas D. Moeglin, (D), Nebraska, 1978, Ng
  Thomas G. Plymate, (D), Minnesota, 1986, Gx
  Paul A. Rollinson, (D), Illinois, 1988, Zn

Missouri University of Science and Technology
Dept of Geology & Geophysics (B,M,D) (2015)
125 McNutt Hall
Rolla, MO 65409-0410
  p. (573) 341-4616
  rocks@mst.edu
  Administrative Assistant: Katherine W. Mattison
Lecturer:
  Patrick S. Mulvany, (D), Missouri (Rolla), 1996, Gg
Adjunct Professor:
  John F. Burst, (D), Missouri, 1950, Eo
  Waldemar M. Dressel, (B), Missouri (Rolla), 1943, Gg
  Charles E. Robertson, (M), Maryland, 1960, Gc
  James E. Vandike, (M), SD Mines, 1979, Hg
  James H. Williams, (D), Missouri (Rolla), 1975, Gg
Emeritus:
  Shelton K. Grant, (D), Utah, 1966, Gz
  Richard D. Hagni, (D), Missouri, 1962, Em

Geza K. Kisvarsanyi, (D), Missouri, 1966, Em
Richard D. Rechtien, (D), Washington, 1964, Ye
Gerald B. Rupert, (D), Missouri, 1964, Ye
Alfred C. Spreng, (D), Wisconsin, 1950, Gr

Moberly Area Comm College - Columbia Campus
Moberly Area Community College - Columbia Campus (2015)
601 Business Loop 70 West
Columbia, MO 65203
p. (573) 234-1067
SandyAnderson@macc.edu
http://www.macc.edu/

Northwest Missouri State University
Dept of Geology-Geography (B,M) (2015)
800 University Drive
Maryville, MO 64468
p. (660) 562-1723
geosci@nwmissouri.edu
Chair:
  C. Renee Rohs, (D), Kansas, 2000, Cg
Professor:
  Gregory D. Haddock, (D), Idaho, 1996, Zi
Associate Professor:
  Theodore L. Goudge, (D), Oklahoma State, 1984, Zy
Assistant Professor:
  James Hickey, (D), Dartmouth, 2006, Ge
  Ming-Chih Hung, (D), Utah, 2003, Zr
  Yanfen Le, (D), Georgia, 2005, Zi
  Leah D. Manos, (M), Tennessee, 1997, Zy
Instructor:
  Jeffrey Bradley, (M), Oklahoma State, 1991, Zg
Emeritus:
  Richard M. Felton, (M), Missouri, 1979, Pg

Ozarks Technical Community College
Physical Sciences (2015)
1001 E. Chestnut Expressway
Springfield, MO 65802
p. (417) 447-8238
ehrichp@otc.edu

Saint Louis University
Earth & Atmospheric Sciences (B,M,D) ☒ (2018)
3642 Lindell Blvd
O'Neil Hall
Room 205
St. Louis, MO 63108
p. (314) 977-3116
robert.herrmann@slu.edu
http://www.slu.edu/department-of-earth-and-atmospheric-sciences-home
Programs: Geoscience with Concentrations in Geology, Geophysics and Environmental Science
Enrollment (2018): B: 69 (10) M: 7 (4) D: 5 (0)
Chair:
  Benjamin de Foy, (D), Cambridge (UK), 1998, As
Professor:
  David J. Crossley, (D), British Columbia, 1973, YdsXy
  Jack Fishman, (D), As
  Daniel M. Hanes, (D), Scripps, 1983, GesGm
  Robert B. Herrmann, (D), St. Louis, 1974, Ys
  Zaitao Pan, (D), Iowa State, 1996, As
  Lupei Zhu, (D), Caltech, 1998, YsGt
Associate Professor:
  John Encarnacion, (D), Michigan, 1994, Gx
  Charles E. Graves, (D), Iowa State, 1988, As
  Robert W. Pasken, (D), St. Louis, 1981, Am
  Linda M. Warren, (D), California (San Diego), 2003, Ygs
Assistant Professor:
  Elizabeth Hasenmueller, (D), Washington Univ (St. Louis), 2011, ClHs
  Wasit Wulamu, (D), Zi

University of Missouri
Dept of Geological Sciences (B,M,D) O☒ (2017)
101 Geology Building
Columbia, MO 65211
p. (573) 882-6785
HuckabeyM@missouri.edu
http://geology.missouri.edu/
Administrative Assistant: Marsha Huckabey
Enrollment (2013): B: 60 (7) M: 13 (11) D: 21 (0)
Chair:
  Kevin L. Shelton, (D), Yale, 1982, EmCs
  Alan Whittington, (D), Open Univ (UK), 1997, GivCp
Director of Geology Field Studies:
  Miriam Barquero-Molina, (D), Texas, 2009, Gct
Professor:
  Mian Liu, (D), Arizona, 1989, Yg
  Kenneth A. MacLeod, (D), Washington, 1992, Pg
  Peter I. Nabelek, (D), SUNY (Stony Brook), 1983, GipCu
  Eric A. Sandvol, (D), New Mexico State, 1995, Ys
Associate Professor:
  Martin S. Appold, (D), Johns Hopkins, 1998, HwEm
  Francisco Gomez, (D), Cornell, 1999, Gt
Assistant Professor:
  John Huntley, (D), Virginia Tech, 2007, PqgPe
  James Schiffbauer, (D), Virginia Tech, 2009, Pgi
Emeritus:
  Michael B. Underwood, (D), Cornell, 1984, GsuGt

University of Missouri, Columbia
School of Natural Resources (B,M,D) (2017)
302 Anheuser-Busch Natural Resources Building
Columbia, MO 65211-7250
p. (573) 882-6446
joses@missouri.edu
http://www.snr.missouri.edu/
Enrollment (2012): B: 165 (15) M: 24 (3) D: 13 (0)
Professor:
  Stephen H. Anderson, (D), North Carolina State, 1985, Sp
  Keith W. Goyne, (D), Penn State, 2003, Sc
  Anthony R. Lupo, (D), Purdue, 1995, As
  Patrick S. Market, (D), St. Louis, 1999, As
  Peter P. Motavalli, (D), Cornell, 1989, So
Extension:
  Patrick E. Guinan, (D), Missouri, 2004, As
Associate Professor:
  Neil I. Fox, (D), Salford (UK), 1998, As
Instructor:
  Eric A. Aldrich, (M), Missouri, 2011, As
Adjunct Professor:
  Frieda Eivazi, (D), Iowa State, 1980, Sc
  Newell R. Kitchen, (D), Colorado State, 1990, So
  Robert J. Kremer, (D), Mississippi State, 1981, Sb
  Robert N. Lerch, (D), Colorado State, 1990, So
  W. Gene Stevens, (D), Mississippi State, 1992, So
  Christopher K. Wikle, (D), Iowa State, As
Emeritus:
  Robert W. Blanchar, (D), Minnesota, 1964, Sc
  James R. Brown, (D), Iowa State, 1963, So
  Clark J. Gantzer, (D), Minnesota, 1980, So
  Ernest C. Kung, (D), Wisconsin, 1963, As
  Randall J. Miles, (D), Texas A&M, 1981, Sd
  Stephen E. Mudrick, (D), MIT, 1973, As

University of Missouri, Kansas City
Dept of Geosciences (B,M,D) ●◯ (2019)
5100 Rockhill Road
Room 420, Robert H. Flarsheim Hall
Kansas City, MO 64110-2499
p. (816) 235-1334
geosciences@umkc.edu
http://cas.umkc.edu/Geosciences/default.asp
t: @UMKC_geosci
Programs: Geology (B); Geography (B); Environmental Science (B)
Certificates: GIS
Enrollment (2018): B: 108 (56) M: 0 (6) D: 26 (2)
Chair:
  Wei Ji, (D), Connecticut, 1991, ZriZy

Professor:
   Jimmy Adegoke, (D), Penn State, 2000, AsZyr
   James B. Murowchick, (D), Penn State, 1984, CgGz
   Tina M. Niemi, (D), Stanford, 1992, Gt
Associate Professor:
   Caroline P. Davies, (D), Arizona State, 2000, GnZce
   Jejung Lee, (D), Northwestern, 2001, GqHw
Assistant Professor:
   Alison Graettinger, (D), Pittsburg, 2012, Gv
   Fengpeng Sun, (D), California (Irvine), 2008, ZoiZc
Emeritus:
   Raymond M. Coveney Jr, (D), Michigan, 1972, EgmCe
   Steven L. Driever, (D), Georgia, 1977, ZnnZn
   Richard J. Gentile, (D), Missouri, 1965, GrsNx
   Syed E. Hasan, (D), Purdue, 1978, Ng

Washington University in St. Louis
Dept of Earth & Planetary Sciences (B,M,D) ☒ (2018)
Campus Box 1169
Rudolph Hall
1 Brookings Drive
St Louis, MO 63130-4899
   p. (314) 935-5610
   slava@wustl.edu
   http://eps.wustl.edu
   f: https://www.facebook.com/WashU.EPSci/
   t: @WUSTL_EPS
   Programs: Geology; Geochemistry; Geophysics; Geobiology
   Administrative Assistant: Katherine M.. Totty
   Enrollment (2017): B: 56 (15) M: 0 (3) D: 31 (5)
Chair:
   Viatcheslav S. Solomatov, (D), Moscow Inst of Physics & Tech, 1990, Yg
Rudolph Professor of Earth & Planeary Sciences:
   Bradley L. Jolliff, (D), SD Mines, 1987, GxzGi
Robert S. Brookings Distinguished Professor:
   Douglas A. Wiens, (D), Northwestern, 1985, YsrYg
James S. McDonnell Distinguished University Professor:
   Raymond E. Arvidson, (D), Brown, 1974, Xg
Professor:
   Jeffrey G. Catalano, (D), Stanford, 2004, Cl
   Robert F. Dymek, (D), Caltech, 1977, GpiGa
   M. Bruce Fegley, (D), MIT, 1980, Xc
   David A. Fike, (D), MIT, 2007, CsmPg
   William B. McKinnon, (D), Caltech, 1981, Xg
   Jill D. Pasteris, (D), Yale, 1980, GbzGe
   Jennifer R. Smith, (D), Pennsylvania, 2001, Ga
   William H. Smith, (D), Princeton, 1966, Xc
   Michael E. Wysession, (D), Northwestern, 1991, Ys
Research Professor:
   Katharina Lodders-Fegley, (D), Max Planck, 1991, Xc
   Alian Wang, (D), Sci/Tech (France), 1987, Ca
Senior Scientist:
   Randy L. Korotev, (D), Wisconsin, 1976, XgCtXm
Associate Professor:
   Alexander S. Bradley, (D), MIT, 2008, CboGg
   Philip Skemer, (D), Yale, 2007, GcCp
Assistant Professor:
   Bronwen L. Konecky, (D), Brown, 2014, PcGn
   Michael J. Krawczynski, (D), MIT, 2011, Cg
   Claire C. Masteller, (D), California (Santa Cruz), 2017, Hg
   Rita Parai, (D), Harvard, 2014, Cs
   Kun Wang, (D), Washington Univ (St. Louis), 2013, XcCsXm
Emeritus:
   Robert E. Criss, (D), Caltech, 1981, Cs
   Harold L. Levin, (D), Washington Univ (St. Louis), 1956, Pi
Research Professor:
   Anne M. Hofmeister, (D), Caltech, 1984, YghXa
Distinguished Scientist:
   Robert D. Hatcher, Jr., (D), Tennessee, 1965, GtcYm
Administrative Officer:
   Robert Gemignani, (M), Notre Dame de Namur Univ, 1998, Zn

Environmental Studies Program (B) (2018)
Box 1169
One Brookings Drive
St. Louis, MO 63130-4899
   p. (314) 935-7047
   enstadmin@levee.wustl.edu
   http://enst.wustl.edu
   Administrative Assistant: Barbara Winston
   Enrollment (2010): B: 161 (59)
Director:
   Jan P. Amend, (D), California (Berkeley), 1995, Co
Professor:
   Raymond E. Arvidson, (D), Brown, 1974, Zn
   Richard Axelbaum, (D), California (Davis), 1988, Ng
   Pratim Biswas, (D), Caltech, 1985, Ng
   Robert Blankenship, (D), California (Berkeley), 1975, Zn
   Robert E. Criss, (D), Caltech, 1981, CsHqXa
   Willem H. Dickhoff, (D), Free (Amsterdam), Zn
   Mike Dudukovic, (D), Illinois Inst Tech, 1971, Zn
   Robert F. Dymek, (D), Caltech, 2006, Gi
   Claude Evans, (D), SUNY (Stony Brook), Zn
   Bruce Fegley, (D), MIT, 1980, Xc
   T.R. Kidder, (D), Harvard, 1988, Zn
   Maxine I. Lipeles, (D), Harvard, 1979, Zn
   William R. Lowry, (D), Stanford, 1988, Zn
   Jill D. Pasteris, (D), Yale, 1980, Zn
   Bruce Petersen, (D), Harvard, Zn
   Robert Pollak, (D), MIT, 1964, Zn
   Tab Rasmussen, (D), Duke, Zn
   Barbara Schaal, (D), Yale, 1974, Zn
   Glenn D. Stone, (D), Arizona, 1988, Zn
   Robert W. Sussman, (D), Duke, 1972, Zn
   Alan R. Templeton, (D), Michigan, 1972, ZicZr
Associate Professor:
   Jon M. Chase, (D), Chicago, Zn
   Clare Palmer, (D), Zn
   Jen R. Smith, (D), Pennsylvania, 2001, Ga
   Jay Turner, (D), Washington (St. Louis), 1993, Zn
Assistant Professor:
   Jeff Catalano, (D), Stanford, 2004, Ca
   Geoff Childs, (D), Indiana, Zn
   Ellen Damschen, (D), North Carolina State, Zn
   Daniel Giammar, (D), Caltech, 2001, Zn
   Young-Shin Jun, Harvard, 2005, Zn
   Tiffany Knight, (D), Pittsburgh, 2003, Pe
   John Orrock, (D), Iowa State, 2004, Zn
Engineering & Science Director:
   Beth Martin, (M), Washington (St. Louis), 1996, Ng
Geology Librarian:
   Clara McLeod, Zn

William Jewell College
Dept of Biology (2015)
Liberty, MO 64068
   p. (816) 781-3806 x230
   allent@william.jewell.edu
   http://www.jewell.edu
Chair:
   Tara Allen
Associate Professor:
   Charles F. J. Newlon, (M), Missouri, 1962, Zg

Montana

Flathead Valley Community College
Geology and Geography (A) (2018)
777 Grandview Drive
Kalispell, MT 59901
   p. (406) 756-3873
   aho@fvcc.edu
   f: https://www.facebook.com/fvccmt/
   t: @fvccmt
   Programs: Transfer programs in geology, geography, environmental science and environmental studies
Associate Professor:
   Anita Ho, (D), Oregon, Gg

Montana State University
Dept of Earth Sciences (B,M,D) O☒ (2019)

P.O. Box 173480
226 Traphagen Hall
Bozeman, MT 59717-3480
p. (406) 994-3331
earth@montana.edu
http://www.montana.edu/wwwes/
Enrollment (2018): B: 228 (45) M: 21 (8) D: 22 (1)

Regents Professor:
Cathy Whitlock, (D), Washington, 1983, Peg
Professor:
Mary Hubbard, (D), MIT, 1988
David R. Lageson, (D), Wyoming, 1980, GctGs
David W. Mogk, (D), Washington, 1984, GpZeGz
Mark L. Skidmore, (D), Alberta, 2001, ClPgGl
David J. Varricchio, (D), Montana State, 1995, Pv
William K. Wyckoff, (D), Syracuse, 1982, Zn
Associate Professor:
Jordy Hendrikx, (D), Canterbury (NZ), 2005, ZnHsAs
Assistant Professor:
Jean Dixon, (D), Dartmouth, 2009, GmScZg
Julia H. Haggerty, (D), Colorado, 2004, Zu
Andrew Laskowski, (D), Arizona, 2016
Jamie McEvoy, (D), Arizona, 2013, Ze
David B. McWethy, (D), Montana State, 2007, PeSb
Madison Myers, (D), Oregon, 2017
Devon Orme, (D), Arizona, 2015
Eric Sproles, (D), Oregon, 2012
Research Assistant Professor:
Frankie Jackson, (D), Montana State, 2007, Pv
Colin Shaw, (D), New Mexico, 2001, GcEmGp
Assistant Teaching Professor:
Christopher Organ, (D), Montana State, 2003
Research Assistant Professor:
David W. Bowen, (D), Colorado, 2001, GroGs
Regents Professor:
John R. Horner, Penn State, 2006, Pv

Dept of Land Resources & Environmental Sciences (B,M,D) (2015)
334 Leon Johnson Hall
P.O. Box 173120
Bozeman, MT 59717-3120
p. (406) 994-7060
lresinfo@montana.edu
http://landresources.montana.edu/
Department Secretary: Peggy Humphrey

Chair:
Tracy M. Sterling, (D)
Professor:
James W. Bauder, (D), Utah State, 1974, Sf
Lisa J. Graumlich, (D), Washington, 1985, Zu
William P. Inskeep, (D), Minnesota, 1985, Sc
Jeffrey S. Jacobsen, (D), Oklahoma State, 1985, So
Gerald A. Nielsen, (D), Wisconsin, 1963, Sd
David M. Ward, (D), Wisconsin, 1975, Zn
Senior Scientist:
Dennis R. Neuman, (M), Montana State, 1972, Zn
Associate Professor:
Richard E. Engel, (D), Minnesota, 1983, Sc
Jon M. Wraith, (D), Utah State, 1989, Sp
Assistant Professor:
Paul B. Hook, (D), Colorado State, 1992, Sf
Rick L. Lawrence, (D), Oregon State, 1998, Zr
Adjunct Professor:
Douglas J. Dollhopf, (D), Montana State, 1975, Zn
Emeritus:
Cliff Montagne, (D), Montana State, 1976, Sd

Montana State University, Billings
Dept of Biological & Physical Sciences (2015)
1500 N. 30th Street
Billings, MT 59101
p. (406) 657-2341 x2028
nsuits@msubillings.edu

Dean:
Tasneem Khaleel, (D), Bangalore, 1970, Pg
Head:
Stanley Wiatr

Professor:
Matt Benacquista, (D), Montana State, 1989, Xy
Thomas T. Zwick, (D), N Colorado, 1977, Ze

Montana Tech of the University of Montana
Dept of Chemistry & Geochemistry (B,M) (2015)
1300 West Park Street
Butte, MT 59701-8997
p. (406) 496-4207
dhobbs@mtech.edu
Department Secretary: Wilma Immonen

Professor:
Douglas A. Coe, (D), Oregon State, 1974, Cg
Douglas A. Drew, (D), Wyoming, 1971, Cg
Donald Stierle, (D), California (Riverside), 1979, Co
Assistant Professor:
Alysia Cox, (D), MIT/WHOI, 2011, CqlCm
John D. Hobbs, (D), New Mexico, 1991, Zn
Research Associate:
Wayne Olmsted, (B), Montana State, 1962, Ca
Andrea Stierle, (D), Montana State, Zn
Instructor:
Stephen R. Parker, (M), Indiana, 1972, Zn
Emeritus:
Frank E. Diebold, (D), Colorado Mines, 1967, Cl
Alexis Volborth, (D), Helsinki, 1954, Ca

Dept of Geological Engineering (B,M) ●☒ (2018)
1300 West Park Street
Butte, MT 59701
p. (406) 496-4262
dconrad@mtech.edu
https://www.mtech.edu/mines-engineering/geological/
Programs: Geological Engineering (B, M); Geotechnical (B); Mining Geology (B); Hydrogeology (B, M); Petroleum Geology (B); Geology (M); Hydrogeological Engineering (M)
Enrollment (2018): B: 37 (9) M: 20 (7)

Chair:
Larry N. Smith, (D), New Mexico, 1988, GsmEo
Professor:
Christopher H. Gammons, (D), Penn State, 1988, CgEmCs
Mary M. MacLaughlin, (D), California (Berkeley), 1997, NrgNt
Diane Wolfgram, (D), California (Berkeley), 1977, EgoNm
Associate Professor:
Glenn D. Shaw, (D), California (Merced), 2009, HwCsHs

Dept of Geophysical Engineering (B,M) ○☒ (2018)
1300 West Park Street
Butte, MT 59701
p. (406) 496-4401
mspeece@mtech.edu
https://www.mtech.edu/mines-engineering/geophysical
f: https://www.facebook.com/GeophysicalEngineeringMontanaTech/
Programs: Geophysical Engineering (B,M); Geoscience (M)
Enrollment (2018): B: 15 (5) M: 6 (4)

Head:
Marvin A. Speece, (D), Wyoming, 1992, YeuYs
Associate Professor:
Xiaobing Zhou, (D), Alaska (Fairbanks), 2002, Zr
Assistant Professor:
Mohamed Khalil, (D), Giessen, 2002, YeuGe
Khalid Miah, (D), Texas, 2008, YexYs
Emeritus:
Curtis A. Link, (D), Houston, 1993, Ye

Rocky Mountain College
Dept of Geology (B) ☒ (2018)
1511 Poly Drive
Billings, MT 59102
p. (406) 657-1101
kalakayt@rocky.edu
http://www.rocky.edu/academics/academic-programs/undergraduate-majors/geology/
f: https://www.facebook.com/rmcgeology/
Programs: Geology

Geology-Petroleum Systems Concentration
Enrollment (2018): B: 11 (5)
Professor:
  Thomas J. Kalakay, (D), Wyoming, 2001, GcpGi
Associate Professor:
  Emily Geraghty Ward, (D), Montana, 2007, ZeGc
  Derek Sjostrom, (D), Dartmouth, 2002, ClGsHs

Salish Kootenai College
Dept of environmental Science (2015)
58138 US Hwy 93
Ronan, MT 59855
  bill_swaney@skc.edu
  http://www.skc.edu

University of Montana
Dept. of Geosciences (B,M,D) (2018)
32 Campus Drive #1296
Missoula, MT 59812-1296
  p. (406) 243-2341
  bendick@mso.umt.edu
  http://www.umt.edu/geosciences
  Programs: Geosciences; Geosciences Earth Education; International Field Geosciences
  Administrative Assistant: Loreene Skeel
  Enrollment (2017): B: 68 (0) M: 16 (0) D: 10 (0)
Chair:
  James R. Staub, (D), South Carolina, 1985, GsrEo
Professor:
  Joel Harper, (D), Wyoming, 1998, GlZc
  Marc S. Hendrix, (D), Stanford, 1992, Gs
  Nancy W. Hinman, (D), California (San Diego), 1987, Cgl
  James W. Sears, (D), Queens, 1979, GctEg
  George D. Stanley, Jr., (D), Kansas, 1977, Pi
Associate Professor:
  Julia Baldwin, (D), MIT, 2003, Gp
  Rebecca Bendick, (D), Colorado, 2000, Yg
  Marco Maneta, (D), Extremadura (Spain), 2006, Hsg
  Andrew Wilcox, (D), Colorado State, 2005, Gm
Assistant Professor:
  Payton Gardner, (D), Utah, 2009, Hw
  Hilary R. Martens, (D), Caltech, 2016, YgsYd
Research Associate:
  Michael Hofmann, (D), Montana, 2005, GosGr

University of Montana Western
Environmental Sciences Dept (B) (2018)
710 South Atlantic Street
Dillon, MT 59725-3598
  p. (406) 683-7134
  rebekah.levine@umwestern.edu
  https://w.umwestern.edu/department/environmental-sciences/
  f: https://www.facebook.com/UMWenvirosciences/
  Programs: Environmental Sciences (B); Environmental Sustainability (B); Earth Science Education (B)
  Enrollment (2018): B: 75 (12)
Chair:
  Rebekah Levine, (D), New Mexico, 2016, GmHsSd
Professor:
  Robert C. Thomas, (D), Washington, 1993, GseZe
  Craig E. Zaspel, (D), Montana State, 1975, Yg
Assistant Professor:
  Arica Crootof, (D), Arizona, 2019, ZcRcZi
  Spruce W. Schoenemann, (D), Washington, 2015, GeCgs
Adjunct Professor:
  Heidi Anderson-Folnagy, (D), Idaho, 2011, Gs
Emeritus:
  R. Stephen Mock, (D), Montana State, 1989, Ca
  Sheila M. Roberts, (D), Calgary, 1996, GeSo

Nebraska

Central Community College
Physical Science (A) (2015)
4500 63rd Street
PO Box 1027
Columbus, NE 68602
  p. (402) 562-1216
  dcondreay@cccneb.edu
  http://www.cccneb.edu/Science-and-Math/
Instructor:
  Denise Condreay, Zg

Chadron State College
Dept of Geosciences (B) O (2018)
1000 Main Street
Chadron, NE 69337
  p. (308) 432-6377
  mleite@csc.edu
  http://www.csc.edu/geoscience
  Programs: Geoscience
  Environmental Geoscience
  Certificates: Water Resources Management
  Administrative Assistant: Stacy Mittleider
  Enrollment (2018): B: 13 (2)
Chair:
  Wendy Jamison, (D)
Professor:
  Michael B. Leite, (D), Wyoming, 1992, Gg
Instructor:
  Jennifer L. Balmat, (M), Chadron State Coll, 2008, GgZe

Creighton University
Dept of Atmospheric Sciences (B,M) (2015)
2500 California Plaza
Omaha, NE 68178
  p. (402) 280-2641
  zehnder@creighton.edu
  Enrollment (2012): B: 10 (0) M: 3 (0)
Professor:
  Joseph A. Zehnder, (D), Chicago, 1986, As
Associate Professor:
  Jon M. Schrage, (D), Purdue, 1998, Am
Assistant Professor:
  Timothy J. Wagner, (D), Wisconsin, 2011, As
Adjunct Professor:
  Richard Ritz, (M), Texas A&M, Am
Emeritus:
  Arthur V. Douglas, (D), Arizona, 1976, Zg

University of Nebraska, Kearney
Dept of Geography (B) (2015)
203 Copeland Hall
Kearney, NE 68849
  p. (308) 865-8355
  combshj@unk.edu
Chair:
  Jason Combs, (D), Nebraska, 2000
Professor:
  Vijay Boken, (D), Manitoba, 1999
  Paul Burger, (D), Oklahoma State, 1997
  Jeremy Dillon, (D), Kansas, 2001
Associate Professor:
  John Bauer, (D), Kansas, 2006
Instructor:
  Nate Eidem, (D), Oregon State, 2011
  Matt Engel, (D), Nebraska, 2007

University of Nebraska, Lincoln
Dept of Earth & Atmospheric Sciences (B,M,D) ● (2018)
126 Bessey Hall
Lincoln, NE 68588-0340
  p. (402) 472-2663
  tfrank2@unl.edu
  http://eas.unl.edu/
  f: https://www.facebook.com/UNLEarthAtmosSci
  Programs: Geology (B); Meteorology-Climatology (B); Earth and Atmospheric Sciences (M,D)
  Administrative Assistant: Janelle Gerry

Administrative Assistant: Tina M. Schinstock
Enrollment (2018): B: 97 (19) M: 38 (18) D: 20 (2)
Chair:
  Tracy D. Frank, (D), Michigan, 1996, GssCs
Professor:
  Mark R. Anderson, (D), Colorado, 1985, As
  Christopher R. Fielding, (D), Durham (UK), 1982, Gs
  Sherilyn C. Fritz, (D), Minnesota, 1985, Pem
  David M. Harwood, (D), Ohio State, 1986, PmGul
  Qi S. Hu, (D), Colorado State, 1992, As
  R. M. Joeckel, (D), Iowa, 1993, GrsGm
  Clinton M. Rowe, (D), Delaware, 1988, As
  David K. Watkins, (D), Florida State, 1984, PmGu
  Vitaly A. Zlotnik, (D), Natl Inst Hydro Engi Geo (Russia), 1979, HwyHq
Associate Professor:
  Caroline M. Burberry, (D), Imperial Coll (UK), 2008, Gct
  Adam L. Houston, (D), Illinois, 2004, Asm
  Richard M. Kettler, (D), Michigan, 1990, Cl
  Ross Secord, (D), Michigan, 2004, PvCs
  Matthew S. Van Den Broeke, (D), Oklahoma, 2011, AsZr
  Peter J. Wagner, (D), Chicago, 1995, PigPq
  Karrie A. Weber, (D), Alabama, 2002, PgCl
Assistant Professor:
  Lynne J. Elkins, (D), MIT/WHOI, 2009, GivCt
  Irina Filina, (D), Texas, 2007, YgGtYe
Research Assistant Professor:
  Mindi L. Searls, (D), Washington, 2007, YgGg
Professor of Practice:
  Mary Anne Holmes, (D), Florida State, 1989, GsOuZn
Emeritus:
  Ronald G. Goble, (D), Toronto, GxEg
  Priscilla C. Grew, (D), California (Berkeley), 1967, GpeZe
  Robert M. Hunt, (D), Columbia, 1971, Pv
  Merlin P. Lawson, (D), Clark, 1973, AmtZr
  David B. Loope, (D), Wyoming, 1981, Gsg
  Darryll T. Pederson, (D), North Dakota, 1971, Hw
  Norman D. Smith, (D), Brown, 1967, Gsm
  Michael R. Voorhies, (D), Wyoming, 1966, Pv
  William J. Wayne, (D), Indiana, 1952, GmlZu

School of Natural Resource Sciences (B,M,D) ●☒ (2018)
Hardin Hall
3310 Holdrege Street
Lincoln, NE 68583-0961
  p. (402) 472-3471
  jcarroll2@unl.edu
  http://snr.unl.edu/
  Programs: Applied Climate Science, Water Science, Environmental Restoration Science.
  Enrollment (2014): M: 4 (0) D: 9 (2)
Professor:
  Ken F. Dewey, (D), Toronto, 1973, Am
  David C. Gosselin, (D), SD Mines, 1987, Ze
  Michael J. Hayes, (D), Missouri, 1994, As
  Qi (Steve) Hu, (D), Colorado State, 1992, As
  Elizabeth A. Walter-Shea, (D), Nebraska, 1987, As
Research Assistant Professor:
  Tsegaye Tadesse, (D), Nebraska, 2002, AsZr
Associate Professor:
  Martha D. Shulski, (D), Minnesota, 2002, As
  Andrew E. Suyker, (D), Nebraska, 2000, As
Emeritus:
  Donald A. Wilhite, (D), Nebraska, 1975, AsZyn
Climate Scientist:
  Deborah J. Bathke, (D), Ohio State, 2004, AsZe

University of Nebraska, Omaha
Dept of Geography and Geology (B,M) (2018)
6001 Dodge Street
DSC 260
Omaha, NE 68182-0199
  p. (402) 554-2662
  rshuster@unomaha.edu
  http://www.unomaha.edu/college-of-arts-and-sciences/geology/
  Programs: Geology, Environmental Science-Earth Science Option, Geography
  Certificates: GIS
  Administrative Assistant: Brenda Todd
  Enrollment (2018): B: 60 (13)
Professor:
  George F. Engelmann, (D), Columbia, 1978, PvGdr
  Harmon D. Maher, Jr., (D), Wisconsin, 1984, GcsGt
Associate Professor:
  Bradley Bereitschaft, (D), North Carolina (Greensboro), 2011, Zuy
  Robert D. Shuster, (D), Kansas, 1985, GiaZe
Assistant Professor:
  Ashlee LD Dere, (D), Penn State, 2014, SdGmCl
  Zac Suriano, (D), Delaware, 2018, Atm
Emeritus:
  Jeffrey S. Peake, (D), Louisiana State, 1977, ZyAtRm
  John F. Shroder, Jr., (D), Utah, 1967, GmlZy

Nevada

College of Southern Nevada - West Charleston
Dept of Physical Sciences (A) (2018)
6375 W. Charleston Blvd.
Las Vegas, NV 89146
  p. (702) 651-7475
  physic@csn.edu
  http://www.csn.edu/pages/2497.asp
Lead Faculty:
  Barbara Graham, (M), ZyAm
  John E. Keller, (D), S Illinois, 2009, GeHw
  Cynthia S. Shroba, (D), Illinois, Gg
Professor:
  Patrick D. Clennan, (M), ZyGe
  Gale D. Martin, (M), Gg
Instructor:
  Douglas Sims, (D), Kingston, 2011, GeSp

Desert Research Institute
Earth & Ecosystems Sciences (2015)
2215 Raggio Parkway
Reno, NV 89512-1095
  p. (775) 673-7300
  bj@dri.edu
Professor:
  John Arnone, (D), Yale, 1988, Pg
  Colleen M. Beck, (D), California (Berkeley), 1979, Sa
  Christian H. Fritsen, (D), S California, 1996, Pg
  Eric McDonald, (D), New Mexico, 1994, Sd
  Alison E. Murray, (D), California (Santa Barbara), 1998, Pg
  David E. Rhode, (D), Washington, 1987, Pe
Staff Geomorphologist:
  Sophie Baker, (M), Dalhousie, 2005, Gm
Associate Research Geomorphologist:
  Steven N. Bacon, (M), Humboldt State, 2003, Gm
Associate Professor:
  Kenneth D. Adams, (D), Nevada (Reno), 1997, Gm
  Thomas F. Bullard, (D), New Mexico, 1995, Gm
  Mary Cablk, (D), Oregon State, 1997, Zr
  Lynn Fenstermaker, (D), Nevada, 2003, Zr
  Giles Marion, (D), California (Berkeley), 1974, Sc
  Kenneth C. McGwire, (D), California (Santa Barbara), 1992, Zy
  David A. Mouat, (D), Oregon State, 1974, Ge
Assistant Professor:
  JoseLuis Antinao, (D), Dalhousie, 2009, Gm
  Donald E. Sabol, Jr, (D), Washington, 1991, Zr
Assistant Research Professor:
  Thomas F. Bullard, (D), New Mexico, 1995, Gma
GIS/Remote Sensing Scientist:
  Timothy B. Minor, (M), California (Santa Barbara), 1982, Zr
Assistant Research Ecologist:
  Richard Jasoni, (D), Texas A&M, 1998, So
Archaeological Technician:
  David Page, (M), Nevada (Reno), 2008, Ga

Great Basin College
Science Dept (A,B) (2018)
1500 College Parkway

Elko, NV 89801
>p. (775) 753-2120
>caroline.meisner@gbcnv.edu
>http://www2.gbcnv.edu/departments/SCI.html
>Programs: Associates of Science - Pattern of Study Geosciences, Associates of Science - Natural Resources, Bachelor's of Arts - Natural Resources

Professor:
> Caroline Bruno Meisner, (M), Oregon State, 2003, ZgSo

Truckee Meadows Community College
Dept of Physical Sciences (2015)
7000 Dandini Boulevard
Reno, NV 89512
>p. (775) 673-7183
>loanderson@tmcc.edu
>http://www.tmcc.edu/physicalsci/

University of Nevada, Las Vegas
Geoscience Department (B,M,D) ●✓🗎 (2019)
4505 S. Maryland Parkway
Box 454010
Las Vegas, NV 89154-4010
>p. (702) 895-3262
>geodept@unlv.edu
>http://geoscience.unlv.edu
>f: https://www.facebook.com/UNLVGeoscience/
>t: @UNLVGeoscience
>Programs: Geology BS, Earht & Environmental Science BS PhD & MS in Geoscience
>Administrative Assistant: Maria I. Rojas
>Administrative Assistant: Elizabeth Y. Smith
>*Enrollment (2016): B: 18 (18) M: 14 (14) D: 0 (3)*

Chair:
> Terry L. Spell, (D), SUNY (Albany), 1991, CcGvi

Professor:
> Brenda J. Buck, (D), New Mexico State, 1996, GbSd
> Andrew D. Hanson, (D), Stanford, 1998, Co
> Ganqing Jiang, (D), Columbia, 2002, GsClGr
> David K. Kreamer, (D), Arizona, 1982, Hw
> Matthew S. Lachniet, (D), Syracuse, 2001, PeCs
> Rodney V. Metcalf, (D), New Mexico, 1990, Gp
> Margaret N. Rees, (D), Kansas, 1984, Gs
> Stephen Rowland, (D), California (Santa Cruz), 1978, Pi
> Wanda J. Taylor, (D), Utah, 1989, Gc
> Michael L. Wells, (D), Cornell, 1991, Gc

Water Resourse Director/ Co-Chair:
> Michael J. Nicholl, (D), Nevada (Reno), 1993

Associate Scientist:
> Kathleen Zanetti, (M), Idaho, 1997, Gg

Assistant Professor:
> Elisabeth M. Hausrath, (D), Penn State, 2007, ScCl

Associate Chair:
> Eugene I. Smith, (D), New Mexico, 1970, Gi

Emeritus:
> Jean S. Cline, (D), Virginia Tech, 1990, EmCe

Associate Reseach Professor:
> Pamela C. Burnley, (D), California (Davis), 1990, GpyGg

University of Nevada, Reno
Center for Neotectonic Studies (2015)
MS 0169
Reno, NV 89557-0169
>p. (775) 784-6067
>wesnousky@unr.edu
>http://neotectonics.seismo.unr.edu/CNSHome.html

Director:
> Steven Wesnousky, (D), Columbia, 1982, Ys

Center for Research in Economic Geology (M,D) ⊠ (2018)
Mail Stop 1169
Reno, NV 89557-1169
>p. (775) 784-1382
>dawnsnell@unr.edu
>Administrative Assistant: Dawn Lee Snell
>*Enrollment (2015): M: 3 (3)*

Director:
> John Muntean, (D), Stanford, Eg

Dept of Geography (B,M,D) ⊠ (2019)
Mail Stop 0154
Reno, NV 89557-0154
>p. (775) 784-6995
>geography@unr.edu
>http://www.unr.edu/geography/

Chair:
> Douglas Boyle, (D), Arizona, Hg

Professor:
> Kate Berry, (D), Colorado, 1993, Zy
> Jill Heaton, (D), Oregon State, 2001, ZniZy
> Scott A. Mensing, (D), California (Berkeley), 1993, Zy
> Paul F. Starrs, (D), California (Berkeley), 1989, Zy

Associate Professor:
> Thomas Albright, (D), Wisconsin, 2007, Zy
> Scott Bassett, (D), Harvard, 2001, Zu
> Stephanie McAfee, (D), Arizona, 2009, Zy

Assistant Professor:
> Jessie Clark, (D), Arizona, 2012, Zu
> Adam Csank, (D), Arizona, 2011, Zy
> Scott Kelley, (D), Arizona State, 2015, Zi
> Kerri Jean Ormerod, (D), Arizona, 2015, Zu

Research Associate:
> Scotty Strachan, (D), Nevada (Reno), 2016, AtPeZy

Lecturer:
> Victoria Randlett, (D), California (Berkeley), 1999, Zu

Adjunct Professor:
> Christopher Ryan, (M), Nevada (Reno), 1998, Zy
> Peter Wigand, (D), Washington State, 1985, Pe

Dept of Geological Sciences and Engineering (B,M,D) O⊠ (2019)
1664 N. Virginia St., MS 0172
Reno, NV 89557-0172
>p. (775) 784-6050
>geology@unr.edu
>http://www.unr.edu/geology
>*Enrollment (2018): B: 145 (0) M: 23 (0) D: 30 (0)*

Research Professor:
> Simon R. Poulson, (D), Penn State, 1990, Cs

Professor:
> John G. Anderson, (D), Columbia, 1976, YsNeYg
> Greg B. Arehart, (D), Michigan, 1992, Eg
> Wendy M. Calvin, (D), Colorado, 1991, XgZrYg
> James R. Carr, (D), Arizona, 1983, NgZuGg
> John N. Louie, Caltech, Ys
> Paula J. Noble, (D), Texas, 1993, PmGnPs
> Scott W. Tyler, (D), Nevada (Reno), 1990, Hw
> Robert J. Watters, (D), Imperial Coll (UK), 1972, Nrg

Associate Professor:
> Stacia Gordon, (D), Minnesota, 2009, Gi

Assistant Professor:
> Ronald J. Breitmeyer, (D), Wisconsin, 2011, GeHwSp
> Wenrong Cao, (D), S California, 2015, Gtc
> Scott W. McCoy, (D), Colorado, 2012, GmNg
> Philipp P. Ruprecht, (D), Washington, 2009, Giv
> Joel S. Scheingross, (D), Caltech, 2016, Gsm

Lecturer:
> John K. McCormack, (D), Nevada (Reno), 1997, Gz

Adjunct Professor:
> Lisa L. Stillings, (D), Penn State, 1994, Cl

Cooperating Faculty:
> Bridget F. Ayling, (D), Australian Nat, 2006, Ce
> Geoffrey Blewitt, (D), Caltech, 1986, Yd
> James E. Faulds, (D), New Mexico, 1989, Gc
> William C. Hammond, (D), Oregon, 2000, Yd
> Rich D. Koehler, (D), Nevada (Reno), 2009, Gg
> Corne Kreemer, (D), SUNY (Stony Brook), 2001, Yd
> John L. Muntean, Stanford, 1998, Em
> Greg Pohll, (D), Nevada (Reno), 1996, Hq
> Steven G. Wesnousky, (D), Columbia, 1982, Ys
> Andrew Zuza, (D), California (Los Angeles), 2016, Gc

Dept of Mining Engineering (B,M,D) (2015)
Mail Stop 0173

Reno, NV 89557-0173
> p. (775) 784-6961
> jameshendrix@unr.edu
> http://www.unr.edu/cos/mining/
> Administrative Assistant: Carla Scott
> *Enrollment (2011): B: 66 (5) M: 8 (0) D: 5 (0)*

Chair:
   Danny L. Taylor, (D), Colorado Mines, 1980, Nm
Professor:
   Jaak Daemen, (D), Minnesota, 1975, Nr
   George Danko, (D), Budapest, 1985, Nm
Associate Professor:
   Carl Nesbitt, (D), Nevada, 1990, Nx
   Thom Seal, (D), Idaho, 2004, Nx
Emeritus:
   Maurice Feunstenau, (D), Nx
   Pierre Mousset-Jones, (D), London, 1988, Nm
Development Technician:
   John D. Leland, (M), Stanford, 1983

Graduate Program of Hydrologic Sciences (M,D) (2016)

1664 N. Virginia Street
MS 0186
Reno, NV 89557-0175
> p. (775) 784-6221
> hydro@unr.edu
> http://www.hydro.unr.edu
> *Enrollment (2015): M: 32 (10) D: 8 (1)*

Research Professor:
   Kumud Acharya, (D), Saitama, HsGeHq
   Kenneth Adams, (D), Nevada (Reno), 1997, Gam
   Braimah Apambire, (D)
   John J. Arnone, (D), Yale, 1988, Sp
   Gayle L. Dana, (D), Nevada (Reno), 1997, Zr
   Joseph Grzymski, (D)
   Roger Jacobson, (D), Penn State, 1973, Cg
   Nick Lancaster, (D), Cambridge, 1977, Gm
   Joseph McConnell, (D), Arizona, 1997, Zn
   Eric McDonald, (D), New Mexico, 1994, Sp
   Alison Murray, (D), California (Santa Barbara), 1998, ObCg
   Daniel Obrist, (D), Nevada (Reno), 2002
   Simon Poulson, (D), Penn State, 1990, Cls
Associate Director:
   Rina Schumer, (D), Nevada (Reno), 2002, HqwGm
Research Professor:
   Christian H. Fritsen, (D), S California, 1995, Og
   Ken Taylor, (D)
Professor:
   Franco Biondi, (D), Arizona, 1994
   Wendy Calvin, (D), Colorado, 1991, YgZr
   George Danko, (D), Hungarian Acad of Sci, 1985, Hq
   Mae Gustin, (D), Arizona, 1988, Cg
   David Kreamer, (D), Arizona, 1982, HwGn
   John Louie, (D), Caltech, 1987, Yg
   Maureen McCarthy, (D)
   Paula Noble, (D), Texas, 1993, Gn
   Anna Panorska, (D), California (Santa Barbara), 1992
   Mark Pinsky, (D)
   Greg Pohll, (D), Nevada (Reno), 1996, Hwq
   Robert G. Qualls, (D), Georgia, 1989, SoHsSf
   Loretta Singletary, (D)
   Scott Tyler, (D), Nevada (Reno), 1990, Hg
   Mark Walker, (D), Cornell, 1998, Zn
Research Hydrologist:
   Brian Andraski, (D), Nevada (Reno)
Senior Scientist:
   Richard Niswonger, (D)
Research Hydrogeologist and Civil Engineer:
   Dave Decker, (D), Nevada (Reno), Hw
Associate Research Professor:
   Marcus Berli, (D), Swiss Fed Inst Tech, Sp
   Li Chen, (D), Chinese Acad of Sci
   Clay A. Cooper, (D), Nevada (Reno), 1999, HySpHq
   Alan Heyvaert, (D), California (Davis), 1998, GnHs
   Justin Huntington, (D), Nevada (Reno), 2011, HgZr
   Richard Jasoni, (D), Texas A&M
   Alexandra Lutz, (D), Nevada (Reno), Hw
   Kenneth McGwire, (D), California (Santa Barbara), 1992, Zr
   Don Sada, (D)
   Rick Susfalk, (D)
   Julian Zhu, (D), Dalhousie
Associate Professor:
   Sudeep Chandra, (D), California (Davis), 2003, Hs
   Keith E. Dennett, (D), Georgia Tech, 1995, Hs
   Eric Marchand, (D), Colorado (Boulder), 2000
   Sherman Swanson, (D), Oregon State, 1983, Hs
   Aleksey S. Telyakovskiy, (D), Wyoming, 2002, Hqy
Associate Research Professor:
   Ronald L. Hershey, (D), Nevada (Reno), 2010, HwCls
Associate Scientist:
   Lisa Shevenell, (D), Nevada (Reno), 1990, Hw
Assistant Research Professor:
   Rishi Parashar, (D), Purdue, 2008, Hq
   Seshadri Rajagopal, (D)
   Casey Schmidt, (D)
Assistant Research Hydrogeologist:
   Rosemary Carroll, (D), Nevada (Reno), 2010, Hq
Assistant Professor:
   Ronald Breitmeyer, (D), Wisconsin, 2011, HwGe
   Adrian Harpold, (D), Cornell, 2010, HqsZr
   Scott McCoy, (D)
   Ben Sullivan, (D)
   Paul Verburg, (D), Wageningen Ag, 1998, SdcSb
   Steve G. Wells, (D), Cincinnati, 1976, Zn
   Yu (Frank) Yang, (D)
Research Associate:
   Ramon Naranjo, (D), Nevada (Reno), 2012, CgHw
   Lisa Stillings, (D), Penn State, 1994, Cm
Assistant Research Professor:
   Tom Bullard, (D), New Mexico, 1995, Gm
Adjunct Professor:
   Jonathan Price, (D)
   Jim Thomas, (D), Nevada (Reno), 1996, Cg
Emeritus:
   Dale Johnson, (D)
   Wally Miller, (D)
   Steve Wheatcraft, (D), Hawaii, 1979, Hw
Cooperating Faculty:
   Chris Benedict
   Jeanne Chambers, (D), Sf
   David Prudic, (D), Nevada (Reno)
   Michael R. Rosen, (D), Texas, 1989, Gn
   Keirith Snyder, (D)
   Mark Weltz, (D)

Great Basin Center for Geothermal Energy (2015)

Mail Stop 0172
1664 N. Virginia St
Reno, NV 89557-0172
> p. (775) 784-7018
> geothermal@unr.edu
> www.gbcge.org

Director:
   Wendy Calvin, (D), Colorado, 1991, ZrYgXg
Professor:
   John Louie, (D), Caltech, 1987, Yse
Research Professor:
   James Faulds, (D), New Mexico, 1989, Gc
Research Associate:
   Nick Hinz, (M), Nevada (Reno), 2004, Gcg

Mackay School of Earth Sciences and Engineering (Director's Office) ☒ (2017)

Mail Stop 0168
Reno, NV 89557-0168
> p. (775) 784-6987
> juliehill@unr.edu
> http://www.mines.unr.edu/Mackay/
> Administrative Assistant: Julie Hill

Director:
   Thom Seal, (D), Idaho, 2004, NxEmNm

Nevada Seismological Lab ☒ (2018)

Mail Stop 0174

Reno, NV 89557-0174
  p. (775) 784-4975
  mainofc@seismo.unr.edu
  http://www.seismo.unr.edu
  Administrative Assistant: Lori McClelland
  Administrative Assistant: Erik Williams
Director:
  Graham Kent, (D), California (San Diego), 1992, Ys
Professor:
  John G. Anderson, (D), Columbia, 1976, Ys
  John N. Louie, (D), Caltech, 1987, YesRn
  Steve Wesnousky, (D), Columbia, 1982, Ys
Research Associate Professor:
  Glenn P. Biasi, (D), Oregon, 1994, Ys
  Ileana Tibuleac, (D), S Methodist, 1999, Ys
Assistant Director, Seismic Network Manager & Development Director:
  Ken D. Smith, (D), Nevada (Reno), 1991, YsGt
Adjunct Research Associate:
  Bill Honjas, (M), Nevada (Reno), Ys
Emeritus Seismic Network Manager:
  David H. von Seggern, (D), Penn State, 1982, YsgEo
Emeritus:
  James N. Brune, (D), Columbia, 1961, Ys
Development Technician:
  Ryan Presser, (A)
Volunteer Adjunct Faculty:
  Aasha Pancha, (D), Nevada (Reno), 2007, Ys
Seismic Systems Analyst:
  David Slater, (B), Calgary, 1990, Ys
Seismic Records Technician:
  Tom Rennie, (D), Nevada (Reno), 2007, Ys
Programmer/Analyst, Seismic Network:
  Gabriel Plank, (B), Cornell, 1994, Ys
Network Seismologist:
  Diane dePolo, (M), Nevada (Reno), 1989, Ys
Development Technician:
  Kent Straley
Associate Engineer:
  John Torrisi, (A)

Western Nevada College
Western Nevada College (2015)
2201 West College Parkway
Carson City, NV 89703
  p. (775) 445-4442
  Winnie.Kortemeier@wnc.edu
  http://www.wnc.edu/academics/division/sme/

New Hampshire

Dartmouth College
Dept of Earth Sciences (B,M,D) ☒ (2018)
227 Fairchild Hall, HB 6105
Hanover, NH 03755
  p. (603) 646-9037
  earth.sciences@dartmouth.edu
  http://www.dartmouth.edu/~earthsci
  f: www.facebook.com/DartmouthEarthSciences
  Programs: Earth Sciences (B,M,D)
  Enrollment (2018): B: 22 (0) M: 10 (0) D: 15 (0)
Professor:
  Xiahong Feng, (D), Case Western, 1991, Cs
  Carl E. Renshaw, (D), Stanford, 1993, HgGc
Associate Professor:
  Robert L. Hawley, (D), Washington, 2005, Gl
  Meredith Kelly, (D), Bern, 2003, Gl
  Mukul Sharma, (D), Rochester, Ge
  Leslie J. Sonder, (D), Harvard, 1986, Gq
Assistant Professor:
  William D. Leavitt, (D), Harvard
  Erich C. Osterberg, (D), Maine, 2007, AsCl
  Marisa Palucis, (D)
  Justin V. Strauss, (D), Harvard, 2015, GsCgGt
Emeritus:
  William Brian Dade, (D), Washington, GshHgEo
  Gary D. Johnson, (D), Iowa State, 1971, GrdYm

Research Professor:
  Brian P. Jackson, (D), Georgia, 1998, CaGeCt

Keene State College
Dept of Geology (B) (2015)
Mail Stop 2001
229 Main Street
Keene, NH 03435-2001
  p. (603) 358-2553
  pnielsen@keene.edu
  http://www.keene.edu/academics/programs/geol/
  Enrollment (2014): B: 17 (4)
Chair:
  Peter A. Nielsen, (D), Alberta, 1977, GxzGc
Associate Professor:
  Steven D. Bill, (D), Case Western, 1982, Pg
Adjunct:
  Carol Leger, (B), Keene State Coll, 2005, Zg
  Dave Obolewicz, (M), Montana Tech, 1978, GgEmAm
Adjunct Professor:
  Charles M. Kerwin, (D), New Hampshire, 2006, Gg
  Edward M. Pokras, (D), Columbia, 1985, GgPmOb

Plymouth State University
Environmental Science and Policy Program (B,M) ⊘ (2019)
17 High Street
Campus Box 48
Plymouth, NH 03264
  p. (603) 536-2573
  ladoner@plymouth.edu
  http://oz.plymouth.edu/esp
  Programs: Environmental Science & Policy BS and MS
  Certificates: GIS (offered through our school's Geography program)
  Enrollment (2018): B: 177 (22) M: 18 (8)
Professor:
  Eric Hoffman, (D), SUNY (Albany), 2000, Am
Undergraduate program coordinator:
  Lisa A. Doner, (D), Colorado, 2001, PcGes
Research Assistant Professor and Director of Research at the Mount Washington Observatory:
  Eric Kelsey, (D), New Hampshire, 2014, Ats
Emeritus:
  Mark P. Turski, (D), Texas, 1994, Eg

University of New Hampshire
Dept of Earth Sciences (B,M,D) ☒ (2018)
214 James Hall
56 College Road
Durham, NH 03824
  p. (603) 862-1718
  earth.sciences@unh.edu
  http://ceps.unh.edu/earth-sciences
  Programs: Earth Sciences; Geochemistry; Geology; Hydrology; Ocean Mapping; Oceanography
  Administrative Assistant: Susan E. Clark
  Enrollment (2018): B: 47 (14) M: 30 (6) D: 13 (0)
Chair:
  Joseph M. Licciardi, (D), Oregon State, 2000, Gl
Research Professor:
  Stephen E. Frolking, (D), New Hampshire, 1993, As
  Cameron P. Wake, (D), New Hampshire, 1993, Gl
Professor:
  Julia G. Bryce, (D), California (Santa Barbara), 1998, CgcCb
  William C. Clyde, (D), Michigan, 1997, PgYmGs
  John E. Hughes-Clarke, (D), Dalhousie, 1988, Og
  Joseph M. Licciardi, (D), Oregon State, 2000, Gl
  Larry A. Mayer, (D), California (San Diego), 1979, Gu
  David C. Mosher, (D), Dalhousie, 1993, Og
  James M. Pringle, (D), MIT/WHOI, 1998, Op
  Ruth K. Varner, (D), New Hampshire, 2000, Cl
Affiliate Professor:
  Andrew Armstrong, (M), Johns Hopkins, 1991, Zn
  Christopher E. Parrish, (D), Wisconsin, Zr
  Douglas C. Vandemark, (M), New Hampshire, 2005, Zr

Research Associate Professor:
   Jack E. Dibb, (D), SUNY (Binghamton), 1988, AcClAs
   Larry G. Ward, (D), South Carolina, 1978, Gu
Associate Professor:
   Michael W. Palace, (D), New Hampshire, Zn
Affiliate Research Associate Professor:
   Mark A. Fahnestock, (D), Caltech, 1991, Gl
Affiliate Associate Professor:
   Joseph Salisbury, Og
   Mary D. Stampone, (D), Delaware, 2009, Zy
Associate Professor:
   Margaret S. Boettcher, (D), MIT/WHOI, 2005, Ygs
   Rosemarie E. Came, (D), MIT/WHOI, 2005, OoZnn
   J. Matthew Davis, (D), New Mexico Tech, 1994, HwYh
   Joel E. Johnson, (D), Oregon State, 2004, Gus
   Jo Laird, (D), Caltech, 1977, Gp
   Anne F. Lightbody, (D), MIT, 2007, HgsZn
   Thomas C. Lippmann, (D), Oregon State, 1992, Onp
Affiliate Research Assistant Professor:
   Alexandra Contosta, (D), New Hampshire, 2011, So
Assistant Professor:
   Robert T. Letscher, (D), Rosenstiel, 2012, OcCbZo
Affiliate Research Associate Professor:
   Brian Calder, (D), Heriot-Watt, 1997, Og
Emeritus:
   Franz E. Anderson, (D), Washington, 1967, Yr
   Francis S. Birch, (D), Princeton, 1969, Yg
   Wallace A. Bothner, (D), Wyoming, 1967, Gc
   Janet W. Campbell, (D), Virginia Tech, 1973, As
   S. Lawrence Dingman, (D), Harvard, 1970, Hq
   Henri E. Gaudette, (D), Illinois, 1963, Cc
   Theodore C. Loder, (D), Alaska, 1971, Oc
Affiliate Research Scientist III:
   Semme Dijkstra, (D), New Brunswick, Og
Affiliate Faculty:
   Rochelle Wigley, (D), Cape Town, 2005, CgGu

Dept of Natural Resources and the Environment (B,M,D) ☒ (2018)
114 James Hall
56 College Rd
Durham, NH 03824-3589
   p. 603-862-1022
   wendy.rose@unh.edu
   https://colsa.unh.edu/natural-resources-environment
   Programs: Environmental Science
   Certificates: GIS certificate
   Enrollment (2018): B: 70 (25) M: 20 (0) D: 20 (0)
Chair:
   Mark Ducey, (D), Zc
Professor:
   Russell Congalton, (D), Zi
   Serita Frey, (D), Sb
   William H. McDowell, (D), HgRw
   Scott Ollinger, (D), Zc
Associate Professor:
   Stuart Grandy, (D), Sc
   Wil Wollheim, (D), Zu
Assistant Professor:
   Jessica Ernakovich, (D), Zc
Research Associate:
   Adam Wymore, (D), Hg

New Jersey

Bergen Community College
Physical Sciences Depart (2015)
400 County Rd 62
Paramus, NJ 07652
   p. (201) 447-7100
   mnotholt@bergen.edu
   http://bergen.edu/academics/academic-divisions-departments/physical-science/

College of New Jersey
Physics Dept (2017)
2000 Pennington Rd.
Ewing, NJ 08628
   p. 609-771-2569
   physics@tcnj.edu
Associate Professor:
   Margaret Benoit, Penn State, 2005, Ys
   Nathan Magee, (D), Penn State, 2006, As
Adjunct Professor:
   Thomas Gillespie, (M), Rutgers, 1986, Gc

Kean University
School of Environmental and Sustainability Sciences (B) (2015)
1000 Morris Avenue
Union, NJ 07083-0411
   p. (908) 737-3737
   fqi@kean.edu
   http://www.kean.edu/KU/College-of-Natural-Applied-Health-Sciences
   Enrollment (2006): B: 160 (0)
Executive Director:
   Paul J. Croft, (D), Rutgers, 1991, AmsZe
Professor:
   Robert Metz, (D), Rensselaer, 1967, Gr
   Shing Yoh, (D), Drexel, 1989, Am
   Constantine S. Zois, (D), Rutgers, 1980, Am
Associate Professor:
   Carrie M. Manfrino, (D), Miami, 1995, GuOb
   Feng Qi, (D), Wisconsin, 2005, ZiyZg
Assistant Professor:
   Kikombo Ngoy, (D), Oregon, 1996, Zy
Lecturer:
   William C. Heyniger, (B), Montclair State, 2011, Am
Secretary:
   Christina Pacia, (N), Zn

Montclair State University
Dept of Earth & Environmental Studies (B,M) (2015)
1 Normal Avenue
Upper Montclair, NJ 07043
   p. (973) 655-4448
   ophorid@montclair.edu
   http://www.csam.montclair.edu/earth/eesweb
Associate Dean:
   Duke U. Ophori, (D), Alberta, 1986, Hw
Chair:
   Jonathan M. Lincoln, (D), Northwestern, 1990, Gr
Associate Dean, Science & Mathematics:
   Michael A. Kruge, (D), California (Berkeley), 1985, Co
Professor:
   Mark J. Chopping, (D), Nottingham, 1998, ZriZe
   Huan E. Feng, (D), SUNY (Stony Brook), 1997, Cm
   Gregory A. Pope, (D), Arizona State, 1994, GmaZy
   Harbans Singh, (D), Rutgers, 1973, Zn
   William Solecki, (D), Rutgers, 1990, Zu
   Rolf Sternberg, (D), Syracuse, 1971, Zn
   Robert W. Taylor, (D), St. Louis, 1971, Zn
   John V. Thiruvathukal, (D), Oregon State, 1968, Yg
Associate Professor:
   Matthew L. Gorring, (D), Cornell, 1997, Gi
Adjunct Professor:
   Kathryn Black, (M), Oklahoma, 1966, Zy
   Matthew S. Tomaso, (M), Texas, 1995, Ga
   Christine Valenti, (M), Montclair State, 1997, Gg
Emeritus:
   Barbara De Beus, Zn
Laboratory Director:
   Yoko Sato, (M), Montclair State, 2000, Gg

New Jersey City University
Dept of Earth and Environmental Sciences (B) (2015)
Rossey Hall - Room 608
2039 Kennedy Boulevard
Jersey City, NJ 07305-1597
   p. (201) 200-3161
   lengland@njcu.edu
   http://www.njcu.edu/dept/geoscience%5Fgeography/
   Enrollment (2006): B: 37 (19)

Chair:
   Deborah Freile, (D), Boston, 1992, GseGg
Professor:
   Martin Abend, (D), Syracuse, 1955, Zy
Research Associate:
   John M. O'Brien, (D), California (Santa Barbara), 1973, Gs
Lecturer:
   William W. Montgomery, (D), W Michigan, 1998, Hw
Adjunct Professor:
   George Papcun, (M), Ze
   Howard Zlotkin, (M), Ze
Emeritus:
   John Marchisin, (M), Montclair State, 1965, Gg

Princeton University
Dept of Geosciences (B,D) ☒ (2019)
113 Guyot Hall
Princeton, NJ 08544-1003
   p. (609) 258-4101
   mrusso@princeton.edu
   http://www.geoweb.princeton.edu
   f: https://www.facebook.com/GeosciencesPU/
   Programs: Environmental Biogeochemistry; Geophysical & Geology; Ocean, Atmosphere & Climate
   Administrative Assistant: Mary Rose Russo
   Enrollment (2018): B: 28 (11) D: 43 (6)
Chair:
   Bess B. Ward, (D), Washington, 1982, ObCb
Associate Chair:
   Thomas S. Duffy, (D), Caltech, 1992, GyYgZm
Professor:
   Gerta Keller, (D), Stanford, 1978, Pm
   Satish C B. Myneni, (D), Ohio State, 1995, Sc
   Tullis C. Onstott, (D), Princeton, 1980, Pg
   Michael Oppenheimer, (D), Chicago, 1970, AsGe
   Allan M. Rubin, (D), Stanford, 1988, Yg
   Jorge L. Sarmiento, (D), Columbia, 1978, Oc
   Daniel M. Sigman, (D), MIT, 1997, ClsPc
   Frederik J. Simons, (D), MIT, 2002, YsdYm
   Jeroen Tromp, (D), Princeton, 1992, Yss
   Gabriel Vecchi, (D), Washington, 2000, ZoOp
Professional Specialist:
   Amal Jayakumar, (D), Goa, 1999, Cm
   Sergey Oleynik, (D), Moscow State (Russia), 1999, Cs
Associate Professor:
   Stephan A. Fueglistaler, (D), ETH (Switzerland), 2002, As
   John A. Higgins, (D), Harvard, 2009, ClPcCs
   Adam C. Maloof, (D), Harvard, 2004, Ym
   R Blair Schoene, (D), MIT, 2006, Cc
Research Scholar:
   Anne Morel-Kraepiel, (D), Princeton, 2001, Em
Assistant Professor:
   Jessica Irving, (D), Trinity Coll (Cambridge), 2009, Yg
   Laure Resplandy, (D), Sorbonne, 2010, Ob
   Xinning Zhang, (D), Caltech, 2010, Cb
Associate Professional Specialist:
   Dawid Szymanowski, (D), ETH Zuirch, 2018, CcGiv
Academic Lab Manager:
   Laurel P. Goodell, (M), Princeton, 1983, ZeGgc
   Danielle M. Schmitt, (M), W Michigan, 1999, Ze
Emeritus:
   Michael L. Bender, (D), Columbia, 1970, Cg
   Lincoln S. Hollister, (D), Caltech, 1966, GptGz
   Francois M M. Morel, (D), Caltech, 1971, Cl
   S. George H. Philander, (D), Harvard, 1970, Op
   Robert A. Phinney, (D), Caltech, 1961, Ys
Post Doctoral Research Fellow:
   Peter Crockford, (D), McGill, 2017, ClsPc
   Romain Darnajoux, (D), Sherbrooke, 2015, Cb
   Jesse Farmer, (D), Columbia, 2017, Cm
   Dario Marconi, (D), Princeton, 2017, Cms
Post Doctoral Research Associate:
   Anne-Sofie Ahm, (D), Copenhagen, 2016, ClsPc
   Etienne Bachmann, (D), Toulouse, 2016, Ys
   Eleanor Berryman, (D), Tech (Berlin), 2016, Gy
   Julius Busecke, (D), Columbia, 2017, Op
   Gregory Davies, (D), Princeton, 2018, Yg
   Michael P. Eddy, (D), MIT, 2016, Cc
   Behrooz Ferdowsi, (D), ETH Zurich, 2014, Yu
   Lucia Gualtieri, (D), Inst de Physique du Globe de Paris (France) & Univ, 2014, Yg
   Jiaze He, (D), North Carolina State, 2015, Yg
   Nadir Jeevanjee, (D), California (Berkeley), 2016, Ap
   Renxing Liang, (D), Oklahoma, 2015, Po
   Enhui Liao, (D), Delaware, 2017, Op
   Ashley Maloney, (D), Washington, 2017, Oc
   Linhan Shen, (D), Caltech, 2017, Cas
   Jingru Sun, (D), Tsinghua, 2018, As
   Sara Thomas, (D), Northwestern, 2017, Cb
   Wenchang Yang, (D), Columbia, 2014, As
Associate Research Scholar:
   Paul Gauthier, (D), Paris, 2010, Sb
   Hom Nath Gharti, (D), Oslo, 2011, YsdYg
   Nicolas Van Oostende, (D), Ghent, 2011, Ob
Undergraduate/Graduate Coordinator:
   Sheryl A. Robas, (A), 1975, Zn

Environmental Engineering & Water Resources Program (B,M,D) ☒ (2018)
E-220 Engineering Quad
Princeton, NJ 08544
   p. (609) 258-4655
   cee@princeton.edu
   https://www.princeton.edu/cee/graduate/programs/eewr/
   Programs: Civil and Environmental Engineering (B,M,D); Geosciences (B,M,D)
   Enrollment (2018): B: 30 (0) M: 4 (0) D: 40 (0)
Director:
   Elie Bou-Zeid, (D), Johns Hopkins, 2005, As
Professor:
   Michael A. Celia, (D), Princeton, 1983, Hw
   Peter R. Jaffe, (D), Vanderbilt, 1981, SbHgw
   James A. Smith, (D), Johns Hopkins, 1981, Hg
   Eric F. Wood, (D), MIT, 1974, HgZrHq
Assistant Professor:
   Catherine A. Peters, (D), Carnegie Mellon, 1992, Hw

Program in Atmospheric & Oceanic Sciences (D) ☒ (2019)
300 Forrestal Road, Sayre Hall
Princeton, NJ 08540-6654
   p. (609) 258-6677
   apval@princeton.edu
   http://www.princeton.edu/aos/
   Enrollment (2018): D: 15 (6)
Professor:
   Denise L. Mauzerall, (D), Harvard, 1996, As
   Michael Oppenheimer, (D), Chicago
   Stephen Pacala, (D), Stanford
   James Smith, (D), Johns Hopkins
   Gabriel Vecchi, (D), Washington
Senior Scientist:
   Isaac Held, (D), Princeton
   Syukuro Manabe, (D), Tokyo, 1958, AstPc
Associate Professor:
   Stephan Fueglistaler, (D), ETH (Switzerland)
   Mark Zondlo, (D), Colorado
Assistant Professor:
   Laure Resplandy, 2010
Lecturer:
   Thomas Delworth, (D), Wisconsin
   Leo Donner, (D), Chicago
   Stephen Garner, (D), MIT
   Stephen Griffies, (D), Pennsylvania
   Robert Hallberg, (D), Washington
   Larry Horowitz, (D), Harvard
   Sonya Legg, (D), Imperial Coll (UK)
   Yi Ming, (D), Princeton
   V. Ramaswamy, (D), SUNY (Albany)
   Rong Zhang, (D), MIT, Op
Emeritus:
   Michael Bender, (D), Columbia
   George Mellor, (D), MIT, 1957, Op
   Isidoro Orlanski, (D), MIT, 1967, AsZoc
   George Philander, (D), Harvard

Jorge L. Sarmiento, (D), Columbia, 1978, Zn

Raritan Valley Community College
Dept of Biology (2015)
118 Lamington Road
Branchburg, NJ 08878
  p. (908) 526-1200
  dtrybuls@raritanval.edu

Rider University
Geological, Environmental, & Marine Sciences (GEMS) (B)
⌂ (2018)
2083 Lawrenceville Road
Lawrenceville, NJ 08648-3099
  p. (609) 896-5092
  browne@rider.edu
  www.rider.edu/gems
  f: www.facebook.com/riderclas/
  t: @RiderCLAS
  Programs: Geosciences (B); Environmental Sciences (B); Marine Sciences (B); Environmental Studies (B); Integrated Sciences and Math (B)
  Enrollment (2017): B: 8 (3)
Chair:
  Kathleen M. Browne, (D), Miami, 1993, GuZe
Professor:
  Jonathan M. Husch, (D), Princeton, 1982, GieXg
  Hongbing Sun, (D), Florida State, 1995, HwCgSc
Associate Professor:
  Daniel L. Druckenbrod, (D), Virginia, 2003, AtZu
  Reed A. Schwimmer, (D), Delaware, 1999, GsOgn
  Gabriela W. Smalley, (D), Maryland, 2002, ObcOp
Adjunct Professor:
  William B. Gallagher, (D), Pennsylvania, 1990, PvgGr
Emeritus:
  Mary Jo Hall, (D), Lehigh, 1981, Gs

Rutgers, The State University of New Jersey
Earth and Planetary Sciences (B,M,D) ⌂ (2019)
Wright Lab
610 Taylor Road
Piscataway, NJ 08854-8066
  p. (848) 445-2044
  gmntn@eps.rutgers.edu
  http://geology.rutgers.edu/
  f: https://www.facebook.com/Earth-and-Planetary-Sciences-at-Rutgers-University-187118004101/
  Programs: Geological Sciences (B.S.), Geology (B.A.), Environmental Geology (B.S.), Planetary Sciences (B.S.)
  Administrative Assistant: Michael Flak, Katanya Myers, Wendy Rodriguez, Tonya Rufus
  Enrollment (2018): B: 30 (10) M: 6 (4) D: 23 (0)
Chair:
  Gregory S. Mountain, (D), Columbia, 1981, GuYrGr
Associate Chair:
  Ying Fan Reinfelder, (D), Utah State, 1992, HgqGe
Undergraduate Director:
  Roy W. Schlische, (D), Columbia, 1990, GctEo
Graduate Program Director:
  James D. Wright, (D), Columbia, 1991, CsOuPm
Distinguished Professor:
  Gail M. Ashley, (D), British Columbia, 1977, GsmSa
  Marie-Pierre Aubry, (D), Marie Curie, Pms
  Paul G. Falkowski, (D), British Columbia, 1975, ObcMPe
  Dennis V. Kent, (D), Columbia, 1974, YmZcGr
  George R. McGhee, Jr., (D), Rochester, 1978, PgqPi
  Kenneth G. Miller, (D), MIT/WHOI, 1982, GurPm
  Yair Rosenthal, (D), WHOI, CmlOc
Professor:
  Craig S. Feibel, (D), Utah, 1988, GrsGa
  Mark D. Feigenson, (D), Princeton, 1982, CgGiv
  Claude T. Herzberg, (D), Edinburgh, 1975, CpGi
  Robert E. Kopp, (D), Caltech, 2007, ZcOoZg
  Vadim Levin, (D), Columbia, 1996, Ysg
  Robert M. Sherrell, (D), MIT/WHOI, 1991, CmOu
  Carl C. Swisher III, (D), California (Berkeley), 1992, Cc
  Martha O. Withjack, (D), Brown, 1977, GctEo
  Nathan Yee, (D), Notre Dame, 2001, PgCo
Associate Professor:
  Juliane Gross, (D), Ruhr (Germany), 2009, Xcm
Assistant Professor:
  Sonia M. Tikoo, (D), MIT, 2014, YmXy
  Jill A. Van Tongeren, (D), Columbia, 2010, GiCgGt
Research Professor:
  Brent D. Turrin, (D), California (Berkeley), 1996, CcGvYm
Assistant Research Professor:
  James V. Browning, (D), Rutgers, 1996, GgsGr
  Linda Godfrey, (D), Cambridge, 1990, CmlCq
  Richard Mortlock, (D), Rutgers, 2017, CgmCs
Co-Director, Rutgers Geology Museum:
  Lauren Neitzke Adamo, (D), Rutgers, 2016, GuPmZe
Research Associate:
  Don Monteverde, (D), Rutgers, 2008, GruGg
  Peter P. Sugarman, (D), Rutgers, 1995, Gr
Lecturer:
  Christopher H. Lepre, (D), Rutgers, 2009, YmSaPv
Distinguished Visiting Professor:
  William A. Berggren, (D), Stockholm, 1962, PmsPe
Emeritus:
  Michael J. Carr, (D), Dartmouth, 1974, Gv
  Richard K. Olsson, (D), Princeton, 1958, PmGr
  Robert E. Sheridan, (D), Columbia, 1968, YrGu
IT Specialist:
  Jason Pappas
Cooperating Faculty:
  Jeremy S. Delaney, (D), Belfast, 1978, Xg

Rutgers, The State University of New Jersey, Newark
Dept of Earth & Environmental Sciences (B,M,D) ●◌ (2017)
101 Warren Street
Smith Hall, room 135
Newark, NJ 07102
  p. (973) 353-5100
  nicky.agate@rutgers.edu
  http://www.ncas.rutgers.edu/ees
  Programs: Geology; Geoscience Engineering; Environmental Sciences
  Certificates: Environmental Geology (G)
  Enrollment (2017): B: 71 (13) M: 3 (0) D: 13 (2)
Chair:
  Lee S. Slater, (D), Lancaster (UK), 1997, YeHw
Professor:
  Yuan Gao, (D), Rhode Island, 1994, As
  Alexander E. Gates, (D), Virginia Tech, 1986, Gc
Associate Research Professor:
  Francisco Artigas, Ohio State, Zi
Associate Professor:
  Evert J. Elzinga, (D), Delaware, 2000, Sc
  Kristina M. Keating, (D), Stanford, 2009, Yug
  Adam B. Kustka, (D), Stony Brook, 2002, OgCmHs
  Dimitrios Ntarlagiannis, (D), Rutgers, 2006, Yu
  Ashaki Rouff, (D), Stony Brook, 2004, Cl
Assistant Professor:
  Mihaela Glamoclija, (D), D'Annunzio (Italy), 2005, PgXgGe
Research Associate:
  Judith Robinson, (D), Rutgers, 2015, Yg
Emeritus:
  Warren Manspeizer, (D), Rutgers, 1964, Gr
  John H. Puffer, (D), Stanford, 1969, GivGe
  Andreas H. Vassiliou, (D), Columbia, 1969, Gz

Stockton University
Dept of Environmental Sciences (B,M) ☒ (2019)
Division of Natural Science and Mathematics
101 Vera King Farris Drive
Galloway, NJ 08205
  p. (609) 652-4620
  jeffrey.webber@stockton.edu
  http://intraweb.stockton.edu/eyos/page.cfm?siteID=183&pageID=23

Professor:
  Tait Chirenje, (D), Florida, Sc
  Weihong Fan, (D), Colorado State, 1993, Zi
  George Zimmermann, (D), Rutgers, 1982, Zn
Associate Professor:
  Daniel Moscovici, (D), Pennsylvania, 2009, Zu
  Matthew Severs, (D), Virginia Tech, 2009, Gg
Assistant Professor:
  Jessica Favorito, (D), Virginia Tech, 2017, So
  Jeffrey Webber, (D), Massachusetts, 2015, Gg
  Emma Witt, (D), Kentucky, 2012, Hg

Dept of Geology (B) ●☒ (2019)
101 Vera King Farris Dr
Galloway, NJ 08205
  p. 609-626-6857
  matthew.severs@stockton.edu
  http://intraweb.stockton.edu/eyos/page.cfm?siteID=183&pageID=33
  Programs: Geology BS and BA
  Enrollment (2018): B: 23 (7)
Associate Professor:
  Matthew Rocky Severs, (D), Virginia Tech, 2007, GxCgEg
Assistant Professor:
  Jessica Favorito, (D), Virginia Tech
  Susanne Moskalski, (D), Delaware, GsOn
  Jeffrey R. Webber, (D), Massachusetts, 2016, GcpGt
  Emma Witt, (D), Hgs
Emeritus:
  Stewart C. Farrell, (D), Massachusetts, 1972, On
  Michael J. Hozik, (D), Massachusetts, 1976, GcYmg

Union County College
Dept of Biology (A) (2015)
1033 Springfield Ave.
Cranford, NJ 07016
  p. (201) 709-7196
  daly@ucc.edu
  Administrative Assistant: Helen Gmitro
Associate Professor:
  Raymond J. Daly, (M), Rutgers, 1975, Pv

William Paterson University
Dept of Environmental Science (B) (2015)
Science Hall
Wayne, NJ 07470
  p. (201) 595-2721
  beckerm2@wpunj.edu
  http://www.wpunj.edu/cosh/departments/environmental-science/
  Enrollment (2011): B: 90 (3)
Chairman:
  Martin A. Becker, (D), Brooklyn Coll, 1997, PgAm
Professor:
  Richard R. Pardi, (D), Pennsylvania, 1983, Cc
Assistant Professor:
  Jennifer R. Callanan, (D), Montclair State, 2008, ScGm
  Karen Swanson, (D), Penn State, 1989, Cc

New Mexico

Eastern New Mexico University
Dept of Physical Sciences (B) ☒ (2018)
1500 S Ave K
STA 33
Portales, NM 88130
  p. (575) 562-2174
  jim.constantopoulos@enmu.edu
  https://www.enmu.edu/academics/degrees-programs/undergraduate-degree/bachelor/environmental-science
  Programs: Environmental Science
  Enrollment (2018): B: 30 (4)
Professor:
  James T. Constantopoulos, (D), Idaho, 1989, GezGx

Mesalands Community College
Mesalands Dinosaur Museum and Natural Sciences Laboratory (A) O☒ (2019)
911 South Tenth Street
Tucumcari, NM 88401
  p. (575) 461-4413
  axelh@mesalands.edu
  http://www.mesalands.edu/
  Programs: Associate of Arts Degree in Natural Sciences, option Paleontology, Associate of Arts Degree in Natural Sciences, option Geology
  Enrollment (2018): A: 5 (2)
Instructor:
  Axel Hungerbuehler, (D), Bristol, 1998, PvgGg

New Mexico Community College
New Mexico Community College (2015)
525 Buena Vista Dr. SE
Albuquerque, NM 87106
  p. (505) 224-3000
  contactcenter@cnm.edu
  http://www.cnm.edu

New Mexico Highlands University
Natural Resources Management Dept (B,M) (2015)
P.O. Box 9000
Las Vegas, NM 87701
  p. (505) 454-3000
  lindlinej@nmhu.edu
  http://www.nmhu.edu/academics/undergraduate/arts_science/natural_resources/
  Enrollment (2012): B: 16 (4) M: 6 (1)
Professor:
  Jennifer Lindline, (D), Bryn Mawr, 1997, GizZe
Associate Professor:
  Michael S. Petronis, (D), New Mexico, 2005, YmGcv

New Mexico Institute of Mining and Technology
Dept of Earth & Environmental Science (B,M,D) O☒ (2019)
801 Leroy Place
Socorro, NM 87801
  p. (575) 835-5634
  geos.dept@npe.nmt.edu
  https://www.nmt.edu/academics/ees/
  Programs: B.S. in Earth Science, B.S. in Environmental Science, M.S. in Geology, M.S. in Geochemistry, M.S. in Geophysics, M.S. in Hydrology, Ph.D. in Earth and Environmental Science, Professional Master of Hydrology
  Certificates: Graduate Certificate Program in Hydrology
  Enrollment (2018): B: 33 (10) M: 36 (16) D: 18 (2)
Professor:
  Susan L. Bilek, (D), California (Santa Cruz), 2001, Ys
  Peter S. Mozley, (D), California (Santa Barbara), 1988, Gs
  Mark A. Person, (D), Johns Hopkins, 1990, HyYhg
  Glenn Spinelli, (D), California (Santa Cruz), 2002, Hw
Senior Volcanologist, NMBG:
  William C. McIntosh, (D), New Mexico Tech, 1990, Cc
Associate Professor:
  Gary Axen, (D), Harvard, 1991, Gct
  Daniel Cadol, (D), Colorado State, 2010, HsGe
  Bruce I. Harrison, (D), New Mexico, 1992, GmSof
  Jolante Van Wijk, (D), Vrije (Amsterdam), 2002, GtoYe
Assistant Professor:
  Ronni Grapenthin, (D), Alaska (Fairbanks), 2012, Gv
  Daniel Jones, (D), Cb
  Ryan Leary, Gs
  Kierran Maher, Washington State, Eg
Associate Research Professor:
  Mark Murray, (D), MIT, Yds
  David B. Reusch, (D), Penn State, 2003, Zn
  Dana S. Ulmer-Scholle, (D), S Methodist, 1992, GsdCl
Map Production Coordinator, NMBG:
  Phillip Miller
Technical Staff Member, Seimologist:
  Charlotte A. Rowe, (D), New Mexico Tech, 2000, Ys

Sr. Geochronologist/Co-Director NM Geochronology Research:
   Matthew T. Heizler, (D), California (Los Angeles), 1993, Cc
Senior Scientist:
   Robert S. Balch, (D), New Mexico Tech, 1997, Yse
Senior Mineralogist/Economic Geologist/ Director XRD Lab/Curator Mineral Museum:
   Virgil L. Lueth, (D), Texas (El Paso), 1988, GzEg
Senior Field Geologist:
   Steven M. Cather, (D), Texas, 1986, Gc
   Daniel Koning, (M), New Mexico, 1999
Senior Economic Geologist:
   Virginia T. McLemore, (D), Texas (El Paso), 1993, Eg
Research Associate, NMBG:
   Matthew Zimmerer, New Mexico Tech
Principal Senior Petroleum Geologist:
   Ronald F. Broadhead, (M), Cincinnati, 1979, GorGs
Principal Senior Environmental Geologist:
   David W. Love, (D), New Mexico, 1980, Ge
Principal Geologist:
   Paul W. Bauer, (D), New Mexico Tech, 1988, Gc
Postdoc Fellow, USGS:
   Jesus Gomez, New Mexico Tech, 2014
Planetary Protection Officer, NASA:
   Catharine A. Conley, (D), Cornell, 1994, Pg
Geophysicist/Field Geologist/Web Information Specialist:
   Shari A. Kelley, (D), S Methodist, 1984, Gt
Geologic Mapping Program Manager:
   J Michael Timmons, (D), New Mexico, 2004, GctGs
Geochemist & Deputy Director Manger of Electron Microprobe Lab:
   Nelia W. Dunbar, (D), New Mexico Tech, 1989, Cg
Emeritus Senior Principal Geophysicist:
   Marshall A. Reiter, (D), Virginia Tech, 1969, Yh
Emeritus Senior Field Geologist:
   Richard Chamberlin, (D), Colorado Mines, 1980, GcvGg
Emeritus Director and State Geologist:
   Charles E. Chapin, (D), Colorado Mines, 1965, Gv
Distinguished Member of Technical Staff:
   Vincent C. Tidwell, (D), New Mexico Tech, 1999, Hwq
Cave & Karst Hydrologist:
   Lewis Land, (D), North Carolina, 1999, Hy
Assistant Professor of Hydrogeology and Applied Geology, Purdue:
   Marty Frisbee, New Mexico Tech, 2010
Assistant Professor:
   Nigel J.F. Blamey, (D), New Mexico Tech, 2000, CaeCl
Adjunct Professor:
   Denis Cohen
   Charles (Jack) Oviatt
   Michael Underwood
Emeritus Senior Environmental Geologist:
   John W. Hawley, (D), Illinois (Urbana), 1962, GeHwGr
Emeritus:
   Antonius J. Budding, (D), Amsterdam
   Andrew R. Campbell, (D), Harvard, 1984, CsEmGz
   Kent C. Condie, (D), California (San Diego), 1965, Ct
   Gerardo W. Gross, (D), Penn State, 1959, Yx
   Jan M. H. Hendrickx, (D), New Mexico State, 1984, Hw
   David B. Johnson, (D), Iowa, 1978, Ps
   Philip R. Kyle, (D), Victoria (NZ), 1976, Giv
   Fred M. Phillips, (D), Arizona, 1981, Hw
   John W. Schlue, (D), California (Los Angeles), 1975, YsgZg
   John L. Wilson, (D), MIT, 1974, Hw

Dept of Mineral Engineering (B,M,D) (2015)
Campus Station
Socorro, NM 87801-9990
   p. (505) 835-5345
   navid.mojtabai@nmt.edu
   http://nmt.edu/academics/mining/index.php
   Department Secretary: Lucero Joanna
Chair:
   Navid Mojtabai, (D), Arizona, 1990, Nr
Professor:
   William X. Chavez, Jr., (D), California (Berkeley), 1984, Em
Associate Professor:
   Cathrine T. Aimone-Martin, (D), Northwestern, 1982, Ng
Assistant Professor:
   Baolin Deng, (D), Johns Hopkins, 1995, Cg
   Randal S. Martin, (D), Washington State, 1992, As
Adjunct Professor:
   William Haneberg, (D), Cincinnati, 1989, Ng
   Per-Anders Persson, (D), Cambridge, 1960, Nr
   Ingar F. Walder, (D), New Mexico Tech, 1991, Cl
Emeritus:
   George B. Griswold, (D), Arizona, Nx
   Kalman I. Oravecz, (D), Witwatersrand, 1967, Nr

New Mexico State University, Alamogordo
New Mexico State University, Alamogordo (2015)
2400 N. Scenic Drive
Alamogordo, NM 88310
   hrnmsua@nmsu.edu
   http://www.nmsua.edu

New Mexico State University, Grants
Dept of Natural Sciences (2015)
1500 N Third St
Grants, NM 87020
   p. (505) 287-6678
   ssgrants@nmsu.edu
   http://www.grants.nmsu.edu

New Mexico State University, Las Cruces
Department of Physics (M,D) ☒ (2018)
MSC 3D
P.O. Box 30001
Las Cruces, NM 88003-8001
   p. (505) 646-3831
   physics@nmsu.edu
   http://physics.nmsu.edu
   Programs: Physics, Geophysics
   Enrollment (2016): D: 2 (2)
Head:
   Heinz Nakote, (D), Arizona State, 1991, Zm
Associate Professor:
   Thomas M. Hearn, (D), Caltech, 1985, Ys
   Boris Kiefer, (D), Michigan, GyYgZm
Assistant Professor:
   Lauren Waszek, (D), Cambridge, 2012, Ys
Emeritus:
   James F. Ni, (D), Cornell, 1984, YsGtYg

Dept of Geological Sciences (B,M) ●☒ (2019)
MSC 3AB, Box 30001
1255 N. Horseshoe
Gardiner Hall, Room 171
Las Cruces, NM 88003
   p. (575) 646-2708
   geology@nmsu.edu
   http://geology.nmsu.edu/
   Programs: Geology (B,M); Geological Sciences (B); Earth and Environmental Systems (B); Earth Science Education (B)
   Administrative Assistant: Lee Hubbard
   Enrollment (2018): B: 38 (9) M: 17 (5)
Head:
   Nancy J. McMillan, (D), S Methodist, 1986, Gi
Professor:
   Jeffrey M. Amato, (D), Stanford, 1995, GcCc
Associate Professor:
   Frank C. Ramos, (D), California (Los Angeles), 2000, CgGg
Assistant Professor:
   Reed J. Burgette, (D), Oregon, 2008, Gtc
   Brian A. Hampton, (D), Purdue, 2006, Gst
   Emily R. Johnson, (D), Oregon, 2008, Gvi

Dept of Plant & Environmental Sciences (A,B,M,D) (2015)
Box 30003
Dept. 3Q
Las Cruces, NM 88003-0003
   p. (505) 646-3405
   esramire@nmsu.edu
   http://aces.nmsu.edu/academics/pes/
   Department Secretary: Paula Ross

Head:
    LeRoy A. Daugherty, (D), Cornell, 1975, Sd
    Richard Pratt, (D)
Professor:
    William C. Lindemann, (D), Minnesota, 1978, Sb
    Bobby D. McCaslin, (D), Minnesota, 1974, Sc
    Theodore W. Sammis, (D), Arizona, 1974, Sp
Assistant Professor:
    Dean Heil, (D), California (Berkeley), 1991, Sc
    Tim L. Jones, (D), Washington State, 1989, Sp
    H. C. Monger, (D), New Mexico State, 1990, Sa

San Juan College

San Juan College (A) (2019)
4601 College Blvd.
Farmington, NM 87402
    p. (505) 566-3325
    burrisj@sanjuancollege.edu
    http://www.sanjuancollege.edu/geology
    Programs: Geology (A)
    Enrollment (2018): A: 20 (4)
Professor:
    John H. Burris, (D), Michigan State, 2004, GgRhGz

University of New Mexico

Dept of Earth & Planetary Sciences (B,M,D) O⊠ (2018)
221 Yale Blvd NE
Northrop Hall, Room 141
MSC03 2040
Albuquerque, NM 87131-0001
    p. (505) 277-4204
    epsdept@unm.edu
    http://epswww.unm.edu/
    Enrollment (2009): B: 184 (36) M: 34 (9) D: 27 (4)
Chair:
    Peter J. Fawcett, (D), Penn State, 1994, Pe
Professor:
    Carl A. Agee, (D), Columbia, 1988, Cp
    Yemane Asmerom, (D), Arizona, 1988, CcPeCg
    Adrian J. Brearley, (D), Manchester (UK), 1984, Gz
    Laura J. Crossey, (D), Wyoming, 1985, Cl
    Maya Elrick, (D), Virginia Tech, 1990, Gs
    Tobias Fischer, (D), Arizona State, 1999, Gv
    Joseph Galewsky, (D), California (Santa Cruz), 1996, Am
    David J. Gutzler, (D), MIT, 1986, As
    Karl E. Karlstrom, (D), Wyoming, 1980, Gt
    James J. Papike, (D), Minnesota, 1964, Ca
    Louis A. Scuderi, (D), California (Los Angeles), 1984, GmPcGs
    Zachary D. Sharp, (D), Michigan, 1987, Cs
    Gary S. Weissmann, (D), California (Davis), 1999, HwGsr
Senior Scientist:
    Nieu-Viorel Atudorei, (D), Lausanne, 1998, Cs
    Victor J. Polyak, (D), Texas Tech, 1998, CcGzm
Associate Professor:
    Brandon Schmandt, (D), Oregon, 2011, Yg
    Chester J. Weiss, (D), Texas A&M, 1998, Ye
Assistant Professor:
    Corinne E. Myers, (D), Kansas, 2013, PeqPi
    Lindsay Lowe Worthington, (D), Texas, 2010, Yg
    Jin Zhang, (D), Illinois, 2014, Gyp
Research Associate:
    Frans J.M. Rietmeijer, (D), Utrecht, 1979, Gp
Lecturer:
    Aurora Pun, (D), New Mexico, 1996, Ca
Adjunct Professor:
    Fraser Goff, (D), California (Santa Cruz), 1977, Cg
    Duane M. Moore, (D), Illinois (Urbana), 1963, Sc
    Thomas E. Williamson, (D), New Mexico, 1993, Pv
Emeritus:
    Roger Y. Anderson, (D), Stanford, 1960, Gn
    Rodney E. Ewing, (D), Stanford, 1974, Gz
    John W. Geissman, (D), Michigan, 1980, Ym
    Rhian Jones, (D), Manchester (UK), 1986, Gz
    Barry S. Kues, (D), Indiana, 1974, Pi
    Leslie M. McFadden, (D), Arizona, 1982, Sd
    Grant A. Meyer, (D), New Mexico, 1993, Gm
    Jane E. Selverstone, (D), MIT, 1985, Gpt
    Lee A. Woodward, (D), Washington, 1962, Gc

Institute of Meteoritics (B,M,D) ⊠ (2017)
MSC03 2050
1 University of New Mexico
Albuquerque, NM 87131
    p. (505) 277-1644
    iom@unm.edu
    meteorite.unm.edu
    Administrative Assistant: Beth Ha
    Enrollment (2017): M: 1 (1) D: 3 (1)
Professor:
    Carl B. Agee, (D), Columbia, 1988, XcGyCp
Senior Scientist III:
    Horton E. Newsom, (D), Arizona, 1981, XgcXm
Senior Scientist III:
    Charles K. Shearer, Jr., (D), Massachusetts, 1983, Gi
Senior Research Scientist III:
    Steven B. Simon, (D), SD Mines, 1988, XmcGi
Assistant Professor:
    Jin Zhang, (D), Illinois, 2014, Gy
Senior Scientist:
    Karen Ziegler, (D), Reading, 1993, CsXc
Research Scientist III:
    Michael N. Spilde, (M), SD Mines, 1987, Gz
Program Manager:
    Shannon Clark, Zn

Water Resources Program (2015)
1915 Roma NE, Room 1044
Albuquerque, NM 87131-1217
    p. (505) 277-5249
    fleckj@unm.edu
    http://www.unm.edu/~wrp/
Director:
    Michael E. Campana, Zn

University of New Mexico, Gallup

Div of Arts and Sciences (2015)
200 College Road
Gallup, NM 87301
    p. (505) 863-7500
    pwatt@unm.edu
    http://www.gallup.unm.edu/

University of New Mexico, Taos

University of New Mexico - Taos (A,B) ● (2015)
1157 County Road 110
Ranchos de Taos, NM 87557
    colnic@unm.edu
    http://www.taos.unm.edu
Adjunct Professor:
    Deborah Ragland, (D), Gg

Western New Mexico University

Dept of Natural Sciences ⊠ (2018)
P.O. Box 680
1000 West College Avenue
Silver City, NM 88062
    p. (575) 538-6352
    corrie.neighbors@wnmu.edu
    http://natsci.wnmu.edu/
Assistant Professor:
    Corrie Neighbors, (D), California (Riverside), 2015, Ysr

New York

Adelphi University

Environmental Studies Program (B,M) O⊠ (2019)
South Avenue
Garden City, NY 11530
    p. (516) 877-4170
    bwygal@adelphi.edu
    http://environmental-studies.adelphi.edu

t: @mammothunter
Enrollment (2018): B: 30 (7) M: 8 (3)
Associate Professor:
  Brian Wygal, (D), Nevada (Reno), 2009, GaRmGe

Adirondack Community College
Science Div (A) (2015)
640 Bay Road
Queensbury, NY 12804
  p. (518) 743-2325
  minkeld@sunyacc.edu
  http://www.sunyacc.edu

Alfred University
Dept of Geology (B) ☒ (2018)
Saxon Drive
Alfred, NY 14802
  p. (607) 871-2208
  fmuller@alfred.edu
  http://ottohmuller.com/ENSweb2008/
  Programs: Geology
  Enrollment (2018): B: 9 (2)
Professor:
  Michele M. Hluchy, (D), Dartmouth, 1988, Cl
  Otto H. Muller, (D), Rochester, 1974, Gc

American Museum of Natural History
Dept of Earth & Planetary Sciences ☒ (2018)
Central Park West at 79th Street
New York, NY 10024-5192
  p. (212)769-5100
  http://www.amnh.org/our-research/physical-sciences/earth-and-planetary-sciences
  Administrative Assistant: Nanette Nicholson
Chair and Curator:
  Denton S. Ebel, (D), Purdue, 1993, XmCgZe
Curator:
  George E. Harlow, (D), Princeton, 1977, Gz
  James D. Webster, (D), Arizona State, 1987, Eg
Senior Museum Specialist:
  Jamie Newman, (M), Brooklyn Coll, Gg
Museum Specialist:
  Sam Alpert, (B), Case Western, 2012, Xm
  Saebyul Choe, (B), Bates, 2014
  Keiji Hammond, (B), Northeastern, 2016
Assistant Curator:
  Nathalie Goodkin, (D), MIT/WHOI, 2007
Specialist:
  Kim V. Fendrich, (M), Arizona, 2016, GzXg
Curator Emeritus:
  Edmond A. Mathez, (D), Washington, 1981, Gx
Special Projects Staff:
  Elizabeth Haussner, (M), Cincinnati, 2016
  Andrea Mason, (M), Brooklyn Coll, 2013
Postdoctoral Fellow:
  Steven Jaret, (D), Stony Brook, 2017
  Nicholas Tailby, (D)
  Natalie Umling, (D), S Carolina, 2017

Div of Paleontology (D) (2015)
Central Park West at 79th Street
New York, NY 10024
  p. (212) 769-5815
  norell@amnh.org
  http://paleo.amnh.org/
  Administrative Assistant: Judy Galkin
  Enrollment (2009): D: 8 (0)
Provost, Professor and Curator:
  Michael J. Novacek, (D), California (Berkeley), 1978, Pv
Dean of the Richard Gilder Grad School, Professor, and Frick Curator:
  John J. Flynn, (D), Columbia, 1983, Pv
Chair, Professor and Curator:
  Mark A. Norell, (D), Yale, 1989, Pv
Professor and Curator:
  Niles Eldredge, (D), Columbia, 1969, Pi
  Neil H. Landman, (D), Yale, 1982, Pi
  John G. Maisey, (D), London, 1974, Pv
  Jin Meng, (D), Columbia, 1991, Pv
Frick Curator Emeritus:
  Richard H. Tedford, (D), California (Berkeley), 1960, Pv
Curator Emeritus:
  Roger L. Batten, (D), Columbia, 1956, Pi
  Eugene S. Gaffney, (D), Columbia, 1969, Pv

Binghamton University
Dept of Geological Sciences and Environmental Studies
(B,M,D) ☒ (2019)
PO Box 6000
Binghamton, NY 13902-6000
  p. (607) 777-2264
  slavetskas@binghamton.edu
  http://geology.binghamton.edu
  Administrative Assistant: Carol Slavetskas
  Enrollment (2012): B: 30 (0) M: 4 (0) D: 12 (0)
Chair:
  Tim K. Lowenstein, (D), Johns Hopkins, 1982, Cl
Professor:
  Joseph R. Graney, (D), Michigan, 1994, ClGeHs
  David M. Jenkins, (D), Chicago, 1980, CpGzp
  H. Richard Naslund, (D), Oregon, 1980, Gi
Associate Professor:
  Peter L. K. Knuepfer, (D), Arizona, 1984, Gt
  Thomas Kulp, (D), Indiana, 2002, Pg
Assistant Professor:
  Alex Nikulin, (D), Rutgers, 2011, Ys
  Molly Patterson, (D), Victoria (NZ), 2015, CmGl
  Jeff Pietras, (D), Wisconsin, 2003, GsEo
Research Associate:
  Alan Jones, (D), Purdue, 1964, Ys
Emeritus:
  Richard E. Andrus, (D), SUNY (Syracuse), 1974, Sf
  Jeffrey S. Barker, (D), Penn State, 1984, Ys
  Donald R. Coates, (D), Columbia, 1956, Gm
  Robert V. Demicco, (D), Johns Hopkins, 1981, Gs
  Steven R. Dickman, (D), California (Berkeley), 1977, Yg
  Thomas W. Donnelly, (D), Princeton, 1959, Gg
  William D. MacDonald, (D), Princeton, 1965, Gc
  Karen M. Salvage, (D), Penn State, 1998, Hw
  James E. Sorauf, (D), Kansas, 1962, Pi
  Francis T. Wu, (D), Caltech, 1966, Ys
Related Staff:
  Michael Hubenthal, (M), Binghamton, 2010, ZeYg

Brooklyn College (CUNY)
Dept of Earth and Environmental Science (B,M) ●☒ (2018)
2900 Bedford Avenue
Brooklyn, NY 11210
  p. (718) 951-5416
  wpowell@brooklyn.cuny.edu
  http://depthome.brooklyn.cuny.edu/geology/
  Enrollment (2009): B: 24 (5) M: 10 (4) D: 3 (1)
Chair:
  Wayne G. Powell, (D), Queen's, 1994, GaCsEm
Chair:
  Jennifer Cherrier, OcCmRw
Professor:
  John A. Chamberlain, (D), Rochester, 1971, Pg
  Zhongqi Cheng, (D), Ohio State, 2001, CaScGe
  Constantin Cranganu, (D), Oklahoma, 1997, GoYhHw
  Peter Groffman, SbCl
  John Marra, (D), Dalhousie, 1977, Ob
  David E. Seidemann, (D), Yale, 1975, Cc
Associate Professor:
  Stephen U. Aja, (D), Washington State, 1989, Cl
  Rebecca Boger, (D), William & Mary, 2002, ZiHs
Assistant Professor:
  Brett Branco, (D), Connecticut, 2007, OnHsOg
  Kennet Flores, (D), Gpc
  Brianne Smith, (D), Hs
Lecturer:
  Matt Garb, (M), Brooklyn Coll, Pg

Laboratory Director:
  Guillermo Rocha, (M), CUNY, 1994, GgeCg

Broome Community College
Dept of Physical Sciences (A) (2015)
Upper Front Street
Box 1017
Binghamton, NY 13902
  p. (607) 778-5000
  smithjj@sunybroome.edu
Professor:
  Bruce K. Oldfield, (M), SUNY (Binghamton), 1988, Gg
Assistant Professor:
  Jason J. Smith, (M), Binghamton, 2009, Ggs

Buffalo State College
Dept of Earth Sciences (B) ●✓ (2019)
1300 Elmwood Avenue
Buffalo, NY 14222
  p. (716) 878-6731
  solargs@buffalostate.edu
  http://www.buffalostate.edu/earthsciences
  Programs: Geology; Earth Sciences
  Administrative Assistant: Cindy Wong
  Enrollment (2018): B: 50 (16)
Acting Chair:
  Daniel E. MacIssac, (D), Xa
Professor:
  Jill K. Singer, (D), Rice, 1986, GsOp
Planetarium Director:
  Kevin K. Williams, (D), Johns Hopkins, 2002, Xg
Associate Professor:
  Elisa T. Bergslien, (D), SUNY (Buffalo), 2002, ClHw
  Gary S. Solar, (D), Maryland, 1999, GcpGt
  Kevin K. Williams, (D), Johns Hopkins, 2002, GmXg
Emeritus:
  John E. Mack, (D), Fordham, 1971, Xg
  Irving Tesmer, (D), Syracuse, Pg

Cayuga Community College
Math and Science (A) (2015)
197 Franklin Street
Auburn, NY 13021
  p. (315) 255-1743
  waters@cayuga-cc.edu
  http://www.cayuga-cc.edu/academics/programs_of_study/math_and_science.php
Professor:
  Abu Z. Badruddin, (D), SUNY (Environ), Zi
  Raymond F. Leszczynski, (M), SUNY (Albany), 1965, GglGm

City College (CUNY)
Dept of Earth & Atmospheric Sciences (B,M) (2015)
New York, NY 10031
  p. (212) 650-6984
  kmcdonald2@ccny.cuny.edu
Chair:
  Jeffrey Steiner, (D), Stanford, 1970, Cp
Professor:
  Stanley Gedzelman, (D), MIT, 1970, As
  Edward E. Hindman, (D), Washington, 1975, As
  Margaret A. Winslow, (D), Columbia, 1979, Gc
Associate Professor:
  Patricia M. Kenyon, (D), Cornell, 1986, Yg
  Federica Raia, (D), Naples, 1997, Giv
  Pengfei Zhang, (D), Utah, 2000, Hw

Colgate University
Department of Geology (B) ●☒ (2018)
13 Oak Drive
Hamilton, NY 13346
  p. (315) 228-7201
  jmcnamara@colgate.edu
  http://www.colgate.edu/academics/departments-and-programs/geology
  Administrative Assistant: Jodi McNamara
  Enrollment (2018): B: 28 (9)
Chair:
  William H. Peck, (D), Wisconsin, 2000, GpiCs
Professor:
  Karen Harpp, (D), Cornell, 1994, GvCgGi
  Amy Leventer, (D), Rice, 1988, Ou
  Constance M. Soja, (D), Oregon, 1985, PiGs
Associate Professor:
  Martin Wong, (D), California (Santa Barbara), 2005, Gtc
Assistant Professor:
  Aubreya Adams, (D), Penn State, 2010, Yg
  Joseph Levy, (D), Brown, 2009, GlsGm
Senior Lecturer:
  Dianne M. Keller, (M), Colgate, 1988, Gz
Emeritus:
  Richard April, (D), Massachusetts, 1978, CgGze
  James McLelland, (D), Chicago, 1961, Gp
  Paul Pinet, (D), Rhode Island, 1972, Ou

Columbia University ⌐
Dept of Earth & Environmental Engineering (2015)
Henry Krumb School of Mines
500 West 120 Street
918 Mudd Bldg
New York, NY 10027
  p. (212) 894-2905
  schlosser@ldeo.columbia.edu
  http://www.eee.columbia.edu/
  Administrative Assistant: Co'Quesie Gilbert
  Department Administrator: Barbara Algin
Acting Chair:
  Nickolas J. Themelis, (D), McGill, 1961, Nx
Professor:
  Paul F. Duby, (D), Columbia, 1962, Nx
  Peter Schlosser, (D), Heidelberg, 1985, Cg
  Ponisseril Somasundaran, (D), California (Berkeley), 1964, Nx
  Tuncel M. Yegulalp, (D), Columbia, 1968, Nm
Associate Professor:
  Ross Bagtzoglou, (D), California (Berkeley), 1990, Hw
Senior Research Scientist:
  Roelof Versteeg, (D), Paris VII, 1991, Yg
Adjunct Professor:
  Vasilis M. Fthenakis, (D), New York, 1991, As
Emeritus:
  Stefan H. Boshkov, (M), Columbia, 1942, Nm
  John T. Kuo, (D), Stanford, 1958, Yg
  Malcolm T. Wane, (M), Columbia, 1954, Nm

Dept of Earth & Environmental Sciences (B,M,D) ☒ (2019)
P.O. Box 1000
61 Route 9W
Palisades, NY 10964
  p. (845) 365-8550
  odland@ldeo.columbia.edu
  http://eesc.columbia.edu
  Programs: Earth and Environmental Sciences
  Certificates: none
Vice Chair:
  Peter B. de Menocal, (D), Columbia, 1991, PeOuCg
Director, Lamont Doherty Earth Observatory:
  Sean Solomon, (D), MIT, 1971, Xy
Dir. Graduate Studies:
  Goran Ekstrom, (D), Harvard, 1987, Ys
Chair:
  Sidney R. Hemming, (D), SUNY (Stony Brook), 1994, CscPe
Professor:
  Wallace S. Broecker, (D), Columbia, 1958, CmOcPe
  Nicholas Christie-Blick, (D), California (Santa Barbara), 1979, Gst
  Joel E. Cohen, (D), Harvard, 1970, Zn
  Hugh Ducklow, (D), Harvard, 1977, ObCb
  Peter M. Eisenberger, (D), Harvard, 1967, Zn
  Steven L. Goldstein, (D), Columbia, 1986, Csg
  Arnold L. Gordon, (D), Columbia, 1965, Op
  Kevin L. Griffin, (D), Duke, 1994, PbeZn
  Peter B. Kelemen, (D), Washington, 1988, GiCg

Jerry F. McManus, (D), Columbia, 1989, PeOu
William H. Menke, (D), Columbia, 1981, YsGvq
John C. Mutter, (D), Columbia, 1982, YrGtYs
Paul E. Olsen, (D), Yale, 1984, PvgGr
Terry A. Plank, (D), Columbia, 1993, GxCg
Lorenzo M. Polvani, (D), MIT, 1988, AmsGq
G. Michael Purdy, (D), Cambridge, 1974, Yr
Peter Schlosser, (D), Heidelberg, 1985, Hw
Adam H. Sobel, (D), MIT, 1998, As
Marc W. Spiegelman, (D), Cambridge, 1989, GqxCg
Martin Stute, (D), Heidelberg, 1989, CsHgGe
Felix Waldhauser, (D), ETH (Switzerland), 1996, YsgGt

Associate Professor:
Sonya Dyhrman, Scripps, 1999, Ob
Arlene M. Fiore, (D), Harvard, 2003, As
Baerbel Hoenisch, (D), Bremen, 2002, PcClm
Meredith Nettles, (D), Harvard, 2005, YsGl
Maria Tolstoy, (D), California (San Diego), 1994, Yr

Assistant Professor:
Ryan P. Abernathey, (D), MIT, 2012, Op

Lecturer:
Alberto Malinverno, (D), Columbia, 1989, Gug
Benjamin S. Orlove, (D), California (Berkeley), 1975, As
Andreas M. Thurnherr, (D), Southampton, 2000, Op
Christopher J. Zappa, (D), Washington, 1999, OpAsZr

Adjunct Professor:
Robert F. Anderson, (D), MIT, 1981, OcCmPe
W. Roger Buck, IV, (D), MIT, 1984, YgGt
John J. Flynn, (D), Columbia, 1983, PvgYm
Alessandra Giannini, (D), Columbia, 2001, AsOpZc
Lisa M. Goddard, (D), Princeton, 1995, AstZc
Arthur L. Lerner-Lam, (D), California (San Diego), 1982, Ys
Douglas G. Martinson, (D), Columbia, 1982, OpGq
Ronald L. Miller, (D), MIT, 1990, As
Mark A. Norell, (D), Yale, 1988, Pv
Dorothy M. Peteet, (D), New York Univ, 1983, PelGn
Andrew W. Robertson, (D), Reading, 1984, As
Joerg Schaefer, (D), ETH Zurich, 2000, Cg
Christopher Small, (D), California (San Diego), 1993, ZrYr
Taro Takahashi, (D), Columbia, 1957, OcCg
Mingfang Ting, (D), Princeton, 1990, As
Spahr C. Webb, (D), California (San Diego), 1984, Yrg
Gisela Winckler, (D), Heidelberg, 1998, Cm

Emeritus:
James D. Hays, (D), Columbia, 1964, PemOu
Paul G. Richards, (D), Caltech, 1970, YsZn
Christopher H. Scholz, (D), MIT, 1967, YxNr
H. James Simpson, Jr., (D), Columbia, 1970, Cm
Lynn R. Sykes, (D), Columbia, 1965, YsGtZn
David Walker, (D), Harvard, 1972, CpGzCg

Dir. Academic Admin & Finance:
Sarah K. Odland, (M), Colorado, 1981, Zn

Asst. Director Climate and Society Program:
Cynthia Thomson, (M), Columbia, 2009, Zn

Lamont-Doherty Earth Observatory (M,D) ☒ (2019)
P.O. Box 1000
61 Route 9W
Palisades, NY 10964
    p. (845) 359-2900
    director@ldeo.columbia.edu
    http://www.ldeo.columbia.edu
    f: https://www.facebook.com/Lamont.Doherty/
    t: @LamontEarth
    Programs: M.S. in Sustainability Science; M.A. in Climate and Society
    Enrollment (2006): M: 40 (0) D: 83 (6)

William B. Ransford Professor of Earth and Planetary Science, Columbia University:
Sean C. Solomon, (D), MIT, 1971, Xgy

Arthur D. Storke Memorial Professor:
Peter Kelemen, (D), Washington, 1987, Gi
Paul E. Olsen, (D), Yale, 1983, Gm

Professor:
Nicholas Christie-Blick, (D), California (Santa Barbara), 1979, Gs
Peter B. deMenocal, (D), Columbia, 1991, Pe
Hugh W. Ducklow, (D), Harvard, 1977

Goran Ekstrom, (D), Harvard, 1987, Yg
Arlene Fiore, (D), Harvard, 2003, As
Steven Goldstein, (D), Columbia, 1986, Cg
Arnold L. Gordon, (D), Columbia, 1965, Op
Kevin Griffin, (D), Duke, 1994, Zn
Galen McKinley, (D), MIT, 2002, Cg
Jerry McManus, (D), Cm
William H. Menke, (D), Columbia, 1981, Ys
John C. Mutter, (D), Columbia, 1982, Yr
Christopher H. Scholz, (D), MIT, 1967, Ys
Adam Sobel, (D), MIT, 1998, As
Marc Spiegelman, (D), Cambridge, 1989, Ys
Renata Wentzcovitch, California (Berkeley), 1988, Ys

Sr. PGI Research Scientist:
Richard Seager, (D), Columbia, 1990, Am

Senior Research Scientist:
Kerstin Lehnert, (D), Albert-Ludwigs Freiburg, 1989, Gi

Research Scientist:
Robert Newton, (D), Columbia, 2001, Ct

Palisades Geophysical Institute/Lamont Research Professor:
Robin E. Bell, (D), Columbia, 1989, Yr

Lamont Research Professor:
Roger W. Buck, (D), MIT, 1984, Yr
Brendon Buckley, (D), Tasmania, 1997, Pe
Suzana Camargo, (D), Tech (Munich), 1992, As
Steven Chillrud, (D), Columbia, 1995, Cg
Rosanne D'Arrigo, (D), Columbia, 1989, Zn
James Davis, (D), MIT, 1986, Yd
James Gaherty, (D), MIT, 1995, Ys
Joaquim Goes, (D), Nagoya, 1996, Gu
David S. Goldberg, (D), Columbia, 1985, Yr
Won-Young Kim, (D), Uppsala, 1986, Ys
Yochanan Kushnir, (D), Oregon State, 1985, As
Braddock Linsley, (D), New Mexico, 1990, Pe
Alberto Malinverno, (D), Columbia, 1989, Go
Douglas G. Martinson, (D), Columbia, 1982, Op
Joerg Schaefer, (D), Swiss Fed Inst, 2000, Pe
Bruce Shaw, (D), Chicago, 1989, Ys
Christopher Small, (D), California (San Diego), 1993, Zr
Michael Steckler, (D), Columbia, 1980, Yv
Ajit Subramaniam, (D), SUNY (Stony Brook), 1995, ObZr
Marco Tedesco, (D), Italian Nat Res Cncl, 2003, GlZr
Andreas Thurnherr, (D), Southampton, 1999, Op
Mingfang Ting, (D), Princeton, 1990, As
Alexander Van Geen, (D), MIT/WHOI, 1989, Cg
Felix Waldhauser, (D), ETH (Switzerland), 1996, Ys
Gisela Winckler, (D), Heidelberg, 1998, Cg
Xiaojun Yuan, (D), California (San Diego), 1994, Op

Heezen Senior Research Scientist:
Suzanne Carbotte, (D), California (Santa Barbara), 1992, Yr

Ewing LDEO Rsrch Professor:
Edward R. Cook, (D), Arizona, 1985, Hw

Ewing Lamont Research Professor:
Robert F. Anderson, (D), MIT, 1981, Cg
Taro Takahashi, (D), Columbia, 1957, Cm

Deputy Director:
Arthur L. Lerner-Lam, (D), California (San Diego), 1982, Ys

Associate Professor:
Baerbel Hoenisch, (D), Alfred Wegener Inst (Germany), 2002, Cm
Meredith Nettles, (D), Harvard, 2005, YsGl

Lamont Associate Research Professor:
Michela Biasutti, (D), Washington, 2003, Am
Benjamin C. Bostick, (D), Stanford, 2002, Sc
Connie Class, (D), Karlsruhe, 1994, Cg
Benjamin Holtzman, (D), Minnesota, 2003, Ys
Michael Kaplan, (D), Colorado, 1999, Gl
Raymond N. Sambrotto, (D), Alaska, 1983, Ob
David Schaff, (D), Stanford, 2001, Ys
Donna Shillington, (D), Wyoming, 2004, Yr
Jason Smerdon, (D), Michigan, 2004, Pe
Susanne Straub, (D), Kiel, 1991, Gv
Christopher Zappa, (D), Washington, 1999, Op

Assistant Professor:
Ryan Abernathy, (D), MIT, 2012, Og
Roisin Commane, (D), Leeds, 2009, Yr

Special Research Scientist:
Pierre E. Biscaye, (D), Yale, 1964, Cm

Enrico Bonatti, (D), Pisa, 1967, Yr
Dake Chen, (D), SUNY (Stony Brook), 1989, Op
James R. Cochran, (D), Columbia, 1977, Yr
Klaus H. Jacob, (D), Goethe (Frankfurt), 1968, Ys
Stanley Jacobs, (B), MIT, 1962, Op
Paul G. Richards, (D), Caltech, 1970, Ys
William B. F. Ryan, (D), Columbia, 1961, Yr
Leonardo Seeber, (B), Columbia, 1964, Ys
William Smethie, (D), Washington, 1979, Cg
Emeritus:
   Mark A. Cane, (D), MIT, 1975, Op
   David Walker, (D), Harvard, 1972, Gx
Special Research Scientist:
   Mikhail Kogan, (D), Inst of Physics (Moscow), 1977, Yd
Senior Research Scientist:
   Andrew Barclay, (D), Oregon, 1998, Ys
Research Scientist:
   Victoria Ferrini, (D), SUNY (Stony Brook), 2004, Ou
   Helga Gomes, (D), Bombay, 1985, Ob
   Gilles Guerin, (D), Columbia, 2000, Gu
   Naomi Henderson, (D), Wisconsin, 1987, Yr
   Timothy Kenna, (D), MIT/WHOI, 2002, Cg
   Frank Nitsche, (D), Alfred Wegener Inst (Germany), 1997, Gu
Lamont Associate Research Professor:
   Laia Andreu-Hayles, (D), Barcelona, 2007, Pe
   Natalie Boelman, (D), Columbia, 2004, Zn
   Timothy Crone, Washington, 2007, Gu
   William Joseph D'Andrea, (D), Brown, 2008, Pe
   Solange Duhamel, (D), Aix-Maraseille II, Pe
   Jonathan E. Nichols, (D), Brown, 2009, Pe
   Pratigya J. Polissar, (D), Massachusetts, 2005, Gg
   Michael Previdi, (M), Rutgers, 2006, As
   Heather M. Savage, (D), Penn State, 2007, Ys
   Beizhan Yan, (D), Rensselaer, 2004, Cg
Lamont Assistant Research Professor:
   Anne Becel, (D), Inst de Physique du Globe de Paris, 2006, Gu
   Einat Lev, (D), MIT, 2009, Gv
Lamont Associate Research Professor:
   Timothy T. Creyts, (D), British Columbia, 2007, Gl

Cornell University
Dept of Earth & Atmospheric Sciences (B,M,D) O☒ (2018)
2122 Snee Hall
Ithaca, NY 14853-1504
   p. (607) 255-3474
   easinfo@cornell.edu
   http://www.eas.cornell.edu/
   Programs: Earth and Atmospheric Sciences (B); Atmospheric Science (B,M,PhD); Geological Sciences (M,PhD); Geological Sciences (MEng)
   Enrollment (2018): B: 43 (13) M: 3 (1) D: 24 (8)
Wold Family Professor in Environmental Balance for Human Sustainability:
   John F.H. Thompson, (D), Toronto, 1982, EmgNx
William and Katherine Snee Professor of Geological Sciences:
   Geoffrey A. Abers, (D), MIT, 1989, YsGt
   Suzanne M. Kay, (D), Brown, 1975, GiCuGz
Sidney Kaufman Professor in Geophysics:
   Larry D. Brown, (D), Cornell, 1976, Ye
J. Preston Levis Professor of Engineering:
   Teresa E. Jordan, (D), Stanford, 1979, GrtPc
Irving Porter Church Professor of Engineering:
   Natalie M. Mahowald, (D), MIT, 1996, As
Hunter R. Rawlings III Professor of Paleontology:
   Warren D. Allmon, (D), Harvard, 1988, PgePi
Charles L. Pack Professor:
   Susan Riha, (D), Washington, 1980, Sf
Professor:
   Richard W. Allmendinger, (D), Stanford, 1979, Gc
   Stephen J. Colucci, (D), SUNY (Albany), 1982, As
   Arthur DeGaetano, (D), Rutgers, 1989, As
   Louis A. Derry, (D), Harvard, 1989, Cl
   Charles H. Greene, (D), Washington, 1985, Ob
   David Hysell, (D), Cornell, 1992, Znr
   Matthew E. Pritchard, (D), Caltech, 2003, YdGvl
   Sara C. Pryor, (D), East Anglia, 1992, AsZn
   William M. White, (D), Rhode Island, 1977, CctCa

Associate Professor:
   Esteban Gazel, (D), Rutgers, 2009, CeGiv
   Katie M. Keranen, (D), Stanford, 2008, Ys
   Rowena B. Lohman, (D), Caltech, 2004, YsZrGt
Assistant Professor:
   Toby R. Ault, (D), Arizona, 2011, Am
Sr. Lecturer:
   Bruce Monger, (D), Hawaii, 1993, Ob
Senior Lecturer:
   Mark Wysocki, (M), Cornell, 1988, As
Adjunct Associate Professor:
   Gregory P. Dietl, (D), North Carolina State, 2002, Peg
Adjunct Professor:
   Martin J. Evans, (D), Wales, 1985, Eo
   Jason Phipps Morgan, (D), Brown, 1985, GvCmPe
   Diego Pol, (D), Columbia, 2005, Pv
   Robert M. Ross, (D), Harvard, 1990, PgZeGs
   Manfred Strecker, (D), Cornell, 1987, Gt
   Martyn Unsworth, (D), Cambridge, 1991, Ye
Emeritus:
   Muawia Barazangi, (D), Columbia, 1971, YsGt
   William A. Bassett, (D), Columbia, 1959, Gz
   Lawrence M. Cathles, (D), Princeton, 1971, GqYgEg
   John L. Cisne, (D), Chicago, 1973, PgsPe
   Bryan L. Isacks, (D), Columbia, 1965, GtYsGm
   Daniel E. Karig, (D), California (San Diego), 1970, YrGl
   Robert W. Kay, (D), Columbia, 1970, Giz
   Warren Knapp, (D), Wisconsin, 1968, As
   Frank H. T. Rhodes, (D), Birmingham, 1950, Pi
   Daniel Wilks, (D), Oregon, 1986, As
Cooperating Faculty:
   Ludmilla Aristilde, (D), California (Berkeley), 2008, Ge
   Rebecca J. Barthelmie, (D), As
   J. Thomas Brenna, (D), Cornell, 1985, Cs
   Oliver H. Gao, (D), California (Davis), 2004, Zn
   Alexander G. Hayes, (D), Caltech, 2011, Xg
   Peter G. Hess, (D), Ac
   Jonathan I. Lunine, (D), Xc
   Sturt W. Manning, (D), Cambridge, 1995, Ga
   Greg C. McLaskey, (D), California (Berkeley), 2011, Ys
   Thomas D. O'Rourke, (D), Illinois, 1975, Ng
   Andy L. Ruina, (D), Brown, 1981, Yx
   Steven W. Squyres, (D), Cornell, 1981, Xg
   Tammo S. Steenhuis, (D), Wisconsin, 1977, Hg
   Scott Steinschneider, (D), Rw
   Yervant Terzian, (D), Xy
   Jefferson W. Tester, (D), MIT, 1971, Ng
   Zellman Warhaft, (D), London, 1975, Zn
   Max Zhang, (D), California (Davis), 2004, Zn

Institute for the Study of the Continents ☒ (2018)
2122 Snee Hall
Ithaca, NY 14853-1504
   p. (607) 255-3474
   easinfo@cornell.edu
   http://www.eas.cornell.edu/
   Enrollment (2010): M: 9 (0) D: 29 (4)
Professor:
   Geoffrey A. Abers, (D), MIT, 1989, Ys
   Richard W. Allmendinger, (D), Stanford, 1979, Gct
   Larry D. Brown, (D), Cornell, 1976, YesGt
   Louis A. Derry, (D), Harvard, 1989, Cg
   David L. Hysell, (D), Cornell, 1992, Znr
   Teresa E. Jordan, (D), Stanford, 1979, GrtZg
   Suzanne M. Kay, (D), Brown, 1975, GiCuGt
   William M. White, (D), Rhode Island, 1977, Ce
Associate Professor:
   Katie M. Keranen, (D), Stanford, 2008, Ys
   Rowena B. Lohman, (D), Caltech, 2004, Yg
   Matthew E. Pritchard, (D), Caltech, 2003, YdGvl
Assistant Professor:
   Greg C. McLaskey, (D), California (Berkeley), 2011, YsNre
Visiting Professor:
   Franklin G. Horowitz, (D), Cornell, 1989, YgZnYe
Emeritus:
   Muawia Barazangi, (D), Columbia, 1971, Ys
   Bryan L. Isacks, (D), Gmt

Robert W. Kay, (D), Columbia, 1970, Gi

Dowling College
Dept of Earth & Marine Sciences (2015)
Oakdale, NY
    asmirnov@dowling.edu

Dutchess Community College
Physical Science (A) (2017)
53 Pendell Road
Poughkeepsie, NY 12601
    p. (845) 431-8550
    rambo@sunydutchess.edu
    http://www.sunydutchess.edu/academics/departments/mathematicsphysicalandcomputersciences/
    Enrollment (2016): A: 5 (2)
Chair:
    Tim Welling, (M), GeSo
Professor:
    Mark McConnaughhay, ZgAmEo
Associate Professor:
    Susan H. Conrad, GsmHs

Graduate School of the City University of New York
PhD Program in Earth & Environmental Sciences (D) ☒ (2018)
365 Fifth Avenue, Room 4306
New York, NY 10016
    p. (212) 817-8240
    ees@gc.cuny.edu
    http://www.gc.cuny.edu/Page-Elements/Academics-Research-Centers-Initiatives/Doctoral-Programs/Earth-and-Environmental-Sciences
    f: https://www.facebook.com/CunyGCEES/
    t: @cunygcees
    Programs: Environmental and Geological Sciences; Geography
    Certificates: GIS
    Administrative Assistant: Judy Li
    Enrollment (2017): D: 99 (11)
Provost:
    William Fritz, (D), Montana (Missoula), Zg
Herbert Kayser Professor:
    Samir Ahmed, (D), Univ Coll (London)
Executive Officer:
    Monica W. Varsanyi, (D), California (Los Angeles)
Distinguished Professor:
    George Hendrey, (D), Washington
Discipline Cooordinator of Geology:
    Nazul Khandaker, (D), Penn State
Professor:
    Terence Agbeyegbe, (D), Essex (UK)
    Sean C. Ahearn, (D), Wisconsin, 1986, Zr
    Jochen Albrecht, (D), Vechta (Germany), Ze
    Teresa Bandosz, (D), Tech (Cracow)
    Stefan Becker, (D), Justus-Liebig, Zn
    Sunil Bhaskaran, (D), New South Wales, 2003, ZfrZi
    Jeffrey Bird, (D), California (Davis), 2001, Sbf
    Anthony Carpi, (D), Cornell
    John A. Chamberlain, (D), Rochester, 1971, Pg
    Zhongqi (Joshua) Cheng, (D), Ohio State, Zg
    Constantin Cranganu, (D), Oklahoma, Zg
    Eric Delson, (D), Columbia, 1973, PvqPs
    Allan Frei, (D), Rutgers
    Alexander Gillerson, (D), Tech (Kazan)
    Ruth Wilson Gilmore, (D), Rutgers
    Hongmian Gong, (D), Georgia
    Yuri Gorokhovich, (D), Grad Ctr (CUNY)
    Kenneth Gould, (D), Northwestern
    Peter Groffman, (D), Georgia, Zg
    Marta Gutman, (D), California (Berkeley)
    Roger A. Hart, (D), Clark
    Gary N. Hemming, (D), SUNY (Stony Brook), Zg
    Reza M. Khanbilvardi, (D), Penn State, 1983, Hy
    Yehuda Klein, (D), California (Berkeley), Hs
    Athanasios Koutavas, (D), Columbia
    Nir Krakauer, (D), Caltech
    Irene S. Leung, (D), California (Berkeley), 1969, Gz
    Tammy Lewis, (D), California (Davis)
    Setha Low, (D), California (Berkeley)
    Allan Ludman, (D), Pennsylvania, 1969, Gg
    Johnny Luo, (D), Columbia, Zg
    Juliana Maantay, (D), Rutgers
    Peter Marcotullio, (D), Columbia
    Steven Markowtiz, (D), Columbia Coll
    John Marra, (D), Dalhousie, Zg
    Kyle McDonald, (D), Michigan, As
    Cecilia M. McHugh, (D), Columbia, 1993, Ou
    Alfredo Morabia, (D), Johns Hopkins, Zg
    Fred Moshary, (D), Columbia
    Stephen Pekar, (D), Rutgers, Zg
    Jonathan Peters, (D), CUNY (Grad Ctr)
    Wayne G. Powell, (D), Queen's, 1994, EmGaCs
    Robert S. Prezant, (D), Delaware (Lewes), 1981, Pg
    Patricia Price, (D), Washington
    Leonid Roytman, (D), Moscow Poly Inst
    Randye Rutberg, (D), Columbia
    David E. Seidemann, (D), Yale, 1976, Cc
    John Seley, (D), Pennsylvania
    William Solecki, (D), Rutgers, Zy
    Gillian Stewart, (D), SUNY (Stony Brook), Oc
    Charles Vorosmarty, (D), New Hampshire, Ng
    John Waldman, (D), Grad Ctr (CUNY), ObHsZc
    William G. Wallace, (D), SUNY (Stony Brook), 1996, Og
    William Wallace, (D), SUNY (Stony Brook), Ge
    Michael Weisberg, (D), Grad Ctr (CUNY), Zy
    Dennis Weiss, (D), New York, 1971, Pg
    Pengfei Zhang, (D), Utah
    Yan Zheng, (D), Columbia, 1999, Cm
    Sharon Zukin, (D), Columbia
Associate Professor:
    Stephen U. Aja, (D), Washington State, 1989, Cl
    Homar S. Barcena, (D), New York
    James Biles, (D), Michigan State
    Karin Block, (D), CUNY (Grad Ctr), Ze
    Rebecca Boger, (D), Virginia Inst of Marine Sci, Zg
    Brett Branco, (D), Connecticut
    Frank S. Buonaiuto Jr., (D), SUNY (Stony Brook), Ze
    Jennifer Cherrier, (D), Florida State, Zg
    Ratan Dahr, (D), Grad Ctr (CUNY)
    Timothy Eaton, (D), Wisconsin, Zg
    Jean Grassman, (D), California (Berkeley)
    Mohamed B. Ibrahim, (D), Alberta, 1985, Zy
    Urs Jans, (D), Swiss Fed Inst Tech
    Peter Kabachnik, (D), California (Los Angeles)
    Cary Karacas, (D), California (Berkeley)
    Patricia M. Kenyon, (D), Cornell, 1986, Yg
    Jacob Mey, (D), Grad Ctr (CUNY)
    Inez Miyares, (D), Arizona State, 1994, Zn
    Hamidreza Norouzi, (D), Grad Ctr (CUNY)
    Gregory O'Mullan, (D), Princeton, Zg
    Rupal Oza
    Michael Piasecki, (D), Michigan
    Laxmi Ramasubramanian, (D), Wisconsin
    Haydee Salmun, (D), Johns Hopkins
    Heather Sloan, (D), Paris, Zz
    Zhengrong Wang, (D), Caltech
    Chuixiang Yi, (D), Nanjing (China)
Associate Scientist:
    Maria Tzortziou, (D), Maryland, As
Assistant Professor:
    Kafui Attoh, (D), Syracuse
    Benjamin Black, Mass, As
    William Blanford, (D), Arizona, Ze
    Christopher Blaszczak-Boxe, (D), CUNY (Grad Ctr), Ze
    James Booth, (D), Washington
    Jean Carmalt, (D), Washington, Ze
    Kennet Flores, (D), Lausann, Zg
    Dianne I. Greenfield, (D), Stony Brook
    Steven Kidder, (D), Caltech, As
    David Lindo Atachati, (D), Georgia, Og
    Marc-Antoine Longpre, (D), Trinity (Dublin), Zg
    Elia Machado, (D), Clark
    Andrew Maroko, (D), Grad Ctr (CUNY)

Mcihael Menser, (D), CUNY (Grad Ctr)
Hari Pant, (D), Dalhousie, Zg
Andrew Reinmann, (D), Boston
Brian Rosa, (D), Manchester
Brianne Smith, (D), Princeton
Filip Stabrowski, (D), California (Berkeley)
Adjunct Professor:
Harold C. Connolly Jr., (D), Gg
Damon Ebel, (D), Purdue
John T. Flynn, (D), Columbia
Vinay Gidwani, (D), California (Berkeley), Zn
Nathalie Goodkin, (D), MIT/WHOI
Julian Gross, (D), Bochum (Germany), Ze
George E. Harlow, (D), Princeton, 1977, Gz
Carsten Kessler, (D), Munster
Neil H. Landman, (D), Yale, 1982, Ps
Edmond A. Mathez, (D), Washington, 1981, Gi
Jin Meng, (D), Columbia, 1991, Pv
Robert P. Nolan, (D), CUNY, 1986, Co
Karl-Heinz Szekielda, (D), Aix-Marseille, Zg
Marco Tedesco, (D)
Distinguished Professor Emeritus:
Martin Paul Schreibman, (D), New York
Emeritus:
Hannes Brueckner, (D), Yale, Ze
Nehru Cherukupalli, (D), Madras
Saul Cohen, (D), Harvard
Otto L. Franke, (D), Karlsruhe, 1962, Gc
Daniel Habib, (D), Penn State, 1965, Pl
William H. Harris, (D), Brown, 1971, Hw
Arthur M. Langer, (D), Columbia, 1965, Gz
Richard S. Liebling, (D), Columbia, 1963, Gz
David C. Locke, (D), Kansas State, 1965, Ca
Peter H. Mattson, (D), Princeton, 1957, Gc
Andrew McIntyre, (D), Columbia, 1967, Pe
Cherukupalli E. Nehru, (D), Madras (India), 1963, Gi
Surenda K. Saxena, (D), Uppsala, 1964, Cg
B. Charlotte Schrieber, (D), Rensselaer, 1974, Gd
Frederick C. Shaw, (D), Harvard, 1965, Pi
David L. Thurber, (D), Columbia, 1964, Cl
College Assistant:
Benjamin DeMott, Zn

Hamilton College
Geosciences Department (B) ☒ (2018)
198 College Hill Road
Clinton, NY 13323
p. (315) 859-4142
dbailey@hamilton.edu
https://my.hamilton.edu/academics/departments?dept=Geosciences
Programs: Geology; Geoarchaeology
Enrollment (2018): B: 10 (7)
Chair:
David G. Bailey, (D), Washington State, 1990, GizGa
Professor:
Cynthia R. Domack, (D), Rice, 1985, Pg
Todd W. Rayne, (D), Wisconsin, 1993, Hy
Barbara J. Tewksbury, (D), Colorado, 1981, Gc
Associate Professor:
Michael L. McCormick, (D), Michigan, 2002, Cb
Assistant Professor:
Catherine C. Beck, (D), Rutgers, 2015, GsnGr

Hartwick College
Dept of Geological and Environmental Sciences (B) (2015)
Johnstone Science Center
1 Hartwick Drive
Oneonta, NY 13820
p. (607) 431-4658
griffingd@hartwick.edu
http://www.hartwick.edu/geology.xml
Administrative Assistant: Nancy Heffernan
Enrollment (2011): B: 28 (9)
Chair:
David H. Griffing, (D), Binghamton, 1994, GduGm

Professor Emeritus:
David Hutchison, (D), West Virginia, 1968, GxgGi
Professor:
Eric L. Johnson, (D), SUNY (Binghamton), 1990, GpcGi
Robert C. Titus, (D), Boston, 1974, Ps
Associate Professor:
Zsuzsanna Balogh-Brunstad, (D), Washington State, 2006, ClHgSo

Hobart & William Smith Colleges
Dept of Geoscience (B) ● ☒ (2018)
300 Pulteney Street
Geneva, NY 14456
p. (315) 781-3586
geoscience@hws.edu
http://www.hws.edu/academics/geoscience/
f: https://www.facebook.com/geoscience.hws
Programs: Geoscience (B)
Enrollment (2018): B: 31 (21)
Professor:
Nan Crystal Arens, (D), Harvard, 1993, Pg
John D. Halfman, (D), Duke, 1987, GeHsGn
Neil Laird, (D), Illinois, 2001, As
Associate Professor:
Tara M. Curtin, (D), Arizona, 2001, PeGs
David Finkelstein, (D), Illinois (Urbana), 1997, CgbCs
David C. Kendrick, (D), Harvard, 1997, PgGg
Nicholas Metz, (D), SUNY (Albany), 2011, As
Technician:
Barbara Halfman, Zn

Hofstra University
Dept of Geology, Environment, and Sustainability (B,M) ☒ (2019)
114 Hofstra University
Hempstead, NY 11549
p. (516) 463-5564
j.b.bennington@hofstra.edu
http://www.hofstra.edu/Academics/Colleges/HCLAS/GEOL/
f: https://www.facebook.com/GESatHU/
Programs: BA / BS Geology, BS Environmental Resources, BA /BS Sustainability Studies, MA Sustainability
Administrative Assistant: Lena Hiller
Enrollment (2018): B: 50 (10) M: 25 (7)
Professor and Chair:
J Bret Bennington, (D), Virginia Tech, 1994, PeGs
Vice Provost for Research and Engagement:
Robert Brinkmann, (D), Wisconsin, 1989, So
Associate Professor:
Emma Christa Farmer, (D), Columbia, 2005, Ou
Director of Sustainability Studies:
Sandra J. Garren, (D), S Florida, 2014, Zi
Assistant Professor:
Jase Bernhardt, Penn State, 2016, Ams
Antonios Marsellos, (D), SUNY (Albany), 2008, GtqZf
Instructor:
Annetta Centrella-Vitale, (M), Stony Brook, 1998
Adina Hakimian, (M), CUNY (Queens), 2012, Gg
Steven C. Okulewicz, (M), CUNY (Brooklyn), 1979, CgGue
Adjunct Professor:
Nehru Cherukupalli, (D), Madras, 1963, Gg
Lillian Hess Tanguay, (D), CUNY, 1993, Gg
Richard Liebling, (D), Columbia, 1963, Gg
Emeritus:
Charles M. Merguerian, (D), Columbia, 1985, Gc
Dennis Radcliffe, Queens, 1966, Gzx

Hudson Valley Community College
Biology, Chemistry, Physics Dept (2015)
80 Vandenburgh Ave.
Troy, NY 12180
p. (518) 629-7453
p.schaefer@hvcc.edu
Assistant Professor:
Ruth H. Major, (M), Syracuse, 1989, GgRh

Hunter College (CUNY)
Dept of Geography (B,M) (2017)
695 Park Avenue
Room 1006 North Building
New York, NY 10021
> p. (212) 772-5265
> imiyares@hunter.cuny.edu
> http://www.geo.hunter.cuny.edu
> Programs: Geography; Environmental Studies; Earth Science Education; Geoinformatics (M); Geography (M)
> Certificates: GIS
> Administrative Assistant: Dana G.. Reimer
> *Enrollment (2006): B: 135 (30) M: 40 (8)*

Chair:
  Allan Frei, (D), Rutgers, 1997, AtHgs
Professor:
  Sean C. Ahearn, (D), Wisconsin, 1986, Zi
  Jochen Albrecht, (D), Vechta (Germany), 1995, ZiyZu
  Hongmian Gong, (D), Georgia, 1997, Zn
  Ines Miyares, (D), Arizona State, 1994, Zen
  Wenge Ni-Meister, (D), Boston, 1997, Zr
  Marianna Pavlovskaya, (D), Clark, 1998, Zi
  William Solecki, (D), Rutgers, 1990, Zu
Associate Professor:
  Frank Buonaiuto, (D), SUNY (Stony Brook), 2003, On
  Mohamed Ibrahim, (D), Alberta, 1985, ZgRwZc
  Rupal Oza, (D), Rutgers, 1999, Zn
  Haydee Salmun, (D), Johns Hopkins, 1989, OpZgAs
Assistant Professor:
  Randye L. Rutberg, (D), Columbia, 2000, CmOcZc
Director, SPARs Lab:
  Thomas Walter, (M), Miami, 1984, ZyeAm
Lecturer:
  Anthony Grande, (M), CUNY (Baruch), 1999, Zn
Adjunct Professor:
  Jack Eichenbaum, (D), Michigan, 1972, Zie
  Edward Linky, (D), Duquesne Law, 1973, Zn
  Teodosia Manecan, (D), Bucharest, 1985, Gp
  Faye Melas, (D), CUNY, 1989, Gs
  Douglas Williamson, (D), CUNY, 2003, Zi
Emeritus:
  Charles A. Heatwole, (D), Michigan State, 1974, Zn
  Karl H. Szekielda, (D), Marseille, 1967, Zrg
GeoScience Lab Tech:
  Amy Jeu, (M), Minnesota, 2002, Zi

Lehman College (CUNY)
Earth, Environmental and Geospatial Sciences (B) ☒ (2019)
250 Bedford Park Boulevard West
Bronx, NY 10468-1589
> p. (718) 960-8660
> heather.sloan@lehman.cuny.edu
> http://www.lehman.edu/academics/eggs/
> Programs: BA in Earth Science, BA in Geography, BS in Environmental Science, MS in GISc
> Certificates: Certificate in Earth Science, Certificate in GIS
> *Enrollment (2014): B: 24 (5)*

Chair:
  Hari Pant, (D), Dalhousie, Cg
Professor:
  Irene S. Leung, (D), California (Berkeley), 1969, Gz
  Juliana Maantay, (D), Rutgers, Zi
Associate Professor:
  Elia Merchado
  Heather Sloan, (D), Paris VI, 1993, Yr
Emeritus:
  Frederick C. Shaw, (D), Harvard, 1965, Ps

Long Island University, Brooklyn Campus
Dept of Physics (2015)
1 University Plaza
Brooklyn, NY 11201-8423
> p. (718) 488-1011
> bkln-admissions@liu.edu
> http://www.liu.edu/Home/Brooklyn

Professor:
  Richard Macomber, (D), Iowa, 1963, Pg
Adjunct Professor:
  Richard A. Jackson, (D), Massachusetts, 1980, Gc
  Alan Siegelberg, (M), Brooklyn, 1977, Gg
Emeritus:
  Samuel R. Kamhi, (D), Columbia, 1963, Gz

Long Island University, C.W. Post Campus
Dept of Earth & Environmental Sciences (B,M) (2015)
720 Northern Boulevard
Brookville, NY 11548-1300
> p. (516) 299-2318
> margaret.boorstein@liu.edu
> Administrative Assistant: Beth Rondot

Chair:
  Margaret F. Boorstein, (D), Columbia, 1977, Zg
Associate Professor:
  Victor DiVenere, (D), Columbia, 1995, Ym
  Lillian Hess-Tanguay, (D), CUNY, 1993, Gs
  E Mark Pires, (D), Michigan State, 1998, Zg
Emeritus:
  Robert S. Harrison, (D), Cambridge, 1965, Zy
  Heinrich Toots, (D), Wyoming, 1965, Pg

Monroe Community College
Geoscience Dept (A) (2016)
1000 E. Henrietta Road
Rochester, NY 14623
> p. (716) 292-2425
> drobertson@monroecc.edu
> http://www.monroecc.edu/depts/geochem/
> Administrative Assistant: Judy Miller
> *Enrollment (2013): A: 7 (0)*

Associate Professor:
  Jessica Barone, (M), Ball State, Ge
  Michael Boester, (M), Zy
  Amanda Colosimo, (M), North Carolina, 2004, Gg
  Daniel E. Robertson, (M), Arizona State, 1986, Eg
Assistant Professor:
  Jonathan Little, (M), ZyGlZi
  Jason Szymanski, (M), GlPe
Instructor:
  Heather Pierce, (M), Connecticut, Zyi

Orange County Community College
Dept of Science, Engineering, and Architecture (A) (2015)
115 South Street
Middletown, NY 10940
> p. (845) 341-4570
> lawrenceobrien@sunyorange.edu
> *Enrollment (2015): A: 3 (1)*

Professor:
  Lawrence E. O'Brien, (M), Michigan, 1972, Gg

Pace University, New York Campus
Dept of Chemistry & Physical Sciences (2015)
1 Pace Plaza
New York, NY 10038
> p. (212) 346-1502
> mshirigarakani@pace.edu
> http://www.pace.edu/dyson/academic-departments-and-programs/chemistry-and-physical-sciences---nyc
> Department Secretary: Pat Calegari

Chair:
  Nigel Yartlett, (D)
Assistant Professor:
  Stephen T. Lofthouse, (M), Hunter (CUNY), 1974, Zg
Adjunct Professor:
  Anatole Dolgoff, (M), Miami (Ohio), 1960, Zg
  William Hansen, (M), Hunter, 1991, As
  John Marchisin, (M), Rutgers, 1965, Zg
  Nathan Reiss, (D), New York, 1973, As

Paleontological Research Institution
Paleontological Research Instituion ☒ (2019)
1259 Trumansburg Road
Ithaca, NY 14850
    p. (607) 273-6623
    allmon@priweb.org
    http://www.priweb.org
    f: https://www.facebook.com/museumoftheearth/
    t: @PRInstitution
Director:
    Warren D. Allmon, (D), Harvard, 1988, Pg
Education Director:
    Robert M. Ross, (D), Harvard, 1990, PgZePc
Director of Collections:
    Gregory Dietl, (D), North Carolina State, 2002, Pg
Director of Publications:
    Jonathan R. Hendricks, (D), Cornell, 2005, Pig

Queens College (CUNY)
School of Earth & Environmental Sciences (B,M,D) (2015)
65-30 Kissena Boulevard
Flushing, NY 11367
    p. (718) 997-3300
    gregory.omullan@qc.cuny.edu
    http://www.qc.edu/EES
    Administrative Assistant: Gladys Sapigao
    Enrollment (2012): B: 150 (19) M: 19 (6) D: 11 (2)
Distinguished Professor, Director & Chair:
    George Hendrey, (D), Washington, 1973, Zg
Professor:
    Nicholas K. Coch, (D), Yale, 1965, GseOu
    N. Gary Hemming, (D), Stony Brook, 1993, CaOcGg
    Allan Ludman, (D), Pennsylvania, 1969, GgtGr
    Steven Markowitz, (D), Columbia, 1981, GbZn
    Cecilia McHugh, (D), Columbia, 1993, Ou
    Alfredo Morabia, (D), Johns Hopkins, 1989, GbZn
    Stephen Pekar, (D), Rutgers, 1999, PeGrOu
    Gillian Stewart, (D), SUNY (Stony Brook), 2005, OcCm
    Yan Zheng, (D), Columbia, 1998, CgHwCm
Associate Professor:
    Jeffrey Bird, (D), California (Davis), 2001, Sbc
    Timothy Eaton, (D), Wisconsin, 2002, HgwGg
Assistant Professor:
    Gregory O'Mullan, (D), Princeton, 2005, PgOb
    Ashaki Rouff, (D), Stony Brook, 2004, Cac
    Chuixiang Yi, (D), Nanjing, 1991, Asm
Emeritus:
    Eugene A. Alexandrov, (D), Columbia, 1959, Eg
    Patrick W. G. Brock, (D), Leeds, 1963, Gg
    Hannes K. Brueckner, (D), Yale, 1968, GgtCg
    Robert M. Finks, (D), Columbia, 1959, Pi
    Daniel Habib, (D), Penn State, 1965, Pl
    Peter H. Mattson, (D), Princeton, 1957, Gc
    Andrew McIntyre, (D), Columbia, 1967, Pe
    B. Charlotte Schreiber, (D), Rensselaer, 1974, Gd
    David H. Speidel, (D), Penn State, 1964, Cgp
    David L. Thurber, (D), Columbia, 1964, Cl

Queensborough Community College
Dept of Biological Sciences and Geology (2015)
222-05 56th Avenue
Bayside, NY 11364
    p. (718) 631-6335
    MGorelick@qcc.cuny.edu
    http://www.qcc.cuny.edu/biologicalsciences/advisors.asp

Rensselaer Polytechnic Institute
Dept of Earth & Environmental Sciences (B,M,D) ☒ (2017)
Science Center 1W19
110 8th Street
Troy, NY 12180-3590
    p. (518) 276-6474
    ees@rpi.edu
    http://www.rpi.edu/dept/ees/
    Programs: Geology (B,M,D); Hydrogeology (B); Environmental Science (B)
    Enrollment (2017): B: 29 (6) M: 1 (3) D: 11 (2)
Head:
    Frank S. Spear, (D), California (Los Angeles), 1976, GpCpGc
Professor:
    Peter A. Fox, (D), Monash, 1985, GqZig
    Steven W. Roecker, (D), MIT, 1981, Yg
    E. Bruce Watson, (D), MIT, 1976, CpcCt
Associate Professor:
    Richard F. Bopp, (D), Columbia, 1979, Co
    Miriam E. Katz, (D), Rutgers, 2001, PmeGu
Assistant Professor:
    Karyn L. Rogers, (D), Washington, 2006, PgCl
    Morgan F. Schaller, (D), Rutgers, 2011, CsPe
Research Associate Professor:
    Daniele J. Cherniak, (D), SUNY (Albany), 1990, Gx
Research Associate:
    Michael R. Ackerson, (D), Renssealear, 2015, CpGiCu
Emeritus:
    M. Brian Bayly, (D), Chicago, 1962, Gc
    Samuel Katz, (D), Columbia, 1955, Yg
    Robert G. La Fleur, (D), Rensselaer, 1961, Hw
    Donald S. Miller, (D), Columbia, 1960, Cc
Laboratory Director:
    Jared W. Singer, (D), Alfred, 2013, CaZm

Skidmore College
Dept of Geosciences (B) ☒ (2019)
815 North Broadway
Saratoga Springs, NY 12866
    p. (518) 580-5190
    afrappie@skidmore.edu
    http://www.https://www.skidmore.edu/geosciences/
    Programs: Geosciences
    Enrollment (2018): B: 14 (3)
Associate Professor:
    Amy Frappier, (D), New Hampshire, 2006, GePe
    Kyle K. Nichols, (D), Vermont, 2002, Gm
Visiting Assistant Professor:
    Margaret Estapa, (D), Maine, 2011, Oc
Assistant Professor:
    Greg Gerbi, (D), MIT/WHOI, OpYg
    Victor Guevara, (D), Virginia Tech, 2017, Gp
Instructor:
    Jennifer Cholnoky, (M), Rensselaer, 2013, Gg

St. Lawrence University
Department of Geology (B) ☒ (2019)
23 Romoda Dr.
Brown Hall
Canton, NY 13617-1475
    p. (315) 229-5851
    astewart@stlawu.edu
    https://www.stlawu.edu/geology
    f: https://www.facebook.com/SLUGeology
    Administrative Assistant: Sherrie Kelly
    Enrollment (2016): B: 35 (17)
Professor:
    Jeffrey R. Chiarenzelli, (D), Kansas, 1989, GzCg
    Antun Husinec, (D), Zagreb, 2002, GsCsGo
Chair:
    Alexander K. Stewart, (D), Cincinnati, 2007, GlmHw
Assistant Professor:
    Judith Nagel-Myers, (D), Muenster, 2006, PisPe
Research Associate:
    George W. Robinson, (D), Queens, 1978, Gz
Visiting Professor:
    Erkan Toraman, (D), Minnesota, 2014, GciGp
Emeritus:
    J. Mark Erickson, (D), North Dakota, 1971, Pgi
Technician:
    Matthew F. Van Brocklin, (M), Akron, 1996, GgZgn

Stony Brook University
Dept of Geosciences (B,M,D) ☒ (2018)

Nicolls Road
Stony Brook, NY 11794-2100
 p. (631) 632-8200
 brian.phillips@stonybrook.edu
 https://www.stonybrook.edu/geosciences
Professor:
 Daniel M. Davis, (D), MIT, 1983, Yg
 Gilbert N. Hanson, (D), Minnesota, 1964, GgZe
 William E. Holt, (D), Arizona, 1989, Ys
 Baosheng Li, (D), SUNY (Stony Brook), 1996, Gy
 Robert C. Lieberman, (D), Columbia, 1969, GyYsx
 Scott M. McLennan, (D), Australian Nat, 1981, Cg
 Hanna Nekvasil, (D), Penn State, 1986, Cp
 Artem Oganov, (D), Univ Coll (London), 2002, Gz
 John B. Parise, (D), James Cook, 1980, Gz
 Brian L. Phillips, (D), Illinois, 1990, GzCl
 Richard J. Reeder, (D), California (Berkeley), 1980, ClGz
 Martin A. Schoonen, (D), Penn State, 1989, Cl
 Donald J. Weidner, (D), MIT, 1972, Gy
 Lianxing Wen, (D), Caltech, 1998, Ys
Associate Professor:
 Timothy Glotch, (D), Arizona State, 2004, XgGz
 E. Troy Rasbury, (D), SUNY (Stony Brook), 1998, Cc
Assistant Professor:
 Michael Sperazza, (D), Montana, 2006, Zn
Curator:
 Stephen C. Englebright, (M), SUNY (Stony Brook), 1975, Ze
Lecturer:
 Christiane W. Stidham, (D), California (Berkeley), 1999, YgGe
Adjunct Professor:
 Robert C. Aller, (D), Yale, 1977, Cm
 Henry J. Bokuniewicz, (D), Yale, 1976, Yr
 J. Kirk Cochran, (D), Yale, 1979, Oc
 Roger D. Flood, (D), MIT, 1978, Gu
 Maureen O'Leary, (D), Johns Hopkins, 1997, Pv
 Michael T. Vaughan, (D), SUNY (Stony Brook), 1979, Yx
Emeritus:
 Garman Harbottle, (D), Columbia, 1949, Cc
 David W. Krause, (D), Michigan, 1982, Pv
 Donald H. Lindsley, (D), Johns Hopkins, 1961, Cp
 Teng-fong Wong, (D), MIT, 1981, Yx
Laboratory Director:
 Owen C. Evans, (D), SUNY (Stony Brook), 1994, Cg

Suffolk County Community College, Ammerman Campus
Dept of Physical Science (A) (2016)
533 College Road
Selden, NY 11784
 p. (631) 451-4338
 butkosd@sunysuffolk.edu
 http://depthome.sunysuffolk.edu/Selden/PhysicalScience/
 Enrollment (2016): A: 27 (0)
Professor:
 Darryl J. Butkos, (M), GgHw
 Michael Inglis, (D), Zg
 Scott A. Mandia, (M), Penn State, 1990, Amt
Associate Professor:
 Matthew Pappas, Zg
Assistant Professor:
 Sean Tvelia, (M), SUNY (Stony Brook), Ggl
Adjunct Professor:
 Jessica Dutton, Zg
 Michael Flanagan, Zg
 Philip Harrington, Zg
 Margaret Lomaga, Zg
 Brian Vorwald, Zg

Sullivan County Community College
Mathematics and Natural Sciences (2015)
112 College Road
Lock Sheldrake, NY 12759
 dlewkiewicz@sunysullivan.edu

SUNY Jefferson
Science (2015)
1220 Coffeen Street
Watertown, NY 13601
 p. (315) 786-2200
 cebeyhoneycutt@sunyjefferson.edu
 http://www.sunyjefferson.edu/academics/programs-study/liberal-arts-sciences-mathematics-science

SUNY Potsdam
Department of Geology (B) ☒ (2018)
220 Timerman Hall
44 Pierrepont Avenue
Potsdam, NY 13676
 p. (315) 267-2286
 rygelmc@potsdam.edu
 http://www.potsdam.edu/academics/AAS/Geology/
 f: https://www.facebook.com/potsdamgeology/
 Programs: Geology, Geographic Information Science
 Administrative Assistant: Roberta Greene
 Enrollment (2018): B: 43 (14)
Professor:
 Michael C. Rygel, (D), Dalhousie, 2005, GsrGo
Assistant Professor:
 Sara E. Bier, (D), Penn State, 2010, GctGg
 Kamal Humagain, (D), Texas Tech, 2016, Zir
 Adam Pearson, (D), Delaware, 2015, Gm
 Page C. Quinton, (D), Missouri, 2016, PcCsPm
 Christian M. Schrader, (D), Georgia, 2009, GiEmGv
Emeritus:
 Robert L. Badger, (D), Virginia Tech, 1989, GipGc
 James D. Carl, (D), Illinois, 1962, Gz
 William T. Kirchgasser, (D), Cornell, 1967, Ps
 Neal R. O'Brien, (D), Illinois, 1963, Gs
 Frank A. Revetta, (D), Rochester, 1970, YgGtg

SUNY, Albany
Dept of Atmospheric and Environmental Sciences (B,M,D) (2015)
1400 Washington Avenue
Albany, NY 12222
 p. (518) 442-4466
 daeschair@albany.edu
 http://www.atmos.albany.edu
Chair:
 Vincent P. Idone, (D), SUNY (Albany), 1982, As
Professor:
 Lance F. Bosart, (D), MIT, 1969, As
 Daniel Keyser, (D), Penn State, 1981, As
 John E. Molinari, (D), Florida State, 1979, As
Senior Research Professor:
 David R. Fitzjarrald, (D), Virginia, 1980, Asm
 Richard R. Perez, (D), SUNY (Albany), 1983, As
 James J. Schwab, (D), Harvard, 1983, Acs
 Christopher J. Walcek, (D), California (Los Angeles), 1983, As
 Wei-Chyung Wang, (D), Columbia, 1973, As
Associate Professor:
 Robert G. Keesee, (D), Colorado, 1979, As
 Christopher D. Thorncroft, (D), Reading, 1988, As
Research Associate:
 Stephen S. Howe, (M), Penn State, 1981, Cs
 David Knight, (D), Washington, 1987, As

SUNY, Buffalo
Dept of Geology (B,M,D) ○◎ (2018)
126 Cooke Hall
Buffalo, NY 14260
 p. (716) 645-3489
 geology@buffalo.edu
 http://www.geology.buffalo.edu
 t: @UBGeology
 Programs: BA, BS Geological Sciences
 MA, MS, PhD Geological Sciences
 Certificates: Graduate: Professional Science Management Certificate
 Administrative Assistant: Alison A. Lagowski
 Enrollment (2018): B: 53 (30) M: 29 (20) D: 15 (5)
Chair:
 Beata M. Csatho, (D), Miskolc (Hungary), 1993, GlZrYg

SUNY Distinguished Teaching Professor:
  Charles E. Mitchell, (D), Harvard, 1983, Pg
Director of Graduate Studies:
  Gregory Valentine, (D), Santa Barbara, 1988, Gv
Professor:
  Richelle Allen-King, (D), Waterloo, 1991, Cg
  Jason P. Briner, (D), Colorado, 2003, Gl
  Marcus I. Bursik, (D), Caltech, 1989, Gv
  Howard R. Lasker, (D), Chicago, 1978, Gu
Associate Professor:
  Tracy K. P. Gregg, (D), Arizona State, 1995, Gv
  Christopher S. Lowry, (D), Wisconsin, 2008, Hw
Assistant Professor:
  Margarete Jadamec, (D), California (Davis), 2009, GtYgZf
  Kimberly Meehan, (D), CUNY City Coll, 2014, Pe
  Erasmus K. Oware, (D), Clemson, 2014, YgHq
  Kristin Poinar, (D), Washington, 2017, Gl
  Ingo Sonder, (D), Wurzburg, 2010, Gv
  Elizabeth K. Thomas, (D), Brown, 2014, Cos
Adjunct Professor:
  Daniel Goode, (D), Princeton, 1998
  Jesse Johnson, (D), Maine, 2002
  Philip Stauffer, (D), California (Santa Cruz), 1999
  Corneilus van der Veen, (D), Utrecht, 1986
Emeritus:
  Mary Alice Coffroth, (D), Miami, 1988, Gu
  Rossman F. Giese, (D), Columbia, 1962, Gz
  Dennis S. Hodge, (D), Wyoming, 1966
  Robert D. Jacobi, (D), Columbia, 1980, GctGs
  Chester C. Langway, (D)
  Michael F. Sheridan, (D), Stanford, 1965, Gv

SUNY, Cortland
Geology Dept (B,M) (2015)
342 Bowers Hall
PO Box 2000
Cortland, NY 13045
  p. (607) 753-2815
  GeologyDept@cortland.edu
  http://www.cortland.edu/geology/
  Administrative Assistant: Susan K. Nevins
  Enrollment (2007): B: 48 (15) M: 8 (6)
Professor:
  David J. Barclay, (D), SUNY (Buffalo), 1998, Glm
  Christopher P. Cirmo, (D), Syracuse, 1994, Hg
  Robert S. Darling, (D), Syracuse, 1992, GzpCp
Associate Professor:
  Christopher A. McRoberts, (D), Syracuse, 1994, Pg
Associate Scientist:
  Gayle C. Gleason, (D), Brown, 1993, Gct
Lecturer:
  Julie L. Barclay, (M), SUNY (Buffalo), 1997, Ggm
Laboratory Director:
  John R. Driscoll, (A), Northwest Electronic, 1974, Zn

SUNY, Fredonia
Department of Geology and Environmental Sciences (B) (2016)
280 Central Avenue
Fredonia, NY 14063-1020
  p. (716) 673-3303
  earth@fredonia.edu
  http://www.fredonia.edu/earth
  Enrollment (2016): B: 40 (7)
Chair:
  Gordon C. Baird, (D), Rochester, 1975, PseGs
Professor:
  Gordon C. Baird, (D), Rochester, 1975, GrPgGs
  Gary G. Lash, (D), Lehigh, 1980, Gr
  Sherri A. Mason, (D), Montana, 2001, ZgGe
Associate Professor:
  Ann K. Deakin, (D), SUNY (Buffalo), 1996, ZiGgZr
Instructor:
  Kimberly Weborg-Benson, (M), Illinois, 1991, GgAsPg

SUNY, Geneseo
Dept of Geological Sciences (B) (2015)
1 College Circle
ISC 235
Geneseo, NY 14454
  p. (585) 245-5291
  lounsbur@geneseo.edu
  http://www.geneseo.edu/geology
  Administrative Assistant: Diane E. Lounsbury
  Enrollment (2015): B: 130 (27)
Chair:
  Benjamin J.C. Laabs, (D), Wisconsin, 2004, GmlGe
Professor:
  Scott D. Giorgis, (D), Wisconsin, 2003, GctYm
  D. Jeffrey Over, (D), Texas Tech, 1990, Ps
Associate Professor:
  Dori J. Farthing, (D), Johns Hopkins, 2001, Gz
  Amy L. Sheldon, (D), Utah, 2002, Hw
Assistant Professor:
  Nicholas H. Warner, (D), Arizona State, 2008, XgGrHg
Emeritus:
  Phillip D. Boger, (D), Ohio State, 1976, Cg
  William J. Brennan, (D), Colorado, 1968, Gc
  Richard B. Hatheway, (D), Cornell, 1969, GxzGp
  James W. Scatterday, (D), Ohio State, 1963, Pg
  Richard A. Young, (D), Washington Univ (St. Louis), 1966, GmXgZr

SUNY, Maritime College
Science Dept (B) (2015)
6 Pennyfield Avenue
Bronx, NY 10465
  p. (718) 409-7380
  kolszewski@sunymaritime.edu
  Enrollment (2009): B: 68 (22)
Chair:
  Kathy Olszewski, (D), SUNY (Stony Brook), 1994, Cg
Associate Professor:
  Marie deAngelis, (D), Washington, 1989, Oc
Assistant Professor:
  Anthony Manzi, (M), Montclair State, Am

SUNY, New Paltz
Geology (B) ☒ (2018)
1 Hawk Drive
New Paltz, NY 12561
  p. (845) 257-3760
  vollmerf@newpaltz.edu
  http://www.newpaltz.edu/geology/
  Programs: Geology
  Environmental Geoscience
  Earth Science Education
  Enrollment (2018): B: 86 (15)
Chair:
  Frederick W. Vollmer, (D), Minnesota, 1985, GctGx
Associate Professor:
  Alexander J. Bartholomew, (D), Cincinnati, 2006, GrPi
  Shafiul H. Chowdhury, (D), W Michigan, 1999, HwGe
  John A. Rayburn, (D), Binghamton, 2004, Gml
Assistant Professor:
  Gordana Garapic, (D), Boston, 2013, Gz
Lecturer:
  Kaustubh Patwardhan, (D), Johns Hopkins, 2009, Ggi
Adjunct Professor:
  Laurel Mutti, (M), Johns Hopkins, 2004, Gg
Emeritus:
  Gilbert J. Brenner, (D), Penn State, 1962, Pl
  Alvin S. Konigsberg, (D), Syracuse, 1969, As
  Constantine Manos, (D), Illinois, 1963, Gs
  Martin S. Rutstein, (D), Brown, 1969, Gz
Related Staff:
  Donald R. Hodder, (B), SUNY (Geneseo), 1985, Gg

SUNY, Oneonta
Dept of Earth and Atmospheric Sciences (B) ☒ (2017)
108 Ravine Parkway
209 Sci. #1
Oneonta, NY 13820-4015

p. (607) 436-3707
Lisa.Hoffman@oneonta.edu
http://www.oneonta.edu/academics/earths/
Programs: Geology; Earth Science; Meteorology; Earth Science Education
Enrollment (2017): B: 91 (0)

Chair:
  Jerome B. Blechman, (D), Wisconsin, 1979, As
Distinguished Teaching Professor:
  James R. Ebert, (D), SUNY (Binghamton), 1984, GrsZe
Associate Professor:
  Keith A. Brunstad, (D), Washington State, 2013, GviEg
  Leigh M. Fall, (D), Texas A&M, 2010, Pgq
  Melissa Godek, (D), Delaware, 2009, AmsZg
  Leslie E. Hasbargen, (D), Minnesota, 2003, GmHgZr
Assistant Professor:
  Christopher Karmosky, (D), Penn State, 2013, AmsZy
Dr.:
  Marta L. Clepper, (D), Kentucky, 2011, ZeGgs
Lecturer:
  Kathryn Metcalf, (M), Arizona, 2012, GctGg
Emeritus:
  P. Jay Fleisher, (D), Washington State, 1967, GlmGe
  Arthur N. Palmer, (D), Indiana, 1969, HgqGm

SUNY, Oswego
Dept of Atmospheric and Geological Sciences (B) ✍ (2018)
394 Shineman Science Center
Oswego, NY 13126
  p. (315) 312-3065
  christine.dallas@oswego.edu
  http://www.oswego.edu/ags
  Programs: Geology; Meteorology
  Enrollment (2015): B: 160 (25)

Professor:
  Alfred J. Stamm, (D), Wisconsin, 1976, Am
  Paul B. Tomascak, (D), Maryland, 1995, CgGzi
  David W. Valentino, (D), Virginia Tech, 1993, GtcYu
Associate Professor:
  Scott Steiger, (D), Texas A&M, 2005, Am
Assistant Professor:
  Rachel J. Lee, (D), Pittsburg, 2013, GvZr
  Steven T. Skubis, (D), SUNY (Albany), 1994, Am
  Justin Stroup, (D), Dartmouth, 2015, GmHwGe
  Michael Veres, (D), Nebraska, 2014, As
Lab/Field Technician:
  Richard Frieman, (M), Buffalo, 2016, Gg

SUNY, Plattsburgh
Center for Earth & Environmental Science (B) ☒ (2019)
101 Broad Street
132 Hudson Hall
Plattsburgh, NY 12901
  p. (518) 564-2028
  cees@plattsburgh.edu
  http://www.plattsburgh.edu/cees
  f: https://www.facebook.com/SUNYPlattsburghCEES/
  Programs: Geology, Environmental Geology, and Earth Science
  Enrollment (2018): B: 18 (3)

Director:
  Edwin A. Romanowicz, (D), Syracuse, 1993, HwGcYu
Professor:
  David A. Franzi, (D), Syracuse, 1984, Gl
  Thomas H. Wolosz, (D), SUNY (Stony Brook), 1984, Pe
Associate Professor:
  Eric Leibensperger, (D), Harvard, 2011, AsOg
Lecturer:
  Keith Gray, (D), Washington State, 2012, Gzc
  Patrick Korths, (M), Syracuse, 2006, RwGg

SUNY, Purchase
Environmental Studies Program (B) ⊘ (2018)
735 Anderson Hill Road
Purchase, NY 10577
  p. (914) 251-6646
  naturalsciences@purchase.edu
  https://www.purchase.edu/academics/environmental-studies/
  Programs: Environmental Studies: Ecology and Policy Concentrations
  Certificates: GIS
  Enrollment (2018): B: 57 (12)

Director of Teaching and Learning Technologies:
  Keith Landa, (D), Michigan, 1989, XgbGg
Professor:
  George P. Kraemer, (D), California (Los Angeles), 1989, ObnCm
Board of Study Coordinator:
  Ryan W. Taylor, (D), Oregon State, 2006, ZiyZg
Assistant Professor:
  Allyson K. Jackson, (D), Oregon State, 2017, ZgcZe
Emeritus:
  James M. Utter, (D), Rutgers, 1971, SfZcRc

SUNY, Stony Brook
School of Marine and Atmospheric Sciences (B,M,D) (2015)
145 Endeavour Hall
Stony Brook, NY 11794-5000
  p. (516) 632-8700
  minghua.zhang@stonybrook.edu
  http://www.somas.stonybrook.edu

Professor:
  Josephine Y. Aller, (D), S California, 1975, ObAs
  Edmund K. M. Chang, (D), Princeton, 1993, As
  J. Kirk Cochran, (D), Yale, 1979, Oc
  Brian Colle, (D), Washington, 1997, As
  David O. Conover, (D), Massachusetts, 1981, Ob
  Roger D. Flood, (D), MIT/WHOI, 1978, Ou
  Marvin A. Geller, (D), MIT, 1969, As
  Christopher Gobler, (D), SUNY (Stony Brook), 1999, Ob
  Sultan Hameed, (D), Manchester, 1968, As
  Darcy J. Lonsdale, (D), Maryland, 1979, Ob
  Glenn R. Lopez, (D), SUNY (Stony Brook), 1976, Ob
  John E. Mak, (D), California (San Diego), 1992, AsCas
  Anne E. McElroy, (D), MIT/WHOI, 1985, ObCb
  Ellen K. Pikitch, (D), Indiana, 1983, Ob
  Mary I. Scranton, (D), MIT/WHOI, 1977, Oc
  R. Lawrence Swanson, (D), Oregon State, 1971, Og
  Gordon T. Taylor, (D), S California, 1983, Obc
  Minghua Zhang, (D), Inst of Atm Physics, 1987, As
Associate Professor:
  Bassem Allam, (D), W Brittany (France), 1998, Ob
  Robert A. Armstrong, (D), Minnesota, 1975, Ocb
  David Black, (D), Ou
  Bruce J. Brownawell, (D), MIT/WHOI, 1986, Oc
  Robert M. Cerrato, (D), Yale, 1980, Ob
  Jackie Collier, (D), Stanford, 1994, Ob
  Michael Frisk, (D), Maryland, 2004
  Marat Khairoutdinov, (D), Oklahoma, 1997, As
  Daniel A. Knopf, (D), Swiss Fed Inst Tech, 2003, As
  Kamazima M. Lwiza, (D), Wales, 1991, Oc
  Bradley Peterson, (D), S Alabama, 1998, Ob
  Joseph Warren, (D), MIT, 2001, Ob
  Robert E. Wilson, (D), Johns Hopkins, 1973, Op
Assistant Science Director, IOCS:
  Demian Chapman, (D), Nova Southeastern, 2007
Assistant Professor:
  Anthony Dvarskas, (D), Maryland, 2007
  Hyemi Kim, (D), Seoul Nat, 2008
  Janet Nye, (D), Maryland, 2008, Ob
  Christopher Wolfe, (D), Oregon State, 2006, Op
  Qingzhi Zhu, (D), Xiamen, 1997, Oc
Faculty Director Semester By The Sea:
  Kurt Bretsch, (D), South Carolina, 2005
Lecturer:
  Lesley Thorne, (D), Duke, 2010
Research Scientist:
  Wuyin Lin, (D), Stony Brook, 2002, As
Research Professor:
  Charles Flagg, (D), MIT, 1977, Op
Engineer:
  Douglas Hill, (D), Columbia, 1977

New York

Ecologist, Author:
　Carl Safina, (D), Rutgers, 1987
Director Riverhead Foundation:
　Robert A. DiGiovanni, Jr., (M), Stony Brook, 2002
Adjunct Professor:
　James Ammerman, (D), Scripps, 1983
　Paul Bowser, (D), Auburn, 1978, Obb
　Carl Brenninkmeijer, (D), Groningen, 1983
　Michael J. Cahill, (D), DePaul, 1978, Zn
　Alistair Dove, Queensland, 1999
　Anga Engel, Oc
　Emmanuelle pales Espinosa, (D), Nante (France), 1999
　Mark Fast, (D), Dalhousie, 2005, Ob
　Scott Ferson, (D), Stony Brook, 1988
　Scott Fowler, (D), 1969
　Roxanne Karimi, (D), Dartmouth, 2007
　Kathryn Kavanagh, (D), James Cook, 1998
　Jeffrey Levinton, (D), Yale, 1971, Ob
　Yangang Liu, (D), Nevada (Reno), 1998
　Stephan Munch, (D), Stony Brook, 2002, Ob
　John Rapaglia, (D), Stony Brook, 2007
　Frank J. Roethel, (D), SUNY (Stony Brook), 1981, Oc
　Jeffrey Tongue, As
　Andrew Vogelmann, (D), Penn State, 1994, As
　Duane E. Waliser, (D), As
　Douglas W. R. Wallace, (D), Dalhousie, 1985, Oc
　Jian Wang, (D), Caltech, 2002, As
Distinguished Professor:
　Cindy Lee, (D), California (San Diego), 1975, OcCo
Emeritus:
　Dong-Ping Wang, (D), Miami, 1975
Affiliated and Joint Faculty:
　Heather L. Lynch, (D), Harvard, 2006
Cooperating Faculty:
　Resit Akcakaya, (D), Stony Brook, 1989
　Stephen Baines, (D), Yale, 1993
　Lee K. Koppelman, (D), New York, 1968, On
　Dianna K. Padilla, (D), Alberta, 1987, Ob
　Sheldon Reaven, (D), California (Berkeley), 1975, Zn

SUNY, The College at Brockport
Dept of the Earth Sciences (B) (2015)
350 New Campus Drive
Brockport, NY 14420-2936
　p. (585) 395-2636
　earthsci@esc.brockport.edu
　http://www.brockport.edu/esc
　Enrollment (2012): B: 117 (16)
Professor:
　Mark R. Noll, (D), Delaware, 1989, Cl
Associate Professor:
　Paul L. Richards, (D), Penn State, 1999, Hg
　Scott M. Rochette, (D), St. Louis, 1998, Am
　James A. Zollweg, (D), Cornell, 1994, Hs
Adjunct Professor:
　David A. Boehm, (M), SUNY (Buffalo), 2003, Gg
　Christine Crafts, (B), SUNY (Brockport), 2000, Am
　Jutta S. Dudley, (D), SUNY (Buffalo), 1998, Gg
　William G. Glynn, (M), Texas A&M, 1984, Gg
　Linda J. Schaffer, (M), SUNY (Brockport), 1988, Ze
Department Secretary:
　Lauri A. Kifer, (A), Monroe Comm Coll, 1997, Zn
Emeritus:
　Robert W. Adams, (D), Johns Hopkins, 1964, Gs
　Whitney J. Autin, (D), Louisiana State, 1989, Gs
　John E. Hubbard, (D), Colorado State, 1968, Hg
　Richard M. Liebe, (D), Iowa, 1962, Ps
　Judy A. Massare, (D), Johns Hopkins, 1983, Pv
　John M. Williams, (B), Goshen, As
On Leave:
　Robert S. Weinbeck, (D), Iowa State, 1980, As
Systems Administrator:
　Thomas M. McDermott, Zn

SUNY, Ulster County Community College
STEM (A) (2017)

Burroughs 105
491 Cottekill Road
Stone Ridge, NY 12484
　p. (845) 687-5230
　schimmrs@sunyulster.edu
　http://people.sunyulster.edu/esc
　Programs: Liberal Arts & Sciences (A); Mathematics & Science (A)
Professor:
　Steven Schimmrich, (M), SUNY (Albany), 1991, GgZgn
Assistant Professor:
　Karen Helgers, (M), SUNY, 1987, GeZgy

Syracuse University
Dept of Earth Sciences (B,M,D) (2017)
204 Heroy Geology Laboratory
Syracuse, NY 13244-1070
　p. (315) 443-2672
　jofitch@syr.edu
　http://earthsciences.syr.edu
　Enrollment (2014): B: 54 (11) M: 13 (4) D: 16 (4)
Chair:
　Laura Lautz, (D), Syracuse, 2005, Hg
　Donald I. Siegel, (D), Minnesota, 1981, Hw
Professor:
　Suzanne L. Baldwin, (D), SUNY (Albany), 1988, Cc
　Paul G. Fitzgerald, (D), Melbourne (Australia), 1988, GtCc
　Linda C. Ivany, (D), Harvard, 1997, PegCs
　Jeffrey Karson, (D), SUNY (Albany), 1977, GtcGv
　Cathryn R. Newton, (D), California (Santa Cruz), 1983, Pg
　Scott D. Samson, (D), Arizona, 1990, Cc
　Christopher A. Scholz, (D), Duke, 1989, Gs
Associate Professor:
　Gregory Hoke, (D), Cornell, 2006, ZgGmt
Assistant Professor:
　Christopher Junium, (D), Penn State, 2010, CsPeCo
　Zunli Lu, (D), Rochester, 2008, Cg
　Robert Moucha, (D), Toronto, 2003, GtYeg
　Jay Thomas, (D), Virginia Tech, 2003, Gi
Instructor:
　Daniel Curewitz, (D), Duke, 1999, OgGct
Emeritus:
　M. E. Bickford, (D), Illinois, 1960, CcGiz
　James C. Brower, (D), Wisconsin, 1964, Pg
Other:
　Bruce Wilkinson, (D), Texas, 1974, Gs

Union College
Geology Dept (B) (2015)
807 Union Street
Schenectady, NY 12308-3107
　p. (518) 388-6770
　geology@union.edu
　http://www.union.edu/academic_depts/geology/
　Administrative Assistant: Deborah A. Klein
　Enrollment (2015): B: 40 (15)
Chair:
　Donald T. Rodbell, (D), Colorado, 1991, GlmGe
Professor:
　John I. Garver, (D), Washington, 1989, GtCcGr
　Kurt T. Hollocher, (D), Massachusetts, 1985, GxCga
Associate Professor:
　Holli M. Frey, (D), Michigan, 2005, GvCag
　David P. Gillikin, (D), Vrije (Brussel), 2005, CsmCl
Lecturer:
　Matthew R. Manon, (D), Michigan, 2008, Gp
　Anouk Verheyden-Gillikin, (D), Vrije (Brussel), 2004
Emeritus:
　George H. Shaw, (D), Washington, 1971, Yx
Related Staff:
　William S. Neubeck, (M), SUNY (Binghamton), 1980, Gm

United States Military Academy
Dept of Geography & Environmental Engineering (B) ●☒ (2018)
745 Brewerton Rd
West Point, NY 10996

p. (914) 938-2300
william.wright@usma.edu
http://www.usma.edu/gene/SitePages/Home.aspx
Programs: Geospatial Information Science; Geography; Environmental Science; Environmental Engineering
Department Secretary: Jean Keller
Enrollment (2018): B: 80 (19)
LTC:
 William C. Wright, (D), Florida, 2017, ZriYd
COL:
 Christopher E. Oxendine, (D), 2014, ZirZf
Assistant Professor:
 Matthew S. O'Banion, (D), Oregon State, 2017, ZrGg
 William C. Wright, (D), Florida, 2017, ZiYd

University of Rochester
Dept of Earth & Environmental Sciences (B,M,D) (2017)
120 Trustee Road
227 Hutchison Hall
Box 270221
Rochester, NY 14620
 p. (585) 275-5713
 ees@earth.rochester.edu
 http://www.earth.rochester.edu
Professor:
 Carmala N. Garzione, (D), Arizona, 2000, Gs
 Gautam Mitra, (D), Johns Hopkins, 1977, GctNr
 Vasilii Petrenko, (D)
 Robert J. Poreda, (D), California (San Diego), 1983, Cg
 John A. Tarduno, (D), Stanford, 1987, YmGtXm
Associate Professor:
 John Kessler, (D)
Associate Scientist:
 Rory D. Cottrell, (D), Rochester, 2000, Ym
 Pennilyn Higgins, (D), Wyoming, 2000, Cs
Assistant Professor:
 Mauricio Ibanez-Mejia, (D)
 Lee Murray, (D)
 Dustin Trail, (D)
 Thomas Weber, (D)
Research Associate:
 Chiara Borrelli, (D), Rensselaer, 2014, PmcCb
Lecturer:
 Karen Berger, (D)
Emeritus:
 Udo Fehn, (D), Tech (Munich), 1973, GeCgOu
 Lawrence W. Lundgren, (D), Yale, 1958, Ge

Utica College
Dept of Geology (B) (2015)
Gordon Science Center
1600 Burrstone Road
Utica, NY 13502
 p. (315) 792-3134
 skanfoush@utica.edu
 https://www.utica.edu/academic/as/geoscience/new/bachelors.cfm
Enrollment (2015): B: 10 (6)
Chair:
 Adam Schoonmaker, (D), SUNY (Albany), 2005, GczGx
Associate Professor:
 Sharon L. Kanfoush, (D), Florida, 2002, GsOuGn
Adjunct Professor:
 Lindsey Geary, (M), Florida State, 2008, Geg
 Tiffany McGivern, (M), Utica Coll, 2014, Geg
Emeritus:
 Herman Muskatt, (D), Syracuse, 1963, GgrPg

Vassar College
Dept of Earth Science & Geography (B) (2015)
Box 735
124 Raymond Avenue
Poughkeepsie, NY 12604-0735
 p. (845) 437-5540
 geo@vassar.edu
 http://earthscienceandgeography.vassar.edu/
 Administrative Assistant: Lois Horst
Enrollment (2015): B: 12 (3)
Chair:
 Mary A. Cunningham, (D), Minnesota, 2001, ZiyZu
Professor:
 Brian J. Godfrey, (D), California (Berkeley), 1984, Zu
 Kirsten M. Menking, (D), California (Santa Cruz), 1995, GmPeGc
 Joseph Nevins, (D), California (Los Angeles), Zu
 Jill S. Schneiderman, (D), Harvard, 1987, Gs
 Jeffrey R. Walker, (D), Dartmouth, 1987, GzScGv
 Yu Zhou, (D), Minnesota, 1995, Zun
Lab Technician and Collections Manager:
 Richard Jones, (M), California (Santa Cruz), 1996
GIS specialist:
 Neil Curri, (B), Zi

York College (CUNY)
Dept of Earth and Physical Sciences (B) O⊘ (2019)
94-20 Guy R. Brewer Blvd
Jamaica, NY 11451
 p. (718)262- 2654
 nkhandaker@york.cuny.edu
 http://www.york.cuny.edu/academics/departments/earth-and-physical-sciences/
 Programs: Geology, Environmental Health Science and K7-12 Earth Science Teacher Certification Program
Enrollment (2018): B: 22 (3)
Chair:
 Timothy Paglione, (D), Boston, Xa
Geology Discipline Coord & CUNY Doctoral Faculty in Earth and Env Sci:
 Nazrul I. Khandaker, (D), Iowa State, 1991, GdeZe
Associate Professor:
 Ratan K. Dhar, (D), CUNY Grad Ctr, 2006, Hw
Assistant Professor:
 Dawn Roberts-Semple, (D), Rutgers, 2012, Gem
Emeritus:
 Stephen Lakatos, (D), Rensselaer, 1971, Cc
 Arthur P. Loring, (D), New York, 1966, Gm
 Stanley Schleifer, (D), CUNY, 1996, Ge

North Carolina

Appalachian State University
Dept of Geography & Planning (B,M) (2015)
323 Rankin Science West
ASU Box 32066
Boone, NC 28608-2066
 p. (828) 262-3000
 schroederk@appstate.edu
 http://www.geo.appstate.edu
Chair:
 James E. Young, (D), Minnesota, 1994, Ze
Professor:
 Michael W. Mayfield, (D), Tennessee, 1984, Hs
 Peter T. Soule, (D), Georgia, 1989, As
 Roger A. Winsor, (D), Illinois, 1975, Zn
Associate Professor:
 Jeff Colby, (D), Colorado, 1995, Zi
 Richard J. Crepeau, (D), California (Irvine), 1995, Zn
 Kathleen Schroeder, (D), Minnesota, 1995, Zn
Assistant Professor:
 Chris Badurek, (D), Buffalo, 2005, Zi
 Rob Brown, (D), Louisiana State, 2001, Zn
 Jana Carp, (D), Illinois (Chicago), 1999, Zn
 Gabrielle Katz, (D), Colorado, 2001, Hg
 Baker Perry, (D), North Carolina, 2006, Zg
Instructor:
 Arthur B. Rex, (M), Appalachian State, 1980, Zi
Lecturer:
 Terence Milstead, (D), Florida State, 2008, Zn
 Saskia van de Gevel, (D), Tennessee, 2008, Zg
Emeritus:
 Neal G. Lineback, (D), Tennessee, 1970, Zy

Dept of Geological and Environmental Sciences (B) O☒ (2019)
PO Box 32067
033 Rankin Science West

Boone, NC 28608-2067
  p. (828) 262-3049
  earth_env_sci@appstate.edu
  http://www.earth.appstate.edu/
  Programs: Geology, Environmental Science, Earth/Environmental Science Secondary Education
  Enrollment (2018): B: 150 (27)

Chair:
  William P. Anderson, (D), North Carolina State, 1999, HwqHs
Professor:
  Ellen A. Cowan, (D), N Illinois, 1988, GsmGu
  Steven J. Hageman, (D), Illinois, 1992, PieGg
  Andrew B. Heckert, (D), New Mexico, 2001, PvgZe
  Cynthia M. Liutkus-Pierce, (D), Rutgers, 2005, Gd
Associate Professor:
  Sarah Carmichael, (D), Johns Hopkins, 2006, GxCgGd
  Gabriele M. Casale, (D), Gct
  Jamie SF Levine, (D), Texas, 2010, Gzp
  Scott T. Marshall, (D), Massachusetts, 2008, YxGc
Assistant Professor:
  William H Armstrong, (D), Colorado, 2017, Glm
  Cole T. Edwards, (D), Ohio State, 2015, GsCs
  Sarah G. Evans, (D), Colorado, 2017, Hwq
Senior Lecturer:
  Laura D. Mallard, (M), Vermont, 2000, GtZe
  Lauren H. Waterworth, (M), Texas A&M, 2002, Gcg
  Brian W. Zimmer, (M), N Arizona, Gg
Lecturer:
  Joey D. Mosteller, (M), Appalachian State, 2009, Gg
  Hannah B. Riegel, (M), Cal State (Northridge), 2015, Ggc

Asheville-Buncombe Technical Comm College
Dept of Chemistry and Physics (A) (2015)
340 Victoria Road
Asheville, NC 28801
  p. (828) 254-1921
  mfender@abtech.edu
  http://www1.abtech.edu/content/arts-and-sciences/chemistryphysics/chemistry-and-physics-overview

Instructor:
  John Bultman, Gg
Adjunct Professor:
  Dan Murphy, Gg

Brevard College
Geology Program (B) ☒ (2019)
1 Brevard College Drive
Brevard, NC 28712
  p. (828) 884-8377
  reynoljh@brevard.edu
  http://www.brevard.edu/reynoljh/
  Administrative Assistant: Beth Banks
Geology Minor Coordinator:
  James H. Reynolds, (D), Dartmouth, 1987, GrvGe

Cape Fear Community College
Dept of Science (2015)
411 N. Front Street
Wilmington, NC 28401
  akolb@cfcc.edu
  http://cfcc.edu/programs/science/
Chair:
  Joy Smoots, (D)

Science Dept (A) (2016)
411 N Front St
Wilmington, NC 28401
  p. (910) 362-7674
  jsmoots@cfcc.edu
  http://cfcc.edu/programs/science/
  Enrollment (2016): A: 5 (0)
Instructor:
  Alvin L. Coleman, (M), Tennessee, 2008, Gg
  Phil Garwood, (D), Edith Cowans (Australia), 1978, Gg

Central Piedmont Community College
Sciences Div (A) (2015)
PO Box 35009
Charlotte, NC 28235
  p. (704) 330-6750
  David.Privette@cpcc.edu
  http://www.cpcc.edu/
Chair:
  Steppen Murphy, (M), S Illinois, Gue
Division Director:
  David Privette, (M), Georgia, 1978, Zy
Instructor:
  Alisa Hylton, (M), Wichita State, Gg

Coastal Carolina University
Dept of Marine Science (2015)
Conway, NC 29528
  jguentze@coastal.edu

Duke University
Div of Earth & Ocean Sciences (B,M,D) (2015)
Nicholas School of the Environment and Earth Sciences
Box 90227
Durham, NC 27708-0227
  p. (919) 684-5847
  bill.chameides@duke.edu
  http://www.nicholas.duke.edu/eos
  Enrollment (2010): B: 37 (15) D: 21 (0)
Chair:
  M. Susan Lozier, (D), Washington, 1989, Op
Professor:
  Paul A. Baker, (D), California (San Diego), 1981, PcCgGg
  Alan E. Boudreau, (D), Washington, 1986, Gi
  Bruce H. Corliss, (D), Rhode Island, 1978, Pm
  Peter K. Haff, (D), Virginia, 1970, Zn
  Robert B. Jackson, (D), Utah State, 1992, Zn
  Emily M. Klein, (D), Columbia, 1988, Gi
  A. Bradshaw Murray, (D), Minnesota, 1995, Gm
  Lincoln F. Pratson, (D), Columbia, 1993, Gs
  Avner Vengosh, (D), Australian Nat, 1990, Hw
Associate Professor:
  Nicolas Cassar, (D), Hawaii, 2003, CbOcb
Associate Scientist:
  Gary S. Dwyer, (D), Duke, 1996, Gs
Assistant Professor:
  Wenhong Li, (D), Georgia Tech, Zn
Lecturer:
  Alexander Glass, (D), Illinois (Urbana), 2006, PiGgZe
Adjunct Professor:
  David J. Erickson, (D), Rhode Island, 1987
  Peter E. Malin, (D), Princeton, 1978, Ys
  Bruce F. Molnia, (D), South Carolina, 1972, ZrGlu
  Daniel D. Richter, (D), Duke, 1980, So
  William H. Schlesinger, (D), Cornell, 1976, Zn
Emeritus:
  Richard T. Barber, (D), Stanford, 1967, Ob
  Duncan Heron, (D), North Carolina, 1958, Grg
  Daniel A. Livingstone, (D), Yale, 1953, Pe
  Ronald D. Perkins, (D), Indiana, 1962, Gdo
  Orrin H. Pilkey, Jr., (D), Florida State, 1962, Ou
Other:
  James S. Clark, (D), Minnesota, 1988, Zn
  Mark N. Feinglos, (D), McGill, 1973, Gz
  Richard F. Kay, (D), Yale, 1973, Zn

East Carolina University
Dept of Geological Sciences (B,M) ○⌁ (2018)
101 Graham Building
Greenville, NC 27858-4353
  p. (252) 328-6360
  culvers@ecu.edu
  http://www.geology.ecu.edu
  f: https://www.facebook.com/ecugeology/
  t: @Ecugeology
  Programs: Geology (B,M)
  Certificates: Hydrogeology and Environmental Geology (G)

Administrative Assistant: Emily Griffin
Administrative Assistant: Lauren Morrison
Enrollment (2018): B: 53 (17) M: 26 (10)
Chair:
   Stephen J. Culver, (D), Wales, 1976, PmGu
Professor:
   David Mallinson, (D), S Florida, 1995, Gu
   Stephen Moysey, (D), Stanford, 2005, Hg
   Catherine A. Rigsby, (D), California (Santa Cruz), 1989, Gs
Teaching Associate Professor:
   Stephen B. Harper, (D), Georgia, 1996, GmgGe
Associate Professor:
   Adriana Heimann, (D), Iowa State, 2006, GzpGi
   Eric M. Horsman, (D), Wisconsin, 2006, GcYgGt
   Eduardo Leorri, (D), Basque (Spain), 2003, GsPm
   Alex K. Manda, (D), Massachusetts, 2009, HwgHq
   Siddhartha Mitra, (D), William & Mary, 1997, Co
   Terri L. Woods, (D), S Florida, 1988, ClGz
Distinguished Research Professor:
   Stanley R. Riggs, (D), Montana, 1967, GusGe
Emeritus:
   Donald W. Neal, (D), W Virginia, 1979, GrdEo

Guilford College
Dept of Geology & Earth Science (B) (2016)
5800 West Friendly Avenue
Greensboro, NC 27410-4173
   p. (336) 316-2263
   ddobson@guilford.edu
   http://www.guilford.edu/academics/departments-and-programs/geology/
   f: https://www.facebook.com/guilfordgeo/
   Enrollment (2016): B: 22 (7)
Professor:
   David M. Dobson, (D), Michigan, 1997, GusPc
   Marlene McCauley, (D), California (Los Angeles), 1986, Cp
Assistant Professor:
   Holly Peterson, (D), British Columbia, 2014, HwGe
Emeritus:
   Cyril H. Harvey, (D), Nebraska, 1960, Gr

North Carolina Ag & Tech State University
Dept of Natural Resources and Environmental Design (B,M)
◎ (2018)
1601 E. Market St
Carver Hall Room 238
Greensboro, NC 27411
   p. (336) 334-7543
   uzo@ncat.edu
   Programs: Environmental Studies
   Certificates: Certificate In Waste Management
   Enrollment (2018): B: 79 (22) M: 13 (7)
Professor:
   Godfrey A. Uzochukwu, (D), Nebraska, 1983, Gz
Assistant Professor:
   Niroj Aryal, (D), Michigan State, 2015, HsgZi

North Carolina Central University
Department of Environmental, Earth and Geospatial Sciences (B,M) ◯-◉ (2018)
1801 Fayetteville Street
Durham, NC 27707-19765
   p. (919) 530-5296
   gvlahovic@nccu.edu
   http://www.nccu.edu/academics/sc/artsandsciences/geospatialscience/index.cfm
   f: https://www.facebook.com/groups/142873172423480/
   t: @DEEGS_NCCU
   Programs: Environmental and Geographic Sciences (B); Earth Science (M)
   Certificates: USGIF Accredited Geospatial Intelligence Certificate (undergraduate and graduate)
   Enrollment (2016): B: 24 (0) M: 20 (0)
Professor:
   John Bang, (D), Texas (El Paso), GeZa
   Gordana Vlahovic, (D), North Carolina, 1999, YsZeRn

Associate Professor:
   Rakesh Malhotra, (D), Georgia, Zri
   Chris McGinn, (D), North Carolina (Greensboro), Zgi
   Timothy Mulrooney, (D), North Carolina (Greensboro), Zir
   Harris Williams, (D), Arizona State, OgZg
   Zhiming Yang, (D), Oklahoma State, Zry
Assistant Professor:
   Carresse Gerald, (D), North Carolina A&T, GeZg

North Carolina State University
Dept of Marine, Earth & Atmospheric Sciences (B,M,D) ◯-◉ (2019)
P.O. Box 8208
Raleigh, NC 27695-8208
   p. (919) 515-3711
   meas-grad-program@ncsu.edu
   http://www.meas.ncsu.edu
   f: https://www.facebook.com/measncsu
   t: @ncsumeas
   Programs: Geology
   Marine Sciences
   Meteorology
   Natural Resources
   Certificates: Climate Change and Society (Graduate)
   Enrollment (2018): B: 215 (0) M: 60 (23) D: 30 (16)
Co-Director of Graduate Programs:
   Elana L. Leithold, (D), Washington, 1987, Gs
Co Director of Graduate Programs:
   Viney P. Aneja, (D), North Carolina State, 1977, Ac
Professor:
   DelWayne Bohnenstiehl, (D), Columbia, 2002, YrsGt
   David J. DeMaster, (D), Yale, 1979, OcCbc
   David B. Eggleston, (D), William & Mary, 1991, Ob
   Ronald V. Fodor, (D), New Mexico, 1972, Gi
   Ruoying He, (D), S Florida, 2002, Op
   Gary M. Lackmann, (D), SUNY (Albany), 1995, As
   Jay F. Levine, (D)
   David A. McConnell, (D), Texas A&M, 1987, ZeRcGc
   Nicholas Meskhidze, (D), Georgia Tech, 2003, As
   Helena Mitasova, (D), Slovak Tech, 1987, Zi
   Matthew Parker, (D), Colorado State, 2002, Am
   Walter Robinson, (D), Columbia, 1985, As
   William J. Showers, (D), Hawaii, 1982, Cs
   Lian Xie, (D), Miami, 1992, As
   Sandra Yuter, (D), Washington, 1996, As
   Yang Zhang, (D), Iowa, 1994, As
Associate Professor:
   Anantha Aiyyer, (D), SUNY (Albany), 2003, As
   Jingpu P. Liu, (D), William & Mary, 2001, Gu
   Chris Osburn, (D), Lehigh, 2000, Cms
   Astrid Schnetzer, (D), Vienna, 2001, Ob
Assistant Professor:
   Stuart P. Bishop, (D), Rhode Island, 2012, OpGq
   Paul K. Byrne, (D), Trinity (Dublin), 2010, XgGcZr
   Lisa Falk, (D), Columbia, 2014, Zg
   Ethan Hyland, (D), Michigan, 2014, GrCsPe
   Markus Petters, (D), Wyoming, 2004, As
   Karl Wegmann, (D), Lehigh, 2008, Gm
Emeritus:
   Satyapal S. Arya, (D), Colorado State, 1968, As
   John C. Fountain, (D), California (Santa Barbara), 1975, Hw
   James P. Hibbard, (D), Cornell, 1988, GctZn
   Daniel Kamykowski, (D), California (San Diego), 1973, ObpOc
   Charles E. Knowles, (D), Texas A&M, 1970, Op
   Sethu S. Raman, (D), Colorado, 1972, As
   Allen J. Riordan, (D), Wisconsin, 1977, As
   Dale A. Russell, (D), Columbia, 1964, Pv
   Fred H. M. Semazzi, (D), Nairobi, 1983, As
   Ping-Tung Shaw, (D), MIT/WHOI, 1982, Op
   Edward F. Stoddard, (D), California (Los Angeles), 1976, Gp
   Charles W. Welby, (D), MIT, 1952, Hw
   Donna L. Wolcott, (D), California (Berkeley), 1972, Ob
   Thomas G. Wolcott, (D), California (Berkeley), 1971, Ob

Dept of Soil Science (B,M,D) (2015)
Box 7619
Raleigh, NC 27695-7619
   p. (919) 515-2655

jeff_mullahey@ncsu.edu
Administrative Assistant: Ashru Shah
Professor:
Aziz Amoozegar, (D), Arizona, 1977, Sp
Stephen W. Broome, (D), North Carolina, 1973, Sf
Donald K. Cassel, (D), California (Davis), 1968, Sp
John L. Havlin, (D), Colorado State, 1983, Sc
Dean L. Hesterberg, (D), California (Riverside), 1988, Sc
Michael T. Hoover, (D), Penn State, 1983, Sd
Greg D. Hoyt, (D), Georgia, 1981, Sb
Daniel W. Israel, (D), Oregon State, 1973, Sb
Harold J. Kleiss, Illinois, 1972, Sd
Deanna L. Osmond, (D), Cornell, 1991, Sb
Wayne P. Robarge, (D), Wisconsin, 1975, Sc
Thomas J. Smyth, (D), North Carolina State, 1981, Sc
Michael J. Vepraskas, (D), Texas A&M, 1980, Sd
Michael G. Wagger, (D), Kansas State, 1983, Sb
Associate Professor:
David A. Crouse, (D), North Carolina State, 1996, Sc
Carl Crozier, (D), North Carolina State, 1992, Sb
David Lindbo, (D), Massachusetts, 1990, Sd
Richard A. McLaughlin, (D), Purdue, 1985, Sc
Assistant Professor:
Alexandria Graves, (D), Virginia Tech, 2003, Sb
Wei Shi, (D), Purdue, Sb
Emeritus:
James W. Gilliam, (D), Mississippi State, 1965, Sb

University of North Carolina, Asheville
Dept of Atmospheric Sciences (B) ●☒ (2018)
CPO 2450
One University Heights
Asheville, NC 28804-3299
p. (828) 251-6149
chennon@unca.edu
https://atms.unca.edu
f: https://www.facebook.com/uncaweather/
t: @uncaweather
Programs: Atmospheric Sciences
Enrollment (2018): B: 36 (11)
Chair:
Chris Hennon, (D), Ohio State, 2003, AstZn
Professor:
Christopher Godfrey, (D), Oklahoma, 2006, AsZnn
Alex Huang, (D), Purdue, 1984, As
Doug Miller, (D), Purdue, 1996, AsZnn

Environmental Studies (B) ☒ (2018)
CPO 2330
Asheville, NC 28804-8511
p. (828) 251-6441
irossell@unca.edu
http://envr.unca.edu
Programs: B.S. Environmental Studies, including concentration in Earth Science.
Administrative Assistant: Debra C. Robbins
Enrollment (2013): B: 144 (44)
Professor:
Delores M. Eggers, (D), North Carolina, 1999, Zn
Kevin K. Moorhead, (D), Florida, 1986, Sbo
Barbara C. Reynolds, (D), Georgia, 2000, Zn
Irene M. Rossell, (D), SUNY (Syracuse), 1995, Zn
Associate Professor:
David P. Gillette, (D), Oklahoma, 2008, Zn
Jackie M. Langille, Tennessee, 2012, GctGe
Jeffrey D. Wilcox, (D), Wisconsin, 2007, HwClGg
Assistant Professor:
Brittani D. McNamee, (D), Idaho, 2013, GzxEg

University of North Carolina, Chapel Hill
Dept of Geological Sciences (B,M,D) ⌁ (2018)
CB 3315, Mitchell Hall
Chapel Hill, NC 27599-3315
p. (919) 966-4516
dcoleman@unc.edu
http://www.geosci.unc.edu
t: @geosciUNC
Programs: BA Earth Science, BS Earth Science, BS Environmental Geoscience
Enrollment (2016): B: 66 (21) M: 4 (5) D: 20 (2)
Chair:
Jonathan M. Lees, (D), Washington, 1989, YsGv
Professor:
Larry K. Benninger, (D), Yale, 1976, ClGuCt
Drew S. Coleman, (D), Kansas, 1991, CcGit
Allen F. Glazner, (D), California (Los Angeles), 1981, GitZf
Jose A. Rial, (D), Caltech, 1979, Ys
Associate Professor:
Laura J. Moore, (D), California (Santa Cruz), 1998, Gme
Tamlin M. Pavelsky, (D), California (Los Angeles), 2008, Hg
Kevin G. Stewart, (D), California (Berkeley), 1987, Gc
Donna M. Surge, (D), Michigan, 2001, Pe
Assistant Professor:
Xiaoming Liu, (D), Maryland, 2013, Cgl
Emeritus:
Joseph Carter, Kansas, Pi
Paul D. Fullagar, (D), Illinois, 1963, CcGaZe
A. Conrad Neumann, (D), Lehigh, 1963, Gu
Daniel A. Textoris, (D), Illinois, 1963, Gd

University of North Carolina, Charlotte
Dept of Geography & Earth Sciences (B,M,D) (2015)
9201 University City Boulevard
Charlotte, NC 28223-0001
p. (704) 687-5973
ges@uncc.edu
http://www.geoearth.uncc.edu/
Enrollment (2010): B: 280 (70) M: 86 (16) D: 32 (8)
Professor:
John F. Bender, (D), SUNY (Stony Brook), 1980, GiCtOu
John A. Diemer, (D), SUNY (Binghamton), 1985, Gs
Associate Professor:
Craig J. Allan, (D), York, 1992, Hg
Andy R. Bobyarchick, (D), SUNY (Albany), 1983, Gc
Scott P. Hippensteel, (D), Delaware, 2000, Gr
Walter Martin, (D), Tennessee, 1984, As
Ross Meentemeyer, (D), North Carolina, 2000, Zi
Assistant Professor:
Manda S. Adams, (D), Wisconsin, 2005, Am
Matt Eastin, (D), Colorado State, 2003, Am
Martha Cary Eppes, (D), New Mexico, 2002, So
Lecturer:
Jake Armour, (M), New Mexico, 2002, GmlGg
Terry Shirley, (M), Penn State, 2003, Asm
Emeritus:
Anne Jefferson, (D), Oregon State, 2006, HgGm
Laboratory Director:
William Garcia, (M), Cincinnati, Pv

University of North Carolina, Pembroke
Geology & Geography Dept (B) ⌁ (2018)
PO Box 1510
Pembroke, NC 28372-1510
p. (910) 775-4024
geo@uncp.edu
https://www.uncp.edu/departments/geology-and-geography
Programs: Geo-Environmental Studies
Certificates: Geospatial Technologies
Enrollment (2018): B: 24 (5)
Chair:
Martin B. Farley, (D), Penn State, 1987, PlZe
Associate Professor:
Jeff B. Chaumba, (D), Georgia, 2009, GxEgGz
Dennis J. Edgell, (D), Kent State, 1992, As
Assistant Professor:
Daren T. Nelson, (D), Utah, 2012, GmHwZe
Instructor:
Jesse Rouse, (D), West Virginia, 2018, Zni
Lecturer:
Amy Gross, (M), North Carolina (Wilmington), 2006, Gg
Nathan E. Phillippi, (M), South Dakota State, 2004, Zgi

Emeritus:
    Suellen Cabe, (D), North Carolina, 1984, Gg
    Thomas E. Ross, (D), Tennessee, 1977, Zy

University of North Carolina, Wilmington
Dept of Earth and Ocean Sciences (B,M) O (2016)
601 South College Road
Wilmington, NC 28403-5944
    p. (910) 962-3490
    lynnl@uncw.edu
    http://www.uncw.edu/earsci/
    Administrative Assistant: Alexis Lee
    Enrollment (2007): B: 78 (21) M: 32 (5)
Chair:
    Lynn A. Leonard, (D), S Florida, 1993, Gu
Professor:
    Michael M. Benedetti, (D), Wisconsin, 2000, GmSaZy
    Nancy R. Grindlay, (D), Rhode Island, 1991, Yg
    Michael S. Smith, (D), Washington Univ (St. Louis), 1990, GzpGa
Associate Professor:
    Lewis J. Abrams, (D), Rhode Island, 1992, Gu
    David E. Blake, (D), Washington State, 1991, Gp
    Douglas W. Gamble, (D), Georgia, 2000, Zy
    Joanne N. Halls, (D), South Carolina, 1996, Zi
    Eric J. Henry, (D), Arizona, 2001, Hw
    Mary E. Hines, (D), Louisiana State, 1992, Zu
Lecturer:
    Roger D. Shew, (M), North Carolina, 1979, Eo
Emeritus:
    Robert T. Argenbright, (D), California (Berkeley), 1990, Zn
    William J. Cleary, (D), South Carolina, 1972, Gu
    James A. Dockal, (D), Iowa, 1980, GdcEm
    W. Burleigh Harris, (D), North Carolina, 1975, GrdGs
    Patricia H. Kelley, (D), Harvard, 1979, Pgi
    Paul A. Thayer, (D), North Carolina, 1967, GdoHg
Laboratory Director:
    Yvonne Marsan, (B), North Carolina (Wilmington), Zi

Wake Technical Community College
Natural Sciences Dept (2015)
9101 Fayetteville Rd
Raleigh, NC 27603
    p. (919) 866-5000
    cdtatum@waketech.edu
    http://www.waketech.edu/programs-courses/credit/natural-sciences

Western Carolina University
Dept of Geosciences & Natural Resources (B) ☒ (2018)
Stillwell Building Room 331
Cullowhee, NC 28723-9047
    p. (828) 227-7367
    mlord@wcu.edu
    http://geology.wcu.edu
    Enrollment (2015): B: 52 (12)
Head:
    Mark L. Lord, (D), North Dakota, 1988, HwGmZe
Whitmire Prof Env Sci:
    Jerry R. Miller, (D), S Illinois, 1990, GmeGf
Director Prog Study Dev Shorelines:
    Robert S. Young, (D), Duke, 1995, GsOn
Associate Provost:
    Brandon E. Schwab, (D), Oregon, 2000, GizGv
Assoc. Dean, Arts & Sciences:
    David A. Kinner, (D), Colorado, 2003, HgZe
Associate Professor:
    Benjamin R. Tanner, (D), Tennessee, 2005, Cos
    Cheryl Waters-Tormey, (D), Wisconsin, 2004, Gct
Assistant Professor:
    Amy Fagan, (D), Notre Dame, 2013, GiXg
    Frank Forcino, (D), Alberta, 2013, PgZeGr
    John P. Gannon, (D), Virginia Tech, 2014, HgSo
Instructor:
    Emily Stafford, (D), Alberta, 2014, PgGg
Emeritus:
    Steven P. Yurkovich, (D), Brown, 1972, Gx

North Dakota

Dickinson State University
Dept of Natural Science (B) (2015)
Dickinson, ND 58601
    p. (701) 227-2114
    Michael.Hastings@dickinsonstate.edu
    http://www.dsu.nodak.edu/
Chair:
    Michael Hastings
Associate Professor:
    Larry D. League, (M), Kansas, 1971, Zy

Minot State University
Dept of Geoscience (B) ☒ (2019)
500 University Avenue West
Minot, ND 58707
    p. (701) 858-3873
    john.webster@minotstateu.edu
    http://www.minotstateu.edu/geology/
    Programs: Geology; Earth Science Education
    Enrollment (2018): B: 19 (3)
Head:
    John R. Webster, (D), Indiana, 1992, Giz
Assistant Professor:
    Joseph Collette, (D), California (Riverside), 2014, PgGsr
    Kathryn Kilroy, (D), Nevada (Reno), 1992, Hg

North Dakota State University
Dept of Geosciences (B) ◎ (2019)
NDSU Dept. 2745
P.O. Box 6050
Fargo, ND 58108-6050
    p. (701) 231-8455
    peter.oduor@ndsu.edu
    http://www.ndsu.edu/geosci
    f: https://www.facebook.com/NDSUGeosciences
    Programs: Geology (B); Geography (minor); Geology (minor); Environmental Geology (minor)
    Enrollment (2018): B: 32 (5)
Chair:
    Bernhardt Saini-Eidukat, (D), Minnesota, 1991, GiCqEg
Professor:
    Kenneth Lepper, (D), Oklahoma State, 2001, CcGml
    Peter Oduor, (D), Missouri (Rolla), 2004, ZiRwZr
Assistant Professor:
    Stephanie S. Day, (D), Minnesota, 2012, GmZiu
    Benjamin J. C. Laabs, (D), Wisconsin, 2004, GlPcGm
    Lydia S. Tackett, (D), S California, 2014, PeGsPi
Lecturer:
    Jessie L. Rock, (M), North Dakota State, 2009, ZgPcg
Emeritus:
    Allan C. Ashworth, (D), Birmingham, 1969, Pe
    Donald P. Schwert, (D), Waterloo, 1978, PeGe

Dept of Soil Science (B,M,D) (2015)
Walster Hall
Fargo, ND 58105
    p. (701) 231-8690
    Thomas.Desutter@ndsu.edu
    Administrative Assistant: Jacinda Wollan
    Enrollment (2014): B: 18 (0) M: 10 (0) D: 4 (0)
Professor:
    Francis Casey, (D), Iowa State, 2000, SpHq
    Dave Franzen
    Robert J. Goos, (D), Colorado, 1980, Sc
Associate Professor:
    Larry J. Cihacek, (D), Iowa State, 1976, Zg
Assistant Professor:
    Amitava Chatterjee, Wyoming
    Aaron Daigh, (D), Iowa State
    Tom DeSutter, (D), Kansas State, So
    Ann-Marie Fortuna
    David G. Hopkins, (D), North Dakota State, 1997, Sd

Abbey Wick, Wyoming

University of North Dakota
Dept of Geography and Geographic Information Science
(B,M) (2019)
221 Centennial Drive
Stop 9020
Grand Forks, ND 58202
    p. (701) 777-4246
    und.geography@und.edu
    http://arts-sciences.und.edu/geography/
    f: https://www.facebook.com/groups/91074880028/
    Programs: Geography (B,M); Geospatial Technologies (minor); Environmental Studies (BA, BS)
    Certificates: GISc (G)
    Enrollment (2018): B: 36 (14) M: 18 (4)
Chair:
    Gregory S. Vandeberg, (D), Kansas State, 2005, GmZiGl
Professor:
    Douglas C. Munski, (D), Illinois, 1978, Ze
    Bradley C. Rundquist, (D), Kansas State, 2000, ZriZy
    Paul Todhunter, (D), California (Los Angeles), 1986, Zy
Associate Professor:
    Enru Wang, (D), Washington, 2005, Zgi
Assistant Professor:
    Christopher Atkinson, (D), Kansas, 2010, AsZi
Instructor:
    Mbongowo J. Mbuh, (D), George Mason, 2015, ZrHs
    Michael A. Niedzielski, (D), Ohio State, 2009, Ziu
    Mikel Smith, (M), N Dakota, 2014, Zn
Emeritus:
    Devon A. Hansen, (D), Utah, 1999, ZnnZn

Harold Hamm School of Geology & Geological Engineering
(B,M,D) (2015)
81 Cornell Street
Stop 8358
Grand Forks, ND 58202
    p. (701) 777-2248
    jolene.marsh@engr.und.edu
    http://engineering.und.edu/geology-and-geological-engineering/
    Enrollment (2014): B: 78 (0) M: 17 (0) D: 10 (0)
Director:
    Joseph H. Hartman, (D), Minnesota, 1984, Pi
Professor:
    William D. Gosnold, (D), S Methodist, 1976, YhCt
    Stephan Nordeng, (D), Michigan, Go
    Dexter Perkins, III, (D), Michigan, 1979, Gp
Associate Professor:
    Philip J. Gerla, (D), Arizona, 1983, Hw
    Ronald K. Matheney, (D), Arizona State, 1989, Cs
    Jaakko Putkonen, (D), Washington, 1997, Gml
Assistant Professor:
    Nels F. Forsman, (D), North Dakota, 1985, GdXgRh
    I-Hsuan Ho, (D), Iowa, Ng
    Dongmei Wang, (D), China, Ng
Emeritus:
    Richard D. LeFever, (D), California (Los Angeles), 1979, GsrGo

Ohio

Ashland University
Dept of Chemistry/Geology/Physics (B) (2019)
401 College Avenue
Ashland, OH 44805
    p. (419) 289-5268
    rcorbin@ashland.edu
    http://www.ashland.edu/departments/geology
    f: https://www.facebook.com/Ashland-University-Sciences-314117456597/
    t: @ausciences
    Programs: Geology (B); Geoscience Technology and Management (B); Geology/Environmental Science (B); Earth Science Education (B)
    Enrollment (2018): B: 16 (2)

Professor:
    Nigel Brush, (D), California (Los Angeles), 1992, GmdGa
Adjunct Professor:
    Tess Holloway, (M), Wright State, 2017, Rn
    Amanda L. Kozak, (M), Wright State, Rn
    Elizabeth Mazzocco, (M), Ohio State, AsRn
    William A. Reinthal, (D), Wisconsin, GzoGg
    Mackenzie Taylor, (M), Miami, 2017, RnGg

Bowling Green State University
Dept of Geology (B,M) (2015)
190 Overman Hall
Bowling Green, OH 43403
    p. (419) 372-2886
    geology@bgsu.edu
    http://www.bgsu.edu/departments/geology/
    Administrative Assistant: Pat A. Wilhelm
    Enrollment (2010): B: 29 (13) M: 23 (2)
Professor:
    James E. Evans, (D), Washington, 1988, GsHsGe
    Charles M. Onasch, (D), Penn State, 1977, Gc
    Robert K. Vincent, (D), Michigan, 1973, Yg
    Peg M. Yacobucci, (D), Harvard, 1999, Pgi
Associate Professor:
    John R. Farver, (D), Brown, 1988, Gy
    Enrique Gomezdelcampo, (D), Tennessee, 2003, Ge
    Peter Gorsevski, (D), Idaho, 2002, ZirGm
    Kurt S. Panter, (D), New Mexico Tech, 1995, Giv
    Jeffrey A. Snyder, (D), Ohio State, 1996, Gm
Lecturer:
    Nichole Elkins, (M), Georgia, 2002, GaZe
    Christopher Pepple, (M), Ze
    Paula J. Steinker, (D), Bowling Green, 1982, Pg
Emeritus:
    Don C. Steinker, (D), California (Berkeley), 1969, Pg
Department IT:
    William Butcher, (B), Rochester, 1972, Gg

Case Western Reserve University
Dept of Earth, Environmental and Planetary Sciences
(B,M,D) (2016)
10900 Euclid Avenue
A.W. Smith #112
Cleveland, OH 44106-7216
    p. (216) 368-3690
    eepsweb@case.edu
    http://eeps.case.edu/
    Enrollment (2016): B: 15 (0) M: 2 (0) D: 5 (0)
Chair:
    James A. Van Orman, (D), MIT, 2000, CgGyCp
Professor:
    Steven A. Hauck, II, (D), Washington Univ (St. Louis), 2001, XyYgXg
    Peter L. McCall, (D), Yale, 1975, Pg
    Peter J. Whiting, (D), California (Berkeley), 1990, Gm
Associate Professor:
    Mulugeta Alene Araya, (D), Turin
    Ralph P. Harvey, (D), Pittsburgh, 1990, Xm
    Beverly Z. Saylor, (D), MIT, 1996, Gs
Assistant Professor:
    Carlo DeMarchi, (D), Georgia Tech
    Zhicheng Jing, (D), Yale, 2010
Adjunct Professor:
    Andrew Dombard, (D), Washington, 2000, Xy
    Joseph T. Hannibal, (D), Kent State, 1990, Pi
    David Saja, (D), Pennsylvania, 1999, Gc
Emeritus:
    Gerald Matisoff, (D), Johns Hopkins, 1978, Cbc
    Samuel M. Savin, (D), Caltech, 1967, Cs

Cedarville University
Dept of Science and Mathematics (B) (2018)
251 North Main Street
Cedarville, OH 45314
    p. (937) 766-7940
    trice@cedarville.edu
    https://www.cedarville.edu/Academic-Programs/Geology.

aspx
f: https://www.facebook.com/cedarvillegeology
Programs: Geology (B); Physical Geology (B); Geoscience (B)
Enrollment (2018): B: 26 (4)
Professor:
　Steven Gollmer, (D), Purdue, GzAmOg
　John H. Whitmore, (D), Loma Linda, 2003, PgGsg
Associate Professor:
　Mark Gathany, (D), Colorado State, ZiuSb
Assistant Professor:
　Thomas L. Rice, (M), Colorado Mines, 1987, GemEo
Adjunct Professor:
　Steve A. Austin, (D), Penn State, 1979, GsEcGd

Central State University
Intl Center for Water Resources Management (B) ☒ (2019)
C.J. McLin Bldg
Wilberforce, OH 45384
　p. (937) 376-6212
　knedunuri@centralstate.edu
　Enrollment (2017): B: 35 (4)
Chairperson and Director:
　Krishna Kumar Nedunuri, (D), Purdue, 1999, HwCaSo
Associate Ditector for Land Grant Research:
　Subramania I. Sritharan, (D), Colorado State, 1984, Hg
Professor:
　Ramanitharan Kandiah, (D), Tulane, 2004, HqSoZi
　Sam Laki, (D), Michigan State, 1992, EgSoHs
　Xiaofang Wei, (D), Indiana State, 2008, ZrOgZe
Associate Professor:
　De Bonne N. Wishart, (D), Rutgers (Newark), 2008, YgCa
　Ning Zhang, (D), West Virginia, 2012, RwEo
Emeritus:
　Samuel Okunade, (D), Kent State, 1986, Gm

Cincinnati Museum Center
Geier Collections and Research Center ☒ (2017)
1301 Western Avenue
Cincinnati, OH 45203
　p. (513) 287-7000
　information@cincymuseum.org
　http://www.cincymuseum.org
Associate Vice President:
　Glenn W. Storrs, (D), Yale, 1986, PvGgr
Curator:
　Brenda Hunda, (D), California (Riverside), 2004, Pi
Research Associate:
　Nigel C. Hughes, (D), Bristol, 1990, Pi
　Takuya Konishi, (D), Alberta, 2009, Pv
　Arnold I. Miller, (D), Chicago, 1986, Pq
　Joshua H. Miller, (D), Chicago, 2009, Pe
　Andrew Webber, (D), Cincinnati, 2007, Pi

Cleveland Museum of Natural History
Dept of Mineralogy ☒ (2018)
1 Wade Oval Drive
University Circle
Cleveland, OH 44106-1767
　p. (216) 231-4600 (Ext. 3229)
　dsaja@cmnh.org
　http://www.cmnh.org/mineralogy
　t: @cmnhmineralogy

Dept of Paleobotany & Paleoecology
1 Wade Oval Drive
Cleveland, OH 44106-1767
　p. (216) 231-4600 (Ext. 3240)
　dsu@cmnh.org
　http://www.cmnh.org/site/ResearchandCollections/InvertebratePaleontology.aspx

Dept of Vertebrate Paleontology (2016)
1 Wade Oval Drive
University Circle
Cleveland, OH 44106-1767
　p. (216) 231-4600
　mryan@cmnh.org
　http://www.cmnh.org/site/ResearchandCollections/VertebratePaleontology.aspx
Head:
　Joseph T. Hannibal, (D), Kent State, 1990, PiEn
Head Technician:
　Lee E. Hall, (B)
Curator:
　Michael J. Ryan, (D)
Collections Manager:
　Amanda R. McGee, (M)

College of Wooster
Dept of Earth Sciences (B) ☒ (2018)
Scovel Hall
944 College Mall
Wooster, OH 44691-2363
　p. (330) 263-2380
　preeder@wooster.edu
　http://www.wooster.edu/academics/areas/geology
　f: https://www.facebook.com/pages/College-of-Wooster-Geology-Department/143144126437
　Programs: Environmental Geoscience or Geology Major
　Administrative Assistant: Patrice Reeder
　Enrollment (2018): B: 28 (15)
Chair:
　Meagen Pollock, (D), Duke, 2007, CuGiv
Professor:
　Gregory C. Wiles, (D), SUNY (Buffalo), 1992, Gl
　Mark A. Wilson, (D), California (Berkeley), 1982, Pi
Associate Professor:
　Shelley Judge, (D), Ohio State, 2007, GtsGc
Visiting Professor:
　Karen E. Alley, (D), Colorado, 2017, GlZr
　Alex Crawford, (D), Colorado, 2017, AtZri

Cuyahoga Community College - Western Campus
Earth Science (A) (2015)
11000 Pleasant Valley Road
Parma, OH 44130
　p. (216) 987-5278
　enroll@tri-c.edu
Assistant Professor:
　Robert Zaleha, Zg
Instructor:
　Carol Fondran, Zg
　Joseph M. Lane, Zg
　Abby N. Norton-Krane, Zg
　Kathryn Sasowsky, Zg
Adjunct Professor:
　Gloria CC Britton, (D), ZgGmZn
　Jennifer Deka, Zg
　John L. Ezerskis, (M), Toledo, 1988, ZgGcHw

Denison University
Dept of Geosciences (B) (2016)
F.W. Olin Science Hall
100 W. College Street
Granville, OH 43023
　p. (740) 587-6217
　hall@denison.edu
　http://www.denison.edu/academics/departments/geosciences/
　Administrative Assistant: Jude Hall
　Enrollment (2013): B: 9 (9)
Professor:
　Tod A. Frolking, (D), Wisconsin, 1985, Zy
　David C. Greene, (D), Nevada (Reno), 1995, GctGe
Associate Professor:
　David H. Goodwin, (D), Arizona, 2003, PiCs
Assistant Professor:
　Erik Klemetti, (D), Oregon State, 2005, GvxGg
　Kate E. Tierney, (D), Ohio State, 2010, CgGrZg
Emeritus:
　Kennard B. Bork, (D), Indiana, 1967, PiRhGs

Robert J. Malcuit, (D), Michigan State, 1973, Gx

Hocking College
GeoEnvironmental Science Program (2015)
3301 Hocking Parkway
Nelsonville, OH 45764-9704
    p. (740) 753-6277
    caudill_m@hocking.edu
    http://www.hocking.edu/programs/geoenvironmental
Professor:
    Michael R. Caudill, (D), Tennessee, 1996, GsSaHw
Instructor:
    Kimberly S. Caudill, (B), Ohio Univ, 1984, GeZuHw

Kent State University
Dept of Geology (B,M,D) O (2015)
221 McGilvrey Hall
Kent, OH 44242
    p. (330) 672-2680
    geology@kent.edu
    http://www.kent.edu/geology/
    Enrollment (2015): B: 154 (23) M: 22 (10) D: 9 (3)
Chair:
    Daniel K. Holm, (D), Harvard, 1992, Gc
Professor:
    Joseph D. Ortiz, (D), Oregon State, 1995, Gs
    Carrie E. Schweitzer, (D), Kent State, 2000, Pi
    Alison J. Smith, (D), Brown, 1991, GnPce
    Neil A. Wells, (D), Michigan, 1984, Gs
Associate Professor:
    David B. Hacker, (D), Kent State, 1998, GcHw
    Anne J. Jefferson, (D), Oregon State, 2006, HgGmCs
Assistant Professor:
    Tathagata Dasgupta, (D), Syracuse, 2010, Cg
    Elizabeth M. Herndon, (D), Penn State, 2012, ClSc
    Christopher J. Rowan, (D), Southampton (U.K.), 2006, Gt
    David M. Singer, (D), Stanford, 2008, Gze
    Jeremy C. Williams, (D), Massachusetts (Boston), 2014, Cg
Emeritus:
    Rodney M. Feldmann, (D), North Dakota, 1967, Pi
    Donald F. Palmer, (D), Princeton, 1968, Yg
    Abdul Shakoor, (D), Purdue, 1982, Ng
Related Staff:
    Merida Keatts, (M), Kent State, 2000, Zn

Kent State University at Stark
Dept of Geology (B) O (2017)
6000 Frank Avenue NW
North Canton, OH 44720
    p. 330-244-3303
    cschweit@kent.edu
    https://www.kent.edu/geology
    Administrative Assistant: Debra Stimer
    Enrollment (2017): B: 20 (0)
Chair:
    Carrie E. Schweitzer, (D), Kent State, 2000, PiGg

Lakeland Community College
Geoscience Dept (A) (2017)
7700 Clocktower Drive
Kirtland, OH 44094
    p. (440) 525-7341
    dpierce@lakelandcc.edu
    http://www.lakelandcc.edu/academic/sh/geol/index.asp
Professor:
    David Pierce, (D), GghsAm

Marietta College
Dept of Petroleum Engineering & Geology (B) ☒ (2018)
215 Fifth Street
Marietta, OH 45750
    p. (740) 376-4775
    sch030@marietta.edu
    http://w3.marietta.edu/departments/Petroleum_Engineering/
    Programs: Geology; Petroleum Engineering; Environmental Engineering
    Enrollment (2017): B: 36 (13)
Professor:
    David L. Jeffery, (D), Texas A&M, 1996, PsGo
Associate Professor:
    Tej P. Gautam, (D), Kent State, 2012, NgGeZi
Instructor:
    Wendy Bartlett, (M), Texas A&M, 1980, Goe
    Veronica Freeman, (M), Texas (Arlington), 1993, Pg
Visiting Professor:
    Paul A. Washington, (D), Connecticut, 1987, Gcs
Administrative Coordinator:
    Susan Hiser, (M), Springfield Coll, 1992, Zn

Miami University
Dept of Geology and Environmental Earth Sciences (B,M,D) O⊘ (2018)
250 S,. Patterson Avenue
118 Shideler Hall
Oxford, OH 45056
    p. (513) 529-3216
    edwardca@MiamiOh.edu
    www.miamioh.edu/geology
    f: https://www.facebook.com/MiamiGeology
    t: @geologymiamioh
    Programs: Geology (B); Environmental Earth Science (B); Earth Science (B)
    Administrative Assistant: Cathy Edwards
    Enrollment (2018): B: 180 (27) M: 11 (3) D: 13 (0)
Janet & Elliot Baines Bicentenntial Professor & Chair:
    Elisabeth Widom, (D), California (Santa Cruz), 1991, CcGve
Professor:
    Michael Brudzinski, (D), Illinois, 2002, Ys
    Brian S. Currie, (D), Arizona, 1998, GstGc
    Yildirm Dilek, (D), California (Davis), 1989, Gt
    Hailiang Dong, (D), Michigan, 1997, Cg
    John F. Rakovan, (D), SUNY (Stony Brook), 1996, GzCl
    Jason Rech, (D), Arizona, 2001, Gm
Associate Professor (Hamilton Campus:
    Mark Krekeler, (D), Illinois (Chicago), 2003, Ge
Associate Professor & Director of IES:
    Jonathan Levy, (D), Wisconsin, 1993, Hw
Assistant Professor:
    Claire McLeod, (D), Durham, 2012, Gx
    Carrie Tyler, (D), Virginia Tech, 2012, Pb
Instructor:
    Tracy Eads, (M), Cincinnati, 1995, Hw
    Jill Mignery, (M), Miami, 2004, Zge
Lecturer (Middletown Campus:
    Tammie Gerke, (D), Cincinnati, 1995, ClGea
Lecturer:
    Todd Dupont, (D), Penn State, 2004, Gl
    Euan Mitchell, (D), New Mexico, 2013
Adjunct Professor:
    Patri Larrea, (D), Zaragoza, 2014
Visiting Professor:
    Hassan Mirnejad, (M), Carleton, 2001, GxCg
Emeritus:
    A. Dwight Baldwin, Jr., (D), Stanford, 1966, Hw
    Mark R. Boardman, (D), North Carolina, 1978, On
    William K. Hart, (D), Case Western, 1982, GivCc
    John M. Hughes, (D), Dartmouth, 1981, Gz
    Robert G. McWilliams, (D), Washington, 1968, Ps
    John K. Pope, (D), Cincinnati, 1966, Pi
    David M. Scotford, (D), Johns Hopkins, 1950, Gp
Geochemistry Lab Manager:
    Amy Wolfe, (D), Pittsburgh, 2010, Cs
Laboratory Director:
    David C. Kuentz, (M), Texas (Arlington), 1986, Ca
Accounting Technician:
    Gail Burger
Director Limper Geology Museum:
    Kendall L. Hauer, (D), Miami (Ohio), 1995, GgZeCg
Cooperating Faculty:
    Jerome Bellian, (D), Texas, 2009, Go
    Carina Colombi, (D), National (San Juan), 2007, Gr

R. Hays Cummins, (D), Texas A&M, 1904, Ge
J Christophe Haley, (D), Johns Hopkins, 1986, Gt
Hassan Mirnejad, (D), Carleton, 2001, Gx
Monica Rakovan, (D), Miami (Ohio), 2011, Hg
Fara Rasoazanamparany, (D), Miami, 2015

Mount Union College
Dept of Geology (B) (2015)
Alliance, OH 44601
    p. (330) 823-3672
    graylm@mountunion.edu
    http://www.mountunion.edu/gy
    Enrollment (2010): B: 8 (2)
Professor:
    Lee M. Gray, (D), Rochester, 1985, PgGg
Department Chair:
    Mark A. McNaught, (D), Rochester, 1991, Gcg
Adjunct Professor:
    Leonard G. Epp, (D), Penn State, 1970, Ob

Muskingum University
Dept of Geology (B) (2016)
163 Stormont Street
New Concord, OH 43762
    p. (740) 826-8306
    svanhorn@muskingum.edu
    http://muskingum.edu/dept/geology/index.html
    Enrollment (2016): B: 33 (6)
Associate Professor:
    Stephen R. Van Horn, (D), Connecticut, 1996, GeZiEo
Associate Professor:
    Eric W. Law, (D), Case Western, 1982, Gp
    David L. Rodland, (D), Virginia Tech, 2003, PieGr

Northwest State Community College
Div of Arts and Sciences (2015)
22600 State Route 34
Archbold, OH 43502
    levans@northweststate.edu
    http://www.northweststate.edu

Oberlin College
Dept of Geology (B) (2016)
52 West Lorain Street
Oberlin, OH 44074-1044
    p. (440) 775-8350
    geology@oberlin.edu
    http://new.oberlin.edu/arts-and-sciences/departments/geology/
    Enrollment (2016): B: 46 (18)
Chair:
    Dennis K. Hubbard, (D), South Carolina, 1977, Gsu
Professor:
    Karla M. Parsons-Hubbard, (D), Rochester, 1993, PgGu
    Steven F. Wojtal, (D), Johns Hopkins, 1982, Gc
Associate Professor:
    F Zeb Page, (D), Michigan, 2005, GpCsg
Assistant Professor:
    Amanda H. Schmidt, (D), Washington, 2010, Gm
Visiting Professor:
    Andrew J. Horst, (D), Stanford, 2013, Gcg
Emeritus:
    Bruce M. Simonson, (D), Johns Hopkins, 1982, GsXmHw
Cooperating Faculty:
    Alison Ricker, (M), Rhode Island, 1977, Zn

Ohio State University
Dept of Civil, Environmental & Geodetic Engineering (B,M,D) ☒ (2018)
470 Hitchcock Hall
2070 Neil Avenue
Columbus, OH 43210
    p. (612) 292-2771
    mackay.49@osu.edu
    http://ceg.osu.edu
    Programs: Civil Engineering
    minor in Survey Engineering
    Enrollment (2016): M: 11 (4) D: 8 (2)
Professor and Chair:
    Allison A. MacKay, (D), MIT, 1998, Cq
Research Professor:
    Charles K. Toth, (D), ZrYdZi
Professor:
    Harvey J. Miller, (D), Ohio State, 1991, Zi
    Alper Yilmaz, (D), C Florida, 2004, ZfrZu
Assistant Professor:
    Rongjun Qin, (D), Zrf
    Lei Wang, (D), Ohio State, Zri

Dept of Geography (B,M,D) (2015)
1036 Derby Hall
154 North Oval Mall
Columbus, OH 43210-1361
    p. (614) 292-2514
    geog_webmaster@osu.edu
    http://www.geography.osu.edu/
Chair:
    Daniel Sui
Professor:
    David H. Bromwich, (D), Wisconsin, 1979, As
    Ellen E. Mosley-Thompson, (D), Ohio State, 1979, As
    Jeffrey C. Rogers, (D), Colorado, 1979, As
Associate Professor:
    Jay S. Holgood, (D), Ohio State, 1984, As
    Jialin Lin, (D), SUNY (Stony Brook), 2001, As
    Bryan G. Mark, (D), Syracuse, 2001, As
Assistant Professor:
    Alvaro Montenegro, (D), Florida State, 2004, OgAsZy
Emeritus:
    A. John Arnfield, (D), McMaster, 1973, As
    John N. Rayner, (D), Canterbury (NZ), 1965, As

Geography (B,M,D) ☒ (2019)
1036 Derby Hall
154 North Oval Mall
Columbus, OH 43210-1361
    p. (614) 292-2514
    geog_webmaster@osu.edu
    https://geography.osu.edu/
    Programs: BA in Geography, BS in Geography, BS in Atmospheric Sciences, MA in Geography, MS in Atmospheric Sciences, PhD in Geography, PhD in Atmospheric Sciences
Associate Professor:
    Alvaro Montenegro, (D), Florida State, 2003, AsOg

School of Earth Sciences (B,M,D) O☒ (2019)
275 Mendenhall Lab
125 South Oval Mall
Columbus, OH 43210-1398
    p. (614) 292-2721
    earthsciences@osu.edu
    http://www.earthsciences.osu.edu
    f: https://www.facebook.com/OSUGeology/
    t: @osuearthscience
    Programs: Earth History, Geodetic Science, Solid Earth Dynamics, and Water, Climate, and the Environment
    Administrative Assistant: Jill Bryant
    Administrative Assistant: Theresa Mooney
    Administrative Assistant: Angeletha M. Rogers
    Enrollment (2011): M: 36 (16) D: 40 (2)
School Director:
    Matthew R. Saltzman, (D), California (Los Angeles), 1996, GsClg
Undergraduate Coordinator:
    Anne E. Carey, (D), Nevada (Reno), 1995, HwCg
Ohio Research Scholar:
    David R. Cole, (D), Penn State, 1980, CgsZa
Ohio Eminent Scholar:
    Michael G. Bevis, (D), Cornell, 1982, Yx
    Frank W. Schwartz, (D), Illinois, 1972, Hw
Distinguished University Scholar:
    William Berry Lyons, (D), Connecticut, 1979, HwCgZg
    CK Shum, (D), Texas, 1982, YdHg

Director, Byrd Polar Research Center:
  Ian M. Howat, (D), California (Santa Cruz), 2006, GlZr
Professor:
  Douglas E. Alsdorf, (D), Cornell, 1996, HgYg
  Loren E. Babcock, (D), Kansas, 1990, Pg
  Michael Barton, (D), Manchester, 1975, GiCpt
  Yu-Ping Chin, (D), Michigan, 1988, CqHw
  Andrea G. Grottoli, (D), Houston, 1998, ObCbOo
  Christopher Jekeli, (D), Ohio State, 1981, Ydv
  Mark A. Kleffner, (D), Ohio State, 1988, PgsGr
  Wendy R. Panero, (D), California (Berkeley), 2001, Yg
  Burkhard A. Schaffrin, (D), Bonn, 1983, Yd
  Lonnie G. Thompson, (D), Ohio State, 1976, GlZgPc
Design Engineer:
  Dana Caccamise, (M)
Senior Scientist:
  Junyi Guo
  John W. Olesik, (D), Wisconsin, 1982, Ca
  Yuchan Yi, (D), Ohio State, 1995, Ydv
Associate Professor:
  Ann Cook, (D), Columbia, 2010, GuYg
  Ozeas S. Costa, Jr, (D), Plymouth, 2002, CblZu
  Thomas Darrah, (D), Rochester, 2009, Cg
  Michael T. Durand, (D), California, 2007, HsZr
  William Ashley Griffith, (D), Stanford, 2008, GcNr
  Motomu Ibaraki, (D), Waterloo, 1994, Hg
  Steven K. Lower, (D), Virginia Tech, 2001, PoZa
  Alan J. Saalfeld, (D), Maryland, 1993, Zi
Senior Research Associate:
  Christopher Gardner, (M)
Research Scientist:
  Paolo Gabrielli, (D), LGGE Grenoble, 2004, CtPeGl
  Susan A. Welch, (D), Delaware, 1997, ClGeCs
Instructional Lab Supervisor:
  Christena Cox, (D), Ohio State, 1994, GeoGc
Assistant Professor:
  Joel D. Barker, (D), Alberta, 2007, CblPe
  Elizabeth Griffith, (D), Stanford, 2008, CsOoCb
  Jill Leonard-Pingel, (D), Scripps Institution of Oceanography, 2012, Pi
  Joachim Moortgat, (D), Radboud, 2006, HwNp
  Audrey Sawyer, (D), Texas, 2011, HwSpHs
  Derek Sawyer, (D), Texas, 2010, YgGoRn
Research Associate:
  Gerald Allen, (D), Ohio State, 2011
  Chunli Dai, (D), Ohio State, 2015, ZrYg
  Renato Frasson, (D), Iowa, 2011, Hg
  Siddharth Gautam, (D), Mumbai, 2009
  Eric Kendrick, (D), Hawaii
  How-wai (Peter) Luk
  Anthony Lutton, (D)
  Myoung-Jong Noh, (D), Inha, 2011, Zr
  Julie Sheets, (D), Ohio State, 1994, Gz
  Stephanie Sherman, (D), Ohio State
Museum Curator :
  Dale M. Gnidovec, (M), Fort Hays State, 1978, PvGr
Adjunct Professor:
  Thomas G. Naymik, (D), Ohio State, 1978, Hw
Visiting Professor:
  David Young, (D), California, 2005, GcpCc
Academy Professor:
  William I. Ausich, (D), Indiana, 1978, PgGs
  Stig M. Bergstrom, (D), Lund (Sweden), 1961, PgGg
  David H. Elliot, (D), Birmingham, 1965, GxtGv
  Lawrence A. Krissek, (D), Oregon State, 1982, Gsu
  Teresa Mensing, (D), Ohio State, 1987, Cg
  Ralph R. B. von Frese, (D), Purdue, 1980, Ye
Emeritus:
  E. Scott Bair, (D), Penn State, 1980, HwGe
  James Bradley
  James W. Collinson, (D), Stanford, 1966, PgGsg
  Charles E. Corbato, (D), California (Los Angeles), 1960, Ye
  Jeffrey J. Daniels, (D), Colorado Mines, 1974, Yg
  James W. Downs, (D), Virginia Tech, 1983, Gz
  Gunter Faure, (D), MIT, 1961, CgHw
  Kenneth A. Foland, (D), Brown, 1972, CcgCs
  Charles E. Herdendorf, (D), Ohio State, 1970, OgGuOn
  Kenneth Jezek, (D), Wisconsin, 1980, YgPcGl
  Garry D. McKenzie, (D), Ohio State, 1968, GmeZe
  Ivan Mueller, (D), Ohio State, 1960, YddYv
  David Nickey, (D), Penn State, 1966
  Hallan C. Noltimier, (D), Newcastle upon Tyne, 1965, YmgGt
  Douglas E. Pride, (D), Illinois, 1969, Eg
  Richard H. Rapp, (D), Ohio State, 1964, Yd
  Rodney T. Tettenhorst, (D), Illinois, 1960, Gz
  Russell O. Utgard, (D), Indiana, 1969, YgEnZe
  Peter N. Webb, (D), Utrecht, 1966, PmRhGr
  Terry J. Wilson, (D), Columbia, 1983, GctGl
Program Coordinator:
  Mike Kositzke
Courtesy Appt.:
  Bryan Mark
Undergraduate Advisor:
  Karen Royce, (D), Ohio State
Space Management Coordinator:
  Dan Dunlap
Related Staff:
  Christina Millan, (D), Ggc

School of Environment and Natural Resources (B,M,D) (2015)
2021 Coffey Road
Room 210 Koffman Hall
Columbus, OH 43210
  p. (614) 292-2265
  white.1917@osu.edu
  http://senr.osu.edu/
  Administrative Assistant: Mary Capoccia
Eminent Scholar:
  Richard P. Dick, (D), Iowa State, 1986, Sb
Associate Director:
  Donald J. Eckert, (D), Ohio State, 1978, Sc
Professor:
  Jerry M. Bigham, (D), North Carolina State, 1977, Sd
  Frank G. Calhoun, (D), Florida, 1971, Sd
  Rattan Lal, (D), Ohio State, 1968, Sp
Associate Professor:
  Nicholas T. Basta, (D), Iowa State, 1989, Sc
  Edward L. McCoy, (D), Oregon State, 1984, Sp
  Brian K. Slater, (D), Wisconsin, 1994, SdZi
Assistant Professor:
  Dawn Ferris, (D), Minnesota, 1997, Sf
Emeritus:
  Warren A. Dick, (D), Iowa State, 1980, SbCbHs

Ohio University
Geological Sciences (B,M) ●⚷ (2018)
316 Clippinger Lab
Athens, OH 45701
  p. (740) 593-1101
  geological.sciences@ohio.edu
  http://www.ohio.edu/geology/
  Programs: BS Geological Sciences, BS Environmental Geology, BA Geological Sciences, Minor Geological Sciences Minor Paleontology
  Enrollment (2018): B: 75 (23) M: 22 (9)
Chair:
  Gregory S. Springer, (D), Colorado State, 2002, GmPcGa
Professor:
  Daniel Hembree, (D), Kansas, 2005, Pi
  Dina L. Lopez, (D), Louisiana State, 1992, CgHwGv
  R. Damian Nance, (D), Cambridge, 1978, Gtc
  Alycia L. Stigall, (D), Kansas, 2004, Pg
Associate Professor:
  Eung Seok Lee, (D), Indiana, 1999, Hw
  Keith A. Milam, (D), Tennessee, 2006, Xg
  Gregory C. Nadon, (D), Toronto, 1991, Gs
Technician & Information Tech:
  Timothy A. Grubb, (A), Washington State Comm Coll, 2002
Administrative Associate:
  Cheri Sheets, (B), Ohio, 1997

Ohio Wesleyan University
Dept of Geology & Geography (B) ☒ (2019)

61 S. Sandusky Street
Delaware, OH 43015
   p. (740) 368-3615
   bsmartin@owu.edu
   http://geo.owu.edu
   Programs: Geology (B.A. and B.S.), Planetary Science Geography, Environmental Science
   Enrollment (2018): B: 26 (10)
Chair:
   Barton S. Martin, (D), Massachusetts, 1991, GxvCg
Director, Environmental Studies Program:
   John B. Krygier, (D), Penn State, 1995, Zi
Professor:
   Keith O. Mann, (D), Iowa, 1987, Pi
Assistant Professor:
   Nathanael S. Amador, (D), Penn State, 2015, ZyGlZr
   Charles C. Trexler, (D), California (Davis), 2018, Gct
Instructor:
   Ashley L. Allen, (M), Wyoming, 2014, Znu
Emeritus:
   Karen H. Fryer, (D), Illinois, 1986, GcpGt
   Richard D. Fusch, (D), Oregon, 1972, Zu
   David H. Hickcox, (D), Oregon, 1979, ZyAm

Shawnee State University
Dept of Natural Sciences (B) ●☒ (2018)
940 Second Street
Portsmouth, OH 45662
   p. (740) 351-3395
   kshoemaker@shawnee.edu
   http://www.shawnee.edu/academics/natural-sciences/majors/geology/
   f: https://www.facebook.com/DeptNaturalSciencesShawneeStateUniversity
   t: @SSU_Science
   Administrative Assistant: Sharon Messer
   Enrollment (2018): B: 16 (9)
Professor:
   Kurt A. Shoemaker, (D), Miami (Ohio), 2004, GxzGm
Assistant Professor:
   Erik B. Larson, (D), Mississippi State, 2014, GmsHg
Other:
   Jeffrey A. Bauer, (D), Ohio State, 1987, Ps

University of Akron
Dept of Geosciences (B,M) O⌐🕮 (2017)
Akron, OH 44325-4101
   p. (330) 972-7630
   butch@uakron.edu
   http://www.uakron.edu/geology/
   Programs: Geology (B,M); Earth Science (B); Environmental Science (B); Geography (B,M); GIS (B); Environmental Geology (M)
   Certificates: Environmental Studies; GIS & Cartography
   Administrative Assistant: Elaine L. Butcher
   Enrollment (2017): B: 125 (37) M: 26 (19)
Professor:
   John A. Peck, (D), Rhode Island, 1995, GseGn
   Ira D. Sasowsky, (D), Penn State, 1992, HwGmo
   David N. Steer, (D), Cornell, 1996, Yes
Professor of Instruction:
   John F. Beltz, (M), Akron, 1992, GgRh
   Meera Chatterjee, (D), Zy
Associate Professor:
   Linda R. Barrett, (D), Michigan State, 1995, SdZri
   John M. Senko, (D), Oklahoma, 2004, Cg
Assistant Professor of Instruction:
   Thomas J. Quick, (M), Akron, 1983, GgCaYe
   Jeremy M. Spencer, (D), Kent State, 2015, AtRnZe
Assistant Professor:
   Shanon P. Donnelly, (D), Indiana, 2009, Zi
   Caleb Holyoke, (D), Brown, 2005, GcnNrYx
   James R. Thomka, (D), Cincinnati, 2015, PggGg

University of Akron - Wayne College
University of Akron - Wayne College (2015)
1901 Smucker Road
Orrville, OH 44667
   p. (330) 972-8934
   WayneCommunityRelations@uakron.edu
   http://www.wayne.uakron.edu/

University of Cincinnati
Dept of Geology (B,M,D) ☒ (2018)
500 Geology/Physics Building
P. O. Box 210013
Cincinnati, OH 45221-0013
   p. (513) 556-3732
   krista.smilek@uc.edu
   http://www.artsci.uc.edu/departments/geology.html
   Programs: Geology
   Enrollment (2018): B: 83 (10) M: 13 (8) D: 17 (2)
Head:
   Lewis Owen, (D), Leicester (UK), 1988, GmlGt
Professor:
   Thomas J. Algeo, (D), Michigan, 1989, Gs
   Carlton E. Brett, (D), Michigan, 1978, Pe
   Attila I. Kilinc, (D), Penn State, 1969, CpGvCu
   Thomas V. Lowell, (D), SUNY (Buffalo), 1986, Gl
Associate Professor:
   Christopher L. Atchison, (D), Ohio State, 2011, Ze
   Brooke E. Crowley, (D), California (Santa Cruz), 2009, CsPev
   Andrew D. Czaja, (D), California (Los Angeles), 2006, PmCl
   Craig Dietsch, (D), Yale, 1985, Gp
   Amy Townsend-Small, (D), Texas, 2006, Co
   Dylan Ward, (D), Colorado, 2010, GmqZi
Assistant Professor:
   Aaron Diefendorf, (D), Penn State, 2010, Co
   Joshua H. Miller, (D), Chicago, 2009, PgvPe
   Reza Soltanian, (D), Wright State, 2015, Hw
   Daniel M. Sturmer, (D), California State (Fullerton), 2003, YeGdo
   Yurena Yanes, (D), La Laguna (Spain), 2005, PeCsPg
Adjunct Professor:
   Brenda R. Hunda, (D), California (Riverside), 2004, Pi
   Glenn W. Storrs, (D), Yale, 1986, Pv
Emeritus:
   Madeleine Briskin, (D), Brown, 1973, Pe
   John E. Grover, (D), Yale, 1972, Gz
   Warren D. Huff, (D), Cincinnati, 1963, GzvGr
   J. Barry Maynard, (D), Harvard, 1972, Cl
   David L. Meyer, (D), Yale, 1971, Pi
   Arnold I. Miller, (D), Chicago, 1986, Pi
   David Nash, (D), Michigan, 1977, Gm
   Paul E. Potter, (D), Chicago, 1952, Gs

University of Dayton
Dept of Geology (B) ● (2016)
300 College Park
SC 179
Dayton, OH 45469-2364
   p. (937) 229-3432
   dgoldman1@udayton.edu
   http://www.udayton.edu/artssciences/geology/
   Administrative Assistant: Darla Titus
   Enrollment (2016): B: 40 (7)
Chair:
   Daniel Goldman, (D), SUNY (Buffalo), 1993, PgqGr
Professor:
   Donald Pair, (D), Syracuse, 1991, Gl
   Michael R. Sandy, (D), London, 1984, Pi
Associate Professor:
   Umesh Haritashya, (D), Indian Inst of Tech, 2005, ZrGm
   Andrea M. Koziol, (D), Chicago, 1988, Cp
   Allen J. McGrew, (D), Wyoming, 1992, GtcGp
   Shuang-Ye Wu, (D), Cambridge, 2000, Zy
Visiting Assistant Professor:
   Zelalem Bedaso, (D), S Florida, 2011, CslGs
Adjunct Professor:
   Sue Klosterman, (M), Wright State, 2005, Gg
   Andrew Rettig, (D), Cincinnati, 2014, Zy
Emeritus:
   Charles J. Ritter, (D), Michigan, 1971, Ct

University of Toledo
Dept of Environmental Sciences (B,M,D) (2015)
2801 W. Bancroft Street
Toledo, OH 43606-3390
   p. (419) 530-2009
   eccarson@wisc.edu
   http://www.utoledo.edu/nsm/envsciences/
   Administrative Assistant: Patricia Hacker
   Enrollment (2011): B: 13 (3) M: 10 (5)
Chair:
   Timothy G. Fisher, (D), Calgary, 1993, GlmGn
Director of the Lake Erie Research Center:
   Carol A. Stepien, (D), S California, 1985, Zn
Associate Professor:
   Richard H. Becker, (D), W Michigan, 2008, ZrGe
   Mark J. Camp, (D), Ohio State, 1974, PiRhEs
   Daryl F. Dwyer, (D), Michigan State, 1986, Zn
   Johan F. Gottgens, (D), Florida, 1992, Zn
   Scott Heckathorn, (D), Illinois, 1995, Zn
   David E. Krantz, (D), South Carolina, 1988, Gs
   James Martin-Hayden, (D), Connecticut, 1994, Hq
   Daryl L. Moorhead, (D), Tennessee, 1985, Zn
   Alison L. Spongberg, (D), Texas A&M, 1994, Co
   Donald J. Stierman, (D), Stanford, 1977, YgGeYs
Assistant Professor:
   Jonathon Bossenbroek, (D), Colorado State (Ft. Collins), 2004, Zn
   Thomas Bridgeman, (D), Michigan, 2001, Zn
   Christine M. Mayer, (D), Illinois, 1998, Zn
   William V. Sigler, (D), Purdue, 1999, Zn
   Michael N. Weintraub, California, 2004, Zn

Wittenberg University
Dept of Geology (B) ☒ (2019)
P.O. Box 720
Springfield, OH 45501-0720
   p. (937) 327-6475
   mzaleha@wittenberg.edu
   http://www.wittenberg.edu/academics/geology
   f: https://www.facebook.com/WittenbergGeology
   Programs: B.S. & B.A. in Geology, B.S. & B.A. in Environmental Science, B.A. in Earth Science Education
   Enrollment (2018): B: 11 (4)
Professor:
   John B. Ritter, (D), Penn State, 1990, Gm
Associate Professor:
   Sarah K. Fortner, (D), Ohio State, 2008, ClZeHg
   Michael J. Zaleha, (D), SUNY (Binghamton), 1994, Gs
Emeritus:
   Katherine L. Bladh, (D), Arizona, 1976, Gi
   Kenneth W. Bladh, (D), Arizona, 1978, GzEm
   Robert W. Morris, (D), Columbia, 1969, PiGg

Wright State University
Dept of Earth and Environmental Science (B,M) (2015)
3640 Colonel Glenn Highway
260 Brehm Lab
Dayton, OH 45435
   p. (937) 775-2201
   david.schmidt@wright.edu
   http://www.wright.edu/ees/
   Enrollment (2010): B: 66 (8) M: 48 (15)
Chair:
   David F. Dominic, (D), West Virginia, 1988, GsrRc
Director of Undergraduate Programs:
   David Schmidt, (D), Ohio State, PiGd
Professor:
   Christopher C. Barton, (D), Yale, 1983, GqYgHq
   C. B. Gregor, (D), Utrecht, 1967, Gs
   Chad Hammerschmidt, (D), Connecticut, 2005, CmOcCb
   Allen G. Hunt, (D), California (Riverside), 1983, GqSpCl
   Robert W. Ritzi, Jr., (D), Arizona, 1989, Hw
Associate Professor:
   Abinash Agrawal, (D), North Carolina, 1990, ClHwPg
   Songlin Cheng, (D), Arizona, 1984, Hw
   Ernest C. Hauser, (D), Wisconsin, 1982, YeGtYu

   William Slattery, (D), CUNY, 1994, ZeGr
   Rebecca Teed, (D), Minnesota, 1999, ZePeGn
   Doyle Watts, (D), Michigan, 1979, YeZr
Director of Sustainability:
   Huntting (Hunt) Brown, (D), Ge
Emeritus:
   Byron Kulander, (D), West Virginia, 1966, Gc
   Benjamin H. Richard, (D), Indiana, 1966, Ye
   Paul J. Wolfe, (D), Case Western, 1966, Ye

Youngstown State University
Dept of Geological & Environmental Sciences (B,M) (2015)
One University Plaza
2120 Moser Hall
Youngstown, OH 44555
   p. (330) 941-3612
   amjacobs@ysu.edu
   http://www.as.ysu.edu/~geology/
   Enrollment (2007): B: 48 (4) M: 21 (1)
Chair:
   Jeffrey C. C., (D), Kent State, 1992, EoNgHy
Director, Env Studies Program:
   Isam E. Amin, (D), Nevada (Reno), 1987, Hw
Professor:
   Raymond E. Beiersdorfer, (D), California (Davis), 1992, Cg
   Alan M. Jacobs, (D), Indiana, 1967, Ge
Assistant Professor:
   Joseph E. Andrew, (D), Kansas, 2002, Gc
   Felicia P. Armstrong, (D), Oklahoma State, 2003, Sb
   Shane V. Smith, (D), Washington State, 2005, Gs
Instructor:
   Harry Bircher, (M), Youngstown State, 1995, Ge
   Brian M. Greene, (D), Kent State, 2001, Ng
   Lawrence P. Gurlea, (M), Pennsylvania, 1971, Cg
Engineering Adjunct:
   Scott C. Martin, (D), Clarkson, 1984, Ge
   Douglas M. Price, (D), Notre Dame, 1988, Cg
Emeritus:
   Ann G. Harris, (M), Miami (Ohio), 1958, Ge
   Ikram U. Khawaja, (D), Indiana, 1969, Ec
   Charles R. Singler, (D), Nebraska, 1969, Gs
Cooperating Faculty:
   Thomas P. Diggins, (D), SUNY, 1997, Pg
   Carl G. Johnston, (D), Cincinnati, 1992, Pg

Zane State College
Oil and Gas Engineering Technology (A) (2015)
1555 Newark Rd
Zanesville, OH 43701
   p. (740) 588-1282
   nwelch@zanestate.edu
   http://www.zanestate.edu
   f: https://www.facebook.com/ZaneStateCollege
   t: @ZaneStateC

Oklahoma

Oklahoma City Community College
Physical Sciences (2015)
7777 South May Avenue
Oklahoma City, OK 73159
   p. (405) 682-1611
   dgregory@occc.edu
   http://www.occc.edu

Oklahoma State University
Boone Pickens School of Geology (B,M,D) O☒ (2018)
105 Noble Research Center
Stillwater, OK 74078-3031
   p. (405) 744-6358
   sandy.earls@okstate.edu
   http://geology.okstate.edu
   Administrative Assistant: Sandy Earls
   Enrollment (2018): B: 60 (17) M: 48 (15) D: 26 (2)

Head:
  Camelia Knapp, (D), Cornell, 2000, GtYgs
Professor:
  Mohamed G. Abdelsalam, (D), Texas (Dallas), 1993, GctYg
  Michael Grammer, (D), Miami, 1991, GsEoGr
  Jay M. Gregg, (D), Michigan State, 1982, Gs
  Todd Halihan, (D), Texas, 2000, HwYuRc
  James Knapp, (D), MIT, 1989, GcYgs
  Jack C. Pashin, (D), Kentucky, 1990, GrcGo
Associate Professor:
  Priyank Jaiswal, (D), Yg
  Daniel A. Lao, (D), Pittsburgh, 2008, GctGg
  James Puckette, (D), Oklahoma State, 1996, Go
  Tracy Quan, (D), MIT/WHOI, 2005, Og
Assistant Professor:
  Ashley Burkett, (D), Indiana State, 2015, Pme
  Ahmed Ismail, (D), Missouri S&T, 2006, Ysu
  Natascha Riedinger, (D), Bremen, 2005, CgGu
  Javier Vilcaez, (D), Tohoku, 2009, NpCl
Visiting Professor:
  Mary E. Hileman, (D), Michigan, 1973, GorGd
Emeritus:
  Arthur Hounslow, (D), Carleton, 1968, Cl
  Douglas Kent, (D), Iowa State, 1969, Hw
  Vernon Scott, (D), Utah, 1975, Ze
  Gary Stewart, (D), Kansas, 1973, Go
  John D. Vitek, (D), Iowa, 1973, Gm
Teaching Assistant Professor:
  Brendan Hanger, (D), Australian Nat, 2014, Ggz

Rogers State University
Dept of Mathematics and Physical Sciences (2015)
1701 W. Will Rogers Blvd.
Claremore, OK 74107
  p. (918) 343-6812
  vwood@rsu.edu
  http://www.rsu.edu

The University of Tulsa
Dept of Geosciences (B,M,D) ☒ (2018)
800 S. Tucker Drive
Tulsa, OK 74104-9700
  p. (918) 631-2517
  dennis-kerr@utulsa.edu
  https://engineering.utulsa.edu/academics/geosciences/
  Programs: Biogeosciences (B); Earth & Environmental Sciences (B); Geology (B); Geophysics (B,M); Geosciences (B,M); Petroleum Engineering (B)
  Administrative Assistant: Beverly A. Phelps
  Enrollment (2018): B: 41 (23) M: 15 (9) D: 6 (0)
Vice President for Research & Dean of Graduate School:
  Janet A. Haggerty, (D), Hawaii, 1982, GudGs
Professor:
  Peter J. Michael, (D), Columbia, 1983, GivGz
  Kerry Sublette, (D), Tulsa, 1985, Ge
Decker Dawson Associate Professor of Applied Geophysics:
  Jingyi Chen, (D), Chinese Acad of Sci, 2005, Yes
Associate Professor:
  Dennis R. Kerr, (D), Wisconsin, 1989, GsrGt
  Steven L. Roche, (D), Colorado Mines, 1997, YesYg
Assistant Professor:
  Junran Li, (D), Virginia, 2008, Gme
  Christine J. Ruhl, (D), Nevada (Reno), 2016, YsGtc
  Bethany P. Theiling, (D), New Mexico, 2012, Cs
Research Associate:
  Robert W. Scott, (D), Kansas, 1967, GsrPi
Applied Associate Professor:
  Winton Cornell, (D), Rhode Island, 1987, GivCa
Emeritus:
  Colin Barker, (D), Oxford, 1965, CgoGo
  J. Bryan Tapp, (D), Oklahoma, 1983, GceZi

Tulsa Community College
Science and Math (A) (2018)
909 S. Boston Avenue
Tulsa, OK 74119
  p. (918) 595-7085
  Kelly.E.Allen@tulsacc.edu
  http://www.tulsacc.edu
  Programs: Geology (A)
  Certificates: GIS
  Enrollment (2018): A: 19 (5)
Professor:
  Kelly E. Allen, (D), Texas Tech, 2000, ZyiZe
Adjunct Professor:
  Douglas Ashe, (M), Oklahoma State, 2018, GgZe
  Claude E. Bolze, (M), Wright State, 1974, GgoZe
  Martin Bregman, (D), New Mexico, 1971, YgGcg
  Christina Opfer, (M), Tulsa, 2015, GgoZe

University of Oklahoma
ConocoPhillips School of Geology & Geophysics (B,M,D)
O (2015)
100 East Boyd
710 Energy Center
Norman, OK 73019-0628
  p. (405) 325-3253
  geology@ou.edu
  http://geology.ou.edu
  Enrollment (2006): B: 87 (0) M: 40 (0) D: 24 (0)
Director:
  R. Douglas Elmore, (D), Michigan, 1981, YmGs
Professor:
  Younane N. Abousleiman, (D), Delaware, 1991, Gg
  Michael H. Engel, (D), Arizona, 1980, Co
  G. Randy Keller, (D), Texas Tech, 1973, Yg
  David London, (D), Arizona State, 1981, CpGiz
  Kurt J. Marfurt, (D), Columbia, 1978, Ys
  Shankar Mitra, (D), Johns Hopkins, 1977, Gc
  R. Paul Philp, (D), Sydney, 1972, Co
  Matthew J. Pranter, (D), Colorado Mines, Go
  Matthew J. Pranter, Gos
  Roger M. Slatt, (D), Alaska, 1970, Go
  Gerilyn S. Soreghan, (D), Arizona, 1992, GsPcZg
  Stephen R. Westrop, (D), Toronto, 1984, Pi
Associate Professor:
  Andrew S. Elwood Madden, (D), Virginia Tech, 2005, ClZaCb
  Megan E. Elwood Madden, (D), Virginia Tech, 2005, ClXgCg
  Richard Lupia, (D), Chicago, 1997, Pm
  John D. Pigott, (D), Northwestern, 1981, GoYeGd
  Barry L. Weaver, (D), Birmingham (UK), 1980, Ct
Assistant Professor:
  Xiaowei Chen, (D), California (San Diego), 2013, YsGt
  Shannon Dulin, (D), Oklahoma, 2014, YmGs
  Michael J. Soreghan, (D), Arizona, 1994, Gs
Adjunct Professor:
  Richard L. Cifelli, (D), Columbia, 1983, Pv
Emeritus:
  Judson L. Ahern, (D), Cornell, 1980, Yg
  James M. Forgotson, Jr., (D), Northwestern, 1956, Go
  M. Charles Gilbert, (D), California (Los Angeles), 1965, Cp
  Charles W. Harper, Jr., (D), Caltech, 1964, Pi
  G. Randy Keller, (D), Texas Tech, 1973, YgGtEo
  David W. Stearns, (D), Texas A&M, 1969, Gc
Cooperating Faculty:
  Neil Suneson, (D), California (Santa Barbara), 1980, Gr
Electron Microprobe Operator:
  George B. Morgan, (D), Oklahoma, 1988, Gi

Mewbourne School of Petroleum & Geological Engineering (B,M,D) (2015)
100 East Boyd Street
Sarkeys Energy Center 1210
Norman, OK 73019-1001
  p. (405) 325-2921
  mpge@ou.edu
  http://mpge.ou.edu/
  f: https://www.facebook.com/OUMPGE
  t: @ou_mpge
  Enrollment (2015): B: 975 (133) M: 101 (25) D: 37 (4)
Director:
  Chandra S. Rai, (D), Hawaii, 1977, Np

Graduate Liaison:
 Deepak Devegowda, (D), Texas A&M, 2008, Np
Director Natural Gas Engineering & Management:
 Suresh Sharma, (D), Oklahoma, 1968, Np
Professor:
 Younane Abousleiman, (D), Delaware, 1991, Nr
 Ramadan Ahmed, (D), Norwegian Inst of Tech, 2001, Np
 Jeff Callard, (D), Louisiana State, 1994, Np
 Ahmed Ghassemi, (D), Oklahoma, 1996, Nr
 Ben Shiau, (D), Oklahoma, 1995
 Carl H. Sondergeld, (D), Cornell, 1977, NpYex
 Musharraf Zaman, (D), Bangladesh Univ of Eng and Tech, 1975, Np
Associate Professor:
 Mashhad Fahes, (D), Imperial Coll (London), 2006, Np
 Ahmad Jamili, (D), Kansas, 2004, Np
 Rouzbeh Moghanloo, (D), Texas, 2012, Np
 Maysam Pournik, (D), Texas A&M, 2008, Np
 Catalin Teodoriu, (D), Ploiesti, Np
 Xingru Wu, (D), Texas, 2006, Np
Assistant Professor:
 Siddharth Misra, (D), Texas, 2015, Np
 Ahmad Sakhaee-Pour, (D), Texas, 2012, Np
Instructor:
 Ilham El-Monier, (D), Texas A&M, 2012, Np
Emeritus:
 Faruk Civan, (D), Oklahoma, 1978, Np
 Roy M. Knapp, (D), Kansas, 1973, Np
 Jean-Claude Roegiers, (D), Minnesota, 1974, Nr
 Subhash N. Shah, (D), New Mexico, 1974, Np

Oregon

Central Oregon Community College
Dept of Science (A) (2015)
2600 NW College Way
Bend, OR 97701
 p. (541) 383-7557
 breynolds@cocc.edu
 http://science.cocc.edu/Programs_Classes/Geology/default.aspx
Associate Professor:
 Robert W. Reynolds, (D), Idaho, 1994, Gv

Oregon State University
College of Earth, Ocean, and Atmospheric Sciences (B,M,D)
● ⊘ (2019)
104 CEOAS Administration Building
Corvallis, OR 97331-5503
 p. (541) 737-1201
 contact@coas.oregonstate.edu
 http://ceoas.oregonstate.edu
 Programs: Earth Science (B); Environmental Sciences (B); Geography and Geospatial Science (B); Geography (M,D); Geology (M,D); Marine Resource Management (M,D); Ocean, Earth, and Atmospheric Sciences (M,D)
 Certificates: GIS (UG); GIS (G); Marine Resource Management (G), Water Conflict Management and Transformation (G)
 Enrollment (2018): B: 706 (98) M: 64 (21) D: 67 (14)
Senior Research:
 Mitchell Lyle, (D), Oregon State, 1978, GgYg
 David Mellinger, Cb
R.S. Yeats Professor of Earthquake Geology and Active Tectonics and Associate Dean for Academic Programs:
 Eric Kirby, (D), MIT, 2001, GtmGc
Exec Dir-Marine Studies Initiative:
 Jack A. Barth, (D), MIT, 1987, On
Director-Budget & Fiscal Planning:
 Sherman H. Bloomer, (D), California (San Diego), 1982, Gi
Director of Ocean Science Program:
 Rob Wheatcroft, (D), Washington, 1990, OgGg
Director of Marine Resource Management Program:
 Flaxen D. Conway, (M), Oregon State, 1986, Gu
Director of Geology Program:
 Adam J. R Kent, (D), Australian Nat, 1995, GiCga
Director of Geography Program:
 Julia A. Jones, (D), Johns Hopkins, 1983, So
Director of Environmental Science Program:
 Laurence Becker, (D), California (Berkeley), 1989, Zn
Professor:
 Jeffrey R. Barnes, (D), Washington, 1983, As
 Kelly Benoit-Bird, (D), Hawaii, 2003, Gu
 Edward J. Brook, (D), MIT/WHOI, 1993, ClPeGl
 Michael E. Campana, (D), Arizona, 1975, Hg
 Lorenzo Ciannelli, (D), Washington, 2002, Og
 Frederick (Rick) Colwell, (D), Virginia Tech, 1986, Zn
 Byron Crump, (D), Washington, 1999
 Shanika de Silva, (D), Open (UK), 1987, Gv
 John H. Dilles, (D), Stanford, 1984, Em
 Gary D. Egbert, (D), Washington, 1987, Yr
 Chris Goldfinger, (D), Oregon State, 1994, Yr
 Miguel A. Goni, (D), Washington, 1992, Gu
 David W. Graham, (D), MIT, 1987, Cm
 Burke R. Hales, Washington, 1995, Oc
 Merrick C. Haller, (D), Delaware, 1999, OnZrOp
 Robert N. Harris, (D), Utah, 1996, Yhr
 Michael Harte, (D), Victoria, 1994, Gu
 Anthony Koppers, (D), Free (Amsterdam), 1988, CgGv
 Michael Kosro, (D), Scripps, 1985, OpZrOn
 Jim Lerczak, (D), Scripps, 2000, Op
 Ricardo Letelier, (D), Hawaii, 1994, ObZr
 Ricardo P. Matano, (D), Princeton, 1991, OpAs
 Andrew J. Meigs, (D), S California, 1995, Gc
 James N. Moum, (D), British Columbia, 1984, Op
 John L. Nabelek, (D), MIT/WHOI, 1984, GtYs
 Jonathan Nash, (D), Oregon State, 2000, Op
 Roger L. Nielsen, (D), S Methodist, 1983, GizGv
 David Noone, (D), Melbourne, 2001, AmZg
 Clare Reimers, (D), Oregon State, 1982, Oc
 Peter Ruggiero, (D), Oregon State, 1997, On
 Roger Samelson, (D), Oregon State, 1987, Op
 Andreas Schmittner, (D), Bern, 1999, AsYr
 Adam Schultz, (D), Washington, 1986, YgmRn
 Eric Skyllingstad, (D), Wisconsin, 1986, Am
 William D. Smyth, (D), Toronto, 1990, Op
 Richard Spinrad, Op
 Yvette H. Spitz, (D), Old Dominion, 1995, Yr
 Frank J. Tepley, III, (D), California (Los Angeles), 1999, GiCs
 Marta E. Torres, (D), Oregon State, 1988, Cm
 Anne M. Trehu, (D), MIT/WHOI, 1982, YrsGt
 Aaron T. Wolf, (D), Wisconsin, 1992, Hg
Senior Research:
 Theodore Durland, (D), Hawaii, 2006, Yg
 Brian Haley, (D), Oregon State, 2004, Cs
 Haruyoshi Matsumoto
 Nicholas Tufillaro, (D), Bryn Mawr, 1990, Cb
Director, Water Resources Graduate Program:
 Mary V. Santelmann, (D), Minnesota, 1988, Zy
Director of Atmospheric Science Program:
 Karen M. Shell, (D), Scripps, 2004, As
Associate Professor:
 Anders Carlson, (D), Oregon State, 2006, GlCt
 Patrick Corcoran, 1989
 Simon P. de Szoeke, (D), Washington, 2004, OpAs
 Hannah Gosnell, (D), Colorado, 2000, Zu
 Alexander Kurapov, (D), St Petersburg, 1994, Yr
 Stephen Lancaster, (D), MIT, 1999, Hs
 Kipp Shearman, (D), Oregon State, 1999, Opn
 Emily L. Shroyer, (D), Oregon State, 2009, On
 Joseph Stoner, (D), Québec (Montréal), 1995, Gsr
 George Waldbusser, (D), Maryland (Baltimore), 2008, Ob
Senior Research:
 Christo Buizert, (D), Copenhagen, 2012, Yr
 Louise A. Copeman, (D), Memorial, 2011, Zn
 Jonathan Fram, (D), California (Berkeley), 2005, Op
 Joseph Haxel
 W. Todd Jarvis, (D), Oregon State, 2006, Hw
 Maria Kavanaugh, Cb
 Jennifer L. McKay, (D), British Columbia, 2004, Cm
 David Rupp, (D), Oregon State, 2005, AsOp
 Jenna Tilt, (D), Washington, 2007, Zu

Assistant Professor:
    Kim S. Bernard, (D), Rhodes, 2007
    Jessica Creveling, (D), Harvard, 2012, GgsGr
    Jennifer Fehrenbacher, (D), N Illinois, 1997, Cm
    Jennifer Hutchings, Gl
    Lauren W. Juranek, Washington, 2007, Cm
    Robert Kennedy, (D), Oregon State, 2004, Zri
    Larry O'Neill, (D), Oregon State, 2007, AmOg
    Alyssa E. Shiel, (D), British Columbia, 2010, ClsCt
    Nick Siler
    Andrew Thurber, (D), California (San Diego), 2010, Cb
    Jamon Van Den Hoek, (D), Wisconsin, 2012, Zur
    James R. Watson, Zy
    Justin Wettstein, (D), Washington, 2007, As
    Greg Wilson, (D), Oregon State, 2013, Onp
    Bo Zhao, (D), Ohio State, 2015, Zi
Senior Instructor:
    Lorene Yokoyama Becker, (M), Wisconsin, 1999, Zi
Instructor:
    Andrea Allan, (D), Oregon State, 2012, As
    Lynette de Silva, (M), Indiana, 2000
    Randall A. Keller, (D), Oregon State, 1996, Gu
    Randall Milstein, (D), Oregon State, 1994
    Rebecca Yalcin, (M), Maine, 2001, Gg
Emeritus:
    A. Jon Kimerling, (D), Wisconsin, 1976, Zir
    Gary P. Klinkhammer, (D), Rhode Island, 1979, CaHsRw
    Robert Lloyd Smith, (D), Oregon State, 1964, OpgZn
    Robert S. Yeats, (D), Washington, 1958, GtYsRg
Dean of College of Earth, Ocean, and Atmospheric Sciences and Professor:
    Roberta Marinelli, (D), S Carolina, 1991, GuOcGe

Portland Community College - Sylvania Campus
Physical Science Dept (A) O (2017)
12000 SW 49th Ave.
Portland, OR 97219
    p. 971-722-8209
    patty.maazouz@pcc.edu
    http://www.pcc.edu/programs/geology/
Instructor:
    Talal Abdulkareem, (D)
    Kali Abel, (M)
    Sharon Delcambre, (D), AsOg
    Gretchen Gebhardt, (M)
    Melinda Hutson, (D), Xm
    Hollie Oakes-Miller, (M), Gg
    Kristy Schepker, (M)
    Steve Todd, Am
    Jonathan Weatherford, (M), Gg

Portland State University
Dept of Geology (B,M,D) (2018)
P.O. Box 751
Portland, OR 97207
    p. (503) 725-3022
    geology@pdx.edu
    http://www.pdx.edu/geology
    t: @PSUGeology
    Programs: Geology
    Enrollment (2018): B: 85 (18) M: 31 (7) D: 2 (0)
Acting Chair:
    Martin J. Streck, (D), Oregon State, 1994, GivCt
Professor:
    Andrew G. Fountain, (D), Washington, 1992, Gl
Associate Professor:
    Kenneth M. Cruikshank, (D), Purdue, 1991, Gc
    Robert Benjamin Perkins, (D), Portland State, 2000, Hq
    Alexander (Alex) M. Ruzicka, (D), Arizona, 1996, Xm
Assistant Professor:
    John T. Bershaw, (D), Rochester, 2011, GsCsHy
    Adam M. Booth, (D), Oregon, 2012, GmZr
    Nancy A. Price, (D), Maine, 2012, GcZeGt
    Ashley Streig, (D), Oregon, 2014, YsGt
Research Associate:
    Richard Hugo, (D), Washington State, 1998, Gy

Adjunct Professor:
    Sheila Alfsen, (D), GgPgOu
    Matthew Brunengo, (D), Portland State, 2012, GgNgGm
    Megan Faust, (D), ZeGr
    Frank D. Granshaw, (D), Portland State, 2011, ZnGl
    Melinda Hutson, (D), Arizona, 1996, Xmc
    William Orr, (D), Michigan State, 1968, GgOg
    Dick Pugh
    Erik Shafer, Portland State, 2015, Gg
    Barry Walker, (D), Oregon State, 2011, GiCgGv
Emeritus:
    Scott F. Burns, (D), Colorado, 1980, Ng
    Michael L. Cummings, (D), Wisconsin, 1978, GgHg
    Paul E. Hammond, (D), Washington, 1963, Gi
    Ansel G. Johnson, (D), Stanford, 1973, Ye
    Richard E. Thoms, (D), California, 1965, Ps
Instructor:
    David Percy, (B), Portland State, 1999, Zi
Cooperating Faculty:
    Christina L. Hulbe, (D), Chicago, 1998, Gl

Rogue Community College
Science Dept-Physical Sciences-Geology (2015)
3345 Redwood Highway
Grants Pass, OR 97527
    p. (541) 245-7527
    jvanbrunt@roguecc.edu
    http://learn.roguecc.edu/science/physical.htm

Southern Oregon University
Dept of Geology (B) (2015)
1250 Siskiyou Boulevard
Ashland, OR 97520
    p. (541) 552-6479
    lane@sou.edu
    Administrative Assistant: Susan Koralek
    Enrollment (2006): B: 45 (5)
Dean, Science:
    Joseph L. Graf, Jr., (D), Yale, 1975, Em
Chair:
    Charles L. Lane, (D), California (Los Angeles), 1987, Hg
Professor:
    Jad A. D'Allura, (D), California (Davis), 1977, Gc
Associate Professor:
    Eric Dittmer, (M), San Jose State, 1972, Gg
Adjunct Professor:
    Vernon J. Crawford, (M), Oregon, 1970, Gg
    Harry W. Smedes, (D), Washington, 1959, Gi
    Richard Ugland, (M), Utah, 1974, Gg
Emeritus:
    Monty A. Elliott, (D), Oregon State, 1971, Gr
    William B. Purdom, (D), Arizona, 1960, Gx

Southwestern Oregon Community College
Dept of Geology (A) (2019)
1988 Newmark
Coos Bay, OR 97420-2912
    p. (541) 888-7216
    rmetzger@socc.edu
    https://www.socc.edu/geology
    Enrollment (2013): A: 2 (0)
Professor:
    Ronald A. Metzger, (D), Iowa, 1991, PmZePs

Tillamook Bay Community College
Associate of Science (2015)
2510 First Street
Tillamook, OR 97141
    bannan@tillamookbay.cc

Treasure Valley Community College
Career and Technical Education ● ⊠ (2018)
650 College Blvd.
Ontario, OR 97914
    p. (541) 881-8866

dtinkler@tvcc.cc
http://www.tvcc.cc.or.us/science/Index.htm
f: https://www.facebook.com/profile.php?id=100005312898088
Instructor:
  Dorothy Tinkler, (D), Texas Tech, 2003, Zi

Umpqua Community College
Dept of Geology Transfer Program (2015)
P.O. Box 967
1140 Umpqua College Rd.
Roseburg, OR 97470
  p. (541) 440-4654
  Jason.Aase@umpqua.edu
  http://www.umpqua.edu/degree-programs/55

University of Oregon
Dept of Earth Sciences (B,M,D) ○◎ (2019)
1272 University of Oregon
Eugene, OR 97403-1272
  p. (541) 346-4573
  pwallace@uoregon.edu
  http://earthsciences.uoregon.edu/
  Programs: Geology; Earth Science; Geophysics; Paleontology; Environmental Geoscience
  Enrollment (2018): B: 0 (35) M: 11 (4) D: 42 (3)
Department Head:
  Paul Wallace, (D), California (Berkeley), 1991, GviCg
Lillis Professor of Volcanology:
  Josef Dufek, (D), Washington, 2006, Gv
Professor:
  Ilya N. Bindeman, (D), Chicago, 1998, CsGvCl
  Rebecca J. Dorsey, (D), Princeton, 1989, Gr
  Eugene D. Humphreys, (D), Caltech, 1985, YsGt
  Mark H. Reed, (D), California (Berkeley), 1977, CgEm
  Alan W. Rempel, (D), Cambridge (UK), 2001, YgNrGl
  Gregory J. Retallack, (D), New England, 1978, PbSoGa
  Joshua J. Roering, (D), California (Berkeley), 2000, Gm
  Douglas R. Toomey, (D), MIT/WHOI, 1987, YsGt
  Ray J. Weldon, (D), Caltech, 1986, Gc
Associate Professor:
  Emilie E. Hooft, (D), MIT, 1996, Yr
  Samantha Hopkins, (D), California (Berkeley), 2005, Pg
  Qusheng Jin, (D), Illinois (Urbana), 2003, Cg
  Carol Paty, (D), Washington, 2006
  Matthew Polizzotto, (D), Stanford, 2007, CqSc
  Dave Sutherland, (D), MIT, 2008, Op
  James Watkins, (D), California (Berkeley), 2010, Cg
Assistant Professor:
  Estelle Chaussard, (D), Miami, 2013, Yd
  Edward B. Davis, (D), California (Berkeley), 2005, PvePq
  Thomas Giachetti, (D), Clermont-Ferrand, 2010, Gvu
  Leif Karlstrom, (D), California (Berkeley), 2011, Yx
  Diego Melgar, (D), Scripps, 2014, Yd
  Valerie J. Sahakian, (D), Scripps, 2015, Yg
  Amanda Thomas, (D), California (Berkeley), 2012, Ys
  Meredith Townsend, (D), Stanford, 2017, Gv
Instructor:
  Marli G. Miller, (D), Washington, 1997, Gc
Emeritus:
  A. Dana Johnston, (D), Minnesota, 1983, Cp
  Marvin A. Kays, (D), Washington Univ (St. Louis), 1960, GpcGt
  William N. Orr, (D), Michigan State, 1967, Pm
  Norman M. Savage, (D), Sydney, 1968, Pi
Office & Business Manager:
  Sandy K. Thoms, (B), Portland State, 1993

Dept of Geography (B,M,D) ◎ (2018)
107 Condon Hall
1251 University of Oregon
Eugene, OR 97403-1251
  p. (541) 346-4555
  uogeog@uoregon.edu
  http://geography.uoregon.edu
  f: https://www.facebook.com/universityoforegongeography/
  Enrollment (2009): B: 167 (55) M: 22 (9) D: 19 (5)

Dean:
  W. Andrew Marcus, (D), Colorado, 1987
Head:
  Dan Gavin, (D), Washington, 2000, ZyPle
Professor:
  Patrick Bartlein, (D), Wisconsin, 1978, Pe
  Amy Lobben, (D), Michigan State, 1999, ZinZn
  Patricia McDowell, (D), Wisconsin, 1980, Gm
  Alexander Murphy, (D), Chicago, 1987, Zn
  Laura Pulido, (D), California (Los Angeles), 1991
  Peter Walker, (M), Oregon, 1966, Zn
Associate Professor:
  Daniel Buck, (D), California (Berkeley), 2002, Zn
  Shaul Cohen, (D), Chicago, 1991, Zn
  Carolyn Fish, (D), Penn State, 2018
  Mark Fonstad, (D), Arizona State, 2000, Zy
  Katharine Meehan, (D), Arizona, 2010, Zn
  Xiaobo Su, (D), Singapore, 2007, Zn
Assistant Professor:
  Leigh Johnson, (D), California (Berkeley), 2011
  Henry Hui Luan, (D), Waterloo, 2016
  Hedda R. Schmidtke, (D), Hamburg, 2004
  Lucas Silva, (D), Guelph, 2011
Instructor:
  Donald Holtgrieve, (D), Oregon, 1973
  Nicholas Kohler, (D), Oregon, 2004, Zi
  Leslie McLees, (D), Oregon, 2012
Emeritus:
  Stanton Cook, (D), California (Berkeley), 1960, Zn
  Carl Johannessen, (D), California (Berkeley), 1959, Zn
  Alvin Urquhart, California (Berkeley), 1962, Zn
  Ronald Wixman, (D), Columbia, 1978, Zn

Western Oregon University
Earth and Physical Science Dept. (B) ☒ (2019)
345 N. Monmouth Ave.
Monmouth, OR 97361
  p. (503) 838-8398
  taylors@wou.edu
  http://www.wou.edu/earthscience
  Programs: B.S. / B.A. Earth Science
  Certificates: Minor/Cert in Geographic Information Science
  Enrollment (2018): B: 50 (10)
Professor:
  Jeffrey A. Myers, (D), Santa Barbara, 1998, Gs
  Stephen B. Taylor, (D), West Virginia, 1999, Gm
  Jeffrey H. Templeton, (D), Oregon State, 1998, GivZe
Instructor:
  Don Ellingson, (M), W Oregon, 1988, AmZgXg
  Matthew Goslin, (M), Oregon State, 1999, Gmg
  Jeremiah Oxford, (M), Oregon State, 2006, Zg
  Grant Smith, (D), Oregon State, 2012, Zg
  Phillip Wade, (M), San Diego State, 1991, Zge

Willamette University
Department of Environmental Science (B) ☒ (2018)
900 State Street
Salem, OR 97301
  p. (503) 370-6587
  spike@willamette.edu
  http://willamette.edu/cla/envs/index.html
  f: https://www.facebook.com/groups/186102921735448/
  Enrollment (2018): B: 99 (0)
Professor:
  Karen Arabas, (D), Penn State, 1997, Zn
Endowed Dempsey Chair:
  Joe Bowersox, (D), Wisconsin, 1995, Zn
Assistant Professor:
  Melinda Butterworth, (D), Arizona
  Katja Meyer, (D), Penn State
  Scott Pike, (D), Georgia, 2000, Gag

Pennsylvania

Allegheny College
Dept of Geology (B) (2015)

520 North Main Street
Meadville, PA 16335
    p. (814) 332-2350
    eboynton@allegheny.edu
    http://www.allegheny.edu/academics/geo/
    Enrollment (2015): B: 20 (0)
Associate Professor:
    Rachel O'Brien, (D), Washington State, 2000, HwGe
Assistant Professor:
    Theresa M. Schwartz, (D), Stanford, 2015, Gs
Visiting Professor:
    Erin M. Birsic, (M), Wisconsin, 2015, Gi
Currently Provost and Dean of the College:
    Ron B. Cole, (D), Rochester, 1993, Gt

Bloomsburg University
Dept of Environmental, Geographical, and Geological Sciences (B) ☒ (2019)
400 East Second Street
Bloomsburg, PA 17815-1301
    p. (570) 389-4108
    mshepard@bloomu.edu
    https://intranet.bloomu.edu/eggs-department
    Programs: Professional Geology, Environmental Geoscience, Geography and Planning. Minors in Hydrology, Spatial Analysis and GIS, Geoscience, Geography.
    Administrative Assistant: Cheryl Smith
    Enrollment (2018): B: 200 (50)
Professor:
    John E. Bodenman, (D), Penn State, 1995, Zu
    Sandra J. Kehoe-Forutan, (D), Queensland, 1991, Znu
    Brett T. McLaurin, (D), Wyoming, 2000, GrsGd
    Michael K. Shepard, (D), Washington Univ (St. Louis), 1994, XgYgZr
    Cynthia Venn, (D), Pittsburgh, 1996, OgClOc
Associate Professor:
    Patricia J. Beyer, (D), Arizona State, 1997, HsZyGm
    Jeffrey C. Brunskill, (D), SUNY (Buffalo), 2005, ZiAmZy
    Jennifer B. Whisner, (D), Tennessee, 2010, GmHw
    S. Christopher Whisner, (D), Tennessee, 2005, GciGz
Assistant Professor:
    Tina Delahunty, (D), Florida, 2002, Zri
    Benjamin Franek, (D), Connecticut, 2013, HsZy
    John G. Hintz, (D), Kentucky, 2005, Zin
    Adrian Van Rythoven, (D), Toronto, 2012, GxzEg
    Dana Xiao, (D), California (Santa Barbara), 2013, Zi

Bryn Mawr College
Dept of Geology (B) ☒ (2019)
101 North Merion Avenue
Bryn Mawr, PA 19010-2899
    p. (610) 526-5115
    aweil@brynmawr.edu
    http://www.brynmawr.edu/geology/
    Programs: Geology
    Enrollment (2018): B: 28 (9)
Professor:
    Arlo B. Weil, (D), Michigan, 2001, GctYm
Associate Professor:
    Donald C. Barber, (D), Colorado, 2001, GesOn
    Selby Cull-Hearth, (D), Washington, 2011, XgGzx
    Pedro J. Marenco, (D), S California, 2007, PgCsPg
Instructor:
    Katherine N. Marenco, (D), S California, 2008, PiePg
Emeritus:
    Maria Luisa B. Crawford, (D), California (Berkeley), 1965, GxzGp
    William A. Crawford, (D), California (Berkeley), 1965, Cg
    Lucian B. Platt, (D), Yale, 1960, Gc
    W. Bruce Saunders, (D), Iowa, 1970, Pi

Bucknell University
Geology and Environmental Geosciencs (B) (2016)
231 O'Leary Center
Lewisburg, PA 17837
    p. (570) 577-1382
    cdaniel@bucknell.edu
    http://www.bucknell.edu/Geology
    f: http://www.facebook.com/BucknellGeology
    Administrative Assistant: Carilee Dill
    Enrollment (2016): B: 36 (7)
Chair:
    Christopher G. Daniel, (D), Rensselaer, 1998, Gp
Professor:
    Mary Beth Gray, (D), Rochester, 1991, Gc
    Carl S. Kirby, (D), Virginia Tech, 1993, Cl
    R. Craig Kochel, (D), Texas, 1980, Gm
    Jeffrey M. Trop, (D), Purdue, 2000, Gs
Associate Professor:
    Ellen K. Herman, (D), Penn State, 2006, Hw
    Robert W. Jacob, (D), Brown, 2006, Yg
Emeritus:
    Jack C. Allen, (D), Princeton, 1962, GizGe
    Edward Cotter, (D), Princeton, 1963, GmPg
Laboratory Director:
    Bradley C. Jordan, (M), Rhode Island, 1983, Gg

California University of Pennsylvania
Dept of Earth Sciences (B) (2015)
250 University Avenue
California, PA 15419
    p. (724) 938-4180
    wickham@calu.edu
    http://www.cup.edu/eberly/earthscience
    Administrative Assistant: Pamela Higinbotham
    Enrollment (2015): B: 195 (25)
Chair:
    Thomas Wickham, (D), Penn State, 2000, Zu
Professor:
    Kyle Fredrick, (D), SUNY (Buffalo), 2008, GgHgGm
Associate Professor:
    Thomas Mueller, (D), Illinois, 1999, Zi
Assistant Professor:
    John Confer, (D), Penn State, 1997, Zn
    Swarndeep S. Gill, (D), Wyoming, 2002, As
    Chad Kauffman, (D), Nebraska, 2000, As
    Susan Ryan, (D), Calgary, 2005, Zg

Carnegie Museum of Natural History
Section of Vertebrate Paleontology (2015)
4400 Forbes Avenue
Pittsburgh, PA 15213
    p. (412) 622-5782
    beardc@CarnegieMNH.org
    http://www.carnegiemnh.org/vp/
Curator:
    K. Christopher Beard, (D), Johns Hopkins, 1989, Pv
    David S. Berman, (D), California (Los Angeles), 1969, PggPv
    Mary R. Dawson, (D), Kansas, 1957, Pv
Assistant Curator:
    Matthew C. Lamanna, (D), Pennsylvania, 2004, PvgPe
Preparator:
    Dan Pickering, (B), Carnegie Mellon, 1983, Zn
    Alan R. Tabrum, (M), SD Mines, 1981, Pv
    Norman Wuerthele, (B), Pittsburgh, 1966, Pv
Curator & Associate Director:
    Zhexi Luo, (D), California (Berkeley), 1987, Pv
Collections Manager:
    Amy C. Henrici, (M), Pittsburgh, 1990, Pv

Clarion University
Dept of Biology and Geosciences (B) ○⋅☒ (2018)
840 Wood Street
Clarion, PA 16214
    p. (814) 393-2317
    cking@clarion.edu
    http://www.clarion.edu/BIGS
    Programs: B.S. Geology, B.S. Environmental Geoscience, Minors in Geology and Environmental Geoscience
    Certificates: GIS
    Enrollment (2018): B: 75 (24)
Professor:
    Yasser M. Ayad, (D), Montreal, 2000, Zi
    Valentine U. James, (D), Texas A&M, ZuRwZn

Anthony J. Vega, (D), Louisiana State, 1994, AtmOp
Craig E. Zamzow, (D), Texas, 1983, Gi
Adjunct Professor:
Sheila Kasar, (D), ZreGe

Delaware County Community College
STEM (A) (2015)
901 Media Line Road
Media, PA 19063
p. (610) 359-5082
jsnyder2@dccc.edu
http://www.dccc.edu/
Professor:
Daniel Childers, (D), Delaware, 2014, ZgOuZi
Associate Professor:
Jennifer L. Snyder, (D), W Michigan, 1998, Zg

Dickinson College
Dept of Earth Sciences (B) ⊠ (2018)
P.O. Box 1773
Carlisle, PA 17013-2896
p. (717) 245-1355
edwardsb@dickinson.edu
https://www.dickinson.edu/earthsciences
f: https://www.facebook.com/pages/category/Product-Service/Dickinson-College-Department-of-Earth-Sciences-108884862477555/
t: @dsonearthsci
Programs: Earth Sciences
Administrative Assistant: Debra Peters
Enrollment (2018): B: 24 (8)
Chair:
Benjamin R. Edwards, (D), British Columbia, 1997, GivCg
Professor:
Marcus M. Key, Jr., (D), Yale, 1988, PiGsa
Associate Professor:
Peter B. Sak, (D), Penn State, 2002, Gcm
Assistant Professor:
Jorden L. Hayes, (D), Wyoming, 2016, Ygs
Alyson M. Thibodeau, (D), Arizona, 2012, CcGaCs
Emeritus:
Jeffery W. Niemitz, (D), S California, 1978, ClOuGn
Noel Potter, Jr., (D), Minnesota, 1969, Gmc
Technician:
Robert Dean, (M), Texas (El Paso), 2004, Gi

Environmental Studies & Environmental Science (2015)
P.O. Box 1773
Carlisle, PA 17013-2896
p. (717) 245-1355
arnoldt@dickinson.edu
Administrative Assistant: Patricia Braught
Associate Professor:
Candie Wilderman, (D), Johns Hopkins, 1984, Hs
Visiting Professor:
Kirsten Hural, (D), Cornell, 1997, Gg

Drexel University
Dept of Biodiversity, Earth & Environmental Science (B,M,D) ⊠ (2018)
2024 MacAlister Hall
3250 Chestnut Street
Philadelphia, PA 19104
p. (215) 571-4651
bees@drexel.edu
http://drexel.edu/bees/
f: https://www.facebook.com/groups/DrexelBEES/
Programs: Geoscience; Environmental Science; Environmental Studies
Enrollment (2016): B: 22 (1) D: 4 (3)
Pilsbry Chair of Malacology:
Gary Rosenberg, (D), Harvard, 1989, Pi
Professor:
David Velinsky, (D), Old Dominion, 1987, OcCms
Associate Professor:
Ted Daeschler, (D), Pennsylvania, 1998, PvGgPg

Assistant Professor:
Marie J. Kurz, (D), Florida, 2013, CqbHw
Amanda Lough, (D), Washington Univ (St. Louis), 2014, Ysg
Jocelyn A. Sessa, (D), Penn State, 2009, PeCsPq
Loyc Vanderkluysen, (D), Hawaii, 2008, GviGz
Elizabeth Watson, (D), California (Berkeley), Zc
Adjunct Professor:
Mitch Cron, (M), Pennsylvania, 2013, Gge
Stephen Huxta, (M), Pennsylvania, 2017, Gg
Carl Mastropaolo, (M), Drexel, Hw

Edinboro University of Pennsylvania
Dept of Geosciences (B) ⊠ (2018)
126 Cooper Hall
230 Scotland Road
Edinboro, PA 16444-0001
p. (814) 732-2529
geosciences@edinboro.edu
http://www.edinboro.edu/academics/schools-and-departments/cshp/departments/geosciences/
f: https://www.facebook.com/pages/Edinboro-University-Geosciences-Department/426407214146067
Programs: Geology (B); Environmental Studies (B); GIS (B)
Enrollment (2017): B: 108 (24)
Chair:
Brian S. Zimmerman, (D), Washington State, 1991, Gzx
Professor:
Baher A. Ghosheh, (D), SUNY (Buffalo), 1988, Zn
David Hurd, (D), Cleveland State, 1997, Zg
Kerry A. Moyer, (D), Penn State, 1993, As
Laurie A. Parendes, (D), Oregon State, 1997, Zn
Joseph F. Reese, (D), Texas, 1995, GctZe
Eric Straffin, (D), Nebraska, 2000, GmsSo
Dale Tshudy, (D), Kent State, 1993, Pi
Associate Professor:
Karen Eisenhart, (D), Colorado, 2004, Zy
Wook Lee, (D), Ohio State, Zi
Assistant Professor:
Richard Deal, (D), South Carolina, 2000, Zi
Tamara Misner, (D), Pittsburg, 2013, GmHs

Elizabethtown College
Dept of Engineering and Physics (B) ⊠ (2019)
One Alpha Drive
Esbenshade Room 160
Elizabethtown, PA 17022-2298
p. (717) 361-1392
mcfaddenj@etown.edu
http://www.etown.edu/PhysicsEngineering.aspx
Administrative Assistant: Jennifer McFadden
Associate Professor:
Michael A. Scanlin, (D), Penn State, Ye
Emeritus:
David Ferruzza, (M), MIT, 1967, AsZn

Franklin and Marshall College
Dept of Earth and Environment (B) ⊠ (2018)
PO Box 3003
Lancaster, PA 17604-3003
p. (717) 358-4133
mbetrone@fandm.edu
www.fandm.edu/earth-environment
f: https://www.facebook.com/FM-Department-of-Earth-Environment
t: @FandMENE
Enrollment (2017): B: 100 (48)
Chair:
Andrew P. deWet, (D), Cambridge, 1989, Ge
Professor:
Carol B. de Wet, (D), Cambridge, 1989, GsdGn
Dorothy J. Merritts, (D), Arizona, 1987, Gm
Stanley A. Mertzman, (D), Case Western, 1971, GizZm
Associate Professor:
Eric Hirsch, (D), Chicago, Zn

Zeshan Ismat, (D), Rochester, 2002, Gc
James E. Strick, (D), Princeton, 1997, RhGe
Robert C. Walter, (D), Case Western, 1981, Cc
Christopher J. Williams, (D), Pennsylvania, 2002, PeSb
Assistant Professor:
Eve Bratman, (D), American, 2009, Zn
Eilzabeth De Santo, (D), Univ Coll (London), Zn
Paul Harnik, (D), Chicago, 2009, Pg
Adjunct Professor:
Suzanna L. Richter, (D), Pennsylvania, 2006, Ge
Visiting Professor:
Timothy D. Bechtel, (D), Brown, 1989, GeYg
Emeritus:
Robert S. Sternberg, (D), Arizona, 1982, YmGa
Roger D. K. Thomas, (D), Harvard, 1970, PgRh
Laboratory Director:
Emily Wilson, (M), Idaho, 2013, CgGi

Gannon University
Earth Science Program (B) (2015)
109 University Square
PMB 3183
Erie, PA 16541
p. (814) 871-7453
olanrewa001@gannon.edu
http://www.gannon.edu/
Assistant Professor:
Johnson Olanrewaju, (D), Penn State, 2002, Cg

Harrisburg Area Community College
Science (A) (2017)
One HACC Drive
Harrisburg, PA 17110
p. 800-222-4222
rkdennis@hacc.edu
http://www.hacc.edu/index.cfm
f: https://www.facebook.com/HACC64
t: @hacc_info
Enrollment (2016): A: 2 (0)
Professor:
James E. Baxter, P.G., (M), Penn State, 1983, GgHwGm

Indiana University of Pennsylvania
Dept of Geography and Regional Planning (2015)
Indiana, PA 15705
p. (724) 357-2250
donaldb@iup.edu
https://www.iup.edu/georegionalplan/

Dept of Geoscience (B) (2018)
111 Walsh Hall
Indiana, PA 15705
p. (724) 357-2379
geoscience-info@iup.edu
http://www.iup.edu/geoscience
Programs: Geology
Earth and Space Science Education
Enrollment (2018): B: 47 (23)
Chair:
Steven A. Hovan, (D), Michigan, 1993, Ou
Professor:
Karen Rose Cercone, (D), Michigan, 1984, Zeg
Jon C. Lewis, (D), Connecticut, 1998, Gc
Associate Professor:
Kenneth S. Coles, (D), Columbia, 1988, XgZeYg
Nicholas Deardorff, (D), Oregon, Gv
Katie Farnsworth, (D), Virginia Inst of Marine Sci, 2003, Ong
Assistant Professor:
Yvonne K. Branan, (D), Michigan Tech, 2007, Gv
Gregory Mount, (D), Florida Atlantic, 2014, Hy
Jonathan P. Warnock, (D), N Illinois, 2013, PemPv
Emeritus:
Joseph C. Clark, (D), Stanford, 1966, Gr
Frank W. Hall, (D), Montana, 1969, Gc
Darlene S. Richardson, (D), Columbia, 1974, Gd
Connie J. Sutton, (M), Indiana (Penn), 1968, Ze

John F. Taylor, (D), Missouri, 1984, PsiGs

Juniata College
Dept of Environmental Science & Studies (2015)
1700 Moore Street
Huntingdon, PA 16652
johnson@juinata.edu
http://www.juniata.edu/departments/environmental/

Dept of Geology (B) ● ☒ (2019)
1700 Moore St
Huntingdon, PA 16652
p. (814) 641-3601
johanesen@juniata.edu
http://www.juniata.edu/academics/departments/geology/
Programs: Geology; Environmental Geology; Earth and Space Science Education
Enrollment (2017): B: 25 (10)
Dept. Chair:
Ryan Mathur, (D), Arizona, 2000, CeYgHw
Associate Professor:
Matthew G. Powell, (D), Johns Hopkins, 2005, PiqPe
Assistant Professor:
Katharine Johanesen, (D), S California, 2011, GzxGc
Research Associate:
Adam J. Ianno, (D), S California, 2015, GgiCu
Emeritus:
Laurence J. Mutti, (D), Harvard, 1978, GxzSc
J. Peter Trexler, (D), Michigan, 1964, Ps
Robert H. Washburn, (D), Columbia, 1966, Gs

Kutztown University of Pennsylvania
Dept of Physical Science (B) ☒ (2018)
Boehm Science Building Room 135
Kutztown, PA 19530
p. (610) 683-4447
simpson@kutztown.edu
http://www.kutztown.edu/acad/geology
Administrative Assistant: Caecilia Holt
Enrollment (2018): B: 35 (11)
Chair:
Edward L. Simpson, (D), Virginia Tech, 1987, Gs
Professor:
Kurt Friehauf, (D), Stanford, 1998, EmCgGx
Laura Sherrod, (D), W Michigan, 2007, YgHwGa
Sarah E. Tindall, (D), Arizona, 2000, Gc
Assistant Professor:
Erin Kraal, (D), California (Santa Cruz), XgGm
Adrienne Oakley, (D), Hawaii, 2009, YrOun
Jacob Sewall, (D), California (Santa Cruz), 2004, GeAs

Lafayette College
Dept of Geology & Environmental Geosciences (B) ☒ (2019)
116 Van Wickle Hall
4 South College Drive
Easton, PA 18042
p. (610) 330-5193
geology@lafayette.edu
http://geology.lafayette.edu
Department Secretary: Rohana Meyerson
Enrollment (2018): B: 34 (7)
Professor:
Dru Germanoski, (D), Colorado State, 1989, Gm
Kira Lawrence, (D), Brown, 2006, Gg
Associate Professor:
Lawrence L. Malinconico, (D), Dartmouth, 1982, YgGcv
David Sunderlin, (D), Chicago, 2004, Gg
Assistant Professor:
Tamara Carley, (D), Vanderbilt, 2014, GviCg
Research Associate:
Mary Ann Malinconico, (D), Columbia, 2002, Eo
Emeritus:
Richard W. Faas, (D), Iowa State, 1964, Gs
Guy L. Hovis, (D), Harvard, 1971, GzxCg

Pennsylvania

Laboratory Director:
  John R. Wilson, (M), Virginia Tech, 2001, Zi

Lehigh University
Dept of Earth & Environmental Sciences (B,M,D) O☒ (2019)
1 W. Packer Ave.
Bethlehem, PA 18015-3001
  p. (610) 758-3660
  nr00@lehigh.edu
  http://www.ees.lehigh.edu
  f: https://www.facebook.com/LehighEES/
  Enrollment (2018): B: 61 (15) M: 8 (3) D: 12 (1)
Department Chair:
  David J. Anastasio, (D), Johns Hopkins, 1988, Gca
Professor:
  Gray E. Bebout, (D), California (Los Angeles), 1989, Gp
  Robert K. Booth, (D), Wyoming, 2003, ZnnPc
  Edward B. Evenson, (D), Michigan, 1972, Gl
  Kenneth P. Kodama, (D), Stanford, 1978, Ym
  Anne S. Meltzer, (D), Rice, 1988, Ys
  Frank J. Pazzaglia, (D), Penn State, 1993, Gm
  Dork Sahagian, (D), Chicago, 1987, PeGvr
  Zicheng Yu, (D), Toronto, 1997, Pe
  Peter K. Zeitler, (D), Dartmouth, 1983, Cc
Associate Professor:
  Benjamin S. Felzer, (D), Brown, 1995, ZcoZn
  Donald P. Morris, (D), Colorado, 1990, Zn
  Stephen C. Peters, (D), Michigan, 2001, Cl
  Joan Ramage, (D), Cornell, 2001, ZrGlm
Assistant Professor:
  Jill I. McDermott, (D), MIT/WHOI, 2015, Cm
Research Associate:
  Bruce D. Idleman, (D), SUNY (Albany), 1990, Cc
  Joshua Stachnik, (D), Wyoming, 2010, Ys
Emeritus:
  Bobb Carson, (D), Washington, 1971, GusGt
  Paul B. Myers, (D), Lehigh, 1960, Hw
Other:
  Bruce R. Hargreaves, (D), California (Berkeley), 1977, Ob

Lock Haven University
Dept of Geology & Physics (B) ☒ (2018)
301 West Church Street
Lock Haven, PA 17745-2390
  p. (570) 484-2048
  mkhalequ@lockhaven.edu
  Programs: Geology
  Administrative Assistant: Denise Rupert
  Enrollment (2018): B: 30 (6)
Professor:
  Loretta D. Dickson, (D), Connecticut, 2006, GizGe
  Md. Khalequzzaman, (D), 1998, HwOnZi
Associate Scientist:
  Thomas C. Wynn, (D), Virginia Tech, 2004, GsPiEo

Mansfield University
Dept of Geosciences (A,B) O☒ (2019)
Belknap Hall
Mansfield, PA 16933
  p. (570) 662-4613
  jdemchak@mansfield.edu
  http://geoggeol.mansfield.edu/
  Enrollment (2016): A: 10 (6) B: 124 (42)
Chair:
  Jennifer Demchak, (D), West Virginia, 2005, Hs
Associate Professor:
  Christopher F. Kopf, (D), Massachusetts, 1999, Gcp
  Lee Stocks, (D), Kent State, 2010, ZyGgm
Assistant Professor:
  Linda Kennedy, (D), North Carolina (Greensboro), 2012, Zy

Mercyhurst University
Dept of Geology (B,M,D) (2015)
501 East 38th Street
Erie, PA 16546
  p. (814) 824-2581
  rbuyce@mercyhurst.edu
  http://mai.mercyhurst.edu
  Enrollment (2012): B: 16 (3)
Director:
  James M. Adovasio, (D), Utah, 1970, GafGs
Professor:
  M. Raymond Buyce, (D), Rensselaer, 1975, GsaOn
Assistant Professor:
  Nicholas Lang, (D), Minnesota, 2006, GvXgGc
  Scott C. McKenzie, (B), Edinboro, 1976, PgXmZe
  Lyman Perscio, (D), New Mexico, 2012, GsSpHg
Adjunct Professor:
  Frank Vento, (D), Pittsburg, 1985, GaSoGm

Millersville University
Dept of Earth Sciences (B,M) (2015)
PO Box 1002
Millersville, PA 17551
  p. (717) 872-3289
  esci@millersville.edu
  http://www.millersville.edu/esci
  Enrollment (2010): B: 23 (4)
Chair:
  Richard D. Clark, (D), Wyoming, 1987, As
Professor:
  Alex J. DeCaria, (D), Maryland, 2000, As
  L. Lynn Marquez, (D), Northwestern, 1998, Cg
  Sepideh Yalda, (D), St. Louis, 1997, As
Associate Professor:
  Sam Earman, (D), New Mexico Tech, 2004, HwCsl
  Ajoy Kumar, (D), Old Dominion, 1996, OpZr
  Jason R. Price, (D), Michigan State, 2003, Gs
  Todd D. Sikora, (D), Penn State, 1996, As
Assistant Professor:
  Robert Vaillancourt, (D), Rhode Island, 1996, Obc
Instructor:
  Joseph Calhoun, (B), Penn State, As
  Mary Ann Schlegel, (M), MIT/WHOI, 1998, Og
Professor:
  Robert S. Ross, (D), Florida State, 1977, As
Emeritus:
  William M. Jordan, (D), Wisconsin, 1965, Gs
  Bernard L. Oostdam, (D), Delaware, 1971, Ou
  Charles K. Scharnberger, (D), Washington (St. Louis), 1971, Gc

Montgomery County Community College
Dept of Science, Technology, Engineering, and Math (A) (2015)
Blue Bell, PA 19422
  p. (215) 641-6446
  rkuhlman@mc3.edu
Professor:
  Robert Kuhlman, (M), Bryn Mawr, 1975, Gg
Instructor:
  George Buchanan, (M), Drexel, 1992, Ng
Adjunct Professor:
  Laurie Martin-Vermilyea, (D), South Carolina, 1992, Ze
  Frank Roberts, (D), Bryn Mawr, 1969, Gp
  Kelly C. Spangler, (M), Drexel, 2003
  Anthony Stevens, (M), Florida, 1981, Ge

Moravian College
Dept of Physics & Earth Science (B) ☒ (2019)
1200 Main Street
Bethlehem, PA 18018-6650
  p. (610) 861-1437
  krieblek@moravian.edu
  http://www.physics.moravian.edu
  Programs: Geology; General Science; Physics
  Administrative Assistant: Ann Sywensky
  Enrollment (2017): B: 1 (1)
Chair:
  Kelly Krieble, (D), Lehigh, 1993, ZnnZn

Pennsylvania State University, Erie
Geoscience Dept (2015)
Erie, PA 16510
>p. (814) 898-6277
>amf11@psu.edu
>http://www.personal.psu.edu/faculty/a/m/amf11/

Chair:
>Anthony M. Foyle

Assistant Professor:
>Eva Tucker, (M), Cincinnati, 1962, Gg

Pennsylvania State University, Monaca
Dept of Geosciences (B) (2015)
100 University Drive
Monaca, PA 15061
>p. (412) 773-3867
>jac7@psu.edu
>http://www.br.psu.edu/default.htm

Assistant Professor:
>John A. Ciciarelli, (D), Penn State, 1971, Gg

Pennsylvania State University, University Park
Dept of Geosciences (B,M,D) ●☒ (2019)
503 Deike Building
University Park, PA 16802-2714
>p. (814) 865-6711
>demian@psu.edu
>http://www.geosc.psu.edu/
>Programs: Geosciences BS, Earth Science & Policy BS, Earth Sciences BS, Geosciences MS, PhD
>Certificates: Sustainability Certificate, MEd certificate
>Administrative Assistant: Amy L. Homan
>*Enrollment (2018): B: 133 (0) M: 20 (0) D: 59 (0)*

Head:
>Demian M. Saffer, (D), California (Santa Cruz), 1999, Hw

Evan Pugh Professor:
>Richard B. Alley, (D), Wisconsin, 1987, Gl
>James F. Kasting, (D), Michigan, 1979, As

Distinguished Professor:
>Katherine H. Freeman, (D), Indiana, 1991, Cos

Director, Earth & Mineral Science Museum:
>Russell W. Graham, (D), Texas, 1976, Pv

Director, Earth & Env Systems Inst:
>Susan L. Brantley, (D), Princeton, 1987, Cg

Director, Astrobiology Research Center:
>Christopher H. House, (D), California (Los Angeles), 1999, Pg

Associate Head, Graduate Programs:
>Mark E. Patzkowsky, (D), Chicago, 1992, Pg

Professor:
>Charles J. Ammon, (D), Penn State, 1991, Ys
>Sridhar Anandakrishnan, (D), Wisconsin, 1990, Ys
>David M. Bice, (D), California (Berkeley), 1989, Gg
>Timothy J. Bralower, (D), California (San Diego), 1986, Pe
>Donald M. Fisher, (D), Brown, 1988, GctZm
>Kevin P. Furlong, (D), Utah, 1981, Gt
>Tanya Furman, (D), MIT, 1989, GiCu
>Peter J. Heaney, (D), Johns Hopkins, 1989, GzClGe
>Klaus Keller, (D), Princeton, 2000, Og
>Lee R. Kump, (D), S Florida, 1986, ClPeCm
>Michael E. Mann, (D), Yale, 1998, As
>Chris Marone, (D), Columbia, 1988, Yx
>Andrew A. Nyblade, (D), Michigan, 1992, Yg
>Peter D. Wilf, (D), Penn, 1998, Pg

Senior Scientist:
>Todd Sowers, (D), Rhode Island, 1991, Cs

Associate Head for Undergraduate Programs:
>Maureen D. Feineman, (D), California (Berkeley), 2004, Cp

Associate Professor:
>Matthew S. Fantle, (D), California (Berkeley), 2005, Cs
>Elizabeth Hajek, (D), Wyoming, 2009, Gs
>Peter C. LaFemina, (D), Miami, 2005, Yd
>Jennifer L. Macalady, (D), California (Davis), 2000, PgClSb
>Eliza Richardson, (D), MIT, 2002, YsZe

Assistant Professor:
>Roman DiBiase, (D), Arizona State, 2011, Gmt
>Bradford Foley, (D), Yale, 2014, YgXyGt
>Sarah Ivory, (D), Arizona, 2013, Pl
>Tess A. Russo, (D), California (Santa Cruz), 2012, Hw
>Christelle Wauthier, (D), Leige, 2011, Zn

Emeritus:
>Shelton S. Alexander, (D), Caltech, 1963, Ys
>Michael A. Arthur, (D), Princeton, 1979, OuClGs
>Hubert L. Barnes, (D), Columbia, 1958, CgEmCe
>Roger J. Cuffey, (D), Indiana, 1966, PiePv
>David H. Eggler, (D), Colorado, 1967, CpGiv
>Terry Engelder, (D), Texas A&M, 1973, Nr
>David (Duff) P. Gold, (D), McGill, 1963, GcgGx
>Earl K. Graham, Jr., (D), Penn State, 1969, Yx
>Roy J. Greenfield, (D), MIT, 1965, YgsYu
>Albert L. Guber, (D), Illinois, 1962, Pe
>Benjamin F. Howell, Jr., (D), Caltech, 1949, Ys
>Derrill M. Kerrick, (D), California (Berkeley), 1968, Gp
>Hiroshi Ohmoto, (D), Princeton, 1969, Cs
>Richard R. Parizek, (D), Illinois, 1961, Hw
>Arthur W. Rose, (D), Caltech, 1958, Cge
>Rudy L. Slingerland, (D), Penn State, 1977, Gs
>Barry Voight, (D), Columbia, 1965, GveGg
>William B. White, (D), Penn State, 1962, CgGzy

Dept of Meteorology and Atmospheric Science (B,M,D) ○☒ (2017)
503 Walker Building
University Park, PA 16802-5013
>p. (814) 865-0478
>meteodept@meteo.psu.edu
>http://www.met.psu.edu
>Programs: Meteorology and Atmospheric Science
>Certificates: Certificate of Achievement in Weather Forecaasting

Professor and Head:
>David J. Stensrud, (D), Penn State, 1992, Am

Professor:
>William H. Brune, (D), Johns Hopkins, 1978, Am
>Eugene E. Clothiaux, (D), Brown, Am
>Kenneth J. Davis, (D), Colorado, 1992, Am
>Jenni L. Evans, (D), Monash, 1990, Am
>Jerry Y. Harrington, (D), Colorado State, 1997, Am
>Gregory S. Jenkins, (D), Am
>James F. Kasting, (D), Michigan, 1979, Am
>Raymond G. Najjar, (D), Princeton, 1990, OcpAt
>Johannes Verlinde, (D), Colorado State, 1992, Am
>George S. Young, (D), Colorado State, Am

Assistant Professor:
>Anthony Didlake, (D), Washington, As
>Melissa Gervais, (D), McGill, As
>Steven J. Greybush, (D), Maryland, 2011, Am
>Matthew R. Kumjian, (D), Oklahoma, 2012, Am
>Sukyoung Lee, (D), Princeton, 1991, Am
>Ying Pan, (D), Penn State, As

Research Assistant:
>William F. Ryan, (M), Maryland, 1990, Am

Director, Meteorology Computing:
>Charles Pavloski, (D), Penn State, 2004, Am

Research Associate:
>Aijun Deng, (D), Penn State, 1999, Am
>Arthur Person, (M), Penn State, 1983, Am
>William S. Syrett, (M), Penn State, 1987, Am

Instructor:
>Frederick J. Gadomski, (M), Penn State, 1983, Am
>Paul Knight, (M), Penn State, 1977, Am

Distinguished Professor:
>J. Michael Fritsch, (D), Colorado State, 1978, Am

Associate Research Professor:
>David R. Stauffer, (D), Penn State, 1990, Am

Associate Professor:
>Hampton N. Shirer, (D), Penn State, 1978, Am

Emeritus:
>Peter R. Bannon, (D), Colorado, 1979, Am
>Craig F. Bohren, (D), Arizona, 1975, Am
>John J. Cahir, (D), Penn State, 1971, Am
>Toby N. Carlson, (D), Imperial Coll (UK), 1965, Am
>John H. E. Clark, (D), Florida State, 1969, Am
>John A. Dutton, (D), Wisconsin, Am
>William M. Frank, (D), Colorado State, 1976, Am

Pennsylvania

Alistair B. Fraser, (D), Imperial Coll (UK), 1968, Am
Charles L. Hosler, (D), Penn State, 1951, Am
Dennis Lamb, (D), Washington, 1970, Am
Nelson L. Seaman, (D), Penn State, 1977, Am
Dennis W. Thomson, (D), Wisconsin, 1968, Am
John C. Wyngaard, (D), Penn State, 1967, Am

Dept of Plant Science (A,B,M,D) (2015)
119 Tyson Building
University Park, PA 16802
   p. (814) 865-6541
   rpm12@psu.edu
   http://plantscience.psu.edu/
Professor:
   Jean-Marc Bollag, (D), Basel, 1959, Sb
   Edward J. Ciolkosz, (D), Wisconsin, 1967, Sd
   Daniel D. Fritton, (D), Iowa State, 1968, Sp
   Sridhar Komarneni, (D), Wisconsin, 1973, Sc
   Gary W. Petersen, (D), Wisconsin, 1965, Sd
Associate Professor:
   Peter J. Landschoot, (D), Rhode Island, 1988, So
   Gregory W. Roth, (D), Penn State, 1987, So
Assistant Professor:
   Rick L. Day, (D), Penn State, 1991, Zu
Research Associate:
   Barry M. Evans, (M), Penn State, 1977, Sd
Adjunct Professor:
   Andrew S. Rogowski, (D), Iowa State, 1964, Sp
   Lawrence A. Schardt, (D), Penn State, 2000, SoHw

Earth and Mineral Sciences Museum & Art Gallery (2015)
116 Deike Building
University Park, PA 16802
   p. (814) 865-6336
   museum@ems.psu.edu
   http://www.ems.psu.edu/outreach/museum
Director:
   Russell W. Graham, (D), Texas, 1976, PveZe

John and Willie Leone Family Dept of Energy and Mineral Engineering (B,M,D) ☒ (2018)
110 Hosler Building
University Park, PA 16802
   p. (814) 865-3437
   eme@ems.psu.edu
   http://www.eme.psu.edu/mnge
Program Chair, and Deike Endowed Chair in Mining Engineering:
   Jeffery L. Kohler, (D), Penn State, 1982, Nm
Assistant Professor:
   Shimin Liu, (D), S Illinois, 2013, Nm

Point Park University

Dept of Environmental Studies (A,B) (2015)
201 Wood Street
Pittsburgh, PA 15222-1994
   p. (412) 392-3900
   jkudlac@pointpark.edu
   http://www.pointpark.edu/Academics/Schools/SchoolofArtsandSciences/Departments/NaturalSciencesandEngineeringTechnology
   Administrative Assistant: Roberta T. Gallick
Head:
   Mark O. Farrell, (D), Carnegie Mellon, 1978, Zn
Professor:
   John J. Kudlac, (D), Pittsburgh, Ng

Shippensburg University

Geography-Earth Science Department (B,M) ●⌂ (2018)
1871 Old Main Drive
Shearer Hall 104
Shippensburg, PA 17257
   p. (717) 477-1685
   tlmyers@ship.edu
   https://www.ship.edu/geo-ess/
   f: https://www.facebook.com/ShipGeoESS/
   Programs: B.Sc. in Geoenvironmental Studies, B.Sc. in Environmental Sustainability, B.Sc. in Geography - GIS, B.Sc. in Geography - Land Use, B.S.Ed. Science Education-Earth and Space Science, B.S.Ed. Science Education-Comprehensive Social Studies, M.Sc. Geoenvironmental Studies
   Certificates: GIS
   Administrative Assistant: Tammy Myers
Chair:
   William L. Blewett, (D), Michigan State, 1991, ZyGmZe
Professor:
   Scott A. Drzyzga, (D), Michigan State, 2007, ZiuGl
   Alison E. Feeney, (D), Michigan State, 1998, ZinZn
   Thomas P. Feeney, (D), Georgia, 1997, ZyGeHw
   Kurtis G. Fuellhart, (D), Penn State, 1998, ZnnZm
   Timothy W. Hawkins, (D), Arizona State, 2004, AsmHg
   Claire A. Jantz, (D), Maryland, 2005, ZuiZc
   Paul G. Marr, (D), Denver, 1996, ZfyZi
   George M. Pomeroy, (D), Akron, 1999, ZuyZn
   Janet S. Smith, (D), Georgia, 1999, Zi
   Christopher J. Woltemade, (D), Wisconsin, 1993, HgRwGm
Associate Professor:
   Michael T. Applegarth, (D), Arizona State, 2001, GmZrSo
   Sean R. Cornell, (D), Cincinnati, 2008, GgsOn
   Kay R. Williams, (D), Georgia, 1995, As
   Joseph T. Zume, (D), Oklahoma, 2007, HgYug
Assistant Professor:
   Russell Hedberg, (D), Penn State, 2018, ZiSoZc

Slippery Rock University

Dept of Geography, Geology, and the Environment (B) (2015)
Slippery Rock, PA 16057
   p. (724) 738-2048
   jack.livingston@sru.edu
   http://academics.sru.edu/gge/
   Administrative Assistant: Bonita L. Vinton
   Enrollment (2006): B: 50 (9)
Chair:
   Jack Livingston
Professor:
   Tamra A. Schiappa, (D), Idaho, 1999, PiGrZe
   Michael J. Zieg, (D), Johns Hopkins, 2001, Gi
Associate Professor:
   Patrick A. Burkhart, (D), Lehigh, 1994, Hg
Assistant Professor:
   Patricia A. Campbell, (D), Pittsburgh, 1994, Gc
   Xianfeng Chen, (D), West Virginia, 2005, Zr
   Julie A. Snow, (D), Rhode Island, 2002, As
   Michael G. Stapleton, (D), Delaware, 1995, So

State Museum of Pennsylvania

Section of Paleontology & Geology (2015)
300 North Street
Harrisburg, PA 17120-0024
   p. (717) 783-9897
   c-sjasinsk@pa.gov
   http://www.statemusempa.org/geologyc.html
   f: https://www.facebook.com/StateMuseumofPA
Acting Curator:
   Steven E. Jasinski, (D), Pennsylvania, pend, PvgPg

Susquehanna University

Dept of Earth & Environmental Sciences (B) (2015)
514 University Ave
Selinsgrove, PA 17870
   p. (570) 372-4216
   straubk@susqu.edu
   http://www.susqu.edu/ees
   Enrollment (2009): B: 29 (7)
Chair:
   Jennifer M. Elick, (D), Tennessee, 1999, Pe
   Katherine H. Straub, (D), Colorado State, 2002, Am
Associate Professor:
   Daniel E. Ressler, (D), Iowa State, 1998, Sp
   Derek J. Straub, (D), As
Assistant Professor:
   Ahmed Lachhab, (D), Iowa, 2006, Hw

Temple University
Earth & Environmental Science (B,M,D) ⊘ (2018)
1901 N. 13th Street
Beury Hall, Rm. 326
Philadelphia, PA 19122-6081
   p. (215) 204-8227
   scox@temple.edu
   http://www.temple.edu/geology
   f: https://www.facebook.com/tu.geology
   Programs: BS, Masters, PhD programs in Geology.
   Administrative Assistant: Shelah Cox

Chair:
   David E. Grandstaff, (D), Princeton, 1974, Cl
Professor:
   Jonathan Nyquist, (D), Wisconsin, 1986, Yg
   Laura Toran, (D), Wisconsin, 1986, Hw
Associate Professor:
   Ilya Buynevich, (D), Boston, 2001, GsPeOn
   Alexandra K. Davatzes, (D), Stanford, 2007, GsXg
   Nicholas Davatzes, (D), Stanford, 2003, GcNrZn
   Dennis O. Terry, (D), Nebraska, 1998, SaPcGr
Assistant Professor:
   Steven Chemtob, (D), 2012
   Bojeong Kim, (D)
   Atsuhiro Muto, (D), 2010, Gl
   Sujith Ravi, (D)
Asst. Lab Manager/Bldg. Coordinator:
   Donald Deigh-Kai, (M), Temple, 2008

Thiel College
Dept of Environmental Science (B) (2015)
Greenville, PA 16125
   p. (412) 589-2821
   areinsel@thiel.edu
Professor:
   James H. Barton, (D), N Colorado, 1977, Zy

University of Pennsylvania
Dept of Earth & Environmental Science (B,M,D) (2015)
240 S. 33rd Street
Philadelphia, PA 19104-6316
   p. (215) 898-5724
   earth@sas.upenn.edu
   http://www.sas.upenn.edu/earth/

Chair:
   Reto Giere, (D), ETH (Switzerland), 1990, GzCgGe
Professor:
   Peter Dodson, (D), Yale, 1974, Pv
   Hermann W. Pfefferkorn, (D), Muenster, 1968, PbePs
Associate Professor:
   David Goldsby, (D)
   Douglas Jerolmack, (D), MIT, 2006, GmYxHq
   Stephen P. Phipps, (D), Princeton, 1984, Gc
   Alain F. Plante, (D), Alberta, 2001, SboCo
Assistant Professor:
   Irina Marinov
   Lauren Sallan, (D), Chicago, 2012, PqvPe
   Jane Willenbring, (D), Dalhousie, 2006, GmCcGl
Lecturer:
   Edward L. Doheny, (D), Indiana, 1967, Ng
   Stanley L. Laskowski, (M), Drexel, 1973, Ge
   Willig B. Sarah, (D), Pennsylvania, 1988, Ge
Emeritus:
   Robert F. Giegengack, Jr., (D), Yale, 1968, Gg
   Arthur H. Johnson, (D), Cornell, 1975, So
Laboratory Manager:
   David R. Vann, (D), Pennsylvania, 1993, GeSfXm
Teaching Faculty:
   Jane Dmochowski, (D), Caltech, 2004, Yg
   Gomaa I. Omar, (D), Pennsylvania, 1985, Cc
Graduate Group Coordinator:
   Joan Buccilli
Director, Professional Masters Programs:
   Yvette Bordeaux, (D), Pennsylvania, 2000, Pg
Department Administrator:
   Arlene Mand, (B), 1971, Zn

Associate Director:
   Maria Andrews, (M)

University of Pittsburgh
Dept of Geology & Environmental Science (B,M,D) ●☒ (2019)
200 SRCC Building
4107 O'Hara Street
Pittsburgh, PA 15260-3332
   p. (412) 624-8780
   gpsgrad@pitt.edu
   http://www.geology.pitt.edu
   Enrollment (2018): B: 218 (79) M: 9 (2) D: 24 (7)

Chair:
   Mark B. Abbott, (D), Minnesota, 1995, Gs
Professor:
   William P. Harbert, (D), Stanford, 1987, Ym
   Michael S. Ramsey, (D), Arizona State, 1996, Zr
   Josef Werne, (D), Northwestern, 2000, CosGn
Associate Professor:
   Rosemary C. Capo, (D), California (Los Angeles), 1990, Cl
   Emily M. Elliott, (D), Johns Hopkins, 2003, Cbs
   Nadine McQuarrie, (D), Arizona, 2001, Gct
   Brian W. Stewart, (D), California (Los Angeles), 1990, CclCq
Assistant Professor:
   Daniel J. Bain, (D), Johns Hopkins, 2004, HgGmZu
   Eitan Shelef
   Brian Thomas
Environmental Reporter, Pittsburgh Post-Gazette:
   S. Don Hopey
Instructor:
   Emily Collins
Lecturer:
   R. Ward Allebach
   Danielle Andrews-Brown, (D)
   Charles E. Jones, (D), Oxford, 1992, GgsPg
Adjunct Professor:
   Matthew C. Lamanna, (D), Pennsylvania, 2004, Pv
   Steven C. Latta
   John S. Pallister, (D), California (Santa Barbara), 1980, GviGg
   Serge Shapiro
   Matthew Watson
Emeritus:
   Thomas Anderson
   Bruce W. Hapke, (D), Cornell, 1962, Xy
   Edward G. Lidiak, (D), Rice, 1963, Gi

University of Pittsburgh, Bradford
Dept of Petroleum Technology (A) (2015)
300 Campus Drive
Bradford, PA 16701-2898
   p. (814) 362-7569
   aap@pitt.edu
   http://www.upb.pitt.edu/academics/petroleumtechnology.aspx
   Administrative Assistant: Janet Shade
   Enrollment (2010): A: 28 (8)

Program Director:
   Assad I. Panah, (D), Oklahoma, 1966, GorGc

West Chester University
Dept of Earth and Space Sciences (B,M) ☒ (2019)
750 South Church Street
Merion Science Center
West Chester, PA 19383
   p. (610) 436-2727
   jhilliker@wcupa.edu
   https://www.wcupa.edu/sciences-mathematics/earthSpaceSciences/
   f: https://www.facebook.com/WCU-Earth-Space-Sciences-143447267219/
   Programs: Geoscience (B,M); Earth Systems (B); Geology (B); Earth and Space Science Education (B)
   Certificates: GIS (B,M)
   Enrollment (2018): B: 120 (28) M: 28 (9)

Chair:
   Joby Hilliker, (D), Penn State, 2002, Am

Professor:
　　Richard M. Busch, (D), Pittsburgh, 1984, GrPgZe
　　Marc R. Gagne, (D), Georgia, 1994, Xa
　　Martin F. Helmke, (D), Iowa State, 2003, HwSdZg
　　Timothy M. Lutz, (D), Pennsylvania, 1979, Gq
　　LeeAnn Srogi, (D), Pennsylvania, 1988, Gp
Associate Professor:
　　Howell Bosbyshell, (D), Bryn Mawr, 2001, GcpGt
　　Cynthia V. Hall, (D), Georgia Tech, 2008, Cg
　　Daria L. Nikitina, (D), Delaware, 2000, Gm
　　Karen Schwarz, (D), Arizona State, XaZe
Assistant Professor:
　　Yong Hoon Kim, (D), South Carolina, 2005, OgZr
　　Christopher Roemmele, (D), Purdue, 2017, ZeGgZg
Adjunct Professor:
　　Vicky Helmke, (M), Iowa State, Gg
　　James "Sandy" Maxwell, (B), Ze
　　Jamie Vann, (M), CgGg

Wilkes University
Dept of Environmental Engineering & Earth Sciences (B)
O☒ (2018)
84 West South Street
Wilkes-Barre, PA 18766
　　p. (570) 408-4610
　　sid.halsor@wilkes.edu
　　http://wilkes.edu/academics/colleges/science-and-engineering/environmental-engineering-earth-sciences/
　　Programs: Geology; Earth & Environmental Sciences; Environmental Engineering
　　Certificates: Sustainability
　　Enrollment (2018): B: 7 (0)
Chair:
　　Marleen Troy, (D), Drexel, 1989, Ht
Professor:
　　Dale A. Bruns, (D), Idaho State, 1981, ZirSf
　　Sid P. Halsor, (D), Michigan Tech, 1989, Giv
　　Kenneth M. Klemow, (D), SUNY (Syracuse), 1982, Pg
　　Prahlad N. Murthy, (D), Texas A&M, 1993, AsZnn
　　Michael A. Steele, (D), Wake Forest, 1988, Pg
　　Brian E. Whitman, (D), Michigan Tech, 1998, HqwSb
Associate Professor:
　　Holly Frederick, (D), Penn State, 1999, Sdb
Assistant Professor:
　　Matthew S. Finkenbinder, (D), Pittsburg, 2015, GsmGl
　　Bobak Karimi, (D), Pittsburgh, 2014, GctYg
Lecturer:
　　Mark A. Kaster, (M), Saint Louis, 1993, Am
　　Julie McMonagle, (M), Lehigh, 1991, Gge
Emeritus:
　　James M. Case, (D), Dalhousie, 1979, Ob
　　Brian T. Redmond, (D), Rensselaer, 1982, GsHwZg

York College of Pennsylvania
Dept of Physical Science (2015)
York, PA 17405
　　p. (717) 846-7788 x333
　　jforesma@ycp.edu
Assistant Professor:
　　William (Bill) Kreiger, (D), Penn State, 1976, GisZg
Adjunct Professor:
　　Ralph Eisenhart, (M), Penn State, 1994, Gg
　　Jeri L. Jones, (B), Catawba, 1977, Ga

Puerto Rico

University of Puerto Rico
Dept of Geology (B,M) O☒ (2018)
PO Box 9000
Mayaguez, PR 00681-9000
　　p. (787) 265-3845
　　lizzette.rodriguez1@upr.edu
　　http://geology.uprm.edu/
　　Programs: Geology
　　Enrollment (2018): B: 118 (15) M: 12 (8)

Professor:
　　Fernando Gilbes, (D), S Florida, 1996, ZrGu
　　James Joyce, (D), Northwestern, 1985, NgGcp
　　Wilson R. Ramirez, (D), Tulane, 2000, GudOu
　　Hernan Santos, (D), Colorado, 1999, PiGdr
Director:
　　Lizzette A. Rodriguez, (D), Michigan Tech, 2007, GviZr
Associate Professor:
　　Lysa Chizmadia, (D), New Mexico, 2004, GzXm
　　Kenneth Stephen Hughes, (D), North Carolina State, 2014, Gcm
　　Alberto Lopez, (D), Northwestern, 2006, YdGtYs
Assistant Professor:
　　Thomas Hudgins, (D), Michigan, 2014, GiCpGv
　　Elizabeth A. Vanacore, (D), Rice, 2008, Ysx
Director of Puerto Rico Seismic Network:
　　Victor Huerfano, (D), Puerto Rico, 2004, Ysg

Rhode Island

Brown University
Dept of Earth, Environmental and Planetary Sciences
(B,M,D) ☒ (2019)
Box 1846, 324 Brook Street
Providence, RI 02912
　　p. (401) 863-3339
　　DEEPS@brown.edu
　　http://www.brown.edu/academics/earth-environmental-planetary-sciences/
　　f: https://www.facebook.com/BrownGeologicalSciences
　　t: @BrownGeoSci
　　Enrollment (2018): B: 0 (12) M: 0 (10) D: 55 (8)
Chair:
　　Greg Hirth, (D), Brown, 1991, GcyYg
Professor:
　　Reid F. Cooper, (D), Cornell, 1983, Gy
　　Karen M. Fischer, (D), MIT, 1988, Ys
　　James W. Head, III, (D), Brown, 1969, XgGvt
　　Timothy D. Herbert, (D), Princeton, 1987, Pe
　　Yongsong Huang, (D), Bristol (UK), 1997, Co
　　Yan Liang, (D), Chicago, 1994, Cp
　　Amanda H. Lynch, (D), Melbourne, 1993, AsGe
　　John F. Mustard, (D), Brown, 1990, Zr
　　Alberto E. Saal, (D), MIT/WHOI, 2000, CgGi
Senior Scientist:
　　David Murray, (D), Oregon State, 1987, Ou
Associate Professor of Research:
　　Steven C. Clemens, (D), Brown, 1990, Ou
Associate Professor:
　　Colleen Dalton, (D), Harvard, 2007, Ys
　　Baylor Fox-Kemper, (D), MIT/WHOI, 2003, OpYnAt
　　Meredith Hastings, (D), Princeton, 2004, As
　　Christian Huber, (D), California (Berkeley), 2009, Gv
　　Ralph E. Milliken, (D), Brown, 2006
　　Stephen Parman, (D), MIT, 2001, CpgCt
　　James M. Russell, (D), Minnesota, 2004, Gn
Assistant Professor:
　　Alexander J. Evans, (D)
　　Brandon C. Johnson, (D), Purdue, 2013, Xg
　　Jung-Eun Lee, (D), California (Berkeley), 2005
Emeritus:
　　Donald W. Forsyth, (D), MIT/WHOI, 1974, YrsGt
　　L. Peter Gromet, (D), Caltech, 1979, Cc
　　John F. Hermance, (D), Toronto, 1967, HqZrHw
　　Paul C. Hess, (D), Harvard, 1968, Gi
　　E. Marc Parmentier, (D), Cornell, 1975, Yg
　　Carle M. Pieters, (D), MIT, 1977, Zr
　　Warren L. Prell, (D), Columbia, 1974, Ou
　　Malcolm J. Rutherford, (D), Johns Hopkins, 1968, Cp
　　Peter H. Schultz, (D), Texas, 1972, Xg
　　Jan A. Tullis, (D), California (Los Angeles), 1971, Gcy
　　Terry E. Tullis, (D), California (Los Angeles), 1971, Yx
　　Thompson Webb, III, (D), Wisconsin, 1971, PelAs
Academic Program Manager:
　　Patricia M. Davey, (B), Rhode Island Coll, 1986, Zn
Academic Department Manager:
　　Dina Egge, (M), 2015

Community College of Rhode Island
Dept of Physics (Geology & Oceanography Div) (A) ☒ (2018)
400 East Avenue
Warwick, RI 02886
  p. (401) 825-2156
  kkortz@ccri.edu
  http://www.ccri.edu/physics/
Professor:
  Karen M. Kortz, (D), Rhode Island, 2009, ZeGg
  Paul White, (D), Gg
Associate Professor:
  Emily Burns, (D), Rhode Island, GgOgZi
  Duayne Rieger, (D), Ys
Assistant Professor:
  Roger M. Hart, (M), 2018, Gg

Providence College
Biology Dept (A,B) ◯ (2019)
1 Cunningham Sq
Providence, RI 02918
  p. (401) 865-2150
  ctoth@providence.edu
  Programs: BA Biology, BS Biology, BA Biology/Secondary Ed, BS Biology/Secondary Ed, BA Biology/Optometry, BS Biology/Optometry, BS Environmental Biology
  Certificates: Neuroscience Certificate
Chair:
  Charles Toth

Roger Williams University
College of Arts & Sciences (B) (2015)
Bristol, RI 02809
  p. (401) 254-3087
  jborden@rwu.edu
  Department Secretary: Valerie Catalano
Head:
  Mark D. Gould, (D), Rhode Island, 1973, Ob
Chair:
  Paul Webb
Professor:
  Thomas Doty, (D), Rhode Island, 1977, Ob
  Richard Heavers, (D), Rhode Island, 1977, Op
  Thomas J. Holstein, (D), Brown, 1969, Zn
  Martine Villalard-Bohnsack, (D), Rhode Island, 1971, Ob
Assistant Professor:
  Tim Scott, (D), SUNY (Stony Brook), 1993, Ob

University of Rhode Island 🏳
Dept of Geosciences (B,M,D) ☒ (2019)
9 East Alumni Ave.,
Kingston, RI 02881
  p. (401) 874-2265
  http://web.uri.edu/geo/
  t: @URI_GEO
  Programs: Geology; Geological Oceanography
  Certificates: Hydrology
  Enrollment (2018): B: 42 (8) M: 6 (0) D: 6 (0)
Associate Dean:
  Anne I. Veeger, (D), Arizona, 1991, HwCl
Chair:
  Brian K. Savage, (D), Caltech, 2004, Ys
Professor:
  Thomas B. Boving, (D), Arizona, 1999, Hw
  David E. Fastovsky, (D), Wisconsin, 1986, PvGsPs
Associate Professor:
  Dawn Cardace, (D), Washington, GpPg
  Simon E. Engelhart, (D), Pennsylvania, OnGmYs
Assistant Professor:
  Ananya Mallik, (D), Rice, 2014, Gx
  Soni M. Pradhanang, (D), HsqHq
Emeritus:
  J. Allan Cain, (D), Northwestern, 1962, Gx
  O Don Hermes, (D), North Carolina, 1967, Gi
  Daniel P. Murray, (D), Brown, 1976, ZeGtp

Graduate School of Oceanography (M,D) (2015)
215 South Ferry Road
Narragansett, RI 02882
  p. (401) 874-6222
  bcorliss@uri.edu
  http://www.gso.uri.edu
  Enrollment (2010): M: 41 (12) D: 31 (8)
Associate Dean:
  John Farrell, (D), Brown, 1991, Ou
  Mark Wimbush, (D), California (San Diego), 1969, Op
Research Professor:
  Theodore J. Smayda, (D), Oslo, 1967, Ob
Professor:
  Robert D. Ballard, (D), Rhode Island, 1974, Ga
  Steven N. Carey, (D), Rhode Island, 1982, Ou
  Jeremy S. Collie, (D), MIT/WHOI, 1985, Ob
  Peter Cornillon, (D), Cornell, 1973, Op
  Steven L. D'Hondt, (D), Princeton, 1989, Ou
  Edward G. Durbin, (D), Rhode Island, 1976, Ob
  Isaac Ginis, (D), Inst Exp Meteor, 1986, Op
  Tetsu Hara, (D), MIT, 1990, Op
  Paul E. Hargraves, (D), William & Mary, 1968, Ob
  David L. Hebert, (D), Dalhousie, 1988, Op
  Christopher Kincaid, (D), Johns Hopkins, 1990, Ou
  John King, (D), Minnesota, 1983, Ou
  Roger Larson, (D), California (San Diego), 1970, Ou
  Margaret Leinen, (D), Rhode Island, 1980, Ou
  John T. Merrill, (D), Colorado, 1976, Oc
  S. Bradley Moran, (D), Dalhousie, 1991, Oc
  Scott W. Nixon, (D), North Carolina, 1970, Ob
  Candace Oviatt, (D), Rhode Island, 1967, Ob
  Hans Thomas Rossby, (D), MIT, 1966, Op
  Lewis Rothstein, (D), Hawaii, 1983, Op
  Haraldur Sigurdsson, (D), Durham (UK), 1970, Ou
  Jennifer Specker, (D), Oregon State, 1980, Ob
  Robert Tyce, (D), California (San Diego), 1976, Ou
  D. Randolph Watts, (D), Cornell, 1973, Op
  Karen Wishner, (D), California (San Diego), 1979, Ob
Associate Professor:
  Brian G. Heikes, (D), Michigan, 1984, As
  Yang Shen, (D), Brown, 1994, Ou
  David C. Smith, (D), California (San Diego), 1994, Ob
Marine Research Scientist:
  Percy Donaghay, (D), Oregon State, 1980, Ob
  Kathleen Donohue, (D), Rhode Island, 1996, Op
  Alfred K. Hanson, Jr., (D), Rhode Island, 1981, Oc
  Barbara K. Sullivan-Watts, (D), Oregon State, 1977, Og
Adjunct Professor:
  Lawrence J. Buckley, (D), New Hampshire, 1975, Oc
  Richard J. Pruell, (D), Rhode Island, 1984, Co
  Charles T. Roman, (D), Delaware, 1981, Ob
Marine Research Scientist:
  Dian J. Gifford, (D), Dalhousie, 1986, Ob
  Robert D. Kenney, (D), Rhode Island, 1984, Ob
Emeritus:
  H. Perry Jeffries, (D), Rutgers, 1959, Ob
  John A. Knauss, (D), California, 1959, Op
  Theodore A. Napora, (D), Yale, 1964, Ob
  Michael E. Pilson, (D), California (San Diego), 1964, Oc
  James G. Quinn, (D), Connecticut, 1967, Oc
  Kenneth A. Rahn, (D), Michigan, 1971, Oc
  Saul B. Saila, (D), Cornell, 1952, Ob
  Jean-Guy E. Schilling, (D), MIT, 1966, OuCuc
  John M. Sieburth, (D), Minnesota, 1954, Ob
  Elijah V. Swift, (D), Johns Hopkins, 1967, Ob

South Carolina

Clemson University
Bob Campbell Geology Museum (A,B,M,D) (2015)
140 Discovery Lane
Clemson, SC 29634-0130
  p. (864) 656-4602
  tsteadm@clemson.edu
  http://www.clemson.edu/geomuseum
Director:
  Todd A. Steadman, (M), Louisiana State, 1987, ZnnZn

Curator:
  David J. Cicimurri, (M), SD Mines, 1998, Pv
  Christian M. Cicimurri, (M), SD Mines, 1999, Pv

Environmental Engineering and Earth Sciences (B,M) (2015)
321 Calhoun Drive
Room 445 Brackett Hall
Clemson, SC 29634-0919
  p. (864) 656-3438
  clemson-eees@lists.clemson.edu
  http://www.clemson.edu/ces/departments/eees/
  Administrative Assistant: Cynthia Rae Gravely
  Enrollment (2013): B: 40 (9) M: 8 (6)
Chair:
  Tanju Karanfil, (D), Michigan, 1995, NgZnn
Professor:
  James W. Castle, (D), Illinois, 1978, GseHw
  Ronald W. Falta, (D), California (Berkeley), 1990, Hq
  Cindy M. Lee, (D), Colorado Mines, 1990, GeClZe
  Lawrence C. Murdoch, (D), Cincinnati, 1991, Hw
  Mark Schlautman, (D), Caltech, 1992, ClHgSc
Assistant Professor:
  Stephen M.J Moysey, (D), Stanford, 2005, HwYuZe
  Brian A. Powell, (D), Clemson, 2004, Cg
  Lindsay C. Shuller-Nickles, (D), Michigan, 2010, Gz
Research Associate:
  Scott E. Brame, (M), Clemson, 1993, Hw
Lecturer:
  Alan B. Coulson, (D), South Carolina, 2009, PgGgCs
Adjunct Professor:
  C. Brannon Andersen, (D), Syracuse, 1994, Cl
  Christian M. Cicimurri, (M), South Dakota, 1999, Pv
  Brian Looney, (D), Minnesota, 1984, Hq
  Vaneaton Price, (D), North Carolina, 1969, Ce
  Tommy Temples, (D), South Carolina, 1996, GoYg
Emeritus:
  Lois B. Krause, (D), Clemson, 1996, Ze
  Fred Molz, (D), Stanford, 1970, Sp
  John R. Wagner, (D), South Carolina, 1993, ZeGm
  Richard D. Warner, (D), Stanford, 1971, Gz

College of Charleston
Dept of Geology & Environmental Geosciences (B,M) (2018)
66 George Street
Charleston, SC 29424
  p. (843) 953-5589
  callahant@cofc.edu
  http://geology.cofc.edu/
  f: https://www.facebook.com/Geology.CofC/
  t: @AquaTimCal
  Programs: Geology
  Enrollment (2018): B: 100 (43)
Chair:
  Timothy J. Callahan, (D), New Mexico Tech, 2001, Hwq
Associate Professor:
  K. Adem Ali, (D), Kent State, 2011, ZrHwNg
  Erin K. Beutel, (D), Northwestern, 2000, Gct
  John Chadwick, (D), Florida, 2002, GiZrGt
  Mitchell W. Colgan, (D), California (Santa Cruz), 1990, PeGe
  M. Scott Harris, (D), Delaware, 2000, OnGam
  Steven C. Jaumé, (D), Columbia, 1994, Ys
  Norman S. Levine, (D), Purdue, 1995, ZiGeNg
  Cassandra R. Runyon, (D), Hawaii, 1988, XgSoZe
  Leslie R. Sautter, (D), South Carolina, 1990, Ob
  Vijay M. Vulava, (D), Swiss Fed Inst Tech, 1998, CgHwSc
Assistant Professor:
  Barbara Beckingham, (D), Maryland, 2011, Cl
  Theodore R. Them, II, (D), Virginia Tech, 2016, CmOcZc
Emeritus:
  James L. Carew, (D), Texas, 1978, PgGd
  Michael P. Katuna, (D), North Carolina, 1974, Gus
  Robert L. Nusbaum, (D), Missouri (Rolla), 1984, Gz
  Alexander W. Ritchie, (D), Texas, 1975, Gc
Laboratory Director:
  Robin Humphreys, (M), Charleston, 2000, Ge

Furman University
Earth and Environmental Sciences (B) (2018)
3300 Poinsett Highway
Greenville, SC 29613
  p. (864) 294-2052
  nina.anthony@furman.edu
  http://ees.furman.edu
  Administrative Assistant: Nina Anthony
  Enrollment (2018): B: 82 (35)
Chair:
  C. Brannon Andersen, (D), Syracuse, 1994, ClGe
Professor:
  Geoffrey Habron, (D), Oregon State, 1999, Zn
  William A. Ranson, (D), Massachusetts, 1979, GxzGp
Associate Professor:
  Weston R. Dripps, (D), Wisconsin, 2003, HwsGe
  Suresh Muthukrishnan, (D), Purdue, 2002, ZiGmZr
Assistant Professor:
  Karen Allen, (D), Georgia, 2016, Zn
  Ruth F. Aronoff, (D), Purdue, 2016, Gc
  Matt Cohen, (D), Arizona State, 2015, Zn
Adjunct Professor:
  Courtney Quinn, (D), Nebraska, 2012, Zn
  Melissa Ranhofer, (D), South Carolina, 2009, Gg
Visiting Research Professor:
  Christopher Romanek, (D), Texas A&M, 1991, Cls
Emeritus:
  John M. Garihan, (D), Penn State, 1973, Gct
  Kenneth A. Sargent, (D), Oklahoma, 1973, Hy
Laboratory Director:
  Lori Nelsen, (M), Furman, 2007, Ca

University of South Carolina
School of the Earth, Ocean & Environment (B,M,D) (2018)
701 Sumter St
EWS 617
Columbia, SC 29208
  p. (803) 777-4535
  khamilton@geol.sc.edu
  http://sc.edu/seoe
  f: https://www.facebook.com/MarineScienceSC/
  Programs: Geological Sciences (B,M,D); Geophysics (B); Marine Science (B,M,D); Environmental Science (B); Environmental Studies (B); Earth & Environmental Resources Management (M,JD);
  Enrollment (2017): B: 582 (111) M: 39 (17) D: 39 (6)
Director:
  Thomas J. Owens, (D), Utah, 1984, Ys
Research Professor:
  Dwayne E. Porter, (D), South Carolina, 1995, OgHsRw
Director of the Belle Baruch Marine Institute:
  James Pinckney, (D), South Carolina, 1992, Obn
Associate Dean:
  Claudia R. Benitez-Nelson, (D), MIT/WHOI, 1999, OcGe
Professor:
  Ron Benner, (D), Georgia, 1984, Ob
  Carol Boggs, (D), Texas, 1979, Zn
  Subrahmanyam Bulusu, (D), Southampton (U.K.), 1998, Op
  Thomas Lekan, (D), Zn
  Joseph Quattro, (D), Rutgers, 1991, Ob
  Tammi Richardson, (D), Dalhousie, 1996, Ob
  Raymond Torres, (D), California (Berkeley), 1997, Hy
  George Voulgaris, (D), Southampton (U.K.), 1992, OnpOg
  Scott M. White, (D), California (Santa Barbara), 2001, YrZr
  Alicia M. Wilson, (D), Johns Hopkins, 1999, Hw
  Neal Woods, (D), Zn
  Gene M. Yogodzinkski, (D), Cornell, 1993, GiCuc
Research Professor, Director of Undergraduate Studies:
  Gwendelyn Geidel, (D), South Carolina, 1982, HwGeCg
Research Associate Professor:
  Jennifer R. Pournelle, (D), California (San Diego), 2003, GaZuSa
Associate Professor:
  David Barbeau, Jr., (D), Arizona, 2003, Gs
  Michael Bizimis, (D), Florida State, 2001, GiCg
  Andrew L. Leier, (D), Arizona, 2005, Gst
  Ryan R. Rykaczewski, (D), Scripps, 2009, ObZoc

Howie Scher, (D), Florida, 2005, GuCl
Alexander E. Yankovsky, (D), Marine Hydrophysical Inst (Ukraine), 1991, OpnAm
Assistant Professor:
Jessica Barnes, (D), Zn
Monica Barra, (D), New York Univ, 2019, Zn
Annie Bourbonnais, (D), Victoria, 2012, Oc
David E. Fuente, (D), North Carolina, 2017, Zn
Conor Harrison, (D), Zn
David Kneas, (D), Zn
Susan Q. Lang, (D), Washington, 2006, Oc
Katherine Ryker, (D), North Carolina State, Ze
Lori A. Ziolkowski, (D), California (Irvine), 2009, CoOcCs
Research Assistant Professor:
Duke Brantley, (D), South Carolina, Ye
Matthew Kimball, (D), Rutgers, 2008, Ob
Erik Smith, (D), Maryland, 2000, HsOb
Emeritus:
Philip Barnes, (D), Zn
Bruce Coull, (D), Lehigh, 1968, Ob
John Mark Dean, (D), Purdue, 1962, Ob
Robert Feller, (D), Washington, 1977, Ob
Madilyn Fletcher, (D), Univ Coll (Wales), 1975, Ob
James N. Kellogg, (D), Princeton, 1981, Yg
Christopher G. Kendall, (D), Imperial Coll (UK), 1966, Gs
Willard S. Moore, (D), SUNY (Stony Brook), 1969, Cc
W. Edwin Sharp, (D), California (Los Angeles), 1964, Gz
Stephen Stancyk, (D)
Pradeep Talwani, (D), Stanford, 1973, Ys
Sarah A. Woodin, (D), Washington, 1972, Ob
Richard Zingmark, (D)

University of South Carolina, Lancaster
University of South Carolina, Lancaster (A) ☒ (2017)
P.O. Box 889
Lancaster, SC 29721
p. (803) 313-7129
martek@mailbox.sc.edu
http://usclancaster.sc.edu
Instructor:
Lynnette Flann Martek, (M), Emporia State, 1994, ZgAm

Winthrop University
Dept of Chemistry, Physics, & Geology (2015)
Sims Science Building
Winthrop University
Rock Hill, SC 29733
p. (803) 323-4949
bolandi@winthrop.edu
http://chem.winthrop.edu
Chair, Environmental Sciences and Studies Program:
Marsha S. Bollinger, (D), South Carolina, 1986, CmOcCc
Professor:
Irene B. Boland, (D), South Carolina, 1996, GtgZe
Associate Professor:
Gwen M. Daley, (D), Virginia Tech, 1999, PqgGs
Scott P. Werts, (D), Johns Hopkins, 2006, ScbPg

Wofford College
Dept of Geology (2015)
Wofford College
429 North Church Street
Spartanburg, SC 29303-3663
p. (864) 597-4527
fergusonta@wofford.edu
http://www.wofford.edu/geology/
Director:
Terry A. Ferguson, (D), Tennessee, 1988, Ga

South Dakota

Black Hills State University
School of Natural Sciences (B) ●☒ (2019)
1200 University Street, Unit 9008
Spearfish, SD 57799-9008
p. (605) 642-6506
Abigail.Domagall@bhsu.edu
https://www.bhsu.edu/Academics/Natural-Sciences/Environmental-Physical-Science
f: https://www.facebook.com/EnvPhysSciBHSU/
Programs: Environmental Physical Science, Earth Science minor, Composite Science Education
Enrollment (2018): B: 35 (7)
Professor:
Mark Gabel, (D), Iowa State, 1982, Pb
Associate Professor:
Abigail M S Domagall, (D), SUNY (Buffalo), 2008, GveZe

Oglala Lakota College
Dept of Math, Science, & Technology (2015)
P.O. Box 490
Kyle, SD 57755
p. (605) 455-6124
hlagarry@olc.edu
http://www.olc.edu/local_links/smet/

South Dakota School of Mines & Technology
Dept of Atmospheric and Environmental Sciences (B,M,D) (2015)
501 E. St. Joseph Street
Rapid City, SD 57701-3995
p. (605) 394-2291
william.capehart@sdsmt.edu
http://www.ias.sdsmt.edu/
Enrollment (2011): B: 15 (0) M: 14 (3)
Associate Professor:
William J. Capehart, (D), Penn State, 1997, AsZoHq
Donna V. Kliche, (D), SD Mines, 2007, AssAs
P. V. Sundareshwar, (D), South Carolina, 2002, ZeCgSb
Assistant Professor:
Adam French, (D), North Carolina State, 2011, Asm
Lisa Kunza, (D), Wyoming, 2012, HsZe
Instructor:
Darren R. Clabo, (M), SD Mines, 2009, AssAs
Emeritus:
Andrew G. Detwiler, (D), SUNY (Albany), 1980, AspAm
John H. Helsdon, (D), SUNY (Albany), 1979, As
Mark R. Hjelmfelt, (D), Chicago, 1980, As
Paul L. Smith, (D), Carnegie Inst, 1960, As

Dept of Geology & Geological Engineering (B,M,D) O☒ (2018)
501 E. Saint Joseph St.
Rapid City, SD 57701-3901
p. (605) 394-2461
laurie.anderson@sdsmt.edu
http://geology.sdsmt.edu
f: https://www.facebook.com/SDSMTGeologyGeologicalEngineering
Programs: Geology (B); Geological Engineering (B); Geology and Geological Engineering (M,D); Paleontology (M)
Certificates: Geospatial Technology (UG,G); Petroleum Systems (G)
Administrative Assistant: Cleo J. Heenan
Enrollment (2018): B: 132 (40) M: 31 (5) D: 13 (1)
Dean of Graduate Education:
Maribeth H. Price, (D), Princeton, 1995, Zir
Department Head; Director Museum of Geology:
Laurie C. Anderson, (D), Wisconsin, 1991, PiePq
Director of Energy Resources Initiative:
Daniel J. Soeder, (M), Bowling Green, 1978, GoeEo
Field Station Director:
Nuri Uzunlar, (D), SD Mines, 1993, EmCeGc
Professor:
Edward F. Duke, (D), Dartmouth, 1984, GpiZr
Timothy L. Masterlark, (D), Wisconsin, 2000, YdhNr
Larry D. Stetler, (D), Washington State, 1993, NxGsm
Senior Scientist:
William M. Roggenthen, (D), Princeton, 1980, NgYg
Associate Professor:
Kurt W. Katzenstein, (D), Nevada (Reno), 2008, NxtZr
Darrin C. Pagnac, (D), California (Riverside), 2005, PvGrZe
J. Foster Sawyer, (D), SD Mines, 2006, GsoHw

Assistant Professor:
  Zeynep O. Baran, (D), Miami, 2012, GcoGt
  Sarah W. Keenan, (D), Tennessee (Knoxville), 2014, PvClb
  Liangping Li, (D), Polytechnic Univ of Valencia (Spain), 2011, HwqHt
  Gokce K. Ustunisik, (D), Cincinnati, 2009, Cp
  Kevin M. Ward, (D), Arizona, 2016, YsGtYg
Coordinator and Instructor:
  Christopher J. Pellowski, (D), SD Mines, 2012, Gg
Associate Director Museum of Geology:
  Sally Y. Shelton, (M), Texas Tech, 1984, PgvGe
Lecturer:
  Curtis V. Price, (M), Dartmouth, 1985, ZifRw
Emeritus:
  Arden D. Davis, (D), SD Mines, 1983, HwNx
  James E. Fox, (D), Wyoming, 1972, Gs
  James E. Martin, (D), Washington, 1979, Ps
  Colin J. Paterson, (D), Otago (NZ), 1978, Eg
  Perry H. Rahn, (D), Penn State, 1965, Ng
  Jack A. Redden, (D), Harvard, 1956, Gp

University of South Dakota
Dept of Earth Sciences & Physics (B) (2015)
414 East Clark Street
Vermillion, SD 57069-2390
  p. (605) 677-5649
  esci@usd.edu
  http://www.usd.edu/earthsciences/
  Enrollment (2011): B: 25 (9)
Chair:
  Timothy H. Heaton, (D), Harvard, 1988, PvOg
Associate Professor:
  Brennan T. Jordan, (D), Oregon State, 2002, GiAm
  Mark R. Sweeney, (D), Washington State, 2004, Gms
Instructor:
  Jeanne M. Fromm, (M), Idaho State, 1995, GgHs

Tennessee

Austin Peay State University
Geosciences Dept (B) (2018)
601 College St
Clarksville, TN 37044
  p. (931) 221-7454
  deibertj@apsu.edu
  http://www.apsu.edu/geosciences
  Programs: Geosciences
  Enrollment (2009): B: 72 (9)
Chair:
  Jack Deibert, (D), Wyoming, GsrGo
Professor:
  Phyllis A. Camilleri, (D), Wyoming, 1994, Gct
  Daniel L. Frederick, (D), Tennessee, 1994, Gsr
  Gregory D. Ridenour, (D), Texas A&M, 1993, HsOgZn
  Robert A. Sirk, (D), Kent State, 1991, RnZyu
Associate Professor:
  Kallina M. Dunkle, (D), Wisconsin, 2012, HwGle
  Christopher Gentry, (D), Indiana State, 2008, Ziy
  Christine Mathenge, (D), Indiana, 2008, Zcy
Assistant Professor:
  Erik L. Haroldson, (D), Wisconsin, 2016, EdGzCg
Emeritus:
  Phillip R. Kemmerly, (D), Oklahoma State, 1973, GmeHg
Laboratory Director:
  Randal P. Roberson, (M), Murray State, 2018, Gg
Other:
  Richard F. Wheeler, (M), Brigham Young, 1980, Ggo

Middle Tennessee State University
Dept of Geosciences (B,M) (2019)
Box 9
Davis Science Building
Room 241
Murfreesboro, TN 37132
  p. (615) 898-2726
  karen.wolfe@mtsu.edu
  http://mtsu.edu/geosciences
  f: https://www.facebook.com/groups/88868852075/
  t: @MTSUGeosciences
  Programs: Geology; Earth Science; Physical Geography; GIS/Remote Sensing
  Administrative Assistant: Karen M. Wolfe
  Enrollment (2018): B: 83 (37) M: 23 (12)
Director:
  Zada Law, (M), Wisconsin, 1980, Zin
Chair:
  Warner Cribb, (D), Ohio State, 1993, GipGz
Professor:
  Mark J. Abolins, (D), Caltech, 1999, Gc
Associate Professor:
  Clay D. Harris, (D), Indiana, 1992, Gse
  Melissa Lobegeier, (D), James Cook, 2001, Pmg
Assistant Professor:
  Jeremy Aber, (D), Kansas State, 2011, Zyi
  Joe D. Collins, (D), Texas (El Paso), 2016, Gms
  Racha El Kadiri, (D), W Michigan, 2014, Hg
  Henrique G. Momm, (D), Mississippi, 2008, ZiHsZr
Instructor:
  Laura R. Collins, (M), Mississippi State, 2005, Gg
Lecturer:
  Alan Brown, (M), Illinois State, 2005, Gg

Motlow State Community College
Dept of Natural Sciences (A) (2015)
PO Box 8500
Lynchburg, TN 37352-8500
  p. (931) 393-1810
  lmayo@mscc.edu
  http://www.mscc.edu/natural_science/
Instructor:
  Lisa L Herring Mayo, (M), Mississippi State, 2000, GgZge

Pellissippi State Community College
Natural and Behavioral Sciences (2015)
10915 Hardin Valley Road
P.O. Box 22990
Knoxville, TN 37801
  p. (865) 694-6685
  jkelley@pstcc.edu
  http://www.pstcc.edu
Adjunct Professor:
  Peter J. Lemiszki, (D), Tennessee, 1992, GcgZi

Roane State Community College - Oak Ridge
Mathematics and Sciences (Geology) (A) (2015)
276 Patton Lane
Harriman, TN 37748
  p. (865) 481-2000
  leea@roanestate.edu
  http://aclee1234.fortunecity.com
Professor:
  Arthur C. Lee, (D), S California, 1994, Ges

Sewanee: University of the South
Dept of Earth and Environmental Systems (B) (2015)
735 University Avenue
Sewanee, TN 37383-1000
  p. (931) 598-1271
  Sherwood@sewanee.edu
  http://www.sewanee.edu/EnvStudies
Chair:
  Scott Torreano, (D), Georgia, 1991, Sf
Professor:
  Martin A. Knoll, (D), Texas (El Paso), 1988, Hw
  Donald B. Potter, Jr., (D), Massachusetts, 1985, Gc
  Stephen A. Shaver, (D), Stanford, 1984, Eg
Associate Professor:
  C. Ken Smith, (D), Florida, 1996, Sf
Adjunct Professor:
  Glendon W. Smalley, (D), Tennessee, 1975, Sf

Tennessee Tech University
Dept of Earth Sciences (B) (2018)
PO Box 5062
Cookeville, TN 38505
    p. (931) 372-3121
    MHarrison@tntech.edu
    https://www.tntech.edu/cas/earth/
    Programs: Geosciences
    Administrative Assistant: Peggy Medlin
    Enrollment (2018): B: 43 (17)
Chair:
    Michael J. Harrison, (D), Illinois (Urbana), 2002, Gc
Professor:
    Evan A. Hart, (D), Tennessee, 2000, Zy
    H. Wayne Leimer, (D), Missouri, 1969, Gz
    Ping-Chi Li, (D), Iowa, 1992, Zi
Associate Professor:
    Joseph Asante, (D), Nevada (Las Vegas), 2012, HwZrGe
    Jeannette Wolak, (D), Montana State, 2011, Gsr
Assistant Professor:
    Lauren Michel, (D), Baylor, 2014, PcSaCl
Adjunct Professor:
    Jason E. Duke, (M), Tennessee Tech, 1995, Zi
Emeritus:
    Larry W. Knox, (D), Indiana, 1974, Pmg

University of Memphis
Center for Earthquake Research & Information (CERI) (M,D) (2018)
3876 Central Avenue, Suite 1
Memphis, TN 38152-3050
    p. (901) 678-2007
    clangstn@memphis.edu
    http://www.memphis.edu/ceri
    Programs: M.S. and PhD concentration in Geophysics
    Enrollment (2018): M: 9 (2) D: 17 (1)
Director:
    Charles A. Langston, (D), Caltech, 1976, Ys
Graduate Coordinator:
    Christine A. Powell, (D), Princeton, 1976, Ys
Professor:
    Randel Tom Cox, (D), Missouri, 1995, GtcGm
Associate Professor:
    Eunseo Choi, (D), Caltech, 2008, Gtq
Research Professor :
    James Dorman, (D), Columbia, 1961, Yse
CERI Founding Director:
    Archibald C. Johnston, (D), Colorado, 1979, Ys
Emeritus:
    Jer-Ming Chiu, (D), Cornell, 1982, Ys
Research Scientist:
    Stephen P. Horton, (D), Nevada (Reno), 1992, Ys
Research Professor :
    Chris Cramer, (D), Stanford, 1976, Ys
    Robert Smalley, Jr., (D), Cornell, 1988, Yd
Assoc. Research Professor:
    Mitchell M. Withers, (D), New Mexico Tech, 1997, Ys
Seismic Network Engineer:
    James Bollwerk, (M), Memphis
Research Equipment Technician II:
    Chris McGoldrick
    David Steiner
Research Associate Technician:
    John Parker
Research Associate II:
    Kent Moran, (D), Memphis
    Holly Withers
Local Technical Support Provider II:
    Robert Debula
    Deshone Marshall
Local Technical Support Provider I:
    James Davis, (D), Memphis, 2013
Director Education & Outreach:
    Gary Patterson, (M), Memphis, Gg
Digital Seismic Systems Supervisor:
    Steve Brewer, (M), Memphis

Assistant Director Administration & Finance:
    Michelle Smith, (B)

Dept of Earth Sciences (B,M,D) (2019)
109 Johnson Hall
488 Patterson Street
Memphis, TN 38152-3550
    p. (901) 678-4571 or 678-2177
    dlarsen@memphis.edu
    http://memphis.edu/earthsciences/
    Programs: Earth Sciences (B, M, D); Geoarchaeology (B); Geography (B, M); Geology (B, M); Archaeology (M); Interdisciplinary Studies (M)
    Certificates: GIS (G)
    Enrollment (2018): B: 63 (16) M: 33 (9) D: 34 (7)
Chair:
    Daniel Larsen, (D), New Mexico, 1994, ClGsHw
Director, Confucius Institute:
    Hsiang-Te Kung, (D), Tennessee, 1980, Zy
Professor:
    Mervin J. Bartholomew, (D), Virginia Tech, 1971, Gtc
    Randel T. Cox, (D), Missouri, 1995, GcmGt
    David H. Dye, (D), Washington, 1980, Ga
    Arleen A. Hill, (D), South Carolina, 2002, Rn
    Esra Ozdenerol, (D), Louisiana State, 2000, Zi
    Jose Pujol, (D), Wyoming, 1985, Ye
    Roy B. Van Arsdale, (D), Utah, 1979, Gcm
Associate Professor:
    Anzhelika Antipova, (D), Louisiana State, 2010, Zg
    Dorian Burnette, (D), Arkansas, 2009, Atm
    Andrew M. Mickelson, (D), Ohio State, 2002, Ga
Assistant Professor:
    Youngsang Kwon, (D), SUNY (Buffalo), 2012, Zi
    Ryan M. Parish, (D), Memphis, 2013, Ga
Visiting Professor:
    Elizabeth Rhenberg, (D), West Virginia, 2015, Pg
Emeritus:
    Phili B. Deboo, (D), Louisiana State, 1963, Pg
    Robert W. Deininger, (D), Rice, 1964, Gx
    James Dorman, (D), Columbia, 1961, Ys
    Archibald C. Johnston, (D), Colorado, 1979, Ys
    David N. Lumsden, (D), Illinois (Urbana), 1965, Gd
Instructor/Coordinator:
    Julie Johnson, (D), Florida Intl, 2012, Gx

University of Tennessee, Chattanooga
Biology, Geology, and Environmental Science (B) (2019)
615 McCallie Ave., Dept. 2653
Chattanooga, TN 37403
    p. (423) 425-4341
    http://www.utc.edu/biology-geology-environmental-science/division-geology/
    f: https://www.facebook.com/GeologyatUTC
    Programs: Geology
    Enrollment (2018): B: 45 (4)
Professor:
    Jonathan W. Mies, (D), North Carolina, 1990, GctHg
Associate Professor:
    Amy Brock-Hon, (D), Nevada (Las Vegas), 2007, GmSc
Assistant Professor:
    Ann E. Holmes, (D), Columbia, 1996, GsrPi
    A K M Azad Hossain, (D), Mississippi, 2008, ZriZg
Adjunct Professor:
    Gregory Brodie, (M), Purdue, 1979, Ge
Geology Laboratory Coordinator:
    Wayne K. Williams, (M), Memphis State, 1980, GdgGo

University of Tennessee, Knoxville
Dept of Earth & Planetary Sciences (B,M,D) (2019)
306 Earth & Planetary Sciences Building
Knoxville, TN 37996-1410
    p. (865) 974-2366
    eps@utk.edu
    http://web.eps.utk.edu/
    f: https://www.facebook.com/UTEPS
    Programs: Geology; Environmental Studies

Enrollment (2018): B: 131 (35) M: 21 (9) D: 27 (3)
Head:
   Michael L. McKinney, (D), Yale, 1985, GePg
Professor:
   Thomas W. Broadhead, (D), Iowa, 1978, Pi
   William M. Dunne, (D), Bristol (UK), 1980, Gc
   Annette S. Engel, (D), Texas, 2004, Cl
   Christopher Fedo, (D), Virginia Tech, 1994, GsXgCg
   Linda Kah, (D), Harvard, 1997, Gs
   Larry D. McKay, (D), Waterloo, 1991, HyGe
   Jeffery E. Moersch, (D), Cornell, 1997, XgZr
   Edmund Perfect, (D), Cornell, 1986, SpHwGq
Associate Professor:
   Devon Burr, (D), Arizona, 2003, XgGm
   Micah Jessup, (D), Virginia Tech, 2007, Gc
   Molly McCanta, (D), Brown, 2004, GiXg
   Colin Sumrall, (D), Texas, 1997, Pi
Associate Scientist:
   Joshua Emery, (D), Arizona, 2003, Xg
Assistant Professor:
   Nick Dygert, (D), Brown, 2015, Giz
   Andrew Steen, (D), North Carolina (Chapel Hill), 2009, CoOb
   Anna Szynkiewicz, (D), Wroclaw, 2004, CslXg
Lecturer:
   Jake Benner, (M), Utah, 2002, GgPg
   William Deane, (M), Tennessee, 1998, Gg
Adjunct Professor:
   Hassina Z. Bilheux, (D), 2003, Zn
   David R. Cole, (D), Penn State, 1980, Ca
   Steven G. Driese, (D), Wisconsin, 1982, SaClGs
   Alice Layton, (D), Purdue, 1987, Zn
   Timothy McCoy, (D), Xm
   David Mittlefehldt, (D), Xm
   Scott Murchie, (D), Xg
   Tommy Phelps, (D), Zn
   Richard T. Williams, II, (D), Virginia Tech, 1979, Yg
Emeritus:
   Theodore C. Labotka, (D), Caltech, 1978, GpCg
   Harry Y. McSween, Jr., (D), Harvard, 1977, XcGi
   Kula C. Misra, (D), W Ontario, 1973, Eg
   Kenneth R. Walker, (D), Yale, 1969, Pe
Other:
   Melanie A. Mayes, (D), Tennessee, 2006, ClGeHw
Cooperating Faculty:
   Janet L. Hopson, (D), Tennessee, 1994, Zn
   Robert Riding, (D), UK, Gs

University of Tennessee, Martin
Dept of Agriculture, Geosciences, and Natural Resources (B) (2015)
256 Brehm Hall
Martin, TN 38238
   p. (731) 881-7260
   mehlhorn@utm.edu
   http://www.utm.edu/departments/caas/agnr/geosciences/
   Enrollment (2013): B: 29 (13)
Professor:
   Paula M. Gale, (D), Arkansas, 1988, SodSc
   Michael A. Gibson, (D), Tennessee, 1988, PgiZe
   Jefferson S. Rogers, (D), Illinois, 1995, Zn
   Robert M. Simpson, (D), Indiana State, 2000, AsmZi
Associate Professor:
   Stan P. Dunagan, (D), Tennessee, 1998, GsSa
Assistant Professor:
   Thomas A. DePriest, (D), Union (Jackson), 2009, Ze
   Benjamin P. Hooks, (D), Maine, 2009, GciNr
Instructor:
   Eleanor E. Gardner, (M), Georgia, GgPg
Emeritus:
   William T. McCutchen, (M), Berea, 1967, Gg
   Robert P. Self, (D), Rice, 1971, Gs
   Helmut C. Wenz, (M), W Michigan, 1968, Zn

Volunteer State Community College
Volunteer State Community College (2015)
1800 Nashville Pike
Gallatin, TN 37006
   p. (615) 230-3294
   Clark.Cropper@volstate.edu
   http://www.volstate.edu

Walters State Community College
Walters State Community College (2015)
500 South Davy Crockett Parkway
Morristown, TN 37813
   p. (423) 585-6764
   http://www.ws.edu

Texas

Alamo Colleges, Palo Alto College
Dept of Geology (A) (2015)
1400 W. Villaret Blvd
San Antonio, TX 78224
   p. (210) 486-3000
   ghagen@alamo.edu
   http://alamo.edu/pac/geology/

Alamo Colleges, San Antonio College
Natural Sciences (2019)
1819 N Main Ave
San Antonio, TX 78212
   p. (210) 486-0840
   dlambert@alamo.edu
   http://alamo.edu/sac/earthsci/
Professor:
   Dean Lambert, (D), Texas, Zy
Adjunct Professor:
   Dwight Jurena, (M), Rennsalear, GgZg
   Ryan E. Rudnicki, (D), Penn State, Zy
   Charles K. Smith, (M), Texas State, Zy

Alvin Community College
Dept of Geology (A) (2015)
3110 Mustang Rd
Alvin, TX 77511
   p. (281) 756-5670
   ddevery@alvincollege.edu
   Enrollment (2012): A: 2 (1)
Dept. Chair of Physical Sciences:
   Dora Devery, (M), Texas Christian, 1979, Gg

Amarillo College
Dept of Physical Science (2016)
P.O. Box 447
Amarillo, TX 79178
   p. (806) 371-5333
   rdhobbs@actx.edu
   https://www.actx.edu/pscience/
   Enrollment (2009): A: 3 (0)
Professor:
   Richard D. Hobbs, (D), Wyoming, 1998, ZrGc
Adjunct Professor:
   David Pertl, (M), W Texas A&M, 1984, Go

Angelo State University
Dept of Physics and Geosciences (B) O☒ (2018)
ASU Station #10904
San Angelo, TX 76909
   p. (325) 942-2242
   joseph.satterfield@angelo.edu
   http://www.angelo.edu/dept/physics/Geosciences/
   Programs: Geosciences (B)
   Enrollment (2018): B: 55 (15)
Professor:
   Joseph I. Satterfield, (D), Rice, 1995, Gc
Assistant Professor:
   Elizabeth Koeman-Shields, (D), Notre Dame, 2015, CgXg
   Heather L. Lehto, (D), S Florida, 2012, GvYsZe

Instructor:
   Jessica Garza, (M), Hawaii, 2011, AmGg
Visiting Assistant Professor:
   Fred L. Wilson, (D), Kansas, 1964, Gm
Adjunct Professor:
   Cary D. Carman, (B), Angelo State, 1999, Hg
   Steven Lyons, (D), Hawaii, 1981, Am
   Robert Purkiss, (M), Texas Tech, 1991, Gg

Austin Community College District
Dept of Earth and Environmental Sciences (A) (2019)
6101 Highland Campus Drive
Austin, TX 78752
   p. (512) 223-7157
   mshepherd@austincc.edu
   http://sites.austincc.edu/ees/
   Programs: Environmental Studies (A); Environmental Technology (A); Geology (A)
   Certificates: Environmental Technology
   Administrative Assistant: Oralia Guerra
   Enrollment (2018): A: 342 (20)
Chair:
   Mark A. Shepherd, (D), Nebraska (Medical Center), 2015, AsObZn
Associate Dean:
   Ronald A. Johns, (D), Texas, 1993, Pi
Professor:
   Robert H. Blodgett, (D), Texas, 1990, RnGsZe
Assistant Professor:
   Peter J. Wehner, (M), Vanderbilt, 1992, Gv
Adjunct Professor:
   Heather L. Beatty, (M), Texas Tech, 1992, Ge
   Peter A. Boone, (D), Texas A&M, 1972, Gro
   Thomas W. Brown, (M), Indiana, 1987, Ges
   Laura E. Chapa, (M), Texas State, 2014, ZgeZr
   M. Jennifer Cooke, (D), Texas, 2005, GmCl
   Leslie M. Davis, (M), Florida State, 1986, Op
   Meredith Y. Denton-Hedrick, (M), Texas A&M, 1992, ZeYeEo
   Maedeh Faraji, (D), Texas, 2007, AsZge
   Kusali R. Gamage, (D), Florida, 2005, HyOuPm
   Katelyn E. Gray, (D), Yale, 2018, PcCb
   Khaled W. Hasan, (D), Texas A&M, 1995, HsZri
   Heather R. Miller, (D), Texas A&M, 2012, Zeg
   Ata U. Rahman, (D), Texas Tech, 1987, GgdGe
   Fabienne M. Rambaud, (M), Texas, 2005, Eg
   Carolyn M. Riess, (M), Texas (El Paso), 1984, GoeGg
   Alina S. Satkoski, (M), Syracuse, 2011, GemGg
   Raymond M. Slade, Jr., (B), SW Texas, 1971, HgsHq
   Jason H. Stephens, (D), Texas, 2014, YrGuYe
   Wenxian Tan, (D), Texas, 2013, Obg
   Anne Turner, (M), Texas, 1986, Hg
Science Laboratory Technician:
   Shannon M. Grace, (B), Texas State, 2010, Zn
   Amie C. Hammond, (B), Texas, 2014, ZeGg
Cooperating Faculty:
   David J. Froehlich, (D), Texas, 1996, Pv

Baylor University
Dept of Environmental science (2015)
Waco, TX 76706
   p. (254) 710-3406
   George_Cobb@baylor.edu
   https://www.baylor.edu/environmentalscience/index.php?id=55209

Dept of Geosciences (B,M,D) (2018)
One Bear Place #97354
101 Bagby Ave.
BSB, 4th Floor, Rm. D409
Waco, TX 76798-7354
   p. (254) 710-2361
   paulette_penney@baylor.edu
   http://www.baylor.edu/geosciences
   f: https://www.facebook.com/baylorgeosciences
   t: @BaylorGeo
   Programs: Geology; Geophysics; Earth Science
   Administrative Assistant: Janelle Atchley
   Administrative Assistant: Jamie J.. Ruth
   Enrollment (2018): B: 39 (23) M: 15 (5) D: 25 (5)
Chair:
   Stacy C. Atchley, (D), Nebraska, 1990, GroGo
W.M. Keck Foundation Professor of Geophysics:
   Robert Jay Pulliam, (D), California (Berkeley), 1991, YsgYe
Keck Foundation Prof. of Geophysics:
   Jay Pulliam, (D), California (Berkeley), 1991, YsgYe
Associate Graduate Dean for Research:
   Steven G. Driese, (D), Wisconsin, 1982, SaClGs
Professor:
   Peter M. Allen, (D), S Methodist, 1977, HgNg
   Vincent S. Cronin, (D), Texas A&M, 1988, GctNg
   Stephen I. Dworkin, (D), Texas, 1991, ClGd
   Stephen Forman, (D), Colorado, CcPc
   Don M. Greene, (D), Oklahoma, 1980, ZyAm
   Lee C. Nordt, (D), Texas A&M, 1996, SdGa
   Kenneth Wilkins, (D), Florida, 1982, Pv
   Joe C. Yelderman, Jr., (D), Wisconsin, 1983, HwgZu
Graduate Program Director:
   Daniel J. Peppe, (D), Yale, 2009, PbcYm
Associate Professor:
   John A. Dunbar, (D), Texas, 1989, YgHg
   William C. Hockaday, (D), Ohio State, 2006, CoaSb
   Joseph D. White, (D), Montana, 1998, Zr
Assistant Professor:
   Kenneth S. Befus, (D), Texas, 2014, GviGv
   Peter B. James, (D), MIT, Ygg
   Scott C. James, (D), California (Irvine), 2001, HwGeEo
   Elizabeth Petsios, (D), S California, 2016, Pie
Emeritus:
   Rena M. Bonem, (D), Oklahoma, 1975, PieOb
   Thomas T. Goforth, (D), S Methodist, 1973, Yg
   Don F. Parker, (D), Texas, 1976, Giv
Luminescence Geochronology Research Specialist:
   Liliana Marin, (M), Illinois (Chicago), 2005, ZiGgCc
Instrumentation Specialist:
   Timothy Meredith
   Ren Zhang, (D), McMaster, 2007, Cs
Laboratory Director:
   Sharon Browning, (M), Gg
Office Manager:
   Paulette Penney, (A), Zn

Blinn College
Agricultural and Natural Science Programs (A) (2017)
902 College Avenue
Brenham, TX 77833
   p. (979) 830-4200
   cl.metz@blinn.edu
   http://www.blinn.edu/natscience/index.htm
Professor:
   Michael Dalman, (M), W Michigan, 1999, Ze
   Cheyl L. Metz, (D), Texas A&M, GgOg

Brookhaven College
Science/Math Div - Geology Dept (2017)
3939 Valley View Lane
Farmers Branch, TX 75244
   p. (972) 860-4758
   LannaBradshaw@dcccd.edu
   http://www.brookhavencollege.edu/instruction/math-science/science/geology/
Chair:
   Lanna K. Bradshaw, (M), TAMU-C, 2000, GgZg

Coastal Bend College
Science Div (A) (2015)
3800 Charco Road
Beeville, TX 78102
   p. (361) 354-2423
   amgarza@coastalbend.edu
   http://www.coastalbend.edu/acdem/science/
   Administrative Assistant:
   Enrollment (2011): A: 3 (0)

Instructor:
Danny Burns, (M), Ball State, 1984, Zgn
Richard Cowart, (D), Texas A&M (Corpus Christi), 2008, GeEo

Collin College - Central Park Campus
Deptartment of Geology (A) (2018)
2200 W. University Drive
McKinney, TX 75071
p. (972) 548-6790
bburkett@collin.edu
http://www.collin.edu/geology/
Enrollment (2016): A: 30 (0)
Professor:
Brett Burkett, (M), SUNY (Buffalo), 2008, Gvg
Related Staff:
Shannon Burkett, (M), SUNY (Buffalo), 2005, Gv

Collin College - Preston Ridge Campus
Dept of Geology and Environmental Science (A) (2017)
9700 Wade Boulevard
Frisco, TX 75035
p. (972) 377-1635
smay@collin.edu
Professor:
Heinrich Goetz, (M), Texas A&M, 1997, Ge
Paul Manganelli, (M), Boston Coll, 1998
S Judson May, (D), New Mexico, 1980, GgoGc

Collin College - Spring Creek Campus
Dept of Geology and Environmental Science (A) (2017)
2800 E. Spring Creek Parkway
Plano, TX 75074
p. (972) 578-5518
dbabcock@collin.edu
http://www.collin.edu/academics/programs/geology.html
Enrollment (2012): A: 10 (0)
Head:
Daphne H. Babcock, (M), Memphis, 1989, Geg
Professor:
Neal Alexandrowicz, (D)
Patrick Gonsulin-Getty, (D)
Instructor:
Stacey Bilich, (B)

Del Mar College
Dept of Natural Sciences (A) (2015)
Corpus Christi, TX 78404
p. (512) 886-1240
jhalcomb@delmar.edu
Professor:
Roger T. Steinberg, (M), Tennessee, 1981, GgoPg
Associate Professor:
Walter V. Kramer, (M), Texas (El Paso), 1970, GgoGx
Emeritus:
Mary S. Thorpe, (M), Baylor, 1966, Ge

El Centro College - Dallas Community College District
Geology Program (A) (2016)
801 Main Street
Dallas, TX 75202
p. (214) 860-2429
nfields@dcccd.edu
http://elcentrocollege.edu/programs/geology
Coordinator:
Nancy Fields, (M), Baylor, Zg
Professor:
Steven McCauley, (M), Mississippi State, GgAm
Bethan Salle, (M), N Colorado, 1999, GgeZg
Adjunct Professor:
Anna F. Banda, (M), Baylor, 2007, GgeGv
David Coffman, (M), Baylor, Gg
Stephanie Coffman, (M), Baylor, Gg
Alice Ruffel, (M), Oklahoma, Gg

El Paso Community College
Dept of Geological Sciences (A) (2015)
P.O. Box 20500
El Paso, TX 79998
p. (915) 831-5161
jvillal6@epcc.edu
http://epcc.edu/InstructionalPrograms/geologicalsciences
Chair:
Joshua Villalobos, (M), Texas (El Paso), 2001, ZeYgu
Coordinator - Valle Verde:
Russell Smith, (M), Florida, Gg
Coordinator - Transmountain Campus:
Kathleen Devaney, (D), California (Los Angeles), 1992, Gg
Coordinator - Northwest Campus:
Deborah Caskey, (M), Texas, Gg
Professor:
Sulaiman Abushagur, (D), Texas (El Paso), 1991, Gg
Instructor:
Robert Rohbaugh, (M), Texas (El Paso)
Adjunct Professor:
Brenda Barnes, (D), Texas (El Paso), Gg
Lawrence Bothern, (M), New Mexico State, Gg
Sabrina Canalda, (M), Texas (El Paso), Gg
Emile Couroux, Gg
Alexandra Falcon, Gg
Musa Hussein, (D), Texas (El Paso), Gg
Adriana Perez, (M), Texas (El Paso), Gg
Kirk Rothemund, Gg

Hardin-Simmons University
Dept of Geology & Environmental Science (B) (2015)
Box 16164
2200 Hickory Street
Abilene, TX 79698-6164
p. (325) 670-1383
ouimette@hsutx.edu
http://www.hsutx.edu/academics/undergraduate/holland/geology
Enrollment (2015): B: 20 (5)
Head:
Mark A. Ouimette, (D), Texas (El Paso), 1994, GieGt
Associate Professor:
Marla Potess, (D), Texas Tech, 2011, Zgu
Steven Rosscoe, (D), Texas Tech, 2008, PmGsPs

Hill College
Div of Mathematics and Sciences (Environmental Science) (2018)
112 Lamar
Hillsboro, TX 76645
p. 254-659-7500 or 817 295-7392
rroberts@hillcollege.edu
http://www.hillcollege.edu/academics/Traditional/Math_Science_EdServices/Geology.html
Programs: Associate of Science

Houston Community College System
Geology Dept (A) (2016)
1010 W. Sam Houston Pkwy.N.
Houston, TX 77043
p. (713) 718-5641
dwight.kranz@hccs.edu
http://learningwebsys.hccs.edu/discipline/geology/
Professor:
Dwight S. Kranz, (M), Texas A&M, 1980, Gg

Kilgore College
Dept of Chemistry and Geology (A) (2019)
1100 Broadway
Kilgore, TX 75662
p. (903) 983-8253
pbuchanan@kilgore.edu
Programs: Geology
Enrollment (2016): A: 5 (2)

Instructor:
  Paul C. Buchanan, (D), Houston, 1995, GgiXm

Lamar University
Dept of Earth and Space Sciences (B) ☒ (2018)
P.O. Box 10031
Beaumont, TX 77710
  p. (409) 880-8236
  jim.jordan@lamar.edu
  http://ess.lamar.edu
Professor:
  Roger W. Cooper, (D), Minnesota, 1978, Gi
  Jim L. Jordan, (D), Rice, 1975, XcgXm
  Donald E. Owen, (D), Kansas, 1963, Gr
  James W. Westgate, (D), Texas, 1988, Pev
Associate Professor:
  Joseph M. Kruger, (D), Arizona, 1991, YgGgZi
Instructor:
  Bennetta Schmidt, (D), Gg
Adjunct Professor:
  Mark Adams, (M), Houston (Clear Lake), Xg
  Cynthia L. Parish, (M), Lamar, 2004, Gg
  Carla M. Tucker, (M), Texas, 1990, Hw
Laboratory Coordinator:
  Karen M. Woods, (B), Lamar, 1991, Zg

Laredo College
Dept of Natural Sciences (2018)
P-41
West End Washington Street
Laredo, TX 78040
  p. (956) 721-5195
  glenn.blaylock@laredo.edu
  http://www.laredo.edu/cms/LCC/Instruction/Divisions/Sciences/Natural_Sciences/Science/
Professor:
  Glenn W. Blaylock, (M), Brigham Young, 1998, Ggv

Lee College
Dept of Physical Sciences (A) (2017)
P. O.Box 818
Baytown, TX 77522
  p. (281) 425-6552
  jdobberstine@lee.edu
  Enrollment (2016): A: 2 (0)
Professor:
  Sharon Gabel, (D), SUNY, 1991, GseGg

Lonestar College - CyFair
Geology Dept (2015)
9191 Barker Cypress Road
Cypress, TX 77433
  p. (281) 290-3919
  michael.r.konvicka@lonestar.edu
  http://www.lonestar.edu/geology-dept-cyfair.htm

Lonestar College - Kingwood
Geology Dept (2015)
20000 Kingwood Drive
Kingwood, TX 77339
  p. (281) 312-1629
  Jean.Whileyman@lonestar.edu
  http://www.lonestar.edu/geology-dept-kingwood.htm

Lonestar College - Montgomery
Geology Dept (2017)
3200 College Park Drive
Conroe, TX 77384
  p. (936) 273-7077
  Michael.J.Sundermann@lonestar.edu
  http://www.lonestar.edu/geology-dept-montgomery.htm
Professor:
  Nathalie N. Brandes, (M), New Mexico Tech
  Cynthia Lawry-Berkins, (D), Texas A&M, 2009, Ges

Lonestar College - North Harris
Geology Dept (2015)
2700 W.W. Thorne Drive
Houston, TX 77073-3499
  p. (281) 618-5685
  tom.hobbs@lonestar.edu
  Enrollment (2007): A: 20 (0)
Head:
  Thomas M. C. Hobbs, (M), Texas (El Paso), 1979, Gg
Professor:
  Peter E. Price, (M), Kentucky, 1979, Zi
Adjunct Professor:
  Penni Major, (M), Gg
  Michelle Mc Mahon, (D), Aberdeen, 1993, Go
  Victor S. Resnic, (D), Nat Pet Inst (Russia), 1971, Go
  Linda C. Tran, (B), Texas A&M, 1999, Zi

Lonestar College - Tomball
Geology Dept (2015)
30555 Tomball Parkway
Tomball, TX 77375
  p. (281) 351-3324
  David.O.Bary@lonestar.edu
  http://www.lonestar.edu/geology-dept-tomball.htm

McLennan Community College
Geology Dept (A) (2015)
1400 College Drive
Waco, TX 76708
  p. (254) 299-8442
  efagner@mclennan.edu
  http://www.mclennan.edu/departments/geol/
Instructor:
  Elaine Alexander, Gg

Midland College
Math and Science Div (A) (2015)
3600 N. Garfield
Midland, TX 79705
  p. (432) 685-4612
  kwaggoner@midland.edu
  http://www.midland.edu/~msd/
  Enrollment (2011): A: 14 (6)
Associate Professor:
  Joan Gawloski, (M), Baylor, GgzGe
  Antony Giles, (M), Sul Ross State, 2006, GgvGi
Assistant Professor:
  Keonho Kim, (D), GgAmPi
Adjunct Professor:
  Karen Waggoner, (D), Texas Tech, GgRhGc

Midwestern State University
Kimbell School of Geosciences (B,M) (2019)
3410 Taft
Wichita Falls, TX 76308
  p. (940) 397-4250
  geology.program@msutexas.edu
  http://www.mwsu.edu/academics/scienceandmath/geosciences/index
  f: https://www.facebook.com/MSUGeosProgram/
  Programs: Geology; Environmental Sciences
  Enrollment (2018): B: 71 (15) M: 16 (2)
Chair & Prothro Distinguished Assoc Professor of Geological Science:
  Jonathan D. Price, (D), Oklahoma, 1998, GiCgGz
Robert L. Bolin Distinguished Professor of Petroleum Geology:
  W. Scott Meddaugh, (D), Harvard, 1982, GoEgCg
Associate Professor and Graduate Advisor:
  Jesse R. Carlucci, (D), Oklahoma, 2012, PeiGs
Assistant Professor:
  Andrew Katumwehe, (D), Oklahoma State, 2016, YuuYe
  Peyton E. Lisenby, (D), Macquarie, 2017, GmeZi
Emeritus:
  Rebecca L. Dodge, (D), Colorado Mines, 1982, Zre
  John Kocurko, (D), Texas Tech, 1972, Gs

North Lake College - Dallas County Community College District
Mathematics and Natural Sciences (A) O (2019)
5001 North MacArthur Blvd
Irving, TX 75038
> p. (972) 273-3500
> mdempsey@dcccd.edu
> https://www.northlakecollege.edu/cd/instruct-divisions/nlc/math-science/pages/default.aspx
> Programs: See link for a full list of course offerings.
> https://www1.dcccd.edu/catalog/coursedescriptions/detail.cfm?loc=DCCCD&heading=Geology

Instructor:
> Leonard Kubicek, Gg

Odessa College
Dept of Geology, Anthropology & Geography (A) (2015)
201 W. University
Odessa, TX 79762
> p. (915) 335-6558
> dedwards@odessa.edu
> Enrollment (2010): A: 1 (1)

Associate Professor:
> Gerald B. McAfee, (M), Sul Ross State, 1966, Pi

Paris Junior College
Dept of Science (2015)
2400 Clarksville Street
Paris, TX 75460
> p. 903-782-0481
> mbarnett@parisjc.edu
> http://www.parisjc.edu/index.php/pjc2/directory-index/C208

Rice University
Center for Computational Earth Science (M,D) (2018)
MS-126
PO Box 1892
Houston, TX 77251-1892
> p. (713) 348-3574
> dmberry@rice.edu
> http://earthscience.rice.edu/centers/ccg/
> Enrollment (2016): M: 7 (2) D: 335 (4)

Associate Director:
> Alan Levander, (D), Stanford, 1984, Ys

Professor:
> Richard G. Gordon, (D), Stanford, 1979, YmGtYd
> Adrian Lenardic, (D), California (Los Angeles), 1995, Gc
> Julia K. Morgan, (D), Cornell, 1993, Gc
> Fenglin Niu, (D), Tokyo, 1997, Ysg
> Dale S. Sawyer, (D), MIT, 1982, Ys
> Colin A. Zelt, (D), British Columbia, 1989, Ys

Assistant Professor:
> Helge Gonnermann, (D), California (Berkeley), 2004, Gv

Dept of Civil and Environmental Engineering (M,D) (2015)
P.O. Box 1892
MS 317
Houston, TX 77251-1892
> p. (713) 348-4951
> rob.griffin@rice.edu
> http://ceve.rice.edu/

Chair:
> Joseph B. Hughes, (D), Iowa, 1992, Zn

Professor:
> Philip B. Bedient, (D), Florida, 1975, Hw
> Arthur A. Few, Jr., (D), Rice, 1969, As
> Mason B. Tomson, (D), Oklahoma State, 1972, Cg
> Calvin H. Ward, (D), Cornell, 1960, Zg
> Mark R. Wiesner, (D), Johns Hopkins, 1985, Zg

Lecturer:
> James B. Blackburn, (D), Texas, 1972, Zu

Adjunct Associate Professor:
> Stanley M. Pier, (D), Purdue, 1952, Co

Adjunct Assistant Professor:
> Charles J. Newell, (D), Rice, 1989, Hw

Adjunct Professor:
> Jean-Yves Bottero, (D), Nancy, 1979, Zn

Cooperating Faculty:
> John Hunter, (M), Indiana Sch Lib Sci, 1974, Zn

Dept of Earth, Environmental and Planetary Sciences (B,M,D) (2018)
MS 126
PO Box 1892
Houston, TX 77251-1892
> p. (713) 348-4880
> geol@rice.edu
> http://earthscience.rice.edu/
> Enrollment (2015): B: 28 (8) M: 45 (18) D: 40 (5)

Chair:
> Cin-Ty A. Lee, (D), Harvard, 2001, CgGiv

Professor:
> Rajdeep Dasgupta, (D), Minnesota, 2006, GxCg
> Gerald R. Dickens, (D), Michigan, 1996, OgCg
> Andre W. Droxler, (D), Miami, 1984, Gs
> Richard G. Gordon, (D), Stanford, 1979, YmGtYd
> Adrian Lenardic, (D), California, 1995, Yg
> Alan R. Levander, (D), Stanford, 1984, Ye
> Caroline A. Masiello, (D), California (Irvine), 1999, Zn
> Julia K. Morgan, (D), Cornell, 1993, GctGv
> Fenglin Niu, (D), Tokyo, 1997, YsGt
> Dale S. Sawyer, (D), MIT, 1982, Yr
> Colin A. Zelt, (D), British Columbia, 1989, Yg

Associate Professor:
> Helge Gonnermann, (D), California (Berkeley), 2004, Gv

Assistant Professor:
> Sylvia G. Dee, (D), S California, 2015, Zo
> Melodie E. French, (D), Texas A&M, 2014, Gc
> Jeffrey A. Nittrouer, (D), Texas, 2010, Gms
> Kirsten L. Siebach, Caltech, 2016, Xg
> Mark Torres, (D), S California, 2015, Cb
> Laurence Y. Yeung, (D), Caltech, 2010, CsAsCl

Adjunct Professor:
> Vitor Abreu, (D), Rice, 1998, Gg
> K. K. Bissada, (D), Washington (St. Louis), 1967, Cg
> Jeffrey J. Dravis, (D), Rice, 1980, Gs
> Paul M. Harris, (D), Miami, 1977, Yr
> N Ross Hill, (D), Virginia, 1978
> Stephen J. Mackwell, (D), Australian Nat, 1985, Yg
> Patrick J. McGovern, (D), MIT, 1996, Xy
> David L. Olgaard, (D), MIT, 1985, Yg
> Stephanie S. Shipp, (D), Rice, 1999, Ze

Emeritus:
> John B. Anderson, (D), Florida State, 1972, Gu
> Albert W. Bally, (D), Zurich, 1953, Gg
> H. C. Clark, (D), Stanford, 1966, YgGeYu
> Dieter Heymann, (D), Amsterdam, 1958, Xm
> William P. Leeman, (D), Oregon, 1974, Gi
> Andreas Luttge, (D), Tubingen, 1990, Cg
> John C. Stormer, Jr., (D), California (Berkeley), 1971, Gi
> Manik Talwani, (D), Columbia, 1959, Yr
> Peter R. Vail, (D), Northwestern, 1959, Gr

Department Administrator:
> Lee Willson, Zn

Saint Mary's University
Dept of Physics and Earth Sciences (2015)
One Camino Santa Maria
San Antonio, TX 78228-8569
> p. (210) 436-3235
> rcardenas@stmarytx.edu
> http://www.stmarytx.edu/acad/physicsandearthscience

Chair:
> Paul Nienaber

Professor:
> David Fitzgerald, (M), Iowa, 1977, Gd

Associate Professor:
> Gene W. Lene, (D), Texas, 1981, Ge

Sam Houston State University
Department of Geography and Geology O (2018)
Huntsville, TX 77341
    p. (936) 294-1073
    fsm002@shsu.edu
    http://www.shsu.edu/academics/geography-geology/gis/graduate.html
    Programs: Geography, Geology, GIS
    Certificates: GIS
Associate Professor:
   Falguni Mukherjee, (D), 2009, Zin

San Antonio Community College
Dept of Chemistry/Earth Sciences/Astronomy (A) (2015)
1300 San Pedro Avenue
CG Rm 207
San Antonio, TX 78212
    p. (210) 486-0045
    tstaggs@alamo.edu
    http://www.alamo.edu/sac/earthsci
Professor:
   Dean P. Lambert, (D), Texas, 1992, Zyi
Associate Professor:
   Anne D. Dietz, (M), Texas A&M (Kingsville), 1989, PgZg
   George R. Stanley, (M), Texas (San Antonio), 2006, ZgRm
   David A. Wood, (D), Arizona, 2000, Znn
Adjunct Professor:
   Thomas Adams, (D), 2011, PvGg
   T Scott Girhard, (M), SW Texas State, ZyAm
   Robert Janusz, (M), Texas (San Antonio), Gg
   Dwight J. Jurena, (M), Rensselaer, 2002, XgYvGz
   Ryan E. Rudnicki, (D), Penn State, 1979, Zyr
   C. Keith Smith, (M), Texas State, 2005, Zy
Cooperating Faculty:
   Steve Dingman, Zn

San Jacinto Community College-Central
Geology Dept (2015)
8060 Spencer Hwy.
Pasadena, TX 77089
    p. (281) 998-6150 x1882
    Karen.Purpera@sjcd.edu
    http://www.sanjac.edu/

San Jacinto Community College-North
Geology Dept (2015)
5800 Uvalde
Houston, TX 77049
    p. (281) 998-6150 x7210
    Kevin.Davis@sjcd.edu
    http://www.sanjac.edu/

San Jacinto Community College-South
Geology Dept (2015)
13735 Beamer Rd.
Houston, TX 77089
    p. (281) 998-6150 x4662
    Joe.Granata@sjcd.edu
    http://www.sanjac.edu/

Southern Methodist University
Roy M. Huffington Dept of Earth Sciences (B,M,D) (2019)
Post Office Box 750395
Dallas, TX 75275-0395
    p. (214) 768-2750
    geol@smu.edu
    http://www.smu.edu/earthsciences
    Programs: Geology, Geophysics
    Administrative Assistant: Stephanie L. Schwob
    Enrollment (2018): B: 31 (19) M: 8 (2) D: 17 (2)
Chair:
   Robert T. Gregory, (D), Caltech, 1981, Cs
Shuler-Foscue Chair in Earth Sciences:
   Zhong Lu, (D), Alaska (Fairbanks), 1996, Zr
Albritton Chair in Earth Sciences:
   Brian W. Stump, (D), California (Berkeley), 1979, Ys
Professor:
   Matthew J. Hornbach, (D), Wyoming, 2005, YrhYe
   Bonnie F. Jacobs, (D), Arizona, 1983, Pl
   James E. Quick, (D), Caltech, 1981, Gv
   Neil J. Tabor, (D), California (Davis), 2002, SdGs
   Crayton J. Yapp, (D), Caltech, 1980, Csl
Hamilton Chair in Earth Sciences:
   Stephen J. Arrowsmith, (D), Leeds, 2004, Ys
Associate Professor:
   Heather R. DeShon, (D), California (Santa Cruz), 2004, Ysr
   M. Beatrice Magnani, (D), Perugia, 2000, YsGtYe
Associate Scientist:
   Christopher T. Hayward, (D), S Methodist, 1997, Ys
   Ian J. Richards, (D), Tennessee, 1994, Cs
Assistant Professor:
   Rita C. Economos, (D), California (Los Angeles), 2009, CsGc
Adjunct Professor:
   Anthony R. Fiorillo, (D), Pennsylvania, 1989, Pv
   Matthew Siegler, (D), California (Los Angeles), 2011, YgXy
   Alisa Winkler, (D), S Methodist, 1990, Pv
   Dale A. Winkler, (D), Texas, 1985, Pv
Emeritus:
   David D. Blackwell, (D), Harvard, 1968, Yh
   James E. Brooks, (D), Washington, 1954, Gr
   Michael J. Holdaway, (D), California (Berkeley), 1963, Gp
   Louis L. Jacobs, (D), Arizona, 1977, Pv
   Robert L. Laury, (D), Wisconsin, 1966, Gd
   John V. Walther, (D), California (Berkeley), 1978, Cg

Stephen F. Austin State University
Dept of Geology (B,M) O (2017)
PO Box 13011 SFA Station
Nacogdoches, TX 75962
    p. (936) 468-3701
    geology@sfasu.edu
    http://www.geology.sfasu.edu/
    Administrative Assistant: Shana Scott
    Enrollment (2010): B: 71 (13) M: 14 (3)
Chair:
   Wesley A. Brown, (D), Texas (El Paso), 2004, YgsGt
Professor:
   R. LaRell Nielson, (D), Utah, 1981, Grs
Associate Professor:
   Chris A. Barker, (D), South Carolina, 1998, Gc
   Kevin W. Stafford, (D), New Mexico Tech, 2008, Hw
Assistant Professor:
   Melinda S. Faulkner, (D), Stephen F. Austin, 2016, GeClEg
   Liane M. Stevens, (D), Montana, 2015, Gpc
Instructor:
   Patricia S. Sharp, (M), Stephen F. Austin, 1978, Gg
Lab Coordinator:
   Wesley L. Turner, (M), Stephen F. Austin, 2016, Gg

Sul Ross State University
Dept of Biology, Geology and Physical Sciences (B,M) O (2019)
Box C-64
Alpine, TX 79832
    p. (432) 837-8112
    measures@sulross.edu
    http://www.sulross.edu/BGPS
    Programs: Geology
    Teacher Certification in Geology
    Certificates: none
    Enrollment (2018): B: 25 (7) M: 13 (4)
Professor:
   Elizabeth A. Measures, (D), Idaho, 1992, Gdq
   Kevin Urbanczyk, (D), Washington State, 1993, GiZi
Assistant Professor:
   Thomas Shiller, (D), Texas Tech, 2017, PvGrd
Lecturer:
   Jesse Kelsch, (M), New Mexico, Gct
Emeritus:
   David M. Rohr, (D), Oregon State, 1978, PiGd

Tarleton State University
Chemistry, Geosciences and Physics (B,M) ☒ (2019)
Box T-540
Stephenville, TX 76402
   p. (254) 968-9143
   rmorgan@tarleton.edu
   http://www.tarleton.edu/CHGP/index.html
   Programs: Geosciences Bachelor's of Science (concentrations in Earth Science, Earth Science with Teacher Certification, Environmental Science, Geology, Petroleum Geology, Hydrogeology, Accelerated bachelor's-to-master's degree), Master's of Science in Geosciences, Master's of Science in Environmental Science
   Certificates: GIS
   Administrative Assistant: Eva Moody
   Enrollment (2018): B: 47 (11) M: 11 (4)
Head:
   Ryan Morgan, (D), Baylor, 2015, Pei
Professor:
   Stephen W. Field, (D), Massachusetts, 1988, Gz
Assistant Professor:
   Catherine Ronck, (D), God
   Christopher Saxon, (D), Oklahoma, Gco
Instructor:
   Joree Burnett, (M), Tarleton State, 2015, HwGe

Tarrant County College, Northeast Campus
Natural Science Dept (A) (2015)
828 Harwood Road
Hurst, TX 76054
   p. (817) 515-6565
   marles.mccurdy@tccd.edu
   https://www.tccd.edu/academics/tcc-catalog/courses-and-programs/geology/
   Enrollment (2011): A: 650 (30)
Associate Professor:
   Meena Balakrishnan, (D), GggGg
   Kevin M. Barrett, (D), Texas State, 2012, AsZgr
Professor:
   Hayden R. Chasteen, (M), NE Louisiana, 1981, GgeGg

Tarrant County College, Southeast Campus
Physical Sciences Dept ☒ (2018)
2100 Southeast Pkwy
Arlington, TX 76018
   p. (817) 515-3238
   thomas.awtry@tccd.edu
   http://www.tccd.edu/academics/tcc-catalog/courses-and-programs/geology/
   Programs: Associate of Science
Associate Professor:
   Chris Baack, (M)
Instructor:
   Samantha Caputi, (M)
   Darren Greenaway, (M)

Temple College
Temple College (2015)
2600 South First Street
Temple, TX 76504
   p. (254) 298-8472
   john.mcclain@templejc.edu
   http://www.templejc.edu/

Texas A&M University
Center For Tectonophysics (M,D) (2017)
3115 TAMU
Department Geology & Geophysics
College of Geosciences
College Station, TX 77843-3115
   p. (979) 845-3296
   chesterf@tamu.edu
   http://tectono.tamu.edu/
   Enrollment (2016): M: 7 (1) D: 13 (1)
Director:
   Frederick Chester, (D), Texas A&M, 1988, GcNrYx
Assistant Director:
   Andreas Kronenberg, (D), Brown, 1983, GtyGc
Professor:
   Judith Chester, (D), Texas A&M, 1992, GcNrGo
   Marcelo J. Sanchez, (D), Politecnica de Catalunya, 2004, NtrNx
Associate Professor:
   Benchuan Duan, (D), California (Riverside), 2006, YsgGq
   David Sparks, (D), Brown, 1992, YdGq
Assistant Professor:
   Patrick M. Fulton, (D), Penn State, 2008, YhHyRn
   Hiroko Kitajima, (D), Texas A&M, 2010, YxSpGc
   Julia S. Reece, (D), Texas, 2011, SpGso

Dept of Atmospheric Sciences (B,M,D) ☒ (2019)
3150 TAMU
College Station, TX 77843-3150
   p. (979) 845-7671
   advising-atmo@geos.tamu.edu
   http://atmo.tamu.edu/
   Programs: Meteorology (B); Atmospheric Sciences (M,D)
   Enrollment (2018): B: 120 (0) M: 20 (0) D: 20 (0)
Department Head:
   Ping Yang, (D), Utah, 1995, As
Instructional Professor:
   Don T. Conlee, (D), Texas A&M, 1994
Distinguished Professor:
   Gerald North, (D), 1966, As
Distinguished Professor :
   Renyi Zhang, (D), MIT, 1993, As
Professor:
   Kenneth P. Bowman, (D), Princeton, 1984, Am
   Sarah D. Brooks, (D), Colorado, 2002, AsZc
   Ping Chang, (D), Princeton, 1988, Op
   Andrew E. Dessler, (D), Harvard, 1994, As
   John Nielsen-Gammon, (D), MIT, 1990, As
   Richard L. Panetta, (D), Wisconsin, 1978, Am
   R. Saravanan, (D), Princeton, 1990, As
   Courtney Schumacher, (D), Washington, 2003, As
   Istvan Szunyogh, (D), Hungarian Acad of Sci, 1994, As
Associate Professor:
   Craig Epifanio, (D), Washington, As
   Robert Korty, (D), MIT, 2005, As
   Gunnar Schade, (D), Johannes-Gutenberg (Germany), 1997, As
Instructional Assistant Professor:
   Tim Logan
Assistant Professor:
   Christopher J. Nowotarski, (D), Penn State, 2013, Ams
   Anita Rapp, (D), Colorado State, 2008, Zr
   Yangyang Xu
Adjunct Professor:
   Alex Dessler, (D), Duke, 1956, As
Emeritus:
   Richard E. Orville, (D), Arizona, 1966, As

Dept of Geography (B,M,D) (2015)
810 Eller O&M Building
3147 TAMU
College Station, TX 77843-3147
   p. (979) 845-7141
   cbruton@geog.tamu.edu
   http://geog.tamu.edu/
   Enrollment (2009): B: 156 (40) M: 16 (8) D: 18 (3)
Head:
   David M. Cairns, (D), Iowa, 1995, Zy
Professor:
   Sarah W. Bednarz, (D), Texas A&M, 1992, Ze
   John R. Giardino, (D), Nebraska, 1979, Gm
   Andrew G. Klein, (D), Cornell, 1997, ZriGl
   Michael R. Waters, (D), Arizona, 1983, Ga
Associate Professor:
   Daniel Z. Sui, (D), Goergia, 1993, Zr
   Vatche P. Tchakerian, (D), California (Los Angeles), 1989, Gm
Assistant Professor:
   Anne Chin, (D), Arizona State, 1994, Gm

Research Associate:
  Jean A. Bowman, (M), Rutgers, 1984, Hg
Emeritus:
  Robert S. Bednarz, (D), Chicago, 1975, Zeu
  Clarissa T. Kimber, (D), Wisconsin, 1969, Zy

Dept of Geology & Geophysics (B,M,D) (2015)
3115 TAMU
College Station, TX 77843-3115
  p. (979) 845-2451
  mcpope@tamu.edu
  http://geoweb.tamu.edu
  Enrollment (2012): B: 278 (41) M: 81 (21) D: 66 (5)
Head:
  John R. Giardino, (D), Nebraska, 1979, GmNg
Regents Professor:
  Mary J. Richardson, (D), MIT, 1980, Op
Professor:
  Tom Blasingame, (D), Texas A&M, 1989, Np
  Richard L. Carlson, (D), Washington, 1976, YrGty
  Frederick M. Chester, (D), Texas A&M, 1988, GcNr
  Judith Chester, (D), Texas A&M, 1992, GcNr
  Mark Everett, (D), Toronto, 1991, Ym
  Richard L. Gibson, Jr., (D), MIT, 1991, Yse
  Ethan L. Grossman, (D), S California, 1982, Csl
  Andrew Hajash, (D), Texas A&M, 1975, Cp
  Bruce Herbert, (D), California (Riverside), 1992, Ge
  Andreas Kronenberg, (D), Brown, 1983, GyNr
  Franco Marcantonio, (D), Columbia, 1994, Ct
  Julie Newman, (D), Rochester, 1993, Gct
  Anne Raymond, (D), Chicago, 1983, Pb
  William W. Sager, (D), Hawaii, 1979, Ou
  David Sparks, (D), Brown, 1992, YgGq
  Yuefeng Sun, (D), Columbia, 1994, GoYe
  Debbie Thomas, (D), North Carolina, 2002, Ou
  Thomas E. Yancey, (D), California (Berkeley), 1971, Pg
  Hongbin Zhan, (D), Nevada (Reno), 1996, Hw
Associate Research Professor:
  Renald Guillemette, (D), Stanford, 1983, Gz
Associate Professor:
  Benchun Duan, (D), California (Riverside), 2006, YsGtYg
  Will Lamb, (D), Wisconsin, 1987, GpCpGz
  Brent Miller, (D), Dalhousie, 1997, Cc
  Julie Newman, (D), Rochester, 1993, Gc
  Thomas Olszewski, (D), Penn State, 2000, Pe
  Michael Pope, (D), Virginia Tech, 1995, GrCc
Assistant Professor:
  Mike Tice, (D), Stanford, 2006, Pg
Assistant Dean :
  Eric Riggs, (D), California (Riverside), 2000, Gy
Lecturer:
  Alfonso Benavides-Iglesias, (D), Texas A&M, 2007, Ys
Emeritus:
  Christopher C. Mathewson, (D), Arizona, 1971, NgHwEg
  John D. Vitek, (D), Iowa, 1973, GmZe
Technical Laboratory Director:
  Michael Heaney, (D), Texas A&M, 1998, Pg
Dean of College of Geosciences:
  Kate Miller, (D), Stanford, 1991, Ys

Dept of Oceanography (B,M,D) (2017)
1204 Eller O&M Blg
MS 3146
College Station, TX 77843-3146
  p. (979) 845-7211
  syvon-lewis@tamu.edu
  http://ocean.tamu.edu
  Programs: Oceanography (B, M, D); Ocean Science and Technology (M)
  Enrollment (2017): B: 8 (0) M: 22 (10) D: 42 (6)
Director, Texas Sea Grant:
  Pamela Plotkin, (D), Texas A&M, 1994, Ob
Interim Dean:
  Debbie Thomas, (D), North Carolina, 2002, Ou
Head:
  Shari A. Yvon-Lewis, (D), Miami, 1994, Oc

Executive Associate Dean for Research:
  Jack G. Baldauf, (D), California (Berkeley), 1984, Ou
Director, Geochemical and Environmental Research Group:
  Anthony Knap, (D), Southampton, Oc
Professor:
  Rainier MW Amon, (D), Texas, 1995, CbOcb
  David A. Brooks, (D), Miami, 1975, Op
  Lisa Campbell, (D), SUNY (Stony Brook), 1985, Ob
  Ping Chang, (D), Princeton, 1988, Op
  Piers Chapman, (D), Univ Coll (N Wales), 1982, OcCm
  Steven F. DiMarco, (D), Texas (Dallas), 1991, OpnOg
  Wilford D. Gardner, (D), MIT/WHOI, 1978, OugZr
  Benjamin S. Giese, (D), Washington, 1989, Op
  Gerardo Gold-Bouchot, (D), Ctr Res Adv Stud (Mexico), 1991, OcCm
  Norman L. Guinasso, (D), Texas A&M, 1984, Opc
  Robert D. Hetland, (D), Florida State, 1999, Op
  Alejandro H. Orsi, (D), Texas A&M, 1993, Op
  Mary Jo Richardson, (D), MIT/WHOI, 1980, Gu
  Gilbert T. Rowe, (D), Duke, 1968, Ob
  Peter H. Santschi, (D), Switzerland (Berne), 1975, Oc
  Niall C. Slowey, (D), MIT, 1991, Ou
Senior Scientist:
  Troy L. Holcombe, (D), Columbia, 1972, OuGcg
  Matthew K. Howard, (D), Texas A&M, 1992, Op
  Ann E. Jochens, (D), Texas A&M, 1977, Op
  Adam Klaus, (D), Hawaii, 1991, Gu
Associate Professor:
  Ayal Anis, (D), Oregon State, 1993, Opn
  Timothy M. Dellapenna, (D), William & Mary, 1999, Ou
  Anja Schulze, (D), Victoria, 2001, Ob
  Achim Stoessel, (D), Hamburg, 1990, OpAmGl
  Daniel C.O Thornton, (D), Queen Mary (London), 1995, Ob
Associate Scientist:
  Steven K. Baum, (D), Texas A&M, 1996, Op
  Jose L. Sericano, (D), Texas A&M, 1993, Cm
Assistant Professor:
  Jessica N. Fitzsimmons, (D), MIT/WHOI, 2013, OcCta
  Henry Potter, (D), Miami, 2014, Op
  Kathryn E. F. Shamberger, (D), Washington, 2011, OcZc
  Jason B. Sylvan, (D), Rutgers, 2008, ObXb
  Chrissy Wiederwohl, (D), Texas A&M, 2012, Op
  Yige Zhang, (D), Yale, 2014, Co
Research Associate:
  Shinichi Kobara, (D), Texas A&M, Zi
  Marion Stoessel, (M), Hamburg, 1985, Op
  Zhankun Wang, (D), Massachusetts (Dartmouth), 2009, Op
Distinguished Professor Emeritus:
  Robert A. Duce, (D), MIT, 1964, Oc
  Worth D. Nowlin, Jr., (D), Texas A&M, 1966, OpZn
Distinguished Professor:
  Gerald R. North, (D), Wisconsin, 1966, Am
Director, Texas Sea Grant College Program:
  Robert R. Stickney, (D), Florida State, 1971, Ob
Emeritus:
  Douglas C. Biggs, (D), MIT/WHOI, 1976, Ob
  George A. Jackson, (D), Caltech, 1976, Op
  Bobby J. Presley, (D), California, 1969, Oc
  Robert H. Stewart, (D), California (San Diego), 1969, Op
Laboratory Director:
  Terry L. Wade, (D), Rhode Island, 1978, OcCao

Dept of Soil & Crop Sciences (B,M,D) (2017)
TAMU 2474
College Station, TX 77843-2474
  p. (979) 845-3603
  cmorgan@tamu.edu
  http://soilcrop.tamu.edu
  Programs: Environmental Plant and Soil Science
  Certificates: none
  Enrollment (2017): B: 32 (15) M: 19 (3) D: 21 (3)
Professor:
  David D. Baltensperger, (D), New Mexico State, 1981, PvCbSb
  Terry Gentry, (D), Arizona, Sb
  Kevin J. McInnes, (D), Kansas State, 1985, Sp
  Cristine L. S. Morgan, (D), Wisconsin, 2003, Spd
  Tony L. Provin, (D), Purdue, 1995, Sc

Paul Schwab, (D), Colorado State, 1981, Sc
Associate Professor:
 Jacqueline A. Aitkenhead-Peterson, (D), New Hampshire, 2000, HsSoGf
 Paul DeLaune, (D), Arkansas, 2002, So
 Youjun Deng, (D), Texas A&M, 2001, GzSc
 Fugen Dou, (D), Texas A&M, 2005, So
 Julie Howe, (D), Wisconsin, 2004, Sc
 Donald McGahan, (D), California (Davis), 2007, Sdc
Assistant Professor:
 Jourdan Bell, Texas A&M, 2014
 Katie Lewis, (D), Texas A&M, 2014, Sco
 Jake Mowrer, (D), Georgia, 2014, Sc
 Haly L. Neely, (D), Texas A&M, 2014, Spd
 Anil Somenhally, (D), Texas A&M, 2010, Sb

Texas A&M University, Commerce
Dept of Biological & Environmental Sciences (B,M) (2015)
Commerce, TX 75429
 p. (903) 886-5378
 DongWon.Choi@tamuc.edu
 http://www.tamu-commerce.edu/biology/
Assistant Professor:
 Haydn A "Chip" Fox, (D), South Carolina, 1994, HgGe

Texas A&M University, Corpus Christi
Dept of Physical and Environmental Sciences (B) (2017)
6300 Ocean Drive
Corpus Christi, TX 78412
 p. (361) 825-6000
 Valeriu.Murgulet@tamucc.edu
 http://geology.tamucc.edu/
 Enrollment (2016): B: 84 (13)
Chair, Dept, Physical and Environmental Sciences:
 Richard Coffin, (D), Delaware, 1986, Ouc
Professor:
 Jennifer M. Smith-Engle, (D), Georgia, 1983, Gs
Associate Professor:
 Thomas H. Naehr, (D), GEOMAR (Kiel), 1996, GuCm
Director, Center for Water Supply Studies:
 Dorina Murgulet, (D), Alabama, 2009, Hw
Assistant Professor:
 Xinping Hu, (D), Old Dominion, 2007, Oc
 Brandi Kiel Reese, (D), Texas A&M, 2011, Obc
Endowed Associate Research Professor:
 James Gibeaut, (D), S Florida, 1991, OnZi
Adjunct Professor:
 Clinton Randall (R Bissell, (M), Oklahoma State, 1984, GosGr
 Erika Locke, (M), California (Los Angeles), 1998, GgoGs
Coordinator, Geology Program:
 Valeriu Murgulet, (D), Alabama, 2010, Cg

Environmental Science Program (B,M) (2018)
6300 Ocean Drive
Corpus Christi, TX 78412
 p. (361) 825-2814
 jennifer.smith-engle@tamucc.edu
 http://sci.tamucc.edu/PENS/index.html
 Programs: Environmental Science
 Enrollment (2018): B: 235 (29) M: 49 (13)
Chair, Physical and Environmental Sciences Dept.:
 Richard Coffin, (D), Delaware, 1986, Oc
Director of National Spill Control School:
 Howard Wood, (M), American Military Univ, 2011, ZeHsZn
Program Coordinator:
 Jennifer M. Smith-Engle, (D), Georgia, 1983, Gs
Endowed Chair for Socioeconomics:
 David Yoskowitz, (D), Texas Tech, 1997, Zn
Endowed Chair for Fisheries and Ocean Health:
 Greg Stunz, (D), Texas A&M, 1999, Zn
Endowed Chair for Ecosystems Studies and Modeling:
 Paul Montagna, (D), South Carolina, 1983, ObCmHs
Director of Center for Coastal Studies:
 Paul Zimba, Mississippi State, 1990, Zn

Professor:
 Fereshteh Billiot, (D), Louisiana State, 2000, Zn
 Cherie McCollough, (D), Texas, 2005, Ze
 Richard McLaughlin, (D), California (Berkeley), 1997, Zn
 Toshiaki Shinoda, (D), Hawaii, 1993, As
Chair, Center for Water Supply Studies:
 Dorina Murgulet, (D), Alabama, 2009, Hw
Associate Professor:
 Darek Bogucki, (D), S California, 1996, Op
 Gregory Buck, (D), Georgia State, 1999, Zn
 Kirk Cammerata, (D), Kentucky, 1987, Zn
 Xinping Hu, (D), Old Dominion, 2007, Cms
 Patrick Larkin, (D), Texas A&M, 1999, Zn
 Riccardo Mozzachiodi, (D), Pisa, 1999, Zn
 Jennifer Pollack, (D), South Carolina, 2006, Ob
 James Silliman, (D), Michigan, 1998, Co
 Michael Starek, (D), Florida, 2008, Zi
 Michael Wetz, (D), Oregon State, 2006, Ob
Assistant Professor:
 Hussain Abdulla, (D), Old Dominion, 2009, Co
 Mohamed Ahmed, (D), W Michigan, 2012, YgZri
 Jeremy Conkle, (D), Louisiana State, 2010, Zn
 Sharon Derrick, (D), Texas A&M, 2001, Zn
 Joseph David Felix, (D), Pittsburg, 2012, Zn
 Chuntao Liu, (D), Wyoming, 2003, As
 Brandi Reese, Texas A&M, 2011, Cb
 Kim Withers, (D), Texas A&M, 1994, Zn
 Lin Zhang, (D), Rhode Island, 2012, Cs
Research Associate:
 Philippe Tissot, (D), Texas A&M, 1994, Zn

Texas A&M University, Kingsville
Dept of Geosciences (B) O (2017)
Campus Box 164
Kingsville, TX 78363
 p. (512) 595-3310
 kftlm00@tamuk.edu
 Enrollment (2016): B: 52 (6)
Professor:
 Thomas L. McGehee, (D), Texas (Dallas), 1987, Cl
Assistant Professor:
 Mark T. Ford, (D), Oregon State, GxzCg
 Brent Hedquist, (D), Arizona State, 2010, ZyAtZi
 Veronica I. Sanchez, (D), Houston, GctGm
 Robert V. Schneider, (D), Texas (El Paso), 1990, GoYes
 Haibin Su, (D), Cincinnati, Zir
 Subbarao Yelisetti, (D), Victoria (Canada), 2014, Ygs
Lecturer:
 Richard M. Parker, (M), Texas A&M, 2000, GgPg

Texas Christian University
Department of Geological Sciences (B,M) ☒ (2018)
TCU Box 298830
2950 West Bowie
Fort Worth, TX 76129
 p. (817) 257-7270
 geology@tcu.edu
 https://geology.tcu.edu/
 Programs: Geology; Applied Geoscience
 Administrative Assistant: Krista Scapelli
 Enrollment (2016): B: 76 (13) M: 30 (6)
TCU Provost and Academic Vice Chancellor:
 R Nowell Donovan, (D), Newcastle upon Tyne, 1972, GdcGt
Chair:
 Helge Alsleben, (D), S California, 2005, GctGg
Professor of Professional Practice:
 Richard A. Denne, (D), Louisiana State, 1990, GoPms
Professor:
 Richard E. Hanson, (D), Columbia, 1983, GxvGt
 John M. Holbrook, (D), Indiana, 1992, Gsr
Associate Professor:
 Arthur B. Busbey, (D), Chicago, 1982, PgvGr
 Xiangyang Xie, (D), Wyoming, 2007, GoEog
Assistant Professor:
 Omar R. Harvey, (D), Texas A&M, 2010, GeHwg

Adjunct Professor:
  Floyd Henk, Jr., (M), Texas Christian, 1981, Go
Emeritus:
  John Breyer, (D), Nebraska, 1977, GsdEo
Professor of Professional Practice:
  Milton Enderlin, (M), Texas Christian, 2010, Nr
  Tamie Morgan, (M), Texas Christian, 1984, Zi

Texas Tech University
Dept of Geosciences (B,M,D) O (2016)
Box 41053
2500 Broadway
Science 125
Lubbock, TX 79409-1053
  p. (806) 834-0497
  alison.winton@ttu.edu
  www.geosciences.ttu.edu
  Enrollment (2015): B: 390 (33) M: 48 (21) D: 23 (0)
Chair:
  Jeffrey A. Lee, (D), Arizona State, 1990, Zy
Horn Professor:
  Sankar Chatterjee, (D), Calcutta, 1970, PvGct
Professor:
  George B. Asquith, (D), Wisconsin, 1966, Go
  Calvin G. Barnes, (D), Oregon, 1982, Gi
  James E. Barrick, (D), Iowa, 1978, Psm
  Gary S. Elbow, (D), Pittsburgh, 1972
  Juske Horita, (D), Texas A&M, 1997
  Thomas M. Lehman, (D), Texas, 1985, Gs
  Moira K. Ridley, (D), Nebraska, 1997, CqlZm
  John L. Schroeder, (D), Texas Tech, 1999, As
  Paul J. Sylvester, (D)
  Aaron S. Yoshinobu, (D), S California, 1999, GctGu
Senior Scientist:
  Melanie A. Barnes, (D), Texas Tech, 2001, CaGie
Associate Professor:
  Eric C. Bruning, (D), Oklahoma, 2008, As
  Perry L. Carter, (D), Ohio State, 1998
  Harold Gurrola, (D), California (San Diego), 1995, Ys
  Callum J. Hetherington, (D), Basel (Switzerland), 2001, GxzCg
  Haraldur R. Karlsson, (D), Chicago, 1988, Cl
  David W. Leverington, (D), Manitoba, 2001, GmZrXg
  Kevin R. Mulligan, (D), Texas A&M, 1997, ZiGmSo
  Seiichi Nagihara, (D), Texas, 1992, Zi
  Dustin E. Sweet, (D), Oklahoma, 2009, GsdGr
  Christopher C. Weiss, (D), Oklahoma, 2004, AssAs
Assistant Professor:
  Brian C. Ancell, (D), Washington, 2006, As
  Guofeng Cao, (D), California (Santa Barbara), 2011, Zi
  Johannes M L Dahl, (D), Ludwig-Maximilians (Germany), 2010, Ams
  Song-Lak Kang, (D), Penn State, 2007, AsHg
Instructor:
  Steven R. Cobb, As
  Linda L. Jones, (M), California (Los Angeles), 1986
  Justin E. Weaver, (M), Texas Tech, 1992, As
Professor:
  Richard E. Peterson, (D), Missouri, 1971, As
Unit Coordinator:
  Alisan C. Sweet, (M), Oklahoma, 2011, Gs
  Debra J. Walker
Senior Technician:
  James M. Browning, (B), Lamar, 1983, Ggx
Senior Business Assistant:
  Alison Winton, (B), Texas Tech, 1992, ZnnZn
Computer Technician:
  Darren W. Hedrick
Academic Advisor:
  Celeste N. Yoshinobu, (M), San Diego State, 1994

Trinity University
Department of Geosciences (B) ☒ (2018)
One Trinity Place #45
San Antonio, TX 78212-7200
  p. (210) 999-7092
  ksurples@trinity.edu
  https://new.trinity.edu/academics/departments/geosciences
  Programs: Geosciences (B); Earth Systems Science (B)
  Enrollment (2018): B: 28 (11)
Chair:
  Kathleen D. Surpless, (D), Stanford, 2001, Gst
Professor:
  Daniel J. Lehrmann, (D), Kansas, 1993, PiGs
  Diane R. Smith, (D), Rice, 1984, Gi
Associate Professor:
  Glenn C. Kroeger, (D), Stanford, 1987, Yg
  Benjamin E. Surpless, (D), Stanford, 1999, GctGi
Assistant Professor:
  Brady A. Ziegler, (D), Virginia Tech, 2018, HwCqGe
Adjunct Professor:
  Leslie F. Bleamaster III, (D), S Methodist, 2003, Xg
Visiting Professor:
  Kurt Knesel, (D), California (Los Angeles), 1998, Gvi
Emeritus:
  Walter Coppinger, (D), Miami (Ohio), 1974, Gcg
  Robert L. Freed, (D), Michigan, 1966, Gz
  Thomas W. Gardner, (D), Cincinnati, 1978, GmtHg

Tyler Junior College
Dept of Geology (2015)
1327 South Baxter Avenue
Tyler, TX 75701
  p. (903) 510-2232
  gbra@tjc.edu
  http://www.tjc.edu/

University of Houston
Allied Geophysical Laboratories (AGL) (B,M,D) O☒ (2018)
SR1 131C
Houston, TX 77204-4231
  p. (713) 743-9150
  rrstewart@uh.edu
  http://www.agl.uh.edu
  Programs: Research and Development in Applied Geophysics especially exploration seismology
Professor:
  Robert R. Stewart, (D), MIT, 1983, YesYu

Dept of Earth and Atmospheric Sciences (B,M,D) O☒ (2018)
Science & Research Building 1
3507 Cullen Blvd, Rm. 312
Houston, TX 77204-5007
  p. (713) 743-3399
  hzhou@uh.edu
  http://www.eas.uh.edu
  Administrative Assistant: Edwina Boateng Kumi
  Administrative Assistant: Jim Parker
  Administrative Assistant: Kirene Ramesar
  Administrative Assistant: Anja Wells
  Enrollment (2016): B: 413 (69) M: 129 (41) D: 128 (10)
Department Chair:
  Hua-Wei Zhou, (D), Caltech, 1989, YseYg
Professor:
  Alan Brandon, (D), Alberta, 1992, GiCac
  John F. Casey, (D), SUNY (Albany), 1980, GtiGu
  John P. Castagna, (D), Texas, 1983, Ye
  Henry S. Chafetz, (D), Texas, 1970, Gds
  Evgeny Chesnokov, (D), Russian Acad of Sci, 1987, YseYx
  Peter Copeland, (D), SUNY (Albany), 1990, CcGt
  Stuart A. Hall, (D), Newcastle, 1976, Ym
  Shuhab D. Khan, (D), Texas (Dallas), 2001, ZrGtYg
  Thomas Lapen, (D), Wisconsin, 2005, Gz
  Aibing Li, (D), Brown, 2000, Ys
  Rosalie F. Maddocks, (D), Kansas, 1965, Pm
  Paul Mann, (D), SUNY (Albany), 1983
  Michael Murphy, (D), California (Los Angeles), 2000, Gct
  Bernhard Rappenglueck, (D), Munich, 1996, As
  Arch M. Reid, (D), Pittsburgh, 1964, GizXm
  William W. Sager, (D), Hawaii, 1983, GtYrm
  Jonathan Snow, (D), MIT/WHOI, 1992, Ca
  Robert Stewart, (D), MIT, 1983, Ye
  John Suppe, (D), Yale, 1969, Gtc

Robert Talbot, (D), Wisconsin, 1981, ActAm
Arthur B. Weglein, (D), CUNY, 1980, Ye
University Distinguished Research Professor:
  Fred Hilterman, (D), Colorado Mines, 1970, Ye
Research Professor:
  Adry Bissada, (D), Washington, 1967, Co
  De-hua Han, (D), Stanford, 1987, Ye
  Leon Thomsen, (D), Columbia, 1969, Ye
Research Associate Professor:
  Donald Van Niewenhuise, (D), South Carolina, 1978, PsGoPe
Associate Professor:
  Regina M. Capuano, (D), Arizona, 1988, HwCqs
  William R. Dupre, (D), Stanford, 1975, On
  Xun Jiang, (D), Caltech, 2006
  Alexander Robinson, (D), California (Los Angeles), 2005, Gc
  Guoquan Wang, (D), Inst of Geology (China), 2001
Research Associate Professor:
  Yongjun Gao, (D), Goettingen, 2004, CstCa
  Virginia Sisson, (D), Princeton, 1985, Gpt
Instructional Assistant Professor:
  Daniel Hauptvogel, (D), CUNY, 2015
  Jennifer N. Lytwyn, (D), Houston, 1993, Gi
Assistant Professor:
  Emily Beverly, (D), Baylor
  Yunsoo Choi, (D), Georgia Tech, 2007
  Qi Fu, (D), Minnesota, 2006, CosCa
  Joel Saylor, (D), Arizona, 2008
  Yuxuan Wang, (D), Harvard, 2005, As
  Julia Wellner, (D), Rice, 2001, GslOu
  Jonny Wu, (D), Royal Holloway Univ U(K), 2010, GtcGo
  Yingcai Zheng, (D), California (Santa Cruz), 2007
Research Scientist:
  Tom Bjorklund, (D), Houston, 2002, Gco
  Martin Cassidy, (D), Houston, 2005, GoCo
  Nikolay Dyaur, (D), Russian Acad of Sci, 1986, Yxs
  Xiangshan Li, (D), Tulane, 2000, Asm
Research Professor:
  James Lawrence, (D), Caltech, 1970, Cl
  Peter Percell, (D), California (Berkeley), 1973, As
Research Associate Professor:
  Dale Bird, (D), Houston, 2004
  Robert Wiley, (D), Colorado Mines, 1980
Research Assistant Professor:
  James Flynn, (M), Houston, 1991
  Charlotte Sjunneskog, (D), Uppsala, 2002
Adjunct Professor:
  Peter Bartok, (M), SUNY (Buffalo), 1972, GoYeNr
Senior Researcher:
  Mike Darnell, (B), Texas A&M, 1974
Postdoctoral Fellow:
  Hao Hu, (D), CAS (China), 2015, Ye
Adjunct:
  Amy Kelly, (D), MIT, 2009
  Gary Morris, (D), Rice, 1995
Researcher:
  Min Sun
  Ewa Szymczyk
  Fuyong Yan
Laboratory Supervisor:
  Minako Righter, (D), Grad Univ for Adv Studies, 2006

University of Houston Downtown
Dept of Natural Sciences (2015)
1 Main Street
Houston, TX 77002
  p. (713) 221-8015
  lyonsp@uhd.edu
Professor:
  Glen K. Merrill, (D), Louisiana State, 1968, Pi
  Penny A. Morris-Smith, (D), California (Berkeley), 1975
Associate Professor:
  Kenneth Johnson, (D), Texas Tech, 1995, GiCac
Lecturer:
  Donald S. Musselwhite, (D), 1995

University of North Texas
Dept of Geography (B,M) (2019)
1155 Union Circle #305279
Denton, TX 76203
  p. (940) 565-2091
  geog@unt.edu
  http://www.geography.unt.edu
  Certificates: GIS
  Administrative Assistant: Tami Deaton
Professor:
  Pinliang Dong, (D), New Brunswick, 2003, Zir
  C. Reid Ferring, (D), Texas (Dallas), 1993, Ga
  Paul F. Hudak, (D), California (Santa Barbara), 1991, Hw
  Joseph R. Oppong, (D), Alberta, 1992, Zn
  Murray Rice, (D), Saskatchewan, 1995, Zn
  Harry F. L. Williams, (D), Simon Fraser, 1989, Gm
  Steve Wolverton, (D), N Texas, 2001, Ga
Associate Professor:
  Waquar Ahmed, (D), Clark, 2007, Zn
  Ipsita Chatterjee, (D), Clark, 2007, Zn
  Matthew Fry, (D), Texas, 2008, Zn
  Kent M. McGregor, (D), Kansas, 1982, AsZg
  Lisa A. Nagaoka, (D), Washington, 2000, Ga
  Feifei Pan, (D), Georgia Tech, 2002, Hw
  Alexandra Ponette-Gonzalez, (D), Yale, 2011, Zyc
  Chetan Tiwari, (D), Iowa, 2008, Zi
Assistant Professor:
  Lu Liang, (D), Zi

University of Texas Rio Grande Valley
School of Earth, Environmental, and Marine Sciences (2015)
Brownsville, TX 78520
  p. (956) 761-2644
  abdullah.rahman@utrgv.edu
  https://www.utrgv.edu/graduate/for-future-students/graduate-programs/program-requirements/ocean-coastal-and-earth-sciences-ms/in

University of Texas, Arlington
Dept of Earth & Environmental Sciences (B,M,D) O (2017)
Box 19049
500 Yates Street
Arlington, TX 76019
  p. (817) 272-2987
  geology@uta.edu
  http://www.uta.edu/ees/
  Enrollment (2014): B: 189 (29) M: 70 (21) D: 28 (3)
Professor:
  Asish Basu, (D), California (Davis), 1975, GxCcs
  Glen Mattioli, (D), Northwestern, 1987, GitYd
  Merlynd K. Nestell, (D), Oregon State, 1966, PmiPs
Associate Professor:
  Qinhong (Max) Hu, (D), Arizona, 1995, GoHwNg
  Andrew Hunt, (D), Liverpool, 1988, GbCg
  Arne M. Winguth, (D), Hamburg, 1997, OpAs
Assistant Professor:
  Majie Fan, (D), Arizona, 2009, GstCs
  Ashley Griffith, (D), Stanford, 2008, Gct
  Liz Griffith, (D), Stanford, 2008, CslCm
  Ashanti Johnson, (D), OgGe
Lecturer:
  Cornelia Winguth, (D), Hamburg, 1998, GugZm
Adjunct Professor:
  John E. Damuth, (D), Columbia, 1973, GusYr
  Galina P. Nestell, (D), VSEGEI (Russia), 1990, PmsPi
Emeritus:
  Brooks Ellwood, (D), Rhode Island, 1977, Yg
  Christopher R. Scotese, (D), Chicago, 1985, Gt
  John S. Wickham, (D), Johns Hopkins, 1969, Gco

University of Texas, Austin
Dept of Marine Science (M,D) ● ☒ (2018)
750 Channel View Drive
Port Aransas, TX 78373-5015
  p. (361) 749-6730

facsearch@utlists.utexas.edu
http://www.utmsi.utexas.edu
f: www.facebook.com/utmsi
Programs: Marine Science
Enrollment (2015): M: 16 (6) D: 16 (1)
Associate Chair:
 Edward J. Buskey, (D), Rhode Island, 1983, Ob
Professor:
 Kenneth H. Dunton, (D), Alaska, 1985, Ob
 Lee A. Fuiman, (D), Michigan, 1983, Ob
 James W. McClelland, (D), Boston, 1998, CgHsCs
 Peter Thomas, (D), Leicester, 1977, Ob
 Tracy A. Villareal, (D), Rhode Island, 1989, Ob
Associate Professor:
 Deana L. Erdner, (D), MIT/WHOI, 1997, Ob
 Andrew J. Esbaugh, (D), Queens, 2005, Ob
 Zhanfei Liu, (D), Stony Brook, 2006, Og
Assistant Professor:
 Brett Baker, (D), Michigan, 2014, Ze
 Brad Erisman, (D), California (San Diego), 2008, Ob
 Amber K. Hardison, (D), William & Mary, 2010, Cm
 Lauren A. Yeager, (D), Florida Intl, 2013, Og
Lecturer:
 Dong-Ha Min, (D), Scripps, 1999, Og
Emeritus:
 Wayne S. Gardner, (D), Wisconsin, 1971, Ob
 Gloria J. Holt, (D), Texas A&M, 1976, Ob

Inst for Geophysics (2017)
JJ Pickle Research Campus
10100 Burnet Road, Bldg. 196 (ROC)
Austin, TX 78758
 p. (512) 471-6156
 utig@ig.utexas.edu
 http://www.ig.utexas.edu/
Associate Director:
 Ian W. D. Dalziel, (D), Edinburgh, 1963, Gt
 Cliff Frohlich, (D), Cornell, 1976, Ys
Research Professor:
 Stephen P. Grand, (D), Caltech, 1986, Ys
 Yosio Nakamura, (D), Penn State, 1963, Ys
 Mrinal K. Sen, (D), Hawaii, 1987, Ye
Professor:
 Sean S. Gulick, (D), Lehigh, 1999, YrGtb
 Paul L. Stoffa, (D), Columbia, 1974, Ye
Research Scientist:
 Gail L. Christeson, (D), MIT/WHOI, 1994, YrsGt
 Craig S. Fulthorpe, (D), Northwestern, 1988, Gu
Senior Scientist:
 James A. Austin, Jr., (D), MIT/WHOI, 1979, Gut
 Nathan L. Bangs, (D), Columbia, 1990, Gu
 John A. Goff, (D), MIT/WHOI, 1990, Yr
 Charles Jackson, (D), Chicago, 1998, ZoOoAt
 Lawrence A. Lawver, (D), California (San Diego), 1976, YrGt
 Paul Mann, (D), SUNY (Albany), 1983, Gt
 Thomas H. Shipley, (D), Rice, 1975, Yr
 Frederick W. Talyor, (D), Cornell, 1979, Pe
Research Scientist:
 Donald D. Blankenship, (D), Wisconsin, 1989, Yg
 Kirk D. McIntosh, (D), California (Santa Cruz), 1992, Yr
 Robert J. Pulliam, (D), California (Berkeley), 1991, Ys
Associate Professor:
 Luc L. Lavier, (D), Columbia, 1999, YgGt
Research Associate:
 John W. Holt, (D), Caltech, 1997, Ye
 David L. Morse, (D), Washington, 1997, Yg
 Robert B. Scott, (D), McGill, 1999, Og
 Roustam K. Seifoullaev, (D), Baku State, 1979, Ye
 Harm Van Avendonk, (D), Scripps, 1998, Yg
Research Professor:
 William E. Galloway, (D), Texas, 1971, GsoGr
Emeritus:
 Milo M. Backus, (D), MIT, 1956, Ye
 Arthur E. Maxwell, (D), Scripps, 1959, Og
Postdoc:
 Christina Holland, (D), S Florida, 2003, Og
 Matthew Hornbach, (D), Wyoming, 2004, Yg

 Timothy Whiteaker, (D), Texas, 2004, Gs
Project Coordinator:
 Patricia E. Ganey-Curry, (B), Texas A&M, 1978, Zn
Program Manager:
 Katherine K. Ellins, (D), Columbia, 1988, Zn
Related Staff:
 Mark Wiederspahn, (B), Bucknell, 1975, Zn

Jackson School of Geosciences (B,M,D) (2017)
Jackson School of Geosciences
2225 Speedway, Stop C1160
Austin, TX 78712-1692
 p. (512) 471-5172
 ckerans@jsg.utexas.edu
 http://www.jsg.utexas.edu
 Enrollment (2012): B: 315 (66) M: 142 (56) D: 157 (12)
Dean, Jackson School of Geosciences:
 Sharon Mosher, (D), Illinois, 1978, Gc
Chair, Department of Geological Sciences:
 Ronald J. Steel, (D), Glasgow (UK), 1970, Gs
Director, Institute for Geophysics:
 Terry Quinn, (D), Brown, 1989, Pe
Director, Bureau of Economic Geology:
 Scott W. Tinker, (D), Colorado, 1996, Go
Acting Director, Energy & Earth Resources Graduate Program:
 William L. Fisher, (D), Kansas, 1961, GsrGo
Professor:
 Jay L. Banner, (D), SUNY (Stony Brook), 1986, Cl
 Christopher J. Bell, (D), California (Berkeley), 1997, Pv
 Philip Bennett, (D), Syracuse, 1988, Hg
 Julia A. Clarke, (D), Yale, 2002, PvgPq
 Mark P. Cloos, (D), California (Los Angeles), 1981, Gcx
 Kerry H. Cook, (D), North Carolina State, 1984, AsPe
 Ian W. D. Dalziel, (D), Edinburgh, 1963, Gt
 Robert E. Dickinson, (D), MIT, 1966, As
 Peter B. Flemings, (D), Cornell, 1990, Gr
 Sergey B. Fomel, (D), Stanford, 2001, YesEo
 Rong Fu, (D), Columbia, 1991, As
 James E. Gardner, (D), Rhode Island, 1993, Gv
 Omar Ghattas, (D), Duke, 1988, ZnYg
 Stephen P. Grand, (D), McGill, 1986, Yg
 Brian Horton, (D), Arizona, 1998, Gs
 Charles Kerans, (D), Carleton, 1982, GsrGo
 Richard A. Ketcham, (D), Texas, 1995, GgCc
 Gary A. Kocurek, (D), Wisconsin, 1980, Gs
 J. Richard Kyle, (D), W Ontario, 1977, EmnCe
 David Mohrig, (D), Washington, 1994, GsmGr
 Timothy B. Rowe, (D), California (Berkeley), 1987, Pv
 Mrinal K. Sen, (D), Hawaii, 1987, Ye
 John M. Sharp, Jr., (D), Illinois, 1974, HqwGe
 Ron J. Steel, (D), Glasgow, 1972, GsoGr
 Daniel Stockli, (D), Stanford, 1999, Gc
 Paul L. Stoffa, (D), Columbia, 1974, Ye
 Robert H. Tatham, (D), Columbia, 1975, Yg
 Clark R. Wilson, (D), California (San Diego), 1975, Yg
 Zong-Liang Yang, (D), Macquarie, 1992, AmsHq
Senior Research Scientist:
 James A. Austin, Jr., (D), MIT/WHOI, 1979, YrGr
 Nathan L. Bangs, (D), Columbia, 1990, Gc
 Donald D. Blankenship, (D), Wisconsin, 1989, Gl
 Gail L. Christeson, (D), MIT, 1994, Yr
 Shirley P. Dutton, (D), Texas, 1986, Gs
 Clifford A. Frohlich, (D), Cornell, 1976, YsGt
 Craig S. Fulthorpe, (D), Northwestern, 1988, GusGr
 John A. Goff, (D), MIT, 1990, Gu
 Bob A. Hardage, (D), Oklahoma State, 1967, Yes
 Susan D. Hovorka, (D), Texas, 1990, Gs
 Michael R. Hudec, (D), Wyoming, 1990, Gc
 Stephen E. Laubach, (D), Illinois, 1986, GcoNp
 Lawrence A. Lawver, (D), Scripps, 1976, Yr
 Robert G. Loucks, (D), Texas, 1976, Gso
 F. Jerry Lucia, (M), Minnesota, 1954, Eo
 Kitty L. Milliken, (D), Texas, 1985, Gd
 Jeffrey G. Paine, (D), Texas, 1991, On
 Stephen C. Ruppel, (D), Tennessee, 1979, GsrGo
 Bridget R. Scanlon, Kentucky, Hg
 Thomas H. Shipley, (D), Rice, 1975, Ys

Frederick W. Taylor, (D), Cornell, 1978, GtmGe
Lesli J. Wood, (D), Colorado State, 1992, Eo
Michael H. Young, (D), Arizona, 1995, SpHwGe
Hongliu Zeng, (D), Texas, 1994, YsGs
Research Scientist:
Peter Eichhubl, (D), California (Santa Barbara), 1997, GcCgNr
Charles S. Jackson, (D), Chicago, 1998, ZoAtOo
Principal Research Scientist:
Patrick Heimbach, (D), Max Planck, 1998, OpGlAs
Associate Professor:
Daniel O. Breecker, (D), New Mexico, 2008, ClsSc
Bayani Cardenas, (D), New Mexico Tech, 2006, Hw
Elizabeth J. Catlos, (D), California (Los Angeles), 2000, GzCg
Marc A. Hesse, (D), Stanford, 2008, GqoYg
Wonsuck Kim, (D), Minnesota, 2007, GsmGr
John Lassiter, (D), California (Berkeley), 1995, Cc
Randall A. Marrett, (D), Cornell, 1990, Gc
Associate Scientist:
Ginny A. Catania, (D), Washington, 2004, Gl
Assistant Professor:
Jaime D. Barnes, (D), New Mexico, 2006, CgsCc
Whitney Behr, (D), S California, 2011, Gc
Joel P. Johnson, (D), MIT, 2007, Gs
Luc L. Lavier, (D), Columbia, 1999, Gt
Jung-Fu Lin, (D), Chicago, 2002, GyYm
Timothy M. Shanahan, (D), Arizona, 2007, PeGsCg
Kyle T. Spikes, (D), Stanford, 2008, Ye
Senior Energy Economist:
Gurcan Gulen, (D), Boston Coll, 1996, Ego
Energy Economist:
Svetlana Ikonnikova, (D), Humboldt (Berlin), 2007, Ego
Research Associate:
Todd Caldwell, (D), Nevada (Reno), 2011, Sp
Sigrid Clift, (B), Texas, 1989, Zg
Brent Elliott, (D), Helsinki, 2001, GzcGt
Andras Fall, (D), Virginia Tech, 2008, CgGcEg
Peter P. Flaig, (D), Alaska (Fairbanks), 2010, GsrEo
Qilong Fu, (D), Regina, 2005, Gsr
Nicholas W. Hayman, (D), Washington, 2003, Ou
Seyyed Abolfazi Hosseini, (D), Tulsa, 2008, Eo
Farzam Javadpour, (D), Calgary, 2006, NpEo
Carey King, (D), Texas, 2004, Eog
Gang Luo, (D), Missouri, Gtc
Lorena G. Moscardelli, (D), Texas, 2007, GmYs
Hardie S. Nance, (M), Texas, 1978, Grc
Maria-Aikaterini Nikolinakou, (D), MIT, 2008, Np
Yuko Okumura, (D), Hawaii, 2005, AsZoOp
Cornel Olariu, (D), Texas (Dallas), 2005, Gr
Mariana Olariu, (D), Texas (Dallas), 2007, GruGs
Christopher Omelon, Cg
Diana C. Sava, (D), Stanford, 2004, Yg
Timothy L. Whiteaker, (D), Texas, 2004, Zi
Brad Wolaver, (D), Texas, 2008, HwGe
Changbing Yang, Hw
Christopher K. Zahm, (D), Colorado Mines, 2002, Yg
Mehdi Zeidouni, (D), Calgary, 2011, Eo
Tongwei Zhang, (D), Chinese Acad of Sci, 1999, Cgs
Senior Lecturer:
Mark A. Helper, (D), Texas, 1985, GctXg
Lecturer:
Hillary C. Olson, (D), Stanford, 1988, Ps
Mary F. Poteet, (D), California (Berkeley), 2001, Pg
Adjunct Professor:
Laurie S. Duncan, (D), Texas
Marcus Gary, (D), Texas, 2009, Hgs
Emeritus:
Daniel Barker, (D), Princeton, 1961, Gi
Robert E. Boyer, (D), Michigan, 1959, Gt
Richard T. Buffler, (D), California (Berkeley), 1967, Gu
William D. Carlson, (D), California (Los Angeles), 1980, GpzCp
Peter T. Flawn, (D), Yale, 1951, Eg
Robert L. Folk, (D), Penn State, 1952, Gd
William E. Galloway, (D), Texas, 1971, Gs
Edward C. Jonas, (D), Illinois, 1954, Gz
Wann Langston, Jr., (D), California (Berkeley), 1952, Pv
Leon E. Long, (D), Columbia, 1959, Cc
Ernest L. Lundelius, (D), Chicago, 1954, PvePg

Earle F. McBride, (D), Johns Hopkins, 1960, Gd
Yosio none Nakamura, (D), Penn State, 1963, YsGt
Douglas Smith, (D), Caltech, 1969, GipGz
James T. Sprinkle, (D), Harvard, 1971, PigGr
Research Scientist:
Fred W. McDowell, (D), Columbia, 1966, Cc
Research Associate Professor:
Sean S. Gulick, (M), Lehigh, 1999, Gc
John W. Holt, (D), Caltech, 1997, GlXyg

University of Texas, Dallas
Dept of Geosciences (B,M,D) O (2019)
Mail Stop ROC21
800 W Campbell Rd
Richardson, TX 75083-3021
  p. (972) 883-2401
  geissman@utdallas.edu
  Programs: No full degrees on line
  Certificates: None
  Administrative Assistant: Gloria J. Eby
  Enrollment (2018): B: 75 (22) M: 45 (11) D: 25 (7)
Professor and Program :
  John Dr. Geissman, (D), Michigan, 1980, GtYm
Director, Center for Lithosphere Studies:
  David Lumley, (D), Stanford, 1995, YsrYn
Professor:
  John F. Ferguson, (D), S Methodist, 1981, Yg
  William I. Manton, (D), Witwatersrand, 1968, Cc
  Robert J. Stern, (D), California (San Diego), 1979, Gt
Associate Professor:
  Tom H. Brikowski, (D), Arizona, 1987, Hw
Assistant Professor:
  Hejun Zhu, (D), Princeton, 2014, Ysx
Research Professor:
  Robert B. Finkelman, (D), Maryland, 1980, GbEcCa
Senior Lecturer:
  William R. Griffin, (D), Texas (Dallas), 2008, Gg
  Ignacio Pujana, (D), Texas (Dallas), 1997, Pm
Retired Professor:
  Richard M. Mitterer, (D), Florida State, 1966, ColOc
  Emile A. Pessagno, Jr., (D), Princeton, 1960, Pm
  Dean C. Presnall, (D), Penn State, 1963, Cp
Retired President and Professor:
  Robert H. Rutford, (D), Minnesota, 1969, Gl
Retired Associate Professor:
  James L. Carter, (D), Rice, 1965, EgmGz
Emeritus:
  Carlos L. V. Aiken, (D), Arizona, 1976, Yv
  George A. McMechan, (M), Toronto, 1972, Ys

Science/Mathematics Education Program (M) (2015)
P. O. Box 830688 FN33
Richardson, TX 75080-9688
  p. (972) 883-2496
  mont@utdallas.edu
  http://www.utdallas.edu/dept/SciMathEd
Professor:
  Thomas R. Butts, (D), Michigan State, 1973, Zn
  Fred L. Fifer, (D), Vanderbilt, 1973, Zn
Associate Professor:
  Cynthia E. Ledbetter, (D), Texas A&M, 1987, Zn
Assistant Professor:
  Homer A. Montgomery, (D), Texas (Dallas), 1988, Pg
  Mary Urquhart, (D), Colorado, 1999, Xy
Instructor:
  Barbara Curry, (M), Texas (Dallas), 1998, Zn

University of Texas, El Paso
Dept of Geological Sciences (B,M,D) O (2019)
500 W. University Avenue
101 Geological Sciences
El Paso, TX 79968-0555
  p. (915) 747-5501
  jdkubicki@utep.edu
  http://science.utep.edu/geology/
  Programs: BA Geology, BS Geology, BS Geophysics, BS

Geology – Secondary Education Minor – 7-12 Science Certification, BS Environmental Science - Biology Concentration, BS Environmental Science - Chemistry Concentration, BS Environmental Science - Geoscience Concentration, BS Environmental Science - Hydroscience Concentration, BS Environmental Science – Secondary Education Minor – 7 – 12 Science Concentration, Fast Track Programs: BS Geology to MS Geology, BS Geology to MBA, Graduate Program: MS Geology, MS Geophysics, MS Environmental Science, PhD Geology
Certificates: Geospatial Information Science and Technology - GIST
Enrollment (2018): B: 171 (0) M: 42 (20) D: 30 (2)
National Seismic Source Facility:
   Galen M. Kaip, (D), Texas (El Paso), 1998, Yx
Professor:
   Elizabeth Y. Anthony, (D), Arizona, 1986, GiCcGv
   Diane I. Doser, (D), Utah, 1984, Ys
   Katherine A. Giles, (D), Arizona, 1991, Go
   Thomas E. Gill, (D), California (Davis), 1995, GmAsCl
   Philip C. Goodell, (D), Harvard, 1970, Ce
   Jose M. Hurtado, (D), MIT, 2002, Gt
   Richard S. Jarvis, (D), Cambridge (UK), 1975, ZyGmAs
   James D. Kubicki, (D), Yale, 1989, ClGe
   Richard P. Langford, (D), Utah, 1989, Gs
   Terry L. Pavlis, (D), Utah, 1982, Gc
   Nicholas E. Pingitore, Jr., (D), Brown, 1973, Cl
   Laura F. Serpa, (D), Cornell, 1986, Ye
   Aaron A. Velasco, (D), California (Santa Cruz), 1993, Yx
Associate Professor:
   Lixin Jin, (D), Michigan, 2007, Ge
   Marianne Karplus, (D), Stanford, 2012, Ysg
   Lin Ma, (D), Michigan, 2008, HgCg
   Deana Pennington, (D), Oregon State, 2002, ZyfZe
Assistant Professor:
   Benjamin Brunner, (D), ETH, 2003, Cs
   Julien Chaput, (D), New Mexico Tech, 2012, GeYes
   Hugo A. Gutierrez-Jurado, (D), New Mexico Tech, 2011, HgsHw
   Jason Ricketts, (D), New Mexico, 2014, GctGi
   Jie Xu, (D), Wisconsin, 2011, ClPoGz
Lecturer:
   Hector Gonzalez-Huizar, (D), Texas (El Paso), 2009, Yg
   Vicki Harder, (D), Texas (El Paso), 1997, Gg
   Musa Hussein, (D), Texas (El Paso), 2007, YgeYm
   Stanley Mubako, (D), S Illinois, 2011, Zi
   Adriana Perez, (D), Texas (El Paso), 2008, Gg
Associate Professor Emeritus:
   William Cornell , (D), California (Los Angeles), Gg
Emeritus:
   Kenneth F. Clark, (D), New Mexico, 1966, Eg
   George Randy Keller , (D), Texas Tech, 1973, Yg
   David V. Le Mone, (D), Michigan State, 1964, Ps
Research Assistant Professor :
   Steven H. Harder, (D), Texas (El Paso), 1986, Yg

University of Texas, Permian Basin
Geology Program (B,M) (2018)
4901 E. University Boulevard
Geology
Odessa, TX 79762-0001
   p. (432) 552-2242
   verma_s@utpb.edu
   Programs: Bachelor in Geology with concentration in 1. Petroleum Geology, 2. General Geology; Masters in Geology: 1. Thesis , 2. Non-Thesis
   Enrollment (2018): B: 20 (0)
Program Head:
   Sumit Verma, (D), Oklahoma, 2015, YeGo
Graduate Program Coordinator:
   Mohamed Zobaa, (D), Missouri S&T, 2011, Pl
Assistant Professor:
   Miles Henderson, (D), Tennessee, 2018, GsdGg
   Joon Heo, (D), Texas A&M, 2012, Hg
Lecturer:
   Robert C. Trentham, (D), Texas (El Paso), 1981, Go

University of Texas, Rio Grande Valley
School of Earth, Environmental and Marine Sciences (B,M)
(2019)
1201 W. University Drive
ESCNE 2.620
Edinburg, TX 78539
   p. (956) 882-5040
   juan.l.gonzalez@utrgv.edu
   https://www.utrgv.edu/seems/
   Programs: BS Environmental Science - Earth and Ocean Sciences Concentration, MS Ocean, Coastal and Earth Sciences, MS Agricultural, Environmental, and Sustainability Sciences
   Enrollment (2018): B: 119 (0) M: 58 (0)
Director:
   David Hicks, (D), Texas (Arlington), 1999, Ob
Professor:
   Carlos Cintra-Buenrostro, (D), Arizona, 2006, PiCsPe
   Abdullah F. Rahman, (D), Zr
Associate Professor:
   Jude Benavides, (D), Rice, 2005, Hsq
   John Breier, (D), Texas, 2006, Cma
   Juan L. Gonzalez, (D), Tulane, 2008, GmsGa
   Elizabeth Heise, (D), Texas A&M, 2001, OnZeGg
Assistant Professor:
   Chu-Lin Cheng, (D), Iowa State, 2009, HgwHq
   Christopher Gabler, (D), Rice, 2012, PeSf
   Cheryl Harrison, (D), California (Santa Cruz), 2012, Op
   James Kang, (D), North Carolina State, 2007, SpcZu
   Brendan Lavy, (D), Texas State, 2017, RnwZg
   Engil I. Pereira, (D), California (Davis), 2014, SbZcSo
Research Associate:
   Lorena Longoria, (B), Texas (Pan Am), 2012, Ob
Lecturer:
   Sarah M. Hardage, (M), New Orleans, 2004, GmZiGg
Adjunct Professor:
   James R. Hinthorne, (D), California (Santa Barbara), 1974, GgzCa

University of Texas, San Antonio
Dept of Geological Sciences (B,M) (2018)
One UTSA Circle
San Antonio, TX 78249-0663
   p. (210) 458-4455
   geosciences@utsa.edu
   http://www.utsa.edu/geosci
   Programs: Geology (B,M); Geoinformatics (M)
   Certificates: GIS (G)
   Enrollment (2017): B: 107 (0) M: 35 (0)
Professor:
   Lance L. Lambert, (D), Iowa, 1992, Pg

Victoria College
Science, Mathematics, & Physical Education (A) (2015)
2200 E. Red River
Victoria, TX 77901
   p. (361)573-3291 (Ext.3432)
   Alisha.Stearman@VictoriaCollege.edu
   https://www.victoriacollege.edu/sciencemathematicsphysicaleducation

Wayland Baptist University
Geology Department (B) (2018)
1900 W. 7th Street
Plainview, TX 79072
   p. (806) 291-1115
   walsht@wbu.edu
   http://www.wbu.edu/academics/schools/math_and_science/geology/default.htm
   f: https://www.facebook.com/Wayland-Baptist-University-Geology-Department-198757676854342/
   Programs: Geology
   Enrollment (2018): B: 4 (1)
Professor:
   Don Parker, (D), GivCg
   Tim R. Walsh, (D), Texas Tech, 2002, GsPmGo

Assistant Professor:
 Mark Bryan, (M), Oklahoma State, 2001, GeoPs

Weatherford College
Weatherford College (2015)
225 College Park Drive
Weatherford, TX 76086
   p. (817) 598-6277
   http://www.wc.edu

West Texas A&M University
Dept of Life, Earth & Environmental Sciences (B,M) (2015)
P. O. Box 60808, WT Station
Canyon, TX 79016-0001
   p. (806) 651-2570
   dsissom@wtamu.edu
   http://www.wtamu.edu/academics/life-earth-environmental-sciences.aspx
   Administrative Assistant: Debi Adams
   Enrollment (2013): B: 13 (2)
Chair:
 David Sissom, (D), Zn
Professor:
 Joseph C. Cepeda, (D), Texas, 1977, GiHw
 David B. Parker, (D), Nebraska, 1996, Sbf
 William J. Rogers, (D), Texas A&M, 1999, Zn
 Gerald E. Schultz, (D), Michigan, 1966, PvGzZg
Associate Professor:
 Gary C. Barbee, (D), Texas A&M, 2004, HwSoZi
Instructor:
 Cindy D. Meador, (M), W Texas A&M, 1989, Zg
 Joe D. Rogers, (M), WTSU (WTAMU), 1987, Ga
 William C. Rogers, (M), WTAMU, 2005, Ge
 Lynn C. Rosa, (M), Oklahoma, 1984, Zge

Western Texas College
Math and Science Dept (2015)
6200 College Avenue
Snyder, TX 79549
   p. (325) 573-8511
   http://wtc.edu/science/index.html
Assistant Professor:
 Troy Lilly, (M), Texas Tech, 1981, GgZg

Wharton County Junior College - Sugarland Campus
Wharton County Junior College - Sugarland Campus (2015)
14004 University Blvd.
Sugarland, TX 77479
   p. (281) 239-1559
   dannyg@wcjc.edu
   http://www.wcjc.edu

Wharton County Junior College - Wharton
Geology (2015)
911 Boling Highway
Wharton, TX 77488
   p. (979) 532-6506
   dannyg@wcjc.edu
   https://www.wcjc.edu/Programs/math-and-science/geology/

Utah

Brigham Young University
Dept of Geography (B) ☒ (2018)
690 SWKT
Provo, UT 84602
   p. (801) 378-3851
   geography@byu.edu
   http://www.geography.byu.edu/
   Programs: Geospatial Science & Technology
   Geospatial Intelligence
   Environmental Systems
   Urban & Regional Planning
   Global Studies
   Travel and Tourism Studies
Chair:
 Ryan R. Jensen, (D), Florida, 2000, Zii
Professor:
 Perry J. Hardin, (D), Utah, 1989, Zr
 Samuel M. Otterstrom, (D), Louisiana State, 1997, Zu
 Matthew J. Shumway, (D), Indiana, 1991, Zn
Associate Professor:
 Matthew F. Bekker, (D), Iowa, 2002, Zy
 Michael J. Clay, (D), California (Davis), 2005
 James A. Davis, (D), Arizona State, 1992, Zn
 Jeffrey O. Durrant, (D), Hawaii, 2001, Ge
 Chad Emmett, (D), Chicago, 1991, Zn
 Clark S. Monson, (D), Hawaii, 2004
 Daniel H. Olsen, (D), Waterloo, 2008
 Brandon Plewe, (D), Buffalo, 1997

Dept of Geological Sciences (B,M) ●◌ (2019)
S 389 ESC
Provo, UT 84602
   p. (801) 422-3918
   geology@byu.edu
   http://www.geology.byu.edu
   Programs: Geology, Earth Space and Science Education
   Enrollment (2018): B: 57 (44) M: 36 (17)
Teaching Professor:
 Randall Skinner, (M), Brigham Young, 1996, Gg
Professor:
 Barry R. Bickmore, (D), Virginia Tech, 2000, ClGz
 Eric H. Christiansen, (D), Arizona State, 1981, GiXgGv
 Ron Harris, (D), London (UK), 1989, Gc
 Jeffrey D. Keith, (D), Wisconsin, 1982, Eg
 Bart J. Kowallis, (D), Wisconsin, 1981, GgCcGz
 John H. McBride, (D), Cornell, 1987, YeGoe
 Stephen T. Nelson, (D), California (Los Angeles), 1991, Cs
 Scott M. Ritter, (D), Wisconsin, 1986, Ps
Research Professor:
 Michael J. Dorais, (D), Georgia, 1987, Gi
 David G. Tingey, (M), Brigham Young, 1989, Zn
Associate Professor:
 Brooks B. Britt, (D), Calgary, 1993, Pg
 Gregory T. Carling, (D), Utah, 2012, HwCts
 Jani Radebaugh, (D), Arizona, 2005, XgGvm
Assistant Professor:
 Samuel M. Hudson, (D), Utah, 2008, GosGr
Visiting Professor:
 R. William Keach II, (M), Cornell, 1986, YeEoGg
Emeritus:
 Myron G. Best, (D), California (Berkeley), Gg
 Dana T. Griffen, (D), Virginia Tech, 1976, Gz
 Lehi F. Hintze, (D), Columbia, 1951, Pg
 Alan L. Mayo, (D), Idaho, 1982, Hw
 Wade E. Miller, (D), California (Berkeley), 1968, Pv
 Thomas H. Morris, (D), Wisconsin, 1986, Gr
 R. Paul Nixon, (D), Brigham Young, 1972, Go
 Morris S. Petersen, (D), Iowa, 1962, Gso
 William R. Phillips, (D), Utah, 1954, Gz

Dept of Plant and Wildlife Sciences (2015)
Provo , UT 84602
   p. 801-422-9389
   pws-grad-secretary@byu.edu
   http://pws.byu.edu/

Salt Lake City Community College-Jordan Campus
Dept of Geosciences (2015)
4600 South Redwood Road
Salt Lake City, UT 84123
   p. (801)957-4150
   adam.dastrup@slcc.edu
   http://www.slcc.edu

Snow College
Dept of Geology (A) (2016)

150 College Ave E
Ephraim, UT 84627
   p. (435) 283-7519
   renee.faatz@snow.edu
   https://www.snow.edu/academics/science_math/geology/index.html
   f: https://www.facebook.com/Snow-College-Geology-616452215106045/
   Enrollment (2016): A: 7 (1)
Chair:
   Renee M. Faatz, (M), Ohio State, 1985, Gg
Assistant Professor:
   Ted L. Olson, (M), Utah, 1976, Ys

Southern Utah University
Dept of Physical Science (B) O (2019)
351 West University Blvd.
Cedar City, UT 84720
   p. (435) 586-7900
   jasonkaiser@suu.edu
   http://www.suu.edu/geology
   f: https://www.facebook.com/SUUGeologyClub
   Programs: B.S. in Geology
   B.A. in Geosciences
   Certificates: GIS
   Administrative Assistant: Rhonda Riley
   Enrollment (2018): B: 44 (8)
Professor:
   Robert L. Eves, (D), Washington State, 1991, CgZeCa
Assistant Professor:
   Jason Kaiser, (D), Oregon State, 2014, GzvCg
   John S. MacLean, (D), Montana, 2009, GctZe
   Grant Shimer, (D), Alaska, Gs

University of Utah
Dept of Atmospheric Sciences (B,M,D) ☒ (2019)
135 S 1460 E, Rm 819
Salt Lake City, UT 84112-0110
   p. (801) 581-6136
   atmos-info@lists.utah.edu
   http://www.atmos.utah.edu
   f: www.facebook.com/UUAtmos
   t: @UofUATMOS
   Programs: Atmospheric Sciences
   Certificates: N/A
   Enrollment (2018): B: 40 (14) M: 11 (6) D: 23 (2)
Chair:
   John Horel, (D), Washington, 1982, AsZf
Professor:
   Timothy J. Garrett, (D), Washington, 2000, As
   Steven Krueger, (D), California (Los Angeles), 1985, As
   John Chun-Han Lin, (D), Harvard, 2003, As
   Gerald Mace, (D), Penn State, 1994, As
   Zhaoxia Pu, (D), Lanzhou, 1997, As
   Jim Steenburgh, (D), Washington, 1995, As
   Edward Zipser, (D), Florida State, 1965, As
Associate Professor:
   Anna Gannet Hallar, (D), Colorado (Boulder), 2003, AspAc
   Kevin D. Perry, (D), Washington, 1995, As
   Thomas Reichler, (D), California (San Diego), 2003, As
   Courtenay Strong, (D), Virginia, 2005, As

Dept of Geography (B,M,D) (2015)
260 S. Central Campus Drive
Rm 270
Salt Lake City, UT 84112-9155
   p. (801) 581-8218
   thomas.kontuly@geog.utah.edu
   http://www.geog.utah.edu
   Administrative Assistant: Susan Van Roosendaal
Chair:
   George F. Hepner, (D), Arizona State, 1979, Zy
Professor:
   Donald R. Currey, (D), Kansas, 1969, Gm
   Thomas M. Kontuly, (D), Pennsylvania, 1978, Zn
   Chung M. Lee, (D), Michigan, 1961, Zu

Assistant Professor:
   Thomas J. Cova, (D), California (Santa Barbara), 1999, Zi
   Richard R. Forster, (D), Cornell, 1997, Zr
Adjunct Professor:
   Jeffrey R. Keaton, (D), Texas A&M, 1988, Zn
   Elliott W. Lips, (M), Colorado State, 1990, Gm
Emeritus:
   Philip C. Emmi, (D), North Carolina, 1979, Zu
   Roger M. Mccoy, (D), Kansas, 1967, Zr
   Merrill K. Ridd, (D), Northwestern, 1963, Zr
Professor-Lecturer:
   Arthur Hampson, (D), Hawaii, 1980, Zy
Assistant Professor:
   Trevor J. Davis, (D), British Columbia, 1999, Zi
Other:
   Fred E. May, (D), Virginia Tech, 1976, Zn

Dept of Geology & Geophysics (B,M,D) ●☒ (2019)
Frederick Albert Sutton Building, Rm 383
115 South 1460 East
Salt Lake City, UT 84112-0102
   p. (801) 581-7062
   gg@utah.edu
   http://www.earth.utah.edu/
   Programs: Geoscience (emphasis in Geology, Environmental Geoscience, or Geophysics)
   Geological Engineering
   Earth Science Composite Teaching
   Earth Science Minor
   Enrollment (2018): B: 135 (40) M: 18 (12) D: 47 (2)
Departmental Chair:
   John M. Bartley, (D), MIT, 1981, GciCc
Professor:
   John R. Bowman, (D), Michigan, 1976, Cs
   Paul D. Brooks, (D), Colorado, 1995, HgCbZg
   Thure E. Cerling, (D), California (Berkeley), 1977, ClsZc
   Marjorie A. Chan, (D), Wisconsin, 1982, Gs
   David A. Dinter, (D), MIT, 1994, GtcGu
   Paul W. Jewell, (D), Princeton, 1989, GmHg
   Cari Johnson, (D), Stanford, 2003, Gs
   William P. Johnson, (D), Colorado, 1993, Ng
   Barbara P. Nash, (D), California (Berkeley), 1971, Gi
   Erich U. Petersen, (D), Michigan, 1983, Em
   Douglas K. Solomon, (D), Waterloo, 1992, Hw
   Michael S. Zhdanov, (D), Moscow State, 1968, Ye
Director, Seismograph Stations:
   Keith Koper, (D), Washington, 1998
Associate Director, Global Change and Sustainability Center:
   Brenda Bowen, (D), Gse
Research Associate Professor:
   Kristine L. Pankow, (D), California, 1999, Ys
Associate Professor:
   Gabriel Bowen, (D), California (Santa Cruz), 2003, Cs
   Michael S. Thorne, (D), Arizona State, 2005, YsxYv
Research Associate Professor:
   James C. Pechmann, (D), Caltech, 1983, YsGt
Research Assistant Professor:
   Diego Fernandez, (D), Buenos Aires, 1991, Cg
Research:
   Alex Gribenko, (D), Yg
Lecturer:
   Holly Godsey, (D), Ze
Assistant Professor:
   Kathleen Allabush, (D), Pe
   Lauren Birgenheier, (D), Nebraska (Lincoln), 2007, Gso
   Randall B. Irmis, (D), California (Berkeley), 2008, Sa
   Sarah Lambart, (D), Cpg
   Fan-Chi Lin, (D), Colorado (Boulder), 2009, Ygs
   Peter C. Lippert, (D), California (Santa Cruz), 2010, YmGtPe
   Lowell Miyagi, (D), California (Berkeley), 2009, Gz
   Jeffrey Moore, (D), California (Berkeley), 2007, NgrGm

Dept of Mining Engineering (B,M,D) (2015)
135 South 1460 East
Room 313
Salt Lake City, UT 84112-0113
   p. (801) 581-7198

mineeng@mines.utah.edu
http://www.mines.utah.edu/mining
Enrollment (2009): B: 48 (11) M: 6 (3) D: 1 (0)
Chair:
  Michael G. Nelson, (D), West Virginia, 1989, Nm
Professor:
  Michael K. McCarter, (D), Utah, 1972, Nm
  William G. Pariseau, (D), Minnesota, 1966, Nr
Adjunct Associate Professor:
  Stephen Bessinger, (D), Nm
Associate Professor:
  Felipe Calizaya, (D), Colorado Mines, 1985, Nm
Adjunct Associate Professor:
  Helmut H. Doelling, (D), Utah, 1964, Gg
  Duane L. Whiting, (B), Utah, 1959, Hw
Adjunct Professor:
  James Donovan, (D), Virginia Tech, 2003, NmrZr
  Krishna P. Sinha, (D), Minnesota, 1979, Nr
  Jeffrey Whyatt, (D), Nm
  Zavis Zavodni, (D), Nm

Energy & Geoscience Institute (M) ● (2016)
423 Wakara Way
Suite 300
Salt Lake City, UT 84108
  p. (801) 581-5126
  egidirector@egi.utah.edu
  http://www.egi.utah.edu
Director:
  Raymond A. Levey, (D), South Carolina, 1981, Eo
Professor:
  Richardson B. Allen, (D), Columbia, 1983, GctZi
  Alastair Fraser, (D), Edinburgh, Gco
  Brian McPherson, (D), Utah, 1996, Ye
  Joseph N. Moore, (D), Penn State, 1975, Gv
  Michal Nemcok, (D), Comenius Univ (Slovakia), 1991, GctGo
  Peter E. Rose, (D), Utah, 1993, Np
  Stuart Simmons, (D), Minnesota, 1986, EmCqHg
  Rasoul Sorkhabi, (D), Japan, 1991, GctGc
  Phillip E. Wannamaker, (D), Utah, 1983, Ye
Scientific Staff :
  Julia Kotulova, (D), Komenius (Slovakia), CoGo
Senior Scientist:
  Bryony Richards-McClung, (D), Go
  Lansing Taylor, (D), Stanford, 1999, Gc
  David Thul, (D), Colorado Mines, 2014, Gg
Associate Professor:
  Glenn W. Johnson, (D), South Carolina, 1997, Gq
  Greg Nash, (D), Utah, Gt
  Marylin Segall, (D), 1991, GeOu
Assistant Professor:
  Shu Jiang , (D), China Geosciences, GoYsNp
  Sudeep Kanungo, (D), Univ Coll (London), Pms
Research Associate:
  Kenneth L. Shaw, (B), British Columbia, 1965, Ye
Instructor:
  William Keach, (M), Cornell, 1986, GoYes
Adjunct Professor:
  Ian Walton, (D), Manchester (UK), 1972, Gq

Utah State University
Department of Geology (B,M,D) ●⌂ (2018)
4505 Old Main Hill
Logan, UT 84322-4505
  p. (435) 797-1273
  geology@usu.edu
  http://geology.usu.edu/
  Programs: Geology; Applied Environmental Geoscience; Earth Science Composite Teaching
  Enrollment (2018): B: 56 (10) M: 21 (4) D: 4 (1)
Head:
  Joel L. Pederson, (D), New Mexico, 1999, Gm
Professor:
  James P. Evans, (D), Texas A&M, 1987, Gc
  Susanne U. Janecke, (D), Utah, 1991, Gt
  W. David Liddell, (D), Michigan, 1979, GsPg
  John W. Shervais, (D), California (Santa Barbara), 1979, Gi
Associate Professor:
  Benjamin J. Burger, Colorado, 2009, PvGs
  Carol M. Dehler, (D), New Mexico, 2001, Gs
  Michelle Fleck, (D), Wyoming, 2001, Ze
  Thomas E. Lachmar, (D), Idaho, 1989, Hw
  Anthony R. Lowry, (D), Utah, 1994, YdsGt
  Tammy M. Rittenour, (D), Nebraska, 2004, GmCcGa
Assistant Professor:
  Alexis K. Ault, (D), Colorado, 2012, GtCc
  Dennis L. Newell, (D), New Mexico, 2007, Cls
Research Associate:
  Kelly K. Bradbury, (D), Utah State, 2012, Gc
Curator of Paleontology, CEU Prehistoric Museum:
  Kenneth Carpenter, (D), Colorado, 1996, PvgPe
Emeritus:
  Donald W. Fiesinger, (D), Calgary, 1975, Gi
  Peter T. Kolesar, (D), California (Riverside), 1973, Cg
  Robert Q. Oaks, Jr., (D), Yale, 1965, GsHwYv

Utah State University Eastern
Dept of Geology (A) ⊠ (2018)
451 East 400 North
Price, UT 84501
  p. (435) 613-5232
  michelle.fleck@usu.edu
  Programs: Associate of Science
  Enrollment (2017): A: 4 (0)
Associate Professor:
  Michelle Cooper Fleck, (D), Wyoming, 2001, GgZy

Utah Valley University
Dept. of Earth Science (B) ○⌂ (2018)
800 West University Parkway
MS 179
Orem, UT 84058
  p. (801) 863-8582
  HORNSDA@uvu.edu
  https://www.uvu.edu/earthscience/index.html
  f: https://www.facebook.com/groups/uvuearthscience/
  Programs: Geology; Geography; Environmental Science; Earth Science Education
  Certificates: GIS
  Enrollment (2018): B: 147 (15)
Professor:
  Michael Bunds, (D), GctRn
  Daniel Horns, (D), NgGeRn
Associate Professor:
  Joel Bradford, (M), Utah, RwGe
  Eddy Cadet, (D), Ge
  James Callison, (D), GeSfo
  Hilary Hungerford, (D), ZuRw
  Daniel Stephen, (D), PiGsPe
  Nathan Toke', (D), Arizona State, 2011, GtmZi
  Weihong Wang, (D), CmGeZi
  Alessandro Zanazzi, (D), CgsPe
Assistant Professor:
  Michael Stearns, (D), California (Santa Barbara), 2014, Gip
  Justin White, ZyiZu

Weber State University
Department of Earth and Environmental Sciences (B) ○⌂ (2019)
1415 Edvalson St. Dept 2507
Ogden, UT 84408-2507
  p. (801) 626-7139
  geosciences@weber.edu
  http://weber.edu/geosciences/
  f: https://www.facebook.com/WSUGeosciences
  Programs: Geology (B); Applied Environmental Geoscience (B); Earth Science Teaching (B)
  Certificates: Geospatial Analysis (GIS)
  Administrative Assistant: Marianne Bischoff
  Enrollment (2018): B: 125 (8)
Chair:
  Richard L. Ford, (D), California (Los Angeles), 1997, GmZeEo

Professor:
  Michael W. Hernandez, (D), Utah, 2004, Zri
  David J. Matty, (D), Rice, 1984, Gig
  Marek Matyjasik, (D), Kent State, 1997, HwGe
  W. Adolph Yonkee, (D), Utah, 1990, Gct
Assistant Professor:
  Elizabeth A. Balgord, (D), Arizona, 2015, Grt
  Carie M. Frantz, (D), S California, 2013, PoCbZn
Adjunct Professor:
  Elise M. Childs, (M), Vanderbilt, 2008, Gg
  Gary S. Davis, (M), Mississippi State, 2007, Gg
  Amanda L. Gentry, (M), Nevada (Las Vegas), 2016, GgsGg
  Gregory B. Nielsen, (D), Utah, 2010, Gg
  Katherine E. Potter, (D), Utah State, 2014, Gig
Emeritus:
  Sydney R. Ash, (D), Reading (UK), 1966, Pb
  Jeffrey G. Eaton, (D), Colorado, 1987, PsGs
  James R. Wilson, (D), Utah, 1976, Gz
Laboratory Director:
  Sara Summers, (M), Notre Dame, 2012, Gg

Vermont

Castleton University
Dept of Natural Sciences (Geology) (B) (2018)
Castleton University
233 South Street
Castleton, VT 05735
  p. (802) 468-1238
  tim.grover@castleton.edu
  http://www.castleton.edu/academics/undergraduate-programs/geology/
  Enrollment (2015): B: 12 (3)
Professor:
  Timothy W. Grover, (D), Oregon, 1988, GptGc
  Helen N. Mango, (D), Dartmouth, 1992, CgGe

Johnson State College
Dept of Environmental & Health Sciences (A,B) (2015)
337 College Hill
Johnson, VT 05656-9464
  p. (802) 635-1325
  Tania.Bacchus@jsc.edu
  Enrollment (2010): B: 37 (11)
Professor:
  Tania S. Bacchus, (D), Maine, 1993, GuAm

Middlebury College
Geology Dept (B) (2016)
276 Bicentennial Hall
Middlebury, VT 05753
  p. (802) 443-5029
  geology_chair@middlebury.edu
  http://www.middlebury.edu/academics/geol
  f: https://www.facebook.com/MiddleburyGeology
  Administrative Assistant: Eileen Brunetto
  Enrollment (2016): B: 30 (15)
Chair:
  David P. West, (D), Maine (Orono), 1993, Gc
Professor:
  Patricia L. Manley, (D), Columbia, 1989, YrGu
  Jeffrey S. Munroe, (D), Wisconsin, 2001, Gln
  Peter C. Ryan, (D), Dartmouth, 1994, Cl
Assistant Professor:
  Will Amidon, (D), Caltech, 2010, CcGmt
  Kristina J. Walowski, (D), Oregon, 2015, Gi
Visiting Professor:
  Thomas O. Manley, (D), Columbia, 1981, Op

Northern Vermont University-Lyndon
Dept of Atmospheric Sciences (B) (2018)
1001 College Road
PO Box 919
Lyndonville, VT 05851
  p. (802) 626-6225
  jason.shafer@northernvermont.edu
  http://meteorology.lyndonstate.edu
  f: https://www.facebook.com/nvuatm/
  t: @lyndonweather
  Enrollment (2018): B: 73 (9)
Chair:
  Janel Hanrahan, (D), Wisconsin, 2010, AsZoRc
Professor:
  Jason Shafer, (D), Utah, 2005, As
Research Associate:
  David Siuta, (D), British Columbia, 2017, AsZn
Adjunct Professor:
  George Loriot, (D), Connecticut, AsZn
Visiting Professor:
  Aaron Preston, (D), Florida State, AsZrAc
Atmospheric Sciences Data Systems Administrator:
  Jason Kaiser, (M), Plymouth State, 2013, AsZn

Norwich University
Dept of Earth and Environmental Sciences (B) (2018)
158 Harmon Drive
Northfield, VT 05663
  p. (802) 485-2304
  rdunn@norwich.edu
  https://www.norwich.edu/programs/geology
  f: https://www.facebook.com/groups/102347383133958/
  Programs: Geology, and Environmental Science
  Enrollment (2018): B: 7 (8)
Professor:
  Richard K. Dunn, (D), Delaware, 1998, GalGs
Associate Professor:
  G. Christopher Koteas, (D), Massachusetts, 2010, GitYh
Assistant Professor:
  Laurie Grigg, (D), Oregon, 2000, PeGnZi
Research Associate:
  George E. Springston, (M), Massachusetts, 1990, GmZiYg
Lecturer:
  Benjamin DeJong, (D), Vermont, 2015, GmHws
Emeritus:
  David S. Westerman, (D), Lehigh, 1972, Gxi

University of Vermont
Dept of Geology (B,M) (2018)
Delehanty Hall
180 Colchester Avenue
Burlington, VT 05405-0122
  p. (802) 656-3396
  geology@uvm.edu
  http://www.uvm.edu/geology
  f: https://www.facebook.com/UVMGeology/
  Programs: Geology
  Administrative Assistant: Robin Hopps
  Administrative Assistant: Srebrenka Mrsic
  Enrollment (2015): B: 48 (0) M: 13 (0)
Chair:
  Andrea Lini, (D), ETH (Switzerland), 1994, CsGn
Professor:
  Paul R. Bierman, (D), Washington, 1993, GmeCc
  John M. Hughes, (D), Dartmouth, 1981, Gz
  Keith A. Klepeis, (D), Texas, 1993, Gct
  Charlotte J. Mehrtens, (D), Chicago, 1979, Gs
Research Associate Professor:
  Andrew W. Schroth, (D), Dartmouth, 2007, ClqCb
Associate Professor:
  Laura E. Webb, (D), Stanford, 1999, GtCcGc
Research Assistant Professor:
  Nicolas Perdrial, (D), Strasbourg (France), 2007, GezCl
Assistant Professor:
  Julia Perdrial, (D), Strasbourg, 2008, ClSc
Lecturer:
  Stephen F. Wright, (D), Minnesota, 1988, Glc
Emeritus:
  David P. Bucke, (D), Oklahoma, 1969, GgdGs
  Barry L. Doolan, (D), SUNY (Binghamton), 1970, Gx
  John C. Drake, (D), Harvard, 1967, Cl

Senior Research Technician:
   Gabriela Mora-Klepeis, (M), Texas, 1992, CacGi

Dept of Plant & Soil Science (B,M,D) (2015)
Jeffords Hall
63 Carrigan Drive
Burlington, VT 05405-1737
   p. (802) 656-2630
   pss@uvm.edu
   http://www.uvm.edu/~pss/

Virginia

Central Virginia Community College
Science, Math, and Engineering (A) (2015)
3506 Wards Road
Lynchburg, VA 24502
   p. (434) 832-7707
   laubj@cvcc.vccs.edu
   http://www.cvcc.vccs.edu/Academics/SME/default.asp
Instructor:
   Mark Tinsley, Zg

College of William & Mary
Dept of Geology (B) ☒ (2018)
PO Box 8795
McGlothlin-Street Hall
Williamsburg, VA 23187-8795
   p. (757) 221-2440
   crroex@wm.edu
   www.wm.edu/as/geology
   Administrative Assistant: Carol Roe
   Enrollment (2018): B: 62 (27)
Chair:
   Gregory S. Hancock, (D), California (Santa Cruz), 1998, Gm
Professor:
   Christopher M. Bailey, (D), Johns Hopkins, 1994, Gc
   Rowan Lockwood, (D), Chicago, 2001, Pei
   R. Heather Macdonald, (D), Wisconsin, 1984, GsZe
   Brent E. Owens, (D), Washington Univ (St. Louis), 1992, Gx
Associate Professor:
   James Kaste, (D), Dartmouth, 2003, ClcSc
Assistant Professor:
   Nicholas Balascio, (D), Massachusetts, 2011, ZyGnl
Research Associate:
   Carl R. Berquist, Jr., (D), William & Mary, 1986, Ou
Lecturer:
   Rebecca Jiron, (D), California (Santa Barbara), 2015, GmtZg
   Linda D. Morse, (B), Virginia Tech, 1983, GeZg
Emeritus:
   Stephen C. Clement, (D), Cornell, 1964, Gz
   P. Geoffrey Feiss, (D), Harvard
   Gerald H. Johnson, (D), Indiana, 1965, Pg

School of Marine Science (M,D) ☒ (2018)
Virginia Institute of Marine Science
P. O. Box 1346
Gloucester Point, VA 23062-1246
   p. (804) 684-7105
   ad-as@vims.edu
   http://www.vims.edu
   Programs: Biological Sciences, Fisheries Science, Aquatic Health Sciences, Physical Sciences (Physical Oceanography, Chemical Oceanography, Marine Geology)
   Certificates: None
   Enrollment (2018): M: 8 (4) D: 10 (3)
Department Chair:
   Carl T. Friedrichs, (D), MIT/WHOI, 1993, On
Research Professor:
   Jian Shen, (D), William & Mary, 1996, Op
Professor:
   Elizabeth A. Canuel, (D), North Carolina, 1992, Oc
   Courtney K. Harris, (D), Virginia, 1999, On
   Steven A. Kuehl, (D), North Carolina State, 1985, Ou
   Harry Wang, (D), Johns Hopkins, 1983, Op

Research Associate Professor:
   William G. Reay, (D), Virginia Tech, 1992, Hg
   Y. Joseph Zhang, (D), Wollongong (Australia), 1996, Opn
Associate Professor:
   Matthew L. Kirwan, (D), Duke, 2007, Gm
Assistant Professor:
   Donglai Gong, (D), Rutgers, 2010, Op
   Christopher J. Hein, (D), Boston, 2012, Ou

George Mason University
Department of Geography and Geoinformation Science (2015)
Fairfax, VA 22030
   p. (703) 993-1210
   ggs@gmu.edu
   https://cos.gmu.edu/ggs/

Dept of Atmospheric, Oceanic, and Earth Sciences (B,M,D) O☒ (2018)
Research Hall 206
4400 University Drive
Fairfax, VA 22030-4444
   p. (703) 993-9587
   ikinter@gmu.edu
   https://cos.gmu.edu/aoes
   Programs: Atmospheric Sciences, Earth Science, Geology, Climate Dynamics
   Administrative Assistant: Maria D'Souza
   Administrative Assistant: Stephanie O'Neill
   Enrollment (2018): B: 69 (30) M: 10 (5) D: 21 (5)
University Professor:
   Edwin K. Schneider, (D), Harvard, 1976, As
   Jagadish Shukla, (D), MIT, 1976, As
Professor:
   Timothy DelSole, (D), Harvard, 1993, As
   Paul A. Dirmeyer, (D), Maryland, 1992, As
   Robert M. Hazen, (D), Harvard, 1975, GzZf
   Linda Hinnov, (D), Johns Hopkins, 1994, GrYg
   Bohua Huang, (D), Maryland, 1992, Op
   Jim Kinter, (D), Princeton, 1984, As
   David M. Straus, (D), Cornell, 1977, As
   Stacey Verardo, (D), CUNY, 1995, Pei
Associate Professor:
   Mark H. Anders, (D), California (Berkeley), 1989, Gc
   Zafer Boybeyi, (D), North Carolina State, 1993, As
   Long S. Chiu, (D), MIT, 1980, AsZrAm
   Barry Klinger, (D), MIT/WHOI, 1992, Op
   Randolph McBride, (D), Louisiana State, 1997, On
   Julia Nord, (D), CUNY (Brooklyn), 1989, Gzg
   Cristiana Stan, (D), Colorado State, 2005, As
Assistant Professor:
   Mbark Baddouh, (D), Wisconsin, 2016, GdzPc
   Natalie Burls, (D), Cape Town, 2010, AsOp
   Geoff Gilleaudeau, (D), Tennessee, 2013, Gs
   Giuseppina Kysar Mattietti, (D), George Washington, 2001, GiZeGa
   Kathy Pegion, (D), George Mason, 2007, AsOp
   Mark Uhen, (D), Michigan, 1996, Pv
Emeritus:
   Richard J. Diecchio, (D), North Carolina, 1980, Gr
   Paul S. Schopf, (D), Princeton, 1978, OpAs

Hampton University
Center for Marine & Coastal Environmental Studies (B) (2015)
100 Queen Street
Hampton, VA 23668
   p. (757) 727-5783
   cas@hamptonu.edu
   Enrollment (2010): B: 32 (9)
Chair:
   George P. Burbanck, (D), Delaware, 1981, On
Professor:
   Benjamin E. Cuker, (D), North Carolina State, 1981, Ob
Associate Professor:
   Robert A. Jordan, (D), Michigan, 1970, Ob
Assistant Professor:
   Deidre M. Gibson, (D), Georgia, 2000, Ob

Adjunct Faculty:
  Emory Morgan, (M), Ze
Related Staff:
  Gary Morgan, (B), North Carolina State, 2007, ZnnZn

James Madison University
Dept of Geology & Environmental Science (B) (2015)
MSC 6903
Memorial Hall
Harrisonburg, VA 22807
  p. (703) 568-6130
  ulansksl@jmu.edu
  http://www.jmu.edu/geology/
  Department Secretary: Sandra Delawder
Head:
  Stanley L. Ulanski, (D), Virginia, 1977, Og
Professor:
  Roddy V. Amenta, (D), Bryn Mawr, 1971, Gc
  Lynn S. Fichter, (D), Michigan, 1972, Gr
  William C. Sherwood, (D), Lehigh, 1961, So
  Steven J. Whitmeyer, (D), Boston, 2004, GctZi
Associate Professor:
  Steven J. Baedke, (D), Indiana, 1998, Hy
  Lewis S. Eaton, (D), Virginia, 1999, Gm
  Eric J. Pyle, (D), Georgia, 1995, Ze
  Kristen E. St. John, (D), Ohio State, 1998, Ou
Emeritus:
  Lance E. Kearns, (D), Delaware, 1977, Gz

Mary Washington College
Dept of Geography (B) (2015)
1301 College Avenue
Fredericksburg, VA 22401-5358
  p. (540) 654-1470
  shanna@umw.edu
  http://www.mwc.edu/geog
  Enrollment (2006): B: 80 (0)
Chair:
  Joseph W. Nicholas, (D), Georgia, 1991, Zy
Associate Professor:
  Donald N. Rallis, (D), Penn State, 1992, Zn
Assistant Professor:
  Dawn S. Bowen, (D), Queen's, 1998, Zn
  Stephen P. Hanna, (D), Kentucky, 1997, Zr
  Farhang Rouhani, (D), Arizona, 2001, Zn

Mountain Empire Community College
Mountain Empire Community College (2015)
3441 Mountain Empire Road
Big Stone Gap, VA 24219
  p. (276) 523-7460
  creynolds@mecc.edu
  http://www.me.vccs.edu/

Northern Virginia Community College - Alexandria
Geology Program (A) (2017)
5000 Dawes Avenue
Alexandria, VA 22311-5097
  p. 703.845.6507
  vzabielski@nvcc.edu
  http://www.nvcc.edu/campuses-and-centers/alexandria/academic-divisions/science/geology.html
Assistant Dean of Geology:
  Victor Zabielski, (D), Brown, 2001, GgOg

Northern Virginia Community College - Annandale
Geology Program (A) (2017)
8333 Little River Turnpike
Annandale, VA 22003
  p. (703) 323-3276
  cbentley@nvcc.edu
  http://www.nvcc.edu/campuses-and-centers/annandale/academic-divisions/math-science--engineering/gol.html
  Programs: Science (B)
  Enrollment (2011): A: 10 (10)

Professor:
  Kenneth Rasmussen, (D), North Carolina, 1989, Gus
Assistant Professor:
  Callan Bentley, (M), Maryland, 2004, Gg
  Shelley Jaye, (M), Wayne State, 1984, Giy

Northern Virginia Community College - Loudoun Campus
Mathematics, Sciences, Technologies and Business Div - Dept of Geology (A) (2018)
21200 Campus Drive
Sterling, VA 20164
  p. (703) 450-2612
  wbour@nvcc.edu
  https://www.nvcc.edu/academics/divisions/mstb/index.html
  Programs: Associates degree in Science
Associate Professor:
  William Bour, (M), George Washington, 1993, Gge
  William Straight, (D), North Carolina State, GgPv
Assistant Professor:
  Okia Ikwuazorm, (M), Gg

Northern Virginia Community College - Woodbridge
Geology Program (A) (2018)
2645 College Drive
Woodbridge, VA 22191
  p. (703) 878-5614
  eburtis@nvcc.edu
  http://www.nvcc.edu/woodbridge/divisions/natural.html
  Enrollment (2014): A: 6 (1)
Instructor:
  Erik Burtis, (M), Montana, GgcGi

Old Dominion University
Dept of Ocean, Earth & Atmospheric Sciences (B,M,D) (2018)
4600 Elkhorn Avenue
Norfolk, VA 23529-0496
  p. (757) 683-4285
  OEASadmin@odu.edu
  http://www.odu.edu/oeas
  Programs: BS in Ocean and Earth Sciences; MS in Ocean and Earth Sciences; PhD in Oceanography
  Certificates: GIS; Ocean Modeling; Marine Science Technology (scheduled for late 2019)
  Enrollment (2018): B: 67 (13)
Chair:
Fred C. Dobbs, (D), Florida State, 1987, Ob
Professor:
  David J. Burdige, (D), California (San Diego), 1983, OcCbo
  Gregory A. Cutter, (D), California (Santa Cruz), 1982, OcCtHs
  H. Rodger Harvey, (D), 1985, Com
  Eileen E. Hofmann, (D), North Carolina State, 1980, Op
  John M. Klinck, (D), Iowa, 1980, Op
  Margaret Mulholland, (D), Maryland, 1998, Ob
Associate Professor:
  Alexander Bochdansky, (D), Memorial, 1997, Ob
  Jennifer Georgen, (D), MIT/WHOI, 2001, Gu
  Nora K. Noffke, (D), Oldenburg, 1997, GsPoGu
  Matthew Schmidt, (D), California, 2005, Ou
  G. Richard Whittecar, Jr., (D), Wisconsin, 1979, GmHwZe
Assistant Professor:
  P. Dreux Chappell, (D), MIT/WHOI, 2009, Oc
Emeritus:
  Larry P. Atkinson, (D), Dalhousie, 1972, Op
  Dennis A. Darby, (D), Wisconsin, 1971, Ou

Patrick Henry Community College
Art, Science, Business, and Technology Dept (A) (2015)
645 Patriot Avenue
Martinsville, VA 24112
  bdooley@ph.vccs.edu
  http://www.ph.vccs.edu/
Assistant Professor:
  Brett Dooley, (M), Virginia Tech, 2005, Gg

Piedmont Virginia Community College
Associate of Science (2015)
501 College Drive
Charlottesville, VA 22902
p. (434) 961-5446
khudson@pvcc.edu
http://www.pvcc.edu/

Radford University
Dept of Geology (B) (2015)
Box 6939
Radford University
Radford, VA 24142
p. (540) 831-5652
geology@radford.edu
http://www.radford.edu/~geol-web
Administrative Assistant: Theresa Gawthrop
Enrollment (2013): B: 62 (12)

Professor:
Rhett B. Herman, (D), Montana State, 1996, YuXaGa
Parvinder S. Sethi, (D), North Carolina State, 1994, Gz
Chester F. Watts, (D), Purdue, 1983, NgrHw
Director, Museum of the Earth Sciences:
Stephen W. Lenhart, (D), Kentucky, 1985, Pg
Associate Professor:
Elizabeth McClellan, (D), Tennessee, GipGd
Jonathan L. Tso, (D), Virginia Tech, 1987, Gcp
Emeritus:
Robert C. Whisonant, (D), Florida State, 1967, GsaGg

Randolph-Macon College
Environmental Studies Program (B) ☒ (2018)
P.O. Box 5005
Ashland, VA 23005
p. (804) 752-3745
mfenster@rmc.edu
http://www.rmc.edu/academics/environmental-studies.aspx
Programs: Environmental Studies major with an Area of Expertise in Geology
Enrollment (2018): B: 31 (15)

Watts Professor of Science:
Michael S. Fenster, (D), Boston, 1995, OnGue
Adjunct Professor:
Charles Saunders, (M), East Carolina, 1990, ZgGrPg

Rappa Hannock Community College
Rappa Hannock Community College (2015)
12745 College Drive
Glenns, VA 23149
p. (804) 435-8970
babdul-malik@rappahannock.edu
http://www.rappahannock.edu

The University of Virginia's College at Wise
Dept of Natural Science (B) ☒ (2018)
1 College Avenue
Wise, VA 24293
p. (276) 328-0203
maw4v@uvawise.edu
https://www.uvawise.edu/academics/department-natural-sciences/
f: https://www.facebook.com/UVaWiseNaturalSciences
t: @UVaWiseNatSci
Programs: BA in Earth Science
Environmental Science, geology track
Administrative Assistant: Alan West

Instructor:
Robert D. VanGundy, (M), North Carolina, 1983, GeHgZg

Tidewater Community College
Geophysical Sciences Dept (A) ☒ (2018)
1700 College Crescent
Virginia Beach, VA 23453
p. (757) 822-7089
RClayton@tcc.edu
Programs: Physical and Historical Geology, Oceanography I & II, Earth Science
Certificates: GIS and Remote Sensing

Professor:
Rodney L. Clayton, (D), NCU, 2016, GgOg
Instructor:
Mike Lyle, Gg
John Waugh, Gg

University of Lynchburg
Environmental Sciences, Studies, and Sustainability (B,M) O☒ (2018)
1501 Lakeside Dr.
Lynchburg, VA 24501
p. (434) 544-8415
haiar@lynchburg.edu
https://www.lynchburg.edu/academics/majors-and-minors/environmental-science/
Programs: Environmental Science; Environmental Studies
Enrollment (2013): B: 40 (10)

Professor:
David R. Perault, (D), Oklahoma, 1998, Ge
Associate Professor:
Brooke Haiar, (D), Oklahoma, 2008, PgGeg

University of Mary Washington
Dept of Earth and Environmental Sciences (B) ●☒ (2017)
1301 College Avenue
Fredericksburg, VA 22401-5358
p. (540) 654-1016
jhayob@umw.edu
http://cas.umw.edu/ees/
Programs: Geology; Environmental Geology; Environmental Science-Natural Track; Environmental Science-Social Track; Interdisciplinary Science Studies
Certificates: GIS
Enrollment (2016): B: 45 (16)

Chair:
Jodie Hayob, (D), Michigan, 1994, GxzGg
Professor:
Grant R. Woodwell, (D), Yale, 1985, Gc
Associate Professor:
Ben O. Kisila, (D), Arkansas, 2002, HsSfZu
Melanie Szulczewski, (D), Wisconsin, 1999, ScClZg
Charles Whipkey, (D), Pittsburgh, 1999, Cg
Instructor:
Sarah A. Morealli, (M), Pittsburgh, 2010, Gg
Emeritus:
Michael L. Bass, (D), Virginia Tech, 1976, Sf
Robert L. McConnell, (D), California (Santa Barbara), 1972, Ge

University of Virginia
Blandy Experimental Farm (2015)
400 Blandy Farm Lane
Boyce, VA 22620
p. (540) 837-1758
Blandy@virginia.edu
http://blandy.virginia.edu/

Director:
David E. Carr, (D), Maryland, 1990, Zn
Associate Director:
Kyle J. Haynes, (D), Louisiana State, 2004, Zn

Dept of Environmental Sciences (B,M,D) ☒ (2018)
Clark Hall
291 McCormick Road
Box 400123
Charlottesville, VA 22904
p. (804) 924-7761
cba4a@virginia.edu
http://www.evsc.virginia.edu
Programs: MA, MS and PhD in Environmental Sciences.
Enrollment (2018): B: 170 (91) M: 17 (6) D: 31 (9)

Chair:
    Michael L. Pace, (D), Georgia, 1981, Zn
Research Full Professor:
    Peter Berg, (D), Tech (Denmark), 1988, So
    William Keene, (M), Virginia, 1981, As
    G. Carleton Ray, (D), Columbia, 1960, Ob
Professor:
    Lawrence Band, (D), California, 1983
    Paolo D'Odorico, (D), Padova, 1998, Hg
    Robert Davis, (D), Delaware, 1988, AsmZy
    Stephan F. J DeWekker, (D), British Columbia, 2002, As
    Scott Doney, (D), M.I.T., 1991, Oc
    Howard E. Epstein, (D), Colorado State, 1997, Zn
    James N. Galloway, (D), California (San Diego), 1972, As
    Janet S. Herman, (D), Penn State, 1982, ClHw
    Alan D. Howard, (D), Johns Hopkins, 1970, GmXg
    Deborah Lawrence, (D), Duke, 1998, ZcoZg
    Manuel Lerdau, (D), Stanford, 1994, Zn
    Stephen A. Macko, (D), Texas, 1981, CosOc
    Karen McGlathery, (D), Cornell, 1992, Ob
    Aaron L. Mills, (D), Cornell, 1975, So
    Herman H. Shugart, Jr., (D), Georgia, 1971, Zg
    David E. Smith, (D), Texas A&M, 1982, Ob
    Vivian E. Thomson, (D), Virginia, 1997, Zn
    Patricia L. Wiberg, (D), Washington, 1987, Og
Research Associate Professor:
    Linda K. Blum, (D), Cornell, 1980, Sb
    David E. Carr, (D), Maryland, 1990, Zn
    Kyle J. Haynes, (D), Louisiana State, 2004, Zn
    Jennie L. Moody, (D), Michigan, 1986, As
    John H. Porter, (D), Virginia, 1988, Zif
Associate Professor:
    Matthew A. Reidenbach, (D), Stanford, 2004, Hg
    Todd M. Scanlon, (D), Virginia, 2002, Hg
    Thomas M. Smith, (D), Tennessee, 1982, Zg
Research Assistant Professor:
    Karen C. Rice, (D), Virginia, 2001, HgCg
Assistant Professor:
    Max Castorani, (D), California, 2014
    Kevin M. Grise, (D), Colorado State, 2011, As
    Sally Pusede, (D), California, 2014, As
    Lauren Simkins
    Xi Yang, (D), Brown, 2014
Emeritus:
    Bruce W. Nelson, (D), Illinois, 1955, Gs
    Wallace E. Reed, (D), Chicago, 1967, Zr
    William F. Ruddiman, (D), Columbia, 1969, Gu

Shenandoah Watershed Study (B,M,D) (2015)
Department of Environmental Sciences
Clark Hall
Charlottesville, VA 22904
    p. (434) 924-3382
    tms2v@virginia.edu
    http://people.virginia.edu/~swas/POST/scripts/overview.php
Associate Professor:
    Todd M. Scanlon, (D), Virginia, 2002, Hgs
Research Scientist:
    Ami L. Riscassi, (D), Virginia, 2009

Virginia Coast Reserve Long Term Ecological Research ☒ (2017)
291 McCormick Road
P.O. Box 400123
Charlottesville, VA 22904-4123
    p. 434-924-0558
    kjm4k@virginia.edu
    http://www.vcrlter.virginia.edu
Professor:
    Karen J. McGlathery, (D), Cornell, 1992, Obn

Virginia Highlands Community College
Virginia Highlands Community College (2015)
100 VHCC Drive
Abingdon, VA 24212
    p. (276)739-2433
    jsurber@vhcc.edu
    http://www.vhcc.edu

Virginia Polytechnic Institute & State University
Dept of Civil & Environmental Engineering (B,M,D) ☒ (2018)
750 Drillfield Drive
200 Patton Hall
Blacksburg, VA 24061-0105
    p. (703) 231-6635
    mwiddows@vt.edu
    http://www.cee.vt.edu/
    Administrative Assistant: Beth Lucas
Graduate Chair:
    Mark A. Widdowson, (D), Auburn, HwSpHq
Professor:
    Stanley Grant, (D), Caltech, Hs
    Jennifer Irish, (D), Delaware, NoOg
Associate Professor:
    Randel Dymond, (D), Penn State, HqsZi
    Erich Hester, (D), North Carolina, Hsw
    Kyle Strom, (D), Iowa, HsGs
Assistant Professor:
    Megan Rippy, (D), Scripps, Rw

Dept of Crop & Soil Environmental Sciences (B,M,D) (2017)
240 Smyth Hall
Blacksburg, VA 24061-0404
    p. (540) 231-6305
    tlthomps@vt.edu
    http://www.cses.vt.edu
    Department Secretary: Nancy Shields
Head:
    John R. Hall, III, (D), Ohio State, 1971, So
Professor:
    Marcus M. Alley, (D), Virginia Tech, 1975, Sc
    James C. Baker, (D), Virginia Tech, 1978, Sd
    Walter L. Daniels, (D), Virginia Tech, 1985, SdGeCg
    Stephen J. Donohue, (D), Purdue, 1974, Sc
    Gregory K. Evanylo, (D), Georgia, 1982, Sc
    Charles Hagedorn, (D), Iowa, 1974, Sb
    Gregory L. Mullins, (D), Purdue, 1985, Sc
    Raymond B. Reneau, (D), Florida, 1969, Sc
    Lucian W. Zelazny, (D), Virginia Tech, 1970, Sc
Associate Professor:
    Duane F. Berry, (D), Michigan State, 1984, Sb
    Matthew J. Eick, (D), Delaware, 1995, Sc
    Naraine Persaud, (D), Florida, 1978, Sp
Assistant Professor:
    John M. Galbraith, (D), Cornell, 1997, Sp
    Carl E. Zipper, (D), Virginia Tech, 1986, Ze
Adjunct Professor:
    Domy C. Adriano, (D), Kansas State, 1970, Sc
    V. C. Baligar, (D), Mississippi State, 1975, Sc
    Pamela J. Thomas, (D), Virginia Tech, 1998, Sd

Dept of Geography (B,M,D) (2015)
115 Major Williams Hall
Blacksburg, VA 24061-0115
    p. (540) 231-7557
    carstens@vt.edu
    Enrollment (2010): B: 170 (49) M: 17 (9) D: 8 (0)
Head:
    Laurence W. Carstensen, (D), North Carolina, 1981, Zi
Professor:
    James B. Campbell, (D), Kansas, 1976, Zr
Associate Professor:
    Lawrence S. Grossman, (D), Australian Nat, 1979, Zn
    Lisa M. Kennedy, (D), Tennessee, Pe
    Resler M. Lynn, (D), Texas State, Zy
Instructor:
    David Carroll, (M), Mississippi State, Am

Dept of Geosciences (B,M,D) (2019)
4044 Derring Hall
Blacksburg, VA 24061
    p. (540) 231-6521
    wstevenh@vt.edu

http://www.geos.vt.edu
Enrollment (2018): B: 66 (26) M: 14 (6) D: 40 (6)
Department Head:
  W. Steven Holbrook, (D), Stanford, 1989, Yur
Research Professor:
  Martin C. Chapman, (D), Virginia Tech, 1998, Ys
  Robert P. Lowell, (D), Oregon State, 1972, Yrh
Professor:
  Robert J. Bodnar, (D), Penn State, 1985, CgEmGv
  Thomas J. Burbey, (D), Nevada (Reno), 1994, HwyZr
  Patricia M. Dove, (D), Princeton, 1991, Cl
  Kenneth A. Eriksson, (D), Witwatersrand, 1977, Gs
  Michael F. Hóchella, Jr., (D), Stanford, 1981, ClGze
  John A. Hole, (D), British Columbia, 1993, Ye
  Scott D. King, (D), Caltech, 1990, Yg
  Richard D. Law, (D), London, 1981, Gc
  Nancy L. Ross, (D), Arizona State, 1985, Gz
  Madeline E. Schreiber, (D), Wisconsin, 1999, Hw
  James A. Spotila, (D), Caltech, 1998, Gtm
  Shuhai Xiao, (D), Harvard, 1998, Pi
Associate Collegiate Professor:
  John A. Chermak, (D), Virginia Tech, 1989, ClGeEd
Associate Professor:
  Mark Caddick, (D), Cambridge, 2005, GpCa
  Benjamin Gill, (D), California (Riverside), 2009, CsPcGs
  Brian W. Romans, (D), Stanford, 2008, GsrEo
  Robert Weiss, (D), Westfalische–Wilhelms, 2005, RnOn
  Ying Zhou, (D), Princeton, 2004, Ys
Assistant Professor:
  Christina Dura, (D), Pennsylvania, 2014, OnRn
  F. Marc Michel, (D), Stony Brook, 2007, ZaGz
  Sterling J. Nesbitt, (D), Columbia, 2009, Pv
  Ryan M. Pollyea, (D), Idaho, 2012, Hq
  D. Sarah Stamps, (D), Purdue, 2013, YdGtRn
  Michelle Stocker, (D), Texas, 2013, Pv
Sr. Research Associate:
  Jing Zhao, (D), Chinese Acad of Sci, 1997
Research Associate:
  Nizhou Han, (D), Iowa State, 1996, Sc
Advanced Instructor:
  Neil E. Johnson, (D), Virginia Tech, 1986, GzEg
Adjunct Professor:
  James S. Beard, (D), California (Davis), 1985, GiClg
  Benedetto DeVivo, (D)
  William S. Henika, (M), Virginia, 1969, Ng
Emeritus:
  F. Donald Bloss, (D), Chicago, 1951, Gz
  G. A. Bollinger, (D), St. Louis, 1967, Ys
  Cahit Coruh, (D), Istanbul, 1970, YesGt
  James R. Craig, (D), Lehigh, 1965, Eg
  Jerry Gibbs, (D), Penn State, Gz
  Gordon C. Grender, (D), Penn State, 1960, Go
  David A. Hewitt, (D), Yale, 1970, Cp
  Wallace D. Lowry, (D), Rochester, 1943, Gc
  J. Fred Read, (D), W Australia, 1971, Gs
  Paul H. Ribbe, (D), Cambridge, 1963, Gz
  J. Donald Rimstidt, (D), Penn State, 1979, Clg
  Edwin S. Robinson, (D), Wisconsin, 1964, Yg
  A. Krishna Sinha, (D), California (Santa Barbara), 1969, Gt
  J. Arthur Snoke, (D), Yale, 1969, Ys
A/P Faculty:
  Gary Glesener, (M), California (Los Angeles), 2016, Ze
Educational Administrator for Outreach:
  S. Llyn Sharp, (M), Virginia Tech, 1990, Ze
Cooperating Faculty:
  Edward Lener, (M), Virginia Tech, 1997, Zn

Virginia Wesleyan College
Dept of Earth and Environmental Sciences (B) (2015)
1584 Wesleyan Dr
Norfolk, VA 23502
  p. 757.455.3200
  emalcolm@vwu.edu
  http://www.vwc.edu/earth-and-envirmental-sciences/
  f: https://www.facebook.com/vwc.ees
  Enrollment (2015): B: 22 (0)

Professor:
  John C. Haley, (D), Johns Hopkins, 1986, GgeZi
Associate Professor:
  Elizabeth Malcolm, (D), Michigan, 2002, AsCgOg
  Garry Noe, (D), California (Riverside), 1982, HsZin

Virginia Western Community College
Geology Dept (A) (2015)
3095 Colonial Avenue
Roanoke, VA 24038
  p. (540) 857-7273
  abalog-szabo@virginiawestern.edu
Professor:
  Anna Balog-Szabo, (D), Virginia Tech, 1996, Gs

Washington & Lee University
Dept of Geology (B) ☒ (2018)
204 West Washington Street
Lexington, VA 24450
  p. (540) 458-8800
  geology@wlu.edu
  http://geology.wlu.edu
  Programs: Geology
  Administrative Assistant: Sarah Barrash. Wilson
  Enrollment (2017): B: 28 (15)
Professor:
  Christopher Connors, (D), Princeton, 1999, GcYeGo
  Lisa Greer, (D), Miami, 2001, GsPe
  David J. Harbor, (D), Colorado State, 1990, Gm
  Elizabeth P. Knapp, (D), Virginia, 1997, Cl
  Jeffrey Rahl, (D), Yale, 2005, Gt
Emeritus:
  Frederick Lyon Schwab, (D), Harvard, 1968, Gdg
  Edgar W. Spencer, (D), Columbia, 1957, GcgZe

Washington

Bellevue College
Earth and Space Sciences Program (A) (2015)
3000 Landerholm Circle SE
Bellevue, WA 98007
  p. (425) 564-3158
  kshort@bellevuecollege.edu
  http://scidiv.bellevuecollege.edu/
Instructor:
  Cary Easterday, (M), GgPg
  Gwyn Jones, (M), Gg
  Deborah Minium, (M), Gg
  Rob Viens, (D), Washington, GgZg

Central Washington University
Dept of Geological Sciences (B,M) ●⚆ (2018)
400 East University Way
MS 7418
Ellensburg, WA 98926-7418
  p. (509) 963-2701
  chair@geology.cwu.edu
  http://www.geology.cwu.edu
  Programs: Geology (B,M); Environmental Geology (B)
  Enrollment (2018): B: 46 (23) M: 18 (9)
Chair:
  Chris Mattinson, (D), Stanford, 2006, GpCcGt
Director, PANGA Laboratory:
  Timothy I. Melbourne, (D), Caltech, 1998, Yd
Professor:
  Wendy A. Bohrson, (D), California (Los Angeles), 1993, GivCc
  Lisa L. Ely, (D), Arizona, 1992, Gm
  Carey A. Gazis, (D), Caltech, 1994, CsHw
  Jeffrey Lee, (D), Stanford, 1990, Gct
Senior Scientist:
  Angela Halfpenny, (D), Liverpool, 2007, Gcz
Associate Professor:
  Anne E. Egger, (D), Stanford, 2010, GtZeRn
  Audrey Huerta, (D), MIT, 1998, Ygd
  Susan Kaspari, (D), Maine, 2007, PcAtGl

Breanyn MacInnes, (D), Washington, 2010, GseGu
GPS Data Analyst:
    Marcelo Santillan, (M), Memphis, 2003, Yd
Assistant Professor:
    Walter Szeliga, (D), Colorado, 2010, YgsYd
Research Associate:
    Paul Winberry, Penn State, 2008, Yg
Science Outreach & Education Coordinator:
    Nick Zentner, (M), Idaho State, 1989, Gg
Lecturer:
    Keegan Fengler, (M), C Washington, Gg
    Winston Norrish, (D), Cincinnati, 1990, GgeGo
Emeritus:
    Robert Bentley, (D), Columbia, 1969, Gi
    Steven Farkas, (D), New Mexico, 1965, Gs
    James R. Hinthorne, (D), California (Santa Barbara), 1974, Ca
    M. Meghan Miller, (D), Stanford, 1987, Yd
    Charles M. Rubin, (D), Caltech, 1990, Gt
Systems Administrator:
    Craig Scrivner, (D), Caltech, 1998

Centralia College
Earth Sciences Program (A) ⊘ (2018)
600 Centralia College Blvd
Centralia, WA 98531
    p. (360) 623-8472
    michelle.harris@centralia.edu
    http://www.centralia.edu/academics/earthscience
    Programs: Geology and Earth Science
    Enrollment (2018): A: 3 (0)
Assistant Professor:
    Michelle Harris, (M), Gg
Emeritus:
    Patrick T. Pringle, (M), Akron, 1982, ZgGvZn

Eastern Washington University
Dept of Geology (B) O☒ (2018)
140 Science Building
Cheney, WA 99004-2439
    p. (509) 359-2286
    geology@ewu.edu
    https://www.ewu.edu/cstem/departments/geology
    Programs: Geology; Environmental Science; Earth and Space Science Education
    Enrollment (2017): B: 33 (19)
Chair:
    Jennifer A. Thomson, (D), Massachusetts, 1992, Gpz
Professor:
    John P. Buchanan, (D), Colorado State, 1985, Hw
    Carmen A. Nezat, (D), Michigan, 2006, ClGeSc
Associate Professor:
    Richard L. Orndorff, (D), Kent State, 1994, Ge
Assistant Professor:
    Chad Pritchard, (D), Washington State, Gct
Lecturer:
    Jeanne Case, (M), California (Riverside), Gg
    Sharen Keattch, (M), Kent State, 1993, Gg
Emeritus:
    Ernest H. Gilmour, (D), Montana, 1967, Pi
    Eugene P. Kiver, (D), Wyoming, 1968, Gl
    Linda B. McCollum, (D), SUNY (Binghamton), 1980, Pg

Edmonds Community College
Geology (A) (2015)
20000 68th Ave W
Lynnwood, WA 98036
    p. (425) 640-1918
    mkelly@edcc.edu
    http://www.edcc.edu/stem/geology/
    Enrollment (2013): A: 4 (0)
Instructor:
    Maria Kelly, Zg
Adjunct Professor:
    Dylan Ahearn, (D), California (Davis)
    Thomas Hamilton, Zg

Everett Community College
Dept of Physical Sciences (A) (2015)
2000 Tower Street
Everett, WA 98201
    p. (425) 388-9429
    sdamp@everettcc.edu
    http://www.everettcc.edu/programs/mathsci/physical/
Instructor:
    Steve Grupp, (M), Colorado Mines, GgOg
Adjunct Professor:
    Alecia Spooner, Zg

Grays Harbor College
Dept of Geology & Oceanography (A) (2015)
Aberdeen, WA 98520
    p. (206) 538-4299
    john.hillier@ghc.edu
Professor:
    John Hillier, (D), Cornell, ZgCa
Emeritus:
    James B. Phipps, (D), Oregon State, 1974, Gu

Green River Community College
Dept of Geology (A) ⌁ (2019)
12401 S.E. 320th
Auburn, WA 98092-3699
    p. (253) 833-9111 (Ext. 4248)
    khoppe@greenriver.edu
    https://www.greenriver.edu/grc/course-descriptions/course-description-detail.aspx?desc=GEOL&deptname=Geology
    Programs: AA, AS; & BAS in Natural Resources in Forest Resource Management
Instructor:
    Kathryn A. Hoppe, (D), Princeton, 1999, ZgOgPg
    Katy Shaw, (M), Washington, ZgOgGg
Adjunct Professor:
    Megan Onufer, (M)

Highline College
Physical Sciences Dept/Geology Program (A) ☒ (2018)
MS 29-3
PO Box 98000
2400 S. 240th St.
Des Moines, WA 98198-9800
    p. (206) 878-3710
    ebaer@highline.edu
    http://geology.highline.edu/
    Enrollment (2018): A: 5 (2)
Instructor:
    Eric M. Baer, (D), California (Santa Barbara), 1995, GgvZg
    Stephaney Puchalski, (D), Indiana, 2011, PgGg
    Jacob Selander, (D), California (Davis)
    Michael Valentine, (D), Massachusetts, 1990, Gt
    Carla Whittington, (M), Indiana, GgvZg

Lower Columbia College
Earth Sciences (Natural Sciences Dept.) (A) (2015)
P.O. Box 3010
Longview, WA 98632
    p. (360) 442-2883
    dcordero@lowercolumbia.edu
    https://lowercolumbia.edu/programs/natural-science.php
Professor:
    David I. Cordero, (M), Portland State, 1997, Zg

North Seattle Community College, North Campus
Earth Science (2015)
9600 College Way North
Seattle, WA 98103
    p. (206) 934-4509
    tfurutani@sccd.ctc.edu
    http://www.northseattle.edu/

Washington

Northwest Indian College
Native Environmental Studies (2015)
2522 Kwina Road
Bellingham, WA 98226
p. (360) 392-4256
toreiro@nwic.edu
http://www.nwic.edu/degrees-and-certificates/bsnes-bachelors-degree

Olympic College
Mathematics, Engineering, Sciences, and Health Div (A) (2015)
Bremerton, WA 98310
p. (360) 475-7777
SMacias@olympic.edu
http://www.olympic.edu/Students/AcadDivDept/MESH/Sciences/Geology/
Instructor:
  Steve E. Macias, (M), Washington, 1996, Gg
Adjunct Professor:
  Katie Howard, Gg

Pacific Lutheran University
Dept of Geosciences (B) (2015)
12180 S. Park
Tacoma, WA 98447
p. (253) 535-7378
geos@plu.edu
http://www.plu.edu/geosciences/
Enrollment (2013): B: 52 (11)
Chair:
  Jill M. Whitman, (D), California (San Diego), 1989, OuGuZe
Professor:
  Duncan Foley, (D), Ohio State, 1978, GeHyZe
Associate Professor:
  Rosemary McKenney, (D), Penn State, 1997, Gm
  Claire E. Todd, (D), Washington, 2007, Glm
Assistant Professor:
  Peter B. Davis, (D), Minnesota, 2008, GcpGz

Peninsula College
Environmental Science (2015)
1502 East Lauridsen Boulevard
Port Angeles, WA 98362
p. (360) 452-9277
jganzhorn@pencol.edu
http://www.pc.ctc.edu

Pierce College
Pierce College (2015)
9401 Farwest Drive SW
Lakewood, WA 98498
p. (253) 964-6676
zayacjm@piercecollege.edu
http://www.piercecollege.edu/departments/physics_planetary_sciences/geology.asp

Seattle Central Community College
Div of Science & Mathematics (2015)
1701 Broadway
Seattle, WA 98122
p. (206) 587-3858
jhull@sccd.ctc.edu
http://seattlecentral.edu/sci-math/
Instructor:
  Katie Gagnon, (D), California (San Diego), 2007, OuZg
  Joseph M. Hull, (D), Rochester, 1988, Gc
Adjunct Professor:
  Michael Harrell, Gg

Shoreline Community College
Geology & Earth Sciences (A) (2015)
16101 Greenwood Avenue North
Seattle, WA 98133
p. (206) 546-4659
eagosta@shoreline.edu
http://www.shoreline.edu/science/geology.htm

Spokane Community College
Dept of Earth Science (2015)
1810 N. Greene Street
Spokane, WA 99217
Andy.Buddington@scc.spokane.edu
http://www.scc.spokane.edu/ArtsSciences/Science/

Tacoma Community College
Dept of Earth Sciences (A) (2018)
6501 South 19th Street
Tacoma, WA 98466
p. (253) 566-5060
rhitz@tacomacc.edu
https://www.tacomacc.edu/academics-programs/programs/geology
Programs: 1) Associate of Science with an Earth Sciences Specialization, 2) Associate of Arts
Enrollment (2018): A: 3 (0)
Professor:
  Ralph B. Hitz, (D), California (Santa Barbara), 1997, PvSaZi
Adjunct Professor:
  Jim McDougall, (D), Zg
  James Peet, (D), Washington
  Michael Valentine, (D), Zg

University of Puget Sound
Geology Dept (B) (2018)
1500 N. Warner Street
Tacoma, WA 98416-1048
p. (253) 879-3814
mvalentine@pugetsound.edu
http://www.pugetsound.edu/academics/departments-and-programs/undergraduate/geology/
f: https://www.facebook.com/upsgeology
Programs: B.S in Geology; B.S.in Natural Science
Administrative Assistant: Leslie Levenson
Enrollment (2018): B: 26 (14)
Chair:
  Michael J. Valentine, (D), Massachusetts, 1990, YmGc
Professor:
  Barry Goldstein, (D), Minnesota, 1985, GlmGs
  Jeffrey H. Tepper, (D), Washington, 1991, GiCgGv
Associate Professor:
  Kena L. Fox-Dobbs, (D), California (Santa Cruz), 2006, PgCsGe
Instructor:
  Ken P. Clark, (M), W Washington, 1989, GgcGi
Emeritus:
  Albert A. Eggers, (D), Dartmouth, 1972, Gv

University of Washington
Dept of Atmospheric Sciences (B,M,D) (2018)
Box 351640
Seattle, WA 98195-1640
p. (206) 543-4250
chair@atmos.washington.edu
http://www.atmos.washington.edu
f: www.facebook.com/UWAtmosSci
t: @UWAtmosSci
Programs: Atmospheric Sciences
Enrollment (2017): B: 73 (18) M: 34 (16) D: 29 (7)
Professor and Chair:
  Dale R. Durran, (D), MIT, 1981, As
Professor:
  Thomas P. Ackerman, (D), Washington, 1976, As
  David S. Battisti, (D), Washington, 1988, As
  Cecilia M. Bitz, (D), Washington, 1997, As
  Christopher S. Bretherton, (D), MIT, 1984, As
  Shuyi S. Chen, (D), Penn State, 1990, As
  Qiang Fu, (D), Utah, 1991, As
  Gregory J. Hakim, (D), SUNY (Albany), 1997, As
  Dennis L. Hartmann, (D), Princeton, 1975, As
  Lyatt Jaegle, (D), Caltech, 1996, As

Daniel A. Jaffe, (D), Washington, 1987, As
Clifford F. Mass, (D), Washington, 1978, As
Joel A. Thornton, (D), California (Berkeley), 2002, As
Robert Wood, (D), Manchester, 1997, As
Research Associate Professor:
    Roger T. Marchand, (D), Virginia Tech, 1997, Ass
Associate Professor:
    Becky Alexander, (D), California (San Diego), 2002, AsCsPe
    Dargan M. Frierson, (D), Princeton, 2005, As
Assistant Professor:
    Daehyun Kim, (D), Seoul Nat, 2010, As
    Abigail L. S. Swann, (D), California (Berkeley), 2010, AsZn
Adjunct Associate Professor:
    Jessica D. Lundquist, (D), Scripps, 2004, As
Affiliate Associate Professor:
    Philip W. Mote, (D), Washington, 1994, As
Adjunct Professor:
    Gerard H. Roe, (D), MIT, 1999, As
    Eric J. Steig, (D), Washington, 1996, CsGlAs
    LuAnne Thompson, (D), MIT, 1990, Og
    Ka-Kit Tung, (D), Harvard, 1977, As
Research Professor Emeritus:
    James E. Tillman, (M), MIT, As
Emeritus:
    Marcia B. Baker, (D), Washington, As
    Robert A. Brown, (D), Washington, 1969, As
    Joost A. Businger, (D), State (Utrecht), As
    Robert J. Charlson, (D), Washington, As
    David S. Covert, (D), Washington, 1974, As
    Thomas C. Grenfell, (D), Washington, 1972, As
    Halstead Harrison, (D), Stanford, 1960, As
    Dean A. Hegg, (D), Washington, 1979, As
    Robert A. Houze, (D), MIT, 1972, As
    Gary A. Maykut, (D), Washington, As
    Peter B. Rhines, (D), Trinity Coll (Cambridge), 1967, Op
    Edward S. Sarachik, (D), Brandeis, As
    John M. Wallace, (D), MIT, 1966, As
    Stephen G. Warren, (D), Harvard, 1973, ApGl
Research Associate Professor:
    Lynn A. McMurdie, (D), Washington, 1989, As
Research Assistant Professor:
    Edward Blanchard, (D), Washington, 2013, As
Affiliate Professor:
    James E. Overland, (D), New York, 1973, As
    Lawrence F. Radke, (D), Washington, 1968, As
    Chidong Zhang, (D), Penn State, 1989, AsmOp
Affiliate Associate Professor:
    Timothy S. Bates, (D), Washington, 1992, As
    Nicholas A. Bond, (D), Washington, 1986, As
    Bradley R. Colman, (D), MIT, 1984, As
    Mark T. Stoelinga, (D), Washington, 1993, As
Affiliate Assistant Professor:
    Bonnie Light, (D), Washington, 2000, As

Dept of Earth & Space Sciences (B,M,D) ●✉ (2018)
070 Johnson Hall
Box 351310
Seattle, WA 98195-1310
    p. (206) 543-1190
    essadv@uw.edu
    http://www.ess.washington.edu
    f: https://www.facebook.com/pages/University-of-Washington-Earth-and-Space-Sciences/132505186787510
    t: @UW_ESS
    Programs: Earth and Space Sciences, Earth and Space Sciences: Geology, Earth and Space Sciences: Biology, Earth and Space Sciences: Environmental, Earth and Space Sciences: Physics, Applied Geosciences, Geological Sciences, Geophysics
    Administrative Assistant: Scott Dakins
    Enrollment (2018): B: 202 (80) M: 22 (14) D: 25 (13)
Chair:
    Kenneth C. Creager, (D), California (San Diego), 1984, Ys
Academic Services:
    Noell Bernard-Kingsley, Ze
Research Professor:
    Howard B. Conway, (D), Canterbury (NZ), 1986, Gl
    Dale P. Winebrenner, (D), Washington, 1985, GlZr
Pacific Northwest Seismic Network:
    Harold Tobin
Affiliate Professor:
    Arthur Frankel, (D), Ys
    Joan Gomberg, (D), YsGt
    Frank I. Gonzalez, (D), Hawaii, 1977, OpRn
    Tony Irving, (D), CgXmGi
    Richard Sack, (D), CgZn
    George Shaw
Affiliate Assistant Professor:
    Ralph Haugerud, (D), Zi
    Brian L. Sherrod, (D), YsGt
Professor:
    George W. Bergantz, (D), Johns Hopkins, 1988, Gi
    J. Michael Brown, (D), Minnesota, 1980, Gy
    Roger Buick, (D), W Australia, 1986, CbXbPm
    David C. Catling, (D), Oxford, 1994, CgAsXb
    Bernard Hallet, (D), California (Los Angeles), 1975, Gml
    Robert H. Holzworth, (D), California (Berkeley), 1977, XyAsZg
    David R. Montgomery, (D), California (Berkeley), 1991, Gm
    Bruce K. Nelson, (D), California (Los Angeles), 1985, CcGie
    Charles A. Nittrouer, (D), Washington, 1978, GuYr
    Eric J. Steig, (D), Washington, 1996, CsGlAs
    Edwin D. Waddington, (D), British Columbia, 1981, Gl
    Peter D. Ward, (D), McMaster, 1976, Pg
    Stephen G. Warren, (D), Harvard, 1973, AsGl
    Robert M. Winglee, (D), Sydney, 1984, XyYx
Research Associate Professor:
    Evan H. Abramson, (D), MIT, 1985, Gy
    Paul A. Bodin, (D), Colorado, 1992, Ys
    Michael P. McCarthy, (D), Washington, 1988, XyYx
    Ronald S. Sletten, (D), Washington, 1995, Sob
Research Assistant Professor:
    Erika M. Harnett, (D), Washington, 2003, XyGev
Affiliate Associate Professor:
    Amit Mushkin, (D), ZrCcGm
Associate Professor:
    Rolf E. Aalto, Gms
    Joshua Bandfield, (D), Arizona State, 2000, Zr
    Derek Booth, Gm
    Juliet G. Crider, (D), Stanford, 1998, GcmNr
    Drew J. Gorman-Lewis, (D), Notre Dame, 2006, PgCl
    Bob Hawley, Gl
    Katharine W. Huntington, (D), MIT, 2006, Gt
    Seth Moran, Gv
    Elizabeth (Liz) Nesbitt, (D), California (Berkeley), 1982, Pim
    Gerard H. Roe, (D), MIT, 1999, AsGml
    David Schmidt, (D), California (Berkeley), 2002, Gt
    Ben Smith, GlZr
    John O.H Stone, (D), Cambridge, 1986, Cca
    Caroline Stromberg, Pbe
    Fangzhen Teng, (D), Maryland, 2005, Cg
    Jeremy Thomas, ZrXa
    Gregory Wilson, (D), Berkeley, 2004, Pvg
Research Assistant Professor:
    Michelle Koutnik, (D), Washington, 2009, Gll
Adjunct Assistant Professor:
    Christian A. Sidor, (D), Chicago, 2000, Pv
Assistant Professor:
    Scott Bennett, (D), California (Davis), 2013, YsGtm
    Tom Brown, CcAt
    Knut Christianson, (D), GlZr
    Trenton Cladouhos, Gcg
    Janice DeCosmo, AtZe
    Alison Duvall, (D), Michigan, 2011, GmcGt
    T.J. Fudge, (D), Gl
    Alexis Licht, (D), Gsn
    Kris Ludwig, Rnc
    Kenichi Matsuoka, Gl
    Tom Neumann, Gl
    Gregg Petrie, Zr
    Dan Shugar, GmlZr
    Erin Wirth, (D), Yale, 2014, YsRn
    Tim Ziemba, Xa
Research Associate:
    Marc Biundo, Ys

Doug Gibbons, ZeYs
Renate Hartog, Ys
Abram Jacobsen, Xg
Scott Kuehner, Ggz
Scientific Instructional Designer:
  Michael Harrell, (D), Washington, Ze
Senior Lecturer:
  Terry W. Swanson, (D), Washington, 1994, CcGe
  Kathy G. Troost, (D), Zun
Lecturer:
  Brian Collins, Gm
  Steven Walters, Zir
Adjunct Professor:
  David S. Battisti, (D), Washington, 1988, As
  John R. Delaney, (D), Arizona, 1977, GuYr
  Harlan Paul Johnson, (D), Washington, 1972, Gu
  Randy LeVeque, (D), Stanford, 1982, GvYs
  John D. Sahr, (D), Cornell, 1990, ZrXym
  William S.D Wilcock, (D), MIT, 1992, Gu
Research Professor Emeritus:
  Stephen D. Malone, (D), Nevada (Reno), 1972, YsGt
  Gary A. Maykut, (D), Washington, 1969, AsGlYr
Research Professor:
  Alan R. Gillespie, (D), Caltech, 1982, GmlZr
Research Associate Professor:
  Robert I. Odom, (D), Washington, 1980, YrGu
Professor:
  Joost Businger, Am
Affiliate Professor:
  Brian Atwater, (D), Delaware, 1980, RnGr
  Charlotte Schreiber, (D), Gsd
Emeritus:
  John B. Adams, (D), Washington, 1961, GcXgZr
  Patricia M. Anderson, (D), Brown, 1982, Ple
  Marcia B. Baker, (D), Washington, 1971, AsYg
  John Booker, (D), California (San Diego), 1968, YmGct
  Joanne (Jody) Bourgeois, (D), Wisconsin, 1980, GsrRh
  Eric S. Cheney, (D), Yale, 1964, Eg
  Darrel S. Cowan, (D), Stanford, 1972, Gct
  Robert S. Crosson, (D), Stanford, 1966, YsGct
  Bernard W. Evans, (D), Oxford (UK), 1959, GpzGi
  I. Stewart (Stu) McCallum, (D), Chicago, 1968, GivGz
  Ronald T. Merrill, (D), California (Berkeley), 1967, Ygm
  George K. Parks, (D), California (Berkeley), 1966, YxXy
  Charles F. Raymond, (D), Caltech, 1969, Gl
  John M. Rensberger, (D), California (Berkeley), 1967, Pvg
  Stewart W. Smith, (D), Caltech, 1961, Ys
  Minze Stuiver, (D), Groningen (Neth), 1958, Cc
  Brian Swanson, Asp
  Joseph A. Vance, (D), Washington, 1957, Gi
Seismology Lab:
  Bill Steele, Ys
Senior Computer Specialist:
  Harvey Greenberg, (M), Cal State (Chico), 1977, Gml
Postdoctoral Researcher:
  Jon Toner, (D)

School of Oceanography (B,M,D) (2015)
Box 357940
Seattle, WA 98195-7940
  p. (206) 543-5060
  admin@ocean.washington.edu
  http://www.ocean.washington.edu/
Professor:
  Knut Aagaard, (D), Washington, 1966, Op
  John A. Baross, (D), Washington, 1972, Ob
  Roy Carpenter, (D), California (San Diego), 1968, No
  William O. Criminale, Jr., (D), Johns Hopkins, 1960, Op
  Eric A. D'Asaro, (D), MIT/WHOI, 1980, Op
  John R. Delaney, (D), Arizona, 1977, Yr
  Jody W. Deming, (D), Maryland, 1981, Ob
  Allan H. Devol, (D), Washington, 1975, Oc
  Steven R. Emerson, (D), Columbia, 1974, OcCmOg
  Charles C. Eriksen, (D), MIT, 1977, Op
  Bruce W. Frost, (D), California (San Diego), 1969, Ob
  Michael C. Gregg, (D), California (San Diego), 1971, Op
  G. Ross Heath, (D), Scripps, 1968, Cm

Barbara M. Hickey, (D), California (San Diego), 1975, Op
H. Paul Johnson, (D), Washington, 1972, Yr
Deborah S. Kelley, (D), Dalhousie, 1990, OuGvp
Marvin D. Lilley, (D), Oregon State, 1983, Ob
Seelye Martin, (D), Johns Hopkins, 1967, Op
James W. Murray, (D), MIT/WHOI, 1973, Oc
Charles A. Nittrouer, (D), Washington, 1978, Ou
Arthur R. M. Nowell, (D), British Columbia, 1975, Ou
Paul D. Quay, (D), Columbia, 1977, Oc
Peter B. Rhines, (D), Cambridge, 1967, Op
Jeffrey E. Richey, (D), California (Davis), 1973, CbHs
Stephen C. Riser, (D), Rhode Island, 1981, Op
Research Professor:
  Mark L. Holmes, (D), Washington, 1975, OuGl
  Ronald L. Shreve, (D), Caltech, 1959, Yr
Associate Professor:
  Virginia E. Armbrust, (D), MIT/WHOI, 1990, Ob
  Susan L. Hautala, (D), Washington, 1992, Op
  Bruce M. Howe, (D), California (San Diego), 1986, Op
  Mitsuhiro Kawase, (D), Princeton, 1986, Op
  Richard G. Keil, (D), Delaware, 1991, Oc
  Evelyn J. Lessard, (D), Rhode Island, 1984, Ob
  Parker MacCready, (D), Washington, Op
  LuAnne Thompson, (D), MIT/WHOI, 1990, Op
  Mark J. Warner, (D), California (San Diego), 1988, Op
  William S. D. Wilcock, (D), MIT/WHOI, 1992, Ou
  Kevin L. Williams, (D), Washington State, 1985, Op
Research Assistant Professor:
  Miles G. Logsdon, (D), Washington, 1997, Zi
  Andrea S. Ogston, (D), Washington, 1997, Ou
Assistant Professor:
  Andrew Barclay, (D), Oregon, 1998, Gu
  Daniel Grunbaum, (D), Cornell, 1992, Ob
  Anitra E. Ingalls, (D), SUNY (Stony Brook), 2002, Oc
  Jeffrey D. Parsons, (D), Illinois, 1998, Ou
  Gabrielle L. Rocap, (D), MIT/WHOI, 2000, Ob
Senior Lecturer:
  Christina M. Emerick, (D), Oregon State, 1985, Ou
Lecturer:
  Richard M. Strickland, (M), Washington, 1975, ObgZg
Adjunct Professor:
  Rose Ann Cattolico, (D), SUNY (Stony Brook), 1973, Zn
  Robert Francis, (D), Washington, 1970, Ob
  Barbara B. Krieger-Brockett, (D), Wayne State, 1976, Ob
  Ronald T. Merrill, (D), California (Berkeley), 1967, Ym
  Edward Sarachik, (D), Brandeis, 1966, As
  Robert C. Spindel, (D), Yale, 1971, Op
Emeritus:
  George C. Anderson, (D), Washington, 1954, Ob
  Karl Banse, (D), Kiel, 1955, Ob
  Joe S. Creager, (D), Texas A&M, 1958, Ou
  Alyn C. Duxbury, (D), Texas A&M, 1963, On
  Terry E. Ewart, (D), Washington, 1965, Op
  Richard H. Gammon, (D), Harvard, 1970, Oc
  Joyce C. Lewin, (D), Yale, 1953, Ob
  Brian T. R. Lewis, (D), Wisconsin, 1970, Yr
  Russell E. McDuff, (D), California (San Diego), 1978, OcGu
  Dean A. McManus, (D), Kansas, 1959, Ou
  Gunnar I. Roden, (M), California (Los Angeles), 1956, Op
  David A. Rothrock, (D), Cambridge, 1968, Op
  Thomas B. Sanford, (D), MIT, 1967, Op
  Richard W. Sternberg, (D), Washington, 1965, Ou
Cooperating Faculty:
  Matthew H. Alford, (D), California (San Diego), 1998, Op
  Edward T. Baker, (D), Washington, 1973, Ou
  Laurie Balistrieri, (M), Washington, 1977, Oc
  John Bullister, (D), California (San Diego), 1984, Oc
  David A. Butterfield, (D), Washington, 1990, Yr
  Glenn A. Cannon, (D), Johns Hopkins, 1969, Op
  Meghan F. Cronin, (D), Rhode Island, 1993, Op
  Brian D. Dushaw, (D), California (San Diego), 1992, Op
  Richard A. Feely, (D), Texas A&M, 1974, Oc
  Don E. Harrison, (D), Harvard, 1977, Op
  Albert Hermann, (D), Washington, 1988, Op
  Robin T. Holcomb, (D), Stanford, 1981, Ou
  Gregory C. Johnson, (D), MIT/WHOI, 1990, Op
  Peter A. Jumars, (D), California (San Diego), 1974, Ob

Kathryn A. Kelly, (D), California (San Diego), 1983, Op
William S. Kessler, (D), Washington, 1989, Op
Craig M. Lee, (D), Washington, 1995, Op
Michael J. McPhaden, (D), California (San Diego), 1980, Og
Curtis D. Mobley, (D), Maryland, 1977, Ob
Harold O. Mofjeld, (D), Washington, 1970, Op
Dennis W. Moore, (D), Harvard, 1968, Op
James H. Morison, (D), Washington, 1980, Op
Jeffrey Napp, (D), California (San Diego), 1986, Ob
Jan Newton, (D), Washington, 1989, Ob
Jeffrey A. Nystuen, (D), California (San Diego), 1985, Op
Joan M. Oltman-Shay, (D), California, 1986, Op
Mary Jane Perry, (D), California (San Diego), 1974, Ob
Thomas Pratt, (N), Gu
Joseph A. Resing, (D), Hawaii, 1997, Oc
Christopher L. Sabine, (D), Hawaii, 1992, Oc
Randy Shuman, (D), Washington, 1978, Ob
Laurenz A. Thomsen, (D), Kiel, 1992, Ob
Cynthia T. Tynan, (D), California (San Diego), 1993, Ob
Rebecca A. Woodgate, (D), Oxford, 1994, Op

Walla Walla Community College
Walla Walla Community College (2015)
500 Tausick Way
Walla Walla, WA 99362
p. (509) 527-4278
steven.may@wwcc.edu
http://www.wwcc.edu

Washington State University
Dept of Crop & Soil Sciences (B,M,D) (2015)
201 Johnson Hall
P.O. Box 646420
Pullman, WA 99164-6420
p. (509) 335-3475
bc.johnson@wsu.edu
http://css.wsu.edu/
Chair:
  Richard Koenig, (D)
Professor:
  David F. Bezdicek, (D), Minnesota, 1967, Sb
  James B. Harsh, (D), California (Berkeley), 1983, Sc
  Shiou Kuo, (D), Maine, 1973, Sc
  Thomas A. Lumpkin, (D), Hawaii, 1978, Zn
  William L. Pan, (D), North Carolina, 1983, Sb
  John P. Reganold, (D), California (Davis), 1980, Soo
Senior Scientist:
  Robert G. Stevens, (D), Colorado State, 1971, Sb
Associate Professor:
  Joan R. Davenport, (D), Guelph, 1985, Sb
  Frank J. Peryea, (D), California, 1983, Sc
Assistant Professor:
  Markus Flury, (D), Swiss Fed Inst Tech, 1994, Sp
Adjunct Professor:
  Alan J. Busacca, (D), California (Davis), 1982, Sa
  Ann C. Kennedy, (D), North Carolina State, 1985, Sb
  Robert I. Papendick, (D), South Dakota State, 1962, Sp
  Jeffery L. Smith, (D), Washington State, 1983, Sb

Peter Hooper GeoAnalytical Lab (B,M,D) (2016)
School of the Environment
1228 Webster Physical Sciences Building
Washington State University
Pullman, WA 99164-2812
p. (509) 335-1626
geolab@mail.wsu.edu
http://environment.wsu.edu/facilities/geolab/
Professor:
  John A. Wolff, (D), London (UK), 1983, Gi
Research Associate:
  Owen K. Neill, (D), Alaska (Fairbanks), 2012, Ca
Emeritus:
  Franklin F. Foit, Jr., (D), Michigan, 1968, Gz
Laboratory Manager:
  Scott Boroughs, (D), Washington State, 2010, GiCa

Research Technologist:
  Charles Knaack, (M), Washington State, 1991, Ca

School of the Environment (B,M,D) ●⌘ (2019)
Webster 1228
P.O. Box 642812
Pullman, WA 99164-2812
p. (509) 335-3009
soe@wsu.edu
http://environment.wsu.edu/
Enrollment (2009): B: 34 (11) M: 30 (9) D: 11 (5)
Professor:
  Stephen Bollens, (D), Washington, 1990, Ob
  Matt Carroll, (D), Washington, 1984, Zn
  C. Kent Keller, (D), Waterloo, 1987, Hw
  Peter B. Larson, (D), Caltech, 1983, Cs
  Dirk Schulze-Makuch, (D), Wisconsin (Milwaukee), 1996, Hw
  Jeffrey D. Vervoort, (D), Cornell, 1994, CcaGt
  John A. Wolff, (D), Imperial Coll (UK), 1983, Giv
Clinical Associate Professor:
  Allyson Beall King, (D), WSU
Associate Professor:
  Sean P. Long , (D), Princeton, 2010, Gc
Clinical Assistant Professor :
  Mike Berger , (D), Oregon, 2004
Assistant Professor:
  Catherine Cooper, (D), Rice, 2005, Yg
Research Associate:
  Scott Boroughs, (D), Washington State, 2010, Gv
  Victor A. Valencia, (D), Arizona, 2005, ZgCcEg
Instructor:
  Kurt Wilkie, (D)
Adjunct Professor:
  Regan L. Patton, (D), Washington State, 1997, Gc
  Stephen P. Reidel, (D), Washington State, 1978, GivGt
Emeritus:
  Keith Blatner , (D), 1983, Zn
  Franklin F. Foit, Jr., (D), Michigan, 1968, Gza
  Andrew Ford, (D), Dartmouth, 1975, Zu
  Eldon H. Franz, (D), Illinois, 1971, Sf
  David R. Gaylord, (D), Wyoming, 1983, Gs
  George Hinman
  Philip E. Rosenberg, (D), Penn State, 1960, GzCg
  A. John Watkinson, (D), Imperial Coll (UK), 1972, Gc
  Gary D. Webster, (D), California (Los Angeles), 1966, Pis

Wenatchee Valley College
Earth Sciences (A) (2015)
1300 Fifth Street
Wenatchee, WA 98801
p. (509) 682-6754
rdawes@wvc.edu
https://www.wvc.edu/directory/departments/geology/
Professor:
  Ralph Dawes, (D), Washington, 1993, GgeAm
Adjunct Professor:
  Kelsay Stanton, (M), W Washington, 2005, GgZe

Western Washington University
Dept of Geology (B,M) O⊠ (2019)
516 High Street
Bellingham, WA 98225-9080
p. (360) 650-3582
geology@wwu.edu
http://geology.wwu.edu/
Programs: Geology; Geophysics
Administrative Assistant: Kate Blizzard
Enrollment (2018): B: 156 (57) M: 38 (10)
Chair:
  Bernard A. Housen, (D), Michigan, 1994, Ym
Director Honors Program:
  Scott R. Linneman, (D), Wyoming, 1990, GmZe
Professor:
  Jackie Caplan-Auerbach, (D), Hawaii, Ys
  Susan M. DeBari, (D), Stanford, 1991, GxZe
  Robert J. Mitchell, (D), Michigan Tech, 1996, Hw

Elizabeth R. Schermer, (D), MIT, 1989, Gt
Associate Professor:
  Colin B. Amos, (D), California (Santa Barbara), 2007, Gc
  Douglas H. Clark, (D), Washington, 1995, Gl
  Pete Stelling, (D), Alaska (Fairbanks), 2003, GxEgGv
Assistant Professor:
  Robyn Dahl, (D), California (Riverside), 2015, PiZe
  Brady Z. Foreman, (D), Wyoming, 2012, Grs
  Sean R. Mulcahy, (D), California (Davis), 2009, GpzGp
  Allison Pfeiffer, (D), California (Santa Cruz), 2017
  Camilo Ponton, (D), MIT, 2012, CsoPc
  Melissa S. Rice, (D), Cornell, 2012, XgGsm
Paleomagnetism Lab Manager:
  Cristina Garcia Lasanta, (D), Zaragoza, 2016, YmGtc
Geochemistry Research Associate:
  Nicole M. McGowan, (D), Macquarie, 2017, CtaCg
Research Associate:
  Russell F. Burmester, (D), Princeton, 1974, Ym
  Eric Grossman, (D), Hawaii, 2001, GusGm
  George Mustoe, (M), W Washington, 1972, PbCaPo
  Charles A. Ross, (D), Yale, 1964, Pm
  Brian Rusk, (D), Oregon, 2003
Senior Instructor:
  Paul Thomas, (M), W Washington, 1997, Gm
Emeritus:
  R. Scott Babcock, (D), Washington, 1970, Cg
  Edwin H. Brown, (D), California (Berkeley), 1966, Gp
  Don J. Easterbrook, (D), Washington, 1962, GlmGe
  David C. Engebretson, (D), Stanford, 1983, Gt
  Thor A. Hansen, (D), Yale, 1978, Pg
  James L. Talbot, (D), Adelaide, 1962, Gc

Whitman College
Dept of Geology (B) ☒ (2018)
345 Boyer
Walla Walla, WA 99362
  p. (509) 527-5225
  baderne@whitman.edu
  http://www.whitman.edu/geology
  Administrative Assistant: Patti Moss
  Enrollment (2018): B: 40 (19)
Chair, Associate Professor:
  Nicholas E. Bader, (D), California (Santa Cruz), 2006, SoHgZi
Professor:
  Kevin R. Pogue, (D), Oregon State, 1993, Gc
  Patrick K. Spencer, (D), Washington, 1984, Pg
Associate Professor:
  Kirsten P. Nicolaysen, (D), MIT, 2001, GiCgGa
Assistant Professor:
  Lyman P. Persico, (D), New Mexico, 2012, Gm
Emeritus:
  Robert J. Carson, (D), Washington, 1970, GmeGl
Geology Technician:
  Elliot Broze, (M), Tromsø, 2017, Gg

Yakima Valley College
Dept of Geology (A) (2015)
P. O. Box 22520
16th Ave. & Nob Hill Blvd.
Yakima, WA 98907
  p. (509) 575-2350 x2366
  dhuycke@yvcc.edu
  http://www.yvcc.edu/FutureStudents/AcademicOptions/Programs/PhysicalSciences/Geology/Pages/default.aspx
Instructor:
  David Huycke, (M), Wyoming, 1979, Gg

West Virginia

Concord University
Dept of Physical Sciences (B) ○⊘ (2018)
1000 Vermillion St.
Athens, WV 24712-1000
  p. (304) 384-5327
  allenj@concord.edu
  https://www.concord.edu/physci/
  f: www.facebook.com/ConcordGeoscience/
  t: @CUGeology
  Programs: Geology
  Enrollment (2018): B: 28 (8)
Chair:
  Joseph L. Allen, (D), Kentucky, 1994, Gct
Professor:
  Alice M. Hawthorne Allen, (D), Indiana, 1998, Xa
Associate Professor:
  Stephen C. Kuehn, (D), Washington State, 2002, CaGv
Assistant Professor:
  Aaron C. Paget, (D), Florida State, 2013, ApOpZr
Research Associate:
  Janine Krippner, (D), Pittsburgh, 2017, GvRc
  Robert Peck, (D), Pennsylvania, 1969, Pg
Instructor:
  Jennifer Phillippe, (M), California (Davis), 2005, GegZe
Adjunct Professor:
  Alyce Lee, (D), Texas A&M, 2009, Og

Fairmont State University
Natural Sciences (B) ⌁ (2019)
1201 Locust Avenue
Fairmont, WV 26554
  p. (304) 367-4393
  dhemler@fairmontstate.edu
  Programs: Earth & Space Science Education
  Enrollment (2017): B: 3 (0)
Coordinator of Geoscience Program:
  Deb Hemler, (D), West Virginia, 1997, Ze
Instructor:
  Todd I. Ensign, (D), West Virginia, 2017, Ze
  Josh Revels, (B), Fairmont State, 2012, Ze
Adjunct Professor:
  Ronald McDowell, (D), Gge
  Barnes Nugent, (M), Gg
Visiting Professor:
  Marcie Raol, (M), West Virginia, 1998, Ze

Marshall University
Dept of Geology (B,M) ☒ (2018)
One John Marshall Drive
Huntington, WV 25755
  p. (304) 696-6720/(304) 696-6756
  geology@marshall.edu
  http://marshall.edu/geology
  Programs: B.S. Geology
  B.A. Geology
  M.S. Physical Science, Geology Concentration
  Enrollment (2018): B: 26 (10)
Chair:
  William L. Niemann, (D), Missouri (Rolla), 1999, Ng
Professor:
  Ronald L. Martino, (D), Rutgers, 1981, GsrPe
Associate Professor:
  Aley El Shazly, (D), Stanford, 1991, GpiCg
Assistant Professor:
  Andrew J. Horst, (D), Syracuse, 2013, GcOgGt

Potomac State College
Dept of Geology & Geography (A) (2015)
101 Fort Avenue
Keyser, WV 26726
  p. (304) 788-6956
  JJNinesteel@mail.wvu.edu
Professor:
  Judy J. Ninesteel, (M), West Virginia, 1995, Hg

West Virginia University
Dept of Geology & Geography (B,M,D) ⊘ (2018)
330 Brooks Hall
P.O. Box 6300
98 Beechurst Ave.
Morgantown, WV 26506-6300
  p. (304) 293-5603

tim.carr@mail.wvu.edu
http://www.geo.wvu.edu
Programs: Geology, Geography, Environmental Sciences
Certificates: GIS
Enrollment (2011): B: 165 (21) M: 45 (10) D: 13 (1)
Chair:
    Timothy R. Carr, (D), Wisconsin, 1981, GoYeGg
Professor:
    Joseph J. Donovan, (D), Penn State, 1992, Hq
    Jason A. Hubbart, (D), Idaho, 2007, HqgHs
    John J. Renton, (D), West Virginia, 1965, Ec
    Jaime Toro, (D), Stanford, 1998, GctGo
    Timothy A. Warner, (D), Purdue, 1992, Zr
    Thomas H. Wilson, (D), West Virginia, 1980, YeGcYg
Associate Professor:
    Kathleen Benison, (D), GrsCm
    Dengliang Gao, (D), Duke, 1997, Ye
    J. Steven Kite, (D), Wisconsin, 1983, GmHsGa
    Dorothy Vesper, (D), Penn State, 2002, Hw
Teaching Assistant Professor:
    Joe Lebold, (D), West Virginia, 2005, PeGrg
Assistant Professor:
    Graham D. M. Andrews, (D), Leicester, 2006, GvtGx
    Shikha Sharma, (D), Lucknow Univ (India), 1998, Csa
    Amy L. Weislogel, (D), Stanford, 2006, Gs
Teaching Assistant Professor:
    Kenneth Brown, (D), Miami, 2015, GgcGe
Adjunct Professor:
    Katherine Lee Avary, (M), North Carolina, 1978, Go
    Bascombe (Mitch) Blake, (D), West Virginia, 2009, GsEcPb
    Alan Brown, (D), Louisiana State, 2002, EoYe
    Katherine R. Bruner, (D), West Virginia, 1991, Gd
    David J. Campagna, (D), Purdue, 1990, GcZrGo
    Blaine Cecil, (D), West Virginia, 1965, Ec
    Phillip Dinterman
    Harry Edenborn, (D), Rutgers, 1982, Scb
    Nick Fedorko, (M), West Virginia, 1998, Ec
    Ronald McDowell, (D), Colorado Mines, 1987, GsPiCe
    Paul F. Ziemkiewicz,
Research Assistant Professor:
    Payam Kavousi Ghahfarohki, (D), West Virginia, 2016, Gog
Emeritus:
    Alan C. Donaldson, (D), Penn State, 1959, Gs
    Thomas W. Kammer, (D), Indiana, 1982, Pi
    Henry W. Rauch, (D), Penn State, 1972, HwGe
    Robert C. Shumaker, (D), Cornell, 1960, Go
    Richard A. Smosna, (D), Illinois, 1973, GsdGo

Wisconsin

Beloit College
R.D. Salisbury Dept of Geology (B) ☒ (2018)
700 College Street
Beloit, WI 53511
    p. (608) 363-2132
    swansons@beloit.edu
    http://www.beloit.edu/geology/
    f: https://www.facebook.com/groups/47867916990/
    Enrollment (2015): B: 20 (11)
Chair:
    Susan K. Swanson, (D), Wisconsin, 2001, HwGm
Associate Professor:
    James R. Rougvie, (D), Texas, 1999, GxpGz
    James J. Zambito, (D), Cincinnati, 2011, PcGsPg
Emeritus:
    Carol Mankiewicz, (D), Wisconsin, 1987, Gs
    Carl V. Mendelson, (D), California (Los Angeles), 1981, PmGs

Lawrence University
Department of Geosciences (B) ☒ (2019)
711 E Boldt Way
Appleton, WI 54911
    p. (920) 832-6731
    knudsena@lawrence.edu
    http://www.lawrence.edu/dept/geology/
    Administrative Assistant: Ellen c. Walsh

Enrollment (2018): B: 16 (5)
Professor:
    Marcia Bjornerud, (D), Wisconsin, 1987, GctRc
Associate Professor:
    Jeffrey J. Clark, (D), Johns Hopkins, 1997, Gm
    Andrew Knudsen, (D), Idaho, 2002, CgGez
Assistant Professor:
    Relena R. Ribbons, (D), Copenhagen, 2017, SbGe
Emeritus:
    John C. Palmquist, (D), Iowa, 1961, GcpGg
    Theodore W. Ross, (D), Washington State, 1969, Gg
    Ronald W. Tank, (D), Indiana, 1962, Ge

Milwaukee Public Museum
Dept of Geology (2015)
800 W. Wells Street
Milwaukee, WI 53233
    p. (414) 278-2741
    sheehan@mpm.edu
    http://www.mpm.edu/collect/geology/geosec-noframes.html
Chair:
    Peter M. Sheehan, (D), California (Berkeley), 1971, Pi
Adjunct Professor:
    Gary D. Rosenberg, (D), California (Los Angeles), 1972, RhPg

Northland College
Dept of Geosciences (B) ● (2017)
1411 Ellis Avenue
Ashland, WI 54806
    p. (715) 682-1852
    tfitz@northland.edu
    Enrollment (2016): B: 39 (8)
Professor:
    Thomas J. Fitz, (D), Delaware, 1999, GgxGc
Assistant Professor:
    Cynthia May, (M), Penn State, 2007, ZiyZn
    David J. Ullman, (D), Wisconsin, 2013, GlPeGm

Saint Norbert College
Geology Dept (B) ☒ (2017)
100 Grant Street
De Pere, WI 54115
    p. (920) 403-3987
    tim.flood@snc.edu
    http://www.snc.edu
    Programs: Geology
    Enrollment (2017): B: 27 (7)
Professor:
    Tim P. Flood, (D), Michigan State, 1987, GvZe
    Nelson R. Ham, (D), Wisconsin, 1994, Gl
Associate Professor:
    Rebecca McKean, (D), Nebraska, 2009, GdPgRh

University of Wisconsin Colleges
Dept of Geography & Geology (A) (2016)
UW Sheboygan
1 University Dr
Sheboygan, WI 53081
    p. (920) 459-6619
    jim.mccluskey@uwc.edu
    http://www.uwc.edu/depts/geography-geology
Chair:
    Karl Byrand, (D), Maryland, 1999, Zny
Chair/CEO UW Marathon County:
    Keith Montgomery, (D), Waterloo, 1986, ZyGg
Professor:
    Norlene Emerson, (D), Wisconsin, 2002, GgPiGs
    Diann Kiesel, (D), Wisconsin, 1998, GgZyGg
    Robert McCallister, (D), Wisconsin, 1996, ZyGgZn
Dean/CEO UW Washington County:
    Alan Paul Price, (D), California (Los Angeles), 1998, ZyGmZg
Associate Professor:
    Iddrisu Adam, (D), Wilfrid Laurier, 2001, ZynZn

Assistant Professor:
  Beth Johnson, (D), N Illinois, 2009, GlRh
  James McCluskey, (D), Rutgers, 1987, ZinZn
  Miclelle Palma, (D), Georgia, 2012, ZnnZn
  Andrew Shears, (D), Kent State, 2011, Zin
  Keith West, (D), Wisconsin (Milwaukee), 2007, ZynZn
Instructor:
  Sanborn Robert, (M), 1991, GgZyn
  Mengist Teklay Berhe
Lecturer:
  Jane Fairchild, (M), Wisconsin, 2003, Am
  Seth Rankin, (M), Wisconsin (Milwaukee), 1974, ZnnZn
Emeritus:
  Thomas Bitner, (M), ZynZn
  Richard Cleek, (M), ZnyZn
  Garret Deckert, (M), ZynZn
  Edwin Dommisee, (M), ZyGgZn
  James Heidt, (M), Zn
  Kenneth Korb, (D)
  Gene E. Musolf, (D), Wisconsin, 1970, Zy
  Shamim Naim, (D)
  Randall Rohe, (D), Colorado, 1978, ZyGgZn
  Leonard Weis, (D)
  Barbara Williams, (D)

University of Wisconsin Oshkosh
Geography Department (B) (2018)
800 Algoma Blvd
Oshkosh, WI 54901
  p. (920) 424-4105
  geography@uwosh.edu
  https://geography.uwosh.edu/
  f: https://www.facebook.com/UWOgeography/
  Programs: Geography
  Certificates: GIS
  Enrollment (2018): B: 32 (14)
Chair:
  Angela G. Subulwa, (D), Kansas, 2009, Zn
Associate Dean:
  Colin Long, (D), Oregon, 2003, PeZyc
Professor:
  Heike C. Alberts, (D), Minnesota, 2003, Zn
  John A. Cross, (D), Illinois, 1979, ZyAst
  Kazimierz J. Zaniewski, (D), Wisconsin (Milwaukee), 1987, Zn
Associate Professor:
  Mamadou Coulibaly, (D), S Illinois, 2004, ZiyHg
Assistant Professor:
  Elizabeth Barron, (D), Rutgers, 2010, Zn

University of Wisconsin, Eau Claire
Dept of Geology (B) ☒ (2018)
157 Phillips Hall
Eau Claire, WI 54702-4004
  p. (715) 836-3732
  steinklm@uwec.edu
  http://www.uwec.edu/academic/geology
  Administrative Assistant: Lorilie M. Steinke
  Enrollment (2011): B: 88 (0)
Chair:
  Kent M. Syverson, (D), Wisconsin, 1992, GlEnGe
Professor:
  Scott K. Clark, (D), Illinois (Urbana), 2007, Ze
  Karen G. Havholm, (D), Texas, 1991, GsZe
  Robert L. Hooper, (D), Washington State, 1983, Gz
  Phillip D. Ihinger, (D), Caltech, 1991, Gx
  J. Brian Mahoney, (D), British Columbia, 1994, GstEg
Assistant Professor:
  Robert Lodge, (D), EmCuGv
  Sarah A. Vitale, (D), Connecticut, 2017, Hq
Senior Lecturer:
  Lori D. Snyder, (M), British Columbia, 1994, Gx

University of Wisconsin, Extension
Dept of Environmental Sciences (2015)
3817 Mineral Point Road
Madison, WI 53705-5100
  p. (608) 262-1705
  elmo.rawling@uwex.edu
Professor:
  John W. Attig, (D), Wisconsin, 1984, Gl
  Thomas J. Evans, (D), Wisconsin, 1994, Gg
  James M. Robertson, (D), Michigan, 1972, Eg
Associate Professor:
  Madeline B. Gotkowitz, (M), New Mexico, 1993, Hw
  David J. Hart, (D), Wisconsin, 2000, Hw
Assistant Professor:
  Eric C. Carson, (D), Wisconsin, 2003, Gl
  Patrick I. McLaughlin, (D), Cincinnati, 2006, Gr
Emeritus:
  Lee Clayton, (D), Illinois, 1965, Gl

University of Wisconsin, Fox Valley
Dept of Geography and Geology (2015)
1478 Midway Rd
Menasha, WI 54952
  p. (920) 832-2600
  joanne.kluessendorf@uwc.edu
  http://uwc.edu/depts/geography-geology/geography-geology

University of Wisconsin, Green Bay
Dept. of Natural & Applied Sciences - Geoscience Program
(B,M) ☒ (2018)
LS-465
2420 Nicolet Drive
Green Bay, WI 54311
  p. (920) 465-2371
  luczajj@uwgb.edu
  http://www.uwgb.edu/geoscience/
  Programs: Geoscience; Earth Science Teaching
  Enrollment (2018): B: 16 (4) M: 4 (2)
Chair of Geoscience:
  John A. Luczaj, (D), Johns Hopkins, 2000, HwGsCl
Professor:
  Kevin J. Fermanich, (D), Wisconsin, 1995, SpRwHt
Associate Professor:
  Ryan Currier, (D), Johns Hopkins, 2011, GizGg
  Steven J. Meyer, (D), Nebraska, 1990, AsmZg
Emeritus:
  Steven I. Dutch, (D), Columbia, 1976, GcgGe
  Ronald D. Stieglitz, (D), Illinois, 1972, GrlHy

University of Wisconsin, Madison
Center for Limnology (M,D) ☒ (2018)
Hasler Lab
680 N. Park Street
Madison, WI 53706
  p. (608)262-3014
  kpelland@wisc.edu
  https://limnology.wisc.edu/
  f: https://www.facebook.com/centerforlimnology/
  t: @WiscLimnology
Chair:
  Kenneth Potter, (D), Johns Hopkins, Ng
Director:
  Jake Vander Zanden, (D), McGill, 1999, On
Professor:
  Calvin B. DeWitt, (D), Michigan, 1963, Ob
  Linda K. Graham, (D), Michigan, 1975, Ob
  John A. Hoopes, (D), MIT, 1965, No
  John E. Kutzbach, (D), Wisconsin, 1966, Pe
  Joy Zedler, (D), Wisconsin, 1968, Ob
Associate Professor:
  Emily Stanley, (D), Arizona State, 1993, Ob
Assistant Professor:
  Sarah Hotchkiss, (D), Minnesota, 1998, Ob
  Carol Lee, (D), Washington, 1998, Ob
  Zheng-yu Liu, (D), MIT, 1991, Op
  Katherine McMahon, (D), California (Berkeley), 2002, Ng
  Chin Wu, (D), MIT, Ng
Emeritus:
  Michael S. Adams, (D), California (Riverside), 1968, Ob

Anders W. Andren, (D), Florida State, 1972, Oc
David E. Armstrong, (D), Wisconsin, 1966, Oc
Steve Carpenter, (D), Wisconsin, 1979, Ob
James F. Kitchell, (D), Colorado, 1970, Ob
William C. Sonzogni, (D), Wisconsin, 1974, Oc

Department of Geography (B,M,D) (2019)
550 N. Park St.
Room 160 Science Hall
Madison, WI 53706
p. (608) 262-2138
exley@wisc.edu
http://www.geography.wisc.edu
f: https://www.facebook.com/UWMadisonGeography/
t: @UWMadisonGeog
Programs: GIS Development (PSM)
Certificates: GIS Fundamentals; Advanced GIS
Enrollment (2018): B: 111 (38) M: 21 (4) D: 43 (9)
Professor:
  Joseph Mason, GmSd
  John W. Williams, (D), Brown, 1999, PeZyPl
Associate Professor:
  Erika Marin-Spiotta, Cb
Assistant Professor:
  Christian Andresen, Zy
  Ken Keefover-Ring, Zg

Dept of Atmospheric & Oceanic Sciences (B,M,D) (2016)
1225 W. Dayton Street
Atmospheric, Oceanic & Space Science Building
Madison, WI 53706-1695
p. (608)262-2828
aos@aos.wisc.edu
http://www.aos.wisc.edu
Enrollment (2011): B: 24 (14) M: 32 (9) D: 34 (5)
Chair:
  Gregory J. Tripoli, (D), Colorado State, 1986, Am
Affiliate:
  Tracey Holloway, (D), Princeton, 2001, AscZn
Professor:
  Steven A. Ackerman, (D), Colorado State, 1987, Zr
  Ankur R. Desai, (D), Penn State, 2006, Asm
  Matthew H. Hitchman, (D), Washington, 1985, As
  Zhengyu Liu, (D), MIT, 1991, Op
  Jonathan E. Martin, (D), Washington, 1992, Am
  Galen McKinley, (D), MIT, 2002, OcpOg
  Michael C. Morgan, (D), MIT, 1994, Am
  Grant W. Petty, (D), Washington, 1990, AsZrAm
  John A. Young, (D), MIT, 1966, AstOp
Senior Scientist:
  Edwin W. Eloranta, (D), Wisconsin, 1972, Yr
Affiliate :
  Chris Kucharik, (D), Wisconsin, 1997, As
Associate Professor:
  Daniel J. Vimont, (D), Washington, 2001, As
Affiliate:
  Samuel Stechmann, (D), Courant Inst, 2008, ZnAs
Assistant Professor:
  Larissa E. Back, (D), Washington, 2007, As
  Tristan S. L'Ecuyer, (D), Colorado State, 2001, As
Adjunct Associate Professor:
  Jeffrey R. Key, (D), Colorado, 1988, ZrAs
Adjunct :
  Andrew Heidinger, (D), Colorado State, 1998, Zr
Adjunct Professor:
  James Kossin, (D), Colorado State, 2000, As
  Steven Platnick, (D), Arizona, Zr
Emeritus:
  Linda M. Keller, (M), Wisconsin, 1971, As
  Francis P. Bretherton, (D), Cambridge, 1961, Zg
  Stefan L. Hastenrath, (D), Bonn, 1959, As
  Donald R. Johnson, (D), Wisconsin, 1965, As
  John E. Kutzbach, (D), Wisconsin, 1966, As
  John M. Norman, (D), Wisconsin, 1971, So
  Robert A. Ragotzkie, (D), Wisconsin, 1953, Ob
  Pao-Kuan Wang, (D), California (Los Angeles), 1978, As

Research Specialist:
  Dierk T. Polzin, (B), Wisconsin, As

Dept of Soil Science (B,M,D) (2015)
1525 Observatory Drive
Madison, WI 53706-1299
p. (608) 262-2633
slspeth@wisc.edu
http://www.soils.wisc.edu
Enrollment (2011): B: 34 (4) M: 16 (2) D: 8 (2)
Chair:
  William L. Bland, (D), Wisconsin, 1984, Sp
Professor:
  Phillip W. Barak, (D), Hebrew, 1988, ScGzRw
  William F. Bleam, (D), Cornell, 1984, Sc
  James G. Bockheim, (D), Washington, 1972, Sf
  William J. Hickey, (D), California (Riverside), 1990, Sb
  King-Jau S. Kung, (D), Cornell, 1984, Sp
  Birl Lowery, (D), Oregon, 1980, Sp
  Frederick W. Madison, (D), Wisconsin, 1972, Sd
  J. Mark Powell, (D), Texas A&M, 1989, So
  Stephen J. Ventura, (D), Wisconsin, 1989, ZrSo
Associate Professor:
  Teri C. Balser, (D), California (Berkeley), 2000, Sb
  Nick J. Balster, (D), Idaho, 1999, Sf
  Carrie A.M Laboski, (D), Minnesota, 2001, So
  Joel A. Pedersen, (D), California (Los Angeles), 2001, So
Assistant Professor:
  Matthew D. Ruark, (D), Purdue, 2006, So
  Douglas J. Soldat, (D), Cornell, 2007, So
Emeritus:
  Larry G. Bundy, (D), Iowa, 1973, So
  Robin F. Harris, (D), Wisconsin, 1972, Sb
  Philip A. Helmke, (D), Wisconsin, 1971, Sc
  Keith A. Kelling, (D), Wisconsin, 1974, So
  Wayne R. Kusssow, (D), Wisconsin, 1966, So
  Kevin McSweeney, (D), Illinois, 1984, SdaSf
  John M. Norman, (D), Wisconsin, 1971, Spo
  E. Jerry Tyler, (D), North Carolina, 1975, Sd

Geological Engineering (B,M,D) (2018)
1415 Engineering Drive
2205 Engineering Hall
Madison, WI 53706-1691
p. (608) 890-2662
likos@wisc.edu
http://www.gle.wisc.edu
Programs: Geological Engineering
Enrollment (2018): B: 92 (0) M: 9 (0) D: 5 (0)
Chair:
  William J. Likos, (D), Colorado Mines, 2000, NgSp
Professor:
  Jean M. Bahr, (D), Stanford, 1987, Hw
  Kurt L. Feigl, (D), Gt
  Laurel Goodwin, California (Berkeley), 1988
  Tracey Holloway, (D), Princeton, As
  Steven P. Loheide, (D), Stanford, 2006, HwSp
  Clifford Thurber, (D), MIT, 1981, Ysx
  Basil Tikoff, (D), Minnesota, 1997, Gc
  Chin Wu, (D), MIT, Onp
Associate Professor:
  Mike Cardiff, Stanford, 2010, Hw
  Dante Fratta, (D), Georgia Tech, 1999, Yx
  James M. Tinjum, (D), Wisconsin, 2003, Ng
Assistant Professor:
  Greeshma Gadikota, (D), Columbia, 2014, GzCqu
  Matt Ginder-Vogel, (D), Stanford, 2006, Sc
  Andrea Hicks, (D), Illinois (Chicago), Zn
  Hiroki Sone, (D), Stanford, 2012, NrYxGt
  Lucas Zoet, (D), Penn State, Gl
Emeritus:
  Tuncer B. Edil, (D), Northwestern, 1973, Ng
  Bezalel C. Haimson, (D), Minnesota, 1968, Nr
  Herbert F. Wang, (D), MIT, 1971, YgHwYx

University of Wisconsin, Milwaukee
Dept of Geosciences (B,M,D) (2017)
P.O. Box 413
Milwaukee, WI 53201-0413
  p. (414) 229-4561
  geosci-office@uwm.edu
  http://www4.uwm.edu/letsci/geosciences/
  f: https://www.facebook.com/UWMgeosci
  Enrollment (2014): B: 75 (17) M: 19 (8) D: 10 (1)
Chair:
  Stephen Q. Dornbos, (D), S California, 2003, PeiPo
Professor:
  Timothy J. Grundl, (D), Colorado Mines, 1988, HwCl
  Mark T. Harris, (D), Johns Hopkins, 1988, GsrGd
  John L. Isbell, (D), Ohio State, 1990, Gs
  Keith A. Sverdrup, (D), Scripps, 1981, YsGt
Associate Professor:
  Barry I. Cameron, (D), N Illinois, 1998, Gvi
  Dyanna M. Czeck, (D), Minnesota, 2001, Gc
  Margaret L. Fraiser, (D), S California, 2005, PiePg
  Lindsay J. McHenry, (D), Rutgers, 2004, GzaXg
  Shangping Xu, (D), Princeton, 2005, Hg
Assistant Professor:
  Julie Bowles, (D), California (San Diego), 2005, Ym
  Erik L. Gulbranson, (D), California (Davis), 2011, CgPe
Adjunct Professor:
  Daniel T. Feinstein, (M), Wisconsin, 1986, Hw
Emeritus:
  Douglas S. Cherkauer, (D), Princeton, 1972, HwGe
  William F. Kean, Jr., (D), Pittsburgh, 1972, Ym
  Norman P. Lasca, (D), Michigan, 1965, GmlSd
  Peter M. Sheehan, (D), California (Berkeley), 1971, Pie

University of Wisconsin, Oshkosh
Geology Dept (B) O☒ (2018)
645 Dempsey Trail
Oshkosh, WI 54901-8649
  p. (920) 424-4460
  mode@uwosh.edu
  http://www.uwosh.edu/geology/
  f: http://www.facebook.com/UWOgeology
  Programs: Geology; Earth Science Secondary Education
  Administrative Assistant: Courtney Maron
  Enrollment (2018): B: 40 (15)
Chair:
  William N. Mode, (D), Colorado, 1980, Gl
Professor:
  Eric E. Hiatt, (D), Colorado, 1997, GsOcGu
  Maureen A. Muldoon, (D), Wisconsin, 1999, Hw
  Timothy S. Paulsen, (D), Illinois, 1997, Gc
  Jennifer M. Wenner, (D), Boston, 2001, Gi
Assistant Professor:
  Benjamin W. Hallett, (D), Rensselaer, 2012, Gpx
  Joseph E. Peterson, (D), N Illinois, 2010, PgGr
Emeritus:
  Norris W. Jones, (D), Virginia Tech, 1968, Gi
  Gene L. La Berge, (D), Wisconsin, 1963, Eg
  James W. McKee, (D), Louisiana State, 1967, Gr
  Brian K. McKnight, (D), Oregon State, 1970, Gdu
  Thomas J. Suszek, (M), Minnesota (Duluth), 1991, Gg
Instrumentation Specialist:
  James A. Amato, (M), Wisconsin (Milwaukee), 2017, GgZi

University of Wisconsin, Parkside
Dept of Geosciences (B) ☒ (2018)
Box 2000
900 Wood Road
Kenosha, WI 53141
  p. (262) 595-2744
  li@uwp.edu
  Programs: Earth Sciences; Environmental Geosciences
  Certificates: NONE
  Enrollment (2018): B: 27 (7)
Chair:
  Li Zhaohui, (D), SUNY (Buffalo), 1994, ClGzHw

Professor:
  John D. Skalbeck, (D), Nevada (Reno), HwYgGg
Assistant Professor:
  Rachel Headley, (D), Washington, 2011, Gl
Research Associate:
  Julie Kindelman, (D), Hs
Emeritus:
  Gerald A. Fowler, PgGg
  Allan F. Schneider, (D), Gl
  James H. Shea, (D), Gs

University of Wisconsin, Platteville
Dept of Geography & Geology (B) (2015)
Platteville, WI 53818
  p. (608) 342-1791
  rawlingj@uwplatt.edu
  http://www.uwplatt.edu/geography/
Professor:
  Charles W. Collins, (D), Nova, 1985, Gm
Associate Professor:
  Richard A. Waugh, (D), Wisconsin, 1995, Gc
Assistant Professor:
  Mari A. Vice, (D), S Illinois, 1993, Gd
Emeritus:
  William A. Broughton, (B), Wisconsin, 1938, Em
  Kenneth A. Shubak, (M), Michigan, 1966, Pi

University of Wisconsin, River Falls
Dept of Plant & Earth Science (B) ☒ (2019)
410 S. Third Street
River Falls, WI 54022
  p. (715) 425-3345
  holly.dolliver@uwrf.edu
  http://www.uwrf.edu/pes/geol/
  Program Assistant: Sue Freiermuth
  Enrollment (2016): B: 50 (0)
Chair:
  Holly A.S. Dolliver, (D), Minnesota, 2007, GmSd
Professor:
  Kerry L. Keen, (D), Minnesota, 1992, HwGse
  Ian S. Williams, (D), California (Santa Barbara), 1981, Yg
Lecturer:
  Kevin Thaisen, (D)

University of Wisconsin, Stevens Point
Dept of Geography and Geology (B,M) (2017)
2001 Fourth Avenue, SCI D332
Stevens Point, WI 54481
  p. (715) 346-2629
  David.Ozsvath@uwsp.edu
  http://www.uwsp.edu/geo/
  Programs: Geoscience (B); Geography (B); Spatial Analysis (B); GIS (M)
  Certificates: GIS
  Administrative Assistant: Mary Clare Sorenson
  Enrollment (2017): B: 63 (27)
Professor:
  Kevin P. Hefferan, (D), Duke, 1992, GctGe
  Neil C. Heywood, (D), Colorado, 1988, Zy
  Eric J. Larson, (D), Oregon State, 2001, Zr
  Karen A. Lemke, (D), Iowa, 1988, GmZy
  David L. Ozsvath, (D), SUNY (Binghamton), 1985, Hw
  Keith W. Rice, (D), Kansas, 1989, Zir
  Michael E. Ritter, (D), Indiana, 1986, Am
Associate Professor:
  Samantha W. Kaplan, (D), Wisconsin, 2003, GnPlGd
  Ismaila Odogba, (D), Louisville, 2009, Zui
Instructor:
  Lisa L. Theo, (M), Wisconsin, 1998, Zy

University of Wisconsin, Superior
Dept of Natural Sciences (B,M) (2015)
PO Box 2000
Belknap & Catlin
Superior, WI 54880

p. (715) 394-8322
natsci@uwsuper.edu
Professor:
William Bajjali, (D), Ottawa, 1994, HwCsg
Associate Professor:
Andy Breckenridge, (D), Minnesota, 2005, GlnGs
Assistant Professor:
Kristin E. Riker-Coleman, (D), Minnesota, 2008, Gg

University of Wisconsin, Whitewater
Geography, Geology, and Environmental Science (B) ☒ (2018)
800 W Main Street
Whitewater, WI 53190
p. (262) 472-1071
geography@uww.edu
http://www.uww.edu/cls/departments/geography-geology-env-sci
Programs: Geography; Geology; Environmental Science
Certificates: GIS
Enrollment (2017): B: 136 (28)
Professor:
Peter Jacobs, (D), Wisconsin, 1994, SdaGm
Dale Splinter, (D), Oklahoma State, 2006, Hs
Associate Professor:
Prajukti Bhattacharyya, (D), Minnesota, 2000, Gce
Eric Compas, (D), Wisconsin, 2008, ZiRw
John Frye, (D), Georgia, 2011, AtZr
Rex Hanger, (D), California (Berkeley), 1992, PigGs
Assistant Professor:
Rocio Duchesne, (D), Montclair State, 2015, Zri
Stephen J. Levas, (D), Ohio State, 2015, CmGe
Jeff Olson, (D), Ohio State, 2013, Ziu

University of Wisconsin-Madison
Department of Geoscience (B,M,D) ●◎ (2018)
1215 West Dayton St
Madison, WI 53706
p. (608) 262-8960
geodept@geology.wisc.edu
http://geoscience.wisc.edu/geoscience/
Programs: Geoscience
Enrollment (2009): B: 58 (17) M: 20 (6) D: 42 (4)
Chair:
D. Charles DeMets, (D), Northwestern, 1988, Ym
Professor:
Alan R. Carroll, (D), Stanford, 1991, Gs
Kurt Feigl, (D), MIT, 1991, Yd
Dante Fratta, (D), Georgia Tech, 1999, Ng
Laurel B. Goodwin, (D), California (Berkeley), 1988, Gc
Clark M. Johnson, (D), Stanford, 1986, CgcCs
D. Clay Kelly, (D), North Carolina, 1999, Pm
Stephen R. Meyers, (D), Northwestern, 2003, PcOoGs
Shanan Peters, (D), Chicago, 2003, Gs
Eric E. Roden, (D), Maryland, 1990, PgSbCl
Bradley S. Singer, (D), Wyoming, 1990, Cc
Clifford H. Thurber, (D), MIT, 1981, YsGv
Basil Tikoff, (D), Minnesota, 1994, Gc
Harold Tobin, (D), California (Santa Cruz), 1995, Yg
John W. Valley, (D), Michigan, 1980, CsGxz
Huifang Xu, (D), Johns Hopkins, 1993, Gz
Senior Scientist:
Brian L. Beard, (D), Wisconsin, 1992, Cc
John Fournelle, (D), Johns Hopkins, 1989, GiZn
Associate Professor:
Michael A. Cardiff, (D), Stanford, 2010, HwYg
Assistant Professor:
Annie Bauer, (D), MIT, 2017, Cca
Chloe Bonamici, (D), Wisconsin, 2013, GxCa
Ken Ferrier, (D), California (Berkeley), 2009, GmZo
Shaun Marcott, (D), Oregon State, 2011, GlPe
Lucas Zoet, (D), Penn State, 2012, GllGm
Emeritus:
Mary P. Anderson, (D), Stanford, 1973, Hw
Jean M. Bahr, (D), Stanford, 1987, Hw
Carl J. Bowser, (D), California (Los Angeles), 1965, Cl
Philip E. Brown, (D), Michigan, 1980, EmGzp

Charles W. Byers, (D), Yale, 1973, Grs
Nikolas I. Christensen, (D), Wisconsin, 1963, Yx
David L. Clark, (D), Iowa, 1957, Pm
Dana H. Geary, (D), Harvard, 1986, Pg
Louis J. Maher, Jr., (D), Minnesota, 1961, Pl
Levi Gordon Medaris, Jr., (D), California, 1966, GipSa
David M. Mickelson, (D), Ohio State, 1971, Glm
Herbert F. Wang, (D), MIT, 1971, YgHwYx
Museum Director:
Richard Slaughter, (D), Iowa, 2001, Pg
Geology Librarian:
Marie Dvorzak, (M), Zn
Related Staff:
David J. Hart, (D), Wisconsin, 2000, Hg
Thomas S. Hooyer, (D), Iowa State, 1999, Hg
David Krabbenhoft, (D), Wisconsin, 1988, ClGeHw
James M. Robertson, (D), Michigan, 1972, Eg
Cooperating Faculty:
Kenneth R. Bradbury, (D), Wisconsin, 1982, Hw

Wyoming

Casper College
Department of Earth and Environmental Sciences (A) ⌐ (2018)
School of Science
125 College Drive
Casper, WY 82601
p. (307) 268-2513
mconnely@caspercollege.edu
http://www.caspercollege.edu/physcience/geology/
Programs: Geology (A); Environmental Science (A); GIS (A)
Certificates: GIS (A)
Enrollment (2017): A: 32 (8)
Klaenhammer Earth Science Chair:
Melissa Connely, (M), Utah State, 2000, GsPev
Instructor:
Karen Sue McCutcheon, (M), Zeg
Jeff Sun, (M), Texas State, Zi
Kent A. Sundell, (D), California (Santa Barbara), 1985, GgvPv
Beth Wisely, (D), Oregon, 2005, YgHwGc

Central Wyoming College
Earth, Energy, Environment (A) (2015)
2660 Peck Avenue
Riverton, WY 82501
p. (307) 855-2000
ssmaglik@cwc.edu
http://www.cwc.edu/academics/programs/earthenviron
f: https://www.facebook.com/cwc.edu/
Enrollment (2015): A: 5 (0)
Professor:
Suzanne M. Smaglik, (M), Colorado Mines, 1987, GgZeCg
Assistant Professor:
Jacki Klancher, (M), Zi

Eastern Wyoming College
Eastern Wyoming College (A) (2015)
3200 West C. Street
Torrington, WY 82240
p. (307) 532-8330
stuart.nelson@ewc.wy.edu
http://www.ewc.wy.edu/
Instructor:
Christopher Wenzel, (D), Wyoming, 2019, SbfHg

Northwest College
Northwest College (2015)
231 West 6th Street
Powell, WY 82435
p. 307.754.6405
Mark.Kitchen@northwestcollege.edu
http://www.nwc.cc.wy.us/

University of Wyoming
Dept of Geology and Geophysics (B,M,D) ●◎ (2019)

Dept. 3006
1000 E. University Ave.
Laramie, WY 82071
p. (307) 766-3386
carrick@uwyo.edu
http://www.uwyo.edu/geolgeophys/
Programs: BS in Geology, BA in Earth Science, BS in Environmental Geology and Geohydrology, MS in Geology, MS in Geophysics, PhD in Geology
Enrollment (2015): B: 157 (26) M: 28 (7) D: 31 (2)

Head:
  Carrick M. Eggleston, (D), Stanford, 1991, ClqCb
Professor:
  Kevin R. Chamberlain, (D), Washington (St. Louis), 1990, CcGt
  Mark T. Clementz, (D), California (Santa Cruz), 2002, Pg
  Carol D. Frost, (D), Cambridge (UK), 1984, CcGit
  Neil F. Humphrey, (D), Washington, 1987, GlmYh
  Barbara E. John, (D), California (Santa Barbara), 1987, Gci
  Subhashis Mallick, (D), Hawaii, 1987, YseYg
  James D. Myers, (D), Johns Hopkins, 1979, Gi
  Bryan N. Shuman, (D), Brown, 2001, PeGne
  Kenneth W.W Sims, (D), California (Berkeley), 1995, CgGvCa
  Ye Zhang, (D), Indiana, 2005, HyGeYu
Senior Scientist:
  Janet Dewey, (M), Auburn, 1993, CaGg
  Susan M. Swapp, (D), Yale, 1982, Gp
Associate Professor:
  Michael J. Cheadle, (D), Cambridge, 1989, Yg
  Po Chen, (D), S California, 2005, YseYg
  Ellen Currano, (D), Penn State, 2008, Pbe
  Kenneth G. Dueker, (D), Oregon, 1994, Ys
  Dario Grana, (D), Stanford, 2013, Ye
  John P. Kaszuba, (D), Colorado Mines, 1997, CgGze
  Brandon McElroy, (D), Texas, 2009, GsmGr
  Clifford S. Riebe, (D), California (Berkeley), 2000, ClGme
Associate Scientist:
  Laura Vietti, (D), Minnesota, 2014, Pv
Assistant Professor:
  James Chapman, (D), Arizona, 2018, Gct
  Kimberly Lau, (D), Stanford, 2017, Cb
  Andrew D. Parsekian, (D), Rutgers, 2011, Yg
  Simone Runyon, (D), Arizona, 2017, EdGip
Adjunct Professor:
  Vladimir Alvarado, (D), Minnesota, 1996, Np
  Eric A. Erslev, (D), Harvard, 1981, GctGg
  Peter H. Hennings, (D), Texas, 1991, Gc
  Ranie Lynds, (D), Wyoming, 2005, Gsr
Emeritus:
  B. Ronald Frost, (D), Washington, 1973, GpiEd
  Robert R. Howell, (D), Arizona, 1980, XgZr
Department Secretary:
  Deborah Prusia

Western Wyoming Community College
Western Wyoming Community College (2015)
2500 College Drive
Rock Springs, WY 82901
p. (307) 382-1662
webmaster@wwcc.wy.edu
http://www.wwcc.wy.edu/academics/geology/default.htm

Algeria

Algerian Petroleum Institute
Algerian Petroleum Institute (B) (2015)
Avenue 1er Novembre
Boumerdes, Skikda, Oran 35000
p. +213 (0) 24 81 90 56
iap@iap.dz
http://www.iap.dz/

Centre Universitaire de Khemis Miliana
Institut de Sciences de la Nature et de la Terre (B) (2015)
Route de Theniet El-Had
Khemis Miliana, Khemis Miliana 44225
p. +213-27-66-42-32
kouben55@hotmail.fr
http://www.cukm.org/

Universite Abou Bekr Belkaid de Tlemcen
Faculté des sciences de la nature et de la vie et sciences de la terre et de l'univers (B) (2015)
BP 119
Tlemcen 13000
p. (+213) 040.91.59.09
webcri@mail.univ-tlemcen.dz
http://snv.univ-tlemcen.dz/

Universite Badji Mokhtar
Faculte des Sciences de la Terre: Amenagement du territoire, Geologie, Hydraulique, Mines (B) (2015)
BP 12
Annaba 23000
d.fst@univ-annaba.dz
http://www.univ-annaba.org/

Universite D'Oran
Faculté des Sciences de la Terre, de la Géographie et de l'Amenagement du Territoire (B) (2015)
Rue du Colonel Lofti
Es-Senia, El-Mnouar, Oran 31000
p. +213 (0) 41 58 19 47
contact@univ-oran.dz
http://www.univ-oran.dz/facultes/F_Terre/index.html

Universite de Annaba
Inst of Natural Sciences (2015)
BP 12
El Hadjar
Annaba
http://www.univ-annaba.org/

Universite de Batna
Dept de Sciences de la Terre (B) (2015)
1, Rue Chahid Boukhlouf Mohamed
El Hadi, Batna 5000
p. (+213) 033.86.06.02
recteur@univ-batna.dz
http://www.univ-batna.dz/

Universite de Jijel
Dept de Geologie (B,M,D) (2015)
Ouled Aissa, Jijel
p. +213 (34) 498016
webmaster@univ-jijel.dz
http://www.univ-jijel.dz/

Universite de Mentouri
Dept des Sciences de la Terre/Geologie (B) (2015)
Campus Ahmed Zouaghi
B.P. 325 Route d Ain El Bey
, Constantine 25000
p. 213 031 90 38 52
univ-constantine@fr.fm
http://www.umc.edu.dz/fst/

Dept de l'Amenagement du territoire (B) (2015)
Campus Ahmed Zouaghi
B.P. 325 Route d Ain El Bey
, Constantine 25000
p. (+213) 31.90.02.07
univ-constantine@fr.fm
http://www.umc.edu.dz/vf/index.php/recherche-scientifique/annuaire-des-laboratoires/110-faculte-des-sciences-de-la-terre-/398-de

Universite des Sciences et de la Technologie
Inst of Earth Sciences (2015)
Houari Boumediene
B.P. 139 Dar El Beida

Eldjazair

Universite des Sciences et de la Technologie Houari Boumediene
Faculte de Sciences de la Tere, Geographie et Amenagement du Territoire (B) (2015)
USTHB-IST BP 32 El Alia
Bab-Ezzouar Alger 16123
    p. (+213) 21247904
    aouabadi@usthb.dz
    http://www.usthb.dz/fst/

Universite Djillali Liabes Sidi Bel Abbes
Faculte des Sciences de la Nature et de la vie (B) (2015)
BP 89
Sidi Bel Abbes 22000
    p. +213 (48) 543018
    mbouziani@univ-sba.dz
    http://www.univ-sba.dz/

Universite Kasdi Merbah Ouargla
Dept de Geologie (B) (2015)
Route de Ghardaia
Ouargla
    p. +213 (29) 712468
    info@ouargla-univ.dz
    http://193.194.92.30/spip/dept/filiere-hydrocarbures-chimie.htm

Dept des Hydrocarbures (B) (2015)
Route de Ghardaia
Ouargla
    p. +213 (29) 712468
    info@ouargla-univ.dz
    http://www.ouargla-univ.dz/index-FR.htm

Universite M'hamed Bouguerra de Boumerdes
Faculte des Hydrocarbures et de la chimie (B) (2015)
Avenue de Ildependance
Boumerdes 35000
    p. +213 (24) 816420
    doyen_fhc@umbb.dz
    http://www.umbb.dz/

Universite Saad Dahlab, Blida
Dept des Sciences de l'Eau et de l'Environnement (B) (2015)
BP 270
Blida 9000
    p. +213 (25) 433625
    contact@univ-blida.dz
    http://www.univ-blida.dz/fac_ingenieur/index.html

Angola

Universidade Agostinho Neto
Faculdade de Geologia (B) (2015)
Av. 4 de Fevereiro
No. 7
2 andar
Luanda C. P. 815
    p. +244-222-333-816
    info@geologia-uan.com
    http://www.geologia-uan.com/

Universidade Independente de Angola
Engenharia dos Recursos Naturais e Ambiente (B) (2016)
Rua da Missao
Bairro Morro Bento II - Corimba
Luanda
    p. (+244) 222 33 89 70
    unia@unia.ao
    http://www.unia.ao/curso.php?cr=6

Vice-Rector:
    Nuno Nascimento Gomes, (M), Lisbon, 1992, GesGu

Universite de Angola
Dept of Geology (2015)
Avenida 4 de Fevereiro 7
Caixa postal 815-C
Luanda

Argentina

Servicio Geologico Minero Argentino
Servicio Geologico Minero Argentino (2015)
Av. Julio Argentino Roca 651. P.B
Capital Federal
    mjanit@mecon.gov.ar
    http://www.segemar.gov.ar/

Universidad de Buenos Aires
Departamento de Ciencias Geológicas - FCEN (B,D) O☒ (2018)
Ciudad Universitaria-Pabellón 2
Intendente Güiraldes 2160
Ciudad Autónoma de Buenos Aires C1428EHA
    p. (5411) 528-58249
    geologia@gl.fcen.uba.ar
    http://www.gl.fcen.uba.ar/
    Programs: Geology, Paleontology

Universidad Nacional de Catamarca
Facultad de Tecnologia Y Ciencias Aplicadas (2015)
Maximio Victoria No. 55
San Fernando del Valle de Catamarca (470, Catamarca CP 4700
    p. (54) 0383-4435 112 int. 112
    daa@tecno.unca.edu.ar
    http://tecno.unca.edu.ar

Universidad Nacional de Cordoba
Facultad de Ciencias Exactas, Fisicas Y Naturales (2015)
Apartado 35-2060 UCR
San Pedro de Montes de Oca
San Jose, Cordoba
    p. +03514332098
    geologia@ucr.ac.cr
    http://www.geologia.ucr.ac.cr/

Universidad Nacional de Jujuy
Instituto de Geologia y Mineria (2015)
San Salvador de Jujuy , Jujuy
    narce@idgym.unju.edu.ar
    http://www.idgym.unju.edu.ar/

Universidad Nacional de la Pampa
Dept of Geology (2015)
Uruguay 151, 6300 Santa Rosa
La Pampa
    susanapaccapelo@exactas.unlpam.edu.ar
    http://www.exactas.unlpam.edu.ar/

Universidad Nacional de la Patagonia San Juan Bosco
Facultad de Ciencias Naturales (2015)
Comodoro Rivadavia, Chubut
    secacademica@unp.edu.ar
    http://www.unp.edu.ar/

Universidad Nacional de La Plata
School of Astronomy and Geophysics (2015)
Paseo del Bosque s/n
B1900FWA
    p. (0221)-423-6593
    academic@fcaglp.unlp.edu.ar
    http://www.fcaglp.unlp.edu.ar/

Universidad Nacional de Río Cuarto
Dept. of Geology (B,M,D) O (2016)
ocampanella@gmail.com
mvillegas@exa.unrc.edu.ar
Rio Cuarto , Cordoba C.P. X5804BYA
    p. (0358) 467-6198
    webgeo@exa.unrc.edu.ar
    http://geo.exa.unrc.edu.ar/
    Enrollment (2016): B: 32 (0) M: 5 (0) D: 2 (0)
Associate Professor:
    Monica B. Villegas, (M), UNRC, 1995, Gsc
Associate Scientist:
    Juan Enrique Otamendi, (D), Nacional de Río Cuarto, 2000, CgaCp
Adjunct Professor:
    Hector Daniel Origlia, (M), Arizona State, 1999, NgZn

Universidad Nacional de Rio Negro
Dept. of Geology (2015)
Building Perito Moreno - Pacheco 460
General Roca , Rio Negro
    sedevallemedio@unrn.edu.ar
    http://www.unrn.edu.ar

Universidad Nacional de Salta
Facultad de Ciencias Naturales (2015)
Salta
    p. (0387) 425-5413
    decnat@unsa.edu.ar
    http://naturales.unsa.edu.ar/

Universidad Nacional de San Juan
Facultad de Ciencias Exactas, Físicas y Naturales (2015)
Parque de Mayo
5400 San Juan
    comunicacionfcefyn1@gmail.com
    http://exactas.unsj.edu.ar/

Universidad Nacional de San Luis
Facultad de Ciencias Exactas, Fisico, Matematicas Y Naturales (2015)
Ejército de los Andes 950
San Luis D5700HHW
    p. +54 (2652) 424027
    sacadfmn@unsl.edu.ar
    http://webfmn.unsl.edu.ar/index.php

Universidad Nacional de Tucuman
Facultad de Ciencias Naturales E Instituto Miguel Lillo (D) (2015)
Miguel Lillo 205
S.M. de Tucuman
San Miguel de Tucuman, Tucuman CP4000
    p. 0381-4239456
    info@csnat.unt.edu.ar
    http://www.csnat.unt.edu.ar/

Universidad Nacional del Comahue
Dept of Geology and Petroleum (2015)
School of Engineering
Buenos Aires 1400
Neuquen 18300 , Patagonia
    p. +54-299-4490300
    sacadfi@uncoma.edu.ar
    http://www.uncoma.edu.ar/

Universidad Nacional del Sur
Departamento de Geologia (2015)
San Juan 670
Primer Piso
Bahia Blanca
Buenos Aires 8000
    p. 54-(0291)-4595147
    secgeo@uns.edu.ar
    http://www.uns.edu.ar/

Australia

Australian National University
Research School of Earth Sciences (M,D) (2015)
Mills Rd.
Canberra, ACT 0200
    p. +61 2 6125 3406
    Director.rses@anu.edu.au
    http://rses.anu.edu.au/
Emeritus:
    Patrick De Deckker, (D), Adelaide, 1981, GuPmGn

Dept of Earth & Marine Sciences (B,M,D) (2015)
47 Daley Road
Canberra, ACT 0200
    p. 02 6125 2056
    ems@anu.edu.au
    http://ems.anu.edu.au/
Reader:
    D C "Bear" McPhail, (D), Princeton, 1991, Cl

Curtin University
Dept of Mining Engineering (B,M,D) (2015)
P. O. Box 597
Kalgoorlie
    e.topal@curtin.edu.au

School of Earth and Planetary Sciences (B,M,D) O (2019)
GPO Box U1987
Perth, WA 6845
    p. +618 92667968
    eps-admin@curtin.edu.au
    https://scieng.curtin.edu.au/schools/school-of-earth-and-planetary-sciences/
    f: https://www.facebook.com/CurtinEPS
    t: @CurtinGeology
    Programs: Geology, Earth Sciences, Surveying, GIS, Petroleum Geoscience, Mineral Exploration and Mining Geology

Dept of Exploration Geophysics (B,M,D) (2017)
GPO Box U1987
Perth, WA 6845
    p. +61 8 9266-3565
    Geophysics-GeneralEnquiries@curtin.edu.au
    http://www.geophysics.curtin.edu.au/
    f: https://www.facebook.com/ExplorationGeophysicsCurtinUniversity
    t: @CurtinGeophys
    Enrollment (2014): B: 27 (13) M: 26 (4) D: 28 (6)
Head of Department:
    Andrej Bona, (D), Calgary, 2002, Yx
Professor:
    Boris Gurevich, (D), Moscow, 1988, Ye
    Anton W. Kepic, (D), British Columbia, 1990, YexYg
    Maxim Lebedev, (D), Mowcow Inst Phys & Tech, 1990, YxNr
Senior Lecturer:
    Vassily Mikhaltsevitch, (D), Kaliningrad State, 1997, Ye
    Andrew Peter Squelch, (D), Nottingham, 1998, NmYeZe
Research Fellow:
    Michael Carson, (D), Univ Coll (Dublin), 2000, Yx
    Aleksandar Dzunic, (M), Belgrade, 1989, YgsYe
Senior Scientist:
    Stephanie Vialle, (D), Paris 7, 2008, YxCqNp
Associate Professor:
    Brett D. Harris, (D), Curtin, 2002, HwYve
    Roman Pevzner, (D), Moscow State, 2004, Yxx
    Milovan Urosevic, (D), Curtin, 1998, YerYx
Lecturer:
    Robert Galvin, (D), Curtin, 2006, Ye
    Stanislav Glubokovskikh, (D), Lomosov (Moscow), 2012, Yxe
    Mahyar Madadi, (D), Inst Adv Studies Basic Sci (Zanjan), 1997, Ye
    Andrew Pethick, (D), Curtin, 2013, YevYg
    Konstantin Tertyshnikov, (D), Curtin, 2014, Ye
    Sasha Ziramov, (M), Belgrade, 2005, Ye

Senior Technical Officer:
   Dominic J. Howman, (B), Curtin, 1994, Ye

Federation University Australia
Dept of Geology (B,M,D) ● (2016)
Federation University
P. O. Box 663
Ballarat, VIC 3353
   p. 613 53279354
   k.dowling@federation.edu.au

Flinders University
School of the Environment (B,M,D) ● (2017)
GPO Box 2100
Adelaide, SA 5001
   p. +61 8 8201 7577
   cheryl.mcdonnell@flinders.edu.au
   http://www.flinders.edu.au/science_engineering/environment/
   Enrollment (2012): B: 44 (14) M: 15 (5) D: 37 (4)
Professor of Hydrogeology:
   Okke Batelaan, (D), Free (Brussels), Hw
Research Fellow:
   Peter Cook, (D), Flinders, Hw
ARC Future Fellow:
   Adrian Werner, (D), Queensland, Hw
Academic Status:
   Nancy Cromar, (D), Napier (UK), Zn
Professor:
   Howard Fallowfield, (D), Dundee (UK), Zn
   Iain Hay, (D), Washington, 1989, Zyy
   Patrick Hesp, (D), Sydney, GmZy
   Andrew Millington, (D), Sussex (UK), ZuyZr
   Craig Simmons, (D), Adelaide, Hw
Director ARA:
   Jorg Hacker, (D), Bonn, ZrAs
Academic Status:
   John Edwards, (D), Adelaide, Zn
   Andew Love, (D), Hw
Associate Professor:
   Erick Bestland, (D), Oregon, GsSa
   Beverley Clarke, (D), Adelaide, Zyn
   Huade Guan, (D), New Mexico Tech, 2005, HgsHg
   Udoy Saikia, (D), Flinders, Zn
DECRA Research Fellow:
   Margaret Shanafield, (D), Nevada, Hw
Sessional Lecturer:
   Harpinder Sandhu, (D), Lincoln (NZ), Zn
   Maria Zotti, Zn
Senior Lecturer:
   David Bass, (D), New England, Zy
   Kirstin Ross, (D), S Australia, Zn
Adjunct Senior Lecturer:
   Simon Benger, (D), Australian Nat, Ziy
   John Hutson, (D), Natal, Spc
   Vincent Post, (D), Amsterdam, Hw
Adjunct Lecturer:
   Stephen Fildes, (M), Adelaide, Zri
   Ben van den Akker, (D), Flinders, Zn
Lecturer:
   Dylan Irvine, (D), Flinders, Hw
   Mark Lethbridge, (D), S Australia, Ziy
   Graziela Miot da Silva, (D), Fed Rio Grande do Sul (Brazil), Gug
   Michael Taylor, (D), Flinders, Zn
   Ilka Wallis, (D), Flinders, HqCg
   Harriet Whiley, (D), Flinders, Zn
Academic Status:
   Gour Dasvarma, (D), Australian Nat, Zn
Adjunct Professor:
   Jim Smith, Zn
Research Fellow:
   Eddie Banks, (D), Flinders, Hw
Adjunct Academic Status:
   Samantha de Ritter, (M), Flinders, Ze
Academic Status:
   Glenn Harrington, (D), Flinders, Hw
   Andrew McGrath, (D), ZrAs

Geoscience Australia
Geoscience Australia ☒ (2017)
GPO Box 378
Canberra, ACT 2601
   p. +61 2 6249 9111
   ref.library@ga.gov.au
   http://www.ga.gov.au/
   t: @GeoAusLibrary

Government of South Australia - Department for State Development
Geological Survey of South Australia (2017)
GPO Box 320
Adelaide, SA 5001
   p. +61 8 8463 3000
   DSD.minerals@sa.gov.au
   http://www.minerals.statedevelopment.sa.gov.au
   t: @PACE_sagov

James Cook University
Geosciences (B,M,D) O☒ (2018)
Building 034
College of Science & Engineering
James Cook University
Townsville, QLD 4811
   p. 07 4781 6947
   eric.roberts@jcu.edu.au
   https://www.jcu.edu.au/college-of-science-and-engineering
   Enrollment (2016): B: 85 (54) M: 6 (2) D: 25 (7)
Professor:
   MIchael I. Bird, (D), Australian Nat, 1988, ClsSa
Senior Scientist:
   Karen E. Joyce, (D), UQ, 2005
Associate Professor:
   Zhaoshan Chang, (D), Washington State, 2003, EmCce
   Eric M. Roberts, (D), Utah, 2005, GsPvCc
   Carl Spandler, (D), Univ, 2005, GxEmCg
Postdoctoral Research Fellow:
   Arianne Ford, (D), James Cook, 2008, GqZiEg
Lecturer:
   James J. Daniell, (D), Sydney, 2010, GumYr
   Jan Marten Huizenga, (D), Vrije (Amsterdam), 1995, GpCgGc
   Christa Placzek, (D), Arizona, 2006, CclGe
   Ioan Sanislav, (D), James Cook, 2009, GctGp
   Peter W. Whitehead, (B), La Trobe, 1986, Ggv
Adjunct Professor:
   Robert Henderson, (D), Victoria (NZ), 1967, GtPiGg
Postdoctoral Research Fellow & Lecturer:
   Hannah L. Hilbert-Wolf, (D), James Cook, 2016, GsCc

Economic Geology Research Centre (A,B,M,D) O☒ (2017)
EGRU
James Cook University
Townsville, QLD 4811
   p. +61 7 4781 4726
   egru@jcu.edu.au
   http://www.jcu.edu.au/economic-geology-research-centre-egru
   Certificates: Professional Development Training

La Trobe University
Environmental Geoscience (B,M,D) ☒ (2018)
Environmental Geoscience
La Trobe University
Victoria 3086
Melbourne, VIC 3086
   p. +61 3 9479 1273
   john.webb@latrobe.edu.au
   http://www.latrobe.edu.au/environmental-geoscience
   Programs: Environmental Geoscience
   Enrollment (2018): B: 39 (5) D: 6 (1)
Associate Professor:
   John A. Webb, (D), Queensland, 1982, GeHwGa
Lecturer:
   David Steart, (D), Victoria, 2003, GgPb

Susan Q. White, (D), La Trobe, 2005, GmgGe
Other:
   Vincent J. Morand, (D), Sydney, 1987, GgcGt

Macquarie University
Dept of Earth and Planetary Sciences (B,M,D) O☒ (2018)
North Ryde
Sydney, NSW 2109
   p. 61-2-98508426
   eps-admin@mq.edu.au
   http://www.eps.mq.edu.au/
   f: https://www.facebook.com/MQeps

Department of Environmental Sciences (B,M,D) (2016)
North Ryde, NSW 2109
   neil.saintilan@mq.edu.au
   http://www.mqu.edu.au/

Monash University
School of Earth, Atmosphere and Environment (B,M,D) ☒ (2018)
PO Box 28E
Clayton, VIC 3800
   p. +61 3 99054884
   earth-atmosphere-environment@monash.edu
   http://www.earth.monash.edu.au/
Professor:
   Peter A. Cawood, (D), Sydney, 1980, GtcGg

New South Wales Resources & Energy
New South Wales Resources & Energy (2017)
NSW Resource & Energy, PO Box 344
Hunter Regional Mail Centre
, NSW 2310
   p. 1300 736 122
   geologicalsurvey.info@trade.nsw.gov.au
   http://www.resourcesandenergy.nsw.gov.au/

Queensland University of Technology
School of Earth, Environmental & Biological Sciences
(B,M,D) ●☒ (2019)
GPO Box 2434
2 George Street
Brisbane, QLD 4001
   p. 61 7 3138 2324
   sef.eebsenquiry@qut.edu.au
   https://www.qut.edu.au/science-engineering/research/research-disciplines/geology-and-geochemistry
   f: https://www.facebook.com/QUTEarthScienceUndergraduatePage/
   Administrative Assistant: Sharon McCann
   Enrollment (2018): B: 120 (20) M: 7 (4) D: 25 (3)
Professor:
   Peter R. Grace, (D), Queensland, 1989, Sb
   David A. Gust, (D), Australian Nat, 1982, GiCpGt
   Balz Kamber, (D), Bern, 1995, GpCgGz
Senior Research Fellow :
   Charlotte M. Allen, (D), Virginia Tech, CcaGi
Research Fellow:
   Henrietta Cathey, (D), Utah, 2006, Giv
Associate Professor:
   Scott E. Bryan, (D), Monash, 1999, GitGs
Senior Lecturer:
   Craig Sloss, (P), Wollongong (Australia), 2005, GsrGm
   Jessica Trofimovs, (D), Monash, 2003, GsvGu
Research Associate:
   David Flannery, (D), New South Wales, 2012, PoCaXg
Lecturer:
   Oliver M. Gaede, (D), W Australia, YeNrYs
   Patrick C. Hayman, (D), Monash, 2009, EgGvi
   David T. Murphy, (D), Queensland, 2002, CgEgCc
   Luke Nothdurft, (D), Queensland Tech, 2009, GdCmGg
   Clemens Scheer, (D), Bonn, 2008, SbGe
   Christoph Schrank, (D), Toronto, 2009, GcqGt
Emeritus:
   John Rigby, Pb

Analytical Laboratory Coordinator :
   Wan-Ping (Sunny) Hu, (D), Canterbury, Ca
Other:
   Shane Russell, (B), Queensland Tech, Ca
Technician - Geology:
   Alex Hepple, (M), Queensland Tech, 2014, ZerGg
Senior Technician - Geology :
   Will Stearman, (B), Queensland Tech, 2011, Geg

Southern Cross University
School of Environment, Science & Engineering (A,B,M,D)
O⌐ (2018)
P. O. Box 157
Lismore, NSW 2480
   p. +61 2 6620 3766
   ese@scu.edu.au
   http://scu.edu.au/environment-science-engineering/
   f: https://www.facebook.com/scu.sese
Professor:
   Bill Boyd, (D), Glasgow, 1982, Pe
   Bradley Eyre, (D), Queensland Tech, Gs
   Peter Harrison, (D), James Cook
Associate Professor:
   Danny Bucher, (D), S Cross, 2001
   Malcolm William Clark, (D), Ca
   Simon Hartley, (B), New England
   Graham Jones, (D), James Cook, Ca
   John Doland Nichols, (D), N Arizona
   Amanda Reichelt-Brushett, (D), S Cross
Lecturer:
   Graeme Palmer, (B), Australian Nat
   Sumith Pathirana, (D), Kent State, Zr
   Kathryn Taffs, (D), Adelaide, Pe
   Michael Whelan, (D), New England, 2013, Zir

University of Adelaide
School of Earth and Environmental Sciences (2015)
Adelaide, SA 5000
   p. +61 8 8303 3999
   graham.heinson@adelaide.edu.au
   http://www.ees.adelaide.edu.au/

Australian School of Petroleum (B,M,D) ● (2017)
Santos Petroleum Engineering Building
North Terrace
Adelaide, SA 5005
   p. 61-8-8313-8000
   admin@asp.adelaide.edu.au
   http://www.asp.adelaide.edu.au/

University of Canberra
Institute for Applied Ecology (2015)
   sarre@aerg.canberra.edu.au
   http://enterprise.canberra.edu.au/WWW/www-crcfe.nsf

School of Education, Science, Technology and Math ☒ (2019)
P. O. Box 1
Belconnen, ACT
   George.Cho@canberra.edu.au
   http://scides.canberra.edu.au/rehs/

University of Melbourne
School of Earth Sciences (B,M,D) (2016)
Melbourne, VIC 3010
   p. +61 3 8344 9866
   head@earthsci.unimelb.edu.au
   http://www.earthsci.unimelb.edu.au

University of New England
Earth Sciences (A,B,M,D) O (2015)
Armidale
NSW
2351
Armidale, NSW 2351

p. +61-2-67732101
geology@une.edu.au
http://www.une.edu.au/about-une/academic-schools/school-of-environmental-and-rural-science/research/life-earth-and-environment/e

Associate Professor:
John R. Paterson, (D), Macquarie, 2005, PigPs

Lecturer:
Phil R. Bell, (D), Alberta, 2011, PgvPg
Luke Milan, (D), GtcGp
Nancy Vickery, (D), New England, GgeYg

Adjunct Professor:
Ian Metcalfe, (D), Leeds (UK), 1976, PmGtg

Emeritus:
Paul Ashley, (D), Macquarie, EgGeEm
Peter Flood, (D), EcGt

University of New South Wales
School of Biological, Earth & Environmental Sciences (B,M,D) (2015)
UNSW
Sydney, NSW 2052
p. 61 2 93852961
bees@unsw.edu.au
http://www.bees.unsw.edu.au/index.html
Enrollment (2013): B: 80 (24) M: 5 (0) D: 14 (4)

Professor:
Michael Archer, (D), Pv
Andy Baker, (D), HwCs
James Goff, (D), Gsm
Suzanne J. Hand, (D), Pv
Martin van Kranendonk, (D), Queens (Canada), GxPeGd

Associate Professor:
David R. Cohen, (D), New South Wales, 1990, Cet
Bryce Kelly, (D), New South Wales, 1995, HwAcGe
Shawn Laffan, (D), Ziy

Dr:
Ian Graham, (D), UTS Sydney, 1995, GxEg

Lecturer:
Catherine Chague-Goff, (D), CgGs

Adjunct Professor:
Derecke Palmer, (D), New South Wales, 2001, Ye

Dr:
Paul G. Lennox, (D), Monash, 1985, GcgZe

Visiting Professor:
Colin R. Ward, (D), New South Wales, 1971, EcGd

University of Newcastle
Discipline of Earth Sciences (B,M,D) O (2015)
School of Environmental & Life Sciences
University Drive
Callaghan, NSW 2308
p. +61 2 4921 8976
name.surname@newcastle.edu.au
http://www.newcastle.edu.au/school/environ-life-science/index.html
Enrollment (2014): B: 132 (121) D: 5 (3)

Associate Professor:
Silvia Frisia, (D), Milan (Italy), 1991, ClPeGd
Phil Geary, (D), UWS, HwSoHw
Gregory Hancock, (D), ZiGmm

Lecturer:
Judy Bailey, (D), Newcastle, 1993, Ec
David Boutelier, (D), Gtc
Alistair Hack, (D), GcCgEm
Anthony Kiem, (D), HqAsZi
Bill Landenberger, (D), Newcastle, 1997, GiCg
Danielle Verdon-Kidd, (D), HqAsZi

University of Queensland
School of Earth Sciences (B,M,D) O (2015)
Faculty of Science
Steele Building
Brisbane, QLD 4072
p. +61 7 3365 1180
enquiries@earth.uq.edu.au
http://www.earth.uq.edu.au/
Enrollment (2015): B: 106 (57) M: 38 (5) D: 60 (6)

University of South Australia
School of Natural and Built Environments (B) O (2016)
GPO Box 2471
Adelaide, SA 5001
Julie.Mills@unisa.edu.au
http://www.unisa.edu.au/it-engineering-and-the-environment/natural-and-built-environments/

Research Fellow:
Juliette Woods, (D), Adelaide, Hwq

University of Sydney
School of Geosciences (A,B,M,D) ☒ (2019)
Room 34
Madsen Building F09
Sydney, NSW 2006
geosciences.enquiries@sydney.edu.au
https://sydney.edu.au/science/schools/school-of-geosciences.html
f: https://www.facebook.com/sydneyunigeo/
t: @sydneyunigeo

Professor:
Jonathan Aitchison, (D), UNE, 1989, GrtPm
Geoffrey L. Clarke, (D), Melbourne, 1987, Gp
Dietmar Muller, (D), Scripps, 1989, Yrg

University of Tasmania
ARC Center of Excellence in Ore Deposits (M,D) O (2017)
Private Bag 79
Hobart, TAS 7001
steve.calladine@utas.edu.au
www.utas.edu.au/codes

School of Earth Sciences (B,M,D) (2015)
Private Bag 79
Hobart, TAS 7001
p. 61 3 6226 2476
Rose.Pongratz@utas.edu.au
http://www.utas.edu.au/earth-sciences/home

Director:
J Bruce Gemmell, (D), Dartmouth, 1987, Eg
Ross R. Large, (D), New England, Eg

Professor:
David R. Cooke, (D), Monash, Eg
Anthony J. Crawford, (D), Melbourne, 1983, Git
Jocelyn McPhie, (D), New England, Gv

Associate Professor:
Ron F. Berry, (D), Flinders, Gc

Lecturer:
Garry J. Davidson, (D), Tasmania, Cs
Peter J. McGoldrick, (D), Melbourne, Ce
Anya Reading, (D), Ysg
Michael Roach, (N), Tasmania, Yg

University of Western Australia
School of Earth & Environment (B,M,D) (2015)
M004
35 Stirling Highway
Crawley, WA 6009
p. 6488 1921
enquiry-see@uwa.edu.au
http://www.see.uwa.edu.au/

University of Wollongong
School of Earth & Environmental Sciences (B,M,D) (2018)
Northfields Avenue
Wollongong, NSW 2522
p. 61 2 4221 4419
anutman@uow.edu.au
http://www.uow.edu.au/science/eesc/

Austria

Geological Survey of Austria
Geologische Bundesanstalt von Osterreich (2015)
Neulinggasse 38, A 1030
Vienna
p. 43-1-712 56 74
office@geologie.ac.at
http://www.geolba.ac.at/

Karl-Franzens-Universitaet Graz
Inst for Earth Science (A,B,M,D) O (2017)
Heinrichstraße 26
Graz 8010
p. +43 316 380-5587
erdwissenschaften@uni-graz.at
http://erdwissenschaften.uni-graz.at/index_en.php
Enrollment (2016): A: 10 (1) B: 40 (35) M: 20 (14) D: 12 (2)

Leopold-Franzens-Universitaet Innsbruck
Inst fuer Geologie (2016)
Innrain 52-f
6020 Innsbruck
p. +43 (0) 512 / 507 - 96125
dekanat-geowiss@uibk.ac.at
http://www.uibk.ac.at/fakultaeten/geo_und_atmosphaeren-wissenschaften/

Montan Universitaet
Dept of Applied Geological Sciences and Geophysics (2015)
Inst. for Prospecting
A-8700 Leoben
geologie@mu-leoben.at

Technische Universitaet Graz
Earth Sciences (2015)
Rechbauerstrasse 12
A-8010 Graz
p. +43(0)316 873 6360
martin.dietzel@tugraz.at

University Leoben
Chair of Geology and Economic Geology (B,M,D) ⊠ (2019)
Montauniversitat Leoben
Peter-Tunnerstrasse 5
Leoben A-8700
p. +43 3842 402 6101
geologie@unileoben.ac.at
http://www.unileoben.ac.at/~buero62/geologie/geologie.html
Programs: Applied Geosciences and Geophysics
Enrollment (2018): B: 300 (0) M: 35 (0) D: 4 (0)
Professor:
  Frank Melcher, (D), Leoben, 1993, GgEgm
Senior Scientist:
  Heinrich Mali, (D), Leoben, GcEgNm
Associate Professor:
  Walter Prochaska, (D), Vienna, Egn
  Gerd Rantitsch, (D), Graz, Ggq
Research Associate:
  Viktor Erlandsson, (M), Gothenburg, 2018, Gg
  Peter Onuk, (D), Leoben, GgCa

University of Innsbruck
Institute of Geology (A,B,M,D) ⊠ (2019)
Innrain 52f
A-6020
Innsbruck, Austria 6020
p. +43 512 507-54300
regina.gratzl@uibk.ac.at
http://www.uibk.ac.at/geologie

University of Salzburg
Inst fuer Geowissenschaften (2015)
Hellbrunner Strasse 34
A-5020 Salzburg
p. +43-662-8044-5200 oder 5400
geo@sbg.ac.at
http://www.uni-salzburg.at/geo

University of Vienna
Dept of Meteorology and Geophysics (B,M,D) ● (2016)
Althanstrasse 14
A-1090 Vienna
p. 0043 1 4277 53701
img-wien@univie.ac.at
http://imgw.univie.ac.at/en/imgw/
Enrollment (2016): B: 40 (30) M: 20 (16) D: 10 (4)
Univ.-Prof.:
  Goetz Bokelmann, (D), Princeton, 1992, Ygs
o.Univ.-Prof.:
  Reinhold Steinacker, (D), Innsbruck, 1975, Ams

Dept of Geography (2017)
Althanstrasse 14
Vienna A-1090
p. 00431427753401
elisabeth.aufhauser@univie.ac.at
http://www.univie.ac.at/Geologie/
Head:
  Thilo Hofmann, (D)

Azerbaijan

Baku State University
The Faculty of Geology (2015)
23 Z. Khalilov Street
370145 Baku
m.mahluga@rambler.ru
http://www.ceebd.co.uk/ceeed/un/az/az003.htm

Bangladesh

Jahangirnagar University
Dept of Geological Sciences (2015)
University Campus, Savar, Dhaka
Dhaka
p. 0088 2 7791045-51 (Ext. 1402)
rabiulju@gmail.com
http://www.juniv.edu/home.php?pg=faculty_science

Rajshahi University
Dept of Geology & Mining (2016)
Motihar, Rajshahi
Rajshahi
p. +880 721-750041
chair.geology@ru.ac.bd
http://www.ru.ac.bd/geol/index.htm

University of Dhaka
Department of Geology (B,M,D) ⊠ (2017)
Curzon Hall Campus
Dhaka 1000
p. (008) 2- 966 1920-73 (Ext. 730
geology@du.ac.bd
http://www.du.ac.bd/academic/department_item/GLG
Enrollment (2017): B: 190 (37) M: 85 (31) D: 2 (0)
Professor:
  Kazi Matin Ahmed, (D), UK, 1994, HwGgHg

Belgium

Faculte Polytechnique de Mons
Dept of Geology (2015)
Fue de Houdain 9
7000 Mons
yves.quinif@fpms.ac.be

Faculte Univ Catholique de Mons
Dept of Geology (2015)
Chaussee de Binche 151
B-7000 Mons

Facultes Univ Notre Dame de la Paix
Dept Geologie (2015)
Rue de Bruxelles 61
B-5000 Namur
vincent.hallet@fundp.ac.be

Geological Survey of Belgium
Service geologique de Belgique (2015)
Department VII
of the Royal Belgian Institute of Natural Sciences (RBINS)
13, Rue Jenner
1000 Brussels
p. +32 (0)2.788.76.61
bgd@natuurwetenschappen.be
http://www.naturalsciences.be/geology/

Ghent University
Dept of Geology (B,M,D) ●⊠ (2018)
Krijgslaan 281-S8
Gent B-9000
p. +32 9 264 45 94
we13@ugent.be
https://www.ugent.be/we/geologie/nl/nl
f: https://www.facebook.com/GeologieUGent
Programs: Geology; Marine and Lacustrine Science and Management; Physical Land Resources
Head:
  Marc De Batist, (D), Ghent, 1989, GnuGs
Professor:
  Stephen Louwye, (D), Pm
  Kristine EMT Walraevens, (D), Ghent, 1987, HwCqYu
Associate Professor:
  Veerle Cnudde, (D), Ghent, 2005, EsHt
  David Van Rooij, (D), Yr
Assistant Professor:
  Sebastien Bertrand, (D), 2001, GsClPc
  Johan De Grave, (D), Cc
  Stijn Dewaele, (D), KU Leuven, 2004, EgGzCg
  Thomas Hermans, (D), HwYeu
  Thijs Vandenbroucke, (D), Ps

Katholieke Universiteit Leuven
Dept of Earth and Environmental Science (B,M,D) (2015)
Department of Earth & Environmental Sciences
Redingenstraat 16
B-3000 Leuven-Heverlee
p. (+32) 016 321450
erik.mathjis@ees.kuleaven.be
http://ees.kuleuven.be/
Professor:
  Patrick Degryse, (D), GaCeGf
  Jan Elsen, (D), Gz
  Philippe Muchez, (D), Univ, 1988, EmCgGd
  Manuel Sintubin, (D), GtcGa
  Robert Speijer, (D), Utrecht, 1994, PmcOo
  Rudy Swennen, Gso
Associate Professor:
  Okke Batelaan, (D), Hw
  Sarah Fowler, (D), Gip
Assistant Professor:
  Marijke Huysmans, (D), Hw

Earth and Environmental Sciences (B,M,D) ⊠ (2017)
Celestijnenlaan 200E PO 2408
  Leuven BE-3001
    p. +32 16 327580
    gert.verstraeten@kuleuven.be
    http://ees.kuleuven.be
    Programs: Geography; Geology

Universite Catholique de Louvain
Dept of Geology (2015)
Bâtiment Mercator
place Louis Pasteur 3
B-1348 Louvain-la-Neuve
p. (32) 10 47 32 97
monique.descampsl@uclouvain.be
http://www.uclouvain.be/geo.html

Universite de Liege
Dept of Geology (B,M,D) (2015)
Batiment B18 (secretariat)
Sart Tilman
Liege B-4000
p. 32 4 366 22 51
TH.Billen@ulg.ac.be
http://www.ulg.ac.be/geolsed/geologie
Chair:
  Frederic P. Boulvain, (D), Brussels, 1990, GdgGs
chargé de cours:
  Nathalie Fagel, (D), GsuGe
  Hans-Balder Havenith, (D), Liège, Ge
Professor:
  Andre-Mathieu Fransolet, (D), Liege, Gz
  Emmanuelle Javaux, (D), Dalhousie, 1999, Pl
  Jacqueline Vander Auwera, (D), Louvain-la-Neuve, 1988, Giv
Emeritus:
  Edouard Poty, (D), Liege, 1981, PiGsPe

Universite Libre de Bruxelles (ULB)
Département des Sciences de la Terre et de l'Environnement (2015)
50, Ave. F. Roosevelt
Brussels 1050
    Julie.Paraire@ulb.ac.be

University of Mons
Dept of Geology and Applied Geology (M,D) ○ (2016)
rue de Houdain, 9
Mons
    gfa@umons.ac.be

Vrije University Brussel
Earth System Sciences (D) ⊠ (2017)
ESSC-WE-VUB
Faculty of Sciences
Pleinlaan 2
Brussels 1050
    p. 0113226293394
    phclaeys@vub.be
    http://we.vub.ac.be/~essc
    f: https://www.facebook.com/amgcvub/
    Enrollment (2017): D: 9 (2)
Head:
  Philippe Claeys, (D), California (Davis), 1993, CgPgXc

Bolivia

Universidad Mayor de San Andres
Dept of Geology (2015)
P. O. Box 12198
Campus Universitario Cota Cota, Calle 27
La Paz
    p. (591-2) 2441983
    webmaster@umsa.bo
    http://www.geologia.umsa.bo/

Botswana

Geological Survey of Botswana
Geological Survey of Botswana (2015)
Khama One Avenue
Plot 1734
Lobatse

p. +267 5330327
http://www.gov.bw/en/Ministries--Authorities/Ministries/Ministry-of-Minerals-Energy-and-Water-Resources-MMWER/Departments1/Depar

University of Botswana
Dept of Geology (B) (2015)
4775 Notwane Rd.
Private Bag UB00704
Gaborone
  p. (267)355 2529
  geology@mopipi.ub.bw
  http://www.ub.bw/home/ac/1/fac/1/dep/79/Geology/

Dept of Environmental Science (B) (2015)
4775 Notwane Road
Private Bag UB 0022
Gaborone
  p. (267) 355-0000
  Sedilamoya.Pansiri@mopipi.ub.bw
  http://www.ub.bw/learning_faculties.cfm?pid=588

Brazil

Federal University of Bahia Geophysics
CPGG (2015)
Salvador, BA
  geofisic@ufba.br
  http://www/pppg.ufba.br/

Geological Survey of Brazil
Servico Geologico do Brasil (2015)
Av. SGAN- Quadra 603 - conjunto J
Parte A - 1º andar
Brasilia - DF 70830-030
  cprmsede@df.cprm.gov.br
  http://www.cprm.gov.br/

Universidad Federal de Rio Grande do Sul
Institute de Geociencia (2015)
Avenida Bento Gonçalves, 9500
Porto Alegre, RS - 91.501-970
  p. +55 51 3308-6337
  igeo@ufrgs.br
  http://www.ufrgs.br/english/the-university/institutes-faculties-and-schools/institute-of-geoscience

Universidade de Brasília
Instituto de Geociências (2015)
Campus Universitário Darcy Ribeiro ICC - Ala Central
CEP 70.910-900 - Brasilia DF
Caixa Postal 04465.
 CEP 70919-970
  p. +61 3307-2433
  igd@unb.br
  http://www.igd.unb.br/

Universidade de Sao Paulo
Inst de Geociencias (2015)
Rua do Lago
562 Cidade Universitaria
05508-080 São Paulo
  p. 63.025.530/0007-08
  gmgigc@usp.br
  http://www.igc.usp.br/

Inst Oceanografico (2015)
Cidade Universitaria, Av. Prof. Luciano Gualberto, Travessa 3 no 380, 05508-900 Sao Paulo - SP

Universidade do Vale do Rio Dos Sinos
Inst de Geociencias (2015)
Av. Unisinos, 950 - Cristo Rei, 93 000 Sao Leopoldo - RS

Universidade Federal da Bahia
Inst de Geociencias (2015)
Rua Augusto Viana s/n - Canela, 40 410 Salvador - BA
  olivia@ufba.br

Universidade Federal de Minas Gerais
Inst de Geociencias (2015)
Av. Antonio Carlos 6.627
 Pampulha, , 31 270 Belo Horizont
  p. 55(31) 3409-5420
  dir@igc.ufmg.br
  http://www.igc.ufmg.br/

Universidade Federal de Ouro Preto
Dept de Geologia (2015)
Campus Morro do Cruzeiro
35 400 Ouro Preto
  p. (31) 3559-1600
  web@degeo.ufop.br
  http://www.degeo.ufop.br/

Universidade Federal de Pernambuco
Centro de Tecnologia - Geociencias (2015)
Av. Agamenon Magalhaes s/n - Santo Amaro, 50 000 Recife - PE
  p. (81) 2126.8105
  secretaria.proacad@ufpe.br
  http://www.ufpe.br/proacad/index.php?option=com_content&view=article&id=150&Itemid=138

Universidade Federal do Ceara
Inst de Geociencias (2015)
Av. Da Universidade, 2853 - Benfica, 60 000 Fortaleza - CE

Universidade Federal do Para
Inst. de Geociencias (B,M,D) ○◐ (2018)
Augusto Correa Street, 01
Campus Guamá
Belem, Pará 66075-110
  p. +55 91 3201 7107
  dirig@ufpa.br
  http://www.ig.ufpa.br/
  Programs: We have at the Geosciences Institute (IG): bachelor's degree in Geophysics, Geology, Meteorology and Oceanography; Master's and Doctoral's programs in Geophysics, Geology and Geochemistry; Professional Masters in Environmental Sciences, Water Resources, Natural Hazards and Risk Management in the Amazon, Environmental Sciences Teaching; Specialization in Water and Environmental Management, Mine Geology and Open-pit mining, Sedimentary Basin Analysis.

Universidade Federal Fluminense
Dept de Geociencias (2015)
Av. Gal. Milton Tavares de Souza
  p. +55 (21) 2629-5951
  gge@vm.uff.br
  http://www.uff.br/degeografia/

Universidade Federal Rural do Rio de Janeiro
Inst de Geociencias (2015)
Km 47 da Antigua Rodovia Rio/Sao Paulo - Seropedico, 23 460 Itaguai - RJ

University of Campinas Geoscience Institute
Inst de Geosciencias (2019)
Caixa Postal: 6152
13083-970
Campinas
  secretaria.prpg@reitoria.unicamp.br
  http://www.ige.unicamp.br/

Bulgaria

Bulgarian Academy of Sciences
Geological Institute (D) (2015)
Acad.G.Bonchev st. bl.24
Sofia 1113
   p. +0359 2 8723 563
   geolinst@geology.bas.bg
   http://www.geology.bas.bg/
Professor Dr, DSc:
   Kristalina Christova Stoykova, (D), Bulgaria Acad Sci, 2008, PemPi
Associate Professor:
   Iliana Boncheva, (D)
   Thomas Noubar Kerestedjian, (D), Bulgaria Acad Sci, 1990, GzeEg

Mining and Geology University
Geology Dept (2015)
Studentski Grad
Sofia 1756
   p. 8060221
   dekangpf@mgu.bg

Ministry of Environment and Water
Ministry of Environment and Water (2015)
22 Maria Louiza Blvd.
Sofia, 1000
   minister@moew.government.bg
   http://www.moew.government.bg/

Sofia University St. Kliment Ohridski
Dept of Geology and Paleontology (2016)
1504 Sofia
15 Tsar Osvobodtel Blvd.
Geology of Fossil, Fuel, Organic Petrology, Organic Geochemistry room 276
   gigeor@gea.uni-sofia.bg
   http://www.uni-sofia.bg/newweb/faculties/geo/departments/geo

Dept of Cartography & GIS (B,M,D) (2015)
1504 Sofia
15 Tsar Osvobodtel Blvd.
GIS, General Physical Geography, Landscape Ecology, Agro-ecology
Sofia, Sofia 1504
   p. 9308261
   popov@gea.uni-sofia.bg
   http://www.gis.gea.uni-sofia.bg/

University of Mining and Geology, St. Ivan Rilski
Faculty of Geology (2015)
Studentski grad
1100 Sofia
   dekangpf@mgu.bg
   http://www.mgu.bg/frame/html

Burkina Faso

Universite de Ouagadougou
L'Unité de Formation et de Recherche en Sciences de la Vie et de la Terre (UFR/SVT) (B) (2015)
03 BP 7021, Ougadougou 03
   p. +226 50-30-70-64/65
   webmaster@univ-ouaga.bf
   http://www.univ-ouaga.bf.html/formations/ufr_SVT/frFormtnsSVTdip1.html

Burundi

Universite du Burundi
Dept of Earth Sciences (B) O (2015)
B.P. 2700
Bujumbura, Burundi
   p. 00257 22 22 55 56
   webmaster@ub.edu.bi
   http://www.ub.edu.bi/

Faculté des Sciences (2015)
B.P. 2700
Bjumbura
   p. 0025722222059
   http://www.ub.edu.bi/ub-fac6.php

Cameroon

Bamenda University of Science and Technology
Dept of Geology (B) (2015)
PO Box 277
Bamenda, NW Province
   p. +237 7726 1789
   http://www.bamendauniversity.com/

Universite de Buea
Dept of Geology and Environmental Science (B) (2015)
PO Box 63
Buea, South West Province
   p. 237-332-2134
   vpktitanji@yahoo.co.uk
   http://ubuea.cm/

Universite de Douala
FACULTY OF SCIENCES (B,M,D) (2015)
BP 2701
Douala, Cameroun
   p. (237) 33 40 75 69
   infos.fs@univ-douala.com
   http://www.facsciences-univ-douala.cm/index.php/contact

Universite de Dschang
Faculte D'Agronomie et des Sciences Agricoles (B,M) (2017)
POB 96
Dschang
   p. (237) 33 45 15 66
   agro.50tenair@gmail.com

Universite de Ngaoundere
School of Geology and Mining (B) (2015)
BP 454
Ngaoundere
   p. +(237) 225 2767
   http://www.univ-ndere.cm/index.php?LANG=EN&ETS=0433A030F35FE61&RUB=E2C0BE24560D78C

Universite de Yaounde 1
Dept des Sciences de la Terre (B) (2015)
BP 812
Yanounde
   p. (237) 222 56 60
   facsciences@uy1.uninet.cm
   http://www.facsciences.uninet.cm/act_aca_dst.html

Canada

Aboriginal Affairs and Northern Development Canada
Mineral Resources Division (2017)
Mineral Resources Division
PO Box 100
Iqaluit, NU X0A 0H0
   p. (867) 975-4293
   nunavutarchives@aandc.gc.ca
   http://nunavutgeoscience.ca/

Acadia University
Dept of Earth and Environmental Science (B,M) ●☒ (2018)
Huggins Science Hall
12 University Avenue HSH 327

Wolfville, NS B4P 2R6
  p. (902) 585-1208
  ees@acadiau.ca
  http://ees.acadiau.ca/
  f: https://www.facebook.com/Acadia-Department-of-Earth-and-Environmental-Science-335568173300748/
  Programs: Geology; Environmental Geoscience; Environmental Science
  Enrollment (2018): B: 45 (16) M: 7 (3)
Head:
  Ian S. Spooner, (D), Calgary, 1994, Ge
Professor:
  Sandra M. Barr, (D), British Columbia, 1973, GitGg
  Nelson O'Driscoll, (D), Ottawa, 2003, CgGgZn
  Peir K. Pufahl, (D), British Columbia, 2001, Gs
  Robert P. Raeside, (D), Calgary, 1982, GptGc
  Clifford R. Stanley, (D), British Columbia, 1988, Cea
Associate Professor:
  Alice Cohen, (D), British Columbia, 2011, ZnnZn
Geology Tech:
  Pam Frail, Gg

Brandon University
Dept of Geology (B,M) ●✉ (2019)
270 18th Street
Brandon, MB R7A 6A9
  p. (204) 727-9677
  somarina@brandonu.ca
  http://www.brandonu.ca/geology
  Programs: Geology
    Mineral Resources
    Energy Resources
    Environmental Geoscience
    Paleontology
  Enrollment (2018): B: 37 (20) M: 3 (1)
Professor:
  Rong-Yu Li, (D), Alberta, 2002, PgmGg
  A. Hamid Mumin, (D), W Ontario, 1994, EgGzt
  Simon A. J Pattison, (D), McMaster, 1992, GsoGd
  Alireza Somarin, (D), New England, 1999, Gig
Emeritus:
  Robert K. Springer, (D), California (Davis), 1971, Gi
  Harvey R. Young, (D), Queen's, 1973, Gd
Instructional Associate:
  Peter J. Adamo, (B), Brandon, 2000, Gg
Laboratory Director:
  Ayat Baig, (M), Lakehead, 2016, EdmEg
Other:
  Paul Alexandre, (D), Loraine, 2000, EgCgGz

Brock University
Dept of Earth Sciences (B,M) ●✉ (2018)
1812 Sir Isaac Brock Way
St. Catharines, ON L2S 3A1
  p. 905 688-5550
  earth@brocku.ca
  http://www.brocku.ca/mathematics-science/departments-and-centres/earth-sciences
  Programs: Earth Sciences; Environmental Geoscience
  Enrollment (2018): B: 65 (21) M: 17 (3)
Professor:
  Uwe Brand, (D), Ottawa, 1979, ClsAs
  Richard J. Cheel, (D), McMaster, 1984, Gs
  Frank Fueten, (D), Toronto, 1989, Gc
  Martin J. Head, (D), Aberdeen, 1990, Pl
  Francine G. McCarthy, (D), Dalhousie, 1992, Pm
  John Menzies, (D), Edinburgh (UK), 1976, GlmGs
Associate Professor:
  Gregory C. Finn, (D), Memorial, 1989, Gx
  Daniel P. McCarthy, (D), Saskatchewan, 1993, Gl
  Mariek Schmidt, (D), Oregon State, 2005, GivXg
Adjunct Professor:
  Paul Budkewitsch, (M), Toronto, 1990, GeZrHq
  Phil McCausland, (D), Gz
  Andrew W. Panko, (D), McMaster, 1985, Ge

Cape Breton University
Math, Physics, Geology (B) (2015)
P.O. Box 5300
Sydney, NS B1P 6L2
  p. (902) 539-5300
  fenton_isenor@cbu.ca
Instructor:
  Fenton M. Isenor, (M), Acadia, 2000, GgNgGm
Emeritus:
  Erwin L. Zodrow, (D), Pi

Capilano University
Geology Dept (A,B) (2015)
2055 Purcell Way
North Vancouver, BC V7J 3H5
  p. (604) 986-1911
  sciences@capilanou.ca
  http://www.capilanou.ca/programs/geology.html
Head:
  Dileep J A Athaide, (M), British Columbia, 1974, GgZge
Professor:
  Jennifer Getsinger, (D), British Columbia

Carleton University
Dept of Earth Sciences (B,M,D) ● (2015)
1125 Colonel By Drive
Ottawa, ON K1S 5B6
  p. (613) 520-5633
  earth.sciences@carleton.ca
  http://earthsci.carleton.ca/
  f: https://www.facebook.com/pages/Department-of-Earth-Sciences-Carleton-University/510369329037382
  t: @ErthSciCarleton
Professor:
  Keith Bell, (D), Oxford, 1964, Cc
  Brian L. Cousens, (D), California (Santa Barbara), 1990, GiCgc
  George R. Dix, (D), Syracuse, 1988, Gsd
  James E. Mungall, (D), McGill, 1993, GiEgCg
  R. Timothy Patterson, (D), California (Los Angeles), 1986, Pm
  Giorgio Ranalli, (D), Illinois, 1970, Yg
  Claudia Schroder-Adams, (D), Dalhousie, 1986, Pm
  George B. Skippen, (D), Johns Hopkins, 1966, Cg
  Richard P. Taylor, (D), Leicester, 1980, Cg
  David H. Watkinson, (D), Penn State, 1965, Em
Associate Professor:
  Gail M. Atkinson, (D), W Ontario, 1993, Ne
  John Blenkinsop, (D), British Columbia, 1972, Cc
  Sharon D. Carr, (D), Carleton, 1990, Gt
  Fred A. Michel, (D), Waterloo, 1982, Hw
Lecturer:
  Ildi Munro, (M), Waterloo, 1975, Pg
Adjunct Professor:
  Robert Berman, (D), British Columbia, 1983, Gp
  Steve L. Cumbaa, (D), Florida, 1975, Pv
  J. Allan Donaldson, (D), Johns Hopkins, 1960, Gs
  T. Scott Ercit, (D), Manitoba, 1986, Gz
  Harold Gibson, (D), Carleton, 1990, Em
  Simon Hanmer, (D), Chelsea (UK), 1977, Gc
  Mark D. Hannington, (D), Toronto, 1989, Em
  Jarmila Kukalova-Peck, (D), Charles (Prague), 1962, Pi
  Dale A. Leckie, (D), McMaster, 1983, Gs
  R. R. Rainbird, (D), Western, 1991, Gs
Emeritus:
  F. K. North, (D), Oxford, 1951, Go

Dalhousie University
Dept of Earth Sciences (B,M,D) ●✉ (2017)
Halifax, NS B3H 3J5
  p. (902) 494-2358
  earth.sciences@dal.ca
  http://earthsciences.dal.ca/index2.html
  Programs: Earth Sciences (minor,B,M,D); Geography (minor)
  Certificates: Geographic Information Science; Materials Science; Science Leadership and Communication

Enrollment (2017): B: 94 (39) M: 24 (9) D: 13 (2)
Chair:
 James M. Brenan, (D), Rensselaer, 1990, CptEm
Professor:
 John C. Gosse, (D), Lehigh, 1994, CcGmRn
 Djordje Grujic, (D), ETH (Switzerland), 1992, Gt
 Mladen Nedimovic, (D), Toronto, 2000, Ysg
 Grant D. Wach, (D), Oxford, 1993, GorGs
Retired:
 Nicholas Culshaw, (D), Ottawa, 1983, Gc
Associate Professor:
 Isabelle Coutand, (D), Rennes, Gg
 Yana Fedortchouk, (D), Victoria, 2006, CpEdGi
 Shannon Sterling, (D), Duke, 2005, HgCbHs
Assistant Professor:
 Lawrence Plug, (D), Alaska (Fairbanks), 2000, Gm
 Owen Sherwood, (D), Dalhousie, 2006, CsOoGo
Research Associate:
 Alan Ruffman, (M), Dalhousie, 1966, Ys
Senior Instructor:
 Mike Young, (M), Queens, 2003, GgcGp
Retired:
 Charles C. Walls, (M), Dalhousie, 1996, Zi
Instructor:
 Richard Cox, (D), Memorial, 1999, GzEdGg
 Chris Greene, (D), Ryerson, 2015, Zi
Honorary Research Associate:
 Prasanta Mukhopadhyay, (D), Jadavpur, 1971, Ec
Adjunct Professor:
 Sandra Barr, (D), British Columbia, 1973, Gx
 D. Brown, (B), Dalhousie
 John Calder, (D), Dalhousie, 1991, Ec
 C. Campbell, (D), Dalhousie
 T. Claire, (D), McMaster
 D Barrie Clarke, (D), Ediburgh, Gi
 M. Deptuck, (D), Dalhousie
 Jarda Dostal, (D), McMaster, 1974, Cg
 A. Dyke, (D), Colorado
 T. J. Fedak, (D), Dalhousie
 R. Fensome, (D), Nottingham
 D. Forbes, (D), British Columbia
 M. Fowler, (D), Newcastle upon Tyne
 D.G. Froese, (D), Calgary
 C. Gerbi, (D), Maine
 D. Gibson, (D), Simon Fraser
 J. Hanley, (D), Toronto
 C. Keen, (D), Cambridge
 Lisa M. Kellman, (D), Quebec (Montreal), 1998, Cg
 Y. Kettanah, (D), Southampton
 E. Kirby, (D), MIT
 T. Lakeman, (D), Alberta
 M. Lavigne, (D), Yale
 K. E. Louden, (D), MIT
 J. Marsh, (D), Maine
 T. Martel, (D), Dalhousie
 Michael Melchin, (D), Western, 1987, Pg
 D. Mosher, (D), Dalhousie
 Peta J. Mudie, (D), Dalhousie, 1980, Pl
 J. E. Mungall, (D), McGill
 J. B. Murphy, (D), McGill
 J. Brendan Murphy, (D), McGill, 1982, Gx
 Michael Parsons, (D), Stanford, Gg
 David Piper, (D), Cantab, 1969, Gs
 P. Pufhal, (D), British Columbia, 2002
 F. W. Richards, (M), Imperial Coll (UK)
 D. Risk, (D), Dalhousie
 Cliff Shaw, (D), W Ontario
 J. Shimeld, (M), Dalhousie
 D. W. Simpson, (D), Australian Nat
 S. Swinden, (D), Memorial
 H. Vincent, (D), Dalhousie
 J. Waldron, (D), Edinburgh
 C. Warron, (D), Oxford
 T. Webster, (D), Dalhousie
 D. M. Whipp, (D), Michigan
 B. Wilson, (D), Wales

Emeritus:
 H.B. S. Cooke, (D), Witwatersrand, 1947, Pv
 Martin R. Gibling, (D), Ottawa, 1978, Gs
 G. Clinton Milligan, (D), Harvard, 1961, Gc
 Peter H. Reynolds, (D), British Columbia
 Patrick J.C. Ryall, (D), Dalhousie, 1974, Yg
 David B. Scott, (D), Dalhousie, 1977, Pm
 Marcos Zentilli, (D), Queen's, 1974, Eg
University Teaching Fellow:
 Anne-Marie Ryan, (D), Dalhousie, 2006, GeZeCg
Administrator:
 Ann Bannon, (B), Dalhousie, 2001, Zn
Cooperating Faculty:
 Christopher Beaumont, (D), Dalhousie, 1973, Ys

Dept of Oceanography (B,M,D) (2018)
Life Sciences Centre
1355 Oxford Street
PO Box 15000
Halifax, NS B3H 4R2
 p. (902) 494-3557
 oceanography@dal.ca
 http://oceanography.dal.ca/
 Enrollment (2015): M: 22 (6) D: 31 (2)
Chair:
 Paul S. Hill, (D), Washington, 1992, Gs
Professor:
 Christopher Beaumont, (D), Dalhousie, 1973, Yg
 Bernard P. Boudreau, (D), Yale, 1985, OcCml
 Katja Fennel, (D), Rostock, 1998, Ob
 Jonathan Grant, (D), South Carolina, 1981, Ob
 Alex E. Hay, (D), British Columbia, 1981, OpnGu
 Dan Kelley, (D), Dalhousie, 1986, Op
 Markus Kienast, (D), British Columbia, 2002, Cs
 Hugh MacIntyre, (D), Deleware, 1996, Ob
 Anna Metaxas, (D), Dalhousie, 1994, Ob
 Jinyu Sheng, (D), Memorial, 1991, Op
 Christopher T. Taggart, (D), McGill, 1986, Ob
 Helmuth Thomas, (D), Rostock, 1997, Oc
 Keith R. Thompson, (D), Liverpool, 1979, Op
 Douglas Wallace, (D), Dalhousie, 1985, Oc
Associate Professor:
 Tetjana Ross, (D), Manitoba, 2003, Op
Assistant Professor:
 Christopher Algar, (D), Dalhousie, 2009, Obc
 David R. Barclay, (D), California (San Diego), 2011, OpNo
 Carolyn Buchwald, (D), MIT/WHOI, 2013, Oc
 Stephanie Kienast, (D), British Columbia, 2002, Gu
 Eric Oliver, (D), Dalhousie, 2011, Op
Lecturer:
 Barry R. Ruddick, (D), MIT, 1977, Op
Adjunct Professor:
 Mark Baumgartner, (D), Oregon, 2002, Ob
 Susanne Craig, (D), Strathclyde, 2000, Op
 Peter Cranford, (D), Dalhousie, 1998, Ob
 Claudio DiBacco, (D), California (San Diego), 1999, Ob
 Dale Ellis, (D), McMaster, 1976, Op
 Kenneth Frank, (D), Toledo, 1978, Ob
 Richard Greatbatch, (D), Cambridge, 1981, Op
 David Greenberg, (D), Liverpool, 1975, Op
 David Hebert, (D), Dalhousie, 1988, Op
 Paul Hines, (D), Bath, 1989, Op
 Bruce D. Johnson, (D), Dalhousie, 1979, Oc
 Sebastian Krastel, (D), Kiel, 1999, Yr
 William K. W. Li, (D), Dalhousie, 1978, Ob
 Keith E. Louden, (D), MIT, 1976, Yr
 Youyu Lu, (D), Victoria, 1997, Op
 Timothy Milligan, (M), Dalhousie, 1997, Gs
 David C. Mosher, (D), Dalhousie, 1993, Ou
 Andreas Oschlies, (D), Kiel, 1994, Ob
 William Perrie, (D), MIT, 1979, OpAm
 David J.W. Piper, (D), Cambridge (UK), 1969, GusGo
 Harold C. Ritchie, (D), McGill, 1982, Am
 Peter C. Smith, (D), MIT/WHOI, 1973, Op
 Ulrich Sommer, (D), Vienna, 1977, Ob
 Toste Tanhua, (D), Goteborg (Sweden), 1997, Oc

Emeritus:
    Anthony J. Bowen, (D), California (San Diego), 1967, On
    John J. Cullen, (D), California (San Diego), 1980, Ob
    Robert O. Fournier, (D), Rhode Island, 1967, Ob
    Marlon R. Lewis, (D), Dalhousie, 1984, Ob
    Eric L. Mills, (D), Yale, 1964, ObZnn
    Robert M. Moore, (D), Southampton, 1977, Oc

Douglas College
Dept of Earth and Environmental Sciences, Faculty of Science and Technology (A) O☒ (2018)
P.O. Box 2503
New Westminster, BC V3L 5B2
    p. (604) 527-5400
    vigourouxcaillibotn@douglascollege.ca
    http://www.douglascollege.ca/programs-courses/faculties/science-technology/earth-and-environmental-sciences
    Programs: Earth and Environmental Science (A); Geological Resources Diploma;
    Environmental Science (A)
    Enrollment (2018): A: 21 (0)
Chair:
    Nathalie Vigouroux-Caillibot, (D), Simon Fraser, 2011, GviCg
    David C. Waddington, (M), Queen's (Canada), 1991, ZgGuEd
Lab Facilitator:
    Denis Beausoleil, (M), Victoria, GgZg
    Corinne Griffing, (D), Simon Fraser, 2018, GgeGm
    Christine Shiels, (M), Saskatchewan, 2017, GgxZg
Instructor:
    Reid Staples, (D), Simon Fraser, 2014, GpxGg
    Selina Tribe, (D), Simon Fraser, 2004, EgGge
    Derek Turner, (D), Simon Fraser, 2014, GemGl
Emeritus:
    Michael C. Wilson, (D), Calgary, 1981, GaPgGs

Lakehead University 📖
Geology (B,M) ●☒ (2018)
955 Oliver Road
Thunder Bay, ON P7B 5E1
    p. (807) 343-8461
    kristine.carey@lakeheadu.ca
    https://www.lakeheadu.ca/academics/departments/geology
    f: https://www.facebook.com/lakeheaduniversity
    t: @mylakehead
    Programs: The Department of Geology offers programs leading to Bachelor and Master of Science degrees. These undergraduate programs are offered in Geology and Earth Science majors. A Geography major combined with a Geology minor and an HBESc in Earth Science are available. Geology and Earth Science programs are also offered concurrently with a BEd from the Department of Undergraduate Studies in Education.
    Certificates: None
    Administrative Assistant: Kristine M. Carey
    Enrollment (2018): B: 60 (26) M: 15 (4)
Chair:
    Peter N. Hollings, (D), Saskatchewan, 1998, Eg
Professor:
    Philip W. Fralick, (D), Toronto, 1985, Gs
    Mary Louise Hill, (D), Princeton, 1985, Gc
Associate Professor:
    Andrew G. Conly, (D), Toronto, 2003, EgCgGz
    Amanda Diochon, (D), Dalhousie, 2009, Ge
    Shannon Zurevinski, (D), Alberta, 2009, Gz
Emeritus:
    Manfred M. Kehlenbeck, (D), Queen's, 1971, Gc
    Stephen A. Kissin, (D), Toronto, 1974, Em
    Edward L. Mercy, (D), Imperial Coll (UK), 1955, Cg
    Roger H. Mitchell, (D), McMaster, 1969, Gi
Geology Technician:
    Kristi Tavener, Zn
    Jonas K. Valiunas, (D), Lakehead, 2017, Zm
Geology Technician:
    Anne Hammond, (B), Lakehead, 1999, Zn

Laurentian University, Sudbury
Harquail School of Earth Sciences (B,M,D) ●☒ (2019)
Ramsey Lake Road
Sudbury, ON P3E 2C6
    p. (705) 675-1151 Ext. 6575
    hes@laurentian.ca
    http://hes.laurentian.ca/
    Administrative Assistant: Roxane J. Mehes
    Enrollment (2016): B: 90 (0) M: 66 (0) D: 21 (0)
Director:
    Doug Tinkham, (D), Alabama, 2002, Gp
Professor:
    Harold L. Gibson, (D), Carleton, 1990, EmGv
    Daniel J. Kontak, (D), Queens, 1985, Eg
    Bruno Lafrance, (D), New Brunswick, 1990, GcEmGt
    Michael Lesher, (D), W Australia, 1984, Em
    Andrew M. McDonald, (D), Carleton, 1992, Gz
    Jeremy P. Richards, (D), Australian Nat, 1990, EmGiCg
    Richard S. Smith, (D), Toronto, Ye
    Elizabeth C. Turner, (D), Queen's (Canada), 1999, GrdEm
Associate Professor:
    Pedro J. Jugo, (D), Alberta, 2003, CpGi
    Matthew I. Leybourne, (D), Ottawa, Cg
    Michael Schindler, (D), Frankfurt, Gz
Assistant Professor:
    Alessandro Ielpi, (D), Siena, 2013, GsmGr
Emeritus:
    Anthony E. Beswick, (D), London, 1965, Gi
    Paul Copper, (D), Imperial Coll (UK), 1965, PieZc
    Richard James, (D), Manchester, 1967, GpiEm
    Reid R. Keays, (D), McMaster, 1968, Em
    Darrel Long, (D), W Ontario, 1976, GsaGl
    Don H. Rousell, (D), Manitoba, 1965, Gc
    Robert E. Whitehead, (D), New Brunswick, 1973, Ce
Other:
    Bruce C. Jago, (D), Toronto, 1990, EgCe
Cooperating Faculty:
    Phillips C. Thurston, (D), Western, 1981, Zg

Manitoba Museum
Dept. of Geology & Paleontology ☒ (2018)
190 Rupert Avenue
Winnipeg, MB R3B 0N2
    p. (204) 956-2830
    gyoung@manitobamuseum.ca
    http://www.manitobamuseum.ca
Curator:
    Graham A. Young, (D), New Brunswick, 1988, PieRc

McGill University
Dept of Earth & Planetary Sciences (B,M,D) ●☒ (2018)
3450 University Street
Room 238
Montreal, QC H3A 0E8
    p. (514) 398-6767
    kristy.thornton@mcgill.ca
    http://www.mcgill.ca/eps
    Programs: Geology (B); Planetary Sciences (B); Earth and Planetary Sciences (M,D)
    Enrollment (2018): B: 21 (7) M: 19 (6) D: 43 (8)
Chair:
    Jeffrey M. McKenzie, (D), Syracuse, 2005, Hw
Professor:
    Don Baker, (D), Penn State, 1985, CuGvCp
    Olivia G. Jensen, (D), British Columbia, 1971, Yg
    Alfonso Mucci, (D), Miami, 1981, CmlOc
    John Stix, (D), Toronto, 1989, Gv
    Anthony E. Williams-Jones, (D), Queen's, 1973, Ce
Associate Professor:
    Galen Halverson, (D), Harvard, 2003, GsrCs
    Yajing Liu, (D), Harvard, 2007, Ygs
    Jeanne Paquette, (D), SUNY (Stony Brook), 1991, Gz
    Christie Rowe, (D), California (Santa Cruz), 2007, GctEm
    Vincent van Hinsberg, (D), Bristol, 2006, EgCg
Assistant Professor:
    Kim Berlo, (D), Bristol (UK), 2006, Gv
    Nicolas B. Cowan, (D), Washington, 2009, AsXyg

Peter Douglas, (D), Yale, 2014, Cs
Natalya Gomez, (D), Harvard, 2013, Og
James Kirkpatrick, (D), Glasgow, 2008, Gct
Nagissa Mahmoudi, (D), McMaster, 2013, Cb
Earth System Science Faculty Lecturer:
William G. Minarik, (D), Rensselaer, 1993, Cp
Adjunct Professor:
Bjorn Sundby, (D), Bergen (Norway), 1966, Cm
Emeritus:
Jafar Arkani-Hamed, (D), MIT, 1969, Xy
Don Francis, (D), MIT, 1974, Gi
Reinhard Hesse, (D), Tech (Munich), 1964, Gs
Andrew J. Hynes, (D), Cambridge, 1972, Gc
Robert F. Martin, (D), Stanford, 1969, GziEm
Colin Stearn, (D), Yale, Ps

Dept of Mining & Materials Engineering (B,M,D) ☒ (2018)
Rm 125 FDA Building
3450 University Street
Montreal, QC H3A 0E8
   p. (514) 398-4986
   roussos.dimitrakopoulos@mcgill.ca
   http://www.mcgill.ca/minmat/mining
   Programs: Mining Engineering
Chair:
   Stephen Yue, (D), Leeds, 1979, Nx
Professor:
   George Demopoulos, (D), McGill, 1982, Nx
   Roussos Dimitrakopoulos, (D), Ecole Polytechnique, 1989, Nm
   Raynald Gauvin, (D)
   Roderick I. Guthrie, (D), London, 1967, Nx
   Ferri Hassani, (D), Nottingham, 1981, Nm
   Hani Mitri, (D), Nottingham, 1981, Nm
Associate Professor:
   Mathieu Brochu, (D), McGill
   Mainul Hasan, (D), McGill, 1987, Nx
   Showan Nazhat, (D)
   Mihriban Pekguleryuz, (D), McGill
Assistant Professor:
   Kirk Bevan, (D), Purdue
   Marta Cerruti, (D)
   Richard Chromik, (D), SUNY
   Nathaniel Quitoriano, (D), MIT
   Jun Song, (D), Princeton
   Kristian Waters, (D)
Lecturer:
   Forence Paray, (D), McGill
Emeritus:
   John J. Jonas, (D), Cambridge, 1960, Nx

Dept of Atmospheric & Oceanic Sciences (B,M,D) ☒ (2017)
805 Sherbrooke Street West
Room 945
Montreal, QC H3A 0B9
   p. (514) 398-3764
   john.gyakum@mcgill.ca
   http://www.mcgill.ca/meteo
   Administrative Assistant: Lucy Nunez
   Enrollment (2017): B: 6 (10) M: 13 (10) D: 31 (5)
Professor:
   Parisa A. Ariya, (D), York, 1996, As
   Peter Bartello, (D), McGill, 1988, As
   John R. Gyakum, (D), MIT, 1981, Am
   Man Kong Yau, (D), MIT, 1977, As
Associate Professor:
   Frederic Fabry, (D), McGill, 1994, As
   Daniel Kirshbaum, (D), Washington, 2005, Asm
   David Straub, (D), Washington, 1990, Op
   Bruno Tremblay, (D), McGill, Op
Assistant Professor:
   Yi Huang, (D), Princeton, 2007, ZrYgAs
   Timothy Merlis, (D), Caltech, 2011, Asm
   Thomas Colin Preston, (D), Duke, 2011, AsZnn
   Andreas Zuend, (D), ETH (Switzerland), 2008, Asm
Adjunct Professor:
   Gilbert Brunet, (D), McGill, 1989, Am
   Ashu Dastoor, (D), IIT

Luc Fillion, (D), McGill, 1991, Ams
Pavlos Kollias, (D), Miami, 2000, ZrAs
Hai Lin, (D), McGill, 1995, As
Emeritus:
   Jacques F. Derome, (D), Michigan, 1968, As
   Henry G. Leighton, (D), Alberta, 1968, As
   Lawrence A. Mysak, (D), Harvard, 1966, Op
   Isztar Zawadzki, (D), McGill, 1972, AsZr

McMaster University
School of Geography & Earth Sciences (B,M,D) ●☒ (2019)
1280 Main Street West
General Science Building
Room 206
Hamilton, ON L8S 4K1
   p. (905) 525-9140 (Ext. 24535)
   geograd@mcmaster.ca
   https://www.science.mcmaster.ca/geo/
   Programs: Honours Earth & Environmental Sciences; Honours Environmental Sciences; Honours Geography & Environmental Sciences; Honours Biology & Environmental Sciences; Environmental Sciences (B); Geographic Information Systems (minor); Earth & Environmental Sciences (D); Geography (D); Earth & Environmental Sciences (M); Geography (M)
   Enrollment (2018): B: 219 (72) M: 40 (17) D: 50 (1)
Director:
   K. Bruce Newbold, (D), McMaster, 1994, Ge
Professor:
   M. Altaf Arain, (D), Arizona, 1997, Hg
   Janok Bhattacharya, (D), McMaster, 1989, Gr
   Sean Carey, (D), McMaster, 2000, Hg
   Vera Chouinard, (D), McMaster, 1986, Zn
   Paulin Coulibaly, (D), Laval, 2000, Hg
   Alan P. Dickin, (D), Oxford, 1981, Cc
   Carolyn H. Eyles, (D), Toronto, 1986, GslGm
   Richard S. Harris, (D), Queen's, 1981, Zu
   Karen Kidd, (D), Alberta, 1996, Gu
   H. Antonio Paez, (D), Tohoku, 2000, Eg
   Edward G. Reinhardt, (D), Carleton, 1996, PmGam
   Darren M. Scott, (D), McMaster, 2000, Zi
   Gregory F. Slater, (D), Toronto, 2001, Cg
   James E. Smith, (D), Waterloo, 1995, Hw
   J. Michael Waddington, (D), York, 1995, Pe
   Allison M. Williams, (D), York, 1997, Zn
   Robert D. Wilton, (D), S California, 1999, Zn
Associate Professor:
   Joseph I. Boyce, (D), Toronto, 1997, Zg
   Sang-Tae Kim, (D), McGill, 2006, Cg
   Suzanne Mills, (D), Saskatchewan, 2007, Zn
   Maureen Padden, (D), ETH Zurich, 2001, CsPc
   Niko Yiannakoulias, (D), Alberta, 2006, Ge
Assistant Professor:
   Luc Bernier, (D), McMaster, 2007, Cg
   Michael Mercier, (D), McMaster, 2004, Zu
Lecturer:
   John MacLachlan, (D), McMaster, 2011, Gs
Adjunct Professor:
   Howard Barker, (D), McMaster, 1991
   Jing Chen, (D), Reading (UK), 1986, AmZry
   Ian G. Droppo, (D), Exeter (UK), 2000
   Susan J. Elliott, (D), McMaster, 1992, Ge
   Michael Pisaric, (D), Queens
   S Martin Taylor, (D), Victoria (Canada), 1974
   Lesley A. Warren, (D), Toronto, 1994, CgHg
University Professor:
   John D. Eyles, (D), London, 1983, Ge
   Henry P. Schwarcz, (D), Caltech, 1960, Cs
Emeritus:
   Brian T. Bunting, (D), London, 1970
   Andrew F. Burghardt, (D), Wisconsin, 1958
   Paul Clifford, (D), London, 1956
   James H. Crocket, (D), MIT
   John J. Drake, (D), McMaster, 1973
   Derek C. Ford, (D), Oxford, 1963, GmHwCc
   Doug Grundy, (D), Manchester (UK), 1966
   Fred L. Hall, (D), MIT, 1975

Leslie J. King, (D), Iowa, 1960
James R. Kramer, (D), Michigan, 1958, Cl
Kao Lee Liaw, (D), Clark, 1972
Robert McNutt, (D), MIT, 1965, CcqHg
Gerry V. Middleton, (D), London (UK), 1954
William A. Morris, (D), 1974, Yg
Yorgos Papageorgiou, (D), Ohio State, 1970
Walter G. Peace, (D), McMaster, 1996, Zu
W. Jack Rink, (D), Florida State, 1990, CcGam
Michael J. Risk, (D), S California, 1971, Pg
Wayne R. Rouse, (D), McGill, 1968
Roger G. Walker, (D), Oxford, 1964
Ming-Ko Woo, (D), British Columbia, 1972, HgZy

Memorial University of Newfoundland
Dept of Earth Sciences (B,M,D) (2015)
Centre for Earth Resources Research, Room ER 4063
Alexander Murray Building
St. John's, NL A1B 3X5
   p. (709) 864-8142
   earthsci@mun.ca
   http://www.mun.ca/earthsciences
Professor:
   Ali E. Aksu, (D), Dalhousie, 1980, Gu
   Elliott T. Burden, (D), Calgary, 1982, PlGso
   Gregory R. Dunning, (D), Memorial, 1984, CcGit
   George A. Jenner, (D), Tasmania, 1982, Gi
   Toby C. J. S. Rivers, (D), Ottawa, 1976, Gp
   Derek H.C. Wilton, (D), Memorial, 1984, EgCae
Associate Professor:
   Tomas J. Calon, (D), Leiden (Neth), 1978, Gc
   Charles A. Hurich, (D), Wyoming, 1988, Ys
   Aphrodite D. Indares, (D), Montreal, 1989, Gp
   Roger A. Mason, (D), Aberdeen, 1978, Gz
   Michael A. Slawinski, (D), Calgary, 1996, Ys
   Paul J. Sylvester, (D), Washington (St. Louis), 1984, Ca
Assistant Professor:
   Alison Leitch, (D), Australian Nat, 1986, Yg
First Year Lab Instructor:
   Roberta (Robbie) Hicks, (M), Gg
Emeritus:
   Jeremy Hall, (D), Glasgow, 1971, YsGtYr
   Richard N. Hiscott, (D), McMaster, 1977, Gs
   Joseph P. Hodych, (D), Toronto, 1971, Ym
   Henry Longerich, (D), Indiana, 1967, Cg
   Michael G. Rochester, (D), Utah, 1959, Yg

Memorial University of Newfoundland, Grenfell Campus
Environmental Science Program (B) ●⊘ (2019)
School of Science and Environment
University Drive
Corner Brook, NL A2H 6P9
   p. (709) 637-6215
   Ian.Warkentin@grenfell.mun.ca
   https://www.grenfell.mun.ca/academics-and-research/
   Pages/school-of-science-and-the-environment/programs/
   Environmental-Science.as
   Programs: Environmental Science
   Enrollment (2018): B: 56 (12)

Mount Allison University
Dept of Geography (B,M) (2015)
144 Main Street
Sackville, NB E4L 1A7
   p. (506) 364-2326
   dlieske@mta.ca
   http://www.mta.ca/departments/geography
Associate Professor:
   Jeffery W. Ollerhead, (D), Guelph, 1994, Yr
Assistant Professor:
   James Xinxia Jiang, (D), Southampton, Gg
Research Associate:
   Thomas A. Clair, (D), McMaster, 1991, Hy

Research Professor:
   David J. Mossman, (D), Otago (NZ), 1970, EmZn
Emeritus:
   Laing Ferguson, (D), Edinburgh, 1960, Pi

Mount Royal University
Dept of Earth and Environmental Sciences (A,B) (2015)
4825 Mount Royal Gate SW
Calgary, AB T3E 6K6
   p. (403) 440-6165
   lstadnyk@mtroyal.ca
   http://www.mtroyal.ab.ca/scitech/earth.shtml
   Administrative Assistant: Leona Stadnyk
Chair:
   Paul Johnston, (D), W Australia, 1986, Pi
Professor:
   John Cox, (D), Aberdeen, 1994, EoGsHw
   Barbara McNicol, (D), Calgary, 1997, Zg
Associate Professor:
   Katherine Boggs, (D), Calgary, 2004, Gt
   Pamela MacQuarrie, (M), Calgary, 1988, Zy
Related Staff:
   Michael Clark, (B), Adams, 1976, Zg

Natural Resources Canada
Ressources Naturelles Canada (2015)
580 Booth Street , 21st Floor
Ottawa, ON K1A 0E4
   debra.tompkinscaron@canada.ca
   http://www.nrcan.gc.ca/
Senior Scientist:
   Sergey V. Samsonov, (D), Western, 2007, YdZrYg
Other:
   Esther Asselin, (M), Laval, 1988, PlGuRh

New Brunswick Dept of Energy and Mines
New Brunswick Dept of Energy and Mines (2017)
Hugh John Flemming Forestry Centre
P. O. Box 6000
Fredericton, NB E3B 5H1
   p. (506) 453-3826
   geoscience@gnb.ca
   http://www.gnb.ca/energy

Queen's University
Dept of Geological Sciences and Geological Engineering (B,M,D) ● (2017)
#240, Bruce Wing, Miller Hall
36 Union St.
Kingston, ON K7L 3N6
   p. (613) 533-2597
   geolundergradassistant@queensu.ca
   www.queensu.ca/geol
   Enrollment (2011): B: 232 (47) M: 46 (20) D: 18 (5)
Head:
   Jean Hutchinson, (D), Toronto, 1992
Professor:
   Mark Diederichs, (D), Toronto, Nr
   Georgia Fotopoulos, (D), Calgary, 2003, Yd
   Laurent Godin, (D), Carleton, 1999, Gct
   Noel P. James, (D), McGill, 1972, Gd
   Guy M. Narbonne, (D), Ottawa, 1981, PigGs
   Gema Olivo, (D), Québec (Montréal), 1995, Eg
   Ronald C. Peterson, (D), Virginia Tech, 1980, Gz
   Victoria H. Remenda, (D), Waterloo, 1993, Hw
Associate Professor:
   Alexander Braun, (D), Goethe, 1999, YgdYv
   John A. Hanes, (D), Toronto, 1979, Cc
   Heather E. Jamieson, (D), Queen's, 1982, Ge
   Daniel Layton-Matthews, (D), Toronto, 2006, EgCte
Research Associate:
   Doug A. Archibald, (D), Queen's, 1982, Cc
Adjunct Professor:
   Rob Harrap, (M), Carleton, 1990, Gc

Emeritus:
   Alan H. Clark, (D), Manchester, 1964, Em
   Robert W. Dalrymple, (D), McMaster, 1977, Gs
   John M. Dixon, (D), Connecticut, 1974, Gct
   Herwart Helmstaedt, (D), New Brunswick, 1968, Gc
   Raymond A. Price, (D), Princeton, 1958, GtcGe

Royal Ontario Museum
Dept of Palaeobiology (2015)
100 Queen's Park
Toronto, ON M5S 2C6
   p. (416) 586-5591
   davidru@rom.on.ca
Curator Of Vertebrate Palaeontology:
   David C. Evans, (D), Toronto, 2007, PvgPq

Department of Earth Sciences ✉ (2019)
100 Queen's Park
Toronto, ON M5S 2C6
   p. (416) 586-5820
   naturalhistory@rom.on.ca
Emeritus:
   Robert I. Gait, (D), Manitoba, 1967, Gz
   Frederick J. Wicks, (D), Oxford, 1969, Gz
Curator:
   Kimberly T. Tait, (D), Arizona, 2007, GzyXm

Royal Tyrrell Museum of Palaeontology
Royal Tyrrell Museum of Palaeontology ✉ (2017)
P.O. Box 7500
Drumheller, AB T0J 0Y0
   p. (403) 823-7707
   tyrrell.info@gov.ab.ca
   http://www.tyrrellmuseum.com/
Executive Director:
   Andrew G. Neuman, (M), Alberta, 1986, Pv
Director, Preservation and Research:
   Donald B. Brinkman, (D), McGill, 1979, Pv
Curator:
   David A. Eberth, (D), Toronto, 1987, Gs
   James D. Gardner, (D), Alberta, 2000, Pv
   Donald Henderson, (D), Bristol, 2000, Pv
   Craig Scott, (D), Alberta, 2008, Pv
Curator:
   Francois Therrien, (D), Johns Hopkins, 2004, Pv
Post-doctoral fellow:
   Caleb Brown, (D), Toronto, 2013, Pv

Saint Francis Xavier University
Dept of Earth Sciences (B,M) (2015)
P.O Box 5000
Antigonish, NS B2G 2W5
   p. (902) 867-5109
   igreen@stfx.ca
   http://sites.stfx.ca/earth_sciences/
   Enrollment (2015): B: 11 (15) M: 13 (3)
Professor:
   Alan J. Anderson, (D), Queen's, 1990, Gx
   Hugo Beltrami, (D), UQAM, 1993, YhAs
   Lisa M. Kellman, (D), Quebec (Montreal), 1997, CsSo
   Michael J. Melchin, (D), Western, 1987, Pi
   J. Brendan Murphy, (D), McGill, 1982, Gt
Associate Professor:
   Dave A. Risk, (D), Dalhousie, 2006, SoZr
Instructor:
   Cindy Murphy, (M), McGill, 1986, Gg
   Colette Rennie, (M), Queen's, 1987, Gg
   Matthew Schumacher, (M), Waterloo, 2006, Zg
   Sid Taylor, (M), Memorial, 1977, Gg

Simon Fraser University
Dept of Earth Sciences (B,M) ✉ (2018)
8888 University Drive
Burnaby, BC V5A 1S6
   p. 778-782-4229
   bcward@sfu.ca
   http://www.sfu.ca/earth-sciences
Professor:
   Diana M. Allen, (D), Carleton, 1996, Hy
   Andrew J. Calvert, (D), Cambridge, 1985, Ys
   Gwenn Flowers, (D), British Columbia, 2000, Gl
   H. Daniel (Dan) Gibson, (D), Carleton, 2003, GcCcGp
   James A. Mac Eachern, (D), Alberta, 1994, Gs
   Dan D. Marshall, (D), Lausanne, 1995, Ca
   Douglas Stead, (D), Nottingham, 1984, Ng
   Derek J. Thorkelson, (D), Carleton, 1992, Gt
   Brent C. Ward, (D), Alberta, 1992, GlmGe
   Glyn Williams-Jones, (D), Open Univ (UK), 2001, GvCeYe
Associate Scientist:
   Dirk Kirste, (D), Calgary, 2001, Cg
Lecturer:
   Kevin Cameron, (M), Memorial, 1986, Gg
   Roberta Donald, (M), British Columbia, 1984, GsZe
Emeritus:
   John J. Clague, (D), British Columbia, 1973, Gl

Universite du Quebec
INRS, centre Eau Terre Environnement (Quebec Geoscience Center) (M,D) (2016)
490 de la Couronne
Quebec, QC G1K 9A9
   p. + (418) 654-4677
   info@ete.inrs.ca
   http://www.ete.inrs.ca
   Administrative Assistant: Pascale Cote
Professor:
   Normand Bergeron, (D), SUNY (Buffalo), 1994, Gm
   Monique Bernier, (D), Zr
   Fateh Chebana, Hq
   Karem Chokmani, Zi
   Pierre Francus, (D), 1997, PeGse
   Bernard Giroux, Yeg
   Erwan Gloaguen, Ye
   Yves Gratton, Op
   Lyal B. Harris, GcYg
   Marc R. LaFleche, (D), Montpellier II, 1991, Ct
   Rene Lefebvre, (D), Laval, 1994, Hw
   Michel Malo, (D), Montreal, 1986, GctGs
   Richard Martel, (D), Laval, 1996, Hw
   Claudio Paniconi, (D), Hw
   Pierre-Simon Ross, GvEg
Senior Scientist:
   Jean H. Bedard, (D), Montreal, 1985, Gi
   Christian Begin, (D), Laval, 1991, PeGmSo
   Benoit Dube, (D), Quebec (Chicoutimi), 1990, Em
   Denis Lavoie, (D), Laval, 1988, Gs
   Yves Michaud, (D), Laval, 1991, Gm
   Michel Parent, (D), W Ontario, 1987, Gl
   Didier Henri Perret, (D), Laval, 1995, NgeZu
Research Associate:
   Eric Boisvert, (M), Quebec (Montreal), 1994, Zn
   Pierre Brouillette, (B), Laval, 1982, Gg
   Kathleen Lauziere, (M), Quebec (Chicoutimi), 1989, Gg
   Marc R. Luzincourt, (B), Quebec (Montreal), 1982, Cs
   Anna Smirnov, (M), Memorial, 1997, Cs
Adjunct Professor:
   Louise Corriveau, (D), McGill, 1989, EmGpg

Universite du Quebec a Chicoutimi
Sciences de la Terre (B,M,D) ●✉ (2019)
555, Boulevard de l'Universite
Chicoutimi, QC G7H 2B1
   p. (418) -545- 5011 (Ext. 5202)
   sue_sc-terre@uqac.ca
   Programs: B.Sc Geology
   B.Sc. geological engineering
   M.Sc. Earth science and geological engineering
   M. Sc. Professional, exploration geology
   Ph.D. Earth sciences
   Enrollment (2018): B: 55 (13) M: 34 (8) D: 6 (1)

Professor:
    Sarah- Jane Barnes, (D), Toronto, 1983, EgGiCt
    Paul Bedard, (D), UQAC, 1992, CaGzEg
    Romain Chesnaux, (D), École Polytechnique (Canada), 2005, Hw
    Damien Gaboury, (D), UQAC, 1999, EmCe
    Ali Saeidi, (D), Lorraine INP (France), 2010, Nrm
    Edward W. Sawyer, (D), Toronto, 1983, GptGi
Assistant Professor:
    Lucie Mattieu, (D), Trinity (Dublin), 2010, EgmCe
Research Associate:
    Silvain Rafini, (D), Universite du Quebec a Montreal, 2008, GcHwEg
Instructor:
    Denis Cote, (M), Quebec (Chicoutimi), 1986, GiEgZi
Lecturer:
    Dominique Genna, (D), Universite du Quebec a Chicoutimi, 2009, GvEn
    Philippe Page, (D), INRS-ETE, 2006, GiCgEm
    Julien Walter, (D), Universite du Quebec a Chicoutimi, 2018, HwqGm
Adjunct Professor:
    Sylvain Raffini, (D), UQAM, 2008, GcHw
Emeritus:
    Guy Archambault, (D), Nr
    Pierre Cousineau, (D), Laval, 1986, Gd
    Jayanta Guha, (D), Jadavpur, 1967, Eg
    Michael D. Higgins, (D), McGill, 1980, Gi
    Alain Rouleau, (D), Waterloo, 1984, Hw
    Denis W. Roy, (D), Princeton, 1976, Gc
On Leave:
    Real Daigneault, (D), Laval, 1991, GcEgZi
Laboratory Director:
    Dany Savard, (M), UQAC, 2010, Ca

Universite du Quebec a Montreal
Département des sciences de la Terre et de l'atmosphère (B,M,D) (2015)
C.P. 8888, succursale Centre-ville
Montreal, QC H3C 3P8
    p. +1 514-987-3000
    dept.sct@uqam.ca
    http://scta.uqam.ca
    Administrative Assistant: France Beauchemin
Professor:
    Florent Barbecot, (D), Hw
    Jean-Pierre Blanchet, (D), Toronto, 1984, Pe
    Gilles Couture, (D), British Columbia, 1987, Yg
    Fiona Ann Darbyshire, (D), Cambridge, 2000, Ys
    Anne de Vernal, (D), Montreal, 1986, Pm
    Alessandro Marco Forte, (D), Toronto, 1989, Yg
    Pierre Gauthier, (D), McGill, 1988, Am
    Eric Girard, (D), McGill, 1999, Am
    Normand Goulet, (D), Queen's, 1976, Gc
    Cherif Hamzaoui, (D), Alziers, 1980, Yg
    Alfred Jaouich, (D), Minnesota, 1975, Sc
    Michel Jebrak, (D), Orleans, 1984, Eg
    Michel Lamothe, (D), W Ontario, 1985, Gl
    Rene Laprise, (D), Toronto, 1988, As
    Marie Larocque, (D), de Poitiers, France, 1998, Hw
    Marc Michel Lucotte, (D), McGill, 1987, Oc
    Daniele Luigi Pinti, (D), France, 1993, Cs
    Martin Roy, (D), Oregon, 1998, Gl
    Ross Stevenson, (D), Arizona, 1989, Cg
    Laxmi Sushama, (D), Melbourne, 1999, Hy
    Julie Mireille Thériault, (D), McGill, 2009, Zn
    Enrico Torlaschi, (D), Ecole Polytechnique (Milan), 1976, Am
    Alain Tremblay, (D), Laval, 1989, Gct
    David Widory, (D), IPGP, 1999, Cs
Associate Professor:
    Sanda Balescu, (D), Libre de Bruxelles, 1988, Pe
    Jean Côté, As
    Bernard Dugas, As
    Stephane Faure, (D), INRS-Georessources, 1995, Gc
    Philippe Gachon, (D), UQAM, 1999, As
    Michel Gauthier, (D), Polytechnique (Montreal), 1982, Eg
    Jean-François Helie, (D), UQAC, 2004, Cs
    Claude Hillaire-Marcel, (D), Paris VI, 1979, Cs
    Jean-Claude Mareschal, (D), Texas A&M, 1975, Yg
    Andre Poirier, (D), UQAM, 2005, Cs
    Gilbert P. Prichonnet, (D), Bordeaux, 1967, Pi
    William W. Shilts
    Gabriel-Constantin Voicu, (D), UQAM, 1995, GcCg

Universite du Quebec a Rimouski
Inst des sciences de la mer de Rimouski (M,D) (2015)
310, Allee des Ursulines
Rimouski, QC G5L 3A1
    p. (418) 723- 8617
    andre_rochon@qc.ca
    http://www.ismer.ca
Director:
    Serge Demers, (D), Laval, 1981, Ob
Professor:
    Celine Audet, (D), Laval, 1985, Ob
    Jean-Claude Brethes, (D), Aix-Marseille, 1978, Ob
    Jean-Pierre Gagne, (D), Montreal, 1993, OcCmo
    Michel Gosselin, (D), Laval, 1990, Ob
    Vladimir G. Koutitonsky, (D), SUNY (Stony Brook), 1985, Op
    Jocelyne Pellerin, (D), Laval, 1982, Ob
    Emilien Pelletier, (D), McGill, 1983, Oc
    Suzanne Roy, (D), Dalhousie, 1986, Ob
    Bjorn Sundby, (D), Bergen (Norway), 1966, Oc
    Bruno Zakardjian, (D), Paris, 1994, Ob

Universite Laval
Dept de geologie (B,M,D) (2015)
Pavillon Pouliot
Faculte des Sciences et de genie
Ste-Foy, QC G1K 7P4
    p. (418) 656-2193
    marc.constantin@ggl.ulaval.ca
    http://www.ggl.ulaval.ca
Head:
    Marc Constantin, (D), Brest, 1995, Gi
    Josee Duchesne, (D), Laval, 1993, Gz
Professor:
    Georges Beaudoin, (D), Ottawa, 1991, EgGzCa
    Richard Fortier, (D), Montreal, 1994, Yg
    Paul W. Glover, (D), East Anglia (UK), 1989, Yx
    Rejean J. Hebert, (D), Brest, 1985, Gi
    Jacques E. Locat, (D), Sherbrooke, 1982, Ng
    Fritz Neuweiler, (D), Berlin, 1995, GsdPg
    Rene Therrien, (D), Waterloo, 1992, HwGe
Eletron Microprobe Specialist:
    Marc Choquette, (D), Laval, 1988, Zn
Associate Scientist:
    Pauline Dansereau, (M), Laval, 1988, Gs
    Andre Levesque, (B), Laval, 1979, Gz
    Pierre Therrien, (M), Laval, 1986, Gq
Research Associate:
    Danielle Cloutier, (D), Laval, GeOg
Adjunct Professor:
    Benoit Fournier, (D), Laval, 1993, NgZmGg
Technician:
    Jean Frenette, (B), Laval, 1985, Gz
    Martin Plante, Cg

University of Alberta
Inst of Geophysical Research (B,M,D) (2015)
Mailstop #615
CEB/Physics
Edmonton, AB T6G 2G7
    p. (780) 492-3521
    msacchi@ualberta.ca
    http://www-geo.phys.ualberta.ca/institute/
    Administrative Assistant: Lee Grimard
Professor:
    Robert A. Creaser, (D), La Trobe, 1992, Cc
    T. Bryant Moodie, (D), Toronto, 1972, Zn
    Robert Rankin, (D), North Wales, 1984, Xy
    Gerhard W. Reuter, (D), McGill, 1985, As
    Wojciech Rozmus, (D), Inst Nuc Res (Poland), Zn
    John C. Samson, (D), Alberta, 1971, Xy

Martin J. Sharp, (D), Aberdeen, Gl
John Shaw, (D), Reading, 1969, Gl
Samuel S. Shen, (D), Wisconsin, As
Bruce R. Sutherland, (D), Toronto, 1994, Zn
Gordon E. Swaters, (D), British Columbia, 1985, Op
Richard D. Sydora, (D), Texas, 1985, Xy
Martyn Unsworth, (D), Cambridge, Zn
John D. Wilson, (D), Guelph, 1980, As
Associate Professor:
  Andrew B. G. Bush, (D), Toronto, As
  Carl Mendoza, (D), Waterloo, 1993, Hw
  Benoit Rivard, (D), Washington, 1990, Zr
  Ben Rostron, (D), Alberta, Zn
  Mauricio D. Sacchi, (D), British Columbia, Ys
Assistant Professor:
  Francis Fenrich, (D), Alberta, 1997, Xy
  Jeff Gu, (D), Harvard, 2001, Ys
  Moritz Heimpel, (D), Johns Hopkins, 1995, Zn
  Vadim Kravchinsky, (D), Irkutsk, 1996, Ym
  Paul Myers, (D), Victoria, 1992, Op
Emeritus:
  Michael E. Evans, (D), Australian Nat, 1969, Ym
  Keith D. Hage, (D), Chicago, 1957, As
  F. Walter Jones, (D), McGill, 1968, Ym
  Edward P. Lozowski, (D), Toronto, 1970, As
  Roger D. Morton, (D), Nottingham, 1959, Em
  Edo Nyland, (D), California (Los Angeles), 1967, Ys
  David Rankin, (D), Alberta, 1960, Ym
  Gordon Rostoker, (D), British Columbia, 1966, Xy
  T.J.T (Tim) Spanos, (D), Alberta, 1977, YgsEo

Dept of Earth & Atmospheric Sciences (B,M,D) ●☒ (2018)
1-26 Earth Sciences Building
Edmonton, AB T6G 2E3
  p. (780) 492-3265
  eas@ualberta.ca
  http://www.ualberta.ca/EAS/
  f: https://www.facebook.com/UofAEarthandAtmospheric-SciencesDepartment
  t: @UofA_EAS
  Programs: Geology, Environmental Earth Sciences, Human Geography, Urban and Regional Planning, Paleontology
  Enrollment (2016): B: 449 (129) M: 0 (17) D: 0 (18)
Chair:
  Stephen T. Johnston, (D), Alberta, 1993, GctGg
Professor & Inaugural Director, Planning Program:
  Sandeep Agrawal, (D), Zu
Professor:
  Robert W. Luth, (D), California (Los Angeles), 1985, Gi
  Martin J. Sharp, (D), Aberdeen, 1982, Gl
Canada Research Chair:
  Thomas Stachel, (D), Wurzburg, 1991, GiCs
Associate Dean (Research):
  Larry M. Heaman, (D), McMaster, 1986, Cc
Associate Chair:
  Murray Gingras, (D), North Carolina, 1987, Go
  Kurt Konhauser, (D), W Ontario, 1993, Pg
Professor:
  Andrew B.G Bush, (D), Toronto, 1995, AsOpPe
  Michael W. Caldwell, (D), McGill, 1995, Pv
  Octavian Catuneanu, (D), Toronto, 1996, Gs
  Thomas Chacko, (D), North Carolina, 1987, Gp
  Robert A. Creaser, (D), La Trobe, 1990, Cc
  Philippe Erdmer, (D), Queen's, 1982, Gc
  Duane Froese, (D), Calgary, 2001, GlmPe
  John Gamon, (D), California (Davis), 1989, Zr
  Christopher DK Herd, (D), New Mexico, 2001, XcGiCp
  Brian Jones, (D), Ottawa, 1974, Ps
  Lindsey Leighton, (D), Michigan, 1999, Pi
  Hans G. Machel, (D), McGill, 1985, Go
  Karlis Muehlenbachs, (D), Chicago, 1971, Cs
  Graham Pearson, (D), Leeds, Cec
  David Potter, (D), Newcastle upon Tyne, Ye
  Gerhard Reuter, (D), McGill, 1985, Am
  Benoit Rivard, (D), Washington (St. Louis), 1990, Zr
  Benjamin J. Rostron, (D), Alberta, 1995, HwGoCg
  G. Arturo Sanchez-Azofeifa, (D), New Hampshire, 1996, ZirZg

Bruce Sutherland, (D), Toronto, 1994, Cm
Martyn Unsworth, (D), Cambridge, GtvYe
John W.F Waldron, (D), Edinburgh, 1981, Gc
John D. Wilson, (D), Guelph, 1980, Am
John-Paul Zonneveld, (D), Alberta, 1999, Go
Associate Chair:
  Tara McGee, (D), Australian Nat, 1996, Zn
Associate Professor:
  Damian Collins, (D), Simon Fraser, 2004, Zn
  Theresa D. Garvin, (D), McMaster, 1999, Zn
  Nicholas Harris, (D), Stanford, Gso
  Jeffrey Kavanaugh, (D), British Columbia, 2001, Ng
  Paul Myers, (D), Victoria, 1996, Og
Assistant Professor:
  Daniel S. Alessi, (D), Notre Dame, 2009, ClGeHw
  Jeff Birchall, (D), Canterbury, 2013, ZuyZg
  Monireh Faramarzi, (D), ETH Zurich, 2010, HgwHq
  Long Li, (D), Lehigh, 2006, CsGpv
  Manish Shirgaokar, (D), Zu
  Amrita Singh, (D), Irvine, 2015, Zu
Distinguished University Professor:
  Nathaniel W. Rutter, (D), Alberta, 1966, Ge
Emeritus:
  Ronald A. Burwash, (D), Minnesota, 1955, Gp
  Ian A. Campbell, (D), Colorado, 1968, Gm
  Brian D. Chatterton, (D), Australian Nat, 1970, Pi
  David M. Cruden, (D), London, 1969, Nr
  John England, (D), Colorado, 1974, Gml
  Leszek A. Kosinski, (D), Warsaw, 1958, Zn
  Arleigh H. Laycock, (D), Minnesota, 1957, Hg
  Edward P. Lozowski, (D), Toronto, 1970, As
  Carl A. Mendoza, (D), Waterloo, 1993, HwSpHq
  O.F. George Sitwell, (D), Toronto, 1968, Zn
  Peter J. Smith, (D), Edinburgh, 1964, Zn
  József Tóth, (D), Utrecht, 1965, Hw
Related Staff:
  Robert Summers, (D), Guelph, 2005, Zn

University of British Columbia
Dept of Earth, Ocean, and Atmospheric Sciences (B,M,D) O (2015)
2020-2207 Main Mall
Vancouver, BC V6T 1Z4
  p. (604) 827-5284
  acairns@eos.ubc.ca
  http://www.eoas.ubc.ca
  Enrollment (2010): B: 361 (105) M: 111 (20) D: 93 (15)
Director, Geological Engineering Program:
  Roger D. Beckie, (D), Princeton, 1992, HwCl
Honorary Professor:
  Mati Raudsepp, (D), Manitoba, 1984, GzeZm
Dean, Science:
  Simon M. Peacock, (D), California (Los Angeles), 1985, GtpYs
Canada Research Chair:
  Roger Francois, (D), British Columbia, 1987, Cm
  Dominique A.M. Weis, (D), Libre de Bruxelles, 1982, CaGeCu
Professor:
  Susan E. Allen, (D), Cambridge, 1988, Op
  Raymond J. Andersen, (D), California (San Diego), 1975, Oc
  Neil Balmforth, (D), Cambridge, 1990, As
  Michael G. Bostock, (D), Australian Nat, 1991, YsGt
  R. Marc Bustin, (D), British Columbia, 1979, Ec
  Gregory M. Dipple, (D), Johns Hopkins, 1991, Gp
  Erik Eberhardt, (D), Saskatchewan, 1998, Ng
  Lee A. Groat, (D), Manitoba, 1988, GzEg
  Oldrich Hungr, (D), Alberta, 1981, Gm
  Mark Jellinek, (D), Australian Nat, 1999, Gg
  Catherine L. Johnson, (D), San Diego, 1994
  Ulrich Mayer, (D), Waterloo, 1999, Nm
  James K. Mortensen, (D), California (Santa Barbara), 1983, Cc
  Douglas W. Oldenburg, (D), California (Santa Barbara), 1974, Yg
  Evgeny Pakhomov, (D), Russian Acad of Sci, 1992, Ob
  James Kelly Russell, (D), Calgary, 1984, GviCg
  James S. Scoates, (D), Wyoming, 1994, GizGv
  Paul L. Smith, (D), McMaster, 1981, Pg
  Douw G. Steyn, (D), British Columbia, 1980, Asm
  Roland B. Stull, (D), Washington, 1975, As
  Curtis Suttle, (D), British Columbia, 1987, Ob

Canada

NSERC Industrial Research Chair in Computational Geoscience:
  Eldad Haber, (D), British Columbia, 1997, YgZn
Director, MDRU:
  Craig J.R. Hart, (D), W Australia, 2004, EmGiCc
Canada Research Chair:
  Maria Maldonado, (D), McGill, 1999, Ob
  Christian Schoof, (D), Oxford, 2002, Gl
Associate Professor:
  Philip Austin, (D), Washington, 1987, As
  Kurt A. Grimm, (D), California (Santa Cruz), 1992, Gs
  Mark S. Johnson, (D), Cornell, 2005, SpHsq
  Lori Kennedy, (D), Texas A&M, 1996, Gc
  Maya G. Kopylova, (D), Moscow, 1990, Gi
  Kristin J. Orians, (D), California (Santa Cruz), 1988, Cm
  Richard A. Pawlowicz, (D), MIT/WHOI, 1994, Op
  Philippe Tortell, (D), Princeton, 2001, Ob
Assistant Professor:
  Kenneth A. Hickey, (D), James Cook, 1995, EgGcg
  Valentina Radic, (D), Alaska (Fairbanks), 2008, GlAsZg
Honorary Research Associate:
  Kevin Kingdon, (M), British Columbia, 1998, Yg
  Michael Maxwell, (D), British Columbia, 1986, Yg
Research Associate:
  Thomas Bissig, (D), Queens, 2001, EmCg
  Farhad Bouzari, (D), Queens, 2003, Eg
  Amanda Bustin, (D), Victoria, 2006, YgNg
  Philip T. Hammer, (D), California (San Diego), 1991, Ys
  Brian Hunt, (D), Tasmania, 2005, Ob
  Bruno Kieffer, (D), Grenoble, 2002, Cs
  Henryk Modzelewski, (D), British Columbia, 2004, As
  Roger Pieters, (D), California (Santa Barbara), Op
  Lin-Ping Song, (D), Sichuan (China), 1996, Ng
  Dave E. Williams, (D), British Columbia, 1987, Oc
Instructor:
  Brett H. Gilley, (M), Simon Fraser, 2003, GsZeRn
  Sara Harris, (D), Oregon State, 1998, Ou
  Tara Ivanochko, (D), British Columbia, 2004, OoZce
  Stuart Sutherland, (D), Leicester (UK), 1992, Pg
Lecturer:
  Francis H. M. Jones, (M), British Columbia, 1987, YeZe
Honorary Lecturer:
  Dileep Athaide, (M), British Columbia, 1975, Gg
Adjunct Professor:
  Robert G. Anderson, (D), Carleton, 1983, Gt
  Stephen Billings, (D), Sydney, 1998, Yg
  Alex Cannon, (D), British Columbia, 2009, As
  Edward Carmack, (D), Washington, 1972, Op
  Michael G. G. Foreman, (D), British Columbia, 1984, Op
  James Haggart, (D), California (Davis), 1984, PgGr
  Catherine J. Hickson, (D), British Columbia, 1987, EgGvZe
  Mark Holzer, (D), Simon Fraser, 1990, As
  R. Lynn Kirlin, (D), Utah State, 1968, As
  Doug McCollor, (D), British Columbia, 2008, Asm
  Barry Narod, (D), British Columbia, 1979, Yx
  Michael Orchard, (D), Hull, 1975, PgGg
  K. Wayne Savigny, (D), Alberta, 1980, Ng
  Barbara H. Scott Smith, (D), Edinburgh, 1977, Gz
  John F. H. Thompson, (D), Toronto, 1984, Em
  Richard E. Thomson, (D), British Columbia, 1971, Op
  Richard Tosdal, (D), California (Santa Barbara), 1988, Em
  Knut von Salzen, (D), Hamburg (Germany), 1997, As
  Chi S. Wong, (D), California (San Diego), 1968, Cm
Emeritus:
  Mary Lou Bevier, (D), California (Santa Barbara), 1982
  Peter M. Bradshaw, (D), Durham (UK), 1965, EgCeGe
  Stephen E. Calvert, (D), California (San Diego), 1964, Cm
  Richard L. Chase, (D), Princeton, 1963, Gu
  Garry K. C. Clarke, (D), Toronto, 1967, Yg
  Ronald M. Clowes, (D), Alberta, 1969, Ys
  Robert M. Ellis, (D), Alberta, 1964, Ys
  William K. Fletcher, (D), Imperial Coll (UK), 1968, Ce
  Michael Healey, (D), Aberdeen, 1969, Ob
  William W. Hsieh, (D), British Columbia, 1981, OpAs
  Alan G. Lewis, (D), Hawaii, 1961, Ob
  Stephen G. Pond, (D), British Columbia, 1965, Op
  R. Doncaster Russell, (D), Toronto, 1954, Yg
  Alastair J. Sinclair, (D), British Columbia, 1964, EmGg
  J. Leslie Smith, (D), British Columbia, 1978, Hw
  Frank J. R. Taylor, (D), Cape Town, 1965, Ob
Undergraduate Program Coordinator:
  Teresa Woodley, Zn
Secretary to the Head:
  Selene Chan, (B), British Columbia, 1995, Zn
Office Support:
  Alicia Warkentin, Zn
Human Resources Manager:
  Cary Thomson, Zn
Graduate Coordinator:
  Audrey Van Slyck
Finance Clerk:
  Anita Lam, Zn
  Kathy Scott, Zn
Director of Resources and Operations:
  Renee Haggart, Zn
Computer Department Manager:
  John Amor, Zn

Soil Science (B,M,D) O (2018)
2357 Main Mall
Vancouver, BC V6T 1Z4
  p. (604) 822-0252
  maja.krzic@ubc.ca
  http://www.landfood.ubc.ca/academics/graduate/soil-science-msc-phd/
  Programs: Soil Science
  Enrollment (2018): B: 4 (4) M: 9 (5) D: 11 (1)
Professor:
  T. Andrew Black, (D), Wisconsin, 1969, Sp
  Christopher Chanway, (D), British Columbia, 1987, Sb
  Sue Grayston, (D), Sheffield, 1988, Sb
  Cindy Prescott, (D), Calgary, 1988, Sb
  Suzanne Simard, (D), Oregon State, 1995, Sb
Associate Professor:
  Maja Krzic, (D), British Columbia, 1997, Sf
Assistant Professor:
  Sean Smukler, (D), California (Davis), 2008, So
Instructor:
  Sandra Brown, (D), British Columbia, 1997, Sp
Emeritus:
  Arthur A. Bomke, (D), Illinois, 1972, Sc
  Leslie M. Lavkulich, (D), Cornell, 1969, Sd
  Hans D. Schreier, (D), British Columbia, 1976, Zg

University of Calgary
Dept of Geoscience (B,M,D) ✉ (2018)
2500 University Drive NW
Calgary, AB T2N 1N4
  p. (403) 220-5841
  youngt@ucalgary.ca
  http://www.geo.ucalgary.ca/
Head:
  Larry R. Lines, (D), British Columbia, 1976, Ye
  Bernhard Mayer, (D), LMU Munich, 1993, CslGe
Professor:
  Benoit Beauchamp, (D), Calgary, 1987, Pm
  Laurence R. Bentley, (D), Princeton, 1990, Hw
  James R. Brown, (D), Uppsala, 1972, Ys
  Kenneth Duckworth, (D), Leeds, 1964, Ye
  Edward D. Ghent, (D), California, 1964, Gp
  Terence M. Gordon, (D), Princeton, 1969, Gq
  Masaki Hayashi, (D), Waterloo, 1997, HwSp
  Charles M. Henderson, (D), Calgary, 1988, Pm
  Ian E. Hutcheon, (D), Carleton, 1977, Cl
  Edward S. Krebes, (D), Alberta, 1980, Ys
  Donald C. Lawton, (D), Auckland, 1979, Ye
  James W. Nicholls, (D), California, 1969, Gi
  Gerald D. Osborn, (D), California, 1972, Gm
  David R. Pattison, (D), Edinburgh, 1985, Gp
  Ronald J. Spencer, (D), Johns Hopkins, 1981, Cg
  Deborah A. Spratt, (D), Johns Hopkins, 1980, Gc
  Robert R. Stewart, (D), MIT, 1983, Ye
  Patrick Wu, (D), Toronto, 1982, Ys

Associate Professor:
   Russell L. Hall, (D), McMaster, 1976, Pi
   Alan R. Hildebrand, (D), Arizona, 1992, XmgXc
   John C. Hopkins, (D), McGill, 1972, Gs
   Gary F. Margrave, (D), Alberta, 1981, Ye
   Cynthia L. Riediger, (D), Waterloo, 1991, Co
   Cathy Ryan, (D), Waterloo, 1994, Hw
Assistant Professor:
   Eva Enkelmann, (D), 2005, CcGmt
   Jean-Michel Maillol, (D), Alberta, 1992, Ym
Senior Instructor:
   Jon W. Jones, (D), Calgary, 1972, Gp
Honorary Professor:
   Brian Norford, (D), Yale, 1959, Gr
Adjunct Professor:
   Philip J. Currie, (D), McGill, 1981, Pv
   Susan L. Gordon, (D), Waterloo, 1999, Hw
   Stephen E. Grasby, (D), Calgary, 1997, Hw
   William D. Gunter, (D), Johns Hopkins, 1974, Cl
   J. Bruce Jamieson, (D), Calgary, 1996, Ne
   Thomas F. Moslow, (D), South Carolina, 1980, Co
   Peter E. Putnam, (D), Calgary, 1985, Gs
   Gerald M. Ross, (D), Carleton, 1983, Gt
   Selim Sayegh, (D), McGill, 1980, Gs
   John-Paul Zonneveld, (D), Alberta, 1999, Pm
Emeritus:
   Finley A. Campbell, (D), Princeton, 1958, Em
   Frederick A. Cook, (D), Cornell, 1980, YeGtYs
   Peter E. Gretener, (D), ETH (Switzerland), 1953, Ng
   Leonard V. Hills, (D), Alberta, 1965, Pl
   Federico F. Krause, (D), Calgary, 1979, GsdGo
   Alfred A. Levinson, (D), Michigan, 1952, Eg
   Alan E. Oldershaw, (D), Liverpool, 1967, Gs
   Philip S. Simony, (D), London, 1963, Gc
   Norman C. Wardlaw, (D), Glasgow, 1960, Go

University of Guelph
School of Environmental Sciences (B,M,D) (2015)
Guelph, ON N1G 2W1
   p. (519) 824-2052
   earnaud@uoguelph.ca
   http://www.ses.uoguelph.ca
Director:
   Jonathan Newman, (D), Albany, 1990, Zn
Director, Controlled Environment Systems Research Facility:
   Michael Dixon, (D), Edinburgh, Zn
Canada Research Chair in Recombinant Antibody Technology:
   Christopher C. Hall, (D), Alberta, Zn
Canada Research Chair in Microbial Ecology:
   Kari Dunfield, (D), Saskatchewan, 2002, SbZn
Associate Dean, Research, OAC:
   Beverley A. Hale, (D), Guelph, 1989, Ct
Associate Dean, Academics, OAC:
   Jonathan Schmidt, (D), Toronto, Zn
Professor:
   Paul Goodwin, (D), California (Davis), Zn
   Andrew Gordon, (D), Alaska (Fairbanks), 1984, Zn
   Ernesto Guzman, (D), California (Davis), Zn
   Tom Hsiang, (D), Washington, Zn
   Hung Lee, (D), McGill, Zn
   Stephen Marshall, (D), Guelph, Zn
   Gard Otis, (D), Kansas, Zn
   Cynthia Scott-Dupree, (D), Simon Fraser, 1986, Zn
   Jack Trevors, (D), Waterloo, Zn
   R. Paul Voroney, (D), Saskatchewan, 1983, Sb
   Claudia Wagner Riddle, (D), Guelph, 1992, Am
Director of the Arboretum:
   Shelley Hunt, (D), Guelph, Zn
Dean, OAC:
   Rob J. Gordon, (D), Guelph, As
Associate Professor:
   Madhur Anand, (D), W Ontario, Zn
   Emmanuelle Arnaud, (D), McMaster, 2002, Gls
   Susan Glasauer, (D), Tech (Munich), 1995, Pg
   Marc Habash, (D), Guelph, Zn
   Rebecca Hallett, (D), Simon Fraser, Zn
   Richard Heck, (D), Saskatchewan, Sd
   John Lauzon, (D), Guelph, So
   Ivan O'Halloran, (D), Saskatchewan, 1986, So
   Paul Sibley, (D), Waterloo, Zn
   Jon Warland, (D), Guelph, 1999, As
Assistant Professor:
   Tim Rennie, Zn
   Laura Van Eerd, (D), Guelph, Zn
   Alan Watson, (M), Guelph, Zn
Instructor:
   Neil Rooney, (D), McGill, Zn
Adjunct Professor:
   Gary Parkin, (D), Guelph, 1994, HwSp
Emeritus:
   George Barron, (D), Iowa State, Zn
   Greg J. Boland, (D), Guelph, Zn
   Ward Chesworth, (D), McMaster, 1967, ClGe
   Les J. Evans, (D), Wales, 1974, Cl
   Austin Fletcher, (D), Alberta, Zn
   Terry J. Gillespie, (D), Guelph, 1968, Am
   Michael J. Goss, (D), Reading, So
   Pieter H. Groenevelt, (D), Wageningen (Neth), 1969, Sp
   Robert Hall, (D), Melbourne, Zn
   Stewart G. Hilts, (D), Toronto, 1981, Zn
   Peter Kevan, (D), Alberta, Zn
   Kenneth M. King, (D), Wisconsin, 1956, As
   I. Peter Martini, (D), McMaster, 1966, Gsl
   Raymond G. McBride, (D), Guelph, 1982, So
   Murray H. Miller, (D), Purdue, 1957, So
   Kaushik Narinder, Zn
   Leonard Ritter, Zn
   Keith Solomon, (D), Illinois, Zn
   Gerry Stephenson, Zn
   George W. Thurtell, (D), Wisconsin, 1965, As
   H. Peter Van Straaten, (D), Goettingen (Germany), 1974, EnScCb

University of Lethbridge
Dept of Geography (B,M) (2015)
4401 University Drive
Lethbridge, AB T1K 3M4
   p. (403) 329-2225
   geography.chair@uleth.ca
   http://home.uleth.ca/geo
   Administrative Assistant: Margaret Cook
Professor:
   Walter E. Aufrecht, (D), Toronto, Zn
   Rene W. Barendregt, (D), Queen's (Canada), 1977, Gm
   Robert J. Rogerson, (D), Macquarie, 1979, Gm
Associate Professor:
   James M. Byrne, (D), Alberta, 1990, Hg
   Hester Jiskoot, (D), Leeds, 2000, GlmZy
   Thomas Johnston, (D), Waterloo, 1988, Ze
   Derek R. Peddle, (D), Waterloo, 1997, Zr
Assistant Professor:
   Craig Coburn, (D), Simon Fraser, 2002, Zy
   Susan Dakin, (D), Waterloo, 2000, Ze
   Stefan Kienzle, (D), Heidelberg, 1993, Zi
   Ivan Townshend, (D), Calgary, 1997, Zn
   Wei Xu, (D), Guelph, 1998, Zi
Adjunct Professor:
   John Dormaar, (D), Alberta, 1961, So
   Ron Hall, Zn
   Larry Herr, (D), Zn
   Daniel L. Johnson, (D), British Columbia, 1983, Zn
   Pano George Karkanis, (D), Uppsala, 1966, So
   Ross McKenzie, (D), Zn
   Anne Smith, (D)
   Derald Smith, (D), Zn
Emeritus:
   Roy J. Fletcher, (D), Clark, 1968, As
   Ian R. MacLachlan, (D), Toronto, 1990, Ze

University of Manitoba
Geological Sciences (B,M,D) (2015)
240 Wallace Building
125 Dysart Road
Winnipeg, MB R3T 2N2

p. (204) 474-9371
Mostafa.Fayek@ad.umanitoba.ca
http://www.umanitoba.ca/geoscience
Administrative Assistant: Brenda Miller
Enrollment (2014): M: 12 (3) D: 8 (2)
Professor:
   Anton Chakhmouradian, (D), St. Petersburg State (Russia), 1997, GziCg
   Nancy Chow, (D), Memorial, 1986, Gs
   Robert J. Elias, (D), Cincinnati, 1979, Pi
   Mostafa Fayek, (D), Saskatchewan, 1996, Cs
   Ian J. Ferguson, (D), Australian Nat, 1988, YemYu
   Andrew Frederiksen, (D), British Columbia, 2001, YsGt
   Norman M. Halden, (D), Glasgow, 1983, Cg
   Frank C. Hawthorne, (D), McMaster, 1973, Gz
   William M. Last, (D), Manitoba, 1980, GsnGo
   Soren Rysgaard, (D), Aarhus (Denmark), 1995, Gl
   Elena Sokolova, (D), Moscow, 1980, Gz
Senior Scholar:
   George S. Clark, (D), Columbia, 1967, Cc
   Barbara L. Sherriff, (D), McMaster, 1988, Gz
   Allan C. Turnock, (D), Johns Hopkins, 1960, Gp
Associate Professor:
   Alfredo Camacho, (D), Australian Nat, 1998, Gt
Assistant Professor:
   Genevieve Ali, (D), Montreal, 2010, Hw
   Zou Zou Kuzyk, (D), Manitoba, 2009, Cg
Research Associate:
   Yassir Abdu, (D), Uppsala, 2004, Gz
Instructor:
   Karen Ferreira, (M)
   William S. Mandziuk, (M), Manitoba, 1989, Gg
   Jeffrey Young, (M), Manitoba, 1992, GcZe
Adjunct Professor:
   Scott Anderson, (D), Dalhousie, 1998
   Christian Bohm, (D), ETH (Switzerland), 1996, Cg
   William Buhay
   David Corrigan, (D), Carleton, Gt
   Jody Deming, (D), Maryland, 1981, Zn
   Michel Houle, (D), Laurentian, 2008, Gv
   Brooke Milne, (D), Ga
   Vince Palace, (D), Manitoba, 1996
   James Reist, (D), Toronto, 1983
   Graham A. Young, (D), New Brunswick, 1988, Pi
Emeritus:
   William C. Brisbin, (D), California (Los Angeles), 1970, Gc
   Robert B. Ferguson, (D), Toronto, 1948, Gz
   Wooil M. Moon, (D), British Columbia, 1976, ZrYnOp
   James T. Teller, (D), Cincinnati, 1970, Gs
Related Staff:
   Neil Ball, (B), Manitoba, 1982, Gz
   Laura Bergen, (M), Manitoba, 2013
   Mark Cooper, (M), Manitoba, 1997, Gz
   Mulu Serzu, (D), Manitoba, 1990, Yg
   Ryan Sharpe, (M), Manitoba, 2012
   Ravinder Sidhu
   Panseok Yang, (D), Memorial, 2002, CaGp
   Misuk Yun, (M), Manitoba, 1986, Cs

University of New Brunswick
Dept of Earth Sciences (B,M,D) (2015)
Fredericton, NB E3B 5A3
   p. (506) 453-4803
   geology@unb.ca
   Administrative Secretary: Merrill Ann Beatty
Chair:
   Cliff S.J. Shaw, (D), W Ontario, 1994, Giv
   Joseph C. White, (D), W Ontario, 1979, Gc
Professor:
   Bruce E. Broster, (D), W Ontario, 1982, Gl
   John Todd Dunn, (D), Alberta, 1983, Gi
   John G. Spray, (D), Cambridge, 1980, GpXgGz
   Paul F. Williams, (D), Sydney, 1969, Gc
Associate Professor:
   Nicholas J. Susak, (D), Princeton, 1981, CglEm
Assistant Professor:
   Karl Butler, (D), British Columbia, 1996, Yg

Research Associate:
   James Whitehead, (D), New Brunswick, 1998, Gg
Honorary Research Professor:
   Henk W. Van De Poll, (D), Swansea, 1970, Gs
Adjunct Professor:
   Richard A. F. Grieve, (D), Toronto, 1970, Xg
   David R. Lentz, (D), Ottawa, 1992, Cg
   Randall F. Miller, (D), Waterloo, 1984, Pg
Emeritus:
   Arnold L. McAllister, (D), McGill, 1950, Em
Geology Librarian:
   Eszter Schwenke, Zn

University of New Brunswick Saint John
Dept of Biological Sciences (B,M,D) ☒ (2018)
PO Box 5050
100 Tucker Park Road
Saint John, NB E2L 4L5
   p. (506) 648-5607
   lwilson@unbsj.ca
   http://www.unbsj.ca/sase/biology/
   Programs: Primarily a Biology department, with a Geology minor as an option.
   Enrollment (2007): B: 2 (0) M: 1 (0)
Professor:
   Lucy A. Wilson, (D), Paris VI, 1986, Ga
Emeritus:
   Alan Logan, (D), Durham (UK), 1962, Ob

University of Ottawa
Dept of Earth and Environmental Sciences (B,M,D) ●☒ (2018)
STEM Complex
Room 342
150 Louis Pasteur Pvt
Ottawa, ON K1N 6N5
   p. (613)-562-5800(Ext.6870)
   cpoirie5@uottawa.ca
   http://www.science.uottawa.ca/est/eng/welcome.html
   Enrollment (2012): B: 78 (18) M: 29 (11) D: 12 (2)
Chair:
   R. William C. Arnott, (D), Alberta, 1987, GsSo
Professor:
   Ian D. Clark, (D), Paris - Sud, 1988, HwGel
   Danielle Fortin, (D), Quebec, 1992, CblGe
   Mark Hannington, (D), Toronto, 1989, Eg
   Keiko Hattori, (D), Tokyo, 1977, CgEmGz
   Michel R. Robin, (D), Waterloo, 1991, Hw
Associate Professor:
   Glenn A. Milne, (D), Toronto, 1998, YgAsZg
   David Schneider, (D), Lehigh, 2000, GtCcGg
Replacement Professor:
   Olivier Nadeau, (D), McGill, 2011, Gx
Assistant Professor:
   Tom A. Al, (D), Waterloo, 1996, HwCgZg
   Pascal Audet, (D), British Columbia, 2008, YgGc
   Jonathan O'Neil, (D), McGill, 2009, Cgc
Lecturer:
   Simone Dumas, (D), Ottawa, 2004, GsOg
Adjunct Professor:
   Frits P. Agterberg, (D), Utrecht, 1961, Gq
   Eric De Kemp, (D), Québec (Chicoutimi), 2000, ZirZg
   Andre Desrochers, (D), Memorial, 1986, Gs
   David Andrew Fisher, (D), Copenhagen, 1978, Gl
   William K. Fyson, (D), Reading (UK), 1960, Gc
   Quentin Gall, (D), Carleton, 1994, GsEmGe
   Richard Goulet, (D), Ottawa, 2001, CgPgGe
   Jeffrey Wayne Hedenquist, (D), Auckland, 1983, Eg
   Donald D. Hogarth, (D), McGill, 1959, Gz
   Dogan Paktunc, (D), Ottawa, 1983, Gz
   Pat E. Rasmussen, (D), Waterloo, 1993, GeAsGb
Emeritus:
   Jan Veizer, (D), Australian Nat, 1971, Cl

University of Regina
Dept of Geology (B,M,D) ●☒ (2018)

3737 Wascana Parkway
Regina, SK S4S 0A2
   p. (306) 585-4147
   geology.office@uregina.ca
   http://www.uregina.ca/science/geology/
   Administrative Assistant: Van Tran
Professor:
   Stephen L. Bend, (D), Newcastle, Gox
   Kathryn Bethune, (D), Queens, GcpGt
   Guoxiang Chi, (D), Queens, Eg
   Ian Coulson, (D), Birmingham (UK), 1996, Gvi
   Hairuo Qing, (D), McGill, 1991, GsCg
Associate Professor:
   Janis Dale, (D), Queens, Gml
   Osman Salad Hersi
   Maria Velez, (D), Amsterdam
Assistant Professor:
   Tsilavo Raharimahefa
Adjunct Professor:
   Kenneth E. Ashton, (D), Gp
   Pier L. Binda, (D), Alberta, 1970, Ge
   Donald M. Kent, (D), Alberta, 1968, Gs
Emeritus:
   Laurence W. Vigrass, (D), Stanford, 1961, Eo

University of Saskatchewan
Dept of Geological Sciences (B,M,D) O (2018)
114 Science Place
Saskatoon, SK S7N 5E2
   p. (306) 966-5683
   sam.butler@usask.ca
   http://artsandscience.usask.ca/geology/
   Programs: Geology; Geophysics; Paleobiology
   Enrollment (2018): B: 132 (47) M: 38 (3) D: 21 (5)
Head:
   Sam Butler, (D), Toronto, 2000, Yg
Professor:
   Kevin M. Ansdell, (D), Saskatchewan, 1992, EgCgGt
   James F. Basinger, (D), Alberta, 1979, Pb
   Luis Buatois, (D), Buenos Aires, 1992, Ps
   Graham George, (D), Sussex, 1983, Cg
   Jim Hendry, (D), Waterloo, 1984, HwClHy
   Chris Holmden, (D), Alberta, 1995, Csc
   Gabriela Mangano, (D), Buenos Aires, 1992, PsGs
   James B. Merriam, (D), York, 1976, Yg
   Igor B. Morozov, (D), Moscow State (Russia), 1985, YsGy
   Yuanming Pan, (D), W Ontario, 1990, Gx
   William P. Patterson, (D), Michigan, 1995, CsPeGn
   Ingrid J. Pickering, (D), Imperial Coll (UK), 1990, CgtZn
   Brian R. Pratt, (D), Toronto, 1989, PgGsr
Associate Professor:
   Matthew B. Lindsay, (D), Waterloo, 2009, ClGg
Assistant Professor:
   Joyce M. McBeth, (D), Manchester, 2007, PgGeSb
   Camille Partin, (D), Gtc
Adjunct Professor:
   Irvine Annesley, (D), Ottawa, 1989, Gi
   Ning Chen, (D), Saskatchewan, 2001, Gz
   David Greenwood, (D), Pg
   Tom Kotzer, (D), Saskatchewan, 1993, CgEg
   Kyle Larson, (D)
   Brett Moldovan, (D), Cg
   Len Wassenaar, (D), Waterloo, 1990, Hw
   Derek A. Wyman, (D), Saskatchewan, 1990, Cg
Emeritus:
   Willi K. Braun, (D), Tubingen, 1958, Ps
   William G. E. Caldwell, (D), Glasgow, 1957, Ps
   Leslie C. Coleman, (D), Princeton, 1955, Gz
   Donald J. Gendzwill, (D), Saskatchewan, 1969, Ye
   Zoltan Hajnal, (D), Manitoba, 1970, YsGtc
   Robin W. Renaut, (D), London, 1982, Gsn
   Mel R. Stauffer, (D), Australian Nat, 1964, Gc

University of Toronto
Dept of Physics, Geophysics Div (B,M,D) (2015)
60 St. George Street
Toronto, ON M5S 1A7
   p. (416) 978-5175
   cliao@physics.utoronto.ca
   http://www.physics.utoronto.ca
Professor:
   Bernd Milkereit, (B), Kiel, 1984, Ye
Emeritus:
   Richard C. Bailey, (D), Cambridge, 1970, Yg
   Richard N. Edwards, (D), Cambridge, 1970, Yr

Dept of Earth Sciences (B,M,D) ● (2017)
Earth Sciences Centre
22 Russell Street
Toronto, ON M5S 3B1
   p. (416) 978-3022
   welcome@es.utoronto.ca
   http://www.es.utoronto.ca
   Business Officer: Silvanna Papaleo
Chair:
   Russell N. Pysklywec, (D), Toronto, 1998, Yg
Professor:
   Nicholas Eyles, (D), Leicester (UK), 1978, Gl
   Grant F. Ferris, (D), Guelph, 1985, Gn
   Henry C. Halls, (D), Toronto, 1970, Ym
   Kenneth W.F. Howard, (D), Birmingham, 1979, Hw
   Andrew D. Miall, (D), Ottawa, 1969, Gso
   Barbara Sherwood Lollar, (D), Waterloo, 1990, Cs
Director, Jack Satterly Geochronology Lab:
   Michael A. Hamilton, (D), Massachusetts, 1993, CcGiCu
Associate Professor:
   Bridget Bergquist, (D), MIT, 2004, Cgb
   Donald W. Davis, (D), Alberta, 1978, Cc
   Grant S. Henderson, (D), W Ontario, 1983, Gz
   Qinya Liu, (D), Caltech, 2006, Ygx
   Daniel J. Schulze, (D), Texas, 1982, Gi
Assistant Professor:
   Carl-Georg Bank, (D), British Columbia, 2002, Zg
   Jörg Bollmann, (D), Swiss Fed Inst Tech, 1995, Pm
   Xu Chu, (D), Yale, 2015, Gx
   Rebecca Ghent, (D), S Methodist, 2002, Zr
   Jochen Halfar, (D), Stanford, 1999, Pe
   Ulrich B. Wortmann, (D), Tech (Munich), Gg
   Zoltan Zajacz, (D), ETH Zurich, 2007, Cu
Research Associate:
   Colin Bray, (D), Oxford, 1980, Cs
Adjunct Professor:
   Jean-Bernard Caron, (D), Toronto, 2005, Pi
   John H. McAndrews, (D), Minnesota, 1964, Pl
Emeritus:
   G M. Anderson, (D), Toronto, 1961, Cp
   Richard C. Bailey, (D), Cambridge, 1970, Yg
   J. J. Fawcett, (D), Manchester, 1961, Gp
   John Gittins, (D), Cambridge (UK), 1959, GipGz
   Anthony J. Naldrett, (D), Queen's, 1964, EmGiCp
   Geoffrey Norris, (D), Cambridge, 1964, Pl
   Pierre-Yves F. Robin, (D), MIT, 1974, Gc
   John C. Rucklidge, (D), Manchester, 1962, Ge
   Walfried M. Schwerdtner, (D), Free Univ Berlin, 1961, GctGq
   Steven D. Scott, (D), Penn State, 1968, GuEmCm
   Edward T. C. Spooner, (D), Manchester (UK), 1976, En
   Peter H. von Bitter, (D), Kansas, 1971, Pm
   John A. Westgate, (D), Alberta, 1964, Gl
   Frederick J. Wicks, (D), Oxford, 1969, Gz

University of Victoria
School of Earth & Ocean Sciences (B,M,D) O (2018)
P.O. Box 1700
3800 Finnerty Road
Victoria, BC V8W 2Y2
   p. (250) 721-6120
   seos@uvic.ca
   http://www.uvic.ca/science/seos/
Associate Director (SCIENCE) NEPTUNE Canada:
   Kim Juniper, (D), Canterbury, 1982, Ob
Associate Dean of Science:
   Kathryn Gillis, (D), Dalhousie, 1987, Gp

Professor:
  Dante Canil, (D), Alberta, 1989, Gi
  Laurence Coogan, (D), Leicester (UK), Gi
  Stanley E. Dosso, (D), British Columbia, 1990, Op
  Adam Monahan, (D), British Columbia, 2000, Zn
  Thomas F. Pedersen, (D), Edinburgh (UK), 1979, CmOcCs
  Vera Pospelova, (D), McGill, 2003, PelOu
  George D. Spence, (D), British Columbia, 1984, Ys
  Verena Tunnicliffe, (D), Yale, 1980, Ob
  Michael J. Whiticar, (D), Kiel, 1978, CosGo
Associate Professor:
  Jay Cullen, (D), Rutgers, 2001, Oc
  John Dower, (D), Victoria, 1994, Ob
  Roberta C. Hamme, (D), Washington, 2003, Oc
  Thomas S. James, (D), Princeton, 1991, YdvYg
  Jody Klymak, (D), Washington, 2001, Op
  Eileen Van Der Flier-Keller, (D), W Ontario, 1985, Cl
  Diana Varela, (D), British Columbia, 1998, Ob
Assistant Professor:
  Colin Goldblatt, (D), East Anglia, 2008, As
  Lucinda Leonard, (D), Victoria, Yg
Adjunct Professor:
  Vaughn Barrie, (D), Wales, 1986, Gu
  John Cassidy, (D), British Columbia, Ys
  James R. Christian, (D), Hawaii, 1995, Obc
  Kenneth L. Denman, (D), British Columbia, 1972, Op
  Richard K. Dewey, (D), British Columbia, 1987, Opg
  Gregory M. Flato, (D), Dartmouth, 1991, As
  John C. Fyfe, (D), McGill, 1987, As
  Richard J. Hebda, (D), British Columbia, 1977, Pl
  Roy D. Hyndman, (D), Australian Nat, 1967, Ys
  Stephen T. Johnston, (D), Alberta, 1993, Gc
  Victor M. Levson, (D), Alberta, 1995, Gl
  David L. Mackas, (D), Dalhousie, 1977, Ob
  Norman McFarlane, (D), Michigan, 1974, As
  Garry C. Rogers, (D), British Columbia, 1983, Ys
  George J. Simandl, (D), Ecole Polytechnique, 1992, En
  Richard Thomson, (D), British Columbia, 1971, Og
  Kelin Wang, (D), W Ontario, 1989, Yg
  Michael Wilmut, (D), Queen's, 1971, Zn
Emeritus:
  Christopher R. Barnes, (D), Ottawa, 1964, PmOgPe
  Ross N. Chapman, (D), British Columbia, 1975, Op
  Christopher J. R. Garrett, (D), Cambridge, 1968, Op
  John T. Weaver, (D), Saskatchewan, 1959, Ym
On Leave:
  Andrew J. Weaver, (D), British Columbia, 1987, Op
Administrative Officer:
  Terry P. Russell, (B), Victoria, 2000, Zn

University of Waterloo

Dept of Earth and Environmental Sciences (B,M,D) ● (2016)
Waterloo, ON N2L 3G1
  p. (519) 888-4567, ext. 32069
  klalbrec@uwaterloo.ca
  https://uwaterloo.ca/earth-environmental-sciences/
  Administrative Assistant: Lorraine Albrecht
Chair:
  Barry G. Warner, (D)
Professor:
  David W. Blowes, (D), Waterloo, 1990, Cg
  Mario Coniglio, (D), Memorial, 1985, Gr
  Maurice B. Dusseault, (D), Alberta, 1977, Ng
  Thomas W. D. Edwards, (D), Waterloo, 1987, Cs
  Stephen George Evans, (D), Alberta, 1983, Ng
  Shaun K. Frape, (D), Queen's, 1979, Cl
  Shoufa Lin, (D), New Brunswick, 1992, Zn
  Carol J. Ptacek, (D), Waterloo, 1992, Cg
  David L. Rudolph, (D), Waterloo, 1989, Hq
  Sherry L. Schiff, (D), Columbia, 1986, Ng
  Edward A. Sudicky, (D), Waterloo, 1983, Hw
  Philippe Van Cappellen, (D)
Associate Professor:
  Anthony E. Endres, (D), British Columbia, 1991, Pe
  Walter Illman, (D)
  Martin Ross, (D)
  Andre Unger, (D), Waterloo, 1995, Hw

Assistant Professor:
  Nandita Basu, (D)
  Carl Guilmette, (D)
  Brian Kendall, (D)
  Lingling Wu, (D)
Lecturer:
  Eric C. Grunsky, (D)
  John Johnson, (D)
  Isabelle McMartin, (D)
  Brent Wolfe, (D)
Adjunct Professor:
  Edward C. Appleyard, (D), Cambridge, 1963, Gp
  Gail Atkinson, (D)
  James F. Barker, (D), Waterloo, 1979, Co
  Steven Berg, (D)
  Alec Blyth, (D)
  Thomas Bullen, (D)
  Lauren Charlet, (D)
  John A. Cherry, (D), Illinois, 1966, Hw
  Peter Condon, (D)
  Rick Devlin
  Michael English, (D)
  Nicholas Eyles, (D)
  John F. Gartner, Zn
  John J. Gibson, (D), Waterloo, 1996, Cs
  Susan Glasuer, (D)
  David Good, (D)
  John Gosse, (D)
  Douglas Gould, (D)
  Norman Halden, (D)
  Daniel Hammarlund, (D)
  Jens Hartman, (D)
  Lin Huang, (D)
  Daniel Hunkeler
  Hyoun-Tae Hwang, (D)
  Richard Jackson, (D)
  Sun-Wook Jeen, (D)
  Michael Krom, (D)
  David R. Lee, (D), Virginia Tech, 1976, Hw
  Yuri Leonenko, (D)
  John Lin, (D)
  Robert Linnen, (D), McGill, 1992, Eg
  Benoit Made, (D)
  Uli Mayer, (D)
  Hossein Memarian, (D)
  John Molson
  Alan V. Morgan, (D), Birmingham, 1970, Gl
  Christopher Neville, (D)
  Dogan Paktunc, (D)
  Sorab Panday, (D)
  Young-Jin Park, (D), Hq
  Gary Parkin, (N)
  Peter Pehme, (D)
  Jennifer Pell, (D)
  Richard Peltier, (D)
  Terry D. Prowse, (D), Canterbury, 1981, Cg
  William Quinton, (D)
  Eric J. Reardon, (D), Penn State, 1974, Cl
  Rashid Rehan, (D)
  Iain Samson, (D)
  Houston C. Saunderson, (N)
  James Sloan, (D)
  James Smith
  John Spoelstra, (D)
  Andrew Stumpf, (D)
  William Taylor, (D)
  Rene Therrien, (D)
  Harvey Thorleifson, (D)
  Martin Thullner, (D)
  Benoit Valley, (D)
  Garth Van der Kamp, (D), Amsterdam, 1973, Hw
  Jan van der Kruk, (D)
  Cees van Staal
  Andrea Vander Woude, (D)
  Owen L. White, (D), Illinois, 1970, Ng
  C. Wolkersdorfer, (D)
  Wenjiao Xiao, (D)

Leiming Zhang, (D)
Emeritus:
   Emil O. Frind, (D), Toronto, 1971, Hq
   Robert W. Gillham, (D), Illinois, 1973, Hq
   Paul Karrow, (D), Illinois, 1957, Gl
Research Professor:
   Ramon Aravena, (D), Waterloo, 1993, Co
Research Associate Professor:
   Will Robertson, (D), Waterloo, 1992, Hw
Other:
   Brewster Conant, Hw

University of Windsor
Dept of Earth and Environmental Sciences (B,M,D) (2015)
401 Sunset Avenue
Windsor, ON N9B 3P4
   p. (519) 253-3000 x2486
   earth@uwindsor.ca
   http://www.uwindsor.ca/ees
   Administrative Assistant: Sharon Horne
Professor:
   Ihsan S. Al-Aasm, (D), Ottawa, 1985, GdClGo
   Aaron Fisk, (D), Manitoba, 1998, ObCs
   V. Chris Lakhan, (D), Toronto, 1982, ZriOn
   Ali Polat, (D), Saskatchewan, 1998, GixCc
   Iain M. Samson, (D), Strathclyde, 1983, CgEmGx
   Frank -. Simpson, (D), Jagellonian (Krakow), 1968, GsrGe
   Alan S. Trenhaile, (D), Wales, 1969, Gm
Associate Professor:
   Maria T. Cioppa, (D), Lehigh, 1997, Ym
   Joel E. Gagnon, (D), McGill, 2006, CaqEg
   Phil A. Graniero, (D), Toronto, 2001, ZinZy
   Cyril G. Rodrigues, (D), Carleton, 1980, Pm
   Christopher Weisener, (D), South Australia, 2003, CoGePg
   Jianwen Yang, (D), Toronto, 1997, HwYg
Instructor:
   Denis Tetrault, (D), Western, 2002, Gg
Emeritus:
   Brian J. Fryer, (D), MIT, 1971, Cla
   Peter P. Hudec, (D), Rensselaer, 1965, NgEmGe
   Terence E. Smith, (D), Wales, 1963, Gi
   David T. A. Symons, (D), Toronto, 1965, YmGtEm
   Andrew Turek, (D), Australian Nat, 1966, Cc

Western University
Dept of Earth Sciences (B,M,D) ○☒ (2018)
Biology & Geology Building
Room 1026
1151 Richmond Street North
London, ON N6A 5B7
   p. (519) 661-3187
   earth-sc@uwo.ca
   http://www.uwo.ca/earth/
   t: @westernuEarth
   Administrative Assistant: Erika Gongora
   Enrollment (2016): B: 123 (15) M: 77 (16) D: 54 (8)
Chair:
   Patricia Corcoran, (D), Dalhousie, 2001, Gd
Robert Hodder Chair:
   Robert Linnen, (D), McGill, 1992, EmCe
NSERC Industrial Research Chair:
   Gail M. Atkinson, (D), W Ontario, 1993, YssYs
Joint appointment with Geography, U.W.O.:
   Desmond Moser, (D), Queens, 1993, GtCcXg
Industrial Research Chair, Joint Appt Physics & Astronomy, U.W.O:
   Gordon R. Osinski, (D), New Brunswick, 2004, XgcGp
Cross-Appt.UWO: Physics&Astronomy:
   Sean R. Shieh, (D), Hawaii, 1998, YxGzZm
Cross-Appt.-Home Dept: Physics & Astronomy, UWO:
   Peter Brown, (D), W Ontario, 1999, Xym
Canada Research Chair, Cross Appt. Biology&Geography, UWO:
   Brian Branfireun, (D), McGill, Ge
Canada Research Chair:
   Frederick J. Longstaffe, (D), McMaster, 1978, Cs
(P.Eng.,P.Geo.):
   Robert A. Schincariol, (D), Ohio State, 1993, HwsNg

Professor:
   Dazhi Jiang, (D), New Brunswick, 1996, GcgGg
   Jisuo Jin, (D), Saskatchewan, 1988, Pi
   A Guy Plint, (D), Oxford, 1980, Gs
   R. Gerhard Pratt, (D), Imperial Coll (UK), 1989, Yeg
   Richard A. Secco, (D), W Ontario, 1988, Gy
Industrial Research Chair:
   Neil Banerjee, (D), Victoria, 2001, EgCgGx
Cross-Appt.UWO: Physics&Astronomy:
   Robert Shcherbakov, (D), Cornell, 2002, YgsYd
Canada Research Chair:
   Audrey Bouvier, (D), École Normale Supérieure de Lyon, 2005, Xc
Associate Professor:
   Burns A. Cheadle, (D), W Ontario, 1986, Go
   Roberta L. Flemming, (D), Queen's (Canada), 1997, GzXmZm
   Katsu Goda, (D), Rn
   Elizabeth A. Webb, (D), W Ontario, 2000, CslSa
Assistant Professor:
   Sheri Molnar, (D), Victoria, 2011, Ys
   Catherine Neish, (D), Arizona, 2008, Xg
   Cameron J. Tsujita, (D), McMaster, 1995, Pi
   Tony Withers, (D), Bristol, 1997, Cp
Research Technician:
   Peter Christoffersen, (M), W Ontario
   Claire Montera
Research Scientist:
   Li Huang, (M), Concordia, 1992, Yr
Research Engineer:
   Matthew Bourassa, (D), Xg
Research Associate:
   Hadi Ghofrani, (D)
W.S. Fyfe Visiting Scientist- in- Residence:
   David Good, (D), McMaster, 1992, Em
Research Associate:
   Brian R. Hart, (D), W Ontario, 1995, Ca
Research Adjunct Professor:
   Natalie Pietrzak-Renaud, (D), W Ontario, 2011, Eg
Associate Curator of Mineralogy, Royal Ontario Museum/Professor, University of Toronto:
   Kim Tait, (D), Arizona, 2007, Xm
Adjunct Professor:
   Keith Baron, (D)
   Melissa Battler
   Rob Carpenter, (D)
   Ed Cloutis, (D), Alberta, 1989, Zi
   Claudia Cochrane, (M), W Ontario, Go
   Yunpeng Dong, (D), Northwest Univ (China), 1997, GtcCg
   Richard Grieve, (D), Toronto, 1970, Xg
   Matt Izawa, (D), W Ontario, 2012, Gz
   Laura Jansen
   Peter Lightfoot
   Phil JA McCausland, (D), Western, 2002, YmXm
   Gero Michel, (D), Harvard, Ys
   Sobhi Nasir, (D), Wuerzburg, 1989
   Sergey Samsonov
   Brenden Smithyman, (D), Ye
   Gordon Southam, (D), Guelph, 1990, Pg
   Yves Thibault, (D), W Ontario, 1990, Gz
   Kristy F. Tiampo, (D), Colorado, 2000, Yd
   Livio Tornabene, (D), Tennessee, 2007, XgZr
   Lisa Van Loon, (D), Ohio State, 2007, Ca
Emeritus:
   Alan Beck, (D), Australian Nat, 1964, Zn
   W. Glen E. Caldwell, (D), Glasgow, 1957, Pm
   William R. Church, (D), Wales, 1961, Gt
   Norman A. Duke, (D), Manitoba, 1983, Eg
   Akio Hayatsu, (D), Toronto, 1965, Cc
   Stephen R. Hicock, (D), W Ontario, 1980, Gl
   Alfred C. Lenz, (D), Princeton, 1959, PiePi
   Robert F. Mereu, (D), W Ontario, 1962, Ys
   H Wayne Nesbitt, (D), Johns Hopkins, 1975, Cl
   H. Currie Palmer, (D), Princeton, 1963, YmGv
   Grant M. Young, (D), Glasgow (UK), 1967, GrCgGt
Laboratory Director:
   Kim Law, Cs
   Wenjun Yong, (D), Yx

Technical Specialist:
  Barry Price
Research Technologist:
  Jon Jacobs, Gg
Research Technician:
  Marc Beauchamp, Ca
  Grace Yau
Lab Technician:
  Steven Wood
Geoscience Collections Curator:
  Alysha McNeil, (D), W Ontario, 2017
Financial Assistant:
  Miyako Maekawa
Computer Technician:
  Bernie Dunn, Ye
Administrative Officer:
  Kristen Harris
Academic Program Coordinator:
  Amy N. Wickham, (M), Connecticut, 2016

York University
Earth and Space Science and Engineering (B,M,D) ● (2015)
4700 Keele Street
Toronto, ON M3J 1P3
  p. (416) 736-5245
  esse@yorku.ca
  http://www.yorku.ca/esse
  Administrative Assistant: Paola Panaro
  Enrollment (2010): B: 100 (38)
Chair:
  Regina Lee, (D), Toronto, 2000
Professor:
  Qiuming Cheng, (D), Ottawa, 1994, Zi
  Christian Haas, (D), Bremen, 1996, Yg
  Gary T. Jarvis, (D), Cambridge, 1978, Yg
  Ian C. McDade, (D), Belfast, 1979, Zr
  Tom McElroy, (D), York, 1985, As
  Spiros Pagiatakis, (D), New Brunswick, 1988, Yd
  Jinjun Shan, (D)
  Peter A. Taylor, (D), Bristol, 1967, As
Associate Professor:
  Costas Armenakis, (D), New Brunswick, 1988, Zr
  Sunil Bisnath, (D), New Brunswick, 2004
  Yongsheng Chen, As
  Michael Daly, (D)
  Baoxin Hu, (D), Boston, 1998, Zr
  Mary Ann Jenkins, (D), Toronto, 1986, As
  Gary P. Klaassen, (D), Toronto, 1983, As
  Brendan Quine, (D)
  Gunho Sohn, (D)
  George Vukovich, (D), As
  James Whiteway, (D), York, 1994, As
Assistant Professor:
  William Colgan, (D), Yg
  Mark Gordon, (D), As
  John E. Moores, (D), Arizona, 2008
Lecturer:
  Hugh Chesser, (M), Toronto, 1987
  Franz Newland, (D), Southampton, 2000, Zr
  Jian-Guo Wang, (D)
Emeritus:
  Keith D. Aldridge, (D), MIT, 1967, YmdYh
  John Miller, (D), Saskatchewan, 1969, Zr
  Gordon G. Shepherd, (D), Toronto, 1956, AsZrAp
  Douglas E. Smylie, (D), Toronto, 1963, Yg
  Anthony M. K. Szeto, (D), Australian Nat, 1982, Yg

Cape Verde

Universidade de Cabo Verde
Dept de Ciencia e Tecnologia (M,D) (2018)
Campus de Palmarejo - CP 279
Palmarejo, Praia - Cabo Verde
Praia
Praca Antonio Lereno, Praia CP 379C
  p. +238 261 99 01

joao.semedo@docente.unicv.edu.cv
http://www.unicv.edu.cv/dct

Faculdade de Engenharia e Ciencias do Mar (B,M,D) ☒ (2017)
Ribeira de Julião
CP 163
Mindelo, São Vicente CP 163
  p. +238 2326561/62
  alexandra.delgado@docente.unicv.edu.cv
  http://www.unicv.edu.cv

Universidade Jean Piaget de Cabo Verde
Ecologia e Desenvolvimento (B) (2015)
Campus Universitario da Cidade da Praia
Praia
  p. +238 629085
  info@unipiaget.cv
  http://www.unipiaget.cv/pdf/cursos/ecdm.pdf

Central African Republic

Universite de Bangui
Dept de Chimie-Biologie-Geologie (B) (2015)
BP 1450
Avenue des Martyrs
Bangui
  p. (236) 61 20 05
  info@univ-bangui.info
  http://www.univ-bangui.org/

Chad

Universite de N'Djamena
Dept de Geologie (B) (2015)
BP 1117
Avenue Mobutu
N'Djamena
  sg@undt.info
  http://www.undt.info/

Chile

Servicio National de Geologia y Mineria de Chile
Servicio National de Geologia y Mineria de Chile (2015)
Avda Sta Maria N° 0104
Providence
  p. 56-2-7375050
  oirs@sernageomin.cl
  http://www.sernageomin.cl/

Universidad Austral de Chile
Inst de Geociencias (2015)
Campus "Isla Teja", Casilla 567, Valdivia
  p. +56 63 2293861
  cienciasdelatierra@uach.cl

Universidad Catolica del Norte
Dept de Ciencias Geologicas (2015)
Avenida Angamos 610, ,
Casilla 1280, Antofagasta
  p. (55) 2355968
  mbembow@ucn.cl
  http://www.ucn.cl/facultades/SitioDeInteres/?cod=2&codItem=110&codPrincipal=1124

Universidad de Chile
Dept de Geologia (2015)
Plaza Ercilla 803, Casilla 13148, Correo 21, Santiago
  colegios@ing.uchile.cl

Dept de Geofisica (2015)
Blanco Encalada 2002
Casilla 2777

Santiago
> p. (56 2)696 6563
> rmunoz@dgf.uchile.cl
> http://www.dgf.uchile.cl/index.html

Universidad de Concepcion
Inst de Geologia Economica Aplicada (2015)
Barrio Universitario - Victor Lamas 1290
Casilla 4107
Concepcion, Region del Bio-Bio
> p. (56) 41 220 44 88
> maravenah@udec.cl
> http://www.udec.cl/postgrado/?q=node/39&codigo=4161

Dept de Geociencia (2015)
Cabina 13 - Barrio Universitario, Casilla 3-C, 4250-1 Concepcion, Region del Bio-Bio

China

Capital Normal University
College of Resources, Environment and Tourism (2015)
West 3rd Ring Road North 105#
Beijing 100048
> p. 86 10 68903321
> info@mail.cnu.edu.cn
> http://www.cnu.edu.cn

China Geological Survey
China Geological Survey (2015)
24 Huangsi Dajie,
Xicheng District
Beijing 100011
> p. +86 10 51632963 51632906
> enwebmaster@mail.cgs.gov.cn
> http://www.cgs.gov.cn/

China University of Geosciences, Beijing
Geosciences (2015)
No.29, Xueyuan Road,
Haidian District
Beijing 100083
> http://www.cugb.edu.cn/EnglishWeb/index.html

China University of Geosciences, Wuhan
Faculty of Geosciences (2015)
No. 388, Lumo Road, Hongshan District
Wuhan 430074
> p. 86-27-87481030
> cugxb@cug.edu.cn
> http://en.cug.edu.cn/cug/index.asp

China University of Mining and Technology, Beijing
College of Geoscience and Surveying Engineering (2015)
Beijing
> dcxy@cumtb.edu.cn
> http://dcxy.cumtb.edu.cn/

Nanjing University
Dept of Earth Sciences (2015)
22 Hankou Road
Nanjing , Jiangsu
> wsj@nju.edu.cn
> http://www.nju.edu.cn/cps/site/njueweb/fg/index.php

University of Hong Kong
Dept of Earth Sciences (B,M,D) ●☒ (2019)
James Lee Science Building
Pokfulam Road
Hong Kong
> p. (852) 2859 1084
> earthsci@hku.hk
> http://www.earthsciences.hku.hk
> Programs: Geology, Earth System Science

Colombia

EAFIT University
Dept of Geology (2015)
Carrera 49 No. 7 Sur - 50
Medellin
> p. 01 8000515 900
> contacto@eafit.edu.co

Escuela de Ingenieria de Antioquia
Dept of Geologic Engineering ☒ (2018)
Calle 25 Sur
#42-63 Envigado
Medellin
> maria.espinosa68@eia.edu.co
> http://www.eia.edu.co/site/index.php/pregrados/programas/ing-geologica.html

Universidad de Santander
Environmental Engineering (B,M) (2015)
Facultad Ingenierias
Cll 70 No. 55-210 Campus Universitario Lagos del Cacique
Bucaramanga, Santander
> p. +57 7 6516500
> nmantilla@udes.edu.co
> http://www.udes.edu.co/programas-profesionales/facultad-ingenierias/ingenieria-ambiental.html

Universidad Industrial de Santander
School of Geology (B,M) O (2016)
Cra 27 Calle 9 Ciudad Universitaria
Bucaramanga, Santander 1
> p. 57-7-6343457
> escgeo@uis.edu.co
> http://geologia.uis.edu.co/eisi/

Director:
> Juan Diego Colegial Gutierrez, (D), Politecnica de Madrid (Spain), 2004, GeNgZr

Professor:
> Luis Carlos Mantilla Figueroa, (D), EgCec

Universidad Nacional de Colombia
Dept de Geociencias (2015)
Carrera 30 No. 45-03, Edificio 224
Calle 45, Cr. 30
Bogota
> lhochoag@unal.edu.co

Costa Rica

Universidad de Costa Rica
Escuela Centroamericana de Geologia (B,M) ●☒ (2018)
Apartado 35-2060 UCR, San Pedro de Montes de Oca, San José
San José
> p. 506 25118128
> geologia@ucr.ac.cr
> http://www.geologia.ucr.ac.cr/
> f: https://www.facebook.com/ECG.UCR/
> t: @geologiaucr
> Programs: Geology (B)

Director:
> Rolando Mora Chinchilla, (M)

Professor:
> Mario Enrique Arias Salguero, (M)
> Allan Astorga Gattgens, (D), Stuttgart, 1996, GesGd
> Elena Badilla Coto, (M)
> Marco Barahona Palomo, (D)
> Lolita Campos Bejarano, (D)
> Guaria Cárdenes Sandí, (D)
> Percy Denyer Chavarría, (D)

Maximiliano Garnier Villarreal, (D), NxYgGg
Lepolt Linkimer Abarca, (D)
Oscar Lucke Castro, (D)
Raúl Mora Amador, (M)
Mauricio Manuel Mora Fernández, (D), Savoie (France), 2003
Stephanie Murillo Maikut, (D)
Luis Guillermo Obando Acuña, (M)
Giovanni Marino Peraldo Huertas, (M)
María Isabel Sandoval Gutiérrez, (D)
Instructor:
 Patrick Durán Leiva, (B)
 María del Pilar Madrigal Quesada, (D)
 Vanessa Rojas Herrera, (A)
 Luis Guillermo Salazar Mondragón, (M)
 Ingrid Vargas Azofeifa, (M)

Croatia

Croatian Geological Survey
Croatian Geological Survey ☒ (2018)
Sachsova 2
P.O.Box 268
Zagreb HR-10000
 p. +38516160888
 ured@hgi-cgs.hr
 http://www.hgi-cgs.hr/
 f: https://www.facebook.com/HGI.CGS/

University of Zagreb
Dept of Geology, Faculty of Science (B,M,D) (2015)
Horvatovac 102a
Zagreb HR-10000
 p. +38514605960
 godsjek@geol.pmf.hr
 http://www.geol.pmf.hr
 Enrollment (2010): B: 121 (27) M: 28 (11) D: 30 (6)
full professor:
 Mladen Juraèiæ, (D), Zagreb, 1987, GueCm
associate professor:
 Dražen Balen, (D), Zagreb, 1999, GxpGi
full professor:
 Darko Tibljaš, (D), Zagreb, 1996, GzScGx
 Vlasta Æosoviæ, (D), Zagreb, 1996, PemPs
emeritus:
 Ivan Gušiæ, (D), Zagreb, 1974, GgPsm

Cyprus

Ministry of Agriculture, Rural Development and Environment
Cyprus Geological Survey Department ☒ (2018)
1 Lefkonos Street
Lefkosia 2064
 p. +357 22409213
 director@gsd.moa.gov.cy
 http://www.moa.gov.cy/gsd
 t: @CY_earthquakes

Czech Republic

Charles University
Inst of Hydrogeology, Engineering Geology and Applied Geophysics (2015)
Albertov 6
128 43 Praha 2
 fischer@natur.cuni.cz
 http://prfdec.natur.cuni.cz/~geophys/ustav/katedra.htm

Institute of Petrology and Structural Geology (B,M,D) (2015)
Albertov 6
Praha 2 128 43
 p. 00420-221951524
 petrol@natur.cuni.cz
 http://petrol.natur.cuni.cz
 Enrollment (2012): B: 3 (5) M: 15 (4) D: 15 (0)

Assoc. Prof.:
 David Dolejs, (D), 2004, GiCpg
Professor:
 Shah Wali Faryad, (D), 1990, GpzGi

Dept of Geophysics (B,M,D) ☒ (2017)
KG MFF UK
V Holesovickach 2
180 00 Prague 8, Czech Republic
Prague, Czech Republic 180 00
 geo@mff.cuni.cz
 http://geo.mff.cuni.cz
 Programs: Solid Earth Geophysics (B); Planetary Science (B)
 Enrollment (2017): B: 5 (0) M: 3 (0) D: 2 (0)

Czech Geological Survey
Cesky Geologicky Ustav (2015)
Klarov 3
118 21 Praha 1
 p. +420 257 089 500
 secretar@geology.cz
 http://www.geology.cz

Geofond (2015)
Kostelni 26
Prague PSC 170 06
 p. +420 234 742 111
 vstrupl@geofond.cz
 http://www.geofond.cz/cz/domu

Masaryk University
Dept of Geological Sciences (B,M,D) O☒ (2018)
Faculty of Science
Kotlarska 2
Brno 611 37
 p. +420 549 49 4322
 geologie@sci.muni.cz
 http://ugv.cz/
 f: https://www.facebook.com/chci.byt.geolog
 Enrollment (2017): B: 134 (30) M: 80 (35) D: 81 (5)
Head:
 Zdenek Losos, (D), Gzy
Professor:
 Ondrej Babek, (D), Gds
 Jiri Kalvoda, (A), PsePm
 Milan Novak, (A), GzpGi
 Antonín Prichystal, (A), GgaGv
Associate Professor:
 Jan Cenpírek, Gz
 Eva Geršlová, Ge
 Martin Ivanov, (A), PgiPe
 Jaromir Leichmann, (A), GxpGi
 Rostislav Melichar, (A), GtcGr
 Slavomír Nehyba, (A), GsdGg
 Radek Skoda, (A), Masaryk (Czech Republic), 2017, GzZmGi
 Marek Slobodnik, (A), GeEg
 Josef Zeman, (D), Comenius Univ (Slovakia), 1987, CgpCe
Assistant Professor:
 Martin Knizek, (D), NgmNr
 Tomas Kuchovsky, (D), Hgy
 Tomas Kumpan, Ps
 Pavel Pracny, Cg
 Adam Ricka, (D), Hgy
Research Associate:
 Renata Copjakova, (D), Gz
Lecturer:
 Nela Dolakova, (A), Plg
Emeritus:
 Rostislav Brzobohaty, (A), PgsPe
 Rudolf Musil, (D), Charles (Prague), 1968, PgvGe

Dept of Mineralogy, Petrology and Geochemistry (2015)
Kotlarska 2
611 37 Brno
 sona@sci.muni.cz

http://www.muni.cz/sci/structure/315020.html

Dept of Geography (2015)
Kotlarska 2
611 37 Brno
dobro@sci.muni.cz
http://www.muni.cz/sci/structure/315030.html

Technical University of Ostrava
Faculty of Mining & Geology (2015)
Vysoka Skola Banska v Ostrava
17.listopadu 15/2172
708-33 Ostrava
p. +420 597 325 456
dekan.hgf@vsb.cz
http://www.hgf.vsb.cz/cs

D.R. of Congo

Marien Ngouabi University
Faculty of Science and Technology (B,M,D) O (2015)
Face Square Général Charles De Gaulle
Bacongo
BP:69
Brazzaville, Congo
p. (+242) 06 623 61 22
jm_ouamba@yahoo.fr
http://www.univ-mngb.net/fs/

Universite de Kinshasa
Dept des Sciences de la Terre (B) (2015)
PO Box 190
Kinshasa XI
p. (+243) 82 333 96 93
rectorat@unikin.cd
www.unikin.cd

Départements et filières Sciences Agronomiques (B) (2015)
PO Box 190
Kinshas XI
p. +243 89 89 20 507
fbkapuku@hotmail.com
http://www.unikin.cd/ogec/

Universite de Lumumbashi
Faculté des sciences (B) (2015)
BP 1825
Lubumbashi
p. +243 263-22-5403
unilu@unilu.net
http://www.unilu.ac.cd/En/Pages/default.aspx

Denmark

Aarhus University
Department of Geoscience (B,M,D) (2016)
Hoegh-Guldbergs Gade 2
DK-8000 Arhus C
geologi@au.dk
http://geo.au.dk

Geological Survey of Denmark and Greenland
Geological Survey of Denmark and Greenland (GEUS) (2016)
Oester Voldgade 10
Copenhagen DK-1350
p. +45 38 20 00
geus@geus.dk
http://www.geus.dk/

Roskilde University
QuadLab, ENSPAC (2015)
Universitetsvej 1
Roskilde DK-4000
p. 45 46743097

storey@ruc.dk
http://www.quadlab.dk/bcms-ui-base/

Technical University of Denmark
Dept of Earth Sciences (2015)
Anker Engelunsved
DK-2800 Lyngby
thho@env.dtu.dk
http://www.igg.dtu.dk/index_z.htm

University of Copenhagen
Inst for Geography and Geology (2015)
Oster Voldgage 5-7
DK-1350 Copenhagen K
p. +4535322400
pab@ign.ku.dk
http://geo.ku.dk/

Dept of Geophysics ☒ (2017)
Juliane Maries Vej 30
Copenhagen 2100 Copenhagen
p. +45 353 20605
losos@sci.muni.cz
http://www.nbi.ku.dk/theinstitute/page52794.htm

Geological Museum (2015)
Oster Voldgade 5-7
Copenhagen K DK-1350
p. +45 35322345
rcp@snm.ku.dk
http://geologi.snm.ku.dk/english/

Djibouti

Insitut de Physique du Globe de Paris, (IPGP)
Dept of Volcanic Systems (B) (2015)
Observatoire Géophysique d'Arta, BP 1888
p. (253) 42 21 92
komorow@ipgp.fr
http://volcano.ipgp.jussieu.fr/djibouti/stationdj.html

Universite de Djibouti
Dept de Geologie/Biologie (B) (2015)
BP 1904
p. +253-250459
webmaster@univ.edu.dj
http://www.univ.edu.dj/facultes.fsti.bg.html

Dominican Republic

Mineterio de Energie y Minas Republica Dominica
Servicio Geologico Nacional Republica Dominica (2016)
Ave Winston Churchill No 75
Edificio J.F. Martinez 3er Piso
Santa Domingo
p. (809) 732 0363
smunoz@sgn.gov.do
http://www.sgn.gov.do

Ecuador

Escuela Politecnica Nacional
Facultad de Geologia, Minas y Petroleos (2015)
Ladron de Guevara E11
253 Quito
p. (593-02) 2-507 - 126
deparamento.geologia@epn.edu.ec
http://www.epn.edu.ec/index.php?option=com_content&view=article&id=1202%3Adepartamento-de-geologia-dg&catid=163&Itemid=342

Escuela Superior Politecnica Del Litoral
Facultad de Ciencias de la Tierra (2015)

Km 30.5, Via Perimentral
Guayaquil EC090150

Universidad de Guayaquil
Facultad de Ciencias Naturales (2015)
Av Raul Gomez Lynx s / n Av Juan Tanca Marengo
Guayaquil
p. 3080777 to 3080758
carmenbonifaz@hotmail.com
http://fccnnugye.com/

Egypt

Ain Shams University
Dept of Geology (B) (2015)
Abbassia 11566
Cairo 11566
p. +20(2)6830963
deans@asunet.shams.edu.eg
http://sci.shams.edu.eg/Departments_Geology_Index.ASP

Dept of Geophysics (B) (2015)
Abbassia 11566
Cairo 11566
p. +20(2)6830963
deans@asunet.shams.edu.eg
http://sci.shams.edu.eg/Departments_Geophysics_Index.ASP

Al Azhar University
Dept of Geology (B) (2015)
Yosief Abbas Street
Cairo 11787
p. +20 (2) 262 3274
k.ubeid@alazhar.edu.ps
http://www.uazhar.edu.eg/bfac/sci/jiolojy.htm

Alexandria University
Dept of Environmental Sciences (B) (2015)
22 Al-Guish Avenue
Alexandria
p. +20 (3) 591 1152
anwar_elfiky@hotmail.com
http://www.alex.edu.eg/dept.jsp?FC=4&CODE=09

Dept of Geology (A,B,M,D) O (2015)
Baghdad Street
Qism Moharram Bek
Alexandria 21511
p. (+203) 3921595
sc-dean@alexu.edu.eg
http://www.sci.alexu.edu.eg/en/Departments/Default.aspx?Dept=4
f: https://www.facebook.com/groups/GSAU.alex/
Enrollment (2014): B: 26 (0)
Professor:
  Galal Mohamed I Galal, (D), Germany, 1994, GgPeGe
  Khalil I. Khalil Ebeid, (D), Germany, 1995, GgEmm
  Kadry Nasser Sediek, (D), USSR, 1991, GgsGd
  Mohamad Nasser Shaaban, (D), U.S.A., 1992, GgsGd
Assistant Lecturer:
  Ahmed Ibrahim Dyab, (M), Egypt, 2013, Gg
Assistant Professor:
  Hossam EL-Din Ahmed EL-S Helba, (D), Germany, 1994, GgZm
  Ahmed Sadek Mansour, (D), Egypt, 1999, Ggs
Instructor:
  Sara Akram M. Mahmoud, (B), Egypt, 2006, Gg
Emeritus Professor:
  Mohamed Waguih El Dakkak, (D), Egypt, 1971, GgsGd
  Galal Abd EL-Ham Ewas, (D), Norway, 1969, GgNr
  Hanafy M. Holail, (D), 1988, Ggs
  Rousine Tanios Toni, (D), Egypt, 1967, GgPg
Emeritus Lecturer:
  Mohamed M. Tamish, (D), West Germany, 1988, GgsCg
Emeritus Professor:
  Mohamed Ahmed Rashed, (D), Moscow (Russia), 1984, GgSo

Oceanography Dept (B) (2015)
22 Al-Guish Avenue
Alexandria
p. +20 (3) 591 1152
h_mitwally@sci.alex.edu.eg
http://www.alex.edu.eg/dept.jsp?FC=4&CODE=07

American University of Cairo
Dept of Petroleum and Energy Engineering (B) (2015)
New Cairo Campus: AUC Avenue
PO Box 74
New Cairo 11835
p. +20.2.2615.1000
amrserag@aucegypt.edu
http://www.aucegypt.edu/academics/dept/peng/Pages/

Assiut University
Dept of Geology (B) (2015)
Assiut Governorate
Assiut City
PO Box 71515
Assiut 71515
p. +20 (88) 235 7007
sci@aun.edu.eg
http://www.aun.edu.eg/fac_sci/depart/ge1.htm

Mining & Metallurgical Engineering Dept (B) (2015)
Assiut Governorate
Assiut City
PO Box 71515
Assiut 71515
p. +20 (88) 235 7007
mamoah@aun.edu.eg
http://www.aun.edu.eg/fac_eng?Dpart/Mining/Mining.htm

Beni Suef University
Geology Dept (B) (2015)
Salah Salem Street
Beni Suef 62511
p. +(20) 822324879, +(20) 2082232
elsherif_zakaria@yahoo.com
http://193.227.1.224/sci/

British University in Egypt
Petroleum Engineering (B) (2015)
Misr Ismalia Road
El Sherouk City 11837

enas.sabry@bue.edu.eg
http://www.bue.edu.eg/

Cairo University
Dept of Geophysics (B) (2015)
University Avenue - Univeristy Square
Giza
p. +20 (2) 572 9584
http://www.freewebtown.com/geophysics2/

Mining, Petroleum, and Metallurgical Engineering (B) (2015)
Univeristy Avenue - University Square
Giza
p. +20 (2) 572 9584
nahed_ecae2003@yahoo.com
http://www.eng.cu.edu.eg/dept/en/mpm/index.htm

Dept of Geology (B) (2015)
University Avenue - University Square
Giza, Cairo
p. +20 (2) 572 9584
portal@cu.edu.eg
http://www.cu.edu.eg/english/

El Mansoura University
Dept of Geology (B,M,D) (2015)

Faculty of Science
at Mansoura
Department of Geology
Mansoura, Dakahliya governorat 35516
p. +20 050 2242388 Ext. 582
ghazala@mans.edu.eg
http://www.mans.edu.eg/facscim/english/GeologyHome/
Enrollment (2012): B: 250 (230) M: 25 (13) D: 15 (7)
Head:
   Hosni H. Ghazala, (D), Mansoura, 1990, Ygx

El Zagazig University
Dept of Geology (B) (2015)
Zagazig
p. +20 (55) 324 577
info@zu.edu.eg
http://www.zu.edu.eg/

Fayoum University
Geology Dept (B,M,D) (2016)
fayoum-Egypt University Zone
ags00@fayoum.edu.eg
Fayoum 63514
p. +20 01005856505
ags00@fayoum.edu.eg
http://www.fayoum.edu.eg/English/Science/Geology/About-Board.aspx
Enrollment (2016): B: 20 (18) M: 4 (0) D: 2 (0)
Professor:
   ahmad gaber shedied, (D), Cairo, 1995, GgHww

Minia University
Geology & Chemistry Dept (B) (2015)
Minia
p. +(208-6) 324 420 #321 443
minia@frcu.eun.rg
http://www.minia.edu.eg

Sohag University
Geology Dept (B) (2015)
EL Kawaser
PO Box 82524
Sohag 82524
p. +(20) 93 4605745
a_abudeif@yahoo.com
http://www.sohag-univ.edu.eg/

South Valley University
Dept of Geology (Aswan Campus) (B) (2015)
Qena 83523
p. +20 (96) 339 756
mohammedsm2003@yahoo.com
http://www.svu.edu.eg/arabic/aswan/sci/en/depart/Geology/Geology.htm

Tanta University
Dept of Geology (B) (2015)
El-Geish Street
Tanta 31527
p. +20 (40) 337 7929
president@tainta.edu.eg
http://www.tanta.edu.eg/ar/Tanta/3elom/geology.html

The Egyptian Geological Survey and Mining Authority
egov@ad.gov.eg
http://www.egsma.gov.eg/

Zagazig University
Dept of Geology (B,M,D) (2015)
Sharkia Governorate
Zagazig City
geology@zu.edu.eg
http://www.zu.edu.eg/

Eritrea

University of Asmara
Dept of Marine Sciences (B) (2015)
PO Box 1220
Asmara
p. 291 1 161926 (Ext. 259)
Zekeria@marine.uoa.edu.er
http://www.uoa.edu.er/academics/dmarine/index.html

Estonia

Geological Survey of Estonia
Eesti Geoloogiakeskus OU (2015)
Juniper tee 82
Tallinn 12618
p. 672 0094
egk@egk.ee
http://www.egk.ee/

University of Tartu
Inst of Geology (B,M,D) ●☒ (2017)
Ravila 14a
Tartu 50411
p. +372 7 375 891
geol@ut.ee
http://www.geoloogia.ut.ee/et
Programs: Geology (B, M, D); Environmental Technology (B)
Enrollment (2017): B: 3 (0) M: 4 (0) D: 2 (0)

Ethiopia

Addis Ababa University
Dept of Planetary and Earth Sciences (2015)
PO Box 1176
Addis Ababa
p. 251-111239462
balem@geo.aau.edu.at
http://www.aau.edu.et/index.php/earth-sciences

School of Civil & Environmental Engineering (B) (2015)
PO Box 1176
Addis Ababa
youngkyunkim@aait.edu.et
http://www.aau.edu.et/index.php/earth-sciences

Arba Minch University
Dept of Geology (B) (2015)
PO Box 21
Arba Minch
p. 251-468814972
yoditayalew@fastmail.fm
http://www.arbaminch-univ.com/

College of Natural Sciences (B) (2015)
PO Box 21
Arba Minch
p. +251-46-8810070
alemayehu.hailemicael@amu.edu.et
http://www.arbamich-univ.com/WTIMeteorlogy.html#

Water & Environmental Engineering (B) (2015)
PO Box 21
Arba Minch
p. +251-46-8810070
nigussie_tg@yahoo.com
http://www.arbaminch-univ.com/WTIWEE.html

Mekelle University
Dept of Earth Science (B) (2015)
P.O.Box: 231
Management Building, Main Campus
Mekelle

p. (+251) 344 40 40 05
cciad.mu@gmail.com
http://www.mu.edu.et/index.php/department-of-earth-science

Institute of Geo-information and Earth Observation Sciences
P.O.Box: 231
Management Building, Main Campus
Makelle
p. +251 914 720398
meseleagw@yahoo.com
http://www.mu.edu.et/index.php/programs/ethiopia-institute-of-technology-mekelle/institute-of-geo-information-and-earth-observat

Semera University
Dept of Geology (B) (2015)
PO Box 132
Semera
p. 251-336660603
semerauniversity@ethionet.et

Wollega University
Dept of Geology (B) (2015)
PO Box 395 Nekemte
Nekemte
p. 251-576615038
wu@ethionet.et
http://www.wuni.edu.et/

Fiji

Ministry of Lands and Mineral Resources
Mineral Resources Dept of Fiji (2015)
Private Mail Bag, GPO
Suva
p. (679) 338 1611
director@mrd.gov.fj
http://www.mrd.gov.fj/

University of the South Pacific
Div of Earth and Environment Science (2015)
Private Mail Bag
GPO Suva
Suva, Fiji Islands
kahsai_k@usp.ac.fj
http://www.sidsnet.org/pacific/usp/earth/

Finland

Aalto University
Dept of civil engineering (B,M,D) ○ (2018)
Rakentajanaukio 4
Espoo
SF-02150 Espoo 00076 AALTO
leena.korkiala-tanttu@aalto.fi
http://civileng.aalto.fi/en/
Programs: Geoengineering, European Mining and Mineral Education Programme

Abo Akademi University
Geology and Mineralogy (B,M,D) O (2016)
Domkyrkotorget 1
Turku Fi-20500
p. (+)358442956429
geologi@abo.fi
http://www.abo.fi/fakultet/geologi
Enrollment (2015): B: 26 (8) M: 50 (4) D: 9 (2)
Professor:
 Olav Eklund, (D), Abo Akademi, 1993, GxeGz

Geological Survey of Finland
Geologian tutkimuskeskus ☒ (2018)
P.O.Box 96

Espoo FI-02151
p. +358 29 503 0000
gtk@gtk.fi
www.gtk.fi
f: https://www.facebook.com/GTK.FI
t: @GTK_FI
Director, Projects and Customers:
 Petri Lintinen, (D)
Director, Operative Units:
 Olli Breilin, (M), Oulu, 1988, GmIGs
Director, Human Resources, Talent Management and Working Environments:
 Helena Tammi, (M)
Director General:
 Mika Nykänen, (M)
Digital Innovations and Corporate data:
 Mikko Eklund, (M)
Director, Science and Innovation:
 Pekka Nurmi, (D)

University of Helsinki
Dept of Geosciences and Geography (B,M,D) (2015)
Gustaf Hällströmink 2
P.O. BOX 68
Helsinki 00014
p. 358-294150827
Mia.Kotilainen@helsinki.fi
http://www.helsinki.fi/geology/
Enrollment (2007): B: 0 (6) M: 26 (15) D: 6 (0)
Head:
 Juha A. Karhu, (D), Helsinki, 1993, Csg

University of Oulu
Dept of Geology (B,M,D) O (2015)
Linnanmaa
FIN-90014
Oulu
 vesa.peuraniemi@oulu.fi
 http://cc.oulu.fi/~geolwww/Geology.htm

Dept of Geophysics (2015)
 Pertti.Kaikkonen@oulu.fi
 http://www.gh.oulu.fi/

France

Bureau de Recherches Geologiques et Minieres
BRGM-Orleans (2015)
3 avenue Claude-Guillemin
BP 36009
Orleans 45060 Cedex 2
 p. +33 (0)2 38 64 34 34
 http://www.brgm.fr/

Catholic University of the West
Environmental Management (2015)
3, place Andre Leroy
BP 808
49008 Anger
 sciences@uco.fr

Centre de Recheches Petrographiques et geochimiques
Center of Petrographic and Geochemical Research (CRPG du CNRS) (2015)
15, rue Notre Dame des Pauvres
BP 20
54501 Vandoeuvre Les Nancy Cedex
 p. + 33 (0)3 83 59 42 02
 rpik@crpg.cnrs-nancy.fr
 http://www.crpg.cnrs-nancy.fr

Ecole des Mines de Paris
Ecole des Mines de Paris (2015)
60-62, Boulevard Saint Michel
75272 PARIS cedex 06
Paris, France 75272
    isabelle.olzenski@mines-paristech.fr
    http://www.ensmp.fr/Eng/ENSMP/aboutENSMP.html

Ecole Nationale Supérieure de Géologie
ENSGéologie (M,D) (2017)
2 Rue du Doyen Marcel Roubault
TSA 70605
VANDOEUVRE-LES-NANCY, Lorraine F-54518
    p. 33 (0)3 83 59 64 15
    ensg-contact@univ-lorraine.fr
    http://www.ensg.univ-lorraine.fr/
Director:
    MONTEL Jean-Marc, (D), Eg

Ecole Nationale Superieure des Mines de Nancy
Ecole des Mines de Nancy (M,D) ● ☒ (2018)
Campus ARTEM
92 rue du Sergent Blandan
CS14234
NANCY 54042
    p. +33 0355662600
    mines-nancy-scolarite-ficm@univ-lorraine.fr
    http://www.mines-nancy.univ-lorraine.fr/content/g%C3%A9oing%C3%A9nierie
    f: https://www.facebook.com/groups/geoingenierie/
    t: @minesnancy
Professor:
    Olivier DECK

Ecole Nationale Superieure des Mines de Saint Etiene
(2015)
158 Cours Fauriel
CS 62362
Saint Etienne 42023
    p. (+33) (0)4 77 420 278
    accueil@ccsti-larotonde.com
    http://www.mines-stetienne.fr/fr

Ecole Normale Superieure de Paris
Dept of Geosciences (M,D) ● (2015)
24 rue Lhomond
Paris 75005
    p. (+33) (0)1 44 32 22 11
    delescluse@geologie.ens.fr
    http://www.geosciences.ens.fr
    Enrollment (2015): M: 20 (10) D: 29 (6)

ENSPM - Institut Francais du Petrole
Centre Exploration (2015)
1-4 avenue de Bois Preau
BP 311
92506 Rueil Malmaison
    vincent.richard@ifpen.fr

French Institute for Research Exploitation of the Sea (IFREMER)
Dept of Marine Geosciences (2015)
BP 70
29263 Plouzane

Higher Natl School of Mines at Saint-Etienne
Departement Geosciences et environnement (A,B,M,D) (2015)
158, cours Fauriel
42023 Saint-Etienne Cedex 02
    guy@emse.fr

Higher Natl School of Mines in Paris (ENSMP)
Deparment of Earth Sciences and environment (2015)
35, rue Saint-Honore
77305 Fontainebleau Cedex
    isabelle.olzenski@mines-paristech.fr

Inst of Physics of the Globe-Paris VII
Lab de Geochimie, Geomateriaux, Geomag, Sismo, Tech, Obser Volcanologiques ☒ (2019)
Tour 14/24 - 2e etage, 4, place Jussieu
75252 Paris Cedex 05
    cartigny@ipgp.fr

Institut de Physique Du Globe De Paris
Institut de Physique Du Globe De Paris (M,D) (2016)
1 rue Jussieu
Paris 75005
    p. 01 83 95 74 00
    accueil@ipgp.fr
    http://www.ipgp.fr/

Institut Polytechnique LaSalle Beauvais
Dept of Geosciences (B,M,D) (2015)
19 Rue Pierre WAGUET - BP 30313
Beauvais 60026 cedex
    p. +33 (0) 3 44 06 89 91
    yannick.vautier@lasalle-beauvais.fr
    http://www.lasalle-beauvais.fr/
    Administrative Assistant: Nathalie Lermurier
Director of the Geosciences Department:
    Yannick Vautier, Inst Géologique Albert-de-Laparent, 1999, GodCo
Director of the Education Program in Geosciences:
    Hervé Leyrit, (D), Gv
Director of School/Companies relations:
    Pascal Barrier, (D), Inst Géologique Albert-de-Laparent, PmGsPs
Professor:
    Olivier Pourret, (D), Rennes I (France), 2006, CgaGe
    Lahcen Zouhri, (D), Lille (France), 2000, HwgGe
Geotechnics:
    Bassam Barakat, (D), Ecole Centrale Paris (France), 1991, NrgGq
Engineer in Mining & Quarry:
    Lucien Corbineau, Inst Géologique Albert-de-Laparent, 2007, EgGgNm
Engineer in Marine Geology:
    Olivier Bain, Inst Géologique Albert-de-Laparent, 2000, OuGgZi
Engineer in Geotechnics:
    Jean-David Vernhes, Polytech Paris UPMC (France), 1998, YgNgr
Engineer in Geology:
    Benoit Proudhon, Inst Géologique Albert-de-Laparent, 1997, GgcGt
Associate Professor:
    Jessica Bonhoure, (D), CREGU-Nancy, 2007, GziCc
    Sadek Brahmi, (D), UPMC Paris VI (France), 1991, Nr
    Claudia Cherubini, (D), Bari (Italia), 2007, HwGe
    Cyril Gagnaison, (D), Sorbonne (France), 2006, PgGsa
    Sébastien Laurent-Charvet, (D), Orléans (France), 2001, GctZe
    Pascale Lutz, (D), Pau (France), 2002, YgxYd
    Mohamed Nasraoui, (D), ENSMP, 1996, EgGzEm
    Elsa Ottavi-Pupier, (D), CRPG-CNRS Nancy, 1996, GvzGi
    Sébastien Potel, (D), Bâle, 2001, GpzGi
    Elodie Saillet, (D), Glasgow (UK), 2009, GctGo
    Renaud Toullec, (D), Bordeaux, 2006, GsoEo
    Ghislain Trullenque, (D), Basel, 2005, GcNrGg

Laboratoire d'Hydrologie et de Geochimie de Strasbourg
1, rue Blessing
67084 Strasbourg Cedex
    p. (+33) (0)3 68 85 04 02
    marie-claie.pierret@unistra.fr
    http://lhyges.unistra.fr/PIERRET-Marie-Claire,250?lang=fr

National Institute of Applied Science
Lab de Mineralogie et Geotechnique (2015)
20, avenue des Buttes de Coesmes

France

35043 Rennes Cedex 7
p. (+33) (0)2 23 23 82 00
olivier.guillou@insa-rennes.fr
http://www.insa-rennes.fr/insa-rennes.html

National Polytechnic Inst. of Grenoble
Ecole Doctorale Terre, Univers, Environnement
Domaine Univ de Grnoble - BP 95
46 avenue Félix Viallet
Cedex 1
Grenoble 38031

Pytheas Institute - Earth Sciences and Astronomy Observatory (PYTHEAS)
Faculte des Sciences de Luminy (2015)
Laboratoires de Geologie Marine et Sedimentologie
70 avenue Leon Lachamp
13288 Marseille
secretariat-direction@osupytheas.fr

Toulouse University
Satellite Geophysics and Oceanography Laboratory (M,D) (2015)
14, Avenue Edouard Belin
31400 Toulouse
p. (+33) (0)5 61 33 29 02
directeur@legos.obs-mip.fr
http://www.legos.obs-mip.fr/Presentation-generale?set_language=en&cl=en

Universite Blaise Pascal (Clermont Ferrand II)
Dept des Sciences de la Terra (2015)
34 Avenue Carnot
63038 Clermont Ferrand Cedex
p. (+33) 04 73 34 67 22
cecile.sergere@univ-bpclermont.fr
http://wwwobs.univ-bpclermont.fr/lmv/cursus/

Université Claude Bernard Lyon 1
Laboratoire de Géologie de Lyon: Terre, Planètes, Environnement (B,M,D) (2016)
Bd du 11 Novembre
Campus de La Doua
Bâtiment Géode
Villeurbanne 69622
p. 0033 (0)472445800
emanuela.mattioli@univ-lyon1.fr
http://lgltpe.ens-lyon.fr/
Administrative Assistant: Marie-Jeanne Barrière
Enrollment (2012): B: 36 (30) M: 16 (16) D: 10 (10)

Professor:
    Pascal Allemand
    Nicolas Coltice
    Fabrice Cordey
    Gilles Cuny
    Isabelle Daniel
    Gilles Dromart
    philippe Gillet
    Stephane Labrosse
    Christophe Lécuyer
    Emanuela Mattioli
    Guillemette Ménot
    Cathy Quantin
    Pierre Thomas
Senior Scientist:
    Thierry Alboussiere
    Vincent Balter
    janne Blichert toft
    Bernard Bourdon
    Razvan Caracas
    Eric Debayle
    Vincent Grossi
    Serge Légendre
    Philippe Herve Leloup
    Bruno Reynard
    yanick Ricard
    Jean Vannier
    Ricard Yanick
Associate Professor:
    Muriel Andréani
    Anne-Marie Aucourt
    Frederic Chambat
    Regis Chirat
    Claude Colombié
    Véronique Daviéro-Gomez
    Renaud Deguen
    Véronique Gardien
    Bernard Gomez
    Vincent Langlois
    Gweltaz Mahéo
    Matthew Makou
    Jean-Emmanuel Martelat
    Davide Olivero
    Vincent Perrier
    Jean-Philippe Perrillat
    Sylvain Pichat
    Bernard Pittet
    Frédéric Quillévéré
    Stéphane Reboulet
    Philippe Sorrel
    Guillaume Suan
    Benoît Tauzin
Associate Scientist:
    Romain Amiot
    Thomas Bodin
    caroline Fitoussi
    Bertrand Lefebvre
    Laurence Lemelle
    Jérémy Martin
    Jan Matas
Emeritus:
    francis Albarede
Related Staff:
    Emmanuelle Albalat
    Ingrid Antheaume
    Florent Arnaud-Godet
    Brigitte Barchasz
    Ghislaine Broillet
    Herve Cardon
    Fabien Dubuffet
    Philippe Fortin
    François Fourel
    Philippe Grandjean
    Naïma Khrouz
    Aline Lamboux
    Gilles Montagnac
    Sophie Passot
    Emmanuel Robert
    Magali Seris
    Philippe Telouk

Universite d'Orleans
Dept des Sciences de la Terre (2015)
BP. 6759
45067 Orleans
scolarite-osuc@univ-orleans.fr

Universite de Bordeaux I
Dept des Sciences de la Terra et de la Mer (2015)
341 Cours de la Liberation
Talence 33400
p. (+33) 05 40 00 88 79
termer@adm.u-bordeaux1.fr
http://www.u-bordeaux1.fr/universite/organisation/composantes-ufr-instituts/ufr-des-sciences-de-la-terre-et-de-la-mer.html

Universite de Bourgogne
UFR Sciences Vie, Terre & Environnement (B,M,D) (2015)
6 boulevard Gabriel
21100 Dijon
dirextion-ufrsvte@u-bourgogne.fr

http://ufr-svte.u-bourgogne.fr/

Universite de Bretagne Occidentale
Departement des Sciences de la Terre (2015)
3 Rue des Archives
Brest 29287
    p. (+33) 02 98 01 61 88
    alain.cottignies@univ-brest.fr
    http://www.univ-brest.fr/ufr-sciences/menu/Les_departements/Sciences_de_la_Terre

Universite de Caen
Dept des Sciences de la Terre (2015)
Esplanade de la Paix
Caen 14000
    p. (+33) 02 31 56 55 87
    isabelle.villette@unicagen.fr
    http://ufrsciences.unicaen.fr/departements/departement-des-sciences-de-la-terre/

Universite de Lorraine - Faculte des Sciences et Technologies
Dept Geosciences (B,M) ●☒ (2018)
Boulevard des Aiguillettes
BP 239
Vandoeuvre Les Nancy, 54506
    p. (+33) (0)3 83 68 47 18
    cecile.fabre@univ-lorraine.fr
    http://www.geologie.uhp-nancy.fr/Php/index.php
    Enrollment (2014): B: 31 (0) M: 51 (0)
Head:
    Cecile FABRE, (D), Earth Sciences, 2000, Gg

Universite de Montpellier
Geosciences Montpellier (M,D) ☒ (2018)
Campus Triolet
CC 060
Montpellier cedex 5 34095
    p. (+33) (0)4 67 14 36 43
    dirgm@gm.univ-montp2.fr
    http://www.gm.univ-montp2.fr/?lang=fr

Universite de Paris Sud (Orsay)
Dept Sciences de la Terre (B,M,D) ●☒ (2018)
15 Rue Georges Clemenceau
Orsay 91400
    p. (+33) (0)1 69 15 49 09
    hermann.zeyen@u-psud.fr
    http://geosciences.geol.u-psud.fr/
    f: https://www.facebook.com/geosciences.paris.sud/
    t: @GEOPS_Orsay
    Programs: Hydro(geo)logy; Ressources; Sedimentary Basins; Planetology; Paleoclimate; Engineering Geology
    Enrollment (2017): B: 65 (42) M: 117 (50) D: 50 (12)

Universite de Paris VI
Lab of Sub-Marine Geodynamics (CEROV) (2015)
Port de la Darse - BP 48
06230 Villefranche-Sur-Mer

Center of Geosynamics Research (2015)
4 Place Jussieu
Paris 75005
    p. (+33) 01 44 27 46 98
    licence.sciterre@upmc.fr
    http://www.upmc.fr/en/education/diplomas/sciences_and_technologies/bachelor_s_degrees/department_of_earth_sciences.html

Universite de Paris VII
Dept des Sciences Physiques de la Terre (2015)
4 place Jussieu
75251 Paris
    p. (+33) (0)1 57 27 57 27
    zarie.rouas@univ-paris-diderot.fr

Universite de Pau
Dept of Geology (A,B,M,D) (2015)
I.P.R.A. - Geologie
Avenue de L'Universite
Pau 64230
    anne-sophie.laloge@univ-pau.fr
    http://www.univ-pau.fr/RECHERCHE/GEOPHY/

Universite de Pau et des Pays de l'Adour
Faculte des Sciences (2015)
Avenue de l'Universite
64000 Pau

Universite de Rennes I
Dept. de Geosciences (2015)
2 Rue du Thabot
35065 Rennes Cedex
    p. (+33) (0)2 23 23 60 76
    florentin.paris@orange.fr
    http://www.geosciences.univ-rennes1.fr/?lang=en

Universite de Strasbourg
Ecole et Observatoire des Sciences de la Terre (B,M,D) ● (2015)
5 rue Descartes
67084 Strasbourg cedex F
    p. (+33) (0)3 68 85 03 53
    eost-contact@unistra.fr
    http://eost.unistra.fr/nouveautes-du-site/
Director:
    Frederic Masson, (D)
Laboratory Director:
    Ulrich Achauer, (D), Karlsruhe, 1990, Ysg

Université Jean Monnet, Saint-Etienne
Département de Géologie - Faculte des Sciences et Techiques (B,M,D) ● (2015)
23 rue du Dr. Paul Michelon
Saint-Etienne Cedex F-42023
    p. (+33) (0)4 77 48 15 85
    veronique.lavastre@univ-st-etienne.fr
    http://portail.univ-st-etienne.fr/bienvenue/presentation/ufr-des-sciences-dpt-geologie-327810.kjsp
Head:
    Bertrand N. MOINE, (D), Macquarie, 2000, CgtGx
Professor:
    Jean-Yves COTTIN, (D), Mus Nat d'Histoire Nat (Paris), 1978
    Damien GUILLAUME, (D)
    Jean-François MOYEN, (D), 2000, GiCtGp
Assistant Professor:
    Marie-Christine GERBE, (D), GisZe

Universite Paul Sabatier (Toulouse III)
Dept Biologie & Geosciences (B,M) ●☒ (2018)
118 Route de Narbonne
31062 Toulouse
    p. (+33) (0)5 82 52 57 21
    fsi.sec@univ-tlse3.fr
    http://www.univ-tlse3.fr/04628859/0/fiche___pagelibre/&RH=ACCUEIL&RF=1237305837890

Université de Franche-Comté
Faculte des Sciences et Techniques (2015)
16, route de Gray
25030 Besancon Cedex
    p. (+33) (0)3 81 66 62 09
    scolarite.ufr-2t@univ-fcomte.fr
    http://sciences.univ-fcomte.fr/pages/fr/menu3795/formations/licence--sciences-de-la-vie-16862-15340.html

France

University of Caen
Dept des Sciences de la Terre (2015)
Lab de Geologie structurals
Esplanadde de la Paix
14032 Caen Cedex

University of Franche-Comte
Lab de Geol Strucurale et appliquee (2015)
Faculte des Sci et Techniques
Place Leclerc
25030 Besancon
    master.geologie@univ-fcomte.fr

Sciences environnementales (2015)
Sciences et techniques
Service Scolarité
16, route de Gray
25030 Besançon cedex
    p. 03 81 66 62 09 / 62 11
    Scolarite.UFR-ST@univ-fcomte.fr
    http://sciences.univ-fcomte.fr/formations/listedesformations.htm

University of Francois Rabelais
Laboratoire de Geologie (2015)
Parc de Grandmot
37200 Tours

University of Lille
Dept of Geology and Laboratory of Oceanology and Geosciences (LOG) (B,M,D) ●☒ (2017)
Cite Scientifique
Batiment SN-5
Villeneuve D'Ascq 59655
    bruno.vendeville@univ-lille1.fr
    http://log.cnrs.fr
    Programs: Geology (B,M,D)

University of Louis Pasteur (Strasbourg 1)
School and Observatory of Earth Sciences (2015)
1, rue Blessing
67084 Strasbourg Cedex

University of Maine 🗐
Faculte des Sci - Lab de Geologie ☒ (2017)
Avenue Olivier Messiasen
72085 Le Mans
    p. (+33) (0)2 43 83 30 00
    http://sciences.univ-lemans.fr/Biologie-Geosciences

University of Montpellier II
Observatoire des sciences de l'univers OREME (2015)
2, place Eugene Bataillon
34095 Montpellier Cedex 05
    eric.servat@msem.univ-montp2.fr

University of Nantes
Departement des Sci de la Terre (2015)
2, rue de la Houssiniere
44072 Nantes Cedex 03
    Christophe.Monnier@univ-nantes.fr

University of Nice
Dept of Earth Science (2015)
Faculte des Sciences
28, avenue Valrose
06034 Nice Cedex
    secretariat-master-PPA@unice.fr

University of Nice-Sopia Antipollis-1
Laboratoire des Sciences de la Terre (2017)
Rue A. Einstein
06560 Valbonne
    dir.recherche@unice.fr
    http://www-geoazur.unice.fr/

University of Orleans
Dept des Sciences de la Terre (2015)
Domaine de la Source
BP 6759
45067 Orleans Cedex 02

University of Perpignan
Centre de Sed et Geochimie marines (2015)
52 Avenue Paul Alduy
66100 Perpignan
    p. (+33) (0)4 68 66 21 39
    facscien@univ-perp.fr

University of Picardy
Dept de Geologie (2015)
Chemin du Thil
80000 Amiens
    p. (+33) (0)4 22 82 76 65
    mohamed.benlahsen@u-picardie.fr
    http://www.u-picardie.fr/jsp/fiche_structure.jsp?STNAV=US&RUBNAV=&CODE=US&LANGUE=0

University of Pierre & Marie Curie (Paris VI)
Dept de Geotectonique (2015)
Tour 26 - ler etage (Boite 219)
4, place Jussieu
75252 Paris Cedex 05

Dept of Living Earth and Environment (2015)
Tour 15 - 4e etage
4, place Jussieu
75252 Paris Cedex 05
    chrystele.sanloup@upmc.fr
    http://www.ipgp.jussieu.fr/

University of Poitiers
Dept des Sciences de la Terre (2015)
40, ave du Recteur Pineau
86022 Poitiers Cedex

University of Provence (Aix-Marseille I)
Lab de Geol Structurale et appliquee (2015)
Centre Saint Charles
3, place Victor Hugo - Case 28
13331 Marseille Cedex 3
    bertrand.martin-garin@univ-amu.fr

University of Reims-Champagne
Dept des Sciences de la Terre (2015)
Moulin de la Housse - BP 1039
51687 Reims Cedex
    p. (+33) (0)3 26 91 34 19
    scolarite.sceiences@univ-reims.fr
    http://www.univ-reims.fr/formation/ufr-instituts-et-ecoles/ufr-sciences-exactes-et-naturelles/presentation,8370.html?

University of Rouen-Upper Normandy
Dept de Geologie (2015)
1 Rue Thomas Becket
76821 Mont-Saint-Aignan
    p. (+33) (0)2 35 14 68 26
    francoise.baillot@univ-rouen.fr

University of Savoy
Laborotoire de Geologie (2015)
Faculte des Sci et Techniques
Campus de Technolac, BP 1104
73011 Chambery Cedex
    Directeur.SceM@univ-smb.fr

University of Science & Technology of Lille
Inst Des Sciences De La Terre (2015)
Flandres - Artois, Cite Scientifique
Bat SN5, BP 36
59655 Villeneuve D'Ascq Cedex
   alain.trentesaux@univ-lille1.fr

University of West Brittany
Earth Sciences (2015)
Groupement de Recherche: GEDO
6, avenue Le Gorgeu
29287 Brest

Gabon

Universite des Sciences et Techniques de Masuku
Dept de Geologie (B) (2015)
BP 901
Franceville
   p. (+241) 01 67 75 78
   http://www.labogabon.net/ustm/facscience/index.html

Universite Omar Bongo
Dept de Geologie (B) (2015)
BP 13 131
Boulevard Leon Mba
Libreville
   p. +241 73 20 45, (241-72) 69 10
   uob@internetgabon.com
   http://www.uob.ga/

Germany

Aachen University of Technology
Inst of Structural Geology, Tectonics and Geomechanics (2015)
Lochnerstr 4-20
D-52056
   ged@ged.rwth-aachen.de
   http://www.ged.rwth-aachen.de

Baden-Wuerttemberg - Landesamt fuer Geologie, Rohstoffe und Bergbau (LGRB)
Regional Government of Freiburg State Office of Geology, Raw Materials and Mining ☒ (2018)
Albertstrasse 5
Freiburg, Baden-Wuerttemberg 79104
   p. ++49 9761 208 3000
   abteilung9@rpf.bwl.de
   http://www.lgrb-bw.de/

Bavarian Environment Agency
Department Geological Survey ☒ (2018)
Buergermeister-Ulrich-Strasse 160
Augsburg, Bavaria 86179
   poststelle@lfu.bayern.de
   https://www.lfu.bayern.de/geologie/
   Programs: None
   Certificates: None

Brandenburg - Landesamt fuer Geowissenschaften und Rohstoffe (LGRB)
State Office for Mining, Geology and Minerals of Brandenburg (2015)
Inselstrabe 26
03046 Cottbus
   info@lbgr.brandenburg.de
   http://www.lbgr.brandenburg.de

Bundesanstalt fur Geowissenschaften und Rohstoffe (BGR)
Federal Institute for Geosciences and Natural Resources ☒ (2018)
Stilleweg 2
Hannover 30655
   poststelle@bgr.de
   http://www.bgr.bund.de

Christian-Albrechts-Universitaet
Geology Dept (2015)
Olshausenstrasse 40/60
2300 Kiel 1

Ernst-Moritz-Arndt Universitaet
Institut fur Geographie und Geologie (2015)
Domstrabe 11
17489 Griesswald
   p. (+49) (0) 3834 86-4502
   geogra@uni-greifswald.de
   http://www.mnf.uni-greifswald.de/institute/geo.html

Freiburg University College
Major Earth and Environmental Sciences (B) (2017)
Bertoldstrasse 17
79085 Freiberg
   p. (+49) 761 203-67342
   studyinfo@ucf.uni-freiburg.de
   http://www.ucf.uni-freiburg.de/

Freie Universitaet Berlin
Inst fuer Palaontologie (2015)
Maltese-Strasse
74-100, Haus D
D-12249 Berlin
   palaeont@zedat.fu-berlin.de
   http://userpage.fu-berlin.de/~palaeont/WELCOME.HTM

Institute of Geological Sciences, Planetary Sciences and Remote Sensing ☒ (2018)
Malteserstr. 74-100
12249 Berlin
   p. (030) 838-70 575
   plansec@zedat.fu-berlin.de
   http://www.geo.fu-berlin.de/geol/fachrichtungen/planet

Friedrich-Schiller-University Jena
Institute for Geosciences (B,M,D) ● (2015)
Burgweg 11
Jena 07749
   p. 0049(0)3641/948600
   geowissenschaften@uni-jena.de
   http://www.geo.uni-jena.de
   Enrollment (2014): B: 17 (43) M: 16 (22) D: 69 (12)
Univ.-Prof. Dr.:
   Sabine Attinger, (D), Hy
   Georg Büchel, (D), Gg
   Christoph Heubeck, (D), Stanford, 1994, GsdPg
   Nina Kukowski, (D), Ygx
   Falko H. Langenhorst, (D), Gz
   Juraj Majzlan, (D), Gz
   Kai U. Totsche, (D), Bayreuth, 1994, HwClSo
   Lothar Viereck, (D)
Jun.-Prof. Dr.:
   Anke Hildebrandt, (D), MIT, 2005, HgSpf
Univ.-Prof. Dr.:
   Kamil M. Ustaszewski, (D), 2004, Gct

Garching Technical University
Inst fur Geologie und Mineralogie (2015)
Lichtenbergerstrasse 4
0-8046 Garching
   rieder@tum.de

Germany

Geologischer Dienst Nordrhein-Westfalen
NRW Geological Survey (2015)
Postfach 10 07 63
De-Greiff-Strasse 195
Krefeld D-47707
  p. +49-2151-897-0
  geoinfo@gd.nrw.de
  http://www.gd.nrw.de/home.php

Geologisches Institut der Universitaet
Dept of Geology (2015)
Plelcherwael 1
87 Wuerzburg

Georg-August University of Goettingen
Dept of Geobiology (B,M,D) ☒ (2018)
Goldschmidstr. 3
Goettingen, Lower Saxony 37077
  p. +49 551 397951
  jreitne@gwdg.de
  http://www.geobiologie.uni-goettingen.de/
  Programs: Bachelor & Master Geosciences/Geobiology
  Enrollment (2018): B: 2 (2) M: 4 (1) D: 10 (4)
Prof.Dr.:
  Joachim Reitner, (D), PggPg
Professor:
  Volker Thiel, Dr., (D), CooCo
Dr.:
  Gernot Arp, Dr., (D), PggPg
  Andreas Reimer, (D), OcCmm
  Jan-Peter Duda, (D), 2014, PgCoGs

Goethe Universitaet
Fachbereich Geowissenschaften (2015)
Altenhoferallee 1
60438 Frankfurt
  p. (+49) (0)69 798 40208
  dekanat-geowiss@em.uni-frankfurt.de
  http://www.uni-frankfurt.de/fb/fb11/index.html

Hamburg - Geological Survey
Ministry of Urban Development and Environmental Protection Agency for the Environment (2015)
Neuenfelder Straße 19
21109 Hamburg
  gla@bsu.hamburg.de
  http://www.geologie.hamburg.de

Helmholtz-Zentrum Potsdam Deutsches Geo-ForschungsZentrum
GeoForschungs Zentrum GFZ (2015)
Telegrafenberg 14473
14473 Potsdam
  p. (+49) 331 288-1045
  presse@gfz-potsdam.de
  http://www.gfz-potsdam.de/startseite/

Hessen - Landesamt fuer Umwelt und Geologie
65203 Wiesbaden, Rheingaustraße 186
  p. 0611-6939-0
  http://www.hlug.de/

Humboldt Universitaet zu Berlin
Palaontologisches Museum (2015)
Invaliden Strasse 43
0-4010 Berlin

Institut fur Palaontologie
Dept of Geology (2015)
O. Weidlich, Fohrenweg 19
W-8504 Stein
  palsek@univie.ac.at

Karlsruhe Institute of Technology
Geophysical Institut (B,M,D) (2015)
Hertzstr. 16
Karlsruhe 76187
  p. +49-721-6084431
  geophysics@gpi.kit.edu
  http://www-gpi.physik.uni-karlsruhe.de/
  Enrollment (2010): B: 15 (0) D: 10 (0)
Professor:
  Thomas Bohlen, (D), Yx
Senior Scientist:
  Thomas Forbriger, (D), Ys
Associate Professor:
  Joachim R.R. Ritter, (D), Universit, 1996, YsGtv
Research Associate:
  Rebecca Harrington, (D), Ys

Institute for Applied Geosciences (2015)
Hertzstr. 16
Karlsruhe
  susanne.winter@kit.edu
  http://www.agw.kit.edu/

Dept of Geology (2015)
Kaiserstrasse 12
76131 Karlsruhe
  p. (+49) 721 608 4219 2/43651
  geophysik@gpi.kit.edu
  http://www.bgu.kit.edu/index.php

Institute for Mineralogy and Geochemistry (2015)
Adenauerring 20b
Karlsruhe D-76131
  p. +49 721 608- 43323
  img@img.uka.de
  http://www.img.kit.edu/

Ludwig-Maximilians-Universitaet Muenchen
Dept of Earth & Enviromental Sciences (B,M,D) (2015)
Theresienstr. 41/III
Munich 80333
  p. 0049/89/21804250
  dingwell@lmu.de
  http://min.geo.uni-muenchen.de/
  Enrollment (2007): B: 173 (28) M: 12 (0) D: 24 (0)
Director:
  Donald Bruce Dingwell, Giv
Professor:
  Alexander Altenbach, Pg
  Wladyslaw Altermann, Gg
  Michael Amler, Pg
  Valerian Bachtadse, Yg
  Hans-Peter Bunge, Yg
  Christina De Campos, Gz
  Karl Thomas Fehr, Gz
  Friedrich Frey, Gz
  Anke Friedrich, (D)
  Anke Friedrich, (D), MIT, 1998, GgtCc
  Helmut Gebrande, (D), LMU Munich (Germany), 1975, YgsEo
  Stuart Gilder, (D), Ym
  Peter Gille, (D), Humboldt (Germany), 1984, GzZm
  Wolfgang Heckel, Gz
  Ernst Hegner, Gg
  Soraya Heuss-Aßbichler, Gz
  Heiner Igel, Yg
  Harald Immel, Pg
  Bernd L. Lammerer, (D), LMU Munich (Germany), 1972, GgtGc
  Reinhold Leinfelder, Pg
  Rocco Malservisi, Yg
  Robert Marschik, (D), EmCg
  Ludwig Masch, Gz
  Wolfgang Moritz, Gz
  Bettina Reichenbacher, (D), Johann Wolfgang Goethe, 1992, PgvPe
  Wolfgang Schmahl, Gz
  Klaus Weber-Diefenbach, Gg

Christian Wolkersdorfer, Gg
Adjunct Professor:
  Frank Trixler, (D), GzZmXc
Emeritus:
  Nikolai Petersen, (D), 1964, Ygm
Other:
  Kirill Aldushin, Gz
  Greta Barbi, Yg
  Robert Barsch, Yg
  Johannes Birner, Gg
  Florian Bleibinhaus, Yg
  Hans Boysen, Gz
  Gilbert Britzke, Yg
  Benoit Cordonnier, Gz
  Alexander Dorfman, Gz
  Thomas Dorfner, Gz
  Barbara Emmer, Yg
  Werner Ertel-Ingrisch, Gz
  Andreas Fichtner, Yg
  Michaela Frei, Gg
  Frantisek Gallovic, Yg
  Alexander Gigler, Gz
  Stefan Grießl, Gz
  Hagen Göttlich, Gz
  Marc Hennemeyer, Gz
  Katja Henßel, Pg
  Kai-Uwe Hess, Gz
  Maria Linda Iaccheri, Gg
  Giampiero Iaffaldano, Yg
  Guntram Jordan, Gz
  Ines Kaiser-Bischoff, Gz
  Thorsten Kowalke, Pg
  Thomas Kunzmann, Gz
  Martin Käser, Yg
  Markus Lackinger, Gz
  Yan Lavallee, Yg
  Maike Lübbe, Gz
  Götz Meisterernst, Gz
  Timo Casjen Merkel, Gz
  Marcus Mohr, Yg
  Lena Müller, Gz
  Dieter Müller-Sohnius, Gz
  Malik Naumann, Pg
  Jens Oeser, (D), Yg
  Sohyun Park, Gz
  Karin Paschert, Gg
  Rossitza Pentcheva, Gz
  Helen Pfuhl, Yg
  Antonio Sebastian Piazzoni, Yg
  Christina Plattner, Yg
  Josep Puente Alvarez de la
  Oliver Riedel, Gz
  Alexander Rocholl, Gz
  Javier Rubio-Sierra, Gz
  Gertrud Rößner, Pg
  Dieter Schmid, Pg
  Julius Schneider, Gz
  Bernhard Schuberth, Yg
  Oliver Spieler, Gz
  Robert Stark, Gz
  Katja Steffens, Gg
  Stefan Strasser, Gz
  Marco Stupazzini, Yg
  Frank Söllner, Gg
  Ferdinand Walther, Gz
  Joachim Wassermann, Yg
  Laura Wehrmann, Pg
  Michael Winklhofer, Yg
  Ayhan Yurtsever, Yg
  Matthias Zeitlhöfler, Gg
  Albert Zink, Gz

Palaeontology & Geobiology, Dept of Earth and Env Sci
(B,M,D) O⊠ (2017)
Richard-Wagner-Strasse 10
Muenchen, Bavaria 80333
  geobiologie@geo.lmu.de
  http://www.palmuc.de
  Programs: Geobiology & Paleobiology (M)
Prof. Dr.:
  Gert Woerheide, (D), Pgi
  William Orsi, (D), PgZn
  Bettina Reichenbacher, (D), PggPv

Marburg Chome University
Geology Dept (2015)
Deutschhausstrabe 10
35032 Marburg
  p. (+49) 06421/28-24257
  edda.walz@geo.uni-marburg.de
  https://www.uni-marburg.de/fb19_alt

Martin-Luther-Universitaet Halle-Wittenberg
Inst for Geosciences and Geography (A,B,M,D) ●◌ (2018)
Von-Seckendorff-Platz 3
D-06120 Halle
  p. ++49-045-55 26010
  direktor@geo.uni-halle.de
  http://www.geo.uni-halle.de
  Programs: Geography (B,M); Geosciences (B,M); Natural Resource Management (B,M); Intl Area Studies (M); Geography Education
Director:
  Christine Fürst, (D), Bonn, 2013, ZucZi

Mecklenburg-Vorpommern - Landesamt fuer Umwelt, Naturschutz und Geologie
State office for the Environment, Nature Conservation and Geology ⊠ (2018)
Goldberger Strasse 12
18273 Gustrow
  p. ++49-3843-777 0
  poststelle@lung.mv-regierung.de
  http://www.lung.mv-regierung.de/

Niedersachsen - Landesamt fuer Bodenforschung
  p. +49 (0)511-643-0
  poststelle-hannover@lbeg.niedersachsen.de
  http://www.lbeg.niedersachsen.de

Rheinisch-Westfaelische Technische Hochschule Aachen
Fachgruppe fuer Geowissenschaften und Geographie (2015)
Lochnerstr. 4-20 (Haus B)
Aachen 52056
  p. 0241 80 96219
  geowiss@rwth-aachen.de
  http://www.fgeo.rwth-aachen.de/

Rheinland-Pfalz - Landesamt fuer Geologie und Bergbau
State Office for Geology and Mining (2015)
Emy-Roeder-Strabe 5
PO Box 100 255
Mainz-Hechtsheim D-55129
  office@lgb-rlp.de
  http://www.lgb-rlp.de/

RWTH Aachen University
Institute of Geology and Palaeontology (2015)
Facultat fur Bergbau
Huttenwesen und Geowissenschaften/Wulwestr 2
D-5100 Aacen
  geo.sek@emr.rwth-aachen.de

Sachsen - Landesamt fuer Umwelt und Geologie (LfUG)
Saxon State Agency for Environment, Agriculture and Geology (2015)

Postfach 54 01 37
Dresden 01 311
  lfulg@smul.sachsen.de
  http://www.smul.sachsen.de/

State Survey for Geology and Mining of Sachsen-Anhalt
State Survey for Geology and Mining of Sachsen-Anhalt ☒ (2019)
Halle, Saxony-Anhalt D-06035
  p. +49 345 52 12 0
  poststelle@lagb.mw.sachsen-anhalt.de
  http://www.lagb.sachsen-anhalt.de

Technical University of Munich
Engineering Geology (B,M,D) ●☒ (2018)
Arcisstr. 21
Munich, Bavaria 80333
  p. +498928925850
  geologie@tum.de
  http://www.eng.geo.tum.de/
  Programs: Geosciences (B); Engineering Geology & Hydrogeology (M); Geothermal Energy (M)
  Enrollment (2017): B: 0 (1) M: 0 (3) D: 11 (0)
Chair-Prof.:
  Kurosch Thuro, (D), Tech (Munich), 1995, NgrGg
Senior Scientist:
  Gerhard Lehrberger, (D), TUM, EgGxg
  Bernhard Lempe, (D), TUM, 2012, GlgNg
Associate Professor:
  Hans Albert Gilg, (D), TUM, 1995, CsEgn
Associate Scientist:
  Katja R. Lokau, (D), TUM, 1999, Ng
  Marion Nickmann, (D), TUM, 2006, Ng
  Bettina Sellmeier, (M), Tech (Munich), Ngr
Research Associate:
  Tamara Breuninger, (M), TUM, 2018, Ngx
  Matthias Brugger, (M), TUM, 2015, NgGg
  Martin Potten, (M), Erlangen (Germany), 2016, NxGg
  Margreta Sonnenwald, (M), TUM, 2016, Ng
  Georg Maximilian Stockinger, (M), TUM, 2015, NrtNx
Laboratory Director:
  Heiko Käsling, (D), TUM, 2009, NrmNg

Technische Hochschule
Institute of Applied Geoscience (2015)
Karolinenplatz 5
64289 Darmstadt
  p. (+49) 6151 16-2171
  herrmann@geo.tu-darmstadt.de
  http://www.geo.tu-darmstadt.de/iag/index.de.jsp

Technische Universitaet Bergakademie Freiberg
Faculty for Geosciences, Geoengineering and Mining (2015)
Bergakademis Freiberg
Postfach 47/Bernhard-Cotta Strasse
Freiberg
  p. +49 (0)3731 / 39 - 3249
  Andrea.Thuemmel@geort.tu-freiberg.de
  http://tu-freiberg.de/fakult3/index.en.html

Technische Universitaet Berlin
Applied Geosciences (2015)
Str. Des 17 Juni 135
W-1000 Berlin 12
  p. +49 30314-7260 5
  http://www.geo.tu-berlin.de/

Technische Universitaet C.W. Braunschweig
Institute of Environmental Geology (2015)
Postfach 3329
Braunschweig D-38023
  geosecret@tu-bs.de
  http://www.tu-braunschweig.de/iug

Inst fuer Geologie und Palaontologie (2015)
Fachbereich fur Physik und Geowissenschaften
Pockelstrasse 14
D-3300 Braunschweig DEN

Inst fuer Geophysik und Meteorologie (2015)
Mendelssohnstr 2-3
D-3300 Braunschweig

Technische Universitaet Clausthal
Institut fuer Geologie und Palaeontologie (2015)
LeibnizstraBe 10
D-38678
Clausthal-Zellerfeld 38678
  Office@geologie.tu-clausthal.de
  http://www.geologie.tu-clausthal.de/

Inst for Petroleum Engineering (2015)
Agricolastrasse 10
38678 Clausthal-Zellerfeld
  p. +49-5323-72-2239
  marion.bischof@tu-clausthal.de
  http://www.ite.tu-clausthal.de/

Technische Universitaet Darmstadt
Institute of Applied Geosciences (B,M,D) (2015)
Schnittspahnstr. 9
Darmstadt D-64287
  p. +49 6151 16 2171
  iag@geo.tu-darmstadt.de
  http://www.geo.tu-darmstadt.de/
  Enrollment (2011): B: 280 (3) M: 54 (0) D: 29 (2)
Acting head:
  Hans-Joachim Kleebe, GzZm
Professor:
  Rafael Ferreiro Maehlmann, Gxp
  Matthias Hinderer, Gs
  Andreas Hoppe, GgZui
  Stephan Kempe, GgCgHs
  Ingo Sass, Ng
  Christoph Schueth, Hw
  Stephan Weinbruch, GzAmZm

Technische Universitat Dresden
Dept of Geosciences (2015)
Mommsenstr 13
0-8027 Dresden
  p. +49 351 463-32863
  doris.salomon@tu-dresden.de
  http://tu-dresden.de/die_tu_dresden/fakultaeten/fakultaet_forst_geo_und_hydrowissenschaften/fachrichtung_geowissenschaften

Thuringen - Landesanstalt fuer Umwelt und Geologie
Thuringen State Institute for Environment and Geology (2015)
Goeschwitzer Str forty-one
07745 Jena
  Poststelle@tlug.thueringen.de
  http://www.tlug-jena.de/de/tlug/

Universitaet Bochum
Institut fuer Geologie, Mineralogie und Geophysik (B,M,D) (2015)
Universitaetsstr. 150
Bochum 44780
  p. +49 234 32 23233
  sabine.sitter@ruhr-uni-bochum.de
  http://www.ruhr-uni-bochum.de/gmg/

Universitaet Bonn
Steinmann-Institut für Geologie, Mineralogie und Paläontologie (B,M,D) ● (2015)
Nussallee 8
Bonn 53115

p. +49 228 73 4803
tmartin@uni-bonn.de
http://www.steinmann.uni-bonn.de/

Universitaet Erlangen-Nuernberg
Geozentrum Nordbayern (2015)
Schlossgarten 5
91054 Erlangen
p. 09131/85-22615
geologie@geol.uni-erlangen.de
http://www.geol.uni-erlangen.de

Institut fur Geographie (2015)
Kochstr. 4/4
91054 Erlangen
p. 09131/85-22633
common@geographie.uni-erlangen.de
http://www.geographie.uni-erlangen.de/

Universitaet Frankfurt
Institut fur Geowissenschaften (2015)
Fachbereisch 17
Senckenberganlage 32-34
6000 Frankfurt
geowissenschaften@em.uni-frankfurt.de
http://www.geo.uni-frankfurt.de/ifg/

Universitaet Freiburg
Institut fur Geo- und Umweltnaturwissenschaften - Geologie (B,M,D) ☒ (2018)
Albertstr. 23-B
Freiburg i. Br., Baden-Wuerttemberg 79104
ulmer@uni-freiburg.de
http://portal.uni-freiburg.de/geowissenschaften
Programs: Geowissenschaften, Geology, Sustainable Materials
Enrollment (2017): B: 140 (50) M: 52 (19) D: 2 (1)

Universitaet Giessen
Inst fuer Geowissenschaften und LithosphSrenforshung (2015)
Senckenbergstrasse 3
35390 Giessen
p. (+49) 641 990
brigitte.becker-lins@geolo.uni-giessen.de
http://www.uni-giessen.de/fbr08/geolith/

Universitaet Goettingen
Geowissenschaftliches Zentrum (2015)
Goldschmidstr. 3
Lower Saxony
Gottingen D-37077
p. +49 551 397951
bhinz@gwdg.de
http://www.uni-goettingen.de/de/125309.html

Universitaet Greifswald
Institute for Geography and Geology (2015)
F.L. Jahn Strasse 17
Greifswald
p. +49 (0)3834 86-4570
geologie@uni-greifswald.de
http://www.mnf.uni-greifswald.de/institute/geo.html

Universitaet Halle-Wittenberg
Institute for Geosciences and Geographie (A,B,M,D) ●☒ (2018)
Von-Seckendorff-Platz 3/4
Halle D-06120
p. +49-0345-55 26055
direktor@geo.uni-halle.de
http://www.geo.uni-halle.de/
Programs: Geography, Geology, International Area Studies, Natural Resource Management, Teachers
Certificates: great variety

Universitaet Hamburg
Center for Earth System Research and Sustainability (M,D) (2015)
Bundesstrasse 53
Hamburg 20146
anke.allner@uni-hamburg.de
http://www.cen.uni-hamburg.de/

Institute of Geophysics (B,M,D) ●☒ (2018)
Bundesstrasse 55
Hamburg 20146
p. +4940428382973
dirk.gajewski@uni-hamburg.de
www.geo.uni-hamburg.de/de/geophysik.html
Programs: BSc Geophysics/Oceanography
MSc Geophysics
Certificates: none
Enrollment (2016): B: 47 (23) M: 14 (11) D: 12 (8)
Professor:
  Dirk J. Gajewski, (D), Karlsruhe, 1987, YesYg

Universitaet Hannover
Institut fur Geologie (2015)
Fachbeleich Erdwissenschaften
Callinstrasse 30
D-30167 Hannover
p. +49-(0)511-762 2343
sekretariat@geowi.uni-hannover.de
http://www.geologie.uni-hannover.de

Universitaet Heidelberg
Insitut fuer Geowissenschaften (2015)
Im Neuenheimer Feld 236
Heidelberg D-69120
p. 06221-54-8291
Elfriede.Hofmann@geow.uni-heidelberg.de
http://www.geow.uni-heidelberg.de/

Facultat fur Chemie und Geowissenschaften (2015)
Im Neuenheimer Feld 234
69120 Heidelberg
p. (+49) 06221/544 844
dcg@urz.uni-heidelberd.de
http://www.chemgeo.uni-hd.de/

Universitaet Leipzig
Institut fur Geophysik und Geologie (2015)
Talstrasse 35
04103 Leipzig
geologie@rz.uni-leipzig.de
http://www.uni-leipzig.de/~geo/

Universitaet Marburg
Inst fuer Geologie und Palaontologie (2015)
Biergenstrasse 12
Lahnberge
D-3550 Marburg-Lahn

Universitaet Muenster
Institut fuer Mineralogie (2015)
Corrensstrabe 24
Muenster 48149
p. +49 251 83-33464
minsek@uni-muenster.de
http://www.uni-muenster.de/Mineralogie/

Institut fuer Geologie und Palaeontologie (B,M,D) ●☒ (2018)
Corrensstr. 24
D-48149 Muenster
p. +49 251-83 33974
sklaus@uni-muenster.de
http://www.uni-muenster.de/GeoPalaeontologie/en/index.html
Enrollment (2016): B: 270 (41) M: 46 (21) D: 25 (3)

Germany

Professor:
   Christine Achten, (D), Frankfurt, 2002, CloEg
   Heinrich Bahlburg, (D), Tech (Berlin), 1986, GsgCc
   Ralph Thomas Becker, (D), Ruhr (Germany), 1990, PisPm
   Ralf Hetzel, (D), Gcm
   Hans Kerp, (D), Utrecht, 1986, PblGg
   Harald Strauss, (D), Göttingen, 1985, CsGg

Universitaet of Kiel
Institute of Geoscience (2015)
Christian-Albrechts-Platz 4
24118 Kiel
   p. (+49) 431 880 3900
   wussow@geophysik.uni-kiel.de
   http://www.ifg.uni-kiel.de/6+M52087573ab0.html

Universitaet Oldenburg
Institute for Chemistry and Biology of the Marine Environment (2015)
Carl-von-Ossietzky-Str. 9-11
Box 2503
Oldenburg 26111
   p. +49-0441-798-5342
   director@icbm.de
   http://www.icbm.uni-oldenburg.de/

Universitaet Potsdam
Institute for Earth and Environmental Sciences (B,M,D) (2015)
Karl-Liebknecht-Str. 24-25
Potsdam 14476
   p. +49 331 977 2116
   sekretariat@geo.uni-potsdam.de
   http://www.geo.uni-potsdam.de/
   Enrollment (2013): B: 319 (14) M: 79 (11) D: 122 (16)

Universitaet Stuttgart
Inst fuer Geol und Palaontogisch (2015)
Fakultaet-7-Geo. und Biowissenschaften
Boblinger Str. 72
D-7000 Stuttgart 10

Universitaet Trier
Regional and Environmental Science (2015)
Behringstrabe 21
D-54296 Trier
   p. (+49) (0) 651-201-4528
   dekanatfb6@uni-trier.de
   http://www.uni-trier.de/index.php?id=2220

Lehrstuhl fuer Geologie (B,M,D) ☒ (2019)
Geowissenschaften (FB VI)
Trier 54286
   p. 0651 201 4647
   ensch@uni-trier.de
   http://www.uni-trier.de/index.php?id=2632
   Programs: Geography; Environmental Sciences; Geoarchaeology
   Enrollment (2018): B: 49 (44) M: 35 (27) D: 4 (0)

Universitaet Tuebingen
Institute for Geoscience (2015)
72076 Tuebingen
Sigwartstr. 10
   Christophe.Pascal@rub.de
   http://www.ifg.uni-tuebingen.de/

Universitaet Würzburg
Institut für Paläontologie (2015)
Pleicherwall 1
D-97070 Würzburg
   p. 49 0931- 31 25 97
   i-palaeontologie@mail.uni-wuerzburg.de
   http://www.palaeontologie.uni-wuerzburg.de

Institut für Geographie und Geologie (2015)
Am Hubland
97074 Würzburg
   p. +49 (0) 931 / 31-85555
   http://www.geographie.uni-wuerzburg.de

Universitaet zu Koeln
Institute for Geology and Mineralogy (2015)
Cologne
   p. +49 221 470-5619
   ellen.stefan@uni-koeln.de
   http://www.geologie.uni-koeln.de/

University of Bayreuth
Bayerisches Geoinstitut (M,D) (2015)
Bayreuth 95440
   p. +49(0)921 55-3700
   bayerisches.geoinstitut@uni-bayreuth.de
   http://www.bgi.uni-bayreuth.de/

University of Bremen
Geosciences (A,B,M,D) ●☒ (2018)
Klagenfurter Strasse 2-4
Bremen, Bremen 28359
   p. +49 421 218 - 65010
   info@geo.uni-bremen.de
   http://www.geo.uni-bremen.de
   Programs: Geosciences (BSc,MSc); Marine Geosciences (MSc); Materials Chemistry and Mineralogy (MSc)
   Enrollment (2015): A: 249 (0) B: 274 (52) M: 249 (62) D: 111 (29)

University of Cologne
Dept of Geology and Mineralogy (B,M,D) ● (2015)
Zuelpicherstr. 49a
50674 Koeln
   p. +49-221-470-5619
   ellen.stefan@uni-koeln.de
   http://www.uni-koeln.de/math-nat-fak/geologie/index.html

Institute of Geophysics and Meteorology (B,M,D) (2015)
Pohligstraße 3
D-50969 Köln
   p. +49 (0)221 470 2552
   sekretar@geo.uni-koeln.de
   http://www.geomet.uni-koeln.de/en/general/home/

Universität Mainz
Institut für Geowissenschaften (B,M,D) ●☒ (2017)
Joh.-J.-Becher-Weg 21
55128 Mainz
   p. +49 6131 39 24373
   igw@uni-mainz.de
   http://www.geowiss.uni-mainz.de/
Head:
   Boris PJ Kaus, (D), Yg
Associate Professor:
   Roman Botcharnikov, (D), Gx
   Jonathan M. Castro, (D), Gv
   Michael Kersten, (D), Ca
   Cees W. Passchier, (D), Gt
   Denis Scholz, (D), Cl
   Bernd R. Schöne, (D), PegPi
   Frank Sirocko, (D), Gs
Adjunct Professor:
   Kirsten I. Grimm, (D), Pm
   Dieter Mertz, (D), Cc
   Thomas Tütken, (D), Pv

West Wilhelms Universitat
Dept of Applied Geology (2015)
Schlossplatz 2
4400 Muenster

Westfälische Wilhelms-Universität Münster
Fachbereich Geowissenschaften (B,M,D) (2016)
Heisenbergstrasse 2
Münster D-48149
 p. +49 251 83-30002
 dekangeo@uni-muenster.de
 http://www.uni-muenster.de/Geowissenschaften/

Ghana

Kwame Nkrumah University of Science and Technology
Petroleum Engineering (B) (2015)
 http://www.knust.edu.gh/ceng/faculties.php

Dept of Geomatic Engineering (B,M,D) O (2017)
geomaticeng@knust.edu.gh
head.geomatic@knust.edu.gh
Kumasi PMB KNUST
 p. +233 (0)3220 60227
 geomaticeng@knust.edu.gh
 http://www.knust.edu.gh/ceng/faculties.php

Geological Engineering (B,M,D) ● (2015)
Geological Engineering Department
College of Engineering
KNUST
Kumasi PMB
 geologicaleng@knust.edu.gh
 http://archive.knust.edu.gh/pages/index.php?siteid=geoleng
Senior Scientist:
 Emmanuel K. Appiah-Adjei, (D), Hohai, 2013, HwgGg
Associate Professor:
 Simon K. Gawu, (D), Witwatersrand, 2004, GgEgZn
Lecturer:
 Bukari Ali, (D), Birmingham (UK), 1996, HwNgYg
 Godfrey C. Amedjoe, (M), Ghana, 2005, GidGc
 Samuel Banash, (M)
 Gordon Foli, (M), 2004
 Solomon S. Gidigasu, (M), Sci & Tech (China), 2013, GeZrNr
 Emmanuel Mensah, (M), Sci & Tech (China), 1998, EgGg

University of Development Studies
Earth and Environmental Science (B) (2015)
PO Box 1350
Tamale
 p. +233 (71) 22422
 edambayi@uds.edu.gh
 http://www.uds.edu.gh/

University of Ghana
Dept of Earth Science (B,M,D) ● (2015)
PO Box LG 58
Legon
Accra
 p. +233-244-116-879
 pmnude@ug.edu.gh
 http://www.ug.edu.gh/index1.php?linkid=185&sublinkid=40&subsublinkid=36
 Enrollment (2013): B: 336 (104) M: 105 (33) D: 2 (1)
Dean of Science:
 Daniel K. Asiedu, (D), GsCgGd
Professor:
 Bruce K. Banoeng-Yakubo, (D), Ghana, GcHw
Senior Scientist:
 Thomas K. Armah, (D), YgeYe
 Jacob M. Kutu, (D), GctGc
 Patrick Asamoah Sakyi, (D), Cg
Head of Department:
 Prosper M. NUDE, (D), Ghana, GipGx
Associate Professor:
 Thomas M. Akabzaa, (D), Ghana, 2004, EgGeNm
 David Atta-Peters, (D), Ghana, Pl
 Johnson Manu, (D), GzEgGg
 Frank K. Nyame, (D), Cg

 Sandow M. Yidana, (D), Hwq
Lecturer:
 Francis Achampong, (D), Ngr
 Chris Y. Anani, (D), Gsd
 Larry-Pax CHEGBELEH, (D), Ng
 Yvonne A.S Loh, (M), Hw
 Issac A. Oppong, (D), Np

University of Mines and Technology
Dept of Geological Engineering (B) (2015)
PO Box 237
Tarkwa
 p. +233 3123 20935
 sps@umat.edu.gh
 http://www.umat.edu/gh/UndergraduatePrograms/geology.html

Dept of Geomatic Engineering (B) (2015)
PO Box 237
Tarkwa
 p. +233(0)362 20324
 gm@umat.edu.gh
 http://www.umat.edu.gh/UndergraduatePrograms/geomatic.html

Dept of Mineral Engineering (B) (2015)
PO Box 237
Tarkwa
 p. +233(0)362 21136
 mr@umat.edu.gh
 http://www.umat.edu.gh/UndergraduatePrograms/mineral.html

Dept of Mining Engineering (B) (2015)
PO Box 237
Tarkwa
 p. +233 3123 20935
 mn@umat.edu.gh
 http://www.umat.edu.gh/UndergraduatePrograms/Mining.html

Greece

Aristotle University of Thessaloniki
School of Geology (2015)
GR-541 24
Thessaloniki 54124
 p. (+30) 2310 99.8450
 info@geo.auth.gr
 http://www.geo.auth.gr/index_en.htm

Greek Institute of Geology & Mineral Exploration
Greek Institute of Geology & Mineral Exploration ☒ (2018)
1, Spirou Louis St.,
Olympic Village
Acharnae P.C. 13677
 dirgen@igme.gr
 http://www.igme.gr

National Technical University of Athens
Dept of Mining Engineering and Metallurgy (2015)
Zographou Campus
15 780 Athens
 secretary@metal.ntua.gr
 http://www.metal.ntua.gr/div-geology/index.html

University of Athens
Faculty of Geology and Geoenvironment (2015)
National & Capodistran Univ. of Athens
Panepistimioupolis, Ilistra
Athens
 secr@geol.uoa.gr
 http://www.geol.uoa.gr/engindex.htm

University of Patras
Dept of Geology (2015)
26110 Patras

Guatemala

Universidad sde San Carlos de Guatemala
Centro de Estudios Superiores de Energia y Minas (2015)
Ciudad Universitaria Zona 12, Guatemala City 01012
 p. (502) 2418-9139 ex. 86211
 usacesem@ing.usac.edu.gt
 http://cesem.ingenieria.usac.edu.gt/

Guinea

Univerité Gamal Abdel Nasser de Conakry
Faculty of Geology and Mining (M,D) ●☒ (2019)
BP 1147
Conakry, Guinée
 p. +224 631 54 48 38
 sekoumoussa@gmail.com
 http://www.uganc.org/index.php/2013-01-12-06-22-44/cere
 Enrollment (2018): M: 30 (0)

Guyana

Guyana Geology and Mines Commission
Guyana Geology and Mines Commission (2015)
Upper Brickdam
Georgetown
 p. 226-5591, 225-2862
 http://www.ggmc.gov.gy/

Haiti

Ecole Nacionale de Geologie Appliquee
Ecole Nacionale de Geologie Appliquee (2015)
B.P. 1560-Varreux, Port-au-Prince

Hungary

Eotvos Lorand University
Insitute of Geography and Earth Sciences (B,M,D) ● (2018)
Pazmany Peter setany 1/C
Budapest H-1117
 p. (+36 1) 381 2191
 altfoldtan@ttk.elte.hu
 http://geosci.elte.hu/en_index.htm
 Enrollment (2014): B: 900 (360) M: 300 (120) D: 40 (18)
Professor:
 Judit Bartholy, (D), Eotvos Lorand, 1978, Am
Associate Professor:
 Gábor Timár, (D), Eotvos Lorand, 2003, YgdZi
Research Professor:
 János Lichtenberger, (D), Hungarian Acad of Sci, 1996, AsZrXy
Professor:
 Kristof Petrovay, (D), Eotvos Lorand, 1992, Xy
 Miklos Kazmer, (D), Eotvos Lorand, 1982, PggRh
Associate Professor:
 Agnes Gorog, (D), Eotvos Lorand, Pmi
 László Lenkey, (D), Vrije (Amsterdam), 1999, YhHwGa
 Balázs Székely, (D), Tuebingen, 2001, GmZri
Associate Scientist:
 Orsolya Ferencz, (D), Budapest, 2001, XyZr
 Anikó Kern, (D), Eotvos Lorand, 2012, ZrAms
 Gábor Molnár, (D), Eotvos Lorand, 2004, ZrYgZi
Assistant Professor:
 László Balázs, (D), Eotvos Lorand, 2009, EoYeg
 Attila Galsa, (D), Eotvos Lorand, 2004, YgHwYe
 Attila Osi, (D), Eotvos Lorand, PviPe
 Emoke Toth, (D), Eotvos Lorand, PmsPb
Research Associate:
 Istvan Szente, (D), Eotvos Lorand, Pig
Professor:
 Ferenc Horváth, (D), Eotvos Lorand, 1971, YgRhGc
 Péter Márton, (D), Eotvos Lorand, 1971, YmgYs

Hungarian Geological Survey
Hungarian Geological Institute (2015)
Stefánia út 14
Budapest HU-1143
 p. 36 1267 1433
 hamort@mgsz.hu

Hungarian Office for Mining and Geology
Hungarian Office for Mining and Geology (2015)
PO Box 95
1590 Budapest 1145
 p. (+36-1) 301-2900
 hivatal@mbfh.hu
 http://www.mbfh.hu

Jozsef Attila University
Dept of Mineralogy, Geochem & Petrology (2015)
P O Box 651
H-6701 Szeged

Dept of Geology & Paleontology (2015)
Egyetem u. 2-6
H-6722 Szeged

Univesity of Szeged
Dept of Physical Geography and Geoinformatics (B,M,D) (2017)
SZTE Termeszeti Foldrajzi Tanszek
Szeged, 6722 Egyetem utca 2. PF:653
Szeged
 p. 0036 62544158
 mezosi@geo.u-szeged.hu
 http://www.geo.u-szeged.hu/ANG/HPfirst.html
 Enrollment (2012): D: 20 (0)
Professor:
 Janos Rakonczai, (D), 1978, Ze
dean of faculty:
 lászló Mucsi, (D), Szeged, 1996, ZirZy
Associate Professor:
 Andrea Farsang, (D), Szeged, 2016, SdZe
 Timea Kiss, (D), Debrecen, 2001, GmZy
 György Sipos, (D), Szeged, 2007, HqYm
 József Szatmári, (D), Szeged, 2007, ZriZy

Iceland

Landmælingar Íslands
National Land Survey of Iceland ☒ (2017)
Stillholt 16-18 300 Akranes
Akranes 300
 p. 4309000
 lmi@lmi.is
 http://www.lmi.is/

University of Iceland
Faculty of Earth Science (2015)
Öskju, Sturlugötu 7
101 Reykjavik
 p. (+354) 5254600
 dagrun@hi.is
 http://english.hi.is/sens/faculty_of_earth_sciences/about_faculty

India

Aligarh Muslim University
Dept of Geology (B,M,D) ● (2015)
Aligarh - U.P.
Aligarh, Uttar Pradesh 202002
 p. 0571-2700615
 lakrao@yahoo.com
 http://www.amu.ac.in/fsc/no4.html

Andhra University
Dept of Geology (2015)
Vishakha-patnam-530 003

College of Science and Technology
Andhra Pradesh
> p. 91-891-2844888
> principal_science@andhrauniversity.info
> http://www.andhrauniversity.info/

Anna University
Dept of Geology (2015)
Guindy, Chennai-600 025,
Tamil Nadu
> p. 044-22358442 / 8444 / 8452
> geonag@gmail.com
> http://www.annauniv.edu/index.php

Annamalai University
Faculty of Marine Sciences (2015)
Annamalainagar-608 025
Tamil Nadu
> p. 91 - 4144 - 238248
> info@annamalaiuniversity.ac.in
> http://annamalaiuniversity.ac.in/marinesciences.htm

Banaras Hindu University
Dept of Geology (B,M,D) ☒ (2017)
Varanasi
U.P. 221005
> hbsrivastava@gmail.com
> http://www.bhu.ac.in/Geology/index.html
> Programs: Geology
> *Enrollment (2017): B: 150 (0) M: 56 (0) D: 50 (0)*

Professor:
> HARI BAHADUR SRIVASTAVA, (D), BHU, 1980, GctGp

Bharathidasan University
Dept of Geology (2015)
School of Geosciences
Tiruchirapalli 620 024
> p. +91 431 2407034
> drmm_bdu@yahoo.co.uk
> http://www.bdu.ac.in/schools/geo_sciences/geology/

Candigarh University
Dept of Geology (2016)
National Highway 95
Chandigarh, Punjab 140413
> p. (+91) 2724481
> chairperson_geology@pu.ac.in
> http://geology.puchd.ac.in/

Cochin University of Science & Technology
Dept of Geology (2015)
Kochi
> p. +91(484)2577550
> akv@cusat.ac.in
> http://www.cusat.ac.in/

Geological Survey of India
Geological Survey of India O☒ (2018)
27, J.L.Nehru Road
Kolkata
Kolkata, West Bengal 700016
> dg.gsi@gov.in
> http://www.gsi.gov.in
> f: https://www.facebook.com/gsipage/
> t: @GeologyIndia
> Certificates: Advanced GIS, Digital Cartography, Geological Mapping, Mineral Exploration, Geophysical Survey, etc

Goa University
Dept of Earth Science (M,D) O☒ (2019)
Taleigao Plateau
Goa University P.O.
Taleigao Plateau
Panaji, Goa 403 206
> p. +91-8669609193
> mkotha@unigoa.ac.in
> http://www.unigoa.ac.in/department.php?adepid=11&mdepid=3
> Programs: Applied Geology
> Certificates: GIS
> *Enrollment (2018): M: 22 (22) D: 1 (1)*

Professor:
> Mahender Kotha, (D), IIT (Bombay), 1987, GsoZi

Associate Professor:
> Anthony V. Viegas, (D), Goa, 1997, GiEmGx

On Contract:
> Poornima Dhavaskar, (M), Geology, 2013, Gge
> Purushottam Verlekar, (M), GrmGg

Assistant Professor:
> Niyati Kalangutkar, (D), Goa, GugGe

Cooperating Faculty:
> Sohini Ganguly, (D), Gi

Guru Nanak Dev University
Dept of Geology (2015)
Amritsar
> http://www.gndu.ac.in/

Indian Institute of Science Education and Research, Kolkata
Dept of Earth Science (B,M,D) ●☒ (2017)
Main Campus
Nadia
Mohanpur (Kolkata), West Bengal 741246
> des.chair@iiserkol.ac.in
> http://earth.iiserkol.ac.in/
> Programs: Geological Sciences

Indian Institute of Technology
Dept of Applied Geology (2015)
Bombay, Powai
Bombay
> ayaz@iitb.ac.in
> http://www.geos.iitb.ac.in/

Indian Institute of Technology, Kharagpur
Dept of Geology & Geophysics (B,M,D) (2015)
Kharagpur, Kharagpur
West Bengal 721302
Kharagpur, West Bengal 721302
> p. +91-3222-282268
> head@gg.iitkgp.ernet.in
> http://www.iitkgp.ac.in/departments/home.php?deptcode=MG

Indian Institute of Technology, Roorkee
Dept of Earth Sciences (M,D) (2015)
Department of Earth Sciences
IITRoorkee
Roorkee, Uttrakhand 247667
> p. +911332285532
> dpkesfes@iitr.ac.in
> http://members.tripod.com/rurkiu/acd-earth/index.html

Indian School of Mines
Dept of Applied Geology (2015)
Dhanbad
West Bengal 826 004
> p. +91-326-2296616
> agl@ismdhanbad.ac.in
> http://www.ismdhanbad.ac.in/depart/geology/index.htm

Institute of Science
Dept of Geology (2015)
Aurangabad 431004
Aurangabad, Maharashtra
> p. +91 (0240) 2400586

director@inosca.org
http://www.inosca.org

Jadavpur University
Dept of Applied Geology (B,M,D) (2015)
Jadavpur
Calcutta
    p. +913324572268
    hod@geology.jdvu.ac.in
    http://www.jaduniv.edu.in/view_department.php?deptid=76

Jmia Millia Islamia
Dept of Geography (2015)
Jamia Nagar
New Delhi-25
    rocketibrahim@yahoo.com
    http://jmi.ac.in/aboutjamia/departments/geography/introduction

Lucknow University
Dept of Geology (2015)
Lucknow 226007 U.P.
    info@lkouniv.ac.in
    http://www.lkouniv.ac.in/dept_geology.htm

Maharaja Sayajirao University of Baroda
Dept of Geology (2015)
Baroda - Gujarat
    p. (919) 426721547
    ischamyal@yahoo.com
    http://www.msubaroda.ac.in/science/index.php

Nagpur University
Dept of Geology (2015)
Rao Bahadur D. Laxminarayan Educational Campus
Law College Square
Amravati Road
Nagpur 440 001
    p. +91-712-253241
    geodeptplacecell@gmail.com
    http://www.nagpuruniversity.org/links/FacultyofScience.htm

Osmania University
Centre of Exploration Geophysics (M,D) O☒ (2019)
Department of Geophysics
Osmania University Campus
Hyderabad, Telangana 500007
    p. (+91) 40-27097116
    head.geophysics@osmania.ac.in
    http://www.osmania.ac.in/
    Programs: Geophysics (M.Sc.)
    Enrollment (2018): M: 29 (29) D: 12 (10)
Head and Chairman Board of Studies:
    Dr Ram Raj Mathur, (D), Osmania, 1989, YemYn
Professor:
    Dr. Veeraiah B, (D), Osmania, 2005, YexYg
Assistant Professor:
    Mrs Manjula Dharmasoth, (M), Osmania (India), 2012, YmeYg
    Dr. Vijay Kumar Dubba, (D), Osmania (India), 2015, YevYs
    Dr Udaya Laxm Gakka, (D), Osmania, 2009, HwYge

Dept of Geology (2015)
Hyderabad 500007 A.P.
    http://www.osmania.ac.in/

Panjab University
Centre for Petroleum and Applied Geology (U.I.E.A.S.T.) (2015)
Chandigarh 160014
    rpatnaik@pu.ac.in
    http://pu.ac.in/

Pondicherry University
Dept of Earth Science (B,M,D) (2015)
The Head
Department of Earth Sciences
Pondicherry University
Pondicherry 605 014
    p. +914132656741
    head.esc@pondiuni.edu.in
    http://www.pondiuni.edu.in/department/department-earth-sciences
Professor:
    Balakrishnan Srinivasan, (D), Jawaharlal Nehru (India), 1986, CcGiz

Presidency College
Dept of Geology (2015)
College Street
Calcutta
    hnb.geol@presidencycollegekolkata.ac.in
    http://www.concentric.net/~slahiri/pc/Department/Geology/geo

Pt. Ravishankar Shukla University
School of Studies in Geology and Water Resource Management (M,D) (2015)
SOS Geology
Pt. Ravishankar Shukla UNiversity
Raipur, Chhattisgarh 492010
    geology@prsu.com
    http://www.prsu.ac.in/
    Enrollment (2013): M: 15 (0) D: 4 (1)
Professor:
    Srikant k. Pande, (D), Nagoya, 1986, GgHyg
Professor:
    Ninad Bodhankar, (D), Raipur, 1992, GcNm
    Kosiyath Rghavan Na Hari, (D), Vikram Univ Ujjain, 1992, GiiCp
    Mohammad Wahdat Yar Khan, (D), AMU, Aligarh, 1979, GdEgCl

University of Baroda
Dept of Geology (2015)
Baroda , Vadodara 390 002
    swanikhil@yahoo.co.in
    http://www.msubaroda.ac.in/deptindex.php?ffac_code=2&fdept_code=6

University of Delhi
Dept of Geology (B,M,D) O⊘ (2018)
Department of Geology
University of Delhi
Delhi
New Delhi 110 007
    p. 27667073
    csdubey@gmail.com
    http://www.geology.du.ac.in/
    Programs: Geology
    Enrollment (2014): B: 105 (1200) M: 60 (600) D: 30 (130)
Director COL:
    C S. Dubey, (D), Delhi, 1993, GexGt

University of Madras
Dept of Geology (2015)
Maraimalai Adigalar Campus
Chennai 600 005
    p. 044 - 22202790
    spmohan50@hotmail.com
    http://www.unom.ac.in/departments/geology/geology.html

Dept of Geology (2015)
Madras, 600 025
Maraimalai Adigalar Campus
    spmohan50@hotmail.com
    http://www.unom.ac.in/departments/geology/geology.html

University of Mysore
Dept of Geology (2015)
Mysore 570 005
    p. +91 821 2419724
    bb@geology.uni-mysore.ac.in
    http://www.uni-mysore.ac.in/geology/#A

University of Pune
Dept of Geology (2015)
Ganeshkhind Road
Pune, Maharashtra 411 007
p. +91-020-25601360
geology@unipune.ac.in
http://www.unipune.ac.in/dept/science/geology/default.htm

Vijayanagara SriKrishnaDevaraya University
Dept of Applied Geology (M,D) (2016)
Dr.C.Venkataiah
Professor of Geology
Vijayanagara SriKrishnaDevaraya University
Bellary, Karnataka 583 104
p. (937) 909-0588
venkataiah.c@gmail.com
http://www.vskub.ac.in

Indonesia

Direktorat Vulkanologi
Volcanological Survey of Indonesia (2015)
Jl. Diponegoro 57
Bandung , West Java
http://portal.vsi.esdm.go.id/

Gadjah Mada University
Dept of Geology (2015)
Jalan Bulaksmur
Yogzakarta 55281
p. (+62) 274 6492340
geografi@geo.ugm.ac.id
http://geo.ugm.ac.id/main/

Indonesian Directorate General of Geology and Mineral Resources
Ministry of Mines and Energy (2015)
pengaduan@esdm.go.id
http://www.esdm.go.id

Institut Teknologi Bandung
Dept of Geology (2015)
Jalan Ganesa, No. 10
Bandung 40132, Jawa Barat
p. 022 2514990
sisfo@fitb.ac.id
http://www.fitb.itb.ac.id/en/

Dept of Geological Engineering (2015)
Jl. Ganesha 10
Bandung, 40135
geologi@gc.itb.ac.id

Trisakti University
Faculty of Earth and Energy Technologies (2015)
Jl. Kiai Tapa, Grogel
Jakarta 11440
p. 5663232 Ext. 8510

Universitas Hasanudin
Dept of Geological Engineering (2015)
Jl. Perintis Kemerdekaan
Ujung Pandang 90245

Universitas Padjadjaran
Dept of Geological Engineering (2015)
Jl. Raya Bandung-Sumedang km.21,
Jatinangor

Universitas Pakuan
Dept of Geological Engineering (2015)
Jl. Pakuan, PO Box 452
Bogor 16143, Jawa Barat
p. 0251-8312206
rektorat@unpak.ac.id
http://www.unpak.ac.id/

Iran

Bu-Ali University
Dept of Geology (A,D) O (2015)
Barati@basu.ac.ir
Hamedan , Hamedan 65174
p. 0098-0811-38381460
Barati@basu.ac.ir
http://www.basu.ac.ir
Administrative Assistant: Meisam Gholipoor

Shahid Bahonar University of Kerman
Department of Geology (B,M,D) ●☒ (2018)
Kerman
contactus@mail.uk.ac.ir
http://www.uk.ac.ir
Programs: Tectonics (B)
Certificates: Morphotectonics (M, D)
Enrollment (2017): B: 22 (25) M: 21 (19) D: 9 (14)
Associate Professor:
  Ahmad Abbasnejad, (D), GemHg
  Reza Derakhshani, (D), GtmGc

University of Tabriz
Dept of Earth Sciences (B,M,D) ●☒ (2018)
29 Bahman Blvd.
University of Tabriz
Tabriz , East Azerbaijan 51666
p. +98 (411) 3335 9894
moazzen@tabrizu.ac.ir
http://www.tabrizu.ac.ir
Programs: Earth Sciences
Certificates: GIS, Mudflow
Enrollment (2018): B: 25 (22) M: 36 (36) D: 18 (18)
Head:
  Ata Allah Nadiri, (D), Iran, 2012, HwGqe
Dean of Faculty:
  Asghar Asghari Moghadam, (D), UK, 1992, HwqHy
Head of the Department:
  Ali Kadkhodai Ilkhchi, (D), Iran, 2008, NpEo
Professor:
  Ali Asghar Calagari, (D), UK, 1997, EgmCg
  Ahmad Jahangiri, (D), India, 2000, GivGg
  Mohsen -. Moayyed, (D), Iran, 2000, GigEg
  Mohssen -. Moazzen, (D), Manchester, 1999, GpCgGz
  Hosseinzadeh Mohammad Reza, (D), Iran, 2005, Egm
Associate Professor:
  Nasir Amel, Iran, 2004, GivGg
  Robab Hajialioghli, (D), Iran, 2007, Gpt
  Reza Vaezi, (D), Iran, 2010, GeHgs
  Behzad Zamani, (D), Iran, 2004, GtYs
Assistant Professor:
  Ebrahim Asghari Ka, (D), Iran, 2001, Ngr
  Ghafour Alavi, (D), Eg
  Ghodrat Barzegari, (D), Iran, Nr
  Mohammad Hassanpour Sedghi, Iran, 2011, YsNeYg
  Fatemeh Mesbahi, (D), Iran, Gt
  Kamal Siahcheshm, (D), Iran, 2011, Egm
Instructor:
  Rahim Jomeiri, (M), Iran, 1992, Yg
  Maqsood Orouji, (M), Ca
  Naser Samimi, (B), Iran, Gg
Lecturer:
  Siavosh Sartipzadeh, (M), Iran, Pm

Ireland

Geological Survey of Ireland
Geological Survey of Ireland (2015)
Beggars Bush

Haddington Road
Dublin 4
    john.butler@gsi.ie
    http://www.gsi.ie/

Geological Survey of Northern Ireland
Geological Survey of Northern Ireland (2015)
Colby House
Stranmillis Court
Belfast BT9 5BF
    gsni@detini.gov.uk
    http://www.bgs.ac.uk/GSNI/

National University of Ireland Galway
Earth and Ocean Sciences (B,D) (2015)
Earth & Ocean Sciences
National University of Ireland
University Road
Galway
    p. + 353 (0)91 492 126
    lorna.larkin@nuigalway.ie
    http://www.nuigalway.ie/eos/
Professor:
    Peter Croot, (D), Otago, Ct
Lecturer:
    Rachel R. Cave, (D), Southampton, Cm
    Eve Daly, (D), National (Ireland), Ygu
    Tiernan Henry, (D), National (Ireland), 2014, HwyEg
    John Murray, (D), Trinity College, PgGs
    Robin Raine, (D), National (Ireland), Yg
    Tyrrell Shane, (D), Univ Coll (Dublin), Gs
    Martin White, (D), Southampton, Op
Emeritus:
    Martin Feely, (D), National (Ireland), 1982, GzxEg

Trinity College
Dept of Geology (2018)
Department of Geology
Museum Building
Trinity College-Dublin
Dublin 2
    p. +353 01 896 1074
    earth@tcd.ie
    http://www.tcd.ie/Geology/
Chair:
    Balz S. Kamber, (D), CgGgp
Associate Professor:
    David Chew, (D), Univ Coll (Dublin), 2001, CcGgt
    Catherine Coxon, GeOu
    Robin Edwards, Ge
    Patrick N. Wyse Jackson, (D), Dublin, 1992, PiRh
Assistant Professor:
    Quentin G. Crowley, Cc
    Seán h. mcClenaghan, Eg
    Chris Nicholas, Eo
    Juan Diego Rodriguez-Blanco, (D), 2006, GzClZm
    Catherine V. Rose, Cg
    Emma L. Tomlinson, Cc

University College Cork
Dept of Geology (B,M,D) (2015)
Donovans Road
Cork
    p. +353 21 4902657
    s.culloty@ucc.ie
    http://www.ucc.ie/ucc/depts/geology/
    Administrative Assistant: Patricia Hegarty
Head:
    John Gamble, Gi
Chair:
    Andy J. Wheeler, (D), Cambridge, 1994, GusGg
Professor:
    Ken Higgs, Pl
Research Associate:
    Tara Davis, Yg

Jim Smith, Pl
Lecturer:
    Bettie M. Higgs, (D), Sheffield, 1977, YgGqRh
    Ed Jarvis, Pl
    Ivor MacCarthy, (D), 1974, Gs
    Pat Meere, (D), National (Ireland), 1992, Gc
    John Reavy, (D), GicGt
Other:
    Mary Lehane, Zn
    Mick O'Callaghan, Zn
    Dan Rose, Zn

University College Dublin
School of Earth Sciences (2015)
Science Centre West
University College Dublin
Belfield, Dublin 4
    p. +353 1 716 2331
    geology@ucd.ie
    http://www.ucd.ie/geology/
Head:
    J. Stephen Daly
Professor:
    Peter D. W. Haughton
    Frank McDermott, (D), Open Univ (UK), 1987, GgCgZg
    Patrick M. Shannon
    John J. Walsh
Associate Professor:
    Christopher J. Bean
    Julian F. Menuge, (D), Cambridge (UK), 1983, EmCcs
    Ian D. Somerville
Lecturer:
    Conrad Childs
    Aggeliki Georgiopopoulou, (D), Southampton (UK), 2006, GusOg
    Ivan Lokmer
    Patrick J. Orr
Adjunct Professor:
    Tom Manzocchi, (D), Goc

Israel

Ben Gurion University of the Negev
Dept of Geological and Env Sciences ☒ (2018)
P.O. Box 653
84105 Beer Sheva
    hatzor@bgu.ac.il
    http://www.bgu.ac.il/geol/

Geological Survey of Israel
Geological Survey of Israel (2015)
30 Malkhe Israel St.
Jerusalem, 95501
    ask_gsi@gsi.gov.il
    http://www.gsi.gov.il/

Hebrew University of Jerusalem
Faculty of Advanced Environmental Studies (2015)
Givat Ram
Jerusalem 91904
    msfeitel@mscc.huji.ac.il

Tel Aviv University
Dept of Geophysical, Atmospheric and Planetary Sciences
(B,M,D) ● (2016)
Ramat Aviv
P O Box 30940
Tel Aviv 69978 69978
    p. 972-3-6408633
    batshevc@tauex.tau.ac.il
    http://geophysics.tau.ac.il/
Full Professor:
    Shmuel Marco, (D), Gg

Geosciences (B,M,D) ● (2015)
Levanon Road

Ramat Aviv
Tel Aviv 6997801
  p. 972-3-6408633
  shmulikm@tau.ac.il
  http://geophysics.tau.ac.il/
Professor:
  Pinhas Alpert, (D), Hebrew, 1980, AsZrAm
  Zvi Ben-Avraham, (D), MIT/WHOI, 1973, YrGt
  Shmulik Marco, (D), Hebrew, 1997, Ggc
  Morris Podolak, (D), Yeshiva, 1974, ZnnZn
  Colin G. Price, (D), Columbia, 1993, As
  Moshe Reshef, (D), Tel Aviv, 1985, Yeg
Senior Scientist:
  Pavel Kishcha, (D), Russian Acad of Sci, 1985, As
Dr:
  Nili Harnik, (D), MIT, 2000, As
Lecturer:
  Gilles Hillel Wust-Bloch, (D), 1990, Ys
  Alon Ziv, (D), Ys
Principal Research Assoc. (Assoc. Professor):
  Lev Eppelbaum, (D), Inst of Geophysics of Georgia, 1989, YvmGt

The Hebrew University of Jerusalem
Institute of Earth Sciences (B,M,D) ●☒ (2017)
Jerusalem 91904
  p. 97226584686
  arimatmon@mail.huji.ac.il
  http://earth.huji.ac.il/
  f: https://www.facebook.com/earthhuji/
  Programs: Geology (B,M); Atmosphere, Oceanography & Climate (B); Environmental Studies (B); Atmospheric Sciences (M); Oceanography (M); Hydrology (M); Environmental Studies (M)
  Enrollment (2017): B: 70 (0)

Italy

Alma Mater Studiorum Università di Bologna
Dipartimento di Scienze Biologiche, Geologiche e Ambientali
(B,M,D) O (2016)
Piazza di Porta San Donato 1
Bologna 40126
  p. ++39 - 051 - 2094238
  alessandro.gargini@unibo.it
  http://www.bigea.unibo.it/it
  Enrollment (2016): B: 120 (30) M: 120 (30) D: 28 (7)

Servizio Geologico d'Italia
Servizio Geologico d'Italia (2015)
Via Vitaliano Brancati
Roma 48-00144
  p. (+39) 0650071
  emergenzeambientali@isprambiente.it
  http://www.isprambiente.gov.it/it/servizi-per-lambiente/il-servizio-geologico-ditalia

Università degli Studi di Padova
Department of Geosciences (B,M,D) ☒ (2018)
Via G. Gradenigo 6
Padova 35131
  p. +390498279110
  geoscienze.direzione@unipd.it
  http://www.geoscienze.unipd.it/
  Enrollment (2017): B: 259 (49) M: 112 (38) D: 17 (7)
Full Professor:
  Gilberto Artioli, (D), Chicago, 1985, GzyZm
  Alessandro Caporali, (D), Ludwig Maximilian (Germany), 1979, YdsZr
  Alberto Carton, (M), Gm
  Giorgio Cassiani, (D), YgHs
  Bernardo Cesare, (D), Gp
  Giulio Di Toro, (D), Gc
  Silvana Martin, Gc
  Fabrizio Nestola, (D), Modena & Reggio Emilia, 2003, Gz
  Giorgio Pennacchioni, Gc
  Cristina Stefani, Gd
  Nicola Surian, Gm
  Massimiliano Zattin, (D), Gs
Associate Professor:
  Claudia Agnini, (D), Padova, 2007, PmeGu
  Lapo Boschi, Yg
  Luca Capraro, (D), Padova, 2002, PelGg
  Andrea D'Alpaos, (D), Hs
  Paolo Fabbri, Hw
  Manuele Faccenda, (D), Gq
  Mario Floris, (D), NgZr
  Alessandro Fontana, (D), Gm
  Eliana Fornaciari, (D), Pgm
  Massimiliano Ghinassi, (D), Gs
  Luca Giusberti, (D), Pm
  Lara Maritan, (D), Gx
  Andrea Marzoli, (D), Cg
  Matteo Massironi, (D), XgZr
  Claudio Mazzoli, Gx
  Stefano Monari, Pg
  Paolo Mozzi, (D), Gm
  Paolo Nimis, Eg
  Nereo Preto, (D), Gs
  Manuel Rigo, (D), Gs
  Gabriella Salviulo, Gz
  Raffaele Sassi, Ggx
  Paolo Scotton, Hg
  Luciano Secco, Gz
  Alberta Silvestri, (D), Gz
  Richard Spiess, Gp
  Annalisa Zaja, Yg
  Dario Zampieri, Gc
Assistant Professor:
  Omar Bartoli, Gp
  Jacopo Boaga, Yg
  Anna Breda, (D), Padova, 2003, GsrGg
  Maria Chiara Dalconi, (D), Gz
  Antonio Galgaro, Ng
  Roberto Gatto, Pg
  Christine Marie Meyzen, (D), Cg
  Leonardo Piccinini, (D), Hg

Università degli Studi di Pavia
Dept of Earth and Environmental Sciences (A,B,M,D) O (2015)
via Ferrata, 1
Pavia 27100
  p. +30 0382 985754
  dvagnini@unipv.it
  http://sciter.unipv.eu/site/home.html
  Enrollment (2014): B: 45 (25) M: 25 (14) D: 15 (5)

Universita Degli Studi di Siena
Centro di GeoTecnologie (2015)
Via Vetri Vecchi 34
San Giovanni Valdarno 52027
  p. +390559119400
  bottacchi@unisi.it
  http://www.geotecnologie.unisi.it

Universita di Bari
Dept Geomineralogico (2015)
Via E. Orabona
4-70125 Bari
  scandale@geomin.uniba.it
  http://www.geomin.uniba.it/frame.htm

Universita di Cagliari
Dept di Geoingegneria e Tecnologie Ambientali (2015)
Via Marengo,3
09124 Cagliari
  p. +39 70 675 52/29
  mazzella@unica.it
  http://geoing.unica.it/digita.htm
Professor:
  Prof. Antonio AM MAZZELLA, Gq

Italy

Universita di Calabria
Dept of Biology Ecology and Earth Science (DIBEST) (D) (2015)
87036 Arcavacata di Rende
Calabria
    crisci@unical.it
    http://www.unical.it

Universita di Camerino
Dept Scienze della Terra (B,M,D) ○☒ (2017)
Via Gentile III da Varano
Camerino, MC 62032
    p. (+39) 737402126
    sst@pec.unicam.it
    http://www.sst.unicam.it/SST/en/course-of-degree
    Programs: Geology (B); Natural Science (B); Geoenvironmental Resources and Risks (M)

Universita di Catania
Inst Scienze della Terra (2015)
Corso Italia 57
95129 Catania
    p. (+39) 095-7195730
    giovali@unict.it
    http://www3.unict.it/idgeg/

Universita di Firenze
Dept Scienze della Terra (B,M,D) ● (2015)
via La Pira 4
50121 Firenze
    direttore@geo.unifi.it
    http://www.dst.unifi.it/

Universita di Genova
Dept Scienze della Terra Ambiente e Vita - DISTAV (B,D) (2015)
Corso Europa 26
16132 Genoa
    p. +39 010 353 8311
    direttore@dipteris.unige.it
    http://www.distav.unige.it/drupalint/index.php
Professor:
    Egidio Armadillo, (D), Ye
Research Associate:
    Donato Belmonte, (D), Cg

Università di Modena e Reggio Emilia
Dept di Scienze Chimiche e Geologiche (B,M,D) (2015)
Largo S. Eufemia, 19
Modena, Italy 41121
    p. 390592055885
    direttore.chimgeo@unimore.it
    http://www.dscg.unimore.it/
    Enrollment (2013): B: 23 (13) M: 14 (4) D: 10 (9)

Universita di Napoli FEDERICO II
Dipartimento di Scienze della Terra, dell'Ambiente e delle Risorse (B,M,D) ● (2016)
L.go S. Marcellino, 10
Naples 80138
    p. +390812538112
    domenico.calcaterra@unina.it
    http://www.distar.unina.it

Universita di Perrugia
Dept Scienze della Terra (2015)
Piazza Università, 1
06100 Perugia
    cclgeol@unipg.it
    http://cclgeol.unipg.it/cclgeol/

Università di Torino
Dept Scienze della Terra (B,M,D) ●☒ (2018)
via Valperga Caluso 35
10125 Torino
    p. 011.6705184
    direzione.scienzeterra@unito.it
    http://www.dst.unito.it/do/home.pl
    Programs: Geological Sciences
    Science in applied Geology
    Enrollment (2018): B: 49 (30) M: 33 (33) D: 10 (0)
Professor:
    Giorgio Carnevale, (D), Pg
    Rodolfo Carosi, (D), Gc
    Daniele Castelli, (D), Gx
    Anna Maria Ferrero, (D), Nr
    Alessandro Pavese, Gz
Associate Professor:
    Rossella Arletti, (D), Gz
    Elena Belluso, (D), Gz
    Piera Benna, (M), Gz
    Alessandro Borghi, (D), Gx
    Marco Bruno, (D), Gz
    Paola Cadoppi, (D), Gc
    Cesare Comina, (D), Yx
    Domenico Antonio De Luca, (D), Hy
    Francesco dela Pierre, (D), Gd
    Massimo Delfino, (D), Pg
    Andrea Festa, (D), Gc
    Maria Gabriella Forno, (M), Gg
    Giandomenico Fubelli, (D), Gm
    Marco Gattiglio, (D), Gc
    Marco Giardino, (D), Gm
    Roberto Giustetto, (D), Gz
    Chiara Teresa Groppo, (D), Gx
    Giuseppe Mandrone, (D), Ng
    Luca Martire, (D), Gd
    Michele Motta, (M), Zy
    Pierluigi Pieruccini, Gm
    Mauro Prencipe, (D), Gz
    Franco Rolfo, (D), Gx
    Piergiorgio Rossetti, (M), Eg
    Sergio Carmelo Vinciguerra, (D), Nr
Assistant Professor:
    Roberto Ajassa, (M), Zy
    Gianni Balestro, (D), Gc
    Carlo Bertok, (D), Gd
    Sabrina Maria Rita Bonetto, (D), Ng
    Corrado Cigolini, (M), Gv
    Diego Coppola, Gv
    Emanuele Costa, (D), Gz
    Anna d'Atri, (D), Gr
    Simona Ferrando, (D), Gx
    Simona Fratianni, (D), Zy
    Rocco Gennari, Gn
    Franco Gianotti, (D), Gg
    Daniele Giordano, (D), Gv
    Salvatore Iaccarino, Gc
    Francesca Lozar, (D), Pg
    Edoardo Martinetto, (D), Pb
    Luciano Masciocco, (M), Ge
    Luigi Motta, (M), Zy
    Marcello Natalicchio, Gs
    Marco Davide Tonon, (M), Ze
    Alberto Vitale Brovarone, Gx
    Elena Zanella, (D), Ym
Emeritus:
    Emiliano Bruno, (M), Gz
    Ezio Callegari, (M), Gx
    Roberto Compagnoni, (M), Gx
    Giovanni Ferraris, (M), Gz
    Giulio Pavia, (M), Pg
    Germano Rigault de la Longrais, (M), Gz

Università di Trieste
Dipartimento di Matematica e Geoscienze (D) (2016)
via Weiss 2
Trieste, Italy I-34128
    p. +39 040 5582055
    dmg@pec.units.it
    http://www.geoscienze.units.it/

Università di Udine
Dipartimento di Georisorse e Territorio (2015)
Dipartimento di Georisorse e Territorio
Via Cotonificio 114
33100 Udine
 ta.tempoindeterminato@uniud.it
 http://udgtls.dgt.uniud.it/

Universita di Urbino
Inst di Geologia Applicata (2015)
Via Muzio Oddi 14
61029 Urbino

Universita Pisa
Dept of Geosciences (D) (2015)
via S. Maria 53
56126 Pisa
 p. +39050847260
 martinelli@dst.unipi.it
 http://www.dst.unipi.it/

University of Ferrara
Dept of Geology ☒ (2017)
Via Cairoli, 32
44121 Ferrara
 p. (+39) 0532 293028
 mob_int@unife.it
 http://www.unife.it/international/education/double-degrees-1/geological-sciences-ferrara-cadiz

University of Milano
Dept of Geology (2015)
Via Festa Del Perdono 7
20126 Milano
 p. (+39) 02 6448 1
 welcome.desk@unimib.it
 http://www.unimib.it/go/46204/Home/English/Academic-Programs/Mathematics-Physics-and-Natural-Sciences/Geological-Sciences-and-Te

University of Parma
Dept of Physics and Earth Sciences (B,M,D) (2019)
Parco Area delle Scienze 157/A
Parma 43100
 p. +39 0521 905326
 dipterr@unipr.it
 http://www.difest.unipr.it
Professor:
 Fulvio Celico, (D), Hgw
Research Associate:
 Andrea Artoni, (D), Gtr
 Tiziano Boschetti, (D), Cgs

University of Siena
Dept of Physical Science, Earth, and Environment (B,M,D) ● (2017)
Strada Laterina, 8
Siena 53100
 p. (+39) 0577 233938
 pec.dsfta@pec.unisipec.it
 http://www.dsfta.unisi.it/it
 f: https://www.facebook.com/dsfta.siena
Professor:
 Mauro Coltorti, (M), GmYg
Associate Professor:
 Luca Maria Foresi, (D), PmGs
 Cecilia Viti, Gz

Jamaica

University of the West Indies Mona Campus
Dept of Geography and Geology (B,M,D) (2015)
University of the West Indies
Mona
Kingston KGN7
 p. 876-927-2728
 geoggeol@uwimona.edu.jm
 http://myspot.mona.uwi.edu/dogg/
Head:
 Simon F. Mitchell, (D), Liverpool, 1993, GsgGs

Japan

Akita University
Deparment of Earth Science and Technology (2015)
1-1 Tagata Gakuen-cho, Akita-shi
Akita

Ehime University
Dept of Earth Sciences (2015)
 hori.rie.mm@ehime-u.ac.jp
 http://www.ehime-u.ac.jp/~cutie/index.html

Geological Survey of Japan
Geological Survey of Japan (2015)
 http://www.gsj.jp/

Hirosaki University
Dept of Earth Science (2015)
 wata@cc.hirosaki-u.ac.jp
 http://sci.hirosaki-u.ac.jp/~earth/index.html

Hiroshima University
Dept of Earth and Planetary Systems Science (2015)
Kagami-yama 1-3-1
Higashi-Hiroshima
Hiroshima 739
 toiawase@geol.sci.hiroshima-u.ac.jp
 http://www.geol.sci.hiroshima-u.ac.jp/index_e.html

Kagoshima University
Faculty of Science (2015)
 koko@sci.kagoshima-u.ac.jp
 http://earth.sci.kagoshima-u.ac.jp/index.html

Kanazawa University
Dept of Earth Sciences (2015)
Kakuma-machi
Kanazawa 920-1192
 fsci-pla-director@edu.kobe-u.ac.jp
 http://earth.s.kanazawa-u.ac.jp/

Kobe University
Earth & Planetary Sciences (2015)
 fsci-pla-director@edu.kobe-u.ac.jp
 http://shidahara1.earth.s.kobe-u.ac.jp/index.html

Kumamoto University
Dept of Earth and Environment (B,M,D) (2015)
2-39-1, Kurokami
Kumamoto City 860-8555
 p. 81-96-342-3411
 tadao@sci.kumamoto-u.ac.jp
 http://www.sci.kumamoto-u.ac.jp/earthsci
Professor:
 Shiro Hasegawa, (D), Pm
 Toshiaki Hasenaka, (D), Gv
 Hiroki Matsuda, (D), Gs
 Tadao Nishiyama, (D), Gp
 Hidetoshi Shibuya, (D), Ymg
 Jun Shimada, (D), Hw
 Akira Yoshiasa, (D), Gz

Nagoya University
Dept of Earth and Planetary Sciences (2015)
 env@post.jimu.nagoya-u.ac.jp
 http://www.eps.nagoya-u.ac.jp/

Shizuoka University
Inst of Geosciences (2015)
Shizuoka 422-8529
  setmasu@ipc.shizuoka.ac.jp
  http://www.sci.shizuoka.ac.jp/~geo/Welcome.html

Tohoku University
Dept of Mineralogy, Petrology and Economic Geology (2015)
Sendai 980-8578
  kaiho@m.tohoku.ac.jp
  http://www.ganko.tohoku.ac.jp/

Tsukuba University
College of Geoscience (2015)
  kankyojoho@ynu.ac.jp
  http://www.geo.tsukuba.ac.jp/

University of Tokyo
Deparment of earth and Planetary Science (2016)
3-1, Hongo 7chome, Bunkyo-ku
Tokyo
  Hirata@eri.u-tokyo.ac.jp
  http://www.eri.u-tokyo.ac.jp/

University of Toyama
Deptt of Earth Sciences (B,M,D) (2015)
3190 Gofuku
Toyama City, Toyama Prefecture 930-8555
  p. 81-76-445-6654
  takeuchi@sci.u-toyama.ac.jp
  http:/www.sci.u-toyama.ac.jp/earth/index-en.html
  Enrollment (2015): B: 208 (0) M: 36 (0) D: 7 (0)
Professor:
  Akira Takeuchi, (D), Osaka, 1979, GtcYe
  Tohru Watanabe, (D), Tokyo, 1991, YxsGv
Associate Professor:
  Shigekazu Kusumoto, (D), Kyoto, 1999, YvdGt

Yokohama National University
Dept of Environment and Natural Sciences (2015)
  kankyojoho@ynu.ac.jp
  http://chigaku.ed.ynu.ac.jp/geology-e.html

Kenya

Jomo Kenyatta University of Agriculture & Tech
Geomatic Engineering and Geospatial Information Systems (B,M,D) ☒ (2017)
PO Box 62000
00200 Nairobi
  p. +254-67-5352391
  gegis@eng.jkuat.ac.ke
  http://www.jkuat.ac.ke/departments/gegis/
  t: @JKUAT_GEGIS
  Programs: Geomatic Engineering; Geospatial Information Sciences
  Enrollment (2014): B: 175 (35) M: 45 (19) D: 8 (0)
Chair:
  Thomas G. Ngigi, (D), Chiba (Japan), 2007, ZriZr
Lecturer:
  Nathan O. Agutu, Jomo Kenyatta, ZgiZr
  Mark Boitt, (M), Stuttgart Univ of Applied Sci, 2010, ZgrZi
  George W. Chege, (M), Jomo Kenyatta, 2014, ZgrZi
  Charles Gaya, (M), Stuttgart Univ of Applied Sci, 2002, ZrgZi
  Andrew Imwati, (D), Nairobi, ZggZi
  Benson K. Kenduiywo, (D), Tech (Darmstadt), 2016, ZriYd
  Fridah K. Kirimi, (M), Jomo Kenyatta, 2010, ZiiZr
  Moffat G. Magondu, (M), Jomo Kenyatta, 2014, ZirZg
  Felix N. Mutua, (D), Tokyo, 2012, ZigZr
  Nancy Mwangi, (M), Jomo Kenyatta, 2008, ZiiZr
  Mercy W. Mwaniki, (M), Jomo Kenyatta, 2010, ZiiZr
  Eunice W. Nduati, (M), Jomo Kenyatta, 2014, ZrrZi
  Patroba A. Odera, (D), Kyoto, 2012, YdZge
  Hunja Waithaka, (D), Hokkaido, YdZgi
  Charles B. Wasomi, (M), Stuttgart Univ of Applied Sci, ZrrZi

Dept of Environmental Studies (B) (2015)
P.O. Box 62000 00200
Nairobi
  pro@jkuat.ac.ke
  http://www.jkuat.ac.ke/

Kenyatta University
School of Environmental Sciences (B,M,D) (2015)
Maseno
  p. +254-057-351620/2
  dean-envs@ku.ac.ke
  http://www.maseno.ac.ke/index3.php?section=schools&page=school-encironment

University of Nairobi
Dept of Geology (B) (2016)
PO Box 30197
Nairobi
  p. +254 20 4449856
  geology@uonbi.ac.ke
  http://geology.uonbi.ac.ke/

Dept of Geography (B) (2015)
PO Box 30917
Nairobi
  p. +254 (020) 318262 (28400)
  samowuor@uonbi.ac.ke
  http://uonbi.ac.ke/departments/?dept_code=HF&&face_code=32

Dept of Meteorology (B,M,D) (2015)
PO Box 30197
Nairobi 00100
  p. 254 020-4449004 (2070)
  dept-meteo@uonbi.ac.ke
  http://www.uonbi.ac.ke

Korea, South

Korea University
Earth and Environmental Sciences (B,M,D) (2015)
Anam-dong, Seongbuk-Gu
Seoul 136-713
  p. 82-2-3290-3170
  sjchoh@korea.ac.kr
  http://ees.korea.ac.kr/
  Enrollment (2015): B: 100 (27) M: 30 (13) D: 8 (4)
Head:
  Meehye Lee, (D), Rhode Island, 1999, AsOgCm
Vice President for Research Affairs:
  Seong-Taek Yun, (D), Korea, 1991, CqHyCl
Vice Dean of Office of Planning and Budget:
  Young Jae Lee, (D), SUNY (Stony Brook), 2005, ScGze
Vice Dean of College of Science:
  Ho Young Jo, (D), Wisconsin, 2003, NgCaGe
Professor:
  Suk-Joo Choh, (D), Texas, 2004, GdsPe
  Seon-Gyu Choi, (D), Waseda, 1983, GzEmg
  Seong-Jae Doh, (D), Rhode Island, 1987, YmgYe
  Jin-Han Ree, (D), SUNY (Albany), 1991, Gct
Associate Professor:
  Scott A. Whattam, (D), Hong Kong, 2002, GiCtg

Latvia

Latvija Valst Geologijas Dienests
Latvia State Geological Survey (2015)
Moscow street 165
Riga, LV-1019
  lvgma@lvgma.gov.lv
  http://mapx.map.vgd.gov.lv/geo3/

University of Latvia
Faculty of Geographical and Earth Sciences (2015)
10 Alberta Str.
202-203 Room
Riga
    zeme@lu.lv
    http://www.lu.lv/e_strukt/fakult/geog/info/

Lebanon

American University of Beirut
Dept of Geology (2015)
Faculty of Arts and Sciences
P.O.Box 11-0236
Riad El-Solh
Beirut 1107 2020
    p. (961)-1-340460/ext. 4160
    arahman@aub.edu.lb

Lesotho

National University of Lesotho
Dept of Geology & Mines (B) (2015)
P.O. Box 750, Maseru
    p. (266) 34 0601
    registrar@nul.ls
    http://www.nul.ls/

Dept of Geography and Environmental Science (B) (2015)
P.O. Roma 180
Maseru 100
    p. +266 2234 0601
    registrar@nul.ls
    http://www.nul.ls/faculties/fost/geography/index.html

Liberia

University of Liberia
Dept of Geology (B) (2015)
Capitol Hill
PO Box 9020
Monrovia 9020
    p. 231-6-422-304
    weekso@fiu.edu
    http://www.universityliberia.org/ul_course_master_list_biology.htm#geology

Dept of Mining Engineering (B) (2015)
Captiol Hill
PO Box 9020
Monrovia 9020
    p. +231 6422304
    weekso@fiu.edu
    http://www.universityliberia.org/ul_course_master_list_mining.htm#mine

Libya

Oil Companies School
Oil Companies School (B) (2015)
    info@ocslibya.com
    http://www.geocites.com/chatalaine/OCS.html

Petroleum Training and Qualifying Institute (PTQI)
Gergarish Road 9KM - Asiahia
Tripoli
    p. +218 21 4833771-5
    info@ptqi.edu.ly
    http://www.ptqi.edu.ly/online/en_home.php#

University of Garyounis
Dept of Geology (B) (2015)
PO Box 1308
Benghazi
    p. +218-61-86304
    uni.office@uob.edu.ly
    http://www.garyounis.edu/

Lithuania

Lithuanian Geological Survey
Lithuanian Geological Survey under the Ministry of Env (2015)
S. Konarskio St. 35
LT-03123 Vilnius
    indre.virbickiene@lgt.lt
    http://www.lgt.lt/

Vilniaus Pedagoginis Universitetas
Dept of Geology (2015)
Studentu 39, 2034 Vilnius
    p. (8 5) 275 89 35
    gmtf.dekanatas@leu.lt
    http://www.leu.lt/lt/gmtf/gmtf_apie_mus/all.html

Vilniaus Universitetas
Dept of Hydrogeology & Engin Geol (2015)
Giurlionio 21/27, 2009, Vilnius
    p. 239 8278
    robert.mokrik@gf.vu.lt
    http://www.vu.lt/en/scientific-report-2012/faculties-and-institutes/faculty-of-natural-sciences#DEPARTMENT_OF_HYDROGEOLOGY_AND_E

Dept of Geology and Mineralogy (B,M,D) ● (2015)
Ciurlionio str. 21/27
LT03101
Vilnius LT03101
    p. +370 5 2398272
    eugenija.rudnickaite@gf.vu.lt
    http://www.geol.gf.vu.lt/lt
    Enrollment (2015): B: 8 (8) M: 11 (11) D: 10 (3)

Luxembourg

Service Geologique du Luxembourg
Service Geologique du Luxembourg (2015)
23, rue du Chemin de Fer
L-8057 Bertrange
    geologie@pch.etat.lu
    http://www.pch.public.lu/administration/organigramme/geo/

Madagascar

Universite d' Antananarivo
Dept of Chemistry (B) (2015)
BP 566
Antananarivo 101
    p. +261 20 22 326 39
    rtianasoa.manoelson@gmail.com
    http://www.univ-antananarivo.mg/

Dept of Biology (B) (2015)
BP 566
Antananarivo 101
    p. +261 20 22 329 39
    arovonjy@yahoo.fr
    http://www.univ-antananarivo.mg/

Dept des Sciences de la Terre (A,B,M,D) (2017)
BP 906
Antananarivo 101 , Madagascar
    p. 00261324205328
    saphirzzf@gmail.com
    http://www.univ-antananarivo.mg/
    Enrollment (2016): A: 144 (138) B: 127 (111) M: 63 (14) D: 1 (0)

Malawi

Mzuu University
Faculty of Environmental Sciences (B) (2015)
Private Bag 201
Mzuzu
p. +(265) 1 320 722/ 320 575
ur@mzuni.ac.mw
http://www.mzuni.ac.mw/index.php?option=com_content&view=article&id=21&Itemid=15

University of Malawi
Dept of Natural Resources Management (B) (2015)
PO Box 219
Lilongwe
p. (265) 01 277 260
david.mkwambisi@bunda.luanar.mw
http://www.bunda.unima.mw/nrm.thm

Chancellor College, Natural Resources and Env Centre (B) (2015)
PO Box 280
Zomba
p. +(265) 1 524 685
geo@chanco.unima.mw
http://www.chanco.unima.mw/department/department.php?DepartmentID=4&Source=Department_of_Geography_and_Earth_Sciences

Malaysia

Minerals and Geoscience Department Malaysia
Minerals and Geoscience Dept Malaysia (2015)
20th Floor, Bangunan Tabung Haji
Jalan Tun Razak
50658 Kuala Lumpur
jmgkll@jmg.gov.my
http://www.jmg.gov.my/

National University of Malaysia
Geology Dept (2015)
43600 Bangi
Selangor
dftsm@ftsm.ukm.my

Universiti Putra Malaysia
Dept of Environmental Sciences (2015)
43400 UPM Serdang
puziah@env.upm.edu.my
http://fsas.upm.edu.my/~sas/envpage/Dept.html

University of Malaya
Dept of Geology (B,M,D) (2015)
University of Malaya
50603 Kuala Lumpur, Malaysia.
Kuala Lumpur, Selangor 50603
p. 03-79674203
ketua_geologi@um.edu.my
http://www.um.edu.my
Head:
   Wan Hasiah Abdullah, (D), Newcastle upon Tyne, 1994, Ec
Professor:
   Teh Guan Hoe, (D), Heidelberg, Ca
   Lee Chai Peng, (D), Liverpool, Pg
   John Kuna Raj, (D), Malaya, 1983, Ng
   Denis N.K Tan, (M), Malaya, Go
Associate Professor:
   Abd Rashid Ahmad, (D), Oxford, Sc
   Azman Abdul Ghani, (D), Liverpool, Gi
   Mohamed Ali Hasan, (M), Ulster, Hg
   Azhar Hj Hussin, (D), London, Gs
   Tajul Anuar Jamaluddin, (D), Wales, 1997, Ng
   Mustaffa Kamal Shuib, (M), London, 1986, Gc
   Samsudin Hj. Taib, (D), Durham, Yg
Lecturer:
   Nuraiteng Tee Abdullah, (D), London, Gr

Ahmad Tajuddin Hj Ibrahim, (D), Newcastle upon Tyne, Ng
Che Noorliza Lat, (M), Nevada (Reno), 1989, Yg
Mat Ruzlin Maulud, (B), Malaya, Zn
Nur Iskandar Taib, (D), Indiana, Gi
Ismail Yusoff, (D), Norwich, Hg

Mali

Universite de Bamako
Sciences de la Terre (B) (2015)
BP 3206
Bamako
p. (223) 222 32 44
phine.romagnoli@unige.ch
http://www.ml.refer.org/u-bamako/spip.php?article.137

Malta

University of Malta
Dept of Geosciences (M,D) O☒ (2018)
Department of Geosciences
University of Malta
Msida Campus
Msida MSD2080
p. (+356) 2340 2362
geo.sci@um.edu.mt
http://www.um.edu.mt/science/geosciences
f: https://www.facebook.com/um.geosciences/
Programs: Geoscience and Physics (B); Geoscience and Mathematics (B); Geoscience (M); Applied Oceanography (M); Petroleum Studies (M)
Enrollment (2018): M: 5 (8) D: 8 (0)
Professor:
   Raymond Ellul, (D), As
   Alfred Micallef, (D), Ap
Associate Professor:
   Alan Deidun, (D)
   Aldo Drago, (D), Southampton, Opg
   Pauline Galea, (D), Victoria (NZ), 1993, YsgZg
   Aaron Micallef, (D), Southampton, 2008, GumOu
Dr:
   Noel Aquilina, (D), Birmingham, Asm
   Sebastiano D'Amico, (D), YsgZg
   Pierre-Sandre Farrugia, (D), Malta
   Anthony Galea, (D), Trieste, Opg
Lecturer:
   Adam Gauci, (D), Malta, 2018, Op

Mauritania

Universite de Nouakchott
Dept de Geologie (B) (2015)
BP 5026
Nouakchott
p. (222) 25 13 82
awa@univ-nkc.mr
http://www.univ-nkc.mr/spip.php?rubrique34

Mauritius

University of Mauritius
Mauritius Radio Telescope (B) (2015)
Reduit
p. (230) 454 1041
nalini@uom.ac.mu
http://www.uom.ac.mu/mrt/mrt2.html

Dept of Chemical and Environmental Engineering (B) (2015)
Reduit
p. (230) 454 1041
d.surroop@uom.ac.mu
http://www.uom.ac.mu/Faculties/FOE/CEE/index.asp

Mexico

Centro de Investigación Científica y de Educación Superior de Ensenada
Earth Sciences Division (M,D) (2016)
Carretera Tijuana-Ensenada # 3918
Zona Playitas
Ensenada, Baja California 22860
    p. (01152-664)1750500
    dir-ct@cicese.mx
    http://www.cicese.mx
    Enrollment (2010): M: 44 (8) D: 20 (2)
Titular researcher:
    Edgardo Canon-Tapia, (D), Hawaii, 1996, GvYmGt
Research Associate:
    Jose G. Acosta, (M), CICESE (Ensenada), 1980, Ne
    Jesus M. Brassea, (M), Cinvestav, 1986, Ye
    Juan M. Espinosa, (M), CICESE (Ensenada), 1983, Ye
    Jose J. Gonzalez, (M), CICESE (Ensenada), 1986, Yd
    Alejandro Hinojosa, (M), CICESE (Ensenada), 1988, Zr
    Luis H. Mendoza, (M), CICESE (Ensenada), 1982, Ne
    Alfonso Reyes, (B), UNAM, Ne
Titular researcher:
    Raul Castro, (D), Nevada (Reno), 1991, Ys
    Juan Contreras, (D), Columbus, 1999, Gt
    Luis A. Delgado-Argote, (D), UNAM, 2000, GitGu
    Francisco Esparza, (D), CICESE (Ensenada), 1991, Ye
    John Fletcher, (D), Utah, 1994, Gpt
    Carlos Flores, (D), Toronto, 1986, Ye
    Jose Frez, (M), California (Los Angeles), 1980, Ys
    Juan Garcia, (D), Oregon State, 1990, Yv
    Ewa Glowacka, (D), Polish Acad of Sci, 1991, Ys
    Enrique Gomez, (D), Toronto, 1981, Ye
Titular Researcher:
    Mario Gonzalez-Escobar, (D), CICESE, 2002, YesYs
Titular researcher:
    Antonio Gonzalez-Fernandez, (D), Complutense, 1996, Gu
    Javier Helenes, (D), Stanford, 1980, Gr
    Thomas Kretzschmar, (D), Tubingen, 1995, Ge
    Margarita Lopez, (D), Toronto, 1985, Cc
    Arturo Martin, (D), Paris XI Orsay, 1988, Gd
    Luis Munguia, (D), California (San Diego), 1982, Ys
    Alejandro Nava, (D), California (San Diego), 1980, Ys
    Marco A. Perez, (D), CICESE, 1995, Ye
    Jose M. Romo, (D), CICESE (Ensenada), 2002, Ye
    Pratap Sahay, (D), Alberta, 1986, Ys
    Rogelio Vazquez, (D), CICESE (Ensenada), 2002, Ye
    Antonio Vidal, (D), CICESE (Ensenada), 2001, Ys
    Bodo Weber, (D), Gp

Ciudad Universitaria
Facultad de Ingenieria (2015)
C.P. 04510
    p. 56 22 08 66
    fainge@servidor.unam.mx
    http://www.ingenieria.unam.mx/

SGM
Servicio Geológico Mexicano (2016)
Blvd. F. Angeles 93.50-4 km.
Hidalgo 42080, Pachuca 42083
    gciadoctec@sgm.gob.mx
    http://www.sgm.gob.mx/

Universidad Autonoma de Baja California Sur
Dept of Marine Geology (B) (2015)
Carretera al Sur km 5.5, Box 19-B
La Paz, BS 23000
    p. 682/2-47-55/2-01-40/2-45-69
    oceanologia.fcm@uabc.edu.mx
Chair:
    Alejandro Alvarez-Arellano, (M), Nacional Auton, 1984, Gs
Professor:
    Rodolfo Cruz-Orozco, (D), Louisiana State, 1974, Gu
    Javier Gaitan-Moran, (M), Intl Inst-Aerospace Sur & Earth Sci,
      1986, Zr
    Carlos A. Galli-Olivier, (D), Utah, 1968, Gd
    Jose I. Peredo-Jaime, (B), Auto de Baja California, 1978, Og
    Luis R. Segura-Vernis, (D), Nacional Auton, 1977, Pi
Associate Professor:
    Alejandro J. Carillo-Chavez, (M), Cincinnati, 1981, Gp
    Efrain Cornejo-Luna, (B), Nacional Auton, 1969, Gq
    Genaro Martinez-Gutierrez, (B), Inst Politecnico Nac, 1982, Gr
    Cesar Martinez-Noriega, (B), Nacional Auton, Ou
    Jaime I. Monroy-Sanchez, (M), 1987, Ge
    Miriam Nunez-Velazco, (B), Tech (Madero), 1980, Go
    Jose A. Perez-Venzor, (B), Autonoma (SLP), 1978, Gi
    Ramon Pimentel-Hernandez, (B), Inst Politecnico Nac, 1980, Yg
    Humberto Rojas-Soriano, (B), Nacional Auton, 1983, Gz
    Paulino Rojo-Garcia, (B), Inst Politecnico Nac, 1983, Gc
Assistant Professor:
    Luis A. Herrera-Gil, (B), Nacional Auton, 1980, Pg
Lecturer:
    Cesar A. Lopez-Ferreira, (B), Nacional Auton, 1979, Hw
Cooperating Faculty:
    Oscar Rodriguez-Plasencia, (B), Escuela Militar de Met, 1959, Am

Universidad Autonoma de Chihuahua
Facultad de Ingenieria (B,M) (2015)
Circuito Universitario Campus 2
Chihuahua, CH 31170
    p. (614) 442-9500
    ancorral@uach.mx
Director:
    Arturo Leal-Bejarano, (B), Auto de Chihuahua, 1976, Zn
Head:
    Arturo Lujan-Lopez, (M), Essex, 1975, Zn
Chair:
    Socorro I. Aguirre-Moriel, (B), ITR (Switzerland), 1979, Nx
    Hector M. Mendoza-Aguilar, (B), Auto de Chihuahua, 1980, Nx
Professor:
    Rafael Chavez-Aguirre, (M), Auto de Chihuahua, 1993, Hw
    Adolfo Chavez-Rodriguez, (D), Arizona, 1987, Hy
    Miguel Franco-Rubio, (M), Nacional Auton, 1978, Eg
    Rafael Madrigal-Rubio, (B), Nacional Auton, 1969, Ng
    Teodulo Mena-Zambrano, (B), Inst Politecnico Nac, 1964, Ng
    Hector Minor-Velazquez, (B), Auto de Chihuahua, 1978, Nx
    Ignacio A. Reyes-Cortes, (B), Texas (El Paso), 1997, Gx
    Manuel Reyes-Cortes, (B), Nacional Auton, 1970, Gi
    Miguel Royo-Ochoa, (M), Auto de Chihuahua, 1997, Hw
    David H. Ruiz-Cisneros, (B), Auto de Chihuahua, 1974, Nx
    Luis M. Trevizo-Cano, (B), Auto de Chihuahua, 1978, Nx

Universidad Autonoma de Nuevo Leon
Facultad de Ciencias de la Tiera (2015)
Hacienda Guadalupe Km. 8 Camino a Cerro Prieto. - Linares, N 67700
    p. (81) 8329 4170 Ext. 4170
    fmedina@fct.uanl.mx
    http://www.fct.uanl.mx/portal/index.php

Universidad Autonoma de San Luis Potosi
Area de Ciencias de la Terra (B,M) (2015)
Facultad de Ingenieri#2.a
Av. Dr. Manuel Nava No. 8
San Luis Potosi#2., SL 78290
    p. (48) 13-82-22
    aaguillonr@uaslp.mx
    Department Secretary: Juan Manuel Torres Aguilera
Director:
    Hector David Atisha-Castillo, (M), Texas, 1978, Zn
Head:
    Joel Milan-Navarro, (M), Nancy (France), 1979, Zr
Professor:
    Jose Refugio Acevedo-Arroyo, (B), Nacional Auton, 1957, Ng
    Luis Garcia-Gutierrez, (M), Stanford, 1951, Eg
    Panfilo R. Martinez-Macias, (M), Nacional Auton, 1995, Gc
    Francisco Javier Orozco-Villasenor, (M), Colorado State, 1983, Eg
    Ramon Ortiz-Aguirre, (M), Madrid, 1979, Hg
    Carlos Francisco Puente-Muniz, (B), Autonoma (SLP), 1979, Hw
    Delfino C. Ruvalcaba-Ruiz, (D), Colorado State, 1982, Em

Juan Manuel Torres-Agulera, (B), Autonoma (SLP), 1980, Gx
Librarian:
   Panfilo R. Martinez-Macias, Zn

Universidad Autonoma de Sonora
Ejecutivo de la Unidad Regional Norte (2015)
Av. Universidad e Irigoyen
Col. Ortiz
Hermosillo

Universidad de Guanajuato
Departamento de Ingeniería en Minas, Metalurgia y Geología
(B) O (2016)
Sede San Matias
Ex Hda. de San Matías s/n
Col. San Javier
Guanajuato, Guanajuato C.P. 36025
   p. ((+473) 73 2 2291 (Ext. 5304)
   juancho@ugto.mx
   http://www.di.ugto.mx/minas/index.php/antecedentes
   Enrollment (2015): B: 7 (7)

Universidad Estatal de Sonora
Escuela Superior de Geociencias (B) O⊠ (2018)
Col. Bugambilias
Hermosillo, SO 83240
   p. 621/5-37-78
   maricela.lopez@ues.mx
   https://www.ues.mx/?p=especiales/ofertaeducativa/malla.aspx&cid=0&sid=13&smid=0&latder=0&params=pa=007_pe=14_tipopa=I
   Programs: minning, metallurgy and geology
   Department Secretary: Manuel Valenzuela-Renteria
   Enrollment (2018): B: 35 (0)
Director:
   Marco A. Gonzalez-Juarez, (B), Sonora, 1977, Gg
   Maricela López, (M), Sonora, 2015, NxGg
Associate Professor:
   Gustavo E. Durazo-Tapia, (B), Sonora, 1981, Em
   Francisco A. Esparza-Yanez, (B), Sonora, 1983, Gg
   David García-Martínez, (D), 2017, GzEmNx
   Raul A. Gongora-Jurado, (B), Auto de Chihuahua, 1977, Nm
   Maria B. Hurtado-De La Ree, (B), Sonora, 1987, Nx
   Leobardo Lopez-Pineda, (B), Nacional Auton, 1980, Yg
   Jesus F. Maytorena-Silva, (B), Sonora, 1981, Gr
   Rogelio Monreal-Saavedra, (D), Texas (Dallas), 1989, Gr
   Gerardo Monteverde-Gutierrez, (B), Sonora, 1980, Nm
   Arnulfo Salazar-Avila, (B), Sonora, 1981, Nx
   Angel Slistan-Grijalva, (B), Sonora, 1981, Nm
Instructor:
   Luis G. Vite-Picazo, (B), Nacional Auton, 1946, Nm

Universidad Nacional Autonoma de Mexico
Ciencias de la Tierra (B) ⊠ (2018)
Ciudad Universitaria,CDMX
Facultad de Ciencias
Ciudad de Mexico 04510
   tierra.coord@ciencias.unam.mx
   Programs: Earth Sciences
   Enrollment (2017): B: 700 (84)
Dr.:
   Blanca Mendoza, (D), Oxford, 1985, Zn

Facultad de Ingeniería (B) (2015)
Av. Universidad 3000, Ciudad Universitaria
Coyoacán, CP 4510
   p. 56 22 08 66
   fainge@servidor.unam.mx
   http://www.ingenieria.unam.mx/paginas/Carreras/ingenieriaMinas/ingMinas_Desc.php
Director:
   Antonio Nieto Antunez, (M), Stanford, 1970, Nm
Professor:
   Angelica Casillas, (B), Inst Politecnico Nac, 1976, Hw
   Esteban Cedillo, (D), Heidelberg, 1988, Em
   Juventino Martinez, (D), Paris VI, 1980, Gt
   Edgardo Meave, (B), Guanajuato, 1947, Gg
   Francisco Medina, (B), Guanajuato, 1982, Zg
   Ricardo Navarro, (B), Inst Politecnico Nac, 1973, Nm
   Salvador Ulloa, (B), Guanajuato, 1947, Eg
   Fernando Vasallo, (D), Lomonosov (USSR), 1982, Em
   Carlos Yanez, (B), Inst Politecnico Nac, 1976, Ce

Inst de Geologia (M,D) (2015)
Apdo. postal 70-296
Ciudad Universitaria
Delegacion Coyoacan, DF 04510
   p. (52) 5622 4314
   igl@geologia.unam.Mx
   http://132.248.20.1/geol.htm
   Administrative Assistant: Ana María Rodriguez Simental
Director:
   Dante J. Moran-Zenteno, (D), Nacional Auton, 1992, Cc
Head:
   Susana A. Alaniz Alvarez, (D), Nacional Auton, 1995, Gc
   Luca Ferrari Pedraglio, (D), Milan, 1992, Gt
   Sergio Cevallos Ferriz, (D), Alberta, 1990, Pg
   Carlos M. Gonzalez Leon, (D), Arizona, 1992, Gr
   Klavdia Oleschko Loutkova, (D), Lomosov (Moscow), 1984, So
   Maria S. Lozano Garcia, (D), d'Aix (Marseilles), 1979, Pe
   Francisco J. Vega Vera, (D), Nacional Auton, 1988, Pg
Chair:
   Gustavo Tolson Jones, (D), Nacional Auton, 1998, Gc
Senior Scientist:
   Shelton Applegate-Pleasants, (D), Chicago, 1961, Pv
   Jorge Aranda, (D), Oregon, 1982, Gi
   Blanca E. Buitron-Sanchez, (D), Nacional Auton, 1974, Pi
   Oscar Carranza-Castaneda, (D), Nacional Auton, 1989, Pv
   Gerardo Carrasco-Nunez, (D), Michigan Tech, 1993, Gv
   Ana L. Carreno, (D), d'Orsay (Paris), 1979, Pm
   Miguel Carrillo Martinez, (D), Paris VI, 1976, Pb
   Liberto De Pablo, (D), Ohio State, 1958, Gz
   Rodolfo Del Arenal-Capetillo, (B), Nacional Auton, 1960, Hw
   Ismael Ferrusquia, (D), Houston, 1971, Gr
   David Flores-Roman, (D), Nacional Auton, 1981, So
   Celestina Gonzalez-Arreola, (D), Nacional Auton, 1989, Pi
   Jose C. Guerrero-Garcia, (D), Texas (Dallas), 1975, Ym
   John D. Keppie, (D), Glasgow, 1964, Gt
   Victor M. Malpica-Cruz, (D), Bordeaux, 1980, Gs
   Enrique Martinez Hernandez, (D), Michigan State, 1979, Pl
   Juventino Martinez-Reyes, (D), Paris VI, 1980, Gt
   Luis M. Mitre-Salazar, (D), Paris IV, 1978, Ge
   Adrian Ortega, (D), Waterloo, 1993, Hw
   Fernando Ortega-Gutierrez, (D), Leeds (UK), 1975, Gp
   Sergio Palacios-Mayorga, (M), Nacional Auton, 1972, So
   Jerjes Pantoja-Alor, (M), Arizona, 1963, Gg
   Maria del Carmen Perrilliat-Montoya, (D), Nacional Auton, 1969, Pi
   Angel Nieto Samaniego, (D), Nacional Auton, 1994, Gc
   Christina D. Siebe Grabach, (D), Hohenheim (Germany), 1993, Sc
   Alicia Silva-Pineda, (D), Nacional Auton, 1980, Pg
   Max Suter-Cargnelutti, (D), Basel, 1978, Gt
   Jordi Tritlla, (D), Barcelona, 1994, Eg
   Ana B. Villasenor-Martinez, (D), Nacional Auton, 1991, Pi
   Reinhard Weber-Gobel, (D), Tubingen, 1967, Pg
Associate Scientist:
   Gerardo J. Aguirre Diaz, (D), Texas, 1993, Gv
   Thierry Calmus, (D), Paris VI, 1983, Gu
   Antoni Camprubi, (D), Barcelona, 1998, Eg
   Alejandro J. Carillo Chavez, (D), Wyoming, 1996, Hg
   Elena Centeno Garcia, (D), Arizona, 1994, Gt
   Rodolfo Corona-Esquivel, (M), Nacional Auton, 1985, Em
   Mariano Elias-Herrera, (M), Nacional Auton, 1982, Gp
   Maria L. Flores-Delgadillo, (M), Nacional Auton, 1987, So
   Gilberto Hernandez-Silva, (D), Nacional Auton, 1983, So
   Rafael Huizar-Alvarez, (D), Franche Comte, 1989, Hw
   Cesar Jacques Ayala, (D), Cincinnati, 1983, Gr
   Marisol Montellano-Ballesteros, (D), California (Berkeley), 1986, Pv
   Amabel M. Oretega Rivera, (D), Queens, 1997, Cc
   Odranoel Quintero, (D), Sorbonne, 1995, Gc
   Jose L. Rodriguez-Castaneda, (M), Pittsburgh, 1984, Gg
   Jaime Roldan Quintana, (M), Iowa, 1976, Gv

Gerardo Sanchez-Rubio, (M), Imperial Coll (UK), 1984, Gv
Jesu Sole, (D), Barcelona, 1996, Cc
Luis F. Vassallo-Morales, (D), Lomonosov, 1981, Em
Maria G. Villasenor-Cabral, (M), Leeds (UK), 1974, Ca
Research Associate:
  Irma Aguilera-Ortiz, (B), Iberoameric, 1971, Ca
  Victor M. Davila-Alcocer, (M), Texas (Dallas), 1986, Ps
  Jose G. Solorio-Munguia, (B), Nacional Auton, 1958, Gz
Adjunct Professor:
  J. Duncan Keppie, (D), Glasgow (Scotland), 1967, Gt
  Luis Silva-Mora, (D), d'Aix (Marseilles), 1979, Gi
Emeritus:
  Gloria Alencaster-Ybarra, (D), Nacional Auton, 1969, Pi
  Zoltan De Cserna, (D), Columbia, 1955, Gt
Geology Librarian:
  Teresa Soledad Medina Malagon, (B), Zn

Inst de Geofisica (2015)
losos@sci.muni.cz
http://tlacaelel.igeofcu.unam.mx/index.eng.html

Mongolia

Gazarchin Institute
Bayanzurkh District
Ulaanbaatar 46
  gazarchin_institute@gazarchin.edu.mn
  http://www.gazarchin.edu.mn/

Mineral Resources Authority of Mongolia
Mineral Resources Authority of Mongolia (2015)
info@mram.gov.mn
http://www.mram.gov.mn/

Mongolian University of Science and Tech
School of Geology and Petroleum Engineering (2015)
Bagatoiruu-46, P. O. Box-520
Ulaanbaatar 46
  p. 976-11-312291
  jepces@must.edu.mn
  http://www.gs.edu.mn

National University of Mongolia
Geology and Geography Faculty (2015)
Ikh Surguuliin gudamj - 1, Baga Toiruu,
Sukhbaatar district,
  Ulaanbaatar
  p. 976-11-311890
  geo@num.edu.mn
  http://geo.num.edu.mn

Morocco

Ecole Nationale de l' Industrie Minerale
Dept de Sciences de la Terre (B) (2015)
Avenue Hadj Ahmed Cherkaoui
BP 753
Agdal, Rabat
  p. (+212) 037 68 02 30
  zaydi@enim.ac.ma
  http://www.enim.ac.ma/formation/sciences_de_la_terre/

Dept de Mines (B) (2015)
Avenue Hadj Ahmed Cherkaoui
BP 753
Adgal, Rabat
  p. (+212) 037 68 02 30
  zaydi@enim.ac.ma
  http://www.enim.ac.ma/formation/mines

Dept de Sciences des Materiaux (B) (2015)
Avenue Hadj Ahmed Cherkaoui
BP 753
Agdal, Rabat
  p. (+212) 037 68 02 30
  zaydi@enim.ac.ma
  http://www.enim.ac.ma/formation/materiaux/

ONAREP - Naional Office of Petroleum Research
ONAREP - Naional Office of Petroleum Research (B) (2015)
PO Box 8030
Rabat 10050
  p. +212 37 28-1616

ONHYM
National Office of Hydrocarbons and Mines (2015)
rh@onhym.com
http://www.onhym.com/Default.aspx?alias=www.onhym.com/EN

School of Mines of Marrakech
School of Mines (B,M,D) (2015)
Rue Machaar, El Harm Quartier Issil-B.P. 38
Marrakech
  p. 212-30-97-79

Universite Abdelmalek Essaadi
Dept de Sciences de la Terre (B) (2015)
  p. +212 539 97 93 16
  presidence@uae.ma
  http://www.uae.ma/portail/FR/

Faculte de Sciences de Tetouen (B) (2015)
BP 2121
Tetouan 93002
  p. +212 (0) 39 97 24 23
  vdap.fs@gmail.com
  http://www.fst.ac.ma/stu.html

Universite Cadi Ayyad
Dept of Geology (2015)
Prince Moulay Abdellah
BP 515
Marakech
  p. +212 (0)5 24 43 46 49
  http://www.fssm.ucam.ac.ma/pages/geologie.php

Universite Hassan 1er - Settat
Dept de Geologie Apliquee (B) (2015)
BP 577
Settat 26000
  p. 05.23.40.07.36
  http://www.fsts.ac.ma/fsts/index/php?option=com_content&task=view&id=37&Itemid=12

Universite Hassan II
Dept of Geology (B) (2015)
9, Rue Tarik Bnou Zia
Anfa Casablanca
  p. 0522 23 06 80
  z.hilmi@fsac.ac.ma
  http://www.fsac.ac.ma/depart/geo/index.html

Universite Hassan II - Mohammedia
Dept de Sciences de la Terre (B) (2015)
279 Cite Yassmina
Mohammedia, Casablanca
  p. +212(33)314635
  presidence@univh2m.ac.ma

Universite Ibn Tofail
Dept de Geologie (B,M,D) ● ☒ (2019)
Abdelhakboua@Yahoo.com
Kenitra, morocco 14 000
  p. +212 061984449
  abdelhakboua@yahoo.com
  http://www.univ-ibntofail.ac.ma/fre/departments.php?esp=4&rub=14&srub=36&srub_=96

Enrollment (2018): M: 27 (35) D: 13 (11)

Universite Ibn Zohr, Agadir
Dept de Sciences de la Terre (B) (2015)
BP 32/S
Agadir 80000
p. +212 (28) 22 71 25
acma2008@esta.ac.ma
http://www.esta.ac.ma/

Universite Mohammed 1er (Oujda)
Ecole Nationale des Sciences Appliquees d'Al Hoceima (ENSAH) (B) (2015)
BP 724
Oujda 60000
p. +212(56)500612
presidence@ump.ma
http://webserver1.ump.ma/ecoles_facultes/ensah

Dept des Sciences de la Terre (B,M,D) (2015)
BP 717
60000 Oujda
p. (212)536500601/02
fso@fso.ump.ma
http://sciences1.univ-oujda.ac.ma/index.htm

Universite Mohammed V
Dept of Geology (2015)
BP 554
Rue Michlifen Agdal
Rbat-Chellah
p. +212 05 37 77 54 71
http://www.fsr.ac.ma/ancien/index.php/departement/giologie.html

Universite Mohammed V (Agdal)
Dept de Genie Minerale (B) (2015)
Avenue Ibnsina
BP 765
Adgal Raba
p. (212 - 537) 77.26.47 - 77.19.0
contact@um5a.ac.ma
http://www.emi.ac.ma

Universite Moulay Ismail, Meknes
Dept de Sciences de la Terre (B) (2015)
Marjane II BP 298
Meknes 5003
p. +212 5 35 53 78 96
doyen@fs-umi.ac.ma
http://ww.umi.ac.ma/

Universite Sidi Mohammed Ben Abdallah
Dept of Geology (2015)
Dhar El Mahraz
BP 42
Atlas-Fes
p. 06 61 35 04 81
dept_geo@fsdmfes.ac.ma
http://www.fsdmfes.ac.ma/Presentation/Departements/Geologie.php

University Mohammed I
Dept of Geology (B,M,D) (2015)
Faculté des Sciences Bd. Mohammed VI
BP 717
Oujda, Oujda-Angad 60000
p. (212)667772215
mbouabdellah2002@yahoo.fr
http://webserver1.ump.ma/

Mozambique

Universidade Eduardo Mondlane
Escola Superior de Ciencias Marinhas e Costeiras (B) (2015)
CP 128
Avenida 1 de Julho
Bairro Chuabo Dembe
Didad de Quelimane, Zambezia
p. +258 24900500/1
valera.dias@uem.mz
http://www.marine.uem.mz/

Laboratorio de Gemologia (B,M) (2015)
Av. Julius Nyerere 3453, Departamento de Física
Campus Universitário
Maputo
p. (+258) 21 497003
uthui@zebra.uem.mz
http://www.fisica.uem.mz/index.php?option=com_content&view=article&id=51&Itemid=57

Universidade Lurio
Faculdade de Engenharia e Ciencias Naturais (B) (2015)
Av. 25 de Setembro
N. 958
Pemba, Cabo Delgado
p. (+258) 27 221238
fecn@unilurio.ac.mz
http://www.unilurio.ac.mz/faculdades_pmb_pt.htm

Namibia

Geological Survey of Namibia in Windhoek
Geological Survey of Namibia in Windhoek (2015)
Private Bag 13297
Windhoek
p. +264-61-2848111
gschneider@mme.gov.na
http://www.mme.gov.na/gsn/

University of Namibia
Dept of Geology (B,M,D) ●☒ (2017)
Private Bag 13301
Windhoek 9000
p. (+264 61) 206 3712
ttjipura@unam.na
http://www.unam.na/faculties/science/departments.html
Programs: Geology (B); Applied Geology in Exploration (M); Hydrogeology (D)
Enrollment (2017): B: 12 (0) M: 15 (0) D: 2 (0)
Chair:
   Ansgar Wanke, Germany, GsCl
Associate Professor:
   Frederick Akalemwa Kamona, (D), Germany, 1991, EgGzCe
   Benjamin S. Mapani, (D), Melbourne (Australia), 1994, GtcGp
Lecturer:
   Martin Harris, (M), Geosciences (Beijing), 2013
   Regmi Kamal, (D), Australian Nat, 2013
   Albertina Nakwafila, (M), Stellenbosch, 2014
   Ester Shalimba, (M), China Geosciences, 2014, Ggp
   Collen Uahengo, (M), Paris Schl Mines, 2013, GgoEg
   Shoopala Uugulu, (M), France, 2012
   Heike Wanke, Germany, Hw

Nepal

Tribhuvan University
Central Dept of Geology (2015)
Gandhi Bhavan, Kirtipur, Kathmandu
info@geology.edu.np
http://www.geology.edu.np/

Netherlands

Delft University of Technology
Department of Geoscience & Engineering (B,M,D) O (2016)
Stevinweg 1
2628 Delft, the Netherlands

p. +31 15 2781328
J.D.Jansen@tudelft.nl
www.tudelft.nl

National Geological Survey of the Netherlands
National Geological Survey of the Netherlands (2015)
Princetonlaan 6
3584 CB Utrecht
p. (+31)30-2564256
http://www.en.geologicalsurvey.nl/

Royal NIOZ
Jaap Sinninghe Damste (D) ☒ (2017)
PO Box 59
Den Burg, Texel 1790 AB
p. +31222369550
Jaap.Damste@nioz.nl
https://www.nioz.nl/en/about/mmb
Programs: Organic Biogeochemistry; Marine Microbiology; Marine Microbial Ecology (w/ Utrecht Univ)

State University of Groningen
Faculty of Spatial Sciences (B,M) ☒ (2018)
P.O. Box 800
Groningen, Netherlands 9700 AV
p. +31 50 363 8891
i.l.veen@rug.nl
http://www.rug.nl/frw
f: http://www.facebook.com/FRWRUG
t: @FRW_RUG
Programs: Geography, urban planning, demography, real estate, education
Enrollment (2014): B: 190 (0) M: 60 (0)

State University of Utrecht
Faculty of Geoscience (B,M,D) (2015)
P.O. Box 80.115
Utrecht 3508 TA
p. (+31) (0)30 253 2044
info@geo.uu.nl
http://www.uu.nl/faculty/geosciences/EN/Contact/Pages

Technical Univ. of Delft
Dept of Geoscience & Engineering, Faculty of Civil Engineering & Geosciences (A,B,M,D) ●☒ (2019)
P. O. Box 5048
Stevinweg 1
2600-GA Delft
p. +31 15 27 86019
m.j.m.ammerlaan@tudelft.nl
https://www.tudelft.nl/en/ceg/about-faculty/departments/geoscience-engineering/
f: https://www.facebook.com/AppliedEarthSciencesTUDelft/
Programs: Applied Geophysics, Environmental Engineering, Geo-Engineering, Geo-Energy Engineering, Geo-Resource Engineering

Universiteit Utrecht
Faculty of Geosciences (B,M) (2016)
Budapestlaan 4b
Utrecht 3584 CD
p. +31 30 253 7890
mscinfo.geo@uu.nl
http://www.geo.uu.nl/

University of Amsterdam
Dept of Geology (2015)
Science Park 904
1098 XH Amsterdam
p. +31 (0)20525 8626
info-science@uva.nl

University of Twente
Faculty of Geo-Information Sciences and Earth Observation (ITC) (M,D) ●⌂ (2019)
P.O. Box 217
Enschede 7500 AE
p. +31 (0)53 487 44 44
info-itc@utwente.nl
https://www.itc.nl
f: https://www.facebook.com/ITC.UTwente
Programs: Master's Geo-information Science and Earth Observation with specialisations in Applied Remote Sensing; Natural Hazards and Disaster Risk Reduction; Geoinformatics; Land Administration; Natural Resources Management; Urban Planning and Management; Water Resources and Environmental Management, Master's Spatial Engineering
Certificates: See: https://www.itc.nl/studyfinder

University of Utrecht
Dept of Earth Sciences ☒ (2017)
Heidelberglaan 2
P.O. Box 80115
Utrecht
C.L.M.Marcelis@uu.nl

Vrije Universiteit Amsterdam
Faculty of Beta Science, Dep. of Earth Sciences (B,M,D) ●☒ (2019)
De Boelelaan 1085-1087
1081 HV Amsterdam 1081 HV
p. +31 (0)20 59 87310
f.bosse@vu.nl
https://science.vu.nl/en/research/earth-sciences/index.aspx
Enrollment (2015): B: 248 (56) M: 67 (45)
Professor:
 Gareth R. Davies, (D), GiCcGf
 Han J. Dolman, (D), CbAmHg
 Sander Houweling, (D), AscAp
 Klaudia Kuiper, (D), Cc
 Ronald R.T. van Balen, (D), GgmGt
 Guido R. van der Werf, (D), Zr
 Wim van Westrenen, (D), CpGz
 Jan Wijbrans, (D), CcGtv
Associate Professor:
 Fraukje M. Brouwer, (D), Utrecht, 2000, GptGc
 Gerald M. Ganssen, (D), ObPe
 Kees Kasse, (D), GmsSd
 Frank J.C. Peeters, (D), PmOg
 Didier Roche, (D), AsZn
 Wouter Schellart, (D), Gt
 Jorien Vonk, (D), Hg
 Pieter Vroon, (D), GvCcl
Assistant Professor:
 Kay J. Beets, Cg
 Janne Koornneef, (D), GvCa
 Kim Naudts, (D), HgPe
 Kei Ogata, (D), Parma (Italy), 2010, GrcGg
 Maarten A. Prins, (D), GsOg
 Monica Sanchez Roman, (D), GdsGg
 Marlies ter Voorde, (D), YgGt
 Jeroen van der Lubbe, (D), Gs
 Nathalie Van der Putten, (D), PecZy
 Ype van der Velde, (D), Hw
 Sander Veraverbeke, (D), Zr
 Martijn C. Westhoff, (D), Hq
Research Associate:
 Antoon Meesters, (D), AmGq
Dr:
 Bernd Andeweg, (D), 2002, GtcGg
 Ron Kaandorp, (D), Gg
 Els Ufkes, (D), PmZgGu

Netherlands Antilles

Royal NIOZ
Marine Geology (2015)

Landsdiep 4
Den Burg 1797 SZ
p. +31(0)222 369 300
Jens.Greinert@nioz.nl
http://www.nioz.nl

Physical Oceanography (2015)
Landsdiep 4
Den Burg 1797 SZ
secretary@nioz.nl
http://www.nioz.nl

New Zealand

Institute of Geological and Nuclear Sciences Ltd (New Zealand)
Institute of Geological and Nuclear Sciences Ltd (New Zealand) (2015)
1 Fairway Drive, Avalon 5010. PO Box 30-368, 5040.
Lower Hutt
p. +64-4-570-1444
G.Alderwick@gns.cri.nz
http://www.gns.cri.nz/
f: https://www.facebook.com/gnsscience

Massey University
Soil and Earth Sciences Group (2015)
Private Bag
Palmerston North
p. 0800 MASSEY (0800 627 739)
contact@massey.ac.nz
http://www.massey.ac.nz/massey/learning/programme-course-paper/programme.cfm?major_code=2044&prog_id=92431

University of Auckland
School of Environment (B,M,D) ☒ (2017)
Private Bag 92019
Auckland Mail Center
Auckland 1142
p. 64 9 3737599
p.kench@auckland.ac.nz
www.env.auckland.ac.nz
Programs: Geography; Earth Sciences; Environmental Science; Environmental Management; Geophysics; Geology

University of Canterbury
Dept of Geological Sciences (B,M,D) ●☒ (2018)
Private Bag 4800
Christchurch 8140
p. (643) 3694384
geology@canterbury.ac.nz
http://www.geol.canterbury.ac.nz/title.html
Enrollment (2014): B: 28 (0) M: 0 (35) D: 0 (20)
Professor Hazard & Disaster Management:
Timothy R H Davies, (D)
Professor:
James W. Cole, (D), Wellington, 1966, Gv
Andy Nicol, (D), Canterbury, 1991, GctGo
Jarg R. Pettinga, (D), Auckland, 1981, Gtc
Senior Lecturer Hazard & Disaster Management:
Thomas Wilson, (D), Canterbury, Gv
Senior Lecturer:
Kari N. Bassett, (D), Minnesota, 1995, GstGd
Catherine Reid, (D), Pg
Associate Professor:
Ben Kennedy, (D), Gv
Lecturer:
Darren Gravley, (D), Gv
Samuel J. Hampton, (D), Canterbury, 2010, GvZe
Alex Nichols, (D)

University of Otago
Dept of Geology (B,M,D) O☒ (2017)
P. O. Box 56
360 Leith Walk
Dunedin 9054
p. +64 3 479-7519
geology@otago.ac.nz
http://www.otago.ac.nz/geology/
Head of Department:
James D. L. White, (D), California (Santa Barbara), 1989, Gvs
Professor:
Dave Craw, (D), Otago, 1981, Eg
R. Ewan Fordyce, (D), Canterbury (NZ), 1981, Pv
David J. Prior, (D), Leeds, 1988, Gct
Senior Lecturer:
Candace E. Martin, (D), Yale, 1990, ClcGz
Virginia G. Toy, (D), Otago, 2008, Gct
Dr:
Christopher M. Moy, (D), Gus
Associate Professor:
Andrew R. Gorman, (D), British Columbia, 2001, YrgYe
Daphne E. Lee, (D), Otago, 1980, Pib
Senior Lecturer:
J. Michael Palin, (D), Yale, 1992, CcGip
Lecturer:
Christina R. Riesselman, (D), Stanford, 2011, Ou
James M. Scott, (D), Otago, 2008, Gp
Steven AF Smith, (D)
Emeritus:
Douglas S. Coombs, (D), Gi
Alan F. Cooper, (D), Otago, 1970, GpiGt
Rick H. Sibson, (D), Imperial Coll (UK), 1977, Gc

University of Waikato
Dept of Earth Sciences (2015)
Private Bag 3105
Hamilton 3240
p. +64 7 838 4625
science@waikato.ac.nz
http://sci.waikato.ac.nz/study/subjects/earth-sciences

Victoria University of Wellington
School of Geography, Environment and Earth Sciences (A,B,M,D) O☒ (2018)
P. O. Box 600
Wellington 6040
p. +6444635337
geo-enquiries@vuw.ac.nz
http://www.vuw.ac.nz/sgees
Programs: Geology, Geophysics, Meteorology, Environmental Science, GIS
Enrollment (2012): B: 153 (107) M: 51 (22)
Professor:
Tim Little, (D), Stanford, 1988, Gct

Dept of Geology (2015)
P. O. Box 600
Wellington
p. +64 4 463 5337
geo-enquiries@vuw.ac.nz
http://www.victoria.ac.nz/sgees/study/postgraduate-study/geology/

Nicaragua

University of Managua
Centro de Investigaciones Geocientificas (2015)
Colonia Miguel Bonilla No. 165, P.O. Box A-131 Cc. Managua, Managua

Niger

Institut de Recherche pour le Developpement au Niger
Institut de Recherche pour le Developpement au Niger (B) (2015)
Avenue de Maradi
Niamey BP 11 416
p. (227) 20 75 38 27

irdniger@ird.fr
http://www.ird.ne/

Universite Abdou Moumouni de Niamey
Dept de Geologie (2015)
B.P. 10662
Niamey
  p. (227) 20 31 50 72
  http://www.secheresse.info/article.php3?id_article=1664
Doyen:
  Abdoulaye M. Alassane

Nigeria

Adekunle Ajasin University
Dept of Earth Sciences (B) (2015)
Akungba Akoko, Ikare Akoko 23401
  p. +234 (34) 246444
  iosazuwa@yahoo.com
  http://www.ajasin.edu.ng/academics/index.php

African University of Science & Technology
Petroleum Engineering (B,M) (2015)
PMB 681
Abuja F.C.T.
Km 10 Airport Road
Garki
  p. +234 9 7800680
  http://aust.edu.ng/

Ahmadu Bello University
Dept of Geology (2015)
Sokoto Road
Samaru-Zaria
Zaria 2222, Kaduna
  p. +234 (69) 50 691
  ybijimi@yahoo.com
  http://www.abu.edu.ng/

American University of Nigeria
School of Arts and Sciences (B) (2015)
Lamido Zubairu Way
PMB 2250, Yola 23455
  p. +234 (805) 5485 702
  dsas@aun.edu.ng
  http://www.abti-american.edu.ng

Caleb University
Dept of Surveying and Geoinformatics (B) (2015)
PMB 21238
Ikeja GPO, Lagos
  p. +234-1-764-7312
  info@calebuniversity.edu.ng
  http://calebuniversity.edu.ng/courses.php?collUid=4

Dept of Environmental Protection and Management (B) (2015)
PMB 21238
Ikeja GPO, Lagos
  p. +234-1-764-7312
  info@calebuniversity.edu.ng
  http://calebuniversity.edu.ng/courses.php?coll_id=4

Crawford University
Dept of Mathematics and Physical Sciences (B) O☒ (2019)
PMB 2001
Faith City, Ogun 112102
  p. +2348023003436
  info@crawforduniversity.edu.ng
  http://www.crawforduniversity.edu.ng
  f: https://www.facebook.com/CrawfordCRU/
  t: @unicrawford
  Programs: Geology and Mineral Sciences

Delta State University
Dept of Geology (B) (2015)
PMB 1
Abraka
  p. +234 (54) 66009
  delsu03@yahoo.com
  htttp://www.delsunigeria.net/
Director:
  S. H. O. Egboh, (D)
Chair:
  E. Adaikpoh, (D)

Enugu State University of Science and Tech
Dept of Geology and Mining (B) (2015)
MB 01660, Enugu
  p. +234 (42) 451319
  dean.fans@esut.edu.ng
  http://www.esut.edu.ng/

Dept of Geography and Meteorology (B) (2015)
MB 01660, Enugu
  p. +234 (42) 451319
  dean.fans@esut.edu.ng
  http://www.esut.edu.ng/

Federal Polytechnic, Bida
Dept of Survey and Geoinformatics (B) (2015)
  http://www.fedpolybidaportal.com/

Dept of Quantitative Surveying (B) (2015)
  http://www.fedpolybidaportal.com/

Federal Polytechnic, Offa
School of Environmental Studies (B) (2015)
  http://www.fedpoffa.edu.ng/

Federal University of Technology
Dept of Applied Geophysics (B,M,D) (2015)
PMB 704
Akure, Ondo
  p. +234 803-595-9029
  joamigun@futa.edu.ng
  http://agp.futa.edu.ng/page.php?pageid=179

Federal University of Technology, Akure
Earth & Mineral Sciences (B) (2015)
PMB 704
Akure
  p. +234 (34) 243744
  sems@futa.edu.ng
  http://www.futa.edu.ng/sems/

Applied Geology (B) ● (2019)
PMB 704
Akure, Ondo
  p. +234 (0) 803 667 2708
  agy@futa.edu.ng
  http://www.futa.edu.ng/agy/
  Enrollment (2015): B: 85 (70)
Professor:
  Yinusa A. Asiwaju-Bello, (D), Leeds, GgHgNg
  Idowu O. Odeyemi, (D), Ibadan, 1976, GgxZr
Doctor:
  O A.G Jegede, (D), Akure, GgNgg
  C T. Okonkwo, (D), Keele, GgxGc
  S A. Opeloye, (D), Bauchi, GgoGs
Mrs:
  Oluwaseyi A. Bamisaye, (D), Akure, 2007, GcZri
Mr:
  Mohammed O. Adepoju, (M), Akure, 2004, GgEg
  Adeshina L. Adisa, (M), Akure, 2012, GgCe
  Isaac O. Ajigo, (M), Jos, 2013, GgNm
  Timothy I. Asowata, (M), Ibadan, 2010, Gge
  Sunday O. Daramola, (M), Akure, 2012, HwGe
  Oladimeji Rilwan Egbeyemi, (M), Ibadan, 2004, GgCg

Emmanuel Eghietseme Igonor, (M), Ibadan, 2007, GgCeGi
Osazuwa A. Ogbahon, (M), Akure, 2011, Ggo
Joshua O. Owoseni, (M), Ibadan, GgHw
Doctor:
  P S. Ola, (D), Akure, GosGg
  Solomon O. Olabode, (D), Akure, GgoGs

Federal University of Technology, Yola
Dept of Surveying & Geoinformatics (B) (2015)
PMB 2076
Yola
    p. +234 (803) 6065646
    survey@futy.edu.ng
    http://www.futy.edu.ng/academic/dsurvey.htm

Dept of Soil Science (B) (2015)
PMB 2076
Yola
    p. +234 805 3372 784
    soilscience@futy.edu.ng
    http://www.futy.edu.ng/academic/dsoil.htm

Dept of Geography (B) (2015)
PMB 2076
Yola
    p. +234 (803) 6065646
    geography@futy.edu.ng
    http://www.futy.edu.ng/academic/dgeography.htm

Dept of Geology (B) (2015)
PMB 2076
Yola
    p. +234 (803) 6065646
    geology@futy.edu.ng
    http://www.futy.edu.ng/academic/dgeology.htm

Gombe State University
Dept of Geography (B) (2015)
    contact@gsu.edu.ng
    http://www.gomsu.org/

Dept of Geology (B) (2015)
    contact@gsu.edu.ng
    http://www.gomsu.org/

Ladoke Akintola University of Technology
Dept of Earth Sciences (B) (2016)
PMB 4000
Ogbomoso, Oyo 4000
    p. +234 8033539911
    contact@lautech.edu.ng
    http://www.lautech.edu.ng/Academics/undergraduates/FPAS/index.html
    Enrollment (2015): B: 13 (13)
Dr:
    Oyelowo Gabriel Bayowa, (D), Obafemi Awolowo, 2013, GggGg
    Abosede Olufunmi Adewoye, (D), Ladoke Akintola, 2014, GeeGe
    Moruffdeen Adedapo Adabanija, (D), Ibadan, 2009, GggGg
Mr:
    Olukayode Adegoke Afolabi, (M), Ibadan, 2000, GxxGx
    Ismaila Abiodun Akinlabi, (M), Ibadan, 2011, GggGg
    Mustapha Taiwo Jimoh, (M), Ibadan, 2006, GxxGx
    Lanre Lateef Kolawole, (M), Newcastle, 1985, GeeGe
    Olusola Christophe Oduneye, (M), Ibadan, 2009, GooGo
    Gbenga Olakunle Ogungbesan, (M), Ibadan, 2006, GooGo

Lagos State Polytechnic
School of Environmental Studies (Quantity Surveying) (B) (2015)
    info@mylaspotech.edu.ng
    http://mylaspotech.net/

National Open University of Nigeria
School of Science & Technology (B) (2015)
PO Box 1866
Bauchi
    p. +234 1-8188849, +234 1-4820720
    sst@noun.edu.ng
    http://www.nou.edu.ng/noun/indexs.jsp

Nnamdi Azikiwe University
Dept of Geology/Meteorology and Env Management (B) (2015)
PMB 5025
Awka, Nnewi 5025
    p. +234 (46) 550018
    emmaojukwu@unizik.edu.ng
    http://www.unizikeduportal.org/

Dept of Geological Sciences (B,M,D) ●⊠ (2017)
Awka Campus
PMB 5025
Awka, Anambra 5025
    p. +234 (46) 550018
    ne.ajaegwu@unizik.edu.ng
    http:wwww.unizikeduportal.org/
    Programs: Geology
Dr.:
    Norbert E. Ajaegwu, (D), Nnamdi Azikiwe (Nigeria), 2014, GorGg

Obafemi Awolowo University
Dept of Geology (B,M,D) ● (2015)
Ile-Ife, Osun State 220002
    p. 08128181062
    geoife@oauife.edu.ng
    http://gly.oauife.edu.ng/
    Enrollment (2014): B: 213 (0) M: 24 (0) D: 12 (0)
Head:
    A. A. Adepelumi, (D), YgsNe
Professor:
    S. A. Adekola, (D), GdoCg
    O. Afolabi, Yg
    J. O. Ajayi, No
    T. R. Ajayi, CgEgGe
    O. A. Alao, (D), Yg
    U. K. Benjamin, GoCg
    S. L. Fadiya, (D), God
    A. O. Ige, Cg
    A. O. Ige, Cg
    E. James-Aworeni, Hg
    J. I. Nwachukwu, GoCg
    O. O. Ocan, GtzGp
    S. B. Ojo, Yg
    V. O. Olarewaju, GozCg
    A. O. Olorunfemi, Go
    M. O. Olorunfemi, Yg
    A. A. Oyawale, GeoCg
    M. A. Rahaman, GpcGi
    B. M. Salami, Hg
    I. A. Tubosun
Associate Professor:
    A Adetunji, (D), EgGi
Assistant Professor:
    Dele E. Falebita, (D), Obafemi Awolowo, YeRnZg
Research Associate:
    M. A. Olayiwola, GugPi
    O. M. Oyebanjo

Dept of Soil Science (B) (2015)
Ile-Ife
    p. +234 (36) 230 290
    midowu@oauife.edu.ng
    http://www.oauife.edu.ng/faculties/agric/soil_sci.html

Department of Geography (B,M,D) O-☖ (2018)
Ile-Ife
    p. +234 (70) 37772355
    sinaayanlade@yahoo.co.uk
    http://www.oauife.edu.ng/faculties/soc_sciences/departments/geo/home.htm
    Programs: Geography and GIS
    Certificates: GIS

Assistant Professor:
  Ayansina Ayanlade, (D), AtZir

Rivers State University of Science & Technology
Inst of Geoscience & Space Technology (IGST) (2015)
P.M.B. 5080
Nkpolu - Oroworukwo
Port Harcourt, Rivers State
info@ust.edu.ng
http://www.ust.edu.ng/

University of Ado - Ekiti
Dept of Agricultural Sciences (B) (2015)
PMB 5363
Ado-Ekiti
  p. +234 (30) 250026
  http://www.unadportal.com/

Geology Dept (B) (2015)
PMB 5363
Ado-Ekiti
  p. +234 (30) 250026
  info@eksu.edu.ng
  http://www.unadportal.com/

University of Benin
Dept of Petroleum Engineering (B) (2015)
PMB 115
Benin City, Edo State
  p. 234 0802 345 5681
  deanengineering@uniben.edu
  http://www.uniben.edu/UniversityOfBenin-PetroleumEngineering.html

Dept of Geology (B,M,D) ● (2015)
P. M. B. 1155
Benin City, Edo State
  p. 234 0802 345 5681
  registrar@uniben.edu
  http://www.uniben.edu/physicalscielcdfaculty.html
Professor:
  Godwin Osarenkhoe Asuen, (D), EcGsg
  Christopher Nnaemeka Akujieze, (D), GgHwGe
  Williams Ogbevire Emofurieta, (D), GzCeGe
  Tony Uzozili Onyeobi, (D), NrgGq
Associate Professor:
  Isaac Okpeseyi Imasuen, (D), W Ontario, Gee
Senior Lecturer:
  Ben Obelenwa Ezenabor, (M), HgZrGe
  Franklin A. Lucas, (D), PmlGr
Lecturer II:
  Aitalokhai Joel Edegbai, (M), GosPi
  Efetobore Gladys Maju-Oyovwikowhe, (M), GszGo
  Sikiru Adeoye Salami, (D), YgeGg
Lecturer I:
  Ovie Odokuma-Alonge, (M), CgGip
Graduate Assistant:
  Tomilola Andre-Obayanju, (B), NgHgGe
Assistant Lecturer:
  Aiyevbekpen Helen Akenzua-Adamcyzk, (M), GeHgGg
  Nosa Samuel Igbinigie, (M), GorGg
  Alexander Ogbamikhumi, (M), YsgGg
  Sunday Erapkpower Okunuwadje, (M), GsoGd
  Edoseghe Edwin Osagiede, (M), GctGx

University of Calabar
Dept of Geology (B) (2015)
PMB 1115 Eta Agbo Road,
Calabar Municipal, Cross River State
  p. (+234) 8173740083
  general@unical.edu.ng
  http://www.unical.edu.ng/pages/programs_courses/sciences.php?nav=departments
Head:
  Nse U. Essien

University of Ibadan
Dept of Geology (B,M,D) (2015)
Ibadan, Oyo State
  p. 02-8101100-8101104
  uigeology@yahoo.com
  http://sci.ui.edu.ng/geowelcome
Professor:
  I. M. Akaegbobi, Cg
  A. A. Elueze, Eg
  Ougbenga. Akindeji. Okunlola, (D), Ibadan, 2001, EgCgGe
  A. I. Olayinka, Yg
  Moshood N. Tijani, (D), Muenster (Germany), 1997, HwCgGe
Lecturer:
  A. E. Abimbola, Cg
  O. C. Adeigbe, (D), Ibadan, 2009, GosGe
  G. O. Adeyemi, NgHg
  O. A. Boboye, Go
  A. T. Bolarinwa, EgGz
  M. E. Nton, Go
  M. A. Oladunjoye, Yg
  O. O. Osinowo, Yg
  I. A. Oyediran, Ng

Dept of Geology (B,M,D) (2016)
Ibadan, Oyo State 200284
  p. +234-8033819066
  ehinola01@yahoo.com
  http://www.ui.edu.ng/?q=departmentofgeology
Professor:
  Olugbenga Ajayi Ehinola, (D), Ibadan, 2002, GoeGs
Head:
  Olugbenga A. Ehinola, (D), Ibadan, 2002, GoeCo
Associate Professor:
  A. S. Olatunji, (D), Ibadan, 2006, CgGeCt

Dept of Petroleum Engineering (B,M,D) (2017)
Oyo Road
Ibadan
  p. +234-2-810-3168
  isehunwa@yahoo.com
  http://www.ui.edu.ng/?q=petengine
  Enrollment (2016): B: 223 (43) M: 17 (17) D: 10 (2)

University of Ilorin
Dept of Geology and Mineral Sciences (B) (2015)
PMB 1515
  p. +234 (31) 221691
  deanfacsci@unilorin.edu.ng
  http://www.unilorin.edu.ng/unilorin/index.php/sciences-dept/geology-mineral-sciences
Head:
  J.I. D. Adekeye, (D), GoCg
Professor:
  S. O. Akande, (D), EgCg
  O. Ogunsanwo, Ng
Lecturer:
  A. Abdurrahman, (M), CeEg
  A. D. Adedcyin, (M), Eg
  O. A. Adekeye, (D), Pm
  S.M. A. Adelana, (M), HgYgGe
  D. A. Alao, NgGz
  O. O. Ige, (M), Ng
  Kehinde A. Olawuyi, (D), Federal Univ Tech (Nigeria), 2015, YeGge
  O. A. Omotoso, (M), Ge
  W. O. Raji, (M), Ye

University of Jos
Dept of Geosciences and Mining (B) (2015)
PMB 2084
Jos 930001
  p. +234 (73) 559 52
  vc@unijos.edu.ng
  http://www.unijos.edu.ng/

University of Lagos
Dept of Metalurgical and Materials Engineering (B) (2015)
Akoka
Yaba, Lagos
p. +234 (1) 493 2600 3
informationunit@unilag.edu.ng
http://www.unilag.edu.ng/index.php?page=home

Dept of Civil and Environmental Engineering (B) (2015)
Akoka
Yaba, Lagos
p. +234 (1) 493 2660 3
informationunit@unilag.edu.ng
http://www.unilag.edu.ng/index.php?page=about_departmentdetail&sno=11

Dept of Petroleum and Gas Engineering (B) (2015)
Akoka
Yaba, Lagos
p. +234 (1) 493 2660 3
informationunit@unilag.edu.ng
http://www.unilag.edu.ng/index.php?page=home

Dept Surveying and Geoinformatics (B) (2015)
Akoka
Yaba, Lagos
p. +234 (1) 493 2660 3
informationunit@unilag.edu.ng
http://www.unilag.edu.ng/index.php?page=about_departmentdetail&sno=16

University of Maiduguri
Dept of Geology (2015)
PMB 1069 FEA
Maiduguri
info@unimaid.edu.ng
http://www.unimaid.edu.ng/root/faculty_of_sciences/dept_geology.html

University of Maiduguri, Borno State
Dept of Geology (B,M,D) (2015)
Bama Road
PMB 1069
Maiduguri
p. +234 (76) 231730
info@unimaid.edu.ng
http://www.unimaid.edu.ng/home.php
Professor:
  I. B. Goni
Assistant Professor:
  Saidu Baba
Lecturer:
  Sani Adamu
  Kyari M. Ajar
  Musa Malana Aji
  Fati Bukar
  Shettima Bukar
  Jalo M. El-Nafaty
  Millitus V. Joseph
  Samaila C. Kali
  Yakubu B. Mohd
  Mohammed Poukar
  Aishatu Sani
  Manaja Mijinyawa Uba
  Sidi M. Waru
  Asabe Kuku Yahaya
  Soloman N. Yusuf
  A. Adam Zarma

University of Nigeria
Dept of Geology (B) (2015)
Nsukka
micah.osilike@unn.edu.ng
http://www.unn.edu.ng/physicalsciences/content/view/700/601/

University of Nigeria Nsukka
Dept of Geology (B) (2015)
anthony.okonta@unn.edu.ng
http://unn.edu.ng/department/geology
Lecturer:
  Alloysuis Okwudili Anyiam
  Luke I. Manah
  Smart Chicka Obiora

Dept of Geoinformatics and Surveying (B) (2015)
batho.okolo@unn.edu.ng
http://www.unn.edu.ng/environmentalstudies/content/view/27/44/

University of Port Harcourt
Dept of Geography & Environmental Management (B) (2015)
East/ West Road
PMB 5323
Choba, Port Harcourt 500001
p. +234 (84) 230890-99
edulms@chimpgroup.com
http://uniport.edu.ng/

Dept of Geology (B) (2015)
East/ West Road
P.M.B. 5323
Choba, Port Harcourt 500001
p. +234 (0)84 817 941
uniport@uniport.edu.ng
http://www.uniport.edu.ng/faculties/science.html

Dept of Petroleum and Gas Engineering (B) (2015)
East/ West Road
PMB 5323
Choba, Port Harcourt 500001
p. +234 (84) 230890-99
dean@eng.uniport.edu.ng
http://uniport.edu.ng/

University of Uyo
Faculty of Science Education (B,M,D) (2015)
PMB 1017
Uyo
p. +234 (85) 200303
vc@uniuyo.edu.ng
http://www.uniuyo.edu.ng/index.htm

Faculty of Social Science (B,M,D) (2015)
PMB 1017
Uyo
p. +234 (85) 200303
vc@uniuyo.edu.ng
http://www.uniuyo.edu.ng/index.htm

Faculty of Engineering (B,M,D) (2016)
PMB 1017
Uyo
p. +234 (85) 200303
vc@uniuyo.edu.ng
http://www.uniuyo.edu.ng/index.htm

Faculty of Agriculture (B,M,D) (2015)
PMB 1017
Uyo, Akwa Ibom State 520001
p. +234 (85) 200303
vc@uniuyo.edu.ng
http://www.uniuyo.edu.ng/index.htm

Faculty of Environmental Studies (B,M,D) (2015)
PMB 1017
Uyo
p. +234 (85) 200303
vc@uniuyo.edu.ng
http://www.uniuyo.edu.ng/index.htm

Western Delta University
Geology and Petroleum Studies (B) (2015)
PMB 10
Oghara
> p. +234 (70) 35400531
> info@wdu.edu.ngg
> http://www.wduniversity.org/

Wukari Jubilee University
Dept of Geography and Environmental Conservation (B) (2015)
PMB 1019
Wukari, Taraba State
> p. +234 (080) 80329138
> wukari_jubilee@yahoo.com
> http://wukarijubileeuniversity.org/

Norway

Norges Geologiske Undersokelse
Geological Survey of Norway ☒ (2018)
Postboks 6315 Torgarden
7491 Trondheim
> ngu@ngu.no
> http://www.ngu.no/no/

Norwegian University of Life Sciences (NMBU)
Section of Soil and Water Sciences, Favulty of Environmental Sciences and NAtural Resource Management (B,M,D)
○☒ (2017)
P.O. Box 5003
1432 AS
Ås 1432
> p. +47 67 23 00 00
> post-mina@nmbu.no
> https://www.nmbu.no/en/faculty/mina
> Programs: Environmental Sciences (B, M); Ecology and Natural Resource Management; Renewable Energy; Sustainable Tourism; Forestry

Professor:
> Trond Børresen, Sp
> Petter Jenssen, Zu
> Tore Krogstad, Sc
> Jan Mulder, Cb
> Gunnhild Riise, Hs
> Trine Sogn, Sb
> Rohrlack Thomas, Hs

Norwegian University of Science and Technology
Dept of Geology and Mineral Resources Engineering
(B,M,D) (2015)
Sem Sælands vei 1
7491 Trondheim
Trondheim
> p. (+47) 73 59 48 00
> iigb-info@ivt.ntnu.no
> http://www.ntnu.no/igb
> f: https://www.facebook.com/geologibergNTNU

Faculty of Engineering Science and Technology (2015)
S. P. Andersens vei 15A
N7034 Trondheim
> p. 73594779
> ingvald.strommen@ntnu.no

University of Bergen
Dept of Earth Science (B,M,D) ○☒ (2017)
Allegaten 41
N-5007 Bergen
> p. +47 55583600
> post@geo.uib.no
> http://www.geo.uib.no/
> f: https://www.facebook.com/earthscienceuib
> t: @uibgeo

University of Oslo
Deptartment of Geosciences (B,M,D) ☒ (2018)
Post Box 1047, Blindern
N-0316 Oslo
> p. +47 22856656
> geosciences@geo.uio.no
> http://www.mn.uio.no/geo/english/
> f: https://www.facebook.com/uiogeo/
> Programs: Two BSc programmes in Geosciences (in Norwegian), Masterprogramme in Geosciences (master's two years)

Mineralogical-Geological Museum (2015)
Sars gate 1
P.O. Box 1172
Blindern, Oslo 0318
> p. +47 228 55050
> postmottak@nhm.uio.no
> http://www.nhm.uio.no/english/

University of Tromso
Geology Dept (2015)
N-9037 Tromso
> matthias.forwick@uit.no
> http://www.ibg.uit.no/geologi/geo_eng_end.html

Oman

Sultan Qaboos University
Dept of Earth Science (B,M,D) ● (2015)
123 Al-Khod
P.O. Box 36
Muscat
> p. +968 2414 6832
> khirbash@squ.edu.om
> http://www.squ.edu.om/earth-sci/tabid/11718/language/en-US
> Enrollment (2013): B: 85 (60) M: 5 (25) D: 2 (1)

Head:
> Salah Al-Khirbash, (D)

Pakistan

COMSATS
Institute of Information Technology (2015)
Park Road, Chak Shahzad
Islamabad
> p. +92-51-9247000-2
> info@ciit.edu.pk
> http://www.comsats.edu.pk/

Geological Survey of Pakistan
> p. (+958) 9211032
> qta@gsp.gov.pk
> http://www.gsp.gov.pk/

Institute of Space Technology
Institute of Space Technology (B,M,D) (2015)
P.O. Box 2750
Islamabad 44000
> p. 92.51.9075100
> info@ist.edu.pk
> http://www.ist.edu.pk/

National University of Sciences and Technology
Institute of Geographic Information System (2015)
RIMMS building, NUST H-12 Campus
Islamabad
> p. +92-51-90854400
> info@igis.nust.edu.pk
> http://igis.nust.edu.pk/

Sind University Pakistan
Dept of Geology (2015)
University of Sindh, Allama I.I. Kazi Campus

Jamshoro
Sindh 76080
    p. 92-22-9213172-9213181-90 Ext:
    dean.science@usindh.edu.pk

University of Engineering and Technology
Geological Engineering (B,M,D) ●☒ (2018)
Lahore, Punjab 54890
    p. 0092-42-99029487
    mzubairab1977@gmail.com
    http://www.uet.edu.pk/faculties/facultiesinfo/geological/index.html?RID=introduction
    Programs: B.Sc. Geological Engineering, M.Sc. Geological Engineering, M.Sc. Geological Sciences, Ph.D. Geological Engineering, Ph.D. Geological Sciences
    Enrollment (2018): B: 50 (50)
Chair:
    Muhammad Zubair Abu Bakar, (D), 2012, NrmNg
Associate Professor:
    Muhammad Farooq Ahmed, (D), Missouri (Rolla), 2013, Nxt

University of Karachi
Dept of Geology (2015)
KARACHI-75270
    p. 99261300-06 Ext: 2295
    vc@uok.edu.pk
    http://uok.edu.pk/faculties/geology/

University of Sargodha
Dept of Earth Science (B,M,D) ● (2015)
Sargodha
    earthscience@uos.edu.pk
    http://www.uos.edu.pk

College of Agriculture (2015)
Sargodha
    agri@uos.edu.pk
    http://www.uos.edu.pk

University of the Punjab
Dept of Geography (2015)
New Campus
Lahore 54590
    chairman@geog.pu.edu.pk
    http://pu.edu.pk/home/department/46/Department-of-Geography

Panama

Universidad de Panama
Facultad de Ciencias Naturales, Exactas y Technologia (2015)
Urbanizacion El Cangrejo, Estafeta Universitaria
    ciencias@ancon.up.ac.pa

Paraguay

Universidad Nacional se Asuncion
Dept de Geologia (2015)
University Campus Km. 11, Asuncion
    facen@facen.una.py

Peru

Geological Survey of Peru
Instituto Geologico Minero y Metalurgico - INGEMMET (2015)
Av. Canada 1470 San Borja
Lima
    p. 051-1-6189800
    webmaster@ingemmet.gob.pe
    http://www.ingemmet.gob.pe/

San Agustin's Natl University
Escuela Academico Profesional de Ingenieria Geologica y Geofisica (2015)

Av. Indraendencia #1015
Arequipa
    p. (054) 244498
    geologia@unsa.edu.pe
    http://www.unsa.edu.pe

Univ Nacional de Ingeneiria
Escuela Academico Profesional de Ingenieria Geologica (2015)
Av. Tupac Amaru S/N
Rimac, Lima
    esanchez@uingemmet.gob.pe
    http://www.uni.edu.pe/sitio/academico/facultades/geologica/

Univ Nacional Mayor San Marcos
Escuela Academico Profesional de Ingenieria Geologica (2016)
Av. Venezuela S/N
Lima
    j_jacay@yahoo.com
    http://www.unmsm.edu.pe

Philippines

University of the Philippines
National Institute of Geological Sciences (B,M,D) ● (2015)
C.P. Garcia corner Velasquez Street
Diliman, Quezon City 1101
    p. (+632) 9296046
    inquire@nigs.org
    http://nigs.org

Poland

AGH University of Science and Technology
Faculty of Geology, Geophysics, and Environmental Protection (2015)
A. Mickiewicza 30 Ave.
30-059 Krakow
    p. +48 12 633 29 36
    aglown@geol.agh.edu.pl
    http://www.wggios.agh.edu.pl/

Faculty of Mining and Geoengineering (2015)
Al. Mickiewicza 30
PL-30059 Krakow
    p. +48 12 617 21 15
    gorn@agh.edu.pl

Jagiellonian University
Institute of Geological Sciences (2015)
Oleandry 2a Str.
30-063 Krakow
    sekretariat.ing@uj.edu.pl
    http://www.ing.uj.edu.pl

Ministry of Environmental Protection, Natural Resources and Forestry
Dept of Geology (2016)
st. Wawelska 52/54
Warsaw 00-922
    p. +48 22 3692449
    Departament.Geologii.i.Koncesji.Geologicznych@mos.gov.pl
    http://www.mos.gov.pl/dg/dga1.htm

Panstwowy Instytut Geologiczny
Polish Geological Institute (2015)
Ul. Rakowiecka 4
PL-00-975 Warszawa
    sekretariat@pgi.gov.pl
    http://www.pgi.gov.pl/

Polish Academy of Sciences
Institute of Geological Sciences (2015)

Al Zwirki i Wigury 93
02-089 Warsaw
    p. (+48 22) 620 33 46, 656 60 63
    ingpanhard@pan.pl

Silesian University of Technology
Faculty of Mining and Geology (B,M,D) ● (2015)
Wydzial Gornictwa i Geologii
Politechnika Slaska
ul.Akademicka 2
Gliwice 44-100
    p. +48 32 2371283
    rg@polsl.pl
    http://www.polsl.pl/Wydzialy/RG/Strony/Witamy.aspx
    Enrollment (2014): B: 2556 (615) M: 327 (283) D: 98 (5)

Univ of Mining & Metallurgy
Dept of Mining and Metallurgical Engineering (2015)
Inst. of Drilling & Oil Exploration
Krakow
    jameshendrix@unr.edu

University of Wroclaw
Inst of Geological Sciences (B,M,D) (2017)
Pl. Maksa Borna 9
50-205 Wroclaw
    p. (+4871) 321-10-76
    sekretariat.ing@uwr.edu.pl
    http://www.ing.uni.wroc.pl/home-page
    Enrollment (2016): D: 14 (8)
Director:
    Krystyna Choma-Moryl, (D), 1983, Ng
Professor:
    Pawel Aleksandrowski, (D), Polish Acad of Sci, 1985, GctYe
    Marek Awdankiewicz, (D), 1997, Gvi
    Piotr Gunia, (D), 1983, Gaz
    Jacek Gurwin, (D), 1997, HwyZf
    Henryk Marszalek, (D), 1994, Hy
    Jacek Puziewicz, (D), 1985, Gix
    Jerzy Sobotka, (D), AGH Univ, 1992, YeuEo
    Andrzej Solecki, (D), 1987, Eg
    Stanislaw Stasko, (D), 1986, Hg
    Jacek Szczepanski, (D), 1999, Ggc
Adjunct Professor:
    Anna Gorecka-Nowak, (D), 1992, Pg
    Maciej Gorka, (D), 2007, AsCa
    Jakub Kierczak, (D), 2007, GeSc
    Antoni Muszer, (D), 1994, Eg
    Anna Pietranik, (D), Wroc, 2006, CgGie
    Marta Rauch, (D), 2002, Gc
    Robert Tarka, (D), 1995, Hg
    Jurand Wojewoda, (D), 1989, Gs

Portugal

Istituto Geologico e Mineiro (IGM)
Geological and Mining Institute of Portugal (2015)
Estrada da Portela
Zambujal
    atendimento@ineti.pt
    http://www.ineti.pt/

Oporto University
Dept of Geology (2015)
Faculdade de Ciencias do Porto
Praca de Gomes Teixeira
4099-002 PORTO
    dg.sec@fc.up.pt
    http://www.fc.up.pt/depts/geo/indexi.html

Universidade da Madeira
Dept de Biologia (B) (2015)
Campus Universitario da Penteada
9000-390 Funchal
    p. +251 291 705 380
    mgouveia@uma.pt
    http://www.uma.pt/Unidades/Biologia/

Universidade de Aveiro
Dept de Geociencias (B,M,D) ☒ (2018)
Campo Universitario Santiago
Aveiro 3810-193
    p. +351 234 370 357
    tavares.rocha@ua.pt
    www.ua.pt/geo
    Programs: Geology, Geological Engineering
Professor:
    Eduardo Ferreira da Silva, (D), CeGbCl

Universidade de Coimbra
Dept de Ciencias da Terra (A,B,M,D) ○☒ (2017)
Rua Silvio Lima, Univ. Coimbra, Pólo II Coimbra
Coimbra 3030-790 Coimbra
    p. +351-239 860 514
    cgeo@ci.uc.pt
    http://www.dct.uc.pt
    Programs: Geology
Professor:
    Rui Pena dos Reis, (D), GorGs
Associate Professor:
    Alcides Castilho Pereira, (D), Coimbra, 1992, GebZr

Universidade de Lisboa
Dept of Geology (A,B,M,D) ●☒ (2018)
Ed. C6, Campo Grande
Campo Grande
Lisboa 1749-016
    p. +35121 750 00 66 / + 351 21 75
    dgeologia@fc.ul.pt
    http://www.fc.ul.pt/en/dg?refer=1
    f: https://www.facebook.com/Departamento-de-Geologia-FCUL-192772970763173
    Programs: Geology
    Certificates: Applied Geology
                   Petroleum Geosciences
    Enrollment (2018): B: 314 (51) M: 87 (13) D: 30 (2)
Professor:
    Maria da Conceição Pombo Freitas, (D), Lisbon, 1996, GusGe
Professor:
    Maria da Conceiç Freitas, (D), GseGu
Associate Professor:
    Mário Cachão, (D), PgeGg
    Nuno Pimentel, (D), GrsGo
Assistant Professor:
    Maria do Rosário Carvalho, (D), HwyZn
    Telmo Bento dos Santos, (D), GptGi

Universidade de Trás-os-Montes e Alto Douro
Departamento de Geologia (B,M,D) (2015)
Apartado 1013
Vila Real 5001-801
    p. +351259350224
    geologia@utad.pt
    http://www.geologia.utad.pt
Professor:
    Maria E.P. Gomes, (D), UTAD, 1996, GiCgGz
Associate Professor:
    Ana M.P. Alencoão, (D), UTAD, 1998, HgGgm
    Artur A.A. Sá, (D), UTAD, 2005, PgiPs
Assistant Professor:
    Maria E.P.S. Abreu, (D), UTAD, 2012, GaaGa
    João C.C.V. Baptista, (D), UTAD, 1998, GmmGa
    Maria R.M. Costa, (D), UTAD, 2000, Hw
    Paulo J.C. Favas, (D), UTAD, 2008, GeCtGg
    José M.M. Lourenço, (D), UTAD, 2006, YeZin
    Alcino S. Oliveira, (D), uTAD, 2002, Hww
    Anabela R.R.C. Oliveira, (D), UTAD, 2011, Gss
    Fernando A.L. Pacheco, (D), UTAD, 2001, Hq
    Luís M.O. Sousa, (D), 2001, EnGgNg
    Rui J.S. Teixeira, (D), UTAD, 2008

Nuno M.O.C.M. Vaz, (D), UTAD, 2010, PmlYg

Universidade do Minho
Dept de Ciencias da Terra (B,M,D) (2015)
Campus de Gualtar
Braga 4710-057
  p. 351 253 604 300
  sec@dct.uminho.pt
  http://www.dct.uminho.pt/index/index.html

Universidade Nova de Lisboa
Dept de Ciencias da Terra (2015)
Qta. da Torre
2829-516 Caparica
  p. (351) 212 948 573
  dct.secretariado@fct.unl.pt
  http://www.dct.fct.unl.pt/

Qatar

University of Qatar
Dept of Geology (B) (2015)
PO Box 2713
Doha
  hamadsaad@qu.edu.qa
  http://www.angelfire.com/ms/GeoQU/
Professor:
  Sobhi J. Nasir, (D), Germany, 1986, Gx
Associate Professor:
  Hamad A. Al-Saad, (D), Egypt, 1996, PsGsu
  Abdulali A. Sadiq, (D), Southampton (UK), 1995, Zr
Assistant Professor:
  Latifa B. AL-Nouimy, (D), Egypt, 1994, Hgw
  Sief A. Alhajari, (D), North Carolina, 1992, Hgw

Reunion

Universite de la Reunion
Faculte de Sciences et Technologie - Sciences de la Terre (B) (2015)
15, avenue Rene Cassin
CS 92003
Saint-Denis, Cedex 9 97744
  p. (262) 0262 938697
  jean-lambert.join@univ-reunion.fr
  http://sciences.univ-reunion.fr/rubrique.php3?id_rubrique=72
Professor:
  Jean-Lambert Join, (D), Hw
Associate Professor:
  Fabrice R. Fontaine, (D), Yg

Laboratoire GeoSciences Reunion (B,M,D) (2015)
15 avenue Rene Cassin
BP 7151
Saint Denis, Messag Cedex 9 97744
  vfamin@univ-reunion.fr
  http://www.geosciencesreunion.fr/
Professor:
  Laurent Michon, (D), GtvGg
Professor:
  Jean-Lambert Join, HwGsHg
Associate Professor:
  Vincent Famin, (D), GcCgGg
  Fabrice R. Fontaine, (D), Ysg
  Claude Smutek, NrZrg

Romania

Alexandru Ioan Cuza
Department of Geology (B,M,D) ☒ (2018)
geology@uaic.ro
Bulevardul Carol I nr. 20A
Iași, Iași 700505
  geology@uaic.ro
  http://geology.uaic.ro

Enrollment (2015): B: 59 (26) M: 40 (18) D: 11 (5)
Head:
  Nicolae Buzgar, (D), A I Cuza Univ of Iasi, 1998, GziGd
Professor:
  Sorin Dorin Baciu, (D), A I Cuza Univ of Iasi, 2001, GcPv
  Dumitru Bulgariu, (D), Cga
  Ovidiu Gabriel Iancu, (D), A I Cuza Univ of Iasi, 1998, GpeCg
  Crina Genoveva Miclaus, (D), A I Cuza Univ of Iasi, 2001, Gsr
  Dan Stumbea, (D), Bolyai Univ of Cluj, 1998, Gez
Associate Professor:
  Daniel Țabară, (D), A I Cuza Univ of Iasi, 2000, Pl
  Andrei Ionut Apopei, (D), A I Cuza Univ of Iasi, 2014, GziGg
  Andrei Buzatu, (D), A I Cuza Univ of Iasi, 2014, GziGg
  Viorel Ionesi, (D), A I Cuza Univ of Iasi, 2005, PmGgHg
Assistant Professor:
  Iuliana Buliga, (D), A I Cuza Univ of Iasi, 2014, Gep
  Traian Gavriloaiei, (D), A I Cuza Univ of Iasi, 1999, CaAsCg
Lecturer:
  Dan Aștefanei, (D), A I Cuza Univ of Iasi, 2012, Cga
  Ciprian Chelariu, (D), A I Cuza Univ of Iasi, 2013, Go
  Claudia Cirimpei, (D), A I Cuza Univ of Iasi, 2010, PgGr
  Tony Cristian Dumitriu, (D), A I Cuza Univ of Iasi, 2015, GcZir
  Mitica Pintilei, (D), A I Cuza Univ of Iasi, 2010, CgeCt
  Bogdan Gabriel Rățoi, (D), A I Cuza Univ of Iasi, 2013, PvGr
  Smaranda Doina Sîrbu, (D), A I Cuza Univ of Iasi, 2002, Ge
  Oana Stan, (D), A I Cuza Univ of Iasi, 2009, GedCb
Visiting Professor:
  Mihai Remus Șaramet, (D), A I Cuza Univ of Iasi, 2004, NpGo
  Mihai Brânzilă, (D), A I Cuza Univ of Iasi, 1998, GgrPg

Babes-Bolyai University
Department of Geology (B,M,D) ☒ (2017)
Str. M. Kogalniceanu nr.1
RO - 400084 Cluj-Napoca
Cluj-Napoca 400084
  p. +40-264-405371
  sorin.filipescu@ubbcluj.ro
  http://bioge.ubbcluj.ro/pages/geologie/
  Programs: Geology; Applied Geology
Chair:
  Sorin Filipescu, (D), Babes-Bolyai, 1996, GrPm
Professor:
  Ioan Bucur, (D), Babes-Bolyai, 1991, PgmGs
  Vlad Codrea, (D), Babes-Bolyai, 1995, PvGo

Institul Geologic al Romaniei
Geological Institute of Romania (2015)
Caransebes, 1 Sector 1
Bucuresti
  office@igr.ro
  http://www.igr.ro/

University of Bucharest
Department of Geophysics - Faculty of Geology and Geophysics (B,M,D) ●☒ (2018)
6 Traian Vuia Street
Sector 2
Bucharest, Bucharest 020956
  p. +40-21-3125024
  bogdan.niculescu@gg.unibuc.ro
  https://gg.unibuc.ro/organizare/departamentul-geofizica/
  Programs: Geological Engineering; Geophysics
  Enrollment (2017): B: 30 (30) M: 20 (20)
Associate Professor:
  Bogdan Mihai NICULESCU, (D), Bucharest, 2002, YgeYu

Russia

Institute of Geology (Karelia, Russia)
Institute of Geology (Karelia, Russia) (D) ☒ (2017)
Pushkinskaya Str. 11
Petrozavodsk 185910
  p. +7 8142 782753

AELITA@krc.karelia.ru
http://igkrc.ru/
Administrative Assistant: Aelita Pervunina
Enrollment (2013): D: 1 (2)

Rwanda

Universite Nationale du Rwanda
Dept of Geology (B) (2015)
Faculty of Science
PO Box 56
Butare
    p. +250 (0) 252 530 122
    info@nur.ac.rw
    http://www.nur.ac.rw/

Dept of Geography (B) (2015)
PO Box 56
Butare
    p. +250 (0) 252 530 122
    info@nur.ac.rw
    http://ww.nur.ac.rw/

Saudi Arabia

King Abdulaziz University
Dept of Geology (2015)
P. O. Box 1744
Building No 55
Jeddah 21441
    jha25@hotmail.com
    http://earthscience.kau.edu.sa/Default.aspx?Site_ID=145&Lng=EN

King Fahd University of Petroleum and Minerals
Geosciences Department (B,M,D) ●☒ (2018)
P. O. Box 5070
KFUPM
Dhahran, Easter Provience 31261
    p. +96638602620
    c-es@kfupm.edu.sa
    https://cpg.kfupm.edu.sa/academics/departments/department-of-geosciences/
    Enrollment (2018): B: 60 (0) M: 53 (0) D: 11 (0)
Chair Professor:
    Luis Gonzalez, (D), Michigan, 1989, CgGn
    Stewart Alan Greenhalgh, (D), Minnesota, 1979, Yse
    John Reijmer, (D), Vrije (Amsterdam), 1991, Gsg
Professor:
    Abdullatif Al-Shuhail, (D), Texas A&M, 1998, Ygs
    Michael A. Kaminski, (D), MIT/WHOI, 1988, GuPm
    Panteleimon Soupios, (D), Aristotle, 2000, Yg
Associate Professor:
    Osman Abdullatif, (D), Khartoum, 1993, Gso
    Khalid Al-Ramadan, (D), Uppsala, 2006, GodGs
    Abdulaziz Al-Shaibani, (D), Texas A&M, 1999, HwGeg
    Sherif Hanafy, (D), Kiel, 2002, Yg
    John Dean Humphrey, (D), Brown, 1987, Gdo
    Mohammad Makkawi, (D), Colorado State, 1998, HwGoq
    Bassam S. Tawabini, (D), King Fahd, 2002, GeHsCa
    Scot Whattam, (D), Hong Kong, GxXc
Assistant Professor:
    Waleed Abdulghani, (D), Go
    Abdullah Al-Shuhail, (D), Calgary, 2011, Ye
    Ammar Elhusseiny, Stanford, Yg
    Ismail Kaka, (D), Carleton (Ottawa), 2006, Ys
    Mohammed Qurban, (D)
    Umair Bin Waheed, (D), KAUST, KSA, 2015, Yg
Lecturer:
    Ayman Al-Lehyani, (M), King Fahd, Yg
    Mutasim Sami Osman, (M), King Fahd, 2014, GosGc

Saudi Geological Survey
Saudi Geological Survey (2015)
P.O. Box: 54141
21514
    p. 966-2-619-5000
    sgs@sgs.org.sa
    http://www.sgs.org.sa/Arabic/Pages/default.aspx

Senegal

Universite Cheikh Anta Diop
Dept de Geologie (B) (2015)
Faculte Des Sciences
BP 5005
Dakar-Fann
    p. +221 869.27.66
    fst@ucad.edu.sn
    http://fst.ucad.sn/index.php?option=com_content&task=view&id=19&Itemid=35

Institut des Sciences de l' Environnement (ISE) (B) ☒ (2019)
BP 5005
Dakar-Fann
    p. +221 869.27.66
    diengdiomaye@yahoo.fr

Sierra Leone

Njala University
Dept of Mining and Metallurgy (B) (2015)
Freetown
    p. +232-22-228788, +232-22-226851
    nuc@sierratel.sl
    http://www.nu-online.com/

Dept of Soil Science (A,B,M,D) (2015)
Freetown
    p. +232-22-228788, +232-22-226851
    nuc@sierratel.sl
    http://ww.nu-online.com/

Dept of Geography & Rural Development (B) (2015)
Freetown
    p. +232-22-228788, +232-22-226851
    nuc@sierratel.sl
    http://www.nu-online.com/

Institute of Env Management and Quality Control (B) (2015)
Freetown
    p. +232-22-228788, +232-22-226851
    nuc@sierratel.sl
    http://www.nu-online.com/

University of Sierra Leone
Dept of Geology (2015)
PO Box 87
Freetown
    aiah_gbakima2000@yahoo.com
    http://www.tusol.org/programmes

Slovakia

Comenius University in Bratislava
Department of Geology and Palaeontology (A,B,M,D) O☒ (2019)
Faculty of Natural Sciences
Ilkovičova 6
Mlynska dolina G
Bratislava 842 15
    p. 00421260296529
    kgp@fns.uniba.sk
    http://geopaleo.fns.uniba.sk/
    Programs: Geology (B); Palaeobiology (B); Dynamic Geology (M); Palaeontology (M,D); Tectonics (D); Sedimentology (D)
    Enrollment (2016): B: 12 (0) M: 1 (0) D: 5 (0)
Professor:
    Michal Kovac, (D), Gsg
    Dusan Plasienka, (D), Gt

Slovenia

Daniela Rehakova, (D), Pms
Aubrecht Roman, (D), Gdr
Associate Professor:
  Jozef Hok, (A), Ggt
  Natalia H. Hudackova, (A), PimPg
  Marianna Kováčová, (A), PlcPe
  Martin Sabol, (D), Comenius Univ (Slovakia), 2000, PvgPs
  Rastislav Vojtko, (D), GctGm
Associate Scientist:
  Matus Hyzny, Pi
  Jan Schlogl, (D), Pig
Assistant Professor:
  Peter Joniak, (D), Pv

Dionyz Stur Institute of Geology
Geology Dept (2015)
Mlynska Dolina 1
81704 Bratislava 11
  p. 02 / 59375111
  secretary@geology.sk
  http://www.geology.sk/new/en

Geologicka Sluzba Slovenskej Republiky
Geological Survey of Slovak Republic (2015)
Mill Valley 1
817 04 Bratislava 11
  secretary@geology.sk
  http://www.sguds.sk

Kosice Technical University
Faculty of Mining, Ecology, Process Control and Geotechnology (2015)
Fakulta BERG
Park Komenskeho 12
040 01 Kosice
  p. +421-55-602 1111
  sekrd.fberg@tuke.sk
  http://www.fberg.tuke.sk/bergweb/index.php?IdLang=0&Selection=4

Slovenia

Geoloski zavod Slovenije
Geological Survey of Slovenia (M,D) (2015)
Dimiceva ulica 14
Ljubljana SI - 1000
  p. +386 1 2809702
  www@geo-zs.si
  http://www.geo-zs.si

University of Ljubljana
Faculty of natural science and engineering (2015)
Aškerčeva c. 12
Ljubljana 1000
  p. 01/470-45-00
  tanja.kocevar@ntf.uni-lj.si
  http://www.uni-lj.si/academies_and_faculties/faculties/2013071111502957/

Oddelek za Geologijo (2015)
Askerceva 12
Ljubljana
  spela.turic@ntf.uni-lj.si
  http://www.ntf.uni-lj.si/og

Somalia

Burao University
Dept of Community Development (B) (2015)
  p. +252 7126481
  info@universityofburao.com
  http://www.buraouniversity.com/rural_and_environmental_studies.htm

Mogadishu University
Somali Centre for Water and Environment (B) (2015)
  p. (252)-5-932454/ 223433/ 658479
  info@mogadishuuniversity.com
  http://www.mogadishuuniversity.com/index.html

Nugaal University
Institute of Geology (B,M) ● (2016)
Lasanod
  p. +252 275 4063 / 7412619
  nugaaluniversity@gmail.com
  http://www.nugaaluniversity.com/index.html

South Africa

Cape Peninsula University of Technology
Dept of Environmental and Occupational Studies (B) (2015)
Cape Town
  p. (+27) 021 959 6230
  vanderwesthuizenh@cput.ac.za
  http://www.cput.ac.za/

Geological Survey of South Africa
Council for Geoscience ● (2015)
Private Bag X112
Pretoria 0001
  p. 0027128411911
  info@geoscience.org.za
  http://www.geoscience.org.za/
Executive Manager:
  Fhatuwani L. Ramagwede, (M), 2011, EgGzEg

Nelson Mandela Metropolitan University
Dept of Geoscience (B,M,D) (2015)
PO Box 77000
Summerstrand, Port Elizabeth 6031
  p. 041 504 2325
  sheila.entress@nmmu.ac.za
  http://geosci.nmmu.ac.za/
  Administrative Assistant: Sheila Entress
Head:
  Nigel Webb, (D)
Professor:
  Vincent Kakembo, Sp
Associate Professor:
  Moctar Doucoure, (D)
  Daniel Mikes, (D)
Post-Doctoral Fellowship:
  Bastien Linol, (D), NMMU, 2013, Gs
Lecturer:
  Callum Anderson, (M), Port Elizabeth, Gzs
  Wilma Britz, (D)
  Gideon Brunsdon, (M)
  Anton de Wit, (D)
  Pakama Syongwana, (D)
  Nicolas Tonnelier, (D)
  Leizel Williams-Bruinders, (M), Zn
Emeritus:
  Peter Booth, (D)
  Anthony Christopher
Related Staff:
  Paul Baldwin
  Willie Deysel

Coastal and Marine Research Institute (B) (2015)
Summerstand Campus (South)
PO Box 77000
Port Elizabeth 6031
  p. +27 41 5042877
  cmr@nmmu.ac.za
  http://www.nmmu.ac.za/default.asp?id=380&bhcp=1

Centre for African Conservation Ecology (M,D) O (2015)
Summerstand Campus (South)
PO Box 77000

Port Elizabeth 6031
p. +27 41 504 2308
Graham.Kerley@nmmu.ac.za
http://ace.nmmu.ac.za/

North-West University
School of Environmental and Health Sciences (B) (2015)
Private Bag x 1290
Potchefstroom 2520
p. (27) 18 299 2528
monica.mosala@nwu.ac.za
http://www.puk.ac.za/opencms/export/PUK/html/fakulteite/natur/geol/index_e.html

Rhodes University
Dept of Geography (B) (2015)
PO Box 94
Grahamstwon 6140
p. +27 (0)46 603 8111
registrar@ru.ac.za
http://oldwww.ru.ac.za/academic/departments/geography/

Institute for Water Research (M,D) (2015)
PO Box 94
Grahamstown 6140
p. +27 46 62224014 / 6222428 / 62
registrar@ru.ac.za
http://www.ru.ac.za/static/institutes/iwr//?request=institutes/iwr/
Professor:
  Denis Hughes, (D), Wales, 1978, Hqs

Dept of Geology (B,M,D) (2015)
PO Box 94
Grahamstown , Eastern Cape Provinc 6140
p. +27 (0)46 603 8309
geolsec@ru.ac.za
http://www.ru.ac.za/geology/
Administrative Assistant: Ashley Goddard
Administrative Assistant: Vuyokazi Nkayi
Head:
  Steve Prevec, Cg
Professor:
  Annette Gotz, Gs
  Peter Horvath, (D), Gp
  Bantubonke Izwe Ntsaluba, Gi
  Briony Proctor
  Hari Tsikos, CgEg
  Yon Yao, Eg
Associate Professor:
  Steffen Buttner, (D), Frankfurt (Germany), 1997, Gcp
Associate Scientist:
  Billy De Klerk, Pg
  Robert Gess, Pg
  Rose Prevec
  Mike Skinner
Emeritus:
  Roger Jacobs
  Julian S. Marsh, (D), Cape Town, 1973, GiCgGv
Laboratory Director:
  Gelu Costin, Gp
Cooperating Faculty:
  John Hepple

Stellenbosch University
Dept of Process Engineering (Chemical and Mineral Processing) (B,M,D) (2015)
Private Bag X1
Matieland 7602
p. +27 21 808-4485
chemeng@sun.ac.za
http://www.chemeng.sun.ac.za/

Centre for Geographical Analysis (B) (2015)
Private Bag X1
Matieland 7602
p. (27) 21 808 3218
rdonaldson@sun.ac.za
http://academic.sun.ac.za/cga/expertise/user.asp?UserID=11

Dept of Earth Sciences (B,M,D) (2015)
Private Bag X1
Matieland, Western Cape 7602
p. +27 (0)21 808 3219
lcon@sun.ac.za
http://www.sun.ac.za/earthSci/
Professor:
  Alakendra N. Roychoudhury, (D), Georgia Tech, 1999, ClmOc

Tshwane University of Technology
Dept of Environmental, Water & Earth Sciences (B,M,D) (2015)
Private Bag x680
Pretoria 0001
p. 012 382 6232
gerberme@tut.ac.za
http://www.tut.ac.za/Students/facultiesdepartments/science/departments/environscience/Pages/default.aspx
Administrative Assistant: Sarah Galebies
Administrative Assistant: Retha Gerber
Enrollment (2015): B: 59 (37) M: 1 (0)
Lecturer:
  Chamunorwa Kambewa, (M), 1998, CgeZe
  Thando Majodina, (M), GgzGe
  Skhumbuzo Sibeko, (B), GggGg
Other:
  Mlindelwa Lupankwa, (D), HywGg

Dept of Environmental Health (B) (2015)
Private Bag X680
Staatsartillerie Road
Pretoria West 0001
p. +27 (0)12 382 5911
vanrooyenps@tut.ac.za
http://www.tut.ac.za/Pages/default.aspx

University of Cape Town
Dept of Oceanography (B) (2015)
Private Bag X3
Rhondebosch 7701
p. +27 (0) 21 650-3277
isabelle.ansorge@uct.ac.za
http://www.sea.uct.ac.za/index.php

Dept of Environmental and Geographical Science (B) (2015)
Shell Environmental & Geographical Science Building
South Lane, Upper Campus
Private Bad X3
Rondebosch 7701
p. (27) 21 - 6502873 / 4
michael.meadows@uct.ac.za
http://www.egs.uct.ac.za/

Dept of Geological Sciences (B,M,D) ● (2016)
Private Bag X3
Rondebosch 7701, Western Cape
p. +27 (0)21-650-2931 
head.geologicalsciences@uct.ac.za
http://www.geology.uct.ac.za/
Administrative Assistant: Lynn Evon
Administrative Assistant: Denise Lesch
Enrollment (2014): B: 30 (30) M: 30 (8) D: 10 (0)
Head:
  Chris Harris, GiCsGg
Professor:
  Steve Richardson, Cg
Senior Scientist:
  Nicholas Laidler
  Petrus J. Le Roux, (D), Cape Town, 2000, CaGia
  Christel Tinguely
Associate Professor:
  John Compton, (D), Harvard, 1986, CmGa

Associate Scientist:
  Fayrooza Rawoot
Assistant Professor:
  Emese M. Bordy, (D), Rhodes, 2001, GsPes
Lecturer:
  Johann Diener, Gp
  Lynnette N. Greyling, (D), Witwatersrand, 2009, EgmCe
  Phillip Janney, (D), California (San Diego), 1996, GiCaXc
  Beth Kahle, YgEoYg
Laboratory Director:
  Kerryn Gray, (M)

University of Fort Hare
Dept of Geology (2015)
Sobukwe Walk 8
Livingstone Hall
Alice, Eastern Cape 5700
  p. +27 (0)40 602 2011
  wkoll@ufh.ac.za
  http://wvw.ufh.ac.za/departments/geology/
Head:
  Oswald Gwavava
Lecturer:
  CJ Gunter
  Vuyokazi Mazomba
Related Staff:
  Luzuko Sigabi

Dept of Geographic Information Systems (B) (2015)
Private Bag X1314
Alice 5700
  p. +27 (0)40 602 2011
  wkoll@ufh.ac.za
  http://www.ufh.ac.za/departments/gis/gishome.html

Dept of Geography, Land Use, and Env Sciences (B) (2015)
Private Bag X1314
Alice 5700
  p. +27 (0)40 602 2011
  wkoll@ufh.ac.za
  http://www.ufh.ac.za/

University of Johannesburg
Dept of Geology (B) (2015)
PO Box 524
Kingsway & University (APK Campus)
Auckland Park 2006
  p. +27 (11) 559-4701
  mdekock@uj.ac.za
  http://www.uj.ac.za/Default.aspx?alias=www.uj.ac.za/geology

Dept of Geography, Environmental Management & Energy Studies (B,M,D) (2015)
PO Box 524
Auckland Park, Gauteng 2006
  p. +27 (0)11 559 2433
  science@uj.ac.za
  http://www.uj.ac.za/Default.aspx?alias=www.uj.ac.za/geography
Dr:
  Isaac T. Rampedi, (D), South Africa, 2010, PgSdHg

Dept of Mining Engineering and Mine Surveying (B) ☒ (2017)
PO Box 17011
Doornfontein 2028
  p. +27 (0)11 559-6628
  hgrobler@uj.ac.za
  http://www.uj.ac.za/Default.aspx?alias=www.uj.ac.za/mining
  Programs: Mine Surveying; Mineral Resource Management; Mining Engineering
  Certificates: Gyroscope Surveying

Dept of Mine Surveying (B) (2015)
  p. +27 (0)11 559-6186
  hgrobler@uj.ac.za
  http://www.uj.ac.za/Default.aspx?alias=www.uj.ac.za/minesurv

University of KwaZulu-Natal
School of Civil Engineering Surveying & Construction (B) (2015)
Centenary Building
King George V Avenue
Durban 4041
  p. (+27) 2603065
  troisc@ukzn.ac.za
  http://geomatics.ukzn.ac.za/HomePage2247.aspx

Centre for Water Resources Research (B,M,D) ☒ (2017)
Room 203
Rabie Saunders Building
Pietermaritzburg, KwaZulu-Natal 3201
  p. +27-(0)33-260 5490
  smithers@ukzn.ac.za
  http://cwrr.ukzn.ac.za/
  Programs: Hydrology; Water Resources; Agricultural Engineering
  Enrollment (2017): B: 55 (0) M: 30 (0) D: 14 (0)

School of Environmental Sciences (B,M,D) ☒ (2019)
Discipline of Geography
King Edward Avenue
Pietermaritzburg Campus
Pietermaritzburg, KwaZulu-Natal 3209
  p. + 27 (0)33 260 5346
  gijsbertsen@ukzn.ac.za
  http://saees.ukzn.ac.za/Homepage.aspx
  f: https://www.facebook.com/FriendsOfUkznAgriculture

Discipline of Geological Sciences, School of Agricultural, Earth and Environmental Sciences (B,M,D) ●☒ (2018)
Private Bag X 54001
Durban, KwaZulu-Natal 4000
  p. +27 (0)31 260 2516
  mccoshp@ukzn.ac.za
  http://www.geology.ukzn.ac.za/
  Programs: Environmental and Engineering Geology (B); Geology and Ore Deposits (B)
  Enrollment (2018): B: 195 (51) M: 14 (1) D: 6 (2)
Professor:
  Tesfaye K. Birke, (D), France, 1999, GcYmGg
  Emmanuel John Carranza, (D), Tech (Delft), 2002, EdCeZr
Associate Professor:
  Andrew Green, (D), KwaZulu-Natal, 2009, GuYgGs
Senior Lecturer:
  Molla Demlie, (D), Ruhr (Germany), 2007, HwGeg
  Jeremy Woodard, (D), Turku (Finland), 2010, CcGiz
Developmental Lecturer:
  Nonkululeko Dladla, (M), KwaZulu-Natal, 2013, GusGr
  Palesa Leuta-Madondo, (M), Western Cape, 2010, CgGig
Lecturer:
  Warwick William Hastie, (D), KwaZulu-Natal, 2013, NrgGc
  Egerton Hingston, (D), Leeds, 2008, NtxNg
  Lauren Hoyer, (D), KwaZulu-Natal, 2016, EgGc
  Philani Mavimbela, (M), Pretoria, 2013, GpzEg
  Saumitra Misra, (D), Calcutta, 1993, GiCgu

University of Limpopo
Dept of Soil Sciences (B) (2015)
Turfloop Campus
Private Bag X1106
Sovenga 0727
  p. +27 (0)15 268 9111
  Funso.Kutu@ul.ac.za
  http://www.ul.ac.za/index.php?Entity=agri_soil_scie

Geography & Environmental Studies Dept (B) (2018)
Turfloop Campus
Private Bag X1106
Sovenga 0727
  p. (+27) (0)15 268 3756

salphy.ramokolo@ul.ac.za
http://www.ul.ac.za/index.php?Entity=agri_geo_environ

School of Physical and Mineral Science (B) (2015)
Turfloop Campus
Private Bag X1106
Sovenga 0727
  p. +27(0) 15 268 3492
  SPMS@ul.ac.za
  http://www.ul.ac.za/index.php?Entity=School%20Main%20Menu&school_id=9
  Administrative Assistant: M. D.. Ramusi
Head:
  K. E. Rammutia
Professor:
  John Dunlevey, (D), Stellenbosch, 1985, GzxEg
  M. Khanyi
  M. A. Letsoalo
  M. A. Mahladisa
  R. M. Makwela
  T, Mobakazi
  T. E. Mosuang
  J. M. T. Mphahlele
  T. T. Netshisaulu
  M. Netsianda
  P. Ntoahee
  O. O. Nubi
  M. Phala
  M. J. Ramusi
  L. Wilsenach
Other:
  J. P. T. Crafford, (N), ZnNgZn

University of Pretoria

Centre of Environmental Studies (M,D) (2015)
Pretora 0002
  p. +27 (0)12 420-3111
  willemferguson@zoology.up.ac.za
  http://www.up.ac.za/centre-environmental-studies

University of Pretoria Natural Hazards Centre (M,D) ☒ (2018)
Department of Geology
University of Pretoria, Private Bag X20, Hatfield,
PRETORIA, 0028
Pretoria 0028
  p. +27 (0)12 420 3613
  andrzej.kijko@up.ac.za
  http://www.up.ac.za/university-of-pretoria-natural-hazard-centre-africa
Professor:
  Andrzej Kijko, (D), AGH Univ, 1979, YsgNe
Research Associate:
  Ansie Smit, (M), Mathematical Statistics, 2010

Dept of Mining Engineering (B,M,D) (2015)
Pretoria 0002
  p. +27 0 12 420-3111
  ssc@up.ac.za
  http://web.up.ac.za/default.asp?ipkCategoryID=2520

Dept of Geography, Geoinformatics and Meteorology (B,M,D) ☒ (2019)
Room 1-3
Geography Building
Cnr Lynnwood and University Roads
Hatfield, Pretoria 0083
  p. +27 (0)12 420 3536
  ggm.support@up.ac.za
  http://web.up.ac.za/ggm
  Programs: Geography, Geoinformatics & Meteorology
  Administrative Assistant: Lunga Ngcongo
  Administrative Assistant: Martha C.. van Aardt
Professor:
  Serena M. Coetzee, (D), Pretoria, 2009, Zi
Associate Professor:
  Greg Breetske, (D)
Lecturer:
  Daniel Darkey
  Nerhene Davis
  Liesl Dyson
  Joos Esterhuizen
  Natalie S. Haussmann
  Michael Loubser, Masters
  P. L. Philemon Tsela
  Victoria Rautenbach, (D)
  Barend van der Merwe, (M), Gm
Related Staff:
  Christel Hansen
  Popi Mahlangu
  Thando Ndarana, (D)
  Erika Pretorius

Dept of Geology (B,M,D) ● (2017)
Pretoria 0002
  p. +27 (0)12 420 2454
  wlady.Altermann@up.ac.za
  http://web.up.ac.za/default.asp?ipkCategoryID=2048

University of the Free State

Department of Geology (B,M,D) ●☒ (2018)
PO Box 339
Bloemfontein, Free State Province 9300
  p. +27 (0)51 401 2515
  roelofsef@ufs.ac.za
  https://www.ufs.ac.za/natagri/departments-and-divisions/geology-home
  Programs: Geology; Environmental Geology; Geochemistry; Mineral Resource Management
  Enrollment (2018): B: 184 (57) M: 42 (6) D: 5 (0)
Prof:
  Frederick Roelofse, (D), Witwatersrand, 2010, Gip
Ms:
  Jarlen J. Beukes, (M), Free State, Gi
  Justine Magson, (M), Free State, 2016, CgGi
  Thendo Mapholi, (B), Free State, Ggp
  Makhadi Rinae, (B), Free State, Gge
Mr:
  Justin Nel, (B), Free State, Gc
  Adriaan Odendaal, (B), Free State, 2010, GgsGr
  Pelele Lehloenya, (M), Free State, Cg
Dr:
  Robert N. Hansen, (D), Stellenbosch, 2014, CgGe
  Matthew Huber, (D), Vienna, EgXm
  Hendrik Minnaar, (D), Free State, Gc
Ms:
  Megan Purchase, (M), Free State, Gx

Institute for Groundwater Studies (B,M,D) (2016)
PO Box 339
Internal Bus 56
Bloemfontein 9300
  p. +27(0)51-4019111
  vermeulend@ufs.ac.za
  http://www.ufs.ac.za/igs

Centre for Environmental Management (B) (2015)
Po Box 338
Bloemfontein 9300
  p. +27(0)51-4019111
  avenantmf@ufs.ac.za

  http://www.ufs.ac.za/faculties/index.php?FCode=04&DCode=106

University of the Western Cape

International Ocean Institute of Southern Africa (B) (2015)
Private Bag X17
Bellville 7535
  p. +27 21 959 3088
  ioi-sa@uwc.ac.za
  http://www.ioisa.org.za/

South Africa

Dept of Earth Science (B) (2015)
Modderdam Road
Private Bag X17
Bellville, Cape Town 7530
p. +27 (0)21 959 2223
wdavids@uwc.ac.za
http://www.uwc.ac.za/Faculties/NS/EarthScience/Pages/default.aspx
Administrative Assistant: Caroline Barnard, Wasielah Davids, Chantal Johannes, Mandy Naidoo
Head:
   Charles Okujeni, (D), Berlin
Chair:
   Jacqueline Goldin, (D), UCT
   Yongxin Xu
Professor:
   Dominic Mazvimavi, (D), Wageningen, 2003
   Jan M. van Bever Donker, (D), Cape Town, 1979, GcpGt
Associate Professor:
   Ebernard Braune
Lecturer:
   Marcelene Andrews
   James Ayuk Ayuk, (M), UWC
   Lewis Jonkey, (M)
   Thokozani Kanyerere
   Mimonitu Opuwari, (D), UWC
   Henok Solomon, (M), UWC
VLIR Coordinator:
   Shauib Dustay
HIVE Manager:
   Yafah Hoosain, (M), UWC
Deputy Head:
   Theo Scheepers, (M), Stellenbosch
Related Staff:
   Janine Becorney
   Shamiel Davids
   Peter Meyer

University of the Witwatersrand

School of Geography, Archaeology and Env Studies (B,M,D) ☒ (2017)
Private Bag 32050
Wits, Johannesburg
p. +27 11 717 6503
donna.koch@wits.ac.za
http://web.wits.ac.za/Academic/Science/Geography/Home.htm
Programs: Geography

School of Mining Engineering (B,M,D) ● (2015)
Private Bag 3
WITS, Johannesburg 2050
p. +27 11-717-7003
bekir.genc@wits.ac.za
http://web.wits.ac.za/Academic/EBE/MiningEng/

School of Geosciences (B,M,D) ● (2016)
Faculty of Science
Private Bag 3
Wits, Johannesburg 2050
p. 27 11 717 6547
sharon.ellis@wits.ac.za
http://web.wits.ac.za/Academic/Science/GeoSciences/Home.htm
Professor:
   Lewis D. Ashwal, (D), Princeton, 1979, GxCgGt
   Roger L. Gibson, (D), Cambridge, 1990, GcpGg
   Kim AA Hein, (D), Tasmania, 1995, GgEmGc
   Judith A. Kinnaird, (D), St. Andrews, 1987, EmGi
Associate Professor:
   Paul A.M. Nex, (D), Univ Coll (Cork), 1997, Eg
Dr:
   Michael QW Jones, (D), Witwatersrand, 1981, YhGtYg
   Grant M. Bybee, (D), Witwatersrand, 2013, Gi
   Katie A. Smart, (D), Alberta, 2011, CsGz
   Zubair A. Jinnah, (D), Witwatersrand, 2011, GsPg

Electron Microprobe Scientist:
   Peter Horvath, (D), Eotvos Lorand, 2002, GpCpGz
Emeritus:
   Carl R. Anhaeusser, (D), Witwatersrand, 1983, GgEgZg

University of Venda

Dept of Geography and Geo-Information Sciences (B) (2015)
University of Venda
Private Bag X5050
Thohoyandou, Limpopo 0950
p. +27 15 962 8593
nthaduleni.nethengwe@univen.ac.za
http://www.univen.ac.za/enviornmental_sciences/dep_geography_geo_sciences.html
Head:
   Nthaduleni S. Nethengwe, (D), West Virginia, RnZiRw
Acting Head:
   T. M. Nelwamondo, (D), UP, 2010, GeZuy
Lecturer:
   E. Kori, (M), Univen, GmZgAt
   M. J. Mokgoebo, (M), Univen
   N. V. Mudau, (D), Univen, 2015, Rm
   A. Muyoki, (D), Howard
   M. Nembudani, (M), Stellenbosch
Related Staff:
   K. H. Mathivha, (M), Univen

Mining and Environmental Geology (A,B,M,D) ☒ (2019)
School of Environmental Science
Private Bag X5050
Thohoyandou , Limpopo Province 0950
p. +27 159628580
ogolaj@univen.ac.za
http://www.univen.ac.za/environmental_sciences/dep_mining_environmental.html
Programs: Geology and Earth Sciences
Enrollment (2018): B: 78 (69) M: 10 (7) D: 4 (1)
Dr.:
   Milton Kataka, (D), Witwatersrand, 2003

GIS Resource Centre (B,M,D) ☒ (2019)
GIS Resource Centre
Private Bag x5050
Thohoyandou, Limpopo 0950
p. +27 15 962 8044
farai.dondofema@univen.ac.za
http://www.univen.ac.za/environmental_sciences/gis_centre.html
t: @chinomukutu3
Programs: GIS, Geography, Remote sensing, Geology, Mining, Hydrology, Environmental Science, Ecology, Urban and Rural planning
Certificates: GIS, Ecology
Enrollment (2018): B: 46 (35) M: 13 (9) D: 4 (1)
Chief Technician:
   Farai Dondofema, (M), Zimbabwe, 2007, ZrHgZi

Institute of Semi-Arid Env and Disaster Management (B) (2015)
p. (+27) 015 962 8513
ndidzum@univen.ac.za
http://www.univen.ac.za/index.php?Entity=Institute%20of%20Semi-Arid%20Environment&Sch=3

Dept of Hydrology and Water Resources (B) (2015)
University of Venda
Private Bag X5050
Thohoyandou, Limpopo 0950
p. +27 015 962 8513
environmental@univen.ac.za
http://www.univen.ac.za/environmental_sciences/dep_hydrology_water.html
Head:
   J. O. Odiyo, (D)
Lecturer:
   J. R. Gumbo, (D)
   P. M. Kundu, (D)

R. Makungo, (M), Venda, 2010, Hqw
Tinyiko Rivers Nkuna, (M), Venda, 2012, HgwAs

University of Zululand
Dept of Geography and Environmental Studies (B) (2015)
Private Bag X1001
KwaDlangezwa 3886
    p. +27 (035) 902 6282
    kamwendog@unizulu.ac.za
    http://www.uzulu.ac.za/scie_geo_env.php

Dept of Hydrology (B,M,D) O (2017)
Private Bag X1001
KwaDlangezwa 3886
    p. +27 (035) 902 6282
    SimonisJ@unizulu.ac.za
    http://www.uzulu.ac.za/scie_hydro.php
    Enrollment (2016): B: 332 (79) M: 15 (2) D: 2 (0)

Spain

Coruna University
University Institute of Geology (D) (2016)
Edificio Servicios Centrales de Investigación
Campus de Elviña s/n
La Coruna 15071
    p. 0034 981167000
    xeoloxia@udc.es
    http://www.udc.es/iux
    Enrollment (2015): D: 8 (80)
Director:
    Juan Ramon Vidal Romaní, (D), Complutense (Madrid), 1983, GmlGc
Professor:
    Antonio Paz Gonzalez, (D), Santiago de Compostela, 1982, Sd
Paleontologist:
    Aurora Grandal d'Anglade, (D), Coruña, 1993, Pv
Associate Professor:
    Elena Pilar de Uña Alvarez, (D), Santiago de Compostela, 1986, Zy
    Cruz Iglesias, (D), Corunna, 2001, GaZmNr
    Jorge Sanjurjo, (D), Corunna, 2005, GeCcGa
    María Teresa Taboada Castro, (D), Santiago de Compostela, 1990, Sd
Associate Scientist:
    Marcos Vaqueiro Rodriguez, (M), Vigo (Spain), GcmYd

Institute of Marine Sciences (CSIC)
Marine Geosciences Departmen (2016)
Passeig Maritim de la Barceloneta
37-49
Barcelona 08003
    p. (+34) 63 230 95 00
    secredir@icm.csic.es
    http://gma.icm.csic.es

Instituto Geologico y Minero de Espana
Instituto Geologico y Minero de Espana (2015)
Rios Rosas, 23
28003 Madrid
    igme@igme.es
    http://www.igme.es

Univ Complutense de Madrid
Facultad de Ciencias Geologicas (B,M,D) O (2018)
C/ Jose Antonio Novais 12
Cuidad Universitaria
Madrid E-28040
    p. 34 91394 4837
    secre.adm@geo.ucm.es
    https://geologicas.ucm.es
    Programs: Geology (Bachelor), Engineering Geology (Bachelor), Environmental Geology (Master), Engineering Geology (Master), Advanced Paleontology (Master), Oil and Ore deposits exploration (Master)
    Certificates: Paleontology (on line), Mineral deposits (on line)
    Enrollment (2018): B: 508 (60) M: 59 (42) D: 70 (8)
Professor:
    Ana María Alonso, (D), Gd
    Eumenio Ancochea, (D), Gv
    Ricardo Arenas, (D), Gp
    José Arribas, (D), Gd
    Juan Luis Arsuaga, (D), Zn
    César Casquet, (D), Gip
    Lourdes Fernández, (D), Gz
    Sixto Fernández, (D), Pg
    Sol López, (D), Gz
    Javier Martín Chivelet, (D), PcGs
    Sergio Rodríguez, (D), PgGg
    Carlos Villaseca, (D), Gi
Vicedean for Research:
    Nieves Meléndez, (D), Ggs
Vicedean for quality:
    María de la Luz García Lorenzo, (D), Cl
Vicedean for Master programs and international affairs:
    María Luisa Canales, (D), Fac CC Geologicas, PmgGg
Vicedean for Bachelor programs :
    Eugenia Arribas, (D), Gd
Vicedean for academic issues:
    Alfonso Muñoz, (D), Ygr
Dean:
    Lorena Ortega, (D), Fac CC Geologicas (Madrid), 1993, EmgGg
Associate Professor:
    Jacobo Abati, (D), Gp
    Saturnino Alba, (D), Gm
    José Antonio Alvarez, (D), Gc
    Pilar Andonaegui, (D), Gg
    Carmen Arias, (D), Gg
    José Manuel Astilleros, (D), Gz
    María Isabel Benito, (D), Gs
    Manuel Bustillo, (D), Eg
    Pedro Castiñeiras, (D), Gp
    María José Comas, (D), Pi
    Elena Crespo, (D), En
    Cristina de Ignacio, (D), Cg
    Raúl de la Horra, (D), Gs
    Lucia de Stefano, (D), Hw
    Gerardo de Vicente, (D), Gt
    Javier Fernández Suárez, (D), Cc
    José María Fernández-Barrenechea, (D), Gz
    María Antonia Fregenal, (D), Gs
    Fernando Gacía Joral, (D), Pi
    María del Carmen Galindo, (D), Cc
    Nuria García, (D), Pv
    Alejandra García Frank, (D), Gg
    Emilia García Romero, (D), Gz
    Laura González, (D), Gr
    José Luis Granja, (D), Yg
    Manuel Hernández, (D), Pv
    Concepción Herrero, (D), Pm
    María Josefa Herrero, (D), Gs
    María José Huertas, (D), Gi
    Juan Miguel Insua, (D), NtGc
    María del Pilar Llanes, (D), Yr
    Francisco Javier Luque, (D), Gz
    José Francisco Martín Duque, (D), Gm
    Gemma Martínez, (D), Pg
    José Jesús Martínez, (D), Gc
    Pedro Martínez, (D), Hw
    Esperanza Montero, (D), Hw
    Belén Muñoz, (D), Gr
    David Orejana, (D), Gig
    Cecilia Pérez-Soba, (D), Gi
    Rubén Piña, (D), GzEm
    Marta Rodríguez, (D), Ggs
    Martín Jesús Rodríguez, (D), Ng
    Ignacio Romeo, (D), GcXg
    Carlos Rossi, (D), GdEo
    Javier Ruiz, (D), Xg
    Nuria Sánchez, (D), Gz
    Sonia Sánchez, (D), Gp
    Yolanda Sánchez, (D), Gs

Juan Ignacio Santisteban, (D), Gr
Esther Sanz, (D), Gd
Miguel Angel Sanz, (D), Zi
Paloma Sevilla, (D), Pi
Meaza Tsige, (D), Ng
David Uribelarrea, (D), Ga
María Josefa Varas, (D), Gd
Cristóbal Viedma, (D), Gz
Fermín Villarroy, (D), Hgy
Elena Vindel, (D), Gz
Assistant Professor:
Laura Domingo, (D), CsPe
Julio Garrote, (D), Gm
Adjunct Professor:
Enrique Aracil, (B), Yu
Sonia Bautista, (M), Ng
María Druet, (D), Yr
Alejandro Faúndez, (B), Ng
Luis Ramón Fernández, (M), Nt
David Jiménez, (B), GcNt
Cristina Martín, (D), Gm
Svetlana Melentijevic, (D), Nx
Luis Eugenio Suárez, (B), Zn

Universidad Autonoma de Madrid
Departamento de Geologia y Geoquimica (M,D) O☒ (2017)
Facultad de ciencias
Campus de Cantoblanco, C/ Francisco Tomás y Valiente, 7
Módulo 06, 6ª planta
Madrid 28049
> p. 91 497 48 00
> directora.geologia@uam.es
> http://www.uam.es/GyG
> Programs: Earth Science; Biology; Chemistry
> Certificates: GIS
> *Enrollment (2017): M: 9 (0) D: 8 (0)*

Universidad de Alicante
Environment and Earth Sciences Dept (B,M,D) (2015)
Facultad de Ciencia
San Vicente de Raspeig
Alicante E-03080
> p. (+34) 96 590 3552
> dctma@ua.es
> http://dctma.ua.es/en/environment-and-earth-sciences-department.html

Universidad de Granada
Dept de Geologia (2015)
Andalusian Institute of Geophysics
Campus Universitario de la Cartuja
18071 Granada
> jaguirre@ugr.es
> http://www.ugr.es/iag/iagpds.html

Universidad de Huelva
Geologia (B,M,D) (2015)
Huelva
> p. 959219809
> secgeo@uhu.es
> http://www.uhu.es/dgeo/

Universidad de Jaen
Departamento de Geologia (B,M,D) O☒ (2018)
Edificio B-3
Campus Universitario
Jaen 23071
> p. +34-953-212295
> jmmolina@ujaen.es
> http://geologia.ujaen.es/

Universidad de Las Palmas de Gran Canaria
Dept de Fisica (B) (2015)
C/Juan de Quesada, nº 30
, Las Palmas de Gran C 35001
> p. (+34) 928 451 000/023
> universidad@ulpgc.es
> http://www.dfis.ulpgc.es/

Facultad de Ciencias del Mar (B,M,D) (2015)
Edificio de Ciencias Básicas
Campus Universitario de Tafira
Las Palmas de Gran Canaria, Las Palmas 35017
> p. +34 928 452 900
> sec_dec_fcm@ulpgc.es
> www.fcm.ulpgc.es

Universidad de Murcia
Dept of Geography and Regional Planning (2015)
Facultad de Sciencia
Calle Santo Cristo 1
Murcia 30100
> p. (868) 88-7446
> jagb@um.es

Universidad de Oviedo
Dept de Geologia (2015)
Campus de Llamaquique
Jesus Arias de Velasco, s/n
33005 Oviedo E-33005
> geodir@geol.uniovi.es
> http://www.geol.uniovi.es/

Universidad de Pais Vascu
Dept de Geologia (2015)
Facultad de Sciencia
37008 Salamanca
> p. 946015491
> sec-centro.fct@ehu.es

Universidad de Palma de Mallorca
Dept de Geologia/Geofisica (2015)
Facultad de Sciencia
07012 Palma

Universidad de Sevilla
Applied Geology to Civil Engineering (D) (2016)
Departamento de Cristalografía y Mineralogía
calle Profesor García González 1
Seville 410012
> p. +34954556318
> igonza@us.es
> http://www.departamento.us.es/dcmqa/
Professor:
Isabel González, (D), GzeSc

Universidad de Valencia
Dept de Geologia (M,D) (2015)
Faculty of Biological Sciences
Building A (2nd and 3rd floor)
C/ Dr. Moliner, 50
Burjassot- (Valencia) 46100
> p. (+34) 96 354 46 02
> dep.geologia@uv.es
> http://www.uv.es/gcologia

Universidad de Valladolid
Dept of Geography (2015)
Facultad de Faculty of Philosophy and Arts
Palacio Santa Cruz/Plaza Santa Cruz B
Valladolid
> p. +34 983 423 005
> belart@fyl.uva.es

Universidad de Zaragoza
Facultad de Ciencias (2015)
50.009 Zaragoza
> seccienz@unizar.es

http://wzar.unizar.es/acad/fac/geolo/adepo.html

Museo de Ciencias Naturales (D) O☒ (2018)
Museo de Ciencias Naturales
Universidad de Zaragoza
Edificio Paraninfo. Plaza Basilio Paraiso
Zaragoza, Aragón 50005
   p. +34 976762096
   museonat@unizar.es
   http://museonat.unizar.es/
   f: https://www.facebook.com/museopaleounizar
   t: @museonat
Professor:
   José Ignacio Canudo, (D), Pv

Universitat Autonoma de Barcelona
Geologia (B,M,D) ●☒ (2018)
Facultat de Ciencies, Edifici C
Campus UAB
Bellaterra, Catalunya 08193
   p. +34 93 581 3022
   d.geologia@uab.es
   http://departaments.uab.cat/geologia/
   Programs: B.S. Geology
   B.S. Environmental Science, MSc. Reservoir Geology and Geophysics, MSc. Paleobiology and Fossil Record, MSc. Resources and Risks, PhD Geology
   Enrollment (2018): B: 284 (46) M: 47 (47) D: 14 (7)
Chair:
   Antoni Teixell Cácharo, (D), 1992, GtcYx
Professor:
   María Luísa Arboleya Cimadevilla, (D), Gc
   Joan Bach Plaza, (D), Gg
   Esteve Cardellach López, (D), GzeCs
   David Gómez Gras, (D), Gd
   Juan Francesc Piniella Febrer, (D), Gz
   Eduard Remacha Grau, (D), Universitat Aut, 1983, Gso
   Antonio Teixell, (D), 1992, Gtc
Associate Professor:
   Lluís Casas Duocastella, (D), GzaYm
   Mercè Corbella Cordomí, (D), GzCl
   Elena Druguet Tantiña, (D), Gc
   Joan Estalrich López, (D), HgGe
   Gúmer García Galán, (D), GiCg
   Rogelio Linares, (D), NgYx
   Oriol Oms Llobet, (D), GrsYm
   Joan Reche Estrada, (D), GpCp
   Enric Vicens Batet, (D), Pg
Assistant Professor:
   Carme Boix, (D), Pm
   Aline Concha, (D), Ge
   Isaac Corral, (D), Gx
   Victor Fondevilla, (D), PgGr
   Marc Furió, (D), Pv
   Ona Margalef, (D), Ge
   Didac Navarro, (D), Gz
   Valentí Oliveras Castro, (D), Gvi
   Joan Poch Serra, (B), Auto Barcelona, 2003, GrsZe
   Felix Sacristan, (M), Ge
   Eduard Saura, (D), Gc
   Bernat Vila, (D), Pv
Lecturer:
   Albert Griera Artigas, (D), GcCl
   Mario Zarroca Hernández, (D), YgNg
Emeritus:
   José Luis Briansó, (D), Gz
   Francisco Martínez Fernández, (D), GpCp

Universitat de Barcelona
Estratigrafia, Paleontologia i Geociencies marines (2015)
c/Marti Franques, s/n
08028
Barcelona
   p. +33 934021384
   secretaria-geologia@ub.edu
   http://www.ub.es/dpep/1welcom3.htm

Dept de Geoquimica, Petrologia I Prospeccio Geologica (2015)
Marti Franques, s/n
Barcelona
8028
   dept-geoquimica@ub.edu
   http://www.ub.es/geoquimi/index.html

Facultat de Ciències de la Terra (Earth Sciences Faculty) (B,M,D) O (2016)
Martí Franqués s/n
Barcelona 08028
   p. +34 934 021 335
   deganat-geologia@ub.edu
   http://www.ub.edu/geologia/en/

Universitat de les Illes Balears
Dept de Ciencies de la Terra (B,M,D) (2015)
Carretera de Valldemossa, km 7.5
Palma, Balearic Islands 07122
   p. +34 971172362
   dct@uib.es
   http://www.uib.es/depart/dctweb/home.htm

University of the Basque Country UPV/EHU
Departamento de Geodinamica (B,M,D) O☒ (2018)
Facultad de Ciencia y Tecnología
Barrio Sarriena, s/n
Leioa, Vizcaya 48940 Leioa
   p. + 34. 94 601 2563
   julia.cuevas@ehu.eus
   http://www.geodinamica.ehu.es/s0001-home1/es/
   Programs: Geology
   Environmental Sciences
   Enrollment (2017): B: 45 (0) D: 2 (0)
Professor:
   Julia Cuevas, (D), 1988, GctGg
Professor:
   Benito Abalos, (D), Basque Country UPV/EHU, 1990, GctGp
   Iñaki Antigüedad, (D), Hw
   Luis Eguiluz, (D), 1988, Ggt
   Jose M. Tubía, (D), 1985, GctYg
Senior Scientist:
   Aitor Aranguren, (D), GcYm
   Nestor Vegas, (D), Pais Vasco UPV/EHU, 2002, GctYm
Associate Professor:
   Jose Julian Esteban, (D), GtCcYg
   Tomas Morales, (D), NgHg
   Pablo Puelles, (D), Gct
   Jesus Angel Uriarte, (D), NgHg
Assistant Professor:
   Arturo Apraiz, (D), Gpt
   Vicente Iribar, (D), HwGm
   Luis Miguel Martinez Torres, (D), Ggt
   Fernando Sarrionandia, (D), Basque Country, 2006, GigGv
Emeritus:
   Hilario Llanos, (D), Hgy

Univesidad de Malaga
Dept de Geography (2015)
El Ejido
29071 Malaga
   ocana@uma.es

Sri Lanka

Sabaragamuwa University of Sri Lanka
Dept of Natural Resources (2015)
P.O Box 02
Belihuloya , Sabaragamuwa Provinc
   p. 0094-45-2280293
   head_nr@sab.ac.lk
   http://www.sab.ac.lk

Sri Lanka Geological Survey & Mines Bureau
Sri Lanka Geological Survey & Mines Bureau (2015)
Senanayake Building, No 4, Galle Road,
Dehiwala
> gsmb@slt.lk
> http://www.gsmb.gov.lk/

University of Moratuwa
Dept of Earth Resources Engineering (2015)
Katubedda, Moratuwa
Western Province
Moratuwa
> p. 0094-11-2650353
> shiromi@earth.mrt.ac.lk
> http://www.ere.mrt.ac.lk/

University of Peradeniya
Dept of Geology (2015)
Central Province
Peradeniya 22000
> p. 0094-81 2394 200/201
> geology@pdn.ac.lk
> http://www.pdn.ac.lk/sci/geology/

Uwa Wellassa University of Sri Lanka
Dept Mineral Resources & Technology (2015)
2nd Mile Post Passara Road
Badulla , Uwa province
> p. 0094-55-2226400
> info@uwu.ac.lk
> http://www.uwu.ac.lk/

Sudan

Cairo University
Dept of Geology (2015)
Khartoum Branch
PO Box 1055
Khartoum

El-Neelain University
Dept of Geology (B) (2015)
Gamhuria Street
Khartoum
> p. +249 (183) 77-441
> http://www.neelain.edu/sd/english.index.htm

Gama'at El Khartoum
Dept of Geology (2015)
PO Box 321
Khartoum
> geology@uofk.edu
> http://science.uofk.edu/index.php?option=com_content&view=article&id=101&Itemid=128&lang=en

International University of Africa
Dept of Geology (B) (2015)
PO Box 2469
Khartoum
> p. (+249) 183 223211
> minerals@iua.edu.sd
> http://www.iua.edu.sd/upldep.htm

Sudan University of Science and Technology
Surveying Engineering Dept (B) (2015)
Southern and Northern Campus
Khartoum
> p. +249183468622
> elhadielnazier@sustech.edu
> http://www.sustech.edu/faculty_en/department.php?coll_no=9&chk=a7b751cd18a66f8cd84b301f24267aab

College of Water & Environmental Engineering (B) (2015)
> p. +24985200512 / 985200511 / 918
> m.ginaya@sustech.edu
> http://www.sustech.edu/faculty_en/index.php?coll_no=13&chk=cdf8953eb5124fa6f08f046043be8bf

College of Petroleum Engineering and Technology (B) (2015)
Southern Campus
Khartoum
> zeinabkhaleel@sustech.edu
> http://www.sustech.edu/faculty_en/index.php?coll_no=23&chk=1d4f14cee52598202e0fc32a12b2dfc2

University of Dongola
Dept of Geology (B) (2015)
PO Box 47
Northern Province, Dongola
> p. +249-241-821-516, +249-241-821
> http://www.uofd.edu.sd/index.php/ar/

University of Gezira
Dept of Geology (2015)
Faculty of Science and Technology
PO Box 20
Wad Medani
> p. (002) 49511825724
> webinfo@uofg.edu.sd
> http://uofg.edu.sd/ENV/

University of Juba
Dept of Geology and Mining (B) (2015)
PO Box 321/1
Khartoum Center, Juba
> p. +249 (83) 222125
> info@juba.edu.sd
> http://www.juba.edu.sd/

Dept of Environmental Studies (B) (2015)
PO Box 321/1
Khartoum Center, Juba
> p. +249 (83) 222125
> info@juba.edu.sd
> http://www.juba.edu.sd/

University of Khartoum
Dept of Geology (B,M,D) (2015)
PO Box 321
Khartoum 11115
> geology@uofk.edu
> http://www.uofk.edu/

Head:
> Ibahim Abdu Mohamed

Professor:
> Saad Eldin Hamad Moha Ali
> Abdelhalim Hassan Elnadi, Goz

Associate Professor:
> Abdelwahab Yousif Abbas
> Salah Bashir Abdalla, Ng
> Osman Mahmoud Abdelatif
> Samia Abdelrahman
> Badr Eldin Khal Agmed
> Fath Elrahman ali Birair
> Omar Elbadri ali Elmaki
> Abdelhafiz Gad Almula
> Mohamed Zaid Awad
> Insaf Sanhoory Babiker
> Almed Sulaiman Daood
> Abdalla Guma Farwa

Instructor:
> Eltayeb Elasha Abdalia
> Sami Osman Ibrahim
> Amro Shaikh Idris Ahmed
> Waleed Elmahdi Siddig
> Saif Eldin Sir Elkhatim

Lecturer:
> Amany Ali Badi

Related Staff:
    Walaa Elnasir Ibrahim

Marine Research Laboratory in Suakin (B) (2015)
PO Box 321
Khartoum 11115
    sbabdalla@hotmail.com
    https://www.fkm.utm.my/marine/mtl/?Album:Visitor

University of Kordofan
Dept of Soil and Water Sciences (B) (2015)
    p. 860008-611-00249
    hadiaabdelatif@kordofan.edu.sd
    http://science.kordofan.edu.sd/

University of Neelain
Dept of Geology (B) (2015)
    science_dean@neelain.edu.sd
    http://www.neelain.edu.sd/

University of the Red Sea
Dept of Geology (B) (2015)
PO Box 24
Red Sea, Port Sudan
    p. +249-311-219-28
    redseauniv44@hotmail.com

Wad El Magbout
Training Centre for Earth Science Technicians (B) (2015)

Suriname

Anton de Kom Universiteit van Suriname
Environmental Sciences Dept (B) O (2016)
Universiteitscomplex Leysweg
P.O.B. 9212
Gebouw 17
Paramaribo, n.a. n.a.
    p. 597465558 (Ext. 308)
    s.carilho@uvs.edu
    http://adekus.uvs.edu/section/sectSection.
    php?secID=5304&lb=Onderwijs%20.%20Technologie%20.%20Milieu
    f: https://www.facebook.com/AdeKUS.official/

Anton de Kom University of Suriname
Dept of Geology & Mining (2015)
Leysweg, P.O. Box 9212, Universiteits Complex, Gebouw VI

Swaziland

University of Swaziland
Dept of Geography, Environmental Science and Planning (B,M,D) (2015)
Private Bag 4
Kwaluseni M201
    p. (+268) 2517-0000
    kwaluseni@uniswa.sz
    http://www.uniswa.sz/academics/science/gep

Sweden

Chalmers University of Technology
Dept of Geology (2015)
S-412 96 Goteborg
Gothenburg
    sofie.hallden@chalmers.se
    http://geo.chalmers.se/index.htm

Karlstad University
Department of Environmental and Life Sciences, Risk and Environmental Studies (B,M,D) ●⌂ (2018)
Karlstad University
Karlstad 65188
    p. +46547001000
    Magnus.Johansson@kau.se
    https://www.kau.se/en/risk-and-environmental-studies
    Programs: Bachelor: Risk and environmental studies
    Master: Risk management in Society

Lund University
Dept of Geology, Historical Geology & Paleontology (2015)
Soluegatan 12
223 62 Lund
    p. (+46) 462221424
    mikael.calner@geol.lu.se
    http://www.science.lu.se/

Stockholm University
Dept of Geological Sciences (B,M,D) ● (2016)
Svante Arrheniusväg 8
106 91 Stockholm
    office@geo.su.se
    http://www.geo.su.se

Sveriges geologiska Undersokning
Geological Survey of Sweden (2016)
Box 670
751 28 Uppsala
    sgu@sgu.se
    http://www.sgu.se/

Umea Universitet
Dept of Ecology and Environmental Science (2015)
Natural Sciences, Johan Bures Road 14, Umeå
S-901 87 Umea
    p. +46 90 786 50 00
    jolina.orrell@emg.umu.se
    http://www.emg.umu.se

University of Gothenberg
Master of Earth Science (2015)
Box 100
S-405 30 Gothenburg
    p. +46 31-786 0000

University of Stockholm
Dept of Geology and Geochemistry (2015)
106 91 Stockholm
    Barbara@geo.su.se
    http://www.geo.su.se/

Uppsala Universitet
Dept of Earth Sciences (B,M,D) ● (2015)
Villavagen 16
UPPSALA SE-752 36
    p. +46-18-4710000
    PREFEKT@geo.uu.se
    http://www.geo.uu.se/default.asp?pageid=1&lan=1

Switzerland

ETH Hoenggerberg
Geophysical Inst ETH (2015)
Sonneggstrasse 5
8092 Zurich
    p. +41 44 633 26 05
    christoph.baerlocher@erdw.ethz.ch
    http://www.geophysics.ethz.ch/

ETH Zurich
Dept of Earth Sciences (2015)
ETH-Zenrum
CH-8092 Zurich
    info@erdw.ethz.ch
    http://www.erdw.ethz.ch/

Federal Office for the Environment (FOEN)
Federal Office for the Environment (FOEN) (2015)
info@bafu.admin.ch
Bern 3003
p. 0041313229311
info@bafu.admin.ch
http://www.bafu.admin.ch/
t: @bafuCH

Université de Lausanne
Dept of Geosciences and Environment (B,M,D) ☒ (2018)
Décanat FGSE
Quartier Mouline
Géopolis
Lausanne, Vaud 1015
p. 0041216923500
doyen.gse@unil.ch
www.unil.ch/gse
Programs: Master of Science in Geology, Master of Science in Environmental Geosciences, Master of Science in Geography, Master of Science in Biogeosciences, Master in Tourism Studies, Master in foundations and practices of sustainability

University of Basel
Geol & Palaeontological Inst (2015)
Bernouillistrasse 32
CH-4056 Basale
joelle.glanzmann@unibas.ch
http://www.unibas.ch/earth/GPI/paleo/index.htm

Inst of Earth Sciences (2015)
Bernoullistrasse 30
Basel 4056
eberhard.parlow@unibas.ch
http://therion.minpet.unibas.ch/minpet/index.html

Dept of Earth Sciences (2015)
Bernoullistr.32
CH-4056 Basel
Joelle.Glanzmann@unibas.ch
http://duw.unibas.ch/

University of Fribourg
Div of Earth Sciences (2017)
Chemin du Musee 6
Perolles
CH-1700 Fribourg
nicole.bruegger@unifr.ch
http://www.unifr.ch/geology/

University of Geneva
Section of Earth Sciences and Environment (2015)
13, rue des Maraichers
Geneva CH-1205
p. 0041223796628
elisabeth.lagut@unige.ch
http://www.unige.ch/sciences/terre/

University of Lausanne
Faculty of Geosciences and Environment (B,M,D) ●☒ (2019)
Geopolis
Lausanne CH-1015
p. +41 21 692 35 00
doyen.gse@unil.ch
www.unil.ch/gse/home.html
Programs: geography, environmental sciences, geology, biogeosciences, tourism studies, foundations and practices of sustainability
Enrollment (2018): B: 388 (0) M: 345 (98) D: 144 (24)

University of Neuchatel
Centre for Hydrogeology and Geothermics (B,M,D) (2016)
Rue Emile-Argand 11
CH-2000 Neuchatel
secretariat.chyn@unine.ch
http://www.unine.ch/chyn

Universität Bern
Institut für Geologie (B,M,D) (2017)
Baltzerstrasse 1+3
Bern CH-3012
p. +41 (0) 31 631 87 61
info@geo.unibe.ch
http://www.geo.unibe.ch

Syria

Damascus University
Dept of Geology (2015)
Jammah Dimasq
Damscus

Taiwan

Academia Sinica
Institute of Earth Sciences ☒ (2018)
128, Section 2, Academia Road, Nangang
Taipei 11529
p. 886-2-2783-9910
jhwang@earth.sinica.edu.tw
http://www.earth.sinica.edu.tw/index_e.php
Distinguished Researcg Fellow:
 Jeen-Hwa Wang, (D), SUNY, 1982, Ysn
Associate Research Fellow:
 Wu-Cheng Chi, (D), California (Berkeley), GtYrs
Assistant Research Fellow:
 Wen-che Yu, (D), Stony Brook, 2007, Ys

Central Geological Survey (MOEA) of Taiwan
Central Geological Survey (MOEA) of Taiwan (2015)
District No. 2, Lane 109
Taipei 235
cgs@moeacgs.gov.tw
http://www.moeacgs.gov.tw/main.jsp

Chinese Culture University
Dept of Geology (2015)
Hanaoka Yangmingshan
Taipei 1111455
p. (02) 2861-0511 rpm 26105
crssge@staff.pccu.edu.tw

National Cheng-Kung University
Dept Earth Sciences (2015)
Tainan
wong56@mail.ncku.edu.tw

National Chung Cheng University
Inst of Seismology (2015)
168 University Rd
Min-Hsiung Chia-Yi
p. (886) -5--2720 x 9 ext: 61201?61209)
seismo@ccu.edu.tw
http://www.eq.ccu.edu.tw

National Taiwan University
Inst of Geology (2015)
245 Choushan Road, Taipei 106-17

Tanzania

ARDHI University
School of Geospatial Sciences and Technolgy (B,M,D) (2015)
P.O.Box 35176
Dar es Salaam
hagai@aru.ac.tz

http://www.aru.ac.tz/page.php?id=63
Enrollment (2012): B: 120 (112) M: 10 (2) D: 6 (0)

Geological Survey of Tanzania
Geological Survey of Tanzania (2015)
P .O Box 903
Dodoma
    madini-do@gst.go.tz
    http://www.gst.go.tz/

University of Dar es Salaam
Dept of Geology (B,M,D) ● (2015)
PO Box 35052
Dar es Salaam
    p. +255 22 2410013
    geology@udsm.ac.tz
    http://www.conas.udsm.ac.tz/geology/
    Enrollment (2015): B: 65 (53) M: 17 (0)
Head:
  Nelson Boniface, (D), Kiel, GcpGz
Professor:
  Makenya A. Maboko, (D), Australian Nat, 1991, GpCgGx
Senior Lecturer:
  Charles Z. Kaaya, (D), Cologne, 1993, GrsGu
  Isaac Muneji Marobhe, (D), Helsinki Univ of Tech (Finland), 1990, YegYx
Associate Professor:
  Shukrani Manya, (D), Dar Es Salaam, 2008, CgGzg
  Hudson H. Nkotagu, (D), Tech (Berlin), 1994, HwgGe
Lecturer:
  Kasanzu Charles, (D), 2014, GgRh
  Emmanuel O. Kazimoto, (D), Kiel, 2014, EgGpCg
  Elisante E. Mshiu, (D), Martin-Luther Halle Wittenberg, 2014, ZrCgZi
  Elisante E. Mshiu, (D), Martin Luther (Germany), 2014, GgCeZr
  Gabriel D. Mulibo, (D), Penn State, 2013, Ysg
  Ferdinand W. Richard, (D), Uppsala, 1999, YsGt

Institute of Marine Sciences (B) (2015)
Mizingani Road
PO Box 35091
Zanzibar
    p. 255-24-2232128/2230741
    director@ims.udsm.ac.tz
    http://www.ims.udsm.ac.tz

University of Dodoma
Dept of Geology and Petroleum Studies (2015)
P.O Box 259
Dodoma
    p. +255 26 2310000
    vc@udom.ac.tz
    http://www.udom.ac.tz/

School of Mines and Petroleum Engineering (2015)
PO Box 259
Dodoma
    p. +255 22 2410013
    vc@udom.ac.tz
    http://www.udom.ac.tz

Thailand

Asian Institute of Technology
Dept of Geology (2015)
P.O. Box 4
Klong Luang
Pathumthani 12120, Bangkok 10501
    p. +66 (0) 2524 6057
    supamas@ait.ac.th
    http://www.set.ait.ac.th/page.php?fol=gte&page=gte

Dept of Mineral Resources
75/10 Rama 6 Road, Phayathai
Bangkok 10400
    pornthip@dmr.go.th
    http://www.dmr.go.th/

Togo

Universite de Lome
Dept de Geologie (B) (2015)
BP 1515
Lomé, TOGO
    p. (+228) 22-25-50-93

Tunisia

Birzeit University
Dept of Geology (2015)
Faculty of Science
P. O. Box 14
7021 Jarzouna Tunis
Bizerte
    p. 216590717
    fsb@fsb.rnu.tn
    http://www.fsb.rnu.tn/index_fr/contact.html

Universite de Carthage
Dept de Geologie (B) (2015)
Jarzouna 7021
    p. +21671841353
    fsb@fsb.rnu.tn
    http://www.fsb.rnu.tn/fsbindex.htm

Universite de Gabes
Dept de Sciences de la Terre (B) (2015)
Cite Riadh
Zerig, Gabes 6072
    p. +216 75 394 800
    mail@fsg.rnu.tn
    http:www/fsg.rnu.tn/PRESENTATION.htm

Inst Superieur des Sciences et Techniques des Eaux de Gabes (B) (2015)
Cite Riadh
Zerig, Gabes 6072
    p. +216 75 394 800
    samir.kamal@isstegb.rnu.tn
    http://www.isstegb.rnu.tn/francais/index.htm

Universite de Sfax
Dept de Sciences de la Terre (B) (2015)
Route de l Aeroport km 0.5
Sfax 3029
    p. 216 74 276 400
    fss@www.fss.rnu.tn
    http://www.fss.rnu.tn

Dept de Genie Materiaux (B) (2015)
B.P:w.3038
Sfax
    p. (216) 74 274 088
    enis@enis.rnu.tn
    http://www.enis.rnu.tn/content/enis00003.htm

Dept de Genie Georessources et Environnement (B) (2015)
B.P:w.3038
Sfax
    p. (216) 74 274 088
    enis@enis.rnu.tn
    http://www.enis.rnu.tn/content/enis00003.htm

Universite de Tunis
Dept of Geology (2015)
Faculty of Science
1 Rue de Beja
Tunis 2092
    p. 21671872600
    http://www.fst.rnu.tn/fr/index.php

Turkey

Canakkale Onsekiz Mart University
Jeoloji Bolumu (2015)
Terzioglu Campus
Canakkale 17020
    sztutkun@comu.edu.tr
    http://jeoloji.comu.edu.tr/

Cukurova Universitesi
Jeoloji Muhendisligi Bolumu (B,M,D) ● (2015)
Faculty of Engineering and Architecture, Department of Geological Engineering
01330 Adana
Balcali
    p. (+ 90 322) 338 67 15
    parlak@cu.edu.tr
    http://jeoloji.cu.edu.tr
Professor:
    Osman Parlak, (D), Geneva, 1996, GipCc

Eskisehir Universitesi
Maden Muhendisligi Bolumu (2015)
Muhendislik Fakultesi
Yunus Emre Kampusu
Eskisehir

Firat University
Dept of Geological Engineering (A,B,M,D) O☒ (2019)
Fırat University, Department of Geological Engineering 23119 Elazıg Turkey
Elazig 23100
    p. 424-2370000-5979
    asasmaz@firat.edu.tr
    http://jeo.muh.firat.edu.tr/tr/node/104
    f: https://www.facebook.com/groups/firatjeoloji/
    Enrollment (2018): A: 1 (1) B: 10 (10) M: 4 (4) D: 2 (2)
Chair:
    Ahmet Sasmaz, (D), Firat, 2006, EgGeCe
Professor:
    Ercan Aksoy, (A), Ggt
    Melahat Beyarslan, (D), GipGx
    Ahmet Feyzi Bingol, (A), GipGv
    Bahattin Cetindag, (A), NgYv
    Zulfu Gurocak, (A), Nrg
    Leyla Kalender, (A), CgcCs
    Ahmet Sagiroglu, (A), Egm
Associate Professor:
    Bunyamin Akgul, (D), GzyGi
    Dicle Bal Akkoca, (D), EnCo
    Hasan Celik, (A), GscEo
    Calibe Koc Tasgin, (A), GsdGg
    Sevcan Kurum, (A), GivGx
Assistant Professor:
    Ayse Didem Kilic, (D), GxpGy
    Esra Ozel Yildirim, (D), Gxi
    Ozlem Oztekin Okan, (D), HgwHs
    Melek Ural, (D), GivCc
Research Assistant:
    Mehmet Kokum, (D), Firat, 2017, GtcGg
Research Assistant:
    Hatice Kara, (M), Cg
    Nevin Ozturk, (D), Firat, 2017, Cg
Research Asistant:
    Elif Akgun, (M), GgcGt
    Onur Alkac, GgsGu
    Gizem Arslan, (M), GxiGp
    Yasemin Aslan, Nr
    Serap Colak Erol, (D), 2014, GgcGt
    Mehmet Ali Erturk, (D), 2016, GiCc
    Mustafa Kanik, (D), 2015, NrSo
    Sibel Kaygili, (D), 2016, GgPgs
    Mahmut Palutoglu, (D), YgsGc
    Mustafa Eren Rizeli, (M), Gxi

    Abdullah Sar, (M), GviGp
    Ismail Yildirim, ScGix

Hacettepe University
Geological Engineering Dept (2015)
Muhendislik Fakultesi
Beytepe Kampusu
    tunay@hacettepe.edu.tr
    http://www.jeo.hun.edu.tr/800x600.htm
Assistant Professor:
    Turker Kurttas, (D), Hacettepe, 1997, HwgZr

Hydrogeological Engineering Programme (B,M,D) ● (2015)
Beytepe Campus
Ankara 06800
    p. +90 (312) 297730
    ekmekci@hacettepe.edu.tr
    http://www.hacettepe.edu.tr/
    Enrollment (2015): B: 235 (42) M: 12 (0) D: 11 (3)
Professor:
    Sakir Simsek, (D), Istanbul, 1982, Hw
    Serdar Bayari, (D), Hacettepe, 1991, Hw
    Mehmet Ekmekci, (D), HwyCs
    Galip Yüce, (D), Hw
    Nur Naciye Özyurt, (D), Hacettepe, Hw
Assistant Professor:
    Levent Tezcan, (D), Hacettepe, 1993, Hw

Istanbul Universitesi
Jeoloji Bolumu (2015)
Muhendislik Fakultesi
Beyazit
    p. 0 (212) 473 70 70 ex. 17600
    koral@istanbul.edu.tr
    http://muhendislik.istanbul.edu.tr/jeoloji/?p=6909

ITU
Jeoloji Bolumu (2015)
Maden Fakultesi
80394, Macka
Istanbul

Middle East Technical University
Inst of Marine Science (2015)
Div. of Marine Geology & Geophysics
P.O. Box 28
33731 Erdemli-Mersin
    p. +90-324 521 3434
    adminims@metu.edu.tr
    http://www.ims.metu.edu.tr/

Dept of Geological Engineering (B,M,D) (2015)
Inonu Bulvari
ODTU
Ankara TR-06531
    p. +90-312-2102682
    gesin@metu.edu.tr
    http://www.geoe.metu.edu.tr/

Mineral Research and Exploration General Directorate of Turkey
    p. +90 312 201 11 51
    mta@mta.gov.tr
    http://www.mta.gov.tr/

Selcuk Universitesi
Faculty of Sciences (2015)
Muhendislik Fakultesi
42040 Konya
    p. (+)903322412484
    fen@selcuk.edu.tr

Uganda

Gulu University
Faculty of Agriculture and Environment (B,M,D) (2016)
P.O. Box 166
Gulu
  p. +256782673491
  d.ongeng@gu.ac.ug
  http://www.gu.ac.ug

Dept of Geography (B) ● (2015)
P.O. Box 166
Gulu
  p. +256772517488
  charles.okumu52@gmail.com
  http://www.gu.ac.ug/index.php?option=com_content&view=category&layout=blog&id=48&Itemid=66
Lecturer:
  Expedito Nuwategeka, (D), Gulu, 2015, ZyyGv

Kabale University
Dept of Env Science and Natural Resources (B) (2015)
PO Box 317
Kabale
  p. 256-4864-22803
  akampabf@gmail.com
  http://www.kabaleuniversity.ac.ag/

Makerere University
Dept of Geology and Petroleum Studies (2015)
P. O. Box 7062
Kampala 041
  p. +256 41 532631-4
  http://mak.ac.ug/
Instructor:
  Kevin Aanyu, (M), Mak
  Wycliff Kawule, (M), ITC
  Robert Mamgbi, (M), CT-Prague
Lecturer:
  Erasmus Barifajio, (D), Mak
  John Mary Kiberu, (D), TUB
  Agnes Alaba Kuterama, (M), ITC
  Andrew Muwanga, (D), Brausnschweig
  Immaculate Nakimera Ssemanda, (D), Mak

Dept of Geography, Geoinformatics & Climatic Sciences (B) (2015)
PO Box 7062
Kampala
  p. +256-41-53126 1
  geog@caes.mak.ac.ug
  http://www.geography.mak.ac.ug/

Institute of Environment and Natural Resources (B) (2015)
PO Box 7062
Kampala 041
  p. 256 414 530134
  fkansiime@muienr.mak.ac.ug
  http://muienr.mak.ac.ug/

Ndejje University
Faculty of Forest Science & Environmental Management (B) (2015)
PO Box 7088
Kampala, Ndejje Hill
  p. +256-0392-730326
  forest@ndejjeuniversity.ac.ug
  http://www.ndejjeuinversity.ac.ug/academics.htm

Ukraine

National Mining University (in Ukraine)
State Mining University of Ukraine (2015)
19 Karl Marx Avenue
Dnepropetrovsk
320600
  dfr@nmuu.dp.ua
  http://www.apex.dp.ua/english/uageo/ukraine.html

United Arab Emirates

United Arab Emirates University
Dept of Geology (B,M,D) ● (2015)
P.O. Box 15551
College of Science, UAE University
Jamma Street
Al-Ain, Abu Dhabi 9713
  p. +971-3-7136380
  Ahmed.murad@uaeu.ac.ae

United Kingdom

Ulster University, Coleraine
School of Geography and Environmental Sciences (A,B,M,D)
○ ⌘ (2018)
Cromore Road
Coleraine BT521SA
  p. 0044(0)2870124401
  a.moore@ulster.ac.uk
  https://www.ulster.ac.uk/faculties/life-and-health-sciences/schools/geography-and-environmental-sciences
  t: @UlsterUniGES
  Programs: Geographic Information Systems; Coastal Zone Management; Environmental Management; Environmental Management and Geographic Information Systems; Environmental Management with Geographic Information Systems; Environmental Pollution and Ecotoxicology; Marine Spatial Planning
  Certificates: Geographic Information Systems; Coastal Zone Management; Environmental Management; Environmental Management and Geographic Information Systems; Environmental Management with Geographic Information Systems; Environmental Pollution and Ecotoxicology; Marine Spatial Planning
  Enrollment (2018): B: 330 (97) M: 110 (42) D: 30 (11)

Aberystwyth University
Dept of Geography & Earth Sciences (B,M,D) ⊠ (2018)
Llandinam Building
Penglais
Aberystwyth SY23 3DB
  p. 01970 622606
  gesstaff@aber.ac.uk
  http://www.aber.ac.uk/en/iges
Professor:
  Ron Fuge, (D), Wales, GbCgGe
Emeritus:
  Michael M J Hambrey, (D), Manchester, 1974, GlsGm

Anglia Ruskin University
Dept of Life Sciences (2015)
Cambridge Campus
East Road
Cambridge CB1 1PT
  p. +44 845 271 3333
  michael.cole@anglia.ac.uk
  http://www.anglia-ruskin.ac.uk/ruskin/en/home/faculties/fst/departments/lifesciences.html

Bangor University
School of Ocean Sciences (B,M,D) (2015)
Menai Bridge,
Isle of Anglesey
LL59 5AB
  p. +44-1248-382854
  oss011@bangor.ac.uk
  http://www.sos.bangor.ac.uk/
Associate Professor:
  Andrew J. Davies, (D), Queen's (Ireland), 2004, ObZir

United Kingdom

Birkbeck College
Dept of Earth and Planetary Sciences (2015)
Department of Earth and Planetary Sciences
University of London - Birkbeck
Malet Street
Bloomsbury, London WC1E 7HX
    p. +44 (0)20 7631 6665
    p.gaunt@bbk.ac.uk
    http://www.bbk.ac.uk/geology/
    Administrative Assistant: Peter Gaunt
Head:
    Gerald Roberts, (D), Durham (UK), 1991, Gct
Professor:
    Charlie Bristow, (D), Leeds (UK), 1987, Gs
    Andy Carter, Zg
    Ian Crawford, Zn
    Hilary Downes, Cg
Instructor:
    Karen Hudson-Edwards, Gez
Lecturer:
    Andy Beard
    Simon Drake, Gv
    Dominic Fortes, Zg
    Steve Hirons, Cl
    Philip Hopley, Pe
    Phillip Pogge von Strandmann, (D), Csl
    Vincent C. H. Tong, Yg
    Charlie Underwood, c.un, Pg

British Antarctic Survey
British Antarctic Survey ☒ (2018)
High Cross, Madingley Road
Cambridge CB3 0ET
    p. +44 (0)1223 221400
    trr@bas.ac.uk
    https://www.bas.ac.uk/team/science-teams/geosciences/
    t: @BAS_News

British Geological Survey
British Geological Survey (2015)
Environmental Science Centre
Nicker Hill
Keyworth
Nottingham NG12 5GG
    p. 0115 936 3100
    enquiries@bgs.ac.uk
    http://www.bgs.ac.uk/

Cardiff University
School of Earth and Ocean Sciences (B,M,D) (2015)
Main Building
Park Place
Cardiff, Wales CF10 3YE
    p. +44 (0)29 2087 4830
    earth-ug@cf.ac.uk
    http://www.cardiff.ac.uk/earth/
Head:
    Ian R. Hall, (D), GuCsGs
    R J. Parkes
Professor:
    J A. Cartwright, (D), Ys
    Dianne Edwards, (D), Pb
    Dianne Edwards, (D), Pb
    I Hall, Zn
    A Harris
    C Harris, Ge
    A C. Kerr, (D), Durham, 1994, Gxi
    Bernard Elgey Leake, (D), Bristol, 1974, GipGz
    R J. Lisle, Gc
    C MacLeod, Gu
    J A. Pearce, Cg
    P N. Pearson, Pe
    D Rickard, Cg
    V P. Wright, Gs
Associate Professor:
    Tiago Alves, (D), Manchester, 2002, GouGt

Commander:
    N Rodgers, (D), Am
Lecturer:
    Rhoda Ballinger, (D), Zn
    Stephen Barker, Pe
    C M. Berry, (D), Pb
    P J. Brabham, (D), Ye
    L C. Cherns, Pg
    Jose Constantine, (D), Hy
    J H. Davies, Yg
    I Fryett, Zn
    TC Hales, (D), Gm
    A Hemsley, Pl
    T Jones, Ge
    C Lear, Ou
    Johan Lissenberg
    Sergio Lourenco, (D), Sp
    R Perkins, Zn
    J Pike, Zn
    H M. Prichard, Gx
    H Sass
    H D. Smith, Zn
    S J. Wakefield, Cm
    C F. Wooldridge, Zn
    Y Yang, Hg

Durham University
Dept of Earth Sciences (B,M,D) ☒ (2018)
Science Laboratories
South Road
Durham DH1 3LE
    p. +44 0191 3342300
    earth.sciences@durham.ac.uk
    http://www.dur.ac.uk/earthsciences/
    f: https://www.facebook.com/DurUniEarthSci
    t: @DurUniEarthSci
    Programs: Geology, Geophysics with Geology, Geoscience, Environmental Geoscience, Earth Science
Professor:
    Mark Allen, Gt
    Andrew Aplin, Go
    Kevin Burton, Cg
    Jon Gluyas, Eo
    Chris Greenwell, (D), CqEoCg
    David Harper, Pg
    Richard Hobbs, Ys
    Robert E. Holdsworth, (D), Leeds, Gc
    Colin G. Macpherson, (D), London, 1994, CgGiCs
    Simon Mathias, Hy
    Kenneth J. McCaffrey, (D), Durham, Gc
    Stefan Nielsen
    Yaoling Niu, (D), Hawaii, 1992, Git
    Christine Peirce, (D), Cambridge, Yg
    Dave Selby, Cc
    Peter Talling, Rn
    Fred Worrall, (D), Cambridge, Cg
Associate Professor:
    James Baldini, Pc
    Richard James Brown, Gv
    Nicola De Paola, Gc
    Darren Grocke, Cs
    Claire Horwell, Rn
    Stuart Jones, Gs
    Ed Llewellin, Gv
    Julie Prytulak, Cg
    Jeroen van Hunen, (D), 2001, Gt
Assistant Professor:
    Martin Smith, Pe
    Fabian Wadsworth, Gv
    Richard Walters, Yd

Edinburgh University
School of Geosciences (B,M,D) ●◎ (2017)
Kings Buildings
West Mains Road
Edinburgh EH9 3JW

p. +44 (0) 131 651 7068
sarah.mcallister@ed.ac.uk
http://www.geos.ed.ac.uk/
f: https://www.facebook.com/geosciences/
t: @geosciencesed
Programs: Geology; Geology and Physical Geography; Geophysics; Environmental GeoSciences; Geoenergy; GIS; Earth Observation; Geoinformation Management
Enrollment (2017): B: 100 (0) M: 1 (0) D: 25 (0)

Director:
    Bruce M. Gittings, (M), Edinburgh, Zi
Head:
    Alexander W. Tudhope, (D), Cs
Chair:
    Chris Dibben, Zn
    Mark D. A. Rounsevell, (D), Zu
    Charles W. J. Withers, (D)
Professor:
    Emily S. Brady, (D), Glas, Ge
    Andrew Curtis, Ys
    Andrew J. Dugmore, (D), ZycGa
    J. Godfrey Fitton, Gi
    Simon L. Harley, Gt
    R. Stuart Haszeldine, (D), Strathclyde, GoRmZo
    Gabriele C. Hegerl, (D), As
    Simon Kelley, (D), London, 1985, CcsZg
    Dick Kroon, Gg
    Ian G. Main, (D), Ys
    Patrick Meir, (D), Edinburgh
    Maurizio Mencuccini, (D), Florence
    John B. Moncrieff, As
    Peter W. Neinow, (D), Cantab, Gl
    Paul Palmer, As
    Jamie R. Pearce, (D), Ge
    Alastair H. F. Robertson, (D), Leicester, 1975, GtgGs
    Hugh D. Sinclair, (D), So
    David Stevenson, (D), As
    Simon F. B. Tett, Am
    Kathryn A. Whaler, (D), Cambridge (UK), 1981, YmRn
    Mathew Williams, (D), East Anglia, 1994, ZcrZu
    Wyn Williams, Ym
    Iain H. Woodhouse, (D), Zi
    Rachel A. Woods, (D), Open, Gs
    Anton M. Ziolkowski, (D), Cambridge (UK), 1971, YesYg
Senior Scientist:
    Nicola Cayzer, (D), Edinburgh, Yg
    Richard Hinton, Cg
Associate Professor:
    Stuart M. V. Gilfillan, (D), Manchester, 2006, CgsCe
Assistant Professor:
    Massimo A. Bollasina, (D), Maryland, 2010, Asp
Research Associate:
    Andrew Bell, Gt
    Ian B. Butler, (D), Cg
    John Craven, Yg
    Jan C. de Hoog, (D), Utrecht, 2001, GxzGv
    Walter Geibert, Cc
    Chris L. Hayward, Ga
    Sian F. Henley, (D), Edinburgh (UK), 2013, CmOcCs
    Mark Naylor, (D), Edinburgh, Gt
    Anthony J. Newton, (D), Edinburgh, Zn
    Laetitia Pichevin, (D), Bordeaux, Cg
    John A. Stevenson, Gv
Instructor:
    Richard L. H. Essery, As
    Raja Ganeshram, Pg
    Andrew T. Mcleod, Zg
    Bryne T. Ngwenya, (D), Reed, Co
    David Reay, (D), As
    Tom Slater, (D), London, Zn
    Samantha Staddon
Lecturer:
    Simon J. Allen
    Mikael Attal, (D), Joseph Fourier, 2003, GmtZy
    Geoffrey Bromily, (D), Ge
    Eliza Calder, Gv
    Mark Chapman, (D), Yg
    Gregory L. Cowie, Co
    Julie Cupples, Zn
    Kyle Dexter
    Ruth Doherty, (D), Edinburgh, As
    Florian Fusseis, Gc
    Daniel Goldberg
    Noel Gourmelen
    Margaret C. Graham, Ge
    Kate V. Heal, Co
    Antonio Ioris, Ge
    Gail D. Jackson, PbGe
    Simon Jung, Og
    Linda Kirstein, Gt
    Eric Laurier, Zn
    Caroline Lehmann, (D)
    Fraser MacDonal
    William A. Mackaness, Zi
    Ondrej Masek, (D), Ng
    Christopher I. McDermott, Hg
    Marc J. Metzger, (D), Wageningen, Ge
    Nina J. Morris
    Simon N. Mudd, (D), Hg
    Caroline Nichol
    Eva Panagiotakopulu, Pe
    Genevieve Patenaude, (D), Ge
    Hugh C. Pumphrey, (D), Mississippi, AscYv
    Kanchana Ruwanpura
    Casey Ryan
    Kate Saunders, (D), GviGz
    Simon J. Shackley, Eg
    Niamh K. Shortt, Eg
    Sran P. Sohi, (D), London, So
    Neil Stuart, Zi
    Daniel Swanton, (D), Durham
    Jenny A. Tait
    Alexander Thomas
    Dan van der Horst, (D), Eg
    Mark Wilkinson, (D)
    Merwether Wilson, Ob
    Ronald M. Wilson, (D)
Emeritus:
    John Grace, As
Teaching Fellow:
    Thomas Challads, Hg
Other:
    Andrew S. Hein, (D), Edinburgh
    Tetsuya Komabayashi, (D), Gz
    Isla Myers-Smith
    Jan Penrose, (D), Toronto
    Marisa Wilson, (D), Oxcon

Exeter University
Camborne School of Mines (2015)
College of Engineering, Mathematics and Physical Sciences
University of Exeter
Penryn Campus
Penryn, Cornwall TR10 9FE
    p. (+44) 1392 661000
    cornwall@exeter.ac.uk
    http://emps.exeter.ac.uk/geology/contact/
Director:
    John Coggan, (D), Newcastle, Nx
    Charlie Moon, (D), Imperial Coll, 1983, CeEmZi
    Neill Wood, (B), Imperial Coll (UK), 1983, YuZen
Professor:
    Hylke J. Glass, (D), NxmEm
    Stephen Hasselbo, Gs
    Kip Jeffrey, Nx
    Bernd Lottermoser, Ca
Lecturer:
    Jens C. Andersen, (D), Copenhagen, GiEmCa
    Ian Bailey, (D), Univ Coll (London), Zn
    Robert Barley
    Christopher Bryan, Nx
    Patrick Foster, Nm
    Sam Hughes, Gt
    Gareth Kennedy, Nm

Kate L. Littler, (D), Univ Coll (London), 2011, PeCl
John Macadam, Ge
Lewis Meyer, (D), Exeter (UK), 2001, Nmr
Kathryn Moore, (D), Bristol, 1999, Ge
Kim Moreton, (D), Exeter Univ, 2008, EnZur
Richard Pascoe, (D), Gz
Duncan Pirrie
Robin Shail, (D), Keele, 1992, GtcEm
Ross Stickland
David Watkins, Hg
Andrew Wetherelt, Nm
Paul Wheeler, Nx
Ben Williamson, (D), London, 1991, EmGv

Heriot-Watt University
Institute of Petroleum Engineering (M,D) (2017)
Research Park, Riccarton
Heriot-Watt University
Edinburgh Campus
Edinburgh EH14 4AS
  p. +44 (0) 131 451 3543
  egis-staff@hw.ac.uk
  https://www.hw.ac.uk/schools/energy-geoscience-infrastructure-society/research/ipe.htm
  f: https://www.facebook.com/hwu.egis
  t: @HWUPetroleum
Director:
  John Ford
  Andy Gardiner
  Sebastian Geiger, (D), ETH (Switzerland), 2004, NpEoYh
  Eric Mackay
  James M. Somerville, Np
  Rink van Dijke
Head:
  Dorrik Stow
Chair:
  John Underhill
Professor:
  Patrick Corbett, Np
  Gary Couples, Np
  David Davies, Ng
  Mahmoud Jamiolahmady, (D), Ins Pet Eng, 2001, Zn
  Colin MacBeth
  Ken Sorbie
Associate Professor:
  Andreas Busch
  Jingsheng Ma, (D), Np
  Uisdean Nicholson
Assistant Professor:
  Hamed Amini
  Elli-Maria (Elma) Charalampidou
  Erkal Ersoy, (D), Eo
  Morteza Haghighat Sefat
Research Associate:
  Achim Ahrens
  Ross Anderson, Eo
  Lorraine Boak
  Matthew Booth
  Rachel Brackenridge
  Rod Burgass
  Antonio Carvalho
  Antonin Chapoy
  Romain Chassagne
  Ilya Fursov
  Alexader Graham
  Sally Hamilton, Gs
  Oleg Ishkov
  Rachel Jamieson
  Marco Lorusso
  Julian Maes
  Pedram Mahzari
  Maria-Daphne Mangriotis, Ys
  Pedro Martinez Garcia
  Robin Shields, Cg
  Mike Singleton
  Oscar Vazquez
  Francesca Watson
  Jinhai Yang
  Zhen Yin
  Zhao Zhang, (D)
  Stephanie Zihms
Instructor:
  Saleh Goodarzian
  Steve McDougall
Lecturer:
  Vasily Demyanov, Zi
  Florian Doster, Hg
  Ahmed H. Elsheikh
  Zeyun Jiang
  Helen Lever, (D), James Cook, 2003, Go
  Helen Lewis
  Khafiz Muradov, (D)
  Gillian E. Pickup, (D), Np
  Asghar Shams, Yg
  Karl D. Stephen
Laboratory Director:
  Jim Buckman
  Mike Christie
Other:
  Bahman Tohidi

Imperial College
Earth Science and Engineering (2018)
Imperial College London
South Kensington Campus
London SW7 2AZ
  p. +44 (0)20 7589 5111
  philip.allen@imperial.ac.uk
  http://www3.imperial.ac.uk/earthscienceandengineering
Director:
  Nigel Brandon, Ge
Head:
  Johannes Cilliers, (D), Cape Town, 1994, Gz
Chair:
  Alastair Fraser, (D), Glasgow, 1995, Go
  Matthew Jackson, (D), Liverpool, 1997
  Howard Johnson, Go
  Peter King, (D), Cambridge, 1982, Np
  Ann H. Muggeridge, (D), Oxford, 1986, NpHwEo
  Yanghua Wang, (D), Imperial Coll, 1997, Yg
Professor:
  Martin J. Blunt, (D), Cambridge, 1988, NpHw
  John Cosgrove, (D), Imperial Coll (London), 1972, Gc
  Sevket Durucan, (D), Nottingham, 1981, NmEcNr
  Sanjeev Gupta, Gs
  Joanna Morgan, (D), Cambridge, 1988, Yg
  Jane Plant, Ge
  Mark Rehkamper, (D), Mainz, 1995, CsaCc
  Mark Sephton, (D), Open, 1996, Xm
  Velisa Vesovic, (D), Imperial Coll, 1998, Np
  Michael Warner, (D), York, 1979, Ys
  Dominik Weiss, (D), Berne, 1998, ClaCq
  Robert W. Zimmerman, (D), California (Berkeley), 1984, NrHwNp
Associate Professor:
  Gerard Gorman, (D), Imperial Coll (London), 2005, Yg
  Samuel Krevor, (D), Eo
  Lidia Lonergan, (D), Oxford, 1991, GtcGs
Research Associate:
  Rebecca Bell, (D), Southampton, 2008, Gt
  Branko Bijeljic, (D), Imperial Coll, 2000, Np
  Raphael Blumenfeld, (D), Cambridge, Zn
  Rhordri Davies, (D), Cardiff, 2007, Yg
  Zita Martins, (D), Leiden, 2007, Xm
  Christopher Pain, Yg
  Randall Perry, Zg
Instructor:
  Jenny Collier, (D), Cambridge, 1989, Yr
  Saskia Goes, (D), California (Santa Cruz), 1995, Yg
  Gary Hampson, (D), Liverpool, 1995, Gd
  Christopher Jackson, (D), Manchester, 2002, Gt
  Anna Korre, Ng
  Jian Guo Liu, (D), Imperial Coll, 2002, Zr
  Stephen Neethling, (D), UMIST, 1999, Gz
  Emma Passmore

Matthew Piggot, Bath, 2001, Hg
Lecturer:
  Ian Bastow, (D), Leeds, 2005, Yg
  Gareth Collins, (D), Imperial Coll (London), 2002, Xm
  Matthew Genge, (D), Univ Coll (London), 1993, Xm
  Kathryn Hadler, (D), Manchester, 2006, Gz
  Cedric John, (D), Potsdam, 2003, Gs
  John-Paul Latham, (D), Imperial Coll, 2003
  Philippa J. Mason, Zr
  Adrian Muxworthy, (D), Oxford, 1998, Ym
  Julie Prytulak, (D), Bristol, 2008, Gg
  Mark Sutton, (D), Wales-Cardiff, 1996, Pg
  Tina van de Flierdt, (D), ETH Zuerich, 2003, Cs
  Alex Whittaker, (D), Edinburgh, 2007, Gt
Visiting Professor:
  Rosalind Coogon, Cs
Emeritus:
  Alain Gringarten, (D), Stanford, 1971, Np
  John Woods, (D), Imperial Coll, 1964

Keele University
Geography, Geology and the Environment (B,M,D) ☒ (2018)
William Smith Building,
Keele University
Keele, Staffordshire ST5 5BG
  p. (+44) 01782 733615
  gge@keele.ac.uk
  http://www.keele.ac.uk/gge/
  f: https://www.facebook.com/KeeleGeology
  t: @KeeleGeology
  Programs: BSc Geology (single & combined honours)
  M. Geology
  MSc Geoscience Research
  Certificates: None
  Enrollment (2014): B: 149 (45) M: 29 (15) D: 10 (2)
Dr:
  Stuart Egan, (D), Keele, Gtc
  Jamie K. Pringle, (D), Heriot-Watt, YuGfa
Associate Professor:
  Ralf Gertisser, (D), Freiburg (Germany), 2001, GivGz
  Ian G. Stimpson, (D), Wales, 1987, YeuRc
Dr:
  Glenda Jones, (D), Keele, NgYg
  Stu Clarke, (D), Keele, GsoGd
  Ralf Halama, (D), Tübingen, GpiCa
  Guido Meinhold, (D), Mainz, GdCa
  Michael Montenari, (D), Tübingen, PmgPs
  Steven L. Rogers, (D), Keele, PgGs

Kingston University
School of Geography, Geology and Environment (2015)
Penrhyn Road Centre
Penrhyn Road
Kingston upon Thames
Surrey KT1 2EE
  p. +44 (0)20 8417 9000
  G.Gillmore@kingston.ac.uk
  http://www.kingston.ac.uk/geolsci/
Director:
  Stuart Downward
Head:
  Gavin Gillmore, (D), Univ Coll (London), Gb
Professor:
  Ian Jarvis, (D), Oxon, 1980, CgGsr
  Peter Treloar, (D), Glasgow, 1978, Gc
  Nigel Walford, (D), London, 1981, Zi
Instructor:
  Peter Hooda, (D), London, 1992, Em
  Annie Hughes, (D), Bristol, 1997, Zi
  Mike Smith, (D), Sheffield, 1999, Zr
Lecturer:
  Andy Adam-Bredford
  Alistair Baird
  Douglas Brown, Zi
  Kerry Brown, (D), Stony Brook, 2004, Ge
  Norman Cheun, Ge
  Tracey Coates
  Hadrian Cook, (D), East Anglia, 1986, Ge
  Peter Garside, (D), Liverpool
  Paul Grant
  Ian Greatbatch, (D), 2008, Zi
  Frances Harris, (D), Kingston, 1995
  Mary Kelly
  David Kidd, (D), St. Andrews, 2005, Ziy
  James Lambert-Smith, (D), Kingston, 2014, Gg
  Andrew Miles, (D), Edinburgh, 2012, Gz
  Stephanie Mills, (D), Witwatersrand, 2006, Gl
  Pamela Murphy
  Colin Ryall
  Neil Thomas
Visiting Professor:
  Rosalind Taylor
Emeritus:
  Richard Moody
  Andy Rankin
Other:
  Andrew Swan

Leicester University
Geology Dept (B,M) (2015)
University Road
Leicester LE1 7RH
  p. +44 (0)116 252 3933
  geology@le.ac.uk
  http://www2.le.ac.uk/departments/geology
Head:
  Richard England, Yg
Professor:
  Sarah Davies, Gs
  Mike Lovell, Yg
  Randy Parrish, Cs
  Mark Purnell, Pg
  Mark Williams, Pg
Lecturer:
  Stewart Fishwick, Yg
  Sarah Gabbott, Pg
  Tom Harvey, Pg
  Gawen Jenkin
  Max Moorkamp, (D), Yg
  Mike Norry, Pe
  Dan Smith, Ge
  Richard Walker, Gs
  Jan Zalasiewicz, Pg
Emeritus:
  Andy D. Saunders, (D), Birmingham (UK), 1976, GiCgGe

Liverpool John Moores University
School of Natural Sciences and Psychology (B,M,D) (2016)
Byrom Street
Liverpool L3 3AF
  p. (+44) 0151 904 6300
  j.r.kirby@ljmu.ac.uk
  https://www.ljmu.ac.uk/research/centres-and-institutes/
  environment-research-group
  f: https://www.facebook.com/groups/LJMU.geography.env.
  sci/
  t: @ljmugeog
  Enrollment (2016): B: 72 (0) M: 3 (0)
Professor:
  Andy Tattersall
Dr:
  Jason Kirby

Nottingham Trent University
Dept of animal, rural, and environmental sciences (2015)
Burton Street
Nottingham NG1 4BU
  p. +44 (0)115 941 8418
  robert.mortimer@ntu.ac.uk
  http://www.ntu.ac.uk

Polytechnic Southwest
Dept of Earth Sciences (2015)
Drake Circus
Plymouth, Devon PL4 8AA
p. (+44) (0)1752 584584
science.environment@plymouth.ac.uk
http://www5.plymouth.ac.uk/schools/school-of-geography-earth-and-environmental-sciences/earth-sciences

Queen's University Belfast
School of Natural and Built Environment (Geography)
(B,M,D) ●☒ (2017)
University Road
Belfast BT7 1NN
p. + 44 (0) 28 90973186
p.warke@qub.ac.uk
http://www.qub.ac.uk/schools/NBE/
f: https://www.facebook.com/QUBGeography/
Programs: Geography (B); Geography with a Language (French or Spanish) (B)
Enrollment (2016): B: 320 (106) M: 6 (6) D: 40 (5)
Head:
    Patricia Warke, (D), Queen's (Ireland), 1994, GmZym
Professor:
    Keith Lilley, (D)
    David Livingstone, (D)
Dr:
    Andrew Newton, (D), 2017, GmlGs
Associate Professor:
    Merav Amir, (D)
    Oliver Dunnett, (D)
    Paul Ell, (D)
    Diarmd Finnegan, (D)
    Nuala Johnson, (D)
    M. Satish Kumar, (D), Jawaharlal Nehru (India), 1991, ZnnZn
    Jennifer McKinley, (D)
    Donal Mullan, (D)
    Helen Roe, (D)
    Alastair Ruffell, (D)
    Ian Shuttleworth, (D)
    Tristan Sturm, (D)
Lecturer:
    Niall Majury, (D), Toronto, 1999

Royal Holloway University of London
Department of Earth Sciences (B,M,D) ☒ (2018)
Department of Earth Sciences
Royal Holloway University of London
Egham Hill
Egham, Surrey TW20 0EX
p. +44 (0) 1784 443 581
info@es.rhul.ac.uk
https://www.royalholloway.ac.uk/earthsciences/home.aspx
t: @RHULEarthSci
Head of Department:
    Jurgen Adam, Gc
Professor:
    Margaret Collinson, Pg
    Howard Falcon-Lang, Pg
    Agust Gudmundsson, Gc
    Robert Hall, (D), Univ Coll (London), 1974, Gt
    Martin King, As
    Dave Mattey, Cs
    Ken R. McClay, (D), Imperial Coll (UK), 1978, GctGo
    Jason Morgan, Yg
    Euan G. Nisbet, (D), Cambridge, 1974, ZgAcGi
    Matthew Thirlwall, Cs
    Paola Vannucchi, Gu
    Dave Waltham, Yg
Research Associate:
    Dave Alderton, Gz
Lecturer:
    Anirban Basu, (D), Illinois (Urbana), CglCs
    Domenico Chiarella, (D), Basilicata (Italy), 2011, Gs
    Kevin C. Clemitshaw, (D), East Anglia (UK), 1986, AsZg
    F. Javier Hernandez Molina, (D), Granada (Spain), 1993, GsuOu
    Saswata Hier-Majumder, Yg
    Christina Manning, Cg
    Nicola Scarselli, (D), Go
    Steve Smith, Ge
    Giulio Solferino, (D), Gg
    Ian M. Watkinson, (D), 2009, GtcRn
Emeritus:
    Andrew Cunningham Scott, Pb
Laboratory Director:
    Nathalie Grassineau, Cs
    David Lowry, Cs

Staffordshire Polytechnic
Geography and the Environment (2015)
    j.w.wheeler@staffs.ac.uk
    http://www.staffs.ac.uk/sands/scis/geology/geology.html#

Dept of Applied Sciences (2015)
College Road, Stroke-on-Trent
Staffordshire ST4 2DE
p. (+44) 1782 294000
r.boast@staffs.ac.uk
http://www.staffs.ac.uk/academic_depts/sciences/subjects/environment/index.jsp

Swansea University Prifysgol Abertawe
Dept of Earth Sciences (2015)
Wallace Building
Swansea University
Singleton Park
Swansea, Wales SA2 8PP
p. +44 (0) 1792 205678 ext 8112
geography@swansea.ac.uk
http://www.swansea.ac.uk/geography/

Teesside University
School of Science and Engineering (B,M) (2016)
Tees Valley
Middlesborough TS1 3BA
p. 01642 738800
sse-admissions@tees.ac.uk
http://www.tees.ac.uk/schools/sse/index.cfm

The Open University
Dept of Environment, Earth & Ecosystems (B,M,D) (2015)
Faculty of Science
The Open University
Walton Hall
Milton Keynes MK7 6AA
p. +44 (0) 1908 652886
Env-Earth-Ecosystems-Enquiries@open.ac.uk
http://www.open.ac.uk/science/environment-earth-ecosystems/
f: https://www.facebook.com/pages/OU-Environment-Earth-Ecosystems/234968859948707
t: @OU_EEE
Head:
    Arlene G. Hunter, (D), Open, 1993, ZeGiCg
Professor:
    Fabrizio Ferrucci, Yg
    David Gowing, (D), Lancaster, 1991, Pb
    Nigel B.W. Harris, (D), Cambridge (UK), 1973, GtiGp
    Walter Oechel, As
Associate Professor:
    Mark A. Brandon, (D), Cambridge (UK), 1995, OgGlOp
    Angela L. Coe, (D), Oxford, GsrCg
    Dave McGarvie, (D), Lancaster, 1985, GviCt
Associate Scientist:
    Peter Sheldon
Research Associate:
    Marie-Laure Bagard
    Gareth Davies, (D), Cranfield, Ze
Reader:
    Neil Edwards, (D)

Instructor:
   Stephen Blake, (D), Lancaster, 1982, Gv
Lecturer:
   Pallavi Anand, (D), Cambridge (UK), 2002, ClPe
   Tom Argles, (D), Oxon, Zi
   Anthony Cohen, (D), Cambridge, 1988, Cs
   Sarah Davies, (D), Sheffield, Ze
   Miranda Dyson, (D), Witwatersrand, 1989, Pg
   Tamsin Edwards, (D)
   Richard Holliman, (D), Open, Ze
   Philip Sexton, (D), Southampton, 2000, Ge
   Carlton Wood
Visiting Professor:
   Stephen Self, (D)
   Edward Youngs, (D), Cambridge (UK), 1972, Sp
Emeritus:
   Phil Potts
   Robert Spicer, (D), Imperial, 1975

The University of Sunderland
Faculty of Applied Science (2015)
Edinburgh Building
City Campus
Chester Rd.
Sunderland SR1 3SD
   p. (+44) 191 515 2000
   john.macintyre@sunderland.ac.uk
   http://www.sunderland.ac.uk/faculties/apsc/ourdepartments/cet/

University College London
Dept of Earth Sciences (B,M,D) ☒ (2018)
Gower Street
London WC1E 6BT
   p. +44 (0)20 7679 2363
   earthsci@ucl.ac.uk
   http://www.ucl.ac.uk/es
Professor:
   Dario Alfe, (D), Intl Sch for Adv Studies (Italy), 1997, ZmGyYh
   Tim Atkinson, (D), Bristol, 1971, PcHwGm
   Paul Bown, (D), Univ Coll (London), 1986, Pm
   David Dobson, (D), Univ Coll (London), 1995, Gz
   Adrian P. Jones, (D), Durham, 1980, GiCgGg
   John M. McArthur, (D), Imperial Coll (UK), 1974, ClHwPs
   Philip Meredith, (D), Imperial Coll, 1983, Nr
   Eric Oelkers, (D), California (Berkeley), 1988, ClGze
   Kevin Pickering, (D), Oxford, 1979, Gs
   Chris Rapley, (D), London, 1976, ZgrGl
   Graham Shields-Zhou, (D), Eidgenössische Tech Hochschule Zürich, 1997, Cg
   Lars Stixrude, (D), California (Berkeley), 1991, Yg
   Julienne Stroeve, (D), Colorado, 1996, GlZr
   Juergen Thurow, (D), Eberhard-Karls, 1987, Gs
   Paul Upchurch, (D), Cambridge, 1993, Pg
   Lidunka Vocadlo, (D), Univ Coll (London), 1993, Gy
   Bridget S. Wade, (D), Edinburgh (UK), PmeCl
   Ian Wood, (D), Univ Coll (London), 1977, Gz
Associate Professor:
   Alex Song, (D), Caltech, Ysg
Lecturer:
   William Graham Burgess, (D), Birmingham, 1987, Hwy
   Anna Ferreira, (D), Oxford, 1005, Ys
   Chris Kilburn, (D), Univ Coll (London), 1984, Yx
   Tom Mitchell, Gc
   Dominic Papineau, (D), Colorado, 2006, Cg
   Phillip Pogge von Strandmann, (D), Open Univ (UK), 2006, Csl
   Pieter Vermeesch, (D), Stanford, 2005, Cc

University College of Swansea
Dept of Geography (2015)
Singleton Park
Swansea SA2 8PP

University of Aberdeen
Dept of Geology and Petroleum Geology (B,M) (2016)
Meston Building
King's College
Aberdeen AB24 3UE
   geology@abdn.ac.uk
   http://www.abdn.ac.uk/geology/
Head:
   David Jolley, (D), PlsGv
Chair:
   Ian Alsop, (D)
   Rob Butler
   Adrian Hartley
   John Howell, (D), Birmingham, 1992, GsoGc
   Andrew Hurst
   Ben Kneller
   David MacDonald
   Randell A. Stephenson, (D), Dalhousie, 1981, YgGtZg
Associate Professor:
   Clare Bond, (D), Gct
Research Associate:
   Steven Andrews
   Robert Daly
   Jyldz Tabyldy Kyzy
   Sam Spinks
   Christian Vallejo
Instructor:
   David Muirhead
Lecturer:
   Stephen Bowden, (D), Newcastle upon Tyne, CoGoCb
   David G. Cornwell, (D), Leicester (UK), 2008, YsGtYe
   Dave Healy
   Malcolm J. Hole
   David Iacopini
   Joyce Neilson
   Colin North
   Nick Schofield

University of Bedfordshire
Dept of Life Sciences (2015)
Park Square
Luton, Bedfordshire LU1 3JU
   p. +44 1234 400400
   international@beds.ac.uk
   http://www.beds.ac.uk/howtoapply/departments/science

University of Birmingham
School of Geography, Earth & Environmental Sciences (B,M,D) (2017)
School of Geography, Earth and Environmental Sciences
University of Birmingham
Edgbaston, Birmingham B15 2TT
   p. +44 (0)121 414 5531
   w.j.bloss@bham.ac.uk
   http://www.birmingham.ac.uk/schools/gees/index.aspx
Director:
   Rob MacKenzie
Head:
   David Hannah, Hg
Chair:
   Eugenia Valsami-Jones, (D), ClGeCa
Professor:
   Stuart Harrad
   Roy M. Harrison, (D), Birmingham (UK), 1989, As
   Jamie Lead
   Alexander Milner
   Tim Reston
   Jon Sadler
   John H. Tellam, (D), Birmingham, 1983, Hw
Associate Professor:
   Paul Anderson
   Bridin Carroll
   Rosie Day, (D), Univ Coll (London), Zn
   Arshad Isakjee
   Lloyd Jenkins
   Stephen M. Jones, (D), Cambridge (UK), 2000, Zg
   Jonathan Oldfield, (D), Zy

United Kingdom

Research Associate:
  Mohamed A. Abdallah
  Salim Alam
  Mohammed Baalousha
  David Beddows
  Ian Boomer
  Leigh R. Crilley, (D), 2013, As
  Mark Cuthbert
  Simon J. Dixon, (D), Gms
  Jonathan Eden
  Sophie Hadfield-Hill
  James Hale
  Stephanie Handley-Sidhu
  Marie Hutton
  Kieran Khamis
  Marcus Kohler
  James Levine
  Yuning Ma
  Paul Martin
  Sara Martinez-Loriente
  Mauro Masiol
  Heiko Moossen
  Catherine L. Muller
  Martin Müller
  Irina Nikolova
  Irina Nikolova
  Matt O'Callaghan
  Isabella Romer Roche
  Zongbo Shi
  Shei Sia Su
  Rick Thomas
  Amey S. Tilak
  Sarah Jane Veevers
  Saskia Warren
  Sebastian Watt
  Jianxin Yin
Deputy Head of School:
  William Bloss
Instructor:
  James Bendle
  Lee Chapman
  Jason Hilton
  Dominique Moran
  Ian Phillips
  Jo Southworth
Lecturer:
  Lauren Andres
  Daniel Arribas-Bel
  Austin Barber
  Nicholas Barrand
  Rebecca Bartlett
  Lesley Batty
  Mike Beazley
  Chris Bradley, (D), Leicester (UK), 1994, Hg
  Xioaming Cai
  Andy Chambers
  Julian Clark
  Juana Maria Delgado-Saborit
  Warren Eastwood
  Steven Emery
  Sara Fregonese
  Guy Harrington
  Alan Hastie
  Alan Herbert
  Phil Jones
  Nick Kettridge
  Stephen Krause
  Mark Ledger
  Peter Lee
  Agnieszka Leszczynski
  Iseult Lynch
  Zena Lynch
  Vlad Mykhnenko
  Patricia Noxolo
  Francis Pope
  Jessica Pykett
  Adam Ramadan
  Joanna Renshaw
  Michael Riley
  Michael Rivett
  John Round
  Ivan Sansom
  Greg Sambrook Smith
  Carl Stevenson
  Emmanouil Tranos
  James Wheeley
  Martin Widman
Emeritus:
  Ian Fairchild, (D), Nottingham, 1978, GsClGm
Birmingham Fellow & Academic Keeper, Lapworth Museum of Geology:
  Richard Butler
Other:
  Tom Dunkley Jones
Related Staff:
  Melanie Bickerton

University of Bristol
School of Earth Sciences (B,M,D) ☒ (2017)
Wills Memorial Building
Queens Road
Bristol BS8 1RJ
  p. +44 117 954 5400
  m.j.walter@bristol.ac.uk
  http://www.bristol.ac.uk/earthsciences/
  f: https://www.facebook.com/School-of-Earth-Sciences-Bristol-University-146277648746662/
  t: @UOBEarthscience
  Programs: Geology; Environmental Geoscience; Geophysics; Palaeobiology
  Enrollment (2013): B: 25 (9) M: 44 (5)
Professor:
  Michael Walter , (D), 1991, GiCpg

University of Cambridge
Dept of Earth Sciences (B,M,D) ●☒ (2018)
Downing Street
Cambridge , Cambridgeshire CB2 3EQ
  p. +44 (0)1223 333400
  satr@cam.ac.uk
  http://www.esc.cam.ac.uk
  f: https://www.facebook.com/pages/Department-of-Earth-Sciences-University-of-Cambridge/131259610400860
  t: @EarthSciCam
  Programs: Earth Sciences
Head:
  Simon Redfern, (D), Cambridge, 1989, Gzy
Professor:
  Michael Bickle, (D), Oxford, 1973, GtYgGo
  Nicholas J. Butterfield, Pg
  Michael Carpenter, Gy
  David Hodell, Ge
  Tim Holland, (D), Oxford, GptGz
  Marian Holness, Gi
  James A. Jackson, (D), Cambridge (UK), 1980, YgGt
  Simon Conway Morris, Pg
  Keith Priestley, YgGt
  Nick Rawlinson, Ygg
  Ekhard Salje, Gy
  Nicky White, YgGt
  Robert White, YgGt
  Eric W. Wolff, (D), Cambridge (UK), 1992, GlPe
  Andy Woods, YgGt
Lecturer:
  David Al-Attar, YdGt
  Alex Copley, YgGt
  Sanne Cottaar, Ys
  Neil Davies, Gs
  Marie Edmonds, Gv
  Ian Farnan, Gz
  Sally Gibson, Gi
  Richard Harrison, Gz
  John Maclennan, Gi
  Kenneth McNamara, Pg

Jerome Anthony Neufeld, GoYgGt
David Norman, Pg
John Rudge, YgGt
Luke C. Skinner, (D), Cambridge (UK), 2005, GeCmOg
Ed Tipper, Cl
Alexandra Turchyn, Ge
Emeritus:
Nigel H. Woodcock, (D), Imperial Coll (UK), 1973, GcsGt

University of Derby
Geographical, Earth and Environmental Sciences (B,M) (2015)
Kedleston Road
Derby DE22 1GAB
  p. +44 (0) 1332 591703
  fehs@derby.ac.uk
  http://www.derby.ac.uk/science/gees/
  Enrollment (2013): B: 120 (30) M: 9 (10)
Head:
  Hugh Rollinson, Cg
Professor:
  Aradhana Mehra, Co
Instructor:
  Jacob Adentunji, Ge
Lecturer:
  Martin Whiteley, Go

University of East Anglia
School of Env Sciences (2015)
Norwich Research Park
Norwich NR4 7TJ
  p. +44 (0)1603 592542
  env.enquiries@uea.ac.uk
  http://www.uea.ac.uk/environmental-sciences
Director:
  Corinne Le Qu, (D), Zg
Professor:
  Jan Alexander, (D), Leeds, GseGd
  Julian E. Andrews, (D), Leicester (UK), 1984, CsGsCl
  Simon Clegg, (D), East Anglia (UK), 1986, AsOcCg
  Brett Day, (D), UEA, 2004, Ge
  Alastair Grant, (D), Wales, 1983, GeObCt
  Karen Heywood, Op
  Kevin Hiscock, (D), Birmingham, 1987, Hwg
  Tim Jickells, (D), Southampton, Ob
  Phillip D. Jones, (D), Newcastle upon Tyne, 1977, AsPeHg
  Andy Jordan, Ge
  Andrew A. Lovett, (D), Wales, 1990, Ziu
  Coling Murrell, As
  Timothy J. Osborn, (D), East Anglia (UK), 1996, At
  Carlos Peres, Sf
  Ian Renfrew, Am
  Ian Renfrew, Am
  Bill Sturges, As
  Roland von Glasow, (D), Max Planck, Yg
  Andrew Watkinson, Ge
  Sir Robert Watson
Assistant Professor:
  Gill Malin, (D), Liverpool (UK), 1983, ObcCb
Research Associate:
  Amy Binner
Instructor:
  Alex Baker, (D), Plymouth Marine Lab, On
  Jenni Barclay, (D), Bristol, Gv
  Paul Burton, (D), Cambridge, Ys
  Mark Chapman, (D), East Anglia, Pm
  Paul Dolman, Zu
  Jan Kaiser, Cs
  Adrian Matthews, (D), Reading, Am
  Claire Reeves, (D), UEA, Cg
  Carol Robinson, Ob
  Parvadha Suntharalingam, Oc
  Cock von Oosterhout, (D), Leiden
  Rachel Warren, Eg
Lecturer:
  Annela Anger-Kraavi, EgGe
  Annela Anger-Kraavi, Eg

Victor Bense, Hg
Alan Bond, (D), Lancaster, Ge
Jason Chilvers, (D), Univ Coll (London), Ge
Pietro Cosentino, Yg
Stephen Dorling, Am
Aldina Franco, (D), UEA, Ge
Robert Hall, (D), Proudman Ocean Lab, On
Tom Hargreaves, Ge
Richard Herd, (D), Lancaster, Gv
Martin Johnson, Cm
Manoj Joshi, (D), Oxford, Zi
Iain Lake, Ge
Irene Lorenzoni, Eg
Nikolai Pedentchouk, (D), Penn State, 2004, Cs
Jane Powell, Ge
Brian Reid, GeCg
Gill Seyfang, Og
Congxiao Shang, (D), Queen Mary, Hw
Peter Simmons, Eg
Trevor Tolhurst, Sb
Jenni Turner, (D), 2008, Gt
Naomi Vaughan, (D), UEA, Am
Charlie Wilson, (D), British Columbia, Ge
Xiaoming Zhai, (D), Dalhousie, Og
Emeritus:
  Peter S. Liss, (D), OcCbAc
Professorial Fellow:
  Kerry Turner, GeEg

University of Exeter
Camborne School of Mines (2015)
Redruth, Cornwall
TR15 3SE
  C.Jeffrey@exeter.ac.uk
  http://www.ex.ac.uk/CSM/

University of Glasgow
School of Geographical and Earth Sciences (2015)
The Gregory Building
East Quadrangle
University Avenue
Glasgow G12 8QQ
  p. +44 (0) 141 330 5436
  GES-General@glasgow.ac.uk
  http://www.gla.ac.uk/schools/ges/
Head:
  Maggie Cusack, Gz
Professor:
  Paul Bishop, (D)
  John Briggs, Zu
  Roderick Brown, Cc
  Gordon Curry, (D), Imperial Coll (UK), 1979, GegPg
  Deborah Dixon
  Jim Hansom, (D), Aberdeen, 1979, OnGmZy
  Trevor Hoey, Gs
  Martin Lee, Gz
  Christopher Philo
Senior Scientist:
  Nicolas Beaudoin
Associate Professor:
  David Forrest, (D), Glasgow, 1996, Zi
Research Associate:
  Enateri Alakpa
  Susan Fitzer, Ob
  Michael Gallagher
  Lydia Hallis
  Ulrich Kelka
  Angela Last
  Paula Lindgren, As
  Alistair Mcgowan, Pg
  Larissa Naylor
  Alan W. Owen, (D), Glasgow, 1977, PgsPg
Instructor:
  Anne Dunlop
  Mhairi Harvey
  Hayden Lorimer

United Kingdom

Rhian Meara
Hester Parr
Vernon Phoenix
Daisy Rood
Lecturer:
Brian Bell, So
David Brown, Gvs
Seamus Coveney, Zi
Tim Dempster, Gz
Jane Drummond, Zi
Derek Fabel, Xg
David Featherstone
Nick Kamenos
Ozan Karaman
Daniel Koehn, Gt
Hannah Mathers, (D), GmlGg
Cheryl Mcgeachan
Simon Naylor
Cristina Persano

University of Leeds
School of Earth and Environment (B,M,D) ☒ (2018)
Maths/Earth and Environment Building
The University of Leeds
Leeds LS2 9JT
   p. +44 113 343 2846
   enquiries@see.leeds.ac.uk
   https://environment.leeds.ac.uk/see
   f: https://www.facebook.com/SchoolofEarthandEnvironment/
   t: @SEELeeds

Director:
Piers M. Forster, (D), Reading (UK), 1994, As
Head:
Simon Bottrell
Chair:
Christopher Collier
Simon Poulton, (D), Cls
Peter Taylor, Ge
Professor:
John Barrett, Ge
Alan Blyth, As
Ian Brooks, (D), UMIST, Zn
Ken Carslaw, (D), East Anglia (UK), 1994, As
Andy Challinor, (D), Leds, 1999, Am
Martyn P. Chipperfield, (D), Cambridge, 1990, AscAp
Surage Dessai, (D), East Anglia
Andy Dougil, (D), Sheffield, So
Paul Field
Quentin Fisher, Np
Paul W.J. Glover, (D), East Anglia (UK), 1989, YxGoa
Andy Gouldson, Ge
Alan Haywood, Pe
Steve Hencher, Ng
Andy Hooper, Ydg
Greg Houseman, (D), Cambridge, Yg
Bill McCaffrey, Gs
Jurgen Neuberg, (D), Colorado, Gv
Jouni Paavola, Ge
Doug Parker, (D), Reading, Am
Jeff Peakall, Gs
Andrew Shepherd
Lindsay C. Stinger, (D), Zn
Graham Stuart, Ys
Paul Wignall, PgGs
Tim Wright, Yd
Bruce W.D. Yardley, (D), Bristol, 1974, CgEgGg
William Young, (D), Ge
Associate Professor:
Doug Angus, Ys
Stephen Arnold, (D), Leeds, As
Wolfgang Buermann, (D), Boston, 2012, Zu
Ian Burke, (D), Southampton, Ge
Robert J. Chapman, (D), Leeds (UK), 1990, EmCe
Steven Dobbie, (D), Dalhousie, As
Luuk Flesken, (D), Wageningen, Ge
Phil Livermore, Ym

Christian Maerz, (D), Bremen, 2008, CmOuEm
Jim McQuaid, (D), Leeds (UK), 1999, As
Daniel Morgan, (D), Open Univ, 2003, Giv
Jon Mound, (D), Toronto, 2001, Gt
Noelle Odling, (D), Queens, Hg
Caroline Peacock, (D), Bristol, Sb
Sebastian Rost, (D), Universit, 2000, Ysg
Sally Russell, (D), Queensland, Ge
Dominick Spracklen, (D), Leeds, 1999, Zn
Julia Steinberger, Zu
Jared West, Hg
Research Associate:
David Banks, Cg
Lauren Gregoire, (D), Bristol
Ruza Ivanovic, (D), Bristol, 2013, OgAsPe
John Marsham, (D), Edinburgh, 2003, As
Instructor:
Nick Dixon, (D), Leeds, 2002, Zu
Tim Foxon, Ge
Chris Green
Dave Hodgson
Jacqueline Houghton
Damian Howells
Geoff Lloyd, (D), Birmingham, 1984
Lecturer:
Emma Bramham, Ym
Roger A. Clark, (D), Leeds, 1982, YseYs
Martin Dallimer, (D), Edinburgh, 2001, Ge
Monica Di Gregorio, (D), London Schl of Econ, Eg
Jen Dyer, (D), Leeds, Ge
Alan Gadian
Clare E. Gordon, (B), Zi
Dabo Guan, Eg
Jason Harvey, (D), 2005, Cg
Mark Hildyard
George Holmes, Ge
Andrea Jackson, (D), Lancaster, As
Julia Leventon, (D), Cent, Ge
Crispin Little, (D), Bristol, 1995, Pg
Piroska Lorinczi, (D), Leeds, 2006, Np
Vernon Manville
Andrew McCair, (D), Cambridge, 1983, Gc
Lucie Middlemiss, (D), Leeds, 2009, Zn
Nigel Mountney, Gs
Rob Newton, (D), Leeds, Ge
Alice OWen, (D), Leeds
Douglas Paton, Gc
Richard Phillips, (D), Oxford, Gt
Claire Quinn, (D), Kings College, Ge
Andrew Ross, (D), Cambridge, Am
Susannah Sallu, Ga
Ivan Savov, (D), S Florida, 2004, Cg
Anne Tallontire
Mark Thomas, (D), Ng
Taija Torvela, (D), Abo Akademi, 2007, GctEg
James van Alstine, Ge
Xianyun Wen, (D), Sichuan (China), 1982, NoAsm
Visiting Professor:
Jane Francis, Pe
Tim Needham, Gc

University of Liverpool
School of Environmental Sciences (B,M,D) (2015)
Jane Herdman Building
4 Brownlow Street
Liverpool, Merseyside L69 3GP
   p. +44 151 794 5146
   envsci@liv.ac.uk
   http://www.liv.ac.uk/environmental-sciences/index.html

Dept of Earth, Ocean, and Ecological Sciences (2015)
Jane Herdman Building
4 Brownlow Street
Liverpool L69 3GP
   p. +44 (0)151 795 4642
   Apboyle@liverpool.ac.uk
   http://www.liv.ac.uk/earth-ocean-and-ecological-sciences/

Head:
    Andreas Rietbrock, YsGv
Professor:
    Daniel Faulkner, Ys
    Richard Holme, Ym
    Nick Kusznir, Yd
    Yan Lavallee, Gv
    Rob Marrs, So
    Jim Marshall, Cs
    Jonathan Sharples, Ocp
    Stan van der Berg, Em
    John Wheeler, (D), GpcGz
    Richard Worden, Gs
Research Associate:
    Katherine Allen, Zi
    James Ball
    Helen Bloomfield, Ob
    Stephen Crowley, Gs
    Jonathan Lauderdale, Og
    Claire Melet
    Andreas Nilsson, Ym
    Vassil M. Roussenov, (D), Sofia (Bulgaria), 1987, OpZo
    Anu Thompson, Oc
    Xiao Wang
    Tsuyuko Yamanaka, Ob
Lecturer:
    Charlotte Jeffrey Abt, Zn
    Andy Biggin, Ym
    Alan P. Boyle, (D), Univ Coll (London), 1982, GzEmZe
    Bryony Caswell, Ob
    Silvio De Angelis, Ys
    Rob Duller, Gvs
    Liz Fisher, Co
    Jonathan Green, Ob
    Mimi Hill, Ym
    Janine Kavanagh, Gv
    Helen Kinvig, Gs
    Harry Leach, (D), Leeds, 1975, Og
    Claire Mahaffey, Oc
    Elisabeth Mariani, Nr
    Kate Parr, Pe
    John Piper, YmGt
    Leonie Robinson, Ob
    Isabelle Ryder, Gt
    Matthew Spencer, Og
    Neil Suttie
    Alessandro Tagliabue, Oc
    Jack Thomson, Ob
Visiting Professor:
    Peter Kokelaar, (D), Wales, 1977, GstGv
Other:
    Sara Henton De Angelis
    Jackie Kendrick
    Rachel Salaun, Oc
    Felix von Aulock

University of Manchester
School of Earth and Environmental Sciences (2015)
Williamson Building
Oxford Road
Manchester M13 9PL
    p. +44 (0) 161 306 9360
    earth.support@manchester.ac.uk
    Administrative Assistant: Steven Olivier
Director:
    Stephen Boult, Hg
Chair:
    Jonathan Redfern, (D), Bristol, 1990, GsoGl
Professor:
    Mike Bowman, (D), Wales, Go
    Andrew T. Chamberlain, (D), Liverpool, 1987, GaSaPv
    Thomas Choularton, (D), Manchester, 1987, As
    Hue Coe, (D), UMIST, 1993, As
    Stephen Flint, (D), Leeds (UK), 1985, GrsGo
    Martin Gallagher, (D), UMIDST, 1986, As
    Jamie Gilmour, (D), Sheffield
    Colin Hughes, (D), Open, 1992, Go
    Francis Livens, Cg
    Jonathan Lloyd, (D), Canterbury, 1993, Ge
    Ian Lyon, (D), 1993, Xc
    Gordon McFiggans, (D), UEA, 2000, As
    Carl Percival, (D), Oxford, 1995, As
    David Poyla, (D), Manchester, 1987, Cm
    Ernest H. Rutter, (D), Imperial Coll (UK), 1970, NrGct
    David Schultz, Am
    Kevin Taylor, Go
    David Vaughan, (D), Oxford, 1971, GzClGe
    Geraint Vaughan, (D), Oxford, 1982, As
    Roy Wogelius, (D), Northwestern, 1990, Cg
Associate Professor:
    Neil C. Mitchell, (D), Oxford, 1989, GumYg
Research Associate:
    Alastair Booth, As
    Gerard Capes, As
    Deborah Chavrit, (D), Nantes (France), 2010, Gi
    Patricia L. Clay, (D), Open Univ (UK), 2010, CgaGg
    Filipa Cox
    Ian Crawford, As
    Sarah Crowther, (D), Oxford, 2003, As
    Christopher Dearden, As
    Patrick Dowey
    Helen Downie, Ge
    Nicholas Edwards, Pg
    Christopher Emersic, (D)
    Torsten Henkel, Cs
    Hazel Jones, As
    Richard Kift, As
    Kimberly Leather, As
    Dantong Liu, As
    Douglas Lowe, As
    William Morgan, As
    Miquel Poyatos-More
    Laura Richards, Ge
    Hugo Ricketts, As
    Athanasios Rizoulis, Ge
    Andrew Smedley, As
    Robert Sparkes, (D), Cambridge, 2012
    Jonathan Taylor, (D), Manchester
    Karen Theis, (D), Manchester, 2008, Zn
    Paul Williams, As
Lecturer:
    Grant Allen, (D), Leicester, 2005, As
    Simon Brocklehurst, (D), MIT, 2002, Gm
    Kate Brodie, (D), Imperial Coll, 1979, Gc
    Rufus Brunt, (D), Leeds, Gs
    Michael Buckley, (D), York, 2008, Ga
    Ray Burgess, (D), Open, Cs
    Victoria Coker, (D), Manchester, 2007, Gz
    Paul Connolly, (D), Manchester, 2006, As
    Stephen Covery-Crump, (D), Univ Coll (London), 1992
    Giles Droop, (D), Oxford, 1979, Gp
    Victoria Egerton, Pg
    David Hodgetts, (D), Keele, 1995, Gsc
    Cathy Hollis, (D), Aberdeen, 1995, Gs
    Merren Jones, (D), Wyoming, 1997, Gst
    Julian Mecklenburgh, Gc
    John Nudds, (D), Dunhelm, 1975, Pg
    Clare Robinson, (D), Lancaster, 1990, Co
    Stefan Schroeder, (D), Bern, 2000, Go
    Bart van Dongen, Utrecht, 2003, Co
Emeritus:
    Christopher Henderson
    Peter Jonas
    Grenville Turner, (D), Oxford, 1962, XcCc

University of Newcastle Upon Tyne
School of Civil Engineering and Geosciences (2016)
Newcastle University
Newcastle upon Tyne
NE1 7RU
    p. +44 (0)191 208 6323
    ceg@ncl.ac.uk
    http://www.ncl.ac.uk/ceg/

Professor of Soil Science / Head of School :
  David Manning
Professor:
  Margaret Carol Bell, Ge
  Andras Bordossy, Zg
  Peter J. Clarke, (D), Oxford (UK), 1997, YdZri
  Stephen Larter, Gg
  Zhenhong Li, Yd
  Phillip Moore, Yd
Research Associate:
  David Alderson, Zi
  Joana Baptista, Ng
  Stephen Blenkinson, Zn
  Bernard Bowler, Go
  Aidan Burton
  Allistair Ford, Gq
  Kirill Palamartchouk, Yd
Instructor:
  James Barhurst, Gs
  Neil Gray, Pg
  Helen Talbot, Co
Lecturer:
  Jamie Amezaga, Ge
  Stuart Barr, Zi
  Colin Davie, (D), Glasgow, 2002, Ngr
  Stuart Edwards, Gq
  Gaetano Elia, gaet, Ng
  David Fairbairn, Gq
  Rachel Gaulton, Zr
  Jean Hall, Ng
  Geoffrey Parkin, Hg
  Nigel Penna, Gq
  Paul Quinn, Hg
Visiting Professor:
  Rick Brassington, Hg
Related Staff:
  Peter Cunningham, Hg

University of Nottingham
Dept of Mineral Resources (2015)
University Park
Nottingham HG7 2RD
  wpadmin@nottingham.ac.uk

University of Oxford
Dept of Earth Sciences (2015)
Department of Earth Sciences
South Parks Road
Oxford, Oxfordshire OX1 3AN
  p. +44 1865 272000
  reception@earth.ox.ac.uk
  http://www.earth.ox.ac.uk/
Head:
  Alex Halliday, Cs
Chair:
  Christopher Ballentine, Xg
  Phillip England, Gt
Professor:
  Martin Brasier, Pg
  Joe Cartwright, Gs
  Shamita Das, Gtv
  Donalf G. Fraser, Cg
  Gideon Henderson, Cg
  Hugh C. Jenkyns, (D), Leicester (UK), 1970, GrsCs
  Samar Khatiwala
  Conall Mac Niocaill, (D), GtgYm
  Tamsin A. Mather, (D), Gv
  Barry Parsons, Yd
  David M. Pyle, (D), Cambridge (UK), 1990, GviRh
  Ros Rickaby, Gz
  Mike Searle, Gt
  Anthony B. Watts, (D), Durham, 1970, GutYv
  Bernard Wood, Gz
  John Woodhouse, Yg
Associate Professor:
  Stuart Robinson, (D), Oxford (UK), 2002, PeGr
  Karin Sigloch, (D), Princeton, 2008, YgsGt
Lecturer:
  Roger Benson, Pg
  Heather Bouman, Cg
  Matt Friedman, Pg
  Lars Hansen, Gz
  Helen Johnson, Og
  Richard Katz, Cg
  Graeme Lloyd, Gs
  Don Porcelli, Cg
  Richard Walker, Zr
Emeritus:
  Dave Waters, (D), Oxford, 1976, GpzGc

University of Plymouth
Dept of Earth Sciences (2018)
Drake Circus
Plymouth, Devon PL4 8AA
  p. +44 (0)1752 584584
  science.technology@plymouth.ac.uk
  http://www.plymouth.ac.uk/schools/sogees
Head:
  Mark Anderson, (D), Wales Cardiff, Gct
Professor:
  Antony Morris, (D), Edinburgh (UK), 1990, YmGtu
  Gregory Price, (D), Reading, 1994, Gsr
Associate Professor:
  Paul D. Cole, (D), 1990, Gvg
  Stephen Grimes, (D), Cardiff, 1998, Cs
  Martin Stokes, (D), Plymouth, 1997, Gm
  Graeme Taylor, Yg
  Matthew Watkinson, (D), Open, 1989, Gs
Lecturer:
  Sarah Boulton, (D), Edinburgh, 2005, Zr
  Arjan Dijkstra, (D), Utrecht, 2001, Gi
  Meriel E.J. FitzPatrick, (D), Plymouth, 1992, PlgGr
  Luca Menegon, (D), Padua, 2006, Gct
  Andrew Merritt, (D), Leeds, 2010, Ng
  Kevin Page, (D), Univ Coll (London), 1988, Gr
  Christopher Smart, (D), Southampton, 1993, Pg
  Colin Wilkins, (D), James Cook, 1991, Eg
Emeritus:
  Malcolm B. Hart, (D), London, 1993, PmGur

University of Portsmouth
School of Earth and Environmental Sciences (2015)
Burnaby Road
Portsmouth, Hampshire PO1 3QL
  p. +44 (0)23 9284 2257
  sees.enquiries@port.ac.uk
  http://www.sci.port.ac.uk/departments/academic/sees
Head:
  Rob Strachan, (D), Keele, 1982, Cc
Professor:
  Andrew Gale, (D), King's College (UK), 1984, PgGrCl
  Jim Smith, (D), Ge
  Craig D. Storey, (D), Leicester, 2002, GptCc
Associate Professor:
  David Giles, (D), Glm
Research Associate:
  Emilie Braund, (D), Graz, 2011, Gz
  Fay Couceiro, (D), On
  Penny Lancaster, (D), Bristol, 2011, Gz
  Robert Loveridge
  Darren Naish
  David Ray, (D)
  Alan Raybould, (D)
  Steve Sweetman, (D), Pg
  Mark Whitton, (D), Portsmouth, 2008, Pg
Instructor:
  Chris Dewdney, (D), Birkbeck, 1983, Yg
  Gary Fones, (D), C Lancashire, Cm
Lecturer:
  John Allen, (D), Southampton, 1996, Og
  Hooshyar Assadullahi
  Philip Benson, (D)

Michelle Bloor, (D), Hg
Dean Bullen, (D), Gt
Anthony Butcher, (D), PlmZe
James Darling, (D), Bristol, 2009, Cs
Mike Fowler, (D), Imperial, Gi
David Franklin, (D), Ym
Martyn Gardiner, (D)
Andy Gibson, (D), Eg
Michelle Hale, (D), Flinders, Og
Nick Koor
David K. Loydell, (D), Aberystwyth, 1989, GrPg
Nicholas Minter, (D), Bristol, 2007, Gs
Derek Rust, (D), Gt
Camen Solana, (D), Gv
Richard Teeuw, (D), Stirring (UK), 1985, AsGmZi
Melvin M. Vopson, (D), C Lancashire, 2002, Yg
Nick Walton, Hw
John Whalley, (D), Gc
Malcolm Whitworth, (D)
Adjunct Professor:
  David Martill, (D), PvgPe

University of Reading
Soil Research Centre (B,M,D) (2015)
Department of Geography and Environmental Science
School of Human and Environmental Sciences
Whiteknights
Reading RG6 6AB
  p. +44 (0)118 378 8911
  shes@reading.ac.uk
  http://www.reading.ac.uk/soil-research-centre
Professor:
  Chris Collins, (D), Sb
Associate Professor:
  Stuart Black, (D), Lancaster, Cc
  Chris Collins, (D), Sc
  McGoff Hazel, (D), Liverpool, Pg
  Steve Robinson, (D), Sf
  Liz Shaw, (D), Sb
Dr:
  Joanna Clark, (D), Leeds, Sc

University of South Wales
Geology Section (2015)
Department of Science
Pontypridd

University of Southampton
School of Ocean and Earth Science (B,M,D) ●☒ (2018)
University of Southampton Waterfront Campus
National Oceanography Centre Southampton
European Way
Southampton, Hampshire SO14 3ZH
  p. +44 (0)23 8059 2011
  soes@soton.ac.uk
  http://www.southampton.ac.uk/oes
  f: https://www.facebook.com/UoSOceanography
  t: @OceanEarthUoS
  Programs: Geology with Physical Geography (B); Geophysical Sciences (B); Oceanography (B); Oceanography with Physical Geography (B); Marine Biology with Oceanography (B); Geology (B,M), Geology with Study Abroad (M); Geophysics (M); Geophysics with study abroad (M); Engineering/Physics/Maths/Geophysics (M); Biology and Marine Biology (M); Marine Biology (M); Marine Biology with study abroad (M); Oceanography (M); Oceanography with French (M); Oceanography with study abroad (M); Foundation Year in Science
Professor:
  Stephen Roberts, (D), Open Univ, 1986, Cg
Head of Physical Oceanography Research Group:
  Alberto Naveira Garabato, (D), Liverpool, 1999, Op
Head of Palaeoceanography and Palaeoclimate Research Group:
  Paul A. Wilson, (D), Cambridge, 1995, GsCg
Head of NOCS Graduate School and Inspire:
  Martin R. Palmer, (D), Leeds, 1984, Cg
Head of Marine Biology & Ecology Research Group:
  Jorg Wiedenmann, (D), Ulm (Germany), 2000, Ob
Head of Marine Biogeochemistry Research Group:
  Christopher Mark Moore, (D), Southampton, 2002, Ob
Head of Geology and Geophysics Research Group:
  Justin Dix, (D), St. Andrews, 1994, GuYuNx
Faculty Director of Graduate School:
  Timothy A. Minshull, (D), Cambridge, 1990, Yr
Director, GAU-Radioanalytical, Professorial Fellow:
  Phillip Warwick, (D), Southampton, 1999, CaGe
Deputy Head of School Research and Enterprise:
  Rachael James, (D), Cambridge, 1995, Oc
Deputy Head of School International:
  Andy Cundy, (D), Southampton, 1994
Deputy Head of School Education:
  Chris Hauton, (D), Southampton, 1995, Ob
Dean, Faculty of Environmental and Life Sciences:
  Rachel A. Mills, (D), Cambridge, 1992, Cm
Professor:
  Nicholas R. Bates, (D), Southampton, 1995, Oc
  Thomas Bibby, (D), Imperial Coll (London), 2003, Ob
  Colin Brownlee, (D), Newcastle upon Tyne, Ob
  Jonathan M. Bull, (D), Edinburgh (UK), 1990, YrGct
  Sybren Drijfhout, (D), Urtecht, 1992, Op
  Gavin Foster, (D), Open, 2000, Csg
  Tim Henstock, (D), Cambridge (UK), 1994, YgrYe
  Alan Kemp, (D), Edinburgh, 1985, Pe
  Maeve Lohan, (D), Southampton, 2003, Ob
  Robert Marsh, (D), Southampton, 2000, Op
  John E.A Marshall, (D), Bristol, 1981, PlGro
  Lisa McNeill, (D), Oregon State, 1998, GtYr
  Duncan A. Purdie, (D), Wales, 1982, Ob
  David Sims, (D), Plymouth, 1994, Ob
  Martin Solan, (D), National (Ireland), 2000, Ob
  Damon A.H Teagle, (D), Cambridge, 1993, CgEmGp
  Toby Tyrrell, (D), Edinburgh (UK), 1993, OgPe
Head of Geochemistry Research Group:
  Juerg M. Matter, (D), Swiss Fed Inst Tech, 2001, Ng
Associate Professor:
  Jonathan Copley, (D), Southampton, 1998, Ob
  Thomas M. Gernon, (D), Bristol, 2007, ZgGvi
  Ivan D. Haigh, (D), Southampton, 2009, On
  Ian C. Harding, (D), Cambridge (UK), 1986, PmlPc
  Antony Jensen, (D), Southampton, 1982, Ob
  Derek Keir, (D), Royal Holloway (London), 2006, Gtv
  Phyllis Lam, (D), Hawaii, 2004, Ob
  Cathy Lucas, (D), Southampton, 1993, Ob
  Catherine A. Rychert, (D), Brown, 2007, Yg
  Rex N. Taylor, (D), Southampton, 1987, GvCa
  Sven Thatje, (D), Bremen, 2003, Ob
  Clive Trueman, (D), Bristol, 1997, Cg
  Jessica H. Whiteside, (D), Columbia, 2006, Pe
Teaching and Research Fellow :
  Hachem Kassem, (D), Southampton, 2016, OnGsZi
SMMI Lecturer:
  Steven Bohaty, (D), California (Santa Cruz), 2006, Og
Senior Tutor, Principle Teaching Fellow:
  Simon R. Boxall, (D), Liverpool, 1985, Op
Senior Tutor:
  Andy J. Barker, (D), Wales (Cardiff), 1983, Gg
Senior Research Fellow:
  Kenneth Collins, (D), Southampton, 1979, Ob
NERC Advanced Senior Research Fellow:
  Tom H.G Ezard, (D), Imperial Coll (London), 2007
Lecturer:
  Maria C. D'Angelo, Ob
  Phillip Fenberg, (D), California (San Diego), 2008, Ob
  Jasmin A. Godbold, (D), Aberdeen, 2009, Ob
  Phillip A. Goodwin, (D), Liverpool, 2007, Pe
  Nicholas Harmon, (D), Brown, 2007, Yg
  Anna Hickman, (D), Southampton, 2007, Og
  Kevin Oliver, (D), East Anglia, 2003, Op
  Marc Rius, (D), Barcelona, 2008, Znn
  Esther Sumner, (D), Bristol, 2009, Gs
  Chuang Xuan, (D), Florida, 2010, Ym
Visiting Professor:
  Ian Townend, (B), Exeter, 1975

Emeritus Fellow:
  Lawrence Hawkins, (D), Southampton, 1985, Ob
Visiting Fellow:
  Claudie Beaulieu, (D), Quebec, 2009, Op
  Laura Grange, (D), Southampton, 2005
  Mathis P. Hain, Princeton, 2013
  Torben Stichel, (D), Christian-Albrechts, 2011
Teaching Fellow, Project Manager for Oman Drilling Project:
  Jude A. Coggon, (D), Durham, 2010, Cg
Senior Research Fellow:
  Matthew Cooper, (D), Cambridge, 1999, Oc
Royal Society University Research Fellow:
  Rosalind M. Coggon, (D), Southampton, 2006, Ce
  Samantha J. Gibbs, (D), Cambridge, 2002, Zg
  Ben A. Ward, (D), Southampton, 2009, Ob

University of St. Andrews
Dept of Earth and Environmental Science (B,M,D) (2015)
Irvine Building
North St.
Fife, Scotland KY16 9AL
  p. (+44) (0)1334 463940
  earthsci@st-andrews.ac.uk
  http://earthsci.st-andrews.ac.uk/
  f: https://www.facebook.com/EarthSciStA/
  t: @EarthSciStA
Director:
  Richard Bates, (D), Wales, Yx
  Ruth Robinson, (D), Penn State, 1997, Gs
Head:
  Tony Prave, (D), Penn State, 1986, GgrGs
Associate Professor:
  Andrea Burke, (D), MIT/WHOI, 2011, ClPc
Assistant Professor:
  Timothy D. Raub, (D), Yale, 2008, YmGgPc
Research Associate:
  Nicky Allison, (D), Edinburgh, 1994, Oc
  Mark Claire, (D), Washington, As
  Catherine Cole, (D), Southampton, 2013, Cm
  Ruth Hindshaw, (D), ETH Zurich, 2011, Cs
  Gareth Izon, (D), Cg
  Coralie Mills
  James Rae, (D), Bristol, Cg
  Vincent Rinterknecht, (D), CcGma
Lecturer:
  Adriam Finch, Gz
  Timothy Hill, (D), Edinburgh, 2003, Ge
  Michael Singer, (D), California (Santa Barbara), 2003, Hg
  John Walden, (D), Wolverhampton Polytechnic, 1990, Ym
  Robert Wilson, (D), W Ontario, 2003, Zn
  Aubrey Zerkle, (D), Penn State, 2006, Co

School of Geography & Geosciences (2015)
Division of Geology
St. Andrews
Fife KY16 9ST
Emeritus:
  Colin K. Ballantyne, (D), Edinburgh (UK), 1981, ZyGml

University of Wales
Dept of Earth and Ocean Sciences (2015)
College of Cardiff
Main Building
Park Place
Cardiff CF1 3YE
  p. +44 (0)29 208 74830
  earth-ug@cf.ac.uk
  http://www.cardiff.ac.uk/earth/
Professor:
  Thomas Blenkinspo
  Ian R. Hall
  Richard Lisle
  Chris MacLeod, Og
  Wolfgang Maier
  Paul Pearson, PmCg
Instructor:
  Stephen Barker
  Huw Davies
  Andrew Kerr, Og
  Jenny Pike, Og
Lecturer:
  Tiego Alves, Og
  Liz Bagshaw
  Rhoda Ballinger, On
  Peter Brabham, Ng
  David Buchs
  Alan Channing, Gd
  Jose Constantine
  T. C. Hales, Gt
  Alan Hemsley, Pl
  Tim Jones
  C. Johan Lissenberg
  Iain MacDonald, Ca
  Rupert Perkins
  Phil Renforth
  David Reynolds
  Henrik Sass, Co
  Simon Wakefield
Emeritus:
  Hazel Prichard, (D), EgGze
Laboratory Director:
  Caroline Lear, Og
Tutorial Fellow:
  Ian Fryett
  Nick Rodgers, Xm

Geography and Earth Sciences (2015)
Llandinam Building
Penglais Campus
Aberystwyth SY23 3DB
  p. +44(0) 1970 622 606
  dges@aber.ac.uk
  http://www.aber.ac.uk/en/iges/
Director:
  Neil Glasser
Head:
  Rhys Jones
Professor:
  Paul A. Brewer, (D), Gm
  John Grattan
  Matthew Hannah
  Alun Hubbard
  Bryn Hubbard
  David Kay
  Henry Lamb
  Richard Lucas
  Mark Macklin
  Alex Maltman, (D), Illinois, 1973
  Nick Pearce
  Andrew D. Thomas, (D), Swansea, 1996, SbGmSf
  Mark Whitehead
  Michael Woods
Instructor:
  Peter Abrahams
  Peter Merriman
  Andrew Mitchell
  Helen Roberts
  Stephen Tooth
Lecturer:
  Charlie Bendall
  Peter Bunting
  Rachel Carr
  Rhys Dafydd Jones
  Sarah Davies
  Carina Fearnley
  Elizabeth Gagen
  Hywel Griffiths
  Kevin Grove
  Andrew Hardy
  Jesse Heley
  Tom Holt
  Gareth Hoskins

Tristram Irvine-Fynn
Cerys Jones
Bill Perkings
Kimberly Peters
George Petropoulos
Mitch Rose
Richard Williams
Sophie Wynne-Jones
Teaching Fellow:
Stefania Amici

School of Ocean Sciences (2015)
Bangor University
Menai Bridge
Anglesey, Bangor LL59 5AB
 p. (01248) 382851
 oss011@bangor.ac.uk
 http://www.bangor.ac.uk/oceansciences/
Director:
 Colin Jago
Professor:
 David Bowers
 Alan Davies
 Michel Kaiser
 Hilary Kennedy
 Chris Richardson
 Tom Rippeth
 James Scourse
 John Simpson
 David Thomas
Associate Professor:
 Simon P. Neill, (D), Strathclyde, 2001, OnpOg
Instructor:
 Jan Geert Hiddink
 Stuart Jenkins
Lecturer:
 Martin Austin
 Jaco H. Baas
 Paul Butler
 Lui Gimenez
 Mattias Green
 Cara Hughes
 Dei G. Huws, (D), Wales, 1992, YruNt
 Suzanna Jackson
 Lewis LeVay
 Shelagh Malham
 Irene Martins
 Ian McCarthy
 Gay Mitchelson-Jacob
 Anna Pienkowski
 Martin Skov
 John Russel Turner
 Katrien van Landeghem
 Stephanie Wilson
Related Staff:
 Timothy Whitton

Uruguay

Direccion Nacional de Mineria y Geologia de Uruguay
Direccion Nacional de Mineria y Geologia de Uruguay (2015)
Hervidero 2861
Montevideo, Montevideo 11800
 p. + 5982 2001951
 infomiem@miem.gub.uy
 http://www.dinamige.gub.uy/

Universidad de la Republica Montevideo
Dept de Geologia (2015)
Avenida 18 de Julio 1968, Montevideo

Universidad de la Republica Oriental del Uruguay (UDELAR)
Instituto de Geología y Paleontología (2015)
Avenida 18 de Julio 1968, Montevideo
CP11 400
Montevideo 10773
 http://www.fcien.edu.uy/menu2/estructura2/ingepa2.html

Deptartamento de Geografia (B) ☒ (2017)
Iguá 4225 Piso 14 Sur C.P: 11400
Montevideo, Montevideo 11400
 p. (598-2) 525 15 52
 geotecno@fcien.edu.uy
 http://geografia.fcien.edu.uy/
Professor:
 Virginia Fernández, (M), Girona, 2001, ZirZn
Assistant Professor:
 Yuri Resnichenko, (M), la República, 2010, ZirZn
Related Staff:
 Carlos Miguel, (B), Valladolid, 4, ZirSf
 Virginia Pedemonte, (B), Fac de Arquitectura, 5, ZirZu

Dept de Suelos y Aguas (2015)
 p. (598 2) 359 82 72
 suelosyaguas@fagro.edu.uy
 http://suelosyaguas.fagro.edu.uy/

Uzbekistan

Institute of Geology and Geophysics
 http://www.ingeo.uz/

Institute of Mineral Resources
 gpniimr@evo.uz
 http://www.gpniimr.uz/

National University of Uzbekistan
Faculty of Geology (2015)
Tashkent
University City 100095
 p. +998712460224
 geology@nuu.uz
 http://nuu.uz/geolog

Tashkent State Technical University
Faculty of Geology (2015)
Uzbekistan, Tashkent, 10095 Universitetskaya street, 2
 p. +998712464600
 TFTU_info@mail.ru

The State Committee for Nature Protection
100 159, Tashkent, pl. Independence, 5
 info@uznature.uz
 http://www.uznature.uz

The State Committee of the Republic of Uzbekistan on Geology and Mineral Resources
11, T.Shevchenko str., Tashkent, Republic of Uzbekistan 100060
 p. +998 (71) 256-8653
 geolcom@bcc.com.uz
 http://www.uzgeolcom.uz

Uzbekistan National Oil and Gas Company - Uzbekneftegaz (UNG)
100047, city Street, Tashkent.
Istiqbol, 21
 p. +998 (71) 233-5757
 kans@uzneftegaz.uz
 http://www.uzneftegaz.uz

Venezuela

Universidad Central de Venezuela
Inst de Ciencias de la Tierra (2015)
Los Chaguaramos, Apdo. 3895, 1010-A Caracas

p. 0212 6366236
coordinv@ciens.ucv.ve
http://www.coordinv.ciens.ucv.ve/investigacion/genci/sitios/35/index.html

Escuela de Geologia, Minas y Geofisica (2015)
Ciudad Universitaria, 47028 Caracas 1041 A

Universidad de la Este, Cumana
Dept de Geologia y Minas (2015)
Apartado Postal 245, Cumana (Estado Sucre)

Universidad de Los Andes
Dept de Geologia y Minas (2015)
Avenida 3, Independencia, La Hechlcera, Merida
ocamacho@ula.ve
http://llama.adm.ula.ve/pingenieria/index.php?option=com_content&view=article&id=313&Itemid=215

Escuela de Ingenieria Geologica (2015)
Av. Tulio Febres C., Merida 5101

Universidad de Oriente, Nucleo Bolivar
Escuela de Ciencias de la Tierra (2015)
Av. Universidad, Campus Universitario La Sabanita, Ciudad Bolivar

Universidad Simon Bolivar
Coordinacion de Ingenieria Geofisica (2015)
Valle de Sarteneja, Baruta, Edo. Miranda 80659, Caracas 1080
p. 9063545
coord-geo@usb.ve
http://www.gc.usb.ve/geocoordweb/index.html

Vietnam

Hanoi University of Mining & Geology
Faculty of Geology (A,B,M,D) O (2015)
Duc Thang Ward - North Tu Liem Distr.
Hanoi
p. +84-4-38387567
diachat@humg.edu.vn
khoadiachat.edu.vn
Enrollment (2015): B: 2800 (420) M: 205 (50) D: 24 (6)
Dean of Faculty:
 Lam Van Nguyen, (D), Hanoi Univ of Mining and Geology, 2002, HwGeHg
Deputy Dean of Faculty:
 Thanh Xuan Ngo, (D), Okayama, 2009, GtcGi

Zambia

Copperbelt University
Mining Dept (B) (2015)
Kitwe
deansot@cbu.ac.zm
http://www.cbu.edu.zm/technology

University of Zambia
Dept of Metallurgy and Mineral Processing (B) (2015)
PO Box 32379
Lusaka
p. +360-21-1-250871
registrar@unza.zm
http://www.unza.zm/index.php?option=com_content&task=view&id=479&Itemid=574

Zimbabwe

Africa University
Faculty of Agriculture and Natural Sciences (B) (2015)
Fairview Rd (Off- Nyanga Rd)
PO Box 1320
Old Mutare, Mutare

p. +2632060075
info@africau.edu
http://www.africau.edu/academic/default.htm

University of Zimbabwe
Dept of Geology (B,M,D) (2015)
Building B047
University of Zimbabwe
P.O. Box MP167
Mount Pleasant, Harare
p. 303211 Ext. 15032
gchipari@science.uz.ac.zw
http://www.uz.ac.zw/index.php/2013-07-09-08-51-40/the-department-of-geology/226-sci/dept-sci/geology-dpt/826-dr-lrm-nhamo
Enrollment (2012): B: 27 (17) M: 1 (0) D: 3 (1)
Chair:
 LRM Nhamo
Lecturer:
 Trendai Jnila, (D)
 Isidro Rafael Vit Manuel, (D)
 Maideyi Lydia Meck, (D)

Dept of Geography & Environmental Science (B,M,D) (2015)
PO Box MP167
Mount Pleasant, Harare
p. +263-04-303211
geography@arts.uz.ac.zw
http://www.uz.ac.zw/science/geography/

Institute of Mining Research (B) (2015)
PO Box MP167
Mount Pleasant, Harare
p. +263-4-336418
speka@science.uz.ac.zw
http://www.uz.ac.zw/

Dept of Mining Engineering (B) (2015)
PO Box MP167
Mount Pleasant, Harare
p. (263) 4 -3335 x ext: 17089)
kudzie@eng.uz.ac.zw
http://www.uz.ac.zw/index.php/mining-about

Dept of Geoinformatics and Surveying (B) (2015)
PO Box MP167
Mount Pleasant, Harare
p. +263 772 318 473
bukalt@eng.uz.ac.zw
http://www.uz.ac.zw/index.php/fac-of-eng/department-of-geoinformatics-and-surveying

Theses and Dissertations, 2016

The following section documents all of the 2016 geoscience dissertations and theses from U.S. and Canadian institutions that were reported to GeoRef Information Services. If you have questions about the data or to make sure your institution's data is included in the future, please contact Monika Long at ml@americangeosciences.org.

Arizona State University
Doctorate
    Miller, Kelly Elizabeth, *The R chondrite record of volatile-rich environments in the early solar system*

Auburn University
Masters
    Ahmed, Nur, *Saltwater intrusion and trace element contamination at the coastal aquifers of the Ganges Delta*
    Cooper, Katherine L., *Ion probe dating of zircons from Wuluke Volcano, NW Tibet, China; constraints on magma evolution*
    Fan, Di, *$^{40}Ar/^{39}Ar$ ages of feldspar and muscovite from the source and detritus of the French Broad River, North Carolina*
    Gunn, James N., III, *$^{40}Ar/^{39}Ar$ age variations among cogenetic feldspars from the Benson Mines, New York*
    Haque, Ziaul, *Petrofacies and detrital geochronology of the upper Pottsville Conglomerate magnafacies, Cahaba Synclinorium, Southern Appalachian fold and thrust belt, Alabama*
    Jahan, Shakura, *Preliminary assessment of the petroleum source rock potential of upper Eocene Kopili Shale, Bengal Basin, Bangladesh*
    Le Blanc, Lainey M., *Geomorphic and sedimentologic analysis of four streams in southeastern Alabama; channel and habitat response to changes in land use*
    Ozsarac, Safak, *Igneous-rock hosted orogenic gold deposit at Hog Mountain, Tallapoosa County, Alabama*
    Saffari Ghandehari, Shahrzad, *Bioremediation of an arsenic-contaminated site, using sulfate-reducing Bacteria, Bay County, Florida*
    Singleton, Robert Frank, III, *Gravity and magnetic modeling of basement beneath the Appalachian Plateau, Valley and Ridge, and Piedmont Provinces, Alabama*
    VanDervoort, Dane S., *Geology of the Wadley South Quadrangle and geochronology of the Dadeville Complex, southernmost Appalachians of east Alabama*

Baylor University
Doctorate
    Agrawal, Mohit, *Multi objective optimization for seismology (MOOS), with application to the Middle East, the Texas Gulf Coast, and the Rio Grande Rift*
    Brownlow, Joshua Wayne, *Influence of hydraulic fracturing on overlying aquifers in the presence of abandoned and converted oil and gas wells; numerical, spatial, uncertainty, and geochemical investigations*
    Xu, Tian, *Application of direct current resistivity method to environmental and hydrological problems*

Masters
    Byram, Ian M., *Depositional and diagenetic controls on reservoir quality and petrophysical responses of the DeBroeck Member of the Rodessa Formation in northwest Louisiana*
    Idleman, Erin E., *Chemostratigraphic analysis and petrophysical modeling of the Upper Jurassic Haynesville Shale in East Texas and west Louisiana*
    Norair, Stephen G., *Streambank erosion assessment; application of dendrogeomorphology, numerical watershed modeling, and model characterization*
    Ntuli, Gift, *Ps receiver function imaging of crustal structure and Moho topography beneath the northeast Caribbean*
    Parizek, Daniel J. M., *Depositional and stratigraphic controls on reservoir distribution within the Rodessa Formation in northwest Louisiana*
    Rasaka, Brandon M., *Correlation of selected earthquakes with seismogenic faults, central Oklahoma*
    Shaw, Kenton A., *Geodetic constraints on deformation of the northeast Caribbean microplates and seismic delay times measured via waveform cross-correlation*
    Worrell, Victoria E., *The seismo-lineament analysis method (SLAM) applied to the South Napa earthquake and antecedent events*

Bachelor
    Abbuhl, Brittany Morgan, *Lateral floral variability in the early Paleocene Nacimiento Formation, San Juan Basin, New Mexico*
    Barton, Mitchell, *Using paleosols from the UNESCO World Heritage Joggins Fossil Cliffs to reconstruct paleoenvironment and paleoclimate of the Carboniferous of Nova Scotia*
    Benton, Nathan, *The diagnostic constraints of shear-wave splitting parameters; an examination of SKS anisotropy detected in and around the Rio Grande rift zone*
    Evans, Zachary, *Early Paleocene climate reconstruction using the isotopic composition of fossil plants*

Bowling Green State University
Masters
    Feitl, Melina G., *Investigation of diatom endemism and species response to climate events using examples from the genera Cyclotella (Lindavia) and Surirella in the Lake El'gygytgyn sediment record*
    Lockshin, Samuel N., *Spatial characterization of Western Interior Seaway paleoceanography using Foraminifera, fuzzy sets and Dempster-Shafer theory*
    Magdic, Matthew James, *Assessment of soil properties in proximity to abandoned oil wells using remote sensing and clay X-ray analysis, Wood County, Ohio*
    Redner, Ellen R., *Magma mixing and evolution at Minna Bluff, Antarctica revealed by amphibole and clinopyroxene analyses*
    Walters, Daryl Georjeanne, *Geospatial analysis of ecological associations and successions in Middle Devonian bioherms of the Great Lakes region*

Brigham Young University
Masters
    Ahern, Alexandra Anne, *Lineations and structural mapping of Io's paterae and mountains; implications for internal stresses*
    Allred, Isaac John, *Spatial trends and facies distribution of the high-energy alluvial Cutler Formation, southeastern Utah*
    Cook, Preston Scott, *Sedimentology and stratigraphy of the Middle Jurassic Preuss Sandstone in northern Utah and eastern Idaho*
    Dailey, Shane Robert, *Geochemistry of the fluorine; and beryllium-rich Spor Mountain Rhyolite, western Utah*
    Goodsell, Timothy Holman, *Trace element inputs from natural and anthropogenic sources in an agricultural watershed, middle Provo River, Utah*
    Little, David A., *Geochemical comparison of ancient and modern eolian dune foresets using principal components analysis*
    Shumway, Jesse Dean, *Facies analysis and depositional environments of the Saints & Sinners Quarry in the Nugget Sandstone of northeastern Utah*
    Spiel, Kinsey Gayle, *Designing an instructional publication on the geology of Capitol Reef National Park*
    Valenza, Jeffery Michael, *Redbeds of the upper Entrada Sandstone, central Utah; facies analysis and regional implications of interfingered sabkha and fluvial terminal splay sediments*

California Institute of Technology
Doctorate
    Case, David Hamilton, *Carbonate-associated microbial ecology at methane seeps; assemblage composition, response to changing environmental conditions, and implications for biomarker longevity*
    Li, Dunzhu, *Some advances in computational geophysics; seismic wave and inverse geodynamic modeling*
    Ma, Yiran, *Imaging the Earth with ambient noise and earthquakes*
    Martens, Hilary Rose, *Using Earth deformation caused by surface mass loading to constrain the elastic structure of the crust and mantle*
    Minchew, Brent M., *Mechanics of deformable glacier beds*
    Nakajima, Miki, *Origin of the Earth and Moon*
    Newcombe, Megan Eve, *Solubility and diffusivity of water in lunar basalt, and chemical zonation in olivine-hosted melt inclusions*
    Prancevic, Jeffrey Paul, *Sediment mobility in steep channels and the transition to landsliding*
    Raven, Morgan Reed, *Organic matter sulfurization in the modern*

ocean

Runnels, Brandon Scott, *A model for energy and morphology of crystalline grain boundaries with arbitrary geometric character*

Scheingross, Joel Simon, *Mechanics of sediment transport and bedrock erosion in steep landscapes*

Siebach, Kirsten Leigh, *Formation and diagenesis of sedimentary rocks in Gale Crater, Mars*

Slotznick, Sarah Pearl, *Coupling textural, magnetic, and modeling techniques to understand Precambrian paleoenvironments*

Sousa, Francis Joseph, *Tectonics of Central and eastern California, Late Cretaceous to modern*

Stevens, Victoria L., *Reconciling geodetic strain and seismicity rate with frequency-magnitude relation of the largest earthquakes*

Masters

Grappone, Joseph Michael, *Investigating the death of the early Paleozoic Moyero River geomagnetic superchron; Middle Ordovician paleomagnetism from Estonia*

California State University at Long Beach
Masters

Blackford, Nolan R., *Testing the relationship between vertical-axis rotation and large-magnitude extension in the central Mojave metamorphic core complex*

Caprile, Jose A., *The geochemical influence of trace element concentrations from marine sedimentary bedrock on freshwater streams in the western Transverse Mountain Ranges*

Kassa, Tesfalidet Ghirmay, *Pore structure of opal-CT and quartz porcelanites, Monterey Formation, California*

Lojasiewicz, Iwo, *A stable isotope study of fluid-rock interactions in the Saddlebag Lake roof pendant, Sierra Nevada, California*

McCarthy, Anne M., *Stable isotope evidence for a massive hydrothermal system related to the Coryell Intrusive Suite, British Columbia*

Moertle, Jasmine A., *Stable isotope evidence for a complex fluid evolution of the northwestern British Columbia Coast Mountains related to terrane accretion*

Njuguna, Wanjiru Margaret, *Mixed siliciclastic-siliceous succession, Miocene Monterey Formation, Point Dume to Paradise Cove, Malibu, California*

Rao, Amar P., *The hydraulic connectivity, perennial warming and relationship to seismicity of the Davis-Schrimpf Seep Field, Salton Trough, California from new and recent temperature time-series*

Zeferjahn, Tanya L., *Submarine groundwater discharge as a freshwater resource for the ancient inhabitants of Rapa Nui*

California State University at Northridge
Masters

Bailey, Claire, *Provenance study of Miocene to Pliocene sand and sandstone intervals recovered on Expedition 317 in the Canterbury Basin, South Island, New Zealand*

Cardona, Jose E., *Constraining the most recent surface rupture on the Garnet Hill strand, San Andreas Fault, Coachella Valley, California*

Cohen, Hannah Rose, *Constraints on the late Miocene/Pliocene reorganization of the San Andreas Fault and basin drainage systems near Tejon Pass, CA, based on stratigraphy, sedimentology, and provenance of the Hungry Valley Formation*

Decker, Meghann Fernley Ilitta, *Triggering mechanisms for a magmatic flare-up of the lower crust in Fiordland, New Zealand, from U-Pb zircon geochronology and O-Hf zircon geochemistry*

Dickey, Nathan W., *Chronology and paleoclimate of late Pleistocene glaciation in the Klamath Mountains, CA*

Gebauer, Samantha Kelly, *Spatio-temporal patterns of lower arc cooling and metamorphism, northern Fiordland, New Zealand*

Johnson, Kyle E., *Facies and depositional history of arc-related, deep-marine volcaniclastic rocks in core recovered on International Ocean Discovery Program Expedition 351, Philippine Sea*

Jones, Kaitlyn L., *The geology of the Singleton Anticline and along-strike change of the San Gorgonio Pass strand, San Andreas Fault from emergent thrust to anticline, near Calimesa, CA*

Joseph, Jeffery, III, *Deformation and fabric transposition in the lower crust during extensional collapse in western Fiordland, New Zealand*

McDonald, Eric Kenneth, *Stratigraphy and detrital zircon U-Pb geochronology of the El Paso Mountains Permian metasedimentary sequence, Southern California*

Moss, Benjamin, *Misorientation analysis with electron backscatter diffraction (EBSD); a novel approach for characterizing hypersolidus to subsolidus fabric development in gabbros from Ocean Drilling Program Hole 735B, Southwest Indian Ridge*

Nissanka Arachchige, Uchitha Suranga, *Formation of intraplate seamount chains by viscous fingering instabilities in the asthenosphere using low Reynolds number miscible fluids with a moving surface boundary*

Rousseau, Nick, *An investigation into the episodic behavior and potential causes of Jurassic magmatism in the northern Sierra Nevada, California*

Wiesenfeld, John Arthur, *Construction of continental crust by deep crustal fractional crystallization and garnet pyroxenite root development; geochemical evidence from Fiordland, New Zealand*

Carleton University
Masters

Blanchard, Jennifer, *Geophysical identification and characterization of mafic-ultramafic intrusions in plume centre regions*

Cunningham, Michael, *Aeromagnetic surveying with unmanned aircraft systems*

Davies, Marissa, *An integrated chemostratigraphic and biostratigraphic framework and benthic foraminifers morphogroup response to paleoenvironmental conditions of the Upper Cretaceous Kanguk Formation, Canadian Arctic Archipelago*

Dziawa, Carolyn, *Quantifying the pressure, temperature, and timing of metamorphism in the Montresor Belt, Nunavut, Canada*

Graff, Jamie Richard, *A history of tectono-magmatism along the Parga Chasma rift system on Venus*

Hache, Michel, *Morphometric analysis of the gull Larus (Aves, Laridae) with implications for small theropod diversity in the Late Cretaceous*

Hayek, Sylvia J., *Seismic basin effects over soft-sediment filled basins in Ottawa, Canada*

Kingdon, Susan Angela, *Pliocene to late Pleistocene magmatism in the Aurora volcanic field, Nevada and California, USA; a petrographic, geochemical, and isotopic study*

Quesnel, Alex, *Cretaceous (Albian to Cenomanian) strata of Eagle Plain, Yukon; subsurface foraminiferal biostratigraphy, regional correlations and paleoenvironments*

Clarkson University
Doctorate

Kumarage, Pubudu, *Simulating transport of gases released from hydrothermal vents and resulting acidification*

Colorado School of Mines
Doctorate

Albinali, Ali, *Analytical solution for anomalous diffusion in fractured nano-porous reservoirs*

Arshad, Hafiz Syed Mahmood, *Numerical analyses of the long term behavior of enhanced geothermal system (EGS)*

Azhari, Amin, *Evaluating the effect of earthquakes on open pit mine slopes*

Cui, Qi, *Stress dependent compaction in tight reservoirs and its impact on long-term production*

Dee, Kato Tsosie, *Dissolved organic carbon (DOC) characteristics in metal-rich waters and implications for copper aquatic toxicity*

Diaz Pantin, Esteban F., *Extended imaging and tomography under two-way operators*

Ding, Dong, *Application of the Lagrangian particle-tracking method to simulating mixing-limited, field-scale biodegradation*

Drennan, Dina Marie, *Biogeochemistry of sulfate reducing bioreactors; how design parameters influence microbial consortia and metal precipitation*

Duan, Yuting, *Elastic wavefield migration and tomography*

Foks, Nathan Leon, *Using adaptive sampling and triangular meshes for the processing and inversion of potential field data*

Gilbert, James M., *Exploring catchment connections with integrated hydrologic models; system interactions and responses to groundwater extraction and climate change in the San Joaquin River basin*

Gulley, Andrew L., *Valuation of ecosystem services impacted by mine site pollution*

Hernandez Bilbao, Eider, *High-resolution chemostratigraphy, sequence stratigraphic correlation, porosity and fracture characterization of the Vaca Muerta Formation, Neuquen Basin, Argentina*

Jefferson, Jennifer L., *Exploring sensitivities of latent heat parameterizations using a coupled, integrated hydrologic model*

Knipper, Kyle, *Improving evapotranspiration estimates in the arid west using multi-platform remote sensing*

Moradi, Ali, *Experimental and numerical analysis of heat transfer in unsaturated soil with an application to soil borehole thermal energy storage (SBTES) systems*

Morse, Michael S., *Field and laboratory investigations of variably* saturated, potential landslides

Parekh, Minal L., *Advancing internal erosion monitoring using seismic methods in field and laboratory studies*

Poeck, Eric O., *Analyzing the potential for unstable mine failures with the calculation of released energy in numerical models*

Williamson, Jacob Lee, *Development and application of field methods for determination of the extent of acid mine drainage contamination and geochemical characteristics of stream sediment recovery*

Wu, Long, *Foreland flexural extension and salt diapir reactivation in oblique extensional systems*

Wu, Xinming, *3D seismic image processing for interpretation*

Masters

Adham, Aidil, *Geomechanics model for wellbore stability analysis in Field "X" North Sumatra Basin*

Al-Khalifa, Aqeel M., *Study of the organic content and mineralogy effects on the acoustic properties of the Niobrara Formation, Denver Basin*

Allen, Nicole K., *Structurally controlled Cu-Zn-(Co-Pb) mineralization in the Neoproterozoic Ombombo Subgroup, Kaololand, Namibia*

Alvarado Blohm, Fernando, *Quantifying the value in geoscientific information using panel data econometrics*

Andrews, Ben P., *Shear-wave splitting and attenuation analysis of downhole microseismic data for reservoir characterization of the Montney Formation, Pouce Coupe, Alberta*

Arias, Elias, *Seismic slope estimation; analyses in 2D/3D*

Baizhanov, Bekdar, *An experimental study of true triaxial stress-induced deformation and permeability anisotropy in sandstones*

Ball, Steffen M., *Hydrothermal nickel sulfide hosted in Neoproterozoic carbonate and evaporitic rocks of the Menda Central Prospect, Democratic Republic of Congo*

Bandler, Aaron J., *Geophysical constraints on critical zone architecture and subsurface hydrology of opposing montane hillslopes*

Bogenschuetz, Nicole, *The effect of the mountain pine beetle on slope stability, soil moisture and root strength*

Borrillo-Hutter, Travis, *Environmental characterization, reclamation, and community monitoring of artisanal and small-scale mining (ASM) impacts to surface waters in Suriname*

Bradford, Emma L., *Parameters influencing compensational behavior of alluvial fans*

Brennan, Stephen W., *Integrated characterization of middle Bakken diagenesis, Williston Basin, North Dakota, U.S.A.*

Bridges, Mason, *Mechanical properties of the Niobrara*

Brown, Hayden Edward, *Spatial and temporal analysis of compensational stacking and gradual migration of an experimental debris-flow fan*

Brown, Nathan David, *Geologic considerations for enhanced oil recovery in Elm Coulee Field, Richland County, Montana, Williston Basin*

Brush, Jennifer F., *Faults and fractures in the Niobrara Formation of Wattenberg Field, Colorado*

Bucker, Wesley Shayne, *Characterization and restoration of inverted features across the southern Taranaki Basin*

Butler, Emily, *P-wave seismic time-lapse analysis of horizontal well completions causing pressure compartmentalization*

Cross, Lauren C., *Structural evolution of the Oil Well Flats detachment system, Canon City Embayment, Colorado*

Dellenbach, Benjamin Arthur, *An outcrop to subsurface stratigraphic analysis of the Niobrara Formation, Sand Wash Basin, Colorado*

Duran, Robert L., *Characterization and analysis of a large rockfall along the Rio Chama, Archuleta County, Colorado*

Durkee, Hannah Marie, *Reservoir characterization and geomechanical evaluation of the Greenhorn Formation in the northern Denver Basin, Colorado*

Ebeling, Katherine Ann, *Toxicity of cadmium, copper, and zinc and their binary mixtures to Daphnia magna in laboratory and field-collected waters*

Edgley, Ryan, *Assessing the efficacy of BMPs to reduce metal loads in the Los Angeles River basin at the watershed scale*

Eldam, Rania, *Evaluating control of slope-aspect on geochemical weathering within the Boulder Creek Critical Zone Observatory*

Eppehimer, Jarred, *Spatio-temporal microseismic analysis of the Woodford Shale, Canadian County, Oklahoma*

Faber, Ethan J., *Development of a landslide risk rating system for small-scale landslides affecting settlements in Guatemala City*

Forrester, Mary Michael, *Ecohydrologic response and atmospheric feedbacks from beetle-induced transpiration losses in the Colorado headwaters*

Gary, Isabel Alida, *Petroleum geology of the B chalk benches of the Niobrara Formation; Wattenberg, Silo, and East Pony Fields, northern Denver Basin, CO/WY*

Graham, Andrew J., *Integrated analysis of Waulsortian-type bioherms of the Lodgepole Formation; Stark County, North Dakota and Bridger Range, Montana*

Grazulis, Alexandra K., *Analysis of stress and geomechanical properties in the Niobrara Formation of Wattenberg Field, Colorado, USA*

Guzman, Mario, *Geology, vein petrography and mineral chemistry of the North Amethyst Deposit, Creede mining district, Creede, Colorado*

Haugen, Benjamin D., *Qualitative and quantitative comparative analyses of 3D LiDAR landslide displacement field measurements*

Hazar, Mehmet, *The tectonostratigraphic evolution, seismic interpretation and 2D section restoration of the offshore eastern Otway Basin, Victoria, Australia*

Heger, Andrew W., *Stratigraphy and reservoir characterization of the Turner Sandstone, southern Powder River basin, Wyoming*

Hinricher, Marcus John, *A geologic comparison of the oil window and dry gas window within the Tow Creek Member of the Niobrara Formation in the Piceance Basin of northwestern Colorado*

Hissem, Anna, *An integrated analysis of the growth history of the Moab-Spanish Valley salt wall and the Moab fault system*

Hodan, Marta D., *Complex fluvio-lacustrine stratigraphic interactions in the Renegade Tongue, Green River Formation, Uinta Basin, Utah*

Horne, Elizabeth, *Kinematics and growth of supra-salt fault systems; a field and subsurface analysis, Salt Valley salt wall, Paradox Basin, Utah*

Hu, Xiexiaomeng, *A coupled geomechanics and flow modeling study for multistage hydraulic fracturing of horizontal wells in enhanced geothermal systems applications*

Hutchinson, Mandi Brooke, *REE enrichment in weathered carbonatite, Bull Hill; Bear Lodge Mountains, Wyoming*

Johnston, Allison, *An integrated geophysical and geochemical approach to characterizing acid mine drainage in a headwater mountain stream in Colorado, USA*

Kamruzzaman, Asm, *Petrophysical rock typing of unconventional shale plays; a case study for the Niobrara Formation of the Denver-Julesburg (DJ) Basin*

Kanno, Cynthia M., *Surface spills at unconventional oil and gas sites; a contaminant transport modeling study for the South Platte alluvial aquifer*

Listiono, Geraldus Adi, *Mixed siliciclastic-carbonate system of the middle member of the Bakken Formation, Williston Basin, North Dakota*

Livo, Kurt, *Mineralogical controls on NMR rock surface relaxivity; a case study of the Fontainebleau Sandstone*

Lockwood, Thomas D., *Outcrop, petrographic, and subsurface analysis of the Sappington Member, Three Forks Formation, and Cottonwood Canyon Member, Lodgepole Formation, central Montana*

Logan, Ryan, *Modeling wildfire impact on hydrologic processes using a precipitation-runoff model*

Lytle, Madison, *The Proterozoic history of the Idaho Springs-Ralston shear zone; evidence for a widespread ca. 1.4 Ga orogenic event in central Colorado*

Mabrey, Alexandria N., *Rock quality index for Niobrara horizontal well drilling and completion optimization, Wattenberg Field, Colorado*

Miller, Savannah R., *An assessment of scaling and information requirements of the classical advection dispersion equation and a temporal-nonlocal advection dispersion equation at the MADE site*

Mueller, Staci K., *Application of time-lapse seismic shear wave inversion to characterize the stimulated rock volume in the Niobrara and Codell reservoirs, Wattenberg Field, CO*

Negri, Jacquelyn A., *Evaluation and validation of multiple predictive models applied to post-wildfire debris-flow hazards*

Panfiloff, Andre, *Experimental evaluation of dynamic elastic properties and anisotropy in shales*

Paskvich, Stephen W., *Phase-shift measurement analyses of time-lapse multicomponent seismic data over a multi-well hydraulic stimulation, Wattenberg Field, Denver Basin*

Quigley, Ashley Kaye, *Setting of volcanogenic massive sulfide* deposits in the Penokean volcanic belt, Great Lakes region, USA

Quigley, Patrick O., *The spectrum of ore deposit types, their alteration and volcanic setting in the Penokean volcanic belt, Great Lakes region, USA*

Ramiro-Ramirez, Sebastian, *Petrographic and petrophysical characterization of the Eagle Ford Shale in La Salle and Gonzales Counties, Gulf Coast region, Texas*

Saxe, Samuel, *Linear modeling and evaluation of controls on flow response in western post-fire watersheds*

Schaiberger, Allison, *Diagenesis in the Upper Cretaceous Eagle Ford Shale, south Texas*

Schwarz, Stephen, *Syn-rift drainages and sedimentary fill architecture; a case study in the Jurassic of the Dampier Sub-basin*

Semmens, Stephen, *An examination of the impact of the natural environment on levee sustainability*

Skewes, Wiley Boulden, *Evolution of magmatic-hydrothermal systems below boiling conditions; evidence from the Copler Au-Cu deposit, east central Turkey*

Stamer, John, *Geologic reservoir characterization of the Codell Sandstone; central Wattenberg Field, Colorado*

Stewart, Joshua C., *Developing remote sensing methods for bedrock mapping of the Front Range Mountains, Colorado*

Stone, Carver Haentjens, *An assessment of risk of hydrocarbon or fracturing fluid migration to fresh water aquifers; case study of Colorado oil and gas fields*

Thunder, Barbara, *The hydro-mechanical analysis of an infiltration-induced landslide along I-70 in Summit County, CO*

Walker, Ella L., *Water use for unconventional energy development in the South Platte River basin of Colorado*

Wescott, Amanda, *Reservoir characterization of the Middle Bakken Member, Fort Berthold region, North Dakota, Williston Basin*

Wieting, Celeste, *Quantifying soil hydraulic property changes with fire severity by laboratory burning*

Zhou, Mengnan, *Optimization of well configuration for a sedimentary enhanced geothermal reservoir*

Columbia University
Doctorate

Chen, Chen, *El Nino Southern Oscillation diversity in a changing climate*

Chen, Jianye, *Evolution and biogeography of frogs and salamanders, inferred from fossils, morphology and molecules*

Doherty, Cathleen Lauren, *Multi-stage evolution of the lithospheric mantle in the West Antarctic rift system; a mantle xenolith study*

Eilon, Zachary Cohen, *New constraints on extensional environments through analysis of teleseisms*

Farmer, Jesse Robert, *Quaternary carbon cycling in the Atlantic Ocean; insights from boron and radiocarbon proxies*

Li, Jiyao, *Seismicity and seismic imaging of the Alaska megathrust fault*

Moulik, Pritwiraj, *Earth's elastic and density structure from diverse seismological observations*

Veitch, Stephen Alexander, *Glacial earthquakes and glacier seismicity in Greenland*

Yanchilina, Anastasia Gennadyevna, *Excess freshwater outflow from the Black Sea-Lake during glacial and deglacial periods and delayed entry of marine water in the early Holocene require evolving sills*

Dalhousie University
Masters

O'Connor, Darragh, *Facies distribution, fluvial architecture, provenance, diagenesis, and reservoir quality of synrift successions from the breakup of Pangea; examples from the Fundy Basin and Orpheus Graben*

Van de Kerckhove, Samantha R., *Tectonic history of the Nepewassi Domain, Central Gneiss Belt, Grenville Province, Ontario; a lithological, structural, metamorphic and geochronological study*

Bachelor

Blume, Oliver, *Synthetic leaching of uranium from South Mountain Batholith granites and Horton Group siltstones of south central Nova Scotia, using a suite of ionic complexes*

Boggild, Kai, *Morphological examination of the NP-28 submarine channel-fan complex in the Amundsen Basin, Arctic Ocean*

Braun, Carmen A., *Chemical stratigraphy of the Windsor Group marine sediments, Shubenacadie Basin, Nova Scotia*

Brittain, Jeremy, *A palynological analysis of late Cretaceous and Paleogene strata from the Eclipse Trough, South Coast section, Bylot Island, Nunavut*

Fitzpatrick, Arthur J., *Land use change in the McIntosh Run watershed, Spryfield, Nova Scotia*

Glas, Nathan, *Using anhydrite crystals and whole rock analysis in a Mississippian salt diapir as a proxy to determine the preservation extent of primary seawater geochemical features through deformation processes*

Hosek, Nicholas, *Trail erosion analysis using close-range structure-from-motion photogrammetry*

Keltie, Erin, *Diamond dissolution in Na_2CO_3, Na_2CO_3-NaCl, and Na_2CO_3-NaF melts at 950C and 0.1 MPa*

McEwan, Robert S., *Visible to near-infrared analysis of granites; a method for charge coupled device spectroscopy*

McLauchlan, Elliot, *The impact of climate change on streamflow in the Blueberry River, northern British Columbia, and the potential influence on oil and gas development*

Steele, Shannon-Morgan M., *Modelling mid-frequency scattering and reverberation in the northern Gulf of Mexico*

Varcoe, Robert, *Does evapotranspiration increase when forests are converted to grasslands?*

Drew University
Doctorate

Dee, Michael, *Roots of Charles Darwin's creativity*

Florida Atlantic University
Masters

Munzenrieder, Cali, *Understanding variability of biogenic gas fluxes from peat soils at high temporal using capacitance moisture probes*

Florida State University
Doctorate

Barton, Cory, *Spatio-temporal evolutions of non-orthogonal equatorial wave modes derived from observations*

Fort Hays State University
Masters

Abrams, Kelsie, *Dental microwear variation in Teleoceras fossiger (Rhinocerotidae) from the Miocene (Hemphillian) of Kansas, with consideration of masticatory processes and enamel microstructure*

Green, John Hunter, *Petrophysical attributes, depositional environment, and diagenetic history of a Mississippian interval from McPherson County, Kansas, USA*

Illinois State University
Masters

Meister, Paul A., *Provenance analysis of granite cobbles in the Tiskilwa Till using U-Pb geochronology*

Indiana State University
Masters

Xu, Jingqi, *Facies and sequence stratigrpahic analyses of the Upper Ordovician shales in northeast Indiana and northwest Ohio*

Indiana University-Purdue University at Indianapolis
Masters

Albert, Ashley Lisbeth, *A laminated carbonate record of late Holocene precipitation-evaporation from Pretty Lake, Lagrange County, Indiana*

Rudloff, Owen M., *A 4600-year record of lake level and hydroclimate variability from an eastern Andean lake in Colombia*

Stamps, Lucas G., *A laminated carbonate record of late Holocene precipitation from Martin Lake, LaGrange County, Indiana*

Kansas State University
Masters

Connolly, Andrew M., *Exploring the relationship between paleobiogeography, deep-diving behavior, and size variation of the Parietal Eye in Mosasaurs*

Lakehead University
Masters

Baig, Ayat, *Characterization of the MAX porphyry Mo deposit using trace element geochemistry in hydrothermal alteration minerals, Trout Lake, B.C.*

D'Angelo, Michael, *Geochemistry, petrography and mineral chemistry of the Guichon Creek and Nicola Batholiths, south central British Columbia*

LaFontaine, Daniel, *Structural and metamorphic control on the Borden gold deposit, Chapleau, Ontario*

McCullough, Monica M., *Sedimentology and paleogeographic*

reconstruction of the strata in and adjacent to the Sudbury impact layer in the northern Paleoproterozoic Animikie Basin

Yip, Christopher Ira, Sedimentology and geochemistry of regressive and transgressive surfaces in the Gunflint Formation, northwestern Ontario

Bachelor

Arnold, Kira, An investigation of the Ney's Lookout lamprophyric dyke at Marathon, ON

Carson, Tracy, A microstructural analysis of the Coffee gold project

Cenedese, Johnathan, Assessing acid generation of mine waste with low-net neutralizing and net-acid producing potentials

Daniels, Jeffrey W., The physics of Archean atmospheric composition

Davis, Sarah, Petrology and geochemistry of the Wolfcamp Lake Basalts

Loma Linda University

Masters

Bird, James Vernon, Jr., Taphonomy of sediments; bioturbation in the Triassic Moenkopi Formation in Southwestern Utah

Massachusetts Institute of Technology

Doctorate

Alpert, Alice Elizabeth, Little Ice Age climate in the western tropical Atlantic inferred from coral geochemical proxies

Amrhein, Daniel Edward, Inferring ocean circulation during the last glacial maximum and last deglaciation using data and models

Nienhuis, Jaap H., Plan-view evolution of wave-dominated deltas

McMaster University

Doctorate

Beatty, Sarah M. B., Dynamic soil water repellency in hydrologic systems

Hendricks, Robert Rombuck, Timing of the emplacement of ancient coastal deposits of Georgia aided by ground penetrating radar and determined by optically stimulated luminescence and electron spin resonance optical dating

Masters

Arcuri, Gabriel, Whole-rock Pb isotope delineation of Archean and Paleoproterozoic crustal terranes in the Grenville Province and adjacent Makkovik Province; evidence for juvenile crustal growth during the Paleoproterozoic

Bennett, Darla, An investigation into the influence of microbes on the cycling of sulfur in a net neutral oxidation reservoir in Sudbury

Brown, Alyson, Paleoenvironmental analysis of Namu Lake, British Columbia

Didemus, Benjamin, Water storage dynamics in peat-filled depressions of the Canadian Shield rock barrens; implications for primary peat formation

Duggan, Sydney, Hydrogeological assessment at the Clarington transformer station using a conventional well cluster with recommendations to establish an advanced groundwater monitoring station

Hodson, Alex, OSL dating of a woodland period occupation at the Hare Hammock ring and mound complex, Bay County, Florida

Ie, Kesia, Oxygen isotope fractionation between hydroxyapatite (HAP)-bound carbonate and water at low temperatures

Kimmerle, Stephanie, Facies analysis and paleodischarge of rivers within a compound incised valley, Cretaceous Ferron Sandstone, Utah

Lee, Rebecca E., Landsystem analysis of three outlet glaciers, southeast Iceland

Rastelli, Jessica, Dissolved organic carbon concentration, patterns and quality at a reclaimed and two natural wetlands, Fort McMurray, Alberta

Spennato, Haley, An assessment of the water table dynamics in a constructed wetland, Fort McMurray, Alberta

Mississippi State University

Doctorate

Lalk, Sarah Paige, Utilization of outdoor resources of enhance understanding of geosciences

Mitchell, Joseph, Addressing questions of prehistoric occupation seasonality at freshwater mussel shell ring sites in the Mississippi Delta; applications in carbonate geochemistry and zooarchaeology

Weremeichik, Jeremy M., Environmental and growth rate effects on trace element incorporation to calcite and aragonite; an experimental study

Masters

Burbach, Curt, Maximizing the informal education of Death Valley National Park ichnofossils

Crabtree, Brandon, Analytical analysis of groundwater availability in Oktibbeha County, Mississippi

Foote, Jeremy Keith, An examination of the hydrological environment in Choctaw County Mississippi since 1995, with a focus on an area surrounding an industrial complex established in 1998

Merritt, Danielle, Estimation of suspended particulate matter concentration in the Mississippi Sound using MODIS imagery

Mitchell, Jonney Luc, Precipitation of aragonite under anoxic conditions; an experimental study

Mullennex, Asa J., Spatial correlation between framework geology and shoreline morphology in Grand Bay, Mississippi

Novak, Aleksandra Vladimirovna, Effects of fluid Mg/Ca and $\delta^{18}O$ on geochemistry of calcium carbonates; studies on inorganic and natural samples

Powell, W. Gabe, Predictive modeling of sinkhole hazards using Synthetic Aperture RADAR interferometry (InSAR) subsidence measurements and local geology

Samai-Odegaarden, Natalie, A petrographic analysis of the microbial thrombolite buildup in the Oxfordian Smackover Formation, Little Cedar Creek Field, Alabama

Smith, Taryn, Surficial geologic mapping of the Vicksburg National Military Park and surrounding areas in Vicksburg, Mississippi

Van Horn, John William, III, Potential of unmanned aerial systems imagery relative to Landsat 8 imagery in the lower Pearl River basin

Missouri University of Science and Technology

Doctorate

Dera, Abdallah Alhadi, Assessment of highway condition using combined geophysical surveys

Lemnifi, Awad Abdussalam Henish, Azimuthal anisotropy beneath north central Africa from shear wave splitting analysis

Obi, Jeremiah Chukwunonso, Geophysical imaging of karst features in Missouri

Masters

Alzaki, Taqi Talib, 2D seismic data and gas chimney interpretation in the South Taranaki Graben, New Zealand

Blue, Aaron Jeffrey, Experimental evaluations of selected sealants to remediate CO_2 leakage

Damasceno, Davi Rodrigues, Numerical analysis of flexural slip during viscoelastic buckle folding

Goetze, B. A., The mantle transition zone beneath South America from stacking of P-to-S receiver functions

Sun, Hao, Data analysis of surfactant-induced wettability alteration in carbonate reservoirs

Wang, Ze, Effect of reservoirs heterogeneity on injection pressure and placement of preformed particle gel for conformance control

Yagci, Fulya Gizem, Reprocessing and structural interpretation of multichannel seismic reflection data from Penobscot, Nova Scotia

Yagci, Gorkem, 3D seismic structural and stratigraphic interpretation of the TUI-3D Field, Taranaki Basin, New Zealand

Zhang, Weicheng, Variable stress orientations in the offshore Nile Delta; the role of salt as a mechanical detachment horizon

Montana State University

Doctorate

Moore-Nall, Anita Louise, Structural controls and chemical characterization of brecciation and uranium-vandium mineralization in the northern Bighorn Basin

Northern Arizona University

Masters

Fisher, Samuel H., Early PaleoAmerican research in the Southwest; an investigation of Clovis activity in Pertrified Forest National Park

Martinez, Daniel J., Assessing site resilience using repeat photography; Photographic analysis of landscape change in Glen Canyon, Arizona

Steen, Douglas P., Late Quaternary paleomagnetism and environmental magnetism at Cascade and Shainin lakes, north-central Brooks Range, Alaska

Ohio State University

Doctorate

Obrycki, John F., Managing soils for environmental science and public health applications

Stevens, Brooke Nan, *Bioaccessibility, bioavailability, and chemical speciation of arsenic in contaminated soils and solid wastes*

Wigmore, Oliver Henry, *Assessing spatiotemporal variability in glacial watershed hydrology; integrating unmanned aerial vehicles and field hydrology, Cordillera Blanca, Peru*

Masters

Rytel, Alexander L., *Hydrous mineral stability in Earth's mantle; implications for water storage and cycling*

Tomashefski, David J., *An erodibility assessment of central Ohio cropland soils*

Vascik, Anne Marie, *Physiographic mapping of Ohio's soil systems*

Bachelor

Bowman, Abbie, *Qanats; understanding world water resources*

Burchwell, Andrew E., *Evidence for using nuclear well logs to access natural gas hydrate reservoirs*

Daniele, John, *Quaternary sediment sources to hemipelagic and contourite dispositional settings off the Iberian Margin, IODP Sites 1385 and 1391*

Dismukes, Evan, *Earth as an analog for mass radius modeling for extra-solar planets*

Domer, George, *Principal component analysis of mineral abundances at IODP Site U1387*

Edgin, Matthew Gordon, *Mineralogical and geochemical analysis of strontium and barium sources in the Point Pleasant Formation*

Gallagher, Connor James, *Restoring surface water-groundwater connectivity in an Appalachian headwater stream on surface mined lands Guy Cove, Kentucky*

Grady, Alexander C., *Geochemical variations in basalts from the south east Indian Ridge*

Gutierrez, Mario Andres, *Upper and middle Miocene ash beds in the central Gulf of Mexico*

Haugh, Ryan A., *Styles of soft-sediment deformation of the lower headwall of the Cape Fear landslide, offshore North Carolina*

Holt, Benjamin D., *Trace element geochemistry of flowback brines from Marcellus Shale*

Koewler, Wesley J., *Stomatal density and index analysis of fossilized Pliocene floral remains from Ellesmere Island, Canada; for the identification of ancient carbon dioxide levels*

Langhorst, Theodore, *Observations of hydraulic controls on the Olentangy River, Columbus, Ohio*

Mave, Megan A., *Tank experiments to quantify fate of microcystin in shallow coastal sediments*

Miller, Laura A., *Rock weathering along small mountainous rivers; Sierra de las Minas, Guatemala*

Miyakawa, Yoko, *Magma evolution along the East Pacific Rise between 11' and 15'*

Perry, Samuel N., *Uranium and thorium storage in $CaSiO_3$ perovskite in the Earth's lower mantle*

Rohan, Tyler J., *Grain size and seismic analysis of IODP Expedition 333, Site C0018 and the pink volcanic ash horizon, Nankai Trough, offshore Japan*

Sethna, Lienne R., *Geochemistry of magmas erupted along the northern East Pacific Rise at 6°-12° implications for mantle source regions and intracrustal evolutionary processes*

Oregon Health and Science University
Doctorate

Lind, Pollyanna, *Geomorphology and sediment dynamics of a humid tropical montane river, Rio Pacuare, Costa Rica*

Oregon State University
Doctorate

Balbas, Andrea M., *Application of cosmogenic nuclides and argon geochronology to paleoclimate, paleomagnetism, and paleohydrology*

Barth, Aaron M., *Geochemical and geostatistical analyses of Quaternary climate variability over millennial-to-orbital timescales*

Bill, Nicholas S., *Late Cenozoic climate, ice-sheet and Earth-surface evolution derived from terrestrial and marine sedimentary archives*

Guerrero, Eduardo Francisco, *Quaternary landscape evolution and the surface expression of plume-lithosphere interactions in the greater Yellowstone area*

Rose, Kelly Kathleen, *Signatures in the subsurface; big & small data approaches for the spatio-temporal analysis of geologic properties & uncertainty reduction*

Masters

Choi, Na Hyung, *Late Pleistocene slip rate along the Panamint Valley fault system, eastern California*

Leydet, David J., *Eastward routing of glacial Lake Agassiz runoff caused the Younger Dryas cold event*

Portland State University
Masters

Hurtado, Heather Ann, *naturally occurring background levels of arsenic in the soils of Southwestern Oregon*

Prescott College
Masters

Jillings, Sarah, *The nature of satisfication and the conditions under which students thrive*

Princeton University
Bachelor

Barker, Ryan, *U-Pb TIMS-TEA geochronology and a new chronostratigraphy for the Canadon Asfalto Basin, central Patagonia*

Beveridge, Alyson K., *Measuring the changing mass of glaciers on the Tibetan Plateau using time-variable gravity from the GRACE mission*

Campbell, Ethan C., *Where three oceans meet; nitrate isotope measurements from the South Atlantic along 34.5°*

Campion, Alison M., *Constraining the timing and amplitude of proposed glacioeustasy during the Late Paleozoic ice age with a continuous carbonate record in Spain*

Christian, Shanna L., *Abandoned oil and gas wells in Pennsylvania; well attributes and effective permeability*

Edwards, Collin, *Mining metagenomic data to understand the lifestyle of atmospheric methane oxidizing bacteria in Antarctic surface soil*

Forden, Atleigh, *Reconstructing fish ecology from otolith geochemistry; past and present*

Liu, Weber, *Analysis of Martian topography via a parameterized spectral approach*

Lowy, Rebecca A., *Arsenic adsorption on Fe-oxides mixed with Mn-oxides; potential of Fe-Mn nanocrystalline coated calcite grains for filtration of arsenic contaminated drinking water*

O'Brien, Evan, *A multiple stressor model of climate change effects on growth and survival of larval Crassostrea gigas*

Simons, Frederik, *Measuring the changing mass of glaciers on the Tibetan Plateau using time-variable gravity from the GRACE mission*

Purdue University
Doctorate

Jang, Won Seok, *Integrated aquifer vulnerability assessment of nitrate contamination in Central Indiana*

Rice University
Doctorate

Chen, Jianxiong, *Frequency-dependent traveltime tomography and full waveform inversion for near-surface seismic refraction data*

Ding, Shuo, *The fate of sulfide during igneous processes of terrestrial planetary bodies*

Erdman, Monica Elizabeth, *Origin and evolution of continental crust; insights from igneous and metamorphic processes in subduction zones*

Fu, Lei, *Migration velocity analysis and waveform inversion with subsurface offset extension*

Hou, Jie, *Accelerating seismic imaging and velocity model building with approximate extended Born inversion*

Liu, Zuolin, *Interactions between biochar, soil, and water*

Nguyen, Chinh Thuc, *Effects of bubble coalescence, permeability, and degassing process on the transition from explosive to effusive eruption of rhyolitic magma*

Masters

Blaser, Austin P., *Controls on magma supply from depth at Kilauea Volcano, Hawai'i*

Demet, Brian Patrick, *Sedimentary processes at ice sheet grounding-zone wedges; comparing planform morphology from the western Ross Sea (Antarctica) to internal stratigraphy from outcrops of the Puget Lowlands (Washington State, U.S.A.)*

Halberstadt, Anna Ruth Weston, *Paleo-ice stream behavior; retreat scenarios and changing controls in the Ross Sea, Antarctica*

Jordan, Brian E., *Mechanisms of late-stage rift deformation; evidence from new 3D seismic data over the west Iberia margin*

Malouta, Alexandra Julia, *Characteristics of serpentinized peridotites from collisional settings; a case study from Domenigoni Valley, California*

Pang, Xiaojiao, *Detection of the low velocity layer atop the 410-km discontinuity beneath northeast China with slowness based CCP stacking*

Tesi Sanjurjo, Mari Anna, *Phases of evolution of the post-rift of the Galicia margin; evidence from the Galicia 3D reflection seismic survey*

Xing, Guangchi, *Effects of shallow density structure on the inversion for crustal shear wave speeds in surface wave tomography*

San Diego State University
Doctorate

Fouad, Geoffrey George, *Flow duration curve prediction for ungauged basins: a data-driven study of the contiguous United States*

Hirakawa, Evan Tyler, *Poro-elasto-plastic off-fault response and dynamics of earthquake faulting*

Withers, Kyle, *Ground motion and variability from 3-D deterministic broadband (0-8 Hz) simulations*

Masters

Jerrett, Andrew Steven, *Paleoseismology of the Imperial Fault at the U.S.-Mexico border, and correlation of regional lake stratigraphy through analysis of oxygen/carbon isotope data*

Shubha Ravikumar, Fnu, *Largest earthquakes in Japan*

Wessel, Kaitlin Nicole, *500 year rupture history of the Imperial Fault at the international border through analysis of faulted Lake Cahuilla sediments, carbon-14 data, and climate data*

Zimmerman, Lucas T., *Utilizing C# to create a Ternary Graph plotting program with rigorous statistical analysis with an analysis of Holocene sands*

San Jose State University
Masters

Emerson, Samuel, *The role of bed shear stress in sediment sorting patterns in a reconstructed gravel bed river*

McKee, Ryan A., *Structure and volcanic evolution of the northern Highland Range, Colorado River extensional corridor, Clark County, Nevada*

Shippensburg University of Pennsylvania
Masters

Galella, Joseph G., *Slackwater sedimentation at Laughlin Milldam, Newville, Pennsylvania*

Stefanic, Michael J., *An alternative explanation for the occurrence of wavellite in Pennsylvania's Blue Ridge*

Simon Fraser University
Doctorate

Medig, Kirsti Pier Rogers, *Sedimentology, geochemistry, and geochronology of unit PR1 of the lower Fifteenmile Group and the Pinguicula Group, Wernecke and Ogilvie Mountains, Yukon, Canada; Mesoproterozoic environments and paleocontinental reconstructions*

Middleton, Mary Ann, *Aquifer-stream connectivity at various scales; application of sediment-water interface temperature and vulnerability assessments of groundwater dependent streams*

Roberts, Nicholas J., *Late Cenozoic geology of La Paz, Bolivia, and its relation to landslide activity*

Zurek, Jeffrey Mark, *Multidisciplinary investigation of the evolution of persistently active basaltic volcanoes*

Masters

Basiru, Morufu Adewale, *In search of Cordilleran point sources to the southern McMurray Sub-basin*

Croquet, Eryne, *Late Wisconsinan paleosols and macrofossils in Chehalis Valley; paleoenvironmental reconstruction and regional significance*

Cullen, Justine Rose, *Optical dating studies of southeastern Patagonian sand wedges in Chile and Argentina*

Jones, Macy Taylor, *Stratigraphy and architecture of shallow-marine strata on an active margin, lower Nanaimo Group, Vancouver Island, BC*

Rathay, Sarah, *Response of a fractured bedrock aquifer to recharge from heavy rainfall events*

Roots, Eric Alexander, *Application of seismic interferometry to imaging a crystalline rock environment at an active VMS mine in Flin Flon, Manitoba, Canada*

Theny, Lucia Maria, *Age, formation and tectonism of the Neoproterozoic Ruddock Creek zinc-lead deposit and host Windermere Supergroup, northern Monashee Mountains, southern Canadian Cordillera*

Van Pelt, Stephanie, *Modeling paleorecharge on the Yucatan Peninsula, Mexico*

Southern Illinois University at Edwardsville
Masters

Adeyemo, Ayomipo E., *River channel changes and stream impoundment on the Provo River, utah; A GIS perspective*

Siech, Katie, *Environmental impacts on aquatic diversity; evaluating marco-invertebrate communities in the Meramec River*

Southern Methodist University
Doctorate

Graf, John F., *Stratigraphy, age, and systematics of two new fossil whales from Angola and stable isotope geochemistry of chicken eggshell, dinosaur eggshell, and paleosol carbonates; early burial diagenesis in Late Cretaceous sediments of the Gobi Basin, Mongolia*

Masters

Campbell, Rachel, *Stratigraphic architecture and reservoir characteristics of slumped distributary mouth bar deposits, Cretaceous Ferron Sandstone, Utah; an analysis of sedimentary fabric and facies utilizing outcrop and core data*

Oldham, Harrison R., *NELE; noise characterization of the northern Mississippi Embayment*

Rauch, Chelsea, *Identification, correlation, and predictive modeling of the line of natural gas expulsion in the Marcellus Shale in northeastern Pennsylvania*

Stanford University
Masters

Charoensawadpong, Panunya, *Building a GIS map to accelerate CO2 utilization for enhanced oil recovery*

Cherry, Jamal, *Optimization strategies for shale gas asset development*

de Chalendar, Jacques, *Impact of Ostwald ripening on residually trapped carbon dioxide*

Elkady, Youssef, *Heat-induced fractures in type I oil shale source rock*

Graveleau, Marguerite, *Pore-scale simulation of mass transfer across immiscible interfaces*

Jiang, Jason Renfeng, *A study of silica nanoparticles for enhanced oil recovery*

McLaughlin, Scott R., *Simulation of CO_2 exsolution for enhanced oil recovery and CO_2 storage*

Xue, Chen, *The application of "OptSpace" algorithm and comparison with "LMaFit" algorithm in three-dimensional seismic data reconstruction via low-rank matrix completion*

Zhang, Ke, *Experimental and numerical investigation of oil recovery from Bakken by miscible CO_2 injection*

State University of New York
Doctorate

Rahman, Shaily, *Cosmogenic silicon-32 reveals extensive authigenic clay formtion in deltaic systems and constrains the marine silica budget*

State University of New York at Binghamton
Doctorate

Janick, Joseph J., *Modeling the chemical sedimentation for the last 30 ka of the Searles Lake Formation, Searles Lake, California*

Schmitkons, Jonathan P., *Sources, transport, deposition, and fate of roadway pollutants in the Binghamton, New York urban corridor*

Masters

Almeida, Kaleo M., *Stability field of the Cl-rich scapolite marialite*

Carpenter, Jeffrey W., *Iron-bearing skarns of the Uhuk Massif, Northwest Territories, Canada*

Lord, Danielle M., *Biological and abiological controls on metal mobilization from shale gas drill cuttings by landfill leachate*

Nelson, Kristina E., *Simulated metal release from drill cuttings associated with natural gas extraction from the Marcellus Shale*

Tufano, Brandon C., *Coupled flexural-dynamic subsidence modeling of a retro-foreland basin, Western Canada Sedimentary Basin*

State University of New York College
Doctorate

Kang, Phil-Goo, *Dynamics and charcteristics of dissolved organic carbon, nitrogen, and sulfur in the Arbutus Lake Watershed in the Adirondack Mountains of New York State*

State University of New York College of Environmental Science and Forestry
Masters

Ng, Nicole, *Spatiotemporal variations of baseflow generation in the*

United State

Stephen F. Austin State University
Doctorate
- Faulkner, Melinda S., *An investigation of hydrogeologic, stratigraphic, and structural controls on Acer grandidentatum communities in a karst landscape, Owl Mountain Province, Fort Hood military installation, Texas*

Masters
- Ayres, Kyle B., *Cone in cone concretions of the Stanley Group in southeastern Oklahoma*
- Eckhoff, Ingrid Jenssen, *Geologic and geochemical characterization of cross-communication potential within the northern Edwards Aquifer system, Texas*
- Ehrhart, Jon T., *Speleogenesis and delineation of megaporosity and karst geohazards through geologic cave mapping and LiDAR analyses associated with infrastructure in Culberson County, Texas*
- Harris, Garrett R., *Evaluation of groundwater quality in Paleozoic aquifers using GIS techniques, central Texas*
- Kakarla, Pawan, *Geotechnical properties of the Travis Peak (Hosston) Formation in East Texas; a compressive and tensile strength analysis using regression analysis*
- Landers, Ashley N., *Speleogenesis of Critchfield bat caves and associated hydrogeology of the northern Edwards Aquifer, Williamson County, Texas*
- Majzoub, Adam F., *Characterization and delineation of karst geohazards along RM652 using electrical resistivity tomography, Culberson County, Texas*
- Stover, Kenneth L., *Seismic interpretation and analysis of the Etouffee Reservoir sands and the surrounding area in Terrebonne Parish, southeast Louisiana*
- Turner, Wesley L., *Characterization of Lower Permian carbonate subaqueous gravity flows in Crockett County, Texas*
- Williamson, Garrett R., *The stratigraphic position of fossil vertebrates from the Pojoaque Member of the Tesuque Formation (middle Miocene, late Barstovian) near Espanola, New Mexico*

SUNY, University at Buffalo
Masters
- Philipps, William Edward, *Improving Holocene glacial chronologies in Wedel Jarlsberg Land, Svalbard and in Uummannaq Fjord, Greenland*
- Venturino, Christian S., *Wrinkle ridges in Syrtis Major, Mars; origin and evolution*

Texas A&M University
Doctorate
- Yuan, Shanshui, *Assessment of multiple approaches for using soil moisture to evaluate drought in the U. S. Great Plains*
- Zamora, Ryan Alexander, *Tropical cyclones in paleoclimate simulation*

Masters
- Elizondo Ugarte, Daniel Fernando, *Subsidiary fault kinematics and displacement transfer at the Mill Creek-Mission Creek Fault stepover, San Bernardino Mountains, California*
- Goggin, Holly Elaine, *Multiple origins of sand-dune topography interactions on Titan*
- Huang, Yibin, *Stream depletion and pumping test interpretation in a horizontally anisotropic aquifer near a stream*
- Radwan, Mohamed Mahmoud Ahmed, *A case study of blast vibration modelling in the Hanason Servtex Quarry, Garden Ridge City, Texas*
- Reed, John Christopher, *An assessment of rockfall activity using dendrogeomorphology; the Amphitheater, Ouray, Colorado, USA*
- Rowley, Taylor, *Evaluating channel migration of the lower Guadalupe River; Seguin, TX to the San Antonio River confluence*
- Sexton, Carolyn Elizabeth, *A dendrogeomorphological assessment of the Donald Duck landslide complex, Grand Mesa, Colorado*
- Ulincy, Amanda Jeanne, *Pleistocene benthic Foraminifera from IODP Site U1344; northern slope of the Bering Sea*

Tulane University School of Science and Eng
Doctorate
- Telfeyan, Katherine Christina, *Analysis of trace element cycling in marsh pore waters of the lower Mississippi River Delta with a case study of vanadium in groundwaters of Texas and Nevada*

Universite Laval
Doctorate
- de Almeida Rodrigues, Andreia, *Concrete deterioration due to sulfide-bearing aggregates*
- Grusson, Youen, *Modelisation de l'evolution hydroclimatique des flux et stocks d'eau verte et d'eau bleue du bassin versant de la Garonne*
- Lechat, Karl Dominique, *Sequestration geologique du CO_2 par carbonatation minerale dans les residus miniers*

Masters
- Banville, David-Roy, *Modelisation cryohydrogeologique tridimensionnelle d'un bassin versant pergelisole; une etude cryohydrogeophysique de proche surface en zone de pergelisol discontinu a Umiujaq au Quebec Nordique*
- Clavet, Charles, *Paleoecologie du Quaternaire dans la region de Wainwright en Alberta*
- Ducharme, Marc-Andre, *Caracterisation du pergelisol; application d'une nouvelle methode afin d'estimer la conductivite thermique a l'aide de la tomodensitometrie*
- Francoeur, Julie, *Etude de l'evolution de la deterioration du beton incorporant des granulats riches en sulfures de fer*
- Langlais, Karine, *Dynamique holocene d'une tourbiere a palses de la region du lac a l'Eau-Claire (Nunavik)*
- Murray, Renaud, *Bilan hydrologique d'un bassin versant dans la region d'Umiujaq au Quebec nordique*
- Philippe, Guillaume, *Impacts d'une variabilite climatique changeante sur la morphologie de berges des chenaux du delta du Gange-Bramapoutre-Meghna et leurs consequences en zones densement peuplees*
- Roy, Nicolas, *Simulations numeriques du transport de methane en provenance de puits de production abandonnes dans des aquiferes peu profonds*

University of Alabama at Tuscaloosa
Doctorate
- Li, Huang, *Integrated modeling and management of groundwater and surface water, Zhangye Basin, northwest China*

Masters
- Bassett, Christine Nicole, *Clams and climate; implications for measuring seasonality in the marine bivalve, Saxidomus gigantea*
- Brenn, Gregory Randall, *Determining the upper mantle seismic structure beneath the northern Transantarctic Mountains, Antarctica, from regional P- and S-wave tomography*
- Essex, Caleb William, *Regional correlation of facies within the Haynesville Formation from onshore Alabama; analysis and implications for provenance and paleostructure*
- McKay, Kathleen Kingry, *Cave air and dripwater variability in Cathedral Caverns, Alabama*
- Merey, Osman, *Kinematic evolution of the Buyuk Menderes Graben in western Turkey inferred from 2-D seismic interpretation and cross section restoration*
- Payne, Taylor N., *$\delta^{15}N$ as a potential proxy for anthropogenic nitrogen loading in Charleston Harbor, South Carolina*
- Tokgil Karaket, Mine, *Structural evolution of the Haymana Basin, central Anatolia, Turkey*

University of Alaska Anchorage
Masters
- Gordaoff, Roberta Michelle, *The house on the hill; A 3800-year-old upland site on Adak Island, the Aleutian islands, Alaska*

University of Alaska at Fairbanks
Doctorate
- Alvizuri, Celso R., *Estimation of full moment tensors, including uncertainties, for earthquakes, volcanic events, and nuclear explosions*
- Barker, Amanda J., *Speciation, transport and mobility of metals in pristine watersheds and contaminated soil systems in Alaska*
- Dixit, Nilesh C., *Tectono-thermal history modeling and reservoir simulation study of the Nenana Basin, central Alaska; implications for regional tectonics and geologic carbon sequestration*
- Kienholz, Christian, *Current state of Alaska's glaciers and evolution of Black Rapids Glacier constrained by observations and modeling*
- Lindgren, Prajna R., *Remote sensing of lake dynamics in Alaska*

Masters
- Cable, William Lambert, *The role of environmental factors in regional and local scale variability in permafrost thermal regime*
- Holloway, Caitlin R., *Paleoethnobotany in interior Alaska*
- Lande, Lauren, *A petrological model for emplacement of the ultramafic Ni-Cu-PGE Alpha Complex, eastern Alaska Range*
- Mejia, Jessica Zimmerman, *Speed-ups and slowdowns; insights into velocity variations measured from 2009-2011 on Yahtse Glacier, Alaska*
- Mixon, Demi C., *The neotectonics, uplift, and accommodation of*

deformation of the Talkeetna Mountains, south-central Alaska
Rosenthal, Jacob L., *Fractured reservoir potential and tectonic development of the Iniskin-Tuxedni region, lower Cook Inlet, Alaska*
Tuzzolino, Amy L., *Compositional characteristics and ages of plutons in the central Ruby Batholith, Alaska; implications for rare-earth-element resources*
Zechmann, Jenna M., *Strategies for applying active seismic subglacial till characterization methods to valley glaciers*

University of Arizona
Doctorate
Truebe, Sarah Anne, *Past climate, modern caves, and future resource management in speleothem paleoclimatology*

University of British Columbia
Doctorate
Boxill, Lois Esther, *The impact of fabric and surface characteristics on the engineering behavior of polymer-amended mature fine tailings*
Cassidy, Alison Elizabeth, *The impacts of High Arctic permafrost disturbances on vegetation and carbon flux dynamics*
Davidson, Sarah, *Modeling disturbance and channel evolution in mountain streams*
Dudill, Ashley Rebecca, *Experimental study of segregation mechanisms in bedload sediment transport*
Granek, Justin, *Application of machine learning algorithms to mineral prospectivity mapping*
Nedilko, Bohdan, *Seismic detection of rockfalls on railway lines*
Roustaei, Ali, *Yield stress fluid flows in uneven geometries; applications to the oil and gas industry*
Sun, Duo, *Storage of carbon dioxide in depleted natural gas reservoirs as gas hydrate*
Tamminga, Aaron, *UAV-based remote sensing of fluvial hydrogeomorphology and aquatic habitat dynamics*
Unglert, Katharina Claudia, *Towards a global classification of volcanic tremor*
van Roggen, Judith, *Revisiting the archaeobotany of Sand Canyon Pueblo; sampling and social context*
Wall, Corey James, *Establishing the age and duration of magmatism in large open-system layered intrusions from the high-precision geochronology of the Neoarchean Stillwater Complex and Paleoproterozoic Bushveld Complex*
Masters
Boucher, Kaleb S., *The structural and fluid evolution of the Efemcukuru epithermal gold deposit, western Turkey*
Brueckman, Christina, *Reliability analysis of discrete fracture network projections from borehole to shaft scale discontinuity data*
Brunetti, Paula, *Magmatic-hydrothermal evolution and post-ore modifications of the Halilaga porphyry Cu-Au deposit, NW Turkey*
Caudle, Dana, *Mineralogy, geochemistry, and geochronology of the KIN Property pegmatites in eastern British Columbia*
Chong, Andrea Denise, *Unintentional contaminant transfer from groundwater to the vadose zone via gas exsolution and ebullition during remediation of volatile organic compounds*
Cook, Natalie Louise, *Establishing the stable isotope and geochemical footprints associated with carbonate-hosted Zn-Pb deposits of the Kootenay Arc*
de Oliveira Gongalves, Adriana, *Analysis of gold extraction processes of artisanal and small-scale gold mining in Portovelo-Zaruma, Ecuador*
Febbo, Gayle Elizabeth, *Structural evolution of the Mitchell Au-Cu-Ag-Mo porphyry deposit, northwestern British Columbia*
Fenwick, Lindsay Alexandra, *Methane and nitrous oxide distributions across the North American Arctic Ocean during summer, 2015*
Gaudet, Matthew A., *The Renard 65 kimberlites; emplacement-related processes in kimberley-type pyroclastic kimberlites*
Laurenzi, Laura, *Investigating metal attenuation processes in mixed sulfide carbonate bearing waste rock*
Leroux, Graham M., *Stratigraphic and petrographic characterization of HS epithermal Au-Ag mineralization at the TV Tower District, Biga Peninsula, NW Turkey*
Liu, Jie, *Chemical composition of coal surface as derived from micro-FTIR and its effects on contact angle*
Liu, Yubo, *Characterisation of block cave mining secondary fragmentation*
MacIver, Michael Ryan, *In situ mineral sediment characterization with light scattering and image analysis*
McMahon, Alexander D., *Influence of turbidity and aeration on the albedo of mountain streams*
Mostaghimi, Nader, *Structural geology and timing of deformation at the Gibraltar copper-molybdenum porphyry deposit, south-central British Columbia*
Panton, Brad, *Numerical modelling of rock anchor pullout and the influence of discrete fracture networks on the capacity of foundation tiedown anchors*
Rich, Shane Daniel, *Geochemical mapping of porphyry deposits and associated alteration through transported overburden*
Shi, Peipei, *Geochemical assessment of the bioavailability of platinum group metals for phytomining*
Ver Hoeve, Thomas James, *Applications of LA-ICP-MS analysis to zircon; assessing downhole fractionation and pre-treatment effects for U-Pb geochronology and trace element variations in accessory minerals from the Bushveld Complex*
Zima, Michael K., *Biogeochemical characterization of the hyporheic zone below the Fraser River near Vancouver, British Columbia*

University of California at Berkeley
Doctorate
Mangiante, David Michael, *Metal and mineral catalyzed organic photochemistry in modern and prebiotic environments*
Oshun, Jasper, *The isotopic evolution of a raindrop through the critical zone*
Reilly, Sean Bryant, *Historical biogeography of reptiles and amphibians from the Lesser Sunda Islands of Indonesia*
Rempe, Daniella Marie, *Controls on critical zone thickness and hydrologic dynamics at the hillslope scale*
Rose, Ian Robert, *True polar wander on convecting planets*
Zhang, Shuai, *First-principles simulations of minerals in Earth and planetary interiors*

University of California at Davis
Doctorate
Mackey, Tyler James, *San, mud, and calcite; microbial landscapes on Antarctic lake beds*
Masters
Levin, Emily, *Paleomagnetic investigation of unlithified sediments from Clear Lake, Northern California and its chronostratigraphic and paleoenvironmental implications*
Weldon, Nicholas Cullum, *Argon diffusion in rhyolite melt at 100 MPa*

University of California at Irvine
Doctorate
Yang, Hongying, *Investigating the impacts of climate, hydrology, and Asian Monsoon intensity on a 13 kyr speleothem record from Laos*

University of California at San Diego
Doctorate
Sibert, Elizabeth Claire, *Ichthyoliths as paleoceanographic and paleoecological proxy and the response of open-ocean fish to Cretaceous and Cenozoic global change*

University of California at Santa Cruz
Doctorate
Shope, James Brandon, *Modeling Pacific Atoll Island shorelines' response to climate change*

University of Florida
Doctorate
Bremner, Paul M., *Three dimensional crustal shear-wave velocity of central Idaho/Eastern Oregon and a verified method for modeling upper mantle anisotropy*

University of Hawaii at Manoa
Doctorate
Acosta-Maeda, Tayro E., *Raman spectroscopy for planetary exploration and characterization of extraterrestrial materials*
Fackrell, Joseph K., *Geochemical evolution of Hawaiian groundwater*
Howell, Samuel M., *Faulting and deformation at divergent and transform plate boundaries*
Janebo, Maria Helena, *Historical explosive eruptions of Hekla and Askja Volcanoes, Iceland; eruption dynamics and source parameters*
Lemelin, Myriam, *The composition of the lunar crust; an in-depth remote sensing view*
Weiss, Jonathan R., *The Bolivian Subandes and beyond; linking wedge deformation processes across multiple timescales*

Masters

Kennedy, Joseph J., *Coupling aircraft and unmanned aerial vehicle remote sensing with simultaneous in-situ coastal measurements to monitor the dynamics of submarine groundwater discharge*

Richardson, Christina, *Geochemical dynamics of nearshore submarine groundwater discharge; Maunalua Bay, O'ahu, Hawai'i*

Robinson, Nichole, *A high-resolution $^{187}Os/^{88}Os$ record for the late Maastrichtian*

Shuler, Christopher K., *Source partitioning of anthropogenic groundwater nitrogen in a mixed-use landscape, Tutuila, American Samoa*

Sigurdardottir, Telma Dis, *Audiomagnetotelluric exploration across Wai'anae Range, O'ahu, Hawai'i*

Tree, Jonathan P., *Mantle potential temperatures of 4.5 to 47 Ma Hawaiian volcanoes using olivine thermometry; implications for melt flux variations*

Watkins, C. Evan, *Constraints on dynamic topography from asymmetric subsidence across the mid-ocean ridges*

Bachelor

McKenzie, Trista, *Quantifying atmospheric fallout of Fukushima-derived radioactive isotopes in the Hawaiian Islands*

McKenzie, Warren, *Aspects of a depression on Elysium Mons, Mars, using HiRISE imagery*

University of Illinois at Chicago
Doctorate

Motamedi, MohammadHosein, *Numerical simulaiton of mechanical reponse of geomaterials from strain hardening to localized failure*

University of Kansas
Doctorate

Grismer, Jesse, *The fragmentation of Gondwanaland; Influence on the historical biogeography and morphological evolution within dragon lizards (Squamata; Agamidae)*

Masters

Alm, Steve Allen, Jr., *A geological and geophysical investigation into the evolution and potential exploitation of a geothermal resource at the Dixie Valley Training Range, Naval Air Station Fallon*

Hallman, Jason Andrew, *Spatial and temporal patterns of Ogallala Formation deposition revealed by U-Pb zircon geochronology*

Jones, Matthew Frazer, *Neoichnology of bats; morphological, ecological, and phylogenetic influences on terrestrial behavior and trackmaking ability within the Chiroptera*

Livers, Amanda Jordan, *Parallel-line beamsteering for enhanced anomaly detection*

Nolan, Jeffery Jordan, *Near-surface void characterization and sensitivity analysis using enhanced processing procedures on passive multi-channel analysis of surface waves (MASW) data*

Pugliano, Tony Michael, *Fundamental stratigraphic, sedimentologic, and petrophysical elements of heterozoan carbonates; grain-rich fining- and shoaling-upward cyclothems and clinothems*

Rawitch, Michael Jess, *Stream CO_2 Degassing: Review of Methods and Laboratory Evaluation of Floating Chambers*

Sweeney, Rafferty Jeremiah, *Sequence stratigraphy and facies distribution in a Miocene carbonate platform; La Rellana Platform area, southeastern Spain*

University of Louisiana at Lafayette
Doctorate

Taylor, Delmetria, *Delineating heavy metal sources with environmental magnetism, x-ray fluorescence, and geospatial analysis; Baton Rouge, Louisiana*

Trabelsi, Racha, *Development of an integrated system to optimize Block 276 production performance*

Masters

Adesiyun, Oludamilola, *High-resolution rock magnetic and paleointensity study of sediments from IODP Site U1389 (West Iberian margin of the North Atlantic Ocean)*

Angelo, Jared Michael, *The use of wavelet energy absorption to estimate hydrocarbon saturation in the North Lissie Field of Wharton County, TX*

Bujard, Jade P., *Geophysical analysis of the Miocene-Pliocene Mangaa Formation for better exploration within the Parihaka 3D survey; Taranaki Basin, New Zealand*

Cai, Xiao, *An analytical method for predicting wellbore temperature profile during drilling gas hydrates reservoirs*

Conlin, Daniel, *Investigating fluid flow in detachment systems through numerical modeling*

Costa, Ayrton, *Geological investigation of Block 17, offshore Angola*

Cozby, Raymond, *Fracture conductivity and its effects on production estimation in shale*

Gunawardana, Prasanna M., *Deep earthquakes spatial distribution; numerical modeling of stress and stored elastic energy distribution within the subducting lithosphere*

Hoang, Tina Hong, *Controls on sediment supply and erosion in the Song Gianh, Central Vietnam*

Juneau, Jacob C., *Understanding how technology-biased and economic-based decisions can impact project profitability; A case study of Creole Field, Cameron, Louisiana*

Li, Hui M. S., *Experimental investigation of geometric effects of core samples on Berea Sandstone geo-mechanical behaviors*

Lipman, Hunter, *Application of inorganic proxies to determine the Paleodepositonal controls on organic matter accumulation for the Marcellus Shale in Northeastern Pennsylvania*

Mason, Shanna, *Characterizing the natural fracture system of the Eagle Ford Shale*

Monlezun, Christian J., *Integrated geospatial and chemical analysis of storm water drainage in Lafayette Parish, Louisiana*

Nguessan, Zegbe, *Horizontal well placement evaluation deep offshore Nigeria*

Perreault, Luke J., *Activation energies of Kerogen in the Eagle Ford Shale estimated from rock eval pyrolysis data; comparison of methods using single and multiple heating rates*

Schaper, Maxwell C., *Fluid-rock interaction within the Picacho Mountains detachment shear zone*

Shane, Timothy E., *Geochemical analysis of parasequences within the productive middle member of the Eagle Ford Estimation at Lozier Canyon near Del Rio, Texas*

University of Maine
Doctorate

Cook, Alden C., *Microstructural analysis of thermoelastic response, nonlinear creep, and pervasive cracking in heterogeneous materials*

Shulman, Deborah J., *Processes leading to the formation of deep granitic shear zones in the Grenville Front tectonic zone, Ontario, Canada*

Masters

Auger, Jeff, *A comparison of global climate reanalysis and climate of South Greenland and the North Atlantic*

Chandler, Emily A., *Sediment accumulations patterns in the Damariscotta River estuary*

Dufour, Christopher G., *Improving the performance of ice sheet modeling through embedded simulation*

Hamley, Catherine M., *Humans and the Falkland Islands warrah; investigating the origins of an extinct endemic canid*

Khan, M. Rakibul Hassan, *Streamflow allocation policy in a changing climate in New England region; a study of hydrologic, ecological, and chemical variability and change*

Osti, Bipush, *GPU-accelerated modelling for geological and ecological processes*

Theodore, Nora, *Epilimnion thickness of lakes in Acadia National Park; trends, controls, and biological implications*

Truong, Chi Khanh Hoang, *Pressurized fast pyrolysis of calcium formate-pretreated biomass*

Weil, Kristen Katherine, *Effects of watershed characteristics on stream vulnerability to urbanization; implications of future land use on streams in Maine, USA*

University of Maryland
Doctorate

Tweedt, Sarah Maureen, *Development and the Early Animal Fossil Record; evoluton and phylogenetic applications*

University of Minnesota at Duluth
Masters

Asp, Kristofer, *An investigation of Ni and Cu isotopic fractionation in basal Duluth Complex Cu-Ni-PGE mineralization, northeastern Minnesota*

Burley, Paul, *Myth as true history; medicine wheels and landmarks as boundary markers of the Lakota World*

Cappio, Laura, *Reconstruction of the Pleistocene environment and climate of the Olorgesailie Basin*

Doyle, Michael, *Geologic and geochemical attributes of the Beaver River diabase and greenstone flow; testing a possible intrusive-volcanic link in the 1.1 GA Midcontinent Rift*

Jasperson, Jenny L., *Seasonal and flood-induced variations in groundwater-surface water exchange dynamics in a shallow aquifer system; Amity Creek, MN*

Leu, Adam R., *Geology and petrology of the Wilder Lake Intrusion, Duluth Complex, northeastern Minnesota*

University of Mississippi
Masters
Dolly, Richard G., *Remobilization of cobalt, copper, nickel, lead, and zinc in a riparian wetland model*

University of Nevada
Masters
Brailo, Courtney M., *A light detecting and ranging (LiDAR) and Global Positioning System (GPS) study of the Truckee Meadows, NV Quaternary fault mapping with ArcGIS, 3D visualization and computational block modeling of the greater Reno area*

University of Nevada at Reno
Doctorate
Warren, Sean N., *Empirical ground support recommendations and weak rock mass classification for underground gold mines in Nevada, USA*
Masters
Richey, Christopher Shaun, *The historical archaeology of ore milling; ideas, environment, and technology*

University of New Hampshire
Masters
Arpino, Meghan Rose, *Effects of high-resolution topography and hydraulics on reach-scale estimates of nitrate transport and retention in the Suncook River, Epsom, NH*
Hodgdon, Taylor, *Developing a chronology for thinning of the Laurentide ice sheet in New Hampshire during the last deglaciation*
Houts, Amanda N., *Reconstructing the glacial history of the Hunafloi Bay region in northwest Iceland using cosmogenic ^{36}Cl surface exposure dating*
Irish, Onni Bowen, *Analysis of CLCS recommendations in light of their relevance to the delineation of a United States extended continental shelf (ECS) in the Arctic*
Lundgren, Dylan, *Assessing bottom sediment methane concentrations and the extent of anaerobic oxidation of methane in the Great Bay estuary, New Hampshire*
Nifong, Kelly L., *Sedimentary environments and depositional history of a paraglacial, estuarine embayment and adjacent inner continental shelf; Portsmouth Harbor, New Hampshire, USA*
Pecanha, Anderson Barbosa da Cruz, *Evaluating the usage of multi-frequency backscatter data as an additional tool for seafloor characterization*
Widlansky, Sarah J., *Paleomagnetism of the Cretaceous Galula Formation and implications for vertebrate paleobiogeography*

University of New Mexico
Doctorate
Frus, Rebecca Jane, *Multidisciplinary work to determine hydrology of arid land springs and how spring waters influence water quality and ecosystem health for desert environments*
Levine, Rebekah, *The influence of beaver activity on modern and Holocene fluvial landscape dynamics in southwestern Montana*
Vander Kaaden, Kathleen, *Experimental investigation into the thermal and magmatic evolution of Mercury*
Masters
Hoots, Charles R., *Seismic imaging of the lithosphere-asthenosphere boundary with a dense broadband array in Central California*
Jackson, Ryan Steele, *Investigation of aqueous processes in the Valle Grande paleo-lake, Valles Caldera as a Martian analog; ChemCam investigation of the John Klein and Cumberland drill holes and tailings, Gale Crater, Mars*
Bachelor
Felicia, Alexandria L., *Preliminary analysis of bifurcations on modern deltas*
Ganter, Will, *Detecting Paleozoic millennial-scale climate changes using pollen analysis of marine limestone-shale rhythmites*
Williams, Jeffrey, *Cl in the early solar system*

University of North Carolina at Wilmington
Masters
Cooke, Kimberly A., *Latitudinal patterns in predation among modern Crepidula species along the east coast of the United States; effects of environmental conditions and taphonomic processes*
Kelly, Bridget Teresa, *Isotopic and taphonomic study of fossil Glycymeris bivalve shells from the North Carolina Pleistocene; interactions with drilling and durophagous predators*
Le, Alyssa, *Introducing novel inputs and agent-based modeling analysis to hydrologic modeling of a poorly gauged basin, western Kenya*
Leathem, Sean P., *Physical effects of reef rugosity on hydrodynamics and sedimentation over transplanted oyster reefs in a shallow tidal creek system*
Mainali, Janardan, *Spatial assessment of climate vulnerability in Nepal; exploring the role of multiple scales and approaches*
Massoll, Jennifer L., *Decadal variations in Western Pacific Warm Pool dynamics as evidenced by Porites corals from Chuuk Atoll, FSM*
Moore, Nicholas O., *Sedimentology, stratigraphy, and provenance of the Late Jurassic Coon Hollow Formation, Pittsburg Landing, Idaho*
Narron, Caroline, *Assessing the impact of climate change on spatiotemporal variations in coastal wetlands of Umm Al Quwain, UAE using geospatial and field techniques*
Neely, Samuel Harrison, *Predator-prey interactions among Pliocene molluscs from the Tjornes Peninsula, Iceland, and East Anglia, England, across the Trans-Arctic invasion*
Sosnowski, Amelia, *Assessing and contextualizing the natural environment in South Sudan's multifaceted conflict using remote sensing*
Stanford, Samantha D., *Relationship of escalation to frequency of drilling predation in Miocene and Pleistocene molluscs of the U. S. Coastal Plain*

University of Oklahoma
Doctorate
Chen, Chen, *Comprehensive analysis of Oklahoma earthquakes; from earthquake monitoring to 3D tomography and relocation*
Lin, Tengfei, *Seismic attribute analysis of depth-migrated data*
Turner, Bryan William, *Utilization of chemostratigraphic proxies for generating and refining sequence stratigraphic frameworks in mudrocks and shales*
Wang, Ting, *An organic geochemical study of Woodford Shale and Woodford-Mississippian tight oil from central Oklahoma*
Masters
Basnet, Shiva, *Controls of pre-existing normal faults on the geometry and location of thrust faults in fold-thrust belts*
Carvajal Garcia, Carlos Patricio, *Atmospheric dust from the Upper Carboniferous Copacabana Formation (Bolivia); a high-resolution record of climate and volcanism from northwestern Gondwana*
Giddens, Emma Lyn, *Pleistocene coral reef destruction in the Florida Keys; paleotempestite evidence from a high resolution LiDAR XRF analysis of Windley Key Quarry, FL*
Hutchinson, Bryce, *Application and limitations of seismic attributes on 2D reconnaissance surveys*
Jacobsen, Rae Travis, *Microseismic location using differential backazimuth angles*
Machado Acosta, Gabriel, *Improving fault images using a directional Laplacian of a Gaussian operator*
Maynard, Sabrina, *Correlation of bioturbated facies, chemostratigraphy, total organic carbon, and sequence stratigraphy in the Woodford Shale of south central Oklahoma*
Mayorga-Gonzalez, Ligia Carolina, *Oil potential of the Asquith Marker, Lewis Shale, greater Green River Basin, Wyoming*
Morgan, Barrett Chance, *Frictional properties of fault gouge in high-velocity/long-distance shear experiments*
Pan, Li, *A gravity and magnetic study of the structure evolution of Sichuan Basin*
Pearson, Catherine Agnes, *Geochemical characterization of the Upper Mississippian Goddard Formation, Springer Group, in the Anadarko Basin of Oklahoma*
Prado Barros, Enkar David, *Seismic facies classification of the lower Strawn Formation, Bend-Arch, Fort Worth Basin, north central Texas*
Qi, Xiao, *An icehouse dust record from the upper Paleozoic (Asselian-Artinskian, Moscovian) Akiyoshi Limestone (Japan)*
Salantur, Burak, *Continuity, connectivity and reservoir characteristics of Desmoinesian fan-delta conglomerates and sandstones, Elk City Field, Anadarko Basin, Oklahoma*
Sexton, Molly Rebecca, *Stratigraphic and textural analysis of nanodiamonds across the Younger Dryas boundary sediments of western Oklahoma*
Smith, Curtis, *Comparative SEM analysis of quartz microtextures on fluvial sand from end-member climate systems*
Villalba, Damian M., *Organic geochemistry of the Woodford Shale, Cherokee Platform, OK and its role in a complex petroleum system*
Zhang, Jing, *Comprehensive reservoir characterization of the Wood-*

ford Shale in parts of Garfield and Kingfisher Counties, Oklahoma

University of Oregon
Doctorate
Cerovski-Darriau, Corina R., *Landslides and landscape evolution over decades to millennia; using tephrochronology, air photos, lidar, and geophysical investigations to reconstruct past landscapes*
Emery, Meaghan Marie, *Assessment of character variation in the crania and teeth of modern artiodactyls for better species diagnosis in the fossil record*
Krogstad, Randy, *Kinematic constraints on tremor and slow slip in Cascadia and implications for fault properties*
Maguffin, Scott Charles, *A biogeochemical study of groundwater arsenic contamination in the southern Willamette Basin, Oregon, USA*
Seligman, Angela Nicole, *Oxygen and hydrogen investigation of volcanic rocks; petrogenesis to paleoclimate*

Masters
Conde, Giselle, *Stomatal index of Ginkgo biloba as a proxy for atmospheric CO_2*
O'Connell, Brennan, *Sedimentology and depositional history of the Miocene-Pliocene southern Bouse Formation, Arizona and California*
Sulak, Daniel J., *Iceberg properties and distributions in three Greenlandic fjords using satellite imagery*

University of Rhode Island
Doctorate
Dzaugis, Mary Elizabeth, *Water radiolysis and chemical production rates near solid-water boundaries*
Phillips, Brennan Theodore, *The influence of hydrothermal plumes on midwater ecology, and a new method to assess pelagic biomass*

Masters
Balcanoff, Jacob, *Determining gas contents of magmas that produce submarine basaltic balloons eruptions; Foerstner 1891 and Socorro 1993*

University of Southern California
Doctorate
Roychaudhuri, Basabdatta, *Spontaneous countercurrent and forced imbibition in gas shales*

University of Texas at Austin
Doctorate
Bush, Meredith Anne, *Stratigraphic signatures of convergent orogenesis from the plte interior; studies form Tibet and the Southern Rocky Mountains*
Fong-Ngern, Rattanaporn, *Deep lacustrine clinoforms; formation, architectures, and deposits*
Frederik, Marina Claudia Geraldina, *Morphology and structure of the accretionary prism offshore north Sumatra, Indonesia and offshore Kodiak Island, USA; a comparison to seek a link between prism formation and hazard potential*
Gao, Ruohan, *Storage, fractionation and melt-crust interaction of basaltic magmas at oceanic and continental settings*
Gooch, Bradley Tyler, *The effects of sedimentary basins on the dynamics of the East Antarctic ice sheet from enhanced groundwater and geothermal heat flow*
Hanna, Romy Darlene, *CM Murchison; Nebular Formation of fine-grained chondrule rims followed by impact processing on the CM parent body*
Hurd, Gregory Samuel, *Revised stratigraphic framework for the Cutoff Formation and implications for deepwater systems modified by large-scale inflections in slope angle below the shelf break*
Licenciado, Sebastian Grabriel Ramirez, *Interactions between sedimentation and deformation in the Kumano Forearc Basin (Japan) and the Malargue Foreland Basin (Argentina)*
Murphy, Brendan Patrick, *Feedbacks among chemical weathering, rock strength and erosion with implications for the climatic control of bedrock river incision*
Parajuli, Sagar Prasad, *New insights into dust aerosol entrainment mechanisms from satellite/ground-based data, climate modeling, and wind-tunnel experiments*
Piliouras, Anastasia, *Ecogeomorphology of fluviodeltaic systems; effects of vegetation on delta morphodynamics*
Pretzlav, Kealie Lynn Goodwin, *Armor development and bedload transport processes during snowmelt and flash floods using laboratory experiments, numerical modeling, and field-based motion-sensor tracers*
Prieto, Maria Isabel, *Sediment gravity-driven vs. bottom-current-controlled processes and interactions; a study of multiple geomor-*
phologic domains in the central Gulf of Mexico and comparison with global systems
Ren, Qi, *Anisotropic seismic characterization of the Eagle Ford Shale; rock-physics modeling, stochastic seismic inversion, and geostatistics*
Sathaye, Kiran, *Geochemical dynamics of Bravo Dome natural CO_2 field*
Thirumalai, Kaustubh Ramesh, *Surface-ocean variability in the northern Gulf of Mexico during the late Holocene*
van der Kolk, Dolores Ann, *Marine-continental transitions in a greenhouse world; reconstructing Late Cretaceous deltas of paleopolar Arctic Alaska and Utah*
Walton, Maureen Anne LeVoir, *Tectonic and sedimentary processes of the Southeast Alaska margin*
Warden, John Gerald, *Microbialites of Lake Clifton, Western Australia; groundwater dependent ecosystems in a threatened environment*
Wu, Guangliang, *Continental extension in orogenic belts; modes of extension, origin of core complexes, and two-phase postorogenic extension*

Masters
Arnost, Daniel Ryan, *Impact of meteoric fluid flow on the thermal evolution of Alpine exhumation; insights from the Gotthard Base Tunnel, Switzerland*
Ayhan, Oguzhan, *Depositional environment, diagenesis and reservoir quality of the middle Bakken Member in the Williston Basin, North Dakota*
Bascon, Luciana de la rocha, *Southern Gulf of Mexico Wilcox source-to-sin; Investigating siliciclastic sedimentation in Mexico deep-water*
Bolhassani, Behnaz, *Model-based cost analysis for pressure and geochemical-based monitoring methods in CO_2-EOR fields; application to field A*
Borgman, Barry Michael, *Characterization of the Cana-Woodford Shale using fractal-based, stochastic inversion; Canadian County, Oklahoma*
Breeden, Benjamin Thomas, III, *Observations on the osteology of Scutellosaurus lawleri Colbert, 1981 (Ornithischia, Thyreophora) on the basis on new specimens from the Lower Jurassic Kayenta Formation of Arizona*
Cacciatore, Christopher Guy, *Interpreting the difference in magnitudes of PETM carbon isotope excursions in paleosol carbonate and paleosol organic matter; oxidation of methane in soils versus elevated soil respiration rates*
Charlton, Timothy C., *A 20,000 year flowstone record of Gulf of Mexico sourced moisture in Texas*
Chatmas, Emily, *The effect of a pre-deposited mobile substrate on terminal fan evolution and channel organization; tank experiments*
Chavez Urbina, Grecia Alexandra, *Eagle Ford Shale; evaluation of companies and well productivity*
Colleps, Cody Lee, *Exhumation of the Lesser Himalaya of northwest India; zircon U-Pb and (U-Th)/He constraints and implications for the Neogene seawater evolution*
Danger, Nick, *Formation and evolution of Pleistocene (MIS 5e) strandplain grainstones along the leeward margin of West Caicos Island, BWI*
Gabb, Kyle Christopher, *A facies-scale chemo-lithostratigraphic composite profile of Del Rio Claystone through Austin Chalk deposition, Late Cretaceous, central Texas, USA*
Halley, Chelsea Jenna, *Laboratory calibration of the CS655 soil moisture sensor*
Jones, Michael, *Implications of geothermal energy production via geopressured gas wells in Texas; merging conceptual understanding of hydrocarbon production and geothermal systems*
Jones, Rebecca Helen, *Comparative paleoenvironmental and architectural analyses of the Anchor Mine Tongue to Upper Sego interval along the Book Cliffs of Utah and the Rangely Anticline of Colorado*
Jung, Eunsil, *Physical modeling of a prograding delta on a mobile substrate; dynamic interactions between progradation and deformation*
Krueworramunee, Kullamard, *Sequence stratigraphy and depositional systems in the Upper Cretaceous (Cenomanian) Woodbine Group, Anderson and Cherokee Counties, Texas*
Lim, Ye Jin, *Experimental investigation of ice-covered deltas; the effects of ice cover on delta morphology*
Marroquin, Selva Mariana, *New records of early Jurassic gladius-bearing coleoids (Prototeuthidina and Loligosepiina) from Alberta, Canada*
McKenzie, Kyle Michael, *Outcrop-derived facies model and cycle architecture of the Tansill G27-G28 high-frequency sequences,*

Rattlesnake Canyon, New Mexico

Mercado, Lauren Kathleen, *The depositional system for the middle and lower Wilcox in Houston Embayment, northern Gulf of Mexico Basin*

O'Brien, Casey Megan, *Chemical and mechanical diagenetic evolution of deformation bands in sandstone*

Osmond, Johnathon Lee, *Fault seal and containment failure analysis of a lower Miocene structure in the San Luis Pass area, offshore Galveston Island, Texas inner shelf*

Prather, Timothy John, *The combined effects of river-flood and tidal processes on the stratigraphy, ichnology, and stratal architecture of the Cretaceous (Campanian) Loyd Delta near Rangely, Colorado, U.S.A.*

Redmond, Lauren Patricia, *Lithofacies, depositional systems, and depositional model of the Mississippian Barnett Formation in the southern Fort Worth Basin*

Regimbal, Kelly Alaine, *Improving resolution of NMO stack using shaping regularization*

Sekhon, Natasha, *Multidecadal rainfall variability in the South Pacific convergence zone using the geochemistry of stalagmites from the Solomon Islands*

Shover, Katherine Rose, *Mass balance of martian sedimentary fans and valleys*

Sierra, Diego Ernesto, *Developing a Vaca Muerta Shale play; an economic assessment approach*

Smith, Faith Martinez, *Does coal mining in West Virginia produce or consume water? A net water balance of seven coal mines in Logan County, West Virginia, an aquifer assessment, and the policies determining water quantities*

Tornabene, Chiara, *Evaluating stable isotope proxies ($\delta^{13}C$, $\delta^{18}O$, and $\delta^{15}N$) for detecting photosymbiosis in fossil corals*

Verma, Rahul, *Hydraulic fracturing sand resource development in the Llano Uplift region, central Texas; resource calculation, favorability analysis, and transportation economics*

Voorhees, Kristopher James, *Anatomy, dimensions, and significance of the penultimate Yates tepee-shelf crest complex, G25 Hairpin HFS, Guadalupe Mountains, New Mexico and Texas*

Yanke, Andrew James, *Feasibility of isotropic inversion in orthorhombic media; the Barrett unconventional model*

Zheng, Hanyue, *Sedimentology and reservoir characterization of the Upper Pennsylvanian Cline Shale, Midland Basin, Texas*

Bachelor

Beam, Jennifer Marie, *The effects of ?incluison orientation on excess compliance and inclusion-based models*

Fryer, Rosemarie C., *Holocene geologic slip rate for the Mission Creek strand of the southern San Andreas Fault, CA*

Palermo, Rose, *Variability in rates and styles of shoreline retreat genertig a sea cliff and wave-cut platform at Sargent Beach, Texas*

Ruthven, Rachel Courtney, *Testing mechanisms and scales of equilibrium using textural and compositional analysis of porphyroblasts in rocks with heterogeneous garnet distributions*

University of Texas at El Paso
Doctorate

Avila, Victor Manuel, *Geophysical constraints on the Hueco and Mesilla Bolsons; structure and geometry*

Molina Sotelo, Castulo, *Geology and mineralization controls surrounding the Palmarejo mining district; a compilation of remote and hands on exploration techniques*

University of Texas at San Antonio
Masters

Allen, Matthew, *New Permian ammonoid localities at Sibley's Last Chance Ranch, Delaware Mountains, Culberson County, West Texas*

Cutler, Eric, *Analysis of cave sediments at three south central Texas caves; potential implications for cave-air CO2 production*

Hoff, Stephen, *The development of a hydrodynamic mixing model to describe the occurrences of the brackish water zone in the Edwards Aquifer, south-central Texas*

Johnson, Joseph, *Integrated stratigraphy and paleoenvironmental reconstruction of a dinosaur track-bearing interval of the Glen Rose Formation at Government Canyon State Natural Area, Bexar County, Texas*

Johnston, William, *3D seismic interpretation of polygonal faulting in Upper Cretaceous sediments, Powder River basin, Wyoming*

Perdomo-Figuroa, Martha, *Quantitative approach to extension and fault characterization within the central and northern Llanos Basin, Colombia*

Sheaffer, Kristine, *Preliminary oxygen isotope analysis of a mixed conodont assemblage from the Type Chappel Limestone (Tournaisian Stage, Mississippian subsystem), San Saba County, central Texas*

Thompson, Reece, *Stable carbon isotope analysis in a south Texas cave investigation sources of CO_2 production*

University of Toledo
Masters

Groat, Lucas, *The physical hydrogeology of the broader historical Irwin Prairie wetland system*

Peters, Joshua, *Geology of Sawyer Quarry Nature Preserve, Lime City, Ohio*

University of Tulsa
Masters

Abdallah, Labiba, *Geochemical characterization of oils from the Mississippian Meramec Formation in eastern Anadarko Basin, Oklahoma*

Adero, Bernard, *The time-reversal imaging in attenuative media*

Boukhamsin, Hani, *Permo-Carboniferous palynological biostratigraphy and paleoenvironments of central and eastern Saudi Arabia*

Campbell, Whitney, *Middle-upper Albian sequence stratigraphy of the interior shelf to shelf margin of the Comanche Shelf, Texas*

Chen, Baorui, *Two dimensional seismic forward modeling in the frequency domain with hybrid boundary conditions*

Hojnacki, Rachel Elaine, *Middle-upper Albian (Cretaceous) sequence stratigraphy and geochemical events of the Fort Stockton Basin, Texas*

Kassie, Ryan, *Inversion of rock properties from pre-stack seismic data using simulated annealing with a rotated coordinate system*

Sessions, Julia, *Early Mississippian St. Joe Group in northeastern Oklahoma to southwestern Missouri; isotope chemostratigraphy*

University of Western Ontario
Doctorate

Harrison, Tanya Nicole, *Martian gully formation and evolution; studies from the local to global scale*

Masters

Adam, Marcus A., *Gold mineralisation at the FAT deposit, Courageous Lake greenstone belt, Northwest Territories, Canada*

Hadden, Shaun Michael, *Anisotropic waveform tomography; application to crosshole data for transversely isotropic media*

Vannelli, Kathleen, *Stratigraphy, sedimentology and paleogeography of the Lower Cretaceous (upper Albian) Peace River, Joli Fou and Pelican Formations, northern Alberta, Canada*

Bachelor

Cookson, Nicolas, *Mineralization styles in the hanging wall and footwall of the Mitchell thrust fault, Mitchell Deposit, KSM, northern British Columbia*

Corlett, Lauren, *Sediment partitioning across a syn-tectonic deltaic complex; Cenomanian Dunvegan Formation, Western Canada*

Digaletos, Maria, *Assessing the performance and zone of influence of the Claudette Mackay-Lassonde Pavilion geothermal system*

Houde, Victoria, *Testing the potential of hackberry carbonate $\delta^{13}C$ and $\delta^{18}O$ as a paleoclimate indicator*

Hutchinson, Scott, *Shock effects and petrological investigation of two newly discovered achondrite meteorites*

Kope, Erin, *Rapid boron geochemistry in whole-rock powders; a fast and cost effective synchrotron tool for uranium exploration*

Kukovica, Jacob, *Statistical analysis of anthropogenic seismicity*

Littleton, Joshua A. H., *Cooling of Earth's core; an experimental perspective*

Maunder, Duncan, *Developing mXRD of ilmenite as a tool for diamond exploration*

Munro, Alexander, *Sm-Nd and Lu-Hf geochronology of ultramafic dykes from the Malanjkhand porphyry copper mine, India*

Smith, Derek, *Analyzing the role of pre-impact lunar topography on impact melt emplacement in comparison to Venusian topography*

Van Zyl, Mauritz, *Raman spectroscopy study of $CaFe_2O_4$ and $CaTi_2O_4$; phases chromite and magnesiochromite*

University of Windsor
Doctorate

Aghbelagh, Yousef Beiraghdar, *Factors controlling the formation of unconformity-related uranium deposits in sedimentary basins; insights from reactive mass transport modeling*

Far, Maryam Shahabi, *The magmatic and volatile evolution of gabbros hosting the Marathon PGE-Cu deposit; evolution of a* *conduit system*

Masters

Ali, Ayat Ruh, *Assessing change and vulnerability of the Guyana coastline with multi-temporal Landsat imagery and survey data*

Friedman, Eyal, *The origin of mantle xenoliths in Tertiary alkaline basalts, British Columbia, Canada; implications for convergent plate margin geodynamic and petrogenetic processes*

Mrad, Carole, *Fluid compartmentalization of Devonian and Mississippian dolostones, Western Canada Sedimentary Basin; evidence from fracture mineralization*

University of Wisconsin at Madison
Doctorate

Johnson, Michael Robert, *An integrated stable isotope record from the late Miocene Pannonian Basin System; the ecology of horses, the life histories of bivalves, and mass-balance modeling*

Lancelle, Chelsea, *Distributed acoustic sensing for imaging near-surface geology and monitoring traffic at Garner Valley, California*

Le Mevel, Helene, *Crustal deformation and magmatic processes at Laguna del Maule volcanic field (Chile); geodetic measurements and numerical models*

Ma, Chao, *Centennial to million-year scale climate cycles in the Late Cretaceous Western Interior Seaway; their implications for geochronology, paleoceanography, and celestial mechanics*

Zhou, Yaoquan, *Oscillatory hydraulic tomography; numerical experiments and laboratory studies*

Masters

Denny, Adam, *Isotopically zoned carbonate cements in early Paleozoic sandstones of the Illinois Basin; $\delta^{18}O$ and $\delta^{13}C$ records of burial and fluid flow*

Hashman, Breana M., *A Mesoarchean microbial iron shuttle, Witwatersrand Basin, South Africa*

Jin, Shiyun, *Incommensurately modulated structures of labradorite feldspars (fAn51) and their petrological implications*

Reddy, Thiruchelvi Rajagobal, *Si isotope variations in the Precambrian rock record explained through experimental studies and development of new methods on femtosecond laser ablation for in situ Si analysis*

Reinisch, Elena C., *Graph theory for analyzing pair-wise data; application to interferometric synthetic aperture radar data*

Sayler, Frances Claire, *Characterization of bedrock secondary porosity using multi-frequency oscillatory flow interference testing*

Shanks, Lindsey Vanni, *On the recurrence of enigmatic nannoplankton blooms in the subtropical South Atlantic during the early Oligocene*

Watkins, W. David, *Local earthquake tomography of the Jalisco, Mexico region*

Zhao, Hangjian, *Evaluating seepage lake drought resilience using stable isotopes of water and groundwater-flow models*

University of Wyoming
Doctorate

Calder, William John, *Interactions among climate change, wildfire, and vegetation shaping landscapes for the last 2000 years*

Graly, Joseph A., *Chemical weathering under the Greenland ice sheet*

Hayes, Jorden L., *Seismic refraction studies of volcanic crust in Costa Rica and of critical zones in the southern Sierra Nevada, California and Laramie Range, Wyoming*

Li, Tao, *Multi-component seismic modeling and robust pre-stack seismic waveform inversion for elastic anisotropic media parameters*

Lukens, Claire E., *Spatial variation in sediment production across mountain catchments; insights from cosmogenic nuclides and tracer thermochronometry*

Marsicek, Jeremiah P., *The spatial and temporal evolution of Holocene climates in North America and Europe*

McLaughlin, J. Fred, *Evolution of the Paleo- and Mesoarchean gneiss of the central Wyoming Province in the western Granite Mountains of Wyoming*

Masters

Fantello, Nadia, *Estimating trapped gas concentrations as bubbles within lake ice using ground-penetrating radar*

Harrington, Joel A., *Temperature distribution and thermal anomalies along a flowline of the Greenland ice sheet*

Hough, Gretchen A., *Sulfur isotopes in bacterially and chemically controlled roll-front deposits*

LaForge, Justin S., *Insights into low-angle normal fault initiation near the brittle-plastic transition; a micro- to macroscale documentation of the Mohave Wash fault system, Chemehuevi Mountains, southeastern CA*

Morey, Jennifer Rose, *Photoelectrochemistry at the mineral/water interface and naturally occurring two-layer solar cells*

Sharma, Hema S., *Azimuthal anisotropy analysis using P-wave multiazimuth seismic data in Rock Springs Uplift, Wyoming, US*

Socianu, Alexa Lea, *Geochemical insights into the distribution of organic matter in the Mowry Shale, southern Powder River basin, Wyoming*

Taylor, Nicholas J., *Utilizing ambient seismic imaging to constrain near-surface weathering and its influence on overlying vegetation*

Wang, Zhan, *Simulation study of miscible CO_2 flooding of Timber Creek*

Utah State University
Masters

Jenkins, David W., *Frictional heating of fault surfaces due to seismic slip; experimental studies on the hematite to magnetite transition and federal and private landownership's effect on oil and gas drilling and production in the southwestern Wyoming checkerboard*

Markowski, Daniel K., *Confirmation of a new geometric and kinematic model of the San Andreas Fault at its southern tip, Durmid Hill, Southern California*

Miller, Roger D., *The Lower Cretaceous Cedar Mountain Formation of eastern Utah; a comparison with the coeval Burro Canyon Formation, including new measured sections on the Uncompahgre Uplift*

Rasmusson, Eric A., *The influence of small displacement faults on seal integrity and lateral movement of fluids*

Varriale, Jerome A., *The MH-2 core from Project Hotspot; description, geologic interpretation, and significance to geothermal exploration in the western Snake River Plain, Idaho*

Vanderbilt University
Masters

Ajayi, Moyosore, *Methane and high volume hydraulic fracturing; quantifying non-point diffuse methane leakage through geochemical surface detection methods*

Cohen, Leland John, *Influences of hardwood riparian vegetation on stream channel geometry in eastern forested environments*

Harmon, Lydia Jane, *Plagioclase, orthopyroxene, clinopyroxene, glass magma-meter and application to Mount Ruapehu, New Zealand and Parana volcanic province, Brazil*

Hermann, Nicholas, *Spatial patterns and driving mechanisms of mid-Holocene moisture in the Western United States; comparison of paleoclimate records and global circulation models*

Patrick, Meagan G., *Stratigraphic evolution of the Ganges-Brahmaputra lower delta plain and its relation to groundwater arsenic distributions*

Tramontano, Samantha, *The importance of minerals and bubbles; (1) the internal trigger test; mapping overpressure regimes for giant magma bodies; (2) developing and incorporating instructional videos and quizzes as a blended and online learning component in an undergraduate optical microscopy curriculum*

Washington University
Doctorate

Haenecour, Pierre, *Stardust in primitive astromaterials; insights into the building blocks and early history of the solar system*

Pratt, Martin James, *Seismic array studies of Antarctica and Madagascar*

Wei, Songqiao, *Seismic studies of the Tonga subduction zone and the Lau back-arc basin*

Wong, Teresa, *Plate tectonics initiation on Earth-like planets; insights from numerical and theoretical analysis of convection-induced lithospheric failure*

Washington University in St Louis
Doctorate

Li, Qingyun, *Calcium carbonate formation in energy-related subsurface environments and engineered systems*

Wesleyan University
Bachelor

Capece, Lena Rose, *The carbon dynamics of East Lake, Newberry Volcano, Oregon*

Cullen, Kate, *Orbital forcing of West and East Antarctic ice sheets 3.8-3.0 Ma; a climate analogue in Weddell Sea ODP Site 697*

Kaufman, Zachary Snow, *Sediment interpretations of ice rafted debris in the Weddell Sea, Antarctica; a 3-3.8 mya record from ODP Site 697*

Moye, Ryan Alexander, *Exploiting vein-architecture to estimate the area of fossil leaves; a multi-site analysis*

Sawyer, Will, *Numerical modeling of pseudotachylyte injection vein formation*

Western Michigan University
Doctorate
    Farag, Abotalib Zaki Abotalib, *An integrated approach for a better understanding of the paleo-hydrology and landscape evolution in the Sahara during the previous wet climatic periods*

Masters
    Beach, Brian A., *Evaluation of an alternative natural surfactant for non-aqueous phase liquid remediation*
    Currie, Bryan J., *Stratigraphy of the Upper Devonian-Lower Mississippian Michigan Basin; review and revision with an emphasis on the Ellsworth petroleum system*
    Erber, Nathan R., *Differentiating Saginaw from Huron-Erie ice lobe meltwater deposits in south central Michigan using longitudinal profiles, grain size analysis, and lithology*
    Ford, Chanse M., *Characterizing groundwater recharge and streamflow using stable isotopes of oxygen and hydrogen*
    Garrett, Jonathan D., *A sequence stratigraphic model for the Silurian A-1 carbonate and modeling of a Niagara-lower Salina reef complex, Michigan Basin, U.S.A.*
    Hinks, Benjamin Davis, *Geochemical and petrological studies on the origin of nickel-copper sulfide mineralization at the Eagle Intrusion in Marquette County, Michigan*
    Hudson, Jeffrey M., *Influence of persulfate on solidification/stabilization characteristics of ISS treatment*
    Ismael, Rozkar Izzaddin, *Slug tests in unconfined aquifers*
    Salim, Rachel Lynn, *Extent of capillary rise in sands and silts*
    Sasso, Andrew Lloyd, *Geochemical and petrological characterizations of peridotite and related rocks in Marquette County, Michigan*
    Suhaimi, Agam Arief, *Pore characterizations and distributions within Niagaran-lower Salina reef complex reservoirs in the Silurian northern Niagaran pinnacle reef trend, Michigan Basin*

Western Washington University
Masters
    Escobar-Burciaga, Ricardo Daniel, *Mineral complexities as evidence for open-system processes in formation of intermediate magmas of the Mount Baker volcanic field, northern Cascade Arc*
    Murphy, Ryan D., *Modeling the effects of forecasted climate change and glacier recession on late summer streamflow in the upper Nooksack River basin*
    Wershow, Harold N., *A Holocene glaciolacustrine record of the Lyman Glacier and implications for glacier fluctuations in the north Cascades, Washington*

Woods Hole Oceanographic Institution
Doctorate
    Boiteau, Rene M., *Molecular determination of marine iron ligands by mass spectrometry*
    Bucholz, Claire E., *Chemical, isotopic, and temporal variations during crustal differentiation; insights from the Dariv igneous complex, western Mongolia*
    Feng, Helen S. H., *Seismic constraints on the processes and consequences of secondary igneous evolution of Pacific oceanic lithosphere*
    Fornace, Kyrstin L., *Late Quaternary climate variability and terrestrial carbon cycling in tropical South America*
    Jackson, Rebecca H., *Dynamics of Greenland's glacial fjords*
    Moulton, Melissa, *Hydrodynamic and morphodynamic responses to surfzone seafloor perturbations*
    Vilasur Swaminathan, Rohith, *Vortices in sinusoidal shear, with applications to Jupiter*

Geological Surveys

Alabama

Geological Survey of Alabama
Geological Survey of Alabama (2018)
420 Hackberry Lane
P.O. Box 869999
Tuscaloosa, AL 35486-6999
  p. (205) 349-2852
  ntew@gsa.state.al.us
  http://www.gsa.state.al.us/
  Programs: None. Research agency.
  Certificates: None. Research agency.
  Enrollment (1995): B: 2 (0)
State Geologist:
  Berry H (Nick) Tew , (D), Alabama, 1999, GroGs
Division Manager, Groundwater Assessment:
  Stephen C. Jones, (M), Alabama, 1996, HgGeCg
Division Manager, Geologic Investigations Division:
  Sandy M. Ebersole, (D), Alabama, 2009, Zi
Division Manager, Energy Investigations Program:
  Denise J. Hills, (M), Delaware, 1998, EoZfYe
Division Manager, Ecosystems Investigations :
  Stuart W. McGregor, (M), Tennessee Tech, 1987, Hs
Deputy Director, GSA:
  Patrick E. O'Neil, (D), Alabama, 1993, Hs
Manager, Geologic Mapping:
  Gene Daniel Irvin, (M), Alabama, 1994, Gg
Manager, Petroleum Systems & Technology:
  David C. Kopaska-Merkel, (D), Kansas, 1983, GsPiZe
Visiting Professor:
  William A. Thomas, (D), Virginia Tech, 1960, GtcGr
Scientific Aid:
  Arthur McLin
GIS Specialist:
  Elizabeth A. Wynn, (M), Alabama, 2008, HsZyi
Geologist & Paleontology Curator:
  T. Lynn Harrell, Jr., (D), Alabama, 2016, PvGg
Geologist:
  Richard E. Carroll, (D), Michigan State, 1992, EcPlGo
  Jamekia Dawson, (B), Alabama, 2015, Hg
  Guohai Jin, (M), Zhejiang, 1989, Np
  Mac McKinney, (B), Alabama, 2007, Hws
  Neil E. Moss, (M), Alabama, 1987, Hw
  Marcella Redden, (M), Alabama, 2004, Gg
  David Tidwell, (M), S Florida, 2005, Gm
  Dane S. VanDervoort, (M), Auburn, 2016, GcpZi
Chemist:
  Rick Wagner, (B), Texas (San Antonio), 1982, Cg
Biologist:
  Rebecca A. Bearden, (B), Auburn, 2007, Hs

Alaska

Alaska Division of Geological & Geophysical Surveys
Department of Natural Resources (2017)
3354 College Road
Fairbanks, AK 99709-3707
  p. (907) 451-5010
  dggspubs@alaska.gov
  http://www.dggs.alaska.gov
  f: http://www.facebook.com/pages/Fairbanks-AK/Alaska-DGGS/346699054500
  t: @akdggs
Director & State Geologist:
  Steven S. Masterman, (M), Alaska (Fairbanks), 1990, EgNg
Petroleum Geologist I:
  David Lepain, (D), Alaska, 1993, EoGos
GeoScientist I:
  Melanie B. Werdon, (D), Alaska (Fairbanks), Eg
Geologist V:
  Janet R. G. Schaefer, (M), Alaska (Fairbanks), Gv
  De Anne S.P. Stevens, (M), Alaska (Fairbanks), Ng
Geologist IV:
  Marwan A. Wartes, (M), Wisconsin, GsoGt
  Gabriel J. Wolken, (D), Gm
Geohydrologist-Geologist IV:
  Ronald P. Daanen, (D), Minnesota (St. Paul), 2004, HyGme
Division Operation Manager:
  Kenneth R. Papp, (M), Alaska (Fairbanks), Gg

Alberta

Alberta Geological Survey
Alberta Geological Survey (2018)
402, Twin Atria Building
4999 - 98 Avenue
Edmonton, AB T6B 2X3
  p. (780) 638-4491
  AGS-Info@aer.ca
  https://ags.aer.ca/

Arizona

Arizona Geological Survey
Arizona Geological Survey (2019)
1955 E 6th St.
PO Box 210184
Tucson, AZ 85721
  p. (520) 621-2352
  fmconway@email.arizona.edu
  http://www.azgs.arizona.edu
  f: https://www.facebook.com/AZ.Geological.Survey
  t: @AZGeology
Director & State Geologist:
  Philip A. Pearthree, (D), Arizona, Ge
Chief, Environmental Geology:
  Ann Youberg, (D), Arizona, Ge
Senior Scientist:
  Michael F. Conway, (D), Michigan Tech, 1993, RcGvg
  Andrew A. Zaffos, (D), Cincinnati, 2014, Zf
Research Geologist:
  Joseph P. Cook, (M), Arizona, 2006, Gm
Research Scientist:
  Carson A. Richardson, (D), Arizona, 2019, Emg
Research Geologist:
  Charles A. Ferguson, (D), Calgary, Gcv
  Brian F. Gootee, (M), Arizona State, Gm
  Brad Johnson, (D), Carleton, 1994, Gc
  Jeri J. Young, (D), Arizona State, Ys

Arkansas

Arkansas Geological Survey
Arkansas Geological Survey (2015)
Vardelle Parham Geology Center
3815 West Roosevelt Road
Little Rock, AR 72204
  p. (501) 296-1877
  ags@arkansas.gov
  http://www.geology.ar.gov/home/
  Administrative Assistant: Laure Hinze
Director & State Geologist:
  Bekki C. White, (M), Centenary, 1993, GoEoGg
Geology Supervisor:
  William D. Hanson, (M), Memphis State, 1991, EgnGs
Geologist Supervisor:
  William L. Prior, (M), Memphis State, 1979, GggEc
Information Systems Analyst:
  James K. Curry, (M), S Methodist, 1978, Zn
Geologist:
  Sandra Chandler, Ze
  Andrew Haner, Zi
  David Johnston, Gg
  Lea Nondorf, Cg

Senior Petroleum Geologist:
   Peng Li, (D), Alabama, 2007, GoEoCo
   M. Ed Ratchford, (D), Idaho, 1994, GocEo
Professional Geologist:
   Richard S. Hutto, (B), Arkansas Tech, 1994, Gg
GIS Analyst:
   Nathan H. Taylor, (B), 2007, ZiyGe
Geology Supervisor:
   Scott Ausbrooks, (B), Arkansas, 2001, YsGeg
   Angela Chandler, (M), Arkansas, 1996, GgrGs
Geologist:
   Ty Johnson, (M), Arkansas, 2008, GgZiGr
   Daniel S. Rains, (M), Arkansas, 2009, GgZi
Deputy Director & Asst State Geologist:
   Mac B. Woodward, (B), S State, 1957, Gog

British Columbia

British Columbia Geological Survey & Dev
British Columbia Geological Survey & Development (2015)
PO Box 9333 Stn Prov Govt
Victoria, BC V8W 9N3
   Geological.Survey@gov.bc.ca
   http://www.empr.gov.bc.ca/MINING/GEOSCIENCE/Pages/default.aspx

California

California Geological Survey
California Geological Survey (2019)
801 K Street
Suite 1200
Sacramento, CA 95814
   p. (916) 445-1825
   cgshq@consrv.ca.gov
   http://www.consrv.ca.gov/CGS/Pages/Index.aspx
State Geologist:
   John G. Parrish, (D)
Supervising Engineering Geologist:
   John Clinkenbeared, (B), Eg
Supervising Engineering Geologist:
   Timothy P. McCrink, (M), New Mexico Tech, 1982, Ng
   William Short, (M), Ng
Senior Sesimologist:
   Hamid Haddidi, (D), Ys
Senior Geologist:
   Chris T. Higgins, (M), California (Davis), 1977, Gg
Senior Engineering Geologist (Specialist):
   Jennifer Thornburg, (M), California (Santa Cruz), Ng
Senior Engineering Geologist:
   David Branum, (B), Ng
   Rui Chen, (D), Edmonton, Nr
   Ron C. Churchill, (D), Minnesota, 1980, Cg
   Tim Dawson, (M), Ng
   Marc Delattre, (M), Ng
   Fred Gius, (M), Gz
   Will Harris, (B), Ng
   Jeremy Lancaster, (M), Ng
   Jeremy Lancaster, (B), Gg
   Donald Lindsay, (M), Ng
   David Longstreth, (M), Ng
   Anne Rosinski, (M), Ng
   Michael A. Silva, (B), California (Davis), 1978, Ng
Senior Engineer:
   Moh J. Huang, (D), Caltech, 1983, Ne
Materials Engineer:
   Badie Rowshandel, (D), Ne
Senior Scientist:
   Badie Rowshandel, (D), Ys
Geologist:
   Lawrence Busch, (A), Eg
Engineering Geologist:
   Jackie Bott, (D), Gc
   Patrick Brand, (M), Ng
   Kevin Doherty, (M), Ng
   Michael Fuller, (B), Ng
   Joshua Goodwin, (M), Gg
   Carlos Guiterrez, (B), Sacramento State, Ng
   Wayne Haydon, (M), Ng
   Cheryl Hayhurst, (B), Ng
   Janis Hernandez, (M), Ng
   Peter Holland, (M), Ng
   Maxime Mareschal, (M), Ng
   Michael McKinney, (M), Gg
   Ante Mlinarevic, (M), Ng
   Matt O'Neal, (B), Gi
   Brian Olson, (M), Ng
   John Oswald, (B), Ng
   Florante Perez, Ng
   Cindy L. Pridmore, (M), San Diego State, 1983, Ng
   Pete Roffers, (M), Ng
   Ron Rubin, (M), Ng
   Gordon Seitz, (D), Ng
   Eleanor Spangler, (M), Ng
   Ellie Spangler, (M), Gc
   Brian Swanson, (M), Ng
   Mark Weigers, (M), Ng
   Chase White, (M), Ng
   Judy Zachariasen, (D), Gg
Civil Engineer:
   Daniel Swensen, (D), Ne
Engineering Geologist:
   Rick I. Wilson, (B), Fresno State, 1987, Ge
Associate Scientist:
   Lijam Hagos, (D), Ys
   Carla Rosa, (M), Ng

Colorado

Colorado Geological Survey
Colorado Geological Survey (2019)
1801 Moly Rd.
Golden, CO 80401
   p. 303-384-2655
   cgs_pubs@mines.edu
   http://www.coloradogeologicalsurvey.org/
   f: https://www.facebook.com/ColoradoGeologicalSurvey
State Geologist:
   Karen Berry, (B), Colorado Mines, NgZu
Senior Mapping Geologist:
   Matt Morgan, (M), Colorado Mines, 2006, GgmXm
Senior Hydrogeologist:
   Peter Barkmann, (M), Montana, 1984, HwGcg
Senior Geothermal Geologist:
   Paul Morgan, (D), Imperial Coll, 2003, YhGtYg
Senior Engineering Geologist:
   Jonathan Lovekin, (M), Colorado Mines, 2007, NgZuGs
Engineering Geologist:
   Jill Carlson, (B), Wesleyan, 1987, NgZu
Senior Engineering Geologist:
   Jonathan White, (B), E Illinois, 1983, NgZuGg
Hydrogeologist:
   Lesley Sebol, (D), Waterloo, 2005, HwGeZg
GIS Hazard Analyst:
   Francis Scot Fitzgerald, (M), Denver, 2011, ZirGg
Geologist:
   Kassandra Lindsey, (M), Portland State, 2015, NgGmZi
   Mike O'Keeffe, (M), New Mexico Tech, 1994, GgeEg
Engineering Geologist:
   Kevin McCoy, (D), Colorado Mines, 2015, NgZiHw
Scientific & Technical Graphic Designer:
   Larry Scott, (B), Univ Arts, 1985, Zy

Connecticut

Dept of Energy and Environmental Protection
Connecticut Geological Survey (2019)
Office of Information Management
79 Elm Street, 6th floor
Hartford, CT 06106-5127
   p. (860) 424-3540
   deep.ctgeosurvey@ct.gov

http://www.ct.gov/deep/geology
State Geologist:
 Margaret A. Thomas, (M), Connecticut, 1983, Gg
Senior Research Associate:
 Randolph P. Steinen, (D), Brown, 1973, Gsg
Civil Engineer:
 Thomas E. Nosal, (M), C Connecticut, 1992, Zn
Research Associate:
 Lindsey Belliveau, (M), Connecticut, 2016, Gmt
 Teresa K. Gagnon, (M), Boston Coll, 1992, Gg
Resource Assistant:
 James M. Bogart, (B), S Connecticut State, 2015, Ge

Delaware
University of Delaware
Delaware Geological Survey (2018)
257 Academy Street
Newark, DE 19716-7501
 p. (302) 831-2833
 delgeosurvey@udel.edu
 http://www.dgs.udel.edu
 Administrative Assistant: Denise T. Heldorfer
 Administrative Assistant: Laura K. Wisk
State Geologist:
 David R. Wunsch, (D), Kentucky, 1992, ClHwNg
Hydrogeologist:
 A. Scott Andres, (M), Lehigh, 1984, Hg
Senior Scientist:
 Peter P. McLaughlin, (D), Louisiana State, 1989, HgPm
Hydrogeologist:
 Thomas E. McKenna, (D), Texas, 1997, Hg
Hydrogeologist:
 Changming He, (D), Nevada (Reno), 2004, GqHg
Associate Scientist:
 Stefanie J. Baxter, (M), Delaware, 1994, On
Research Associate:
 John A. Callahan, (M), Delaware, 2014, Zri
 Jaime L. Tomlinson, (M), Delaware, 2006, Zn
Emeritus:
 John H. Talley, (M), Franklin and Marshall, 1974, As
Scientist:
 Kelvin W. Ramsey, (D), Delaware, 1988, As
 William S. Schenck, (M), Delaware, 1997, ZiGi
GIS Specialist:
 Lillian T. Wang, (M), Delaware, 2005, Zi

Florida
Florida Geological Survey
Florida Dept of Environmental Protection (2018)
Commonwealth Building
3000 Commonwealth Blvd, Suite 1
Tallahassee, FL 32303
 p. (850) 617-0300
 jonathan.arthur@FloridaDEP.gov
 https://floridadep.gov/fgs
State Geologist:
 Jonathan D. Arthur, (D), Florida State, 1994, HwCgGe
Assistant State Geologist:
 Guy H. Means, (M), Florida State, 2009, PgGe

Georgia
Georgia Dept of Natural Resources
Georgia Environmental Protection Div (2016)
Environmental Protection Division
2 Martin Luther King Jr. Dr., Suite 1152
East Tower
Atlanta, GA 30334-9004
 p. (404) 657-5947
 askepd@gaepd.org
 http://epd.georgia.gov/
Senior Scientist:
 James Kennedy, (D), Texas A&M, 1981

Hawaii
Dept of Land & Natural Resources
Commission on Water Resource Management (2018)
Kalanimoku Building
P.O. Box 621
1151 Punchbowl Street, #227
Honolulu, HI 96809
 p. (808) 587-0214
 dlnr.cwrm@hawaii.gov
 http://dlnr.hawaii.gov/cwrm

Idaho
University of Idaho
Idaho Geological Survey (2018)
875 Perimeter Dr. MS 3014
University of Idaho
Moscow, ID 83844-3014
 p. (208) 885-7991
 igs@uidaho.edu
 http://www.idahogeology.org/
Senior Geologist:
 Dennis M. Feeney, (M), W Washington, 2008, GgiYg
Associate Scientist:
 Virginia S. Gillerman, (D), California (Berkeley), 1982, Eg
 Reed S. Lewis, (D), Oregon State, 1990, GgiEg
Related Staff:
 Michael Ratchford, (D), Idaho, 1994, GcoEg

Illinois
Illinois State Geological Survey
Energy & Earth Resource Center (2015)
615 E. Peabody Drive.
Champaign, IL 61820-6964
 p. (217) 244-2430
 finley@isgs.uiuc.edu
Center Director:
 Robert J. Finley, (D), South Carolina, 1975, Eo
Scientist:
 Latif A. Khan, (D), Tech, 1971, Nx
Senior Scientist:
 Scott M. Frailey, (D), Missouri (Rolla), 1989, Np
 Massoud Rostam-Abadi, (D), Wayne State, 1982, Zn
Scientist:
 Mei-In (Melissa) Chou, (D), Michigan State, 1977, Co
 Sheng-Fu Joseph Chou, (D), Michigan State, 1977, Co
 Joseph A. Devera, (M), S Illinois, 1985, Gr
 Ivan G. Krapac, (M), Illinois, 1987, Ca
 Zakaria Lasemi, (D), Miami, 1990, En
 Hannes E. Leetaru, (D), Illinois, 1997, Eo
 Donald G. Mikulic, (D), Oregon State, 1979, Ps
 Beverly Seyler, (M), SUNY, 1978, Eo
Associate Scientist:
 Cheri A. Chenoweth, (B), Illinois, 1979, Ec
 Joan E. Crockett, (B), Illinois, 1983, EoZg
 John P. Grube, (M), Colorado Mines, 1984, Eo
 Bryan G. Huff, (M), Illinois, 1984, Eo
 Rex A. Knepp, (M), Go
Assistant Scientist:
 F. Brett Denny, (B), Missouri (Rolla), 1985, Gr
 Christopher P. Korose, (B), Illinois, 1995, Ec
 Vinodkumar A. Patel, (B), Inst of Tech, 1973, Zn
Emeritus:
 Pam Cookus, Zn
Assistant Scientist:
 Kathleen M. Henry, (B), Illinois State, 1982, Gg

Illinois State Geological Survey (2017)
615 East Peabody Drive
Champaign, IL 61820-6964
 p. 217-333-4747
 info@isgs.illinois.edu
 https://www.isgs.illinois.edu/
 f: https://www.facebook.com/ILGeoSurvey

t: @ILGeoSurvey
Administrative Assistant: Tamra S. Montgomery
Director:
Richard C. Berg, (D), Illinois, 1979, GleGm
Senior Geologist:
Anne L. Ellison, (M), Ge
Geoscience Information Stewardship:
Mark Yacucci, (M), Illinois State
Associate Geologist:
Scott D. Elrick, (M), California (Riverside), 1999, EcGsPg
Senior Geologist:
Zakaria Lasemi, (D), Illinois, Eg
Senior Geochemist:
Samuel V. Panno, (M), S Illinois, 1978, ClGgHw
Head Geochemistry:
Randall Locke, (M), Cl
Chief Scientist:
Steven E. Brown, (M), Wisconsin, 1990, Glm
Associate Quaternary Geologist:
David A. Grimley, (D), Illinois, 1996, GmlSc
Senior Scientist:
Richard A. Cahill, (M), Maryland, 1974, CaGeCt
Associate Quaternary Geologist:
Andrew C. Phillips, (D), Illinois (Chicago), 1993, GsmZi
Adjunct Assistant Professor:
Ethan Theuerkauf, (D), North Carolina, 2016, GmuGs
Assistant Petroleum Geologist:
Nathan D. Webb, (M), Illinois (Urbana), 2009, EoGo
Illinois State Geologist:
E. Donald McKay III, (D), Illinois, 1977, GlsGr
Wetlands Geology Specialist:
Colleen Long, (M), North Carolina (Chapel Hill), 2012, Gg
Jessica Monson, (M), Gg
Wetlands Geologist and Head:
James J. Miner, (M), Gg
Wetlands Geologist:
Jessica R. Ackerman, Zn
Team Leader/Associate Geologist:
Dale R. Schmidt, (M), Ge
Team Leader:
D. Adomaitis, Ge
Senior Petroleum Geologist:
Hannes E. Leetaru, (D), Illinois, 1997, Gor
Senior Paleontologist:
Joseph A. Devera, (M), Pg
Senior Geophysicist:
Timothy H. Larson, (D)
Remote Sensing Data Manager:
Janet Carmarca
Program Manager, Illinois Height Modernization Project:
Sheena K. Beaverson, Zr
Principal Engineering Geologist:
R. A. Bauer, (M), Ng
Media and Information Technology Administrator:
Daniel Byers
Map Standards Coordinator:
Jennifer Carrell, Zg
Hydrogeologist and Assistant Section Head:
Yu-Feng Forrest Lin, (D), Wisconsin, 2002, Hg
Geospatial Applications Developer/GIS Specialist:
Melony Barrett, Zi
Geologic Specialist:
Zohreh Askari, (M), Azad, 1997, Gsc
Alan R. Myers, (B), EcGo
Jennifer M. Obrad, (M), EcGo
Enviromental Data Coordinator:
Clint Beccue, Ge
Associate Wetlands Geologist:
Steven Benton, Gg
Keith W. Carr, (M), Gg
Eric T. Plankell, (M), Hg
Geoff Pociask, (M), Hw
Associate Quaternary Geologist:
Olivier J. Caron, (D), Québec à Montréal, Ng
Associate Geologist:
Cheri Chenoweth, GoEc
Christopher P. Korose, (M), EcGo

Associate Geohydrologist:
Jason F. Thomason, (D), Hw
Associate Geochemist:
Shari E. Effert-Fanta, (M), Cas
Associate Engineering Geologist:
Andrew Anderson, Ng
Greg A. Kientop, (M), Ge
Associate Economic Geologist:
F. Brett Denny, (M), GzEg
Associate Director Advanced Energy Technology Initiative:
Sallie Greenberg, (D), Illinois (Urbana), 2013, ClZg
Assistant Wetlands Geologist:
Kathleen E. Bryant, (M), Gg
Assistant Section Head, Environmental Assessments:
Mark Collier, Ge
Assistant Section Head:
B. Brandon Curry, (D), Illinois (Urbana), 1995, Gel
Assistant Geologist:
James Damico, (M)
Craig R. Decker, (B)
Scott R. Ellis, (B), Ge
Bradley Ettlie, (B), Ge
Jared Freiburg, (M), GgzGx
James W. Geiger, (B), Ge
Nathan P. Grigsby, (B)
Matthew P. Spaeth, (M), Ge
Assistant Geochemist:
Peter M. Berger, (M), Illinois (Urbana), 2008, Cg
Other:
Hong Wang, Illinois (Urbana), 1996, Cg

University of Illinois
Illinois State Geological Survey/Prairie Research Institute (2018)
615 E. Peabody Dr.
Champaign, IL 61820-6964
  p. (217) 333-4747
  info@isgs.illinois.edu
  http://www.isgs.edu
  f: https://www.facebook.com/ILGeoSurvey/
  t: @ILGeoSurvey
State Geologist:
Richard Berg, (D), Illinois (Urbana-Champaign), 1979, GlRwZu
Energy & Minerals Group:
Steven Whittaker, (D), Saskatchewan
Principle Geologist:
B. Brandon Curry, Illinois, Gl
Principal Scientist:
Randall A. Locke II, (M), CgHw
Principal Geologist:
Scott D. Elrick, (M), California (Riverside), 1998, EcGgs
Assoociate Geologist:
Geoffrey Pociask, (M), Wisconsin, Hs
Head:
Anne L. Ellison, (M), Ge
Hannes Leetaru, (D), Illinois, Go
Jason Thomason, (D), Iowa State, Gl
Mark A. Yacucci, (M), Zi
Senior Scientist:
Yongqi Lu, (D), Zm
Chief Scientist:
Steven E. Brown, (M), Gl

Illinois State Water Survey (2018)
2204 Griffith Drive
Champaign, IL 61820
  p. (217) 333-2210
  pattih@illinois.edu
  http://www.isws.illinois.edu/
  Programs: N/A
  Certificates: N/A
Head, Groundwater Science Section:
Walton R. Kelly, (D), Virginia, 1993, Hw
Senior Scientist:
Steven D. Wilson, (M), Illinois, 1988, Hw
Associate Professional Scientist:
George S. Roadcap, (D), Illinois, 2004, Hw

Associate Scientist:
  Daniel Abrams, (D), Indiana, 2013, Hw
  Daniel Hadley, (M), N Arizona, 2014, Hw
  Devin Mannix, (M), S Illinois, 2013, Hw

Indiana

Indiana University
Indiana Geological and Water Survey (2017)
611 North Walnut Grove Avenue
Bloomington, IN 47405
  p. (812) 855-7636
  igsinfo@indiana.edu
  http://igws.indiana.edu/
State Geologist & Director:
  Todd A. Thompson, (D), Indiana, 1987, GsmGu
Senior Scientist:
  Maria Mastalerz, (D), Silesian Tech, 1988, Ec
Asitant Director for Research:
  Lee J. Florea, (D), S Florida, 2006, HqGeCl
Associate Scientist:
  Sally L. Letsinger, (D), Indiana, 2001, HqZir
Research Associate:
  Patrick I. McLaughlin, (D), Cincinnati, 2006, GrClGs
Reservoir Geologist:
  Cristian R. Medina, (M), Indiana, 2007, GoeHw
Research Geophysicist/Hydrologist:
  Kevin M. Ellett, (M), California (Davis), 2002, YeHw
Research Geologist:
  José Luis Antinao, (D), Dalhousie, 2009, CcGlr
  Robert J. Autio, (M), Indiana, 1992, HgGle
  Alyssa M. Bancroft, (D), Ohio State, 2014, PmGr
  Tracy D. Branam, (M), Indiana, 1991, CasEm
  Christopher Dintaman, (M), Indiana, 1997, Hw
  Agnieszka Drobniak, (D), Sci Tech (Krakow), 2002, Ec
  Jayson Eldridge, (B), California (Davis), 2012, EoGr
  Nancy R. Hasenmueller, (M), Ohio State, 1969, Gre
  Rebecca A. Meyer, Ec
  Gary J. Motz, (M), Akron, 2010, PqgPi
  Shawn C. Naylor, (M), Montana, 2006, Hw
  John A. Rupp, (M), E Washington, 1980, Go
Quaternary Geologist:
  Henry M. Loope, (D), Wisconsin, 2013, Gls

Iowa

Iowa Dept of Natural Resources
Iowa Geological and Water Survey (2015)
109 Trowbridge Hall
Iowa City, IA 52242-1319
  p. (319) 335-1575
  MaryPat.Heitman@dnr.iowa.gov
  http://www.igsb.uiowa.edu
  Administrative Assistant: Mary Pat. Heitman
Research Geologist:
  Richard A. Langel, (M), Iowa, 1996, Gg
State Geologist:
  Robert D. Libra, Gg
Section Supervisor, Geology and Groundwater Studies:
  J. Michael Gannon, (M), Arizona, Gg
Section Supervisor, Geographic Information:
  Chris Ensminger
Research Geologist:
  Mary R. Howes, Gg
  Lynette S. Seigley, (M), Iowa, Gg
Natural Resource Biologist:
  Jacklyn Gautsch, (B), Wisconsin, Zn
GIS Technician:
  Chris Kahle, (M), Kansas, Zi
Geologist 3, Research Geologist:
  Paul Hiaibao Liu, (D), Nebraska, Gg
  Robert M. McKay, (B), Tulane, Gg
  Deborah J. Quade, (M), Iowa, 1992, GleGm
  Robert Rowden, (M), Iowa, Gg
  Keith Schilling, (M), Iowa State, Gg
  Stephanie Tassier-Surine, (M), Massachusetts, Gg
  Paul E. VanDorpe, (M), Wayne State, Gg
Geologist 3, Remote Sensing Analyst:
  James D. Giglierano, (M), Purdue, GgZri
  Pete Kollasch, (M), Iowa, GgZri
Geologist 3, GIS Analyst:
  Kathryne Clark, (M), New Mexico, GgZi
Geologist 3, Geographic Information System Analyst:
  Calvin Wolter, (B), Arizona, GgZi
Geologist 2, Research Geologist:
  Michael Bounk, (M), Iowa, Gg
  Chad Fields, (M), N Iowa, Gg
Geologist 3, NRGIS Library Manager and GIS Analyst:
  Casey Kohrt, (B), Iowa State, GgZi

Kansas

University of Kansas
Kansas Geological Survey (2017)
1930 Constant Avenue
West Campus
Lawrence, KS 66047-3724
  p. (785) 864-3965
  http://www.kgs.ku.edu/
  f: www.facebook.com/KansasGeologicalSurvey
  t: @ksgeology
Director:
  Rolfe D. Mandel, (D), Kansas, 1991, Ga
Section Chief, Senior Scientist:
  James J. Butler, Jr., (D), Stanford, 1987, Hw
  Richard D. Miller, (D), Leoben, 2007, Ye
Research Project Director:
  Kenneth A. Nelson, (B), Kansas, 1993, Zy
Manager, Wichita Well Sample Library:
  Mike Dealy, (B), Fort Hays State, 1979, Gg
Manager, Geohydrology Support Services:
  Blake Brownie Wilson, (M), Kansas State, 1993, Zi
Senior Scientific Fellow:
  Evan K. Franseen, (D), Wisconsin, 1989, Gs
  Lynn W. Watney, (D), Kansas, 1985, Gr
  Donald O. Whittemore, (D), Penn State, 1973, Hw
Section Chief, Senior Scientist:
  Greg A. Ludvigson, (D), Iowa, 1988, GrCsPc
Emeritus, Senior Scientist:
  Ricardo A. Olea, (D), Kansas, 1982, GqoEc
Associate Scientist:
  Geoffrey C. Bohling, (D), Kansas, 1999, Hw
  Gaisheng Liu, (D), Alabama, 2004, Hy
  Kerry D. Newell, (D), Kansas, 1996, Go
Research Project Manager:
  Eileen Battles, (B), Kansas, 1993, Zy
Petroleum Engineer:
  Yehven I. Holubnyak, (M), North Dakota, 2008, Np
Associate Researcher Senior:
  Robert S. Sawin, (M), Kansas State, 1977, Gg
Assistant Researcher Senior:
  Edward Reboulet, (M), Boise State, 2003, Hy
Assistant Research Professor:
  Julian Ivanov, (D), Kansas, 2002, Yg
Emeritus, Senior Scientist:
  Pieter Berendsen, (D), California (Riverside), 1971, Cg
  Robert W. Buddemeier, (D), Washington, 1969, Hy
  Tim R. Carr, (D), Wisconsin, 1981, Gg
  John C. Davis, (D), Wyoming, 1967, GqoGd
Emeritus, Senior Scientific Fellow:
  Lawrence L. Brady, (D), Kansas, 1971, Gg
  John H. Doveton, (D), Edinburgh, 1969, Go
Emeritus, Director:
  Rex C. Buchanan, (M), Wisconsin, 1982, Ze
Emeritus, Associate Scientist:
  Truman Waugh, (B), Washburn, 1963, Ca
Emeritus, Assistant Director:
  Lawrence H. Skelton, (M), Wichita State, 1991, Gge
Research Engineer Senior:
  Brett Bennet, (B), Kansas, 1982, Ng
Outreach Manager, Geologist:
  Susan G. Stover, (M), Kansas, 1993, Gg

Data Resources Library Manager, Geologist:
  Daniel R. Suchy, (D), McGill, 1992, Gg
Assistant Scientist:
  Tandis Bidgoli, (D), Kansas, 2014, Gc
  Andrea Brookfield, (D), Waterloo, 2009, Hy
  Jon J. Smith, (D), Kansas, 2007, Gg

Kentucky

University of Kentucky
Kentucky Geological Survey (2017)
228 Mining & Minerals Resources Building
504 Rose Street
Lexington, KY 40506-0107
  p. (859) 257-5500
  jerryw@uky.edu
  http://www.uky.edu/KGS
Interim State Geologist and Director:
  Gerald A. Weisenfluh, (D), South Carolina, 1982, Ec
Head, Western Kentucky Office:
  David A. Williams, (M), E Kentucky, 1979, Eg
Head, Water Resources Section:
  Charles J. Taylor, (M), Kentucky, 1992, HwgCl
Head, Geoscience Information Management:
  Doug C. Curl, (M), Tennessee, 1998, ZfGgc
Head, Geologic Hazards Section:
  Zhenming Wang, (D), Kentucky, 1998, Ys
Head, Energy & Minerals:
  David C. Harris, (M), SUNY (Stony Brook), 1982, Go
Manager, Well Sample & Core Library:
  Patrick J. Gooding, (M), E Kentucky, 1983, Eo
Geologist:
  Rick Bowersox, (D), S Florida, 2006, Go
Senior Scientist:
  J. Richard Bowersox, (D), S Florida, 2006, GoePq
Geologist:
  Stephen F. Greb, (D), Kentucky, 1992, GsEcGs
Geologist:
  John Hickman, (D), Kentucky, 2011, EoGtc
Adjunct Professor:
  Gerald A. Weisenfluh, (D), South Carolina, 1982, EcZi
Technology Transfer Officer:
  Michael J. Lynch, (B), E Kentucky, 1975, Zn
Manager, Administration:
  Kathryn E. Ellis, (M), Kentucky, 2012, Zn
Hydrogeologist:
  E. Glynn Beck, (M), East Carolina, 1997, Hw
  James C. Currens, (M), E Kentucky, 1978, Hg
  Junfeng Zhu, (D), Arizona, 2005, Hw
Geologist:
  Matt Crawford, (M), E Kentucky, 2001, Zn
  Bart Davidson, (M), E Kentucky, 1986, Hg
  Cortland F. Eble, (D), West Virginia, 1988, Pl
  Brandon C. Nuttall, (B), E Kentucky, 1971, Eo
  Thomas M. Parris, (D), California (Santa Barbara), 1998, Eo
  Thomas N. Sparks, (M), Duke, 1979, Gg

Louisiana

Louisiana State University
Basin Research Energy Section (2015)
Louisiana Geological Survey/LSU
208 Howe Russell Geoscience Complex
Baton Rouge, LA 70803-4101
  p. (225) 578-8328
  bharde1@lsu.edu
  http://www.bri.lsu.edu
  Office Coordinator: Cherri B. Webre
Associate Professor:
  Ronald K. Zimmerman, (D), Louisiana State, 1966, Go
Assistant Professor:
  Clayton F. Breland, (D), Tennessee, 1980, Ye
  John B. Echols, (D), Louisiana State, 1966, Go
Research Associate:
  Brian J. Harder, (B), Louisiana State, 1981, Eo
  Bobby L. Jones, (B), Louisiana State, 1953, Go
  Phillip W. Lemay, (B), Centenary, 1999, Gg
  Michael B. Miller, (M), North Carolina, 1982, Go
  Lloyd R. Milner, (B), Louisiana State, 1985, Gg
  Patrick M. O'Neill, (B), Louisiana State, 1985, Sf
Computer Analyst:
  Reed J. Bourgeois, (B), Louisiana State, 1983, Zn
Accountant Technician:
  Carla Domingue, Zn

Louisiana Geological Survey (2015)
3079-Energy
Coast and Environment Bldg.
Baton Rouge, LA 70803
  p. (225) 578-5320
  hammer@lsu.edu
  http://www.lgs.lsu.edu
State Geologist and Professor:
  Chacko J. John, (D), Delaware, 1977, GosGe
Assistant Professor:
  Douglas A. Carlson, (D), Wisconsin (Milwaukee), 2001, Hw
  Marty R. Horn, (D), Texas (Arlington), 1996, EoGgr
GIS Coordinator:
  R Hampton Peele, (M), Louisiana State, 2000, ZirZg
Computer Analyst:
  Reed J. Bourgeois, (B), Nicholls State, 1985, Zn
Research Associate:
  Brian J. Harder, (B), Louisiana State, 1981, Eo
  Paul V. Heinrich, (M), Illinois, 1982, GsaGm
  Bobby Jones
  Richard P. McCulloh, (M), Texas, 1977, Gg
  Lloyd R. Milner, (B), Louisiana State, 1985, Gg
  Patrick M. O'Neill, (B), Louisiana State, 1985, Sf
  Robert L. Paulsell, (B), Louisiana State, 1987, ZyGg
  Lisa G. Pond, (B), Louisiana State, 1987, Zg
  Arren Schulingkamp, Gg
Cartographic Manager:
  John I. Snead, (B), Louisiana State, 1978, Gm
Assistant Director:
  John E. Johnston, III, (M), Texas, 1977, EoGe
Office Coordinator:
  Melissa H. Esnault, (B), Louisiana Tech, Zn

Maine

Dept of Agriculture, Conservation, and Forestry
Maine Geological Survey (2018)
93 State House Station
Augusta, ME 04333-0093
  p. (207) 287-2801
  mgs@maine.gov
  http://www.maine.gov/dacf/mgs/
  Administrative Assistant: Tammara Roberts
State Geologist:
  Robert G. Marvinney, (D), Syracuse, 1986, Gg
Senior Geologist:
  Lindsay Spigel, (D), Wisconsin, 2006, Gm
Physical Geologist:
  Henry N. Berry IV, (D), Massachusetts, 1989, Gg
Marine Geologist:
  Stephen M. Dickson, (D), Maine, 1999, OnGuOu
  Peter Slovinsky, (M), South Carolina, 2001, Gu
Hydrogeologist:
  Ryan Gordon, (D), Syracuse, 2014, HwGm
  Daniel B. Locke, (B), Maine, 1982, Hw
Director, Earth Resources Information:
  Christian Halsted, (B), Maine, 1995, Zi
GIS Coordinator:
  Amber Whittaker, (M), New Mexico, 2006, ZiGcCg

Manitoba

Manitoba Geological Survey
Manitoba Growth, Enterprise and Trade (2018)
360-1395 Ellice Avenue
Winnipeg, MB R3G 3P2
  p. 1-800-223-5215

christian.bohm@gov.mb.ca
http://www.manitoba.ca/iem/index.html
Director:
 Alisa Ramrattan
Senior Scientist:
 Christian Bohm, (D), ETH (Switzerland), 1996, GgCgEg
 Michelle P.B. Nicolas, (M), Manitoba, 1997, GrsGo

Maryland
Maryland Department of Natural Resources
Maryland Geological Survey (2018)
2300 St. Paul Street
Baltimore, MD 21218-5210
 p. (410) 554-5500
 mgsinfo.dnr@maryland.gov
 http://www.mgs.md.gov/
State Geologist:
 Richard A. Ortt, Jr., (B), Johns Hopkins, 1991, NtGeYr
Hydrogeologist, Program Chief:
 David C. Andreasen, (B), Maryland, 1985, Hw
Hydrogeologist:
 Jon Achmad, Hw
Geologist/Program Chief:
 Stephen Van Ryswick, (B), Maryland, 2002, Ge
Senior Scientist:
 David K. Brezinski, (D), Pittsburgh, 1984, Gr
Hydrogeologist:
 Andrew Staley, Hw
 Tiffany J. VanDerwerker, (M), Virginia Tech, Hw
Geologist:
 Christopher B. Connallon, (M), Michigan State, 2015, GmZi
 William D. Junkin, (M), California (Santa Barbara), 2018, GvrGg
 Rebecca H. Kavage-Adams, (M), Virginia Tech, 2002, GmZiGe
 Katherine A. Knippler, (B), Susquehanna, 2001, Ge
 Heather A. Quinn, (M), Florida, GgHg
 Elizabeth R. Sylvia, (M), Towson, 2013, Ge
Education Specialist:
 Dale W. Shelton, (B), Towson, 1986, ZeGg

Massachusetts
Massachusetts Geological Survey
Dept of Geosciences (2018)
Univ of Massachusetts (Amherst)
627 North Pleasant Street
Amherst, MA 01003
 p. (413) 545-2286
 sbmabee@geo.umass.edu
 Programs: N/A
 Certificates: N/A
State Geologist:
 Stephen B. Mabee, (D), Massachusetts, 1992, Hw

Minnesota
University of Minnesota
Minnesota Geological Survey (2018)
2609 West Territorial Road
Saint Paul, MN 55114-1009
 p. 612-626-2969
 mgs@umn.edu
 https://www.mngs.umn.edu/index.html
 f: https://www.facebook.com/MinnesotaGeologicalSurvey
Director:
 Harvey Thorleifson, (D), Colorado, 1989, Gl
Geologist:
 Terrence Boerboom, (M), Minnesota (Duluth), 1987, GivGc
Geologist:
 Jennifer Horton, (M), Toledo, 2015, Gl
Geologist:
 Jacqueline Hamilton, (M), Minnesota, 2007, ZiGe
 Andrew J. Retzler, (M), Idaho State, 2013, GrsPi

Mississippi
Mississippi Office of Geology
Geospatial Resources Div (2015)
P.O. Box 2279
700 N. State St
Jackson, MS 39225-2279
 p. (601) 961-5500
 barbara_yassin@deq.state.ms.us
 www.deq.state.ms.us
Geologist:
 Steven D. Champlin, (B), Alabama, 1976, Go
GIS Analyst:
 Barbara E. Yassin, (B), Illinois State, 1989, ZyiGm
Geologist:
 Peter S. Hutchins, (B), Millsaps, 1990, Zn

Mississippi Dept of Environmental Quality (2016)
P.O. Box 2279
Jackson, MS 39225-2279
 p. (601) 961-5500
 mbograd@mdeq.ms.gov
 http://www.deq.state.ms.us/
State Geologist:
 Michael B. E. Bograd, (M), Mississippi, 2002, Gg

Missouri
Missouri Dept of Natural Resources
Dam & Reservoir Safety (2015)
PO Box 250
Buehler Bldg/111 Fairgrounds Rd
Rolla, MO 65402
 p. (573) 368-2175
 bob.clay@dnr.mo.gov
Professional Staff:
 Robert Clay, (M), Oklahoma State, 1977, Zn
Professional Staff:
 Glenn D. Lloyd, (B), Zn
Other:
 Paul Simon, (B), MST
 Ryan Stack, (B), MST

Div of Geology and Land Survey (2015)
PO Box 250
111 Fairgrounds Road
Rolla, MO 65402-0250
 p. (573) 368-2100
 joe.gillman@dnr.mo.gov
 http://www.dnr.state.mo.us/geology.htm
 Administrative Assistant: Tami L.. Allison
Division Director:
 Mimi R. Garstang, (B), SW Missouri State, Ge
Deputy Director & Assistant State Geologist:
 James W. Duley, (B), C Missouri, 1975, Hy

Geological Survey Program (2018)
PO Box 250
111 Fairgrounds Rd
Rolla, MO 65402-0250
 p. (573) 368-2100
 gspgeol@dnr.mo.gov
 http://www.dnr.mo.gov/geology
Professional Staff:
 Joe Gillman, (B), Missouri State, 1992, Gg
Senior Scientist:
 Pat Mulvany, (D), Missouri S&T, 1996
Unit Chief:
 Justin Davis, (M), Missouri S&T, 2012
 Jeremiah Jackson, (M), Missouri State, 2011, Gg
 Brenna McDonald, (B), SE Missouri State, 1997, Ge
 Cheryl M. Seeger, (D), Missouri S&T, 2003, Gig
 Chris Vierrether, (M), Missouri S&T, 1988, GgoGe
 Vicki Voigt, (B), Missouri S&T, 2010, Gg
Section Chief:
 Larry Pierce, (M), Zn
 Kyle Rollins, (B), Missouri State, 1986, ZyGg

Sherri Stoner, (B), Missouri State, 1991
Program Director:
 Amber Steele, (M), Missouri, 2011, Sd
Geologist:
 Peter Bachle, (B), Missouri S&T, 1997, Gg
 Fletcher Bone, (B), C Missouri, 2007, Gg
 David L. Bridges, (D), Missouri S&T, 2011, GgiPg
 John H. Corley, (M), Missouri, 2014, Geg
 Jeff Crews, (M), Missouri S&T, 2004, Hy
 Trevor Ellis, (B), Missouri S&T, 2010, Gg
 Kyle Ganz, (M), Missouri S&T, 2013
 Airin J. Haselwander, (B), Missouri S&T, 2011, Gg
 Terry Hawkins, (B), Brigham Young, 1988, Geg
 Lisa Lori, (M), Missouri S&T, 2016, Gp
 Brad Mitchell, (M), Missouri S&T, 2010
 Matt Parker, (B), Missouri S&T, 1993
 John Pate, (B), Tennessee, 2007, Ge
 Kirsten Schaefer, (B), Illinois State, 2016
 Michael A. Siemens, (M), Wichita State, 1985, GgHgZi
 Molly Starkey, (M), Missouri State, 2011
Geological Tech:
 Cecil Boswell, Zn
 Eric Hohl, Zn
 Karen Loveland, Zn
 Dan Nordwald, Zn
 Patrick Scheel, Zn
 Fred Shaw, Zn
Environmental Specialist:
 Andrew Combs
Deputy Director:
 Jerry L. Prewett, (B), Missouri State, 1992, GgHw

Water Resources Program (2018)
PO Box 250
111 Fairgrounds Rd
Rolla, MO 65401-0250
 p. (573) 368-2175
 mowaters@dnr.mo.gov
 http://www.dnr.state.mo.us/mgs/
Geologist:
 Scott Kaden, Hw
Associate Scientist:
 Robert Bacon, Hs
Professional Staff:
 Charles Du Charme, (B), Hs

Montana

Montana Tech of The University of Montana
Montana Bureau of Mines & Geology (A,B,M,D) (2018)
1300 West Park Street
Butte, MT 59701-8997
 p. (406) 496-4180
 jmetesh@mtech.edu
 http://www.mbmg.mtech.edu
 f: https://www.facebook.com/MontanaGeology/
 Administrative Assistant: Margaret Delaney
State Geologist:
 John J. Metesh, (D), Montana, 2003, HwCa
Assistant Director RET:
 Marvin R. Miller, (M), Indiana, 1965, Hw
Research Division Chief:
 Thomas W. Patton, (M), Montana Tech, 1987, Hw
Program Manager Geologist:
 Catherine McDonald, (M), California (Davis), 1992, Grs
Assistant Research Geologist:
 Jeff Lonn, (M), Montana, 1985, GgtGc
Senior Hydrogeologist:
 Jon C. Reiten, (M), North Dakota, 1983, Hw
Geologist:
 Susan M. Vuke, (M), Montana, 1982, GgrGs
Professor:
 Colleen G. Elliott, (D), New Brunswick, 1988, GctGg
Senior Research Hydrogeologist:
 John R. Wheaton, (M), Montana, 1987, HwEc
Program Manager Hydrogeologist:
 Ginette Abdo, (M), Penn State, 1989, Hw

 John I. La Fave, (M), Texas, 1987, Hw
Hydrogeologist:
 Andrew L. Bobst, (M), Binghamton, 2000, HwCgEo
 Gary Icopini, (D), Michigan State, 2000, ClHwCb
Geologist:
 Michael C. Stickney, (M), Montana, 1980, YsGtm
Publications Editor:
 Susan A. Barth, (M), Montana Tech, 2009, Zn
Assistant Research Hydrogeologist:
 Camela Carstarphen, (M), Oregon State, 1991, Hw
Assistant Research Geologist:
 Phyllis Hargrave, (M), Montana Tech, 1990, Gg
Sr. Hydrogeologist:
 Kirk B. Waren, (M), Wright State, 1988, HwsZe
Hydrogeologist:
 Terence E. Duaime, (B), Montana Tech, 1978, Hws
Senior Research Geologist, Museum Curator:
 Richard B. Berg, (D), Montana, 1964, EnGz
Seismic Analyst:
 Deborah Smith
Research Assistant III:
 Jaqueline R. Timmer, (B), Montana Tech, 1990, Ca
Hydrogeologist:
 Daniel D. Blythe, (B), Montana State, 2006, Hws
GIS Specialist:
 Ken L. Sandau, (A), Montana Tech, 1999, Zi
 Paul R. Thale, (M), Montana State, 1994, Zi
Geologic Cartographer:
 Susan M. Smith, (B), Montana State, 1970, Zn
Director of Sponsored Programs:
 Joanne Lee
Chemist:
 Ashley Huft, (B), Montana Tech, 2008, Ca
Associate Research Professor:
 Steve F. McGrath, (M), Montana Tech, 1992, CaEgCe
Assistant Hydrogeologist:
 Mary K. Sutherland, (M), Montana, 2009, Hws
Computer Software Eng/Applications:
 Luke Buckley, (B), Montana Tech, 1995, Zn

Nebraska

Unversity of Nebraska - Lincoln
Conservation & Survey Div (D) (2015)
Conservation & Survey Division
3310 Holdrege Street
616 Hardin Hall
Lincoln, NE 68583-0996
 p. (402) 472-3471
 mkuzila1@unl.edu
 http://snr.unl.edu/csd/
Director:
 Mark S. Kuzila, (D), Nebraska, 1988, Sd
Professor:
 Xun-Hong Chen, (D), Wyoming, 1994, Hq
 David C. Gosselin, (D), SD Mines, 1987, CgHw
 James W. Merchant, (D), Kansas, 1984, Zru
 Jozsef Szilagyi, (D), California (Davis), 1997, HqAs
Senior Scientist:
 Susan Olafsen-Lackey, (B), SD Mines, 1982, Hw
 Steven S. Sibray, (M), New Mexico, 1977, HwGg
Associate Professor:
 Paul Hanson, (D), Nebraska, 2005, Gm
 Matt Joeckel, (D), Iowa, 1993, GsrGg
Research Associate:
 Leslie M. Howard, (M), Nebraska, 1989, Zyi
Emeritus:
 Marvin P. Carlson, (D), Nebraska, 1969, Gt
 Robert F. Diffendal, Jr., (D), Nebraska, 1971, GrPiRh
 Duane A. Eversoll, (M), Nebraska, 1977, Ng
 Anatoly Gitelson, (D), Inst Radio Tech (Russia), 1972, Zn
 James W. Goeke, (M), Colorado State, 1970, HwGmYe
 Donald C. Rundquist, (D), Nebraska, 1977, Zri
 James Swinehart, (D), Gg

Nevada

University of Nevada 1
Nevada Bureau of Mines and Geology (2018)
Mail Stop 178
University of Nevada
Reno, NV 89557-0088
   p. (775) 682-8766
   jfaulds@unr.edu
   http://www.nbmg.unr.edu/
   f: https://www.facebook.com/Nevada-Bureau-of-Mines-and-Geology-106397989390636/
   Programs: N/A
   Certificates: N/A
   Administrative Assistant: Alex Nesbitt
State Geologist/Professor:
   James E. Faulds, (D), New Mexico, 1989, Gc
Professor:
   Geoffrey Blewitt, (D), Caltech, 1996, Yd
   William C. Hammond, (D), Oregon, 2000, Yd
   Christopher D. Henry, (D), Texas, 1975, GtvCc
   Corne Kreemer, (D), SUNY (Stony Brook), 2001, YdRn
Geologic Mapping Specialist:
   Seth Dee, (M), Oregon, 2006, Gg
Director, Center for Research in Economic Geology:
   John Muntean, (D), Stanford, 1998, Eg
Associate Professor:
   Bridget Ayling, (D), Australia, 2006, CgYhGo
   Craig M. dePolo, (D), Nevada (Reno), 1998, Ng
Assistant Professor:
   Rich Koehler, (D), Nevada (Reno), 2009, GmtNg
   Mike Ressel, (D), Nevada (Reno), 2005, EgGiCc
   Andrew V. Zuza, (D), California (Los Angeles), 2016, Gct
Adjunct Professor:
   Mark F. Coolbaugh, (D), Nevada (Reno), 2003, Eg
Emeritus:
   John W. Bell, (M), Arizona State, 1974, NgGm
   Stephen B. Castor, (D), Nevada (Reno), 1972, Eg
   Larry J. Garside, (M), Nevada (Reno), 1968, Gg
   Liang-Chi Hsu, (D), California (Los Angeles), 1966, Cp
   Daphne D. LaPointe, (M), Montana, 1977, Gg
   Paul J. Lechler, (D), Nevada (Reno), 1995, Cg
   Jonathan G. Price, (D), California (Berkeley), 1977, Gg
   Alan R. Ramelli, (M), Nevada (Reno), 1988, Ng
   Lisa Shevenell, (D), Nevada, 1990, Cg
   Joseph V. Tingley, (M), Nevada (Reno), 1963, Em
Geologic Information Specialist:
   David A. Davis, (M), Nevada (Reno), 1990, Gg

New Hampshire

New Hampshire Geological Survey
New Hampshire Dept of Environmental Services (2015)
29 Hazen Drive
P.O.Box 95
Concord, NH 03302-0095
   p. (603) 271-1975
   geology@des.nh.gov
   http://des.nh.gov/organization/commissioner/gsu/
State Geologist:
   Frederick H. Chormann, Jr., (M), New Hampshire, 1985, HyZiGm
Geoscience Program Specialist:
   Gregory A. Barker, (B), Rhode Island, 1985, GglZi
Fluvial Geomorphology Specialist:
   Shane Csiki, (D), Illinois (Urbana), 2014, ZyGm
Outreach Coordinator:
   Lee Wilder, (M), New Hampshire, 1972, GgZee
Hydrologist:
   Jeremy D. Nicoletti, (B), James Madison, 2009, GgHg
   Neil F. Olson, (M), Idaho State, 2010, GeHy

New Jersey

New Jersey Geological and Water Survey
New Jersey Geological and Water Survey ☐ (2018)
PO Box 420, Mail Code:29-01
Trenton, NJ 08625-0420
   p. (609) 292-1185
   njgsweb@dep.nj.gov
   www.njgeology.org
   Administrative Assistant: Tenika Jacobs
State Geologist:
   Jeffrey L. Hoffman, (M), Princeton, 1981, HqwRw
Bureau Chief:
   David L. Pasicznyk, (B), Temple, 1979, Yg
Research Scientist 1:
   Steven E. Spayd, (M), UMD New Jersey, 2004, HwGb
   Peter J. Sugarman, (D), Rutgers, 1995, Gr
Supervising Geologist:
   Eric W. Roman, (M), Temple, 1999, Hw
Supervising Env Specialist:
   Helen L. Rancan, (M), Stevens, 1995, Hs
Supervising Env Engineer:
   Richard Shim-Chim, (B), Toronto, 1973, Hw
Research Scientist 2:
   James T. Boyle, (M), New Mexico Tech, 1984, Hw
   Donald H. Monteverde, (D), Rutgers, 2008, Grg
Research Scientist 1:
   Scott D. Stanford, (D), Rutgers, 2001, Gml
GIS Specialist 1:
   Zehdreh Allen-Lafayette, (M), Syracuse, 1994, Zi
   Mark A. French, (B), Rutgers, 1989, Hq
   Ted J. Pallis, (M), Montclair State, 1994, Zi
   Ronald Pristas, (B), Penn State, 1990, Zi
Principal Environmental Specialist:
   Raymond T. Bousenberry, (M), New Jersey Inst Tech, 2007, Ge
   Steven E. Domber, (M), Wisconsin, 2000, Hw
   Gregg M. Steidl, (B), Rutgers, 1995, Ge
Research Scientist 2:
   John H. Dooley, (M), New Mexico Tech, 1983, CgGfe
GIS Specialist 2:
   Michael W. Girard, (B), Bloomsburg, 1996, Zi
Section Chief:
   William P. Graff, (B), Clark, 1979, Zn
Investigator:
   Walter Marzulli, Zy

New Mexico

New Mexico Institute of Mining & Technology
New Mexico Bureau of Geology & Mineral Resources (2015)
801 Leroy Place
Socorro, NM 87801-4796
   p. (575) 835-5420
   Nelia.Dunbar@nmt.edu
   http://geoinfo.nmt.edu/
Director and State Geologist:
   L Greer Price, (M), Washington, 1974, Gg
Senior Volcanologist:
   William C. McIntosh, (D), New Mexico Tech, 1990, Cc
Field Geologist:
   Bruce Allen, (D), New Mexico, 1993, Gm
Senior Field Geologist:
   Steven M. Cather, (D), Texas, 1986, Gs
Principal Senior Geologist:
   Paul W. Bauer, (D), New Mexico Tech, 1988, Gp
Adjunct Faculty:
   George S. Austin, (D), Iowa, 1971, En
   David W. Love, (D), New Mexico, 1980, Ge
Senior Mining Engineer:
   Robert W. Eveleth, (B), New Mexico Tech, 1969, Nm
Senior Field Geologist:
   Richard M. Chamberlin, (D), Colorado Mines, 1980, Gg
Senior Env Geologist Emeritus:
   John W. Hawley, (D), Illinois, 1962, Ge
Senior Chemist:
   Lynn A. Brandvold, (M), North Dakota State, 1964, Ca
Principal Senior Geophysicist:
   Marshall A. Reiter, (D), Virginia Tech, 1969, Yh
Geologist:
   James M. Barker, (M), California (Santa Barbara), 1972, En
Emeritus Director & State Geologist:
   Peter Scholle, (D)

Emeritus:
  Charles E. Chapin, (D), Colorado Mines, 1966, Gt
  Ibrahim H. Gundiler, (D), New Mexico Tech, 1975, Nx
Webmaster/Geologist:
  Adam S. Read, (M), New Mexico, 1997, Gg
Senior Lab Associate:
  Lisa Peters, (M), Texas (El Paso), 1987, Gg
Senior Geophysicist, Field Geologist:
  Shari Kelley, (D), S Methodist, 1984, Yg
Principal Senior Petroleum Geologist:
  Ronald F. Broadhead, (M), Cincinnati, 1979, Go
Principal Senior Hydrogeologist:
  Peggy Johnson, (M), New Mexico Tech, 1990, Hw
Minerals Outreach Liason:
  Virginia McLemore, (D), Texas (El Paso), 1994, Em
Mineralogist/Economic Geologist:
  Virgil W. Lueth, (D), Texas, 1988, Gg
Manager, Digital Cartography Lab:
  Glen Jones, (B), New Mexico Tech, 1989, Zn
Hydrologist; Hydrogeology Program Manager:
  Stacy Timmons, (M), Oregon State, 2002, Gg
Hydrogeologist:
  Lewis A. Land, (D), North Carolina, 1999, Hg
GIS Cartographer:
  David J. McCraw, (M), New Mexico, 1985, Zy
Geological Librarian:
  Maureen Wilks, (D), New Mexico Tech, 1991, Gp
Geochronologist:
  Matthew T. Heizler, (D), California (Los Angeles), 1993, Cc
Field Geologist:
  Geoffrey Rawling, (D), New Mexico Tech, 2002, Gg
Economic Geologist:
  Douglas Bland, (M), Wyoming, 1982, Eg
Deputy Director; Geochemist:
  Nelia W. Dunbar, (D), New Mexico Tech, 1989, Gi
Deputy Director & Manager, Geologic Mapping Program:
  Michael Timmons, (D), New Mexico, 2004, Gg
Chemistry Lab Manager:
  Bonnie A. Frey, (M), New Mexico Tech, 2002, Cg
Related Staff:
  Gretchen K. Hoffman, (M), Arizona, 1979, Ec

New York

New York State Geological Survey
3000 Cultural Education Center
Madison Avenue
Albany, NY 12230
  p. (518) 473-6262
  djornov@mail.nysed.gov
  http://www.nysm.nysed.gov/nysgs/
  Administrative Assistant: Donna Jornov
Curator of Sedimentary Geology:
  Charles Ver Straeten, (D), Rochester, 1996, Grs
Curator of Geology:
  Marian V. Lupulescu, (D), IASI (Romania), 1987, GzxEm
Senior Scientist:
  Andrew Kozlowski, (D), W Michigan, 2002, Gl
  Langhorne Smith, (D), Virginia Tech, 1996, Go
Museum Scientist 1:
  Brian Bird, (D), GlZi
Project Geologist:
  James Leone, (B), SUNY (Albany), 2006, EoGoCg
  Brian Slater, (M), SUNY (Albany), 2007, GoEoGr
Education Specialist:
  Kathleen Bonk, (B), SUNY (Albany), 2009, GoCg
  Brandon L. Graham
State Paleontologist and Paleontology Curator:
  Ed Landing, (D), Michigan, 1979, PscCc
Emeritus:
  Michael A. Hawkins, (B), USNY Regents, 1972, Gz
Paleontology Collections Technician:
  Frank Mannolini
Technician:
  Michael Pascussi, (A), Schenectady Comm Coll, 1999, Eo
Related Staff:
  Linda A. VanAller-Hernick, (B), St Rose, 1974, Pgb

Cooperating Faculty:
  Barry Floyd, (M), Rensselaer, 1987, Zn

North Carolina

North Carolina Geological Survey
Dept of Environmental Quality (2018)
1612 Mail Service Center
Raleigh, NC 27699-1612
  p. 919-707-9210
  kenneth.b.taylor@ncdenr.gov
  http://portal.ncdenr.org/lr/geological_home
  Administrative Assistant: Joyce Sanford
State Geologist:
  Kenneth B. Taylor, (D), Saint Louis, 1991, YsGgEo
Senior Geologist:
  Philip Bradley, (M), North Carolina State, 1996, Gcg
Senior Geologist :
  Bart Cattanach, (M), North Carolina State, 1998, Gc
Senior Geologist:
  Kathleen M. Farrell, (D), Louisiana State, 1989, Gs
  Jeffrey C. Reid, (D), Georgia, 1981
Engineering Geologist:
  Richard M. Wooten, (M), Georgia, 1980, RnGc
Project Geologist:
  Randy Bechtel, (M), SUNY (Buffalo), ZeGg
Geologist I:
  Nick Bozdog, (B), W Carolina, 2005, Gg
  Kenny Gay, (M), East Carolina, 1980, GzsPi
  Sierra J. Isard, (M), Iowa, 2014, Gcg

North Dakota

North Dakota Geological Survey
North Dakota Geological Survey (2018)
1016 E. Calgary Ave.
600 East Boulevard Avenue
Bismarck, ND 58505-0840
  p. (701) 328-8000
  emurphy@nd.gov
  https://www.dmr.nd.gov/ndgs/
State Geologist:
  Edward C. Murphy, (M), North Dakota, 1983, Zn
Paleontologist:
  Jeff Person, Pg
Geologist:
  Ned Kruger, Gg
  Lorraine A. Manz, (D), London, 1982, GmlGg
  Tim Nesheim, (M), Iowa, 2009, Gg

Northern Territory

Northern Territory Government Minerals and Energy
48-50 Smith St
Paspalis Centrepoint Building
GPO Box 3000
Darwin, NT 0800
  p. +61 8 8999 5511
  minerals@nt.gov.au
  http://www.minerals.nt.gov.au

Northwest Territories Geological Survey
Industry Tourism and Investment - Government of the Northwest Territories (2018)
P.O. Box 1320
Yellowknife, NT X1A 2L9
  p. (867) 767-9211
  NTGS@gov.nt.ca
  http://www.nwtgeoscience.ca/

Nova Scotia

Geological Survey of Canada
Atlantic Div (2015)

Bedford Institute of Oceanography
1 Challenger Drive
P.O. Box 1006
Dartmouth, NS B2Y 4A2
p. (902) 426-4386
Pat.Dennis@NRCan-RNCan.gc.ca
http://gsc.nrcan.gc.ca/org/atlantic/index_e.php
Senior Scientist:
Kumiko Azetsu-Scott, (D), Dalhousie, 1992, Ocp

Nova Scotia Natural Resources
Nova Scotia Natural Resources (2015)
P.O. Box 698
Founders Square
Halifax, NS B3J 3M8
p. (902) 424-5935
http://www.gov.ns.ca/natr/

Ohio

Ohio Dept of Natural Resources
Div of Geological Survey (2018)
2045 Morse Road Bldg. C-1
Columbus, OH 43229
p. (614) 265-6576
geo.survey@dnr.state.oh.us
http://www.ohiodnr.com/geosurvey/
Programs: N/A
Certificates: N/A
Geologist:
Michael P. Solis, (M), Kentucky, 2010, GgcGt
Senior Scientist:
Frank Fugitt, (B), Ohio Univ, GrHw
Geologist:
Daniel R. Blake, (M), Wright State, 2013, Ysg
Geologic Assistant:
Madge R. Fitak, (B), Mount Union, 1972, Gg

Oklahoma

University of Oklahoma
Oklahoma Geological Survey (2015)
100 East Boyd
Energy Center
Suite N-131
Norman, OK 73019-0628
p. (405) 325-3031
ogs@ou.edu
http://www.ogs.ou.edu/homepage.php
Geologist:
Thomas M. Stanley, (D), Kansas, 2000, PiGsc
Adjunct Professor:
Kyle E. Murray, (D), Colorado Mines, 2003, HwZiGo
Seismologist:
Amberlee Darold, (M), Oregon, 2012, Ys
Geologist:
Julie M. Chang, (D), Texas (El Paso), 2006, GgZnn
Brittany Pritchett, Go

Ontario

Ontario Geological Survey
Ontario Geological Survey (2018)
933 Ramsey Lake Road
Sudbury, ON P3E6B5
p. (705) 670-5758
johanne.roux@ontario.ca
http://www.mndm.gov.on.ca/en/mines-and-minerals
f: www.facebook.com/OGSgeology
t: @OGSgeology
Director:
Renee-Luce. Simard, (D), Dalhousie, 2005, GgEgZe
Senior Manager, Earth Resources & Geoscience Mapping Section:
John Hechler, (M), Laurentian
Senior Manager, Resident Geologist Program:
Mark Smyk

Senior Manager, Geoservices Section:
John Beals

Oregon

Oregon Dept of Geology & Mineral Industries
Coastal Field Office (2015)
PO Box 1033
Newport, OR 97365
p. (541) 574-6658
Jonathan.Allan@dogami.state.or.us
Regional Geologist:
Rob Witter, (D), Oregon, 1999, Zg
Geologist:
George R. Priest, (D), Oregon State, 1980, Gg
Coastal Section Supervisor:
Jonathan C. Allan, (D), Canterbury (NZ), 1998, Gm

Oregon Dept of Geology and Mineral Industries
Oregon Dept of Geology and Mineral Industries (2017)
800 NE Oregon Street
Suite 965
Portland, OR 97232-2162
p. (971) 673-1555
dogami-info@oregon.gov
http://www.oregongeology.org
State Geologist:
Brad Avy, (M), Alaska
Chief Scientist, Deputy Director:
Ian P. Madin, (M), Oregon State, Gg
Reclamationist:
Ben Mundie, Ge
Industrial Minerals Geologist:
Clark Niewendorp, (M), Eg
Hydrogeologist:
Robert Brinkmann, (B), Colorado State, 1983, HwGoEm
Hydrocarbon and Metallic Ore Geologist:
Robert Houston, (M), Oregon, Ge
Geotechnical Engineer:
Yumei Wang, (M), California (Berkeley), 1988, Ne
Engineering Geologist:
Bill Burns, (M), Portland State, 1999, Ng
Eastern Oregon Regional Geologist:
Jason McClaughry, (M), WSU, 2003, Gg
Coastal Geomorphologist:
Jonathan C. Allan, (D), Canterbury (NZ), 1998, Gm

Pennsylvania

Pennsylvania Bureau of Topographic & Geologic Survey
DCNR-Pennsylvania Geological Survey (2018)
3240 Schoolhouse Road
Middletown, PA 17057-3534
p. (717) 702-2017
ra-nr-topo_maindesk@pa.gov
http://www.dcnr.pa.gov/Geology/Pages
f: https://www.facebook.com/PennsylvaniaGeology
Administrative Assistant: Connie Cross
Administrative Assistant: Audrey Kissinger
State Geologist and Bureau Director:
Gale C. Blackmer, (D), Penn State, 1992, Gc
Geologist Manger:
Kristin M. Carter, (M), Lehigh, 1993, GoHw
Geologist Manager:
Michael E. Moore, (B), Penn State, 1975, HwZi
Stuart O. Reese, (M), Tennessee, 1986, Hw
Senior Geologic Scientist:
Rose-Anna Behr, (M), New Mexico Tech, 1999, Gcr
Aaron D. Bierly, (B), Pitt (Johnstown), GgZn
Helen L. Delano, (M), SUNY (Binghamton), 1979, GgmNg
Clifford H. Dodge, (M), Northwestern, 1976, GrEcRh
Antonette K. Markowski, (M), S Illinois, 1990, Eoc
Victoria V. Neboga, (M), Kiev, 1985, Hw
Caron E. Pawlicki, (M), Pittsburgh, 1986, ZiRc
Zachery Schagrin, (M), West Chester, Gug
Katherine W. Schmid, (M), Pittsburgh, 2005, GoCc

James R. Shaulis, (M), Penn State, 1985, EcZe
Geologist Supervisor:
   Brian J. Dunst, (B), Indiana (Penn), 1982, EgoHw
   Kristen Hand, (B), Nicholls State, HwGg
   Stephen G. Shank, (D), Penn State, 1993, GziGp
   Simeon Suter, (B), Millersville, 1982, HwGg
Geologic Scientist:
   Robin Anthony, (B), Case Western, Go
   Craig Ebersole, (B), Juniata
   Leonard J. Lentz, (M), North Carolina State, 1983, Ec
   John C. Neubaum, (M), American Military Univ, 2001, EcnGg
Librarian:
   Jody L. Smale, (M), Clarion, 2010, Zn
IT Technician:
   Mark A. Dornes
IT Generalist:
   David Fletcher
IT Adminstrator:
   Sandipkumar P. Patel, (B)
Clerk Typist:
   Lynn J. Levino, Go
   Jody L. Rebuck, (B), Messiah
   Renee Speicher

Puerto Rico

Puerto Rico Bureau of Geology
Dept of Natural & Environmental Resources (2015)
Apartado 9066600
Puerta de Tierra Station
San Juan, PR 00906-6600
   p. (787) 722-2526
Director:
   Vanessa del S. Rodriguez, Zn

Quebec

Ministère de l'Énergie et des Ressources naturelles Québec
5700 4eme avenue ouest
local A-409
Quebec, QC G1H 6R1
   p. 1-866-248-6936
   services.clientele@mern.gouv.qc.ca
   http://www.mern.gouv.qc.ca/

Queensland

Queensland Env and Resource Management
GPO Box 2454
Brisbane, QLD 4001
info@derm.qld.gov.au
http://www.derm.qld.gov.au/

Rhode Island

University of Rhode Island
Rhode Island Geological Survey (D) (2015)
9 East Alumni Ave.
314 Woodward Hall
University of Rhode Island
Kingston, RI 02881
   p. 401.874.2191
   rigsurv@etal.uri.edu
   http://www.uri.edu/cels/geo/GEO_risurvey.html
State Geologist (Research Professor Emeritus):
   Jon C. Boothroyd, (D), South Carolina, 1974, GslOn
Professor:
   Thomas Boving, (D), Arizona, 1999, Hg
   David Fastovsky, (D), Wisconsin, 1986, PvGsPs
Assistant Professor:
   Dawn Cardace, (D), Washington Univ (St. Louis), 2006, Pg
   Simon Engelhard, (D), Pennsylvania, 2010, Gg
   Brian Savage, (D), Caltech, 2004, Ys

Lecturer:
   Elizabeth Laliberte, (D), Rhode Island, 1997, Og
Research Associate:
   Bryan A. Oakley, (D), Rhode Island, 2012, Gsl

Saskatchewan

Saskatchewan Energy and Resources
Saskatchewan Ministry Energy and Resources (2018)
1000 - 2103 - 11th Avenue
Regina, SK S4P 3Z8
   p. (306) 787-9580
   dmoadmin.econ@gov.sk.ca
   http://www.saskatchewan.ca/

South Carolina

South Carolina Dept of Natural Resources
Geological Survey (2017)
5 Geology Road
Columbia, SC 29212
   p. 803.896.7931
   scgs@dnr.sc.gov
   http://www.dnr.sc.gov/geology/
Director:
   Charles W. Clendenin, Jr., (D), Witwatersrand, 1989, Eg
Geologist III:
   William R. Doar, III, (D), South Carolina, 2014, GrdGp
Chief Geologist:
   C. Scott Howard, (D), Delaware
Research Associate:
   Andy Wykel, (B), USC, 2017, Gg
Geologist II:
   Katherine E. Luciano, (M), Coll of Charleston, GusGm
Program Manager-Drilling:
   Joe Koch, Zg
Digitizer II:
   Matt Henderson, (B), South Carolina, Zi

South Dakota

South Dakota Dept of Environment and Natural Resources
Geological Survey Program (2017)
Akeley-Lawrence Science Center
University of South Dakota
414 East Clark Street
Vermillion, SD 57069-2390
   p. (605) 677-5227
   derric.iles@usd.edu
   http://www.sdgs.usd.edu/
State Geologist:
   Derric L. Iles, (M), Iowa State, 1977, HwGg
Geologist:
   Darren J. Johnson, (M), Washington, 2009, GgEo
Geologist:
   Nicholas J. Krohe, (M), Idaho State, 2016, Gg

Tennessee

Tennessee Geological Survey
Dept of Environment & Conservation (2018)
William R. Snodgrass TN Tower
312 Rosa L. Parks Ave., 12th Floor
Nashville, TN 37243
   p. (615) 532-1502
   Ronald.Zurawski@tn.gov
   https://www.tn.gov/environment/program-areas/tennessee-geological-survey.html
State Geologist:
   Ronald P. Zurawski, (M), Vanderbilt, 1973, GgEg
Environmental Consultant:
   Albert B. Horton, (M), Vanderbilt, 1981, GgZi
   Peter J. Lemiszki, (D), Tennessee, 1992, GcZiGg
   Barry W. Miller, (M), Tennessee, 1989, EcZiGg

Environmental Scientist:
  Vince Antonacci, (M), Ball State, 1987, GgZi
  Ronald J. Clendening, (B), Tennessee Tech, 1986, GgcGm
Admin. Services Asst. 2 :
  Carolyn A. Patton

Texas

University of Texas at Austin, Jackson School of Geosciences
Bureau of Economic Geology (2015)
University Station, Box X
Austin, TX 78713-8924
  p. (512) 471-1534
  begmail@beg.utexas.edu
  http://www.beg.utexas.edu
Director:
  Scott W. Tinker, (D), Colorado, 1996, Gor
Associate Director:
  Jay P. Kipper, (M), Trinity (San Antonio), 1983, Zn
  Eric C. Potter, (M), Oregon State, 1975, Eo
Senior Research Scientist:
  Shirley P. Dutton, (D), Texas, 1986, Gd
  Bob A. Hardage, (D), Oklahoma State, 1967, Yg
  Susan D. Hovorka, (D), Texas, 1990, Gse
  Michael R. Hudec, (D), Wyoming, 1990, GcEo
  Martin P. A. Jackson, (D), Cape Town, 1976, Gc
  Stephen E. Laubach, (D), Illinois, 1986, GcNrGz
  Bob Loucks, (D), Texas, 1976, Go
  Bridget R. Scanlon, (D), Kentucky, 1985, Hw
Research Scientist Associate V:
  Tucker F. Hentz, (M), Kansas, 1982, GrsGo
Research Scientist:
  William A. Ambrose, (M), Texas, 1983, Gs
  Peter Eichhubl, (D), California (Santa Barbara), 1997, GcCgNr
  Jean-Philippe Nicot, (D), Texas, 1998, HwCg
Senior Scientist:
  Ian Duncan, (D), British Columbia, 1982, Gg
  F. Jerry Lucia, (M), Minnesota, 1954, Go
  Stephen C. Ruppel, (D), Tennessee, 1979, Gd
Senior Research Scientist:
  Kitty L. Milliken
Research Scientist:
  Sergey Fomel, (D), Stanford, 2002, Yx
  Jeffrey G. Paine, (D), Washington, 1991, Gr
  Julia Stowell Gale, (D), Exeter (UK), 1987, Gc
  Hongliu Zeng, (D), Texas, 1994, Gs
Associate Professor:
  Charles Kerans, (D), Carleton, 1982, Gd
Research Scientist:
  Tim Dooley, (D), London (UK), 1994, Gc
Associate Scientist:
  Katherine D. Romanak, (D), Texas, 1997, CgScGe
Research Scientist Associate IV:
  Robert C. Reedy, (M), New Mexico Tech, 1996, Hgw
Research Scientist Associate:
  John R. Andrews, (B), North Carolina, 1990, Zi
Research Associate:
  Bruce Cutright
  Qilong Fu, (D), Regina, 2005, Gso
  H, Scott Hamlin, (D), Texas, Gr
  Ursula Hammes, (D), Colorado, 1992, Gg
  Farzam Javadpour, (D)
  Timothy A. Meckel, (D)
  Osareni C. Ogiesoba, (D)
  Katherine D. Romanak, (D)
  Diana Sava, (D), Stanford, 2004, Yg
  Changbing Yang, (D)
  Christopher K. Zahm, (D)
  Beverly Blakeney DeJarnett, (M), Penn State, 1986, GrsGx
  Edward W. Collins, (M), Stephen F. Austin, 1978, Ge
  Micheal V. DeAngelo, (M), Texas (El Paso), 1988, Ye
  Xavier Janson, (D), Miami, 2002, Gr
Research Scientist Associate:
  Seay Nance, (D), Texas, 1988, Gg

Research Scientist Associate IV:
  Robert M. Reed, (D), Texas, 1999, GxoGc
IT Manager:
  Ron Russell, (M), Oklahoma, 1993, Zn
Associate Director:
  Michael H. Young, Arizona, 1995, Ge
512-471-7135:
  George Bush, (B), Texas Tech
Research Scientist Associate IV:
  Thomas A. Tremblay, (M), Texas, 1992, Zy
Research Scientist Associate:
  Caroline Breton, (B), Texas, 2001, Zi
  Dallas B. Dunlap, (B), Texas, 1997, Ye
  Tiffany Hepner, (M), S Florida, 2000, Og
Project Manager:
  Rebecca C. Smyth, (M), Texas, 1995, Hg
  Ramon H. Trevino, (M), Texas (Arlington), 1988, GsoGe
Related Staff:
  Joseph S. Yeh, (B), Fu-Jen Catholic, 1977, Gm

Utah

Utah Geological Survey
Dept of Natural Resources (2019)
1594 West North Temple, Ste 3110
Box 146100
Salt Lake City, UT 84114-6100
  p. (801) 537-3300
  http://geology.utah.gov
  f: https://www.facebook.com/UTGeologicalSurvey/
  t: @utahgeological
State Geologist:
  Bill Keach, (M), Cornell, 1986, Yg
Curator Core Research Center:
  Peter J. Nielsen, (M), Brigham Young, 1992, Eo

Environmental Sciences Program (2015)
1594 West North Temple, Ste 3110
PO Box 146100
Salt Lake City, UT 84114-6100
  p. (801) 537-3389
  pamperri@utah.gov
  http://geology.utah.gov
Manager:
  Mike Lowe, (M), Utah State, 1985, Hw
Geologist:
  Charles E. Bishop, (B), Utah, 1986, Hw
  Hugh A. Hurlow, (D), Washington, 1982, Hw
  Janae Wallace, (M), N Arizona, 1993, Hw
Senior Geologist:
  James I. Kirkland, (D), Colorado, 1990, Pg
Senior Scientist:
  David B. Madsen, (D), Missouri, 1973, Ga
Research Associate:
  Martha C. Hayden, (B), Utah, 1978, Pv
Geotechnician:
  Alison Corey, Zn

Geologic Hazards Program (2015)
1594 West North Temple
P O Box 146100
Salt Lake City, UT 84114-6100
  p. (801) 537-3300
  rickallis@utah.gov
  http://geology.utah.gov/ghp/
Program Manager:
  Steve D. Bowman, (D), Nevada (Reno), 2002, NgZn
Senior Scientist:
  William R. Lund, (B), Idaho, 1970, Ng
Paleoseismologist:
  Christopher B. DuRoss, (M), Utah, 2004, Ng
Landslide Geologist:
  Gregg S. Beukelman, (M), Boise State, 1997, NgZi
Hazards Mapping Geologist:
  Jessica Castleton, (B), Weber State, 2005, Ng
  Adam McKean, (M), Brigham Young, 2011, Gmg

Hazards Geologist:
   Tyler R. Knudsen, (M), Nevada (Las Vegas), 2005, Ng
   Gregory N. McDonald, (B), Utah, 1992, Ng
Hazard Mapping Geologist:
   Ben A. Erickson, (M), Utah, 2011, Ng
Debris Flow/Landslide Geologist:
   Richard E. Giraud, (M), Idaho, 1986, Ng
Geologist:
   Mike Hylland, (M), Oregon State, 1990, PgYs
GIS Analyst:
   Corey Unger, (B), Weber State, 2001, Zi

Geologic Mapping Program (2015)
1594 West North Temple, Ste 3110
PO Box 146100
Salt Lake City, UT 84114-6100
   p. (801) 537-3355
   grantwillis@utah.gov
   http://geology.utah.gov
Senior Scientist:
   Robert Biek, (M), N Illinois, 1988, Gg
Adjunct Associate Professor:
   James I. Kirkland, (D), Colorado, 1990, PvsPi
Senior Geologist:
   Donald Clark, (M), N Illinois, 1987, Gg
   Jonathan K. King, (M), Wyoming, 1984, Gg
   Douglas A. Sprinkel, (M), Utah State, 1977, GosGg
Manager:
   Grant C. Willis, (M), Brigham Young, 1983, Gg
GIS Analyst:
   Basia Matyjasik, (B), Warsaw, 1988, Gg
Senior GIS Analyst:
   Kent D. Brown, Zn

Vermont

Agency of Natural Resources, Dept of Environmental Conservation
Vermont Geological Survey (2018)
1 National Life Drive
Main 2
Montpelier, VT 05620-3902
   p. (802) 522-5210
   marjorie.gale@vermont.gov
   https://dec.vermont.gov/geological-survey
State Geologist:
   Marjorie H. Gale, (M), Vermont, 1980, GgcGe
Geologist -Environmental Scientist:
   Jonathan Kim, (D), Buffalo, 1995, GetCg

Victoria

Geological Survey of Victoria, Australia
Victoria - Dept of Economic Development, Jobs, Transport and Resources (2016)
GPO Box 2392
Melbourne, VIC 3001
   p. +61 3 94528906
   gsv.info@ecodev.vic.gov.au
   http://www.earthresources.vic.gov.au/earth-resources

Victoria - Dept of Sustainability and Environment
8 Nicholson Street
East Melbourne, VIC 3002
   p. +61 3 5332 5000
   peter.walsh@parliament.vic.gov.au
   http://www.dse.vic.gov.au/

Virginia

Division of Geology and Mineral Resources
Virginia Dept of Mines, Minerals & Energy (2018)
Fontaine Research Park
900 Natural Resources Drive
Suite 500
Charlottesville, VA 22903
   p. (434) 951-6341
   DGMRInfo@dmme.virginia.gov
   http://dmme.virginia.gov/DGMR/divisiongeologymineral-resources.shtml
State Geologist:
   David B. Spears, (M), Virginia Tech, 1983, GcEg
Manager, Geologic Mapping:
   Matthew J. Heller, (M), North Carolina State, 1996, GgRnGe
Manager, Economic Geology:
   William L. Lassetter, (M), Nevada (Reno), 1996, EgHg

University of Virginia
Virginia State Climatology Office (2015)
291 McCormick Road
P.O. Box 400123
Charlottesville, VA 22904-4123
   p. 434-924-0548
   climate@virginia.edu
   http://climate.virginia.edu/home.htm
Director:
   Patrick J. Michaels, (D), Wisconsin, 1979, As

Washington

Washington Geological Survey
Washington Dept of Natural Resources (2019)
1111 Washington Street, SE, MS 47007
Olympia, WA 98504-7007
   p. (360) 902-1450
   geology@dnr.wa.gov
   http://www.dnr.wa.gov/geology/
Geology Librarian:
   Stephanie Earls, (M), Washington, 2010, Gge
State Geologist:
   David K. Norman, (M), Utah, 1980, Gg

West Virginia

West Virginia Geological & Economic Survey
Mont Chateau Research Center
1 Mont Chateau Road
Morgantown, WV 26508-8079
   p. (304) 594-2331
   info@geosrv.wvnet.edu
   http://www.wvgs.wvnet.edu
   f: https://www.facebook.com/WVGeoSurvey
State Geologist:
   Michael Ed. Hohn, (D), Indiana, 1976, Eg
Adjunct Professor:
   Paula J. Hunt, (M), Purdue, 1988, GgHw

Wisconsin

University of Wisconsin, Extension
Wisconsin Geological and Natural History Survey (2018)
3817 Mineral Point Road
Madison, WI 53705-5100
   p. (608) 262-1705
   jill.pongetti@wgnhs.uwex.edu
   http://wgnhs.uwex.edu/
   f: https://www.facebook.com/WGNHS
   t: @WGNHS
Director and State Geologist:
   Kenneth R. Bradbury, (D), Wisconsin, 1982, Hw
Hydrogeologist:
   David J. Hart, (D), Wisconsin, 2000, HwYg
Geologist:
   Eric C. Carson, (D), Wisconsin, 2003, Gl
   J. Elmo Rawling, (D), Gl
   Esther K. Stewart, (M), Idaho State, 2008, Gg
State Geologist and Director:
   James M. Robertson, (D), Michigan, 1972, Eg
Professor:
   Thomas J. Evans, (D), Wisconsin, 1994, Eg

Emeritus:
  Bruce A. Brown, (D), Manitoba, 1984, GcEg
  Ron G. Hennings, (M), Wisconsin, 1977
  Fred W. Madison
  Roger M. Peters, (B), Wisconsin, 1969, Gg
Samples Coordinator:
  Carsyn Ames, (M), Iowa State, 2018
Project Geologist:
  Stefanie Dodge, (B), Wisconsin (Milwaukee), Gg
Outreach Manager:
  M. Carol L. McCartney, (D), Wisconsin, 1979, RcGee
Office Manager:
  Jill Pongetti, (B)
Hydrogeologist:
  Mike J. Parsen, (M), Hw
GIS Specialist:
  Steve Mauel
  Caroline Rose, (B), Wisconsin
  Kathy Roushar
Geotechnician:
  Peter M. Chase, (B), Wisconsin (Milwaukee), 1985, HwYgRw
Geologist:
  Irene D. Lippelt, (B), Manitoba, 1978, Gg
Editor:
  Linda Deith, (B)
Assistant Director:
  Peter Schoephoester, (M), Wisconsin (Parkside)
Archivist:
  Brad Gottschalk, (B), Lawrence
Administrative Manager:
  Sushmita Lotlikar, (M), Wisconsin, 2014
Related Staff:
  Amber Boudreau
  Grace E. Graham, (B), Beloit College, 2013, Hw

Wyoming

Wyoming State Geological Survey
Wyoming State Geological Survey (2018)
P.O. Box 1347
Laramie, WY 82073
  p. (307) 766-2286
  wsgs-info@wyo.gov
  http://www.wsgs.wyo.gov/
Director:
  Erin A. Campbell, (D), Wyoming, 1997, Gc
Geologist:
  Robert Gregory, (M), En
  Ranie Lynds, (D), EoGro
  Wayne Sutherland, (M), Eng

Yukon

Yukon Geological Survey
Yukon Geological Survey (2015)
  suzanne.roy@gov.yk.ca
  http://www.geology.gov.yk.ca/

U.S. Federal Agencies & International Organizations

Agencies and Intl Organizations

United Nations Education, Scientific, and Cultural Organization
International Geoscience Programme (2015)
Division of Ecological and Earth Sciences
1 rue Miollis
Paris, Cedex 15 F-75732
    p. +33 (0)1 45 68 10 00
    m.alaawah@unesco.org
    http://www.unesco.org/new/en/natural-sciences/environment/earth-sciences/international-geoscience-programme/
IGCP Chairperson:
    Patricia Vickers-Rich, (D), Columbia, 1972
Team Leader, Hydrogeology:
    Gil Mahe
Team Leader, Global Change and Evolution of Life:
    Guy Narbonne
Team Leader, Geohazards:
    Andrej Gosar
Team Leader, Geodynamic:
    George Gibson
Team Leader, Earth Resources:
    Robert Moritz
IGCP/SIDA Representative:
    Vivi Vajda, (D), Lund, 1998

Geological Survey of India
Geological Survey of India (2018)
27, J.L.Nehru Road
Kolkata, West Bengal 700016
    dg.gsi@gov.in
    http://www.gsi.gov.in
    f: https://www.facebook.com/gsipage/
    t: @GeologyIndia
    Certificates: Advanced GIS, Digital Cartography, Geological Mapping, Mineral Exploration, Geophysical Survey, etc

Goa University
Dept of Earth Science (M,D) (2019)
Taleigao Plateau
Goa University P.O.
Taleigao Plateau
Panaji, Goa 403 206
    p. +91-8669609193
    mkotha@unigoa.ac.in
    http://www.unigoa.ac.in/department.php?adepid=11&mdepid=3
    Programs: Applied Geology
    Certificates: GIS
    Enrollment (2018): M: 22 (22) D: 1 (1)
Professor:
    Mahender Kotha, (D), IIT (Bombay), 1987, GsoZi
Associate Professor:
    Anthony V. Viegas, (D), Goa, 1997, GiEmGx
On Contract:
    Poornima Dhavaskar, (M), Geology, 2013, Gge
    Purushottam Verlekar, (M), GrmGg
Assistant Professor:
    Niyati Kalangutkar, (D), Goa, GugGe
Cooperating Faculty:
    Sohini Ganguly, (D), Gi

Ministry of Environmental Protection, Natural Resources and Forestry
Dept of Geology (2016)
st. Wawelska 52/54
Warsaw 00-922
    p. +48 22 3692449
    Departament.Geologii.i.Koncesji.Geologicznych@mos.gov.pl
    http://www.mos.gov.pl/dg/dga1.htm

Sveriges geologiska Undersokning
Geological Survey of Sweden (2016)
Box 670
751 28 Uppsala
    sgu@sgu.se
    http://www.sgu.se/

Argonne National Laboratory
Chemical Technology Div (2015)
9700 South Cass Avenue
Building 205
Argonne, IL 60439
    p. (630) 252-4383
    ebunel@anl.gov
Senior Scientist:
    Milton Blander, (D), Yale, 1953, Xc
    Paul A. Fenter, (D), Pennsylvania, 1990, ClZm
Associate Scientist:
    Allen J. Bakel, (D), Oklahoma, 1990, Co
    Ronald P. Chiarello, (D), Northeastern, 1990, Zm
    Donald G. Graczyk, (D), Wisconsin, 1975, Cs
    Ben D. Holt, (D), Illinois Inst Tech, 1969, Ca
    James J. Mazer, (D), Northwestern, 1987, Cl
    David J. Wronkiewicz, (D), New Mexico Tech, 1989, Cg
Adjunct Professor:
    Neil C. Sturchio, (D), Washington (St. Louis), 1983, Cg
Related Staff:
    Alice M. Essling, (B), St. Xavier, 1956, Ca
    Edmund A. Huff, (M), Chicago, 1957, Ca
    Francis J. Markun, (B), Lewis, 1961, Cc
    Florence P. Smith, (B), Southern, 1968, Ca

Geosciences & Information Technology Section (2015)
9700 South Cass Avenue
Argonne, IL 60439
    p. (630) 252-6034
    djmiller@anl.gov
    Department Secretary: Sue Baumann
Manager, Geosciences & Information Tech:
    Lisa A. Durham, (M), Purdue, 1989, Hw
Assistant System Engineer:
    Cheong-yip R. Yuen, (D), Wisconsin (Milwaukee), 1986, Ge
Associate Scientist:
    Robert L. Johnson, (D), Cornell, 1991, Zn
Research Associate:
    John Ditmars, (D), Caltech, 1971, Zn
    Jennifer Herbert, (B), N Illinois, 1997, Gg
    Zhenhua Jiang, (D), Duke, 1992, Zn
    David S. Miller, (D), Johns Hopkins, 1995, Ge
    Terri L. Patton, (M), NE Illinois, 1989, Cl
    John J. Quinn, (M), Minnesota, 1992, Hq
Geology Librarian:
    Swati Wagh, Zn

Environmental Research Div (2015)
9700 South Cass Avenue
Argonne, IL 60439
    p. (630) 252-3879
    bmlesht@anl.gov
    http://www.anl.gov/ER/
Senior Scientist:
    Jeffrey S. Gaffney, (D), California (Riverside), 1975, As
    Raymond M. Miller, (D), Illinois State, 1975, Sb
    Marvin L. Wesely, (D), Wisconsin, 1970, As
Associate Scientist:
    Jacqueline C. Burton, (D), Tennessee, 1978, Cg
    Richard L. Coulter, (D), Penn State, 1976, As
    Paul V. Doskey, (D), Wisconsin, 1982, As
    Julie D. Jastrow, (D), Chicago, 1994, Sb
    Lorraine M. LaFreniere, (D), Wisconsin, 1980, Zn
    In Young Lee, (D), California (Los Angeles), 1975, As
    Barry M. Lesht, (D), Chicago, 1977, Op

Nancy A. Marley, (D), Florida State, 1984, As
William T. Meyer, (D), Imperial Coll (UK), 1973, Cg
Robert A. Sedivy, (M), Georgia Tech, 1979, Hw
Jack D. Shannon, (D), Oklahoma, 1975, Am
Douglas L. Sisterson, (M), Wyoming, 1975, As
Mohamed Sultan, (D), Washington (St. Louis), 1984, Cg
John L. Walker, (D), Imperial Coll (UK), 1964, Cg
Y Eugene Yan, (D), Ohio State, 1998, Gg
Research Associate:
Richard H. Becker, (M), Washington (St. Louis), Zr
Clyde B. Dennis, (M), Florida, 1979, Zn
Richard L. Hart, (B), Illinois Inst Tech, 1970, As
Timothy J. Martin, (B), Beloit, 1974, As
Barney W. Nashold, (M), Illinois (Chicago Circle), 1976, Zn
Kent A. Orlandini, (B), Illinois, 1957, Cc
Candace M. Rose, (B), Benedictine Coll, 1985, Zn

Energy Systems Div (2015)
9700 South Cass Avenue
Bldg 362 E-340
Argonne, IL 60439-4815
p. (630) 252-3392
energy_systems@anl.gov
Administrative Assistant: Barbara Sullivan
Head:
Donald O. Johnson, (D), Illinois, 1972, Gr
Senior Scientist:
Lyle D. McGinnis, (D), Illinois, 1965, Yg
Scientist:
Kenneth L. Brubaker, (D), Wisconsin, 1972, As
Dorland E. Edgar, (D), Purdue, 1976, Gm
Steven F. Miller, (B), Knox, 1984, Gg
R. Eric Zimmerman, (D), Northwestern, 1972, Ng
Research Associate:
Paul C. Heigold, (D), Illinois, 1969, Yg
Theresa C. Scholtz, (B), N Illinois, 1985, Ge
Michael D. Thompson, (M), N Illinois, 1989, Yg
Related Staff:
John F. Schneider, (M), N Illinois, 1977, Ca
Linda M. Shem, (M), Northwestern, 1991, Ge
Patrick L. Wilkey, (M), Illinois, 1976, Ng

Chemistry Div (2015)
9700 South Cass Avenue
Argonne, IL 60439
p. (630) 972-3570
ebunel@anl.gov
Senior Scientist:
Dieter M. Gruen, (D), Chicago, 1951, Zm
Michael J. Pellin, (D), Illinois, 1978, Zm
Associate Scientist:
Wallis F. Calaway, (D), Indiana, 1975, Zm

Bureau of Land Management
Colorado State Office (2015)
2850 Youngfield Street
Lakewood, CO 80215
p. (303) 239-3600
hhankins@blm.gov

California State Office (2018)
2800 Cottage Way
Suite W-1623
Sacramento, CA 95825
p. (916) 978-4400
mailto:BLM_CA_Web_SO@blm.gov
https://www.blm.gov/office/california-state-office
f: https://www.blm.gov/media/social-media

Arizona State Office (2015)
One North Central Ave
Suite 800
Phoenix, AZ 85004
p. (602) 417-9500
egomez@blm.gov

Alaska State Office (2015)
222 West Seventh Ave
#13
Anchorage, AK 99513
p. (907) 271-3212
dlassuy@blm.gov

National Training Center (2015)
9828 North 31st Avenue
Phoenix, AZ 85051
p. (602) 906-5500
dwilkins@blm.gov

National Operations Center (2015)
PO Box 25047
Denver, CO 80225
p. (303) 236-8857
lgraham@blm.gov

Eastern States Office (2015)
7450 Boston Boulevard
Springfield, VA 22153
p. (703) 440-1600
es_general_web@blm.gov

Idaho State Office (2015)
1387 South Vinnell Way
Boise, ID 83709
p. (208) 373-4000
sellis@blm.gov

Montana State Office (2015)
5001 Southgate Drive
Billings, MT 59101
p. (406) 896-5012
kiszler@blm.gov

Nevada State Office (2015)
1340 Financial Blvd
Reno, NV 89502
p. (775) 861-6400
jswickard@blm.gov

New Mexico State Office (2015)
301 Dinosaur Trail
Santa Fe, NM 87502
p. (505) 954-2098
jjuen@blm.gov

Oregon State Office (2015)
333 SW 1st Avenue
Portland, OR 97204
p. (503) 808-6001
b2jackso@blm.gov

Utah State Office (2015)
440 West 200 South
Suite 500
Salt Lake City, UT 84101
p. (801) 539-4001
utsomail@blm.gov

Wyoming State Office (2015)
PO Box 1828
Cheyenne, WY 82003
p. (307) 775-6256
jcamargo@blm.gov

Headquarters Directorate (2015)
1849 C Street NW
Rm 5665
Washington, DC 20240
p. (202) 208-3801
director@blm.gov

Dept of Agriculture
Natural Resources Conservation Service (2016)

14th and Independence Ave., SW
Washington, DC 20250
 p. (202) 720-7246
 jason.weller@wdc.usda.gov
 http://www.nrcs.usda.gov

Dept of Commerce
National Oceanic & Atmospheric Administration (2015)
Silver Spring Metro Center 3
1315 East-West Highway
Silver Spring, MD 20910-3282
 p. (202) 482-3436
 edward.horton@noaa.gov
 http://www.noaa.gov

National Inst of Standards & Technology (2018)
100 Bureau Drive
Stop 1070
Gaithersburg, MD 20899-3460
 p. (301) 975-2000
 inquiries@nist.gov
 https://www.nist.gov/

Dept of Defense
Space & Naval Warfare Systems Command (2015)
4301 Pacific Highway
San Diego, CA 92110-3127
 p. (619) 524-7053
 http://www.spawar.navy.mil/

Air Force Center for Environmental Excellence (2015)
3207 North Road
Brooks Air Force Base, TX 78235-5363
 p. (210) 536-2162

U.S. Army Corps of Engineers (2015)
441 G Street, NW
Washington, DC 20314
 p. (202) 761-0001
 http://www.usace.army.mil/working.html

U.S. Army Chemical & Biological Defense Command (2015)
The Pentagon
Washington, DC 20310
 p. (703) 695-0363
 http://www.army.mil/csa/

U.S. Army Research Laboratory Command (2015)
The Pentagon
Washington, DC 20310
 p. (703) 695-0363
 http://www.army.mil/csa/

Defense Threat Reduction Agency (2015)
8725 John T. Kingman Road
MS 6201
Ft. Belvoir, VA 22060-6201
 p. (703) 767-4883
 dtra.publicaffairs@dtra.mil
 http://www.dtra.mil

Office of Naval Research (2015)
Ballston Towers #1
800 North Quincy Street
Arlington, VA 22203
 p. (703) 696-4767
 http://www.onr.navy.mil/

Naval Intelligence Command (2015)
4251 Suitland Road
Washington, DC 20395-5720
 p. (301) 669-4000
 http://www.nmic.navy.mil/

Air Weather Service (2015)
106 Peacekeeper Drive
Offutt Air Force Base, NE 68113-4039
 p. (402) 294-5749

Space Command (2015)
150 Vandenberg Street
Peterson Air Force Base, CO 80914-4020
 p. (719) 554-3001
 http://www.af.mil/sites/afspc.shtml

U.S. Special Operations Command (Air Force) (2015)
100 Bartley Street
Hurlburt Field, FL 32544-5273
 p. (850) 884-2323
 http://www.af.mil/sites/afsoc.html

Naval Oceanography Command (2015)
U.S. Naval Observatory
34th Street & Massachusetts Avenue, NW
Washington, DC 20007
 p. (202) 762-1020
 http://www.oceanographer.navy.mil

Dept of Energy
Assistant Secretary for Fossil Energy (2015)
Forrestal Building
1000 Independence Avenue, SW
Washington, DC 20585-0301
 p. (202) 586-6660
 FECommunications@hq.doe.gov
 http://www.fe.doe.gov/

Office of Oil & Gas (2015)
Forrestal Building
1000 Independence Avenue, SW
Washington, DC 20585-0640
 p. (202) 586-6012

Asst Secretary for Energy Efficiency & Renewable Energy (2015)
Forrestal Building
1000 Independence Avenue, SW
Washington, DC 20585-0121
 p. (202) 586-9220
 http://www.eren.doe.gov/

Assistant Secretary for Environmental Management (2017)
Forrestal Building
1000 Independence Avenue, SW
Washington, DC 20585-0113
 p. (202) 586-7709
 EM.WebContentManager@em.doe.gov
 http://www.em.doe.gov/

Office of Resource Management (2015)
Forrestal Building
1000 Independence Avenue, SW
Washington, DC 20585-0620
 p. (202) 586-3521

Assistant Secretary for Environment, Safety & Health (2015)
Forrestal Building
1000 Independence Avenue, SW
Washington, DC 20585-0119
 p. (202) 586-6151
 http://www.eh.doe.gov/

Energy Information Administration (2015)
Forrestal Building
1000 Independence Avenue, SW
Washington, DC 20585-0601
 p. (202) 586-8800
 howard.gruenspecht@eia.gov
 http://www.eia.doe.gov/

Office of Energy Projects (2015)
Federal Energy Regulatory Commission
888 First Street, NE
Washington, DC 20426

p. (202) 219-2700

Office of Coal, Nuclear, Electric & Alternate Fuels (2015)
COMSAT Building
950 L'Enfant Plaza, SW
Washington, DC 20024
p. (202) 287-7990

Coal & Power Systems (2015)
Forrestal Building
1000 Independence Avenue, SW
Washington, DC 20585-0320
p. (202) 586-1650

Office of Nuclear Energy, Science & Technology (2015)
Forrestal Building
1000 Independence Avenue, SW
Washington, DC 20585-0117
p. (202) 586-6630
http://www.ne.doe.gov/

Dept of Health & Human Services
Agency for Toxic Substances & Disease Registry (2015)
1600 Clifton Road
Atlanta, GA 30333
p. (404) 639-7000
http://www.atsdr.cdc.gov/atsdrhome.html

Centers for Disease Control & Prevention (2015)
1600 Clifton Rd.
Atlanta, GA 30333
p. (404) 639-3535
http://www.cdc.gov

Dept of Labor
Mine Safety & Health Administration (2015)
Balston Tower #3
4015 Wilson Boulevard
Arlington, VA 22203
p. (703) 235-1385
http://www.msha.gov

Dept of State
Intl Boundary & Water Commission, US & Mexico (2015)
The Commons
4171 North Mesa, Suite C-310
El Paso, TX 79902-1441
p. 1-800-262-8857
john.merino@ibwc.state.gov

Dept of the Interior
Office of Surface Mining, Reclamation & Enforcement (2015)
1951 Constitution Ave. N.W.
Washington, DC 20240
p. (202) 208-2565
osm-getinfo@osmre.gov
http://www.osmre.gov/

U.S. Fish & Wildlife Service (2015)
1849 C. Street N.W.
Washington, DC 20240
p. (202) 208-4545
http://www.fws.gov/

National Park Service (2015)
1849 C. Street N.W.
Washington, DC 20240
p. (202) 208-4621
http://www.nps.gov

Office of the Secretary (2015)
1849 C. Street N.W.
Washington, DC 20240
p. (202) 208-3100
feedback@ios.doi.gov

http://www.doi.gov

Bureau of Land Management (2015)
Office of Public Affairs
1849 C Street, Room 406-LS
Washington, DC 20240
p. (202) 208-3801
director@blm.gov
http://www.blm.gov

Bureau of Indian Affairs (2015)
1849 C. Street N.W.
Washington, DC 20240
p. (202) 208-7163
http://www.doi.gov/bureau-indian-affairs.html

Bureau of Reclamation (2015)
1849 C. Street N.W.
Washington, DC 20240
p. (202) 513-0501
http://www.usbr.gov

Dept of the Treasury
Internal Revenue Service (2015)
1111 Constitution Avenue, NW
Washington, DC 20224
p. (202) 622-9511
http://www.irs.ustreas.gov

Dept of Transportation
Federal Aviation Administration (2015)
800 Independence Avenue, SW
Washington, DC 20591
p. (202) 267-3484
http://www.faa.gov

U.S. Coast Guard (2015)
2100 Second Street, SW
Washington, DC 20593
p. (202) 366-4000
http://www.uscg.mil

Environmental Protection Agency
Office of Underground Storage Tanks (2015)
Crystal Gateway One
1235 Jefferson Davis Highway
Arlington, VA 22202
p. (703) 603-9900
hoskinson.carolyn@epa.gov

Office of Emergency & Remedial Response (Superfund/Oil Programs) (2015)
Crystal Gateway One
1235 Jefferson Davis Highway
Arlington, VA 22202
p. (703) 603-8960
oilinfo@epamail.epa.gov
http://www.epa.gov/superfund

Office of Ground Water & Drinking Water (2015)
1200 Pennsylvania Avenue, NW
Washington, DC 20460
p. (202) 260-5400
grevatt.peter@epa.gov

Office of Wetlands, Oceans & Watersheds (2015)
Fairchild Building
499 South Capitol Street, SW
Washington, DC 20003
p. (202) 260-7166
best-wong.benita@epa.gov

Assistant Administrator for Water Programs (2017)
1200 Pennsylvania Avenue, NW
Washington, DC 20460
p. (202) 260-5700

Beauvais.joel@Epa.gov
https://www.epa.gov/aboutepa/about-office-water

Office of Wastewater Management (2015)
1200 Pennsylvania Avenue, NW
Washington, DC 20460
p. (202) 564-0748
ow-general@epa.gov

Office of Atmospheric Programs (2015)
501 Third Street, NW
Washington, DC 20001
p. (202) 564-9140
McCabe.janet@Epa.gov

Office of Solid Waste (2015)
1301 Constitutiion Ave. NW
Arlington, VA 22202
p. (202) 566-0200
aastanislaus@epa.gov
http://www2.epa.gov/aboutepa/about-office-solid-waste-and-emergency-response-oswer

Assistant Administrator for Environmental Information (2015)
1200 Pennsylvania Avenue, NW
Washington, DC 20460
p. (202) 564-6665
Nelson.kimberly@Epa.gov

Office of Science Policy (2016)
1200 Pennsylvania Ave., NW
MC-8104R
Washington, DC 20460
p. (202) 564-6705
hauchman.fred@epa.gov

Office of Science & Technology (2015)
1200 Pennsylvania Avenue, NW
Washington, DC 20460
p. (202) 260-5400
Southerland.elizabeth@Epa.gov

National Center for Environmental Research (2015)
Ronald Reagan Building
1300 Pennsylvania Avenue, NW
Washington, DC 20004
p. (202) 564-6825
johnson.jim@epa.gov

Office of Research and Development (2018)
USEPA-ORD-8101R
1200 Pennsylvania Avenue, NW
Washington, DC 20460
p. 202-564-3512
frithsen.jeff@epa.gov
http://www.epa.gov/research

Safety, Health & Environmental Management Div (2015)
1200 Pennsylvania Avenue, NW
Washington, DC 20460
p. (202) 564-1640
dfe@epa.gov

National Air & Radiation Environmental Laboratory (2015)
540 South Morris Avenue
Montgomery, AL 36115-2601
p. (334) 270-3404
griggs.john@epa.gov

Federal Emergency Management Agency
Office of the Director (2015)
500 C Street, SW
Washington, DC 20472
p. (202) 646-4600
http://www.fema.gov/

Jet Propulsion Laboratory
Earth & Space Sciences Div (2015)
Geology & Planetology Section
Pasadena, CA 91109
p. (818) 354-3440
daniel.j.mccleese@jpl.nasa.gov
Department Administrative Manager: Murray Geller
Lead Scientist:
  Bruce E. Banerdt, (D), S California, 1983, Xy
Section Manager:
  Ronald G. Blom, (D), California (Santa Barbara), 1987, Zr
Section Member:
  Michael J. Abrams, (M), Caltech, 1973, Zr
  Ronald E. Alley, (M), Northwestern, 1972, Zr
  Diana L. Blaney, (D), Hawaii, 1990, Xg
  Bonnie J. Buratti, (D), Cornell, 1983, Zr
  Robert E. Crippen, (D), California (Santa Barbara), 1989, Zr
  Joy A. Crisp, (D), Princeton, 1984, Gv
  Thomas C. Duxbury, (M), Purdue, 1966, Zn
  Diane L. Evans, (D), Washington, 1981, Gm
  Tom G. Farr, (D), Washington, 1981, Gm
  Ken E. Herkenhoff, (D), Caltech, 1989, Xg
  Simon J. Hook, (D), Durham (UK), 1989, Zr
  Erik R. Ivins, (M), California (Los Angeles), 1976, Gt
  Anne B. Kahle, (D), California (Los Angeles), 1975, Zr
  Harold R. Lang, (D), Calgary, 1983, Gr
  Kyle C. McDonald, (D), Michigan, 1991, Zn
  Robert M. Nelson, (D), Pittsburgh, 1978, Xg
  Eni G. Njoku, (D), MIT, 1976, Zr
  Frank D. Palluconi, (M), Penn State, 1963, Zr
  David C. Pieri, (D), Cornell, 1979, Gt
  Jeffrey J. Plant, (D), Washington (St. Louis), 1991, Zn
  Jeffrey Plescia, (D), S California, 1985, Zn
  Carol A. Raymond, (D), Columbia, 1989, Yr
  Suzanne E. Smrekar, (D), S Methodist, 1990, Xy
  Linda J. Spilker, (D), California (Los Angeles), 1992, XyZr
  Ellen R. Stofan, (D), Brown, 1989, Xg
  Glenn J. Veeder, (D), Caltech, 1974, Zn
Section Member:
  Matthew P. Golombek, (D), Massachusetts, 1981, XgyGt
Senior Scientist:
  Ronald S. Sanders, (D), Brown, 1970, Gr

Lawrence Livermore National Laboratory
Atmospheric, Earth and Energy (2015)
7000 East Avenue
L-203
Livermore, CA 94550
p. (925) 423-1848
antoun1@llnl.gov
https://www-pls.llnl.gov/?url=about_pls-atmospheric_earth_and_energy_division
Administrative Assistant: Laura Long
Head:
  Kenneth J. Jackson, (D), California (Berkeley), 1983, Cl
Senior Scientist:
  Roger D. Aines, (D), Caltech, 1984, Cg
  Bill L. Bourcier, (D), Oregon State, 1983, Cg
  Thomas A. Buscheck, (D), California (Berkeley), 1984, Ng
  Steven Carle, (D), California (Davis), 1996, Hw
  Charles R. Carrigan, (D), California (Los Angeles), 1977, Yg
  Susan A. Carroll, (D), Northwestern, 1988, Cl
  M Lee Davisson, (M), California (Davis), 1992, Cg
  Quingyun Duan, (D), Arizona, 1991, Hg
  Robert C. Finkel, (D), California (San Diego), 1974, Cg
  Samuel (Julio) J. Friedmann, (D), MIT, Gg
  Richard B. Knapp, (D), Arizona, 1978, Ng
  Gayle A. Pawloski, (B), Cal State (Hayward), 1979, Gg
  Abelardo L. Ramirez, (M), Purdue, 1979, Ng
  Sarah K. Roberts, (M), Arizona State, 1990, Cl
  Andrew F B Tompson, (D), Princeton, 1985, HwSpCc
  Jeffrey L. Wagoner, (M), California (Riverside), 1977, Gs
  Ananda M. Wijesinghe, (D), MIT, 1978, Nr
  Sun Yunwei, (D), Israel Inst of Tech, 1995, Ng
  Mavrik Zavarin, Yr

Los Alamos National Laboratory
Chemistry Div (B,M,D) (2015)
P.O. Box 1663
Los Alamos, NM 87545
 p. (505) 667-4457
 chemistry@lanl.gov
 http://pearl1.lanl.gov/external/default.htm
Division Leader:
 Alexander J. Gancarz, (D), Caltech, 1976, Cc
Staff:
 Kent D. Abney, (D), Colorado State, 1987, Zn
 Stephen F. Agnew, (D), Washington State, 1981, Zn
 Moses Attrep, (D), Arkansas, 1965, Ze
 Timothy M. Benjamin, (D), Caltech, 1979, Ca
 Scott M. Bowen, (D), New Mexico, 1983, Ca
 James R. Brainard, (D), Indiana, 1979, Zn
 Jeffrey C. Bryan, (D), Washington, 1988, Zn
 Carol J. Burns, (D), California (Berkeley), 1987, Zn
 Timothy P. Burns, (D), Nebraska, 1986, Co
 Gilbert W. Butler, (D), California (Berkeley), 1967, Cc
 Edwin P. Chamberlain, (D), Texas A&M, 1971, Cc
 David L. Clark, (D), Indiana, 1986, Zn
 Dean A. Cole, (D), Iowa, 1985, Co
 David B. Curtis, (D), Oregon State, 1974, Cc
 Paul R. Dixon, (D), Yale, 1989, Zn
 Robert J. Donahoe, (D), North Carolina State, 1985, Zn
 Stephen K. Doorn, (D), Northwestern, 1989, Zn
 Clarence J. Duffy, (D), British Columbia, 1977, Cg
 Deward W. Efurd, (D), Arkansas, 1975, Ct
 Phillip G. Eller, (D), Ohio State, 1971, Zn
 June T. Fabryka-Martin, (D), Arizona, 1988, Hg
 Bryan L. Fearey, (D), Iowa State, 1986, Cc
 David L. Finnegan, (D), Maryland, 1984, As
 John R. Fitzpatrick, (M), New Mexico, 1983, Zn
 Malcolm M. Fowler, (D), Washington (St. Louis), 1972, As
 Sammy R. Garcia, (B), New Mexico Highlands, 1972, Ca
 Russell E. Gritzo, (B), New Mexico State, 1983, Zn
 Richard C. Heaton, (D), Illinois, 1973, Zn
 Sara B. Helmick, (B), Southwestern, 1955, Zn
 David R. Janecky, (D), Minnesota, 1982, Cl
 Kung King-Hsi, (D), Cornell, 1989, Zn
 Scott Kinkead, (D), Idaho, 1983, Zn
 Gregory J. Kubas, (D), Northwestern, 1970, As
 Pat J. Langston-Unkefer, (D), Texas A&M, 1978, Co
 Patrick A. Longmire, (D), New Mexico, 1991, Zn
 Michael MacInnes, (M), Wisconsin, 1969, Zn
 Allen S. Mason, (D), Miami, 1974, As
 Charles M. Miller, (D), Stanford, 1980, Cc
 Geoffrey G. Miller, (D), Rensselaer, 1984, Cc
 Terrance L. Morgan, (M), Colorado State, 1984, Ng
 David Morris, (D), North Carolina State, 1984, Cc
 Eugene J. Mroz, (D), Maryland, 1976, As
 Michael T. Murrell, (D), California (San Diego), 1980, Cc
 Allen E. Ogard, (D), Chicago, 1957, Hw
 Jose A. Olivares, (D), Iowa State, 1985, Cs
 Hain Oona, (D), Arizona, 1979, Zn
 Kevin C. Ott, (D), Caltech, 1983, Zn
 Edward S. Patera, (D), Arizona State, 1982, Cg
 Richard E. Perrin, (B), Denver, 1952, Cc
 Eugene J. Peterson, (D), Arizona State, 1976, Ct
 Dennis Phillips, (D), Hawaii, 1976, Co
 Jane Poths, (D), Chicago, 1982, Cc
 Pamela Z. Rogers, (D), California (Berkeley), 1981, Cl
 Donald J. Rokop, (D), Lake Forest, 1985, Cc
 Robert S. Rundberg, (D), CUNY, 1978, Gq
 Nancy N. S. Sauer, (D), Iowa State, 1986, Zn
 Norman C. Schroeder, (D), Iowa State, 1985, Cc
 Louis A. Silks, (B), Suffolk, 1978, Zn
 Paul H. Smith, (D), California (Berkeley), 1987, Zn
 Zita V. Svitra, (D), Roosevelt, 1967, Zn
 Basil I. Swanson, (D), Northwestern, 1969, Zn
 C. Drew Tait, (D), North Carolina State, 1984, Cp
 Wayne A. Taylor, (B), New Mexico, 1978, Cl
 Kimberly W. Thomas, (D), California (Berkeley), 1978, Ct
 Joseph L. Thompson, (D), Penn State, 1963, Ct
 Ines Triay, (D), Miami, 1985, Hg
 Clifford J. Unkefer, (D), Minnesota, 1981, Zn
 David J. Vieira, (D), California (Berkeley), 1978, Zn
 Jerry B. Wilhelmy, (D), California (Berkeley), 1969, Zn
 Kurt Wolfsberg, (D), Washington (St. Louis), 1959, Cc
 William H. Woodruff, (D), Purdue, 1972, Zn
 Mary Anne Yates, (D), Carnegie Mellon, 1976, Zn
Emeritus:
 Ernest A. Bryant, (D), Washington (St. Louis), 1956, Cg
 Merle E. Bunker, (D), Indiana, 1950, Zn
 William R. Daniels, (D), New Mexico, 1965, Ct
 Donald L. Hull, (M), Iowa State, 1974, Zn

Earth & Environmental Sciences Div (2016)
P.O. Box 1663
Los Alamos, NM 87545
 p. (505) 667-3644
 wallacet@lanl.gov
 Enrollment (1996): B: 17 (17) M: 6 (6)
Technical Staff:
 James E. Bossert, (D), Colorado State, 1990, As
Director:
 Terry C. Wallace, Jr., (D), Caltech, 1983, Ys
Technical Staff:
 Kay H. Birdsell, (M), Colorado, 1985, Hqy
 James W. Carey, (D), Harvard, 1990, CpYxNr
 Lianjie Huang, (D), Paris, 1994, Yg
 Kenneth H. Wohletz, (D), Arizona State, 1980, YxGvRm
Technical Staff:
 Paul Aamodt, (M), Nevada, 1991, Zn
 Douglas Alde, (D), Illinois, 1979, Zn
 M. James Aldrich, (D), New Mexico, 1972, Gc
 Fairley J. Barnes, (D), New Mexico State, 1986, Zn
 Naomi M. Becker, (D), Wisconsin, 1991, Hg
 James D. Blacic, (D), California (Los Angeles), 1971, Nr
 Rainer Bleck, (D), Penn State, 1968, Am
 Christopher R. Bradley, (D), MIT, 1993, Yg
 Thomas L. Brake, (M), New Mexico State, 1985, Nr
 David E. Broxton, (M), New Mexico, 1976, Gx
 Wendee M. Brunish, (D), Illinois, 1981, Zn
 Gilles Y. Bussod, (D), California (Los Angeles), 1988, Yx
 Katherine Campbell, (D), New Mexico, 1979, Gq
 Theodore C. Carney, (M), Colorado State, 1981, Ng
 Gregory L. Cole, (D), Arizona, 1990, Zg
 Steffanie Coonley, (M), New Mexico, 1984, Zn
 Keeley R. Costigan, (D), Colorado State, 1992, As
 William Cottingame, (D), Texas, 1984, As
 James L. Craig, (B), Nevada, 1974, Gg
 Bruce M. Crowe, (D), California (Santa Barbara), 1974, Gv
 Zora V. Dash, (M), New Mexico, 1988, Zn
 Deborah J. Daymon, (M), Idaho, 1994, Zn
 Kalpak Dighe, (D), Clemson, 1994, Zn
 John C. Dinsmoor, (B), Colorado Mines, 1989, Nm
 Alison M. Dorries, (D), Harvard, 1986, Zn
 Donald S. Dreesen, (B), New Mexico, 1968, Np
 David V. Duchane, (D), Michigan, 1978, Zn
 Michael H. Ebinger, (D), Purdue, 1988, So
 C L Edwards, (D), New Mexico Tech, 1975, Yg
 C. James Elliott, (D), Yale, Zn
 Scott M. Elliott, (D), California, 1983, As
 Perry D. Farley, (D), Oklahoma, 1981, Zn
 George T. Farmer, (D), Cincinnati, 1968, Hw
 David N. Fogel, (M), Cal State, 1997, Zn
 Carl W. Gable, (D), Harvard, 1989, Gt
 Edward S. Gaffney, (D), Caltech, 1973, Yg
 Anthony F. Gallegos, (D), Colorado State, 1970, Pg
 Jamie N. Gardner, (D), California (Davis), 1985, Gg
 Fraser Goff, (D), California (Santa Cruz), 1977, Cg
 Jeffrey C. Hansen, (B), Weber State, 1971, Zn
 Charles D. Harrington, (D), Indiana, 1970, Gm
 Hans E. Hartse, (D), New Mexico Tech, 1991, Ys
 Ward L. Hawkins, (B), Nevada (Reno), 1977, Zg
 Grant Heiken, (D), California (Santa Barbara), 1972, Gv
 Donald D. Hickmott, (D), MIT, 1988, Gp
 Steve T. Hildebrand, (D), Texas (Dallas), 1993, Ys
 Emil F. Homuth, (D), Washington, 1974, Ye
 Paul A. Johnson, (M), Arizona, 1984, Yg
 Eric M. Jones, (D), Wisconsin, 1970, As

Hemendra N. Kalia, (D), Missouri Sch of Mines, 1970, Nm
Jim C. Kao, (D), Illinois, 1985, As
Danny Katzman, (M), New Mexico, 1991, HyGem
Elizabeth Keating, (D), Wisconsin, 1995, Gg
Sharad Kelkar, (M), Texas, 1979, Np
C. F. Keller, Jr., (D), Indiana, 1969, As
Richard G. Kovach, (M), Naval Postgrad Sch, 1979, Nm
Donathon J. Krier, (M), New Mexico, 1980, Gv
Thomas D. Kunkle, (D), Hawaii, 1978, Xy
Edward M. Kwicklis, (M), Colorado, 1987, Gg
Chung Chieng A. Lai, (D), Texas A&M, 1984, Am
Schon S. Levy, (M), Texas, 1975, Gi
Peter C. Lichtner, (D), Mainz, 1974, Zn
Rodman Linn, (D), New Mexico State, 1997, Ng
Lynn McDonald, (M), Cal State, Zn
Maureen A. McGraw, (D), California (Berkeley), 1996, Zn
Laurie A. McNair, (D), Carnegie Mellon, 1995, Zn
Wayne R. Meadows, (B), Cal State (Bakersfield), 1974, Yg
Theodore Mockler, (M), Carnegie Mellon, 1986, Zn
Orrin B. Myers, (D), Colorado State, 1992, Pg
Balu Nadiga, (D), Caltech, 1992, Zn
Brent D. Newman, (D), New Mexico Tech, 1996, Gg
John W. Nyhan, (D), Colorado State, 1972, So
Ronald D. Oliver, (B), Oregon Inst of Tech, 1972, Yg
Howard J. Patton, (D), MIT, 1978, Yg
Frank V. Perry, (D), California, 1988, Gv
William S. Phillips, (D), MIT, 1985, Yg
Eugene W. Pokorny, (B), Missouri, 1979, Nm
William M. Porch, (D), Washington, 1971, Yg
Allyn R. Pratt, (M), Boise State, 1977, Ge
George Randall, (D), SUNY (Binghamton), Yg
Steen Rasmussen, (D), Tech (Denmark), 1985, Yg
Jon M. Reisner, (D), Iowa State, As
Steven L. Reneau, (D), California (Berkeley), 1988, Gm
Douglas O. Revelle, (D), Michigan, 1974, As
Peter Roberts, (D), MIT, 1989, Ys
Bruce A. Robinson, (D), MIT, 1985, Zn
R. Roussel-Dupre, (D), Colorado, 1979, Zn
Thomas J. Shankland, (D), Harvard, 1966, Yx
Catherine H. Smith, (D), New Mexico State, 1995, Ca
Wendy E. Soll, (D), MIT, 1991, Hw
Everett P. Springer, (D), Utah State, 1983, Hq
Lee Steck, (D), California, Gg
Robert P. Swift, (D), Washington, 1969, Nr
E.M.D. Symbalisty, (D), Chicago, 1984, Zn
Steven R. Taylor, (D), MIT, 1980, Ys
James Tencate, (D), Texas, 1992, Zn
Bryan J. Travis, (D), Florida State, 1974, Hq
David T. Vaniman, (D), California (Santa Cruz), 1976, Gx
Richard G. Warren, (M), New Mexico, 1972, Gi
Douglas J. Weaver, (M), Nevada, 1995, Zn
Thomas A. Weaver, (D), Chicago, 1973, Yg
Rodney W. Whitaker, (D), Indiana, 1976, As
Earl M. Whitney, (D), Utah, Zn
Judith L. Winterkamp, (B), Texas, 1977, Zn
Giday WoldeGabriel, (D), Case Western, 1987, Gx
Andrew V. Wolfsberg, (D), Stanford, 1993, Hw
George A. Zyvoloski, (D), California (Santa Barbara), 1975, Ng
Project Leader:
  Mark T. Peters, (D), Chicago, 1992, Yg
Technical Staff:
  James N. Albright, (D), Chicago, 1969, Yg
  Donald W. Brown, (M), California (Los Angeles), 1961, Nr
  Leigh S. House, (D), Columbia, 1982, Ys
  Larry Allan Jones, (B), Nebraska, 1972, Ng
Technical Staff:
  W. Scott Baldridge, (D), Caltech, 1979, GicGt
Other:
  Christina "Tina" Behr-Andres, (D), Michigan Tech, 1992, Zn

National Aeronautics & Space Administration
George C. Marshall Space Flight Center (2015)
Marshall Space Flight Center, AL 35812-0001
  p. (256) 544-1910
  patrick.e.scheuermann@nasa.gov

Goddard Space Flight Center (2015)
Greenbelt Road
Greenbelt, MD 20771-0001
  p. (301) 286-5121
  stephanie.s.keene@nasa.gov

NASA Headquarters (2015)
Washington, DC 20546-0001
  p. (202) 358-2345
  info-center@hq.nasa.gov
  http://www.hq.nasa.gov/

Ames Research Center (2015)
Building BN200
Moffett Field, CA 94035-1000
  p. (650) 604-5111
  michael.s.mewhinney@nasa.gov
Chair:
  Penelope Boston, (D), Colorado (Boulder), 1985

Lyndon B. Johnson Space Center (2015)
Houston, TX 77058-3696
  p. (281) 483-5309
  ellen.ochoa-1@nasa.gov

National Oceanic and Atmospheric Admin
National Centers for Environmental Information (2018)
David Skaggs Research Center
325 Broadway
Boulder, CO 80303
  p. (303) 497-6826
  ncei.info@noaa.gov
  http://www.ngdc.noaa.gov/
  f: http://www.facebook.com/NOAANCEIoceangeo
  t: @NOAANCEIocngeo

National Science Foundation
Atmospheric and Geospace Sciences Div (2015)
4201 Wilson Boulevard
Arlington, VA 22230
  p. (703) 292-8520
  pshepson@nsf.gov
  http://www.nsf.gov/div/index.jsp?div=AGS
Division Director:
  Paul B. Shepson
Section Head for NCAR/Facilities Section:
  Stephan P. Nelson
Section Head for Geospace Section:
  Richard A. Behnke
Section Head for Atmosphere Section:
  David J. Verardo
Program Director for Solar Terrestrial Research Program:
  Therese M. Jorgensen
Program Director for Physical and Dynamic Meteorology Program:
  A. Gannet Hallar
  Chungu Lu
  Bradley F. Smull
Program Director for Paleoclimate Program:
  Candace O. Major
Program Director for Magnetospheric Physics Program:
  Raymond J. Walker
Program Director for Geospace Facilities:
  Robert M. Robinson
Program Director for Education and Cross Disciplinary Activities Program:
  Linda George
Program Director for Climate and Large Scale Dynamics Program & Carbon and Water in Earth Systems Program:
  Eric T. DeWeaver
Program Director for Climate and Large Scale Dynamics Program:
  Anjuli Bamzai
Program Director for Atmospheric Chemistry Program:
  Peter Milne
  Anne-Marie Schmoltner
Program Director for Aeronomy Program:
  Anja Stromme

Program Coodinator for NCAR/Facilities Section:
  Sarah L. Ruth
Facilities Program Manager for NCAR/Facilities Section:
  Linnea M. Avallone
Assistant Program Director for Atmosphere Section:
  Nicholas F. Anderson

Ocean Sciences Division (2016)
4201 Wilson Boulevard
Arlington, VA 22230
  p. (703) 292-8580
  rwmurray@nsf.gov
  http://www.nsf.gov/div/index.jsp?div=OCE
Division Director:
  Richard W. Murray
Section Head for Integrative Programs Section:
  Bauke Houtman
Program Director for Ship Operations Program:
  Rose Dufour
Program Director for Physical Oceanography Program:
  Eric C. Itsweire
  Baris Mete Uz
Program Director for Oceanographic Instrumentation and Technical Service Programs:
  James Holik
Program Director for Ocean Observatories Initiative:
  Jean M. McGovern
Program Director for Ocean Education Programs:
  Elizabeth L. Rom
Program Director for Ocean Drilling Programs:
  James F. Allan
Program Director for Marine Geology and Geophysics Program:
  Candace O. Major
  Barbara Ransom
Program Director for Chemical Oceanography Program:
  Donald Rice
Program Director for Biological Oceanography Program:
  David L. Garrison
Program Director:
  John Walter
Program Directof for Ocean Drilling Programs:
  Thomas Janacek
Associate Program Director for Oceanographic Technology and Interdisciplinary Coordination Program:
  Kandace S. Binkley

Office of the Director (2015)
4201 Wilson Boulevard
Arlington, VA 22230
  p. (703) 292-5111
  fcordova@nsf.gov
  http://www.nsf.gov
  f: https://www.facebook.com/US.NSF
  t: @NSF

Earth Sciences Div (2017)
4201 Wilson Boulevard
Arlington, VA 22230
  p. (703) 292-8550
  cfrost@nsf.gov
  http://www.nsf.gov/div/index.jsp?div=EAR
  t: @NSF_EAR
Program Director for Hydrologic Sciences:
  Thomas Torgersen
Program Director for Instrumentation and Facilities:
  David Lambert
Section Head:
  Gregory J. Anderson
Section Head :
  Lina Patino, (D)
Section for Antarctic Sciences:
  Scott G. Borg, (D), Arizona State, 1984, Git
Program Director for Tectonics Program:
  David M. Fountain
  Stephen S. Harlan
Program Director for Petrology and Geochemistry Program:
  Sonia Esperanca

  Jennifer Wade
Program Director for Integrated Earth Systems:
  Leonard E. Johnson
Program Director for Instrumentation and Facilities:
  Russell C. Kelz
Program Director for Geophysics Program:
  Robin Reichlin
Program Director for Geobiology and low-T Geochemistry:
  Jonathan Wynn
Program Director for Geobiology and Low Temp Geochemistry:
  Enriqueta C. Barrera
Program Director for Earthscope:
  Margaret Benoit

Directorate for Geosciences (2018)
2415 Eisenhower Avenue
Room C8000
Alexandria, VA 22314
  p. (703) 292-8500
  sborg@nsf.gov
  https://www.nsf.gov/dir/index.jsp?org=GEO
  f: https://www.facebook.com/US.NSF
  t: @NSF
Program Director for International Activities:
  Maria Uhle

Nuclear Regulatory Commission
NRC Headquarters (2015)
One White Flint North Building
11555 Rockville Pike
Rockville, MD 20852
  p. (301) 415-8200
  http://www.nrc.gov/

Oak Ridge National Laboratory
Environmental Sciences Div (2015)
P.O. Box 2008
Mail Stop 6035
Oak Ridge, TN 37831
  p. (865) 574-7374
  envsci@ornl.gov
  http://www.esd.ornl.gov
Director:
  Stephen G. Hildebrand, (D), Michigan, 1973, Zn
Professor:
  Baohua Gu, (D), California (Berkeley), 1991, CbqSb
Senior Scientist:
  Marshall Adams, (D), North Carolina, 1974, Ob
  Jeff Amthor, (D), Yale, 1987, Sf
  Tom Ashwood, (M), Murray State, 1975, Ge
  Mark Bevelhimer, (D), Tennessee, 1990, Hs
  Terence J. Blasing, (D), Wisconsin, 1975, As
  Thomas Boden, (M), Miami, 1985, As
  Craig Brandt, (M), Tennessee, 1988, Ge
  Scott C. Brooks, (D), Virginia, 1994, Cl
  Robert S. Burlage, (D), Tennessee, 1990, Sb
  Meng-Dawn Cheng, (D), Illinois, 1986, As
  Robert B. Cook, (D), Columbia, 1981, Cg
  William E. Doll, (D), Wisconsin, 1983, Yg
  Thomas O. Early, (D), Washington (St. Louis), 1969, Cl
  Philip M. Jardine, (D), Virginia Tech, 1985, Sc
  Liyuan Liang, (D), Caltech, 1988, Cl
  Steven E. Lindberg, (D), Florida State, 1979, Cl
  John F. McCarthy, (D), Rhode Island, 1975, Co
  Gerilynn R. Moline, (D), Wisconsin, 1992, Hw
  Tony V. Palumbo, (D), North Carolina State, 1980, Sb
  Tommy J. Phelps, (D), Wisconsin, 1985, Sb
  Ellen D. Smith, (M), Wisconsin, 1979, Hg
  Brian P. Spalding, (D), Cornell, 1976, Sc
  Robert S. Turner, (D), Pennsylvania, 1983, Cl
  David B. Watson, (M), New Mexico Tech, 1983, Hw
  Olivia M. West, (D), MIT, 1991, Ng
Associate Scientist:
  Tammy Beaty, Zi
  Mary Anna Bogle, (M), Miami, 1975, Ge
  Norman D. Farrow, (B), Oregon State, 1974, Hg

Deputy Director:
    Gary K. Jacobs, (D), Penn State, 1981, Cg

Energy Div (2015)
P.O. Box 2008
Oak Ridge, TN 37831-6187
    p. (615) 574-5510
    wullschlegsd@ornl.gov
    Administrative Assistant: Teresa D. Ferguson
Director:
    Robert B. Shelton, (D), S Illinois, 1970, Zn
Chair:
    Donald W. Lee, (D), Michigan, 1977, Hw
Senior Scientist:
    Richard H. Ketelle, (M), Tennessee, 1977, Ge
    Russell Lee, (D), McMaster, 1978, Zy
    William P. Staub, (D), Iowa State, 1969, Ng
Research Associate:
    Arthur C. Curtis, (M), Colorado State, 1993, Ge
    Robert O. Johnson, (D), Tennessee, 1984, Hg
    Richard R. Lee, (M), Temple, 1982, Gr
    John D. Tauxe, (D), Texas, 1994, Hw

Pacific Northwest National Laboratory
Applied Geology & Geochemistry (2015)
Environmental Technology Div
Wayne J. Martin, Manager
PO Box 999 MSIN K6-81
Richland, WA 99352
    p. (509) 376-5952
    Christopher.Brown@pnnl.gov
    http://www.pnl.gov/agg
    Administrative Assistant: Charissa J. Chou
Head:
    George R. Holdren, (D), Johns Hopkins, 1977, Cl
Staff Scientist:
    Christopher J. Murray, (D), Stanford, 1992, Gq
Senior Research Engineer:
    Mark D. White, (D), Colorado State, 1986, Hy
Lab Fellow:
    Bernard P. McGrail, (D), Columbia Southern, 1996, Cg
Senior Scientist:
    Douglas B. Barnett, (M), E Washington, 1985, Em
    Bruce N. Bjornstad, (M), E Washington, 1980, Gs
    Kirk J. Cantrell, (D), Georgia Tech, 1989, Cl
    Amy P. Gamerdinger, (D), Cornell, 1989, Cl
    Tyler J. Gilmore, (M), Idaho, 1987, Hg
    Floyd N. Hodges, (D), Texas, 1975, Cg
    Duane G. Horton, (D), Illinois, 1983, Gz
    Kenneth M. Krupka, (D), Penn State, 1984, Cl
    George V. Last, (M), Washington State, 1997, Ge
    Jonathan W. Lindberg, (M), Washington State, 1995, Ec
    Shas V. Mattigod, (D), Washington State, 1976, Cl
    Alan C. Rohay, (D), Washington, 1982, Ys
    Herbert T. Schaef, (M), Texas Tech, 1991, Cg
    R. Jeffrey Serne, (B), Washington, 1969, CgSoCl
    Mark D. Sweeney, (B), C Washington, 1985, Yg
    Bruce A. Williams, (B), Colorado Mines, 1980, Ng
Associate Scientist:
    Yi-Ju Chien, (M), Stanford, 1998, Gq
    Jonathan P. Icenhower, (D), Oklahoma, 1995, Cl
    David C. Lanigan, (B), Michigan Tech, 1980, Hg
    Virginia L. Legore, (B), Oregon State, 1974, Cg
    Clark W. Lindenmeier, (M), E Washington, 1995, Cg
    Paul F. Martin, (B), Pacific Lutheran, 1982, Cg
    Kent E. Parker, (M), Washington State, 1995, Cg
Research Associate:
    Alexandra B. Amonette, (M), CUNY, 1976, Cg
    Deborah S. Burke, (B), E Washington, 1992, Sc
    Elsa A. Camacho, (B), Heritage, 1996, Cg
    Matthew J. O'Hara, (B), Montana, 1996, Cg
    Robert D. Orr, (B), N State, 1994, Ca

Sandia National Laboratory
Geoscience and Environment (2015)
1515 Eubank SE
P.O. Box 5800
Albuquerque, NM 87185
    gobrela@sandia.gov
    http://www.sandia.gov/
Research Scientist:
    Thomas Dewers, (D), Indiana, 1990, Yx

Smithsonian Institution
Smithsonian Astrophysical Observatory (2017)
60 Garden Street
Cambridge, MA 02138
    p. (617) 495-7100
    calcock@cfa.harvard.edu
    https://www.cfa.harvard.edu/sao
    f: https://www.facebook.com/HarvardSmithsonianCenter-ForAstrophysics

U.S. Geological Survey
Office of the Director (2015)
12201 Sunrise Valley Drive, MS 100
Reston, VA 20192
    p. (703) 648-7411
    abwade@usgs.gov
    www.usgs.gov

Regional Director, Northwest (2015)
909 1st Avenue
Seattle, WA 98104
    p. (206) 220-4600
    ddlynch@usgs.gov
    http://www.usgs.gov

Regional Director, Midwest (2015)
1451 Green Road
Ann Arbor, MI 48105
    p. (737) 214-7207
    lcarl@usgs.gov
    http://www.usgs.gov

Regional Director, Northeast (2015)
12201 Sunrise Valley Drive, MS 953
Reston, VA 20192
    p. (703) 648-6660
    druss@usgs.gov
    http://www.usgs.gov

Regional Director, Southeast (2015)
1770 Corporate Drive, Suite 500
Norcross, GA 30093
    p. (770) 409-7701
    jdweaver@usgs.gov
    http://www.usgs.gov

Director, Office of Science Quality & Integrity (2015)
12201 Sunrise Valley Drive, MS 911
Reston, VA 20192
    p. (703) 648-6601
    athornhill@usgs.gov
    http://www.usgs.gov

Assoc Director, Office of Communications & Publishing (2015)
12201 Sunrise Valley Drive, MS 119
Reston, VA 20192
    p. (703) 648-5750
    bwainman@usgs.gov
    http://www.usgs.gov

Assoc Director, Office of Budget, Planning, & Integration (2015)
12201 Sunrise Valley Drive, MS 105
Reston, VA 20192
    p. (703) 648-4443
    cburzyk@usgs.gov
    http://www.usgs.gov

Associate Director, Administration & Enterprise Info (2017)
12201 Sunrise Valley Drive, MS 201

Reston, VA 20192
> p. (703) 648-7261
> timothy_quinn@ios.doi.gov
> http://www.usgs.gov

Associate Director, Core Science Systems (M) (2017)
12201 Sunrise Valley Drive, MD 108
Reston, VA 20192
> p. (703) 648-5747
> kgallagher@usgs.gov
> http://www.usgs.gov

Associate Director, Water Resources (D) (2018)
12201 Sunrise Valley Drive, MS 436
Reston, VA 20192
> p. (703) 648-4557
> dcline@usgs.gov
> water.usgs.gov

Associate Director, Climate and Land-Use Change (2015)
12201 Sunrise Valley Drive, MD 409
Reston, VA 20192
> p. (703) 648-5215
> sryker@usgs.gov
> http://www.usgs.gov

Natural Hazards Mission Area (2016)
12201 Sunrise Valley Drive, MS 111
Reston, VA 20192
> p. (703) 648-6600
> applegate@usgs.gov
> http://www.usgs.gov/natural_hazards/

Regional Safety Manager (2018)
PO Box 596
Brightwood, OR 97011
> p. (503) 622-4432
> wsimonds@usgs.gov
> http://www.usgs.gov

Associate Director, Ecosystems (M) (2016)
12201 Sunrise Valley Drive, MS 300
Reston, VA 20192
> p. (703) 648-4050
> akinsinger@usgs.gov
> http://www.usgs.gov

Associate Director, Human Capital (2015)
12201 Sunrise Valley Drive, MS 201
Reston, VA 20192
> p. (703) 648-7261
> dwade@usgs.gov
> http://www.usgs.gov

SPECIALTY CODES

Specialty codes are used to indicate the research or teaching specialties of faculty members listed in the directory. Bold numbers are total number of individuals for each major category. Numbers in parentheses are individual specialty totals.

GEOLOGY 6573
Gg	General Geology (1326)
Ga	Archaeological Geology (136)
Ge	Environmental Geology (753)
Gm	Geomorphology (639)
Gl	Glaciology and Glacial Geology (329)
Gu	Marine Geology (260)
Gz	Mineralogy & Crystallography (568)
Gn	Paleolimnology (68)
Go	Petroleum Geology (320)
Gx	General Petrology (221)
Gi	Igneous Petrology (565)
Gp	Metamorphic Petrology (288)
Gd	Sedimentary Petrology (198)
Gs	Sedimentology (883)
Gr	Physical Stratigraphy (357)
Gc	Structural Geology (793)
Gt	Tectonics (736)
Gv	Volcanology (367)
Gq	Mathematical Geology (68)
Gy	Mineral Physics (78)
Gb	Medical Geology (21)
Gf	Forensic Geology (12)

ECONOMIC GEOLOGY 627
Eg	General Economic Geology (279)
Ec	Coal (60)
Em	Metallic Ore Deposits (153)
En	Industrial Minerals (32)
Eo	Oil and Gas (125)
Es	Construction Materials (SSG) (2)
Ed	Ore Deposits (Other) (11)

GEOCHEMISTRY 2241
Cg	General Geochemistry (646)
Ca	Analytical Geochemistry (191)
Cp	Experimental Petrology/Phase Equilibria (99)
Ce	Exploration Geochemistry (62)
Cc	Geochronology & Radioisotopes (337)
Cl	Low-temperature Geochemistry (398)
Cm	Marine Geochemistry (199)
Co	Organic Geochemistry (136)
Cs	Stable Isotopes (348)
Ct	Trace Element Distribution (81)
Cb	Biogeochemistry (84)
Cu	High-temperature Geochemistry (25)
Cq	Aqueous Geochemistry (36)

GEOPHYSICS 1707
Yg	General Geophysics (628)
Yx	Experimental Geophysics (98)
Ye	Exploration Geophysics (241)
Yd	Geodesy (97)
Ym	Geomagnetism & Paleomagnetism (161)
Yv	Gravity (30)
Yh	Heat Flow (30)
Ys	Seismology (552)
Yr	Marine Geophysics (166)
Yn	Nonlinear Geophysics (7)
Yu	Near-surface Geophysics (44)

PALEONTOLOGY 1593
Pg	General Paleontology (525)
Ps	Paleostratigraphy (115)
Pm	Micropaleontology (176)
Pb	Paleobotany (67)
Pl	Palynology (66)
Pq	Quantitative Paleontology (26)
Pv	Vertebrate Paleontology (257)
Pi	Invertebrate Paleontology (256)
Pe	Paleoecology (343)
Po	Geomicrobiology (14)
Pc	Paleoclimatology (83)

HYDROLOGY 2000
Hg	General Hydrology (338)
Hw	Ground Water/Hydrogeology (711)
Hq	Quantitative Hydrology (120)
Hs	Surface Waters (181)
Hy	Geohydrology (105)
Ht	Materials Transport (5)

SOIL SCIENCE 1385
Sp	Soil Physics/Hydrology (109)
Sc	Soil Chemistry/Mineralogy (185)
Sd	Pedology/Classification/Morphology (82)
Sf	Forest Soils/Rangelands/Wetlands (57)
Sb	Soil Biology/Biochemistry (116)
Sa	Paleopedology/Archeology (30)
So	Other Soil Science (228)

ENGINEERING GEOLOGY 626
Ng	General Engineering Geology (288)
Ne	Earthquake Engineering (24)
Nx	Mining Tech/Extractive Metallurgy (42)
Nm	Mining Engineering (89)
Np	Petroleum Engineering (68)
Nr	Rock Mechanics (115)
No	Ocean Engineering/Mining (29)
Nt	Geotechnical Engineering (13)
Nx	Geological Engineering (17)

OCEANOGRAPHY 1537
Og	General Oceanography (224)
Ob	Biological Oceanography (450)
Oc	Chemical Oceanography (253)
Ou	Geological Oceanography (153)
Op	Physical Oceanography (402)
On	Shore and Nearshore Processes (125)
Oo	Paleooceanography (21)

ASTRONOMICAL SCIENCES 370
Xc	Cosmochemistry (67)
Xg	Extraterrestrial Geology (154)
Xy	Extraterrestrial Geophysics (89)
Xm	Meteorites & Tektites (70)
Xa	Astrophysics (30)
Xp	Heliophysics (2)
Xb	Astrobiology (9)

ATMOSPHERIC SCIENCES 1179
As	Atmospheric Sciences (876)
Am	Meteorology (332)
Ac	Atmospheric Chemistry (20)
Ap	Atmospheric Physics (19)
At	Climatology (58)

GEOSCIENCE & SOCIETY 157
Rn	Natural Hazards (58)
Rm	Manmade Hazards (7)
Rw	Water Quality/Use (34)
Rc	Geoscience Communication (19)
Rh	History of Geoscience (42)

OTHER 3289
Zg	General Earth Sciences (407)
Ze	Earth Science Education (360)
Zy	Physical Geography (338)
Zr	Remote Sensing (454)
Zm	Material Science (57)
Zu	Land Use/Urban Geology (122)
Zi	Geographic Information Systems (532)
Zf	Geoinformatics (26)
Zc	Global Change (77)
Za	Nanogeoscience (6)
Zo	Climate Modeling (31)
Zn	Not Elsewhere Classified (1310)

TOTAL 23,284

Faculty Specialty Index

GEOLOGY
General Geology

Abbott, Lon, University of Colorado
Abousleiman, Younane N., University of Oklahoma
Abreu, Vitor, Rice University
Abushagur, Sulaiman, El Paso Community College
Adabanija, Moruffdeen A., Ladoke Akintola University of Technology
Adamo, Peter J., Brandon University
Adepoju, Mohammed O., Federal University of Technology, Akure
Adinolfi, Bryan, Southern Connecticut State University
Adisa, Adeshina L., Federal University of Technology, Akure
Agbogun, Henry, Fort Hays State University
Ahola, John, Bristol Community College
Ajigo, Isaac O., Federal University of Technology, Akure
Akgun, Elif, Firat University
Akinlabi, Ismaila A., Ladoke Akintola University of Technology
Aksoy, Ercan, Firat University
Akujieze, Christopher N., University of Benin
Alexander, Dane, Western Michigan University
Alexander, Elaine, McLennan Community College
Alfsen, Sheila, Portland State University
Alkac, Onur, Firat University
Allison, Alivia J., Emporia State University
Altermann, Wladyslaw, Ludwig-Maximilians-Universitaet Muenchen
Amaach, Noureddin, College of Staten Island/CUNY
Amato, James A., University of Wisconsin, Oshkosh
Anderson, Paula, University of Arkansas, Fayetteville
Andonaegui, Pilar, Univ Complutense de Madrid
Andronicos, Chris, Purdue University
Anhaeusser, Carl R., University of the Witwatersrand
Antonacci, Vince, Tennessee Geological Survey
Arias, Carmen, Univ Complutense de Madrid
Ashe, Douglas, Tulsa Community College
Asiwaju-Bello, Yinusa A., Federal University of Technology, Akure
Asowata, Timothy I., Federal University of Technology, Akure
Athaide, Dileep J., Capilano University
Athaide, Dileep, University of British Columbia
Athey, Jennifer E., Alaska Div of Geological & Geophysical Surveys
Bach Plaza, Joan, Universitat Autonoma de Barcelona
Bachle, Peter, Missouri Dept of Natural Resources
Baer, Eric M., Highline College
Bagley, Evan, University of Southern Mississippi
Baker, Cathy, Arkansas Tech University
Balakrishnan, Meena, Tarrant County College, Northeast Campus
Ballero, Deniz Z., Georgia State Univ, Perimeter College, Online
Bally, Albert W., Rice University
Balmat, Jennifer L., Chadron State College
Balsley, Christopher, Southern Connecticut State University
Banda, Anna F., El Centro College - Dallas Comm Coll District
Barclay, Julie L., SUNY, Cortland
Barker, Andy J., University of Southampton
Barker, Gregory A., New Hampshire Geological Survey
Barminski, Robert, Hartnell College
Barnes, Brenda, El Paso Community College
Bassett, Damon, Missouri State University
Baxter, P.G., James E., Harrisburg Area Community College
Bayowa, Oyelowo G., Ladoke Akintola University of Technology
Beatty, Lynne, Johnson County Community College
Beausoleil, Denis, Douglas College
Beltz, John F., University of Akron
Bender, Hallock J., Mt. San Antonio College
Benner, Jake, University of Tennessee, Knoxville
Bennett, Sara, Western Illinois University
Benson, Donna M., Mesa Community College
Benson, Robert G., Adams State University
Bentley, Callan, Northern Virginia Community College - Annandale
Benton, Steven, Illinois State Geological Survey
Berry IV, Henry N., Dept of Agriculture, Conservation, and Forestry
Best, Myron G., Brigham Young University
Beyer, Adam, Nicholls State University
Bice, David M., Pennsylvania State University, University Park
Biek, Robert, Utah Geological Survey
Bierly, Aaron D., Pennsylvania Bur of Topo & Geologic Survey
Birner, Johannes, Ludwig-Maximilians-Universitaet Muenchen
Blaylock, Glenn W., Laredo College
Boehm, David A., SUNY, The College at Brockport
Bograd, Michael B., Mississippi Office of Geology
Bohm, Christian, Manitoba Geological Survey
Bolze, Claude E., Tulsa Community College
Bone, Fletcher, Missouri Dept of Natural Resources
Boroughs, Terry J., American River College
Bothern, Lawrence, El Paso Community College
Bouker, Polly A., Georgia State Univ, Perimeter College, Newton Campus
Bounk, Michael, Iowa Dept of Natural Resources
Bour, William, Northern Virginia Comm Coll - Loudoun Campus
Bouse, Robin, El Camino College
Bowen, Esther E., Argonne National Laboratory
Bowles, Zack, Mesa Community College
Bowring, Samuel A., Massachusetts Institute of Technology
Bozdog, Nick, North Carolina Geological Survey
Bradshaw, Lanna K., Brookhaven College
Brady, Lawrence L., University of Kansas
Branciforte, Chloe, Mesa Community College
Brânzilă, Mihai, Alexandru Ioan Cuza
Bridges, David L., Missouri Dept of Natural Resources
Bries Korpik, Jill, Inver Hills Community College
Brill, Jr., Richard C., Honolulu Community College
Brock, Patrick W. G., Queens College (CUNY)
Brouillette, Pierre, Universite du Quebec
Brown, Alan, Middle Tennessee State University
Brown, Kenneth, West Virginia University
Brown, Summer, University of Kentucky
Browning, James M., Texas Tech University
Browning, James V., Rutgers, The State Univ of New Jersey
Browning, Sharon, Baylor University
Broze, Elliot, Whitman College
Brueckner, Hannes K., Queens College (CUNY)
Brunengo, Matthew, Portland State University
Bruschke, Freddi Jo, California State University, Fullerton
Bryant, Kathleen E., Illinois State Geological Survey
Bryant, Marita, Bates College
Buchanan, Paul C., Kilgore College
Buchwald, Caryl E., Carleton College
Bucke, David P., University of Vermont
Bulger, Daniel E., Georgia State Univ, Perimeter Coll, Decatur Campus
Bultman, John, Asheville-Buncombe Technical Community College
Burd, Aurora, Antelope Valley College
Burns, Emily, Community College of Rhode Island
Burris, John H., San Juan College
Burtis, Erik, Northern Virginia Community College - Woodbridge
Butcher, William, Bowling Green State University
Butkos, Darryl J., Suffolk County Comm Coll, Ammerman Campus
Büchel, Georg, Friedrich-Schiller-University Jena
Cabe, Suellen, University of North Carolina, Pembroke
Caldwell, Andy, Front Range Community College - Larimer
Caldwell, Marianne O., Hillsborough Community College
Callahan, Caitlin N., Grand Valley State University
Cameron, Kevin, Simon Fraser University
Canalda, Sabrina, El Paso Community College
Carr, Keith W., Illinois State Geological Survey
Carr, Tim R., University of Kansas
Case, Jeanne, Eastern Washington University
Caskey, Deborah, El Paso Community College
Cervato, Cinzia C., Iowa State University of Science & Technology
Chadima, Sarah A., South Dakota Dept of Env and Nat Res
Chamberlin, Richard M., New Mexico Institute of Mining & Technology
Chandler, Angela, Arkansas Geological Survey
Chang, Julie M., University of Oklahoma
Charles, Kasanzu, University of Dar es Salaam
Chasteen, Hayden R., Tarrant County College, Northeast Campus
Chernosky, Jr., Joseph V., University of Maine
Cherukupalli, Nehru, Hofstra University
Childs, Elise M., Weber State University
Cholnoky, Jennifer, Skidmore College
Christensen, Wesley P., South Dakota Dept of Env and Nat Res
Ciciarelli, John A., Pennsylvania State University, Monaca
Clark, Donald, Utah Geological Survey
Clark, Kathryne, Iowa Dept of Natural Resources

Clark, Ken P., University of Puget Sound
Clayton, Rodney L., Tidewater Community College
Clendening, Ronald J., Tennessee Geological Survey
Coffman, David, El Centro College - Dallas Comm Coll District
Coffman, Stephanie, El Centro College - Dallas Comm Coll District
Colak Erol, Serap, Firat University
Coleman, Alvin L., Cape Fear Community College
Collins, Laura R., Middle Tennessee State University
Colosimo, Amanda, Monroe Community College
Connolly Jr., Harold C., Grad Sch of the City Univ of New York
Conquy, Xenia, Broward College
Cooper, Jennifer, Southern Connecticut State University
Cooper, Jonathon L., Carleton College
Cornell, Kristie, University of Louisiana at Lafayette
Cornell, Sean R., Shippensburg University
Cornell , William , University of Texas, El Paso
Couroux, Emile, El Paso Community College
Coutand, Isabelle, Dalhousie University
Craig, James L., Los Alamos National Laboratory
Crain, John R., College of the Sequoias
Crawford, Vernon J., Southern Oregon University
Creveling, Jessica, Oregon State University
Criswell, James, Cape Fear Community College
Cron, Mitch, Drexel University
Cronoble, James M., Metropolitan State College of Denver
Cummings, Michael L., Portland State University
Curtis, Lynn A., Broward College
Cvetko Tešoviæ, Blanka, University of Zagreb
Damir, Buckoviæ, University of Zagreb
Davis, David A., University of Nevada
Davis, Gary S., Weber State University
Dawes, Ralph, Wenatchee Valley College
Day, Damon P., Mt. San Antonio College
Deakin, Joann, Cochise College
Dealy, Mike, University of Kansas
Deane, William, University of Tennessee, Knoxville
Dee, Seth, University of Nevada
Deering, Chad, Michigan Technological University
Delano, Helen L., Pennsylvania Bur of Topo & Geologic Survey
Devaney, Kathleen, El Paso Community College
Devery, Dora, Alvin Community College
Devlahovich, Vincent A., College of the Canyons
DeVries-Zimmerman, Suzanne, Hope College
Dhavaskar, Poornima, Goa University
Di Fiori, Sara, El Camino College
Dinklage, William, Santa Barbara City College
Dittmer, Eric, Southern Oregon University
Dodge, Stefanie, University of Wisconsin, Extension
Doelling, Helmut H., University of Utah
Doherty, David J., Wayne State University
Donnelly, Thomas W., Binghamton University
Dooley, Brett, Patrick Henry Community College
Dougan, Bernie,
Doyle, Joseph, Bridgewater State University
Dressel, Waldemar M., Missouri University of Science and Technology
Drobny, Gerry, Cosumnes River College
Dudley, Jutta S., SUNY, The College at Brockport
Duncan, Ian, Univ of Texas at Austin, Jackson Sch of Geosciences
Dushman, Beth, American River College
Dyab, Ahmed I., Alexandria University
Earls, Stephanie, Washington Geological Survey
Easterday, Cary, Bellevue College
Edson, Carol, Las Positas College
Egbeyemi, Oladimeji R., Federal University of Technology, Akure
Eguiluz, Luis, Univ of the Basque Country UPV/EHU
Eisenhart, Ralph, York College of Pennsylvania
El Dakkak, Mohamed W., Alexandria University
Ellis, Trevor, Missouri Dept of Natural Resources
Emerson, Norlene, University of Wisconsin Colleges
Engelhard, Simon, University of Rhode Island
Erlandsson, Viktor, University Leoben
Escartin, Javier, Woods Hole Oceanographic Institution
Escuer, Joan, Universitat Autonoma de Barcelona
Esparza-Yanez, Francisco A., Universidad Estatal de Sonora
Evans, Thomas J., University of Wisconsin, Extension
Ewas, Galal A., Alexandria University
Faatz, Renee M., Snow College

FABRE, Cecile, Univ de Lorraine - Faculte des Sci et Tech
Fagnan, Brian A., South Dakota Dept of Env and Nat Res
Fahrenbach, Mark D., South Dakota Dept of Env and Nat Res
Falcon, Alexandra, El Paso Community College
Farquharson, Phil, MiraCosta College
Farrell, Michael, Cuyamaca College
Feeney, Dennis M., University of Idaho
Fengler, Keegan , Central Washington University
Fields, Chad, Iowa Dept of Natural Resources
Figg, Sean, Palomar College
Filkorn, Harry, Los Angeles Pierce College
Finley, Mark, Heartland Community College
Fitak, Madge R., Ohio Dept of Natural Resources
Fitchett, Rebekah, Northeastern Illinois University
Fitz, Thomas J., Northland College
Fleck, Michelle C., Utah State University Eastern
Flower, Hilary, Saint Petersburg College, Clearwater
Flowers Falls, Emily, Washington & Lee University
Forno, Maria Gabriella, Università di Torino
Frail, Pam, Acadia University
Fredrick, Kyle, California University of Pennsylvania
Freed, Andrew M., Purdue University
Frei, Michaela, Ludwig-Maximilians-Universitaet Muenchen
Freiburg, Jared, Illinois State Geological Survey
Friedmann, Samuel (Julio) J., Lawrence Livermore National Laboratory
Friedrich, Anke, Ludwig-Maximilians-Universitaet Muenchen
Frieman, Richard, SUNY, Oswego
Fromm, Jeanne M., University of South Dakota
Frost, Gina M., San Joaquin Delta College
Fuerst, Alice, Metropolitan Community College-Kansas City
Gagnon, Teresa K., Dept of Energy and Environmental Protection
Galal, Galal M., Alexandria University
Gale, Marjorie H., Agency of Natural Res, Dept of Env Conservation
Gannon, J. Michael, Iowa Dept of Natural Resources
García Frank, Alejandra, Univ Complutense de Madrid
Gardner, Eleanor E., University of Tennessee, Martin
Gardner, Jamie N., Los Alamos National Laboratory
Garrett, Maureen, Arizona Western College
Garside, Larry J., University of Nevada
Garwood, Phil, Cape Fear Community College
Gawloski, Joan, Midland College
Gawu, Simon K., Kwame Nkrumah Univ of Science and Tech
Gentry, Amanda L., Weber State University
Gianotti, Franco, Università di Torino
Giegengack, Jr., Robert F., University of Pennsylvania
Giglierano, James D., Iowa Dept of Natural Resources
Giles, Antony, Midland College
Gillman, Joe, Missouri Dept of Natural Resources
Glynn, William G., SUNY, The College at Brockport
Gonzalez-Juarez, Marco A., Universidad Estatal de Sonora
Goodwin, Joshua, California Geological Survey
Gould, Joseph C., Saint Petersburg College, Clearwater
Grace, Cathy A., University of Mississippi
Greene, Mott, University of Washington
Griffin, William R., University of Texas, Dallas
Griffing, Corinne, Douglas College
Gross, Amy, University of North Carolina, Pembroke
Grover, Jeffrey A., Cuesta College
Grupp, Steve, Everett Community College
Gušiæ, Ivan, University of Zagreb
Hagadorn, James W., Denver Museum of Nature & Science
Hakimian, Adina, Hofstra University
Hale, Leslie J., Smithsonian Inst / Natl Museum of Natural History
Haley, John C., Virginia Wesleyan College
Hall, Clarence A., University of California, Los Angeles
Hall, Tracy, Georgia Highlands College
Hamdan, Abeer, Phoenix College
Hamecher, Emily A., California State University, Fullerton
Hammes, Ursula, Univ of Texas at Austin, Jackson Sch of Geosciences
Hams, Jacquelyn E., Los Angeles Valley College
Hand, Linda M., College of San Mateo
Hanger, Brendan, Oklahoma State University
Hanna, Ruth L., Las Positas College
Hanson, Gilbert N., Stony Brook University
Harder, Vicki, University of Texas, El Paso
Hargrave, Phyllis, Montana Tech of The University of Montana
Harma, Roberta L., Moorpark College

Faculty by Specialty

Harone, Imad, College of Staten Island/CUNY
Harrell, Michael, Seattle Central Community College
Harris, Ann, Eastern Kentucky University
Harris, Michelle, Centralia College
Harrison, Linda, Western Michigan University
Hart, George, Louisiana State University
Hart, Roger M., Community College of Rhode Island
Hartley, Susan, Lake Superior College
Haselwander, Airin J., Missouri Dept of Natural Resources
Hauer, Kendall L., Miami University
Hayes, Garry F., California State University, Stanislaus
Hayes, Garry F., Modesto Junior College
Heck, Frederick R., Ferris State University
Heerschap, Lauren, Fort Lewis College
Hegner, Ernst, Ludwig-Maximilians-Universitaet Muenchen
Heidlauf, Lisa, Wheaton College
Hein, Kim A., University of the Witwatersrand
Helba, Hossam EL-Din A., Alexandria University
Heller, Matthew J., Division of Geology and Mineral Resources
Helmke, Vicky, West Chester University
Henley, Barbara D., Mercer University
Henry, Kathleen M., Illinois State Geological Survey
Hepburn, J. Christopher, Boston College
Herbert, Jennifer, Argonne National Laboratory
Hernandez, Larry, MiraCosta College
Herring Mayo, Lisa L., Motlow State Community College
Herzig, Chuck, El Camino College
Hess Tanguay, Lillian, Hofstra University
Hetherington, Jean, Diablo Valley College
Hickcox, Charles W., Emory University
Hicks, Roberta (Robbie), Memorial University of Newfoundland
Higgins, Chris T., California Geological Survey
Hill, Catherine R., Arizona Western College
Hill, Chris, Grossmont College
Hinthorne, James R., University of Texas, Rio Grande Valley
Hinz, Nicholas, University of Nevada
Ho, Anita, Flathead Valley Community College
Hobbs, Thomas M., Lonestar College - North Harris
Hochstaedter, Alfred, Monterey Peninsula College
Hodder, Donald R., SUNY, New Paltz
Hok, Jozef, Comenius University in Bratislava
Holail, Hanafy M., Alexandria University
Holmes, Stevie L., South Dakota Dept of Env and Nat Res
Hood, Teresa A., University of Miami
Hood, William C., Colorado Mesa University
Hopkins, David M., University of Alaska, Fairbanks
Hoppe, Andreas, Technische Universitaet Darmstadt
Horn, John, Metropolitan Community College-Kansas City
Horton, Albert B., Tennessee Geological Survey
House, Martha, Pasadena City College
Howard, Katie, Olympic College
Howes, Mary R., Iowa Dept of Natural Resources
Hunt, Paula J., West Virginia Geological & Economic Survey
Hural, Kirsten, Dickinson College
Hussein, Musa, El Paso Community College
Hutto, Richard S., Arkansas Geological Survey
Huxta, Stephen, Drexel University
Huycke, David, Yakima Valley College
Hylland, Michael D., Utah Geological Survey
Hylton, Alisa, Central Piedmont Community College
Hynes, Joanna, California State University, Fullerton
Iaccheri, Maria Linda, Ludwig-Maximilians-Universitaet Muenchen
Ianno, Adam J., Juniata College
Idstein, Peter J., University of Kentucky
Igonor, Emmanuel E., Federal University of Technology, Akure
Ikwuazorm, Okia, Northern Virginia Comm Coll - Loudoun Campus
Irvin, Gene D., Geological Survey of Alabama
Isenor, Fenton M., Cape Breton University
Jackson, Hiram, Cosumnes River College
Jackson, Jeremiah, Missouri Dept of Natural Resources
Jacobs, Jon, Western University
Jaecks, Glenn, American River College
Janusz, Robert, San Antonio Community College
Jegede, O A., Federal University of Technology, Akure
Jellinek, Mark, University of British Columbia
Jensen, Ann R., South Dakota Dept of Env and Nat Res
Jiang, James Xinxia, Mount Allison University

Johnson, Darren J., South Dakota Dept of Env and Nat Res
Johnson, Edward, College of Staten Island/CUNY
Johnson, Kurt, Alaska Div of Geological & Geophysical Surveys
Johnson, Ty, Arkansas Geological Survey
Johnston, David, Arkansas Geological Survey
Johnston, Paul L., Emporia State University
Jones, Charles E., University of Pittsburgh
Jones, Gwyn, Bellevue College
Jordan, Bradley C., Bucknell University
Jurena, Dwight, Alamo Colleges, San Antonio College
Kaandorp, Ron, Vrije Universiteit Amsterdam
Kaldor, Michael, Miami-Dade College (Wolfson Campus)
Kapp, Jessica, University of Arizona
Karwoski, Todd, University of Maryland
Katz, Cindi, Grad Sch of the City Univ of New York
Kaye, John M., Mississippi State University
Kaygili, Sibel, Firat University
Keane, Christopher M., American Geosciences Institute
Keating, Elizabeth, Los Alamos National Laboratory
Keattch, Sharen, Eastern Washington University
Keene, Deborah A., University of Alabama
Kelley, Neil P., Vanderbilt University
Kempe, Stephan, Technische Universitaet Darmstadt
Kerwin, Charles M., Keene State College
Kerwin, Michael W., University of Denver
Ketcham, Richard A., University of Texas, Austin
Khalil Ebeid, Khalil I., Alexandria University
Kiesel, Diann, University of Wisconsin Colleges
Kim, Keonho, Midland College
King, Jonathan K., Utah Geological Survey
Kirkby, Kent C., University of Minnesota, Twin Cities
Klaus, James S., University of Miami
Klee, Thomas M., Hillsborough Community College
Klosterman, Sue, University of Dayton
Knowlton, Kelly, Northwestern State University
Kodosky, Larry, Oakland Community College
Koehler, Rich D., University of Nevada, Reno
Koenig, Brian, College of the Desert
Kohrt, Casey, Iowa Dept of Natural Resources
Kolkas, Mosbah, College of Staten Island/CUNY
Kollasch, Pete, Iowa Dept of Natural Resources
Kowallis, Bart J., Brigham Young University
Kraft, Kaatje, Mesa Community College
Kraft, Kaatje,
Kramer, J. Curtis, University of the Pacific
Kramer, Kate, McHenry County College
Kramer, Walter V., Del Mar College
Kranz, Dwight S., Houston Community College System
Krohe, Nicholas J., South Dakota Dept of Env and Nat Res
Kroon, Dick, Edinburgh University
Kruger, Ned, North Dakota Geological Survey
Kubicek, Leonard, North Lake College - Dallas County Comm Coll District
Kuehner, Scott, University of Washington
Kuhlman, Robert, Montgomery County Community College
Kuhnhenn, Gary L., Eastern Kentucky University
Kukoè, Duje, University of Zagreb
Kumpf, Amber C., Muskegon Community College
Kuntz, Mark R., Elgin Community College
Kutis, Michael, Ball State University
Kwicklis, Edward M., Los Alamos National Laboratory
Lacy, Tor, Cerritos College
Lambert-Smith, James, Kingston University
Lammerer, Bernd L., Ludwig-Maximilians-Universitaet Muenchen
Lancaster, Jeremy, California Geological Survey
Langel, Richard A., Iowa Dept of Natural Resources
Langhorst, Glenn, Fond du Lac Tribal and Community College
LaPointe, Daphne D., University of Nevada
Larter, Stephen, University of Newcastle Upon Tyne
Lauziere, Kathleen, Universite du Quebec
Lawrence, Kira, Lafayette College
Leite, Michael B., Chadron State College
Lemay, Phillip W., Louisiana State University
Leszczynski, Raymond F., Cayuga Community College
Levy, Melissa H., American River College
Lewis, Chris, City College of San Francisco
Lewis, Mary, Contra Costa College
Lewis, Reed S., University of Idaho

Liauw, Henri L., Broward College
Libra, Robert D., Iowa Dept of Natural Resources
Liebling, Richard, Hofstra University
Lilly, Troy, Western Texas College
Lippelt, Irene D., University of Wisconsin, Extension
Liu, Paul Hiaibao, Iowa Dept of Natural Resources
Locke, Erika, Texas A&M University, Corpus Christi
Lombard, Armand J., Mesa Community College
Long, Colleen, Illinois State Geological Survey
Lonn, Jeff, Montana Tech of The University of Montana
Lovett, Cole, Lake Michigan College
Lu, Yuehan, University of Alabama
Ludman, Allan, Queens College (CUNY)
Ludman, Allan, Grad Sch of the City Univ of New York
Lueth, Virgil W., New Mexico Institute of Mining & Technology
Lužar-Oberiter, Borna, University of Zagreb
Lyle, Mike, Tidewater Community College
Lyle, Mitchell, Oregon State University
Lyman, John C., Bakersfield College
Mabee, Stephen B., University of Massachusetts, Amherst
Macdonald, Francis, University of California, Santa Barbara
Macias, Steve E., Olympic College
MacLachlan, James, Metropolitan State College of Denver
Madin, Ian P., Oregon Dept of Geology and Mineral Industries
Magee, Robert, Virginia Wesleyan College
Mahaffee, Tina, Middle Georgia College
Mahlen, Nancy J., SUNY, Geneseo
Mahmoud, Sara A., Alexandria University
Majodina, Thando, Tshwane University of Technology
Major, Penni, Lonestar College - North Harris
Major, Ruth H., Hudson Valley Community College
Mandziuk, William S., University of Manitoba
Mansour, Ahmed S., Alexandria University
Mapholi, Thendo, University of the Free State
Marchand, Gerard, Mount Holyoke College
Marchisin, John, New Jersey City University
Marco, Shmuel, Tel Aviv University
Marco, Shmulik, Tel Aviv University
Marshall, Thomas R., South Dakota Dept of Env and Nat Res
Martin, Gale D., College of Southern Nevada - West Charleston Campus
Martinez Torres, Luis Miguel, Univ of the Basque Country UPV/EHU
Martinuš, Maja, University of Zagreb
Marty, Kevin, Imperial Valley College
Martz, Todd, University of California, San Diego
Marvinney, Robert G., Dept of Agriculture, Conservation, and Forestry
Mata, Scott, California State University, Fullerton
Mattox, Tari, Grand Rapids Community College
Matyjasik, Basia, Utah Geological Survey
Mauldin-Kinney, Virginia L., Georgia State Univ, Perimeter Coll, Dunwoody Camp
May, S J., Collin College - Preston Ridge Campus
Mayer, Larry, Woods Hole Oceanographic Institution
McArthur, Russell , City College of San Francisco
McCall, Rosemary, College of Staten Island/CUNY
McCauley, Steven, El Centro College - Dallas Comm Coll District
McClaughry, Jason, Oregon Dept of Geology and Mineral Industries
McCoy, Floyd W., Windward Community College
McCulloh, Richard P., Louisiana State University
McCutchen, William T., University of Tennessee, Martin
McDermott, Frank, University College Dublin
McDowell, Robin J., Georgia State Univ, Perimeter Coll, Dunwoody Camp
McDowell, Ronald, Fairmont State University
McKay, Robert M., Iowa Dept of Natural Resources
McKinney, Michael, California Geological Survey
McMonagle, Julie, Wilkes University
Meave, Edgardo, Univ Nac Autonoma de Mexico
Meister, Paul A., Illinois State University
Melcher, Frank, University Leoben
Meldahl, Keith H., MiraCosta College
Meléndez, Nieves, Univ Complutense de Madrid
Metz, Cheyl L., Blinn College
Metzler, Christopher V., MiraCosta College
Meyer, Jeffrey W., Santa Barbara City College
Michel, Suzanne, Cuyamaca College
Millan, Christina, Ohio State University
Millen, Timothy M., Elgin Community College
Miller, David, California State University, Bakersfield
Miller, Steven F., Argonne National Laboratory
Miller-Hicks, Bryan, Cuyamaca College
Milner, Lloyd R., Louisiana State University
Milner, Lloyd R., Louisiana State University
Miner, James J., Illinois State Geological Survey
Minium, Deborah, Bellevue College
Moll, Nancy E., College of the Desert
Monet, Julie, California State University, Chico
Monson, Jessica, Illinois State Geological Survey
Montayne, Simone, Alaska Div of Geological & Geophysical Surveys
Montwill, Gail F., Santiago Canyon College
Morand, Vincent J., La Trobe University
Morealli, Sarah A., University of Mary Washington
Morgan, Matt, Colorado Geological Survey
Morris, Billy, Georgia Highlands College
Mosteller, Joey D., Appalachian State University
Mshiu, Elisante E., University of Dar es Salaam
Mulvany, Patrick S., Missouri University of Science and Technology
Munn, Barbara J., California State University, Sacramento
Murphy, Cindy, Saint Francis Xavier University
Murphy, Dan, Asheville-Buncombe Technical Community College
Muskatt, Herman, Utica College
Mutti, Laurel, SUNY, New Paltz
Nagy-Shadman, Elizabeth, Pasadena City College
Nance, Seay, Univ of Texas at Austin, Jackson Sch of Geosciences
Nesheim, Tim, North Dakota Geological Survey
Newman, Brent D., Los Alamos National Laboratory
Newman, Jamie, American Museum of Natural History
Nicoletti, Jeremy D., New Hampshire Geological Survey
Nielsen, Gregory B., Weber State University
Nixon III, Roy (Rick) A., Georgia State Univ, Perimeter College, Online
Norman, David K., Washington Geological Survey
Norrish, Winston, Central Washington University
Nugent, Barnes, Fairmont State University
O'Brien, Lawrence E., Orange County Community College
O'Keeffe, Mike, Colorado Geological Survey
Oakes-Miller, Hollie, Portland Community College - Sylvania Campus
Obolewicz, Dave, Keene State College
Odendaal, Adriaan, University of the Free State
Odeyemi, Idowu O., Federal University of Technology, Akure
Ogbahon, Osazuwa A., Federal University of Technology, Akure
Okonkwo, C T., Federal University of Technology, Akure
Olabode, Solomon O., Federal University of Technology, Akure
Oldfield, Bruce K., Broome Community College
Oldham, Richard L., American River College
Onuk, Peter, University Leoben
Opeloye, S A., Federal University of Technology, Akure
Opfer, Christina, Tulsa Community College
Opperman, William, Broward College
Orr, William , Portland State University
Oswald, Peter J., University of Alaska, Anchorage
Owoseni, Joshua O., Federal University of Technology, Akure
Pande, Srikant k., Pt. Ravishankar Shukla University
Pantoja-Alor, Jerjes, Univ Nac Autonoma de Mexico
Papp, Kenneth R., Alaska Div of Geological & Geophysical Surveys
Parish, Cynthia L., Lamar University
Parker, Richard M., Texas A&M University, Kingsville
Parnella, Bill, University of Delaware
Parrick, Brittany, Ohio Dept of Natural Resources
Parsons, Michael, Dalhousie University
Paschert, Karin, Ludwig-Maximilians-Universitaet Muenchen
Patterson, Gary, University of Memphis
Patton, Jason A., Arkansas Tech University
Patton, Terri, Argonne National Laboratory
Patwardhan, Kaustubh, SUNY, New Paltz
Pawloski, Gayle A., Lawrence Livermore National Laboratory
Pellowski, Christopher J., South Dakota School of Mines & Technology
Perez, Adriana, El Paso Community College
Perez, Adriana, University of Texas, El Paso
Perkis, Bill, Gogebic Community College
Pesavento, Jim, Palomar College
Peters, Lisa, New Mexico Institute of Mining & Technology
Peters, Roger M., University of Wisconsin, Extension
Peterson, Joseph E., Elgin Community College
Phillips, Michael, Illinois Valley Community College
Pierce, David, Lakeland Community College
Pirie, Diane H., Florida International University

Faculty by Specialty

Pokras, Edward M., Keene State College
Polissar, Pratigya J., Columbia University
Poole, T. Craig, Fresno City College
Posiloviæ, Hrvoje, University of Zagreb
Pound, Kate S., Saint Cloud State University
Powers, Elizabeth, California State University, Bakersfield
Prave, Tony, University of St. Andrews
Prewett, Jerry L., Missouri Dept of Natural Resources
Price, Jonathan G., University of Nevada
Price, L G., New Mexico Institute of Mining & Technology
Prichystal, Antonín, Masaryk University
Priesendorf, Carl, Metropolitan Community College-Kansas City
Priest, George R., Oregon Dept of Geology & Mineral Industries
Priewisch, Alexandra, Fresno City College
Prior, William L., Arkansas Geological Survey
Prothero, Donald, Los Angeles Pierce College
Proudhon, Benoit, Institut Polytechnique LaSalle Beauvais (ex-IGAL)
Prytulak, Julie, Imperial College
Purkiss, Robert, Angelo State University
Quick, Thomas J., University of Akron
Quinn, Heather A., Maryland Department of Natural Resources
Ragland, Deborah, University of New Mexico, Taos
Rahman, Ata U., Austin Community College District
Rains, Daniel S., Arkansas Geological Survey
Ranhofer, Melissa, Furman University
Rantitsch, Gerd, University Leoben
Rashed, Mohamed A., Alexandria University
Rath, Carolyn, California State University, Fullerton
Rathburn, Sara L., Colorado State University
Rawling, Geoffrey, New Mexico Institute of Mining & Technology
Ray, Waverly, Cuyamaca College
Raymer, Janet, Metropolitan Community College-Kansas City
Read, Adam S., New Mexico Institute of Mining & Technology
Redden, Marcella, Geological Survey of Alabama
Reesman, Arthur L., Vanderbilt University
Reif, Samantha, Lincoln Land Community College
Rennie, Colette, Saint Francis Xavier University
Repka, James, Saddleback Community College
Richards, Bill, North Idaho College
Riegel, Hannah B., Appalachian State University
Riker-Coleman, Kristin E., University of Wisconsin, Superior
Rinae, Makhadi, University of the Free State
Riordan, Jean, Alaska Div of Geological & Geophysical Surveys
Rivera, Mark, University of Alaska, Anchorage
Roberson, Randal P., Austin Peay State University
Robert, Sanborn, University of Wisconsin Colleges
Robinson, Sarah, United States Air Force Academy
Rocha, Guillermo, Brooklyn College (CUNY)
Roche, James E., Louisiana State University
Rodgers, Jim, Wyoming State Geological Survey
Rodríguez, Marta, Univ Complutense de Madrid
Rodriguez-Castaneda, Jose L., Univ Nac Autonoma de Mexico
Ross, Theodore W., Lawrence University
Rothemund, Kirk, El Paso Community College
Rounds, Steven W., California State University, Sacramento
Rowden, Robert, Iowa Dept of Natural Resources
Ruetz, Joseph W., Stanford University
Ruffel, Alice, El Centro College - Dallas Comm Coll District
Rumrill, Julie, Southern Connecticut State University
Ruppert, Kelly R., California State University, Fullerton
Sacramentogrilo, Isabelle, San Diego State University
Salle, Bethan, El Centro College - Dallas Comm Coll District
Samimi, Naser, University of Tabriz
Samra, Charles, Wyoming State Geological Survey
Sassi, Raffaele, Università degli Studi di Padova
Sato, Yoko, Montclair State University
Sawin, Robert S., University of Kansas
Schafer, Carl M., Macomb Community College, Center Campus
Scheffler, Joanna, Mesa Community College
Schenck, William S., University of Delaware
Schilling, Keith, Iowa Dept of Natural Resources
Schimmrich, Steven, SUNY, Ulster County Community College
Schmidt, Bennetta, Lamar University
Schulingkamp, Arren, Louisiana State University
Schulte, Kimberly D., Georgia State Univ, Perimeter College, Online
Schultz, Jan, Santa Barbara City College
Sediek, Kadry N., Alexandria University

Seigley, Lynette S., Iowa Dept of Natural Resources
Severs, Matthew, Stockton University
Shaaban, Mohamad N., Alexandria University
Shade, Harry, West Valley College
Shafer, Erik, Portland State University
Shakun, Jeremy D., Boston College
Shalimba, Ester, University of Namibia
Shannon, Jeremy, Michigan Technological University
Sharp, Patricia S., Stephen F. Austin State University
shediied, ahmad g., Fayoum University
Shiels, Christine, Douglas College
Shimizu, Melinda, Mesa Community College
Shinn, Eugene, University of South Florida
Shorey, Christian V., Colorado School of Mines
Shroba, Cynthia S., College of Southern Nevada - West Charleston Campus
Sibeko, Skhumbuzo, Tshwane University of Technology
Sicard, Karri, Alaska Div of Geological & Geophysical Surveys
Siegelberg, Alan, Long Island University, Brooklyn Campus
Siemens, Michael A., Missouri Dept of Natural Resources
Simard, Renee-Luce ., Ontario Geological Survey
Skelton, Lawrence H., University of Kansas
Skinner, Randall, Brigham Young University
Sledzinski, Grazyna, Wayne State University
Smaglik, Suzanne M., Central Wyoming College
Smith, Jason J., Broome Community College
Smith, Jon J., University of Kansas
Smith, Mike, Front Range Community College - Larimer
Smith, Russell, El Paso Community College
Smithson, Jayne, Contra Costa College
Snyder, Daniel, Middle Georgia College
Snyder, Noah, Boston College
Solferino, Giulio, Royal Holloway University of London
Solis, Michael P., Ohio Dept of Natural Resources
Sparks, Thomas N., University of Kentucky
Speed, Don, Phoenix College
Sperone, Felice, Wayne State University
Stakes, Debra, Cuesta College
Stanton, Kathryn, Sacramento City College
Stanton, Kelsay, Wenatchee Valley College
Steart, David, La Trobe University
Steck, Lee, Los Alamos National Laboratory
Steffens, Katja, Ludwig-Maximilians-Universitaet Muenchen
Steinberg, Roger T., Del Mar College
Stewart, Esther K., University of Wisconsin, Extension
Stover, Susan G., University of Kansas
Straight, William, Northern Virginia Comm Coll - Loudoun Campus
Straub, Kyle M., Tulane University
Stumpf, Andrew, Illinois State University
Suchy, Daniel R., University of Kansas
Summers, Sara, Weber State University
Sumrall, Jonathan B., Fort Hays State University
Sundell, Ander, College of Western Idaho
Sundell, Kent A., Casper College
Sunderlin, David, Lafayette College
Suszek, Thomas J., University of Wisconsin, Oshkosh
Swyrtek, Sheila M., Mott Community College
Szczepanski, Jacek, University of Wroclaw
Söllner, Frank, Ludwig-Maximilians-Universitaet Muenchen
Tajik, Atieh, Georgia State Univ
Tamish, Mohamed M., Alexandria University
Tapanila, Lori, Idaho State University
Tassier-Surine, Stephanie, Iowa Dept of Natural Resources
Taylor, Carolyn, Mesa Community College
Taylor, Penny M., Mount Holyoke College
Taylor, Sid, Saint Francis Xavier University
Taylor, Witt, Bentley University
Tetrault, Denis, University of Windsor
Thole, Jeffrey T., Macalester College
Thomas, Margaret A., Dept of Energy and Environmental Protection
Thompson, Nils W., Georgia State Univ, Perimeter College, Online
Thompson, Nils W., Georgia State Univ, Perimeter College, Clarkston
Thorleifson, Harvey, University of Minnesota, Twin Cities
Thornberry-Ehrlich, Trista L., Colorado State University
Thul, David, University of Utah
Timmons, Michael, New Mexico Institute of Mining & Technology
Timmons, Stacy, New Mexico Institute of Mining & Technology

Tolley, William, Southern Connecticut State University
Tomiæ, Vladimir, University of Zagreb
Toni, Rousine T., Alexandria University
Towery, Brooke L., Pensacola Junior College
Tucker, Eva, Pennsylvania State University, Erie
Turbeville, John, MiraCosta College
Turner, Wesley L., Stephen F. Austin State University
Turner, III, Henry, University of Arkansas, Fayetteville
Tvelia, Sean, Suffolk County Comm Coll, Ammerman Campus
Uahengo, Collen, University of Namibia
Ugland, Richard, Southern Oregon University
Ullrich, Alexander D., Georgia State Univ, Perimeter Coll, Dunwoody
Urquhart, Joanne, Bowdoin College
Valenti, Christine D., Georgia State Univ, Perimeter College, Online
Valenti, Christine, Montclair State University
van Balen, Ronald R., Vrije Universiteit Amsterdam
Van Brocklin, Matthew F., St. Lawrence University
VanDorpe, Paul E., Iowa Dept of Natural Resources
Vickery, Nancy, University of New England
Vidoviæ, Jelena, University of Zagreb
Viens, Rob, Bellevue College
Vierrether, Chris, Missouri Dept of Natural Resources
Vig, Pradeep K., Kaskaskia College
Vinton, Bonita L., Slippery Rock University
Voigt, Vicki, Missouri Dept of Natural Resources
Vuke, Susan M., Montana Tech of The University of Montana
Wacker, Michael, Florida International University
Wadleigh, Hank, Cypress College
Waggoner, Karen, Midland College
Wakefield, Kelli, Mesa Community College
Waugh, John, Tidewater Community College
Weatherford, Jonathan, Portland Community College - Sylvania Campus
Webber, Jeffrey, Stockton University
Weber, Diane, Illinois Central College
Weber-Diefenbach, Klaus, Ludwig-Maximilians-Universitaet Muenchen
Weborg-Benson, Kimberly, SUNY, Fredonia
Weeden, Lori, University of Massachusetts, Lowell
Wein, Agatha, Cuyamaca College
Werhner, Matthew J., Hillsborough Community College
West, Robert, East Los Angeles College
Wheeler, Richard F., Austin Peay State University
White, Paul, Community College of Rhode Island
Whitehead, James, University of New Brunswick
Whitehead, Peter W., James Cook University
Whitehill, Matthew, Lake Superior College
Whittier, Michael, California State University, Stanislaus
Whittington, Carla, Highline College
Wilbur, Bryan, Pasadena City College
Wilder, Lee, New Hampshire Geological Survey
Willahan, Duane, Gavilan College
Williams, Curtis J., Cypress College
Williams, James H., Missouri University of Science and Technology
Willis, Grant C., Utah Geological Survey
Willis, Marc, Fullerton College
Willsey, Shawn P., College of Southern Idaho
Wilson, Jeffrey A., University of Michigan
Wolfe, Ben, Metropolitan Community College-Kansas City
Wolkersdorfer, Christian, Ludwig-Maximilians-Universitaet Muenchen
Wolter, Calvin, Iowa Dept of Natural Resources
Wood, Jacqueline, Delgado Community College
Woodall, Debra W., Daytona State College
Worcester, Peter, Eastern Kentucky University
Wortmann, Ulrich B., University of Toronto
Wykel, Andy, South Carolina Dept of Natural Resources
Wypych, Alicja, Alaska Div of Geological & Geophysical Surveys
Yalcin, Kaplan, Oregon State University
Yalcin, Rebecca, Oregon State University
Yan, Y E., Argonne National Laboratory
Yoshida, Glenn, Los Angeles Southwest College
Young, Mike, Dalhousie University
Zabel, Garrett E., Colorado Mountain College
Zabielski, Victor, Northern Virginia Community College - Alexandria
Zachariasen, Judy, California Geological Survey
Zanetti, Kathleen, University of Nevada, Las Vegas
Zawiskie, John M., Wayne State University
Zayac, John M., Los Angeles Pierce College
Zeitlhöfler, Matthias, Ludwig-Maximilians-Universitaet Muenchen
Zentner, Nick, Central Washington University
Zimmer, Brian W., Appalachian State University
Zurawski, Ronald P., Tennessee Geological Survey

Archaeological Geology

Abreu, Maria E., Universidade de Trás-os-Montes e Alto Douro
Adams, Kenneth, University of Nevada, Reno
Adovasio, James M., Mercyhurst University
Ballard, Robert D., University of Rhode Island
Bowers, Peter, University of Alaska, Fairbanks
Buckley, Michael, University of Manchester
Chamberlain, Andrew T., University of Manchester
Degryse, Patrick, Katholieke Universiteit Leuven
Dunn, Richard K., Norwich University
Dye, David H., University of Memphis
Elkins, Nichole, Bowling Green State University
Fadem, Cynthia M., Earlham College
Farrand, William, University of Michigan
Ferguson, Terry A., Wofford College
Ferring, C. Reid, University of North Texas
Garrison, Ervan G., University of Georgia
Greenlee, Diana M., University of Louisiana, Monroe
Gundersen, James N., Wichita State University
Gunia, Piotr, University of Wroclaw
Haynes, Jr., C. Vance, University of Arizona
Hayward, Chris L., Edinburgh University
Holliday, Vance, University of Arizona
Holliday, Vance T., University of Arizona
Iglesias, Cruz, Coruna University
Jones, Jeri L., York College of Pennsylvania
Kelley, Alice R., University of Maine
Kvamme, Kenneth L., Boston University
Lee-Gorishti, Yolanda, Southern Connecticut State University
Madsen, David B., Utah Geological Survey
Mandel, Rolfe D., University of Kansas
Mandel, Rolfe, University of Kansas
Manning, Sturt W., Cornell University
Mason, Owen, University of Alaska, Fairbanks
Mickelson, Andrew M., University of Memphis
Milne, Brooke, University of Manitoba
Nagaoka, Lisa A., University of North Texas
Ortmann, Anthony L., Murray State University
Page, David, Desert Research Institute
Parish, Ryan M., University of Memphis
Pike, Scott, Willamette University
Pope, Richard, University of Derby
Pournelle, Jennifer R., University of South Carolina
Powell, Wayne G., Brooklyn College (CUNY)
Reitz, Elizabeth J., University of Georgia
Rogers, Joe D., West Texas A&M University
Sallu, Susannah, University of Leeds
Schiffman, Robert A., Bakersfield College
Smith, Jen R., Washington University in St. Louis
Smith, Jennifer R., Washington University in St. Louis
Stafford, C. Russell, Indiana State University
Swyrtek, Sheila, Charles Stewart Mott Community College
Thieme, Donald, Valdosta State University
Tomaso, Matthew S., Montclair State University
Towner, Ronald, University of Arizona
Uribelarrea, David, Univ Complutense de Madrid
Venter, Marcie, Murray State University
Vento, Frank, Mercyhurst University
Waters, Michael R., Texas A&M University
Wilson, Lucy A., University of New Brunswick Saint John
Wilson, Michael C., Douglas College
Wolverton, Steve, University of North Texas
Wygal, Brian, Adelphi University

Environmental Geology

Abbasnejad, Ahmad, Shahid Bahonar University of Kerman
Aden, Douglas J., Ohio Dept of Natural Resources
Adentunji, Jacob, University of Derby
Adewoye, Abosede O., Ladoke Akintola University of Technology
Adomaitis, D., Illinois State Geological Survey
Akenzua-Adamcyzk, Aiyevbekpen H., University of Benin
Albee-Scott, Steven R., Jackson College
Amezaga, Jamie, University of Newcastle Upon Tyne
Apotsos, Alex, Williams College

Faculty by Specialty

Argyilan, Erin, Indiana University Northwest
Aristilde, Ludmilla, Cornell University
Ashwood, Tom, Oak Ridge National Laboratory
Astorga Gattgens, Allan, Universidad de Costa Rica
Babcock, Daphne H., Collin College - Spring Creek Campus
Baker-Treloar, Elizabeth, University of San Diego
Bang, John, North Carolina Central University
Barber, Donald C., Bryn Mawr College
Barone, Jessica, Monroe Community College
Barrett, John, University of Leeds
Bartholemew, Paul, University of New Haven
Bartolucci, Valerio, Broward College
Beatty, Heather L., Austin Community College District
Beccue, Clint, Illinois State Geological Survey
Bechtel, Timothy D., Franklin and Marshall College
Beebe, Alex, University of South Alabama
Bell, Margaret C., University of Newcastle Upon Tyne
Bilanovic, Dragoljub D., Bemidji State University
Binda, Pier L., University of Regina
Bircher, Harry, Youngstown State University
Bogart, James M., Dept of Energy and Environmental Protection
Bogle, Mary Anna, Oak Ridge National Laboratory
Bond, Alan, University of East Anglia
Bousenberry, Raymond T., New Jersey Geological and Water Survey
Brady, Emily S., Edinburgh University
Brake, Sandra S., Indiana State University
Brandon, Nigel, Imperial College
Brandt, Craig, Oak Ridge National Laboratory
Branfireun, Brian, Western University
Breitmeyer, Ronald J., University of Nevada, Reno
Brodie, Gregory, University of Tennessee, Chattanooga
Bromily, Geoffrey, Edinburgh University
Brown, Huntting (Hunt), Wright State University
Brown, Kerry, Kingston University
Brown, Thomas W., Austin Community College District
Bryan, Mark, Wayland Baptist University
Budkewitsch, Paul, Brock University
Buliga, Iuliana, Alexandru Ioan Cuza
Burke, Ian, University of Leeds
Cadet, Eddy, Utah Valley University
Callison, James, Utah Valley University
Cathcart, Eric, University of San Diego
Caudill, Kimberly S., Hocking College
Chapman, LeeAnna Y., University of San Diego
Chappell, James R., Colorado State University
Chaput, Julien, University of Texas, El Paso
Chaudhuri, Sambhudas, Kansas State University
Chernoff, Barry, Wesleyan University
Cheun, Norman, Kingston University
Chhetri, Parveen, California State University, Dominguez Hills
Chilvers, Jason, University of East Anglia
Cloutier, Danielle, Universite Laval
Colegial Gutierrez, Juan D., Universidad Industrial de Santander
Collier, Mark, Illinois State Geological Survey
Collins, Edward W., Univ of Texas at Austin, Jackson Sch of Geosciences
Concha, Aline, Universitat Autonoma de Barcelona
Congleton, John D., University of West Georgia
Constantopoulos, James T., Eastern New Mexico University
Cook, Hadrian, Kingston University
Corley, John H., Missouri Dept of Natural Resources
Cowart, Richard, Coastal Bend College
Cox, Christena, Ohio State University
Coxon, Catherine, Trinity College
Cummins, R. Hays, Miami University
Curry, B. B., Illinois State Geological Survey
Curry, B. B., Northern Illinois University
Curry, Gordon, University of Glasgow
Curtis, Arthur C., Oak Ridge National Laboratory
Dallimer, Martin, University of Leeds
Davis, R. Laurence, University of New Haven
Davis, Steven J., University of California, Irvine
Day, Brett, University of East Anglia
Denniston, Rhawn F., Cornell College
deWet, Andrew P., Franklin and Marshall College
Dimmick, Charles W., Central Connecticut State University
Diochon, Amanda, Lakehead University
Downie, Helen, University of Manchester
Dubey, C S., University of Delhi
Duncan, Ian J., University of Texas, Austin
Durrant, Jeffrey O., Brigham Young University
Dyer, Jen, University of Leeds
Dymond, Salli F., University of Minnesota, Duluth
Eastler, Thomas E., University of Maine - Farmington
Edwards, Robin, Trinity College
Elliott, Susan J., McMaster University
Ellis, Scott R., Illinois State Geological Survey
Ellison, Anne L., University of Illinois
Ellison, Anne L., Illinois State Geological Survey
Ettlie, Bradley, Illinois State Geological Survey
Eyles, John D., McMaster University
Faulkner, Melinda S., Stephen F. Austin State University
Favas, Paulo J., Universidade de Trás-os-Montes e Alto Douro
Fehn, Udo, University of Rochester
Field, Cathryn K., Bowdoin College
Flesken, Luuk, University of Leeds
Foley, Duncan, Pacific Lutheran University
Foret, Jim, University of Louisiana at Lafayette
Foxon, Tim, University of Leeds
Franco, Aldina, University of East Anglia
Frappier, Amy, Skidmore College
Galicki, Stan, Millsaps College
Gallen, Sean, Colorado State University
Garrison, Trent, Northern Kentucky University
Garstang, Mimi R., Missouri Dept of Natural Resources
Geary, Lindsey, Utica College
Geiger, James W., Illinois State Geological Survey
Gerald, Carresse, North Carolina Central University
Geršlová, Eva, Masaryk University
Gidigasu, Solomon S., Kwame Nkrumah Univ of Science and Tech
Glenn, Craig R., University of Hawai'i, Manoa
Goetz, Heinrich, Collin College - Preston Ridge Campus
Gomes, Nuno N., Universidade Independente de Angola
Gomezdelcampo, Enrique, Bowling Green State University
Gouldson, Andy, University of Leeds
Graham, Margaret C., Edinburgh University
Grant, Alastair, University of East Anglia
Halfman, John D., Hobart & William Smith Colleges
Hanes, Daniel M., Saint Louis University
Hargreaves, Tom, University of East Anglia
Harris, Ann G., Youngstown State University
Harris, C, Cardiff University
Harris, Jr., Stanley E., Southern Illinois University Carbondale
Harvey, Omar R., Texas Christian University
Havenith, Hans-Balder, Universite de Liege
Havig, Jeff, University of Minnesota, Twin Cities
Hawkins, Terry, Missouri Dept of Natural Resources
Hawley, John W., New Mexico Institute of Mining & Technology
Hawley, John W., New Mexico Institute of Mining and Technology
Helgers, Karen, SUNY, Ulster County Community College
Herbert, Bruce, Texas A&M University
Hickey, James, Northwest Missouri State University
Hill, Timothy, University of St. Andrews
Hodell, David, University of Cambridge
Holmes, George, University of Leeds
Hoppie, Bryce W., Minnesota State University
Houston, Robert, Oregon Dept of Geology and Mineral Industries
Howell, Dave, Leicester University
Hubeny, J B., Salem State University
Hudson-Edwards, Karen, Birkbeck College
Humphreys, Robin, College of Charleston
Ioris, Antonio, Edinburgh University
Isaacson, Carl, Bemidji State University
Jackman, Toni K., Wichita State University
Jacobs, Alan M., Youngstown State University
Jamieson, Heather E., Queen's University
Jin, Lixin, University of Texas, El Paso
Jones, T, Cardiff University
Jordan, Andy, University of East Anglia
Jovanovic, Vladimir, College of Staten Island/CUNY
Kairies-Beatty, Candace L., Winona State University
Keller, John E., College of Southern Nevada - West Charleston Campus
Kennedy, Taylor, Orange Coast College
Ketelle, Richard H., Oak Ridge National Laboratory
Kientop, Greg A., Illinois State Geological Survey

Kierczak, Jakub, University of Wroclaw
Kim, Jonathan, Agency of Natural Res, Dept of Env Conservation
Kirchner, James W., University of California, Berkeley
Knippler, Katherine A., Maryland Department of Natural Resources
Kolawole, Lanre L., Ladoke Akintola University of Technology
Krekeler, Mark, Miami University
Kretzschmar, Thomas, Centro de Inv Científica y de Ed Sup de Ensenada
Lake, Iain, University of East Anglia
Laskowski, Stanley L., University of Pennsylvania
Last, George V., Pacific Northwest National Laboratory
Lawry-Berkins, Cynthia, Lonestar College - Montgomery
Lee, Arthur C., Roane State Community College - Oak Ridge
Lee, Cindy M., Clemson University
Lene, Gene W., Saint Mary's University
Leventon, Julia, University of Leeds
Lloyd, Jonathan, University of Manchester
Love, David W., New Mexico Institute of Mining and Technology
Lundgren, Lawrence W., University of Rochester
Macadam, John, Exeter University
Manser, Nathan, Michigan Technological University
Margalef, Ona, Universitat Autonoma de Barcelona
Martin, Scott C., Youngstown State University
Masciocco, Luciano, Università di Torino
Matthews, Robert A., University of California, Davis
May, Michael, Western Kentucky University
Mayer, Margaret, Dine' College
McConnell, Robert L., University of Mary Washington
McCullough, Jr., Edgar J., University of Arizona
McDonald, Brenna, Missouri Dept of Natural Resources
McGivern, Tiffany, Utica College
McKinney, Michael L., University of Tennessee, Knoxville
Metzger, Marc J., Edinburgh University
Meyer, Brian, Georgia State Univ
Meyer, W. Craig, Los Angeles Pierce College
Miller, David S., Argonne National Laboratory
Mitre-Salazar, Luis M., Univ Nac Autonoma de Mexico
Moloney, Marguerite M., Nicholls State University
Monroy-Sanchez, Jaime I., Univ Autonoma de Baja California Sur
Moore, Kathryn, Exeter University
Morse, Linda D., College of William & Mary
Mouat, David A., Desert Research Institute
Mundie, Ben, Oregon Dept of Geology and Mineral Industries
Nelwamondo, T. M., University of Venda
Newbold, K. B., McMaster University
Newton, Rob, University of Leeds
Null, E. Jan, San Francisco State University
O'Connor, Yuet-Ling, Pasadena City College
Olson, Neil F., New Hampshire Geological Survey
Omotoso, O. A., University of Ilorin
Orndorff, Richard L., Eastern Washington University
Oyawale, A. A., Obafemi Awolowo University
Paavola, Jouni, University of Leeds
Panko, Andrew W., Brock University
Pate, John, Missouri Dept of Natural Resources
Patenaude, Genevieve, Edinburgh University
Pearce, Jamie R., Edinburgh University
Pearthree, Philip A., Arizona Geological Survey
Penzo, Michael A., Bridgewater State University
Perault, David R., University of Lynchburg
Perdrial, Nicolas, University of Vermont
Pereira, Alcides C., Universidade de Coimbra
Peterson, Jon W., Hope College
Phillippe, Jennifer, Concord University
Piotrowski, Alexander, University of Cambridge
Pisani-Gareau, Tara, Boston College
Plant, Jane, Imperial College
Pohopien, Kazimierz M., Mt. San Antonio College
Pope, Jeanette K., DePauw University
Poudel, Durga, University of Louisiana at Lafayette
Powell, Jane, University of East Anglia
Pratt, Allyn R., Los Alamos National Laboratory
Quinn, Claire, University of Leeds
Rasmussen, Pat E., University of Ottawa
Reeves, Donald Matt, University of Alaska, Anchorage
Reid, Brian, University of East Anglia
Rezaie-Boroon, Mohammad H., California State University, Los Angeles
Rice, Thomas L., Cedarville University
Richards, Laura, University of Manchester
Richter, Suzanna L., Franklin and Marshall College
Rizoulis, Athanasios, University of Manchester
Roberts, Ray L., Hill College
Roberts, Sheila M., University of Montana Western
Roberts-Semple, Dawn, York College (CUNY)
Rocheford, MaryKathryn (Kat), Lake Superior State University
Rogers, William C., West Texas A&M University
Rucklidge, John C., University of Toronto
Ruhl, Laura S., University of Arkansas at Little Rock
Russell, Sally, University of Leeds
Rutter, Nathaniel W., University of Alberta
Ryan, Anne-Marie, Dalhousie University
Sacristan, Felix, Universitat Autonoma de Barcelona
Sanjurjo, Jorge, Coruna University
Sarah, Willig B., University of Pennsylvania
Satkoski, Alina A., Austin Community College District
Schleifer, Stanley, York College (CUNY)
Schmidt, Dale R., Illinois State Geological Survey
Schoenemann, Spruce W., University of Montana Western
Scholtz, Theresa C., Argonne National Laboratory
Scott, Robert B., University of Texas, Austin
Segall, Marylin, University of Utah
Sewall, Jacob, Kutztown University of Pennsylvania
Sexton, Philip, The Open University
Sharma, Mukul, Dartmouth College
Shem, Linda M., Argonne National Laboratory
Sims, Douglas, College of Southern Nevada - West Charleston Campus
Sîrbu, Smaranda D., Alexandru Ioan Cuza
Skinner, Luke C., University of Cambridge
Slobodnik, Marek, Masaryk University
Smith, Dan, Leicester University
Smith, Jim, University of Portsmouth
Smith, Steve, Royal Holloway University of London
Snider, Henry I., Eastern Connecticut State University
Spaeth, Matthew P., Illinois State Geological Survey
Spahr, Paul, Ohio Dept of Natural Resources
Spooner, Ian S., Acadia University
Stahle, David W., University of Arkansas, Fayetteville
Stan, Oana, Alexandru Ioan Cuza
Stearman, Will, Queensland University of Technology
Steidl, Gregg M., New Jersey Geological and Water Survey
Stevens, Anthony, Montgomery County Community College
Stinchcomb, Gary E., Murray State University
Stumbea, Dan, Alexandru Ioan Cuza
Sturm, Diana, University of South Alabama
Sublette, Kerry, The University of Tulsa
Sylvia, Elizabeth R., Maryland Department of Natural Resources
Tank, Ronald W., Lawrence University
Tawabini, Bassam S., King Fahd University of Petroleum and Minerals
Taylor, Peter, University of Leeds
Thorpe, Mary S., Del Mar College
Triplett, Laura, Gustavus Adolphus College
Turchyn, Alexandra, University of Cambridge
Turner, Derek, Douglas College
Turner, Kerry, University of East Anglia
Vaezi, Reza, University of Tabriz
van Alstine, James, University of Leeds
Van Horn, Stephen R., Muskingum University
Van Ryswick, Stephen, Maryland Department of Natural Resources
VanGundy, Robert D., The University of Virginia's College at Wise
Vann, David R., University of Pennsylvania
Visser, Jenneke M., University of Louisiana at Lafayette
Wallace, William, Grad Sch of the City Univ of New York
Walsh, Maud M., Louisiana State University
Watkinson, Andrew, University of East Anglia
Webb, John A., La Trobe University
Welling, Tim, Dutchess Community College
Wilson, Charlie, University of East Anglia
Wilson, Rick I., California Geological Survey
Witkowski, Christine, Middlesex Community College
Wunsch, David R., University of Delaware
Yiannakoulias, Niko, McMaster University
Youberg, Ann, Arizona Geological Survey
Young, Michael H., Univ of Texas at Austin, Jackson Sch of Geosciences
Young, William, University of Leeds
Yuen, Cheong-yip R., Argonne National Laboratory

Geomorphology

Aalto, Rolf E., University of Washington
Adams, Kenneth D., Desert Research Institute
Adams, Peter N., University of Florida
Alba, Saturnino, Univ Complutense de Madrid
Allan, Jonathan C., Oregon Dept of Geology and Mineral Industries
Allen, Bruce, New Mexico Institute of Mining & Technology
Allen, Phillip, Frostburg State University
Anders, Alison M., University of Illinois, Urbana-Champaign
Anderson, Robert S., University of California, Santa Cruz
Anderson, Robert, University of Colorado
Anderson, Suzanne P., University of Colorado
Antinao, JoseLuis, Desert Research Institute
Applegarth, Michael T., Shippensburg University
Armour, Jake, University of North Carolina, Charlotte
Aslan, Andres, Colorado Mesa University
Attal, Mikael, Edinburgh University
Bacon, Steven N., Desert Research Institute
Baker, Sophie, Desert Research Institute
Baker, Victor R., University of Arizona
Baptista, João C., Universidade de Trás-os-Montes e Alto Douro
Barendregt, Rene W., University of Lethbridge
Beget, James E., University of Alaska, Fairbanks
Belliveau, Lindsey, Dept of Energy and Environmental Protection
Benedetti, Michael M., University of North Carolina, Wilmington
Bergeron, Normand, Universite du Quebec
Berta, Susan, Indiana State University
Bierman, Paul R., University of Vermont
Blisniuk, Kimberly, San Jose State University
Bookhagen, Bodo, University of California, Santa Barbara
Booth, Adam M., Portland State University
Booth, Derek, University of Washington
Bradley, William C., University of Colorado
Breilin, Olli, Geological Survey of Finland
Brewer, Paul A., University of Wales
Brock-Hon, Amy, University of Tennessee, Chattanooga
Brocklehurst, Simon, University of Manchester
Brush, Nigel, Ashland University
Bull, William B., University of Arizona
Bullard, Thomas F., Desert Research Institute
Bullard, Tom, University of Nevada, Reno
Burbank, Doug, University of California, Santa Barbara
Burbank, Douglas W., University of California, Santa Barbara
Burke, Raymond M., Humboldt State University
Campbell, Ian A., University of Alberta
Carson, Robert J., Whitman College
Carton, Alberto, Università degli Studi di Padova
Carver, Gary A., Humboldt State University
Castillon, David A., Missouri State University
Chenoweth, M. S., University of Louisiana, Monroe
Chin, Anne, Texas A&M University
Clark, James A., Wheaton College
Clark, Jeffrey J., Lawrence University
Clark, Marin, University of Michigan
Coates, Donald R., Binghamton University
Colgan, Patrick M., Grand Valley State University
Collins, Brian, University of Washington
Collins, Charles W., University of Wisconsin, Platteville
Collins, Joe D., Middle Tennessee State University
Coltorti, Mauro, University of Siena
Connallon, Christopher B., Maryland Department of Natural Resources
Constantine, Jose A., Williams College
Cook, Joseph P., Arizona Geological Survey
Cooke, M. J., Austin Community College District
Cornwell, Kevin J., California State University, Sacramento
Cotter, Edward, Bucknell University
Crosby, Benjamin T., Idaho State University
Currey, Donald R., University of Utah
Dale, Janis, University of Regina
Daly, Julia F., University of Maine - Farmington
Daniels, J. Michael, University of Denver
Day, Stephanie S., North Dakota State University
DeJong, Benjamin, Norwich University
Dethier, David P., Williams College
DiBiase, Roman , Pennsylvania State University, University Park
Dietrich, William E., University of California, Berkeley
Dixon, Jean, Montana State University
Dixon, John C., University of Arkansas, Fayetteville
Dixon, Simon J., University of Birmingham
Dogwiler, Toby J., Illinois State University
Dolliver, Holly A., University of Wisconsin, River Falls
Dort, Jr., Wakefield, University of Kansas
Durbin, James, University of Southern Indiana
Duvall, Alison, University of Washington
Eaton, Lewis S., James Madison University
Edgar, Dorland E., Argonne National Laboratory
Ely, Lisa L., Central Washington University
England, John, University of Alberta
Evans, Diane L., Jet Propulsion Laboratory
Fagherazzi, Sergio, Boston University
Farr, Tom G., Jet Propulsion Laboratory
Ferrier, Ken, University of Wisconsin-Madison
Finnegan, Noah J., University of California, Santa Cruz
Fontana, Alessandro, Università degli Studi di Padova
Ford, Derek C., McMaster University
Ford, Richard L., Weber State University
Fubelli, Giandomenico, Università di Torino
Furbish, David J., Vanderbilt University
Gabet, Emmanuel, San Jose State University
Garcia, Antonio F., California Polytechnic State University
Garcia, Marcelo H., University of Illinois, Urbana-Champaign
Gardner, Thomas W., Trinity University
Garrote, Julio, Univ Complutense de Madrid
Gasparini, Nicole, Tulane University
Germanoski, Dru, Lafayette College
Ghoneim, Eman, Boston University
Giardino, John R., Texas A&M University
Giardino, Marco, Università di Torino
Gill, Thomas E., University of Texas, El Paso
Gillam, Mary L., Fort Lewis College
Gillespie, Alan R., University of Washington
Gonzalez, Juan L., University of Texas, Rio Grande Valley
Gootee, Brian F., Arizona Geological Survey
Gordon, Steven J., United States Air Force Academy
Gosse, John, University of Kansas
Gran, Karen B., University of Minnesota, Duluth
Greenberg, Harvey, University of Washington
Grimley, David A., Illinois State Geological Survey
Guccione, Margaret J., University of Arkansas, Fayetteville
Hales, TC, Cardiff University
Hallet, Bernard, University of Washington
Hancock, Gregory S., College of William & Mary
Hansen, Edward C., Hope College
Hanson, Lindley S., Salem State University
Hanson, Paul, Unversity of Nebraska - Lincoln
Harbor, David J., Washington & Lee University
Harbor, Jon M., Purdue University
Hardage, Sarah M., University of Texas, Rio Grande Valley
Harper, Stephen B., East Carolina University
Harrington, Charles D., Los Alamos National Laboratory
Harrison, Bruce I., New Mexico Institute of Mining and Technology
Hasbargen, Leslie E., SUNY, Oneonta
Heinzel, Chad E., University of Northern Iowa
Heitmuller, Franklin, University of Southern Mississippi
Hesp, Patrick A., Louisiana State University
Hesp, Patrick, Flinders University
Higgins, Charles G., University of California, Davis
Hooke, Roger L., University of Maine
Hooks, W. Gary, University of Alabama
Hopkins, Kenneth D., University of Northern Colorado
Howard, Alan D., University of Virginia
Hubbard, Trent, Alaska Div of Geological & Geophysical Surveys
Hungr, Oldrich, University of British Columbia
Hyatt, James A., Eastern Connecticut State University
Isacks, Bryan L., Cornell University
Jerolmack, Douglas, University of Pennsylvania
Jewell, Paul W., University of Utah
Jiron, Rebecca, College of William & Mary
Johnson, Sarah E., Northern Kentucky University
Kasse, Kees, Vrije Universiteit Amsterdam
Kavage-Adams, Rebecca H., Maryland Department of Natural Resources
Kehew, Alan E., Western Michigan University
Keller, Edward A., University of California, Santa Barbara
Kelsey, Harvey M., Humboldt State University

Kemmerly, Phillip R., Austin Peay State University
Kendrick, Katherine J., University of California, Riverside
Kenny, Ray, Fort Lewis College
Kirwan, Matthew L., College of William & Mary
Kiss, Timea, Univesity of Szeged
Kite, J. Steven, West Virginia University
Knott, Jeffrey R., California State University, Fullerton
Kochel, R. Craig, Bucknell University
Koehler, Rich, University of Nevada
Kori, E., University of Venda
Kowaleski, Douglas, University of Massachusetts, Amherst
Kowalewski, Douglas E., Worcester State University
Laabs, Benjamin J., SUNY, Geneseo
Lamb, Michael P., California Institute of Technology
Lancaster, Nick, University of Nevada, Reno
Larsen, Isaac J., University of Massachusetts, Amherst
Larson, Erik B., Shawnee State University
Lasca, Norman P., University of Wisconsin, Milwaukee
Lehre, Andre K., Humboldt State University
Lemke, Karen A., University of Wisconsin, Stevens Point
Leverington, David W., Texas Tech University
Levine, Rebekah, University of Montana Western
Li, Junran, The University of Tulsa
Lifton, Nathaniel A., Purdue University
Lininger, Katherine, University of Colorado
Linneman, Scott R., Western Washington University
Lips, Elliott W., University of Utah
Lisenby, Peyton E., Midwestern State University
Long, Ann D., University of Illinois, Urbana-Champaign
Loring, Arthur P., York College (CUNY)
MacGregor, Kelly, Macalester College
Madej, Mary Ann, U.S. Geological Survey
Manz, Lorraine A., North Dakota Geological Survey
Marchant, David R., Boston University
Marchetti, David W., Western Colorado University
Marshall, Jeffrey S., California State Polytechnic University, Pomona
Marshall, Jill A., University of Arkansas, Fayetteville
Martín, Cristina, Univ Complutense de Madrid
Martín Duque, José Francisco, Univ Complutense de Madrid
Mason, Joseph, University of Wisconsin, Madison
Mathers, Hannah, University of Glasgow
McCoy, Scott W., University of Nevada, Reno
McDowell, Patricia, University of Oregon
McKean, Adam, Utah Geological Survey
McKenney, Rosemary, Pacific Lutheran University
McKenzie, Garry D., Ohio State University
McMillan, Margaret E., University of Arkansas at Little Rock
Menking, Kirsten M., Vassar College
Merritts, Dorothy J., Franklin and Marshall College
Meyer, Grant A., University of New Mexico
Meyer, Jessica R., University of Iowa
Michaud, Yves, Universite du Quebec
Miller, Jerry R., Western Carolina University
Misner, Tamara, Edinboro University of Pennsylvania
Montgomery, David R., University of Washington
Moon, Seulgi, University of California, Los Angeles
Moore, Andrew, Earlham College
Moore, Laura J., University of North Carolina, Chapel Hill
Morgan, Daniel J., Vanderbilt University
Moscardelli, Lorena G., University of Texas, Austin
Mossa, Joann, University of Florida
Mount, Jeffrey F., University of California, Davis
Mozzi, Paolo, Università degli Studi di Padova
Murray, A. Bradshaw, Duke University
Musselman, Zachary A., Millsaps College
Mylroie, John E., Mississippi State University
Namikas, Steven, Louisiana State University
Napieralski, Jacob A., University of Michigan, Dearborn
Nash, David, University of Cincinnati
Nelson, Daren T., University of North Carolina, Pembroke
Nelson, Robert S., Illinois State University
Neubeck, William S., Union College
Newton, Andrew, Queen's University Belfast
Newton, Robert M., Smith College
Nichols, Kyle K., Skidmore College
Nielsen, Dennis N., Winona State University
Nikitina, Daria L., West Chester University
Nittrouer, Jeffrey A., Rice University
Okunade, Samuel, Central State University
Olsen, Paul E., Columbia University
ONeal, Michael, University of Delaware
Orme, Amalie, California State University, Northridge
Osborn, Gerald D., University of Calgary
Oskin, Michael, University of California, Davis
Ouimet, William, University of Connecticut
Oviatt, Charles G., Kansas State University
Owen, Lewis, University of Washington
Owen, Lewis, University of Cincinnati
Paradise, Thomas R., University of Arkansas, Fayetteville
Parker, Gary, University of Illinois, Urbana-Champaign
Pasternack, Gregory B., University of California, Davis
Patton, Peter C., Wesleyan University
Pavlowsky, Robert T., Missouri State University
Pazzaglia, Frank J., Lehigh University
Pearson, Adam, SUNY Potsdam
Pederson, Joel L., Utah State University
Pelletier, Jon D., University of Arizona
Perron, Taylor, Massachusetts Institute of Technology
Persico, Lyman P., Whitman College
Pierce, Jennifer L., Boise State University
Pieruccini, Pierluigi, Università di Torino
Pitlick, John, University of Colorado
Pizzuto, James E., University of Delaware
Plug, Lawrence, Dalhousie University
Polk, Jason, Western Kentucky University
Pope, Gregory A., Montclair State University
Potter, Jr., Noel, Dickinson College
Putkonen, Jaakko, University of North Dakota
Rayburn, John A., SUNY, New Paltz
Rech, Jason, Miami University
Reed, Denise J., University of New Orleans
Refsnider, Kurt, Prescott College
Reneau, Steven L., Los Alamos National Laboratory
Rhoads, Bruce, University of Illinois, Urbana-Champaign
Rhodes, Dallas D., Humboldt State University
Rice-Snow, R. Scott, Ball State University
Rittenour, Tammy M., Utah State University
Ritter, John B., Wittenberg University
Robinson, Cordula, Boston University
Rockwell, Thomas K., San Diego State University
Roering, Joshua J., University of Oregon
Rogerson, Robert J., University of Lethbridge
Roth, Danica, Colorado School of Mines
Savina, Mary E., Carleton College
Sawyer, Carol F., University of South Alabama
Schaller, Mirjam, University of Michigan
Schiefer, Erik, Northern Arizona University
Schmidt, Amanda H., Oberlin College
Schmutz, Phillip P., University of West Florida
Scuderi, Louis A., University of New Mexico
Shepherd, Stephanie L., Auburn University
Shreve, Ronald L., University of California, Los Angeles
Shroder, Jr., John F., University of Nebraska, Omaha
Shugar, Dan, University of Washington
Sklar, Leonard, San Francisco State University
Smith, Richard, University of Arkansas, Fayetteville
Snead, John I., Louisiana State University
Snyder, Jeffrey A., Bowling Green State University
Souch, Catherine J., Indiana University, Indianapolis
Spigel, Lindsay, Dept of Agriculture, Conservation, and Forestry
Springer, Gregory S., Ohio University
Springston, George E., Norwich University
Stanford, Scott D., New Jersey Geological and Water Survey
Stine, Scott W., California State University, East Bay
Stokes, Martin, University of Plymouth
Straffin, Eric, Edinboro University of Pennsylvania
Strasser, Jeffrey C., Augustana College
Stroup, Justin, SUNY, Oswego
Surian, Nicola, Università degli Studi di Padova
Sweeney, Mark R., University of South Dakota
Székely, Balázs, Eotvos Lorand University
Taylor, Stephen B., Western Oregon University
Tchakerian, Vatche P., Texas A&M University
Ten Brink, Norman W., Grand Valley State University

Theuerkauf, Ethan, Illinois State Geological Survey
Thomas, Paul, Western Washington University
Thorson, Robert M., University of Connecticut
Tidwell, David, Geological Survey of Alabama
Tomkin, Jonathan H., University of Illinois, Urbana-Champaign
Toy, Terrence J., University of Denver
Tranel, Lisa M., Illinois State University
Trenhaile, Alan S., University of Windsor
Tucker, Gregory E., University of Colorado
van der Merwe, Barend, University of Pretoria
van Dijk, Deanna, Calvin College
Vandeberg, Gregory S., University of North Dakota
Vidal Romaní, Juan Ramon, Coruna University
Vitek, John D., Oklahoma State University
Vitek, John D., Texas A&M University
Wakabayashi, John, California State University, Fresno
Walters, James C., University of Northern Iowa
Walther, Suzanne, University of San Diego
Ward, Dylan, University of Cincinnati
Warke, Patricia, Queen's University Belfast
Wayne, William J., University of Nebraska, Lincoln
Webb, Robert H., University of Arizona
Wegmann, Karl, North Carolina State University
Weirich, Frank H., University of Iowa
Werner, Bradley T., University of California, San Diego
West, Nicole, Central Michigan University
Whisner, Jennifer B., Bloomsburg University
White, Susan Q., La Trobe University
Whiting, Peter J., Case Western Reserve University
Whittecar, Jr., G. Richard, Old Dominion University
Wickert, Andrew D., University of Minnesota, Twin Cities
Wilcock, Peter W., Johns Hopkins University
Wilcox, Andrew, University of Montana
Willenbring, Jane, University of Pennsylvania
Williams, Harry F. L., University of North Texas
Williams, Kevin K., Buffalo State College
Wilson, Fred L., Angelo State University
Wilson, Greg C., Grand Valley State University
Wilson, Monte D., Boise State University
Wohl, Ellen E., Colorado State University
Wolken, Gabriel J., Alaska Div of Geological & Geophysical Surveys
Wood, Spencer H., Boise State University
Yeh, Joseph S., Univ of Texas at Austin, Jackson Sch of Geosciences
Young, Richard A., SUNY, Geneseo

Glaciology and Glacial Geology

Aber, James S., Emporia State University
Aizen, Vladimir, University of Idaho
Alley, Karen E., College of Wooster
Alley, Richard B., Pennsylvania State University, University Park
Andrews, John T., University of Colorado
Angle, Michael P., Ohio Dept of Natural Resources
Armstrong, William H., Appalachian State University
Arnaud, Emmanuelle, University of Guelph
Attig, John W., University of Wisconsin, Extension
Barclay, David J., SUNY, Cortland
Bartholomaus, Timothy, University of Idaho
Bassis, Jeremy, University of Michigan
Berg, Richard C., Illinois State Geological Survey
Berg, Richard, University of Illinois
Berthold, Angela, University of Minnesota
Bevis, Kenneth A., Hanover College
Bezada Diaz, Maximiliano, University of Minnesota
Bird, Brian, New York State Geological Survey
Birkel, Sean, University of Maine
Blake, Kevin, Eastern Michigan University
Blankenship, Donald D., University of Texas, Austin
Booth, Adam, Imperial College
Borns, Jr., Harold W., University of Maine
Breckenridge, Andy, University of Wisconsin, Superior
Brigham-Grette, Julie, University of Massachusetts, Amherst
Briner, Jason P., SUNY, Buffalo
Bromley, Gordon R., University of Maine
Broster, Bruce E., University of New Brunswick
Brown, Steven E., University of Illinois
Brown, Steven E., Illinois State Geological Survey
Brugger, Keith A., University of Minnesota, Morris
Campbell, Seth W., University of Maine

Carlson, Anders, Oregon State University
Carson, Eric C., University of Wisconsin, Extension
Catania, Ginny A., University of Texas, Austin
Christianson, Knut, University of Washington
Clague, John J., Simon Fraser University
Clark, Douglas H., Western Washington University
Clark, Peter U., Oregon State University
Clayton, Lee, University of Wisconsin, Extension
Clebnik, Sherman M., Eastern Connecticut State University
Conway, Howard B., University of Washington
Cotter, James F., University of Minnesota, Morris
Creyts, Timothy T., Columbia University
Crossen, Kristine J., University of Alaska, Anchorage
Csatho, Beata M., SUNY, Buffalo
Curry, B. B., University of Illinois
Das, Sarah B., Woods Hole Oceanographic Institution
Davis, P. Thompson, Bentley University
Denton, George H., University of Maine
Doughty, Alice, Bates College
Douglass, Daniel C., Northeastern University
Dupont, Todd, Miami University
Dupont, Todd, University of California, Irvine
Easterbrook, Don J., Western Washington University
Enderlin, Ellyn, Boise State University
Enderlin, Ellyn M., University of Maine
Evenson, Edward B., Lehigh University
Eyles, Nicholas, University of Toronto
Fahnestock, Mark A., University of New Hampshire
Fisher, David A., University of Ottawa
Fisher, Timothy G., University of Toledo
Fleisher, P. Jay, SUNY, Oneonta
Flowers, Gwenn, Simon Fraser University
Fountain, Andrew G., Portland State University
Franzi, David A., SUNY, Plattsburgh
Froese, Duane, University of Alberta
Fudge, T.J., University of Washington
Goldstein, Barry, University of Puget Sound
Gowan, Angela S., University of Minnesota
Hall, Brenda L., University of Maine
Ham, Nelson R., Saint Norbert College
Hambrey, Michael M., Aberystwyth University
Hamilton, Gordon S., University of Maine
Harper, Joel, University of Montana
Hawley, Bob, University of Washington
Hawley, Robert L., Dartmouth College
Headley, Rachel, University of Wisconsin, Parkside
Hicock, Stephen R., Western University
Hock, Regine M., University of Alaska, Fairbanks
Holt, John W., University of Texas, Austin
Hooke, Roger L., University of Minnesota, Twin Cities
Horton, Jennifer, University of Minnesota
Howat, Ian M., Ohio State University
Hughes, Terence J., University of Maine
Hughes III, Richard O., Crafton Hills College
Hulbe, Christina L., Portland State University
Humphrey, Neil F., University of Wyoming
Hutchings, Jennifer, Oregon State University
Iverson, Neal R., Iowa State University of Science & Technology
Jiskoot, Hester, University of Lethbridge
Johnson, Beth, University of Wisconsin Colleges
Johnson, James B., Colorado Mesa University
Joughin, Ian, University of Washington
Kaplan, Michael, Columbia University
Karrow, Paul, University of Waterloo
Kelly, Meredith, Dartmouth College
Kiver, Eugene P., Eastern Washington University
Knaeble, Alan, University of Minnesota
Koutnik, Michelle, University of Washington
Kozlowski, Andrew, New York State Geological Survey
Laabs, Benjamin J., North Dakota State University
Lamothe, Michel, Universite du Quebec a Montreal
Lampkin, Derrick, University of Maryland
Lea, Peter D., Bowdoin College
Lempe, Bernhard, Technical University of Munich
Leonard, Eric M., Colorado College
Levson, Victor M., University of Victoria
Levy, Joseph, Colgate University

Licciardi, Joseph M., University of New Hampshire
Licht, Kathy J., Indiana Univ / Purdue Univ, Indianapolis
Loope, Henry M., Indiana University
Lowell, Thomas V., University of Cincinnati
Lusardi, Barb, University of Minnesota
Malone, Andrew, University of Illinois at Chicago
Marcott, Shaun, University of Wisconsin-Madison
Matsuoka, Kenichi, University of Washington
McCarthy, Daniel P., Brock University
McKay III, E. Donald, Illinois State Geological Survey
Menzies, John, Brock University
Meyer, Gary, University of Minnesota
Mickelson, David M., University of Wisconsin-Madison
Miller, Gifford H., University of Colorado
Mills, Stephanie, Kingston University
Mode, William N., University of Wisconsin, Oshkosh
Mooers, Howard, University of Minnesota, Duluth
Morgan, Alan V., University of Waterloo
Munroe, Jeffrey S., Middlebury College
Muto, Atsuhiro, Temple University
Nash, Thomas A., Ohio Dept of Natural Resources
Neinow, Peter W., Edinburgh University
Neumann, Tom, University of Washington
Nguyen, Maurice, University of Minnesota
Pair, Donald, University of Dayton
Parent, Michel, Universite du Quebec
Poinar, Kristin, SUNY, Buffalo
Putnam, Aaron E., University of Maine
Quade, Deborah J., Iowa Dept of Natural Resources
Radic, Valentina, University of British Columbia
Rawling, J. E., University of Wisconsin, Extension
Raymond, Charles F., University of Washington
Retelle, Michael J., Bates College
Rexius, James E., Schoolcraft College
Ridge, John C., Tufts University
Rignot, Eric, University of California, Irvine
Rodbell, Donald T., Union College
Roy, Martin, Universite du Quebec a Montreal
Rutford, Robert H., University of Texas, Dallas
Rysgaard, Soren, University of Manitoba
Schneider, Allan F., University of Wisconsin, Parkside
Schoof, Christian, University of British Columbia
Schroeder, Dustin M., Stanford University
Schulz, Layne D., South Dakota Dept of Env and Nat Res
Sharp, Martin J., University of Alberta
Shaw, John, University of Alberta
Smith, Ben , University of Washington
Spencer, Matt, Lake Superior State University
Staley, Amie, University of Minnesota
Stansell, Nathan D., Northern Illinois University
Stearns, Leigh, University of Kansas
Stewart, Alexander K., St. Lawrence University
Stroeve, Julienne, University College London
Swanger, Kate, University of Massachusetts, Lowell
Syverson, Kent M., University of Wisconsin, Eau Claire
Szymanski, Jason, Monroe Community College
Taylor, Lawrence D., Albion College
Tedesco, Marco, Columbia University
Thackray, Glenn D., Idaho State University
Thomason, Jason, University of Illinois
Thompson, Lonnie G., Ohio State University
Thorleifson, Harvey, University of Minnesota
Todd, Claire E., Pacific Lutheran University
Totten, Stanley M., Hanover College
Tulaczyk, Slawek, University of California, Santa Cruz
Ullman, David J., Northland College
Waddington, Edwin D., University of Washington
Wagner, Kaleb, University of Minnesota
Wake, Cameron P., University of New Hampshire
Ward, Brent C., Simon Fraser University
Werner, Alan, Mount Holyoke College
Westgate, John A., University of Toronto
Wilch, Thomas I., Albion College
Wiles, Gregory C., College of Wooster
Winebrenner, Dale P., University of Washington
Wolff, Eric W., University of Cambridge
Wright, Stephen F., University of Vermont
Zoet, Lucas, University of Wisconsin, Madison
Zoet, Lucas, University of Wisconsin-Madison

Marine Geology

Abrams, Lewis J., University of North Carolina, Wilmington
Aksu, Ali E., Memorial University of Newfoundland
Allen, Katherine A., University of Maine
Aluwihare, Lihini I., University of California, San Diego
Anderson, John B., Rice University
Austin, Jr., James A., University of Texas, Austin
Bacchus, Tania S., Johnson State College
Bangs, Nathan L., University of Texas, Austin
Barclay, Andrew, University of Washington
Barrie, Vaughn, University of Victoria
Becel, Anne, Columbia University
Belknap, Daniel F., University of Maine
Benoit-Bird, Kelly, Oregon State University
Bentley, Samuel J., Louisiana State University
Brooks, Gregg R., Eckerd College
Brown, Kevin M., University of California, San Diego
Browne, Kathleen M., Rider University
Buffler, Richard T., University of Texas, Austin
Calmus, Thierry, Univ Nac Autonoma de Mexico
Carlin, Joe, California State University, Fullerton
Carson, Bobb, Lehigh University
Chase, Richard L., University of British Columbia
Cleary, William J., University of North Carolina, Wilmington
Coffroth, Mary Alice, SUNY, Buffalo
Conway, Flaxen D., Oregon State University
Cook, Ann, Ohio State University
Crone, Timothy, Columbia University
Cruz-Orozco, Rodolfo, Univ Autonoma de Baja California Sur
Curray, Joseph R., University of California, San Diego
Damuth, John E., University of Texas, Arlington
Daniell, James J., James Cook University
Das, Indrani, Columbia University
De Deckker, Patrick, Australian National University
Delaney, John R., University of Washington
Dix, Justin, University of Southampton
Dladla, Nonkululeko, University of KwaZulu-Natal
Dobson, David M., Guilford College
Donnelly, Jeffrey, Woods Hole Oceanographic Institution
Driscoll, Neal W., University of California, San Diego
Dugan, Brandon, Colorado School of Mines
Eberli, Gregor P., University of Miami
Edwards, Margo H., University of Hawai'i, Manoa
Flood, Roger D., Stony Brook University
Fornari, Daniel J., Woods Hole Oceanographic Institution
Freitas, Maria da Conceição P., Universidade de Lisboa
Fryer, Patricia B., University of Hawai'i, Manoa
Fulthorpe, Craig S., University of Texas, Austin
Gee, Jeffrey S., University of California, San Diego
Georgen, Jennifer, Old Dominion University
Georgen, Jennifer, Florida State University
Georgiopopoulou, Aggeliki, University College Dublin
Giosan, Liviu, Woods Hole Oceanographic Institution
Glickson, Deborah, National Acad of Sci, Eng, and Medicine
Goes, Joaquim, Columbia University
Goff, John A., University of Texas, Austin
Goni, Miguel A., Oregon State University
Gonzalez-Fernandez, Antonio, Centro de Inv Científica y de Ed Sup de Ensenada
Goodman-Tchernov, Beverly, McMaster University
Grabowski, Jonathan, Northeastern University
Green, Andrew, University of KwaZulu-Natal
Greene, H. Gary, Moss Landing Marine Laboratories
Grossman, Eric, Western Washington University
Guerin, Gilles, Columbia University
Haggerty, Janet A., The University of Tulsa
Hall, Ian R., Cardiff University
Harte, Michael, Oregon State University
Hine, Albert C., University of South Florida
Humphris, Susan E., Woods Hole Oceanographic Institution
Jacobson, Gary L., Grossmont College
Johnson, Harlan P., University of Washington
Johnson, Joel E., University of New Hampshire
Juraèiæ, Mladen, University of Zagreb
Kalangutkar, Niyati, Goa University

Kaminski, Michael A., King Fahd University of Petroleum and Minerals
Katuna, Michael P., College of Charleston
Kearney, Micheal S., University of Maryland
Keller, Randall A., Oregon State University
Kelley, Joseph T., University of Maine
Kidd, Karen, McMaster University
Kienast, Stephanie, Dalhousie University
Klasik, John A., California State Polytechnic University, Pomona
Klaus, Adam, Texas A&M University
Laine, Edward P., Bowdoin College
Lasker, Howard R., SUNY, Buffalo
Leonard, Lynn A., University of North Carolina, Wilmington
Lindo Atichati, David, College of Staten Island/CUNY
Liu, Jingpu P., North Carolina State University
Lonsdale, Peter F., University of California, San Diego
Luciano, Katherine E., South Carolina Dept of Natural Resources
MacLeod, C, Cardiff University
Malinverno, Alberto , Columbia University
Mallinson, David, East Carolina University
Manfrino, Carrie M., Kean University
Marinelli, Roberta, Oregon State University
Mayer, Larry A., University of New Hampshire
McCoy, Jr., Floyd W., University of Hawai'i, Manoa
Meadows, Guy A., Michigan Technological University
Mekik, Figen A., Grand Valley State University
Meylan, Maurice A., University of Southern Mississippi
Micallef, Aaron, University of Malta
Miller, Kenneth G., Rutgers, The State Univ of New Jersey
Miot da Silva, Graziela, Flinders University
Mitchell, Neil C., University of Manchester
Moberly, Ralph, University of Hawai'i, Manoa
Mountain, Gregory S., Rutgers, The State Univ of New Jersey
Moy, Christopher M., University of Otago
Naehr, Thomas H., Texas A&M University, Corpus Christi
Neitzke Adamo, Lauren, Rutgers, The State Univ of New Jersey
Neumann, A. Conrad, University of North Carolina, Chapel Hill
Nicholson, David, Woods Hole Oceanographic Institution
Nitsche, Frank, Columbia University
Nittrouer, Charles A., University of Washington
Norris, Richard D., University of California, San Diego
Oehlert, Amanda M., University of Miami
Olayiwola, M. A., Obafemi Awolowo University
Patton, Jason R., Humboldt State University
Peterson, Larry C., University of Miami
Phipps, James B., Grays Harbor College
Pietro, Kathryn R., Woods Hole Oceanographic Institution
Piper, David J., Dalhousie University
Pratt, Thomas, University of Washington
Purkis, Sam J., University of Miami
Ramirez, Wilson R., University of Puerto Rico
Raphael, Constantine N., Eastern Michigan University
Rasmussen, Kenneth, Northern Virginia Community College - Annandale
Rea, David K., University of Michigan
Reed, Donald L., San Jose State University
Richardson, Mary Jo, Texas A&M University
Richaud, Mathieu, California State University, Fresno
Riggs, Stanley R., East Carolina University
Rodolfo, Kelvin S., University of Illinois at Chicago
Ross, David A., Woods Hole Oceanographic Institution
Ruddiman, William F., University of Virginia
Schagrin, Zachery, Pennsylvania Bur of Topo & Geologic Survey
Scher, Howie, University of South Carolina
Schwartz, David, Cabrillo College
Scott, Steven D., University of Toronto
Skarke, Adam, Mississippi State University
Slagle, Angela, Columbia University
Slovinsky, Peter, Dept of Agriculture, Conservation, and Forestry
Tinto, Kirsteen, Columbia University
Tucholke, Brian E., Woods Hole Oceanographic Institution
Van Andel, Tjeerd H., Stanford University
Vannucchi, Paola, Royal Holloway University of London
Ward, Larry G., University of New Hampshire
Watts, Anthony B., University of Oxford
Wehmiller, John F., University of Delaware
Wheeler, Andy J., University College Cork
Wilcock, William S., University of Washington
Winguth, Cornelia, University of Texas, Arlington

Mineralogy & Crystallography

Abdu, Yassir, University of Manitoba
Ague, Jay J., Yale University
Akgul, Bunyamin, Firat University
Alderton, Dave, Royal Holloway University of London
Aldushin, Kirill, Ludwig-Maximilians-Universitaet Muenchen
Altaner, Stephen P., University of Illinois, Urbana-Champaign
Anderson, Callum, Nelson Mandela Metropolitan University
Apopei, Andrei I., Alexandru Ioan Cuza
Arletti, Rossella, Università di Torino
Artioli, Gilberto, Università degli Studi di Padova
Astilleros, José Manuel, Univ Complutense de Madrid
Ball, Neil, University of Manitoba
Bassett, William A., Cornell University
Beane, Rachel J., Bowdoin College
Belluso, Elena, Università di Torino
Benna, Piera, Università di Torino
Bermanec, Vladimir, University of Zagreb
Bish, David L., Indiana University, Bloomington
Bladh, Kenneth W., Wittenberg University
Blanchard, Frank N., University of Florida
Bloss, F. D., Virginia Polytechnic Institute & State University
Bonhoure, Jessica, Institut Polytechnique LaSalle Beauvais (ex-IGAL)
Boyle, Alan P., University of Liverpool
Boysen, Hans, Ludwig-Maximilians-Universitaet Muenchen
Braund, Emilie, University of Portsmouth
Brearley, Adrian J., University of New Mexico
Briansó, José Luis, Universitat Autonoma de Barcelona
Brown, Jr., Gordon E., Stanford University
Bruno, Emiliano, Università di Torino
Bruno, Marco, Università di Torino
Burnham, Charles, Fort Lewis College
Burnham, Charles W., Harvard University
Burns, Peter C., University of Notre Dame
Buzatu, Andrei, Alexandru Ioan Cuza
Buzgar, Nicolae, Alexandru Ioan Cuza
Cardellach López, Esteve, Universitat Autonoma de Barcelona
Carl, James D., SUNY Potsdam
Carrigan, Charles W., Olivet Nazarene University
Casas Duocastella, Lluís, Universitat Autonoma de Barcelona
Catlos, Elizabeth J., University of Texas, Austin
Cenpírek, Jan, Masaryk University
Chakhmouradian, Anton, University of Manitoba
Chen, Ning, University of Saskatchewan
Chiarenzelli, Jeffrey R., St. Lawrence University
Chizmadia, Lysa, University of Puerto Rico
Choi, Seon-Gyu, Korea University
Cilliers, Johannes, Imperial College
Clement, Stephen C., College of William & Mary
Cody, Robert, Iowa State University of Science & Technology
Cofer, Harland E., Georgia Southwestern State University
Coker, Victoria, University of Manchester
Cole, Kevin C., Grand Valley State University
Coleman, Leslie S., University of Saskatchewan
Collins, Lorence G., California State University, Northridge
Cooper, Mark, University of Manitoba
Copjakova, Renata, Masaryk University
Corbella Cordomí, Mercè, Universitat Autonoma de Barcelona
Cordonnier, Benoit, Ludwig-Maximilians-Universitaet Muenchen
Costa, Emanuele, Università di Torino
Cox, Richard, Dalhousie University
Cusack, Maggie, University of Glasgow
Dalconi, Maria C., Università degli Studi di Padova
Darling, Robert S., SUNY, Cortland
De Campos, Christina, Ludwig-Maximilians-Universitaet Muenchen
De Nault, Kenneth J., University of Northern Iowa
De Pablo, Liberto, Univ Nac Autonoma de Mexico
Dempster, Tim, University of Glasgow
Deng, Youjun, Texas A&M University
Denicourt, Raymond, University of South Florida, Tampa
Denny, F. B., Illinois State Geological Survey
Dobson, David, University College London
Dollase, Wayne A., University of California, Los Angeles
Dorfman, Alexander, Ludwig-Maximilians-Universitaet Muenchen
Dorfner, Thomas, Ludwig-Maximilians-Universitaet Muenchen
Downs, James W., Ohio State University
Duchesne, Josee, Universite Laval

Dunlevey, John, University of Limpopo
Dutrow, Barbara L., Louisiana State University
Elliott, Brent, University of Texas, Austin
Elsen, Jan, Katholieke Universiteit Leuven
Emofurieta, Williams O., University of Benin
Enders, Merritt S., Colorado School of Mines
Ercit, T. Scott, Carleton University
Ertel-Ingrisch, Werner, Ludwig-Maximilians-Universitaet Muenchen
Ewing, Rodney C., Stanford University
Ewing, Rodney E., University of New Mexico
Farnan, Ian, University of Cambridge
Farthing, Dori J., SUNY, Geneseo
Farzaneh, Akbar, University of Tabriz
Feather, Russell, Smithsonian Inst / Natl Museum of Natural History
Feely, Martin, National University of Ireland Galway
Fehr, Karl Thomas, Ludwig-Maximilians-Universitaet Muenchen
Feinglos, Mark N., Duke University
Fendrich, Kim V., American Museum of Natural History
Ferguson, Robert B., University of Manitoba
Fernández, Lourdes, Univ Complutense de Madrid
Fernández-Barrenechea, José María, Univ Complutense de Madrid
Ferraris, Giovanni, Università di Torino
Field, Stephen W., Tarleton State University
Finch, Adriam, University of St. Andrews
Flemming, Roberta L., Western University
Foit, Jr., Franklin F., Washington State University
Fransolet, Andre-Mathieu, Universite de Liege
Freed, Robert L., Trinity University
Frenette, Jean, Universite Laval
Frey, Friedrich, Ludwig-Maximilians-Universitaet Muenchen
Frueh, Alfred J., University of Connecticut
Gadikota, Greeshma, University of Wisconsin, Madison
Gait, Robert I., Royal Ontario Museum
Gann, Delbert E., Millsaps College
Garapic, Gordana, SUNY, New Paltz
García Romero, Emilia, Univ Complutense de Madrid
García-Martínez, David, Universidad Estatal de Sonora
Garvie, Laurence, Arizona State University
Gay, Kenny, North Carolina Geological Survey
Gibbs, Jerry, Virginia Polytechnic Institute & State University
Giere, Reto, University of Pennsylvania
Giese, Rossman F., SUNY, Buffalo
Gigler, Alexander, Ludwig-Maximilians-Universitaet Muenchen
Gille, Peter, Ludwig-Maximilians-Universitaet Muenchen
Gius, Fred, California Geological Survey
Giustetto, Roberto, Università di Torino
Gollmer, Steven, Cedarville University
González, Isabel , Universidad de Sevilla
Grant, Shelton K., Missouri University of Science and Technology
Gray, Keith, SUNY, Plattsburgh
Grießl, Stefan, Ludwig-Maximilians-Universitaet Muenchen
Griffen, Dana T., Brigham Young University
Groat, Lee A., University of British Columbia
Grover, John E., University of Cincinnati
Guggenheim, Stephen J., University of Illinois at Chicago
Guillemette, Renald, Texas A&M University
Gunter, Mickey E., University of Idaho
Göttlich, Hagen, Ludwig-Maximilians-Universitaet Muenchen
Hadler, Kathryn, Imperial College
Hall, Anne M., Emory University
Hansen, Lars, University of Oxford
Harlow, George E., Grad Sch of the City Univ of New York
Harlow, George E., American Museum of Natural History
Harrison, Richard, University of Cambridge
Hawkins, David, Wellesley College
Hawkins, Michael A., New York State Geological Survey
Hawthorne, Frank C., University of Manitoba
Hazen, Robert M., Carnegie Institution for Science
Hazen, Robert M., George Mason University
Heaney, Peter J., Pennsylvania State University, University Park
Heckel, Wolfgang, Ludwig-Maximilians-Universitaet Muenchen
Heimann, Adriana, East Carolina University
Henderson, Grant S., University of Toronto
Hennemeyer, Marc, Ludwig-Maximilians-Universitaet Muenchen
Hess, Kai-Uwe, Ludwig-Maximilians-Universitaet Muenchen
Heuss-Aßbichler, Soraya, Ludwig-Maximilians-Universitaet Muenchen
Hogarth, Donald D., University of Ottawa

Hollabaugh, Curtis L., University of West Georgia
Hooper, Robert L., University of Wisconsin, Eau Claire
Horton, Duane G., Pacific Northwest National Laboratory
Hovis, Guy L., Lafayette College
Huff, Warren D., University of Cincinnati
Hughes, John M., Miami University
Hughes, John M., University of Vermont
Hummer, Daniel R., Southern Illinois University Carbondale
Izawa, Matt, Western University
Joesten, Raymond, University of Connecticut
Johanesen, Katharine, Juniata College
Johnson, Neil E., Virginia Polytechnic Institute & State University
Jonas, Edward C., University of Texas, Austin
Jones, Rhian, University of New Mexico
Jordan, Guntram, Ludwig-Maximilians-Universitaet Muenchen
Kaiser, Jason, Southern Utah University
Kaiser-Bischoff, Ines, Ludwig-Maximilians-Universitaet Muenchen
Kamhi, Samuel R., Long Island University, Brooklyn Campus
Kampf, Anthony R., Los Angeles County Museum of Natural History
Kearns, Lance E., James Madison University
Keller, Dianne M., Colgate University
Kerestedjian, Thomas N., Bulgarian Academy of Sciences
Kleebe, Hans-Joachim, Technische Universitaet Darmstadt
Komabayashi, Tetsuya, Edinburgh University
Krol, Michael A., Bridgewater State University
Kunzmann, Thomas, Ludwig-Maximilians-Universitaet Muenchen
Lackinger, Markus, Ludwig-Maximilians-Universitaet Muenchen
Lancaster, Penny, University of Portsmouth
Langenhorst, Falko H., Friedrich-Schiller-University Jena
Langer, Arthur M., Grad Sch of the City Univ of New York
Lapen, Thomas, University of Houston
Leavens, Peter B., University of Delaware
Lee, Martin, University of Glasgow
Leimer, H. Wayne, Tennessee Tech University
Leitz, Robert E., Metropolitan State College of Denver
Leung, Irene S., Lehman College (CUNY)
Leung, Irene S., Grad Sch of the City Univ of New York
Levesque, Andre, Universite Laval
Levine, Jamie S., Appalachian State University
Lian, Jie, University of Michigan
Liebling, Richard S., Grad Sch of the City Univ of New York
López, Sol, Univ Complutense de Madrid
Losh, Steven, Minnesota State University
Losos, Zdenek, Masaryk University
Lueth, Virgil L., New Mexico Institute of Mining and Technology
Lupulescu, Marian V., New York State Geological Survey
Luque, Francisco Javier, Univ Complutense de Madrid
Lübbe, Maike, Ludwig-Maximilians-Universitaet Muenchen
Majzlan, Juraj, Friedrich-Schiller-University Jena
Manu, Johnson, University of Ghana
Marshall, Craig, University of Kansas
Martin, Robert F., McGill University
Masch, Ludwig, Ludwig-Maximilians-Universitaet Muenchen
Mason, Roger A., Memorial University of Newfoundland
McCausland, Phil, Brock University
McCormack, John K., University of Nevada, Reno
McCormick, George R., University of Iowa
McCurry, Michael O., Idaho State University
McDonald, Andrew M., Laurentian University, Sudbury
McHenry, Lindsay J., University of Wisconsin, Milwaukee
McNamee, Brittani D., University of North Carolina, Asheville
Meisteremst, Götz, Ludwig-Maximilians-Universitaet Muenchen
Merkel, Timo Casjen, Ludwig-Maximilians-Universitaet Muenchen
Miles, Andrew, Kingston University
Miyagi, Lowell, University of Utah
Moritz, Wolfgang, Ludwig-Maximilians-Universitaet Muenchen
Müller, Lena, Ludwig-Maximilians-Universitaet Muenchen
Müller-Sohnius, Dieter, Ludwig-Maximilians-Universitaet Muenchen
Navarro, Didac, Universitat Autonoma de Barcelona
Neethling, Stephen, Imperial College
Nestola, Fabrizio, Università degli Studi di Padova
Nord, Julia, George Mason University
Novak, Milan , Masaryk University
Nusbaum, Robert L., College of Charleston
Oganov, Artem, Stony Brook University
Paktunc, Dogan, University of Ottawa
Paquette, Jeanne, McGill University

Faculty by Specialty

Parise, John B., Stony Brook University
Park, Sohyun, Ludwig-Maximilians-Universitaet Muenchen
Pascoe, Richard, Exeter University
Pavese, Alessandro, Università di Torino
Peacor, Donald R., University of Michigan
Pentcheva, Rossitza, Ludwig-Maximilians-Universitaet Muenchen
Peterson, Ronald C., Queen's University
Phillips, Brian L., Stony Brook University
Phillips, William R., Brigham Young University
Piña, Rubén, Univ Complutense de Madrid
Piniella Febrer, Juan Francesc, Universitat Autonoma de Barcelona
Post, Jeffrey E., Smithsonian Inst / Natl Museum of Natural History
Potter, Lee S., University of Northern Iowa
Prencipe, Mauro, Università di Torino
Radcliffe, Dennis, Hofstra University
Rakovan, John F., Miami University
Raudsepp, Mati, University of British Columbia
Redfern, Simon, University of Cambridge
Reinthal, William A., Ashland University
Reuss, Robert L., Tufts University
Ribbe, Paul H., Virginia Polytechnic Institute & State University
Rickaby, Ros, University of Oxford
Riedel, Oliver, Ludwig-Maximilians-Universitaet Muenchen
Rigault de la Longrais, Germano, Università di Torino
Robinson, George W., St. Lawrence University
Robinson, Paul D., Southern Illinois University Carbondale
Rocholl, Alexander, Ludwig-Maximilians-Universitaet Muenchen
Rodriguez-Blanco, Juan Diego, Trinity College
Rojas-Soriano, Humberto, Univ Autonoma de Baja California Sur
Rosenberg, Philip E., Washington State University
Ross, Nancy L., Virginia Polytechnic Institute & State University
Rossman, George R., California Institute of Technology
Rouse, Roland C., University of Michigan
Rubio-Sierra, Javier, Ludwig-Maximilians-Universitaet Muenchen
Rutstein, Martin S., SUNY, New Paltz
Salviulo, Gabriella, Università degli Studi di Padova
Sánchez, Nuria, Univ Complutense de Madrid
Schindler, Michael, Laurentian University, Sudbury
Schmahl, Wolfgang, Ludwig-Maximilians-Universitaet Muenchen
Schneider, Julius, Ludwig-Maximilians-Universitaet Muenchen
Schroeder, Paul A., University of Georgia
Scott Smith, Barbara H., University of British Columbia
Secco, Luciano, Università degli Studi di Padova
Sethi, Parvinder S., Radford University
Shank, Stephen G., Pennsylvania Bur of Topo & Geologic Survey
Sharp, Thomas G., Arizona State University
Sharp, W. Edwin, University of South Carolina
Sheets, Julie, Ohio State University
Shen, Guoyin, Carnegie Institution for Science
Sherriff, Barbara L., University of Manitoba
Shim, Sang-Heon, Arizona State University
Shuller-Nickles, Lindsay C., Clemson University
Silvestri, Alberta, Università degli Studi di Padova
Simmons, William B., University of New Orleans
Simmons, William B., University of Michigan
Singer, David M., Kent State University
Skoda, Radek, Masaryk University
Smith, Michael S., University of North Carolina, Wilmington
Smyth, Joseph R., University of Colorado
Snow, Eleanour, University of South Florida, Tampa
Sokolova, Elena, University of Manitoba
Solorio-Munguia, Jose G., Univ Nac Autonoma de Mexico
Spieler, Oliver, Ludwig-Maximilians-Universitaet Muenchen
Spilde, Michael N., University of New Mexico
Stark, Robert, Ludwig-Maximilians-Universitaet Muenchen
Stefano, Christopher J., Michigan Technological University
Stewart, Dion C., Georgia State Univ, Perimeter College, Alpharetta
Strasser, Stefan, Ludwig-Maximilians-Universitaet Muenchen
Tait, Kimberly T., Royal Ontario Museum
Tettenhorst, Rodney T., Ohio State University
Thibault, Yves, Western University
Tibljaš, Darko, University of Zagreb
Tomašiæ, Nenad, University of Zagreb
Trixler, Frank, Ludwig-Maximilians-Universitaet Muenchen
Uzochukwu, Godfrey A., North Carolina Agricultural & Tech State Univ
Vassiliou, Andreas H., Rutgers, The State Univ of New Jersey, Newark
Vaughan, David, University of Manchester
Viedma, Cristóbal, Univ Complutense de Madrid
Vindel, Elena, Univ Complutense de Madrid
Viti, Cecilia, University of Siena
Walker, Jeffrey R., Vassar College
Walther, Ferdinand, Ludwig-Maximilians-Universitaet Muenchen
Warner, Richard D., Clemson University
Webb, Christine R., Smithsonian Inst / Natl Museum of Natural History
Weinbruch, Stephan, Technische Universitaet Darmstadt
Wenk, Hans-Rudolf, University of California, Berkeley
Wicks, Frederick J., Royal Ontario Museum
Wicks, Frederick J., University of Toronto
Wilson, James R., Weber State University
Winchell, Robert E., California State University, Long Beach
Wise, Michael A., Smithsonian Inst / Natl Museum of Natural History
Wise, Michael A., University of New Orleans
Wood, Bernard, University of Oxford
Wood, Ian, University College London
Wylie, Ann G., University of Maryland
Xu, Huifang, University of Wisconsin-Madison
Yoshiasa, Akira, Kumamoto University
Yurtsever, Ayhan, Ludwig-Maximilians-Universitaet Muenchen
Zimmerman, Brian S., Edinboro University of Pennsylvania
Zink, Albert, Ludwig-Maximilians-Universitaet Muenchen
Zurevinski, Shannon, Lakehead University
Šeavnièar, Stjepan, University of Zagreb
Žigoveèki Gobac, Željka, University of Zagreb

Paleolimnology
Anderson, Roger Y., University of New Mexico
Axford, Yarrow L., Northwestern University
Brady, Kristina, University of Minnesota, Twin Cities
Brant, Lynn A., University of Northern Iowa
Brenner, Mark, University of Florida
Buchheim, Paul, Loma Linda University
Davies, Caroline P., University of Missouri, Kansas City
De Batist, Marc, Ghent University
Ferris, Grant F., University of Toronto
Gennari, Rocco, Università di Torino
Glaser, Paul H., University of Minnesota, Twin Cities
Heyvaert, Alan, University of Nevada, Reno
Hillman, Aubrey, University of Louisiana at Lafayette
Kaplan, Samantha W., University of Wisconsin, Stevens Point
Kirby, Matthew E., California State University, Fullerton
Lisiecki, Lorraine, University of California, Santa Barbara
McGlue, Michael M., University of Kentucky
Noren, Anders, University of Minnesota, Twin Cities
Porter, Susannah, University of California, Santa Barbara
Rosen, Michael R., University of Nevada, Reno
Russell, James M., Brown University
Shapley, Mark, University of Minnesota, Twin Cities
Smith, Alison J., Kent State University
Steinman, Byron A., University of Minnesota, Duluth
Stevens (Landon), Lora R., California State University, Long Beach
Stone, Alexander, University of Minnesota, Twin Cities
Theissen, Kevin, University of Saint Thomas
Waters, Matthew N., Auburn University

Petroleum Geology
Abdulghani, Waleed, King Fahd University of Petroleum and Minerals
Adeigbe, O. C., University of Ibadan
Adekeye, J.I. D., University of Ilorin
Ajaegwu, Norbert E., Nnamdi Azikiwe University
Al-Ramadan, Khalid, King Fahd University of Petroleum and Minerals
Ali, Hendratta N., Fort Hays State University
Alves, Tiago, Cardiff University
Andrews, Richard D., University of Oklahoma
Anthony, Robin, Pennsylvania Bur of Topo & Geologic Survey
Aplin, Andrew, Durham University
Arnold, Dan, Heriot-Watt University
Asquith, George B., Texas Tech University
Avary, Katherine L., West Virginia University
Bartlett, Wendy, Marietta College
Bartok, Peter, University of Houston
Bellian, Jerome, Miami University
Bend, Stephen L., University of Regina
Benjamin, U. K., Obafemi Awolowo University
Bissell, Clinton R., Texas A&M University, Corpus Christi
Boboye, O. A., University of Ibadan

Bonk, Kathleen, New York State Geological Survey
Bowersox, J. Richard, University of Kentucky
Bowersox, John R., University of Kentucky
Bowersox, Rick, University of Kentucky
Bowler, Bernard, University of Newcastle Upon Tyne
Bowman, Mike, University of Manchester
Broadhead, Ronald F., New Mexico Institute of Mining and Technology
Broadhead, Ronald F., New Mexico Institute of Mining & Technology
Burton, Bradford R., Western Colorado University
Carr, Timothy R., University of Kansas
Carr, Timothy R., West Virginia University
Carter, Kristin M., Pennsylvania Bur of Topo & Geologic Survey
Cassidy, Martin, University of Houston
Champlin, Steven D., Mississippi Office of Geology
Cheadle, Burns A., Western University
Chelariu, Ciprian, Alexandru Ioan Cuza
Chenoweth, Cheri, Illinois State Geological Survey
Cochrane, Claudia, Western University
Cranganu, Constantin, Brooklyn College (CUNY)
Curtis, John B., Colorado School of Mines
Dallegge, Todd A., University of Northern Colorado
Denne, Richard A., Texas Christian University
Doveton, John H., University of Kansas
Echols, John B., Louisiana State University
Edegbai, Aitalokhai J., University of Benin
Ehinola, Olugbenga A., University of Ibadan
Ehinola, Olugbenga A., University of Ibadan
Eide, Elizabeth A., National Acad of Sci, Eng, and Medicine
Elnadi, Abdelhalim H., University of Khartoum
Fadiya, S. L., Obafemi Awolowo University
Fakhari, Mohammad D., Ohio Dept of Natural Resources
Finley, Robert J., University of Illinois, Urbana-Champaign
Folorunso, I. O., University of Ilorin
Forgotson, Jr., James M., University of Oklahoma
Fraser, Alastair, Imperial College
Giles, Katherine A., University of Texas, El Paso
Gillespie, Janice, California State University, Bakersfield
Gingras, Murray, University of Alberta
Graham, Stephan A., Stanford University
Grender, Gordon C., Virginia Polytechnic Institute & State University
Hanks, Catherine L., University of Alaska, Fairbanks
Harris, David C., University of Kentucky
Harun, Nina, Alaska Div of Geological & Geophysical Surveys
Haszeldine, R. S., Edinburgh University
Henk, Jr., Floyd, Texas Christian University
Herrmann, Leo A., Louisiana Tech University
Hileman, Mary E., Oklahoma State University
Hofmann, Michael, University of Montana
Hu, Qinhong (Max), University of Texas, Arlington
Hudson, Samuel M., Brigham Young University
Hughes, Colin, University of Manchester
Igbinigie, Nosa S., University of Benin
Jiang, Shu, University of Utah
John, Chacko J., Louisiana State University
Johnson, Howard, Imperial College
Jones, Bobby L., Louisiana State University
Kavousi Ghahfarohki, Payam, West Virginia University
Keach, William, University of Utah
Knepp, Rex A., Illinois State Geological Survey
Leetaru, Hannes E., Illinois State Geological Survey
Leetaru, Hannes E., University of Illinois, Urbana-Champaign
Leetaru, Hannes, University of Illinois
Lever, Helen, Heriot-Watt University
Levino, Lynn J., Pennsylvania Bur of Topo & Geologic Survey
Li, Peng, Arkansas Geological Survey
Loucks, Bob, Univ of Texas at Austin, Jackson Sch of Geosciences
Lucia, F. Jerry, Univ of Texas at Austin, Jackson Sch of Geosciences
Machel, Hans G., University of Alberta
Malinverno, Alberto, Columbia University
Manzocchi, Tom, University College Dublin
Mazzullo, Salvatore J., Wichita State University
Mc Mahon, Michelle, Lonestar College - North Harris
Meddaugh, W. S., Midwestern State University
Medina, Cristian R., Indiana University
Miller, Michael B., Louisiana State University
Minzoni, Marcello, University of Alabama
Neufeld, Jerome A., University of Cambridge

Newell, Kerry D., University of Kansas
Nixon, R. Paul, Brigham Young University
Nordeng, Stephan, University of North Dakota
North, F. K., Carleton University
Nton, M. E., University of Ibadan
Nunez-Velazco, Miriam, Univ Autonoma de Baja California Sur
Nwachukwu, J. I., Obafemi Awolowo University
Oduneye, Olusola C., Ladoke Akintola University of Technology
Ogungbesan, Gbenga O., Ladoke Akintola University of Technology
Ola, P S., Federal University of Technology, Akure
Olarewaju, V. O., Obafemi Awolowo University
Olorunfemi, A. O., Obafemi Awolowo University
Osman, Mutasim S., King Fahd University of Petroleum and Minerals
Panah, Assad I., University of Pittsburgh, Bradford
Pashin, Jack C., Mississippi State University
Pena dos Reis, Rui, Universidade de Coimbra
Pertl, David, Amarillo College
Pigott, John D., University of Oklahoma
Pisel, Jesse, Wyoming State Geological Survey
Pranter, Matthew J., University of Oklahoma
Pritchett, Brittany, University of Oklahoma
Puckette, James, Oklahoma State University
Ratchford, M. E., Arkansas Geological Survey
Resnic, Victor S., Lonestar College - North Harris
Richards-McClung, Bryony, University of Utah
Riess, Carolyn M., Austin Community College District
Ronck, Catherine, Tarleton State University
Rupp, John A., Indiana University
Scarselli, Nicola, Royal Holloway University of London
Schmid, Katherine W., Pennsylvania Bur of Topo & Geologic Survey
Schneider, Robert V., Texas A&M University, Kingsville
Schroeder, Stefan, University of Manchester
Sharman, Glenn R., University of Arkansas, Fayetteville
Shumaker, Robert C., West Virginia University
Simpson, Altus, Cypress College
Slater, Brian, New York State Geological Survey
Slatt, Roger M., University of Oklahoma
Smith, Langhorne, New York State Geological Survey
Soeder, Daniel J., South Dakota School of Mines & Technology
Sonnenberg, Stephen A., Colorado School of Mines
Sprinkel, Douglas A., Utah Geological Survey
Stewart, Gary, Oklahoma State University
Stright, Lisa, Colorado State University
Sullivan, Raymond, San Francisco State University
Sun, Yuefeng, Texas A&M University
Tan, Denis N., University of Malaya
Taylor, Kevin, University of Manchester
Temples, Tommy, Clemson University
Tinker, Scott W., University of Texas, Austin
Tinker, Scott W., Univ of Texas at Austin, Jackson Sch of Geosciences
Toner, Rachel, Wyoming State Geological Survey
Trentham, Robert C., University of Texas, Permian Basin
Van Kooten, Gerald K., Calvin College
Vautier, Yannick, Institut Polytechnique LaSalle Beauvais (ex-IGAL)
Wach, Grant D., Dalhousie University
Wardlaw, Norman C., University of Calgary
Weimer, Paul, University of Colorado
White, Bekki C., Arkansas Geological Survey
Whiteley, Martin, University of Derby
Woodward, Mac B., Arkansas Geological Survey
Xie, Xiangyang, Texas Christian University
Zimmerman, Ronald K., Louisiana State University
Zonneveld, John-Paul, University of Alberta

General Petrology
Afolabi, Olukayode A., Ladoke Akintola University of Technology
Ague, Jay J., Yale University
Aird, Hannah M., California State University, Chico
Anderson, Alan J., Saint Francis Xavier University
Arslan, Gizem, Firat University
Ashwal, Lewis D., University of the Witwatersrand
Balen, Dražen, University of Zagreb
Barr, Sandra, Dalhousie University
Basu, Asish, University of Texas, Arlington
Bickel, Charles E., San Francisco State University
Biševac, Vanja, University of Zagreb
Blackburn, William H., University of Kentucky
Bonamici, Chloe, University of Wisconsin-Madison

Faculty by Specialty

Borghi, Alessandro, Università di Torino
Botcharnikov, Roman, Universität Mainz
Brady, John B., Smith College
Brandriss, Mark E., Smith College
Brown, Michael, University of Maryland
BrownLee, Sarah J., Wayne State University
Broxton, David E., Los Alamos National Laboratory
Cain, J. Allan, University of Rhode Island
Callegari, Ezio, Università di Torino
Carmichael, Sarah, Appalachian State University
Castelli, Daniele, Università di Torino
Chaumba, Jeff B., University of North Carolina, Pembroke
Cherniak, Daniele J., Rensselaer Polytechnic Institute
Chu, Xu, University of Toronto
Compagnoni, Roberto, Università di Torino
Compton, Robert R., Stanford University
Cordell, Ann S., University of Florida
Corral, Isaac, Universitat Autonoma de Barcelona
Crawford, Maria Luisa B., Bryn Mawr College
Dasgupta, Rajdeep, Rice University
de Hoog, Jan C., Edinburgh University
DeBari, Susan M., Western Washington University
Deininger, Robert W., University of Memphis
Dobrzhinetskaya, Larissa F., University of California, Riverside
Doolan, Barry L., University of Vermont
Eklund, Olav, Abo Akademi University
Elliot, David H., Ohio State University
Encarnacion, John, Saint Louis University
Erickson, Rolfe C., Sonoma State University
Ferrando, Simona, Università di Torino
Ferreiro Maehlmann, Rafael, Technische Universitaet Darmstadt
Finn, Gregory C., Brock University
Ford, Mark T., Texas A&M University, Kingsville
Frank, Charles O., Southern Illinois University Carbondale
Ganguly, Jiba, University of Arizona
Goble, Ronald G., University of Nebraska, Lincoln
Gonzales, David A., Fort Lewis College
Graham, Ian, University of New South Wales
Gray, Norman H., University of Connecticut
Groppo, Chiara Teresa, Università di Torino
Hacker, Bradley R., University of California, Santa Barbara
Haileab, Bereket, Carleton College
Hanson, Richard E., Texas Christian University
Hatheway, Richard B., SUNY, Geneseo
Hayob, Jodie, University of Mary Washington
Hetherington, Callum J., Texas Tech University
Hollocher, Kurt T., Union College
Hutchison, David, Hartwick College
Ihinger, Phillip D., University of Wisconsin, Eau Claire
Jercinovic, Michael J., University of Massachusetts, Amherst
Jimoh, Mustapha T., Ladoke Akintola University of Technology
Johnson, Julie, University of Memphis
Jolliff, Bradley L., Washington University in St. Louis
Kerr, A C., Cardiff University
Keskinen, Mary J., University of Alaska, Fairbanks
Kilic, Ayse Didem, Firat University
Leichmann, Jaromir, Masaryk University
Lewy, Robert, Bakersfield College
Malcuit, Robert J., Denison University
Mallik, Ananya, University of Rhode Island
Manning, Craig E., University of California, Los Angeles
Maritan, Lara, Università degli Studi di Padova
Martin, Barton S., Ohio Wesleyan University
Mathez, Edmond A., American Museum of Natural History
Mazzoli, Claudio, Università degli Studi di Padova
McFadden, Rory, Gustavus Adolphus College
McLeod, Claire, Miami University
Mirnejad, Hassan, Miami University
Moecher, David P., University of Kentucky
Moore, Daniel K., Brigham Young University - Idaho
Moore, Gordon M., University of California, Davis
Murphy, J. Brendan, Dalhousie University
Mutti, Laurence J., Juniata College
Nadeau, Olivier, University of Ottawa
Nasir, Sobhi J., University of Qatar
Nesse, William D., University of Northern Colorado
Nielsen, Peter A., Keene State College

Noblett, Jeffrey B., Colorado College
Owens, Brent E., College of William & Mary
Ozel Yildirim, Esra , Firat University
Pan, Yuanming, University of Saskatchewan
Peterson, Virginia L., Grand Valley State University
Plank, Terry A., Columbia University
Plymate, Thomas G., Missouri State University
Prichard, H M., Cardiff University
Ptacek, Anton D., San Diego State University
Purchase, Megan, University of the Free State
Purdom, William B., Southern Oregon University
Ranson, William A., Furman University
Reed, Robert M., Univ of Texas at Austin, Jackson Sch of Geosciences
Reyes-Cortes, Ignacio A., Universidad Autonoma de Chihuahua
Rizeli, Mustafa Eren, Firat University
Roadifer, Jack E., Colorado Mesa University
Rolfo, Franco, Università di Torino
Rougvie, James R., Beloit College
Severs, Matthew R., Stockton University
Sharma, Shiv K., University of Hawai'i, Manoa
Shoemaker, Kurt A., Shawnee State University
Snyder, Lori D., University of Wisconsin, Eau Claire
Spandler, Carl, James Cook University
Stelling, Pete, Western Washington University
Thy, Peter, University of California, Davis
Torres-Aguilera, Juan Manuel, Universidad Autonoma de San Luis Potosi
Van Buer, Nicholas J., California State Polytechnic University, Pomona
van Kranendonk, Martin, University of New South Wales
Van Rythoven, Adrian, Bloomsburg University
Vaniman, David T., Los Alamos National Laboratory
Vanko, David A., Towson University
Vitale Brovarone, Alberto, Università di Torino
Walker, David, Columbia University
Warren, Jessica, University of Delaware
Weaver, Stephen G., Colorado College
Welsh, James L., Gustavus Adolphus College
Westerman, David S., Norwich University
Whattam, Scot , King Fahd University of Petroleum and Minerals
Whelan, Peter M., University of Minnesota, Morris
WoldeGabriel, Giday, Los Alamos National Laboratory
Wortel, Matthew J., University of Iowa
Yurkovich, Steven P., Western Carolina University

Igneous Petrology
Adams, Gerald E., Columbia College Chicago
Albee, Arden L., California Institute of Technology
Allen, Jack C., Bucknell University
Amedjoe, Godfrey C., Kwame Nkrumah Univ of Science and Tech
Amel, Nasir, University of Tabriz
Andersen, Jens C., Exeter University
Anderson, James Lawford, Boston University
Annesley, Irvine, University of Saskatchewan
Anthony, Elizabeth Y., University of Texas, El Paso
Aranda, Jorge, Univ Nac Autonoma de Mexico
Badger, Robert L., SUNY Potsdam
Bailey, David G., Hamilton College
Baldridge, W. Scott, Los Alamos National Laboratory
Barboni, Melanie, Arizona State University
Barker, Daniel, University of Texas, Austin
Barnes, Calvin G., Texas Tech University
Barr, Sandra M., Acadia University
Barth, Andrew P., Indiana Univ / Purdue Univ, Indianapolis
Barton, Michael, Ohio State University
Beard, James S., Virginia Polytechnic Institute & State University
Bedard, Jean H., Universite du Quebec
Bender, E. E., Orange Coast College
Bender, John F., University of North Carolina, Charlotte
Benimoff, Alan I., College of Staten Island/CUNY
Bentley, Robert, Central Washington University
Bergantz, George W., University of Washington
Beswick, Anthony E., Laurentian University, Sudbury
Beukes, Jarlen J., University of the Free State
Beyarslan, Melahat, Firat University
Bingol, Ahmet Feyzi, Firat University
Birsic, Erin M., Allegheny College
Bizimis, Michael, University of South Carolina
Blackerby, Bruce A., California State University, Fresno
Bladh, Katherine L., Wittenberg University

Bloomer, Sherman H., Oregon State University
Boerboom, Terrence, University of Minnesota
Bohrson, Wendy A., Central Washington University
Borg, Scott G., National Science Foundation
Boroughs, Scott, Washington State University
Boudreau, Alan E., Duke University
Brandon, Alan, University of Houston
Brandriss, Mark E., Williams College
Brophy, James G., Indiana University, Bloomington
Browne, Brandon L., Humboldt State University
Brueseke, Matthew E., Kansas State University
Bryan, Scott E., Queensland University of Technology
Bybee, Grant M., University of the Witwatersrand
Byerly, Gary R., Louisiana State University
Canil, Dante, University of Victoria
Carmichael, Ian S., University of California, Berkeley
Casquet, César, Univ Complutense de Madrid
Castillo, Paterno R., University of California, San Diego
Cathey, Henrietta, Queensland University of Technology
Cecil, M. R., California State University, Northridge
Cepeda, Joseph C., West Texas A&M University
Chadwick, John, College of Charleston
Chavrit, Deborah, University of Manchester
Cheney, John T., Amherst College
Chesner, Craig A., Eastern Illinois University
Christiansen, Eric H., Brigham Young University
Clague, David A., University of Hawai'i, Manoa
Claiborne, Lily L., Vanderbilt University
Clark, Russell G., Albion College
Clarke, D B., Dalhousie University
Clausen, Benjamin L., Loma Linda University
Clemens-Knott, Diane, California State University, Fullerton
Coleman, Robert G., Stanford University
Condit, Christopher D., University of Massachusetts, Amherst
Constantin, Marc, Universite Laval
Coogan, Laurence, University of Victoria
Coombs, Douglas S., University of Otago
Cooper, Kari M., University of California, Davis
Cooper, Roger W., Lamar University
Cornell, Winton, The University of Tulsa
Cote, Denis, Universite du Quebec a Chicoutimi
Cottle, John, University of California, Santa Barbara
Cousens, Brian L., Carleton University
Crawford, Anthony J., University of Tasmania
Cribb, Warner, Middle Tennessee State University
Currier, Ryan, University of Wisconsin, Green Bay
Davies, Gareth R., Vrije Universiteit Amsterdam
Dean, Robert, Dickinson College
Delgado-Argote, Luis A., Centro de Inv Científica y de Ed Sup de Ensenada
Dick, Henry J B., Woods Hole Oceanographic Institution
Dickson, Loretta D., Lock Haven University
Dijkstra, Arjan, University of Plymouth
Dingwell, Donald Bruce, Ludwig-Maximilians-Universitaet Muenchen
Dolejs, David, Charles University
Dorais, Michael J., Brigham Young University
Downs, Robert T., University of Arizona
Dunbar, Nelia W., New Mexico Institute of Mining & Technology
Dunn, John Todd, University of New Brunswick
Dygert, Nick, University of Tennessee, Knoxville
Dymek, Robert F., Washington University in St. Louis
Edwards, Benjamin R., Dickinson College
Elders, Wilfred A., University of California, Riverside
Elkins, Lynne J., University of Nebraska, Lincoln
Erturk, Mehmet Ali, Firat University
Fagan, Amy, Western Carolina University
Fernandez, Louis A., California State University, San Bernardino
Fiesinger, Donald W., Utah State University
Fitton, J. G., Edinburgh University
Fleming, Thomas H., Southern Connecticut State University
Flower, Martin F., University of Illinois at Chicago
Fodor, Ronald V., North Carolina State University
Fournelle, John, University of Wisconsin-Madison
Fowler, Mike, University of Portsmouth
Fowler, Sarah, Katholieke Universiteit Leuven
Fox, Lydia K., University of the Pacific
Francis, Don, McGill University

Furman, Tanya, Pennsylvania State University, University Park
Gaetani, Glenn A., Woods Hole Oceanographic Institution
Gamble, John, University College Cork
Ganguly, Sohini, Goa University
Garcia, Michael O., University of Hawai'i, Manoa
García Galán, Gúmer, Universitat Autonoma de Barcelona
Garrison, Jennifer M., California State University, Los Angeles
Geist, Dennis J., University of Idaho
GERBE, Marie-Christine, Université Jean Monnet, Saint-Etienne
Gertisser, Ralf, Keele University
Ghani, Azman Abdul, University of Malaya
Gibson, David, University of Maine - Farmington
Gibson, Sally, University of Cambridge
Gill, James B., University of California, Santa Cruz
Gittins, John, University of Toronto
Glazner, Allen F., University of North Carolina, Chapel Hill
Gomes, Maria E., Universidade de Trás-os-Montes e Alto Douro
Gordon, Stacia , University of Nevada, Reno
Gorring, Matthew L., Montclair State University
Green, John C., University of Minnesota, Duluth
Greenberg, Jeffrey K., Wheaton College
Grove, Timothy L., Massachusetts Institute of Technology
Grunder, Anita L., Oregon State University
Gualda, Guilherme, Vanderbilt University
Gust, David A., Queensland University of Technology
Haggerty, Stephen E., Florida International University
Halsor, Sid P., Wilkes University
Hammer, Julia E., University of Hawai'i, Manoa
Hammersley, Lisa, California State University, Sacramento
Hammond, Paul E., Portland State University
Hannah, Judith L., Colorado State University
Hanson, Sarah L., Adrian College
Hari, Kosiyath R., Pt. Ravishankar Shukla University
Harris, Chris, University of Cape Town
Hart, William K., Miami University
Hearn Jr, B. Carter, Smithsonian Inst / Natl Museum of Natural History
Hebert, Rejean J., Universite Laval
Hermes, O D., University of Rhode Island
Hess, Paul C., Brown University
Hidalgo, Paulo J., Georgia State Univ
Higgins, Michael D., Universite du Quebec a Chicoutimi
Hirschmann, Marc M., University of Minnesota, Twin Cities
Hirt, William H., College of the Siskiyous
Holness, Marian, University of Cambridge
Howarth, Geoffrey, University of Georgia
Hudgins, Thomas, University of Puerto Rico
Huertas, María José, Univ Complutense de Madrid
Husch, Jonathan M., Rider University
Huysken, Kristin, Indiana University Northwest
Irvine, T. Neil, Carnegie Institution for Science
Jagoutz, Oliver, Massachusetts Institute of Technology
Jahangiri, Ahmad, University of Tabriz
Janney, Phillip, University of Cape Town
Jaye, Shelley, Northern Virginia Community College - Annandale
Jenner, George A., Memorial University of Newfoundland
Johnson, Kenneth, University of Houston Downtown
Johnson, Kevin T. M., University of Hawai'i, Manoa
Jones, Adrian P., University College London
Jones, Norris W., University of Wisconsin, Oshkosh
Jordan, Brennan T., University of South Dakota
Kay, Robert W., Cornell University
Kay, Suzanne M., Cornell University
Kelemen, Peter B., Woods Hole Oceanographic Institution
Kelemen, Peter B., Columbia University
Kempton, Pamela D., Kansas State University
Kent, Adam J., Oregon State University
Klein, Emily M., Duke University
Kopylova, Maya G., University of British Columbia
Koteas, G. Christopher, Norwich University
Kreiger, William (Bill), York College of Pennsylvania
Kurum, Sevcan, Firat University
Kyle, Philip R., New Mexico Institute of Mining and Technology
Kysar Mattietti, Giuseppina, George Mason University
Lackey, Jade Star, Pomona College
Landenberger, Bill, University of Newcastle
Lange, Rebecca A., University of Michigan
Le Roux, Veronique, Woods Hole Oceanographic Institution

Faculty by Specialty

Leake, Bernard E., Cardiff University
Leeman, William P., Rice University
Lehnert, Kerstin, Columbia University
Lesher, Charles E., University of California, Davis
Levy, Schon S., Los Alamos National Laboratory
Lidiak, Edward G., University of Pittsburgh
Lindline, Jennifer, New Mexico Highlands University
Luth, Robert W., University of Alberta
Lux, Daniel R., University of Maine
Lytwyn, Jennifer N., University of Houston
Maclennan, John, University of Cambridge
Mahood, Gail A., Stanford University
Mana, Sara, Salem State University
Manduca, Cathryn A., Carleton College
Marsh, Julian S., Rhodes University
Mathez, Edmond A., Grad Sch of the City Univ of New York
Mattioli, Glen, University of Texas, Arlington
Matty, David J., Weber State University
Mayborn, Kyle R., Western Illinois University
McCallum, I. Stewart (Stu), University of Washington
McCanta, Molly, University of Tennessee, Knoxville
McClellan, Elizabeth, Radford University
McMillan, Nancy J., New Mexico State University, Las Cruces
Medaris, Jr., Levi G., University of Wisconsin-Madison
Memeti, Valbone, California State University, Fullerton
Mertzman, Stanley A., Franklin and Marshall College
Michael, Peter J., The University of Tulsa
Miller, Calvin F., Vanderbilt University
Miller, James, University of Minnesota, Duluth
Miller, Jonathan S., San Jose State University
Mills, James G., DePauw University
Misra, Saumitra, University of KwaZulu-Natal
Mitchell, Roger H., Lakehead University
Moayyed, Mohsen -., University of Tabriz
Morgan, Daniel, University of Leeds
Morgan, George B., University of Oklahoma
Morse, Stearns A., University of Massachusetts, Amherst
MOYEN, Jean-François, Université Jean Monnet, Saint-Etienne
Mungall, James E., Carleton University
Mustart, David A., San Francisco State University
Myers, James D., University of Wyoming
Nabelek, Peter I., University of Missouri
Nash, Barbara P., University of Utah
Naslund, H. Richard, Binghamton University
Naumann, Terry R., University of Alaska, Anchorage
Neal, Clive R., University of Notre Dame
Nehru, Cherukupalli E., Grad Sch of the City Univ of New York
Nelson, Stephen A., Tulane University
Nelson, Wendy, Towson University
Nicholls, James W., University of Calgary
Nicholson, Kirsten N., Ball State University
Nicolaysen, Kirsten P., Whitman College
Nielsen, Roger L., Oregon State University
Niu, Yaoling, Durham University
Ntsaluba, Bantubonke I., Rhodes University
NUDE, Prosper M., University of Ghana
O'Neal, Matt, California Geological Survey
Orejana, David, Univ Complutense de Madrid
Ouimette, Mark A., Hardin-Simmons University
Page, Philippe, Universite du Quebec a Chicoutimi
Panter, Kurt S., Bowling Green State University
Parker, Don, Wayland Baptist University
Parker, Don F., Baylor University
Parlak, Osman, Cukurova Universitesi
Patino-Douce, Marta, University of Georgia
Pérez-Soba, Cecilia, Univ Complutense de Madrid
Perez-Venzor, Jose A., Univ Autonoma de Baja California Sur
Perfit, Michael R., University of Florida
Petrinec, Zorica, University of Zagreb
Philpotts, Anthony R., University of Connecticut
Polat, Ali, University of Windsor
Potter, Katherine E., Weber State University
Price, Jonathan D., Midwestern State University
Puffer, John H., Rutgers, The State Univ of New Jersey, Newark
Puziewicz, Jacek, University of Wroclaw
Raia, Federica, City College (CUNY)
Ratajeski, Kent, University of Kentucky
Reagan, Mark K., University of Iowa
Reavy, John, University College Cork
Reid, Arch M., University of Houston
Reidel, Stephen P., Washington State University
Reyes-Cortes, Manuel, Universidad Autonoma de Chihuahua
Roden, Michael F., University of Georgia
Roelofse, Frederick, University of the Free State
Rooney, Tyrone, Michigan State University
Ross, Martin E., Northeastern University
Ruprecht, Philipp P., University of Nevada, Reno
Saini-Eidukat, Bernhardt, North Dakota State University
Sarrionandia, Fernando, Univ of the Basque Country UPV/EHU
Saunders, Andy D., Leicester University
Schmidt, Mariek, Brock University
Schrader, Christian M., SUNY Potsdam
Schulze, Daniel J., University of Toronto
Schwab, Brandon E., Western Carolina University
Schwab, Brandon E., Humboldt State University
Schwartz, Joshua J., California State University, Northridge
Scoates, James S., University of British Columbia
Seaman, Sheila J., University of Massachusetts, Amherst
Seeger, Cheryl M., Missouri Dept of Natural Resources
Sen, Gautam, Florida International University
Shaw, Cliff S., University of New Brunswick
Shearer, Jr., Charles K., University of New Mexico
Shervais, John W., Utah State University
Shuster, Robert D., University of Nebraska, Omaha
Sigurdson, David, California State University, Dominguez Hills
Silva-Mora, Luis, Univ Nac Autonoma de Mexico
Sinton, John M., University of Hawai'i, Manoa
Sirbescu, Mona, Central Michigan University
Size, William B., Emory University
Sjoblom, Megan, Brigham Young University - Idaho
Smedes, Harry W., Southern Oregon University
Smith, Diane R., Trinity University
Smith, Douglas, University of Texas, Austin
Smith, Eugene I., University of Nevada, Las Vegas
Smith, Matthew C., University of Florida
Smith, Terence E., University of Windsor
Somarin, Alireza, Brandon University
Spera, Frank J., University of California, Santa Barbara
Springer, Robert K., Brandon University
Stachel, Thomas, University of Alberta
Stearns, Michael, Utah Valley University
Stern, Charles R., University of Colorado
Stewart, Michael A., University of Illinois, Urbana-Champaign
Stork, Allen L., Western Colorado University
Stormer, Jr., John C., Rice University
Streck, Martin J., Portland State University
Stull, Robert J., California State University, Los Angeles
Szymanski, David, Bentley University
Tacinelli, John C., Rochester Community & Technical College
Taib, Nur Iskandar, University of Malaya
Templeton, Jeffrey H., Western Oregon University
Tepley, III, Frank J., Oregon State University
Tepper, Jeffrey H., University of Puget Sound
Thakurta, Joyashish, Western Michigan University
Thomas, Jay, Syracuse University
Ural, Melek, Firat University
Urbanczyk, Kevin, Sul Ross State University
Van Tongeren, Jill A., Rutgers, The State Univ of New Jersey
Vance, Joseph A., University of Washington
Vance, Robert K., Georgia Southern University
Vander Auwera, Jacqueline, Universite de Liege
Vetter, Scott K., Centenary College of Louisiana
Viegas, Anthony V., Goa University
Villaseca, Carlos, Univ Complutense de Madrid
von der Handt, Anette, University of Minnesota, Twin Cities
Wadsworth, William B., Whittier College
Walker, Barry, Portland State University
Walowski, Kristina J., Middlebury College
Walter, Michael, University of Bristol
Wanless, Dorsey, Boise State University
Warren, Richard G., Los Alamos National Laboratory
Webster, John R., Minot State University
Weiblen, Paul W., University of Minnesota, Twin Cities
Weiland, Thomas J., Georgia Southwestern State University

Wendlandt, Richard F., Colorado School of Mines
Wenner, Jennifer M., University of Wisconsin, Oshkosh
Whattam, Scott A., Korea University
White, Craig M., Boise State University
White, John C., Eastern Kentucky University
Whitney, James A., University of Georgia
Whittington, Alan, University of Missouri
Wiese, Katryn, City College of San Francisco
Wirth, Karl R., Macalester College
Wittke, James, Northern Arizona University
Wobus, Reinhard A., Williams College
Wolf, Michael B., Augustana College
Wolff, John A., Washington State University
Wolff, John A., Washington State University
Worcester, Peter A., Hanover College
Wulff, Andrew H., Whittier College
Yogodzinski, Gene M., University of South Carolina
Young, Davis A., Calvin College
Zamzow, Craig E., Clarion University
Zieg, Michael J., Slippery Rock University

Metamorphic Petrology

Abati, Jacobo, Univ Complutense de Madrid
Allen, Gary C., University of New Orleans
Appleyard, Edward C., University of Waterloo
Apraiz, Arturo, Univ of the Basque Country UPV/EHU
Arenas, Ricardo, Univ Complutense de Madrid
Ashton, Kenneth E., University of Regina
Baldwin, Julia, University of Montana
Bartoli, Omar, Università degli Studi di Padova
Bauer, Paul W., New Mexico Institute of Mining & Technology
Bebout, Gray E., Lehigh University
Berg, Christopher A., Orange Coast College
Berg, Christopher A., University of West Georgia
Berman, Robert, Carleton University
Blake, David E., University of North Carolina, Wilmington
Brouwer, Fraukje M., Vrije Universiteit Amsterdam
Brown, Edwin H., Western Washington University
Burnley, Pamela C., University of Nevada, Las Vegas
Burwash, Ronald A., University of Alberta
Caddick, Mark, Virginia Polytechnic Institute & State University
Cardace, Dawn, University of Rhode Island
Carillo-Chavez, Alejandro J., Univ Autonoma de Baja California Sur
Carlson, William D., University of Texas, Austin
Castiñeiras, Pedro, Univ Complutense de Madrid
Cesare, Bernardo, Università degli Studi di Padova
Chacko, Thomas, University of Alberta
Clarke, Geoffrey L., University of Sydney
Cooper, Alan F., University of Otago
Costin, Gelu, Rhodes University
Cruz-Uribe, Alicia M., University of Maine
Daniel, Christopher G., Bucknell University
Davidson, Cameron, Carleton College
Dawson, Jane P., Iowa State University of Science & Technology
DeAngelis, Michael T., University of Arkansas at Little Rock
Devore, George W., Florida State University
Diener, Johann, University of Cape Town
Dietsch, Craig, University of Cincinnati
Dipple, Gregory M., University of British Columbia
dos Santos, Telmo B., Universidade de Lisboa
Droop, Giles, University of Manchester
Duke, Edward F., South Dakota School of Mines & Technology
Dunn, Steven R., Mount Holyoke College
Dymek, Robert F., Washington University in St. Louis
Eckert, James O., Yale University
El Shazly, Aley, Marshall University
Elias-Herrera, Mariano, Univ Nac Autonoma de Mexico
Evans, Bernard W., University of Washington
Faryad, Shah Wali, Charles University
Fawcett, J. J., University of Toronto
Fletcher, John, Centro de Inv Científica y de Ed Sup de Ensenada
Flores, Kennet, Brooklyn College (CUNY)
Foster Jr., Charles T., University of Iowa
Frost, B. Ronald, University of Wyoming
Ghent, Edward D., University of Calgary
Giaramita, Mario J., California State University, Stanislaus
Gillis, Kathryn, University of Victoria
Goodge, John W., University of Minnesota, Duluth
Grew, Edward S., University of Maine
Grew, Priscilla C., University of Nebraska, Lincoln
Grover, Timothy W., Castleton University
Grover, Timothy W., University of Northern Colorado
Guevara, Victor, Skidmore College
Hacker, Bradley R., University of California, Santa Barbara
Hajialioghli, Robab, University of Tabriz
Halama, Ralf, Keele University
Hallett, Benjamin W., University of Wisconsin, Oshkosh
Henry, Darrell J., Louisiana State University
Hepburn, J. Christopher, Boston College
Hickmott, Donald D., Los Alamos National Laboratory
Hoisch, Thomas D., Northern Arizona University
Holdaway, Michael J., Southern Methodist University
Holland, Tim, University of Cambridge
Hollister, Lincoln S., Princeton University
Horvath, Peter, Rhodes University
Horvath, Peter, University of the Witwatersrand
Huizenga, Jan Marten, James Cook University
Iancu, Ovidiu G., Alexandru Ioan Cuza
Indares, Aphrodite D., Memorial University of Newfoundland
James, Richard, Laurentian University, Sudbury
Johnson, Eric L., Hartwick College
Jones, Jon W., University of Calgary
Kamber, Balz, Queensland University of Technology
Kays, Marvin A., University of Oregon
Kerrick, Derrill M., Pennsylvania State University, University Park
La Tour, Timothy E., Georgia State Univ
Labotka, Theodore C., University of Tennessee, Knoxville
Laird, Jo, University of New Hampshire
Lamb, Will, Texas A&M University
Law, Eric W., Muskingum University
Leech, Mary L., San Francisco State University
Liou, Juhn G., Stanford University
Lori, Lisa, Missouri Dept of Natural Resources
Maboko, Makenya A., University of Dar es Salaam
Magloughlin, Jerry F., Colorado State University
Mahan, Kevin H., University of Colorado
Manecan, Teodosia, Hunter College (CUNY)
Manon, Matthew R., Union College
Martínez Fernández, Francisco, Universitat Autonoma de Barcelona
Mattinson, Chris, Central Washington University
Mavimbela, Philani, University of KwaZulu-Natal
McLelland, James, Colgate University
Menold, Carrie A., Albion College
Metcalf, Rodney V., University of Nevada, Las Vegas
Metzger, Ellen P., San Jose State University
Moazzen, Mohssen -., University of Tabriz
Mogk, David W., Montana State University
Morton, Douglas M., University of California, Riverside
Mulcahy, Sean R., Western Washington University
Nishiyama, Tadao, Kumamoto University
Ortega-Gutierrez, Fernando, Univ Nac Autonoma de Mexico
Page, F Zeb, Oberlin College
Palin, Richard, Colorado School of Mines
Panish, Peter T., University of Massachusetts, Amherst
Parkinson, Christopher D., University of New Orleans
Pattison, David R., University of Calgary
Peck, William H., Colgate University
Penniston-Dorland, Sarah, University of Maryland
Perkins, III, Dexter, University of North Dakota
Pervunina, Aelita, Institute of Geology (Karelia, Russia)
Plummer, Charles C., California State University, Sacramento
Potel, Sébastien, Institut Polytechnique LaSalle Beauvais (ex-IGAL)
Radakovich, Amy, University of Minnesota
Raeside, Robert P., Acadia University
Rahaman, M. A., Obafemi Awolowo University
Reche Estrada, Joan, Universitat Autonoma de Barcelona
Redden, Jack A., South Dakota School of Mines & Technology
Rietmeijer, Frans J., University of New Mexico
Rivers, Toby C. J. S., Memorial University of Newfoundland
Roberts, Frank, Montgomery County Community College
Roselle, Gregory T., Brigham Young University - Idaho
Rosenfeld, John L., University of California, Los Angeles
Rumble, III, Douglas, Carnegie Institution for Science
Sánchez, Sonia, Univ Complutense de Madrid
Sawyer, Edward W., Universite du Quebec a Chicoutimi

Schiffman, Peter, University of California, Davis
Scotford, David M., Miami University
Scott, James M., University of Otago
Selverstone, Jane E., University of New Mexico
Sisson, Virginia, University of Houston
Sorensen, Sorena S., Smithsonian Inst / Natl Museum of Natural History
Spear, Frank S., Rensselaer Polytechnic Institute
Spiess, Richard, Università degli Studi di Padova
Spray, John G., University of New Brunswick
Srogi, LeeAnn, West Chester University
Staples, Reid, Douglas College
Stevens, Liane M., Stephen F. Austin State University
Stoddard, Edward F., North Carolina State University
Storey, Craig D., University of Portsmouth
Stout, James H., University of Minnesota, Twin Cities
Stowell, Harold H., University of Alabama
Swapp, Susan M., University of Wyoming
Thomson, Jennifer A., Eastern Washington University
Tinkham, Doug, Laurentian University, Sudbury
Turnock, Allan C., University of Manitoba
Walsh, Emily O., University of Iowa
Walsh, Emily O., Cornell College
Waters, Dave, University of Oxford
Weber, Bodo, Centro de Inv Científica y de Ed Sup de Ensenada
Wheeler, John, University of Liverpool
Whitney, Donna L., University of Minnesota, Twin Cities
Wilks, Maureen, New Mexico Institute of Mining & Technology
Wintsch, Robert P., Indiana University, Bloomington
Woodland, Bertram G., Field Museum of Natural History
Zhai, Xiaoming, College of Lake County

Sedimentary Petrology
Adekola, S. A., Obafemi Awolowo University
Al-Aasm, Ihsan S., University of Windsor
Alberstadt, Leonard P., Vanderbilt University
Alonso, Ana María, Univ Complutense de Madrid
Arribas, Eugenia, Univ Complutense de Madrid
Arribas, José, Univ Complutense de Madrid
Babek, Ondrej, Masaryk University
Baddouh, Mbark, George Mason University
Basu, Abhijit, Indiana University, Bloomington
Benson, Donald J., University of Alabama
Bertog, Janet, Northern Kentucky University
Bertok, Carlo, Università di Torino
Boulvain, Frederic P., Universite de Liege
Brengman, Latisha A., University of Minnesota, Duluth
Bruner, Katherine R., West Virginia University
Budd, David A., University of Colorado
Carozzi, Albert V., University of Illinois, Urbana-Champaign
Chafetz, Henry S., University of Houston
Channing, Alan, University of Wales
Choh, Suk-Joo, Korea University
Corcoran, Patricia, Western University
Cousineau, Pierre, Universite du Quebec a Chicoutimi
dela Pierre, Francesco, Università di Torino
Dockal, James A., University of North Carolina, Wilmington
Donovan, R N., Texas Christian University
Dutton, Shirley P., Univ of Texas at Austin, Jackson Sch of Geosciences
Fitzgerald, David, Saint Mary's University
Folk, Robert L., University of Texas, Austin
Forsman, Nels F., University of North Dakota
Franseen, Evan K., University of Kansas
Galli-Olivier, Carlos A., Univ Autonoma de Baja California Sur
Gómez Gras, David, Universitat Autonoma de Barcelona
Gonzalez, Maria M., Central Michigan University
Griffing, David H., Hartwick College
Grigsby, Jeffry D., Ball State University
Hampson, Gary, Imperial College
Horton, Jr., Robert A., California State University, Bakersfield
Humphrey, John D., King Fahd University of Petroleum and Minerals
James, Noel P., Queen's University
Kaczmarek, Stephen E., Western Michigan University
Kerans, Charles, Univ of Texas at Austin, Jackson Sch of Geosciences
Khan, Mohammad Wahdat Y., Pt. Ravishankar Shukla University
Khandaker, Nazrul I., York College (CUNY)
Kirkland, Brenda L., Mississippi State University
Laury, Robert L., Southern Methodist University
Liutkus-Pierce, Cynthia M., Appalachian State University

Lowe, Donald R., Stanford University
Lumsden, David N., University of Memphis
Martin, Arturo, Centro de Inv Científica y de Ed Sup de Ensenada
Martire, Luca, Università di Torino
McBride, Earle F., University of Texas, Austin
McKean, Rebecca, Saint Norbert College
McKnight, Brian K., University of Wisconsin, Oshkosh
Measures, Elizabeth A., Sul Ross State University
Meinhold, Guido, Keele University
Melim, Leslie A., Western Illinois University
Milliken, Kitty L., University of Texas, Austin
Montañez, Isabel P., University of California, Davis
Moore, Bruce R., University of Kentucky
Moshier, Stephen O., Wheaton College
Muntean, Thomas, Adrian College
Nothdurft, Luke, Queensland University of Technology
Ortega-Ariza, Diana, Columbus State University
Pedone, Vicki A., California State University, Northridge
Perkins, Ronald D., Duke University
Portner, Ryan, San Jose State University
Randazzo, Anthony F., University of Florida
Richardson, Darlene S., Indiana University of Pennsylvania
Roman, Aubrecht, Comenius University in Bratislava
Rossi, Carlos, Univ Complutense de Madrid
Ruppel, Stephen C., Univ of Texas at Austin, Jackson Sch of Geosciences
Saja, David B., Cleveland Museum of Natural History
Sanchez Roman, Monica, Vrije Universiteit Amsterdam
Sandberg, Philip A., University of Illinois, Urbana-Champaign
Sanz, Esther, Univ Complutense de Madrid
Savrda, Charles E., Auburn University
Schreiber, B. Charlotte, Queens College (CUNY)
Schrieber, B. Charlotte, Grad Sch of the City Univ of New York
Schwab, Frederick L., Washington & Lee University
Siegrist, jr, Henry G., University of Guam
Stefani, Cristina, Università degli Studi di Padova
Sutterlin, Peter G., Wichita State University
Suttner, Lee J., Indiana University, Bloomington
Sutton, Sally J., Colorado State University
Teixeira, Bernardo, University of Louisiana at Lafayette
Textoris, Daniel A., University of North Carolina, Chapel Hill
Thayer, Paul A., University of North Carolina, Wilmington
Uddin, Ashraf, Auburn University
Varas, María Josefa, Univ Complutense de Madrid
Vice, Mari A., University of Wisconsin, Platteville
Videtich, Patricia E., Grand Valley State University
Walker, Theodore R., University of Colorado
Williams, Wayne K., University of Tennessee, Chattanooga
Young, Harvey R., Brandon University
Zeigler, E. Lynn, Georgia State Univ, Perimeter College, Clarkston

Sedimentology
Abbott, Mark B., University of Pittsburgh
Abbott, Patrick L., San Diego State University
Abdullatif, Osman, King Fahd University of Petroleum and Minerals
Adams, Robert W., SUNY, The College at Brockport
Alexander, Jan, University of East Anglia
Alexander, Jane L., College of Staten Island/CUNY
Algeo, Thomas J., University of Cincinnati
Allison, Mead A., Tulane University
Alvarez-Arellano, Alejandro, Univ Autonoma de Baja California Sur
Ambrose, William A., Univ of Texas at Austin, Jackson Sch of Geo
Ambrose, William, University of Texas, Austin
Anani, Chris Y., University of Ghana
Andersen, David W., San Jose State University
Anderson, Thomas B., Sonoma State University
Anderson-Folnagy, Heidi, University of Montana Western
Anfinson, Owen, Sonoma State University
Anglen, Jeff, California State University, Fresno
Arnott, R. William C., University of Ottawa
Ashley, Gail M., Rutgers, The State Univ of New Jersey
Asiedu, Daniel K., University of Ghana
Askari, Zohreh, Illinois State Geological Survey
Austin, Steve A., Cedarville University
Autin, Whitney J., SUNY, The College at Brockport
Bahlburg, Heinrich, Universitaet Muenster
Balog-Szabo, Anna, Virginia Western Community College
Barbeau, Jr., David, University of South Carolina
Barhurst, James, University of Newcastle Upon Tyne

Bartley, Julie K., Gustavus Adolphus College
Bassett, Kari N., University of Canterbury
Beck, Catherine C., Hamilton College
Behl, Richard J., California State University, Long Beach
Bekker, Andrey, University of California, Riverside
Belt, Edward S., Amherst College
Benito, María Isabel, Univ Complutense de Madrid
Berger, Wolfgang H., University of California, San Diego
Bergmann, Kristin, Massachusetts Institute of Technology
Bershaw, John T., Portland State University
Bertrand, Sebastien, Ghent University
Best, James L., University of Illinois, Urbana-Champaign
Bestland, Erick, Flinders University
Bhattacharya, Janok, University of Houston
Birgenheier, Lauren, University of Utah
Bjornstad, Bruce N., Pacific Northwest National Laboratory
Blake, Bascombe (Mitch), West Virginia University
Blum, Michael D., University of Kansas
Bodenbender, Brian E., Hope College
Boothroyd, Jon C., University of Rhode Island
Bordy, Emese M., University of Cape Town
Borowski, Walter S., Eastern Kentucky University
Bourgeois, Joanne (Jody), University of Washington
Bowen, Brenda , University of Utah
Brand, Brittany, Boise State University
Brandon, Christine M., Bridgewater State University
Breda, Anna, Università degli Studi di Padova
Breyer, John, Texas Christian University
Bristow, Charlie, Birkbeck College
Broda, James E., Woods Hole Oceanographic Institution
Brunt, Rufus, University of Manchester
Burns, Diane M., Eastern Illinois University
Busby, Cathy, University of California, Santa Barbara
Buyce, M. Raymond, Mercyhurst University
Buynevich, Ilya, Temple University
Caputo, Mario V., San Diego State University
Carrapa, Barbara, University of Arizona
Carroll, Alan R., University of Wisconsin-Madison
Cartwright, Joe, University of Oxford
Caruthers, Andrew, Western Michigan University
Cassel, Elizabeth J., University of Idaho
Castle, James W., Clemson University
Cather, Steven M., New Mexico Institute of Mining & Technology
Catuneanu, Octavian, University of Alberta
Caudill, Michael R., Hocking College
Celik, Hasan, Firat University
Chan, Marjorie A., University of Utah
Cheel, Richard J., Brock University
Chiarella, Domenico, Royal Holloway University of London
Chow, Nancy, University of Manitoba
Chowns, Timothy M., University of West Georgia
Christie-Blick, Nicholas, Columbia University
Clarke, Stu, Keele University
Clift, Peter, Louisiana State University
Coch, Nicholas K., Queens College (CUNY)
Coe, Angela L., The Open University
Colburn, Ivan P., California State University, Los Angeles
Cole, Rex D., Colorado Mesa University
Connely, Melissa, Casper College
Conrad, Susan H., Dutchess Community College
Cope, Tim D., DePauw University
Corsetti, Frank A., University of Southern California
Cowan, Clinton A., Carleton College
Cowan, Ellen A., Appalachian State University
Cox, Ronadh, Williams College
Crowley, Stephen, University of Liverpool
Cullen, James L., Salem State University
Currie, Brian S., Miami University
Dade, William B., Dartmouth College
Dalrymple, Robert W., Queen's University
Dansereau, Pauline, Universite Laval
Davatzes, Alexandra K., Temple University
Davies, Neil, University of Cambridge
Davies, Sarah, Leicester University
Davis, Jr., Richard A., University of South Florida, Tampa
de la Horra, Raúl, Univ Complutense de Madrid
de Wet, Carol B., Franklin and Marshall College
Dehler, Carol M., Utah State University
Deibert, Jack, Austin Peay State University
Demicco, Robert V., Binghamton University
Deocampo, Daniel M., Georgia State Univ
Desrochers, Andre, University of Ottawa
Diemer, John A., University of North Carolina, Charlotte
Dix, George R., Carleton University
Dominic, David F., Wright State University
Donald, Roberta, Simon Fraser University
Donaldson, Alan C., West Virginia University
Donaldson, J. Allan, Carleton University
Dravis, Jeffrey J., Rice University
Droste, John B., Indiana University, Bloomington
Droxler, Andre W., Rice University
Dumas, Simone, University of Ottawa
Dunagan, Stan P., University of Tennessee, Martin
Dutta, Prodip K., Indiana State University
Dutton, Shirley P., University of Texas, Austin
Dwyer, Gary S., Duke University
Eberth, David A., Royal Tyrrell Museum of Palaeontology
Edmonds, Douglas A., Indiana University, Bloomington
Edwards, Cole T., Appalachian State University
Egenhoff, Sven O., Colorado State University
Elliott, Jr., William S., University of Southern Indiana
Elrick, Maya, University of New Mexico
Enos, Paul, University of Kansas
Eriksson, Kenneth A., Virginia Polytechnic Institute & State University
Ethridge, Frank G., Colorado State University
Evans, James E., Bowling Green State University
Eyles, Carolyn H., McMaster University
Eyre, Bradley, Southern Cross University
Faas, Richard W., Lafayette College
Fagel, Nathalie, Universite de Liege
Fairchild, Ian, University of Birmingham
Fan, Majie, University of Texas, Arlington
Farkas, Steven, Central Washington University
Farrell, Kathleen M., North Carolina Geological Survey
Fedo, Christopher, University of Tennessee, Knoxville
Fielding, Christopher R., University of Nebraska, Lincoln
Fillmore, Robert P., Western Colorado University
Finkenbinder, Matthew S., Wilkes University
Finzel, Emily, University of Iowa
Fischer, Alfred G., University of Southern California
Fisher, William L., University of Texas, Austin
Flaig, Peter P., University of Texas, Austin
Fletcher, Charles H., University of Hawai'i, Manoa
Fouke, Bruce W., University of Illinois, Urbana-Champaign
Fox, James E., South Dakota School of Mines & Technology
Fox, William T., Williams College
Fralick, Philip W., Lakehead University
Frank, Tracy D., University of Nebraska, Lincoln
Franseen, Evan K., University of Kansas
Frazier, William J., Columbus State University
Fregenal, María Antonia, Univ Complutense de Madrid
Freile, Deborah, New Jersey City University
Freitas, Maria d., Universidade de Lisboa
Fu, Qilong, University of Texas, Austin
Fu, Qilong, Univ of Texas at Austin, Jackson Sch of Geosciences
Gabel, Sharon, Lee College
Gaines, Robert R., Pomona College
Gall, Quentin, University of Ottawa
Galli, Kenneth G., Boston College
Galloway, William E., University of Texas, Austin
Galloway, William E., University of Texas, Austin
Gani, M. Royhan, Western Kentucky University
Garrison, Robert E., University of California, Santa Cruz
Garzione, Carmala N., University of Rochester
Gaylord, David R., Washington State University
Gerhard, Lee C., University of Kansas
Ghinassi, Massimiliano, Università degli Studi di Padova
Gianniny, Gary, Fort Lewis College
Gibling, Martin R., Dalhousie University
Gilleaudeau, Geoff, George Mason University
Gillespie, Robb, Western Michigan University
Gilley, Brett H., University of British Columbia
Glumac, Bosiljka, Smith College
Goff, James, University of New South Wales

Faculty by Specialty

Goldstein, Robert H., University of Kansas
Goodbred, Jr., Steven L., Vanderbilt University
Gore, Pamela J. W., Emory University
Gore, Pamela J. W., Georgia State Univ, Perimeter College, Clarkston
Gorte, Mary C., Delta College
Gotz, Annette, Rhodes University
Grammer, Michael, Oklahoma State University
Greb, Stephen F., University of Kentucky
Greb, Stephen F., University of Kentucky
Greene, Todd, California State University, Chico
Greer, Lisa, Washington & Lee University
Gregg, Jay M., Oklahoma State University
Gregor, C. B., Wright State University
Grimm, Kurt A., University of British Columbia
Grippo, Alessandro, Santa Monica College
Grotzinger, John P., California Institute of Technology
Grove, Karen, San Francisco State University
Guo, Junhua, California State University, Bakersfield
Gupta, Sanjeev, Imperial College
Hajek, Elizabeth, Pennsylvania State University, University Park
Hall, Mary Jo, Rider University
Halverson, Galen, McGill University
Hamilton, Sally, Heriot-Watt University
Hampton, Brian A., New Mexico State University, Las Cruces
Harris, Clay D., Middle Tennessee State University
Harris, Mark T., University of Wisconsin, Milwaukee
Harris, Nicholas, University of Alberta
Harrison, III, William B., Western Michigan University
Hasselbo, Stephen, Exeter University
Havholm, Karen G., University of Wisconsin, Eau Claire
Hayes, Miles O., University of New Orleans
Haywick, Douglas W., University of South Alabama
Hazel, Jr., Joseph E., Northern Arizona University
Heinrich, Paul V., Louisiana State University
Henderson, Miles, University of Texas, Permian Basin
Hendrix, Marc S., University of Montana
Hernandez Molina, F. J., Royal Holloway University of London
Herrero, María Josefa, Univ Complutense de Madrid
Hess-Tanguay, Lillian, Long Island University, C.W. Post Campus
Hesse, Reinhard, McGill University
Heubeck, Christoph, Friedrich-Schiller-University Jena
Hiatt, Eric E., University of Wisconsin, Oshkosh
Hickson, Thomas A., University of Saint Thomas
Higgins, Sean M., Columbia University
Hilbert-Wolf, Hannah L., James Cook University
Hill, Paul S., Dalhousie University
Hill, Philip R., University of Victoria
Hinderer, Matthias, Technische Universitaet Darmstadt
Hiscott, Richard N., Memorial University of Newfoundland
Hodgetts, David, University of Manchester
Hoey, Trevor, University of Glasgow
Holbrook, John M., Texas Christian University
Hollis, Cathy, University of Manchester
Holmes, Ann E., University of Tennessee, Chattanooga
Holmes, Mary Anne, University of Nebraska, Lincoln
Hopkins, John C., University of Calgary
Horton, Brian, University of Texas, Austin
Houston, William (Bill), Lake Superior State University
Hovorka, Susan D., University of Texas, Austin
Hovorka, Susan D., Univ of Texas at Austin, Jackson Sch of Geosciences
Howard, Jeffrey L., Wayne State University
Howell, John, University of Aberdeen
Hubbard, Dennis K., Oberlin College
Huntoon, Jacqueline E., Michigan Technological University
Hurtgen, Matthew T., Northwestern University
Husinec, Antun, St. Lawrence University
Hussin, Azhar Hj, University of Malaya
Ielpi, Alessandro, Laurentian University, Sudbury
Ingersoll, Raymond V., University of California, Los Angeles
Isbell, John L., University of Wisconsin, Milwaukee
Jackson, Chester W., Georgia Southern University
Jackson, William T., University of South Alabama
Jaeger, John M., University of Florida
Janson, Xavier, University of Texas, Austin
Jiang, Ganqing, University of Nevada, Las Vegas
Jinnah, Zubair A., University of the Witwatersrand
Joeckel, Matt, Unversity of Nebraska - Lincoln

John, Cedric, Imperial College
Johnson, Cari, University of Utah
Johnson, Joel P., University of Texas, Austin
Jones, David S., Amherst College
Jones, Lawrence S., Colorado Mesa University
Jones, Merren, University of Manchester
Jones, Stuart, Durham University
Jordan, William M., Millersville University
Kah, Linda, University of Tennessee, Knoxville
Kamola, Diane, University of Kansas
Kanfoush, Sharon L., Utica College
Kendall, Christopher G., University of South Carolina
Kent, Donald M., University of Regina
Kerans, Charles, University of Texas, Austin
Kerr, Dennis R., The University of Tulsa
Kidwell, Susan M., University of Chicago
Kilibarda, Zoran, Indiana University Northwest
Kim, Wonsuck, University of Texas, Austin
Kinvig, Helen, University of Liverpool
Koc Tasgin, Calibe, Firat University
Kocurek, Gary A., University of Texas, Austin
Kocurko, John, Midwestern State University
Kokelaar, Peter, University of Liverpool
Kopaska-Merkel, David C., Geological Survey of Alabama
Kotha, Mahender, Goa University
Kovac, Michal, Comenius University in Bratislava
Kovaèiæ, Marijan, University of Zagreb
Kraft, John C., University of Delaware
Krantz, David E., University of Toledo
Kraus, Mary J., University of Colorado
Krause, Federico F., University of Calgary
Krissek, Lawrence A., Ohio State University
Kurtanjek, Dražen, University of Zagreb
Langford, Richard P., University of Texas, El Paso
Last, William M., University of Manitoba
Lavoie, Denis, Universite du Quebec
Leary, Ryan, New Mexico Institute of Mining and Technology
Leckie, Dale A., Carleton University
LeFever, Richard D., University of North Dakota
Lehman, Thomas M., Texas Tech University
Leier, Andrew L., University of South Carolina
Leithold, Elana L., North Carolina State University
Leorri, Eduardo, East Carolina University
Levin, Naomi, Johns Hopkins University
Levy, Laura B., Humboldt State University
Licht, Alexis, University of Washington
Liddell, W. David, Utah State University
Link, Paul K., Idaho State University
Linn, Anne M., National Acad of Sci, Eng, and Medicine
Linol, Bastien, Nelson Mandela Metropolitan University
Little, William W., Brigham Young University - Idaho
Lloyd, Graeme, University of Oxford
Lock, Brian E., University of Louisiana at Lafayette
Long, Darrel, Laurentian University, Sudbury
Loope, David B., University of Nebraska, Lincoln
Loucks, Robert G., University of Texas, Austin
Ludvigson, Gregory A., University of Kansas
Lynds, Ranie, University of Wyoming
Mac Eachern, James A., Simon Fraser University
MacCarthy, Ivor, University College Cork
Macdonald, R. Heather, College of William & Mary
MacInnes, Breanyn, Central Washington University
Macintyre, Ian G., Smithsonian Inst / Natl Museum of Natural History
MacLachlan, John, McMaster University
Mahoney, J. Brian, University of Wisconsin, Eau Claire
Maju-Oyovwikowhe, Efetobore G., University of Benin
Malpica-Cruz, Victor M., Univ Nac Autonoma de Mexico
Manker, John P., Georgia Southwestern State University
Mankiewicz, Carol, Beloit College
Manos, Constantine, SUNY, New Paltz
Marjanac, Tihomir, University of Zagreb
Marsaglia, Kathleen M., California State University, Northridge
Martini, I. Peter, University of Guelph
Martino, Ronald L., Marshall University
Marzolf, John E., Southern Illinois University Carbondale
Matsuda, Hiroki, Kumamoto University
McCaffrey, Bill, University of Leeds

McCarthy, Paul, University of Alaska, Fairbanks
McCrone, Alistair W., Humboldt State University
McDowell, Ronald, West Virginia University
McElroy, Brandon, University of Wyoming
McNeill, Donald F., University of Miami
Mehrtens, Charlotte J., University of Vermont
Melas, Faye, Hunter College (CUNY)
Merrill, Robert D., California State University, Fresno
Meyers, Jamie A., Winona State University
Miall, Andrew D., University of Toronto
Miclaus, Crina G., Alexandru Ioan Cuza
Middleton, Larry T., Northern Arizona University
Milligan, Timothy, Dalhousie University
Minter, Nicholas, University of Portsmouth
Minzoni, Rebecca T., University of Alabama
Mitchell, Simon F., University of the West Indies Mona Campus
Mohrig, David, University of Texas, Austin
Monecke, Katrin, Wellesley College
Morton, Robert, University of South Florida, Tampa
Moskalski, Susanne, Stockton University
Mount, Jeffrey F., University of California, Davis
Mountney, Nigel, University of Leeds
Mozley, Peter S., New Mexico Institute of Mining and Technology
Mrinjek, Ervin, University of Zagreb
Munoz, Samuel, Northeastern University
Myers, Jeffrey A., Western Oregon University
Myrow, Paul M., Colorado College
Nadon, Gregory C., Ohio University
Nalin, Ronald, Loma Linda University
Natalicchio, Marcello, Università di Torino
Neal, William J., Grand Valley State University
Nehyba, Slavomír, Masaryk University
Nelson, Bruce W., University of Virginia
Neuweiler, Fritz, Universite Laval
Nick, Kevin, Loma Linda University
Noffke, Nora K., Old Dominion University
O'Brien, John M., New Jersey City University
O'Brien, Neal R., SUNY Potsdam
Oakley, Bryan A., University of Rhode Island
Oaks, Jr., Robert Q., Utah State University
OConnell, Suzanne B., Wesleyan University
Okunuwadje, Sunday E., University of Benin
Oldershaw, Alan E., University of Calgary
Oliveira, Anabela R., Universidade de Trás-os-Montes e Alto Douro
Oppo, Davide, University of Louisiana at Lafayette
Ortiz, Joseph D., Kent State University
Paola, Christopher, University of Minnesota, Twin Cities
Pattison, Simon A., Brandon University
Peakall, Jeff, University of Leeds
Pearson, Eugene F., University of the Pacific
Peck, John A., University of Akron
Perscio, Lyman, Mercyhurst University
Peters, Shanan, University of Wisconsin-Madison
Petersen, Morris S., Brigham Young University
Phillips, Andrew C., Illinois State Geological Survey
Pickering, Kevin, University College London
Pietras, Jeff, Binghamton University
Pikelj, Kristina, University of Zagreb
Piper, David, Dalhousie University
Platt, Brian F., University of Mississippi
Plink-Bjorklund, Piret, Colorado School of Mines
Plint, A G., Western University
Potter, Paul E., University of Cincinnati
Potter-McIntyre, Sally, Southern Illinois University Carbondale
Powell, Ross D., Northern Illinois University
Pratson, Lincoln F., Duke University
Preto, Nereo, Università degli Studi di Padova
Price, Greogry, University of Plymouth
Price, Jason R., Millersville University
Prins, Maarten A., Vrije Universiteit Amsterdam
Pufahl, Peir K., Acadia University
Putnam, Peter E., University of Calgary
Qing, Hairuo, University of Regina
Rainbird, R. R., Carleton University
Ramirez, Pedro C., California State University, Los Angeles
Ramsey, Kelvin W., University of Delaware
Rankey, Gene, University of Kansas

Read, J. Fred, Virginia Polytechnic Institute & State University
Reams, Max W., Olivet Nazarene University
Redfern, Jonathan, University of Manchester
Redmond, Brian T., Wilkes University
Rees, Margaret N., University of Nevada, Las Vegas
Reid, Pam, University of Miami
Reid, Steven K., Morehead State University
Reijmer, John, King Fahd University of Petroleum and Minerals
Remacha Grau, Eduard, Universitat Autonoma de Barcelona
Renaut, Robin W., University of Saskatchewan
Ridgway, Kenneth D., Purdue University
Riding, Robert, University of Tennessee, Knoxville
Rigo, Manuel, Università degli Studi di Padova
Rigsby, Catherine A., East Carolina University
Ritts, Malinda, University of Idaho
Roberts, Eric M., James Cook University
Roberts, Harry H., Louisiana State University
Robinson, Ruth, University of St. Andrews
Rogers, Raymond R., Macalester College
Romans, Brian W., Virginia Polytechnic Institute & State University
Runkel, Anthony, University of Minnesota, Twin Cities
Runkel, Anthony, University of Minnesota
Ruppel, Stephen C., University of Texas, Austin
Rygel, Michael C., SUNY Potsdam
Saltzman, Matthew R., Ohio State University
Sánchez, Yolanda, Univ Complutense de Madrid
Sawyer, J. F., South Dakota School of Mines & Technology
Sayegh, Selim, University of Calgary
Saylor, Beverly Z., Case Western Reserve University
Scheingross, Joel S., University of Nevada, Reno
Schieber, Juergen, Indiana University, Bloomington
Schneiderman, Jill S., Vassar College
Scholz, Christopher A., Syracuse University
Schreiber, Charlotte, University of Washington
Schwartz, Theresa M., Allegheny College
Schwimmer, Reed A., Rider University
Scott, Robert W., The University of Tulsa
Self, Robert P., University of Tennessee, Martin
Setterholm, Dale, University of Minnesota
Shane, Tyrrell, National University of Ireland Galway
Shaw, John B., University of Arkansas, Fayetteville
Shea, James H., University of Wisconsin, Parkside
Shimer, Grant, Southern Utah University
Siewers, Fredrick D., Western Kentucky University
Simms, Alexander, University of California, Santa Barbara
Simonson, Bruce M., Oberlin College
Simpson, Edward L., Kutztown University of Pennsylvania
Simpson, Frank -., University of Windsor
Singer, Jill K., Buffalo State College
Singler, Charles R., Youngstown State University
Sirocko, Frank, Universität Mainz
Slingerland, Rudy L., Pennsylvania State University, University Park
Sloss, Craig, Queensland University of Technology
Smith, Jon, University of Kansas
Smith, Larry N., Montana Tech of the University of Montana
Smith, Michael E., Northern Arizona University
Smith, Norman D., University of Nebraska, Lincoln
Smith, Shane V., Youngstown State University
Smith-Engle, Jennifer M., Texas A&M University, Corpus Christi
Smosna, Richard A., West Virginia University
Soreghan, Gerilyn S., University of Oklahoma
Soreghan, Michael J., University of Oklahoma
Soster, Frederick M., DePauw University
Southard, John B., Massachusetts Institute of Technology
Staub, James R., University of Montana
Steel, Ron J., University of Texas, Austin
Steel, Ronald J., University of Texas, Austin
Steenberg, Julia, University of Minnesota
Steinen, Randolph P., Dept of Energy and Environmental Protection
Stoner, Joseph, Oregon State University
Strauss, Justin V., Dartmouth College
Sumner, Dawn Y., University of California, Davis
Sumner, Esther, University of Southampton
Surpless, Kathleen D., Trinity University
Sweet, Alisan C., Texas Tech University
Sweet, Dustin E., Texas Tech University
Swennen, Rudy, Katholieke Universiteit Leuven

Faculty by Specialty

Swett, Keene, University of Iowa
Teller, James T., University of Manitoba
Thomas, Robert C., University of Montana Western
Thompson, Todd A., Indiana University
Thurow, Juergen, University College London
Totten, Matthew W., Kansas State University
Toullec, Renaud, Institut Polytechnique LaSalle Beauvais (ex-IGAL)
Trevino, Ramon H., Univ of Texas at Austin, Jackson Sch of Geosciences
Triplehorn, Donald M., University of Alaska, Fairbanks
Trofimovs, Jessica, Queensland University of Technology
Trop, Jeffrey M., Bucknell University
Törnqvist, Torbjörn E., Tulane University
Ulmer-Scholle, Dana S., New Mexico Institute of Mining and Technology
Underwood, Michael B., University of Missouri
Van De Poll, Henk W., University of New Brunswick
van der Lubbe, Jeroen, Vrije Universiteit Amsterdam
Villegas, Monica B., Universidad Nacional de Rio Cuarto
Voice, Peter J., Western Michigan University
Wagoner, Jeffrey L., Lawrence Livermore National Laboratory
Walker, Richard, Leicester University
Walsh, Tim R., Wayland Baptist University
Walton, Anthony W., University of Kansas
Wang, Ping, University of South Florida, Tampa
Wanke, Ansgar, University of Namibia
Wanless, Harold R., University of Miami
Warme, John E., Colorado School of Mines
Wartes, Marwan A., Alaska Div of Geological & Geophysical Surveys
Washburn, Robert H., Juniata College
Watkinson, Matthew, University of Plymouth
Weislogel, Amy L., West Virginia University
Wellner, Julia, University of Houston
Wells, Neil A., Kent State University
Whalen, Michael T., University of Alaska, Fairbanks
Whisonant, Robert C., Radford University
Whiteaker, Timothy, University of Texas, Austin
Wilkinson, Bruce H., University of Michigan
Wilkinson, Bruce, Syracuse University
Wilson, Carol A., Louisiana State University
Wilson, Paul A., University of Southampton
Winkelstern, Ian Z., Grand Valley State University
Wittkop, Chad, Minnesota State University
Wizevich, Michael, Central Connecticut State University
Wojewoda, Jurand, University of Wroclaw
Wolak, Jeannette, Tennessee Tech University
Wood, Lesli, Colorado School of Mines
Woodruff, Jonathan D., University of Massachusetts, Amherst
Woods, Adam D., Santiago Canyon College
Woods, Adam D., California State University, Fullerton
Woods, Rachel A., Edinburgh University
Worden, Richard, University of Liverpool
Wright, V P., Cardiff University
Wynn, Thomas C., Lock Haven University
Yang, Wan, Wichita State University
Yeager, Kevin M., University of Kentucky
Yellen, Brian, University of Massachusetts, Amherst
Young, Robert S., Western Carolina University
Zaleha, Michael J., Wittenberg University
Zattin, Massimiliano, Università degli Studi di Padova
Zeng, Hongliu, Univ of Texas at Austin, Jackson Sch of Geosciences

Physical Stratigraphy
Abdullah, Nuraiteng Tee, University of Malaya
Aitchison, Jonathan, University of Sydney
Alvarez, Walter, University of California, Berkeley
Anderson, Raymond R., University of Iowa
Archer, Allen W., Kansas State University
Aschoff, Jennifer, University of Alaska, Anchorage
Atchley, Stacy C., Baylor University
Baird, Gordon C., SUNY, Fredonia
Balgord, Elizabeth A., Weber State University
Bart, Philip J., Louisiana State University
Bartholomew, Alexander J., SUNY, New Paltz
Benison, Kathleen, West Virginia University
Bhattacharya, Janok, McMaster University
Blakeney DeJarnett, Beverly, Univ of Texas at Austin, Jackson Sch of Geo
Blakey, Ronald C., Northern Arizona University
Boone, Peter A., Austin Community College District
Bowen, David W., Montana State University

Brezinski, David K., Maryland Department of Natural Resources
Brooks, James E., Southern Methodist University
Busch, Richard M., West Chester University
Byers, Charles W., University of Wisconsin-Madison
Catacosinos, Paul A., Delta College
Chipping, David H., California Polytechnic State University
Clark, Joseph C., Indiana University of Pennsylvania
Colombi, Carina, Miami University
Coniglio, Mario, University of Waterloo
Cramer, Bradley D., University of Iowa
d'Atri, Anna, Università di Torino
Dalton, Richard F., New Jersey Geological and Water Survey
Denny, F. Brett, Illinois State Geological Survey
Devera, Joseph A., Illinois State Geological Survey
Diecchio, Richard J., George Mason University
Diffendal, Jr., Robert F., Unversity of Nebraska - Lincoln
Doar, III, William R., South Carolina Dept of Natural Resources
Dodge, Clifford H., Pennsylvania Bur of Topo & Geologic Survey
Dorsey, Rebecca J., University of Oregon
Drzewiecki, Peter A., Eastern Connecticut State University
Ebert, James R., SUNY, Oneonta
Elliott, Monty A., Southern Oregon University
Evanoff, Emmett, University of Northern Colorado
Evans, Kevin R., Missouri State University
Feibel, Craig S., Rutgers, The State Univ of New Jersey
Ferrusquia, Ismael, Univ Nac Autonoma de Mexico
Fichter, Lynn S., James Madison University
Filipescu, Sorin, Babes-Bolyai University
Flemings, Peter B., University of Texas, Austin
Flint, Stephen, University of Manchester
Foreman, Brady Z., Western Washington University
Freeman, Rebecca, University of Kentucky
Fritz, William J., College of Staten Island/CUNY
Fugitt, Frank, Ohio Dept of Natural Resources
Gardulski, Anne F., Tufts University
Gentile, Richard J., University of Missouri, Kansas City
González, Laura, Univ Complutense de Madrid
Gonzalez Leon, Carlos M., Univ Nac Autonoma de Mexico
Hamlin, H, Scott, Univ of Texas at Austin, Jackson Sch of Geosciences
Harris, W. Burleigh, University of North Carolina, Wilmington
Harvey, Cyril H., Guilford College
Hasenmueller, Nancy R., Indiana University
Hattin, Donald E., Indiana University, Bloomington
Heckel, Philip H., University of Iowa
Heermance, Richard V., California State University, Northridge
Helenes, Javier, Centro de Inv Científica y de Ed Sup de Ensenada
Hentz, Tucker F., Univ of Texas at Austin, Jackson Sch of Geosciences
Heron, Duncan, Duke University
Hinnov, Linda, George Mason University
Hippensteel, Scott P., University of North Carolina, Charlotte
Holland, Steven M., University of Georgia
Hyland, Ethan, North Carolina State University
Jacques Ayala, Cesar, Univ Nac Autonoma de Mexico
Janson, Xavier, Univ of Texas at Austin, Jackson Sch of Geosciences
Jenkyns, Hugh C., University of Oxford
Joeckel, R. M., University of Nebraska, Lincoln
Johnson, Donald O., Argonne National Laboratory
Johnson, Gary D., Dartmouth College
Johnson, William C., University of Kansas
Jordan, Robert R., University of Delaware
Jordan, Teresa E., Cornell University
Kaaya, Charles Z., University of Dar es Salaam
King, Norman R., University of Southern Indiana
King, Jr., David T., Auburn University
Kulp, Mark A., University of New Orleans
Lang, Harold R., Jet Propulsion Laboratory
Langenheim, Jr., Ralph L., University of Illinois, Urbana-Champaign
Lash, Gary G., SUNY, Fredonia
Lee, Richard R., Oak Ridge National Laboratory
Lincoln, Jonathan M., Montclair State University
Loydell, David K., University of Portsmouth
Lozinsky, Richard P., Fullerton College
Ludvigson, Greg A., University of Kansas
Lundblad, Steven P., University of Hawai'i, Hilo
Mancini, Ernest A., University of Alabama
Manger, Walter, University of Arkansas, Fayetteville
Manspeizer, Warren, Rutgers, The State Univ of New Jersey, Newark

Martinez-Gutierrez, Genaro, Univ Autonoma de Baja California Sur
Maytorena-Silva, Jesus F., Universidad Estatal de Sonora
McDonald, Catherine, Montana Tech of The University of Montana
McKee, James W., University of Wisconsin, Oshkosh
McLaughlin, Patrick I., University of Wisconsin, Extension
McLaughlin, Patrick I., Indiana University
McLaughlin, Jr., Peter P., University of Delaware
McLaurin, Brett T., Bloomsburg University
Meckel, Timothy A., University of Texas, Austin
Merriam, Daniel F., Wichita State University
Metz, Robert, Kean University
Monreal-Saavedra, Rogelio, Universidad Estatal de Sonora
Monteverde, Don, Rutgers, The State Univ of New Jersey
Monteverde, Donald H., New Jersey Geological and Water Survey
Morris, Thomas H., Brigham Young University
Muñoz, Belén, Univ Complutense de Madrid
Nance, Hardie S., University of Texas, Austin
Neal, Donald W., East Carolina University
Nicolas, Michelle P., Manitoba Geological Survey
Nielson, R. LaRell, Stephen F. Austin State University
Norford, Brian, University of Calgary
Ogata, Kei, Vrije Universiteit Amsterdam
Olariu, Cornel, University of Texas, Austin
Olariu, Mariana, University of Texas, Austin
Oms Llobet, Oriol, Universitat Autonoma de Barcelona
Owen, Donald E., Lamar University
Page, Kevin, University of Plymouth
Paine, Jeffrey G., Univ of Texas at Austin, Jackson Sch of Geosciences
Parcell, William C., Wichita State University
Pashin, Jack C., Oklahoma State University
Peterson, Gary L., San Diego State University
Peterson, Larry C., University of Miami
Pimentel, Nuno, Universidade de Lisboa
Poch Serra, Joan, Universitat Autonoma de Barcelona
Pope, Michael, Texas A&M University
Retzler, Andrew J., University of Minnesota
Reynolds, James H., Brevard College
Sadler, Peter M., University of California, Riverside
Sanders, Ronald S., Jet Propulsion Laboratory
Santisteban, Juan Ignacio, Univ Complutense de Madrid
Scott, Thomas M., University of South Florida, Tampa
Simmons, Lizanne V., Santiago Canyon College
Snyder, Walter S., Boise State University
Spreng, Alfred C., Missouri University of Science and Technology
Stieglitz, Ronald D., University of Wisconsin, Green Bay
Sugarman, Peter J., New Jersey Geological and Water Survey
Sugarman, Peter P., Rutgers, The State Univ of New Jersey
Suneson, Neil, University of Oklahoma
Swenson, John B., University of Minnesota, Duluth
Tew, Berry H., Geological Survey of Alabama
Tierney, Kate, University of Iowa
Turner, Elizabeth C., Laurentian University, Sudbury
Vail, Peter R., Rice University
Valenzuela-Renteria, Manuel, Universidad Estatal de Sonora
Van de Water, Peter, California State University, Fresno
Ver Straeten, Charles, New York State Geological Survey
Verlekar, Purushottam, Goa University
Vondra, Carl F., Iowa State University of Science & Technology
Watney, Lynn W., University of Kansas
Watney, W. Lynn, University of Kansas
Weimer, Robert J., Colorado School of Mines
Witzke, Brian J., University of Iowa
Young, Grant M., Western University
Zachry, Doy L., University of Arkansas, Fayetteville

Structural Geology
Bodhankar, Ninad, Pt. Ravishankar Shukla University
Abalos, Benito, Univ of the Basque Country UPV/EHU
Abdelsalam, Mohamed G., Oklahoma State University
Abolins, Mark J., Middle Tennessee State University
Adam, Jurgen, Royal Holloway University of London
Adams, John B., University of Washington
Akciz, Sinan, California State University, Fullerton
Alaniz Alvarez, Susana A., Univ Nac Autonoma de Mexico
Aldrich, M. James, Los Alamos National Laboratory
Aleksandrowski, Pawel, University of Wroclaw
Allard, Stephen T., Winona State University
Allen, Joseph L., Concord University

Allen, Richardson B., University of Utah
Allison, David T., University of South Alabama
Allmendinger, Richard W., Cornell University
Alsleben, Helge, Texas Christian University
Alvarez, José Antonio, Univ Complutense de Madrid
Amato, Jeffrey M., New Mexico State University, Las Cruces
Amenta, Roddy V., James Madison University
Amos, Colin B., Western Washington University
Anastasio, David J., Lehigh University
Anders, Mark H., George Mason University
Anderson, James L., University of Hawai'i, Hilo
Anderson, Mark, University of Plymouth
Andrew, Joseph E., Youngstown State University
Aranguren, Aitor, Univ of the Basque Country UPV/EHU
Arboleya Cimadevilla, María Luísa, Universitat Autonoma de Barcelona
Armstrong, Phillip A., California State University, Fullerton
Aronoff, Ruth F., Furman University
Arrowsmith, Ramon, Arizona State University
Axen, Gary, New Mexico Institute of Mining and Technology
Aydin, Atilla, Stanford University
Babaie, Hassan A., Georgia State Univ
Baciu, Sorin D., Alexandru Ioan Cuza
Bailey, Christopher M., College of William & Mary
Baird, Graham, University of Northern Colorado
Balestro, Gianni, Università di Torino
Bamisaye, Oluwaseyi A., Federal University of Technology, Akure
Bangs, Nathan L., University of Texas, Austin
Banoeng-Yakubo, Bruce K., University of Ghana
Baran, Zeynep O., South Dakota School of Mines & Technology
Barineau, Clinton I., Columbus State University
Barker, Chris A., Stephen F. Austin State University
Barnes, Charles W., Northern Arizona University
Barquero-Molina, Miriam, University of Missouri
Bartley, John M., University of Utah
Bauer, Paul W., New Mexico Institute of Mining and Technology
Bayly, M. Brian, Rensselaer Polytechnic Institute
Behr, Rose-Anna, Pennsylvania Bur of Topo & Geologic Survey
Behr, Whitney, University of Texas, Austin
Berry, Ron F., University of Tasmania
Bethune, Kathryn, University of Regina
Beutel, Erin K., College of Charleston
Bhattacharyya, Prajukti, University of Wisconsin, Whitewater
Bidgoli, Tandis, University of Kansas
Bier, Sara E., SUNY Potsdam
Bilodeau, William L., California Lutheran University
Birke, Tesfaye K., University of KwaZulu-Natal
Bishop, Kim, California State University, Los Angeles
Bjorklund, Tom, University of Houston
Bjornerud, Marcia, Lawrence University
Blackmer, Gale C., Pennsylvania Bur of Topo & Geologic Survey
Bobyarchick, Andy R., University of North Carolina, Charlotte
Bond, Clare, University of Aberdeen
Boniface, Nelson, University of Dar es Salaam
Bosbyshell, Howell, West Chester University
Bothner, Wallace A., University of New Hampshire
Bott, Jackie, California Geological Survey
Bradbury, Kelly K., Utah State University
Bradley, Michael, Eastern Michigan University
Bradley, Philip, North Carolina Geological Survey
Brandon, Mark T., Yale University
Brennan, William J., SUNY, Geneseo
Brisbin, William C., University of Manitoba
Brodie, Kate, University of Manchester
Brown, Bruce A., University of Wisconsin, Extension
Bunds, Michael, Utah Valley University
Burberry, Caroline M., University of Nebraska, Lincoln
Burchfiel, B. C., Massachusetts Institute of Technology
Burger, H. Robert, Smith College
Burks, Rachel J., Towson University
Burmeister, Kurtis C., University of the Pacific
Buttner, Steffen, Rhodes University
Bykerk-Kauffman, Ann, California State University, Chico
Byrne, Timothy, University of Connecticut
Cadoppi, Paola, Università di Torino
Calon, Tomas J., Memorial University of Newfoundland
Camilleri, Phyllis A., Austin Peay State University
Campagna, David J., West Virginia University

Campbell, Erin A., Wyoming State Geological Survey
Campbell, Patricia A., Slippery Rock University
Carlson, Diane H., California State University, Sacramento
Carosi, Rodolfo, Università di Torino
Casale, Gabriele M., Appalachian State University
Cashman, Susan M., Humboldt State University
Caskey, John, San Francisco State University
Cather, Steven M., New Mexico Institute of Mining and Technology
Cattanach, Bart, North Carolina Geological Survey
Çemen, Ibrahim, University of Alabama
Chamberlin, Richard, New Mexico Institute of Mining and Technology
Chapman, Alan, Macalester College
Chapman, James, University of Wyoming
Chester, Frederick M., Texas A&M University
Chester, Frederick, Texas A&M University
Chester, Judith, Texas A&M University
Cladouhos, Trenton, University of Washington
Clayton, Robert W., Brigham Young University - Idaho
Cloos, Mark P., University of Texas, Austin
Colby, Thomas A., University of Arkansas at Little Rock
Coleman, Jr., Paul J., University of California, Los Angeles
Connelly, Jeffrey B., University of Arkansas at Little Rock
Connors, Christopher, Washington & Lee University
Cooke, Michele L., University of Massachusetts, Amherst
Coppinger, Walter, Trinity University
Cosgrove, John, Imperial College
Cowan, Darrel S., University of Washington
Cox, Randel T., University of Memphis
Crespi, Jean M., University of Connecticut
Crider, Juliet G., University of Washington
Cronin, Vincent S., Baylor University
Crowley, Peter D., Amherst College
Cruikshank, Kenneth M., Portland State University
Cuevas, Julia, Univ of the Basque Country UPV/EHU
Culshaw, Nicholas, Dalhousie University
Czeck, Dyanna M., University of Wisconsin, Milwaukee
D'Allura, Jad A., Southern Oregon University
Daigneault, Real, Universite du Quebec a Chicoutimi
Davatzes, Nicholas, Temple University
Davis, George H., University of Arizona
Davis, Peter B., Pacific Lutheran University
Dawers, Nancye H., Tulane University
De Paola, Nicola, Durham University
Di Toro, Giulio, Università degli Studi di Padova
DiPietro, Joseph A., University of Southern Indiana
Dixon, John M., Queen's University
Dooley, Tim, Univ of Texas at Austin, Jackson Sch of Geosciences
Douglas, Bruce, Indiana University, Bloomington
Draper, Grenville, Florida International University
Druguet Tantiña, Elena, Universitat Autonoma de Barcelona
Duebendorfer, Ernest M., Northern Arizona University
Dumitriu, Tony C., Alexandru Ioan Cuza
Dunne, George C., California State University, Northridge
Dunne, William M., University of Tennessee, Knoxville
Dunning, Jeremy D., Indiana University, Bloomington
Dutch, Steven I., University of Wisconsin, Green Bay
Eichhubl, Peter, University of Texas, Austin
Eichhubl, Peter, Univ of Texas at Austin, Jackson Sch of Geosciences
Elliott, Colleen G., Montana Tech of The University of Montana
Erdmer, Philippe, University of Alberta
Erslev, Eric A., University of Wyoming
Erslev, Eric A., Colorado State University
Eusden, J. Dykstra, Bates College
Evans, James P., Utah State University
Evans, Mark, Central Connecticut State University
Famin, Vincent, Universite de la Reunion
Faulds, James E., University of Nevada
Faulds, James, University of Nevada, Reno
Faulds, James E., University of Nevada, Reno
Faure, Stephane, Universite du Quebec a Montreal
Fayon, Annia K., University of Minnesota, Twin Cities
Ferguson, Charles A., Arizona Geological Survey
Festa, Andrea, Università di Torino
Fischer, Mark P., Northern Illinois University
Fisher, Donald M., Pennsylvania State University, University Park
Forte, Adam, Louisiana State University
Franke, Otto L., Grad Sch of the City Univ of New York

Fraser, Alastair, University of Utah
French, Melodie E., Rice University
Fryer, Karen H., Ohio Wesleyan University
Fryxell, Joan E., California State University, San Bernardino
Fueten, Frank, Brock University
Fusseis, Florian, Edinburgh University
Fyson, William K., University of Ottawa
Gale, Julia F., University of Texas, Austin
Gani, Nahid, Western Kentucky University
Garihan, John M., Furman University
Gates, Alexander E., Rutgers, The State Univ of New Jersey, Newark
Gattiglio, Marco, Università di Torino
Gibson, H. Daniel (Dan), Simon Fraser University
Gibson, Roger L., University of the Witwatersrand
Gibson, Ronald C., Golden West College
Gillespie, Thomas, College of New Jersey
Gilotti, Jane A., University of Iowa
Giorgis, Scott D., SUNY, Geneseo
Girty, Gary H., San Diego State University
Gleason, Gayle C., SUNY, Cortland
Godin, Laurent, Queen's University
Gold, David (Duff) P., Pennsylvania State University, University Park
Goodwin, Laurel B., University of Wisconsin-Madison
Gottardi, Raphael, University of Louisiana at Lafayette
Goulet, Normand, Universite du Quebec a Montreal
Gray, Mary Beth, Bucknell University
Greene, David C., Denison University
Griera Artigas, Albert, Universitat Autonoma de Barcelona
Griffith, Ashley, University of Texas, Arlington
Griffith, William A., Ohio State University
Groshong, Jr., Richard H., University of Alabama
Gross, Michael, Florida International University
Groszos, Mark S., Valdosta State University
Gudmundsson, Agust, Royal Holloway University of London
Guitierrez-Alonso, Gabriel, Florida International University
Gulick, Sean S., University of Texas, Austin
Guth, Peter L., United States Naval Academy
Hack, Alistair, University of Newcastle
Hacker, David B., Kent State University
Halfpenny, Angela, Central Washington University
Hall, Frank W., Indiana University of Pennsylvania
Hanley, Thomas B., Columbus State University
Hanmer, Simon, Carleton University
Hannula, Kimberly, Fort Lewis College
Harrap, Rob, Queen's University
Harris, Lyal B., Universite du Quebec
Harris, Ron, Brigham Young University
Harrison, Michael J., Tennessee Tech University
Hawkins, John F., Auburn University
Hefferan, Kevin P., University of Wisconsin, Stevens Point
Heimsath, Arjun, Arizona State University
Helmstaedt, Herwart, Queen's University
Helper, Mark A., University of Texas, Austin
Hendrix, Thomas E., Grand Valley State University
Hennings, Peter H., University of Wyoming
Hetherington, Eric D., College of the Sequoias
Hetzel, Ralf, Universitaet Muenster
Hibbard, James P., North Carolina State University
Hill, Mary Louise, Lakehead University
Hinz, Nick, University of Nevada, Reno
Hirth, Greg, Brown University
Hirth, Gregory, Woods Hole Oceanographic Institution
Hoffman, Paul F., Harvard University
Holdsworth, Robert E., Durham University
Holm, Daniel K., Kent State University
Holst, Timothy, University of Minnesota, Duluth
Holyoke, Caleb, University of Akron
Hooks, Benjamin P., University of Tennessee, Martin
Horsman, Eric M., East Carolina University
Horst, Andrew J., Marshall University
Horst, Andrew J., Oberlin College
Hozik, Michael J., Stockton University
Hudec, Michael R., Univ of Texas at Austin, Jackson Sch of Geosciences
Hudec, Michael R., University of Texas, Austin
Hudleston, Peter J., University of Minnesota, Twin Cities
Hughes, Kenneth S., University of Puerto Rico
Hull, Joseph M., Seattle Central Community College

Hynes, Andrew J., McGill University
Iaccarino, Salvatore, Università di Torino
Isard, Sierra J., North Carolina Geological Survey
Ismat, Zeshan, Franklin and Marshall College
Jackson, Martin P. A., Univ of Texas at Austin, Jackson Sch of Geo
Jackson, Richard A., Long Island University, Brooklyn Campus
Jacobi, Robert D., SUNY, Buffalo
Jessup, Micah, University of Tennessee, Knoxville
Jiang, Dazhi, Western University
Jiménez, David, Univ Complutense de Madrid
Jirsa, Mark, University of Minnesota
John, Barbara E., University of Wyoming
Johnson, Arvid M., Purdue University
Johnson, Brad, Arizona Geological Survey
Johnson, Scott E., University of Maine
Johnston, Scott, California Polytechnic State University
Johnston, Stephen T., University of Alberta
Johnston, Stephen T., University of Victoria
Jones, Gustavo Tolson, Univ Nac Autonoma de Mexico
Kalakay, Thomas J., Rocky Mountain College
Karabinos, Paul, Williams College
Karimi, Bobak, Wilkes University
Kath, Randal L., University of West Georgia
Kehlenbeck, Manfred M., Lakehead University
Kelsch, Jesse, Sul Ross State University
Kelty, Thomas, California State University, Long Beach
Kennedy, Lori, University of British Columbia
Kirkpatrick, James, McGill University
Klepeis, Keith A., University of Vermont
Klimczak, Christian, University of Georgia
Knapp, James, Oklahoma State University
Konig, Ron, University of Arkansas, Fayetteville
Kopf, Christopher F., Mansfield University
Kruckenberg, Seth C., Boston College
Kruckenberg, Seth, Boston College
Krugh, W C., California State University, Bakersfield
Kuiper, Yvette D., Colorado School of Mines
Kulander, Byron, Wright State University
Kutu, Jacob M., University of Ghana
Lafrance, Bruno, Laurentian University, Sudbury
Lageson, David R., Montana State University
Langille, Jackie M., University of North Carolina, Asheville
Lao, Daniel A., Oklahoma State University
Lapusta, Nadia, California Institute of Technology
Laubach, Stephen E., University of Texas, Austin
Laubach, Stephen E., Univ of Texas at Austin, Jackson Sch of Geosciences
Laurent-Charvet, Sébastien, Institut Polytechnique LaSalle Beauvais
Law, Richard D., Virginia Polytechnic Institute & State University
Lee, Jeffrey, Central Washington University
Lemiszki, Peter J., Tennessee Geological Survey
Lemiszki, Peter J., Pellissippi State Community College
Lenardic, Adrian, Rice University
Lennox, Paul G., University of New South Wales
Lewis, Jon C., Indiana University of Pennsylvania
Lincoln, Beth Z., Albion College
Lisle, R J., Cardiff University
Little, Tim, Victoria University of Wellington
Livaccari, Richard F., Colorado Mesa University
Long , Sean P., Washington State University
Loveland, Andrea M., Wyoming State Geological Survey
Lowry, Wallace D., Virginia Polytechnic Institute & State University
Luther, Amy, Louisiana State University
Ma, Chong, Auburn University
MacDonald, William D., Binghamton University
MacLean, John S., Southern Utah University
Maher, Jr., Harmon D., University of Nebraska, Omaha
Mali, Heinrich, University Leoben
Malo, Michel, Universite du Quebec
Malone, David H., Illinois State University
Markley, Michelle J., Mount Holyoke College
Marrett, Randall A., University of Texas, Austin
Marshak, Stephen, University of Illinois, Urbana-Champaign
Martin, Silvana, Università degli Studi di Padova
Martínez, José Jesús, Univ Complutense de Madrid
Martinez-Macias, Panfilo R., Universidad Autonoma de San Luis Potosi
Marty, Kevin G., Imperial Valley College
Mattson, Peter H., Grad Sch of the City Univ of New York

Mattson, Peter H., Queens College (CUNY)
McCaffrey, Kenneth J., Durham University
McCair, Andrew, University of Leeds
McClay, Ken R., Royal Holloway University of London
McGill, George E., University of Massachusetts, Amherst
McIntosh, Kirk D., University of Texas, Austin
McKay, Matthew P., Missouri State University
McNaught, Mark A., Mount Union College
McNulty, Brendan A., California State University, Dominguez Hills
McQuarrie, Nadine, University of Pittsburgh
Mecklenburgh, Julian, University of Manchester
Meere, Pat, University College Cork
Meigs, Andrew J., Oregon State University
Menegon, Luca, University of Plymouth
Merguerian, Charles M., Hofstra University
Metcalf, Kathryn, SUNY, Oneonta
Mezger, Jochen E., University of Alaska, Fairbanks
Mies, Jonathan W., University of Tennessee, Chattanooga
Miller, Elizabeth L., Stanford University
Miller, Marli G., University of Oregon
Miller, Robert B., San Jose State University
Milligan, G. Clinton, Dalhousie University
Minnaar, Hendrik, University of the Free State
Miranda, Elena A., California State University, Northridge
Mitchell, Tom, University College London
Mitra, Gautam, University of Rochester
Mitra, Shankar, University of Oklahoma
Mookerjee, Matty, Sonoma State University
Moore, J. Casey, University of California, Santa Cruz
Morgan, Julia K., Rice University
Morgan, Julia K., Rice University
Morgan, Sven S., Iowa State University of Science & Technology
Morrow, Robert H., South Carolina Dept of Natural Resources
Mosher, Sharon, University of Texas, Austin
Muller, Otto H., Alfred University
Murphy, Michael, University of Houston
Needham, Tim, University of Leeds
Nel, Justin, University of the Free State
Nemcok, Michal, University of Utah
Neuhauser, Kenneth R., Fort Hays State University
Newman, Julie, Texas A&M University
Nicol, Andy, University of Canterbury
Northrup, Clyde J., Boise State University
Nourse, Jonathan A., California State Polytechnic University, Pomona
O'Hara, Kieran D., University of Kentucky
O'Meara, Stephanie A., Colorado State University
Onasch, Charles M., Bowling Green State University
Osagiede, Edoseghe E., University of Benin
Palmquist, John C., Lawrence University
Paterson, Scott R., University of Southern California
Paton, Douglas, University of Leeds
Patton, Regan L., Washington State University
Paulsen, Timothy S., University of Wisconsin, Oshkosh
Pavlis, Terry L., University of Texas, El Paso
Pawley, Alison, University of Manchester
Pearson, David M., Idaho State University
Pennacchioni, Giorgio, Università degli Studi di Padova
Perry, Bruce, California State University, Long Beach
Petrie, Elizabeth S., Western Colorado University
Phipps, Stephen P., University of Pennsylvania
Platt, John P., University of Southern California
Platt, Lucian B., Bryn Mawr College
Pogue, Kevin R., Whitman College
Pollard, David D., Stanford University
Potter, Jr., Donald B., Sewanee: University of the South
Price, Nancy A., Portland State University
Prior, David J., University of Otago
Pritchard, Chad, Eastern Washington University
Puelles, Pablo, Univ of the Basque Country UPV/EHU
Quintero, Odranoel, Univ Nac Autonoma de Mexico
Raffini, Sylvain, Universite du Quebec a Chicoutimi
Rafini, Silvain, Universite du Quebec a Chicoutimi
Ratchford, Michael , University of Idaho
Rauch, Marta, University of Wroclaw
Reber, Jacqueline, Iowa State University of Science & Technology
Ree, Jin-Han, Korea University
Reese, Joseph F., Edinboro University of Pennsylvania

Reinen, Linda A., Pomona College
Resor, Phillip G., Wesleyan University
Reynolds, Stephen J., Arizona State University
Ricketts, Jason, University of Texas, El Paso
Ritchie, Alexander W., College of Charleston
Roberts, Gerald, Birkbeck College
Robertson, Charles E., Missouri University of Science and Technology
Robin, Pierre-Yves F., University of Toronto
Robinson, Alexander, University of Houston
Robinson, Delores, University of Alabama
Robinson, Kevin, San Diego State University
Robinson, Peter, University of Massachusetts, Amherst
Rodgers, David W., Idaho State University
Rogers, Robert D., California State University, Stanislaus
Rojo-Garcia, Paulino, Univ Autonoma de Baja California Sur
Romeo, Ignacio, Univ Complutense de Madrid
Rousell, Don H., Laurentian University, Sudbury
Rowe, Christie, McGill University
Roy, Denis W., Universite du Quebec a Chicoutimi
Rusmore, Margaret E., Occidental College
Saillet, Elodie, Institut Polytechnique LaSalle Beauvais (ex-IGAL)
Saja, David, Case Western Reserve University
Sak, Peter B., Dickinson College
Saleeby, Jason B., California Institute of Technology
Samaniego, Angel Nieto, Univ Nac Autonoma de Mexico
Sanchez, Veronica I., Texas A&M University, Kingsville
Sanislav, Ioan, James Cook University
Satterfield, Joseph I., Angelo State University
Saura, Eduard, Universitat Autonoma de Barcelona
Saxon, Christopher, Tarleton State University
Scharnberger, Charles K., Millersville University
Schlische, Roy W., Rutgers, The State Univ of New Jersey
Schmidt, Keegan L., Lewis-Clark State College
Schoonmaker, Adam, Utica College
Schrank, Christoph, Queensland University of Technology
Schwerdtner, Walfried M., University of Toronto
Sears, James W., University of Montana
Shaw, Colin, Montana State University
Shuib, Mustaffa Kamal, University of Malaya
Sibson, Rick H., University of Otago
Siddoway, Christine S., Colorado College
Simony, Philip S., University of Calgary
Singleton, John, Colorado State University
Skehan, James W., Boston College
Skehan, S.J., James W., Boston College
Skemer, Philip, Washington University in St. Louis
Solar, Gary S., Buffalo State College
Sorkhabi, Rasoul, University of Utah
Spangler, Ellie, California Geological Survey
Spencer, Edgar W., Washington & Lee University
Spratt, Deborah A., University of Calgary
SRIVASTAVA, HARI B., Banaras Hindu University
Stauffer, Mel R., University of Saskatchewan
Stearns, David W., University of Oklahoma
Stewart, Kevin G., University of North Carolina, Chapel Hill
Stinson, Amy L., Irvine Valley College
Stinson, Amy L., Santiago Canyon College
Stockli, Daniel, University of Texas, Austin
Stowell Gale, Julia, Univ of Texas at Austin, Jackson Sch of Geosciences
Strayer, Luther M., California State University, East Bay
Sullivan, Walter, Colby College
Suneson, Neil H., University of Oklahoma
Surpless, Benjamin E., Trinity University
Talbot, James L., Western Washington University
Tapp, J. B., The University of Tulsa
Tate, Garrett W., Vanderbilt University
Taylor, Lansing, University of Utah
Taylor, Wanda J., University of Nevada, Las Vegas
Tewksbury, Barbara J., Hamilton College
Teyssier, Christian P., University of Minnesota, Twin Cities
Thompson, Margaret D., Wellesley College
Tikoff, Basil, University of Wisconsin-Madison
Tikoff, Basil, University of Wisconsin, Madison
Timmons, J M., New Mexico Institute of Mining and Technology
Tindall, Sarah E., Kutztown University of Pennsylvania
Titus, Sarah J., Carleton College
Tobisch, Othmar T., University of California, Santa Cruz

Toraman, Erkan, St. Lawrence University
Toro, Jaime, West Virginia University
Torvela, Taija, University of Leeds
Toy, Virginia G., University of Otago
Treloar, Peter, Kingston University
Tremblay, Alain, Universite du Quebec a Montreal
Trexler, Charles C., Ohio Wesleyan University
Trudgill, Bruce D., Colorado School of Mines
Trullenque, Ghislain, Institut Polytechnique LaSalle Beauvais (ex-IGAL)
Tso, Jonathan L., Radford University
Tubía, Jose M., Univ of the Basque Country UPV/EHU
Tull, James F., Florida State University
Tullis, Jan A., Brown University
Twiss, Robert J., University of California, Davis
Ustaszewski, Kamil M., Friedrich-Schiller-University Jena
Van Arsdale, Roy B., University of Memphis
van Bever Donker, Jan M., University of the Western Cape
van der Pluijm, Ben, University of Michigan
VanDervoort, Dane S., Geological Survey of Alabama
Vaqueiro Rodriguez, Marcos, Coruna University
Vegas, Nestor, Univ of the Basque Country UPV/EHU
Vogl, Jim, University of Florida
Voicu, Gabriel-Constantin, Universite du Quebec a Montreal
Vojtko, Rastislav, Comenius University in Bratislava
Vollmer, Frederick W., SUNY, New Paltz
Waag, Charles J., Boise State University
Waldron, John W., University of Alberta
Walker, J. Douglas, University of Kansas
Washington, Paul A., Marietta College
Waters-Tormey, Cheryl, Western Carolina University
Waterworth, Lauren H., Appalachian State University
Watkinson, A. John, Washington State University
Waugh, Richard A., University of Wisconsin, Platteville
Webber, Jeffrey R., Stockton University
Weber, John C., Grand Valley State University
Weil, Arlo B., Bryn Mawr College
Weldon, Ray J., University of Oregon
Wells, Michael L., University of Nevada, Las Vegas
West, David P., Middlebury College
Whalley, John, University of Portsmouth
Whisner, S. Christopher, Bloomsburg University
White, Joseph C., University of New Brunswick
Whitmeyer, Steven J., James Madison University
Wickham, John S., University of Texas, Arlington
Wilkerson, M. S., University of Illinois, Urbana-Champaign
Wilkerson, M. Scott, DePauw University
Williams, Michael L., University of Massachusetts, Amherst
Williams, Paul F., University of New Brunswick
Wilson, Terry J., Ohio State University
Winslow, Margaret A., City College (CUNY)
Withjack, Martha O., Rutgers, The State Univ of New Jersey
Wojtal, Steven F., Oberlin College
Woodcock, Nigel H., University of Cambridge
Woodward, Lee A., University of New Mexico
Woodwell, Grant R., University of Mary Washington
Workman-Ford, Kerry, California State University, Fresno
Yikilmaz, Burak, University of California, Davis
Yonkee, W. A., Weber State University
Yoshinobu, Aaron S., Texas Tech University
Young, David, Ohio State University
Young, Jeffrey, University of Manitoba
Zampieri, Dario, Università degli Studi di Padova
Zimmerman, Jr., Jay, Southern Illinois University Carbondale
Zuza, Andrew, University of Nevada, Reno
Zuza, Andrew V., University of Nevada

Tectonics
Abercrombie, Rachel, Boston University
Allen, Mark, Durham University
Anderson, Robert G., University of British Columbia
Andeweg, Bernd, Vrije Universiteit Amsterdam
Arkle, Jenny C., Augustana College
Artoni, Andrea, University of Parma
Ault, Alexis K., Utah State University
Avouac, Jean-Philippe, California Institute of Technology
Barnhart, William, University of Iowa
Barth, Nicolas, University of California, Riverside
Bartholomew, Mervin J., University of Memphis

Bell, Andrew, Edinburgh University
Bell, Rebecca, Imperial College
Bennett, Scott E., U.S. Geological Survey
Bickle, Michael, University of Cambridge
Bieler, David B., Centenary College of Louisiana
Bird, G. Peter, University of California, Los Angeles
Boggs, Katherine, Mount Royal University
Boland, Irene B., Winthrop University
Boutelier, David, University of Newcastle
Boyer, Robert E., University of Texas, Austin
Bradshaw, John, University of Canterbury
Bullen, Dean, University of Portsmouth
Burgette, Reed J., New Mexico State University, Las Cruces
Burgmann, Roland, University of California, Berkeley
Busby, Cathy J., University of California, Davis
Camacho, Alfredo, University of Manitoba
Cao, Wenrong, University of Nevada, Reno
Carlson, Marvin P., Unversity of Nebraska - Lincoln
Carr, Sharon D., Carleton University
Casey, John F., University of Houston
Cawood, Peter A., Monash University
Centeno Garcia, Elena, Univ Nac Autonoma de Mexico
Chapin, Charles E., New Mexico Institute of Mining & Technology
Chi, Wu-Cheng, Academia Sinica
Choi, Eunseo, University of Memphis
Church, William R., Western University
Cole, Ron B., Allegheny College
Contreras, Juan, Centro de Inv Científica y de Ed Sup de Ensenada
Corrigan, David, University of Manitoba
Cowgill, Eric S., University of California, Davis
Cox, Randel T., University of Memphis
Cunningham, Dickson, Eastern Connecticut State University
Dallmeyer, R. David, University of Georgia
Dalziel, Ian W. D., University of Texas, Austin
Das, Shamita, University of Oxford
Davis, Gregory A., University of Southern California
De Cserna, Zoltan, Univ Nac Autonoma de Mexico
de Vicente, Gerardo, Univ Complutense de Madrid
DeCelles, Peter G., University of Arizona
Derakhshani, Reza, Shahid Bahonar University of Kerman
DeVecchio, Duane, Arizona State University
Dewey, John, University of California, Davis
Dickinson, William R., University of Arizona
Dilek, Yildirm, Miami University
Dinter, David A., University of Utah
Dolan, James F., University of Southern California
Dong, Yunpeng, Western University
Dooley, Tim P., University of Texas, Austin
Driscoll, Peter E., Carnegie Institution for Science
Ducea, Mihai N., University of Arizona
Dumond, Gregory, University of Arkansas, Fayetteville
Egan, Stuart, Keele University
Egger, Anne E., Central Washington University
Engebretson, David C., Western Washington University
England, Phillip, University of Oxford
Esteban, Jose Julian, Univ of the Basque Country UPV/EHU
Feigl, Kurt L., University of Wisconsin, Madison
Ferrari Pedraglio, Luca, Univ Nac Autonoma de Mexico
Fitzgerald, Paul G., Syracuse University
Fosdick, Julie C., Indiana University, Bloomington
Foster, David A., University of Florida
Furlong, Kevin P., Pennsylvania State University, University Park
Gable, Carl W., Los Alamos National Laboratory
Gans, Phillip B., University of California, Santa Barbara
Garver, John I., Union College
Gehrels, George E., University of Arizona
Geissman, John D., University of Texas, Dallas
Gerbi, Christopher C., University of Maine
German, Chris, Woods Hole Oceanographic Institution
Gifford, Jennifer N., University of Mississippi
Gomez, Francisco, University of Missouri
Gries, John C., Wichita State University
Grujic, Djordje, Dalhousie University
Guenthner, Willy, University of Illinois, Urbana-Champaign
Hales, T. C., University of Wales
Haley, J C., Miami University
Hall, Robert, Royal Holloway University of London

Hansen, Vicki L., University of Minnesota, Duluth
Haq, Saad, Purdue University
Harley, Simon L., Edinburgh University
Harms, Tekla A., Amherst College
Harris, Nigel B., The Open University
Hatcher, Jr., Robert D., Washington University in St. Louis
Hemphill-Haley, Mark, Humboldt State University
Henderson, Robert, James Cook University
Henry, Christopher D., University of Nevada
Hughes, Sam, Exeter University
Huntington, Katharine W., University of Washington
Hurtado, Jose M., University of Texas, El Paso
Insel, Nadja, Northeastern Illinois University
Isacks, Bryan L., Cornell University
Ivins, Erik R., Jet Propulsion Laboratory
Jackson, Christopher, Imperial College
Jacobson, Carl E., Iowa State University of Science & Technology
Jadamec, Margarete, SUNY, Buffalo
Janecke, Susanne U., Utah State University
John, Hickman B., University of Kentucky
Judge, Shelley, College of Wooster
Kapp, Paul A., University of Arizona
Karlstrom, Karl E., University of New Mexico
Karson, Jeffrey, Syracuse University
Kato, Terence T., California State University, Chico
Keir, Derek, University of Southampton
Keller, Edward A., University of California, Santa Barbara
Kelley, Shari A., New Mexico Institute of Mining and Technology
Keppie, J. Duncan, Univ Nac Autonoma de Mexico
Keppie, John D., Univ Nac Autonoma de Mexico
Kirby, Eric, Oregon State University
Kirstein, Linda, Edinburgh University
Kleinspehn, Karen L., University of Minnesota, Twin Cities
Klemperer, Simon L., Stanford University
Knapp, Camelia, Oklahoma State University
Knuepfer, Peter L. K., Binghamton University
Koehn, Daniel, University of Glasgow
Kokum, Mehmet, Firat University
Koons, Peter O., University of Maine
Kronenberg, Andreas, Texas A&M University
Lamb, Melissa A., University of Saint Thomas
Lavier, Luc L., University of Texas, Austin
Lin, Jian, Woods Hole Oceanographic Institution
Lonergan, Lidia, Imperial College
Loveless, Jack, Smith College
Luo, Gang, University of Texas, Austin
Mac Niocaill, Conall, University of Oxford
Mallard, Laura A., Appalachian State University
Malone, Shawn J., Ball State University
Mann, Paul, University of Texas, Austin
Mapani, Benjamin S., University of Namibia
Marsellos, Antonios, Hofstra University
Martinez, Juventino, Univ Nac Autonoma de Mexico
Martinez-Reyes, Juventino, Univ Nac Autonoma de Mexico
May, Daniel J., University of New Haven
McClelland, William C., University of Iowa
McGill, Sally F., California State University, San Bernardino
McGrew, Allen J., University of Dayton
McNeill, Lisa, University of Southampton
McPherson, Robert C., Humboldt State University
McRivette, Michael, Albion College
Melichar, Rostislav, Masaryk University
Melosh, Henry J., University of Arizona
Mesbahi, Fatemeh, University of Tabriz
Michon, Laurent, Universite de la Reunion
Mickus, Kevin L., Missouri State University
Milan, Luke, University of New England
Molnar, Peter, University of Colorado
Moore, Gregory F., University of Hawai'i, Manoa
Moores, Eldridge M., University of California, Davis
Moser, Desmond, Western University
Moucha, Robert, Syracuse University
Mound, Jon, University of Leeds
Mueller, Karl J., University of Colorado
Murphy, J. Brendan, Saint Francis Xavier University
Möller, Andreas, University of Kansas
Nabelek, John L., Oregon State University

Faculty by Specialty

Nadin, Elisabeth S., University of Alaska, Fairbanks
Nance, R. Damian, Ohio University
Nash, Greg, University of Utah
Naylor, Mark, Edinburgh University
Ngo, Thanh X., Hanoi University of Mining & Geology
Niemi, Nathan A., University of Michigan
Niemi, Tina M., University of Missouri, Kansas City
Ocan, O. O., Obafemi Awolowo University
Onderdonk, Nate, California State University, Long Beach
Partin, Camille, University of Saskatchewan
Passchier, Cees W., Universität Mainz
Pavlis, Terry L., University of New Orleans
Peacock, Simon M., University of British Columbia
Peterman, Emily M., Bowdoin College
Pettinga, Jarg R., University of Canterbury
Phillips, Richard, University of Leeds
Pieri, David C., Jet Propulsion Laboratory
Plasienka, Dusan, Comenius University in Bratislava
Pluhar, Christopher J., California State University, Fresno
Price, Raymond A., Queen's University
Rahl, Jeffrey, Washington & Lee University
Regalla, Christine, Boston University
Reusch, Douglas N., University of Maine - Farmington
Richter, Frank M., University of Chicago
Robertson, Alastair H., Edinburgh University
Roeske, Sarah M., University of California, Davis
Ross, Gerald M., University of Calgary
Rowan, Christopher J., Kent State University
Rowley, David B., University of Chicago
Royden, Leigh H., Massachusetts Institute of Technology
Rubin, Charles M., Central Washington University
Rust, Derek, University of Portsmouth
Ryder, Isabelle, University of Liverpool
Sager, William W., University of Houston
Sample, James C., Northern Arizona University
Schellart, Wouter, Vrije Universiteit Amsterdam
Schermer, Elizabeth R., Western Washington University
Schmidt, David, University of Washington
Schneider, David, University of Ottawa
Scotese, Christopher R., University of Texas, Arlington
Searle, Mike, University of Oxford
Shail, Robin, Exeter University
Shaw, John, Harvard University
Sinha, A. Krishna, Virginia Polytechnic Institute & State University
Sintubin, Manuel, Katholieke Universiteit Leuven
Spotila, James A., Virginia Polytechnic Institute & State University
Srimal, Neptune, Florida International University
Stanley, Jessica, University of Idaho
Steltenpohl, Mark G., Auburn University
Stern, Robert J., University of Texas, Dallas
Stock, Joann M., California Institute of Technology
Strecker, Manfred, Cornell University
Streepey Smith, Meg, Earlham College
Suppe, John, University of Houston
Suter-Cargnelutti, Max, Univ Nac Autonoma de Mexico
Takeuchi, Akira, University of Toyama
Taylor, Brian, University of Hawai'i, Manoa
Taylor, Frederick W., University of Texas, Austin
Taylor, Michael H., University of Kansas
Teixell, Antonio, Universitat Autonoma de Barcelona
Teixell Cácharo, Antoni, Universitat Autonoma de Barcelona
Thigpen, Ryan, University of Kentucky
Thomas, William A., Geological Survey of Alabama
Thorkelson, Derek J., Simon Fraser University
Toke', Nathan, Utah Valley University
Turner, Jenni, University of East Anglia
Umhoefer, Paul J., Northern Arizona University
Unsworth, Martyn, University of Alberta
Valentine, Michael, Highline College
Valentino, David W., SUNY, Oswego
van Hunen, Jeroen, Durham University
Van Wijk, Jolante, New Mexico Institute of Mining and Technology
vanKeken, Peter E., Carnegie Institution for Science
Wallace, Laura, University of Texas, Austin
Watkinson, Ian M., Royal Holloway University of London
Webb, Laura E., University of Vermont
Wernicke, Brian P., California Institute of Technology

Whipple, Kelin, Arizona State University
Whittaker, Alex, Imperial College
Willis, Julie B., Brigham Young University - Idaho
Wong, Martin, Colgate University
Wright, James (Jim) E., University of Georgia
Wu, Jonny, University of Houston
Yeats, Robert S., Oregon State University
Yin, An, University of California, Los Angeles
Yule, J. Douglas, California State University, Northridge
Zamani, Behzad, University of Tabriz

Volcanology

, Anthony F., Universite de la Reunion
Aguirre Diaz, Gerardo J., Univ Nac Autonoma de Mexico
Ancochea, Eumenio, Univ Complutense de Madrid
Anderson, Steven W., University of Northern Colorado
Anderson, Jr., Alfred T., University of Chicago
Andrews, Benjamin, Smithsonian Inst / Natl Museum of Natural History
Andrews, Graham D., West Virginia University
Awdankiewicz, Marek, University of Wroclaw
Banik, Tenley, Illinois State University
Barclay, Jenni, University of East Anglia
Befus, Kenneth S., Baylor University
Blake, Stephen, The Open University
Boroughs, Scott, Washington State University
Branan, Yvonne K., Indiana University of Pennsylvania
Branney, Mike, Leicester University
Brown, David, University of Glasgow
Brown, Richard J., Durham University
Brunstad, Keith A., SUNY, Oneonta
Burkett, Brett, Collin College - Central Park Campus
Burkett, Shannon, Collin College - Central Park Campus
Bursik, Marcus I., SUNY, Buffalo
Calder, Eliza, Edinburgh University
Cameron, Barry I., University of Wisconsin, Milwaukee
Cameron, Cheryl, Alaska Div of Geological & Geophysical Surveys
Camp, Victor E., San Diego State University
Canon-Tapia, Edgardo, Centro de Inv Científica y de Ed Sup de Ensenada
Carley, Tamara, Lafayette College
Carr, Michael J., Rutgers, The State Univ of New Jersey
Carrasco-Nunez, Gerardo, Univ Nac Autonoma de Mexico
Castro, Jonathan M., Universität Mainz
Chapin, Charles E., New Mexico Institute of Mining and Technology
Chapman, Marshall, Morehead State University
Cigolini, Corrado, Università di Torino
Clarke, Amanda, Arizona State University
Cole, James W., University of Canterbury
Cole, Paul D., University of Plymouth
Connor, Charles B., University of South Florida, Tampa
Coppola, Diego, Università di Torino
Coulson, Ian, University of Regina
Crisp, Joy A., Jet Propulsion Laboratory
Crowe, Bruce M., Los Alamos National Laboratory
de Silva, Shanika, Oregon State University
Deardorff, Nicholas, Indiana University of Pennsylvania
Domagall, Abigail M., Black Hills State University
Drake, Simon, Birkbeck College
Dufek, Josef, University of Oregon
Dufek, Josef, Georgia Institute of Technology
Duffield, Wendell A., Northern Arizona University
Duller, Rob, University of Liverpool
Edmonds, Marie, University of Cambridge
Eggers, Albert A., University of Puget Sound
Escobar-Wolf, Rudiger, Michigan Technological University
Fagents, Sarah, University of Hawai'i, Manoa
Fischer, Tobias, University of New Mexico
Flood, Tim P., Saint Norbert College
Frey, Holli M., Union College
Gansecki, Cheryl A., University of Hawai'i, Hilo
Gardner, James E., University of Texas, Austin
Genareau, Kimberly, University of Alabama
Genna, Dominique, Universite du Quebec a Chicoutimi
Giachetti, Thomas, University of Oregon
Giordano, Daniele, Università di Torino
Godchaux, Martha M., Mount Holyoke College
Gonnermann, Helge, Rice University
Graettinger, Alison, University of Missouri, Kansas City
Grapenthin, Ronni, New Mexico Institute of Mining and Technology

Gravley, Darren, University of Canterbury
Green, Jack, California State University, Long Beach
Gregg, Tracy K. P., SUNY, Buffalo
Gutmann, James T., Wesleyan University
Hampton, Samuel J., University of Canterbury
Harpp, Karen, Colgate University
Harwood, Richard D., Black Hawk College
Hasenaka, Toshiaki, Kumamoto University
Hausback, Brian, California State University, Sacramento
Head, Elisabet M., Northeastern Illinois University
Heiken, Grant, Los Alamos National Laboratory
Heliker, Christina C., University of Hawai'i, Hilo
Herd, Richard, University of East Anglia
Holm, Richard F., Northern Arizona University
Hon, Ken, University of Hawai'i, Hilo
Houghton, Bruce F., University of Hawai'i, Manoa
Houle, Michel, University of Manitoba
Huber, Christian, Brown University
Hughes, Scott S., Idaho State University
Iverson, Richard, University of Washington
Jackson, Marie D., University of Utah
Johnson, Emily R., New Mexico State University, Las Cruces
Junkin, William D., Maryland Department of Natural Resources
Kavanagh, Janine, University of Liverpool
Kennedy, Ben, University of Canterbury
Kieffer, Susan W., University of Illinois, Urbana-Champaign
Klemetti, Erik, Denison University
Knesel, Kurt, Trinity University
Kobs-Nawotniak, Shannon E., Idaho State University
Koornneef, Janne, Vrije Universiteit Amsterdam
Krier, Donathon J., Los Alamos National Laboratory
Krippner, Janine, Concord University
Lang, Nicholas, Mercyhurst University
Larsen, Jessica F., University of Alaska, Fairbanks
Lavallee, Yan, University of Liverpool
Le Mevel, Helene, Carnegie Institution for Science
Lee, Rachel J., SUNY, Oswego
Lehto, Heather L., Angelo State University
LeMasurier, Wesley E., University of Colorado, Denver
Lev, Einat , Columbia University
LeVeque, Randy, University of Washington
Leyrit, Hervé, Institut Polytechnique LaSalle Beauvais (ex-IGAL)
Llewellin, Ed, Durham University
Lockwood, John P., University of Hawai'i, Hilo
Magsino, Sammantha L., National Acad of Sci, Eng, and Medicine
Manga, Michael, University of California, Berkeley
Maria, Tony, University of Southern Indiana
Mather, Tamsin A., University of Oxford
Matoza, Robin, University of California, Santa Barbara
McGarvie, Dave, The Open University
McPhie, Jocelyn, University of Tasmania
Michelfelder, Gary, Missouri State University
Moore, Joseph N., University of Utah
Moran, Seth, University of Washington
Morton, Ronald, University of Minnesota, Duluth
Neuberg, Jurgen, University of Leeds
O'Sullivan, Katie, California State University, Bakersfield
Oliveras Castro, Valentí, Universitat Autonoma de Barcelona
Ort, Michael H., Northern Arizona University
Ottavi-Pupier, Elsa, Institut Polytechnique LaSalle Beauvais (ex-IGAL)
Pallister, John S., University of Pittsburgh
Perry, Frank V., Los Alamos National Laboratory
Phipps Morgan, Jason, Cornell University
Putirka, Keith D., California State University, Fresno
Pyle, David M., University of Oxford
Quick, James E., Southern Methodist University
Reynolds, Robert W., Central Oregon Community College
Rhodes, J. Michael, University of Massachusetts, Amherst
Riggs, Nancy, Northern Arizona University
Rodriguez, Lizzette A., University of Puerto Rico
Roldan Quintana, Jaime, Univ Nac Autonoma de Mexico
Roman, Diana C., Carnegie Institution for Science
Rose, William I., Michigan Technological University
Ross, Pierre-Simon, Universite du Quebec
Russell, James K., University of British Columbia
Sanchez-Rubio, Gerardo, Univ Nac Autonoma de Mexico
Sar, Abdullah, Firat University
Saunders, Kate, Edinburgh University
Schaefer, Janet R. G., Alaska Div of Geological & Geophysical Surveys
Sennert, Sally K., Smithsonian Inst / Natl Museum of Natural History
Shea, Thomas, University of Hawai'i, Manoa
Sheridan, Michael F., SUNY, Buffalo
Smith, Alan L., California State University, San Bernardino
Solana, Camen, University of Portsmouth
Sonder, Ingo, SUNY, Buffalo
Soule, S. Adam, Woods Hole Oceanographic Institution
Stevenson, John A., Edinburgh University
Stix, John, McGill University
Straub, Susanne, Columbia University
Swanson, Donald A., University of Hawai'i, Manoa
Taylor, Rex N., University of Southampton
Teasdale, Rachel, California State University, Chico
Townsend, Meredith, University of Oregon
Ukstins Peate, Ingrid, University of Iowa
Valentine, Gregory, SUNY, Buffalo
Van Ry, Michael, Orange Coast College
Vanderkluysen, Loyc, Drexel University
Venzke, Edward, Smithsonian Inst / Natl Museum of Natural History
Vigouroux-Caillibot, Nathalie, Douglas College
Voight, Barry, Pennsylvania State University, University Park
Vroon, Pieter, Vrije Universiteit Amsterdam
Wadsworth, Fabian, Durham University
Wallace, Paul, University of Oregon
Waters, Laura E., Sonoma State University
Webber, Karen L., University of Michigan
Webber, Karen L., University of New Orleans
Wehner, Peter J., Austin Community College District
White, James D., University of Otago
Williams, Stanley N., Arizona State University
Williams-Jones, Glyn, Simon Fraser University
Wilson, Thomas, University of Canterbury
Wulff, Andrew, Western Kentucky University

Mathematical Geology

Agterberg, Frits P., University of Ottawa
Barton, Christopher C., Wright State University
Bolton, Edward W., Yale University
Campbell, Katherine, Los Alamos National Laboratory
Cathles, Lawrence M., Cornell University
Chien, Yi-Ju, Pacific Northwest National Laboratory
Chou, Charissa J., Pacific Northwest National Laboratory
Cornejo-Luna, Efrain, Univ Autonoma de Baja California Sur
Davis, John C., University of Kansas
de Hoop, Maarten, Purdue University
Doveton, John H., University of Kansas
Edwards, Stuart, University of Newcastle Upon Tyne
Faccenda, Manuele, Università degli Studi di Padova
Fairbairn, David, University of Newcastle Upon Tyne
Ford, Allistair, University of Newcastle Upon Tyne
Ford, Arianne, James Cook University
Fox, Peter A., Rensselaer Polytechnic Institute
Gordon, Terence M., University of Calgary
Harbaugh, John W., Stanford University
Harding, Chris, Iowa State University of Science & Technology
He, Changming, University of Delaware
Hesse, Marc A., University of Texas, Austin
Hunt, Allen G., Wright State University
Johnson, Glenn W., University of Utah
Journel, Andre G., Stanford University
Lee, Jejung, University of Missouri, Kansas City
Lutz, Timothy M., West Chester University
MAZZELLA, Prof. Antonio A., Universita di Cagliari
Murray, Christopher J., Pacific Northwest National Laboratory
Olea, Ricardo A., University of Kansas
Penna, Nigel, University of Newcastle Upon Tyne
Rogova, Galina L., SUNY, Buffalo
Rundberg, Robert S., Los Alamos National Laboratory
Sonder, Leslie J., Dartmouth College
Spiegelman, Marc W., Columbia University
Sun, Alexander, University of Texas, Austin
Therrien, Pierre, Universite Laval
Tuttle, Samuel , Mount Holyoke College
Walton, Ian, University of Utah

Mineral Physics

Abramson, Evan H., University of Washington
Bass, Jay D., University of Illinois, Urbana-Champaign
Berryman, Eleanor, Princeton University
Brown, J. Michael, University of Washington
Bukowinski, Mark S., University of California, Berkeley
Carpenter, Michael, University of Cambridge
Chen, Jiuhua, Florida International University
Cohen, Ronald E., Carnegie Institution for Science
Cooper, Reid F., Brown University
Duffy, Thomas S., Princeton University
Farver, John R., Bowling Green State University
Fenter, Paul, University of Illinois at Chicago
Fischer, Rebecca A., Harvard University
Goncharov, Alexander F., Carnegie Institution for Science
Heinz, Dion L., University of Chicago
Hugo, Richard, Portland State University
Jackson, Jennifer M., California Institute of Technology
Jacobsen, Steven D., Northwestern University
Jeanloz, Raymond, University of California, Berkeley
Karato, Shun-ichiro, Yale University
Kavner, Abby, University of California, Los Angeles
Kiefer, Boris, New Mexico State University, Las Cruces
Knittle, Elise, University of California, Santa Cruz
Kronenberg, Andreas, Texas A&M University
Lee, Kanani K., Yale University
Li, Baosheng, Stony Brook University
Lieberman, Robert C., Stony Brook University
Lin, Jung-Fu, University of Texas, Austin
Liu, Zhenxian, Carnegie Institution for Science
Militzer, Burkhard, University of California, Berkeley
Riggs, Eric, Texas A&M University
Salje, Ekhard, University of Cambridge
Saxena, Surendra K., Florida International University
Secco, Richard A., Western University
Stracher, Glenn B., East Georgia State College
Struzhkin, Viktor V., Carnegie Institution for Science
Tschauner, Oliver, University of Nevada, Las Vegas
Tyburczy, James A., Arizona State University
Vocadlo, Lidunka, University College London
Weidner, Donald J., Stony Brook University
Williams, Quentin, University of California, Santa Cruz
Zhang, Jin, University of New Mexico

Medical Geology
Buck, Brenda J., University of Nevada, Las Vegas
Finkelman, Robert B., University of Texas, Dallas
Fuge, Ron, Aberystwyth University
Gillmore, Gavin, Kingston University
Hunt, Andrew, University of Texas, Arlington
Markowitz, Steven, Queens College (CUNY)
Morabia, Alfredo, Queens College (CUNY)
Pasteris, Jill D., Washington University in St. Louis

Forensic Geology
Williams, Thomas, University of Idaho

ECONOMIC GEOLOGY
General Economic Geology
Adedcyin, A. D., University of Ilorin
Adetunji, A, Obafemi Awolowo University
Akabzaa, Thomas M., University of Ghana
Akande, S. O., University of Ilorin
Alavi, Ghafour, University of Tabriz
Alexandre, Paul, Brandon University
Alexandrov, Eugene A., Queens College (CUNY)
Allard, Gilles O., University of Georgia
Anger-Kraavi, Annela, University of East Anglia
Ansdell, Kevin M., University of Saskatchewan
Arehart, Greg B., University of Nevada, Reno
Ashley, Paul, University of New England
Banerjee, Neil, Western University
Barnes, Sarah- J., Universite du Quebec a Chicoutimi
Barton, Mark D., University of Arizona
Beaudoin, Georges, Universite Laval
Bland, Douglas, New Mexico Institute of Mining & Technology
Bolarinwa, A. T., University of Ibadan
Bornhorst, Theodore J., Michigan Technological University

Bouzari, Farhad, University of British Columbia
Bradshaw, Peter M., University of British Columbia
Brimhall, George H., University of California, Berkeley
Burt, Donald M., Arizona State University
Busch, Lawrence, California Geological Survey
Bustillo, Manuel, Univ Complutense de Madrid
Calagari, Ali Asghar, University of Tabriz
Camprubi, Antoni, Univ Nac Autonoma de Mexico
Carter, James L., University of Texas, Dallas
Castor, Stephen B., University of Nevada
Chang, Zhaoshan, Colorado School of Mines
Cheney, Eric S., University of Washington
Cheng, Yanbo, James Cook University
Chi, Guoxiang, University of Regina
Clark, Kenneth F., University of Texas, El Paso
Clendenin, Jr., Charles W., South Carolina Dept of Natural Resources
Clinkenbeared, John, California Geological Survey
Conly, Andrew G., Lakehead University
Cook, Robert B., Auburn University
Cooke, David R., University of Tasmania
Coolbaugh, Mark F., University of Nevada
Corbineau, Lucien, Institut Polytechnique LaSalle Beauvais (ex-IGAL)
Coron, Cynthia R., Southern Connecticut State University
Corral, Isaac, James Cook University
Coveney Jr, Raymond M., University of Missouri, Kansas City
Craig, James R., Virginia Polytechnic Institute & State University
Craw, Dave, University of Otago
Crawford, Thomas J., University of West Georgia
Dewaele, Stijn, Ghent University
Di Gregorio, Monica, University of Leeds
Donovan, John F., Winona State University
Duke, Norman A., Western University
Dunst, Brian J., Pennsylvania Bur of Topo & Geologic Survey
Elueze, A. A., University of Ibadan
Evans, Thomas J., University of Wisconsin, Extension
Flawn, Peter T., University of Texas, Austin
Franco-Rubio, Miguel, Universidad Autonoma de Chihuahua
Garcia-Gutierrez, Luis, Universidad Autonoma de San Luis Potosi
Gauthier, Michel, Universite du Quebec a Montreal
Gemmell, J B., University of Tasmania
Gibson, Andy, University of Portsmouth
Gillerman, Virginia S., Boise State University
Gillerman, Virginia S., University of Idaho
Greyling, Lynnette N., University of Cape Town
Guan, Dabo, University of Leeds
Guha, Jayanta, Universite du Quebec a Chicoutimi
Gulen, Gurcan, University of Texas, Austin
Hannington, Mark, University of Ottawa
Hanson, William D., Arkansas Geological Survey
Harris, DeVerle P., University of Arizona
Hayman, Patrick C., Queensland University of Technology
Hedenquist, Jeffrey W., University of Ottawa
Hickey, Kenneth A., University of British Columbia
Hickson, Catherine J., University of British Columbia
Hohn, Michael E., West Virginia Geological & Economic Survey
Hollenbaugh, Kenneth M., Boise State University
Holley, Elizabeth, Colorado School of Mines
Hollings, Peter N., Lakehead University
Hoyer, Lauren, University of KwaZulu-Natal
Huber, Matthew, University of the Free State
Ikonnikova, Svetlana, University of Texas, Austin
Jago, Bruce C., Laurentian University, Sudbury
Jean-Marc, MONTEL, Ecole Nationale Supérieure de Géologie (ENSG)
Jebrak, Michel, Universite du Quebec a Montreal
Jessey, David R., California State Polytechnic University, Pomona
Kamilli, Robert J., University of Arizona
Kamona, Frederick A., University of Namibia
Karginoglu, Yusuf , Firat University
Kazimoto, Emmanuel O., University of Dar es Salaam
Keith, Jeffrey D., Brigham Young University
Kelly, William C., University of Michigan
Kesler, Stephen E., University of Michigan
Kish, Stephen A., Florida State University
Kontak, Daniel J., Laurentian University, Sudbury
La Berge, Gene L., University of Wisconsin, Oshkosh
Laki, Sam, Central State University
Large, Ross R., University of Tasmania

Lasemi, Zakaria, Illinois State Geological Survey
Lassetter, William L., Division of Geology and Mineral Resources
Layton-Matthews, Daniel, Queen's University
Lehrberger, Gerhard, Technical University of Munich
Levinson, Alfred A., University of Calgary
Linnen, Robert, University of Waterloo
Lorenzoni, Irene, University of East Anglia
Maher, Kierran, New Mexico Institute of Mining and Technology
Mantilla Figueroa, Luis C., Universidad Industrial de Santander
Masterman, Steven S., Alaska Div of Geological & Geophysical Surveys
Mattieu, Lucie, Universite du Quebec a Chicoutimi
mcClenaghan, Seán h., Trinity College
McLemore, Virginia T., New Mexico Institute of Mining and Technology
Mensah, Emmanuel, Kwame Nkrumah Univ of Science and Tech
Misra, Kula C., University of Tennessee, Knoxville
Mohammad Reza, Hosseinzadeh, University of Tabriz
Morton, Penelope, University of Minnesota, Duluth
Mumin, A. Hamid, Brandon University
Muntean, John, University of Nevada
Muntean, John, University of Nevada, Reno
Muszer, Antoni, University of Wroclaw
Nasraoui, Mohamed, Institut Polytechnique LaSalle Beauvais (ex-IGAL)
Nex, Paul A., University of the Witwatersrand
Niewendorp, Clark, Oregon Dept of Geology and Mineral Industries
Nimis, Paolo, Università degli Studi di Padova
Okunlola, Ougbenga. A., University of Ibadan
Olivo, Gema, Queen's University
Orozco-Villasenor, Francisco Javier, Univ Autonoma de San Luis Potosi
Paez, H. A., McMaster University
Paterson, Colin J., South Dakota School of Mines & Technology
Pietrzak-Renaud, Natalie, Western University
Potra, Adriana, University of Arkansas, Fayetteville
Prichard, Hazel, University of Wales
Pride, Douglas E., Ohio State University
Prochaska, Walter, University Leoben
Ramagwede, Fhatuwani L., Geological Survey of South Africa
Rambaud, Fabienne M., Austin Community College District
Ressel, Mike, University of Nevada
Ridley, John R., Colorado State University
Robertson, Daniel E., Monroe Community College
Robertson, James M., University of Wisconsin-Madison
Rossetti, Piergiorgio, Università di Torino
Sagiroglu, Ahmet, Firat University
Sasmaz, Ahmet, Firat University
Schardt, Christian, University of Minnesota, Duluth
Seedorff, Eric, University of Arizona
Shackley, Simon J., Edinburgh University
Shaver, Stephen A., Sewanee: University of the South
Shortt, Niamh K., Edinburgh University
Siahcheshm, Kamal, University of Tabriz
Simmons, Peter, University of East Anglia
Solecki, Andrzej, University of Wroclaw
Stucker, James D., Ohio Dept of Natural Resources
Talley, John H., University of Delaware
Titley, Spencer R., University of Arizona
Tribe, Selina, Douglas College
Tritlla, Jordi, Univ Nac Autonoma de Mexico
Turski, Mark P., Plymouth State University
Twelker, Evan, Alaska Div of Geological & Geophysical Surveys
Ulloa, Salvador, Univ Nac Autonoma de Mexico
van der Horst, Dan, Edinburgh University
van Hinsberg, Vincent, McGill University
Warren, Rachel, University of East Anglia
Webster, James D., American Museum of Natural History
Werdon, Melanie B., Alaska Div of Geological & Geophysical Surveys
Wilkins, Colin, University of Plymouth
Williams, David A., University of Kentucky
Wilton, Derek H., Memorial University of Newfoundland
Wolfgram, Diane, Montana Tech of the University of Montana
Yao, Yon, Rhodes University
Yates, Martin G., University of Maine
Yellich, John, Western Michigan University
Zentilli, Marcos, Dalhousie University

Coal
Abdullah, Wan Hasiah, University of Malaya
Asuen, Godwin O., University of Benin
Bailey, Judy, University of Newcastle
Bustin, R. Marc, University of British Columbia
Calder, John, Dalhousie University
Cardott, Brian J., University of Oklahoma
Carroll, Richard E., Geological Survey of Alabama
Cecil, Blaine, West Virginia University
Chenoweth, Cheri A., Illinois State Geological Survey
Drobniak, Agnieszka, Indiana University
Elrick, Scott D., University of Illinois
Elrick, Scott D., Illinois State Geological Survey
Fedorko, Nick, West Virginia University
Flood, Peter, University of New England
Hoffman, Gretchen K., New Mexico Institute of Mining & Technology
Hower, James C., University of Kentucky
Huggett, William, Southern Illinois University Carbondale
Kehoe, Kelsey, Wyoming State Geological Survey
Khawaja, Ikram U., Youngstown State University
Korose, Christopher P., Illinois State Geological Survey
Korose, Christopher P., Illinois State Geological Survey
Lentz, Leonard J., Pennsylvania Bur of Topo & Geologic Survey
Lindberg, Jonathan W., Pacific Northwest National Laboratory
Mastalerz, Maria D., Indiana University, Bloomington
Mastalerz, Maria, Indiana University
Meyer, Rebecca A., Indiana University
Miller, Barry W., Tennessee Geological Survey
Mukhopadhyay, Prasanta, Dalhousie University
Myers, Alan R., Illinois State Geological Survey
Neubaum, John C., Pennsylvania Bur of Topo & Geologic Survey
Obrad, Jennifer M., Illinois State Geological Survey
Renton, John J., West Virginia University
Rimmer, Susan M., Southern Illinois University Carbondale
Shaulis, James R., Pennsylvania Bur of Topo & Geologic Survey
Ward, Colin R., University of New South Wales
Weisenfluh, Gerald A., University of Kentucky
Wright, Chris, Ohio Dept of Natural Resources

Metallic Ore Deposits
Barnett, Douglas B., Pacific Northwest National Laboratory
Bissig, Thomas, University of British Columbia
Broughton, William A., University of Wisconsin, Platteville
Brown, Philip E., University of Wisconsin-Madison
Campbell, Finley A., University of Calgary
Cedillo, Esteban, Univ Nac Autonoma de Mexico
Chang, Zhaoshan, James Cook University
Chapman, Robert J., University of Leeds
Chavez, Jr., William X., New Mexico Institute of Mining and Technology
Clark, Alan H., Queen's University
Cline, Jean S., University of Nevada, Las Vegas
Corona-Esquivel, Rodolfo, Univ Nac Autonoma de Mexico
Corriveau, Louise, Universite du Quebec
Crowe, Douglas E., University of Georgia
Dilles, John H., Oregon State University
Dube, Benoit, Universite du Quebec
Durazo-Tapia, Gustavo E., Universidad Estatal de Sonora
Einaudi, Marco T., Stanford University
Fifarek, Richard, Southern Illinois University Carbondale
Friehauf, Kurt, Kutztown University of Pennsylvania
Gaboury, Damien, Universite du Quebec a Chicoutimi
Gibson, Harold, Carleton University
Gibson, Harold L., Laurentian University, Sudbury
Good, David, Western University
Graf, Jr., Joseph L., Southern Oregon University
Guilbert, John M., University of Arizona
Hagni, Richard D., Missouri University of Science and Technology
Hannington, Mark D., Carleton University
Hart, Craig J., University of British Columbia
Hooda, Peter, Kingston University
Keays, Reid R., Laurentian University, Sudbury
Kinnaird, Judith A., University of the Witwatersrand
Kissin, Stephen A., Lakehead University
Kisvarsanyi, Geza K., Missouri University of Science and Technology
Koestel, Mark, Mt. San Antonio College
Kyle, J. Richard, University of Texas, Austin
Lesher, Michael, Laurentian University, Sudbury
Linnen, Robert, Western University
Lodge, Robert, University of Wisconsin, Eau Claire
Macfarlane, Andrew W., Florida International University
McAllister, Arnold L., University of New Brunswick
McLemore, Virginia, New Mexico Institute of Mining & Technology

Melchiorre, Erik, California State University, San Bernardino
Menuge, Julian F., University College Dublin
Monecke, Thomas, Colorado School of Mines
Morel-Kraepiel, Anne, Princeton University
Morton, Roger D., University of Alberta
Mossman, David J., Mount Allison University
Muchez, Philippe, Katholieke Universiteit Leuven
Muntean, John L., University of Nevada, Reno
Naldrett, Anthony J., University of Toronto
Newberry, Rainer J., University of Alaska, Fairbanks
Ortega, Lorena, Univ Complutense de Madrid
Petersen, Erich U., University of Utah
Powell, Wayne G., Grad Sch of the City Univ of New York
Richards, Jeremy P., Laurentian University, Sudbury
Richardson, Carson A., Arizona Geological Survey
Ripley, Edward M., Indiana University, Bloomington
Romberger, Samuel B., Colorado School of Mines
Ruvalcaba-Ruiz, Delfino C., Universidad Autonoma de San Luis Potosi
Salaun, Pascal, University of Liverpool
Shelton, Kevin L., University of Missouri
Simmons, Stuart, University of Utah
Sinclair, Alastair J., University of British Columbia
Spry, Paul G., Iowa State University of Science & Technology
Thompson, John F., Cornell University
Thompson, John F. H., University of British Columbia
Tingley, Joseph V., University of Nevada
Tosdal, Richard, University of British Columbia
Uzunlar, Nuri, South Dakota School of Mines & Technology
van der Berg, Stan, University of Liverpool
Vasallo, Fernando, Univ Nac Autonoma de Mexico
Vassallo-Morales, Luis F., Univ Nac Autonoma de Mexico
Watkinson, David H., Carleton University
Wheeler, Greg, California State University, Sacramento
Williamson, Ben, Exeter University

Industrial Minerals
Austin, George S., New Mexico Institute of Mining & Technology
Bal Akkoca, Dicle, Firat University
Barker, James M., New Mexico Institute of Mining & Technology
Berg, Richard B., Montana Tech of The University of Montana
Calengas, Peter L., Western Illinois University
Crespo, Elena, Univ Complutense de Madrid
Gregory, Robert, Wyoming State Geological Survey
Krukowski, Stanley T., University of Oklahoma
Lasemi, Zakaria, Illinois State Geological Survey
McClellan, Guerry H., University of Florida
Moreton, Kim, Exeter University
Power, W. Robert, Georgia State Univ
Simandl, George J., University of Victoria
Sousa, Luís M., Universidade de Trás-os-Montes e Alto Douro
Spooner, Edward T. C., University of Toronto
Sutherland, Wayne, Wyoming State Geological Survey
Van Straaten, H. Peter, University of Guelph

Oil and Gas
Anderson, Ross, Heriot-Watt University
Balázs, László, Eotvos Lorand University
Brown, Alan, West Virginia University
Burst, John F., Missouri University of Science and Technology
C., Jeffrey C., Youngstown State University
Cox, John, Mount Royal University
Crockett, Joan E., Illinois State Geological Survey
Deisher, Jeffrey, Ohio Dept of Natural Resources
Eldridge, Jayson, Indiana University
Ersoy, Erkal, Heriot-Watt University
Evans, Martin J., Cornell University
Finley, Robert J., Illinois State Geological Survey
Gillis, Robert, Alaska Div of Geological & Geophysical Surveys
Gluyas, Jon, Durham University
Gooding, Patrick J., University of Kentucky
Grube, John P., Illinois State Geological Survey
Harder, Brian J., Louisiana State University
Herriott, Trystan, Alaska Div of Geological & Geophysical Surveys
Hickman, John, University of Kentucky
Hills, Denise J., Geological Survey of Alabama
Hooks, Chris H., Geological Survey of Alabama
Horn, Marty R., Louisiana State University
Hosseini, Seyyed Abolfazi, University of Texas, Austin
Huff, Bryan G., Illinois State Geological Survey
Hulett, Sam, Ohio Dept of Natural Resources
Johnston, III, John E., Louisiana State University
King, Carey, University of Texas, Austin
Krevor, Samuel, Imperial College
Leetaru, Hannes E., Illinois State Geological Survey
Leighton, Morris W., University of Illinois, Urbana-Champaign
Leone, James, New York State Geological Survey
Lepain, David, Alaska Div of Geological & Geophysical Surveys
Levey, Raymond A., University of Utah
Lucia, F. J., University of Texas, Austin
Lynds, Ranie, Wyoming State Geological Survey
Malinconico, Mary Ann, Lafayette College
Markowski, Antonette K., Pennsylvania Bur of Topo & Geologic Survey
Nicholas, Chris, Trinity College
Nielsen, Peter J., Utah Geological Survey
Nuttall, Brandon C., University of Kentucky
Oldershaw, Michael, Bakersfield College
Parris, Thomas M., University of Kentucky
Pascussi, Michael, New York State Geological Survey
Potter, Eric C., Univ of Texas at Austin, Jackson Sch of Geosciences
Seyler, Beverly, Illinois State Geological Survey
Shew, Roger D., University of North Carolina, Wilmington
Tew, Nick, University of Alabama
Vanden Bergy, Michael, Utah Geological Survey
Vigrass, Laurence W., University of Regina
Waid, Christopher, Ohio Dept of Natural Resources
Webb, Nathan D., Illinois State Geological Survey
Wood, Lesli J., University of Texas, Austin
Zeidouni, Mehdi, University of Texas, Austin

Construction Materials (SSG)
Cnudde, Veerle, Ghent University

Ore Deposits (Other)
Baig, Ayat, Brandon University
Carranza, Emmanuel John, University of KwaZulu-Natal
Haroldson, Erik L., Austin Peay State University
Mantei, Erwin J., Missouri State University
Runyon, Simone, University of Wyoming

GEOCHEMISTRY
General Geochemistry
Atefanei, Dan, Alexandru Ioan Cuza
Abimbola, A. E., University of Ibadan
Aines, Roger D., Lawrence Livermore National Laboratory
Ajayi, T. R., Obafemi Awolowo University
Akaegbobi, I. M., University of Ibadan
Allen, Douglas, Salem State University
Allen-King, Richelle, SUNY, Buffalo
Amonette, Alexandra B., Pacific Northwest National Laboratory
Anderson, Robert F., Columbia University
April, Richard, Colgate University
Atekwana, Eliot A., University of Delaware
Ayers, John C., Vanderbilt University
Ayling, Bridget, University of Nevada
Babcock, R. S., Western Washington University
Bales, Roger C., University of Arizona
Banks, David, University of Leeds
Barker, Colin, The University of Tulsa
Barnes, Hubert L., Pennsylvania State University, University Park
Barnes, Jaime D., University of Texas, Austin
Basu, Anirban, Royal Holloway University of London
Beets, Kay J., Vrije Universiteit Amsterdam
Beiersdorfer, Raymond E., Youngstown State University
Belmonte, Donato, Universita di Genova
Bender, Michael L., Princeton University
Berendsen, Pieter, University of Kansas
Berger, Peter M., Illinois State Geological Survey
Bergquist, Bridget, University of Toronto
Berkelhammer, Max, University of Illinois at Chicago
Bernhard, Joan M., Woods Hole Oceanographic Institution
Bernier, Luc, McMaster University
Bird, Dennis K., Stanford University
Bissada, K. K., Rice University
Blake, Ruth E., Yale University
Blowes, David W., University of Waterloo
Bodnar, Robert J., Virginia Polytechnic Institute & State University

Boger, Phillip D., SUNY, Geneseo
Bohm, Christian, University of Manitoba
Bolge, Louise, Columbia University
Boschetti, Tiziano, University of Parma
Bouman, Heather, University of Oxford
Bourcier, Bill L., Lawrence Livermore National Laboratory
Brantley, Susan L., Pennsylvania State University, University Park
Broecker, Wallace S., Columbia University
Brooks, Paul D., University of Arizona
Brounce, Maryjo, University of California, Riverside
Bryant, Ernest A., Los Alamos National Laboratory
Bryce, Julia G., University of New Hampshire
Bulgariu, Dumitru, Alexandru Ioan Cuza
Burton, Jacqueline C., Argonne National Laboratory
Burton, Kevin, Durham University
Buseck, Peter R., Arizona State University
Butler, Ian B., Edinburgh University
Cai, Yue, Columbia University
Camacho, Elsa A., Pacific Northwest National Laboratory
Catling, David C., University of Washington
Centorbi, Tracey, University of Maryland
Chague-Goff, Catherine, University of New South Wales
Chatterjee, Nilanjan, Massachusetts Institute of Technology
Chen, Chu-Yung, University of Illinois, Urbana-Champaign
Chillrud, Steven, Columbia University
Chorover, Jonathan D., University of Arizona
Churchill, Ron C., California Geological Survey
Claeys, Philippe, Vrije University Brussel
Clark, Jordan, University of California, Santa Barbara
Class, Connie, Columbia University
Clay, Patricia L., University of Manchester
Coe, Douglas A., Montana Tech of the University of Montana
Coggon, Jude A., University of Southampton
Cole, David R., Ohio State University
Collier, James D., Fort Lewis College
Cook, Robert B., Oak Ridge National Laboratory
Cowman, Tim C., South Dakota Dept of Env and Nat Res
Crawford, William A., Bryn Mawr College
Darrah, Thomas, Ohio State University
Dasgupta, Tathagata, Kent State University
Davisson, M L., Lawrence Livermore National Laboratory
Dawson, M. Robert, Iowa State University of Science & Technology
Day, James, University of California, San Diego
de Ignacio, Cristina, Univ Complutense de Madrid
Deng, Baolin, New Mexico Institute of Mining and Technology
Derry, Louis A., Cornell University
Ding, Kang, University of Minnesota, Twin Cities
Dong, Hailiang, Miami University
Dooley, John H., New Jersey Geological and Water Survey
Dostal, Jarda, Dalhousie University
Downes, Hilary, Birkbeck College
Drew, Douglas A., Montana Tech of the University of Montana
Duffy, Clarence J., Los Alamos National Laboratory
Dunbar, Nelia W., New Mexico Institute of Mining and Technology
Durham, William, Massachusetts Institute of Technology
Eby, G. Nelson, University of Massachusetts, Lowell
Eiler, John M., California Institute of Technology
Erhardt, Andrea M., University of Kentucky
Evans, Owen C., Stony Brook University
Eves, Robert L., Southern Utah University
Fajkoviæ, Hana, University of Zagreb
Fall, Andras, University of Texas, Austin
Farley, Kenneth A., California Institute of Technology
Farmer, G. Lang, University of Colorado
Faure, Gunter, Ohio State University
Feigenson, Mark D., Rutgers, The State Univ of New Jersey
Fernandez, Diego, University of Utah
Finkel, Robert C., Lawrence Livermore National Laboratory
Finkelstein, David, Hobart & William Smith Colleges
Flegal, Russell, University of California, Santa Cruz
Flowers, Rebecca M., University of Colorado
Foustoukos, Dionysis, Carnegie Institution for Science
Fraser, Donalf G., University of Oxford
Frey, Bonnie A., New Mexico Institute of Mining & Technology
Gambrell, Robert P., Louisiana State University
Gammons, Christopher H., Montana Tech of the University of Montana
Garbesi, Karina, California State University, East Bay
George, Graham, University of Saskatchewan
Ghiorso, Mark , University of Washington
Gilfillan, Stuart M., Edinburgh University
Gleason, James D., University of Michigan
Goff, Fraser, University of New Mexico
Goff, Fraser, Los Alamos National Laboratory
Goldstein, Steven, Columbia University
Gonzalez, Luis , King Fahd University of Petroleum and Minerals
Gosselin, David C., Unversity of Nebraska - Lincoln
Gouldey, Jeremy C., Grand Valley State University
Goulet, Richard, University of Ottawa
Grant, Willard H., Emory University
Gulbranson, Erik L., University of Wisconsin, Milwaukee
Gurlea, Lawrence P., Youngstown State University
Gustin, Mae, University of Nevada, Reno
Gutierrez, Melida, Missouri State University
Gysi, Alex, Colorado School of Mines
Halden, Norman M., University of Manitoba
Hall, Cynthia V., West Chester University
Hansen, Robert N., University of the Free State
Hart, Stanley R., Woods Hole Oceanographic Institution
Harvey, Jason, University of Leeds
Hattori, Keiko, University of Ottawa
Hemming, Sidney, Columbia University
Henderson, Gideon, University of Oxford
Hervig, Richard, Arizona State University
Hickey-Vargas, Rosemary, Florida International University
Hinman, Nancy W., University of Montana
Hinton, Richard, Edinburgh University
Hodges, Floyd N., Pacific Northwest National Laboratory
Hudson, Robert J. M., University of Illinois, Urbana-Champaign
Ige, A. O., Obafemi Awolowo University
Irving, Tony, University of Washington
Izon, Gareth, University of St. Andrews
Jacobs, Gary K., Oak Ridge National Laboratory
Jacobson, Roger, University of Nevada, Reno
Jarvis, Ian, Kingston University
Jerde, Eric, Morehead State University
Jin, Qusheng, University of Oregon
John, Bratton F., Wayne State University
Johnson, Clark M., University of Wisconsin-Madison
Johnston, David T., Harvard University
Kalender, Leyla, Firat University
Kallemeyn, Gregory, University of California, Los Angeles
Kamber, Balz S., Trinity College
Kambewa, Chamunorwa , Tshwane University of Technology
Kamenov, George D., University of Florida
Kaplan, Isaac R., University of California, Los Angeles
Kara, Hatice, Firat University
Kaszuba, John P., University of Wyoming
Katz, Richard, University of Oxford
Kellman, Lisa M., Dalhousie University
Kenna, Timothy, Columbia University
Kim, Sang-Tae, McMaster University
Kiro Feldman, Yael, Columbia University
Kirste, Dirk, Simon Fraser University
Knudsen, Andrew, Lawrence University
Koeman-Shields, Elizabeth, Angelo State University
Kolesar, Peter T., Utah State University
Koppers, Anthony, Oregon State University
Koretsky, Carla, Western Michigan University
Kotzer, Tom, University of Saskatchewan
Krawczynski, Michael J., Washington University in St. Louis
Ku, Teh-Lung, University of Southern California
Kuzyk, Zou Zou, University of Manitoba
Labidi, Jabrane, University of Maryland
Lange, Eric, Ball State University
Langmuir, Charles H., Harvard University
Le Roex, Anton, University of Cape Town
Lechler, Paul J., University of Nevada
Lee, Cin-Ty A., Rice University
Legore, Virginia L., Pacific Northwest National Laboratory
Lehloenya, Pelele, University of the Free State
Lentz, David R., University of New Brunswick
Lerman, Abraham, Northwestern University
Leuta-Madondo, Palesa, University of KwaZulu-Natal
Leybourne, Matthew I., Laurentian University, Sudbury

Faculty by Specialty

Faculty by Specialty

Lindenmeier, Clark W., Pacific Northwest National Laboratory
Liu, Xiaoming, University of North Carolina, Chapel Hill
Livens, Francis, University of Manchester
Locke II, Randall A., University of Illinois
Longerich, Henry, Memorial University of Newfoundland
Lopez, Dina L., Ohio University
Loyd, Sean, California State University, Fullerton
Lu, Zunli, Syracuse University
Luttge, Andreas, Rice University
Lyons, Timothy W., University of California, Riverside
Macpherson, Colin G., Durham University
Magson, Justine, University of the Free State
Mango, Helen N., Castleton University
Manning, Christina, Royal Holloway University of London
Manya, Shukrani, University of Dar es Salaam
Marquez, L. Lynn, Millersville University
Martin, Paul F., Pacific Northwest National Laboratory
Martin, Scot T., Harvard University
Marzoli, Andrea, Università degli Studi di Padova
McClelland, James W., University of Texas, Austin
McDonough, William F., University of Maryland
McGrail, Bernard P., Pacific Northwest National Laboratory
McGuire, Jennifer, University of Saint Thomas
McIntosh, Jennifer C., University of Arizona
McKibben, Michael A., University of California, Riverside
McKinley, Galen, Columbia University
McLennan, Scott M., Stony Brook University
Meduniæ, Gordana, University of Zagreb
Meixner, Thomas, University of Arizona
Mensing, Teresa, Ohio State University
Mercy, Edward L., Lakehead University
Meyer, William T., Argonne National Laboratory
Meyzen, Christine M., Università degli Studi di Padova
Michel, Jacqueline, University of New Orleans
Milkov, Alexei, Colorado School of Mines
MOINE, Bertrand N., Université Jean Monnet, Saint-Etienne
Moldovan, Brett, University of Saskatchewan
Mortlock, Richard , Rutgers, The State Univ of New Jersey
Mukasa, Samuel, University of Minnesota, Twin Cities
Mukhopadhyay, Sujoy, University of California, Davis
Murgulet, Valeriu, Texas A&M University, Corpus Christi
Murowchick, James B., University of Missouri, Kansas City
Murphy, David T., Queensland University of Technology
Nagy, Kathryn L., University of Illinois at Chicago
Naranjo, Ramon, University of Nevada, Reno
Nondorf, Lea, Arkansas Geological Survey
Nyame, Frank K., University of Ghana
O'Driscoll, Nelson, Acadia University
O'Hara, Matthew J., Pacific Northwest National Laboratory
O'Neil, Jonathan, University of Ottawa
Odokuma-Alonge, Ovie, University of Benin
Odom, LeRoy A., Florida State University
Okulewicz, Steven C., Hofstra University
Olanrewaju, Johnson, Gannon University
Olatunji, A. S., University of Ibadan
Olsen, Khris B., Pacific Northwest National Laboratory
Olszewski, Kathy, SUNY, Maritime College
Omelon, Christopher, University of Texas, Austin
Otamendi, Juan E., Universidad Nacional de Rio Cuarto
Ozturk, Nevin, Firat University
Palinkaš, Ladislav, University of Zagreb
Palmer, Martin R., University of Southampton
Pant, Hari, Lehman College (CUNY)
Papineau, Dominic, University College London
Parker, Kent E., Pacific Northwest National Laboratory
Parris, Thomas M., University of Kentucky
Patera, Edward S., Los Alamos National Laboratory
Pearce, J A., Cardiff University
Peate, David W., University of Iowa
Peng, Yongbo, Louisiana State University
Piccoli, Philip M., University of Maryland
Pichevin, Laetitia, Edinburgh University
Pickering, Ingrid J., University of Saskatchewan
Pietranik, Anna, University of Wroclaw
Pintilei, Mitica, Alexandru Ioan Cuza
Plank, Terry, Columbia University
Plante, Martin, Universite Laval
Pommier, Anne, University of California, San Diego
Porcelli, Don, University of Oxford
Poreda, Robert J., University of Rochester
Pourmand, Ali, University of Miami
Pourret, Olivier, Institut Polytechnique LaSalle Beauvais (ex-IGAL)
Powell, Brian A., Clemson University
Pracny, Pavel, Masaryk University
Prevec, Steve, Rhodes University
Price, Douglas M., Youngstown State University
Prohiæ, Esad, University of Zagreb
Prowse, Terry D., University of Waterloo
Prytulak, Julie, Durham University
Ptacek, Carol J., University of Waterloo
Rae, James, University of St. Andrews
Ramos, Frank C., New Mexico State University, Las Cruces
Reed, Mark H., University of Oregon
Reeves, Claire, University of East Anglia
Reiners, Peter, University of Arizona
Richardson, Steve, University of Cape Town
Rickard, D, Cardiff University
Riedinger, Natascha, Oklahoma State University
Roberts, Stephen, University of Southampton
Rohs, C. Renee, Northwest Missouri State University
Rollinson, Hugh, University of Derby
Romanak, Katherine D., Univ of Texas at Austin, Jackson Sch of Geo
Rose, Arthur W., Pennsylvania State University, University Park
Rose, Catherine V., Trinity College
Rudnick, Roberta L., University of Maryland
Ruiz, Joaquin, University of Arizona
Ruttenberg, Kathleen C., University of Hawai'i, Manoa
Saal, Alberto E., Brown University
Sack, Richard, University of Washington
Sakyi, Patrick A., University of Ghana
Salters, Vincent J., Florida State University
Samson, Iain M., University of Windsor
Savov, Ivan, University of Leeds
Saxena, Surenda K., Grad Sch of the City Univ of New York
Schaef, Herbert T., Pacific Northwest National Laboratory
Schaefer, Joerg, Columbia University
Schauble, Edwin A., University of California, Los Angeles
Schlosser, Peter, Columbia University
Schrag, Daniel P., Harvard University
Schubert, Brian, University of Louisiana at Lafayette
Seitz, Jeffery C., California State University, East Bay
Senko, John M., University of Akron
Serne, R. Jeffrey, Pacific Northwest National Laboratory
Shahar, Anat, Carnegie Institution for Science
Shevenell, Lisa, University of Nevada
Shields, Robin, Heriot-Watt University
Shields-Zhou, Graham, University College London
Sims, Kenneth W., University of Wyoming
Skinner, Brian J., Yale University
Skippen, George B., Carleton University
Slater, Gregory F., McMaster University
Smethie, William, Columbia University
Speidel, David H., Queens College (CUNY)
Spencer, Ronald J., University of Calgary
Stack, Andrew, Georgia Institute of Technology
Stebbins, Jonathan F., Stanford University
Steele, Kenneth F., University of Arkansas, Fayetteville
Stevenson, Ross, Universite du Quebec a Montreal
Strmiæ Palinkaš, Sabina, University of Zagreb
Stubbins, Aron, Northeastern University
Sturchio, Neil C., Argonne National Laboratory
Sultan, Mohamed, Argonne National Laboratory
Susak, Nicholas J., University of New Brunswick
Taylor, Richard P., Carleton University
Teagle, Damon A., University of Southampton
Tefend, Karen S., University of West Georgia
Teng, Fangzhen, University of Washington
Thomas, Jim, University of Nevada, Reno
Tierney, Kate E., Denison University
Tomascak, Paul B., SUNY, Oswego
Tomson, Mason B., Rice University
Trueman, Clive, University of Southampton
Tsikos, Hari, Rhodes University
Van Geen, Alexander, Columbia University

Van Orman, James A., Case Western Reserve University
Vann, Jamie, West Chester University
Varekamp, Johan C., Wesleyan University
Vulava, Vijay M., College of Charleston
Wagner, Rick, Geological Survey of Alabama
Walker, John L., Argonne National Laboratory
Walther, John V., Southern Methodist University
Wang, Hong, Illinois State Geological Survey
Wang, Yang, Florida State University
Warburton, David L., Florida Atlantic University
Warren, Lesley A., McMaster University
Watkins, James, University of Oregon
Whipkey, Charles, University of Mary Washington
White, William B., Pennsylvania State University, University Park
Wigley, Rochelle, University of New Hampshire
Williams, Jeremy C., Kent State University
Wilson, Emily, Franklin and Marshall College
Winckler, Gisela, Columbia University
Wogelius, Roy, University of Manchester
Wooden, Joseph, Stanford University
Worrall, Fred, Durham University
Wronkiewicz, David J., Argonne National Laboratory
Wyman, Derek A., University of Saskatchewan
Yan, Beizhan, Columbia University
Yang, Qiang, Columbia University
Yardley, Bruce W., University of Leeds
Young, Nicolas, Columbia University
Zanazzi, Alessandro, Utah Valley University
Zeman, Josef, Masaryk University
Zhang, Tongwei, University of Texas, Austin
Zheng, Yan, Queens College (CUNY)
Zolotov, Mikhail Y., Arizona State University

Analytical Geochemistry
Aguilera-Ortiz, Irma, Univ Nac Autonoma de Mexico
Ash, Richard, University of Maryland
Barnes, Melanie A., Texas Tech University
Beauchamp, Marc, Western University
Bedard, Paul, Universite du Quebec a Chicoutimi
Benjamin, Timothy M., Los Alamos National Laboratory
Bhattacharyya, Sidhartha, University of Alabama
Blamey, Nigel J., New Mexico Institute of Mining and Technology
Bowen, Scott M., Los Alamos National Laboratory
Branam, Tracy D., Indiana University
Brandvold, Lynn A., New Mexico Institute of Mining & Technology
Cahill, Richard A., Illinois State Geological Survey
Castaneda, Isla, University of Massachusetts, Amherst
Catalano, Jeff, Washington University in St. Louis
Chappaz, Anthony, Central Michigan University
Cheng, Zhongqi, Brooklyn College (CUNY)
Clark, Malcolm W., Southern Cross University
Cole, David R., University of Tennessee, Knoxville
De Carlo, Eric H., University of Hawai'i, Manoa
Dewey, Janet, University of Wyoming
Effert-Fanta, Shari E., Illinois State Geological Survey
Elam, Tim, University of Washington
Essling, Alice M., Argonne National Laboratory
Frew, Nelson M., Woods Hole Oceanographic Institution
Gabitov, Rinat, Mississippi State University
Gagnon, Joel E., University of Windsor
Garcia, Sammy R., Los Alamos National Laboratory
Gavriloaiei, Traian, Alexandru Ioan Cuza
Halbig, Joseph B., University of Hawai'i, Hilo
Hart, Brian R., Western University
Hemming, N. G., Queens College (CUNY)
Hinthorne, James R., Central Washington University
Hoe, Teh Guan, University of Malaya
Holt, Ben D., Argonne National Laboratory
Hu, Wan-Ping (Sunny), Queensland University of Technology
Huff, Edmund A., Argonne National Laboratory
Huff, Ashley, Montana Tech of The University of Montana
Jackson, Brian P., Dartmouth College
Jedrysek, Mariusz O., University of Wroclaw
Johnson, Carl G., Woods Hole Oceanographic Institution
Jones, Graham, Southern Cross University
Kersten, Michael, Universität Mainz
Klinkhammer, Gary P., Oregon State University
Knaack, Charles, Washington State University
Knudstrup, Renee, Salem State University
Krapac, Ivan G., Illinois State Geological Survey
Kuehn, Stephen C., Concord University
Kuentz, David C., Miami University
Landry, Peter B., Woods Hole Oceanographic Institution
Le Roux, Petrus J., University of Cape Town
Locke, David C., Grad Sch of the City Univ of New York
Lottermoser, Bernd, Exeter University
Lundstrom, Elizabeth, University of Minnesota, Twin Cities
MacDonald, Iain, University of Wales
Marshall, Dan D., Simon Fraser University
McGrath, Steve F., Montana Tech of The University of Montana
McMurtry, Gary M., University of Hawai'i, Manoa
Messo, Charles W., University of Dar es Salaam
Mock, R. Stephen, University of Montana Western
Mora-Klepeis, Gabriela, University of Vermont
Mujumba, Jean K., University of Dar es Salaam
Navarre-Sitchler, Alexis, Colorado School of Mines
Neill, Owen K., Washington State University
Nelsen, Lori, Furman University
Olesik, John W., Ohio State University
Olmsted, Wayne, Montana Tech of the University of Montana
Orouji, Maqsood, University of Tabriz
Orr, Robert D., Pacific Northwest National Laboratory
Papike, James J., University of New Mexico
Pike, Steven M., Woods Hole Oceanographic Institution
Pun, Aurora, University of New Mexico
Ragland, Paul C., Florida State University
Ranville, James, Colorado School of Mines
Rose, Timothy, Smithsonian Inst / Natl Museum of Natural History
Rouff, Ashaki, Queens College (CUNY)
Russell, Shane, Queensland University of Technology
Savard, Dany, Universite du Quebec a Chicoutimi
Schneider, John F., Argonne National Laboratory
Shen, Linhan, Princeton University
Shimizu, Nobumichi, Woods Hole Oceanographic Institution
Singer, Jared W., Rensselaer Polytechnic Institute
Smith, Catherine H., Los Alamos National Laboratory
Smith, Florence P., Argonne National Laboratory
Snow, Jonathan, University of Houston
Student, James J., Central Michigan University
Sylvester, Paul J., Memorial University of Newfoundland
Thomas, Donald M., University of Hawai'i, Manoa
Thompson, Christopher J., Pacific Northwest National Laboratory
Timmer, Jaqueline R., Montana Tech of The University of Montana
Underwood, Ben, Indiana University, Bloomington
Van Loon, Lisa, Western University
Villasenor-Cabral, Maria G., Univ Nac Autonoma de Mexico
Voelker, Bettina, Colorado School of Mines
Volborth, Alexis, Montana Tech of the University of Montana
Wang, Alian, Washington University in St. Louis
Warwick, Phillip, University of Southampton
Waugh, Truman, University of Kansas
Weis, Dominique A., University of British Columbia
Williams, Kim R., Colorado School of Mines
Yang, Panseok, University of Manitoba

Experimental Petrology/Phase Equilibria
Ackerson, Michael R., Rensselaer Polytechnic Institute
Agee, Carl A., University of New Mexico
Anderson, G M., University of Toronto
Asimow, Paul D., California Institute of Technology
Brenan, James M., Dalhousie University
Candela, Philip A., University of Maryland
Carey, James W., Los Alamos National Laboratory
Cottrell, Elizabeth, University of Maryland
Cottrell, Elizabeth, Smithsonian Inst / Natl Museum of Natural History
Davis, Fred A., University of Minnesota, Duluth
Eggler, David H., Pennsylvania State University, University Park
Ernst, W. Gary, Stanford University
Fedortchouk, Yana, Dalhousie University
Fei, Yingwei, University of Maryland
Fei, Yingwei, Carnegie Institution for Science
Feineman, Maureen D., Pennsylvania State University, University Park
Frank, Mark R., Northern Illinois University
Gilbert, M. Charles, University of Oklahoma
Hajash, Andrew, Texas A&M University
Herzberg, Claude T., Rutgers, The State Univ of New Jersey

Hewitt, David A., Virginia Polytechnic Institute & State University
Hsu, Liang-Chi, University of Nevada
Jenkins, David M., Binghamton University
Johnston, A. Dana, University of Oregon
Kilinc, Attila I., University of Cincinnati
Koster Van Groos, August F., University of Illinois at Chicago
Koziol, Andrea M., University of Dayton
Lambart, Sarah, University of Utah
Lazar, Codi, California State University, San Bernardino
Liang, Yan, Brown University
Lindsley, Donald H., Stony Brook University
London, David, University of Oklahoma
Lundstrom, Craig C., University of Illinois, Urbana-Champaign
McCauley, Marlene, Guilford College
Minarik, William G., McGill University
Mosenfelder, Jed, University of Minnesota, Twin Cities
Munoz, James L., University of Colorado
Mysen, Bjorn O., Carnegie Institution for Science
Nekvasil, Hanna, Stony Brook University
Newton, Robert C., University of California, Los Angeles
Parman, Stephen, Brown University
Presnall, Dean C., University of Texas, Dallas
Robert, Genevieve, Bates College
Rutherford, Malcolm J., Brown University
Seewald, Jeffrey S., Woods Hole Oceanographic Institution
Simmons, Craig, Colorado School of Mines
Spera, Frank J., University of California, Santa Barbara
Steiner, Jeffrey, City College (CUNY)
Stolper, Edward M., California Institute of Technology
Tait, C. Drew, Los Alamos National Laboratory
Ustunisik, Gokce K., South Dakota School of Mines & Technology
van Westrenen, Wim, Vrije Universiteit Amsterdam
Walker, David, Columbia University
Watson, E. Bruce, Rensselaer Polytechnic Institute
Windom, Kenneth E., Iowa State University of Science & Technology
Withers, Tony, Western University
Wyllie, Peter J., California Institute of Technology

Exploration Geochemistry
Abdurrahman, A., University of Ilorin
Ayling, Bridget F., University of Nevada, Reno
Borojeviæ Šoštariæ, Sibila, University of Zagreb
Closs, L. Graham, Colorado School of Mines
Coggon, Rosalind M., University of Southampton
Cohen, David R., University of New South Wales
da Silva, Eduardo F., Universidade de Aveiro
Dreschoff, Gisela, University of Kansas
Fletcher, William K., University of British Columbia
Gazel, Esteban, Cornell University
Goodell, Philip C., University of Texas, El Paso
Mathur, Ryan, Juniata College
McGoldrick, Peter J., University of Tasmania
Moon, Charlie, Exeter University
Pearson, Graham, University of Alberta
Price, Vaneaton, Clemson University
Stanley, Clifford R., Acadia University
White, William M., Cornell University
Whitehead, Robert E., Laurentian University, Sudbury
Williams-Jones, Anthony E., McGill University
Yanez, Carlos, Univ Nac Autonoma de Mexico

Geochronology & Radioisotopes
Allen, Charlotte M., Queensland University of Technology
Amidon, Will, Middlebury College
Ankney, Meagan, Northwestern University
Antinao, José Luis, Indiana University
Archibald, Doug A., Queen's University
Asmerom, Yemane, University of New Mexico
Baksi, Ajoy K., Louisiana State University
Baldwin, Suzanne L., Syracuse University
Bauer, Annie, University of Wisconsin-Madison
Baxter, Ethan, Boston College
Beard, Brian L., University of Wisconsin-Madison
Bell, Keith, Carleton University
Bickford, M. E., Syracuse University
Black, Stuart, University of Reading
Blackburn, Terrence, University of California, Santa Cruz
Blenkinsop, John, Carleton University

Blusztajn, Jurek, Woods Hole Oceanographic Institution
Blythe, Ann, Occidental College
Brown, Roderick, University of Glasgow
Brown, Tom, University of Washington
Butler, Gilbert W., Los Alamos National Laboratory
Carlson, Richard W., Carnegie Institution for Science
Chamberlain, Edwin P., Los Alamos National Laboratory
Chamberlain, Kevin R., University of Wyoming
Cheng, Hai, University of Minnesota, Twin Cities
Chew, David, Trinity College
Clark, George S., University of Manitoba
Coleman, Drew S., University of North Carolina, Chapel Hill
Copeland, Peter, University of Houston
Cowart, James B., Florida State University
Creaser, Robert A., University of Alberta
Crowley, Jim, Boise State University
Crowley, Quentin G., Trinity College
Curtice, Joshua M., Woods Hole Oceanographic Institution
Curtis, David B., Los Alamos National Laboratory
Curtis, Garniss H., University of California, Berkeley
Davis, Donald W., University of Toronto
De Grave, Johan, Ghent University
DePaolo, Donald J., University of California, Berkeley
Dickin, Alan P., McMaster University
Dulai, Henrietta, University of Hawai'i, Manoa
Dunning, Gregory R., Memorial University of Newfoundland
Dutton, Andrea, University of Florida
Eddy, Michael P., Princeton University
Edwards, R. Lawrence, University of Minnesota, Twin Cities
Enkelmann, Eva, University of Calgary
Fearey, Bryan L., Los Alamos National Laboratory
Fenton, Cassandra, Colorado Mesa University
Fernández Suárez, Javier, Univ Complutense de Madrid
Foland, Kenneth A., Ohio State University
Forman, Stephen, Baylor University
Frost, Carol D., University of Wyoming
Fullagar, Paul D., University of North Carolina, Chapel Hill
Galindo, María del Carmen, Univ Complutense de Madrid
Gallup, Christina D., University of Minnesota, Duluth
Gancarz, Alexander J., Los Alamos National Laboratory
Gaschnig, Richard M., University of Massachusetts, Lowell
Gaudette, Henri E., University of New Hampshire
Geibert, Walter, Edinburgh University
Georgiev, Svetoslav, Colorado State University
Goehring, Brent M., Tulane University
Gosse, John C., Dalhousie University
Gromet, L. P., Brown University
Hall, Chris, University of Michigan
Hames, Willis E., Auburn University
Hamilton, Michael A., University of Toronto
Hanan, Barry B., San Diego State University
Hanes, John A., Queen's University
Harbottle, Garman, Stony Brook University
Harrison, T. Mark, University of California, Los Angeles
Hayatsu, Akio, Western University
Heaman, Larry M., University of Alberta
Heatherington, Ann L., University of Florida
Heizler, Matthew T., New Mexico Institute of Mining & Technology
Heizler, Matthew T., New Mexico Institute of Mining and Technology
Hodges, Kip V., Arizona State University
Horan, Mary F., Carnegie Institution for Science
Hourigan, Jeremy, University of California, Santa Cruz
Hughen, Konrad A., Woods Hole Oceanographic Institution
Idleman, Bruce D., Lehigh University
Jacobsen, Stein B., Harvard University
Jull, A. J. Timothy, University of Arizona
Karner, Daniel B., Sonoma State University
Kelley, Simon, Edinburgh University
Kimbrough, David L., San Diego State University
Konter, Jasper G., University of Hawai'i, Manoa
Krummenacher, Daniel, San Diego State University
Kuiper, Klaudia, Vrije Universiteit Amsterdam
Kurz, Mark D., Woods Hole Oceanographic Institution
Lakatos, Stephen, York College (CUNY)
Lassiter, John, University of Texas, Austin
Lepper, Kenneth, North Dakota State University
Lively, Rich, University of Minnesota

Long, Leon E., University of Texas, Austin
Lopez, Margarita, Centro de Inv Científica y de Ed Sup de Ensenada
Maneiro, Kathryn A., Wheaton College
Manton, William I., University of Texas, Dallas
Markun, Francis J., Argonne National Laboratory
McDowell, Fred W., University of Texas, Austin
McIntosh, William C., New Mexico Institute of Mining and Technology
McLean, Noah, University of Kansas
McNutt, Robert, McMaster University
McWilliams, Michael O., Stanford University
Menninga, Clarence, Calvin College
Mertz, Dieter, Universität Mainz
Miller, Brent, Texas A&M University
Miller, Charles M., Los Alamos National Laboratory
Miller, Donald S., Rensselaer Polytechnic Institute
Miller, Geoffrey G., Los Alamos National Laboratory
Min, Kyoungwon, University of Florida
Mock, Timothy D., Carnegie Institution for Science
Monteleone, Brian D., Woods Hole Oceanographic Institution
Moore, Willard S., University of South Carolina
Moran-Zenteno, Dante J., Univ Nac Autonoma de Mexico
Morris, David, Los Alamos National Laboratory
Mortensen, James K., University of British Columbia
Mueller, Paul A., University of Florida
Mukasa, Samuel B., University of Michigan
Murrell, Michael T., Los Alamos National Laboratory
Nelson, Bruce K., University of Washington
Noakes, John E., University of Georgia
Omar, Gomaa I., University of Pennsylvania
Oretega Rivera, Amabel M., Univ Nac Autonoma de Mexico
Orlandini, Kent A., Argonne National Laboratory
Osmond, John K., Florida State University
Palin, J. Michael, University of Otago
Pardi, Richard R., William Paterson University
Perrin, Richard E., Los Alamos National Laboratory
Placzek, Christa, James Cook University
Polyak, Victor J., University of New Mexico
Poths, Jane, Los Alamos National Laboratory
Puchtel, Igor, University of Maryland
Rasbury, E. Troy, Stony Brook University
Renne, Paul, University of California, Berkeley
Rhodes, Edward J., University of California, Los Angeles
Rink, W. J., McMaster University
Rinterknecht, Vincent, University of St. Andrews
Rioux, Matt, University of California, Santa Barbara
Rokop, Donald J., Los Alamos National Laboratory
Rubin, Kenneth H., University of Hawai'i, Manoa
Samson, Scott D., Syracuse University
Schmitz, Mark D., Boise State University
Schneider, Robert J., Woods Hole Oceanographic Institution
Schoene, R B., Princeton University
Schroeder, Norman C., Los Alamos National Laboratory
Seidemann, David E., Grad Sch of the City Univ of New York
Seidemann, David E., Brooklyn College (CUNY)
Selby, Dave, Durham University
Shea, Erin, University of Alaska, Anchorage
Shirey, Steven B., Carnegie Institution for Science
Silver, Leon T., California Institute of Technology
Singer, Bradley S., University of Wisconsin-Madison
Sole, Jesu, Univ Nac Autonoma de Mexico
Southon, John, University of California, Irvine
Spell, Terry L., University of Nevada, Las Vegas
Spencer, Joel Q., Kansas State University
Srinivasan, Balakrishnan, Pondicherry University
Stewart, Brian W., University of Pittsburgh
Stone, John O., University of Washington
Strachan, Rob, University of Portsmouth
Stuiver, Minze, University of Washington
Swanson, Karen, William Paterson University
Swanson, Terry W., University of Washington
Swisher III, Carl C., Rutgers, The State Univ of New Jersey
Szymanowski, Dawid, Princeton University
Tera, Fouad, Carnegie Institution for Science
Thibodeau, Alyson M., Dickinson College
Tomlinson, Emma L., Trinity College
Turek, Andrew, University of Windsor
Turrin, Brent D., Rutgers, The State Univ of New Jersey

Van Schmus, W. Randall, University of Kansas
Vermeesch, Pieter, University College London
Vervoort, Jeffrey D., Washington State University
Walker, Richard J., University of Maryland
Walter, Robert C., Franklin and Marshall College
Wampler, J. Marion, Georgia State Univ
White, William M., Cornell University
Widom, Elisabeth, Miami University
Wielicki, Matthew, University of Alabama
Wielicki, Michelle, University of Alabama
Wijbrans, Jan, Vrije Universiteit Amsterdam
Wolfsberg, Kurt, Los Alamos National Laboratory
Woodard, Jeremy, University of KwaZulu-Natal
Yang, Gang, Colorado State University
Yin, Qing-zhu, University of California, Davis
Zeitler, Peter K., Lehigh University
Zimmerman, Aaron, Colorado State University
Zou, Haibo, Auburn University
Zreda, Marek, University of Arizona

Low-temperature Geochemistry
Achten, Christine, Universitaet Muenster
Agrawal, Abinash, Wright State University
Aharon, Paul, University of Alabama
Ahm, Anne-Sofie, Princeton University
Aja, Stephen U., Grad Sch of the City Univ of New York
Aja, Stephen U., Brooklyn College (CUNY)
Alessi, Daniel S., University of Alberta
Alexander, Clark R., Georgia Institute of Technology
Amend, Jan, University of Southern California
Anand, Pallavi, The Open University
Andersen, C. B., Clemson University
Andersen, C. Brannon, Furman University
Angino, Ernest E., University of Kansas
Arrhenius, Gustaf, University of California, San Diego
Baker, Leslie L., University of Idaho
Balogh-Brunstad, Zsuzsanna, Hartwick College
Banner, Jay L., University of Texas, Austin
Beckingham, Barbara, College of Charleston
Benner, Shawn, Boise State University
Benninger, Larry K., University of North Carolina, Chapel Hill
Bergslien, Elisa T., Buffalo State College
Bertine, Kathe K., San Diego State University
Bickmore, Barry R., Brigham Young University
Bird, MIchael I., James Cook University
Bischoff, William D., Wichita State University
Blanton, Jackson O., Georgia Institute of Technology
Bogner, Jean E., University of Illinois at Chicago
Bowser, Carl J., University of Wisconsin-Madison
Brabander, Daniel J., Wellesley College
Brand, Uwe, Brock University
Breecker, Daniel O., University of Texas, Austin
Brook, Edward J., Oregon State University
Brooks, Scott C., Oak Ridge National Laboratory
Burke, Andrea, University of St. Andrews
Burke, Roger A., University of Georgia
Cantrell, Kirk J., Pacific Northwest National Laboratory
Cao, Hongsheng, Wichita State University
Capo, Rosemary C., University of Pittsburgh
Carroll, Susan A., Lawrence Livermore National Laboratory
Casey, William H., University of California, Davis
Catalano, Jeffrey G., Washington University in St. Louis
Cerling, Thure E., University of Utah
Chermak, John A., Virginia Polytechnic Institute & State University
Chesworth, Ward, University of Guelph
Christina, Thomas D., Georgia Institute of Technology
Clark, Jordan F., University of California, Santa Barbara
Cody, Anita M., Iowa State University of Science & Technology
Crockford, Peter, Princeton University
Crossey, Laura J., University of New Mexico
Datta, Saugata, Kansas State University
Derry, Louis A., Cornell University
Diebold, Frank E., Montana Tech of the University of Montana
Donahoe, Rona J., University of Alabama
Dostie, Philip, Bates College
Dove, Patricia M., Virginia Polytechnic Institute & State University
Drake, John C., University of Vermont
Dworkin, Stephen I., Baylor University

Early, Thomas O., Oak Ridge National Laboratory
Eggleston, Carrick M., University of Wyoming
Elliott, W. Crawford, Georgia State Univ
Elswick, Erika R., Indiana University, Bloomington
Elwood Madden, Andrew S., University of Oklahoma
Elwood Madden, Megan E., University of Oklahoma
Engel, Annette S., University of Tennessee, Knoxville
Evans, Les J., University of Guelph
Fein, Jeremy B., University of Notre Dame
Fenter, Paul A., Argonne National Laboratory
Ferrell, Jr., Ray E., Louisiana State University
Fortner, Sarah K., Wittenberg University
Frape, Shaun K., University of Waterloo
Frisia, Silvia, University of Newcastle
Fryer, Brian J., University of Windsor
Gamerdinger, Amy P., Pacific Northwest National Laboratory
García Lorenzo, María de la Luz, Univ Complutense de Madrid
Gerke, Tammie, Miami University
Graf, Donald L., University of Illinois, Urbana-Champaign
Grandstaff, David E., Temple University
Graney, Joseph R., Binghamton University
Graustein, William C., Yale University
Greenberg, Sallie, Illinois State Geological Survey
Gunter, William D., University of Calgary
Haas, Johnson R., Western Michigan University
Hannigan, Robyn, University of Massachusetts, Boston
Hanor, Jeffrey S., Louisiana State University
Harrison, Wendy J., Colorado School of Mines
Hasenmueller, Elizabeth, Saint Louis University
Heikoop, Jeffrey, University of South Florida, Tampa
Herman, Janet S., University of Virginia
Herndon, Elizabeth M., Kent State University
Higgins, John A., Princeton University
Hirons, Steve, Birkbeck College
Hixon, Amy E., University of Notre Dame
Hluchy, Michele M., Alfred University
Hochella, Jr., Michael F., Virginia Polytechnic Institute & State University
Holdren, George R., Pacific Northwest National Laboratory
Hounslow, Arthur, Oklahoma State University
Hui, Alice, Indiana University, Bloomington
Hutcheon, Ian E., University of Calgary
Icenhower, Jonathan P., Pacific Northwest National Laboratory
Icopini, Gary, Montana Tech of The University of Montana
Ito, Emi, University of Minnesota, Twin Cities
Jacinthe, Pierre-Andre, Indiana Univ / Purdue Univ, Indianapolis
Jackson, Kenneth J., Lawrence Livermore National Laboratory
Jacobson, Andrew D., Northwestern University
Jahnke, Richard, Georgia Institute of Technology
Janecky, David R., Los Alamos National Laboratory
Karlsson, Haraldur R., Texas Tech University
Kaste, James, College of William & Mary
Kelly, Jacque L., Georgia Southern University
Kettler, Richard M., University of Nebraska, Lincoln
Kirby, Carl S., Bucknell University
Knapp, Elizabeth P., Washington & Lee University
Kneeshaw, Tara A., Grand Valley State University
Krabbenhoft, David, University of Wisconsin-Madison
Kramer, James R., McMaster University
Krupka, Kenneth M., Pacific Northwest National Laboratory
Ku, Timothy C., Wesleyan University
Kubicki, James D., University of Texas, El Paso
Kump, Lee R., Pennsylvania State University, University Park
Kurtz, Andrew, Boston University
Langman, Jeffrey, University of Idaho
Larsen, Daniel, University of Memphis
Lawrence, James, University of Houston
Leckie, James O., Stanford University
Liang, Liyuan, Oak Ridge National Laboratory
Lindberg, Steven E., Oak Ridge National Laboratory
Lindsay, Matthew B., University of Saskatchewan
Locke, Randall, Illinois State Geological Survey
Long, Austin, University of Arizona
Long, David T., Michigan State University
Lowenstein, Tim K., Binghamton University
Macalady, Donald L., Colorado School of Mines
Martin, Candace E., University of Otago
Martin, Ellen E., University of Florida

Martin, Jonathan B., University of Florida
Martini, Anna M., Amherst College
Mattigod, Shas V., Pacific Northwest National Laboratory
Mayes, Melanie A., University of Tennessee, Knoxville
Maynard, J. Barry, University of Cincinnati
Mazer, James J., Argonne National Laboratory
McArthur, John M., University College London
McGee, David, Massachusetts Institute of Technology
McGehee, Thomas L., Texas A&M University, Kingsville
McPhail, D C "Bear", Australian National University
Merino, Enrique, Indiana University, Bloomington
Moore, Joel, Towson University
Morel, Francois M M., Princeton University
Munk, LeeAnn, University of Alaska, Anchorage
Nelson, Marc A., University of Arkansas, Fayetteville
Nesbitt, H W., Western University
Neumann, Klaus, Ball State University
Newell, Dennis L., Utah State University
Nezat, Carmen A., Eastern Washington University
Niemitz, Jeffery W., Dickinson College
Noll, Mark R., SUNY, The College at Brockport
Norton, Stephen A., University of Maine
Nzengung, Valentine A., University of Georgia
Oelkers, Eric, University College London
Olsen, Amanda A., University of Maine
Oster, Jessica L., Vanderbilt University
Panno, Samuel V., Illinois State Geological Survey
Parnell, Roderic A., Northern Arizona University
Patton, Terri L., Argonne National Laboratory
Perdrial, Julia, University of Vermont
Peters, Stephen C., Lehigh University
Pichler, Thomas, University of South Florida, Tampa
Pingitore, Jr., Nicholas E., University of Texas, El Paso
Poulson, Simon, University of Nevada, Reno
Poulton, Simon, University of Leeds
Rademacher, Laura K., University of the Pacific
Raven, Morgan, University of California, Santa Barbara
Reardon, Eric J., University of Waterloo
Reeder, Richard J., Stony Brook University
Rhodes, Amy L., Smith College
Riebe, Clifford S., University of Wyoming
Rimstidt, J. Donald, Virginia Polytechnic Institute & State University
Roberts, Sarah K., Lawrence Livermore National Laboratory
Rogers, Pamela Z., Los Alamos National Laboratory
Romanek, Christopher, Furman University
Rouff, Ashaki, Rutgers, The State Univ of New Jersey, Newark
Roychoudhury, Alakendra N., Stellenbosch University
Runnells, Donald D., University of Colorado
Rustad, James R., University of California, Davis
Ryan, Peter C., Middlebury College
Schlautman, Mark, Clemson University
Scholz, Denis, Universität Mainz
Schoonen, Martin A., Stony Brook University
Schoonmaker, Jane E., University of Hawai'i, Manoa
Schroth, Andrew W., University of Vermont
Shiel, Alyssa E., Oregon State University
Sigman, Daniel M., Princeton University
Sjostrom, Derek, Rocky Mountain College
Skidmore, Mark L., Montana State University
Stillings, Lisa L., University of Nevada, Reno
Stoessell, Ronald K., University of New Orleans
Stone, Alan T., Johns Hopkins University
Sturchio, Neil C., University of Delaware
Suarez, Celina, University of Arkansas, Fayetteville
Sverjensky, Dimitri A., Johns Hopkins University
Swart, Peter K., University of Miami
Taillefert, Martial, Georgia Institute of Technology
Taylor, Wayne A., Los Alamos National Laboratory
Telmer, Kevin, University of Victoria
Templeton, Alexis, University of Colorado
Thurber, David L., Queens College (CUNY)
Thurber, David L., Grad Sch of the City Univ of New York
Tipper, Ed, University of Cambridge
Toner, Brandy M., University of Minnesota, Twin Cities
Turner, Robert S., Oak Ridge National Laboratory
Valsami-Jones, Eugenia, University of Birmingham
Van Der Flier-Keller, Eileen, University of Victoria

Varner, Ruth K., University of New Hampshire
Veizer, Jan, University of Ottawa
Velbel, Michael A., Michigan State University
Voglesonger, Kenneth M., Northeastern Illinois University
Walder, Ingar F., New Mexico Institute of Mining and Technology
Walker, Charles T., California State University, Long Beach
Wallace, Adam F., University of Delaware
Walter, Lynn M., University of Michigan
Wang, Jianwei, Louisiana State University
Wasylenki, Laura E., Indiana University, Bloomington
Wasylenki, Laura, Northern Arizona University
Wehmiller, John F., University of Delaware
Weiss, Dominik, Imperial College
Welch, Susan A., Ohio State University
West, A. Joshua (Josh), University of Southern California
Whittemore, Donald O., University of Kansas
Widanagamage, Inoka, University of Mississippi
Wood, James R., Michigan Technological University
Woods, Terri L., East Carolina University
Wunsch, David R., University of Delaware
Xu, Jie, University of Texas, El Paso
Yuretich, Richard F., University of Massachusetts, Amherst
Zhaohui, Li, University of Wisconsin, Parkside

Marine Geochemistry
Hoenisch, Baerbel, Columbia University
Adkins, Jess F., California Institute of Technology
Aller, Robert C., Stony Brook University
Aller, Robert C., SUNY, Stony Brook
Barry, Peter, Woods Hole Oceanographic Institution
Basak, Chandranath, California State University, Bakersfield
Berelson, William M., University of Southern California
Betzer, Peter R., University of South Florida
Biscaye, Pierre E., Columbia University
Bollinger, Marsha S., Winthrop University
Breier, John, University of Texas, Rio Grande Valley
Broecker, Wallace S., Columbia University
Byrne, Robert H., University of South Florida
Cahill, Kevin, Woods Hole Oceanographic Institution
Calvert, Stephen E., University of British Columbia
Cave, Rachel R., National University of Ireland Galway
Charette, Matthew A., Woods Hole Oceanographic Institution
Church, Thomas M., University of Delaware
Coble, Paula G., University of South Florida
Cole, Catherine, University of St. Andrews
Compton, John, University of Cape Town
Cuomo, Carmela, University of New Haven
Dai, Minhan, Woods Hole Oceanographic Institution
De Carlo, Eric H., University of Hawai'i, Manoa
Dickson, Andrew G., University of California, San Diego
Doney, Scott C., Woods Hole Oceanographic Institution
Dutkiewicz, Stephanie, Massachusetts Institute of Technology
Farmer, Jesse, Princeton University
Fehrenbacher, Jennifer, Oregon State University
Feng, Huan E., Montclair State University
Filippelli, Gabriel M., Indiana Univ / Purdue Univ, Indianapolis
Fones, Gary, University of Portsmouth
Francois, Roger, University of British Columbia
Fulweiler, Robinson, Boston University
Galy, Valier, Woods Hole Oceanographic Institution
Gerwick, William H., University of California, San Diego
Glazer, Brian, University of Hawai'i, Manoa
Godfrey, Linda, Rutgers, The State Univ of New Jersey
Gospodinova, Kalina D., Woods Hole Oceanographic Institution
Goudreau, Joanne, Woods Hole Oceanographic Institution
Graham, David W., Oregon State University
Grassian, Vicki, University of California, San Diego
Hammerschmidt, Chad, Wright State University
Hammond, Douglas E., University of Southern California
Hansel, Colleen, Woods Hole Oceanographic Institution
Hardison, Amber K., University of Texas, Austin
Heath, G. Ross, University of Washington
Henderson, Paul, Woods Hole Oceanographic Institution
Henley, Sian F., Edinburgh University
Horner, Tristan, Woods Hole Oceanographic Institution
Hu, Xinping, Texas A&M University, Corpus Christi
Ingall, Ellery D., Georgia Institute of Technology
Jayakumar, Amal, Princeton University

Johnson, Martin, University of East Anglia
Juranek, Lauren W., Oregon State University
Kastner, Miriam, University of California, San Diego
Klein, Frieder, Woods Hole Oceanographic Institution
Latimer, Jennifer C., Indiana State University
Levas, Stephen J., University of Wisconsin, Whitewater
Li, Yuan-Hui, University of Hawai'i, Manoa
Long, Matthew, Woods Hole Oceanographic Institution
Luther, III, George W., University of Delaware
Mackenzie, Fred T., University of Hawai'i, Manoa
Maerz, Christian, University of Leeds
Malahoff, Alexander, University of Hawai'i, Manoa
Marconi, Dario, Princeton University
McCorkle, Daniel C., Woods Hole Oceanographic Institution
McDermott, Jill I., Lehigh University
McGillis, Wade, Columbia University
McKay, Jennifer L., Oregon State University
McManus, Jerry, Columbia University
Mills, Rachel A., University of Southampton
Montoya, Joseph, Georgia Institute of Technology
Moore, Bradley S., University of California, San Diego
Mucci, Alfonso, McGill University
Murray, Richard, Boston University
Nuester, Jochen, California State University, Chico
Orians, Kristin J., University of British Columbia
Osburn, Chris, North Carolina State University
Owen, Robert M., University of Michigan
Patterson, Molly, Binghamton University
Pedersen, Thomas F., University of Victoria
Peucker-Ehrenbrink, Bernhard, Woods Hole Oceanographic Institution
Popp, Brian N., University of Hawai'i, Manoa
Poyla, David, University of Manchester
Ravizza, Gregory, University of Hawai'i, Manoa
Ries, Justin, Northeastern University
Rosenthal, Yair, Rutgers, The State Univ of New Jersey
Russell, Ann D., University of California, Davis
Rutberg, Randye L., Hunter College (CUNY)
Sansone, Francis J., University of Hawai'i, Manoa
Schoonmaker, Jane E., University of Hawai'i, Manoa
Sericano, Jose L., Texas A&M University
Seyfried, Jr., William E., University of Minnesota, Twin Cities
Sherrell, Robert M., Rutgers, The State Univ of New Jersey
Simpson, Jr., H. James, Columbia University
Smith, Joseph P., United States Naval Academy
Stillings, Lisa, University of Nevada, Reno
Sulanowska, Margaret, Woods Hole Oceanographic Institution
Sundby, Bjorn, McGill University
Sutherland, Bruce, University of Alberta
Takahashi, Taro, Columbia University
Tan, Chunyang, University of Minnesota, Twin Cities
Them, II, Theodore R., College of Charleston
Thompson, Geoffrey, Woods Hole Oceanographic Institution
Tivey, Margaret K., Woods Hole Oceanographic Institution
Torres, Marta E., Oregon State University
Tripati, Aradhna, University of California, Los Angeles
Trull, Thomas, Woods Hole Oceanographic Institution
Ullman, William J., University of Delaware
Valentine, David, University of California, Santa Barbara
Wakefield, S J., Cardiff University
Wang, Weihong, Utah Valley University
Wang, Z. Aleck, Woods Hole Oceanographic Institution
Wankel, Scott, Woods Hole Oceanographic Institution
Weiss, Ray F., University of California, San Diego
Winckler, Gisela, Columbia University
Windom, Herbert L., Georgia Institute of Technology
Wong, Chi S., University of British Columbia
Zheng, Yan, Grad Sch of the City Univ of New York

Organic Geochemistry
Abdulla, Hussain, Texas A&M University, Corpus Christi
Amend, Jan P., Washington University in St. Louis
Anderson, Ken B., Southern Illinois University Carbondale
Aravena, Ramon, University of Waterloo
Bada, Jeffrey L., University of California, San Diego
Bakel, Allen J., Argonne National Laboratory
Banfield, Jillian, University of California, Berkeley
Barker, James F., University of Waterloo
Bianchi, Thomas S., University of Florida

Bissada, Adry, University of Houston
Blair, Neal E., Northwestern University
Bopp, Richard F., Rensselaer Polytechnic Institute
Bowden, Stephen, University of Aberdeen
Brassell, Simon C., Indiana University, Bloomington
Burns, Timothy P., Los Alamos National Laboratory
Chen, Bob, University of Massachusetts, Boston
Chou, Mei-In (Melissa), Illinois State Geological Survey
Chou, Sheng-Fu Joseph, Illinois State Geological Survey
Cody, George D., Carnegie Institution for Science
Cole, Dean A., Los Alamos National Laboratory
Cowie, Gregory L., Edinburgh University
Diefendorf, Aaron, University of Cincinnati
Engel, Michael H., University of Oklahoma
Farrington, John W., Woods Hole Oceanographic Institution
Feakins, Sarah, University of Southern California
Fernandez, Loretta, Northeastern University
Filley, Timothy R., Purdue University
Fisher, Liz, University of Liverpool
Francis, Robert D., California State University, Long Beach
Freeman, Katherine H., Pennsylvania State University, University Park
Fu, Qi, University of Houston
Hanson, Andrew D., University of Nevada, Las Vegas
Hartnett, Hilairy E., Arizona State University
Harvey, H. Rodger, Old Dominion University
Hayes, John M., Indiana University, Bloomington
Heal, Kate V., Edinburgh University
Hites, Ronald A., Indiana University, Bloomington
Hockaday, William C., Baylor University
Huang, Yongsong, Brown University
Kenig, Fabien, University of Illinois at Chicago
Kotulova, Julia, University of Utah
Kruge, Michael A., Montclair State University
Langston-Unkefer, Pat J., Los Alamos National Laboratory
Love, Gordon D., University of California, Riverside
MacCarthy, Patrick, Colorado School of Mines
Macko, Stephen A., University of Virginia
McCarthy, John F., Oak Ridge National Laboratory
Mehra, Aradhana, University of Derby
Meyers, Philip A., University of Michigan
Mitra, Siddhartha, East Carolina University
Mitterer, Richard M., University of Texas, Dallas
Moldowan, J. Michael, Stanford University
Moslow, Thomas F., University of Calgary
Nelson, Robert K., Woods Hole Oceanographic Institution
Ngwenya, Bryne T., Edinburgh University
Nolan, Robert P., Grad Sch of the City Univ of New York
Perdue, Edward Michael, Georgia Institute of Technology
Petsch, Steven, University of Massachusetts, Amherst
Phillips, Dennis, Los Alamos National Laboratory
Philp, R. Paul, University of Oklahoma
Pier, Stanley M., Rice University
Pratt, Lisa M., Indiana University, Bloomington
Pruell, Richard J., University of Rhode Island
Reddy, Christopher M., Woods Hole Oceanographic Institution
Repeta, Daniel J., Woods Hole Oceanographic Institution
Riediger, Cynthia L., University of Calgary
Robinson, Clare, University of Manchester
Rohrssen, Megan, Central Michigan University
Sass, Henrik, University of Wales
Sepúlveda, Julio C., University of Colorado
Sessions, Alex L., California Institute of Technology
Silliman, James, Texas A&M University, Corpus Christi
Spongberg, Alison L., University of Toledo
Steen, Andrew, University of Tennessee, Knoxville
Stierle, Donald, Montana Tech of the University of Montana
Summons, Roger, Massachusetts Institute of Technology
Talbot, Helen, University of Newcastle Upon Tyne
Tanner, Benjamin R., Western Carolina University
Thiel, Dr., Volker, Georg-August University of Goettingen
Thomas, Elizabeth K., SUNY, Buffalo
Tierney, Jessica, University of Arizona
Townsend-Small, Amy, University of Cincinnati
van Dongen, Bart, University of Manchester
Van Vleet, Edward S., University of South Florida
Venkatesan, M I., University of California, Los Angeles
Visscher, Pieter T., University of Connecticut

Visscher, Pieter, University of Connecticut
Voorhees, Kent J., Colorado School of Mines
Wakeham, Stuart G., Georgia Institute of Technology
Weisener, Christopher, University of Windsor
Werne, Josef, University of Pittsburgh
Whelan, Jean K., Woods Hole Oceanographic Institution
Whiticar, Michael J., University of Victoria
Zerkle, Aubrey, University of St. Andrews
Zhang, Yige, Texas A&M University
Zhuang, Guangsheng, Louisiana State University
Zimmerman, Andrew, University of Florida
Ziolkowski, Lori A., University of South Carolina

Stable Isotopes

Anderson, Thomas F., University of Illinois, Urbana-Champaign
Anderson, William T., Florida International University
Andrews, Julian E., University of East Anglia
Atudorei, Nieu-Viorel, University of New Mexico
Bao, Huiming, Louisiana State University
Bedaso, Zelalem, University of Dayton
Berke, Melissa, University of Notre Dame
Bindeman, Ilya N., University of Oregon
Blum, Joel D., University of Michigan
Bowen, Gabriel, University of Utah
Bowman, John R., University of Utah
Bray, Colin, University of Toronto
Brenna, J. T., Cornell University
Brunner, Benjamin, University of Texas, El Paso
Buczynska, Anna, Northern Illinois University
Burgess, Ray, University of Manchester
Burns, Stephen J., University of Massachusetts, Amherst
Campbell, Andrew R., New Mexico Institute of Mining and Technology
Cao, Xiaobin, Louisiana State University
Clayton, Robert N., University of Chicago
Cohen, Anthony, The Open University
Conroy, Jessica, University of Illinois, Urbana-Champaign
Coogon, Rosalind, Imperial College
Criss, Robert E., Washington University in St. Louis
Criss, Robert E., Washington University in St. Louis
Crowley, Brooke E., University of Cincinnati
Darling, James, University of Portsmouth
Davidson, Garry J., University of Tasmania
Denniston, Rhawn F., University of Iowa
Dettman, David, University of Arizona
Deuser, Werner G., Woods Hole Oceanographic Institution
Dodd, Justin P., Northern Illinois University
Domingo, Laura, Univ Complutense de Madrid
Dorale, Jeffrey A., University of Iowa
Douglas, Peter, McGill University
Eastoe, Chris J., University of Arizona
Economos, Rita C., Southern Methodist University
Edwards, Thomas W. D., University of Waterloo
Elder, Kathryn L., Woods Hole Oceanographic Institution
Fantle, Matthew S., Pennsylvania State University, University Park
Farquhar, James, University of Maryland
Fayek, Mostafa, University of Manitoba
Feng, Xiahong, Dartmouth College
Fike, David A., Washington University in St. Louis
Fisher, David E., University of Miami
Foster, Gavin, University of Southampton
Fricke, Henry C., Colorado College
Gagnon, Alan R., Woods Hole Oceanographic Institution
Gao, Yongjun, University of Houston
Gazis, Carey A., Central Washington University
Gibson, John J., University of Waterloo
Gilbert, Kathleen W., Wellesley College
Gilg, Hans A., Technical University of Munich
Gill, Benjamin, Virginia Polytechnic Institute & State University
Gillikin, David P., Union College
Goldstein, Steven L., Columbia University
Gonzalez, Luis A., University of Kansas
Graczyk, Donald G., Argonne National Laboratory
Grassineau, Nathalie, Royal Holloway University of London
Gregory, Robert T., Southern Methodist University
Griffith, Elizabeth, Ohio State University
Griffith, Liz, University of Texas, Arlington
Grimes, Stephen, University of Plymouth
Grocke, Darren, Durham University

Grossman, Ethan L., Texas A&M University
Guo, Weifu, Woods Hole Oceanographic Institution
Haley, Brian, Oregon State University
Halliday, Alex, University of Oxford
Helie, Jean-François, Universite du Quebec a Montreal
Hemming, Sidney R., Columbia University
Henkel, Torsten, University of Manchester
Higgins, Pennilyn, University of Rochester
Hillaire-Marcel, Claude, Universite du Quebec a Montreal
Hilton, David R., University of California, San Diego
Hindshaw, Ruth, University of St. Andrews
Holk, Gregory J., California State University, Long Beach
Holmden, Chris, University of Saskatchewan
Horton, Travis, University of Canterbury
Howe, Stephen S., SUNY, Albany
Hren, Michael, University of Connecticut
Ingram, B. Lynn, University of California, Berkeley
Johnson, Beverly J., Bates College
Junium, Christopher, Syracuse University
Kaiser, Jan, University of East Anglia
Kaplan, Isaac R., University of California, Los Angeles
Karhu, Juha A., University of Helsinki
Kaufman, Alan J., University of Maryland
Keigwin, Lloyd D., Woods Hole Oceanographic Institution
Kellman, Lisa M., Saint Francis Xavier University
Kieffer, Bruno, University of British Columbia
Kienast, Markus, Dalhousie University
Kirtland Turner, Sandra, University of California, Riverside
Knauth, L. Paul, Arizona State University
Kohn, Matthew, Boise State University
Krantz, David E., University of Delaware
Kreutz, Karl J., University of Maine
Krishnamurthy, R. V., Western Michigan University
Lambert, W. J., University of Alabama
Larson, Peter B., Washington State University
Law, Kim, Western University
Leavitt, Steven W., University of Arizona
Lefticariu, Liliana, Southern Illinois University Carbondale
Li, Long, University of Alberta
Lini, Andrea, University of Vermont
Lohmann, Kyger C., University of Michigan
Longstaffe, Frederick J., Western University
Lowry, David, Royal Holloway University of London
Luzincourt, Marc R., Universite du Quebec
Marshall, Jim, University of Liverpool
Masterson, Andrew, Northwestern University
Matheney, Ronald K., University of North Dakota
Mattey, Dave, Royal Holloway University of London
Mayer, Bernhard, University of Calgary
Michalski, Greg, Purdue University
Mix, Alan C., Oregon State University
Muehlenbachs, Karlis, University of Alberta
Munizzi, Jordan S., University of Kentucky
Nelson, Stephen T., Brigham Young University
Nielsen, Sune G., Woods Hole Oceanographic Institution
O'Neil, James, University of Michigan
Ohmoto, Hiroshi, Pennsylvania State University, University Park
Oleynik, Sergey, Princeton University
Olivares, Jose A., Los Alamos National Laboratory
Padden, Maureen, McMaster University
Parai, Rita, Washington University in St. Louis
Parrish, Randy, Leicester University
Passey, Benjamin, University of Michigan
Patterson, William P., University of Saskatchewan
Pedentchouk, Nikolai, University of East Anglia
Pinti, Daniele Luigi, Universite du Quebec a Montreal
Pogge von Strandmann, Phillip, University College London
Pogge von Strandmann, Phillip, Birkbeck College
Poirier, Andre, Universite du Quebec a Montreal
Ponton, Camilo, Western Washington University
Popp, Brian N., University of Hawai'i, Manoa
Poulson, Simon R., University of Nevada, Reno
Ravelo, Christina, University of California, Santa Cruz
Rehkamper, Mark, Imperial College
Richards, Ian J., Southern Methodist University
Rye, Danny M., Yale University
Savin, Samuel M., Case Western Reserve University

Schaller, Morgan F., Rensselaer Polytechnic Institute
Schimmelmann, Arndt, Indiana University, Bloomington
Schwarcz, Henry P., McMaster University
Sedlacek, Alexa, University of Northern Iowa
Shahar, Anat, University of Maryland
Sharma, Shikha, West Virginia University
Sharp, Zachary D., University of New Mexico
Sherwood, Owen, Dalhousie University
Sherwood Lollar, Barbara, University of Toronto
Shieh, Yuch-Ning, Purdue University
Showers, William J., North Carolina State University
Smart, Katie A., University of the Witwatersrand
Smirnov, Anna, Universite du Quebec
Sowers, Todd, Pennsylvania State University, University Park
Steig, Eric J., University of Washington
Steig, Eric J., University of Washington
Strauss, Harald, Universitaet Muenster
Stute, Martin, Columbia University
Syrup, Krista A., Moraine Valley Community College
Szynkiewicz, Anna, University of Tennessee, Knoxville
Taylor, Hugh P., California Institute of Technology
Theiling, Bethany P., The University of Tulsa
Thirlwall, Matthew, Royal Holloway University of London
Tudhope, Alexander W., Edinburgh University
Valley, John W., University of Wisconsin-Madison
van de Flierdt, Tina, Imperial College
Wanamaker, Alan D., Iowa State University of Science & Technology
Wang, Jianhua, Carnegie Institution for Science
Webb, Elizabeth A., Western University
Welp, Lisa, Purdue University
Wenner, David B., University of Georgia
White, James W. C., University of Colorado
Widory, David, Universite du Quebec a Montreal
Wolfe, Amy, Miami University
Wright, James D., Rutgers, The State Univ of New Jersey
Yapp, Crayton J., Southern Methodist University
Yeung, Laurence Y., Rice University
Young, Edward D., University of California, Los Angeles
Yun, Misuk, University of Manitoba
Zhang, Lin, Texas A&M University, Corpus Christi
Zhang, Ren, Baylor University
Ziegler, Karen, University of New Mexico
Zierenberg, Robert A., University of California, Davis

Trace Element Distribution
Coale, Kenneth H., Moss Landing Marine Laboratories
Condie, Kent C., New Mexico Institute of Mining and Technology
Croot, Peter, National University of Ireland Galway
Cullers, Robert L., Kansas State University
Daniels, William R., Los Alamos National Laboratory
Efurd, Deward W., Los Alamos National Laboratory
Frey, Frederick A., Massachusetts Institute of Technology
Gabrielli, Paolo, Ohio State University
Hale, Beverley A., University of Guelph
Kar, Aditya, Fort Valley State University
Kyte, Frank T., University of California, Los Angeles
LaFleche, Marc R., Universite du Quebec
Liese, Homer C., University of Connecticut
Marcantonio, Franco, Texas A&M University
McGowan, Nicole M., Western Washington University
Newton, Robert, Columbia University
Nicolescu, Stefan, Yale University
Peterson, Eugene J., Los Alamos National Laboratory
Ritter, Charles J., University of Dayton
Ryan, Jeffrey G., University of South Florida, Tampa
Seifert, Karl E., Iowa State University of Science & Technology
Simonetti, Tony, University of Notre Dame
Thomas, Kimberly W., Los Alamos National Laboratory
Thompson, Joseph L., Los Alamos National Laboratory
Von Reden, Karl F., Woods Hole Oceanographic Institution
Weaver, Barry L., University of Oklahoma
Wildeman, Thomas R., Colorado School of Mines

Biogeochemistry
Amon, Rainier M., Texas A&M University
Anbar, Ariel D., Arizona State University
Babbin, Andrew, Massachusetts Institute of Technology
Barker, Joel D., Ohio State University

Bradley, Alexander S., Washington University in St. Louis
Buick, Roger, University of Washington
Cassar, Nicolas, Duke University
Costa, Jr, Ozeas S., Ohio State University
Darnajoux, Romain, Princeton University
Dolman, Han J., Vrije Universiteit Amsterdam
Elliott, Emily M., University of Pittsburgh
Fortin, Danielle, University of Ottawa
Fowle, David A., University of Kansas
Gu, Baohua, Oak Ridge National Laboratory
Jaisi, Deb, University of Delaware
Jones, Daniel, New Mexico Institute of Mining and Technology
Kavanaugh, Maria, Oregon State University
Lau, Kimberly, University of Wyoming
Mahmoudi, Nagissa, McGill University
Marin-Spiotta, Erika, University of Wisconsin, Madison
Matisoff, Gerald, Case Western Reserve University
McCormick, Michael L., Hamilton College
Mellinger, David, Oregon State University
Mulder, Jan, Norwegian University of Life Sciences (NMBU)
Pearson, Ann, Harvard University
Reddy, K R., University of Florida
Reese, Brandi, Texas A&M University, Corpus Christi
Richardson, Justin, University of Massachusetts, Amherst
Richey, Jeffrey E., University of Washington
Roberts, Jennifer A., University of Kansas
Sanford, Robert A., University of Illinois, Urbana-Champaign
Smith, Robert, University of Idaho
Thomas, Sara, Princeton University
Thurber, Andrew, Oregon State University
Torres, Mark, Rice University
Tufillaro, Nicholas, Oregon State University
Winnick, Matthew, University of Massachusetts, Amherst
Zhang, Xinning, Princeton University

High-temperature Geochemistry
Baker, Don, McGill University
Jackson, Matt, University of California, Santa Barbara
Le Voyer, Marion, Smithsonian Inst / Natl Museum of Natural History
Pollock, Meagen, College of Wooster
Reid, Mary R., Northern Arizona University
Zajacz, Zoltan, University of Toronto
Zhang, Youxue, University of Michigan

Aqueous Geochemistry
Chin, Yu-Ping, Ohio State University
Clark, Elyse, Fitchburg State University
Cox, Alysia, Montana Tech of the University of Montana
Greenwell, Chris, Durham University
Hon, Rudolph, Boston College
Kurz, Marie J., Drexel University
MacKay, Allison A., Ohio State University
O'Shea, Bethany, University of San Diego
Polizzotto, Matthew, University of Oregon
Ridley, Moira K., Texas Tech University
Saccocia, Peter J., Bridgewater State University
Yun, Seong-Taek, Korea University

GEOPHYSICS
General Geophysics
Adams, Aubreya, Colgate University
Adepelumi, A. A., Obafemi Awolowo University
Afolabi, O., Obafemi Awolowo University
Ahern, Judson L., University of Oklahoma
Ahmed, Mohamed, Texas A&M University, Corpus Christi
Aihaiti, Muhetaer, Carnegie Institution for Science
Al-Lehyani, Ayman, King Fahd University of Petroleum and Minerals
Al-Shuhail, Abdullatif, King Fahd University of Petroleum and Minerals
Alao, O. A., Obafemi Awolowo University
Albright, James N., Los Alamos National Laboratory
Angelopoulos, Vassilis, University of California, Los Angeles
Armah, Thomas K., University of Ghana
Audet, Pascal, University of Ottawa
Aurnou, Jonathan M., University of California, Los Angeles
Bachtadse, Valerian, Ludwig-Maximilians-Universitaet Muenchen
Bailey, Richard C., University of Toronto
Barbi, Greta, Ludwig-Maximilians-Universitaet Muenchen

Barklage, Mitchell, Northwestern University
Barsch, Robert, Ludwig-Maximilians-Universitaet Muenchen
Bastow, Ian, Imperial College
Beaumont, Christopher, Dalhousie University
Becker, Alex, University of California, Berkeley
Behn, Mark, Boston College
Bendick, Rebecca, University of Montana
Bercovici, David, Yale University
Beresnev, Igor A., Iowa State University of Science & Technology
Billings, Stephen, University of British Columbia
Bina, Craig R., Northwestern University
Birch, Francis S., University of New Hampshire
Blankenship, Donald D., University of Texas, Austin
Bleibinhaus, Florian, Ludwig-Maximilians-Universitaet Muenchen
Bloxham, Jeremy, Harvard University
Boaga, Jacopo, Università degli Studi di Padova
Boehler, Reinhard, Carnegie Institution for Science
Boettcher, Margaret S., University of New Hampshire
Bokelmann, Goetz, University of Vienna
Boschi, Lapo, Università degli Studi di Padova
Bradley, Christopher R., Los Alamos National Laboratory
Braun, Alexander, Queen's University
Bregman, Martin, Tulsa Community College
Briggs, Derek E., Yale University
Britzke, Gilbert, Ludwig-Maximilians-Universitaet Muenchen
Brodholt, John, University College London
Brooks, Debra A., Santiago Canyon College
Brown, Wesley A., Stephen F. Austin State University
Buck, IV, W. R., Columbia University
Bunge, Hans-Peter, Ludwig-Maximilians-Universitaet Muenchen
Bustin, Amanda, University of British Columbia
Butler, Karl, University of New Brunswick
Butler, Sam, University of Saskatchewan
Calvin, Wendy, University of Nevada, Reno
Campbell, David L., University of Iowa
Carlson, Barry A., Delta College
Carpenter, Philip J., Northern Illinois University
Carrigan, Charles R., Lawrence Livermore National Laboratory
Cassiani, Giorgio, Università degli Studi di Padova
Cayzer, Nicola, Edinburgh University
Chapman, Mark, Edinburgh University
Chase, Clement G., University of Arizona
Cheadle, Michael J., University of Wyoming
Chow, Paul, Carnegie Institution for Science
Clark, H. C., Rice University
Clarke, Garry K. C., University of British Columbia
Coakley, Bernard, University of Alaska, Fairbanks
Colgan, William, York University
Comas, Xavier, Florida Atlantic University
Cooper, Catherine , Washington State University
Copley, Alex, University of Cambridge
Cosentino, Pietro, University of East Anglia
Couture, Gilles, Universite du Quebec a Montreal
Craig, Mitchell S., California State University, East Bay
Craven, John, Edinburgh University
Daly, Eve, National University of Ireland Galway
Daniels, Jeffrey J., Ohio State University
Davies, Gregory, Princeton University
Davies, J H., Cardiff University
Davies, Rhordri, Imperial College
Davis, Daniel M., Stony Brook University
Davis, Tara, University College Cork
Dengler, Lorinda, Humboldt State University
Dewdney, Chris, University of Portsmouth
Dickman, Steven R., Binghamton University
Dieterich, James H., University of California, Riverside
Dmochowski, Jane, University of Pennsylvania
Doll, William E., Oak Ridge National Laboratory
Dorman, Leroy M., University of California, San Diego
Dunbar, John A., Baylor University
Durland, Theodore, Oregon State University
Dzunic, Aleksandar, Curtin University
Ebel, John E., Boston College
Edwards, C L, Los Alamos National Laboratory
Ehlers, Todd A., University of Michigan
Ekstrom, Goran, Columbia University
Elhusseiny, Ammar , King Fahd University of Petroleum and Minerals

Ellwood, Brooks, University of Texas, Arlington
Emmer, Barbara, Ludwig-Maximilians-Universitaet Muenchen
England, Richard, Leicester University
Fehler, Michael, Massachusetts Institute of Technology
Ferguson, John F., University of Texas, Dallas
Ferrucci, Fabrizio, The Open University
Fialko, Yuri A., University of California, San Diego
Fichtner, Andreas, Ludwig-Maximilians-Universitaet Muenchen
Filina, Irina, University of Nebraska, Lincoln
Fishwick, Stewart, Leicester University
Flesch, Lucy M., Purdue University
Foley, Bradford, Pennsylvania State University, University Park
Fontaine, Fabrice R., Universite de la Reunion
Forte, Alessandro Marco, Universite du Quebec a Montreal
Fortier, Richard, Universite Laval
Gaffney, Edward S., Los Alamos National Laboratory
Gaidos, Eric J., University of Hawai'i, Manoa
Gallovic, Frantisek, Ludwig-Maximilians-Universitaet Muenchen
Galsa, Attila, Eotvos Lorand University
Gebrande, Helmut, Ludwig-Maximilians-Universitaet Muenchen
Ghazala, Hosni H., El Mansoura University
Goes, Saskia, Imperial College
Goforth, Thomas T., Baylor University
Gonzalez-Huizar, Hector, University of Texas, El Paso
Goodliffe, Andrew, University of Alabama
Gorman, Gerard, Imperial College
Graham, Gina R., Alaska Div of Geological & Geophysical Surveys
Gramsch, Stephen, Carnegie Institution for Science
Grand, Stephen P., University of Texas, Austin
Granja, José Luis, Univ Complutense de Madrid
Greenfield, Roy J., Pennsylvania State University, University Park
Gregg, Patricia, University of Illinois, Urbana-Champaign
Gribenko, Alex, University of Utah
Grindlay, Nancy R., University of North Carolina, Wilmington
Gualtieri, Lucia, Princeton University
Haas, Christian, York University
Haber, Eldad, University of British Columbia
Hamzaoui, Cherif, Universite du Quebec a Montreal
Hanafy, Sherif , King Fahd University of Petroleum and Minerals
Hardage, Bob A., Univ of Texas at Austin, Jackson Sch of Geosciences
Harder, Steven H., University of Texas, El Paso
Harmon, Nicholas, University of Southampton
Harry, Dennis L., Colorado State University
Hay, Carling, Boston College
Hayes, Jorden L., Dickinson College
He, Jiaze, Princeton University
Heigold, Paul C., Argonne National Laboratory
Hellio, Gabrielle, University of Southampton
Henson, Harvey, Southern Illinois University Carbondale
Henstock, Tim, University of Southampton
Henyey, Thomas L., University of Southern California
Hier-Majumder, Saswata, Royal Holloway University of London
Higgs, Bettie M., University College Cork
Hofmeister, Anne M., Washington University in St. Louis
Hornbach, Matthew, University of Texas, Austin
Horowitz, Franklin G., Cornell University
Horváth, Ferenc, Eotvos Lorand University
Houseman, Greg, University of Leeds
Huang, Lianjie, Los Alamos National Laboratory
Huerta, Audrey, Central Washington University
Hussein, Musa, University of Texas, El Paso
Iaffaldano, Giampiero, Ludwig-Maximilians-Universitaet Muenchen
Igel, Heiner, Ludwig-Maximilians-Universitaet Muenchen
Irving, Jessica, Princeton University
Ishii, Miaki, Harvard University
Ito, Garrett T., University of Hawai'i, Manoa
Ivanov, Julian, University of Kansas
Jackson, James A., University of Cambridge
Jacob, Robert W., Bucknell University
Jaiswal, Priyank, Oklahoma State University
James, Peter B., Baylor University
Jarvis, Gary T., York University
Jayawickreme, Dushmantha, Southern Connecticut State University
Jensen, Olivia G., McGill University
Jezek, Kenneth, Ohio State University
Johnson, Kaj, Indiana University, Bloomington
Johnson, Paul A., Los Alamos National Laboratory
Johnson, Verner C., Colorado Mesa University
Jomeiri, Rahim, University of Tabriz
Jurdy, Donna M., Northwestern University
Kafka, Alan L., Boston College
Kahle, Beth, University of Cape Town
Kaplinski, Matthew A., Northern Arizona University
Katz, Samuel, Rensselaer Polytechnic Institute
Kaus, Boris P., Universität Mainz
Kavner, Abby, University of California, Los Angeles
Keach, Bill, Utah Geological Survey
Keller, G. Randy, University of Oklahoma
Keller , George R., University of Texas, El Paso
Kelley, Shari, New Mexico Institute of Mining & Technology
Kellogg, James N., University of South Carolina
Kellogg, Louise H., University of California, Davis
Kenyon, Patricia M., City College (CUNY)
Kenyon, Patricia M., Grad Sch of the City Univ of New York
King, Scott D., Virginia Polytechnic Institute & State University
Kingdon, Kevin, University of British Columbia
Kinsland, Gary L., University of Louisiana at Lafayette
Korenaga, Jun, Yale University
Kroeger, Glenn C., Trinity University
Kruger, Joseph M., Lamar University
Kruse, Sarah E., University of South Florida, Tampa
Kukowski, Nina, Friedrich-Schiller-University Jena
Kuo, John T., Columbia University
Käser, Martin, Ludwig-Maximilians-Universitaet Muenchen
Lat, Che Noorliza, University of Malaya
Lathrop, Daniel, University of Maryland
Lavallee, Yan, Ludwig-Maximilians-Universitaet Muenchen
Lavier, Luc L., University of Texas, Austin
Leitch, Alison, Memorial University of Newfoundland
Lenardic, Adrian, Rice University
Leonard, Lucinda, University of Victoria
Lewis, Stephen D., California State University, Fresno
Lin, Fan-Chi, University of Utah
Lin, Guoqing, University of Miami
Liu, Lanbo, University of Connecticut
Liu, Lijun, University of Illinois, Urbana-Champaign
Liu, Mian, University of Missouri
Liu, Qinya, University of Toronto
Liu, Yajing, McGill University
Lohman, Rowena B., Cornell University
Lopez-Pineda, Leobardo, Universidad Estatal de Sonora
Louie, John, University of Nevada, Reno
Lovell, Mike, Leicester University
Lozos, Julian, California State University, Northridge
Luttrell, Karen, Louisiana State University
Lutz, Pascale, Institut Polytechnique LaSalle Beauvais (ex-IGAL)
Mackwell, Stephen J., Rice University
Mahatsente, Rezene, University of Alabama
Malinconico, Lawrence L., Lafayette College
Malservisi, Rocco, Ludwig-Maximilians-Universitaet Muenchen
Mao, Ho-kwang (David), University of Chicago
Mareschal, Jean-Claude, Universite du Quebec a Montreal
Martens, Hilary R., University of Montana
Mavko, Gerald M., Stanford University
Maxwell, Michael, University of British Columbia
McElwaine, Jim, Durham University
McGinnis, Lyle D., Argonne National Laboratory
Meade, Brendan, Harvard University
Meadows, Wayne R., Los Alamos National Laboratory
Meert, Joseph G., University of Florida
Meng, Yue, Carnegie Institution for Science
Merriam, James B., University of Saskatchewan
Merrill, Ronald T., University of Washington
Milne, Glenn A., University of Ottawa
Minchew, Brent, Massachusetts Institute of Technology
Mitrovica, Jerry X., Harvard University
Mohr, Marcus, Ludwig-Maximilians-Universitaet Muenchen
Montesi, Laurent G., University of Maryland
Moorkamp, Max, Leicester University
Morgan, F D., Massachusetts Institute of Technology
Morgan, Jason, Royal Holloway University of London
Morgan, Joanna, Imperial College
Morris, William A., McMaster University
Morrison, H. Frank, University of California, Berkeley

Morse, David L., University of Texas, Austin
Muñoz, Alfonso, Univ Complutense de Madrid
Murphy, Vincent, Boston College
NICULESCU, Bogdan M., University of Bucharest
Niu, Qifei i., Boise State University
Nunn, Jeffrey A., Louisiana State University
Nur, Amos M., Stanford University
Nyblade, Andrew A., Pennsylvania State University, University Park
Nyquist, Jonathan, Temple University
O'Farrell, Keely A., University of Kentucky
Oeser, Jens, Ludwig-Maximilians-Universitaet Muenchen
Ogiesoba, Osareni C., University of Texas, Austin
Ojo, S. B., Obafemi Awolowo University
Oladunjoye, M. A., University of Ibadan
Olayinka, A. I., University of Ibadan
Oldenburg, Douglas W., University of British Columbia
Olgaard, David L., Rice University
Oliver, Ronald D., Los Alamos National Laboratory
Olorunfemi, M. O., Obafemi Awolowo University
Olson, Peter L., Johns Hopkins University
Osinowo, O. O., University of Ibadan
Oware, Erasmus K., SUNY, Buffalo
Paillet, Fred, University of Arkansas, Fayetteville
Pain, Christopher, Imperial College
Palmer, Donald F., Kent State University
Palutoglu, Mahmut, Firat University
Panero, Wendy R., Ohio State University
Park, Changyong, Carnegie Institution for Science
Parmentier, E. Marc, Brown University
Parsekian, Andrew D., University of Wyoming
Pasicznyk, David L., New Jersey Geological and Water Survey
Patton, Howard J., Los Alamos National Laboratory
Peavy, Samuel T., Georgia Southwestern State University
Peirce, Christine, Durham University
Pelton, John R., Boise State University
Peters, Mark T., Los Alamos National Laboratory
Petersen, Nikolai, Ludwig-Maximilians-Universitaet Muenchen
Pfuhl, Helen, Ludwig-Maximilians-Universitaet Muenchen
Phillips, William S., Los Alamos National Laboratory
Piazzoni, Antonio Sebastian, Ludwig-Maximilians-Univ Muenchen
Pimentel-Hernandez, Ramon, Univ Autonoma de Baja California Sur
Plattner, Christina, Ludwig-Maximilians-Universitaet Muenchen
Pollack, Henry N., University of Michigan
Porch, William M., Los Alamos National Laboratory
Porter, Ryan, Northern Arizona University
Priestley, Keith, University of Cambridge
Pysklywec, Russell N., University of Toronto
Raine, Robin, National University of Ireland Galway
Ranalli, Giorgio, Carleton University
Randall, George, Los Alamos National Laboratory
Rasmussen, Steen, Los Alamos National Laboratory
Ravat, Dhananjay, University of Kentucky
Rawlinson, Nick, University of Cambridge
Rempel, Alan W., University of Oregon
Revetta, Frank A., SUNY Potsdam
Rice, James R., Harvard University
Richard, Robert, University of California, Los Angeles
Richards, Mark A., University of California, Berkeley
Richardson, Randall M., University of Arizona
Roach, Michael, University of Tasmania
Roberts, Mark L., Woods Hole Oceanographic Institution
Robinson, Edwin S., Virginia Polytechnic Institute & State University
Robinson, Judith, Rutgers, The State Univ of New Jersey, Newark
Rochester, Michael G., Memorial University of Newfoundland
Roecker, Steven W., Rensselaer Polytechnic Institute
Romanovsky, Vladimir, University of Alaska, Fairbanks
Rothman, Daniel H., Massachusetts Institute of Technology
Rubin, Allan M., Princeton University
Rudge, John, University of Cambridge
Rudolph, Maxwell L., University of California, Davis
Russell, Christopher T., University of California, Los Angeles
Russell, R. Doncaster, University of British Columbia
Russo, Raymond, University of Florida
Ryall, Patrick J., Dalhousie University
Rychert, Catherine A., University of Southampton
Sabra, Karim, Georgia Institute of Technology
Sahakian, Valerie J., University of Oregon

Salami, Sikiru A., University of Benin
Sammis, Charles G., University of Southern California
Sauck, William A., Western Michigan University
Sava, Diana, Univ of Texas at Austin, Jackson Sch of Geosciences
Sava, Diana C., University of Texas, Austin
Sawyer, Derek, Ohio State University
Schmandt, Brandon, University of New Mexico
Schriver, David, University of California, Los Angeles
Schubert, Gerald, University of California, Los Angeles
Schuberth, Bernhard, Ludwig-Maximilians-Universitaet Muenchen
Schultz, Adam, Oregon State University
Searls, Mindi L., University of Nebraska, Lincoln
Segall, Paul, Stanford University
Serzu, Mulu, University of Manitoba
Shams, Asghar, Heriot-Watt University
Shcherbakov, Robert, Western University
Sherrod, Laura, Kutztown University of Pennsylvania
Siegler, Matthew, Southern Methodist University
Sigloch, Karin, University of Oxford
Simila, Gerald W., California State University, Northridge
Simmons, M. G., Massachusetts Institute of Technology
Simms, Janet E., Mississippi State University
Sinogeikin, Stanislav, Carnegie Institution for Science
Sit, Stefany , University of Illinois at Chicago
Sleep, Norman H., Stanford University
Smylie, Douglas E., York University
Solomatov, Viatcheslav S., Washington University in St. Louis
Sonett, Charles P., University of Arizona
Soupios, Panteleimon , King Fahd University of Petroleum and Minerals
Spanos, T.J.T (Tim), University of Alberta
Sparks, David, Texas A&M University
Stearns, Richard G., Vanderbilt University
Stein, Carol A., University of Illinois at Chicago
Stephenson, Randell A., University of Aberdeen
Stidham, Christiane W., Stony Brook University
Stierman, Donald J., University of Toledo
Stixrude, Lars, University College London
Stupazzini, Marco, Ludwig-Maximilians-Universitaet Muenchen
Suckale, Jenny, Stanford University
Sweeney, Mark D., Pacific Northwest National Laboratory
Szeliga, Walter, Central Washington University
Szeto, Anthony M. K., York University
Taib, Samsudin Hj., University of Malaya
Tatham, Robert H., University of Texas, Austin
Taylor, Graeme, University of Plymouth
ten Brink, Uri S., Woods Hole Oceanographic Institution
ter Voorde, Marlies, Vrije Universiteit Amsterdam
Thiruvathukal, John V., Montclair State University
Thompson, Michael D., Argonne National Laboratory
Timár, Gábor, Eotvos Lorand University
Tobin, Harold, University of Wisconsin-Madison
Tong, Vincent C., Birkbeck College
Tsai, Victor, California Institute of Technology
Tsoflias, George P., University of Kansas
Utgard, Russell O., Ohio State University
Van Avendonk, Harm, University of Texas, Austin
van Keken, Peter J., University of Michigan
Velli, Marco, University of California, Los Angeles
Vernhes, Jean-David, Institut Polytechnique LaSalle Beauvais (ex-IGAL)
Versteeg, Roelof, Columbia University
Vincent, Robert K., Bowling Green State University
von Glasow, Roland, University of East Anglia
Vopson, Melvin M., University of Portsmouth
Waheed, Umair B., King Fahd University of Petroleum and Minerals
Waltham, Dave, Royal Holloway University of London
Wang, Chi-Yuen, University of California, Berkeley
Wang, Herbert F., University of Wisconsin, Madison
Wang, Herbert F., University of Wisconsin-Madison
Wang, Kelin, University of Victoria
Wang, Yanghua, Imperial College
Warren, Linda M., Saint Louis University
Wassermann, Joachim, Ludwig-Maximilians-Universitaet Muenchen
Weaver, Thomas A., Los Alamos National Laboratory
Weeraratne, Dayanthie, California State University, Northridge
Wettlaufer, John S., Yale University
White, Nicky, University of Cambridge
White, Robert, University of Cambridge

Whitman, Dean, Florida International University
Williams, Ian S., University of Wisconsin, River Falls
Williams, II, Richard T., University of Tennessee, Knoxville
Wilson, Clark R., University of Texas, Austin
Winberry, Paul, Central Washington University
Winklhofer, Michael, Ludwig-Maximilians-Universitaet Muenchen
Wisely, Beth, Casper College
Wishart, De Bonne N., Central State University
Woodhouse, John, University of Oxford
Woods, Andy, University of Cambridge
Worthington, Lindsay L., University of New Mexico
Xiao, Yuming, Carnegie Institution for Science
Yelisetti, Subbarao, Texas A&M University, Kingsville
Yuen, David A., University of Minnesota, Twin Cities
Zahm, Christopher K., University of Texas, Austin
Zaja, Annalisa, Università degli Studi di Padova
Zakharova, Natalia, Central Michigan University
Zarroca Hernández, Mario, Universitat Autonoma de Barcelona
Zaspel, Craig E., University of Montana Western
Zelt, Colin A., Rice University
Zha, Changsheng, Carnegie Institution for Science
Zhang, Chi, University of Kansas
Zoback, Mark D., Stanford University

Experimental Geophysics
Anderson, Orson L., University of California, Los Angeles
Askari, Roohollah, Michigan Technological University
Bates, Richard, University of St. Andrews
Bevis, Michael G., Ohio State University
Bohlen, Thomas, Karlsruhe Institute of Technology
Bona, Andrej, Curtin University
Bridges, Frank, University of California, Santa Cruz
Bussod, Gilles Y., Los Alamos National Laboratory
Carson, Michael, Curtin University
Christensen, Nikolas I., University of Wisconsin-Madison
Comina, Cesare, Università di Torino
Dewers, Thomas, Sandia National Laboratory
Dyaur, Nikolay, University of Houston
Evans, J. B., Massachusetts Institute of Technology
Fomel, Sergey, Univ of Texas at Austin, Jackson Sch of Geosciences
Fratta, Dante, University of Wisconsin, Madison
Glover, Paul W., Universite Laval
Glover, Paul W., University of Leeds
Glubokovskikh, Stanislav, Curtin University
Graham, Jr., Earl K., Pennsylvania State University, University Park
Gross, Gerardo W., New Mexico Institute of Mining and Technology
Kaip, Galen M., University of Texas, El Paso
Karlstrom, Leif, University of Oregon
Key, Kerry, University of California, San Diego
Kilburn, Chris, University College London
Kitajima, Hiroko, Texas A&M University
Kohlstedt, David L., University of Minnesota, Twin Cities
Lebedev, Maxim, Curtin University
Manghnani, Murli H., University of Hawai'i, Manoa
Mao, Ho-kwang, Carnegie Institution for Science
Marone, Chris, Pennsylvania State University, University Park
Marshall, Hans-Peter, Boise State University
Marshall, Scott T., Appalachian State University
Mei, Shenghua, University of Minnesota, Twin Cities
Narod, Barry, University of British Columbia
Olhoeft, Gary R., Colorado School of Mines
Parks, George K., University of Washington
Pevzner, Roman, Curtin University
Ruina, Andy L., Cornell University
Schmitt, Douglas R., Purdue University
Scholz, Christopher H., Columbia University
Shankland, Thomas J., Los Alamos National Laboratory
Shaw, George H., Union College
Shieh, Sean R., Western University
Shu, Jinfu, Carnegie Institution for Science
Somayazulu, Maddury, Carnegie Institution for Science
Tullis, Terry E., Brown University
Vanorio, Tiziana, Stanford University
Vaughan, Michael T., Stony Brook University
Velasco, Aaron A., University of Texas, El Paso
Vialle, Stephanie, Curtin University
Watanabe, Tohru, University of Toyama
Wohletz, Kenneth H., Los Alamos National Laboratory
Wong, Teng-fong, Stony Brook University
Yong, Wenjun, Western University
Zebker, Howard A., Stanford University
Zimmerman, Mark, University of Minnesota, Twin Cities

Exploration Geophysics
Al-Shuhail, Abdullah, King Fahd University of Petroleum and Minerals
Alumbaugh, David L., University of California, Berkeley
Armadillo, Egidio, Universita di Genova
B, Dr. V., Osmania University
Backus, Milo M., University of Texas, Austin
Becker, Alex, University of California, Berkeley
Biondi, Biondo L., Stanford University
Black, Ross A., University of Kansas
Brabham, P J., Cardiff University
Brantley, Duke, University of South Carolina
Brassea, Jesus M., Centro de Inv Científica y de Ed Sup de Ensenada
Breland, Clayton F., Louisiana State University
Brown, Larry D., Cornell University
Carmichael, Robert S., University of Iowa
Castagna, John P., University of Houston
Chandler, Val W., University of Minnesota, Twin Cities
Chandler, Val, University of Minnesota
Chen, Jingyi, The University of Tulsa
Claerbout, Jon F., Stanford University
Cook, Frederick A., University of Calgary
Corbato, Charles E., Ohio State University
Coruh, Cahit, Virginia Polytechnic Institute & State University
DeAngelo, Micheal V., Univ of Texas at Austin, Jackson Sch of Geo
Donaldson, Paul R., Boise State University
Dubba, Dr. Vijay K., Osmania University
Duckworth, Kenneth, University of Calgary
Dunlap, Dallas B., Univ of Texas at Austin, Jackson Sch of Geosciences
Dunn, Bernie, Western University
Ellett, Kevin M., Indiana University
Esparza, Francisco, Centro de Inv Científica y de Ed Sup de Ensenada
Espinosa, Juan M., Centro de Inv Científica y de Ed Sup de Ensenada
Falebita, Dele E., Obafemi Awolowo University
Ferguson, Ian J., University of Manitoba
Flores, Carlos, Centro de Inv Científica y de Ed Sup de Ensenada
Fomel, Sergey B., University of Texas, Austin
Gaede, Oliver M., Queensland University of Technology
Gajewski, Dirk J., Universitaet Hamburg
Galvin, Robert, Curtin University
Gao, Dengliang, West Virginia University
Gendzwill, Donald J., University of Saskatchewan
Giroux, Bernard, Universite du Quebec
Gloaguen, Erwan, Universite du Quebec
Gomez, Enrique, Centro de Inv Científica y de Ed Sup de Ensenada
Gonzalez-Escobar, Mario, Centro de Inv Científica y de Ed Sup de Ensenada
Grana, Dario, University of Wyoming
Gurevich, Boris, Curtin University
Han, De-hua, University of Houston
Hardage, Bob A., University of Texas, Austin
Harris, James B., Millsaps College
Hauser, Ernest C., Wright State University
Herrmann, Felix J., Georgia Institute of Technology
Hilterman, Fred, University of Houston
Hinze, William J., Purdue University
Hole, John A., Virginia Polytechnic Institute & State University
Holt, John W., University of Texas, Austin
Homuth, Emil F., Los Alamos National Laboratory
Howman, Dominic J., Curtin University
Hu, Hao, University of Houston
Jiracek, George R., San Diego State University
Johnson, Ansel G., Portland State University
Jones, Francis H., University of British Columbia
Keach II, R. William, Brigham Young University
Kepic, Anton W., Curtin University
Khalil, Mohamed, Montana Tech of the University of Montana
Lawton, Donald C., University of Calgary
Lee, Ki Ha, University of California, Berkeley
Levander, Alan R., Rice University
Li, Yaoguo, Colorado School of Mines
Lines, Larry R., University of Calgary
Link, Curtis A., Montana Tech of the University of Montana
Louie, John N., University of Nevada, Reno
Lourenço, José M., Universidade de Trás-os-Montes e Alto Douro

Madadi, Mahyar, Curtin University
Margrave, Gary F., University of Calgary
Marobhe, Isaac M., University of Dar es Salaam
Mathur, Dr R., Osmania University
McBride, John H., Brigham Young University
McPherson, Brian, University of Utah
Miah, Khalid, Montana Tech of the University of Montana
Mikhaltsevitch, Vassily, Curtin University
Milkereit, Bernd, University of Toronto
Miller, Richard D., University of Kansas
Morrison, Huntly Frank, University of California, Berkeley
Olawuyi, Kehinde A., University of Ilorin
Olsen, Kim B., San Diego State University
Palmer, Derecke, University of New South Wales
Perez, Marco A., Centro de Inv Científica y de Ed Sup de Ensenada
Pethick, Andrew, Curtin University
Potter, David, University of Alberta
Pratt, R. Gerhard, Western University
Pujol, Jose, University of Memphis
Raef, Abdelmoneam E., Kansas State University
Raji, W. O., University of Ilorin
Rechtien, Richard D., Missouri University of Science and Technology
Rector, James W., University of California, Berkeley
Reshef, Moshe, Tel Aviv University
Richard, Benjamin H., Wright State University
Roche, Steven L., The University of Tulsa
Romo, Jose M., Centro de Inv Científica y de Ed Sup de Ensenada
Rupert, Gerald B., Missouri University of Science and Technology
Sarwar, A. K. Mostofa, University of New Orleans
Sava, Paul, Colorado School of Mines
Sbar, Marc, University of Arizona
Scanlin, Michael A., Elizabethtown College
Seifoullaev, Roustam K., University of Texas, Austin
Sen, Mrinal K., University of Texas, Austin
Serpa, Laura F., University of Texas, El Paso
Sexton, John L., Southern Illinois University Carbondale
Shaw, Kenneth L., University of Utah
Slater, Lee S., Rutgers, The State Univ of New Jersey, Newark
Smith, Richard S., Laurentian University, Sudbury
Smithyman, Brenden, Western University
Snieder, Roel, Colorado School of Mines
Sobotka, Jerzy, University of Wroclaw
Speece, Marvin A., Montana Tech of the University of Montana
Spikes, Kyle T., University of Texas, Austin
Sprenke, Kenneth F., University of Idaho
Steer, David N., University of Akron
Sternberg, Ben K., University of Arizona
Stewart, Robert, University of Houston
Stewart, Robert R., University of Houston
Stewart, Robert R., University of Calgary
Stimpson, Ian G., Keele University
Stoffa, Paul L., University of Texas, Austin
Sturmer, Daniel M., University of Cincinnati
Swidinsky, Andrei, Colorado School of Mines
Tertyshnikov, Konstantin, Curtin University
Thomsen, Leon, University of Houston
Tsvankin, Ilya D., Colorado School of Mines
Turpening, Roger M., Michigan Technological University
Unsworth, Martyn, Cornell University
Urosevic, Milovan, Curtin University
Vazquez, Rogelio, Centro de Inv Científica y de Ed Sup de Ensenada
Verma, Sumit, University of Texas, Permian Basin
von Frese, Ralph R., Ohio State University
Wannamaker, Phillip E., University of Utah
Watts, Doyle, Wright State University
Weglein, Arthur B., University of Houston
Weiss, Chester J., University of New Mexico
Wilson, Thomas H., West Virginia University
Wolfe, Paul J., Wright State University
Zhang, Bo, University of Alabama
Zhang, Rui, University of Louisiana at Lafayette
Zhdanov, Michael S., University of Utah
Ziolkowski, Anton M., Edinburgh University
Ziramov, Sasha, Curtin University

Geodesy

Al-Attar, David, University of Cambridge
Bennett, Richard, University of Arizona
Blewitt, Geoffrey, University of Nevada, Reno
Blewitt, Geoffrey, University of Nevada
Caporali, Alessandro, Università degli Studi di Padova
Chaussard, Estelle, University of Oregon
Clarke, Peter J., University of Newcastle Upon Tyne
Crossley, David J., Saint Louis University
Davis, James , Columbia University
Elliott, Julie, Purdue University
Evans, Eileen L., California State University, Northridge
Feigl, Kurt, University of Wisconsin-Madison
Flake, Rex , Central Washington University
Fotopoulos, Georgia, Queen's University
Funning, Gareth J., University of California, Riverside
Gonzalez, Jose J., Centro de Inv Científica y de Ed Sup de Ensenada
Hammond, William C., University of Nevada
Hammond, William C., University of Nevada, Reno
Herring, Thomas A., Massachusetts Institute of Technology
Hiroji, Anand D., University of Southern Mississippi
Hooper, Andy, University of Leeds
James, Thomas S., University of Victoria
Jekeli, Christopher, Ohio State University
King, Robert W., Massachusetts Institute of Technology
Kogan, Mikhail, Columbia University
Kreemer, Corne, University of Nevada
Kreemer, Corne, University of Nevada, Reno
Kusznir, Nick, University of Liverpool
LaFemina, Peter C., Pennsylvania State University, University Park
Li, Zhenhong, University of Newcastle Upon Tyne
Lopez, Alberto, University of Puerto Rico
Lowry, Anthony R., Utah State University
Masterlark, Timothy L., South Dakota School of Mines & Technology
Melbourne, Timothy I., Central Washington University
Melgar, Diego, University of Oregon
Miller, M. Meghan, Central Washington University
Moore, Phillip, University of Newcastle Upon Tyne
Mueller, Ivan, Ohio State University
Murray, Mark, New Mexico Institute of Mining and Technology
Newman, Andrew V., Georgia Institute of Technology
Odera, Patroba A., Jomo Kenyatta University of Agriculture & Tech
Pagiatakis, Spiros, York University
Palamartchouk, Kirill, University of Newcastle Upon Tyne
Parsons, Barry, University of Oxford
Pritchard, Matthew E., Cornell University
Rapp, Richard H., Ohio State University
Rundle, John, University of California, Davis
Samsonov, Sergey V., Natural Resources Canada
Santillan, Marcelo, Central Washington University
Schaffrin, Burkhard A., Ohio State University
Shum, CK, Ohio State University
Smalley, Jr., Robert, University of Memphis
Smith-Konter, Bridget R., University of Hawai'i, Manoa
Sparks, David, Texas A&M University
Spinler, Joshua C., University of Arkansas at Little Rock
Stamps, D. S., Virginia Polytechnic Institute & State University
Stegman, Dave, University of California, San Diego
Tiampo, Kristy F., Western University
Waithaka, Hunja, Jomo Kenyatta University of Agriculture & Technology
Walters, Richard, Durham University
Wdowinski, Shimon, Florida International University
Wells, David E., University of Southern Mississippi
Wright, Tim, University of Leeds
Yi, Yuchan, Ohio State University

Geomagnetism & Paleomagnetism

Aldridge, Keith D., York University
Backus, George E., University of California, San Diego
Banerjee, Subir K., University of Minnesota, Twin Cities
Biedermann, Andrea, University of Minnesota, Twin Cities
Biggin, Andy, University of Liverpool
Bilardello, Dario, University of Minnesota, Twin Cities
Bogue, Scott W., Occidental College
Booker, John, University of Washington
Bowles, Julie, University of Wisconsin, Milwaukee
Braginsky, Stanislav I., University of California, Los Angeles
Bramham, Emma , University of Leeds
Brown, Laurie L., University of Massachusetts, Amherst
Burmester, Russell F., Western Washington University
Channell, James E., University of Florida

Chen, Xiaojia, Carnegie Institution for Science
Cioppa, Maria T., University of Windsor
Clough, Gene A., Bates College
Coe, Robert S., University of California, Santa Cruz
Constable, Catherine G., University of California, San Diego
Cottrell, Rory D., University of Rochester
DeMets, D. Charles, University of Wisconsin-Madison
Dharmasoth, Mrs M., Osmania University
DiVenere, Victor, Long Island University, C.W. Post Campus
Doh, Seong-Jae, Korea University
Dulin, Shannon, University of Oklahoma
Ellwood, Brooks B., Louisiana State University
Elmore, R. Douglas, University of Oklahoma
Evans, David A., Yale University
Evans, Michael E., University of Alberta
Everett, Mark, Texas A&M University
Feinberg, Joshua, University of Minnesota, Twin Cities
Ferre', Eric C., University of Louisiana at Lafayette
Franklin, David, University of Portsmouth
Fraser-Smith, Antony C., Stanford University
Fu, Roger, Harvard University
Garcia Lasanta, Cristina, Western Washington University
Geissman, John W., University of Michigan
Geissman, John W., University of New Mexico
Gilder, Stuart, Ludwig-Maximilians-Universitaet Muenchen
Gordon, Richard G., Rice University
Guerrero-Garcia, Jose C., Univ Nac Autonoma de Mexico
Hall, Frank R., University of New Orleans
Hall, Stuart A., University of Houston
Halls, Henry C., University of Toronto
Harbert, William P., University of Pittsburgh
Hill, Mimi, University of Liverpool
Hodych, Joseph P., Memorial University of Newfoundland
Holme, Richard, University of Liverpool
Housen, Bernard A., Western Washington University
Jackson, Michael, University of Minnesota, Twin Cities
Jones, F. Walter, University of Alberta
Kean, Jr., William F., University of Wisconsin, Milwaukee
Kelso, Paul R., Lake Superior State University
Kent, Dennis V., Rutgers, The State Univ of New Jersey
Kodama, Kenneth P., Lehigh University
Kravchinsky, Vadim, University of Alberta
Larson, Edwin E., University of Colorado
Lepre, Christopher H., Rutgers, The State Univ of New Jersey
Lima, Eduardo A., Massachusetts Institute of Technology
Lippert, Peter C., University of Utah
Livermore, Phil, University of Leeds
Lund, Steven P., University of Southern California
Maillol, Jean-Michel, University of Calgary
Maloof, Adam C., Princeton University
Marshall, Monte, San Diego State University
Márton, Péter, Eotvos Lorand University
McCausland, Phil J., Western University
Merrill, Ronald T., University of Washington
Morris, Antony, University of Plymouth
Moskowitz, Bruce M., University of Minnesota, Twin Cities
Muxworthy, Adrian, Imperial College
Negrini, Robert M., California State University, Bakersfield
Nilsson, Andreas, University of Liverpool
Noltimier, Hallan C., Ohio State University
Opdyke, Neil D., University of Florida
Palmer, H. C., Western University
Papitashvili, Vladimir O., National Science Foundation
Pares, Josep M., University of Michigan
Park, Stephen K., University of California, Riverside
Parker, Robert L., University of California, San Diego
Paty, Carol M., Georgia Institute of Technology
Petronis, Michael S., New Mexico Highlands University
Piper, John, University of Liverpool
Plattner, Alain, University of Alabama
Rankin, David, University of Alberta
Raub, Timothy D., University of St. Andrews
Richter, Carl, University of Louisiana at Lafayette
Roberts, Paul H., University of California, Los Angeles
Shibuya, Hidetoshi, Kumamoto University
Smirnov, Aleksey V., Michigan Technological University
Srnka, Len, University of California, San Diego
Sternberg, Robert S., Franklin and Marshall College
Stone, David B., University of Alaska, Fairbanks
Symons, David T., University of Windsor
Tarduno, John A., University of Rochester
Tauxe, Lisa, University of California, San Diego
Tikoo, Sonia M., Rutgers, The State Univ of New Jersey
Valentine, Michael J., University of Puget Sound
Van der Voo, Rob, University of Michigan
Verosub, Kenneth L., University of California, Davis
Volk, Michael, University of Minnesota, Twin Cities
Walden, John, University of St. Andrews
Weaver, John T., University of Victoria
Weiss, Benjamin, Massachusetts Institute of Technology
Whaler, Kathryn A., Edinburgh University
Williams, Wyn, Edinburgh University
Xuan, Chuang, University of Southampton
Zanella, Elena, Università di Torino

Gravity
Aiken, Carlos L., University of Texas, Dallas
Eppelbaum, Lev, Tel Aviv University
Garcia, Juan, Centro de Inv Científica y de Ed Sup de Ensenada
Grannell, Roswitha B., California State University, Long Beach
Kusumoto, Shigekazu, University of Toyama
Steckler, Michael, Columbia University

Heat Flow
Beltrami, Hugo, Saint Francis Xavier University
Blackwell, David D., Southern Methodist University
Fulton, Patrick M., Texas A&M University
Harris, Robert N., Oregon State University
Hsieh, Wen-Pin, Academia Sinica
Huang, Shaopeng, University of Michigan
Jones, Michael Q., University of the Witwatersrand
Lenkey, László, Eotvos Lorand University
Morgan, Paul, Colorado Geological Survey
Morgan, Paul, Northern Arizona University
Reiter, Marshall A., New Mexico Institute of Mining and Technology
Sclater, John G., University of California, San Diego
Smith, Douglas L., University of Florida
Starrfield, Sumner, Arizona State University

Seismology
Abers, Geoffrey A., Cornell University
Achauer, Ulrich, Universite de Strasbourg
Agnew, Duncan C., University of California, San Diego
Alexander, Shelton S., Pennsylvania State University, University Park
Allen, Clarence R., California Institute of Technology
Allen, Richard M., University of California, Berkeley
Ammon, Charles J., Pennsylvania State University, University Park
Ampuero, Jean-Paul, California Institute of Technology
Anandakrishnan, Sridhar, Pennsylvania State University, University Park
Anderson, John G., University of Nevada, Reno
Angus, Doug, University of Leeds
Archuleta, Ralph J., University of California, Santa Barbara
Arrowsmith, Stephen J., Southern Methodist University
Assatourians, Karen, Western University
Aster, Richard C., Colorado State University
Atkinson, Gail M., Western University
Ausbrooks, Scott, Arkansas Geological Survey
Austermann, Jacqueline, Columbia University
Bachmann, Etienne, Princeton University
Bagley, Brian, University of Minnesota, Twin Cities
Balch, Robert S., New Mexico Institute of Mining and Technology
Barazangi, Muawia, Cornell University
Barclay, Andrew, Columbia University
Barker, Jeffrey S., Binghamton University
Beaumont, Christopher, Dalhousie University
Beck, Susan L., University of Arizona
Beghein, Caroline, University of California, Los Angeles
Ben-Zion, Yehuda, University of Southern California
Benavides-Iglesias, Alfonso, Texas A&M University
Bennett, Scott, University of Washington
Benoit, Margaret, College of New Jersey
Beroza, Gregory C., Stanford University
Bezada, Max, University of Minnesota, Twin Cities
Biasi, Glenn P., University of Nevada, Reno
Bilek, Susan L., New Mexico Institute of Mining and Technology
Biundo, Marc, University of Washington

Blake, Daniel R., Ohio Dept of Natural Resources
Blakely, Robert F., Indiana University, Bloomington
Bodin, Paul A., University of Washington
Bollinger, G. A., Virginia Polytechnic Institute & State University
Bostock, Michael G., University of British Columbia
Bozdag, Ebru, Colorado School of Mines
Braile, Lawrence W., Purdue University
Brodsky, Emily, University of California, Santa Cruz
Brown, James R., University of Calgary
Brudzinski, Michael, Miami University
Brumbaugh, David S., Northern Arizona University
Brune, James N., University of Nevada, Reno
Burdick, Scott, Wayne State University
Burton, Paul, University of East Anglia
Byrnes, Joseph, University of Minnesota, Twin Cities
Calvert, Andrew J., Simon Fraser University
Cann, Johnson R., Woods Hole Oceanographic Institution
Caplan-Auerbach, Jackie, Western Washington University
Cardimona, Steve, Mendocino College
Cartwright, J A., Cardiff University
Cassidy, John, University of Victoria
Castro, Raul, Centro de Inv Científica y de Ed Sup de Ensenada
Chapman, Martin C., Virginia Polytechnic Institute & State University
Chen, Po, University of Wyoming
Chen, Wang-Ping, University of Illinois, Urbana-Champaign
Chen, Xiaowei, University of Oklahoma
Chesnokov, Evgeny, University of Houston
Chiu, Jer-Ming, University of Memphis
Christensen, Douglas, University of Alaska, Fairbanks
Cicerone, Robert D., Bridgewater State University
Cipar, John J., Boston College
Clark, Roger A., University of Leeds
Clayton, Robert W., California Institute of Technology
Clowes, Ronald M., University of British Columbia
Cochran, Elizabeth, University of California, Riverside
Collins, John A., Woods Hole Oceanographic Institution
Conder, James A., Southern Illinois University Carbondale
Cormier, Vernon F., University of Connecticut
Cornwell, David G., University of Aberdeen
Cottaar, Sanne , University of Cambridge
Cramer, Chris, University of Memphis
Creager, Kenneth C., University of Washington
Crosson, Robert S., University of Washington
Curtis, Andrew, Edinburgh University
D'Amico, Sebastiano, University of Malta
Dalton, Colleen, Brown University
Darbyshire, Fiona Ann, Universite du Quebec a Montreal
Darold, Amberlee, University of Oklahoma
Davis, Paul M., University of California, Los Angeles
Day, Steven M., San Diego State University
De Angelis, Silvio, University of Liverpool
Denolle, Marine, Harvard University
dePolo, Diane, University of Nevada, Reno
DeShon, Heather R., Southern Methodist University
Dorman, James, University of Memphis
Doser, Diane I., University of Texas, El Paso
Douilly, Roby, University of California, Riverside
Dreger, Douglas S., University of California, Berkeley
Duan, Benchuan, Texas A&M University
Duan, Benchun, Texas A&M University
Dueker, Kenneth G., University of Wyoming
Dunham, Eric M., Stanford University
Ebel, John E., Boston College
Eilon, Zach, University of California, Santa Barbara
Ekstrom, Goran, Columbia University
Ellis, Robert M., University of British Columbia
Ellsworth, William L., Stanford University
Farrell, Jamie, University of Utah
Faulkner, Daniel, University of Liverpool
Ferreira, Anna, University College London
Finley, William R., University of Louisiana at Lafayette
Fischer, Karen M., Brown University
Fontaine, Fabrice R., Universite de la Reunion
Forbriger, Thomas, Karlsruhe Institute of Technology
Ford, Heather, University of California, Riverside
Fox, Jeff, Ohio Dept of Natural Resources
Frankel, Arthur, University of Washington
Frazer, L. N., University of Hawai'i, Manoa
Frederiksen, Andrew, University of Manitoba
Frez, Jose, Centro de Inv Científica y de Ed Sup de Ensenada
Frohlich, Cliff, University of Texas, Austin
Frohlich, Clifford A., University of Texas, Austin
Fujita, Kazuya, Michigan State University
Gabrielov, Andrei, Purdue University
Gaherty, James, Columbia University
Galea, Pauline, University of Malta
Gao, Haiying, University of Massachusetts, Amherst
Garces, Milton A., University of Hawai'i, Manoa
Garnero, Edward, Arizona State University
Gharti, Hom Nath, Princeton University
Ghosh, Abhijit, University of California, Riverside
Gibson, Jr., Richard L., Texas A&M University
Gilbert, Hersh, Purdue University
Gilpin, Bernard J., Golden West College
Glowacka, Ewa, Centro de Inv Científica y de Ed Sup de Ensenada
Gomberg, Joan, University of Washington
Grand, Stephen P., University of Texas, Austin
Greenhalgh, Stewart A., King Fahd University of Petroleum and Minerals
Gu, Jeff, University of Alberta
Gurnis, Michael C., California Institute of Technology
Gurrola, Harold, Texas Tech University
Haddidi, Hamid, California Geological Survey
Hager, Bradford H., Massachusetts Institute of Technology
Hagerty, Michael, Boston College
Hagos, Lijam, California Geological Survey
Hajnal, Zoltan, University of Saskatchewan
Hall, Jeremy, Memorial University of Newfoundland
Hamburger, Michael W., Indiana University, Bloomington
Hamilton, Warren, Colorado School of Mines
Hammer, Philip T., University of British Columbia
Hansen, Samantha E., University of Alabama
Harrington, Rebecca, Karlsruhe Institute of Technology
Harris, Jerry M., Stanford University
Hartog, Renate, University of Washington
Hartse, Hans E., Los Alamos National Laboratory
Hassanpour Sedghi, Mohammad, University of Tabriz
Hauksson, Egill, California Institute of Technology
Hawman, Robert B., University of Georgia
Hayward, Christopher T., Southern Methodist University
Hearn, Thomas M., New Mexico State University, Las Cruces
Helmberger, Donald V., California Institute of Technology
Herrmann, Robert B., Saint Louis University
Hildebrand, Steve T., Los Alamos National Laboratory
Hobbs, Richard, Durham University
Holland, Austin A., University of Oklahoma
Holt, William E., Stony Brook University
Holtzman, Benjamin, Columbia University
Honjas, Bill, University of Nevada, Reno
Horton, Stephen P., University of Memphis
House, Leigh S., Los Alamos National Laboratory
Houston, Heidi B., University of Southern California
Howell, Jr., Benjamin F., Pennsylvania State University, University Park
Huerfano, Victor, University of Puerto Rico
Humphreys, Eugene D., University of Oregon
Hurich, Charles A., Memorial University of Newfoundland
Hyndman, Roy D., University of Victoria
Ismail, Ahmed, Oklahoma State University
Jackson, David D., University of California, Los Angeles
Jacob, Klaus H., Columbia University
Jasbinsek, John J., California Polytechnic State University
Jaumé, Steven C., College of Charleston
Ji, Chen, University of California, Santa Barbara
Johnson, Jeffrey , Boise State University
Johnson, Lane R., University of California, Berkeley
Johnson, Roy A., University of Arizona
Johnston, Archibald C., University of Memphis
Jones, Alan, Binghamton University
Jones, Craig H., University of Colorado
Jordan, Thomas H., University of Southern California
Kafka, Alan L., Boston College
Kaka, Ismail, King Fahd University of Petroleum and Minerals
Kanamori, Hiroo, California Institute of Technology
Karplus, Marianne, University of Texas, El Paso
Kent, Graham, University of Nevada, Reno

Keranen, Katie M., Cornell University
Kijko, Andrzej, University of Pretoria
Kim, Won-Young, Columbia University
Knopoff, Leon, University of California, Los Angeles
Kovach, Robert L., Stanford University
Krebes, Edward S., University of Calgary
Langston, Charles A., University of Memphis
Laske, Gabi, University of California, San Diego
Lavallee, Daniel, University of California, Santa Barbara
Lay, Thorne, University of California, Santa Cruz
Lees, Jonathan M., University of North Carolina, Chapel Hill
Lekic, Vedran, University of Maryland
Lerner-Lam, Arthur L., Columbia University
Levander, Alan, Rice University
Levin, Vadim, Rutgers, The State Univ of New Jersey
Li, Aibing, University of Houston
Li, Yong-Gang, University of Southern California
Liberty, Lee M., Boise State University
Linde, Alan T., Carnegie Institution for Science
Liner, Christopher, University of Arkansas, Fayetteville
Lizarralde, Daniel, Woods Hole Oceanographic Institution
Lohman, Rowena B., Cornell University
Long, Maureen D., Yale University
Lorenzo, Juan M., Louisiana State University
Lough, Amanda, Drexel University
Louie, John N., University of Nevada, Reno
Louie, John, University of Nevada, Reno
Lumley, David, University of Texas, Dallas
Ma, Shuo, San Diego State University
Magnani, M. Beatrice, Southern Methodist University
Main, Ian G., Edinburgh University
Malin, Peter E., Duke University
Mallick, Subhashis, University of Wyoming
Malone, Stephen D., University of Washington
Mangriotis, Maria-Daphne , Heriot-Watt University
Marfurt, Kurt J., University of Oklahoma
Masters, T. Guy, University of California, San Diego
McCarthy, Christine, Columbia University
McClelland, Lori, University of Nevada, Reno
McGuire, Jeffrey J., Woods Hole Oceanographic Institution
McLaskey, Greg C., Cornell University
McMechan, George A., University of Texas, Dallas
Meltzer, Anne S., Lehigh University
Meng, Lingsen, University of California, Los Angeles
Menke, William H., Columbia University
Mereu, Robert F., Western University
Michel, Gero, Western University
Mikesell, Dylan, Boise State University
Miller, Kate, Texas A&M University
Minster, J. Bernard H., University of California, San Diego
Molnar, Sheri, Western University
Morozov, Igor B., University of Saskatchewan
Moulis, Anastasia, Boston College
Mulibo, Gabriel D., University of Dar es Salaam
Munguia, Luis, Centro de Inv Científica y de Ed Sup de Ensenada
Nakamura, Yosio, University of Texas, Austin
Nakamura, Yosio n., University of Texas, Austin
Nava, Alejandro, Centro de Inv Científica y de Ed Sup de Ensenada
Nedimovic, Mladen, Dalhousie University
Neighbors, Corrie, Western New Mexico University
Nettles, Meredith, Columbia University
Ni, James F., New Mexico State University, Las Cruces
Nikulin, Alex, Binghamton University
Nissen-Meyer, Targe, University of Oxford
Niu, Fenglin, Rice University
Nowack, Robert L., Purdue University
Nyland, Edo, University of Alberta
Ogbamikhumi, Alexander, University of Benin
Oglesby, David D., University of California, Riverside
Okal, Emile A., Northwestern University
Okaya, David A., University of Southern California
Oliver, Adolph A., Chabot College
Olson, Ted L., Snow College
Owens, Thomas J., University of South Carolina
Pancha, Aasha, University of Nevada, Reno
Pankow, Kristine L., University of Utah
Panning, Mark, University of Florida

Park, Jeffrey J., Yale University
Pavlis, Gary L., Indiana University, Bloomington
Pechmann, James C., University of Utah
Peng, Zhigang, Georgia Institute of Technology
Pennington, Wayne D., Michigan Technological University
Persaud, Patricia, Louisiana State University
Phinney, Robert A., Princeton University
Plank, Gabriel, University of Nevada, Reno
Polet, Jascha, California State Polytechnic University, Pomona
Powell, Christine A., University of Memphis
Pride, Steven, University of California, Berkeley
Pulliam, Jay, Baylor University
Pulliam, Robert J., University of Texas, Austin
Pulliam, Robert J., Baylor University
Reading, Anya, University of Tasmania
Rector, III, James W., University of California, Berkeley
Rennie, Tom, University of Nevada, Reno
Revenaugh, Justin, University of Minnesota, Twin Cities
Rial, Jose A., University of North Carolina, Chapel Hill
Richard, Ferdinand W., University of Dar es Salaam
Richards, Paul G., Columbia University
Richardson, Eliza, Pennsylvania State University, University Park
Rieger, Duayne, Community College of Rhode Island
Rietbrock, Andreas, University of Liverpool
Ritsema, Jeroen, University of Michigan
Ritter, Joachim R., Karlsruhe Institute of Technology
Roberts, Peter, Los Alamos National Laboratory
Rogers, Garry C., University of Victoria
Rohay, Alan C., Pacific Northwest National Laboratory
Romanowicz, Barbara A., University of California, Berkeley
Rost, Sebastian, University of Leeds
Rowe, Charlotte A., New Mexico Institute of Mining and Technology
Rowshandel, Badie, California Geological Survey
Ruff, Larry J., University of Michigan
Ruffman, Alan, Dalhousie University
Ruhl, Christine J., The University of Tulsa
Sacchi, Mauricio D., University of Alberta
Sacks, I. Selwyn, Carnegie Institution for Science
Sahay, Pratap, Centro de Inv Científica y de Ed Sup de Ensenada
Sandvol, Eric A., University of Missouri
Savage, Brian K., University of Rhode Island
Savage, Heather M., Columbia University
Sawyer, Dale S., Rice University
Schaff, David, Columbia University
Schlue, John W., New Mexico Institute of Mining and Technology
Scholz, Christopher H., Columbia University
Schutt, Derek L., Colorado State University
Schwab, Fred, University of California, Los Angeles
Schwartz, Susan, University of California, Santa Cruz
Seeber, Leonardo, Columbia University
Serpa, Laura F., University of New Orleans
Shaw, Bruce, Columbia University
Shearer, Peter M., University of California, San Diego
Sheehan, Anne F., University of Colorado
Sherrod, Brian L., University of Washington
Shipley, Thomas H., University of Texas, Austin
Shragge, Jeffrey, Colorado School of Mines
Simons, Frederik J., Princeton University
Simons, Mark, California Institute of Technology
Slater, David, University of Nevada, Reno
Slawinski, Michael A., Memorial University of Newfoundland
Smith, Ken D., University of Nevada, Reno
Smith, Stewart W., University of Washington
Smythe, William, University of California, Los Angeles
Snoke, J. Arthur, Virginia Polytechnic Institute & State University
Song, Alex, University College London
Song, Xiaodong, University of Illinois, Urbana-Champaign
Spence, George D., University of Victoria
Spetzler, Hartmut A., University of Colorado
Spiegelman, Marc, Columbia University
Stachnik, Joshua, Lehigh University
Steele, Bill, University of Washington
Steeples, Don W., University of Kansas
Steidl, Jamison H., University of California, Santa Barbara
Stein, Seth, Northwestern University
Stephen, Ralph A., Woods Hole Oceanographic Institution
Stickney, Michael C., Montana Tech of The University of Montana

Street, Ronald L., University of Kentucky
Streig, Ashley, Portland State University
Stuart, Graham, University of Leeds
Stump, Brian W., Southern Methodist University
Sverdrup, Keith A., University of Wisconsin, Milwaukee
Sykes, Lynn R., Columbia University
Talwani, Pradeep, University of South Carolina
Tanimoto, Toshiro, University of California, Santa Barbara
Tape, Carl, University of Alaska, Fairbanks
Taylor, Kenneth B., North Carolina Geological Survey
Taylor, Steven R., Los Alamos National Laboratory
Teng, Ta-liang, University of Southern California
Thomas, Amanda, University of Oregon
Thorne, Michael S., University of Utah
Thurber, Clifford H., University of Wisconsin-Madison
Thurber, Clifford, University of Wisconsin, Madison
Tibuleac, Ileana, University of Nevada, Reno
Toksoz, M N., Massachusetts Institute of Technology
Toomey, Douglas R., University of Oregon
Tromp, Jeroen, Princeton University
Tura, Ali, Colorado School of Mines
Van Avendonk, Harm J., University of Texas, Austin
van der Hilst, Robert, Massachusetts Institute of Technology
van der Lee, Suzan, Northwestern University
Vanacore, Elizabeth A., University of Puerto Rico
Vidal, Antonio, Centro de Inv Científica y de Ed Sup de Ensenada
Vidale, John E., University of Southern California
Vlahovic, Gordana, North Carolina Central University
von Seggern, David H., University of Nevada, Reno
Wagner, Lara S., Carnegie Institution for Science
Waite, Gregory P., Michigan Technological University
Waldhauser, Felix, Columbia University
Wallace, Jr., Terry C., Los Alamos National Laboratory
Wang, Jeen-Hwa, Academia Sinica
Wang, Zhenming, University of Kentucky
Ward, Kevin M., South Dakota School of Mines & Technology
Warner, Michael, Imperial College
Waszek, Lauren, New Mexico State University, Las Cruces
Wen, Lianxing, Stony Brook University
Wentzcovitch, Renata, Columbia University
Wesnousky, Steve, University of Nevada, Reno
Wesnousky, Steven G., University of Nevada, Reno
Wesnousky, Steven, University of Nevada, Reno
Wiens, Douglas A., Washington University in St. Louis
Williams, Erik, University of Nevada, Reno
Wirth, Erin, University of Washington
Withers, Mitchell M., University of Memphis
Wolf, Lorraine W., Auburn University
Wolny, Dave, Colorado Mesa University
Woolery, Edward W., University of Kentucky
Wu, Francis T., Binghamton University
Wu, Patrick, University of Calgary
Wust-Bloch, Gilles H., Tel Aviv University
Wysession, Michael E., Washington University in St. Louis
Young, Jeri J., Arizona Geological Survey
Yu, Wen-che, Academia Sinica
Zandt, George, University of Arizona
Zelt, Colin A., Rice University
Zeng, Hongliu, University of Texas, Austin
Zhan, Zhongwen, California Institute of Technology
Zhou, Hua-Wei, University of Houston
Zhou, Ying, Virginia Polytechnic Institute & State University
Zhu, Hejun, University of Texas, Dallas
Zhu, Lupei, Saint Louis University
Ziv, Alon, Tel Aviv University

Marine Geophysics
Anderson, Franz E., University of New Hampshire
Archer, David, University of Chicago
Austin, Jr., James A., University of Texas, Austin
Becker, Keir, University of Miami
Behn, Mark D., Woods Hole Oceanographic Institution
Bell, Robin E., Columbia University
Ben-Avraham, Zvi, Tel Aviv University
Bohnenstiehl, DelWayne, North Carolina State University
Bokuniewicz, Henry J., Stony Brook University
Bonatti, Enrico, Columbia University
Bowin, Carl O., Woods Hole Oceanographic Institution

Buck, Roger W., Columbia University
Buizert, Christo, Oregon State University
Bull, Jonathan M., University of Southampton
Butterfield, David A., University of Washington
Canales Cisneros, Juan Pablo, Woods Hole Oceanographic Institution
Cande, Steven C., University of California, San Diego
Carbotte, Suzanne, Columbia University
Carlson, Richard L., Texas A&M University
Chamberlain, C. Page, Stanford University
Christeson, Gail L., University of Texas, Austin
Cochran, James R., Columbia University
Collier, Jenny, Imperial College
Commane, Roisin, Columbia University
Constable, Steven C., University of California, San Diego
Deans, Jeremy, University of Southern Mississippi
Delaney, John R., University of Washington
Druet, María, Univ Complutense de Madrid
Dunn, Robert A., University of Hawai'i, Manoa
Edwards, Richard N., University of Toronto
Egbert, Gary D., Oregon State University
Eloranta, Edwin W., University of Wisconsin, Madison
Evans, Robert L., Woods Hole Oceanographic Institution
Farges, Francois, Stanford University
Forsyth, Donald W., Brown University
Freyberg, David L., Stanford University
Fricker, Helen A., University of California, San Diego
Goff, John A., University of Texas, Austin
Goldberg, David S., Columbia University
Goldfinger, Chris, Oregon State University
Gorelick, Steven, Stanford University
Gorman, Andrew R., University of Otago
Gulick, Sean S., University of Texas, Austin
Harris, Paul M., Rice University
Harrison, Christopher G., University of Miami
Henderson, Naomi, Columbia University
Hildebrand, John A., University of California, San Diego
Hooft, Emilie E., University of Oregon
Hornbach, Matthew J., Southern Methodist University
Huang, Li, Western University
Huws, Dei G., University of Wales
Ji, Peng, Columbia University
Johnson, H. Paul, University of Washington
Karig, Daniel E., Cornell University
Kingslake, Jonathan, Columbia University
Kitanidis, Peter K., Stanford University
Krastel, Sebastian , Dalhousie University
Kummerow, Christian D., Colorado State University
Kurapov, Alexander, Oregon State University
Lawver, Lawrence A., University of Texas, Austin
Lewis, Brian T. R., University of Washington
Llanes, María del Pilar, Univ Complutense de Madrid
Louden, Keith E., Dalhousie University
Lowell, Robert P., Virginia Polytechnic Institute & State University
Luyendyk, Bruce P., University of California, Santa Barbara
MacAyeal, Douglas R., University of Chicago
Mack, Seymour, California State University, Fresno
Madsen, John A., University of Delaware
Manley, Patricia L., Middlebury College
Martinez, Fernando, University of Hawai'i, Manoa
McClain, James S., University of California, Davis
McIntosh, Kirk D., University of Texas, Austin
Minshull, Timothy A., University of Southampton
Mittelstaedt, Eric L., University of Idaho
Monismith, Stephen G., Stanford University
Muller, Dietmar, University of Sydney
Mutter, John C., Columbia University
Mutter, John C., Columbia University
Nilsson, Anders, Stanford University
Oakley, Adrienne, Kutztown University of Pennsylvania
Odom, Robert I., University of Washington
Ollerhead, Jeffery W., Mount Allison University
Orcutt, John A., University of California, San Diego
Purdy, G. Michael, Columbia University
Raymond, Carol A., Jet Propulsion Laboratory
Rodriguez Simental, Ana María, Univ Nac Autonoma de Mexico
Ryan, William B. F., Columbia University
Sandwell, David T., University of California, San Diego

Sawyer, Dale S., Rice University
Schouten, Hans, Woods Hole Oceanographic Institution
Sheridan, Robert E., Rutgers, The State Univ of New Jersey
Shillington, Donna, Columbia University
Shipley, Thomas H., University of Texas, Austin
Shreve, Ronald L., University of Washington
Silver, Eli A., University of California, Santa Cruz
Sloan, Heather, Lehman College (CUNY)
Sohn, Robert A., Woods Hole Oceanographic Institution
Sorlien, Christopher C., University of California, Santa Barbara
Spitz, Yvette H., Oregon State University
Spormann, Alfred, Stanford University
Stephens, Jason H., Austin Community College District
Swift, Stephen A., Woods Hole Oceanographic Institution
Talwani, Manik, Rice University
Thompson, David W. J., Colorado State University
Tivey, Maurice A., Woods Hole Oceanographic Institution
Tolstoy, Maria, Columbia University
Trehu, Anne M., Oregon State University
Van Rooij, David, Ghent University
Wattrus, Nigel J., University of Minnesota, Duluth
Webb, Spahr, Columbia University
Webb, Spahr C., Columbia University
Wessel, Paul, University of Hawai'i, Manoa
Wetzel, Laura R., Eckerd College
White, Scott M., University of South Carolina
Wooding, Frank B., Woods Hole Oceanographic Institution
Xu, Li, Woods Hole Oceanographic Institution
Zavarin, Mavrik, Lawrence Livermore National Laboratory
Zhu, Wenlu, University of Maryland

Near-surface Geophysics
Aracil, Enrique, Univ Complutense de Madrid
Bradford, John, Colorado School of Mines
Clement, William P., University of Massachusetts, Amherst
Ferdowsi, Behrooz, Princeton University
Herman, Rhett B., Radford University
Holbrook, W. S., Virginia Polytechnic Institute & State University
Katumwehe, Andrew, Midwestern State University
Keating, Kristina M., Rutgers, The State Univ of New Jersey, Newark
Lamb, Andrew P., University of Arkansas, Fayetteville
McGrath, Daniel, Colorado State University
Ntarlagiannis, Dimitrios, Rutgers, The State Univ of New Jersey, Newark
Pringle, Jamie K., Keele University
Wood, Neill, Exeter University

PALEONTOLOGY
General Paleontology
Adrain, Tiffany S., University of Iowa
Allmon, Warren D., Cornell University
Allmon, Warren D., Paleontological Research Institution
Altenbach, Alexander, Ludwig-Maximilians-Universitaet Muenchen
Amler, Michael, Ludwig-Maximilians-Universitaet Muenchen
Anderson, Wayne I., University of Northern Iowa
Arens, Nan Crystal, Hobart & William Smith Colleges
Arnone, John, Desert Research Institute
Arp, Dr., Gernot, Georg-August University of Goettingen
Ausich, William I., Ohio State University
Awramik, Stanley M., University of California, Santa Barbara
Babcock, Loren E., Ohio State University
Bailey, Jack B., Western Illinois University
Bailey, Jake, University of Minnesota, Twin Cities
Bajraktareviæ, Zlatan, University of Zagreb
Baumiller, Tomasz K., University of Michigan
Beatty, William L., Winona State University
Beck, Charles, University of Michigan
Belasky, Paul, Ohlone College
Bell, Phil R., University of New England
Benson, Roger, University of Oxford
Bergstrom, Stig M., Ohio State University
Berman, David S., Carnegie Museum of Natural History
Berry, David R., California State Polytechnic University, Pomona
Bill, Steven D., Keene State College
Blackwelder, Patricia L., University of Miami
Bordeaux, Yvette, University of Pennsylvania
Bosak, Tanja, Massachusetts Institute of Technology
Branstrator, Jon W., Earlham College
Brasier, Martin, University of Oxford
Bray, Emily, University of Colorado
Brink, Alyson, University of Southern Mississippi
Britt, Brooks B., Brigham Young University
Brower, James C., Syracuse University
Brzobohaty, Rostislav, Masaryk University
Bucur, Ioan, Babes-Bolyai University
Budd, Ann F., University of Iowa
Busbey, Arthur B., Texas Christian University
Bush, Andrew M., University of Connecticut
Butterfield, Nicholas J., University of Cambridge
Cachão, Mário, Universidade de Lisboa
Caldwell, Roy, University of California, Berkeley
Cardace, Dawn, University of Rhode Island
Carew, James L., College of Charleston
Carlson, Sandra J., University of California, Davis
Carnevale, Giorgio, Università di Torino
Caron, Jean-Bernard, Royal Ontario Museum
Carter, Burchard D., Georgia Southwestern State University
Chamberlain, John A., Brooklyn College (CUNY)
Chamberlain, John A., Grad Sch of the City Univ of New York
Cheetham, Alan H., Smithsonian Inst / Natl Museum of Natural History
Cherns, L C., Cardiff University
Chin, Karen, University of Colorado
Christie, Max L., University of Illinois, Urbana-Champaign
Cirimpei, Claudia, Alexandru Ioan Cuza
Cisne, John L., Cornell University
Clementz, Mark T., University of Wyoming
Clyde, William C., University of New Hampshire
Cohen, Phoebe A., Williams College
Collette, Joseph, Minot State University
Collins, Laurel S., Florida International University
Collinson, James W., Ohio State University
Collinson, Margaret, Royal Holloway University of London
Colman, Albert, University of Chicago
Conley, Catharine A., New Mexico Institute of Mining and Technology
Coorough Burke, Patricia J., Milwaukee Public Museum
Coulson, Alan B., Clemson University
Cowen, Richard, University of California, Davis
Cramer, Howard R., Emory University
Crane, Peter R., University of Chicago
Curran, H. Allen, Smith College
Darroch, Simon A., Vanderbilt University
De Klerk, Billy, Rhodes University
Deboo, Phili B., University of Memphis
Delfino, Massimo, Università di Torino
Derstler, Kraig L., University of New Orleans
Devera, Joseph A., Illinois State Geological Survey
Dietl, Gregory, Paleontological Research Institution
Dietz, Anne D., San Antonio Community College
Diggins, Thomas P., Youngstown State University
Dodd, J. Robert, Indiana University, Bloomington
Domack, Cynthia R., Hamilton College
Droser, Mary L., University of California, Riverside
Duda, Jan-Peter, Georg-August University of Goettingen
Dupraz, Christophe, University of Connecticut
Dyson, Miranda, The Open University
Edwards, Nicholas, University of Manchester
Egerton, Victoria, University of Manchester
Erickson, J. Mark, St. Lawrence University
Falcon-Lang, Howard, Royal Holloway University of London
Fall, Leigh M., SUNY, Oneonta
Farmer, Jack D., Arizona State University
Felton, Richard M., Northwest Missouri State University
Fernández, Sixto, Univ Complutense de Madrid
Ferriz, Sergio Cevallos, Univ Nac Autonoma de Mexico
Fio, Karmen, University of Zagreb
Fisher, Frank, University of Colorado
Flessa, Karl W., University of Arizona
Flood, Beverley, University of Minnesota, Twin Cities
Fondevilla, Victor, Universitat Autonoma de Barcelona
Foote, Michael J., University of Chicago
Forcino, Frank, Western Carolina University
Fornaciari, Eliana, Università degli Studi di Padova
Fournier, Gregory, Massachusetts Institute of Technology
Fowler, Gerald A., University of Wisconsin, Parkside
Fox, David L., University of Minnesota, Twin Cities
Fox-Dobbs, Kena L., University of Puget Sound

Freeman, Veronica, Marietta College
Friedman, Matt, University of Oxford
Fritsen, Christian H., Desert Research Institute
Gabbott, Sarah, Leicester University
Gagnaison, Cyril, Institut Polytechnique LaSalle Beauvais (ex-IGAL)
Gahn, Forest J., Brigham Young University - Idaho
Gaidos, Eric J., University of Hawai'i, Manoa
Gale, Andrew, University of Portsmouth
Gallegos, Anthony F., Los Alamos National Laboratory
Ganeshram, Raja, Edinburgh University
Garb, Matt, Brooklyn College (CUNY)
Gatto, Roberto, Università degli Studi di Padova
Geary, Dana H., University of Wisconsin-Madison
Gess, Robert, Rhodes University
Gibson, Michael A., University of Tennessee, Martin
Glamoclija, Mihaela, Rutgers, The State Univ of New Jersey, Newark
Glasauer, Susan, University of Guelph
Goldman, Daniel, University of Dayton
Gorecka-Nowak, Anna, University of Wroclaw
Gorman-Lewis, Drew J., University of Washington
Gray, Lee M., Mount Union College
Gray, Neil, University of Newcastle Upon Tyne
Green, Keith E., Cypress College
Greenwood, David, University of Saskatchewan
Haggart, James, University of British Columbia
Haiar, Brooke, University of Lynchburg
Hansen, Thor A., Western Washington University
Harnik, Paul, Franklin and Marshall College
Harper, David, Durham University
Harries, Peter J., University of South Florida, Tampa
Harvey, Tom, Leicester University
Hawkins, Andrew, Towson University
Hazel, McGoff, University of Reading
Heaney, Michael, Texas A&M University
Henderson, Wayne G., California State University, Fullerton
Hendy, Ingrid, University of Michigan
Henßel, Katja, Ludwig-Maximilians-Universitaet Muenchen
Herrera-Gil, Luis A., Univ Autonoma de Baja California Sur
Hickman, Carole S., University of California, Berkeley
Hintze, Lehi F., Brigham Young University
Hopkins, Samantha, University of Oregon
House, Christopher H., Pennsylvania State University, University Park
Hughes, Nigel C., University of California, Riverside
Hylland, Mike, Utah Geological Survey
Immel, Harald, Ludwig-Maximilians-Universitaet Muenchen
Ivanov, Martin, Masaryk University
Jackson, Jeremy B. C., University of California, San Diego
Johnson, Gerald H., College of William & Mary
Johnston, Carl G., Youngstown State University
Jones, Dan, University of Minnesota, Twin Cities
Jones, Douglas S., University of Florida
Jorge, Maria Luisa, Vanderbilt University
Kauffman, Erle G., Indiana University, Bloomington
Kazmer, Miklos, Eotvos Lorand University
Kelley, Patricia H., University of North Carolina, Wilmington
Kendrick, David C., Hobart & William Smith Colleges
Kern, J. Philip, San Diego State University
Khaleel, Tasneem, Montana State University, Billings
Kirkland, James I., Utah Geological Survey
Kirschvink, Joseph L., California Institute of Technology
Kleffner, Mark A., Ohio State University
Klemow, Kenneth M., Wilkes University
Koehl, Mimi A. R., University of California, Berkeley
Konhauser, Kurt, University of Alberta
Kowalke, Thorsten, Ludwig-Maximilians-Universitaet Muenchen
Kozdon, Reinhard, Columbia University
Kulp, Thomas, Binghamton University
Lambert, Lance L., University of Texas, San Antonio
Laporte, Leo F., University of California, Santa Cruz
Leinfelder, Reinhold, Ludwig-Maximilians-Universitaet Muenchen
Li, Rong-Yu, Brandon University
Lieberman, Bruce S., University of Kansas
Lindberg, David R., University of California, Berkeley
Linsley, David, Colgate University
Little, Crispin, University of Leeds
Love, Renee, University of Idaho
Lozar, Francesca, Università di Torino

Lucas-Clark, Joyce, City College of San Francisco
Macalady, Jennifer L., Pennsylvania State University, University Park
MacLeod, Kenneth A., University of Missouri
Macomber, Richard, Long Island University, Brooklyn Campus
Mandra, York T., San Francisco State University
Marenco, Pedro J., Bryn Mawr College
Martínez, Gemma, Univ Complutense de Madrid
McBeth, Joyce M., University of Saskatchewan
McCall, Peter L., Case Western Reserve University
McCollum, Linda B., Eastern Washington University
McGhee, Jr., George R., Rutgers, The State Univ of New Jersey
Mcgowan, Alistair, University of Glasgow
McHugh, Julia, Colorado Mesa University
McKenzie, Scott C., Mercyhurst University
McMenamin, Mark, Mount Holyoke College
McNamara, Kenneth, University of Cambridge
McRoberts, Christopher A., SUNY, Cortland
Mead, James I., Northern Arizona University
Means, Guy H., Florida Geological Survey
Melchin, Michael, Dalhousie University
Meyer Dombard, DArcy, University of Illinois at Chicago
Mezga, Aleksandar, University of Zagreb
Miller, Joshua H., University of Cincinnati
Miller, Randall F., University of New Brunswick
Mitchell, Charles E., SUNY, Buffalo
Mock, Thomas, University of East Anglia
Moe-Hoffman, Amy P., Mississippi State University
Monaco, Pat, University of Colorado
Monari, Stefano, Università degli Studi di Padova
Montgomery, Homer A., University of Texas, Dallas
Moro, Alan, University of Zagreb
Morris, Simon C., University of Cambridge
Munro, Ildi, Carleton University
Murray, Alison E., Desert Research Institute
Murray, John, National University of Ireland Galway
Musil, Rudolf, Masaryk University
Myers, Orrin B., Los Alamos National Laboratory
Naumann, Malik, Ludwig-Maximilians-Universitaet Muenchen
Nealson, Kenneth, University of Southern California
Newman, Dianne K., California Institute of Technology
Newton, Cathryn R., Syracuse University
Norman, David, University of Cambridge
Nudds, John, University of Manchester
O'Mullan, Gregory, Queens College (CUNY)
Olcott Marshall, Alison, University of Kansas
Ono, Shuhei, Massachusetts Institute of Technology
Onstott, Tullis C., Princeton University
Orchard, Michael, University of British Columbia
Orphan, Victoria, California Institute of Technology
Orsi, William, Ludwig-Maximilians-Universitaet Muenchen
Owen, Alan W., University of Glasgow
Parsons-Hubbard, Karla M., Oberlin College
Patzkowsky, Mark E., Pennsylvania State University, University Park
Pavia, Giulio, Università di Torino
Peck, Robert, Concord University
Peng, Lee Chai, University of Malaya
Perez-Huerta, Alberto, University of Alabama
Person, Jeff, North Dakota Geological Survey
Peter, Mark, Ohio Dept of Natural Resources
Peterson, Joseph E., University of Wisconsin, Oshkosh
Petuch, Edward J., Florida Atlantic University
Pezelj, Đurđica, University of Zagreb
Plotnick, Roy E., University of Illinois at Chicago
Poteet, Mary F., University of Texas, Austin
Pratt, Brian R., University of Saskatchewan
Prezant, Robert S., Grad Sch of the City Univ of New York
Pruss, Sara B., Smith College
Puchalski, Stephaney, Highline College
Purnell, Mark, Leicester University
Rampedi, Isaac T., University of Johannesburg
Rathburn, Anthony, Indiana State University
Raup, David M., University of Chicago
Reichenbacher, Bettina, Ludwig-Maximilians-Universitaet Muenchen
Reid, Catherine, University of Canterbury
Reitner, Joachim, Georg-August University of Goettingen
Rhenberg, Elizabeth, University of Memphis
Risk, Michael J., McMaster University

Robbins, Lisa, University of South Florida, Tampa
Roden, Eric E., University of Wisconsin-Madison
Rodríguez, Sergio, Univ Complutense de Madrid
Rogers, Karyn L., Rensselaer Polytechnic Institute
Rogers, Kristina C., University of Minnesota, Twin Cities
Rogers, Raymond, University of Minnesota, Twin Cities
Rogers, Steven L., Keele University
Rosenfeld, Carla, University of Minnesota, Twin Cities
Ross, Robert M., Paleontological Research Institution
Ross, Robert M., Cornell University
Runnegar, Bruce, University of California, Los Angeles
Rößner, Gertrud, Ludwig-Maximilians-Universitaet Muenchen
Sá, Artur A., Universidade de Trás-os-Montes e Alto Douro
Santelli, Cara M., University of Minnesota, Twin Cities
Sanudo, Sergio, University of Southern California
Scatterday, James W., SUNY, Geneseo
Schiffbauer, James, University of Missouri
Schmid, Dieter, Ludwig-Maximilians-Universitaet Muenchen
Schopf, J. William, University of California, Los Angeles
Schrenk, Matt, Michigan State University
Scudder, Sylvia J., University of Florida
Seibt, Ulrike, University of California, Los Angeles
Shapiro, Russell S., California State University, Chico
Shelton, Sally Y., South Dakota School of Mines & Technology
Shropshire, K. Lee, University of Northern Colorado
Silva-Pineda, Alicia, Univ Nac Autonoma de Mexico
Slatkin, Montgomery, University of California, Berkeley
Slaughter, Richard, University of Wisconsin-Madison
Smart, Christopher, University of Plymouth
Smith, Dena M., University of Colorado
Smith, Paul L., University of British Columbia
Sorauf, James E., University of South Florida, Tampa
Southam, Gordon, Western University
Spencer, Patrick K., Whitman College
Stafford, Emily, Western Carolina University
Steele, Michael A., Wilkes University
Steinker, Don C., Bowling Green State University
Steinker, Paula J., Bowling Green State University
Stetter, Karl O., University of California, Los Angeles
Stifel, Peter B., University of Maryland
Stigall, Alycia L., Ohio University
Stratton, James F., Eastern Illinois University
Superchi-Culver, Tonia, University of Colorado
Sutherland, Stuart, University of British Columbia
Sutton, Mark, Imperial College
Sweetman, Steve, University of Portsmouth
Tesmer, Irving, Buffalo State College
Thomas, Roger D. K., Franklin and Marshall College
Thomka, James R., University of Akron
Thompson, Jann W. M., Smithsonian Inst / Natl Museum of Nat History
Thompson, Joel B., Eckerd College
Tice, Mike, Texas A&M University
Toots, Heinrich, Long Island University, C.W. Post Campus
Toth, Natalie, Denver Museum of Nature & Science
Underwood, Charlie, Birkbeck College
Upchurch, Paul, University College London
Valentine, James W., University of California, Berkeley
Van Alstine, James B., University of Minnesota, Morris
Van Iten, Heyo, Hanover College
VanAller-Hernick, Linda A., New York State Geological Survey
Vega Vera, Francisco J., Univ Nac Autonoma de Mexico
Vicens Batet, Enric, Universitat Autonoma de Barcelona
Wake, David B., University of California, Berkeley
Wake, Marvalee H., University of California, Berkeley
Wallace, Steve, University of Colorado
Ward, Peter D., University of Washington
Weber, Karrie A., University of Nebraska, Lincoln
Weber-Gobel, Reinhard, Univ Nac Autonoma de Mexico
Wehrmann, Laura, Ludwig-Maximilians-Universitaet Muenchen
Weiss, Dennis, Grad Sch of the City Univ of New York
West, Ronald R., Kansas State University
White, Lisa D., San Francisco State University
Whitmore, John H., Cedarville University
Whitton, Mark, University of Portsmouth
Wignall, Paul, University of Leeds
Wilf, Peter D., Pennsylvania State University, University Park
Williams, Mark, Leicester University
Woerheide, Prof. Dr. Gert, Ludwig-Maximilians-Universitaet Muenchen
Yacobucci, Peg M., Bowling Green State University
Yancey, Thomas E., Texas A&M University
Yee, Nathan, Rutgers, The State Univ of New Jersey
Zalasiewicz, Jan, Leicester University

Paleostratigraphy

Al-Saad, Hamad A., University of Qatar
Arden, Daniel D., Georgia Southwestern State University
Baarli, B. Gudveig, Williams College
Baird, Gordon C., SUNY, Fredonia
Barrick, James E., Texas Tech University
Bauer, Jeffrey A., Shawnee State University
Berggren, William A., Woods Hole Oceanographic Institution
Braun, Willi K., University of Saskatchewan
Buatois, Luis, University of Saskatchewan
Caldwell, William G. E., University of Saskatchewan
Cohen, Andrew S., University of Arizona
Davila-Alcocer, Victor M., Univ Nac Autonoma de Mexico
Eaton, Jeffrey G., Weber State University
Ettensohn, Frank R., University of Kentucky
Finney, Stanley C., California State University, Long Beach
Grubb, Barbara, Mt. San Antonio College
Isaacson, Peter E., University of Idaho
Jeffery, David L., Marietta College
Johnson, David B., New Mexico Institute of Mining and Technology
Johnson, Markes E., Williams College
Jones, Brian, University of Alberta
Kalvoda, Jiri, Masaryk University
Kirchgasser, William T., SUNY Potsdam
Kumpan, Tomas, Masaryk University
Kurtz, Vincent E., Missouri State University
Landing, Ed, New York State Geological Survey
Landman, Neil H., Grad Sch of the City Univ of New York
Le Mone, David V., University of Texas, El Paso
Leatham, W. Britt, California State University, San Bernardino
Lewis, Gerald L., Pasadena City College
Liebe, Richard M., SUNY, The College at Brockport
Mangano, Gabriela, University of Saskatchewan
Martin, James E., South Dakota School of Mines & Technology
Maurrasse, Florentin J., Florida International University
McWilliams, Robert G., Miami University
Mikulic, Donald G., Illinois State Geological Survey
Miller, James F., Missouri State University
Miller, Richard H., San Diego State University
Murphy, Michael A., University of California, Riverside
Ogg, James G., Purdue University
Olson, Hillary C., University of Texas, Austin
Over, D. Jeffrey, SUNY, Geneseo
Ritter, Scott M., Brigham Young University
Rosario, Jose, California State University, East Bay
Sageman, Bradley B., Northwestern University
Shaw, Frederick C., Lehman College (CUNY)
Smith, Gerald R., University of Michigan
Stearn, Colin, McGill University
Stevens, Calvin H., San Jose State University
Taylor, John F., Indiana University of Pennsylvania
Thoms, Richard E., Portland State University
Titus, Robert C., Hartwick College
Tobin, Tom S., University of Alabama
Trexler, J. Peter, Juniata College
Van Niewenhuise, Donald, University of Houston
Vandenbroucke, Thijs, Ghent University
Zinsmeister, William J., Purdue University

Micropaleontology

Adekeye, O. A., University of Ilorin
Agnini, Claudia, Università degli Studi di Padova
Arnold, Anthony J., Florida State University
Aubry, Marie-Pierre, Rutgers, The State Univ of New Jersey
Bancroft, Alyssa M., Indiana University
Barnes, Christopher R., University of Victoria
Barrier, Pascal, Institut Polytechnique LaSalle Beauvais (ex-IGAL)
Beauchamp, Benoit, University of Calgary
Berggren, William A., Rutgers, The State Univ of New Jersey
Boix, Carme, Universitat Autonoma de Barcelona
Bollmann, Jörg, University of Toronto
Borrelli, Chiara, University of Rochester

Bown, Paul, University College London
Burke, Collette D., Wichita State University
Burkett, Ashley, Oklahoma State University
Buzas, Martin A., Smithsonian Inst / Natl Museum of Natural History
Caldwell, W. Glen E., Western University
Canales, María Luisa, Univ Complutense de Madrid
Carreno, Ana L., Univ Nac Autonoma de Mexico
Chapman, Mark, University of East Anglia
Ciesielski, Paul F., University of Florida
Clark, David L., University of Wisconsin-Madison
Corliss, Bruce H., Duke University
Culver, Stephen J., East Carolina University
Curry, William B., Woods Hole Oceanographic Institution
Czaja, Andrew D., University of Cincinnati
de Vernal, Anne, Universite du Quebec a Montreal
Eicher, Don L., University of Colorado
Eicher, Donald, University of Colorado
Finger, Ken, University of California, Berkeley
Fluegeman, Richard H., Ball State University
Foresi, Luca Maria, University of Siena
Frankel, Larry, University of Connecticut
Fritts, Paul J., California State University, Long Beach
Giusberti, Luca, Università degli Studi di Padova
Goldstein, Susan T., University of Georgia
Gorog, Agnes, Eotvos Lorand University
Grimm, Kirsten I., Universität Mainz
Harding, Ian C., University of Southampton
Hart, Malcolm B., University of Plymouth
Harwood, David M., University of Nebraska, Lincoln
Hasegawa, Shiro, Kumamoto University
Hemphill-Haley, Eileen, Humboldt State University
Henderson, Charles M., University of Calgary
Herrero, Concepción, Univ Complutense de Madrid
Huber, Brian T., Smithsonian Inst / Natl Museum of Natural History
Hunt, Gene, Smithsonian Inst / Natl Museum of Natural History
Ingle, Jr., James C., Stanford University
Ionesi, Viorel, Alexandru Ioan Cuza
Ishman, Scott E., Southern Illinois University Carbondale
Kanungo, Sudeep, University of Utah
Katz, Miriam E., Rensselaer Polytechnic Institute
Keller, Gerta, Princeton University
Kelly, D. Clay, University of Wisconsin-Madison
Klapper, Gilbert, Northwestern University
Knox, Larry W., Tennessee Tech University
Leadbetter, Jared R., California Institute of Technology
Leckie, R. Mark, University of Massachusetts, Amherst
Lewandowski, Katherine, Eastern Illinois University
Lobegeier, Melissa, Middle Tennessee State University
Louwye, Stephen, Ghent University
Lucas, Franklin A., University of Benin
Lundin, Robert F., Arizona State University
Lupia, Richard, University of Oklahoma
Maddocks, Rosalie F., University of Houston
Martin, Ronald E., University of Delaware
McCarthy, Francine G., Brock University
McCartney, Kevin, University of Maine, Presque Isle
Mendelson, Carl V., Beloit College
Metcalfe, Ian, University of New England
Metzger, Ronald A., Southwestern Oregon Community College
Montenari, Michael, Keele University
Nathan , Stephen, Eastern Connecticut State University
Nathan, Stephen, University of Massachusetts, Amherst
Nestell, Galina P., University of Texas, Arlington
Nestell, Merlynd K., University of Texas, Arlington
Noble, Paula J., University of Nevada, Reno
Olsson, Richard K., Rutgers, The State Univ of New Jersey
Oppo, Delia W., Woods Hole Oceanographic Institution
Orr, William N., University of Oregon
Patterson, R. Timothy, Carleton University
Pearson, Paul, University of Wales
Peeters, Frank J., Vrije Universiteit Amsterdam
Pessagno, Jr., Emile A., University of Texas, Dallas
Puckett, Mark, University of Southern Mississippi
Pujana, Ignacio, University of Texas, Dallas
Rathburn, Anthony E., California State University, Bakersfield
Rehakova, Daniela, Comenius University in Bratislava
Reinhardt, Edward G., McMaster University
Roberts, Charles, Mt. San Antonio College
Robinson, Edward, Florida International University
Rodrigues, Cyril G., University of Windsor
Ross, Charles A., Western Washington University
Rosscoe, Steven, Hardin-Simmons University
Sartipzadeh, Siavosh, University of Tabriz
Scherer, Reed P., Northern Illinois University
Schroder-Adams, Claudia, Carleton University
Scott, David B., Dalhousie University
Sen Gupta, Barun K., Louisiana State University
Siesser, William G., Vanderbilt University
Sloan, Doris, University of California, Berkeley
Sloan, Jon R., California State University, Northridge
Speijer, Robert, Katholieke Universiteit Leuven
Stott, Lowell D., University of Southern California
Thomas, Ellen, Yale University
Thomas, Ellen, Wesleyan University
Toth, Emoke, Eotvos Lorand University
Ufkes, Els , Vrije Universiteit Amsterdam
Vaz, Nuno M., Universidade de Trás-os-Montes e Alto Douro
von Bitter, Peter H., University of Toronto
Wade, Bridget S., University College London
Walker, Sally E., University of Georgia
Watkins, David K., University of Nebraska, Lincoln
Webb, Peter N., Ohio State University
Wise, Jr, Sherwood W., Florida State University
Zonneveld, John-Paul, University of Calgary

Paleobotany
Ash, Sydney R., Weber State University
Basinger, James F., University of Saskatchewan
Beck, John H., Boston College
Berry, C M., Cardiff University
Boyce, Charles K., University of Chicago
Burnham, Robyn J., University of Michigan
Carrillo Martinez, Miguel, Univ Nac Autonoma de Mexico
Currano, Ellen, University of Wyoming
Dilcher, David L., University of Florida
DiMichele, William A., Smithsonian Inst / Natl Museum of Nat History
Donoghue, Michael J., Yale University
Edwards, Dianne, Cardiff University
Erwin, Diane, University of California, Berkeley
Gabel, Mark, Black Hills State University
Gastaldo, Robert A., Colby College
Gowing, David, The Open University
Griffin, Kevin L., Columbia University
Heath, Kathleen M., Indiana State University
Hu, Shusheng, Yale University
Jackson, Gail D., Edinburgh University
Jarzen, David M., University of Florida
Kerp, Hans, Universitaet Muenster
Knoll, Andrew H., Harvard University
Looy, Cynthia, University of California, Berkeley
Manchester, Steven R., University of Florida
Martinetto, Edoardo, Università di Torino
McElwain, Jenny, Field Museum of Natural History
Miller, Ian, Denver Museum of Nature & Science
Mustoe, George, Western Washington University
Peppe, Daniel J., Baylor University
Pfefferkorn, Hermann W., University of Pennsylvania
Punyasena, Surangi W., University of Illinois, Urbana-Champaign
Raymond, Anne, Texas A&M University
Retallack, Gregory J., University of Oregon
Rigby, John, Queensland University of Technology
Scott, Andrew C., Royal Holloway University of London
Stromberg, Caroline, University of Washington
Taggart, Ralph E., Michigan State University
Taylor, Edith, University of Kansas
Thomasson, Joseph R., Fort Hays State University
Tiffney, Bruce H., University of California, Santa Barbara
Tyler, Carrie, Miami University
Wing, Scott L., Smithsonian Inst / Natl Museum of Natural History

Palynology
Abar, Daniel, Alexandru Ioan Cuza
Anderson, Patricia M., University of Washington
Asselin, Esther, Natural Resources Canada
Atta-Peters, David, University of Ghana

Brenner, Gilbert J., SUNY, New Paltz
Brush, Grace S., Johns Hopkins University
Burden, Elliott T., Memorial University of Newfoundland
Butcher, Anthony, University of Portsmouth
Calcote, Randy, University of Minnesota, Twin Cities
Darrell, James H., Georgia Southern University
Davis, Owen K., University of Arizona
Dolakova, Nela, Masaryk University
Eble, Cortland F., University of Kentucky
Farley, Martin B., University of North Carolina, Pembroke
FitzPatrick, Meriel E., University of Plymouth
Fowell, Sarah J., University of Alaska, Fairbanks
Habib, Daniel, Grad Sch of the City Univ of New York
Habib, Daniel, Queens College (CUNY)
Head, Martin J., Brock University
Hebda, Richard J., University of Victoria
Hemsley, A, Cardiff University
Hemsley, Alan, University of Wales
Higgs, Ken, University College Cork
Hills, Leonard V., University of Calgary
Ivory, Sarah, Pennsylvania State University, University Park
Jacobs, Bonnie F., Southern Methodist University
Jarvis, Ed, University College Cork
Javaux, Emmanuelle, Universite de Liege
Jolley, David, University of Aberdeen
Kováčová, Marianna, Comenius University in Bratislava
Leyden, Barbara W., University of South Florida, Tampa
Maher, Jr., Louis J., University of Wisconsin-Madison
Marshall, John E., University of Southampton
Martinez Hernandez, Enrique, Univ Nac Autonoma de Mexico
McAndrews, John H., University of Toronto
Mudie, Peta J., Dalhousie University
Norris, Geoffrey, University of Toronto
O'Keefe, Jennifer, Morehead State University
Rich, Fredrick J., Georgia Southern University
Rueger, Bruce F., Colby College
Smith, Jim, University College Cork
Stefanova, Ivanka, University of Minnesota, Twin Cities
Strother, Paul L., Boston College
Strother, Paul K., Boston College
Warny, Sophie, Louisiana State University
Zobaa, Mohamed, University of Texas, Permian Basin

Quantitative Paleontology
Daley, Gwen M., Winthrop University
Huntley, John, University of Missouri
Marshall, Charles R., University of California, Berkeley
Miller, Arnold I., Cincinnati Museum Center
Motz, Gary J., Indiana University
Parker, William C., Florida State University
Sallan, Lauren, University of Pennsylvania

Vertebrate Paleontology
Adams, Thomas, San Antonio Community College
Applegate-Pleasants, Shelton, Univ Nac Autonoma de Mexico
Archer, Michael, University of New South Wales
Baltensperger, David D., Texas A&M University
Barnosky, Anthony D., University of California, Berkeley
Bartels, William S., Albion College
Beard, K. Christopher, Carnegie Museum of Natural History
Behrensmeyer, Anna K., Smithsonian Inst / Natl Museum of Natural His
Bell, Christopher J., University of Texas, Austin
Bloch, Jonathan, University of Florida
Bolt, John R., Field Museum of Natural History
Brand, Leonard R., Loma Linda University
Brinkman, Donald B., Royal Tyrrell Museum of Palaeontology
Brochu, Christopher A., University of Iowa
Brown, Caleb, Royal Tyrrell Museum of Palaeontology
Burger, Benjamin J., Utah State University
Caldwell, Michael W., University of Alberta
Campbell, Kenneth E., Los Angeles County Museum of Natural History
Canudo, José Ignacio, Universidad de Zaragoza
Carlson, Keith J., Gustavus Adolphus College
Carpenter, Kenneth, University of Colorado
Carpenter, Kenneth, Utah State University
Carrano, Matthew T., Smithsonian Inst / Natl Museum of Natural History
Carranza-Castaneda, Oscar, Univ Nac Autonoma de Mexico
Chatterjee, Sankar, Texas Tech University

Chiappe, Luis M., Los Angeles County Museum of Natural History
Cicimurri, Christian M., Clemson University
Cicimurri, David J., Clemson University
Cifelli, Richard L., University of Oklahoma
Clarke, Julia A., University of Texas, Austin
Clemens, William A., University of California, Berkeley
Codrea, Vlad, Babes-Bolyai University
Cooke, H.B. S., Dalhousie University
Coombs, Margery C., Amherst College
Covert, Herbert, University of Colorado
Cumbaa, Steve L., Carleton University
Currie, Philip J., University of Calgary
Curry Rogers, Kristina A., Macalester College
Daeschler, Ted, Drexel University
Daly, Raymond J., Union County College
Davis, Edward B., University of Oregon
Dawson, Mary R., Carnegie Museum of Natural History
Dawson, Mary, University of Colorado
Delson, Eric, Grad Sch of the City Univ of New York
DeSantis, Larisa R., Vanderbilt University
Dodson, Peter, University of Pennsylvania
Druckenmiller, Patrick S., University of Alaska, Fairbanks
Dundas, Robert G., California State University, Fresno
Dzanh, Trihn, University of Colorado
Eberle, Jaelyn J., University of Colorado
Elliott, David K., Northern Arizona University
Emry, Robert, Smithsonian Inst / Natl Museum of Natural History
Engelmann, George F., University of Nebraska, Omaha
Esperante, Raul, Loma Linda University
Evans, David C., Royal Ontario Museum
Fastovsky, David E., University of Rhode Island
Fastovsky, David, University of Rhode Island
Fiorillo, Anthony R., Southern Methodist University
Flynn, John J., Columbia University
Flynn, John J., American Museum of Natural History
Fordyce, R. Ewan, University of Otago
Froehlich, David J., Austin Community College District
Furió, Marc, Universitat Autonoma de Barcelona
Gaffney, Eugene S., American Museum of Natural History
Gallagher, William B., Rider University
García, Nuria, Univ Complutense de Madrid
Garcia, William, University of North Carolina, Charlotte
Gardner, James D., Royal Tyrrell Museum of Palaeontology
Gauthier, Jacques, Yale University
Gauthier, Jacques A., Yale University
Gillette, David D., Northern Arizona University
Gnidovec, Dale M., Ohio State University
Goodwin, Mark B., University of California, Berkeley
Gottfried, Michael D., Michigan State University
Graham, Russell W., Pennsylvania State University, University Park
Grandal d'Anglade, Aurora, Coruna University
Grande, Lance, Field Museum of Natural History
Gray, Robert S., Santa Barbara City College
Green, Harry W., University of California, Berkeley
Hammer, William R., Augustana College
Hand, Suzanne J., University of New South Wales
Hargrave, Jennifer E., University of Louisiana at Lafayette
Harrell, Jr., T. L., Geological Survey of Alabama
Harris, Judith A., University of Colorado
Hayden, Martha C., Utah Geological Survey
Heaton, Timothy H., University of South Dakota
Heckert, Andrew B., Appalachian State University
Henderson, Donald, Royal Tyrrell Museum of Palaeontology
Henrici, Amy C., Carnegie Museum of Natural History
Hernández, Manuel, Univ Complutense de Madrid
Hitz, Ralph B., Tacoma Community College
Holroyd, Pat, University of California, Berkeley
Holtz, Jr., Thomas R., University of Maryland
Horner, John R., Montana State University
Hungerbuehler, Axel, Mesalands Community College
Hunt, Robert M., University of Nebraska, Lincoln
Indeck, Jeff, University of Colorado
Jackson, Frankie, Montana State University
Jacobs, Louis L., Southern Methodist University
Jasinski, Steven E., State Museum of Pennsylvania
Johnson, Ned K., University of California, Berkeley
Joniak, Peter , Comenius University in Bratislava

Keenan, Sarah W., South Dakota School of Mines & Technology
Kirkland, James I., Utah Geological Survey
Knell, Michael J., Southern Connecticut State University
Koch, Paul L., University of California, Santa Cruz
Konishi, Takuya, Cincinnati Museum Center
Krause, David W., Stony Brook University
Krishtalka, Leonard, University of Kansas
Lamanna, Matthew C., University of Pittsburgh
Lamanna, Matthew C., Carnegie Museum of Natural History
Lamb, James P., University of West Alabama
Langston, Jr., Wann, University of Texas, Austin
Lidicker, Jr., William Z., University of California, Berkeley
Lindsay, Everett H., University of Arizona
Lundelius, Ernest L., University of Texas, Austin
Luo, Zhexi, Carnegie Museum of Natural History
Lyson, Tyler, Denver Museum of Nature & Science
MacFadden, Bruce J., University of Florida
Maisey, John G., American Museum of Natural History
Makovicky, Peter J., Field Museum of Natural History
Marcot, Jonathan, University of Colorado
Martill, David, University of Portsmouth
Martin, James E., University of Louisiana at Lafayette
Massare, Judy A., SUNY, The College at Brockport
McDonald, Greg, University of Colorado
McFadden, Bruce J., University of Florida
McLeod, Samuel A., Los Angeles County Museum of Natural History
Meng, Jin, Grad Sch of the City Univ of New York
Meng, Jin, American Museum of Natural History
Merck, Jr., John W., University of Maryland
Miller, Wade E., Brigham Young University
Montellano-Ballesteros, Marisol, Univ Nac Autonoma de Mexico
Morales, Michael A., Emporia State University
Motani, Ryosuke, University of California, Davis
Naples, Virginia, Northern Illinois University
Nesbitt, Sterling J., Virginia Polytechnic Institute & State University
Neuman, Andrew G., Royal Tyrrell Museum of Palaeontology
Njau, Jackson K., Indiana University, Bloomington
Norell, Mark A., Columbia University
Norell, Mark A., American Museum of Natural History
Norris, Christopher A., Yale University
Novacek, Michael J., American Museum of Natural History
O'Leary, Maureen, Stony Brook University
Olsen, Paul E., Columbia University
Osi, Attila, Eotvos Lorand University
Padian, Kevin, University of California, Berkeley
Pagnac, Darrin C., South Dakota School of Mines & Technology
Parham, James, California State University, Fullerton
Patton, James L., University of California, Berkeley
Pol, Diego, Cornell University
Polly, P D., Indiana University, Bloomington
Rățoi, Bogdan G., Alexandru Ioan Cuza
Ray, Clayton E., Smithsonian Inst / Natl Museum of Natural History
Rensberger, John M., University of Washington
Rieppel, Olivier C., Field Museum of Natural History
Robinson, Peter, University of Colorado
Rowe, Timothy B., University of Texas, Austin
Russell, Dale A., North Carolina State University
Sabol, Martin, Comenius University in Bratislava
Sagebiel, J. C., San Bernardino County Museum
Samonds, Karen, Northern Illinois University
Sankey, Julia, California State University, Stanislaus
Schiebout, Judith A., Louisiana State University
Schultz, Gerald E., West Texas A&M University
Schwartz, Hilde, University of California, Santa Cruz
Schwimmer, David R., Columbus State University
Scott, Craig, Royal Tyrrell Museum of Palaeontology
Sears, Karen, University of Colorado
Secord, Ross, University of Nebraska, Lincoln
Semken, Jr., Holmes A., University of Iowa
Sertich, Joseph, Denver Museum of Nature & Science
Shaw, Christopher A., Los Angeles County Museum of Natural History
Shiller, Thomas, Sul Ross State University
Sidor, Christian A., University of Washington
Smith, Kathlyn M., Georgia Southern University
Steadman, David W., University of Florida
Stearley, Ralph F., Calvin College
Stocker, Michelle, Virginia Polytechnic Institute & State University

Storrs, Glenn W., University of Cincinnati
Storrs, Glenn W., Cincinnati Museum Center
Sumida, Stuart S., California State University, San Bernardino
Tabrum, Alan R., Carnegie Museum of Natural History
Tedford, Richard H., American Museum of Natural History
Therrien, Francois, Royal Tyrrell Museum of Palaeontology
Tütken, Thomas, Universität Mainz
Uhen, Mark, George Mason University
VanRegenmorter, John M., Grand Valley State University
Varricchio, David J., Montana State University
Vietti, Laura, University of Wyoming
Vila, Bernat, Universitat Autonoma de Barcelona
Voorhies, Michael R., University of Nebraska, Lincoln
Vrba, Elisabeth S., Yale University
Whatley, Robin L., Columbia College Chicago
Wilkins, Kenneth, Baylor University
Williamson, Thomas E., University of New Mexico
Wilson, Gregory, University of Washington
Wilson, Laura E., Fort Hays State University
Winkler, Alisa, Southern Methodist University
Winkler, Dale A., Southern Methodist University
Wood, Aaron R., Iowa State University of Science & Technology
Woodburne, Michael O., University of California, Riverside
Wuerthele, Norman, Carnegie Museum of Natural History
Wyss, Andre R., University of California, Santa Barbara
Zakrzewski, Richard J., Fort Hays State University

Invertebrate Paleontology
Adrain, Jonathan M., University of Iowa
Alencaster-Ybarra, Gloria, Univ Nac Autonoma de Mexico
Anderson, Laurie C., South Dakota School of Mines & Technology
Bailey, Richard H., Northeastern University
Batten, Roger L., American Museum of Natural History
Becker, Ralph Thomas, Universitaet Muenster
Bishop, Gale A., Georgia Southern University
Blake, Daniel B., University of Illinois, Urbana-Champaign
Bonem, Rena M., Baylor University
Bonuso, Nicole, California State University, Fullerton
Bork, Kennard B., Denison University
Brandt, Danita S., Michigan State University
Briggs, Derek E., Yale University
Broadhead, Thomas W., University of Tennessee, Knoxville
Brown, Lewis M., Lake Superior State University
Buitron-Sanchez, Blanca E., Univ Nac Autonoma de Mexico
Buss, Leo W., Yale University
Butts, Susan H., Yale University
Cairns, Stephen, Smithsonian Inst / Natl Museum of Natural History
Caldwell, Roy L., University of California, Berkeley
Camp, Mark J., University of Toledo
Caron, Jean-Bernard, University of Toronto
Carter, Joseph, University of North Carolina, Chapel Hill
Chatelain, Edward E., Valdosta State University
Chatterton, Brian D., University of Alberta
Cintra-Buenrostro, Carlos, University of Texas, Rio Grande Valley
Clark, Murlene W., University of South Alabama
Comas, María José, Univ Complutense de Madrid
Copper, Paul, Laurentian University, Sudbury
Cuffey, Roger J., Pennsylvania State University, University Park
Dahl, Robyn, Western Washington University
Day, James E., Illinois State University
Deline, Brad, University of West Georgia
Dutro, Thomas, Smithsonian Inst / Natl Museum of Natural History
Eldredge, Niles, American Museum of Natural History
Elias, Robert J., University of Manitoba
Erwin, Douglas H., Smithsonian Inst / Natl Museum of Natural History
Evanoff, Emmett, University of Colorado
Feldmann, Rodney M., Kent State University
Ferguson, Laing, Mount Allison University
Finks, Robert M., Queens College (CUNY)
Finnegan, Seth, University of California, Berkeley
Fisher, Daniel C., University of Michigan
Fraiser, Margaret L., University of Wisconsin, Milwaukee
Full, Robert J., University of California, Berkeley
Gacía Joral, Fernando, Univ Complutense de Madrid
Gilmour, Ernest H., Eastern Washington University
Glass, Alexander, Duke University
Gonzalez-Arreola, Celestina, Univ Nac Autonoma de Mexico
Goodwin, David H., Denison University

Hageman, Steven J., Appalachian State University
Hall, Russell L., University of Calgary
Hanger, Rex, University of Wisconsin, Whitewater
Hannibal, Joseph T., Case Western Reserve University
Hannibal, Joseph T., Cleveland Museum of Natural History
Harper, Jr., Charles W., University of Oklahoma
Hartman, Joseph H., University of North Dakota
Hegna, Thomas A., Western Illinois University
Hembree, Daniel, Ohio University
Hendricks, Jonathan R., Paleontological Research Institution
Hudackova, Natalia H., Comenius University in Bratislava
Hughes, Nigel C., Cincinnati Museum Center
Hunda, Brenda, Cincinnati Museum Center
Hunda, Brenda R., University of Cincinnati
Hyzny, Matus, Comenius University in Bratislava
Jablonski, David, University of Chicago
James, Matthew J., Sonoma State University
Jin, Jisuo, Western University
Johns, Ronald A., Austin Community College District
Johnston, Paul, Mount Royal University
Kammer, Thomas W., West Virginia University
Key, Jr., Marcus M., Dickinson College
Kosloski, Mary, University of Iowa
Kues, Barry S., University of New Mexico
Kukalova-Peck, Jarmila, Carleton University
Labandeira, Conrad C., Smithsonian Inst / Natl Museum of Natural Hist
LaBarbera, Michael C., University of Chicago
Landman, Neil H., American Museum of Natural History
Lee, Daphne E., University of Otago
Leggitt, Leroy, Loma Linda University
Lehrmann, Daniel J., Trinity University
Leighton, Lindsey, University of Alberta
Lenz, Alfred C., Western University
Leonard-Pingel, Jill, Ohio State University
Levin, Harold L., Washington University in St. Louis
Lewis, Ronald D., Auburn University
Lidgard, Scott H., Field Museum of Natural History
Lindberg, David R., University of California, Berkeley
Lockley, Martin G., University of Colorado, Denver
LoDuca, Steven T., Eastern Michigan University
Mann, Keith O., Ohio Wesleyan University
Marenco, Katherine N., Bryn Mawr College
Mason, Charles E., Morehead State University
McAfee, Gerald B., Odessa College
Melchin, Michael J., Saint Francis Xavier University
Merrill, Glen K., University of Houston Downtown
Meyer, David L., University of Cincinnati
Miller, Arnold I., University of Cincinnati
Miller, William C., Humboldt State University
Morris, Robert W., Wittenberg University
Nagel-Myers, Judith, St. Lawrence University
Narbonne, Guy M., Queen's University
Nesbitt, Elizabeth (Liz), University of Washington
Nitecki, Matthew H., Field Museum of Natural History
Oleinik, Anton, Florida Atlantic University
Paterson, John R., University of New England
Perrilliat-Montoya, Maria del Carmen, Univ Nac Autonoma de Mexico
Petsios, Elizabeth, Baylor University
Pietsch, Carlie, San Jose State University
Pojeta, John, Smithsonian Inst / Natl Museum of Natural History
Pope, John K., Miami University
Portell, Roger W., University of Florida
Poty, Edouard, Universite de Liege
Powell, Matthew G., Juniata College
Prichonnet, Gilbert P., Universite du Quebec a Montreal
Rhodes, Frank H. T., Cornell University
Ritterbush, Linda A., California Lutheran University
Robison, Richard A., University of Kansas
Rodland, David L., Muskingum University
Rohr, David M., Sul Ross State University
Rosenberg, Gary, Drexel University
Rowell, Albert J., University of Kansas
Rowland, Stephen, University of Nevada, Las Vegas
Sandy, Michael R., University of Dayton
Santos, Heman, University of Puerto Rico
Saunders, W. Bruce, Bryn Mawr College
Savage, Norman M., University of Oregon

Schiappa, Tamra A., Slippery Rock University
Schlogl, Jan , Comenius University in Bratislava
Schmidt, David, Wright State University
Schweitzer, Carrie E., Kent State University
Schweitzer, Carrie E., Kent State University at Stark
Segura-Vernis, Luis R., Univ Autonoma de Baja California Sur
Selden, Paul A., University of Kansas
Sevilla, Paloma, Univ Complutense de Madrid
Shaw, Frederick C., Grad Sch of the City Univ of New York
Sheehan, Peter M., University of Wisconsin, Milwaukee
Sheehan, Peter M., Milwaukee Public Museum
Shubak, Kenneth A., University of Wisconsin, Platteville
Soja, Constance M., Colgate University
Sorauf, James E., Binghamton University
Spinosa, Claude, Boise State University
Sprinkle, James T., University of Texas, Austin
Squires, Richard L., California State University, Northridge
Stanley, Thomas M., University of Oklahoma
Stanley, Jr., George D., University of Montana
Stephen, Daniel, Utah Valley University
Stock, Carl W., University of Alabama
Sumrall, Colin, University of Tennessee, Knoxville
Szente, Istvan, Eotvos Lorand University
Szlavecz, Katalin, Johns Hopkins University
Tapanila, Leif, Idaho State University
Tshudy, Dale, Edinboro University of Pennsylvania
Tsujita, Cameron J., Western University
Villasenor-Martinez, Ana B., Univ Nac Autonoma de Mexico
Visaggi, Christy, Georgia State Univ
Votaw, Robert, Indiana University Northwest
Wagner, Peter J., University of Nebraska, Lincoln
Wagner, Peter J., Field Museum of Natural History
Waller, Thomas R., Smithsonian Inst / Natl Museum of Natural History
Webber, Andrew, Cincinnati Museum Center
Webster, Gary D., Washington State University
Webster, Mark, University of Chicago
Westrop, Stephen R., University of Oklahoma
Wilson, Mark A., College of Wooster
Wyse Jackson, Patrick N., Trinity College
Xiao, Shuhai, Virginia Polytechnic Institute & State University
Young, Graham A., University of Manitoba
Young, Graham A., Manitoba Museum
Zachos, Louis G., University of Mississippi
Zodrow, Erwin L., Cape Breton University
Zumwalt, Gary S., Louisiana Tech University

Paleoecology
Allabush, Kathleen, University of Utah
Almendinger, James E., University of Minnesota, Twin Cities
Andreu-Hayles, Laia, Columbia University
Arkle, Kelsey M., Augustana College
Ashworth, Allan C., North Dakota State University
Baker, Richard G., University of Iowa
Balescu, Sanda, Universite du Quebec a Montreal
Barker, Stephen, Cardiff University
Bartlein, Patrick, University of Oregon
Begin, Christian, Universite du Quebec
Bennington, J Bret, Hofstra University
Billups, Katharina, University of Delaware
Blanchet, Jean-Pierre, Universite du Quebec a Montreal
Bottjer, David J., University of Southern California
Boyd, Bill, Southern Cross University
Bralower, Timothy J., Pennsylvania State University, University Park
Brett, Carlton E., University of Cincinnati
Briskin, Madeleine, University of Cincinnati
Buckley, Brendon, Columbia University
Caissie, Beth E., Iowa State University of Science & Technology
Camill III, Philip, Bowdoin College
Capraro, Luca, Università degli Studi di Padova
Carlucci, Jesse R., Midwestern State University
Chapin, F. Stuart, University of California, Berkeley
Charles, Christopher D., University of California, San Diego
Cobb, Kim M., Georgia Institute of Technology
Cohen, Anne L., Woods Hole Oceanographic Institution
Cole, Julia E., University of Arizona
Colgan, Mitchell W., College of Charleston
Curtin, Tara M., Hobart & William Smith Colleges
Curtis, Jason H., University of Florida

D'Andrea, William J., Columbia University
de Menocal, Peter B., Columbia University
Dekens, Petra, San Francisco State University
deMenocal, Peter B., Columbia University
Dietl, Gregory P., Cornell University
Dornbos, Stephen Q., University of Wisconsin, Milwaukee
Duhamel, Solange, Columbia University
Edlund, Mark B., University of Minnesota, Twin Cities
Elick, Jennifer M., Susquehanna University
Endres, Anthony E., University of Waterloo
Engstrom, Daniel R., University of Minnesota, Twin Cities
Evans, Michael N., University of Maryland
Fariduddin, Mohammad, Northeastern Illinois University
Fawcett, Peter J., University of New Mexico
Francis, Jane, University of Leeds
Francus, Pierre, Universite du Quebec
Fritts, Harold C., University of Arizona
Fritz, Sherilyn C., University of Nebraska, Lincoln
Gabler, Christopher, University of Texas, Rio Grande Valley
Goodwin, Phillip A., University of Southampton
Grigg, Laurie, Norwich University
Guber, Albert L., Pennsylvania State University, University Park
Halfar, Jochen, University of Toronto
Hasiotis, Stephen T., University of Kansas
Hays, James D., Columbia University
Haywood, Alan, University of Leeds
He, Helen, University of East Anglia
Herbert, Timothy D., Brown University
Herrmann, Achim, Louisiana State University
Hill, Tessa M., University of California, Davis
Hopley, Philip, Birkbeck College
Hu, Feng-Sheng, University of Illinois, Urbana-Champaign
Hughes, Malcolm K., University of Arizona
Huybers, Peter, Harvard University
Ivany, Linda C., Syracuse University
Johnson, Claudia C., Indiana University, Bloomington
Johnson, Robert G., University of Minnesota, Twin Cities
Johnson, Thomas C., University of Massachusetts, Amherst
Jones, Douglas S., University of Florida
Kemp, Alan, University of Southampton
Kennedy, Lisa M., Virginia Polytechnic Institute & State University
Knight, Tiffany, Washington University in St. Louis
Koutavas, Athanasios, College of Staten Island/CUNY
Kutzbach, John E., University of Wisconsin, Madison
Lachniet, Matthew S., University of Nevada, Las Vegas
Lea, David W., University of California, Santa Barbara
Lebold, Joe, West Virginia University
Linsley, Braddock, Columbia University
Littler, Kate L., Exeter University
Livingstone, Daniel A., Duke University
Lohmann, George P., Woods Hole Oceanographic Institution
Long, Colin, University of Wisconsin Oshkosh
Lozano Garcia, Maria S., Univ Nac Autonoma de Mexico
Mann, Daniel H., University of Alaska, Fairbanks
Marchal, Olivier, Woods Hole Oceanographic Institution
Martin, Anthony J., Emory University
Martin, Pamela, University of Chicago
Mayewski, Paul A., University of Maine
McIntyre, Andrew, Queens College (CUNY)
McIntyre, Andrew, Grad Sch of the City Univ of New York
McManus, Jerry F., Woods Hole Oceanographic Institution
McManus, Jerry F., Columbia University
McWethy, David B., Montana State University
Medina Elizalde, Martin A., Auburn University
Meehan, Kimberly, SUNY, Buffalo
Miller, Joshua H., Cincinnati Museum Center
Miller, Keith B., Kansas State University
Miller, Molly F., Vanderbilt University
Mishler, Brent, University of California, Berkeley
Morgan, Ryan, Tarleton State University
Myers, Corinne E., University of New Mexico
Myrbo, Amy, University of Minnesota, Twin Cities
Nichols, Jonathan E., Columbia University
Norry, Mike, Leicester University
Olszewski, Thomas, Texas A&M University
Ortegren, Jason, University of West Florida
Panagiotakopulu, Eva, Edinburgh University

Parr, Kate, University of Liverpool
Pearson, P N., Cardiff University
Pekar, Stephen, Queens College (CUNY)
Peteet, Dorothy M., Columbia University
Phelps, William, Riverside City College
Poli, Maria-Serena, Eastern Michigan University
Pospelova, Vera, University of Victoria
Poulsen, Christopher J., University of Michigan
Power, Mary E., University of California, Berkeley
Quinn, Terry, University of Texas, Austin
Raymo, Maureen, Columbia University
Rhode, David E., Desert Research Institute
Rindsberg, Andrew K., University of West Alabama
Robinson, Stuart, University of Oxford
Royer, Dana, Wesleyan University
Sahagian, Dork, Lehigh University
Salzer, Matthew W., University of Arizona
Sauer, Peter, Indiana University, Bloomington
Schaefer, Joerg, Columbia University
Schellenberg, Stephen A., San Diego State University
Schelske, Claire L., University of Florida
Schwert, Donald P., North Dakota State University
Schöne, Bernd R., Universität Mainz
Sessa, Jocelyn A., Drexel University
Severinghaus, Jeffrey P., University of California, San Diego
Shaak, Graig D., University of Florida
Shanahan, Timothy M., University of Texas, Austin
Shroat-Lewis, Rene A., University of Arkansas at Little Rock
Shuman, Bryan N., University of Wyoming
Sinha, Ashish, California State University, Dominguez Hills
Sloan, Lisa C., University of California, Santa Cruz
Smerdon, Jason, Columbia University
Smith, Dena M., University of Colorado
Smith, Martin, Durham University
Speer, James, Indiana State University
Sremac, Jasenka, University of Zagreb
Stoykova, Kristalina C., Bulgarian Academy of Sciences
Surge, Donna M., University of North Carolina, Chapel Hill
Swetnam, Thomas W., University of Arizona
Tackett, Lydia S., North Dakota State University
Taffs, Kathryn, Southern Cross University
Talyor, Frederick W., University of Texas, Austin
Thompson, William G., Woods Hole Oceanographic Institution
Trouet, Valerie, University of Arizona
Uno, Kevin, Columbia University
Van der Putten, Nathalie, Vrije Universiteit Amsterdam
Verardo, Stacey, George Mason University
Vermeij, Geerat J., University of California, Davis
Waddington, J. M., McMaster University
Walker, Kenneth R., University of Tennessee, Knoxville
Warnock, Jonathan P., Indiana University of Pennsylvania
Webb, III, Thompson, Brown University
Weldeab, Syee, University of California, Santa Barbara
Westgate, James W., Lamar University
Whiteside, Jessica H., University of Southampton
Whitlock, Cathy, Montana State University
Wigand, Peter, University of Nevada, Reno
Williams, Christopher J., Franklin and Marshall College
Williams, John W., University of Wisconsin, Madison
Wolosz, Thomas H., SUNY, Plattsburgh
Woodhouse, Connie, University of Arizona
Yanes, Yurena, University of Cincinnati
Yu, Zicheng, Lehigh University
Æosoviæ, Vlasta, University of Zagreb

Geomicrobiology
Chan, Clara S., University of Delaware
Flannery, David, Queensland University of Technology
Frantz, Carie M., Weber State University
Liang, Renxing, Princeton University
Lower, Steven K., Ohio State University
Osburn, Magdalena, Northwestern University

Paleoclimatology
Andrus, C. Fred T., University of Alabama
Atkinson, Tim, University College London
Baker, Paul A., Duke University
Baldini, James, Durham University

Cleaveland, Malcolm K., University of Arkansas, Fayetteville
Cook, Tim, University of Massachusetts, Amherst
Cotton, Jennifer, California State University, Northridge
Doner, Lisa A., Plymouth State University
Gray, Katelyn E., Austin Community College District
Hoenisch, Baerbel, Columbia University
Johnson, Thomas, University of Minnesota, Duluth
Kaspari, Susan, Central Washington University
Koffman, Bess, Colby College
Konecky, Bronwen L., Washington University in St. Louis
Martín Chivelet, Javier, Univ Complutense de Madrid
McKay, Nicholas, Northern Arizona University
Meyers, Stephen R., University of Wisconsin-Madison
Michel, Lauren, Tennessee Tech University
Quinton, Page C., SUNY Potsdam
Railsback, L. Bruce, University of Georgia
Salacup, Jeff, University of Massachusetts, Amherst
Sullivan, Donald G., University of Denver
Touchan, Ramzi, University of Arizona
Williams, Alton P., Columbia University
Winsor, Kelsey, Northern Arizona University
Zachos, James, University of California, Santa Cruz
Zambito, James J., Beloit College

HYDROLOGY
General Hydrology
Adelana, S.M. A., University of Ilorin
AL-Nouimy, Latifa B., University of Qatar
Alencoão, Ana M., Universidade de Trás-os-Montes e Alto Douro
Alhajari, Sief A., University of Qatar
Allan, Craig J., University of North Carolina, Charlotte
Allen, Peter M., Baylor University
Alsdorf, Douglas E., Ohio State University
Andres, A. S., University of Delaware
Arain, M. A., McMaster University
Autio, Robert J., Indiana University
Bain, Daniel J., University of Pittsburgh
Barnard, Holly R., University of Colorado
Baron, Dirk, California State University, Bakersfield
Becker, Naomi M., Los Alamos National Laboratory
Bennett, Philip, University of Texas, Austin
Bense, Victor, University of East Anglia
Bloor, Michelle, University of Portsmouth
Bork, Karrigan, University of the Pacific
Borsa, Adrian, University of California, San Diego
Boult, Stephen, University of Manchester
Boving, Thomas, University of Rhode Island
Bowman, Jean A., Texas A&M University
Boyle, Douglas, University of Nevada, Reno
Bradley, Chris, University of Birmingham
Brassington, Rick, University of Newcastle Upon Tyne
Brooks, Paul D., University of Utah
Brown, David L., California State University, Chico
Burkhart, Patrick A., Slippery Rock University
Byrne, James M., University of Lethbridge
Campana, Michael E., Oregon State University
Carey, Sean, McMaster University
Carillo Chavez, Alejandro J., Univ Nac Autonoma de Mexico
Carman, Cary D., Angelo State University
Celico, Fulvio, University of Parma
Challads, Thomas, Edinburgh University
Chaubey, Indrajeet, Purdue University
Cheng, Chu-Lin, University of Texas, Rio Grande Valley
Cianfrani, Christina, Hampshire College
Cirmo, Christopher P., SUNY, Cortland
Coulibaly, Paulin, McMaster University
Crawford, Nicholas, Western Kentucky University
Cunningham, Peter, University of Newcastle Upon Tyne
Currens, James C., University of Kentucky
D'Odorico, Paolo, University of Virginia
Dahlke, Helen E., University of California, Davis
Darby, Jeannie, University of California, Davis
Davidson, Bart, University of Kentucky
Dawson, Jamekia, Geological Survey of Alabama
Doster, Florian, Heriot-Watt University
Dowling, Carolyn B., Ball State University
Duan, Quingyun, Lawrence Livermore National Laboratory
Duex, Timothy W., University of Louisiana at Lafayette
Eaton, Timothy, Queens College (CUNY)
El Kadiri, Racha, Middle Tennessee State University
Enright, Richard L., Bridgewater State University
Entekhabi, Dara, Massachusetts Institute of Technology
Eshleman, Keith N., University of Maryland
Estalrich López, Joan, Universitat Autonoma de Barcelona
Ezenabor, Ben O., University of Benin
Fabryka-Martin, June T., Los Alamos National Laboratory
Faramarzi, Monireh, University of Alberta
Farrow, Norman D., Oak Ridge National Laboratory
Fedele, Juan J., Saint Cloud State University
Foufoula-Georgiou, Efi, University of Minnesota, Twin Cities
Fox, Haydn A., Texas A&M University, Commerce
Frasson, Renato, Ohio State University
Gannon, John P., Western Carolina University
Gary, Marcus, University of Texas, Austin
Gilmore, Tyler J., Pacific Northwest National Laboratory
Grieneisen, Michael L., University of California, Davis
Groves, Christopher, Western Kentucky University
Guan, Huade, Flinders University
Gutierrez-Jurado, Hugo A., University of Texas, El Paso
Hagan, Robert M., University of California, Davis
Hamann, Hillary, University of Denver
Hannah, David, University of Birmingham
Hanson, Blaine R., University of California, Davis
Hart, David J., University of Wisconsin-Madison
Harter, Thomas, University of California, Davis
Hasan, Mohamed Ali, University of Malaya
Hauswirth, Scott, California State University, Northridge
Heimann, William H., Fort Hays State University
Heitz, Leroy F., University of Guam
Heo, Joon, University of Texas, Permian Basin
Hernes, Peter J., University of California, Davis
Hildebrandt, Anke, Friedrich-Schiller-University Jena
Hill, Mary C., University of Kansas
Hooyer, Thomas S., University of Wisconsin-Madison
Hornberger, George, Vanderbilt University
Hubbard, John E., SUNY, The College at Brockport
Huntington, Justin, University of Nevada, Reno
Ibaraki, Motomu, Ohio State University
Inamdar, Shreeram, University of Delaware
Jacobs, Katharine L., University of Arizona
James-Aworeni, E., Obafemi Awolowo University
Jefferson, Anne, University of North Carolina, Charlotte
Jefferson, Anne J., Kent State University
Johannesson, Karen H., Tulane University
Johnson, Robert L., Argonne National Laboratory
Johnson, Robert O., Oak Ridge National Laboratory
Jones, Stephen C., Geological Survey of Alabama
Kandel, Hari, Lake Superior State University
Kang, Peter, University of Minnesota, Twin Cities
Kasenow, Michael, Eastern Michigan University
Katz, Gabrielle, Appalachian State University
Kelly, Walton R., Illinois State University
Kesel, Richard H., Louisiana State University
Keyantash, John, California State University, Dominguez Hills
Kilroy, Kathyrn, Minot State University
Kinner, David A., Western Carolina University
Kuchovsky, Tomas, Masaryk University
Land, Lewis A., New Mexico Institute of Mining & Technology
Lane, Charles L., Southern Oregon University
Lanigan, David C., Pacific Northwest National Laboratory
Lautz, Laura, Syracuse University
Lavanchy, G. Thomas, University of Denver
Laycock, Arleigh H., University of Alberta
Lee, Michael D., California State University, East Bay
Liess, Stefan, University of Minnesota, Twin Cities
Lightbody, Anne F., University of New Hampshire
Lin, Yu-Feng F., Illinois State Geological Survey
Liu, Gaisheng, University of Kansas
Llanos, Hilario, Univ of the Basque Country UPV/EHU
Ma, Lin, University of Texas, El Paso
Masteller, Claire C., Washington University in St. Louis
McColl, Kaighin, Harvard University
McDermott, Christopher I., Edinburgh University

McDowell, William H., University of New Hampshire
McElwee, Carl D., University of Kansas
McKenna, Thomas E., University of Delaware
McLaughlin, Peter P., University of Delaware
McNamara, James P., Boise State University
Megdal, Sharon B., University of Arizona
Mehnert, Edward, Illinois State University
Molotch, Noah, University of Colorado
Monohan, Carrie, California State University, Chico
Moysey, Stephen, East Carolina University
Mudd, Simon N., Edinburgh University
Naudts, Kim, Vrije Universiteit Amsterdam
Ninesteel, Judy J., Potomac State College
Nkuna, Tinyiko R., University of Venda for Science & Technology
O'Reilly, Catherine M., Illinois State University
Odling, Noelle, University of Leeds
Ortiz-Aguirre, Ramon, Universidad Autonoma de San Luis Potosi
Oztekin Okan, Ozlem, Firat University
Palmer, Arthur N., SUNY, Oneonta
Papuga, Shirley, Wayne State University
Parkin, Geoffrey, University of Newcastle Upon Tyne
Pavelsky, Tamlin M., University of North Carolina, Chapel Hill
Pettijohn, J. C., University of Illinois, Urbana-Champaign
Piccinini, Leonardo, Università degli Studi di Padova
Piggot, Matthew, Imperial College
Plankell, Eric T., Illinois State Geological Survey
Prestegaard, Karen L., University of Maryland
Puente, Carlos E., University of California, Davis
Quinn, Paul, University of Newcastle Upon Tyne
Rakovan, Monica, Miami University
Reay, William G., College of William & Mary
Reedy, Robert C., Univ of Texas at Austin, Jackson Sch of Geosciences
Reidenbach, Matthew A., University of Virginia
Reinfelder, Ying Fan, Rutgers, The State Univ of New Jersey
Renshaw, Carl E., Dartmouth College
Rice, Karen C., University of Virginia
Richards, Paul L., SUNY, The College at Brockport
Ricka, Adam, Masaryk University
Root, Tara L., Florida Atlantic University
Rouse, Joseph D., University of Guam
Saar, Martin O., University of Minnesota, Twin Cities
Salami, B. M., Obafemi Awolowo University
Sandoval, Samuel, University of California, Davis
Sandoval Solis, Samuel, University of California, Davis
Scanlon, Bridget R., University of Texas, Austin
Scanlon, Todd M., University of Virginia
Scanlon, Todd M., University of Virginia
Schilling, Keith E., University of Iowa
Scotton, Paolo, Università degli Studi di Padova
Screaton, Elizabeth J., University of Florida
Singer, Michael, University of St. Andrews
Slade, Jr., Raymond M., Austin Community College District
Smith, Ellen D., Oak Ridge National Laboratory
Smith, James A., Princeton University
Smith, Michael, Geological Survey of Alabama
Smith, Ronald M., Pacific Northwest National Laboratory
Smyth, Rebecca C., Univ of Texas at Austin, Jackson Sch of Geosciences
Sorooshian, Soroosh, University of California, Irvine
Sorooshian, Soroosh, University of Arizona
Spane, Frank A., Pacific Northwest National Laboratory
Sritharan, Subramania I., Central State University
Stafford, James, Wyoming State Geological Survey
Stasko, Stanislaw, University of Wroclaw
Steele, Ken, University of Arkansas, Fayetteville
Steenhuis, Tammo S., Cornell University
Sterling, Shannon, Dalhousie University
Tarka, Robert, University of Wroclaw
Thorne, Paul D., Pacific Northwest National Laboratory
Triay, Ines, Los Alamos National Laboratory
Turner, Anne, Austin Community College District
Tyler, Scott, University of Nevada, Reno
Van Stan, John, Georgia Southern University
Vandike, James E., Missouri University of Science and Technology
Vermeul, Vincent R., Pacific Northwest National Laboratory
Villarroy, Fermín, Univ Complutense de Madrid
Vivoni, Enrique, Arizona State University
Vonk, Jorien, Vrije Universiteit Amsterdam

Wallender, Wes W., University of California, Davis
Wallender, Wesley W., University of California, Davis
Watkins, David, Exeter University
West, Jared, University of Leeds
White, Jeffrey R., Indiana University, Bloomington
Witt, Emma, Stockton University
Wolf, Aaron T., Oregon State University
Woltemade, Christopher J., Shippensburg University
Woo, Ming-Ko, McMaster University
Wood, Eric F., Princeton University
Wymore, Adam, University of New Hampshire
Xu, Shangping, University of Wisconsin, Milwaukee
Yan, Eugene, University of Illinois at Chicago
Yang, Y, Cardiff University
Ye, Ming, Florida State University
You, Jinsheng, University of Nebraska, Lincoln
Yusoff, Ismail, University of Malaya
Zaccaria, Daniele, University of California, Davis
Zhang, Minghua, University of California, Davis
Ziemer, Robert R., Humboldt State University
Zume, Joseph T., Shippensburg University

Ground Water/Hydrogeology

Simsek, Sakir, Hacettepe University
Abdo, Ginette, Montana Tech of The University of Montana
Abrams, Daniel, University of Illinois
Ahmed, Kazi M., University of Dhaka
Al, Tom A., University of Ottawa
Al-Shaibani, Abdulaziz, King Fahd University of Petroleum and Minerals
Alexander, Scott, University of Minnesota, Twin Cities
Alexander, Jr., E. Calvin, University of Minnesota, Twin Cities
Ali, Bukari, Kwame Nkrumah Univ of Science and Tech
Ali, Genevieve, University of Manitoba
Amin, Isam E., Youngstown State University
Anderson, Mary P., University of Wisconsin-Madison
Anderson, William P., Appalachian State University
Andreasen, David C., Maryland Department of Natural Resources
Andres, A. S., University of Delaware
Antigüedad, Iñaki, Univ of the Basque Country UPV/EHU
Appiah-Adjei, Emmanuel K., Kwame Nkrumah Univ of Science and Tech
Appold, Martin S., University of Missouri
Arthur, Jonathan D., Florida Geological Survey
Asante, Joseph, Tennessee Tech University
Asghari Moghadam, Asghar, University of Tabriz
Bagtzoglou, Ross, Columbia University
Bahr, Jean M., University of Wisconsin-Madison
Bahr, Jean M., University of Wisconsin, Madison
Bair, E. Scott, Ohio State University
Bajjali, William, University of Wisconsin, Superior
Baker, Andy, University of New South Wales
Baldwin, Jr., A. Dwight, Miami University
Banks, Eddie, Flinders University
Barbecot, Florent, Universite du Quebec a Montreal
Barbee, Gary C., West Texas A&M University
Barkmann, Peter, Colorado Geological Survey
Barrash, Warren, Boise State University
Batelaan, Okke, Katholieke Universiteit Leuven
Batelaan, Okke, Flinders University
Bayari, Serdar, Hacettepe University
Beck, E. G., University of Kentucky
Becker, Matthew, California State University, Long Beach
Beckie, Roger D., University of British Columbia
Bedient, Philip B., Rice University
Bennett, Steven W., Western Illinois University
Bentley, Laurence R., University of Calgary
Bergamaschi, Brian, California State University, Sacramento
Besancon, James, Wellesley College
Bethke, Craig M., University of Illinois, Urbana-Champaign
Bishop, Charles E., Utah Geological Survey
BlackEagle, Cory, Eastern Kentucky University
Blythe, Daniel D., Montana Tech of The University of Montana
Bobst, Andrew L., Montana Tech of The University of Montana
Bohling, Geoff, University of Kansas
Bohling, Geoffrey C., University of Kansas
Bolster, Diogo, University of Notre Dame
Boutt, David, University of Massachusetts, Amherst
Boving, Thomas B., University of Rhode Island
Boyle, James T., New Jersey Geological and Water Survey

Bradbury, Kenneth R., University of Wisconsin, Extension
Bradbury, Kenneth R., University of Wisconsin-Madison
Brahana, John V., University of Arkansas, Fayetteville
Brame, Scott E., Clemson University
Bratcher, Susan, California State University, Fresno
Bray, Erin, California State University, Northridge
Breitmeyer, Ronald, University of Nevada, Reno
Brikowski, Tom H., University of Texas, Dallas
Brinkmann, Robert, Oregon Dept of Geology and Mineral Industries
Brookfield, Andrea, University of Kansas
Brusseau, Mark L., University of Arizona
Buchanan, John P., Eastern Washington University
Burbey, Thomas J., Virginia Polytechnic Institute & State University
Burgess, William G., University College London
Burnett, Joree, Tarleton State University
Butler, Jr., James J., University of Kansas
Callahan, Timothy J., College of Charleston
Capuano, Regina M., University of Houston
Cardenas, Bayani, University of Texas, Austin
Cardiff, Michael A., University of Wisconsin-Madison
Cardiff, Mike, University of Wisconsin, Madison
Carey, Anne E., Ohio State University
Carle, Steven, Lawrence Livermore National Laboratory
Carling, Gregory T., Brigham Young University
Carlson, Catherine A., Eastern Connecticut State University
Carlson, Douglas A., Louisiana State University
Carstarphen, Camela, Montana Tech of The University of Montana
Carvalho, Maria d., Universidade de Lisboa
Casillas, Angelica, Univ Nac Autonoma de Mexico
Cassidy, Daniel P., Western Michigan University
Celia, Michael A., Princeton University
Chase, Peter M., University of Wisconsin, Extension
Chavez-Aguirre, Rafael, Universidad Autonoma de Chihuahua
Cheng, Songlin, Wright State University
Cherkauer, Douglas S., University of Wisconsin, Milwaukee
Cherry, John A., University of Waterloo
Cherubini, Claudia, Institut Polytechnique LaSalle Beauvais (ex-IGAL)
Chesnaux, Romain, Universite du Quebec a Chicoutimi
Chowdhury, Shafiul H., SUNY, New Paltz
Clarey, Timothy L., Delta College
Clark, Ian D., University of Ottawa
Conant, Brewster, University of Waterloo
Cook, Edward R., Columbia University
Cook, Peter, Flinders University
Costa, Maria R., Universidade de Trás-os-Montes e Alto Douro
Croxen III, Fred W., Arizona Western College
Davidson, Gregg R., University of Mississippi
Davis, Arden D., South Dakota School of Mines & Technology
Davis, J. Matthew, University of New Hampshire
Davis, Ralph K., University of Arkansas, Fayetteville
de Stefano, Lucia, Univ Complutense de Madrid
Decker, Dave, University of Nevada, Reno
Del Arenal-Capetillo, Rodolfo, Univ Nac Autonoma de Mexico
Demlie, Molla, University of KwaZulu-Natal
Denizman, Can, Valdosta State University
Devlin, J. F., University of Kansas
Dhar, Ratan K., York College (CUNY)
Dimova, Natasha T., University of Alabama
Dintaman, Christopher, Indiana University
Domber, Steven E., New Jersey Geological and Water Survey
Doran, Peter, Louisiana State University
Doss, Paul K., University of Southern Indiana
Douglas, Ellen, University of Massachusetts, Boston
Dripps, Weston R., Furman University
Druhan, Jennifer, University of Illinois, Urbana-Champaign
Duaime, Terence E., Montana Tech of The University of Montana
Dunkle, Kallina M., Austin Peay State University
Durham, Lisa A., Argonne National Laboratory
Durham, Lisa, Argonne National Laboratory
Eads, Tracy, Miami University
Earman, Sam, Millersville University
Ekmekci, Mehmet, Hacettepe University
El-Kadi, Aly I., University of Hawai'i, Manoa
Ellis, Andre, California State University, Los Angeles
Erickson, Melinda L., University of Minnesota, Twin Cities
Esling, Steven P., Southern Illinois University Carbondale
Evans, Sarah G., Appalachian State University

Fabbri, Paolo, Università degli Studi di Padova
Fairley, Jerry P., University of Idaho
Famiglietti, James, University of California, Irvine
Farmer, George T., Los Alamos National Laboratory
Feinstein, Daniel T., University of Wisconsin, Milwaukee
Ferre, Paul A., University of Arizona
Fisher, Andrew T., University of California, Santa Cruz
Fitzpatrick, Stephan, Georgia State Univ, Perimeter College, Clarkston
Flowers, George C., Tulane University
Fogg, Graham E., University of California, Davis
Fountain, John C., North Carolina State University
Frisbee, Marty, Purdue University
Fryar, Alan E., University of Kentucky
Gakka, Dr U., Osmania University
Gardner, Payton, University of Montana
Garven, G, Tufts University
Ge, Shemin, University of Colorado
Geary, Phil, University of Newcastle
Geidel, Gwendelyn, University of South Carolina
Gemperline, Johanna M., Maryland Department of Natural Resources
Gerla, Philip J., University of North Dakota
Godsey, Sarah E., Idaho State University
Goeke, James W., Unversity of Nebraska - Lincoln
Gordon, Ryan, Dept of Agriculture, Conservation, and Forestry
Gordon, Susan L., University of Calgary
Gorelick, Steven M., Stanford University
Gotkowitz, Madeline B., University of Wisconsin, Extension
Gouzie, Douglas R., Missouri State University
Graham, Grace E., University of Wisconsin, Extension
Grasby, Stephen E., University of Calgary
Grismer, Mark E., University of California, Davis
Grundl, Timothy J., University of Wisconsin, Milwaukee
Gurdak, Jason, San Francisco State University
Gurwin, Jacek, University of Wroclaw
Habana, Nathan C., University of Guam
Hadley, Daniel, University of Illinois
Haggerty, Roy D., Oregon State University
Halihan, Todd, Oklahoma State University
Hampton, Duane R., Western Michigan University
Hand, Kristen, Pennsylvania Bur of Topo & Geologic Survey
Hargis, David, University of Arizona
Harrington, Glenn, Flinders University
Harris, Brett D., Curtin University
Harris, William H., Grad Sch of the City Univ of New York
Hart, David J., University of Wisconsin, Extension
Harter, Thomas L., University of California, Davis
Harter, Thomas, University of California, Davis
Hatch, Christine, University of Massachusetts, Amherst
Hayashi, Masaki, University of Calgary
Hays, Phillip D., University of Arkansas, Fayetteville
Heiss, James, University of Massachusetts, Lowell
Helmke, Martin F., West Chester University
Hendrickx, Jan M., New Mexico Institute of Mining and Technology
Hendry, Jim, University of Saskatchewan
Henry, Eric J., University of North Carolina, Wilmington
Henry, Tiernan, National University of Ireland Galway
Herman, Ellen K., Bucknell University
Hermans, Thomas, Ghent University
Hershey, Ronald L., University of Nevada, Reno
Hibbs, Barry, California State University, Los Angeles
Hiscock, Kevin, University of East Anglia
Hoff, Jean L., Saint Cloud State University
Hoover, Karin A., California State University, Chico
Horner, Tim C., California State University, Sacramento
Howard, Kenneth W., University of Toronto
Howe III, Thomas R., Western Michigan University
Hu, Bill X., Florida State University
Hudak, Paul F., University of North Texas
Huizar-Alvarez, Rafael, Univ Nac Autonoma de Mexico
Hurlow, Hugh, Utah Geological Survey
Hurlow, Hugh A., Utah Geological Survey
Huysmans, Marijke, Katholieke Universiteit Leuven
Iles, Derric L., South Dakota Dept of Env and Nat Res
Iqbal, Mohammad Z., University of Northern Iowa
Iribar, Vicente , Univ of the Basque Country UPV/EHU
Irvine, Dylan, Flinders University
James, Scott C., Baylor University

Jannik, Nancy O., Winona State University
Jarvis, W. T., Oregon State University
Jenson, John W., University of Guam
Johnson, Peggy, New Mexico Institute of Mining & Technology
Johnson, Thomas M., University of Illinois, Urbana-Champaign
Join, Jean-Lambert, Universite de la Reunion
Juster, Thomas C., University of South Florida, Tampa
Kaden, Scott, Missouri Dept of Natural Resources
Keen, Kerry L., University of Wisconsin, River Falls
Keller, C. Kent, Washington State University
Kelly, Bryce, University of New South Wales
Kelly, Walton R., University of Illinois
Kent, Douglas, Oklahoma State University
Khalequzzaman, Md., Lock Haven University
Kim, Yongsang (Barry), University of Guam
Kirk, Scott, Ohio Dept of Natural Resources
Knoll, Martin A., Sewanee: University of the South
Kreamer, David K., University of Nevada, Las Vegas
Kreamer, David, University of Nevada, Reno
Kurttas, Turker, Hacettepe University
La Fave, John I., Montana Tech of The University of Montana
La Fleur, Robert G., Rensselaer Polytechnic Institute
Lachhab, Ahmed, Susquehanna University
Lachmar, Thomas E., Utah State University
LaFreniere, Lorraine, Argonne National Laboratory
Lahiri, Chayan, Adams State University
Lambert, Carolyn D., University of Northern Colorado
Larocque, Marie, Universite du Quebec a Montreal
Laton, W. R., California State University, Fullerton
Lee, David R., University of Waterloo
Lee, Donald W., Oak Ridge National Laboratory
Lee, Eung Seok, Ohio University
Lee, Ming-Kuo, Auburn University
Lefebvre, Rene, Universite du Quebec
Lemke, Lawrence D., Central Michigan University
Lenczewski, Melissa E., Northern Illinois University
Leonhart, Leo S., University of Arizona
Levy, Jonathan, Miami University
Li, Liangping, South Dakota School of Mines & Technology
Locke, Daniel B., Dept of Agriculture, Conservation, and Forestry
Loh, Yvonne A., University of Ghana
Loheide, Steven P., University of Wisconsin, Madison
Lopez-Ferreira, Cesar A., Univ Autonoma de Baja California Sur
Lord, Mark L., Western Carolina University
Love, Andrew, Flinders University
Lowe, Mike, Utah Geological Survey
Lowry, Christopher S., SUNY, Buffalo
Luczaj, John A., University of Wisconsin, Green Bay
Luhmann, Andrew J., Wheaton College
Lutz, Alexandra, University of Nevada, Reno
Lyons, William B., Ohio State University
Mabee, Stephen B., Massachusetts Geological Survey
MacNish, Robert, University of Arizona
Macpherson, Gwendolyn L., University of Kansas
Makkawi, Mohammad, King Fahd University of Petroleum and Minerals
Malzone, Jonathan, Eastern Kentucky University
Manda, Alex K., East Carolina University
Mannix, Devin, University of Illinois
Marino, M. A., University of California, Davis
Marino, Miguel A., University of California, Davis
Martel, Richard, Universite du Quebec
Martínez, Pedro, Univ Complutense de Madrid
Mastropaolo, Carl, Drexel University
Matyjasik, Marek, Weber State University
May, James, Mississippi State University
Mayer, James R., University of West Georgia
Mayo, Alan L., Brigham Young University
McCurdy, Maureen, Louisiana Tech University
McKenzie, Jeffrey M., McGill University
McKinney, Mac, Geological Survey of Alabama
Mendoza, Carl, University of Alberta
Mendoza, Carl A., University of Alberta
Metesh, John J., Montana Tech of The University of Montana
Michael, Holly A., University of Delaware
Michel, Fred A., Carleton University
Milewski, Adam, University of Georgia
Miller, Marvin R., Montana Tech of The University of Montana

Misra, Debsmita, University of Alaska, Fairbanks
Mitchell, Robert J., Western Washington University
Moline, Gerilynn R., Oak Ridge National Laboratory
Montero, Esperanza, Univ Complutense de Madrid
Montgomery, William W., New Jersey City University
Moore, Michael E., Pennsylvania Bur of Topo & Geologic Survey
Moortgat, Joachim, Ohio State University
Moran, Jean E., California State University, East Bay
Moss, Neil E., Geological Survey of Alabama
Moysey, Stephen M., Clemson University
Muldoon, Maureen A., University of Wisconsin, Oshkosh
Murdoch, Lawrence C., Clemson University
Murgulet, Dorina, Texas A&M University, Corpus Christi
Murray, Kent S., University of Michigan, Dearborn
Murray, Kyle E., University of Oklahoma
Myers, Paul B., Lehigh University
Nadiri, Ata Allah, University of Tabriz
Narasimhan, T. N., University of California, Berkeley
Naylor, Shawn C., Indiana University
Naymik, Thomas G., Ohio State University
Neboga, Victoria V., Pennsylvania Bur of Topo & Geologic Survey
Nedunuri, Krishna K., Central State University
Nelson, Craig, Ohio Dept of Natural Resources
Neuman, Shlomo P., University of Arizona
Newcomer, Darrell R., Pacific Northwest National Laboratory
Newell, Charles J., Rice University
Nguyen, Lam V., Hanoi University of Mining & Geology
Nicot, Jean-Philippe, Univ of Texas at Austin, Jackson Sch of Geosciences
Nkotagu, Hudson H., University of Dar es Salaam
Noonan, Mathew T., South Dakota Dept of Env and Nat Res
O'Brien, Arnold L., University of Massachusetts, Lowell
O'Brien, Rachel, Allegheny College
O'Reilly, Andrew M., University of Mississippi
Oberdorfer, June A., San Jose State University
Ogard, Allen E., Los Alamos National Laboratory
Olafsen-Lackey, Susan, Unversity of Nebraska - Lincoln
Oliveira, Alcino S., Universidade de Trás-os-Montes e Alto Douro
Olyphant, Greg A., Indiana University, Bloomington
Ophori, Duke U., Montclair State University
Opper, Carl, Saint Petersburg College, Clearwater
Ortega, Adrian, Univ Nac Autonoma de Mexico
Osborn, Stephen G., California State Polytechnic University, Pomona
Ozsvath, David L., University of Wisconsin, Stevens Point
Pan, Feifei, University of North Texas
Paniconi, Claudio, Universite du Quebec
Parizek, Richard R., Pennsylvania State University, University Park
Parkin, Gary, University of Guelph
Parsen, Mike J., University of Wisconsin, Extension
Patton, Thomas W., Montana Tech of The University of Montana
Pederson, Darryll T., University of Nebraska, Lincoln
Peters, Catherine A., Princeton University
Peterson, Eric W., Illinois State University
Peterson, Holly, Guilford College
Pfannkuch, Hans O., University of Minnesota, Twin Cities
Phillips, Fred M., New Mexico Institute of Mining and Technology
Pleasants, Mark S., Ohio Dept of Natural Resources
Pociask, Geoff, Illinois State Geological Survey
Pohll, Greg, University of Nevada, Reno
Post, Vincent, Flinders University
Prestegaard, Karen L., University of Maryland
Price, Rene, Florida International University
Puente-Muniz, Carlos Francisco, Univ Autonoma de San Luis Potosi
Puleo, Jack, University of Delaware
Quinn, John, Argonne National Laboratory
Raab, James, Ohio Dept of Natural Resources
Rains, Mark C., University of South Florida, Tampa
Rauch, Henry W., West Virginia University
Reese, Stuart O., Pennsylvania Bur of Topo & Geologic Survey
Reeve, Andrew S., University of Maine
Reeves, Donald Matthew M., Western Michigan University
Reichard, James S., Georgia Southern University
Reiten, Jon C., Montana Tech of The University of Montana
Remenda, Victoria H., Queen's University
Remson, Irwin, Stanford University
Riemersma, Peter E., Grand Valley State University
Rios-Sanchez, Miriam, Bemidji State University
Ritzi, Jr., Robert W., Wright State University

Roadcap, George S., University of Illinois
Robbins, Gary A., University of Connecticut
Robertson, Wendy, Central Michigan University
Robertson, Will, University of Waterloo
Robin, Michel R., University of Ottawa
Roman, Eric W., New Jersey Geological and Water Survey
Romanowicz, Edwin A., SUNY, Plattsburgh
Ronayne, Michael J., Colorado State University
Rose, Seth E., Georgia State Univ
Rostron, Benjamin J., University of Alberta
Rouleau, Alain, Universite du Quebec a Chicoutimi
Royo-Ochoa, Miguel, Universidad Autonoma de Chihuahua
Russo, Tess A., Pennsylvania State University, University Park
Ryan, Cathy, University of Calgary
Saffer, Demian M., Pennsylvania State University, University Park
Saint, Prem K., California State University, Fullerton
Salvage, Karen M., Binghamton University
Samuelson, Alan C., Ball State University
Sanders, Laura L., Northeastern Illinois University
Sanford, William E., Colorado State University
Sasowsky, Ira D., University of Akron
Sawyer, Audrey, Ohio State University
Scanlon, Bridget R., Univ of Texas at Austin, Jackson Sch of Geosciences
Scheidt, Brian, Mineral Area College
Schincariol, Robert A., Western University
Schlosser, Peter, Columbia University
Schmitz, Darrel W., Mississippi State University
Schreiber, Madeline E., Virginia Polytechnic Institute & State University
Schueth, Christoph, Technische Universitaet Darmstadt
Schulmeister, Marcia K., Emporia State University
Schulze-Makuch, Dirk, Washington State University
Schwartz, Frank W., Ohio State University
Scott, Christopher A., University of Arizona
Sebol, Lesley, Colorado Geological Survey
Sedivy, Robert, Argonne National Laboratory
Sedivy, Robert A., Argonne National Laboratory
Sendlein, Lyle V. A., University of Kentucky
Seyoum, Wondwosen M., Illinois State University
Shanafield, Margaret, Flinders University
Shang, Congxiao, University of East Anglia
Shaw, Glenn D., Montana Tech of the University of Montana
Sheldon, Amy L., SUNY, Geneseo
Shevenell, Lisa, University of Nevada, Reno
Shim-Chim, Richard, New Jersey Geological and Water Survey
Shimada, Jun, Kumamoto University
Sibray, Steven S., Unversity of Nebraska - Lincoln
Siegel, Donald I., Syracuse University
Simmons, Craig, Flinders University
Simpkins, William W., Iowa State University of Science & Technology
Singha, Kamini, Colorado School of Mines
Skalbeck, John D., University of Wisconsin, Parkside
Smith, J. Leslie, University of British Columbia
Smith, James E., McMaster University
Smith, James E., University of Arizona
Smith, Karen, Argonne National Laboratory
Soll, Wendy E., Los Alamos National Laboratory
Solomon, Douglas K., University of Utah
Soltanian, Reza, University of Cincinnati
Spangler, Daniel P., University of Florida
Spayd, Steven E., New Jersey Geological and Water Survey
Spinelli, Glenn, New Mexico Institute of Mining and Technology
Springer, Abraham E., Northern Arizona University
Stafford, Kevin W., Stephen F. Austin State University
Staley, Andrew, Maryland Department of Natural Resources
Stewart, Mark T., University of South Florida, Tampa
Stotler, Randy, University of Kansas
Strack, Otto D., University of Minnesota, Twin Cities
Sudicky, Edward A., University of Waterloo
Suen, C. J., California State University, Fresno
Sukop, Michael C., Florida International University
Sun, Hongbing, Rider University
Suter, Simeon, Pennsylvania Bur of Topo & Geologic Survey
Sutherland, Mary K., Montana Tech of The University of Montana
Swanson, Susan K., Beloit College
Tabidian, M. Ali, California State University, Northridge
Taboga, Karl, Wyoming State Geological Survey
Tagne, Gilles V., Wheaton College
Tauxe, John D., Oak Ridge National Laboratory
Taylor, Charles J., University of Kentucky
Tellam, John H., University of Birmingham
Tezcan, Levent, Hacettepe University
Therrien, Rene, Universite Laval
Thomason, Jason F., Illinois State Geological Survey
Tick, Geoffrey, University of Alabama
Tidwell, Vincent C., New Mexico Institute of Mining and Technology
Tijani, Moshood N., University of Ibadan
Tilton, Eric E., University of Colorado
Tipping, Robert, University of Minnesota
Tompson, Andrew F., Lawrence Livermore National Laboratory
Toran, Laura, Temple University
Tóth, József, University of Alberta
Totsche, Kai U., Friedrich-Schiller-University Jena
Tucker, Carla M., Lamar University
Tyler, Scott W., University of Nevada, Reno
Unger, Andre, University of Waterloo
Upchurch, Sam B., University of South Florida, Tampa
Vacher, H. Leonard, University of South Florida, Tampa
Valerio, Mitch, Ohio Dept of Natural Resources
Van Der Hoven, Steve, Illinois State University
Van der Kamp, Garth, University of Waterloo
van der Velde, Ype, Vrije Universiteit Amsterdam
VanDerwerker, Tiffany J., Maryland Department of Natural Resources
Veeger, Anne I., University of Rhode Island
Veeger, Anne, University of Rhode Island
Vengosh, Avner, Duke University
Vesper, Dorothy, West Virginia University
Wallace, Janae, Utah Geological Survey
Walraevens, Kristine E., Ghent University
Walter, Julien, Universite du Quebec a Chicoutimi
Walton, Nick, University of Portsmouth
Wang, Dong, California State University, Fresno
Wanke, Heike, University of Namibia
Waren, Kirk B., Montana Tech of The University of Montana
Wassenaar, Len, University of Saskatchewan
Watson, David B., Oak Ridge National Laboratory
Weissmann, Gary S., University of New Mexico
Welby, Charles W., North Carolina State University
Welhan, John A., Idaho State University
Werner, Adrian, Flinders University
Wheatcraft, Steve, University of Nevada, Reno
Wheaton, John R., Montana Tech of The University of Montana
Whiting, Duane L., University of Utah
Whittemore, Donald O., University of Kansas
Wicks, Carol M., Louisiana State University
Widdowson, Mark A., Virginia Polytechnic Institute & State University
Wilcox, Jeffrey D., University of North Carolina, Asheville
Wilson, Alicia M., University of South Carolina
Wilson, John L., New Mexico Institute of Mining and Technology
Wilson, Lorne G., University of Arizona
Wilson, Steven D., University of Illinois
Wolaver, Brad, University of Texas, Austin
Wolfsberg, Andrew V., Los Alamos National Laboratory
Woods, Juliette, University of South Australia
Yan, Eugene, Argonne National Laboratory
Yang, Changbing, University of Texas, Austin
Yang, Jianwen, University of Windsor
Yelderman, Jr., Joe C., Baylor University
Yidana, Sandow M., University of Ghana
Yüce, Galip, Hacettepe University
Zarnetske, Jay, Michigan State University
Zhan, Hongbin, Texas A&M University
Zhang, Pengfei, City College (CUNY)
Zhang, You-Kuan, University of Iowa
Zheng, Chunmiao, University of Alabama
Zhu, Junfeng, University of Kentucky
Ziegler, Brady A., Trinity University
Zimmer, Margaret, University of California, Santa Cruz
Zlotnik, Vitaly A., University of Nebraska, Lincoln
Zouhri, Lahcen, Institut Polytechnique LaSalle Beauvais (ex-IGAL)
Zreda, Marek G., University of Arizona
Özyurt, Nur N., Hacettepe University

Quantitative Hydrology
Benson, David A., Colorado School of Mines
Birdsell, Kay H., Los Alamos National Laboratory

Bywater-Reyes, Sharon, University of Northern Colorado
Carroll, Rosemary, University of Nevada, Reno
Chebana, Fateh, Universite du Quebec
Chen, Xun-Hong, Unversity of Nebraska - Lincoln
Covington, Matthew, University of Arkansas, Fayetteville
Danko, George, University of Nevada, Reno
Dingman, S. Lawrence, University of New Hampshire
Donovan, Joseph J., West Virginia University
Duckstein, Lucien, University of Arizona
Dymond, Randel, Virginia Polytechnic Institute & State University
Evans, David G., California State University, Sacramento
Falta, Ronald W., Clemson University
Florea, Lee J., Indiana University
Flores, Alejandro N., Boise State University
French, Mark A., New Jersey Geological and Water Survey
Frind, Emil O., University of Waterloo
Gillham, Robert W., University of Waterloo
Gupta, Hoshin V., University of Arizona
Harpold, Adrian, University of Nevada, Reno
Hermance, John F., Brown University
Hoffman, Jeffrey L., New Jersey Geological and Water Survey
Holt, Robert M., University of Mississippi
Hubbart, Jason A., West Virginia University
Hughes, Denis, Rhodes University
Kandiah, Ramanitharan, Central State University
Kiem, Anthony, University of Newcastle
Letsinger, Sally L., Indiana University
Li, Dan, Boston University
Loague, Keith, Stanford University
Looney, Brian, Clemson University
Maddock, III, Thomas, University of Arizona
Martin-Hayden, James, University of Toledo
Meko, David M., University of Arizona
Michaud, Jene D., University of Hawai'i, Hilo
Pacheco, Fernando A., Universidade de Trás-os-Montes e Alto Douro
Pangle, Luke, Georgia State Univ
Parashar, Rishi, University of Nevada, Reno
Park, Young-Jin, University of Waterloo
Perkins, Robert B., Portland State University
Pohll, Greg, University of Nevada, Reno
Pollyea, Ryan M., Virginia Polytechnic Institute & State University
Puente, Carlos E., University of California, Davis
Quinn, John J., Argonne National Laboratory
Restrepo, Jorge I., Florida Atlantic University
Rudolph, David L., University of Waterloo
Salvucci, Guido D., Boston University
Schumer, Rina, University of Nevada, Reno
Sharp, Jr., John M., University of Texas, Austin
Sipos, György, Univesity of Szeged
Springer, Everett P., Los Alamos National Laboratory
Stockton, Charles W., University of Arizona
Szilagyi, Jozsef, Unversity of Nebraska - Lincoln
Telyakovskiy, Aleksey S., University of Nevada, Reno
Thorbjarnarson, Kathryn W., San Diego State University
Travis, Bryan J., Los Alamos National Laboratory
Verdon-Kidd, Danielle, University of Newcastle
Vitale, Sarah A., University of Wisconsin, Eau Claire
Wallis, Ilka, Flinders University
Westhoff, Martijn C., Vrije Universiteit Amsterdam
Whitman, Brian E., Wilkes University
Yeh, Tian-Chyi J., University of Arizona
Zhang, Yong, University of Alabama

Surface Waters
Acharya, Kumud, University of Nevada, Reno
Aitkenhead-Peterson, Jacqueline A., Texas A&M University
Aryal, Niroj, North Carolina Agricultural & Tech State University
Bacon, Robert, Missouri Dept of Natural Resources
Bauer, Carl J., University of Arizona
Bearden, Rebecca A., Geological Survey of Alabama
Benavides, Jude, University of Texas, Rio Grande Valley
Bevelhimer, Mark, Oak Ridge National Laboratory
Beyer, Patricia J., Bloomsburg University
Blanken, Peter D., University of Colorado
Blersch, Stacey S., Columbus State University
Blumentritt, Dylan, Winona State University
Cadol, Daniel, New Mexico Institute of Mining and Technology
Chandra, Sudeep, University of Nevada, Reno

Christian, Alan D., University of Massachusetts, Boston
Clark, Robert A., University of Arizona
D'Alpaos, Andrea, Università degli Studi di Padova
Demchak, Jennifer, Mansfield University
Dennett, Keith E., University of Nevada, Reno
Doran, Peter T., University of Illinois at Chicago
Du Charme, Charles, Missouri Dept of Natural Resources
Duan, Shuiwang, University of Maryland
Durand, Michael T., Ohio State University
Franek, Benjamin, Bloomsburg University
Franz, Kristie, Iowa State University of Science & Technology
Goodrich, David C., University of Arizona
Grable, Judy, Valdosta State University
Grant, Stanley, Virginia Polytechnic Institute & State University
Gulliver, John S., University of Minnesota, Twin Cities
Hamlet, Alan, University of Notre Dame
Harris, Randa R., University of West Georgia
Hasan, Khaled W., Austin Community College District
Hawkins, R. B., University of Arizona
Helz, George, University of Maryland
Hester, Erich, Virginia Polytechnic Institute & State University
Ince, Simon, University of Arizona
Kindelman, Julie, University of Wisconsin, Parkside
Kisila, Ben O., University of Mary Washington
Klein, Yehuda, Grad Sch of the City Univ of New York
Knight, Allen W., University of California, Davis
Kunza, Lisa, South Dakota School of Mines & Technology
Lancaster, Stephen, Oregon State University
Lansey, Kevin E., University of Arizona
Ledford, Sarah, Georgia State Univ
Lund, Jay R., University of California, Davis
Maneta, Marco, University of Montana
Maxwell, Reed M., Colorado School of Mines
Mayfield, Michael W., Appalachian State University
McGregor, Stuart W., Geological Survey of Alabama
McIsaac, Gregory F., University of Illinois, Urbana-Champaign
Nearing, Grey S., University of Alabama
Noe, Garry, Virginia Wesleyan College
O'Neil, Patrick E., Geological Survey of Alabama
Pociask, Geoffrey, University of Illinois
Pradhanang, Soni M., University of Rhode Island
Rancan, Helen L., New Jersey Geological and Water Survey
Richard, Gigi A., Colorado Mesa University
Ridenour, Gregory D., Austin Peay State University
Riise, Gunnhild, Norwegian University of Life Sciences (NMBU)
Royer, Todd, Indiana University, Bloomington
Scott, Verne H., University of California, Davis
Shuttleworth, W. James, University of Arizona
Smerdon, Ernest T., University of Arizona
Smith, Brianne, Brooklyn College (CUNY)
Smith, Erik, University of South Carolina
Smith, Laurence C., University of California, Los Angeles
Smith, Sean, University of Maine
Splinter, Dale, University of Wisconsin, Whitewater
Stieglitz, Marc, Georgia Institute of Technology
Strom, Kyle, Virginia Polytechnic Institute & State University
Swanson, Sherman, University of Nevada, Reno
Thomas, Rohrlack, Norwegian University of Life Sciences (NMBU)
Troch, Peter A., University of Arizona
Valdes, Juan B., University of Arizona
Van Meter, Kimberly, University of Illinois at Chicago
Wampler, Peter J., Grand Valley State University
White, Jeffrey, Indiana University, Bloomington
Wilderman, Candie, Dickinson College
Wynn, Elizabeth A., Geological Survey of Alabama
Zollweg, James A., SUNY, The College at Brockport

Geohydrology
Allen, Diana M., Simon Fraser University
Attinger, Sabine, Friedrich-Schiller-University Jena
Baedke, Steven J., James Madison University
Beddows, Patricia A., Northwestern University
Brookfield, Andrea, University of Kansas
Buddemeier, Robert W., University of Kansas
Burkart, Michael R., Iowa State University of Science & Technology
Castro, Maria Clara, University of Michigan
Chavez-Rodriguez, Adolfo, Universidad Autonoma de Chihuahua
Chormann, Jr., Frederick H., New Hampshire Geological Survey

Clair, Thomas A., Mount Allison University
Constantine, Jose, Cardiff University
Cooper, Clay A., University of Nevada, Reno
Crews, Jeff, Missouri Dept of Natural Resources
Daanen, Ronald P., Alaska Div of Geological & Geophysical Surveys
De Luca, Domenico Antonio, Università di Torino
Dowd, John F., University of Georgia
Drake, Lon D., University of Iowa
Duley, James W., Missouri Dept of Natural Resources
Fogg, Graham E., University of California, Davis
Gamage, Kusali R., Austin Community College District
Gartner, Janette, Bentley University
Gierke, John S., Michigan Technological University
Hardy, Douglas R., University of Massachusetts, Amherst
Hong, Sung-ho, Murray State University
Hyndman, David W., Michigan State University
Katzman, Danny, Los Alamos National Laboratory
Khanbilvardi, Reza M., Grad Sch of the City Univ of New York
Knight, Rosemary J., Stanford University
Koch, Magaly, Boston University
Land, Lewis, New Mexico Institute of Mining and Technology
Leap, Darrell I., Purdue University
Liu, Gaisheng, University of Kansas
Lovell, Mark D., Brigham Young University - Idaho
Lupankwa, Mlindelwa, Tshwane University of Technology
Marszalek, Henryk, University of Wroclaw
Mathias, Simon, Durham University
Mayer, Alex S., Michigan Technological University
McIntosh, Jennifer, University of Arizona
McKay, Larry D., University of Tennessee, Knoxville
McKenna, Thomas E., University of Delaware
Mount, Gregory, Indiana University of Pennsylvania
Ng, Crystal, University of Minnesota, Twin Cities
Noyes, Joanne M., South Dakota Dept of Env and Nat Res
Oshun, Jasper, Humboldt State University
Person, Mark A., New Mexico Institute of Mining and Technology
Rayne, Todd W., Hamilton College
Reboulet, Edward, University of Kansas
Rovey, Charles W., Missouri State University
Sargent, Kenneth A., Furman University
Sushama, Laxmi, Universite du Quebec a Montreal
Torres, Raymond, University of South Carolina
White, Mark D., Pacific Northwest National Laboratory
Zhang, Ye, University of Wyoming

Materials Transport
Troy, Marleen, Wilkes University

SOIL SCIENCE
Soil Physics/Hydrology
Amoozegar, Aziz, North Carolina State University
Anderson, Stephen H., University of Missouri, Columbia
Arnone, John J., University of Nevada, Reno
Baker, John M., University of Minnesota, Twin Cities
Berli, Marcus, University of Nevada, Reno
Black, T. Andrew, University of British Columbia
Bland, William L., University of Wisconsin, Madison
Boast, Charles W., University of Illinois, Urbana-Champaign
Brown, Sandra, University of British Columbia
Børresen, Trond, Norwegian University of Life Sciences (NMBU)
Caldwell, Todd, University of Texas, Austin
Casey, Francis, North Dakota State University
Cassel, Donald K., North Carolina State University
Cushman, John, Purdue University
Ellsworth, Timothy R., University of Illinois, Urbana-Champaign
Fermanich, Kevin J., University of Wisconsin, Green Bay
Feyereisen, Gary, University of Minnesota, Twin Cities
Flury, Markus, Washington State University
Fritton, Daniel D., Pennsylvania State University, University Park
Galbraith, John M., Virginia Polytechnic Institute & State University
Grattan, Stephen R., University of California, Davis
Groenevelt, Pieter H., University of Guelph
Gupta, Satish C., University of Minnesota, Twin Cities
Hook, James E., University of Georgia
Hopmans, Jan W., University of California, Davis
Horton, Robert, Iowa State University of Science & Technology
Hsiao, Theodore C., University of California, Davis

Hutson, John, Flinders University
Jawitz, James W., University of Florida
Johnson, Mark S., University of British Columbia
Jones, Tim L., New Mexico State University, Las Cruces
Kakembo, Vincent, Nelson Mandela Metropolitan University
Kanemasu, Edward T., University of Georgia
Kang, James, University of Texas, Rio Grande Valley
Kirkham, Mary Beth, Kansas State University
Kluitenberg, Gerard J., Kansas State University
Knappenberger, Thorsten J., Auburn University
Kung, King-Jau S., University of Wisconsin, Madison
Lal, Rattan, Ohio State University
Lourenco, Sergio, Cardiff University
Lowery, Birl, University of Wisconsin, Madison
McCoy, Edward L., Ohio State University
McDonald, Eric, University of Nevada, Reno
McInnes, Kevin J., Texas A&M University
Molz, Fred, Clemson University
Moncrief, John F., University of Minnesota, Twin Cities
Morgan, Cristine L., Texas A&M University
Mulla, David J., University of Minnesota, Twin Cities
Neely, Haly L., Texas A&M University
Nielsen, Donald R., University of California, Davis
Niu, Guo-Yue, University of Texas, Austin
Nkedi-Kizza, Peter, University of Florida
Norman, John M., University of Wisconsin, Madison
Papendick, Robert I., Washington State University
Perfect, Edmund, University of Tennessee, Knoxville
Persaud, Naraine, Virginia Polytechnic Institute & State University
Radcliffe, David E., University of Georgia
Reece, Julia S., Texas A&M University
Ressler, Daniel E., Susquehanna University
Rogowski, Andrew S., Pennsylvania State University, University Park
Sammis, Theodore W., New Mexico State University, Las Cruces
Selim, Hussein M., Louisiana State University
Sharma, Vasudha, University of Minnesota, Twin Cities
Simmons, F. William, University of Illinois, Urbana-Champaign
Spokas, Kurt, University of Minnesota, Twin Cities
Stone, Loyd, Kansas State University
Thomas, John, University of Florida
Venterea, Rodney T., University of Minnesota, Twin Cities
Wang, Zhi (Luke), California State University, Fresno
Warrick, Arthur W., University of Arizona
Welch, Steve M., Kansas State University
Wierenga, Peter J., University of Arizona
Wierenga, Peter J., University of Arizona
Wraith, Jon M., Montana State University
Yeh, Jim, University of Arizona
Young, Michael H., University of Texas, Austin
Youngs, Edward, The Open University

Soil Chemistry/Mineralogy
Adriano, Domy C., Virginia Polytechnic Institute & State University
Adriano, Domy C., University of Georgia
Ahmad, Abd Rashid, University of Malaya
Alley, Marcus M., Virginia Polytechnic Institute & State University
Appel, Christopher (Chip) S., California Polytechnic State University
Artiola, Janick F., University of Arizona
Baligar, V. C., Virginia Polytechnic Institute & State University
Barak, Phillip W., University of Wisconsin, Madison
Basta, Nicholas T., Ohio State University
Bertsch, Paul M., University of Georgia
Blanchar, Robert W., University of Missouri, Columbia
Bleam, William F., University of Wisconsin, Madison
Bloom, Paul R., University of Minnesota, Twin Cities
Bomke, Arthur A., University of British Columbia
Bostick, Benjamin C., Columbia University
Burke, Deborah S., Pacific Northwest National Laboratory
Callanan, Jennifer R., William Paterson University
Casey, William H., University of California, Davis
Chirenje, Tait, Stockton University
Chorover, Jon, University of Arizona
Chorover, Jonathan, University of Arizona
Clark, Joanna, University of Reading
Collins, Chris, University of Reading
Crouse, David A., North Carolina State University
Curry, Joan E., University of Arizona
Dahlgran, Randy, University of California, Davis

Donohue, Stephen J., Virginia Polytechnic Institute & State University
Dowdy, Robert H., University of Minnesota, Twin Cities
Dudley, Lynn M., Florida State University
Eckert, Donald J., Ohio State University
Edenborn, Harry, West Virginia University
Eick, Matthew J., Virginia Polytechnic Institute & State University
Eivazi, Frieda, University of Missouri, Columbia
Elzinga, Evert J., Rutgers, The State Univ of New Jersey, Newark
Engel, Richard E., Montana State University
Evangelou, V. P., Iowa State University of Science & Technology
Evanylo, Gregory K., Virginia Polytechnic Institute & State University
Fendorf, Scott E., Stanford University
Fernandez, Fabian G., University of Minnesota, Twin Cities
Gamble, Audrey V., Auburn University
Gascho, Gary J., University of Georgia
Gaston, Lewis A., Louisiana State University
Ginder-Vogel, Matt, University of Wisconsin, Madison
Goos, Robert J., North Dakota State University
Goyne, Keith W., University of Missouri, Columbia
Grandy, Stuart, University of New Hampshire
Guertal, Elizabeth A., Auburn University
Han, Nizhou, Virginia Polytechnic Institute & State University
Harris, Jr., Willie G., University of Florida
Harsh, James B., Washington State University
Hassett, John J., University of Illinois, Urbana-Champaign
Havlin, John L., North Carolina State University
He, Zhenli, University of Florida
Heil, Dean, New Mexico State University, Las Cruces
Helmke, Philip A., University of Wisconsin, Madison
Hemzacek Laukant, Jean M., Northeastern Illinois University
Hesterberg, Dean L., North Carolina State University
Hettiarachchi, Ganga, Kansas State University
Howe, Julie, Texas A&M University
Huluka, Gobena E., Auburn University
Inskeep, William P., Montana State University
Jaouich, Alfred, Universite du Quebec a Montreal
Jardine, Philip M., Oak Ridge National Laboratory
Jones, Robert L., University of Illinois, Urbana-Champaign
Kabengi, Nadine, Georgia State Univ
Kaiser, Daniel E., University of Minnesota, Twin Cities
Kissel, David E., University of Georgia
Komarneni, Sridhar, Pennsylvania State University, University Park
Koskinen, William C., University of Minnesota, Twin Cities
Krogstad, Tore, Norwegian University of Life Sciences (NMBU)
Kuo, Shiou, Washington State University
Laird, David A., Iowa State University of Science & Technology
Lamb, John A., University of Minnesota, Twin Cities
Lee, Young Jae, Korea University
Lewis, Katie, Texas A&M University
Li, Yuncong, University of Florida
Ma, Lena Q., University of Florida
Malzer, Gary L., University of Minnesota, Twin Cities
Marion, Giles, Desert Research Institute
Massey, Michael S., California State University, East Bay
McCaslin, Bobby D., New Mexico State University, Las Cruces
McLaughlin, Richard A., North Carolina State University
Miao, Yuxin, University of Minnesota, Twin Cities
Miller, William P., University of Georgia
Moore, Duane M., University of New Mexico
Mowrer, Jake, Texas A&M University
Mullins, Gregory L., Virginia Polytechnic Institute & State University
Mulvaney, Richard L., University of Illinois, Urbana-Champaign
Mylavarapu, S R., University of Florida
Myneni, Satish C B., Princeton University
Nair, Vimala D., University of Florida
O'Connor, George A., University of Florida
Olson, Kenneth R., University of Illinois, Urbana-Champaign
Pagliari, Paulo H., University of Minnesota, Twin Cities
Peck, Theodore R., University of Illinois, Urbana-Champaign
Peryea, Frank J., Washington State University
Pierzynski, Gary M., Kansas State University
Plank, Owen C., University of Georgia
Provin, Tony L., Texas A&M University
Quade, Jay, University of Arizona
Rehm, George W., University of Minnesota, Twin Cities
Reneau, Raymond B., Virginia Polytechnic Institute & State University
Robarge, Wayne P., North Carolina State University

Rosen, Carl J., University of Minnesota, Twin Cities
Roy, William R., University of Illinois, Urbana-Champaign
Russelle, Michael P., University of Minnesota, Twin Cities
Schmitt, Michael A., University of Minnesota, Twin Cities
Schwab, Paul, Texas A&M University
Segars, William P., University of Georgia
Seyfferth, Angelia L., University of Delaware
Shuman, Larry M., University of Georgia
Siebe Grabach, Christina D., Univ Nac Autonoma de Mexico
Sims, Albert L., University of Minnesota, Twin Cities
Smyth, Thomas J., North Carolina State University
Spalding, Brian P., Oak Ridge National Laboratory
Strawn, Daniel G., University of Idaho
Stucki, Joseph W., University of Illinois, Urbana-Champaign
Szulczewski, Melanie, University of Mary Washington
Thien, Steve J., Kansas State University
Thompson, Michael L., Iowa State University of Science & Technology
Toor, Gurpal, University of Florida
Trumbore, Susan E., University of California, Irvine
Werts, Scott P., Winthrop University
Wilson, Melissa L., University of Minnesota, Twin Cities
Wright, Alan, University of Florida
Yildirim, Ismail, Firat University
Zelazny, Lucian W., Virginia Polytechnic Institute & State University

Pedology/Classification/Morphology

Anderson, James L., University of Minnesota, Twin Cities
Baker, James C., Virginia Polytechnic Institute & State University
Barrett, Linda R., University of Akron
Bell, James C., University of Minnesota, Twin Cities
Bettis III, E. A., University of Iowa
Bigham, Jerry M., Ohio State University
Birkeland, Peter W., University of Colorado
Burras, Lee, Iowa State University of Science & Technology
Calhoun, Frank G., Ohio State University
Ciolkosz, Edward J., Pennsylvania State University, University Park
Collins, Mary E., University of Florida
Cooper, Terence H., University of Minnesota, Twin Cities
Daniels, Walter L., Virginia Polytechnic Institute & State University
Darmody, Robert G., University of Illinois, Urbana-Champaign
Daugherty, LeRoy A., New Mexico State University, Las Cruces
Dere, Ashlee L., University of Nebraska, Omaha
Evans, Barry M., Pennsylvania State University, University Park
Farsang, Andrea, Univesity of Szeged
Frederick, Holly, Wilkes University
Heck, Richard, University of Guelph
Hoover, Michael T., North Carolina State University
Hopkins, David G., North Dakota State University
Jacobs, Peter, University of Wisconsin, Whitewater
Jelinski, Nicolas, University of Minnesota, Twin Cities
Johnson-Maynard, Jodi, University of Idaho
Kleiss, Harold J., North Carolina State University
Kuzila, Mark S., Unversity of Nebraska - Lincoln
Lavkulich, Leslie M., University of British Columbia
Lindbo, David, North Carolina State University
Madison, Frederick W., University of Wisconsin, Madison
Mbila, Monday O., Alabama A&M University
McDaniel, Paul A., University of Idaho
McDonald, Eric, Desert Research Institute
McFadden, Leslie M., University of New Mexico
McGahan, Donald, Texas A&M University
McSweeney, Kevin, University of Wisconsin, Madison
Miles, Randall J., University of Missouri, Columbia
Miller, Gerald A., Iowa State University of Science & Technology
Montagne, Cliff, Montana State University
Nater, Edward A., University of Minnesota, Twin Cities
Nielsen, Gerald A., Montana State University
Nordt, Lee C., Baylor University
Paz Gonzalez, Antonio, Coruna University
Petersen, Gary W., Pennsylvania State University, University Park
Post, Donald F., University of Arizona
Ransom, Michel D., Kansas State University
Ryder, Roy, University of South Alabama
Sandor, Jonathan A., Iowa State University of Science & Technology
Shaw, Joey N., Auburn University
Slater, Brian K., Ohio State University
Steele, Amber , Missouri Dept of Natural Resources
Taboada Castro, María T., Coruna University

Tabor, Neil J., Southern Methodist University
Thomas, Pamela J., Virginia Polytechnic Institute & State University
Tyler, E. Jerry, University of Wisconsin, Madison
Vepraskas, Michael J., North Carolina State University
Verburg, Paul, University of Nevada, Reno
West, Larry T., University of Georgia
Yoo, Kyungsoo, University of Minnesota, Twin Cities

Forest Soils/Rangelands/Wetlands
Amthor, Jeff, Oak Ridge National Laboratory
Andrus, Richard E., Binghamton University
Balster, Nick J., University of Wisconsin, Madison
Bass, Michael L., University of Mary Washington
Bauder, James W., Montana State University
Bockheim, James G., University of Wisconsin, Madison
Broome, Stephen W., North Carolina State University
Chambers, Jeanne, University of Nevada, Reno
Clark, Mark W., University of Florida
Clark, Melissa, Indiana University, Bloomington
Comerford, Nicholas B., University of Florida
David, Mark B., University of Illinois, Urbana-Champaign
Dietze, Michael, Boston University
Ferris, Dawn, Ohio State University
Fick, Walter H., Kansas State University
Franz, Eldon H., Washington State University
Gunderson, Lance, Emory University
Hallock, Brent G., California Polytechnic State University
Hook, Paul B., Montana State University
Inglett, Patrick, University of Florida
Krzic, Maja, University of British Columbia
Laingen, Christopher R., Eastern Illinois University
Meretsky, Vicky J., Indiana University, Bloomington
O'Neill, Patrick M., Louisiana State University
Osborne, Todd, University of Florida
Owensby, Clenton E., Kansas State University
Peres, Carlos, University of East Anglia
Posler, Gerry L., Kansas State University
Riha, Susan, Cornell University
Robinson, Steve, University of Reading
Smalley, Glendon W., Sewanee: University of the South
Smith, C. Ken, Sewanee: University of the South
Taskey, Ronald D., California Polytechnic State University
Torreano, Scott, Sewanee: University of the South
Utter, James M., SUNY, Purchase
Wegner, John, Emory University

Soil Biology/Biochemistry
Allan, Deborah L., University of Minnesota, Twin Cities
Armstrong, Felicia P., Youngstown State University
Balser, Teri C., University of Wisconsin, Madison
Berry, Duane F., Virginia Polytechnic Institute & State University
Bezdicek, David F., Washington State University
Bird, Jeffrey, Grad Sch of the City Univ of New York
Bird, Jeffrey, Queens College (CUNY)
Blum, Linda K., University of Virginia
Bollag, Jean-Marc, Pennsylvania State University, University Park
Burlage, Robert S., Oak Ridge National Laboratory
Chanway, Christopher, University of British Columbia
Cheng, H. H., University of Minnesota, Twin Cities
Collins, Chris, University of Reading
Crozier, Carl, North Carolina State University
Davenport, Joan R., Washington State University
Dick, Richard P., Ohio State University
Dick, Warren A., Ohio State University
Dunfield, Kari, University of Guelph
Feng, Yucheng, Auburn University
Frey, Serita, University of New Hampshire
Fultz, Lisa M., Louisiana State University
Gauthier, Paul, Princeton University
Gentry, Terry, Texas A&M University
Gerba, Charles P., University of Arizona
Gilliam, James W., North Carolina State University
Grace, Peter R., Queensland University of Technology
Graham, Jr., James H., University of Florida
Graves, Alexandria, North Carolina State University
Grayston, Sue, University of British Columbia
Groffman, Peter, Brooklyn College (CUNY)
Gutknecht, Jessica L., University of Minnesota, Twin Cities

Hagedorn, Charles, Virginia Polytechnic Institute & State University
Harris, Glendon H., University of Georgia
Harris, Robin F., University of Wisconsin, Madison
Hartel, Peter G., University of Georgia
Hickey, William J., University of Wisconsin, Madison
Hladik, Christine, Georgia Southern University
Hoyt, Greg D., North Carolina State University
Ishii, Satoshi, University of Minnesota, Twin Cities
Israel, Daniel W., North Carolina State University
Jaffe, Peter R., Princeton University
Jastrow, Julie D., Argonne National Laboratory
Kennedy, Ann C., Washington State University
Knudsen, Guy, University of Idaho
Kremer, Robert J., University of Missouri, Columbia
Lindemann, William C., New Mexico State University, Las Cruces
Matson, Pamela A., Stanford University
Miller, Raymond M., Argonne National Laboratory
Molina, Jean-Alex E., University of Minnesota, Twin Cities
Moorhead, Kevin K., University of North Carolina, Asheville
Morra, Matthew J., University of Idaho
Ogram, Andrew V., University of Florida
Oikawa, Patricia, California State University, East Bay
Osmond, Deanna L., North Carolina State University
Palumbo, Tony V., Oak Ridge National Laboratory
Pan, William L., Washington State University
Parker, David B., West Texas A&M University
Peacock, Caroline, University of Leeds
Pereira, Engil I., University of Texas, Rio Grande Valley
Phelps, Tommy J., Oak Ridge National Laboratory
Plante, Alain F., University of Pennsylvania
Prasad, Rishi, Auburn University
Prescott, Cindy, University of British Columbia
Ribbons, Relena R., Lawrence University
Rice, Chuck W., Kansas State University
Ruehr, Thomas A., California Polytechnic State University
Sadowsky, Michael J., University of Minnesota, Twin Cities
Scheer, Clemens, Queensland University of Technology
Shaw, Liz, University of Reading
Shi, Wei, North Carolina State University
Silva, Lucas C., University of California, Davis
Simard, Suzanne, University of British Columbia
Smith, Jeffery L., Washington State University
Sogn, Trine, Norwegian University of Life Sciences (NMBU)
Somenhally, Anil, Texas A&M University
Stevens, Robert G., Washington State University
Thomas, Andrew D., University of Wales
Tolhurst, Trevor, University of East Anglia
Torrents, Alba, University of Maryland
Voroney, R. Paul, University of Guelph
Wagger, Michael G., North Carolina State University
Wenzel, Christopher, Eastern Wyoming College
Wilkie, Ann C., University of Florida

Paleopedology/Archeology
Beck, Colleen M., Desert Research Institute
Busacca, Alan J., Washington State University
Driese, Steven G., Baylor University
Driese, Steven G., University of Tennessee, Knoxville
Follmer, Leon R., University of Illinois, Urbana-Champaign
Herrmann, Edward W., Indiana University, Bloomington
Irmis, Randall B., University of Utah
Monger, H. C., New Mexico State University, Las Cruces
Terry, Dennis O., Temple University

Other Soil Science
Adamsen, Floyd, University of Arizona
Adee, Eric, Kansas State University
Bader, Nicholas E., Whitman College
Balser, Teri , University of Florida
Bell, Brian, University of Glasgow
Berg, Peter, University of Virginia
Bonczek, James, University of Florida
Brinkmann, Robert, Hofstra University
Brown, James R., University of Missouri, Columbia
Budke, William, Ventura College
Bundy, Larry G., University of Wisconsin, Madison
Cabrera, Miguel L., University of Georgia
Cambardella, Cynthia, Iowa State University of Science & Technology

Ciampitti, Ignacio, Kansas State University
Claassen, Mark, Kansas State University
Coleman, Tommy L., Alabama A&M University
Contosta, Alexandra, University of New Hampshire
Cramer, Gary, Kansas State University
Cruse, Richard M., Iowa State University of Science & Technology
Dahlgren, Randy A., University of California, Davis
Daroub, Samira H., University of Florida
DeLaune, Paul, Texas A&M University
DeSutter, Tom, North Dakota State University
Dingus, Delmar D., California Polytechnic State University
Dormaar, John, University of Lethbridge
Dou, Fugen, Texas A&M University
Dougil, Andy, University of Leeds
Drescher, Andrew, University of Minnesota, Twin Cities
Duncan, Stewart, Kansas State University
Eberle, Bill, Kansas State University
Ebinger, Michael H., Los Alamos National Laboratory
Ehler, Stan, Kansas State University
Eppes, Martha C., University of North Carolina, Charlotte
Favorito, Jessica, Stockton University
Ferre, Paul (Ty), University of Arizona
Fjell, Dale, Kansas State University
Flores-Delgadillo, Maria L., Univ Nac Autonoma de Mexico
Flores-Roman, David, Univ Nac Autonoma de Mexico
Fritz, Allan, Kansas State University
Gale, Paula M., University of Tennessee, Martin
Gantzer, Clark J., University of Missouri, Columbia
Geisseler, Daniel, University of California, Davis
Gerber, Stefan, University of Florida
Gordon, Barney, Kansas State University
Goss, Michael J., University of Guelph
Haag, Lucas, Kansas State University
Hall, III, John R., Virginia Polytechnic Institute & State University
Halverson, Larry, Iowa State University of Science & Technology
Henning, Stanley J., Iowa State University of Science & Technology
Hernandez-Silva, Gilberto, Univ Nac Autonoma de Mexico
Hochmuth, George J., University of Florida
Holman, John, Kansas State University
Hopmans, Jan W., University of California, Davis
Horwath, William R., University of California, Davis
Houlton, Ben, University of California, Davis
Inglett, Kanika Sharma, University of Florida
Jackson, Louise E., University of California, Davis
Jacobsen, Jeffrey S., Montana State University
Janssen, Keith, Kansas State University
Jasoni, Richard, Desert Research Institute
Johnson, Arthur H., University of Pennsylvania
Johnson, Jane M., University of Minnesota, Twin Cities
Jones, Julia A., Oregon State University
Jugulam, Mithila, Kansas State University
Karkanis, Pano G., University of Lethbridge
Karlen, Douglas L., Iowa State University of Science & Technology
Kaspar, Thomas A., Iowa State University of Science & Technology
Keeney, Dennis R., Iowa State University of Science & Technology
Kelling, Keith A., University of Wisconsin, Madison
Killorn, Randy J., Iowa State University of Science & Technology
Kitchen, Newell R., University of Missouri, Columbia
Kolka, Randall K., University of Minnesota, Twin Cities
Kramer, Marc, University of Florida
Kusssow, Wayne R., University of Wisconsin, Madison
Laboski, Carrie A., University of Wisconsin, Madison
Landschoot, Peter J., Pennsylvania State University, University Park
Lauzon, John, University of Guelph
Lerch, Robert N., University of Missouri, Columbia
Liang, George, Kansas State University
Liebens, Johan, University of West Florida
Loutkova, Klavdia Oleschko, Univ Nac Autonoma de Mexico
Loynachan, Thomas E., Iowa State University of Science & Technology
Mackowiak, Cheryl, University of Florida
Mallarino, Antonio W., Iowa State University of Science & Technology
Marrs, Rob, University of Liverpool
McBride, Raymond G., University of Guelph
Mengel, David, Kansas State University
Miller, Murray H., University of Guelph
Mills, Aaron L., University of Virginia
Min, Doo-Hong, Kansas State University
Mishra, Umakant, Argonne National Laboratory
Moorberg, Colby, Kansas State University
Moorman, Thomas B., Iowa State University of Science & Technology
Morgan, Kelly, University of Florida
Morris, Geoffrey, Kansas State University
Motavalli, Peter P., University of Missouri, Columbia
Nelson, Nathan, Kansas State University
Norman, John M., University of Wisconsin, Madison
Nyhan, John W., Los Alamos National Laboratory
O'Geen, Toby, University of California, Davis
O'Halloran, Ivan, University of Guelph
Obour, Augustine, Kansas State University
Obreza, Thomas A., University of Florida
Palacios-Mayorga, Sergio, Univ Nac Autonoma de Mexico
Parikh, Sanjai, University of California, Davis
Pedersen, Joel A., University of Wisconsin, Madison
Perumal, Ram, Kansas State University
Peterson, Dallas, Kansas State University
Picardal, Flynn W., Indiana University, Bloomington
Polito, Thomas A., Iowa State University of Science & Technology
Powell, J. Mark, University of Wisconsin, Madison
Prasad, Vara, Kansas State University
Presley, DeAnn, Kansas State University
Qualls, Robert G., University of Nevada, Reno
Randall, Gyles W., University of Minnesota, Twin Cities
Rechcigl, John E., University of Florida
Reganold, John P., Washington State University
Regehr, David, Kansas State University
Rice, Pamela J., University of Minnesota, Twin Cities
Richards, James H., University of California, Davis
Richter, Daniel D., Duke University
Risk, Dave A., Saint Francis Xavier University
Rolston, Dennis E., University of California, Davis
Roozeboom, Kraig, Kansas State University
Roth, Gregory W., Pennsylvania State University, University Park
Ruark, Matthew D., University of Wisconsin, Madison
Ruiz-Diaz, Dorivar, Kansas State University
Santos, Eduardo, Kansas State University
Sartain, Jerry B., University of Florida
Sassenrath, Gretchen, Kansas State University
Schaap, Marcel, University of Arizona
Schapaugh, Bill T., Kansas State University
Schardt, Lawrence A., Pennsylvania State University, University Park
Schlegel, Alan, Kansas State University
Schumann, Arnold W., University of Florida
Scow, Kate, University of California, Davis
Sherwood, William C., James Madison University
Shoup, Doug, Kansas State University
Shroyer, Jim, Kansas State University
Shuford, James W., Alabama A&M University
Silveira, Maria, University of Florida
Silvertooth, Jeffrey C., University of Arizona
Sinclair, Hugh D., Edinburgh University
Sletten, Ronald S., University of Washington
Smith, Terry L., California Polytechnic State University
Smukler, Sean, University of British Columbia
Sohi, Sran P., Edinburgh University
Soldat, Douglas J., University of Wisconsin, Madison
Southard, Randal J., University of California, Davis
Stahlman, Phillip, Kansas State University
Stapleton, Michael G., Slippery Rock University
Stevens, W. G., University of Missouri, Columbia
Strock, Jeffrey S., University of Minnesota, Twin Cities
Stubler, Craig, California Polytechnic State University
Szecsody, James E., Pacific Northwest National Laboratory
Tesso, Tesfaye, Kansas State University
Thompson, Curtis, Kansas State University
Tomlinson, Peter, Kansas State University
Tubana, Brenda S., Louisiana State University
Vanderlip, Richard, Kansas State University
Voss, Regis D., Iowa State University of Science & Technology
Walworth, James, University of Arizona
Wesley Wood, Charley, University of Florida
Whitney, D.A., Kansas State University
Wilson, P. Christopher, University of Florida
Zhang, Guorong, Kansas State University

ENGINEERING GEOLOGY
General Engineering Geology
, Ebrahim A., University of Tabriz
Abdalla, Salah B., University of Khartoum
Acevedo-Arroyo, Jose Refugio, Univ Autonoma de San Luis Potosi
Achampong, Francis, University of Ghana
Adams, Herbert G., California State University, Northridge
Adeyemi, G. O., University of Ibadan
Aimone-Martin, Cathrine T., New Mexico Institute of Mining and Tech
Alao, D. A., University of Ilorin
Anderson, Andrew, Illinois State Geological Survey
Andre-Obayanju, Tomilola, University of Benin
Axelbaum, Richard, Washington University in St. Louis
Aydin, Adnan, University of Mississippi
Baptista, Joana, University of Newcastle Upon Tyne
Bauer, R. A., Illinois State Geological Survey
Bautista, Sonia, Univ Complutense de Madrid
Bell, David, University of Canterbury
Bell, John W., University of Nevada
Bennet, Brett, University of Kansas
Berry, Karen, Colorado Geological Survey
Beukelman, Gregg S., Utah Geological Survey
Biswas, Pratim, Washington University in St. Louis
Bonetto, Sabrina Maria Rita, Università di Torino
Bowman, Steve D., Utah Geological Survey
Brabham, Peter, University of Wales
Brady, Roland H., California State University, Fresno
Brand, Patrick, California Geological Survey
Branum, David, California Geological Survey
Breuninger, Tamara, Technical University of Munich
Brugger, Matthias, Technical University of Munich
Brunkal, Holly, Western Colorado University
Buchanan, George, Montgomery County Community College
Burns, Bill, Oregon Dept of Geology and Mineral Industries
Burns, Scott F., Portland State University
Buscheck, Thomas A., Lawrence Livermore National Laboratory
Carlson, Jill, Colorado Geological Survey
Carney, Theodore C., Los Alamos National Laboratory
Caron, Olivier J., Illinois State Geological Survey
Carr, James R., University of Nevada, Reno
Castleton, Jessica, Utah Geological Survey
Cato, Kerry, California State University, San Bernardino
Cetindag, Bahattin, Firat University
CHEGBELEH, Larry-Pax, University of Ghana
Choma-Moryl, Krystyna, University of Wroclaw
Davie, Colin, University of Newcastle Upon Tyne
Davies, David, Heriot-Watt University
Dawson, Tim, California Geological Survey
Delattre, Marc, California Geological Survey
dePolo, Craig M., University of Nevada
Doheny, Edward L., University of Pennsylvania
Doherty, Kevin, California Geological Survey
DuRoss, Christopher B., Utah Geological Survey
Dusseault, Maurice B., University of Waterloo
Eberhardt, Erik, University of British Columbia
Edil, Tuncer B., University of Wisconsin, Madison
Elia, Gaetano, University of Newcastle Upon Tyne
Erickson, Ben A., Utah Geological Survey
Evans, Stephen G., University of Waterloo
Eversoll, Duane A., Unversity of Nebraska - Lincoln
Faúndez, Alejandro, Univ Complutense de Madrid
Ferriz, Horacio, California State University, Stanislaus
Floris, Mario, Università degli Studi di Padova
Fournier, Benoit, Universite Laval
Fratta, Dante, University of Wisconsin-Madison
Fuller, Michael, California Geological Survey
Galgaro, Antonio, Università degli Studi di Padova
Gautam, Tej P., Marietta College
Giraud, Richard E., Utah Geological Survey
Greene, Brian M., Youngstown State University
Gretener, Peter E., University of Calgary
Guiterrez, Carlos, California Geological Survey
Hall, Jean, University of Newcastle Upon Tyne
Haneberg, William C., University of Kentucky
Haneberg, William, New Mexico Institute of Mining and Technology
Harris, Will, California Geological Survey
Hasan, Syed E., University of Missouri, Kansas City
Haydon, Wayne, California Geological Survey
Hayhurst, Cheryl, California Geological Survey
Hencher, Steve, University of Leeds
Henika, William S., Virginia Polytechnic Institute & State University
Hernandez, Janis, California Geological Survey
Ho, I-Hsuan, University of North Dakota
Holland, Peter, California Geological Survey
Holzer, Thomas, Stanford University
Horns, Daniel, Utah Valley University
Hudec, Peter P., University of Windsor
Ibrahim, Ahmad Tajuddin Hj, University of Malaya
Ige, O. O., University of Ilorin
Jamaluddin, Tajul Anuar, University of Malaya
Jo, Ho Young, Korea University
Johnpeer, Gary D., Cerritos College
Johnson, William P., University of Utah
Jones, Glenda, Keele University
Jones, Larry Allan, Los Alamos National Laboratory
Joyce, James, University of Puerto Rico
Karanfil, Tanju, Clemson University
Kavanaugh, Jeffrey, University of Alberta
Knapp, Richard B., Lawrence Livermore National Laboratory
Knizek, Martin, Masaryk University
Knudsen, Tyler R., Utah Geological Survey
Korre, Anna, Imperial College
Kreylos, Oliver, University of California, Davis
Krohn, James P., Los Angeles Pierce College
Kudlac, John J., Point Park University
Lancaster, Jeremy, California Geological Survey
Likos, William J., University of Wisconsin, Madison
Linares, Rogelio, Universitat Autonoma de Barcelona
Lindsay, Donald, California Geological Survey
Lindsey, Kassandra, Colorado Geological Survey
Linn, Rodman, Los Alamos National Laboratory
Locat, Jacques E., Universite Laval
Lokau, Katja R., Technical University of Munich
Longstreth, David, California Geological Survey
Lovekin, Jonathan, Colorado Geological Survey
Lund, William R., Utah Geological Survey
Lupogo, Keneth, University of Dar es Salaam
Madrigal-Rubio, Rafael, Universidad Autonoma de Chihuahua
Mandrone, Giuseppe, Università di Torino
Mareschal, Maxime, California Geological Survey
Martel, Stephen J., University of Hawai'i, Manoa
Martin, Beth, Washington University in St. Louis
Masek, Ondrej, Edinburgh University
Mathewson, Christopher C., Texas A&M University
Matter, Juerg M., University of Southampton
McCoy, Kevin, Colorado Geological Survey
McCrink, Timothy P., California Geological Survey
McDonald, Gregory N., Utah Geological Survey
McMahon, Katherine, University of Wisconsin, Madison
Mena-Zambrano, Teodulo, Universidad Autonoma de Chihuahua
Merifield, Paul M., University of California, Los Angeles
Merritt, Andrew, University of Plymouth
Mlinarevic, Ante, California Geological Survey
Moeglin, Thomas D., Missouri State University
Mogilevskaya, Sonia, University of Minnesota, Twin Cities
Moore, Jeffrey, University of Utah
Morales, Tomas, Univ of the Basque Country UPV/EHU
Morgan, Terrance L., Los Alamos National Laboratory
Nickmann, Marion, Technical University of Munich
Niemann, William L., Marshall University
Nieto, Alberto S., University of Illinois, Urbana-Champaign
O'Rourke, Thomas D., Cornell University
Ogunsanwo, O., University of Ilorin
Olson, Brian, California Geological Survey
Origlia, Hector D., Universidad Nacional de Rio Cuarto
Oswald, John, California Geological Survey
Oyediran, I. A., University of Ibadan
Patrick, David M., University of Southern Mississippi
Perez, Florante, California Geological Survey
Perret, Didier H., Universite du Quebec
Pestrong, Raymond, San Francisco State University
Pipkin, Bernard W., University of Southern California
Poeter, Eileen P., Colorado School of Mines

Potter, Kenneth, University of Wisconsin, Madison
Poulton, Mary M., University of Arizona
Poulton, Mary, University of Arizona
Pridmore, Cindy L., California Geological Survey
Rahn, Perry H., South Dakota School of Mines & Technology
Raj, John K., University of Malaya
Ramelli, Alan R., University of Nevada
Ramirez, Abelardo L., Lawrence Livermore National Laboratory
Rockaway, John D., Northern Kentucky University
Rodríguez, Martín Jesús, Univ Complutense de Madrid
Roffers, Pete, California Geological Survey
Roggenthen, William M., South Dakota School of Mines & Technology
Rosa, Carla, California Geological Survey
Rosinski, Anne, California Geological Survey
Rubin, Ron, California Geological Survey
Santi, Paul M., Colorado School of Mines
Sass, Ingo, Technische Universitaet Darmstadt
Savigny, K. Wayne, University of British Columbia
Schiff, Sherry L., University of Waterloo
Seitz, Gordon, California Geological Survey
Sellmeier, Bettina, Technical University of Munich
Shakoor, Abdul, Kent State University
Short, William, California Geological Survey
Silva, Michael A., California Geological Survey
Song, Lin-Ping, University of British Columbia
Sonnenwald, Margreta, Technical University of Munich
Spangler, Eleanor, California Geological Survey
Staub, William P., Oak Ridge National Laboratory
Stead, Douglas, Simon Fraser University
Stevens, De Anne S., Alaska Div of Geological & Geophysical Surveys
Swanson, Brian, California Geological Survey
Tester, Jefferson W., Cornell University
Thomas, Mark, University of Leeds
Thornburg, Jennifer, California Geological Survey
Thuro, Kurosch, Technical University of Munich
Tinjum, James M., University of Wisconsin, Madison
Tsige, Meaza, Univ Complutense de Madrid
Turner, A. Keith, Colorado School of Mines
Uriarte, Jesus Angel, Univ of the Basque Country UPV/EHU
Urzua, Alfredo, Boston College
Villeneuve, Marlene C., University of Canterbury
Vorosmarty, Charles, Grad Sch of the City Univ of New York
Walton, Gabriel, Colorado School of Mines
Wang, Dongmei, University of North Dakota
Watts, Chester F., Radford University
Weigers, Mark, California Geological Survey
West, Olivia M., Oak Ridge National Laboratory
West, Terry R., Purdue University
White, Chase, California Geological Survey
White, Jonathan, Colorado Geological Survey
White, Owen L., University of Waterloo
Wilkey, Patrick L., Argonne National Laboratory
Williams, Bruce A., Pacific Northwest National Laboratory
Williams, John W., San Jose State University
Wu, Chin, University of Wisconsin, Madison
Yarbrough, Lance D., University of Mississippi
Yunwei, Sun, Lawrence Livermore National Laboratory
Zhou, Wendy W., Colorado School of Mines
Zimmerman, R. Eric, Argonne National Laboratory
Zyvoloski, George A., Los Alamos National Laboratory

Earthquake Engineering
Acosta, Jose G., Centro de Inv Científica y de Ed Sup de Ensenada
Atkinson, Gail M., Carleton University
Heaton, Thomas H., California Institute of Technology
Huang, Moh J., California Geological Survey
Jamieson, J. Bruce, University of Calgary
Kutter, Bruce, University of California, Davis
Mendoza, Luis H., Centro de Inv Científica y de Ed Sup de Ensenada
Michaels, Paul, Boise State University
Oommen, Thomas, Michigan Technological University
Reyes, Alfonso, Centro de Inv Científica y de Ed Sup de Ensenada
Rowshandel, Badie, California Geological Survey
Swensen, Daniel, California Geological Survey
Wang, Yumei, Oregon Dept of Geology and Mineral Industries

Mining Tech/Extractive Metallurgy
Aguirre-Moriel, Socorro I., Universidad Autonoma de Chihuahua
Bryan, Christopher, Exeter University
Coggan, John, Exeter University
Demopoulos, George, McGill University
Doyle, Fiona M., University of California, Berkeley
Duby, Paul F., Columbia University
Evans, James W., University of California, Berkeley
Feunstenau, Maurice, University of Nevada, Reno
Fuerstenau, Douglas W., University of California, Berkeley
Glass, Hylke J., Exeter University
Griswold, George B., New Mexico Institute of Mining and Technology
Gundiler, Ibrahim H., New Mexico Institute of Mining & Technology
Guthrie, Roderick I., McGill University
Hasan, Mainul, McGill University
Hiskey, J. Brent, University of Arizona
Hurtado-De La Ree, Maria B., Universidad Estatal de Sonora
Jeffrey, Kip, Exeter University
Jonas, John J., McGill University
Khan, Latif A., Illinois State Geological Survey
Kinabo, Crispin P., University of Dar es Salaam
Lee, Jaeheon, University of Arizona
López, Maricela, Universidad Estatal de Sonora
Mendoza-Aguilar, Hector M., Universidad Autonoma de Chihuahua
Minor-Velazquez, Hector, Universidad Autonoma de Chihuahua
Nesbitt, Carl, University of Nevada, Reno
Ruiz-Cisneros, David H., Universidad Autonoma de Chihuahua
Salazar-Avila, Arnulfo, Universidad Estatal de Sonora
Sastry, Kalanadh V. S., University of California, Berkeley
Seal, Thom, University of Nevada, Reno
Smith, Karl A., University of Minnesota, Twin Cities
Somasundaran, Ponisseril, Columbia University
Themelis, Nickolas J., Columbia University
Trevizo-Cano, Luis M., Universidad Autonoma de Chihuahua
Wheeler, Paul, Exeter University
Yue, Stephen, McGill University

Mining Engineering
MASSACCI, Prof. Giorgio, Universita di Cagliari
Bandopadhyay, Sukumar, University of Alaska, Fairbanks
Bessinger, Stephen, University of Utah
Boshkov, Stefan H., Columbia University
Brune, Jürgen F., Colorado School of Mines
Calizaya, Felipe, University of Utah
Chatterjee, Snehamoy, Michigan Technological University
Chen, Gang, University of Alaska, Fairbanks
Dagdelen, Kadri, Colorado School of Mines
Danko, George, University of Nevada, Reno
Dimitrakopoulos, Roussos, McGill University
Dinsmoor, John C., Los Alamos National Laboratory
Donovan, James, University of Utah
Durucan, Sevket, Imperial College
Eveleth, Robert W., New Mexico Institute of Mining & Technology
Foster, Patrick, Exeter University
Ganguli, Rajive, University of Alaska, Fairbanks
Gongora-Jurado, Raul A., Universidad Estatal de Sonora
Hafez, Sabry ., University of Alaska, Fairbanks
Hassani, Ferri, McGill University
Honkaer, Rick, University of Kentucky
Kalia, Hemendra N., Los Alamos National Laboratory
Kaunda, Rennie, Colorado School of Mines
Kennedy, Gareth, Exeter University
Kim, Kwangmin, University of Arizona
Kohler, Jeffery L., Pennsylvania State University, University Park
Kovach, Richard G., Los Alamos National Laboratory
Kuchta, Mark, Colorado School of Mines
Liu, Shimin, Pennsylvania State University, University Park
Lusk, Braden , University of Kentucky
MANCA, Prof. Pierpaolo P., Universita di Cagliari
Mayer, Ulrich, University of British Columbia
McCarter, Michael K., University of Utah
Meyer, Lewis, Exeter University
Miller, Hugh, Colorado School of Mines
Mitri, Hani, McGill University
Momayez, Moe, University of Arizona
Monteverde-Gutierrez, Gerardo, Universidad Estatal de Sonora
Mousset-Jones, Pierre, University of Nevada, Reno
Navarro, Ricardo, Univ Nac Autonoma de Mexico
Nelson, Michael G., University of Utah
Nieto Antunez, Antonio, Univ Nac Autonoma de Mexico

Novak, Thomas, University of Kentucky
Petr, Vilem, Colorado School of Mines
Pokorny, Eugene W., Los Alamos National Laboratory
Silva-Castro, Jhon, University of Kentucky
Slistan-Grijalva, Angel, Universidad Estatal de Sonora
Sottile, Jr., Joseph, University of Kentucky
Squelch, Andrew P., Curtin University
Taylor, Danny L., University of Nevada, Reno
Tenorio, Victor O., University of Arizona
Tharp, Thomas M., Purdue University
Vite-Picazo, Luis G., Universidad Estatal de Sonora
Wala, Andrew M., University of Kentucky
Wane, Malcolm T., Columbia University
Wedding, William C., University of Kentucky
Wetherelt, Andrew, Exeter University
Whyatt, Jeffrey, University of Utah
Yegulalp, Tuncel M., Columbia University
Zavodni, Zavis, University of Utah
Zhang, Jinhong, University of Arizona
Zipf, Karl, Colorado School of Mines

Petroleum Engineering
Șaramet, Mihai R., Alexandru Ioan Cuza
Ahmed, Ramadan, University of Oklahoma
Alvarado, Vladimir, University of Wyoming
Bijeljic, Branko, Imperial College
Blasingame, Tom, Texas A&M University
Blunt, Martin J., Imperial College
Callard, Jeff, University of Oklahoma
Civan, Faruk, University of Oklahoma
Cooper, George A., University of California, Berkeley
Corbett, Patrick, Heriot-Watt University
Couples, Gary, Heriot-Watt University
Devegowda, Deepak, University of Oklahoma
Dreesen, Donald S., Los Alamos National Laboratory
El-Monier, Ilham, University of Oklahoma
Fahes, Mashhad, University of Oklahoma
Fisher, Quentin, University of Leeds
Frailey, Scott M., Illinois State Geological Survey
Geiger, Sebastian, Heriot-Watt University
Gringarten, Alain, Imperial College
Holubnyak, Yehven I., University of Kansas
Jamili, Ahmad, University of Oklahoma
Javadpour, Farzam, University of Texas, Austin
Jin, Guohai, Geological Survey of Alabama
Kadkhodai Ilkhchi, Ali, University of Tabriz
Kelkar, Sharad, Los Alamos National Laboratory
King, Peter, Imperial College
Knapp, Roy M., University of Oklahoma
Lorinczi, Piroska, University of Leeds
Ma, Jingsheng, Heriot-Watt University
Misra, Siddharth, University of Oklahoma
Moghanloo, Rouzbeh, University of Oklahoma
Muggeridge, Ann H., Imperial College
Nikolinakou, Maria-Aikaterini, University of Texas, Austin
Oppong, Issac A., University of Ghana
Patzek, Tad W., University of California, Berkeley
Pickup, Gillian E., Heriot-Watt University
Pournik, Maysam, University of Oklahoma
Rai, Chandra S., University of Oklahoma
Rose, Peter E., University of Utah
Sakhaee-Pour, Ahmad, University of Oklahoma
Shah, Subhash N., University of Oklahoma
Sharma, Suresh, University of Oklahoma
Somerville, James M., Heriot-Watt University
Sondergeld, Carl H., University of Oklahoma
Teodoriu, Catalin, University of Oklahoma
Vesovic, Velisa, Imperial College
Wu, Xingru, University of Oklahoma
Zaman, Musharraf, University of Oklahoma

Rock Mechanics
Abousleiman, Younane, University of Oklahoma
Abu Bakar, Muhammad Z., University of Engineering and Technology
Agioutantis, Zach, University of Kentucky
Archambault, Guy, Universite du Quebec a Chicoutimi
Asbury, Brian, Colorado School of Mines
Aslan, Yasemin, Firat University
Barakat, Bassam, Institut Polytechnique LaSalle Beauvais (ex-IGAL)
Barzegari, Ghodrat, University of Tabriz
Blacic, James D., Los Alamos National Laboratory
Brahmi, Sadek, Institut Polytechnique LaSalle Beauvais (ex-IGAL)
Brake, Thomas L., Los Alamos National Laboratory
Brown, Donald W., Los Alamos National Laboratory
Chen, Rui, California Geological Survey
Cruden, David M., University of Alberta
Cundall, Peter A., University of Minnesota, Twin Cities
Daemen, Jaak, University of Nevada, Reno
Detournay, Emmanuel M., University of Minnesota, Twin Cities
Diederichs, Mark, Queen's University
Enderlin, Milton, Texas Christian University
Engelder, Terry, Pennsylvania State University, University Park
Ferrero, Anna Maria, Università di Torino
Germanovich, Leonid, Georgia Institute of Technology
Ghassemi, Ahmed, University of Oklahoma
Gordon, Robert B., Yale University
Gurocak, Zulfu, Firat University
Guzina, Bojan B., University of Minnesota, Twin Cities
Haimson, Bezalel C., University of Wisconsin, Madison
Hastie, Warwick W., University of KwaZulu-Natal
Kanik, Mustafa, Firat University
Kemeny, John M., University of Arizona
Kim, Eunhye, Colorado School of Mines
Kulatilake, Pinnaduwa H. S. W., University of Arizona
Käsling, Heiko, Technical University of Munich
Labuz, Joseph F., University of Minnesota, Twin Cities
MacLaughlin, Mary M., Montana Tech of the University of Montana
Mariani, Elisabeth, University of Liverpool
Meredith, Philip, University College London
Mojtabai, Navid, New Mexico Institute of Mining and Technology
Nelson, Priscilla P., Colorado School of Mines
Onyeobi, Tony U., University of Benin
Oravecz, Kalman I., New Mexico Institute of Mining and Technology
Ozbay, M. Ugur, Colorado School of Mines
Pariseau, William G., University of Utah
Pec, Matej, Massachusetts Institute of Technology
Perry, Kyle, University of Kentucky
Persson, Per-Anders, New Mexico Institute of Mining and Technology
Roegiers, Jean-Claude, University of Oklahoma
Rutter, Ernest H., University of Manchester
Saeidi, Ali, Universite du Quebec a Chicoutimi
Sinha, Krishna P., University of Utah
Smutek, Claude, Universite de la Reunion
Sone, Hiroki, University of Wisconsin, Madison
Stockinger, Georg M., Technical University of Munich
Swift, Robert P., Los Alamos National Laboratory
Unrug, Kot F., University of Kentucky
Vinciguerra, Sergio, Leicester University
Vinciguerra, Sergio Carmelo, Università di Torino
Watters, Robert J., University of Nevada, Reno
Wijesinghe, Ananda M., Lawrence Livermore National Laboratory
Zimmerman, Robert W., Imperial College

Ocean Engineering/Mining
Ajayi, J. O., Obafemi Awolowo University
Cambazoglu, Mustafa K., University of Southern Mississippi
Carpenter, Roy, University of Washington
Gibson, Carl H., University of California, San Diego
Hodgkiss, Jr., William S., University of California, San Diego
Hoopes, John A., University of Wisconsin, Madison
Irish, Jennifer, Virginia Polytechnic Institute & State University
Nootz, Gero, University of Southern Mississippi
Wen, Xianyun, University of Leeds

Geotechnical Engineering
Fernández, Luis Ramón, Univ Complutense de Madrid
Hingston, Egerton, University of KwaZulu-Natal
Insua, Juan Miguel, Univ Complutense de Madrid
Ortt, Jr., Richard A., Maryland Department of Natural Resources
Sanchez, Marcelo J., Texas A&M University

Geological Engineering
Garnier Villarreal, Maximiliano, Universidad de Costa Rica
Higgins, Jerry D., Colorado School of Mines
Katzenstein, Kurt W., South Dakota School of Mines & Technology
Melentijevic, Svetlana, Univ Complutense de Madrid
Potten, Martin, Technical University of Munich

Stetler, Larry D., South Dakota School of Mines & Technology

OCEANOGRAPHY
General Oceanography
Abernathy, Ryan, Columbia University
Allen, John, University of Portsmouth
Alves, Tiego, University of Wales
Barros, Tony, Miami-Dade College (Wolfson Campus)
Becker, Janet M., University of Hawai'i, Manoa
Bohaty, Steven, University of Southampton
Brandes, Jay, Georgia Institute of Technology
Brandon, Mark A., The Open University
Brown, Erik T., University of Minnesota, Duluth
Buchanan, Donald G., San Bernardino Valley College
Calder, Brian, University of New Hampshire
Chamberlin, William S., Fullerton College
Chandler, Cyndy, Woods Hole Oceanographic Institution
Ciannelli, Lorenzo, Oregon State University
Cooney, Michael, University of Hawai'i, Manoa
Curewitz, Daniel, Syracuse University
Deen, Patricia A., Palomar College
DeJesus, Roman, Fullerton College
Denton, George, University of South Florida
Dickens, Gerald R., Rice University
Dijkstra, Semme, University of New Hampshire
Dong, Charles, El Camino College
Falkowski, Paul G., University of Hawai'i, Manoa
Field, Richard T., University of Delaware
Fritsen, Christian H., University of Nevada, Reno
Froelich, Philip, Florida State University
Gomez, Natalya, McGill University
Gordon, Elizabeth S., Fitchburg State University
Grossman, Walter, San Bernardino Valley College
Hale, Michelle, University of Portsmouth
Hepner, Tiffany, Univ of Texas at Austin, Jackson Sch of Geosciences
Herdendorf, Charles E., Ohio State University
Hernes, Peter, University of California, Davis
Hickman, Anna, University of Southampton
Holland, Christina, University of Texas, Austin
Holliday, Joseph W., El Camino College
Horton, Radley, Columbia University
Hoyt, William H., University of Northern Colorado
Hughes-Clarke, John E., University of New Hampshire
Ivanovic, Ruza, University of Leeds
Johnson, Ashanti, University of Texas, Arlington
Johnson, Helen, University of Oxford
Judkins, Heather L., Saint Petersburg College, Clearwater
Jung, Simon, Edinburgh University
Keller, Klaus, Pennsylvania State University, University Park
Kerr, Andrew, University of Wales
Kim, Yong Hoon, West Chester University
King, Andrew, University of Illinois at Chicago
Kustka, Adam B., Rutgers, The State Univ of New Jersey, Newark
Laliberte, Elizabeth, University of Rhode Island
Lauderdale, Jonathan, University of Liverpool
Leach, Harry, University of Liverpool
Lear, Caroline, University of Wales
Ledbetter, Michael T., University of Arkansas at Little Rock
Lee, Alyce, Concord University
Lee, Dong, Columbia University
Lee, Stephen C., Los Angeles Pierce College
Leinen, Margaret, University of California, San Diego
Lenz, Petra H., University of Hawai'i, Manoa
Lindo Atachati, David, Grad Sch of the City Univ of New York
Liu, Zhanfei, University of Texas, Austin
Ma, Yanxia, Louisiana State University
MacLeod, Chris, University of Wales
Maxwell, Arthur E., University of Texas, Austin
McPhaden, Michael J., University of Washington
McWilliams, James, University of California, Los Angeles
Min, Dong-Ha, University of Texas, Austin
Montenegro, Alvaro, Ohio State University
Moore, Dennis, University of Hawai'i, Manoa
Mosher, David C., University of New Hampshire
Muir, William, San Bernardino Valley College
Myers, Paul, University of Alberta
Noyes, Jim, El Camino College
Peredo-Jaime, Jose I., Univ Autonoma de Baja California Sur
Pike, Jenny, University of Wales
Porter, Dwayne E., University of South Carolina
Quan, Tracy, Oklahoma State University
Radulski, Robert, Southern Connecticut State University
Rappe, Michael S., University of Hawai'i, Manoa
Roesler, Collin, Bowdoin College
Salisbury, Joseph, University of New Hampshire
Schlegel, Mary A., Millersville University
Schneider, Niklas, University of Hawai'i, Manoa
Schroeder, William W., Dauphin Island Sea Lab
Scott, Robert B., University of Texas, Austin
Seyfang, Gill, University of East Anglia
Spencer, Matthew, University of Liverpool
Steger, John M., Miami Dade College (Kendall Campus)
Sullivan-Watts, Barbara K., University of Rhode Island
Swanson, R. L., SUNY, Stony Brook
Thomas, Megan D., United States Naval Academy
Thompson, LuAnne, University of Washington
Thomson, Richard, University of Victoria
Thurman, Harold V., Mt. San Antonio College
Trujillo, Alan P., Palomar College
Tyrrell, Toby, University of Southampton
Ulanski, Stanley L., James Madison University
Veeramony, Jayaram, Mississippi State University
Venn, Cynthia, Bloomsburg University
Venti, Nicholas, University of Massachusetts, Amherst
Walker, Nan D., Louisiana State University
Wallace, William G., Grad Sch of the City Univ of New York
Wang, Lei, Columbia University
Westervelt, Daniel, Columbia University
Wheatcroft, Rob, Oregon State University
Wiberg, Patricia L., University of Virginia
Wiesenburg, Denis A., University of Southern Mississippi
Williams, Harris, North Carolina Central University
Williams, Ric, University of Liverpool
Wiltshire, John C., University of Hawai'i, Manoa
Wolcott, Ray, Cuyamaca College
Wolff, George, University of Liverpool
Wu, Yutian, Columbia University
Yeager, Lauren A., University of Texas, Austin
Yin, Jianjun, University of Arizona
Yon, Lisa, Palomar College
Zhai, Xiaoming, University of East Anglia

Biological Oceanography
Adams, Marshall, Oak Ridge National Laboratory
Adams, Michael S., University of Wisconsin, Madison
Algar, Christopher, Dalhousie University
Allam, Bassem, SUNY, Stony Brook
Allen, Eric E., University of California, San Diego
Aller, Josephine Y., SUNY, Stony Brook
Anderson, George C., University of Washington
Apprill, Amy, Woods Hole Oceanographic Institution
Armbrust, Virginia E., University of Washington
Atkinson, Marlin J., University of Hawai'i, Manoa
Audet, Celine, Universite du Quebec a Rimouski
Auster, Peter, University of Connecticut
Azam, Farooq, University of California, San Diego
Baltz, Donald M., Louisiana State University
Banse, Karl, University of Washington
Barber, Bruce, University of South Florida
Barber, Richard T., Duke University
Barlow, Jay P., University of California, San Diego
Baross, John A., University of Washington
Bartlett, Douglas H., University of California, San Diego
Baumann, Hannes, University of Connecticut
Baumann, Zofia, University of Connecticut
Baumgartner, Mark, Dalhousie University
Benfield, Mark C., Louisiana State University
Benner, Ron, University of South Carolina
Bibby, Thomas, University of Southampton
Bienfang, Paul K., University of Hawai'i, Manoa
Biggs, Douglas C., Texas A&M University
Blake, Norman J., University of South Florida
Bloomfield, Helen, University of Liverpool
Boar, Rosalind, University of East Anglia

Bochdansky, Alexander, Old Dominion University
Bollens, Stephen, Washington State University
Boudrias, Michel A., University of San Diego
Bowen, Jennifer, University of Massachusetts, Boston
Bowser, Paul, SUNY, Stony Brook
Breitbart, Mya, University of South Florida
Brethes, Jean-Claude, Universite du Quebec a Rimouski
Briggs, John C., University of South Florida
Brownlee, Colin, University of Southampton
Bruno, Barbara, University of Hawai'i, Manoa
Bucklin, Ann, University of Connecticut
Burton, Ronald S., University of California, San Diego
Buskey, Edward J., University of Texas, Austin
Campbell, Lisa, Texas A&M University
Carlile, Amy L., University of New Haven
Carney, Robert S., Louisiana State University
Carpenter, Steve, University of Wisconsin, Madison
Carr, Mark, University of California, Santa Cruz
Case, James M., Wilkes University
Caswell, Bryony, University of Liverpool
Cerrato, Robert M., SUNY, Stony Brook
Checkley, David M., University of California, San Diego
Christian, James R., University of Victoria
Church, Matthew, University of Hawai'i, Manoa
Collie, Jeremy S., University of Rhode Island
Collier, Jackie, SUNY, Stony Brook
Collins, Kenneth, University of Southampton
Conover, David O., SUNY, Stony Brook
Copley, Jonathan, University of Southampton
Coull, Bruce, University of South Carolina
Cranford, Peter, Dalhousie University
Croll, Don, University of California, Santa Cruz
Cuba, Thomas, University of South Florida
Cuker, Benjamin E., Hampton University
Cullen, John J., Dalhousie University
D'Angelo, Maria C., University of Southampton
D'Elia, Christopher, University of South Florida
Daly, Kendra L., University of South Florida
Dam, Hans G., University of Connecticut
Davies, Andrew J., Bangor University
Dayton, Paul K., University of California, San Diego
Dean, John Mark, University of South Carolina
Demers, Serge, Universite du Quebec a Rimouski
Deming, Jody W., University of Washington
Denton, Gary R. W., University of Guam
DeWitt, Calvin B., University of Wisconsin, Madison
DiBacco, Claudio, Dalhousie University
Dieterle, Dwight A., University of South Florida
Dobbs, Fred C., Old Dominion University
Donaghay, Percy, University of Rhode Island
Doty, Thomas, Roger Williams University
Dower, John, University of Victoria
Drazen, Jeffrey C., University of Hawai'i, Manoa
Ducklow, Hugh, Columbia University
Duffy, Tara , Northeastern University
Dunton, Kenneth H., University of Texas, Austin
Durbin, Edward G., University of Rhode Island
Dyhrman, Sonya, Columbia University
Edgcomb, Virginia, Woods Hole Oceanographic Institution
Eggleston, David B., North Carolina State University
Epp, Leonard G., Mount Union College
Erdner, Deana L., University of Texas, Austin
Erisman, Brad, University of Texas, Austin
Esbaugh, Andrew J., University of Texas, Austin
Falkowski, Paul G., Rutgers, The State Univ of New Jersey
Fast, Mark, SUNY, Stony Brook
Felbeck, Horst, University of California, San Diego
Feller, Robert, University of South Carolina
Fenberg, Phillip, University of Southampton
Fennel, Katja, Dalhousie University
Fisher, Jr., Thomas R., University of Maryland
Fisk, Aaron, University of Windsor
Fitzer, Susan, University of Glasgow
Fletcher, Madilyn, University of South Carolina
Fournier, Robert O., Dalhousie University
Francis, Robert, University of Washington
Frank, Kenneth, Dalhousie University
Franks, Peter J. S., University of California, San Diego
Frost, Bruce W., University of Washington
Fuiman, Lee A., University of Texas, Austin
Gaasterland, Terry, University of California, San Diego
Gallagher, Eugene, University of Massachusetts, Boston
Ganssen, Gerald M., Vrije Universiteit Amsterdam
Gardner, Wayne S., University of Texas, Austin
Gates, Ruth D., University of Hawai'i, Manoa
Gibson, Deidre M., Hampton University
Gifford, Dian J., University of Rhode Island
Gilbert, Patricia M., University of Maryland
Gobler, Christopher, SUNY, Stony Brook
Godbold, Jasmin A., University of Southampton
Goetze, Erica, University of Hawai'i, Manoa
Gomes, Helga, Columbia University
Gosselin, Michel, Universite du Quebec a Rimouski
Gould, Mark D., Roger Williams University
Graham, Linda K., University of Wisconsin, Madison
Graham, Michael, Moss Landing Marine Laboratories
Graham, William (., University of Southern Mississippi
Grant, Jonathan, Dalhousie University
Green, Jonathan, University of Liverpool
Greene, Charles H., Cornell University
Greenhow, Danielle, University of Southern Mississippi
Greer, Adam, University of Southern Mississippi
Grigg, Richard W., University of Hawai'i, Manoa
Grottoli, Andrea G., Ohio State University
Grunbaum, Daniel, University of Washington
Hargraves, Paul E., University of Rhode Island
Hargreaves, Bruce R., Lehigh University
Harvey, James T., Moss Landing Marine Laboratories
Hastings, Philip A., University of California, San Diego
Hauton, Chris, University of Southampton
Hawkins, Lawrence, University of Southampton
Healey, Michael, University of British Columbia
Heil, Cynthia, University of South Florida
Hessler, Robert R., University of California, San Diego
Hicks, David, University of Texas, Rio Grande Valley
Holland, Nicholas D., University of California, San Diego
Holt, Gloria J., University of Texas, Austin
Hood, Raleigh, University of Maryland
Hopkins, Thomas S., Dauphin Island Sea Lab
Hopkins, Thomas L., University of South Florida
Hotchkiss, Sarah, University of Wisconsin, Madison
Humm, Harold J., University of South Florida
Hunsucker, Kelli, Florida Institute of Technology
Hunt, Brian, University of British Columbia
Jeffries, H. Perry, University of Rhode Island
Jensen, Antony, University of Southampton
Jickells, Tim, University of East Anglia
Johnson, Kevin, Florida Institute of Technology
Jordan, Robert A., Hampton University
Juhl, Andrew, Columbia University
Jumars, Peter A., University of Washington
Juniper, Kim, University of Victoria
Kamykowski, Daniel, North Carolina State University
Kana, Todd M., University of Maryland
Karl, David M., University of Hawai'i, Manoa
Kaufmann, Ronald S., University of San Diego
Kelley, Christopher, University of Hawai'i, Manoa
Kelly, John, University of New Haven
Kemp, Paul, University of Hawai'i, Manoa
Kenney, Robert D., University of Rhode Island
Kimball, Matthew, University of South Carolina
Kirschenfeld, Taylor, University of West Florida
Kitchell, James F., University of Wisconsin, Madison
Kooyman, Gerald L., University of California, San Diego
Kraemer, George P., SUNY, Purchase
Krieger-Brockett, Barbara B., University of Washington
Lam, Phyllis, University of Southampton
Landry, Michael R., University of California, San Diego
LaRock, Paul A., Louisiana State University
Lee, Carol, University of Wisconsin, Madison
Leichter, James J., University of California, San Diego
Lessard, Evelyn J., University of Washington
Letelier, Ricardo, Oregon State University
Levin, Lisa A., University of California, San Diego

Faculty by Specialty

Levinton, Jeffrey, SUNY, Stony Brook
Lewin, Joyce C., University of Washington
Lewis, Alan G., University of British Columbia
Lewis, Marlon R., Dalhousie University
Li, William K., Dalhousie University
Lilley, Marvin D., University of Washington
Lin, Senjie, University of Connecticut
Logan, Alan, University of New Brunswick Saint John
Lohan, Maeve, University of Southampton
Longoria, Lorena, University of Texas, Rio Grande Valley
Lonsdale, Darcy J., SUNY, Stony Brook
Lopez, Glenn R., SUNY, Stony Brook
Lucas, Cathy, University of Southampton
MacIntyre, Hugh, Dalhousie University
Mackas, David L., University of Victoria
Maldonado, Maria, University of British Columbia
Malin, Gill, University of East Anglia
Marra, John, Brooklyn College (CUNY)
Martiny, Adam, University of California, Irvine
Matassa, Catherine, University of Connecticut
McCarthy, James T., Harvard University
McElroy, Anne E., SUNY, Stony Brook
McGlathery, Karen J., University of Virginia
McGlathery, Karen, University of Virginia
McGowan, John A., University of California, San Diego
McManus, George B., University of Connecticut
Mendelssohn, Irving A., Louisiana State University
Metaxas, Anna, Dalhousie University
Meylan, Anne, University of South Florida
Miller, Douglas C., University of Delaware
Mills, Eric L., Dalhousie University
Milroy, Scott P., University of Southern Mississippi
Mobley, Curtis D., University of Washington
Monger, Bruce, Cornell University
Montagna, Paul, Texas A&M University, Corpus Christi
Moore, Christopher M., University of Southampton
Moran, Dawn, Woods Hole Oceanographic Institution
Mueller, Erich M., University of South Alabama
Mulholland, Margaret, Old Dominion University
Munch, Stephan, SUNY, Stony Brook
Murray, Alison, University of Nevada, Reno
Murtugudde, Raghuram G., University of Maryland
Napora, Theodore A., University of Rhode Island
Napp, Jeffrey, University of Washington
Newman, William A., University of California, San Diego
Newton, Jan, University of Washington
Nixon, Scott W., University of Rhode Island
Nosal, Andrew, University of San Diego
Nye, Janet, SUNY, Stony Brook
Ohman, Mark D., University of California, San Diego
Oliver, John S., Moss Landing Marine Laboratories
Oliver, Matthew J., University of Delaware
Oschlies, Andreas, Dalhousie University
Oviatt, Candace, University of Rhode Island
Padilla, Dianna K., SUNY, Stony Brook
Pakhomov, Evgeny, University of British Columbia
Palenik, Brian, University of California, San Diego
Palevsky, Hilary, Wellesley College
Paul, John H., University of South Florida
Peebles, Ernst B., University of South Florida
Pellerin, Jocelyne, Universite du Quebec a Rimouski
Perry, Mary Jane, University of Washington
Peterson, Bradley, SUNY, Stony Brook
Pikitch, Ellen K., SUNY, Stony Brook
Pinckney, James, University of South Carolina
Plotkin, Pamela, Texas A&M University
Pollack, Jennifer, Texas A&M University, Corpus Christi
Potts, Donald C., University of California, Santa Cruz
Powers, Sean, University of South Alabama
Prairie, Jennifer, University of San Diego
Purdie, Duncan A., University of Southampton
Quattro, Joseph, University of South Carolina
Rabalais, Nancy N., Louisiana State University
Ragotzkie, Robert A., University of Wisconsin, Madison
Ray, G. Carleton, University of Virginia
Redalje, Donald G., University of Southern Mississippi
Reese, Brandi K., Texas A&M University, Corpus Christi

Resplandy, Laure, Princeton University
Richardson, Tammi, University of South Carolina
Robinson, Carol, University of East Anglia
Robinson, Leonie, University of Liverpool
Rocap, Gabrielle L., University of Washington
Roman, Charles T., University of Rhode Island
Roughgarden, Joan, Stanford University
Rouse, Gregory W., University of California, San Diego
Rowe, Gilbert T., Texas A&M University
Roy, Suzanne, Universite du Quebec a Rimouski
Ruttan, Lore, Emory University
Rykaczewski, Ryan R., University of South Carolina
Saila, Saul B., University of Rhode Island
Sambrotto, Raymond N., Columbia University
Sautter, Leslie R., College of Charleston
Schnetzer, Astrid, North Carolina State University
Schulze, Anja, Texas A&M University
Scott, Tim, Roger Williams University
Searcy, Steven, University of San Diego
Selph, Karen E., University of Hawai'i, Manoa
Shaw, Richard F., Louisiana State University
Shuman, Randy, University of Washington
Shumway, Sandra, University of Connecticut
Sibert, John R., University of Hawai'i, Manoa
Sieburth, John M., University of Rhode Island
Silver, Mary W., University of California, Santa Cruz
Sims, David, University of Southampton
Smalley, Gabriela W., Rider University
Smayda, Theodore J., University of Rhode Island
Smith, Craig R., University of Hawai'i, Manoa
Smith, David C., University of Rhode Island
Smith, David E., University of Virginia
Smith, Jennifer E., University of California, San Diego
Solan, Martin, University of Southampton
Sommer, Ulrich, Dalhousie University
Sousa, Wayne P., University of California, Berkeley
Specker, Jennifer, University of Rhode Island
Stakes, Debra S., Stanford University
Stanley, Emily, University of Wisconsin, Madison
Steppe, Cecily N., United States Naval Academy
Steward, Grieg F., University of Hawai'i, Manoa
Stickney, Robert R., Texas A&M University
Stoecker, Diane, University of Maryland
Stoner, Betsy, Bentley University
Strickland, Richard M., University of Washington
Subramaniam, Ajit, Columbia University
Sugihara, George, University of California, San Diego
Suttle, Curtis, University of British Columbia
Swain, Geoffrey, Florida Institute of Technology
Swift, Elijah V., University of Rhode Island
Sylvan, Jason B., Texas A&M University
Taggart, Christopher T., Dalhousie University
Talley, Drew, University of San Diego
Tan, Wenxian, Austin Community College District
Taylor, Frank J. R., University of British Columbia
Taylor, Gordon T., SUNY, Stony Brook
Thatje, Sven, University of Southampton
Thomas, Florence, University of Hawai'i, Manoa
Thomas, Peter, University of Texas, Austin
Thompson, Diane, Boston University
Thomsen, Laurenz A., University of Washington
Thomson, Jack, University of Liverpool
Thornton, Daniel C., Texas A&M University
Tlusty, Michael, University of Massachusetts, Boston
Toonen, Robert, University of Hawai'i, Manoa
Torres, Joseph J., University of South Florida
Tortell, Philippe, University of British Columbia
Treude, Tina, University of California, Los Angeles
Tunnicliffe, Verena, University of Victoria
Turner, Robert E., Louisiana State University
Tynan, Cynthia T., University of Washington
Urban-Rich, Juanita, University of Massachusetts, Boston
Vacquier, Victor D., University of California, San Diego
Vaillancourt, Robert, Millersville University
Van Oostende, Nicolas, Princeton University
Varela, Diana, University of Victoria
Vargo, Gabriel A., University of South Florida

Vaudrey, Jamie, University of Connecticut
Villalard-Bohnsack, Martine, Roger Williams University
Villareal, Tracy A., University of Texas, Austin
Waldbusser, George, Oregon State University
Waldman, John, Grad Sch of the City Univ of New York
Walsh, John J., University of South Florida
Ward, Ben A., University of Southampton
Ward, Bess B., Princeton University
Ward, J. Evan, University of Connecticut
Warren, Joseph, SUNY, Stony Brook
Wells, Randy, University of South Florida
Welschmeyer, Nicholas A., Moss Landing Marine Laboratories
Weng, Kevin, University of Hawai'i, Manoa
Wetz, Michael, Texas A&M University, Corpus Christi
Wiedenmann, Jorg, University of Southampton
Wilson, Merwether, Edinburgh University
Wishner, Karen, University of Rhode Island
Wolcott, Donna L., North Carolina State University
Wolcott, Thomas G., North Carolina State University
Woodin, Sarah A., University of South Carolina
Yamanaka, Tsuyuko, University of Liverpool
Yandle, Tracy, Emory University
Young, Richard E., University of Hawai'i, Manoa
Zakardjian, Bruno, Universite du Quebec a Rimouski
Zedler, Joy, University of Wisconsin, Madison
Zehr, Jonathan P., University of California, Santa Cruz
Zhang, Huan, University of Connecticut

Chemical Oceanography
Allison, Nicky, University of St. Andrews
Andersen, Raymond J., University of British Columbia
Anderson, Robert F., Columbia University
Andersson, Andreas, University of California, San Diego
Andren, Anders W., University of Wisconsin, Madison
Andrews, Allen H., University of Hawai'i, Manoa
Annett, Amber, University of Southampton
Armstrong, David E., University of Wisconsin, Madison
Armstrong, Robert A., SUNY, Stony Brook
Azetsu-Scott, Kumiko, Geological Survey of Canada
Bacon, Michael P., Woods Hole Oceanographic Institution
Balistrieri, Laurie, University of Washington
Barbeau, Katherine A., University of California, San Diego
Bates, Nicholas R., University of Southampton
Benitez-Nelson, Claudia R., University of South Carolina
Benway, Heather, Woods Hole Oceanographic Institution
Bishop, James K., University of California, Berkeley
Boudreau, Bernard P., Dalhousie University
Bourbonnais, Annie, University of South Carolina
Boyle, Edward A., Massachusetts Institute of Technology
Brownawell, Bruce J., SUNY, Stony Brook
Bruland, Kenneth W., University of California, Santa Cruz
Buchwald, Carolyn, Dalhousie University
Buckley, Lawrence J., University of Rhode Island
Buesseler, Ken O., Woods Hole Oceanographic Institution
Bullister, John, University of Washington
Burdige, David J., Old Dominion University
Canuel, Elizabeth A., College of William & Mary
Chan, Kwan M., California State University, Long Beach
Chapman, Piers, Texas A&M University
Chappell, P. D., Old Dominion University
Cherrier, Jennifer, Brooklyn College (CUNY)
Cochran, J. K., Stony Brook University
Cochran, J. Kirk, SUNY, Stony Brook
Coffin, Richard, Texas A&M University, Corpus Christi
Cooper, Matthew, University of Southampton
Cornwell, Jeffery C., University of Maryland
Crutzen, Paul, University of California, San Diego
Culberson, Charles H., University of Delaware
Cullen, Jay, University of Victoria
Cutter, Gregory A., Old Dominion University
deAngelis, Marie, SUNY, Maritime College
Delaney, Margaret L., University of California, Santa Cruz
DeMaster, David J., North Carolina State University
Devol, Allan H., University of Washington
Doney, Scott, University of Virginia
Druffel, Ellen R. M., University of California, Irvine
Drysdale, Jessica, Woods Hole Oceanographic Institution
Duce, Robert A., Texas A&M University

Emerson, Steven R., University of Washington
Emile-Geay, Julien, University of Southern California
Engel, Anga, SUNY, Stony Brook
Estapa, Margaret, Skidmore College
Fanning, Kent A., University of South Florida
Feely, Richard A., University of Washington
Fenical, William H., University of California, San Diego
Fisher, Nicholas S., SUNY, Stony Brook
Fitzgerald, William F., University of Connecticut
Fitzsimmons, Jessica N., Texas A&M University
Fogel, Marilyn L., University of Delaware
Fox, Austin, Florida Institute of Technology
Fox, Lewis, Broward College
Fredricks, Helen, Woods Hole Oceanographic Institution
Gagne, Jean-Pierre, Universite du Quebec a Rimouski
Gammon, Richard H., University of Washington
Garcia-Rubio, Luis H., University of South Florida
Gieskes, Joris M., University of California, San Diego
Glover, David M., Woods Hole Oceanographic Institution
Gold-Bouchot, Gerardo, Texas A&M University
Gosselin, Kelsey, Woods Hole Oceanographic Institution
Granger, Julie, University of Connecticut
Hales, Burke R., Oregon State University
Hamme, Roberta C., University of Victoria
Hanson, Jr., Alfred K., University of Rhode Island
Hayes, Christopher T., University of Southern Mississippi
Haymet, Anthony D., University of California, San Diego
Ho, David, University of Hawai'i, Manoa
Hollander, David J., University of South Florida
Hong, Gi-Hoon, Wayne State University
Hu, Xinping, Texas A&M University, Corpus Christi
Huebert, Barry J., University of Hawai'i, Manoa
Hughes, Chambers, University of California, San Diego
Ingalls, Anitra E., University of Washington
James, Rachael, University of Southampton
Johnson, Bruce D., Dalhousie University
Keil, Richard G., University of Washington
Kiene, Ronald P., University of South Alabama
Knap, Anthony, Texas A&M University
Kujawinski, Elizabeth B., Woods Hole Oceanographic Institution
Lang, Susan Q., University of South Carolina
LaVigne, Michéle, Bowdoin College
Lee, Cindy, SUNY, Stony Brook
Letscher, Robert T., University of New Hampshire
Lima, Ivan D., Woods Hole Oceanographic Institution
Liss, Peter S., University of East Anglia
Loder, Theodore C., University of New Hampshire
Longnecker, Krista, Woods Hole Oceanographic Institution
Lott, Dempsey E., Woods Hole Oceanographic Institution
Lucotte, Marc Michel, Universite du Quebec a Montreal
Lund, David, University of Connecticut
Lwiza, Kamazima M., SUNY, Stony Brook
Mahaffey, Claire, University of Liverpool
Maloney, Ashley, Princeton University
Marchitto, Thomas M., University of Colorado
Martin, William R., Woods Hole Oceanographic Institution
Mason, Robert P., University of Connecticut
Matsumoto, Katsumi, University of Minnesota, Twin Cities
McDuff, Russell E., University of Washington
McIlvin, Matt, Woods Hole Oceanographic Institution
McKinley, Galen, University of Wisconsin, Madison
McNichol, Ann P., Woods Hole Oceanographic Institution
Measures, Christopher, University of Hawai'i, Manoa
Merrill, John T., University of Rhode Island
Moffett, James W., University of Southern California
Moore, Robert M., Dalhousie University
Moran, S. Bradley, University of Rhode Island
Murray, James W., University of Washington
Najjar, Raymond G., Pennsylvania State University, University Park
Nuzzio, Donald B., University of Delaware
Paytan, Adina, Stanford University
Pellenbarg, Robert, College of the Desert
Pelletier, Emilien, Universite du Quebec a Rimouski
Pennock, Jonathan R., Dauphin Island Sea Lab
Pilson, Michael E., University of Rhode Island
Precedo, Laura, Broward College
Presley, Bobby J., Texas A&M University

Pyrtle, Ashanti J., University of South Florida
Quay, Paul D., University of Washington
Quinn, James G., University of Rhode Island
Rahn, Kenneth A., University of Rhode Island
Ravelo, Ana C., University of California, Santa Cruz
Reimer, Andreas, Georg-August University of Goettingen
Reimers, Clare, Oregon State University
Resing, Joseph A., University of Washington
Rheuban, Jennie, Woods Hole Oceanographic Institution
Roethel, Frank J., SUNY, Stony Brook
Russell, Joellen, University of Arizona
Sabine, Christopher L., University of Washington
Saito, Mak A., Woods Hole Oceanographic Institution
Salaun, Rachel, University of Liverpool
Sandwith, Zoe, Woods Hole Oceanographic Institution
Santschi, Peter H., Texas A&M University
Sarmiento, Jorge L., Princeton University
Sayles, Frederick L., Woods Hole Oceanographic Institution
Schwartz, Matthew C., University of West Florida
Scranton, Mary I., SUNY, Stony Brook
Shamberger, Kathryn E., Texas A&M University
Sharp, Jonathan H., University of Delaware
Sharples, Jonathan, University of Liverpool
Shiller, Alan M., University of Southern Mississippi
Sholkovitz, Edward R., Woods Hole Oceanographic Institution
Simjouw, Jean-Paul, University of New Haven
Skoog, Annelie, University of Connecticut
Sonzogni, William C., University of Wisconsin, Madison
Soule, Melissa, Woods Hole Oceanographic Institution
Stewart, Gillian, Queens College (CUNY)
Stewart, Gillian, Grad Sch of the City Univ of New York
Sundby, Bjorn, Universite du Quebec a Rimouski
Suntharalingam, Parvadha, University of East Anglia
Swarr, Gretchen, Woods Hole Oceanographic Institution
Sylva, Sean, Woods Hole Oceanographic Institution
Tagliabue, Alessandro, University of Liverpool
Takahashi, Taro, Columbia University
Tanhua, Toste, Dalhousie University
Thomas, Helmuth, Dalhousie University
Thompson, Anu, University of Liverpool
Timmermans, Mary-Louise, Yale University
Tobias, Craig, University of Connecticut
Ullman, William J., University of Delaware
Van Mooy, Benjamin, Woods Hole Oceanographic Institution
Velinsky, David, Drexel University
Veron, Alain J., University of Delaware
Vlahos, Penny, University of Connecticut
Wade, Terry L., Texas A&M University
Wallace, Douglas, Dalhousie University
Wallace, Douglas W. R., SUNY, Stony Brook
Williams, Dave E., University of British Columbia
Windsor, Jr., John G., Florida Institute of Technology
Yao, Wensheng, University of South Florida
Yvon-Lewis, Shari A., Texas A&M University
Zafiriou, Oliver C., Woods Hole Oceanographic Institution
Zhu, Qingzhi, SUNY, Stony Brook

Geological Oceanography

Agardy, Rachael, Central Michigan University
Aiello, Ivano, Moss Landing Marine Laboratories
Alexander, Clark R., Georgia Southern University
Alt, Jeffrey C., University of Michigan
Arthur, Michael A., Pennsylvania State University, University Park
Bain, Olivier, Institut Polytechnique LaSalle Beauvais (ex-IGAL)
Baker, Edward T., University of Washington
Baldauf, Jack G., Texas A&M University
Berquist, Jr., Carl R., College of William & Mary
Billups, Katharina, University of Delaware
Black, David, SUNY, Stony Brook
Busch, William H., University of New Orleans
Bush, David M., University of West Georgia
Byrne, Timothy, University of Connecticut
Carey, Steven N., University of Rhode Island
Clemens, Steven C., Brown University
Coffin, Richard, Texas A&M University, Corpus Christi
Cook, Mea S., Williams College
Creager, Joe S., University of Washington
D'Hondt, Steven L., University of Rhode Island
Darby, Dennis A., Old Dominion University
Dellapenna, Timothy M., Texas A&M University
Diercks, Arne R., University of Southern Mississippi
Dudley, Walter, University of Hawai'i, Manoa
Emerick, Christina M., University of Washington
Farmer, Emma C., Hofstra University
Farrell, John, University of Rhode Island
Ferrini, Victoria, Columbia University
Finney, Bruce P., University of Alaska, Fairbanks
Flood, Roger D., SUNY, Stony Brook
Gagnon, Katie, Seattle Central Community College
Gardner, Wilford D., Texas A&M University
Hallock-Muller, Pamela, University of South Florida
Harris, Sara, University of British Columbia
Hawkins, James W., University of California, San Diego
Hayman, Nicholas W., University of Texas, Austin
Hein, Christopher J., College of William & Mary
Holcomb, Robin T., University of Washington
Holcombe, Troy L., Texas A&M University
Holmes, Mark L., University of Washington
Honjo, Susumu, Woods Hole Oceanographic Institution
Hovan, Steven A., Indiana University of Pennsylvania
Kelley, Deborah S., University of Washington
Kincaid, Christopher, University of Rhode Island
King, John, University of Rhode Island
Kominz, Michelle A., Western Michigan University
Kuehl, Steven A., College of William & Mary
Larson, Roger, University of Rhode Island
Lear, C, Cardiff University
Leinen, Margaret, University of Rhode Island
Leventer, Amy, Colgate University
Locker, Stanley D., University of South Florida
Longworth, Brett, Woods Hole Oceanographic Institution
Manganini, Steven J., Woods Hole Oceanographic Institution
Martinez-Noriega, Cesar, Univ Autonoma de Baja California Sur
McHugh, Cecilia M., Grad Sch of the City Univ of New York
McHugh, Cecilia, Queens College (CUNY)
McManus, Dean A., University of Washington
Moore, Theodore C., University of Michigan
Mosher, David C., Dalhousie University
Mottl, Michael J., University of Hawai'i, Manoa
Murray, David, Brown University
Muza, Jay P., Broward College
Naar, David F., University of South Florida
Nittrouer, Charles A., University of Washington
Nowell, Arthur R. M., University of Washington
Ogston, Andrea S., University of Washington
Oostdam, Bernard L., Millersville University
Parsons, Jeffrey D., University of Washington
Pilkey, Jr., Orrin H., Duke University
Pinet, Paul, Colgate University
Prell, Warren L., Brown University
Quinn, Terrence, University of South Florida
Riesselman, Christina R., University of Otago
Sager, William W., Texas A&M University
Schilling, Jean-Guy E., University of Rhode Island
Schmidt, Matthew, Old Dominion University
Shen, Yang, University of Rhode Island
Sigurdsson, Haraldur, University of Rhode Island
Slowey, Niall C., Texas A&M University
Smith, Stephen V., University of Hawai'i, Manoa
Snoeckx, Hilde, University of West Florida
Sommerfield, Christopher K., University of Delaware
St. John, Kristen E., James Madison University
Stanley, Daniel J., Smithsonian Inst / Natl Museum of Natural History
Sternberg, Richard W., University of Washington
Thomas, Debbie, Texas A&M University
Trembanis, Art, University of Delaware
Tyce, Robert, University of Rhode Island
Uchupi, Elazar, Woods Hole Oceanographic Institution
Wallace, Davin, University of Southern Mississippi
Whitman, Jill M., Pacific Lutheran University
Wilcock, William S., University of Washington
Zarillo, Gary, Florida Institute of Technology
Zeebe, Richard E., University of Hawai'i, Manoa

Physical Oceanography

Aagaard, Knut, University of Washington

Abernathey, Ryan P., Columbia University
Alford, Matthew H., University of Washington
Allen, Susan E., University of British Columbia
Andrefouet, Serge, University of South Florida
Anis, Ayal, Texas A&M University
Armi, Laurence, University of California, San Diego
Arnone, Robert (Bob), University of Southern Mississippi
Atkinson, Larry P., Old Dominion University
Barclay, David R., Dalhousie University
Baum, Steven K., Texas A&M University
Beaulieu, Claudie, University of Southampton
Bishop, Stuart P., North Carolina State University
Bogucki, Darek, Texas A&M University, Corpus Christi
Bohlen, Walter F., University of Connecticut
Boicourt, William, University of Maryland
Bordoni, Simona, California Institute of Technology
Bostater, Charles, Florida Institute of Technology
Bourke, Robert H., Naval Postgraduate School
Bowman, Malcolm J., SUNY, Stony Brook
Boxall, Simon R., University of Southampton
Boyd, John, University of Michigan
Bracco, Annalisa, Georgia Institute of Technology
Breaker, Laurence, Moss Landing Marine Laboratories
Brooks, David A., Texas A&M University
Buckingham, Michael J., University of California, San Diego
Buijsman, Maarten, University of Southern Mississippi
Bulusu, Subrahmanyam, University of South Carolina
Busecke, Julius, Princeton University
Cane, Mark A., Columbia University
Cannon, Glenn A., University of Washington
Carder, Kendall L., University of South Florida
Carmack, Edward, University of British Columbia
Carter, Glenn, University of Hawai'i, Manoa
Carton, James, University of Maryland
Cessi, Paola, University of California, San Diego
Chang, Ping, Texas A&M University
Chao, Shenn-Yu, University of Maryland
Chapman, Ross N., University of Victoria
Chen, Dake, Columbia University
Chiu, Ching-Sang, Naval Postgraduate School
Chu, Peter C., Naval Postgraduate School
Cole, Rick, University of South Florida
Collins, Curtis A., Naval Postgraduate School
Cornillon, Peter, University of Rhode Island
Craig, Susanne , Dalhousie University
Crease, James, University of Delaware
Criminale, Jr., William O., University of Washington
Cronin, Meghan F., University of Washington
D'Asaro, Eric A., University of Washington
Davis, Leslie M., Austin Community College District
de Szoeke, Simon P., Oregon State University
Denman, Kenneth L., University of Victoria
Dever, Edward P., Oregon State University
Dewey, Richard K., University of Victoria
Di Lorenzo, Emanuele, Georgia Institute of Technology
Dierssen, Heidi, University of Connecticut
DiMarco, Steven F., Texas A&M University
Donohue, Kathleen, University of Rhode Island
Donovan, Jeff C., University of South Florida
Dorman, Clive E., San Diego State University
Dosso, Stanley E., University of Victoria
Drago, Aldo, University of Malta
Drijfhout, Sybren, University of Southampton
Dushaw, Brian D., University of Washington
Dutrieux, Pierre, Columbia University
Edson, James, University of Connecticut
Ellis, Dale, Dalhousie University
English, David C., University of South Florida
Eriksen, Charles C., University of Washington
Ewart, Terry E., University of Washington
Fedorov, Alexey V., Yale University
Ferrari, Raffaele, Massachusetts Institute of Technology
Firing, Eric, University of Hawai'i, Manoa
Flagg, Charles, SUNY, Stony Brook
Flament, Pierre J., University of Hawai'i, Manoa
Flierl, Glenn R., Massachusetts Institute of Technology
Flynn, Russell L., Cypress College

Follows, Michael, Massachusetts Institute of Technology
Foreman, Michael G., University of Victoria
Foreman, Michael G. G., University of British Columbia
Fox-Kemper, Baylor, Brown University
Fram, Jonathan, Oregon State University
Galea, Anthony, University of Malta
Galperin, Boris, University of South Florida
Garfield, Newell (Toby), San Francisco State University
Garrett, Christopher J. R., University of Victoria
Garwood, Roland W., Naval Postgraduate School
Gauci, Adam, University of Malta
Gerbi, Greg, Skidmore College
Giese, Benjamin S., Texas A&M University
Gille, Sarah T., University of California, San Diego
Ginis, Isaac, University of Rhode Island
Gnanadesikan, Anand, Johns Hopkins University
Gong, Donglai, College of William & Mary
Gonzalez, Frank I., University of Washington
Gordon, Arnold L., Columbia University
Gordon, Arnold L., Columbia University
Gratton, Yves, Universite du Quebec
Greatbatch, Richard , Dalhousie University
Greenberg, David, Dalhousie University
Gregg, Michael C., University of Washington
Guinasso, Norman L., Texas A&M University
Haine, Thomas W., Johns Hopkins University
Hara, Tetsu, University of Rhode Island
Harrison, Cheryl, University of Texas, Rio Grande Valley
Harrison, Don E., University of Washington
Hautala, Susan L., University of Washington
Hay, Alex E., Dalhousie University
He, Ruoying, North Carolina State University
Heavers, Richard, Roger Williams University
Hebert, David, Dalhousie University
Hebert, David L., University of Rhode Island
Heimbach, Patrick, University of Texas, Austin
Hendershott, Myrl C., University of California, San Diego
Herbers, Thomas H., Naval Postgraduate School
Hermann, Albert, University of Washington
Hetland, Robert D., Texas A&M University
Heywood, Karen, University of East Anglia
Hickey, Barbara M., University of Washington
Hines, Paul, Dalhousie University
Hofmann, Eileen E., Old Dominion University
Howard, Matthew K., Texas A&M University
Howden, Stephan, University of Southern Mississippi
Howe, Bruce M., University of Washington
Hsieh, William W., University of British Columbia
Huang, Bohua, George Mason University
Ierley, Glenn R., University of California, San Diego
Jackson, George A., Texas A&M University
Jacobs, Stanley, Columbia University
Jochens, Ann E., Texas A&M University
Johnson, Gregory C., University of Washington
Justic, Dubravko, Louisiana State University
Kaplan, Alexey, Columbia University
Kawase, Mitsuhiro, University of Washington
Kelley, Dan, Dalhousie University
Kelly, Kathryn A., University of Washington
Kessler, William S., University of Washington
Klinck, John M., Old Dominion University
Klinger, Barry, George Mason University
Kloosterziel, Rudolf C., University of Hawai'i, Manoa
Klymak, Jody, University of Victoria
Knauss, John A., University of Rhode Island
Knowles, Charles E., North Carolina State University
Kosro, Michael, Oregon State University
Koutitonsky, Vladimir G., Universite du Quebec a Rimouski
Kumar, Ajoy, Millersville University
Kuperman, William A., University of California, San Diego
Lebow, Ruth Y., Los Angeles Pierce College
Lee, Craig M., University of Washington
Lerczak, Jim, Oregon State University
Lesht, Barry, University of Illinois at Chicago
Lesht, Barry M., Argonne National Laboratory
Liao, Enhui, Princeton University
Liu, Zheng-yu, University of Wisconsin, Madison

Faculty by Specialty

Liu, Zhengyu, University of Wisconsin, Madison
Lozier, M. Susan, Duke University
Lu, Youyu, Dalhousie University
Lukas, Roger, University of Hawai'i, Manoa
Luther, Douglas S., University of Hawai'i, Manoa
Luther, Mark E., University of South Florida
MacCready, Parker, University of Washington
Magaard, Lorenz, University of Hawai'i, Manoa
Malanotte-Rizzoli, Paola M., Massachusetts Institute of Technology
Manley, Thomas O., Middlebury College
Marsh, Robert, University of Southampton
Marshall, John C., Massachusetts Institute of Technology
Martin, Seelye, University of Washington
Martinson, Douglas G., Columbia University
Martinson, Douglas G., Columbia University
Maslowski, Wieslaw, Naval Postgraduate School
Matano, Ricardo P., Oregon State University
Maul, George A., Florida Institute of Technology
McCreary, Julian P., University of Hawai'i, Manoa
McManus, Margaret Anne, University of Hawai'i, Manoa
Mellor, George, Princeton University
Melville, W. Kendall, University of California, San Diego
Merrifield, Mark A., University of Hawai'i, Manoa
Mitchum, Gary T., University of South Florida
Mofjeld, Harold O., University of Washington
Moore, Dennis W., University of Washington
Moore, Jefferson K., University of California, Irvine
Morison, James H., University of Washington
Moum, James N., Oregon State University
Muller, Andrew C., United States Naval Academy
Muller, Peter, University of Hawai'i, Manoa
Munk, Walter H., University of California, San Diego
Myers, Paul, University of Alberta
Mysak, Lawrence A., McGill University
Nash, Jonathan, Oregon State University
Naveira Garabato, Alberto, University of Southampton
Nechaev, Dmitri, University of Southern Mississippi
Nowlin, Jr., Worth D., Texas A&M University
Nystuen, Jeffrey A., University of Washington
O'Donnell, James, University of Connecticut
Oliver, Eric, Dalhousie University
Oliver, Kevin, University of Southampton
Oltman-Shay, Joan M., University of Washington
Orsi, Alejandro H., Texas A&M University
Paduan, Jeffrey D., Naval Postgraduate School
Pawlowicz, Richard A., University of British Columbia
Perrie, William, Dalhousie University
Perry, David J., Chabot College
Petruncio, Emil T., United States Naval Academy
Philander, S. George H., Princeton University
Pieters, Roger, University of British Columbia
Pinkel, Robert, University of California, San Diego
Pond, Stephen G., University of British Columbia
Potter, Henry, Texas A&M University
Powell, Brian, University of Hawai'i, Manoa
Powell, Thomas M., University of California, Berkeley
Primeau, Francois, University of California, Irvine
Pringle, James M., University of New Hampshire
Qiu, Bo, University of Hawai'i, Manoa
Radko, Timour, Naval Postgraduate School
Rhines, Peter B., University of Washington
Richards, Kelvin J., University of Hawai'i, Manoa
Richardson, Mary J., Texas A&M University
Riser, Stephen C., University of Washington
Roden, Gunnar I., University of Washington
Roemmich, Dean H., University of California, San Diego
Ross, Tetjana, Dalhousie University
Rossby, Hans T., University of Rhode Island
Rothrock, David A., University of Washington
Rothstein, Lewis, University of Rhode Island
Rouse, Jr., Lawrence J., Louisiana State University
Roussenov, Vassil M., University of Liverpool
Ruddick, Barry R., Dalhousie University
Rudnick, Daniel L., University of California, San Diego
Salmon, Richard L., University of California, San Diego
Salmun, Haydee, Hunter College (CUNY)
Samelson, Roger, Oregon State University

Sanford, Lawrence P., University of Maryland
Sanford, Thomas B., University of Washington
Schopf, Paul S., George Mason University
Schulz, William J., United States Naval Academy
Send, Uwe, University of California, San Diego
Shaw, Ping-Tung, North Carolina State University
Shearman, Kipp, Oregon State University
Shen, Jian, College of William & Mary
Sheng, Jinyu, Dalhousie University
Smith, Peter C., Dalhousie University
Smith, Robert L., Oregon State University
Smyth, William D., Oregon State University
Southam, John R., University of Miami
Spindel, Robert C., University of Washington
Spinrad, Richard, Oregon State University
Stanton, Timothy P., Naval Postgraduate School
Stewart, Robert H., Texas A&M University
Stoessel, Achim, Texas A&M University
Stoessel, Marion, Texas A&M University
Stramski, Dariusz, University of California, San Diego
Straub, David, McGill University
Sutherland, Dave, University of Oregon
Swaters, Gordon E., University of Alberta
Talley, Lynne D., University of California, San Diego
Thompson, Andrew F., California Institute of Technology
Thompson, Keith R., Dalhousie University
Thompson, LuAnne, University of Washington
Thomson, Richard E., University of British Columbia
Thurnherr, Andreas, Columbia University
Thurnherr, Andreas M., Columbia University
Timmermann, Axel, University of Hawai'i, Manoa
Tokmakian, Robin T., Naval Postgraduate School
Tremblay, Bruno, McGill University
Tziperman, Eli, Harvard University
Veronis, George, Yale University
Von Schwind, Joseph J., Naval Postgraduate School
Wang, Harry, College of William & Mary
Wang, Zhankun, Texas A&M University
Warner, Mark J., University of Washington
Watts, D. Randolph, University of Rhode Island
Weaver, Andrew J., University of Victoria
Webster, Ferris, University of Delaware
Weisberg, Robert H., University of South Florida
Wells, Neil C., University of Southampton
White, Martin, National University of Ireland Galway
Whitney, Michael, University of Connecticut
Wiederwohl, Chrissy, Texas A&M University
Wiggert, Jerry D., University of Southern Mississippi
Williams, Kevin L., University of Washington
Wilson, Robert E., SUNY, Stony Brook
Wimbush, Mark, University of Rhode Island
Winguth, Arne M., University of Texas, Arlington
Wolfe, Christopher, SUNY, Stony Brook
Woodgate, Rebecca A., University of Washington
Wu, Jin, University of Delaware
Wunsch, Carl I., Massachusetts Institute of Technology
Wunsch, Carl, Harvard University
Yankovsky, Alexander E., University of South Carolina
Young, William R., University of California, San Diego
Yuan, Xiaojun, Columbia University
Zappa, Christopher J., Columbia University
Zappa, Christopher, Columbia University
Zhang, Rong, Princeton University
Zhang, Y. J., College of William & Mary

Shore and Nearshore Processes
Ashton, Andrew, Woods Hole Oceanographic Institution
Baker, Alex, University of East Anglia
Ballinger, Rhoda, University of Wales
Barth, Jack A., Oregon State University
Baxter, Stefanie J., University of Delaware
Boardman, Mark R., Miami University
Bowen, Anthony J., Dalhousie University
Branco, Brett, Brooklyn College (CUNY)
Briggs, Tiffany Roberts M., Florida Atlantic University
Buonaiuto, Frank, Hunter College (CUNY)
Burbanck, George P., Hampton University
Camann, Eleanor J., Red Rocks Community College

Couceiro, Fay, University of Portsmouth
Dickson, Stephen M., Dept of Agriculture, Conservation, and Forestry
Dunbar, Robert B., Stanford University
Dupre, William R., University of Houston
Dura, Christina, Virginia Polytechnic Institute & State University
Duxbury, Alyn C., University of Washington
Engelhart, Simon E., University of Rhode Island
Farnsworth, Katie, Indiana University of Pennsylvania
Farrell, Stewart C., Stockton University
Fenster, Michael S., Randolph-Macon College
FitzGerald, Duncan M., Boston University
Friedrichs, Carl T., College of William & Mary
Gibeaut, James, Texas A&M University, Corpus Christi
Giese, Graham S., Woods Hole Oceanographic Institution
Griggs, Garry, University of California, Santa Cruz
Guza, Robert T., University of California, San Diego
Haigh, Ivan D., University of Southampton
Hall, Robert, University of East Anglia
Haller, Merrick C., Oregon State University
Hansom, Jim, University of Glasgow
Harris, Courtney K., College of William & Mary
Harris, M. Scott, College of Charleston
Heise, Elizabeth, University of Texas, Rio Grande Valley
Helmuth, Brian, Northeastern University
Jackson, Chester M., Georgia Southern University
Kassem, Hachem, University of Southampton
Kemp, Andrew, Tufts University
Kennedy, Andrew, University of Notre Dame
Kineke, Gail C., Boston College
Koppelman, Lee K., SUNY, Stony Brook
Lippmann, Thomas C., University of New Hampshire
MacMahan, Jamie, Naval Postgraduate School
McBride, Randolph, George Mason University
McManus, Margaret A., University of California, Santa Cruz
Neill, Simon P., University of Wales
Oakley, Bryan, Eastern Connecticut State University
Otvos, Ervin G., University of Southern Mississippi
Oxner-Jones, D. M., Ohio Dept of Natural Resources
Ozkan-Haller, Tuba, Oregon State University
Paine, Jeffrey G., University of Texas, Austin
Rosen, Peter S., Northeastern University
Ruggiero, Peter, Oregon State University
Seibel, Erwin, San Francisco State University
Shroyer, Emily L., Oregon State University
Siedlecki, Samantha, University of Connecticut
Sommerfield, Christopher K., University of Delaware
Theuerkauf, Ethan, University of Illinois at Chicago
Thornton, Edward B., Naval Postgraduate School
Vander Zanden, Jake, University of Wisconsin, Madison
Voulgaris, George, University of South Carolina
Westerink, Joannes J., University of Notre Dame
Wilson, Greg, Oregon State University
Winant, Clinton D., University of California, San Diego
Wu, Chin, University of Wisconsin, Madison

Paleooceanography
Came, Rosemarie E., University of New Hampshire
Ivanochko, Tara, University of British Columbia
Lynch-Stieglitz, Jean, Georgia Institute of Technology

ASTRONOMICAL SCIENCES
Cosmochemistry
Agee, Carl B., University of New Mexico
Alexander, Conel M., Carnegie Institution for Science
Bermingham, Katherine R., University of Maryland
Blake, Geoffrey A., California Institute of Technology
Blander, Milton, Argonne National Laboratory
Bodenheimer, Peter, University of California, Santa Cruz
Bouvier, Audrey, Western University
Boynton, William V., University of Arizona
Brown, Robert H., University of Arizona
Burnett, Donald S., California Institute of Technology
Ciesla, Fred, University of Chicago
Dauphas, Nicolas, University of Chicago
Davis, Andrew, University of Chicago
Drake, Michael J., University of Arizona
Fegley, Bruce, Washington University in St. Louis
Fegley, M. Bruce, Washington University in St. Louis

Greenwood, James P., Wesleyan University
Gross, Juliane, Rutgers, The State Univ of New Jersey
Grossman, Lawrence, University of Chicago
Herd, Christopher D., University of Alberta
Humayun, Munir, Florida State University
Huntress, Jr., Wesley T., Carnegie Institution for Science
Jordan, Jim L., Lamar University
Korycansky, Don, University of California, Santa Cruz
Kring, David, University of Arizona
Lauretta, Dante, University of Arizona
Lewis, John S., University of Arizona
Lin, Douglas, University of California, Santa Cruz
Lodders-Fegley, Katharina, Washington University in St. Louis
Lunine, Jonathan I., Cornell University
Lunine, Jonathan I., University of Arizona
Lyon, Ian, University of Manchester
MacPherson, Glenn J., Smithsonian Inst / Natl Museum of Natural History
McKeegan, Kevin D., University of California, Los Angeles
McSween, Jr., Harry Y., University of Tennessee, Knoxville
Mojzsis, Stephen J., University of Colorado
Nittler, Larry R., Carnegie Institution for Science
Smith, William H., Washington University in St. Louis
Swindle, Timothy, University of Arizona
Telus, Myriam, University of California, Santa Cruz
Turner, Grenville, University of Manchester
Vogt, Steven, University of California, Santa Cruz
Wang, Kun, Washington University in St. Louis

Extraterrestrial Geology
Adams, Mark, Lamar University
Anderson, Jennifer L., Winona State University
Arvidson, Raymond E., Washington University in St. Louis
Asphaug, Erik, University of California, Santa Cruz
Baker, Victor R., University of Arizona
Ballentine, Christopher, University of Oxford
Batygin, Konstantin, California Institute of Technology
Blaney, Diana L., Jet Propulsion Laboratory
Bleamaster III, Leslie F., Trinity University
Bourassa, Matthew, Western University
Burr, Devon, University of Tennessee, Knoxville
Byrne, Paul K., North Carolina State University
Calvin, Wendy M., University of Nevada, Reno
Christensen, Philip R., Arizona State University
Coles, Kenneth S., Indiana University of Pennsylvania
Craddock, Robert A., Smithsonian Institution / National Air & Space Mus
Cull-Hearth, Selby, Bryn Mawr College
Delaney, Jeremy S., Rutgers, The State Univ of New Jersey
Dietz, Richard D., University of Northern Colorado
Ehlmann, Bethany L., California Institute of Technology
Elachi, Charles, California Institute of Technology
Emery, Joshua, University of Tennessee, Knoxville
Fabel, Derek, University of Glasgow
Fishman, George, Bentley University
Gilmore, Martha S., Wesleyan University
Glotch, Timothy, Stony Brook University
Golombek, Matthew P., Jet Propulsion Laboratory
Grant, John A., Smithsonian Institution / National Air & Space Museum
Grieve, Richard, Western University
Grieve, Richard A. F., University of New Brunswick
Grosfils, Eric B., Pomona College
Hayes, Alexander G., Cornell University
Head, III, James W., Brown University
Herkenhoff, Ken E., Jet Propulsion Laboratory
Horgan, Briony, Purdue University
Howell, Robert R., University of Wyoming
Hynek, Brian M., University of Colorado
Jacobsen, Abram, University of Washington
Jakosky, Bruce M., University of Colorado
Johnson, Brandon C., Brown University
Jurena, Dwight J., San Antonio Community College
Karuntillake, Suniti, Louisiana State University
Knutson, Heather, California Institute of Technology
Korotev, Randy L., Washington University in St. Louis
Kraal, Erin, Kutztown University of Pennsylvania
Landa, Keith, SUNY, Purchase
Leake, Martha A., Valdosta State University
Li, Lin, Indiana Univ / Purdue Univ, Indianapolis
Lorenz, Ralph, University of Arizona

Mack, John E., Buffalo State College
Massironi, Matteo, Università degli Studi di Padova
Maxwell, Ted A., Smithsonian Institution / National Air & Space Museum
McEwen, Alfred S., University of Arizona
McGowan, Eileen, University of Massachusetts, Amherst
McKinnon, William B., Washington University in St. Louis
Milam, Keith A., Ohio University
Minton, David, Purdue University
Moersch, Jeffery E., University of Tennessee, Knoxville
Mouginis-Mark, Peter J., University of Hawai'i, Manoa
Murchie, Scott, University of Tennessee, Knoxville
Neish, Catherine, Western University
Nelson, Robert M., Jet Propulsion Laboratory
Newsom, Horton E., University of New Mexico
Osinski, Gordon R., Western University
Parsons, Reid A., Fitchburg State University
Piatex, Jennifer L., Central Connecticut State University
Radebaugh, Jani, Brigham Young University
Rice, Melissa S., Western Washington University
Ruiz, Javier , Univ Complutense de Madrid
Runyon, Cassandra R., College of Charleston
Schultz, Peter H., Brown University
Shepard, Michael K., Bloomsburg University
Siebach, Kirsten L., Rice University
Solomon, Sean C., Columbia University
Squyres, Steven W., Cornell University
Stofan, Ellen R., Jet Propulsion Laboratory
Strom, Robert G., University of Arizona
Tornabene, Livio, Western University
Turtle, Elizabeth, University of Arizona
Warner, Nicholas H., SUNY, Geneseo
Watters, Thomas R., Smithsonian Institution / National Air & Space Mus
Williams, Kevin K., Buffalo State College
Wordsworth, Robin, Harvard University
Zimbelman, James R., Smithsonian Institution / Nat Air & Space Mus

Extraterrestrial Geophysics
Angelopoulos, Vassilis, University of California, Los Angeles
Arkani-Hamed, Jafar, McGill University
Asphaug, Erik I., University of Arizona
Banerdt, Bruce E., Jet Propulsion Laboratory
Benacquista, Matt, Montana State University, Billings
Bord, Don, University of Michigan, Dearborn
Brown, Michael E., California Institute of Technology
Brown, Peter, Western University
Coleman, Paul J., University of California, Los Angeles
Dombard, Andrew, Case Western Reserve University
Dombard, Andrew J., University of Illinois at Chicago
Fenrich, Francis, University of Alberta
Ferencz, Orsolya, Eotvos Lorand University
Garrick-Bethell, Ian, University of California, Santa Cruz
Giacalone, Joe, University of Arizona
Goldreich, Peter M., California Institute of Technology
Greenberg, Richard J., University of Arizona
Hapke, Bruce W., University of Pittsburgh
Harnett, Erika M., University of Washington
Hauck, II, Steven A., Case Western Reserve University
Holzworth, Robert H., University of Washington
Hood, Lonnie L., University of Arizona
Hubbard, William B., University of Arizona
Jewitt, David, University of California, Los Angeles
Jokipii, Jack R., University of Arizona
Kivelson, Margaret G., University of California, Los Angeles
Kunkle, Thomas D., Los Alamos National Laboratory
Larsen, Kristine, Central Connecticut State University
Lucey, Paul G., University of Hawai'i, Manoa
Margot, Jean-Luc, University of California, Los Angeles
Matzke, David, University of Michigan, Dearborn
McCarthy, Michael P., University of Washington
McGovern, Patrick J., Rice University
McPherron, Robert L., University of California, Los Angeles
Muhleman, Duane O., California Institute of Technology
Newman, William I., University of California, Los Angeles
Nimmo, Francis, University of California, Santa Cruz
Paige, David A., University of California, Los Angeles
Petrovay, Kristof, Eotvos Lorand University
Rankin, Robert, University of Alberta
Robinson, R. Craig, Central Connecticut State University
Rostoker, Gordon, University of Alberta
Russell, Christopher T., University of California, Los Angeles
Samson, John C., University of Alberta
Schlichting, Hilke, University of California, Los Angeles
Schmerr, Nicholas C., University of Maryland
Seager, Sara, Massachusetts Institute of Technology
Smrekar, Suzanne E., Jet Propulsion Laboratory
Solomon, Sean , Columbia University
Spilker, Linda J., Jet Propulsion Laboratory
Steele, Andrew, Carnegie Institution for Science
Stevenson, David J., California Institute of Technology
Sydora, Richard D., University of Alberta
Terzian, Yervant, Cornell University
Urquhart, Mary, University of Texas, Dallas
Walker, Raymond J., University of California, Los Angeles
Winglee, Robert M., University of Washington
Zuber, Maria T., Massachusetts Institute of Technology

Meteorites & Tektites
Albin, Edward, Georgia State Univ, Perimeter College, Online
Alpert, Sam, American Museum of Natural History
Binzel, Richard P., Massachusetts Institute of Technology
Cartwright, Julia, University of Alabama
Collins, Gareth, Imperial College
Corrigan, Catherine, Smithsonian Inst / Natl Museum of Natural History
Dunn, Tasha L., Colby College
Ebel, Denton S., American Museum of Natural History
French, Bevan, Smithsonian Inst / Natl Museum of Natural History
Genge, Matthew, Imperial College
Glass, Billy P., University of Delaware
Hartman, Ron N., Mt. San Antonio College
Harvey, Ralph P., Case Western Reserve University
Heymann, Dieter, Rice University
Hildebrand, Alan R., University of Calgary
Hutson, Melinda, Portland Community College - Sylvania Campus
Hutson, Melinda, Portland State University
Jurewicz, Amy , Arizona State University
Karner, James M., University of Utah
Krot, Alexander N., University of Hawai'i, Manoa
Martins, Zita, Imperial College
McCoy, Timothy, University of Tennessee, Knoxville
McCoy, Timothy J., Smithsonian Inst / Natl Museum of Natural History
Melosh, H. J., Purdue University
Mittlefehldt, David, University of Tennessee, Knoxville
Rodgers, Nick, University of Wales
Rubin, Alan E., University of California, Los Angeles
Ruzicka, Alexander (Alex) M., Portland State University
Sephton, Mark, Imperial College
Simon, Steven B., University of New Mexico
Sipiera, Paul P., Mercer University
Swindle, Timothy D., University of Arizona
Tait, Kim, Western University
Taylor, G. Jeffrey, University of Hawai'i, Manoa
Wadhwa, Meenakshi, Field Museum of Natural History
Warren, Paul, University of California, Los Angeles
Wasson, John T., University of California, Los Angeles

Astrophysics
Boss, Alan P., Carnegie Institution for Science
Butler, R. Paul, Carnegie Institution for Science
Chambers, John E., Carnegie Institution for Science
Cruzen, Shawn, Columbus State University
Desch, Steven, Arizona State University
Gagne, Marc R., West Chester University
Groppi, Christopher, Arizona State University
Hawthorne Allen, Alice M., Concord University
Horst, Sarah, Johns Hopkins University
Jacobson, Seth A., Northwestern University
Kulkarni, Shrinivas R., California Institute of Technology
MacIssac, Daniel E., Buffalo State College
Morgan, Siobahn M., University of Northern Iowa
Paglione, Timothy, York College (CUNY)
Puckett, Andrew, Columbus State University
Scannapieco, Evan, Arizona State University
Schwarz, Karen, West Chester University
Sheppard, Scott S., Carnegie Institution for Science
Weinberger, Alycia J., Carnegie Institution for Science
Williams, Rosa, Columbus State University

Ziemba, Tim, University of Washington

Heliophysics
Moldwin, Mark B., University of Michigan

ATMOSPHERIC SCIENCES
Atmospheric Sciences
Ackerman, Thomas P., University of Washington
Adames-Corraliza, Angel F., University of Michigan
Adegoke, Jimmy, University of Missouri, Kansas City
Agee, Ernest M., Purdue University
Aiyyer, Anantha, North Carolina State University
Aldrich, Eric A., University of Missouri, Columbia
Alexander, Becky, University of Washington
Allan, Andrea, Oregon State University
Allard, Jason, Valdosta State University
Allen, Grant, University of Manchester
Allen, Robert J., University of California, Riverside
Alpert, Pinhas, Tel Aviv University
Anastasio, Cort, University of California, Davis
Ancell, Brian C., Texas Tech University
Anderson, Bruce, Boston University
Anderson, James G., Harvard University
Anderson, Mark R., University of Nebraska, Lincoln
Antonescu, Adrian, University of Manchester
Aquilina, Noel, University of Malta
Arakawa, Akio, University of California, Los Angeles
Ariya, Parisa A., McGill University
Arnfield, A. John, Ohio State University
Arnold, David L., Frostburg State University
Arnold, Stephen, University of Leeds
Arritt, Raymond W., Iowa State University of Science & Technology
Arya, Satyapal S., North Carolina State University
Atkinson, Christopher, University of North Dakota
Austin, Philip, University of British Columbia
Back, Larissa E., University of Wisconsin, Madison
Baker, Marcia B., University of Washington
Baldwin, Michael, Purdue University
Balmforth, Neil, University of British Columbia
Barlow, Mathew, University of Massachusetts, Lowell
Barnes, Elizabeth A., Colorado State University
Barnes, Jeffrey R., Oregon State University
Barrett, Bradford S., United States Naval Academy
Barrett, Kevin M., Tarrant County College, Northeast Campus
Bartello, Peter, McGill University
Barthelmie, Rebecca J., Cornell University
Bates, Timothy S., University of Washington
Bathke, Deborah J., University of Nebraska, Lincoln
Battisti, David S., University of Washington
Beard, Kenneth V., University of Illinois, Urbana-Champaign
Bell, Michael M., Colorado State University
Bennartz, Ralf, Vanderbilt University
Bergin, Michael H., Georgia Institute of Technology
Bierly, Gregory, Indiana State University
Bitz, Cecilia M., University of Washington
Black, Benjamin, Grad Sch of the City Univ of New York
Black, Robert X., Georgia Institute of Technology
Blanchard, Edward, University of Washington
Blasing, Terence J., Oak Ridge National Laboratory
Blechman, Jerome B., SUNY, Oneonta
Blyth, Alan, University of Leeds
Boden, Thomas, Oak Ridge National Laboratory
Boering, Kristie, University of California, Berkeley
Bollasina, Massimo A., Edinburgh University
Bond, Nicholas A., University of Washington
Booth, Alastair, University of Manchester
Bosart, Lance F., SUNY, Albany
Bossert, James E., Los Alamos National Laboratory
Bou-Zeid, Elie, Princeton University
Boybeyi, Zafer, George Mason University
Bradley, Raymond S., University of Massachusetts, Amherst
Bretherton, Christopher S., University of Washington
Bromwich, David H., Ohio State University
Brooks, Sarah D., Texas A&M University
Brown, Robert A., University of Washington
Brubaker, Kenneth L., Argonne National Laboratory
Bruning, Eric C., Texas Tech University
Budikova, Dagmar, Illinois State University
Burls, Natalie, George Mason University
Bush, Andrew B., University of Alberta
Bush, Andrew B. G., University of Alberta
Businger, Joost A., University of Washington
Calhoun, Joseph, Millersville University
Camargo, Suzana, Columbia University
Campbell, Janet W., University of New Hampshire
Cannon, Alex, University of British Columbia
Capehart, William J., South Dakota School of Mines & Technology
Capes, Gerard, University of Manchester
Carslaw, Ken, University of Leeds
Case Hanks, Anne T., University of Louisiana, Monroe
Castro, Mark S., University of Maryland
Cess, Robert D., SUNY, Stony Brook
Chameides, William L., Duke University
Chang, Edmund K., SUNY, Stony Brook
Chang, Hai-ru, Georgia Institute of Technology
Chang, Young-Soo, Argonne National Laboratory
Charlevoix, Donna J., University of Illinois, Urbana-Champaign
Charlson, Robert J., University of Washington
Chen, Shuyi S., University of Washington
Chen, Tsing-Chang, Iowa State University of Science & Technology
Chen, Yi-Leng, University of Hawai'i, Manoa
Chen, Yongsheng, York University
Cheng, Meng-Dawn, Oak Ridge National Laboratory
Chipperfield, Martyn P., University of Leeds
Chiu, Christine, Colorado State University
Chiu, Long S., George Mason University
Choularton, Thomas, University of Manchester
Chu, Pao-Shin, University of Hawai'i, Manoa
Chuang, Patrick Y., University of California, Santa Cruz
Clabo, Darren R., South Dakota School of Mines & Technology
Claire, Mark, University of St. Andrews
Clark, Richard D., Millersville University
Clarke, Antony D., University of Hawai'i, Manoa
Clegg, Simon, University of East Anglia
Clemitshaw, Kevin C., Royal Holloway University of London
Cobb, Steven R., Texas Tech University
Coe, Hue, University of Manchester
Colle, Brian, SUNY, Stony Brook
Collett, Jr., Jeffrey L., Colorado State University
Collins, William D., University of California, Berkeley
Collis, Scott, Argonne National Laboratory
Colman, Bradley R., University of Washington
Colucci, Stephen J., Cornell University
Connolly, Paul, University of Manchester
Cook, Kerry H., University of Texas, Austin
Corcoran, William, Missouri State University
Costigan, Keeley R., Los Alamos National Laboratory
Côté, Jean, Universite du Quebec a Montreal
Cottingame, William, Los Alamos National Laboratory
Coulter, Richard L., Argonne National Laboratory
Covert, David S., University of Washington
Cowan, Nicolas B., McGill University
Crawford, Ian, University of Manchester
Crawford, James, Georgia Institute of Technology
Crilley, Leigh R., University of Birmingham
Cronin, Timothy W., Massachusetts Institute of Technology
Crowther, Sarah, University of Manchester
Curry, Judith A., Georgia Institute of Technology
Cziczo, Dan, Massachusetts Institute of Technology
Davis, Robert, University of Virginia
de Foy, Benjamin, Saint Louis University
Dearden, Christopher, University of Manchester
DeCaria, Alex J., Millersville University
DeConto, Robert, University of Massachusetts, Amherst
DeGaetano, Arthur, Cornell University
Delcambre, Sharon, Portland Community College - Sylvania Campus
DelSole, Timothy, George Mason University
Dempsey, David P., San Francisco State University
Deng, Yi, Georgia Institute of Technology
Denning, A. Scott, Colorado State University
Derome, Jacques F., McGill University
Desai, Ankur R., University of Wisconsin, Madison
Dessler, Alex, Texas A&M University
Dessler, Andrew E., Texas A&M University

Detwiler, Andrew G., South Dakota School of Mines & Technology
DeWekker, Stephan F., University of Virginia
Dickinson, Robert E., University of Texas, Austin
Didlake, Anthony, Pennsylvania State University, University Park
Dirmeyer, Paul A., George Mason University
Dobbie, Steven, University of Leeds
Dobler, Scott, Western Kentucky University
Doherty, Ruth, Edinburgh University
Dominguez, Francina, University of Illinois, Urbana-Champaign
Doskey, Paul V., Argonne National Laboratory
Dugas, Bernard, Universite du Quebec a Montreal
Dunn, Allison L., Worcester State University
Durkee, Josh, Western Kentucky University
Durran, Dale R., University of Washington
Dutcher, Allen, University of Nebraska, Lincoln
Edgell, Dennis J., University of North Carolina, Pembroke
Elliott, Scott M., Los Alamos National Laboratory
Ellul, Raymond , University of Malta
Epifanio, Craig, Texas A&M University
Essery, Richard L., Edinburgh University
Esslinger, Kelly L., Arizona Western College
Fabry, Frederic, McGill University
Faloona, Ian C., University of California, Davis
Fan, Xingang, Western Kentucky University
Faraji, Maedeh, Austin Community College District
Feng, Song, University of Arkansas, Fayetteville
Feng, Yan, Argonne National Laboratory
Fernando, Joe, University of Notre Dame
Ferruzza, David, Elizabethtown College
Few, Jr., Arthur A., Rice University
Fink, Uwe, University of Arizona
Finnegan, David L., Los Alamos National Laboratory
Fiore, Arlene M., Columbia University
Fiore, Arlene, Columbia University
Fischer, Emily, Colorado State University
Fishman, Jack, Saint Louis University
Fitzjarrald, David R., SUNY, Albany
Flato, Gregory M., University of Victoria
Fletcher, Roy J., University of Lethbridge
Forster, Piers M., University of Leeds
Fovell, Robert, University of California, Los Angeles
Fowler, Malcolm M., Los Alamos National Laboratory
Fox, Neil I., University of Missouri, Columbia
Frame, Jeffrey, University of Illinois, Urbana-Champaign
Frederick, John E., University of Chicago
French, Adam, South Dakota School of Mines & Technology
Frierson, Dargan M., University of Washington
Frolking, Stephen E., University of New Hampshire
Fthenakis, Vasilis M., Columbia University
Fu, Qiang, University of Washington
Fu, Rong, University of Texas, Austin
Fueglistaler, Stephan A., Princeton University
Fuhrmann, Christopher, Mississippi State University
Fultz, Dave, University of Chicago
Fung, Inez, University of California, Berkeley
Fyfe, John C., University of Victoria
Gachon, Philippe, Universite du Quebec a Montreal
Gaffney, Jeffrey S., Argonne National Laboratory
Gallagher, Martin, University of Manchester
Galloway, James N., University of Virginia
Gallus, William A., Iowa State University of Science & Technology
Gao, Yuan, Rutgers, The State Univ of New Jersey, Newark
Garcia, Oswaldo, San Francisco State University
Garrett, Timothy J., University of Utah
Gedzelman, Stanley, City College (CUNY)
Geller, Marvin A., SUNY, Stony Brook
Gervais, Melissa, Pennsylvania State University, University Park
Ghate, Virendra P., Argonne National Laboratory
Ghil, Michael, University of California, Los Angeles
Giannini, Alessandra, Columbia University
Gill, Swarndeep S., California University of Pennsylvania
Gluhovsky, Alexander, Purdue University
Goddard, Lisa M., Columbia University
Godfrey, Christopher, University of North Carolina, Asheville
Goldblatt, Colin, University of Victoria
Gonzalez, Alex, Iowa State University of Science & Technology
Goodman, Paul, University of Arizona

Goodrich, Greg, Western Kentucky University
Gordon, Mark, York University
Gordon, Rob J., University of Guelph
Gorka, Maciej, University of Wroclaw
Grace, John, Edinburgh University
Granger, Darryl E., Purdue University
Graves, Charles E., Saint Louis University
Grenfell, Thomas C., University of Washington
Griffis, Timothy J., University of Minnesota, Twin Cities
Grise, Kevin M., University of Virginia
Grotjahn, Richard , University of California, Davis
Guinan, Patrick E., University of Missouri, Columbia
Gutzler, David J., University of New Mexico
Hage, Keith D., University of Alberta
Hakim, Gregory J., University of Washington
Hallar, Anna G., University of Utah
Hameed, Sultan, SUNY, Stony Brook
Hanrahan, Janel, Northern Vermont University-Lyndon
Hansen, William, Pace University, New York Campus
Hanson, Howard, Florida Atlantic University
Harnik, Nili, Tel Aviv University
Harrison, Halstead, University of Washington
Harrison, Roy M., University of Birmingham
Harshvardhan, Purdue University
Hart, Richard L., Argonne National Laboratory
Hartmann, Dennis L., University of Washington
Hastenrath, Stefan L., University of Wisconsin, Madison
Hastings, Meredith, Brown University
Hawkins, Timothy W., Shippensburg University
Hayes, Michael J., University of Nebraska, Lincoln
Heald, Colette, Massachusetts Institute of Technology
Hegerl, Gabriele C., Edinburgh University
Hegg, Dean A., University of Washington
Heikes, Brian G., University of Rhode Island
Helsdon, John H., South Dakota School of Mines & Technology
Hence, Deanna A., University of Illinois, Urbana-Champaign
Henderson, Gina R., United States Naval Academy
Hennon, Chris, University of North Carolina, Asheville
Hickmon, Nicki, Argonne National Laboratory
Hindman, Edward E., City College (CUNY)
Hirschboeck, Katherine K., University of Arizona
Hitchman, Matthew H., University of Wisconsin, Madison
Hjelmfelt, Mark R., South Dakota School of Mines & Technology
Holberg, Jay B., University of Arizona
Holgood, Jay S., Ohio State University
Holloway, Tracey, University of Wisconsin, Madison
Holzer, Mark, University of British Columbia
Horel, John, University of Utah
Horton, Daniel E., Northwestern University
Houston, Adam L., University of Nebraska, Lincoln
Houweling, Sander, Vrije Universiteit Amsterdam
Houze, Robert A., University of Washington
Hoyos, Carlos, Georgia Institute of Technology
Hu, Qi S., University of Nebraska, Lincoln
Hu, Qi (Steve), University of Nebraska, Lincoln
Huang, Alex, University of North Carolina, Asheville
Huey, Gregory L., Georgia Institute of Technology
Hugli, Wilbur G., University of West Florida
Hunten, Donald M., University of Arizona
Hurrell, James, Colorado State University
Hutyra, Lucy, Boston University
Idone, Vincent P., SUNY, Albany
Ingersoll, Andrew P., California Institute of Technology
Jablonowski, Christiane, University of Michigan
Jackson, Andrea, University of Leeds
Jacob, Daniel J., Harvard University
Jaegle, Lyatt, University of Washington
Jaffe, Daniel A., University of Washington
Jain, Atul K., University of Illinois, Urbana-Champaign
Jenkins, Mary Ann, York University
Jewett, Brian, University of Illinois, Urbana-Champaign
Johnson, Donald R., University of Wisconsin, Madison
Jones, Eric M., Los Alamos National Laboratory
Jones, Hazel, University of Manchester
Jones, Phillip D., University of East Anglia
Kaiser, Jason, Northern Vermont University-Lyndon
Kang, Song-Lak, Texas Tech University

Kao, Jim C., Los Alamos National Laboratory
Karamperidou, Christina, University of Hawai'i, Manoa
Kasting, James F., Pennsylvania State University, University Park
Kauffman, Chad, California University of Pennsylvania
Keables, Michael J., University of Denver
Keeling, Ralph F., University of California, San Diego
Keene, William, University of Virginia
Keesee, Robert G., SUNY, Albany
Keller, Linda M., University of Wisconsin, Madison
Keller, Jr., C. F., Los Alamos National Laboratory
Kennel, Charles F., University of California, San Diego
Keyser, Daniel, SUNY, Albany
Khairoutdinov, Marat, SUNY, Stony Brook
Kidder, Steven, Grad Sch of the City Univ of New York
Kieu, Chanh Q., Indiana University, Bloomington
Kift, Richard, University of Manchester
Kim, Daehyun, University of Washington
Kim, Hyemi, Georgia Institute of Technology
Kim, Saewung, University of California, Irvine
King, Kenneth M., University of Guelph
King, Martin, Royal Holloway University of London
Kinter, Jim, George Mason University
Kirkpatrick, Cody, Indiana University, Bloomington
Kirlin, R. Lynn, University of British Columbia
Kirshbaum, Daniel, McGill University
Kishcha, Pavel, Tel Aviv University
Klaassen, Gary P., York University
Kliche, Donna V., South Dakota School of Mines & Technology
Kluver, Daria, Central Michigan University
Knapp, Warren, Cornell University
Knight, David, SUNY, Albany
Knopf, Daniel A., SUNY, Stony Brook
Koehler, Thomas, United States Air Force Academy
Konigsberg, Alvin S., SUNY, New Paltz
Korty, Robert, Texas A&M University
Kossin, James, University of Wisconsin, Madison
Kota, Jozsef, University of Arizona
Kotamarthi, Rao, Argonne National Laboratory
Kreidenweis, Sonia M., Colorado State University
Krueger, Steven, University of Utah
Kuang, Zhiming, Harvard University
Kubas, Gregory J., Los Alamos National Laboratory
Kucharik, Chris, University of Wisconsin, Madison
Kung, Ernest C., University of Missouri, Columbia
Kursinski, Robert, University of Arizona
Kushnir, Yochanan, Columbia University
Kutzbach, John E., University of Wisconsin, Madison
L'Ecuyer, Tristan S., University of Wisconsin, Madison
Lackmann, Gary M., North Carolina State University
Laird, Neil, Hobart & William Smith Colleges
Laprise, Rene, Universite du Quebec a Montreal
Lasher-Trapp, Sonia, University of Illinois, Urbana-Champaign
Lazarus, Steven M., Florida Institute of Technology
Leather, Kimberly, University of Manchester
Lee, In Young, Argonne National Laboratory
Lee, Meehye, Korea University
Leibensperger, Eric, SUNY, Plattsburgh
Leighton, Henry G., McGill University
Leppert, Ken, University of Louisiana, Monroe
Lew, Jeffrey, University of California, Los Angeles
Li, Tim, University of Hawai'i, Manoa
Li, Xiangshan, University of Houston
Lichtenberger, János, Eotvos Lorand University
Light, Bonnie, University of Washington
Lin, Hai, McGill University
Lin, Jialin, Ohio State University
Lin, John C., University of Utah
Lin, Wuyin, SUNY, Stony Brook
Lindgren, Paula, University of Glasgow
Liou, Kuo Nan, University of California, Los Angeles
Liu, Chuntao, Texas A&M University, Corpus Christi
Liu, Dantong, University of Manchester
Liu, Jiping, Georgia Institute of Technology
Liu, Ping, SUNY, Stony Brook
Lombardo, Kelly, University of Connecticut
Lora, Juan, Yale University
Loriot, George, Northern Vermont University-Lyndon

Lowe, Douglas, University of Manchester
Lozowski, Edward P., University of Alberta
Lundquist, Jessica D., University of Washington
Luo, Chao, Georgia Institute of Technology
Lupo, Anthony R., University of Missouri, Columbia
Lynch, Amanda H., Brown University
Lyons, Lawrence, University of California, Los Angeles
Lyons, Walter A., University of Northern Colorado
Maasch, Kirk A., University of Maine
Mace, Gerald, University of Utah
Magee, Nathan, College of New Jersey
Magnusdottir, Gudrun, University of California, Irvine
Mahowald, Natalie M., Cornell University
Mak, John E., SUNY, Stony Brook
Mak, Mankin, University of Illinois, Urbana-Champaign
Malcolm, Elizabeth, Virginia Wesleyan College
Maloney, Eric, Colorado State University
Manabe, Syukuro, Princeton University
Mann, Michael E., Pennsylvania State University, University Park
Manning, Andrew, University of East Anglia
Marchand, Roger T., University of Washington
Mark, Bryan G., Ohio State University
Market, Patrick S., University of Missouri, Columbia
Marley, Nancy A., Argonne National Laboratory
Marsham, John, University of Leeds
Martin, Randal S., New Mexico Institute of Mining and Technology
Martin, Timothy J., Argonne National Laboratory
Martin, Walter, University of North Carolina, Charlotte
Mason, Allen S., Los Alamos National Laboratory
Mass, Clifford F., University of Washington
Mauzerall, Denise L., Princeton University
Mawalagedara, Rachindra, Iowa State University of Science & Tech
Maykut, Gary A., University of Washington
Mayor, Shane D., California State University, Chico
Mazzocco, Elizabeth, Ashland University
McCollor, Doug, University of British Columbia
McDonald, Kyle, Grad Sch of the City Univ of New York
McElroy, Michael B., Harvard University
McElroy, Tom, York University
McFarlane, Norman, University of Victoria
McFarquhar, Greg M., University of Illinois, Urbana-Champaign
McFiggans, Gordon, University of Manchester
McGregor, Kent M., University of North Texas
McMurdie, Lynn A., University of Washington
McQuaid, Jim, University of Leeds
McWilliams, James C., University of California, Los Angeles
Mechoso, Carlos R., University of California, Los Angeles
Merlis, Timothy, McGill University
Meskhidze, Nicholas, North Carolina State University
Metz, Nicholas, Hobart & William Smith Colleges
Meyer, Steven J., University of Wisconsin, Green Bay
Michaels, Patrick J., University of Virginia
Miller, Doug, University of North Carolina, Asheville
Miller, Ronald L., Columbia University
Millet, Dylan B., University of Minnesota, Twin Cities
Mitchell, Jonathan, University of California, Los Angeles
Modzelewski, Henryk, University of British Columbia
Molina, Mario J., University of California, San Diego
Molinari, John E., SUNY, Albany
Monahan, Edward C., University of Connecticut
Moncrieff, John B., Edinburgh University
Montenegro, Alvaro, Ohio State University
Monteverdi, John P., San Francisco State University
Moody, Jennie L., University of Virginia
Morgan, William, University of Manchester
Mosley-Thompson, Ellen E., Ohio State University
Mote, Philip W., University of Washington
Moyer, Elisabeth, University of Chicago
Moyer, Kerry A., Edinboro University of Pennsylvania
Mroz, Eugene J., Los Alamos National Laboratory
Mudrick, Stephen E., University of Missouri, Columbia
Mullen, Steven L., University of Arizona
Murphy, Todd, University of Louisiana, Monroe
Murrell, Coling, University of East Anglia
Murthy, Prahlad N., Wilkes University
Nakamura, Noboru, University of Chicago
Nappo, Carmen, Georgia Institute of Technology

Nathan, Terrence R., University of California, Davis
Neelin, J. David, University of California, Los Angeles
Nenes, Athanasios, Georgia Institute of Technology
Nesbitt, Stephen W., University of Illinois, Urbana-Champaign
Newman, Steven B., Central Connecticut State University
Nielsen-Gammon, John, Texas A&M University
Niyogi, Dev, Purdue University
Nogueira, Ricardo, Georgia State Univ
Norris, Joel R., University of California, San Diego
North, Gerald, Texas A&M University
Nugent, Alison D., University of Hawai'i, Manoa
O'Brien, Travis A., University of California, Davis
O'Gorman, Paul, Massachusetts Institute of Technology
Obrist, Daniel, University of Massachusetts, Lowell
Oechel, Walter, The Open University
Ojala, Carl F., Eastern Michigan University
Okalebo, Jane, University of Nebraska, Lincoln
Okumura, Yuko, University of Texas, Austin
Oppenheimer, Michael, Princeton University
Orlanski, Isidoro, Princeton University
Orlove, Benjamin S., Columbia University
Orville, Richard E., Texas A&M University
Osterberg, Erich C., Dartmouth College
Overland, James E., University of Washington
Overpeck, Jonathan T., University of Arizona
Palmer, Paul, Edinburgh University
Pan, Ying, Pennsylvania State University, University Park
Pan, Zaitao, Saint Louis University
Pani, Eric A., University of Louisiana, Monroe
Paulson, Suzanne, University of California, Los Angeles
Pegion, Kathy, George Mason University
Percell, Peter, University of Houston
Percival, Carl, University of Manchester
Perez, Richard R., SUNY, Albany
Perry, Kevin D., University of Utah
Peterson, Richard E., Texas Tech University
Petters, Markus, North Carolina State University
Petty, Grant W., University of Wisconsin, Madison
Pierce, Jeff, Colorado State University
Pierrehumbert, Raymond T., University of Chicago
Plumb, Raymond A., Massachusetts Institute of Technology
Polzin, Dierk T., University of Wisconsin, Madison
Powell, Mark L., Los Angeles Pierce College
Prather, Kimberly A., University of California, San Diego
Prather, Michael J., University of California, Irvine
Preston, Aaron, Northern Vermont University-Lyndon
Preston, Thomas C., McGill University
Previdi, Michael, Columbia University
Price, Colin G., Tel Aviv University
Prinn, Ronald G., Massachusetts Institute of Technology
Pryor, Sara C., Cornell University
Pu, Zhaoxia, University of Utah
Pumphrey, Hugh C., Edinburgh University
Pusede, Sally, University of Virginia
Radke, Lawrence F., University of Washington
Raman, Sethu S., North Carolina State University
Ramanathan, V., University of California, San Diego
Ramsey, Kelvin W., University of Delaware
Randall, David A., Colorado State University
Rappenglueck, Bernhard, University of Houston
Rasmussen, Kristen L., Colorado State University
Rauber, Robert M., University of Illinois, Urbana-Champaign
Ravishankara, A.R., Colorado State University
Rawlins, Michael, University of Massachusetts, Amherst
Ray, Pallav, Florida Institute of Technology
Rayner, John N., Ohio State University
Reay, David, Edinburgh University
Reichler, Thomas, University of Utah
Reisner, Jon M., Los Alamos National Laboratory
Reiss, Nathan, Pace University, New York Campus
Reuter, Gerhard W., University of Alberta
Revelle, Douglas O., Los Alamos National Laboratory
Richter, David, University of Notre Dame
Ricketts, Hugo, University of Manchester
Riemer, Nicole, University of Illinois, Urbana-Champaign
Riordan, Allen J., North Carolina State University
Ritsche, Michael, Argonne National Laboratory

Robertson, Andrew W., Columbia University
Robinson, Walter, North Carolina State University
Robinson, Walter A., University of Illinois, Urbana-Champaign
Roche, Didier, Vrije Universiteit Amsterdam
Roe, Gerard H., University of Washington
Rogers, Jeffrey C., Ohio State University
Ross, Robert S., Millersville University
Rowe, Clinton M., University of Nebraska, Lincoln
Rupp, David, Oregon State University
Russell, Armistead G., Georgia Institute of Technology
Russell, Lynn M., University of California, San Diego
Rutledge, Steven A., Colorado State University
Saltzman, Eric S., University of California, Irvine
Samson, Perry, University of Michigan
Sandel, Bill R., University of Arizona
Sarachik, Edward S., University of Washington
Sarachik, Edward, University of Washington
Saravanan, R., Texas A&M University
Schade, Gunnar, Texas A&M University
Schlosser, C. Adam, Massachusetts Institute of Technology
Schmittner, Andreas, Oregon State University
Schneider, Edwin K., George Mason University
Schneider, Tapio, California Institute of Technology
Schroeder, John L., Texas Tech University
Schumacher, Courtney, Texas A&M University
Schumacher, Russ, Colorado State University
Seeley, Mark W., University of Minnesota, Twin Cities
Selin, Noelle, Massachusetts Institute of Technology
Semazzi, Fred H. M., North Carolina State University
Serreze, Mark, University of Colorado
Shafer, Jason, Northern Vermont University-Lyndon
Shell, Karen M., Oregon State University
Shellito, Lucinda, University of Northern Colorado
Shen, Samuel S., University of Alberta
Shen, Xinhua, University of Northern Iowa
Shepherd, Gordon G., York University
Shepherd, Mark A., Austin Community College District
Shepson, Paul B., Purdue University
Sherman-Morris, Kathleen M., Mississippi State University
Shinoda, Toshiaki, Texas A&M University, Corpus Christi
Shirley, Terry, University of North Carolina, Charlotte
Shukla, Jagadish, George Mason University
Shulski, Martha D., University of Nebraska, Lincoln
Sikora, Todd D., Millersville University
Simpson, Robert M., University of Tennessee, Martin
Sisterson, Doug, Argonne National Laboratory
Sisterson, Douglas L., Argonne National Laboratory
Siuta, David, Northern Vermont University-Lyndon
Smedley, Andrew, University of Manchester
Smith, Paul L., South Dakota School of Mines & Technology
Smith, Phillip J., Purdue University
Smith, Ronald B., Yale University
Snow, Julie A., Slippery Rock University
Snyder, Peter K., University of Minnesota, Twin Cities
Snyder, Richard L., University of California, Davis
Sobel, Adam, Columbia University
Sobel, Adam H., Columbia University
Sokolik, Irina, Georgia Institute of Technology
Solomon, Susan, Massachusetts Institute of Technology
Somerville, Richard C., University of California, San Diego
Soule, Peter T., Appalachian State University
Srivastava, Ramesh C., University of Chicago
Sriver, Ryan, University of Illinois, Urbana-Champaign
St. John, James C., Georgia Institute of Technology
Stan, Cristiana, George Mason University
Staten, Paul W., Indiana University, Bloomington
Steenburgh, Jim, University of Utah
Stevens, Philip S., Indiana University, Bloomington
Stevenson, David, Edinburgh University
Steyn, Douw G., University of British Columbia
Stickel, Robert, Georgia Institute of Technology
Stoelinga, Mark T., University of Washington
Stone, Peter H., Massachusetts Institute of Technology
Straub, Derek J., Susquehanna University
Straus, David M., George Mason University
Strobel, Darrell F., Johns Hopkins University
Strong, Courtenay, University of Utah

Stull, Roland B., University of British Columbia
Sturges, Bill, University of East Anglia
Stutz, Jochen P., University of California, Los Angeles
Sun, Jingru, Princeton University
Sun, Wen-Yih, Purdue University
Suyker, Andrew E., University of Nebraska, Lincoln
Swann, Abigail L., University of Washington
Swanson, Brian, University of Washington
Szunyogh, Istvan, Texas A&M University
Tadesse, Tsegaye, University of Nebraska, Lincoln
Takle, Eugene S., Iowa State University of Science & Technology
Talley, John H., University of Delaware
Tatarskii, Viatcheslav, Georgia Institute of Technology
Taylor, Elwynn, Iowa State University of Science & Technology
Taylor, Peter A., York University
Teeuw, Richard, University of Portsmouth
Thorncroft, Christopher D., SUNY, Albany
Thorne, Richard M., University of California, Los Angeles
Thornton, Joel A., University of Washington
Thurtell, George W., University of Guelph
Tillman, James E., University of Washington
Ting, Mingfang, Columbia University
Tomasko, Martin G., University of Arizona
Tongue, Jeffrey, SUNY, Stony Brook
Trapp, Robert J., University of Illinois, Urbana-Champaign
Tung, Ka-Kit, University of Washington
Tung, Wen-wen, Purdue University
Turco, Richard, University of California, Los Angeles
Twine, Tracy E., University of Minnesota, Twin Cities
Tzortziou, Maria, Grad Sch of the City Univ of New York
Van Den Broeke, Matthew S., University of Nebraska, Lincoln
van den Heever, Sue, Colorado State University
van Leeuwen, Peter Jan, Colorado State University
Vanos, Jennifer K., University of California, San Diego
Vaughan, Geraint, University of Manchester
Veres, Michael, SUNY, Oswego
Vimont, Daniel J., University of Wisconsin, Madison
Vincent, Dayton G., Purdue University
Vogelmann, Andrew, SUNY, Stony Brook
von Salzen, Knut, University of British Columbia
von Salzen, Knut, University of Victoria
Vukovich, George, York University
Wagner, Timothy J., Creighton University
Walcek, Christopher J., SUNY, Albany
Waliser, Duane E., SUNY, Stony Brook
Wallace, John M., University of Washington
Walter-Shea, Elizabeth A., University of Nebraska, Lincoln
Wang, Hsiang-Jui, Georgia Institute of Technology
Wang, Jian, SUNY, Stony Brook
Wang, Pao-Kuan, University of Wisconsin, Madison
Wang, Wei-Chyung, SUNY, Albany
Wang, Yuhang, Georgia Institute of Technology
Wang, Yuqing, University of Hawai'i, Manoa
Wang, Yuxuan, University of Houston
Wang, Zhuo, University of Illinois, Urbana-Champaign
Warland, Jon, University of Guelph
Warren, Stephen G., University of Washington
Waugh, Darryn W., Johns Hopkins University
Wax, Charles L., Mississippi State University
Weare, B. C., University of California, Davis
Weaver, Justin E., Texas Tech University
Weber, Rodney J., Georgia Institute of Technology
Webster, Peter J., Georgia Institute of Technology
Weinbeck, Robert S., SUNY, The College at Brockport
Weiss, Christopher C., Texas Tech University
Wennberg, Paul O., California Institute of Technology
Wesely, Marvin L., Argonne National Laboratory
Wettstein, Justin, Oregon State University
Whitaker, Rodney W., Los Alamos National Laboratory
Whiteway, James, York University
Wikle, Christopher K., University of Missouri, Columbia
Wilhelmson, Robert B., University of Illinois, Urbana-Champaign
Wilhite, Donald A., University of Nebraska, Lincoln
Wilks, Daniel, Cornell University
Williams, John M., SUNY, The College at Brockport
Williams, Kay R., Shippensburg University
Williams, Paul, University of Manchester
Willoughby, Hugh E., Florida International University
Wilson, John D., University of Alberta
Wine, Paul H., Georgia Institute of Technology
Wofsy, Steven C., Harvard University
Wood, Kim, Mississippi State University
Wood, Robert, University of Washington
Wu, Shiliang, Michigan Technological University
Wu, Xiaoqing, Iowa State University of Science & Technology
Wu, Yutian, Purdue University
Wuebbles, Donald J., University of Illinois, Urbana-Champaign
Wysocki, Mark, Cornell University
Xie, Lian, North Carolina State University
Yalda, Sepideh, Millersville University
Yang, Ping, Texas A&M University
Yang, Wenchang, Princeton University
Yarger, Douglas N., Iowa State University of Science & Technology
Yau, Man Kong, McGill University
Yi, Chuixiang, Queens College (CUNY)
Young, John A., University of Wisconsin, Madison
Yu, Jin-Yi, University of California, Irvine
Yung, Yuk L., California Institute of Technology
Yuter, Sandra, North Carolina State University
Zaitchik, Benjamin, Johns Hopkins University
Zawadzki, Isztar, McGill University
Zehnder, Joseph A., Creighton University
Zender, Charles, University of California, Irvine
Zeng, Ning, University of Maryland
Zhang, Chidong, University of Washington
Zhang, Henian, Georgia Institute of Technology
Zhang, Minghua, SUNY, Stony Brook
Zhang, Renyi, Texas A&M University
Zhang, Xi, University of California, Santa Cruz
Zhang, Yang, North Carolina State University
Zhu, Ping, Florida International University
Zhuang, Qianlai, Purdue University
Zipser, Edward, University of Utah
Zuend, Andreas, McGill University

Meteorology
Adams, Manda S., University of North Carolina, Charlotte
Allen, John, Central Michigan University
Ault, Toby R., Cornell University
Babb, David M., Pennsylvania State University, University Park
Bahrmann, Chad, Pennsylvania State University, University Park
Bannon, Peter R., Pennsylvania State University, University Park
Bartholy, Judit, Eotvos Lorand University
Baxter, Martin, Central Michigan University
Bernhardt, Jase, Hofstra University
Biasutti, Michela, Columbia University
Billings, Brian J., Saint Cloud State University
Blackwell, Keith G., University of South Alabama
Bleck, Rainer, Los Alamos National Laboratory
Bohren, Craig F., Pennsylvania State University, University Park
Bowman, Kenneth P., Texas A&M University
Brown, Michael E., Mississippi State University
Brown, Paul W., University of Arizona
Brune, William H., Pennsylvania State University, University Park
Brunet, Gilbert, McGill University
Businger, Joost, University of Washington
Businger, Steven, University of Hawai'i, Manoa
Cahir, John J., Pennsylvania State University, University Park
Carlson, Richard E., Iowa State University of Science & Technology
Carlson, Toby N., Pennsylvania State University, University Park
Carroll, David, Virginia Polytechnic Institute & State University
Challinor, Andy, University of Leeds
Chang, Chih-Pei, Naval Postgraduate School
Chen, Hway-Jen, Naval Postgraduate School
Chen, Jing, McMaster University
Clark, John H., Pennsylvania State University, University Park
Clothiaux, Eugene E., Pennsylvania State University, University Park
Colby, Frank P., University of Massachusetts, Lowell
Cook, David R., Argonne National Laboratory
Corona, Thomas J., Metropolitan State College of Denver
Coulter, Richard L., Argonne National Laboratory
Crafts, Christine, SUNY, The College at Brockport
Creasey, Robert L., Naval Postgraduate School
Croft, Paul J., Kean University
Croft, Paul J., Mississippi State University

Cullen, Heidi, Georgia Institute of Technology
Czarnetzki, Alan C., University of Northern Iowa
Dahl, Johannes M., Texas Tech University
Davidson, Kenneth L., Naval Postgraduate School
Davis, Kenneth J., Pennsylvania State University, University Park
Deng, Aijun, Pennsylvania State University, University Park
Dewey, Ken F., University of Nebraska, Lincoln
Dixon, P. Grady, Fort Hays State University
Dorling, Stephen, University of East Anglia
Duncan Tabb, Neva, Saint Petersburg College, Clearwater
Durkee, Philip A., Naval Postgraduate School
Dutton, John A., Pennsylvania State University, University Park
Dyer, Jamie, Mississippi State University
Eastin, Matt, University of North Carolina, Charlotte
Ellingson, Don, Western Oregon University
Elsberry, Russell L., Naval Postgraduate School
Emanuel, Kerry A., Massachusetts Institute of Technology
Evans, Jenni L., Pennsylvania State University, University Park
Fairchild, Jane, University of Wisconsin Colleges
Farrell, Brian F., Harvard University
Ferger, Marisa, Pennsylvania State University, University Park
Fillion, Luc, McGill University
Finley, Jason P., Los Angeles Pierce College
Fitzpatrick, Patrick J., Mississippi State University
Flory, David M., Iowa State University of Science & Technology
Flynn, Wendilyn, University of Northern Colorado
Frank, William M., Pennsylvania State University, University Park
Fraser, Alistair B., Pennsylvania State University, University Park
Frederickson, Paul A., Naval Postgraduate School
Fritsch, J. Michael, Pennsylvania State University, University Park
Fullmer, James W., Southern Connecticut State University
Gadomski, Frederick J., Pennsylvania State University, University Park
Galewsky, Joseph, University of New Mexico
Garza, Jessica, Angelo State University
Gauthier, Pierre, Universite du Quebec a Montreal
Gillespie, Terry J., University of Guelph
Girard, Eric, Universite du Quebec a Montreal
Godek, Melissa, SUNY, Oneonta
Greybush, Steven J., Pennsylvania State University, University Park
Griswold, Jennifer S., University of Hawai'i, Manoa
Guest, Peter S., Naval Postgraduate School
Gunter, William S., Columbus State University
Gutowski, William J., Iowa State University of Science & Technology
Gyakum, John R., McGill University
Hacker, Joshua P., Naval Postgraduate School
Hamill, Paul, McHenry County College
Haney, Christa M., Mississippi State University
Haney, Robert, Naval Postgraduate School
Hansen, Anthony R., Saint Cloud State University
Harr, Patrick A., Naval Postgraduate School
Harrington, Jerry Y., Pennsylvania State University, University Park
Hehr, John G., University of Arkansas, Fayetteville
Heyniger, William C., Kean University
Hilliker, Joby, West Chester University
Hoffman, Eric, Plymouth State University
Hosler, Charles L., Pennsylvania State University, University Park
Illari, Lodovica, Massachusetts Institute of Technology
Jenkins, Gregory S., Pennsylvania State University, University Park
Jin, Fei Fei, University of Hawai'i, Manoa
Jordan, Mary S., Naval Postgraduate School
Karmosky, Christopher, SUNY, Oneonta
Kaster, Mark A., Wilkes University
Kasting, James F., Pennsylvania State University, University Park
Keeler, Jason, Central Michigan University
Kimball, Sytske K., University of South Alabama
Knight, Paul, Pennsylvania State University, University Park
Kubesh, Rodney, Saint Cloud State University
Kumjian, Matthew R., Pennsylvania State University, University Park
Lai, Chung Chieng A., Los Alamos National Laboratory
Lamb, Dennis, Pennsylvania State University, University Park
Lander, Mark A., University of Guam
Lawson, Merlin P., University of Nebraska, Lincoln
Lee, Sukyoung, Pennsylvania State University, University Park
Leonard, Meredith L., Los Angeles Valley College
Lerach, David G., University of Northern Colorado
Lindzen, Richard S., Massachusetts Institute of Technology
Liu, Yongqiang, Georgia Institute of Technology

Lyons, Steven, Angelo State University
Mandia, Scott A., Suffolk County Comm Coll, Ammerman Campus
Manzi, Anthony, SUNY, Maritime College
Martin, Jonathan E., University of Wisconsin, Madison
Matthews, Adrian, University of East Anglia
McGauley, Michael G., Miami Dade College (Kendall Campus)
Meesters, Antoon, Vrije Universiteit Amsterdam
Mendenhall, Larry, Mt. San Antonio College
Mercer, Andrew, Mississippi State University
Montgomery, Michael T., Naval Postgraduate School
Moore, Richard W., Naval Postgraduate School
Morgan, Michael C., University of Wisconsin, Madison
Morschauser, Lindsey, Mississippi State University
Mower, Richard, Central Michigan University
Murphree, James Thomas, Naval Postgraduate School
Murray, Andrew, University of South Alabama
Nielsen, Kurt E., Naval Postgraduate School
Nietfeld, Daniel, Creighton University
Ning, Liang, University of Massachusetts, Amherst
Noone, David, Oregon State University
Nordstrom, Greg, Mississippi State University
North, Gerald R., Texas A&M University
Nowotarski, Christopher J., Texas A&M University
Nuss, Wendell A., Naval Postgraduate School
O'Neill, Larry, Oregon State University
Panetta, Richard L., Texas A&M University
Park, Myung-Sook, Naval Postgraduate School
Parker, Doug, University of Leeds
Parker, Matthew, North Carolina State University
Pasken, Robert W., Saint Louis University
Pavloski, Charles, Pennsylvania State University, University Park
Penny, Andrew, Naval Postgraduate School
Person, Arthur, Pennsylvania State University, University Park
Polvani, Lorenzo M., Columbia University
Priest, Eric, College of Lake County
Renard, Robert J., Naval Postgraduate School
Renfrew, Ian, University of East Anglia
Reuter, Gerhard, University of Alberta
Ritchie, Harold C., Dalhousie University
Ritter, Michael E., University of Wisconsin, Stevens Point
Ritz, Richard, Creighton University
Rochette, Scott M., SUNY, The College at Brockport
Rockwood, Anthony A., Metropolitan State College of Denver
Rodgers, N, Cardiff University
Rodriguez-Plasencia, Oscar, Univ Autonoma de Baja California Sur
Ross, Andrew, University of Leeds
Russell, William H., Los Angeles Pierce College
Ryan, William F., Pennsylvania State University, University Park
Sanabia, Elizabeth R., United States Naval Academy
Scarnato, Barbara V., Naval Postgraduate School
Schaffer, David L., Wichita State University
Schrage, Jon M., Creighton University
Schultz, David, University of Manchester
Seager, Richard, Columbia University
Seaman, Nelson L., Pennsylvania State University, University Park
Shannon, Jack D., Argonne National Laboratory
Shirer, Hampton N., Pennsylvania State University, University Park
Skubis, Steven T., SUNY, Oswego
Skyllingstad, Eric, Oregon State University
Smith, David R., United States Naval Academy
Smith, Dwight E., United States Naval Academy
Stamm, Alfred J., SUNY, Oswego
Stauffer, David R., Pennsylvania State University, University Park
Steiger, Scott, SUNY, Oswego
Steinacker, Reinhold, University of Vienna
Stensrud, David J., Pennsylvania State University, University Park
Straub, Katherine H., Susquehanna University
Syrett, William, Pennsylvania State University, University Park
Terwey, Wes, University of South Alabama
Tett, Simon F., Edinburgh University
Thomson, Dennis W., Pennsylvania State University, University Park
Todd, Steve, Portland Community College - Sylvania Campus
Torlaschi, Enrico, Universite du Quebec a Montreal
Tripoli, Gregory J., University of Wisconsin, Madison
Underwood, Stephen J., Georgia Southern University
Vaughan, Naomi, University of East Anglia
Verlinde, Johannes, Pennsylvania State University, University Park

Vernon, James Y., Los Angeles Pierce College
Wagner Riddle, Claudia, University of Guelph
Wallace, Tim, Mississippi State University
Wang, Bin, University of Hawai'i, Manoa
Wang, Qing, Naval Postgraduate School
Wash, Carlyle H., Naval Postgraduate School
Weisman, Robert A., Saint Cloud State University
Williams, Aaron, University of South Alabama
Williams, Forrest, Naval Postgraduate School
Williams, Roger T., Naval Postgraduate School
Wilson, John D., University of Alberta
Wyngaard, John C., Pennsylvania State University, University Park
Wysong, Jr., James F., Hillsborough Community College
Yang, Zong-Liang, University of Texas, Austin
Yoh, Shing, Kean University
Young, George S., Pennsylvania State University, University Park
Zhao, Jingxia, University of Hawai'i, Manoa
Zois, Constantine S., Kean University
Zunkel, Paul, Emporia State University

Atmospheric Chemistry
Aneja, Viney P., North Carolina State University
Dibb, Jack E., University of New Hampshire
Hess, Peter G., Cornell University
Schwab, James J., SUNY, Albany
Talbot, Robert, University of Houston

Atmospheric Physics
Atreya, Sushil, University of Michigan
Jeevanjee, Nadir, Princeton University
Malakar, Nabin, Worcester State University
Micallef, Alfred, University of Malta
Paget, Aaron C., Concord University
Torri, Giuseppe, University of Hawai'i, Manoa
Warren, Stephen G., University of Washington

Climatology
Ayanlade, Ayansina, Obafemi Awolowo University
Bradley, Alice, Williams College
Burnette, Dorian, University of Memphis
Crawford, Alex, College of Wooster
Daoust, Mario, Missouri State University
DeCosmo, Janice, University of Washington
Diem, Jeremy E., Georgia State Univ
Druckenbrod, Daniel L., Rider University
Feldl, Nicole, University of California, Santa Cruz
Foster, Stuart A., Western Kentucky University
Frei, Allan, Hunter College (CUNY)
Frye, John, University of Wisconsin, Whitewater
Hirschboeck, Katherine K., University of Arizona
Kelsey, Eric, Plymouth State University
Osborn, Timothy J., University of East Anglia
Payne, Ashley E., University of Michigan
Schultze, Steven R., University of South Alabama
Skinner, Christopher, University of Massachusetts, Lowell
Spencer, Jeremy M., University of Akron
Stine, Alexander, San Francisco State University
Strachan, Scotty, University of Nevada, Reno
Suriano, Zac, University of Nebraska, Omaha
Tian, Di, Auburn University
Tudhope, Alexander, University of Washington
Vega, Anthony J., Clarion University
Yow, Donald M., Eastern Kentucky University

GEOSCIENCE & SOCIETY
Natural Hazards
Aranda, Angela, California State University, Fullerton
Atwater, Brian, University of Washington
Blodgett, Robert H., Austin Community College District
Goda, Katsu, Western University
Haney, Jennifer, Bloomsburg University
Hill, Arleen A., University of Memphis
Holloway, Tess, Ashland University
Horwell, Claire, Durham University
Johanson, Erik, Florida Atlantic University
Kozak, Amanda L., Ashland University
Lavy, Brendan, University of Texas, Rio Grande Valley
Ludwig, Kris, University of Washington

Nethengwe, Nthaduleni S., University of Venda
Sirk, Robert A., Austin Peay State University
Talling, Peter, Durham University
Taylor, Mackenzie, Ashland University
Vidale, Jon, University of Washington
Weiss, Robert, Virginia Polytechnic Institute & State University
Wittke, Seth, Wyoming State Geological Survey
Wooten, Richard M., North Carolina Geological Survey

Manmade Hazards
Keil, Chris, Wheaton College

Water Quality/Use
Bouwer, Edward, Johns Hopkins University
Bradford, Joel, Utah Valley University
Keller, Troy, Columbus State University
Korths, Patrick, SUNY, Plattsburgh
Kuwabara, James, City College of San Francisco
Muller, Marc, University of Notre Dame
Rippy, Megan, Virginia Polytechnic Institute & State University
Steinschneider, Scott, Cornell University
Trainor-Guitton, Whitney, Colorado School of Mines
Zhang, Ning, Central State University

Geoscience Communication
Boss, Stephen K., University of Arkansas, Fayetteville
Conway, Michael F., Arizona Geological Survey
McCartney, M. Carol L., University of Wisconsin, Extension

History of Geoscience
Case, Stephen, Olivet Nazarene University
Dezzani, Raymond, University of Idaho
Drobnyk, John W., Southern Connecticut State University
Oreskes, Naomi, Harvard University
Rosenberg, Gary D., Milwaukee Public Museum
Strick, James E., Franklin and Marshall College

OTHER
General Earth Sciences
Agutu, Nathan O., Jomo Kenyatta University of Agriculture & Tech
Admire, Amanda R., Humboldt State University
Ahrens, Donald C., Modesto Junior College
Antipova, Anzhelika, University of Memphis
Balch, Jennifer K., University of Colorado
Bank, Carl-Georg, University of Toronto
Barron, Robert, Michigan Technological University
Beach Davis, Janet, Heartland Community College
Benjamin, Patricia A., Worcester State University
Berry, Leonard, Florida Atlantic University
Boger, Rebecca, Grad Sch of the City Univ of New York
Boitt, Mark, Jomo Kenyatta University of Agriculture & Technology
Booher, Gary, El Camino College
Boorstein, Margaret F., Long Island University, C.W. Post Campus
Bordossy, Andras, University of Newcastle Upon Tyne
Boyce, Joseph I., McMaster University
Bradley, Jeffrey, Northwest Missouri State University
Branlund, Joy, Southwestern Illinois College - Sam Wolf Granite City
Bretherton, Francis P., University of Wisconsin, Madison
Bries-Korpik, Jill, Century College
Britton, Gloria C., Cuyahoga Community College - Western Campus
Brod, Joseph D., Hillsborough Community College
Bruening, Susan, Eastern Connecticut State University
Burns, Danny, Coastal Bend College
Busa, Mark, Middlesex Community College
Carlson, Galen R., California State University, Fullerton
Carlson, Heath, Eastern Connecticut State University
Carrell, Jennifer, Illinois State Geological Survey
Carter, Andy, Birkbeck College
Champion, Kyle M., Hillsborough Community College
Chapa, Laura E., Austin Community College District
Chege, George W., Jomo Kenyatta University of Agriculture & Tech
Chen, jiquan, Michigan State University
Cheng, Zhongqi (Joshua), Grad Sch of the City Univ of New York
Cherrier, Jennifer, Grad Sch of the City Univ of New York
Childers, Daniel, Delaware County Community College
Cihacek, Larry J., North Dakota State University
Clark, Michael, Mount Royal University
Cleveland, Natasha, Frederick Community College
Clift, Sigrid, University of Texas, Austin

Faculty by Specialty

Cole, Gregory L., Los Alamos National Laboratory
Condreay, Denise, Central Community College
Cordero, David I., Lower Columbia College
Cornebise, Michael W., Eastern Illinois University
Cranganu, Constantin, Grad Sch of the City Univ of New York
Davis, James A., Eastern Illinois University
de Wit, Cary, University of Alaska, Fairbanks
Deka, Jennifer, Cuyahoga Community College - Western Campus
DeLima, Lynn-Ann, Eastern Connecticut State University
Dennison, Robert, Heartland Community College
Dod, Bruce D., Mercer University
Dolgoff, Anatole, Pace University, New York Campus
Douglas, Arthur V., Creighton University
Douthat, Jr., James R., Hillsborough Community College
Dutton, Jessica, Suffolk County Comm Coll, Ammerman Campus
Eaton, Timothy, Grad Sch of the City Univ of New York
Erski, Theodore, McHenry County College
Ezerskis, John L., Cuyahoga Community College - Western Campus
Falk, Lisa, North Carolina State University
Fatherree, James W., Hillsborough Community College
Fielding, Lynn, El Camino College
Fields, Nancy, El Centro College - Dallas Comm Coll District
Flanagan, Michael, Suffolk County Comm Coll, Ammerman Campus
Flores, Kennet, Grad Sch of the City Univ of New York
Fondran, Carol, Cuyahoga Community College - Western Campus
Fortes, Dominic, Birkbeck College
Foss, Donald J., College of Marin
Fritz, William, Grad Sch of the City Univ of New York
Gernon, Thomas M., University of Southampton
Gibbs, Samantha J., University of Southampton
Goetz, Andrew R., University of Denver
Goulden, Michael L., University of California, Irvine
Groffman, Peter, Grad Sch of the City Univ of New York
Hamilton, Thomas, Edmonds Community College
Harrington, Philip, Suffolk County Comm Coll, Ammerman Campus
Haverluk, Terry W., United States Air Force Academy
Hawkins, Ward L., Los Alamos National Laboratory
Hayes, Van E., Hillsborough Community College
Hemming, Gary N., Grad Sch of the City Univ of New York
Hendrey, George, Queens College (CUNY)
Hillier, John, Grays Harbor College
Hobbs, Chasidy, University of West Florida
Hoke, Gregory, Syracuse University
Hoppe, Kathryn A., Green River Community College
Hurd, David, Edinboro University of Pennsylvania
Ibrahim, Mohamed, Hunter College (CUNY)
Ide, Kayo, University of California, Los Angeles
Imwati, Andrew, Jomo Kenyatta University of Agriculture & Technology
Inglis, Michael, Suffolk County Comm Coll, Ammerman Campus
Jackson, Allyson K., SUNY, Purchase
Johnson, Jean M., Dalton State Community College
Johnson, Kathleen, University of California, Irvine
Jones, Stephen M., University of Birmingham
Kane, Mustapha, Florida Gateway College
Keefover-Ring, Ken, University of Wisconsin, Madison
Keeling, David J., Western Kentucky University
Kellner, Patricia, Los Angeles Harbor College
Kelly, Maria, Edmonds Community College
Khade, Vishnu R., Eastern Connecticut State University
Koch, Joe, South Carolina Dept of Natural Resources
Lane, Joseph M., Cuyahoga Community College - Western Campus
Le Qu, Corinne, University of East Anglia
Leger, Carol, Keene State College
Leighty, Robert S., Mesa Community College
Lerch, Derek, Feather River College
Levasseur, Emile, Eastern Connecticut State University
Locke, James L., College of Marin
Lofthouse, Stephen T., Pace University, New York Campus
Lomaga, Margaret, Suffolk County Comm Coll, Ammerman Campus
Longpre, Marc-Antoine, Grad Sch of the City Univ of New York
Luo, Johnny, Grad Sch of the City Univ of New York
Mack, John, Los Angeles Harbor College
Mackenzie, Kristen, Denver Museum of Nature & Science
MacNeil, James E., Hillsborough Community College
Marchisin, John, Pace University, New York Campus
Marlowe, Brian W., Hillsborough Community College
Marra, John, Grad Sch of the City Univ of New York
Martek, Lynnette F., University of South Carolina, Lancaster
Mason, Sherri A., SUNY, Fredonia
McConnaughhay, Mark, Dutchess Community College
McDougall, Jim, Tacoma Community College
McGinn, Chris, North Carolina Central University
Mcleod, Andrew T., Edinburgh University
McNicol, Barbara, Mount Royal University
Meador, Cindy D., West Texas A&M University
Medina, Francisco, Univ Nac Autonoma de Mexico
Meisner, Caroline B., Great Basin College
Mignery, Jill, Miami University
Mitra, Chandana, Auburn University
Morabia, Alfredo, Grad Sch of the City Univ of New York
Moreno, Rafael, Metropolitan State College of Denver
Morton, Bruce, Eastern Connecticut State University
Motyka, James, Eastern Connecticut State University
Munasinghe, Tissa, Los Angeles Harbor College
Nagel, Athena, Mississippi State University
Narey, Martha A., University of Denver
Neumann, Patricia, El Camino College
Neves, Douglas, El Camino College
Newlon, Charles F. J., William Jewell College
Nisbet, Euan G., Royal Holloway University of London
North, Leslie, Western Kentucky University
Norton-Krane, Abby N., Cuyahoga Community College - Western
O'Brien, Suzanne R., Bridgewater State University
O'Mullan, Gregory, Grad Sch of the City Univ of New York
Olney, Jessica L., Hillsborough Community College
Olney, Matthew P., Hillsborough Community College
Osborn, Joe, Century College
Oxford, Jeremiah, Western Oregon University
Oymayan, Avo, Florida Gateway College
Pant, Hari, Grad Sch of the City Univ of New York
Pappas, Matthew, Suffolk County Comm Coll, Ammerman Campus
Pedley, Kate, University of Canterbury
Pekar, Stephen, Grad Sch of the City Univ of New York
Peprah, Ebenezer, El Camino College
Perry, Baker, Appalachian State University
Perry, Randall, Imperial College
Pettengill, Gordon H., Massachusetts Institute of Technology
Phillippi, Nathan E., University of North Carolina, Pembroke
Pires, E M., Long Island University, C.W. Post Campus
Pittman, Jason, Folsom Lake College
Pond, Lisa G., Louisiana State University
Potess, Marla, Hardin-Simmons University
Pringle, Patrick T., Centralia College
Rapley, Chris, University College London
Renfrew, Melanie, Los Angeles Harbor College
Ritter, Paul, Heartland Community College
Rock, Jessie L., North Dakota State University
Rosa, Lynn C., West Texas A&M University
Roth, Leonard T., Hillsborough Community College
Ryan, Susan, California University of Pennsylvania
Sasowsky, Kathryn, Cuyahoga Community College - Western Campus
Saunders, Charles, Randolph-Macon College
Saunders, Ralph, California State University, Dominguez Hills
Schreier, Hans D., University of British Columbia
Schulze, Karl, Waubonsee Community College
Schumacher, Matthew, Saint Francis Xavier University
Shaw, Katy, Green River Community College
Shugart, Jr., Herman H., University of Virginia
Sloan, Heather, Grad Sch of the City Univ of New York
Smilnak, Roberta A., Metropolitan State College of Denver
Smith, Grant, Western Oregon University
Smith, H. Dixon, Metropolitan State College of Denver
Smith, Steven J., University of Northern Iowa
Smith, Thomas M., University of Virginia
Snyder, Jennifer L., Delaware County Community College
Soash, Norman E., Hillsborough Community College
Spooner, Alecia, Everett Community College
Springer, Kathleen B., San Bernardino County Museum
Stermer, Ed, Illinois Central College
Stone, Jim, Aims Community College
Sumrall, Jeanne L., Fort Hays State University
Szekielda, Karl-Heinz, Grad Sch of the City Univ of New York
Tarr, Alexander, Worcester State University
Thomas, Ray G., University of Florida

Thurston, Phillips C., Laurentian University, Sudbury
Tidwell, Allan, Chipola College
Tinsley, Mark, Central Virginia Community College
Tongdee, Poetchanaporn, Hillsborough Community College
Travers, Steven, Heartland Community College
Valencia, Victor A., Washington State University
Valentine, Michael, Tacoma Community College
Van Cleave, Kevan A., Hillsborough Community College
van de Gevel, Saskia, Appalachian State University
Veblen, Thomas T., University of Colorado
Voorhees, David H., Waubonsee Community College
Vorwald, Brian, Suffolk County Comm Coll, Ammerman Campus
Waddington, David C., Douglas College
Wade, Phillip, Western Oregon University
Walter, Nathan A., University of West Georgia
Wang, Enru, University of North Dakota
Ward, Calvin H., Rice University
Warger, Jane, Chaffey College
Webb, Craig A., Mt. San Antonio College
White, Susan, Los Angeles Harbor College
Wiesner, Mark R., Rice University
Williams, Mark W., University of Colorado
Wills, William V., Hillsborough Community College
Winterbottom, Wesley, Eastern Connecticut State University
Witter, Rob, Oregon Dept of Geology & Mineral Industries
Woods, Karen M., Lamar University
XU, Yingfeng, University of Louisiana at Lafayette
Yacucci, Mark, Heartland Community College
Zaleha, Robert, Cuyahoga Community College - Western Campus

Earth Science Education
Albach, Suzanne M., Washtenaw Community College
Albrecht, Jochen, Grad Sch of the City Univ of New York
Atchison, Christopher L., University of Cincinnati
Attrep, Moses, Los Alamos National Laboratory
Aylward, Linda, Illinois Central College
Baker, Brett, University of Texas, Austin
Barnett, Michael, Boston College
Bechtel, Randy, North Carolina Geological Survey
Bednarski, Marsha, Central Connecticut State University
Bednarz, Robert S., Texas A&M University
Bednarz, Sarah W., Texas A&M University
Bergwerff, Kenneth A., Calvin College
Bernard-Kingsley, Noell, University of Washington
Black, Alice (Jill), Missouri State University
Blanford, William , Grad Sch of the City Univ of New York
Blaszczak-Boxe, Christopher, Grad Sch of the City Univ of New York
Block, Karin, Grad Sch of the City Univ of New York
Brueckner, Hannes, Grad Sch of the City Univ of New York
Buchanan, Rex C., University of Kansas
Buonaiuto Jr., Frank S., Grad Sch of the City Univ of New York
Burns, Sandra, Central Connecticut State University
Butcher, Patricia M., California State University, Fullerton
Carden-Jessen, Melanie E., Missouri State University
Carmalt, Jean, Grad Sch of the City Univ of New York
Cercone, Karen Rose, Indiana University of Pennsylvania
Chandler, Sandra, Arkansas Geological Survey
Clark, Scott K., University of Wisconsin, Eau Claire
Clary, Renee M., Mississippi State University
Clepper, Marta L., SUNY, Oneonta
Coburn, Daniel, Southern Connecticut State University
Cooney, Timothy M., University of Northern Iowa
Crowder, Margaret, Western Kentucky University
d'Alessio, Matthew, California State University, Northridge
Dakin, Susan, University of Lethbridge
Dalman, Michael, Blinn College
Davies, Gareth, The Open University
Davies, Sarah, The Open University
De Kock, Walter E., University of Northern Iowa
de Ritter, Samantha, Flinders University
Dehne, Kevin T., Delta College
DeKraker, Dan, Cerritos College
Denton-Hedrick, Meredith Y., Austin Community College District
DePriest, Thomas A., University of Tennessee, Martin
Dieckmann, Melissa S., Eastern Kentucky University
Douglass, David N., Pasadena City College
Elkins, Joe T., University of Northern Colorado
Emerson, Cheryl R., Illinois Central College

Englebright, Stephen C., Stony Brook University
Ensign, Todd I., Fairmont State University
Falkena, Lee, University of Illinois at Chicago
Faust, Megan, Portland State University
Ferguson, Julie , University of California, Irvine
Flanagan, Timothy, Berkshire Community College
Fleck, Michelle, Utah State University
Frankic, Anamarija, University of Massachusetts, Boston
Geraghty Ward, Emily, Rocky Mountain College
Gibbons, Doug, University of Washington
Glesener, Gary, Virginia Polytechnic Institute & State University
Godsey, Holly, University of Utah
Goodell, Laurel P., Princeton University
Gosselin, David C., University of Nebraska, Lincoln
Gray, Kyle R., University of Northern Iowa
Gross, Julian, Grad Sch of the City Univ of New York
Hamane, Angel, California State University, Los Angeles
Hammond, Amie C., Austin Community College District
Harrell, Michael, University of Washington
Heid, Kelly L., Grand Valley State University
Hemler, Deb, Fairmont State University
Hepple, Alex, Queensland University of Technology
Herman, Theodore C., West Valley College
Holliman, Richard, The Open University
Hubenthal, Michael, Binghamton University
Hunter, Arlene G., The Open University
Jahanyar, Alireza, University of Tabriz
Johnston, Thomas, University of Lethbridge
Kornreich Wolf, Susan, Augustana College
Kortz, Karen M., Community College of Rhode Island
Krause, Lois B., Clemson University
Krockover, Gerald H., Purdue University
Kusnick, Judith E., California State University, Sacramento
LaDue, Nicole D., Northern Illinois University
Libarkin, Julie C., Michigan State University
Longoria, Jose F., Florida International University
Ludwikoski, David J., Comm College of Baltimore County, Catonsville
MacLachlan, Ian R., University of Lethbridge
Martin-Vermilyea, Laurie, Montgomery County Community College
Mattox, Stephen R., Grand Valley State University
Maxwell, James "Sandy", West Chester University
McCollough, Cherie, Texas A&M University, Corpus Christi
McConnell, David A., North Carolina State University
McCutcheon, Karen Sue, Casper College
McEvoy, Jamie , Montana State University
McGeary, Susan, University of Delaware
McNeal, Karen S., Auburn University
Messina, Paula, San Jose State University
Miller, Heather R., Austin Community College District
Miyares, Ines, Hunter College (CUNY)
Morgan, Emory, Hampton University
Morton, Allan E., Central Arizona College
Mulkey, Sean, Illinois Central College
Munski, Douglas C., University of North Dakota
Murray, Daniel P., University of Rhode Island
Nakagawa, Masami, Colorado School of Mines
O'Neal, Claire J., University of Delaware
Oxley, Meghan, University of Washington
Papcun, George, New Jersey City University
Pepple, Christopher, Bowling Green State University
Petit, Martin A., Illinois Central College
Phillips, Paul, Fort Hays State University
Preston, William L., California Polytechnic State University
Pyle, Eric J., James Madison University
Rakonczai, Janos, Univesity of Szeged
Raol, Marcie, Fairmont State University
Refenes, James L., Concordia University
Revels, Josh, Fairmont State University
Roemmele, Christopher, West Chester University
Russ, Jean M., College of Southern Maryland
Ryker, Katherine, University of South Carolina
Schaffer, Linda J., SUNY, The College at Brockport
Schmitt, Danielle M., Princeton University
Scott, Vernon, Oklahoma State University
Semken, Steven, Arizona State University
Sharp, S. L., Virginia Polytechnic Institute & State University
Shelton, Dale W., Maryland Department of Natural Resources

Shepardson, Daniel P., Purdue University
Shipp, Stephanie S., Rice University
Skinner, Lisa, Northern Arizona University
Slattery, William, Wright State University
Smay, Jessica J., San Jose City College
Spurr, Aaron, University of Northern Iowa
Straw, Byron, University of Northern Colorado
Sundareshwar, P. V., South Dakota School of Mines & Technology
Sutton, Connie J., Indiana University of Pennsylvania
Taylor, Ta-Shana A., University of Miami
Teed, Rebecca, Wright State University
Thompson, Kenneth W., Emporia State University
Tonon, Marco Davide, Università di Torino
Villalobos, Joshua, El Paso Community College
Viskupic, Karen, Boise State University
Wagner, John R., Clemson University
White, Lisa D., University of California, Berkeley
Wood, Howard, Texas A&M University, Corpus Christi
Wright, Carrie L., University of Southern Indiana
Young, James E., Appalachian State University
Zipper, Carl E., Virginia Polytechnic Institute & State University
Zlotkin, Howard, New Jersey City University
Zwick, Thomas T., Montana State University, Billings

Physical Geography
Aay, Henry, Calvin College
Abend, Martin, New Jersey City University
Aber, Jeremy, Middle Tennessee State University
Adam, Iddrisu, University of Wisconsin Colleges
Ajassa, Roberto, Università di Torino
Albright, Thomas , University of Nevada, Reno
Allen, Kelly E., Tulsa Community College
Amador, Nathanael S., Ohio Wesleyan University
Andresen, Christian, University of Wisconsin, Madison
Applegate, Toby, University of Massachusetts, Amherst
Aubert, John E., American River College
Balascio, Nicholas, College of William & Mary
Ballantyne, Colin K., University of St. Andrews
Barnett, Roger T., University of the Pacific
Barton, James H., Thiel College
Bascom, Johnathan, Calvin College
Bass, David, Flinders University
Bates, Mary, College of the Canyons
Battles, Eileen, University of Kansas
Bein, F. L., Indiana University, Indianapolis
Bekker, Matthew F., Brigham Young University
Berry, Kate, University of Nevada, Reno
Bitner, Thomas, University of Wisconsin Colleges
Black, Kathryn, Montclair State University
Blackburn, William, Western Kentucky University
Blewett, William L., Shippensburg University
Boester, Michael, Monroe Community College
Brothen, Jerry, El Camino College
Brothers, Timothy S., Indiana University, Indianapolis
Butzow, Dean G., Lincoln Land Community College
Cairns, David M., Texas A&M University
Camille, Michael A., University of Louisiana, Monroe
Campbell, Glenn A., Eastern Kentucky University
Carter, Greg, University of Southern Mississippi
Chacko, Elizabeth, George Washington University
Chaddock, Lisa, Cuyamaca College
Chatterjee, Meera, University of Akron
Clarke, Beverley, Flinders University
Clennan, Patrick D., College of Southern Nevada - West Charleston
Coburn, Craig, University of Lethbridge
Comrie, Andrew, University of Arizona
Coveney, Eamonn, Fort Hays State University
Croft, Gary M., San Bernardino Valley College
Cross, John A., University of Wisconsin Oshkosh
Csank, Adam, University of Nevada, Reno
Csiki, Shane, New Hampshire Geological Survey
Curry, Janel M., Calvin College
Davidson, Fiona M., University of Arkansas, Fayetteville
Dawsey, Cyrus B., Auburn University
de Uña Alvarez, Elena P., Coruna University
Deckert, Garret, University of Wisconsin Colleges
DeWeese, Georgina G., University of West Georgia
Dexter, Leland R., Northern Arizona University
Dommissee, Edwin, University of Wisconsin Colleges
Dubsky, Scott , United States Air Force Academy
Dugmore, Andrew J., Edinburgh University
Duncan, Ian, City College of San Francisco
Ebiner, Matt, El Camino College
Eisenhart, Karen, Edinboro University of Pennsylvania
Engstrom, Vanessa, San Bernardino Valley College
Feeney, Thomas P., Shippensburg University
Fonstad, Mark, University of Oregon
Fratianni, Simona, Università di Torino
Frolking, Tod A., Denison University
Gamble, Douglas W., University of North Carolina, Wilmington
Gaubatz, Piper, University of Massachusetts, Amherst
Gavin, Dan, University of Oregon
Girhard, T S., San Antonio Community College
Goudge, Theodore L., Northwest Missouri State University
Graham, Barbara, College of Southern Nevada - West Charleston Campus
Greene, Don M., Baylor University
Gripshover, Margaret , Western Kentucky University
Hall, Luke D., Ventura College
Hampson, Arthur, University of Utah
Harley, Grant, University of Idaho
Harrison, Robert S., Long Island University, C.W. Post Campus
Hart, Evan A., Tennessee Tech University
Harty, John P., Johnson County Community College
Hay, Iain, Flinders University
Hazen, Helen, University of Denver
Hedquist, Brent, Texas A&M University, Kingsville
Heibel, Todd, San Bernardino Valley College
Hepner, George F., University of Utah
Hess, Darrel E., City College of San Francisco
Heywood, Neil C., University of Wisconsin, Stevens Point
Hickcox, David H., Ohio Wesleyan University
Hindle, Tobin, Florida Atlantic University
Holland, Edward C., University of Arkansas, Fayetteville
Holt, David, University of Southern Mississippi
Houston, Serin D., Mount Holyoke College
Howard, Leslie M., Unversity of Nebraska - Lincoln
Ibrahim, Mohamed B., Grad Sch of the City Univ of New York
Jarvis, Richard S., University of Texas, El Paso
Jeon, Sung Bae, Eastern Kentucky University
Johnson, Mark O., Worcester State University
Jones, Alice, Eastern Kentucky University
Jordan, Karen J., University of South Alabama
Karpilo, Jr, Ronald J., Colorado State University
Kebbede, Girma, Mount Holyoke College
Kennedy, Linda, Mansfield University
Key, Doug, Palomar College
Khan, Belayet H., Eastern Illinois University
Kiage, Lawrence M., Georgia State Univ
Kilcoyne, John R., Metropolitan State College of Denver
Kimber, Clarissa T., Texas A&M University
Kung, Hsiang-Te, University of Memphis
Kuzera, Kristopher , University of Denver
Laity, Julie E., California State University, Northridge
Lambert, Dean P., San Antonio Community College
Lambert, Dean, Alamo Colleges, San Antonio College
League, Larry D., Dickinson State University
Lee, Jeffrey A., Texas Tech University
Lee, Russell, Oak Ridge National Laboratory
Lineback, Neal G., Appalachian State University
Little, Jonathan, Monroe Community College
Liu, Jing, Santa Monica College
LoVetere, Crystal, Cerritos College
Lynn, Resler M., Virginia Polytechnic Institute & State University
MacQuarrie, Pamela, Mount Royal University
Maher, John, Johnson County Community College
Maio, Chris, University of Alaska, Fairbanks
Mann, Dan, University of Alaska, Fairbanks
Manos, Leah D., Northwest Missouri State University
Markwith, Scott H., Florida Atlantic University
Marzulli, Walter, New Jersey Geological and Water Survey
McAfee, Stephanie , University of Nevada, Reno
McCallister, Robert, University of Wisconsin Colleges
McCoy, William D., University of Massachusetts, Amherst
McCraw, David J., New Mexico Institute of Mining & Technology
McGwire, Kenneth C., Desert Research Institute

Mensing, Scott A., University of Nevada, Reno
Millette, Thomas L., University of Massachusetts, Amherst
Minnich, Richard A., University of California, Riverside
Montgomery, Keith, University of Wisconsin Colleges
Montoya, Judith, Northern Arizona University
Morris, Pete, Santa Monica College
Motta, Luigi, Università di Torino
Motta, Michele, Università di Torino
Musolf, Gene E., University of Wisconsin Colleges
Nelson, Kenneth A., University of Kansas
Ngoy, Kikombo, Kean University
Nicholas, Joseph W., Mary Washington College
Norwood, James, Auburn University
Nuwategeka, Expedito, Gulu University
Ogbuchiekwe, Edmund Jekwu, San Bernardino Valley College
Oldfield, Jonathan, University of Birmingham
Olson, Kimberly, American River College
Paulsell, Robert L., Louisiana State University
Peake, Jeffrey S., University of Nebraska, Omaha
Pennington, Deana, University of Texas, El Paso
Pesses, Michael, Antelope Valley College
Phillips, Nathan, Boston University
Pierce, Heather, Monroe Community College
Pinnt, Todd, Arizona Western College
Ponette-Gonzalez, Alexandra, University of North Texas
Precht, Francis L., Frostburg State University
Price, Alan P., University of Wisconsin Colleges
Price, Marie D., George Washington University
Privette, David, Central Piedmont Community College
Rahman, Abu, Antelope Valley College
Reese, Andy, University of Southern Mississippi
Rettig, Andrew, University of Dayton
Robinson, Michael A., Santa Barbara City College
Rodgers, John C., Mississippi State University
Rohe, Randall, University of Wisconsin Colleges
Rollins, Kyle, Missouri Dept of Natural Resources
Rose, Leanna S., University of West Georgia
Ross, Thomas E., University of North Carolina, Pembroke
Rudnicki, Ryan E., San Antonio Community College
Rudnicki, Ryan E., Alamo Colleges, San Antonio College
Ryan, Christopher, University of Nevada, Reno
Sandlin, Stephen H., San Bernardino Valley College
Santelmann, Mary V., Oregon State University
Schafer, Tom, Fort Hays State University
Schmidt, Lisa, San Bernardino Valley College
Scott, Larry, Colorado Geological Survey
Scott, Thomas A., University of California, Riverside
Sharkey, Debra, Cosumnes River College
Smith, C. K., San Antonio Community College
Smith, Charles K., Alamo Colleges, San Antonio College
Smith, Steven C., American River College
Solecki, William , Grad Sch of the City Univ of New York
Stahmann, Paul, McHenry County College
Stampone, Mary D., University of New Hampshire
Starrs, Paul F., University of Nevada, Reno
Stern, Herschel I., MiraCosta College
Stevens, Stan, University of Massachusetts, Amherst
Stocks, Lee, Mansfield University
Swarts, Stanley W., Northern Arizona University
Theo, Lisa L., University of Wisconsin, Stevens Point
Thomsen, Charles E., American River College
Todhunter, Paul, University of North Dakota
Tremblay, Thomas A., Univ of Texas at Austin, Jackson Sch of Geo
van Leeuwen, Willem, University of Arizona
Veverka, Laura, Metropolitan Community College-Kansas City
Vogel, Eve, University of Massachusetts, Amherst
Walter, Thomas, Hunter College (CUNY)
Watson, James R., Oregon State University
Webb, Craig, Eastern Kentucky University
Weisberg, Michael, Grad Sch of the City Univ of New York
Welford, Mark R., Georgia Southern University
West, Keith, University of Wisconsin Colleges
White, Justin, Utah Valley University
Wilkie, Richard W., University of Massachusetts, Amherst
Wilkins, David E., Boise State University
Wilson, Jeffey S., Indiana University, Indianapolis
Wu, Shuang-Ye, University of Dayton

Wyckoff, John W., University of Colorado, Denver
Yassin, Barbara E., Mississippi Office of Geology
Ye, Hengchun, California State University, Los Angeles
Zhou, Yuyu, Iowa State University of Science & Technology

Remote Sensing

Aanstoos, James V., Iowa State University of Science & Technology
Abdalati, Waleed, University of Colorado
Abrams, Michael J., Jet Propulsion Laboratory
Ackerman, Steven A., University of Wisconsin, Madison
Ahearn, Sean C., Grad Sch of the City Univ of New York
Ali, K. Adem, College of Charleston
Alley, Ronald E., Jet Propulsion Laboratory
Bandfield, Joshua, University of Washington
Barros, Jose Antonio, Florida International University
Beaverson, Sheena K., Illinois State Geological Survey
Becker, Richard H., University of Toledo
Becker, Richard H., Argonne National Laboratory
Berlin, Graydon L., Northern Arizona University
Bernier, Monique, Universite du Quebec
Binford, Micheal W., University of Florida
Blom, Ronald G., Jet Propulsion Laboratory
Boulton, Sarah, University of Plymouth
Broadfoot, Lyle A., University of Arizona
Buratti, Bonnie J., Jet Propulsion Laboratory
Cablk, Mary, Desert Research Institute
Callahan, John A., University of Delaware
Calvin, Wendy, University of Nevada, Reno
Campbell, Bruce A., Smithsonian Institution / National Air & Space Mus
Campbell, James B., Virginia Polytechnic Institute & State University
Carn, Simon, Michigan Technological University
Cetin, Haluk, Murray State University
Chen, Shu-Hua, University of California, Davis
Chen, Xianfeng, Slippery Rock University
Chopping, Mark J., Montclair State University
Crippen, Robert E., Jet Propulsion Laboratory
Dai, Chunli, Ohio State University
Dana, Gayle L., University of Nevada, Reno
Dash, Padmanava, Mississippi State University
Delahunty, Tina, Bloomsburg University
Di Girolamo, Larry, University of Illinois, Urbana-Champaign
Dodge, Rebecca L., Midwestern State University
Dondofema, Farai, University of Venda for Science & Technology
Duchesne, Rocio, University of Wisconsin, Whitewater
Easson, Gregory L., University of Mississippi
El-Baz, Farouk, Boston University
Engstrom, Ryan, George Washington University
Fenstermaker, Lynn, Desert Research Institute
Ferris, Michael H., California State University, Dominguez Hills
Fildes, Stephen , Flinders University
Flynn, Luke P., University of Hawai'i, Manoa
Forster, Richard R., University of Utah
Frankenberg, Christian, California Institute of Technology
Friedl, Mark, Boston University
Friedl, Mark A., Boston University
Frost, Eric G., San Diego State University
Gaitan-Moran, Javier, Univ Autonoma de Baja California Sur
Gammack-Clark, James, Florida Atlantic University
Gamon, John, University of Alberta
Gaulton, Rachel, University of Newcastle Upon Tyne
Gaya, Charles , Jomo Kenyatta University of Agriculture & Technology
Ghent, Rebecca, University of Toronto
Gibson, Glen, United States Air Force Academy
Gilbes, Fernando, University of Puerto Rico
Gimmestad, Gary G., Georgia Institute of Technology
Giraldo, Mario A., California State University, Northridge
Glenn, Nancy, Boise State University
Goetz, Alexander, University of Colorado
Hacker, Jorg, Flinders University
Hanna, Stephen P., Mary Washington College
Hardin, Perry J., Brigham Young University
Haritashya, Umesh, University of Dayton
Hay, Rodrick, California State University, Dominguez Hills
Heidinger, Andrew, University of Wisconsin, Madison
Hernandez, Michael W., Weber State University
Hinojosa, Alejandro, Centro de Inv Cientifica y de Ed Sup de Ensenada
Hobbs, Richard D., Amarillo College
Hook, Simon J., Jet Propulsion Laboratory

Hossain, A K M Azad, University of Tennessee, Chattanooga
Hu, Baoxin, York University
Hu, Zhiyong, University of West Florida
Huang, Yi, McGill University
Humes, Karen, University of Idaho
Hung, Ming-Chih, Northwest Missouri State University
Hutchinson, Charles F., University of Arizona
Jaworski, Eugene, Eastern Michigan University
Ji, Wei, University of Missouri, Kansas City
Jin, Yufang, University of California, Davis
Johnson, Elias, Missouri State University
Kahle, Anne B., Jet Propulsion Laboratory
Kasar, Sheila , Clarion University
Kenduiywo, Benson K., Jomo Kenyatta University of Agriculture & Tech
Kennedy, Robert, Oregon State University
Kern, Anikó, Eotvos Lorand University
Kessler, Fritz, Frostburg State University
Key, Jeffrey R., University of Wisconsin, Madison
Khan, Shuhab D., University of Houston
Klein, Andrew G., Texas A&M University
Klemas, Victor, University of Delaware
Kollias, Pavlos, McGill University
Kudela, Raphael M., University of California, Santa Cruz
Lakhan, V. Chris, University of Windsor
Larson, Eric J., University of Wisconsin, Stevens Point
Lawrence, Rick L., Montana State University
Lee, Keenan, Colorado School of Mines
Lee, Zhongping, University of Massachusetts, Boston
Liu, Jian G., Imperial College
Lu, Zhong, Southern Methodist University
Lulla, Kamlesh, University of Delaware
Lyon, Ronald J. P., Stanford University
Malhotra, Rakesh, North Carolina Central University
Mason, Philippa J., Imperial College
Mbuh, Mbongowo J., University of North Dakota
Mccoy, Roger M., University of Utah
McDade, Ian C., York University
McGrath, Andrew, Flinders University
McGwire, Kenneth, University of Nevada, Reno
Merchant, James W., Unversity of Nebraska - Lincoln
Meyer, Franz J., University of Alaska, Fairbanks
Miao, Xin, Missouri State University
Middleton, Carrie A., Dept of Interior Office of Surface Mining Reclamation and Enforcement
Milan-Navarro, Joel, Universidad Autonoma de San Luis Potosi
Miller, John, York University
Millette, Thomas L., Mount Holyoke College
Minor, Timothy B., Desert Research Institute
Molnár, Gábor, Eotvos Lorand University
Molnia, Bruce F., Duke University
Moon, Wooil M., University of Manitoba
Mshiu, Elisante E., University of Dar es Salaam
Muller-Karger, Frank E., University of South Florida
Mushkin, Amit, University of Washington
Mustard, John F., Brown University
Myneni, Ranga , Boston University
Nduati, Eunice W., Jomo Kenyatta University of Agriculture & Tech
Newland, Franz, York University
Ngigi, Thomas G., Jomo Kenyatta University of Agriculture & Tech
Ni-Meister, Wenge, Hunter College (CUNY)
Njoku, Eni G., Jet Propulsion Laboratory
Noh, Myoung-Jong, Ohio State University
O'Banion, Matthew S., United States Military Academy
Palluconi, Frank D., Jet Propulsion Laboratory
Parrish, Christopher E., University of New Hampshire
Pathirana, Sumith, Southern Cross University
Peddle, Derek R., University of Lethbridge
Peltzer, Gilles, University of California, Los Angeles
Petrie, Gregg, University of Washington
Pieters, Carle M., Brown University
Platnick, Steven, University of Wisconsin, Madison
Powell, Rebecca , University of Denver
Price, Kevin, Kansas State University
Qin, Rongjun, Ohio State University
Quisenberry, Dan R., Mercer University
Rahman, Abdullah F., University of Texas, Rio Grande Valley
Ramage, Joan, Lehigh University

Ramsey, Michael S., University of Pittsburgh
Ramspott, Matthew E., Frostburg State University
Rapp, Anita, Texas A&M University
Reed, Wallace E., University of Virginia
Ridd, Merrill K., University of Utah
Rivard, Benoit, University of Alberta
Roberts, Charles E., Florida Atlantic University
Rundquist, Bradley C., University of North Dakota
Rundquist, Donald C., Unversity of Nebraska - Lincoln
Sadiq, Abdulali A., University of Qatar
Sahr, John D., University of Washington
Schaaf, Crystal, Boston University
Schaaf, Crystal, University of Massachusetts, Boston
Schenk, Anton, SUNY, Buffalo
Seirer, Jami J., Fort Hays State University
Small, Christopher, Columbia University
Smith, Mike, Kingston University
Smith, Peter, University of Arizona
Strahler, Alan, Boston University
Strain, Priscilla L., Smithsonian Institution / National Air & Space Mus
Sui, Daniel Z., Texas A&M University
Sultan, Mohamed, Western Michigan University
Szatmári, József, Univesity of Szeged
Thomas, Jeremy, University of Washington
Torres Parisian, Cathleen, University of Minnesota, Twin Cities
Toth, Charles K., Ohio State University
Tullis, Jason A., University of Arkansas, Fayetteville
Ustin, Susan L., University of California, Davis
Ustin, Susan L., University of California, Davis
van der Werf, Guido R., Vrije Universiteit Amsterdam
Vandemark, Douglas C., University of New Hampshire
Velicogna, Isabella, University of California, Irvine
Ventura, Stephen J., University of Wisconsin, Madison
Veraverbeke, Sander, Vrije Universiteit Amsterdam
Viertel, Dave, Eastern Illinois University
Walker, Richard, University of Oxford
Wang, Lei, Ohio State University
Warner, Timothy A., West Virginia University
Wasomi, Charles B., Jomo Kenyatta University of Agriculture & Tech
Watson, Kelly, Eastern Kentucky University
Wei, Xiaofang, Central State University
Weng, Qihao, Indiana State University
White, Joseph D., Baylor University
Woo, David, California State University, East Bay
Woodcock, Curtis, Boston University
Woodcock, Curtis E., Boston University
Wright, William C., United States Military Academy
Yan, Xiao-Hai, University of Delaware
Yang, Zhiming, North Carolina Central University
Yool, Stephen R., University of Arizona
Yu, Qian, University of Massachusetts, Amherst
Zhang, Caiyun, Florida Atlantic University
Zhou, Xiaobing, Montana Tech of the University of Montana
Zhou, Xiaolu, Georgia Southern University

Material Science
Alfe, Dario, University College London
Calaway, Wallis F., Argonne National Laboratory
Chiarello, Ronald P., Argonne National Laboratory
Chrzan, Daryl, University of California, Berkeley
deFontaine, Didier, University of California, Berkeley
DeJonghe, Lutgard, University of California, Berkeley
Devine, Thomas M., University of California, Berkeley
Eberhart, Mark E., Colorado School of Mines
Ferrari, Mauro, University of California, Berkeley
Glaeser, Andreas, University of California, Berkeley
Gronsky, Ronald, University of California, Berkeley
Gruen, Dieter M., Argonne National Laboratory
Haller, Eugene, University of California, Berkeley
Hayden, Geoffrey W., Mercer University
Hojjatie, Barry, Valdosta State University
Lu, Yongqi, University of Illinois
Morris, Jr., J. W., University of California, Berkeley
Nakote, Heinz, New Mexico State University, Las Cruces
Navrotsky, Alexandra, University of California, Davis
Pellin, Michael J., Argonne National Laboratory
Pylypenko, Svitlana, Colorado School of Mines
Ritchie, Robert O., University of California, Berkeley

Sands, Timothy, University of California, Berkeley
Strobel, Timothy A., Carnegie Institution for Science
Valiunas, Jonas K., Lakehead University
Voller, Vaughan R., University of Minnesota, Twin Cities
Weber, Eicke, University of California, Berkeley

Land Use/Urban Geology

Agrawal, Sandeep , University of Alberta
Bassett, Scott, University of Nevada, Reno
Bereitschaft, Bradley, University of Nebraska, Omaha
Birchall, Jeff, University of Alberta
Blackburn, James B., Rice University
Bodenman, John E., Bloomsburg University
Bonine, Michael E., University of Arizona
Bremer, Keith A., Fort Hays State University
Briggs, John, University of Glasgow
Buermann, Wolfgang, University of Leeds
Clark, Jessie , University of Nevada, Reno
Day, Rick L., Pennsylvania State University, University Park
Dixon, Nick, University of Leeds
Dolman, Paul, University of East Anglia
Emmi, Philip C., University of Utah
Ford, Andrew, Washington State University
Fusch, Richard D., Ohio Wesleyan University
Fürst, Christine, Martin-Luther-Universitaet Halle-Wittenberg
Godfrey, Brian J., Vassar College
Gosnell, Hannah, Oregon State University
Graff, Thomas O., University of Arkansas, Fayetteville
Graumlich, Lisa J., Montana State University
Haggerty, Julia H., Montana State University
Harris, Richard S., McMaster University
Hines, Mary E., University of North Carolina, Wilmington
Hintz, Rashauna, University of Arkansas, Fayetteville
Holl, Karen D., University of California, Santa Cruz
Hungerford, Hilary, Utah Valley University
James, Valentine U., Clarion University
Jantz, Claire A., Shippensburg University
Jenssen, Petter, Norwegian University of Life Sciences (NMBU)
Kaushal, Sujay, University of Maryland
Larson, David J., California State University, East Bay
Lee, Chung M., University of Utah
Liao, Haifeng (Felix), University of Idaho
Marston, Sallie A., University of Arizona
Mercier, Michael , McMaster University
Millington, Andrew, Flinders University
Moscovici, Daniel, Stockton University
Nevins, Joseph, Vassar College
Odogba, Ismaila, University of Wisconsin, Stevens Point
Ormerod, Kerri Jean , University of Nevada, Reno
Otterstrom, Samuel M., Brigham Young University
Peace, Walter G., McMaster University
Platt, Rutherford H., University of Massachusetts, Amherst
Pomeroy, George M., Shippensburg University
Ptak, Tom, University of Idaho
Randlett, Victoria, University of Nevada, Reno
Riebesell, John, University of Michigan, Dearborn
Rounsevell, Mark D., Edinburgh University
Shirgaokar, Manish, University of Alberta
Singh, Amrita, University of Alberta
Skillen, James, Calvin College
Solecki, William, Hunter College (CUNY)
Solecki, William, Montclair State University
Stark, Robin, Northwestern University
Steinberger, Julia, University of Leeds
Sullivan, Jack B., University of Maryland
Thomas, Valerie, Georgia Institute of Technology
Tilt, Jenna, Oregon State University
Troost, Kathy G., University of Washington
Van Den Hoek, Jamon, Oregon State University
Wickham, Thomas, California University of Pennsylvania
Wollheim, Wil, University of New Hampshire
Zhou, Yu, Vassar College

Geographic Information Systems

Adhikari, Sanchayeeta, California State University, Northridge
Ahearn, Sean C., Hunter College (CUNY)
Albrecht, Jochen, Hunter College (CUNY)
Alderson, David, University of Newcastle Upon Tyne
Algeo, Catherine, Western Kentucky University
Allen, Katherine, University of Liverpool
Allen-Lafayette, Zehdreh, New Jersey Geological and Water Survey
Ambinakudige, Shrinidhi, Mississippi State University
Amer, Reda, Tulane University
Andrews, John R., Univ of Texas at Austin, Jackson Sch of Geosciences
Anthony-Zajanc, Kate, Idaho State University
Argles, Tom, The Open University
Artigas, Francisco, Rutgers, The State Univ of New Jersey, Newark
Ayad, Yasser M., Clarion University
Badruddin, Abu Z., Cayuga Community College
Badurek, Chris, Appalachian State University
Bailey, Keiron D., University of Arizona
Barr, Stuart, University of Newcastle Upon Tyne
Barrett, Melony, Illinois State Geological Survey
Bauer, Emily, University of Minnesota
Beaty, Tammy, Oak Ridge National Laboratory
Becker, Lorene Y., Oregon State University
Benger, Simon, Flinders University
Benson, Jane L., Murray State University
Betchwars, Corey, University of Minnesota
Bloechle, Amber, University of West Florida
Bloomgren, Bruce, University of Minnesota
Boger, Rebecca, Brooklyn College (CUNY)
Boroushaki, Soheil, California State University, Northridge
Bottenberg, H. Carrie, Idaho State University
Bowlick, Forrest J., University of Massachusetts, Amherst
Boyack, Diana L., Idaho State University
Breton, Caroline, Univ of Texas at Austin, Jackson Sch of Geosciences
Brown, Douglas, Kingston University
Bruns, Dale A., Wilkes University
Brunskill, Jeffrey C., Bloomsburg University
Burton, Christopher G., Auburn University
Busby, Michael R., Murray State University
Cao, Guofeng, Texas Tech University
Carstensen, Laurence W., Virginia Polytechnic Institute & State University
Carter, David, Department of Interior Office of Surface Mining Reclamation and Enforcement
Cary, Kevin, Western Kentucky University
Cheng, Qiuming, York University
Cheung, Wing, Palomar College
Cheung, Wing H., Palomar College
Chokmani, Karem, Universite du Quebec
Chopra, Prame N., Australian National University
Christopherson, Gary, University of Arizona
Cicha, Jarrod, University of Minnesota
Cloutis, Ed, Western University
Coetzee, Serena M., University of Pretoria
Colby, Jeff, Appalachian State University
Compas, Eric, University of Wisconsin, Whitewater
Congalton, Russell, University of New Hampshire
Cothren, Jackson D., University of Arkansas, Fayetteville
Coulibaly, Mamadou, University of Wisconsin Oshkosh
Cova, Thomas J., University of Utah
Coveney, Seamus, University of Glasgow
Cunningham, Mary A., Vassar College
Curri, Neil, Vassar College
Curry, Susan, University of Florida
Dai, Dajun, Georgia State Univ
Davis, Trevor J., University of Utah
De Kemp, Eric, University of Ottawa
Deakin, Ann K., SUNY, Fredonia
Deal, Richard, Edinboro University of Pennsylvania
Delparte, Donna M., Idaho State University
Demyanov, Vasily, Heriot-Watt University
Diver, Kim, Wesleyan University
Dogwiler, Toby, Missouri State University
Dong, Pinliang, University of North Texas
Donnelly, Shanon P., University of Akron
Douglass, Gordon, Utah Geological Survey
Drake, Luke, California State University, Northridge
Drummond, Jane, University of Glasgow
Drzyzga, Scott A., Shippensburg University
Duke, Jason E., Tennessee Tech University
Earle, Robert, American River College
Eichenbaum, Jack, Hunter College (CUNY)
Fan, Chao, University of Idaho

Fan, Weihong, Stockton University
Feeney, Alison E., Shippensburg University
Fernández, Virginia, Universidad de la Republica Oriental del Urugua
Fitzgerald, Francis S., Colorado Geological Survey
Forrest, David, University of Glasgow
Garren, Sandra J., Hofstra University
Gathany, Mark, Cedarville University
Gauthier, Donald J., Los Angeles Valley College
Gentry, Christopher, Austin Peay State University
Girard, Michael W., New Jersey Geological and Water Survey
Gittings, Bruce M., Edinburgh University
Gopal, Sucharita, Boston University
Gordon, Clare E., University of Leeds
Gorsevski, Peter, Bowling Green State University
Grala, Katarzynz, Mississippi State University
Graniero, Phil A., University of Windsor
Greatbatch, Ian, Kingston University
Greene, Chris, Dalhousie University
Grunwald, Sabine -., University of Florida
Haddock, Gregory D., Northwest Missouri State University
Halls, Joanne N., University of North Carolina, Wilmington
Halsted, Christian, Dept of Agriculture, Conservation, and Forestry
Hamilton, Jacqueline, University of Minnesota
Hancock, Gregory, University of Newcastle
Haner, Andrew, Arkansas Geological Survey
Hansen, William J., Worcester State University
Haugerud, Ralph, University of Washington
Hedberg, Russell, Shippensburg University
Henderson, Matt, South Carolina Dept of Natural Resources
Herried, Brad, University of Minnesota, Twin Cities
Hick, Steven R., University of Denver
Hill, Malcolm D., Northeastern University
Hintz, John G., Bloomsburg University
Hong, Jessie, University of West Georgia
Hotz, Helenmary, University of Massachusetts, Boston
Howard, Hugh H., American River College
Huang, Jane, Fitchburg State University
Huffman, F. T., Eastern Kentucky University
Hughes, Annie, Kingston University
Humagain, Kamal, SUNY Potsdam
Jennings, Nathan, American River College
Jensen, Ryan R., Brigham Young University
Jeu, Amy, Hunter College (CUNY)
Joshi, Manoj, University of East Anglia
Juntunen, Thomas, University of Minnesota, Twin Cities
Kahle, Chris, Iowa Dept of Natural Resources
Kambesis, Patricia, Western Kentucky University
Kar, Bandana, University of Southern Mississippi
Kelleher, Cole, University of Minnesota, Twin Cities
Kelley, Scott , University of Nevada, Reno
Kidd, David, Kingston University
Kienzle, Stefan, University of Lethbridge
Kimerling, A. Jon, Oregon State University
Kirimi, Fridah K., Jomo Kenyatta University of Agriculture & Tech
Klancher, Jacki, Central Wyoming College
Kobara, Shinichi, Texas A&M University
Kohler, Nicholas, University of Oregon
Kostelnick, John, Illinois State University
Krizek, Jeffrey, San Bernardino Valley College
Kronenfeld, Barry J., Eastern Illinois University
Krygier, John B., Ohio Wesleyan University
Kwon, Youngsang, University of Memphis
Laffan, Shawn, University of New South Wales
LaRue, Michelle, University of Minnesota, Twin Cities
Law, Zada, Middle Tennessee State University
Le, Yanfen, Northwest Missouri State University
Lee, Wook, Edinboro University of Pennsylvania
Lethbridge, Mark, Flinders University
Levine, Norman S., College of Charleston
Li, Gary, California State University, East Bay
Li, Jing, University of Denver
Li, Ping-Chi, Tennessee Tech University
Liang, Lu, University of North Texas
Lisichenko, Richard, Fort Hays State University
Liu, Weibo, Florida Atlantic University
Lobben, Amy, University of Oregon
Logsdon, Miles G., University of Washington

Longbrake, David, University of Denver
Lovett, Andrew A., University of East Anglia
Lukinbeal, Christopher, University of Arizona
Luo, Jun, Missouri State University
Lupo, Tom, American River College
Maantay, Juliana, Lehman College (CUNY)
Maas, Regan, California State University, Northridge
Machovina, Brett , United States Air Force Academy
Mackaness, William A., Edinburgh University
Magondu, Moffat G., Jomo Kenyatta University of Agriculture & Tech
Marcano, Eugenio J., Mount Holyoke College
Marin, Liliana, Baylor University
Marsan, Yvonne, University of North Carolina, Wilmington
Marzen, Luke J., Auburn University
May, Cynthia, Northland College
Mayer, Jordan, University of Minnesota
McCluskey, James, University of Wisconsin Colleges
Meentemeyer, Ross, University of North Carolina, Charlotte
Meng, Qingmin, Mississippi State University
Metcalf, Meredith, Eastern Connecticut State University
Micander, Rachel, University of Nevada
Miguel, Carlos, Universidad de la Republica Oriental del Uruguay
Miller, Harvey J., Ohio State University
Mitasova, Helena, North Carolina State University
Momm, Henrique G., Middle Tennessee State University
Morgan, John D., University of West Florida
Morgan, Tamie, Texas Christian University
Morris, John A., Mississippi State University
Mubako, Stanley, University of Texas, El Paso
Mucsi, lászló, Univesity of Szeged
Mueller, Thomas, California University of Pennsylvania
Mukherjee, Falguni, Sam Houston State University
Mulligan, Kevin R., Texas Tech University
Mulrooney, Timothy, North Carolina Central University
Muthukrishnan, Suresh, Furman University
Mutua, Felix N., Jomo Kenyatta University of Agriculture & Technology
Mwangi, Nancy , Jomo Kenyatta University of Agriculture & Technology
Mwaniki, Mercy W., Jomo Kenyatta University of Agriculture & Tech
Nagihara, Seiichi, Texas Tech University
Nemon, Amy, Western Kentucky University
Niedzielski, Michael A., University of North Dakota
Nimako, Solomon Nana Kwaku, San Bernardino Valley College
Obermeyer, Nancy J., Indiana State University
Oduor, Peter, North Dakota State University
Olsen, Patrick, University of Idaho
Olson, Jeff, University of Wisconsin, Whitewater
Oxendine, Christopher E., United States Military Academy
Ozdenerol, Esra, University of Memphis
Palladino, Steve D., Ventura College
Pallis, Ted J., New Jersey Geological and Water Survey
Palmer, Evan, United States Air Force Academy
Pavlovskaya, Marianna, Hunter College (CUNY)
Pawlicki, Caron E., Pennsylvania Bur of Topo & Geologic Survey
Pedemonte, Virginia, Universidad de la Republica Oriental del Uruguay
Peele, R Hampton, Louisiana State University
Percy, David, Portland State University
Porter, Claire, University of Minnesota, Twin Cities
Porter, John H., University of Virginia
Portillo, Danny, United States Air Force Academy
Price, Curtis V., South Dakota School of Mines & Technology
Price, Maribeth H., South Dakota School of Mines & Technology
Price, Peter E., Lonestar College - North Harris
Pristas, Ronald, New Jersey Geological and Water Survey
Proctor, Elizabeth, City College of San Francisco
Qi, Feng, Kean University
Qiu, Xiaomin, Missouri State University
Raber, George, University of Southern Mississippi
Resnichenko, Yuri, Universidad de la Republica Oriental del Uruguay
Rex, Arthur B., Appalachian State University
Rice, Keith W., University of Wisconsin, Stevens Point
Riney, Kaylin, South Carolina Dept of Natural Resources
Robinson, Lori, University of Minnesota
Robinson, Michael, Georgia Southern University
Roof, Steven, Hampshire College
Rowley, Rex J., Illinois State University
Saalfeld, Alan J., Ohio State University
Sadd, James L., Occidental College

Sanchez-Azofeifa, G. Arturo, University of Alberta
Sandau, Ken L., Montana Tech of The University of Montana
Sanz, Miguel Angel, Univ Complutense de Madrid
Schenck, William S., University of Delaware
Scott, Darren M., McMaster University
Seong, Jeong C., University of West Georgia
Shears, Andrew, University of Wisconsin Colleges
Sleezer, Richard O., Emporia State University
Smith, Betty E., Eastern Illinois University
Smith, Janet S., Shippensburg University
Starek, Michael , Texas A&M University, Corpus Christi
Stuart, Neil, Edinburgh University
Su, Haibin, Texas A&M University, Kingsville
Sun, Jeff, Casper College
Sun, Yifei, California State University, Northridge
Taylor, Nathan H., Arkansas Geological Survey
Taylor, Ryan W., SUNY, Purchase
Templeton, Alan R., Washington University in St. Louis
Thale, Paul R., Montana Tech of The University of Montana
Thayn, Jonathan B., Illinois State University
Tinkler, Dorothy, Treasure Valley Community College
Tiwari, Chetan, University of North Texas
Tong, Daoqin, University of Arizona
Tran, Linda C., Lonestar College - North Harris
Tu, Wei, Georgia Southern University
Unger, Corey, Utah Geological Survey
VanHorn, Jason E., Calvin College
Veisze, Paul M., American River College
Vincent, Paul C., Valdosta State University
Vlcan, Jennifer, University of Nevada
Walford, Nigel, Kingston University
Walls, Charles C., Dalhousie University
Walters, Steven, University of Washington
Wang, Lillian T., University of Delaware
Weber, Keith, Idaho State University
Weiss, Alfred W., Waubonsee Community College
Wen, Yuming, University of Guam
Whelan, Michael, Southern Cross University
White, Scott, Fort Lewis College
Whiteaker, Timothy L., University of Texas, Austin
Whittaker, Amber, Dept of Agriculture, Conservation, and Forestry
Williamson, Douglas, Hunter College (CUNY)
Wilson, Blake B., University of Kansas
Wilson, John R., Lafayette College
Wilson, Roy R., Eastern Connecticut State University
Woodhouse, Iain H., Edinburgh University
Wright, William C., United States Military Academy
Wulamu, Wasit, Saint Louis University
Xiao, Dana, Bloomsburg University
Xie, Zhixiao, Florida Atlantic University
Xu, Wei, University of Lethbridge
Yacucci, Mark A., University of Illinois
Yan, Jun, Western Kentucky University
Ye, Gordon, City College of San Francisco
Yin, Zhi-Yong, University of San Diego
Zhang, Guiming, University of Denver
Zhang, Qiaofeng (Robin), Murray State University
Zhao, Bo, Oregon State University

Geoinformatics
Bhaskaran, Sunil, Grad Sch of the City Univ of New York
Curl, Doug C., University of Kentucky
Limp, William (., University of Arkansas, Fayetteville
Mansur, Adam, Smithsonian Inst / Natl Museum of Natural History
Marr, Paul G., Shippensburg University
O'Dean, Emily, University of Nevada
Sui, Daniel, University of Arkansas, Fayetteville
Yilmaz, Alper, Ohio State University
Zaffos, Andrew A., Arizona Geological Survey

Global Change
Boschmann, E. Eric, University of Denver
Crimmins, Michael A., University of Arizona
Crootof, Arica, University of Montana Western
Ducey, Mark, University of New Hampshire
El Masri, Bassil, Murray State University
Ernakovich, Jessica, University of New Hampshire
Felzer, Benjamin S., Lehigh University

Gilligan, Jonathan M., Vanderbilt University
Hicke, Jeffrey, University of Idaho
Kaufman, Darrell S., Northern Arizona University
Kopp, Robert E., Rutgers, The State Univ of New Jersey
Lawrence, Deborah, University of Virginia
Mathenge, Christine, Austin Peay State University
Nyantakyi-Frimpong, Hanson, University of Denver
Oches, Rick, Bentley University
Ollinger, Scott, University of New Hampshire
Rood, Richard B., University of Michigan
Rooney-Varga, Juliette, University of Massachusetts, Lowell
Sutton, Paul C., University of Denver
Taylor, Matthew, University of Denver
Tierney, Sean, University of Denver
Trigoso, Erika, University of Denver
Watson, Elizabeth, Drexel University
Williams, Mathew, Edinburgh University

Nanogeoscience
Michel, F. M., Virginia Polytechnic Institute & State University

Climate Modeling
Abatzoglou, John, University of Idaho
Cheng, Linyin, University of Arkansas, Fayetteville
Dee, Sylvia G., Rice University
Gutowski, William J., Iowa State University of Science & Technology
Hegewisch, Katherine, University of Idaho
Jackson, Charles S., University of Texas, Austin
Jackson, Charles, University of Texas, Austin
Sun, Fengpeng, University of Missouri, Kansas City
Vecchi, Gabriel, Princeton University

Not Elsewhere Classified
(Deep) McGregor, Mary, Montgomery College
Aamodt, Paul, Los Alamos National Laboratory
Abney, Kent D., Los Alamos National Laboratory
Abt, Charlotte J., University of Liverpool
Ackerman, Jessica R., Illinois State Geological Survey
Acosta, Patricia E., Williams College
Adams, Debi, West Texas A&M University
Agnew, Stephen F., Los Alamos National Laboratory
Aguilera, Juan Manuel Torres, Universidad Autonoma de San Luis Potosi
Ahmed, Waquar, University of North Texas
Alavi, Hedy, Johns Hopkins University
Alberts, Heike C., University of Wisconsin Oshkosh
Albrecht, Lorraine, University of Waterloo
Albright, Katia, University of Nevada, Reno
Alde, Douglas, Los Alamos National Laboratory
Algin, Barbara, Columbia University
Ali-Bray, Julie, Mt. San Antonio College
Allen, Ashley L., Ohio Wesleyan University
Allen, Karen, Furman University
Allison, Tami L., Missouri Dept of Natural Resources
Ambruster, W. Scott, University of Alaska, Fairbanks
Amelung, Falk, University of Miami
Amor, John, University of British Columbia
Anand, Madhur, University of Guelph
Antar, Ali A., Central Connecticut State University
Anthony, Leona M., University of Hawai'i, Manoa
Anthony, Nina, Furman University
Arabas, Karen, Willamette University
Argenbright, Robert T., University of North Carolina, Wilmington
Armstrong, Andrew, University of New Hampshire
Arsuaga, Juan Luis, Univ Complutense de Madrid
Arundale, Wendy H., University of Alaska, Fairbanks
Arvidson, Raymond E., Washington University in St. Louis
Ashour-Abdalla, Maha, University of California, Los Angeles
Atisha-Castillo, Hector David, Universidad Autonoma de San Luis Potosi
Au, Whitlow W L., University of Hawai'i, Manoa
Aufrecht, Walter E., University of Lethbridge
Ayers, Joseph , Northeastern University
Bailey, Ian, Exeter University
Baldizon, Ileana, Miami-Dade College (Wolfson Campus)
Ball, Elizabeth, University of Nevada, Reno
Ball, William P., Johns Hopkins University
Ballinger, Rhoda, Cardiff University
Banks, Beth, Brevard College
Bannon, Ann, Dalhousie University
Barmore, Garrett, University of Nevada, Reno

Barnbaum, Cecilia S., Valdosta State University
Barnes, Fairley J., Los Alamos National Laboratory
Barnes, Jessica, University of South Carolina
Barnes, Philip, University of South Carolina
Barnes, Randal J., University of Minnesota, Twin Cities
Barra, Monica, University of South Carolina
Barron, Elizabeth, University of Wisconsin Oshkosh
Barron, George, University of Guelph
Barth, Susan A., Montana Tech of The University of Montana
Bartl, Simona, Moss Landing Marine Laboratories
Baskin, Perry A., Valdosta State University
Bass, Jerry, University of Southern Mississippi
Basso, Bruno, Michigan State University
Baumann, Sue, Argonne National Laboratory
Beatty, Merrill Ann, University of New Brunswick
Beck, Alan, Western University
Becker, Laurence, Oregon State University
Becker, Stefan, Grad Sch of the City Univ of New York
Becker, Udo, University of Michigan
Bederman, Sanford H., Georgia State Univ
Behr-Andres, Christina "Tina", Los Alamos National Laboratory
Bellew, Angela, University of Iowa
Berchem, Jean, University of California, Los Angeles
Betancourt, Julio, University of Arizona
Bezdecny, Kris, California State University, Los Angeles
Bienvenu, Nadean S., University of Louisiana at Lafayette
Bilheux, Hassina Z., University of Tennessee, Knoxville
Billiot, Fereshteh, Texas A&M University, Corpus Christi
Bischoff, Marianne, Weber State University
Bishoff, Carolyn, University of Minnesota, Twin Cities
Black, Annjeannette , Comm College of Baltimore County, Catonsville
Blankenship, Robert, Washington University in St. Louis
Blatner , Keith , Washington State University
Blenkinson, Stephen, University of Newcastle Upon Tyne
Blumenfeld, Raphael, Imperial College
Boelman, Natalie, Columbia University
Boggs, Carol, University of South Carolina
Boisvert, Eric, Universite du Quebec
Boland, Greg J., University of Guelph
Boland, John J., Johns Hopkins University
Booth, Robert K., Lehigh University
Bosh, Amanda, Massachusetts Institute of Technology
Bossenbroek, Jonathon , University of Toledo
Boswell, Cecil, Missouri Dept of Natural Resources
Bottero, Jean-Yves, Rice University
Bourgeois, Reed J., Louisiana State University
Bourgeois, Reed J., Louisiana State University
Bowen, Dawn S., Mary Washington College
Bowen, Jennifer, Northeastern University
Bowen, Robert, University of Massachusetts, Boston
Bowersox, Joe, Willamette University
Bozzato, Edda, Laurentian University, Sudbury
Bradley, Michael D., University of Arizona
Brainard, James R., Los Alamos National Laboratory
Bratman, Eve, Franklin and Marshall College
Braught, Patricia, Dickinson College
Brewer, Margene, Calvin College
Bridgeman, Thomas, University of Toledo
Brink, Benita A., Adams State University
Brooks, Ian, University of Leeds
Brown, Bradford D., University of Idaho
Brown, Kent D., Utah Geological Survey
Brown, Rob, Appalachian State University
Brunetto, Eileen, Middlebury College
Brunish, Wendee M., Los Alamos National Laboratory
Bryan, Jeffrey C., Los Alamos National Laboratory
Bryant, Anita M., University of West Georgia
Buck, Daniel, University of Oregon
Buck, Gregory, Texas A&M University, Corpus Christi
Buckley, Luke, Montana Tech of The University of Montana
Bullamore, Henry W., Frostburg State University
Bunker, Merle E., Los Alamos National Laboratory
Burk, Sue, University of Maryland
Burns, Carol J., Los Alamos National Laboratory
Burns, John W., Mt. San Antonio College
Busse, Friedrich H., University of California, Los Angeles
Butts, Thomas R., University of Texas, Dallas

Byrand, Karl, University of Wisconsin Colleges
Cahill, Michael J., SUNY, Stony Brook
Cahoy, Kerri, Massachusetts Institute of Technology
Cailliet, Gregor M., Moss Landing Marine Laboratories
Calegari, Pat, Pace University, New York Campus
Cammerata, Kirk , Texas A&M University, Corpus Christi
Campana, Michael E., University of New Mexico
Cantarero, Debra A., Pasadena City College
Capoccia, Mary, Ohio State University
Carey, Kristine M., Lakehead University
Carey, Tara, Black Hawk College
Carp, Jana, Appalachian State University
Carr, David E., University of Virginia
Carroll, Matt, Washington State University
Carroll, Tracy, Missouri State University
Catalano, Valerie, Roger Williams University
Catau, John C., Missouri State University
Cattolico, Rose Ann, University of Washington
Caupp, Craig L., Frostburg State University
Chan, Selene, University of British Columbia
Chaney, Phil L., Auburn University
Chase, Anne, University of Arizona
Chase, Jon M., Washington University in St. Louis
Chatterjee, Ipsita, University of North Texas
Cheek, William H., Missouri State University
Chen, Kai Loon, Johns Hopkins University
Cheney, Donald, Northeastern University
Chief, Karletta, University of Arizona
Childs, Geoff, Washington University in St. Louis
Choquette, Marc, Universite Laval
Chouinard, Vera, McMaster University
Christopher, Micol, Mt. San Antonio College
Church, Warren, Columbus State University
Clark, David L., Los Alamos National Laboratory
Clark, James S., Duke University
Clark, Robert O., Northern Arizona University
Clark, Shannon, University of New Mexico
Clay, Robert, Missouri Dept of Natural Resources
Cleek, Richard, University of Wisconsin Colleges
Cochran, David, University of Southern Mississippi
Coffin, Richard B., University of Delaware
Cohen, Alice, Acadia University
Cohen, Joel E., Columbia University
Cohen, Matt, Furman University
Cohen, Shaul, University of Oregon
Colby, Bonnie C., University of Arizona
Colby, Sarah, Northern Arizona University
Coleman, Gary D., University of Maryland
Collins, Damian, University of Alberta
Collins, Lisa J., Northwestern University
Colwell, Frederick (Rick), Oregon State University
Confer, John, California University of Pennsylvania
Conkle, Jeremy, Texas A&M University, Corpus Christi
Connon, CAPT (Ret) Brian D., University of Southern Mississippi
Cook, Stanton, University of Oregon
Cook, Steve, Oregon State University
Cookus, Pam, Illinois State Geological Survey
Coonley, Steffanie, Los Alamos National Laboratory
Copeman, Louise A., Oregon State University
Corcoran, Deborah, Missouri State University
Corey, Alison, Utah Geological Survey
Costea, Lidia, Northeastern Illinois University
Costello, Margaret, California State University, Long Beach
Cote, Pascale, Universite du Quebec
Cox, Shelah, Temple University
Crafford, J. P., University of Limpopo
Crawford, Ian, Birkbeck College
Crawford, Matt, University of Kentucky
Crepeau, Richard J., Appalachian State University
Cromar, Nancy, Flinders University
Crumbly, Isaac J., Fort Valley State University
Cupples, Julie, Edinburgh University
Curry, Barbara, University of Texas, Dallas
Curry, James K., Arkansas Geological Survey
D'Arrigo, Rosanne, Columbia University
Damschen, Ellen, Washington University in St. Louis
Dash, Zora V., Los Alamos National Laboratory

Dasvarma, Gour, Flinders University
Daugherty, Carolyn M., Northern Arizona University
Davey, Patricia M., Brown University
Davis, James A., Brigham Young University
Day, Rosie, University of Birmingham
Daymon, Deborah J., Los Alamos National Laboratory
De Beus, Barbara, Montclair State University
De Santo, Eilzabeth, Franklin and Marshall College
de Wit, Julien, Massachusetts Institute of Technology
Dean, Jeffrey S., University of Arizona
Deaton, Tami, University of North Texas
DeGraff, Jerry, California State University, Fresno
Delawder, Sandra, James Madison University
Demanet, Laurent, Massachusetts Institute of Technology
Deming, Jody, University of Manitoba
DeMott, Benjamin, Grad Sch of the City Univ of New York
Dennis, Clyde B., Argonne National Laboratory
Dennis, Pam, University of Florida
Derrick, Sharon, Texas A&M University, Corpus Christi
Deschaine, Sylvia, Bates College
Dessler, Alexander, University of Arizona
Detrich, William, Northeastern University
Dibben, Chris, Edinburgh University
Dickerson, Richard E., University of California, Los Angeles
Dickhoff, Willem H., Washington University in St. Louis
Dighe, Kalpak, Los Alamos National Laboratory
Dill, Carilee, Bucknell University
Dingman, Steve, San Antonio Community College
Ditmars, John, Argonne National Laboratory
Divine, Aaron, Northern Arizona University
Dixon, Clifton, University of Southern Mississippi
Dixon, Michael, University of Guelph
Dixon, Paul R., Los Alamos National Laboratory
Doherty, Cheryl, University of Delaware
Dollhopf, Douglas J., Montana State University
Domingue, Carla, Louisiana State University
Donahoe, Robert J., Los Alamos National Laboratory
Doorn, Stephen K., Los Alamos National Laboratory
Dorries, Alison M., Los Alamos National Laboratory
Driever, Steven L., University of Missouri, Kansas City
Driscoll, John R., SUNY, Cortland
Duchane, David V., Los Alamos National Laboratory
Dudukovic, Mike, Washington University in St. Louis
Duff, John, University of Massachusetts, Boston
Duffey, Patricia, Fort Hays State University
Dunton, Karen K., Illinois State University
Duxbury, Thomas C., Jet Propulsion Laboratory
Dvorzak, Marie, University of Wisconsin-Madison
Dwyer, Daryl F., University of Toledo
Earls, Sandy, Oklahoma State University
Ebert, David, Moss Landing Marine Laboratories
Eby, Gloria J., University of Texas, Dallas
Eby, Stephanie, Northeastern University
Edwards, John , Flinders University
Eggers, Delores M., University of North Carolina, Asheville
Eisenberger, Peter M., Columbia University
Eller, Phillip G., Los Alamos National Laboratory
Ellins, Katherine K., University of Texas, Austin
Elliott, C. James, Los Alamos National Laboratory
Ellis, Hugh, Johns Hopkins University
Ellis, Kathryn E., University of Kentucky
Emmett, Chad, Brigham Young University
Epstein, Howard E., University of Virginia
Esnault, Melissa H., Louisiana State University
Esser, Corinne, University of California, Davis
Evans, Claude, Washington University in St. Louis
Evans, Krista M., Missouri State University
Fadiman, Maria, Florida Atlantic University
Fallowfield, Howard, Flinders University
Fares, Mary A., Valdosta State University
Farley, Perry D., Los Alamos National Laboratory
Farrell, Mark O., Point Park University
Favor, Michael, University of Michigan, Dearborn
Felix, Joseph D., Texas A&M University, Corpus Christi
Ferguson, Teresa D., Oak Ridge National Laboratory
Ferreira, Maryanne F., Woods Hole Oceanographic Institution
Ferren, Richard L., Berkshire Community College
Fifer, Fred L., University of Texas, Dallas
Fish, Jennifer M., University of California, Santa Cruz
Fitzpatrick, John R., Los Alamos National Laboratory
Fitzsimmons, Kevin, University of Arizona
Flaherty, Frank A., Valdosta State University
Fletcher, Austin, University of Guelph
Flores-Garcia, Mari C., California State University, Northridge
Floyd, Barry, New York State Geological Survey
Fogel, David N., Los Alamos National Laboratory
Foley, John P., Montana Tech of The University of Montana
Foreman, Taylor, Denver Museum of Nature & Science
Foster , Alec , Illinois State University
Foster, Michael, Moss Landing Marine Laboratories
Foti, Pamela, Northern Arizona University
Fox, Laurel R., University of California, Santa Cruz
Freiermuth, Sue, University of Wisconsin, River Falls
Frisk, DeAnn M., Iowa State University of Science & Technology
Fry, Matthew, University of North Texas
Fryett, I, Cardiff University
Fuellhart, Kurtis G., Shippensburg University
Fuente, David E., University of South Carolina
Fuentes, Edna L., University of Illinois at Chicago
Gagosian, Carol, Wellesley College
Gallick, Roberta T., Point Park University
Ganey-Curry, Patricia E., University of Texas, Austin
Gao, Mengsheng, University of Florida
Gao, Oliver H., Cornell University
Garcia, Jan, Occidental College
Gartner, John F., University of Waterloo
Garvin, Theresa D., University of Alberta
Gautsch, Jacklyn, Iowa Dept of Natural Resources
Gehrels, Tom, University of Arizona
Geller, Jonathan, Moss Landing Marine Laboratories
Geller, Murray, Jet Propulsion Laboratory
Gemignani, Robert, Washington University in St. Louis
Gerhardt, Hannes, University of West Georgia
Gerry, Janelle, University of Nebraska, Lincoln
Ghattas, Omar, University of Texas, Austin
Ghosheh, Baher A., Edinboro University of Pennsylvania
Giammar, Daniel, Washington University in St. Louis
Gibbs, Gaynell, Louisiana State University
Gibson, Deana, Missouri State University
Gidwani, Vinay, Grad Sch of the City Univ of New York
Gifford-Gonzalez, Dianne, University of California, Santa Cruz
Gilbert, Co'Quesie, Columbia University
Gillespie, Thomas, Emory University
Gillette, David P., University of North Carolina, Asheville
Gitelson, Anatoly, Unversity of Nebraska - Lincoln
Glaser, Brian, Black Hawk College
Glenn, Ed, University of Arizona
Gmitro, Helen, Union County College
Goldhamer, David A., University of California, Davis
Gong, Hongmian, Hunter College (CUNY)
Good, Daniel B., Georgia Southern University
Goodin, Ruth, University of Miami
Goodwin, Paul, University of Guelph
Gordon, Andrew, University of Guelph
Gottgens, Johan F., University of Toledo
Gouhier, Tarik, Northeastern University
Grace, Shannon M., Austin Community College District
Graff, William P., New Jersey Geological and Water Survey
Grande, Anthony, Hunter College (CUNY)
Granshaw, Frank D., Portland State University
Grattan, Stephen R., University of California, Davis
Gravely, Cynthia Rae, Clemson University
Gray, Sarah, University of San Diego
Green, Brittany, Kansas State University
Greene, Pauline R., University of Louisiana at Lafayette
Greene, Roberta, SUNY Potsdam
Griffin, Kevin, Columbia University
Griffith, Caitlin, University of Arizona
Gritzo, Russell E., Los Alamos National Laboratory
Grossman, Lawrence S., Virginia Polytechnic Institute & State University
Guerra, Oralia, Austin Community College District
Guerrieri, Mary, University of Arizona
Guikema, Seth, Johns Hopkins University
Guzman, Ernesto, University of Guelph

Habash, Marc, University of Guelph
Habron, Geoffrey, Furman University
Hacker, Patricia, University of Toledo
Haddad, Brent, University of California, Santa Cruz
Haff, Peter K., Duke University
Hafner, James A., University of Massachusetts, Amherst
Haggart, Renee, University of British Columbia
Halfman, Barbara, Hobart & William Smith Colleges
Hall, Christopher C., University of Guelph
Hall, I, Cardiff University
Hall, Jude, Denison University
Hall, Robert, University of Guelph
Hall, Ron, University of Lethbridge
Halladay, Sylvia R., Ohio Dept of Natural Resources
Hallett, Rebecca, University of Guelph
Hamilton, George, Berkshire Community College
Hammersley, Charles, Northern Arizona University
Hammond, Anne, Lakehead University
Haney, Mary, University of South Florida, Tampa
Hanke, Steve H., Johns Hopkins University
Hankins, Katherine B., Georgia State Univ
Hansen, Devon A., University of North Dakota
Hansen, Jeffrey C., Los Alamos National Laboratory
Hardy, Shaun J., Carnegie Institution for Science
Harris, Virginia M., Wesleyan University
Harrison, Conor, University of South Carolina
Hawley, Rebecca D., Northern Arizona University
Haynes, Kyle J., University of Virginia
Haynes, Kyle J., University of Virginia
Heaton, Jill, University of Nevada, Reno
Heaton, Richard C., Los Alamos National Laboratory
Heatwole, Charles A., Hunter College (CUNY)
Heckathorn, Scott, University of Toledo
Heenan, Cleo J., South Dakota School of Mines & Technology
Hegarty, Patricia, University College Cork
Heidt, James, University of Wisconsin Colleges
Heimpel, Moritz, University of Alberta
Heitman, Mary P., Iowa Dept of Natural Resources
Helmick, Sara B., Los Alamos National Laboratory
Hendrikx, Jordy, Montana State University
Henshel, Diane S., Indiana University, Bloomington
Herr, Larry, University of Lethbridge
Hicks, Andrea, University of Wisconsin, Madison
Higinbotham, Pamela, California University of Pennsylvania
Hildebrand, Stephen G., Oak Ridge National Laboratory
Hill, Christopher, Massachusetts Institute of Technology
Hill, Julie, University of Nevada, Reno
Hiller, Lena, Hofstra University
Hilpert, Markus, Johns Hopkins University
Hilts, Stewart G., University of Guelph
Hinderaker, Pam, Iowa State University of Science & Technology
Hirsch, Eric, Franklin and Marshall College
Hiser, Susan, Marietta College
Hobbs, Benjamin F., Johns Hopkins University
Hobbs, John D., Montana Tech of the University of Montana
Hockey, Thomas A., University of Northern Iowa
Hodges, Jackie, Fort Valley State University
Hohl, Eric, Missouri Dept of Natural Resources
Holbrook, Amanda, Morehead State University
Holdren, John P., Harvard University
Hollenbeck, Diane, Metropolitan State College of Denver
Holstein, Thomas J., Roger Williams University
Holt, Caecilia, Kutztown University of Pennsylvania
Homan, Amy L., Pennsylvania State University, University Park
Hommel, Demian, Oregon State University
Hopson, Janet L., University of Tennessee, Knoxville
Horne, Sharon, University of Windsor
Hrepic, Zdeslav, Columbus State University
Hsiang, Tom, University of Guelph
Hsiao, Theodore C., University of California, Davis
Huang, Norden E., University of Delaware
Huckabey, Marsha, University of Missouri
Huffman, Debra E., University of South Florida
Huffman, Robert L., Mercer University
Hughes, Joseph B., Rice University
Hughes, Randall, Northeastern University
Hull, Donald L., Los Alamos National Laboratory

Humphrey, Peggy, Montana State University
Hunt, Kathy, University of Maryland
Hunt, Shelley, University of Guelph
Hunter, John, Rice University
Huntoon, Laura, University of Arizona
Hutchins, Peter S., Mississippi Office of Geology
Hysell, David, Cornell University
Hysell, David L., Cornell University
Iantria, Linnea, Missouri State University
Immonen, Wilma, Montana Tech of the University of Montana
Ioannides, Dimitri, Missouri State University
Ivy, Russell L., Florida Atlantic University
Jackson, Robert B., Duke University
Jacobs, Tenika, New Jersey Geological and Water Survey
Jamiolahmady, Mahmoud, Heriot-Watt University
Janetos, Tony, Boston University
Jarcho, Kari A., University of Minnesota, Twin Cities
Jensen, Scott W., South Dakota Dept of Env and Nat Res
Jiang, Zhenhua, Argonne National Laboratory
Joanna, Lucero, New Mexico Institute of Mining and Technology
Johannessen, Carl, University of Oregon
Johnson, Daniel L., University of Lethbridge
Johnson, Emily P., Boston University
Johnson, Jeanne L., Louisiana State University
Johnson, Judy L., University of Alaska, Fairbanks
Johnson, Robert L., Argonne National Laboratory
Jones, Glen, New Mexico Institute of Mining & Technology
Jones, Gwilym, Northeastern University
Jones, Minnie O., University of Illinois at Chicago
Jones, III, John P., University of Arizona
Jornov, Donna, New York State Geological Survey
Joseph, Miranda, University of Arizona
Juanes, Ruben, Massachusetts Institute of Technology
Jun, Young-Shin, Washington University in St. Louis
Jurmanovich, Barb, Delta College
Juszczyk, Carmen, University of Colorado
Jutla, Rajinder S., Missouri State University
Karl, Tami S., Florida State University
Kay, Richard F., Duke University
Keala, Lori, Pomona College
Keating, Martha E., Bentley University
Keaton, Jeffrey R., University of Utah
Keatts, Merida, Kent State University
Kehoe-Forutan, Sandra J., Bloomsburg University
Keller, Jean, United States Military Academy
Kelly, Kimberly, Montgomery College
Kelly, Sherrie, St. Lawrence University
Kennedy, Christina B., Northern Arizona University
Kennel, Charles F., University of California, Los Angeles
Kevan, Peter, University of Guelph
Keyser, Jan, University of South Alabama
Khurana, Krishan, University of California, Los Angeles
Kidder, T.R., Washington University in St. Louis
Kidwell, Jen, Stanford University
Kifer, Lauri A., SUNY, The College at Brockport
Kile, Susan, Eastern Illinois University
Kim, Stacy, Moss Landing Marine Laboratories
Kim, Wan, Carnegie Institution for Science
Kimberly, Shaw, Columbus State University
Kimbro, David, Northeastern University
King-Hsi, Kung, Los Alamos National Laboratory
Kinkead, Scott, Los Alamos National Laboratory
Kipper, Jay P., Univ of Texas at Austin, Jackson Sch of Geosciences
Kneas, David, University of South Carolina
Kohlstedt, Sally G., University of Minnesota, Twin Cities
Komoto, Cary, Normandale Community College
Kontuly, Thomas M., University of Utah
Koralek, Susan, Southern Oregon University
Kosinski, Leszek A., University of Alberta
Kostov, Svilen, Georgia Southwestern State University
Kressler, Sharon J., University of Minnesota, Twin Cities
Krieble, Kelly, Moravian College
Kruse, Jennifer, Gustavus Adolphus College
Kumar, M. Satish, Queen's University Belfast
Kuntz, Kara, Fort Hays State University
LaBella, Joel, Wesleyan University
LaDochy, Steve, California State University, Los Angeles

LaFreniere, Lorraine M., Argonne National Laboratory
Lagowski, Alison A., SUNY, Buffalo
Lam, Anita, University of British Columbia
Lane, Mark, Palomar College
Larkin, Patrick, Texas A&M University, Corpus Christi
Larock, B. E., University of California, Davis
Larson, Harold P., University of Arizona
Laurier, Eric, Edinburgh University
Layton, Alice, University of Tennessee, Knoxville
Leal-Bejarano, Arturo, Universidad Autonoma de Chihuahua
Lebofsky, Larry A., University of Arizona
Ledbetter, Cynthia E., University of Texas, Dallas
Lee, Alexis, University of North Carolina, Wilmington
Lee, Hung, University of Guelph
Lehane, Mary, University College Cork
Lekan, Thomas, University of South Carolina
Leland, John, University of Nevada, Reno
Lemmond, Peter C., Woods Hole Oceanographic Institution
Lener, Edward, Virginia Polytechnic Institute & State University
Lerdau, Manuel, University of Virginia
Lermurier, Nathalie, Institut Polytechnique LaSalle Beauvais (ex-IGAL)
Leveque, Connie, University of Maine, Presque Isle
Li, Wenhong, Duke University
Lichtner, Peter C., Los Alamos National Laboratory
Lin, Hsing K., University of Alaska, Fairbanks
Lin, Shoufa, University of Waterloo
Lin, Xiaomao, Kansas State University
Lindley, Stacy, California State University, Sacramento
Linky, Edward, Hunter College (CUNY)
Lipeles, Maxine I., Washington University in St. Louis
Lloyd, Glenn D., Missouri Dept of Natural Resources
Loeb, Valerie, Moss Landing Marine Laboratories
Long, Laura, Lawrence Livermore National Laboratory
Long, Lisa, Ohio Dept of Natural Resources
Longmire, Patrick A., Los Alamos National Laboratory
Lotterhos, Kathleen, Northeastern University
Lounsbury, Diane E., SUNY, Geneseo
Loveland, Karen, Missouri Dept of Natural Resources
Lovseth, John, Principia College
Lowry, William R., Washington University in St. Louis
Loxsom, Fred, Eastern Connecticut State University
Lucas, Beth, Virginia Polytechnic Institute & State University
Lujan-Lopez, Arturo, Universidad Autonoma de Chihuahua
Lumpkin, Thomas A., Washington State University
Lynch, Michael J., University of Kentucky
Maciha, Mark J., Northern Arizona University
MacInnes, Michael, Los Alamos National Laboratory
Mahler, Robert L., University of Idaho
Maier, Raina M., University of Arizona
Malagon, Teresa Soledad Medina, Univ Nac Autonoma de Mexico
Malega, Ron, Missouri State University
Mand, Arlene, University of Pennsylvania
Mangel, Mark S., University of California, Santa Cruz
Marks, Dennis W., Valdosta State University
Marshall, Stephen, University of Guelph
Martinez-Macias, Panfilo R., Universidad Autonoma de San Luis Potosi
Masiello, Caroline A., Rice University
Mattison, Katherine W., Missouri University of Science and Technology
Maulud, Mat Ruzlin, University of Malaya
May, Diane M., Missouri State University
May, Fred E., University of Utah
Mayer, Christine M., University of Toledo
Maynard, David, California State University, San Bernardino
McArthur, Russell, San Francisco State University
McAtee, Mike, California State Polytechnic University, Pomona
McConnell, Joseph, University of Nevada, Reno
McCoy, Sue, Cosumnes River College
McDermott, Thomas M., SUNY, The College at Brockport
McDonald, Kyle C., Jet Propulsion Laboratory
McDonald, Lynn, Los Alamos National Laboratory
McFadden, Jennifer, Elizabethtown College
McGee, Tara, University of Alberta
McGraw, Maureen A., Los Alamos National Laboratory
McKenzie, Connie, Louisiana Tech University
McKenzie, Ross, University of Lethbridge
McLaughlin, Richard, Texas A&M University, Corpus Christi
McLeod, Clara, Washington University in St. Louis

McMillan, Robert S., University of Arizona
McNair, Laurie A., Los Alamos National Laboratory
McNamara, Jodi, Colgate University
Medlin, Peggy, Tennessee Tech University
Meehan, Katharine, University of Oregon
Mehes, Roxane J., Laurentian University, Sudbury
Mendoza, Blanca, Univ Nac Autonoma de Mexico
Messer, Sharon, Shawnee State University
Meyer, Christopher, Smithsonian Inst / Natl Museum of Natural History
Meyer, Judith, Missouri State University
Meyerson, Rohana, Lafayette College
Middlemiss, Lucie, University of Leeds
Miller, Judy, Monroe Community College
Miller, Mark, University of Southern Mississippi
Miller, Ted R., South Dakota Dept of Env and Nat Res
Milligan, Richard, Georgia State Univ
Mills, Suzanne, McMaster University
Milner, Jennifer, Augustana College
Milstead, Terence, Appalachian State University
Mitchneck, Beth A., University of Arizona
Mittleider, Stacy, Chadron State College
Miyares, Inez, Grad Sch of the City Univ of New York
Mockler, Theodore, Los Alamos National Laboratory
Moe-Hoffman, Amy P., Mississippi State University
Mollner, Daniel, Gustavus Adolphus College
Momen, Nasim, Boston University
Monahan, Adam, University of Victoria
Montgomery, Tamra S., Illinois State Geological Survey
Moodie, T. Bryant, University of Alberta
Moody, Eva, Tarleton State University
Mooney, Phillip, Sonoma State University
Moore, Clyde, Louisiana State University
Moorhead, Daryl L., University of Toledo
Moran-Taylor, Michelle, University of Denver
Morehouse, Barbara, University of Arizona
Morgan, Gary, Hampton University
Morin, Paul, University of Minnesota, Twin Cities
Morris, Brenda, Palomar College
Morris, Donald P., Lehigh University
Morrison, Lauren, East Carolina University
Moss, Patti, Whitman College
Mozzachiodi, Riccardo, Texas A&M University, Corpus Christi
Mueller, Amy, Northeastern University
Murphy, Alexander, University of Oregon
Murphy, Edward C., North Dakota Geological Survey
Myers, Clifford D., Berkshire Community College
Myers, Tammy, Shippensburg University
Nadiga, Balu, Los Alamos National Laboratory
Nakatsuka, James, University of California, Los Angeles
Narinder, Kaushik, University of Guelph
Nashold, Barney W., Argonne National Laboratory
Nesbitt, Alex, University of Nevada
Neuman, Dennis R., Montana State University
Nevins, Susan K., SUNY, Cortland
Newman, Jonathan, University of Guelph
Newton, Anthony J., Edinburgh University
Newton, Seth A., Geological Survey of Alabama
Nichols, Terry E., University of Arkansas, Fayetteville
Nicholson, Nanette, American Museum of Natural History
Nielsen, Mary, University of South Dakota
Noll, Michael G., Valdosta State University
Nordwald, Dan, Missouri Dept of Natural Resources
Norman, Catherine, Johns Hopkins University
Nosal, Thomas E., Dept of Energy and Environmental Protection
O'Callaghan, Mick, University College Cork
O'Day, Sandra, Central Connecticut State University
O'Melia, Charles R., Johns Hopkins University
O'Neil, Jennifer, University of Nevada, Reno
Occhiuzzi, Tony, Mesa Community College
Odland, Sarah K., Columbia University
Oglesby, Elizabeth, University of Arizona
Okafor, Florence A., Alabama A&M University
Olson, Derek, Naval Postgraduate School
Olson, James R., South Dakota Dept of Env and Nat Res
Oona, Hain, Los Alamos National Laboratory
Oppong, Joseph R., University of North Texas
Orrock, John, Washington University in St. Louis

Otis, Gard, University of Guelph
Ott, Kevin C., Los Alamos National Laboratory
Ouellette, Vicki L., University of Northern Colorado
Owens, Tamera, Muskegon Community College
Oza, Rupal, Hunter College (CUNY)
Pace, Michael L., University of Virginia
Pacia, Christina, Kean University
Palace, Michael W., University of New Hampshire
Palm, Risa I., Georgia State Univ
Palma, Miclelle, University of Wisconsin Colleges
Palmer, Christina, California State University, San Bernardino
Palmer, Clare, Washington University in St. Louis
Papaleo, Silvana, University of Toronto
Parendes, Laurie A., Edinboro University of Pennsylvania
Parker, Jim, University of Houston
Parker, Joan, Moss Landing Marine Laboratories
Parker, Marjorie, Bowdoin College
Parker, Stephen R., Montana Tech of the University of Montana
Parrish, Pia, San Diego State University
Pasteris, Jill D., Washington University in St. Louis
Patel, Vinodkumar A., Illinois State Geological Survey
Patino-Douce, Alberto E., University of Georgia
Patronas, Dennis, Dauphin Island Sea Lab
Patterson, Jodi T., Utah Geological Survey
Patterson, Mark, Northeastern University
Penney, Paulette, Baylor University
Pepper, Ian L., University of Arizona
Peri, Francesco, University of Massachusetts, Boston
Perkins, R, Cardiff University
Petersen, Bruce, Washington University in St. Louis
Peterson, Peter A., Iowa State University of Science & Technology
Phelps, Tommy, University of Tennessee, Knoxville
Pickering, Dan, Carnegie Museum of Natural History
Pickett, Nicki, Broward College
Pierce, Larry, Missouri Dept of Natural Resources
Pike, J, Cardiff University
Plane, David A., University of Arizona
Plant, Jeffrey J., Jet Propulsion Laboratory
Plescia, Jeffrey, Jet Propulsion Laboratory
Pociask, Walter, Wayne State University
Podolak, Morris, Tel Aviv University
Pollak, Robert, Washington University in St. Louis
Polovina, Jeffrey J., University of Hawai'i, Manoa
Polsky, Colin, Florida Atlantic University
Powers, Roger W., University of Alaska, Fairbanks
Poynton, Helen, University of Massachusetts, Boston
Prewitt, Charles, University of Arizona
Prichard, Terry L., University of California, Davis
Pulkkinen, Tuija, University of Michigan
Pundsack, Jonathan, University of Minnesota, Twin Cities
Purtle, Jennifer M., University of Arkansas, Fayetteville
Quinn, Courtney, Furman University
Quintero, Sylvia, University of Arizona
Radil, Steven, University of Idaho
Rallis, Donald N., Mary Washington College
Rankin, Seth, University of Wisconsin Colleges
Rasmussen, Tab, Washington University in St. Louis
Ravela, Sai, Massachusetts Institute of Technology
Reaven, Sheldon, SUNY, Stony Brook
Rennie, Tim, University of Guelph
Reusch, David B., New Mexico Institute of Mining and Technology
Reynolds, Barbara C., University of North Carolina, Asheville
Rice, Benjamin, Northwestern University
Rice, Murray, University of North Texas
Ricker, Alison, Oberlin College
Ridgwell, Andy, University of California, Riverside
Ridky, Alice M., Colby College
Rieke, George H., University of Arizona
Riley, James, University of Arizona
Riley, Rhonda, Southern Utah University
Riordan, Lindsay, Tufts University
Ritter, Leonard, University of Guelph
Rius, Marc, University of Southampton
Robas, Sheryl A., Princeton University
Robbins, Debra C., University of North Carolina, Asheville
Roberts, A. Lynn, Johns Hopkins University
Robinson, Bruce A., Los Alamos National Laboratory
Robinson, David, The Open University
Robinson, Mark, Arizona State University
Robinson, William, University of Massachusetts, Boston
Rodriguez, Vanessa del S., Puerto Rico Bureau of Geology
Roe, Carol, College of William & Mary
Roemer, Elizabeth, University of Arizona
Rogers, Jefferson S., University of Tennessee, Martin
Rogers, William J., West Texas A&M University
Roinstad, Lori L., South Dakota Dept of Env and Nat Res
Rojas, Adena, Johns Hopkins University
Rollinson, Paul A., Missouri State University
Rondot, Beth, Long Island University, C.W. Post Campus
Rooney, Neil, University of Guelph
Rose, Candace M., Argonne National Laboratory
Rose, Dan, University College Cork
Rosengaus, Rebeca, Northeastern University
Ross, Kirstin, Flinders University
Ross, Paula, New Mexico State University, Las Cruces
Rossell, Irene M., University of North Carolina, Asheville
Rostam-Abadi, Massoud, Illinois State Geological Survey
Rostron, Ben, University of Alberta
Rouhani, Farhang, Mary Washington College
Rouse, Jesse, University of North Carolina, Pembroke
Roussel-Dupre, R., Los Alamos National Laboratory
Rowland, Scott K., University of Hawai'i, Manoa
Rozmus, Wojciech, University of Alberta
Rumstay, Kenneth S., Valdosta State University
Rupert, Denise, Lock Haven University
Russell, Ron, Univ of Texas at Austin, Jackson Sch of Geosciences
Russell, Terry P., University of Victoria
Russell, Theresa J., University of Arkansas, Fayetteville
Russo, Mary Rose, Princeton University
Ruzicka, Jaromir, University of Hawai'i, Manoa
Saikia, Udoy, Flinders University
Saku, James C., Frostburg State University
Sammarco, Paul W., Louisiana State University
Sanchez, Charles, University of Arizona
Sandhu, Harpinder, Flinders University
Santander, Erma, University of Arizona
Sapigao, Gladys, Queens College (CUNY)
Sarin, Manmohan, University of Delaware
Sarmiento, Jorge L., Princeton University
Sauer, Nancy N. S., Los Alamos National Laboratory
Sawyer, Suzanne, Utah Geological Survey
Scapelli, Krista, Texas Christian University
Scarpino, Samuel, Northeastern University
Schaal, Barbara, Washington University in St. Louis
Schauss, Kim E., University of Southern Indiana
Scheel, Patrick, Missouri Dept of Natural Resources
Scheidemen, Kathy J., University of California, Santa Barbara
Schlesinger, William H., Duke University
Schlumpberger, Debbie, Saint Cloud State University
Schmidt, Jonathan, University of Guelph
Schoenberger, Erica, Johns Hopkins University
Schroeder, Kathleen, Appalachian State University
Schwankl, Larry J., University of California, Davis
Schwenke, Eszter, University of New Brunswick
Schwob, Stephanie L., Southern Methodist University
Scott, Kathy, University of British Columbia
Scott-Dupree, Cynthia, University of Guelph
Scyphers, Steven, Northeastern University
Sears, Heather, Northeastern University
Seifu, Abiye, Columbus State University
Shade, Janet, University of Pittsburgh, Bradford
Shah, Ashru, North Carolina State University
Sharma, Govind, Alabama A&M University
Shaw, Fred, Missouri Dept of Natural Resources
Shchukin, Eugene D., Johns Hopkins University
Sheley, Christina, Indiana University, Bloomington
Shelton, Robert B., Oak Ridge National Laboratory
Shields, Nancy, Virginia Polytechnic Institute & State University
Shock, Everett, Arizona State University
Showman, Adam, University of Arizona
Shumway, Matthew J., Brigham Young University
Sibley, Paul, University of Guelph
Sigler, William V., University of Toledo
Silks, Louis A., Los Alamos National Laboratory

Simon, Kathy, California Polytechnic State University
Simonetti, Stephanie, University of Notre Dame
Singh, Hanumant, Northeastern University
Singh, Harbans, Montclair State University
Sissom, David, West Texas A&M University
Sitwell, O.F. George, University of Alberta
Skeel, Loreene, University of Montana
Slater, Tom, Edinburgh University
Slavetskas, Carol, Binghamton University
Smale, Jody L., Pennsylvania Bur of Topo & Geologic Survey
Smith, Derald, University of Lethbridge
Smith, Donald R., University of California, Santa Cruz
Smith, Elizabeth Y., University of Nevada, Las Vegas
Smith, Erik, University of Saint Thomas
Smith, H D., Cardiff University
Smith, Jim, Flinders University
Smith, K. L., Wichita State University
Smith, Mikel, University of North Dakota
Smith, Paul H., Los Alamos National Laboratory
Smith, Peter J., University of Alberta
Smith, Susan M., Montana Tech of The University of Montana
Solomon, Keith, University of Guelph
Somers, Jr., Arnold E., Valdosta State University
Sorenson, Mary Clare, University of Wisconsin, Stevens Point
Southwell, Benjamin, Lake Superior State University
Sperazza, Michael, Stony Brook University
Spracklen, Dominick, University of Leeds
Stadnyk, Leona, Mount Royal University
Standridge, Debbie, Georgia Southwestern State University
Starr, Richard, Moss Landing Marine Laboratories
Steadman, Todd A., Clemson University
Stechmann, Samuel, University of Wisconsin, Madison
Steller, Diana, Moss Landing Marine Laboratories
Stephenson, Gerry, University of Guelph
Stepien, Carol A., University of Toledo
Sternberg, Rolf, Montclair State University
Stevens, Joan M., California Polytechnic State University
Stierle, Andrea, Montana Tech of the University of Montana
Stimer, Debra, Kent State University at Stark
Stinger, Lindsay C., University of Leeds
Stone, Glenn D., Washington University in St. Louis
Strangeway, Robert J., University of California, Los Angeles
Strong, Ellen, Smithsonian Inst / Natl Museum of Natural History
Stunz, Greg, Texas A&M University, Corpus Christi
Su, Xiaobo, University of Oregon
Suárez, Luis Eugenio, Univ Complutense de Madrid
Subulwa, Angela G., University of Wisconsin Oshkosh
Sullivan, Barbara, Argonne National Laboratory
Summers, Robert, University of Alberta
Sussman, Robert W., Washington University in St. Louis
Sutherland, Bruce R., University of Alberta
Svitra, Zita V., Los Alamos National Laboratory
Swanson, Basil I., Los Alamos National Laboratory
Swetnam, Thomas W., University of Arizona
Swindall, Diane, University of California, Davis
Symbalisty, E.M.D., Los Alamos National Laboratory
Sywensky, Ann, Moravian College
Taney, R. Marieke, Northern Arizona University
Tanji, K. K., University of California, Davis
Tavener, Kristi, Lakehead University
Taylor, David J., University of Chicago
Taylor, Michael, Flinders University
Taylor, Robert W., Montclair State University
Tencate, James, Los Alamos National Laboratory
Teplitski, Max, University of Florida
Ter-Simonian, Vardui, University of Southern California
Theis, Karen, University of Manchester
Thériault, Julie Mireille, Universite du Quebec a Montreal
Thompson, Glennis, University of California, Berkeley
Thomson, Cary, University of British Columbia
Thomson, Cynthia, Columbia University
Thomson, Vivian E., University of Virginia
Tingey, David G., Brigham Young University
Tissot, Philippe, Texas A&M University, Corpus Christi
Todd, Brenda, University of Nebraska, Omaha
Tomlinson, Jaime L., University of Delaware
Torcellini, Paul, Eastern Connecticut State University
Townshend, Ivan, University of Lethbridge
Tracy, Matthew, United States Air Force Academy
Trevors, Jack, University of Guelph
Triplehorn, Judy, University of Alaska, Fairbanks
Trost, G K., University of Arkansas, Fayetteville
Trout, Jennifer, Western Michigan University
Trussell, Geoffrey, Northeastern University
Tuller, Markus, University of Arizona
Turner, Jay, Washington University in St. Louis
Tyning, Thomas F., Berkshire Community College
Unkefer, Clifford J., Los Alamos National Laboratory
Unsworth, Martyn, University of Alberta
Urquhart, Alvin, University of Oregon
van den Akker, Ben, Flinders University
Van Eerd, Laura, University of Guelph
van Norden, Maxim F., University of Southern Mississippi
Van Otten, George A., Northern Arizona University
Van Roosendaal, Susan, University of Utah
Varadi, Ferenc D., University of California, Los Angeles
Varady, Robert G., University of Arizona
Veeder, Glenn J., Jet Propulsion Laboratory
Vieira, David J., Los Alamos National Laboratory
Vollmer, Steve, Northeastern University
Wada, Ikuko, University of Minnesota, Twin Cities
Wadhwa, Meenakshi, Arizona State University
Wagh, Swati, Argonne National Laboratory
Waite, Cynthia H., University of Southern California
Walker, Mark, University of Nevada, Reno
Walker, Peter, University of Oregon
Walker, Raymond J., University of California, Los Angeles
Walraven, Brenda, Mercer University
Walrod, Amanda G., University of Arkansas, Fayetteville
Walsh, Christopher, University of Maryland
Walsh, Daniel E., University of Alaska, Fairbanks
Walsh, Ellen c., Lawrence University
Warburg, Helena F., Williams College
Ward, David M., Montana State University
Warhaft, Zellman, Cornell University
Warkentin, Alicia, University of British Columbia
Waterstone, Marvin, University of Arizona
Watson, Alan, University of Guelph
Wauthier, Christelle, Pennsylvania State University, University Park
Weaver, Douglas J., Los Alamos National Laboratory
Webre, Cherri B., Louisiana State University
Weinstein, Charles E., Berkshire Community College
Weintraub, Michael N., University of Toledo
Wells, Steve G., University of Nevada, Reno
Welton, Leicha, University of Alaska, Fairbanks
Wenz, Helmut C., University of Tennessee, Martin
West, Alan, The University of Virginia's College at Wise
Wetzel, Dan L., University of Alaska, Fairbanks
Whiley, Harriet, Flinders University
White, George W., Frostburg State University
Whitney, Earl M., Los Alamos National Laboratory
Wiederspahn, Mark, University of Texas, Austin
Wilder, Margaret, University of Arizona
Wilhelmy, Jerry B., Los Alamos National Laboratory
William, Nancy, Kansas State University
Williams, Allison M., McMaster University
Williams-Bruinders, Leizel, Nelson Mandela Metropolitan University
Willson, Lee, Rice University
Wilmut, Michael, University of Victoria
Wilson, Robert, University of St. Andrews
Wilson, Sarah B., Washington & Lee University
Wilson, Thomas B., University of Arizona
Wilton, Robert D., McMaster University
Winsor, Roger A., Appalachian State University
Winston, Barbara, Washington University in St. Louis
Winterkamp, Judith L., Los Alamos National Laboratory
Winton, Alison, Texas Tech University
Wisdom, Jack, Massachusetts Institute of Technology
Wise, Dana, Southern Illinois University Carbondale
Withers, Kim, Texas A&M University, Corpus Christi
Wixman, Ronald, University of Oregon
Wolfe, Karen M., Middle Tennessee State University
Wollan, Jacinda, North Dakota State University
Wong, Cindy, Buffalo State College

Wood, Carolyn F., Dauphin Island Sea Lab
Woodard, Gary C., University of Arizona
Woodley, Teresa, University of British Columbia
Woodruff, William H., Los Alamos National Laboratory
Woods, Neal, University of South Carolina
Wooldridge, C F., Cardiff University
Wright, Kathyrn, Western Michigan University
Wu, David T., Colorado School of Mines
Wyckoff, William K., Montana State University
Yarbrough, Robert A., Georgia Southern University
Yates, Mary Anne, Los Alamos National Laboratory
Yelle, Roger, University of Arizona
Yoakum, Barbara, United States Naval Academy
Yoskowitz, David , Texas A&M University, Corpus Christi
Young, Donald W., University of Arizona
Young, Patrick, Arizona State University
Young, Priscilla E., South Dakota Dept of Env and Nat Res
Yutzy, Gale, Frostburg State University
Zajac, Roman N., University of New Haven
Zamstein, Lavi, Columbus State University
Zaniewski, Kazimierz J., University of Wisconsin Oshkosh
Zeiger, Elaine, Field Museum of Natural History
Zelizer, Nora, Princeton University
Zhang, Max, Cornell University
Zimba, Paul, Texas A&M University, Corpus Christi
Zimmermann, George, Stockton University
Zotti, Maria, Flinders University
Zurick, David, Eastern Kentucky University
Zwiefelhofer, Luke, Winona State University

Faculty Index

A

Aştefanei, Dan, 004023201455 astdan@uaic.ro,
　Alexandru Ioan Cuza – Cga
Aagaard, Knut, (206) 543-8942 aagaard@apl.washington.edu,
　University of Washington – Op
Aalto, Rolf E., +44 (0) 1392-26-3344 aalto@uw.edu,
　University of Washington – Gms
Aamodt, Paul, (505) 665-1331 plaamod@lanl.gov,
　Los Alamos National Laboratory – Zn
Aanstoos, James V., 515-294-1032 aanstoos@iastate.edu,
　Iowa State University of Science & Technology – ZrAs
Aay, Henry, (616) 526-7033 aay@calvin.edu,
　Calvin College – Zyu
Abalos, Benito, benito.abalos@ehu.eus,
　University of the Basque Country UPV/EHU – GctGp
Abatzoglou, John, 208-885-6239 jabatzoglou@uidaho.edu,
　University of Idaho – ZoyZc
Abbasnejad, Ahmad, abbasnejadahmad@yahoo.com,
　Shahid Bahonar University of Kerman – GemHg
Abbott, Lon, (303) 492-6172 lon.abbott@colorado.edu,
　University of Colorado – GgtGm
Abbott, Mark B., (412) 624-1408 mabbott1@pitt.edu,
　University of Pittsburgh – Gs
Abbott, Patrick L., (619) 594-5591 pabbott@mail.sdsu.edu,
　San Diego State University – Gs
Abdalati, Waleed, Waleed.Abdalati@colorado.edu,
　University of Colorado – Zr
Abdelsalam, Mohamed G., 405-744-6358 mohamed.abdel_salam@okstate.edu,
　Oklahoma State University – GctYg
Abdo, Ginette, 406-496-4152 gabdo@mtech.edu,
　Montana Tech of The University of Montana – Hw
Abdu, Yassir, (204) 474-7356 abdu@cc.umanitoba.ca,
　University of Manitoba – Gz
Abdulghani, Waleed, +96638602848 wmaghani@kfupm.edu.sa,
　King Fahd University of Petroleum and Minerals – Go
Abdulla, Hussain, 361-825-6050 Hussain.Abdulla@tamucc.edu,
　Texas A&M University, Corpus Christi – Co
Abdullah, Nuraiteng Tee, 03-79674229 naiteng@um.edu.my,
　University of Malaya – Gr
Abdullah, Wan Hasiah, 03-79674232 wanhasia@um.edu.my,
　University of Malaya – Ec
Abdullatif, Osman, +96638601479 osmanabd@kfupm.edu.sa,
　King Fahd University of Petroleum and Minerals – Gso
Abdurrahman, A., aabdrrahman@unilorin.edu.ng,
　University of Ilorin – CeEg
Abend, Martin, (201) 200-3161 mabend@njcu.edu,
　New Jersey City University – Zy
Aber, James S., jaber@emporia.edu,
　Emporia State University – Gl
Abernathey, Ryan P., 845-365-8185 rpa@ldeo.columbia.edu,
　Columbia University – Op
Abernathy, Ryan, rpa@ldeo.columbia.edu,
　Columbia University – Og
Abers, Geoffrey A., 607-255-3879 abers@cornell.edu,
　Cornell University – YsGt
Abimbola, A. E., af.abimbola@mail.ui.edu.ng,
　University of Ibadan – Cg
Abney, Kent D., (505) 665-3894 Los Alamos National Laboratory – Zn
Abolins, Mark J., (615)904-8372 Mark.Abolins@mtsu.edu,
　Middle Tennessee State University – Gc
Abousleiman, Younane, 405-325-2900 yabousle@ou.edu,
　University of Oklahoma – Nr
Abrams, Daniel, (217) 244-1520 dbabrams@illinois.edu,
　University of Illinois – Hw
Abrams, Lewis J., 910-962-2350 abramsl@uncw.edu,
　University of North Carolina, Wilmington – Gu
Abrams, Michael J., (818) 354-0937 Jet Propulsion Laboratory – Zr
Abramson, Evan H., 206-616-4388 evana@uw.edu,
　University of Washington – Gy
Abreu, Maria E., msabreu@utad.pt,
　Universidade de Trás-os-Montes e Alto Douro – GaaGa
Abt, Charlotte J., +440151 794 5178　　Chj@liverpool.ac.uk,
　University of Liverpool – Zn
Abu Bakar, Muhammad Z., 0092-42-99029487 mzubairab1977@gmail.com,
　University of Engineering and Technology – NrmNg
Abushagur, Sulaiman, 915-831-2539 sulaiman@epcc.edu,
　El Paso Community College – Gg
Acharya, Kumud, (702) 862-5371 kumud.acharya@dri.edu,
　University of Nevada, Reno – HsGeHq
Achauer, Ulrich, ulrich.achauer@unistra.fr,
　Universite de Strasbourg – Ysg
Achten, Christine, +49-251-83-33941 achten@uni-muenster.de,
　Universitaet Muenster – CloEg
Ackerman, Jessica R., 217-333-4258 jracker@illinois.edu,
　Illinois State Geological Survey – Zn
Ackerman, Steven A., (608)263-3647 stevea@ssec.wisc.edu,
　University of Wisconsin, Madison – Zr
Ackerman, Thomas P., (206) 221-2767 ackerman@atmos.washington.edu,
　University of Washington – As
Ackerson, Michael R., ackerm4@rpi.edu,
　Rensselaer Polytechnic Institute – CpGiCu
Acosta, Patricia E., (413) 597-2221 patricia.e.acosta@williams.edu,
　Williams College – Zn
Adabanija, Moruffdeen A., +2348037786592 maadabanija@lautech.edu.ng,
　Ladoke Akintola University of Technology – GggGg
Adam, Iddrisu, iddrisu.adam@uwc.edu,
　University of Wisconsin Colleges – ZynZn
Adam, Jurgen, +44 1784 414258 Jurgen.Adam@rhul.ac.uk,
　Royal Holloway University of London – Gc
Adames-Corraliza, Angel F., afadames@umich.edu,
　University of Michigan – Ast
Adamo, Peter J., 204-727-9683 adamop@brandonu.ca,
　Brandon University – Gg
Adams, Aubreya, 315-228-7202 aadams@colgate.edu,
　Colgate University – Yg
Adams, Debi, 806-651-2570 dadams@wtamu.edu,
　West Texas A&M University – Zn
Adams, Gerald E., 312-344-7540 gadams@colum.edu,
　Columbia College Chicago – Gi
Adams, Herbert G., (818) 677-2575 herb.adams@csun.edu,
　California State University, Northridge – Ng
Adams, John B., adams@ess.washington.edu,
　University of Washington – GcXgZr
Adams, Kenneth, 775.673.7345 ken.adams@dri.edu,
　University of Nevada, Reno – Gam
Adams, Manda S., 704-687-5984 Manda.Adams@uncc.edu,
　University of North Carolina, Charlotte – Am
Adams, Mark, capt.mark.adams@gmail.com,
　Lamar University – Xg
Adams, Marshall, (865) 574-7335 sma@ornl.gov,
　Oak Ridge National Laboratory – Ob
Adams, Michael S., (608) 263-5994 University of Wisconsin, Madison – Ob
Adams, Peter N., (352) 846-0825 adamsp@ufl.edu,
　University of Florida – Gm
Adams, Thomas, 210-486-0045 tadams67@alamo.edu,
　San Antonio Community College – PvGg
Adamsen, Floyd, (602) 437-1702 f.j.adamsen@gmail.com,
　University of Arizona – So
Adedcyin, A. D., deloadedoyin@yahoo.com,
　University of Ilorin – Eg
Adee, Eric, (785) 354-7236 eadee@ksu.edu,
　Kansas State University – So
Adegoke, Jimmy, (816) 235-2978 adegokej@umkc.edu,
　University of Missouri, Kansas City – AsZyr
Adeigbe, O. C., oc.adeigbe@mail.ui.edu.ng,
　University of Ibadan – GosGe
Adekola, S. A., 234-708-928-5170 adekoladsolo@gmail.com,
　Obafemi Awolowo University – GdoCg
Adelana, S.M. A., adelana@gmx.net,
　University of Ilorin – HgYgGe
Aden, Douglas J., (614) 265-6579 douglas.aden@dnr.ohio.gov,
　Ohio Dept of Natural Resources – GeRnGl
Adentunji, Jacob, j.adetunji@derby.ac.uk,
　University of Derby – Ge
Adepelumi, A. A., 234-8128181062 aadepelu@oauife.edu.ng,
　Obafemi Awolowo University – YgsNe
Adepoju, Mohammed O., 08034722855 moadepoju@futa.edu.ng,

417

Faculty Index -A

Federal University of Technology, Akure – GgEg
Adetunji, A, 234-8137433622 Obafemi Awolowo University – EgGi
Adewoye, Abosede O., +2347032045714, +2347036602812 aoadewoye55@lautech.edu.ng,
Ladoke Akintola University of Technology – GeeGe
Adeyemi, G. O., go.adeyemi@mail.ui.edu.ng,
University of Ibadan – NgHg
Adhikari, Sanchayeeta, 818-677-5630 sadhikari@csun.edu,
California State University, Northridge – Zi
Adinolfi, Bryan, adinolfib1@southernct.edu,
Southern Connecticut State University – GgZg
Adisa, Adeshina L., 08029458886 adeshina4me2000@gmail.com,
Federal University of Technology, Akure – GgCe
Adkins, Jess F., (626) 395-8550 jess@gps.caltech.edu,
California Institute of Technology – Cm
Admire, Amanda R., 707-826-3111 ara11@humboldt.edu,
Humboldt State University – Zg
Adomaitis, D., 217-244-8872 Dadomait@illinois.edu,
Illinois State Geological Survey – Ge
Adovasio, James M., 814-824-2581 adovasio@mercyhurst.edu,
Mercyhurst University – GafGs
Adrain, Jonathan M., (319) 335-1539 jonathan-adrain@uiowa.edu,
University of Iowa – Pi
Adrain, Tiffany S., (319) 335-1822 tiffany-adrain@uiowa.edu,
University of Iowa – Pg
Adriano, Domy C., (803) 725-2752 University of Georgia – Sc
Afolabi, O., 234-803-512-4838 toltobi@yahoo.com,
Obafemi Awolowo University – Yg
Afolabi, Olukayode A., +2348052158722 oaafolabi@lautech.edu.ng,
Ladoke Akintola University of Technology – GxxGx
Agardy, Rachael, 989-774-7596 agard1rv@cmich.edu,
Central Michigan University – Ou
Agbogun, Henry, 785-628-4574 hmagobogun@fhsu.edu,
Fort Hays State University – GgYg
Agee, Carl A., (505) 277-1644 agee@unm.edu,
University of New Mexico – Cp
Agee, Ernest M., (765) 494-3282 eagee@purdue.edu,
Purdue University – As
Agioutantis, Zach, 859-257-953 zach.agioutantis@uky.edu,
University of Kentucky – NrmEc
Agnew, Duncan C., (858) 534-2590 dagnew@ucsd.edu,
University of California, San Diego – Ys
Agnew, Stephen F., (505) 665-1764 Los Alamos National Laboratory – Zn
Agnini, Claudia, 39-049-827918 claudia.agnini@unipd.it,
Università degli Studi di Padova – PmeGu
Agrawal, Abinash, (937) 775-3455 abinash.agrawal@wright.edu,
Wright State University – ClHwPg
Agrawal, Sandeep , 780-492-1230 sagrawal@ualberta.ca,
University of Alberta – Zu
Agterberg, Frits P., 613-996-2374 Frits.Agterberg@nrcan-rncan.gc.ca,
University of Ottawa – Gq
Ague, Jay J., 203-432-3171 jay.ague@yale.edu,
Yale University – Gzx
Aharon, Paul, (205) 348-2528 paharon@ua.edu,
University of Alabama – ClmCq
Ahearn, Sean C., (212) 772-5327 Graduate School of the City University of New York – Zr
Ahern, Judson L., (405) 325-4480 jahern@ou.edu,
University of Oklahoma – Yg
Ahm, Anne-Sofie, 609-258-4101 aahm@princeton.edu,
Princeton University – ClsPc
Ahmad, Abd Rashid, 03-79674156 abrashid@um.edu.my,
University of Malaya – Sc
Ahmed, Kazi M., +880 1711846840 kmahmed@du.ac.bd,
University of Dhaka – HwGgHg
Ahmed, Mohamed, 361-825-3278 Mohamed.Ahmed@tamucc.edu,
Texas A&M University, Corpus Christi – YgZri
Ahmed, Muhammad F., mfanr5@mst.edu,
University of Engineering and Technology – Nxt
Ahmed, Waqar, 940-565-2721 waquar.ahmed@unt.edu,
University of North Texas – Zn
Ahola, John, John.Ahola@bristolcc.edu,
Bristol Community College – Gg
Ahrens, Donald C., (209) 575-6300 Modesto Junior College – Zg
Aiello, Ivano, (831) 771-4400 iaiello@mlml.calstate.edu,
Moss Landing Marine Laboratories – OuGmu
Aiken, Carlos L., (972) 883-2450 aiken@utdallas.edu,
University of Texas, Dallas – Yv
Aines, Roger D., (925) 423-7184 aines1@llnl.gov,
Lawrence Livermore National Laboratory – Cg
Aird, Hannah M., 530-898-6369 haird@csuchico.edu,
California State University, Chico – GxEg
Aitchison, Jonathan, geoscience.headofschool@sydney.edu.au,
University of Sydney – GrtPm
Aitkenhead-Peterson, Jacqueline A., (979) 845-3682 jacqui_a-p@tamu.edu,
Texas A&M University – HsSoGf
Aiyyer, Anantha, 919-515-7973 aaiyyer@ncsu.edu,
North Carolina State University – As
Aizen, Vladimir, 208-885-5888 aizen@uidaho.edu,
University of Idaho – GlHg
Aja, Stephen U., (718) 951-5405 Graduate School of the City University of New York – Cl
Aja, Stephen U., 7189515000 x2881 suaja@brooklyn.cuny.edu,
Brooklyn College (CUNY) – Cl
Ajaegwu, Norbert E., (+234) 8023649729 ne.ajaegwu@unizik.edu.ng,
Nnamdi Azikiwe University – GorGg
Ajassa, Roberto, roberto.ajassa@unito.it,
Università di Torino – Zy
Ajayi, J. O., 234-803-401-4357 owoajayi@oauife.edu.ng,
Obafemi Awolowo University – No
Ajayi, T. R., 234-803-725-8924 traajayi@oauife.edu.ng,
Obafemi Awolowo University – CgEgGe
Ajigo, Isaac O., 08032107038 ioajigo@futa.edu.ng,
Federal University of Technology, Akure – GgNm
Akabzaa, Thomas M., 233-246325685 akabzaa@ug.edu.gh,
University of Ghana – EgGeNm
Akaegbobi, I. M., izumike2002@yahoo.com,
University of Ibadan – Cg
Akciz, Sinan, (657) 278-4371 sakciz@fullerton.edu,
California State University, Fullerton – GcRn
Akenzua-Adamcyzk, Aiyevbekpen H., aiyevbekpen.akenzua@uniben.edu,
University of Benin – GeHgGg
Akgul, Bunyamin, 00904242370000-5988 bakgul@firat.edu.tr,
Firat University – GzyGi
Akgun, Elif , 00904242370000-5972 efiratligil@firat.edu.tr,
Firat University – GgcGt
Akinlabi, Ismaila A., +2348050225113 iaakinlabi@lautech.edu.ng,
Ladoke Akintola University of Technology – GggGg
Aksoy, Ercan, 00904242370000-5974 Firat University – Ggt
Aksu, Ali E., (709) 737-8385 aaksu@sparky2.esd.mun.ca,
Memorial University of Newfoundland – Gu
Akujieze, Christopher N., krisjiaku@yahoo.com,
University of Benin – GgHwGe
Al, Tom A., 613-562-5800 6966 tom.al@uottawa.ca,
University of Ottawa – HwCgZg
Al-Aasm, Ihsan S., (519) 253-3000 ext 2494 alaasm@uwindsor.ca,
University of Windsor – GdClGo
Al-Attar, David, +44 (0) 1223 348935 da380@cam.ac.uk,
University of Cambridge – YdGt
Al-Lehyani, Ayman, +96638601661 allehyani@kfupm.edu.sa,
King Fahd University of Petroleum and Minerals – Yg
Al-Ramadan, Khalid, +96638607175 ramadank@kfupm.edu.sa,
King Fahd University of Petroleum and Minerals – GodGs
Al-Saad, Hamad A., hamadsaad@qu.edu.qa,
University of Qatar – PsGsu
Al-Shaibani, Abdulaziz, +96638604002 shaibani@kfupm.edu.sa,
King Fahd University of Petroleum and Minerals – HwGeg
Al-Shuhail, Abdullah, +966138602538 shuhail@kfupm.edu.sa,
King Fahd University of Petroleum and Minerals – Ye
Al-Shuhail, Abdullatif, +96638603584 ashuhail@kfupm.edu.sa,
King Fahd University of Petroleum and Minerals – Ygs
Alao, O. A., 234-805-466-7314 olade77@yahoo.com,
Obafemi Awolowo University – Yg
Alavi, Hedy, (410) 516-7091 alavi@jhu.edu,
Johns Hopkins University – Zn
Alba, Saturnino, sdeal@baucm.es,
Univ Complutense de Madrid – Gm
Albach, Suzanne M., 734-677-5111 salbach@wccnet.edu,
Washtenaw Community College – ZegGe
Albee, Arden L., 626.395.6260 aalbee@caltech.edu,
California Institute of Technology – GizXg
Albee-Scott, Steven R., (517) 796-8526 albeescsteven@jccmi.edu,
Jackson College – GePeSf
Alberstadt, Leonard P., 615-322-2160 leonard.p.alberstadt@vanderbilt.edu,

Vanderbilt University – Gd
Alberts, Heike C., (920) 424-4105 alberts@uwosh.edu,
  University of Wisconsin Oshkosh – Zn
Albin, Edward, ealbin@gsu.edu,
  Georgia State University, Perimeter College, Online – Xm
Albrecht, Jochen, (212) 772-5221 jochen@hunter.cuny.edu,
  Hunter College (CUNY) – ZiyZu
Albrecht, Lorraine, klalbrec@uwaterloo.ca,
  University of Waterloo – Zn
Albright, James N., (505) 667-4318 j_albright@lanl.gov,
  Los Alamos National Laboratory – Yg
Albright, Katia, (775) 682-8370 kalbright@unr.edu,
  University of Nevada, Reno – Zn
Albright, Thomas , (775) 784-6673 talbright@unr.edu,
  University of Nevada, Reno – Zy
Alde, Douglas, (505) 667-0488 dxa@lanl.gov,
  Los Alamos National Laboratory – Zn
Alderson, David, +44 (0) 191 208 7121 david.alderson@ncl.ac.uk,
  University of Newcastle Upon Tyne – Zi
Alderton, Dave, +44 1784 443585 D.Alderton@rhul.ac.uk,
  Royal Holloway University of London – Gz
Aldrich, Eric A., 573-882-6301 aldriche@missouri.edu,
  University of Missouri, Columbia – As
Aldrich, M. James, (505) 667-1495 jaldrich@lanl.gov,
  Los Alamos National Laboratory – Gc
Aldridge, Keith D., (416) 399-0124 keith@yorku.ca,
  York University – YmdYh
Aldushin, Kirill, 089/2180 4337 kirill.aldushin@lrz.uni-muenchen.de,
  Ludwig-Maximilians-Universitaet Muenchen – Gz
Aleksandrowski, Pawel, pawel.aleksandrowski@uwr.edu.pl,
  University of Wroclaw – GctYe
Alessi, Daniel S., (780) 492-8019 alessi@ualberta.ca,
  University of Alberta – ClGeHw
Alexander, Becky, (206) 543-0164 beckya@atmos.washington.edu,
  University of Washington – AsCsPe
Alexander, Clark R., 912-598-2329 clark.alexander@skio.usg.edu,
  Georgia Southern University – OuGsOn
Alexander, Conel M., (202) 478-8478 calexander@carnegiescience.edu,
  Carnegie Institution for Science – Xc
Alexander, Dane, (269) 387-5486 dane.alexander@wmich.edu,
  Western Michigan University – GgZnn
Alexander, Elaine, (254) 299-8442 ealexander@mclennan.edu,
  McLennan Community College – Gg
Alexander, Jan, +44 (0)1603 59 3759 j.alexander@uea.ac.uk,
  University of East Anglia – GseGd
Alexander, Jane L., 718-982-3013 jane.alexander@csi.cuny.edu,
  College of Staten Island/CUNY – GsClt
Alexander, Scott, 612-626-4164 alexa107@umn.edu,
  University of Minnesota, Twin Cities – Hw
Alexander, Shelton S., (814) 863-7246 shel@geosc.psu.edu,
  Pennsylvania State University, University Park – Ys
Alexander, Jr., E. Calvin, (612) 624-3517 alexa001@umn.edu,
  University of Minnesota, Twin Cities – HwCcl
Alexandre, Paul, 204-727-9693 alexandrep@brandonu.ca,
  Brandon University – EgCgGz
Alfe, Dario, +44 202 7679 32361 d.alfe@ucl.ac.uk,
  University College London – ZmGyYh
Alford, Matthew H., 206-221-3257 malford@apl.washington.edu,
  University of Washington – Op
Alfsen, Sheila, sheila.alfsen@pdx.edu,
  Portland State University – GgPgOu
Algar, Christopher, (902) 494-7192 calgar@dal.ca,
  Dalhousie University – Obc
Algeo, Catherine, (270) 745-5922 katie.algeo@wku.edu,
  Western Kentucky University – Zi
Algeo, Thomas J., (513) 556-4195 thomas.algeo@uc.edu,
  University of Cincinnati – Gs
Algin, Barbara, (212) 854-2905 ba110@columbia.edu,
  Columbia University – Zn
Ali, Bukari, +233 20 330 7976 bukariali@yahoo.co.uk,
  Kwame Nkrumah University of Science and Technology – HwNgYg
Ali, Genevieve, 204-474-7266 Genevieve.Ali@umanitoba.ca,
  University of Manitoba – Hw
Ali, Hendratta N., (785) 628-4608 hnali@fhsu.edu,
  Fort Hays State University – GogYs
Ali, K. Adem, (843) 953-0877 alika@cofc.edu,
  College of Charleston – ZrHwNg

Alkac, Onur, 00904242370000-5970 oalkac@firat.edu.tr,
  Firat University – GgsGu
Allabush, Kathleen, 801-581-7062 k.ritterbush@utah.edu,
  University of Utah – Pe
Allam, Bassem, (632) 632-8745 bassem.allam@stonybrook.edu,
  SUNY, Stony Brook – Ob
Allan, Andrea, (541) 737-3427 aallan@coas.oregonstate.edu,
  Oregon State University – As
Allan, Craig J., 704-687-5999 cjallan@email.uncc.edu,
  University of North Carolina, Charlotte – Hg
Allan, Deborah L., (612) 625-3158 dallan@soils.umn.edu,
  University of Minnesota, Twin Cities – Sb
Allan, Jonathan C., (541) 574-6658 Oregon Dept of Geology and Mineral
  Industries – Gm
Allan, Jonathan C., (541) 574-5568 jonathan.allan@dogami.state.or.us,
  Oregon Dept of Geology & Mineral Industries – Gm
Allard, Gilles O., (706) 542-2420 goallar@uga.edu,
  University of Georgia – Eg
Allard, Jason, (229) 333-5752 jmallard@valdosta.edu,
  Valdosta State University – AsZg
Allard, Stephen T., 507-457-2739 sallard@winona.edu,
  Winona State University – GctGx
Allen, Ashley L., (740) 368-3624 alallen@owu.edu,
  Ohio Wesleyan University – Znu
Allen, Bruce, 505-366-2531 allenb@gis.nmt.edu,
  New Mexico Institute of Mining & Technology – Gm
Allen, Charlotte M., +61 7 3138 0177 cm.allen@qut.edu.au,
  Queensland University of Technology – CcaGi
Allen, Clarence R., (626) 395-6904 allen@gps.caltech.edu,
  California Institute of Technology – Ys
Allen, Diana M., (778) 782-3967 dallen@sfu.ca,
  Simon Fraser University – Hy
Allen, Douglas, dallen@salemstate.edu,
  Salem State University – CgHsCt
Allen, Eric E., (858) 534-2570 eallen@ucsd.edu,
  University of California, San Diego – Ob
Allen, Gary C., 828-747-7818 gallen@uno.edu,
  University of New Orleans – Gp
Allen, Grant, +44 0161 306-6851 Grant.Allen@manchester.ac.uk,
  University of Manchester – As
Allen, John, 989-774-1923 allen4jt@cmich.edu,
  Central Michigan University – Ams
Allen, John, +44 023 9284 2257 john.allen@port.ac.uk,
  University of Portsmouth – Og
Allen, Joseph L., 304-384-5238 allenj@concord.edu,
  Concord University – Gct
Allen, Karen, (864)294-2504 karen.allen@furman.edu,
  Furman University – Zn
Allen, Katherine, +44 0151 795 4646   K.A.Allen@liverpool.ac.uk,
  University of Liverpool – Zi
Allen, Katherine A., Katherine.Allen@maine.edu,
  University of Maine – Gu
Allen, Kelly E., 918-595-7085 Kely.E.Allen@tulsacc.edu,
  Tulsa Community College – ZyiZe
Allen, Mark, +44 191 33 42344 m.b.allen@durham.ac.uk,
  Durham University – Gt
Allen, Peter M., (254)710-2189 peter_allen@baylor.edu,
  Baylor University – HgNg
Allen, Phillip, 301/687-4891 ppallen@frostburg.edu,
  Frostburg State University – GmePe
Allen, Richard M., (510) 642-1275 rallen@seismo.berkeley.edu,
  University of California, Berkeley – Ys
Allen, Richardson B., (801) 581-7574 pallen@egi.utah.edu,
  University of Utah – GctZi
Allen, Robert J., (951) 827-4870 robert.allen@ucr.edu,
  University of California, Riverside – As
Allen, Susan E., (604) 822-2828 sallen@eos.ubc.ca,
  University of British Columbia – Op
Allen-King, Richelle, (716) 645-4287 richelle@buffalo.edu,
  SUNY, Buffalo – Cg
Allen-Lafayette, Zehdreh, (609) 292-2576 zehdreh.allen-lafayette@dep.nj.gov,
  New Jersey Geological and Water Survey – Zi
Aller, Josephine Y., (631) 632-8655 josephine.aller@stonybrook.edu,
  SUNY, Stony Brook – ObAs
Aller, Robert C., (631) 632-8746 robert.aller@stonybrook.edu,
  SUNY, Stony Brook – Cm
Alley, Karen E., 330-287-1956 kalley@wooster.edu,

College of Wooster – GlZr
Alley, Marcus M., (540) 231-9777 malley@vt.edu,
  Virginia Polytechnic Institute & State University – Sc
Alley, Richard B., (814) 863-1700 rba6@psu.edu,
  Pennsylvania State University, University Park – Gl
Alley, Ronald E., (818) 354-0751 ron@lithos.jpl.nasa.gov,
  Jet Propulsion Laboratory – Zr
Allison, Alivia J., 620-341-5984 aalliso2@emporia.edu,
  Emporia State University – GgaGm
Allison, David T., 251-460-6381 dallison@southalabama.edu,
  University of South Alabama – Gc
Allison, Mead A., (504) 862-3270 meadallison@tulane.edu,
  Tulane University – Gs
Allison, Nicky, +44 01334 463952 na9@st-andrews.ac.uk,
  University of St. Andrews – Oc
Allison, Tami L., tami.allison@dnr.mo.gov,
  Missouri Dept of Natural Resources – Zn
Allmon, Warren D., 6072736623x14 wda1@cornell.edu,
  Cornell University – PgePi
Allmon, Warren D., (607) 273-6623 (Ext. 14) wda1@cornell.edu,
  Paleontological Research Institution – Pg
Almendinger, James E., 651-433-5953 dinger@smm.org,
  University of Minnesota, Twin Cities – Pe
Alonso, Ana María, alonsoza@ucm.es,
  Univ Complutense de Madrid – Gd
Alpert, Pinhas, pinhas@post.tau.ac.il,
  Tel Aviv University – AsZrAm
Alpert, Sam, (212) 769-5383 salpert@amnh.org,
  American Museum of Natural History – Xm
Alsdorf, Douglas E., 614 247-6908 alsdorf.1@osu.edu,
  Ohio State University – HgYg
Alsleben, Helge, (817) 257-5455 h.alsleben@tcu.edu,
  Texas Christian University – GctGg
Alt, Jeffrey C., (734) 764-8380 jalt@umich.edu,
  University of Michigan – Ou
Altaner, Stephen P., (217) 244-1244 altaner@illinois.edu,
  University of Illinois, Urbana-Champaign – Gz
Altenbach, Alexander, 089/2180 6598 a.altenbach@lrz.uni-muenchen.de,
  Ludwig-Maximilians-Universitaet Muenchen – Pg
Altermann, Wladyslaw, 089/2180 6552 wlady.altermann@iaag.geo.uni-muenchen.de,
  Ludwig-Maximilians-Universitaet Muenchen – Gg
Alumbaugh, David L., (510) 215-4227 dalumbaugh@berkeley.edu,
  University of California, Berkeley – Ye
Aluwihare, Lihini I., (858) 822-4886 laluwihare@ucsd.edu,
  University of California, San Diego – Gu
Alvarado, Vladimir, 307-766-6464 valvarad@uwyo.edu,
  University of Wyoming – Np
Alvarez, José Antonio, jaalvare@ucm.es,
  Univ Complutense de Madrid – Gc
Alvarez, Walter, (510) 642-2602 platetec@berkeley.edu,
  University of California, Berkeley – Gr
Alves, Tiego, +44(0)29 208 76754 AlvesT@cf.ac.uk,
  University of Wales – Og
Aly, Mohamed H., 479-575-4524 aly@uark.edu,
  University of Arkansas, Fayetteville – Gg
Amador, Nathanael S., 740-368-3619 nsamador@owu.edu,
  Ohio Wesleyan University – ZyGlZr
Amato, James A., (920) 424-2268 amatoj@uwosh.edu,
  University of Wisconsin, Oshkosh – GgZi
Amato, Jeffrey M., (575) 646-3017 amato@nmsu.edu,
  New Mexico State University, Las Cruces – GcCc
Ambinakudige, Shrinidhi, (662) 268-1032 x210 shrinidhi@geosci.msstate.edu,
  Mississippi State University – ZiyZr
Ambrose, William, 512-471-0258 william.ambrose@beg.utexas.edu,
  University of Texas, Austin – GsrEo
Ambruster, W. Scott, (907) 474-7161 ffwsa@aurora.alaska.edu,
  University of Alaska, Fairbanks – Zn
Amedjoe, Godfrey C., +233 24 5961 073 chiri.amedjoe@gmail.com,
  Kwame Nkrumah University of Science and Technology – GidGc
Amel, Nasir, +98 (411) 339 2696 amel@tabrizu.ac.ir,
  University of Tabriz – GivGg
Amelung, Falk, 305 421-4949 famelung@rsmas.miami.edu,
  University of Miami – Zn
Amend, Jan, (213) 740-0652 janamend@usc.edu,
  University of Southern California – ClmCo
Amend, Jan P., (314) 935-8651 amend@wustl.edu,
  Washington University in St. Louis – Co
Amenta, Roddy V., (540) 568-6674 amentarv@jmu.edu,
  James Madison University – Gc
Amer, Reda, 504-862-3220 ramer1@tulane.edu,
  Tulane University – ZirGe
Amezaga, Jaime, +44 (0) 191 208 4876 jaime.amezaga@ncl.ac.uk,
  University of Newcastle Upon Tyne – Ge
Amidon, Will, 802-443-5980 wamidon@middlebury.edu,
  Middlebury College – CcGmt
Amin, Isam E., (330) 941-2293 ieamin@ysu.edu,
  Youngstown State University – Hw
Amler, Michael, 089/2180 6602 amler@lrz.uni-muenchen.de,
  Ludwig-Maximilians-Universitaet Muenchen – Pg
Ammon, Charles J., (814) 865-2310 cammon@geosc.psu.edu,
  Pennsylvania State University, University Park – Ys
Amon, Rainier M., 409 740 4719 amonr@tamug.edu,
  Texas A&M University – CbOcb
Amonette, Alexandra B., (509) 376-5019 Pacific Northwest National Laboratory – Cg
Amoozegar, Aziz, (919) 515-3967 North Carolina State University – Sp
Amor, John, (604) 822-6933 jamor@eos.ubc.ca,
  University of British Columbia – Zn
Amos, Colin B., 360-650-3587 colin.amos@wwu.edu,
  Western Washington University – Gc
Ampuero, Jean-Paul, 626.395.6958 ampuero@gps.caltech.edu,
  California Institute of Technology – Ys
Amthor, Jeff, (865) 576-2773 Oak Ridge National Laboratory – Sf
Anand, Madhur, (519) 824-4120 Ext.56254 manand@uoguelph.ca,
  University of Guelph – Zn
Anand, Pallavi, +44 (0) 1908 652225 x 52225 pallavi.anand@open.ac.uk,
  The Open University – ClPe
Anandakrishnan, Sridhar, (814) 863-6742 sak@essc.psu.edu,
  Pennsylvania State University, University Park – Ys
Anastasio, Cort, 530-754-6095 canastasio@ucdavis.edu,
  University of California, Davis – As
Anastasio, David J., (610) 758-5117 dja2@lehigh.edu,
  Lehigh University – Gca
Anbar, Ariel D., (480) 965-0767 anbar@asu.edu,
  Arizona State University – CbsXb
Ancell, Brian C., 806-834-3143 brian.ancell@ttu.edu,
  Texas Tech University – As
Ancochea, Eumenio, anco@ucm.es,
  Univ Complutense de Madrid – Gv
Anders, Alison M., 217-244-3917 amanders@illinois.edu,
  University of Illinois, Urbana-Champaign – Gm
Anders, Mark H., mander44@gmu.edu,
  George Mason University – Gc
Andersen, C. Brannon, (864) 294-3366 brannon.andersen@furman.edu,
  Furman University – ClGe
Andersen, David W., (408) 924-5014 david.andersen@sjsu.edu,
  San Jose State University – GsCl
Andersen, Jens C., +44 01326 371836 j.c.andersen@exeter.ac.uk,
  Exeter University – GiEmCa
Andersen, Raymond J., (604) 822-4511 randersn@eos.ubc.ca,
  University of British Columbia – Oc
Anderson, Alan J., (902) 867-2309 aanderso@stfx.ca,
  Saint Francis Xavier University – Gx
Anderson, Andrew, 217-244-0995 acandrsn@illinois.edu,
  Illinois State Geological Survey – Ng
Anderson, Bruce, brucea@bu.edu,
  Boston University – AsOp
Anderson, Callum, 27 41 504 2811 callum.anderson@nmmu.ac.za,
  Nelson Mandela Metropolitan University – Gzs
Anderson, G M., (416) 978-2062 University of Toronto – Cp
Anderson, James G., (617) 495-5922 anderson@huarp.harvard.edu,
  Harvard University – As
Anderson, James L., 808-932-7553 jamesa@hawaii.edu,
  University of Hawai'i, Hilo – GctYd
Anderson, James L., (612) 625-0279 jandersn@soils.umn.edu,
  University of Minnesota, Twin Cities – Sd
Anderson, James Lawford, lawford@bu.edu,
  Boston University – Gi
Anderson, Jennifer L., 507-457-2457 jlanderson@winona.edu,
  Winona State University – XgYgZe
Anderson, John B., (713) 348-4652 johna@rice.edu,
  Rice University – Gu
Anderson, John G., (775) 784-4265 jga@seismo.unr.edu,

University of Nevada, Reno – Ys
Anderson, John G., (775) 784-1954 jga@unr.edu,
University of Nevada, Reno – YsNeYg
Anderson, Ken B., 618-453-7389 kanderson@geo.siu.edu,
Southern Illinois University Carbondale – Co
Anderson, Laurie C., (605) 394-1290 Laurie.Anderson@sdsmt.edu,
South Dakota School of Mines & Technology – PiePq
Anderson, Mark, +44 1752 584768 M.Anderson@plymouth.ac.uk,
University of Plymouth – Gct
Anderson, Mark R., (402) 472-6656 manderson4@unl.edu,
University of Nebraska, Lincoln – As
Anderson, Mary P., (914) 365-8335 andy@geology.wisc.edu,
University of Wisconsin-Madison – Hw
Anderson, Orson L., (310) 825-2386 University of California, Los Angeles – Yx
Anderson, Patricia M., pata@uw.edu,
University of Washington – Ple
Anderson, Paula, 479-575-3355 pea001@uark.edu,
University of Arkansas, Fayetteville – Gg
Anderson, Raymond R., (319) 335-1589 Raymond.Anderson@dnr.iowa.gov,
University of Iowa – Gr
Anderson, Robert, 303-735-4684 Robert.S.Anderson@Colorado.EDU,
University of Colorado – Gm
Anderson, Robert F., boba@ldeo.columbia.edu,
Columbia University – Cg
Anderson, Robert F., (845) 365-8508 boba@ldeo.columbia.edu,
Columbia University – OcCmPe
Anderson, Robert G., (604) 822-2449 University of British Columbia – Gt
Anderson, Robert S., (831) 459-3342 randerson@es.ucsc.edu,
University of California, Santa Cruz – Gm
Anderson, Ross, +44 (0) 131 451 3798 r.anderson@hw.ac.uk,
Heriot-Watt University – Eo
Anderson, Stephen H., 573-882-6303 andersons@missouri.edu,
University of Missouri, Columbia – Sp
Anderson, Steven W., (970) 351-2973 steven.anderson@unco.edu,
University of Northern Colorado – GvZeXg
Anderson, Suzanne P., (303) 492-7071 suzanne.anderson@colorado.edu,
University of Colorado – GmCl
Anderson, Thomas B., (775) 747-1438 tom.anderson@sonoma.edu,
Sonoma State University – GsrGd
Anderson, Thomas F., 12173442177 tfanders@illinois.edu,
University of Illinois, Urbana-Champaign – CslCm
Anderson, Wayne I., (319) 273-3125 wayne.anderson@uni.edu,
University of Northern Iowa – Pg
Anderson, William B., (828) 262-7540 andersonwp@appstate.edu,
Appalachian State University – HwqHs
Anderson, William T., (305) 348-2693 andersow@fiu.edu,
Florida International University – CsPcPcl
Anderson, Jr., Alfred T., (773) 702-8138 University of Chicago – Gv
Anderson-Folnagy, Heidi, (406) 683-7134 heidi.anderson@umwestern.edu,
University of Montana Western – Gs
Andersson, Andreas, (858) 822-2486 aandersson@ucsd.edu,
University of California, San Diego – Oc
Andeweg, Bernd, 31 20 5987339 bernd.andeweg@vu.nl,
Vrije Universiteit Amsterdam – GtcGg
Andonaegui, Pilar, andonaeg@ucm.es,
Univ Complutense de Madrid – Gg
Andreasen, David C., (410) 260-8814 david.andreasen@maryland.gov,
Maryland Department of Natural Resources – Hw
Andres, A. S., (302) 831-2833 asandres@udel.edu,
University of Delaware – HwZi
Andres, A. S., 302-831-0599 asandres@udel.edu,
University of Delaware – Hg
Andresen, Christian, candresen@wisc.edu,
University of Wisconsin, Madison – Zy
Andrews, Benjamin, (202) 633-1818 andrewsb@si.edu,
Smithsonian Inst / Natl Museum of Natural Hist– Gv
Andrews, Graham D., (304) 293-2192 graham.andrews@mail.wvu.edu,
West Virginia University – GvtGx
Andrews, John R., 512-471-1534 john.andrews@beg.utexas.edu,
University of Texas at Austin, Jackson School of Geosciences – Zi
Andrews, John T., (303) 492-5183 john.t.andrews@colorado.edu,
University of Colorado – GluGs
Andrews, Julian E., +44 (0)1603 59 2536 j.andrews@uea.ac.uk,
University of East Anglia – CsGsCl
Andrews, Richard D., 405-325-3991 rdandrews@ou.edu,
University of Oklahoma – Go

Andronicos, Chris, (765) 494-5982 candroni@purdue.edu,
Purdue University – Gg
Andrus, C. Fred T., (205) 348-5177 fandrus@ua.edu,
University of Alabama – PcGaCs
Andrus, Richard E., (607) 777-2453 Binghamton University – Sf
Aneja, Viney P., 91951557808 viney_aneja@ncsu.edu,
North Carolina State University – Ac
Anfinson, Owen, anfinson@sonoma.edu,
Sonoma State University – GsdGg
Angelopoulos, Vassilis, (310) 794-7090 vassilis@ucla.edu,
University of California, Los Angeles – XyAsYm
Anger-Kraavi, Annela, +44 (0)1603 59 2633 a.anger-kraavi@uea.ac.uk,
University of East Anglia – Eg
Angino, Ernest E., (785) 843-7503 University of Kansas – Cl
Angle, Michael P., 614 265 6602 mike.angle@dnr.state.oh.us,
Ohio Dept of Natural Resources – GlgHw
Angus, Doug, +4401133431326 d.angus@leeds.ac.uk,
University of Leeds – Ys
Anhaeusser, Carl R., 011-7176581 carl.anhaeusser@wits.ac.za,
University of the Witwatersrand – GgEgZg
Anis, Ayal, (409) 740-4987 anisa@tamug.tamu.edu,
Texas A&M University – Opn
Ankney, Meagan, Meagan@earth.northwestern.edu,
Northwestern University – CcqCa
Annett, Amber, +44 023 80596041 A.L.Annett@soton.ac.uk,
University of Southampton – Oc
Ansdell, Kevin M., (306) 966-5698 kevin.ansdell@usask.ca,
University of Saskatchewan – EgCgGt
Antar, Ali A., (860) 832-2931 antar@ccsu.edu,
Central Connecticut State University – Zn
Anthony, Elizabeth Y., (915) 747-5483 eanthony@utep.edu,
University of Texas, El Paso – GiCcGv
Anthony, Leona M., (808) 956-8763 leonaa@hawaii.edu,
University of Hawai'i, Manoa – Zn
Anthony, Nina, 864-294-2052 nina.anthony@furman.edu,
Furman University – Zn
Anthony, Robin, 412-442-4295 robanthony@pa.gov,
Pennsylvania Bureau of Topographic & Geologic Survey – Go
Anthony-Zajanc, Kate, anthcath@isu.edu,
Idaho State University – Zi
Antinao, José Luis , (812) 855-1366 jantinao@indiana.edu,
Indiana University – CcGlr
Antinao, JoseLuis, 775-673-7450 JoseLuis.antinao@dri.edu,
Desert Research Institute – Gm
Antipova, Anzhelika, 901-678-2178 antipova@memphis.edu,
University of Memphis – Zg
Antonacci, Vince, (615) 532-1507 Vince.Antonacci@tn.gov,
Tennessee Geological Survey – GgZi
Antonescu, Adrian, +44 0161 306-3911 bogdan.antonescu@manchester.ac.uk,
University of Manchester – As
Aplin, Andrew, +44 191 33 42332 Durham University – Go
Apopei, Andrei I., 0040747760718 andrei.apopei@uaic.ro,
Alexandru Ioan Cuza – GziGg
Apotsos, Alex, 413-597-5082 alex.apotsos@williams.edu,
Williams College – Ge
Appel, Christopher (Chip) S., (805) 756-1691 cappel@calpoly.edu,
California Polytechnic State University – Sc
Appiah-Adjei, Emmanuel K., +233 20 7934 556 ekappiah-adjei.soe@knust.edu.gh,
Kwame Nkrumah University of Science and Technology – HwgGg
Applegarth, Michael T., (717) 477-1712 mtappl@ship.edu,
Shippensburg University – GmZrSo
Applegate, Toby, 413-545-1535 tapplega@geo.umass.edu,
University of Massachusetts, Amherst – Zy
Appold, Martin S., (573) 882-0701 appoldm@missouri.edu,
University of Missouri – HwEm
Apprill, Amy, (505) 289-2649 aapprill@whoi.edu,
Woods Hole Oceanographic Institution – Ob
Apraiz, Arturo, arturo.apraiz@ehu.eus,
University of the Basque Country UPV/EHU – Gpt
April, Richard, (315) 228-7212 rapril@colgate.edu,
Colgate University – CgGze
Aquilina, Noel , noel.aquilina@um.edu.mt,
University of Malta – Asm
Arabas, Karen, 503-370-6666 karabas@willamette.edu,
Willamette University – Zn
Aracil, Enrique, earacil@ucm.es,

Faculty Index -A

Faculty Index -A

Univ Complutense de Madrid – Yu
Arain, M. A., (905) 525-9140 (Ext. 27941) arainm@mcmaster.ca,
  McMaster University – Hg
Arakawa, Akio, (310) 825-9874 aar@atmos.ucla.edu,
  University of California, Los Angeles – As
Aranda, Angela, (657) 278-3551 angperez@fullerton.edu,
  California State University, Fullerton – Rn
Arboleya Cimadevilla, María Luísa, ++935811951 marialuisa.arboleya@uab.cat,
  Universitat Autonoma de Barcelona – Gc
Archambault, Guy, garchambault@uqac.ca,
  Universite du Quebec a Chicoutimi – Nr
Archer, Allen W., (785) 532-2244 aarcher@ksu.edu,
  Kansas State University – Gr
Archer, David, (773) 702-0823 University of Chicago – Yr
Archer, Michael, m.archer@unsw.edu.au,
  University of New South Wales – Pv
Archibald, Doug A., (613) 545-6594 Queen's University – Cc
Archuleta, Ralph J., (805) 893-8441 archuleta@geol.ucsb.edu,
  University of California, Santa Barbara – Ys
Arehart, Greg B., (775) 784-6470 arehart@unr.edu,
  University of Nevada, Reno – Eg
Arenas, Ricardo, rarenas@ucm.es,
  Univ Complutense de Madrid – Gp
Arens, Nan Crystal, 315-781-3930 arens@hws.edu,
  Hobart & William Smith Colleges – Pg
Argenbright, Robert T., (910) 962-3498 argenbrightr@uncw.edu,
  University of North Carolina, Wilmington – Zn
Argles, Tom, tom.argles@open.ac.uk,
  The Open University – Zi
Argyilan, Erin, 219-980-7124 eargyila@iun.edu,
  Indiana University Northwest – GeHwAm
Arias, Carmen, cariasf@ucm.es,
  Univ Complutense de Madrid – Gg
Ariya, Parisa A., (514) 398-3615 parisa.ariya@mcgill.ca,
  McGill University – As
Arkani-Hamed, Jafar, 514-398-6767 jafar@physics.utoronto.ca,
  McGill University – Xy
Arkle, Jenny C., 309-794-7844 jennyarkle@augustana.edu,
  Augustana College – GtmCc
Arkle, Kelsey M., 309-794-7487 kelseyarkle@augustana.edu,
  Augustana College – PeGsg
Arletti, Rossella, rossella.arletti@unito.it,
  Università di Torino – Gz
Armadillo, Egidio, +390103538085 egidio@dipteris.unige.it,
  Universita di Genova – Ye
Armbrust, Virginia E., (206) 616-1783 armbrust@ocean.washington.edu,
  University of Washington – Ob
Armi, Laurence, (858) 534-6843 larmi@ucsd.edu,
  University of California, San Diego – Op
Armour, Jake, 704-687-5968 jarmour@uncc.edu,
  University of North Carolina, Charlotte – GmlGg
Armstrong, Andrew, 603-862-4559 andya@ccom.unh.edu,
  University of New Hampshire – Zn
Armstrong, David E., (608) 262-2470 University of Wisconsin, Madison
  – Oc
Armstrong, Felicia P., 330-941-1385 fparmstrong@ysu.edu,
  Youngstown State University – Sb
Armstrong, Phillip A., (657) 278-3169 parmstrong@fullerton.edu,
  California State University, Fullerton – GcYg
Armstrong, Robert A., (609) 258-5260 robert.armstrong@stonybrook.edu,
  SUNY, Stony Brook – Ocb
Arnaud, Emmanuelle, 519-824-4120 x58087 earnaud@uoguelph.ca,
  University of Guelph – Gls
Arnold, Anthony J., (850) 644-4228 arnold@gly.fsu.edu,
  Florida State University – Pm
Arnold, Dan, +44 (0) 131 451 8298 d.arnold@hw.ac.uk,
  Heriot-Watt University – Go
Arnold, David L., 301/687-4053 dlarnold@frostburg.edu,
  Frostburg State University – As
Arnold, Stephen, +44(0) 113 34 37245 s.arnold@leeds.ac.uk,
  University of Leeds – As
Arnone, John, (775) 673-7445 jarnone@dri.edu,
  Desert Research Institute – Pg
Arnone, Robert (Bob), 228.688.6268 robert.arnone@usm.edu,
  University of Southern Mississippi – Op
Arnott, R. William C., (613)562-5800 6854 warnott@uottawa.ca,
  University of Ottawa – GsSo

Aronoff, Ruth F., (864)294-3363 ruth.aronoff@furman.edu,
  Furman University – Gc
Arp, Dr., Gernot, +49 (0)551 397986 garp@gwdg.de,
  Georg-August University of Goettingen – PggPg
Arrhenius, Gustaf, (858) 534-2961 arrhenius@ucsd.edu,
  University of California, San Diego – ClGzCm
Arribas, Eugenia, earribas@ucm.es,
  Univ Complutense de Madrid – Gd
Arritt, Raymond W., (515) 294-9870 rwarritt@iastate.edu,
  Iowa State University of Science & Technology – As
Arrowsmith, Ramon, (480) 965-3541 ramon.arrowsmith@asu.edu,
  Arizona State University – GcmGt
Arrowsmith, Stephen J., sarrowsmith@smu.edu,
  Southern Methodist University – Ys
Arslan, Gizem, 00904242370000-5972 gturus@firat.edu.tr,
  Firat University – GxiGp
Arsuaga, Juan Luis, azara@ucm.es,
  Univ Complutense de Madrid – Zn
Arthur, Jonathan D., (850) 617-0320 jonathan.arthur@FloridaDEP.gov,
  Florida Geological Survey – HwCgGe
Arthur, Michael A., (814) 863-6054 maa6@psu.edu,
  Pennsylvania State University, University Park – OuClGs
Artigas, Francisco, 201-460-2801 Francisco.Artigas@rutgers.edu,
  Rutgers, The State University of New Jersey, Newark – Zi
Artiola, Janick F., (520) 621-3516 jartiola@email.arizona.edu,
  University of Arizona – ScHsCa
Artioli, Gilberto, 39-049-8279162 gilberto.artioli@unipd.it,
  Università degli Studi di Padova – GzyZm
Arundale, Wendy H., (907) 474-7039 University of Alaska, Fairbanks – Zn
Arvidson, Raymond E., (314) 935-5679 arvidson@wunder.wustl.edu,
  Washington University in St. Louis – Xg
Arya, Satyapal S., (919) 515-7002 pal_arya@ncsu.edu,
  North Carolina State University – As
Aryal, Niroj, (336) 285-3832 naryal@ncat.edu,
  North Carolina Agricultural & Tech State University – HsgZi
Asante, Joseph, (931) 372-3576 jasante@tntech.edu,
  Tennessee Tech University – HwZrGe
Asbury, Brian, 303-273-3123 basbury@mines.edu,
  Colorado School of Mines – Nr
Aschoff, Jennifer, (907) 786-1442 jaschoff@uaa.alaska.edu,
  University of Alaska, Anchorage – GrdGo
Asghari Moghadam, Asghar, +98 (411) 339 2703 moghadam@tabrizu.ac.ir,
  University of Tabriz – HwqHy
Ash, Richard, (301) 405-7504 rdash@umd.edu,
  University of Maryland – Ca
Ashley, Gail M., (848) 445-2221 gmashley@eps.rutgers.edu,
  Rutgers, The State University of New Jersey – GsmSa
Ashley, Paul, +61-2-67732348 pashley@une.edu.au,
  University of New England – EgGeEm
Ashour-Abdalla, Maha, (310) 825-8881 mabdalla@igpp.ucla.edu,
  University of California, Los Angeles – Zn
Ashton, Andrew, 508-289-3751 aashton@whoi.edu,
  Woods Hole Oceanographic Institution – On
Ashwal, Lewis D., +27 11 717 6652 lewis.ashwal@wits.ac.za,
  University of the Witwatersrand – GxCgGt
Ashwood, Tom, (865) 574-7542 Oak Ridge National Laboratory – Ge
Ashworth, Allan C., 701-231-7919 allan.ashworth@ndsu.edu,
  North Dakota State University – Pe
Asimow, Paul D., (626) 395-4133 asimow@gps.caltech.edu,
  California Institute of Technology – CpGzy
Asiwaju-Bello, Yinusa A., 08036672708 Federal University of Technology,
  Akure – GgHgNg
Askari, Roohollah, (906) 487-2029 raskari@mtu.edu,
  Michigan Technological University – YxxNp
Askari, Zohreh, (217) 300-1819 askari@illinois.edu,
  Illinois State Geological Survey – Gsc
Aslan, Andres, (970) 248-1614 aaslan@coloradomesa.edu,
  Colorado Mesa University – Gm
Aslan, Yasemin, 00904242370000-5959 yaslan@hotmail.com,
  Firat University – Nr
Asmerom, Yemane, (505) 277-4204 asmerom@unm.edu,
  University of New Mexico – CcPeCg
Asowata, Timothy I., 08065348350 Federal University of Technology,
  Akure – Gge
Asphaug, Erik, (831) 459-2260 easphaug@pmc.ucsc.edu,
  University of California, Santa Cruz – Xg
Asphaug, Erik I., (831) 566-7877 asphaug@lpl.arizona.edu,

University of Arizona – XygXa
Asquith, George B., 806-834-0497 george.asquith@ttu.edu,
   Texas Tech University – Go
Assatourians, Karen, 519-661-2111, ext. 84715 kassatou@uwo.ca,
   Western University – Ys
Asselin, Esther, (418) 654-2612 esther.asselin@canada.ca,
   Natural Resources Canada – PlGuRh
Aster, Richard C., (970) 491-7606 rick.aster@colostate.edu,
   Colorado State University – YsGv
Astilleros, José Manuel, jmastill@ucm.es,
   Univ Complutense de Madrid – Gz
Atchison, Christopher L., (513) 556-3613 christopher.atchison@uc.edu,
   University of Cincinnati – Ze
Atchley, Stacy C., (254) 710-2196 stacy_atchley@baylor.edu,
   Baylor University – GroGo
Atekwana, Eliot A., 302-8312569 eatekwan@udel.edu,
   University of Delaware – CgsCq
Athaide, Dileep, (604) 984-1771 dathaide@eos.ubc.ca,
   University of British Columbia – Gg
Athaide, Dileep J., 604-986-1911, loc 7552 dathaide@capilanou.ca,
   Capilano University – GgZge
Athey, Jennifer E., (907) 451-5028 jennifer.athey@alaska.gov,
   Alaska Division of Geological & Geophysical Surveys – Gg
Atkinson, Gail M., 519-661-2111 x.84207 gatkins6@uwo.ca,
   Western University – YssYs
Atkinson, Gail M., gma@ccs.carleton.ca,
   Carleton University – Ne
Atkinson, Larry P., (757) 683-4926 latkinso@odu.edu,
   Old Dominion University – Op
Atkinson, Marlin J., 235-2224 mja@hawaii.edu,
   University of Hawai'i, Manoa – Ob
Atkinson, Tim, +44 020 7679 37711 t.atkinson@ucl.ac.uk,
   University College London – PcHwGm
Atkinson, Jr., William W., (303) 492-6103 william.atkinson@colorado.edu,
   University of Colorado – EmCg
Atreya, Sushil, (734) 936-0489 atreya@umich.edu,
   University of Michigan – ApZy
Attal, Mikael, +44 (0) 131 650 8533 mikael.attal@ed.ac.uk,
   Edinburgh University – GmtZy
Attig, John W., (608) 262-6131 jwattig@wisc.edu,
   University of Wisconsin, Extension – Gl
Attinger, Sabine, 0049(0)3641/948651 sabine.attinger@ufz.de,
   Friedrich-Schiller-University Jena – Hy
Attrep, Moses, (505) 667-0088 Los Alamos National Laboratory – Ze
Atudorei, Nieu-Viorel, atudorei@unm.edu,
   University of New Mexico – Cs
Atwater, Brian, 206-553-2927 atwater@uw.edu,
   University of Washington – RnGr
Au, Whitlow W L., 808-247-5026 wau@hawaii.edu,
   University of Hawai'i, Manoa – Zn
Aubert, John E., (916) 484-8637 aubertj@arc.losrios.edu,
   American River College – Zy
Aubry, Marie-Pierre, (732) 445-0822 aubry@eps.rutgers.edu,
   Rutgers, The State University of New Jersey – Pms
Audet, Celine, (418) 723-1986 (Ext. 1744) celine_audet@uqar.qc.ca,
   Universite du Quebec a Rimouski – Ob
Audet, Pascal, (613)562-5800 2344 pascal.audet@uottawa.ca,
   University of Ottawa – YgGc
Aufrecht, Walter E., (403) 329-2485 aufrecht@uleth.ca,
   University of Lethbridge – Zn
Ault, Alexis K., aault@email.arizona.edu,
   Utah State University – GtCc
Ault, Toby R., 607-255-1509 tra38@cornell.edu,
   Cornell University – Am
Aurnou, Jonathan M., (310) 825-2054 aurnou@epss.ucla.edu,
   University of California, Los Angeles – YgmYx
Ausbrooks, Scott, 501 683-0119 scott.ausbrooks@arkansas.gov,
   Arkansas Geological Survey – YsGeg
Ausich, William I., (614) 292-3353 ausich.1@osu.edu,
   Ohio State University – PgGs
Auster, Peter, 860-405-9118 peter.auster@uconn.edu,
   University of Connecticut – Ob
Austermann, Jacqueline, jackya@ldeo.columbia.edu,
   Columbia University – Ys
Austin, George S., (505) 835-5230 george@gis.nmt.edu,
   New Mexico Institute of Mining & Technology – En
Austin, Philip, (604) 822-2175 paustin@eos.ubc.ca,
   University of British Columbia – As
Austin, Jr., James A., (512) 471-0450 jamie@ig.utexas.edu,
   University of Texas, Austin – Gut
Autin, Whitney J., 585-395-5738 dirtguy@esc.brockport.edu,
   SUNY, The College at Brockport – Gs
Autio, Robert J., (812) 856-9104 rjautio@indiana.edu,
   Indiana University – HgGle
Avary, Katherine L., 304 594 2331 avary@geosrv.wvnet.edu,
   West Virginia University – Go
Avouac, Jean-Philippe, 626.395.2350 avouac@gps.caltech.edu,
   California Institute of Technology – GtYs
Awdankiewicz, Marek, marek.awdankiewicz@uwr.edu.pl,
   University of Wroclaw – Gvi
Awramik, Stanley M., (805) 893-3830 Awramik@geol.ucsb.edu,
   University of California, Santa Barbara – Pg
Axelbaum, Richard, rla@me.wustl.edu,
   Washington University in St. Louis – Ng
Axen, Gary, 575.835.5178 gaxen@ees.nmt.edu,
   New Mexico Institute of Mining and Technology – Gct
Axford, Yarrow L., 847.467.2268 yarrow@earth.northwestern.edu,
   Northwestern University – GnPmi
Ayad, Yasser M., yayad@clarion.edu,
   Clarion University – Zi
Ayanlade, Ayansina, (803) 777-2355 sinaayanlade@yahoo.co.uk,
   Obafemi Awolowo University – AtZir
Aydin, Adnan, (662) 915-1342 aaydin@olemiss.edu,
   University of Mississippi – NgrYe
Aydin, Atilla, (650) 725-8708 aydin@pangea.stanford.edu,
   Stanford University – Gc
Ayers, John C., 615-322-2158 john.c.ayers@vanderbilt.edu,
   Vanderbilt University – CgGx
Ayers, Joseph , 7815817370 x309 lobster@neu.edu,
   Northeastern University – Zn
Ayling, Bridget, (775) 682-8768 bayling@unr.edu,
   University of Nevada – CgYhGo
Aylward, Linda, (309) 694-5256 Illinois Central College – Ze
Azam, Farooq, (858) 534-6850 fazam@ucsd.edu,
   University of California, San Diego – Ob
Azetsu-Scott, Kumiko, (902) 426-8572 kumiko.azetsu-scott@mar.dfo-mpo.gc.ca,
   Geological Survey of Canada – Ocp

B

Baarli, B. Gudveig, (413) 597-2329 gudveig.baarli@williams.edu,
   Williams College – PssPi
Babaie, Hassan A., (404) 413-5766 hbabaie@gsu.edu,
   Georgia State University – GcZf
Babb, David M., (814) 863-3918 dmb16@psu.edu,
   Pennsylvania State University, University Park – Am
Babbin, Andrew, (617) 253-2181 babbin@mit.edu,
   Massachusetts Institute of Technology – Cb
Babcock, Daphne H., (972) 578-5518 dbabcock@collin.edu,
   Collin College - Spring Creek Campus – Geg
Babcock, Loren E., (614) 292-2103 babcock.5@osu.edu,
   Ohio State University – Pg
Babcock, R. S., (360) 650-3592 babcock@wwu.edu,
   Western Washington University – Cg
Babek, Ondrej, +420 549 49 3163 Masaryk University – Gds
Bacchus, Tania S., (802) 635-1329 Tania.Bacchus@jsc.edu,
   Johnson State College – GuAm
Bach Plaza, Joan, ++935811272 joan.bach@uab.cat,
   Universitat Autonoma de Barcelona – Gg
Bachle, Peter, (573) 368-2472 peter.bachle@dnr.mo.gov,
   Missouri Dept of Natural Resources – Gg
Bachmann, Etienne, 609-258-4101 etienneb@princeton.edu,
   Princeton University – Ys
Bachtadse, Valerian, 089/2180 4237 valerian@geophysik.uni-muenchen.de,
   Ludwig-Maximilians-Universitaet Muenchen – Yg
Baciu, Sorin D., 0040740404546 dsbaciu@gmail.com,
   Alexandru Ioan Cuza – GcPv
Back, Larissa E., 608-262-0776 lback@wisc.edu,
   University of Wisconsin, Madison – As
Backus, George E., (858) 534-2468 gebackus@ucsd.edu,
   University of California, San Diego – YmsOp
Bacon, Robert, (573) 526-0807 Missouri Dept of Natural Resources – Hs
Bacon, Steven N., 775-673-7473 Steve.Bacon@dri.edu,
   Desert Research Institute – Gm
Bada, Jeffrey L., (858) 534-4258 jbada@ucsd.edu,

University of California, San Diego – Co
Baddouh, Mbark, (703) 993-3892 mbaddouh@gmu.edu,
George Mason University – GdzPc
Bader, Nicholas E., (509) 527-5113 baderne@whitman.edu,
Whitman College – SoHgZi
Badger, Robert L., badgerrl@potsdam.edu,
SUNY Potsdam – GipGc
Badruddin, Abu Z., 315 2948610 ext.231 badruddin@cayuga-cc.edu,
Cayuga Community College – Zi
Badurek, Chris, (828) 262-7054 badurekca@appstate.edu,
Appalachian State University – Zi
Baedke, Steven J., (540) 568-6156 baedkesj@jmu.edu,
James Madison University – Hy
Baer, Eric M., (206) 592-3513 ebaer@highline.edu,
Highline College – GgvZg
Bagley, Brian, 612-624-8539 bagl0025@umn.edu,
University of Minnesota, Twin Cities – YsZn
Bagley, Evan, 601-266-4729 Evan.Bagley@usm.edu,
University of Southern Mississippi – GgEo
Bagtzoglou, Ross, (212) 854-3154 Columbia University – Hw
Bahlburg, Heinrich, +49-251-83-33935 bahlbur@uni-muenster.de,
Universitaet Muenster – GsgCc
Bahr, Jean M., (608) 262-2396 jmbahr@geology.wisc.edu,
University of Wisconsin, Madison – Hw
Bahr, Jean M., (608) 262-5513 jmbahr@geology.wisc.edu,
University of Wisconsin-Madison – Hw
Bahrmann, Chad, 814-865-9500 cbahrmann@psu.edu,
Pennsylvania State University, University Park – Am
Baig, Ayat, 204-227-7499 baiga@branedonu.ca,
Brandon University – EdmEg
Bailey, Christopher M., (757) 221-2445 cmbail@wm.edu,
College of William & Mary – Gc
Bailey, David G., (315) 859-4142 dbailey@hamilton.edu,
Hamilton College – GizGa
Bailey, Ian, +4401326 259322 i.bailey@exeter.ac.uk,
Exeter University – Zn
Bailey, Jack B., (309) 298-1481 JB-Bailey@wiu.edu,
Western Illinois University – Pg
Bailey, Jake, (612) 624-1603 baileyj@umn.edu,
University of Minnesota, Twin Cities – Pg
Bailey, Judy, 61 02 4921 5415 Judy.Bailey@newcastle.edu.au,
University of Newcastle – Ec
Bailey, Keiron D., kbailey@email.arizona.edu,
University of Arizona – Zi
Bailey, Richard C., (416) 978-3231 bailey@physics.utoronto.ca,
University of Toronto – Yg
Bailey, Richard H., (617) 373-3181 r.bailey@neu.edu,
Northeastern University – Pi
Bain, Daniel J., (412) 624-8780 dbain@pitt.edu,
University of Pittsburgh – HgGmZu
Bain, Olivier, +33(0)3 44069304 olivier.bain@lasalle-beauvais.fr,
Institut Polytechnique LaSalle Beauvais (ex-IGAL) – OuGgZi
Bair, E. Scott, 614-592-3900 bair.1@osu.edu,
Ohio State University – HwGe
Baird, Gordon C., (716) 673-3840 baird@fredonia.edu,
SUNY, Fredonia – GrPgGs
Baird, Graham, 970-351-2830 Graham.Baird@unco.edu,
University of Northern Colorado – GcgGz
Bajjali, William, (715) 394 8056 wbajjali@uwsuper.edu,
University of Wisconsin, Superior – HwCsg
Bajraktareviæ, Zlatan, +38514606098 zbajrak@geol.pmf.hr,
University of Zagreb – PgvPm
Bakel, Allen J., (630) 252-5486 Argonne National Laboratory – Co
Baker, Alex, +44 (0)1603 59 1529 alex.baker@uea.ac.uk,
University of East Anglia – On
Baker, Andy, a.baker@unsw.edu.au,
University of New South Wales – HwCs
Baker, Brett, 361-749-6774 brett_baker@utexas.edu,
University of Texas, Austin – Ze
Baker, Cathy, (479) 968-0661 cbaker@atu.edu,
Arkansas Tech University – Gg
Baker, Don, 514-398-7485 don.baker@mcgill.ca,
McGill University – CuGvCp
Baker, Edward T., (206) 526-6251 baker@pmel.noaa.gov,
University of Washington – Ou
Baker, James C., (540) 231-9785 suebrown@vt.edu,
Virginia Polytechnic Institute & State University – Sd

Baker, John M., (612) 625-4249 jbaker@soils.umn.edu,
University of Minnesota, Twin Cities – Sp
Baker, Leslie L., 885-6192 lbaker@uidaho.edu,
University of Idaho – CIXgGe
Baker, Marcia B., mbbaker@u.washington.edu,
University of Washington – As
Baker, Marcia B., 206-685-9697 mbbaker@uw.edu,
University of Washington – AsYg
Baker, Paul A., (919) 684-6450 pbaker@duke.edu,
Duke University – PcCgGg
Baker, Richard G., (319) 335-1827 dick-baker@uiowa.edu,
University of Iowa – PebPc
Baker, Sophie, 775-673-7434 Sophie.baker@dri.edu,
Desert Research Institute – Gm
Baker, Victor R., (520) 621-7875 baker@email.arizona.edu,
University of Arizona – GmXgRh
Baker-Treloar, Elizabeth, (619) 260-6822 ebaker@sandiego.edu,
University of San Diego – Geu
Baksi, Ajoy K., (225) 388-3422 abaksi@geol.lsu.edu,
Louisiana State University – Cc
Bal Akkoca, Dicle, 00904242370000-5982 dbal@firat.edu.tr,
Firat University – EnCo
Balakrishnan, Meena, 817-515-6360 meena.balakrishnan@tccd.edu,
Tarrant County College, Northeast Campus – GggGg
Balascio, Nicholas, 757-221-2880 nbalascio@wm.edu,
College of William & Mary – ZyGnl
Balázs, László, balazslaszlo@windowslive.com,
Eotvos Lorand University – EoYeg
Balch, Robert S., 575-835-5305 balch@prrc.nmt.edu,
New Mexico Institute of Mining and Technology – Yse
Baldauf, Jack G., (979) 845-3651 jbaldauf@tamu.edu,
Texas A&M University – Ou
Baldini, James, +44 191 33 42334 james.baldini@durham.ac.uk,
Durham University – Pc
Baldizon, Ileana, (305) 237-3658 ileana.baldizo@mdc.edu,
Miami-Dade College (Wolfson Campus) – Zn
Baldridge, W. Scott, (505) 667-4338 sbaldridge@lanl.gov,
Los Alamos National Laboratory – GicGt
Baldwin, Julia, 406-243-5778 jbaldwin@mso.umt.edu,
University of Montana – Gp
Baldwin, Michael, baldwin@purdue.edu,
Purdue University – As
Baldwin, Suzanne L., (315) 443-2672 sbaldwin@syr.edu,
Syracuse University – Cc
Baldwin, Jr., A. Dwight, dbbaldwin@comcast.net,
Miami University – Hw
Balen, Dražen, +38514605967 drbalen@geol.pmf.hr,
University of Zagreb – GxpGi
Bales, Roger C., rbales@eng.ucmerced.edu,
University of Arizona – Cg
Balescu, Sanda, 514-987-3000 #3126 sanda.balescu@club-internet.fr,
Universite du Quebec a Montreal – Pe
Balestro, Gianni, gianni.balestro@unito.it,
Università di Torino – Gc
Balgord, Elizabeth A., 801-626-6225 balgord@email.arizona.edu,
Weber State University – Grt
Balistrieri, Laurie, (206) 543-8966 balistri@ocean.washington.edu,
University of Washington – Oc
Ball, Elizabeth, (775) 682-7351 ball@mines.unr.edu,
University of Nevada, Reno – Zn
Ball, Neil, (204) 474-8075 na_ball@umanitoba.ca,
University of Manitoba – Gz
Ball, William P., (410) 516-5434 bball@jhu.edu,
Johns Hopkins University – Zn
Ballantyne, Colin K., ckb@st-andrews.ac.uk,
University of St. Andrews – ZyGml
Ballard, Robert D., (401) 874-6115 bballard@gso.uri.edu,
University of Rhode Island – Ga
Ballentine, Christopher, +44 (1865) 272000 (72938) chrisb@earth.ox.ac.uk,
University of Oxford – Xg
Ballero, Deniz Z., 678-212-7567 dballero@gsu.edu,
Georgia State University, Perimeter College, Online – GgPm
Ballinger, Rhoda, +44(0)29 208 76671 BallingerRC@cf.ac.uk,
University of Wales – On
Ballinger, Rhoda, ballingerRC@cf.ac.uk,
Cardiff University – Zn
Bally, Albert W., (713) 348-6063 geol@rice.edu,

Rice University – Gg
Balmat, Jennifer L., (308) 432-6483 jbalmat@csc.edu,
    Chadron State College – GgZe
Balmforth, Neil, (604) 822-9835 njb@math.ubc.ca,
    University of British Columbia – As
Balog-Szabo, Anna, 540-857-7222 abalog-szabo@virginiawestern.edu,
    Virginia Western Community College – Gs
Balogh-Brunstad, Zsuzsanna, (607) 431-4734 balogh_brunz@hartwick.edu,
    Hartwick College – ClHgSo
Balser, Teri , 352-294-3157 tcbalser@ufl.edu,
    University of Florida – Soo
Balser, Teri C., (608) 262-0132 tcbalser@wisc.edu,
    University of Wisconsin, Madison – Sb
Balsley, Christopher, 203-392-6647 balsleyc1@southernct.edu,
    Southern Connecticut State University – Gg
Balster, Nick J., (608) 263-5719 njbalster@wisc.edu,
    University of Wisconsin, Madison – Sf
Baltensperger, David D., (979) 845-3041 dbaltensperger@tamu.edu,
    Texas A&M University – PvCbSb
Baltz, Donald M., (225) 388-6512 Louisiana State University – Ob
Bamisaye, Oluwaseyi A., 07031533939 adunseyi@gmail.com,
    Federal University of Technology, Akure – GcZri
Bancroft, Alyssa M., (812) 856-5313 ambancro@indiana.edu,
    Indiana University – PmGr
Banda, Anna F., (903) 802-0186 aperry@dcccd.edu,
    El Centro College - Dallas Community College District – GgeGv
Bandfield, Joshua, joshband@uw.edu,
    University of Washington – Zr
Bandopadhyay, Sukumar, (907) 474-6876 sbandopadhyay@alaska.edu,
    University of Alaska, Fairbanks – Nm
Banerdt, Bruce E., (818) 354-5413 Jet Propulsion Laboratory – Xy
Banerjee, Neil, (519) 661-2111 x.83727 nbanerj3@uwo.ca,
    Western University – EgCgGx
Banerjee, Subir K., (612) 624-5722 banerjee@tc.umn.edu,
    University of Minnesota, Twin Cities – Ym
Banfield, Jillian, 510-642-9488 jill@seismo.berkeley.edu,
    University of California, Berkeley – CoGz
Bang, John, (919) 530-6569 jjbang@nccu.edu,
    North Carolina Central University – GeZa
Bangs, Nathan L., (512) 471-0424 nathan@ig.utexas.edu,
    University of Texas, Austin – Gu
Banik, Tenley, 309-438-8922 tjbanik@ilstu.edu,
    Illinois State University – GviCc
Bank, Carl-Georg, (416) 978-4381 bank@geology.utoronto.ca,
    University of Toronto – Zg
Banks, Beth, 828-884-8164 Beth.Banks@brevard.edu,
    Brevard College – Zn
Banks, David, +44(0) 113 34 35244 d.banks@see.leeds.ac.uk,
    University of Leeds – Cg
Banks, Eddie, eddie.banks@flinders.edu.au,
    Flinders University – Hw
Banner, Jay L., (512) 471-5016 banner@mail.utexas.edu,
    University of Texas, Austin – Cl
Bannon, Ann, ann.bannon@dal.ca,
    Dalhousie University – Zn
Bannon, Peter R., (814) 863-1309 bannon@ems.psu.edu,
    Pennsylvania State University, University Park – Am
Banse, Karl, (206) 543-5079 banse@ocean.washington.edu,
    University of Washington – Ob
Bao, Huiming, (225) 578-3419 bao@lsu.edu,
    Louisiana State University – CsAcCu
Baptista, Joana, +44 (0) 191 208 5899 joana.baptista@ncl.ac.uk,
    University of Newcastle Upon Tyne – Ng
Baptista, João C., jbaptist@utad.pt,
    Universidade de Trás-os-Montes e Alto Douro – GmmGa
Barak, Phillip W., (608) 890-0689 phillip.barak@wisc.edu,
    University of Wisconsin, Madison – ScGzRw
Barakat, Bassam, +33(0)3 44068973 bassam.barakat@lasalle-beauvais.fr,
    Institut Polytechnique LaSalle Beauvais (ex-IGAL) – NrgGq
Baran, Zeynep O., (605) 394-2461 Zeynep.Baran@sdsmt.edu,
    South Dakota School of Mines & Technology – GcoGt
Barazangi, Muawia, 607 255-6411 mb44@cornell.edu,
    Cornell University – YsGt
Barbeau, Katherine A., (858) 822-4339 kbarbeau@ucsd.edu,
    University of California, San Diego – Oc
Barbeau, Jr., David, dbarbeau@geol.sc.edu,
    University of South Carolina – Gs

Barbecot, Florent, 514-987-3000 #7786 barbecot.florent@uqam.ca,
    Universite du Quebec a Montreal – Hw
Barbee, Gary C., (806) 651-2294 gbarbee@wtamu.edu,
    West Texas A&M University – HwSoZi
Barber, Donald C., (610) 526-5110 dbarber@brynmawr.edu,
    Bryn Mawr College – GesOn
Barber, Richard T., (919) 728-2111 rbarber@duke.edu,
    Duke University – Ob
Barbi, Greta, 089/2180 4234 barbi@gophysik.uni-muenchen.de,
    Ludwig-Maximilians-Universitaet Muenchen – Yg
Barboni, Melanie, (480) 965-0612 melanie.barboni@asu.edu,
    Arizona State University – Gi
Barclay, Andrew, barclay@ldeo.columbia.edu,
    Columbia University – Ys
Barclay, Andrew, 206-543-8956 barclay@ldeo.columbia.edu,
    University of Washington – Gu
Barclay, David J., (607) 753-2921 david.barclay@cortland.edu,
    SUNY, Cortland – Glm
Barclay, David R., (902) 494-4164 dbarclay@dal.ca,
    Dalhousie University – OpNo
Barclay, Jenni, +44 (0)1603 59 3887 j.barclay@uea.ac.uk,
    University of East Anglia – Gv
Barclay, Julie L., julie.barclay@cortland.edu,
    SUNY, Cortland – Ggm
Barendregt, Rene W., (403) 329-2530 barendregt@uleth.ca,
    University of Lethbridge – Gm
Barhurst, James, +44 (0) 191 208 5431 james.bathurst@ncl.ac.uk,
    University of Newcastle Upon Tyne – Gs
Barineau, Clinton I., 706-507-8092 barineau_clinton@columbusstate.edu,
    Columbus State University – GctGg
Barker, Andy J., +44 (0)23 80593641 A.J.Barker@soton.ac.uk,
    University of Southampton – Gg
Barker, Chris A., 936 468-2340 cbarker@sfasu.edu,
    Stephen F. Austin State University – Gc
Barker, Colin, (918) 631-3014 colin-barker@utulsa.edu,
    The University of Tulsa – CgoGo
Barker, Daniel, (512) 471-5502 danbarker@mail.utexas.edu,
    University of Texas, Austin – Gi
Barker, Gregory A., (603) 271-7332 gbarker@des.state.nh.us,
    New Hampshire Geological Survey – GglZi
Barker, James M., (505) 825-5114 jbark@nmt.edu,
    New Mexico Institute of Mining & Technology – En
Barker, Jeffrey S., (607) 777-2522 jbarker@binghamton.edu,
    Binghamton University – Ys
Barker, Joel D., (740) 725-6097 barker.246@osu.edu,
    Ohio State University – CblPe
Barker, Stephen, barkers3@cardiff.ac.uk,
    Cardiff University – Pe
Barkmann, Peter, 303-384-2642 barkmann@mines.edu,
    Colorado Geological Survey – HwGcg
Barlow, Jay P., (858) 546-7178 jbarlow@ucsd.edu,
    University of California, San Diego – Ob
Barlow, Mathew, (978) 934-3908 mathew_barlow@uml.edu,
    University of Massachusetts, Lowell – As
Barminski, Robert, 831-770-7056 rbarminski@aol.com,
    Hartnell College – GgOg
Barmore, Garrett, 775-784-4528 gbarmore@unr.edu,
    University of Nevada, Reno – Zn
Barnard, Holly R., Holly.Barnard@colorado.edu,
    University of Colorado – Hg
Barnbaum, Cecilia S., (229) 249-2645 cbarnbau@valdosta.edu,
    Valdosta State University – Zn
Barnes, Calvin G., 806-834-7389 cal.barnes@ttu.edu,
    Texas Tech University – Gi
Barnes, Charles W., (928) 774-6079 chuck.barnes@nau.edu,
    Northern Arizona University – GcXg
Barnes, Christopher R., (250) 721-8847 crbarnes@uvic.ca,
    University of Victoria – PmOgPe
Barnes, Elizabeth A., eabarnes@atmos.colostate.edu,
    Colorado State University – As
Barnes, Fairley J., (505) 667-4933 fyb@lanl.gov,
    Los Alamos National Laboratory – Zn
Barnes, Hubert L., (814) 865-7573 barnes@psu.edu,
    Pennsylvania State University, University Park – CgEmCe
Barnes, Jaime D., jdbarnes@jsg.utexas.edu,
    University of Texas, Austin – CgsCc
Barnes, Jeffrey R., (541) 737-5685 barnes@coas.oregonstate.edu,

Oregon State University – As
Barnes, Jessica, jebarnes@mailbox.sc.edu,
    University of South Carolina – Zn
Barnes, Melanie A., (806) 834-7965 melanie.barnes@ttu.edu,
    Texas Tech University – CaGie
Barnes, Philip, pbarnes@environ.sc.edu,
    University of South Carolina – Zn
Barnes, Randal J., (612) 625-5828 University of Minnesota, Twin Cities – Zn
Barnes, Sarah- J., 418 545 5011 sjbarnes@uqac.ca,
    Universite du Quebec a Chicoutimi – EgGiCt
Barnett, Douglas B., (509) 376-3416 brent.barnett@pnl.gov,
    Pacific Northwest National Laboratory – Em
Barnett, Michael, 617-552-8300 barnetge@bc.edu,
    Boston College – Ze
Barnett, Roger T., rbarnett@pacific.edu,
    University of the Pacific – Zy
Barnhart, William, (319) 384-4732 william-barnhart-1@uiowa.edu,
    University of Iowa – Gt
Barnosky, Anthony D., barnosky@berkeley.edu,
    University of California, Berkeley – Pv
Baron, Dirk, (661) 654-3044 California State University, Bakersfield – HgGeZe
Barone, Jessica, 585-292-2448 jbarone@monroecc.edu,
    Monroe Community College – Ge
Baross, John A., (206) 543-0833 University of Washington – Ob
Barquero-Molina, Miriam, (573) 882-9557 barqueromolinam@missouri.edu,
    University of Missouri – Gct
Barr, Sandra M., (902) 585-1340 sandra.barr@acadiau.ca,
    Acadia University – GitGg
Barr, Stuart, +44 (0) 191 208 6449 stuart.barr@ncl.ac.uk,
    University of Newcastle Upon Tyne – Zi
Barra, Monica, (803) 576-8340 mbarra@seie.sc.edu,
    University of South Carolina – Zn
Barrash, Warren, (208) 426-1229 wbarrash@boisestate.edu,
    Boise State University – HwYxGe
Barrett, Bradford S., 410-293-6567 bbarrett@usna.edu,
    United States Naval Academy – As
Barrett, John, +44(0) I13 34 32394 j.r.barrett@leeds.ac.uk,
    University of Leeds – Ge
Barrett, Kevin M., (817) 515-6352 kevin.barrett@tccd.edu,
    Tarrant County College, Northeast Campus – AsZgr
Barrett, Linda R., 330 972-7620 barrett@uakron.edu,
    University of Akron – SdZri
Barrett, Melony, 217-333-7917 mebarret@illinois.edu,
    Illinois State Geological Survey – Zi
Barrick, James E., (806) 834-2717 jim.barrick@ttu.edu,
    Texas Tech University – Psm
Barrie, Vaughn, (250) 363-6424 University of Victoria – Gu
Barrier, Pascal, +33(0)3 44068975 pascal.barrier@lasalle-beauvais.fr,
    Institut Polytechnique LaSalle Beauvais (ex-IGAL) – PmGsPs
Barron, Elizabeth, (920) 424-4105 barrone@uwosh.edu,
    University of Wisconsin Oshkosh – Zn
Barron, George, gbarron@uoguelph.ca,
    University of Guelph – Zn
Barron, Robert, (906) 487-2096 rjbarron@mtu.edu,
    Michigan Technological University – ZgGzi
Barros, Tony, (305) 237-3754 tbarros@mdc.edu,
    Miami-Dade College (Wolfson Campus) – Og
Barry, Peter, pbarry@whoi.edu,
    Woods Hole Oceanographic Institution – Cm
Barsch, Robert, 089/2180 4201 barsch@geophysik.uni-muenchen.de,
    Ludwig-Maximilians-Universitaet Muenchen – Yg
Bart, Philip J., (225) 388-3109 pbart@geol.lsu.edu,
    Louisiana State University – Gr
Bartello, Peter, (514) 398-8075 peter.bartello@mcgill.ca,
    McGill University – As
Bartels, William S., (517) 629-0313 wbartels@albion.edu,
    Albion College – Pv
Barth, Andrew P., (317) 274-1243 ibsz100@iupui.edu,
    Indiana University / Purdue University, Indianapolis – Gi
Barth, Jack A., (541) 737-1607 barth@coas.oregonstate.edu,
    Oregon State University – On
Barth, Nicolas, (951) 827-3138 nic.barth@ucr.edu,
    University of California, Riverside – Gt
Barth, Susan A., (406) 496-4687 sbarth@mtech.edu,
    Montana Tech of The University of Montana – Zn

Barthelmie, Rebecca J., rb737@cornell.edu,
    Cornell University – Ass
Bartholomaus, Timothy, (208) 885-6217 tbartholomaus@uidaho.edu,
    University of Idaho – Gl
Bartholomew, Alexander J., 845-257-3765 barthola@newpaltz.edu,
    SUNY, New Paltz – GrPi
Bartholomew, Mervin J., (901) 678-1613 jbrthlm1@memphis.edu,
    University of Memphis – Gtc
Bartholy, Judit, +36 20 3722945 bartholy@caesar.elte.hu,
    Eotvos Lorand University – Am
Bartl, Simona, 831-771-4400 sbartl@mlml.calstate.edu,
    Moss Landing Marine Laboratories – Zn
Bartlein, Patrick, (541) 346-4967 bartlein@uoregon.edu,
    University of Oregon – Pe
Bartlett, Douglas H., (858) 534-5233 dbartlett@ucsd.edu,
    University of California, San Diego – Ob
Bartlett, Wendy, 740-376-4782 bartletw@marietta.edu,
    Marietta College – Goe
Bartley, John M., (801) 58--7162 john.bartley@utah.edu,
    University of Utah – GciCc
Bartley, Julie K., 507-933-7541 jbartley@gustavus.edu,
    Gustavus Adolphus College – GsPgo
Bartok, Peter, 713-444-7751 peter@bartokinc.com,
    University of Houston – GoYeNr
Bartolucci, Valerio, (954) 201-6678 vbartolu@broward.edu,
    Broward College – Ge
Barton, Christopher C., (727) 215-5538 chris.barton@wright.edu,
    Wright State University – GqYgHq
Barton, James H., (412) 589-2821 Thiel College – Zy
Barton, Mark D., (520) 621-8529 mdbarton@email.arizona.edu,
    University of Arizona – Eg
Barton, Michael, (614) 292-3132 barton.2@osu.edu,
    Ohio State University – GiCpt
Barzegari, Ghodrat, Gbarzegari@gmail.com,
    University of Tabriz – Nr
Basak, Chandranath, (661) 654-3400 cbasak@csub.edu,
    California State University, Bakersfield – CmOoCs
Bascom, Johnathan, (616) 526-7053 jbascom@calvin.edu,
    Calvin College – Zy
Basinger, James F., (306) 966-5684 jim.basinger@usask.ca,
    University of Saskatchewan – Pb
Baskin, Perry A., (229) 259-5052 pbaskin@valdosta.edu,
    Valdosta State University – Zn
Bass, David, david.bass@flinders.edu.au,
    Flinders University – Zy
Bass, Jay D., (217) 333-1018 jaybass@illinois.edu,
    University of Illinois, Urbana-Champaign – Gy
Bass, Jerry, 601 266-4732 joby@usm.edu,
    University of Southern Mississippi – Zn
Bass, Michael L., mbass@umw.edu,
    University of Mary Washington – Sf
Bassett, Damon, (417) 836-4897 dbassett@missouristate.edu,
    Missouri State University – Gg
Bassett, Kari N., (03) 369-4495 kari.bassett@canterbury.ac.nz,
    University of Canterbury – GstGd
Bassett, Scott, (775) 784-1434 sbassett@unr.edu,
    University of Nevada, Reno – Zu
Bassett, William A., wab7@cornell.edu,
    Cornell University – Gz
Bassis, Jeremy, (734) 615-3606 jbassis@umich.edu,
    University of Michigan – GlOgZg
Basta, Nicholas T., 614-292-6282 basta.4@osu.edu,
    Ohio State University – Sc
Bastow, Ian, +44 20 759 42974 i.bastow@imperial.ac.uk,
    Imperial College – Yg
Basu, Abhijit, 8128556654/5581 basu@indiana.edu,
    Indiana University, Bloomington – Gd
Basu, Anirban, +44 1784 414083 anirban.basu@rhul.ac.uk,
    Royal Holloway University of London – CglCs
Basu, Asish, (817) 272-2987 abasu@uta.edu,
    University of Texas, Arlington – GxCcs
Batelaan, Okke, okke.batelaan@ees.kuleuven.be,
    Katholieke Universiteit Leuven – Hw
Bates, Mary, (661) 362-5054 Mary.Bates@canyons.edu,
    College of the Canyons – Zyg
Bates, Nicholas R., N.R.Bates@soton.ac.uk,
    University of Southampton – Oc

Bates, Richard, +44 01334 463997 crb@st-andrews.ac.uk,
University of St. Andrews – Yx
Bates, Timothy S., (206) 526-6248 Tim.Bates@noaa.gov,
University of Washington – As
Bathke, Deborah J., dbathke2@unl.edu,
University of Nebraska, Lincoln – AsZe
Battisti, David S., (206) 543-2019 battisti@uw.edu,
University of Washington – As
Battisti, David S., david@atmos.washington.edu,
University of Washington – As
Battles, Eileen, (785) 864-2129 battles@kgs.ku.edu,
University of Kansas – Zy
Batygin, Konstantin, 626-395-2920 kbatygin@gps.caltech.edu,
California Institute of Technology – Xg
Bauder, James W., (406) 994-5685 jbauder@montana.edu,
Montana State University – Sf
Bauer, Annie, annie.bauer@wisc.edu,
University of Wisconsin-Madison – Cca
Bauer, Emily, (612) 626-0909 bauer010@umn.edu,
University of Minnesota – Zi
Bauer, Jeffrey A., jbauer@shawnee.edu,
Shawnee State University – Ps
Bauer, Paul W., (505) 835-5106 bauer@nmbg.nmt.edu,
New Mexico Institute of Mining & Technology – Gp
Bauer, R. A., 217-244-2394 rabauer@illinois.edu,
Illinois State Geological Survey – Ng
Baum, Steven K., (979) 845-0793 sbaum@ocean.tamu.edu,
Texas A&M University – Op
Baumann, Hannes, 860-405-9297 hannes.baumann@uconn.edu,
University of Connecticut – Ob
Baumann, Zofia, (860) 405-9281 zofia.baumann@uconn.edu,
University of Connecticut – ObcCm
Baumgartner, Mark, mbaumgartner@whoi.edu,
Dalhousie University – Ob
Baumiller, Tomasz K., (734) 764-7543 tomaszb@umich.edu,
University of Michigan – Pg
Bautista, Sonia, sbautist@ucm.es,
Univ Complutense de Madrid – Ng
Baxter, Ethan, 617-552-3640 ethan.baxter@bc.edu,
Boston College – Ccg
Baxter, Martin, 989-774-2055 baxte1ma@cmich.edu,
Central Michigan University – Ams
Baxter, Stefanie J., 302-831-1576 steff@udel.edu,
University of Delaware – On
Baxter, P.G., James E., 717-780-2377 jebaxter@hacc.edu,
Harrisburg Area Community College – GgHwGm
Bayari, Serdar, +90 (312) 2977740 serdar@hacettepe.edu.tr,
Hacettepe University – Hw
Bayly, M. Brian, (518) 283-1721 bayly.brian@gmail.com,
Rensselaer Polytechnic Institute – Gc
Bayowa, Oyelowo G., +2348030820291 obayowa@lautech.edu.ng,
Ladoke Akintola University of Technology – GggGg
Beach Davis, Janet, (309) 268-8513 janet.beach-davis@heartland.edu,
Heartland Community College – Zg
Beane, Rachel J., (207) 725-3160 rbeane@bowdoin.edu,
Bowdoin College – GzxZe
Beard, Brian L., 608-262-1806 beardb@geology.wisc.edu,
University of Wisconsin-Madison – Cc
Beard, James S., (276) 634-4170 jim.beard@vmnh.virginia.gov,
Virginia Polytechnic Institute & State University – GiClg
Beard, K. Christopher, (412) 622-5782 beardc@CarnegieMNH.org,
Carnegie Museum of Natural History – Pv
Beard, Kenneth V., (217) 333-1676 k-beard@uiuc.edu,
University of Illinois, Urbana-Champaign – As
Bearden, Rebecca A., 205-247-3623 rbearden@gsa.state.al.us,
Geological Survey of Alabama – Hs
Beatty, Heather L., heather.beatty@austincc.edu,
Austin Community College District – Ge
Beatty, Lynne, 9134698500 x3785 lbeatty@jccc.edu,
Johnson County Community College – GgZy
Beatty, Merrill Ann, (506) 453-4803 mbeatty@unb.ca,
University of New Brunswick – Zn
Beatty, William L., 507-474-5789 wbeatty@winona.edu,
Winona State University – PgiGs
Beaty, Tammy, (865) 574-0119 taw@ornl.gov,
Oak Ridge National Laboratory – Zi
Beauchamp, Marc, 519-661-2111 ext 88104 mbeauch6@uwo.ca,
Western University – Ca
Beaudoin, Georges, (418) 656-3141 beaudoin@ggl.ulaval.ca,
Universite Laval – EgGzCa
Beaulieu, Claudie, clbeauli@ucsc.edu,
University of Southampton – Op
Beaumont, Christopher, (902) 494-3779 chris.beaumont@dal.ca,
Dalhousie University – Yg
Beausoleil, Denis, 604-777-6019 beausoleild@douglascollege.ca,
Douglas College – GgZg
Beaverson, Sheena K., 217-244-9306 sbeavers@illinois.edu,
Illinois State Geological Survey – Zr
Bebout, Gray E., (610) 758-5831 geb0@lehigh.edu,
Lehigh University – Gp
Beccue, Clint, 217-265-5161 cbeccue@illinois.edu,
Illinois State Geological Survey – Ge
Becel, Anne, annebcl@ldeo.columbia.edu,
Columbia University – Gu
Bechtel, Randy, 919-707-9204 randy.bechtel@ncdenr.gov,
North Carolina Geological Survey – ZeGg
Bechtel, Timothy D., 717 291-4133 timothy.bechtel@fandm.edu,
Franklin and Marshall College – GeYg
Beck, Catherine C., 315-859-4847 ccbeck@hamilton.edu,
Hamilton College – GsnGr
Beck, Charles, (734) 763-5089 chbeck@umich.edu,
University of Michigan – Pg
Beck, Colleen M., (702) 795-8077 colleen@dri.edu,
Desert Research Institute – Sa
Beck, E. G., 270.827.3414 x23 ebeck@uky.edu,
University of Kentucky – Hw
Beck, John H., 617-552-8300 john.beck@bc.edu,
Boston College – Pb
Beck, Susan L., (520) 621-8628 slbeck@email.arizona.edu,
University of Arizona – YsGt
Becker, Alex, (510) 643-9181 University of California, Berkeley – Yg
Becker, Janet M., (808) 956-6514 jbecker@soest.hawaii.edu,
University of Hawai'i, Manoa – Og
Becker, Keir, (305) 421-4661 kbecker@rsmas.miami.edu,
University of Miami – YrGu
Becker, Laurence, (541) 737-9504 beckerla@geo.oregonstate.edu,
Oregon State University – Zn
Becker, Lorene Y., 541-737-6993 beckelo@geo.oregonstate.edu,
Oregon State University – Zi
Becker, Matthew, 562-985-8983 mbecker3@csulb.edu,
California State University, Long Beach – Hwg
Becker, Naomi M., (505) 667-2165 nmb@lanl.gov,
Los Alamos National Laboratory – Hg
Becker, Ralph Thomas, +49-251-83-33951 rbecker@uni-muenster.de,
Universitaet Muenster – PisPm
Becker, Richard H., (630) 252-7595 Argonne National Laboratory – Zr
Becker, Richard H., (419) 530-4571 richard.becker@utoledo.edu,
University of Toledo – ZrGe
Becker, Stefan, 718-960-1120 stefan.becker@lehman.cuny.edu,
Graduate School of the City University of New York – Zn
Becker, Udo, (734) 615-6894 ubecker@umich.edu,
University of Michigan – Zn
Beckie, Roger D., (604) 822-6462 rbeckie@eos.ubc.ca,
University of British Columbia – HwCl
Beckingham, Barbara, (843) 953-0483 beckinghamba@cofc.edu,
College of Charleston – Cl
Bedard, Jean H., (418) 654-2671 jbedard@nrcan.gc.ca,
Universite du Quebec – Gi
Bedard, Paul, 4185455011, 2276 pbedard@uqac.ca,
Universite du Quebec a Chicoutimi – CaGzEg
Bedaso, Zelalem, (937) 229-2393 zbedaso1@udayton.edu,
University of Dayton – CslGs
Beddows, Patricia A., (847) 491-7460 patricia@earth.northwestern.edu,
Northwestern University – HyGma
Bedient, Philip B., (713) 348-4953 bedient@rice.edu,
Rice University – Hw
Bednarski, Marsha, (860) 832-2943 bednarskim@ccsu.edu,
Central Connecticut State University – Ze
Bednarz, Robert S., r-bednarz@tamu.edu,
Texas A&M University – Zeu
Bednarz, Sarah W., (979) 845-1579 s-bednarz@tamu.edu,
Texas A&M University – Ze
Beebe, Alex, dbeebe@southalabama.edu,
University of South Alabama – Ge

Beets, Kay J., +31 20 59 87357 c.j.beets@vu.nl,
  Vrije Universiteit Amsterdam – Cg
Befus, Kenneth S., (254)710-2192 Kenneth_Befus@baylor.edu,
  Baylor University – GviGv
Beget, James E., ffjeb1@uaf.edu,
  University of Alaska, Fairbanks – Gm
Beghein, Caroline, (310) 825-0742 cbeghein@ucla.edu,
  University of California, Los Angeles – Ysg
Begin, Christian, (418) 654-2648 christian.begin@canada.ca,
  Universite du Quebec – PeGmSo
Behl, Richard J., (562) 985-5850 behl@csulb.edu,
  California State University, Long Beach – GsPeGd
Behn, Mark, 617 552-2180 mark.behn@bc.edu,
  Boston College – YgrGt
Behn, Mark D., (508) 289-3637 mbehn@whoi.edu,
  Woods Hole Oceanographic Institution – Yr
Behr, Rose-Anna, (717) 702-2035 rosbehr@pa.gov,
  Pennsylvania Bureau of Topographic & Geologic Survey – Gcr
Behr, Whitney, 512-232-1941 behr@utexas.edu,
  University of Texas, Austin – Gc
Behr-Andres, Christina "Tina", (505) 667-3644 behr-andres@lanl.gov,
  Los Alamos National Laboratory – Zn
Behrensmeyer, Anna K., (202) 633-1307 Smithsonian Inst / Natl Museum of Natural Hist– Pv
Beiersdorfer, Raymond E., (330) 941-1753 rebeiersdorfer@ysu.edu,
  Youngstown State University – Cg
Bein, F. L., (317) 274-1100 Indiana University, Indianapolis – Zy
Bekker, Andrey, (951) 827-4611 andrey.bekker@ucr.edu,
  University of California, Riverside – GsrCg
Bekker, Matthew F., (801) 422-1961 matthew_bekker@byu.edu,
  Brigham Young University – Zy
Belasky, Paul, 510-676-0945 pbelasky@ohlone.edu,
  Ohlone College – PgqGs
Belknap, Daniel F., (207) 581-2159 belknap@maine.edu,
  University of Maine – GusGa
Bell, Andrew, +44 (0) 131 650 4918 a.bell@ed.ac.uk,
  Edinburgh University – Gt
Bell, Brian, +44 01413306898 Brian.Bell@glasgow.ac.uk,
  University of Glasgow – So
Bell, Christopher J., (512) 471-7301 cjbell@mail.utexas.edu,
  University of Texas, Austin – Pv
Bell, David, (03) 3642-717 david.bell@canterbury.ac.nz,
  University of Canterbury – Ng
Bell, James C., (651) 328-1359 bellx007@umn.edu,
  University of Minnesota, Twin Cities – SdfZe
Bell, John W., (775) 784-1939 jbell@unr.edu,
  University of Nevada – NgGm
Bell, Keith, keith_bell@carleton.ca,
  Carleton University – Cc
Bell, Margaret C., +44 (0) 191 208 7936 margaret.bell@ncl.ac.uk,
  University of Newcastle Upon Tyne – Ge
Bell, Michael M., mmbell@colostate.edu,
  Colorado State University – As
Bell, Rebecca, +44 20 759 40903 rebecca.bell@imperial.ac.uk,
  Imperial College – Gt
Bell, Robin E., robinb@ldeo.columbia.edu,
  Columbia University – Yr
Bellew, Angela, (319) 335-1819 angela-bellew@uiowa.edu,
  University of Iowa – Zn
Bellian, Jerome, jerry.bellian@whiting.com,
  Miami University – Go
Belliveau, Lindsey, 860-424-3581 lindsey.belliveau@ct.gov,
  Dept of Energy and Environmental Protection – Gmt
Belluso, Elena, elena.belluso@unito.it,
  Università di Torino – Gz
Belmonte, Donato, +39 010 353 8136 Universita di Genova – Cg
Belt, Edward S., (413) 542-2712 esbelt@amherst.edu,
  Amherst College – Gs
Beltrami, Hugo, 902-867-2326 hbeltram@stfx.ca,
  Saint Francis Xavier University – YhAs
Beltz, John F., 330 972-6687 jfb4@uakron.edu,
  University of Akron – GgRh
Ben-Avraham, Zvi, zviba@tau.ac.il,
  Tel Aviv University – YrGt
Ben-Zion, Yehuda, (213) 740-6734 ybz@earth.usc.edu,
  University of Southern California – Ys
Benacquista, Matt, (406) 657-2341 Montana State University, Billings – Xy

Benavides, Jude, 956 883-5938 jude.benavides@utrgv.edu,
  University of Texas, Rio Grande Valley – Hsq
Benavides-Iglesias, Alfonso, a.benavides@geos.tamu.edu,
  Texas A&M University – Ys
Bend, Stephen L., (306) 585-4021 University of Regina – Gox
Bender, E. E., (714) 432-5681 ebender@occ.cccd.edu,
  Orange Coast College – GitGz
Bender, John F., 704-687- 5956 jfbender@email.uncc.edu,
  University of North Carolina, Charlotte – GiCtOu
Bender, Michael L., (609) 258-5807 bender@princeton.edu,
  Princeton University – Cg
Bendick, Rebecca, 406-243-5774 bendick@mso.umt.edu,
  University of Montana – Yg
Benedetti, Michael M., (910) 962-7650 benedettim@uncw.edu,
  University of North Carolina, Wilmington – GmSaZy
Benfield, Mark C., (225) 388-6372 mbenfie@lsu.edu,
  Louisiana State University – Ob
Benger, Simon, simon.benger@flinders.edu.au,
  Flinders University – Ziy
Benimoff, Alan I., 718-982-2835 alan.benimoff@csi.cuny.edu,
  College of Staten Island/CUNY – GizGp
Benison, Kathleen , 304-293-5603 West Virginia University – GrsCm
Benitez-Nelson, Claudia R., 777-0018 cbnelson@geol.sc.edu,
  University of South Carolina – OcGe
Benito, María Isabel, mibenito@ucm.es,
  Univ Complutense de Madrid – Gs
Benjamin, Patricia A., 508-929-8606 pbenjamin@worcester.edu,
  Worcester State University – Zg
Benjamin, Timothy M., (505) 667-5154 Los Alamos National Laboratory – Ca
Benjamin, U. K., 234-8038312732 uzochukwu.benjamin@oauife.edu.ng,
  Obafemi Awolowo University – GoCg
Benna, Piera, piera.benna@unito.it,
  Università di Torino – Gz
Benner, Jake, (865) 974-2366 jbenner@utk.edu,
  University of Tennessee, Knoxville – GgPg
Benner, Ron, benner@mailbox.sc.edu,
  University of South Carolina – Ob
Bennet, Brett, 785-864-2117 bb@kgs.ku.edu,
  University of Kansas – Ng
Bennett, Philip, (512) 471-3587 pbennett@mail.utexas.edu,
  University of Texas, Austin – Hg
Bennett, Richard, (520) 621-2324 rab@geo.arizona.edu,
  University of Arizona – YdGt
Bennett, Sara, 309/298-1905 SC-Bennett@wiu.edu,
  Western Illinois University – Gg
Bennett, Scott, scottekb@uw.edu,
  University of Washington – YsGtm
Bennett, Steven W., (309) 298-1256 SW-Bennett1@wiu.edu,
  Western Illinois University – Hw
Benninger, Larry K., (919) 962-0699 lbenning@email.unc.edu,
  University of North Carolina, Chapel Hill – ClGuCt
Bennington, J Bret, 516 463-5568 geojbb@hofstra.edu,
  Hofstra University – PeGs
Benoit-Bird, Kelly, 541-737-2063 kbenoit@coas.oregonstate.edu,
  Oregon State University – Gu
Bense, Victor, +44 (0)1603 59 1297 v.bense@uea.ac.uk,
  University of East Anglia – Hg
Benson, David A., 303-273-3806 dbenson@mines.edu,
  Colorado School of Mines – Hq
Benson, Donna M., (480) 461-7247 donnabenson@mesacc.edu,
  Mesa Community College – Gg
Benson, Jane L., (270) 809-3106 jbenson1@murraystate.edu,
  Murray State University – Zi
Benson, Robert G., (719) 587-7921 rgbenson@adams.edu,
  Adams State University – Gg
Benson, Roger, +44 (1865) 272000 roger.benson@earth.ox.ac.uk,
  University of Oxford – Pg
Bentley, Callan, (703) 323-3276 cbentley@nvcc.edu,
  Northern Virginia Community College - Annandale – Gg
Bentley, Laurence R., (403) 220-4512 University of Calgary – Hw
Bentley, Samuel J., 225-578-5735 sjb@lsu.edu,
  Louisiana State University – Gu
Benton, Steven, 217-244-0082 s-benton@illinois.edu,
  Illinois State Geological Survey – Gg
Benway, Heather, (508) 289-2838 hbenway@whoi.edu,
  Woods Hole Oceanographic Institution – OcCmOo

Berchem, Jean, (310) 206-6484 jberchem@igpp.ucla.edu,
University of California, Los Angeles – Zn
Bercovici, David, (203) 432-3168 david.bercovici@yale.edu,
Yale University – Yg
Bereitschaft, Bradley, 402-554-2674 bbereitschaft@unomaha.edu,
University of Nebraska, Omaha – Zuy
Berelson, William M., (213) 740-5828 berelson@usc.edu,
University of Southern California – Cm
Berendsen, Pieter, 785-864-2141 pieterb@kgs.ku.edu,
University of Kansas – Cg
Beresnev, Igor A., (515) 294-7529 beresnev@iastate.edu,
Iowa State University of Science & Technology – YgsYe
Berg, Christopher A., (714)432-0202 x21229 cberg5@occ.cccd.edu,
Orange Coast College – GpZeGt
Berg, Christopher A., (678) 839-4059 cberg@westga.edu,
University of West Georgia – Gpi
Berg, Peter, (804) 924-1318 pb8n@virginia.edu,
University of Virginia – So
Berg, Richard, 217-244-2776 rberg@illinois.edu,
University of Illinois – GlRwZu
Berg, Richard B., (406) 496-4172 dberg@mtech.edu,
Montana Tech of The University of Montana – EnGz
Berg, Richard C., (217) 244-2776 rberg@illinois.edu,
Illinois State Geological Survey – GleGm
Bergantz, George W., 206-685-4972 bergantz@uw.edu,
University of Washington – Gi
Berger, Peter M., 217-333-7078 pmberger@illinois.edu,
Illinois State Geological Survey – Cg
Berger, Wolfgang H., (858) 822-2545 wberger@ucsd.edu,
University of California, San Diego – Gs
Bergeron, Normand, (418) 654-3703 normand.bergeron@ete.inrs.ca,
Universite du Quebec – Gm
Berggren, William A., 848-445-8523 wberggren@whoi.edu,
Rutgers, The State University of New Jersey – PmsPe
Berggren, William A., (508) 289-2593 wberggren@whoi.edu,
Woods Hole Oceanographic Institution – Ps
Bergin, Michael H., (404) 894-9723 mike.bergin@ce.gatech.edu,
Georgia Institute of Technology – Is
Bergmann, Kristin, (617) 253-9852 kdberg@mit.edu,
Massachusetts Institute of Technology – GsCl
Bergquist, Bridget, bergquist@es.utoronto.ca,
University of Toronto – Cgb
Bergslien, Elisa T., 716-878-3793 bergslet@buffalostate.edu,
Buffalo State College – ClHw
Bergstrom, Stig M., (614) 292-4473 bergstrom.1@osu.edu,
Ohio State University – PgGg
Bergwerff, Kenneth A., (616) 526-6371 kbergwer@calvin.edu,
Calvin College – Ze
Berke, Melissa, Melissa.Berke.1@nd.edu,
University of Notre Dame – Cs
Berkelhammer, Max, 312-413-8271 berkelha@uic.edu,
University of Illinois at Chicago – Cg
Berli, Marcus, 702.862.5452 markus.berli@dri.edu,
University of Nevada, Reno – Sp
Berlin, Graydon L., lenn.berlin@nau.edu,
Northern Arizona University – Zr
Berlo, Kim, (514) 398-5884 kim.berlo@mcgill.ca,
McGill University – Gv
Berman, David S., (412) 622-3248 bermand@carnegiemnh.org,
Carnegie Museum of Natural History – PggPv
Bermanec, Vladimir, +38514605972 vberman@public.carnet.hr,
University of Zagreb – GzEnGi
Bermingham, Katherine R., (301) 405-2707 kberming@umd.edu,
University of Maryland – XcCcg
Bernard-Kingsley, Noell, 206-616-8511 noelleon@uw.edu,
University of Washington – Ze
Bernhard, Joan M., 508-289-3480 jbernhard@whoi.edu,
Woods Hole Oceanographic Institution – Cg
Bernhardt, Jase, 516-463-5731 jase.e.bernhardt@hofstra.edu,
Hofstra University – Ams
Bernier, Luc, (905) 525-9140 (Ext. 26364) berniejm@mcmaster.ca,
McMaster University – Cg
Bernier, Monique, 418 654-2585 monique.bernier@ete.inrs.ca,
Universite du Quebec – Zr
Beroza, Gregory C., (650) 723-4958 beroza@stanford.edu,
Stanford University – Ys
Berquist, Jr., Carl R., (757) 221-2448 crberq@wm.edu,
College of William & Mary – Ou
Berry, C M., Chris.Berry@earth.cf.ac.uk,
Cardiff University – Pb
Berry, David R., 909-869-3455 drberry@csupomona.edu,
California State Polytechnic University, Pomona – PgGo
Berry, Duane F., (540) 231-9792 duberry@vt.edu,
Virginia Polytechnic Institute & State University – Sb
Berry, Karen, 303-384-2640 kaberry@mines.edu,
Colorado Geological Survey – NgZu
Berry, Kate, (775) 784-6344 kberry@unr.edu,
University of Nevada, Reno – Zy
Berry, Leonard, 561 297-2935 berry@fau.edu,
Florida Atlantic University – Zg
Berry, Ron F., 61 3 6226 2456 Ron.Berry@utas.edu.au,
University of Tasmania – Gc
Berry IV, Henry N., 207-287-7179 henry.n.berry@maine.gov,
Dept of Agriculture, Conservation, and Forestry – Gg
Berryman, Eleanor, 609-258-4101 eb14@princeton.edu,
Princeton University – Gy
Bershaw, John T., (503) 725-3778 bershaw@pdx.edu,
Portland State University – GsCsHy
Berta, Susan, 812-237-2260 sberta@indstate.edu,
Indiana State University – GmZr
Berthold, Angela, (612) 626-6744 berth084@d.umn.edu,
University of Minnesota – Gl
Bertine, Kathe K., (619) 594-6369 kbertine@geology.sdsu.edu,
San Diego State University – Cl
Bertog, Janet, 859-572-1523 bertogj@nku.edu,
Northern Kentucky University – Gd
Bertok, Carlo, carlo.bertok@unito.it,
Università di Torino – Gd
Bertsch, Paul M., (803) 725-2752 University of Georgia – Sc
Besancon, James, (781) 283-3030 jbesancon@wellesley.edu,
Wellesley College – Hww
Best, James L., 217-244-1839 jimbest@illinois.edu,
University of Illinois, Urbana-Champaign – Gs
Best, Myron G., best_myron_g@byu.edu,
Brigham Young University – Ig
Beswick, Anthony E., (705) 675-1151 Laurentian University, Sudbury – Gi
Betancourt, Julio, jlbetanc@usgs.gov,
University of Arizona – Zn
Betchwars, Corey, (612) 625-3507 betch038@umn.edu,
University of Minnesota – Zi
Bettis III, E. A., (319) 335-1831 art-bettis@uiowa.edu,
University of Iowa – SdGmSa
Betzer, Peter R., (727) 553-1130 pbetzer@marine.usf.edu,
University of South Florida – Cm
Beukelman, Gregg S., (801) 903-8138 greggbeukelman@msn.com,
Utah Geological Survey – NgZi
Beukes, Jarlen J., (051) 401-2318 beukesj@ufs.ac.za,
University of the Free State – Gi
Beutel, Erin K., (843) 953-5591 beutele@cofc.edu,
College of Charleston – Gct
Bevelhimer, Mark, (865) 576-0266 mb2@ornl.gov,
Oak Ridge National Laboratory – Hs
Bevis, Kenneth A., (812) 866-7307 bevis@hanover.edu,
Hanover College – GlmGe
Bevis, Michael G., 614 247 5071 bevis.6@osu.edu,
Ohio State University – Yx
Beyarslan, Melahat, 00904242370000-5985 mbeyarslan@firat.edu.tr,
Firat University – GipGx
Beyer, Adam, adam.beyer@nicholls.edu,
Nicholls State University – Gg
Beyer, Patricia J., (570) 389-4570 pbeyer@bloomu.edu,
Bloomsburg University – HsZyGm
Bezada, Max, 614-624-3280 mbezada@umn.edu,
University of Minnesota, Twin Cities – Ys
Bezada Diaz, Maximiliano, 612-625-3507 mbezadad@umn.edu,
University of Minnesota – Gl
Bezdecny, Kris, (323) 343-2400 kbezdec@calstatela.edu,
California State University, Los Angeles – ZniZn
Bezdicek, David F., (509) 335-3644 bezdicek@wsu.edu,
Washington State University – Sb
Bhaskaran, Sunil, 718-289-5566 sunil.bhaskaran@bcc.cuny.edu,
Graduate School of the City University of New York – ZfrZi
Bhattacharya, Janok, bhattaj@mcmaster.ca,
University of Houston – Gsr

Bhattacharya, Janok, (905) 525-9140 (Ext. 23528) bhattaj@mcmaster.ca, McMaster University – Gr

Bhattacharyya, Prajukti, (262) 472-5257 bhattacj@uww.edu, University of Wisconsin, Whitewater – Gce

Bhattacharyya, Sidhartha, 205-348-6427 bhatt001@crimson.ua.edu, University of Alabama – CatGe

Bianchi, Thomas S., 352-392-6138 tbianchi@ufl.edu, University of Florida – Co

Biasi, Glenn P., (775) 784-4576 glenn@seismo.unr.edu, University of Nevada, Reno – Ys

Biasutti, Michela, biasutti@ldeo.columbia.edu, Columbia University – Am

Bibby, Thomas, +44 (0)23 8059 6446 tsb@noc.soton.ac.uk, University of Southampton – Ob

Bice, David M., 814-865-4477 dbice@geosc.psu.edu, Pennsylvania State University, University Park – Gg

Bickel, Charles E., (415) 338-1963 bickel@sfsu.edu, San Francisco State University – Gx

Bickford, M. E., (315) 443-2672 mebickfo@syr.edu, Syracuse University – CcGiz

Bickle, Michael, +44 (0) 1223 333484 mb72@esc.cam.ac.uk, University of Cambridge – GtYgGo

Bickmore, Barry R., (801) 422-4680 barry_bickmore@byu.edu, Brigham Young University – ClGz

Bidgoli, Tandis, (785) 864-3315 bidgoli@kgs.ku.edu, University of Kansas – GcoCc

Biek, Robert, (801) 537-3356 bobbiek@utah.gov, Utah Geological Survey – Gg

Bieler, David B., (318) 869-5234 dbieler@centenary.edu, Centenary College of Louisiana – GtrYe

Bienfang, Paul K., 808-956-7402 bienfang@soest.hawaii.edu, University of Hawai'i, Manoa – Ob

Bienvenu, Nadean S., (337) 482-6647 nadean@louisiana.edu, University of Louisiana at Lafayette – Zn

Bier, Sara E., (315) 267-3482 bierse@potsdam.edu, SUNY Potsdam – GctGg

Bierly, Aaron D., 717-702-2034 aabierly@pa.gov, Pennsylvania Bureau of Topographic & Geologic Survey – GgZn

Bierly, Gregory, 812-237-3225 Gregory.Bierly@indstate.edu, Indiana State University – Asm

Bierman, Paul R., (802) 656-4411 pbierman@uvm.edu, University of Vermont – GmeCc

Biggin, Andy, +440151 794 3460 A.Biggin@liverpool.ac.uk, University of Liverpool – Ym

Biggs, Douglas C., (979) 845-3423 d-biggs@tamu.edu, Texas A&M University – Ob

Bigham, Jerry M., (614) 292-9066 bigham.1@osu.edu, Ohio State University – Sd

Bijeljic, Branko, +44 20 759 46420 b.bijeljic@imperial.ac.uk, Imperial College – Np

Bilanovic, Dragoljub D., 218-755-2801 dbilanovic@bemidjistate.edu, Bemidji State University – GePgCa

Bilardello, Dario, (612) 624-5049 dario@umn.edu, University of Minnesota, Twin Cities – Ym

Bilek, Susan L., 575.835.6510 sbilek@nmt.edu, New Mexico Institute of Mining and Technology – Ys

Bilheux, Hassina Z., (865) 384-9630 bilheuxhn@ornl.gov, University of Tennessee, Knoxville – Zn

Bill, Steven D., (603) 358-2552 sbill@keene.edu, Keene State College – Pg

Billings, Brian J., 320-308-3298 bjbillings@stcloudstate.edu, Saint Cloud State University – Am

Billings, Stephen, sbillings@skyri.com, University of British Columbia – Yg

Billiot, Fereshteh, 361-825-6067 Fereshteh.Billiot@tamucc.edu, Texas A&M University, Corpus Christi – Zn

Billups, Katharina, 302-645-4249 kbillups@udel.edu, University of Delaware – Pe

Bilodeau, William L., (805) 493-3264 bilodeau@callutheran.edu, California Lutheran University – GcsGg

Bina, Craig R., (847) 491-5097 craig@earth.northwestern.edu, Northwestern University – YgZmGy

Bindeman, Ilya N., 541-346-3817 bindeman@uoregon.edu, University of Oregon – CsGvCl

Binford, Micheal W., (352) 392-0494 mbinford@geog.ufl.edu, University of Florida – Zr

Bingol, Ahmet Feyzi, 00904242370000-5973 fbingol@firat.edu.tr, Firat University – GipGv

Binzel, Richard P., (617) 253-6486 rpb@mit.edu, Massachusetts Institute of Technology – Xm

Biondi, Biondo L., (650) 723-9831 biondo@sep.stanford.edu, Stanford University – Ye

Birch, Francis S., (603) 862-1718 fsb@cisunix.unh.edu, University of New Hampshire – Yg

Birchall, Jeff, (780) 248-5758 jeff.birchall@ualberta.ca, University of Alberta – ZuyZg

Bircher, Harry, 330-549-9051 harrybircher@buckeyecivildesign.com, Youngstown State University – Ge

Bird, Brian, bbird@mail.nysed.gov, New York State Geological Survey – GlZi

Bird, Dennis K., (650) 723-1664 bird@pangea.stanford.edu, Stanford University – Cg

Bird, G. Peter, (310) 825-1126 pbird@epss.ucla.edu, University of California, Los Angeles – GtYsRn

Bird, Jeffrey, 718-997-3332 jeffrey.bird@qc.cuny.edu, Graduate School of the City University of New York – Sbf

Bird, MIchael I., 61742421137 michael.bird@jcu.edu.au, James Cook University – ClsSa

Birdsell, Kay H., (505) 665-0260 khb@lanl.gov, Los Alamos National Laboratory – Hqy

Birgenheier, Lauren, (801) 585-3158 lauren.birgenheier@utah.edu, University of Utah – Gso

Birke, Tesfaye K., (031) 260-8781 birket@ukzn.ac.za, University of KwaZulu-Natal – GcYmGg

Birkel, Sean, sean.birkel@maine.edu, University of Maine – Gl

Birkeland, Peter W., (303) 492-6985 birkelap@stripe.colorado.edu, University of Colorado – Sd

Birner, Johannes, 089/2180 6565 johannes.birner@web.de, Ludwig-Maximilians-Universitaet Muenchen – Gg

Birsic, Erin M., (814) 332-2872 ebirsic@allegheny.edu, Allegheny College – Gi

Biscaye, Pierre E., biscaye@ldeo.columbia.edu, Columbia University – Cm

Bischoff, Marianne, 801-626-7139 mariannebischoff@weber.edu, Weber State University – Zn

Bischoff, William D., (316) 978-6659 Wichita State University – Cl

Bish, David L., (812) 855-2039 bish@indiana.edu, Indiana University, Bloomington – GzySc

Bishoff, Carolyn, 612-625-0317 cbishoff@umn.edu, University of Minnesota, Twin Cities – Zn

Bishop, Charles E., (801) 537-3361 nrugs.cbishop@state.ut.us, Utah Geological Survey – Hw

Bishop, Gale A., (912) 681-5361 Georgia Southern University – Pi

Bishop, James K., (510) 495-2457 bishop@atmos.berkeley.edu, University of California, Berkeley – Oc

Bishop, Kim, 323 343-2409 Kbishop@calstatela.edu, California State University, Los Angeles – Gc

Bishop, Stuart P., 919-515-7894 spbishop@ncsu.edu, North Carolina State University – OpGq

Bissada, Adry, 713-743-4026 kbissada@uh.edu, University of Houston – Co

Bissig, Thomas, 604-822-5503 tbissig@eos.ubc.ca, University of British Columbia – EmCg

Biswas, Pratim, pratim.biswas@seas.wustl.edu, Washington University in St. Louis – Ng

Bitz, Cecilia M., (206) 543-1339 bitz@atmos.washington.edu, University of Washington – As

Biundo, Marc, mbiundo@uw.edu, University of Washington – Ys

Bizimis, Michael, 803-777-5565 mbizimis@geol.sc.edu, University of South Carolina – GiCg

Biševac, Vanja, +38514605969 vabisevac@geol.pmf.hr, University of Zagreb – GxpGz

Bjorklund, Tom, 281-589-6846 tbjorklund@uh.edu, University of Houston – Gco

Bjornerud, Marcia, (920) 832-7015 marcia.bjornerud@lawrence.edu, Lawrence University – GctRc

Bjornstad, Bruce N., (509) 373-6948 bruce.bjornstad@pnl.gov, Pacific Northwest National Laboratory – Gs

Blacic, James D., (505) 667-6815 jblacic@lanl.gov, Los Alamos National Laboratory – Nr

Black, Alice (Jill), (417) 836-5300 ablack@missouristate.edu, Missouri State University – ZeAsHw

Black, Annjeannette, 443-840-4560 ablack@ccbcmd.edu,
Community College of Baltimore County, Catonsville – Zn
Black, Benjamin, 212-650-7027 Graduate School of the City University of New York – As
Black, David, (631) 632-8676 david.black@stonybrook.edu,
SUNY, Stony Brook – Ou
Black, Kathryn, (973) 655-4448 Montclair State University – Zy
Black, Robert X., (404) 894-1756 rob.black@eas.gatech.edu,
Georgia Institute of Technology – As
Black, Ross A., (785) 864-2740 black@ku.edu,
University of Kansas – Ye
Black, Stuart, s.black@rdg.ac.uk,
University of Reading – Cc
Black, T. Andrew, (604) 822-2730 University of British Columbia – Sp
Blackburn, James B., (713) 524-1012 blackbur@rice.edu,
Rice University – Zu
Blackburn, Terrence, terryb@ucsc.edu,
University of California, Santa Cruz – Cc
Blackburn, William, (270) 745-8849 will.blackburn@wku.edu,
Western Kentucky University – Zy
Blackmer, Gale C., (717) 702-2032 gblackmer@pa.gov,
Pennsylvania Bureau of Topographic & Geologic Survey – Gc
Blackwelder, Patricia L., 305 421-4677 pblackwelder@rsmas.miami.edu,
University of Miami – Pg
Blackwell, Keith G., 251-460-6302 kblackwell@southalabama.edu,
University of South Alabama – Am
Bladh, Katherine L., klbladh@wittenberg.edu,
Wittenberg University – Gi
Blair, Neal E., (847) 491-8790 n-blair@northwestern.edu,
Northwestern University – CoSbOc
Blake, Bascombe (Mitch), (304) 594-2331 blake@geosrv.wvnet.edu,
West Virginia University – GsEcPb
Blake, Daniel B., (217) 333-3836 dblake@illinois.edu,
University of Illinois, Urbana-Champaign – Pi
Blake, Daniel R., 740 201 2763 daniel.blake@dnr.state.oh.us,
Ohio Dept of Natural Resources – Ysg
Blake, David E., (910) 962-3387 blaked@uncw.edu,
University of North Carolina, Wilmington – Gp
Blake, Geoffrey A., (626) 395-6296 gab@gps.caltech.edu,
California Institute of Technology – Xc
Blake, Norman J., (727) 553-1521 nblake@marine.usf.edu,
University of South Florida – Ob
Blake, Ruth E., (203) 432-3191 ruth.blake@yale.edu,
Yale University – Cg
Blake, Stephen, stephen.blake@open.ac.uk,
The Open University – Gv
Blakely, Robert F., (812) 855-1339 Indiana University, Bloomington – Ys
Blakeney DeJarnett, Beverly, 713-896-6740 bev.dejarnett@beg.utexas.edu,
University of Texas at Austin, Jackson School of Geosciences – GrsGx
Blakey, Ronald C., (928) 523-2740 ronald.blakey@nau.edu,
Northern Arizona University – Gr
Blamey, Nigel J., 905-688-3365 nblamey@brocku.ca,
New Mexico Institute of Mining and Technology – CaeCl
Blanchard, Edward, (206) 543-5219 ed@atmos.washington.edu,
University of Washington – As
Blanchard, Frank N., (352) 392-2296 University of Florida – Gz
Blanchet, Jean-Pierre, 514-987-3000 #3316 blanchet.jean-pierre@uqam.ca,
Universite du Quebec a Montreal – Pe
Bland, Douglas, (505) 466-6696 dmbland@comcast.net,
New Mexico Institute of Mining & Technology – Eg
Bland, William L., (608) 262-0221 wlbland@wisc.edu,
University of Wisconsin, Madison – Sp
Blander, Milton, (630) 252-4548 Argonne National Laboratory – Xc
Blaney, Diana L., (818) 354-5419 Jet Propulsion Laboratory – Xg
Blanford, William, 718-997-3303 Graduate School of the City University of New York – Ze
Blanken, Peter D., blanken@colorado.edu,
University of Colorado – HsAsm
Blankenship, Donald D., (512) 471-0489 blank@ig.utexas.edu,
University of Texas, Austin – Yg
Blankenship, Robert, blankship@wustl.edu,
Washington University in St. Louis – Zn
Blasing, Terence J., (865) 574-7368 blasingtj@ornl.gov,
Oak Ridge National Laboratory – As
Blasingame, Tom, (979) 845-2292 t-blasingame@spindletop.tamu.edu,
Texas A&M University – Np
Blaszczak-Boxe, Christopher, 718-270-6449 Graduate School of the City University of New York – Ze
Blatner, Keith, 509-335-4499 blatner@wsu.edu,
Washington State University – Zn
Blaylock, Glenn W., 956-764-5715 glenn.blaylock@laredo.edu,
Laredo College – Ggv
Bleam, William F., (608) 262-9956 wfbleam@wisc.edu,
University of Wisconsin, Madison – Sc
Bleamaster III, Leslie F., (210) 999-7740 lbleamas@trinity.edu,
Trinity University – Xg
Blechman, Jerome B., (607) 436-3322 jerome.blechman@oneonta.edu,
SUNY, Oneonta – As
Bleck, Rainer, (505) 665-9150 bleck@lanl.gov,
Los Alamos National Laboratory – Am
Bleibinhaus, Florian, 089/2180 4202 bleibi@geophysik.uni-muenchen.de,
Ludwig-Maximilians-Universitaet Muenchen – Yg
Blenkinson, Stephen, +44 (0)191 208 7933 stephen.blenkinsop@ncl.ac.uk,
University of Newcastle Upon Tyne – Zn
Blenkinsop, John, jblenkin@ccs.carleton.ca,
Carleton University – Cc
Blersch, Stacey S., 706-507-8068 blersch_stacey@columbusstate.edu,
Columbus State University – HsgRw
Blewett, William L., (717) 477-1513 wlblew@ship.edu,
Shippensburg University – ZyGmZe
Blewitt, Geoffrey, (775) 682-8778 gblewitt@unr.edu,
University of Nevada, Reno – Yd
Blisniuk, Kimberly, 408 924-5045 kimberly.blisniuk@sjsu.edu,
San Jose State University – Gmt
Bloch, Jonathan, jbloch@flmnh.ufl.edu,
University of Florida – Pv
Block, Karin, 212-650-8543 Graduate School of the City University of New York – Ze
Blodgett, Robert H., (512) 223-4276 rblodget@austincc.edu,
Austin Community College District – RnGsZe
Bloechle, Amber, (850) 857-6121 abloechle@uwf.edu,
University of West Florida – Zi
Blom, Ronald G., (818) 354-4681 Jet Propulsion Laboratory – Zr
Bloom, Paul R., (612) 625-4711 pbloom@soils.umn.edu,
University of Minnesota, Twin Cities – Sc
Bloomer, Sherman H., (541) 737-4811 sherman.bloomer@oregonstate.edu,
Oregon State University – Gi
Bloomfield, Helen, +440151 795 4652   H.J.Bloomfield@liverpool.ac.uk,
University of Liverpool – Ob
Bloomgren, Bruce, (612) 626-4108 bloom005@umn.edu,
University of Minnesota – Zi
Bloor, Michelle, +44 023 92 842295 michelle.bloor@port.ac.uk,
University of Portsmouth – Hg
Bloxham, Jeremy, (617) 495-9517 jeremy_bloxham@harvard.edu,
Harvard University – Yg
Blum, Joel D., (734) 764-1435 jdblum@umich.edu,
University of Michigan – CsaCl
Blum, Linda K., (434) 924-0560 lkb2e@virginia.edu,
University of Virginia – Sb
Blum, Michael D., 785-864-4974 University of Kansas – GsrGm
Blumenfeld, Raphael, r.blumenfeld@imperial.ac.uk,
Imperial College – Zn
Blumentritt, Dylan, 507.457.5234 dblumentritt@winona.edu,
Winona State University – HsGm
Blunt, Martin J., 442075946500 m.blunt@imperial.ac.uk,
Imperial College – NpHw
Blusztajn, Jurek, (508) 289-2692 jblusztajn@whoi.edu,
Woods Hole Oceanographic Institution – CcGi
Blyth, Alan, +44(0) 113 34 31632 a.m.blyth@leeds.ac.uk,
University of Leeds – As
Blythe, Ann, 323 259-2553 ablythe@oxy.edu,
Occidental College – Cc
Blythe, Daniel D., 406.496.4379 dblythe@mtech.edu,
Montana Tech of The University of Montana – Hws
Boaga, Jacopo, jacopo.boaga@unipd.it,
Università degli Studi di Padova – Yg
Boar, Rosalind, +44 (0)1603 59 3103 r.boar@uea.ac.uk,
University of East Anglia – Ob
Boardman, Mark R., (513) 529-3230 boardman@MiamiOh.edu,
Miami University – On
Boast, Charles W., (217) 333-1278 University of Illinois, Urbana-Champaign – Sp
Boboye, O. A., oa.boboye@mail.ui.edu.ng,
University of Ibadan – Go

Bobst, Andrew L., 406.496.4409 abobst@mtech.edu,
 Montana Tech of The University of Montana – HwCgEo
Bobyarchick, Andy R., 704-687-5998 arbobyar@email.uncc.edu,
 University of North Carolina, Charlotte – Gc
Bochdansky, Alexander, 757-6834933 abochdan@odu.edu,
 Old Dominion University – Ob
Bockheim, James G., (608) 263-5903 bockheim@wisc.edu,
 University of Wisconsin, Madison – Sf
Boden, Thomas, (865) 241-4842 tab@ornl.gov,
 Oak Ridge National Laboratory – As
Bodenbender, Brian E., (616) 395-7541 bodenbender@hope.edu,
 Hope College – GsPiGe
Bodenheimer, Peter, (831) 459-2064 peter@ucolick.org,
 University of California, Santa Cruz – Xc
Bodenman, John E., (570) 389-4697 boden@bloomu.edu,
 Bloomsburg University – Zu
Bodin, Paul A., 206-616-7315 bodin@uw.edu,
 University of Washington – Ys
Bodnar, Robert J., (540) 231-7455 rjb@vt.edu,
 Virginia Polytechnic Institute & State University – CgEmGv
Boerboom, Terrence, (612) 626-3369 boerb001@umn.edu,
 University of Minnesota – GivGc
Boering, Kristie, (510) 642-3472 boering@cchem.berkeley.edu,
 University of California, Berkeley – As
Boester, Michael, mboester@monroecc.edu,
 Monroe Community College – Zy
Boettcher, Margaret S., 603-862-0580 margaret.boettcher@unh.edu,
 University of New Hampshire – Ygs
Bogart, James M., james.bogart@ct.gov,
 Dept of Energy and Environmental Protection – Ge
Boger, Rebecca, 718-951-5416 rboger@brooklyn.cuny.edu,
 Graduate School of the City University of New York – Zg
Boger, Rebecca, 718-951-5000 x2159 rboger@brooklyn.cuny.edu,
 Brooklyn College (CUNY) – ZiHs
Boggs, Carol, boggscl@mailbox.sc.edu,
 University of South Carolina – Zn
Boggs, Katherine, (403) 440-6645 kboggs@mtroyal.ca,
 Mount Royal University – Gt
Bogle, Mary Anna, (865) 574-7824 amb@ornl.gov,
 Oak Ridge National Laboratory – Ge
Bogner, Jean E., jbogner@uic.edu,
 University of Illinois at Chicago – ClSbHy
Bograd, Michael B., (601) 961-5528 mbograd@mdeq.ms.gov,
 Mississippi Office of Geology – Gg
Bogucki, Darek, 361-825-825-2836 Darek.Bogucki@tamucc.edu,
 Texas A&M University, Corpus Christi – Op
Bogue, Scott W., (323) 259-2563 bogue@oxy.edu,
 Occidental College – Ym
Bohaty, Steven, +44 (0)23 8059 3040 S.Bohaty@noc.soton.ac.uk,
 University of Southampton – Og
Bohlen, Thomas, +49-721-6084416 thomas.bohlen@kit.edu,
 Karlsruhe Institute of Technology – Yx
Bohlen, Walter F., 860-405-9176 walter.bohlen@uconn.edu,
 University of Connecticut – Op
Bohling, Geoff, 785-864-2093 geoff@kgs.ku.edu,
 University of Kansas – Hw
Bohm, Christian, (204) 945-6549 christian.bohm@gov.mb.ca,
 Manitoba Geological Survey – GgCgEg
Bohnenstiehl, DelWayne, (919) 515-7449 drbohnen@ncsu.edu,
 North Carolina State University – YrsGt
Bohren, Craig F., (814) 865-2951 bohren@ems.psu.edu,
 Pennsylvania State University, University Park – Am
Bohrson, Wendy A., (509) 963-2835 bohrson@geology.cwu.edu,
 Central Washington University – GivCc
Boicourt, William, (410) 221-8426 boicourt@hpl.umces.edu,
 University of Maryland – Op
Boisvert, Eric, (418) 654-3705 eboisvert@nrcan.gc.ca,
 Universite du Quebec – Zn
Boitt, Mark , mboitt@jkuat.ac.ke,
 Jomo Kenyatta University of Agriculture & Technology – ZgrZi
Bokuniewicz, Henry J., (631) 632-8674 henry.bokuniewicz@stonybrook.edu,
 SUNY, Stony Brook – Yr
Boland, Greg J., (519) 824-4120 x52735 gboland@uoguelph.ca,
 University of Guelph – Zn
Boland, Irene B., 803-323-4949 bolandi@winthrop.edu,
 Winthrop University – GtgZe
Boland, John J., (410) 516-7103 jboland@jhu.edu,
 Johns Hopkins University – Zn
Bolarinwa, A. T., at.bolarinwa@mail.ui.edu.ng,
 University of Ibadan – EgGz
Bollag, Jean-Marc, (814) 863-0843 jmbollag@psu.edu,
 Pennsylvania State University, University Park – Sb
Bollasina, Massimo A., +44 (0) 131 651 3464 massimo.bollasina@ed.ac.uk,
 Edinburgh University – Asp
Bollens, Stephen, 360-546-9116 sbollens@vancouver.wsu.edu,
 Washington State University – Ob
Bollinger, G. A., (540) 231-6521 Virginia Polytechnic Institute & State University – Ys
Bollinger, Marsha S., 803-323-4944 bollingerm@winthrop.edu,
 Winthrop University – CmOcCc
Bollmann, Jörg, (416) 978-2061 bollmann@geology.utoronto.ca,
 University of Toronto – Pm
Bolster, Diogo, Diogo.Bolster.5@nd.edu,
 University of Notre Dame – Hw
Bolt, John R., (312) 665-7629 jbolt@fieldmuseum.org,
 Field Museum of Natural History – Pv
Bolton, Edward W., (203) 432-3149 edward.bolton@yale.edu,
 Yale University – GqHqCq
Bolze, Claude E., claude.bolze@tulsacc.edu,
 Tulsa Community College – GgoZe
Bomke, Arthur A., (604) 822-6534 University of British Columbia – Sc
Bona, Andrej, +61 8 9266 7194 A.Bona@curtin.edu.au,
 Curtin University – Yx
Bonamici, Chloe, bonamici@wisc.edu,
 University of Wisconsin-Madison – GxCa
Bonczek, James, 352-294-3112 bonczek@ufl.edu,
 University of Florida – SoHsg
Bond, Alan, +44 (0)1603 59 3402 alan.bond@uea.ac.uk,
 University of East Anglia – Ge
Bond, Clare, +44 (0)1224 273492 clare.bond@abdn.ac.uk,
 University of Aberdeen – Gct
Bond, Nicholas A., (206) 526-6459 Nicholas.Bond@noaa.gov,
 University of Washington – As
Bone, Fletcher, (573) 368-2183 fletcher.bone@dnr.mo.gov,
 Missouri Dept of Natural Resources – Gg
Bonem, Rena M., (254) 710-2187 rena_bonem@baylor.edu,
 Baylor University – PieOb
Bonetto, Sabrina Maria Rita, sabrina.bonetto@unito.it,
 Università di Torino – Ng
Bonhoure, Jessica, +33(0)3 44068994 jessica.bonhoure@lasalle-beauvais.fr,
 Institut Polytechnique LaSalle Beauvais (ex-IGAL) – GziCc
Boniface, Nelson, +255222410013 nelson.boniface@udsm.ac.tz,
 University of Dar es Salaam – GcpGz
Bonine, Michael E., kebonine@u.arizona.edu,
 University of Arizona – Zu
Bonk, Kathleen, 518-473-9988 kbonk@mail.nysed.gov,
 New York State Geological Survey – GoCg
Bonuso, Nicole, (657) 278-8451 nbonuso@fullerton.edu,
 California State University, Fullerton – Pie
Booher, Gary, (310) 660-3593 gbooher@elcamino.edu,
 El Camino College – Zg
Booker, John, 206-543-0489 jrbooker@uw.edu,
 University of Washington – YmGct
Bookhagen, Bodo, (805) 893-3568 bodo@geog.ucsb.edu,
 University of California, Santa Barbara – Gm
Boone, Peter A., pboone@austincc.edu,
 Austin Community College District – Gro
Boorstein, Margaret F., (516) 299-2318 maboorst@liu.edu,
 Long Island University, C.W. Post Campus – Zg
Booth, Adam, +44 20 759 46528 a.booth@imperial.ac.uk,
 Imperial College – Gl
Booth, Adam M., 503-725-3320 boothad@pdx.edu,
 Portland State University – GmZr
Booth, Alastair, murray.booth@manchester.ac.uk,
 University of Manchester – As
Booth, Derek, 206-914-5031 dbooth@uw.edu,
 University of Washington – Gm
Booth, Robert K., rkb205@lehigh.edu,
 Lehigh University – ZnnPc
Boothroyd, Jon C., (401) 874-2265 jon_boothroyd@uri.edu,
 University of Rhode Island – GslOn
Bopp, Richard F., (518) 276-3075 boppr@rpi.edu,
 Rensselaer Polytechnic Institute – Co
Bord, Don, (313) 593-5483 dbord@umich.edu,

University of Michigan, Dearborn – Xy
Bordeaux, Yvette, (215) 898-9191 bordeaux@sas.upenn.edu,
　University of Pennsylvania – Pg
Bordoni, Simona, 626.395.2672 bordoni@gps.caltech.edu,
　California Institute of Technology – Op
Bordossy, Andras, +44 (0) 191 208 6319 andras.bardossy@ncl.ac.uk,
　University of Newcastle Upon Tyne – Zg
Bordy, Emese M., 021-650-2901 emese.bordy@uct.ac.za,
　University of Cape Town – GsPes
Borg, Scott G., 703-292-8500 sborg@nsf.gov,
　National Science Foundation – Git
Borghi, Alessandro, alessandro.borghi@unito.it,
　Università di Torino – Gx
Bork, Karrigan, 916-733-2818 kbork@PACIFIC.EDU,
　University of the Pacific – Hg
Bork, Kennard B., (928) 554-4942 bork@denison.edu,
　Denison University – PiRhGs
Bornhorst, Theodore J., 906-487-2721 tjb@mtu.edu,
　Michigan Technological University – EgCgGx
Borns, Jr., Harold W., (207) 581-2196 borns@maine.edu,
　University of Maine – Gl
Borojeviæ Šoštariæ, Sibila, +38514605961 sborojsost@geol.pmf.hr,
　University of Zagreb – CegCc
Boroughs, Scott, geoentoptics@wsu.edu,
　Washington State University – Gv
Boroughs, Scott, 509-335-1626 scott.boroughs@wsu.edu,
　Washington State University – GiCa
Boroughs, Terry J., (916) 331-8596 BorougT@arc.losrios.edu,
　American River College – GgXgCg
Boroushaki, Soheil, 818-677-4715 soheil.boroushaki@csun.edu,
　California State University, Northridge – Zi
Borowski, Walter S., (859) 622-1277 w.borowski@eku.edu,
　Eastern Kentucky University – GsrOu
Borrelli, Chiara, (585) 275-7884 cborrelli@ur.rochester.edu,
　University of Rochester – PmcCb
Borsa, Adrian, (858) 534-6895 aborsa@ucsd.edu,
　University of California, San Diego – Hg
Bosak, Tanja, 617-324-3959 tbosak@mit.edu,
　Massachusetts Institute of Technology – Pg
Bosart, Lance F., (518) 442-4564 bosart@atmos.albany.edu,
　SUNY, Albany – As
Bosbyshell, Howell, (610) 436-2805 hbosbyshell@wcupa.edu,
　West Chester University – GcpGt
Boschmann, E. Eric, 303-871-4387 Eric.Boschmann@du.edu,
　University of Denver – Zc
Bosh, Amanda, 617-253-4115 asbosh@mit.edu,
　Massachusetts Institute of Technology – Zn
Boss, Alan P., (202) 478-8858 aboss@carnegiescience.edu,
　Carnegie Institution for Science – Xa
Boss, Stephen K., (479) 575-7134 sboss@uark.edu,
　University of Arkansas, Fayetteville – RcYgGg
Bossenbroek, Jonathon , 419-530-8376 jonathon.bossenbroek@utoledo.edu,
　University of Toledo – Zn
Bossert, James E., (505) 667-6268 bossert@lanl.gov,
　Los Alamos National Laboratory – As
Bostater, Charles, (321) 674-8096 bostater@fit.edu,
　Florida Institute of Technology – Op
Bostick, Benjamin C., bostick@ldeo.columbia.edu,
　Columbia University – Sc
Bostock, Michael G., (604) 822-2082 bostock@eos.ubc.ca,
　University of British Columbia – YsGt
Boston, Penelope, 604 224 1141 penelope.j.boston@nasa.gov,
　National Aeronautics & Space Administration –
Boswell, Cecil, 573-368-2146 cecil.boswell@dnr.mo.gov,
　Missouri Dept of Natural Resources – Zn
Bothern, Lawrence, 915-831-5088 lbothern@epcc.edu,
　El Paso Community College – Gg
Bothner, Wallace A., (603) 862-1718 wally.bothner@unh.edu,
　University of New Hampshire – Gc
Bottenberg, H. Carrie, 208-282-3538 bottcarr@isu.edu,
　Idaho State University – ZiGtc
Bottero, Jean-Yves, bottero@arbois.cerege.fr,
　Rice University – Zn
Bottjer, David J., (213) 740-6100 dbottjer@usc.edu,
　University of Southern California – Pe
Bou-Zeid, Elie, (609) 258-5429 ebouzeid@princeton.edu,
　Princeton University – As

Boudreau, Alan E., (919) 684-5646 boudreau@duke.edu,
　Duke University – Gi
Boudreau, Bernard P., (902) 494-8895 bernie.boudreau@dal.ca,
　Dalhousie University – OcCml
Boudrias, Michel A., (619) 260-4794 boum@sandiego.edu,
　University of San Diego – Ob
Bouker, Polly A., (770) 278-1320 pbouker@gsu.edu,
　Georgia State University, Perimeter College, Newton Campus – Gg
Boult, Stephen, +44 0161 275-3867 s.boult@manchester.ac.uk,
　University of Manchester – Hg
Boulton, Sarah, +44 1752 584762 sarah.boulton@plymouth.ac.uk,
　University of Plymouth – Zr
Boulvain, Frederic P., 32 4 366 22 52 fboulvain@ulg.ac.be,
　Universite de Liege – GdgGs
Bouman, Heather, +44 (1865) 272019 Heather.Bouman@earth.ox.ac.uk,
　University of Oxford – Cg
Bounk, Michael, Michael.Bounk@dnr.iowa.gov,
　Iowa Dept of Natural Resources – Gg
Bour, William, (703) 450-2612 wbour@nvcc.edu,
　Northern Virginia Community College - Loudoun Campus – Gge
Bourassa, Matthew, 519-661-2111 ext 80370 mbouras@uwo.ca,
　Western University – Xg
Bourbonnais, Annie, (803) 777-7042 abourbonnais@seoe.sc.edu,
　University of South Carolina – Oc
Bourcier, Bill L., (925) 423-3745 bourcier1@llnl.gov,
　Lawrence Livermore National Laboratory – Cg
Bourgeois, Joanne (Jody), (206) 685-2443 jbourgeo@uw.edu,
　University of Washington – GsrRh
Bourgeois, Reed J., (225) 388-8879 reed@lgs.bri.lsu.edu,
　Louisiana State University – Zn
Bourke, Robert H., (831) 656-2673 rbourke@nps.edu,
　Naval Postgraduate School – Op
Bouse, Robin, (310) 660-3593 rbouse@elcamino.edu,
　El Camino College – Gg
Bousenberry, Raymond T., (609) 984-6587 raymond.bousenberry@dep.nj.gov,
　New Jersey Geological and Water Survey – Ge
Boutt, David, (413) 545-2724 dboutt@geo.umass.edu,
　University of Massachusetts, Amherst – HwqHy
Bouvier, Audrey, 519-661-2111 x.88516 audrey.bouvier@uwo.ca,
　Western University – Xc
Bouwer, Edward, (410) 516-7437 bouwer@jhu.edu,
　Johns Hopkins University – RwHws
Bouzari, Farhad, (604) 822-1874 fbouzari@eos.ubc.ca,
　University of British Columbia – Eg
Boving, Thomas, (401) 874-7053 boving@uri.edu,
　University of Rhode Island – Hg
Bowden, Stephen, +44 (0)1224 273467 s.a.bowden@abdn.ac.uk,
　University of Aberdeen – CoGoCb
Bowen, Anthony J., (902) 494-7082 tony.bowen@dal.ca,
　Dalhousie University – On
Bowen, Brenda , (801) 585-5326 brenda.bowen@utah.edu,
　University of Utah – Gse
Bowen, David W., dbowen@montana.edu,
　Montana State University – GroGs
Bowen, Dawn S., (540) 654-1491 dbowen@mwc.edu,
　Mary Washington College – Zn
Bowen, Esther E., (630) 252-7553 ebowen@anl.gov,
　Argonne National Laboratory – Gg
Bowen, Gabriel, (801) 585-7925 gabe.bowen@utah.edu,
　University of Utah – Cs
Bowen, Jennifer, 7815817370 x346 je.bowen@northeastern.edu,
　Northeastern University – Zn
Bowen, Jennifer, jennifer.bowen@umb.edu,
　University of Massachusetts, Boston – ObPgZn
Bowen, Scott M., (505) 667-4313 Los Alamos National Laboratory – Ca
Bowersox, J. Richard, (859) 323-0536 j.r.bowersox@uky.edu,
　University of Kentucky – GoePq
Bowersox, Joe, 503-370-6220 jbowerso@willamette.edu,
　Willamette University – Zn
Bowersox, John R., (859) 257-1147 j.r.bowersox@uky.edu,
　University of Kentucky – God
Bowersox, Rick, 859-323-0536, j.r.bowersox@uky.edu
　University of Kentucky – Go
Bowin, Carl O., (508) 289-2572 cbowin@whoi.edu,
　Woods Hole Oceanographic Institution – YrGtg
Bowler, Bernard, +44 (0) 191 208 5931 bernard.bowler@ncl.ac.uk,
　University of Newcastle Upon Tyne – Go

Bowles, Julie, 414-229-6110 bowlesj@uwm.edu,
  University of Wisconsin, Milwaukee – Ym
Bowles, Zack, zbowles@mesacc.edu,
  Mesa Community College – Gg
Bowlick, Forrest J., (413)-577-3816 fbowlick@umass.edu,
  University of Massachusetts, Amherst – ZieZg
Bowman, Jean A., (409) 862-6544 jbowman@tamu.edu,
  Texas A&M University – Hg
Bowman, John R., (801) 581-7250 john.bowman@utah.edu,
  University of Utah – Cs
Bowman, Kenneth P., (979) 845-4060 k-bowman@tamu.edu,
  Texas A&M University – Am
Bowman, Malcolm J., (631) 632-8669 malcolm.bowman@stonybrook.edu,
  SUNY, Stony Brook – Op
Bowman, Mike, michael.bowman@manchester.ac.uk,
  University of Manchester – Go
Bowman, Steve D., (801) 537-3304 stevebowman@utah.gov,
  Utah Geological Survey – NgGe
Bown, Paul, +44 020 7679 32431 p.bown@ucl.ac.uk,
  University College London – Pm
Bowring, Samuel A., sbowring@mit.edu,
  Massachusetts Institute of Technology – Gg
Bowser, Carl J., (608) 262-8955 bowser@geology.wisc.edu,
  University of Wisconsin-Madison – Cl
Bowser, Paul, prb4@cornell.edu,
  SUNY, Stony Brook – Obb
Boxall, Simon R., +44 (0)23 80592744 simon.boxall@soton.ac.uk,
  University of Southampton – Op
Boyack, Diana L., (208) 282-3137 boyadian@isu.edu,
  Idaho State University – ZiGg
Boybeyi, Zafer, (703) 993-1560 zboybeyi@gmu.edu,
  George Mason University – As
Boyce, Charles K., (773) 834-7640 University of Chicago – Pb
Boyce, Joseph I., (905) 525-9140 (Ext. 24188) boycej@mcmaster.ca,
  McMaster University – Zg
Boyd, Bill, william.boyd@scu.edu.au,
  Southern Cross University – Pe
Boyd, John, (734) 764-3338 jpboyd@umich.edu,
  University of Michigan – OpAs
Boyer, Robert E., (512) 471-7228 reboyer@mail.utexas.edu,
  University of Texas, Austin – Gt
Boyle, Alan P., +44 0151 794 5154 apboyle@liverpool.ac.uk,
  University of Liverpool – GzEmZe
Boyle, Douglas, 775-784-6995 University of Nevada, Reno – Hg
Boyle, Edward A., (617) 253-3388 eaboyle@mit.edu,
  Massachusetts Institute of Technology – Oc
Boyle, James T., (609) 984-6587 jim.boyle@dep.nj.gov,
  New Jersey Geological and Water Survey – Hw
Boynton, William V., (520) 621-6941 wboynton@lpl.arizona.edu,
  University of Arizona – XcZr
Boysen, Hans, 089/2180 4333 boysen@lmu.de,
  Ludwig-Maximilians-Universitaet Muenchen – Gz
Bozdag, Ebru, 303-273-3578 bozdag@mines.edu,
  Colorado School of Mines – Ys
Bozdog, Nick, 828-296-4635 nick.bozdog@ncdenr.gov
  North Carolina Geological Survey – Gg
Bozzato, Edda, 7056751151, ext. 2272 ebozzato@laurentian.ca,
  Laurentian University, Sudbury – ZnnZn
Brabander, Daniel J., 781-283-3056 dbraband@wellesley.edu,
  Wellesley College – Cl
Brabham, P J., brabham@cf.ac.uk,
  Cardiff University – Ye
Brabham, Peter, +44(0)29 208 74334 Brabham@cardiff.ac.uk,
  University of Wales – Ng
Bracco, Annalisa, (404) 894-1749 annalisa@eas.gatech.edu,
  Georgia Institute of Technology – OpZnn
Bradbury, Kelly K., kellykbradbury@gmail.com,
  Utah State University – Gc
Bradbury, Kenneth R., (608) 263-7921 ken.bradbury@uwex.edu,
  University of Wisconsin-Madison – Hw
Bradford, Joel, (801) 863-7263 bradfojo@uvu.edu,
  Utah Valley University – RwGe
Bradford, John, jbradford@mines.edu,
  Colorado School of Mines – Yu
Bradley, Alexander S., (314) 935-6333 abradley@levee.wustl.edu,
  Washington University in St. Louis – CboGg
Bradley, Alice, (413) 597-4649 alice.c.bradley@williams.edu,
  Williams College – AtOn
Bradley, Chris, +44 (0)121 414 8097 c.bradley@bham.ac.uk,
  University of Birmingham – Hg
Bradley, Christopher R., (505) 665-6713 cbradley@lanl.gov,
  Los Alamos National Laboratory – Yg
Bradley, Jeffrey, (660) 562-1818 jbradle@nwmissouri.edu,
  Northwest Missouri State University – Zg
Bradley, Michael, (313) 487-0218 Eastern Michigan University – Gc
Bradley, Michael D., (520) 621-3865 mdb@hwr.arizona.edu,
  University of Arizona – Zn
Bradley, Philip, 919-707-9241 pbradley@ncdenr.gov,
  North Carolina Geological Survey – Gcg
Bradley, Raymond S., (413) 545-2120 rbradley@geo.umass.edu,
  University of Massachusetts, Amherst – As
Bradshaw, John, (03) 3667-001 x7487 john.bradshaw@canterbury.ac.nz,
  University of Canterbury – Gt
Bradshaw, Lanna K., 972/860-4713 LannaBradshaw@dcccd.edu,
  Brookhaven College – GgZg
Bradshaw, Peter M., (604) 681-8600 pbradshaw@firstpointminerals.com,
  University of British Columbia – EgCeGe
Brady, Emily S., +44 (0) 131 650 9137 Emily.Brady@ed.ac.uk,
  Edinburgh University – Ge
Brady, John B., (413) 585-3953 jbrady@smith.edu,
  Smith College – GxzZe
Brady, Kristina, 612-626-7889 brad0311@umn.edu,
  University of Minnesota, Twin Cities – Gn
Brady, Lawrence L., (785) 864-2159 lbrady@kgs.ku.edu,
  University of Kansas – Ag
Brady, Roland H., (559) 278-2391 roland_brady@csufresno.edu,
  California State University, Fresno – NgZuGa
Braginsky, Stanislav I., (310) 794-1331 jbragin@ucla.edu,
  University of California, Los Angeles – Ym
Brahana, John V., 479-575-3355 brahana@uark.edu,
  University of Arkansas, Fayetteville – HwGe
Brahmi, Sadek, +33(0)3 44068978 sadek.brahmi@lasalle-beauvais.fr,
  Institut Polytechnique LaSalle Beauvais (ex-IGAL) – Nr
Braile, Lawrence W., (765) 494-5979 braile@purdue.edu,
  Purdue University – Ys
Brainard, James R., (505) 667-0150 Los Alamos National Laboratory – Zn
Brake, Sandra S., (812) 237-2270 sandra.brake@indstate.edu,
  Indiana State University – GeEm
Brake, Thomas L., (702) 794-7828 brake_thomas_l@lanl.gov,
  Los Alamos National Laboratory – Nr
Bralower, Timothy J., 814-863-1240 bralower@psu.edu,
  Pennsylvania State University, University Park – Pe
Brame, Scott E., (864) 656-7167 scott@clemson.edu,
  Clemson University – Hw
Bramham, Emma, +44(0) 113 34 35595 E.K.Bramham@leeds.ac.uk,
  University of Leeds – Ym
Branam, Tracy D., (812) 855-8390 tbranam@indiana.edu,
  Indiana University – CasEm
Branan, Yvonne K., ykbranan@iup.edu,
  Indiana University of Pennsylvania – Gv
Brand, Brittany, 208-426-4154 brittanybrand@boisestate.edu,
  Boise State University – Gsv
Brand, Leonard R., (909) 558-4530 lbrand@llu.edu,
  Loma Linda University – Pv
Brand, Uwe, (905) 688-5550 x3529 uwe.brand@brocku.ca,
  Brock University – ClsAs
Brandes, Jay, (912) 598-2361 jay.brandes@skio.usg.edu,
  Georgia Institute of Technology – Og
Brandon, Alan, 713-743-2359 abrandon@uh.edu,
  University of Houston – GiCac
Brandon, Christine M., (508) 531-2125 christine.brandon@bridgew.edu,
  Bridgewater State University – Gs
Brandon, Mark A., 01908 655172 mark.brandon@open.ac.uk,
  The Open University – OgGlOp
Brandon, Mark T., (203) 432-3135 mark.brandon@yale.edu,
  Yale University – Gc
Brandon, Nigel, +44 20 759 45704 n.brandon@imperial.ac.uk,
  Imperial College – Ge
Brandriss, Mark E., mbrandri@smith.edu,
  Williams College – Gi
Brandriss, Mark E., (413) 585-3585 mbrandri@smith.edu,
  Smith College – Gx
Brandt, Craig, (865) 574-1921 fcb@ornl.gov,
  Oak Ridge National Laboratory – Ge

Brandt, Danita S., (517) 355-6595 brandt@msu.edu,
    Michigan State University – PiGr
Brandvold, Lynn A., (505) 835-5517 lynnb@nmt.edu,
    New Mexico Institute of Mining & Technology – Ca
Branfireun, Brian, bbranfir@uwo.ca,
    Western University – Ge
Branlund, Joy, (618) 797-7451 joy.branlund@swic.edu,
    Southwestern Illinois College - Sam Wolf Granite City Campus – Zg
Branney, Mike, +440116 252 3647 mjb26@le.ac.uk,
    Leicester University – Gv
Branstrator, Jon W., (765) 983-1339 jonb@earlham.edu,
    Earlham College – Pg
Brant, Lynn A., (319) 273-6160 lynn.brant@uni.edu,
    University of Northern Iowa – Gn
Brantley, Duke, brantley@mailbox.sc.edu,
    University of South Carolina – Ye
Brantley, Susan L., (814) 863-1739 brantley@eesi.psu.edu,
    Pennsylvania State University, University Park – Cg
Brânzilă, Mihai, 004023201495 mib@uaic.ro,
    Alexandru Ioan Cuza – GgrPg
Brasier, Martin, +44 (1865) 272074 martin.brasier@earth.ox.ac.uk,
    University of Oxford – Pg
Brassea, Jesus M., jbrassea@cicese.mx,
    Centro de Investigación Científica y de Educación Superior de Ensenada – Ye
Brassell, Simon C., (812) 855-3786 simon@indiana.edu,
    Indiana University, Bloomington – Cob
Brassington, Rick, +44 (0) 192 576 6754 rick.brassington@ncl.ac.uk,
    University of Newcastle Upon Tyne – Hg
Braught, Patricia, (717) 245-1355 braught@dickinson.edu,
    Dickinson College – Zn
Braun, Alexander, (613) 533-6621 braun@queensu.ca,
    Queen's University – YgdYv
Braund, Emilie, +44 023 9284 2257 emilie.bruand@port.ac.uk,
    University of Portsmouth – Gz
Bray, Colin, (416) 978-6516 cjbray@quartz.geology.utoronto.ca,
    University of Toronto – Cs
Bray, Erin, 818-677-4668 ebray@csun.edu,
    California State University, Northridge – HwsGm
Breaker, Laurence, 831-771-4400 lbreaker@mlml.calstate.edu,
    Moss Landing Marine Laboratories – Op
Brearley, Adrian J., (505) 277-4163 brearley@unm.edu,
    University of New Mexico – Gz
Breckenridge, Andy, abrecken@uwsuper.edu,
    University of Wisconsin, Superior – GlnGs
Breecker, Daniel O., 512-471-6166 breecker@jsg.utexas.edu,
    University of Texas, Austin – ClsSc
Breier, John, 956 882-6625 john.breier@utrgv.edu,
    University of Texas, Rio Grande Valley – Cma
Breilin, Olli, +358 (0)29 503 5200 olli.breilin@gtk.fi,
    Geological Survey of Finland – GmlGs
Breitbart, Mya, (727) 553-3520 mya@marine.usf.edu,
    University of South Florida – Ob
Breitmeyer, Ronald, 775.682.6049 rbreitmeyer@unr.edu,
    University of Nevada, Reno – HwGe
Breland, Clayton F., (225) 388-8300 clayton@lgs.bri.lsu.edu,
    Louisiana State University – Ye
Bremer, Keith A., (785) 628-4644 kabremer@fhsu.edu,
    Fort Hays State University – ZuiZn
Brenan, James M., 902-494-2355 jbrenan@dal.ca,
    Dalhousie University – CptEm
Brengman, Latisha A., 218-726-7586 lbrengma@d.umn.edu,
    University of Minnesota, Duluth – Gd
Brenna, J. T., (607) 255-9182 jtb4@cornell.edu,
    Cornell University – CsaCo
Brenner, Mark, (352) 392-9617 brenner@ufl.edu,
    University of Florida – Gn
Bretherton, Christopher S., (206) 685-7414 breth@washington.edu,
    University of Washington – As
Bretherton, Francis P., (608)262-7497 fbretherton@charter.net,
    University of Wisconsin, Madison – Zg
Brethes, Jean-Claude, (418) 724-1779 cjmichau@globetrotter.net,
    Universite du Quebec a Rimouski – Ob
Breton, Caroline, 512-471-0322 cari.breton@beg.utexas.edu,
    University of Texas at Austin, Jackson School of Geosciences – Zi
Brett, Carlton E., (513) 556-4556 carlton.brett@uc.edu,
    University of Cincinnati – Pe

Breuninger, Tamara, ‭+49 89 28925885‬ tamara.breuninger@tum.de,
    Technical University of Munich – Ngx
Brewer, Margene, (616) 526-8415 mkb25@calvin.edu,
    Calvin College – ZnGgZn
Brewer, Paul A., +44 (0)1970 622 586 pqb@aber.ac.uk,
    University of Wales – Gm
Breyer, John, j.breyer@tcu.edu,
    Texas Christian University – GsdEo
Brezinski, David K., (410) 554-5526 david.brezinski@maryland.gov,
    Maryland Department of Natural Resources – Gr
Briansó, José Luis, joseluis.brianso@uab.cat,
    Universitat Autonoma de Barcelona – Gz
Bridgeman, Thomas, 419-530-8373 Thomas.Bridgeman@utoledo.edu,
    University of Toledo – Zn
Bridges, David L., (573) 368-2656 david.bridges@dnr.mo.gov,
    Missouri Dept of Natural Resources – GgiPg
Bridges, Frank, (831) 459-2893 bridges@cats.ucsc.edu,
    University of California, Santa Cruz – Yx
Bries Korpik, Jill, (651) 450-3726 jkorpik@inverhills.mnscu.edu,
    Inver Hills Community College – Gg
Bries-Korpik, Jill, (651) 779-3434 jill.brieskorpik@century.edu,
    Century College – Zg
Briggs, Derek E., (203) 432-8590 derek.briggs@yale.edu,
    Yale University – Yg
Briggs, John, +44 0141 330 8744 John.Briggs@glasgow.ac.uk,
    University of Glasgow – Zu
Briggs, Tiffany Roberts M., (561) 297-4669 briggst@fau.edu,
    Florida Atlantic University – On
Brigham-Grette, Julie, (413) 577-2270 juliebg@geo.umass.edu,
    University of Massachusetts, Amherst – GlPcGu
Brikowski, Tom H., (972) 883-6242 brikowi@utdallas.edu,
    University of Texas, Dallas – Hw
Brill, Jr., Richard C., (808) 845-9488 brill@hawaii.edu,
    Honolulu Community College – Gg
Brimhall, George H., 510.642.5868 brimhall@eps.berkeley.edu,
    University of California, Berkeley – Eg
Briner, Jason P., 716-645-4326 jbriner@buffalo.edu,
    SUNY, Buffalo – Gl
Brink, Alyson, 601-266-4729 Alyson.Brink@usm.edu,
    University of Southern Mississippi – PgvGg
Brink, Benita A., (719) 587-7426 babrink@adams.edu,
    Adams State University – Zn
Brinkman, Donald B., (403) 823-7707 Royal Tyrrell Museum of Palaeontology – Pv
Brinkmann, Robert, (541) 967-2039 Oregon Dept of Geology and Mineral Industries – HwGoEm
Brinkmann, Robert, 516-463-7348 robert.brinkmann@hofstra.edu,
    Hofstra University – So
Brisbin, William C., (204) 474-9454 w_brisbin@umanitoba.ca,
    University of Manitoba – Gc
Briskin, Madeleine, madeleine.briskin@uc.edu,
    University of Cincinnati – Pe
Bristow, Charlie, +44 020 3073 8025 c.bristow@ucl.ac.uk,
    Birkbeck College – Gs
Britt, Brooks B., 801-422-7316 brooks_britt@byu.edu,
    Brigham Young University – Pg
Britton, Gloria C., (216) 987-5228 gloria.britton@tri-c.edu,
    Cuyahoga Community College - Western Campus – ZgGmZn
Britzke, Gilbert, 089/2180 4209 gilbert.brietzke@geophysik.uni-muenchen.de,
    Ludwig-Maximilians-Universitaet Muenchen – Yg
Broadfoot, Lyle A., (520) 621-4303 broadfoot@vega.lpl.arizona.edu,
    University of Arizona – Zr
Broadhead, Ronald F., (505) 835-5202 ron@gis.nmt.edu,
    New Mexico Institute of Mining & Technology – Go
Broadhead, Thomas W., (865) 974-1151 twbroadhead@utk.edu,
    University of Tennessee, Knoxville – Pi
Brochu, Christopher A., (319) 353-1808 chris-brochu@uiowa.edu,
    University of Iowa – Pv
Brock, Patrick W. G., (718) 997-3328 patrick.brock@qc.cuny.edu,
    Queens College (CUNY) – Gg
Brock-Hon, Amy, (423) 425-4409 amy-brock-hon@utc.edu,
    University of Tennessee, Chattanooga – GmSc
Brocklehurst, Simon, +44 0161 275-3037 Simon.H.Brocklehurst@manchester.ac.uk,
    University of Manchester – Gm
Brod, Joseph D., 813-744-8018 jbrod@hccfl.edu,

Hillsborough Community College – Zg
Broda, James E., (508) 289-2466 jbroda@whoi.edu,
    Woods Hole Oceanographic Institution – Gs
Brodholt, John, +44 020 7679 32622 j.brodholt@ucl.ac.uk,
    University College London – Yg
Brodie, Gregory, (423) 425-5915 gregory-brodie@utc.edu,
    University of Tennessee, Chattanooga – Ge
Brodie, Kate, +44 0161 275-3948 Kate.Brodie@manchester.ac.uk,
    University of Manchester – Gc
Brodsky, Emily, (831) 459-1854 brodsky@ucsc.edu,
    University of California, Santa Cruz – YsgHw
Broecker, Wallace S., broecker@ldeo.columbia.edu,
    Columbia University – Cg
Broecker, Wallace S., (845) 365-8413 broecker@ldeo.columbia.edu,
    Columbia University – CmOcPe
Bromily, Geoffrey, +44 (0) 131 650 8519 geoffrey.bromiley@ed.ac.uk,
    Edinburgh University – Ge
Bromley, Gordon R., gordon.bromley@maine.edu,
    University of Maine – Gl
Bromwich, David H., (614) 688-5314 bromwich.1@osu.edu,
    Ohio State University – As
Brook, Edward J., (541) 737-8197 brooke@geo.oregonstate.edu,
    Oregon State University – ClPeGl
Brookfield, Andrea, 785-864-2199 andrea@kgs.ku.edu,
    University of Kansas – Hy
Brooks, David A., (979) 845-5527 dbrooks@ocean.tamu.edu,
    Texas A&M University – Op
Brooks, Debra A., (714) 564-4788 Santiago Canyon College – Yg
Brooks, Gregg R., (727) 864-8992 brooksgr@eckerd.edu,
    Eckerd College – Gu
Brooks, Ian, +44(0) 113 34 36743 i.m.brooks@leeds.ac.uk,
    University of Leeds – Zn
Brooks, Paul D., (520) 621-3424 brooks@hwr.arizona.edu,
    University of Arizona – Cg
Brooks, Paul D., (801) 585-2858 paul.brooks@utah.edu,
    University of Utah – HgCbZg
Brooks, Sarah D., (979) 845-5632 sbrooks@tamu.edu,
    Texas A&M University – AsZc
Brooks, Scott C., (865) 574-6398 3sb@ornl.gov,
    Oak Ridge National Laboratory – Cl
Broome, Stephen W., (919) 515-2643 North Carolina State University – Sf
Brophy, James G., (812) 855-6417 brophy@indiana.edu,
    Indiana University, Bloomington – Gi
Broster, Bruce E., (506) 453-4804 broster@unb.ca,
    University of New Brunswick – Gl
Brothen, Jerry, (310) 660-3593 jbrothen@elcamino.edu,
    El Camino College – Zy
Brothers, Timothy S., (317) 274-1101 tbrother@iupui.edu,
    Indiana University, Indianapolis – Zy
Brouillette, Pierre, (418) 654-2567 pbrouill@nrcan.gc.ca,
    Universite du Quebec – Gg
Brounce, Maryjo, (951) 827-3151 maryjo.brounce@ucr.edu,
    University of California, Riverside – Cg
Brouwer, Fraukje M., +31-20-598 7335 fraukje.brouwer@vu.nl,
    Vrije Universiteit Amsterdam – GptGc
Brower, James C., (315) 443-2672 Syracuse University – Pg
Brown, Alan, abrown11@houston.oilfield.slb.com,
    West Virginia University – EoYe
Brown, Bradford D., (208) 722-6701 University of Idaho – Zn
Brown, Bruce A., (608) 263-3201 babrown1@wisc.edu,
    University of Wisconsin, Extension – GcEg
Brown, Caleb, (403) 823-7707 Royal Tyrrell Museum of Palaeontology – Pv
Brown, Cathe, (202) 633-1788 brownc@si.edu,
    Smithsonian Inst / Natl Museum of Natural Hist– Gzi
Brown, David, +44 01413307410 David.Brown@glasgow.ac.uk,
    University of Glasgow – Gvs
Brown, David L., (530) 898-4035 dlbrown@csuchico.edu,
    California State University, Chico – Hg
Brown, Donald W., (505) 667-1926 dwb@lanl.gov,
    Los Alamos National Laboratory – Nr
Brown, Douglas, +44 208 417 2245 Doug.Brown@kingston.ac.uk,
    Kingston University – Zi
Brown, Edwin H., (360) 650-3597 ehbrown@wwu.edu,
    Western Washington University – Gp
Brown, Erik T., (218) 726-8891 etbrown@d.umn.edu,
    University of Minnesota, Duluth – Og

Brown, Huntting (Hunt), 937 775-4996 hunt.brown@wright.edu,
    Wright State University – Ge
Brown, J. Michael, 206-616-6058 jmbrown5@uw.edu,
    University of Washington – Gy
Brown, James R., (403) 220-7484 jbrown@geo.ucalgary.ca,
    University of Calgary – Ys
Brown, Kenneth, 304-293-4107 kenneth.brown1@mail.wvu.edu,
    West Virginia University – GgcGe
Brown, Kent D., 801-537-3350 kentbrown@utah.gov,
    Utah Geological Survey – Zn
Brown, Kerry, K.Brown@kingston.ac.uk,
    Kingston University – Ge
Brown, Kevin M., (858) 534-5368 kmbrown@ucsd.edu,
    University of California, San Diego – Gu
Brown, Larry D., (607) 255-6346 ldb7@cornell.edu,
    Cornell University – Ye
Brown, Larry D., ldb7@cornell.edu,
    Cornell University – YesGt
Brown, Laurie L., (413) 545-0245 lbrown@geo.umass.edu,
    University of Massachusetts, Amherst – Ymg
Brown, Lewis M., (906) 635-2155 lbrown@lssu.edu,
    Lake Superior State University – PiZe
Brown, Michael, (301) 405-4080 mbrown@umd.edu,
    University of Maryland – GxpGt
Brown, Michael E., (626) 395-8423 mbrown@gps.caltech.edu,
    California Institute of Technology – Xy
Brown, Michael E., (662) 325-2906 mebrown@ra.msstate.edu,
    Mississippi State University – Am
Brown, Paul W., (520) 621-1319 pbrown@ag.arizona.edu,
    University of Arizona – Am
Brown, Peter, 5196612111 ext.86458 pbrown@uwo.ca,
    Western University – Xym
Brown, Philip E., (608) 262-5954 pebrown@wisc.edu,
    University of Wisconsin-Madison – EmGzp
Brown, Richard J., +44 (0) 191 33 42303 richard.brown3@durham.ac.uk,
    Durham University – Gv
Brown, Rob, (828) 262-7222 brownrn@appstate.edu,
    Appalachian State University – Zn
Brown, Robert A., rabrown@atmos.washington.edu,
    University of Washington – As
Brown, Robert H., (520) 626-9045 rhb@lpl.arizona.edu,
    University of Arizona – Xc
Brown, Roderick, +44 01413305460 Roderick.Brown@glasgow.ac.uk,
    University of Glasgow – Cc
Brown, Sandra, 604-822-5965 University of British Columbia – Sp
Brown, Steven E., 217-333-5143 steebrow@illinois.edu,
    University of Illinois – Gl
Brown, Summer, 859-218-0873 summer.brown@uky.edu,
    University of Kentucky – GgZei
Brown, Thomas W., tbrown1@austincc.edu,
    Austin Community College District – Ges
Brown, Tom, (925) 423-8507 tabrown@llnl.gov,
    University of Washington – CcAt
Brown, Wesley A., (936) 468-2422 brownwa1@sfasu.edu,
    Stephen F. Austin State University – YgsGt
Brown, Jr., Gordon E., (650) 723-9168 gordon@pangea.stanford.edu,
    Stanford University – Gz
Brownawell, Bruce J., (631) 632-8658 bruce.brownawell@stonybrook.edu,
    SUNY, Stony Brook – Oc
Browne, Brandon L., (707) 826-3950 brandon.browne@humboldt.edu,
    Humboldt State University – GivZe
Browne, Kathleen M., (609) 896-5408 browne@rider.edu,
    Rider University – GuZe
Browning, James M., 806-834-7782 james.browning@ttu.edu,
    Texas Tech University – Ggx
Browning, James V., 848-445-3368 jvb@eps.rutgers.edu,
    Rutgers, The State University of New Jersey – GgsGr
Browning, Sharon, (254)710-2159 sharon_browning@baylor.edu,
    Baylor University – Ge
Brownlee, Colin, +44 (0)1752 633347 cb1e11@soton.ac.uk,
    University of Southampton – Ob
BrownLee, Sarah J., (313) 577-6223 sarah.brownlee@wayne.edu,
    Wayne State University – GxcGy
Broxton, David E., 505-667-2492 broxton@lanl.gov,
    Los Alamos National Laboratory – Gx
Broze, Elliot, brozeea@whitman.edu,
    Whitman College – Gg

Brubaker, Kenneth L., (630) 252-7630 Argonne National Laboratory – As
Brudzinski, Michael, 513-529-9758 brudzimr@MiamiOh.edu,
 Miami University – Ys
Brueckner, Hannes, 718-997-3303 hannes.brueckner@gc.cuny.edu,
 Graduate School of the City University of New York – Ze
Brueckner, Hannes K., (845) 521-4637 hannes@ldeo.columbia.edu,
 Queens College (CUNY) – GgtCg
Bruening, Susan, bruenings@easternct.edu,
 Eastern Connecticut State University – Zg
Brueseke, Matthew E., (785) 532-6724 brueseke@ksu.edu,
 Kansas State University – GitGv
Brugger, Keith A., (320) 589-6310 bruggeka@mrs.umn.edu,
 University of Minnesota, Morris – GllGm
Brugger, Matthias, +49 89 289 - 25886 m.brugger@tum.de,
 Technical University of Munich – NgGg
Bruland, Kenneth W., (831) 459-4587 bruland@cats.ucsc.edu,
 University of California, Santa Cruz – OcCtm
Brumbaugh, David S., (928) 523-7191 david.brumbaugh@nau.edu,
 Northern Arizona University – Ys
Brune, James N., (775) 784-4974 brune@seismo.unr.edu,
 University of Nevada, Reno – Ys
Brune, Jürgen F., 303.273.3704 jbrune@mines.edu,
 Colorado School of Mines – Nm
Brune, William H., (814) 865-3286 whb2@psu.edu,
 Pennsylvania State University, University Park – Am
Brunengo, Matthew, 503-725-3391 mbruneng@pdx.edu,
 Portland State University – GgNgGm
Bruner, Katherine R., (304) 293-5603 bruner@geo.wvu.edu,
 West Virginia University – Gd
Brunet, Gilbert, 514 421-4617 gilbert.brunet@canada.ca,
 McGill University – Am
Brunetto, Eileen, (802) 443-5970 efahey@middlebury.edu,
 Middlebury College – Zn
Bruning, Eric C., 806-834-3120 eric.bruning@ttu.edu,
 Texas Tech University – As
Brunish, Wendee M., (505) 667-5724 wb@lanl.gov,
 Los Alamos National Laboratory – Zn
Brunkal, Holly, 970.943.2180 hbrunkal@western.edu,
 Western Colorado University – NgGm
Brunner, Benjamin, 915-747-5507 bbrunner@utep.edu,
 University of Texas, El Paso – Cs
Bruno, Barbara, (808) 956-0901 barb@hawaii.edu,
 University of Hawai'i, Manoa – Ob
Bruno, Emiliano, emiliano.bruno@unito.it,
 Università di Torino – Gz
Bruns, Dale A., 570-408-4603 dale.bruns@wilkes.edu,
 Wilkes University – ZirSf
Brunskill, Jeffrey C., 570-389-4355 jbrunski@bloomu.edu,
 Bloomsburg University – ZiAmZy
Brunstad, Keith A., (607) 436-3066 keith.brunstad@oneonta.edu,
 SUNY, Oneonta – GviEg
Brunt, Rufus, +44 0161 306-6816 rufus.brunt@manchester.ac.uk,
 University of Manchester – Gs
Bruschke, Freddi Jo, 657-278-3551 fbruschke@fullerton.edu,
 California State University, Fullerton – Gg
Brush, Grace S., (410) 516-7107 gbrush@jhu.edu,
 Johns Hopkins University – PleGn
Brush, Nigel, (419) 289-5271 nbrush@ashland.edu,
 Ashland University – GmdGa
Brusseau, Mark L., (520) 621-3244 brusseau@ag.arizona.edu,
 University of Arizona – Hw
Bryan, Christopher, +44 01326 259482 C.G.Bryan@exeter.ac.uk,
 Exeter University – Nx
Bryan, Scott E., +61 7 3138 4827 scott.bryan@qut.edu.au,
 Queensland University of Technology – GitGs
Bryant, Anita M., (678) 839-4051 abryant@westga.edu,
 University of West Georgia – Zn
Bryant, Ernest A., (505) 667-2422 Los Alamos National Laboratory – Cg
Bryant, Kathleen E., 217-244-9045 kebryant@illinois.edu,
 Illinois State Geological Survey – Gg
Bryant, Marita, (207) 786-6452 mbryant@bates.edu,
 Bates College – Gg
Bryce, Julia G., (603) 862-3139 julie.bryce@unh.edu,
 University of New Hampshire – CgcCb
Brzobohaty, Rostislav, +420 549 49 3326 rosta@sci.muni.cz,
 Masaryk University – PgsPe
Buatois, Luis, luis.buatois@usask.ca,
 University of Saskatchewan – Ps
Buchanan, Donald G., 9093844399 ext. 5467 dbuchanan@sbccd.cc.ca.us,
 San Bernardino Valley College – Og
Buchanan, George, gbuchana@mc3.edu,
 Montgomery County Community College – Ng
Buchanan, John P., (509) 359-7493 jbuchanan@ewu.edu,
 Eastern Washington University – Hw
Buchanan, Paul C., (903) 983-8253 pbuchanan@kilgore.edu,
 Kilgore College – GgiXm
Buchanan, Rex C., (785) 864-2106 rex@kgs.ku.edu,
 University of Kansas – Ze
Buchheim, Paul, (909) 558-4530 pbuchheim@llu.edu,
 Loma Linda University – GnsPe
Buchwald, Carolyn, (902) 494-3666 cbuchwald@dal.ca,
 Dalhousie University – Oc
Buck, Brenda J., (702) 895-1694 buckb@unlv.nevada.edu,
 University of Nevada, Las Vegas – GbSd
Buck, Daniel, (541) 346-2353 danielb@uoregon.edu,
 University of Oregon – Zn
Buck, Gregory, 361-825-3717 Gregory.Buck@tamucc.edu,
 Texas A&M University, Corpus Christi – Zn
Buck, Roger W., buck@ldeo.columbia.edu,
 Columbia University – Yr
Buck, IV, W. R., (845) 365-8592 buck@ldeo.columbia.edu,
 Columbia University – YgGt
Bucke, David P., (802) 899-3584 david.bucke@uvm.edu,
 University of Vermont – GgdGs
Buckingham, Michael J., (858) 534-7977 mbuckingham@ucsd.edu,
 University of California, San Diego – Op
Buckley, Brendon, bmb@ldeo.columbia.edu,
 Columbia University – Pe
Buckley, Lawrence J., (401) 874-6671 lbuckley@gso.uri.edu,
 University of Rhode Island – Oc
Buckley, Luke, (406) 496-4677 lbuckley@mtech.edu,
 Montana Tech of The University of Montana – Zn
Buckley, Michael, +44(0)161 306 5175 M.Buckley@manchester.ac.uk,
 University of Manchester – Ga
Bucklin, Ann, 860-405-9260 ann.bucklin@uconn.edu,
 University of Connecticut – Ob
Bucur, Ioan, +40-264-405371 ibucur@bioge.ubbcluj.ro,
 Babes-Bolyai University – PgmGs
Buczynska, Anna, 815-753-7945 abuczynska@niu.edu,
 Northern Illinois University – Csa
Budd, Ann F., (319) 335-1818 ann-budd@uiowa.edu,
 University of Iowa – Pgi
Budd, David A., (303) 492-3988 david.budd@colorado.edu,
 University of Colorado – GdsEo
Buddemeier, Robert W., (785) 864-2112 buddrw@kgs.ku.edu,
 University of Kansas – Hy
Budikova, Dagmar, (309) 438-2546 dbudiko@ilstu.edu,
 Illinois State University – AsZyi
Buermann, Wolfgang, +44(0) 113 34 34958 w.buermann@leeds.ac.uk,
 University of Leeds – Zu
Buesseler, Ken O., (508) 289-2309 kbuesseler@whoi.edu,
 Woods Hole Oceanographic Institution – Oc
Buick, Roger, 206-543-1913 buick@uw.edu,
 University of Washington – CbXbPm
Buijsman, Maarten, (228) 688-2385 maarten.buijsman@usm.edu,
 University of Southern Mississippi – Op
Buizert, Christo, 541-737-1572 buizertc@science.oregonstate.edu,
 Oregon State University – Yr
Bukowinski, Mark S., (510) 642-0977 University of California, Berkeley – Gy
Bulgariu, Dumitru, (023) 220-1656 dbulgariu@yahoo.com,
 Alexandru Ioan Cuza – Cga
Bulger, Daniel E., 678-891-2415 dbulger1@gsu.edu,
 Georgia State University, Perimeter College, Decatur Campus – GgZn
Buliga, Iuliana, 0040744297849 iuliana.buliga@uaic.ro,
 Alexandru Ioan Cuza – Gep
Bull, Jonathan M., +44 (0)23 8059 3078 bull@soton.ac.uk,
 University of Southampton – YrGct
Bull, William B., (520) 621-6024 University of Arizona – Gm
Bullamore, Henry W., (301) 687-4413 hbullamore@frostburg.edu,
 Frostburg State University – Zn
Bullard, Thomas F., (775) 673-7420 tbullard@dri.edu,
 Desert Research Institute – Gm
Bullen, Dean, +44 023 92 842289 dean.bullen@port.ac.uk,

University of Portsmouth – Gt
Bullister, John, (206) 526-6741 bullister@pmel.noaa.gov,
University of Washington – Oc
Bultman, John, (828) 254-1921 ext 319 jbultman@abtech.edu,
Asheville-Buncombe Technical Community College – Gg
Bulusu, Subrahmanyam, 803-777-2572 sbulusu@geol.sc.edu,
University of South Carolina – Op
Bunds, Michael, (801) 863-6306 Michael.Bunds@uvu.edu,
Utah Valley University – GctRn
Bundy, Larry G., (608) 263-2889 lgbundy@wisc.edu,
University of Wisconsin, Madison – So
Bunge, Hans-Peter, 089/2180 4225 bunge@lmu.de,
Ludwig-Maximilians-Universitaet Muenchen – Yg
Buonaiuto, Frank, 212-650-3092 frank.buonaiuto@hunter.cuny.edu,
Hunter College (CUNY) – On
Buratti, Bonnie J., (818) 354-7427 Jet Propulsion Laboratory – Zr
Burbanck, George P., (757) 727-5783 george.burbanck@hamptonu.edu,
Hampton University – On
Burbank, Doug, (805) 893-7858 burbank@ucsb.edu,
University of California, Santa Barbara – GmtGs
Burberry, Caroline M., (402) 472-7157 cburberry2@unl.edu,
University of Nebraska, Lincoln – Gct
Burbey, Thomas J., (540) 231-6696 tjburbey@vt.edu,
Virginia Polytechnic Institute & State University – HwyZr
Burchfiel, B. C., bcburch@mit.edu,
Massachusetts Institute of Technology – Gc
Burden, Elliott T., (709) 864-8388 etburden@mun.ca,
Memorial University of Newfoundland – PlGso
Burdick, Scott, (313) 577-6412 sburdick@wayne.edu,
Wayne State University – Ysg
Burdige, David J., (757) 683-4930 dburdige@odu.edu,
Old Dominion University – OcCbo
Burger, Benjamin J., 435.722.1778 benjamin.burger@usu.edu,
Utah State University – PvGs
Burger, H. Robert, rburger@smith.edu,
Smith College – GcYeZi
Burgess, Ray, +44 0161 275-3958 ray.burgess@manchester.ac.uk,
University of Manchester – Cs
Burgess, William G., +44 020 7679 37820 william.burgess@ucl.ac.uk,
University College London – Hwy
Burgette, Reed J., 575-646-3782 burgette@nmsu.edu,
New Mexico State University, Las Cruces – Gtc
Burgmann, Roland, (510) 643-9545 burgmann@seismo.berkeley.edu,
University of California, Berkeley – GtYd
Burk, Sue, 301-405-6244 University of Maryland – Zn
Burkart, Michael R., mburkart@iastate.edu,
Iowa State University of Science & Technology – Hy
Burke, Andrea, +44 01334 463910 ab276@st-andrews.ac.uk,
University of St. Andrews – ClPc
Burke, Collette D., (316) 978-3140 Wichita State University – Pm
Burke, Deborah S., (509) 372-2483 deborah.burke@pnl.gov,
Pacific Northwest National Laboratory – Sc
Burke, Ian, +44(0) 113 34 37532 lab 33965 i.t.burke@leeds.ac.uk,
University of Leeds – Ge
Burke, Raymond M., (707) 826-4292 rmb2@humboldt.edu,
Humboldt State University – Gm
Burke, Roger A., (706) 542-2652 University of Georgia – Cl
Burkett, Ashley, 405-744-6358 ashley.burkett@okstate.edu,
Oklahoma State University – Pme
Burkett, Brett, (976) 548-6510 bburkett@collin.edu,
Collin College - Central Park Campus – Gvg
Burkett, Shannon, (972) 548-6611 sburkett@collin.edu,
Collin College - Central Park Campus – Gv
Burkhart, Patrick A., (724) 738-2502 patrick.burkhart@sru.edu,
Slippery Rock University – Hg
Burks, Rachel J., (410) 704-3005 rburks@towson.edu,
Towson University – Gc
Burlage, Robert S., (865) 574-7321 Oak Ridge National Laboratory – Sb
Burls, Natalie, 703-993-5756 nburls@gmu.edu,
George Mason University – AsOp
Burmeister, Kurtis C., (209) 946-2398 kburmeister@pacific.edu,
University of the Pacific – GctEg
Burmester, Russell F., (360) 650-3654 russ.burmester@wwu.edu,
Western Washington University – Ym
Burnett, Donald S., (626) 395-6117 burnett@gps.caltech.edu,
California Institute of Technology – Xc
Burnett, Joree, 254-968-1868 jburnett@tarleton.edu,
Tarleton State University – HwGe
Burnette, Dorian, 901-678-4452 djbrntte@memphis.edu,
University of Memphis – Atm
Burnham, Charles, burnham_c@fortlewis.edu,
Fort Lewis College – Gz
Burnham, Robyn J., (734) 764-0489 rburnham@umich.edu,
University of Michigan – Pb
Burnley, Pamela C., 702-895-2536 pamela.burnley@unlv.edu,
University of Nevada, Las Vegas – GpyGg
Burns, Bill, (971) 673-1555 Oregon Dept of Geology and Mineral Industries – Ng
Burns, Carol J., (505) 665-1765 Los Alamos National Laboratory – Zn
Burns, Danny, 361-354-2405 deburns@coastalbend.edu,
Coastal Bend College – Zgn
Burns, Diane M., (217) 581-2626 dmburns@eiu.edu,
Eastern Illinois University – Gsr
Burns, Emily, eburns@ccri.edu,
Community College of Rhode Island – GgOgZi
Burns, Peter C., (574) 631-5380 peter.burns.50@nd.edu,
University of Notre Dame – Gz
Burns, Sandra, (860) 832-2934 burns@ccsu.edu,
Central Connecticut State University – Ze
Burns, Scott F., (503) 725-3389 burnss@pdx.edu,
Portland State University – Ng
Burns, Stephen J., (413) 545-0142 sburns@geo.umass.edu,
University of Massachusetts, Amherst – Cs
Burns, Timothy P., (505) 667-4600 Los Alamos National Laboratory – Co
Burr, Devon, (865) 974-2366 dburr1@utk.edu,
University of Tennessee, Knoxville – XgGm
Burras, Lee, (515) 294-0559 lburras@iastate.edu,
Iowa State University of Science & Technology – Sd
Burris, John H., (505) 566-3325 burrisj@sanjuancollege.edu,
San Juan College – GgRhGz
Bursik, Marcus I., (716) 645-4265 mib@buffalo.edu,
SUNY, Buffalo – Gv
Burst, John F., (573) 341-4616 Missouri University of Science and Technology – Eo
Burt, Donald M., (480) 965-6180 donald.burt@asu.edu,
Arizona State University – EgCgGz
Burtis, Erik, (703) 878-5614 eburtis@nvcc.edu,
Northern Virginia Community College - Woodbridge – GgcGi
Burton, Bradford R., 970.943.2252 bburton@western.edu,
Western Colorado University – GocGt
Burton, Christopher G., 334-844-3418 cgb0038@abuurn.edu,
Auburn University – Zi
Burton, Jacqueline C., (630) 252-8795 jcburton@anl.gov,
Argonne National Laboratory – Cg
Burton, Kevin, +44 (0) 191 33 44298 kevin.burton@durham.ac.uk,
Durham University – Cg
Burton, Paul, +44 (0)1603 59 2982 p.burton@uea.ac.uk,
University of East Anglia – Ys
Burton, Ronald S., (858) 822-5784 rburton@ucsd.edu,
University of California, San Diego – Ob
Burwash, Ronald A., (780) 492-3085 ronald.burwash@telus.net,
University of Alberta – Gp
Busa, Mark, 860-343-5779 MBusa@mxcc.commnet.edu,
Middlesex Community College – ZgEgYg
Busacca, Alan J., (509) 335-1859 busacca@wsu.edu,
Washington State University – Sa
Busbey, Arthur B., (817) 257-7301 a.busbey@tcu.edu,
Texas Christian University – PgvGr
Busby, Cathy, (805) 893-4068 University of California, Santa Barbara – Gs
Busby, Cathy J., cjbusby@ucdavis.edu,
University of California, Davis – Gt
Busby, Michael R., (270) 809-3370 mbusby@murraystate.edu,
Murray State University – Zi
Busch, Richard M., (610) 436-2716 rbusch@wcupa.edu,
West Chester University – GrPgZe
Busch, William H., (503) 392-3341 whbusch@gmail.com,
University of New Orleans – Ou
Buscheck, Thomas A., (925) 423-9390 buscheck1@llnl.gov,
Lawrence Livermore National Laboratory – Ng
Buseck, Peter R., (480) 965-3945 pbuseck@asu.edu,
Arizona State University – Cg
Busecke, Julius, 609-258-4101 jbusecke@princeton.edu,
Princeton University – Op
Bush, Andrew B., (780) 492-0351 andrew.bush@ualberta.ca,

Bush, Andrew M., (860) 486-9358 andrew.bush@uconn.edu,
University of Connecticut – PgqPi
Bush, Andrew B. G., (780) 492-0351 andrew.bush@ualberta.ca,
University of Alberta – As
Bush, David M., (678) 839-4057 dbush@westga.edu,
University of West Georgia – Ou
Businger, Joost, 206-543-4250 University of Washington – Am
Buskey, Edward J., (361) 749-3102 ed.buskey@utexas.edu,
University of Texas, Austin – Ob
Buss, Leo W., 203-432-3869 leo.buss@yale.edu,
Yale University – Pi
Busse, Friedrich H., (310) 825-7698 University of California, Los Angeles – Zn
Bussod, Gilles Y., (505) 667-7220 gbussod@lanl.gov,
Los Alamos National Laboratory – Yx
Bustillo, Manuel, bustillo@ucm.es,
Univ Complutense de Madrid – Eg
Bustin, R. Marc, (604) 822-6179 mbustin@eos.ubc.ca,
University of British Columbia – Ec
Butcher, Anthony, +44 023 92 842486 anthony.butcher@port.ac.uk,
University of Portsmouth – PlmZe
Butcher, Patricia M., (657) 278-3561 pbutcher@fullerton.edu,
California State University, Fullerton – Zeg
Butkos, Darryl J., (631) 451-4354 butkosd@sunysuffolk.edu,
Suffolk County Community College, Ammerman Campus – GgHw
Butler, Gilbert W., (505) 667-6005 Los Alamos National Laboratory – Cc
Butler, Ian B., +44 (0) 131 650 5885 ian.butler@ed.ac.uk,
Edinburgh University – Cg
Butler, Karl, (506) 453-4804 University of New Brunswick – Yg
Butler, R. Paul, (202) 478-8866 pbutler@carnegiescience.edu,
Carnegie Institution for Science – Xa
Butler, Sam, (306) 966-5702 sam.butler@usask.ca,
University of Saskatchewan – Yg
Butler, Jr., James J., (785) 864-2116 jbutler@kgs.ku.edu,
University of Kansas – Hw
Butterfield, David A., (206) 526-6722 dab3@u.washington.edu,
University of Washington – Yr
Butterfield, Nicholas J., +44 (0) 1223 333379 njb1005@esc.cam.ac.uk,
University of Cambridge – Pg
Buttner, Steffen, +27 (0)46-603-8775 s.buettner@ru.ac.za,
Rhodes University – Gcp
Butts, Susan H., (203) 432-3037 susan.butts@yale.edu,
Yale University – PiGsPe
Butzow, Dean G., 217-786-4923 dean.butzow@llcc.edu,
Lincoln Land Community College – Zy
Buyce, M. Raymond, rbuyce@mercyhurst.edu,
Mercyhurst University – GsaOn
Buynevich, Ilya, (215) 204-3635 coast@temple.edu,
Temple University – GsPeOn
Buzas, Martin A., (202) 633-1313 Smithsonian Inst / Natl Museum of Natural Hist– Pm
Buzatu, Andrei, 0040232201463 andrei.buzatu@uaic.ro,
Alexandru Ioan Cuza – GziGg
Buzgar, Nicolae, 0040232201462 nicolae.buzgar@uaic.ro,
Alexandru Ioan Cuza – GziGd
Bybee, Grant M., 011 717 6633 grant.bybee@wits.ac.za,
University of the Witwatersrand – Gi
Büchel, Georg, +49(0)3641 948640 georg.buechel@uni-jena.de,
Friedrich-Schiller-University Jena – Gg
Byerly, Gary R., (225) 578-5318 glbyer@lsu.edu,
Louisiana State University – Gi
Byers, Charles W., cwbyers@wisc.edu,
University of Wisconsin-Madison – Grs
Bykerk-Kauffman, Ann, 530-898-6305 abykerk-kauffman@csuchico.edu,
California State University, Chico – GcZe
Byrand, Karl, (920) 459-6619 karl.byrand@uwc.edu,
University of Wisconsin Colleges – Zny
Byrne, James M., (403) 329-2002 byme@uleth.ca,
University of Lethbridge – Hg
Byrne, Paul K., 919-513-2578 paul.byrne@ncsu.edu,
North Carolina State University – XgGcZr
Byrne, Robert H., (727) 553-1508 byrne@marine.usf.edu,
University of South Florida – Cm
Byrne, Timothy, (860) 486-3142 tim.byrne@uconn.edu,
University of Connecticut – Gc
Byrne, Timothy, 860-486-8144 tim.byrne@uconn.edu,
University of Connecticut – Ou
Byrnes, Joseph, jsbyrnes@umn.edu,
University of Minnesota, Twin Cities – Ys
Bywater-Reyes, Sharon, 970-351-1086 sharon.bywaterreyes@unco.edu,
University of Northern Colorado – HqsHt

C

C., Jeffrey C., (330) 941-3612 jcdick@ysu.edu,
Youngstown State University – EoNgHy
Cablk, Mary, (775) 673-7371 mcablk@dri.edu,
Desert Research Institute – Zr
Cabrera, Miguel L., (706) 542-1242 University of Georgia – So
Cachão, Mário, mcachao@fc.ul.pt,
Universidade de Lisboa – PgeGg
Caddick, Mark, 540-231-9919 caddick@vt.edu,
Virginia Polytechnic Institute & State University – GpCa
Cadet, Eddy, (801) 863-8881 cadeted@uvu.edu,
Utah Valley University – Ge
Cadol, Daniel, 575-835-5645 dcadol@ees.nmt.edu,
New Mexico Institute of Mining and Technology – HsGe
Cadoppi, Paola, paola.cadoppi@unito.it,
Università di Torino – Gc
Cahill, Kevin, (508) 289-2925 kcahill@whoi.edu,
Woods Hole Oceanographic Institution – Cm
Cahill, Michael J., (631) 588-8778 mjc@germanocahill.com,
SUNY, Stony Brook – Zn
Cahill, Richard A., 217-244-2532 racahill49@gmail.com,
Illinois State Geological Survey – CaGeCt
Cahir, John J., (814) 863-8358 cahir@ems.psu.edu,
Pennsylvania State University, University Park – Am
Cahoy, Kerri, 617-324-6005 kcahoy@mit.edu,
Massachusetts Institute of Technology – Zn
Cai, Yue, cai@ldeo.columbia.edu,
Columbia University – Cg
Cairns, David M., (979) 845-2783 cairns@geog.tamu.edu,
Texas A&M University – Zy
Cairns, Stephen, cairnss@si.edu,
Smithsonian Inst / Natl Museum of Natural Hist– Pi
Caissie, Beth E., 515-294-7528 bethc@iastate.edu,
Iowa State University of Science & Technology – Pe
Calagari, Ali Asghar, +98 (411) 339 2699 calagari@tabrizu.ac.ir,
University of Tabriz – EgmCg
Calaway, Wallis F., (630) 972-3586 Argonne National Laboratory – Zm
Calcote, Randy, calco001@umn.edu,
University of Minnesota, Twin Cities – Ple
Calder, Brian, 603-862-0526 brc@ccom.unh.edu,
University of New Hampshire – Og
Calder, Eliza, +44 (0) 131 650 4910 ecalder@staffmail.ed.ac.uk,
Edinburgh University – Gv
Calder, John, (902) 424-2778 Dalhousie University – Ec
Caldwell, Andy, 970-204-8228 andrew.caldwell@frontrange.edu,
Front Range Community College - Larimer – Gg
Caldwell, Marianne O., mcaldwell@hccfl.edu,
Hillsborough Community College – Ggs
Caldwell, Michael W., 780-492-3458 mw.caldwell@ualberta.ca,
University of Alberta – Pv
Caldwell, Roy, (510) 642-1391 rlcaldwell@berkeley.edu,
University of California, Berkeley – Pg
Caldwell, Todd, 512-471-2003 todd.caldwell@beg.utexas.edu,
University of Texas, Austin – Sp
Caldwell, W. Glen E., 519-661-3187 gcaldwel@uwo.ca,
Western University – Pm
Caldwell, William G. E., (519) 661-3857 University of Saskatchewan – Ps
Calegari, Pat, (212) 346-1502 Pace University, New York Campus – Zn
Calengas, Peter L., (309) 298-1151 PL-Calengas@wiu.edu,
Western Illinois University – En
Calhoun, Frank G., (330) 263-3722 scalhoun@coas.oregonstate.edu,
Ohio State University – Sd
Calhoun, Joseph, 717-872-3289 jcalhoun@hearst.com,
Millersville University – As
Calizaya, Felipe, (801) 581-5422 felipe.calizaya@utah.edu,
University of Utah – Nm
Callahan, Caitlin N., 616-331-3601 callahac@gvsu.edu,
Grand Valley State University – GgZg
Callahan, John A., 302-831-3584 john.callahan@udel.edu,
University of Delaware – Zri
Callahan, Timothy J., (843) 953-8278 callahant@cofc.edu,

College of Charleston – Hwq
Callanan, Jennifer R., (973) 720-3979 callananj@wpunj.edu,
    William Paterson University – ScGm
Callard, Jeff, callard@ou.edu,
    University of Oklahoma – Np
Callison, James, (801) 863-8679 JCallison@uvu.edu,
    Utah Valley University – GeSfo
Calon, Tomas J., (709) 737-8398 tcalon@sparky2.esd.mun.ca,
    Memorial University of Newfoundland – Gc
Calvert, Andrew J., (778) 782-5511 acalvert@sfu.ca,
    Simon Fraser University – Ys
Calvert, Stephen E., (604) 822-5210 calvert@eos.ubc.ca,
    University of British Columbia – Cm
Calvin, Wendy, 775.784.1785 wcalvin@unr.edu,
    University of Nevada, Reno – YgZr
Camacho, Alfredo, (204) 474-7413 camacho@cc.umanitoba.ca,
    University of Manitoba – Gt
Camacho, Elsa A., (509) 376-5473 elsa.camacho@pnl.gov,
    Pacific Northwest National Laboratory – Cg
Camann, Eleanor J., 303-914-6290 eleanor.camann@rrcc.edu,
    Red Rocks Community College – OnZeGs
Camargo, Suzana, suzana@ldeo.columbia.edu,
    Columbia University – As
Cambardella, Cynthia, (515) 294-2921 cindy.cambardella@ars.usda.gov,
    Iowa State University of Science & Technology – So
Cambazoglu, Mustafa K., 228.688.3009 kemal.cambazoglu@usm.edu,
    University of Southern Mississippi – No
Came, Rosemarie E., 603-862-1720 rosemarie.came@unh.edu,
    University of New Hampshire – OoZnn
Cameron, Barry I., 414-229-3136 bcameron@uwm.edu,
    University of Wisconsin, Milwaukee – Gvi
Cameron, Cheryl, (907) 451-5012 cheryl.cameron@alaska.gov,
    Alaska Division of Geological & Geophysical Surveys – Gv
Cameron, Kevin, (778) 782-4703 kjc@sfu.ca,
    Simon Fraser University – Gg
Camill III, Philip, 207-721-5149 pcamill@bowdoin.edu,
    Bowdoin College – PebSb
Camille, Michael A., (318) 342-1750 University of Louisiana, Monroe – Zy
Camilleri, Phyllis A., (931) 221-7317 camillerip@apsu.edu,
    Austin Peay State University – Gct
Cammerata, Kirk , 361-825-2468 Kirk.Cammerata@tamucc.edu,
    Texas A&M University, Corpus Christi – Zn
Camp, Mark J., (419) 530-2398 mark.camp@utoledo.edu,
    University of Toledo – PiRhEs
Camp, Victor E., (619) 594-7170 vcamp@mail.sdsu.edu,
    San Diego State University – Gv
Campagna, David J., 757-603-3221 davidjcampagna@gmail.com,
    West Virginia University – GcZrGo
Campana, Michael E., (541) 737-2413 michael.campana@oregonstate.edu,
    Oregon State University – Hg
Campbell, Andrew R., (575) 835-5327 campbell@nmt.edu,
    New Mexico Institute of Mining and Technology – CsEmGz
Campbell, Bruce A., 202-633-2472 Smithsonian Institution / National Air & Space Museum – Zr
Campbell, David L., (319) 335-1314 david-l-campbell@uiowa.edu,
    University of Iowa – Yg
Campbell, Erin A., 307-766-2286 erin.campbell@wyo.gov,
    Wyoming State Geological Survey – Gc
Campbell, Finley E., (403) 220-6801 University of Calgary – Em
Campbell, Glenn A., (859) 622-6474 glenn.campbell@eku.edu,
    Eastern Kentucky University – Zy
Campbell, Ian A., ian.campbell@ualberta.ca,
    University of Alberta – Gm
Campbell, James B., (540) 231-5841 Virginia Polytechnic Institute & State University – Zr
Campbell, Katherine, (505) 667-2799 ksc@lanl.gov,
    Los Alamos National Laboratory – Gq
Campbell, Kenneth E., (213) 763-3425 kcampbell@nhm.org,
    Los Angeles County Museum of Natural History – PvGgr
Campbell, Lisa, (979) 845-5706 lisacampbell@tamu.edu,
    Texas A&M University – Ob
Campbell, Patricia A., (724) 738-4405 patricia.campbell@sru.edu,
    Slippery Rock University – Gc
Campbell, Seth W., seth.campbell@maine.edu,
    University of Maine – Gl
Canalda, Sabrina, 915-831- 2617 mren@epcc.edu,
    El Paso Community College – Gg

Canales, María Luisa, 34 913944819 mcanales@ucm.es,
    Univ Complutense de Madrid – PmgGg
Canales Cisneros, Juan Pablo, (508) 289-2893 jpcanales@whoi.edu,
    Woods Hole Oceanographic Institution – Yr
Cande, Steven C., (858) 534-1552 scande@ucsd.edu,
    University of California, San Diego – Yr
Candela, Philip A., (301) 405-2783 candela@geol.umd.edu,
    University of Maryland – CpgEg
Cane, Mark A., mcane@ldeo.columbia.edu,
    Columbia University – Op
Canil, Dante, (250) 472-4180 dcanil@uvic.ca,
    University of Victoria – Gi
Cantarero, Debra A., (626) 585-7138 dacantarero@pasadena.edu,
    Pasadena City College – Zn
Cantrell, Kirk J., (509) 376-2136 kirk.cantrell@pnl.gov,
    Pacific Northwest National Laboratory – Cl
Canudo, José Ignacio, jicanudo@unizar.es,
    Universidad de Zaragoza – Pv
Canuel, Elizabeth A., (804) 684-7134 ecanuel@vims.edu,
    College of William & Mary – Oc
Cao, Guofeng, 806-834-8920 guofeng.cao@ttu.edu,
    Texas Tech University – Zi
Cao, Hongsheng, 316-978-3140 Wichita State University – Cl
Cao, Wenrong, 775-784-1770 wenrongc@unr.edu,
    University of Nevada, Reno – Gtc
Cao, Xiaobin, xcao@lsu.edu,
    Louisiana State University – Cs
Capehart, William J., (605) 394-2291 william.capehart@sdsmt.edu,
    South Dakota School of Mines & Technology – AsZoHq
Capes, Gerard, Gerard.Capes@manchester.ac.uk,
    University of Manchester – As
Caplan-Auerbach, Jackie, 360-650-4153 caplanj@wwu.edu,
    Western Washington University – Ys
Capo, Rosemary C., (412) 624-8873 rcapo@pitt.edu,
    University of Pittsburgh – Cl
Capoccia, Mary, (614) 292-8522 capoccia.6@osu.edu,
    Ohio State University – Zn
Caporali, Alessandro, 39-049-8279122 alessandro.caporali@unipd.it,
    Università degli Studi di Padova – YdsZr
Capraro, Luca, 39-049-8279182 luca.capraro@unipd.it,
    Università degli Studi di Padova – PelGg
Capuano, Regina M., 713-743-2957 capuano@uh.edu,
    University of Houston – HwCqs
Caputo, Mario V., (909) 214-7742 mvcaputo@earthlink.net,
    San Diego State University – GsdOn
Carbotte, Suzanne, carbotte@ldeo.columbia.edu,
    Columbia University – Yr
Cardace, Dawn, 401-874-9384 cardace@uri.edu,
    University of Rhode Island – GpPg
Cardellach López, Esteve, ++34935813091 esteve.cardellach@uab.cat,
    Universitat Autonoma de Barcelona – GzeCs
Carden-Jessen, Melanie E., (417) 836-3231 mcardenjessen@missouristate.edu,
    Missouri State University – Ze
Cardenas, Bayani, 512-471-9425 cardenas@mail.utexas.edu,
    University of Texas, Austin – Hw
Cardiff, Michael A., (608) 262-8960 cardiff@wisc.edu,
    University of Wisconsin-Madison – HwYg
Cardiff, Mike, cardiff@wisc.edu,
    University of Wisconsin, Madison – Hw
Cardimona, Steve, (707) 468-3219 scardimo@mendocino.edu,
    Mendocino College – Ys
Cardott, Brian J., (405) 325-8065 bcardott@ou.edu,
    University of Oklahoma – EcoGg
Carew, James L., (843) 953-5592 carewj@cofc.edu,
    College of Charleston – PgGd
Carey, Anne E., (614) 292-2375 carey.145@osu.edu,
    Ohio State University – HwCg
Carey, James W., (505) 667-5540 bcarey@lanl.gov,
    Los Alamos National Laboratory – CpYxNr
Carey, Kristine M., (807) 343-8461 kristine.carey@lakeheadu.ca,
    Lakehead University – Zn
Carey, Sean, 905-525-9140 (Ext. 20134) careysk@mcmaster.ca,
    McMaster University – Hg
Carey, Steven N., (401) 874-6209 scarey@gso.uri.edu,
    University of Rhode Island – Ou
Carey, Tara, 309-796-5274 careyt@bhc.edu,
    Black Hawk College – Zn

Carle, Steven, (925) 423-5039 carle1@llnl.gov,
   Lawrence Livermore National Laboratory – Hw
Carley, Tamara, 610-330-5199 carleyt@lafayette.edu,
   Lafayette College – GviCg
Carlile, Amy L., 203-479-4257 acarlile@newhaven.edu,
   University of New Haven – ObZn
Carlin, Joe, (657) 278-3054 jcarlin@fullerton.edu,
   California State University, Fullerton – GuOuGs
Carling, Gregory T., (801) 422-2622 greg.carling@byu.edu,
   Brigham Young University – HwCts
Carlson, Anders, 541-737-3625 acarlson@coas.oregonstate.edu,
   Oregon State University – GlCt
Carlson, Barry A., bacarlso@delta.edu,
   Delta College – Yg
Carlson, Catherine A., 860-465-5218 carlsonc@easternct.edu,
   Eastern Connecticut State University – Hwg
Carlson, Diane H., carlsondh@csus.edu,
   California State University, Sacramento – Gc
Carlson, Douglas A., 225/578-3671 dcarlson@lsu.edu,
   Louisiana State University – Hw
Carlson, Galen R., 657-278-3882 gcarlson@exchange.fullerton.edu,
   California State University, Fullerton – Zg
Carlson, Heath, carlsonh@easternct.edu,
   Eastern Connecticut State University – Zg
Carlson, Jill, 303-384-2643 carlson@mines.edu,
   Colorado Geological Survey – NgZu
Carlson, Keith J., (507) 933-7333 jcarlson@gac.edu,
   Gustavus Adolphus College – Pv
Carlson, Marvin P., (402) 472-3471 rockdrmpc@hotmail.com,
   Unversity of Nebraska - Lincoln – Gt
Carlson, Richard E., (515) 294-9868 richard@iastate.edu,
   Iowa State University of Science & Technology – Am
Carlson, Richard L., (979) 845-1398 carlson@geo.tamu.edu,
   Texas A&M University – YrGty
Carlson, Richard W., (202) 478-8474 rcarlson@carnegiescience.edu,
   Carnegie Institution for Science – Ccg
Carlson, Sandra J., (530) 752-2834 sjcarlson@ucdavis.edu,
   University of California, Davis – Pg
Carlson, Toby N., (814) 863-1582 tnc@psu.edu,
   Pennsylvania State University, University Park – Am
Carlson, William D., (512) 471-4770 wcarlson@jsg.utexas.edu,
   University of Texas, Austin – GpzCp
Carlucci, Jesse R., (940) 397-4448 jesse.carlucci@msutexas.edu,
   Midwestern State University – PeiGs
Carmack, Edward, (250) 363-6585 carmack@dfo-mpo.gc.ca,
   University of British Columbia – Op
Carmalt, Jean, 212-237-8195 Graduate School of the City University of New York – Ze
Carman, Cary D., 325-944-4600 cdcarman@usgs.gov,
   Angelo State University – Hg
Carmichael, Ian S., (510) 642-2577 ian@eps.berkeley.edu,
   University of California, Berkeley – Gi
Carmichael, Robert S., (319) 337-6499 robert-carmichael@uiowa.edu,
   University of Iowa – YexGg
Carmichael, Sarah, (828) 262-8471 carmichaelsk@appstate.edu,
   Appalachian State University – GxCgGd
Carn, Simon, 906/487-1756 scarn@mtu.edu,
   Michigan Technological University – ZrGv
Carnevale, Giorgio, giorgio.carnevale@unito.it,
   Università di Torino – Pg
Carney, Robert S., (225) 388-6511 Louisiana State University – Ob
Carney, Theodore C., (505) 667-3415 tedc@lanl.gov,
   Los Alamos National Laboratory – Ng
Caron, Jean-Bernard, jcaron@rom.on.ca,
   Royal Ontario Museum – PgiPg
Caron, Jean-Bernard, (416) 586-5753 jcaron@rom.on.ca,
   University of Toronto – Pi
Caron, Olivier J., 217-300-0198 caron@illinois.edu,
   Illinois State Geological Survey – Ng
Carosi, Rodolfo, rodolfo.carosi@unito.it,
   Università di Torino – Gc
Carp, Jana, (828) 262-7091 carpje@appstate.edu,
   Appalachian State University – Zn
Carpenter, Kenneth, (435) 613-5752 ken.carpenter@usu.edu,
   Utah State University – PvgPe
Carpenter, Michael, +44 (0) 1223 333483 mc43@esc.cam.ac.uk,
   University of Cambridge – Gy

Carpenter, Philip J., (815) 753-1523 pjcarpenter@niu.edu,
   Northern Illinois University – YgHwNg
Carpenter, Roy, (206) 543-8535 rcarp@u.washington.edu,
   University of Washington – No
Carr, David E., 540-837-1758 blandy@virginia.edu,
   University of Virginia – Zn
Carr, James R., (775) 784-4244 carr@unr.edu,
   University of Nevada, Reno – NgZuGg
Carr, Keith W., 217-265-0267 kw-carr@illinois.edu,
   Illinois State Geological Survey – Gg
Carr, Mark, (831) 459-3958 carr@biology.ucsc.edu,
   University of California, Santa Cruz – Ob
Carr, Michael J., carr@rutgers.edu,
   Rutgers, The State University of New Jersey – Gv
Carr, Timothy R., (785) 864-2135 tcarr@kgs.ku.edu,
   University of Kansas – Go
Carr, Timothy R., 304-293-9660 tim.carr@mail.wvu.edu,
   West Virginia University – GoYeGg
Carrano, Matthew T., (202) 633-1506 Smithsonian Inst / Natl Museum of Natural Hist– Pv
Carranza, Emmanuel John, (031) 260-2803 carranzae@ukzn.ac.za,
   University of KwaZulu-Natal – EdCeZr
Carrapa, Barbara, 520-621-4910 bcarrapa@email.arizona.edu,
   University of Arizona – Gs
Carrell, Jennifer, 217-244-2764 jcarrell@illinois.edu,
   Illinois State Geological Survey – Zg
Carrigan, Charles R., (925) 422-3941 carrigan1@llnl.gov,
   Lawrence Livermore National Laboratory – Yg
Carrigan, Charles W., (815) 939-5346 CCarriga@olivet.edu,
   Olivet Nazarene University – GzCgGc
Carroll, Alan R., (608) 262-2368 acarroll@geology.wisc.edu,
   University of Wisconsin-Madison – Gs
Carroll, David, (540) 231-5469 carrolld@vt.edu,
   Virginia Polytechnic Institute & State University – Am
Carroll, Matt, 509-335-2235 carroll@wsu.edu,
   Washington State University – Zn
Carroll, Richard E., 205-247-3551 rcarroll@gsa.state.al.us,
   Geological Survey of Alabama – EcPlGo
Carroll, Rosemary, rosemary.carroll@dri.edu,
   University of Nevada, Reno – Hq
Carroll, Susan A., (925) 423-7552 Lawrence Livermore National Laboratory – Cl
Carroll, Tracy, (417) 836-5800 TracyCarroll@MissouriState.edu,
   Missouri State University – Zn
Carslaw, Ken, +44(0) 113 34 31597 k.s.carslaw@leeds.ac.uk,
   University of Leeds – As
Carson, Bobb, bc00@lehigh.edu,
   Lehigh University – GusGt
Carson, Eric C., 608-890-1998 eccarson@wisc.edu,
   University of Wisconsin, Extension – Gl
Carson, Michael, +618 9266-4973 Michael.carson@curtin.edu.au,
   Curtin University – Yx
Carson, Robert J., (509) 527-5224 carsonrj@whitman.edu,
   Whitman College – GmeGl
Carstarphen, Camela, 406-496-4633 ccarstarphen@mtech.edu,
   Montana Tech of The University of Montana – Hw
Carstensen, Laurence W., (540) 231-2600 Virginia Polytechnic & State University – Zi
Carter, Andy, +44 020 7679 2418 a.carter@ucl.ac.uk,
   Birkbeck College – Zg
Carter, Burchard D., (229) 931-2325 burchard.carter@gsw.edu,
   Georgia Southwestern State University – Pg
Carter, David, (303) 293-5014 dcarter@osmre.gov,
   Department of Interior Office of Surface Mining Reclamation and Enforcement – Zir
Carter, Glenn, (808) 956-9267 gscarter@hawaii.edu,
   University of Hawai'i, Manoa – Op
Carter, Greg, (228) 214-3305 greg.carter@usm.edu,
   University of Southern Mississippi – Zyi
Carter, James L., 972 883-2455 jcarter@utdallas.edu,
   University of Texas, Dallas – EgmGz
Carter, Joseph, clams@email.unc.edu,
   University of North Carolina, Chapel Hill – Pi
Carter, Kristin M., 412.442.4234 krcarter@pa.gov,
   Pennsylvania Bureau of Topographic & Geologic Survey – GoHw
Carton, Alberto, +390498279141 alberto.carton@unipd.it,
   Università degli Studi di Padova – Gm

Carton, James, (301) 405-5365 University of Maryland – Op
Cartwright, J A., cartwrightJA@cf.ac.uk,
    Cardiff University – Ys
Cartwright, Joe, +44 (1865) 272000 joe.cartwright@earth.ox.ac.uk,
    University of Oxford – Gs
Caruthers, Andrew, (269) 387-8633 andrew.caruthers@wmich.edu,
    Western Michigan University – GsCsGr
Carvalho, Maria d., mdrcarvalho@fc.ul.pt,
    Universidade de Lisboa – HwyZn
Carver, Gary A., (907) 487-4551 Humboldt State University – Gmt
Cary, Kevin, (270) 745-2981 kevin.cary@wku.edu,
    Western Kentucky University – Zi
Casas Duocastella, Lluís, ++34935868365 Lluis.Casas@uab.cat,
    Universitat Autonoma de Barcelona – GzaYm
Case, James M., james.case@wilkes.edu,
    Wilkes University – Ob
Case, Jeanne, 509-359-4288 jdcase@ewu.edu,
    Eastern Washington University – Gg
Case, Stephen, (815) 939-5681 scase@olivet.edu,
    Olivet Nazarene University – RhXg
Case Hanks, Anne T., (318) 342-1822 casehanks@ulm.edu,
    University of Louisiana, Monroe – AsZe
Casey, Francis, francis.casey@ndsu.edu,
    North Dakota State University – SpHq
Casey, John F., 713-743-2824 jfcasey@uh.edu,
    University of Houston – GtiGu
Casey, William H., (916) 752-3211 University of California, Davis – Cl
Cashman, Susan M., (707) 826-3114 smc1@humboldt.edu,
    Humboldt State University – Gc
Caskey, Deborah, 915-831-8905 dcaskey1@epcc.edu,
    El Paso Community College – Gg
Caskey, John, (415) 405-0353 caskey@sfsu.edu,
    San Francisco State University – Gc
Casquet, César, casquet@ucm.es,
    Univ Complutense de Madrid – Gip
Cassar, Nicolas, 919 681-8865 nicolas.cassar@duke.edu,
    Duke University – CbOcb
Cassel, Donald K., (919) 515-1457 North Carolina State University – Sp
Cassel, Elizabeth J., (208) 885-4289 ecassel@uidaho.edu,
    University of Idaho – Gs
Cassiani, Giorgio, +390498279189 giorgio.cassiani@unipd.it,
    Università degli Studi di Padova – YgHs
Cassidy, Daniel P., (269) 387-5324 daniel.cassidy@wmich.edu,
    Western Michigan University – Hw
Cassidy, John, cassidy@pgc.nrcan.gc.ca,
    University of Victoria – Ys
Cassidy, Martin, 713-616-5853 mcassidy@uh.edu,
    University of Houston – GoCo
Castagna, John P., 713-743-8699 jpcastagna@uh.edu,
    University of Houston – Ye
Castaneda, Isla, 413 577-1124 isla@geo.umass.edu,
    University of Massachusetts, Amherst – Ca
Castelli, Daniele, daniele.castelli@unito.it,
    Università di Torino – Gx
Castillo, Paterno R., (858) 534-0383 pcastillo@ucsd.edu,
    University of California, San Diego – Giu
Castillon, David A., (417) 836-5801 DavidCastillon@MissouriState.edu,
    Missouri State University – Gm
Castiñeiras, Pedro, castigar@ucm.es,
    Univ Complutense de Madrid – Gp
Castle, James W., (864) 656-5015 jcastle@clemson.edu,
    Clemson University – GseHw
Castleton, Jessica, (801) 537-3381 jessicacastleton@utah.gov,
    Utah Geological Survey – Ng
Castor, Stephen B., 775-682-8766 scastor@unr.edu,
    University of Nevada – Eg
Castro, Maria Clara, 734-615-3812 mccastro@umich.edu,
    University of Michigan – Hy
Castro, Mark S., (301) 689-3115 castro@al.umces.edu,
    University of Maryland – As
Castro, Raul, raul@cicese.mx,
    Centro de Investigación Científica y de Educación Superior de Ensenada – Ys
Caswell, Bryony, +440151 795 4390   B.A.Caswell@liverpool.ac.uk,
    University of Liverpool – Ob
Catalano, Jeff, 314-935-6015 catalano@levee.wustl.edu,
    Washington University in St. Louis – Ca

Catania, Ginny A., (512) 471-0403 gcatania@utig.ig.utexas.edu,
    University of Texas, Austin – Gl
Catau, John C., (417) 836-5801 johncatau@missouristate.edu,
    Missouri State University – Zn
Cathcart, Eric, (619) 260-4099 cathcart@sandiego.edu,
    University of San Diego – Geg
Cather, Steven M., (505) 835-5153 steve@gis.nmt.edu,
    New Mexico Institute of Mining & Technology – Gs
Cathey, Henrietta, +61 7 3138 0416 henrietta.cathey@qut.edu.au,
    Queensland University of Technology – Giv
Cathles, Lawrence M., (607) 255-2844 lmc19@cornell.edu,
    Cornell University – GqYgEg
Catling, David C., (206) 543-8653 dcatling@uw.edu,
    University of Washington – CgAsXb
Catlos, Elizabeth J., 512-471-4762 ejcatlos@gmail.com,
    University of Texas, Austin – GzCg
Cato, Kerry, (909) 537-5409 kerry.cato@csusb.edu,
    California State University, San Bernardino – Ng
Cattanach, Bart, 828-296-4634 bart.cattanach@ncdenr.gov,
    North Carolina Geological Survey – Gc
Cattolico, Rose Ann, (206) 543-9363 racat@u.washington.edu,
    University of Washington – Zn
Catuneanu, Octavian, (780) 492-6569 octavian.catuneanu@ualberta.ca,
    University of Alberta – Gs
Caudill, Kimberly S., 740-753-6289 caudillk7007@hocking.edu,
    Hocking College – GeZuHw
Caudill, Michael R., 740-753-6277 caudillm@hocking.edu,
    Hocking College – GsSaHw
Caupp, Craig L., (301) 687-4755 ccaupp@frostburg.edu,
    Frostburg State University – ZnHs
Cave, Rachel R., +353 (0)91 492 351 rachel.cave@nuigalway.ie,
    National University of Ireland Galway – Cm
Cawood, Peter A., (+44) (0)1334 463911 peter.cawood@monash.edu,
    Monash University – GtcGg
Cayzer, Nicola, +44 (0) 131 650 8527 Nicola.Cayzer@ed.ac.uk,
    Edinburgh University – Yg
Cecil, Blaine, 703 648 6415 bcecil@usgs.gov,
    West Virginia University – Ec
Cecil, M. R., (818) 677-7009 robinson.cecil@csun.edu,
    California State University, Northridge – GiCcGz
Celia, Michael A., (609) 258-5425 Princeton University – Hw
Celik, Hasan, 00904242370000-5980 hasancelik@firat.edu.tr,
    Firat University – GscEo
Çemen, Ibrahim, 205-348-8019 icemen@ua.edu,
    University of Alabama – Gct
Cenpírek, Jan, +420 549 49 6450 jcemp@sci.muni.cz,
    Masaryk University – Gz
Centorbi, Tracey, (301) 405-6965 tlcento@umd.edu,
    University of Maryland – Cg
Cepeda, Joseph C., 806-651-2272 jcepeda@mail.wtamu.edu,
    West Texas A&M University – GiHw
Cercone, Karen Rose, (724) 357-5623 kcercone@iup.edu,
    Indiana University of Pennsylvania – Zeg
Cerling, Thure E., (801) 581-5558 thure.cerling@utah.edu,
    University of Utah – ClsZc
Cerrato, Robert M., (631) 632-8666 robert.cerrato@stonybrook.edu,
    SUNY, Stony Brook – Ob
Cervato, Cinzia C., (515) 294-7583 cinzia@iastate.edu,
    Iowa State University of Science & Technology – GgZe
Cesare, Bernardo, +390498279148 bernardo.cesare@unipd.it,
    Università degli Studi di Padova – Gp
Cessi, Paola, (858) 534-0622 pcessi@ucsd.edu,
    University of California, San Diego – Op
Cetin, Haluk, (270) 809-2085 hcetin@murraystate.edu,
    Murray State University – ZrGe
Cetindag, Bahattin, 00904242370000-5957 bcetindag@firat.edu.tr,
    Firat University – NgYv
Chacko, Elizabeth, (202) 994-6185 echack@gwu.edu,
    George Washington University – Zy
Chacko, Thomas, (780) 492-5395 tom.chacko@ualberta.ca,
    University of Alberta – Gp
Chaddock, Lisa, 6196604000 x3033 Lisa.Chaddock@gcccd.edu,
    Cuyamaca College – Zy
Chadima, Sarah A., 605-677-6166 sarah.chadima@usd.edu,
    South Dakota Dept of Environment and Natural Resources – Gg
Chadwick, John, (843) 953-5950 chadwickj@cofc.edu,
    College of Charleston – GiZrGt

Chafetz, Henry S., (713) 743-3227 hchafetz@uh.edu,
University of Houston – Gds

Chague-Goff, Catherine, c.chague-goff@unsw.edu.au,
University of New South Wales – CgGs

Chakhmouradian, Anton, (204) 474-7278 anton.chakhmouradian@umanitoba.ca,
University of Manitoba – GziCg

Challads, Thomas, +44 (0) 131 650 8543 tchallan@staffmail.ed.ac.uk,
Edinburgh University – Hg

Challinor, Andy, +44(0) 113 34 33194 a.j.challinor@leeds.ac.uk,
University of Leeds – Am

Chamberlain, Andrew T., +44 (0)161 306 4176 andrew.chamberlain@manchester.ac.uk,
University of Manchester – GaSaPv

Chamberlain, Edwin P., (718) 951-5416 Los Alamos National Laboratory – Cc

Chamberlain, John A., (718) 951-5926 Graduate School of the City University of New York – Pg

Chamberlain, John A., 7189515000 x2885 johnc@brooklyn.cuny.edu,
Brooklyn College (CUNY) – Pg

Chamberlain, Kevin R., (307) 766-2914 kchamber@uwyo.edu,
University of Wyoming – CcGt

Chamberlin, Richard, (575) 835-5310 richard@nmbg.nmt.edu,
New Mexico Institute of Mining and Technology – GcvGg

Chamberlin, William S., (714) 992-7443 schamberlin@fullcoll.edu,
Fullerton College – OgbZe

Chambers, Jeanne, 775-784-5329 chambers@unr.edu,
University of Nevada, Reno – SfHg

Chambers, John E., (202) 478-8855 jchambers@carnegiescience.edu,
Carnegie Institution for Science – Xa

Chameides, William L., 613-8004 bill.chameides@duke.edu,
Duke University – AsGe

Champion, Kyle M., 813-253-7326 Hillsborough Community College – Zg

Champlin, Steven D., (601) 961-5506 stephen_champlin@deq.state.ms.us,
Mississippi Office of Geology – Go

Chan, Clara S., 302-831-1819 cschan@udel.edu,
University of Delaware – PoObGz

Chan, Kwan M., (562) 985-4817 kmchan@csulb.edu,
California State University, Long Beach – Oc

Chan, Marjorie A., (801) 581-6551 marjorie.chan@utah.edu,
University of Utah – Gs

Chan, Selene, (604) 822-2034 schan@eos.ubc.ca,
University of British Columbia – Zn

Chandler, Angela, (501) 683-0111 angela.chandler@arkansas.gov,
Arkansas Geological Survey – GgrGs

Chandler, Cyndy, (508) 289-2765 cchandler@whoi.edu,
Woods Hole Oceanographic Institution – Og

Chandler, Sandra, (501) 683-0125 sandra.chandler@arkansas.gov,
Arkansas Geological Survey – Ze

Chandler, Val, (612) 626-4976 chand004@umn.edu,
University of Minnesota – Ye

Chandler, Val W., 612-627-4780 (Ext. 203) chand004@umn.edu,
University of Minnesota, Twin Cities – Ye

Chandra, Sudeep, 775.784.6221 sudeep@unr.edu,
University of Nevada, Reno – Hs

Chaney, Phil L., (334) 844-3420 chanepl@auburn.edu,
Auburn University – ZnnZn

Chang, Chih-Pei, 831-656-2840 cpchang@nps.edu,
Naval Postgraduate School – Am

Chang, Edmund K., (631) 632-6170 kar.chang@stonybrook.edu,
SUNY, Stony Brook – As

Chang, Hai-ru, hrc@eas.gatech.edu,
Georgia Institute of Technology – As

Chang, Julie M., 405-325-7055 jmchang@ou.edu,
University of Oklahoma – GgZnn

Chang, Ping, (979) 845-8196 pchang@ocean.tamu.edu,
Texas A&M University – Op

Chang, Young-Soo, (630) 252-4076 changy@anl.gov,
Argonne National Laboratory – As

Chang, Zhaoshan, 303-384-2127 chang@mines.edu,
Colorado School of Mines – Eg

Chang, Zhaoshan, +61 7 47816434 zhaoshan.chang@jcu.edu.au,
James Cook University – EmCce

Channell, James E., 352-392-3658 jetc@ufl.edu,
University of Florida – YmGt

Channing, Alan, +44(0)29 208 76213 ChanningA@cardiff.ac.uk,
University of Wales – Gd

Chanway, Christopher, 604–822–3716 christopher.chanway@ubc.ca,
University of British Columbia – Sb

Chao, Shenn-Yu, (410) 221-8427 chao@hpl.umces.edu,
University of Maryland – Op

Chapa, Laura E., (512) 820-9079 laura.chapa@austincc.edu,
Austin Community College District – ZgeZr

Chapin, Charles E., (505) 835-5613 chapin@gis.nmt.edu,
New Mexico Institute of Mining & Technology – Gt

Chapin, Charles E., 505-344-5817 chapin@gis.nmt.edu,
New Mexico Institute of Mining and Technology – Gv

Chapin, F. Stuart, (510) 642-1003 University of California, Berkeley – Pe

Chapman, Alan, chapman@macalester.edu,
Macalester College – Gct

Chapman, LeeAnna Y., 6192604600x2441 leeannachapman@sandiego.edu,
University of San Diego – GeZe

Chapman, Mark, +44 (0)1603 59 3114 mark.chapman@uea.ac.uk,
University of East Anglia – Pm

Chapman, Mark, +44 (0) 131 650 8521 mchapman@staffmail.ed.ac.uk,
Edinburgh University – Yg

Chapman, Marshall, (606) 783-5397 m.chapman@moreheadstate.edu,
Morehead State University – Gv

Chapman, Martin C., (540) 231-5036 mcc@vt.edu,
Virginia Polytechnic Institute & State University – Ys

Chapman, Piers, (979) 845-9399 piers.chapman@tamu.edu,
Texas A&M University – OcCm

Chapman, Robert J., +44(0) 113 34 33190 r.j.chapman@leeds.ac.uk,
University of Leeds – EmCe

Chapman, Ross N., (250) 472-4340 chapman@uvic.ca,
University of Victoria – Op

Chappaz, Anthony, 989-774-4388 chapp1a@cmich.edu,
Central Michigan University – Cat

Chappell, James R., (970) 491-5147 Jim_Chappell@partner.nps.gov,
Colorado State University – Ge

Chappell, P. D., 757-683-4937 pdchappe@odu.edu,
Old Dominion University – Oc

Chaput, Julien, 915-747-7003 jachaput@utep.edu,
University of Texas, El Paso – GeYes

Charette, Matthew A., (508) 289-3205 mcharette@whoi.edu,
Woods Hole Oceanographic Institution – Cm

Charles, Christopher D., (858) 534-5911 ccharles@ucsd.edu,
University of California, San Diego – Pe

Charles, Kasanzu, Kcharls16@yahoo.com,
University of Dar es Salaam – GgRh

Charlevoix, Donna J., (217) 244-9575 charlevo@atmos.uiuc.edu,
University of Illinois, Urbana-Champaign – As

Charlson, Robert J., charlson@chem.washington.edu,
University of Washington – As

Chase, Anne, (520) 621-6004 achase@email.arizona.edu,
University of Arizona – Zn

Chase, Clement G., (520) 621-2417 cgchase@email.arizona.edu,
University of Arizona – Yg

Chase, Jon M., jchase@biology2.wustl.edu,
Washington University in St. Louis – Zn

Chase, Peter M., (608) 265.6003 peter.chase@wgnhs.uwex.edu,
University of Wisconsin, Extension – HwYgRw

Chase, Richard L., (604) 822-3086 rchase@eos.ubc.ca,
University of British Columbia – Gu

Chasteen, Hayden R., 817-515-6694 hayden.chasteen@tccd.edu,
Tarrant County College, Northeast Campus – GgeGg

Chatelain, Edward E., (229) 333-5758 echatela@valdosta.edu,
Valdosta State University – PiGlPv

Chatterjee, Ipsita, 940-565-2372 Ipsita.Chatterjee@unt.edu,
University of North Texas – Zn

Chatterjee, Meera, 330 972-2394 meera@uakron.edu,
University of Akron – Zy

Chatterjee, Nilanjan, (617) 253-1995 nchat@mit.edu,
Massachusetts Institute of Technology – Cg

Chatterjee, Sankar, 806-834-4590 sankar.chatterjee@ttu.edu,
Texas Tech University – PvGct

Chatterjee, Snehamoy, 906-487-2516 schatte1@mtu.edu,
Michigan Technological University – NmGqNx

Chatterton, Brian D., (780) 492-3085 brian.chatterton@ualberta.ca,
University of Alberta – Pi

Chaubey, Indrajeet, (765) 494-3258 ichaubey@purdue.edu,
Purdue University – Hg

Chaudhuri, Sambhudas, (785) 532-2246 ksuncsc@ksu.edu,
Kansas State University – GeoCc

Chaumba, Jeff B., (910) 522-5787 jeff.chaumba@uncp.edu,
University of North Carolina, Pembroke – GxEgGz

Chaussard, Estelle, 541-346-4573 estelle@uoregon.edu,
University of Oregon – Yd

Chavrit, Deborah, +44 0161 275-0760 deborah.chavrit@manchester.ac.uk,
University of Manchester – Gi

Cheadle, Burns A., 519-661-2111 x.89009 bcheadle@uwo.ca,
Western University – Go

Cheadle, Michael J., (307) 766-3206 cheadle@uwyo.edu,
University of Wyoming – Yg

Chebana, Fateh, fateh.chebana@ete.inrs.ca,
Universite du Quebec – Hq

Checkley, David M., (858) 534-4228 dcheckley@ucsd.edu,
University of California, San Diego – Ob

Cheek, William H., (417) 836-5801 billcheek@missouristate.edu,
Missouri State University – Zn

Cheel, Richard J., (905) 688-5550 (Ext. 3512) RCheel@brocku.ca,
Brock University – Gs

Chelariu, Ciprian, 0040232201486 ciprian.chelariu@uaic.ro,
Alexandru Ioan Cuza – Go

Chen, Bob, bob.chen@umb.edu,
University of Massachusetts, Boston – CoOc

Chen, Gang, (907)474-6875 gchen@alaska.edu,
University of Alaska, Fairbanks – Nm

Chen, Hway-Jen, 831-656-3788 hjchen@nps.edu,
Naval Postgraduate School – Am

Chen, Jing, chenj@geog.utoronto.ca,
McMaster University – AmZry

Chen, Jingyi, 918-631-2517 jingyi-chen@utulsa.edu,
The University of Tulsa – Yes

Chen, jiquan, (419) 932-1517 jqchen@msu.edu,
Michigan State University – Zg

Chen, Jiuhua, 305-348-3030 chenj@fiu.edu,
Florida International University – Gyz

Chen, Po, 307-766-3086 pchen@uwyo.edu,
University of Wyoming – YseYg

Chen, Shu-Hua, 530-752-1822 shachen@ucdavis.edu,
University of California, Davis – ZrAsm

Chen, Shuyi S., (206) 685-1736 shuyic@uw.edu,
University of Washington – As

Chen, Tsing-Chang, (515) 294-9874 tmchen@iastate.edu,
Iowa State University of Science & Technology – As

Chen, Wang-Ping, (217) 333-2744 wpchen@illinois.edu,
University of Illinois, Urbana-Champaign – YsGt

Chen, Xianfeng, (724) 738-2385 xianfeng.chen@sru.edu,
Slippery Rock University – Zr

Chen, Xiaowei, 508-289-3820 xiaowei.fengr@gmail.com,
University of Oklahoma – YsGt

Chen, Xun-Hong, (402) 472-0772 xchen2@unl.edu,
Unversity of Nebraska - Lincoln – Hq

Chen, Yi-Leng, (808) 956-2570 yileng@hawaii.edu,
University of Hawai'i, Manoa – Asm

Chen, Yongsheng, (416)736-2100 #40124 yochen@yorku.ca,
York University – As

Cheney, Donald, d.cheney@neu.edu,
Northeastern University – Zn

Cheney, Eric S., 206-543-1163 vaalbara@uw.edu,
University of Washington – Eg

Cheney, John T., (413) 542-2311 jtcheney@amherst.edu,
Amherst College – Gi

Cheng, Chu-Lin, 956 665-2464 chulin.cheng@utrgv.edu,
University of Texas, Rio Grande Valley – HgwHq

Cheng, H. H., (612) 625-1793 hcheng@soils.umn.edu,
University of Minnesota, Twin Cities – Sb

Cheng, Hai, 86-29-83395119 cheng021@xjtu.edu.cn,
University of Minnesota, Twin Cities – CcPe

Cheng, Linyin, 479-575-5569 lc032@uark.edu,
University of Arkansas, Fayetteville – Zo

Cheng, Meng-Dawn, (423) 241-5918 ucn@ornl.gov,
Oak Ridge National Laboratory – As

Cheng, Qiuming, (416)736-2100 #22842 qiuming@yorku.ca,
York University – Zi

Cheng, Songlin, 937 775-3455 songlin.cheng@wright.edu,
Wright State University – Hw

Cheng, Yanbo, 07 47816808 yanbo.cheng1@jcu.edu.au,
James Cook University – EgGgEm

Cheng, Zhongqi, 7189515000 x2647 zcheng@brooklyn.cuny.edu,
Brooklyn College (CUNY) – CaScGe

Chenoweth, Cheri, 217-244-4610 cchenowe@illinois.edu,
Illinois State Geological Survey – GoEc

Chenoweth, M. S., (318) 342-1887 chenoweth@ulm.edu,
University of Louisiana, Monroe – GmZir

Cherkauer, Douglas S., (262) 628-3672 aquadoc@uwm.edu,
University of Wisconsin, Milwaukee – HwGe

Chermak, John A., 540-239-4504 jchermak@vt.edu,
Virginia Polytechnic Institute & State University – ClGeEd

Cherniak, Daniele J., (518) 276-3358 chernd@rpi.edu,
Rensselaer Polytechnic Institute – Gx

Chernoff, Barry, 860 6852452 bchernoff@wesleyan.edu,
Wesleyan University – Ge

Cherrier, Jennifer, 718-951-5000, ext. 2927 jennifer.cherrier18@brooklyn.cuny.edu,
Graduate School of the City University of New York – Zg

Cherubini, Claudia, +33(0)3 44068977 claudia.cherubini@lasalle-beauvais.fr,
Institut Polytechnique LaSalle Beauvais (ex-IGAL) – HwGe

Cherukupalli, Nehru, 516 463-6545 geonec@hofstra.edu,
Hofstra University – Gg

Chesnaux, Romain, 418-545-5011 ext: 5426 rchesnaux@uqac.ca,
Universite du Quebec a Chicoutimi – Hw

Chesner, Craig A., (217) 581-2626 cachesner@eiu.edu,
Eastern Illinois University – GivGz

Chesnokov, Evgeny, 713-743-2579 emchesnokov@uh.edu,
University of Houston – YseYx

Chester, Frederick, (979) 845-3296 chesterf@tamu.edu,
Texas A&M University – GcNrYx

Chester, Judith, (979) 845-1380 chesterj@geo.tamu.edu,
Texas A&M University – GcNr

Chesworth, Ward, (519) 824-4120 (Ext. 52457) wcheswor@uoguelph.ca,
University of Guelph – ClGe

Cheun, Norman, +44 020 8417 2811 K.W.Cheung@kingston.ac.uk,
Kingston University – Ge

Cheung, Wing, 7607441150 x3652 wcheung@palomar.edu,
Palomar College – ZirZy

Chew, David, + 353 1 8963481 chewd@tcd.ie,
Trinity College – CcGgt

Chhetri, Parveen, (310) 243-3377 pchhetri@csudh.edu,
California State University, Dominguez Hills – Geg

Chi, Wu-Cheng, 886-2-2783-9910 ext 510 wchi@sinica.edu.tw,
Academia Sinica – GtYrs

Chiappe, Luis M., (213) 863-3323 lchiappe@nhm.org,
Los Angeles County Museum of Natural History – Pv

Chiarella, Domenico, +44 1784 443890 domenico.chiarella@rhul.ac.uk,
Royal Holloway University of London – Gs

Chiarello, Ronald P., (630) 252-9327 Argonne National Laboratory – Zm

Chiarenzelli, Jeffrey R., 315-229-5202 jchiarenzelli@stlawu.edu,
St. Lawrence University – GzCg

Chief, Karletta, (520) 626-5598 kchief@email.arizona.edu,
University of Arizona – Zn

Chien, Yi-Ju, (509) 373-4822 Yi-Hu.chien@pnl.gov,
Pacific Northwest National Laboratory – Gq

Childers, Daniel, (610) 359-5242 dchilder@dccc.edu,
Delaware County Community College – ZgOuZi

Chilvers, Jason, +44 (0)1603 59 3130 jason.chilvers@uea.ac.uk,
University of East Anglia – Ge

Chin, Anne, (409) 845-7141 Texas A&M University – Gm

Chin, Karen, 303-735-3074 karen.chin@colorado.edu,
University of Colorado – Pg

Chin, Yu-Ping, (614) 292-6953 chin.15@osu.edu,
Ohio State University – CqHw

Chipperfield, Martyn P., +44(0) 113 34 36459 m.chipperfield@leeds.ac.uk,
University of Leeds – AscAp

Chipping, David H., 805) 528-0914 dchippin@calpoly.edu,
California Polytechnic State University – Grt

Chirenje, Tait, 609-652-4588 tait.chirenje@stockton.edu,
Stockton University – Sc

Chiu, Ching-Sang, (831) 656-3239 chiu@nps.edu,
Naval Postgraduate School – Op

Chiu, Christine, christine.chiu@colostate.edu,
Colorado State University – As

Chiu, Jer-Ming, (901) 678-2007 jerchiu@memphis.edu,
University of Memphis – Ys

Chiu, Long S., (703) 993-1984 lchiu@gmu.edu,
George Mason University – AsZrAm

Chizmadia, Lysa, (787) 265-3845 University of Puerto Rico – GzXm

Choh, Suk-Joo, 82-2-3290-3180 sjchoh@korea.ac.kr,
    Korea University – GdsPe
Choi, Eunseo, (901) 678-4923 echoi2@memphis.edu,
    University of Memphis – Gtq
Choi, Seon-Gyu, 82-2-3290-3174 seongyu@korea.ac.kr,
    Korea University – GzEmg
Chokmani, Karem, karem.chokmani@ete.inrs.ca,
    Universite du Quebec – Zi
Cholnoky, Jennifer, (518) 580-8127 jcholnok@skidmore.edu,
    Skidmore College – Gg
Choma-Moryl, Krystyna, krystyna.choma-moryl@uwr.edu.pl,
    University of Wroclaw – Ng
Chopping, Mark J., 973-655-4448 choppingm@mail.montclair.edu,
    Montclair State University – ZriZe
Chormann, Jr., Frederick H., (603) 271-1975 frederick.chormann@des.nh.gov,
    New Hampshire Geological Survey – HyZiGm
Chorover, Jon, 520-626-5635 chorover@cals.arizona.edu,
    University of Arizona – Sc
Chorover, Jonathan, (520) 621-7228 chorover@cals.arizona.edu,
    University of Arizona – ScClo
Chorover, Jonathan D., (520) 626-5635 chorover@cals.arizona.edu,
    University of Arizona – Cg
Chou, Charissa J., (509) 372-3804 Charissa.chou@pnl.gov,
    Pacific Northwest National Laboratory – Gq
Chou, Mei-In (Melissa), 217-244-0312 chou@isgs.uiuc.edu,
    Illinois State Geological Survey – Co
Chou, Sheng-Fu Joseph, 217-244-2744 jchou@isgs.uiuc.edu,
    Illinois State Geological Survey – Co
Chouinard, Vera, (905) 525-9140 (Ext. 23518) chouinar@mcmaster.ca,
    McMaster University – Zn
Choularton, Thomas, +44 0161 306-3950 choularton@manchester.ac.uk,
    University of Manchester – As
Chow, Nancy, (204) 474-6451 n_chow@umanitoba.ca,
    University of Manitoba – Gs
Chowns, Timothy M., (678) 839-4052 tchowns@westga.edu,
    University of West Georgia – GsrGg
Christensen, Douglas, doug@giseis.alaska.edu,
    University of Alaska, Fairbanks – Ys
Christensen, Nikolas I., (608) 265-4469 chris@geology.wisc.edu,
    University of Wisconsin-Madison – Yx
Christensen, Philip R., (480) 965-7105 phil.christensen@asu.edu,
    Arizona State University – Xg
Christensen, Wesley P., 605-677-6149 wes.christensen@usd.edu,
    South Dakota Dept of Environment and Natural Resources – Gg
Christeson, Gail L., 512-471-0463 gail@ig.utexas.edu,
    University of Texas, Austin – Yr
Christian, Alan D., 617-287-6639 alan.christian@umb.edu,
    University of Massachusetts, Boston – HsGmCs
Christian, James R., jim.christian@canada.ca,
    University of Victoria – Obc
Christiansen, Eric H., 801422 2113 eric_christiansen@byu.edu,
    Brigham Young University – GiXgGv
Christianson, Knut, knut@uw.edu,
    University of Washington – GlZr
Christie, Max L., 217-333-3540 mlc7@illinois.edu,
    University of Illinois, Urbana-Champaign – PgGr
Christie-Blick, Nicholas, (845) 365-8180 ncb@ldeo.columbia.edu,
    Columbia University – Gst
Christie-Blick, Nicholas, ncb@ldeo.columbia.edu,
    Columbia University – Gs
Chrzan, Daryl, (510) 643-1624 dcchrzan@berkeley.edu,
    University of California, Berkeley – Zm
Chu, Pao-Shin, chu@hawaii.edu,
    University of Hawai'i, Manoa – As
Chu, Peter C., (831) 656-3688 pcchu@nps.edu,
    Naval Postgraduate School – OpZr
Chuang, Patrick Y., (831) 459-1501 pchuang@pmc.ucsc.edu,
    University of California, Santa Cruz – As
Church, Matthew, (808) 956-8779 mjchurch@hawaii.edu,
    University of Hawai'i, Manoa – Ob
Church, Thomas M., (302) 831-2558 tchurch@udel.edu,
    University of Delaware – Cm
Church, Warren, 706-507-8093 church_warren@columbusstate.edu,
    Columbus State University – ZnPcZc
Church, William R., (519) 661-3192 wrchurch@uwo.ca,
    Western University – Gt
Churchill, Ron C., (916) 327-0745 California Geological Survey – Cg

Ciampitti, Ignacio, (785) 532-6940 ciampitti@ksu.edu,
    Kansas State University – So
Cianfrani, Christina, ccNS@hampshire.edu,
    Hampshire College – Hg
Ciannelli, Lorenzo, 541-737-3142 lciannelli@coas.oregonstate.edu,
    Oregon State University – Og
Cicerone, Robert D., (508) 531-2713 rcicerone@bridgew.edu,
    Bridgewater State University – YsgXy
Cicha, Jarrod, 612-626-4468 cich0060@morris.umn.edu,
    University of Minnesota – Zi
Ciciarelli, John A., (412) 773-3867 jac7@psu.edu,
    Pennsylvania State University, Monaca – Gg
Cicimurri, Christian M., (864)656-4602 cmcici@clemson.edu,
    Clemson University – Pv
Cicimurri, Christian M., (864) 650-8456 cmcici@clemson.edu,
    Clemson University – Pv
Cicimurri, David J., (864) 656-4601 dcheech@clemson.edu,
    Clemson University – Pv
Ciesielski, Paul F., (352) 392-2231 pciesiel@ufl.edu,
    University of Florida – Pm
Ciesla, Fred, (773) 702-8169 University of Chicago – Xcg
Cifelli, Richard L., (405) 325-4712 rlc@ou.edu,
    University of Oklahoma – Pv
Cigolini, Corrado, corrado.cigolini@unito.it,
    Università di Torino – Gv
Cihacek, Larry J., (701) 231-8572 North Dakota State University – Zg
Cilliers, Johannes, +44 20 759 47360 j.j.cilliers@imperial.ac.uk,
    Imperial College – Gz
Cintra-Buenrostro, Carlos, 956 882-5746 carlos.cntra@utrgv.edu,
    University of Texas, Rio Grande Valley – PiCsPe
Ciolkosz, Edward J., (814) 865-1530 f8i@psu.edu,
    Pennsylvania State University, University Park – Sd
Cioppa, Maria T., 519-253-3000 ext. 2502 mcioppa@uwindsor.ca,
    University of Windsor – Ym
Cipar, John J., 617-552-8300 cipar@bc.edu,
    Boston College – Ys
Cirimpei, Claudia, (023) 220-1496 claudia.cirimpei@yahoo.com,
    Alexandru Ioan Cuza – PgGr
Cirmo, Christopher P., (607) 753-2924 cirmoc@cortland.edu,
    SUNY, Cortland – Hg
Cisne, John L., john.cisne@cornell.edu,
    Cornell University – PgsPe
Civan, Faruk, (405) 325-6778 fcivan@ou.edu,
    University of Oklahoma – Np
Claassen, Mark, (785) 532-6101 mclaasse@ksu.edu,
    Kansas State University – So
Clabo, Darren R., 605-394-1996 Darren.Clabo@sdsmt.edu,
    South Dakota School of Mines & Technology – AssAs
Cladouhos, Trenton, 206 729-2400 ttcladouhos@gmail.com,
    University of Washington – Gcg
Claerbout, Jon F., (650) 723-3717 Stanford University – Ye
Claeys, Philippe, (322) 629-3391 phclaeys@vub.ac.be,
    Vrije University Brussel – CgPgXc
Clague, David A., (808) 967-8819 clague@mbari.org,
    University of Hawai'i, Manoa – Giv
Clague, John J., (778) 782-4657 Simon Fraser University – Gl
Claiborne, Lily L., (615) 343-4515 lily.claiborne@vanderbilt.edu,
    Vanderbilt University – GivCg
Claire, Mark, +44 01334 463688 mc229@st-andrews.ac.uk,
    University of St. Andrews – As
Clarey, Timothy L., (989) 686-9252 tlclarey@delta.edu,
    Delta College – HwGcPv
Clark, Alan H., (613) 533-6187 Queen's University – Em
Clark, David L., (505) 665-0005 Los Alamos National Laboratory – Zn
Clark, David L., (608) 262-4972 dlc@geology.wisc.edu,
    University of Wisconsin-Madison – Pm
Clark, Donald, (801) 537-3344 donclark@utah.gov,
    Utah Geological Survey – Gg
Clark, Douglas H., (360) 650-7939 doug.clark@wwu.edu,
    Western Washington University – Gl
Clark, George S., (204) 474-7343 gs_clark@umanitoba.ca,
    University of Manitoba – Cc
Clark, H. C., (713) 527-4887 hcclark@owlnet.rice.edu,
    Rice University – YgGeYu
Clark, Ian D., 613 562-5800 Ext 6834 idclark@uottawa.ca,
    University of Ottawa – HwGel
Clark, James A., james.clark@wheaton.edu,

Wheaton College – GmYgZi
Clark, James S., (919) 660-7402 jimclark@duke.edu,
    Duke University – Zn
Clark, Jeffrey J., (920) 832-6733 Lawrence University – Gm
Clark, Jessie , (775) 784-6345 jessieclark@unr.edu,
    University of Nevada, Reno – Zu
Clark, Joanna, j.m.clark@reading.ac.uk,
    University of Reading – Sc
Clark, John H., 814-863-1581 clark@ems.psu.edu,
    Pennsylvania State University, University Park – Am
Clark, Jordan, (805) 893-7838 University of California, Santa Barbara – Cg
Clark, Joseph C., (724) 357 5622 Indiana University of Pennsylvania – Gr
Clark, Kathryne, Kathryne.Clark@dnr.iowa.gov,
    Iowa Dept of Natural Resources – GgZi
Clark, Ken P., (253) 879-3138 kclark@pugetsound.edu,
    University of Puget Sound – GgcGi
Clark, Kenneth F., 915-581-8371 clark@utep.edu,
    University of Texas, El Paso – Eg
Clark, Malcolm W., malcolm.clark@scu.edu.au,
    Southern Cross University – Ca
Clark, Marin, 734-615-0484 marinkc@umich.edu,
    University of Michigan – Gm
Clark, Mark W., (352) 392-1803 (Ext. 316) clarkmw@ufl.edu,
    University of Florida – Sf
Clark, Melissa, (812) 855-4556 Indiana University, Bloomington – SfZg
Clark, Michael, (403) 440-8944 mdclark@mtroyal.ca,
    Mount Royal University – Zg
Clark, Murlene W., 251-460-6381 mclark@southalabama.edu,
    University of South Alabama – PiGrg
Clark, Peter U., (541) 737-1247 clarkp@geo.oregonstate.edu,
    Oregon State University – Gl
Clark, Richard D., 717-872-3930 Richard.clark@millersville.edu,
    Millersville University – As
Clark, Robert A., (520) 621-3842 clark@hwr.arizona.edu,
    University of Arizona – Hs
Clark, Robert O., (928) 523-1321 robert.clark@nau.edu,
    Northern Arizona University – Zn
Clark, Roger A., +44(0) 113 34 35221 r.a.clark@leeds.ac.uk,
    University of Leeds – YseYs
Clark, Russell G., (517) 629-0312 rgclark@albion.edu,
    Albion College – GiZi
Clark, Scott K., (715) 836-2958 clarksco@uwec.edu,
    University of Wisconsin, Eau Claire – Ze
Clark, Shannon, (505) 277-1644 skclark@unm.edu,
    University of New Mexico – Zn
Clarke, Amanda, 480-965-6590 amanda.clarke@asu.edu,
    Arizona State University – Gv
Clarke, Antony D., (808) 956-6215 tclarke@soest.hawaii.edu,
    University of Hawai'i, Manoa – As
Clarke, Beverley, beverley.clarke@flinders.edu.au,
    Flinders University – Zyn
Clarke, Garry K. C., (604) 822-3602 gclarke@eos.ubc.ca,
    University of British Columbia – Yg
Clarke, Geoffrey L., +61293512919 geoffrey.clarke@usyd.edu.au,
    University of Sydney – Gp
Clarke, Julia A., (512) 232-7563 julia_clarke@jsg.utexas.edu,
    University of Texas, Austin – PvgPq
Clarke, Peter J., +44 (0) 191 208 6351 peter.clarke@ncl.ac.uk,
    University of Newcastle Upon Tyne – YdZri
Clarke, Stu, (+44) 01782 733171 s.m.clarke@keele.ac.uk,
    Keele University – GsoGd
Clary, Renee M., (662) 268-1032 x215 rclary@geosci.msstate.edu,
    Mississippi State University – ZePgGe
Class, Connie, class@ldeo.columbia.edu,
    Columbia University – Cg
Clausen, Benjamin L., (909) 558-4548 bclausen@llu.edu,
    Loma Linda University – GiCtYg
Clay, Patricia L., +44 0161 275-0407 patricia.clay@manchester.ac.uk,
    University of Manchester – CgaGg
Clay, Robert, 573/368-2177 bob.clay@dnr.mo.gov,
    Missouri Dept of Natural Resources – Zn
Clayton, Lee, (608) 263-6839 lclayton@wisc.edu,
    University of Wisconsin, Extension – Gl
Clayton, Robert N., (773) 702-7777 University of Chicago – Cs
Clayton, Robert W., (626) 395-6909 clay@gps.caltech.edu,
    California Institute of Technology – Ys
Clayton, Robert W., 208-496-1906 claytonr@byui.edu,
    Brigham Young University - Idaho – GcYgNg
Clayton, Rodney L., 757-822-7089 RClayton@tcc.edu,
    Tidewater Community College – GgOg
Cleary, William J., 910-962-2320 clearyw@uncw.edu,
    University of North Carolina, Wilmington – Gu
Cleaveland, Malcolm K., (479) 575-3355 mcleavel@uark.edu,
    University of Arkansas, Fayetteville – PcZc
Clebnik, Sherman M., (860) 465-4323 clebniks@easternct.edu,
    Eastern Connecticut State University – Gl
Clegg, Simon, +44 (0)1603 59 3185 s.clegg@uea.ac.uk,
    University of East Anglia – AsOcCg
Clemens, Steven C., (401) 863-1964 steven_clemens@brown.edu,
    Brown University – Ou
Clemens, William A., (510) 642-6675 University of California, Berkeley
    – Pv
Clemens-Knott, Diane, (657) 278-2369 dclemensknott@fullerton.edu,
    California State University, Fullerton – Gig
Clement, William P., (413) 545-5910 wclement@geo.umass.edu,
    University of Massachusetts, Amherst – Yue
Clementz, Mark T., (307) 766-6048 mclement1@uwyo.edu,
    University of Wyoming – Pg
Clemitshaw, Kevin C., +44 1784 414026 k.clemitshaw@rhul.ac.uk,
    Royal Holloway University of London – AsZg
Clendenin, Jr., Charles W., (803) 896-7702 ClendeninB@dnr.sc.gov,
    South Carolina Dept of Natural Resources – Eg
Clendening, Ronald J., (615) 532-1504 ron.clendening@tn.gov,
    Tennessee Geological Survey – GgcGm
Clennan, Patrick D., 702-651-7501 patrick.clennan@csn.edu,
    College of Southern Nevada - West Charleston Campus – ZyGe
Clepper, Marta L., 607-436-3736 marta.clepper@oneonta.edu,
    SUNY, Oneonta – ZeGgs
Cleveland, Natasha, 301-846-2563 ncleveland@frederick.edu,
    Frederick Community College – Zge
Clift, Peter, 225-578-2153 pclift@lsu.edu,
    Louisiana State University – GsuGo
Clift, Sigrid, 512-471-0320 sigrid.clift@beg.utexas.edu,
    University of Texas, Austin – Zg
Cline, Jean S., (702) 575-9968 jean.cline@unlv.edu,
    University of Nevada, Las Vegas – EmCe
Cloos, Mark P., (512) 471-4170 cloos@jsg.utexas.edu,
    University of Texas, Austin – Gcx
Closs, L. Graham, (303) 273-3856 lcloss@mines.edu,
    Colorado School of Mines – Ce
Clothiaux, Eugene E., 814-865-2915 cloth@meteo.psu.edu,
    Pennsylvania State University, University Park – Am
Clough, Gene A., (207) 786-6396 gclough@bates.edu,
    Bates College – Ym
Cloutier, Danielle, (418) 656-7679 danielle.cloutier@ggl.ulaval.ca,
    Universite Laval – GeOg
Cloutis, Ed, 204-786-9386 e.cloutis@uwinnipeg.ca,
    Western University – Zi
Clowes, Ronald M., (604) 822-4138 clowes@eos.ubc.ca,
    University of British Columbia – Ys
Clyde, William C., 603-862-3148 will.clyde@unh.edu,
    University of New Hampshire – PgYmGs
Cnudde, Veerle, veerle.cnudde@ugent.be,
    Ghent University – EsHt
Coakley, Bernard, 907-474-5385 bernard.coakley@gi.alaska.edu,
    University of Alaska, Fairbanks – Yg
Coale, Kenneth H., (831) 771-4400 coale@mlml.calstate.edu,
    Moss Landing Marine Laboratories – CtcOc
Cobb, Kim M., (404) 894-3895 kcobb@eas.gatech.edu,
    Georgia Institute of Technology – PeCsOg
Cobb, Steven R., 806-834-1395 steve.cobb@ttu.edu,
    Texas Tech University – As
Coble, Paula G., (727) 553-1631 pcoble@marine.usf.edu,
    University of South Florida – Cm
Coburn, Craig, (403) 317-2818 craig.coburn@uleth.ca,
    University of Lethbridge – Zy
Coburn, Daniel, 203-392-5835 coburnd1@southernct.edu,
    Southern Connecticut State University – Ze
Coch, Nicholas K., (718) 997-3326 nicholas.coch@qc.cuny.edu,
    Queens College (CUNY) – GseOu
Cochran, David, 601-266-6014 david.cochran@usm.edu,
    University of Southern Mississippi – Zn
Cochran, Elizabeth, cochran@ucr.edu,
    University of California, Riverside – Ys

Cochran, J. K., (631) 632-8746 kcochran@notes.cc.sunysb.edu,
  Stony Brook University – Oc
Cochran, J. Kirk, (631) 632-8733 kirk.cochran@stonybrook.edu,
  SUNY, Stony Brook – Oc
Cochran, James R., jrc@ldeo.columbia.edu,
  Columbia University – Yr
Codrea, Vlad, +40-264-405300 ext 5205 vcodrea@bioge.ubbcluj.ro,
  Babes-Bolyai University – PvGo
Cody, Anita M., amcody@iastate.edu,
  Iowa State University of Science & Technology – Cl
Cody, George D., 202-478-8980 gcody@carnegiescience.edu,
  Carnegie Institution for Science – Co
Cody, Robert, rdcody@iastate.edu,
  Iowa State University of Science & Technology – Gz
Coe, Douglas A., (406) 496-4207 dcoe@mtech.edu,
  Montana Tech of the University of Montana – Cg
Coe, Hue, +44 0161 306-9362 hugh.coe@manchester.ac.uk,
  University of Manchester – As
Coe, Robert S., (831) 459-2393 rcoe@pmc.ucsc.edu,
  University of California, Santa Cruz – Ym
Coetzee, Serena M., 27 (0)12 420 3823 serena.coetzee@up.ac.za,
  University of Pretoria – Zi
Coffin, Richard, 361-825-2814 Richard.Coffin@tamucc.edu,
  Texas A&M University, Corpus Christi – Ouc
Coffroth, Mary Alice, 716-645-4871 coffroth@buffalo.edu,
  SUNY, Buffalo – Gu
Coggan, John, +44 01326 371824 J.Coggan@exeter.ac.uk,
  Exeter University – Nx
Coggon, Jude A., +44 (0)23 80596539 jude.coggon@soton.ac.uk,
  University of Southampton – Cg
Cohen, Alice, 902-585-1126 alice.cohen@acadiau.ca,
  Acadia University – ZnnZn
Cohen, Andrew S., (520) 621-4691 cohen@email.arizona.edu,
  University of Arizona – Ps
Cohen, Anne L., (508) 289-2958 acohen@whoi.edu,
  Woods Hole Oceanographic Institution – PecOo
Cohen, Anthony, anthony.cohen@open.ac.uk,
  The Open University – Cs
Cohen, David R., 612 9385 8084 d.cohen@unsw.edu.au,
  University of New South Wales – Cet
Cohen, Joel E., (212) 327-8883 cohen@rockvax.rockefeller.edu,
  Columbia University – Zn
Cohen, Matt, (864) 294 - 2505 matthew.cohen@furman.edu,
  Furman University – Zn
Cohen, Phoebe A., 413-597-2358 phoebe.a.cohen@williams.edu,
  Williams College – Pgg
Cohen, Ronald E., 202-478-8937 rcohen@carnegiescience.edu,
  Carnegie Institution for Science – Gy
Cohen, Shaul, (541) 346-4500 scohen@uoregon.edu,
  University of Oregon – Zn
Coker, Victoria, +44 0161 275-3803 vicky.coker@manchester.ac.uk,
  University of Manchester – Gz
Colak Erol, Serap, 00904242370000-5995 serapcolak@firat.edu.tr,
  Firat University – GgcGt
Colburn, Ivan P., (323) 343-2413 California State University, Los Angeles
  – Gs
Colby, Bonnie C., (520) 621-4775 bcolby@email.arizona.edu,
  University of Arizona – Zn
Colby, Frank P., (978) 934-3906 frank_colby@uml.edu,
  University of Massachusetts, Lowell – Am
Colby, Jeff, (828) 262-7126 colbyj@appstate.edu,
  Appalachian State University – Zi
Colby, Sarah, (928) 523-4561 Northern Arizona University – Zn
Colby, Thomas A., tacolby@ualr.edu,
  University of Arkansas at Little Rock – GcNg
Cole, Catherine, +44 01334 464018 csc5@st-andrews.ac.uk,
  University of St. Andrews – Cm
Cole, David R., (865) 574-5473 cole.618@osu.edu,
  University of Tennessee, Knoxville – Ca
Cole, David R., (614) 688-7407 cole.618@osu.edu,
  Ohio State University – CgsZa
Cole, Dean A., (505) 665-0832 Los Alamos National Laboratory – Co
Cole, Gregory L., (505) 667-1858 gcole@lanl.gov,
  Los Alamos National Laboratory – Zg
Cole, James W., (03) 3642-766 jim.cole@canterbury.ac.nz,
  University of Canterbury – Gv
Cole, Julia E., (520) 626-2341 jecole@email.arizona.edu,
  University of Arizona – PeCsAs
Cole, Kevin C., 616-331-3791 colek@gvsu.edu,
  Grand Valley State University – Gz
Cole, Paul D., +44 1752 585985 paul.cole@plymouth.ac.uk,
  University of Plymouth – Gvg
Cole, Rex D., (970) 248-1599 rcole@coloradomesa.edu,
  Colorado Mesa University – GsrGo
Cole, Rick, (727) 553-1522 rickcole@rdsea.com,
  University of South Florida – Op
Cole, Ron B., (814) 332-3393 rcole@allegheny.edu,
  Allegheny College – Gt
Colegial Gutierrez, Juan D., (316) 237-2685 colegial@uis.edu.co,
  Universidad Industrial de Santander – GeNgZr
Coleman, Alvin L., (910) 362-7365 acoleman@cfcc.edu,
  Cape Fear Community College – Gg
Coleman, Drew S., 919-962-0705 dcoleman@email.unc.edu,
  University of North Carolina, Chapel Hill – CcGit
Coleman, Paul J., (310) 825-1776 pcoleman@igpp.ucla.edu,
  University of California, Los Angeles – Xy
Coleman, Robert G., (650) 723-9205 coleman@pangea.stanford.edu,
  Stanford University – Gi
Coleman, Tommy L., (205) 851-5462 tcoleman@aamu.edu,
  Alabama A&M University – So
Coleman, Jr., Paul J., (310) 825-1776 University of California, Los Angeles
  – Gc
Coles, Kenneth S., (724) 357-5626 kcoles@iup.edu,
  Indiana University of Pennsylvania – XgZeYg
Colgan, Mitchell W., (843) 953-7171 colganm@cofc.edu,
  College of Charleston – PeGe
Colgan, Patrick M., (616) 313-3201 colganp@gvsu.edu,
  Grand Valley State University – GmlPe
Colgan, William, (416)736-2100 #77703 colgan@yorku.ca,
  York University – Yg
Colle, Brian, (631) 632-3174 brian.colle@stonybrook.edu,
  SUNY, Stony Brook – As
Collett, Jr., Jeffrey L., collett@atmos.colostate.edu,
  Colorado State University – As
Collette, Joseph, 701-858-4142 joseph.collette@minotstateu.edu,
  Minot State University – PgGsr
Collie, Jeremy S., (401) 874-6859 jcollie@gso.uri.edu,
  University of Rhode Island – Ob
Collier, Jackie, (631) 632-8696 jackie.collier@stonybrook.edu,
  SUNY, Stony Brook – Ob
Collier, James D., (970) 247-7129 collier_j@fortlewis.edu,
  Fort Lewis College – Cg
Collier, Jenny, +44 20 759 46547 r.coggon@imperial.ac.uk,
  Imperial College – Yr
Collier, Mark, 217-300-1171 mcollier@illinois.edu,
  Illinois State Geological Survey – Ge
Collins, Brian, 206-685-1910 bcollins@uw.edu,
  University of Washington – Gm
Collins, Curtis A., (831) 656-3271 collins@nps.edu,
  Naval Postgraduate School – Op
Collins, Damian, 780-492-3197 damian.collins@ualberta.ca,
  University of Alberta – Zn
Collins, Edward W., (512) 471-6247 eddie.collins@beg.utexas.edu,
  University of Texas at Austin, Jackson School of Geosciences – Ge
Collins, Gareth, +44 20 759 41518 g.collins@imperial.ac.uk,
  Imperial College – Xm
Collins, Joe D., (615) 898-2376 joe.collins@mtsu.edu,
  Middle Tennessee State University – Gms
Collins, John A., (508) 289-2733 jcollins@whoi.edu,
  Woods Hole Oceanographic Institution – Ys
Collins, Kenneth, +44 (0)23 8059 6010 kjc@noc.soton.ac.uk,
  University of Southampton – Ob
Collins, Laura R., (615) 494-8635 laura.collins@mtsu.edu,
  Middle Tennessee State University – Gg
Collins, Laurel S., (305) 348-1732 collinsl@fiu.edu,
  Florida International University – Pg
Collins, Lisa J., (847) 491-3238 earth@northwestern.edu,
  Northwestern University – ZnnZn
Collins, Lorence G., lorencec@sysmatrix.net,
  California State University, Northridge – Gzx
Collins, Mary E., (352) 392-1951 mec@ufl.edu,
  University of Florida – Sd
Collins, William D., (510) 495-2407 wdcollins@berkeley.edu,
  University of California, Berkeley – As

Collinson, James W., collinson.1@osu.edu,
  Ohio State University – PgGsg
Collinson, Margaret, +44 1784 443607 M.Collinson@rhul.ac.uk,
  Royal Holloway University of London – Pg
Collis, Scott, 630-252-0550 scollis@anl.gov,
  Argonne National Laboratory – As
Colman, Albert, (773) 824-1278 University of Chicago – PgCs
Colman, Bradley R., Brad.Colman@noaa.gov,
  University of Washington – As
Colosimo, Amanda, (585) 292-2421 acolosimo@monroecc.edu,
  Monroe Community College – Gg
Coltorti, Mauro, +390577233814 mauro.coltorti@unisi.it,
  University of Siena – GmYg
Colucci, Stephen J., (607) 255-1752 sjc25@cornell.edu,
  Cornell University – As
Colwell, Frederick (Rick), (541) 737-5220 rcolwell@coas.oregonstate.edu,
  Oregon State University – Zn
Comas, María José, mjcomas@ucm.es,
  Univ Complutense de Madrid – Pi
Comas, Xavier, 561 297-3256 xcomas@fau.edu,
  Florida Atlantic University – Yg
Comerford, Nicholas B., (850) 875-7100 nbc@ufl.edu,
  University of Florida – Sf
Comina, Cesare, cesare.comina@unito.it,
  Università di Torino – Yx
Compton, John, (021) 650-2927 john.compton@uct.ac.za,
  University of Cape Town – CmGa
Comrie, Andrew, (520) 621-1585 comrie@email.arizona.edu,
  University of Arizona – Zy
Conder, James A., (618) 453-7352 conder@geo.siu.edu,
  Southern Illinois University Carbondale – Ysr
Condie, Kent C., (575) 835-5531 kcondie@nmt.edu,
  New Mexico Institute of Mining and Technology - Ct
Condit, Christopher D., (413) 545-0272 ccondit@geo.umass.edu,
  University of Massachusetts, Amherst – Gi
Condreay, Denise, 402-562-1216 dcondreay@cccneb.edu,
  Central Community College – Zg
Confer, John, 724-938-4211 confer@calu.edu,
  California University of Pennsylvania – Zn
Congalton, Russell, russ.congalton@unh.edu,
  University of New Hampshire – Zi
Congleton, John D., (678) 839-4066 jconglet@westga.edu,
  University of West Georgia – GeZiHs
Conkle, Jeremy, 361-825-2682 Jeremy.Conkle@tamucc.edu,
  Texas A&M University, Corpus Christi – Zn
Conley, Catharine A., 202-358-3912 cassie.conley@nasa.gov,
  New Mexico Institute of Mining and Technology – Pg
Conly, Andrew G., (807) 343-8463 andrew.conly@lakeheadu.ca,
  Lakehead University – EgCgGz
Connallon, Christopher B., (410) 554 5545 christopher.connallon@maryland.gov,
  Maryland Department of Natural Resources – GmZi
Connelly, Jeffrey B., (501) 569-3546 jbconnelly@ualr.edu,
  University of Arkansas at Little Rock – GcNg
Connely, Melissa, 307-268-2017 mconnely@caspercollege.edu,
  Casper College – GsPev
Connolly, Paul, paul.connolly@manchester.ac.uk,
  University of Manchester – As
Connolly Jr., Harold C., 856-256-5261 hconnolly@gc.cuny.edu,
  Graduate School of the City University of New York – Gg
Connon, CAPT (Ret) Brian D., (228) 688-3720 brian.connon@usm.edu,
  University of Southern Mississippi – ZnnOg
Connor, Charles B., (813) 974-0325 cconnor@cas.usf.edu,
  University of South Florida, Tampa – Gv
Connors, Christopher, (540) 458-8170 connorsc@wlu.edu,
  Washington & Lee University – GcYeGo
Conover, David O., David.Conover@stonybrook.edu,
  SUNY, Stony Brook – Ob
Conrad, Susan H., 845-431- 8534 conrad@sunydutchess.edu,
  Dutchess Community College – GsmHs
Conroy, Jessica, 217-244-4855 jconro@illinois.edu,
  University of Illinois, Urbana-Champaign – Cs
Constable, Catherine G., (858) 534-3183 cconstable@ucsd.edu,
  University of California, San Diego – Ymg
Constable, Steven C., (858) 534-2409 sconstable@ucsd.edu,
  University of California, San Diego – Yr
Constantin, Marc, (418) 656-3139 marc.constantin@ggl.ulaval.ca,
  Universite Laval – Gi
Constantine, Jose, constantineja@cf.ac.uk,
  Cardiff University – Hy
Constantine, Jose A., 413-597-3298 jconstantine@williams.edu,
  Williams College – GmZi
Constantopoulos, James T., (575) 562-2651 jim.constantopoulos@enmu.edu,
  Eastern New Mexico University – GezGx
Contosta, Alexandra, alix.contosta@unh.edu,
  University of New Hampshire – So
Conway, Flaxen D., (541) 737-1339 fconway@coas.oregonstate.edu,
  Oregon State University – Gu
Conway, Howard B., (206) 685-8085 hcon@uw.edu,
  University of Washington – Gl
Conway, Michael F., 520.621.2352 fmconway@email.arizona.edu,
  Arizona Geological Survey – RcGvg
Coogan, Laurence, lacoogan@uvic.ca,
  University of Victoria – Gi
Coogon, Rosalind, +44 20 759 46547 r.coggon@imperial.ac.uk,
  Imperial College – Cs
Cook, Ann, (614) 247-6085 cook.1129@osu.edu,
  Ohio State University – GuYg
Cook, David R., (630) 252-5840 drcook@anl.gov,
  Argonne National Laboratory – AmsSo
Cook, Edward R., drdendro@ldeo.columbia.edu,
  Columbia University – Hw
Cook, Frederick A., (250) 537-8892 fcook@ucalgary.ca,
  University of Calgary – YeGtYs
Cook, Hadrian, +44 020 8417 67756 H.Cook@kingston.ac.uk,
  Kingston University – Ge
Cook, Joseph P., 520-621-2470 joecook@email.arizona.edu,
  Arizona Geological Survey – Gm
Cook, Kerry H., 512-232-7931 kc@jsg.utexas.edu,
  University of Texas, Austin – AsPe
Cook, Mea S., 413-597-4541 mea.s.cook@williams.edu,
  Williams College – OuGe
Cook, Peter, peter.cook@flinders.edu.au,
  Flinders University – Hw
Cook, Robert B., (334) 844-4282 cookrob@auburn.edu,
  Auburn University – Eg
Cook, Steve, (541) 737-0962 cooks@geo.oregonstate.edu,
  Oregon State University – Zn
Cook, Tim, 413-545-1831 tcook@geo.umass.edu,
  University of Massachusetts, Amherst – Pc
Cooke, David R., 61 3 6226 7605 d.cooke@utas.edu.au,
  University of Tasmania – Eg
Cooke, Michele L., (413) 547-3142 cooke@geo.umass.edu,
  University of Massachusetts, Amherst – Gc
Cookus, Pam, 217-244-2486 cookus@isgs.uiuc.edu,
  Illinois State Geological Survey – Zn
Coolbaugh, Mark F., mfc@unr.edu,
  University of Nevada – Eg
Coombs, Douglas S., +64 3 479-7505 doug.coombs@otago.ac.nz,
  University of Otago – Gi
Coombs, Margery C., mcoombs@amherst.edu,
  Amherst College – Pv
Cooney, Michael, (808) 956-7337 mcooney@hawaii.edu,
  University of Hawai'i, Manoa – Og
Cooney, Timothy M., (319) 273-2918 timothy.cooney@uni.edu,
  University of Northern Iowa – Ze
Coonley, Steffanie, (505) 665-2330 scoonley@lanl.gov,
  Los Alamos National Laboratory – Zn
Cooper, Alan F., +64 3 479-7515 alan.cooper@stonebow.otago.ac.nz,
  University of Otago – GpiGt
Cooper, Catherine, 509-335-1501 cmcooper@wsu.edu,
  Washington State University – Yg
Cooper, Clay A., (775) 673-7372 clay.cooper@dri.edu,
  University of Nevada, Reno – HySpHq
Cooper, Jennifer, 203-392-5842 cooperj1@southernct.edu,
  Southern Connecticut State University – GgcGi
Cooper, Jonathon L., (507) 222-4401 jlcooper@carleton.edu,
  Carleton College – Gg
Cooper, Kari M., (530) 754-8826 kmcooper@ucdavis.edu,
  University of California, Davis – Gi
Cooper, Mark, (204) 474-8075 mark_cooper@umanitoba.ca,
  University of Manitoba – Gz
Cooper, Matthew, +44 (0)23 80592062 matthew.cooper@noc.soton.ac.uk,
  University of Southampton – Oc

Cooper, Reid F., (401) 863-2160 Reid_Cooper@Brown.edu,
  Brown University – Gy
Cooper, Roger W., (409) 880-8239 roger.cooper@lamar.edu,
  Lamar University – Gi
Cooper, Terence H., (612) 625-7747 tcooper@soils.umn.edu,
  University of Minnesota, Twin Cities – Sd
Coorough Burke, Patricia J., (414) 278-6155 coorough@mpm.edu,
  Milwaukee Public Museum – PgiPe
Cope, Tim D., (765) 658-6443 tcope@depauw.edu,
  DePauw University – GsZiGr
Copeland, Peter, 713-743-3649 copeland@uh.edu,
  University of Houston – CcGt
Copeman, Louise A., 541-737-6840 Oregon State University – Zn
Copjakova, Renata, +549 49 3073 copjakova@sci.muni.cz,
  Masaryk University – Gz
Copley, Alex, +44 (0) 1223 748937 acc41@cam.ac.uk,
  University of Cambridge – YgGt
Copley, Jonathan, +44 (0)23 8059 6621 jtc@noc.soton.ac.uk,
  University of Southampton – Ob
Copper, Paul, pcopper@laurentian.ca,
  Laurentian University, Sudbury – PieZc
Coppinger, Walter, (406) 287-2288 wcopping@trinity.edu,
  Trinity University – Gcg
Coppola, Diego, diego.coppola@unito.it,
  Università di Torino – Gv
Corbato, Charles E., 614-451-6537 ccorbato@gmail.com,
  Ohio State University – Ye
Corbella Cordomí, Mercè, ++34935811973 merce.corbella@uab.cat,
  Universitat Autonoma de Barcelona – GzCl
Corbett, Patrick, +44 (0) 131 451 3171 p.corbett@hw.ac.uk,
  Heriot-Watt University – Np
Corbineau, Lucien, +33(0)3 44068962 lucien.corbineau@lasalle-beauvais.fr,
  Institut Polytechnique LaSalle Beauvais (ex-IGAL) – EgGgNm
Corcoran, Deborah, 417-836-6889 debcorcoran@missouristate.edu,
  Missouri State University – Zn
Corcoran, Patricia, 519-661-2111, ext. 86836 pcorcor@uwo.ca,
  Western University – Gd
Corcoran, William, (417) 836-5801 williamcorcoran@missouristate.edu,
  Missouri State University – As
Cordell, Ann S., 3523921721 x491 cordell@flmnh.ufl.edu,
  University of Florida – Gxa
Cordero, David I., 360-442-2883 dcordero@lowercolumbia.edu,
  Lower Columbia College – Zg
Cordonnier, Benoit, 089/2180 4271 cordonnier@min.uni-muenchen.de,
  Ludwig-Maximilians-Universitaet Muenchen – Gz
Corey, Alison, (801) 537-3122 nrugs.acorey@state.ut.us,
  Utah Geological Survey – Zn
Corley, John H., (573) 368-2132 john.corley@dnr.mo.gov,
  Missouri Dept of Natural Resources – Geg
Corliss, Bruce H., (919) 684-2951 bruce.corliss@duke.edu,
  Duke University – Pm
Cormier, Vernon F., (860) 486-3547 vernon.cormier@uconn.edu,
  University of Connecticut – YsGt
Cornebise, Michael W., (217) 581-7486 mwcornebise@eiu.edu,
  Eastern Illinois University – Zg
Cornell, Kristie, 337-482-1455 kcornell@louisiana.edu,
  University of Louisiana at Lafayette – Gg
Cornell, Sean R., 717-477-1310 srcornell@ship.edu,
  Shippensburg University – GgsOn
Cornell , William , Cornell@utep.edu,
  University of Texas, El Paso – Gg
Cornell, Winton, (918) 631-3248 winton-cornell@utulsa.edu,
  The University of Tulsa – GivCa
Cornillon, Peter, (401) 874-6283 pcornillon@gso.uri.edu,
  University of Rhode Island – Op
Cornwell, David G., +44 (0)1224 273448 d.cornwell@abdn.ac.uk,
  University of Aberdeen – YsGtYe
Cornwell, Jeffery C., (410) 221-8445 cornwell@hpl.umces.edu,
  University of Maryland – Oc
Cornwell, Kevin J., (916) 278-6667 cornwell@csus.edu,
  California State University, Sacramento – GmeHs
Coron, Cynthia R., (203) 392-5840 coronc1@southernct.edu,
  Southern Connecticut State University – Eg
Corona, Thomas J., (303) 556-8525 Metropolitan State College of Denver – Am
Corral, Isaac, +61 7 47815681 isaac.corralcalleja@jcu.edu.au,
  James Cook University – EgGvCg

Corrigan, Catherine, (202) 633-1855 corriganc@si.edu,
  Smithsonian Inst / Natl Museum of Natural Hist– XmcXg
Corriveau, Louise, (418) 654-2672 louise.corriveau@canada.ca,
  Universite du Quebec – EmGpg
Corsetti, Frank A., (213) 740-6123 fcorsetti@earth.usc.edu,
  University of Southern California – Gs
Coruh, Cahit, (386) 447-1289 coruh@vt.edu,
  Virginia Polytechnic Institute & State University – YesGt
Cosentino, Pietro, p.cosentino@uea.ac.uk,
  University of East Anglia – Yg
Cosgrove, John, +44 20 759 46466 j.cosgrove@imperial.ac.uk,
  Imperial College – Gc
Costa, Emanuele, emanuele.costa@unito.it,
  Università di Torino – Gz
Costa, Jr, Ozeas S., (419) 755-4128 costa.47@osu.edu,
  Ohio State University – CblZu
Costea, Lidia, (773) 442-6050 L-Costea@neiu.edu,
  Northeastern Illinois University – ZnnZn
Costello, Margaret, (562) 985-4809 mcostel2@csulb.edu,
  California State University, Long Beach – Zn
Costigan, Keeley R., (505) 665-4788 krc@lanl.gov,
  Los Alamos National Laboratory – As
Costin, Gelu, +27 (0)46-603-8316 g.costin@ru.ac.za,
  Rhodes University – Gp
Cote, Denis, (418) 545-5011 dcote@uqac.ca,
  Universite du Quebec a Chicoutimi – GiEgZi
Côté, Jean, 514 987-3000 #2351 cote.jean@uqam.ca,
  Universite du Quebec a Montreal – As
Cote, Pascale, (418) 654-2601 pacote@nrcan.gc.ca,
  Universite du Quebec – Zn
Cothren, Jackson D., (479) 575-6790 jcothre@uark.edu,
  University of Arkansas, Fayetteville – Zir
Cottaar, Sanne , sc845@cam.ac.uk,
  University of Cambridge – Ys
Cotter, James F., (320) 589-6312 cotterjf@mrs.umn.edu,
  University of Minnesota, Morris – Gl
Cottingame, William, (505) 667-8339 wcottingame@lanl.gov,
  Los Alamos National Laboratory – As
Cottle, John, (805) 893-3471 cottle@geol.ucsb.edu,
  University of California, Santa Barbara – Gi
Cotton, Jennifer, (818) 677-7978 jennifer.cotton@csun.edu,
  California State University, Northridge – Pc
Cottrell, Elizabeth, cottrell@si.edu,
  University of Maryland – CpgGx
Cottrell, Elizabeth, (202) 633-1859 cottrelle@si.edu,
  Smithsonian Inst / Natl Museum of Natural Hist– Cp
Cottrell, Rory D., rory@earth.rochester.edu,
  University of Rochester – Ym
Couceiro, Fay, +44 023 92 842294 fay.couceiro@port.ac.uk,
  University of Portsmouth – On
Coulibaly, Mamadou, (920) 424-4105 coulibalym@uwosh.edu,
  University of Wisconsin Oshkosh – ZiyHg
Coulibaly, Paulin, (905) 525-9140 (Ext. 23354) couliba@mcmaster.ca,
  McMaster University – Hg
Coull, Bruce, bccoull@sc.edu,
  University of South Carolina – Ob
Coulson, Alan B., (864) 656-1897 acoulso@clemson.edu,
  Clemson University – PgGgCs
Coulter, Richard L., (631) 252-5833 rcoulter@anl.gov,
  Argonne National Laboratory – Am
Couples, Gary, +44 (0) 131 451 3123 g.couples@hw.ac.uk,
  Heriot-Watt University – Np
Cousens, Brian L., 001-613-520-2600 x4436 brian.cousens@carleton.ca,
  Carleton University – GiCgc
Cousineau, Pierre, (418) 545-5011 pcousine@uqac.ca,
  Universite du Quebec a Chicoutimi – Gd
Couture, Gilles, 514-987-3000 #8905 couture.gilles@uqam.ca,
  Universite du Quebec a Montreal – Yg
Cova, Thomas A., (801) 581-7930 tom.cova@geog.utah.edu,
  University of Utah – Zi
Coveney, Eamonn, ewcoveney@fhsu.edu,
  Fort Hays State University – Zyn
Coveney, Seamus, +44 0141 330 7750 Seamus.Coveney@glasgow.ac.uk,
  University of Glasgow – Zi
Coveney Jr, Raymond M., (816) 235-1334 coveneyr@umkc.edu,
  University of Missouri, Kansas City – EgmCe
Covert, David S., (206) 685-7461 dcovert@atmos.washington.edu,

University of Washington – As
Covington, Matthew, 479-575-3876 mcoving@uark.edu,
University of Arkansas, Fayetteville – HqGmg
Cowan, Clinton A., (507) 222-7021 ccowan@carleton.edu,
Carleton College – GsPe
Cowan, Darrel S., (206) 543-4033 darrel@uw.edu,
University of Washington – Gct
Cowan, Ellen A., (828) 262-2260 cowanea@appstate.edu,
Appalachian State University – GsmGu
Cowan, Nicolas B., (514) 398-1967 nicolas.cowan@mcgill.ca,
McGill University – AsXyg
Cowart, James B., (850) 644-5784 jcowart@mailer.fsu.edu,
Florida State University – Cc
Cowart, Richard, (361) 644-3049 recowart@coastalbend.edu,
Coastal Bend College – GeEo
Cowen, Richard, richard@blueaokfarm.com,
University of California, Davis – Pg
Cowgill, Eric S., (530) 754-6574 escowgill@ucdavis.edu,
University of California, Davis – Gtc
Cowie, Gregory L., +44 (0) 131 650 8502 Dr.Greg.Cowie@ed.ac.uk,
Edinburgh University – Co
Cowman, Tim C., 605-677-6151 tim.cowman@usd.edu,
South Dakota Dept of Environment and Natural Resources – CgGg
Cox, Alysia, (406) 496-4185 acox@mtech.edu,
Montana Tech of the University of Montana – CqlCm
Cox, Christena, 614-292-0138 cox.1@osu.edu,
Ohio State University – GeoGc
Cox, John, (403) 440-6160 jcox@mtroyal.ca,
Mount Royal University – EoGsHw
Cox, Randel T., (901) 678-4361 randycox@memphis.edu,
University of Memphis – GcmGt
Cox, Randel T., (901) 678-4870 randycox@memphis.edu,
University of Memphis – GtcGm
Cox, Richard, 902 494 3362 Richard.Cox@dal.ca,
Dalhousie University – GzEdGg
Cox, Ronadh, (413) 597-2297 ronadh.cox@williams.edu,
Williams College – GsXg
Cox, Shelah, 215 204-8227 scox@temple.edu,
Temple University – Zn
Coxon, Catherine, + 353 1 8962235 cecoxon@tcd.ie,
Trinity College – GeOu
Craddock, Robert A., 202-633-2473 Smithsonian Institution / National Air & Space Museum – Xg
Crafford, J. P., +27 (0)15 268 2217 krappie.crafford@ul.ac.za,
University of Limpopo – ZnNgZn
Craig, James L., (505) 665-7996 jlcraig@lanl.gov,
Los Alamos National Laboratory – Gg
Craig, James R., (540) 231-5222 jrcraig@vt.edu,
Virginia Polytechnic Institute & State University – Eg
Craig, Mitchell S., (510) 885-3425 mitchell.craig@csueastbay.edu,
California State University, East Bay – Yg
Craig, Susanne , (902) 494-4381 susanne.craig@dal.ca,
Dalhousie University – Op
Crain, John R., (209) 730-3812 College of the Sequoias – Gg
Cramer, Bradley D., (319) 335-0704 bradley-cramer@uiowa.edu,
University of Iowa – GrPsGs
Cramer, Chris, (901) 678-2007 ccramer@memphis.edu,
University of Memphis – Ys
Cramer, Gary, (620) 662-9021 gcramer@ksu.edu,
Kansas State University – So
Crane, Peter R., (773) 922-9410 University of Chicago – Pg
Cranford, Peter, (902) 426-3277 cranfordp@mar.dfo-mpo.gc.ca,
Dalhousie University – Ob
Cranganu, Constantin, 7189515000 x2878 cranganu@brooklyn.cuny.edu,
Brooklyn College (CUNY) – GoYhHw
Craven, John, +44 (0) 131 650 7887 John.Craven@ed.ac.uk,
Edinburgh University – Yg
Craw, Dave, +64 3 479-7529 dave.craw@otago.ac.nz,
University of Otago – Eg
Crawford, Alex, 330-263-1980 acrawford@wooster.edu,
College of Wooster – AtZri
Crawford, Anthony J., tony.crawford@utas.edu.au,
University of Tasmania – Git
Crawford, Ian, +44 020 3073 8026 i.crawford@bbk.ac.uk,
Birkbeck College – Zn
Crawford, Ian, +44 0161 306-6850 I.Crawford@manchester.ac.uk,
University of Manchester – As

Crawford, James, 757-864-7231 j.h.crawford@larc.nasa.gov,
Georgia Institute of Technology – As
Crawford, Maria Luisa B., (610) 526-5111 mcrawfor@brynmawr.edu,
Bryn Mawr College – GxzGp
Crawford, Matt, 859.323.0510 mcrawford@uky.edu,
University of Kentucky – Zn
Crawford, Nicholas, (270) 745-5889 nicholas.crawford@wku.edu,
Western Kentucky University – Hg
Crawford, Thomas J., (678) 839-4062 University of West Georgia – Eg
Crawford, Vernon J., (541) 552-6479 crawford@sou.edu,
Southern Oregon University – Gg
Crawford, William A., (610) 526-5112 Bryn Mawr College – Cg
Creager, Kenneth C., 206-685-2803 kcc@uw.edu,
University of Washington – Ys
Crease, James, (302) 645-4240 University of Delaware – Op
Creaser, Robert A., (780) 492-2942 robert.creaser@ualberta.ca,
University of Alberta – Cc
Creasey, Robert L., 831-656-3178 creasey@nps.edu,
Naval Postgraduate School – Am
Crepeau, Richard J., (828) 262-7052 crepeaurj@appstate.edu,
Appalachian State University – Zn
Crespi, Jean M., (860) 486-0601 crespi@geol.uconn.edu,
University of Connecticut – Gc
Crespo, Elena, ecrespo@ucm.es,
Univ Complutense de Madrid – En
Creveling, Jessica, 541-737-2112 crevelij@oregonstate.edu,
Oregon State University – GgsGr
Crews, Jeff, (573) 368-2356 jeff.crews@dnr.mo.gov,
Missouri Dept of Natural Resources – Hy
Creyts, Timothy T., tcreyts@ldeo.columbia.edu,
Columbia University – Gl
Cribb, Warner, (615) 898-2379 warner.cribb@mtsu.edu,
Middle Tennessee State University – GipGz
Crider, Juliet G., 206-543-8715 criderj@uw.edu,
University of Washington – GcmNr
Crilley, Leigh R., +44 (0)121 414 5523 l.crilley@bham.ac.uk,
University of Birmingham – As
Criminale, Jr., William O., (206) 543-9506 lascala@amath.washington.edu,
University of Washington – Op
Crimmins, Michael A., (520) 626-4244 crimmins@email.arizona.edu,
University of Arizona – Zc
Crippen, Robert E., (818) 354-2475 robert.e.crippen@jpl.nasa.gov,
Jet Propulsion Laboratory – Zr
Crisp, Joy A., (818) 354-9036 Jet Propulsion Laboratory – Gv
Criss, Robert E., (314) 935-7441 criss@levee.wustl.edu,
Washington University in St. Louis – Cs
Criss, Robert E., criss@wustl.edu,
Washington University in St. Louis – CsHqXa
Criswell, James, (910) 392-7536 jcriswell@cfcc.edu,
Cape Fear Community College – Gg
Crockett, Joan E., (217) 244-2388 crockett@illinois.edu,
Illinois State Geological Survey – EoZg
Crockford, Peter, 609-258-4101 pwc2@princeton.edu,
Princeton University – ClsPc
Croft, Gary M., gcroft@valleycollege.edu,
San Bernardino Valley College – Zy
Croft, Paul J., (908) 737-3737 pcroft@kean.edu,
Kean University – AmsZe
Croll, Don, (831) 459-3610 dcroll@cats.ucsc.edu,
University of California, Santa Cruz – Ob
Cromar, Nancy, nancy.cromar@flinders.edu.au,
Flinders University – Zn
Cronin, Meghan F., (206) 526-6449 Meghan.F.Cronin@noaa.gov,
University of Washington – Op
Cronin, Timothy W., twcronin@mit.edu,
Massachusetts Institute of Technology – As
Cronin, Vincent S., (254) 710-2174 vince_cronin@baylor.edu,
Baylor University – GctNg
Cronoble, James M., (303) 556-3070 Metropolitan State College of Denver – Gg
Croot, Peter, + 353 (0)91 492 194 peter.croot@nuigalway.ie,
National University of Ireland Galway – Ct
Crootof, Arica, 406-683-7075 arica.crootof@umwestern.edu,
University of Montana Western – ZcRcZi
Crosby, Benjamin T., (208) 282-2949 crosby@isu.edu,
Idaho State University – GmHsZy
Cross, John A., (920) 424-4105 cross@uwosh.edu,

University of Wisconsin Oshkosh – ZyAst
Crossen, Kristine J., (907) 786-6838 kjcrossen@uaa.alaska.edu,
　University of Alaska, Anchorage – Gl
Crossey, Laura J., (505) 277-4204 lcrossey@unm.edu,
　University of New Mexico – Cl
Crossley, David J., (314) 977-3153 crossley@eas.slu.edu,
　Saint Louis University – YdsXy
Crosson, Robert S., crosson@uw.edu,
　University of Washington – YsGct
Crouse, David A., (919) 515-7302 North Carolina State University – Sc
Crowder, Margaret, (270) 745-5973 margaret.crowder@wku.edu,
　Western Kentucky University – Ze
Crowe, Bruce M., (702) 794-7206 bmc@lanl.gov,
　Los Alamos National Laboratory – Gv
Crowe, Douglas E., 706-542-2382 crowe@gly.uga.edu,
　University of Georgia – Em
Crowley, Brooke W., (513) 556-7181 brooke.crowley@uc.edu,
　University of Cincinnati – CsPev
Crowley, Jim, jimcrowley@boisestate.edu,
　Boise State University – Ccg
Crowley, Peter D., (413) 542-2715 pdcrowley@amherst.edu,
　Amherst College – Gc
Crowley, Quentin G., + 353 1 8962403 crowleyq@tcd.ie,
　Trinity College – Cc
Crowley, Stephen, +44 0151 794 5163　　Sfcrow@liverpool.ac.uk,
　University of Liverpool – Gs
Crowther, Sarah, +44 0161 275-0407 Sarah.Crowther@manchester.ac.uk,
　University of Manchester – As
Croxen III, Fred W., 928-344-7586 fred.croxen@azwestern.edu,
　Arizona Western College – HwPvZi
Crozier, Carl, (919) 793-4428 North Carolina State University – Sb
Cruden, David M., (780) 492-3085 dave.cruden@ualberta.ca,
　University of Alberta – Nr
Cruikshank, Kenneth M., (503) 725-3383 cruikshankk@pdx.edu,
　Portland State University – Gc
Crumbly, Isaac J., (912) 825-6454 Fort Valley State University – Zn
Cruse, Richard M., (515) 294-7850 rmc@iastate.edu,
　Iowa State University of Science & Technology – So
Crutzen, Paul, pcrutzen@ucsd.edu,
　University of California, San Diego – Oc
Cruz-Uribe, Alicia M., (207) 581-4494 alicia.cruzuribe@maine.edu,
　University of Maine – GpCpt
Cruzen, Shawn,　706-649-1785 cruzen_shawn@columbusstate.edu,
　Columbus State University – Xa
Csank, Adam, 775-784-6663　acsank@unr.edu,
　University of Nevada, Reno – Zy
Csatho, Beata M., (716) 645-4325 bcsatho@buffalo.edu,
　SUNY, Buffalo – GlZrYg
Csiki, Shane, (603) 271-2876 shane.csiki@des.nh.gov,
　New Hampshire Geological Survey – ZyGm
Cuffey, Roger J., (814) 865-1293 rcuffey@psu.edu,
　Pennsylvania State University, University Park – PiePv
Cuker, Benjamin E., (757) 727-5783 benjamin.cuker@hamptonu.edu,
　Hampton University – Ob
Cullen, Jay, 250-721-6120 jcullen@uvic.ca,
　University of Victoria – Oc
Cullen, John J., (902) 494-4667 john.cullen@dal.ca,
　Dalhousie University – Ob
Cullers, Robert L., (785) 532-2240 rcullers@ksu.edu,
　Kansas State University – Ct
Culshaw, Nicholas, (902) 494-3501 Dalhousie University – Gc
Culver, Stephen J., (252) 328-6360 culvers@ecu.edu,
　East Carolina University – PmGu
Cummings, Michael L., (503) 725-3395 cummingsm@pdx.edu,
　Portland State University – GgHg
Cummins, R. Hays, (513) 529-1338 cumminrh@MiamiOh.edu,
　Miami University – Ge
Cundall, Peter A., (612) 626-0369 pacundall@aol.com,
　University of Minnesota, Twin Cities – Nr
Cunningham, Dickson, 860-465-4321 cunninghamw@easternct.edu,
　Eastern Connecticut State University – GtcGx
Cunningham, Mary A., 845-437-5547 macunningham@vassar.edu,
　Vassar College – ZiyZu
Cunningham, Peter, +44 (0) 191 208 7932 peter.cunningham@ncl.ac.uk,
　University of Newcastle Upon Tyne – Hg
Cuomo, Carmela, ccuomo@newhaven.edu,
　University of New Haven – CmmGu

Cupples, Julie, +44 (0) 131 651 4315 Julie.Cupples@ed.ac.uk,
　Edinburgh University – Zn
Curewitz, Daniel, 315-443-2672 dcurewit@syr.edu,
　Syracuse University – OgGct
Curl, Doug C., 859-323-0519 doug@uky.edu,
　University of Kentucky – ZfGgc
Curran, H. Allen, (413) 585-3943 acurran@smith.edu,
　Smith College – Pg
Currano, Ellen, (307) 766-2819 ecurrano@uwyo.edu,
　University of Wyoming – Pbe
Curray, Joseph R., (858) 459-2130 jcurray@ucsd.edu,
　University of California, San Diego – Gu
Currens, James C., 859-323-0526 currens@uky.edu,
　University of Kentucky – Hg
Currey, Donald R., (801) 581-8218 University of Utah – Gm
Curri, Neil, 845-437-7708 necurri@vassar.edu,
　Vassar College – Zi
Currie, Brian S., (513) 529-7578 curriebs@miamioh.edu,
　Miami University – GstGc
Currie, Philip J., (403) 823-7707 University of Calgary – Pv
Currier, Ryan, (920) 465-2582 currierr@uwgb.edu,
　University of Wisconsin, Green Bay – GizGg
Curry, B. B., (217) 244-5787 b-curry@illinois.edu,
　Northern Illinois University – GelPs
Curry, Gordon, +44 01413305444 gordon.curry@glasgow.ac.uk,
　University of Glasgow – GegPg
Curry, James K., (501) 682-6992 james.curry@arkansas.gov,
　Arkansas Geological Survey – Zn
Curry, Janel M., (616) 526-6869 Calvin College – Zy
Curry, Joan E., (520) 626-5081 curry@ag.arizona.edu,
　University of Arizona – Sc
Curry, Judith A., (404) 894-3948 judith.curry@eas.gatech.edu,
　Georgia Institute of Technology – AsZrn
Curry, Susan, (352) 294-3147 scurry@ufl.edu,
　University of Florida – ZiSo
Curry, William B., (508) 289-2591 wcurry@whoi.edu,
　Woods Hole Oceanographic Institution – Pm
Curry Rogers, Kristina A., 651-696-6288 krogers@macalester.edu,
　Macalester College – Pv
Curtice, Joshua M., (508) 289-2618 jcurtice@whoi.edu,
　Woods Hole Oceanographic Institution – Cc
Curtin, Tara M., (315) 781-3928 curtin@hws.edu,
　Hobart & William Smith Colleges – PeGs
Curtis, Andrew, +44 (0) 131 650 8515 Andrew.Curtis@ed.ac.uk,
　Edinburgh University – Ys
Curtis, Arthur C., (615) 576-7642 Oak Ridge National Laboratory – Ge
Curtis, David B., (505) 667-4845 Los Alamos National Laboratory – Cc
Curtis, Garniss H., (510) 845-0333 University of California, Berkeley – Cc
Curtis, Jason H., (352) 392-2296 jcurtis@che.ufl.edu,
　University of Florida – Pe
Curtis, John B., jbcurtis@mines.edu,
　Colorado School of Mines – GoCo
Curtis, Lynn A., (954) 201-6942 lcurtis@broward.edu,
　Broward College – GgPg
Cusack, Maggie, +44　01413305491 Maggie.Cusack@glasgow.ac.uk,
　University of Glasgow – Gz
Cushman, John, (765) 494-8040 jcushman@purdue.edu,
　Purdue University – Sp
Cutter, Gregory A., (757) 683-4929 gcutter@odu.edu,
　Old Dominion University – OcCtHs
Cvetko Tešoviæ, Blanka, +38514606107 blankacvetko@yahoo.com,
　University of Zagreb – GgPsm
Czaja, Andrew D., 513-556-3574 andrew.czaja@uc.edu,
　University of Cincinnati – PmCl
Czarnetzki, Alan C., (319) 273-2152 alan.czarnetzki@uni.edu,
　University of Northern Iowa – Am
Czeck, Dyanna M., 414-229-3948 dyanna@uwm.edu,
　University of Wisconsin, Milwaukee – Gc
Cziczo, Dan, (617) 324-4882 djcziczo@mit.edu,
　Massachusetts Institute of Technology – As

D

d'Alessio, Matthew, (818) 677-3647 matthew.dalessio@csun.edu,
　California State University, Northridge – ZeGt
D'Allura, Jad A., (541) 552-6480 dallura@sou.edu,
　Southern Oregon University – Gc
D'Alpaos, Andrea, +390498279117 andrea.dalpaos@unipd.it,

Università degli Studi di Padova – Hs
D'Amico, Sebastiano, +356 2340 3101 sebdamico@gmail.com,
   University of Malta – YsgZg
D'Andrea, William J., dandrea@ldeo.columbia.edu,
   Columbia University – Pe
D'Asaro, Eric A., (206) 545-2982 dasaro@apl.washington.edu,
   University of Washington – Op
d'Atri, Anna, anna.datri@unito.it,
   Università di Torino – Gr
D'Hondt, Steven L., (401) 792-6808 dhondt@gso.uri.edu,
   University of Rhode Island – Ou
D'Odorico, Paolo, (434) 924-7241 pd6v@virginia.edu,
   University of Virginia – Hg
da Silva, Eduardo F., eafsilva@ua.pt,
   Universidade de Aveiro – CeGbCl
Daanen, Ronald P., 907-451-5965 ronald.daanen@alaska.gov,
   Alaska Division of Geological & Geophysical Surveys – HyGme
Dade, William B., 603-646-0286 William.B.Dade@Dartmouth.edu,
   Dartmouth College – GsHgEo
Daemen, Jaak, (775) 7844309 daemen@mines.unr.edu,
   University of Nevada, Reno – Nr
Daeschler, Ted, 215 299 1133 ebd29@drexel.edu,
   Drexel University – PvGgPg
Dagdelen, Kadri, (303) 273-3711 kdagdele@mines.edu,
   Colorado School of Mines – Nm
Dahl, Johannes M., 806-834-6197 johannes.dahl@ttu.edu,
   Texas Tech University – Ams
Dahl, Robyn, 360-650-7207 Robyn.Dahl@wwu.edu,
   Western Washington University – PiZe
Dahlgran, Randy, 530-752-2814 radahlgren@ucdavis.edu,
   University of California, Davis – Sc
Dahlke, Helen E., (530) 302-5358 hdahlke@ucdavis.edu,
   University of California, Davis – Hgs
Dai, Chunli, (614) 292-2269 dai.56@osu.edu,
   Ohio State University – ZrYg
Dai, Dajun, 404 413 5750 geoddd@langate.gsu.edu,
   Georgia State University – Zi
Daigneault, Real, (418) 545-5011 rdaignea@uqac.ca,
   Universite du Quebec a Chicoutimi – GcEgZi
Dakin, Susan, (403) 329-2279 susan.dakin@uleth.ca,
   University of Lethbridge – Ze
Dalconi, Maria C., +390498279163 mariachiara.dalconi@unipd.it,
   Università degli Studi di Padova – Gz
Dale, Janis, (306) 585-4840 University of Regina – Gml
Daley, Gwen M., 803-323-4973 daleyg@winthrop.edu,
   Winthrop University – PqgGs
Dallegge, Todd A., 970-351-1086 Todd.Dallegge@unco.edu,
   University of Northern Colorado – GosGr
Dallimer, Martin, +44(0) 113 34 35279 M.Dallimer@leeds.ac.uk,
   University of Leeds – Ge
Dallmeyer, R. David, (706) 542-7448 dallmeyr@uga.edu,
   University of Georgia – Gt
Dalman, Michael, (979) 830-4206 mdalman@blinn.edu,
   Blinn College – Ze
Dalrymple, Robert W., (613) 533-6186 dalrympl@queensu.ca,
   Queen's University – Gs
Dalton, Colleen, 401-863-5875 Colleen_Dalton@brown.edu,
   Brown University – Ys
Dalton, Richard F., (609) 292-2576 richard.dalton@dep.nj.gov,
   New Jersey Geological and Water Survey – Gr
Daly, Eve, +353 (0)91 492 183 eve.daly@nuigalway.ie,
   National University of Ireland Galway – Ygu
Daly, Julia F., (207) 778-7403 dalyj@maine.edu,
   University of Maine - Farmington – GmHsGu
Daly, Kendra L., (727) 553-1041 kdaly@marine.usf.edu,
   University of South Florida – Ob
Daly, Raymond J., daly@ucc.edu,
   Union County College – Pv
Dalziel, Ian W. D., (512) 471-0431 ian@ig.utexas.edu,
   University of Texas, Austin – Gt
Dam, Hans G., (860) 405-9098 hans.dam@uconn.edu,
   University of Connecticut – Ob
Damir, Buckoviæ, +38514606106 buckovic@geol.pmf.hr,
   University of Zagreb – GgPms
Damschen, Ellen, damschen@wustl.edu,
   Washington University in St. Louis – Zn
Damuth, John E., (972) 623-7368 damuth@uta.edu,
   University of Texas, Arlington – GusYr
Dana, Gayle L., (775) 674-7538 gdana@dri.edu,
   University of Nevada, Reno – Zr
Daniel, Christopher G., (570) 577-1133 cdaniel@bucknell.edu,
   Bucknell University – Gp
Daniell, James J., james.daniell@jcu.edu.au,
   James Cook University – GumYr
Daniels, J. Michael, (303) 871-7531 j.michael.daniels@du.edu,
   University of Denver – GmSp
Daniels, Jeffrey J., 614-292-6704 daniels.9@osu.edu,
   Ohio State University – Yg
Daniels, Walter L., (540) 231-7175 wdaniels@vt.edu,
   Virginia Polytechnic Institute & State University – SdGeCg
Daniels, William R., (505) 667-4546 Los Alamos National Laboratory – Ct
Danko, George, (775) 784-4284 danko@mines.unr.edu,
   University of Nevada, Reno – Hq
Dansereau, Pauline, dansero@ggl.ulaval.ca,
   Universite Laval – Gs
Daoust, Mario, 417-836-5301 mariodaoust@missouristate.edu,
   Missouri State University – AtZy
Daramola, Sunday O., 08060256588 sunday.daramola@gmail.com,
   Federal University of Technology, Akure – HwGe
Darby, Dennis A., (757) 683-4701 ddarby@odu.edu,
   Old Dominion University – Ou
Darby, Jeannie, (916) 752-5670 University of California, Davis – Hg
Darbyshire, Fiona Ann, 514-987-3000 #5054 darbyshire.fiona_ann@uqam.ca,
   Universite du Quebec a Montreal – Ys
Darling, James, +44 023 92 842247 james.darling@port.ac.uk,
   University of Portsmouth – Cs
Darling, Robert S., (607) 753-2923 darlingr@cortland.edu,
   SUNY, Cortland – GzpCp
Darmody, Robert G., (217) 333-9489 University of Illinois, Urbana-
   Champaign - Sd
Darnajoux, Romain, 609-258-4101 romaind@princeton.edu,
   Princeton University – Cb
Darold, Amberlee, 405/325-8611 Amberlee.P.Darold-1@ou.edu,
   University of Oklahoma – Ys
Daroub, Samira H., (561) 993-1500 sdaroub@ufl.edu,
   University of Florida – So
Darrah, Thomas, (614) 688-2132 darrah.24@osu.edu,
   Ohio State University – Cg
Darrell, James H., (912) 478-5361 JDarrell@GeorgiaSouthern.edu,
   Georgia Southern University – Pl
Darroch, Simon A., simon.a.darroch@vanderbilt.edu,
   Vanderbilt University – PgiPe
Das, Sarah B., 508-289-2464 sdas@whoi.edu,
   Woods Hole Oceanographic Institution – GlAtZg
Dasgupta, Rajdeep, 713.348.2664 Rajdeep.Dasgupta@rice.edu,
   Rice University – GxCg
Dasgupta, Tathagata, 330-672-4104 tdasgupt@kent.edu,
   Kent State University – Cg
Dash, Padmanava, 662-325-3915 pd175@msstate.edu,
   Mississippi State University – ZrHg
Dash, Zora V., (505) 667-1923 zvd@lanl.gov,
   Los Alamos National Laboratory – Zn
Dasvarma, Gour, gour.dasvarma@flinders.edu.au,
   Flinders University – Zn
Datta, Saugata, (785) 532-2241 sdatta@ksu.edu,
   Kansas State University – ClHwGe
Daugherty, Carolyn M., carolyn.daugherty@nau.edu,
   Northern Arizona University – Zn
Daugherty, LeRoy A., (505) 646-3406 New Mexico State University, Las
   Cruces – Sd
Dauphas, Nicolas, (773) 702-2930 University of Chicago – Xc
Davatzes, Alexandra K., (215) 204-3907 alix@temple.edu,
   Temple University – GsXg
Davatzes, Nicholas, (215) 204-2837 davatzes@temple.edu,
   Temple University – GcNrZn
Davenport, Joan R., (509) 786-9384 jdavenp@tricity.wsu.edu,
   Washington State University – Sb
Davey, Patricia M., 863-2449 patricia_davey@brown.edu,
   Brown University – Zn
David, Mark B., (217) 333-4308 dmnicol@illinois.edu,
   University of Illinois, Urbana-Champaign – Sf
Davidson, Bart, 859-323-0524 bdavidson@uky.edu,
   University of Kentucky – Hg
Davidson, Cameron, (507) 222-7144 cdavidso@carleton.edu,

Carleton College – GptCc
Davidson, Fiona M., (479) 575-3879 fdavidso@comp.uark.edu,
University of Arkansas, Fayetteville – Zy
Davidson, Garry J., 61 3 6226 2815 Garry.Davidson@utas.edu.au,
University of Tasmania – Cs
Davidson, Gregg R., (662) 915-5824 davidson@olemiss.edu,
University of Mississippi – Hw
Davidson, Kenneth L., 831-656-2309 kldavids@nps.edu,
Naval Postgraduate School – Am
Davie, Colin, +44 (0) 191 208 6458 colin.davie@ncl.ac.uk,
University of Newcastle Upon Tyne – Ngr
Davies, Andrew J., andrew.j.davies@bangor.ac.uk,
Bangor University – ObZir
Davies, David, +44 (0) 131 451 3569 d.davies@hw.ac.uk,
Heriot-Watt University – Ng
Davies, Gareth, +44(0)7780864555 gareth.davies@open.ac.uk,
The Open University – Ze
Davies, Gareth R., g.r.davies@vu.nl,
Vrije Universiteit Amsterdam – GiCcGf
Davies, Gregory, 609-258-4101 gd3@princeton.edu,
Princeton University – Yg
Davies, J H., huw@earth.cf.ac.uk,
Cardiff University – Yg
Davies, Neil, +44 (0) 1223 333453 nsd27@cam.ac.uk,
University of Cambridge – Gs
Davies, Rhodri, +44 20 759 45722 rhodri.davies@imperial.ac.uk,
Imperial College – Yg
Davies, Sarah, sarah.davies@open.ac.uk,
The Open University – Ze
Davies, Sarah, +440116 252 3624 sjd27@le.ac.uk,
Leicester University – Gs
Davis, Andrew, (773) 702-8164 University of Chicago – Xc
Davis, Arden D., (605) 394-2527 Arden.Davis@sdsmt.edu,
South Dakota School of Mines & Technology – HwNx
Davis, Daniel M., 631-632-8217 daniel.davis@sunysb.edu,
Stony Brook University – Yg
Davis, David A., (775) 682-8767 ddavis@unr.edu,
University of Nevada – Gg
Davis, Donald W., (416) 946-0365 dond@geology.utoronto.ca,
University of Toronto – Cc
Davis, Edward B., 541-346-3461 edavis@uoregon.edu,
University of Oregon – PvePq
Davis, Fred A., (218) 726-8331 fdavis@d.umn.edu,
University of Minnesota, Duluth – CpGi
Davis, George H., (520) 621-1856 gdavis@u.arizona.edu,
University of Arizona – Gc
Davis, Gregory A., (213) 740-6726 gdavis@usc.edu,
University of Southern California – GtcGe
Davis, J. Matthew, (603) 862-1718 matt.davis@unh.edu,
University of New Hampshire – HwYh
Davis, James , jdavis@ldeo.columbia.edu,
Columbia University – Yd
Davis, James A., (217) 581-5528 jadavis2@eiu.edu,
Eastern Illinois University – ZgAm
Davis, James A., (801) 378-3852 james_davis@byu.edu,
Brigham Young University – Zn
Davis, John C., john.davis5@mchsi.com,
University of Kansas – GqoGd
Davis, Kenneth J., (814) 863-8601 davis@met.psu.edu,
Pennsylvania State University, University Park – Am
Davis, Leslie M., ldavis1@austincc.edu,
Austin Community College District – Op
Davis, Owen K., (520) 621-7953 odavis@email.arizona.edu,
University of Arizona – Pl
Davis, P. Thompson, (781) 891-3479 pdavis@bentley.edu,
Bentley University – GlmZc
Davis, Paul M., (310) 825-1343 pdavis@epss.ucla.edu,
University of California, Los Angeles – YsmYg
Davis, Peter B., 253-538-5770 davispb@plu.edu,
Pacific Lutheran University – GcpGz
Davis, R. Laurence, (203) 932-7108 rldavis@newhaven.edu,
University of New Haven – GemHg
Davis, Ralph K., (479) 575-4515 ralphd@comp.uark.edu,
University of Arkansas, Fayetteville – HwGgg
Davis, Robert, (804) 924-0579 red3u@virginia.edu,
University of Virginia – AsmZy
Davis, Steven J., (949) 824-1821 sjdavis@uci.edu,
University of California, Irvine – GeCsZu
Davis, Tara, +353 21 4903696 t.davis@ucc.ie,
University College Cork – Yg
Davis, Trevor J., (801) 587-9019 University of Utah – Zi
Davis, Jr., Richard A., (813) 974-2773 rdavis@chuma.cas.usf.edu,
University of South Florida, Tampa – Gs
Davisson, M L., (925) 423-5993 davisson2@llnl.gov,
Lawrence Livermore National Laboratory – Cg
Dawers, Nancye H., (504) 865-5198 ndawers@tulane.edu,
Tulane University – GcoGm
Dawes, Ralph, 509-682-6754 rdawes@wvc.edu,
Wenatchee Valley College – GgeAm
Dawsey, Cyrus B., (334) 844-3418 dawsecb@auburn.edu,
Auburn University – Zy
Dawson, Jamekia, 205-247-3611 jdawson@gsa.state.al.us,
Geological Survey of Alabama – Hg
Dawson, Jane P., (515) 294-6302 jpdawson@iastate.edu,
Iowa State University of Science & Technology – Gp
Dawson, Mary R., (412) 622-3246 dawsonm@carnegiemnh.org,
Carnegie Museum of Natural History – Pv
Day, Brett, +44 (0)1603 59 1413 brett.day@uea.ac.uk,
University of East Anglia – Ge
Day, James, (858) 534-5431 jmdday@ucsd.edu,
University of California, San Diego – Cg
Day, James E., (309) 438-8678 jeday@ilstu.edu,
Illinois State University – PisPm
Day, Rick L., (814) 863-1615 r4d@psu.edu,
Pennsylvania State University, University Park – Zu
Day, Rosie, +44 (0)121 41 48096 r.j.day@bham.ac.uk,
University of Birmingham – Zn
Day, Stephanie S., 701-231-8837 stephanie.day@ndsu.edu,
North Dakota State University – GmZiu
Day, Steven M., (619) 594-2663 sday@mail.sdsu.edu,
San Diego State University – Ys
Daymon, Deborah J., (505) 667-9021 deba@lanl.gov,
Los Alamos National Laboratory – Zn
Dayton, Paul K., (858) 534-6740 pdayton@ucsd.edu,
University of California, San Diego – Ob
De Angelis, Silvio, +44-151-794-5161 S.De-Angelis@liverpool.ac.uk,
University of Liverpool – Ys
De Batist, Marc, marc.debatist@ugent.be,
Ghent University – GnuGs
De Beus, Barbara, (973) 655-4448 debeusb@mail.montclair.edu,
Montclair State University – Zn
De Campos, Christina, 089/2180 4264 campos@min.uni-muenchen.de,
Ludwig-Maximilians-Universitaet Muenchen – Gz
De Carlo, Eric H., (808) 956-6473 edecarlo@soest.hawaii.edu,
University of Hawai'i, Manoa – Ca
De Carlo, Eric H., (808) 956-5924 edecarlo@soest.hawaii.edu,
University of Hawai'i, Manoa – Cm
De Deckker, Patrick, +61-2-6125 2070 patrick.dedeckker@anu.edu.au,
Australian National University – GuPmGn
de Foy, Benjamin, (314) 977-3122 bdefoy@slu.edu,
Saint Louis University – As
De Grave, Johan, johan.degrave@ugent.be,
Ghent University – Cc
de Hoog, Jan C., +44 (0) 131 650 8525 ceesjan.dehoog@ed.ac.uk,
Edinburgh University – GxzGv
de Hoop, Maarten, (765) 496-6439 mdehoop@math.purdue.edu,
Purdue University – Gq
de Ignacio, Cristina, cris@ucm.es,
Univ Complutense de Madrid – Cg
De Kemp, Eric, 613-947-3738 edekemp@nrcan.gc.ca,
University of Ottawa – ZirZg
De Klerk, Billy, +27 (0)46-622-2312 b.deklerk@ru.ac.za,
Rhodes University – Pg
De Kock, Walter E., (319) 266-6577 University of Northern Iowa – Ze
de la Horra, Raúl, rhorraba@ucm.es,
Univ Complutense de Madrid – Gs
de Menocal, Peter B., (845) 365-8483 peter@ldeo.columbia.edu,
Columbia University – PeOuCg
De Nault, Kenneth J., (319) 273-2033 kenneth.denault@uni.edu,
University of Northern Iowa – Gz
De Santo, Eilzabeth, (717) 358-4555 edesanto@fandm.edu,
Franklin and Marshall College – Zn
de Silva, Shanika, (541) 737-1212 desilvsh@geo.oregonstate.edu,
Oregon State University – Gv

de Stefano, Lucia, luciads@ucm.es,
    Univ Complutense de Madrid – Hw
de Szoeke, Simon P., 541-737-8391 sdeszoek@coas.oregonstate.edu,
    Oregon State University – OpAs
de Uña Alvarez, Elena P., 00 34 988387137 edeuna@uvigo.es,
    Coruna University – Zy
de Vernal, Anne, 514-987-3000 #8599 devernal.anne@uqam.ca,
    Universite du Quebec a Montreal – Pm
de Vicente, Gerardo, gdv@ucm.es,
    Univ Complutense de Madrid – Gt
de Wet, Carol B., 717-358-4388 cdewet@fandm.edu,
    Franklin and Marshall College – GsdGn
de Wit, Cary, (907) 474-7141 cwdewit@alaska.edu,
    University of Alaska, Fairbanks – Zg
de Wit, Julien, (617) 258-0209 jdewit@mit.edu,
    Massachusetts Institute of Technology – Zn
Deakin, Ann K., (716) 673-3303 deakin@fredonia.edu,
    SUNY, Fredonia – ZiGgZr
Deakin, Joann, 520-249-9042 deakinj@cochise.edu,
    Cochise College – Ggg
Deal, Richard, (814) 732-1733 rdeal@edinboro.edu,
    Edinboro University of Pennsylvania – Zi
Dealy, Mike, (316) 943-2343 mdealy@kgs.ku.edu,
    University of Kansas – Gg
Dean, Jeffrey S., (520) 621-2320 jdean@ltrr.arizona.edu,
    University of Arizona – Zn
Dean, John Mark, jmdean36@gmail.com,
    University of South Carolina – Ob
Dean, Robert, 717-245-1109 deanr@dickinson.edu,
    Dickinson College – Gi
Deane, William, (865) 974-2366 wdeane@utk.edu,
    University of Tennessee, Knoxville – Gg
deAngelis, Marie, 718-409-7380 mdeangelis@sunymaritime.edu,
    SUNY, Maritime College – Oc
DeAngelis, Michael T., 569-3542 mtdeangelis@ualr.edu,
    University of Arkansas at Little Rock – GpCp
DeAngelo, Micheal V., (512) 232-3373 michael.deangelo@beg.utexas.edu,
    University of Texas at Austin, Jackson School of Geosciences – Ye
Deans, Jeremy, 601-266-4729 Jeremy.Deans@usm.edu,
    University of Southern Mississippi – YrGc
Dearden, Christopher, +44 0161 306-3911 christopher.dearden@manchester.ac.uk,
    University of Manchester – As
Deardorff, Nicholas, 724-357-2379 n.deardorff@iup.edu,
    Indiana University of Pennsylvania – Gv
Deaton, Tami, (940) 565-2091 tami.deaton@unt.edu,
    University of North Texas – Zn
DeBari, Susan M., (360) 650-3588 debari@wwu.edu,
    Western Washington University – GxZe
Deboo, Phili B., (901) 678-4424 pdeboo@memphis.edu,
    University of Memphis – Pg
DeCaria, Alex J., (717) 871-4739 alex.decaria@millersville.edu,
    Millersville University – As
DeCelles, Peter G., (520) 621-4910 decelles@email.arizona.edu,
    University of Arizona – GtsGc
Decker, Dave, 775.673.7353 dave.decker@dri.edu,
    University of Nevada, Reno – Hw
DeConto, Robert, (413) 545-3426 deconto@geo.umass.edu,
    University of Massachusetts, Amherst – As
DeCosmo, Janice, 206-221-6178 jdecosmo@uw.edu,
    University of Washington – AtZe
Dee, Seth, (775) 682-7704 sdee@unr.edu,
    University of Nevada – Gg
Dee, Sylvia G., 713.348.4889 sylvia.dee@rice.edu,
    Rice University – Zo
Deen, Patricia A., 7607441150 ext. 2512 pdeen@palomar.edu,
    Palomar College – OgZg
Deering, Chad, 906/487-1187 cddeerin@mtu.edu,
    Michigan Technological University – Ggi
DeGaetano, Arthur, (607) 255-0385 atd2@cornell.edu,
    Cornell University – As
Dehler, Carol M., (435) 797-0764 carol.dehler@usu.edu,
    Utah State University – Gs
Dehne, Kevin T., (989) 686-9326 ktdehne@delta.edu,
    Delta College – Ze
Deibert, Jack, (931) 221-6318 deibertj@apsu.edu,
    Austin Peay State University – GsrGo

Deininger, Robert W., (901) 678-2177 University of Memphis – Gx
Deisher, Jeffrey , 740 2012765 jeffrey.deischer@dnr.state.oh.us,
    Ohio Dept of Natural Resources – Eo
DeJesus, Roman, 714-992-7462 rdejesus@fullcoll.edu,
    Fullerton College – Og
DeJong, Benjamin, bdejong@norwich.edu,
    Norwich University – GmHws
Deka, Jennifer, (216) 987-5827 Jennifer.Deka@tri-c.edu,
    Cuyahoga Community College - Western Campus – Zg
Dekens, Petra, (415) 338-6015 dekens@sfsu.edu,
    San Francisco State University – Pe
DeKraker, Dan, (562) 860-2451 x2668 ddekraker@cerritos.edu,
    Cerritos College – ZeOpGg
Delahunty, Tina, (570) 389-5181 tdelahun@bloomu.edu,
    Bloomsburg University – Zri
Delaney, Jeremy S., jsd@eps.rutgers.edu,
    Rutgers, The State University of New Jersey – XgmGx
Delaney, John R., 206-543-4830 jdelaney@ocean.washington.edu,
    University of Washington – GuYr
Delaney, Margaret L., (831) 459-4736 delaney@cats.ucsc.edu,
    University of California, Santa Cruz – Oc
Delano, Helen L., 717.702.2031 hdelano@pa.gov,
    Pennsylvania Bureau of Topographic & Geologic Survey – GgmNg
DeLaune, Paul, 940-552-9941x207 pbdelaune@ag.tamu.edu,
    Texas A&M University – So
Delawder, Sandra, delawdsa@jmu.edu,
    James Madison University – Zn
Deline, Brad, (678) 839-4061 bdeline@westga.edu,
    University of West Georgia – Pi
Dellapenna, Timothy M., (409) 740-4952 dellapet@tamug.tamu.edu,
    Texas A&M University – Ou
Delparte, Donna M., 208-282-4419 delparte@isu.edu,
    Idaho State University – Zi
DelSole, Timothy, (703) 993-5715 tdelsole@gmu.edu,
    George Mason University – As
Delson, Eric, (718) 960-8405 eric.delson@lehman.cuny.edu,
    Graduate School of the City University of New York – PvqPs
Demanet, Laurent, (617) 253-2614 laurent@math.mit.edu,
    Massachusetts Institute of Technology – Zn
DeMaster, David J., (919) 515-7026 david_demaster@ncsu.edu,
    North Carolina State University – OcCbc
Demchak, Jennifer, 570-662-4613 jdemchak@mansfield.edu,
    Mansfield University – Hs
deMenocal, Peter B., peter@ldeo.columbia.edu,
    Columbia University – Pe
Demers, Serge, (418) 723-1986 (Ext. 1483) Universite du Quebec a Rimouski – Ob
DeMets, D. Charles, (608) 262-8598 chuck@geology.wisc.edu,
    University of Wisconsin-Madison – Ym
Demicco, Robert V., (607) 777-2604 demicco@binghamton.edu,
    Binghamton University – Gs
Deming, Jody, jdeming@u.washington.edu,
    University of Manitoba – Zn
Deming, Jody W., (206) 543-0845 jdeming@u.washington.edu,
    University of Washington – Ob
Demlie, Molla, (031) 260-7380 demliem@ukzn.ac.za,
    University of KwaZulu-Natal – HwGeg
Demopoulos, George, (514) 398-4755 McGill University – Nx
DeMott, Benjamin, 212-817-8244 bdemott@gc.cuny.edu,
    Graduate School of the City University of New York – Zn
Dempsey, David P., (415) 338-7716 ddempsey@norte.sfsu.edu,
    San Francisco State University – As
Dempster, Tim, +4401413305445 Tim.Dempster@glasgow.ac.uk,
    University of Glasgow – Gz
Demyanov, Vasily, +44 (0) 131 451 8298 v.demyanov@hw.ac.uk,
    Heriot-Watt University – Zi
Deng, Aijun, (814) 863-8253 axd157@psu.edu,
    Pennsylvania State University, University Park – Am
Deng, Baolin, (505) 835-5505 New Mexico Institute of Mining and Technology – Cg
Deng, Yi, 404-385-1821 yi.deng@eas.gatech.edu,
    Georgia Institute of Technology – AsZoAm
Deng, Youjun, 979-862-8476 yjd@tamu.edu,
    Texas A&M University – GzSc
Dengler, Lorinda, (707) 826-3115 lad1@humboldt.edu,
    Humboldt State University – Yg
Denicourt, Raymond, (813) 974-2236 University of South Florida, Tampa

– Gz

Denizman, Can, (229) 333-5752 Valdosta State University – Hw

Denman, Kenneth L., (250) 363-8230 denmank@ec.gc.ca,
University of Victoria – Op

Denne, Richard A., 817-257-4423 r.denne@tcu.edu,
Texas Christian University – GoPms

Dennett, Keith E., (775) 784-4056 kdennett@unr.edu,
University of Nevada, Reno – Hs

Denning, A. Scott, (970) 491-8359 scott.denning@colostate.edu,
Colorado State University – AsZg

Dennis, Clyde B., (630) 252-5999 cbdennis@anl.gov,
Argonne National Laboratory – Zn

Dennison, Robert, (309) 268-8646 robert.dennison@heartland.edu,
Heartland Community College – Zgn

Denniston, Rhawn F., (319) 895-4306 rdenniston@cornellcollege.edu,
Cornell College – GeCs

Denny, F. B., 618-985-3394 x240 fdenny@illinois.edu,
Illinois State Geological Survey – GzEg

Denny, F. Brett, 618-985-3394 denny@isgs.uiuc.edu,
Illinois State Geological Survey – Gr

Denolle, Marine, (617) 496-2708 mdenolle@fas.harvard.edu,
Harvard University – Ys

Denton, Gary R. W., (671) 735-2690 gdenton@triton.uog.edu,
University of Guam – Ob

Denton, George H., (207) 581-2193 debbies@maine.edu,
University of Maine – Gl

Denton-Hedrick, Meredith Y., mdentonh@austincc.edu,
Austin Community College District – ZeYeEo

Deocampo, Daniel M., 404413-5750 deocampo@gsu.edu,
Georgia State University – Gsn

DePaolo, Donald J., (510) 642-7686 DJDePaolo@lbl.gov,
University of California, Berkeley – Cc

dePolo, Craig M., (775) 682-8770 eq_dude@sbcglobal.net,
University of Nevada – Ng

dePolo, Diane, (775) 784-4976 diane@seismo.unr.edu,
University of Nevada, Reno – Ys

DePriest, Thomas A., (731) 881-7441 tdepriest@utm.edu,
University of Tennessee, Martin – Ze

Derakhshani, Reza, derakhshani@uk.ac.ir,
Shahid Bahonar University of Kerman – GtmGc

Dere, Ashlee L., 402 554 3317 adere@unomaha.edu,
University of Nebraska, Omaha – SdGmCl

Derome, Jacques F., (514) 398-6079 jacques.derome@mcgill.ca,
McGill University – As

Derrick, Sharon, 361-825-3637 Sharon.Derrick@tamucc.edu,
Texas A&M University, Corpus Christi – Zn

Derry, Louis A., lad9@cornell.edu,
Cornell University – Cg

Derstler, Kraig L., (504) 280-6799 kderstle@uno.edu,
University of New Orleans – PgiPv

Desai, Ankur R., (608) 265-9201 desai@aos.wisc.edu,
University of Wisconsin, Madison – Asm

DeSantis, Larisa R., 615-343-7831 larisa.desantis@vanderbilt.edu,
Vanderbilt University – Pve

Desch, Steven, (480) 965-7742 steve.desch@asu.edu,
Arizona State University – Xa

Deschaine, Sylvia, (207) 786-6490 sdescha2@bates.edu,
Bates College – Zn

DeShon, Heather R., 214-768-2916 Southern Methodist University – Ysr

Desrochers, Andre, 613-562-5838 adesro@uottawa.ca,
University of Ottawa – Gs

Dessler, Alexander, (520) 621-4589 dessler@vega.lpl.arizona.edu,
University of Arizona – Zn

Dessler, Andrew E., 979-862-1427 adessler@tamu.edu,
Texas A&M University – As

DeSutter, Tom, 701-231-8690 Thomas.Desutter@ndsu.edu,
North Dakota State University – So

Dethier, David P., (413) 597-2078 david.p.dethier@williams.edu,
Williams College – Gm

Detournay, Emmanuel M., (612) 625-5522 detou001@tc.umn.edu,
University of Minnesota, Twin Cities – NrpNm

Detrich, William, (617) 373-4495 iceman@neu.edu,
Northeastern University – Zn

Dettman, David, (520) 621-4618 dettman@email.arizona.edu,
University of Arizona – Cs

Detwiler, Andrew G., (605) 394-1995 andrew.detwiler@sdsmt.edu,
South Dakota School of Mines & Technology – AspAm

Deuser, Werner G., (508) 289-2551 wdeuser@whoi.edu,
Woods Hole Oceanographic Institution – CsGsu

Devaney, Kathleen, 915-831-5161 kdevaney@epcc.edu,
El Paso Community College – Gg

DeVecchio, Duane, (480) 727-2636 duane.devecchio@asu.edu,
Arizona State University – GtmGc

Devegowda, Deepak, deepak.devegowda@ou.edu,
University of Oklahoma – Np

Dever, Edward P., 541-737-2749 edever@coas.oregonstate.edu,
Oregon State University – Op

Devera, Joseph A., 618-985-3394 devera@isgs.uiuc.edu,
Illinois State Geological Survey – Gr

Devera, Joseph A., 618-985-3394 x241 j-devera@illinois.edu,
Illinois State Geological Survey – Pg

Devery, Dora, (281) 756-5670 ddevery@alvincollege.edu,
Alvin Community College – Gg

Devlahovich, Vincent A., (661) 362-3658 Vincent.Devlahovich@canyons.edu,
College of the Canyons – GgZye

Devlin, J. F., 785-864-4994 jfdevlin@ku.edu,
University of Kansas – Hw

Devol, Allan H., (206) 543-1292 devol@u.washington.edu,
University of Washington – Oc

DeVries-Zimmerman, Suzanne, (616) 395-7297 zimmerman@hope.edu,
Hope College – Gge

Dewaele, Stijn, +32 9 264 4618 stijndg.dewaele@ugent.be,
Ghent University – EgGzCg

Dewdney, Chris, +44 023 92 842417 chris.dewdney@port.ac.uk,
University of Portsmouth – Yg

DeWeese, Georgina G., (678) 839-4065 gdeweese@westga.edu,
University of West Georgia – Zy

DeWekker, Stephan F., (434) 924-3324 sfd3d@virginia.edu,
University of Virginia – As

Dewers, Thomas, 505-845-0631 tdewers@sandia.gov,
Sandia National Laboratory – Yx

deWet, Andrew P., (717) 291-3815 andy.dewet@fandm.edu,
Franklin and Marshall College – Ge

Dewey, Janet, (307) 223-2265 jdewey2@uwyo.edu,
University of Wyoming – CaGg

Dewey, John, jfdewey@ucdavis.edu,
University of California, Davis – Gt

Dewey, Richard K., (250) 472-4009 rdewey@uvic.ca,
University of Victoria – Opg

Dezzani, Raymond, 208-885-7360 dezzani@uidaho.edu,
University of Idaho – RhZnn

Dhar, Ratan K., 718-262-2889 rdhar@york.cuny.edu,
York College (CUNY) – Hw

Dharmasoth, Mrs M., +914027097116 manju.geophysics@gmail.com,
Osmania University – YmeYg

Di Fiori, Sara, (310) 660-3593 sdifiori@elcamino.edu,
El Camino College – GgOg

Di Girolamo, Larry, (217) 333-3080 gdi@illinois.edu,
University of Illinois, Urbana-Champaign – Zr

Di Gregorio, Monica, +44(0) 113 34 31592 m.digregorio@leeds.ac.uk,
University of Leeds – Eg

Di Lorenzo, Emanuele, (404) 894-3994 edl@gatech.edu,
Georgia Institute of Technology – OpZnn

Di Toro, Giulio, +390498279105 giulio.ditoro@unipd.it,
Università degli Studi di Padova – Gc

DiBacco, Claudio, (902) 426-9778 Claudio.DiBacco@dfo-mpo.gc.ca,
Dalhousie University – Ob

Dibb, Jack E., 603-862-3063 jack.dibb@unh.edu,
University of New Hampshire – AcClAs

Dibben, Chris, +44 (0) 131 650 2552 chris.dibben@ed.ac.uk,
Edinburgh University – Yn

DiBiase, Roman , 814-865-7388 rdibiase@psu.edu,
Pennsylvania State University, University Park – Gmt

Dick, Henry J B., (508) 289-2590 hdick@whoi.edu,
Woods Hole Oceanographic Institution – GitGc

Dick, Richard P., 614-247-7605 dick.78@osu.edu,
Ohio State University – Sb

Dick, Warren A., (330) 263-3877 dick.5@osu.edu,
Ohio State University – SbCbHs

Dickens, Gerald R., 713.348.5130 jerry@rice.edu,
Rice University – CgCg

Dickerson, Richard E., (310) 825-5864 University of California, Los Angeles – Zn

Dickhoff, Willem H., (314) 935-4169 wimd@howdy.wustl.edu,

455

Washington University in St. Louis – Zn
Dickin, Alan P., (905) 525-9140 (Ext. 24365) dickin@mcmaster.ca, McMaster University – Cc
Dickinson, Robert E., 512-232-7933 robted@jsg.utexas.edu, University of Texas, Austin – As
Dickinson, William R., (520) 299-5220 wrdickin@dakotacom.net, University of Arizona – Gt
Dickman, Steven R., (607) 777-2857 dickman@binghamton.edu, Binghamton University – Yg
Dickson, Andrew G., (858) 822-2990 adickson@ucsd.edu, University of California, San Diego – Cm
Dickson, Loretta D., (570) 484-2068 ldickson@lockhaven.edu, Lock Haven University – GizGe
Dickson, Stephen M., (207) 287-7174 stephen.m.dickson@maine.gov, Dept of Agriculture, Conservation, and Forestry – OnGuOu
Didlake, Anthony, 814-863-4636 acd171@psu.edu, Pennsylvania State University, University Park – As
Diebold, Frank E., (406) 496-4215 Montana Tech of the University of Montana – Cl
Diecchio, Richard J., (703) 993-5394 rdiecchi@gmu.edu, George Mason University – Gr
Dieckmann, Melissa S., (859) 622-1274 melissa.dieckmann@eku.edu, Eastern Kentucky University – ZeCl
Diederichs, Mark, 613-533-6504 diederim@queensu.ca, Queen's University – Nr
Diefendorf, Aaron, (513) 556-3787 aaron.diefendorf@uc.edu, University of Cincinnati – Co
Diem, Jeremy E., (404) 413-5770 jdiem@gsu.edu, Georgia State University – At
Diemer, John A., 704-687-5994 jadiemer@email.uncc.edu, University of North Carolina, Charlotte – Gs
Diener, Johann, 021-650-2925 johann.diener@uct.ac.za, University of Cape Town – Gp
Diercks, Arne R., 662.915.2301 arne.diercks@usm.edu, University of Southern Mississippi – Ou
Dierssen, Heidi, 860-405-9239 heidi.dierssen@uconn.edu, University of Connecticut – Op
Dieterich, James H., (951) 827-2976 james.dieterich@ucr.edu, University of California, Riverside – Yg
Dieterle, Dwight A., (727) 553-1114 ddieterle@marine.usf.edu, University of South Florida – Ob
Dietl, Gregory, 607-273-6623 gpd3@cornell.edu, Paleontological Research Institution – Pg
Dietrich, William E., (510) 642-2633 bill@geomorph.berkeley.edu, University of California, Berkeley – Gm
Dietsch, Craig, (513) 556-2547 craig.dietsch@uc.edu, University of Cincinnati – Gp
Dietz, Anne D., 210-486-0470 adietz@alamo.edu, San Antonio Community College – PgZg
Dietz, Richard D., (970) 351-2950 richard.dietz@unco.edu, University of Northern Colorado – Xg
Dietze, Michael, dietze@bu.edu, Boston University – SfZr
Diffendal, Jr., Robert F., 402-472-7546 rfd@unl.edu, Unversity of Nebraska - Lincoln – GrPiRh
Diggins, Thomas P., 330-941-3605 tpdiggins@ysu.edu, Youngstown State University – Pg
Dighe, Kalpak, (505) 665-1701 kdighe@lanl.gov, Los Alamos National Laboratory – Zn
Dijkstra, Arjan, +44 1752 584774 arjan.dijkstra@plymouth.ac.uk, University of Plymouth – Gi
Dijkstra, Semme, 603-862-0525 semmed@ccom.unh.edu, University of New Hampshire – Og
Dilcher, David L., 3523921721 x460 dilcher@flmnh.ufl.edu, University of Florida – Pb
Dilcher, David L., (352) 392-1721 dilcher@flmnh.ufl.edu, University of Florida – Pb
Dilek, Yildirm, (513) 529-2212 dileky@MiamiOh.edu, Miami University – Gt
Dilles, John H., (541) 737-1245 dillesj@geo.oregonstate.edu, Oregon State University – Em
DiMarco, Steven F., (979) 862-4168 sdimarco@tamu.edu, Texas A&M University – OpnOg
DiMichele, William A., (202) 633-1319 Smithsonian Inst / Natl Museum of Natural Hist– Pb
Dimmick, Charles W., (860) 832-2936 dimmick@ccsu.edu, Central Connecticut State University – Ge

Dimova, Natasha T., 205-348-0256 ntdimova@ua.edu, University of Alabama – HwCcm
Ding, Kang, 612-626-1860 mlcd@umn.edu, University of Minnesota, Twin Cities – Cg
Dingman, S. Lawrence, (603) 862-1718 University of New Hampshire – Hq
Dingus, Delmar D., (805) 756-2753 ddingus@calpoly.edu, California Polytechnic State University – So
Dingwell, Donald Bruce, 089/21804136 dingwell@min.uni-muenchen.de, Ludwig-Maximilians-Universitaet Muenchen – Giv
Dinklage, William, 805-730-4114 wsdinklage@sbcc.edu, Santa Barbara City College – Gge
Dinsmoor, John C., (702) 295-6189 john_dinsmoor@lanl.gov, Los Alamos National Laboratory – Nm
Dintaman, Christopher, (812) 856-5654 cdintama@indiana.edu, Indiana University – Hw
Dinter, David A., (801) 581-7937 david.dinter@utah.edu, University of Utah – GtcGu
Diochon, Amanda, (807) 343-8444 adiochon@lakeheadu.ca, Lakehead University – Ge
DiPietro, Joseph A., (812) 465-7041 dipietro@usi.edu, University of Southern Indiana – GcpGm
Dipple, Gregory M., (604) 827-0653 gdipple@eoas.ubc.ca, University of British Columbia – Gp
Dirmeyer, Paul A., (703) 993-5363 pdirmeye@gmu.edu, George Mason University – As
Ditmars, John, (630) 252-5953 Argonne National Laboratory – Zn
Dittmer, Eric, (541) 552-6496 dittmer@sou.edu, Southern Oregon University – Gg
DiVenere, Victor, (516) 299-2318 divenere@liu.edu, Long Island University, C.W. Post Campus – Ym
Diver, Kim, 860-685-2610 kdiver@wesleyan.edu, Wesleyan University – ZiyZg
Divine, Aaron, (928) 523-7835 Aaron.Divine@nau.edu, Northern Arizona University – ZnnZn
Dix, George R., (613) 520-2600 george.dix@carleton.ca, Carleton University – Gsd
Dix, Justin, +44 (0)23 8059 3057 j.k.dix@soton.ac.uk, University of Southampton – GuYuNx
Dixon, Clifton, 601-266-4731 c.dixon@usm.edu, University of Southern Mississippi – Zn
Dixon, Jean, (406) 994-3342 jean.dixon@montana.edu, Montana State University – GmScZg
Dixon, John C., (479) 422-2115 jcdixon@uark.edu, University of Arkansas, Fayetteville – GmXgSo
Dixon, John M., john.dixon@queensu.ca, Queen's University – Gct
Dixon, Michael, (519) 824-4120 Ext.52555 dixon@ces.uoguelph.ca, University of Guelph – Zn
Dixon, Nick, +44(0) 113 34 34931 fuensd@leeds.ac.uk, University of Leeds – Zu
Dixon, P. Grady, (785) 628-4536 pgdixon@fhsu.edu, Fort Hays State University – AmZyAs
Dladla, Nonkululeko, (031) 260-2801 dladlan2@ukzn.ac.za, University of KwaZulu-Natal – GusGr
Dmochowski, Jane, 215-5735388 janeed@sas.upenn.edu, University of Pennsylvania – Yg
Doar, III, William R., (803) 609-7065 doarw@dnr.sc.gov, South Carolina Dept of Natural Resources – GrdGp
Dobbie, Steven, +44(0) 113 34 36725 j.s.e.dobbie@leeds.ac.uk, University of Leeds – As
Dobbs, Fred C., (757) 683-4301 fdobbs@odu.edu, Old Dominion University – Ob
Dobler, Scott, (270) 745-7078 scott.dobler@wku.edu, Western Kentucky University – As
Dobrzhinetskaya, Larissa F., (951) 827-2028 larissa@ucrac1.ucr.edu, University of California, Riverside – Gx
Dobson, David, +44 020 7679 32398 d.dobson@ucl.ac.uk, University College London – Gz
Dobson, David M., (336) 316-2278 ddobson@guilford.edu, Guilford College – GusPc
Dockal, James A., (563) 845-1034 dockal@uncw.edu, University of North Carolina, Wilmington – GdcEm
Dodd, J. Robert, (516) 632-8204 dodd@indiana.edu, Indiana University, Bloomington – Pg
Dodd, Justin P., 815-753-7949 jdodd@niu.edu, Northern Illinois University – Cs

Dodge, Clifford H., 717.702.2036 cdodge@pa.gov,
Pennsylvania Bureau of Topographic & Geologic Survey – GrEcRh
Dodge, Rebecca L., rebecca.dodge@msutexas.edu,
Midwestern State University – Zre
Dodge, Stefanie, (414) 324-5676 stefanie.dodge@uwex.edu,
University of Wisconsin, Extension – Gg
Dodson, Peter, (215) 898-8784 dodsonp@vet.upenn.edu,
University of Pennsylvania – Pv
Doh, Seong-Jae, 82-2-3290-3173 sjdoh@korea.ac.kr,
Korea University – YmgYe
Doheny, Edward L., (215) 898-6085 University of Pennsylvania – Ng
Doherty, Cheryl, (302) 831-2569 sparrow@udel.edu,
University of Delaware – Zn
Doherty, David J., (313) 577-2506 ah0654@wayne.edu,
Wayne State University – GgEoGx
Doherty, Ruth, +44 (0) 131 650 6759 ruth.doherty@ed.ac.uk,
Edinburgh University – As
Dolakova, Nela, +420 549 49 3542 nela@sci.muni.cz,
Masaryk University – Plg
Dolan, James F., (213) 740-8599 dolan@usc.edu,
University of Southern California – Gt
Dolejs, David, +420-221951525 Charles University – GiCpg
Dolgoff, Anatole, (212) 346-1502 Pace University, New York Campus – Zg
Doll, William E., d8e@ornl.gov,
Oak Ridge National Laboratory – Yg
Dollase, Wayne A., (310) 825-5823 dollase@ucla.edu,
University of California, Los Angeles – Gz
Dollhopf, Douglas J., (406) 944-5594 dollhopf@montana.edu,
Montana State University – Zn
Dolliver, Holly A., holly.dolliver@uwrf.edu,
University of Wisconsin, River Falls – GmSd
Dolman, Han J., +31 20 5987358 han.dolman@vu.nl,
Vrije Universiteit Amsterdam – CbAmHg
Dolman, Paul, +44 (0)1603 59 3175 p.dolman@uea.ac.uk,
University of East Anglia – Zu
Domack, Cynthia R., (315) 859-4710 cdomack@hamilton.edu,
Hamilton College – Pg
Domagall, Abigail M., (605) 642-6506 abigail.domagall@bhsu.edu,
Black Hills State University – GveZe
Dombard, Andrew J., (312) 996-9206 adombard@uic.edu,
University of Illinois at Chicago – Xy
Domber, Steven E., (609) 984-6587 steven.domber@dep.nj.gov,
New Jersey Geological and Water Survey – Hw
Domingo, Laura, ldomingo@ucm.es,
Univ Complutense de Madrid – CsPe
Domingue, Carla, (225) 388-8407 carla@lgs.bri.lsu.edu,
Louisiana State University – Zn
Dominguez, Francina, 217-265-5483 francina@illinois.edu,
University of Illinois, Urbana-Champaign – AsHg
Dominic, David F., 937 775-2201 david.dominic@wright.edu,
Wright State University – GsrRc
Donaghay, Percy, (401) 874-6944 donaghay@gso.uri.edu,
University of Rhode Island – Ob
Donahoe, Robert J., (505) 667-7603 Los Alamos National Laboratory – Zn
Donahoe, Rona J., 205-348-1879 rdonahoe@geo.ua.edu,
University of Alabama – Cl
Donald, Roberta, (778) 782-4925 robbie_donald@sfu.ca,
Simon Fraser University – GsZe
Donaldson, Alan C., (304) 293-5603 adonalds@wvu.edu,
West Virginia University – Gs
Donaldson, Paul R., (208) 426-3639 pdonalds@boisestate.edu,
Boise State University – Ye
Dondofema, Farai, +27 15 962 80044 farai.dondofema@univen.ac.za,
University of Venda for Science & Technology – ZrHgZi
Doner, Lisa A., 603-535-2245 ladoner@plymouth.edu,
Plymouth State University – PcGes
Doney, Scott, 434-924-0570 scd5c@virginiae.edu,
University of Virginia – Oc
Dong, Charles, (310) 660-3593 El Camino College – Og
Dong, Hailiang, (513) 529-2517 dongh@MiamiOh.edu,
Miami University – Cg
Dong, Pinliang, (940) 565-2377 pinliang.dong@unt.edu,
University of North Texas – Zir
Dong, Yunpeng, ydong265@uwo.ca,
Western University – GtcCg
Donnelly, Jeffrey, 508-289-2994 jdonnelly@whoi.edu,
Woods Hole Oceanographic Institution – Gu

Donnelly, Shanon P., 330 972-7630 sd51@uakron.edu,
University of Akron – Zi
Donoghue, Michael J., 203-432-1935 michael.donoghue@yale.edu,
Yale University – Pbg
Donohue, Kathleen, (401) 874-6615 kdonohue@gso.uri.edu,
University of Rhode Island – Op
Donohue, Stephen J., (540) 231-9740 donohue@vt.edu,
Virginia Polytechnic Institute & State University – Sc
Donovan, James, (801) 585-3029 james.donovan@utah.edu,
University of Utah – NmrZr
Donovan, Jeff C., (727) 553-1116 jdonovan@marine.usf.edu,
University of South Florida – Op
Donovan, Joseph J., (304) 293-5603 donovan@geo.wvu.edu,
West Virginia University – Hq
Donovan, R N., (817) 257-7214 r.donovan@tcu.edu,
Texas Christian University – GdcGt
Doolan, Barry L., (802) 656-0248 bdoolan@uvm.edu,
University of Vermont – Gx
Dooley, Brett, bdooley@ph.vccs.edu,
Patrick Henry Community College – Gg
Dooley, John H., (609) 292-2576 john.dooley@dep.state.nj.us,
New Jersey Geological and Water Survey – CgGfe
Dooley, Tim, 512-471-8261 tim.dooley@beg.utexas.edu,
University of Texas at Austin, Jackson School of Geosciences – Gc
Dorais, Michael J., (801) 422-1347 dorais@byu.edu,
Brigham Young University – Gi
Dorale, Jeffrey A., (319) 335-0822 jeffrey-dorale@uiowa.edu,
University of Iowa – Cs
Doran, Peter, pdoran@lsu.edu,
Louisiana State University – Hws
Doran, Peter T., (225) 578-3955 pdoran@lsu.edu,
University of Illinois at Chicago – HswAm
Dorfman, Alexander, 089/2180 4275 Ludwig-Maximilians-Universitaet Muenchen – Gz
Dorfner, Thomas, 089/2180 4278 dorfner@min.uni-muenchen.de,
Ludwig-Maximilians-Universitaet Muenchen – Gz
Dorling, Stephen, +44 (0)1603 59 2533 s.dorling@uea.ac.uk,
University of East Anglia – Am
Dormaar, John, (403) 327-4561 University of Lethbridge – So
Dorman, Clive E., (619) 594-5707 cdorman@mail.sdsu.edu,
San Diego State University – OpAmt
Dorman, James, (901) 678-2007 dorman@comcast.net,
University of Memphis – Yse
Dorman, James, (901) 678-4753 dorman@comcast.net,
University of Memphis – Ys
Dorman, Leroy M., (858) 534-2406 ldorman@ucsd.edu,
University of California, San Diego – Yg
Dornbos, Stephen Q., (414) 229-6630 sdornbos@uwm.edu,
University of Wisconsin, Milwaukee – PeiPo
Dorries, Alison M., (505) 665-6952 adorries@lanl.gov,
Los Alamos National Laboratory – Zn
Dorsey, Rebecca J., (541) 346-4431 rdorsey@uoregon.edu,
University of Oregon – Gr
Dort, Jr., Wakefield, (785) 864-4974 University of Kansas – Gm
dos Santos, Telmo B., tmsantos@fc.ul.pt,
Universidade de Lisboa – GptGi
Doser, Diane I., 915-747-5851 doser@utep.edu,
University of Texas, El Paso – Ys
Doskey, Paul V., (630) 252-7662 pvdoskey@anl.gov,
Argonne National Laboratory – As
Doss, Paul K., (812) 465-7132 pdoss@usi.edu,
University of Southern Indiana – HwGeZu
Dosso, Stanley E., (250) 472-4341 sdosso@uvic.ca,
University of Victoria – Op
Dostal, Jarda, (902) 420-5747 Dalhousie University – Cg
Doster, Florian, +44 (0)131 451 3171 f.doster@hw.ac.uk,
Heriot-Watt University – Hg
Dostie, Philip, (207) 786-6485 pdostie@bates.edu,
Bates College – Cl
Doty, Thomas, (401) 254-3066 Roger Williams University – Ob
Dou, Fugen, (409) 752-2741 ext. 2223 f-dou@aesrg.tamu.edu,
Texas A&M University – So
Doughty, Alice, 207-786-6113 adoughty@bates.edu,
Bates College – Gl
Dougil, Andy, +44(0) 113 34 36782 a.j.dougill@leeds.ac.uk,
University of Leeds – So
Douglas, Arthur V., (402) 280-2464 sonora@creighton.edu,

Creighton University – Zg
Douglas, Bruce, 812-855-3848 douglasb@indiana.edu,
  Indiana University, Bloomington – Gc
Douglas, Ellen, ellen.douglas@umb.edu,
  University of Massachusetts, Boston – HwqHy
Douglas, Peter, 514-398-3677 peter.douglas@mcgill.ca,
  McGill University – Cs
Douglass, Daniel C., (617) 373-4381 d.douglass@northeastern.edu,
  Northeastern University – GlSdPc
Douglass, David N., (626) 585-7036 dndouglass@pasadena.edu,
  Pasadena City College – Ze
Douglass, Gordon, 801-538-4810 gdouglass@utah.gov,
  Utah Geological Survey – Zi
Douilly, Roby, (951) 827-3180 roby.douilly@ucr.edu,
  University of California, Riverside – YsGtYg
Dove, Patricia M., (540) 231-2444 pdove@vt.edu,
  Virginia Polytechnic Institute & State University – Cl
Doveton, John H., (785) 864-2100 doveton@kgs.ku.edu,
  University of Kansas – Go
Doveton, John H., 785-864-3965 doveton@kgs.ku.edu,
  University of Kansas – Gq
Dowd, John F., (706) 542-2383 Jdowd@uga.edu,
  University of Georgia – Hy
Dowdy, Robert H., (612) 625-7058 bdowdy@soils.umn.edu,
  University of Minnesota, Twin Cities – Sc
Dower, John, 250-721-6120 dower@uvic.ca,
  University of Victoria – Ob
Dowling, Carolyn B., 765-285-8274 cbdowling@bsu.edu,
  Ball State University – HgGeCg
Downes, Hilary, +44 020 3073 8027 h.downes@ucl.ac.uk,
  Birkbeck College – Cg
Downie, Helen, helen.downie@manchester.ac.uk,
  University of Manchester – Ge
Downs, Robert T., (520) 626-8092 rdowns@email.arizona.edu,
  University of Arizona – Gi
Drake, John C., (802) 656-0244 jdrake@uvm.edu,
  University of Vermont – Cl
Drake, Lon D., (319) 335-1826 University of Iowa – Hy
Drake, Luke, 818-677-3508 luke.drake@csun.edu,
  California State University, Northridge – Zi
Drake, Michael J., (520) 621-6962 drake@lpl.arizona.edu,
  University of Arizona – Xc
Drake, Simon, +440203 073 8024 drakesimon1@gmail.com,
  Birkbeck College – Gv
Draper, Grenville, (305) 348-3087 draper@fiu.edu,
  Florida International University – GctRh
Drazen, Jeffrey C., (808) 956-6567 jdrazen@hawaii.edu,
  University of Hawai'i, Manoa – Ob
Dreesen, Donald S., (505) 667-1913 dreesen@lanl.gov,
  Los Alamos National Laboratory – Np
Dreger, Douglas S., (510) 643-1719 dreger@seismo.berkeley.edu,
  University of California, Berkeley – Ys
Drescher, Andrew, (612) 625-2374 University of Minnesota, Twin Cities
  – So
Dreschoff, Gisela, (785) 864-4517 University of Kansas – Ce
Dressel, Waldemar M., (573) 341-4616 Missouri University of Science and
  Technology – Gg
Drew, Douglas A., (404) 496-4202 ddrew@mtech.edu,
  Montana Tech of the University of Montana – Cg
Driese, Steven G., 254-710-2361 steven_driese@baylor.edu,
  University of Tennessee, Knoxville – SaClGs
Driese, Steven G., (254) 710-2194 steven_driese@baylor.edu,
  Baylor University – SaClGs
Driever, Steven L., (816) 235-2971 drievers@umkc.edu,
  University of Missouri, Kansas City – ZnnZn
Drijfhout, Sybren, +44 (0)23 8059 6202 S.S.Drijfhout@soton.ac.uk,
  University of Southampton – Op
Dripps, Weston R., (864) 294-3392 weston.dripps@furman.edu,
  Furman University – HwsGe
Driscoll, John R., (607) 753-2926 driscollj@cortland.edu,
  SUNY, Cortland – Zn
Driscoll, Neal W., (858) 822-5026 ndriscoll@ucsd.edu,
  University of California, San Diego – Gu
Driscoll, Peter E., 202-478-8827 pdriscoll@carnegiescience.edu,
  Carnegie Institution for Science – GtYm
Drobniak, Agnieszka, (812) 855-2687 agdrobni@indiana.edu,
  Indiana University – Ec

Drobny, Gerry, (916) 691-7204 Cosumnes River College – Gg
Droop, Giles, +44 0161 275-3809 giles.droop@manchester.ac.uk,
  University of Manchester – Gp
Droser, Mary L., 951-827-3797 mary.droser@ucr.edu,
  University of California, Riverside – Pg
Droxler, Andre W., 713.348.4885 andre@rice.edu,
  Rice University – Gs
Druckenbrod, Daniel L., 609-895-5422 ddruckenbrod@rider.edu,
  Rider University – AtZu
Druckenmiller, Patrick S., 907-474-6954 ffpsd@uaf.edu,
  University of Alaska, Fairbanks – Pv
Druet, María, mdruetve@ucm.es,
  Univ Complutense de Madrid – Yr
Druffel, Ellen R. M., (949) 824-2116 edruffel@uci.edu,
  University of California, Irvine – Oc
Druguet Tantiña, Elena, ++935813730 elena.druguet@uab.cat,
  Universitat Autonoma de Barcelona – Gc
Druhan, Jennifer, (217) 300-0142 jdruhan@illinois.edu,
  University of Illinois, Urbana-Champaign – HwCs
Drummond, Jane, +4401413304208 Jane.Drummond@glasgow.ac.uk,
  University of Glasgow – Zi
Drysdale, Jessica, (508) 289-3739 jdrysdale@whoi.edu,
  Woods Hole Oceanographic Institution – Oc
Drzewiecki, Peter A., (860) 465-4322 drzewieckip@easternct.edu,
  Eastern Connecticut State University – GrsGd
Drzyzga, Scott A., (717) 477-1307 sadrzy@ship.edu,
  Shippensburg University – ZiuGl
Duaime, Terence E., (406) 496-4157 tduaime@mtech.edu,
  Montana Tech of The University of Montana – Hws
Duan, Benchuan, (979) 845-3297 bduan@tamu.edu,
  Texas A&M University – YsgGq
Duan, Quingyun, (925) 422-7704 duan2@llnl.gov,
  Lawrence Livermore National Laboratory – Hg
Duan, Shuiwang, (301) 405-2407 sduan@umd.edu,
  University of Maryland – HsZu
Dubba, Dr. Vijay K., +919550627034 ddvkumar7@gmail.com,
  Osmania University – YevYs
Dube, Benoit, (418) 654-2669 bdube@nrcan.gc.ca,
  Universite du Quebec – Em
Dubey, C S., (981) 107-4867 csdubey@gmail.com,
  University of Delhi – GexGt
Dubsky, Scott, 719-333-3080 scott.dubsky@usafa.edu,
  United States Air Force Academy – ZyGm
Duby, Paul F., (212) 854-2928 pfd1@columbia.edu,
  Columbia University – Nx
Duce, Robert A., (979) 845-5756 rduce@ocean.tamu.edu,
  Texas A&M University – Oc
Ducea, Mihai N., (520) 621-5171 ducea@email.arizona.edu,
  University of Arizona – Gt
Ducey, Mark, 603-862-4429 mark.ducey@unh.edu,
  University of New Hampshire – Zc
Duchane, David V., (505) 667-9893 duchane@lanl.gov,
  Los Alamos National Laboratory – Zn
Duchesne, Josee, (418) 656-2177 duchesne@ggl.ulaval.ca,
  Universite Laval – Gz
Duchesne, Rocio, duchesnr@uww.edu,
  University of Wisconsin, Whitewater – Zri
Ducklow, Hugh, 845-365-8167 hducklow@ldeo.columbia.edu,
  Columbia University – ObCb
Duckstein, Lucien, (520) 621-2274 duckstein@engref.fr,
  University of Arizona – Hq
Duckworth, Kenneth, (403) 220-7375 University of Calgary – Ye
Duda, Jan-Peter, +49 551 397960 jan-peter.duda@geo.uni-goettingen.de,
  Georg-August University of Goettingen – PgCoGs
Dudley, Lynn M., 850-644-4214 dudley@gly.fsu.edu,
  Florida State University – Sc
Dudley, Walter, 808-933-3905 dudley@hawaii.edu,
  University of Hawai'i, Manoa – OuRn
Dudukovic, Mike, dudu@che.wustl.edu,
  Washington University in St. Louis – Zn
Duebendorfer, Ernest M., (928) 523-7510 ernie.d@nau.edu,
  Northern Arizona University – Gct
Dueker, Kenneth G., (307) 766-2657 dueker@uwyo.edu,
  University of Wyoming – Ys
Duex, Timothy W., (337) 482-6222 twd7693@louisiana.edu,
  University of Louisiana at Lafayette – HgGex
Dufek, Josef, 541-346-4788 jdufek@uoregon.edu,

University of Oregon – Gv
Dufek, Josef, (404) 984-9472 josef.dufek@eas.gatech.edu,
 Georgia Institute of Technology – GvZnn
Duff, John, john.duff@umb.edu,
 University of Massachusetts, Boston – ZnnZn
Duffey, Patricia, 785-628-5389 pduffey@fhsu.edu,
 Fort Hays State University – Zn
Duffield, Wendell A., (928) 523-4852 Northern Arizona University – Gv
Duffy, Clarence J., (505) 667-5154 Los Alamos National Laboratory – Cg
Duffy, Tara , (617) 373-6323 t.duffy@neu.edu,
 Northeastern University – Ob
Duffy, Thomas S., (609) 258-6769 duffy@princeton.edu,
 Princeton University – GyYgZm
Dugan, Brandon, (713) 348-5088 dugan@mines.edu,
 Colorado School of Mines – GuYrHw
Dugas, Bernard, 514 987-3000 #5319 dugas.bernard@uqam.ca,
 Universite du Quebec a Montreal – As
Dugmore, Andrew J., +44 (0) 131 650 8156 andrew.dugmore@ed.ac.uk,
 Edinburgh University – ZycGa
Duhamel, Solange, sduhamel@ldeo.columbia.edu,
 Columbia University – Pe
Duke, Edward F., (605) 394-2388 Edward.Duke@sdsmt.edu,
 South Dakota School of Mines & Technology – GpiZr
Duke, Jason E., (931) 372-3121 jduke@tntech.edu,
 Tennessee Tech University – Zi
Duke, Norman A., (519) 661-3199 nduke@uwo.ca,
 Western University – Eg
Dulai, Henrietta, (808) 956-0720 hdulaiov@hawaii.edu,
 University of Hawai'i, Manoa – CcHw
Duley, James W., 573-368-2105 bill.duley@dnr.mo.state,
 Missouri Dept of Natural Resources – Hy
Dulin, Shannon, (405) 325-5488 sdulin@ou.edu,
 University of Oklahoma – YmGs
Duller, Rob, +44-151-794-5179 Robert.Duller@liverpool.ac.uk,
 University of Liverpool – Gvs
Dumas, Simone, (613)562-5800 6230 dumas@uottawa.ca,
 University of Ottawa – GsOg
Dumitriu, Tony C., 0040756584797 dtonycristian@yahoo.com,
 Alexandru Ioan Cuza – GcZir
Dumond, Gregory, 479-575-3411 gdumond@uark.edu,
 University of Arkansas, Fayetteville – GtcGg
Dunagan, Stan P., (731) 881-7437 sdunagan@utm.edu,
 University of Tennessee, Martin – GsSa
Dunbar, John A., (254) 710-2191 john_dunbar@baylor.edu,
 Baylor University – YgHg
Dunbar, Nelia W., (575) 835-5783 nelia@nmbg.nmt.edu,
 New Mexico Institute of Mining & Technology – Gi
Dunbar, Robert B., (650) 725-6830 dunbar@pangea.stanford.edu,
 Stanford University – On
Duncan, Ian, 512-471-5117 ian.duncan@beg.utexas.edu,
 University of Texas at Austin, Jackson School of Geosciences – Gg
Duncan, Ian, rduncan@ccsf.edu,
 City College of San Francisco – Zy
Duncan, Ian J., 512-471-5117 ian.duncan@beg.utexas.edu,
 University of Texas, Austin – GeCe
Duncan, Stewart, 785532277 sduncan@ksu.edu,
 Kansas State University – So
Duncan Tabb, Neva, 721-791-2758 Duncantabb.Neva@spcollege.edu,
 Saint Petersburg College, Clearwater – Am
Dundas, Robert G., (559) 278-6984 rdundas@csufresno.edu,
 California State University, Fresno – PvGr
Dunfield, Kari, (519) 824-4120 x58088 dunfield@uoguelph.ca,
 University of Guelph – SbZn
Dunkle, Kallina M., (931) 221-7451 dunklek@apsu.edu,
 Austin Peay State University – HwGle
Dunlap, Dallas B., (512) 471-4858 dallas.dunlap@beg.utexas.edu,
 University of Texas at Austin, Jackson School of Geosciences – Ye
Dunlevey, John, +27 (0)15 268 3483 john.dunlevey@ul.ac.za,
 University of Limpopo – GzxEg
Dunn, Allison L., (508) 929-8641 adunn@worcester.edu,
 Worcester State University – AsZyHg
Dunn, Bernie, 519-850-2362 bdunn@uwo.ca,
 Western University – Ye
Dunn, John Todd, (506) 453-4804 University of New Brunswick – Gi
Dunn, Richard K., (802) 485-2304 rdunn@norwich.edu,
 Norwich University – GalGs
Dunn, Robert A., (808) 956-3728 dunnr@hawaii.edu,
 University of Hawai'i, Manoa – Yrs
Dunn, Steven R., 413-538-2531 sdunn@mtholyoke.edu,
 Mount Holyoke College – Gp
Dunn, Tasha L., (207) 859-5808 tldunn@colby.edu,
 Colby College – XmGiz
Dunne, George C., (818) 677-2511 george.dunne@csun.edu,
 California State University, Northridge – Gc
Dunne, William M., (423) 974-4161 wdunne@utk.edu,
 University of Tennessee, Knoxville – Gc
Dunning, Gregory R., (709) 864-8481 gdunning@mun.ca,
 Memorial University of Newfoundland – CcGit
Dunning, Jeremy D., (812) 856-4448 dunning@indiana.edu,
 Indiana University, Bloomington – Gc
Dunst, Brian J., 412-442-4230 bdunst@pa.gov,
 Pennsylvania Bureau of Topographic & Geologic Survey – EgoHw
Dunton, Karen K., (309) 438-7640 kkedunt@ilstu.edu,
 Illinois State University – Zn
Dunton, Kenneth H., (361) 749-6744 ken.dunton@utexas.edu,
 University of Texas, Austin – Ob
Dupont, Todd, (513) 529-9734 dupontt@miamioh.edu,
 Miami University – Gl
Dupont, Todd, 949-824-1133 tdupont@uci.edu,
 University of California, Irvine – Gl
Dupraz, Christophe, 860-486-1394 christophe.dupraz@uconn.edu,
 University of Connecticut – Pg
Dupre, William R., 713-743-4987 wdupre@uh.edu,
 University of Houston – On
Dura, Christina, tinadura@vt.edu,
 Virginia Polytechnic Institute & State University – OnRn
Durand, Michael T., (614)247-4835 durand.8@osu.edu,
 Ohio State University – HsZr
Durbin, Edward G., (401) 874-6850 edurbin@gso.uri.edu,
 University of Rhode Island – Ob
Durbin, James, 812-465-1208 jdurbin@usi.edu,
 University of Southern Indiana – Gm
Durham, Lisa, 630-252-3170 ladurham@anl.gov,
 Argonne National Laboratory – Hw
Durham, William, 617-253-5810 wbdurham@mit.edu,
 Massachusetts Institute of Technology – Cg
Durkee, Josh, joshua.durkee@wku.edu,
 Western Kentucky University – AsZy
Durkee, Philip A., 831-656-2517 durkee@nps.edu,
 Naval Postgraduate School – AmZr
Durland, Theodore, 541-737-5058 tdurland@coas.oregonstate.edu,
 Oregon State University – Yg
DuRoss, Christopher B., (801) 537-3348 christopherduross@utah.gov,
 Utah Geological Survey – Ng
Durran, Dale R., (206) 543-7440 drdee@uw.edu,
 University of Washington – As
Durrant, Jeffrey O., (801) 422-4116 jodurrant@byu.edu,
 Brigham Young University – Ge
Durucan, Sevket, +44 20 759 47354 s.durucan@imperial.ac.uk,
 Imperial College – NmEcNr
Dushaw, Brian D., (206) 685-4198 dushaw@apl.washington.edu,
 University of Washington – Op
Dutch, Steven I., (920) 465-2371 dutchs@uwgb.edu,
 University of Wisconsin, Green Bay – GcgGe
Dutcher, Allen, adutcher1@unl.edu,
 University of Nebraska, Lincoln – As
Dutkiewicz, Stephanie, 617-253-8039 stephd@mit.edu,
 Massachusetts Institute of Technology – Cm
Dutrieux, Pierre, pierred@ldeo.columbia.edu,
 Columbia University – Op
Dutro, Thomas, (202) 633-1322 Smithsonian Inst / Natl Mus of Natl Hist– Pi
Dutrow, Barbara L., (225) 388-2525 dutrow@geol.lsu.edu,
 Louisiana State University – Gz
Dutta, Prodip K., (812) 237-2268 Prodip.Dutta@indstate.edu,
 Indiana State University – Gs
Dutton, Andrea, 352-392-3626 adutton@ufl.edu,
 University of Florida – CcGsPc
Dutton, Jessica, duttonj@sunysuffolk.edu,
 Suffolk County Community College, Ammerman Campus – Zg
Dutton, John A., 814-865-1534 dutton@ems.psu.edu,
 Pennsylvania State University, University Park – Am
Dutton, Shirley P., 512-471-0329 shirley.dutton@beg.utexas.edu,
 University of Texas, Austin – Gs

Duvall, Alison, 206-211-8311 aduvall@uw.edu,
  University of Washington – GmcGt
Duxbury, Thomas C., (818) 354-4301 Jet Propulsion Laboratory – Zn
Dvorzak, Marie, (608) 262-8956 mdvorzak@geology.wisc.edu,
  University of Wisconsin-Madison – Zn
Dworkin, Stephen I., (254) 710-2186 steve_dworkin@baylor.edu,
  Baylor University – ClGd
Dwyer, Daryl F., 419-5302661 daryl.dwyer@utoledo.edu,
  University of Toledo – Zn
Dwyer, Gary S., (919) 681-8164 gsd3@duke.edu,
  Duke University – Gs
Dyab, Ahmed I., 002-03-3921595 seawolf_5000@yahoo.com,
  Alexandria University – Gg
Dyaur, Nikolay, 713-743-6539 ndyaur@uh.edu,
  University of Houston – Yxs
Dye, David H., 901-678-3330 daviddye@memphis.edu,
  University of Memphis – Ga
Dyer, Jamie, (662) 268-1032 Ext 220 jamie.dyer@msstate.edu,
  Mississippi State University – Am
Dyer, Jen, +44(0) 113 34 39086 j.dyer@leeds.ac.uk,
  University of Leeds – Ge
Dygert, Nick, 865-974-6022 ndygert1@utk.edu,
  University of Tennessee, Knoxville – Giz
Dyhrman, Sonya, 845-365-8165 sdyhrman@ldeo.columbia.edu,
  Columbia University – Ob
Dymek, Robert F., (314) 935-5344 bob_d@levee.wustl.edu,
  Washington University in St. Louis – GpiGa
Dymond, Randel, (540) 231-9962 dymond@vt.edu,
  Virginia Polytechnic Institute & State University – HqsZi
Dymond, Salli F., (218) 726-6582 dymon003@umn.edu,
  University of Minnesota, Duluth – Ge
Dyson, Miranda, +44 (0)1908 653398 x 53398 miranda.dyson@open.ac.uk,
  The Open University – Pg
Dzunic, Aleksandar, +61 8 9266 2297 A.Dzunic@curtin.edu.au,
  Curtin University – YgsYe

E

Eads, Tracy, 513-529-3515 eadsts@miamioh.edu,
  Miami University – Hw
Earls, Sandy, 405-744-6358 sandy.earls@okstate.edu,
  Oklahoma State University – Zn
Earls, Stephanie, (360) 902-1473 stephanie.earls@dnr.wa.gov,
  Washington Geological Survey – Gge
Early, Thomas O., eot@ornl.gov,
  Oak Ridge National Laboratory – Cl
Earman, Sam, (717) 871-4336 searman@millersville.edu,
  Millersville University – HwCsl
Easson, Gregory L., (662) 915-5995 geasson@olemiss.edu,
  University of Mississippi – Zr
Easterbrook, Don J., 360 319-5389 don.easterbrook@wwu.edu,
  Western Washington University – GlmGe
Easterday, Cary, cary.easterday@bellevuecollege.edu,
  Bellevue College – GgPg
Eastin, Matt, 704-687-5914 mdeastin@uncc.edu,
  University of North Carolina, Charlotte – Am
Eastler, Thomas E., 207-778-7401 eastler@maine.edu,
  University of Maine - Farmington – GeZrGg
Eastoe, Chris J., (520) 791-7430 eastoe@email.arizona.edu,
  University of Arizona – CsGg
Eaton, Jeffrey G., jeaton@weber.edu,
  Weber State University – PsGs
Eaton, Lewis S., (540) 568-3339 eatonls@jmu.edu,
  James Madison University – Gm
Eaton, Timothy, 718-997-3335 timothy.eaton@qc.cuny.edu,
  Graduate School of the City University of New York – Zg
Eaton, Timothy, 718-997-3327 timothy.eaton@qc.cuny.edu,
  Queens College (CUNY) – HgwGg
Ebel, Denton S., (212) 769-5381 debel@amnh.org,
  American Museum of Natural History – XmCgZe
Ebel, John E., (617) 552-3399 ebel@bc.edu,
  Boston College – Ys
Ebel, John E., 617-552-8319 ebel@bc.edu,
  Boston College – Yg
Eberhardt, Erik, (604) 827-5573 eeberhardt@eos.ubc.ca,
  University of British Columbia – Ng
Eberhart, Mark E., (303) 273-3726 meberhar@mines.edu,
  Colorado School of Mines – Zm

Eberle, Bill, (785) 532-6101 weberle@ksu.edu,
  Kansas State University – SoZu
Eberle, Jaelyn J., 303-492-8069 jaelyn.eberle@colorado.edu,
  University of Colorado – Pv
Eberle, Jaelyn J., jaelyn.eberle@colorado.edu,
  University of Colorado – Pve
Eberli, Gregor P., (305) 421-4678 geberli@rsmas.miami.edu,
  University of Miami – GusYx
Ebert, David, 831-771-4400 debert@mlml.calstate.edu,
  Moss Landing Marine Laboratories – Zn
Ebert, James R., (607) 436-3065 james.ebert@oneonta.edu,
  SUNY, Oneonta – GrsZe
Eberth, David A., 403-823-7707 Royal Tyrrell Museum of Palaeontology – Gs
Ebiner, Matt, (310) 660-3593 mebiner@elcamino.edu,
  El Camino College – Zy
Ebinger, Michael H., (505) 667-3147 mhe@lanl.gov,
  Los Alamos National Laboratory – So
Eble, Cortland F., 8592575500 x 149 eble@uky.edu,
  University of Kentucky – Pl
Eble, Cortland F., 859-323-0540 eble@uky.edu,
  University of Kentucky – Pl
Eby, G. Nelson, (978) 934-3907 nelson_eby@uml.edu,
  University of Massachusetts, Lowell – CgGiCt
Eby, Gloria J., (972) 883-24004 gloria.eby@utdallas.edu,
  University of Texas, Dallas – Zn
Eby, Stephanie, (617) 373-4380 s.eby@neu.edu,
  Northeastern University – Zn
Echols, John B., (225) 388-8573 john@lgs.bri.lsu.edu,
  Louisiana State University – Go
Eckert, Donald J., (614) 292-9048 eckert.1@osu.edu,
  Ohio State University – Sc
Eckert, James O., (203) 432-3181 xwn6w3i02@sneakemail.com,
  Yale University – Gp
Economos, Rita C., 214-768-3313 reconomos@smu.edu,
  Southern Methodist University – CsGc
Eddy, Michael P., (609) 258-4101 meddy@princeton.edu,
  Princeton University – Cc
Edegbai, Aitalokhai J., aitalokhai.edegbai@uniben.edu,
  University of Benin – GosPi
Edgar, Dorland E., (630) 252-7596 Argonne National Laboratory – Gm
Edgcomb, Virginia, 508-289-3734 vedgcomb@whoi.edu,
  Woods Hole Oceanographic Institution – ObZc
Edgell, Dennis J., (910) 521-6479 dennis.edgell@uncp.edu,
  University of North Carolina, Pembroke – As
Edil, Tuncer B., (608) 262-3225 edil@engr.wisc.edu,
  University of Wisconsin, Madison – Ng
Edlund, Mark B., mbedlund@smm.org,
  University of Minnesota, Twin Cities – Pe
Edmonds, Douglas A., (812) 855-4512 edmondsd@indiana.edu,
  Indiana University, Bloomington – Gsm
Edmonds, Marie, +44 (0) 1223 333463 medm06@esc.cam.ac.uk,
  University of Cambridge – Gv
Edson, Carol, 925-424-1336 cedson@laspositascollege.edu,
  Las Positas College – Gg
Edson, James, james.edson@uconn.edu,
  University of Connecticut – Op
Edwards, Benjamin R., (717) 254-8934 edwardsb@dickinson.edu,
  Dickinson College – GivCg
Edwards, C L, (505) 667-8464 cledwards@lanl.gov,
  Los Alamos National Laboratory – Yg
Edwards, Dianne, +44(0)29 208 74264 edwardsd2@cardiff.ac.uk,
  Cardiff University – Pb
Edwards, Dianne, edwardsd2@cf.ac.uk,
  Cardiff University – Pb
Edwards, Margo H., (808) 956-5232 margo@soest.hawaii.edu,
  University of Hawai'i, Manoa – Gu
Edwards, Nicholas, +44 0161 275-3810 nicholas.edwards@manchester.ac.uk,
  University of Manchester – Pg
Edwards, R. Lawrence, 612-626-0207 edwar001@umn.edu,
  University of Minnesota, Twin Cities – Cc
Edwards, Richard N., (416) 978-2267 edwards@core.physics.utoronto.ca,
  University of Toronto – Yr
Edwards, Robin, robin.edwards@tcd.ie,
  Trinity College – Ge
Edwards, Stuart, +44 (0) 191 208 3986 stuart.edwards@ncl.ac.uk,

University of Newcastle Upon Tyne – Gq
Effert-Fanta, Shari E., 217-244-2192 sfanta@illinois.edu, Illinois State Geological Survey – Cas
Efurd, Deward W., (505) 667-2437 Los Alamos National Laboratory – Ct
Egan, Stuart, (+44) 01782 733174 s.s.egan@keele.ac.uk, Keele University – Gtc
Egbert, Gary D., (541) 737-2947 egbert@coas.oregonstate.edu, Oregon State University – Yr
Egbeyemi, Oladimeji R., 08167781509 oregbeyemi@futa.edu.ng, Federal University of Technology, Akure – GgCg
Egenhoff, Sven O., (970) 491-0749 sven@warnercnr.colostate.edu, Colorado State University – Gsu
Egerton, Victoria, victoria.egerton@manchester.ac.uk, University of Manchester – Pg
Eggers, Delores M., (828) 251-6654 eggers@unca.edu, University of North Carolina, Asheville – Zn
Eggler, David H., (814) 571-1960 dhe1@psu.eu, Pennsylvania State University, University Park – CpGiv
Eggleston, Carrick M., (307) 766-6769 carrick@uwyo.edu, University of Wyoming – ClqCb
Eggleston, David B., (919) 515-7840 eggleston@ncsu.edu, North Carolina State University – Ob
Eguiluz, Luis, luis.eguiluz@ehu.eus, University of the Basque Country UPV/EHU – Ggt
Ehinola, Olugbenga A., +2348033819066 oa.ehinola@mail.ui.edu.ng, University of Ibadan – GoeCo
Ehler, Stan, (785) 532-6101 swehler@ksu.edu, Kansas State University – So
Ehlers, Todd A., (734) 763-5112 tehlers@umich.edu, University of Michigan – Yg
Ehlmann, Bethany L., 626.395.6720 ehlmann@caltech.edu, California Institute of Technology – XgZrGz
Eichenbaum, Jack, (718) 961-8406 jaconet@aol.com, Hunter College (CUNY) – Zie
Eicher, Don L., (303) 492-6187 University of Colorado – Pm
Eichhubl, Peter, (512) 475-8829 peter.eichhubl@beg.utexas.edu, University of Texas at Austin, Jackson School of Geosciences – Gc-CgNr
Eick, Matthew J., (540) 231-8943 eick@vt.edu, Virginia Polytechnic Institute & State University – Sc
Eide, Elizabeth A., (202) 334-2392 eeide@nas.edu, National Academy of Sciences, Engineering, and Medicine – GogGt
Eiler, John M., (626) 395-6239 eiler@gps.caltech.edu, California Institute of Technology – Cg
Eilon, Zach, eilon@geol.ucsb.edu, University of California, Santa Barbara – Ys
Einaudi, Marco T., (650) 723-0575 marco@pangea.stanford.edu, Stanford University – Em
Eisenberger, Peter M., (917) 568-6915 petere@ldeo.columbia.edu, Columbia University – Zn
Eisenhart, Karen, (814) 732-2398 keisenhart@edinboro.edu, Edinboro University of Pennsylvania – Zy
Eisenhart, Ralph, (717) 846-7788 York College of Pennsylvania – Gg
Eivazi, Frieda, 573-681-5459 University of Missouri, Columbia – Sc
Eklund, Olav, +358442956429 olav.eklund@abo.fi, Abo Akademi University – GxeGz
Ekmekci, Mehmet, +90 312 2977730 ekmekci@hacettepe.edu.tr, Hacettepe University – HwyCs
Ekstrom, Goran, ekstrom@ldeo.columbia.edu, Columbia University – Yg
Ekstrom, Goran, (845) 365-8427 ekstrom@ldeo.columbia.edu, Columbia University – Ys
El Dakkak, Mohamed W., 002-03-3921595 waguihm@yahoo.com, Alexandria University – GgsGd
El Kadiri, Racha, 615-484-7641 racha.elkadiri@mtsu.edu, Middle Tennessee State University – Hg
El Masri, Bassil, (270) 809-3110 belmasri@murraystate.edu, Murray State University – Zc
El Shazly, Aley, (304) 696-6756 elshazly@marshall.edu, Marshall University – GpiCg
El-Baz, Farouk, (617) 353-5081 farouk@bu.edu, Boston University – Zr
El-Kadi, Aly I., (808) 956-6331 elkadi@hawaii.edu, University of Hawai'i, Manoa – Hw
El-Monier, Ilham, ilham.el-monier@ou.edu, University of Oklahoma – Np
Elachi, Charles, 626.395.8858 charles.elachi@jpl.nasa.gov, California Institute of Technology – Xg
Elam, Tim, 206-685-3092 wtelam@apl.uw.edu, University of Washington – Cag
Elder, Kathryn L., (508) 289-2513 kelder@whoi.edu, Woods Hole Oceanographic Institution – Cs
Elders, Wilfred A., wilfred.elders@ucr.edu, University of California, Riverside – Gi
Eldredge, Niles, (212) 769-5723 American Museum of Natural History – Pi
Eldridge, Jayson, (812) 856-2382 jayeldri@indiana.edu, Indiana University – EoGr
Elhusseiny, Ammar, +966138602625 ammar.elhusseiny@kfupm.edu.sa, King Fahd University of Petroleum and Minerals – Yg
Elia, Gaetano, +44 (0) 191 208 7934 University of Newcastle Upon Tyne – Ng
Elias, Robert J., (204) 474-8862 eliasrj@ms.umanitoba.ca, University of Manitoba – Pi
Elick, Jennifer M., 570-372-4214 elick@susqu.edu, Susquehanna University – Pe
Elkins, Joe T., 970-351-3060 joe.elkins@unco.edu, University of Northern Colorado – ZeGgCl
Elkins, Lynne J., (402) 472-7563 lelkins@unl.edu, University of Nebraska, Lincoln – GivCt
Eller, Phillip G., (505) 667-5830 Los Alamos National Laboratory – Zn
Ellett, Kevin M., 812-856-3671 kmellett@indiana.edu, Indiana University – YeHw
Ellingson, Don, 503-838-8865 ellingd@wou.edu, Western Oregon University – AmZgXg
Ellins, Katherine K., (512) 232-3251 kellins@ig.utexas.edu, University of Texas, Austin – Zn
Elliot, David H., (614) 292-5076 elliot.1@osu.edu, Ohio State University – GxtGv
Elliott, Brent, 512-471-1812 brent.elliott@beg.utexas.edu, University of Texas, Austin – GzcGt
Elliott, C. James, (505) 667-9517 cje@lanl.gov, Los Alamos National Laboratory – Zn
Elliott, Colleen G., (406) 496-4143 celliott@mtech.edu, Montana Tech of The University of Montana – GctGg
Elliott, David K., (928) 523-7188 david.elliott@nau.edu, Northern Arizona University – Pv
Elliott, Emily M., (412) 624-8780 eelliott@pitt.edu, University of Pittsburgh – Cbs
Elliott, Julie, (765) 496-7206 julieelliott@purdue.edu, Purdue University – Yd
Elliott, Monty A., (541) 552-6482 Southern Oregon University – Gr
Elliott, Scott M., (505) 667-0949 sme@lanl.gov, Los Alamos National Laboratory – As
Elliott, Susan J., (519) 888-4567 (Ext. 33923) elliotts@uwaterloo.ca, McMaster University – Ge
Elliott, W. Crawford, (404) 413-5756 wcelliott@gsu.edu, Georgia State University – Clc
Elliott, Jr., William S., (812) 228-5053 wselliott@usi.edu, University of Southern Indiana – GsrCl
Ellis, Andre, aellis3@calstatela.edu, California State University, Los Angeles – HwCg
Ellis, Hugh, (901) 678-2007 hugh.ellis@jhu.edu, Johns Hopkins University – Zn
Ellis, Kathryn E., (859) 323-0504 kathryn.ellis@uky.edu, University of Kentucky – Zn
Ellis, Robert M., (604) 822-6574 rellis@eos.ubc.ca, University of British Columbia – Ys
Ellis, Scott R., 217-265-5105 ellis3@illinois.edu, Illinois State Geological Survey – Ge
Ellis, Trevor, 573-368-2153 Trevor.Ellis@dnr.mo.gov, Missouri Dept of Natural Resources – Gg
Ellison, Anne L., 217-244-2502 aerdmann@illinois.edu, University of Illinois – Ge
Ellsworth, Timothy R., (217) 333-2055 University of Illinois, Urbana-Champaign – Sp
Ellsworth, William L., (650) 723-9390 wellsworth@stanford.edu, Stanford University – Ys
Ellwood, Brooks B., (225) 578-3416 ellwood@lsu.edu, Louisiana State University – Ym
Elmore, R. Douglas, (405) 325-3253 delmore@ou.edu, University of Oklahoma – YmGs
Elnadi, Abdelhalim H., geology@uofk.edu, University of Khartoum – Goz
Eloranta, Edwin W., (608)262-7327 eloranta@lidar.ssec.wisc.edu,

University of Wisconsin, Madison – Yr
Elrick, Maya, (505) 277-5077 dolomite@unm.edu,
University of New Mexico – Gs
Elrick, Scott D., (217) 333-3222 elrick@illinois.edu,
University of Illinois – EcGgs
Elsberry, Russell L., 831-656-2373 elsberry@nps.edu,
Naval Postgraduate School – Am
Elsen, Jan, jan.elsen@ees.kuleuven.be,
Katholieke Universiteit Leuven – Gz
Elswick, Erika R., (812) 855-2493 eelswick@indiana.edu,
Indiana University, Bloomington – Cl
Elueze, A. A., aa.elueze@mail.ui.edu.ng,
University of Ibadan – Eg
Elwood Madden, Andrew S., (405) 325-3253 amadden@ou.edu,
University of Oklahoma – ClZaCb
Ely, Lisa L., (509) 963-2177 ely@cwu.edu,
Central Washington University – Gm
Elzinga, Evert J., 973-353-5238 elzinga@newark.rutgers.edu,
Rutgers, The State University of New Jersey, Newark – Sc
Emanuel, Kerry A., (617) 253-2462 emanuel@mit.edu,
Massachusetts Institute of Technology – AmRn
Emerick, Christina M., (206) 543-2491 tina@ocean.washington.edu,
University of Washington – Ou
Emerson, Cheryl R., (309) 694-5373 Illinois Central College – Ze
Emerson, Norlene, (608) 647-6186 x109 norlene.emerson@uwc.edu,
University of Wisconsin Colleges – GgPiGs
Emerson, Steven R., (206) 543-0428 emerson@u.washington.edu,
University of Washington – OcCmOg
Emery, Joshua, (865) 974-2366 jemery2@utk.edu,
University of Tennessee, Knoxville – Xg
Emile-Geay, Julien, 213-740-2945 julieneg@usc.edu,
University of Southern California – Oc
Emmer, Barbara, 089/2180 4231 Ludwig-Maximilians-Universitaet Muenchen – Yg
Emmett, Chad, (801) 422-7886 chad_emmett@byu.edu,
Brigham Young University – Zn
Emmi, Philip C., (801) 581-5562 pcemmi@geog.utah.edu,
University of Utah – Zu
Emofurieta, Williams O., wemofu@uniben.edu,
University of Benin – GzCeGe
Emry, Robert, (202) 633-1323 Smithsonian Inst / Natl Museum of Natural Hist– Pv
Encarnacion, John, (314) 977-3119 encarnjp@eas.slu.edu,
Saint Louis University – Gx
Enderlin, Milton, 817 257 5318 m.enderlin@tcu.edu,
Texas Christian University – Nr
Enders, Merritt S., 303-273-3066 menders@mines.edu,
Colorado School of Mines – Gzg
Endres, Anthony E., (519) 888-4567 (Ext. 3552) University of Waterloo – Pe
Engebretson, David C., (360) 650-3595 david.engebretson@wwu.edu,
Western Washington University – Gt
Engel, Anga, mpeschke@geomar.de,
SUNY, Stony Brook – Oc
Engel, Annette S., 865-974-2366 aengel1@utk.edu,
University of Tennessee, Knoxville – Cl
Engel, Michael H., (405) 325-4435 ab1635@ou.edu,
University of Oklahoma – Co
Engel, Richard E., (406) 994-5295 rengel@montana.edu,
Montana State University – Sc
Engelder, Terry, (814) 865-3620 engelder@geosc.psu.edu,
Pennsylvania State University, University Park – Nr
Engelhard, Simon, (401) 874-2187 engelhart@uri.edu,
University of Rhode Island – Gg
Engelmann, George F., 402-554-4804 gengelmann@unomaha.edu,
University of Nebraska, Omaha – PvGdr
England, John, (780) 492-5673 john.england@ualberta.ca,
University of Alberta – Gml
England, Phillip, +44 (1865) 272000 philip@earth.ox.ac.uk,
University of Oxford – Gt
England, Richard, +440116 252 3522 hodgeology@le.ac.uk,
Leicester University – Yg
Englebright, Stephen C., (631) 632-8230 Stony Brook University – Ze
English, David C., (727) 553-1503 denglish@marine.usf.edu,
University of South Florida – Op
Engstrom, Daniel R., (612) 433-5953 (Ext. 18) dre@umn.edu,
University of Minnesota, Twin Cities – Pe

Engstrom, Ryan, rengstro@gwu.edu,
George Washington University – Zr
Enos, Paul, (785) 864-9714 enos@ku.edu,
University of Kansas – GsdGu
Enright, Richard L., (508) 531-1390 enright@bridgew.edu,
Bridgewater State University – HgZrEg
Ensign, Todd I., 304-367-8438 tensign@fairmontstate.edu,
Fairmont State University – Ze
Entekhabi, Dara, (617) 253-9698 darae@mit.edu,
Massachusetts Institute of Technology – Hg
Epifanio, Craig, cepi@tamu.edu,
Texas A&M University – As
Eppelbaum, Lev, +97236405086 levap@post.tau.ac.il,
Tel Aviv University – YvmGt
Eppes, Martha C., 704-687-5993 meppes@uncc.edu,
University of North Carolina, Charlotte – So
Epstein, Howard E., (434) 924-4308 hee2b@virginia.edu,
University of Virginia – Zn
Erdmer, Philippe, (403) 492-2676 philippe.erdmer@ualberta.ca,
University of Alberta – Gc
Erdner, Deana L., (361) 749-6719 derdner@utexas.edu,
University of Texas, Austin – Ob
Erhardt, Andrea M., 859-257-6931 andrea.erhardt@uky.edu,
University of Kentucky – CgsCm
Erickson, Ben A., (801) 537-3379 benerickson@utah.gov,
Utah Geological Survey – Ng
Erickson, J. Mark, (315) 379-5198 St. Lawrence University – Pgi
Erickson, Melinda L., 763-783-3231 merickso@usgs.gov,
University of Minnesota, Twin Cities – Hw
Erickson, Rolfe C., (707) 664-2334 rolfe.erickson@sonoma.edu,
Sonoma State University – Gx
Eriksen, Charles C., (206) 543-6528 eriksen@u.washington.edu,
University of Washington – Op
Eriksson, Kenneth A., (540) 231-4680 kaeson@vt.edu,
Virginia Polytechnic Institute & State University – Gs
Erisman, Brad, 361-749-6833 berisman@utexas.edu,
University of Texas, Austin – Ob
Ernst, W. Gary, (650) 723-2750 ernst@pangea.stanford.edu,
Stanford University – Cp
Erski, Theodore, (815) 455-8992 terski@mchenry.edu,
McHenry County College – Zg
Erslev, Eric A., (970) 231-2654 eric.erslev@colostate.edu,
Colorado State University – GctGg
Ersoy, Erkal, +44 (0)131 451 4777 erkal.ersoy@pet.hw.ac.uk,
Heriot-Watt University – Eo
Ertel-Ingrisch, Werner, 089/2180 4275 ertel@min.uni-muenchen.de,
Ludwig-Maximilians-Universitaet Muenchen – Gz
Erturk, Mehmet Ali, 00904242370000-5981 maerturk@firat.edu.tr,
Firat University – GiCc
Erwin, Diane, (510) 642-3921 dmerwin@berkeley.edu,
University of California, Berkeley – Pb
Erwin, Douglas H., (202) 633-1324 Smithsonian Inst / Natl Museum of Natural Hist– Pi
Esbaugh, Andrew J., (361) 749-6835 a.esbaugh@austin.utexas.edu,
University of Texas, Austin – Ob
Escobar-Wolf, Rudiger, 906/487-2128 rpescoba@mtu.edu,
Michigan Technological University – Gv
Escuer, Joan, joan.escuer@uab.cat,
Universitat Autonoma de Barcelona – Gg
Eshleman, Keith N., (301) 689-7170 eshleman@al.umces.edu,
University of Maryland – Hg
Esling, Steven P., (618) 453-7376 esling@siu.edu,
Southern Illinois University Carbondale – HwGl
Esnault, Melissa H., (225) 578-5320 mesnau1@lsu.edu,
Louisiana State University – Zn
Esparza, Francisco, fesparz@cicese.mx,
Centro de Investigación Científica y de Educación Superior de Ensenada – Ye
Esser, Corinne, (530) 752-3668 caesser@ucdavis.edu,
University of California, Davis – Zn
Essery, Richard L., +44 (0) 131 651 9093 Richard.Essery@ed.ac.uk,
Edinburgh University – As
Essling, Alice M., (630) 252-3493 Argonne National Laboratory – Ca
Esslinger, Kelly L., kelly.esslinger@azwestern.edu,
Arizona Western College – AsOpGg
Estalrich López, Joan, ++935811270 joan.estalrich@uab.cat,
Universitat Autonoma de Barcelona – HgGe

Estapa, Margaret, 518-589-5477 mestapa@skidmore.edu,
Skidmore College – Oc
Esteban, Jose Julian, jj.esteban@ehu.es,
University of the Basque Country UPV/EHU – GtCcYg
Ethridge, Frank G., (970) 491-6195 fredpet@cnr.colostate.edu,
Colorado State University – Gs
Ettensohn, Frank R., (859) 257-1401 f.ettensohn@uky.edu,
University of Kentucky – PsGdPi
Ettlie, Bradley, 217-265-6543 ettlie76@illinois.edu,
Illinois State Geological Survey – Ge
Eusden, J. Dykstra, (207) 786-6015 deusden@bates.edu,
Bates College – Gc
Evangelou, V. P., (515) 294-9237 Iowa State University of Science & Technology – Sc
Evanoff, Emmett, 970-351-2647 Emmett.Evanoff@unco.edu,
University of Northern Colorado – GrPgGs
Evans, Barry M., (814) 863-3531 bmel@psu.edu,
Pennsylvania State University, University Park – Sd
Evans, Bernard W., 206-543-1163 bwevans@uw.edu,
University of Washington – GpzGi
Evans, Claude, (314) 935-6684 Washington University in St. Louis – Zn
Evans, David A., (203) 432-3127 david.evans@yale.edu,
Yale University – YmGt
Evans, David C., davide@rom.on.ca,
Royal Ontario Museum – PvgPq
Evans, David G., (916) 278-7840 dave_evans@csus.edu,
California State University, Sacramento – HqYg
Evans, Diane L., (818) 354-2418 Jet Propulsion Laboratory – Gm
Evans, Eileen L., (818) 677-5026 eileen.evans@csun.edu,
California State University, Northridge – YdGtYg
Evans, J. B., (617) 253-2856 brievans@mit.edu,
Massachusetts Institute of Technology – Yx
Evans, James E., (419) 372-2414 evansje@bgsu.edu,
Bowling Green State University – GsHsGe
Evans, James P., (435) 797-1267 james.evans@usu.edu,
Utah State University – Gc
Evans, Jenni L., (814) 865-3240 evans@meteo.psu.edu,
Pennsylvania State University, University Park – Am
Evans, Kevin R., (417) 836-5590 kevinevans@missouristate.edu,
Missouri State University – GrsGt
Evans, Krista M., (417) 836-5688 KristaEvans@MissouriState.edu,
Missouri State University – Zn
Evans, Les J., (519) 824-4120 (Ext. 53017) levans@lrs.uoguelph.ca,
University of Guelph – Cl
Evans, Mark, evansmaa@ccsu.edu,
Central Connecticut State University – GciGz
Evans, Michael E., (780) 492-5517 evans@phys.ualberta.ca,
University of Alberta – Ym
Evans, Michael N., (301) 405-8763 mnevans@umd.edu,
University of Maryland – PeAsCs
Evans, Owen C., (631) 632-8061 owen.evans@sunysb.edu,
Stony Brook University – Cg
Evans, Robert L., (508) 289-2673 revans@whoi.edu,
Woods Hole Oceanographic Institution – Yx
Evans, Thomas J., (608) 263-4125 tevans@wisc.edu,
University of Wisconsin, Extension – Gg
Evanylo, Gregory K., (540) 231-9739 gevanylo@vt.edu,
Virginia Polytechnic Institute & State University – Sc
Eveleth, Robert W., (505) 835-5325 beveleth@gis.nmt.edu,
New Mexico Institute of Mining & Technology – Nm
Evenson, Edward B., (610) 758-3659 ebe0@lehigh.edu,
Lehigh University – Gl
Everett, Mark, (979) 862-2129 everett@geo.tamu.edu,
Texas A&M University – Ym
Eversoll, Duane A., (402) 472-7524 deversoll2@unl.edu,
Unversity of Nebraska - Lincoln – Ng
Eves, Robert L., (435) 586-1934 eves@suu.edu,
Southern Utah University – CgZeCa
Ewart, Terry E., (206) 543-1327 ewart@apl.washington.edu,
University of Washington – Op
Ewas, Galal A., 002-03-3921595 goueiss@gmail.com,
Alexandria University – GgNr
Ewing, Rodney C., (650) 497-6203 rewing1@stanford.edu,
Stanford University – GzeCl
Eyles, Carolyn H., (905) 525-9140 (Ext. 24077) eylesc@mcmaster.ca,
McMaster University – GslGm
Eyles, John D., (905) 525-9140 (Ext. 23152) eyles@mcmaster.ca,
McMaster University – Ge
Eyles, Nicholas, 416-287-7195 eyles@utsc.utoronto.ca,
University of Toronto – Gl
Eyre, Bradley, bradley.eyre@scu.edu.au,
Southern Cross University – Gs
Ezerskis, John L., (216) 268-0493 john.ezerskis@tri-c.edu,
Cuyahoga Community College - Western Campus – ZgGcHw

F

Faatz, Renee M., (435) 283-7519 renee.faatz@snow.edu,
Snow College – Gg
Fabbri, Paolo, +390498279124 paolo.fabbri@unipd.it,
Università degli Studi di Padova – Hw
Fabel, Derek, +4401413305473 Derek.Fabel@glasgow.ac.uk,
University of Glasgow – Xg
FABRE, Cecile, cecile.fabre@univ-lorraine.fr,
Universite de Lorraine - Faculte des Sciences et Technologies – Gg
Fabry, Frederic, (514) 398-7733 frederic.fabry@staff.mcgill.ca,
McGill University – As
Fabryka-Martin, June T., (505) 665-2300 Los Alamos National Laboratory – Hg
Faccenda, Manuele, +390498279159 manuele.faccenda@unipd.it,
Università degli Studi di Padova – Gq
Fadem, Cynthia M., (765) 983-1231 fademcy@earlham.edu,
Earlham College – GaSoCs
Fadiman, Maria, 561 297-3314 mfadiman@fau.edu,
Florida Atlantic University – Zn
Fadiya, S. L., 234-803-332-0230 fadiyalawrence@yahoo.co.uk,
Obafemi Awolowo University – God
Fagan, Amy, 828 227 3820 alfagan@wcu.edu,
Western Carolina University – GiXg
Fagel, Nathalie, +32 4 3662209 Nathalie.Fagel@ulg.ac.be,
Universite de Liege – GsuGe
Fagents, Sarah, 808-956-3163 fagents@higp.hawaii.edu,
University of Hawai'i, Manoa – Gv
Fagherazzi, Sergio, (617) 353-2092 sergio@bu.edu,
Boston University – GmOnGs
Fagnan, Brian A., 605-394-6652 brian.fagnan@state.sd.us,
South Dakota Dept of Environment and Natural Resources – Gg
Fahes, Mashhad, mashhad.fahes@ou.edu,
University of Oklahoma – Np
Fahnestock, Mark A., 603-862-0322 mark.fahnestock@unh.edu,
University of New Hampshire – Gl
Fahrenbach, Mark D., 605-394-6830 mark.fahrenbach@state.sd.us,
South Dakota Dept of Environment and Natural Resources – Gg
Fairbairn, David, +44 (0) 191 208 6353 david.fairbairn@ncl.ac.uk,
University of Newcastle Upon Tyne – Gq
Fairchild, Ian, +44 (0)121 414 4181 i.j.fairchild@bham.ac.uk,
University of Birmingham – GsClGm
Fairchild, Jane, jane.fairchild@uwc.edu,
University of Wisconsin Colleges – Am
Fairley, Jerry P., (208) 885-9259 jfairley@uidaho.edu,
University of Idaho – HwGv
Fajkoviæ, Hana, +38514605969 hanaf@geol.pmf.hr,
University of Zagreb – CgGeCm
Fakhari, Mohammad D., 614 6584 mohammad.fakhari@dnr.state.oh.us,
Ohio Dept of Natural Resources – GocNx
Falcon-Lang, Howard, +44 1784 414039 Howard.Falcon-Lang@rhul.ac.uk,
Royal Holloway University of London – Pg
Falebita, Dele E., 234-703-298-7836 delefale@oauife.edu.ng,
Obafemi Awolowo University – YeRnZg
Falk, Lisa, 919-515-8458 esfalk@ncsu.edu,
North Carolina State University – Zg
Falkena, Lee, 312-996-4499 lfalkena@uic.edu,
University of Illinois at Chicago – ZeGig
Falkowski, Paul G., falco@imcs.rutgers.edu,
University of Hawai'i, Manoa – Og
Falkowski, Paul G., (848) 932-6555 (Ext. 370) falko@eps.rutgers.edu,
Rutgers, The State University of New Jersey – ObCmPe
Fall, Andras, (512) 471-8334 andras.fall@beg.utexas.edu,
University of Texas, Austin – CgGcEg
Fall, Leigh M., 607 436-2615 leigh.fall@oneonta.edu,
SUNY, Oneonta – Pgq
Fallowfield, Howard, howard.fallowfield@flinders.edu.au,
Flinders University – Zn
Falta, Ronald W., (864) 656-0125 faltar@clemson.edu,
Clemson University – Hq

Famiglietti, James, (949)824-9434 jfamigli@uci.edu,
    University of California, Irvine – Hw
Fan, Chao, 208-885-0949 cfan@uidaho.edu,
    University of Idaho – Zir
Fan, Majie, (817) 272-2987 mfan@uta.edu,
    University of Texas, Arlington – GstCs
Fan, Weihong, weihong.fan@stockton.edu,
    Stockton University – Zi
Fanning, Kent A., (727) 553-1594 kaf@marine.usf.edu,
    University of South Florida – OcGu
Fantle, Matthew S., 814-863-9968 mfantle@geosc.psu.edu,
    Pennsylvania State University, University Park – Cs
Faraji, Maedeh, maedeh.faraji@austincc.edu,
    Austin Community College District – AsZge
Fares, Mary A., (229) 333-5755 mfares@valdosta.edu,
    Valdosta State University – Zn
Fariduddin, Mohammad, 773/442-6059 M-Fariduddin@neiu.edu,
    Northeastern Illinois University – Pe
Farley, Kenneth A., (626) 395-6005 farley@gps.caltech.edu,
    California Institute of Technology – Cg
Farley, Martin B., 910-521-6478 martin.farley@uncp.edu,
    University of North Carolina, Pembroke – PlZe
Farley, Perry D., (505) 667-2415 dfarley@lanl.gov,
    Los Alamos National Laboratory – Zn
Farmer, Emma C., 516 463-5568 geoecf@hofstra.edu,
    Hofstra University – Ou
Farmer, G. Lang, 303-492-6534 farmer@colorado.edu,
    University of Colorado – Cg
Farmer, George T., (505) 665-0225 gfarmer@lanl.gov,
    Los Alamos National Laboratory – Hw
Farmer, Jack D., (480) 965-6748 jfarmer@asu.edu,
    Arizona State University – Pg
Farmer, Jesse, 609-258-4101 jess.farmer@princeton.edu,
    Princeton University – Cm
Farnan, Ian, +44 (0) 1223 333431 ifarnan@esc.cam.ac.uk,
    University of Cambridge – Gz
Farnsworth, Katie, 724-357-3406 katie.farnsworth@iup.edu,
    Indiana University of Pennsylvania – Ong
Farquhar, James, (301) 405-5034 jfarquha@umd.edu,
    University of Maryland – Cs
Farquharson, Phil, pfarquharson@miracosta.edu,
    MiraCosta College – Gg
Farr, Tom G., (818) 354-9057 tom.farr@jpl.nasa.gov,
    Jet Propulsion Laboratory – Gm
Farrand, William, (734) 764-1435 wfarrand@umich.edu,
    University of Michigan – Ga
Farrell, Brian F., (617) 495-2998 farrell@seas.harvard.edu,
    Harvard University – Am
Farrell, Jamie, 801-581-7856 jamie.farrell@utah.edu,
    University of Utah – Ys
Farrell, John, (401) 874-6561 jfarrell@gso.uri.edu,
    University of Rhode Island – Ou
Farrell, Kathleen M., (919) 733-7353 kathleen.farrell@ncdenr.gov,
    North Carolina Geological Survey – Gs
Farrell, Mark O., (412) 392-3879 Point Park University – Zn
Farrell, Michael, michaelfarrell@myefolio.com,
    Cuyamaca College – Gg
Farrell, Stewart C., (609) 652-4245 Stockton University – On
Farrington, John W., 5082741926 (cell Phone) jfarrington@whoi.edu,
    Woods Hole Oceanographic Institution – ComOc
Farrow, Norman D., ndf@ornl.gov,
    Oak Ridge National Laboratory – Hg
Farsang, Andrea, +3662544156 farsang@geo.u-szeged.hu,
    Univesity of Szeged – SdZe
Farthing, Dori J., 585-245-5298 farthing@geneseo.edu,
    SUNY, Geneseo – Gz
Farver, John R., (419) 372-7203 jfarver@bgnet.bgsu.edu,
    Bowling Green State University – Gy
Faryad, Shah Wali, +420221951521 faryad@natur.cuni.cz,
    Charles University – GpzGi
Fastovsky, David, (401) 874-2185 defastov@uri.edu,
    University of Rhode Island – PvGsPs
Fatherree, James W., 813-253-7906 jfatherree@hccfl.edu,
    Hillsborough Community College – Zg
Faulds, James, (775) 682-8751 jfaulds@unr.edu,
    University of Nevada, Reno – Gc
Faulds, James E., (775) 682-6650 jfaulds@unr.edu,
    University of Nevada – Gc
Faulds, James E., (775) 682-8751 jfaulds@unr.edu,
    University of Nevada, Reno – Gc
Faulkner, Daniel, +440151 794 5169   Faulkner@liverpool.ac.uk,
    University of Liverpool – Ys
Faulkner, Melinda S., (936) 468-2236 mgshaw@sfasu.edu,
    Stephen F. Austin State University – GeClEg
Faúndez, Alejandro, afaundez@ucm.es,
    Univ Complutense de Madrid – Ng
Faure, Stephane, 514-987-3000 #2369 faure.stephane@uqam.ca,
    Universite du Quebec a Montreal – Gc
Faust, Megan, megfaust@pdx.edu,
    Portland State University – ZeGr
Favas, Paulo J., 351259350220 pjcf@utad.pt,
    Universidade de Trás-os-Montes e Alto Douro – GeCtGg
Favor, Michael, (313) 593-5235 University of Michigan, Dearborn – Zn
Favorito, Jessica, jessica.favorito@stockton.edu,
    Stockton University – So
Fawcett, J. J., (416) 978-3027 fawcett@quartz.geology.utoronto.ca,
    University of Toronto – Gp
Fawcett, Peter J., (505) 277-3867 fawcett@unm.edu,
    University of New Mexico – Pe
Fayek, Mostafa, (204) 474-7982 fayek@cc.umanitoba.ca,
    University of Manitoba – Cs
Fayon, Annia K., 612-626-9805 fayon001@umn.edu,
    University of Minnesota, Twin Cities – Gc
Feakins, Sarah, (213) 740-7168 feakins@usc.edu,
    University of Southern California – CoGuCs
Fearey, Bryan L., (505) 665-2423 Los Alamos National Laboratory – Cc
Feather, Russell, (202) 633-1793 featherr@si.edu,
    Smithsonian Inst / Natl Museum of Natural Hist– Gz
Fedele, Juan J., 320-308-1049 jjfedele@stcloudstate.edu,
    Saint Cloud State University – Hg
Fedo, Christopher, (865) 974-2366 cfedo@utk.edu,
    University of Tennessee, Knoxville – GsXgCg
Fedorov, Alexey V., (203) 432-3153 alexey.fedorov@yale.edu,
    Yale University – Op
Fedortchouk, Yana, 902 494 8432 yana@dal.ca,
    Dalhousie University – CpEdGi
Feely, Martin, +353 (0)91 492 129 martin.feely@nuigalway.ie,
    National University of Ireland Galway – GzxEg
Feely, Richard A., (206) 526-6214 feely@pmel.noaa.gov,
    University of Washington – Oc
Feeney, Alison E., (717) 477-1319 aefeen@ship.edu,
    Shippensburg University – ZinZn
Feeney, Dennis M., (208) 885-5203 dmfeeney@uidaho.edu,
    University of Idaho – GgiYg
Feeney, Thomas P., (717) 477-1297 tpfeen@ship.edu,
    Shippensburg University – ZyGeHw
Fegley, Bruce, (314) 935-4852 bfegley@levee.wustl.edu,
    Washington University in St. Louis – Xc
Fehler, Michael, 617-253-3589 fehler@mit.edu,
    Massachusetts Institute of Technology – Yg
Fehn, Udo, (585) 244-4868 udo.fehn@rochester.edu,
    University of Rochester – GeCgOu
Fehr, Karl Thomas, 089/2180 4256 fehr@min.uni-muenchen.de,
    Ludwig-Maximilians-Universitaet Muenchen – Gz
Fehrenbacher, Jennifer, 541-737-6285 fehrenje@coas.oregonstate.edu,
    Oregon State University – Cm
Fei, Yingwei, fei@gl.ciw.edu,
    University of Maryland – CpGxCg
Fei, Yingwei, 202-478-8936 yfel@carnegiescience.edu,
    Carnegie Institution for Science – Cp
Feibel, Craig S., (848) 445-2721 feibel@eps.rutgers.edu,
    Rutgers, The State University of New Jersey – GrsGa
Feigenson, Mark D., (848) 445-3149 feigy@eps.rutgers.edu,
    Rutgers, The State University of New Jersey – CgGiv
Feigl, Kurt, 608-262-0176 feigl@wisc.edu,
    University of Wisconsin-Madison – Yd
Feigl, Kurt L., (608) 262-0176 feigl@wisc.edu,
    University of Wisconsin, Madison – Gt
Fein, Jeremy B., (574) 631-6101 fein.1@nd.edu,
    University of Notre Dame – Cl
Feinberg, Joshua, (612) 624-8429 feinberg@umn.edu,
    University of Minnesota, Twin Cities – YmGze
Feineman, Maureen D., (814) 863-6649 mdf12@psu.edu,
    Pennsylvania State University, University Park – Cp

Feinglos, Mark N., (919) 668-1367 mark.feinglos@duke.edu,
Duke University – Gz

Feinstein, Daniel T., (414) 962-2582 dtfeinst@usgs.gov,
University of Wisconsin, Milwaukee – Hw

Felbeck, Horst, (858) 534-6647 hfelbeck@ucsd.edu,
University of California, San Diego – Ob

Feldl, Nicole, 831-459-3693 nfeldl@ucsc.edu,
University of California, Santa Cruz – At

Feldmann, Rodney M., (330) 672-2506 rfeldman@kent.edu,
Kent State University – Pi

Felix, Joseph D., 361-825-4180 Joseph.Felix@tamucc.edu,
Texas A&M University, Corpus Christi – Zn

Feller, Robert, feller@biol.sc.edu,
University of South Carolina – Ob

Felton, Richard M., (660) 562-1569 rfelton@nwmissouri.edu,
Northwest Missouri State University – Pg

Felzer, Benjamin S., (610) 758-3536 bsf208@lehigh.edu,
Lehigh University – ZcoZn

Fenberg, Phillip, +44 (0)23 80592729 P.B.Fenberg@soton.ac.uk,
University of Southampton – Ob

Fendorf, Scott E., (650) 723-5238 fendorf@pangea.stanford.edu,
Stanford University – Sc

Fendrich, Kim V, kfendrich@amnh.org,
American Museum of Natural History – GzXg

Feng, Huan E., (973) 655-7549 fengh@mail.montclair.edu,
Montclair State University – Cm

Feng, Song, (479) 575-4748 songfeng@uark.edu,
University of Arkansas, Fayetteville – AsZyAm

Feng, Xiahong, 603-646-1712 xiahong.feng@dartmouth.edu,
Dartmouth College – Cs

Feng, Yan, 630-252-2550 yfeng@anl.gov,
Argonne National Laboratory – As

Feng, Yucheng, (334) 844-3967 yfeng@auburn.edu,
Auburn University – SbCb

Fengler, Keegan , 509.963.2719 keegan@geology.cwu.edu,
Central Washington University – Gg

Fenical, William H., (858) 534-2133 wfenical@ucsd.edu,
University of California, San Diego – Oc

Fennel, Katja, (902) 494-4526 katja.fennel@dal.ca,
Dalhousie University – Ob

Fenrich, Francis, (780) 492-2149 frances@space.ualberta.ca,
University of Alberta – Xy

Fenster, Michael S., (804)752-3745 mfenster@rmc.edu,
Randolph-Macon College – OnGue

Fenstermaker, Lynn, (702) 862-5412 lynn@dri.edu,
Desert Research Institute – Zr

Fenter, Paul, (630) 252-7053 fenter@anl.gov,
University of Illinois at Chicago – Gy

Fenton, Cassandra, 970-248-1077 cfenton@coloradomesa.edu,
Colorado Mesa University – CcGmCl

Ferdowsi, Behrooz, 609-258-4101 behrooz@princeton.edu,
Princeton University – Yu

Ferencz, Orsolya, orsi@sas.elte.hu,
Eotvos Lorand University – XyZr

Ferger, Marisa, (814) 863-4229 mferger@psu.edu,
Pennsylvania State University, University Park – Am

Ferguson, Charles A., 520.621.2470 caf@email.arizona.edu,
Arizona Geological Survey – Gcv

Ferguson, Ian J., (204) 474-9154 ij_ferguson@umanitoba.ca,
University of Manitoba – YemYu

Ferguson, John F., (972) 883-2410 ferguson@utdallas.edu,
University of Texas, Dallas – Yg

Ferguson, Julie , (949)824-9411 julie.ferguson@uci.edu,
University of California, Irvine – Ze

Ferguson, Terry A., (864) 597-4527 fergusonta@wofford.edu,
Wofford College – Ga

Fermanich, Kevin J., (920) 465-2240 fermanik@uwgb.edu,
University of Wisconsin, Green Bay – SpRwHt

Fernandez, Diego, (801) 587-9366 diego.fernandez@utah.edu,
University of Utah – Cg

Fernandez, Fabian G., 612-625-7460 fabiangf@umn.edu,
University of Minnesota, Twin Cities – ScbSo

Fernandez, Loretta, 617.373.5461 l.fernandez@neu.edu,
Northeastern University – Co

Fernandez, Louis A., (909) 537-5024 California State University, San Bernardino – Gi

Fernández, Lourdes, lfdiaz@ucm.es,
Univ Complutense de Madrid – Gz

Fernández, Virginia, (598) 2525 1552 vivi@fcien.edu.uy,
Universidad de la Republica Oriental del Uruguay (UDELAR) – ZirZn

Fernández Suárez, Javier, jfsuarez@ucm.es,
Univ Complutense de Madrid – Cc

Ferrari, Raffaele, (617) 253-1291 raffaele@mit.edu,
Massachusetts Institute of Technology – OpAsOb

Ferre, Paul A., (520) 621-2952 ty@hwr.arizona.edu,
University of Arizona – Hw

Ferre, Paul (Ty), (520) 621-6082 ty@hwr.arizona.edu,
University of Arizona – So

Ferre', Eric C., 337-482-2228 eric.ferre@louisiana.edu,
University of Louisiana at Lafayette – YmGc

Ferreira, Anna, +44 020 7679 37704 a.ferreira@ucl.ac.uk,
University College London – Ys

Ferreira, Maryanne F., (508) 289-2266 mferreira@whoi.edu,
Woods Hole Oceanographic Institution – Zn

Ferrell, Jr., Ray E., (225) 388-5306 rferrell@lsu.edu,
Louisiana State University – Cl

Ferren, Richard L., (413) 236-4553 rferren@berkshirecc.edu,
Berkshire Community College – Zn

Ferrero, Anna Maria, anna.ferrero@unito.it,
Università di Torino – Nr

Ferring, C. Reid, (940) 565-2993 reid.ferring@unt.edu,
University of North Texas – Ga

Ferrini, Victoria, ferrini@ldeo.columbia.edu,
Columbia University – Ou

Ferris, Dawn, 419-755-3909 dferris@sciencesocieties.org,
Ohio State University – Sf

Ferris, Grant F., 416-978-0526 ferris@quartz.geology.utoronto.ca,
University of Toronto – Gn

Ferris, Michael H., 310 243-3405 mferris@csudh.edu,
California State University, Dominguez Hills – Zr

Ferriz, Horacio, 209 667-3874 hferriz@csustan.edu,
California State University, Stanislaus – NgHgYe

Ferruzza, David, (not) li-sted ferruzzad@etown.edu,
Elizabethtown College – AsZn

Festa, Andrea, andrea.festa@unito.it,
Università di Torino – Gc

Few, Jr., Arthur A., (713) 527-4003 (Ext. 3601) few@rice.edu,
Rice University – As

Feyereisen, Gary, 612-625-0968 gfeyer@umn.edu,
University of Minnesota, Twin Cities – Sp

Fialko, Yuri A., (858) 822-5028 yfialko@ucsd.edu,
University of California, San Diego – YgdGt

Fichter, Lynn S., (540) 568-6531 fichtels@jmu.edu,
James Madison University – Gr

Fichtner, Andreas, 089/2180 4230 andreas.fichtner@geophysik.uni-muenchen.de,
Ludwig-Maximilians-Universitaet Muenchen – Yg

Fick, Walter H., (785) 532-7223 whfick@ksu.edu,
Kansas State University – Sf

Field, Cathryn K., cfield@bowdoin.edu,
Bowdoin College – Ge

Field, Richard T., 302-831-2695 University of Delaware – Og

Field, Stephen W., (254) 968-9887 field@tarleton.edu,
Tarleton State University – Gz

Fielding, Christopher R., (402) 472-9801 cfielding2@unl.edu,
University of Nebraska, Lincoln – Gs

Fielding, Lynn, (310) 660-3593 lfielding@elcamino.edu,
El Camino College – Zg

Fields, Chad, Chad.Fields@dnr.iowa.gov,
Iowa Dept of Natural Resources – Gg

Fields, Nancy, 214-860-2429 nfields@dcccd.edu,
El Centro College - Dallas Community College District – Zg

Fiesinger, Donald W., don.fiesinger@usu.edu,
Utah State University – Gi

Figg, Sean, (760)744-1150 ext 2513 sfigg@palomar.edu,
Palomar College – Gg

Fike, David A., (314) 935-6607 dfike@levee.wustl.edu,
Washington University in St. Louis – CsmPg

Fildes, Stephen , stephen.fildes@flinders.edu.au,
Flinders University – Zri

Filina, Irina, (402) 472-2077 ifilina2@unl.edu,
University of Nebraska, Lincoln – YgGtYe

Filipescu, Sorin, +40-264-405300 ext 5206 sorin.filipescu@ubbcluj.ro,
Babes-Bolyai University – GrPm

Filippelli, Gabriel M., (317) 274-3795 gfilippe@iupui.edu,
  Indiana University / Purdue University, Indianapolis – CmGb
Filkorn, Harry, filkornh@piercecollege.edu,
  Los Angeles Pierce College – GgPi
Filley, Timothy R., (765) 494-6581 filley@purdue.edu,
  Purdue University – Co
Fillion, Luc, 514-421-4770 luc.fillion@ec.gc.ca,
  McGill University – Ams
Fillmore, Robert P., (970) 943-2650 rfillmore@western.edu,
  Western Colorado University – Gs
Finch, Adriam, +44 (0) 1334 462384 aaf1@st-and.ac.uk,
  University of St. Andrews – Gz
Finger, Ken, 510-6432559 kfinger@berkeley.edu,
  University of California, Berkeley – PmZnn
Fink, Uwe, (520) 2736 uwefink@lpl.arizona.edu,
  University of Arizona – As
Finkel, Robert C., (925) 422-2044 finkel1@llnl.gov,
  Lawrence Livermore National Laboratory – Cg
Finkelman, Robert B., (972) 883-2459 bobf@utdallas.edu,
  University of Texas, Dallas – GbEcCa
Finkelstein, David, (315) 781-4443 finkelstein@hws.edu,
  Hobart & William Smith Colleges – CgbCs
Finkenbinder, Matthew S., (570) 408-3871 matthew.finkenbinder@wilkes.edu,
  Wilkes University – GsmGl
Finks, Robert M., (718) 997-3305 Queens College (CUNY) – Pi
Finley, Jason P., finleyjp@piercecollege.edu,
  Los Angeles Pierce College – Am
Finley, Mark, (309) 268-8642 mark.finley@heartland.edu,
  Heartland Community College – Gg
Finley, Robert J., 217-244-8389 finley@isgs.uiuc.edu,
  Illinois State Geological Survey – Eo
Finley, William R., 337-482-6468 WFINLEY@ROCMANDO.COM,
  University of Louisiana at Lafayette – YsZi
Finn, Gregory C., (905) 688-5550 x3528 greg.finn@brocku.ca,
  Brock University – Gx
Finnegan, David L., (505) 667-6548 Los Alamos National Laboratory – As
Finnegan, Noah J., 831-459-5110 nfinnegan@pmc.ucsc.edu,
  University of California, Santa Cruz – Gm
Finnegan, Seth, sethf@berkeley.edu,
  University of California, Berkeley – Pig
Finney, Bruce P., (907) 474-7724 University of Alaska, Fairbanks – Ou
Finney, Stanley C., (562) 985-8637 scfinney@csulb.edu,
  California State University, Long Beach – Psi
Finzel, Emily, (319) 335-0405 emily-finzel@uiowa.edu,
  University of Iowa – GstEo
Fio, Karmen, +38514606088 karmen.fio@gmail.com,
  University of Zagreb – PgGg
Fiore, Arlene, amfiore@ldeo.columbia.edu,
  Columbia University – As
Fiore, Arlene M., (845) 365-8550 amfiore@ldeo.columbia.edu,
  Columbia University – As
Fiorillo, Anthony R., (214) 421-3466 Southern Methodist University – Pv
Firing, Eric, (808) 956-7894 efiring@hawaii.edu,
  University of Hawai'i, Manoa – Op
Fischer, Alfred G., (213) 740-5821 fisher@usc.edu,
  University of Southern California – Gs
Fischer, Emily, evf@rams.colostate.edu,
  Colorado State University – As
Fischer, Karen M., (401) 863-1360 Karen_Fischer@Brown.edu,
  Brown University – Ys
Fischer, Mark P., (815) 753-0523 mfischer@niu.edu,
  Northern Illinois University – GcoHy
Fischer, Rebecca A., 617-384-6992 rebeccafischer@fas.harvard.edu,
  Harvard University – Gy
Fischer, Tobias, (505) 277-0683 fischer@unm.edu,
  University of New Mexico – Gv
Fish, Jennifer M., 831-459-1235 jmsfish@ucsc.edu,
  University of California, Santa Cruz – Zn
Fisher, Andrew T., (831) 459-5598 afisher@ucsc.edu,
  University of California, Santa Cruz – Hw
Fisher, Daniel C., (734) 764-0488 dcfisher@umich.edu,
  University of Michigan – Pi
Fisher, David A., 613-996-7623 fisher@nrcan-mcan.gc.ca,
  University of Ottawa – Gl
Fisher, David E., dfisher@miami.edu,
  University of Miami – Cs
Fisher, Donald M., (814) 865-3206 fisher@geosc.psu.edu,
  Pennsylvania State University, University Park – GctZm
Fisher, Liz, +440151 795 4390   E.H.Fisher@liverpool.ac.uk,
  University of Liverpool – Co
Fisher, Nicholas S., (631) 632-8649 nicholas.fisher@stonybrook.edu,
  SUNY, Stony Brook – Oc
Fisher, Quentin , +44(0) 113 34 31920 q.j.fisher@leeds.ac.uk,
  University of Leeds – Np
Fisher, Timothy G., (419) 530-2009 timothy.fisher@utoledo.edu,
  University of Toledo – GlmGn
Fisher, William L., (512) 471-5600 wfisher@mail.utexas.edu,
  University of Texas, Austin – GsrGo
Fisher, Jr., Thomas R., (410) 221-8432 fisher@hpl.umces.edu,
  University of Maryland – ObZiCl
Fishman, George , (781) 891-2721 gfishman@bentley.edu,
  Bentley University – XgbXc
Fishman, Jack, (314) 977-3132 jfishma2@slu.edu,
  Saint Louis University – As
Fishwick, Stewart, +44 0116 252 3810 sf130@le.ac.uk,
  Leicester University – Yg
Fisk, Aaron, 519-253-3000 x4740 afisk@uwindsor.ca,
  University of Windsor – ObCs
Fitak, Madge R., 614-265-6585 madge.fitak@dnr.state.oh.us,
  Ohio Dept of Natural Resources – Gg
Fitchett, Rebekah, 773-442-6052 R-Fitchett@neiu.edu,
  Northeastern Illinois University – GgCg
Fitton, J. G., +44 (0) 131 650 8529 Godfrey.Fitton@ed.ac.uk,
  Edinburgh University – Gi
Fitz, Thomas J., (715) 682-1852 tfitz@northland.edu,
  Northland College – GgxGc
Fitzer, Susan, +44 01413305442 Susan.Fitzer@glasgow.ac.uk,
  University of Glasgow – Ob
Fitzgerald, David, (210) 436-3235 Saint Mary's University – Gd
FitzGerald, Duncan M., (617) 353-2530 dunc@bu.edu,
  Boston University – OnGsu
Fitzgerald, Francis S., 303-384-2644 ffitzger@mines.edu,
  Colorado Geological Survey – ZirGg
Fitzgerald, Paul G., (315) 443-2672 pgfitzge@syr.edu,
  Syracuse University – GtCc
Fitzgerald, William F., 860-405-9158 william.fitzgerald@uconn.edu,
  University of Connecticut – Oc
Fitzjarrald, David R., (518) 437-8735 dfitzjarrald@albany.edu,
  SUNY, Albany – Asm
Fitzpatrick, John R., (505) 667-4761 Los Alamos National Laboratory – Zn
FitzPatrick, Meriel E., +44 1752 584769 m.e.fitzpatrick@plymouth.ac.uk,
  University of Plymouth – PlgGr
Fitzpatrick, Patrick J., (228) 688-1157 fitz@gri.msstate.edu,
  Mississippi State University – Am
Fitzpatrick, Stephan, (678) 891-3773 sfitzpatrick1@gsu.edu,
  Georgia State University, Perimeter College, Clarkston Campus – HwCgSo
Fitzsimmons, Jessica N., 979-845-5137 jessfitz@tamu.edu,
  Texas A&M University – OcCta
Fitzsimmons, Kevin, (520) 626-3324 kevfitz@ag.arizona.edu,
  University of Arizona – Zn
Fjell, Dale, (785) 532-5833 dfjell@ksu.edu,
  Kansas State University – So
Flagg, Charles, (631) 632-3184 charles.flagg@stonybrook.edu,
  SUNY, Stony Brook – Op
Flaherty, Frank A., (229) 333-5665 flaherty@valdosta.edu,
  Valdosta State University – Zn
Flaig, Peter P., peter.flaig@beg.utexas.edu,
  University of Texas, Austin – GsrEo
Flake, Rex , 509.963.1114 rex@geology.cwu.edu,
  Central Washington University – Yd
Flament, Pierre J., (808) 956-6663 pflament@hawaii.edu,
  University of Hawai'i, Manoa – Op
Flanagan, Michael, flanagam@sunysuffolk.edu,
  Suffolk County Community College, Ammerman Campus – Zg
Flanagan, Timothy, (413) 236-4503 tflanaga@berkshirecc.edu,
  Berkshire Community College – Zei
Flannery, David, +61 7 3138-1615 david.flannery@qut.edu.au,
  Queensland University of Technology – PoCaXg
Flato, Gregory M., (250) 363-8223 greg.flato@ec.gc.ca,
  University of Victoria – As
Flawn, Peter T., (512) 471-1825 pflawn@po.utexas.edu,
  University of Texas, Austin – Eg
Fleck, Michelle, (435) 613-5232 michelle.fleck@usu.edu,

Utah State University – Ze
Flegal, Russell, (831) 459-2093 rflegal@es.ucsc.edu,
University of California, Santa Cruz – Cg
Fleisher, P. Jay, (607) 286-7541 fleishpj@oneonta.edu,
SUNY, Oneonta – GlmGe
Fleming, Thomas H., 203-392-5837 flemingt1@southernct.edu,
Southern Connecticut State University – GiCgc
Flemings, Peter B., 512-471-6156 flemings@ig.utexas.edu,
University of Texas, Austin – Gr
Flemming, Roberta L., (519) 661-3143 rflemmin@uwo.ca,
Western University – GzXmZm
Flesch, Lucy M., (765) 494-0263 lmflesch@purdue.edu,
Purdue University – Yg
Fleskens, Luuk, +31(0) 317 485467 l.fleskens@leeds.ac.uk,
University of Leeds – Ge
Flessa, Karl W., (520) 621-7336 kflessa@email.arizona.edu,
University of Arizona – PgeGe
Fletcher, Austin, rfletche@uoguelph.ca,
University of Guelph – Zn
Fletcher, Charles H., (808) 956-2582 fletcher@soest.hawaii.edu,
University of Hawai'i, Manoa – GsOnZu
Fletcher, John, jfletche@cicese.mx,
Centro de Investigación Científica y de Educación Superior de Ensenada – Gpt
Fletcher, William K., (604) 822-2392 wfletcher@eos.ubc.ca,
University of British Columbia – Ce
Flierl, Glenn R., (617) 253-4692 glenn@lake.mit.edu,
Massachusetts Institute of Technology – Op
Flint, Stephen, +44 0161 306-6971 stephen.flint@manchester.ac.uk,
University of Manchester – GrsGo
Flood, Beverley, 612-624-1603 floo0017@umn.edu,
University of Minnesota, Twin Cities – Pg
Flood, Peter, +61-2-67732329 pflood@une.edu.au,
University of New England – EcGt
Flood, Roger D., (631) 632-6971 roger.flood@stonybrook.edu,
SUNY, Stony Brook – Ou
Flood, Tim P., (920) 403-1356 tim.flood@snc.edu,
Saint Norbert College – GvZe
Florea, Lee J., (812) 855-1376 lflorea@indiana.edu,
Indiana University – HqGeCl
Flores, Alejandro N., lejoflores@boisestate.edu,
Boise State University – HqAs
Flores, Kennet, 718-951-5000 x3253 KEFlores@brooklyn.cuny.edu,
Brooklyn College (CUNY) – Gpc
Flores-Garcia, Mari C., (818) 677-3541 mari.flores@csun.edu,
California State University, Northridge – Zn
Floris, Mario, +390498279121 mario.floris@unipd.it,
Università degli Studi di Padova – NgZr
Flory, David M., (515) 294-0264 flory@iastate.edu,
Iowa State University of Science & Technology – AmsZn
Flower, Hilary, Flower.Hilary@spcollege.edu,
Saint Petersburg College, Clearwater – Gg
Flower, Martin F., +1 312 391-7632 flower@uic.edu,
University of Illinois at Chicago – GiCuGt
Flowers, George C., (504) 862-3192 flowers@tulane.edu,
Tulane University – Hw
Flowers, Gwenn, (778)782-6638 gflowers@sfu.ca,
Simon Fraser University – Gl
Flowers, Rebecca M., (303) 492-5135 rebecca.flowers@colorado.edu,
University of Colorado – CgcGt
Flowers Falls, Emily, (540) 458-8868 fallse@wlu.edu,
Washington & Lee University – Gg
Floyd, Barry, 518-473-9988 esogis@mail.nysed.gov,
New York State Geological Survey – Zn
Fluegeman, Richard H., (765) 285-8267 rfluegem@bsu.edu,
Ball State University – PmGro
Flynn, John J., (212) 769-5806 American Museum of Natural History – Pv
Flynn, Luke P., (808) 956-3154 flynn@higp.hawaii.edu,
University of Hawai'i, Manoa – Zr
Flynn, Wendilyn, 970-351-1071 wendilyn.flynn@unco.edu,
University of Northern Colorado – Ams
Fodor, Ronald V., (919) 515-7177 ron_fodor@ncsu.edu,
North Carolina State University – Gi
Fogel, David N., (505) 665-3305 fogel@lanl.gov,
Los Alamos National Laboratory – Zn
Fogg, Graham E., (916) 752-6810 University of California, Davis – Hy
Foit, Jr., Franklin F., (509) 335-3093 Washington State University – Gz

Foland, Kenneth A., foland.1@osu.edu,
Ohio State University – CcgCs
Foley, Bradford, (814) 863-3591 bjf5382@psu.edu,
Pennsylvania State University, University Park – YgXyGt
Foley, Duncan, (253) 535-7568 foleyd@plu.edu,
Pacific Lutheran University – GeHyZe
Foley, John P., (406) 496-4414 jfoley@mtech.edu,
Montana Tech of The University of Montana – Zn
Folk, Robert L., (512) 471-5294 rlfolk@mail.utexas.edu,
University of Texas, Austin – Gd
Follmer, Leon R., (217) 359-2090 lfollmer@illinois.edu,
University of Illinois, Urbana-Champaign – SaGlm
Follows, Michael, (617) 253-5939 mick@mit.edu,
Massachusetts Institute of Technology – Op
Folorunso, I. O., foloisgood@yahoo.com,
University of Ilorin – GoCg
Fomel, Sergey, 512-475-9573 sergey.fomel@beg.utexas.edu,
University of Texas at Austin, Jackson School of Geosciences – Yx
Fondevilla, Victor, victor.fondevilla@uab.cat,
Universitat Autonoma de Barcelona – PgGr
Fondran, Carol, (216) 987-5227 Carol.Fondran@tri-c.edu,
Cuyahoga Community College - Western Campus – Zg
Fones, Gary, +44 023 92 842252 gary.fones@port.ac.uk,
University of Portsmouth – Cm
Fonstad, Mark, fonstad@uoregon.edu,
University of Oregon – Zy
Fontaine, Fabrice R., +262262938207 fabrice.fontaine@univ-reunion.fr,
Universite de la Reunion – Yg
Fontana, Alessandro, +390498279118 alessandro.fontana@unipd.it,
Università degli Studi di Padova – Gm
Foote, Michael J., (773) 702-4320 mfoote@midway.uchicago.edu,
University of Chicago – Pg
Forbriger, Thomas, +49-721-6084593 thomas.forbriger@kit.edu,
Karlsruhe Institute of Technology – Ys
Forcino, Frank, (828) 227-2888 flforcino@wcu.edu,
Western Carolina University – PgZeGr
Ford, Allistair, +44 (0) 191 208 7121 alistair.ford@ncl.ac.uk,
University of Newcastle Upon Tyne – Gq
Ford, Andrew, 509-335-7846 forda@wsu.edu,
Washington State University – Zu
Ford, Arianne, Arianne.Ford@jcu.edu.au,
James Cook University – GqZiEg
Ford, Derek C., (905) 525-9140 x20132 dford@mcmaster.ca,
McMaster University – GmHwCc
Ford, Heather, (951) 827-3194 heather.ford@ucr.edu,
University of California, Riverside – Ys
Ford, Mark T., 361-593-2767 mark.ford@tamuk.edu,
Texas A&M University, Kingsville – GxzCg
Ford, Richard L., (801) 626-6942 rford@weber.edu,
Weber State University – GmZeEo
Fordyce, R. Ewan, +64 3 479-7510/7535 ewan.fordyce@otago.ac.nz,
University of Otago – Pv
Foreman, Brady Z., 360-650-2546 brady.foreman@wwu.edu,
Western Washington University – Grs
Foreman, Michael G., 250-363-6306 mike.foreman@dfo-mpo.gc.ca,
University of Victoria – OpZo
Foreman, Taylor, (303) 370-6367 Taylor.Foreman@dmns.org,
Denver Museum of Nature & Science – Zn
Foresi, Luca Maria, luca.foresi@unisi.it,
University of Siena – PmGs
Foret, Jim, 337-482-6064 jaf3663@louisiana.edu,
University of Louisiana at Lafayette – Ge
Forgotson, Jr., James M., (405) 325-4451 jforgot@ou.edu,
University of Oklahoma – Go
Forman, Stephen, (254)710-2495 steve_forman@baylor.edu,
Baylor University – CcPc
Fornaciari, Eliana, +390498279185 eliana.fornaciari@unipd.it,
Università degli Studi di Padova – Pgm
Fornari, Daniel J., (508) 289-2857 dfornari@whoi.edu,
Woods Hole Oceanographic Institution – Gu
Forno, Maria Gabriella, gabriella.forno@unito.it,
Università di Torino – Gg
Forrest, David, +4401413305401 david.forrest@glasgow.ac.uk,
University of Glasgow – Zi
Forsman, Nels F., (701) 777-4349 nels.forsman@engr.und.edu,
University of North Dakota – GdXgRh
Forster, Piers M., +44(0) 113 34 36476 p.m.forster@leeds.ac.uk,

University of Leeds – As
Forster, Richard R., (801) 581-3611 rick.forster@geog.utah.edu,
University of Utah – Zr
Forsyth, Donald W., (401) 863-1699 donald_forsyth@brown.edu,
Brown University – YrsGt
Forte, Adam, aforte8@lsu.edu,
Louisiana State University – Gcm
Forte, Alessandro Marco, 514-987-3000 #5607 forte.alessandro@uqam.ca,
Universite du Quebec a Montreal – Yg
Fortes, Dominic, andrew.fortes@ucl.ac.uk,
Birkbeck College – Zg
Fortier, Richard, (418) 656-2746 richard.fortier@ggl.ulaval.ca,
Universite Laval – Yg
Fortin, Danielle, 613-562-5800 6423 dfortin@uottawa.ca,
University of Ottawa – CblGe
Fortner, Sarah K., (937) 327-7328 sfortner@wittenberg.edu,
Wittenberg University – ClZeHg
Fosdick, Julie C., (812) 855-6109 jfosdick@indiana.edu,
Indiana University, Bloomington – Gt
Foss, Donald J., (415) 485-9523 jcprice@metro.net,
College of Marin – Zg
Foster , Alec , (309) 438-0922 alfost2@ilstu.edu,
Illinois State University – Zn
Foster, David A., (352) 392-2231 dafoster@ufl.edu,
University of Florida – GtCc
Foster, Gavin, +44 (0)23 8059 3786 Gavin.Foster@noc.soton.ac.uk,
University of Southampton – Csg
Foster, Patrick, +44 01326 371828 P.J.Foster@exeter.ac.uk,
Exeter University – Nm
Foster, Stuart A., (270) 745-5983 stuart.foster@wku.edu,
Western Kentucky University – At
Foster Jr., Charles T., (319) 335-1801 tom-foster@uiowa.edu,
University of Iowa – GptGg
Foti, Pamela, (928) 523-6196 Pam.foti@nau.edu,
Northern Arizona University – Zn
Fotopoulos, Georgia, (613) 533-6639 georgia.fotopoulos@queensu.ca,
Queen's University – Yd
Foufoula-Georgiou, Efi, (612) 626-0369 University of Minnesota, Twin Cities – Hg
Fouke, Bruce W., (217) 244-5431 fouke@illinois.edu,
University of Illinois, Urbana-Champaign – GsdGb
Fountain, Andrew G., (503) 725-3386 andrew@pdx.edu,
Portland State University – Gl
Fountain, John C., 919-515-3717 fountain@ncsu.edu,
North Carolina State University – Hw
Fournelle, John, (608) 438-7480 johnf@geology.wisc.edu,
University of Wisconsin-Madison – GiZn
Fournier, Benoit, (418) 656-3930 benoit.fournier@ggl.ulaval.ca,
Universite Laval – NgZmGg
Fournier, Gregory, 617-324-6164 g4nier@mit.edu,
Massachusetts Institute of Technology – Pg
Fournier, Robert O., (902) 494-3666 robert.fournier@dal.ca,
Dalhousie University – Ob
Fovell, Robert, (310) 206-9956 fovell@atmos.ucla.edu,
University of California, Los Angeles – As
Fowell, Sarah J., (907) 474-7810 sjfowell@alaska.edu,
University of Alaska, Fairbanks – Pl
Fowle, David A., (785) 864-1955 fowle@ku.edu,
University of Kansas – Cb
Fowler, Gerald A., fowler@uwp.edu,
University of Wisconsin, Parkside – PgGg
Fowler, Malcolm M., (505) 667-5439 Los Alamos National Laboratory – As
Fowler, Mike, +44 023 92 842293 mike.fowler@port.ac.uk,
University of Portsmouth – Gi
Fowler, Sarah, sarah.fowler@ees.kuleuven.be,
Katholieke Universiteit Leuven – Gip
Fox, Austin, (321) 674-7463 afox@fit.edu,
Florida Institute of Technology – Oc
Fox, David L., 612-624-6361 dlfox@umn.edu,
University of Minnesota, Twin Cities – Pg
Fox, James E., (605) 394-2461 James.Fox@sdsmt.edu,
South Dakota School of Mines & Technology – Gs
Fox, Jeff, 740 2012762 jeff.fox@dnr.state.oh.us,
Ohio Dept of Natural Resources – YsmZe
Fox, Laurel R., (831) 459-2533 fox@biology.ucsc.edu,
University of California, Santa Cruz – Zn

Fox, Lewis, 954-201-6674 lfox@broward.edu,
Broward College – Oc
Fox, Lydia K., (209) 946-2481 lkfox@pacific.edu,
University of the Pacific – GizZe
Fox, Neil I., 573-882-2144 foxn@missouri.edu,
University of Missouri, Columbia – As
Fox, Peter A., (518) 727-4862 pfox@cs.rpi.edu,
Rensselaer Polytechnic Institute – GqZig
Fox-Dobbs, Kena L., 253-879-2458 kena@pugetsound.edu,
University of Puget Sound – PgCsGe
Fox-Kemper, Baylor, 401-863-3979 baylor@brown.edu,
Brown University – OpYnAt
Foxon, Tim, +44(0) 113 34 37910 t.j.foxon@leeds.ac.uk,
University of Leeds – Ge
Frail, Pam, (902) 585-1513 pam.frail@acadiau.ca,
Acadia University – Gg
Frailey, Scott M., 217-244-2430 sfrailey@illinois.edu,
Illinois State Geological Survey – Np
Fraiser, Margaret L., 414-229-3827 mfraiser@uwm.edu,
University of Wisconsin, Milwaukee – PiePg
Fralick, Philip W., (807) 343-8288 philip.fralick@lakeheadu.ca,
Lakehead University – Gs
Fram, Jonathan, 541-737-3966 jfram@coas.oregonstate.edu,
Oregon State University – Op
Frame, Jeffrey, 217-244-9575 frame@illinois.edu,
University of Illinois, Urbana-Champaign – Asm
Francis, Don, donald.francis@mcgill.ca,
McGill University – Gi
Francis, Robert, (206) 543-7345 University of Washington – Ob
Francis, Robert D., (562) 985-4929 jfrancis@csulb.edu,
California State University, Long Beach – Co
Franco, Aldina, +44 (0)1603 59 2721 a.franco@uea.ac.uk,
University of East Anglia – Ge
Franek, Benjamin, (570) 389-4567 bfranek@bloomu.edu,
Bloomsburg University – HsZy
Frank, Charles O., (618) 453-7365 frank@geo.siu.edu,
Southern Illinois University Carbondale -- Gx
Frank, Kenneth, (902) 426-3498 frankk@mar.dfo-mpo.gc.ca,
Dalhousie University – Ob
Frank, Mark R., (815) 753-8395 mfrank@niu.edu,
Northern Illinois University – Cp
Frank, Tracy D., (402) 472-9799 tfrank2@unl.edu,
University of Nebraska, Lincoln – GssCs
Frank, William M., frank@ems.psu.edu,
Pennsylvania State University, University Park – Am
Franke, Otto L., (212) 650-6984 Graduate School of the City University of New York – Gc
Frankel, Arthur, 206-553-0626 afrankel@uw.edu,
University of Washington – Ys
Frankenberg, Christian, 626-395-6143 cfranken@caltech.edu,
California Institute of Technology – Zr
Frankic, Anamarija, anamarija.frankic@umb.edu,
University of Massachusetts, Boston – Ze
Franklin, David, +44 023 92 843540 david.franklin@port.ac.uk,
University of Portsmouth – Ym
Franks, Peter J. S., (858) 534-7528 pfranks@ucsd.edu,
University of California, San Diego – Ob
Franseen, Evan K., (785) 864-2072 evanf@kgs.ku.edu,
University of Kansas – Gs
Fransolet, Andre-Mathieu, 32 4 366 22 06 amfransolet@ulg.ac.be,
Universite de Liege – Gz
Frantz, Carie M., (801) 626-6181 cariefrantz@weber.edu,
Weber State University – PoCbZn
Franz, Eldon H., (206) 300-9259 franz@wsu.edu,
Washington State University – Sf
Franz, Kristie, (515) 294-7454 kfranz@iastate.edu,
Iowa State University of Science & Technology – Hs
Franzi, David A., (518) 564-4033 franzida@plattsburgh.edu,
SUNY, Plattsburgh – Gl
Frappier, Amy, 518-580-8371 afrappie@skidmore.edu,
Skidmore College – GePe
Fraser, Alastair, 44-20-7594-6530 afraser@egi.utah.edu,
University of Utah – Gco
Fraser, Alistair B., abf1@psu.edu,
Pennsylvania State University, University Park – Am
Fraser, Donalf G., +44 (1865) 272033 don@earth.ox.ac.uk,
University of Oxford – Cg

Fraser-Smith, Antony C., (650) 723-3684 Stanford University – Ym
Frasson, Renato , frasson.1@osu.edu,
　Ohio State University – Hg
Fratta, Dante, 608-265-5644 fratta@wisc.edu,
　University of Wisconsin-Madison – Ng
Frazer, L. N., (808) 956-3724 neil@soest.hawaii.edu,
　University of Hawai'i, Manoa – Ys
Frederick, Daniel L., (931) 221-7455 frederickd@apsu.edu,
　Austin Peay State University – Gsr
Frederick, Holly, 570-408-4880 holly.frederick@wilkes.edi,
　Wilkes University – Sdb
Frederick, John E., (773) 702-3237 University of Chicago – As
Frederickson, Paul A., 831-595-5212 pafreder@nps.edu,
　Naval Postgraduate School – Am
Frederiksen, Andrew, (204) 474-9460 andrew_frederiksen@umanitoba.ca,
　University of Manitoba – YsGt
Fredrick, Kyle, (724) 938-4180 fredrick@calu.edu,
　California University of Pennsylvania – GgHgGm
Fredricks, Helen, 508-289-3678 hfredricks@whoi.edu,
　Woods Hole Oceanographic Institution – Oc
Freed, Andrew M., freed@purdue.edu,
　Purdue University – Gg
Freed, Robert L., (210) 999-7092 bfreed@trinity.edu,
　Trinity University – Gz
Freeman, Katherine H., (814) 863-8177 khf4@psu.edu,
　Pennsylvania State University, University Park – Cos
Freeman, Rebecca, (859) 257-3758 rebecca.freeman@uky.edu,
　University of Kentucky – GrPi
Freeman, Veronica, 740-376-4779 freemanv@marietta.edu,
　Marietta College – Pg
Fregenal, María Antonia, mariana@ucm.es,
　Univ Complutense de Madrid – Gs
Frei, Allan, (212) 772-5322 afrei@hunter.cuny.edu,
　Hunter College (CUNY) – AtHgs
Frei, Michaela, 089/2180 6590 michaela.frei@iaag.geo.uni-muenchen.de,
　Ludwig-Maximilians-Universitaet Muenchen – Gg
Freiburg, Jared, 217-244-2495 freiburg@illinois.edu,
　Illinois State Geological Survey – GgzGx
Freiermuth, Sue, (715) 425-3345 susan.m.freiermuth@uwrf.edu,
　University of Wisconsin, River Falls – Zn
Freile, Deborah, (201) 200-3161 dfreile@njcu.edu,
　New Jersey City University – GseGg
Freitas, Maria d., cfreitas@fc.ul.pt,
　Universidade de Lisboa – GseGu
Freitas, Maria da Conceição P., +351217500352 cfreitas@fc.ul.pt,
　Universidade de Lisboa – GusGe
French, Adam, (605) 394-1649 adam.french@sdsmt.edu,
　South Dakota School of Mines & Technology – Asm
French, Bevan, (202) 633-1326 Smithsonian Inst / Natl Museum of Natural Hist– Xm
French, Mark A., (609) 984-6587 mark.french@dep.state.nj.us,
　New Jersey Geological and Water Survey – Hq
French, Melodie E., 713.348.5088 mefrench@rice.edu,
　Rice University – Gc
Frenette, Jean, 418-656-8123 frenette@ggl.ulaval.ca,
　Universite Laval – Gz
Frew, Nelson M., (508) 289-2489 nfrew@whoi.edu,
　Woods Hole Oceanographic Institution – Ca
Frey, Bonnie A., (505) 835-5160 bfrey@nmt.edu,
　New Mexico Institute of Mining & Technology – Cg
Frey, Frederick A., (617) 253-2818 fafrey@mit.edu,
　Massachusetts Institute of Technology – CtGvi
Frey, Friedrich, 089/21804332 f.frey@lmu.de,
　Ludwig-Maximilians-Universitaet Muenchen – Gz
Frey, Holli M., 518-388-6418 freyh@union.edu,
　Union College – GvCag
Frey, Serita, serita.frey@unh.edu,
　University of New Hampshire – Sb
Fricke, Henry C., (719) 389-6514 hfricke@coloradocollege.edu,
　Colorado College – Cs
Fricker, Helen A., (858) 534-6145 hafricker@ucsd.edu,
　University of California, San Diego – Yr
Friedl, Mark, friell@bu.edu,
　Boston University – Zr
Friedl, Mark A., (617) 353-4807 Boston University – Zr
Friedman, Matt, +44 (1865) 272035 matt.friedman@earth.ox.ac.uk,
　University of Oxford – Pg

Friedmann, Samuel (Julio) J., (925) 423-0585 friedmann2@llnl.gov,
　Lawrence Livermore National Laboratory – Gg
Friedrich, Anke, friedrich@iaag.geo.uni-muenchen.de,
　Ludwig-Maximilians-Universitaet Muenchen – GgtCc
Friedrichs, Carl T., (804) 684-7303 cfried@vims.edu,
　College of William & Mary – On
Friehauf, Kurt, (610) 683-4446 friehauf@kutztown.edu,
　Kutztown University of Pennsylvania – EmCgGx
Frieman, Richard, richard.frieman@oswego.edu,
　SUNY, Oswego – Gg
Frierson, Dargan M., (206) 685-7364 dargan@atmos.washington.edu,
　University of Washington – As
Frind, Emil O., (519) 888-4567 (Ext. 3959) University of Waterloo – Hq
Frisbee, Marty, (765) 494-8678 mfrisbee@purdue.edu,
　Purdue University – Hws
Frisia, Silvia, +61 2 4921 5402 Silvia.Frisia@newcastle.edu.au,
　University of Newcastle – ClPeGd
Frisk, DeAnn M., (515) 294-4477 dfrisk@iastate.edu,
　Iowa State University of Science & Technology – Zn
Fritsch, J. Michael, (814) 863-1842 fritsch@ems.psu.edu,
　Pennsylvania State University, University Park – Am
Fritsen, Christian H., (775) 673-7487 cfritsen@dri.edu,
　Desert Research Institute – Pg
Fritton, Daniel D., (814) 865-1143 ddf@psu.edu,
　Pennsylvania State University, University Park – Sp
Fritts, Harold C., (520) 621-1608 hfritts@ltrr.arizona.edu,
　University of Arizona – Pe
Fritz, Allan, (785) 532-7245 akf@ksu.edu,
　Kansas State University – So
Fritz, Sherilyn C., (402) 472-6431 sfritz2@unl.edu,
　University of Nebraska, Lincoln – Pem
Fritz, William, 718-982-2440 william.fritz@csi.cuny.edu,
　Graduate School of the City University of New York – Zg
Fritz, William J., 718-982-2400 william.fritz@csi.cuny.edu,
　College of Staten Island/CUNY – Grv
Froehlich, David J., (512) 223-4894 eohippus@austincc.edu,
　Austin Community College District – PvZn
Froelich, Philip, 850-644-4331 froelich@ocean.fsu.edu,
　Florida State University – Og
Froese, Duane, (780) 492-1968 duane.froese@ualberta.ca,
　University of Alberta – GlmPe
Frohlich, Cliff, (512) 471-0460 cliff@ig.utexas.edu,
　University of Texas, Austin – Ys
Frolking, Stephen E., (603) 862-0244 steve.frolking@unh.edu,
　University of New Hampshire – As
Frolking, Tod A., (740) 587-6222 frolking@denison.edu,
　Denison University – Zy
Fromm, Jeanne M., (605) 582-3416 Jeanne.Fromm@usd.edu,
　University of South Dakota – GgHs
Frost, B. Ronald, (307) 399-5585 rfrost@uwyo.edu,
　University of Wyoming – GpiEd
Frost, Bruce W., 206-543-7186 frost@ocean.washington.edu,
　University of Washington – Ob
Frost, Carol D., frost@uwyo.edu,
　University of Wyoming – CcGit
Frost, Eric G., (619) 594-5003 eric.frost@sdsu.edu,
　San Diego State University – Zr
Frost, Gina M., 209 954 5380 gfrost@deltacollege.edu,
　San Joaquin Delta College – Gg
Fry, Matthew, 940-369-7576 mfry@unt.edu,
　University of North Texas – Zn
Fryar, Alan E., (859) 257-4392 alan.fryar@uky.edu,
　University of Kentucky – HwGeCl
Frye, John , fryej@uww.edu,
　University of Wisconsin, Whitewater – AtZr
Fryer, Brian J., 519-253-3000 ext, 3750 bfryer@uwindsor.ca,
　University of Windsor – Cla
Fryer, Karen H., khfryer@owu.edu,
　Ohio Wesleyan University – GcpGt
Fryer, Patricia B., (808) 956-3146 pfryer@soest.hawaii.edu,
　University of Hawai'i, Manoa – GuiGt
Fryett, I, fryettI@cf.ac.uk,
　Cardiff University – Zn
Fryxell, Joan E., (909) 537-5311 jfryxell@csusb.edu,
　California State University, San Bernardino – Gct
Fthenakis, Vasilis M., (516) 282-2830 Columbia University – As
Fu, Qi, (713) 743-3660 qfu3@uh.edu,

University of Houston – CosCa
Fu, Qiang, (206) 685-2070 qfu@atmos.washington.edu,
 University of Washington – As
Fu, Qilong, 512-232-9372 qilong.fu@beg.utexas.edu,
 University of Texas, Austin – Gsr
Fu, Roger, 617-384-6991 rogerfu@fas.harvard.edu,
 Harvard University – YmXym
Fu, Rong, 512-232-7932 rongfu@jsg.utexas.edu,
 University of Texas, Austin – As
Fubelli, Giandomenico, giandomenico.fubelli@unito.it,
 Università di Torino – Gm
Fudge, T.J., 206-543-0162 tjfudge@uw.edu,
 University of Washington – Gl
Fueglistaler, Stephan A., (609) 258-8238 stf@princeton.edu,
 Princeton University – As
Fuellhart, Kurtis G., (717) 477-1309 kgfuel@ship.edu,
 Shippensburg University – ZnnZn
Fuente, David E., 803 777-2757 fuente@seoe.sc.edu,
 University of South Carolina – Zn
Fuentes, Edna L., (312) 996-6123 eriver15@uic.edu,
 University of Illinois at Chicago – Zn
Fueten, Frank, (905) 688-5550 (Ext. 3856) FFueten@brocku.ca,
 Brock University – Gc
Fuge, Ron, +44 (0)1970 622 642 rrf@aber.ac.uk,
 Aberystwyth University – GbCgGe
Fugitt, Frank, (614) 265-6759 frank.fugitt@dnr.state.oh.us,
 Ohio Dept of Natural Resources – GrHw
Fuhrmann, Christopher, (662) 268-1032 Ext 219 cmf396@msstate.edu,
 Mississippi State University – As
Fuiman, Lee A., (361) 749-6775 lee.fuiman@utexas.edu,
 University of Texas, Austin – Ob
Fujita, Kazuya, 517-355-0142 fujita@msu.edu,
 Michigan State University – Ys
Full, Robert J., (510) 642-9896 rjfull@garnet.berkeley.edu,
 University of California, Berkeley – Pi
Fullagar, Paul D., fullagar@unc.edu,
 University of North Carolina, Chapel Hill – CcGaZe
Fullmer, James W., (203) 392-5841 fullmerj1@southernct.edu,
 Southern Connecticut State University – Am
Fulthorpe, Craig S., (512) 471-0459 craig@ig.utexas.edu,
 University of Texas, Austin – Gu
Fulton, Patrick M., (979) 862-2493 pfulton@tamu.edu,
 Texas A&M University – YhHyRn
Fultz, Lisa M., 225-578-1344 lfutz@agcenter.lsu.edu,
 Louisiana State University – Sb
Fulweiler, Robinson, 617-358-5466 rwf@bu.edu,
 Boston University – Cm
Fung, Inez, (510) 643-9367 inez@atmos.berkeley.edu,
 University of California, Berkeley – As
Funning, Gareth J., gareth.funning@ucr.edu,
 University of California, Riverside – YdgYs
Furbish, David J., 615-322-2137 david.j.furbish@vanderbilt.edu,
 Vanderbilt University – GmHg
Furió, Marc, marc.furio@uab.cat,
 Universitat Autonoma de Barcelona – Pv
Furlong, Kevin P., (814) 863-0567 kevin@geodyn.psu.edu,
 Pennsylvania State University, University Park – Gt
Furman, Tanya, (814) 865-5782 furman@psu.edu,
 Pennsylvania State University, University Park – GiCu
Fusch, Richard D., (740) 368-3616 rdfusch@owu.edu,
 Ohio Wesleyan University – Zu
Fusseis, Florian, +44 (0) 131 650 6755 ffusseis@staffmail.ed.ac.uk,
 Edinburgh University – Gc
Fyfe, John C., (250) 363-8236 john.fyfe@ec.gc.ca,
 University of Victoria – As
Fürst, Christine, +345 5526017 christine.fuerst@geo.uni-halle.de,
 Martin-Luther-Universitaet Halle-Wittenberg – ZucZi
Fyson, William K., 613-745-6645 wfyson@uottawa.ca,
 University of Ottawa – Gc

G

Gaasterland, Terry, (858) 822-4600 tgaasterland@ucsd.edu,
 University of California, San Diego – Ob
Gabbott, Sarah, +440116 252 3636 sg21@le.ac.uk,
 Leicester University – Pg
Gabel, Mark, (605) 642-9035 Black Hills State University – Pb
Gabel, Sharon, 281-425-6335 sgabel@lee.edu,
 Lee College – GseGg
Gabet, Emmanuel, 408 924-5035 manny.gabet@sjsu.edu,
 San Jose State University – Gm
Gabitov, Rinat, (662) 268-1032 Ext 218 rg850@msstate.edu,
 Mississippi State University – Cas
Gable, Carl W., (505) 665-3533 gable@lanl.gov,
 Los Alamos National Laboratory – Gt
Gabler, Christopher, 956 882 7656 christopher.gabler@utrgv.edu,
 University of Texas, Rio Grande Valley – PeSf
Gaboury, Damien, 418-545-5011 dgaboury@uqac.ca,
 Universite du Quebec a Chicoutimi – EmCe
Gabrielli, Paolo, 614 292 6664 gabrielli.1@osu.edu,
 Ohio State University – CtPeGl
Gabrielov, Andrei, (765) 496-2868 gabriea@purdue.edu,
 Purdue University – Ys
Gachon, Philippe, 514-987-3000 #2601 gachon.philippe@uqam.ca,
 Universite du Quebec a Montreal – As
Gacía Joral, Fernando, fgjoral@ucm.es,
 Univ Complutense de Madrid – Pi
Gadomski, Frederick J., (814) 863-4229 gadomski@mail.meteo.psu.edu,
 Pennsylvania State University, University Park – Am
Gaede, Oliver M., +61 7 3138 2535 oliver.gaede@qut.edu.au,
 Queensland University of Technology – YeNrYs
Gaetani, Glenn A., (508) 289-3724 ggaetani@whoi.edu,
 Woods Hole Oceanographic Institution – Gi
Gaffney, Edward S., (505) 665-6387 gaffney@lanl.gov,
 Los Alamos National Laboratory – Yg
Gaffney, Eugene S., (212) 769-5801 American Museum of Natural History – Pv
Gaffney, Jeffrey S., (630) 252-5178 gaffney@anl.gov,
 Argonne National Laboratory – As
Gagnaison, Cyril, +33(0)3 44068997 cyril.gagnaison@lasalle-beauvais.fr,
 Institut Polytechnique LaSalle Beauvais (ex-IGAL) – PgGsa
Gagne, Jean-Pierre, 418-723-1986 ext 1870 jean-pierre_gagne@uqar.ca,
 Universite du Quebec a Rimouski – OcCmo
Gagne, Marc R., (610) 436-3014 mgagne@wcupa.edu,
 West Chester University – Xa
Gagnon, Alan R., (508) 289-2961 agagnon@whoi.edu,
 Woods Hole Oceanographic Institution – Cs
Gagnon, Joel E., (519) 253-3000 x2496 jgagnon@uwindsor.ca,
 University of Windsor – CaqEg
Gagnon, Katie, 206-516-3161 KGagnon@sccd.ctc.edu,
 Seattle Central Community College – OuZg
Gagnon, Teresa K., (860) 424-3680 teresa.gagnon@ct.gov,
 Dept of Energy and Environmental Protection – Gg
Gagosian, Carol, 781-283-3151 cgagosian@wellesley.edu,
 Wellesley College – Zn
Gaherty, James, gaherty@ldeo.columbia.edu,
 Columbia University – Ys
Gahn, Forest J., (208) 496-1900 gahnf@byui.edu,
 Brigham Young University - Idaho – PggGs
Gaidos, Eric J., (808) 956-7897 gaidos@hawaii.edu,
 University of Hawai'i, Manoa – Yg
Gaines, Robert R., (909) 621-8674 robert.gaines@pomona.edu,
 Pomona College – GsClPi
Gajewski, Dirk J., +4940428382975 dirk.gajewski@uni-hamburg.de,
 Universitaet Hamburg – YesYg
Gakka, Dr U., +914027097116 udayalaxmi.g@gmail.com,
 Osmania University – HwYge
Galal, Galal M., 002-03-3921595 galalgalal2004@yahoo.com,
 Alexandria University – GgPeGe
Galbraith, John M., (540) 231-9784 ttcf@vt.edu,
 Virginia Polytechnic Institute & State University – Sp
Gale, Andrew, +44 023 92 846127 andy.gale@port.ac.uk,
 University of Portsmouth – PgGrCl
Gale, Julia F., (512) 232-7957 julia.gale@beg.utexas.edu,
 University of Texas, Austin – GctGo
Gale, Marjorie H., (802) 522-5210 marjorie.gale@vermont.gov,
 Agency of Natural Resources, Dept of Env Conservation – GgcGe
Gale, Paula M., (731) 881-7326 pgale@utm.edu,
 University of Tennessee, Martin – SodSc
Galea, Anthony, anthony.galea@um.edu.mt,
 University of Malta – Pi
Galea, Pauline, +356 2340 3034 pauline.galea@um.edu.mt,
 University of Malta – YsgZg
Galewsky, Joseph, 505-277-4204 galewsky@unm.edu,
 University of New Mexico – Am

Galgaro, Antonio, +390498279123 antonio.galgaro@unipd.it,
Università degli Studi di Padova – Ng
Galicki, Stan, (601) 974-1340 galics@millsaps.edu,
Millsaps College – Ged
Galindo, María del Carmen, cgalindo@ucm.es,
Univ Complutense de Madrid – Cc
Gall, Quentin, (613) 234-0188 qgall@rogers.com,
University of Ottawa – GsEmGe
Gallagher, Eugene, eugene.gallagher@umb.edu,
University of Massachusetts, Boston – Ob
Gallagher, Martin, +44 0161 306-3937 martin.gallagher@manchester.ac.uk,
University of Manchester – As
Gallagher, William B., 609-896-5000 ext. 7784 wgallagher@rider.edu,
Rider University – PvgGr
Gallegos, Anthony F., (505) 665-0862 agallegos@lanl.gov,
Los Alamos National Laboratory – Pg
Gallen, Sean, sean.gallen@colostate.edu,
Colorado State University – GemGt
Galli, Kenneth G., (617) 552-4504 kenneth.galli@bc.edu,
Boston College – GsdGg
Gallovic, Frantisek, 089/2180 4209 frantisek.gallovic@geophysik.uni-muenchen.de,
Ludwig-Maximilians-Universitaet Muenchen – Yg
Galloway, James N., (804) 924-1303 jng@virginia.edu,
University of Virginia – As
Galloway, William E., (512) 471-5673 galloway@austin.utexas.edu,
University of Texas, Austin – GsoGr
Galloway, William E., (512) 471-0494 galloway@ischool.utexas.edu,
University of Texas, Austin – Gs
Gallup, Christina D., (218) 726-8984 cgallup@d.umn.edu,
University of Minnesota, Duluth – Cc
Gallus, William A., (515) 294-2270 wgallus@iastate.edu,
Iowa State University of Science & Technology – As
Galperin, Boris, (727) 553-1101 bgalperin@marine.usf.edu,
University of South Florida – Op
Galsa, Attila, gali@pangea.elte.hu,
Eotvos Lorand University – YgHwYe
Galy, Valier, (508) 289-2340 vgaly@whoi.edu,
Woods Hole Oceanographic Institution – Cm
Gamage, Kusali R., kgamage@austincc.edu,
Austin Community College District – HyOuPm
Gamble, Audrey V., (334) 844-4100 avg0001@auburn.edu,
Auburn University – Sc
Gamble, Douglas W., (910) 962-3778 gambled@uncw.edu,
University of North Carolina, Wilmington – Zy
Gamble, John, +353 21 4903955 j.gamble@ucc.ie,
University College Cork – Gi
Gambrell, Robert P., (225) 388-6426 Louisiana State University – Cg
Gamerdinger, Amy P., (509) 373-3077 amy.gamerdinger@pnl.gov,
Pacific Northwest National Laboratory – Cl
Gammack-Clark, James, 561 297-0314 jgammack@fau.edu,
Florida Atlantic University – Zri
Gammon, Richard H., (206) 543-1609 gammon@macmail.chem.washington.edu,
University of Washington – Oc
Gammons, Christopher H., (406) 496-4763 cgammons@mtech.edu,
Montana Tech of the University of Montana – CgEmCs
Gamon, John, 780-492-0345 gamon@ualberta.ca,
University of Alberta – Zr
Gancarz, Alexander J., (505) 667-2606 Los Alamos National Laboratory – Cc
Ganeshram, Raja, +44 (0) 131 650 7364 R.Ganeshram@ed.ac.uk,
Edinburgh University – Pg
Ganey-Curry, Patricia E., (512) 471-0408 patty@ig.utexas.edu,
University of Texas, Austin – Zn
Ganguli, Rajive, (907) 474-7212 rganguli@alaska.edu,
University of Alaska, Fairbanks – Nm
Ganguly, Jiba, 520-621-6006 ganguly@email.arizona.edu,
University of Arizona – Gx
Gani, M. Royhan, (270) 745-5977 royhan.gani@wku.edu,
Western Kentucky University – GsrGg
Gani, Nahid, nahid.gani@wku.edu,
Western Kentucky University – Gc
Gann, Delbert E., 601-974-1341 gannde@millsaps.edu,
Millsaps College – Gz
Gannon, J. Michael, Mike.Gannon@dnr.iowa.gov,
Iowa Dept of Natural Resources – Gg

Gannon, John P., (828) 227-3813 jpgannon@wcu.edu,
Western Carolina University – HgSo
Gans, Phillip B., (805) 893-2642 University of California, Santa Barbara – Gt
Gansecki, Cheryl A., (808) 932-7549 gansecki@hawaii.edu,
University of Hawai'i, Hilo – GviZe
Ganssen, Gerald M., +31 20 59 87369 g.m.ganssen@vu.nl,
Vrije Universiteit Amsterdam – ObPe
Gantzer, Clark J., 573-882-0611 University of Missouri, Columbia – So
Gao, Dengliang, 304-293-3310 Dengliang.Gao@mail.wvu.edu,
West Virginia University – Ye
Gao, Haiying, 413 577-1250 haiyinggao@geo.umass.edu,
University of Massachusetts, Amherst – Ys
Gao, Mengsheng, 352-392-1951 msgao@ufl.edu,
University of Florida – Zn
Gao, Oliver H., hg55@cornell.edu,
Cornell University – Zn
Gao, Yongjun, 713-743-4382 yongjungao@uh.edu,
University of Houston – CstCa
Gao, Yuan, 973-353-1139 yuangaoh@newark.rutgers.edu,
Rutgers, The State University of New Jersey, Newark – As
Garbesi, Karina, 510-885-3172 karina.garbesi@csueastbay.edu,
California State University, East Bay – Cg
Garces, Milton A., (808) 325-1558 milton@higp.hawaii.edu,
University of Hawai'i, Manoa – Ys
Garcia, Antonio F., (805) 756-2430 afgarcia@calpoly.edu,
California Polytechnic State University – Gmg
Garcia, Jan, (323) 259-2823 jangarcia@oxy.edu,
Occidental College – Zn
Garcia, Juan, jgarcia@cicese.mx,
Centro de Investigación Científica y de Educación Superior de Ensenada – Yv
Garcia, Marcelo H., (217) 356-9482 mhgarcia@illinois.edu,
University of Illinois, Urbana-Champaign – Gmu
Garcia, Michael O., (808) 956-6641 garcia@soest.hawaii.edu,
University of Hawai'i, Manoa – Gi
García, Nuria, nugarcia@ucm.es,
Univ Complutense de Madrid – Pv
Garcia, Oswaldo, (415) 338-2061 ogarcia@sfsu.edu,
San Francisco State University – As
Garcia, Sammy R., (505) 667-4151 Los Alamos National Laboratory – Ca
Garcia, William, 704-687-5982 wjgarcia@uncc.edu,
University of North Carolina, Charlotte – Pv
García Frank, Alejandra, agfrank@ucm.es,
Univ Complutense de Madrid – Gg
García Galán, Gúmer, ++935812835 gumer.galan@uab.cat,
Universitat Autonoma de Barcelona – GiCg
Garcia Lasanta, Cristina, (360) 650-3835 Cristina.GarciaLasanta@wwu.edu,
Western Washington University – YmGtc
García Lorenzo, María de la Luz, mglorenzo@ucm.es,
Univ Complutense de Madrid – Cl
Garcia-Rubio, Luis H., (727) 553-1246 garcia@marine.usf.edu,
University of South Florida – Oc
Gardner, Eleanor E., 731-881-7444 egardne3@utm.edu,
University of Tennessee, Martin – GgPg
Gardner, James D., (403) 823-7707 Royal Tyrrell Museum of Palaeontology – Pv
Gardner, James E., (512) 471-0953 gardner@mail.utexas.edu,
University of Texas, Austin – Gv
Gardner, Jamie N., (505) 667-1799 jgardner@lanl.gov,
Los Alamos National Laboratory – Gg
Gardner, Payton, (406) 243-2458 payton.gardner@umontana.edu,
University of Montana – Hw
Gardner, Thomas W., tgardner@trinity.edu,
Trinity University – GmtHg
Gardner, Wayne S., (361) 749-6823 wayne.gardner@utexas.edu,
University of Texas, Austin – Ob
Gardner, Wilford D., (979) 845-7211 wgardner@ocean.tamu.edu,
Texas A&M University – OugZr
Gardulski, Anne F., 617-627-2891 anne.gardulski@tufts.edu,
Tufts University – GrdGg
Garfield, Newell (Toby), (415) 338-3713 garfield@sfsu.edu,
San Francisco State University – Op
Garihan, John M., (864) 294-2052 jack.garihan@furman.edu,
Furman University – Gct
Garnero, Edward, (480) 965-7653 garnero@asu.edu,
Arizona State University – Ys

Garren, Sandra J., 516-463-5565 sandra.j.garren@hofstra.edu,
    Hofstra University – Zi
Garrett, Christopher J. R., (250) 721-7702 garrett@uvphys.phys.uvic.ca,
    University of Victoria – Op
Garrett, Timothy J., 801-581-5768 tim.garrett@utah.edu,
    University of Utah – As
Garrick-Bethell, Ian, 831-459-1277 igarrick@ucsc.edu,
    University of California, Santa Cruz – Xy
Garrison, Ervan G., (706) 542-1097 egarriso@uga.edu,
    University of Georgia – GamGn
Garrison, Jennifer M., (323) 343-2412 jgarris@calstatela.edu,
    California State University, Los Angeles – GiCcGv
Garrison, Robert E., (831) 459-5563 University of California, Santa Cruz
    – Gs
Garrison, Trent, (859) 572-6683 garrisont@nku.edu,
    Northern Kentucky University – GeHwEc
Garrote, Julio, juliog@ucm.es,
    Univ Complutense de Madrid – Gm
Garside, Larry J., 775-784-6693 lgarside@unr.edu,
    University of Nevada – Gg
Garstang, Mimi R., 573-368-2101 mimi.garstang@dnr.mo.gov,
    Missouri Dept of Natural Resources – Ge
Gartner, Janette, (781) 891-2901 jgartner@bentley.edu,
    Bentley University – Hy
Garven, G, (617) 627-3795 grant.garven@tufts.edu,
    Tufts University – HwqHy
Garver, John I., (518) 388-6770 garverj@union.edu,
    Union College – GtCcGr
Garvie, Laurence, (480) 965-3361 lgarvie@asu.edu,
    Arizona State University – GzXc
Garvin, Theresa D., (780) 492-4593 theresa.garvin@ualberta.ca,
    University of Alberta – Zn
Garwood, Phil, 910 392 7111 pgarwood@cfcc.edu,
    Cape Fear Community College – Gg
Garwood, Roland W., garwood@nps.edu,
    Naval Postgraduate School – Op
Garza, Jessica, 325-486-6987 jessica.garza@angelo.edu,
    Angelo State University – AmGg
Garzione, Carmala N., (585) 273-4572 garzione@earth.rochester.edu,
    University of Rochester – Gs
Gaschnig, Richard M., (978) 934-3706 richard_gaschnig@uml.edu,
    University of Massachusetts, Lowell – CctCa
Gascho, Gary J., (912) 386-3329 University of Georgia – Sc
Gasparini, Nicole, 504-862-3197 ngaspari@tulane.edu,
    Tulane University – GmHs
Gastaldo, Robert A., (207) 859-5807 ragastal@colby.edu,
    Colby College – PbGs
Gaston, Lewis A., (504) 388-1323 lagaston@agcenter.lsu.edu,
    Louisiana State University – Sc
Gates, Alexander E., 973-353-5034 agates@rutgers.edu,
    Rutgers, The State University of New Jersey, Newark – Gc
Gates, Ruth D., (808) 236-7420 rgates@hawaii.edu,
    University of Hawai'i, Manoa – Ob
Gathany, Mark, 937-766-3823 mgathany@cedarville.edu,
    Cedarville University – ZiuSb
Gattiglio, Marco, marco.gattiglio@unito.it,
    Università di Torino – Gc
Gatto, Roberto, +390498279172 roberto.gatto@unipd.it,
    Università degli Studi di Padova – Pg
Gaubatz, Piper, (413) 545-0768 gaubatz@geo.umass.edu,
    University of Massachusetts, Amherst – Zy
Gauci, Adam, adam.gauci@um.edu.mt,
    University of Malta – Op
Gaudette, Henri E., (603) 862-1718 University of New Hampshire – Cc
Gaulton, Rachel, +44 (0) 191 208 6577 rachel.gaulton@ncl.ac.uk,
    University of Newcastle Upon Tyne – Zr
Gautam, Tej P., 740-376-4371 tej.gautam@marietta.edu,
    Marietta College – NgGeZi
Gauthier, Donald J., (818) 778-5514 gauthidj@lavc.edu,
    Los Angeles Valley College – ZiyAm
Gauthier, Jacques, (203) 432-3150 jacques.gauthier@yale.edu,
    Yale University – Pv
Gauthier, Michel, 514-987-3000 #4560 gauthier.michel@uqam.ca,
    Universite du Quebec a Montreal – Eg
Gauthier, Paul, (609) 258-7442 ppg@princeton.edu,
    Princeton University – Sb
Gauthier, Pierre, 514-987-3000 #3304 gauthier.pierre@uqam.ca,
    Universite du Quebec a Montreal – Am
Gautsch, Jacklyn, 319-335-1761 jackie.gautsch@dnr.iowa.gov,
    Iowa Dept of Natural Resources – Zn
Gavin, Dan, (541) 346-5787 dgavin@uoregon.edu,
    University of Oregon – ZyPle
Gavriloaiei, Traian, 0040232201462 tgavrilo@uaic.ro,
    Alexandru Ioan Cuza – CaAsCg
Gawloski, Joan, 432-685-4630 jgawloski@midland.edu,
    Midland College – GgzGe
Gawu, Simon K., +233 244 067804 skygawu@yahoo.com,
    Kwame Nkrumah University of Science and Technology – GgEgZn
Gay, Kenny, 9197337353 x28 kenny.gay@ncdenr.gov,
    North Carolina Geological Survey – GzsPi
Gaya, Charles , cogaya@jkuat.ac.ke,
    Jomo Kenyatta University of Agriculture & Technology – ZrgZi
Gaylord, David R., (509) 335-8127 gaylordd@wsu.edu,
    Washington State University – Gs
Gazel, Esteban, egazel@cornell.edu,
    Cornell University – CeGiv
Gazis, Carey A., 509.963-2820 cgazis@geology.cwu.edu,
    Central Washington University – CsHw
Ge, Shemin, (303) 492-8323 ges@colorado.edu,
    University of Colorado – HwEoYg
Geary, Dana H., dana@geology.wisc.edu,
    University of Wisconsin-Madison – Pg
Geary, Lindsey, (315)792-3134 Utica College – Geg
Geary, Phil, 61 02 4921 6726 phil.geary@newcastle.edu.au,
    University of Newcastle – HwSoHw
Gebrande, Helmut, 089/2180 4325 gebrande@geophysik.uni-muenchen.de,
    Ludwig-Maximilians-Universitaet Muenchen – YgsEo
Gedzelman, Stanley, (212) 650-6470 City College (CUNY) – As
Gee, Jeffrey S., (858) 534-4707 jsgee@ucsd.edu,
    University of California, San Diego – Gu
Gehrels, George E., (520) 349-4702 ggehrels@email.arizona.edu,
    University of Arizona – Gt
Gehrels, Tom, (520) 621-6970 tgehrels@lpl.arizona.edu,
    University of Arizona – Zn
Geibert, Walter, +44 (0) 131 651 7704 Walter.Geibert@ed.ac.uk,
    Edinburgh University – Cc
Geidel, Gwendelyn, (803) 777-7171 geidel@environ.sc.edu,
    University of South Carolina – HwGeCg
Geiger, James W., 217-265-8989 jgeiger@illinois.edu,
    Illinois State Geological Survey – Ge
Geiger, Sebastian, s.geiger@hw.ac.uk,
    Heriot-Watt University – NpEoYh
Geissman, John D., 972-883-2403 geissman@utdallas.edu,
    University of Texas, Dallas – GtYm
Geissman, John W., (505) 277-3433 jgeiss@unm.edu,
    University of Michigan – Ym
Geist, Dennis J., (208) 885-6491 dgeist@uidaho.edu,
    University of Idaho – Giv
Geller, Jonathan, 831-771-4400 geller@mlml.calstate.edu,
    Moss Landing Marine Laboratories – Zn
Geller, Marvin A., (631) 632-8701 marvin.geller@stonybrook.edu,
    SUNY, Stony Brook – As
Gemignani, Robert, (314) 935-4614 rgemigna@levee.wustl.edu,
    Washington University in St. Louis – Zn
Gemmell, J B., 61 3 6226 2893 bruce.gemmell@utas.edu.au,
    University of Tasmania – Eg
Gemperline, Johanna M., (410) 554-5552 johanna.gemperline@maryland.gov,
    Maryland Department of Natural Resources – Hw
Genareau, Kimberly, 205-348-1878 kdg@ua.edu,
    University of Alabama – Gvi
Gendzwill, Donald J., don.gendzwill@usask.ca,
    University of Saskatchewan – Ye
Genge, Matthew, +44 20 759 46499 m.genge@imperial.ac.uk,
    Imperial College – Xm
Genna, Dominique, gennadomi@hotmail.com,
    Universite du Quebec a Chicoutimi – GvEn
Gentile, Richard J., (816) 235-2974 gentiler@umkc.edu,
    University of Missouri, Kansas City – GrsNx
Gentry, Amanda L., (303) 319-0695 amandagentry@weber.edu,
    Weber State University – GgsGg
Gentry, Christopher, 931-221-7478 gentryc@apsu.edu,
    Austin Peay State University – Ziy
Gentry, Terry, 979-845-5323 tjgentry@tamu.edu,
    Texas A&M University – Sb

George, Graham, 306-966-5722 g.george@usask.ca,
 University of Saskatchewan – Cg
Georgen, Jennifer, 850-645-4987 georgen@gly.fsu.edu,
 Florida State University – Gu
Georgen, Jennifer, 757-683-5198 jgeorgen@odu.edu,
 Old Dominion University – Gu
Georgiev, Svetoslav, (970)-491-3789 svetoslav.georgiev@colostate.edu,
 Colorado State University – CcGo
Georgiopopoulou, Aggeliki, (+353) 1 716 2062 aggie.georg@ucd.ie,
 University College Dublin – GusOg
Geraghty Ward, Emily, emily.ward@rocky.edu,
 Rocky Mountain College – ZeGc
Gerald, Carresse, (919) 530-7117 cgerald6@nccu.edu,
 North Carolina Central University – GeZg
Gerba, Charles P., (520) 621-6906 gerba@ag.arizona.edu,
 University of Arizona – Sb
GERBE, Marie-Christine, 33-477485123 gerbe@univ-st-etienne.fr,
 Université Jean Monnet, Saint-Etienne – GisZe
Gerber, Stefan, 352294-3174 sgerber@ufl.edu,
 University of Florida – So
Gerbi, Christopher C., 207 581-2153 University of Maine – Gt
Gerbi, Greg, 518-580-5127 ggerbi@skidmore.edu,
 Skidmore College – OpYg
Gerhard, Lee C., 78538643965 leegtn37@gmail.com,
 University of Kansas – GsoGd
Gerhardt, Hannes, (678) 839-4064 hgerhard@westga.edu,
 University of West Georgia – Zn
Gerke, Tammie, 513-727-3268 gerketl@miamioh.edu,
 Miami University – ClGea
Gerla, Philip J., (701) 777-3305 phil.gerla@engr.und.edu,
 University of North Dakota – Hw
German, Chris, 508-289-2853 cgerman@whoi.edu,
 Woods Hole Oceanographic Institution – Gt
Germanoski, Dru, (610) 330-5196 germanod@lafayette.edu,
 Lafayette College – Gm
Germanovich, Leonid, (404) 894-2284 leonid@ce.gatech.edu,
 Georgia Institute of Technology – NrYg
Gernon, Thomas M., +44 (0)23 8059 2670 thomas.gernon@noc.soton.ac.uk,
 University of Southampton – ZgGvi
Gerry, Janelle, 402-472-2663 University of Nebraska, Lincoln – Zn
Gertisser, Ralf, (+44) 01782 733181 r.gertisser@keele.ac.uk,
 Keele University – GivGz
Gervais, Melissa, mmg62@psu.edu,
 Pennsylvania State University, University Park – As
Gerwick, William H., (858) 534-0578 wgerwick@ucsd.edu,
 University of California, San Diego – Cm
Geršlová, Eva, +420 549 49 4027 gerslova@sci.muni.cz,
 Masaryk University – Ge
Gess, Robert, +27 (0)82-7595848 robg@imaginet.co.za,
 Rhodes University – Pg
Ghani, Azman Abdul, 03-79674234 azmangeo@um.edu.my,
 University of Malaya – Gi
Gharti, Hom Nath, (609) 258-2605 hgharti@princeton.edu,
 Princeton University – YsdYg
Ghassemi, Ahmed, ahmad.ghassemi@ou.edu,
 University of Oklahoma – Nr
Ghate, Virendra P., (630) 252-1609 vghate@anl.gov,
 Argonne National Laboratory – AsmZr
Ghattas, Omar, (512) 232-4304 omar@ices.utexas.edu,
 University of Texas, Austin – ZnYg
Ghazala, Hosni H., (109) 688-7904 ghazala@mans.edu.eg,
 El Mansoura University – Ygx
Ghent, Edward D., (403) 220-5847 ghent@geo.ucalgary.ca,
 University of Calgary – Gp
Ghent, Rebecca, (416) 978-0597 ghentr@geology.utoronto.ca,
 University of Toronto – Zr
Ghil, Michael, (310) 206-2285 ghil@atmos.ucla.edu,
 University of California, Los Angeles – AsYdOp
Ghinassi, Massimiliano, +390498279181 massimiliano.ghinassi@unipd.it,
 Università degli Studi di Padova – Gs
Ghiorso, Mark, 206-550-1850 ghiorso@uw.edu,
 University of Washington – CgGx
Ghoneim, Eman, ghoneime@uncw.edu,
 Boston University – Gm
Ghosh, Abhijit, (951) 827-4493 aghosh@ucr.edu,
 University of California, Riverside – YsGt
Ghosheh, Baher A., (814) 732-2207 ghosheh@edinboro.edu,
 Edinboro University of Pennsylvania – Zn
Giacalone, Joe, 520-626-8365 giacalone@lpl.arizona.edu,
 University of Arizona – Xy
Giachetti, Thomas, tgiachet@uoregon.edu,
 University of Oregon – Gvu
Giannini, Alessandra, (845) 680-4473 alesall@iri.columbia.edu,
 Columbia University – AsOpZc
Gianniny, Gary, (970) 247-7254 gianniny_g@fortlewis.edu,
 Fort Lewis College – Gs
Gianotti, Franco, franco.gianotti@unito.it,
 Università di Torino – Gg
Giaramita, Mario J., (209) 667-3558 mgiaramita@csustan.edu,
 California State University, Stanislaus – Gpz
Giardino, John R., (979) 845-3224 giardino@geo.tamu.edu,
 Texas A&M University – GmNg
Giardino, John R., (409) 867-9067 rickg@tamu.edu,
 Texas A&M University – Gm
Giardino, Marco, marco.giardino@unito.it,
 Università di Torino – Gm
Gibbons, Doug, 206-685-8180 dgibbons@uw.edu,
 University of Washington – ZeYs
Gibbs, Gaynell, ocean@lsu.edu,
 Louisiana State University – Zn
Gibbs, Samantha J., +44 (0)23 80592003 Samantha.Gibbs@noc.soton.ac.uk,
 University of Southampton – Zg
Gibeaut, James, 361-825-2060 James.Gibeaut@tamucc.edu,
 Texas A&M University, Corpus Christi – OnZi
Gibling, Martin R., mgibling@is.dal.ca,
 Dalhousie University – Gs
Gibson, Andy, +44 023 92 842654 andy.gibson@port.ac.uk,
 University of Portsmouth – Eg
Gibson, Carl H., (858) 534-3184 cgibson@ucsd.edu,
 University of California, San Diego – No
Gibson, David, (207) 778-7402 dgibson@maine.edu,
 University of Maine - Farmington – GigGz
Gibson, Deana, (417) 836-5801 deanagibson@missouristate.edu,
 Missouri State University – Zn
Gibson, Deidre M., 757.727.5883 deidre.gibson@hamptonu.edu,
 Hampton University – Ob
Gibson, Glen, 719-333-3080 glen.gibson@usafa.edu,
 United States Air Force Academy – Zri
Gibson, H. Daniel (Dan), (778) 782-7057 hdgibson@sfu.ca,
 Simon Fraser University – GcCcGp
Gibson, Harold L., 7056751151 x2371 hgibson@laurentian.ca,
 Laurentian University, Sudbury – EmGv
Gibson, Michael A., (731) 881-7435 mgibson@utm.edu,
 University of Tennessee, Martin – PgiZe
Gibson, Roger L., 011 717 6553 roger.gibson@wits.ac.za,
 University of the Witwatersrand – GcpGg
Gibson, Ronald C., (714) 895-8194 rgibson@gwc.cccd.edu,
 Golden West College – Gc
Gibson, Sally, +44 (0) 1223 333401 sally@esc.cam.ac.uk,
 University of Cambridge – Gi
Gibson, Jr., Richard L., (979) 862-8653 gibson@tamu.edu,
 Texas A&M University – Yse
Gidigasu, Solomon S., +233 27 7807 707 ssrgidigasu.soe@knust.edu.gh,
 Kwame Nkrumah University of Science and Technology – GeZrNr
Gidwani, Vinay, 612-625-1397 Graduate School of the City University of New York – Zn
Giegengack, Jr., Robert F., (215) 898-5191 gieg@sas.upenn.edu,
 University of Pennsylvania – Gg
Giere, Reto, giere@sas.upenn.edu,
 University of Pennsylvania – GzCgGe
Gierke, John S., (906) 487-2535 jsgierke@mtu.edu,
 Michigan Technological University – Hy
Giese, Benjamin S., (979) 845-2306 bgiese@ocean.tamu.edu,
 Texas A&M University – Op
Giese, Graham S., (508) 289-2297 ggiese@whoi.edu,
 Woods Hole Oceanographic Institution – On
Giese, Rossman F., (716)645-4263 glgclay@buffalo.edu,
 SUNY, Buffalo – Gz
Gieskes, Joris M., (858) 534-4257 jgieskes@ucsd.edu,
 University of California, San Diego – Oc
Gifford, Dian J., (401) 874-6690 dgifford@uri.edu,
 University of Rhode Island – Ob
Gifford, Jennifer N., 662-915-2079 jngiffor@olemiss.edu,
 University of Mississippi – GtxCg

Gifford-Gonzalez, Dianne, (831) 459-2633 University of California, Santa Cruz – Zn
Gigler, Alexander, 089/2180 4185 Ludwig-Maximilians-Universitaet Muenchen – Gz
Giglierano, James D., James.Giglierano@dnr.iowa.gov, Iowa Dept of Natural Resources – GgZri
Gilbert, Co'Quesie, (619) 534-2470 cag65@columbia.edu, Columbia University – Zn
Gilbert, Hersh, (765) 496-9518 hersh@purdue.edu, Purdue University – Ys
Gilbert, Kathleen W., 781-283-3086 kgilbert@wellesley.edu, Wellesley College – Cs
Gilbert, M. Charles, (405) 325-3253 mcgilbert@ou.edu, University of Oklahoma – Cp
Gilbert, Patricia M., (410) 221-8422 University of Maryland – Ob
Gilbes, Fernando, 787 2653845 gilbes@cacique.uprm.edu, University of Puerto Rico – ZrGu
Gilder, Stuart, 089/2180 4239 stuart.gilder@geophysik.uni-muenchen.de, Ludwig-Maximilians-Universitaet Muenchen – Ym
Giles, Antony, (432) 685-5580 agiles@midland.edu, Midland College – GgvGi
Giles, David, +44 023 92 842248 david.giles@port.ac.uk, University of Portsmouth – Glm
Giles, Katherine A., 915-747-7075 kagiles@utep.edu, University of Texas, El Paso – Go
Gilfillan, Stuart M., +44 (0) 131 651 3462 stuart.gilfillan@ed.ac.uk, Edinburgh University – CgsCe
Gilg, Hans A., +49 89 289 - 25855 agilg@tum.de, Technical University of Munich – CsEgn
Gill, Benjamin, 540-231-7485 bcgill@vt.edu, Virginia Polytechnic Institute & State University – CsPcGs
Gill, Fiona, +44(0) 113 34 35190 f.gill@leeds.ac.uk, University of Leeds – Ca
Gill, James B., (831) 459-3842 jgill@pmc.ucsc.edu, University of California, Santa Cruz – Gi
Gill, Swarndeep S., (724) 938-1677 gill@calu.edu, California University of Pennsylvania – As
Gill, Thomas E., (915) 747-5168 tegill@utep.edu, University of Texas, El Paso – GmAsCl
Gillam, Mary L., gillam@rmi.net, Fort Lewis College – Gm
Gille, Peter, 089/2180 4355 peter.gille@lrz.uni-muenchen.de, Ludwig-Maximilians-Universitaet Muenchen – GzZm
Gille, Sarah T., (858) 822-4425 sgille@ucsd.edu, University of California, San Diego – Op
Gilleaudeau, Geoff, (703) 993-3289 ggilleau@gmu.edu, George Mason University – Gs
Gillerman, Virginia S., (208) 332-4420 vgillerm@uidaho.edu, University of Idaho – Eg
Gillerman, Virginia S., (208) 426-4002 vgillerm@boisestate.edu, Boise State University – Eg
Gillespie, Alan R., arg3@uw.edu, University of Washington – GmlZr
Gillespie, Janice, 661-654-3040 jgillespie@csub.edu, California State University, Bakersfield – Go
Gillespie, Robb, (269) 387-5364 robb.gillespie@wmich.edu, Western Michigan University – GsoGm
Gillespie, Terry J., (519) 824-4120 (Ext. 54276) tgillesp@uoguelph.ca, University of Guelph – Am
Gillespie, Thomas, (609) 771-2569 College of New Jersey – Gc
Gillespie, Thomas, thomas.gillespie@emory.edu, Emory University – Zn
Gillette, David P., (828) 251-6366 dgillett@unca.edu, University of North Carolina, Asheville – Zn
Gilley, Brett H., bgilley@eoas.ubc.ca, University of British Columbia – GsZeRn
Gilliam, James W., (919) 515-2040 North Carolina State University – Sb
Gilligan, Jonathan M., (615) 322-2420 jonathan.gilligan@vanderbilt.edu, Vanderbilt University – ZcRwZu
Gillikin, David P., (518) 388-6679 gillikind@union.edu, Union College – CsmCl
Gillis, Kathryn, 250-721-6120 kgillis@uvic.ca, University of Victoria – Gp
Gillis, Robert, (907) 451-5024 robert.gillis@alaska.gov, Alaska Division of Geological & Geophysical Surveys – Eo
Gillman, Joe, (573) 368-2101 joe.gillman@dnr.mo.gov, Missouri Dept of Natural Resources – Gg

Gillmore, Gavin, +44 020 8417 2518 G.Gillmore@kingston.ac.uk, Kingston University – Gb
Gilmore, Martha S., (860) 685-3129 mgilmore@wesleyan.edu, Wesleyan University – XgGmZr
Gilmore, Tyler J., (509) 376-2370 tyler.gilmore@pnl.gov, Pacific Northwest National Laboratory – Hg
Gilmour, Ernest H., (509) 359-7480 egilmour@ewu.edu, Eastern Washington University – Pi
Gilotti, Jane A., (319) 335-1097 jane-gilotti@uiowa.edu, University of Iowa – GctGp
Gilpin, Bernard J., (714) 895-8233 bgilpin@gwc.cccd.edu, Golden West College – Ys
Gimmestad, Gary G., (404) 493-1331 gary.gimmestad@gmail.com, Georgia Institute of Technology – ZrAs
Ginder-Vogel, Matt, 608-262-0768 mgindervogel@wisc.edu, University of Wisconsin, Madison – Sc
Gingras, Murray, 780-492-1963 mgringras@ualberta.ca, University of Alberta – Go
Ginis, Isaac, (401) 874-6484 iginis@gso.uri.edu, University of Rhode Island – Op
Giordano, Daniele, daniele.giordano@unito.it, Università di Torino – Gv
Giorgis, Scott D., (585) 245-5293 giorgis@geneseo.edu, SUNY, Geneseo – GctYm
Giosan, Liviu, 508-289-2257 lgiosan@whoi.edu, Woods Hole Oceanographic Institution – GusPc
Giraldo, Mario A., (818) 677-4431 mario.giraldo@csun.edu, California State University, Northridge – ZriZg
Girard, Eric, 514-987-3000 #3325 girard.eric@uqam.ca, Universite du Quebec a Montreal – Am
Girard, Michael W., (609) 292-2576 mike.girard@dep.state.nj.us, New Jersey Geological and Water Survey – Zi
Giraud, Richard E., (801) 537-3351 richardgiraud@utah.gov, Utah Geological Survey – Ng
Girhard, T S., 486-0045 tgirhard@alamo.edu, San Antonio Community College – ZyAm
Giroux, Bernard, bernard.giroux@ete.inrs.ca, Universite du Quebec – Yeg
Girty, Gary H., (619) 594-2552 ggirty@mail.sdsu.edu, San Diego State University – Gc
Gitelson, Anatoly, gitelson@calmit.unl.edu, Unversity of Nebraska - Lincoln – Zn
Gittings, Bruce M., +44 (0) 131 650 2558 bruce@ed.ac.uk, Edinburgh University – Zi
Gittins, John, (416) 483-9345 j.gittins@utoronto.ca, University of Toronto – GipGz
Giusberti, Luca, +390498279183 Università degli Studi di Padova – Pm
Giustetto, Roberto, roberto.giustetto@unito.it, Università di Torino – Gz
Glamoclija, Mihaela, (973) 353-2509 m.glamoclija@rutgers.edu, Rutgers, The State University of New Jersey, Newark – PgXgGe
Glasauer, Susan, (519) 824-4120 xt52453 glasauer@uoguelph.ca, University of Guelph – Pg
Glaser, Brian, 309-796-5238 glaserb@bhc.edu, Black Hawk College – Zn
Glaser, Paul H., 612-624-8395 glase001@umn.edu, University of Minnesota, Twin Cities – Gn
Glass, Alexander, (919) 684-6167 alex.glass@duke.edu, Duke University – PiGgZe
Glass, Billy P., (302) 449-2464 bglass@udel.edu, University of Delaware – XmcGz
Glass, Hylke J., +44 01326 371823 h.j.glass@exeter.ac.uk, Exeter University – NxmEm
Glazer, Brian, 808-956-6658 glazer@hawaii.edu, University of Hawai'i, Manoa – Cm
Glazner, Allen F., (919) 962-0689 afg@unc.edu, University of North Carolina, Chapel Hill – GitZf
Gleason, Gayle C., (607) 753-2816 gleasong@cortland.edu, SUNY, Cortland – Gct
Gleason, James D., 734-764-9523 jdgleaso@umich.edu, University of Michigan – Cg
Glenn, Craig R., (808) 956-2200 glenn@soest.hawaii.edu, University of Hawai'i, Manoa – GeCmGs
Glenn, Ed, (520) 626-2664 eglenn@ag.arizona.edu, University of Arizona – Zn
Glenn, Nancy, 208.221.1245 nancyglenn@boisestate.edu, Boise State University – ZrNgZi

Faculty Index -G

Glickson, Deborah, (202) 334-2024 dglickson@nas.edu,
  National Academy of Sciences, Engineering, and Medicine – GutGg
Gloaguen, Erwan, erwan.gloaguen@ete.inrs.ca,
  Universite du Quebec – Ye
Glotch, Timothy, (631) 632-1168 timothy.glotch@stonybrook.edu,
  Stony Brook University – XgGz
Glover, David M., (508) 289-2656 dglover@whoi.edu,
  Woods Hole Oceanographic Institution – Oc
Glover, Paul W., 418-656-5180 paul.glover@ggl.ulaval.ca,
  Universite Laval – Yx
Glover, Paul W., +44(0) 113 34 35213 p.w.j.glover@leeds.ac.uk,
  University of Leeds – YxGoa
Glowacka, Ewa, glowacka@cicese.mx,
  Centro de Investigación Científica y de Educación Superior de Ensenada – Ys
Glubokovskikh, Stanislav, +618 9266-7190 stanislav.glubokovskikh@curtin.edu.au,
  Curtin University – Yxe
Gluhovsky, Alexander, (765) 494-0670 aglu@purdue.edu,
  Purdue University – As
Glumac, Bosiljka, (413) 585-3680 bglumac@smith.edu,
  Smith College – Gs
Gluyas, Jon, +44 (0) 191 33 42302 j.g.gluyas@durham.ac.uk,
  Durham University – Eo
Glynn, William G., wglynn@brockport.edu,
  SUNY, The College at Brockport – Gg
Gnanadesikan, Anand, (410) 516-0722 gnanades@jhu.edu,
  Johns Hopkins University – OpcAs
Gnidovec, Dale M., (614) 292-6896 gnidovec.1@osu.edu,
  Ohio State University – PvGr
Gobler, Christopher, (631) 632-5043 Christopher.Gobler@stonybrook.edu,
  SUNY, Stony Brook – Ob
Goda, Katsu, kgoda2@uwo.ca,
  Western University – Rn
Godbold, Jasmin A., +44 (0)23 80593639 J.A.Goldbold@soton.ac.uk,
  University of Southampton – Ob
Godchaux, Martha M., (208) 882-9062 Mount Holyoke College – Gv
Goddard, Lisa M., (845) 680-4430 goddard@iri.columbia.edu,
  Columbia University – AstZc
Godek, Melissa, (607) 436-3375 melissa.godek@oneonta.edu,
  SUNY, Oneonta – AmsZg
Godfrey, Brian J., 845-437-5544 godfrey@vassar.edu,
  Vassar College – Zu
Godfrey, Christopher, 828-232-5160 cgodfrey@unca.edu,
  University of North Carolina, Asheville – AsZnn
Godfrey, Linda, godfrey@marine.rutgers.edu,
  Rutgers, The State University of New Jersey – CmlCq
Godin, Laurent, (613) 533-3223 godinl@queensu.ca,
  Queen's University – Gct
Godsey, Holly, (801) 587-7865 University of Utah – Ze
Godsey, Sarah E., 208-282-3170 godsey@isu.edu,
  Idaho State University – HwGm
Goehring, Brent M., (504) 862-3196 bgoehrin@tulane.edu,
  Tulane University – CcGlm
Goeke, James W., (308) 530-4437 jgoeke@unl.edu,
  Unversity of Nebraska - Lincoln – HwGmYe
Goes, Joaquim, jig@ldeo.columbia.edu,
  Columbia University – Gu
Goes, Saskia, +44 20 759 46434 s.goes@imperial.ac.uk,
  Imperial College – Yg
Goetz, Alexander, 303-492-5086 goetz@cses.colorado.edu,
  University of Colorado – Zr
Goetz, Andrew R., (303) 871-2674 agoetz@du.edu,
  University of Denver – Zgu
Goetz, Heinrich, (972) 377-1079 hgoetz@collin.edu,
  Collin College - Preston Ridge Campus – Ge
Goetze, Erica, (808) 956-7156 egoetze@hawaii.edu,
  University of Hawai'i, Manoa – Ob
Goff, Fraser, (505) 667-8060 fraser@lanl.gov,
  Los Alamos National Laboratory – Cg
Goff, James, j.goff@unsw.edu.au,
  University of New South Wales – Gsm
Goff, John A., 512-471-0476 goff@ig.utexas.edu,
  University of Texas, Austin – Gu
Goforth, Thomas T., (254) 710-2183 tom_goforth@baylor.edu,
  Baylor University – Yg
Golabi, Mohammad H., 671-735-2143 mgolabi@triton.uog.edu,
  University of Guam – Spc
Gold, David (Duff) P., (814) 865-9993 gold@ems.psu.edu,
  Pennsylvania State University, University Park – GcgGx
Gold-Bouchot, Gerardo, (979) 845-9826 ggold@tamu.edu,
  Texas A&M University – OcCm
Goldberg, David S., goldberg@ldeo.columbia.edu,
  Columbia University – Yr
Goldblatt, Colin, 250-721-6120 czg@uvic.ca,
  University of Victoria – As
Goldfinger, Chris, (541) 737-9622 gold@coas.oregonstate.edu,
  Oregon State University – Yr
Goldhamer, David A., (559) 646-6500 dagoldhamer@ucdavis.edu,
  University of California, Davis – Zn
Goldman, Daniel, (937) 229-5637 dgoldman1@udayton.edu,
  University of Dayton – PgqGr
Goldreich, Peter M., (626) 395-6193 pmg@gps.caltech.edu,
  California Institute of Technology – Xy
Goldstein, Barry, (253) 879-3822 goldstein@pugetsound.edu,
  University of Puget Sound – GlmGs
Goldstein, Robert H., (785) 864-2738 gold@ku.edu,
  University of Kansas – Gs
Goldstein, Steven, steveg@ldeo.columbia.edu,
  Columbia University – Cg
Goldstein, Steven L., (845) 365-8787 steveg@ldeo.columbia.edu,
  Columbia University – Csg
Goldstein, Susan T., (706) 542-2397 sgoldst@uga.edu,
  University of Georgia – Pm
Gollmer, Steven, 937-766-7764 gollmers@cedarville.edu,
  Cedarville University – GzAmOg
Golombek, Matthew P., (818) 393-7948 mgolombek@jpl.nasa.gov,
  Jet Propulsion Laboratory – XgyGt
Gomberg, Joan, 206-616-5581 gomberg@usgs.gov,
  University of Washington – YsGt
Gomes, Helga, helga@ldeo.columbia.edu,
  Columbia University – Ob
Gomes, Maria E., 351259350261 mgomes@utad.pt,
  Universidade de Trás-os-Montes e Alto Douro – GiCgGz
Gomes, Nuno N., +244924987900 ngomes999@gmail.com,
  Universidade Independente de Angola – GesGu
Gomez, Enrique, egomez@cicese.mx,
  Centro de Investigación Científica y de Educación Superior de Ensenada – Ye
Gomez, Francisco, (573) 882-9744 fgomez@missouri.edu,
  University of Missouri – Gt
Gomez, Natalya, 514-398-4885 natalya.gomez@mcgill.ca,
  McGill University – Og
Gómez Gras, David, ++935813093 david.gomez@uab.cat,
  Universitat Autonoma de Barcelona – Gd
Gomezdelcampo, Enrique, 419 372 2886 egomezd@bgsu.edu,
  Bowling Green State University – Ge
Goncharov, Alexander F., 202-478-8947 agoncharov@carnegiescience.edu,
  Carnegie Institution for Science – GyYxh
Gong, Donglai, 804-684-7529 gong@vims.edu,
  College of William & Mary – Op
Gong, Hongmian, (212) 772-4658 gong@hunter.cuny.edu,
  Hunter College (CUNY) – Zn
Goni, Miguel A., (541) 737-0578 mgoni@coas.oregonstate.edu,
  Oregon State University – Gu
Gonnermann, Helge, (713) 348-6263 Helge.M.Gonnermann@rice.edu,
  Rice University – Gv
Gonzales, David A., (970) 247-7378 gonzales_d@fortlewis.edu,
  Fort Lewis College – Gx
Gonzalez, Alex, (515) 294-8729 agon@iastate.edu,
  Iowa State University of Science & Technology – As
Gonzalez, Frank I., 206-290-0903 figonzal@uw.edu,
  University of Washington – OpRn
González, Isabel , +34954556317 igonza@us.es,
  Universidad de Sevilla – GzeSc
Gonzalez, Jose J., javier@cicese.mx,
  Centro de Investigación Científica y de Educación Superior de Ensenada – Yd
Gonzalez, Juan L., (956) 665-3523 juan.l.gonzalez@utrgv.edu,
  University of Texas, Rio Grande Valley – GmsGa
González, Laura, lgacebron@ucm.es,
  Univ Complutense de Madrid – Gr
Gonzalez, Luis, +966138607981 gonzalez@kfupm.edu.sa,
  King Fahd University of Petroleum and Minerals – CgGn

Gonzalez, Luis A., 785-864-2743 lgonzlez@ku.edu,
University of Kansas – CsGd
Gonzalez, Maria M., 989-774-2295 gonza1mm@cmich.edu,
Central Michigan University – Gd
Gonzalez Leon, Carlos M., cmgleon@servidor.unam.mx,
Universidad Nacional Autonoma de Mexico – Gr
Gonzalez-Huizar, Hector, 915-747-5305 hectorg@utep.edu,
University of Texas, El Paso – Yg
Good, Daniel B., (912) 478-5361 DanGood@GeorgiaSouthern.edu,
Georgia Southern University – Zn
Good, David, dgood3@uwo.ca,
Western University – Em
Goodbred, Jr., Steven L., (615) 343-6424 steven.goodbred@vanderbilt.edu,
Vanderbilt University – GsOn
Goodell, Laurel P., (609) 258-1043 laurel@princeton.edu,
Princeton University – ZeGgc
Goodell, Philip C., 915-747-5593 goodell@utep.edu,
University of Texas, El Paso – Ce
Goodge, John W., (218) 726-7491 jgoodge@d.umn.edu,
University of Minnesota, Duluth – Gp
Goodin, Ruth, 305-284-4253 ruthgoodin@miami.edu,
University of Miami – Zn
Gooding, Patrick J., (859) 389-8810 gooding@uky.edu,
University of Kentucky – Eo
Goodliffe, Andrew, amg@ua.edu,
University of Alabama – YgeGt
Goodman, Paul, 520-621-8484 pgoodman@email.arizona.edu,
University of Arizona – As
Goodman-Tchernov, Beverly, 972-4-8288424 bgoodman@univ.haifa.ac.il,
McMaster University – Gu
Goodrich, David C., (520) 670-6380 goodrich@tucson.ars.ag.gov,
University of Arizona – Hs
Goodrich, Greg, gregory.goodrich@wku.edu,
Western Kentucky University – Ast
Goodwin, David H., (740) 587-5621 goodwind@denison.edu,
Denison University – PiCs
Goodwin, Laurel B., (608) 265-4234 laurel@geology.wisc.edu,
University of Wisconsin-Madison – Gc
Goodwin, Mark B., (505) 835-5178 mark@berkeley.edu,
University of California, Berkeley – Pv
Goodwin, Paul, (519) 824-4120 Ext.52754 pgoodwin@uoguelph.ca,
University of Guelph – Zn
Goodwin, Phillip A., +44 (0)23 80596161 P.A.Goodwin@soton.ac.uk,
University of Southampton – Pe
Goos, Robert J., (701) 231-8581 North Dakota State University – Sc
Gootee, Brian F., 602.708.8846 bgootee@email.arizona.edu,
Arizona Geological Survey – Gm
Gopal, Sucharita, suchi@bu.edu,
Boston University – Zi
Gordon, Andrew, (519) 824-4120 ext 52415 agordon@uoguelph.ca,
University of Guelph – Zn
Gordon, Arnold L., agordon@ldeo.columbia.edu,
Columbia University – Op
Gordon, Arnold L., (845) 365-8325 agordon@ldeo.columbia.edu,
Columbia University – Op
Gordon, Barney, (785) 532-6101 bgordon@ksu.edu,
Kansas State University – So
Gordon, Clare E., +44(0) 113 34 35210 c.e.gordon@leeds.ac.uk,
University of Leeds – Zi
Gordon, Mark, (416)736-2100 mgordon@yorku.ca,
York University – As
Gordon, Richard G., (713) 348-5279 rgg@rice.edu,
Rice University – YmGtYd
Gordon, Rob J., (519) 824-4120 Ext.52285 rjgordon@uoguelph.ca,
University of Guelph – As
Gordon, Robert B., (203) 432-3125 robert.gordon@yale.edu,
Yale University – Nr
Gordon, Ryan, 207-287-7178 ryan.gordon@maine.gov,
Dept of Agriculture, Conservation, and Forestry – HwGm
Gordon, Stacia , 775-784-6476 staciag@unr.edu,
University of Nevada, Reno – Gi
Gordon, Steven J., (719) 333-3067 usafa.dfeg@usafa.edu,
United States Air Force Academy – Gm
Gordon, Terence M., (403) 220-8301 tmg@geo.ucalgary.ca,
University of Calgary – Gq
Gore, Pamela J. W., 678-891-3754 pgore@gsu.edu,
Georgia State University, Perimeter College, Clarkston Campus – GsZeGn
Gorecka-Nowak, Anna, anna.gorecka-nowak@uwr.edu.pl,
University of Wroclaw – Pg
Gorelick, Steven, (650) 725-2950 gorelick@geo.stanford.edu,
Stanford University – Yr
Gorka, Maciej, maciej.gorka@uwr.edu.pl,
University of Wroclaw – AsCa
Gorman, Andrew R., +64 3 479-7516 andrew.gorman@otago.ac.nz,
University of Otago – YrgYe
Gorman, Gerard, +44 20 759 49985 g.gorman@imperial.ac.uk,
Imperial College – Yg
Gorman-Lewis, Drew J., 206-543-3541 dgormanl@uw.edu,
University of Washington – PgCl
Gorog, Agnes, gorog@ludens.elte.hu,
Eotvos Lorand University – Pmi
Gorring, Matthew L., (973) 655-5409 gorringm@mail.montclair.edu,
Montclair State University – Gi
Gorsevski, Peter, (419) 372-7201 peterg@bgsu.edu,
Bowling Green State University – ZirGm
Gosnell, Hannah, (541) 737-1222 gosnell@colorado.edu,
Oregon State University – Zu
Gosnold, William D., (701) 777-2631 will.gosnold@engr.und.edu,
University of North Dakota – YhCt
Gospodinova, Kalina D., 508-289-3212 kgospodinova@whoi.edu,
Woods Hole Oceanographic Institution – Cm
Goss, Michael J., (519) 824-4120 (Ext. 2491) mgoss@uoguelph.ca,
University of Guelph – So
Gosse, John, 902-494-6632 jcgosse@is.dal.CA,
University of Kansas – Gm
Gosselin, David C., 402-472-8919 dgosselin2@unl.edu,
University of Nebraska, Lincoln – Ze
Gosselin, Kelsey, 508-289-3664 kgosselin@whoi.edu,
Woods Hole Oceanographic Institution – Oc
Gosselin, Michel, 4187241761 x1284 michel_gosselin@uqar.ca,
Universite du Quebec a Rimouski – Ob
Gotkowitz, Madeline B., (608) 262-1580 mbgotkow@wisc.edu,
University of Wisconsin, Extension – Hw
Gottardi, Raphael, 337-482-6177 rxg0121@louisiana.edu,
University of Louisiana at Lafayette – Gcp
Gottfried, Michael D., 517-432-5480 gottfrie@msu.edu,
Michigan State University – Pv
Gottgens, Johan F., (419) 530-8451 johan.gottgens@utoledo.edu,
University of Toledo – Zn
Gotz, Annette, +27 (0)46-603-8313 a.gotz@ru.ac.za,
Rhodes University – Gs
Goudge, Theodore L., (660) 562-1798 tgoudge@nwmissouri.edu,
Northwest Missouri State University – Zy
Goudreau, Joanne, (508) 289-2560 jgoudreau@whoi.edu,
Woods Hole Oceanographic Institution – Cm
Gouhier, Tarik, 7815817370 x302 t.gouhier@neu.edu,
Northeastern University – Zn
Gould, Joseph C., gould.joe@spcollege.edu,
Saint Petersburg College, Clearwater – Gg
Gould, Mark D., (401) 254-3087 Roger Williams University – Ob
Goulden, Michael L., (949) 824-1983 mgoulden@uci.edu,
University of California, Irvine – Zg
Gouldey, Jeremy C., 616-331-2995 gouldjer@gvsu.edu,
Grand Valley State University – CgPeCg
Gouldson, Andy, +44(0) 113 34 36417 a.gouldson@leeds.ac.uk,
University of Leeds – Ge
Goulet, Normand, (514) 987-3375 r27254@er.uqam.ca,
Universite du Quebec a Montreal – Gc
Goulet, Richard, 613-943-9922 Richard.Goulet@cnsc-ccsn.gc.ca,
University of Ottawa – CgPgGe
Gouzie, Douglas R., (417) 836-5228 douglasgouzie@missouristate.edu,
Missouri State University – HwGeCl
Gowan, Angela S., (612) 626-6451 gowa0001@umn.edu,
University of Minnesota – Gl
Gowing, David, +44 (0)1908 659468 x 59468 david.gowing@open.ac.uk,
The Open University – Pb
Goyne, Keith W., 573-882-0090 University of Missouri, Columbia – Sc
Grable, Judy, (229) 333-5752 Valdosta State University – Hs
Grabowski, Jonathan, 7815817370 x337 j.grabowski@neu.edu,
Northeastern University – GuEg
Grace, Cathy A., (662)915-1799 cag@olemiss.edu,
University of Mississippi – Gg
Grace, John, +44 (0) 131 650 5400 jgrace@ed.ac.uk,

Edinburgh University – As
Grace, Peter R., 61 7 3138 2610 pr.grace@qut.edu.au,
  Queensland University of Technology – Sb
Grace, Shannon M., (512) 223-4891 sgrace@austincc.edu,
  Austin Community College District – Zn
Graczyk, Donald G., (630) 252-3489 Argonne National Laboratory – Cs
Graettinger, Alison, 816-235-6701 graettingera@umkc.edu,
  University of Missouri, Kansas City – Gv
Graf, Jr., Joseph L., (541) 552-6861 graf@sou.edu,
  Southern Oregon University – Em
Graff, Thomas O., (479) 575-3878 tgraff@comp.uark.edu,
  University of Arkansas, Fayetteville – Zu
Graff, William P., (609) 292-2576 bill.graff@dep.nj.gov,
  New Jersey Geological and Water Survey – Zn
Graham, Barbara, 702-651-4173 barbara.graham@csn.edu,
  College of Southern Nevada - West Charleston Campus – ZyAm
Graham, David W., (541) 737-4140 dgraham@coas.oregonstate.edu,
  Oregon State University – Cm
Graham, Gina R., 907-451-5031 gina.graham@alaska.gov,
  Alaska Division of Geological & Geophysical Surveys – Yg
Graham, Grace E., (608) 263-4125 grace.graham@wgnhs.uwex.edu,
  University of Wisconsin, Extension – Hw
Graham, Ian, i.graham@unsw.edu.au,
  University of New South Wales – GxEg
Graham, Margaret C., +44 (0) 131 650 4767 Margaret.Graham@ed.ac.uk,
  Edinburgh University – Ge
Graham, Michael, (831) 771-4400 mgraham@mlml.calstate.edu,
  Moss Landing Marine Laboratories – Ob
Graham, Russell W., (814) 865-6336 graham@ems.psu.edu,
  Pennsylvania State University, University Park – Pv
Graham, Stephan A., (650) 723-0507 graham@pangea.stanford.edu,
  Stanford University – Go
Graham, William (., (228) 688-3177 monty.graham@usm.edu,
  University of Southern Mississippi – Ob
Graham, Jr., Earl K., (814) 865-2273 graham@ems.psu.edu,
  Pennsylvania State University, University Park – Yx
Graham, Jr., James H., (863) 956-1151 jhgraham@ufl.edu,
  University of Florida – Sb
Grala, Katarzynz, (662) 268-1032 Ext 222 kg160@msstate.edu,
  Mississippi State University – Zi
Grammer, Michael, 405-744-6358 michael.grammer@okstate.edu,
  Oklahoma State University – GsEoGr
Gran, Karen B., 218-726-7406 kgran@d.umn.edu,
  University of Minnesota, Duluth – Gms
Grana, Dario, 307-766-3449 dgrana@uwyo.edu,
  University of Wyoming – Ye
Grand, Stephen P., (512) 471-3005 steveg@maestro.geo.utexas.edu,
  University of Texas, Austin – Ys
Grandal d'Anglade, Aurora, 00 34 981 167000 xeaurora@udc.es,
  Coruna University – Pv
Grande, Anthony, 212-772-5265 agrande@hunter.cuny.edu,
  Hunter College (CUNY) – Zn
Grande, Lance, (312) 665-7632 lgrande@fieldmuseum.org,
  Field Museum of Natural History – Pv
Grandstaff, David E., (215) 204-8228 grand@temple.edu,
  Temple University – Cl
Grandy, Stuart, stuart.grandy@unh.edu,
  University of New Hampshire – Sc
Graney, Joseph R., (607) 777-6347 jgraney@binghamton.edu,
  Binghamton University – ClGeHs
Granger, Darryl E., (765) 494-0043 dgranger@purdue.edu,
  Purdue University – As
Granger, Julie, 860-405-9094 julie.granger@uconn.edu,
  University of Connecticut – Cs
Graniero, Phil A., (519) 253-3000 x2485 graniero@uwindsor.ca,
  University of Windsor – ZinZy
Grannell, Roswitha B., (562) 985-4927 grannell@csulb.edu,
  California State University, Long Beach – Yv
Granshaw, Frank D., (503) 725-3391 fgransha@pdx.edu,
  Portland State University – ZnGl
Grant, Alastair, +44 (0)1603 59 2537 a.grant@uea.ac.uk,
  University of East Anglia – GeObCt
Grant, John A., 202-633-2474 Smithsonian Institution / National Air & Space Museum – Xg
Grant, Jonathan, (902) 494-2021 jon.grant@dal.ca,
  Dalhousie University – Ob
Grant, Shelton K., (573) 341-4616 Missouri University of Science and Technology – Gz
Grant, Stanley, 703-361-5606 stanleyg@vt.edu,
  Virginia Polytechnic Institute & State University – Hs
Grapenthin, Ronni, 575-835-5924 rg@nmt.edu,
  New Mexico Institute of Mining and Technology – Gv
Grassian, Vicki, (858) 534-2499 vhgrassian@ucsd.edu,
  University of California, San Diego – Cm
Grassineau, Nathalie, +44 1784 443810 Nathalie.Grassineau@rhul.ac.uk,
  Royal Holloway University of London – Cs
Grattan, Stephen R., (530) 752-1130 srgrattan@ucdavis.edu,
  University of California, Davis – Zn
Grattan, Stephen R., 530-752-4618 srgrattan@ucdavis.edu,
  University of California, Davis – Spp
Gratton, Yves, yves.gratton@ete.inrs.ca,
  Universite du Quebec – Op
Graumlich, Lisa J., (406) 994-5320 dalylisa@montana.edu,
  Montana State University – Zu
Graustein, William C., (203) 287-2853 william.graustein@yale.edu,
  Yale University – Cl
Gravely, Cynthia Rae, (864) 656-3438 gravelc@clemson.edu,
  Clemson University – Zn
Graves, Alexandria, 919-513-0635 alexandria_graves@ncsu.edu,
  North Carolina State University – Sb
Graves, Charles E., (314) 977-3121 gravesce@slu.edu,
  Saint Louis University – As
Gravley, Darren, +64 3 3667001 Ext 45683 darren.gravley@canterbury.ac.nz,
  University of Canterbury – Gv
Gray, Katelyn E., katelyn.gray@austincc.edu,
  Austin Community College District – PcCb
Gray, Keith, (518) 564-4032 kgray003@plattsburgh.edu,
  SUNY, Plattsburgh – Gzc
Gray, Kyle R., 319-273-2809 kyle.gray@uni.edu,
  University of Northern Iowa – ZeGg
Gray, Lee M., (330) 823-3605 graylm@mountunion.edu,
  Mount Union College – PgGg
Gray, Mary Beth, (570) 577-1146 mbgray@bucknell.edu,
  Bucknell University – Gc
Gray, Neil, +44 (0) 191 208 4887 neil.gray@ncl.ac.uk,
  University of Newcastle Upon Tyne – Pg
Gray, Norman H., dlfox@umn.edu,
  University of Connecticut – Gx
Gray, Robert S., (805) 965-0581 gray@sbcc.net,
  Santa Barbara City College – PvGzx
Gray, Sarah, (619) 260-4098 sgray@sandiego.edu,
  University of San Diego – Zn
Grayston, Sue, 604–822–5928 sue.grayston@ubc.ca,
  University of British Columbia – Sb
Greatbatch, Ian, +44 020 8417 2879 I.Greatbatch@kingston.ac.uk,
  Kingston University – Zi
Greatbatch, Richard , rgreatbatch@geomar.de,
  Dalhousie University – Op
Greb, Stephen F., 8592575500 x136 greb@uky.edu,
  University of Kentucky – GsEcGr
Greb, Stephen F., (859) 323-0542 greb@uky.edu,
  University of Kentucky – GsEcGs
Green, Andrew, (031) 260-2516 greena1@ukzn.ac.za,
  University of KwaZulu-Natal – GuYgGs
Green, Brittany, (785) 532-6101 bdgreen@ksu.edu,
  Kansas State University – Zn
Green, Harry W., (510) 642-3059 University of California, Berkeley – Pv
Green, Jack, (562) 985-4198 jgreen3@csulb.edu,
  California State University, Long Beach – Gv
Green, John C., (218) 726-7208 jgreen@d.umn.edu,
  University of Minnesota, Duluth – GieGv
Green, Jonathan, +44 0151 795 4385 Jonathan.Green@liverpool.ac.uk,
  University of Liverpool – Ob
Greenberg, David, (902) 426-2431 greenbergd@mar.dfo-mpo.gc.ca,
  Dalhousie University – Op
Greenberg, Harvey, (206) 685-7981 hgreen@uw.edu,
  University of Washington – Gml
Greenberg, Jeffrey K., jeffrey.greenberg@wheaton.edu,
  Wheaton College – GicGe
Greenberg, Richard J., (520) 621-6940 greenberg@lpl.arizona.edu,
  University of Arizona – Xy
Greenberg, Sallie, (217) 244-4068 sallieg@illinois.edu,
  Illinois State Geological Survey – ClZg
Greene, Brian M., 412-395-7323 Brian.Greene@usace.army.mil,

Youngstown State University – Ng
Greene, Charles H., chg2@cornell.edu,
  Cornell University – Ob
Greene, Chris, 902-494-7018 csgreene@dal.ca,
  Dalhousie University – Zi
Greene, David C., (740) 587-6476 greened@denison.edu,
  Denison University – GctGe
Greene, Don M., (254) 710-2193 don_greene@baylor.edu,
  Baylor University – ZyAm
Greene, Mott, 206-713-0917 mgreene@uw.edu,
  University of Washington – GgAmOg
Greene, Pauline R., (337) 482-6468 geology@louisiana.edu,
  University of Louisiana at Lafayette – Zn
Greene, Roberta, (315) 267-2286 greenera@potsdam.edu,
  SUNY Potsdam – Zn
Greene, Todd, 530-898-5546 tjgreene@csuchico.edu,
  California State University, Chico – Gsr
Greenfield, Roy J., (814) 237-1810 rkg4@psu.edu,
  Pennsylvania State University, University Park – YgsYu
Greenhalgh, Stewart A., +966138608716 greenhalgh@kfupm.edu.sa,
  King Fahd University of Petroleum and Minerals – Yse
Greenhow, Danielle, (228) 688-2309 danielle.greenhow@usm.edu,
  University of Southern Mississippi – Ob
Greenlee, Diana M., (318) 926-3314 greenlee@ulm.edu,
  University of Louisiana, Monroe – GaCo
Greenwell, Chris, +44 (0) 191 33 42324 chris.greenwell@durham.ac.uk,
  Durham University – CqEoCg
Greenwood, James P., 860 685-2545 jgreenwood@wesleyan.edu,
  Wesleyan University – Xc
Greer, Adam, 228.688.2325 adam.greer@usm.edu,
  University of Southern Mississippi – Ob
Greer, Lisa, 540-458-8871 greerl@wlu.edu,
  Washington & Lee University – GsPe
Gregg, Jay M., 405-744-6358 jay.gregg@okstate.edu,
  Oklahoma State University – Gs
Gregg, Michael C., (206) 543-1353 gregg@apl.washington.edu,
  University of Washington – Op
Gregg, Patricia, 217-333-3540 pgregg@illinois.edu,
  University of Illinois, Urbana-Champaign – YgCg
Gregg, Tracy K. P., (716) 645-4328 tgregg@buffalo.edu,
  SUNY, Buffalo – Gv
Gregor, C. B., 937 775-3455 Wright State University – Gs
Gregory, Robert, (307) 766-2286 Ext. 237 robert.gregory@wyo.gov,
  Wyoming State Geological Survey – En
Gregory, Robert T., (214) 768-3075 Southern Methodist University – Cs
Grender, Gordon C., (540) 231-6521 Virginia Polytechnic Institute & State University – Go
Grenfell, Thomas C., tcg@atmos.washington.edu,
  University of Washington – As
Gretener, Peter E., (403) 220-5849 University of Calgary – Ng
Grew, Edward S., (207) 581-2169 esgrew@maine.edu,
  University of Maine – Gpz
Grew, Priscilla C., (402) 472-2095 pgrew1@unl.edu,
  University of Nebraska, Lincoln – GpeZe
Greybush, Steven J., 814-867-4926 sjg213@psu.edu,
  Pennsylvania State University, University Park – Am
Greyling, Lynnette N., 021-650-294886 l.greyling@uct.ac.za,
  University of Cape Town – EgmCe
Gribenko, Alex, (801) 585-6484 alex.gribenko@utah.edu,
  University of Utah – Yg
Grieneisen, Michael L., mgrien@ucdavis.edu,
  University of California, Davis – Hg
Griera Artigas, Albert, ++935811035 albert.griera@uab.cat,
  Universitat Autonoma de Barcelona – GcCl
Gries, John C., (316) 978-3140 Wichita State University – Gt
Grieve, Richard, rgrieve2@uwo.ca,
  Western University – Xg
Grieve, Richard A. F., (613) 995-5372 geology@unb.ca,
  University of New Brunswick – Xg
Grießl, Stefan, 089/2180 4188 stefan.griessl@uni.muenchen.de,
  Ludwig-Maximilians-Universitaet Muenchen – Gz
Griffen, Dana T., 801-422-2305 dana_griffen@byu.edu,
  Brigham Young University – Gz
Griffin, Kevin, griff@ldeo.columbia.edu,
  Columbia University – Zn
Griffin, Kevin L., (845) 365-8371 griff@ldeo.columbia.edu,
  Columbia University – PbeZn

Griffin, William R., (972) 883-2430 griffin@utdallas.edu,
  University of Texas, Dallas – Gg
Griffing, Corinne, 604-527-5217 griffingc@douglascollege.ca,
  Douglas College – GgeGm
Griffing, David H., 607-431-4629 griffingd@hartwick.edu,
  Hartwick College – GduGm
Griffis, Timothy J., (612) 625-3117 tgriffis@soils.umn.edu,
  University of Minnesota, Twin Cities – As
Griffith, Ashley, (817) 272-2987 wagriff@uta.edu,
  University of Texas, Arlington – Gct
Griffith, Caitlin, (520) 626-3806 griffith@lpl.arizona.edu,
  University of Arizona – Zn
Griffith, Elizabeth, griffith.906@osu.edu,
  Ohio State University – CsOoCb
Griffith, Liz, (817) 272-2987 lgriff@uta.edu,
  University of Texas, Arlington – CslCm
Griffith, William A., griffith.233@osu.edu,
  Ohio State University – GcNr
Grigg, Laurie, 802 485 3323 lgrigg@norwich.edu,
  Norwich University – PeGnZi
Grigg, Richard W., (808) 956-7186 rgrigg@soest.hawaii.edu,
  University of Hawai'i, Manoa – Ob
Griggs, Garry, 831-459-5006 griggs@ucsc.edu,
  University of California, Santa Cruz – On
Grigsby, Jeffry D., (765) 285-8270 jgrigsby@bsu.edu,
  Ball State University – GdClGo
Grimes, Stephen, +44 1752 584759 stephen.grimes@plymouth.ac.uk,
  University of Plymouth – Cs
Grimley, David A., 217-244-7324 dgrimley@illinois.edu,
  Illinois State Geological Survey – GmlSc
Grimm, Kurt A., (604) 822-9258 kgrimm@eos.ubc.ca,
  University of British Columbia – Gs
Grindlay, Nancy R., (910) 962-2352 grindlayn@uncw.edu,
  University of North Carolina, Wilmington – Yg
Gringarten, Alain, +44 20 759 47440 a.gringarten@imperial.ac.uk,
  Imperial College – Np
Grippo, Alessandro, grippo_alessandro@smc.edu,
  Santa Monica College – Gs
Grise, Kevin M., (434) 924-0433 kmg3r@virginia.edu,
  University of Virginia – As
Grismer, Mark E., (530) 752-3245 megrismer@ucdavis.edu,
  University of California, Davis – Hw
Griswold, George B., (505) 299-6192 New Mexico Institute of Mining and Technology – Nx
Griswold, Jennifer S., smalljen@hawaii.edu,
  University of Hawai'i, Manoa – AmZr
Gritzo, Russell E., (505) 667-0481 Los Alamos National Laboratory – Zn
Groat, Lee A., (604) 822-1289 groat@mail.ubc.ca,
  University of British Columbia – GzEg
Grocke, Darren, +44 (0) 191 33 42282 Durham University – Cs
Groenevelt, Pieter H., (519) 824-4120 (Ext. 53585) pgroenev@lrs.uoguelph.ca,
  University of Guelph – Sp
Groffman, Peter , 719-951-5000, ext. 5416 peter.groffman@brooklyn.cuny.edu,
  Graduate School of the City University of New York – Zg
Groffman, Peter, 212-413-3143 Peter.Groffman@brooklyn.cuny.edu,
  Brooklyn College (CUNY) – SbCl
Gromet, L. P., 401-863-1920 Peter_Gromet@brown.edu,
  Brown University – Cc
Groppi, Christopher, (480) 965-6436 cgroppi@asu.edu,
  Arizona State University – XaZrn
Groppo, Chiara Teresa, chiara.groppo@unito.it,
  Università di Torino – Gx
Grosfils, Eric B., (909) 621-8673 egrosfils@pomona.edu,
  Pomona College – XgGvq
Groshong, Jr., Richard H., 205-348-1882 rhgroshon@cs.com,
  University of Alabama – Gc
Gross, Amy, 910-521-6588 amy.gross@uncp.edu,
  University of North Carolina, Pembroke – Gg
Gross, Gerardo W., grossgw@nmt.edu,
  New Mexico Institute of Mining and Technology – Yx
Gross, Julian, 848-445-3619 jgross@eps.rutgers.edu,
  Graduate School of the City University of New York – Ze
Gross, Michael, (305) 348-3932 grossm@fiu.edu,
  Florida International University – Gc
Grossman, Eric, (360) 650-4697 Eric.Grossman@wwu.edu,
  Western Washington University – GusGm
Grossman, Ethan L., (979) 845-0637 e-grossman@tamu.edu,

Texas A&M University – Csl
Grossman, Lawrence, (773) 702-8153 University of Chicago – Xc
Grossman, Lawrence S., (540) 231-5116 Virginia Polytechnic Institute & State University – Zn
Grossman, Walter, wgrossman@yahoo.com, San Bernardino Valley College – Og
Groszos, Mark S., (229) 333-5664 msgroszo@valdosta.edu, Valdosta State University – GctEg
Grotjahn, Richard, 530-752-2246 Grotjahn@ucdavis.edu, University of California, Davis – As
Grottoli, Andrea G., 614 292 5782 grottoli.1@osu.edu, Ohio State University – ObCbOo
Grotzinger, John P., 626.395.6785 grotz@gps.caltech.edu, California Institute of Technology – GsPeXg
Grove, Karen, (415) 338-2617 kgrove@sfsu.edu, San Francisco State University – Gs
Grove, Timothy L., (617) 253-2878 tlgrove@mit.edu, Massachusetts Institute of Technology – Gi
Grover, Jeffrey A., 8055463100 x2759 jgrover@cuesta.edu, Cuesta College – Gg
Grover, John E., john.grover@uc.edu, University of Cincinnati – Gz
Grover, Timothy W., (802) 468-1289 tim.grover@castleton.edu, Castleton University – GptGc
Grover, Timothy W., 970-351-2398 timothy.grover@unco.edu, University of Northern Colorado – GpzGt
Groves, Christopher, (270) 745-5974 chris.groves@wku.edu, Western Kentucky University – Hg
Grube, John P., 217-244-1716 grube@isgs.uiuc.edu, Illinois State Geological Survey – Eo
Gruen, Dieter M., (630) 972-3513 Argonne National Laboratory – Zm
Grujic, Djordje, (902) 494-2208 dgrujic@is.dal.ca, Dalhousie University – Gt
Grunbaum, Daniel, (206) 221-6594 grunbaum@ocean.washington.edu, University of Washington – Ob
Grunder, Anita L., (541) 737-5189 grundera@geo.oregonstate.edu, Oregon State University – Gi
Grundl, Timothy J., 414-229-4765 grundl@uwm.edu, University of Wisconsin, Milwaukee – HwCl
Grunwald, Sabine -., (352) 294-3145 sabgru@ufl.edu, University of Florida – ZiSoZr
Grupp, Steve, 425-388-9450 sgrupp@everettcc.edu, Everett Community College – GgOg
Gu, Baohua, 865-574-7286 gub1@ornl.gov, Oak Ridge National Laboratory – CbqSb
Gu, Jeff, (780) 492-2292 jgu@phys.ualberta.ca, University of Alberta – Ys
Gualda, Guilherme, 615-322-2976 g.gualda@vanderbilt.edu, Vanderbilt University – GivGz
Gualtieri, Lucia, 609-258-4101 luciag@princeton.edu, Princeton University – Yg
Guan, Dabo, +44(0) 113 34 37432 d.guan@leeds.ac.uk, University of Leeds – Eg
Guan, Huade, huade.guan@flinders.edu.au, Flinders University – HgsHg
Guccione, Margaret J., (479) 575-3354 guccione@comp.uark.edu, University of Arkansas, Fayetteville – GmaGg
Gudmundsson, Agust, +44 1784 276345 Agust.Gudmundsson@rhul.ac.uk, Royal Holloway University of London – Gc
Guenthner, Willy, wrg@illinois.edu, University of Illinois, Urbana-Champaign – GtzCg
Guerra, Oralia, (512) 223-7483 oguerra1@austincc.edu, Austin Community College District – Zn
Guerrieri, Mary, (520) 621-2828 mary@lpl.arizona.edu, University of Arizona – Zn
Guertal, Elizabeth A., eguertal@acesag.auburn.edu, Auburn University – Sc
Guest, Peter S., 831-656-2451 pguest@nps.edu, Naval Postgraduate School – Am
Guevara, Victor, vguevara@skidmore.edu, Skidmore College – Gp
Guggenheim, Stephen J., (312) 996-3263 xtal@uic.edu, University of Illinois at Chicago – Gz
Guha, Jayanta, 418 545 5222 jguha@uqac.c, Universite du Quebec a Chicoutimi – Eg
Guilbert, John M., (520) 621-6024 j.guilbert@comcast.net, University of Arizona – Em

Guillemette, Renald, (979) 845-6301 guillemette@geo.tamu.edu, Texas A&M University – Gz
Guinan, Patrick E., 573-882-5909 guinanp@missouri.edu, University of Missouri, Columbia – As
Guinasso, Norman L., (979) 862-2323 norman@geos.tamu.edu, Texas A&M University – Opc
Gulbranson, Erik L., 414-229-1153 gulbrans@uwm.edu, University of Wisconsin, Milwaukee – CgPe
Gulen, Gurcan, (713) 654-5404 gurcan.gulen@beg.utexas.edu, University of Texas, Austin – Ego
Gulick, Sean S., (512) 471-3262 sean@ig.utexas.edu, University of Texas, Austin – YrGtb
Gulick, Sean S., 512-471-0483 sean@ig.utexas.edu, University of Texas, Austin – Gc
Gulliver, John S., (612) 625-4080 gulli003@tc.umn.edu, University of Minnesota, Twin Cities – Hs
Gundersen, James N., (316) 978-3140 Wichita State University – Ga
Gunderson, Lance, 404 727 8108 lgunder@emory.edu, Emory University – Sf
Gundiler, Ibrahim H., (505) 835-5730 gundiler@gis.nmt.edu, New Mexico Institute of Mining & Technology – Nx
Gunia, Piotr, piotr.gunia@uwr.edu.pl, University of Wroclaw – Gaz
Gunter, Mickey E., (208) 885-6015 mgunter@uidaho.edu, University of Idaho – GzEnGb
Gunter, William D., (403) 472-4406 University of Calgary – Cl
Gunter, William S., 706-507-8094 gunter_william@columbusstate.edu, Columbus State University – Ams
Guo, Junhua, jguo1@csub.edu, California State University, Bakersfield – Gs
Guo, Weifu, (508) 289-3380 wguo@whoi.edu, Woods Hole Oceanographic Institution – CslPe
Gupta, Hoshin V., (520) 275-5534 hoshin.gupta@hwr.arizona.edu, University of Arizona – HqZgAs
Gupta, Sanjeev, +44 20 759 46527 s.gupta@imperial.ac.uk, Imperial College – Gs
Gupta, Satish C., (612) 625-1241 sgupta@soils.umn.edu, University of Minnesota, Twin Cities – Sp
Gurdak, Jason, (415) 338-6869 jgurdak@sfsu.edu, San Francisco State University – HwCqAt
Gurevich, Boris, +61 8 9266-7359 B.Gurevich@curtin.edu.au, Curtin University – Ye
Gurlea, Lawrence P., lisobar@aol.com, Youngstown State University – Cg
Gurnis, Michael C., (626) 395-6979 gurnis@gps.caltech.edu, California Institute of Technology – Ys
Gurocak, Zulfu, 00904242370000-5991 zgurocak@firat.edu.tr, Firat University – Nrg
Gurrola, Harold, 806-834-8625 harold.gurrola@ttu.edu, Texas Tech University – Ys
Gurwin, Jacek, jacek.gurwin@uwr.edu.pl, University of Wroclaw – HwyZf
Gust, David A., 61 7 3138 2217 d.gust@qut.edu.au, Queensland University of Technology – GiCpGt
Gustin, Mae, 775.784.4203 mgustin@cabnr.unr.edu, University of Nevada, Reno – Cg
Guth, Peter L., (410) 293-6560 pguth@usna.edu, United States Naval Academy – GcZiy
Guthrie, Roderick I., (514) 398-4755 McGill University – Nx
Gutierrez, Melida, (417) 836-5967 mgutierrez@missouristate.edu, Missouri State University – CgHy
Gutierrez-Jurado, Hugo A., 915-747-5159 hagutierrez@utep.edu, University of Texas, El Paso – HgsHw
Gutknecht, Jessica L., 612-626-8435 jgutknec@umn.edu, University of Minnesota, Twin Cities – Sb
Gutmann, James T., (860) 685-2258 jgutmann@wesleyan.edu, Wesleyan University – GviGz
Gutowski, William J., (515) 294-5632 gutowski@iastate.edu, Iowa State University of Science & Technology – Am
Gutzler, David J., (505) 277-3328 gutzler@unm.edu, University of New Mexico – As
Guza, Robert T., (858) 534-0585 rguza@ucsd.edu, University of California, San Diego – On
Guzina, Bojan B., (612) 626-0789 guzina@wave.ce.umn.edu, University of Minnesota, Twin Cities – Nr
Guzman, Ernesto, (519) 824-4120 Ext.53609 eguzman@uoguelph.ca, University of Guelph – Zn

Gušiæ, Ivan, +38514606102 ivangusic@yahoo.com,
 University of Zagreb – GgPsm
Gyakum, John R., (514) 398-6076 john.gyakum@mcgill.ca,
 McGill University – Am
Gysi, Alex, (303) 273-3828 agysi@mines.edu,
 Colorado School of Mines – Cg
Göttlich, Hagen, 089/2180 5615 goettlich@ennab.de,
 Ludwig-Maximilians-Universitaet Muenchen – Gz

H

Haag, Lucas, (785) 462-6281 lhaag@ksu.edu,
 Kansas State University – So
Haas, Christian, (416)736-2100 #77705 haasc@yorku.ca,
 York University – Yg
Haas, Johnson R., (269) 387-2878 johnson.haas@wmich.edu,
 Western Michigan University – Cl
Habana, Nathan C., (671) 735-2693 nhabana@triton.uog.edu,
 University of Guam – Hw
Habash, Marc, (519) 824-4120 Ext.52748 mhabash@uoguelph.ca,
 University of Guelph – Zn
Haber, Eldad, (604) 822-4525 haber@eos.ubc.ca,
 University of British Columbia – YgZn
Habib, Daniel, (718) 997-3333 Graduate School of the City University of New York – Pl
Habron, Geoffrey, 864-294-3413 geoffrey.habron@furman.edu,
 Furman University – Zn
Hacker, Bradley R., (805) 893-7952 hacker@geol.ucsb.edu,
 University of California, Santa Barbara – Gp
Hacker, David B., 330-672-8831 dhacker@kent.edu,
 Kent State University – GcHw
Hacker, Jorg, jorg.hacker@flinders.edu.au,
 Flinders University – ZrAs
Hacker, Joshua P., 831-656-2722 jphacker@nps.edu,
 Naval Postgraduate School – Am
Hacker, Patricia, 419-530-5058 patricia.hacker@utoledo.edu,
 University of Toledo – Zn
Haddad, Brent, (831) 459-4149 bhaddad@cats.ucsc.edu,
 University of California, Santa Cruz – Zn
Haddock, Gregory D., (660) 562-1719 haddock@nwmissouri.edu,
 Northwest Missouri State University – Zi
Hadler, Kathryn, +44 20 759 47198 k.hadler@imperial.ac.uk,
 Imperial College – Gz
Hadley, Daniel, drhadley@illinois.edu,
 University of Illinois – Hw
Hafez, Sabry ., (907) 474- 6917 ssabour@alaska.edu,
 University of Alaska, Fairbanks – Nm
Haff, Peter K., (919) 684-5902 haff@duke.edu,
 Duke University – Zn
Hafner, James A., (413) 545-0778 hafner@geo.umass.edu,
 University of Massachusetts, Amherst – Zn
Hagadorn, James W., (303) 370-6058 james.hagadorn@dmns.org,
 Denver Museum of Nature & Science – GgsPg
Hagan, Robert M., (530) 752-0453 University of California, Davis – Hg
Hagedorn, Charles, (540) 231-4895 chagedor@vt.edu,
 Virginia Polytechnic Institute & State University – Sb
Hageman, Steven J., (828) 262-6609 hagemansj@appstate.edu,
 Appalachian State University – PieGg
Hager, Bradford H., (617) 253-0126 bhhager@mit.edu,
 Massachusetts Institute of Technology – Ys
Hagerty, Michael, 617-552-8300 hagertmb@bc.edu,
 Boston College – Ys
Haggart, James, jim.haggart@canada.ca,
 University of British Columbia – PgGr
Haggart, Renee, (604) 822-2789 rhaggart@eos.ubc.ca,
 University of British Columbia – Zn
Haggerty, Janet A., (918) 631-2304 janet-haggerty@utulsa.edu,
 The University of Tulsa – GudGs
Haggerty, Julia H., (406) 994-6904 julia.haggerty@montana.edu,
 Montana State University – Zu
Haggerty, Roy D., (541) 737-0663 Roy.Haggerty@oregonstate.edu,
 Oregon State University – Hw
Haggerty, Stephen E., (305) 348-7338 haggerty@fiu.edu,
 Florida International University – GiEgGz
Hagni, Richard D., (573) 341-4657 rhagni@umr.edu,
 Missouri University of Science and Technology – Em
Haigh, Ivan D., +44 (023) 80596501 i.d.haigh@soton.ac.uk,
 University of Southampton – On

Haileab, Bereket, (507) 222-5746 bhaileab@carleton.edu,
 Carleton College – GxzGi
Haimson, Bezalel C., (608) 262-2563 bhaimson@wisc.edu,
 University of Wisconsin, Madison – Nr
Haine, Thomas W., (410) 516-7048 thomas.haine@jhu.edu,
 Johns Hopkins University – OpgAs
Hajash, Andrew, (979) 845-0642 hajash@geo.tamu.edu,
 Texas A&M University – Cp
Hajek, Elizabeth, hajek@psu.edu,
 Pennsylvania State University, University Park – Gs
Hajnal, Zoltan, (306) 966-5694 zoltan.hajnal@usask.ca,
 University of Saskatchewan – YsGtc
Hakim, Gregory J., (206) 685-2439 hakim@atmos.washington.edu,
 University of Washington – As
Hakimian, Adina, 516-463-6545 adina.i.hakimian@hofstra.edu,
 Hofstra University – Gg
Halama, Ralf, +44 (0) 1782 7 34960 r.halama@keele.ac.uk,
 Keele University – GpiCa
Halbig, Joseph B., halbig@wazoo.com,
 University of Hawai'i, Hilo – Ca
Halden, Norman M., (204) 474-6910 nm_halden@umanitoba.ca,
 University of Manitoba – Cg
Hale, Beverley A., (519) 824-4120 (Ext. 53434) bhale@uoguelph.ca,
 University of Guelph – Ct
Hale, Leslie J., (202) 633-1796 halel@si.edu,
 Smithsonian Inst / Natl Museum of Natural Hist– GgZnRh
Hale, Michelle, +44 023 92 842290 michelle.hale@port.ac.uk,
 University of Portsmouth – Gg
Hales, Burke R., (541) 737-8121 bhales@coas.oregonstate.edu,
 Oregon State University – Oc
Hales, T. C., +44(0)29 208 74329 HalesT@cardiff.ac.uk,
 University of Wales – Gt
Hales, TC, halest@cf.ac.uk,
 Cardiff University – Gm
Haley, Brian, 541-737-2649 bhaley@coas.oregonstate.edu,
 Oregon State University – Cs
Haley, J C., jachaley@vwc.edu,
 Miami University – Gts
Haley, John C., 757-455-3407 jchaley@vwc.edu,
 Virginia Wesleyan College – GgeZi
Halfar, Jochen, (905) 828-5419 jochen.halfar@utoronto.ca,
 University of Toronto – Pe
Halfman, John D., (315) 781-3918 halfman@hws.edu,
 Hobart & William Smith Colleges – GeHsGn
Halfpenny, Angela, 509-963-2826 angela.halfpenny@cwu.edu,
 Central Washington University – Gcz
Halihan, Todd, 405-744-6358 todd.halihan@okstate.edu,
 Oklahoma State University – HwYuRc
Hall, Anne M., (404) 727-2863 ahall04@emory.edu,
 Emory University – Gz
Hall, Brenda L., (207) 581-2191 brendah@maine.edu,
 University of Maine – Gl
Hall, Chris, (734) 764-6391 cmhall@umich.edu,
 University of Michigan – Cc
Hall, Christopher C., (519) 824-4120 Ext.52740 jchall@uoguelph.ca,
 University of Guelph – Zn
Hall, Clarence A., (310) 825-1010 hall@epss.ucla.edu,
 University of California, Los Angeles – GgtGe
Hall, Cynthia V., (610) 436-1003 chall@wcupa.edu,
 West Chester University – Cg
Hall, Frank R., (504) 280-1105 frhall@mac.com,
 University of New Orleans – Ym
Hall, Ian R., +44(0)29 208 76212 hall@cardiff.ac.uk,
 Cardiff University – GuCsGs
Hall, Jean, +44 (0) 191 208 8783 jean.hall@ncl.ac.uk,
 University of Newcastle Upon Tyne – Ng
Hall, Jeremy, (709) 864-7569 jeremyh@mun.ca,
 Memorial University of Newfoundland – YsGtYr
Hall, Jude, (740) 587-6217 hall@denison.edu,
 Denison University – Zn
Hall, Luke D., (805) 642-3211 lhall@vcccd.net,
 Ventura College – Zy
Hall, Robert, +44 1784 443897 robert.hall@rhul.ac.uk,
 Royal Holloway University of London – Gt
Hall, Robert, +44 (0)1603 59 2550 robert.hall@uea.ac.uk,
 University of East Anglia – On
Hall, Russell L., (403) 220-6678 University of Calgary – Pi

Hall, Stuart A., 713-743-3416 sahgeo@uh.edu,
  University of Houston – Ym
Hall, Tracy, 678-872-8415 thall@highlands.edu,
  Georgia Highlands College – Gg
Hall, III, John R., (504) 231-6305 jrhall3@vt.edu,
  Virginia Polytechnic Institute & State University – So
Halladay, Sylvia R., 614 265 6624 sylvia.halladay@dnr.state.oh.us,
  Ohio Dept of Natural Resources – Zn
Hallar, Anna G., (801) 587-7238 gannet.hallar@utah.edu,
  University of Utah – AspAc
Haller, Merrick C., (541) 737-9141 merrick.haller@oregonstate.edu,
  Oregon State University – OnZrOp
Hallet, Bernard, 206-685-2409 hallet@uw.edu,
  University of Washington – Gml
Hallett, Benjamin W., 920-424-0868 hallettb@uwosh.edu,
  University of Wisconsin, Oshkosh – Gpx
Hallett, Rebecca, (519) 824-4120 Ext.54488 rhallett@uoguelph.ca,
  University of Guelph – Zn
Halliday, Alex, +44 (1865) 272969 alex.halliday@earth.ox.ac.uk,
  University of Oxford – Cs
Hallock, Brent G., (805)756-2436 bhallock@calpoly.edu,
  California Polytechnic State University – Sf
Hallock-Muller, Pamela, (727) 553-1567 pmuller@usf.edu,
  University of South Florida – OuGsPm
Halls, Henry C., 905-828-5363 hlhalls@utm.utoronto.ca,
  University of Toronto – Ym
Halls, Joanne N., (910) 962-7614 hallsj@uncw.edu,
  University of North Carolina, Wilmington – Zi
Halsor, Sid P., (570) 408-4611 sid.halsor@wilkes.edu,
  Wilkes University – Giv
Halsted, Christian, (207) 287-7175 christian.h.halsted@maine.gov,
  Dept of Agriculture, Conservation, and Forestry – Zi
Halverson, Galen, (514) 398-4894 galen.halverson@mcgill.ca,
  McGill University – GsrCs
Halverson, Larry, (515) 294-0495 larryh@iastate.edu,
  Iowa State University of Science & Technology – So
Ham, Nelson R., (920) 403-3977 nelson.ham@snc.edu,
  Saint Norbert College – Gl
Hamane, Angel, ahamane2@calstatela.edu,
  California State University, Los Angeles – Ze
Hamann, Hillary, (303) 871-3977 hillary.hamann@du.edu,
  University of Denver – HgZy
Hambrey, Michael M., +44 (0)1970 621 860, mjh@aber.ac.uk,
  Aberystwyth University – GlsGm
Hamburger, Michael W., (812) 855-2934 hamburg@indiana.edu,
  Indiana University, Bloomington – YsGtv
Hamdan, Abeer, abeer.hamdan@phoenixcollege.edu,
  Phoenix College – Gg
Hamecher, Emily A., 626-660-8314 ehamecher@fullerton.edu,
  California State University, Fullerton – GgzGi
Hameed, Sultan, (631) 632-8319 sultan.hameed@stonybrook.edu,
  SUNY, Stony Brook – As
Hames, Willis E., (334) 844-4881 hameswe@auburn.edu,
  Auburn University – CcGpt
Hamill, Paul, (815) 455-8698 phamill@mchenry.edu,
  McHenry County College – AmZg
Hamilton, George, (413) 499-4660 bhamilton@berkshirecc.edu,
  Berkshire Community College – Zn
Hamilton, Gordon S., (207) 581-3446 gordon.hamilton@maine.edu,
  University of Maine – Gl
Hamilton, Jacqueline, (612) 626-8292 stub0035@umn.edu,
  University of Minnesota – ZiGe
Hamilton, Michael A., (416) 946-7424 mahamilton@es.utoronto.ca,
  University of Toronto – CcGiCu
Hamilton, Sally, +44 (0)131 451 3198 s.hamilton@hw.ac.uk,
  Heriot-Watt University – Gs
Hamilton, Thomas, 4256401339x7067 thomas.hamilton@edcc.edu,
  Edmonds Community College – Zg
Hamilton, Warren, 303-384-2047 whamilton@mines.edu,
  Colorado School of Mines – Ys
Hamlet, Alan, Alan.Hamlet.1@nd.edu,
  University of Notre Dame – Hs
Hamlin, H, Scott, 512-4759527 scott.hamlin@beg.utexas.edu,
  University of Texas at Austin, Jackson School of Geosciences – Gr
Hamme, Roberta C., (250) 472-4014 rhamme@uvic.ca,
  University of Victoria – Oc
Hammer, Julia E., 808-956-5996 jhammer@hawaii.edu,
  University of Hawai'i, Manoa – GiCpGv
Hammer, Philip T., (604) 822-5703 phammer@eos.ubc.ca,
  University of British Columbia – Ys
Hammer, William R., (309) 794-7487 williamhammer@augustana.edu,
  Augustana College – Pv
Hammerschmidt, Chad, 937 775-3457 chad.hammerschmidt@wright.edu,
  Wright State University – CmOcCb
Hammersley, Charles, (928) 523-6655 charles.hammersley@nau.edu,
  Northern Arizona University – Zn
Hammersley, Lisa, 916-278-7200 hammersley@csus.edu,
  California State University, Sacramento – Gig
Hammes, Ursula, 512-471-1891 ursula.hammes@beg.utexas.edu,
  University of Texas at Austin, Jackson School of Geosciences – Gg
Hammond, Amie C., ahammond@austincc.edu,
  Austin Community College District – ZeGg
Hammond, Anne, (807) 343-8677 anne.hammond@lakeheadu.ca,
  Lakehead University – Zn
Hammond, Douglas E., (213) 740-5837 dhammond@usc.edu,
  University of Southern California – Cm
Hammond, Paul E., (503) 725-3387 hammondp@pdx.edu,
  Portland State University – Gi
Hammond, William C., (775) 784-6436 whammond@unr.edu,
  University of Nevada, Reno – Yd
Hampson, Arthur, (801) 585-5698 spike.hampson@geog.utah.edu,
  University of Utah – Zy
Hampson, Gary , +44 20 759 46475 g.j.hampson@imperial.ac.uk,
  Imperial College – Gd
Hampton, Brian A., 575-646-2997 bhampton@nmsu.edu,
  New Mexico State University, Las Cruces – Gst
Hampton, Duane R., (269) 387-5496 duane.hampton@wmich.edu,
  Western Michigan University – HwSpHg
Hampton, Samuel J., +64 3 3667001 Ext 6770 samuel.hampton@canterbury.ac.nz,
  University of Canterbury – GvZe
Hams, Jacquelyn E., (818) 778-5566 hamsje@lavc.edu,
  Los Angeles Valley College – GgOuZe
Hamzaoui, Cherif, 514-987-3000 #6837 hamzaoui.cherif@uqam.ca,
  Universite du Quebec a Montreal – Yg
Han, De-hua, 713-743-9293 dhan@uh.edu,
  University of Houston – Ye
Han, Nizhou, (540) 231-2403 nhan@vt.edu,
  Virginia Polytechnic Institute & State University – Sc
Hanafy, Sherif , +9661368603765 sherif.mahmoud@kfupm.edu.sa,
  King Fahd University of Petroleum and Minerals – Yg
Hanan, Barry B., (619) 594-6710 bhanan@mail.sdsu.edu,
  San Diego State University – Cc
Hancock, Gregory, 61 02 4921 5090 Greg.Hancock@newcastle.edu.au,
  University of Newcastle – ZiGmm
Hancock, Gregory S., (757) 221-2446 gshanc@wm.edu,
  College of William & Mary – Gm
Hand, Kristen, 717–702–2046 khand@pa.gov,
  Pennsylvania Bureau of Topographic & Geologic Survey – HwGg
Hand, Linda M., 650 574-6633 hand@smccd.edu,
  College of San Mateo – GgPgOg
Hand, Suzanne J., s.hand@unsw.edu.au,
  University of New South Wales – Pv
Haneberg, William, (505) 255-8005 New Mexico Institute of Mining and Technology – Ng
Haneberg, William C., (859) 257-5500 bill.haneberg@uky.edu,
  University of Kentucky – NgZi
Haner, Andrew, (501) 683-0153 andrew.haner@arkansas.gov,
  Arkansas Geological Survey – Zi
Hanes, Daniel M., (314) 977-3703 dan.hanes@slu.edu,
  Saint Louis University – GesGm
Hanes, John A., (613) 533-6188 hanes@queensu.ca,
  Queen's University – Cc
Haney, Christa M., 662-268-1032 Ext 224 meloche@geosci.msstate.edu,
  Mississippi State University – Am
Haney, Jennifer, (570) 389-5305 jhaney@bloomu.edu,
  Bloomsburg University – Rn
Hanger, Brendan, 405-744-6358 brendan.hanger@okstate.edu,
  Oklahoma State University – Ggz
Hanger, Rex, hangerr@uww.edu,
  University of Wisconsin, Whitewater – PigGs
Hanke, Steve H., (410) 516-7183 hanke@jhu.edu,
  Johns Hopkins University – Zn
Hankins, Katherine B., 404 413-5775 khankins@gsu.edu,

Georgia State University – Zn
Hanks, Catherine L., 907-474-5562 chanks@gi.alaska.edu,
 University of Alaska, Fairbanks – Go
Hanna, Ruth L., 925-424-1319 rhanna@laspositascollege.edu,
 Las Positas College – Gg
Hanna, Stephen P., (540) 654-1490 shanna@umw.edu,
 Mary Washington College – Zr
Hannah, David, +44 (0)121 41 46925 d.m.hannah@bham.ac.uk,
 University of Birmingham – Hg
Hannah, Judith L., (970) 491-1329 jhannah@warnercnr.colostate.edu,
 Colorado State University – GiCc
Hannibal, Joseph T., 216-231-4600 jhanniba@cmnh.org,
 Cleveland Museum of Natural History – PiEn
Hannigan, Robyn, robyn.hannigan@umb.edu,
 University of Massachusetts, Boston – ClmCt
Hannington, Mark, 613-562-5292 Mark.Hannington@uottawa.ca,
 University of Ottawa – Eg
Hannula, Kimberly, (970) 247-7463 hannula_k@fortlewis.edu,
 Fort Lewis College – GctGp
Hanor, Jeffrey S., (225) 388-3418 hanor@geol.lsu.edu,
 Louisiana State University – Cl
Hanrahan, Janel, 802-626-6370 janel.hanrahan@northernvermont.edu,
 Northern Vermont University-Lyndon – AsZoRc
Hansel, Colleen, (508) 289-3738 chansel@whoi.edu,
 Woods Hole Oceanographic Institution – Cm
Hansen, Anthony R., 320-308-2009 arhansen@stcloudstate.edu,
 Saint Cloud State University – Am
Hansen, Devon A., devon.hansen@und.edu,
 University of North Dakota – Znn
Hansen, Jeffrey C., (505) 667-5043 jchansen@lanl.gov,
 Los Alamos National Laboratory – Zn
Hansen, Lars, +44 (1865) 272000 lars.hansen@earth.ox.ac.uk,
 University of Oxford – Gz
Hansen, Robert N., +27 (0)51 401 2712 hansenr@ufs.ac.za,
 University of the Free State – CgGe
Hansen, Samantha E., (205) 348-7089 shansen@ua.edu,
 University of Alabama – YsGt
Hansen, Thor A., thor.hansen@wwu.edu,
 Western Washington University – Pg
Hansen, Vicki L., (218) 726-8628 vhansen@d.umn.edu,
 University of Minnesota, Duluth – GtXgGc
Hansen, William, (212) 346-1502 Pace University, New York Campus – As
Hansen, William J., (508) 929-8608 whansen@worcester.edu,
 Worcester State University – ZiyZr
Hansom, Jim, +4401413305406 jim.hansom@glasgow.ac.uk,
 University of Glasgow – OnGmZy
Hanson, Andrew D., (702) 895-1092 University of Nevada, Las Vegas – Co
Hanson, Blaine R., (530) 752-1130 brhanson@ucdavis.edu,
 University of California, Davis – Hg
Hanson, Gilbert N., (631) 632-8210 gilbert.hanson@stonybrook.edu,
 Stony Brook University – GgZe
Hanson, Howard, 561 297 2460 hphanson@fau.edu,
 Florida Atlantic University – AsOg
Hanson, Lindley S., lhanson@salemstate.edu,
 Salem State University – GmgOn
Hanson, Paul, (402) 472-7762 phanson2@unl.edu,
 Unversity of Nebraska - Lincoln – Gm
Hanson, Richard E., (817) 257-7996 r.hanson@tcu.edu,
 Texas Christian University – GxvGt
Hanson, Sarah L., 517-264-3944 slhanson@adrian.edu,
 Adrian College – Giz
Hanson, William D., 501 683-0115 doug.hanson@arkansas.gov,
 Arkansas Geological Survey – EgnGs
Hanson, Jr., Alfred N., (401) 874-6899 akhanson@gso.uri.edu,
 University of Rhode Island – Oc
Hapke, Bruce W., (412) 624-8876 hapke@pitt.edu,
 University of Pittsburgh – Xy
Haq, Saad, (765) 496-7206 haq@purdue.edu,
 Purdue University – Gt
Hara, Tetsu, (401) 874-6509 thara@uri.edu,
 University of Rhode Island – Op
Harbaugh, John W., (650) 723-3365 harbaugh@pangea.stanford.edu,
 Stanford University – Gq
Harbert, William P., (412) 624-8874 harbert@pitt.edu,
 University of Pittsburgh – Ym
Harbor, David J., 540 458 8871 harbord@wlu.edu,
 Washington & Lee University – Gm

Harbor, Jon M., (765) 494-4753 jharbor@purdue.edu,
 Purdue University – GmlGe
Harbottle, Garman, garman@bnl.gov,
 Stony Brook University – Cc
Hardage, Bob A., 512-471-0300 bob.hardage@beg.utexas.edu,
 University of Texas, Austin – Yes
Hardage, Sarah M., 956 665 3023 sarah.fearnleyhardage@utrgv.edu,
 University of Texas, Rio Grande Valley – GmZiGg
Harder, Brian J., (225) 578-8533 bharde1@lsu.edu,
 Louisiana State University – Eo
Harder, Steven H., (915) 747-5746 harder@utep.edu,
 University of Texas, El Paso – Yg
Harder, Vicki, 915-747-5501 vmharder@utep.edu,
 University of Texas, El Paso – Gg
Hardin, Perry J., (801) 378-6062 perry_hardin@byu.edu,
 Brigham Young University – Zr
Harding, Chris, (515) 294-7521 charding@iastate.edu,
 Iowa State University of Science & Technology – Gq
Harding, Ian C., +44 (0)23 80592071 ich@noc.soton.ac.uk,
 University of Southampton – PmlPc
Hardison, Amber K., (361) 749-6705 amber.hardison@utexas.edu,
 University of Texas, Austin – Cm
Hardy, Douglas R., (802) 649-1829 dhardy@geo.umass.edu,
 University of Massachusetts, Amherst – Hy
Hardy, Shaun J., (202) 478-7960 shardy@carnegiescience.edu,
 Carnegie Institution for Science – Zn
Hargis, David, 619-521-0165 dhargis@hargis.com,
 University of Arizona – Hwq
Hargrave, Jennifer E., 337-482-0678 jhargrave@louisiana.edu,
 University of Louisiana at Lafayette – PvGrZe
Hargrave, Phyllis, 406-496-4606 phargrave@mtech.edu,
 Montana Tech of The University of Montana – Gg
Hargraves, Paul E., (401) 874-6241 pharg@gso.uri.edu,
 University of Rhode Island – Ob
Hargreaves, Bruce R., (610) 758-3683 brh0@lehigh.edu,
 Lehigh University – Ob
Hargreaves, Tom, +44 (0)1603 59 3116 tom.hargreaves@uea.ac.uk,
 University of East Anglia – Ge
Hari, Kosiyath R., krharigeology@gmail.com,
 Pt. Ravishankar Shukla University – GiiCp
Haritashya, Umesh, 937-229-2939 Umesh.Haritashya@notes.udayton.edu,
 University of Dayton – ZrGm
Harley, Grant, 208-885-0950 gharley@uidaho.edu,
 University of Idaho – ZynZn
Harley, Simon L., +44 (0) 131 650 4839 Simon.Harley@ed.ac.uk,
 Edinburgh University – Gt
Harlow, George E., (212) 769-5378 Graduate School of the City University of New York – Gz
Harmon, Nicholas, +44 (0)23 80594783 N.Harmon@noc.soton.ac.uk,
 University of Southampton – Yg
Harms, Tekla A., (413) 542-2711 taharms@amherst.edu,
 Amherst College – Gt
Harnett, Erika M., 206-543-0212 eharnett@uw.edu,
 University of Washington – XyGev
Harnik, Nili, 03-640-6359 harnik@tau.ac.il,
 Tel Aviv University – As
Harnik, Paul, (717) 358-5946 paul.harnik@fandm.edu,
 Franklin and Marshall College – Pg
Haroldson, Erik L., 931221-7449 haroldsone@apsu.edu,
 Austin Peay State University – EdGzCg
Harper, David, +44 (0) 191 33 47143 david.harper@durham.ac.uk,
 Durham University – Pg
Harper, Joel, (406) 243-2341 joel.harper@umontana.edu,
 University of Montana – GlZc
Harper, Stephen B., (252) 328-6773 harpers@ecu.edu,
 East Carolina University – GmgGe
Harper, Jr., Charles W., (405) 325-7725 charper@gcn.ou.edu,
 University of Oklahoma – Pi
Harpold, Adrian, (775) 784-6759 aharpold@cabnr.unr.edu,
 University of Nevada, Reno – HqsZr
Harpp, Karen, (315) 228-7211 kharpp@colgate.edu,
 Colgate University – GvCgGi
Harr, Patrick A., 831-656-3787 paharr@nps.edu,
 Naval Postgraduate School – Am
Harrap, Rob, 613-533-2553 harrap@queensu.ca,
 Queen's University – Gc
Harrell, Michael, 206-543-0367 mdh666@uw.edu,

University of Washington – Ze
Harrell, Michael, 206-344-4392 MHarrell@sccd.ctc.edu,
    Seattle Central Community College – Gg
Harrell, Jr., T. L., 205-247-3559 TLHarrell@gsa.sate.al.us,
    Geological Survey of Alabama – PvGg
Harries, Peter J., (813) 974-4974 harries@chuma.cas.usf.edu,
    University of South Florida, Tampa – Pg
Harrington, Charles D., (505) 667-0078 charrington@lanl.gov,
    Los Alamos National Laboratory – Gm
Harrington, Glenn, glenn.harrington@flinders.edu.au,
    Flinders University – Hw
Harrington, Rebecca, +49-721-6084625 rebecca.harrington@kit.edu,
    Karlsruhe Institute of Technology – Ys
Harris, Ann, ann.harris@eku.edu,
    Eastern Kentucky University – GgZy
Harris, Ann G., 330-941-3613 agharris@cc.ysu.edu,
    Youngstown State University – Ge
Harris, Brett D., +618 9266-3089 B.Harris@curtin.edu.au,
    Curtin University – HwYve
Harris, C, harrisC@cf.ac.uk,
    Cardiff University – Ge
Harris, Chris, 021-650-4886 chris.harris@uct.ac.za,
    University of Cape Town – GiCsGg
Harris, Clay D., 615-904-8019 clay.harris@mtsu.edu,
    Middle Tennessee State University – Gse
Harris, Courtney K., (804) 684-7194 ckharris@vims.edu,
    College of William & Mary – On
Harris, David C., (859) 323-0545 dcharris@uky.edu,
    University of Kentucky – Go
Harris, DeVerle P., (520) 621-6024 dharris@geo.arizona.edu,
    University of Arizona – Eg
Harris, Glendon H., (706) 542-2968 University of Georgia – Sb
Harris, James B., 601-974-1343 harrijb@millsaps.edu,
    Millsaps College – Ye
Harris, Jerry M., (650) 723-0496 Stanford University – Ys
Harris, M. Scott, (843) 953-0864 harriss@cofc.edu,
    College of Charleston – OnGam
Harris, Mark T., (414) 229-5483 mtharris@uwm.edu,
    University of Wisconsin, Milwaukee – GsrGd
Harris, Michelle, (360) 623-8472 michelle.harris@centralia.edu,
    Centralia College – Gg
Harris, Nicholas, (780) 492-0356 nharris@ualberta.ca,
    University of Alberta – Gso
Harris, Nigel B., n.b.w.harris@open.ac.uk,
    The Open University – GtiGp
Harris, Randa R., (678) 839-4056 rharris@westga.edu,
    University of West Georgia – Hs
Harris, Richard S., (905) 525-9140 (Ext. 27216) harrisr@mcmaster.ca,
    McMaster University – Zu
Harris, Robert N., (541) 737-4370 rharris@coas.oregonstate.edu,
    Oregon State University – Yhr
Harris, Robin F., (608) 263-5691 rfharris@wisc.edu,
    University of Wisconsin, Madison – Sb
Harris, Ron, (801) 422-9264 rharris@byu.edu,
    Brigham Young University – Gc
Harris, Sara, (604) 822-5674 sara@eos.ubc.ca,
    University of British Columbia – Ou
Harris, Virginia M., (860) 685-2244 vharris@wesleyan.edu,
    Wesleyan University – Zn
Harris, W. Burleigh, (910) 962-3492 harrisw@uncw.edu,
    University of North Carolina, Wilmington – GrdGs
Harris, William H., (718) 951-5416 Graduate School of the City University of New York – Hw
Harris, Jr., Stanley E., (618) 453-3351 Southern Illinois University Carbondale – Ge
Harris, Jr., Willie G., (352) 294-3110 apatite@ufl.edu,
    University of Florida – Scd
Harrison, Bruce I., (575) 835-5864 bruce@nmt.edu,
    New Mexico Institute of Mining and Technology – GmSof
Harrison, Cheryl, 956 882-8456 cheryl.harrison@utrgv.edu,
    University of Texas, Rio Grande Valley – Op
Harrison, Christopher G., 305 421-4610 charrison@rsmas.miami.edu,
    University of Miami – Yr
Harrison, Conor, cmharris@mailbox.sc.edu,
    University of South Carolina – Zn
Harrison, Don E., (206) 526-6225 harrison@pmel.noaa.gov,
    University of Washington – Op

Harrison, Halstead, harrison@atmos.washington.edu,
    University of Washington – As
Harrison, Linda , (269) 387-8642 linda.harrison@wmich.edu,
    Western Michigan University – GgPmZn
Harrison, Michael J., (931) 372-3751 mharrison@tntech.edu,
    Tennessee Tech University – Gc
Harrison, Richard, +44 (0) 1223 333380 rjh40@esc.cam.ac.uk,
    University of Cambridge – Gz
Harrison, Robert S., (516) 299-2318 Long Island University, C.W. Post Campus – Zy
Harrison, Roy M., +44 (0)121 41 43494 r.m.harrison@bham.ac.uk,
    University of Birmingham – As
Harrison, T. Mark, (310) 825-7970 tmh@argon.ess.ucla.edu,
    University of California, Los Angeles – Cc
Harrison, Wendy J., (303) 273-3821 wharriso@mines.edu,
    Colorado School of Mines – Cl
Harrison, III, William B., (269) 387-8691 william.harrison_iii@wmich.edu,
    Western Michigan University – GsrPg
Harry, Dennis L., (970) 491-2714 dharry@warnercnr.colostate.edu,
    Colorado State University – Yg
Harsh, James B., (509) 335-3650 harsh@wsu.edu,
    Washington State University – Sc
Harshvardhan, (765) 494-0693 harsh@purdue.edu,
    Purdue University – As
Hart, Brian R., bhart@uwo.ca,
    Western University – Ca
Hart, Craig J., (604) 822-5149 chart@eoas.ubc.ca,
    University of British Columbia – EmGiCc
Hart, David J., (608) 262-2307 dave.hart@wgnhs.uwex.edu,
    University of Wisconsin, Extension – HwYg
Hart, David J., 608-262-1580 djhart@wisc.edu,
    University of Wisconsin-Madison – Hg
Hart, Evan A., (931) 372-3121 EHart@TnTech.edu,
    Tennessee Tech University – Zy
Hart, Malcolm B., +441752584761 mhart@plymouth.ac.uk,
    University of Plymouth – PmGur
Hart, Richard L., (630) 252-5839 rlhart@anl.gov,
    Argonne National Laboratory – As
Hart, Roger M., rmhart1@ccri.edu,
    Community College of Rhode Island – Gg
Hart, Stanley R., (520) 625-4543 shart@whoi.edu,
    Woods Hole Oceanographic Institution – CgGvCp
Hart, William K., (513) 529-3217 hartwk@miamioh.edu,
    Miami University – GivCc
Harte, Michael, mharte@coas.oregonstate.edu,
    Oregon State University – Gu
Hartel, Peter G., (706) 542-0898 University of Georgia – Sb
Harter, Thomas, 530-752-2709 thharter@ucdavis.edu,
    University of California, Davis – Hw
Harter, Thomas, 530-400-1784 ThHarter@ucdavis.edu,
    University of California, Davis – Hg
Harter, Thomas L., (530) 752-2709 thharter@ucdavis.edu,
    University of California, Davis – Hw
Hartley, Susan, (218)279-2661 s.hartley@lsc.edu,
    Lake Superior College – Gg
Hartman, Joseph H., (701) 777-5055 joseph.hartman@engr.und.edu,
    University of North Dakota – Pi
Hartmann, Dennis L., (206) 543-7460 dennis@atmos.washington.edu,
    University of Washington – As
Hartnett, Hilairy E., (480) 965-5593 h.hartnett@asu.edu,
    Arizona State University – ComOc
Hartog, Renate, 206-685-7079 jrhartog@uw.edu,
    University of Washington – Ys
Hartse, Hans E., (505) 664-8495 Los Alamos National Laboratory – Ys
Harty, John P., jharty1@jccc.edu,
    Johnson County Community College – Zy
Harun, Nina, 907-451-5085 nina.harun@alaska.gov,
    Alaska Division of Geological & Geophysical Surveys – Go
Harvey, Cyril H., (336) 316-2238 charvey@guilford.edu,
    Guilford College – Gr
Harvey, H. Rodger, 757-683-5657 rharvey@odu.edu,
    Old Dominion University – Com
Harvey, James T., harvey@mlml.calstate.edu,
    Moss Landing Marine Laboratories – Ob
Harvey, Jason, +44(0) 113 34 34033 feejh@leeds.ac.uk,
    University of Leeds – Cg
Harvey, Omar R., 817-257-4272 omar.harvey@tcu.edu,

Texas Christian University – GeHwg
Harvey, Ralph P., (216) 368-0198 rph@case.edu,
 Case Western Reserve University – Xm
Harvey, Tom, +440116 252 3644 thph2@le.ac.uk,
 Leicester University – Pg
Harwood, David M., (402) 472-2648 dharwood1@unl.edu,
 University of Nebraska, Lincoln – PmGul
Harwood, Richard D., (309) 796-5271 harwoodr@bhc.edu,
 Black Hawk College – Gv
Hasan, Khaled W., khaled.hasan@austincc.edu,
 Austin Community College District – HsZri
Hasan, Mainul, (514) 398-4755 McGill University – Nx
Hasan, Mohamed Ali, 03-79674203/4141 alihasan@um.edu.my,
 University of Malaya – Hg
Hasan, Syed E., (816) 235-2976 hasans@umkc.edu,
 University of Missouri, Kansas City – Ng
Hasbargen, Leslie E., (607) 436-2741 leslie.hasbargen@oneonta.edu,
 SUNY, Oneonta – GmHgZr
Haselwander, Airin J., (573) 368-2196 airin.haselwander@dnr.mo.gov,
 Missouri Dept of Natural Resources – Gg
Hasenmueller, Elizabeth, (314) 977-7518 elizabeth.hasenmueller@slu.edu,
 Saint Louis University – ClHs
Hasenmueller, Nancy R., (812) 855-1360 hasenmue@indiana.edu,
 Indiana University – Gre
Hasiotis, Stephen T., (785) 864-4941 hasiotis@ku.edu,
 University of Kansas – Pe
Hassani, Ferri, (514) 398-4755 McGill University – Nm
Hassanpour Sedghi, Mohammad, +98 (411) 339 2697 University of Tabriz – YsNeYg
Hasselbo, Stephen, +44 01326 253651 S.P.Hesselbo@exeter.ac.uk,
 Exeter University – Gs
Hassett, John J., (217) 333-9472 University of Illinois, Urbana-Champaign – Sc
Hastenrath, Stefan L., (608)262-3659 slhasten@wisc.edu,
 University of Wisconsin, Madison – As
Hastie, Warwick W., 031-2602519 HastieW@ukzn.ac.za,
 University of KwaZulu-Natal – NrgGc
Hastings, Meredith, 401-863-3658 Meredith_Hastings@brown.edu,
 Brown University – As
Hastings, Philip A., (858) 822-2913 phastings@ucsd.edu,
 University of California, San Diego – Ob
Haszeldine, R. S., +44 (0) 131 650 8549 stuart.haszeldine@ed.ac.uk,
 Edinburgh University – GoRmZo
Hatch, Christine, 413 577-2245 chatch@geo.umass.edu,
 University of Massachusetts, Amherst – HwgSp
Hatcher, Jr., Robert D., (865) 974-6565 bobmap@utk.edu,
 Washington University in St. Louis – GtcYm
Hatheway, Richard B., hatheway@geneseo.edu,
 SUNY, Geneseo – GxzGp
Hattin, Donald E., (812) 855-8232 hattin@indiana.edu,
 Indiana University, Bloomington – Gr
Hattori, Keiko, 613-562-5800 6866 khattori@uottawa.ca,
 University of Ottawa – CgEmGz
Hauck, II, Steven A., (216) 368-3675 hauck@case.edu,
 Case Western Reserve University – XyYgXg
Hauer, Kendall L., (513) 529-3220 hauerkl@miamioh.edu,
 Miami University – GgZeCg
Haugerud, Ralph, 206-713-7453 haugerud@uw.edu,
 University of Washington – Zi
Hauksson, Egill, (626) 395-6954 hauksson@gps.caltech.edu,
 California Institute of Technology – YsGtYg
Hausback, Brian, (916) 278-6521 hausback@csus.edu,
 California State University, Sacramento – Gv
Hauser, Ernest C., 937 775-3443 ernest.hauser@wright.edu,
 Wright State University – YeGtYu
Hausrath, Elisabeth M., 702-895-1134 elisabeth.hausrath@unlv.edu,
 University of Nevada, Las Vegas – ScCl
Hauswirth, Scott, (818) 677-4880 scott.hauswirth@csun.edu,
 California State University, Northridge – HgCg
Hautala, Susan L., (206) 543-0596 susanh@ocean.washington.edu,
 University of Washington – Op
Hauton, Chris, +44 (0)23 80595784 ch10@noc.soton.ac.uk,
 University of Southampton – Ob
Havenith, Hans-Balder, 32 4 3662035 HB.Havenith@ulg.ac.be,
 Universite de Liege – Ge
Haverluk, Terry W., (719) 333-8746 usafa.dfeg@usafa.edu,
 United States Air Force Academy – Zg

Havholm, Karen G., (715) 836-2945 havholkg@uwec.edu,
 University of Wisconsin, Eau Claire – GsZe
Havig, Jeff, jhavig@umn.edu,
 University of Minnesota, Twin Cities – GeCg
Havlin, John L., (919) 515-2655 havlin@ncsu.edu,
 North Carolina State University – Sc
Hawkins, Andrew, 410-704-3187 dhawkins@towson.edu,
 Towson University – Pg
Hawkins, David, 781-283-3554 dhawkins@wellesley.edu,
 Wellesley College – Gzi
Hawkins, James W., (858) 534-2161 jhawkins@ucsd.edu,
 University of California, San Diego – OuGip
Hawkins, John F., (334) 844-4894 jfh0005@auburn.edu,
 Auburn University – GcZeGt
Hawkins, Lawrence, +44 (0)23 80593426 leh@noc.soton.ac.uk,
 University of Southampton – Ob
Hawkins, Michael A., (518) 486-2011 michael.hawkins@nysed.gov,
 New York State Geological Survey – Gz
Hawkins, R. B., (520) 621-7273 rhawkins@ag.arizona.edu,
 University of Arizona – Hs
Hawkins, Terry, (573) 368-2164 terry.hawkins@dnr.mo.gov,
 Missouri Dept of Natural Resources – Geg
Hawkins, Timothy W., (717) 477-1662 twhawk@ship.edu,
 Shippensburg University – AsmHg
Hawkins, Ward L., (505) 667-5835 whawkins@lanl.gov,
 Los Alamos National Laboratory – Zg
Hawley, Bob, robert.l.hawley@dartmouth.edu,
 University of Washington – Gl
Hawley, John W., (505) 255-4847 hgeomatters@gmail.com,
 New Mexico Institute of Mining and Technology – GeHwGr
Hawley, Rebecca D., (928) 523-1251 d.hawley@nau.edu,
 Northern Arizona University – Zn
Hawley, Robert L., 603-646-2373 Robert.Hawley@dartmouth.edu,
 Dartmouth College – Gl
Hawman, Robert B., (706) 542-2398 rob@seismo.gly.uga.edu,
 University of Georgia – YseYu
Hawthorne, Frank C., (204) 474-8861 frank_hawthorne@umanitoba.ca,
 University of Manitoba – Gz
Hawthorne Allen, Alice M., 304-384-6273 amhallen@concord.edu,
 Concord University – Xa
Hay, Alex E., (902) 494-6657 alex.hay@dal.ca,
 Dalhousie University – OpnGu
Hay, Carling, carling.hay@bc.edu,
 Boston College – YgZoOn
Hay, Rodrick, (310) 243-2547 rhay@csudh.edu,
 California State University, Dominguez Hills – Zr
Hayashi, Masaki, (403) 220-2794 hayashi@ucalgary.ca,
 University of Calgary – HwSp
Hayden, Geoffrey W., (912) 752-2597 Mercer University – Zm
Hayden, Martha C., (801) 537-3311 nrugs.mhayden@state.ut.us,
 Utah Geological Survey – Pv
Hayes, Alexander G., (607) 255-1712 agh4@cornell.edu,
 Cornell University – XgZrGs
Hayes, Christopher T., (228) 688-3469 christopher.t.hayes@usm.edu,
 University of Southern Mississippi – Oc
Hayes, Garry F., (209) 575-6294 Modesto Junior College – Gg
Hayes, Garry F., ghayes@csustan.edu,
 California State University, Stanislaus – Gg
Hayes, John M., (914) 365-8470 jhayes@whoi.edu,
 Indiana University, Bloomington – Co
Hayes, Jorden L., (717) 245-8303 hayesjo@dickinson.edu,
 Dickinson College – Ygs
Hayes, Michael J., (402) 472-4271 mhayes2@unl.edu,
 University of Nebraska, Lincoln – As
Hayes, Miles O., 504-280-6325 mhayes@uno.edu,
 University of New Orleans – Gs
Hayes, Van E., 813-253-7685 vhayes@hccfl.edu,
 Hillsborough Community College – Zg
Hayman, Nicholas W., 512-471-7721 hayman@ig.utexas.edu,
 University of Texas, Austin – Ou
Hayman, Patrick C., +61 7 3138 5259 patrick.hayman@qut.edu.au,
 Queensland University of Technology – EgGvi
Haymet, Anthony D., (858) 534-2827 thaymet@ucsd.edu,
 University of California, San Diego – OcZgn
Haynes, Kyle J., (540) 837-1758 kjh8w@virginia.edu,
 University of Virginia – Zn
Haynes, Jr., C. Vance, (520) 621-6307 University of Arizona – Ga

Hayob, Jodie, (540) 654-1425 jhayob@umw.edu,
   University of Mary Washington – GxzGg
Hays, James D., (845) 365-8403 jimhays@ldeo.columbia.edu,
   Columbia University – PemOu
Hays, Phillip D., (479) 575-7343 pdhays@uark.edu,
   University of Arkansas, Fayetteville – HwCsGg
Hayward, Chris L., +44 (0) 131 650 5827 chris.hayward@ed.ac.uk,
   Edinburgh University – Ga
Haywick, Douglas W., (334) 460-6381 dhaywick@jaguar1.usouthal.edu,
   University of South Alabama – Gs
Haywood, Alan, +44(0) 113 34 38657 earamh@leeds.ac.uk,
   University of Leeds – Pe
Hazel, McGoff, H.J.McGoff@rdg.ac.uk,
   University of Reading – Pg
Hazel, Jr., Joseph E., (928) 523-9145 joseph.hazel@nau.edu,
   Northern Arizona University – Gs
Hazen, Helen, 303-871-2672 helen.hazen@du.edu,
   University of Denver – Zy
Hazen, Robert M., (703) 993-2163 rhazen@gmu.edu,
   George Mason University – GzZf
Hazen, Robert M., (202) 478-8962 rhazen@carnegiescience.edu,
   Carnegie Institution for Science – Gz
He, Changming, (302) 831-4917 hchm@udel.edu,
   University of Delaware – GqHg
He, Helen, +44 (0)1603 59 2091 yi.he@uea.ac.uk,
   University of East Anglia – Pe
He, Jiaze, 609-258-4101 jiazeh@princeton.edu,
   Princeton University – Yg
He, Ruoying, 919-513-0943 ruoying_he@ncsu.edu,
   North Carolina State University – Op
He, Zhenli, 772-468-3922 zhe@ufl.edu,
   University of Florida – Sc
Head, Elisabet M., 773-442-6055 E-Head@neiu.edu,
   Northeastern Illinois University – GvZrCg
Head, Martin J., (905) 688-5550 (Ext. 5216) mjhead@brocku.ca,
   Brock University – Pl
Head, III, James W., (401) 863-2526 james_head_iii@brown.edu,
   Brown University – XgGvt
Headley, Rachel, headley@uwp.edu,
   University of Wisconsin, Parkside – Gl
Heal, Kate V., +44 (0) 131 650 5420 K.Heal@ed.ac.uk,
   Edinburgh University – Co
Heald, Colette, (617) 324-5666 heald@mit.edu,
   Massachusetts Institute of Technology – As
Healey, Michael, (604) 822-4705 mhealey@eos.ubc.ca,
   University of British Columbia – Ob
Heaman, Larry M., (780) 492-2778 larry.heaman@ualberta.ca,
   University of Alberta – Cc
Heaney, Michael , (979) 845-7841 heaney@geo.tamu.edu,
   Texas A&M University – Pg
Heaney, Peter J., (814) 865-6821 pjheaney@psu.edu,
   Pennsylvania State University, University Park – GzClGe
Hearn, Thomas M., (505) 646-5076 thearn@nmsu.edu,
   New Mexico State University, Las Cruces – Ys
Hearn Jr, B. Carter, (202) 633-1756 hearnc@si.edu,
   Smithsonian Inst / Natl Museum of Natural Hist– Gic
Heath, G. Ross, (206) 543-3153 rheath@u.washington.edu,
   University of Washington – Cm
Heath, Kathleen M., (812) 237-3004 kheath@indstate.edu,
   Indiana State University – Pb
Heatherington, Ann L., (352) 392-6220 aheath@ufl.edu,
   University of Florida – CcGti
Heaton, Jill, (775) 784-8056 jheaton@unr.edu,
   University of Nevada, Reno – ZniZy
Heaton, Richard C., (505) 667-1141 Los Alamos National Laboratory – Zn
Heaton, Thomas H., (626) 395-6897 heaton@gps.caltech.edu,
   California Institute of Technology – Ne
Heaton, Timothy H., (605) 677-6122 Timothy.Heaton@usd.edu,
   University of South Dakota – PvOg
Heatwole, Charles A., (212) 772-5323 cah@geo.hunter.cuny.edu,
   Hunter College (CUNY) – Zn
Heavers, Richard, (401) 254-3095 Roger Williams University – Op
Hebda, Richard J., (250) 387-5493 University of Victoria – Pl
Hebert, David, (902) 426-1216 david.hebert@dfo-mpo.gc.ca,
   Dalhousie University – Op
Hebert, David L., (401) 874-6610 herbert@gso.uri.edu,
   University of Rhode Island – Op
Hebert, Rejean J., 418-656-3137 herbert@ggl.ulaval.ca,
   Universite Laval – Gi
Heck, Frederick R., (231) 591-2588 heckf@ferris.edu,
   Ferris State University – GgZg
Heck, Richard, (519) 824-4120 (Ext. 52450) rheck@uoguelph.ca,
   University of Guelph – Sd
Heckathorn, Scott, 419-530-4328 scott.heckathorn@utoledo.edu,
   University of Toledo – Zn
Heckel, Philip H., (319) 335-1804 philip-heckel@uiowa.edu,
   University of Iowa – GrPsGd
Heckel, Wolfgang, 089/2180 4331 heckl@lmu.de,
   Ludwig-Maximilians-Universitaet Muenchen – Gz
Heckert, Andrew B., (828) 262-7609 heckertab@appstate.edu,
   Appalachian State University – PvgZe
Hedberg, Russell, 717-477-1515 rchedberg@ship.edu,
   Shippensburg University – ZiSoZc
Hedquist, Brent, 361-593-3586 brent.hedquist@tamuk.edu,
   Texas A&M University, Kingsville – ZyAtZi
Heenan, Cleo J., (605) 394-2461 Cleo.Heenan@sdsmt.edu,
   South Dakota School of Mines & Technology – Zn
Heermance, Richard V., (818) 677-4357 richard.heermance@csun.edu,
   California State University, Northridge – GrCc
Heerschap, Lauren, 970-247-7620 Heerschap_L@fortlewis.edu,
   Fort Lewis College – GgEgGt
Hefferan, Kevin P., (715) 346-4453 kheffera@uwsp.edu,
   University of Wisconsin, Stevens Point – GctGe
Hegarty, Patricia, +353 21 4902533 p.hegarty@ucc.ie,
   University College Cork – Zn
Hegerl, Gabriele C., +44 (0) 131 651 9092 Gabi.Hegerl@ed.ac.uk,
   Edinburgh University – As
Hegewisch, Katherine, 208-885-6622 khegewisch@uidaho.edu,
   University of Idaho – Zo
Hegg, Dean A., (206) 543-1984 deanhegg@atmos.washington.edu,
   University of Washington – As
Hegna, Thomas A., (309) 298-1366 ta-hegna@wiu.edu,
   Western Illinois University – PigPg
Hegner, Ernst, 089/2180 4274 hegner@lmu.de,
   Ludwig-Maximilians-Universitaet Muenchen – Gg
Hehr, John G., (479) 575-7428 jghehr@uark.edu,
   University of Arkansas, Fayetteville – AmZy
Heibel, Todd, 909-384-8638 theibel@sbccd.cc.ca.us,
   San Bernardino Valley College – Zyi
Heid, Kelly L., 616-331-3783 heidke@gvsu.edu,
   Grand Valley State University – Ze
Heidinger, Andrew, 608-263-6757 heidinger@ssec.wisc.edu,
   University of Wisconsin, Madison – Zr
Heidlauf, Lisa, lisa.heidlauf@wheaton.edu,
   Wheaton College – Gg
Heigold, Paul C., (630) 252-7861 Argonne National Laboratory – Yg
Heiken, Grant, (505) 667-8477 heiken@lanl.gov,
   Los Alamos National Laboratory – Gv
Heikes, Brian G., (401) 874-6810 zagar@notos.gso.uri.edu,
   University of Rhode Island – As
Heil, Dean, (505) 646-3405 New Mexico State University, Las Cruces – Sc
Heimann, Adriana, 252 328 5206 heimanna@ecu.edu,
   East Carolina University – GzpGi
Heimann, William H., (785) 625-5663 whheimann@fhsu.edu,
   Fort Hays State University – HgwGm
Heimbach, Patrick, (512) 232-7694 heimbach@utexas.edu,
   University of Texas, Austin – OpGlAs
Heimpel, Moritz, (780) 492-3519 mheimpel@phys.ualberta.ca,
   University of Alberta – Zn
Heimsath, Arjun, 480-965-5585 arjun.heimsath@asu.edu,
   Arizona State University – Gc
Hein, Christopher J., 804-684-7533 hein@vims.edu,
   College of William & Mary – Ou
Hein, Kim A., 011 717 6623 kim.ncube0hein@wits.ac.za,
   University of the Witwatersrand – GgEmGc
Heinrich, Paul V., 225-578-4398 heinric@lsu.edu,
   Louisiana State University – GsaGm
Heinz, Dion L., (773) 702-3046 heinz@uchicago.edu,
   University of Chicago – GyzYx
Heinzel, Chad E., (319) 273-6168 chad.heinzel@uni.edu,
   University of Northern Iowa – GmaSa
Heise, Elizabeth, 956 882-6769 elizabeth.heise@utrgv.edu,
   University of Texas, Rio Grande Valley – OnZeGg
Heiss, James, (978) 934-6304 james_heiss@uml.edu,

University of Massachusetts, Lowell – Hw
Heitman, Mary P., MaryPat.Heitman@dnr.iowa.gov,
 Iowa Dept of Natural Resources – Zn
Heitmuller, Franklin, 601-266-5423 franklin.heitmuller@usm.edu,
 University of Southern Mississippi – GmsGe
Heitz, Leroy F., lheitz@triton.uog.edu,
 University of Guam – Hg
Heizler, Matthew T., (575) 835-5244 matt@nmt.edu,
 New Mexico Institute of Mining and Technology – Cc
Helba, Hossam EL-Din A., 002-03-3921595 kahralran2009@Yahoo.com,
 Alexandria University – GgZm
Helenes, Javier, jhelenes@cicese.mx,
 Centro de Investigación Científica y de Educación Superior de Ensenada – Gr
Helgers, Karen, (845) 687-5224 helgersk@sunyulster.edu,
 SUNY, Ulster County Community College – GeZgy
Helie, Jean-François, 514-987-3000 #2413 helie.jean-francois@uqam.ca,
 Universite du Quebec a Montreal – Cs
Heliker, Christina C., (808) 967-8807 cheliker@usgs.gov,
 University of Hawai'i, Hilo – Gv
Heller, Matthew J., (434) 951-6351 matt.heller@dmme.virginia.gov,
 Division of Geology and Mineral Resources – GgRnGe
Hellio, Gabrielle, +44 023 80599258 G.C.Hellio@soton.ac.uk,
 University of Southampton – Yg
Helmberger, Donald V., (626) 395-6998 helm@gps.caltech.edu,
 California Institute of Technology – Ys
Helmick, Sara B., (505) 667-9583 Los Alamos National Laboratory – Zn
Helmke, Martin F., 610-436-3565 mhelmke@wcupa.edu,
 West Chester University – HwSdZg
Helmke, Philip A., (608) 263-4947 pahelmke@wisc.edu,
 University of Wisconsin, Madison – Sc
Helmke, Vicky, (610) 738-0308 vhelmke@wcupa.edu,
 West Chester University – Gg
Helmstaedt, Herwart, (613) 533-6175 Queen's University – Gc
Helmuth, Brian, 7815817370 x307 b.helmuth@northeastern.edu,
 Northeastern University – OnZcRc
Helper, Mark A., (512) 471-1009 helper@jsg.utexas.edu,
 University of Texas, Austin – GctXg
Helsdon, John H., (605) 394-2291 John.Helsdon@sdsmt.edu,
 South Dakota School of Mines & Technology – As
Hembree, Daniel, (740) 597-1495 hembree@ohio.edu,
 Ohio University – Pi
Hemler, Deb, (304) 367-4393 dhemler@fairmontstate.edu,
 Fairmont State University – Ze
Hemming, Gary N., 718-997-3335 hemming@qc.cuny.edu,
 Graduate School of the City University of New York – Zg
Hemming, Sidney, sidney@ldeo.columbia.edu,
 Columbia University – Cg
Hemming, Sidney R., (845) 365-8417 sidney@ldeo.columbia.edu,
 Columbia University – CscPe
Hemphill-Haley, Eileen, emh21@humboldt.edu,
 Humboldt State University – Pm
Hemphill-Haley, Mark, (707) 826-3933 mah54@humboldt.edu,
 Humboldt State University – Gt
Hemsley, A, hemsleyAR@cf.ac.uk,
 Cardiff University – Pl
Hemsley, Alan, +44(0)29 208 75367 HemsleyAR@cardiff.ac.uk,
 University of Wales – Pl
Hemzacek Laukant, Jean M., 773/442-6056 J-Hemzacek@neiu.edu,
 Northeastern Illinois University – Sc
Hence, Deanna A., 217-244-6014 dhence@illinois.edu,
 University of Illinois, Urbana-Champaign – AsmAp
Hencher, Steve, S.R.Hencher@leeds.ac.uk,
 University of Leeds – Ng
Hendershott, Myrl C., (858) 534-5705 mhendershott@ucsd.edu,
 University of California, San Diego – Op
Henderson, Charles M., (403) 220-6170 henderson@geo.ucalgary.ca,
 University of Calgary – Pm
Henderson, Donald, (403) 823-7707 Royal Tyrrell Museum of Palaeontology – Pv
Henderson, Gideon, gideon.henderson@earth.ox.ac.uk,
 University of Oxford – Cg
Henderson, Gina R., 410-293-6555 ghenders@usna.edu,
 United States Naval Academy – As
Henderson, Grant S., (416) 978-6041 henders@geology.utoronto.ca,
 University of Toronto – Gz
Henderson, Matt, (803) 896-7931 hendersonm@dnr.sc.gov,
 South Carolina Dept of Natural Resources – Zi
Henderson, Miles, (432) 552-2245 henderson_m@utpb.edu,
 University of Texas, Permian Basin – GsdGg
Henderson, Naomi, naomi@ldeo.columbia.edu,
 Columbia University – Yr
Henderson, Paul, (508) 289-3466 phenderson@whoi.edu,
 Woods Hole Oceanographic Institution – Cm
Henderson, Robert, 747814746 bob.henderson@jcu.edu.au,
 James Cook University – GtPiGg
Henderson, Wayne G., (657) 278-3561 whenderson@fullerton.edu,
 California State University, Fullerton – PgGg
Hendrey, George, 718-997-3325 george.hendrey@qc.cuny.edu,
 Queens College (CUNY) – Zg
Hendricks, Jonathan R., 6072736623 (Ext. 20) jrh42@cornell.edu,
 Paleontological Research Institution – Pig
Hendrickx, Jan M., (575) 835-5892 hendrick@nmt.edu,
 New Mexico Institute of Mining and Technology – Hw
Hendrikx, Jordy, (406) 994-6918 jordy.hendrikx@montana.edu,
 Montana State University – ZnHsAs
Hendrix, Marc S., 406-243-5867 marc.hendrix@umontana.edu,
 University of Montana – Gs
Hendrix, Thomas E., 616-331-3728 Grand Valley State University – Gc
Hendry, Jim, (306) 966-5720 jim.hendry@usask.ca,
 University of Saskatchewan – HwClHy
Hendy, Ingrid, 734615-6892 ihendy@umich.edu,
 University of Michigan – Pg
Henika, William S., (540) 231-4298 bhenika@vt.edu,
 Virginia Polytechnic Institute & State University – Ng
Henkel, Torsten, +44 0161 275-3810 Torsten.Henkel@manchester.ac.uk,
 University of Manchester – Cs
Henley, Sian F., +44 (0) 131 650 7010 s.f.henley@ed.ac.uk,
 Edinburgh University – CmOcCs
Hennemeyer, Marc, 089/2180 4340 marc@hennemeyer.de,
 Ludwig-Maximilians-Universitaet Muenchen – Gz
Henning, Stanley J., (515) 294-7846 sjhennin@iastate.edu,
 Iowa State University of Science & Technology – So
Hennings, Peter H., (307) 766-3386 University of Wyoming – Gc
Hennon, Chris, (828) 232-5159 chennon@unca.edu,
 University of North Carolina, Asheville – AstZn
Henrici, Amy C., (412) 622-3265 henricia@carnegiemnh.org,
 Carnegie Museum of Natural History – Pv
Henry, Christopher D., (775) 682-8753) chenry@unr.edu,
 University of Nevada – GtvCc
Henry, Darrell J., (225) 388-2693 dhenry@geol.lsu.edu,
 Louisiana State University – Gp
Henry, Eric J., 910-962-7622 henrye@uncw.edu,
 University of North Carolina, Wilmington – Hw
Henry, Kathleen M., (217) 244-8994 kmhenry@illinois.edu,
 Illinois State Geological Survey – Gg
Henry, Tiernan, +353 (0)91 495 096 tiernan.henry@nuigalway.ie,
 National University of Ireland Galway – HwyEg
Henshel, Diane S., (812) 855-4556 dhenshel@indiana.edu,
 Indiana University, Bloomington – Zn
Henson, Harvey, 618-536-6666 henson@geo.siu.edu,
 Southern Illinois University Carbondale – Yg
Henstock, Tim, +44 (0)23 80596491 then@noc.soton.ac.uk,
 University of Southampton – YgrYe
Hentz, Tucker F., (512) 471-7281 tucker.hentz@beg.utexas.edu,
 University of Texas at Austin, Jackson School of Geosciences – GrsGo
Henyey, Thomas L., (213) 740-5832 henyey@usc.edu,
 University of Southern California – Yg
Henßel, Katja, 089/2180 6624 k.henssel@lrz.uni-muenchen.de,
 Ludwig-Maximilians-Universitaet Muenchen – Pg
Heo, Joon, heo_j@utpb.edu,
 University of Texas, Permian Basin – Hg
Hepburn, J. Christopher, 617-552-3642 john.hepburn@bc.edu,
 Boston College – Gg
Hepner, George F., (801) 581-6021 george.hepner@geog.utah.edu,
 University of Utah – Zy
Hepner, Tiffany, (512) 475-9572 tiffany.hepner@beg.utexas.edu,
 University of Texas at Austin, Jackson School of Geosciences – Og
Hepple, Alex, 3138 5051 a.hepple@qut.edu.au,
 Queensland University of Technology – ZerGg
Herbers, Thomas H., (831) 656-2917 herbers@nps.edu,
 Naval Postgraduate School – Op
Herbert, Bruce, (979) 845-2405 herbert@geo.tamu.edu,
 Texas A&M University – Ge

Herbert, Jennifer, (630) 252-0493 Argonne National Laboratory – Gg
Herbert, Timothy D., (401) 863-1207 timothy_herbert@brown.edu,
    Brown University – Pe
Herd, Christopher D., (780) 492-5798 herd@ualberta.ca,
    University of Alberta – XcGiCp
Herd, Richard, +44 (0)1603 59 3667 r.herd@uea.ac.uk,
    University of East Anglia – Gv
Herdendorf, Charles E., 440-934-1514 herdendorf.1@osu.edu,
    Ohio State University – OgGuOn
Herkenhoff, Ken E., (818) 354-3539 ken.e.herkenhoff@jpl.nasa.gov,
    Jet Propulsion Laboratory – Xg
Herman, Ellen K., (570) 577-3088 ekh008@bucknell.edu,
    Bucknell University – Hw
Herman, Janet S., (434) 924-0553 jsh5w@virginia.edu,
    University of Virginia – ClHw
Herman, Rhett B., 540 831-5441 rherman@radford.edu,
    Radford University – YuXaGa
Hermance, John F., (508) 252-3116 john_hermance@brown.edu,
    Brown University – HqZrHw
Hermann, Albert, 206-526-6495 albert.j.hermann@noaa.gov,
    University of Washington – Opb
Hermans, Thomas, thomas.hermans@ugent.be,
    Ghent University – HwYeu
Hernandez, Larry, (760) 757-2121, x6329 lhernandez@miracosta.edu,
    MiraCosta College – Gg
Hernández, Manuel, hdezfdez@ucm.es,
    Univ Complutense de Madrid – Pv
Hernandez, Michael W., 801-626-8186 mhernandez@weber.edu,
    Weber State University – Zri
Hernandez Molina, F. J., javier.hernandez-molina@rhul.ac.uk,
    Royal Holloway University of London – GsuOu
Herndon, Elizabeth M., (330) 672-2680 eherndo1@kent.edu,
    Kent State University – ClSc
Hernes, Peter, 530-752-7827 pjhernes@ucdavis.edu,
    University of California, Davis – Og
Hernes, Peter J., 530-754-4327 pjhernes@ucdavis.edu,
    University of California, Davis – Hg
Heron, Duncan, (919) 684-5321 duncan.heron@duke.edu,
    Duke University – Grg
Herried, Brad, (612) 625-6501 herri147@umn.edu,
    University of Minnesota, Twin Cities – Zi
Herring, Thomas A., (617) 253-5941 tah@mit.edu,
    Massachusetts Institute of Technology – Yd
Herring Mayo, Lisa L., 931-393-2136 lmayo@mscc.edu,
    Motlow State Community College – GgZge
Herriott, Trystan, (907) 451-5011 trystan.herriott@alaska.gov,
    Alaska Division of Geological & Geophysical Surveys – Eo
Herrmann, Achim, 225-578-3016 aherrmann@lsu.edu,
    Louisiana State University – PeGd
Herrmann, Edward W., (812) 856-0587 edherrma@indiana.edu,
    Indiana University, Bloomington – Sa
Herrmann, Felix J., +1 (404) 385-7069 felix.herrmann@gatech.edu,
    Georgia Institute of Technology – YesZn
Herrmann, Robert B., (314) 977-3120 rbh@eas.slu.edu,
    Saint Louis University – Ys
Hershey, Ronald L., (775) 673-7393 ron.hershey@dri.edu,
    University of Nevada, Reno – HwCls
Hervig, Richard, (480) 965-8427 hervig@asu.edu,
    Arizona State University – Zri
Herzberg, Claude T., (848) 445-3154 herzberg@eps.rutgers.edu,
    Rutgers, The State University of New Jersey – CpGi
Herzig, Chuck, (310) 660-3593 cherzig@elcamino.edu,
    El Camino College – GgOg
Hesp, Patrick, patrick.hesp@flinders.edu.au,
    Flinders University – GmZy
Hesp, Patrick A., (225) 578-6244 pahesp@lsu.edu,
    Louisiana State University – Gm
Hess, Darrel E., (415) 239-3104 dhess@ccsf.edu,
    City College of San Francisco – Zy
Hess, Kai-Uwe, 089/2180 4275 hess@min.uni-muenchen.de,
    Ludwig-Maximilians-Universitaet Muenchen – Gz
Hess, Paul C., (401) 863-1929 Paul_Hess@Brown.edu,
    Brown University – Gi
Hess, Peter G., 607-255-2495 peter.hess@cornell.edu,
    Cornell University – Acs
Hess Tanguay, Lillian, 516 463-6545 geolht@hofstra.edu,
    Hofstra University – Gg
Hess-Tanguay, Lillian, (516) 299-2318 lhess@liu.edu,
    Long Island University, C.W. Post Campus – Gs
Hesse, Marc A., (512) 471-0768 mhesse@jsg.utexas.edu,
    University of Texas, Austin – GqoYg
Hesse, Reinhard, 514-398-3627 reinhard.hesse@mcgill.ca,
    McGill University – Gs
Hessler, Robert R., (858) 534-2665 rhessler@ucsd.edu,
    University of California, San Diego – Ob
Hester, Erich, (540) 231-9758 ehester@vt.edu,
    Virginia Polytechnic Institute & State University – Hsw
Hesterberg, Dean L., (919) 515-2636 North Carolina State University – Sc
Hetherington, Callum J., (806) 834-3110 callum.hetherington@ttu.edu,
    Texas Tech University – GxzCg
Hetherington, Eric D., (209) 730-3812 College of the Sequoias – Gc
Hetherington, Jean, (925) 685-1230 (Ext. 462) jhetheri@dvc.edu,
    Diablo Valley College – Gg
Hetland, Robert D., (979) 458-0096 rhetland@ocean.tamu.edu,
    Texas A&M University – Op
Hettiarachchi, Ganga, (785) 532-7209 ganga@ksu.edu,
    Kansas State University – Sc
Hetzel, Ralf, +49-251-83-33908 rahetzel@uni-muenster.de,
    Universitaet Muenster – Gcm
Heubeck, Christoph, 0049(0)3641/948620 christoph.heubeck@uni-jena.de,
    Friedrich-Schiller-University Jena – GsdPg
Heuss-Aßbichler, Soraya, 089/21804252 soraya@min.uni-muenchen.de,
    Ludwig-Maximilians-Universitaet Muenchen – Gz
Hewitt, David A., (540) 231-6521 dhewitt@vt.edu,
    Virginia Polytechnic Institute & State University – Cp
Heymann, Dieter, (713) 348-4890 dieter@owlnet.rice.edu,
    Rice University – Xm
Heyniger, William C., (908) 737-3660 wheynige@kean.edu,
    Kean University – Am
Heyvaert, Alan, 775.673.7322 alan.heyvaert@dri.edu,
    University of Nevada, Reno – GnHs
Heywood, Karen, +44 (0)1603 59 2555 k.heywood@uea.ac.uk,
    University of East Anglia – Op
Heywood, Neil C., (715) 346-4452 nheywood@uwsp.edu,
    University of Wisconsin, Stevens Point – Zy
Hiatt, Eric E., (920) 424-7001 hiatt@uwosh.edu,
    University of Wisconsin, Oshkosh – GsOcGu
Hibbard, James P., (919) 515-7242 jphibbar@ncsu.edu,
    North Carolina State University – GctZn
Hibbs, Barry, (323) 343-2414 bhibbs@calstatela.edu,
    California State University, Los Angeles – Hw
Hick, Steven R., 303-871-2535 shick@du.edu,
    University of Denver – Zi
Hickcox, Charles W., (404) 727-0118 geocwh@emory.edu,
    Emory University – Gg
Hicke, Jeffrey, 208-885-6240 jhicke@uidaho.edu,
    University of Idaho – ZcyZn
Hickey, Barbara M., (206) 543-4737 bhickey@u.washington.edu,
    University of Washington – Op
Hickey, James, (660) 562-1817 jhickey@nwmissouri.edu,
    Northwest Missouri State University – Ge
Hickey, Kenneth A., (778) 384-7074 khickey@eos.ubc.ca,
    University of British Columbia – EgGcg
Hickey, William J., (608) 262-9018 wjhickey@wisc.edu,
    University of Wisconsin, Madison – Sb
Hickey-Vargas, Rosemary, (305) 348-3471 hickey@fiu.edu,
    Florida International University – Cg
Hickman, Anna, +44 (0)23 80592132 A.Hickman@noc.soton.ac.uk,
    University of Southampton – Og
Hickman, Carole S., (510) 642-3429 University of California, Berkeley – Pg
Hickman, John, (859) 323-0541 jhickman@uky.edu,
    University of Kentucky – EoGtc
Hickmon, Nicki, 630-252-7662 nhickmon@anl.gov,
    Argonne National Laboratory – As
Hickmott, Donald D., (505) 667-8753 dhickmott@lanl.gov,
    Los Alamos National Laboratory – Gp
Hicks, Andrea, hicks5@wisc.edu,
    University of Wisconsin, Madison – Zn
Hicks, David, 956 882 5055 david.hicks@utrgv.edu,
    University of Texas, Rio Grande Valley – Ob
Hicks, Roberta (Robbie), (709) 737-8349 rhicks@mun.ca,
    Memorial University of Newfoundland – Gg
Hickson, Catherine J., (604) 761-5573 ttgeo@telus.net,

University of British Columbia – EgGvZe
Hickson, Thomas A., (651) 962-5241 tahickson@stthomas.edu,
University of Saint Thomas – GsmGe
Hicock, Stephen R., shicock@uwo.ca,
Western University – Gl
Hidalgo, Paulo J., (404) 413-5789 phidalgo@gsu.edu,
Georgia State University – GivGz
Hier-Majumder, Saswata, Saswata.Hier-Majumder@rhul.ac.uk,
Royal Holloway University of London – Yg
Higgins, Charles G., paulaH88@hotmail.com ,
University of California, Davis – Gm
Higgins, Chris T., (916) 322-9997 California Geological Survey – Gg
Higgins, Jerry D., (303) 273-3817 jhiggins@mines.edu,
Colorado School of Mines – NxgNt
Higgins, John A., (609) 258-2756 jahiggin@princeton.edu,
Princeton University – ClPcCs
Higgins, Michael D., 4185455011 x 5052 mhiggins@uqac.ca,
Universite du Quebec a Chicoutimi – Gi
Higgins, Pennilyn, pennilyn.higgins@rochester.edu,
University of Rochester – Cs
Higgs, Bettie M., +353 860476789 b.higgs@ucc.ie,
University College Cork – YgGqRh
Higgs, Ken, +353 21 4902290 k.higgs@ucc.ie,
University College Cork – Pl
Higinbotham, Pamela, 724-938-4180 higinbotham@calu.edu,
California University of Pennsylvania – Zn
Hilbert-Wolf, Hannah L., hannah.hilbertwolf@my.jcu.edu.au,
James Cook University – GsCc
Hildebrand, Alan R., 403-220-2291 ahildebr@ucalgary.ca,
University of Calgary – XmgXc
Hildebrand, John A., (858) 534-4069 jhildebrand@ucsd.edu,
University of California, San Diego – Yr
Hildebrand, Stephen G., hildebrandsg@ornl.gov,
Oak Ridge National Laboratory – Zn
Hildebrand, Steve T., (505) 667-4318 hildebrand@lanl.gov,
Los Alamos National Laboratory – Ys
Hildebrandt, Anke, 349-3641-948651 hildebrandt.a@uni-jena.de,
Friedrich-Schiller-University Jena – HgSpf
Hileman, Mary E., (405) 744-4341 mary.hileman@okstate.edu,
Oklahoma State University – GorGd
Hill, Arleen A., 901-678-2589 aahill@memphis.edu,
University of Memphis – Rn
Hill, Catherine R., 928344-7719 catherine.hill@azwestern.edu,
Arizona Western College – GgOgHw
Hill, Chris, chris.hill@gcccd.edu,
Grossmont College – GgOg
Hill, Christopher, 617-253-6430 cnh@mit.edu,
Massachusetts Institute of Technology – Zn
Hill, Julie, (775) 784-6987 juliehill@unr.edu,
University of Nevada, Reno – Zn
Hill, Malcolm D., (617) 373-4377 m.hill@neu.edu,
Northeastern University – ZiGze
Hill, Mary C., 785-864-2728 mchill@ku.edu,
University of Kansas – HgqHw
Hill, Mary Louise, (807) 343-8319 mary.louise.hill@lakeheadu.ca,
Lakehead University – Gc
Hill, Mimi, +44 0151 794 3462, M.Hill@liverpool.ac.uk,
University of Liverpool – Ym
Hill, Paul S., (902) 494-2266 paul.hill@dal.ca,
Dalhousie University – Gs
Hill, Philip R., philip.hill@canada.ca,
University of Victoria – Gsu
Hill, Tessa M., (530) 752-0179 tmhill@ucdavis.edu,
University of California, Davis – Pe
Hill, Timothy, +44 01334 464013 tch2@st-andrews.ac.uk,
University of St. Andrews – Ge
Hillaire-Marcel, Claude, 514-987-3000 #3376 hillaire-marcel.claude@uqam.ca,
Universite du Quebec a Montreal – Cs
Hiller, Lena, 516 463-5564 geolzh@hofstra.edu,
Hofstra University – Zn
Hillier, John, jhillier@ghc.edu,
Grays Harbor College – ZgCa
Hilliker, Joby, (610) 436-2213 jhilliker@wcupa.edu,
West Chester University – Am
Hillman, Aubrey, (337) 482-5162 aubrey.hillman@louisiana.edu,
University of Louisiana at Lafayette – GneGa
Hills, Denise J., (205) 247-3694 dhills@gsa.state.al.us,
Geological Survey of Alabama – EoZfYe
Hills, Leonard V., (403) 220-5848 University of Calgary – Pl
Hilterman, Fred, 713-850-7600 x3318 fhilterman@uh.edu,
University of Houston – Ye
Hilton, David R., (858) 822-0639 drhilton@ucsd.edu,
University of California, San Diego – Cs
Hilts, Stewart G., (519) 824-4120 (Ext. 52448) shilts@uoguelph.ca,
University of Guelph – Zn
Hindle, Tobin, 561 297-2846 thindle@fau.edu,
Florida Atlantic University – ZyiZc
Hindman, Edward E., (212) 650-6469 City College (CUNY) – As
Hindshaw, Ruth, +44 01334 463936 rh71@st-andrews.ac.uk,
University of St. Andrews – Cs
Hine, Albert C., (727) 553-1161 hine@usf.edu,
University of South Florida – Gus
Hines, Mary E., (910) 962-3012 hinese@uncw.edu,
University of North Carolina, Wilmington – Zu
Hines, Paul, phines50@gmail.com,
Dalhousie University – Op
Hingston, Egerton, (031) 260-2805 hingstone@ukzn.ac.za,
University of KwaZulu-Natal – NtxNg
Hinman, Nancy W., (406) 243-5277 nancy.hinman@umontana.edu,
University of Montana – Cgl
Hinnov, Linda, 703-993-3082 lhinnov@gmu.edu,
George Mason University – GrYg
Hinojosa, Alejandro, alhinc@cicese.mx,
Centro de Investigación Científica y de Educación Superior de Ensenada – Zr
Hinton, Richard, +44 (0) 131 650 8548 Richard.Hinton@ed.ac.uk,
Edinburgh University – Cg
Hintz, John G., (570)389-4140 jhintz@bloomu.edu,
Bloomsburg University – Zin
Hintz, Rashauna, 479-575-3355 rmicken@uark.edu,
University of Arkansas, Fayetteville – Zu
Hintze, Lehi F., 801 422-6361 lehi.hintze@gmail.com,
Brigham Young University – Pg
Hinz, Nicholas, (775)784-1446 nhinz@unr.edu,
University of Nevada – Gg
Hinze, William J., wjh730@comcast.net,
Purdue University – YevGt
Hippensteel, Scott P., 704-687-5992 shippens@email.uncc.edu,
University of North Carolina, Charlotte – Gr
Hiroji, Anand D., (228) 688-1510 anand.hiroji@usm.edu,
University of Southern Mississippi – Yd
Hirons, Steve, +44 020 3073 8028 s.hirons@ucl.ac.uk,
Birkbeck College – Cl
Hirsch, Eric, eric.hirsch@fandm.edu,
Franklin and Marshall College – Zn
Hirschboeck, Katherine K., (520) 621-6466 katie@ltrr.arizona.edu,
University of Arizona – AtHsRn
Hirschmann, Marc M., 612-625-6698 hirsc022@umn.edu,
University of Minnesota, Twin Cities – Gi
Hirt, William H., (530) 938-5255 hirt@siskiyous.edu,
College of the Siskiyous – Gi
Hirth, Greg, (401) 863-7063 Greg_Hrth@Brown.edu,
Brown University – GcyYg
Hirth, Gregory, 508-289-2776 Greg_Hirth@brown.edu,
Woods Hole Oceanographic Institution – Gc
Hiscock, Kevin, +44 (0)1603 59 3104 k.hiscock@uea.ac.uk,
University of East Anglia – Hwg
Hiscott, Richard N., (709) 737-8394 rhiscott@sparky2.esd.mun.ca,
Memorial University of Newfoundland – Gs
Hiser, Susan, 740-376-4775 sch030@marietta.edu,
Marietta College – Zn
Hiskey, J. Brent, (520) 621-6185 jbh@engr.arizona.edu,
University of Arizona – Nx
Hitchman, Matthew H., (608)262-4653 matt@aos.wisc.edu,
University of Wisconsin, Madison – As
Hites, Ronald A., (812) 855-0193 hitesr@indiana.edu,
Indiana University, Bloomington – Co
Hitz, Ralph B., (253) 566-5299 rhitz@tacomacc.edu,
Tacoma Community College – PvSaZi
Hixon, Amy E., (574) 631-1872 ahixon@nd.edu,
University of Notre Dame – ClqCc
Hjelmfelt, Mark R., (605) 394-2291 Mark.Hjelmfelt@sdsmt.edu,
South Dakota School of Mines & Technology – As
Hladik, Christine, (912) 478-0338 chladik@georgiasouthern.edu,

Georgia Southern University – SbZr
Hluchy, Michele M., (607) 871-2838 Alfred University – Cl
Hlusko, Leslea, (510) 643-8838 hlusko@berkeley.edu,
 University of California, Berkeley – Znn
Ho, Anita, (406) 756-3873 aho@fvcc.edu,
 Flathead Valley Community College – Gg
Ho, David, (808) 956-3311 ho@hawaii.edu,
 University of Hawai'i, Manoa – OcCb
Ho, I-Hsuan, (701) 777-6156 ihsuan.ho@engr.und.edu,
 University of North Dakota – Ng
Hobbs, Benjamin F., (410) 516-4681 bhobbs@jhu.edu,
 Johns Hopkins University – Zn
Hobbs, Chasidy, (850) 474-2735 chobbs@uwf.edu,
 University of West Florida – ZgeZy
Hobbs, John D., dhobbs@mtech.edu,
 Montana Tech of the University of Montana – Zn
Hobbs, Richard, +44 (0) 191 33 44295 r.w.hobbs@durham.ac.uk,
 Durham University – Ys
Hobbs, Richard D., (806) 371-5333 rdhobbs@actx.edu,
 Amarillo College – ZrGc
Hobbs, Thomas M., 281-618-5796 Tom.Hobbs@nhmccd.edu,
 Lonestar College - North Harris – Gg
Hochella, Jr., Michael F., (540) 231-6227 hochella@vt.edu,
 Virginia Polytechnic Institute & State University – ClGze
Hochmuth, George J., 352-392-1803 318 hoch@ufl.edu,
 University of Florida – So
Hochstaedter, Alfred, (831) 646-4149 ahochstaedter@mpc.edu,
 Monterey Peninsula College – GgOgZg
Hock, Regine M., 907-474-7691 regine.hock@gi.alaska.edu,
 University of Alaska, Fairbanks – Gl
Hockaday, William C., (254)7102639 william_hockaday@baylor.edu,
 Baylor University – CoaSb
Hockey, Thomas A., (319) 273-2065 thomas.hockey@uni.edu,
 University of Northern Iowa – Zn
Hodder, Donald R., (914) 257-3757 SUNY, New Paltz – Gg
Hodell, David, +44 (0) 1223 330270 dhod07@esc.cam.ac.uk,
 University of Cambridge – Ge
Hodges, Floyd N., (509) 376-4627 floyd.hodges@pnl.gov,
 Pacific Northwest National Laboratory – Cg
Hodges, Kip V., 480-965-5331 kvhodges@asu.edu,
 Arizona State University – CcGtXg
Hodgetts, David, David.Hodgetts@manchester.ac.uk,
 University of Manchester – Gsc
Hodgkiss, Jr., William S., (858) 534-1798 whodgkiss@ucsd.edu,
 University of California, San Diego – No
Hodych, Joseph P., (709) 864-7567 jhodych@mun.ca,
 Memorial University of Newfoundland – Ym
Hoe, Teh Guan, 03-79674231 tehgh@um.edu.my,
 University of Malaya – Ca
Hoenisch, Baerbel, (845) 365-8828 hoenisch@ldeo.columbia.edu,
 Columbia University – PcClm
Hoey, Trevor, +4401413307736 Trevor.Hoey@glasgow.ac.uk,
 University of Glasgow – Gs
Hoff, Jean L., (320) 308-5914 jhoff@stcloudstate.edu,
 Saint Cloud State University – HwZeGg
Hoffman, Gretchen K., (505) 835-5640 gretchen@gis.nmt.edu,
 New Mexico Institute of Mining & Technology – Ec
Hoffman, Jeffrey L., (609) 292-1185 jeffrey.l.hoffman@dep.nj.gov,
 New Jersey Geological and Water Survey – HqwRw
Hoffman, Paul F., (617) 496-6380 paulfhoffman@gmail.com,
 Harvard University – Gc
Hofmann, Eileen E., (757) 683-5334 ehofmann@odu.edu,
 Old Dominion University – Op
Hofmann, Michael, (406) 243-5855 michael.hofmann@umontana.edu,
 University of Montana – GosGr
Hofmeister, Anne M., (314) 935-7440 hofmeist@levee.wustl.edu,
 Washington University in St. Louis – YghXa
Hogarth, Donald D., 613-731-8090 dhogarth@uottawa.ca,
 University of Ottawa – Gz
Hohl, Eric, 573-368-2168 eric.hohl@dnr.mo.gov,
 Missouri Dept of Natural Resources – Zn
Hohn, Michael E., 304 594-2331 hohn@geosrv.wvnet.edu,
 West Virginia Geological & Economic Survey – Eg
Hoisch, Thomas D., (928) 523-1904 thomas.hoisch@nau.edu,
 Northern Arizona University – Gp
Hojjatie, Barry, (229) 333-5753 bhojjati@valdosta.edu,
 Valdosta State University – Zm

Hok, Jozef, jozef.hok@uniba.sk,
 Comenius University in Bratislava – Ggt
Hoke, Gregory, (315) 443-1903 gdhoke@syr.edu,
 Syracuse University – ZgGmt
Holail, Hanafy M., 002-03-3921595 hanafyholail@hotmail.com,
 Alexandria University – Ggs
Holberg, Jay B., (520) 621-4571 holberg@vega.lpl.arizona.edu,
 University of Arizona – As
Holbrook, Amanda, (606) 783-2381 a.holbrook@moreheadstate.edu,
 Morehead State University – Zn
Holbrook, John M., 817-257-6275 john.holbrook@tcu.edu,
 Texas Christian University – Gsr
Holbrook, W. S., 540-231-6521 wstevenh@vt.edu,
 Virginia Polytechnic Institute & State University – Yur
Holcomb, Robin T., (206) 543-5274 rholcomb@ocean.washington.edu,
 University of Washington – Ou
Holcombe, Troy L., 979 845-3528 or 979 690-7398 tholcombe@ocean.tamu.edu,
 Texas A&M University – OuGcg
Holdaway, Michael J., (214) 692-2750 Southern Methodist University – Gp
Holdren, George R., (509) 376-2242 rich.holdren@pnl.gov,
 Pacific Northwest National Laboratory – Cl
Holdren, John P., 617-495-1498 john_holdren@hks.harvard.edu,
 Harvard University – Zn
Holdsworth, Robert E., 0191 3742529 r.e.holdsworth@durham.ac.uk,
 Durham University – Gc
Hole, John A., (540) 231-3858 hole@vt.edu,
 Virginia Polytechnic Institute & State University – Ye
Holgood, Jay S., (614) 292-3999 hobgood.1@osu.edu,
 Ohio State University – As
Holk, Gregory J., 562-985-5006 gholk@csulb.edu,
 California State University, Long Beach – CsGxEm
Holl, Karen D., (831) 459-3668 kholl@cats.ucsc.edu,
 University of California, Santa Cruz – Zu
Hollabaugh, Curtis L., (678) 839-4050 chollaba@westga.edu,
 University of West Georgia – GzCg
Holland, Austin A., 405-325-8497 austin.holland@ou.edu,
 University of Oklahoma – YsZnn
Holland, Edward C., 479-575-6635 echollan@uark.edu,
 University of Arkansas, Fayetteville – Zy
Holland, Nicholas D., (858) 534-2085 nholland@ucsd.edu,
 University of California, San Diego – Ob
Holland, Steven M., (706) 542-0424 stratum@gly.uga.edu,
 University of Georgia – Gr
Holland, Tim, +44 (0) 1223 333466 tjbh@esc.cam.ac.uk,
 University of Cambridge – GptGz
Hollander, David J., (727) 553-1019 davidh@usf.edu,
 University of South Florida – Oc
Hollenbaugh, Kenneth M., (208) 426-3700 khollenb@boisestate.edu,
 Boise State University – Eg
Holley, Elizabeth, 303-273-3409 eholley@mines.edu,
 Colorado School of Mines – Eg
Holliday, Joseph W., 3106603593 xt. 3371 jholliday@elcamino.edu,
 El Camino College – Og
Holliday, Vance, vthollid@email.arizona.edu,
 University of Arizona – Ga
Holliday, Vance T., (520) 621-4734 vthollid@email.arizona.edu,
 University of Arizona – GamSa
Holliman, Richard, +44(0) 1908 654 646 x 54646 richard.holliman@open.ac.uk,
 The Open University – Ze
Hollings, Peter N., (807) 343-8329 peter.hollings@lakeheadu.ca,
 Lakehead University – Eg
Hollis, Cathy, +44 0161 306-6583 Cathy.Hollis@manchester.ac.uk,
 University of Manchester – Gs
Hollister, Lincoln S., (609) 258-4106 linc@princeton.edu,
 Princeton University – GptGz
Hollocher, Kurt T., (518) 388-6518 hollochk@union.edu,
 Union College – GxCga
Holloway, Tess, thollow2@ashland.edu,
 Ashland University – Rn
Holloway, Tracey, (608) 262-5356 taholloway@wisc.edu,
 University of Wisconsin, Madison – AscZn
Holloway, Tracey, taholloway@wisc.edu,
 University of Wisconsin, Madison – As
Holm, Daniel K., (330) 672-4094 dholm@kent.edu,
 Kent State University – Gc
Holm, Richard F., richard.holm@nau.edu,

Northern Arizona University – Gv
Holman, John, (620) 276-8286 jholman@ksu.edu,
   Kansas State University – So
Holmden, Chris, (306) 966-5697 chris.holmden@usask.ca,
   University of Saskatchewan – Csc
Holme, Richard, +44 0151 794 5254   R.T.Holme@liverpool.ac.uk,
   University of Liverpool – Ym
Holmes, Ann E., (423) 425-1704 ann-holmes@utc.edu,
   University of Tennessee, Chattanooga – GsrPi
Holmes, George, +44(0) 113 34 31163 G.Holmes@leeds.ac.uk,
   University of Leeds – Ge
Holmes, Mark L., 425 487-2639 mholmes@uw.edu,
   University of Washington – OuGl
Holmes, Mary Anne, 402-588-2492 mholmes2@unl.edu,
   University of Nebraska, Lincoln – GsOuZn
Holmes, Stevie L., 605-677-6147 stevie.holmes@usd.edu,
   South Dakota Dept of Environment and Natural Resources – Gg
Holness, Marian, +44 (0) 1223 333434 marian@esc.cam.ac.uk,
   University of Cambridge – Gi
Holroyd, Pat, (510) 642-3733 pholroyd@berkeley.edu,
   University of California, Berkeley – Pv
Holst, Timothy, tholst@d.umn.edu,
   University of Minnesota, Duluth – GcYn
Holstein, Thomas J., (401) 254-3097 Roger Williams University – Zn
Holt, Ben D., (630) 252-4347 Argonne National Laboratory – Ca
Holt, Caecilia, 610 683-4445 holt@kutztown.edu,
   Kutztown University of Pennsylvania – Zn
Holt, David, 228-214-3255 david.h.holt@usm.edu,
   University of Southern Mississippi – Zyi
Holt, Gloria J., (361) 749-6716 joanholt@utexas.edu,
   University of Texas, Austin – Ob
Holt, John W., 512-471-0487 jack@ig.utexas.edu,
   University of Texas, Austin – GlXyg
Holt, Robert M., (662) 915-6687 rmholt@olemiss.edu,
   University of Mississippi – Hq
Holt, William E., 631-632-8215 william.holt@sunysb.edu,
   Stony Brook University – Ys
Holtz, Jr., Thomas R., (301) 405-4084 tholtz@umd.edu,
   University of Maryland – PvgPg
Holtzman, Benjamin, benh@ldeo.columbia.edu,
   Columbia University – Ys
Holubnyak, Yehven I., (785) 864-2070 eugene@kgs.ku.edu,
   University of Kansas – Np
Holyoke, Caleb, (330) 972-7635 cholyoke@uakron.edu,
   University of Akron – GcNrYx
Holzer, Mark, (604) 822-0531 mholzer@langara.bc.ca,
   University of British Columbia – As
Holzworth, Robert H., (206) 685-7410 bobholz@uw.edu,
   University of Washington – XyAsZg
Homan, Amy L., (814) 865-2622 aqs3@psu.edu,
   Pennsylvania State University, University Park – Zn
Hommel, Demian, 541-737-5070 hommeld@geo.oregonstate.edu,
   Oregon State University – Zn
Homuth, Emil F., (702) 794-7351 fred_homuth@notes.ymp.gov,
   Los Alamos National Laboratory – Ye
Hon, Ken, (808) 974-7302 kenhon@hawaii.edu,
   University of Hawai'i, Hilo – Gvi
Hon, Rudolph, (617) 552-3656 hon@bc.edu,
   Boston College – Cq
Hong, Gi-Hoon, (313) 577-2506 gihoonh@gmail.com,
   Wayne State University – OcCcm
Hong, Jessie, (678) 839-5466 jhong@westga.edu,
   University of West Georgia – ZieZg
Hong, Sung-ho, 270-809-6319 shong4@murraystate.edu,
   Murray State University – HyGe
Honjas, Bill, (775) 784-6613 bhonjas@optimsoftware.com ,
   University of Nevada, Reno – Ys
Honjo, Susumu, (508) 289-2589 shonjo@whoi.edu,
   Woods Hole Oceanographic Institution – Ou
Honkaer, Rick, 859-257-1108 rick.honaker@uky.edu,
   University of Kentucky – Nm
Hood, Lonnie L., (520) 621-6936 lon@lpl.arizona.edu,
   University of Arizona – XyAs
Hood, Raleigh, (305) 361-4668 rhood@umces.edu,
   University of Maryland – Ob
Hood, Teresa A., 305-284-8647 t.hood@miami.edu,
   University of Miami – Gg

Hood, William C., (970) 241-8020 Colorado Mesa University – Gg
Hooda, Peter, +44 020 8417 2155 p.hooda@kingston.ac.uk,
   Kingston University – Em
Hooft, Emilie E., (541) 346-4762 emilie@uoregon.edu,
   University of Oregon – Yr
Hook, James E., (912) 386-3182 University of Georgia – Sp
Hook, Paul B., (406) 944-3724 paul@intermountainaquatics.com,
   Montana State University – Sf
Hook, Simon J., (818) 354-0974 simon@lithos.jpl.nasa.gov,
   Jet Propulsion Laboratory – Zr
Hooke, Roger L., (207) 581-2203 rhooke@acadia.net,
   University of Maine – Gm
Hooks, Benjamin P., (731) 881-7430 bhooks@utm.edu,
   University of Tennessee, Martin – GciNr
Hooks, Chris H., (205) 247-3721 chooks@gsa.state.al.us,
   Geological Survey of Alabama – EoGxo
Hooks, W. Gary, (205) 348-1877 University of Alabama – Gm
Hooper, Andy, +44(0) 113 34 37723 a.hooper@leeds.ac.uk,
   University of Leeds – Ydg
Hooper, Robert L., (715) 836-4932 hooperrl@uwec.edu,
   University of Wisconsin, Eau Claire – Gz
Hoover, Karin A., 530-898-6269 khoover@csuchico.edu,
   California State University, Chico – Hw
Hoover, Michael T., (919) 515-7305 North Carolina State University – Sd
Hooyer, Thomas S., 608-263-4175 hooyer@uwm.edu,
   University of Wisconsin-Madison – Hg
Hopkins, David G., (701) 231-8948 North Dakota State University – Sd
Hopkins, David M., (907) 474-7565 University of Alaska, Fairbanks – Gg
Hopkins, John C., (403) 220-5842 hopkins@geo.ucalgary.ca,
   University of Calgary – Gs
Hopkins, Kenneth D., (970) 351-2853 kenneth.hopkins@unco.edu,
   University of Northern Colorado – Gml
Hopkins, Samantha, 541-346-5976 shopkins@uoregon.edu,
   University of Oregon – Pg
Hopkins, Thomas S., (205) 348-1791 Dauphin Island Sea Lab – Ob
Hopley, Philip, +44 020 3073 8029 p.hopley@ucl.ac.uk,
   Birkbeck College – Pe
Hopmans, Jan W., (916) 752-3060 University of California, Davis – So
Hopmans, Jan W., jwhopmans@ucdavis.edu,
   University of California, Davis – So
Hopmans, Jan W., (530) 752-3060 jwhopmans@ucdavis.edu,
   University of California, Davis – Sp
Hoppe, Kathryn A., 2538339111 x 4323 khoppe@greenriver.edu,
   Green River Community College – ZgOgPg
Hoppie, Bryce W., (507) 389-2315 bryce.hoppie@mnsu.edu,
   Minnesota State University – GeHw
Hopson, Janet L., 865-946-1460 hopsonj@ornl.gov,
   University of Tennessee, Knoxville – Zn
Horan, Mary F., 202-478-8481 mhoran@carnegiescience.edu,
   Carnegie Institution for Science – Cc
Horel, John, 801-581-7091 john.horel@utah.edu,
   University of Utah – AsZf
Horgan, Briony, (765) 496-2290 briony@purdue.edu,
   Purdue University – XgZr
Horn, John, (816) 604-3132 john.horn@mcckc.edu,
   Metropolitan Community College-Kansas City – GgZy
Horn, Marty R., 225-578-2681 mhorn@lsu.edu,
   Louisiana State University – EoGgr
Hornbach, Matthew J., 214-768-2389 mhornbach@smu.edu,
   Southern Methodist University – YrhYe
Hornberger, George, (615) 343-1144 george.m.hornberger@vanderbilt.edu,
   Vanderbilt University – HgwHs
Horne, Sharon, (519) 253-3000 ext, 2528 shorne@uwindsor.ca,
   University of Windsor – Zn
Horner, John R., (406) 994-3982 jhorner@montana.edu,
   Montana State University – Pv
Horner, Tim C., (916) 278-5635 hornertc@csus.edu,
   California State University, Sacramento – HwGs
Horner, Tristan, (508) 289-3825 thorner@whoi.edu,
   Woods Hole Oceanographic Institution – Cm
Horns, Daniel, (801) 863-8582 hornsda@uvu.edu,
   Utah Valley University – NgGeRn
Horowitz, Franklin G., +16072754955 frank.horowitz@cornell.edu,
   Cornell University – YgZnYe
Horsman, Eric M., 252 3285265 horsmane@ecu.edu,
   East Carolina University – GcYgGt
Horst, Andrew J., 440-775-8714 ahirst@oberlin.edu,

Oberlin College – Gcg
Horst, Andrew J., (304) 696-5435 horst@marshall.edu,
 Marshall University – GcOgGt
Horst, Sarah, 410-516-5286 sarah.horst@jhu.edu,
 Johns Hopkins University – XaAcXb
Horton, Albert B., (615) 532-1509 Albert.Horton@tn.gov,
 Tennessee Geological Survey – GgZi
Horton, Brian, 512-471-5172 horton@mail.utexas.edu,
 University of Texas, Austin – Gs
Horton, Daniel E., 847-467-6185 danethan@earth.northwestern.edu,
 Northwestern University – AsPeZn
Horton, Duane G., (509) 376-6868 duane.horton@pnl.gov,
 Pacific Northwest National Laboratory – Gz
Horton, Jennifer, (612) 626-4067 jmhorton@umn.edu,
 University of Minnesota – Gl
Horton, Radley, hortonr@ldeo.columbia.edu,
 Columbia University – Og
Horton, Robert, (515) 294-7843 rhorton@iastate.edu,
 Iowa State University of Science & Technology – Sp
Horton, Stephen P., (901) 678-2007 shorton@memphis.edu,
 University of Memphis – Ys
Horton, Travis, +64 3 3667001 Ext 7734 travis.horton@canterbury.ac.nz,
 University of Canterbury – Cs
Horton, Jr., Robert A., 661-654-3059 rhorton@csub.edu,
 California State University, Bakersfield – GdsGo
Horváth, Ferenc, frankh@ludens.elte.hu,
 Eotvos Lorand University – YgRhGc
Horvath, Peter, +27 (0)46-603-8312 p.horvath@ru.ac.za,
 Rhodes University – Gp
Horvath, Peter, 011 717 6539 peter.horvath@wits.ac.za,
 University of the Witwatersrand – GpCpGz
Horwath, William R., (530) 754-6029 wrhorwath@ucdavis.edu,
 University of California, Davis – Sob
Horwell, Claire, +44 (0) 191 33 42253 claire.horwell@durham.ac.uk,
 Durham University – Rn
Hosler, Charles L., (814) 863-8358 hosler@ems.psu.edu,
 Pennsylvania State University, University Park – Am
Hossain, A K M Azad, 423-425-4404 azad-hossain@utc.edu,
 University of Tennessee, Chattanooga – ZriZg
Hosseini, Seyyed Abolfazi, 512-471-1534 seyyed.hosseini@beg.utexas.edu,
 University of Texas, Austin – Eo
Houghton, Bruce F., (808) 956-2561 bhought@soest.hawaii.edu,
 University of Hawai'i, Manoa – Gv
Hounslow, Arthur, 405-372-2328 Oklahoma State University – Cl
Hourigan, Jeremy, 831-459-2879 hourigan@ucsc.edu,
 University of California, Santa Cruz – Cc
House, Christopher H., (814) 865-8802 chouse@geosc.psu.edu,
 Pennsylvania State University, University Park – Pg
House, Leigh S., house@lanl.gov,
 Los Alamos National Laboratory – Ys
Houseman, Greg, +44(0) 113 34 35206 g.a.houseman@leeds.ac.uk,
 University of Leeds – Yg
Housen, Bernard A., (360) 650-3588 bernieh@wwu.edu,
 Western Washington University – Ym
Houston, Adam L., ahouston2@unl.edu,
 University of Nebraska, Lincoln – Asm
Houston, Heidi B., 310 691 9152 houstonh@usc.edu,
 University of Southern California – YsGt
Houston, Robert, (541) 967-2039 Oregon Dept of Geology and Mineral Industries – Ge
Houston, Serin D., 413-538-2055 shouston@mtholyoke.edu,
 Mount Holyoke College – Zyn
Houston, William (Bill), 906-635-2155 whouston1@lssu.edu,
 Lake Superior State University – Gsg
Houweling, Sander, +31 20 59 83687 s.houweling@vu.nl,
 Vrije Universiteit Amsterdam – AscAp
Houze, Robert A., (206) 543-6922 houze@atmos.washington.edu,
 University of Washington – As
Hovan, Steven A., (724) 357-5625 hovan@iup.edu,
 Indiana University of Pennsylvania – Ou
Hovis, Guy L., (610) 258-3907 hovisguy@lafayette.edu,
 Lafayette College – GzxCg
Hovorka, Susan D., 512-471-4863 susan.hovorka@beg.utexas.edu,
 University of Texas, Austin – Gs
Howard, Alan D., (804) 924-0563 ah6p@virginia.edu,
 University of Virginia – GmXg
Howard, Hugh H., 916 484-8805 howardh@arc.losrios.edu,
 American River College – Zi
Howard, Jeffrey L., 313-577-3258 jhoward@wayne.edu,
 Wayne State University – GsScGa
Howard, Katie, 360-475-7700 khoward@llion.org,
 Olympic College – Gg
Howard, Kenneth W., 416-287-7233 University of Toronto – Hw
Howard, Leslie M., (402) 472-9192 thoward@unl.edu,
 Unversity of Nebraska - Lincoln – Zyi
Howard, Matthew K., (979) 862-4169 mkhoward@tamu.edu,
 Texas A&M University – Op
Howarth, Geoffrey, Howarth@uga.edu,
 University of Georgia – Gi
Howat, Ian M., (614) 292-4839 howat.4@osu.edu,
 Ohio State University – GlZr
Howden, Stephan, (228) 688-5284 stephan.howden@usm.edu,
 University of Southern Mississippi – Op
Howe, Bruce M., (206) 543-9141 billhowe@cs.washington.edu,
 University of Washington – Op
Howe, Julie, jhowe@tamu.edu,
 Texas A&M University – Sc
Howe, Stephen S., (518) 442-5053 showe@albany.edu,
 SUNY, Albany – Cs
Howe III, Thomas R., (269) 387-5492 thomas.r.howe@wmich.edu,
 Western Michigan University – HwGem
Howell, Dave, +440116 252 3804 dah29@le.ac.uk,
 Leicester University – Ge
Howell, John, +44 (0)1224 272606 john.howell@abdn.ac.uk,
 University of Aberdeen – GsoGc
Howell, Robert R., (307) 766-6296 rhowell@uwyo.edu,
 University of Wyoming – XgZr
Howell, Jr., Benjamin F., (814) 863-0886 howellbf@aol.com,
 Pennsylvania State University, University Park – Ys
Hower, James C., (859) 257-0261 james.hower@uky.edu,
 University of Kentucky – Ec
Howes, Mary R., mary.howes@dnr.iowa.gov,
 Iowa Dept of Natural Resources – Gg
Howman, Dominic J., +61 8 9266-2329 D.Howman@curtin.edu.au,
 Curtin University – Ye
Hoyer, Lauren, 031-2601613 HoyerL@ukzn.ac.za,
 University of KwaZulu-Natal – EgGc
Hoyos, Carlos, carlos.hoyos@eas.gatech.edu,
 Georgia Institute of Technology – As
Hoyt, Greg D., (919) 684-3562 North Carolina State University – Sb
Hoyt, William H., (970) 351-2487 william.hoyt@unco.edu,
 University of Northern Colorado – OgGsZe
Hozik, Michael J., 609 652-4277 HozikM@Stockton.Edu,
 Stockton University – GcYmg
Hren, Michael, 860-486-9511 michael.hren@uconn.edu,
 University of Connecticut – CslCo
Hrepic, Zdeslav, 706-569-3052 hrepic_zdeslav@columbusstate.edu,
 Columbus State University – Zn
Hsiang, Tom, (519) 824-4120 Ext.52753 thsiang@uoguelph.ca,
 University of Guelph – Zn
Hsiao, Theodore C., (530) 752-0691 tchsiao@ucdavis.edu,
 University of California, Davis – SpAmZn
Hsieh, Wen-Pin, 886-227839910-509 wphsieh@earth.sinica.edu.tw,
 Academia Sinica – YhGy
Hsieh, William W., (250) 388-0508 whsieh@eos.ubc.ca,
 University of British Columbia – OpAs
Hsu, Liang-Chi, (775) 682-8746 lihsu@unr.edu,
 University of Nevada – Cp
Hu, Baoxin, (416)736-2100 #20557 baoxin@yorku.ca,
 York University – Zr
Hu, Bill X., 850-644-3943 hu@gly.fsu.edu,
 Florida State University – Hw
Hu, Feng-Sheng, (217) 244-2982 fhu@illinois.edu,
 University of Illinois, Urbana-Champaign – Pe
Hu, Hao, hhu5@central.uh.edu,
 University of Houston – Ye
Hu, Qi S., 402-472-6642 QHU2@unl.edu,
 University of Nebraska, Lincoln – As
Hu, Qi (Steve), qhu2@unl.edu,
 University of Nebraska, Lincoln – As
Hu, Qinhong (Max), (817) 272-5398 maxhu@uta.edu,
 University of Texas, Arlington – GoHwNg
Hu, Shusheng, 203-432-3790 shusheng.hu@yale.edu,
 Yale University – PblGs

Hu, Wan-Ping (Sunny), +61 7 3138 7314 sunny.hu@qut.edu.au,
  Queensland University of Technology – Ca
Hu, Xinping, 361-825-3395 Xinping.Hu@tamucc.edu,
  Texas A&M University, Corpus Christi – Cms
Hu, Zhiyong, 850-474-3494 zhu@uwf.edu,
  University of West Florida – Zr
Huang, Alex, (828) 232-5157 University of North Carolina, Asheville – As
Huang, Bohua, (703) 993-6084 bhuang@gmu.edu,
  George Mason University – Op
Huang, Li, (519) 661-3188 LHuang3@uwo.ca,
  Western University – Yr
Huang, Lianjie, (505) 665-1108 ljh@lanl.gov,
  Los Alamos National Laboratory – Yg
Huang, Moh J., (916) 322-9304 California Geological Survey – Ne
Huang, Norden E., (301) 286-8879 University of Delaware – Zn
Huang, Shaopeng, (734) 763-3169 shaopeng@umich.edu,
  University of Michigan – YhZcGt
Huang, Yi, 514-398-8217 yi.huang@mcgill.ca,
  McGill University – ZrYgAs
Huang, Yongsong, Yongsong_Huang@Brown.edu,
  Brown University – Co
Hubbard, Dennis K., 440-775-8346 dennis.hubbard@oberlin.edu,
  Oberlin College – Gsu
Hubbard, Trent, (907) 451-5009 trent.hubbard@alaska.gov,
  Alaska Division of Geological & Geophysical Surveys – GmNgZi
Hubbard, William B., (520) 621-6942 hubbard@lpl.arizona.edu,
  University of Arizona – XyYvAs
Hubbart, Jason A., (304) 293-2472 jason.hubbart@mail.wvu.edu,
  West Virginia University – HqgHs
Hubenthal, Michael, (607) 777-4612 hubenth@iris.edu,
  Binghamton University – ZeYg
Hubeny, J B., bhubeny@salemstate.edu,
  Salem State University – GeOnCs
Huber, Brian T., (202) 633-1328 Smithsonian Inst / Natl Museum of Natural Hist– Pm
Huber, Christian, christian_huber@brown.edu,
  Brown University – Gv
Huber, Matthew, +27 (0)51 401 3944 huberms@ufs.ac.za,
  University of the Free State – EgXm
Huckabey, Marsha, (573) 882-2040 huckabeym@missouri.edu,
  University of Missouri – Zn
Hudackova, Natalia H., natalia.hudackova@uniba.sk,
  Comenius University in Bratislava – PimPg
Hudak, Paul F., (940) 565-4312 paul.hudak@unt.edu,
  University of North Texas – Hw
Hudec, Michael R., 512-471-1428 michael.hudec@beg.utexas.edu,
  University of Texas, Austin – Gc
Hudec, Peter P., (519) 253-3000 x2491 hudec@uwindsor.ca,
  University of Windsor – NgEmGe
Hudgins, Thomas, (787) 265-3845 thomas.hudgins@upr.edu,
  University of Puerto Rico – GiCpGv
Hudleston, Peter J., 612-625-0046 hudle001@umn.edu,
  University of Minnesota, Twin Cities – Gc
Hudson, Robert J. M., (217) 333-7641 rjhudson@uiuc.edu,
  University of Illinois, Urbana-Champaign – Cg
Hudson, Samuel M., sam.hudson@byu.edu,
  Brigham Young University – GosGr
Hudson-Edwards, Karen, +44 0203 073 8030 k.hudson-edwards@bbk.ac.uk,
  Birkbeck College – Gez
Huebert, Barry J., (808) 258-4673 huebert@hawaii.edu,
  University of Hawai'i, Manoa – OcCg
Huerfano, Victor, (787) 265-3845 victor.huerfano@upr.edu,
  University of Puerto Rico – Ysg
Huerta, Audrey, 509.963.2718 huerta@geology.cwu.edu,
  Central Washington University – Ygd
Huertas, María José, huertas@ucm.es,
  Univ Complutense de Madrid – Gi
Huey, Gregory L., (404) 894-5541 greg.huey@eas.gatech.edu,
  Georgia Institute of Technology – As
Huff, Bryan G., 217-244-2509 huff@isgs.uiuc.edu,
  Illinois State Geological Survey – Eo
Huff, Edmunt A., (630) 252-3633 Argonne National Laboratory – Ca
Huff, Warren D., (513) 556-3731 warren.huff@uc.edu,
  University of Cincinnati – GzvGr
Huffman, Debra E., (727) 553-3930 debrah@usf.edu,
  University of South Florida – Zn
Huffman, F. T., (859) 622-6968 tyler.huffman@eku.edu,
  Eastern Kentucky University – Zi
Huffman, Robert L., (912) 752-2704 Mercer University – Zn
Huft, Ashley, (406) 496-4789 ahuft@mtech.edu,
  Montana Tech of The University of Montana – Ca
Huggett, William, 618-453-7392 huggett@geo.siu.edu,
  Southern Illinois University Carbondale – Ec
Hughen, Konrad A., (508) 289-3353 khughen@whoi.edu,
  Woods Hole Oceanographic Institution – Cc
Hughes, Annie, +44 020 8417 2603 Ku08925@kingston.ac.uk,
  Kingston University – Zi
Hughes, Chambers, (858) 534-7121 chughes@ucsd.edu,
  University of California, San Diego – Oc
Hughes, Colin, colin.hughes@manchester.ac.uk,
  University of Manchester – Go
Hughes, Denis, +27 46 6224014 d.hughes@ru.ac.za,
  Rhodes University – Hqs
Hughes, John M., (802) 656-9443 jmhughes@uvm.edu,
  University of Vermont – Gz
Hughes, John M., John.M.Hughes@uvm.edu,
  Miami University – Gz
Hughes, Joseph B., (713)348-5603 david.v.hughes@rice.edu,
  Rice University – Zn
Hughes, Kenneth S., (787) 265-3845 kenneth.hughes@upr.edu,
  University of Puerto Rico – Gcm
Hughes, Malcolm K., (520) 621-6470 mhughes@ltrr.arizona.edu,
  University of Arizona – Pe
Hughes, Nigel C., (951) 827-3098 nigel.hughes@ucr.edu,
  University of California, Riverside – Pg
Hughes, Nigel C., nigel.hughes@ucr.edu,
  Cincinnati Museum Center – Pi
Hughes, Randall, 7815817370 x314 rhughes@neu.edu,
  Northeastern University – Zn
Hughes, Sam, +44 07727 096492 S.P.Hughes@exeter.ac.uk,
  Exeter University – Gt
Hughes, Scott S., (530) 533-1933 hughscot@isu.edu,
  Idaho State University – GvXgGi
Hughes III, Richard O., 909-389-3237 rihughes@craftonhills.edu,
  Crafton Hills College – GlZe
Hughes-Clarke, John E., 603-862-5505 jhc@ccom.unh.edu,
  University of New Hampshire – Og
Hugli, Wilbur G., 850-474-3470 whugli@uwf.edu,
  University of West Florida – As
Hugo, Richard, 503-725-3356 hugo@pdx.edu,
  Portland State University – Gy
Hui, Alice, akhui@umail.iu.edu,
  Indiana University, Bloomington – Cl
Huizenga, Jan Marten, (074) 781-4597 jan.huizenga@jcu.edu.au,
  James Cook University – GpCgGc
Hulbe, Christina L., chulbe@pdx.edu,
  Portland State University – Gl
Hulett, Sam, 614 265 6573 sam.hulett@dnr.state.oh.us,
  Ohio Dept of Natural Resources – Eo
Hull, Donald L., (505) 667-4151 Los Alamos National Laboratory – Zn
Hull, Joseph M., jhull@sccd.ctc.edu,
  Seattle Central Community College – Gc
Huluka, Gobena E., (334) 844-4100 hulukgo@auburn.edu,
  Auburn University – Sc
Humagain, Kamal, (315) 267-3356 humagak@potsdam.edu,
  SUNY Potsdam – Zir
Humayun, Munir, 850-644-5860 humayun@magnet.fsu.edu,
  Florida State University – XcCgGi
Humes, Karen, 208-885-6506 khumes@uidaho.edu,
  University of Idaho – ZriZn
Hummer, Daniel R., 618-453-7386 daniel.hummer@siu.edu,
  Southern Illinois University Carbondale – GzyCu
Humphrey, John D., +966 13 860 7982 humphrey@kfupm.edu.sa,
  King Fahd University of Petroleum and Minerals – Gdo
Humphrey, Neil F., (307) 314-2332 neil@uwyo.edu,
  University of Wyoming – GlmYh
Humphrey, Peggy, (406) 994-5718 peggyh@montana.edu,
  Montana State University – Zn
Humphreys, Eugene D., (541) 852-0091 genehumphreys@gmail.com,
  University of Oregon – YsGt
Humphreys, Robin, (843) 953-7424 humphreysr@cofc.edu,
  College of Charleston – Ge
Humphris, Susan E., (508) 289-3451 shumphris@whoi.edu,
  Woods Hole Oceanographic Institution – GuCm

Hunda, Brenda, (513) 455-7160 bhunda@cincymuseum.org, Cincinnati Museum Center – Pi
Hung, Ming-Chih, (660) 562-1797 mhung@nwmissouri.edu, Northwest Missouri State University – Zr
Hungerbuehler, Axel, (575) 461-3446 axelh@mesalands.edu, Mesalands Community College – PvgGg
Hungerford, Hilary, (801) 863-7160 hilary.hungerford@uvu.edu, Utah Valley University – ZuRw
Hungr, Oldrich, (604) 822-8471 ohungr@eos.ubc.ca, University of British Columbia – Gm
Hunsucker, Kelli, (321) 674-8437 khunsucker@fit.edu, Florida Institute of Technology – Ob
Hunt, Allen G., (937) 775-3116 allen.hunt@wright.edu, Wright State University – GqSpCl
Hunt, Andrew, 817 272 2987 hunt@uta.edu, University of Texas, Arlington – GbCg
Hunt, Brian, (604) 822-9135 bhunt@eos.ubc.ca, University of British Columbia – Ob
Hunt, Gene, (202) 633-1331 Smithsonian Inst / Natl Museum of Natural Hist– Pm
Hunt, Kathy, bhunt@ipst.umd.edu, University of Maryland – Zn
Hunt, Paula J., (304) 594-2331 phunt@geosrv.wvnet.edu, West Virginia Geological & Economic Survey – GgHw
Hunt, Robert M., (402) 472-4604 rhunt2@unl.edu, University of Nebraska, Lincoln – Pv
Hunt, Shelley, (519) 824-420 Ext.53065 shunt@uoguelph.ca, University of Guelph – Zn
Hunten, Donald M., (520) 621-4002 dhunten@lpl.arizona.edu, University of Arizona – As
Hunter, Arlene G., +44 (0) 1908 655400 x55400 a.g.hunter@open.ac.uk, The Open University – ZeGiCg
Huntington, Justin, 775.673.7670 justin.huntington@dri.edu, University of Nevada, Reno – HgZr
Huntington, Katharine W., 206-543-1750 kate1@uw.edu, University of Washington – Gt
Huntley, John, (573) 884-8083 University of Missouri – PqgPe
Huntoon, Jacqueline E., (906) 487-2440 jeh@mtu.edu, Michigan Technological University – GsZe
Huntoon, Laura, huntoon@email.arizona.edu, University of Arizona – Zn
Huntress, Jr., Wesley T., (202) 478-8910 whuntress@carnegiescience.edu, Carnegie Institution for Science – Xc
Hural, Kirsten, (717) 245-1632 Dickinson College – Gg
Hurd, David, (814) 732-2493 dhurd@edinboro.edu, Edinboro University of Pennsylvania – Zg
Hurich, Charles A., (709) 737-2384 churich@mun.ca, Memorial University of Newfoundland – Ys
Hurlow, Hugh, 801-537-3385 hughhurlow@utah.gov, Utah Geological Survey – Hw
Hurrell, James, james.hurrell@colostate.edu, Colorado State University – As
Hurtado, Jose M., (915) 747-5669 jhurtado@utep.edu, University of Texas, El Paso – E
Hurtgen, Matthew T., (847) 491-7539 matt@earth.northwestern.edu, Northwestern University – GsPgGr
Husch, Jonathan M., (609) 896-5330 husch@rider.edu, Rider University – GieXg
Husinec, Antun, (315) 229-5248 ahusinec@stlawu.edu, St. Lawrence University – GsCsGo
Hussein, Musa, (915) 747-5424 mjhussein@utep.edu, University of Texas, El Paso – YgeYm
Hussin, Azhar Hj, 03-79674203 azharhh@um.edu.my, University of Malaya – Gs
Hutcheon, Ian E., (403) 220-6744 ian@earth.geo.ucalgary.ca, University of Calgary – Cl
Hutchings, Jennifer, (541) 737-4453 jhutchings@coas.oregonstate.edu, Oregon State University – Gl
Hutchins, Peter S., (601) 961-5505 peter_hutchins@deq.state.ms.us, Mississippi Office of Geology – Zn
Hutchinson, Charles F., chuck@ag.arizona.edu, University of Arizona – Zr
Hutchison, David, 607-431-4730 hutchisond@hartwick.edu, Hartwick College – GxgGi
Hutson, John , john.hutson@flinders.edu.au, Flinders University – Spc
Hutson, Melinda, 971-722-4146 mhutson@pcc.edu, Portland Community College - Sylvania Campus – Xm
Hutson, Melinda, 503-725-3372 mhutson@pdx.edu, Portland State University – Xmc
Hutto, Richard S., (501) 683-0151 richard.hutto@arkansas.gov, Arkansas Geological Survey – Gg
Hutyra, Lucy, lrhutyra@bu.edu, Boston University – AsZr
Huws, Dei G., 01248 382523 oss082@bangor.ac.uk, University of Wales – YruNt
Huybers, Peter, 617-495-8391 phuybers@fas.harvard.edu, Harvard University – Pe
Huycke, David, 509-574-4817 dhuycke@yvcc.edu, Yakima Valley College – g
Huysken, Kristin, (219) 980-6739 khuysken@iun.edu, Indiana University Northwest – GizGc
Huysmans, Marijke, marijke.huysmans@ees.kuleuven.be, Katholieke Universiteit Leuven – Hw
Hyatt, James A., (860) 465-5789 hyattj@easternct.edu, Eastern Connecticut State University – Gml
Hyland, Ethan, 207-999-9047 ehyland@unity.ncsu.edu, North Carolina State University – GrCsPe
Hylland, Michael D., (801) 537-3382 mikehylland@utah.gov, Utah Geological Survey – Gg
Hylton, Alisa, (704)330-6297 Alisa.Hylton@cpcc.edu, Central Piedmont Community College – Gg
Hyndman, David W., 517-353-4442 hyndman@msu.edu, Michigan State University – Hy
Hyndman, Roy D., (250) 363-6428 University of Victoria – Ys
Hynek, Brian M., 303-735-4312 brian.hynek@colorado.edu, University of Colorado – Xg
Hynes, Andrew J., andrew.hynes@mcgill.ca, McGill University – Gc
Hynes, Joanna , (657) 278-5464 jfantozzi@fullerton.edu, California State University, Fullerton – Gg
Hysell, David, dlh37@cornell.edu, Cornell University – Znr

I

Iaccheri, Maria Linda, 089/2180 4274 Ludwig-Maximilians-Universitaet Muenchen – Gg
Iaffaldano, Giampiero, 089/2180 4220 giampiero@geophysik.uni-muenchen.de, Ludwig-Maximilians-Universitaet Muenchen – Yg
Iancu, Ovidiu G., 0040232201455 ogiancu@uaic.ro, Alexandru Ioan Cuza – GpeCg
Ianno, Adam J., (814) 641-6662 iannoa@juniata.edu, Juniata College – GgiCu
Iantria, Linnea, (417) 836-4486 liantria@missouristate.edu, Missouri State University – Zn
Ibaraki, Motomu, (614) 292-7528 ibaraki.1@osu.edu, Ohio State University – Hg
Ibrahim, Ahmad Tajuddin Hj, 03-79674230 ahmadt@um.edu.my, University of Malaya – Ng
Ibrahim, Mohamed, 212-772-5267 mibrahim@hunter.cuny.edu, Hunter College (CUNY) – ZgRwZc
Icenhower, Jonathan P., (509) 372-0078 jon.icenhower@pnl.gov, Pacific Northwest National Laboratory – Cl
Icopini, Gary, 406-496-4841 gicopini@mtech.edu, Montana Tech of The University of Montana – ClHwCb
Ide, Kayo, (310) 206-6484 University of California, Los Angeles – Zg
Idleman, Bruce D., bdi2@lehigh.edu, Lehigh University – Cc
Idone, Vincent P., (518) 442-4577 vpi@atmos.albany.edu, SUNY, Albany – As
Idstein, Peter J., 859-257-2770 peter.idstein@uky.edu, University of Kentucky – Gg
Ielpi, Alessandro, (705) 675-1151 aielpi@laurentian.ca, Laurentian University, Sudbury – GsmGr
Ierley, Glenn R., gierley@ucsd.edu, University of California, San Diego – Op
Igel, Heiner, 089/2180 4204 heiner.igel@lmu.de, Ludwig-Maximilians-Universitaet Muenchen – Yg
Iglesias, Cruz, 0034981167000 cruzi@udc.es, Coruna University – GaZmNr
Ignor, Emmanuel E., 08066943346 eeigonor@futa.edu.ng, Federal University of Technology, Akure – GgCeGi
Ihinger, Phillip D., (715) 836-2158 ihinger@uwec.edu,

University of Wisconsin, Eau Claire – Gx
Ikonnikova, Svetlana, 512-232-9364 s.ikonnikova@mail.utexas.edu,
 University of Texas, Austin – Ego
Iles, Derric L., (605) 677-6148 derric.iles@usd.edu,
 South Dakota Dept of Environment and Natural Resources – HwGg
Illari, Lodovica, (617) 253-2286 illari@mit.edu,
 Massachusetts Institute of Technology – Am
Imasuen, Isaac O., okpeseyi.imasuen@uniben.edu,
 University of Benin – Gee
Immel, Harald, 089/2180 6615 h.immel@lrz.uni-muenchen.de,
 Ludwig-Maximilians-Universitaet Muenchen – Pg
Immonen, Wilma, (406) 496-4182 wimmonen@mtech.edu,
 Montana Tech of the University of Montana – Zn
Inamdar, Shreeram, 302-831-8877 inamdar@udel.edu,
 University of Delaware – Hg
Ince, Simon, (520) 621-5082 University of Arizona – Hs
Indares, Aphrodite D., (709) 737-2456 afin@sparky2.esd.mun.ca,
 Memorial University of Newfoundland – Gp
Ingall, Ellery D., (404) 894-3883 ellery.ingall@eas.gatech.edu,
 Georgia Institute of Technology – CmlOc
Ingalls, Anitra E., 206-221-6748 aingalls@u.washington.edu,
 University of Washington – Oc
Ingersoll, Andrew P., 626 395-6167 api@gps.caltech.edu,
 California Institute of Technology – As
Ingersoll, Raymond V., (310) 825-8634 ringer@epss.ucla.edu,
 University of California, Los Angeles – Gst
Ingle, Jr., James C., (650) 723-3366 ingle@pangea.stanford.edu,
 Stanford University – Pm
Inglett, Kanika Sharma, 352-294-3164 University of Florida – So
Inglett, Patrick, 352-392-1803 pinglett@ufl.edu,
 University of Florida – Sf
Inglis, Michael, 631-451-4120 ingism@sunysuffolk.edu,
 Suffolk County Community College, Ammerman Campus – Zg
Ingram, B. Lynn, (510) 643-1474 ingram@eps.berkeley.edu,
 University of California, Berkeley – Cs
Insel, Nadja, 773-442-6058 N-Insel@neiu.edu,
 Northeastern Illinois University – Gt
Inskeep, William P., (406) 994-5077 binskeep@montana.edu,
 Montana State University – Sc
Insua, Juan Miguel, insuarev@ucm.es,
 Univ Complutense de Madrid – NtGc
Ioannides, Dimitri, 417-836-5801 DIoannides@MissouriState.edu,
 Missouri State University – Zn
Ionesi, Viorel, vioion@uaic.ro,
 Alexandru Ioan Cuza – PmGgHg
Ioris, Antonio, +44 (0) 131 651 9090 a.ioris@ed.ac.uk,
 Edinburgh University – Ge
Iqbal, Mohammad Z., (319) 273-2998 m.iqbal@uni.edu,
 University of Northern Iowa – Hw
Iribar, Vicente , vicente.iribar@ehu.eus,
 University of the Basque Country UPV/EHU – HwGm
Irish, Jennifer, (540) 231-2298 jirish@vt.edu,
 Virginia Polytechnic Institute & State University – NoOg
Irmis, Randall B., (801) 581-6555 irmis@umnh.utah.edu,
 University of Utah – Sa
Irvin, Gene D., 205/247-3542 dirvin@gsa.state.al.us,
 Geological Survey of Alabama – Gg
Irvine, Dylan, dylan.irvine@flinders.edu.au,
 Flinders University – Hw
Irvine, T. Neil, 202-478-8950 nirvine@carnegiescience.edu,
 Carnegie Institution for Science – Gi
Irving, Jessica, 609 258-4536 jirving@princeton.edu,
 Princeton University – Yg
Irving, Tony, irvingaj@uw.edu,
 University of Washington – CgXmGi
Isaacson, Carl, 218-755-4104 cisaacson@bemidjistate.edu,
 Bemidji State University – Ge
Isaacson, Peter E., (208) 885-7969 isaacson@uidaho.edu,
 University of Idaho – Psi
Isacks, Bryan L., bli1@cornell.edu,
 Cornell University – Gmt
Isacks, Bryan L., 607-316-8266 bli1@cornell.edu,
 Cornell University – GtYsGm
Isard, Sierra J., 828-296-4629 sierra.isard@ncdenr.gov,
 North Carolina Geological Survey – Gcg
Isbell, John L., 414-229-2877 jisbell@uwm.edu,
 University of Wisconsin, Milwaukee – Gs

Isenor, Fenton M., Fenton_Isenor@cbu.ca,
 Cape Breton University – GgNgGm
Ishii, Miaki, 617-384-8066 ishii@eps.harvard.edu,
 Harvard University – Yg
Ishii, Satoshi, 612-624-7902 ishi0040@umn.edu,
 University of Minnesota, Twin Cities – SbPg
Ishman, Scott E., (618) 453-7377 sishman@siu.edu,
 Southern Illinois University Carbondale – Pmc
Ismail, Ahmed, 405-744-6358 ahmed.ismail@okstate.edu,
 Oklahoma State University – Ysu
Ismat, Zeshan, 717-358-4485 zeshan.ismat@fandm.edu,
 Franklin and Marshall College – Gc
Israel, Daniel W., (919) 515-2388 North Carolina State University – Sb
Ito, Emi, 612-624-7881 eito@umn.edu,
 University of Minnesota, Twin Cities – ClGnCs
Ito, Garrett T., (808) 956-9717 gito@hawaii.edu,
 University of Hawai'i, Manoa – YgrGt
Ivanochko, Tara, (604) 827-3179 tivanochko@eos.ubc.ca,
 University of British Columbia – OoZce
Ivanov, Julian, (785) 864-2089 jivanov@kgs.ku.edu,
 University of Kansas – Yg
Ivanov, Martin, +420 549 49 4600 mivanov@sci.muni.cz,
 Masaryk University – PgiPe
Ivanovic, Ruza, +44(0) 113 34 34945 r.ivanovic@leeds.ac.uk,
 University of Leeds – OgAsPe
Ivany, Linda C., (315) 443-3626 lcivany@syr.edu,
 Syracuse University – PegCs
Iverson, Neal R., (515) 294-8048 niverson@iastate.edu,
 Iowa State University of Science & Technology – Gl
Iverson, Richard , 360-993-8920 riverson@usgs.gov,
 University of Washington – Gv
Ivins, Erik R., (818) 354-4785 Jet Propulsion Laboratory – Gt
Ivory, Sarah, sarah_ivory@psu.edu,
 Pennsylvania State University, University Park – Pl
Izon, Gareth, +44 01334 463936 gji3@st-andrews.ac.uk,
 University of St. Andrews – Cg

J

Jablonowski, Christiane, (734) 763-6238 cjablono@umich.edu,
 University of Michigan – AsmZo
Jablonski, David, (773) 702-8163 djablons@midway.uchicago.edu,
 University of Chicago – Pi
Jacinthe, Pierre-Andre, pjacinth@iupui.edu,
 Indiana University / Purdue University, Indianapolis – Cl
Jackson, Allyson K., (914) 251-6646 allyson.jackson@purchase.edu,
 SUNY, Purchase – ZgcZe
Jackson, Andrea, +44(0) 113 34 36728 a.v.jackson@leeds.ac.uk,
 University of Leeds – As
Jackson, Brian P., (603) 646-1272 brian.jackson@dartmouth.edu,
 Dartmouth College – CaGeCt
Jackson, Charles, (512) 471-0401 charles@ig.utexas.edu,
 University of Texas, Austin – ZoOoAt
Jackson, Chester M., (912) 598-2328 cjackson@georgiasouthern.edu,
 Georgia Southern University – OnGs
Jackson, Chester W., (912) 478-0174 cjackson@georgiasouthern.edu,
 Georgia Southern University – GsmGu
Jackson, Christopher, +44 20 759 47450v c.jackson@imperial.ac.uk,
 Imperial College – Gt
Jackson, David D., (310) 825-0421 djackson@ucla.edu,
 University of California, Los Angeles – YsdYg
Jackson, Frankie, (406) 994-6642 frankiej@montana.edu,
 Montana State University – Pv
Jackson, Gail D., +44 (0) 131 650 5436 G.Jackson@ed.ac.uk,
 Edinburgh University – PbGe
Jackson, George A., (979) 845-0405 gjackson@ocean.tamu.edu,
 Texas A&M University – Op
Jackson, Hiram, (916) 691-7605 Cosumnes River College – Gg
Jackson, James A., +44 (0) 1223 337197 jaj2@cam.ac.uk,
 University of Cambridge – YgGt
Jackson, Jennifer M., (626) 395-6780 jackson@gps.caltech.edu,
 California Institute of Technology – Gy
Jackson, Jeremiah, 573-368-2182 jeremiah.jackson@dnr.mo.gov,
 Missouri Dept of Natural Resources – Gg
Jackson, Jeremy B. C., (858) 518-7613 jbjackson@ucsd.edu,
 University of California, San Diego – PgePi
Jackson, Kenneth J., (925) 422-6053 jackson8@llnl.gov,
 Lawrence Livermore National Laboratory – Cl

Jackson, Louise E., 530-754-9116 lejackson@ucdavis.edu,
University of California, Davis – So

Jackson, Marie D., 801-581-7062 m.d.jackson@utah.edu,
University of Utah – Gv

Jackson, Martin P. A., (512) 475-9548 martin.jackson@beg.utexas.edu,
University of Texas at Austin, Jackson School of Geosciences – Gc

Jackson, Matt, jackson@geol.ucsb.edu,
University of California, Santa Barbara – Cu

Jackson, Michael, 612-624-5274 jacks057@umn.edu,
University of Minnesota, Twin Cities – Ym

Jackson, Richard A., (718) 460-1476 Long Island University, Brooklyn Campus – Gc

Jackson, Robert B., 919 660 7408 jackson@duke.edu,
Duke University – Zn

Jackson, William T., wjackson@southalabama.edu,
University of South Alabama – GsCc

Jacob, Daniel J., (617) 495-1794 djj@io.harvard.edu,
Harvard University – As

Jacob, Klaus H., jacob@ldeo.columbia.edu,
Columbia University – Ys

Jacob, Robert W., (570) 577-1791 rwj003@bucknell.edu,
Bucknell University – Yg

Jacobi, Robert D., (716) 645-3489 rdjacobi@buffalo.edu,
SUNY, Buffalo – GctGs

Jacobs, Alan M., 330-941-2933 amjacobs@ysu.edu,
Youngstown State University – Ge

Jacobs, Jon, 519-661-2111 ext 86752 jjacobs@uwo.ca,
Western University – Gg

Jacobs, Katharine L., (520) 626-0684 jacobsk@email.arizona.edu,
University of Arizona – Hg

Jacobs, Louis L., (214) 768-2773 Southern Methodist University – Pv

Jacobs, Peter, jacobsp@uww.edu,
University of Wisconsin, Whitewater – SdaGm

Jacobs, Tenika, (609) 984-6587 tenika.jacobs@dep.state.nj.us,
New Jersey Geological and Water Survey – Zn

Jacobsen, Abram, 360-734-5019 abramj@uw.edu,
University of Washington – Xg

Jacobsen, Jeffrey S., (406) 994-4605 jefj@montana.edu,
Montana State University – So

Jacobsen, Stein B., (617) 495-5233 jacobsen@neodymium.harvard.edu,
Harvard University – Cc

Jacobsen, Steven D., 847-467-1825 s-jacobsen@northwestern.edu,
Northwestern University – GyzzZm

Jacobson, Andrew D., (847) 491-3132 adj@earth.northwestern.edu,
Northwestern University – ClPeCa

Jacobson, Carl E., cejac@iastate.edu,
Iowa State University of Science & Technology – GtcGp

Jacobson, Roger, (775) 673-7364 roger@dri.edu,
University of Nevada, Reno – Cg

Jacobson, Seth A., 847-491-3459 sethajacobson@earth.northwestern.edu,
Northwestern University – XacXg

Jadamec, Margarete, 716-645-4262 mjadamec@buffalo.edu,
SUNY, Buffalo – GtYgZf

Jaecks, Glenn, jaecksg@arc.losrios.edu,
American River College – GgOgPg

Jaeger, John M., (352) 846-1381 jmjaeger@ufl.edu,
University of Florida – Gs

Jaegle, Lyatt, (206) 685-2679 jaegle@atmos.washington.edu,
University of Washington – As

Jaffe, Daniel A., (425) 352-5357 djaffe@u.washington.edu,
University of Washington – As

Jaffe, Peter R., (609) 258-4653 jaffe@princeton.edu,
Princeton University – SbHgw

Jago, Bruce C., (705) 675-1151 x7227 bjago@laurentian.ca,
Laurentian University, Sudbury – EgCe

Jagoutz, Oliver, (617) 324-5514 jagoutz@mit.edu,
Massachusetts Institute of Technology – GitGc

Jahangiri, Ahmad, +98 (411) 339 2695 jahangiri@tabriu.ac.ir,
University of Tabriz – GivGg

Jain, Atul K., (217) 333-2128 jain1@illinois.edu,
University of Illinois, Urbana-Champaign – As

Jaisi, Deb, 302-831-1376 jaisi@udel.edu,
University of Delaware – Cb

Jaiswal, Priyank, 405-744-6358 priyank.jaiswal@okstate.edu,
Oklahoma State University – Yg

Jakosky, Bruce M., 303-492-8004 bruce.jakosky@colorado.edu,
University of Colorado – Xg

Jamaluddin, Tajul Anuar, 03-79674152 taj@um.edu.my,
University of Malaya – Ng

James, Matthew J., (707) 664-2301 james@sonoma.edu,
Sonoma State University – PiRhPg

James, Noel P., (613) 533-6170 jamesn@queensu.ca,
Queen's University – Gd

James, Peter B., (254)710-2351 P_James@baylor.edu,
Baylor University – Ygg

James, Rachael, +44 (0)23 80599005 R.H.James@soton.ac.uk,
University of Southampton – Oc

James, Richard, 7056751151 x2271 rjames@laurentian.ca,
Laurentian University, Sudbury – GpiEm

James, Scott C., (254) 710-2534 sc_james@baylor.edu,
Baylor University – HwGeEo

James, Thomas S., (250) 363-6403 thomas.james@canada.ca,
University of Victoria – YdvYg

James, Valentine U., (814) 393-1938 vjames@clarion.edu,
Clarion University – ZuRwZn

James-Aworeni, E., 234-705-326-6216 dawnjames@gmail.com,
Obafemi Awolowo University – Hg

Jamieson, Heather E., (613) 545-6181 Queen's University – Ge

Jamieson, J. Bruce, (403) 288-0803 University of Calgary – Ne

Jamili, Ahmad, ahmad.jamili@ou.edu,
University of Oklahoma – Np

Jamiolahmady, Mahmoud, +44 (0) 131 451 3122 m.jamiolahmady@pet.hw.ac.uk,
Heriot-Watt University – Zn

Janecke, Susanne U., (435) 797-3877 susanne.janecke@usu.edu,
Utah State University – Gt

Janecky, David R., (505) 667-7603 Los Alamos National Laboratory – Cl

Janetos, Tony, ajanetos@bu.edu,
Boston University – Zn

Janney, Phillip, +27-21-650-2929 phil.janney@uct.ac.za,
University of Cape Town – GiCaXc

Janson, Xavier, 512-475-9524 xavier.janson@beg.utexas.edu,
University of Texas at Austin, Jackson School of Geosciences – Gr

Janssen, Keith, (785) 532-6101 kjanssen@ksu.edu,
Kansas State University – So

Jantz, Claire A., (717) 477-1399 cajant@ship.edu,
Shippensburg University – ZuiZc

Janusz, Robert, 210-486-0045 San Antonio Community College – Gg

Jaouich, Alfred, 514-987-3000 #3378 jaouich.alfred@uqam.ca,
Universite du Quebec a Montreal – Sc

Jarcho, Kari A., (612) 625-5251 kjarcho@umn.edu,
University of Minnesota, Twin Cities – Zn

Jardine, Philip M., ipj@ornl.gov,
Oak Ridge National Laboratory – Sc

Jarvis, Ed, +353 21 4902698 e.jarvis@ucc.ie,
University College Cork – Pl

Jarvis, Gary T., (416)736-2100 #77710 jarvis@yorku.ca,
York University – Yg

Jarvis, Ian, +44 020 8417 2526 i.jarvis@kingston.ac.uk,
Kingston University – CgGsr

Jarvis, Richard S., (915) 747-5263 rsjarvis@utep.edu,
University of Texas, El Paso – ZyGmAs

Jarvis, W. T., 541-737-4032 todd.jarvis@oregonstate.edu,
Oregon State University – Hw

Jarzen, David M., 3523921721 x 245 dilcher@flmnh.ufl.edu,
University of Florida – Pb

Jasbinsek, John J., (805) 756-2013 jjasbins@calpoly.edu,
California Polytechnic State University – Ysg

Jasinski, Steven E., (717) 783-9897 c-sjasinsk@pa.gov,
State Museum of Pennsylvania – PvgPg

Jasoni, Richard, 775-673-7472 richard.jasoni@dri.edu,
Desert Research Institute – So

Jastrow, Julie D., (630) 252-3226 jdjastrow@anl.gov,
Argonne National Laboratory – Sb

Jaumé, Steven C., (843) 953-1802 jaumes@cofc.edu,
College of Charleston – Ys

Javadpour, Farzam, (512) 232-8068 farzam.javadpour@beg.utexas.edu,
University of Texas, Austin – NpEo

Javaux, Emmanuelle, 32 4 366 54 22 EJ.Javaux@ulg.ac.be,
Universite de Liege – Pl

Jawitz, James W., 352-392-1951 (203) jawitz@ufl.edu,
University of Florida – SpHsw

Jayakumar, Amal, (609) 258-6294 ajayakum@princeton.edu,
Princeton University – Cm

Jayawickreme, Dushmantha, (203) 392-7286 jayawickred1@southernct.edu,
  Southern Connecticut State University – YgSpNg
Jaye, Shelley, (703) 425-5180 sjaye@nvcc.edu,
  Northern Virginia Community College - Annandale – Giy
Jean-Marc, MONTEL, +33 3 83 59 64 00 jean-marc.montel@univ-lorraine.fr,
  Ecole Nationale Supérieure de Géologie (ENSG) – Eg
Jeanloz, Raymond, (510) 642-2639 jeanloz@uclink.berkeley.edu,
  University of California, Berkeley – Gy
Jebrak, Michel, 514-987-3000 #3986 jebrak.michel@uqam.ca,
  Universite du Quebec a Montreal – Eg
Jedrysek, Mariusz O., mariusz.jedrysek@uwr.edu.pl,
  University of Wroclaw – CaGe
Jeevanjee, Nadir, 609-258-4101 nadirj@princeton.edu,
  Princeton University – Ap
Jefferson, Anne, 704-687-5977 ajefferson@uncc.edu,
  University of North Carolina, Charlotte – HgGm
Jefferson, Anne J., (330) 672-2746 ajeffer9@kent.edu,
  Kent State University – HgGmCs
Jeffery, David L., 740-376-4844 jefferyd@marietta.edu,
  Marietta College – PsGo
Jeffrey, Kip, +4401326259442 C.Jeffrey@exeter.ac.uk,
  Exeter University – Nx
Jeffries, H. Perry, (401) 874-6222 jeffries@uri.edu,
  University of Rhode Island – Ob
Jekeli, Christopher, 614 292 7117 jekeli.1@osu.edu,
  Ohio State University – Ydv
Jelinski, Nicolas, 612-626-9936 jeli0026@umn.edu,
  University of Minnesota, Twin Cities – SdCs
Jellinek, Mark, mjellinek@eos.ubc.ca,
  University of British Columbia – Gg
Jenkins, David M., (607) 777-2736 dmjenks@binghamton.edu,
  Binghamton University – CpGzp
Jenkins, Gregory S., 814-865-0479 gsj1@psu.edu,
  Pennsylvania State University, University Park – Am
Jenkins, Mary Ann, (416)736-2100 #22992 maj@yorku.ca,
  York University – As
Jenkyns, Hugh C., +44 (1865) 272023 hugh.jenkyns@earth.ox.ac.uk,
  University of Oxford – GrsCs
Jenner, George A., (709) 737-8387 gjenner@sparky2.esd.mun.ca,
  Memorial University of Newfoundland – Gi
Jensen, Ann R., 605-677-6159 ann.jensen@usd.edu,
  South Dakota Dept of Environment and Natural Resources – Gg
Jensen, Antony, +44 (0)23 80593428 acj@noc.soton.ac.uk,
  University of Southampton – Ob
Jensen, Olivia G., 514-398-3587 olivia.jensen@mcgill.ca,
  McGill University – Yg
Jensen, Ryan R., (801) 422-5386 rjensen@byu.edu,
  Brigham Young University – Zii
Jensen, Scott W., 605-677-6869 scott.jensen@usd.edu,
  South Dakota Dept of Environment and Natural Resources – Zn
Jenson, John W., (671) 735-2689 jjenson@triton.uog.edu,
  University of Guam – HwGel
Jercinovic, Michael J., (413) 545-2431 mjj@geo.umass.edu,
  University of Massachusetts, Amherst – Gx
Jerde, Eric, (606) 783-5406 e.jerde@moreheadstate.edu,
  Morehead State University – Cg
Jerolmack, Douglas, 215-746-2823 sediment@sas.upenn.edu,
  University of Pennsylvania – GmYxHq
Jessey, David R., 909-869-3457 drjessey@csupomona.edu,
  California State Polytechnic University, Pomona – Eg
Jessup, Micah, (865) 974-2366 mjessup@utk.edu,
  University of Tennessee, Knoxville – Gc
Jeu, Amy, (212) 772-4019 ajeu@hunter.cuny.edu,
  Hunter College (CUNY) – Zi
Jewell, Paul W., (801) 581-6636 paul.jewell@utah.edu,
  University of Utah – GmHg
Jewett, Brian, 217-333-3957 bjewett@illinois.edu,
  University of Illinois, Urbana-Champaign – As
Jewitt, David, (310) 825-3880 jewitt@epss.ucla.edu,
  University of California, Los Angeles – XyZnn
Jezek, Kenneth, 614-292-7973 jezek.1@osu.edu,
  Ohio State University – YgPcGl
Ji, Chen, (805) 893-2923 ji@geol.ucsb.edu,
  University of California, Santa Barbara – Ys
Ji, Peng, pengji@ldeo.columbia.edu,
  Columbia University – Yr
Ji, Wei, (816) 235-1334 jiwei@umkc.edu,
  University of Missouri, Kansas City – ZriZy
Jiang, Dazhi, 519-661-3192 djiang3@uwo.ca,
  Western University – GcgGg
Jiang, Ganqing, (702) 895-2708 ganqing.jiang@unlv.edu,
  University of Nevada, Las Vegas – GsClGr
Jiang, James Xinxia, (506) 364-2326 Mount Allison University – Gg
Jiang , Shu, 801-585-9816 sjiang@egi.utah.edu,
  University of Utah – GoYsNp
Jiang, Zhenhua, (410) 436-6864 Argonne National Laboratory – Zn
Jickells, Tim, +44 (0)1603 59 3117 t.jickells@uea.ac.uk,
  University of East Anglia – Ob
Jiménez, David, davidj03@ucm.es,
  Univ Complutense de Madrid – GcNt
Jimoh, Mustapha T., +2348052472942 mtjimoh@lautech.edu.ng,
  Ladoke Akintola University of Technology – GxxGx
Jin, Fei Fei, 808-956-4645 jff@hawaii.edu,
  University of Hawai'i, Manoa – AmsZo
Jin, Guohai, 205-247-3560 gjin@gsa.state.al.us,
  Geological Survey of Alabama – Np
Jin, Jisuo, (519) 661-4061 jjin@uwo.ca,
  Western University – Pi
Jin, Lixin, (915) 747-5559 ljin2@utep.edu,
  University of Texas, El Paso – Ge
Jin, Qusheng, 541-346-4999 qjin@uoregon.edu,
  University of Oregon – Cg
Jin, Yufang, (530) 601 9805 jin.yufang@gmail.com,
  University of California, Davis – Zr
Jinnah, Zubair A., (011) 717-6554 zubair.jinnah@wits.ac.za,
  University of the Witwatersrand – GsPg
Jiracek, George R., (619) 594-5160 gjiracek@mail.sdsu.edu,
  San Diego State University – Ye
Jiron, Rebecca, 757-221-2484 rljiron@wm.edu,
  College of William & Mary – GmtZg
Jirsa, Mark, (612) 626-4028 jirsa001@umn.edu,
  University of Minnesota – Gc
Jiskoot, Hester, (403) 329-2739 hester.jiskoot@uleth.ca,
  University of Lethbridge – GlmZy
Jo, Ho Young, 82-2-3290-3179 hyjo@korea.ac.kr,
  Korea University – NgCaGe
Jochens, Ann E., (979) 845-6714 ajochens@ocean.tamu.edu,
  Texas A&M University – Op
Joeckel, Matt, (402) 472-7520 rjoeckel3@unl.edu,
  Unversity of Nebraska - Lincoln – GsrGg
Joesten, Raymond, raymond.joesten@uconn.edu,
  University of Connecticut – GzZmGp
Johanesen, Katharine, (814) 641-3601 johanesen@juniata.edu,
  Juniata College – GzxGc
Johannesson, Karen H., 504-862-3193 kjohanne@tulane.edu,
  Tulane University – HgCg
Johanson, Erik, (561) 297-4153 ejohanson@fau.edu,
  Florida Atlantic University – RnZy
John, Barbara E., (307) 223-1951 bjohn@uwyo.edu,
  University of Wyoming – Gci
John, Bratton F., (313) 577-2506 jbratton@limno.com,
  Wayne State University – Cg
John, Cedric, +44 20 759 46461 cedric.john@imperial.ac.uk,
  Imperial College – Gs
John, Chacko J., (225) 578-8681 cjohn@lsu.edu,
  Louisiana State University – GosGe
John, Hickman B., 859-323-0541 jhickman@uky.edu,
  University of Kentucky – Gto
Johnpeer, Gary D., (714) 255-9819 gjohnpeer@earthlink.net,
  Cerritos College – Ng
Johns, Ronald A., (512) 223-7491 rjohns@austincc.edu,
  Austin Community College District – Pi
Johnson, Ansel G., (503) 725-3381 Portland State University – Ye
Johnson, Arthur H., ahj@sas.upenn.edu,
  University of Pennsylvania – So
Johnson, Ashanti, (817) 272-2987 ashanti@uta.edu,
  University of Texas, Arlington – OgGe
Johnson, Beth , beth.johnson@uwc.edu,
  University of Wisconsin Colleges – GlRh
Johnson, Beverly J., 207 786-6062 bjohnso3@bates.edu,
  Bates College – CsbCl
Johnson, Brad, 520-621-2470 bradjohnson@email.arizona.edu,
  Arizona Geological Survey – Gc

Johnson, Brandon C., (401) 863-5163 brandon_johnson@brown.edu, Brown University – Xg
Johnson, Bruce D., (902) 494-3259 bruce.johnson@dal.ca, Dalhousie University – Oc
Johnson, Cari, (801) 585-3782 cari.johnson@utah.edu, University of Utah – Gs
Johnson, Carl G., (508) 289-2304 cjohnson@whoi.edu, Woods Hole Oceanographic Institution – Ca
Johnson, Clark M., (828) 265-8680 clarkj@geology.wisc.edu, University of Wisconsin-Madison – CgcCs
Johnson, Claudia C., 812-855-0646 claudia@indiana.edu, Indiana University, Bloomington – Pe
Johnson, Daniel L., (403) 327-4561 University of Lethbridge – Zn
Johnson, Darren J., 605-677-6162 darren.j.johnson@usd.edu, South Dakota Dept of Environment and Natural Resources – GgEo
Johnson, David B., (575) 835-5635 djohnson@nmt.edu, New Mexico Institute of Mining and Technology – Ps
Johnson, Donald O., (630) 252-3392 Argonne National Laboratory – Gr
Johnson, Donald R., (608)262-2538 donj@ssec.wisc.edu, University of Wisconsin, Madison – As
Johnson, Elias, (417) 836-5801 Missouri State University – Zr
Johnson, Emily P., (617) 353-9709 Boston University – Zn
Johnson, Emily R., 575-646-3795 erj@nsmu.edu, New Mexico State University, Las Cruces – Gvi
Johnson, Eric L., 1-607-4314658 johnsone@hartwick.edu, Hartwick College – GpcGi
Johnson, Gary D., (603) 636-2371 gary.d.johnson@dartmouth.edu, Dartmouth College – GrdYm
Johnson, Glenn W., (801) 581-6151 gjohnson@egi.utah.edu, University of Utah – Gq
Johnson, Gregory C., (206) 526-6806 University of Washington – Op
Johnson, H. Paul, (206) 543-8474 johnson@ocean.washington.edu, University of Washington – Yr
Johnson, Helen, +44 01865 272142 Helen.Johnson@earth.ox.ac.uk, University of Oxford – Og
Johnson, Howard, +44 20 759 46461 h.d.johnson@imperial.ac.uk, Imperial College – Go
Johnson, Jane M., (320) 589-3411 University of Minnesota, Twin Cities – So
Johnson, Jean M., (706) 272-2666 jmjohnson@daltonstate.edu, Dalton State Community College – Zg
Johnson, Jeanne L., (225) 354-9562 jeannej@lsu.edu, Louisiana State University – Zn
Johnson, Jeffrey , 208-426-2959 jeffrey.b.johnson@gmail.com, Boise State University – YsGv
Johnson, Joel E., 603-862-1718 joel.johnson@unh.edu, University of New Hampshire – Gus
Johnson, Joel P., 512-232-5288 joelj@jsg.utexas.edu, University of Texas, Austin – Gs
Johnson, Judy L., (907) 474-7388 jljohnson21@alaska.edu, University of Alaska, Fairbanks – ZnnZn
Johnson, Julie, 901-678-4217 jjhnsn79@memphis.edu, University of Memphis – Gx
Johnson, Kaj, 812-855-3612 kajjohns@indiana.edu, Indiana University, Bloomington – Yg
Johnson, Kathleen, 949-824-6174 kathleen.johnson@uci.edu, University of California, Irvine – Zg
Johnson, Kenneth, (713) 222-5375 johnsonk@uhd.edu, University of Houston Downtown – GiCac
Johnson, Kevin, 321-674-7186 johnson@fit.edu, Florida Institute of Technology – Ob
Johnson, Kevin T. M., 808-956-3444 kjohnso2@hawaii.edu, University of Hawai'i, Manoa – Gi
Johnson, Kurt, (907) 696-0079 kurt.johnson@alaska.gov, Alaska Division of Geological & Geophysical Surveys – Gg
Johnson, Lane R., LRJohnson@lbl.gov, University of California, Berkeley – Ys
Johnson, Mark S., (778) 999-2102 mark.johnson.2102@gmail.com, University of British Columbia – SpHsq
Johnson, Markes E., 413-597-2329 markes.e.johnson@williams.edu, Williams College – PseGm
Johnson, Martin, +44 (0)1603 59 1299 martin.johnson@uea.ac.uk, University of East Anglia – Cm
Johnson, Ned K., (510) 642-3059 University of California, Berkeley – Pv
Johnson, Neil E., 540-231-1785 johnsonne@vt.edu, Virginia Polytechnic Institute & State University – GzEg
Johnson, Paul A., (507) 933-7442 Los Alamos National Laboratory – Yg
Johnson, Peggy, (575) 835-5819 peggy@nmbg.nmt.edu, New Mexico Institute of Mining & Technology – Hw
Johnson, Robert G., 612-626-0853 johns088@umn.edu, University of Minnesota, Twin Cities – Pe
Johnson, Robert L., (630) 252-7004 rljohnson@anl.gov, Argonne National Laboratory – Zn
Johnson, Robert O., (507) 933-7442 Oak Ridge National Laboratory – Hg
Johnson, Roy A., (520) 621-4890 johnson6@email.arizona.edu, University of Arizona – Ys
Johnson, Sarah E., 859-572-6907 johnsonsa@nku.edu, Northern Kentucky University – GmNgZi
Johnson, Scott E., (207) 581-2142 johnsons@maine.edu, University of Maine – Gc
Johnson, Thomas, tcj@d.umn.edu, University of Minnesota, Duluth – PcGn
Johnson, Thomas C., (218) 341-4599 tcj@d.umn.edu, University of Massachusetts, Amherst – PeCoOu
Johnson, Thomas M., (217) 244-2002 tmjohnsn@illinois.edu, University of Illinois, Urbana-Champaign – HwCls
Johnson, Ty, (501) 683-0153 ty.johnson@arkansas.gov, Arkansas Geological Survey – GgZiGr
Johnson, Verner C., (970) 248-1672 vjohnson@coloradomesa.edu, Colorado Mesa University – Yg
Johnson, William C., 785-864-5548 wcj@ku.edu, University of Kansas – Grm
Johnson, William P., (801) 581-5033 william.johnson@utah.edu, University of Utah – Ng
Johnston, A. Dana, (541) 346-5588 adjohn@uoregon.edu, University of Oregon – Cp
Johnston, Archibald C., (901) 678-2007 ajohnstn@memphis.edu, University of Memphis – Ys
Johnston, Carl G., 330-941-7151 cgjohnston@ysu.edu, Youngstown State University – Pg
Johnston, David, (501) 683-0126 david.johnston@arkansas.gov, Arkansas Geological Survey – Gg
Johnston, David T., 617-496-5024 johnston@eps.harvard.edu, Harvard University – CgPg
Johnston, Paul, (403) 440-6174 pajohnston@mtroyal.ca, Mount Royal University – Pi
Johnston, Paul L., 620-342-4098 jillpaulj@hotmail.com, Emporia State University – Gg
Johnston, Scott, 805-756-1650 scjohnst@calpoly.edu, California Polytechnic State University – GcpCc
Johnston, Stephen T., (780) 492-5249 stjohnst@ualberta.ca, University of Alberta – GctGg
Johnston, Thomas, (403) 329-2534 johnston@uleth.ca, University of Lethbridge – Ze
Johnston, III, John E., (225) 578-8657 hammer@lsu.edu, Louisiana State University – EoGe
Join, Jean-Lambert, +262262938697 jean-lambert.join@univ-reunion.fr, Universite de la Reunion – Hw
Jokipii, Jack R., (520) 621-4256 jokipii@lpl.arizona.edu, University of Arizona – Xy
Jolley, David, +44 (0)1224 273450 d.jolley@abdn.ac.uk, University of Aberdeen – PlsGv
Jolliff, Bradley L., (314) 935-5622 blj@levee.wustl.edu, Washington University in St. Louis – GxzGi
Jomeiri, Rahim, +98 (411) 339 2704 University of Tabriz – Yg
Jonas, John J., (514) 398-4755 McGill University – Nx
Jones, Adrian P., +44 020 7679 32415 adrian.jones@ucl.ac.uk, University College London – GiCgGg
Jones, Alan, (607) 777-2518 Binghamton University – Ys
Jones, Alice, (859) 622-1424 Alice.Jones@eku.edu, Eastern Kentucky University – Zyn
Jones, Bobby L., (225) 388-8328 Louisiana State University – Go
Jones, Brian, (780) 492-3074 brian.jones@ualberta.ca, University of Alberta – Ps
Jones, Charles E., (412) 624-6347 cejones@pitt.edu, University of Pittsburgh – GgsPg
Jones, Craig H., (303) 492-6994 craig.jones@colorado.edu, University of Colorado – YsGtYg
Jones, Dan, dsjones@umn.edu, University of Minnesota, Twin Cities – Pg
Jones, David S., (413) 542-2714 djones@amherst.edu, Amherst College – GsClGr
Jones, Douglas S., (352) 273-1902 dsjones@flmnh.ufl.edu, University of Florida – PeiPg

Jones, Douglas S., 3523921721 x485 dsjones@flmnh.ufl.edu,
University of Florida – Pg
Jones, Eric M., (505) 667-6386 honais@lanl.gov,
Los Alamos National Laboratory – As
Jones, F. Walter, (780) 492-0667 wjones@phys.ualberta.ca,
University of Alberta – Ym
Jones, Francis H., (604) 822-2138 fjones@eoas.ubc.ca,
University of British Columbia – YeZe
Jones, Glen, (505) 835-5627 glen@gis.nmt.edu,
New Mexico Institute of Mining & Technology – Zn
Jones, Glenda, +44 1782 734309 g.m.jones@keele.ac.uk,
Keele University – NgYg
Jones, Jon W., (403) 220-5024 University of Calgary – Gp
Jones, Julia A., (541) 737-1224 jonesj@geo.oregonstate.edu,
Oregon State University – So
Jones, Larry Allan, (505) 667-0142 ljones@lanl.gov,
Los Alamos National Laboratory – Ng
Jones, Lawrence S., (970) 248-1708 lajones@coloradomesa.edu,
Colorado Mesa University – GsmGr
Jones, Merren, Merren.A.Jones@manchester.ac.uk,
University of Manchester – Gst
Jones, Michael Q., (011) 717-6628 michael.jones@wits.ac.za,
University of the Witwatersrand – YhGtYg
Jones, Minnie O., (312) 996-3154 mojones@uic.edu,
University of Illinois at Chicago – Zn
Jones, Norris W., (920) 424-4460 jonesnw@uwosh.edu,
University of Wisconsin, Oshkosh – Gi
Jones, Phillip D., +44 (0)1603 59 2090 p.jones@uea.ac.uk,
University of East Anglia – AsPeHg
Jones, Rhian, 505-277-4204 rjones@unm.edu,
University of New Mexico – Gz
Jones, Robert L., (217) 333-9490 University of Illinois, Urbana-Champaign – Sc
Jones, Stephen C., (205) 247-3601 sjones@gsa.state.al.us,
Geological Survey of Alabama – HgGeCg
Jones, Stephen M., +44 (0)121 41 46155 s.jones.4@bham.ac.uk,
University of Birmingham – Zg
Jones, Stuart, +44 (0) 191 33 42319 stuart.jones@durham.ac.uk,
Durham University – Gs
Jones, T, jonesTP@cf.ac.uk,
Cardiff University – Ge
Jones, Tim L., (505) 646-3405 New Mexico State University, Las Cruces – Sp
Jones, III, John P., jpjones@email.arizona.edu,
University of Arizona – Zn
Jordan, Andy, +44 (0)1603 59 2552 a.jordan@uea.ac.uk,
University of East Anglia – Ge
Jordan, Bradley C., (570) 577-3024 jordan@bucknell.edu,
Bucknell University – Gg
Jordan, Brennan T., (605) 677-6143 Brennan.Jordan@usd.edu,
University of South Dakota – GiAm
Jordan, Guntram, 089/2180 4353 guntram.jodan@lrz.uni-muenchen.de,
Ludwig-Maximilians-Universitaet Muenchen – Gz
Jordan, Jim L., (409) 880-8211 jim.jordan@lamar.edu,
Lamar University – XcgXm
Jordan, Karen J., 251-460-6381 kjordan@southalabama.edu,
University of South Alabama – Zyr
Jordan, Mary S., 831-656-7571 jordan@nps.edu,
Naval Postgraduate School – Am
Jordan, Robert A., (757) 727-5783 robert.jordan@hamptonu.edu,
Hampton University – Ob
Jordan, Robert R., (302) 831-6415 rrjordan@udel.edu,
University of Delaware – Gr
Jordan, Teresa E., (607) 255-3596 tej1@cornell.edu,
Cornell University – GrtZg
Jordan, Thomas H., (213) 821-1237 tjordan@usc.edu,
University of Southern California – Ys
Jorge, Maria Luisa, (615) 322-2160 malu.jorge@vanderbilt.edu,
Vanderbilt University – Pg
Jornov, Donna, djornov@mail.nysed.gov,
New York State Geological Survey – Zn
Joshi, Manoj, +44 (0)1603 59 3647 m.joshi@uea.ac.uk,
University of East Anglia – Zi
Joughin, Ian, 206-221-3177 ian@apl.washington.edu,
University of Washington – GlZrAt
Journel, Andre G., (650) 723-1594 journel@pangea.stanford.edu,
Stanford University – Gq

Joyce, James, (787) 265-3845 james.joyce@upr.edu,
University of Puerto Rico – NgGcp
Juanes, Ruben, 617-253-7191 juanes@mit.edu,
Massachusetts Institute of Technology – Zn
Judge, Shelley, (330) 263-2297 sjudge@wooster.edu,
College of Wooster – GtsGc
Judkins, Heather L., Judkins.Heather@spcollege.edu,
Saint Petersburg College, Clearwater – Og
Jugo, Pedro J., 7056751151 x2106 pjugo@laurentian.ca,
Laurentian University, Sudbury – CpGi
Jugulam, Mithila, (785) 532-2755 mithila@ksu.edu,
Kansas State University – So
Juhl, Andrew, andyjuhl@ldeo.columbia.edu,
Columbia University – Ob
Jull, A. J. Timothy, (520) 621-6816 jull@u.arizona.edu,
University of Arizona – CclCg
Jung, Simon, +44 (0) 131 650 4837 Simon.Jung@ed.ac.uk,
Edinburgh University – Og
Juniper, Kim, 250-721-6120 kjuniper@uvic.ca,
University of Victoria – Ob
Junium, Christopher, 315-443-8969 ckjunium@syr.edu,
Syracuse University – CsPeCo
Junkin, William D., (410) 554-5543 william.junkin@maryland.gov,
Maryland Department of Natural Resources – GvrGg
Juntunen, Thomas, (612) 626-0505 junt0015@umn.edu,
University of Minnesota, Twin Cities – Zir
Juračiæ, Mladen, +38514606099 mjuracic@geol.pmf.hr,
University of Zagreb – GueCm
Juranek, Lauren W., 541-737-2368 ljuranek@coas.oregonstate.edu,
Oregon State University – Cm
Jurdy, Donna M., (847) 491-7163 donna@earth.northwestern.edu,
Northwestern University – YgXyg
Jurena, Dwight, djurena@alamo.edu,
Alamo Colleges, San Antonio College – GgZg
Jurena, Dwight J., 210-486-0472 djurena@alamo.edu,
San Antonio Community College – XgYvGz
Jurewicz, Amy, (480) 773-1796 amy.jurewicz@asu.edu,
Arizona State University – Xm
Jurmanovich, Barb, 989-686-9445 Delta College – Zn
Juster, Thomas C., (813) 974-9691 University of South Florida, Tampa – Hw
Juszczyk, Carmen, (303) 492-2330 CarmenJ@Colorado.EDU,
University of Colorado – Zn
Jutla, Rajinder S., (417) 836-5298 rajinderjutla@missouristate.edu,
Missouri State University – Zn

K

Kaandorp, Ron, ron.kaandorp@falw.vu.nl,
Vrije Universiteit Amsterdam – Gg
Kabengi, Nadine, 404-413-5207 kabengi@gsu.edu,
Georgia State University – ScClGe
Kaczmarek, Stephen E., (269) 387-5479 stephen.kaczmarek@wmich.edu,
Western Michigan University – GdCgGz
Kaden, Scott, 573-368-2194 nrkades@mail.dnr.state.mo.us,
Missouri Dept of Natural Resources – Hw
Kadkhodai Ilkhchi, Ali, +98 (411) 339 2724 University of Tabriz – NpEo
Kafka, Alan L., (617) 552-3650 kafka@bc.edu,
Boston College – Yg
Kafka, Alan L., 617-552-8300 kafka@bc.edu,
Boston College – Ys
Kah, Linda, (865) 974-2366 lckah@utk.edu,
University of Tennessee, Knoxville – Gs
Kahle, Anne B., (818) 354-7265 anne@aster.jpl.nasa.gov,
Jet Propulsion Laboratory – Zr
Kahle, Beth, 021-650-2900 beth.kahle@uct.ac.za,
University of Cape Town – YgEoYg
Kahle, Chris, Chris.Kahle@dnr.iowa.gov,
Iowa Dept of Natural Resources – Zi
Kaip, Galen M., (915) 747-6817 gkaip@utep.edu,
University of Texas, El Paso – Yx
Kairies-Beatty, Candace L., (507) 474-5789 ckairiesbeatty@winona.edu,
Winona State University – GeClRw
Kaiser, Daniel E., 612-624-3482 dekaiser@umn.edu,
University of Minnesota, Twin Cities – Sc
Kaiser, Jan, +44 (0)1603 59 3393 j.kaiser@uea.ac.uk,
University of East Anglia – Cs
Kaiser, Jason, 802-626-6685 jason.kaiser@northernvermont.edu,

Northern Vermont University-Lyndon – AsZn
Kaiser, Jason, (435) 865-8275 jasonkaiser@suu.edu,
　Southern Utah University – GzvCg
Kaiser-Bischoff, Ines, 089/2180 4314 kaiser-bischoff@lmu.de,
　Ludwig-Maximilians-Universitaet Muenchen – Gz
Kaka, Ismail, +96638603879 skaka@kfupm.edu.sa,
　King Fahd University of Petroleum and Minerals – Ys
Kakembo, Vincent, 27 41 504 4516 Vincent.Kakembo@nmmu.ac.za,
　Nelson Mandela Metropolitan University – Sp
Kalakay, Thomas J., 406-657-1101 kalakayt@rocky.edu,
　Rocky Mountain College – GcpGi
Kalangutkar, Niyati, (866) 960-9198 Goa University – GugGe
Kaldor, Michael, (305) 237-3025 michael.kaldor@mdc.edu,
　Miami-Dade College (Wolfson Campus) – Gg
Kalender, Leyla, 00904242370000-5984 leylakalender@firat.edu.tr,
　Firat University – CgcCs
Kalia, Hemendra N., (702) 295-5767 hemendra_kalia@notes.ymp.gov,
　Los Alamos National Laboratory – Nm
Kallemeyn, Gregory, (310) 825-3202 University of California, Los Angeles – Cg
Kalvoda, Jiri, +420 549 49 4756 dino@sci.muni.cz,
　Masaryk University – PsePm
Kamber, Balz, +61 7 3138 2261 balz.kamber@qut.edu.au,
　Queensland University of Technology – GpCgGz
Kamber, Balz S., + 353 1 8962957 kamberbs@tcd.ie,
　Trinity College – CgGgp
Kambesis, Patricia, (270) 745-5984 pat.kambesis@wku.edu,
　Western Kentucky University – ZigHw
Kambewa, Chamunorwa , 078 321 6710 kambewac@tut.ac.za,
　Tshwane University of Technology – CgeZe
Kamenov, George D., (352) 846-3599 kamenov@ufl.edu,
　University of Florida – CgaEm
Kamhi, Samuel R., (718) 460-1476 Long Island University, Brooklyn Campus – Gz
Kamilli, Robert J., (520) 670-5576 bkamilli@usgs.gov,
　University of Arizona – EgmCg
Kaminski, Michael A., +966138602344 kaminski@kfupm.edu.sa,
　King Fahd University of Petroleum and Minerals – GuPm
Kammer, Thomas W., (304) 293-5603 tkammer@wvu.edu,
　West Virginia University – Pi
Kamola, Diane, (785) 864-2724 dlkamola@ku.edu,
　University of Kansas – Gs
Kamona, Frederick A., afkamona@unam.na,
　University of Namibia – EgGzCe
Kampf, Anthony R., (213) 763-3328 akampf@nhm.org,
　Los Angeles County Museum of Natural History – Gz
Kamykowski, Daniel, (919) 859-3428 dkamyko@ncsu.edu,
　North Carolina State University – ObpOc
Kana, Todd M., (410) 221-8481 kana@hpl.umces.edu,
　University of Maryland – Ob
Kanamori, Hiroo, (626) 395-6914 hiroo@gps.caltech.edu,
　California Institute of Technology – Ysg
Kandel, Hari, 906-635-2618 hkandel@lssu.edu,
　Lake Superior State University – HgZi
Kandiah, Ramanitharan, (937) 376-6260 rkandiah@centralstate.edu,
　Central State University – HqSoZi
Kane, Mustapha, 386.754. 4452 mustapha.kane@fgc.edu,
　Florida Gateway College – Zg
Kanemasu, Edward T., (706) 542-2151 University of Georgia – Sp
Kanfoush, Sharon L., (315) 792-3134 skanfoush@utica.edu,
　Utica College – GsOuGn
Kang, James, 956 665 3526 jihoon.kang@utrgv.edu,
　University of Texas, Rio Grande Valley – SpcZu
Kang, Peter, pkkang@umn.edu,
　University of Minnesota, Twin Cities – Hg
Kang, Song-Lak, (806) 834-1139 song-lak.kang@ttu.edu,
　Texas Tech University – AsHg
Kanik, Mustafa, 00904242370000-5964 Firat University – NrSo
Kanungo, Sudeep, 801-585-7852 skanungo@egi.utah.edu,
　University of Utah – Pms
Kao, Jim C., (505) 667-9226 Los Alamos National Laboratory – As
Kaplan, Alexey, alexeyk@ldeo.columbia.edu,
　Columbia University – Op
Kaplan, Isaac R., (310) 825-5076 University of California, Los Angeles – Cg
Kaplan, Isaac R., (310) 825-5706 irkaplan@ucla.edu,
　University of California, Los Angeles – Csg

Kaplan, Michael, mkaplan@ldeo.columbia.edu,
　Columbia University – Gl
Kaplan, Samantha W., (715) 346-4149 skaplan@uwsp.edu,
　University of Wisconsin, Stevens Point – GnPlGd
Kaplinski, Matthew A., (928) 523-9145 Northern Arizona University – Yg
Kapp, Jessica, 520-626-5701 jkapp@email.arizona.edu,
　University of Arizona – Gg
Kapp, Paul A., (520) 626-8763 pkapp@email.arizona.edu,
　University of Arizona – GtcGm
Kar, Aditya, (912) 825-6844 Fort Valley State University – Ct
Kar, Bandana, 601-266-5786 bandana.kar@usm.edu,
　University of Southern Mississippi – Zi
Kara, Hatice, 00904242370000-5965 haticekara@firat.edu.tr,
　Firat University – Cg
Karabinos, Paul, (413) 597-2079 paul.m.karabinos@williams.edu,
　Williams College – Gc
Karamperidou, Christina, 808-956-2565 ckaramp@hawaii.edu,
　University of Hawai'i, Manoa – AsZoc
Karanfil, Tanju, (864) 656-1005 tkaranf@clemson.edu,
　Clemson University – NgZnn
Karato, Shun-ichiro, (203) 432-3147 shun-ichiro.karato@yale.edu,
　Yale University – Gy
Karginoglu, Yusuf , 00904242370000-5989 Firat University – Eg
Karhu, Juha A., 358-294150834 Juha.Karhu@helsinki.fi,
　University of Helsinki – Csg
Karig, Daniel E., dek9@cornell.edu,
　Cornell University – YrGl
Karimi, Bobak, 570-408-4698 bobak.karimi@wilkes.edu,
　Wilkes University – GctYg
Karkanis, Pano G., (403) 317-0156 Res. (403) 330-9445 Cel. karkanis@telusplanet.net,
　University of Lethbridge – So
Karl, David M., (808) 956-8964 dkarl@soest.hawaii.edu,
　University of Hawai'i, Manoa – Ob
Karl, Tami S., (850) 644-5861 karl@gly.fsu.edu,
　Florida State University – Zn
Karlen, Douglas L., (515) 294-3336 Doug.Karlen@ars.usda.gov,
　Iowa State University of Science & Technology – So
Karlsson, Haraldur R., 806-834-7978 hal.karlsson@ttu.edu,
　Texas Tech University – Cl
Karlstrom, Karl E., (505) 277-4346 kek1@unm.edu,
　University of New Mexico – Gt
Karlstrom, Leif, (541) 346-4323 leif@uoregon.edu,
　University of Oregon – Yx
Karmosky, Christopher, Christopher.Karmosky@oneonta.edu,
　SUNY, Oneonta – AmsZy
Karner, Daniel B., (707) 664-2854 karner@sonoma.edu,
　Sonoma State University – Cc
Karner, James M., 801-585-5976 j.karner@utah.edu,
　University of Utah – Xm
Karpilo, Jr, Ronald J., (970) 225-3500 ron_karpilo@partner.nps.gov,
　Colorado State University – Zy
Karplus, Marianne, (915) 747-5413 mkarplus@utep.edu,
　University of Texas, El Paso – Ysg
Karson, Jeffrey, 315-443-7976 jakarson@syr.edu,
　Syracuse University – GtcGv
Karuntillake, Suniti, 225-578-2337 sunitiw@lsu.edu,
　Louisiana State University – Xg
Karwoski, Todd, (301) 405-0084 karwoski@geol.umd.edu,
　University of Maryland – Gt
Kasar, Sheila , 814-393-2718 skazar@clarion.edu,
　Clarion University – ZreGe
Kaspar, Thomas A., (515) 294-8873 Tom.Kaspar@ars.usda.gov,
　Iowa State University of Science & Technology – So
Kaspari, Susan, kaspari@geology.cwu.edu,
　Central Washington University – PcAtGl
Kasse, Kees, +31 20 59 87381 c.kasse@vu.nl,
　Vrije Universiteit Amsterdam – GmsSd
Kassem, Hachem, +44 (0)23 8059 6205 hachem.kassem@soton.ac.uk,
　University of Southampton – OnGsZi
Kaste, James, 757-221-2591 jmkaste@wm.edu,
　College of William & Mary – ClcSc
Kaster, Mark A., 570-408-5046 mark.kaster@wilkes.edu,
　Wilkes University – Am
Kasting, James F., (814) 865-3207 jfk4@psu.edu,
　Pennsylvania State University, University Park – As
Kastner, Miriam, (858) 534-2065 mkastner@ucsd.edu,

University of California, San Diego – CmlCm
Kaszuba, John P., 307-766-3392 jkaszub1@uwyo.edu,
University of Wyoming – CgGze
Kath, Randal L., (678) 839-4063 rkath@westga.edu,
University of West Georgia – GcNgHg
Kato, Terence T., 530-898-5262 tkato@csuchico.edu,
California State University, Chico – GtcGp
Katumwehe, Andrew, (940) 397-4031 andrew.katumwehe@msutexas.edu,
Midwestern State University – YuuYe
Katuna, Michael P., katunam@cofc.edu,
College of Charleston – Gus
Katz, Cindi, 212-817-8728 ckatz@gc.cuny.edu,
Graduate School of the City University of New York – Gg
Katz, Gabrielle, (828) 262-3000 katzgl@appstate.edu,
Appalachian State University – Hg
Katz, Miriam E., (518) 276-8521 katzm@rpi.edu,
Rensselaer Polytechnic Institute – PmeGu
Katz, Richard, +44-1865-282122 richard.katz@earth.ox.ac.uk,
University of Oxford – Cg
Katzenstein, Kurt W., (605) 394-2461 kurt.katzenstein@sdsmt.edu,
South Dakota School of Mines & Technology – NxtZr
Katzman, Danny, (505) 667-0599 katzman@lanl.gov,
Los Alamos National Laboratory – HyGem
Kauffman, Chad, 724 938-5760 kauffman@cup.edu,
California University of Pennsylvania – As
Kauffman, Erle G., (812) 855-5154 Indiana University, Bloomington – Pg
Kaufman, Alan J., (301) 405-0395 kaufman@geol.umd.edu,
University of Maryland – Csg
Kaufman, Darrell S., (928) 523-7192 darrell.kaufman@nau.edu,
Northern Arizona University – Zc
Kaufmann, Ronald S., (619) 260-5904 kaufmann@sandiego.edu,
University of San Diego – Ob
Kaunda, Rennie, 303-273-3772 rkaunda@mines.edu,
Colorado School of Mines – Nm
Kaushal, Sujay, (301) 405-0454 skaushal@umd.edu,
University of Maryland – ZuGeHs
Kavage-Adams, Rebecca H., (410) 554-5553 rebecca.adams@maryland.gov,
Maryland Department of Natural Resources – GmZiGe
Kavanagh, Janine, +44-151-794-5150 Janine.Kavanagh@liverpool.ac.uk,
University of Liverpool – Gv
Kavanaugh, Jeffrey, 780-492-1740 jeff.kavanaugh@ualberta.ca,
University of Alberta – Ng
Kavanaugh, Maria, mkavanau@coas.oregonstate.edu,
Oregon State University – Cb
Kavner, Abby, (310) 206-3675 akavner@ucla.edu,
University of California, Los Angeles – GyYg
Kavousi Ghahfarohki, Payam, 304-293-5603 pkavousi@mail.wvu.edu,
West Virginia University – Gog
Kawase, Mitsuhiro, (206) 543-0766 kawase@ocean.washington.edu,
University of Washington – Op
Kay, Richard F., (919) 684-2143 richard.kay@duke.edu,
Duke University – Zn
Kay, Robert W., (607) 255-3461 rwk6@cornell.edu,
Cornell University – Gi
Kay, Suzanne M., (607) 255-4701 smk16@cornell.edu,
Cornell University – GiCuGt
Kay, Suzanne M., (607) 227-5143 smk16@cornell.edu,
Cornell University – GiCuGz
Kaye, John M., (662) 325-3915 Mississippi State University – Gg
Kaygili, Sibel, 00904242370000-5962 skaygili@firat.edu.tr,
Firat University – GgPgs
Kays, Marvin A., (541) 346-4578 makays@oregon.uoregon.edu,
University of Oregon – GpcGt
Kazimoto, Emmanuel O., +255222410013 ekazimoto@udsm.ac.tz,
University of Dar es Salaam – EgGpCg
Kazmer, Miklos, +36-1-372-2500 ext. 8627 mkazmer@gmail.com,
Eotvos Lorand University – PggRh
Keables, Michael J., (303) 871-2653 michael.keables@du.edu,
University of Denver – AstAm
Keach, Bill, (801) 537-3301 Utah Geological Survey – Yg
Keach, William, 801-585-1717 bkeach@egi.utah.edu,
University of Utah – GoYes
Keach II, R. William, 801-857-7728 bkeach@byu.edu,
Brigham Young University – YeEoGg
Keala, Lori, (909) 621-8675 lkeala@pomona.edu,
Pomona College – Zn
Kean, Jr., William F., 414-229-5231 wkean@uwm.edu,

University of Wisconsin, Milwaukee – Ym
Keane, Christopher M., (703) 379-2480 keane@americangeosciences.org,
American Geosciences Institute – GgmZf
Kearney, Micheal S., (301) 405-4057 kearneym@umd.edu,
University of Maryland – Gu
Kearns, Lance E., 540-568-6130 kearnsle@jmu.edu,
James Madison University – Gz
Keating, Elizabeth, (505) 665-6714 ekeating@lanl.gov,
Los Alamos National Laboratory – Gg
Keating, Kristina M., 973-353-1263 kmkeat@newark.rutgers.edu,
Rutgers, The State University of New Jersey, Newark – Yug
Keating, Martha E., (781) 891-2980 mkeating@bentley.edu,
Bentley University – Zn
Keaton, Jeffrey R., (801) 581-8218 University of Utah – Zn
Keattch, Sharen , 509-359-7358 skeattch@ewu.edu,
Eastern Washington University – Gg
Keatts, Merida, 330-672-2897 mkeatts@kent.edu,
Kent State University – Zn
Keays, Reid R., (705) 675-1151 Laurentian University, Sudbury – Em
Kebbede, Girma, 413-538-2004 gkebbede@mtholyoke.edu,
Mount Holyoke College – Zy
Keeler, Jason, 989-774-3179 keele1j@cmich.edu,
Central Michigan University – Am
Keeling, David J., (270) 745-4555 david.keeling@wku.edu,
Western Kentucky University – Zg
Keeling, Ralph F., (858) 534-7582 rkeeling@ucsd.edu,
University of California, San Diego – As
Keen, Kerry L., (715) 425-3729 kerry.l.keen@uwrf.edu,
University of Wisconsin, River Falls – HwGse
Keenan, Sarah W., (605) 394-2461 Sarah.Keenan@sdsmt.edu,
South Dakota School of Mines & Technology – PvClb
Keene, Deborah A., 205-348-3334 dakeene@ua.edu,
University of Alabama – Gga
Keene, William, (804) 924-0586 wck@virginia.edu,
University of Virginia – As
Keeney, Dennis R., (515) 294-8066 drkeeney@iastate.edu,
Iowa State University of Science & Technology – So
Keesee, Robert G., (518) 442-4566 rgk@atmos.albany.edu,
SUNY, Albany – As
Kehew, Alan E., (269) 387-5495 alan.kehew@wmich.edu,
Western Michigan University – GmlHw
Kehoe, Kelsey, (307) 766-2286 Ext. 233 kelsey.kehoe@wyo.gov,
Wyoming State Geological Survey – EcGo
Kehoe-Forutan, Sandra J., (570) 389-4106 skehoe@bloomu.edu,
Bloomsburg University – Znu
Keigwin, Lloyd D., (508) 289-2784 lkeigwin@whoi.edu,
Woods Hole Oceanographic Institution – CsGul
Keil, Chris, (630) 752-7271 chris.keil@wheaton.edu,
Wheaton College – Rm
Keil, Richard G., (206) 616-1947 rickkeil@ocean.washington.edu,
University of Washington – Oc
Keir, Derek, +44 (023) 8059 6614 D.Keir@soton.ac.uk,
University of Southampton – Gtv
Keith, Jeffrey D., 801-422-2189 jeff_keith@byu.edu,
Brigham Young University – Eg
Kelemen, Peter, peterk@ldeo.columbia.edu,
Columbia University – Gi
Kelemen, Peter B., (845) 365-8728 peterk@ldeo.columbia.edu,
Columbia University – GiCg
Kelkar, Sharad, (505) 667-4639 kelkar@vega.lanl.gov,
Los Alamos National Laboratory – Np
Kelleher, Cole, 612-626-0505 University of Minnesota, Twin Cities – Zi
Keller, C. Kent, (509) 335-3040 ckkeller@wsu.edu,
Washington State University – Hw
Keller, Dianne M., (315) 228-7893 dkeller@colgate.edu,
Colgate University – Gz
Keller, Edward A., (805) 893-4207 keller@geol.ucsb.edu,
University of California, Santa Barbara – Gm
Keller, G. Randy, (405) 325-3031 grkeller@ou.edu,
University of Oklahoma – Yg
Keller, G. Randy, (405) 255-0608 grkeller@ou.edu,
University of Oklahoma – YgGtEo
Keller , George R., keller@utep.edu,
University of Texas, El Paso – Yg
Keller, Gerta, (609) 258-4117 gkeller@princeton.edu,
Princeton University – Pm
Keller, Jean, (845) 938-4185 United States Military Academy – Zn

Keller, John E., 702-651-5887 john.keller@csn.edu,
  College of Southern Nevada - West Charleston Campus – GeHw
Keller, Klaus, (814) 865-6718 kkeller@geosc.psu.edu,
  Pennsylvania State University, University Park – Og
Keller, Linda M., (608) 265-2209 lmkeller@wisc.edu,
  University of Wisconsin, Madison – As
Keller, Randall A., 541-737-7648 kellerr@geo.oregonstate.edu,
  Oregon State University – Gu
Keller, Troy, 706-507-8099 keller_troy@columbusstate.edu,
  Columbus State University – RwZc
Keller, Jr., C. F., (505) 667-0920 cfk@lanl.gov,
  Los Alamos National Laboratory – As
Kelley, Alice R., (207) 581-2056 akelley@maine.edu,
  University of Maine – GalGm
Kelley, Christopher, (808) 956-7437 ckelley@hawaii.edu,
  University of Hawai'i, Manoa – Ob
Kelley, Dan, (902) 494-1694 dan.kelley@dal.ca,
  Dalhousie University – Op
Kelley, Deborah S., (206) 685-9556 dskelley@uw.edu,
  University of Washington – OuGvp
Kelley, Joseph T., (207) 581-2162 jtkelley@maine.edu,
  University of Maine – Gua
Kelley, Neil P., neil.p.kelley@vanderbilt.edu,
  Vanderbilt University – GgOgPg
Kelley, Patricia H., (910) 448-0526 kelleyp@uncw.edu,
  University of North Carolina, Wilmington – Pgi
Kelley, Scott, 775-784-6705 scottkelley@unr.edu,
  University of Nevada, Reno – Zi
Kelley, Shari, (575) 661-6171 sakelley@nmbg.nmt.edu,
  New Mexico Institute of Mining & Technology – Yg
Kelley, Shari A., 505-412-9269 sakelley@nmbg.nmt.edu,
  New Mexico Institute of Mining and Technology – Gt
Kelley, Simon, +44 (0)131 650 2537 simon.kelley@ed.ac.uk,
  Edinburgh University – CcsZg
Kelling, Keith A., (608) 263-2795 kkelling@wisc.edu,
  University of Wisconsin, Madison – So
Kellman, Lisa M., (902) 867-5086 lkellman@stfx.ca,
  Saint Francis Xavier University – CsSo
Kellner, Patricia, pakellner@socal.rr.com,
  Los Angeles Harbor College – Zg
Kellogg, James N., (803) 777-4501 kellogg@geol.sc.edu,
  University of South Carolina – Yg
Kellogg, Louise H., (530) 752-3690 kellogg@ucdavis.edu,
  University of California, Davis – Yg
Kelly, Bryce, bryce.kelly@unsw.edu.au,
  University of New South Wales – HwAcGe
Kelly, D. Clay, (608) 262-1698 ckelly@geology.wisc.edu,
  University of Wisconsin-Madison – Pm
Kelly, Jacque L., (912) 478-8677 JKelly@GeorgiaSouthern.edu,
  Georgia Southern University – Cl
Kelly, John, 203-479-4822 jkelly@newhaven.edu,
  University of New Haven – ObZn
Kelly, Kathryn A., (206) 543-9810 kkelly@apl.washington.edu,
  University of Washington – Op
Kelly, Kimberly, (240) 567-5227 Kimberly.Kelly@montgomerycollege.edu,
  Montgomery College – Zn
Kelly, Maria, (425) 640-1918 mkelly@edcc.edu,
  Edmonds Community College – Zg
Kelly, Meredith, 603-646-9647 Meredith.Kelly@dartmouth.edu,
  Dartmouth College – Gl
Kelly, Sherrie, 315-229-5851 skelly@stlawu.edu,
  St. Lawrence University – Zn
Kelly, Walton R., (217) 898-0657 wkelly@illinois.edu,
  University of Illinois – Hw
Kelly, William C., (734) 764-1435 billkell@umich.edu,
  University of Michigan – Eg
Kelsch, Jesse, (432) 837-8657 jkelsch@sulross.edu,
  Sul Ross State University – Gct
Kelsey, Eric, ekelsey2@plymouth.edu,
  Plymouth State University – Ats
Kelsey, Harvey M., (707) 826-3991 hmk1@humboldt.edu,
  Humboldt State University – Gmt
Kelso, Paul R., (906) 635-2158 pkelso@lssu.edu,
  Lake Superior State University – YmGct
Kelty, Thomas, (562) 985-4589 t.kelty@csulb.edu,
  California State University, Long Beach – Gc
Kemeny, John M., (520) 621-4448 kemeny@email.arizona.edu,
  University of Arizona – Nr
Kemmerly, Phillip R., (931) 221-7471 kemmerlyp@apsu.edu,
  Austin Peay State University – GmeHg
Kemp, Alan, +44 (0)23 80592788 aesk@noc.soton.ac.uk,
  University of Southampton – Pe
Kemp, Andrew, (617) 627-0869 andrew.kemp@tufts.edu,
  Tufts University – OnPmOg
Kemp, Paul, (808) 956-6220 paulkemp@hawaii.edu,
  University of Hawai'i, Manoa – Ob
Kempton, Pamela D., (785) 532-6724 pkempton@ksu.edu,
  Kansas State University – GiCtc
Kendrick, David C., 315-781-3929 kendrick@hws.edu,
  Hobart & William Smith Colleges – PgGg
Kendrick, Katherine J., (951) 276-4418 kendrick@gps.caltech.edu,
  University of California, Riverside – Gm
Kenduiywo, Benson K., bkenduiywo@jkuat.ac.ke,
  Jomo Kenyatta University of Agriculture & Technology – ZriYd
Kenig, Fabien, (312) 996-3020 fkenig@uic.edu,
  University of Illinois at Chicago – CobOo
Kenna, Timothy, tkenna@ldeo.columbia.edu,
  Columbia University – Cg
Kennedy, Ann C., (509) 335-1554 akennedy@wsu.edu,
  Washington State University – Sb
Kennedy, Ben, +64 3 3667001 Ext 7775 ben.kennedy@canterbury.ac.nz,
  University of Canterbury – Gv
Kennedy, Christina B., tina.kennedy@nau.edu,
  Northern Arizona University – Zn
Kennedy, Gareth, +44 01326 371876 G.A.Kennedy@exeter.ac.uk,
  Exeter University – Nm
Kennedy, Linda, 570-662-4609 lkennedy@mansfield.edu,
  Mansfield University – Zy
Kennedy, Lisa M., (540) 231-1422 kennedy1@vt.edu,
  Virginia Polytechnic Institute & State University – Pe
Kennedy, Lori, (604) 822-1811 lkennedy@eos.ubc.ca,
  University of British Columbia – Gc
Kennedy, Robert, (541) 737-6332 rkennedy@coas.oregonstate.edu,
  Oregon State University – Zri
Kennedy, Taylor, tkennedy1@occ.cccd.edu,
  Orange Coast College – Ge
Kennel, Charles F., (310) 825-4018 University of California, Los Angeles – Zn
Kennel, Charles F., (858) 822-6424 ckennel@ucsd.edu,
  University of California, San Diego – As
Kenney, Robert D., (401) 874-6664 rkenney@uri.edu,
  University of Rhode Island – Ob
Kenny, Ray, (970) 247-7462 kenny_r@fortlewis.edu,
  Fort Lewis College – Gm
Kent, Adam J., (541) 737-1205 adam.kent@geo.oregonstate.edu,
  Oregon State University – GiCga
Kent, Dennis V., 848-445-7049 dvk@rutgers.edu,
  Rutgers, The State University of New Jersey – YmZcGr
Kent, Douglas, 405-377-0166 Oklahoma State University – Hw
Kent, Graham, (775) 784-4977 gkent@seismo.unr.edu,
  University of Nevada, Reno – Ys
Kenyon, Patricia M., (212) 650-6472 City College (CUNY) – Yg
Kepic, Anton W., +61 8 9266-7503 A.Kepic@curtin.edu.au,
  Curtin University – YexYg
Keppie, J. Duncan, 56224303 duncan@servidor.unam.mx,
  Universidad Nacional Autonoma de Mexico – Gt
Kerans, Charles, (512) 471-1368 charles.kerans@beg.utexas.edu,
  University of Texas at Austin, Jackson School of Geosciences – Gd
Kerans, Charles, (512) 471-3519 ckerans@jsg.utexas.edu,
  University of Texas, Austin – GsrGo
Kerestedjian, Thomas N., +359 2 979 2244 thomas@geology.bas.bg,
  Bulgarian Academy of Sciences – GzeEg
Kern, Anikó, anikoc@nimbus.elte.hu,
  Eotvos Lorand University – ZrAms
Kerp, Hans, +49-251-83-23966 kerp@uni-muenster.de,
  Universitaet Muenster – PblGg
Kerr, A C., kerra@cf.ac.uk,
  Cardiff University – Gxi
Kerr, Andrew, +44(0)29 208 74578 KerrA@cardiff.ac.uk,
  University of Wales – Og
Kerr, Dennis R., 918 631 3020 dennis-kerr@utulsa.edu,
  The University of Tulsa – GsrGt
Kerrick, Derrill M., (814) 865-7574 kerrick@geosc.psu.edu,
  Pennsylvania State University, University Park – Gp

Kerwin, Charles M., 603.358.2405 ckerwin@keene.edu,
  Keene State College – Gg
Kerwin, Michael W., 303-871-3998 mkerwin@du.edu,
  University of Denver – GgeZc
Kesel, Richard H., (225) 578-5880 gakesel@lsu.edu,
  Louisiana State University – Hg
Keskinen, Mary J., (907) 474-7769 mjkeskinen@alaska.edu,
  University of Alaska, Fairbanks – Gxz
Kesler, Stephen E., (734) 763-5057 skesler@umich.edu,
  University of Michigan – Eg
Kessler, Fritz, (301) 687-4266 fkessler@frostburg.edu,
  Frostburg State University – Zr
Kessler, William S., (206) 526-6221 kessler@pmel.noaa.gov,
  University of Washington – Op
Ketcham, Richard A., (512) 471-6942 ketcham@jsg.utexas.edu,
  University of Texas, Austin – GgCc
Kettler, Richard M., (402) 472-0882 rkettler1@unl.edu,
  University of Nebraska, Lincoln – Cl
Kevan, Peter, pkevan@uoguelph.ca,
  University of Guelph – Zn
Key, Doug, 7607441150 ext.2515 dkey@palomar.edu,
  Palomar College – Zy
Key, Jeffrey R., (608) 263-2605 jkey@ssec.wisc.edu,
  University of Wisconsin, Madison – ZrAs
Key, Kerry, (858) 822-2975 kkey@ucsd.edu,
  University of California, San Diego – Yx
Key, Jr., Marcus M., (717) 245-1448 key@dickinson.edu,
  Dickinson College – PiGsa
Keyantash, John, (310) 243-2363 jkeyantash@csudh.edu,
  California State University, Dominguez Hills – HgAsZe
Keyser, Daniel, (518) 442-4559 dkeyser@albany.edu,
  SUNY, Albany – As
Khairoutdinov, Marat, (631) 632-6339 marat.khairoutdinov@stonybrook.edu,
  SUNY, Stony Brook – As
Khalequzzaman, Md., (570) 484-2075 mkhalequ@lockhaven.edu,
  Lock Haven University – HwOnZi
Khalil, Mohamed, 406-496-4716 mkhalil@mtech.edu,
  Montana Tech of the University of Montana – YeuGe
Khalil Ebeid, Khalil I., 01227431425 kebeid@yahoo.com,
  Alexandria University – GgEmm
Khan, Belayet H., 217-581-6246 bhkhan@eiu.edu,
  Eastern Illinois University – ZyAm
Khan, Latif A., 217-244-2383 info@isgs.illinois.edu,
  Illinois State Geological Survey – Nx
Khan, Mohammad Wahdat Y., (982) 719-7331 mwykhan@rediffmail.com,
  Pt. Ravishankar Shukla University – GdEgCl
Khan, Shuhab D., (713) 743-5404 sdkhan@uh.edu,
  University of Houston – ZrGtYg
Khanbilvardi, Reza M., (215) 650-8009 Graduate School of the City University of New York – Hy
Khandaker, Nazrul I., (718) 262-2079 nkhandaker@york.cuny.edu,
  York College (CUNY) – GdeZe
Khawaja, Ikram U., john@cis.ysu.edu,
  Youngstown State University – Ec
Khurana, Krishan, (310) 825-8240 University of California, Los Angeles – Zn
Kiage, Lawrence M., (404) 413-5777 geolkk@langate.gsu.edu,
  Georgia State University – ZyPlZr
Kidd, David, +44 020 8417 62541 david.kidd@kingston.ac.uk,
  Kingston University – Ziy
Kidd, Karen, 905-525-9140 (Ext. 23550) karenkidd@mcmaster.ca,
  McMaster University – Gu
Kidder, Steven, 212-650-8431 skidder@ccny.cuny.edu,
  Graduate School of the City University of New York – As
Kidder, T.R., trkidder@wustl.edu,
  Washington University in St. Louis – Zn
Kidwell, Jen, (650) 723-0891 jparis@stanford.edu,
  Stanford University – Zn
Kidwell, Susan M., (773) 702-3008 skidwell@uchicago.edu,
  University of Chicago – GsPe
Kiefer, Boris, 575 646 1932 bkiefer@nmsu.edu,
  New Mexico State University, Las Cruces – GyYgZm
Kieffer, Bruno, bkieffer@eos.ubc.ca,
  University of British Columbia – Cs
Kieffer, Susan W., (217) 244-6206 skieffer@illinois.edu,
  University of Illinois, Urbana-Champaign – Gv
Kienast, Markus, (902) 494-8338 markus.kienast@dal.ca,
  Dalhousie University – Cs
Kienast, Stephanie, (902) 494-2203 stephanie.kienast@dal.ca,
  Dalhousie University – Gu
Kiene, Ronald P., (251) 861-7526 rkiene@jaguar1.usouthal.edu,
  University of South Alabama – Oc
Kientop, Greg A., 217-265-6581 gkientop@illinois.edu,
  Illinois State Geological Survey – Ge
Kienzle, Stefan, stefan.kienzle@uleth.ca,
  University of Lethbridge – Zi
Kiesel, Diann, (608) 355-5223 diann.kiesel@uwc.edu,
  University of Wisconsin Colleges – GgZyGg
Kieu, Chanh Q., (812) 856-5704 ckieu@indiana.edu,
  Indiana University, Bloomington – As
Kifer, Lauri A., (585) 395-2636 lmulley@brockport.edu,
  SUNY, The College at Brockport – Zn
Kift, Richard, +44 0161 306-8770 Richard.Kift@manchester.ac.uk,
  University of Manchester – As
Kijko, Andrzej, +27 12 420 3613 andrzej.kijko@up.ac.za,
  University of Pretoria – YsgNe
Kilburn, Chris, +44 020 7679 37194 c.kilburn@ucl.ac.uk,
  University College London – Yx
Kilcoyne, John R., (303) 556-4258 Metropolitan State College of Denver – Zy
Kile, Susan, 217-581-2626 skkile@eiu.edu,
  Eastern Illinois University – Zn
Kilibarda, Zoran, (219) 980-6753 zkilibar@iun.edu,
  Indiana University Northwest – GsmGd
Kilic, Ayse Didem, 00904242370000-5969 adkilic@firat.edu.tr,
  Firat University – GxpGy
Kilinc, Attila I., (513) 556-5967 attila.kilinc@uc.edu,
  University of Cincinnati – CpGvCu
Killorn, Randy J., (515) 294-1923 rkillorn@iastate.edu,
  Iowa State University of Science & Technology – So
Kilroy, Kathryn, 701-858-3114 kathryn.kilroy@minotstateu.edu,
  Minot State University – Hg
Kim, Daehyun, (206) 221-8935 daehyun@atmos.washington.edu,
  University of Washington – As
Kim, Eunhye, 303-273-3428 ekim1@mines.edu,
  Colorado School of Mines – NrmNt
Kim, Hyemi, (404) 894-1738 hyemi.kim@eas.gatech.edu,
  Georgia Institute of Technology – As
Kim, Jonathan, (802) 522-5401 jon.kim@vermont.gov,
  Agency of Natural Resources, Dept of Env Conservation – GetCg
Kim, Keonho, 432-685-4739 kkim@midland.edu,
  Midland College – GgAmPi
Kim, Kwangmin, kimkm@email.arizona.edu,
  University of Arizona – NmrNx
Kim, Saewung, (949)824-4531 saewungk@uci.edu,
  University of California, Irvine – As
Kim, Sang-Tae, (905) 525-9140 (Ext. 26494) sangtae@mcmaster.ca,
  McMaster University – Cg
Kim, Stacy, 831-771-4400 skim@mlml.calstate.edu,
  Moss Landing Marine Laboratories – Zn
Kim, Wan, (202) 478-8871 wkim@carnegiescience.edu,
  Carnegie Institution for Science – Zn
Kim, Won-Young, wykim@ldeo.columbia.edu,
  Columbia University – Ys
Kim, Wonsuck, 512-471-4203 delta@jsg.utexas.edu,
  University of Texas, Austin – GsmGr
Kim, Yong Hoon, (610) 436-2203 ykim@wcupa.edu,
  West Chester University – OgZr
Kim, Yongsang (Barry), (671) 735-2693 kimys@triton.uog.edu,
  University of Guam – Hw
Kimball, Matthew, matt@belle.baruch.sc.edu,
  University of South Carolina – Ob
Kimber, Clarissa T., (409) 845-7141 Texas A&M University – Zy
Kimberly, Shaw, 706-507-8344 Columbus State University – Zn
Kimbro, David, 7815817370 x310 d.kimbro@neu.edu,
  Northeastern University – Zn
Kimbrough, David L., (619) 594-1385 dkimbrough@mail.sdsu.edu,
  San Diego State University – Cc
Kimerling, A. Jon, kimerlia@geo.oregonstate.edu,
  Oregon State University – Zir
Kincaid, Christopher, (401) 874-6571 kincaid@gso.uri.edu,
  University of Rhode Island – Ou
Kineke, Gail C., (617) 552-3655 gail.kineke.1@bc.edu,
  Boston College – On

King, Andrew, alking@uic.edu,
University of Illinois at Chicago – Og
King, Carey, 512-471-5468 cking@jsg.utexas.edu,
University of Texas, Austin – Eog
King, John, (401) 874-6594 jking@gso.uri.edu,
University of Rhode Island – Ou
King, Jonathan K., (801) 537-3354 jonking@utah.gov,
Utah Geological Survey – Gg
King, Kenneth M., (519) 824-4120 (Ext. 52453) kenmking@rogers.com,
University of Guelph – As
King, Martin, +44 1784 414038 M.King@rhul.ac.uk,
Royal Holloway University of London – As
King, Norman R., (812) 464-1794 nking@usi.edu,
University of Southern Indiana – Gr
King, Peter, +44 20 759 47362 peter.king@imperial.ac.uk,
Imperial College – Np
King, Robert W., (617) 253-7064 rwk@chandler.mit.edu,
Massachusetts Institute of Technology – Yd
King, Scott D., 540-231-6521 Virginia Polytechnic Institute & State University – Yg
King, Jr., David T., (334) 844-4882 kingdat@auburn.edu,
Auburn University – GrZn
Kinkead, Scott, (505) 665-1760 Los Alamos National Laboratory – Zn
Kinnaird, Judith A., 27117176583 judith.kinnaird@wits.ac.za,
University of the Witwatersrand – EmGi
Kinner, David A., (828) 227-3821 dkinner@wcu.edu,
Western Carolina University – HgZe
Kinsland, Gary L., (337) 288-6421 glkinsland@louisiana.edu,
University of Louisiana at Lafayette – YgeGt
Kinter, Jim, (703) 993-5700 jkinter@gmu.edu,
George Mason University – As
Kinvig, Helen, +440151 795 4657 H.Kinvig@liverpool.ac.uk,
University of Liverpool – Gs
Kipper, Jay P., (512) 475-9505 jay.kipper@beg.utexas.edu,
University of Texas at Austin, Jackson School of Geosciences – Zn
Kirby, Carl S., (570) 577-1385 kirby@bucknell.edu,
Bucknell University – Cl
Kirby, Eric, (541) 737-5169 kirbye@geo.oregonstate.edu,
Oregon State University – GtmGc
Kirby, Matthew E., (657) 278-2158 mkirby@fullerton.edu,
California State University, Fullerton – GnPeGg
Kirchgasser, William T., (315) 267-2296 kirchgwt@potsdam.edu,
SUNY Potsdam – Ps
Kirchner, James W., (510) 643-8559 kirchner@seismo.berkeley.edu,
University of California, Berkeley – Ge
Kirimi, Fridah K., fkirimi@jkuat.ac.ke,
Jomo Kenyatta University of Agriculture & Technology – ZiiZr
Kirk, Scott, 614 265 6742 scott.kirk@dnr.state.oh.us,
Ohio Dept of Natural Resources – Hw
Kirkby, Kent C., (612) 624-1392 kirkby@tc.umn.edu,
University of Minnesota, Twin Cities – Gg
Kirkham, Mary Beth, (785) 532-0422 mbk@ksu.edu,
Kansas State University – Sp
Kirkland, Brenda J., 662-268-1032 Ext 228 blk39@msstate.edu,
Mississippi State University – GdEoPo
Kirkland, James I., (801) 537-3307 nrugs.jkirklan@state.ut.us,
Utah Geological Survey – Pg
Kirkpatrick, Cody, (812) 855-3481 codykirk@indiana.edu,
Indiana University, Bloomington – As
Kirkpatrick, James, (514) 398-7442 james.kirkpatrick@mcgill.ca,
McGill University – Gct
Kirlin, R. Lynn, (250) 721-8681 lkirlin@comcast.net,
University of British Columbia – As
Kiro Feldman, Yael, ykiro@ldeo.columbia.edu,
Columbia University – Cg
Kirschenfeld, Taylor, 850-474-2746 University of West Florida – Ob
Kirschvink, Joseph L., (626) 395-6136 kirschvink@caltech.edu,
California Institute of Technology – Pg
Kirshbaum, Daniel, 514-398-3347 daniel.kirshbaum@mcgill.ca,
McGill University – Asm
Kirste, Dirk, 778-782-5365 dkirste@sfu.ca,
Simon Fraser University – Cg
Kirstein, Linda, +44 (0) 131 650 4838 Linda.Kirstein@ed.ac.uk,
Edinburgh University – Gt
Kirtland Turner, Sandra, (951) 827-3191 sandra.kirtlandturner@ucr.edu,
University of California, Riverside – Cs
Kirwan, Matthew L., 804-684-7054 kirwan@vims.edu,
College of William & Mary – Gm
Kish, Stephen A., (850) 644-2064 kish@gly.fsu.edu,
Florida State University – Eg
Kishcha, Pavel, 972-3-6407411 pavelk@post.tau.ac.il,
Tel Aviv University – As
Kisila, Ben O., (540) 654-1107 bkisila@umw.edu,
University of Mary Washington – HsSfZu
Kiss, Timea, +3662544156 kisstimi@gmail.com,
Univesity of Szeged – GmZy
Kissel, David E., (706) 542-0900 University of Georgia – Sc
Kissin, Stephen A., (807) 343-8220 stephen.kissin@lakeheadu.ca,
Lakehead University – Em
Kisvarsanyi, Geza K., (573) 341-4616 Missouri University of Science and Technology – Em
Kitajima, Hiroko, 979.458.2717 kitaji@tamu.edu,
Texas A&M University – YxSpGc
Kitchen, Newell R., 573-882-1138 kitchenn@missouri.edu,
University of Missouri, Columbia – So
Kite, J. Steven, (304) 293-5603 steve.kite@mail.wvu.edu,
West Virginia University – GmHsGa
Kivelson, Margaret G., (310) 825-3435 mkivelson@igpp.ucla.edu,
University of California, Los Angeles – Xy
Kiver, Eugene P., (509) 359-7959 eugene.kiver@ewu.edu,
Eastern Washington University – Gl
Klaassen, Gary P., (416)736-2100 #77727 gklaass@yorku.ca,
York University – As
Klancher, Jacki, 307-855-2205 jklanche@cwc.edu,
Central Wyoming College – Zi
Klapper, Gilbert, (847) 732-1859 g-klapper@northwestern.edu,
Northwestern University – Pm
Klasik, John A., 909-869-3453 jaklasik@csupomona.edu,
California State Polytechnic University, Pomona – Gu
Klaus, Adam, (979) 845-3055 aklaus@odpemail.tamu.edu,
Texas A&M University – Gu
Klaus, James S., 305-284-3426 j.klaus@miami.edu,
University of Miami – Gg
Klee, Thomas M., (813) 253-7259 tklee@hccfl.edu,
Hillsborough Community College – Gg
Kleffner, Mark A., (419) 995-8208 kleffner.1@osu.edu,
Ohio State University – PgsGr
Klein, Andrew G., (979) 845-7179 klein@geog.tamu.edu,
Texas A&M University – ZriGl
Klein, Emily M., (919) 684-5965 ek4@duke.edu,
Duke University – Gi
Klein, Frieder, fklein@whoi.edu,
Woods Hole Oceanographic Institution – Cm
Klein, Yehuda, 718-951-5000 ext 5131 yklein@brooklyn.cuny.edu,
Graduate School of the City University of New York – Hs
Kleinspehn, Karen L., 612-624-0537 klein004@umn.edu,
University of Minnesota, Twin Cities – Gt
Kleiss, Harold J., (919) 515-2643 North Carolina State University – Sd
Klemas, Victor, (302) 831-8256 klemas@udel.edu,
University of Delaware – Zr
Klemetti, Erik, 740-587-5788 klemettie@denison.edu,
Denison University – GvxGg
Klemow, Kenneth M., (570) 408-4758 kenneth.klemow@wilkes.edu,
Wilkes University – Pg
Klemperer, Simon L., (650) 723-8214 sklemp@stanford.edu,
Stanford University – GtYes
Klepeis, Keith A., (802) 656-0247 kklepeis@uvm.edu,
University of Vermont – Gct
Kliche, Donna V., 605-394-1957 Donna.Kliche@sdsmt.edu,
South Dakota School of Mines & Technology – AssAs
Klimczak, Christian, klimczak@uga.edu,
University of Georgia – GcmGt
Klinck, John M., (757) 683-6005 Klinck@ccpo.odu.edu,
Old Dominion University – Op
Klinger, Barry, (703) 993-9227 bklinger@gmu.edu,
George Mason University – Op
Klinkhammer, Gary P., (541) 737-5209 gklinkhammer@coas.oregonstate.edu,
Oregon State University – CaHsRw
Kloosterziel, Rudolf C., 808-956-7668 rudolf@soest.hawaii.edu,
University of Hawai'i, Manoa – Op
Klosterman, Sue, (937) 229-2661 Sue.Klosterman@notes.udayton.edu,
University of Dayton – Gg
Kluitenberg, Gerard J., (785) 532-7215 gjk@ksu.edu,

Kansas State University – Sp
Kluver, Daria, 989-774-1249 kluve1db@cmich.edu,
 Central Michigan University – Asm
Klymak, Jody, 250-721-6120 jklymak@uvic.ca,
 University of Victoria – Op
Knaack, Charles, (509) 335-6742 knaack@wsu.edu,
 Washington State University – Ca
Knaeble, Alan, (612) 626-2495 knaeb001@umn.edu,
 University of Minnesota – Gl
Knap, Anthony, 979-862-2323 ext 111 tknap@tamu.edu,
 Texas A&M University – Oc
Knapp, Camelia, 405-744-6358 camelia.knapp@okstate.edu,
 Oklahoma State University – GtYgs
Knapp, Elizabeth P., (540) 458-8867 knappe@wlu.edu,
 Washington & Lee University – Cl
Knapp, James, 405-744-6358 james.knapp@okstate.edu,
 Oklahoma State University – GcYgs
Knapp, Richard B., (925) 423-3328 knapp4@llnl.gov,
 Lawrence Livermore National Laboratory – Ng
Knapp, Roy M., 405-325-6829 knapp@ou.edu,
 University of Oklahoma – Np
Knapp, Warren, (607) 255-3034 wwk2@cornell.edu,
 Cornell University – As
Knappenberger, Thorsten J., (334) 844-4100 tjk0013@auburn.edu,
 Auburn University – Sp
Knauss, John A., (401) 874-6141 jknauss@gso.uri.edu,
 University of Rhode Island – Op
Knauth, L. Paul, knauth@asu.edu,
 Arizona State University – Cs
Kneeshaw, Tara A., 616-331-8996 kneeshta@gvsu.edu,
 Grand Valley State University – ClGe
Knell, Michael J., (203) 392-5836 knellm1@southernct.edu,
 Southern Connecticut State University – PvGsPe
Knepp, Rex A., 217-244-2422 knepp@isgs.uiuc.edu,
 Illinois State Geological Survey – Go
Knesel, Kurt, 210-999-7606 kknesel@trinity.edu,
 Trinity University – Gvi
Knight, Allen W., (530) 752-0453 aknight557@aol.com,
 University of California, Davis – Hs
Knight, David, (518) 442-4204 knight@atmos.albany.edu,
 SUNY, Albany – As
Knight, Paul, (814) 863-4229 knight@mail.meteo.psu.edu,
 Pennsylvania State University, University Park – Am
Knight, Rosemary J., (650) 723-4746 Stanford University – Hy
Knight, Tiffany, tknight@biology2.wustl.edu,
 Washington University in St. Louis – Pe
Knippler, Katherine A., (410) 554-5543   katherine.knippler@maryland.gov,
 Maryland Department of Natural Resources – Ge
Knittle, Elise, (831) 459-4949 eknittle@pmc.ucsc.edu,
 University of California, Santa Cruz – Gy
Knizek, Martin, +420 549 49 6298 kniza@sci.muni.cz,
 Masaryk University – NgmNr
Knoll, Andrew H., (617) 495-9306 aknoll@harvard.edu,
 Harvard University – Pb
Knoll, Martin A., (931) 598-1713 mknoll@sewanee.edu,
 Sewanee: University of the South – Hw
Knopf, Daniel A., (631) 632-3092 Daniel.Knopf@stonybrook.edu,
 SUNY, Stony Brook – As
Knopoff, Leon, (310) 825-1885 knopoff@physics.ucla.edu,
 University of California, Los Angeles – Ys
Knott, Jeffrey R., (657) 278-5547 jknott@fullerton.edu,
 California State University, Fullerton – Gmg
Knowles, Charles E., (919) 515-3700 ernie_knowles@ncsu.edu,
 North Carolina State University – Op
Knox, Larry W., (931) 372-3523 lknox@tntech.edu,
 Tennessee Tech University – Pmg
Knudsen, Andrew, (920) 832-6731 knudsena@lawrence.edu,
 Lawrence University – CgGez
Knudsen, Guy, gknudsen@uidaho.edu,
 University of Idaho – Sb
Knudsen, Tyler R., (435) 865-9036 tylerknudsen@utah.gov,
 Utah Geological Survey – Ng
Knudstrup, Renee, rknudstrup@salemstate.edu,
 Salem State University – Ca
Knuepfer, Peter L. K., (607) 777-2389 Binghamton University – Gt
Knutson, Heather, 626.395.4268 hknutson@caltech.edu,
 California Institute of Technology – Xg

Kobara, Shinichi, 979 845 4089 shinichi@tamu.edu,
 Texas A&M University – Zi
Kobs-Nawotniak, Shannon E., (208) 282-3365 kobsshan@isu.edu,
 Idaho State University – Gv
Koc Tasgin, Calibe, 00904242370000-5976 calibekoc@firat.edu.tr,
 Firat University – GsdGg
Koch, Joe, (803) 896-4167 kochj@dnr.sc.gov,
 South Carolina Dept of Natural Resources – Zg
Koch, Magaly, mkoch@bu.edu,
 Boston University – HyZri
Koch, Paul L., (831) 459-5861 pkoch@pmc.ucsc.edu,
 University of California, Santa Cruz – Pv
Kochel, R. Craig, (570) 577-3032 kochel@bucknell.edu,
 Bucknell University – Gm
Kocurek, Gary A., (512) 471-5855 garyk@mail.utexas.edu,
 University of Texas, Austin – Gs
Kocurko, John, 940-397-4250 Midwestern State University – Gs
Kodama, Kenneth P., (610) 758-3663 kpk0@lehigh.edu,
 Lehigh University – Ym
Kodosky, Larry, 248-232-4538 lgkodosk@oaklandcc.edu,
 Oakland Community College – GgvEg
Koehl, Mimi A. R., (510) 642-8103 University of California, Berkeley – Pg
Koehler, Rich, (775) 682-8763 rkoehler@unr.edu,
 University of Nevada – GmtNg
Koehler, Thomas , (719) 333-8712 thomas.koehler@usafa.edu,
 United States Air Force Academy – As
Koehn, Daniel, Daniel.Koehn@glasgow.ac.uk,
 University of Glasgow – Gt
Koeman-Shields, Elizabeth, 325-486-6767 elizabeth.koeman-shields@angelo.edu,
 Angelo State University – CgXg
Koenig, Brian, bkoenig@collegeofthedesert.edu,
 College of the Desert – Gge
Kogan, Mikhail, (845) 365-8882 kogan@ldeo.columbia.edu,
 Columbia University – Yd
Kohler, Jeffery L., 814-865-9834 JK9@psu.edu,
 Pennsylvania State University, University Park – Nm
Kohler, Nicholas, (541) 346-4160 nicholas@uoregon.edu,
 University of Oregon – Zi
Kohlstedt, David L., (612) 626-1544 dlkohl@umn.edu,
 University of Minnesota, Twin Cities – YxGzi
Kohlstedt, Sally G., 612-624-9368 sgk@umn.edu,
 University of Minnesota, Twin Cities – Zn
Kohn, Matthew, mattkohn@boisestate.edu,
 Boise State University – CsGpPv
Kokum, Mehmet, 00904242370000-5963 mkokum@firat.edu.tr,
 Firat University – GtcGg
Kolawole, Lanre L., +2348032277598 llkolawole@lautech.edu.ng,
 Ladoke Akintola University of Technology – GeeGe
Kolesar, Peter T., peter.t.kolesar@gmail.com,
 Utah State University – Cg
Kolka, Randall K., (218) 326-7100 University of Minnesota, Twin Cities – So
Kolkas, Mosbah, mossbah.kolkas@csi.cuny.edu,
 College of Staten Island/CUNY – Gg
Kollias, Pavlos, 514-398-1500 pavlos.kollias@stonybrook.edu,
 McGill University – ZrAs
Komabayashi, Tetsuya, +44 (0) 131 650 8518 tetsuya.komabayashi@ed.ac.uk,
 Edinburgh University – Gz
Komarneni, Sridhar, (814) 865-1542 komarneni@psu.edu,
 Pennsylvania State University, University Park – Sc
Kominz, Michelle A., (269) 387-5340 michelle.kominz@wmich.edu,
 Western Michigan University – OuYgGu
Komoto, Cary, 952-358-9007 cary.komoto@normandale.edu,
 Normandale Community College – ZnnZn
Konecky, Bronwen L., (415) 420-8083 bronwen.konecky@colorado.edu,
 Washington University in St. Louis – PcGn
Konhauser, Kurt, 780-492-2571 kurtk@ualberta.ca,
 University of Alberta – Pg
Konig, Ron, 479-575-3355 University of Arkansas, Fayetteville – Gc
Konishi, Takuya, 513-556-9726 konishta@ucmail.uc.edu,
 Cincinnati Museum Center – Pv
Kontak, Daniel J., 7056751151 x2352 dkontak@laurentian.ca,
 Laurentian University, Sudbury – Eg
Konter, Jasper G., 808-956-8705 jkonter@hawaii.edu,
 University of Hawai'i, Manoa – CcGiCt
Kontuly, Thomas M., (801) 581-8218 thomas.kontuly@geog.utah.edu,

Koons, Peter O., (207)-581-2158 peter.koons@maine.edu,
  University of Maine – Gt
Koornneef, Janne, +31 20 59 81824 j.m.koornneef@vu.nl,
  Vrije Universiteit Amsterdam – GvCa
Kooyman, Gerald L., (858) 534-2091 gkooyman@ucsd.edu,
  University of California, San Diego – Ob
Kopaska-Merkel, David C., (205) 247-3695 davidkm@gsa.state.al.us,
  Geological Survey of Alabama – GsPiZe
Kopf, Christopher F., (570) 662-4615 ckopf@mansfield.edu,
  Mansfield University – Gcp
Kopp, Robert E., (732) 200-2705 robert.kopp@rutgers.edu,
  Rutgers, The State University of New Jersey – ZcOoZg
Koppers, Anthony, 541-737-5425 akoppers@coas.oregonstate.edu,
  Oregon State University – CgGv
Kopylova, Maya G., (604) 822-0865 mkopylova@eos.ubc.ca,
  University of British Columbia – Gi
Koralek, Susan, koralek@sou.edu,
  Southern Oregon University – Zn
Korenaga, Jun, (203) 432-7381 jun.korenaga@yale.edu,
  Yale University – YgsCg
Koretsky, Carla, (269) 387-4372 carla.koretsky@wmich.edu,
  Western Michigan University – CglHs
Kori, E., +27 15 962 8565 edmore.kori@univen.ac.za,
  University of Venda – GmZgAt
Kornreich Wolf, Susan, (309) 794-7369 susanwolf@augustana.edu,
  Augustana College – Ze
Korose, Christopher P., 217-333-7256 korose@illinois.edu,
  Illinois State Geological Survey – Ec
Korotev, Randy L., (314) 935-5637 korotev@wustl.edu,
  Washington University in St. Louis – XgCtXm
Korre, Anna, +44 20 759 47372 a.korre@imperial.ac.uk,
  Imperial College – Ng
Korths, Patrick, 518-564-2028 pkort001@plattsburgh.edu,
  SUNY, Plattsburgh – RwGg
Korty, Robert, 979-847-9090 korty@tamu.edu,
  Texas A&M University – As
Kortz, Karen M., kkortz@ccri.edu,
  Community College of Rhode Island – ZeGg
Korycansky, Don, (831) 459-5843 University of California, Santa Cruz – Xc
Koskinen, William C., (612) 625-4276 koskinen@soils.umn.edu,
  University of Minnesota, Twin Cities – Sc
Kosloski, Mary, (319) 335-0893 mary-kosloski@uiowa.edu,
  University of Iowa – Pig
Kosro, Michael, (541) 737-3079 kosro@coas.oregonstate.edu,
  Oregon State University – OpZrOn
Kossin, James, 608-265-5356 kossin@ssec.wisc.edu,
  University of Wisconsin, Madison – As
Kostelnick, John, 309-438-7679 jckoste@ilstu.edu,
  Illinois State University – Zi
Koster Van Groos, August F., (312) 996-8678 kvg@uic.edu,
  University of Illinois at Chicago – Cp
Kostov, Svilen, 229-931-2321 skostov@gsw.edu,
  Georgia Southwestern State University – Zn
Kota, Jozsef, (520) 621-4396 kota@lpl.arizona.edu,
  University of Arizona – As
Kotamarthi, Rao, 630-252-7164 vrkotamarthi@anl.gov,
  Argonne National Laboratory – As
Koteas, G. Christopher, 802 485 3321 gkoteas@norwich.edu,
  Norwich University – GiYh
Kotha, Mahender, +91-832-6519329 mkotha@unigoa.ac.in,
  Goa University – GsoZi
Kotulova, Julia, jkotulova@egi.utah.edu,
  University of Utah – CoGo
Koutavas, Athanasios, 718-982-2972 tom.koutavas@csi.cuny.edu,
  College of Staten Island/CUNY – PeOg
Koutitonsky, Vladimir G., (418) 724-1986 (Ext. 1763) vgk@uqar.qc.ca,
  Universite du Quebec a Rimouski – Op
Koutnik, Michelle, 206-221-5041 mkoutnik@uw.edu,
  University of Washington – Gll
Kováčová, Marianna, marianna.kovacova@uniba.sk,
  Comenius University in Bratislava – PlcPe
Kovac, Michal, +421260296555 kovacm@uniba.sk,
  Comenius University in Bratislava – Gsg
Kovach, Richard G., (702) 295-6180 rkovach@lanl.gov,
  Los Alamos National Laboratory – Nm
Kovach, Robert L., (650) 723-4827 Stanford University – Ys
Kovaèiæ, Marijan, +38514605963 mkovacic@geol.pmf.hr,
  University of Zagreb – GsdGx
Kowaleski, Douglas, 857-234-9339 dkowal@geo.umass.edu,
  University of Massachusetts, Amherst – Gm
Kowalewski, Douglas E., 508-929-8646 douglas.kowalewski@worcester.edu,
  Worcester State University – GmlZy
Kowalke, Thorsten, 089/2180 6733 t.kowalke@lrz.uni-muenchen.de,
  Ludwig-Maximilians-Universitaet Muenchen – Pg
Kowallis, Bart J., (801) 422-2467 bkowallis@byu.edu,
  Brigham Young University – GgCcGz
Kozak, Amanda L., akozak@ashland.edu,
  Ashland University – Rn
Koziol, Andrea M., (937) 229-2954 Andrea.Koziol@notes.udayton.edu,
  University of Dayton – Cp
Kozlowski, Andrew, 518 486-2012 akozlows@mail.nysed.gov,
  New York State Geological Survey – Gl
Kraal, Erin, 484-646-5859 kraal@kutztown.edu,
  Kutztown University of Pennsylvania – XgGm
Krabbenhoft, David, (608) 821-3843 dpkrabbe@usgs.gov,
  University of Wisconsin-Madison – ClGeHw
Kraemer, George P., 914-251-6640 george.kraemer@purchase.edu,
  SUNY, Purchase – ObnCm
Kraft, Kaatje, 480-654-7382 vanderhoeven@mesacc.edu,
  Mesa Community College – Gg
Kramer, J. Curtis, (209) 946-2482 ckramer@pacific.edu,
  University of the Pacific – Gg
Kramer, James R., kramer@mcmaster.ca,
  McMaster University – Cl
Kramer, Kate, (815) 479-7877 kkramer@mchenry.edu,
  McHenry County College – GgZg
Kramer, Marc, 352-294-3165 mgkramer@ufl.edu,
  University of Florida – So
Kramer, Walter V., (361) 698-1385 wkramer@delmar.edu,
  Del Mar College – GgoGx
Krantz, David E., (419) 530-2662 david.krantz@utoledo.edu,
  University of Toledo – Gs
Kranz, Dwight S., (713) 718-5641 dwight.kranz@hccs.edu,
  Houston Community College System – Gg
Krapac, Ivan G., 217-333-6442 krapac@isgs.uiuc.edu,
  Illinois State Geological Survey – Ca
Krastel, Sebastian, skrastel@geophysik.uni-kiel.de,
  Dalhousie University – Yr
Kraus, Mary J., 303-492-7251 mary.kraus@colorado.edu,
  University of Colorado – GsSa
Krause, David W., (303) 370-6379 david.krause@sunysb.edu,
  Stony Brook University – Pv
Krause, Federico F., (403) 220-5845 fkrause@ucalgary.ca,
  University of Calgary – GsdGo
Krause, Lois B., (864) 656-7653 Clemson University – Ze
Kravchinsky, Vadim, (780) 492-5591 vkrav@phys.ualberta.ca,
  University of Alberta – Ym
Krawczynski, Michael J., 314-935-6328 mikekraw@levee.wustl.edu,
  Washington University in St. Louis – Cg
Kreamer, David, 702.895.3553 dave.kreamer@unlv.edu,
  University of Nevada, Reno – HwGn
Krebes, Edward S., (403) 220-5028 University of Calgary – Ys
Kreemer, Corne, (775) 682-8780 kreemer@unr.edu,
  University of Nevada, Reno – Yd
Kreidenweis, Sonia M., sonia@atmos.colostate.edu,
  Colorado State University – As
Kreiger, William (Bill), (717) 815-1379 wkreiger@ycp.edu,
  York College of Pennsylvania – GisZg
Krekeler, Mark, 513-785-3106 krekelmp@miamioh.edu,
  Miami University – Ge
Kremer, Robert J., 573-882-6408 University of Missouri, Columbia – Sb
Kressler, Sharon J., 612-625-5068 kress004@umn.edu,
  University of Minnesota, Twin Cities – Zn
Kretzschmar, Thomas, tkretzsc@cicese.mx,
  Centro de Investigación Científica y de Educación Superior de Ensenada – Ge
Kreutz, Karl J., (207)-581-3011 karl.kreutz@maine.edu,
  University of Maine – Cs
Krevor, Samuel, s.krevor@imperial.ac.uk,
  Imperial College – Eo
Krieble, Kelly, (610) 861-1437 krieblek@moravian.edu,
  Moravian College – ZnnZn

Krieger-Brockett, Barbara B., (206) 543-2216 krieger@cheme.washington.edu,
University of Washington – Ob
Krier, Donathon J., (505) 665-7834 krier@lanl.gov,
Los Alamos National Laboratory – Gv
Kring, David, (281) 486-2119 kring@lpi.usra.edu,
University of Arizona – XcGiXg
Krippner, Janine, 304-384-5327 Concord University – GvRc
Krishnamurthy, R. V., (269) 387-5501 r.v.krishnamurthy@wmich.edu,
Western Michigan University – Cs
Krishtalka, Leonard, (785) 864-4540 krishtalka@ku.edu,
University of Kansas – Pv
Krissek, Lawrence A., (614) 292-1924 krissek.1@osu.edu,
Ohio State University – Gsu
Krockover, Gerald H., (765) 494-5795 hawk1@purdue.edu,
Purdue University – Ze
Kroeger, Glenn C., (210) 999-7607 gkroeger@trinity.edu,
Trinity University – Yg
Krohe, Nicholas J., 605-677-3923 Nick.Krohe@usd.edu,
South Dakota Dept of Environment and Natural Resources – Gg
Krohn, James P., krohnjp@piercecollege.edu,
Los Angeles Pierce College – NgGg
Kronenberg, Andreas, (979) 845-0132 a-kronenberg@geos.tamu.edu,
Texas A&M University – GtyGc
Kronenfeld, Barry J., (217) 581-7014 bjkronenfeld@eiu.edu,
Eastern Illinois University – Ziy
Kroon, Dick, +44 (0) 131 651 7089 D.Kroon@ed.ac.uk,
Edinburgh University – Gg
Krot, Alexander N., 808-956-3900 sasha@higp.hawaii.edu,
University of Hawai'i, Manoa – Xm
Kruckenberg, Seth, 617-552-8300 seth.kruckenberg@bc.edu,
Boston College – Gc
Kruckenberg, Seth C., 617-552-3647 seth.kruckenberg@bc.edu,
Boston College – Gct
Krueger, Steven, 801-581-3903 steve.krueger@utah.edu,
University of Utah – As
Kruge, Michael A., 973-655-7668 krugem@mail.montclair.edu,
Montclair State University – Co
Kruger, Joseph M., (409) 880-8233 joseph.kruger@lamar.edu,
Lamar University – YgGgZi
Kruger, Ned , ekadrmas@nd.gov,
North Dakota Geological Survey – Gg
Krugh, W C., 661-654-3126 wkrugh@csub.edu,
California State University, Bakersfield – GcmCc
Krukowski, Stanley T., 405-325-3031 skrukowski@ou.edu,
University of Oklahoma – En
Krupka, Kenneth M., (509) 376-4412 ken.krupka@pnl.gov,
Pacific Northwest National Laboratory – Cl
Kruse, Jennifer, (507) 933-7333 jkruse@gustavus.edu,
Gustavus Adolphus College – Zn
Kruse, Sarah E., (813) 974-7341 University of South Florida, Tampa – Yg
Krygier, John B., (740) 368-3622 jbkrygie@owu.edu,
Ohio Wesleyan University – Zi
Krzic, Maja, 604-822-0252 maja.krzic@ubc.ca,
University of British Columbia – Sf
Ku, Teh-Lung, (213) 740-5826 rku@usc.edu,
University of Southern California – CgcGn
Ku, Timothy C., (860)685-2265 tcku@wesleyan.edu,
Wesleyan University – Cl
Kuang, Zhiming, 617-495-2354 kuang@eps.harvard.edu,
Harvard University – As
Kubas, Gregory J., (505) 667-5846 Los Alamos National Laboratory – As
Kubesh, Rodney, 320-308-4217 rjkubesh@stcloudstate.edu,
Saint Cloud State University – Am
Kubicek, Leonard, (972) 273-3508 lenkubicek@dcccd.edu,
North Lake College - Dallas County Community College District – Gg
Kubicki, James D., 915-747-5501 University of Texas, El Paso – ClGe
Kucharik, Chris, 608-890-3021 kucharik@wisc.edu,
University of Wisconsin, Madison – As
Kuchovsky, Tomas, +420 549 49 5452 tomas@sci.muni.cz,
Masaryk University – Hgy
Kuchta, Mark, (303) 273-3306 mkuchta@mines.edu,
Colorado School of Mines – Nm
Kudela, Raphael M., (831) 459-3290 kudela@cats.ucsc.edu,
University of California, Santa Cruz – Zr
Kudlac, John J., (412) 392-3423 Point Park University – Ng
Kuehl, Steven A., (804) 684-7118 kuehl@vims.edu,
College of William & Mary – Ou

Kuehn, Stephen C., 304-384-6322 sckuehn@concord.edu,
Concord University – CaGv
Kuehner, Scott, 206-543-8393 kuehner@uw.edu,
University of Washington – Ggz
Kuentz, David C., (513) 529-5992 kuentzdc@MiamiOh.edu,
Miami University – Ca
Kues, Barry S., (505) 277-3626 bkues@unm.edu,
University of New Mexico – Pi
Kuhlman, Robert, rkuhlman@mc3.edu,
Montgomery County Community College – Gg
Kuhnhenn, Gary L., (859) 622-8140 gary.kuhnhenn@eku.edu,
Eastern Kentucky University – GgZe
Kuiper, Klaudia, k.f.kuiper@vu.nl,
Vrije Universiteit Amsterdam – Cc
Kuiper, Yvette D., 303-273-3105 ykuiper@mines.edu,
Colorado School of Mines – Gc
Kujawinski, Elizabeth B., (508) 289-3493 ekujawinski@whoi.edu,
Woods Hole Oceanographic Institution – Oc
Kukoè, Duje, +38514606111 duje.kukoc@geol.pmf.hr,
University of Zagreb – Gg
Kukowski, Nina, 0049(0)3641/948680 nina.kukowski@uni-jena.de,
Friedrich-Schiller-University Jena – Ygx
Kulander, Byron, byron.kulander@wright.edu,
Wright State University – Gc
Kulatilake, Pinnaduwa H. S. W., (520) 621-6064 kulatila@u.arizona.edu,
University of Arizona – Nr
Kulkarni, Shrinivas R., (626) 395-4010 srk@astro.caltech.edu,
California Institute of Technology – Xa
Kulp, Mark A., (504) 280-1170 mkulp@uno.edu,
University of New Orleans – GrmGs
Kulp, Thomas, tkulp@binghamton.edu,
Binghamton University – Pg
Kumar, Ajoy, (717) 871-2432 ajoy.kumar@millersville.edu,
Millersville University – OpZr
Kumar, M. Satish, +442890973479 s.kumar@qub.ac.uk,
Queen's University Belfast – ZnnZn
Kumjian, Matthew R., 814-863-1581 kumjian@psu.edu,
Pennsylvania State University, University Park – Am
Kummerow, Christian D., kummerow@atmos.colostate.edu,
Colorado State University – Yr
Kump, Lee R., (814) 863-1274 lkump@psu.edu,
Pennsylvania State University, University Park – ClPeCm
Kumpan, Tomas, +420 549 49 3314 kumpan@sci.muni.cz,
Masaryk University – Ps
Kumpf, Amber C., (231) 777-0289 amber.kumpf@muskegoncc.edu,
Muskegon Community College – GgYrOu
Kung, Ernest C., 573-882-5909 University of Missouri, Columbia – As
Kung, Hsiang-Te, (901) 678-4538 hkung@memphis.edu,
University of Memphis – Zy
Kung, King-Jau S., (608) 262-6530 kskung@wisc.edu,
University of Wisconsin, Madison – Sp
Kunkle, Thomas D., (505) 667-1259 Los Alamos National Laboratory – Xy
Kuntz, Kara, (785) 628-5804 kkuntz@fhsu.edu,
Fort Hays State University – Zn
Kuntz, Mark R., (847) 697-1000 mkuntz@elgin.edu,
Elgin Community College – Gg
Kunza, Lisa, (605) 394-2449 lisa.kunza@sdsmt.edu,
South Dakota School of Mines & Technology – HsZe
Kunzmann, Thomas, 089/2180 4292 kunzmann@min.uni-muenchen.de,
Ludwig-Maximilians-Universitaet Muenchen – Gz
Kuo, Shiou, (206) 840-4573 skuo@wsu.edu,
Washington State University – Sc
Kuperman, William A., (858) 534-3158 wkuperman@ucsd.edu,
University of California, San Diego – Op
Kurapov, Alexander, 541-737-2865 kurapov@coas.oregonstate.edu,
Oregon State University – Yr
Kursinski, Robert, (520) 626-3338 kursinsk@atmo.arizona.edu,
University of Arizona – As
Kurtanjek, Dražen, +38514605965 dkurtan@inet.hr,
University of Zagreb – GsdGg
Kurttas, Turker, +90-312-2977760 kurttast@gmail.com,
Hacettepe University – HwgZr
Kurtz, Andrew, kurtz@bu.edu,
Boston University – Clg
Kurtz, Vincent E., (417) 836-5801 Missouri State University – Ps
Kurum, Sevcan, 00904242370000-5992 skurum@firat.edu.tr,
Firat University – GivGx

Kurz, Marie J., mk3483@drexel.edu,
Drexel University – CqbHw
Kurz, Mark D., (508) 289-2888 mkurz@whoi.edu,
Woods Hole Oceanographic Institution – Cc
Kushnir, Yochanan, kushnir@ldeo.columbia.edu,
Columbia University – As
Kusnick, Judith E., (916) 278-4692 California State University, Sacramento – Ze
Kusssow, Wayne R., (608) 263-3631 wrkussow@wisc.edu,
University of Wisconsin, Madison – So
Kustka, Adam B., 973-353-5509 kustka@newark.rutgers.edu,
Rutgers, The State University of New Jersey, Newark – OgCmHs
Kusumoto, Shigekazu, 81-76-445-6653 kusu@sci.u-toyama.ac.jp,
University of Toyama – YvdGt
Kusznir, Nick, +44-151-794-5182 N.Kusznir@liverpool.ac.uk,
University of Liverpool – Yd
Kutis, Michael, 765-285-2487 mkutis@bsu.edu,
Ball State University – Gg
Kutzbach, John E., (608)262-0392 jek@facstaff.wisc.edu,
University of Wisconsin, Madison – As
Kuwabara, James, 415-452-7776 kuwabara@usgs.gov,
City College of San Francisco – RwOgCg
Kuzera, Kristopher, 303-871-3378 kristopher.kuzera@du.edu,
University of Denver – ZyHg
Kuzila, Mark S., (402) 472-7537 mkuzila@unl.edu,
Unversity of Nebraska - Lincoln – Sd
Kuzyk, Zou Zou, 204-272-1535 umkuzyk@cc.umanitoba.ca,
University of Manitoba – Cg
Kvamme, Kenneth L., (617) 353-3415 Boston University – Ga
Kwicklis, Edward M., (505) 665-7408 kwicklis@lanl.gov,
Los Alamos National Laboratory – Gg
Kwon, Youngsang, 901-678-2979 ykwon@memphis.edu,
University of Memphis – Zi
Kyle, J. Richard, (512) 471-4351 rkyle@jsg.utexas.edu,
University of Texas, Austin – EmnCe
Kyle, Philip R., (575) 835-5995 kyle@nmt.edu,
New Mexico Institute of Mining and Technology – Giv
Kysar Mattietti, Giuseppina, (703) 993-9269 gkysar@gmu.edu,
George Mason University – GiZeGa
Kyte, Frank T., (310) 825-2015 kyte@igpp.ucla.edu,
University of California, Los Angeles – Ct
Käser, Martin, 089/2180 4138 martin.kaeser@geophysik.uni-muenchen.de,
Ludwig-Maximilians-Universitaet Muenchen – Yg
Käsling, Heiko, +49 89 289 25831 heiko.kaesling@tum.de,
Technical University of Munich – NrmNg

L

L'Ecuyer, Tristan S., 608-262-2828 University of Wisconsin, Madison – As
La Berge, Gene L., (920) 424-4460 laberge@uwosh.edu,
University of Wisconsin, Oshkosh – Eg
La Fave, John I., 406-496-4306 jlafave@mtech.edu,
Montana Tech of The University of Montana – Hw
La Tour, Timothy E., 404413-5767 tlatour@gsu.edu,
Georgia State University – Gp
Laabs, Benjamin J., (701) 231-6197 benjamin.laabs@ndsu.edu,
North Dakota State University – GlPcGm
Laabs, Benjamin J., 585-245-5305 laabs@geneseo.edu,
SUNY, Geneseo – GmlGe
Labandeira, Conrad C., (202) 633-1336 Smithsonian Inst / Natl Museum of Natural Hist– Pi
LaBarbera, Michael C., (773) 702-8092 mlabarbe@uchicago.edu,
University of Chicago – Pi
LaBella, Joel, (860)685-2242 jlabella@wesleyan.edu,
Wesleyan University – Zn
Laboski, Carrie A., (608) 263-2795 laboski@wisc.edu,
University of Wisconsin, Madison – So
Labotka, Theodore C., (865) 974-2366 tlabotka@utk.edu,
University of Tennessee, Knoxville – GpCg
Labuz, Joseph F., (612) 625-9060 jlabuz@umn.edu,
University of Minnesota, Twin Cities – Nr
Lachhab, Ahmed, 570-374-4215 lachhab@susqu.edu,
Susquehanna University – Hw
Lachmar, Thomas E., (435) 797-1247 tom.lachmar@gmail.com,
Utah State University – Hw
Lachniet, Matthew S., 702-895-4388 matthew.lachniet@unlv.edu,
University of Nevada, Las Vegas – PeCs
Lackey, Jade Star, (909) 621-8677 jadestar.lackey@pomona.edu,
Pomona College – GipCs
Lackinger, Markus, 089/2180 Ludwig-Maximilians-Universitaet Muenchen – Gz
Lackmann, Gary M., 919-515-1439 gary@ncsu.edu,
North Carolina State University – As
Lacy, Tor, tlacy@cerritos.edu,
Cerritos College – Gg
LaDochy, Steve, (323) 343-3222 sladoch@calstatela.edu,
California State University, Los Angeles – Zn
LaDue, Nicole D., (815) 753-7935 nladue@niu.edu,
Northern Illinois University – Ze
LaFemina, Peter C., 814-865-7326 pfemina@geosc.psu.edu,
Pennsylvania State University, University Park – Yd
Laffan, Shawn, shawn.laffan@unsw.edu.au,
University of New South Wales – Ziy
LaFleche, Marc R., (418) 654-2670 marc.richer-lafleche@ete.inrs.ca,
Universite du Quebec – Ct
Lafrance, Bruno, 7056751151 x2264 blafrance@laurentian.ca,
Laurentian University, Sudbury – GcEmGt
LaFreniere, Lorraine, (630) 252-7969 lafreniere@anl.gov,
Argonne National Laboratory – HwGoPs
Lageson, David R., (406) 994-6913 lageson@montana.edu,
Montana State University – GctGs
Lagowski, Alison A., (716) 645-4856 aal@buffalo.edu,
SUNY, Buffalo – Zn
Lahiri, Chayan, (719) 587-7357 chayanlahiri@adams.edu,
Adams State University – HwZr
Lai, Chung Chieng A., (505) 665-6635 cal@lanl.gov,
Los Alamos National Laboratory – Am
Laine, Edward P., edlaine@bowdoin.edu,
Bowdoin College – Gu
Laingen, Christopher R., (217) 581-2999 crlaingen@eiu.edu,
Eastern Illinois University – SfZn
Laird, David A., (515) 294-1581 dalaird@iastate.edu,
Iowa State University of Science & Technology – Sc
Laird, Jo, (603) 862-1718 jl@cisunix.unh.edu,
University of New Hampshire – Gp
Laird, Neil, 315-781-3603 laird@hws.edu,
Hobart & William Smith Colleges – As
Laity, Julie E., (818) 677-3532 julie.laity@csun.edu,
California State University, Northridge – Zy
Lakatos, Stephen, (718) 262-2589 York College (CUNY) – Cc
Lake, Iain, +44 (0)1603 59 3744 i.lake@uea.ac.uk,
University of East Anglia – Ge
Lakhan, V. Chris, (519) 253-3000 x2183 lakan@uwindsor.ca,
University of Windsor – ZriOn
Laki, Sam, (937) 376-6272 slaki@centralstate.edu,
Central State University – EgSoHs
Lal, Rattan, 614-292-9069 lal.1@osu.edu,
Ohio State University – Sp
Laliberte, Elizabeth, (401) 874-5512 elalib@mail.uri.edu,
University of Rhode Island – Og
Lam, Anita, 604-822-2736 alam@eos.ubc.ca,
University of British Columbia – Zn
Lam, Phyllis, +44 (0)23 8059 8388 P.Lam@southampton.ac.uk,
University of Southampton – Ob
Lamanna, Matthew C., (412) 624-8780 University of Pittsburgh – Pv
Lamanna, Matthew C., (412) 578-2696 lamannam@carnegiemnh.org,
Carnegie Museum of Natural History – PvgPe
Lamb, Andrew P., 479-575-8685 aplamb@uark.edu,
University of Arkansas, Fayetteville – YumYv
Lamb, Dennis, (814) 883-0174 lno@psu.edu,
Pennsylvania State University, University Park – Am
Lamb, James P., 205 652 3725 jlamb@uwa.edu,
University of West Alabama – PvePg
Lamb, John A., (612) 625-1772 jlamb@soils.umn.edu,
University of Minnesota, Twin Cities – Sc
Lamb, Melissa A., (651) 962-5242 malamb@stthomas.edu,
University of Saint Thomas – GtcGs
Lamb, Michael P., 626.395.3612 mpl@gps.caltch.edu,
California Institute of Technology – Gm
Lamb, Will, (979) 845-3075 lamb@geo.tamu.edu,
Texas A&M University – GpCpGz
Lambart, Sarah, 801-581-7062 sarah.lambart@utah.edu,
University of Utah – Cpg
Lambert, Carolyn D., 970-351-2647 Carolyn.Lambert@unco.edu,
University of Northern Colorado – Hw

Lambert, Dean, dlambert@alamo.edu,
   Alamo Colleges, San Antonio College – Zy
Lambert, Dean P., 210-486-0471 dlambert@alamo.edu,
   San Antonio Community College – Zyi
Lambert, Lance L., (210) 458-4455 Lance.Lambert@utsa.edu,
   University of Texas, San Antonio – Pg
Lambert, W. J., 205-348-4404 jlambert@ua.edu,
   University of Alabama – CsPeCg
Lambert-Smith, James, J.S.Lambert-Smith@kingston.ac.uk,
   Kingston University – Gg
Lamothe, Michel, 514-987-3000 #3361 lamothe.michel@uqam.ca,
   Universite du Quebec a Montreal – Gl
Lampkin, Derrick, 301 405 7797 dlampkin@umd.edu,
   University of Maryland – GlAsZr
Lancaster, Nick, (775) 673-7304 nick@dri.edu,
   University of Nevada, Reno – Gm
Lancaster, Penny, +44 023 9284 2272 penny.lancaster@port.ac.uk,
   University of Portsmouth – Gz
Lancaster, Stephen, 541-737-9258 lancasts@geo.oregonstate.edu,
   Oregon State University – Hs
Land, Lewis, 575-887-5508 lland@nmbg.nmt.edu,
   New Mexico Institute of Mining and Technology – Hy
Land, Lewis A., 505-887-5505 lland@gis.nmt.edu,
   New Mexico Institute of Mining & Technology – Hg
Landa, Keith, (914) 251-6450 keith.landa@purchase.edu,
   SUNY, Purchase – XgbGg
Landenberger, Bill, 61 02 4921 6366 bill.landenberger@newcastle.edu.au,
   University of Newcastle – GiCg
Lander, Mark A., (671) 735-2685 mlander@triton.uog.edu,
   University of Guam – Am
Landing, Ed, (518) 473-8071 elanding@mail.nysed.gov,
   New York State Geological Survey – PscCc
Landman, Neil H., (212) 769-5723 Graduate School of the City University
   of New York – Ps
Landman, Neil H., (212) 769-5712 American Museum of Natural History
   – Pi
Landry, Michael R., (858) 534-4702 mlandry@ucsd.edu,
   University of California, San Diego – Ob
Landry, Peter B., (508) 289-3443 plandry@whoi.edu,
   Woods Hole Oceanographic Institution – Ca
Landschoot, Peter J., (814) 863-1017 pcl11@psu.edu,
   Pennsylvania State University, University Park – So
Lane, Charles L., (541) 552-6114 lane@sou.edu,
   Southern Oregon University – Hg
Lane, Joseph M., (216) 987-5227 Joseph.Lane@tri-c.edu,
   Cuyahoga Community College - Western Campus – Zg
Lane, Mark, 7607441150 xt. 2951 mlane@palomar.edu,
   Palomar College – Zn
Lang, Harold R., (304) 293-5603 harold@lithos.jpl.nasa.gov,
   Jet Propulsion Laboratory – Gr
Lang, Nicholas, 814-824-3646 nlang@mercyhurst.edu,
   Mercyhurst University – GvXgGc
Lang, Susan Q., 803-777-8832 slang@geol.sc.edu,
   University of South Carolina – Oc
Lange, Eric, (765) 285-8272 eslange@bsu.edu,
   Ball State University – Cg
Lange, Rebecca A., (734) 764-7421 becky@umich.edu,
   University of Michigan – Gi
Langel, Richard A., 319-335-4102 richard.langel@dnr.iowa.gov,
   Iowa Dept of Natural Resources – Gg
Langenheim, Jr., Ralph L., rlangenh@illinois.edu,
   University of Illinois, Urbana-Champaign – Gr
Langenhorst, Falko H., 0049(0)3641/948730 falko.langenhorst@uni-jena.de,
   Friedrich-Schiller-University Jena – Gz
Langer, Arthur M., (718) 951-4793 Graduate School of the City University
   of New York – Gz
Langford, Richard P., (915) 747-5968 langford@utep.edu,
   University of Texas, El Paso – Gs
Langhorst, Glenn, 218-879-0719 glang@fdltcc.edu,
   Fond du Lac Tribal and Community College – Gg
Langille, Jackie M., (828) 251-6453 jlangill@unca.edu,
   University of North Carolina, Asheville – GctGe
Langman, Jeffrey, (208) 885-0310 jlangman@uidaho.edu,
   University of Idaho – ClHws
Langmuir, Charles H., (617) 384-9948 langmuir@eps.harvard.edu,
   Harvard University – Cg

Langston, Charles A., (901) 678-4869 clangstn@memphis.edu,
   University of Memphis – Ys
Langston, Jr., Wann, (512) 471-7736 wannl@mail.utexas.edu,
   University of Texas, Austin – Pv
Langston-Unkefer, Pat J., (505) 665-2556 Los Alamos National Laboratory
   – Co
Lanigan, David C., (509) 376-9308 david.lanigan@pnl.gov,
   Pacific Northwest National Laboratory – Hg
Lansey, Kevin E., (520) 621-2512 lansey@engr.arizona.edu,
   University of Arizona – Hs
Lao, Daniel A., (405) 744-6358 daniel.lao_davila@okstate.edu,
   Oklahoma State University – GctGg
Lapen, Thomas, 713-743-6122 tjlapen@uh.edu,
   University of Houston – Gz
LaPointe, Daphne D., (775) 682-8772 dlapoint@unr.edu,
   University of Nevada – Gg
Laporte, Leo F., laporte@ucsc.edu,
   University of California, Santa Cruz – Pg
Laprise, Rene, 514-987-3000 #3302 laprise.rene@uqam.ca,
   Universite du Quebec a Montreal – As
Lapusta, Nadia, (626) 395-2277 lapusta@caltech.edu,
   California Institute of Technology – GcYs
Large, Ross R., 61 3 6226 2819 Ross.Large@utas.edu.au,
   University of Tasmania – Eg
Larkin, Patrick, 361-825-3258 Patrick.Larkin@tamucc.edu,
   Texas A&M University, Corpus Christi – Zn
LaRock, Paul A., (225) 388-6307 Louisiana State University – Ob
Larocque, Marie, 514-987-3000 #1515 larocque.marie@uqam.ca,
   Universite du Quebec a Montreal – Hw
Larsen, Daniel, (901) 678-4358 dlarsen@memphis.edu,
   University of Memphis – ClGsHw
Larsen, Isaac J., 413-545-0538 ilarsen@geo.umass.edu,
   University of Massachusetts, Amherst – Gm
Larsen, Jessica F., 907-474-7992 jflarsen@alaska.edu,
   University of Alaska, Fairbanks – Gv
Larsen, Kristine, (860) 832-2938 larsen@ccsu.edu,
   Central Connecticut State University – Xy
Larson, David J., 510-885-3132 david.larson@csueastbay.edu,
   California State University, East Bay – Zu
Larson, Edwin E., 303-492-6172 University of Colorado – Ym
Larson, Eric J., 715-346-4098 Eric.Larsen@uwsp.edu,
   University of Wisconsin, Stevens Point – Zr
Larson, Erik B., (740) 351-3144 elarson@shawnee.edu,
   Shawnee State University – GmsHg
Larson, Harold P., (520) 621-6943 hplarson@u.arizona.edu,
   University of Arizona – Zn
Larson, Peter B., (509) 335-3095 plarson@wsu.edu,
   Washington State University – Cs
Larson, Roger, (401) 874-6165 rlar@uri.edu,
   University of Rhode Island – Ou
Larter, Stephen, +44 (0) 191 208 5956 alex.leathard@ncl.ac.uk,
   University of Newcastle Upon Tyne – Gg
LaRue, Michelle, larue010@umn.edu,
   University of Minnesota, Twin Cities – Zi
Lasca, Norman P., (414) 229-4602 nplasca@uwm.edu,
   University of Wisconsin, Milwaukee – GmlSd
Lasemi, Zakaria, 217-244-6944 lasemi@isgs.uiuc.edu,
   Illinois State Geological Survey – En
Lash, Gary G., (716) 673-3842 lash@fredonia.edu,
   SUNY, Fredonia – Gr
Lasher-Trapp, Sonia, 217-244-4250 slasher@illinois.edu,
   University of Illinois, Urbana-Champaign – As
Laske, Gabi, (858) 534-8774 glaske@ucsd.edu,
   University of California, San Diego – Ys
Lasker, Howard R., 716-645-4870 hlasker@buffalo.edu,
   SUNY, Buffalo – Gu
Laskowski, Stanley L., (215) 573-3164 slaskows@sas.upenn.edu,
   University of Pennsylvania – Ge
Lassetter, William L., (434) 951-6361 william.lassetter@dmme.virginia.gov,
   Division of Geology and Mineral Resources – EgHg
Lassiter, John, (512) 471-4002 lassiter1@mail.utexas.edu,
   University of Texas, Austin – Cc
Last, George V., (509) 376-3961 george.last@pnl.gov,
   Pacific Northwest National Laboratory – Ge
Last, William M., (204) 474-8361 wm_last@umanitoba.ca,
   University of Manitoba – GsnGo
Lat, Che Noorliza, 03-79674157 noorliza@um.edu.my,

University of Malaya – Yg
Lathrop, Daniel, (301) 405-1594 lathrop@umd.edu,
  University of Maryland – Yg
Latimer, Jennifer C., (812) 237-2254 jen.latimer@indstate.edu,
  Indiana State University – CmGeb
Laton, W. R., (657) 278-7514 wlaton@fullerton.edu,
  California State University, Fullerton – HwGeHy
Lau, Kimberly, Kimberly.Lau@uwyo.edu,
  University of Wyoming – Cb
Laubach, Stephen E., (512) 471-6303 steve.laubach@beg.utexas.edu,
  Univ of Texas at Austin, Jackson School of Geosciences – GcNrGz
Lauderdale, Jonathan, J.Lauderdale@liverpool.ac.uk,
  University of Liverpool – Og
Laurent-Charvet, Sébastien, +33(0)3 44068995 sebastien.laurent-charvet@lasalle-beauvais.fr,
  Institut Polytechnique LaSalle Beauvais (ex-IGAL) – GctZe
Lauretta, Dante, (520) 626-1138 lauretta@lpl.arizona.edu,
  University of Arizona – XcmXg
Laurier, Eric, +44 (0) 131 651 4303 Eric.Laurier@ed.ac.uk,
  Edinburgh University – Zn
Lautz, Laura, 315-443-1196 lklautz@syr.edu,
  Syracuse University – Hg
Lauziere, Kathleen, (418) 654-2658 klauzier@nrcan.gc.ca,
  Universite du Quebec – Gg
Lauzon, John, (519) 824-4120 (Ext. 52459) lauzonj@uoguelph.ca,
  University of Guelph – So
Lavallee, Daniel, 450 465 5392 daniel.lavallee@mail.mcgill.ca,
  University of California, Santa Barbara – YsnAs
Lavallee, Yan, +440151 794 5183, Yan.Lavallee@liverpool.ac.uk,
  University of Liverpool – Gv
Lavallee, Yan, 089/2180 4221 yanlavallee@hotmail.com,
  Ludwig-Maximilians-Universitaet Muenchen – Yg
Lavanchy, G. Thomas, 303-871-7521 thomas.lavanchy@du.edu,
  University of Denver – Hg
Lavier, Luc L., 512-471-0455 luc@ig.utexas.edu,
  University of Texas, Austin – Gt
LaVigne, Michéle, 207-798-4283 mlavign@bowdoin.edu,
  Bowdoin College – OcPe
Lavkulich, Leslie M., (604) 822-3477 University of British Columbia – Sd
Lavoie, Denis, (418) 654-2571 delavoie@nrcan.gc.ca,
  Universite du Quebec – Gs
Lavy, Brendan, 956 665 5280 brendadn.lavy@utrgv.edu,
  University of Texas, Rio Grande Valley – RnwZi
Law, Eric W., (740) 826-8242 ericlaw@muskingum.edu,
  Muskingum University – Gp
Law, Kim, 519-661-2111 ext 83881 krlaw@uwo.ca,
  Western University – Cs
Law, Richard D., (540) 231-6685 rdlaw@vt.edu,
  Virginia Polytechnic Institute & State University – Gc
Law, Zada, (615) 494-8805 zada.law@mtsu.edu,
  Middle Tennessee State University – Zin
Lawrence, Deborah, (434) 924-0581 dl3c@virginia.edu,
  University of Virginia – ZcoZg
Lawrence, James, jimrslawrence@gmail.com,
  University of Houston – Cl
Lawrence, Kira, (610) 330-5194 lawrenck@lafayette.edu,
  Lafayette College – Gg
Lawrence, Rick L., (406) 994-5409 Montana State University – Zr
Lawry-Berkins, Cynthia, 936-273-7407 cynthia.lawryberkins@lonestar.edu,
  Lonestar College - Montgomery – Ges
Lawson, Merlin P., (402) 202-5392 mlawson1@unl.edu,
  University of Nebraska, Lincoln – AmtZr
Lawton, Donald C., (403) 220-5718 University of Calgary – Ye
Lawver, Lawrence A., 512-471-0433 lawver@ig.utexas.edu,
  University of Texas, Austin – Yr
Lay, Thorne, (831) 459-3164 tlay@pmc.ucsc.edu,
  University of California, Santa Cruz – Ys
Layton, Alice, 865-974-8072 alayton@utk.edu,
  University of Tennessee, Knoxville – Zn
Layton-Matthews, Daniel, (613) 533-6338 dlayton@queensu.ca,
  Queen's University – EgCte
Lazar, Codi, 909-537-5586 clazar@csusb.edu,
  California State University, San Bernardino – Cp
Lazarus, Steven M., 321-674-2160 slazarus@fit.edu,
  Florida Institute of Technology – As
Le, Yanfen, 660-562-1525 le@nwmissouri.edu,
  Northwest Missouri State University – Zi
Le Mevel, Helene, 202-478-8842 hlemevel@carnegiescience.edu,
  Carnegie Institution for Science – Gv
Le Mone, David V., 915-747-5501 lemone@utep.edu,
  University of Texas, El Paso – Ps
Le Qu, Corinne, +44 (0)1603 59 2840 c.lequere@uea.ac.uk,
  University of East Anglia – Zg
Le Roex, Anton, 021-650-2902 anton.leroex@uct.ac.za,
  University of Cape Town – Cg
Le Roux, Petrus J., 021-650-4139 petrus.leroux@uct.ac.za,
  University of Cape Town – CaGia
Le Roux, Veronique, (508) 289-3549 vleroux@whoi.edu,
  Woods Hole Oceanographic Institution – Gi
Le Voyer, Marion, (626) 786-3716 levoyermarion@gmail.com,
  Smithsonian Inst / Natl Museum of Natural Hist– Cu
Lea, David W., (805) 893-8665 lea@geol.ucsb.edu,
  University of California, Santa Barbara – PeCm
Lea, Peter D., (207) 725-3439 plea@bowdoin.edu,
  Bowdoin College – Gl
Leach, Harry, +44 0151 794 4097 leach@liverpool.ac.uk,
  University of Liverpool – Og
Leadbetter, Jared R., 626.395.4182 jleadbetter@caltech.edu,
  California Institute of Technology – Pm
Leake, Bernard E., 00442920876421 leakeb@cf.ac.uk,
  Cardiff University – GipGz
Leake, Martha A., (229) 333-5756 mleake@valdosta.edu,
  Valdosta State University – Xg
Leap, Darrell I., (765) 494-3699 mountains2oceans@comcast.net,
  Purdue University – Hy
Lear, C, carrie@earth.cf.ac.uk,
  Cardiff University – Ou
Lear, Caroline, +44(0)29 208 79004 LearC@cardiff.ac.uk,
  University of Wales – Og
Leary, Ryan, ryan.leary@nmt.edu,
  New Mexico Institute of Mining and Technology – Gs
Leatham, W. Britt, (909) 537-5322 bleatham@csusb.edu,
  California State University, San Bernardino – PsmOg
Leather, Kimberly, Kimberley.Leather@manchester.ac.uk,
  University of Manchester – As
Leavens, Peter B., 302-831-8106 pbl@udel.edu,
  University of Delaware – Gz
Leavitt, Steven W., (520) 621-6468 sleavitt@ltrr.arizona.edu,
  University of Arizona – Csc
Lebedev, Maxim, +61 8 9266 3519 m.lebedev@curtin.edu.au,
  Curtin University – YxNr
Lebofsky, Larry A., (520) 621-6947 lebofsky@lpl.arizona.edu,
  University of Arizona – ZngXm
Lebold, Joe, (304) 293-0749 joe.lebold@mail.wvu.edu,
  West Virginia University – PeGrg
Lechler, Paul J., (775) 682-8773 plechler@unr.edu,
  University of Nevada – Cg
Leckie, R. Mark, (413) 545-1948 mleckie@geo.umass.edu,
  University of Massachusetts, Amherst – PmGru
Ledford, Sarah, 404-413-5780 sledford@gsu.edu,
  Georgia State University – HsRw
Lee, Alexis, (910) 962-3736 leea@uncw.edu,
  University of North Carolina, Wilmington – Zn
Lee, Alyce, 304-256-0270 alee@concord.edu,
  Concord University – Og
Lee, Arthur C., (865) 220-9145 leea@roanestate.edu,
  Roane State Community College - Oak Ridge – Ges
Lee, Chung M., (801) 581-8218 chunglee@geog.utah.edu,
  University of Utah – Zu
Lee, Cin-Ty A., 713.348.5084 ctlee@rice.edu,
  Rice University – CgGiv
Lee, Cindy, (632) 220-2101 cindy.lee@stonybrook.edu,
  SUNY, Stony Brook – OcCo
Lee, Cindy M., (864) 656-0672 lc@clemson.edu,
  Clemson University – GeClZe
Lee, Craig M., (206) 685-7656 craig@apl.washington.edu,
  University of Washington – Op
Lee, Daphne E., +64 3 479-7525 daphne.lee@otago.ac.nz,
  University of Otago – Pib
Lee, Eung Seok, (740) 593-1101 leee1@ohio.edu,
  Ohio University – Hw
Lee, Hung, (519) 824-4120 Ext.53828 hlee@uoguelph.ca,
  University of Guelph – Zn
Lee, In Young, (630) 252-8724 Argonne National Laboratory – As

Lee, Jaeheon, 520-626-4967 jaeheon@email.arizona.edu,
  University of Arizona – Nx
Lee, Jeffrey, (509) 963-2801 jeff@geology.cwu.edu,
  Central Washington University – Gct
Lee, Jeffrey A., 806-834-8228 jeff.lee@ttu.edu,
  Texas Tech University – Zy
Lee, Jejung, leej@umkc.edu,
  University of Missouri, Kansas City – GqHw
Lee, Kanani K., (203) 432-4354 kanani.lee@yale.edu,
  Yale University – Gy
Lee, Keenan, (303) 273-3808 klee@mines.edu,
  Colorado School of Mines – Zr
Lee, Martin, +4401413302634 Martin.Lee@glasgow.ac.uk,
  University of Glasgow – Gz
Lee, Meehye, 82-2-3290-3178 meehye@korea.ac.kr,
  Korea University – AsOgCm
Lee, Michael D., 510-885-3155 michael.lee@csueastbay.edu,
  California State University, East Bay – Hg
Lee, Ming-Kuo, (334) 844-4898 leeming@auburn.edu,
  Auburn University – HwGe
Lee, Rachel J., (315) 312-5506 rachel.lee@oswego.edu,
  SUNY, Oswego – GvZr
Lee, Sukyoung, (814) 863-1587 slg9@psu.edu,
  Pennsylvania State University, University Park – Am
Lee, Wook, (814) 732-2291 wlee@edinboro.edu,
  Edinboro University of Pennsylvania – Zi
Lee, Young Jae, 82-2-3290-3181 youngjlee@korea.ac.kr,
  Korea University – ScGze
Lee, Zhongping, zhongping.lee@umb.edu,
  University of Massachusetts, Boston – ZrOg
Lee-Gorishti, Yolanda, 203-392-6647 yolanda.lee-gorishti@uconn.edu,
  Southern Connecticut State University – GaZe
Leech, Mary L., (415) 338-1144 leech@sfsu.edu,
  San Francisco State University – GptGz
Leeman, William P., (713) 348-4892 leeman@rice.edu,
  Rice University – Gi
Lees, Jonathan M., (919) 962-0695 jonathan.lees@unc.edu,
  University of North Carolina, Chapel Hill – YsGv
Leetaru, Hannes, 217 333-5058 hleetaru@illinois.edu,
  University of Illinois – Go
Lefebvre, Rene, (418) 654-2651 rene.lefebvre@ete.inrs.ca,
  Universite du Quebec – Hw
LeFever, Richard D., (701) 777-3014 richard.lefever@engr.und.edu,
  University of North Dakota – GsrGo
Lefticariu, Liliana, (618) 453-7373 lefticariu@geo.siu.edu,
  Southern Illinois University Carbondale – CsbGe
Leger, Carol, 603.358.2570 cleger@keene.edu,
  Keene State College – Zg
Leggitt, Leroy, lleggitt@llu.edu,
  Loma Linda University – Pi
Legore, Virginia L., (509) 376-5019 virginia.legore@pnl.gov,
  Pacific Northwest National Laboratory – Cg
Lehane, Mary, +353 21 4902764 m.lehane@ucc.ie,
  University College Cork – Zn
Lehloenya, Pelele, (051) 401-3783 lehloenyapb@ufs.ac.za,
  University of the Free State – Cg
Lehman, Thomas M., 806-834-3148 tom.lehman@ttu.edu,
  Texas Tech University – Gs
Lehnert, Kerstin, lehnert@ldeo.columbia.edu,
  Columbia University – Gi
Lehrberger, Gerhard, +49 89 289 25832 lehrberger@tum.de,
  Technical University of Munich – EgGxg
Lehre, Andre K., (707) 616-8573 akl1@humboldt.edu,
  Humboldt State University – GmHs
Lehrmann, Daniel J., 210-999-7654 dlehrmann@trinity.edu,
  Trinity University – PiGs
Lehto, Heather L., (325) 486-6990 heather.lehto@angelo.edu,
  Angelo State University – GvYsZe
Leibensperger, Eric, (518) 564-4104 eleib003@plattsburgh.edu,
  SUNY, Plattsburgh – AsOg
Leichmann, Jaromir, +420 549 49 5559 leichman@sci.muni.cz,
  Masaryk University – GxpGi
Leichter, James J., (858) 822-5330 jleichter@ucsd.edu,
  University of California, San Diego – Ob
Leier, Andrew L., 803-777-9941 aleier@geol.sc.edu,
  University of South Carolina – Gst
Leighton, Henry G., (514) 398-3766 henry.leighton@mcgill.ca,
  McGill University – As
Leighton, Lindsey, 780-492-3983 lleighto@ualberta.ca,
  University of Alberta – Pi
Leighty, Robert S., (480) 461-7021 rleighty@mesacc.edu,
  Mesa Community College – Zg
Leimer, H. Wayne, (931) 372-3522 hwleimer@tntech.edu,
  Tennessee Tech University – Gz
Leinen, Margaret, (401) 874-6222 mleinen@nsf.gov,
  University of Rhode Island – Ou
Leinen, Margaret, 858-534-2827 mleinen@ucsd.edu,
  University of California, San Diego – OgPe
Leinfelder, Reinhold, 089/2180 6629 r.leinfelder@lrz.uni-muenchen.de,
  Ludwig-Maximilians-Universitaet Muenchen – Pg
Leitch, Alison, (709) 737-3306 aleitch@mun.ca,
  Memorial University of Newfoundland – Yg
Leite, Michael B., (308) 432-6377 mleite@csc.edu,
  Chadron State College – Gg
Leithold, Elana L., (919) 515-7282 lonnie_leithold@ncsu.edu,
  North Carolina State University – Gs
Leitz, Robert E., (303) 556-3072 Metropolitan State College of Denver –
  Gz
Lekan, Thomas, lekan@sc.edu,
  University of South Carolina – Zn
Lekic, Vedran, (301) 405-4086 ved@umd.edu,
  University of Maryland – Ysg
Leland, John, 775-784-6670 jleland@unr.edu,
  University of Nevada, Reno – Zn
Lemay, Phillip W., (225) 388-6922 philip@lgs.bri.lsu.edu,
  Louisiana State University – Gg
Lemiszki, Peter J., (865) 594-5596 peter.lemiszki@tn.gov,
  Pellissippi State Community College – GcgZi
Lemke, Karen A., (715) 346-2709 klemke@uwsp.edu,
  University of Wisconsin, Stevens Point – GmZy
Lemke, Lawrence D., (989) 774-1144 l.d.lemke@cmich.edu,
  Central Michigan University – HwGer
Lemmond, Peter C., (508) 289-2457 plemmond@whoi.edu,
  Woods Hole Oceanographic Institution – Zn
Lempe, Bernhard, +49 89289 25862 lempe@tum.de,
  Technical University of Munich – GlgNg
Lenardic, Adrian, (713) 348-4883 ajns@rice.edu,
  Rice University – Yg
Lenczewski, Melissa E., (815) 753-7937 lenczewski@niu.edu,
  Northern Illinois University – HwGeCo
Lene, Gene W., (210) 436-3011 glene@stmarytx.edu,
  Saint Mary's University – Ge
Lener, Edward, (540) 231-9249 lener@vt.edu,
  Virginia Polytechnic Institute & State University – Zn
Lenhart, Stephen W., slenhart@radford.edu,
  Radford University – Pg
Lennox, Paul G., + 61 2 9385 8096 p.lennox@unsw.edu.au,
  University of New South Wales – GcgZe
Lentz, David R., (506) 547-2070 University of New Brunswick – Cg
Lentz, Leonard J., 717–702–2040 lelentz@pa.gov,
  Pennsylvania Bureau of Topographic & Geologic Survey – Ec
Lenz, Alfred C., (519) 661-3195 aclenz@uwo.ca,
  Western University – PiePi
Lenz, Petra H., 808-956-8003 petra@pbrc.hawaii.edu,
  University of Hawai'i, Manoa – Og
Leonard, Eric M., (719) 389-6513 eleonard@coloradocollege.edu,
  Colorado College – Gl
Leonard, Lynn A., 910-962-2338 lynnl@uncw.edu,
  University of North Carolina, Wilmington – Gu
Leonard, Meredith L., 818.778.5595 leonarml@lavc.edu,
  Los Angeles Valley College – AmZei
Leonard-Pingel, Jill, leonard-pingel.1@osu.edu,
  Ohio State University – Pi
Leone, James, 518-473-9988 jleone@mail.nysed.gov,
  New York State Geological Survey – EoGoCg
Leonhart, Leo S., (520) 881-7300 University of Arizona – Hw
Leorri, Eduardo, 252 737 2529 leorrie@ecu.edu,
  East Carolina University – GsPm
Lepain, David, 907-451-5085 david.lepain@alaska.gov,
  Alaska Division of Geological & Geophysical Surveys – EoGos
Lepper, Kenneth, (701) 231-6746 ken.lepper@ndsu.edu,
  North Dakota State University – CcGml
Leppert, Ken, (318) 342-1918 University of Louisiana, Monroe – As
Lepre, Christopher H., 848-445-3410 clepre@eps.rutgers.edu,

Rutgers, The State University of New Jersey – YmSaPv
Lerach, David G., (970) 351-2853 David.Lerach@unco.edu,
    University of Northern Colorado – Ams
Lerch, Derek, dlerch@frc.edu,
    Feather River College – Zg
Lerch, Robert N., 573-882-1402 University of Missouri, Columbia – So
Lerczak, Jim, 541-737-6128 jlerczak@coas.oregonstate.edu,
    Oregon State University – Op
Lerdau, Manuel, (434) 924-3325 mtl5g@virginia.edu,
    University of Virginia – Zn
Lerman, Abraham, 847-491-7385 alerman@northwestern.edu,
    Northwestern University – CgGnCm
Lermurier, Nathalie, +33(0)3 44062540 nathalie.lermurier@lasalle-beauvais.fr,
    Institut Polytechnique LaSalle Beauvais (ex-IGAL) – Zn
Lerner-Lam, Arthur L., lerner@ldeo.columbia.edu,
    Columbia University – Ys
Lerner-Lam, Arthur L., (845) 365-8356 lerner@ldeo.columbia.edu,
    Columbia University – Ys
Lesher, Charles E., (530) 752-9779 celesher@ucdavis.edu,
    University of California, Davis – Gi
Lesher, Michael, 7056751151 x2276 MLesher@laurentian.ca,
    Laurentian University, Sudbury – Em
Lesht, Barry, 312-413-3176 blesht@uic.edu,
    University of Illinois at Chicago – Op
Lesht, Barry M., (630) 252-4208 bmlesht@anl.gov,
    Argonne National Laboratory – Op
Lessard, Evelyn J., (206) 543-8795 elessard@u.washington.edu,
    University of Washington – Ob
Leszczynski, Raymond F., 3152948613 ext. 2313 Ray@cayuga-cc.edu,
    Cayuga Community College – GglGm
Letelier, Ricardo, (541) 737-3890 letelier@coas.oregonstate.edu,
    Oregon State University – ObZr
Lethbridge, Mark, mark.lethbridge@flinders.edu.au,
    Flinders University – Ziy
Letscher, Robert T., 603-862-1008 robert.letscher@unh.edu,
    University of New Hampshire – OcCbZo
Letsinger, Sally L., (812) 855-1356 sletsing@indiana.edu,
    Indiana University – HqZir
Leung, Irene S., (718) 960-8572 Graduate School of the City University of New York – Gz
Leuta-Madondo, Palesa, (031) 260-2517 leutap@ukzn.ac.za,
    University of KwaZulu-Natal – CgGig
Lev, Einat , einatlev@ldeo.columbia.edu,
    Columbia University – Gv
Levander, Alan, (713) 348-6064 alan@rice.edu,
    Rice University – Ys
Levas, Stephen J., 262-472-6200 levass@uww.edu,
    University of Wisconsin, Whitewater – CmGe
Levasseur, Emile , levasseure@easternct.edu,
    Eastern Connecticut State University – Zg
Leventer, Amy, (315) 228-7214 aleventer@colgate.edu,
    Colgate University – Ou
Leventon, Julia, +44(0) 113 34 31635 J.Leventon@leeds.ac.uk,
    University of Leeds – Ge
LeVeque, Randy, 206.685.3037 rjl@uw.edu,
    University of Washington – GvYs
Lever, Helen, +44 (0) 131 451 4057 h.lever@hw.ac.uk,
    Heriot-Watt University – Go
Leverington, David W., 806-834-5310 david.leverington@ttu.edu,
    Texas Tech University – GmZrXg
Levesque, Andre, levesque.andre@ggl.ulaval.ca,
    Universite Laval – Gz
Levey, Raymond A., (801) 585-3826 rlevey@egi.utah.edu,
    University of Utah – Eo
Levin, Harold L., (314) 935-5647 LEVIN@LEVEE.WUSTL.EDU,
    Washington University in St. Louis – Pi
Levin, Lisa A., (858) 534-3579 llevin@ucsd.edu,
    University of California, San Diego – Ob
Levin, Naomi, 410-516-4317 nlevin3@jhu.edu,
    Johns Hopkins University – Gs
Levin, Vadim, 848-445-5415 vlevin@eps.rutgers.edu,
    Rutgers, The State University of New Jersey – Ysg
Levine, Norman S., (843) 953-5308 levinen@cofc.edu,
    College of Charleston – ZiGeNg
Levine, Rebekah, 406-683-7134 rebekah.levine@umwestern.edu,
    University of Montana Western – GmHsSd
Levino, Lynn J., 412-442-4299 Pennsylvania Bureau of Topographic & Geologic Survey – Go
Levinson, Alfred A., (403) 220-5846 University of Calgary – Eg
Levinton, Jeffrey, (631) 632-8602 jeffrey.levinton@stonybrook.edu,
    SUNY, Stony Brook – Ob
Levson, Victor M., (250) 952-0391 vic.levson@gems9.gov.bc.ca,
    University of Victoria – Gl
Levy, Jonathan, (513) 529-1947 levyj@MiamiOh.edu,
    Miami University – Hw
Levy, Joseph, jlevy@colgate.edu,
    Colgate University – GlsGm
Levy, Laura B., 707-826-3165 laura.levy@humboldt.edu,
    Humboldt State University – GslPc
Levy, Melissa H., (916) 484-8684 levym@arc.losrios.edu,
    American River College – GgZgy
Levy, Schon S., (505) 667-9504 sslevy@lanl.gov,
    Los Alamos National Laboratory – Gi
Lew, Alan A., (928) 523-6567 alan.lew@nau.edu,
    Northern Arizona University – Zn
Lew, Jeffrey, (310) 825-3023 lew@atmos.ucla.edu,
    University of California, Los Angeles – As
Lewandowski, Katherine, (217) 581-7270 kjlewandowski@eiu.edu,
    Eastern Illinois University – Pmc
Lewis, Alan G., (604) 822-3626 alewis@eos.ubc.ca,
    University of British Columbia – Ob
Lewis, Brian T. R., (206) 543-7419 blewis@u.washington.edu,
    University of Washington – Yr
Lewis, Chris, cjlewis@ccsf.edu,
    City College of San Francisco – GgEmSo
Lewis, Gerald L., (626) 585-7137 gllewis@pasadena.edu,
    Pasadena City College – Ps
Lewis, John S., (360) 873-8781 jsl@u.arizona.edu,
    University of Arizona – Xcm
Lewis, Jon C., (724) 357-5624 jclewis@iup.edu,
    Indiana University of Pennsylvania – Gc
Lewis, Katie, 806-746-6101 krothlisberger@tamu.edu,
    Texas A&M University – Sco
Lewis, Marlon R., (902) 494-3513 marlon.lewis@dal.ca,
    Dalhousie University – Ob
Lewis, Mary, 5102357800 x.4284 mlewis@contracosta.edu,
    Contra Costa College – Gg
Lewis, Reed S., (208) 885-7472 reedl@uidaho.edu,
    University of Idaho – GgiEg
Lewis, Ronald D., (334) 844-4886 lewisrd@auburn.edu,
    Auburn University – Pi
Lewis, Stephen D., (559) 278-6956 slewis@csufresno.edu,
    California State University, Fresno – Yg
Leybourne, Matthew I., 7056751151 x2263 mleybourne@laurentian.ca,
    Laurentian University, Sudbury – Cg
Leyden, Barbara W., (813) 974-0324 University of South Florida, Tampa – Pl
Leyrit, Hervé, +33(0)3 44068998 herve.leyrit@lasalle-beauvais.fr,
    Institut Polytechnique LaSalle Beauvais (ex-IGAL) – Gv
Li, Aibing, 713-743-2878 ali2@uh.edu,
    University of Houston – Ys
Li, Baosheng, 631 632-9642 baosheng.li@sunysb.edu,
    Stony Brook University – Gy
Li, Chusi, (812) 855-1558 cli@indiana.edu,
    Indiana University, Bloomington – GiEm
Li, Dan, lidan@bu.edu,
    Boston University – HqAs
Li, Gary, 510-885-3165 gary.li@csueastbay.edu,
    California State University, East Bay – Zi
Li, Jing, 303-871-4687 jing.li145@du.edu,
    University of Denver – Zi
Li, Junran, 918-631-2517 junran-li@utulsa.edu,
    The University of Tulsa – Gme
Li, Liangping, (605) 394-2461 Liangping.Li@sdsmt.edu,
    South Dakota School of Mines & Technology – HwqHt
Li, Lin, ll3@iupui.edu,
    Indiana University / Purdue University, Indianapolis – XgZrg
Li, Long, 780-492-9288 long4@ualberta.ca,
    University of Alberta – CsGpv
Li, Peng, (501) 683-0118 peng.li@arkansas.gov,
    Arkansas Geological Survey – GoEoCo
Li, Ping-Chi, (931) 372-3752 PLI@TnTech.edu,
    Tennessee Tech University – Zi
Li, Rong-Yu, 204-727-9684 lir@brandonu.ca,

Brandon University – PgmGg
Li, Tim, 808-956-9427 timli@hawaii.edu,
University of Hawai'i, Manoa – Asm
Li, Wenhong, 919 684-5015 wl66@duke.edu,
Duke University – Zn
Li, William K., (902) 426-6349 lib@mar.dfo-mpo.gc.ca,
Dalhousie University – Ob
Li, Xiangshan, 713-743-0742 xli10@uh.edu,
University of Houston – Asm
Li, Yaoguo, (303) 273-3510 ygli@mines.edu,
Colorado School of Mines – Yev
Li, Yong-Gang, (213) 740-3556 hylirenko@marshall.usc.edu,
University of Southern California – Ys
Li, Yuan-Hui, 808-956-6297 yhli@soest.hawaii.edu,
University of Hawai'i, Manoa – Cm
Li, Yuncong, (305) 246-7000 yunli@ufl.edu,
University of Florida – Sco
Li, Zhenhong, +44 (0) 191 208 5704 zhenhong.li@ncl.ac.uk,
University of Newcastle Upon Tyne – Yd
Liang, George, (785) 532-6101 gliang@ksu.edu,
Kansas State University – So
Liang, Liyuan, 865-241-3933 2ll@ornl.gov,
Oak Ridge National Laboratory – Cl
Liang, Lu, 940.369.5198 lu.liang@unt.edu,
University of North Texas – Zi
Liang, Renxing, 609-258-4101 Princeton University – Po
Liang, Yan, (401) 863-9477 Yan_Liang@Brown.edu,
Brown University – Cp
Liao, Enhui, 609-258-4101 enhui.liao@princeton.edu,
Princeton University – Op
Liao, Haifeng (Felix), 208-885-6452 hliao@uidaho.edu,
University of Idaho – ZuiZn
Liauw, Henri L., hliauwap@broward.edu,
Broward College – Gg
Libarkin, Julie C., (517) 355-8369 libarkin@msu.edu,
Michigan State University – Zen
Liberty, Lee M., (208) 426-1166 lml@cgiss.boisestate.edu,
Boise State University – Ys
Libra, Robert D., Robert.Libra@dnr.iowa.gov,
Iowa Dept of Natural Resources – Gg
Licciardi, Joseph M., 603-862-1718 joe.licciardi@unh.edu,
University of New Hampshire – Gl
Licciardi, Joseph M., (603) 862-3135 joe.licciardi@unh.edu,
University of New Hampshire – Gl
Licht, Alexis, licht@uw.edu,
University of Washington – Gsn
Licht, Kathy J., (317) 278-1343 Indiana University / Purdue University, Indianapolis – Gl
Lichtenberger, János, lityi@sas.elte.hu,
Eotvos Lorand University – AsZrXy
Lichtner, Peter C., (505) 667-3420 lichtner@lanl.gov,
Los Alamos National Laboratory – Zn
Liddell, W. David, (435) 797-1261 dave.liddell@usu.edu,
Utah State University – GsPg
Lidgard, Scott H., (312) 665-7625 slidgard@fieldmuseum.org,
Field Museum of Natural History – Pi
Lidiak, Edward G., (412) 624-8871 egl@pitt.edu,
University of Pittsburgh – Gi
Lidicker, Jr., William Z., (510) 642-3059 University of California, Berkeley – Pv
Liebe, Richard M., (585) 395-5100 (Ext. 7524) rliebe@weather.brockport.edu,
SUNY, The College at Brockport – Ps
Liebens, Johan, 850-474-2065 liebens@uwf.edu,
University of West Florida – So
Lieberman, Bruce S., (785) 864-2741 blieber@ku.edu,
University of Kansas – Pgi
Lieberman, Robert C., (631) 632-8214 robert.liebermann@sunysb.edu,
Stony Brook University – GyYsx
Liebling, Richard, 516 463-6545 georsl@hofstra.edu,
Hofstra University – Gg
Liebling, Richard S., (212) 772-5412 Graduate School of the City University of New York – Gz
Lifton, Nathaniel A., (765) 494-0754 nlifton@purdue.edu,
Purdue University – GmCc
Light, Bonnie, (206) 543-9824 bonnie@apl.washington.edu,
University of Washington – As
Lightbody, Anne F., 603-862--0711 anne.lightbody@unh.edu,

University of New Hampshire – HgsZn
Likos, William J., 608-890-2662 likos@wisc.edu,
University of Wisconsin, Madison – NgSp
Lilley, Marvin D., (206) 543-0859 lilley@ocean.washington.edu,
University of Washington – Ob
Lilly, Troy, 325-574-7922 tlilly@wtc.edu,
Western Texas College – GgZg
Lima, Eduardo A., 617-324-2829 limaea@mit.edu,
Massachusetts Institute of Technology – Ym
Lima, Ivan D., ilima@whoi.edu,
Woods Hole Oceanographic Institution – Oc
Limp, William (., (479) 575-7909 flimp@uark.edu,
University of Arkansas, Fayetteville – Zfi
Lin, Douglas, (831) 459-2732 lin@lick.ucsc.edu,
University of California, Santa Cruz – Xc
Lin, Fan-Chi, (801) 581-4373 fanchi.lin@utah.edu,
University of Utah – Ygs
Lin, Guoqing, glin@rsmas.miami.edu,
University of Miami – Ys
Lin, Hsing K., (907) 474-6347 hklin@alaska.edu,
University of Alaska, Fairbanks – Zn
Lin, Jialin, (614) 292-6634 lin.789@osu.edu,
Ohio State University – As
Lin, Jian, (508) 289-2576 jlin@whoi.edu,
Woods Hole Oceanographic Institution – Gt
Lin, John C., 801-581-7530 john.lin@utah.edu,
University of Utah – As
Lin, Jung-Fu, 512-471-8054 afu@jsg.utexas.edu,
University of Texas, Austin – GyYm
Lin, Senjie, (860) 405-9168 senjie.lin@uconn.edu,
University of Connecticut – Ob
Lin, Shoufa, (519) 888-4567 (Ext. 6557) shoufa@uwaterloo.ca,
University of Waterloo – Zn
Lin, Wuyin, (631) 632-3141 Wuyin.Lin@stonybrook.edu,
SUNY, Stony Brook – As
Lin, Xiaomao, (785) 532-6816 xlin@ksu.edu,
Kansas State University – Zn
Lin, Yu-Feng F., 217-333-0235 yflin@illinois.edu,
Illinois State Geological Survey – Hg
Linares, Rogelio, ++935811259 rogelio.linares@uab.cat,
Universitat Autonoma de Barcelona – NgYx
Lincoln, Beth Z., (517) 629-0331 blincoln@albion.edu,
Albion College – Gc
Lincoln, Jonathan M., (973) 655-7273 lincolnj@mail.montclair.edu,
Montclair State University – Gr
Lindberg, David R., (510) 642-3926 University of California, Berkeley – Pi
Lindberg, David R., drl@berkeley.edu,
University of California, Berkeley – Pg
Lindberg, Jonathan W., (509) 376-5005 jon.lindberg@pnl.gov,
Pacific Northwest National Laboratory – Ec
Lindberg, Steven E., partnerships@ornl.gov,
Oak Ridge National Laboratory – Cl
Lindbo, David, (919) 793-4428 North Carolina State University – Sd
Linde, Alan T., alinde@carnegiescience.edu,
Carnegie Institution for Science – Ys
Lindemann, William C., (505) 646-3405 New Mexico State University, Las Cruces – Sb
Lindenmeier, Clark W., (509) 376-8419 clark.lindenmeier@pnl.gov,
Pacific Northwest National Laboratory – Cl
Lindgren, Paula, 441413305442 Paula.Lindgren@glasgow.ac.uk,
University of Glasgow – As
Lindley, Stacy, (916) 278-6337 stacy.lindley@csus.edu,
California State University, Sacramento – Zn
Lindline, Jennifer, (505) 426-2046 lindlinej@nmhu.edu,
New Mexico Highlands University – GizZe
Lindo Atachati, David, 718-982-2919 david.lindo@csi.cuny.edu,
Graduate School of the City University of New York – Og
Lindsay, Everett H., (520) 621-6024 ehlind@geo.arizona.edu,
University of Arizona – Pv
Lindsay, Matthew B., (306) 966-5693 matt.lindsay@usask.ca,
University of Saskatchewan – ClGg
Lindsey, Kassandra, 303-384-2660 kolindsey@mines.edu,
Colorado Geological Survey – NgGmZi
Lindsley, Donald H., (631) 632-8195 donald.lindsley@stonybrook.edu,
Stony Brook University – Cp
Lindzen, Richard S., (617) 253-2432 rlindzen@mit.edu,
Massachusetts Institute of Technology – Am

Lineback, Neal G., (828) 262-3000 linebackng@appstate.edu,
 Appalachian State University – Zy
Liner, Christopher, 479-575-5667 liner@uark.edu,
 University of Arkansas, Fayetteville – YseGg
Lines, Larry R., (403) 220-5841 University of Calgary – Ye
Lini, Andrea, (802) 656-0245 andrea.lini@uvm.edu,
 University of Vermont – CsGn
Lininger, Katherine, katherine.lininger@colorado.edu,
 University of Colorado – Gm
Link, Curtis A., (406) 496-4165 clink@mtech.edu,
 Montana Tech of the University of Montana – Ye
Link, Paul K., (208) 282-3365 linkpaul@isu.edu,
 Idaho State University – Gs
Linky, Edward, (212) 772-5265 linky.edward@epamail.epa.gov,
 Hunter College (CUNY) – Zn
Linn, Anne M., (202) 334-2744 alinn@nas.edu,
 National Academy of Sciences, Engineering, and Medicine – Gs
Linn, Rodman, (505) 665-6254 rrl@lanl.gov,
 Los Alamos National Laboratory – Ng
Linneman, Scott R., (360) 650-3446 Scott.Linneman@wwu.edu,
 Western Washington University – GmZe
Linnen, Robert, 519-661-2111 x89207 rlinnen@uwo.ca,
 Western University – EmCe
Linol, Bastien, bastien.aeon@gmail.com,
 Nelson Mandela Metropolitan University – Gs
Liou, Juhn G., (650) 723-2716 liou@pangea.stanford.edu,
 Stanford University – Gp
Lipeles, Maxine I., (314) 935-5482 milipeles@seas.wustl.edu,
 Washington University in St. Louis – Zn
Lippelt, Irene D., (608) 262-7430 irene.lippelt@wgnhs.uwex.edu,
 University of Wisconsin, Extension – Gg
Lippert, Peter C., (801) 581-4599 pete.lippert@utah.edu,
 University of Utah – YmGtPe
Lippmann, Thomas C., 603-862-4450 lippmann@ccom.unh.edu,
 University of New Hampshire – Onp
Lips, Elliott W., (801) 581-8218 elliott.lips@geog.utah.edu,
 University of Utah – Gm
Lisenby, Peyton E., (940) 397-4475 peyton.lisenby@msutexas.edu,
 Midwestern State University – GmeZi
Lisichenko, Richard, 785-628-4159 rlisiche@fhsu.edu,
 Fort Hays State University – Zie
Lisiecki, Lorraine, (805) 893-4437 lisiecki@geol.ucsb.edu,
 University of California, Santa Barbara – Gn
Lisle, R J., lisle@cf.ac.uk,
 Cardiff University – Gc
Liss, Peter S., +44 (0)1603 59 2563 p.liss@uea.ac.uk,
 University of East Anglia – OcCbAc
Little, Crispin, +44(0) 113 34 36621 earctsl@leeds.ac.uk,
 University of Leeds – Pg
Little, Jonathan, jlittle@monroecc.edu,
 Monroe Community College – ZyGlZi
Little, Tim, +64 4 463 6198 Tim.Little@vuw.ac.nz,
 Victoria University of Wellington – Gct
Little, William W., (208) 496-7679 littlew@byui.edu,
 Brigham Young University - Idaho – GsdGm
Littler, Kate L., 01326255725 k.littler@exeter.ac.uk,
 Exeter University – PeCl
Liu, Chuntao, 361-825-3845 Chuntao.Liu@tamucc.edu,
 Texas A&M University, Corpus Christi – As
Liu, Dantong, dantong.liu@manchester.ac.uk,
 University of Manchester – As
Liu, Gaisheng, 785-864-2115 gliu@kgs.ku.edu,
 University of Kansas – Hy
Liu, Jian G., +44 20 759 46418 j.g.liu@imperial.ac.uk,
 Imperial College – Zr
Liu, Jing, liu_jing@smc.edu,
 Santa Monica College – ZyiZr
Liu, Jingpu P., 919-515-3711 jpliu@ncsu.edu,
 North Carolina State University – Gu
Liu, Jiping, jiping.liu@eas.gatech.edu,
 Georgia Institute of Technology – As
Liu, Lanbo, (860) 486-1388 lanbo.liu@uconn.edu,
 University of Connecticut – Yg
Liu, Lijun, (217) 300-0378 ljliu@illinois.edu,
 University of Illinois, Urbana-Champaign – YgGtm
Liu, Mian, (573) 882-3784 LiuM@missouri.edu,
 University of Missouri – Yg

Liu, Paul Hiaibao, Paul.Liu@dnr.iowa.gov,
 Iowa Dept of Natural Resources – Gg
Liu, Ping, (631) 632-3195 Ping.Liu@stonybrook.edu,
 SUNY, Stony Brook – As
Liu, Qinya, 416-978-5434 University of Toronto – Ygx
Liu, Shimin, (814) 863-4491 szl3@psu.edu,
 Pennsylvania State University, University Park – Nm
Liu, Weibo, 561 297-4965 liuw@fau.edu,
 Florida Atlantic University – Zi
Liu, Xiaoming, (919) 962-0675 xiaomliu@unc.edu,
 University of North Carolina, Chapel Hill – Cgl
Liu, Yajing, (514) 398-4085 yajing.liu@mcgill.ca,
 McGill University – Ygs
Liu, Yongqiang, (706) 559-4240 yliu@fs.fed.us,
 Georgia Institute of Technology – Am
Liu, Zhanfei, 361-749-6772 zhanfei.liu@utexas.edu,
 University of Texas, Austin – Og
Liu, Zhengyu, (608)262-0777 zliu3@wisc.edu,
 University of Wisconsin, Madison – Op
Liu, Zhenxian, (314) 882-3784 zxliu@bnl.gov,
 Carnegie Institution for Science – Gy
Liutkus-Pierce, Cynthia M., 828-262-6933 liutkuscm@appstate.edu,
 Appalachian State University – Gd
Livaccari, Richard F., (970) 248-1081 rlivacca@coloradomesa.edu,
 Colorado Mesa University – Gc
Lively, Rich, (612) 626-3103 lively@umn.edu,
 University of Minnesota – Cc
Livens, Francis, +44 0161 275-4647 francis.livens@manchester.ac.uk,
 University of Manchester – Cg
Livermore, Phil, +44(0) 113 34 30379 p.w.livermore@leeds.ac.uk,
 University of Leeds – Ym
Livingstone, Daniel A., (919) 684-3264 livingst@duke.edu,
 Duke University – Pe
Lizarralde, Daniel, 508-289-2942 dlizarralde@whoi.edu,
 Woods Hole Oceanographic Institution – Ys
Llanes, María del Pilar, pllanes@ucm.es,
 Univ Complutense de Madrid – Yr
Llewellin, Ed, +44 (0) 191 33 42336 ed.llewellin@durham.ac.uk,
 Durham University – Gv
Lloyd, Glenn D., 573/368-2176 glenn.lloyd@dnr.mo.gov,
 Missouri Dept of Natural Resources – Zn
Lloyd, Graeme, +44 (1865) 272056 graeme.lloyd@earth.ox.ac.uk,
 University of Oxford – Gs
Lloyd, Jonathan, +44 0161 275-7155 Jon.Lloyd@manchester.ac.uk,
 University of Manchester – Ge
Loague, Keith, (650) 723-0847 keith@pangea.stanford.edu,
 Stanford University – Hq
Lobben, Amy, (541) 346-4566 lobben@uoregon.edu,
 University of Oregon – ZinZn
Lobegeier, Melissa, 615-898-2403 Melissa.Lobegeier@mtsu.edu,
 Middle Tennessee State University – Pmg
Locat, Jacques E., (418) 656-2179 locat@ggl.ulaval.ca,
 Universite Laval – Ng
Lock, Brian E., (337) 482-6823 belock@louisiana.edu,
 University of Louisiana at Lafayette – Gs
Locke, Daniel B., (207) 287-7171 daniel.b.locke@maine.gov,
 Dept of Agriculture, Conservation, and Forestry – Hw
Locke, David C., (718) 997-3271 Graduate School of the City University
 of New York – Ca
Locke, Erika, erika.locke@tamucc.edu,
 Texas A&M University, Corpus Christi – GgoGs
Locke, James L., (415) 485-9526 jim@marin.edu,
 College of Marin – Zg
Locke, Randall, (217) 333-3866 rlocke@illinois.edu,
 Illinois State Geological Survey – Cl
Locker, Stanley D., (727) 553-1502 stan@marine.usf.edu,
 University of South Florida – Ou
Lockley, Martin G., (303) 556-4884 University of Colorado, Denver – Pi
Lockwood, John P., (808) 967-8579 jplockwood@volcanologist.com,
 University of Hawai'i, Hilo – Gvg
Lockwood, Rowan, (757) 221-2878 rxlock@wm.edu,
 College of William & Mary – Pei
Lodders-Fegley, Katharina, (314) 935-4851 lodders@levee.wustl.edu,
 Washington University in St. Louis – Xc
Lodge, Robert, 715-836-4361 lodgerw@uwec.edu,
 University of Wisconsin, Eau Claire – EmCuGv
Loeb, Valerie, 831-771-4400 loeb@mlml.calstate.edu,

Moss Landing Marine Laboratories – Zn
Lofthouse, Stephen T., (212) 346-1760 slofthouse@fsmail.pace.edu,
　Pace University, New York Campus – Zg
Logan, Alan, (506) 648-5715 logan@unbsj.ca,
　University of New Brunswick Saint John – Ob
Logsdon, Miles G., (206) 543-5334 mlog@u.washington.edu,
　University of Washington – Zi
Lohan, Maeve, +44 (0)23 80596449 M.Lohan@soton.ac.uk,
　University of Southampton – Ob
Loheide, Steven P., (608) 265-5277 loheide@wisc.edu,
　University of Wisconsin, Madison – HwSp
Lohman, Rowena B., rbl62@cornell.edu,
　Cornell University – YsZrGt
Lohmann, George P., (508) 289-2840 glohmann@whoi.edu,
　Woods Hole Oceanographic Institution – Pe
Lohmann, Kyger C., (734) 763-2298 kacey@umich.edu,
　University of Michigan – CsGsCl
Lokau, Katja R., +49 89 289 25857 katja.lokau@tum.de,
　Technical University of Munich – Ng
Lomaga, Margaret, lomagam@sunysuffolk.edu,
　Suffolk County Community College, Ammerman Campus – Zg
Lombardo, Kelly, 860-405-9256 kelly.lombardo@uconn.edu,
　University of Connecticut – Asm
London, David, (405) 325-3253 dlondon@ou.edu,
　University of Oklahoma – CpGiz
Lonergan, Lidia, +44 20 759 46465 l.lonergan@imperial.ac.uk,
　Imperial College – GtcGs
Long, Ann D., (217) 244-6172 annlong@illinois.edu,
　University of Illinois, Urbana-Champaign – Gm
Long, Austin, (520) 621-8888 along@geo.arizona.edu,
　University of Arizona – Cl
Long, Colin, (920) 424-4105 longco@uwosh.edu,
　University of Wisconsin Oshkosh – PeZyc
Long, Colleen, longcm@illinois.edu,
　Illinois State Geological Survey – Gg
Long, David T., (517) 353-9618 long@msu.edu,
　Michigan State University – ClGeb
Long, Leon E., (512) 459-7838 leonlong@jsg.utexas.edu,
　University of Texas, Austin – Cc
Long, Lisa, 614 265 6590 lisa.long@dnr.state.oh.usm,
　Ohio Dept of Natural Resources – Zn
Long, Matthew, 508-289-2798 mlong@whoi.edu,
　Woods Hole Oceanographic Institution – Cm
Long, Maureen D., (203) 432-5031 maureen.long@yale.edu,
　Yale University – Ysg
Long , Sean P., 509-335-8868 sean.p.long@wsu.edu,
　Washington State University – Gc
Longbrake, David, David.Longbrake@du.edu,
　University of Denver – Zi
Longerich, Henry, (709) 737-8380 henry@sparky2.esd.mun.ca,
　Memorial University of Newfoundland – Cg
Longnecker, Krista, (508) 289-2824 klongnecker@whoil.edu,
　Woods Hole Oceanographic Institution – Oc
Longoria, Jose F., (786) 210-0590 longoria@fiu.edu,
　Florida International University – ZeGoe
Longoria, Lorena, 956 665-5124 lorena.longoria@utrgv.edu,
　University of Texas, Rio Grande Valley – Ob
Longpre, Marc-Antoine, 718-997-3259 mlongpre@qc.cuny.edu,
　Graduate School of the City University of New York – Zg
Longstaffe, Frederick J., (519) 661-3177 flongsta@uwo.ca,
　Western University – Cs
Longworth, Brett, 508-289-3559 blongworth@whoi.edu,
　Woods Hole Oceanographic Institution – Ou
Lonn, Jeff, (406) 496-4177 jlonn@mtech.edu,
　Montana Tech of The University of Montana – GgtGc
Lonsdale, Darcy J., (631) 632-8712 darcy.lonsdale@stonybrook.edu,
　SUNY, Stony Brook – Ob
Lonsdale, Peter F., (858) 534-2855 plonsdale@ucsd.edu,
　University of California, San Diego – Gu
Looney, Brian, 803-725-3692 brian02.looney@srnl.doe.gov,
　Clemson University – Hq
Loope, David B., (402) 472-2647 dloope1@unl.edu,
　University of Nebraska, Lincoln – Gsg
Loope, Henry M., (812) 856-3117 hloope@indiana.edu,
　Indiana University – Gls
Looy, Cynthia, 510-642-1607 looy@berkeley.edu,
　University of California, Berkeley – Pb

Lopez, Alberto, (787) 265-3845 alberto.lopez3@upr.edu,
　University of Puerto Rico – YdGtYs
Lopez, Dina L., (740) 593-9435 lopezd@ohio.edu,
　Ohio University – CgHwGv
Lopez, Glenn R., (631) 632-8660 glenn.lopez@stonybrook.edu,
　SUNY, Stony Brook – Ob
Lopez, Margarita, marlopez@cicese.mx,
　Centro de Investigación Científica y de Educación Superior de Ensenada – Cc
López, Maricela , (662) 370-6262 maricela.lopez@gmail.com,
　Universidad Estatal de Sonora – NxGg
López, Sol, antares@ucm.es,
　Univ Complutense de Madrid – Gz
Lora, Juan, 203432-2627 Juan.Lora@yale.edu,
　Yale University – As
Lord, Mark L., (828) 227-2271 mlord@wcu.edu,
　Western Carolina University – HwGmZe
Lorenz, Ralph, (520) 621-5585 rlorenz@lpl.arizona.edu,
　University of Arizona – Xg
Lorenzo, Juan M., (225) 388-2497 juan@geol.lsu.edu,
　Louisiana State University – Ys
Lorenzoni, Irene, +44 (0)1603 59 3173 i.lorenzoni@uea.ac.uk,
　University of East Anglia – Eg
Lori, Lisa, 573-368-2152 lisa.lori@dnr.mo.gov,
　Missouri Dept of Natural Resources – Gp
Lorinczi, Piroska, +44(0) 113 34 39245 earpl@leeds.ac.uk,
　University of Leeds – Np
Loring, Arthur P., (718) 262-2079 York College (CUNY) – Gm
Loriot, George, 802-626-6378 george.loriot@northernvermont.edu,
　Northern Vermont University-Lyndon – AsZn
Losh, Steven, (507) 389-6323 steven.losh@mnsu.edu,
　Minnesota State University – GzxGo
Losos, Zdenek, ++420 549 49 5623 losos@sci.muni.cz,
　Masaryk University – Gzy
Lott, Dempsey E., (508) 289-2929 dlott@whoi.edu,
　Woods Hole Oceanographic Institution – Oc
Lotterhos, Kathleen, (781) 581-7370 k.lotterhos@neu.edu,
　Northeastern University – Zn
Lottermoser, Bernd, +44 01326 255973 B.Lottermoser@exeter.ac.uk,
　Exeter University – Ca
Loucks, Bob, (512) 471-0366 bob.loucks@beg.utexas.edu,
　University of Texas at Austin, Jackson School of Geosciences – Go
Louden, Keith E., (902) 494-3452 keith.louden@dal.ca,
　Dalhousie University – Yr
Lough, Amanda, (215) 895 6456 amanda.c.lough@drexel.edu,
　Drexel University – Ysg
Louie, John, (775) 784-4219 louie@unr.edu,
　University of Nevada, Reno – Yg
Louie, John N., (775) 299-3835 louie@seismo.unr.edu,
　University of Nevada, Reno – Ys
Lounsbury, Diane E., (585) 245-5291 lounsbur@geneseo.edu,
　SUNY, Geneseo – Zn
Lourenço, José M., 00351259350280 martinho@utad.pt,
　Universidade de Trás-os-Montes e Alto Douro – YeZin
Lourenco, Sergio, lourencosd@cf.ac.uk,
　Cardiff University – Sp
Love, David W., (505) 835-5146 dave@gis.nmt.edu,
　New Mexico Institute of Mining & Technology – Ge
Love, Gordon D., glove@ucr.edu,
　University of California, Riverside – Cog
Love, Renee, (208) 885-4079 rlove@uidaho.edu,
　University of Idaho – Pg
Lovekin, Jonathan, 303-384-2654 jlovekin@mines.edu,
　Colorado Geological Survey – NgZuGs
Loveland, Andrea M., (307) 766-2286 andrea.loveland@wyo.gov,
　Wyoming State Geological Survey – Gct
Loveland, Karen, 573-368-2142 karen.loveland@dnr.mo.gov,
　Missouri Dept of Natural Resources – Zn
Loveless, Jack, (413) 585-2657 jloveles@smith.edu,
　Smith College – Gtc
Lovell, Mark D., (208) 496-7680 lovellm@byui.edu,
　Brigham Young University - Idaho – HyZiGo
Lovell, Mike, +440116 252 3798 mike.lovell@le.ac.uk,
　Leicester University – Yg
LoVetere, Crystal, clovetere@cerritos.edu,
　Cerritos College – Zy
Lovett, Andrew A., +44 (0)1603 59 3126 a.lovett@uea.ac.uk,

University of East Anglia – Ziu
Lovett, Cole, (269) 927-8744 lovett@lakemichigancollege.edu,
Lake Michigan College – Gg
Lovseth, John, 618.374.5291 john.lovseth@principia.edu,
Principia College – Zni
Lowe, Donald R., (650) 725-3040 lowe@pangea.stanford.edu,
Stanford University – Gd
Lowe, Douglas, Douglas.Lowe@manchester.ac.uk,
University of Manchester – As
Lowe, Mike, (801) 537-3389 nrugs.mlowe@state.ut.us,
Utah Geological Survey – Hw
Lowell, Robert P., (540) 231-6004 rlowell@vt.edu,
Virginia Polytechnic Institute & State University – Yrh
Lowell, Thomas V., (513) 556-4165 thomas.lowell@uc.edu,
University of Cincinnati – Gl
Lowenstein, Tim K., (607) 777-4254 lowenst@binghamton.edu,
Binghamton University – Cl
Lower, Steven K., 614 292-1571 lower.9@osu.edu,
Ohio State University – PoZa
Lowery, Birl, (608) 262-2752 blowery@wisc.edu,
University of Wisconsin, Madison – Sp
Lowry, Anthony R., (435) 797-7096 tony.lowry@usu.edu,
Utah State University – YdsGt
Lowry, Christopher S., 716-645-4266 cslowry@buffalo.edu,
SUNY, Buffalo – Hw
Lowry, David, +44 1784 443105 D.Lowry@rhul.ac.uk,
Royal Holloway University of London – Cs
Lowry, Wallace D., (540) 231-6521 Virginia Polytechnic Institute & State University – Gc
Lowry, William R., (314) 935-5821 Washington University in St. Louis – Zn
Loxsom, Fred, loxsomf@easternct.edu,
Eastern Connecticut State University – Zn
Loyd, Sean, 657-278-4537 sloyd@fullerton.edu,
California State University, Fullerton – CgGe
Loydell, David K., +44 023 92 842698 david.loydell@port.ac.uk,
University of Portsmouth – GrPg
Loynachan, Thomas E., (515) 290-7154 teloynac@iastate.edu,
Iowa State University of Science & Technology – So
Lozier, M. Susan, (919) 681-8199 s.lozier@duke.edu,
Duke University – Op
Lozinsky, Richard P., (714) 992-7445 rlozinsky@fullcoll.edu,
Fullerton College – GreGm
Lozos, Julian, (818) 677-7977 julian.lozos@csun.edu,
California State University, Northridge – YgGt
Lozowski, Edward P., (403) 492-0348 University of Alberta – As
Lozowski, Edward P., (780) 492-0385 edward.lozowski@ualberta.ca,
University of Alberta – As
Lu, Yongqi, 217 244-4985 yongqilu@illinois.edu,
University of Illinois – Zm
Lu, Youyu, (902) 426-7780 youyu.lu@ec.gc.ca,
Dalhousie University – Op
Lu, Yuehan, 205-348-1882 yuehanlu@as.ua.edu,
University of Alabama – GgOuGe
Lu, Zhong, 214-768-0101 Southern Methodist University – Zr
Lu, Zunli, 315-443-0281 zunlilu@syr.edu,
Syracuse University – Cg
Lucas, Beth, (540) 231-4595 blucas06@vt.edu,
Virginia Polytechnic Institute & State University – Zn
Lucas, Cathy, +44 (0)23 80596617 cathy.lucas@noc.soton.ac.uk,
University of Southampton – Ob
Lucas, Franklin A., frank.lucas@uniben.edu,
University of Benin – PmlGr
Lucas-Clark, Joyce, 415-452-5046 jluclark@comcast.net,
City College of San Francisco – PgGg
Lucey, Paul G., (808) 956-3137 lucey@higp.hawaii.edu,
University of Hawai'i, Manoa – Xy
Lucia, F. J., 512-471-1534 jerry.lucia@beg.utexas.edu,
University of Texas, Austin – Eo
Lucia, F. Jerry, (512) 471-7367 jerry.lucia@beg.utexas.edu,
University of Texas at Austin, Jackson School of Geosciences – Go
Luciano, Katherine E., (843) 953-6843 lucianok@dnr.sc.gov,
South Carolina Dept of Natural Resources – GusGm
Lucotte, Marc Michel, 514-987-3000 #3767 lucotte.marc_michel@uqam.ca,
Universite du Quebec a Montreal – Oc
Luczaj, John A., (920) 465-5139 luczajj@uwgb.edu,
University of Wisconsin, Green Bay – HwGsCl

Ludman, Allan, (718) 997-3271 Graduate School of the City University of New York – Gg
Ludman, Allan, (718) 997-3300 allan.ludman@qc.cuny.edu,
Queens College (CUNY) – GgtGr
Ludvigson, Greg A., (785) 864-2734 gludvigson@ku.edu,
University of Kansas – GrCsPc
Ludwig, Kris, kaludwig@usgs.gov,
University of Washington – Rnc
Ludwikoski, David J., 443-840-4216 dludwikoski@ccbcmd.edu,
Community College of Baltimore County, Catonsville – ZegZn
Lueth, Virgil L., 575-835-5140 vwlueth@nmt.edu,
New Mexico Institute of Mining and Technology – GzEg
Luhmann, Andrew J., 630 752 7476 andrew.luhmann@wheaton.edu,
Wheaton College – HwGmCl
Lukas, Roger, (808) 956-7896 rlukas@soest.hawaii.edu,
University of Hawai'i, Manoa – Op
Lukinbeal, Christopher, clukinbe@email.arizona.edu,
University of Arizona – Zi
Lumley, David, (972) 883-2424 David.Lumley@utdallas.edu,
University of Texas, Dallas – YsrYn
Lumpkin, Thomas A., (509) 335-2726 christie.lumpkin@wsu.edu,
Washington State University – Zn
Lumsden, David N., (901) 678-2774 dlumsden@memphis.edu,
University of Memphis – Gd
Lund, David, 860-405-9331 david.lund@uconn.edu,
University of Connecticut – Oc
Lund, Jay R., (916) 752-5671 University of California, Davis – Hs
Lund, Steven P., (213) 740-5835 slund@usc.edu,
University of Southern California – Ym
Lund, William R., (435) 865-9034 billlund@utah.gov,
Utah Geological Survey – Ng
Lundblad, Steven P., 808-932-7548 slundbla@hawaii.edu,
University of Hawai'i, Hilo – Gra
Lundelius, Ernest L., (512) 232-5513 erniel@utexas.edu,
University of Texas, Austin – PvePg
Lundquist, Jessica D., (206) 685-7594 jdlund@uw.edu,
University of Washington – As
Lundstrom, Craig C., (217) 244-6293 lundstro@illinois.edu,
University of Illinois, Urbana-Champaign – Cpg
Lundstrom, Elizabeth, elundstr@umn.edu,
University of Minnesota, Twin Cities – Ca
Lunine, Jonathan I., (520) 621-2789 jlunine@lpl.arizona.edu,
University of Arizona – Xc
Lunine, Jonathan I., 607-255-5911 jlunine@astro.cornell.edu,
Cornell University – Xc
Luo, Chao, chao.luo@eas.gatech.edu,
Georgia Institute of Technology – As
Luo, Gang, 512-475-6613 gangluo66@gmail.com,
University of Texas, Austin – Gtc
Luo, Johnny, 212-650-8936 luo@sci.ccny.cuny.edu,
Graduate School of the City University of New York – Zg
Luo, Jun, (417) 836-4273 junluo@missouristate.edu,
Missouri State University – Zi
Luo, Zhexi, (412) 622-6578 calleryb@CarnegieMNH.org,
Carnegie Museum of Natural History – Pv
Lupankwa, Mlindelwa, 012 382 6213 lupankwam@tut.ac.za,
Tshwane University of Technology – HywGg
Lupia, Richard, (405) 325-7229 rlupia@ou.edu,
University of Oklahoma – Pm
Lupo, Anthony R., (573) 884-1638 lupoa@missouri.edu,
University of Missouri, Columbia – As
Lupulescu, Marian V., (518) 474-1432 marian.lupulescu@nysed.gov,
New York State Geological Survey – GzxEm
Luque, Francisco Javier, jluque@ucm.es,
Univ Complutense de Madrid – Gz
Lusardi, Barb, (612) 626-4791 lusar001@umn.edu,
University of Minnesota – Gl
Lusk, Braden, 859-257-1105 braden.lusk@uky.edu,
University of Kentucky – Nm
Luth, Robert W., (780) 492-2740 robert.luth@ualberta.ca,
University of Alberta – Gi
Luther, Amy, (225) 578-2337 aluther@lsu.edu,
Louisiana State University – GcZe
Luther, Douglas S., (808) 956-5875 dluther@soest.hawaii.edu,
University of Hawai'i, Manoa – Op
Luther, Mark E., (727) 553-1528 mluther@marine.usf.edu,
University of South Florida – Op

Luther, III, George W., (302) 645-4208 luther@udel.edu,
 University of Delaware – Cm
Luttge, Andreas, (713) 348-6304 aluttge@rice.edu,
 Rice University – Cg
Luttrell, Karen, (225) 578-5620 kluttrell@lsu.edu,
 Louisiana State University – Yg
Lutz, Alexandra, 775.673.7418 alexandra.lutz@dri.edu,
 University of Nevada, Reno – Hw
Lutz, Pascale, +33(0)3 44068990 pascale.lutz@lasalle-beauvais.fr,
 Institut Polytechnique LaSalle Beauvais (ex-IGAL) – YgxYd
Lutz, Timothy M., (610) 436-3498 tlutz@wcupa.edu,
 West Chester University – Gq
Lux, Daniel R., (207) 581-4494 dlux@maine.edu,
 University of Maine – Gi
Luyendyk, Bruce P., (805) 893-2827 University of California, Santa Barbara – Yr
Luzincourt, Marc R., (418) 654-3715 mluzinco@nrcan.gc.ca,
 Universite du Quebec – Cs
Lužar-Oberiter, Borna, +38514606115 bluzar@geol.pmf.hr,
 University of Zagreb – Gg
Lwiza, Kamazima M., (631) 632-7309 kamazima.lwiza@stonybrook.edu,
 SUNY, Stony Brook – Oc
Lübbe, Maike, 089/2180 4337 Meike.Luebbe@lrz.uni-muenchen.de,
 Ludwig-Maximilians-Universitaet Muenchen – Gz
Lyle, Mike, 757-822-7189 tclylem@tcc.edu,
 Tidewater Community College – Gg
Lyle, Mitchell, 541-737-3427 mlyle@coas.oregonstate.edu,
 Oregon State University – GgYg
Lynch, Michael J., (859) 323-0561 mike.lynch@uky.edu,
 University of Kentucky – Zn
Lynch-Stieglitz, Jean, (404) 894-3944 jean@eas.gatech.edu,
 Georgia Institute of Technology – OoPcCs
Lynds, Ranie, (307) 766-2286 Ext. 235 ranie.lynds@wyo.gov,
 Wyoming State Geological Survey – EoGro
Lynds, Ranie, (307) 766-2286 ranie.lynds@wyo.gov,
 University of Wyoming – Gsr
Lynn, Resler M., (540) 231-5790 resler@vt.edu,
 Virginia Polytechnic Institute & State University – Zy
Lyon, Ian, Ian.Lyon@manchester.ac.uk,
 University of Manchester – Xc
Lyon, Ronald J. P., (650) 725-8077 lyon@pangea.stanford.edu,
 Stanford University – Zr
Lyons, Lawrence, (310) 206-7876 larry@atmos.ucla.edu,
 University of California, Los Angeles – As
Lyons, Steven, Steve.Lyons@noaa.gov,
 Angelo State University – Am
Lyons, Timothy W., (951) 827-3106 timothy.lyons@ucr.edu,
 University of California, Riverside – CgOcCl
Lyons, William B., (614) 688-3241 lyons.142@osu.edu,
 Ohio State University – HwCgZg
Lyson, Tyler, 303-370-6328 Tyler.Lyson@dmns.org,
 Denver Museum of Nature & Science – Pv
Lytwyn, Jennifer N., 713-743-3296 jlytwyn@uh.edu,
 University of Houston – Gi

M

Ma, Chong, (334) 844-4203 chongma@auburn.edu,
 Auburn University – GctGg
Ma, Jingsheng, +44 (0)131 451 8296 jingsheng.ma@pet.hw.ac.uk,
 Heriot-Watt University – Np
Ma, Lena Q., (352) 392-9063 lqma@ufl.edu,
 University of Florida – Sc
Ma, Lin, 915-747-5218 lma@utep.edu,
 University of Texas, El Paso – HgCg
Ma, Shuo, 619-594-3091 sma@mail.sdsu.edu,
 San Diego State University – Ys
Ma, Yanxia, yma@lsu.edu,
 Louisiana State University – Og
Maantay, Juliana, 718-960-8574 maantay@aol.com,
 Lehman College (CUNY) – Zi
Maas, Regan, 818-677-3515 regan.maas@csun.edu,
 California State University, Northridge – Zi
Maasch, Kirk A., (207) 581-2197 kirk@iceage.umeqs.maine.edu,
 University of Maine – As
Mabee, Stephen B., (413) 545-4814 sbmabee@geo.umass.edu,
 Massachusetts Geological Survey – Hw
Maboko, Makenya A., +255222410013 mmaboko@uccmail.co.tz,
 University of Dar es Salaam – GpCgGx
Mac Eachern, James A., 778-782-5388 jmaceach@sfu.ca,
 Simon Fraser University – Gs
Mac Niocaill, Conall, +44 (0)1865 282135 conallm@earth.ox.ac.uk,
 University of Oxford – GtgYm
Macadam, John, J.D.Macadam@exeter.ac.uk,
 Exeter University – Ge
Macalady, Jennifer L., (814) 865-6330 jlm80@psu.edu,
 Pennsylvania State University, University Park – PgClSb
MacAyeal, Douglas R., (773) 702-8027 drm7@midway.uchicago.edu,
 University of Chicago – Yr
MacCarthy, Ivor, +353 21 4902152 i.maccarthy@ucc.ie,
 University College Cork – Gs
MacCarthy, Patrick, pmaccart@mines.edu,
 Colorado School of Mines – Co
MacCready, Parker, (206) 685-9588 parker@ocean.washington.edu,
 University of Washington – Op
Macdonald, Francis, francism@geol.ucsb.edu,
 University of California, Santa Barbara – Gg
MacDonald, Iain, +44(0)29 208 77302 McdonaldI1@cardiff.ac.uk,
 University of Wales – Ca
Macdonald, R. Heather, (757) 221-2443 rhmacd@wm.edu,
 College of William & Mary – GsZe
MacDonald, William D., (607) 777-2863 Binghamton University – Gc
Macdougall, J. Douglas, jdmacdougall@ucsd.edu,
 University of California, San Diego –
Mace, Gerald, 801-585-9489 jay.mace@utah.edu,
 University of Utah – As
MacFadden, Bruce J., 3523921721 x496 bmacfadd@flmnh.ufl.edu,
 University of Florida – Pv
Macfarlane, Andrew W., (305) 348-3980 macfarla@fiu.edu,
 Florida International University – Em
MacGregor, Kelly, (651) 696-6441 macgregor@macalester.edu,
 Macalester College – Gml
Machel, Hans G., (780) 492-5659 hans.machel@ualberta.ca,
 University of Alberta – Go
Machovina, Brett , 719-333-3080 brett.machovina@usafa.edu,
 United States Air Force Academy – Zir
Macias, Steve E., (360) 475-7711 smacias@olympic.edu,
 Olympic College – Gg
Maciha, Mark J., (928) 523-8242 Mark.Maciha@nau.edu,
 Northern Arizona University – ZnnZn
MacInnes, Breanyn, 509-963-2827 macinnes@geology.cwu.edu,
 Central Washington University – GseGu
MacInnes, Michael, (505) 665-2154 Los Alamos National Laboratory – Zn
MacIntyre, Hugh, (902) 494-2932 hugh.macintyre@dal.ca,
 Dalhousie University – Ob
Macintyre, Ian G., (202) 633-1339 Smithsonian Inst / Natl Museum of Natural Hist– Gs
MacIssac, Daniel E., (716) 878-6731 solargs@buffalostate.edu,
 Buffalo State College – Xa
Mack, John E., (716) 878-5006 mackje@buffalostate.edu,
 Buffalo State College – Xg
Mackaness, William A., +44 (0) 131 650 8163 william.mackaness@ed.ac.uk,
 Edinburgh University – Zi
Mackas, David L., (250) 363-6442 mackas@ios.bc.ca,
 University of Victoria – Ob
MacKay, Allison A., 614-247-7652 mackay.49@osu.edu,
 Ohio State University – Cq
Mackenzie, Fred T., (808) 395-7907 fredm@soest.hawaii.edu,
 University of Hawai'i, Manoa – CmGes
Mackenzie, Kristen, 303-370-8372 Kristen.MacKenzie@dmns.org,
 Denver Museum of Nature & Science – Zg
Macko, Stephen A., (434) 924-6849 sam8f@virginia.edu,
 University of Virginia – CosOc
Mackowiak, Cheryl, 850-875-7126 Echo13@ufl.edu,
 University of Florida – So
MacLachlan, Ian R., (403) 329-2076 maclachlan@uleth.ca,
 University of Lethbridge – Ze
MacLachlan, John, (905) 525-9140 (Ext. 24195) maclacjc@mcmaster.ca,
 McMaster University – Gs
MacLaughlin, Mary M., (406) 496-4655 mmaclaughlin@mtech.edu,
 Montana Tech of the University of Montana – NrgNt
MacLean, John S., 435.586.1937 johnmaclean@suu.edu,
 Southern Utah University – GctZe
Maclennan, John, +44 (0) 1223 761602 jmac05@esc.cam.ac.uk,
 University of Cambridge – Gi

MacLeod, C, macleod@cf.ac.uk,
Cardiff University – Gu
MacLeod, Chris, +44(0)29 208 74332 MacLeod@cardiff.ac.uk,
University of Wales – Og
MacLeod, Kenneth A., (573) 884-3118 macleodk@missouri.edu,
University of Missouri – Pg
MacMahan, Jamie, jhMacMah@nps.edu,
Naval Postgraduate School – Onp
MacNish, Robert, (520) 621-5082 macnish@hwr.arizona.edu,
University of Arizona – Hw
Macomber, Richard, (718) 460-1476 Long Island University, Brooklyn Campus – Pg
Macpherson, Colin G., 0191 3342283 colin.macpherson@durham.ac.uk,
Durham University – CgGiCs
MacPherson, Glenn J., (202) 633-1803 macphers@si.edu,
Smithsonian Inst / Natl Museum of Natural Hist– XcGi
Macpherson, Gwendolyn L., (785) 864-2742 glmac@ku.edu,
University of Kansas – Hw
MacQuarrie, Pamela, (403) 440-6176 pmacquarrie@mtroyal.ca,
Mount Royal University – Zy
Madadi, Mahyar, +61 8 9266-2324 Mahyar.Madadi@curtin.edu.au,
Curtin University – Ye
Maddock, III, Thomas, (520) 621-7120 maddock@hwr.arizona.edu,
University of Arizona – Hq
Maddocks, Rosalie F., 713-743-2772 rmaddocks@uh.edu,
University of Houston – Pm
Madej, Mary Ann, mary_ann_madej@usgs.gov,
U.S. Geological Survey – Gme
Madin, Ian P., (971) 673-1555 Oregon Dept of Geology and Mineral Industries – Gg
Madison, Frederick W., (608) 263-4004 fredmad@wisc.edu,
University of Wisconsin, Madison – Sd
Madsen, David B., (801) 537-3314 nrugs.dmadsen@state.ut.us,
Utah Geological Survey – Ga
Madsen, John A., (302) 831-1608 jmadsen@udel.edu,
University of Delaware – YrOuZe
Maerz, Christian, +44 (0) 1133431504 c.maerz@leeds.ac.uk,
University of Leeds – CmOuEm
Magaard, Lorenz, 808-956-7509 lorenz@hawaii.edu,
University of Hawai'i, Manoa – Op
Magloughlin, Jerry F., (970) 491-1812 jerrym@cnr.colostate.edu,
Colorado State University – Gpz
Magnani, M. Beatrice, 214-768-1751 Southern Methodist University – YsGtYe
Magnusdottir, Gudrun, (949) 824-3520 gudrun@uci.edu,
University of California, Irvine – As
Magondu, Moffat G., mmagondu@jkuat.ac.ke,
Jomo Kenyatta University of Agriculture & Technology – ZirZg
Magsino, Sammantha L., (202) 334-2744 smagsino@nas.edu,
National Academy of Sciences, Engineering, and Medicine – GvNg
Magson, Justine, +27 (0)51 401 2373 markramj1@ufs.ac.za,
University of the Free State – CgGi
Mahaffee, Tina, 478-934-3400 tmahaffee@mgc.edu,
Middle Georgia College – Gg
Mahaffey, Claire, +44 0151 794 4090   Claire.Mahaffey@liverpool.ac.uk,
University of Liverpool – Oc
Mahan, Kevin H., kevin.mahan@colorado.edu,
University of Colorado – Gpc
Mahatsente, Rezene, 205-348-5772 rmahatsente@ua.edu,
University of Alabama – YgmYv
Maher, John, 9134698500 x5953 jmaher1@jccc.edu,
Johnson County Community College – Zy
Maher, Kierran, 575-835-6354 kmaher@nmt.edu,
New Mexico Institute of Mining and Technology – Eg
Maher, Jr., Harmon D., 402-554-4807 harmon_maher@unomaha.edu,
University of Nebraska, Omaha – GcsGt
Maher, Jr., Louis J., (608) 262-9595 maher@geology.wisc.edu,
University of Wisconsin-Madison – Pl
Mahlen, Nancy J., (585) 245-5016 mahlen@geneseo.edu,
SUNY, Geneseo – GgpCc
Mahler, Robert L., (208) 885-7025 University of Idaho – Zn
Mahmoud, Sara A., 002-03-3921595 geo_soso2006@yahoo.com,
Alexandria University – Gg
Mahmoudi, Nagissa, nagissa.mahmoudi@mcgill.ca,
McGill University – Cb
Mahoney, J. Brian, (715) 836-4952 mahonej@uwec.edu,
University of Wisconsin, Eau Claire – GstEg

Mahood, Gail A., (650) 723-1429 gail@pangea.stanford.edu,
Stanford University – Gi
Mahowald, Natalie M., 607-255-5166 nmm63@cornell.edu,
Cornell University – As
Maier, Raina M., (520) 621-7231 rmaier@ag.arizona.edu,
University of Arizona – Zn
Maillol, Jean-Michel, (403) 220-8393 maillol@ucalgary.ca,
University of Calgary – Ym
Main, Ian G., +44 (0) 131 650 4911 Ian.Main@ed.ac.uk,
Edinburgh University – Ys
Maio, Chris, (907) 474-5651 cvmaio@alaska.edu,
University of Alaska, Fairbanks – Zy
Maisey, John G., (212) 769-5811 American Museum of Natural History – Pv
Majodina, Thando, (012) 382-6283 majodinato@tut.ac.za,
Tshwane University of Technology – GgzGe
Major, Penni, 281-765-7865 penny.westerfeld@nhmccd.edu,
Lonestar College - North Harris – Gg
Major, Ruth H., 518-629-7131 Hudson Valley Community College – GgRh
Maju-Oyovwikowhe, Efetobore G., efetobore.maju-oyovwikowhe@uniben.edu,
University of Benin – GszGo
Majzlan, Juraj, +49(0)3641 948700 juraj.majzlan@uni-jena.de,
Friedrich-Schiller-University Jena – Gz
Mak, John E., (631) 632-8673 john.mak@stonybrook.edu,
SUNY, Stony Brook – AsCas
Mak, Mankin, (217) 333-8071 mak@atmos.uiuc.edu,
University of Illinois, Urbana-Champaign – As
Makkawi, Mohammad, +966138602621 makkawi@kfupm.edu.sa,
King Fahd University of Petroleum and Minerals – HwGoq
Makovicky, Peter J., (312) 665-7633 pmakovicky@fieldmuseum.org,
Field Museum of Natural History – Pv
Makungo, R., +27 15 962 8577 rachel.makungo@univen.ac.za,
University of Venda for Science & Technology – Hqw
Malahoff, Alexander, (808) 956-6802 malahoff@hawaii.edu,
University of Hawai'i, Manoa – Cm
Malakar, Nabin, nmalakar@worcester.edu,
Worcester State University – Ap
Malanotte-Rizzoli, Paola M., (617) 253-2451 rizzoli@mit.edu,
Massachusetts Institute of Technology – Op
Malcolm, Elizabeth, 757-233-8751 emalcolm@vwc.edu,
Virginia Wesleyan College – AsCgOg
Malcuit, Robert J., malcuit@denison.edu,
Denison University – Gx
Malega, Ron, 417-836-4556 rmalega@missouristate.edu,
Missouri State University – Zn
Malhotra, Rakesh, rmalhotra@nccu.edu,
North Carolina Central University – Zri
Malhotra, Renu, (520) 626-5899 renu@lpl.arizona.edu,
University of Arizona –
Malin, Gill, +44 (0)1603 59 2531 g.malin@uea.ac.uk,
University of East Anglia – ObcCb
Malin, Peter E., (919) 681-8889 p.malin@auckland.ac.nz,
Duke University – Ys
Malinconico, Lawrence L., (610) 330-5195 malincol@lafayette.edu,
Lafayette College – YgGcv
Malinverno, Alberto , (845) 365-8577 alberto@ldeo.columbia.edu,
Columbia University – Gug
Malinverno, Alberto, alberto@ldeo.columbia.edu,
Columbia University – Go
Mallarino, Antonio W., (515) 294-6200 apmallar@iastate.edu,
Iowa State University of Science & Technology – So
Mallick, Subhashis, 307-766-2884 smallick@uwyo.edu,
University of Wyoming – YseYg
Mallik, Ananya, 401)-874-2671 ananya_mallik@uri.edu,
University of Rhode Island – Gx
Mallinson, David, (252) 328-1344 mallinsond@ecu.edu,
East Carolina University – Gu
Malo, Michel, michel.malo@ete.inrs.ca,
Universite du Quebec – GctGs
Malone, Andrew, 312-413-1868 amalone@uic.edu,
University of Illinois at Chicago – GlAt
Malone, David H., (309) 438-7649 dhmalon@ilstu.edu,
Illinois State University – GcsEg
Malone, Shawn J., 765-285-8263 sjmalone@bsu.edu,
Ball State University – GtxGm
Malone, Stephen D., 206-685-3811 smalone@uw.edu,
University of Washington – YsGt

Maloney, Ashley, 609-258-4101 Princeton University – Oc
Maloney, Eric, emaloney@atmos.colostate.edu,
 Colorado State University – As
Maloof, Adam C., (609) 258-4101 maloof@princeton.edu,
 Princeton University – Ym
Malservisi, Rocco, 089/2180 4202 malservisi@rsmas.miami.edu,
 Ludwig-Maximilians-Universitaet Muenchen – Yg
Malzer, Gary L., (612) 625-6728 gmalzer@soils.umn.edu,
 University of Minnesota, Twin Cities – Sc
Malzone, Jonathan, (859) 622-1278 Jonathan.Malzone@eku.edu,
 Eastern Kentucky University – HwgGm
Mana, Sara, smana@salemstate.edu,
 Salem State University – GicGv
Manabe, Syukuro, (609) 258-2790 manabe@princeton.edu,
 Princeton University – AstPc
MANCA, Prof. Pierpaolo P., 070-675-5529 ppmanca@unica.it,
 Universita di Cagliari – Nm
Manchester, Steven R., (352) 392-1721 steven@flmnh.ufl.edu,
 University of Florida – PbGs
Mancini, Ernest A., (205) 348-4319 emancini@as.ua.edu,
 University of Alabama – Gro
Mand, Arlene, (215) 573-3164 amand@sas.upenn.edu,
 University of Pennsylvania – Zn
Manda, Alex K., (252) 328-9403 mandaa@ecu.edu,
 East Carolina University – HwgHq
Mandel, Rolfe, (785) 864-2171 mandel@ku.edu,
 University of Kansas – Gam
Mandia, Scott A., 631-451-4104 mandias@sunysuffolk.edu,
 Suffolk County Community College, Ammerman Campus – Amt
Mandra, York T., (415) 338-2061 ytmandra@sfsu.edu,
 San Francisco State University – Pg
Mandrone, Giuseppe, giuseppe.mandrone@unito.it,
 Università di Torino – Ng
Manduca, Cathryn A., (507) 222-7096 cmanduca@carleton.edu,
 Carleton College – Gi
Mandziuk, William S., (204) 474-7826 mandziu0@cc.umanitoba.ca,
 University of Manitoba – Gg
Manecan, Teodosia, (212) 772-5265 tmanecan@hunter.cuny.edu,
 Hunter College (CUNY) – Gp
Maneiro, Kathryn A., 630 752 7228 kathryn.maneiro@wheaton.edu,
 Wheaton College – CcGic
Maneta, Marco, 406-243-2454 marco.maneta@mso.umt.edu,
 University of Montana – Hsg
Manfrino, Carrie M., (908) 737-3697 cmanfrin@kean.edu,
 Kean University – GuOb
Manga, Michael, (510) 643-8532 manga@seismo.berkeley.edu,
 University of California, Berkeley – GvXyHw
Manganini, Steven J., (508) 289-2778 smanganini@whoi.edu,
 Woods Hole Oceanographic Institution – Ou
Mangano, Gabriela, 306-966-5730 gabriela.mangano@usask.ca,
 University of Saskatchewan – PsGs
Mangel, Mark S., (831) 459-5785 msmangel@cats.ucsc.edu,
 University of California, Santa Cruz – Zn
Manger, Walter, 479-575-3355 wmanger@uark.edu,
 University of Arkansas, Fayetteville – Gr
Manghnani, Murli H., (808) 956-7825 murli@soest.hawaii.edu,
 University of Hawai'i, Manoa – Yx
Mango, Helen N., (802) 468-1478 helen.mango@castleton.edu,
 Castleton University – CgGe
Mangriotis, Maria-Daphne , +44 (0) 131 451 3565 m.mangriotis@hw.ac.uk,
 Heriot-Watt University – Ys
Manley, Patricia L., (802) 443-5430 patmanley@middlebury.edu,
 Middlebury College – YrGu
Manley, Thomas O., (802) 443-3114 tmanley@middlebury.edu,
 Middlebury College – Op
Mann, Dan, dhmann@alaska.edu,
 University of Alaska, Fairbanks – Zy
Mann, Daniel H., (907) 474-5872 University of Alaska, Fairbanks – Pe
Mann, Keith O., (740) 368-3620 komann@owu.edu,
 Ohio Wesleyan University – Pi
Mann, Michael E., (814) 863-4075 mann@meteo.psu.edu,
 Pennsylvania State University, University Park – As
Mann, Paul, (512) 471-0452 paulm@ig.utexas.edu,
 University of Texas, Austin – Gt
Manning, Andrew, a.manning@uea.ac.uk,
 University of East Anglia – As
Manning, Christina, +44 1784 443835 c.manning@es.rhul.ac.uk,
 Royal Holloway University of London – Cg
Manning, Craig E., (310) 206-3290 manning@epss.ucla.edu,
 University of California, Los Angeles – GxCp
Manning, Sturt W., 607-255-8650 sm456@cornell.edu,
 Cornell University – Ga
Mannix, Devin, (217) 244-2419 mannix@illinois.edu,
 University of Illinois – Hw
Manon, Matthew R., (518) 388-8015 manonm@union.edu,
 Union College – Gp
Manos, Leah D., (660) 562-1385 lmanos@nwmissouri.edu,
 Northwest Missouri State University – Zy
Manser, Nathan, 906/487-2894 Michigan Technological University – GeNm
Mansour, Ahmed S., 002-03-3921595 ah_sadek@hotmail.com,
 Alexandria University – Ggs
Manspeizer, Warren, (973) 353-5100 mansp@andromeda.rutgers.edu,
 Rutgers, The State University of New Jersey, Newark – Gr
Mansur, Adam, mansura@si.edu,
 Smithsonian Inst / Natl Museum of Natural Hist– Zf
Manton, William I., (972) 883-2441 manton@utdallas.edu,
 University of Texas, Dallas – Cc
Manz, Lorraine A., 701 328 8000 lmanz@nd.gov,
 North Dakota Geological Survey – GmlGg
Manzi, Anthony, 718-409-7371 amanzi@sunymaritime.edu,
 SUNY, Maritime College – Am
Manzocchi, Tom, (+353) 1 716 2605 tom.manzocchi@ucd.ie,
 University College Dublin – Goc
Mao, Ho-kwang, 202-478-8960 hmao@carnegiescience.edu,
 Carnegie Institution for Science – Yx
Mao, Ho-kwang (David), (773) 702-8136 University of Chicago – Yg
Mapani, Benjamin S., +264-61-2063745 bmapani@unam.na,
 University of Namibia – GtcGp
Mapholi, Thendo, +27 (0)51 401 2724 mapholit@ufs.ac.za,
 University of the Free State – Ggp
Marcano, Eugenio J., (413) 538-3237 emarcano@mtholyoke.edu,
 Mount Holyoke College – ZiSoZn
Marcantonio, Franco, 979-845-9240 marcantonio@geo.tamu.edu,
 Texas A&M University – Ct
Marchal, Olivier, (508) 289-3374 omarchal@whoi.edu,
 Woods Hole Oceanographic Institution – Pe
Marchand, Gerard, (413) 538-3173 gmarchan@mtholyoke.edu,
 Mount Holyoke College – Gga
Marchand, Roger T., (206) 685-3757 rojmarch@u.washington.edu,
 University of Washington – Ass
Marchant, David R., (617) 353-3236 marchant@bu.edu,
 Boston University – Gm
Marchetti, David W., (970) 943-2367 dmarchetti@western.edu,
 Western Colorado University – GmClc
Marchisin, John, (212) 346-1502 Pace University, New York Campus – Zg
Marchisin, John, (201) 200-3161 jmarchisin@njcu.edu,
 New Jersey City University – Gg
Marchitto, Thomas M., 303-492-7739 tom.marchitto@colorado.edu,
 University of Colorado – Oc
Marco, Shmuel, 972-3-6405689 shmulik.marco@gmail.com,
 Tel Aviv University – Ga
Marco, Shmulik, shmulikm@tau.ac.il,
 Tel Aviv University – Ggc
Marconi, Dario, 609-258-4101 dmarconi@princeton.edu,
 Princeton University – Cms
Marenco, Katherine N., (610) 526-5627 kmarenco@brynmawr.edu,
 Bryn Mawr College – PiePg
Marenco, Pedro J., (610) 526-7580 pmarenco@brynmawr.edu,
 Bryn Mawr College – PgCsPg
Mareschal, Jean-Claude, 514-987-3000 #6864 mareschal.jean-claude@uqam.ca,
 Universite du Quebec a Montreal – Yg
Marfurt, Kurt J., (405) 325-5669 kmarfurt@ou.edu,
 University of Oklahoma – Ys
Margalef, Ona, ona.margalef@uab.cat,
 Universitat Autonoma de Barcelona – Ge
Margot, Jean-Luc, (310) 206-8345 jlm@epss.ucla.edu,
 University of California, Los Angeles – XyYdv
Margrave, Gary F., (403) 220-4606 University of Calgary – Ye
Maria, Tony, 812-461-5326 ahmaria@usi.edu,
 University of Southern Indiana – Gv
Mariani, Elisabeth, +440151 794 5180 Mariani@liverpool.ac.uk,
 University of Liverpool – Nr

Marin, Liliana, (254)710-2648 Liliana_Marin@baylor.edu, Baylor University – ZiGgCc

Marinelli, Roberta, 541-737-5195 marinelr@oregonstate.edu, Oregon State University – GuOcGe

Marino, Miguel A., (530) 752-0684 mamarino@ucdavis.edu, University of California, Davis – Hw

Marion, Giles, (775) 673-7349 gmarion@dri.edu, Desert Research Institute – Sc

Maritan, Lara, +390498279143 lara.maritan@unipd.it, Università degli Studi di Padova – Gx

Marjanac, Tihomir, +38514606109 marjanac@geol.pmf.hr, University of Zagreb – GsgYg

Mark, Bryan G., (614) 247-6180 mark.9@osu.edu, Ohio State University – As

Market, Patrick S., 573-882-1496 marketp@missouri.edu, University of Missouri, Columbia – As

Markley, Michelle J., (413) 538-2814 mmarkley@mtholyoke.edu, Mount Holyoke College – GcgGt

Markowitz, Steven, 718-6704180 Queens College (CUNY) – GbZn

Markowski, Antonette K., (717) 702-2038 amarkowski@pa.gov, Pennsylvania Bureau of Topographic & Geologic Survey – Eoc

Marks, Dennis W., 229-244-5159 Valdosta State University – Zn

Markun, Francis J., (630) 252-7521 Argonne National Laboratory – Cc

Markwith, Scott H., 561 297-2102 smarkwit@fau.edu, Florida Atlantic University – Zy

Marley, Nancy A., (630) 252-5014 namarley@anl.gov, Argonne National Laboratory – As

Marlowe, Brian W., 813-253-7685 bmarlowe2@hccfl.edu, Hillsborough Community College – Zg

Marobhe, Isaac M., 255222410013 marobhe@udsm.ac.tz, University of Dar es Salaam – YegYx

Marone, Chris, (814) 865-7964 cjm38@psu.edu, Pennsylvania State University, University Park – Yx

Marquez, L. Lynn, (717) 871-4339 lynn.marquez@millersville.edu, Millersville University – Cg

Marr, Paul G., (717) 477-1656 pgmarr@ship.edu, Shippensburg University – ZfyZi

Marra, John, (718) 951-5000 x2594 jfm7780@brooklyn.cuny.edu, Brooklyn College (CUNY) – Ob

Marrett, Randall A., (512) 471-2113 marrett@mail.utexas.edu, University of Texas, Austin – Gc

Marrs, Rob, +44 0151 795 5172 Calluna@liverpool.ac.uk, University of Liverpool – So

Marsaglia, Kathleen M., (818) 677-6309 kathie.marsaglia@csun.edu, California State University, Northridge – Gs

Marsan, Yvonne, 910-962-7288 marsany@uncw.edu, University of North Carolina, Wilmington – Zi

Marschik, Robert, 089/2180 6527 marschik@lmu.de, Ludwig-Maximilians-Universitaet Muenchen – EmCg

Marsellos, Antonios, (516) 463-5567 antonios.e.marsellos@hofstra.edu, Hofstra University – GtqZf

Marsh, Julian S., +27 (0)46-603-7394 goonie.marsh@ru.ac.za, Rhodes University – GiCgGv

Marsh, Robert, +44 (0)23 8059 6214 rma@noc.soton.ac.uk, University of Southampton – Op

Marshak, Stephen, (217) 333-7705 smarshak@illinois.edu, University of Illinois, Urbana-Champaign – Gct

Marshall, Charles R., 510- 642-1821 crmarshall@berkeley.edu, University of California, Berkeley – PqgPi

Marshall, Craig, (785) 864-6029 cpmarshall@ku.edu, University of Kansas – GzZmXb

Marshall, Dan D., 778-782-5474 Simon Fraser University – Ca

Marshall, Hans-Peter, hpmarshall@boisestate.edu, Boise State University – YxGl

Marshall, Jeffrey S., 909-869-3461 marshall@csupomona.edu, California State Polytechnic University, Pomona – GmtHs

Marshall, Jill A., 479-575-2420 jillm@uark.edu, University of Arkansas, Fayetteville – Gm

Marshall, Jim, +44-151-794-5177 Isotopes@liverpool.ac.uk, University of Liverpool – Cs

Marshall, John C., (617) 253-9615 jmarsh@mit.edu, Massachusetts Institute of Technology – Op

Marshall, John E., +44 (0)23 80592015 jeam@noc.soton.ac.uk, University of Southampton – PlGro

Marshall, Monte, mmarshall@mail.sdsu.edu, San Diego State University – YmGtc

Marshall, Scott T., 828-265-8680 marshallst@appstate.edu, Appalachian State University – YxGc

Marshall, Stephen, (519) 824-4120 Ext.52720 samarsha@uoguelph.ca, University of Guelph – Zn

Marshall, Thomas R., 605-677-6164 tom.marshall@usd.edu, South Dakota Dept of Environment and Natural Resources – GgEo

Marsham, John, +44(0) 113 34 36422 j.marsham@leeds.ac.uk, University of Leeds – As

Marston, Sallie A., marston@email.arizona.edu, University of Arizona – Zu

Martel, Richard, (418) 654-2683 rene.martel@ete.inrs.ca, Universite du Quebec – Hw

Martel, Stephen J., (808) 956-7797 martel@soest.hawaii.edu, University of Hawai'i, Manoa – Ng

Martens, Hilary R., (406) 243-6855 hilary.martens@umontana.edu, University of Montana – YgsYd

Martill, David, +44 023 92 842256 david.martill@port.ac.uk, University of Portsmouth – PvgPe

Martin, Anthony J., (740) 368-3621 geoam@learnlink.emory.edu, Emory University – Pe

Martin, Arturo, amartin@cicese.mx, Centro de Investigación Científica y de Educación Superior de Ensenada – Gd

Martin, Barton S., (740) 368-3621 bsmartin@owu.edu, Ohio Wesleyan University – GxvCg

Martin, Beth, 314-935-4136 martin@wustl.edu, Washington University in St. Louis – Ng

Martin, Candace E., +64 3 479-7526 candace.martin@otago.ac.nz, University of Otago – ClcGz

Martín, Cristina, crismartin@ucm.es, Univ Complutense de Madrid – Gm

Martin, Ellen E., (352) 392-2141 eemartin@ufl.edu, University of Florida – Cl

Martin, Gale D., 702-651- 3141 Gale.Martin@csn.edu, College of Southern Nevada - West Charleston Campus – Gg

Martin, James E., (337) 482-6468 geology@louisiana.edu, University of Louisiana at Lafayette – PvGg

Martin, James E., (605) 394-2461 James.Martin@sdsmt.edu, South Dakota School of Mines & Technology – Ps

Martin, Jonathan B., (352) 392-6219 jbmartin@ufl.edu, University of Florida – ClHg

Martin, Jonathan E., (608)262-9845 jon@aos.wisc.edu, University of Wisconsin, Madison – Am

Martin, Pamela, (773) 834-5245 University of Chicago – Pe

Martin, Paul F., (509) 376-2519 paul.martin@pnl.gov, Pacific Northwest National Laboratory – Cg

Martin, Randal S., (505) 835-5469 New Mexico Institute of Mining and Technology – As

Martin, Robert F., (514) 324-2579 robert.martin@mcgill.ca, McGill University – GziEm

Martin, Ronald E., (302) 831-2569 daddy@udel.edu, University of Delaware – PmePg

Martin, Scot T., 617-495-7620 smartin@seas.harvard.edu, Harvard University – Cg

Martin, Scott C., 330-941-3026 scmartin@ysu.edu, Youngstown State University – Ge

Martin, Seelye, (206) 543-6438 seelye@ocean.washington.edu, University of Washington – Op

Martin, Silvana, +390498279104 silvana.martin@unipd.it, Università degli Studi di Padova – Gc

Martin, Timothy J., (630) 252-8708 tjmartin@anl.gov, Argonne National Laboratory – As

Martin, Walter, 704-687-5954 wemartin@uncc.edu, University of North Carolina, Charlotte – As

Martin, William R., (508) 289-2836 wmartin@whoi.edu, Woods Hole Oceanographic Institution – Oc

Martín Chivelet, Javier, j.m.chivelet@ucm.es, Univ Complutense de Madrid – PcGs

Martin-Hayden, James, (419) 530-2634 jhayden@geology.utoledo.edu, University of Toledo – Hq

Martinez, Fernando, (808) 956-6882 martinez@soest.hawaii.edu, University of Hawai'i, Manoa – Yr

Martínez, Gemma, gemmamar@ucm.es, Univ Complutense de Madrid – Pg

Martínez Fernández, Francisco, ++935811513 francisco.martinez@uab.cat, Universitat Autonoma de Barcelona – GpCp

Martini, Anna M., (413) 542-2067 ammartini@amherst.edu, Amherst College – Cl

Martini, I. Peter, (519) 824-4120 x52488 pmartini@uoguelph.ca, University of Guelph – Gsl
Martino, Ronald L., (304) 696-2715 martinor@marshall.edu, Marshall University – GsrPe
Martins, Zita, +44 20 759 49982 z.martins@imperial.ac.uk, Imperial College – Xm
Martinson, Douglas G., dgm@ldeo.columbia.edu, Columbia University – Op
Martinson, Douglas G., (845) 365-8830 dgm@ldeo.columbia.edu, Columbia University – OpGq
Martinuš, Maja, +38514606089 majamarti@geol.pmf.hr, University of Zagreb – Gg
Martiny, Adam, 949-824-9713 amartiny@uci.edu, University of California, Irvine – Ob
Martire, Luca, luca.martire@unito.it, Università di Torino – Gd
Marty, Kevin, 760-355-5761 kevin.marty@imperial.edu, Imperial Valley College – Gg
Martz, Todd, (858) 534-7466 trmartz@ucsd.edu, University of California, San Diego – Gg
Marvinney, Robert G., (207) 287-2804 robert.g.marvinney@maine.gov, Dept of Agriculture, Conservation, and Forestry – Gg
Marzen, Luke J., (334) 844-3462 marzelj@auburn.edu, Auburn University – Zir
Marzolf, John E., (618) 453-7372 marzolf@geo.siu.edu, Southern Illinois University Carbondale – Gs
Marzoli, Andrea, +390498279154 andrea.marzoli@unipd.it, Università degli Studi di Padova – Cg
Marzulli, Walter, (609) 292-2576 walt.marzulli@dep.state.nj.us, New Jersey Geological and Water Survey – Zy
Masch, Ludwig, 089/2180 4273 masch@min.uni-muenchen.de, Ludwig-Maximilians-Universitaet Muenchen – Gz
Masciocco, Luciano, luciano.masciocco@unito.it, Università di Torino – Ge
Masek, Ondrej, +44 (0) 131 650 5095 ondrej.masek@ed.ac.uk, Edinburgh University – Ng
Masiello, Caroline A., 713.348.5234 masiello@rice.edu, Rice University – Zn
Maslowski, Wieslaw, (831) 656-3162 maslowski@nps.edu, Naval Postgraduate School – Op
Mason, Allen S., (505) 667-4140 Los Alamos National Laboratory – As
Mason, Charles E., (606) 783-2166 c.mason@moreheadstate.edu, Morehead State University – PiGrs
Mason, Owen, (907) 474-6293 University of Alaska, Fairbanks – Ga
Mason, Philippa J., +44 20 759 46528 p.j.mason@imperial.ac.uk, Imperial College – Zr
Mason, Robert P., (860) 405-9129 robert.mason@uconn.edu, University of Connecticut – OcCtm
Mason, Roger A., (709) 737-4385 rmason@mun.ca, Memorial University of Newfoundland – Gz
Mason, Sherri A., 716.673.3292 mason@fredonia.edu, SUNY, Fredonia – ZgGe
Mass, Clifford F., cliff@atmos.washington.edu, University of Washington – As
Massare, Judy A., (585) 395-2419 jmassare@brockport.edu, SUNY, The College at Brockport – Pv
Massey, Michael S., mike.massey@csueastbay.edu, California State University, East Bay – ScGeCg
Massironi, Matteo, +390498279116 matteo.massironi@unipd.it, Università degli Studi di Padova – XgZr
Mastalerz, Maria, (812) 855-5412 mmastale@indiana.edu, Indiana University – Ec
Mastalerz, Maria D., 812-855-7416 mmastale@indiana.edu, Indiana University, Bloomington – Ec
Masterlark, Timothy L., (605) 394-2461 Timothy.Masterlark@sdsmt.edu, South Dakota School of Mines & Technology – YdhNr
Masterman, Steven S., (907) 451-5007 steve.masterman@alaska.gov, Alaska Division of Geological & Geophysical Surveys – EgNg
Masters, T. Guy, (858) 534-4122 tmasters@ucsd.edu, University of California, San Diego – YsGy
Masterson, Andrew, andym@earth.northwestern.edu, Northwestern University – CslCa
Mata, Scott, (657) 278-7096 smata@fullerton.edu, California State University, Fullerton – Gg
Matano, Ricardo P., (541) 737-2212 rmatano@coas.oregonstate.edu, Oregon State University – OpAs
Matassa, Catherine, catherine.matassa@uconn.edu, University of Connecticut – Ob
Matheney, Ronald K., (701) 777-4569 ronald.matheney@engr.und.edu, University of North Dakota – Cs
Mathenge, Christine, 931-221-6434 mathengec@apsu.ecu, Austin Peay State University – Zcy
Mather, Tamsin A., +44 (0)1865 282125 tamsin.mather@earth.ox.ac.uk, University of Oxford – Gv
Mathers, Hannah, hannah.mathers@glasgow.ac.uk, University of Glasgow – GmlGg
Mathewson, Christopher C., (979) 845-2488 mathewson@geo.tamu.edu, Texas A&M University – NgHwEg
Mathez, Edmond A., (212) 769-5379 Graduate School of the City University of New York – Gi
Mathez, Edmond A., 212-769-5459 mathez@amnh.org, American Museum of Natural History – Gx
Mathias, Simon, s.a.mathias@durham.ac.uk, Durham University – Hy
Mathur, Dr R., +919347303173 ramrajmathur@gmail.com, Osmania University – YemYn
Mathur, Ryan, 814-641-3725 mathur@juniata.edu, Juniata College – CeYgHw
Matisoff, Gerald, (216) 368-3690 gerald.matisoff@case.edu, Case Western Reserve University – Cbc
Matoza, Robin, matoza@geol.ucsb.edu, University of California, Santa Barbara – Gv
Matson, Pamela A., (650) 725-6812 matson@pangea.stanford.edu, Stanford University – Sb
Matsumoto, Katsumi, 612-624-0275 katsumi@umn.edu, University of Minnesota, Twin Cities – Oc
Matsuoka, Kenichi, +47 77 75 06 45 matsuoka@npolar.no, University of Washington – Gl
Matter, Juerg M., +44 (0)23 80593042 J.Matter@southampton.ac.uk, University of Southampton – Ng
Mattey, Dave, +44 1784 443587 D.Mattey@rhul.ac.uk, Royal Holloway University of London – Cs
Matthews, Adrian, +44 (0)1603 59 3733 a.j.matthews@uea.ac.uk, University of East Anglia – Am
Matthews, Robert A., (916) 752-0179 University of California, Davis – Ge
Mattieu, Lucie, 418-545-5011 x2538 lucie.mathieu1@uqac.ca, Universite du Quebec a Chicoutimi – EgmCe
Mattigod, Shas V., (509) 376-4311 shas.mattigod@pnl.gov, Pacific Northwest National Laboratory – Cl
Mattinson, Chris, 509.963.1628 mattinson@geology.cwu.edu, Central Washington University – GpCcGt
Mattioli, Glen, (817) 272-2987 mattioli@uta.edu, University of Texas, Arlington – GitYd
Mattison, Katherine W., (573) 341-4616 kmattisn@umr.edu, Missouri University of Science and Technology – Zn
Mattox, Stephen R., (616) 331-3734 mattoxs@gvsu.edu, Grand Valley State University – ZeGv
Mattox, Tari, 616-234-2119 tmattox@grcc.edu, Grand Rapids Community College – GgvCg
Mattson, Peter H., (718) 997-3300 Graduate School of the City University of New York – Gc
Mattson, Peter H., (718) 997-3335 Queens College (CUNY) – Gc
Matty, David J., 801-626-7195 dmatty@weber.edu, Weber State University – Gig
Matyjasik, Basia, (801) 537-3122 basiamatyjasik@utah.gov, Utah Geological Survey – Gg
Matyjasik, Marek, (801) 626-7726 mmatyjasik@weber.edu, Weber State University – HwGe
Matzke, David, (313) 593-5036 dmatzke@umich.edu, University of Michigan, Dearborn – Xy
Maul, George A., 321-674-8096 gmaul@fit.edu, Florida Institute of Technology – Op
Mauldin-Kinney, Virginia L., vmauldin-kinney1@gsu.edu, Georgia State University, Perimeter College, Dunwoody – GgXg
Maulud, Mat Ruzlin, 03-79674144 matruzlin@um.edu.my, University of Malaya – Zn
Maurrasse, Florentin J., (305) 348-2350 maurrass@fiu.edu, Florida International University – PsGsCm
Mauzerall, Denise L., (609) 258-2498 mauzeral@princeton.edu, Princeton University – As
Mavimbela, Philani, (031) 260-7081 mavimbelap@ukzn.ac.za, University of KwaZulu-Natal – GpzEg
Mavko, Gerald M., (650) 723-9438 Stanford University – Yg
Mawalagedara, Rachindra, 515-294-8633 rmawala@iastate.edu,

Iowa State University of Science & Technology – As
Maxwell, Arthur E., 512-471-0411 art@ig.utexas.edu,
   University of Texas, Austin – Og
Maxwell, James "Sandy", (610) 436-2240 jmaxwell@wcupa.edu,
   West Chester University – Ze
Maxwell, Michael, (604) 296-4218 Michael_Maxwell@golder.com,
   University of British Columbia – Yg
Maxwell, Reed M., 303-384-2456 rmaxwell@mines.edu,
   Colorado School of Mines – Hs
May, Cynthia, (715) 682-1499 cmay@northland.edu,
   Northland College – ZiyZn
May, Daniel J., (203) 932-7262 dmay@newhaven.edu,
   University of New Haven – GtxGe
May, Diane M., dianemay@missouristate.edu,
   Missouri State University – Zn
May, Fred E., (801) 581-8218 pscem.fmay@state.ut.us,
   University of Utah – Zn
May, James, (662) 325-5806 Mississippi State University – Hw
May, Michael, (502) 745-4555 michael.may@wku.edu,
   Western Kentucky University – Ge
May, S J., (972) 377-1635 smay@collin.edu,
   Collin College - Preston Ridge Campus – GgoGc
Mayborn, Kyle R., (309) 298-1577 KR-Mayborn@wiu.edu,
   Western Illinois University – Gic
Mayer, Alex S., (906) 487-3372 asmayer@mtu.edu,
   Michigan Technological University – Hy
Mayer, Bernhard, (403) 220-5389 bmayer@ucalgary.ca,
   University of Calgary – CslGe
Mayer, Christine M., (419) 530-8377 christine.mayer@utoledo.edu,
   University of Toledo – Zn
Mayer, James R., (678) 839-4055 jmayer@westga.edu,
   University of West Georgia – Hw
Mayer, Jordan, 612-626-2672 jmayer@umn.edu,
   University of Minnesota – Zi
Mayer, Larry A., 603-862-2615 larry.mayer@unh.edu,
   University of New Hampshire – Gu
Mayer, Margaret, 928-724-6722 Dine' College – GeZn
Mayer, Ulrich, (604) 822-1539 umayer@eos.ubc.ca,
   University of British Columbia – Nm
Mayes, Melanie A., 865-574-7336 mayesma@ornl.gov,
   University of Tennessee, Knoxville – ClGeHw
Mayewski, Paul A., (207) 581-3019 paul.mayewski@maine.edu,
   University of Maine – Pe
Mayfield, Michael W., (828) 262-7058 mayfldmw@appstate.edu,
   Appalachian State University – Hs
Maynard, David, (909) 537-4321 dmaynard@csusb.edu,
   California State University, San Bernardino – ZnnZn
Maynard, J. Barry, maynarjb@uc.edu,
   University of Cincinnati – Cl
Mayo, Alan L., 801-422-2338 alan_mayo@byu.edu,
   Brigham Young University – Hw
Mayor, Shane D., 530-898-6337 sdmayor@csuchico.edu,
   California State University, Chico – AsZrAp
Mazer, James J., (630) 252-7362 Argonne National Laboratory – Cl
MAZZELLA, Prof. Antonio A., 070-675-5542 mazzella@unica.it,
   Universita di Cagliari – Gq
Mazzocco, Elizabeth, emazzocc@ashland.edu,
   Ashland University – AsRn
Mazzoli, Claudio, +390498279144 claudio.mazzoli@unipd.it,
   Università degli Studi di Padova – Gx
Mazzullo, Salvatore J., (316) 978-3140 Wichita State University – Go
Mbila, Monday O., 256-372-4185 monday.mbila@aamu.edu,
   Alabama A&M University – Sd
Mbuh, Mbongowo J., 701-777-4587 mbongowo.mbuh@und.edu,
   University of North Dakota – ZrHs
Mc Mahon, Michelle, 281-765-7865 mmckubota@aol.com,
   Lonestar College - North Harris – Go
McAfee, Gerald B., (432) 335-6558 bmcafee@odessa.edu,
   Odessa College – Pi
McAfee, Stephanie , (775) 784-6999 smcafee@unr.edu,
   University of Nevada, Reno – Zy
McAllister, Arnold L., (506) 453-4804 University of New Brunswick – Em
McAndrews, John H., (416) 586-5609 University of Toronto – Pl
McArthur, John M., +44 020 7679 2376 j.mcarthur@ucl.ac.uk,
   University College London – ClHwPs
McArthur, Russell, (415) 338-1755 San Francisco State University – Zn
McArthur, Russell , rmcarthur@ccsf.edu,
   City College of San Francisco – Gg
McAtee, Mike, 909-869-3648 lmmcatee@csupomona.edu,
   California State Polytechnic University, Pomona – Zn
McBeth, Joyce M., joyce.mcbeth@usask.ca,
   University of Saskatchewan – PgGeSb
McBride, Earle F., (512) 471-1905 efmcbride@mail.utexas.edu,
   University of Texas, Austin – Gd
McBride, John H., (801) 422-5219 john_mcbride@byu.edu,
   Brigham Young University – YeGoe
McBride, Randolph, (703) 993-1642 rmcbride@gmu.edu,
   George Mason University – On
McBride, Raymond G., (519) 824-4120 (Ext. 52492) rmcbride@uoguelph.ca,
   University of Guelph – So
McCaffrey, Bill, +44(0) 113 34 36625 w.d.mccaffrey@leeds.ac.uk,
   University of Leeds – Gs
McCaffrey, Kenneth J., 0191 3742523 k.j.mccaffrey@durham.ac.uk,
   Durham University – Gc
McCair, Andrew, +44(0) 113 34 35219 a.m.mccaig@leeds.ac.uk,
   University of Leeds – Gc
McCall, Peter L., (216) 368-3676 plm4@case.edu,
   Case Western Reserve University – Pg
McCallister, Robert, (608) 758-6556 robert.mccallister@uwc.edu,
   University of Wisconsin Colleges – ZyGgZn
McCallum, I. Stewart (Stu), mccallum@uw.edu,
   University of Washington – GivGz
McCanta, Molly, 865-974-6003   mmccanta@utk.edu,
   University of Tennessee, Knoxville – GiXg
McCarter, Michael K., (801) 581-8603 k.mccarter@utah.edu,
   University of Utah – Nm
McCarthy, Christine, mccarthy@ldeo.columbia.edu,
   Columbia University – Ys
McCarthy, Daniel P., (905) 688-5550 (Ext. 3864) dmccarthy2@brocku.ca,
   Brock University – Gl
McCarthy, Francine G., (905) 688-5550 (Ext. 4286) Francine@brocku.ca,
   Brock University – Pm
McCarthy, James T., 617-495-2330 jmccarthy@oeb.harvard.edu,
   Harvard University – Ob
McCarthy, Michael P., 206-685-2543 mccarthy@uw.edu,
   University of Washington – XyYx
McCarthy, Paul, (907) 474-6894 pjmccarthy@alaska.edu,
   University of Alaska, Fairbanks – GsSaPc
McCartney, Kevin, (207) 768-9482 kevin.mccartney@umpi.edu,
   University of Maine, Presque Isle – Pm
McCartney, M. Carol R., (608) 263-7393 carol.mccartney@wgnhs.uwex.edu,
   University of Wisconsin, Extension – RcGee
McCaslin, Bobby D., (505) 646-3405 New Mexico State University, Las Cruces – Sc
McCauley, Marlene, (336) 316-2236 mmccauley@guilford.edu,
   Guilford College – Cp
McCauley, Steven, 214-860-2398 smccauley@dcccd.edu,
   El Centro College - Dallas Community College District – GgAm
McCausland, Phil, pmccausl@brocku.ca,
   Brock University – Gz
McCausland, Phil J., pmccausl@uwo.ca,
   Western University – YmXm
McClain, James S., (916) 752-7093 University of California, Davis – Yr
McClaughry, Jason, (541) 523-3133 Oregon Dept of Geology and Mineral Industries – Gg
McClay, Ken R., +44 1784 443618 k.mcclay@rhul.ac.uk,
   Royal Holloway University of London – GctGo
McClellan, Elizabeth, emcclellan@radford.edu,
   Radford University – GipGd
McClellan, Guerry H., (352) 392-2231 guerry@ufl.edu,
   University of Florida – En
McClelland, James W., (361) 749-6756 jimm@utexas.edu,
   University of Texas, Austin – CgHsCs
McClelland, Lori, (775)784-4977 mcclella@unr.edu,
   University of Nevada, Reno – Ys
McClelland, William C., 319-335-1827 bill-mcclelland@uiowa.edu,
   University of Iowa – GtCcGc
mcClenaghan, Seán h., + 353 1 8961585 mcclens@tcd.ie,
   Trinity College – Eg
McCluskey, James, (715) 261-6362 jim.mccluskey@uwc.edu,
   University of Wisconsin Colleges – ZinZn
McColl, Kaighin, kmccoll@seas.harvard.edu,
   Harvard University – Hg
McCollough, Cherie, 361-825-3166 Cherie.McCollough@tamucc.edu,

Texas A&M University, Corpus Christi – Ze
McCollum, Linda B., (509) 359-7473 lmccollum@ewu.edu,
　　Eastern Washington University – Pg
McConnaughhay, Mark, 845-431-8536 mcconn@sunydutchess.edu,
　　Dutchess Community College – ZgAmEo
McConnell, David A., (919) 515-0381 damcconn@ncsu.edu,
　　North Carolina State University – ZeRcGc
McConnell, Joseph, (775) 673-7348 jmcconn@dri.edu,
　　University of Nevada, Reno – Zn
McConnell, Robert L., rmconnel@umw.edu,
　　University of Mary Washington – Ge
McCorkle, Daniel C., (508) 289-2949 dmccorkle@whoi.edu,
　　Woods Hole Oceanographic Institution – Cm
McCormack, John K., (775) 784-4518 mccormac@mines.unr.edu,
　　University of Nevada, Reno – Gz
McCormick, George R., 419-798-4812 erie-@msn.com,
　　University of Iowa – GzEmCa
McCormick, Michael L., 315-859-4832 mmccormi@hamilton.edu,
　　Hamilton College – Cb
McCoy, Edward L., (330) 263-3884 mccoy.13@osu.edu,
　　Ohio State University – Sp
McCoy, Floyd W., 236-9115 fmccoy@hawaii.edu,
　　Windward Community College – GgaGu
McCoy, Kevin, 303-384-2632 kemccoy@mines.edu,
　　Colorado Geological Survey – NgZiHw
Mccoy, Roger M., (801) 581-8218 University of Utah – Zr
McCoy, Scott W., 775-682-7205 scottmccoy@unr.edu,
　　University of Nevada, Reno – GmNg
McCoy, Timothy, 202-633-2206 mccoyt@si.edu,
　　University of Tennessee, Knoxville – Xm
McCoy, William D., (413) 545-1535 wdmccoy@geo.umass.edu,
　　University of Massachusetts, Amherst – Zy
McCoy, Jr., Floyd W., (808) 235-7318 University of Hawai'i, Manoa – Gu
McCraw, David J., (505) 835-5594 djmc@mailhost.nmt.edu,
　　New Mexico Institute of Mining & Technology – Zy
McCreary, Julian P., (808) 956-2216 jay@soest.hawaii.edu,
　　University of Hawai'i, Manoa – Op
McCrink, Timothy P., (916) 324-2549 California Geological Survey – Ng
McCulloh, Richard P., (225) 578-5327 mccullo@lsu.edu,
　　Louisiana State University – Ng
McCullough, Jr., Edgar J., (520) 621-6024 University of Arizona – Ge
McCurdy, Maureen, (318) 257-3165 mm@coes.latech.edu,
　　Louisiana Tech University – Hw
McCurry, Michael O., (208) 282-3960 mccumich@isu.edu,
　　Idaho State University – Gz
McCutcheon, Karen Sue, 307-472 5359 kmccutcheon@caspercollege.edu,
　　Casper College – Zeg
McDade, Ian C., (416)736-2100 #22859 mcdade@yorku.ca,
　　York University – Zr
McDaniel, Paul A., (208) 885-7012 University of Idaho – Sd
McDermott, Christopher I., +44 (0) 131 650 5931 cmcdermo@staffmail.ed.ac.uk,
　　Edinburgh University – Hg
McDermott, Frank, (+353) 1 716 2327 frank.mcdermott@ucd.ie,
　　University College Dublin – GgCgZg
McDermott, Jill I., (610) 758-3683 jill.mcdermott@lehigh.edu,
　　Lehigh University – Cm
McDermott, Thomas M., (585) 395-5718 SUNY, The College at Brockport – Zn
McDonald, Andrew M., 7056751151 x2266 amcdonald@laurentian.ca,
　　Laurentian University, Sudbury – Gz
McDonald, Brenna, (573) 368-2163 brenna.mcdonald@dnr.mo.gov,
　　Missouri Dept of Natural Resources – Ge
McDonald, Catherine, (406) 496-4883 kmcdonald@mtech.edu,
　　Montana Tech of The University of Montana – Grs
McDonald, Eric, (775) 673-7302 emcdonald@dri.edu,
　　University of Nevada, Reno – Sp
McDonald, Gregory N., (801) 537-3383 gregmcdonald@utah.gov,
　　Utah Geological Survey – Ng
McDonald, Kyle, 212-650-6452 kmcdonald@ccny.cuny.edu,
　　Graduate School of the City University of New York – As
McDonald, Kyle C., (818) 354-3440 kyle.mcdonald@jpl.nasa.gov,
　　Jet Propulsion Laboratory – Zn
McDonald, Lynn, (505) 667-1582 lmcdonald@lanl.gov,
　　Los Alamos National Laboratory – Zn
McDonough, William F., (301) 405-5561 mcdonoug@geol.umd.edu,
　　University of Maryland – Cg

McDougall, Jim, 253-566-5060 JMcDougall@tacomacc.edu,
　　Tacoma Community College – Zg
McDowell, Fred W., (512) 471-1672 mcdowell@mail.utexas.edu,
　　University of Texas, Austin – Cc
McDowell, Patricia, (541) 346-4567 pmcd@uoregon.edu,
　　University of Oregon – Gm
McDowell, Robin J., 770-274-5466 rmcdowell@gsu.edu,
　　Georgia State University, Perimeter College, Dunwoody Campus – Gge
McDowell, Ronald, (304) 594-2331 mcdowell@geosrv.wvnet.edu,
　　West Virginia University – GsPiCe
McDowell, Ronald, rmcdowell@fairmontstate.edu,
　　Fairmont State University – Gge
McDowell, William H., (603) 862-2249 bill.mcdowell@unh.edu,
　　University of New Hampshire – HgRw
McDuff, Russell E., (360) 453-7877 mcduff@uw.edu,
　　University of Washington – OcGu
McElroy, Anne E., (631) 632-8488 anne.mcelroy@stonybrook.edu,
　　SUNY, Stony Brook – ObCb
McElroy, Brandon, (307) 766-3601 bmcelroy@uwyo.edu,
　　University of Wyoming – GsmGr
McElroy, Michael B., (617) 495-9261 mbm@io.harvard.edu,
　　Harvard University – As
McElroy, Tom, (416) 736-2100 x22113 tmcelroy@yorku.ca,
　　York University – As
McElwain, Jenny, (312) 665-7635 jennifer.mcelwain@gmail.com,
　　Field Museum of Natural History – Pb
McElwaine, Jim, james.mcelwaine@durham.ac.uk,
　　Durham University – Yg
McElwee, Carl D., (785) 843-4164 cmcelwee@ku.edu,
　　University of Kansas – HgYg
McEvoy, Jamie , (406) 994-4069 jamie.mcevoy@montana.edu,
　　Montana State University – Ze
McEwen, Alfred, (520) 621-4573 mcewen@lpl.arizona.edu,
　　University of Arizona – Xg
McFadden, Bruce J., (352) 846-2000 bmacfadd@flmnh.ufl.edu,
　　University of Florida – Pv
McFadden, Jennifer, 717-361-1392 mcfaddenj@etown.edu,
　　Elizabethtown College – ZnnZn
McFadden, Leslie M., (505) 277-6121 lmcfadnm@unm.edu,
　　University of New Mexico – Sd
McFadden, Rory, 507-933-7333 rmcfadde@gustavus.edu,
　　Gustavus Adolphus College – GxcGz
McFarlane, Norman, (250) 363-8227 norm.mcfarlane@ec.gc.ca,
　　University of Victoria – As
McFarquhar, Greg M., (217) 265-5458 mcfarq@illinois.edu,
　　University of Illinois, Urbana-Champaign – Asm
McFiggans, Gordon, +44 0161 306-3954 Gordon.B.Mcfiggans@manchester.ac.uk,
　　University of Manchester – As
McGahan, Donald, (254) 968-9701 mcgahan@tarleton.edu,
　　Texas A&M University – Sdc
McGarvie, Dave, +44 (0) 131 549 7140 x 71140 dave.mcgarvie@open.ac.uk,
　　The Open University – GviCt
McGauley, Michael G., 305-237-2687 mmcgaule@mdc.edu,
　　Miami Dade College (Kendall Campus) – AmOp
McGeary, Susan, (302) 831-8174 smcgeary@udel.edu,
　　University of Delaware – Ze
McGee, David, (617) 324-3545 davidmcg@mit.edu,
　　Massachusetts Institute of Technology – Cl
McGee, Tara, 780-492-3042 tmcgee@ualberta.ca,
　　University of Alberta – Zn
McGehee, Thomas L., (512) 595-3590 kftlm00@tamuk.edu,
　　Texas A&M University, Kingsville – Cl
McGhee, Jr., George R., 848-445-8523 mcghee@eps.rutgers.edu,
　　Rutgers, The State University of New Jersey – PgqPi
McGill, George E., (413) 545-0140 University of Massachusetts, Amherst – Gc
McGill, Sally F., (909) 537-5347 smcgill@csusb.edu,
　　California State University, San Bernardino – GtYdGm
McGillis, Wade, mcgillis@ldeo.columbia.edu,
　　Columbia University – Cm
McGinn, Chris, (919) 530-6269 cmcginn@nccu.edu,
　　North Carolina Central University – Zgi
McGinnis, Lyle D., (630) 252-8722 Argonne National Laboratory – Yg
McGivern, Tiffany, (315)792-3134 Utica College – Geg
McGlathery, Karen, (804) 924-0558 kjm4k@virginia.edu,
　　University of Virginia – Ob

McGlue, Michael M., 859-257-1952 michael.mcglue@uky.edu, University of Kentucky – GnrGm
McGoldrick, Peter J., 61 3 6226 7209 University of Tasmania – Ce
Mcgowan, Alistair, +4401413305449 Alistair.McGowan@glasgow.ac.uk, University of Glasgow – Pg
McGowan, Eileen, 413-586-8305 emcgowan@geo.umass.edu, University of Massachusetts, Amherst – Xg
McGowan, John A., (858) 534-2074 jmcgowan@ucsd.edu, University of California, San Diego – Ob
McGowan, Nicole M., 360-650-3835 Nicole.McGowan@wwu.edu, Western Washington University – CtaCg
McGrail, Bernard P., (509) 376-9193 pete.mcgrail@pnl.gov, Pacific Northwest National Laboratory – Cg
McGrath, Andrew, andrew.mcgrath@flinders.edu.au, Flinders University – ZrAs
McGrath, Steve F., 406-496-4767 smcgrath@mtech.edu, Montana Tech of The University of Montana – CaEgCe
McGraw, Maureen A., (505) 665-8128 mcgraw@lanl.gov, Los Alamos National Laboratory – Zn
McGregor, Kent M., (940) 565-2380 kent.mcgregor@unt.edu, University of North Texas – AsZg
McGregor, Stuart W., 205-247-3629 smcgregor@gsa.state.al.us, Geological Survey of Alabama – Hs
McGrew, Allen J., (937) 229-5638 amcgrew1@udayton.edu, University of Dayton – GtcGp
McGuire, Jeffrey J., (508) 289-3290 jmcguire@whoi.edu, Woods Hole Oceanographic Institution – Ys
McGuire, Jennifer, 651-962-5254 jtmcguire@stthomas.edu, University of Saint Thomas – CgHsGb
McGwire, Kenneth, 775.673.7324 ken.mcgwire@dri.edu, University of Nevada, Reno – Zr
McHenry, Lindsay J., (414) 229-3951 lmchenry@uwm.edu, University of Wisconsin, Milwaukee – GzaXg
McHugh, Cecilia, (718) 997-3330 cecilia.mchugh@qc.cuny.edu, Queens College (CUNY) – Ou
McHugh, Cecilia M., (718) 997-3322 Graduate School of the City University of New York – Ou
McHugh, Julia, 970-248-1993 jumchugh@coloradomesa.edu, Colorado Mesa University – Pgv
McIlvin, Matt, 508-289-2884 mmcilvin@whoi.edu, Woods Hole Oceanographic Institution – Oc
McInnes, Kevin J., (979) 845-5986 k-mcinnes@tamu.edu, Texas A&M University – Sp
McIntosh, Jennifer, 520-626-2282 mcintosh@hwr.arizona.edu, University of Arizona – Hy
McIntosh, Kirk D., 512-471-0480 kirk@ig.utexas.edu, University of Texas, Austin – Gc
McIntosh, William C., (505) 835-5324 mcintosh@nmt.edu, New Mexico Institute of Mining & Technology – Cc
McIntyre, Andrew, (718) 997-3329 Queens College (CUNY) – Pe
McIsaac, Gregory F., (217) 333-9411 University of Illinois, Urbana-Champaign – Hs
McKay, Jennifer L., 541-737-4054 mckay@coas.oregonstate.edu, Oregon State University – Cm
McKay, Larry D., (865) 974-5498 lmckay@utk.edu, University of Tennessee, Knoxville – HyGe
McKay, Matthew P., (417) 836-5318 matthewmckay@missouristate.edu, Missouri State University – Gct
McKay, Nicholas, 928-523-1918 Nicholas.Mckay@nau.edu, Northern Arizona University – PcZo
McKay, Robert M., Robert.McKay@dnr.iowa.gov, Iowa Dept of Natural Resources – Gg
McKay III, E. Donald, (217) 356-7338 emckay@illinois.edu, Illinois State Geological Survey – GlsGr
McKean, Adam, (801) 537-3386 adammckean@utah.gov, Utah Geological Survey – Gmg
McKean , Rebecca, 920-403-3227 rebecca.mckean@snc.edu, Saint Norbert College – GdPgRh
McKee, James W., (414) 424-4460 mckee@athenet.net, University of Wisconsin, Oshkosh – Ce
McKeegan, Kevin D., (310) 825-3580 mckeegan@epss.ucla.edu, University of California, Los Angeles – XcCsc
McKenna, Thomas E., 302-831-8257 mckennat@udel.edu, University of Delaware – Hg
McKenna, Thomas E., 302-831-2833 mckennat@udel.edu, University of Delaware – Hy
McKenney, Rosemary, 253-535-8726 mckennra@plu.edu, Pacific Lutheran University – Gm
McKenzie, Connie, mckenzie@coes.latech.edu, Louisiana Tech University – Zn
McKenzie, Jeffrey M., (514) 398-6767 jeffrey.mckenzie@mcgill.ca, McGill University – Hw
McKenzie, Phyllis, (202) 633-1860 mckenzie@si.edu, Smithsonian Inst / Natl Museum of Natural Hist–
McKenzie, Ross, ross.mckenzie@agric.gov.ab.ca, University of Lethbridge – Zn
McKenzie, Scott C., (814) 824-2382 smckenzie@mercyhurst.edu, Mercyhurst University – PgXmZe
McKibben, Michael A., (951) 827-3444 michael.mckibben@ucr.edu, University of California, Riverside – Cg
McKinley, Galen, (608) 262-4817 galen@aos.wisc.edu, University of Wisconsin, Madison – OcpOg
McKinley, Galen, mckinley@ldeo.columbia.edu, Columbia University – Cg
McKinney, Mac, 205-247-3549 mmckinney@gsa.state.al.us, Geological Survey of Alabama – Hws
McKinney, Michael L., (865) 974-6359 mmckinne@utk.edu, University of Tennessee, Knoxville – GePg
McKinnon, William B., (314) 935-5604 mckinnon@wustl.edu, Washington University in St. Louis – Xg
McKnight, Brian K., (920) 233-4595 briankmcknight@yahoo.com, University of Wisconsin, Oshkosh – Gdu
McLaskey, Greg C., gcm8@cornell.edu, Cornell University – YsNr
McLaughlin, Patrick I., 812-855-1350 pimclaug@iu.edu, Indiana University – GrClGs
McLaughlin, Patrick I., (608) 262-8658 pimclaughlin@wisc.edu, University of Wisconsin, Extension – Gr
McLaughlin, Peter P., 302-831-8263 ppmclau@udel.edu, University of Delaware – HgPm
McLaughlin, Richard, 361-825-2010 Richard.Mclaughlin@tamucc.edu, Texas A&M University, Corpus Christi – Zn
McLaughlin, Richard A., (919) 515-7306 North Carolina State University – Sc
McLaughlin, Jr., Peter P., 302-831-2833 ppmclau@udel.edu, University of Delaware – Gr
McLaurin, Brett T., 570-389-4142 bmclauri@bloomu.edu, Bloomsburg University – GrsGd
McLean, Noah, noahmc@ku.edu, University of Kansas – CcGq
McLemore, Virginia, (505) 835-5521 ginger@gis.nmt.edu, New Mexico Institute of Mining & Technology – Em
McLennan, Scott M., (631) 632-8194 scott.mclennan@sunysb.edu, Stony Brook University – Cg
Mcleod, Andrew T., +44 (0) 131 650 5434 Andy.McLeod@ed.ac.uk, Edinburgh University – Zg
McLeod, Claire, 513-529-9662 mcleodcl@miamioh.edu, Miami University – Gx
McLeod, Clara, cpmcleod@wustl.edu, Washington University in St. Louis – Zn
McLeod, Samuel A., (213) 763-3325 smcleod@ref.usc.edu, Los Angeles County Museum of Natural History – Pv
McManus, Dean A., (206) 543-0587 mcmanus@u.washington.edu, University of Washington – Ou
McManus, George B., 860-405-9164 george.mcmanus@uconn.edu, University of Connecticut – Ob
McManus, Jerry, jmcmanus@ldeo.columbia.edu, Columbia University – Cm
McManus, Jerry F., (508) 289-3328 jmcmanus@whoi.edu, Woods Hole Oceanographic Institution – Pe
McManus, Jerry F., (845) 365-8722 jmcmanus@ldeo.columbia.edu, Columbia University – PeOu
McManus, Margaret A., (831) 459-4736 University of California, Santa Cruz – On
McManus, Margaret Anne, (808) 956-8623 mamc@hawaii.edu, University of Hawai'i, Manoa – Op
McMechan, George A., (972) 883-2419 mcmec@utdallas.edu, University of Texas, Dallas – Ys
McMenamin, Mark, 413-538-2280 mmcmenam@mtholyoke.edu, Mount Holyoke College – PgGst
McMillan, Margaret E., (501) 569-3024 memcmillan@ualr.edu, University of Arkansas at Little Rock – Gm
McMillan, Nancy J., 575-646-5000 nmcmilla@nmsu.edu, New Mexico State University, Las Cruces – Gi

McMillan, Robert S., (520) 621-6968 bob@lpl.arizona.edu,
University of Arizona – ZnnZn
McMonagle, Julie, (570) 408-4604 julie.mcmonagle@wilkes.edu,
Wilkes University – Gge
McMullin, David W., (902) 585-1276 david.mcmullin@acadiau.ca,
Acadia University –
McMurdie, Lynn A., (206) 685-9405 mcmurdie@atmos.washington.edu,
University of Washington – As
McMurtry, Gary M., (808) 956-6858 garym@soest.hawaii.edu,
University of Hawai'i, Manoa – CaGvu
McNair, Laurie A., (505) 665-3328 mcnair@lanl.gov,
Los Alamos National Laboratory – Zn
McNamara, James P., (208) 426-1354 jmcnamar@boisestate.edu,
Boise State University – Hg
McNamara, Jodi, (315) 228-7201 jmcnamara@colgate.edu,
Colgate University – Zn
McNamara, Kenneth, +44 (0) 1223 333410 kmcn07@esc.cam.ac.uk,
University of Cambridge – Pg
McNamee, Brittani D., (828) 350-4554 bmcnamee@unca.edu,
University of North Carolina, Asheville – GzxEg
McNaught, Mark A., (330) 829-8226 mcnaugma@mountunion.edu,
Mount Union College – Gcg
McNeal, Karen S., (334) 844-4282 ksm0041@auburn.edu,
Auburn University – ZeGeCm
McNeill, Donald F., (305) 284-3360 dmcneill@rsmas.miami.edu,
University of Miami – GsrGu
McNeill, Lisa, +44 (0)23 80593640 lcmn@noc.soton.ac.uk,
University of Southampton – GtYr
McNichol, Ann P., (508) 289-3394 amcnichol@whoi.edu,
Woods Hole Oceanographic Institution – Oc
McNicol, Barbara, (403) 440-6175 bmcnicol@mtroyal.ca,
Mount Royal University – Zg
McNulty, Brendan A., (310) 243-3412 bmcnulty@csudh.edu,
California State University, Dominguez Hills – Gc
McNutt, Robert, (905) 525-9140 mcnutt@mcmaster.ca,
McMaster University – CcqHg
McPhaden, Michael J., (206) 526-6783 mcphaden@pmel.noaa.gov,
University of Washington – Og
McPhail, D C "Bear", +61 2 6125 2776 bear@ems.anu.edu.au,
Australian National University – Cl
McPherron, Robert L., (310) 825-1882 rmcpherr@igpp.ucla.edu,
University of California, Los Angeles – Xy
McPherron, Robert L., (818) 887-6220 rmcpherron@igpp.ucla.edu,
University of California, Los Angeles – Xy
McPherson, Brian, 801-581-5634 bmcpherson@egi.utah.edu,
University of Utah – Ye
McPherson, Robert C., (707) 826-5828 rm4@humboldt.edu,
Humboldt State University – Gt
McPhie, Jocelyn, 61 3 6226 2892 Jocelyn.McPhie@utas.edu.au,
University of Tasmania – Gv
McQuaid, Jim, +44(0) 113 34 36724 j.b.mcquaid@leeds.ac.uk,
University of Leeds – As
McQuarrie, Nadine, (412) 624-8870 nmcq@pitt.edu,
University of Pittsburgh – Gct
McRivette, Michael, 517-629-0276 mmcrivette@albion.edu,
Albion College – GtZir
McRoberts, Christopher A., (607) 753-2925 mcroberts@cortland.edu,
SUNY, Cortland – Pg
McSween, Jr., Harry Y., (865) 974-6359 mcsween@utk.edu,
University of Tennessee, Knoxville – XcGi
McSweeney, Kevin, (608) 262-0331 kmcsween@wisc.edu,
University of Wisconsin, Madison – SdaSf
McWethy, David B., (406) 994-6915 dmcwethy@montana.edu,
Montana State University – PeSb
McWilliams, James, (310) 206-2829 jcm@atmos.ucla.edu,
University of California, Los Angeles – OgAs
McWilliams, Michael O., (650) 723-3718 mcwilliams@stanford.edu,
Stanford University – CcYme
McWilliams, Robert G., mcwillrg@miamioh.edu,
Miami University – Ps
Meade, Brendan, 617-495-8921 meade@fas.harvard.edu,
Harvard University – Yg
Meador, Cindy D., 806-651-2582 cmeador@wtamu.edu,
West Texas A&M University – Zg
Meadows, Guy A., 906/487-1106 gmeadows@mtu.edu,
Michigan Technological University – Gu
Meadows, Wayne R., (505) 665-0291 wmeadows@lanl.gov,
Los Alamos National Laboratory – Yg
Means, Guy H., (850) 617-0312 Guy.Means@FloridaDEP.gov,
Florida Geological Survey – PgGe
Measures, Christopher, (808) 956-8693 chrism@soest.hawaii.edu,
University of Hawai'i, Manoa – Oc
Measures, Elizabeth A., (432) 837-8117 measures@sulross.edu,
Sul Ross State University – Gdq
Mechoso, Carlos R., (310) 825-3057 mechoso@atmos.ucla.edu,
University of California, Los Angeles – As
Meckel, Timothy A., 512-471-4306 tip.meckel@beg.utexas.edu,
University of Texas, Austin – Gr
Mecklenburgh, Julian, +440161 275-3821 Julian.Mecklenburgh@manchester.ac.uk,
University of Manchester – Gc
Medaris, Jr., Levi G., (608) 833-4258 medaris@geology.wisc.edu,
University of Wisconsin-Madison – GipSa
Meddaugh, W. S., (940) 397-4469 scott.meddaugh@mwsu.edu,
Midwestern State University – GoEgCg
Medina, Cristian R., 855-9992 crmedina@indiana.edu,
Indiana University – GoeHw
Medina Elizalde, Martin A., (334) 844-4966 mam0199@auburn.edu,
Auburn University – PeHyCl
Medlin, Peggy, (931) 372-3121 pmedlin@tntech.edu,
Tennessee Tech University – Zn
Meduniæ, Gordana, +38514605909 gpavlovi@inet.hr,
University of Zagreb – CgScGe
Meehan, Katharine, 541-346-4521 meehan@uoregon.edu,
University of Oregon – Zn
Meehan, Kimberly, (716)645-4293 kcmeehan@buffalo.edu,
SUNY, Buffalo – Pe
Meentemeyer, Ross, 704-687-5944 rkmeente@email.uncc.edu,
University of North Carolina, Charlotte – Zi
Meere, Pat, +353 21 4903056 p.meere@ucc.ie,
University College Cork – Gc
Meert, Joseph G., (352) 392-2231 jmeert@ufl.edu,
University of Florida – YgGt
Meesters, Antoon, +31 20 59 87362 a.g.c.a.meesters@vu.nl,
Vrije Universiteit Amsterdam – AmGq
Megdal, Sharon B., (520) 792-9591 smegdal@ag.arizona.edu,
University of Arizona – Hg
Mehes, Roxane J., (705) 673-6575 rmehes@laurentian.ca,
Laurentian University, Sudbury – Zn
Mehra, Aradhana, +44 01332 591133 a.mehra@derby.ac.uk,
University of Derby – Co
Mehrtens, Charlotte J., (802) 656-0243 cmehrten@uvm.edu,
University of Vermont – Gs
Mei, Shenghua, 612-626-0572 meixx002@umn.edu,
University of Minnesota, Twin Cities – Yx
Meigs, Andrew J., (541) 737-1214 meigsa@geo.oregonstate.edu,
Oregon State University – Gc
Meinhold, Guido, +44 1782 733617 g.meinhold@keele.ac.uk,
Keele University – GdCa
Meisner, Caroline B., (775) 340-7341 caroline.meisner@gbcnv.edu,
Great Basin College – ZgSo
Meister, Paul A., (309) 438-7479 pameist@ilstu.edu,
Illinois State University – GgHgZn
Meisterernst, Götz, 089/2180 4356 goetz.meisterernst@lrz.uni-muenchen.de,
Ludwig-Maximilians-Universitaet Muenchen – Gz
Meixner, Thomas, 520-626-1532 tmeixner@email.arizona.edu,
University of Arizona – CgRw
Mekik, Figen A., (616) 331-3020 mekikf@gvsu.edu,
Grand Valley State University – GuCm
Meko, David M., (520) 621-3457 dmeko@ltrr.arizona.edu,
University of Arizona – Hq
Melas, Faye, (212) 772-5265 ffmelas@yahoo.com,
Hunter College (CUNY) – Gs
Melbourne, Timothy I., (509) 963-2799 tim@geology.cwu.edu,
Central Washington University – Yd
Melcher, Frank, frank.melcher@unileoben.ac.at,
University Leoben – GgEgm
Melchin, Michael, (902) 867-5177 Dalhousie University – Pg
Melchiorre, Erik, (909) 537-7754 emelch@csusb.edu,
California State University, San Bernardino – EmHwCl
Meldahl, Keith H., (760) 757-2121, x6412 kmeldahl@miracosta.edu,
MiraCosta College – Gg
Meléndez, Nieves, nievesml@ucm.es,
Univ Complutense de Madrid – Ggs

Melgar, Diego, (541) 346-3488 dmelgar@uoregon.edu,
University of Oregon – Yd

Melichar, Rostislav, +420 549 49 5812 melda@sci.muni.cz,
Masaryk University – GtcGr

Melim, Leslie A., (309) 298-1377 LA-Melim@wiu.edu,
Western Illinois University – Gd

Mellinger, David, 541-867-0372 David.Mellinger@oregonstate.edu,
Oregon State University – Cb

Mellor, George, (609) 258-6570 glmellor@princeton.edu,
Princeton University – Op

Melosh, H. J., (765) 494-3290 jmelosh@purdue.edu,
Purdue University – XmgXy

Melosh, Henry J., (520) 621-2806 jmelosh@lpl.arizona.edu,
University of Arizona – Gt

Meltzer, Anne S., (610) 758-3673 asm3@lehigh.edu,
Lehigh University – Ys

Melville, W. Kendall, (858) 534-0478 kmelville@ucsd.edu,
University of California, San Diego – Op

Memeti, Valbone, (657) 278-2036 vmemeti@fullerton.edu,
California State University, Fullerton – GivCa

Mendelssohn, Irving A., (225) 388-6425 Louisiana State University – Ob

Mendoza, Carl, (780) 492-2664 carl.mendoza@ualberta.ca,
University of Alberta – Hw

Mendoza, Luis H., lmendoza@cicese.mx,
Centro de Inv Científica y de Educación Superior de Ensenada – Ne

Menegon, Luca, +44 1752 584931 luca.menegon@plymouth.ac.uk,
University of Plymouth – Gct

Meng, Jin, (212) 496-3337 American Museum of Natural History – Pv

Meng, Lingsen, meng@epss.ucla.edu,
University of California, Los Angeles – Ys

Meng, Qingmin, 662-268-1032 Ext 240 Mississippi State University – Zig

Mengel, David, (785) 532-2166 dmengel@ksu.edu,
Kansas State University – So

Menke, William H., menke@ldeo.columbia.edu,
Columbia University – Ys

Menke, William H., (845) 365-8438 menke@ldeo.columbia.edu,
Columbia University – YsGvq

Menking, Kirsten M., 845-437-5545 kimenking@vassar.edu,
Vassar College – GmPeGc

Menninga, Clarence, menn@calvin.edu,
Calvin College – CcGg

Menold, Carrie A., 517-629-0312 cmenold@albion.edu,
Albion College – Gpz

Mensah, Emmanuel, +233 24 4186 193 emmamensah6@yahoo.com,
Kwame Nkrumah University of Science and Technology – EgGg

Mensing, Scott A., (775) 784-6346 smensing@unr.edu,
University of Nevada, Reno – Zy

Mensing, Teresa, (740) 725-6234 mensing.1@osu.edu,
Ohio State University – Cg

Menuge, Julian F., (+353) 1 716 2141 j.f.menuge@ucd.ie,
University College Dublin – EmCcs

Menzies, John, (905) 688-5550 x3865 jmenzies@brocku.ca,
Brock University – GlmGs

Mercer, Andrew, 662-268-1032 Ext 231 mercer@gri.msstate.edu,
Mississippi State University – Ams

Merchant, James W., (402) 472-7531 jmerchant@unl.edu,
Unversity of Nebraska - Lincoln – Zru

Mercier, Michael , (905) 525-9140 (Ext. 27597) merciema@mcmaster.ca,
McMaster University – Zu

Merck, Jr., John W., (301) 405-2808 jmerck@umd.edu,
University of Maryland – Pv

Meredith, Philip, +44 020 7679 37824 p.meredith@ucl.ac.uk,
University College London – Nr

Meretsky, Vicky J., (812) 855-5971 meretsky@indiana.edu,
Indiana University, Bloomington – Sf

Mereu, Robert F., (519) 661-3605 a424@uwo.ca,
Western University – Ys

Merguerian, Charles M., 516 463-5567 geocmm@hofstra.edu,
Hofstra University – Gc

Merifield, Paul M., (310) 794-5019 pmerifie@ucla.edu,
University of California, Los Angeles – Ng

Merino, Enrique, (812) 855-5088 merino@indiana.edu,
Indiana University, Bloomington – Cl

Merkel, Timo Casjen, 089/2180 4337 casjen.merkel@lrz.uni-muenchen.de,
Ludwig-Maximilians-Universitaet Muenchen – Gz

Merlis, Timothy, 514-398-3140 timothy.merlis@mcgill.ca,
McGill University – Asm

Merriam, James B., (306) 966-5716 jim.merriam@usask.ca,
University of Saskatchewan – Yg

Merrifield, Mark A., (808) 956-6161 markm@soest.hawaii.edu,
University of Hawai'i, Manoa – Op

Merrill, Glen K., (713) 221-8168 merrillg@uhd.edu,
University of Houston Downtown – Pi

Merrill, John T., (401) 874-6715 jmerrill@boreas.gso.uri.edu,
University of Rhode Island – Oc

Merrill, Ronald T., (206) 543-6686 University of Washington – Ym

Merrill, Ronald T., rtm6@uw.edu,
University of Washington – Ygm

Merritt, Andrew, +44 1752 584702 andrew.merritt@plymouth.ac.uk,
University of Plymouth – Ng

Merritts, Dorothy J., (717) 291-4398 dorothy.merritts@fandm.edu,
Franklin and Marshall College – Gm

Mertzman, Stanley A., (717) 291-3818 stan.mertzman@fandm.edu,
Franklin and Marshall College – GizZm

Meskhidze, Nicholas, 919-515-7243 nicholas_meskhidze@ncsu.edu,
North Carolina State University – As

Messer, Sharon, (740) 351-3456 smesser@shawnee.edu,
Shawnee State University – Zn

Messina, Paula, (408) 924-5027 paula.messina@sjsu.edu,
San Jose State University – ZeGm

Metaxas, Anna, (902) 494-3021 metaxas@dal.ca,
Dalhousie University – Ob

Metcalf, Kathryn, (607) 436-3067 Kathryn.Metcalf@oneonta.edu,
SUNY, Oneonta – GctGg

Metcalf, Meredith, (860) 465-4370 metcalfm@easternct.edu,
Eastern Connecticut State University – ZirHw

Metcalf, Rodney V., (702) 895-4442 kim.metcalf@unlv.edu,
University of Nevada, Las Vegas – Gp

Metcalfe, Ian, +61-2-67726297 imetcal2@une.edu.au,
University of New England – PmGtg

Metesh, John J., (406) 496-4159 jmetesh@mtech.edu,
Montana Tech of The University of Montana – HwCa

Metz, Cheyl L., 979 209-7461 cl.metz@blinn.edu,
Blinn College – GgOg

Metz, Nicholas, (315) 781-3615 metz@hws.edu,
Hobart & William Smith Colleges – As

Metz, Robert, (908) 737-3687 rmetz@kean.edu,
Kean University – Gr

Metzger, Ellen P., (408) 924-5048 ellen.metzger@sjsu.edu,
San Jose State University – GpZe

Metzger, Marc J., +44 (0) 131 651 4446 mmetzger@staffmail.ed.ac.uk,
Edinburgh University – Ge

Metzger, Ronald A., (541) 888-7216 rmetzger@socc.edu,
Southwestern Oregon Community College – PmZePs

Metzler, Christopher V., (760) 944-4449, x7738 cmetzler@miracosta.edu,
MiraCosta College – Gg

Meyer, Brian, bmeyer2@gsu.edu,
Georgia State University – GesHw

Meyer, David L., (513) 556-4530 david.meyer@uc.edu,
University of Cincinnati – Pi

Meyer, Franz J., 907-474-7767 fmeyer@gi.alaska.edu,
University of Alaska, Fairbanks – Zr

Meyer, Gary, (612) 626-1741 meyer015@umn.edu,
University of Minnesota – Gl

Meyer, Grant A., (505) 277-4204 gmeyer@unm.edu,
University of New Mexico – Gm

Meyer, Jeffrey W., (805) 965-0531 (Ext. 4270) meyerj@sbcc.edu,
Santa Barbara City College – GgzGx

Meyer, Jessica R., (319) 335-1831 jessica-meyer@uiowa.edu,
University of Iowa – Gm

Meyer, Judith, (417) 836-5604 judithmeyer@missouristate.edu,
Missouri State University – Zn

Meyer, Lewis, +44 01326 253766 l.h.i.meyer@exeter.ac.uk,
Exeter University – Nmr

Meyer, Rebecca A., (812) 855-2687 reameyer@indiana.edu,
Indiana University – Ec

Meyer, Steven J., (920) 465-5022 meyers@uwgb.edu,
University of Wisconsin, Green Bay – AsmZg

Meyer, W. Craig, (818) 710-4241 meyerwc@piercecollege.edu,
Los Angeles Pierce College – GeuPm

Meyer, William T., (630) 969-6586 meyer@iastate.edu,
Argonne National Laboratory – Cg

Meyer Dombard, DArcy, 312-996-2423 drmd@uic.edu,
University of Illinois at Chicago – Pg

Meyers, Jamie A., jmeyers@winona.edu,
  Winona State University – Gsd
Meyers, Philip A., (734) 764-0597 pameyers@umich.edu,
  University of Michigan – CoGnu
Meyers, Stephen R., 608-890-2574 smeyers@geology.wisc.edu,
  University of Wisconsin-Madison – PcOoGs
Meylan, Anne, (727) 896-8626 University of South Florida – Ob
Meylan, Maurice A., (601) 266-4527 mmeylan@otr.usm.edu,
  University of Southern Mississippi – Gu
Meyzen, Christine M., +390498279153 christine.meyzen@unipd.it,
  Università degli Studi di Padova – Cg
Mezga, Aleksandar, +38514606116 amezga@geol.pmf.hr,
  University of Zagreb – Pg
Mezger, Jochen E., (907) 474-7809 jemezger@alaska.edu,
  University of Alaska, Fairbanks – GcpCc
Miah, Khalid, (406) 496-4888 kmiah@mtech.edu,
  Montana Tech of the University of Montana – YexYs
Miall, Andrew D., (416) 978-8841 miall@es.utoronto.ca,
  University of Toronto – Gso
Miao, Xin, (417) 836-5173 xinmiao@missouristate.edu,
  Missouri State University – Zri
Miao, Yuxin, 612-625-8101 ymiao@umn.edu,
  University of Minnesota, Twin Cities – ScZf
Micallef, Aaron, +356 23403662 aaron.micallef@um.edu.mt,
  University of Malta – GumOu
Micander, Rachel, 775-682-6351 rmicander@unr.edu,
  University of Nevada – Zie
Michael, Holly A., (302) 831-4197 hmichael@udel.edu,
  University of Delaware – Hwq
Michael, Peter J., (918) 631-3017 pjm@utulsa.edu,
  The University of Tulsa – GivGz
Michaels, Patrick J., (804) 924-0549 pmichaels@cato.org,
  University of Virginia – As
Michaels, Paul, (208) 426-1929 pm@cgiss.boisestate.edu,
  Boise State University – Ne
Michalski, Greg, (765) 494-3704 gmichals@purdue.edu,
  Purdue University – CsGeAs
Michaud, Jene D., 808-974-7411 jene@hawaii.edu,
  University of Hawai'i, Hilo – HqwGm
Michaud, Yves, (418) 654-2647 Universite du Quebec – Gm
Michel, F. M., 540-231-3299 mfrede2@vt.edu,
  Virginia Polytechnic Institute & State University – ZaGz
Michel, Fred A., fmichel@ccs.carleton.ca,
  Carleton University – Hw
Michel, Jacqueline, 504-280-6325 jmichel@uno.edu,
  University of New Orleans – Cg
Michel, Lauren, (931) 372-3188 lmichel@tntech.edu,
  Tennessee Tech University – PcSaCl
Michel, Suzanne, 6196447454 x3028 Suzanne.Michel@gcccd.edu,
  Cuyamaca College – Gg
Michelfelder, Gary, (417) 836-3171 garymichelfelder@missouristate.edu,
  Missouri State University – GviCg
Mickelson, Andrew M., 901-678-4505 amicklsn@memphis.edu,
  University of Memphis – Ga
Mickelson, David M., (608) 262-7863 mickelson@geology.wisc.edu,
  University of Wisconsin-Madison – Glm
Mickus, Kevin L., (417) 836-6375 kevinmickus@missouristate.edu,
  Missouri State University – GtYv
Miclaus, Crina G., 00402324095 crina_miclaus@yahoo.co.uk,
  Alexandru Ioan Cuza – Gsr
Middlemiss, Lucie, +44(0) 113 34 35246 l.k.middlemiss@leeds.ac.uk,
  University of Leeds – Zn
Middleton, Carrie A., (303) 293-5019 cmiddleton@osmre.gov,
  Department of Interior Office of Surface Mining Reclamation and Enforcement – ZrGfZi
Middleton, Larry T., (928) 523-2429 larry.middleton@nau.edu,
  Northern Arizona University – Gs
Mies, Jonathan W., (423) 425-4606 Jonathan-Mies@utc.edu,
  University of Tennessee, Chattanooga – GctHg
Mignery, Jill, 513-529-03225 henryjm@MiamiOh.edu,
  Miami University – Zge
Miguel, Carlos, (598) 25251552 cmiguel@fcien.edu.uy,
  Universidad de la Republica Oriental del Uruguay (UDELAR) – ZirSf
Mikesell, Dylan, 208-426-1404 dylanmikesell@boisestate.edu,
  Boise State University – YsuYx
Mikhaltsevitch, Vassily, +61 8 9266-4976 V.Mikhaltsevitch@curtin.edu.au,
  Curtin University – Ye

Mikulic, Donald G., 217-244-2518 mikulic@isgs.uiuc.edu,
  Illinois State Geological Survey – Ps
Milam, Keith A., (740) 593-1106 milamk@ohio.edu,
  Ohio University – Xg
Milan, Luke, +61267732019 lmilan@une.edu.au,
  University of New England – GtcGp
Miles, Randall J., 573-882-6607 University of Missouri, Columbia – Sd
Milewski, Adam, (706) 542-2652 milewski@uga.edu,
  University of Georgia – Hw
Militzer, Burkhard, militzer@seismo.berkeley.edu,
  University of California, Berkeley – Gy
Milkereit, Bernd, (416) 978-2466 bm@physics.utoronto.ca,
  University of Toronto – Ye
Milkov, Alexei, 303-273-3887 amilkov@mines.edu,
  Colorado School of Mines – Cg
Millan, Christina, 614-292-0863 millan.2@osu.edu,
  Ohio State University – Ggc
Millen, Timothy M., 847-697-1000 tmillen@elgin.edu,
  Elgin Community College – Gg
Miller, Arnold I., (513) 556-4022 arnold.miller@uc.edu,
  Cincinnati Museum Center – Pq
Miller, Barry W., (865) 594-5599 barry.miller@tn.gov,
  Tennessee Geological Survey – EcZiGg
Miller, Brent, (979) 458-3671 bvmiller@geo.tamu.edu,
  Texas A&M University – Cc
Miller, Calvin F., 615-322-2232 calvin.miller@vanderbilt.edu,
  Vanderbilt University – Gi
Miller, Charles M., (505) 667-8415 Los Alamos National Laboratory – Cc
Miller, David, dmiller37@csub.edu,
  California State University, Bakersfield – GgtGc
Miller, David S., (630) 252-7191 Argonne National Laboratory – Ge
Miller, Donald S., (518) 478-0758 milled2@rpi.edu,
  Rensselaer Polytechnic Institute – Cc
Miller, Doug, 828-232-5158 dmiller@unca.edu,
  University of North Carolina, Asheville – AsZnn
Miller, Douglas C., (302) 645-4277 dmiller@udel.edu,
  University of Delaware – Ob
Miller, Elizabeth L., (650) 723-1149 miller@pangea.stanford.edu,
  Stanford University – Gc
Miller, Geoffrey G., (506) 643-2361 Los Alamos National Laboratory – Cc
Miller, Gerald A., (515) 291-3442 soil@iastate.edu,
  Iowa State University of Science & Technology – SdoGm
Miller, Gifford H., 303-492-6962 gmiller@colorado.edu,
  University of Colorado – GlCc
Miller, Harvey J., (801) 585-3972 miller.81@osu.edu,
  Ohio State University – Zi
Miller, Heather R., heather.miller@austincc.edu,
  Austin Community College District – Zeg
Miller, Hugh, (303) 273-3558 hmiller@mines.edu,
  Colorado School of Mines – Nmg
Miller, Ian, 303-370-8351 Ian.Miller@dmns.org,
  Denver Museum of Nature & Science – PbGg
Miller, James, mille066@d.umn.edu,
  University of Minnesota, Duluth – GiEd
Miller, James F., (417) 836-5801 Missouri State University – Ps
Miller, Jerry R., (828) 227-2269 jmiller@wcu.edu,
  Western Carolina University – GmeGf
Miller, John, (416)736-5245 jrmiller@yorku.ca,
  York University – Zr
Miller, Jonathan S., (408) 924-5015 jonathan.miller@sjsu.edu,
  San Jose State University – GiCc
Miller, Joshua H., josh.miller@uc.edu,
  Cincinnati Museum Center – Pe
Miller, Joshua H., (513) 556-6704 joshua.miller@uc.edu,
  University of Cincinnati – PgvPe
Miller, Judy, jmiller264@monroecc.edu,
  Monroe Community College – Zn
Miller, Kate, (979) 845-3651 kcmiller@tamu.edu,
  Texas A&M University – Ys
Miller, Keith B., (785) 532-2250 kbmill@ksu.edu,
  Kansas State University – Pe
Miller, Kenneth G., (848) 445-3622 kgm@eps.rutgers.edu,
  Rutgers, The State University of New Jersey – GurPm
Miller, M. Meghan, (509) 963-2825 meghan@cwu.edu,
  Central Washington University – Yd
Miller, Mark, 601-266-4729 m.m.miller@usm.edu,
  University of Southern Mississippi – Zn

Miller, Marli G., (541) 346-4410 millerm@uoregon.edu,
University of Oregon – Gc
Miller, Marvin R., (406) 496-4155 mmiller@mtech.edu,
Montana Tech of The University of Montana – Hw
Miller, Michael B., (225) 388-3412 byron@lgs.bri.lsu.edu,
Louisiana State University – Go
Miller, Molly F., 615-322-3528 molly.miller@vanderbilt.edu,
Vanderbilt University – PeGs
Miller, Murray H., (519) 824-4120 (Ext. 53758) jmmiller7@sympatico.ca,
University of Guelph – So
Miller, Randall F., (506) 643-2361 University of New Brunswick – Pg
Miller, Raymond M., (630) 252-3395 rmmiller@anl.gov,
Argonne National Laboratory – Sb
Miller, Richard D., (785) 864-2091 rmiller@kgs.ku.edu,
University of Kansas – Ye
Miller, Richard D., (785) 864-3965 rmiller@kgs.ku.edu,
University of Kansas – Ye
Miller, Richard H., (619) 594-5118 rmiller@geology.sdsu.edu,
San Diego State University – Ps
Miller, Robert B., (408) 924-5025 robert.b.miller@sjsu.edu,
San Jose State University – Gct
Miller, Ronald L., (212) 678-5577 rmiller@giss.nasa.gov,
Columbia University – As
Miller, Steven F., (506) 643-2361 Argonne National Laboratory – Gg
Miller, Ted R., 605-677-6867 ted.miller@usd.edu,
South Dakota Dept of Environment and Natural Resources – Zn
Miller, Wade E., 801 422-2321 wem@geology.byu.edu,
Brigham Young University – Pv
Miller, William C., (707) 826-3110 wm1@humboldt.edu,
Humboldt State University – Pi
Miller, William P., (601) 325-2912 University of Georgia – Sc
Miller-Hicks, Bryan, Bryan.Miller-Hicks@gcccd.edu,
Cuyamaca College – Gg
Millet, Dylan B., 612-626-3259 dbm@umn.edu,
University of Minnesota, Twin Cities – As
Millette, Thomas L., (413) 538-2813 tmillett@mtholyoke.edu,
Mount Holyoke College – ZriGm
Milligan, Richard, 404 413-5778 rmilligan@gsu.edu,
Georgia State University – Zn
Milligan, Timothy, (902) 426-3273 milligan@dfo-mpo.gc.ca,
Dalhousie University – Gs
Milliken, Kitty L., (512) 471-6082 kittym@mail.utexas.edu,
University of Texas, Austin – Gd
Millington, Andrew, andrew.millington@flinders.edu.au,
Flinders University – ZuyZr
Mills, Aaron L., (804) 924-0564 alm7d@virginia.edu,
University of Virginia – So
Mills, Eric L., (902) 471-2016 e.mills@dal.ca,
Dalhousie University – ObZnn
Mills, James G., 765-658-4669 jmills@depauw.edu,
DePauw University – Gi
Mills, Rachel A., +44 (0)23 80592678 Rachel.Mills@soton.ac.uk,
University of Southampton – Cm
Mills, Stephanie, +44 020 8417 2950 S.Mills@kingston.ac.uk,
Kingston University – Gl
Mills, Suzanne, (905) 525-9140 smills@mcmaster.ca,
McMaster University – Zn
Milne, Glenn A., 613-562-5800 6424 gamilne@uottawa.ca,
University of Ottawa – YgAsZg
Milner, Jennifer, (309) 794-7318 JenniferMilner@augustana.edu,
Augustana College – Zn
Milner, Lloyd R., (225) 578-3410 lmilne1@lsu.edu,
Louisiana State University – Gg
Milroy, Scott P., (228) 688-7128 scott.milroy@usm.edu,
University of Southern Mississippi – Ob
Milstead, Terence, (828) 262-7057 milsteadtm@appstate.edu,
Appalachian State University – Zn
Min, Dong-Ha, 512-475-9290 dongha@austin.utexas.edu,
University of Texas, Austin – Og
Min, Doo-Hong, dmin@ksu.edu,
Kansas State University – So
Min, Kyoungwon, (352) 392-2720 kmin@ufl.edu,
University of Florida – Cc
Minarik, William G., 514-398-2596 william.minarik@mcgill.ca,
McGill University – Cp
Minchew, Brent, (617) 324-3704 minchew@mit.edu,
Massachusetts Institute of Technology – Yg

Miner, James J., 217-244-5786 miner@illinois.edu,
Illinois State Geological Survey – Gg
Minium, Deborah, (425) 564-5120 deborah.minium@bellevuecollege.edu,
Bellevue College – Gg
Minnaar, Hendrik, +27 (0)51 401 2372 minnaarh@ufs.ac.za,
University of the Free State – Gc
Minnich, Richard A., (951) 827-5515 richard.minnich@ucr.edu,
University of California, Riverside – Zy
Minor, Timothy B., (775) 673-7477 tminor@dri.edu,
Desert Research Institute – Zr
Minshull, Timothy A., +44 (0)23 80596569 tmin@noc.soton.ac.uk,
University of Southampton – Yr
Minster, J. Bernard H., (858) 945-0693 jbminster@ucsd.edu,
University of California, San Diego – Ys
Minter, Nicholas, +44 023 92 842288 nic.minter@port.ac.uk,
University of Portsmouth – Gs
Minton, David, (765) 494-3292 daminton@purdue.edu,
Purdue University – Xg
Minzoni, Marcello, (205)348-0768 marcello.minzoni@ua.edu,
University of Alabama – GosEo
Minzoni, Rebecca T., (205)348-6050 rebecca.minzoni@ua.edu,
University of Alabama – GsPme
Miot da Silva, Graziela, graziela.miotdasilva@flinders.edu.au,
Flinders University – Gug
Miranda, Elena A., 818-677-4671 elena.miranda@csun.edu,
California State University, Northridge – Gc
Mirnejad, Hassan , 513-529-3216 mirnejh@miamioh.edu,
Miami University – GxCg
Mirnejad, Hassan, mirnejh@miamioh.edu,
Miami University – Gx
Mishler, Brent, (510) 642-6810 bmishler@berkeley.edu,
University of California, Berkeley – Pe
Mishra, Umakant, 630-252-1108 umishra@anl.gov,
Argonne National Laboratory – So
Misner, Tamara, 814-732-1352 tmisner@edinboro.edu,
Edinboro University of Pennsylvania – GmHs
Misra, Debsmita, (907) 474-5339 dmisra@alaska.edu,
University of Alaska, Fairbanks – HwZrNg
Misra, Kula C., (865) 974-6020 kmisra@utk.edu,
University of Tennessee, Knoxville – Eg
Misra, Saumitra, (031) 260-2800 misras@ukzn.ac.za,
University of KwaZulu-Natal – GiCgu
Misra, Siddharth, 405-325-6787 misra@ou.edu,
University of Oklahoma – Np
Mitasova, Helena, 919-513-1327 helena_mitasova@ncsu.edu,
North Carolina State University – Zi
Mitchell, Charles E., (716) 645-4290 cem@buffalo.edu,
SUNY, Buffalo – Pg
Mitchell, Jonathan, (310) 825-2970 mitch@epss.ucla.edu,
University of California, Los Angeles – AsXg
Mitchell, Neil C., neil.mitchell@manchester.ac.uk,
University of Manchester – GumYg
Mitchell, Robert J., (360) 650-3591 robert.mitchell@geol.wwu.edu,
Western Washington University – Hw
Mitchell, Roger H., (807) 343-8287 roger.mitchell@lakeheadu.ca,
Lakehead University – Gi
Mitchell, Simon F., 876-927-2728 geoggeol@uwimona.edu.jm,
University of the West Indies Mona Campus – GsgGs
Mitchell, Tom, +44 (0)20 7679 7361 tom.mitchell@ucl.ac.uk,
University College London – Gc
Mitchneck, Beth A., bethm@email.arizona.edu,
University of Arizona – Zn
Mitchum, Gary T., (727) 553-3941 gmitchum@marine.usf.edu,
University of South Florida – Op
Mitra, Chandana, (334) 844-4229 czm0033@auburn.edu,
Auburn University – ZgAsZi
Mitra, Gautam, (585) 275-5816 gautam.mitra@rochester.edu,
University of Rochester – GctNr
Mitra, Shankar, (405) 325-4462 smitra@ou.edu,
University of Oklahoma – Gc
Mitra, Siddhartha, 252 328 6611 mitras@ecu.edu,
East Carolina University – Co
Mitri, Hani, (514) 398-4755 McGill University – Nm
Mitrovica, Jerry X., 617-496-2732 jxm@eps.harvard.edu,
Harvard University – Yg
Mittelstaedt, Eric L., (208) 885-2045 emittelstaedt@uidaho.edu,
University of Idaho – Yr

Mitterer, Richard M., mitterer@utdallas.edu, University of Texas, Dallas – ColOc
Mittlefehldt, David, (281) 483-5043 david.w.mittlefehldt@nasa.gov, University of Tennessee, Knoxville – Xm
Mix, Alan C., 541-737-5212 amix@coas.oregonstate.edu, Oregon State University – Cs
Miyagi, Lowell, (801) 581-6619 lowell.miyagi@utah.edu, University of Utah – Gz
Miyares, Ines, (212) 772-5443 imiyares@hunter.cuny.edu, Hunter College (CUNY) – Zen
Moayyed, Mohsen -., +98 (413) 339 2616 moayyed@tabrizu.ac.ir, University of Tabriz – GigEg
Moazzen, Mohssen -., +98 (411) 339 2679 moazzen@tabrizu.ac.ir, University of Tabriz – GpCgGz
Moberly, Ralph, (808) 956-8765 ralph@soest.hawaii.edu, University of Hawai'i, Manoa – GutGs
Mobley, Curtis D., (206) 230-8166 University of Washington – Ob
Mock, R. Stephen, (406) 683-7261 steve.mock@umwestern.edu, University of Montana Western – Ca
Mock, Thomas, +44 (0)1603 59 2566 t.mock@uea.ac.uk, University of East Anglia – Pg
Mock, Timothy D., 202-478-8466 tmock@carnegiescience.edu, Carnegie Institution for Science – Cc
Mockler, Theodore, (505) 667-4318 mockler@lanl.gov, Los Alamos National Laboratory – Zn
Mode, William N., (920) 424-7004 mode@uwosh.edu, University of Wisconsin, Oshkosh – Gl
Modzelewski, Henryk, (604) 822-3591 hmodzelewski@eos.ubc.ca, University of British Columbia – As
Moe-Hoffman, Amy P., apm105@msstate.edu, Mississippi State University – PgZe
Moe-Hoffman, Amy P., 662-268-1032 Ext 234 paleomoe@gmail.com, Mississippi State University – Zn
Moecher, David P., (859) 257-6939 moker@uky.edu, University of Kentucky – GxtCg
Moeglin, Thomas D., (417) 836-5800 Missouri State University – Ng
Moersch, Jeffery E., (865) 974-2366 jmoersch@utk.edu, University of Tennessee, Knoxville – XgZr
Moffett, James W., 213-740-5626 jmoffett@usc.edu, University of Southern California – OcCba
Mofjeld, Harold O., (206) 526-6819 mofjeld@pmel.noaa.gov, University of Washington – Op
Moghanloo, Rouzbeh, rouzbeh.gm@ou.edu, University of Oklahoma – Np
Mogilevskaya, Sonia, 612-625-4810 mogil003@umn.edu, University of Minnesota, Twin Cities – Ng
Mogk, David W., (406) 600-4071 mogk@montana.edu, Montana State University – GpZeGz
Mohammad Reza, Hosseinzadeh, +98 (411) 339 2697 University of Tabriz – Egm
Mohr, Marcus, 089/2180 4230 marcus.mohr@geophysik.uni-muenchen.de, Ludwig-Maximilians-Universitaet Muenchen – Yg
Mohrig, David, (512) 471-2282 mohrig@jsg.utexas.edu, University of Texas, Austin – GsmGr
MOINE, Bertrand N., 33-477481513 moineb@univ-st-etienne.fr, Université Jean Monnet, Saint-Etienne – CgtGx
Mojzsis, Stephen J., (303) 492-5014 stephen.mojzsis@colorado.edu, University of Colorado – XcCcGp
Moldowan, J. Michael, (650) 725-0913 moldowan@pangea.stanford.edu, Stanford University – Co
Moldwin, Mark B., (734) 647-3370 mmoldwin@umich.edu, University of Michigan – Xp
Molina, Jean-Alex E., (651) 647-9865 jamolina@umn.edu, University of Minnesota, Twin Cities – SbCgSo
Molina, Mario J., (858) 534-1696 mjmolina@ucsd.edu, University of California, San Diego – As
Molinari, John E., (518) 442-4562 molinari@atmos.albany.edu, SUNY, Albany – As
Moll, Nancy E., 760-776-7272 nmoll@collegeofthedesert.edu, College of the Desert – Gge
Mollner, Daniel, (507) 933-7569 Gustavus Adolphus College – Zn
Molnár, Gábor, molnar@sas.elte.hu, Eotvos Lorand University – ZrYgZi
Molnar, Peter, 303-492-4936 peter.molnar@colorado.edu, University of Colorado – Gt
Molnar, Sheri, 519-661-2111, ext.87031 smolnar8@uwo.ca, Western University – Ys

Molnia, Bruce F., (703) 648-4120 bmolnia@usgs.gov, Duke University – ZrGlu
Moloney, Marguerite M., (985) 448-4878 marguerite.moloney@nicholls.edu, Nicholls State University – GeZnn
Molotch, Noah, Noah.Molotch@colorado.edu, University of Colorado – Hg
Momayez, Moe, (520) 621-6580 moe.momayez@arizona.edu, University of Arizona – NmYxNr
Momen, Nasim, (617) 353-5679 Boston University – Zn
Momm, Henrique G., (615) 904-8372 henrique.momm@mtsu.edu, Middle Tennessee State University – ZiHsZr
Monahan, Adam, 250-721-6120 monahana@uvic.ca, University of Victoria – Zn
Monahan, Edward C., (860) 405-9110 edward.monahan@uconn.edu, University of Connecticut – AsOp
Monari, Stefano, +390498279171 stefano.monari@unipd.it, Università degli Studi di Padova – Pg
Moncrief, John F., (612) 625-2771 moncr001@umn.ed, University of Minnesota, Twin Cities – Sp
Moncrieff, John B., +44 (0) 131 650 5402 J.Moncrieff@ed.ac.uk, Edinburgh University – As
Monecke, Katrin, kmonecke@wellesley.edu, Wellesley College – Gsn
Monecke, Thomas, 303-273-3841 tmonecke@mines.edu, Colorado School of Mines – Em
Monet, Julie, 530-898-3460 jmonet@csuchico.edu, California State University, Chico – GgNg
Monger, Bruce, bcm3@cornell.edu, Cornell University – Ob
Monger, H. C., (505) 646-3405 New Mexico State University, Las Cruces – Sa
Monson, Jessica, 217-265-6895 jlbm@illinois.edu, Illinois State Geological Survey – Gg
Montagna, Paul, 361-825-2040 paul.montagna@tamucc.edu, Texas A&M University, Corpus Christi – ObCmHs
Montagne, Cliff, (406) 599-7755 montagne@montana.edu, Montana State University – Sd
Montañez, Isabel P., (530) 754-7823 ipmontanez@ucdavis.edu, University of California, Davis – Gd
Montayne, Simone, (907) 451-5036 simone.montayne@alaska.gov, Alaska Division of Geological & Geophysical Surveys – Gg
Monteleone, Brian D., (508) 289-2405 bmonteleone@whoi.edu, Woods Hole Oceanographic Institution – Cc
Montenari, Michael, (+44) 01782 733162 m.montenari@keele.ac.uk, Keele University – PmgPs
Montenegro, Alvaro, (614) 688-5451 montenegro.8@osu.edu, Ohio State University – OgAsZy
Montero, Esperanza, emontero@ucm.es, Univ Complutense de Madrid – Hw
Montesi, Laurent G., (301) 405-7534 montesi@umd.edu, University of Maryland – YgXyGt
Monteverde, Don, dmonte@eps.rutgers.edu, Rutgers, The State University of New Jersey – GruGg
Monteverde, Donald H., (609) 292-2576 don.monteverde@dep.state.nj.us, New Jersey Geological and Water Survey – Grg
Monteverdi, John P., (415) 338-7728 montever@sfsu.edu, San Francisco State University – As
Montgomery, David R., 206-685-2560 bigdirt@uw.edu, University of Washington – Gm
Montgomery, Homer A., 972-883-2496 mont@utdallas.edu, University of Texas, Dallas – Pg
Montgomery, Keith, (715) 845-9602 keith.montgomery@uwc.edu, University of Wisconsin Colleges – ZyGg
Montgomery, Michael T., (831) 656-2296 mtmontgo@nps.edu, Naval Postgraduate School – AmsZg
Montgomery, Tamra S., 217-333-5105 tmntgmry@illinois.edu, Illinois State Geological Survey – Zn
Montgomery, William W., (201) 200-3161 wmontgomery@njcu.edu, New Jersey City University – Hw
Montoya, Joseph, (404) 385-0479 Georgia Institute of Technology – Cm
Montoya, Judith, (928) 523-8523 judith.montoya@nau.edu, Northern Arizona University – Zy
Montwill, Gail F., (909) 946-1796 Santiago Canyon College – Gg
Moodie, T. Bryant, (780) 492-5742 bryant.moodie@ualberta.ca, University of Alberta – Zn
Moody, Eva, (254) 968-9143 moody@tarleton.edu,

Tarleton State University – Zn
Moody, Jennie L., (804) 924-0592 jlm8h@virginia.edu, University of Virginia – As
Mooers, Howard, (218) 726-7239 hmooers@d.umn.edu, University of Minnesota, Duluth – GleHw
Mookerjee, Matty, (707) 664-2002 matty.mookerjee@sonoma.edu, Sonoma State University – GcYgZi
Moon, Charlie, +44 01326 371822 c.j.moon@exeter.ac.uk, Exeter University – CeEmZi
Moon, Seulgi, sgmoon@g.ucla.edu, University of California, Los Angeles – Gm
Moon, Wooil M., (204) 474-9833 wmoon@cc.umanitoba.ca, University of Manitoba – ZrYnOp
Mooney, Phillip, (707) 664-2328 mooneyp@sonoma.edu, Sonoma State University – ZnGc
Moorberg, Colby, (785) 532-7207 Kansas State University – So
Moore, Andrew, (765) 983-1672 moorean@earlham.edu, Earlham College – Gm
Moore, Bradley S., (858) 822-6650 bsmoore@ucsd.edu, University of California, San Diego – Cm
Moore, Christopher M., +44 (0)23 80594801 cmm297@noc.soton.ac.uk, University of Southampton – Ob
Moore, Daniel K., (208) 496-1902 moored@byui.edu, Brigham Young University - Idaho – GxiCp
Moore, Dennis, dmoore@pmel.noaa.gov, University of Hawai'i, Manoa – Og
Moore, Dennis W., (734) 763-0202 dennis.w.moore@noaa.gov, University of Washington – Op
Moore, Duane M., 505-277-4204 dewey33@unm.edu, University of New Mexico – Sc
Moore, Gordon M., (530) 752-5829 gomo@ucdavis.edu, University of California, Davis – Gx
Moore, Gregory F., (808) 956-6854 gmoore@hawaii.edu, University of Hawai'i, Manoa – Gt
Moore, J. Casey, (831) 459-2574 jcmoore@pmc.ucsc.edu, University of California, Santa Cruz – Gc
Moore, Jefferson K., 949-824-5391 jkmoore@uci.edu, University of California, Irvine – Op
Moore, Jeffrey, (801) 585-0491 jeff.moore@utah.edu, University of Utah – NgrGm
Moore, Joel, (410) 704-4245 moore@towson.edu, Towson University – ClsSc
Moore, Joseph N., (801) 585-6931 jmoore@egi.utah.edu, University of Utah – Gv
Moore, Kathryn, +44 (0)1326 255693 K.Moore@exeter.ac.uk, Exeter University – Ge
Moore, Laura J., (919) 962-5960 moorelj@email.unc.edu, University of North Carolina, Chapel Hill – Gme
Moore, Michael E., 717.702.2024 michmoore@pa.gov, Pennsylvania Bureau of Topographic & Geologic Survey – HwZi
Moore, Phillip, +44 (0) 191 208 5040 philip.moore@ncl.ac.uk, University of Newcastle Upon Tyne – Yd
Moore, Richard W., 831-656-1041 rwmoor1@nps.edu, Naval Postgraduate School – Am
Moore, Robert M., (902) 494-3871 robert.moore@dal.ca, Dalhousie University – Oc
Moore, Theodore C., (734) 763-0202 tedmoore@umich.edu, University of Michigan – OuPmc
Moores, Eldridge M., (916) 752-0352 University of California, Davis – Gt
Moorhead, Daryl L., (419) 530-2017 daryl.moorhead@utoledo.edu, University of Toledo – Zn
Moorhead, Kevin K., (828) 232-5183 moorhead@unca.edu, University of North Carolina, Asheville – Sbo
Moorkamp, Max, +440116 252 3632 mm489@le.ac.uk, Leicester University – Yg
Moorman, Thomas B., (515) 294-2308 tom.moorman@ars.usda.gov, Iowa State University of Science & Technology – So
Moortgat, Joachim, (614) 688-2140 moortgat.1@osu.edu, Ohio State University – HwNp
Mora-Klepeis, Gabriela, 802-656-0246 gmora@uvm.edu, University of Vermont – CacGi
Morabia, Alfredo, 718-670-4180 Queens College (CUNY) – GbZn
Morabia, Alfredo, 718-650-4222 amorabia@qc.cuny.edu, Graduate School of the City University of New York – Zg
Morales, Michael A., (620) 794-0191 mmorales@emporia.edu, Emporia State University – PvGrg
Morales, Tomas, tomas.morales@ehu.eus,
University of the Basque Country UPV/EHU – NgHg
Moran, Dawn, (508) 289-4918 dmoran@whoi.edu, Woods Hole Oceanographic Institution – Ob
Moran, Jean E., 510-885-2491 jean.moran@csueastbay.edu, California State University, East Bay – HwCl
Moran, S. Bradley, (401) 874-6530 moran@gso.uri.edu, University of Rhode Island – Oc
Moran, Seth, 360-993-8934 smoran@usgs.gov, University of Washington – Gv
Moran-Taylor, Michelle, 303-871-2513 michelle.moran-taylor@du.edu, University of Denver – Zn
Moran-Zenteno, Dante J., dante@tonatiuh.igeofcu.unam.mx, Universidad Nacional Autonoma de Mexico – Cc
Morand, Vincent J., +61 3 9479 5641 v.morand@latrobe.edu.au, La Trobe University – GgcGt
Morealli, Sarah A., 540 654-1402 smoreall@umw.edu, University of Mary Washington – Gg
Morehouse, Barbara, morehoub@email.arizona.edu, University of Arizona – Zn
Morel, Francois M M., (609) 258-2416 morel@princeton.edu, Princeton University – Cl
Morel-Kraepiel, Anne, 609-258-7415 kraepiel@Princeton.EDU, Princeton University – Em
Moreno, Rafael, (303) 556-8477 Metropolitan State College of Denver – Zg
Moreton, Kim, (+44) (0) 7854825368 k.moreton@exeter.ac.uk, Exeter University – EnZur
Morgan, Cristine L., (979) 845-3603 cmorgan@tamu.edu, Texas A&M University – Spd
Morgan, Daniel, +44(0) 113 34 35202 d.j.morgan@leeds.ac.uk, University of Leeds – Giv
Morgan, Daniel J., (615) 343-3141 dan.morgan@vanderbilt.edu, Vanderbilt University – GmCc
Morgan, Emory, (757) 727-5783 emory.morgan@hampton.edu, Hampton University – Ze
Morgan, F D., (617) 253-7857 fdmorgan@mit.edu, Massachusetts Institute of Technology – Yg
Morgan, Gary, 757.727.5783 gary.morgan@hamptonu.edu, Hampton University – ZnnZn
Morgan, George B., (405) 325-2642 gmorgan@ou.edu, University of Oklahoma – Gi
Morgan, Jason, +44 1784 443606 Jason.Morgan@rhul.ac.uk, Royal Holloway University of London – Yg
Morgan, Joanna, +44 20 759 46423 j.v.morgan@imperial.ac.uk, Imperial College – Yg
Morgan, John D., 850-474-2224 jmorgan3@uwf.edu, University of West Florida – Zi
Morgan, Julia K., (713) 348-6330 morganj@rice.edu, Rice University – GctGv
Morgan, Kelly, 239-658-3400 conserv@ufl.edu, University of Florida – So
Morgan, Matt, 303-384-2632 mmorgan@mines.edu, Colorado Geological Survey – GgmXm
Morgan, Michael C., (608)265-8159 morgan@aurora.aos.wisc.edu, University of Wisconsin, Madison – Am
Morgan, Paul, 303-384-2648 morgan@mines.edu, Colorado Geological Survey – YhGtYg
Morgan, Paul, paul.morgan@nau.edu, Northern Arizona University – Yh
Morgan, Ryan, 254-968-9894 rmorgan@tarleton.edu, Tarleton State University – Pei
Morgan, Siobahn M., (319) 273-2389 siobahn.morgan@uni.edu, University of Northern Iowa – Xa
Morgan, Sven S., 515-294-1837 smorgan@iastate.edu, Iowa State University of Science & Technology – Gc
Morgan, Tamie, (817) 257-7743 tamie.morgan@tcu.edu, Texas Christian University – Zi
Morgan, Terrance L., (505) 667-0837 Los Alamos National Lab – Ng
Morgan, William, +44 0161 306-6586 Will.Morgan@manchester.ac.uk, University of Manchester – As
Morin, Paul, 612-626-0505 lpaul@umn.edu, University of Minnesota, Twin Cities – Zn
Morison, James H., (206) 543-1394 morison@apl.washington.edu, University of Washington – Op
Moritz, Wolfgang, 089/2180 4336 w.moritz@lrz.uni-muenchen.de, Ludwig-Maximilians-Universitaet Muenchen – Gz
Moro, Alan, +38514606093 amoro@geol.pmf.hr,

University of Zagreb – PgsGg
Morozov, Igor B., (306) 966-2761 igor.morozov@usask.ca,
  University of Saskatchewan – YsGy
Morra, Matthew J., (208) 885-6315 University of Idaho – Sb
Morris, Antony, +44 1752 584766 a.morris@plymouth.ac.uk,
  University of Plymouth – YmGtu
Morris, Billy, 706-368-7528 bmorris@highlands.edu,
  Georgia Highlands College – Gg
Morris, Brenda, (760) 744-1150 ext. 2512 bmorris@palomar.edu,
  Palomar College – Zn
Morris, David, (314) 935-6926 Los Alamos National Laboratory – Cc
Morris, Donald P., (610) 758-5175 dpm2@lehigh.edu,
  Lehigh University – Zn
Morris, Geoffrey, (785) 532-3397 gpmorris@ksu.edu,
  Kansas State University – So
Morris, John A., 662-268-1032 Ext 235 jam16@msstate.edu,
  Mississippi State University – ZirAm
Morris, Simon C., +44 (0) 1223 333414 sc113@esc.cam.ac.uk,
  University of Cambridge – Pg
Morris, Thomas H., (801) 422-3761 tom_morris@byu.edu,
  Brigham Young University – Gr
Morris, William A., (905) 525-9140 (Ext. 20116) morriswa@mcmaster.ca,
  McMaster University – Yg
Morrison, Lauren, 252 328 6360 morrisonl14@ecu.edu,
  East Carolina University – Zn
Morrow, Robert H., 803.896.1214 morrowr@dnr.sc.gov,
  South Carolina Dept of Natural Resources – GcxGt
Morschauser, Lindsey, 662-268-1032 Ext 243 lcm193@msstate.edu,
  Mississippi State University – Am
Morse, David L., (512) 232-3241 University of Texas, Austin – Yg
Morse, Linda D., (757) 221-2444 ldmors@wm.edu,
  College of William & Mary – GeZg
Morse, Stearns A., (413) 545-0175 tm@geo.umass.edu,
  University of Massachusetts, Amherst – Gi
Mortensen, James K., (604) 822-6208 jmortensen@eos.ubc.ca,
  University of British Columbia – Cc
Mortlock, Richard , (848) 445-3423 rmortloc@eps.rutgers.edu,
  Rutgers, The State University of New Jersey – CgmCs
Morton, Allan E., (602) 426-4351 Central Arizona College – Ze
Morton, Bruce, mortonb@easternct.edu,
  Eastern Connecticut State University – Zg
Morton, Douglas M., (951) 276-6397 scamp@ucrac1.ucr.edu,
  University of California, Riverside – Gp
Morton, Penelope, pmorton@d.umn.edu,
  University of Minnesota, Duluth – EgCg
Morton, Robert, (813) 974-2773 University of South Florida, Tampa – Gs
Morton, Roger D., (780) 492-3265 University of Alberta – Em
Morton, Ronald, rmorton@d.umn.edu,
  University of Minnesota, Duluth – GvEg
Moscardelli, Lorena G., 512-471-4971 lorena.moscardelli@beg.utexas.edu,
  University of Texas, Austin – GmYs
Moscovici, Daniel, (215) 688-2910 daniel.moscovici@stockton.edu,
  Stockton University – Zu
Mosenfelder, Jed, jmosenfe@umn.edu,
  University of Minnesota, Twin Cities – Cp
Moser, Desmond, 519-661-4214 dmoser22@uwo.ca,
  Western University – GtCcXg
Mosher, David C., 603-862-5493 dmosher@ccom.unh.edu,
  University of New Hampshire – Og
Mosher, David C., (902) 426-3149 mosher@agc.bio.ns.ca,
  Dalhousie University – Ou
Mosher, Sharon, (512) 471-4135 mosher@mail.utexas.edu,
  University of Texas, Austin – Gc
Moshier, Stephen O., (630) 752-5856 stephen.moshier@wheaton.edu,
  Wheaton College – Gd
Moskalski, Susanne, susanne.moskalski@stockton.edu,
  Stockton University – GsOn
Moskowitz, Bruce M., 612-624-1547 bmosk@umn.edu,
  University of Minnesota, Twin Cities – Ym
Mosley-Thompson, Ellen E., (614) 292-2580 thompson.4@osu.edu,
  Ohio State University – As
Moslow, Thomas F., (403) 269-6911 University of Calgary – Co
Moss, Neil E., 205-247-3557 nmoss@gsa.state.al.us,
  Geological Survey of Alabama – Hw
Moss, Patti, 509-527-5225 mosspm@whitman.edu,
  Whitman College – Zn
Mossa, Joann, (352) 294-7510 mossa@ufl.edu,
  University of Florida – GmHsZy
Mossman, David J., (506) 364-2326 dmossman@mta.ca,
  Mount Allison University – EmZn
Motani, Ryosuke, 530-754-6284 rmotani@ucdavis.edu,
  University of California, Davis – Pv
Motavalli, Peter P., 573-884-3212 motavallip@missouri.edu,
  University of Missouri, Columbia – So
Mote, Philip, 541-737-5694 pmote@coas.oregonstate.edu,
  Oregon State University –
Mote, Philip W., pmote@coas.oregonstate.edu,
  University of Washington – As
Mottl, Michael J., (808) 956-7006 mmottl@soest.hawaii.edu,
  University of Hawai'i, Manoa – OuCgm
Motyka, James, motykaj@easternct.edu,
  Eastern Connecticut State University – Zg
Motz, Gary J., (812) 856-3500 garymotz@indiana.edu,
  Indiana University – PqgPi
Mouat, David A., (775) 673-7402 dmouat@dri.edu,
  Desert Research Institute – Ge
Moucha, Robert, 315-443-6239 rmoucha@syr.edu,
  Syracuse University – GtYeg
Mouginis-Mark, Peter J., (808) 956-3147 University of Hawai'i, Manoa – Xg
Moulis, Anastasia, 617-552-8300 anastasia.macherides@bc.edu,
  Boston College – Ys
Moum, James N., 541-737-2553 jmoum@coas.oregonstate.edu,
  Oregon State University – Op
Mound, Jon, +44(0) 113 34 35216 earjem@leeds.ac.uk,
  University of Leeds – Gt
Mount, Gregory, 724-357-7662 gregory.mount@iup.edu,
  Indiana University of Pennsylvania – Hy
Mount, Jeffrey F., (916) 752-7092 University of California, Davis – Gs
Mount, Jeffrey F., jfmount@ucdavis.edu,
  University of California, Davis – Gm
Mountain, Gregory S., 848-445-0817 gmtn@eps.rutgers.edu,
  Rutgers, The State University of New Jersey – GuYrGr
Mountney, Nigel, +44(0) 113 34 35249 n.p.mountney@leeds.ac.uk,
  University of Leeds – Gs
Mousset-Jones, Pierre, (775) 784-6959 mousset@mines.unr.edu,
  University of Nevada, Reno – Nm
Mower, Richard, 989-774-3821 mower1rn@cmich.edu,
  Central Michigan University – Am
Mowrer, Jake, 979-845-5366 jake.mowrer@tamu.edu,
  Texas A&M University – Sc
Moy, Christopher M., 64-3-479-5279 chris.moy@otago.ac.nz,
  University of Otago – Gus
MOYEN, Jean-François, (+33)477481510 jean.francois.moyen@univ-st-etienne.fr,
  Université Jean Monnet, Saint-Etienne – GiCtGp
Moyer, Elisabeth, (773) 834-2992 University of Chicago – As
Moyer, Kerry A., (814) 732-2454 kmoyer@edinboro.edu,
  Edinboro University of Pennsylvania – As
Moysey, Stephen, (864) 353-1517 moyseys18@ecu.edu,
  East Carolina University – Hg
Moysey, Stephen M., (864) 656-5019 smoysey@clemson.edu,
  Clemson University – HwYuZe
Mozley, Peter S., (575) 835-5311 mozley@nmt.edu,
  New Mexico Institute of Mining and Technology – Gs
Mozzachiodi, Riccardo, 361-825-3634 Riccardo.Mozzachiodi@tamucc.edu,
  Texas A&M University, Corpus Christi – Zn
Mozzi, Paolo, +390498279190 paolo.mozzi@unipd.it,
  Università degli Studi di Padova – Gm
Mrinjek, Ervin, +38514606057 ervin.mrinjek@zg.t-com.hr,
  University of Zagreb – GscGg
Mroz, Eugene J., (505) 667-7758 Los Alamos National Laboratory – As
Mshiu, Elisante E., *255768384430 mshiutz@gmail.com,
  University of Dar es Salaam – ZrCgZi
Mshiu, Elisante E., mshiutz@udsm.ac.tz,
  University of Dar es Salaam – GgCeZr
Mubako, Stanley, 915-747-7372 stmubako@utep.edu,
  University of Texas, El Paso – Zi
Mucci, Alfonso, (514) 398-4892 alfonso.mucci@mcgill.ca,
  McGill University – CmlOc
Muchez, Philippe, 32 16 327584 philippe.muchez@kuleuven.be,
  Katholieke Universiteit Leuven – EmCgGd
Mucsi, lászló, +3662544156 mucsi@geo.u-szeged.hu,
  Univesity of Szeged – ZirZy

Mudd, Simon N., +44 (0) 131 650 2535 simon.m.mudd@ed.ac.uk, Edinburgh University – Hg
Mudrick, Stephen E., (573) 882-6721 University of Missouri, Columbia – As
Muehlenbachs, Karlis, (780) 492-2827 karlis.muehlenbachs@ualberta.ca, University of Alberta – Cs
Mueller, Amy, 617.373.8131 a.mueller@northeastern.edu, Northeastern University – Zn
Mueller, Erich M., (334) 460-7136 em256@cornell.edu, University of South Alabama – Ob
Mueller, Ivan, 860-726-2069 mueller.3@osu.edu, Ohio State University – YddYv
Mueller, Karl J., (303) 492-7336 karl.mueller@colorado.edu, University of Colorado – GtcGm
Mueller, Paul A., (352) 392-2231 pamueller@ufl.edu, University of Florida – CcGit
Mueller, Thomas, (724) 938-4255 mueller@calu.edu, California University of Pennsylvania – Zi
Muggeridge, Ann H., +44 20 759 47379 a.muggeridge@imperial.ac.uk, Imperial College – NpHwEo
Muhleman, Duane O., (626) 395-6186 dom@gps.caltech.edu, California Institute of Technology – Xy
Muir, William, 909-387-1603 wmuir@sbccd.cc.ca.us, San Bernardino Valley College – Og
Mukasa, Samuel B., (734) 936-3227 mukasa@umich.edu, University of Michigan – Cc
Mukherjee, Falguni, (936) 294-1073 fsm002@shsu.edu, Sam Houston State University – Zin
Mukhopadhyay, Sujoy, (530) 752-4711 sujoy@ucdavis.edu, University of California, Davis – Cg
Mulcahy, Sean R., 360-650-3645 sean.mulcahy@wwu.edu, Western Washington University – GpzGp
Muldoon, Maureen A., (920) 424-4461 muldoon@uwosh.edu, University of Wisconsin, Oshkosh – Hw
Mulholland, Margaret, (757) 683-3972 mmulholl@odu.edu, Old Dominion University – Ob
Mulibo, Gabriel D., gmbelwa@yahoo.com, University of Dar es Salaam – Ysg
Mulla, David J., (612) 625-6721 mulla003@umn.edu, University of Minnesota, Twin Cities – Sp
Mullen, Steven L., (520) 621-6842 mullen@atmo.arizona.edu, University of Arizona – As
Muller, Andrew C., (410) 293-6569 amuller@usna.edu, United States Naval Academy – OpnOu
Muller, Dietmar, (029) 351-3244 d.muller@usyd.edu.au, University of Sydney – Yrg
Muller, Marc, mmuller1@nd.edu, University of Notre Dame – Rw
Muller, Otto H., (607) 871-2208 Alfred University – Gc
Muller, Peter, (808) 956-8081 pmuller@soest.hawaii.edu, University of Hawai'i, Manoa – Op
Muller-Karger, Frank E., (727) 553-3335 University of South Florida – Zr
Mulligan, Kevin R., 806-834-0391 kevin.mulligan@ttu.edu, Texas Tech University – ZiGmSo
Mullins, Gregory L., (540) 231-4383 gmullins@vt.edu, Virginia Polytechnic Institute & State University – Sc
Mulrooney, Timothy, (919) 530-6269 tmulroon@nccu.edu, North Carolina Central University – Zir
Mulvaney, Richard L., (217) 333-9467 University of Illinois, Urbana-Champaign – Sc
Mulvany, Patrick S., (573) 341-4616 mulvany@umr.edu, Missouri University of Science and Technology – Gg
Mumin, A. Hamid, 204-727-9685 mumin@brandonu.ca, Brandon University – EgGzt
Munasinghe, Tissa, munasit@lahc.edu, Los Angeles Harbor College – Zg
Mundie, Ben, (541) 967-2039 Oregon Dept of Geology and Mineral Industries – Ge
Mungall, James E., 613 558 9337 jamesmungall@cunet.carleton.ca, Carleton University – GiEgCg
Munguia, Luis, lmunguia@cicese.mx, Centro de Investigación Científica y de Educación Superior de Ensenada – Ys
Munizzi, Jordan S., (859) 257-6222 jmunizzi@uky.edu, University of Kentucky – Cs
Munk, LeeAnn, 907 786-6895 lamunk@uaa.alaska.edu, University of Alaska, Anchorage – Cle

Munk, Walter H., (858) 534-2877 wmunk@ucsd.edu, University of California, San Diego – Op
Munn, Barbara J., (916) 278-6811 California State University, Sacramento – Ggp
Muñoz, Alfonso, amunoz@ucm.es, Univ Complutense de Madrid – Ygr
Munroe, Jeffrey S., (802) 443-3446 jmunroe@middlebury.edu, Middlebury College – Gln
Munski, Douglas C., (701) 777-4591 douglas.munski@und.edu, University of North Dakota – Ze
Muntean, John, (775) 682-8748 munteanj@unr.edu, University of Nevada – Eg
Muntean, Thomas, 517-264-3943 tmuntean@adrian.edu, Adrian College – Gde
Murchie, Scott, 240-228-6235 scott.murchie@jhuapl.edu, University of Tennessee, Knoxville – Xg
Murdoch, Lawrence C., (864) 656-2597 lmurdoc@clemson.edu, Clemson University – Hw
Murgulet, Dorina, (361) 825-2309 Dorina.Murgulet@tamucc.edu, Texas A&M University, Corpus Christi – Hw
Murgulet, Valeriu, 361-825-6023 Valeriu.Murgulet@tamucc.edu, Texas A&M University, Corpus Christi – Cg
Murowchick, James B., (816) 235-2979 murowchickj@umkc.edu, University of Missouri, Kansas City – CgGz
Murphree, James Thomas, 831-656-2723 murphree@nps.edu, Naval Postgraduate School – Am
Murphy, Alexander, (541) 346-4571 abmurphy@uoregon.edu, University of Oregon – Zn
Murphy, Cindy, (902) 867-2299 cmurphy@stfx.ca, Saint Francis Xavier University – Gg
Murphy, David T., 61 07 31382329 david.murphy@qut.edu.au, Queensland University of Technology – CgEgCc
Murphy, Edward C., (701) 328-8002 emurphy@nd.gov, North Dakota Geological Survey – Zn
Murphy, J. Brendan, (902) 867-2481 Dalhousie University – Gx
Murphy, Michael, 713-743-3564 mmurphy@central.uh.edu, University of Houston – Gct
Murphy, Michael A., michael.murphy@ucr.edu, University of California, Riverside – Ps
Murphy, Todd, (318) 342-3428 murphy@ulm.edu, University of Louisiana, Monroe – As
Murphy, Vincent, 617-552-8300 Boston College – Yg
Murray, A. Bradshaw, (919) 684-5847 abmurray@duke.edu, Duke University – Gm
Murray, Alison, 775.673.7361 alison.murray@dri.edu, University of Nevada, Reno – ObCg
Murray, Andrew, 251-460-7325 amurray@southalabama.edu, University of South Alabama – Am
Murray, Christopher J., (509) 376-5848 chris.murray@pnl.gov, Pacific Northwest National Laboratory – Gq
Murray, Daniel P., (401) 874-2265 dpmurray@uri.edu, University of Rhode Island – ZeGtp
Murray, David, (401) 863-3531 David_Murray@Brown.EDU, Brown University – Ou
Murray, James W., (206) 543-4730 jmurray@u.washington.edu, University of Washington – Oc
Murray, John, + 353 (0)91 495 095 john.murray@nuigalway.ie, National University of Ireland Galway – PgGs
Murray, Kent S., (313) 436-9129 kmurray@umich.edu, University of Michigan, Dearborn – Hw
Murray, Kyle E., (405) 325-7502 kyle.murray@ou.edu, University of Oklahoma – HwZiGo
Murray, Mark, 575.835.6930 murray@ees.nmt.edu, New Mexico Institute of Mining and Technology – Yds
Murray, Richard, (617) 353-6532 rickm@bu.edu, Boston University – Cm
Murrell, Coling, +44 (0)1603 59 2959 j.c.murrell@uea.ac.uk, University of East Anglia – As
Murrell, Michael T., (505) 667-0967 Los Alamos National Laboratory – Cc
Murthy, Prahlad N., 570-408-4617 prahlad.murthy@wilkes.edu, Wilkes University – AsZnn
Murtugudde, Raghuram G., (301) 314-2622 ragu@umd.edu, University of Maryland – ObZr
Mushkin, Amit, 972(02)5314254 mushkin@uw.edu, University of Washington – ZrCcGm
Musil, Rudolf, +420 549 49 5997 rudolf@sci.muni.cz, Masaryk University – PgvGe

Muskatt, Herman, 792-3028 ewelch@utica.edu,
    Utica College – GgrPg
Musolf, Gene E., (715) 845-9602 University of Wisconsin Colleges – Zy
Musselman, Zachary A., (601) 974-1344 musseza@millsaps.edu,
    Millsaps College – Gm
Mustard, John F., (401) 863-1264 john_mustard@brown.edu,
    Brown University – Zr
Mustart, David A., (415) 338-7729 mustart@sfsu.edu,
    San Francisco State University – GizGv
Mustoe, George, george.mustoe@wwu.edu,
    Western Washington University – PbCaPo
Muthukrishnan, Suresh, (864) 294-3361 suresh.muthukrishnan@furman.edu,
    Furman University – ZiGmZr
Muto, Atsuhiro, 215-204-3699 amuto@temple.edu,
    Temple University – Gl
Mutter, John C., jcm@ldeo.columbia.edu,
    Columbia University – Yr
Mutter, John C., (646) 269-6942 jcm7@columbia.edu,
    Columbia University – YrGtYs
Mutti, Laurel, muttil@newpaltz.edu,
    SUNY, New Paltz – Gg
Mutti, Laurence J., (814) 641-3067 mutti@juniata.edu,
    Juniata College – GxzSc
Mutua, Felix N., fnmutua@jkuat.ac.ke,
    Jomo Kenyatta University of Agriculture & Technology – ZigZr
Muxworthy, Adrian, +44 20 759 46442 adrian.muxworthy@imperial.ac.uk,
    Imperial College – Ym
Muza, Jay P., (954) 736-8231 jmuza@broward.edu,
    Broward College – OuPmOo
Mwangi, Nancy , nwmwangi@jkuat.ac.ke,
    Jomo Kenyatta University of Agriculture & Technology – ZiiZr
Myers, Alan R., 217-300-2570 amyers76@illinois.edu,
    Illinois State Geological Survey – EcGo
Myers, Clifford D., (413) 236-4601 cmyers@berkshirecc.edu,
    Berkshire Community College – Zn
Myers, Corinne E., 505.277.4204 cemyers@unm.edu,
    University of New Mexico – PeqPi
Myers, James D., (307) 766-2203 magma@uwyo.edu,
    University of Wyoming – Gi
Myers, Jeffrey A., 503-838-8365 myersj@wou.edu,
    Western Oregon University – Gs
Myers, Orrin B., (505) 665-3742 obm@lanl.gov,
    Los Alamos National Laboratory – Pg
Myers, Paul, 780-492-6706 pmyers@ualberta.ca,
    University of Alberta – Og
Myers, Paul B., (610) 758-3665 pbm1@lehigh.edu,
    Lehigh University – Hw
Myers, Tammy, (717) 477-1685 tlmyers@ship.edu,
    Shippensburg University – Zn
Mylavarapu, S R., 352-392-1951 ext 202 raom@ufl.edu,
    University of Florida – Sc
Müller, Lena, 089/2180 4340 lena-mueller@lrz.uni-muenchen.de,
    Ludwig-Maximilians-Universitaet Muenchen – Gz
Müller-Sohnius, Dieter, 089/2180 4259 mueso@min.uni-muenchen.de,
    Ludwig-Maximilians-Universitaet Muenchen – Gz
Mylroie, John E., 662-268-1032 Ext 237 mylroie@geosci.msstate.edu,
    Mississippi State University – GmHyGs
Myneni, Satish C B., (609) 258-5848 smyneni@princeton.edu,
    Princeton University – Sc
Myrbo, Amy, 612-626-7889 amyrbo@umn.edu,
    University of Minnesota, Twin Cities – PeZe
Myrow, Paul M., (719) 389-6790 pmyrow@coloradocollege.edu,
    Colorado College – Gsr
Mysak, Lawrence A., (514) 398-3768 lawrence.mysak@mcgill.ca,
    McGill University – Op
Mysen, Bjorn O., (202) 478-8975 bmysen@carnegiescience.edu,
    Carnegie Institution for Science – Cp
Möller, Andreas, 785864358 amoller@ku.edu,
    University of Kansas – Gt

N

Naar, David F., (727) 553-1637 dnaar@usf.edu,
    University of South Florida – Ou
Nabelek, John L., 541-737-2757 nabelek@coas.oregonstate.edu,
    Oregon State University – GtYs
Nabelek, Peter I., (573) 884-6463 nabelekp@missouri.edu,
    University of Missouri – GipCu

Nadeau, Olivier, 613-558-8395 onadeau@uottawa.ca,
    University of Ottawa – Gx
Nadiga, Balu, (505) 667-9466 balu@lanl.gov,
    Los Alamos National Laboratory – Zn
Nadin, Elisabeth S., (907) 474-5181 enadin@alaska.edu,
    University of Alaska, Fairbanks – Gtc
Nadiri, Ata Allah, +989143003725 nadiri@tabrizu.ac.ir,
    University of Tabriz – HwGqe
Nadon, Gregory C., (740) 593-4212 nadon@ohio.edu,
    Ohio University – Gs
Naehr, Thomas H., (361) 825-2470 Thomas.Naehr@tamucc.edu,
    Texas A&M University, Corpus Christi – GuCm
Nagaoka, Lisa A., (940) 565-2510 lisa.nagaoka@unt.edu,
    University of North Texas – Ga
Nagel, Athena, (662) 268-1032 ext 238 amo58@msstate.edu,
    Mississippi State University – ZgiZr
Nagel-Myers, Judith, 315-229-5239 jnagel@stlawu.edu,
    St. Lawrence University – PisPe
Nagihara, Seiichi, 806-834-4481 seiichi.nagihara@ttu.edu,
    Texas Tech University – Zi
Nagy, Kathryn L., 312-355-3276 klnagy@uic.edu,
    University of Illinois at Chicago – Cg
Nagy-Shadman, Elizabeth, 626 585-3369 eanagy-shadman@pasadena.edu,
    Pasadena City College – Gg
Nair, Vimala D., (352) 392-1803 Ext. 324 vdn@ufl.edu,
    University of Florida – Sc
Najjar, Raymond G., (814) 863-1586 najjar@meteo.psu.edu,
    Pennsylvania State University, University Park – OcpAt
Nakagawa, Masami, 303-384-2132 mnakagaw@mines.edu,
    Colorado School of Mines – Ze
Nakamura, Noboru, (773) 702-3802 nnn@bethel.uchicago.edu,
    University of Chicago – As
Nakamura, Yosio, (512) 471-0428 yosio@ig.utexas.edu,
    University of Texas, Austin – Ys
Nakote, Heinz, (575) 646-7627 hnakotte@nmsu.edu,
    New Mexico State University, Las Cruces – Zm
Naldrett, Anthony J., +441243531823 ajnaldrett@yahoo.com,
    University of Toronto – EmGiCp
Nalin, Ronald, (909) 558-7151 rnalin@llu.edu,
    Loma Linda University – Gsr
Namikas, Steven, (225) 578-6142 snamik1@lsu.edu,
    Louisiana State University – Gms
Nance, Hardie S., 512-471-6285 seay.nance@beg.utexas.edu,
    University of Texas, Austin – Grc
Nance, R. Damian, (740) 593-1107 nance@ohio.edu,
    Ohio University – Gtc
Nance, Seay, 512-471-6285 seay.nance@beg.utexas.edu,
    University of Texas at Austin, Jackson School of Geosciences – Gg
Napieralski, Jacob A., 313.593.5157 jnapiera@umich.edu,
    University of Michigan, Dearborn – GmlZi
Naples, Virginia, 815-753-7820 vlnaples@niu.edu,
    Northern Illinois University – Pv
Napora, Theodore A., (401) 874-6222 University of Rhode Island – Ob
Napp, Jeffrey, (206) 526-4148 jeff.napp@noaa.gov,
    University of Washington – Ob
Naranjo, Ramon, 775-887=7659 rnaranjo@usgs.gov,
    University of Nevada, Reno – CgHw
Narbonne, Guy M., (613) 533-6168 narbonne@queensu.ca,
    Queen's University – PiGGs
Narey, Martha A., mnarey@du.edu,
    University of Denver – Zg
Narod, Barry, (604) 822-2267 narod@eos.ubc.ca,
    University of British Columbia – Yx
Nash, Barbara P., (801) 581-8587 barb.nash@utah.edu,
    University of Utah – Gi
Nash, David, david.nash@uc.edu,
    University of Cincinnati – Gm
Nash, Greg, 801-585-9986 gnash@egi.utah.edu,
    University of Utah – Gt
Nash, Jonathan, 541-737-4573 nash@coas.oregonstate.edu,
    Oregon State University – Op
Nash, Thomas A., 614 265 6582 thomas.nash@dnr.state.oh.us,
    Ohio Dept of Natural Resources – Gl
Nashold, Barney W., (630) 252-7698 bwnashold@anl.gov,
    Argonne National Laboratory – Zn
Nasir, Sobhi J., 00974-4852207 nasir54@qu.edu.qa,
    University of Qatar – Gx

Naslund, H. Richard, (607) 777-4313 Binghamton University – Gi
Nasraoui, Mohamed, +33 (0)344063813 mohmed.nasraoui@lasalle-beauvais.fr,
   Institut Polytechnique LaSalle Beauvais (ex-IGAL) – EgGzEm
Natalicchio, Marcello, marcello.natalicchio@unito.it,
   Università di Torino – Gs
Nater, Edward A., (612) 625-1725 enater@umn.edu,
   University of Minnesota, Twin Cities – Sd
Nathan , Stephen, 860-465-5579 nathans@easternct.edu,
   Eastern Connecticut State University – PmEo
Nathan, Stephen, 413-967-5448 snathan@geo.umass.edu,
   University of Massachusetts, Amherst – PmOg
Nathan, Terrence R., (530) 752-1609 trnathan@ucdavis.edu,
   University of California, Davis – Asm
Naudts, Kim, +31 20 59 87354 Vrije Universiteit Amsterdam – HgPe
Naumann, Malik, 089/2180 6714 Ludwig-Maximilians-Universitaet Muenchen – Pg
Naumann, Terry R., trnaumann@uaa.alaska.edu,
   University of Alaska, Anchorage – Gi
Navarre-Sitchler, Alexis, 303-384-2219 asitchle@mines.edu,
   Colorado School of Mines – Ca
Navarro, Didac, didac.navarro@uab.cat,
   Universitat Autonoma de Barcelona – Gz
Naveira Garabato, Alberto, +44 (0)23 80592680 acng@noc.soton.ac.uk,
   University of Southampton – Op
Navrotsky, Alexandra, (530) 752-3292 anavrotsky@ucdavis.edu,
   University of California, Davis – Zm
Naylor, Mark, +44 (0) 131 650 4918 Mark.Naylor@ed.ac.uk,
   Edinburgh University – Gt
Naylor, Shawn C., 812-855-2504 snaylor@indiana.edu,
   Indiana University – Hw
Naymik, Thomas G., 614-433-9280 tnaymik@geosyntec.com,
   Ohio State University – Hw
Nduati, Eunice W., enduati@jkuat.ac.ke,
   Jomo Kenyatta University of Agriculture & Technology – ZrrZi
Neal, Clive R., (574) 631-8328 neal.1@nd.edu,
   University of Notre Dame – Gi
Neal, Donald W., (252) 328-4392 neald@ecu.edu,
   East Carolina University – GrdEo
Neal, William J., (616) 331-3381 nealw@gvsu.edu,
   Grand Valley State University – GsdGr
Nealson, Kenneth, (213) 821-2271 knealson@usc.edu,
   University of Southern California – Pg
Nearing, Grey L., (205) 348-6268 gsnearing@ua.edu,
   University of Alabama – HsqZr
Neboga, Victoria V., 717–702–2026 vneboga@pa.gov,
   Pennsylvania Bureau of Topographic & Geologic Survey – Hw
Nechaev, Dmitri, (228) 688-2573 dmitri.nechaev@usm.edu,
   University of Southern Mississippi – Op
Nedimovic, Mladen, 1 902 494 4524 mladen@Dal.Ca,
   Dalhousie University – Ysg
Nedunuri, Krishna K., (937) 376-6455 knedunuri@centralstate.edu,
   Central State University – HwCaSo
Needham, Tim, +44(0) 113 34 39104 D.T.Needham@leeds.ac.uk,
   University of Leeds – Gc
Neelin, J. David, (310) 206-3734 neelin@atmos.ucla.edu,
   Università di California, Los Angeles – As
Neely, Haly L., 979-845-3041 hneely@tamu.edu,
   Texas A&M University – Spd
Neethling, Stephen, +44 20 759 49341 s.neethling@imperial.ac.uk,
   Imperial College – Gz
Negrini, Robert M., (661) 654-2185 rnegrini@csub.edu,
   California State University, Bakersfield – YmgGn
Nehru, Cherukupalli E., (718) 951-5417 Graduate School of the City University of New York – Gi
Nehyba, Slavomír, +420 549 49 6067 slavek@sci.muni.cz,
   Masaryk University – GsdGg
Neighbors, Corrie, (505) 538-6352 neighborsc@wnmu.edu,
   Western New Mexico University – Ysr
Neill, Owen K., (509) 335-6770 owen.neill@wsu.edu,
   Washington State University – Ca
Neill, Simon P., 01248 383258 s.p.neill@bangor.ac.uk,
   University of Wales – OnpOg
Neinow, Peter W., +44 (0) 131 650 9139 Peter.Nienow@ed.ac.uk,
   Edinburgh University – Gl
Neish, Catherine, 519-661-2111, ext.83188 cneish@uwo.ca,
   Western University – Xg

Neitzke Adamo, Lauren, (848) 932-7243 lauren.adamo@rutgers.edu,
   Rutgers, The State University of New Jersey – GuPmZe
Nekvasil, Hanna, 631-632-8201 hanna.nekvasil@sunysb.edu,
   Stony Brook University – Cp
Nel, Justin, (051) 401-3402 nelwj@ufs.ac.za,
   University of the Free State – Gc
Nelsen, Lori, (864)294-3481 lori.nelsen@furman.edu,
   Furman University – Ca
Nelson, Bruce K., (206) 543-4434 bnelson@uw.edu,
   University of Washington – CcGie
Nelson, Bruce W., bwn@virginia.edu,
   University of Virginia – Gs
Nelson, Craig, 614 265 6603 craig.nelson@dnr.state.oh.us,
   Ohio Dept of Natural Resources – Hw
Nelson, Daren T., (910) 775-4589 daren.nelson@uncp.edu,
   University of North Carolina, Pembroke – GmHwZe
Nelson, Kenneth A., (785) 864-2164 nelson@kgs.ku.edu,
   University of Kansas – Zy
Nelson, Marc A., (501) 575-3964 manelson@comp.uark.edu,
   University of Arkansas, Fayetteville – Cl
Nelson, Michael G., (801) 585-3064 nelsonelson@aol.com,
   University of Utah – Nm
Nelson, Nathan, (785) 532-5115 nonelson@ksu.edu,
   Kansas State University – So
Nelson, Priscilla P., 303-273-3700 pnelson@mines.edu,
   Colorado School of Mines – NrgNm
Nelson, Robert K., (508) 289-2656 rnelson@whoi.edu,
   Woods Hole Oceanographic Institution – Co
Nelson, Robert M., (818) 354-1797 Jet Propulsion Laboratory – Xg
Nelson, Robert S., (309)438-7808 rsnelso@ilstu.edu,
   Illinois State University – Gm
Nelson, Stephen A., (504) 862-3194 snelson@tulane.edu,
   Tulane University – GivRn
Nelson, Stephen T., 801-422-8688 steve_nelson@byu.edu,
   Brigham Young University – Cs
Nelson, Wendy, (410) 704-3133 wrnelson@towson.edu,
   Towson University – GiCc
Nelwamondo, T. M., +27 15 962 8582 tshililo.nelwamondo@univen.ac.za,
   University of Venda – GeZuy
Nemcok, Michal, +421-2-5463-0337 mnemcok@egi.utah.edu,
   University of Utah – GctGo
Nemon, Amy, (270) 745-6952 amy.nemon@wku.edu,
   Western Kentucky University – Zig
Nenes, Athanasios, (404) 894-9225 nenes@eas.gatech.edu,
   Georgia Institute of Technology – As
Nesbitt, Alex, (775) 682-8762 anesbitt@unr.edu,
   University of Nevada – Zn
Nesbitt, Carl, 775-784-8287 carln@unr.edu,
   University of Nevada, Reno – Nx
Nesbitt, Elizabeth (Liz), (206) 543-5949 lnesbitt@uw.edu,
   University of Washington – Pim
Nesbitt, H W., (519) 661-3194 hwn@uwo.ca,
   Western University – Cl
Nesbitt, Stephen W., (217) 244-3740 snesbitt@illinois.edu,
   University of Illinois, Urbana-Champaign – As
Nesbitt, Sterling J., 540-231-6330 sjn2104@vt.edu,
   Virginia Polytechnic Institute & State University – Pv
Nesheim, Tim, tonesheim@nd.gov,
   North Dakota Geological Survey – Gg
Nesse, William D., (970) 351-2830 w.d.nesse@gmail.com,
   University of Northern Colorado – Gxz
Nestell, Galina P., (817) 272-2987 gnestell@uta.edu,
   University of Texas, Arlington – PmsPi
Nestola, Fabrizio, +390498279160 fabrizio.nestola@unipd.it,
   Università degli Studi di Padova – Gz
Nethengwe, Nthaduleni S., +27 15 962 8593 nthaduleni.nethengwe2@univen.ac.za,
   University of Venda – RnZiRw
Nettles, Meredith, (845) 365-8613 nettles@ldeo.columbia.edu,
   Columbia University – YsGl
Nettles, Meredith, nettles@ldeo.columbia.edu,
   Columbia University – YsGl
Neubaum, John C., (717) 702-2039 jneubaum@pa.gov,
   Pennsylvania Bureau of Topographic & Geologic Survey – EcnGg
Neubeck, William S., (518) 388-6519 neubeckw@union.edu,
   Union College – Gm
Neuberg, Jurgen, +44(0) 113 34 36769 j.neuberg@leeds.ac.uk,

Faculty Index -N

University of Leeds – Gv
Neufeld, Jerome A., +44 (0) 1223 765709 jn271@cam.ac.uk, University of Cambridge – GoYgGt
Neuhauser, Kenneth R., (785) 628-5349 kneuhaus@fhsu.edu, Fort Hays State University – GceGs
Neuman, Andrew G., (403) 823-7707 Royal Tyrrell Museum of Palaeontology – Pv
Neuman, Dennis R., (406) 994-4822 dneuman@montana.edu, Montana State University – Zn
Neuman, Shlomo P., (520) 621-7114 neuman@hwr.arizona.edu, University of Arizona – HwqGq
Neumann, A. Conrad, (919) 962-0190 neumann@marine.unc.edu, University of North Carolina, Chapel Hill – Gu
Neumann, Klaus, 765-285-8262 kneumann@bsu.edu, Ball State University – Cl
Neumann, Patricia, (310) 660-3593 pneumann@elcamino.edu, El Camino College – Zg
Neumann, Tom, 301-614-5923 thomas.neumann@nasa.gov, University of Washington – Gl
Neuweiler, Fritz, (418) 656-7479 fritz.neuweiler@ggl.ulaval.ca, Universite Laval – GsdPg
Neves, Douglas, (310) 660-3593 dneves@elcamino.edu, El Camino College – Zg
Nevins, Joseph, 845-437-7823 jonevins@vassar.edu, Vassar College – Zu
Nevins, Susan K., (607) 753-2815 GeologyDept@cortland.edu, SUNY, Cortland – Zn
Newberry, Rainer J., ffrn@uaf.edu, University of Alaska, Fairbanks – Em
Newbold, K. B., (905) 525-9140 (Ext. 27948) newbold@mcmaster.ca, McMaster University – Ge
Newcomer, Darrell R., (509) 376-1054 darrell.newcomer@pnl.gov, Pacific Northwest National Laboratory – Hw
Newell, Charles J., (713) 663-6600 Rice University – Hw
Newell, Dennis L., dennis.newell@usu.edu, Utah State University – Cls
Newell, Kerry D., (785) 864-2183 dnewell@kgs.ku.edu, University of Kansas – Go
Newland, Franz, (416)736-2100 #22095 franz.newland@lassonde.yorku.ca, York University – Zr
Newman, Andrew V., (404) 894-3976 anewman@gatech.edu, Georgia Institute of Technology – YdsYg
Newman, Brent D., (505) 667-3021 bnewman@lanl.gov, Los Alamos National Laboratory – Gg
Newman, Dianne K., (636) 395-3543 dkn@caltech.edu, California Institute of Technology – Pg
Newman, Jamie, 212-769-5386 jamien@amnh.org, American Museum of Natural History – Gg
Newman, Jonathan, (519) 824-4120 Ext.52147 jnewma01@uoguelph.ca, University of Guelph – Zn
Newman, Julie, (979) 845-9283 newman@geo.tamu.edu, Texas A&M University – Gct
Newman, Steven B., (860) 832-2940 ccsuwxman@gmail.com, Central Connecticut State University – As
Newman, William A., (858) 534-2150 wnewman@ucsd.edu, University of California, San Diego – Ob
Newman, William I., (310) 825-3912 win@ucla.edu, University of California, Los Angeles – Xy
Newsom, Horton E., (505) 277-0375 newsom@unm.edu, University of New Mexico – XgcXm
Newton, Anthony J., +44 (0) 131 650 2546 Anthony.Newton@ed.ac.uk, Edinburgh University – Zn
Newton, Cathryn R., 315-443-2672 crnewton@syr.edu, Syracuse University – Pg
Newton, Jan, (206) 616-3641 newton@ocean.washington.edu, University of Washington – Ob
Newton, Rob, +44(0) 113 34 31631 d.oneill@leeds.ac.uk, University of Leeds – Ge
Newton, Robert, bnewton@ldeo.columbia.edu, Columbia University – Ct
Newton, Robert C., (310) 206-2917 rcnewton@ucla.edu, University of California, Los Angeles – CpGp
Newton, Robert M., (413) 585-3946 rnewton@smith.edu, Smith College – Gm
Newton, Seth A., 205-247-3708 snewton@gsa.state.al.us, Geological Survey of Alabama – ZnGg
Nex, Paul A., 011 717 6563 Paul.Nex@wits.ac.za, University of the Witwatersrand – Eg
Nezat, Carmen A., (509) 359-7959 cnezat@ewu.edu, Eastern Washington University – ClGeSc
Ng, Crystal, 612-624-9243 gcng@umn.edu, University of Minnesota, Twin Cities – Hy
Ngigi, Thomas G., tgngigi@jkuat.ac.ke, Jomo Kenyatta University of Agriculture & Technology – ZriZr
Ngo, Thanh X., +84-9782 24246 ngoxuanthanh@humg.edu.vn, Hanoi University of Mining & Geology – GtcGi
Ngoy, Kikombo, (908) 737-3718 kngoy@kean.edu, Kean University – Zy
Nguyen, Maurice, 612-626-8739 nguyenm@umn.edu, University of Minnesota – Gl
Ngwenya, Bryne T., +44 (0) 131 650 8507 Bryne.Ngwenya@ed.ac.uk, Edinburgh University – Co
Ni, James F., (505) 646-1920 jni@nmsu.edu, New Mexico State University, Las Cruces – YsGtYg
Ni-Meister, Wenge, (212) 772-5321 wenge.ni-meister@hunter.cuny.edu, Hunter College (CUNY) – Zr
Nicholas, Chris, +353 1 8962176 nicholyj@tcd.ie, Trinity College – Eo
Nicholas, Joseph W., (540) 654-1470 jnicola@mwc.edu, Mary Washington College – Zy
Nicholls, James W., (403) 220-7127 University of Calgary – Gi
Nichols, Jonathan E., jnichols@ldeo.columbia.edu, Columbia University – Pe
Nichols, Kyle K., 518-580-5194 knichols@skidmore.edu, Skidmore College – Gm
Nichols, Terry E., (501) 575-7317 tenichol@comp.uark.edu, University of Arkansas, Fayetteville – Zn
Nicholson, David, (508) 289-3547 dnicholson@whoi.edu, Woods Hole Oceanographic Institution – Gu
Nicholson, Kirsten N., 765-285-8268 knichols@bsu.edu, Ball State University – Gi
Nicholson, Nanette, 212-769-5390 nnicholson@amnh.org, American Museum of Natural History – Zn
Nick, Kevin, (909) 558-4530 knick@llu.edu, Loma Linda University – GsbGr
Nickmann, Marion, +49 89 289 25853 nickmann@tum.de, Technical University of Munich – Ng
Nicol, Andy, andy.nicol@canterbury.ac.nz, University of Canterbury – GctGo
Nicolas, Michelle P., (204) 945-6571 michelle.nicolas@gov.mb.ca, Manitoba Geological Survey – GrsGo
Nicolaysen, Kirsten P., 509-527-4934 nicolakp@whitman.edu, Whitman College – GiCgGa
Nicolescu, Stefan, 203-432-3141 stefan.nicolescu@yale.edu, Yale University – CtGpz
Nicoletti, Jeremy D., (603) 271-5762 jeremy.nicoletti@des.nh.gov, New Hampshire Geological Survey – GgHg
Nicot, Jean-Philippe, (512) 471-6246 jp.nicot@beg.utexas.edu, University of Texas at Austin, Jackson School of Geosciences – HwCg
NICULESCU, Bogdan M., +40-21-3125024 bogdan.niculescu@gg.unibuc.ro, University of Bucharest – YgeYu
Niedzielski, Michael A., michael.niedzielski@und.edu, University of North Dakota – Ziu
Nielsen, Donald R., (530) 753-5760 drnielsen@ucdavis.edu, University of California, Davis – Sp
Nielsen, Gerald A., (406) 994-5075 nielsenmontana@aol.com, Montana State University – Sd
Nielsen, Gregory B., (801) 626-6394 gnielsen@weber.edu, Weber State University – Gg
Nielsen, Kurt E., 831-656-2295 nielsen@nps.edu, Naval Postgraduate School – Am
Nielsen, Mary, 605-677-5649 Mary.Nielsen@usd.edu, University of South Dakota – Zn
Nielsen, Peter A., (603) 358-2553 pnielsen@keene.edu, Keene State College – GxzGc
Nielsen, Peter J., (801) 537-3359 peternielsen@utah.gov, Utah Geological Survey – Eo
Nielsen, Roger L., (541) 737-1235 nielsenr@geo.oregonstate.edu, Oregon State University – GizGv
Nielsen, Sune G., (508) 289-2837 snielsen@whoi.edu, Woods Hole Oceanographic Institution – Cs
Nielsen-Gammon, John, (979) 862-2248 n-g@tamu.edu, Texas A&M University – As

Niemann, William L., (304) 696-6721 niemann@marshall.edu, Marshall University – Ng

Niemi, Nathan A., 734-764-6377 naniemi@umich.edu, University of Michigan – Gtc

Niemi, Tina M., (816) 235-5342 niemit@umkc.edu, University of Missouri, Kansas City – Gt

Niemitz, Jeffery W., niemitz@dickinson.edu, Dickinson College – ClOuGn

Niewendorp, Clark, (971) 673-1555 Oregon Dept of Geology and Mineral Industries – Eg

Nikitina, Daria L., (610) 436-3103 dnikitina@wcupa.edu, West Chester University – Gm

Nikolinakou, Maria-Aikaterini, 512-475-6613 mariakat@austin.utexas.edu, University of Texas, Austin – Np

Nikulin, Alex, (607) 777-2518 anikulin@binghamton.edu, Binghamton University – Ys

Nimis, Paolo, +390498279161 paolo.nimis@unipd.it, Università degli Studi di Padova – Eg

Nimmo, Francis, 831-459-1783 fnimmo@ucsc.edu, University of California, Santa Cruz – Xy

Ninesteel, Judy J., (304) 788-6956 JJNinesteel@mail.wvu.edu, Potomac State College – Hg

Ning, Liang, 814-8760987 lun115@psu.edu, University of Massachusetts, Amherst – Am

Nisbet, Euan G., +44 1784 443809 e.nisbet@rhul.ac.uk, Royal Holloway University of London – ZgAcGi

Nissen-Meyer, Tarje, +44 (1865) 282149 tarje.nissen-meyer@earth.ox.ac.uk, University of Oxford – Ys

Nitecki, Matthew H., (312) 665-7093 mnitecki@fmnh.org, Field Museum of Natural History – Pi

Nitsche, Frank, fnitsche@ldeo.columbia.edu, Columbia University – Gu

Nittler, Larry R., (202) 478-8460 lnittler@carnegiescience.edu, Carnegie Institution for Science – Xc

Nittrouer, Charles A., 206-543-5099 nittroue@ocean.washington.edu, University of Washington – GuYr

Nittrouer, Jeffrey A., 713.348.4886 nittrouer@rice.edu, Rice University – Gms

Niu, Fenglin, 713.348.4122 niu@rice.edu, Rice University – YsGt

Niu, Guo-Yue, niu@geo.utexas.edu, University of Texas, Austin – SpAs

Niu, Yaoling, +44 (0) 191 33 42311 yaoling.niu@durham.ac.uk, Durham University – Git

Nixon, R. Paul, 801-422-4657 paul_nixon@byu.edu, Brigham Young University – Go

Nixon, Scott W., (401) 874-6803 University of Rhode Island – Ob

Nixon III, Roy (Rick) A., rnixon3@gsu.edu, Georgia State University, Perimeter College, Online – Gg

Niyogi, Dev, (765) 494-9531 climate@purdue.edu, Purdue University – As

Njau, Jackson K., (812) 856-3170 jknjau@indiana.edu, Indiana University, Bloomington – Pv

Njoku, Eni G., (818) 354-3693 eni.g.njoku@jpl.nasa.gov, Jet Propulsion Laboratory – Zr

Nkedi-Kizza, Peter, (352) 392-1951 kizza@ufl.edu, University of Florida – Sp

Nkotagu, Hudson H., nkotaguh@yahoo.com, University of Dar es Salaam – HwgGe

Nkuna, Tinyiko R., +27 15 962 8017 tinyiko.nkuna@univen.ac.za, University of Venda for Science & Technology – HgwAs

Noakes, John E., (706) 542-1395 University of Georgia – Cc

Noble, Paula J., 775-784-6211 noblepj@unr.edu, University of Nevada, Reno – PmGnPs

Noblett, Jeffrey B., (719) 389-6516 jnoblett@coloradocollege.edu, Colorado College – Gxv

Noe, Garry, (757) 455-3284 gnoe@vwc.edu, Virginia Wesleyan College – HsZin

Noffke, Nora K., (757) 683-3313 nnoffke@odu.edu, Old Dominion University – GsPoGu

Nogueira, Ricardo, 404-413-5791 rnogueira@gsu.edu, Georgia State University – Asm

Noh, Myoung-Jong, (614) 292-5445 noh.56@osu.edu, Ohio State University – Zr

Nolan, Robert P., (718) 951-4242 Graduate School of the City University of New York – Co

Noll, Mark R., (585) 395-5717 mnoll@esc.brockport.edu, SUNY, The College at Brockport – Cl

Noll, Michael G., (229) 333-7143 mgnoll@valdosta.edu, Valdosta State University – Zn

Noltimier, Hallan C., (614) 292-9796 noltimier.2@osu.edu, Ohio State University – YmgGt

Nondorf, Lea, (501) 683-0110 lea.nondorf@arkansas.gov, Arkansas Geological Survey – Cg

Noonan, Mathew T., 605-677-6152 matthew.noonan@usd.edu, South Dakota Dept of Environment and Natural Resources – Hw

Noone, David, (541) 737-3629 dcn@coas.oregonstate.edu, Oregon State University – AmZg

Nootz, Gero, 228.688.3091 gero.nootz@usm.edu, University of Southern Mississippi – No

Nord, Julia, (703) 993-3395 jnord@gmu.edu, George Mason University – Gzg

Nordeng, Stephan, (701) 777-3455 stephan.nordeng@engr.und.edu, University of North Dakota – Go

Nordstrom, Greg, 662-268-1032 Ext 248 gjn2@msstate.edu, Mississippi State University – Am

Nordt, Lee C., (254) 710-4288 lee_nordt@baylor.edu, Baylor University – SdGa

Nordwald, Dan, 573-368-2451 dan.nordwald@dnr.mo.gov, Missouri Dept of Natural Resources – Zn

Norell, Mark A., (212) 769-5804 American Museum of Natural History – Pv

Noren, Anders, 612-626-3298 noren021@umn.edu, University of Minnesota, Twin Cities – Gn

Norford, Brian, (403) 292-7000 University of Calgary – Gr

Norman, David, +44 (0) 1223 333426 dn102@esc.cam.ac.uk, University of Cambridge – Pg

Norman, David K., (360) 902-1439 dave.norman@dnr.wa.gov, Washington Geological Survey – Gg

Norman, John M., (608)262-4576 jmnorman@wisc.edu, University of Wisconsin, Madison – So

Norman, John M., (608) 262-2633 jmnorman@wisc.edu, University of Wisconsin, Madison – Spo

Norris, Christopher A., 203-432-3748 christopher.norris@yale.edu, Yale University – Pv

Norris, Geoffrey, (416) 978-4851 norris@quartz.geology.utoronto.ca, University of Toronto – Pl

Norris, Joel R., (858) 822-4420 jnorris@ucsd.edu, University of California, San Diego – As

Norris, Richard D., (858) 822-1868 rnorris@ucsd.edu, University of California, San Diego – Gu

Norrish, Winston, 509.963.2192 norrishw@geology.cwu.edu, Central Washington University – GgeGo

Norry, Mike, +440116 252 3803 nah@le.ac.uk, Leicester University – Pe

North, Gerald R., (979) 845-8083 g-north@tamu.edu, Texas A&M University – Am

North, Leslie, leslie.north@wku.edu, Western Kentucky University – Zg

Northrup, Clyde J., (208) 426-1581 cjnorth@boisestate.edu, Boise State University – Gc

Norton, Stephen A., (207) 581-2156 norton@maine.edu, University of Maine – Cl

Norton-Krane, Abby N., (216) 987-5227 Abby.Norton-Krane@tri-c.edu, Cuyahoga Community College - Western Campus – Zg

Norwood, James, 334-844-3414 jan0003@auburn.edu, Auburn University – Zy

Nosal, Andrew, 6192604600 x2438 anosal@sandiego.edu, University of San Diego – Ob

Nosal, Thomas E., 86004243590 thomas.nosal@ct.gov, Dept of Energy and Environmental Protection – Zn

Nothdurft, Luke, +61 7 3138 1531 l.nothdurft@qut.edu.au, Queensland University of Technology – GdCmGg

Nourse, Jonathan A., 909-869-3460 janourse@csupomona.edu, California State Polytechnic University, Pomona – GctEm

Novacek, Michael J., (212) 769-5805 American Museum of Natural History – Pv

Novak, Milan, +420 549 49 6188 mnovak@sci.muni.cz, Masaryk University – GzpGi

Novak, Thomas, 859-257-3818 thomas.novak@uky.edu, University of Kentucky – Nm

Nowack, Robert L., (765) 494-5978 nowack@purdue.edu, Purdue University – Ys

Nowell, Arthur R. M., (206) 543-6605 nowell@cofs.washington.edu,

University of Washington – Ou
Nowlin, Jr., Worth D., (979) 846-6747 wnowlin@tamu.edu,
 Texas A&M University – OpZn
Nowotarski, Christopher J., (979) 845-3305 cjnowotarski@tamu.edu,
 Texas A&M University – Ams
Noyes, Jim, (310) 660-3593 tnoyes@elcamino.edu,
 El Camino College – Og
Noyes, Joanne M., (605) 394-6972 joanne.noyes@state.sd.us,
 South Dakota Dept of Environment and Natural Resources – HyGg
Ntarlagiannis, Dimitrios, 973-353-5189 dimntar@newark.rutgers.edu,
 Rutgers, The State University of New Jersey, Newark – Yu
Nton, M. E., aa.elueze@mail.ui.edu.ng,
 University of Ibadan – Go
Ntsaluba, Bantubonke I., +27 (0)46-603-8309 b.ntsaluba@ru.ac.za,
 Rhodes University – Gi
Nudds, John, +44 0161 275-7861 john.nudds@manchester.ac.uk,
 University of Manchester – Pg
NUDE, Prosper M., +233-244-116879 pmnude@ug.edu.gh,
 University of Ghana – GipGx
Nuester, Jochen, jnuester@csuchico.edu,
 California State University, Chico – Cm
Nugent, Alison D., 808-956-2878 anugent@hawaii.edu,
 University of Hawai'i, Manoa – AsZnn
Nugent, Barnes, barnes@geosrv.wvnet.edu,
 Fairmont State University – Gg
Null, E. Jan, (415) 338-2061 San Francisco State University – Ge
Nunn, Jeffrey A., (225) 388-6657 jeff@geol.lsu.edu,
 Louisiana State University – Yg
Nur, Amos M., (650) 723-9526 Stanford University – Yg
Nusbaum, Robert L., (843) 953-5596 nusbaumr@cofc.edu,
 College of Charleston – Gz
Nuss, Wendell A., 831-656-2308 nuss@nps.edu,
 Naval Postgraduate School – Am
Nuttall, Brandon C., 859-323-0544 bnuttall@uky.edu,
 University of Kentucky – Eo
Nuwategeka, Expedito, +256 782 889 985 nuwategeka@gmail.com,
 Gulu University – ZyyGv
Nwachukwu, J. I., 234-803-725-8122 jnwachuk@oauife.edu.ng,
 Obafemi Awolowo University – GoCg
Nyantakyi-Frimpong, Hanson, 303-871-4175 hnyanta2@du.edu,
 University of Denver – Zc
Nyblade, Andrew A., (814) 863-8341 andy@geosc.psu.edu,
 Pennsylvania State University, University Park – Yg
Nye, Janet, (631) 632-3187 janet.nye@stonybrook.edu,
 SUNY, Stony Brook – Ob
Nyhan, John W., (505) 667-3163 Los Alamos National Laboratory – So
Nyland, Edo, (780) 492-5502 edo@phys.ualberta.ca,
 University of Alberta – Ys
Nyquist, Jonathan, (215) 204-7484 nyq@temple.edu,
 Temple University – Yg
Nystuen, Jeffrey A., (206) 543-1343 nystuen@apl.washington.edu,
 University of Washington – Op
Nzengung, Valentine A., (706) 202-4296 vnzengun@uga.edu,
 University of Georgia – ClGeCq

O

O'Banion, Matthew S., 845-938-2326 matthew.obanion@westpoint.edu,
 United States Military Academy – ZrGg
O'Brien, Arnold L., (978) 934-3902 arnold_obrien@uml.edu,
 University of Massachusetts, Lowell – Hw
O'Brien, John M., (201) 200-3161 jobrien@njcu.edu,
 New Jersey City University – Gs
O'Brien, Lawrence E., (914) 341-4570 lobrien@sunyorange.edu,
 Orange County Community College – Gg
O'Brien, Neal R., (315) 267-2287 obriennr@potsdam.edu,
 SUNY Potsdam – Gs
O'Brien, Rachel, (814) 332-2875 robrien@allegheny.edu,
 Allegheny College – HwGe
O'Brien, Suzanne R., (508) 531-1390 s6obrien@bridgew.edu,
 Bridgewater State University – Zg
O'Brien, Travis A., 530- 830-2640 tob@ucdavis.edu,
 University of California, Davis – AstAp
O'Callaghan, Mick, +353 21 4902662 mick.ocallaghan@ucc.ie,
 University College Cork – Zn
O'Connor, George A., (352) 392-1803 (Ext. 329) gao@ufl.edu,
 University of Florida – Sc
O'Connor, Yuet-Ling, yloconnor@pasadena.edu,
 Pasadena City College – Ge
O'Dean, Emily, 775-682-9780 eodean@unr.edu,
 University of Nevada – Zf
O'Donnell, James, 860-405-9171 james.odonnell@uconn.edu,
 University of Connecticut – Op
O'Driscoll, Nelson, 902-585-1679 nelson.odriscoll@acadiau.ca,
 Acadia University – CgGgZn
O'Farrell, Keely A., 859-323-4876 k.ofarrell@uky.edu,
 University of Kentucky – Ygd
O'Geen, Toby, 530-752-2155 atogeen@ucdavis.edu,
 University of California, Davis – So
O'Gorman, Paul, (617) 452-3382 pog@mit.edu,
 Massachusetts Institute of Technology – As
O'Halloran, Ivan, (519) 674-1635 iohallo@ridgetownc.uoguelph.ca,
 University of Guelph – So
O'Hara, Kieran D., (859) 269-5161 geokoh@uky.edu,
 University of Kentucky – GcCg
O'Hara, Matthew J., (509) 373-1671 matt.ohara@pnl.gov,
 Pacific Northwest National Laboratory – Cg
O'Keefe, Jennifer, (606) 783-2349 j.okeefe@moreheadstate.edu,
 Morehead State University – PlEcZe
O'Keeffe, Mike, 303-384-2637 okeeffe@mines.edu,
 Colorado Geological Survey – GgeEg
O'Leary, Maureen, 631-444-3730 maureen.oleary@sunysb.edu,
 Stony Brook University – Pv
O'Meara, Stephanie A., (970) 491-6655 stephanie_o'meara@partner.nps.gov,
 Colorado State University – Gc
O'Melia, Charles R., (410) 516-7102 omelia@jhu.edu,
 Johns Hopkins University – Zn
O'Mullan, Gregory, 718-997-3452 gomullan@qc.cuny.edu,
 Queens College (CUNY) – PgOb
O'Neal, Claire J., 302-831-8909 cjoneal@udel.edu,
 University of Delaware – E
O'Neil, James, (734) 764-1435 jro@umich.edu,
 University of Michigan – Cs
O'Neil, Jennifer, 775-682-8747 joneil@unr.edu,
 University of Nevada, Reno – Zn
O'Neil, Jonathan, 613-562-5800 6273 Jonathan.oneil@uottawa.ca,
 University of Ottawa – Cgc
O'Neil, Patrick E., 205-247-3586 poneil@gsa.state.al.us,
 Geological Survey of Alabama – Hs
O'Neill, Larry, (541) 737-2064 loneill@coas.oregonstate.edu,
 Oregon State University – AmOg
O'Neill, Patrick M., (504) 388-2681 pat@lgs.bri.lsu.edu,
 Louisiana State University – Sf
O'Neill, Patrick M., (225) 578-8590 poneil2@lsu.edu,
 Louisiana State University – Sf
O'Reilly, Andrew M., 662-915-2483 aoreilly@olemiss.edu,
 University of Mississippi – Hwq
O'Reilly, Catherine M., 309-438-3493 cmoreil@ilstu.edu,
 Illinois State University – HgCgGg
O'Rourke, Thomas D., (607) 255-6470 tdo1@cornell.edu,
 Cornell University – Nge
O'Shea, Bethany, (619) 260-4243 bethoshea@sandiego.edu,
 University of San Diego – CqGeCt
O'Sullivan, Katie, (661) 654-3991 kosullivan@csub.edu,
 California State University, Bakersfield – GviXg
Oakes-Miller, Hollie, hollie.oakesmiller@pcc.edu,
 Portland Community College - Sylvania Campus – Gg
Oakley, Adrienne, 484-646-4334 oakley@kutztown.edu,
 Kutztown University of Pennsylvania – YrOun
Oakley, Bryan, (860) 465-0418 oakleyb@easternct.edu,
 Eastern Connecticut State University – OnGml
Oakley, Bryan A., boakley@my.uri.edu,
 University of Rhode Island – Gsl
Oaks, Jr., Robert Q., (435) 752-0867 boboaks@comcast.net,
 Utah State University – GsHwYv
Oberdorfer, June A., (408) 924-5026 june.oberdorfer@sjsu.edu,
 San Jose State University – HwGeHy
Obermeyer, Nancy J., 812-237-4351 njobie28@kiva.net,
 Indiana State University – Zi
Obolewicz, Dave, dobolewicz@keene.edu,
 Keene State College – GgEmAm
Obour, Augustine, (785) 625-3425 aobour@ksu.edu,
 Kansas State University – So
Obrad, Jennifer M., 217-333-8741 jobrad@illinois.edu,
 Illinois State Geological Survey – EcGo

Obreza, Thomas A., 352-392-1951 ext 243 obreza@ufl.edu,
University of Florida – So
Obrist, Daniel, (978) 934-3900 Daniel_Obrist@uml.edu,
University of Massachusetts, Lowell – As
Ocan, O. O., 234-803-705-8796 Obafemi Awolowo University – GtzGp
Occhiuzzi, Tony, 4807528888 x 81369 tocchiuzzi.cds@tuhsd.k12.az.us,
Mesa Community College – Zn
Oches, Rick, (781) 891-2937 roches@bentley.edu,
Bentley University – ZcePe
OConnell, Suzanne B., (860) 685-2262 soconnell@wesleyan.edu,
Wesleyan University – GsuGe
Odendaal, Adriaan, +27 (0)51 401 9928 odendaalai@ufs.ac.za,
University of the Free State – GgsGr
Odera, Patroba A., podera@jkuat.ac.ke,
Jomo Kenyatta University of Agriculture & Technology – YdZge
Odeyemi, Idowu O., 08033436110 Federal University of Technology, Akure – GgxZr
Odland, Sarah K., 845-365-8633 odland@ldeo.columbia.edu,
Columbia University – Zn
Odling, Noelle, +44(0) 113 34 32806 n.e.odling@leeds.ac.uk,
University of Leeds – Hg
Odogba, Ismaila, 715-346-4451 iodogba@uwsp.edu,
University of Wisconsin, Stevens Point – Zui
Odokuma-Alonge, Ovie, ovie.odokuma-alonge@uniben.edu,
University of Benin – CgGip
Odom, LeRoy A., (850) 644-6706 odom@magnet.fsu.edu,
Florida State University – Cg
Odom, Robert I., 206-685-3788 odom@apl.washington.edu,
University of Washington – YrGu
Oduneye, Olusola C., +2348066744769 ocoduneye@lautech.edu.ng,
Ladoke Akintola University of Technology – GooGo
Oduor, Peter, (701) 231-7145 Peter.Oduor@ndsu.edu,
North Dakota State University – ZiRwZr
Oechel, Walter, walter.oechel@open.ac.uk,
The Open University – As
Oehlert, Amanda M., 305-421-4662 aoehlert@rsmas.miami.edu,
University of Miami – Gu
Oelkers, Eric, +44 020 7679 37870 e.oelkers@ucl.ac.uk,
University College London – ClGze
Oeser, Jens, 089/21804208 jens.oeser@lmu.de,
Ludwig-Maximilians-Universitaet Muenchen – Yg
Oganov, Artem, 631-632-1429 artem.oganov@sunysb.edu,
Stony Brook University – Gz
Ogard, Allen E., (505) 667-6344 Los Alamos National Laboratory – Hw
Ogata, Kei, +31 20 59 87288 k.ogata@vu.nl,
Vrije Universiteit Amsterdam – GrcGg
Ogbahon, Osazuwa A., 08064253825 oaogbahon@futa.edu.ng,
Federal University of Technology, Akure – Ggo
Ogbamikhumi, Alexander, alexander.ogbamikhumi@uniben.edu,
University of Benin – YsgGg
Ogg, James G., (765) 494-8681 jogg@purdue.edu,
Purdue University – PsGsYm
Ogiesoba, Osareni C., 512-471-6250 osareni.ogiesoba@beg.utexas.edu,
University of Texas, Austin – Ygs
Oglesby, David D., (951) 827-2036 david.oglesby@ucr.edu,
University of California, Riverside – Ys
Oglesby, Elizabeth, eoglesby@email.arizona.edu,
University of Arizona – Zn
Ogram, Andrew V., 3523921951 (Ext 211) aogram@ufl.edu,
University of Florida – Sb
Ogston, Andrea S., (206) 543-0768 ogston@ocean.washington.edu,
University of Washington – Ou
Ogungbesan, Gbenga O., +2348053432408 googungbesan@lautech.edu.ng,
Ladoke Akintola University of Technology – GooGo
Ogunsanwo, O., femiogunsanwo2003@yahoo.com,
University of Ilorin – Ng
Ohman, Mark D., (858) 534-2754 mohman@ucsd.edu,
University of California, San Diego – Ob
Ohmoto, Hiroshi, (814) 865-4074 ohmoto@geosc.psu.edu,
Pennsylvania State University, University Park – Cs
Oikawa, Patricia, patty.oikawa@csueastbay.edu,
California State University, East Bay – SbCoSf
Ojala, Carl F., (313) 487-0218 Eastern Michigan University – As
Ojo, S. B., 234-803-703-7226 sbojo@oauife.edu.ng,
Obafemi Awolowo University – Yg
Okafor, Florence A., (256) 372-4926 florence.okafor@aamu.edu,
Alabama A&M University – ZnEn

Okal, Emile A., (847) 491-3194 emile@earth.northwestern.edu,
Northwestern University – YsgYr
Okalebo, Jane, jane.akalebo@gmail.com,
University of Nebraska, Lincoln – As
Okaya, David A., (213) 740-7452 okaya@earth.usc.edu,
University of Southern California – Ys
Okulewicz, Steven C., (516) 463-6545 geosco@hofstra.edu,
Hofstra University – CgGue
Okumura, Yuko, 512-471-0383 yukoo@ig.utexas.edu,
University of Texas, Austin – AsZoOp
Okunade, Samuel, (937) 376-6455 Central State University – Gm
Okunlola, Ougbenga. A., +2348033814070 o.okunlola@mail.ui.edu.ng,
University of Ibadan – EgCgGe
Okunuwadje, Sunday E., sunday.okunuwadje@uniben.edu,
University of Benin – GsoGd
Ola, P S., 08035927594 psola@futa.edu.ng,
Federal University of Technology, Akure – GosGg
Olabode, Solomon O., 08033783498 Federal University of Technology, Akure – GgoGs
Oladunjoye, M. A., ma.oladunjoye@mail.ui.edu.ng,
University of Ibadan – Yg
Olafsen-Lackey, Susan, (402) 371-6512 slackey@unl.edu,
Unversity of Nebraska - Lincoln – Hw
Olanrewaju, Johnson, 814-871-7453 olanrewa001@gannon.edu,
Gannon University – Cg
Olarewaju, V. O., 234-803-403-2030 volarewa@oauife.edu.ng,
Obafemi Awolowo University – GozCg
Olariu, Cornel, 512-471-1519 cornelo@jsg.utexas.edu,
University of Texas, Austin – Gr
Olariu, Mariana, iulialet@yahoo.com,
University of Texas, Austin – GruGs
Olatunji, A. S., +2348055475865 as.olatunji@mail.ui.edu.ng,
University of Ibadan – CgGeCt
Olawuyi, Kehinde A., +2348107331222 gideonola2001@yahoo.com,
University of Ilorin – YeGge
Olayinka, A. I., aa.elueze@mail.ui.edu.ng,
University of Ibadan – Yg
Olayiwola, M. A., 234-806-448-7416 mayiwola@oauife.edu.ng,
Obafemi Awolowo University – GugPi
Olcott Marshall, Alison, (785) 864-1917 olcott@ku.edu,
University of Kansas – PgCoPo
Oldenburg, Douglas W., (604) 822-5406 doldenburg@eos.ubc.ca,
University of British Columbia – Yg
Oldershaw, Alan E., (403) 220-3258 oldersha@geo.ucalgary.ca,
University of Calgary – Gs
Oldfield, Jonathan, +44 (0)121 414 2943 j.d.oldfield@bham.ac.uk,
University of Birmingham – Zy
Oldham, Richard L., 568-3100 (12147) OldhamR@arc.losrios.edu,
American River College – Gg
Olea, Ricardo A., (703) 648-6414 rolea@usgs.gov,
University of Kansas – GqoEc
Oleinik, Anton, (561) 297-3297 aoleinik@fau.edu,
Florida Atlantic University – Pi
Olesik, John W., (614) 292-6954 olesik.2@osu.edu,
Ohio State University – Ca
Oleynik, Sergey, (609) 258-2390 soleynik@princeton.edu,
Princeton University – Cs
Olhoeft, Gary R., (303) 273-3458 golhoeft@mines.edu,
Colorado School of Mines – YxmYu
Olivares, Jose A., (505) 665-2643 Los Alamos National Laboratory – Cs
Oliveira, Alcino S., soliveir@utad.pt,
Universidade de Trás-os-Montes e Alto Douro – Hww
Oliver, Adolph A., (415) 786-6865 Chabot College – Ys
Oliver, Kevin, +44 (0)23 80596490 K.Oliver@noc.soton.ac.uk,
University of Southampton – Op
Oliver, Matthew J., 302-645-4079 moliver@udel.edu,
University of Delaware – ObZrOg
Oliver, Ronald D., (702) 794-7095 ron_oliver@lanl.gov,
Los Alamos National Laboratory – Yg
Oliveras Castro, Valentí, ++935813092 valenti.oliveras@uab.cat,
Universitat Autonoma de Barcelona – Gvi
Olivo, Gema, 613 533-6998 olivo@queensu.ca,
Queen's University – Eg
Ollerhead, Jeffery W., (506) 364-2428 jollerhead@mta.ca,
Mount Allison University – Yr
Ollinger, Scott, scott.ollinger@unh.edu,
University of New Hampshire – Zc

Olmsted, Wayne, (406) 496-4151 Montana Tech of the University of Montana – Ca
Olney, Jessica L., 813-253-7647 jolney2@hccfl.edu, Hillsborough Community College – ZgGg
Olney, Matthew P., molney@hccfl.edu, Hillsborough Community College – ZgGg
Olorunfemi, A. O., 234-803-392-1712 akinrewa@oauife.edu.ng, Obafemi Awolowo University – Go
Olorunfemi, M. O., 234-903-719-2169 mlorunfe@oauife.edu.ng, Obafemi Awolowo University – Yg
Olsen, Amanda A., 207 581-2194 amanda.olsen@umit.maine.edu, University of Maine – Cl
Olsen, Khris B., (509) 376-4114 kb.olsen@pnl.gov, Pacific Northwest National Laboratory – Cg
Olsen, Kim B., 619-594-2649 kbolsen@mail.sdsu.edu, San Diego State University – Ye
Olsen, Patrick, 208-885-0810 polsen@uidaho.edu, University of Idaho – Zi
Olsen, Paul E., kolsen@ldeo.columbia.edu, Columbia University – Gm
Olsen, Paul E., (845) 365-8491 polsen@ldeo.columbia.edu, Columbia University – PvgGr
Olson, Hillary C., (512) 653-8356 holson@austin.utexas.edu, University of Texas, Austin – Ps
Olson, James R., 605-677-6866 jim.olson@usd.edu, South Dakota Dept of Environment and Natural Resources – Zn
Olson, Jeff, olsonjl@uww.edu, University of Wisconsin, Whitewater – Ziu
Olson, Kenneth R., (217) 333-9639 University of Illinois, Urbana-Champaign – Sc
Olson, Kimberly, olsonk@arc.losrios.edu, American River College – Zye
Olson, Neil F., (603) 271-2875 neil.olson@des.nh.gov, New Hampshire Geological Survey – GeHy
Olson, Peter L., (410) 516-7707 olson@jhu.edu, Johns Hopkins University – Yg
Olson, Ted L., 435 283-7533 ted.olson@snow.edu, Snow College – Ys
Olsson, Richard K., 848-445-3043 olsson@eps.rutgers.edu, Rutgers, The State University of New Jersey – PmGr
Olszewski, Kathy, (718) 409-7366 kolszewski@sunymaritime.edu, SUNY, Maritime College – Cg
Olszewski, Thomas, 979-845-2465 tomo@geo.tamu.edu, Texas A&M University – Pe
Oltman-Shay, Joan M., (425) 644-9660 University of Washington – Op
Olyphant, Greg A., (812) 855-5154 olyphant@indiana.edu, Indiana University, Bloomington – Hw
Omar, Gomaa I., (215) 898-6908 gomar@sas.upenn.edu, University of Pennsylvania – Cc
Omelon, Christopher, 512-471-5440 omelon@jsg.utexas.edu, University of Texas, Austin – Cg
Omotoso, O. A., deletoso2002@yahoo.com, University of Ilorin – Ge
Oms Llobet, Oriol, ++34935811218 joseporiol.oms@uab.cat, Universitat Autonoma de Barcelona – GrsYm
Onasch, Charles M., (419) 372-7197 conasch@bgnet.bgsu.edu, Bowling Green State University – Gc
Onderdonk, Nate, (562) 985-2654 nate.onderdonk@csulb.edu, California State University, Long Beach – Gtm
ONeal, Michael, 302-831-8273 michael@udel.edu, University of Delaware – Gm
Ono, Shuhei, (617) 253-0474 sono@mit.edu, Massachusetts Institute of Technology – PgCs
Onstott, Tullis C., (609) 258-6898 tullis@princeton.edu, Princeton University – Pg
Oommen, Thomas, (906) 487-2045 toommen@mtu.edu, Michigan Technological University – NeZiNg
Oona, Hain, (505) 667-5685 Los Alamos National Laboratory – Zn
Oostdam, Bernard L., 717-871-9928 boostdam@hotmail.com, Millersville University – Ou
Opdyke, Neil D., (352) 392-6127 drno@ufl.edu, University of Florida – Ym
Opeloye, S A., 08060981972 Federal University of Technology, Akure – GgoGs
Opfer, Christina, christina.opfer@tulsacc.edu, Tulsa Community College – GgoZe
Ophori, Duke U., (973) 655-7558 ophorid@mail.montclair.edu, Montclair State University – Hw
Oppenheimer, Michael, (609) 258-2338 omichael@princeton.edu, Princeton University – AsGe
Opper, Carl, 727-791-2536 opper.carl@spcollege.edu, Saint Petersburg College, Clearwater – Hw
Opperman, William, wopperma@broward.edu, Broward College – Gg
Oppo, Davide, 337-482-6676 davide.oppo@louisiana.edu, University of Louisiana at Lafayette – Gs
Oppo, Delia W., 508-289-2681 doppo@whoi.edu, Woods Hole Oceanographic Institution – Pm
Oppong, Joseph R., (940) 565-2181 oppong@unt.edu, University of North Texas – Zn
Orchard, Michael, 604-669-0909 morchard@nrcan.gc.ca, University of British Columbia – PgGg
Orcutt, John A., (858) 534-2887 jorcutt@ucsd.edu, University of California, San Diego – YrsYd
Orejana, David, dorejana@ucm.es, Univ Complutense de Madrid – Gig
Oreskes, Naomi, 617-495-3480 oreskes@fas.harvard.edu, Harvard University – Rh
Orians, Kristin J., (604) 822-6571 korians@eoas.ubc.ca, University of British Columbia – Cm
Origlia, Hector D., +5435846761998 doriglia@exa.unrc.edu.ar, Universidad Nacional de Rio Cuarto – NgZn
Orlandini, Kent A., (630) 252-4236 Argonne National Laboratory – Cc
Orlanski, Isidoro, (609) 258-1319 isidoro.orlanski@noaa.gov, Princeton University – AsZoc
Orlove, Benjamin S., bsorlove@ucdavis.edu, Columbia University – As
Orme, Amalie, (818) 677-3532 amalie.orme@csun.edu, California State University, Northridge – Gm
Ormerod, Kerri Jean , (775) 784-6347 kormerod@unr.edu, University of Nevada, Reno – Zu
Orndorff, Richard L., (509) 359-2855 rorndorff@ewu.edu, Eastern Washington University – Ge
Orphan, Victoria, (626) 395-1786 vorphan@gps.caltech.edu, California Institute of Technology – Pg
Orr, Robert D., (209) 376-2481 robert.orr@pnl.gov, Pacific Northwest National Laboratory – Ca
Orr, William , orrw@pdx.edu, Portland State University – GgOg
Orr, William N., (541) 346-4577 worr@darkwing.uoregon.edu, University of Oregon – Pm
Orrock, John, orrock@wustl.edu, Washington University in St. Louis – Zn
Orsi, Alejandro H., (979) 845-4014 aorsi@ocean.tamu.edu, Texas A&M University – Op
Ort, Michael H., (928) 523-9363 michael.ort@nau.edu, Northern Arizona University – Gv
Ortega, Lorena, 34 913944837 decgeo@ucm.es, Univ Complutense de Madrid – EmgGg
Ortega-Ariza, Diana, 706-507-8090 ortegaariza_diana@columbusstate.edu, Columbus State University – GdgGr
Ortegren, Jason, 850-474-2491 jortegren@uwf.edu, University of West Florida – PeZu
Ortiz, Joseph D., 330-672-2225 jortiz@kent.edu, Kent State University – Gs
Ortmann, Anthony L., 270-809-6755 aortmann@murraystate.edu, Murray State University – GaYg
Ortt, Jr., Richard A., (410) 554-5503 richard.ortt@maryland.gov, Maryland Department of Natural Resources – NtGeYr
Orville, Richard E., (979) 845-7671 rorville@tamu.edu, Texas A&M University – As
Osagiede, Edoseghe E., edoseghe.osagiede@uniben.edu, University of Benin – GctGx
Osborn, Gerald D., (403) 220-6448 University of Calgary – Gm
Osborn, Joe, (651) 779-3434 joe.osborn@century.edu, Century College – Zg
Osborn, Stephen G., (909)869-3009 sgosborn@csupomona.edu, California State Polytechnic University, Pomona – HwClSp
Osborn, Timothy J., +44 (0)1603 59 2089 t.osborn@uea.ac.uk, University of East Anglia – At
Osborne, Todd, 352-392-1803 osbornet@ufl.edu, University of Florida – Sf
Osburn, Chris, 919-515-0382 chris_osburn@ncsu.edu, North Carolina State University – Cms

Osburn, Magdalena , 847-491-4254 maggie@earth.northwestern.edu,
Northwestern University – PoCbs
Oschlies, Andreas, aoschlies@geomar.de,
Dalhousie University – Ob
Oshun, Jasper, 707-826-3112 jo1196@humboldt.edu,
Humboldt State University – HyGmg
Osi, Attila, hungaros@freemail.hu,
Eotvos Lorand University – PviPe
Osinski, Gordon R., 519-661-2111 ext.84208 gosinbski@uwo.ca,
Western University – XgcGp
Oskin, Michael, (530) 752-3993 meoskin@ucdavis.edu,
University of California, Davis – GmtGc
Osman, Mutasim S., +966138603184 mutasimsami@kfupm.edu.sa,
King Fahd University of Petroleum and Minerals – GosGc
Osmond, Deanna L., (919) 515-7303 deanna_osmond@ncsu.edu,
North Carolina State University – Sb
Osmond, John K., (850) 644-5860 osmond@gly.fsu.edu,
Florida State University – Cc
Oster, Jessica L., 615-322-1461 jessica.l.oster@vanderbilt.edu,
Vanderbilt University – ClsPc
Osterberg, Erich C., erich.c.osterberg@dartmouth.edu,
Dartmouth College – AsCl
Otamendi, Juan E., +543584676198 jotamendi@exa.unrc.edu.ar,
Universidad Nacional de Rio Cuarto – CgaCp
Otis, Gard, (519) 814-4120 Ext.52478 gotis@uoguelph.ca,
University of Guelph – Zn
Ott, Kevin C., (505) 667-4600 Los Alamos National Laboratory – Zn
Ottavi-Pupier, Elsa, +33 (0)3 44069358 elsa.ottavi-pupier@lasalle-beauvais.fr,
Institut Polytechnique LaSalle Beauvais (ex-IGAL) – GvzGi
Otterstrom, Samuel M., (801) 422-7751 samuel_otterstrom@byu.edu,
Brigham Young University – Zu
Otvos, Ervin G., 228-388-7200 ervin.otvos@usm.edu,
University of Southern Mississippi – OnGsm
Ouellette, Vicki L., (970) 351-2647 vicki.ouellette@unco.edu,
University of Northern Colorado – Zn
Ouimet, William, 860-486-3772 william.ouimet@uconn.edu,
University of Connecticut – GmZy
Ouimette, Mark A., (325) 670-1383 ouimette@hsutx.edu,
Hardin-Simmons University – GieGt
Over, D. Jeffrey, (585) 245-5294 over@geneseo.edu,
SUNY, Geneseo – Ps
Overland, James E., james.e.overland@noaa.gov,
University of Washington – As
Overpeck, Jonathan T., 520-626-4345 jto@email.arizona.edu,
University of Arizona – As
Oviatt, Candace, (401) 874-6132 coviatt@gso.uri.edu,
University of Rhode Island – Ob
Oviatt, Charles G., (505) 304-6515 joviatt@ksu.edu,
Kansas State University – Gm
Oware, Erasmus K., (716)645-4260 erasmuso@buffalo.edu,
SUNY, Buffalo – YgHq
Owen, Alan W., +4401413305461 alan.owen@glasgow.ac.uk,
University of Glasgow – PgsPg
Owen, Donald E., (409) 880-8234 donald.owen@lamar.edu,
Lamar University – Gr
Owen, Lewis, (513) 556-4203 lewis.owen@uc.edu,
University of Cincinnati – GmlGt
Owen, Lewis, 513-410-2339 Lewis.Owen@uc.edu,
University of Washington – GmCcGt
Owen, Robert M., (734) 763-4593 rowen@umich.edu,
University of Michigan – CmGu
Owens, Brent E., (757) 221-1813 beowen@wm.edu,
College of William & Mary – Gx
Owens, Tamera, (231) 777-0289 tamera.owens@muskegoncc.edu,
Muskegon Community College – Zn
Owens, Thomas J., 803-777-4530 owens@seis.sc.edu,
University of South Carolina – Ys
Owensby, Clenton E., (785) 532-7232 owensby@ksu.edu,
Kansas State University – Sf
Owoseni, Joshua O., 08060719411 Federal University of Technology,
Akure – GgHw
Oxendine, Christopher E., 845-938-4354 christopher.oxendine@westpoint.edu,
United States Military Academy – ZirZf
Oxford, Jeremiah, 503-838-8680 oxfordj@wou.edu,
Western Oregon University – Zg
Oxley, Meghan, 206-543-8904 what@uw.edu,
University of Washington – Ze
Oxner-Jones, D. M., 740 201 2761 dalton.jones@dnr.state.oh.us,
Ohio Dept of Natural Resources – OnGgZg
Oyawale, A. A., 234-803-331-2150 aoyawale@oauife.edu.ng,
Obafemi Awolowo University – GeoCg
Oyediran, I. A., ia.oyediran@mail.ui.edu.ng,
University of Ibadan – Ng
Oza, Rupal, 212-650-3035 rupal.oza@hunter.cuny.edu,
Hunter College (CUNY) – Zn
Ozbay, M. Ugur, (303) 273-3710 mozbay@mines.edu,
Colorado School of Mines – Nr
Ozdenerol, Esra, 901-678-2787 eozdenrl@memphis.edu,
University of Memphis – Zi
Ozel Yildirim, Esra , 00904242370000-5953 eozel@firat.edu.tr,
Firat University – Gxi
Ozkan-Haller, Tuba, 541-737-9170 ozkan@coas.oregonstate.edu,
Oregon State University – On
Ozsvath, David L., (715) 346-2287 dozsvath@uwsp.edu,
University of Wisconsin, Stevens Point – Hw
Oztekin Okan, Ozlem, 00904242370000-5983 ooztekin@firat.edu.tr,
Firat University – HgwHs
Ozturk, Nevin, 00904242370000-5956 nevinozturk@firat.edu.tr,
Firat University – Cg

P

Paavola, Jouni, +44(0) 113 34 36787 j.paavola@leeds.ac.uk,
University of Leeds – Ge
Pace, Michael L., (434) 924-6541 mlp5fy@virginia.edu,
University of Virginia – Ge
Pacheco, Fernando A., fpacheco@utad.pt,
Universidade de Trás-os-Montes e Alto Douro – Hq
Pacia, Christina, (908) 737-3738 cpacia@kean.edu,
Kean University – Zn
Padden, Maureen, (905) 525-9140 (Ext. 20118) paddenm@mcmaster.ca,
McMaster University – CsPc
Padian, Kevin, (510) 642-7434 kpadian@berkeley.edu,
University of California, Berkeley – Pv
Paduan, Jeffrey D., (831) 656-3350 paduan@nps.edu,
Naval Postgraduate School – Op
Paez, H. A., 905-525-9140 (Ext. 26099) paezha@mcmaster.ca,
McMaster University – Eg
Page, David, 775-673-7110 dave.page@dri.edu,
Desert Research Institute – Ga
Page, F Zeb, (440) 775-6701 zeb.page@oberlin.edu,
Oberlin College – GpCsg
Page, Kevin, +44 1752 584750 kevin.page@plymouth.ac.uk,
University of Plymouth – Gr
Page, Philippe, philippe_page@uqac.ca,
Universite du Quebec a Chicoutimi – GiCgEm
Paget, Aaron C., 304-384-6006 apaget@concord.edu,
Concord University – ApOpZr
Pagiatakis, Spiros, (416)736-2100 #77757 spiros@yorku.ca,
York University – Yd
Pagliari, Paulo H., 507-752-5065 pagli005@umn.edu,
University of Minnesota, Twin Cities – Sc
Paglione, Timothy, 718-262-2654 tpaglione@york.cuny.edu,
York College (CUNY) – Xa
Pagnac, Darrin C., 605-394-2469 Darrin.Pagnac@sdsmt.edu,
South Dakota School of Mines & Technology – PvGrZe
Paige, David A., (310) 825-4268 dap@epss.ucla.edu,
University of California, Los Angeles – Xy
Paillet, Fred, 479-575-3355 fredp@cox.net,
University of Arkansas, Fayetteville – YgHw
Pain, Christopher, +44 20 759 49322 c.pain@imperial.ac.uk,
Imperial College – Yg
Paine, Jeffrey G., 512-471-1260 jeff.paine@beg.utexas.edu,
University of Texas, Austin – On
Paine, Jeffrey G., (512) 475-9524 jeff.paine@beg.utexas.edu,
University of Texas at Austin, Jackson School of Geosciences – Gr
Pair, Donald, (937) 229-2936 Don.Pair@notes.udayton.edu,
University of Dayton – Gl
Pakhomov, Evgeny, (604) 827-5564 epakhomov@eos.ubc.ca,
University of British Columbia – Ob
Paktunc, Dogan, 613-947-7061 Dogan.Packtunc@nrcan-rncan.gc.ca,
University of Ottawa – Gz
Palace, Michael W., 603-862-4193 mike.palace@unh.edu,
University of New Hampshire – Zn

Palamartchouk, Kirill, +44 191 208 6421 kirill.palamartckirill.palamart-chouk@ncl.ac.uk,
University of Newcastle Upon Tyne – Yd
Palenik, Brian, (858) 534-7505 bpalenik@ucsd.edu,
University of California, San Diego – Ob
Palin, J. Michael, +64 3 479-9083 michael.palin@otago.ac.nz,
University of Otago – CcGip
Palin, Richard, 303-273-3819 rmpalin@mines.edu,
Colorado School of Mines – Gp
Palinkaš, Ladislav, +38514605971 lpalinka@geol.pmf.hr,
University of Zagreb – CgeCp
Palladino, Steve D., (805) 654-6400 (Ext. 1365) Ventura College – Zi
Pallis, Ted J., (609) 984-6587 ted.pallis@dep.nj.gov,
New Jersey Geological and Water Survey – Zi
Pallister, John S., (360) 993-8964 jpallist@usgs.gov,
University of Pittsburgh – GviGg
Palluconi, Frank D., (818) 354-8362 frank.d.palluconi@jpl.nasa.gov,
Jet Propulsion Laboratory – Zr
Palm, Risa I., (404) 413-2574 risapalm@gsu.edu,
Georgia State University – Zn
Palma, Miclelle, (262) 521-5542 michelle.palma@uwc.edu,
University of Wisconsin Colleges – ZnnZn
Palmer, Arthur N., (607) 436-3064 arthur.palmer@oneonta.edu,
SUNY, Oneonta – HgqGm
Palmer, Christina, (909) 537-5336 cpalmer@csusb.edu,
California State University, San Bernardino – Zn
Palmer, Clare, cpalmer@wustl.edu,
Washington University in St. Louis – Gz
Palmer, Derecke, 612 9385 8719 d.palmer@unsw.edu.au,
University of New South Wales – Ye
Palmer, Donald F., (330) 672-2680 dpalmer@kent.edu,
Kent State University – Yg
Palmer, Evan, 719-333-3080 ronald.palmer@usafa.edu,
United States Air Force Academy – Zig
Palmer, H. C., (519) 661-2111 x86749 cpalmer@uwo.ca,
Western University – YmGv
Palmer, Martin R., +44 (0)23 80596607 pmrp@noc.soton.ac.uk,
University of Southampton – Cg
Palmer, Paul, +44 (0) 131 650 7724 paul.palmer@ed.ac.uk,
Edinburgh University – As
Palmquist, John C., (414) 832-6732 palmquij@lawrence.edu,
Lawrence University – GcpGg
Palumbo, Tony V., (423) 576-8002 avp@ornl.gov,
Oak Ridge National Laboratory – Sb
Palutoglu, Mahmut, 00904242370000-5954 mpalutoglu@firat.edu.tr,
Firat University – YgsGc
Pan, Feifei, 940-369-5109 feifei/pan@unt.edu,
University of North Texas – Hw
Pan, William L., (509) 335-3611 wlpan@wsu.edu,
Washington State University – Sb
Pan, Ying, yyp5033@psu.edu,
Pennsylvania State University, University Park – As
Pan, Yuanming, (706) 966-5699 yuanming.pan@usask.ca,
University of Saskatchewan – Gx
Pan, Zaitao, (314) 977-3114 panz@eas.slu.edu,
Saint Louis University – As
Panagiotakopulu, Eva, +44 (0) 131 650 2531 Eva.P@ed.ac.uk,
Edinburgh University – Pe
Panah, Assad I., (814) 362-7569 aap@pitt.edu,
University of Pittsburgh, Bradford – GorGc
Pancha, Aasha, (775) 784-4254 pancha@seismo.unr.edu,
University of Nevada, Reno – Ys
Panero, Wendy R., 614 292 6290 panero.1@osu.edu,
Ohio State University – Yg
Panetta, Richard L., (979) 845-1386 r-panetta@tamu.edu,
Texas A&M University – Am
Pangle, Luke, lpangle@gsu.edu,
Georgia State University – HqCs
Pani, Eric A., (318) 342-1878 University of Louisiana, Monroe – As
Paniconi, Claudio, claudio.paniconi@ete.inrs.ca,
Universite du Quebec – Hw
Panish, Peter T., (413) 545-2593 panish@geo.umass.edu,
University of Massachusetts, Amherst – Gp
Pankow, Kristine L., 801-585-6484 pankow@seis.utah.edu,
University of Utah – Ys
Panning, Mark, (352) 392-2634 mpanning@ufl.edu,
University of Florida – Ys

Panno, Samuel V., (217) 244-2456 s-panno@illinois.edu,
Illinois State Geological Survey – ClGgHw
Pant, Hari, 718-960-5859 hari.pant@hunter.cuny.edu,
Graduate School of the City University of New York – Zg
Panter, Kurt S., (419) 372-7337 kpanter@bgnet.bgsu.edu,
Bowling Green State University – Giv
Paola, Christopher, 612-624-8025 cpaola@umn.edu,
University of Minnesota, Twin Cities – Gs
Papaleo, Silvanna, spapaleo@zircon.geology.utoronto.ca,
University of Toronto – Zn
Papcun, George, 12012993161 gpapcun@njcu.edu,
New Jersey City University – Ze
Papendick, Robert I., (509) 335-1552 papendick@wsu.edu,
Washington State University – Sp
Papike, James J., (505) 277-1633 jpapike@unm.edu,
University of New Mexico – Ca
Papineau, Dominic, d.papineau@ucl.ac.uk,
University College London – Cg
Papitashvili, Vladimir O., 703-292-8033 vpapita@nsf.gov,
National Science Foundation – YmXpa
Papp, Kenneth R., (907) 696-0079 kenneth.papp@alaska.gov,
Alaska Division of Geological & Geophysical Surveys – Gg
Pappas, Matthew, 631-451-4301 pappasm@sunysuffolk.edu,
Suffolk County Community College, Ammerman Campus – Zg
Papuga, Shirley, (313) 577-9436 shirley.papuga@wayne.edu,
Wayne State University – Hgs
Paquette, Jeanne, (514) 398-4402 jeanne.paquette@mcgill.ca,
McGill University – Gz
Paradise, Thomas R., (479) 575-3159 paradise@uark.edu,
University of Arkansas, Fayetteville – GmZyRn
Parai, Rita, 314-935-3974 rparai@levee.wustl.edu,
Washington University in St. Louis – Cs
Parashar, Rishi, 775.673-7496 rishi.parashar@dri.edu,
University of Nevada, Reno – Hq
Parcell, William C., (316)978- 3140 Wichita State University – Gr
Pardi, Richard R., (201) 595-2695 William Paterson University – Cc
Parendes, Laurie A., (814) 732-2840 lparendes@edinboro.edu,
Edinboro University of Pennsylvania – Zn
Parent, Michel, (418) 654-2657 Universite du Quebec – Gl
Pares, Josep M., (734) 615-0472 jmpares@umich.edu,
University of Michigan – Ym
Parham, James, (657) 278-2043 jparham@fullerton.edu,
California State University, Fullerton – PvgGg
Parise, John B., (631) 632-8196 john.parise@sunysb.edu,
Stony Brook University – Gz
Pariseau, William G., (801) 581-5164 w.pariseau@utah.edu,
University of Utah – Nr
Parish, Cynthia L., cparish@gt.rr.com,
Lamar University – Gg
Parish, Ryan M., 901-678-2606 rmparish@memphis.edu,
University of Memphis – Ga
Parizek, Richard R., (814) 865-3012 parizek@ems.psu.edu,
Pennsylvania State University, University Park – Hw
Park, Jeffrey J., (203) 432-3172 jeffrey.park@yale.edu,
Yale University – YsZgc
Park, Myung-Sook, 831-656-2858 mpark@nps.edu,
Naval Postgraduate School – Am
Park, Sohyun, 089/2180 4333 soohyun.park@physik.uni-muenchen.de,
Ludwig-Maximilians-Universitaet Muenchen – Gz
Park, Stephen K., (951) 827-4501 magneto@ucrmt.ucr.edu,
University of California, Riverside – Ym
Parker, David B., 806-651-4099 dparker@wtamu.edu,
West Texas A&M University – Sbf
Parker, Don, 806-291-1121 Don.parker@wbu.edu,
Wayland Baptist University – GivCg
Parker, Don F., (254) 710-2192 don_parker@baylor.edu,
Baylor University – Giv
Parker, Doug, +44(0) 113 34 36739 d.j.parker@leeds.ac.uk,
University of Leeds – Am
Parker, Gary, 217-244-6161 parkerg@illinois.edu,
University of Illinois, Urbana-Champaign – Gm
Parker, Jim, 713-743-6750 jlparker9@uh.edu,
University of Houston – Zn
Parker, Joan, 831-771-4400 parker@mlml.calstate.edu,
Moss Landing Marine Laboratories – Zn
Parker, Kent E., (509) 373-6337 kent.parker@pnl.gov,
Pacific Northwest National Laboratory – Cg

Parker, Marjorie, (207) 725-3628 mparker@bowdoin.edu,
  Bowdoin College – Zn
Parker, Matthew, 919-513-4367 mdparker@ncsu.edu,
  North Carolina State University – Am
Parker, Richard M., 361-593-3590 Richard.Parker@tamuk.edu,
  Texas A&M University, Kingsville – GgPg
Parker, Robert L., (858) 534-2475 rlparker@ucsd.edu,
  University of California, San Diego – Ym
Parker, Stephen R., (406) 496-4185 sparker@mtech.edu,
  Montana Tech of the University of Montana – Zn
Parker, William C., (850) 644-1568 parker@gly.fsu.edu,
  Florida State University – PqeGq
Parkin, Gary, (519) 824-4120 (Ext. 52452) gparkin@uoguelph.ca,
  University of Guelph – HwSp
Parkin, Geoffrey, +44 (0) 191 208 6146 geoff.parkin@ncl.ac.uk,
  University of Newcastle Upon Tyne – Hg
Parkinson, Christopher D., 504-280-6792 cparkins@uno.edu,
  University of New Orleans – Gp
Parks, George K., (510) 643-5512 parks@ssl.berkeley.edu,
  University of Washington – YxXy
Parlak, Osman, 03223387081 parlak@cu.edu.tr,
  Cukurova Universitesi – GipCc
Parman, Stephen, (401) 863-3352 Stephen_Parman@brown.edu,
  Brown University – CpgCt
Parmentier, E. Marc, (401) 863-1700 EM_Parmentier@Brown.edu,
  Brown University – Yg
Parnell, Roderic A., (928) 523-3329 roderic.parnell@nau.edu,
  Northern Arizona University – Cl
Parnella, Bill, 302-831-3156 parnella@udel.edu,
  University of Delaware – Hg
Parr, Kate, +44 0151 795 4640 Kate.Parr@liverpool.ac.uk,
  University of Liverpool – Pe
Parrick, Brittany, 614 265 6581 Brittany.Parrick@dnr.state.oh.us,
  Ohio Dept of Natural Resources – Gg
Parris, Thomas M., 859-323-0527 mparris@uky.edu,
  University of Kentucky – Eo
Parris, Thomas M., 859-257-1147 mparris@uky.edu,
  University of Kentucky – Cg
Parrish, Christopher E., 603-862-3438 cparrish@ccom.unh.edu,
  University of New Hampshire – Zr
Parrish, Pia, (619) 594-5587 pparrish@mail.sdsu.edu,
  San Diego State University – Zn
Parrish, Randy, +440116 252 3315 rrp@nigl.nerc.ac.uk,
  Leicester University – Cs
Parsekian, Andrew D., (307) 766-3603 aparseki@uwyo.edu,
  University of Wyoming – Yg
Parsen, Mike J., 608629419 mike.parsen@wgnhs.uwex.edu,
  University of Wisconsin, Extension – Hw
Parsons, Barry, +44 (1865) 272017 barry.parsons@earth.ox.ac.uk,
  University of Oxford – Yd
Parsons, Jeffrey D., (206) 221-6627 parsons@ocean.washington.edu,
  University of Washington – Ou
Parsons-Hubbard, Karla M., (440) 775-8353 karla.hubbard@oberlin.edu,
  Oberlin College – PgGu
Partin, Camille, camille.partin@usask.ca,
  University of Saskatchewan – Gtc
Paschert, Karin, 089/2180 6573 karin.paschert@lmu.de,
  Ludwig-Maximilians-Universitaet Muenchen – Gg
Pascoe, Richard, +44 01326 371838 R.D.Pascoe@exeter.ac.uk,
  Exeter University – Gz
Pascussi, Michael, (518) 486-4820 mpascucc@mail.nysed.gov,
  New York State Geological Survey – Eo
Pashin, Jack C., 405-744-6358 jack.pashin@okstate.edu,
  Oklahoma State University – GrcGo
Pashin, Jack C., (205) 349-2852 jpashin@gsa.state.al.us,
  Mississippi State University – GoEcGs
Pasicznyk, David L., (609) 984-6587 dave.pasicznyk@dep.state.nj.us,
  New Jersey Geological and Water Survey – Yg
Pasken, Robert W., (314) 977-3125 paskenrw@slu.edu,
  Saint Louis University – Am
Passey, Benjamin, (410) 340-9473 passey@umich.edu,
  University of Michigan – CsaCl
Pasteris, Jill D., (314) 935-5434 pasteris@levee.wustl.edu,
  Washington University in St. Louis – GbzGe
Pasteris, Jill D., pasteris@wustl.edu,
  Washington University in St. Louis – Zn
Pasternack, Gregory B., (530) 754-9243 gpast@ucdavis.edu,
  University of California, Davis – GmHg
Pate, John, (573) 368-2148 john.pate@dnr.mo.gov,
  Missouri Dept of Natural Resources – Ge
Patel, Vinodkumar A., 217-244-0639 patel@isgs.uiuc.edu
  Illinois State Geological Survey – Zn
Patenaude, Genevieve, +44 (0) 131 651 4472 Genevieve.Patenaude@ed.ac.uk,
  Edinburgh University – Ge
Patera, Edward S., (505) 667-4457 Los Alamos National Laboratory – Cg
Paterson, Colin J., (605) 394-2461 Colin.Paterson@sdsmt.edu,
  South Dakota School of Mines & Technology – Eg
Paterson, John R., +61-2-67732101 jpater20@une.edu.au,
  University of New England – PigPs
Paterson, Scott R., (213) 740-6103 paterson@usc.edu,
  University of Southern California – Ge
Patino-Douce, Alberto E., (706) 542-2394 alpatino@uga.edu,
  University of Georgia – ZnnZn
Patino-Douce, Marta, (706) 542-2399 mapatino@uga.edu,
  University of Georgia – Gi
Paton, Douglas, +44(0) 113 34 35238 D.A.Paton@leeds.ac.uk,
  University of Leeds – Gc
Patrick, David M., (601) 266-4530 University of Southern Mississippi – Ng
Patronas, Dennis, dpatronas@disl.org,
  Dauphin Island Sea Lab – Zn
Patterson, Gary, (901) 678-5264 glpttrsn@memphis.edu,
  University of Memphis – Gg
Patterson, Jodi T., (801) 537-3310 jpatters@utah.gov,
  Utah Geological Survey – Zn
Patterson, Mark, (781) 581-7370 m.patterson@neu.edu,
  Northeastern University – Zn
Patterson, Molly , (607) 777-2831 patterso@binghamton.edu,
  Binghamton University – CmGl
Patterson, R. Timothy, tpatters@ccs.carleton.ca,
  Carleton University – Pm
Patterson, William P., (306) 966-5691 bill.patterson@usask.ca,
  University of Saskatchewan – CsPeGn
Pattison, David R., (403) 220-3263 pattison@geo.ucalgary.ca,
  University of Calgary – Gp
Pattison, Simon A., 204-727-7468 pattison@brandonu.ca,
  Brandon University – GsoGd
Patton, Howard J., (505) 667-1003 patton@lanl.gov,
  Los Alamos National Laboratory – Yg
Patton, James L., (510) 642-3059 University of California, Berkeley – Pv
Patton, Jason A., 479-968-0676 jpatton@atu.edu,
  Arkansas Tech University – GgeGc
Patton, Jason R., Jason.Patton@humboldt.edu,
  Humboldt State University – GuOuGg
Patton, Peter C., (860) 685-2268 ppatton@wesleyan.edu,
  Wesleyan University – Gm
Patton, Regan L., (509) 534-3670 rpatton@wsu.edu,
  Washington State University – Gc
Patton, Terri, 630-252-3294 tlpatton@anl.gov,
  Argonne National Laboratory – Gg
Patton, Thomas W., 406-496-4153 tpatton@mtech.edu,
  Montana Tech of The University of Montana – Hw
Patwardhan, Kaustubh, 845-257-3738 patwardk@newpaltz.edu,
  SUNY, New Paltz – Ggi
Paty, Carol M., (404) 894-2860 carol.paty@eas.gatech.edu,
  Georgia Institute of Technology – YmgZn
Patzkowsky, Mark E., (814) 863-1959 brachio@geosc.psu.edu,
  Pennsylvania State University, University Park – Pg
Paul, John H., (727) 553-1168 jpaul@marine.usf.edu,
  University of South Florida – Ob
Paulsell, Robert L., (225) 578-8655 rpaulsell@lsu.edu,
  Louisiana State University – ZyGg
Paulsen, Timothy S., (920) 424-7002 paulsen@uwosh.edu,
  University of Wisconsin, Oshkosh – Gc
Paulson, Suzanne, (310) 206-4442 paulson@atmos.ucla.edu,
  University of California, Los Angeles – As
Pavelsky, Tamlin M., (919) 962-4239 pavelsky@unc.edu,
  University of North Carolina, Chapel Hill – Hg
Pavese, Alessandro, alessandro.pavese@unito.it,
  Università di Torino – Gz
Pavlis, Gary L., (812) 855-5141 pavlis@indiana.edu,
  Indiana University, Bloomington – YsGtYe
Pavlis, Terry L., (504) 280-6797 tpavlis@uno.edu,
  University of New Orleans – Gt
Pavlis, Terry L., 915-747-5570 tlpavlis@utep.edu,

University of Texas, El Paso – Gc
Pavloski, Charles, (814) 865-0478 pavloski@essc.psu.edu,
 Pennsylvania State University, University Park – Am
Pavlovskaya, Marianna, (212) 772-5320 mpavlov@hunter.cuny.edu,
 Hunter College (CUNY) – Zi
Pavlowsky, Robert T., (417) 836-8473 robertpavlowsky@missouristate.edu,
 Missouri State University – Gm
Pawley, Alison, +44 0161 275-3944 alison.pawley@manchester.ac.uk,
 University of Manchester – Gc
Pawlicki, Caron E., (717) 702-2042 coneil@pa.gov,
 Pennsylvania Bureau of Topographic & Geologic Survey – ZiRc
Pawloski, Gayle A., (925) 423-0437 pawloski1@llnl.gov,
 Lawrence Livermore National Laboratory – Gg
Pawlowicz, Richard A., (604) 822-1356 rpawlowicz@eos.ubc.ca,
 University of British Columbia – Op
Payne, Ashley E., 734-647-1991 aepayne@umich.edu,
 University of Michigan – At
Paytan, Adina, (650) 724-4073 apaytan@pangea.stanford.edu,
 Stanford University – Oc
Paz Gonzalez, Antonio, 34 981167000 tucho@udc.es,
 Coruna University – Sd
Pazzaglia, Frank J., (610) 758-3667 fjp3@lehigh.edu,
 Lehigh University – Gm
Peace, Walter G., peacew@mcmaster.ca,
 McMaster University – Zu
Peacock, Caroline, +44(0) 113 34 37877 C.L.Peacock@leeds.ac.uk,
 University of Leeds – Sb
Peacock, Simon M., simon.peacock@ubc.ca,
 University of British Columbia – GtpYs
Peacor, Donald R., (734) 764-1452 drpeacor@umich.edu,
 University of Michigan – Gz
Peakall, Jeff, +44(0) 113 34 35205 j.peakall@leeds.ac.uk,
 University of Leeds – Gs
Peake, Jeffrey S., jpeake@unomaha.edu,
 University of Nebraska, Omaha – ZyAtRm
Pearce, Jamie R., +44 (0) 131 650 2294 jamie.pearce@ed.ac.uk,
 Edinburgh University – Ge
Pearson, Adam, 315 267-2951 pearsoaj@potsdam.edu,
 SUNY Potsdam – Gm
Pearson, Ann, (617) 384-8392 pearson@eps.harvard.edu,
 Harvard University – Cb
Pearson, David M., 208-282-3486 peardavi@isu.edu,
 Idaho State University – Gc
Pearson, Eugene F., (209) 946-2926 epearson@pacific.edu,
 University of the Pacific – GsPgZg
Pearson, Graham, (780) 492-4156 gdpearson@ualberta.ca,
 University of Alberta – Cec
Pearson, P N., PearsonP@cf.ac.uk,
 Cardiff University – Pe
Pearson, Paul, +44(0)29 208 74579 PearsonP@cardiff.ac.uk,
 University of Wales – PmCg
Pearthree, Philip A., 520-621-2470 pearthre@email.arizona.edu,
 Arizona Geological Survey – Ge
Peate, David W., (319) 335-0567 david-peate@uiowa.edu,
 University of Iowa – CgGi
Peavy, Samuel T., 229-931-2330 speavy@canes.gsw.edu,
 Georgia Southwestern State University – Yg
Pec, Matej, 617-324-7279 mpec@mit.edu,
 Massachusetts Institute of Technology – Nr
Pechmann, James C., (801) 581-3858 pechmann@seis.utah.edu,
 University of Utah – YsGt
Peck, John A., (330) 972-7659 jpeck@uakron.edu,
 University of Akron – GseGn
Peck, Robert, 304-384-5327 Concord University – Pg
Peck, Theodore R., (217) 333-9486 University of Illinois, Urbana-Champaign – Sc
Peck, William H., (315) 228-7698 wpeck@colgate.edu,
 Colgate University – GpiCs
Peddle, Derek R., derek.peddle@uleth.ca,
 University of Lethbridge – Zr
Pedemonte, Virginia, (598) 25251552 vpedemonte@fcien.edu.uy,
 Universidad de la Republica Oriental del Uruguay (UDELAR) – ZirZu
Pedentchouk, Nikolai, +44 (0)1603 59 3395 n.pedentchouk@uea.ac.uk,
 University of East Anglia – Cs
Pedersen, Joel A., (608) 263-4971 joelpedersen@wisc.edu,
 University of Wisconsin, Madison – So
Pedersen, Thomas F., tfp@uvic.ca,
 University of Victoria – CmOcCs
Pederson, Darryll T., (402) 472-7563 dpederson2@unl.edu,
 University of Nebraska, Lincoln – Hw
Pederson, Joel L., (435) 797-7097 joel.pederson@usu.edu,
 Utah State University – Gm
Pedley, Kate, +64 3 3667001 Ext 3892 kate.pedley@canterbury.ac.nz,
 University of Canterbury – Zg
Pedone, Vicki A., (818) 677-3541 vicki.pedone@csun.edu,
 California State University, Northridge – GdCs
Peebles, Ernst B., (727) 553-3983 epeebles@mail.usf.edu,
 University of South Florida – ObCs
Peele, R Hampton, (225) 328-6651 hampton@lsu.edu,
 Louisiana State University – ZirZg
Peeters, Frank J., +31 20 5987419 f.j.c.peeters@vu.nl,
 Vrije Universiteit Amsterdam – PmOg
Pegion, Kathy, 703-993-5727 kpegion@gmu.edu,
 George Mason University – AsOp
Peirce, Christine, 0191 3742515 christine.peirce@durham.ac.uk,
 Durham University – Yg
Pekar, Stephen, (718) 997-3305 stephen.pekar@qc.cuny.edu,
 Queens College (CUNY) – PeGrOu
Pellenbarg, Robert, rpellenbarg@collegeofthedesert.edu,
 College of the Desert – OcGu
Pellerin, Jocelyne, (418) 724-1704 jocelyne_pellerin@uqar.uquebec.ca,
 Universite du Quebec a Rimouski – Ob
Pelletier, Emilien, (418) 723-1986 (Ext. 1764) emilien_pelletier@uqar.ca,
 Universite du Quebec a Rimouski – Oc
Pelletier, Jon D., (520) 626-2126 jdpellet@email.arizona.edu,
 University of Arizona – Gm
Pellin, Michael J., (630) 972-3510 Argonne National Laboratory – Zm
Pellowski, Christopher J., (605) 394-2465 Christopher.Pellowski@sdsmt.edu,
 South Dakota School of Mines & Technology – Gg
Pelton, John R., (208) 426-3640 jpelton@boisestate.edu,
 Boise State University – YgeYs
Peltzer, Gilles, (310) 206-2156 peltzer@epss.ucla.edu,
 University of California, Los Angeles – ZrGt
Pena dos Reis, Rui, 965042654 penareis@dct.uc.pt,
 Universidade de Coimbra – GorGs
Peng, Lee Chai, 03-79674233 leecp@um.edu.my,
 University of Malaya – Pg
Peng, Yongbo, 225-578-3413 ypeng@lsu.edu,
 Louisiana State University – Cg
Peng, Zhigang, 404-894-0231 zhigang.peng@eas.gatech.edu,
 Georgia Institute of Technology – YsgZn
Penna, Nigel, +44 (0) 191 208 8747 nigel.penna@ncl.ac.uk,
 University of Newcastle Upon Tyne – Gq
Pennacchioni, Giorgio, +390498279106 giorgio.pennacchioni@unipd.it,
 Università degli Studi di Padova – Gc
Penney, Paulette, 254710-2361 paulette_penney@baylor.edu,
 Baylor University – Zn
Pennington, Deana, 915-747-5867 ddpennington@utep.edu,
 University of Texas, El Paso – ZyfZe
Pennington, Wayne D., (906) 487-2531 wayne@mtu.edu,
 Michigan Technological University – YsNe
Penniston-Dorland, Sarah, (301) 405-4087 sarahpd@geol.umd.edu,
 University of Maryland – Gp
Pennock, Jonathan R., (334) 861-7531 jonathan.pennock@unh.edu,
 Dauphin Island Sea Lab – Oc
Penny, Andrew, 831-656-3101 abpenny@nps.edu,
 Naval Postgraduate School – Am
Pentcheva, Rossitza, 089/2180 4352 pentcheva@lrz.uni-muenchen.de,
 Ludwig-Maximilians-Universitaet Muenchen – Gz
Penzo, Michael A., (508) 531-1390 mpenzo@bridgew.edu,
 Bridgewater State University – Ge
Peppe, Daniel J., (254) 710-2629 daniel_peppe@baylor.edu,
 Baylor University – PbcYm
Pepper, Ian L., (520) 621-7234 ipepper@ag.arizona.edu,
 University of Arizona – Zn
Peprah, Ebenezer, (310) 660-3593 epeprah@elcamino.edu,
 El Camino College – Zg
Perault, David R., (434) 544-8370 perault@lynchburg.edu,
 University of Lynchburg – Ge
Percell, Peter, 713-743-6724 ppercell@uh.edu,
 University of Houston – As
Percival, Carl, C.Percival@manchester.ac.uk,
 University of Manchester – As
Percy, David, (503) 725-3373 Portland State University – Zi

Perdrial, Julia, (802) 656-0665 julia.perdrial@uvm.edu,
　　University of Vermont – ClSc
Perdrial, Nicolas, Nicolas.Perdrial@uvm.edu,
　　University of Vermont – GezCl
Pereira, Alcides C., +351-239 860 563 apereira@dct.uc.pt,
　　Universidade de Coimbra – GebZr
Pereira, Engil I., 956 665 2220 engil.pereira@utrgv.edu,
　　University of Texas, Rio Grande Valley – SbZcSo
Peres, Carlos, +44 (0)1603 59 2549 c.peres@uea.ac.uk,
　　University of East Anglia – Sf
Perez, Adriana, 915-747-5501 aperez13@utep.edu,
　　University of Texas, El Paso – Gg
Perez, Adriana, 915-831-8875 aperez28@epcc.edu,
　　El Paso Community College – Gg
Perez, Florante, (213) 620-5026 California Geological Survey – Ng
Perez, Marco A., mperez@cicese.mx,
　　Centro de Investigación Científica y de Educación Superior de Ensenada – Ye
Perez, Richard R., (518) 437-8751 perez@asrc.cestm.albany.edu,
　　SUNY, Albany – As
Perez-Huerta, Alberto, (205) 348-8382 aphuerta@ua.edu,
　　University of Alabama – PgClGz
Pérez-Soba, Cecilia, pesoa@ucm.es,
　　Univ Complutense de Madrid – Gi
Perfect, Edmund, (865) 974-6017 eperfect@utk.edu,
　　University of Tennessee, Knoxville – SpHwGq
Perfit, Michael R., (352) 392-2128 mperfit@ufl.edu,
　　University of Florida – Giu
Peri, Francesco, Francesco.Peri@umb.edu,
　　University of Massachusetts, Boston – Zn
Perkins, Ronald D., (919) 684-3376 rperkins@duke.edu,
　　Duke University – Gdo
Perkins, III, Dexter, (701) 777-2991 dexter.perkins@engr.und.edu,
　　University of North Dakota – Gp
Perkis, Bill, bill.perkis@gogebic.edu,
　　Gogebic Community College – Gg
Perret, Didier H., (418) 654-2686 didier.perret@canada.ca,
　　Universite du Quebec – NgeZu
Perrie, William, (902) 426-3985 william.perrie@dfo-mpo.gc.ca,
　　Dalhousie University – OpAm
Perrin, Richard E., (505) 667-4755 Los Alamos National Laboratory – Cc
Perron, Taylor, (617) 253-5735 perron@mit.edu,
　　Massachusetts Institute of Technology – Gm
Perry, Baker, (828) 262-7597 perrylb@appstate.edu,
　　Appalachian State University – Zg
Perry, David J., (415) 786-6887 Chabot College – Op
Perry, Frank V., (505) 667-1033 fperry@lanl.gov,
　　Los Alamos National Laboratory – Gv
Perry, Kevin D., (801) 581-6138 kevin.perry@utah.edu,
　　University of Utah – As
Perry , Kyle, 859-257-0133 kyle.perry@uky.edu,
　　University of Kentucky – Nr
Perry, Randall, +44 20 759 46425 r.perry@imperial.ac.uk,
　　Imperial College – Zg
Persaud, Naraine, (540) 231-3817 npers@vt.edu,
　　Virginia Polytechnic Institute & State University – Sp
Persaud, Patricia, 225-578-5676 ppersaud@lsu.edu,
　　Louisiana State University – Ys
Perscio, Lyman, 814-824-2076 lpersico@mercyhurst.edu,
　　Mercyhurst University – GsSpHg
Persico, Lyman P., 509-527-5157 persiclp@whitman.edu,
　　Whitman College – Gm
Person, Arthur, (814) 863-8568 Pennsylvania State University, University Park – Am
Person, Jeff, jjperson@nd.gov,
　　North Dakota Geological Survey – Pg
Person, Mark A., (575) 835-6506 mperson@ees.nmt.edu,
　　New Mexico Institute of Mining and Technology – HyYhg
Perumal, Ram, (785) 625-3425 perumal@ksu.edu,
　　Kansas State University – So
Pervunina, Aelita, +78142766039 aelita@krc.karelia.ru,
　　Institute of Geology (Karelia, Russia) – GpvEm
Peryea, Frank J., (509) 663-8181 Washington State University – Sc
Pesavento, Jim, 7607441150 x2516 jpesavento@palomar.edu,
　　Palomar College – GgZg
Pessagno, Jr., Emile A., (972) 883-2444 pessagno@utdallas.edu,
　　University of Texas, Dallas – Pm

Pesses, Michael, mpesses@avc.edu,
　　Antelope Valley College – Zyi
Pestrong, Raymond, (415) 338-2080 rayp@sfsu.edu,
　　San Francisco State University – Ng
Peteet, Dorothy M., (845) 365-8420 peteet@ldeo.columbia.edu,
　　Columbia University – PelGn
Peter, Mark, 614 265 6529 mark.peter@dnr.state.oh.us,
　　Ohio Dept of Natural Resources – PgZe
Peterman, Emily M., 207-725-3846 epeterma@bowdoin.edu,
　　Bowdoin College – Gtp
Peters, Catherine A., (609) 258-5645 cap@phoenix.princeton.edu,
　　Princeton University – Hw
Peters, Lisa, (505) 835-5217 lisa@nmt.edu,
　　New Mexico Institute of Mining & Technology – Gg
Peters, Mark T., (202) 586-9279 Mark_Peters@notes.ymp.gov,
　　Los Alamos National Laboratory – Yg
Peters, Roger M., roger.peter@uwex.edu,
　　University of Wisconsin, Extension – Gg
Peters, Shanan, 608-262-5987 peters@geology.wisc.edu,
　　University of Wisconsin-Madison – Gs
Peters, Stephen C., 610/758-3660 scp2@lehigh.edu,
　　Lehigh University – Cl
Petersen, Bruce, (314) 935-5643 petersen@wustl.edu,
　　Washington University in St. Louis – Zn
Petersen, Erich U., (801) 581-7238 erich.petersen@utah.edu,
　　University of Utah – Em
Petersen, Gary W., (814) 865-1540 gwp2@psu.edu,
　　Pennsylvania State University, University Park – Sd
Petersen, Nikolai, 0049/89/2180-4233 petersen@geophysik.uni-muenchen.de,
　　Ludwig-Maximilians-Universitaet Muenchen – Ygm
Peterson, Bradley, Bradley.Peterson@stonybrook.edu,
　　SUNY, Stony Brook – Ob
Peterson, Dallas, (785) 532-0405 Kansas State University – So
Peterson, Eric W., (309) 438-7865 ewpeter@ilstu.edu,
　　Illinois State University – Hws
Peterson, Eugene J., (505) 667-5182 Los Alamos National Laboratory – Ct
Peterson, Gary L., (619) 594-5594 gpeterson@geology.sdsu.edu,
　　San Diego State University – Gr
Peterson, Holly, 336-316-2263 petersonhe@guilford.edu,
　　Guilford College – HwGe
Peterson, Jon W., peterson@hope.edu,
　　Hope College – Ge
Peterson, Joseph E., (847) 697-1000 jpeterson@elgin.edu,
　　Elgin Community College – Gg
Peterson, Joseph E., 920-424-4463 petersoj@uwosh.edu,
　　University of Wisconsin, Oshkosh – PgGr
Peterson, Larry C., 305-284-6821 l.peterson@miami.edu,
　　University of Miami – Gr
Peterson, Larry C., 305 421-4692 lpeterson@rsmas.miami.edu,
　　University of Miami – GuPcm
Peterson, Peter A., (515) 294-9652 pap@iastate.edu,
　　Iowa State University of Science & Technology – Zn
Peterson, Richard E., 806-834-3418 richard.peterson@ttu.edu,
　　Texas Tech University – As
Peterson, Ronald C., (613) 533-6180 peterson@queensu.ca,
　　Queen's University – Gz
Peterson, Virginia L., (616) 331-2811 petersvi@gvsu.edu,
　　Grand Valley State University – GxcGt
Pethick, Andrew, +618 9266-2297 Andrew.Pethick@curtin.edu.au,
　　Curtin University – YevYg
Petit, Martin A., (309) 694-5327 Illinois Central College – Ze
Petr, Vilem, 303-273-3222 vpetr@mines.edu,
　　Colorado School of Mines – Nm
Petrie, Elizabeth S., (970) 943-2117 epetrie@western.edu,
　　Western Colorado University – GcoYe
Petrie, Gregg, 206-552-6385 gregg.petrie@ieee.org,
　　University of Washington – Zr
Petrinec, Zorica, +38514605969 zoricap@geol.pmf.hr,
　　University of Zagreb – GipGz
Petronis, Michael S., (505) 454-3513 mspetro@nmhu.edu,
　　New Mexico Highlands University – YmGcv
Petrovay, Kristof, +36 20 3722500/6621 k.petrovay@astro.elte.hu,
　　Eotvos Lorand University – Xy
Petruncio, Emil T., 410-293-6564 petrunci@usna.edu,
　　United States Naval Academy – OpZr
Petsch, Steven, (413) 545-4413 spetsch@geo.umass.edu,
　　University of Massachusetts, Amherst – Col

Petsios, Elizabeth, (254)710-2388 elizabeth_petsios@baylor.edu, Baylor University – Pie
Pettengill, Gordon H., (617) 253-4281 ghp@space.mit.edu, Massachusetts Institute of Technology – Zg
Petters, Markus, 919-515-7144 markus_petters@ncsu.edu, North Carolina State University – As
Pettijohn, J. C., 217-333-3540 jcpettij@illinois.edu, University of Illinois, Urbana-Champaign – Hg
Pettinga, Jarg R., (03) 3667-001 x7716 jarg.pettinga@canterbury.ac.nz, University of Canterbury – Gtc
Petty, Grant W., (608)263-3265 gpetty@aos.wisc.edu, University of Wisconsin, Madison – AsZrAm
Petuch, Edward J., (561) 297-2398 epetuch@fau.edu, Florida Atlantic University – Pg
Peucker-Ehrenbrink, Bernhard, (508) 289-2518 bpeucker@whoi.edu, Woods Hole Oceanographic Institution – CmlHs
Pevzner, Roman, +618 9266-9805 R.Pevzner@curtin.edu.au, Curtin University – Yxx
Pezelj, Đurđica, +38514606117 durpezelj@yahoo.com, University of Zagreb – PgbPm
Pfannkuch, Hans O., (612) 624-1620 h2olafpf@umn.edu, University of Minnesota, Twin Cities – HwGeRh
Pfefferkorn, Hermann W., (215) 898-5156 hpfeffer@sas.upenn.edu, University of Pennsylvania – PbePs
Pfuhl, Helen, 089/2180 4230 helen.pfuhl@geophysik.uni-muenchen.de, Ludwig-Maximilians-Universitaet Muenchen – Yg
Phelps, Tommy, 865-574-7290 phelpstj1@ornl.gov, University of Tennessee, Knoxville – Zn
Phelps, William, 951-222-8350 william.phelps@rcc.edu, Riverside City College – PeOpPg
Philander, S. George H., (609) 258-5683 gphlder@princeton.edu, Princeton University – Op
Phillippe, Jennifer, 304-384-5157 jphillippe@concord.edu, Concord University – GegZe
Phillippi, Nathan E., 910-521-6588 nathan.phillippi@uncp.edu, University of North Carolina, Pembroke – Zgi
Phillips, Andrew C., (217) 333-2513 aphillps@illinois.edu, Illinois State Geological Survey – GsmZi
Phillips, Brian L., (631) 632-6853 brian.phillips@stonybrook.edu, Stony Brook University – GzCl
Phillips, Dennis, (505) 667-4253 Los Alamos National Laboratory – Co
Phillips, Fred M., (575) 835-5540 phillips@nmt.edu, New Mexico Institute of Mining and Technology – Hw
Phillips, Michael, (815) 224-0394 mike_phillips@ivcc.edu, Illinois Valley Community College – GgeGm
Phillips, Paul, (785) 628-5969 pphillip@fhsu.edu, Fort Hays State University – Zen
Phillips, Richard, +44(0) 113 34 31728 R.J.Phillips@leeds.ac.uk, University of Leeds – Gt
Phillips, William R., (801) 378-4545 Brigham Young University – Gz
Phillips, William S., (505) 667-8106 wsp@lanl.gov, Los Alamos National Laboratory – Yg
Philp, R. Paul, (405) 325-4469 pphilp@ou.edu, University of Oklahoma – Co
Philpotts, Anthony R., (860) 486-1394 anthony.philpotts@uconn.edu, University of Connecticut – Gi
Phinney, Robert A., (609) 258-4118 rphinney@princeton.edu, Princeton University – Ys
Phipps, Stephen P., (215) 898-4602 sphipps@sas.upenn.edu, University of Pennsylvania – Gc
Phipps Morgan, Jason, jp369@cornell.edu, Cornell University – GvCmPe
Piazzoni, Antonio Sebastian, 089/2180 4230 antonio.piazzoni@geophysik.uni-muenchen.de, Ludwig-Maximilians-Universitaet Muenchen – Yg
Picardal, Flynn W., (812) 855-0732 picardal@indiana.edu, Indiana University, Bloomington – So
Piccinini, Leonardo, +390498279124 leonardo.piccinini@unipd.it, Università degli Studi di Padova – Hg
Piccoli, Philip M., (301) 405-6966 piccoli@umd.edu, University of Maryland – Cg
Pichevin, Laetitia, +44 (0) 131 650 5980 Laetitia.Pichevin@ed.ac.uk, Edinburgh University – Cg
Pichler, Thomas, (813) 974-0321 pichler@cas.usf.edu, University of South Florida, Tampa – Cl
Pickering, Ingrid J., (306) 966-5706 ingrid.pickering@usask.ca, University of Saskatchewan – CgtZn

Pickering, Kevin, +44 020 7679 31325 kt.pickering@ucl.ac.uk, University College London – Gs
Pickett, Nicki, 954-201-6677 npickett@broward.edu, Broward College – Zn
Pickup, Gillian E., +44 (0) 131 451 3168 g.pickup@hw.ac.uk, Heriot-Watt University – Np
Pier, Stanley M., (713) 522-0633 Rice University – Co
Pierce, David, 440-525-7341 dpierce@lakelandcc.edu, Lakeland Community College – GgHsAm
Pierce, Heather, (585) 292-2426 hpierce@monroecc.edu, Monroe Community College – Zyi
Pierce, Jeff, jeffrey.pierce@colostate.edu, Colorado State University – As
Pierce, Jennifer L., (208) 426-5380 jenpierce@boisestate.edu, Boise State University – GmRcZc
Pierce, Larry, (573) 368-2191 larry.pierce@dnr.mo.gov, Missouri Dept of Natural Resources – Zn
Pieri, David C., (818) 354-6299 Jet Propulsion Laboratory – Gt
Pierrehumbert, Raymond T., (773) 702-8811 University of Chicago – As
Pieruccini, Pierluigi, pierluigi.pieruccini@unito.it, Università di Torino – Gm
Pierzynski, Gary M., 785-532-6101 gmp@ksu.edu, Kansas State University – Sc
Pieters, Carle M., (401) 863-2417 Carle_Pieters@Brown.EDU, Brown University – Zr
Pieters, Roger, (604) 822-4297 rpieters@eos.ubc.ca, University of British Columbia – Op
Pietranik, Anna, anna.pietranik@uwr.edu.pl, University of Wroclaw – CgGie
Pietras, Jeff, (607) 777-3348 jpietras@binghamton.edu, Binghamton University – GsEo
Pietro, Kathryn R., (508) 289-3862 kpietro@whoi.edu, Woods Hole Oceanographic Institution – Gu
Pietrzak-Renaud, Natalie, 519-473-3766 npietrz@uwo.ca, Western University – Eg
Pietsch, Carlie, 408 924-5279 carlie.pietsch@sjsu.edu, San Jose State University – PiePc
Pigott, John D., (405) 325-4498 jpigott@ou.edu, University of Oklahoma – GoYeGd
Pike, J, PikeJ@cf.ac.uk, Cardiff University – Zn
Pike, Jenny, +44(0)29 208 75181 PikeJ@cardiff.ac.uk, University of Wales – Og
Pike, Scott, (503) 370-6587 spike@willamette.edu, Willamette University – Gag
Pike, Steven M., (508) 289-2350 spike@whoi.edu, Woods Hole Oceanographic Institution – Ca
Pikelj, Kristina, +38514606113 kpikelj@geol.pmf.hr, University of Zagreb – Gs
Pikitch, Ellen K., (631) 632-9599 ellen.pikitch@stonybrook.edu, SUNY, Stony Brook – Ob
Pilkey, Jr., Orrin H., (919) 684-4238 opilkey@geo.duke.edu, Duke University – Ca
Pilson, Michael E., (401) 874-6104 pilson@gso.uri.edu, University of Rhode Island – Oc
Pimentel, Nuno , pimentel@fc.ul.pt, Universidade de Lisboa – GrsGo
Pinckney, James, 803 777 7133 pinckney@sc.edu, University of South Carolina – Obn
Pinet, Paul, (315) 228-7656 ppinet@colgate.edu, Colgate University – Ou
Pingitore, Jr., Nicholas E., 915-747-5754 npingitore@utep.edu, University of Texas, El Paso – Cl
Piniella Febrer, Juan Francesc, ++34935813088 juan.piniella@uab.cat, Universitat Autonoma de Barcelona – Gz
Pinkel, Robert, (858) 534-2056 rpinkel@ucsd.edu, University of California, San Diego – Op
Pinnt, Todd, todd.pinnt@azwestern.edu, Arizona Western College – Zyi
Pinti, Daniele Luigi, 514-987-3000 #2572 pinti.daniele@uqam.ca, Universite du Quebec a Montreal – Cs
Pintilei, Mitica, 0040232401494 mpintilei@gmail.com, Alexandru Ioan Cuza – CgeCt
Piotrowski, Alexander, +44 (0) 1223 333473 apio04@esc.cam.ac.uk, University of Cambridge – Ge
Piper, David J., (902) 426-6580 david.piper@canada.ca, Dalhousie University – GusGo

Piper, John, +44 0151 794 3461 Sg04@liverpool.ac.uk, University of Liverpool – YmGt
Pipkin, Bernard W., (213) 740-6106 pipkin@usc.edu, University of Southern California – Ng
Pires, E M., (516) 299-2318 mark.pires@liu.edu, Long Island University, C.W. Post Campus – Zg
Pirie, Diane H., (305) 348-2876 Florida International University – Gg
Pisani-Gareau, Tara, (617) 552-0843 pisanoga@bc.edu, Boston College – Ge
Pisel, Jesse, (.307) 766-2286 Ext. 225 jesse.pisel@wyo.gov, Wyoming State Geological Survey – GoEoGr
Pitlick, John, pitlick@colorado.edu, University of Colorado – Gm
Pittman, Jason, (916) 608-6668 PittmaJ@flc.losrios.edu, Folsom Lake College – ZgiZi
Pizzuto, James E., (302) 831-2710 pizzuto@udel.edu, University of Delaware – Gm
Placzek, Christa, 0747 814 756 christa.placzek@jcu.edu.au, James Cook University – CclGe
Plane, David A., plane@email.arizona.edu, University of Arizona – Zn
Plank, Gabriel, (775) 784-7039 gabe@seismo.unr.edu, Universitat of Nevada, Reno – Ys
Plank, Owen C., (706) 542-9072 University of Georgia – Sc
Plank, Terry, tplank@ldeo.columbia.edu, Columbia University – Cg
Plank, Terry A., (845) 365-8410 tplank@ldeo.columbia.edu, Columbia University – GxCg
Plankell, Eric T., 217-265-8029 eplankel@illinois.edu, Illinois State Geological Survey – Hg
Plant, Jane, +44 20 759 47416 jane.plant@imperial.ac.uk, Imperial College – Ge
Plant, Jeffrey J., (818) 393-3799 plant@jpl.nasa.gov, Jet Propulsion Laboratory – Zn
Plante, Alain F., (215) 898-9269 aplante@sas.upenn.edu, University of Pennsylvania – SboCo
Plante, Martin, (418) 656-8121 martin.plante@ggl.ulaval.ca, Universite Laval – Cg
Plasienka, Dusan, 0042160296529 plasienka@fns.uniba.sk, Comenius University in Bratislava – Gt
Platnick, Steven, steven.platnick@nasa.gov, University of Wisconsin, Madison – Zr
Platt, Brian F., 662-915-5440 bfplatt@olemiss.edu, University of Mississippi – GsPe
Platt, John P., (213) 821-1194 jplatt@usc.edu, University of Southern California – GctGp
Platt, Rutherford H., (413) 545-2499 rplatt@geo.umass.edu, University of Massachusetts, Amherst – Zu
Plattner, Alain, 205-348-0387 amplattner@ua.edu, University of Alabama – YmuYg
Plattner, Christina, 089/21804220 christina.plattner@geophysik.uni-muenchen.de, Ludwig-Maximilians-Universitaet Muenchen – Yg
Pleasants, Mark S., 614 265 6748 mark.pleasants@dnr.state.oh.us, Ohio Dept of Natural Resources – HwSp
Plescia, Jeffrey, (202) 358-0295 Jet Propulsion Laboratory – Zn
Plink-Bjorklund, Piret, (303) 384-2042 pplink@mines.edu, Colorado School of Mines – Gs
Plint, A G., (519) 661-3179 gplint@uwo.ca, Western University – Gs
Plotkin, Pamela, 979-845-3902 plotkin@tamu.edu, Texas A&M University – Ob
Plotnick, Roy E., (312) 996-2111 plotnick@uic.edu, University of Illinois at Chicago – PgiGq
Plug, Lawrence, 902-494-1200 lplug@is.dal.ca, Dalhousie University – Gm
Pluhar, Christopher J., (559) 278-1128 cpluhar@csufresno.edu, California State University, Fresno – GtYmNg
Plumb, Raymond A., (617) 253-6281 plumb@mit.edu, Massachusetts Institute of Technology – As
Plummer, Charles C., plummercc@csus.edu, California State University, Sacramento – Gp
Poch Serra, Joan, ++935811085 joan.poch@uab.cat, Universitat Autonoma de Barcelona – GrsZe
Pociask, Geoff, 217-265-8212 pociask@illinois.edu, Illinois State Geological Survey – Hw
Pociask, Geoffrey , 217 265-8029 pociask@illinois.edu, University of Illinois – Hs
Pociask, Walter, (313) 577-2506 wpociask@gmail.com, Wayne State University – Zn
Podolak, Morris, (052) 838-0976 morris@post.tau.ac.il, Tel Aviv University – ZnnZn
Poeter, Eileen P., epoeter@mines.edu, Colorado School of Mines – Ng
Pogge von Strandmann, Phillip, +44 020 7679 33637 p.strandmann@ucl.ac.uk, University College London – Csl
Pogue, Kevin R., (509) 527-5955 pogue@whitman.edu, Whitman College – Gc
Pohll, Greg, 775-682-6349 greg.pohll@dri.edu, University of Nevada, Reno – Hwq
Pohll, Greg, (775) 674-7523 greg.pohll@dri.edu, University of Nevada, Reno – Hqw
Poinar, Kristin, (716)645-4286 kpoinar@buffalo.edu, SUNY, Buffalo – Gl
Poirier, Andre, 514-987-3000 #1718 Universite du Quebec a Montreal – Cs
Pojeta, John, (202) 633-1347 Smithsonian Inst / Natl Museum of Natural Hist– Pi
Pokorny, Eugene W., (702) 295-7496 eugene_pokorny@lanl.gov, Los Alamos National Laboratory – Nm
Pokras, Edward M., epokras@keene.edu, Keene State College – GgPmOb
Polat, Ali, 519-253-3000 x2495 polat@uwindsor.ca, University of Windsor – GixCc
Polet, Jascha, 909-869-3459 jpolet@csupomona.edu, California State Polytechnic University, Pomona – Ysg
Polito, Thomas A., (515) 294-0513 tpolito@iastate.edu, Iowa State University of Science & Technology – So
Polizzotto, Matthew, (541) 346-5217 mpolozzo@uoregon.edu, University of Oregon – CqSc
Polk, Jason, jason.polk@wku.edu, Western Kentucky University – Gm
Pollack, Henry N., (734) 763-0084 hpollack@umich.edu, University of Michigan – Yg
Pollack, Jennifer, 361-825-2041 Jennifer.Pollack@tamucc.edu, Texas A&M University, Corpus Christi – Ob
Pollak, Robert, pollak@wustl.edu, Washington University in St. Louis – Zn
Pollard, David D., (650) 723-4679 dpollard@pangea.stanford.edu, Stanford University – Gc
Pollock, Meagen, (330) 263-2202 mpollock@wooster.edu, College of Wooster – CuGiv
Polly, P D., (812) 855-7994 pdpolly@indiana.edu, Indiana University, Bloomington – PvgPq
Pollyea, Ryan M., 540-231-7929 rpollyea@vt.edu, Virginia Polytechnic Institute & State University – Hq
Polovina, Jeffrey J., (808) 983-5390 jeffrey.polovina@noaa.gov, University of Hawai'i, Manoa – Zn
Polsky, Colin, (954) 236-1088 cpolsky@fau.edu, Florida Atlantic University – Znc
Polvani, Lorenzo M., (212) 854-7331 lmp3@columbia.edu, Columbia University – AmsGq
Polyak, Victor J., (505) 277-4204 polyak@unm.edu, University of New Mexico – CcGzm
Polzin, Dierk T., dtpolzin@wisc.edu, University of Wisconsin, Madison – As
Pomeroy, George M., (717) 477-1776 gmpome@ship.edu, Shippensburg University – ZuyZn
Pommier, Anne, (858) 822-5025 pommier@ucsd.edu, University of California, San Diego – Cg
Pond, Lisa G., (225) 578-0401 lgpond@lsu.edu, Louisiana State University – Zg
Pond, Stephen G., (604) 822-2205 spond@eos.ubc.ca, University of British Columbia – Op
Ponette-Gonzalez, Alexandra, 940-565-4012 alexandra.ponette@unt.edu, University of North Texas – Zyc
Ponton, Camilo, (360) 650-3648 Camilo.Ponton@wwu.edu, Western Washington University – CsoPc
Poole, T. Craig, (559) 442-4600 craig.poole@fresnocitycollege.edu, Fresno City College – Gg
Pope, Gregory A., (973) 655-7385 popeg@mail.montclair.edu, Montclair State University – GmaZy
Pope, Jeanette K., 765-658-4105 jpope@depauw.edu, DePauw University – Ge
Pope, Michael , (979) 845-4376 mcpope@geo.tamu.edu,

Texas A&M University – GrCc
Pope, Richard, +44 (0)1332 591751 r.j.pope@derby.ac.uk,
University of Derby – Gam
Popp, Brian N., (808) 956-6206 popp@soest.hawaii.edu,
University of Hawai'i, Manoa – Cs
Porcelli, Don, +44 (1865) 282121 don.porcelli@earth.ox.ac.uk,
University of Oxford – Cg
Porch, William M., (505) 667-0971 wporch@lanl.gov,
Los Alamos National Laboratory – Yg
Poreda, Robert J., (585) 275-0051 robert.poreda@rochester.edu,
University of Rochester – Cg
Portell, Roger W., 3523921721 x258 portell@flmnh.ufl.edu,
University of Florida – PisPv
Porter, Claire, 612-626-0505 porte254@umn.edu,
University of Minnesota, Twin Cities – Zi
Porter, Dwayne E., 803-777-4615 porter@sc.edu,
University of South Carolina – OgHsRw
Porter, John H., (804) 924-8999 jhp7e@virginia.edu,
University of Virginia – Zif
Porter, Ryan, 928-523-2429 Ryan.Porter@nau.edu,
Northern Arizona University – YgsGt
Porter, Susannah, (805) 893-8954 porter@geol.ucsb.edu,
University of California, Santa Barbara – Gn
Portillo, Danny, (719) 333-3080 United States Air Force Academy – Zi
Portner, Ryan, ryan.portner@sjsu.edu,
San Jose State University – Gdv
Posiloviæ, Hrvoje, +38514606087 posilovic@geol.pmf.hr,
University of Zagreb – Gg
Posler, Gerry L., (785) 532-6101 gposler@ksu.edu,
Kansas State University – Sf
Pospelova, Vera, (250) 721-6314 vpospe@uvic.ca,
University of Victoria – PelOu
Post, Donald F., (520) 621-1262 postdf@ag.arizona.edu,
University of Arizona – Sd
Post, Jeffrey E., (202) 633-1814 postj@si.edu,
Smithsonian Inst / Natl Museum of Natural Hist– Gz
Post, Vincent , vincent.post@flinders.edu.au,
Flinders University – Hw
Poteet, Mary F., 512-471-5209 mpoteet@jsg.utexas.edu,
University of Texas, Austin – Pg
Potel, Sébastien, +33(0)3 44068999 sebastien.potel@lasalle-beauvais.fr,
Institut Polytechnique LaSalle Beauvais (ex-IGAL) – GpzGi
Potess, Marla, (325) 670-1395 marla.potess@hsutx.edu,
Hardin-Simmons University – Zgu
Poths, Jane, (505) 667-1506 Los Alamos National Laboratory – Cc
Potra, Adriana, 479-575-6419 potra@uark.edu,
University of Arkansas, Fayetteville – EgCcGg
Potten, Martin, +49 89 289 - 25885 martin.potten@tum.de,
Technical University of Munich – NxGg
Potter, David, (780) 2481518 dkpotter@ualberta.ca,
University of Alberta – Ye
Potter, Eric C., (512) 471-7090 eric.potter@beg.utexas.edu,
University of Texas at Austin, Jackson School of Geosciences – Eo
Potter, Henry, hpotter@tamu.edu,
Texas A&M University – Op
Potter, Lee S., 319-273-2759 lspotter@earthlink.net,
University of Northern Iowa – Gz
Potter, Paul E., (513) 556-3732 University of Cincinnati – Gs
Potter, Jr., Donald B., (931) 598-1479 bpotter@sewanee.edu,
Sewanee: University of the South – Gc
Potter, Jr., Noel, pottern@dickinson.edu,
Dickinson College – Gmc
Potts, Donald C., (831) 459-4417 potts@biology.ucsc.edu,
University of California, Santa Cruz – Ob
Poty, Edouard, 32 4 366 52 83 e.poty@ulg.ac.be,
Universite de Liege – PiGsPe
Poudel, Durga, 337-482-6163 ddpoudel@louisiana.edu,
University of Louisiana at Lafayette – Ge
Poulsen, Christopher J., (734) 615-2236 poulsen@umich.edu,
University of Michigan – Pe
Poulson, Simon, 775.784.1833 poulson@mines.unr.edu,
University of Nevada, Reno – Cls
Poulson, Simon R., (775)784-1104 poulson@mines.unr.edu,
University of Nevada, Reno – Cs
Poulton, Mary, 520-621-8391 mpoulton@email.arizona.edu,
University of Arizona – Ng
Poulton, Simon, +44(0) 113 34 35237 s.poulton@leeds.ac.uk,
University of Leeds – Cls
Pound, Kate S., 320-308-2014 kspound@stcloudstate.edu,
Saint Cloud State University – GgZeGd
Pourmand, Ali, apourmand@rsmas.miami.edu,
University of Miami – Cga
Pourret, Olivier, +33(0)3 44068979 olivier.pourret@lasalle-beauvais.fr,
Institut Polytechnique LaSalle Beauvais (ex-IGAL) – CgaGe
Powell, Brian, (808) 956-6724 powellb@hawaii.edu,
University of Hawai'i, Manoa – Op
Powell, Brian A., 864656-1004 bpowell@clemson.edu,
Clemson University – Cg
Powell, Christine A., (901) 678-8455 capowell@memphis.edu,
University of Memphis – Ys
Powell, J. Mark, (608) 890-0070 jmpowel2@wisc.edu,
University of Wisconsin, Madison – So
Powell, Jane, +44 (0)1603 59 2822 j.c.powell@uea.ac.uk,
University of East Anglia – Ge
Powell, Matthew G., (814) 641-3602 powell@juniata.edu,
Juniata College – PiqPe
Powell, Rebecca , 303-871-2667 Rebecca.L.Powell@du.edu,
University of Denver – Zr
Powell, Ross D., (815) 753-7952 rpowell@niu.edu,
Northern Illinois University – GslZc
Powell, Thomas M., (510) 642-7455 zackp@socrates.berkeley.edu,
University of California, Berkeley – Op
Powell, Wayne G., (718) 951-5761 wpowell@brooklyn.cuny.edu,
Graduate School of the City University of New York – EmGaCs
Powell, Wayne G., (718) 951-5416 wpowell@brooklyn.cuny.edu,
Brooklyn College (CUNY) – GaCsEm
Power, Mary E., (510) 643-7776 University of California, Berkeley – Pe
Powers, Roger W., (907) 474-6188 University of Alaska, Fairbanks – Zn
Powers, Sean, spowers@southalabama.edu,
University of South Alabama – Ob
Poyla, David, +44 0161 275-3818 david.polya@manchester.ac.uk,
University of Manchester – Cm
Poynton, Helen, helen.poynton@umb.edu,
University of Massachusetts, Boston – Zn
Pracny, Pavel, +420 549 49 8523 pracny@sci.muni.cz,
Masaryk University – Cg
Pradhanang, Soni M., (401) 874-5980 spradhanang@uri.edu,
University of Rhode Island – HsqHq
Prairie, Jennifer, (619) 260-8820 jcprairie@sandiego.edu,
University of San Diego – Ob
Pranter, Matthew J., (405) 325-3253 matthew.pranter@ou.edu,
University of Oklahoma – Go
Pranter, Matthew J., 405-325-4451 matthew.pranter@ou.edu,
University of Oklahoma – Gos
Prasad, Rishi, (334) 844-4100 rzp0050@auburn.edu,
Auburn University – Sb
Prasad, Vara , (785) 532-3746 vara@ksu.edu,
Kansas State University – So
Prather, Kimberly A., (858) 822-5312 kprather@ucsd.edu,
University of California, San Diego – AsCmHg
Prather, Michael J., (949) 824-5838 mprather@uci.edu,
University of California, Irvine – As
Pratson, Lincoln F., (919) 681-8077 lincoln.pratson@duke.edu,
Duke University – Gs
Pratt, Allyn R., (505) 667-4308 pratt_allyn_r@lanl.gov,
Los Alamos National Laboratory – Ge
Pratt, Brian R., (306) 966-5725 brian.pratt@usask.ca,
University of Saskatchewan – PgGsr
Pratt, Lisa M., (812) 855-9203 prattl@indiana.edu,
Indiana University, Bloomington – Co
Pratt, R. Gerhard, 519-661-2111 ext. 86690 gpratt2@uwo.ca,
Western University – Yeg
Prave, Tony, +44 (0)1334462381 ap13@st-andrews.ac.uk,
University of St. Andrews – GgrGs
Precedo, Laura, 954-201-6674 lprecedo@broward.edu,
Broward College – Oc
Precht, Francis L., (301) 687-4440 fprecht@frostburg.edu,
Frostburg State University – Zy
Prell, Warren L., (401) 863-3221 Warren_Prell@Brown.edu,
Brown University – Ou
Prencipe, Mauro, mauro.prencipe@unito.it,
Università di Torino – Gz
Prescott, Cindy, 604–822–4701 cindy.prescott@ubc.ca,
University of British Columbia – Sb

Presley, Bobby J., (979) 845-5136 bpresley@ocean.tamu.edu, Texas A&M University – Oc
Presley, DeAnn, 785532128 deann@ksu.edu, Kansas State University – So
Presnall, Dean C., (972) 883-2444 presnall@utdallas.edu, University of Texas, Dallas – Cp
Prestegaard, Karen L., (301) 405-6982 kpresto@geol.umd.edu, University of Maryland – Hgw
Preston, Aaron, 802-626-6496 aaron.preston@northernvermont.edu, Northern Vermont University-Lyndon – AsZrAc
Preston, Thomas C., 514-398-3766 thomas.preston@mcgill.ca, McGill University – AsZnn
Preston, William L., (805) 756-2210 wpreston@calpoly.edu, California Polytechnic State University – Ze
Preto, Nereo, +390498279174 nereo.preto@unipd.it, Università degli Studi di Padova – Gs
Prevec, Steve, +27 (0)46-603-8309 s.prevec@ru.ac.za, Rhodes University – Cg
Previdi, Michael, mprevidi@ldeo.columbia.edu, Columbia University – As
Prewett, Jerry L., (573) 368-2101 jerry.prewett@dnr.mo.gov, Missouri Dept of Natural Resources – GgHw
Prezant, Robert S., (718) 997-4105 RPrezant@QC1.QC.EDU, Graduate School of the City University of New York – Pg
Price, Alan P., (262) 521-5498 paul.price@uwc.edu, University of Wisconsin Colleges – ZyGmZg
Price, Colin G., +97236406029 cprice@flash.tau.ac.il, Tel Aviv University – As
Price, Curtis V., (605) 394-2461 Curtis.Price@sdsmt.edu, South Dakota School of Mines & Technology – ZifRw
Price, Douglas M., 330-941-3019 dmprice@ysu.edu, Youngstown State University – Cg
Price, Greogry, +44 1752 584771 G.Price@plymouth.ac.uk, University of Plymouth – Gsr
Price, Jason R., 717-872-3005 Jason.Price@millersville.edu, Millersville University – Gs
Price, Jonathan D., (940) 397-4288 jonathan.price@mwsu.edu, Midwestern State University – GiCgGz
Price, Jonathan G., 775/682-8766 jprice@unr.edu, University of Nevada – Gg
Price, Kevin, (785) 532-6101 kpprice@ksu.edu, Kansas State University – Zr
Price, L G., (575) 835-5752 greer@nmbg.nmt.edu, New Mexico Institute of Mining & Technology – Gg
Price, Maribeth H., (605) 394-2461 Maribeth.Price@sdsmt.edu, South Dakota School of Mines & Technology – Zir
Price, Marie D., (202) 994-6187 George Washington University – Zy
Price, Nancy A., 503-725-3398 naprice@pdx.edu, Portland State University – GcZeGt
Price, Peter E., 281-765-7764 Peter.E.Price@nhmccd.edu, Lonestar College - North Harris – Zi
Price, Raymond A., (613) 533-6542 pricera@queensu.ca, Queen's University – GtcGe
Price, Rene, (305) 348-3119 pricer@fiu.edu, Florida International University – HwClRw
Prichard, Hazel, +44(0)29 208 74323 prichard@cardiff.ac.uk, University of Wales – EgGze
Prichard, Terry L., (559) 468-2086 tlprichard@ucdavis.edu, University of California, Davis – Zn
Prichonnet, Gilbert P., 514-987-3000 #3383 prichonnet.gilbert@uqam.ca, Universite du Quebec a Montreal – Pi
Prichystal, Antonín, +420 549 49 6699 prichy@sci.muni.cz, Masaryk University – GgaGv
Pride, Douglas E., 614-329-0941 pride.1@osu.edu, Ohio State University – Eg
Pride, Steven, (510) 495-2823 SRPride@lbl.gov, University of California, Berkeley – Ys
Pridmore, Cindy L., (916) 925-6902 California Geological Survey – Ng
Priesendorf, Carl, (816) 604-2549 carl.priesendorf@mcckc.edu, Metropolitan Community College-Kansas City – GgZyGe
Priest, Eric, (847) 543-2585 epriest@clcillinois.edu, College of Lake County – AmZe
Priest, George R., (541) 574-6642 george.priest@dogami.state.or.us, Oregon Dept of Geology & Mineral Industries – Gg
Priestley, Keith, +44 (0) 1223 337195 University of Cambridge – YgGt
Priewisch, Alexandra, 505-504-3675 alexandra.priewisch@fresnocitycollege.edu, Fresno City College – Gg
Primeau, Francois, 949-824-9435 fprimeau@uci.edu, University of California, Irvine – Op
Pringle, James M., (603) 862-5000 jpringle@cisunix.unh.edu, University of New Hampshire – Op
Pringle, Jamie K., (+44) 01782 733163 j.k.pringle@keele.ac.uk, Keele University – YuGfa
Pringle, Patrick T., 360-623-8584 ppringle@centralia.edu, Centralia College – ZgGvZr
Prinn, Ronald G., 617-253-2452 rprinn@mit.edu, Massachusetts Institute of Technology – As
Prins, Maarten A., +31 20 59 83635 m.a.prins@vu.nl, Vrije Universiteit Amsterdam – GsOg
Prior, David J., +64 3 479-5279 david.prior@otago.ac.nz, University of Otago – Gct
Prior, William L., 501 683-0117 bill.prior@arkansas.gov, Arkansas Geological Survey – GggEc
Pristas, Ronald, (609) 984-6587 ron.pristas@dep.state.nj.us, New Jersey Geological and Water Survey – Zi
Pritchard, Chad, 509-359-7026 cpritchard@ewu.edu, Eastern Washington University – Gct
Pritchard, Matthew E., (607) 255-4870 mp337@cornell.edu, Cornell University – YdGvl
Pritchard, Matthew E., mp337@cornell.edu, Cornell University – YdGvl
Pritchett, Brittany, 405-325-7331 brittanyp@ou.edu, University of Oklahoma – Go
Privette, David, (704)330-6750 David.Privette@cpcc.edu, Central Piedmont Community College – Zy
Prohiæ, Esad, +38514605963 eprohic@geol.pmf.hr, University of Zagreb – CglGe
Proudhon, Benoit, +33(0)3 44068996 benoit.proudhon@lasalle-beauvais.fr, Institut Polytechnique LaSalle Beauvais (ex-IGAL) – GgcGt
Provin, Tony L., (979) 862-4955 t-provin@tamu.edu, Texas A&M University – Sc
Pruell, Richard J., (401) 782-3000 University of Rhode Island – Co
Pruss, Sara B., (413) 585-3948 spruss@smith.edu, Smith College – Pg
Pryor, Sara C., 607-255-3376 sp2279@cornell.edu, Cornell University – AsZn
Prytulak, Julie, +44 20 759 46474 j.prytulak@imperial.ac.uk, Imperial College – Gg
Prytulak, Julie, julie.prytulak@durham.ac.uk, Durham University – Cg
Ptak, Tom, 208-885-6238 tptak@uidaho.edu, University of Idaho – Zu
Pu, Zhaoxia, 801-585-3864 zhaoxia.pu@utah.edu, University of Utah – As
Puchalski, Stephaney, spuchalski@highline.edu, Highline College – PgGg
Puchtel, Igor, (301) 405-4054 ipuchtel@umd.edu, University of Maryland – CcgGi
Puckett, Andrew, 706-507-8098 puckett_andrew@columbusstate.edu, Columbus State University – Xa
Puckett, Mark, 601-266-4729 Mark.Puckett@usm.edu, University of Southern Mississippi – PmiGg
Puckette, James, 405-744-6358 jim.puckette@okstate.edu, Oklahoma State University – Go
Puelles, Pablo, pablo.puelles@ehu.eus, University of the Basque Country UPV/EHU – Gct
Puente, Carlos E., (916) 752-0689 University of California, Davis – Hg
Pufahl, Peir K., 902-585-1858 peir.pufahl@acadiau.ca, Acadia University – Gs
Puffer, John H., (973) 353-5100 jpuffer@andromeda.rutgers.edu, Rutgers, The State University of New Jersey, Newark – GivGe
Pujana, Ignacio, (972) 883-2461 pujana@utdallas.edu, University of Texas, Dallas – Pm
Pujol, Jose, (901) 678-4827 jpujol@memphis.edu, University of Memphis – Ye
Puleo, Jack, 302-831-2440 jpuleo@udel.edu, University of Delaware – Hw
Pulkkinen, Tuija, 734-615-3583 tuija@umich.edu, University of Michigan – Zn
Pulliam, Jay, (254) 710-2183 jay_pulliam@baylor.edu, Baylor University – YsgYe
Pulliam, Robert J., (512) 471-6156 jay@ig.utexas.edu, University of Texas, Austin – Ys

Pulliam, Robert J., (254) 710-2183 jay_pulliam@baylor.edu,
   Baylor University – YsgYe
Pumphrey, Hugh C., +44 (0) 131 650 6026 hugh.pumphrey@ed.ac.uk,
   Edinburgh University – AscYv
Pun, Aurora, (505) 277-5629 apun@unm.edu,
   University of New Mexico – Ca
Pundsack, Jonathan, 612-626-0505 pundsack@umn.edu,
   University of Minnesota, Twin Cities – Zn
Punyasena, Surangi W., (217) 244-8049 punyasena@life.illinois.edu,
   University of Illinois, Urbana-Champaign – PblPe
Purchase, Megan, +27 (0)51 401 7158 purchasemd@ufs.ac.za,
   University of the Free State – Gx
Purdie, Duncan A., +44 (0)23 80592263 duncan.purdie@noc.soton.ac.uk,
   University of Southampton – Ob
Purdom, William B., (541) 552-6494 purdom@sou.edu,
   Southern Oregon University – Gx
Purdy, G. Michael, (845) 365-8348 mpurdy@ldeo.columbia.edu,
   Columbia University – Yr
Purkis, Sam J., 305-421-4351 spurkis@rsmas.miami.edu,
   University of Miami – Gu
Purkiss, Robert, 325-486-6987 robert.purkiss@angelo.edu,
   Angelo State University – Gg
Purnell, Mark, +440116 252 3645 map2@le.ac.uk,
   Leicester University – Pg
Purtle, Jennifer M., (501) 575-7317 jms14@comp.uark.edu,
   University of Arkansas, Fayetteville – Zn
Pusede, Sally, (434) 924-4544 sep6a@virginia.edu,
   University of Virginia – As
Putirka, Keith D., (559) 278-4524 kputirka@csufresno.edu,
   California State University, Fresno – GviGp
Putkonen, Jaakko, (701) 777-3213 jaakko.putkonen@engr.und.edu,
   University of North Dakota – Gml
Putnam, Aaron E., 207-581-2186 aaron.putnam@maine.edu,
   University of Maine – Gll
Putnam, Peter E., (403) 215-7850 University of Calgary – Gs
Puziewicz, Jacek, jacek.puziewicz@uwr.edu.pl,
   University of Wroclaw – Gix
Pyle, Eric J., 540-568-7115 pyleej@jmu.edu,
   James Madison University – Ze
Pylypenko, Svitlana, (303) 384-2140 spylypen@mines.edu,
   Colorado School of Mines – Zmn
Pyrtle, Ashanti J., apyrtle@seas.marine.usf.edu,
   University of South Florida – Oc
Pysklywec, Russell N., (416) 978-4852 russ@geology.utoronto.ca,
   University of Toronto – Yg

Q

Qi, Feng, (908) 737-3702 fqi@kean.edu,
   Kean University – ZiyZg
Qin, Rongjun, qin.324@osu.edu,
   Ohio State University – Zrf
Qiu, Bo, (808) 956-4098 bo@soest.hawaii.edu,
   University of Hawai'i, Manoa – Op
Qiu, Xiaomin, (417) 836-3219 qiu@missouristate.edu,
   Missouri State University – Zi
Quade, Deborah J., -319-337-9785 deborah.quade@dnr.iowa.gov,
   Iowa Dept of Natural Resources – GleGm
Quade, Jay, 520-626-3223 quadej@email.arizona.edu,
   University of Arizona – Sc
Qualls, Robert G., (775) 327-5014 qualls@unr.edu,
   University of Nevada, Reno – SoHsSf
Quan, Tracy, 405-744-6358 tracy.quan@okstate.edu,
   Oklahoma State University – Og
Quattro, Joseph, Quattro@biol.sc.edu,
   University of South Carolina – Ob
Quay, Paul D., (206) 545-8061 pdquay@u.washington.edu,
   University of Washington – Oc
Quick, Thomas J., (330) 972-6935 tquick@uakron.edu,
   University of Akron – GgCaYe
Quinn, Claire, +44(0) 113 34 38700 c.h.quinn@leeds.ac.uk,
   University of Leeds – Ge
Quinn, Courtney, (864)294-3655 courtney.quinn@furman.edu,
   Furman University – Zn
Quinn, Heather A., (410) 554-5522 heather.quinn@maryland.gov,
   Maryland Department of Natural Resources – GgHg
Quinn, James G., (401) 874-6219 jgquinn@gso.uri.edu,
   University of Rhode Island – Oc
Quinn, John, (630) 252-5357 quinnj@anl.gov,
   Argonne National Laboratory – Hw
Quinn, Paul, +44 (0) 191 208 5773 p.f.quinn@ncl.ac.uk,
   University of Newcastle Upon Tyne – Hg
Quinn, Terry, 512-471-0377 quinn@ig.utexas.edu,
   University of Texas, Austin – Pe
Quintero, Sylvia, (520) 621-6025 squinter@email.arizona.edu,
   University of Arizona – Zn
Quinton, Page C., (315) 267-2815 quintopc@potsdam.edu,
   SUNY Potsdam – PcCsPm

R

Rățoi, Bogdan G., 0040742955966 bog21rat@gmail.com,
   Alexandru Ioan Cuza – PvGr
Raab, James, 614 265 6747 jim.raab@dnr.state.oh.us,
   Ohio Dept of Natural Resources – Hw
Rabalais, Nancy N., (225) 851-2800 Louisiana State University – Ob
Raber, George, 601-266-5807 george.raber@usm.edu,
   University of Southern Mississippi – Zi
Radakovich, Amy, (612) 626-4434 rada0042@d.umn.edu,
   University of Minnesota – Gp
Radcliffe, David E., (706) 542-0897 University of Georgia – Sp
Radcliffe, Dennis, 516-463-6545 dennis.radcliffe@hofstra.edu,
   Hofstra University – Gzx
Radebaugh, Jani, 801 422-9127 jani.radebaugh@byu.edu,
   Brigham Young University – XgGvm
Rademacher, Laura K., (209) 946-7351 lrademacher@pacific.edu,
   University of the Pacific – ClHySo
Radic, Valentina, 604-827-1446 vradic@eos.ubc.ca,
   University of British Columbia – GlAsZg
Radil, Steven, 208-885-5058 sradil@uidaho.edu,
   University of Idaho – Zn
Radke, Lawrence F., radke@ucar.edu,
   University of Washington – As
Radko, Timour, (831) 656-3318 tradko@nps.edu,
   Naval Postgraduate School – Op
Rae, James, +44 01334 464948 jwbr@st-andrews.ac.uk,
   University of St. Andrews – Cg
Raef, Abdelmoneam E., (785) 532-2240 abraef@ksu.edu,
   Kansas State University – YesZr
Raeside, Robert P., (902) 585-1323 rob.raeside@acadiau.ca,
   Acadia University – GptGc
Rafini, Silvain, 514-987-3000 poste 2369 silvain.rafini@gmail.com,
   Universite du Quebec a Chicoutimi – GcHwEg
Ragland, Paul C., (850) 644-5018 ragland@magnet.fsu.edu,
   Florida State University – Ca
Ragotzkie, Robert A., ragotzkie@wisc.edu,
   University of Wisconsin, Madison – Ob
Rahaman, M. A., 234-800-337-7441 mrahaman@oauife.edu.ng,
   Obafemi Awolowo University – GpcGi
Rahl, Jeffrey, 540-458-8101 rahlj@wlu.edu,
   Washington & Lee University – Gt
Rahman, Abdullah F., 956 761-2644 abdullah.rahman@utrgv.edu,
   University of Texas, Rio Grande Valley – Zr
Rahn, Kenneth A., (401) 874-6713 krahn@uri.edu,
   University of Rhode Island – Oc
Rahn, Perry H., (605) 394-2527 Perry.Rahn@sdsmt.edu,
   South Dakota School of Mines & Technology – Ng
Rai, Chandra S., (405) 325-6866 crai@ou.edu,
   University of Oklahoma – Np
Railsback, L. Bruce, (706) 542-3453 rlsbk@gly.uga.edu,
   University of Georgia – PcClGd
Raine, Robin, +353 (0)91 492271 robin.raine@nuigalway.ie,
   National University of Ireland Galway – Yg
Rains, Daniel S., 479-601-7521 daniel.rains@arkansas.gov,
   Arkansas Geological Survey – GgZi
Rains, Mark C., 974-0323 University of South Florida, Tampa – Hw
Raj, John K., 03-79674225 jkraj@um.edu.my,
   University of Malaya – Ng
Raji, W. O., lanreraji24@unilorin.edu.ng,
   University of Ilorin – Ye
Rakovan, John F., (513) 529-3245 rakovajf@miamioh.edu,
   Miami University – GzCl
Rakovan, Monica, rakovamt@miamioh.edu,
   Miami University – Hg
Rallis, Donald N., (540) 654-1492 drallis@mwcgw.mwc.edu,
   Mary Washington College – Zn

Ramage, Joan, (610) 758-6410 jmr204@lehigh.edu, Lehigh University – ZrGlm

Ramagwede, Fhatuwani L., 0027128411911 info@geoscience.org.za, Geological Survey of South Africa – EgGzEg

Raman, Sethu S., (919) 515-7144 sethu_raman@ncsu.edu, North Carolina State University – As

Ramanathan, V., (858) 534-0219 vramanathan@ucsd.edu, University of California, San Diego – As

Rambaud, Fabienne M., frambaud@austincc.edu, Austin Community College District – Eg

Ramelli, Alan R., (775) 784-4151 ramelli@unr.edu, University of Nevada – Ng

Ramirez, Abelardo L., (925) 422-6919 ramirez3@llnl.gov, Lawrence Livermore National Laboratory – Ng

Ramirez, Pedro C., (323) 343-2417 pramire@calstatela.edu, California State University, Los Angeles – Gs

Ramirez, Wilson R., (787) 265-3845 wilson.ramirez1@upr.edu, University of Puerto Rico – GudOu

Ramos, Frank C., 575-646-2511 framos@nmsu.edu, New Mexico State University, Las Cruces – CgGg

Rampedi, Isaac T., +27115592429 isaacr@uj.ac.za, University of Johannesburg – PgSdHg

Ramsey, Kelvin W., (302) 831-3586 kwramsey@udel.edu, University of Delaware – As

Ramsey, Kelvin W., 302-831-2833 kwramsey@udel.edu, University of Delaware – Gs

Ramsey, Michael S., (412) 624-8772 ramsey@ivis.eps.pitt.edu, University of Pittsburgh – Zr

Ramspott, Matthew E., 301/687-4412 meramspott@frostburg.edu, Frostburg State University – Zr

Ranalli, Giorgio, granalli@ccs.carleton.ca, Carleton University – Yg

Rancan, Helen L., 609-984-6587 helen.rancan@dep.state.nj.us, New Jersey Geological and Water Survey – Hs

Randall, David A., randall@atmos.colostate.edu, Colorado State University – As

Randall, George, (505) 667-9483 Los Alamos National Laboratory – Yg

Randall, Gyles W., (507) 835-3620 University of Minnesota, Twin Cities – So

Randazzo, Anthony F., (352) 392-2634 afrgeohazards@bellsouth.net, University of Florida – Gd

Randlett, Victoria, (775) 327-5078 randlett@unr.edu, University of Nevada, Reno – Zu

Ranhofer, Melissa, (864)294-3647 melissa.ranhofer@furman.edu, Furman University – Gg

Rankey, Gene, (785) 864-6028 grankey@ku.edu, University of Kansas – GsrGu

Rankin, Robert, (780) 492-5082 rankin@phys.ualberta.ca, University of Alberta – Xy

Ransom, Michel D., (785) 532-7203 mdransom@ksu.edu, Kansas State University – SdcGm

Ranson, William A., (864) 294-3364 bill.ranson@furman.edu, Furman University – GxzGp

Ranville, James, (303) 273-3004 jranvill@mines.edu, Colorado School of Mines – Ca

Raol, Marcie, (304)367-8379 Marcie.Rice@fairmontstate.edu, Fairmont State University – Ze

Rapley, Chris, +44 020 310 86320 christopher.rapley@ucl.ac.uk, University College London – ZgrGl

Rapp, Anita, arapp@tamu.edu, Texas A&M University – Zr

Rappe, Michael S., (808) 236-7464 rappe@hawaii.edu, University of Hawai'i, Manoa – Og

Rappenglueck, Bernhard, 713-743-2469 brappenglueck@uh.edu, University of Houston – As

Rasbury, E. Troy, 631-632-1488 troy.rasbury@sunysb.edu, Stony Brook University – Cc

Rashed, Mohamed A., 002-03-3921595 Rashedmohamed@yahoo.com, Alexandria University – GgSo

Rasmussen, Craig, (520) 621-7223 crasmuss@cals.arizona.edu, University of Arizona – GmCl

Rasmussen, Kenneth, (703) 323-2139 krasmussen@nvcc.edu, Northern Virginia Community College - Annandale – Gus

Rasmussen, Kristen L., kristenr@rams.colostate.edu, Colorado State University – As

Rasmussen, Pat E., (613) 868-8609 pat.rasmussen@canada.gc.ca, University of Ottawa – GeAsGb

Rasmussen, Steen, (505) 665-0052 steen@lanl.gov, Los Alamos National Laboratory – Yg

Rasmussen, Tab, (314) 935-4844 dtrasmus@wustl.edu, Washington University in St. Louis – Zn

Ratajeski, Kent, (859) 257-4444 kent.ratajeski@email.uky.edu, University of Kentucky – Gi

Ratchford, M. E., (501) 683-0118 ed.ratchford@arkansas.gov, Arkansas Geological Survey – GocEo

Ratchford, Michael , 208-885-7991 eratchford@uidaho.edu, University of Idaho – GcoEg

Rath, Carolyn, (657) 278-7096 crath@fullerton.edu, California State University, Fullerton – Gg

Rathburn, Anthony, 812-237-2269 trathburn@gmail.com, Indiana State University – PgOu

Rathburn, Anthony E., (661) 654-3281 arathburn@csub.edu, California State University, Bakersfield – PmOoCb

Rathburn, Sara L., (970) 491-6956 sara.rathburn@colostate.edu, Colorado State University – Ggm

Raub, Timothy D., +44 01334 464012 timraub@st-andrews.ac.uk, University of St. Andrews – YmGgPc

Rauber, Robert M., (217) 333-2835 r-rauber@illinois.edu, University of Illinois, Urbana-Champaign – As

Rauch, Henry W., (304) 293-2187 hrauch@wvu.edu, West Virginia University – HwGe

Rauch, Marta, marta.rauch@uwr.edu.pl, University of Wroclaw – Gc

Raudsepp, Mati, (604) 822-6396 mraudsepp@eoas.ubc.ca, University of British Columbia – GzeZm

Ravat, Dhananjay, (859) 257-4726 dhananjay.ravat@uky.edu, University of Kentucky – YgmYv

Ravela, Sai, 617-253-0997 ravela@mit.edu, Massachusetts Institute of Technology – Zn

Ravelo, Ana C., 831-459-3722 acr@pmc.ucsc.edu, University of California, Santa Cruz – OcGe

Raven, Morgan, raven@geol.ucsb.edu, University of California, Santa Barbara – Cl

Ravizza, Gregory, 808-956-2916 gravizza@soest.hawaii.edu, University of Hawai'i, Manoa – Cm

Rawling, Geoffrey, (505) 366-2535 geoff@gis.nmt.edu, New Mexico Institute of Mining & Technology – Gg

Rawling, J. E., (608) 263-6839 elmo.rawling@wgnhs.uwex.edu, University of Wisconsin, Extension – Gl

Rawlins, Michael, 413 545-0659 rawlins@geo.umass.edi, University of Massachusetts, Amherst – As

Rawlinson, Nick, nr441@cam.ac.uk, University of Cambridge – Ygg

Ray, G. Carleton, (804) 924-0551 cr@virginia.edu, University of Virginia – Ob

Ray, Pallav, 321-674-7191 pray@fit.edu, Florida Institute of Technology – AstZo

Ray, Waverly, 6196447454 x3018 Waverly.Ray@gcccd.edu, Cuyamaca College – Gg

Rayburn, John A., 845-257-3767 rayburnj@newpaltz.edu, SUNY, New Paltz – Gml

Raymond, Anne, (979) 845-0644 raymond@geo.tamu.edu, Texas A&M University – Pb

Raymond, Carol A., (818) 354-8690 Jet Propulsion Laboratory – Yr

Raymond, Charles F., 206-685-9697 cfr2@uw.edu, University of Washington – Gl

Rayne, Todd W., (315) 859-4698 trayne@hamilton.edu, Hamilton College – Hy

Rea, David K., (734) 936-0521 davidrea@umich.edu, University of Michigan – Gu

Read, Adam S., (505) 366-2533 adamread@gis.nmt.edu, New Mexico Institute of Mining & Technology – Gg

Read, J. Fred, (540) 231-5124 jread@vt.edu, Virginia Polytechnic Institute & State University – Gs

Reading, Anya, 61 3 6226 2477 anya.reading@utas.edu.au, University of Tasmania – Ysg

Reagan, Mark K., (319) 335-1802 mark-reagan@uiowa.edu, University of Iowa – Gi

Reams, Max W., (815) 939-5394 mreams@olivet.edu, Olivet Nazarene University – GsmPg

Reaven, Sheldon, (631) 632-8765 sheldon.reaven@stonybrook.edu, SUNY, Stony Brook – Zn

Reavy, John, +353 21 4904574 j.reavy@ucc.ie, University College Cork – GicGt

Reay, David, +44 (0) 131 650 7723 david.reay@ed.ac.uk, Edinburgh University – As
Reay, William G., 804-684-7119 wreay@vims.edu, College of William & Mary – Hg
Reber, Jacqueline, (515) 294-7513 jreber@iastate.edu, Iowa State University of Science & Technology – Gct
Reboulet, Edward, 785-864-2173 reboulet@kgs.ku.edu, University of Kansas – Hy
Rech, Jason, 513-529-1935 rechja@MiamiOh.edu, Miami University – Gm
Rechcigl, John E., 813-633-4111 rechcigl@ufl.edu, University of Florida – So
Reche Estrada, Joan, ++935811781 joan.reche@uab.cat, Universitat Autonoma de Barcelona – GpCp
Rechtien, Richard D., (573) 341-4616 Missouri University of Science and Technology – Ye
Rector, James W., (510) 643-7820 University of California, Berkeley – Ye
Redalje, Donald G., (228) 688-1174 donald.redalje@usm.edu, University of Southern Mississippi – Ob
Redden, Jack A., (605) 394-2461 South Dakota School of Mines & Technology – Gp
Redden, Marcella, 205-247-3654 mmcintyre@gsa.state.al.us, Geological Survey of Alabama – Gg
Reddy, Christopher M., (508) 289-2316 creddy@whoi.edu, Woods Hole Oceanographic Institution – Co
Reddy, K R., (352) 295-3154 krr@ufl.edu, University of Florida – CbSbc
Redfern, Jonathan, +440161 275-3773 jonathan.redfern@manchester.ac.uk, University of Manchester – GsoGl
Redfern, Simon, +44 (0) 1223 333475 satr@cam.ac.uk, University of Cambridge – Gzy
Redmond, Brian T., (570) 696-2906 brian.redmond@wilkes.edu, Wilkes University – GsHwZg
Ree, Jin-Han, 82-2-3290-3175 reejh@korea.ac.kr, Korea University – Gct
Reece, Julia S., 979-458-2728 jreece@geos.tamu.edu, Texas A&M University – SpGso
Reed, Denise J., djreed@uno.edu, University of New Orleans – GmOn
Reed, Donald L., (408) 924-5036 donald.reed@sjsu.edu, San Jose State University – GuYg
Reed, Mark H., (541) 346-5587 mhreed@uoregon.edu, University of Oregon – CgEm
Reed, Robert M., (512) 471-0356 rob.reed@beg.utexas.edu, University of Texas at Austin, Jackson School of Geosciences – GxoGc
Reed, Wallace E., wer@virginia.edu, University of Virginia – Zr
Reeder, Richard J., (631) 632-8208 rjreeder@stonybrook.edu, Stony Brook University – ClGz
Reedy, Robert C., (512) 471-7244 bob.reedy@beg.utexas.edu, University of Texas at Austin, Jackson School of Geosciences – Hgw
Rees, Margaret N., (702) 895-3890 peg.rees@unlv.edu, University of Nevada, Las Vegas – Gs
Reese, Andy, 601-266-4729 andy.reese@usm.edu, University of Southern Mississippi – Zy
Reese, Brandi, 361-825-3022 Brandi.Reese@tamucc.edu, Texas A&M University, Corpus Christi – Cb
Reese, Joseph F., (814) 732-2529 jreese@edinboro.edu, Edinboro University of Pennsylvania – GctZe
Reese, Stuart O., 717.702.2028 sreese@pa.gov, Pennsylvania Bureau of Topographic & Geologic Survey – Hw
Reesman, Arthur L., 615-322-2976 Vanderbilt University – Gg
Reeve, Andrew S., (207) 581-2353 asreeve@maine.edu, University of Maine – Hw
Reeves, Claire, +44 (0)1603 59 3625 c.reeves@uea.ac.uk, University of East Anglia – Cg
Reeves, Donald Matt, (907) 786-1372 dmreeves2@uaa.alaska.edu, University of Alaska, Anchorage – GeHgZn
Reeves, Donald Matthew M., (269) 387-5493 matt.reeves@wmich.edu, Western Michigan University – HwqNr
Refenes, James L., (734) 995-7594 james.refenes@cuaa.edu, Concordia University – Ze
Refsnider, Kurt, (928) 350-2256 kurt.refsnider@prescott.edu, Prescott College – GmlPe
Regalla, Christine, cregalla@bu.edu, Boston University – GtcGm
Reganold, John P., (509) 335-8856 reganold@wsu.edu, Washington State University – Soo
Regehr, David, (785) 532-6101 dregehr@ksu.edu, Kansas State University – So
Rehakova, Daniela, +421260296700 rehakova@fns.uniba.sk, Comenius University in Bratislava – Pms
Rehkamper, Mark, +44 20 759 46391 markrehk@imperial.ac.uk, Imperial College – CsaCc
Rehm, George W., (612) 625-6210 grehm@soils.umn.edu, University of Minnesota, Twin Cities – Sc
Reichard, James S., (912) 478-1153 JReich@GeorgiaSouthern.edu, Georgia Southern University – Hw
Reichenbacher, Bettina, 089/2180 6603 b.reichenbacher@lrz.uni-muenchen.de, Ludwig-Maximilians-Universitaet Muenchen – PgvPe
Reichler, Thomas, 801-585-0040 thomas.reichler@utah.edu, University of Utah – As
Reid, Arch M., (919) 386-1711 archreid@gmail.com, University of Houston – GizXm
Reid, Brian, +44 (0)1603 59 2357 b.reid@uea.ac.uk, University of East Anglia – GeCg
Reid, Catherine, 3667001 Ext 7764 catherine.reid@canterbury.ac.nz, University of Canterbury – Pg
Reid, Mary R., (928) 523-7200 mary.reid@nau.edu, Northern Arizona University – CucGi
Reid, Pam, 305 421-4606 preid@rsmas.miami.edu, University of Miami – Gs
Reid, Steven K., 606-783-5293 s.reid@morehead-st.edu, Morehead State University – GsoGe
Reidel, Stephen P., (509) 376-9932 sreidel@wsu.edu, Washington State University – GivGt
Reidenbach, Matthew A., (434) 243-4937 mar5jj@virginia.edu, University of Virginia – Hg
Reif, Samantha, 217-786-2764 Samantha.Reif@llcc.edu, Lincoln Land Community College – Gg
Reijmer, John, +966138602626 reijmer@kfupm.edu.sa, King Fahd University of Petroleum and Minerals – Gsg
Reimer, Andreas, +49 (0)551 392164 areimer@gwdg.de, Georg-August University of Goettingen – OcCmm
Reimers, Clare, 541-737-0220 creimers@coas.oregonstate.edu, Oregon State University – Oc
Reinen, Linda A., (909) 621-8672 lreinen@pomona.edu, Pomona College – Gct
Reiners, Peter, 520-626-2236 reiners@email.arizona.edu, University of Arizona – Cg
Reinfelder, Ying Fan, 848-445-2044 yingfan@eps.rutgers.edu, Rutgers, The State University of New Jersey – HgqGe
Reinhardt, Edward G., (905) 525-9140 (Ext. 27594) ereinhar@mcmaster.ca, McMaster University – PmGam
Reintha, William A., wreintha@ashland.edu, Ashland University – GzoGg
Reisner, Jon M., (505) 665-1889 reisner@lanl.gov, Los Alamos National Laboratory – As
Reiss, Nathan, (212) 346-1502 Pace University, New York Campus – As
Reiten, Jon C., 406-657-2630 jreiten@mtech.edu, Montana Tech of The University of Montana – Hw
Reiter, Marshall A., (575) 835-5306 mreiter@nmt.edu, New Mexico Institute of Mining and Technology – Yh
Reitner, Joachim, +49 (0)551 397950 jreitne@gwdg.de, Georg-August University of Goettingen – PggPg
Reitz, Elizabeth J., (706) 542-1464 ereitz@arches.uga.edu, University of Georgia – Ga
Remacha Grau, Eduard, ++935811603 eduard.remacha@uab.cat, Universitat Autonoma de Barcelona – Gso
Remenda, Victoria H., (613) 533-6594 remendav@queensu.ca, Queen's University – Hw
Rempel, Alan W., (541) 346-6316 rempel@uoregon.edu, University of Oregon – YgNrGl
Remson, Irwin, (650) 723-9191 Stanford University – Hw
Renard, Robert J., 831-375-2354 bobandotty@aol.com, Naval Postgraduate School – Am
Renaut, Robin W., (306) 966-5705 robin.renaut@usask.ca, University of Saskatchewan – Gsn
Reneau, Raymond B., (540) 231-9779 reneau@vt.edu, Virginia Polytechnic Institute & State University – Sc
Reneau, Steven L., (505) 665-3151 sreneau@lanl.gov, Los Alamos National Laboratory – Gm
Renfrew, Ian, +44 (0)1603 59 2557 i.renfrew@uea.ac.uk,

University of East Anglia – Am
Renfrew, Melanie, renfremp@lahc.edu,
  Los Angeles Harbor College – Zg
Renne, Paul, (510) 644-1350 prenne@bgc.org,
  University of California, Berkeley – Cc
Rennie, Colette, (902) 867-2299 crennie@stfx.ca,
  Saint Francis Xavier University – Gg
Rennie, Tim, (613) 258-8336 Ext. 61286 trennie@kemptvillec.uoguelph.ca,
  University of Guelph – Zn
Rennie, Tom, (775) 682-6088 tom@seismo.unr.edu,
  University of Nevada, Reno – Ys
Rensberger, John M., rensb@uw.edu,
  University of Washington – Pvg
Renshaw, Carl E., (603) 646-3365 carl.renshaw@dartmouth.edu,
  Dartmouth College – HgGc
Renton, John J., (304) 293-5603 jrenton@wvu.edu,
  West Virginia University – Ec
Repeta, Daniel J., (508) 289-2635 drepeta@whoi.edu,
  Woods Hole Oceanographic Institution – Co
Repka, James, (949) 582-4694 jrepka@saddleback.edu,
  Saddleback Community College – GgmZe
Reshef, Moshe, 972-36406880 mosher@post.tau.ac.il,
  Tel Aviv University – Yeg
Resing, Joseph A., (206) 526-6184 resing@pmel.noaa.gov,
  University of Washington – Oc
Resnic, Victor S., 281-618-5800 rvictor@mcleodusa.net,
  Lonestar College - North Harris – Go
Resnichenko, Yuri, (598) 2525 1552 yresni@fcien.edu.uy,
  Universidad de la Republica Oriental del Uruguay (UDELAR) – ZirZn
Resor, Phillip G., 860 6853139 presor@wesleyan.edu,
  Wesleyan University – Gct
Resplandy, Laure, 609-258-4101 laurer@princeton.edu,
  Princeton University – Ob
Ressel, Mike, (775) 682-7844 mressel@unr.edu,
  University of Nevada – EgGiCc
Ressler, Daniel E., 570-372-4216 resslerd@susqu.edu,
  Susquehanna University – Sp
Restrepo, Jorge I., (561) 297-2795 restrepo@fau.edu,
  Florida Atlantic University – Hqw
Retallack, Gregory J., (541) 346-4558 gregr@uoregon.edu,
  University of Oregon – PbSoGa
Retelle, Michael J., (207) 786-6155 mretelle@bates.edu,
  Bates College – Gl
Rettig, Andrew, 937-229-2261 arettig1@udayton.edu,
  University of Dayton – Zy
Retzler, Andrew J., (612) 626-3895 aretzler@umn.edu,
  University of Minnesota – GrsPi
Reusch, David B., 575-835-5404 dreusch@ees.nmt.edu,
  New Mexico Institute of Mining and Technology – Zn
Reusch, Douglas N., 207-778-7463 info@reuschlaw.de,
  University of Maine - Farmington – GtCgOu
Reuss, Robert L., 617-627-3494 bert.reuss@tufts.edu,
  Tufts University – Gzi
Reuter, Gerhard, (780) 492-0358 gerhard.reuter@ualberta.ca,
  University of Alberta – Am
Revelle, Douglas O., (505) 667-1256 revelle@lanl.gov,
  Los Alamos National Laboratory – As
Revenaugh, Justin, 612-624-7553 justinr@umn.edu,
  University of Minnesota, Twin Cities – Ys
Revetta, Frank A., (315) 267-3441 revettfa@potsdam.edu,
  SUNY Potsdam – YgGtg
Rex, Arthur B., (828) 262-6911 rexab@appstate.edu,
  Appalachian State University – Zi
Rexius, James E., 734-462-4400 jrexius@schoolcraft.edu,
  Schoolcraft College – Gl
Reyes, Alfonso, reyeszca@cicese.mx,
  Centro de Investigación Científica y de Educación Superior de Ensenada – Ne
Reynolds, Barbara C., (828) 232-5048 kreynolds@unca.edu,
  University of North Carolina, Asheville – Zn
Reynolds, James H., (828) 884-8377 reynoljh@brevard.edu,
  Brevard College – GrvGe
Reynolds, Robert W., (541) 383-7557 Central Oregon Community College – Gv
Reynolds, Stephen J., (480) 965-9049 sreynolds@asu.edu,
  Arizona State University – GctZe
Rezaie-Boroon, Mohammad H., (323) 343-2406 mrezaie@calstatela.edu,
  California State University, Los Angeles – GeCg
Rhenberg, Elizabeth, (901) 678-2177 erhenbrg@memphis.edu,
  University of Memphis – Pg
Rheuban, Jennie, (508) 289-3782 jrheuban@whoi.edu,
  Woods Hole Oceanographic Institution – Oc
Rhines, Peter B., (206) 543-0593 rhines@atmos.washington.edu,
  University of Washington – Op
Rhoads, Bruce, brhoads@illinois.edu,
  University of Illinois, Urbana-Champaign – Gm
Rhode, David E., (775) 673-7310 dave@dri.edu,
  Desert Research Institute – Pe
Rhodes, Amy L., (413) 585-3947 arhodes@smith.edu,
  Smith College – ClGe
Rhodes, Dallas D., 912-601-1600 dallas.rhodes@humboldt.edu,
  Humboldt State University – GmZre
Rhodes, Edward J., (310) 825-3880 erhodes@epss.ucla.edu,
  University of California, Los Angeles – CcGma
Rhodes, Frank H. T., (607) 255-6233 Cornell University – Pi
Rhodes, J. Michael, (413) 545-2841 jmrhodes@geo.umass.edu,
  University of Massachusetts, Amherst – GviCa
Rial, Jose A., (919) 966-4553 jose_rial@unc.edu,
  University of North Carolina, Chapel Hill – Ys
Ribbe, Paul H., (540) 231-6880 ribbe@vt.edu,
  Virginia Polytechnic Institute & State University – Gz
Ribbons, Relena R., relena.r.ribbons@lawrence.edu,
  Lawrence University – SbGe
Rice, Benjamin, 847-491-8190 ben@earth.northwestern.edu,
  Northwestern University – Zn
Rice, Chuck W., (785) 532-7217 cwrice@ksu.edu,
  Kansas State University – Sb
Rice, James R., (617) 495-3445 rice@esag.harvard.edu,
  Harvard University – Ygu
Rice, Karen C., (434) 243-3429 kcr4y@virginia.edu,
  University of Virginia – HgCg
Rice, Keith W., 715-346-4454 krice@uwsp.edu,
  University of Wisconsin, Stevens Point – Zir
Rice, Melissa S., (360) 650-3592 melissa.rice@wwu.edu,
  Western Washington University – XgGsm
Rice, Murray, 940-565-3861 murray.rice@unt.edu,
  University of North Texas – Zn
Rice, Pamela J., (612) 625-1909 price@soils.umn.edu,
  University of Minnesota, Twin Cities – So
Rice, Thomas L., (937) 766-6140 trice@cedarville.edu,
  Cedarville University – GemEo
Rice-Snow, R. Scott, 765-285-8269 ricesnow@bsu.edu,
  Ball State University – Gm
Rich, Fredrick J., (912) 478-0849 frich@georgiasouthern.edu,
  Georgia Southern University – Pl
Richard, Benjamin H., benjamin.richard@wright.edu,
  Wright State University – Ye
Richard, Gigi A., 970-248-1689 grichard@coloradomesa.edu,
  Colorado Mesa University – Hsg
Richard, Robert, (310) 825-6663 rrichard@igpp.ucla.edu,
  University of California, Los Angeles – Yg
Richards, Bill, (208) 769-3477 bill.richards@nic.edu,
  North Idaho College – GgZyGz
Richards, James H., (530) 752-0170 jhrichards@ucdavis.edu,
  University of California, Davis – So
Richards, Jeremy P., (705) 675-1151 jrichards2@laurentian.ca,
  Laurentian University, Sudbury – EmGiCg
Richards, Kelvin J., 808-956-5399 rkelvin@hawaii.edu,
  University of Hawai'i, Manoa – Op
Richards, Laura, +44 0161 306-0361 laura.richards@manchester.ac.uk,
  University of Manchester – Ge
Richards, Mark A., Mark_Richards@berkeley.edu,
  University of California, Berkeley – Yg
Richards, Paul G., (845) 365-8389 richards@ldeo.columbia.edu,
  Columbia University – YsZn
Richards, Paul L., prichard@brockport.edu,
  SUNY, The College at Brockport – Hg
Richards-McClung, Bryony, 801-585-0599 bmcclung@egi.utah.edu,
  University of Utah – Go
Richardson, Carson A., carichardson@email.arizona.edu ,
  Arizona Geological Survey – Emg
Richardson, Eliza, (814) 863-2507 eliza@psu.edu,
  Pennsylvania State University, University Park – YsZe
Richardson, Justin, 413-545-4840 JBRichardson@umass.edu,

551

University of Massachusetts, Amherst – CbSc
Richardson, Mary J., (979) 845-7966 mrichardson@ocean.tamu.edu,
    Texas A&M University – Op
Richardson, Randall M., (520) 621-4950 rmr@email.arizona.edu,
    University of Arizona – Yg
Richardson, Steve, 021-650-2921 steve.richardson@uct.ac.za,
    University of Cape Town – Cg
Richardson, Tammi, richardson@biol.sc.edu,
    University of South Carolina – Ob
Richaud, Mathieu, (559) 278-4557 mathieu@csufresno.edu,
    California State University, Fresno – GusCs
Richey, Jeffrey E., (206) 543-7339 jrichey@u.washington.edu,
    University of Washington – CbHs
Richter, Carl, (337) 482-5353 richter@louisiana.edu,
    University of Louisiana at Lafayette – Ym
Richter, Daniel D., 919-613-8031 drichter@duke.edu,
    Duke University – So
Richter, David, David.Richter.26@nd.edu,
    University of Notre Dame – As
Richter, Frank M., (773) 702-8118 richter@geosci.uchicago.edu,
    University of Chicago – Gt
Richter, Suzanna L., (717) 358-5843 suzanna.richter@fandm.edu,
    Franklin and Marshall College – Ge
Ricka, Adam, +420 549 49 6605 ricka@sci.muni.cz,
    Masaryk University – Hgy
Rickaby, Ros, +44 (1865) 272034 rosalind.rickaby@earth.ox.ac.uk,
    University of Oxford – Gz
Rickard, D, rickard@cf.ac.uk,
    Cardiff University – Cg
Ricketts, Hugo, +44 0161 306-3911 h.ricketts@manchester.ac.uk,
    University of Manchester – As
Ricketts, Jason, (915) 747-5599 jricketts@utep.edu,
    University of Texas, El Paso – GctGi
Ridd, Merrill K., (801) 581-7939 merrill.ridd@geog.utah.edu,
    University of Utah – Zr
Ridenour, Gregory D., (931) 221-7454 ridenourg@apsu.edu,
    Austin Peay State University – HsOgZn
Ridge, John C., (617) 627-3494 jack.ridge@tufts.edu,
    Tufts University – GlnGg
Ridgway, Kenneth D., (765) 494-3269 ridge@purdue.edu,
    Purdue University – Gs
Ridgwell, Andy, (951) 827-3186 andy@seao2.org,
    University of California, Riverside – Zn
Riding, Robert, 865-974-2366 rriding@utk.edu,
    University of Tennessee, Knoxville – Gs
Ridky, Alice M., (207) 859-5800 amridky@colby.edu,
    Colby College – Zn
Ridley, John R., (970) 491-5943 jridley@colostate.edu,
    Colorado State University – Eg
Ridley, Moira K., (806) 834-0627 moira.ridley@ttu.edu,
    Texas Tech University – CqlZm
Riebe, Clifford S., 307-766-3965 criebe@uwyo.edu,
    University of Wyoming – ClGme
Riebesell, John, (313) 593-5132 jriebese@umich.edu,
    University of Michigan, Dearborn – Zu
Riedel, Oliver, 089/2180 4335 oliver.riedl@lrz.uni-muenchen.de,
    Ludwig-Maximilians-Universitaet Muenchen – Gz
Riediger, Cynthia L., (403) 220-8783 riediger@geo.ucalgary.ca,
    University of Calgary – Co
Riedinger, Natascha, 405-744-6358 natascha.riedinger@okstate.edu,
    Oklahoma State University – CgGu
Rieke, George H., (520) 621-2832 grieke@as.arizona.edu,
    University of Arizona – Zn
Riemer, Nicole, 217-244-2844 nriemer@illinois.edu,
    University of Illinois, Urbana-Champaign – As
Riemersma, Peter E., 616-331-3553 riemersp@gvsu.edu,
    Grand Valley State University – Hw
Rieppel, Olivier C., (312) 665-7630 orieppel@fieldmuseum.org,
    Field Museum of Natural History – Pv
Ries, Justin, 7815817370 x342 j.ries@neu.edu,
    Northeastern University – Cm
Riess, Carolyn M., (512) 468-1832 criess@austincc.edu,
    Austin Community College District – GoeGg
Riesselman, Christina R., +64 3 479-7505 christina.riesselman@otago.ac.nz,
    University of Otago – Ou
Rietbrock, Andreas, +44-151-794-5181 A.Rietbrock@liverpool.ac.uk,
    University of Liverpool – YsGv

Rietmeijer, Frans J., (505) 277-5733 fransjmr@unm.edu,
    University of New Mexico – Gp
Rigby, John, 61 7 3138 1638 j.rigby@qut.edu.au,
    Queensland University of Technology – Pb
Riggs, Eric, 9798453651 emriggs@geos.tamu.edu,
    Texas A&M University – Gy
Riggs, Nancy, (928) 523-9362 nancy.riggs@nau.edu,
    Northern Arizona University – Gv
Riggs, Stanley R., (252) 328-6015 riggss@ecu.edu,
    East Carolina University – GusGe
Rignot, Eric, (949) 824-3739 erignot@uci.edu,
    University of California, Irvine – GlOpZr
Rigo, Manuel, +390498279175 manuel.rigo@unipd.it,
    Università degli Studi di Padova – Gs
Rigsby, Catherine A., (252) 328-4297 rigsbyc@ecu.edu,
    East Carolina University – Gs
Riha, Susan, (607) 255-1729 sjr4@cornell.edu,
    Cornell University – Sf
Riker-Coleman, Kristin E., 715-394-8410 krikerco@uwsuper.edu,
    University of Wisconsin, Superior – Gg
Riley, James, (520) 626-6681 jjriley@ag.arizona.edu,
    University of Arizona – Zn
Riley, Rhonda, riley@suu.edu,
    Southern Utah University – Zn
Rimmer, Susan M., (618) 453-7369 srimmer@siu.edu,
    Southern Illinois University Carbondale – EcCgGo
Rimstidt, J. Donald, (540) 231-6894 jdr02@vt.edu,
    Virginia Polytechnic Institute & State University – Clg
Rinae, Makhadi, +27 (0)51 401 9008 makhadi@ufs.ac.za,
    University of the Free State – Gge
Rindsberg, Andrew K., 205 652 3416 arindsberg@uwa.edu,
    University of West Alabama – PeGePi
Riney, Kaylin, 803.896.7931 rineyk@dnr.sc.gov,
    South Carolina Dept of Natural Resources – Zi
Rink, W. J., 850 229-1443 rinkwj@mcmaster.ca,
    McMaster University – CcGam
Rinterknecht, Vincent, +33 (0)1 45 07 55 81 vincent.rinterknecht@lgp.cnrs.fr,
    University of St. Andrews – CcGma
Riordan, Allen J., (919) 515-7973 al_riordan@ncsu.edu,
    North Carolina State University – As
Riordan, Jean, (907) 696-0079 jean.riordan@alaska.gov,
    Alaska Division of Geological & Geophysical Surveys – Gg
Riordan, Lindsay, (617) 627-3494 lindsay.riordan@tufts.edu,
    Tufts University – Zn
Rios-Sanchez, Miriam, 218-755-2563 mriossanchez@bemidjistate.edu,
    Bemidji State University – HwZrGg
Rioux, Matt, rioux@geol.ucsb.edu,
    University of California, Santa Barbara – Cc
Ripley, Edward M., (812) 855-1196 ripley@indiana.edu,
    Indiana University, Bloomington – Em
Rippy, Megan, 703-361-5606 mrippy@vt.edu,
    Virginia Polytechnic Institute & State University – Rw
Riser, Stephen C., (206) 543-1187 riser@uw.edu,
    University of Washington – Op
Risk, Dave A., drisk@stfx.ca,
    Saint Francis Xavier University – SoZr
Ritchie, Alexander W., (843) 953-5591 ritchiea@cofc.edu,
    College of Charleston – Gc
Ritchie, Harold C., (902) 494-5192 hritchie@phys.ocean.dal.ca,
    Dalhousie University – Am
Ritsche, Michael, 630-252-1554 mtritsche@anl.gov,
    Argonne National Laboratory – As
Ritsema, Jeroen, (734) 615-6405 jritsema@umich.edu,
    University of Michigan – Ysg
Rittenour, Tammy M., (435) 213-5756 tammy.rittenour@usu.edu,
    Utah State University – GmCcGa
Ritter, Charles J., (937) 229-2953 University of Dayton – Ct
Ritter, Joachim R., +49-721-60844539 joachim.ritter@kit.edu,
    Karlsruhe Institute of Technology – YsGtv
Ritter, John B., (937) 327-7332 jritter@wittenberg.edu,
    Wittenberg University – Gm
Ritter, Leonard, (519) 824-4120 Ext.52980 lritter@uoguelph.ca,
    University of Guelph – Zn
Ritter, Michael E., 715-346-4449 mritter@uwsp.edu,
    University of Wisconsin, Stevens Point – Am
Ritter, Paul, (309) 268-8640 paul.ritter@heartland.edu,
    Heartland Community College – Zg

Ritter, Scott M., 801-4224239 scott_ritter@byu.edu,
   Brigham Young University – Ps
Ritterbush, Linda A., ritterbu@clunet.edu,
   California Lutheran University – Pi
Ritts, Malinda, (208) 885-1179 mritts@uidaho.edu,
   University of Idaho – GsCg
Ritz, Richard, (402) 280-2461 richard.ritz@afwa.af.mil,
   Creighton University – Am
Ritzi, Jr., Robert W., 937 775-3455 robert.ritzi@wright.edu,
   Wright State University – Hw
Rius, Marc, +44 (0)23 8059 3275 m.rius@soton.ac.uk,
   University of Southampton – Znn
Rivard, Benoit, (780) 492-0345 benoit.rivard@ualberta.ca,
   University of Alberta – Zr
Rivera, Mark, 907-786-1235 marivera@uaa.alaska.edu,
   University of Alaska, Anchorage – Gg
Rivers, Toby C. J. S., (709) 737-8392 trivers@sparky2.esd.mun.ca,
   Memorial University of Newfoundland – Gp
Rizeli, Mustafa Eren, 00904242370000-5961 merizeli@firat.edu.tr,
   Firat University – Gxi
Rizoulis, Athanasios, +44 0161 275-0311 A.Rizoulis@manchester.ac.uk,
   University of Manchester – Ge
Roach, Michael, 61 3 6226 2474 University of Tasmania – Yg
Roadcap, George S., (217) 333-7951 roadcap@illinois.edu,
   University of Illinois – Hw
Robarge, Wayne P., (919) 515-1454 North Carolina State University – Sc
Robas, Sheryl A., (609) 258-6144 srobas@princeton.edu,
   Princeton University – Zn
Robbins, Debra C., (828) 251-6441 drobbins@unca.edu,
   University of North Carolina, Asheville – Zn
Robbins, Gary A., (860) 486-2448 gary.robbins@uconn.edu,
   University of Connecticut – Hw
Roberson, Randal P., (931) 221-1004 robersonr@apsu.edu,
   Austin Peay State University – Gg
Robert, Genevieve, (207) 786-6105 grobert@bates.edu,
   Bates College – CpGiz
Robert, Sanborn, (262) 335-5263 robert.sanborn@uwc.edu,
   University of Wisconsin Colleges – GgZyn
Roberts, A. Lynn, (410) 516-4387 lroberts@jhu.edu,
   Johns Hopkins University – Zn
Roberts, Charles E., (561) 297-3254 croberts@fau.edu,
   Florida Atlantic University – Zri
Roberts, Eric M., +61747816947 eric.roberts@jcu.edu.au,
   James Cook University – PsPvCc
Roberts, Frank, (504) 388-2964 Montgomery County Community College
   – Gp
Roberts, Gerald, +44 020 3073 8033 gerald.roberts@ucl.ac.uk,
   Birkbeck College – Gct
Roberts, Harry H., (225) 388-2964 harry@antares.esl.lsu.edu,
   Louisiana State University – Gs
Roberts, Jennifer A., (785) 864-4997 jenrob@ku.edu,
   University of Kansas – Cbl
Roberts, Mark L., (508) 289-3654 mroberts@whoi.edu,
   Woods Hole Oceanographic Institution – Yg
Roberts, Paul H., (310) 206-2707 roberts@math.ucla.edu,
   University of California, Los Angeles – Ym
Roberts, Peter, (505) 667-1199 proberts@lanl.gov,
   Los Alamos National Laboratory – Ys
Roberts, Ray L., 817295-7392 rroberts@hillcollege.edu,
   Hill College – Ge
Roberts, Sarah K., (925) 423-4112 roberts28@llnl.gov,
   Lawrence Livermore National Laboratory – Cl
Roberts, Sheila M., (406) 683-7017 sheila.roberts@umwestern.edu,
   University of Montana Western – GeSo
Roberts, Stephen, +44 ()023 80593246 steve.roberts@noc.soton.ac.uk,
   University of Southampton – Cg
Roberts-Semple, Dawn, 718-262-2775 drobertssemple@york.cuny.edu,
   York College (CUNY) – Gem
Robertson, Alastair H., +44 (0) 131 650 8546 alastair.robertson@ed.ac.uk,
   Edinburgh University – GtgGs
Robertson, Andrew W., 845-680-4491 awr@iri.columbia.edu,
   Columbia University – As
Robertson, Charles E., (573) 341-4616 Missouri University of Science and
   Technology – Gc
Robertson, Daniel E., 585-292-2422 drobertson@monroecc.edu,
   Monroe Community College – Eg
Robertson, James M., (608) 263-7384 james.robertson@uwex.edu,
   University of Wisconsin, Extension – Eg
Robertson, Wendy, 989-774-7517 wendy.robertson@cmich.edu,
   Central Michigan University – HwCg
Robin, Michel R., 613-562-5800 Ext 6852 mrobin@uottawa.ca,
   University of Ottawa – Hw
Robin, Pierre-Yves F., 905 828-5419 University of Toronto – Gc
Robinson, Alexander, 713-743-2547 acrobinson@uh.edu,
   University of Houston – Gc
Robinson, Bruce A., (505) 667-1910 robinson@lanl.gov,
   Los Alamos National Laboratory – Zn
Robinson, Carol, +44 (0)1603 59 3174 carol.robinson@uea.ac.uk,
   University of East Anglia – Ob
Robinson, Clare, +44 0161 275-3296 A.Rizoulis@manchester.ac.uk,
   University of Manchester – Co
Robinson, Cordula, cordula@crsa.bu.edu,
   Boston University – Gm
Robinson, David, +44 (0)1908 653493 x 53493 david.robinson@open.ac.uk,
   The Open University – Zn
Robinson, Delores, dmr@ua.edu,
   University of Alabama – Gc
Robinson, Edward, (305) 348-3572 draper@fiu.edu,
   Florida International University – Pm
Robinson, Edwin S., (540) 231-6521 esrobinson@vt.edu,
   Virginia Polytechnic Institute & State University – Yg
Robinson, George W., grobinson@stlawu.edu,
   St. Lawrence University – Gz
Robinson, Judith, 973-353-1976 judy.robinson@rutgers.edu,
   Rutgers, The State University of New Jersey, Newark – Yg
Robinson, Kevin, (619) 594-1386 rockrobinson@gmail.com,
   San Diego State University – Gc
Robinson, Leonie, +44 0151 795 4387 Leonie.Robinson@liverpool.ac.uk,
   University of Liverpool – Ob
Robinson, Lori, (612) 626-7429 robin126@umn.edu,
   University of Minnesota – Zi
Robinson, Mark, (480) 727-9691 mark.s.robinson@asu.edu,
   Arizona State University – Zn
Robinson, Michael, (912) 598-3310 mike.robinson@skio.usg.edu,
   Georgia Southern University – ZiOn
Robinson, Michael A., marobinson3@sbcc.edu,
   Santa Barbara City College – ZyiAm
Robinson, Paul D., (618) 453-7373 robinson@geo.siu.edu,
   Southern Illinois University Carbondale – Gz
Robinson, Peter, (303) 492-5108 peter.robinson@colorado.edu,
   University of Colorado – Pv
Robinson, Peter, (413) 545-2286 probinson@geo.umass.edu,
   University of Massachusetts, Amherst – Gc
Robinson, R. Craig, (860) 832-2950 Central Connecticut State University
   – Xy
Robinson, Ruth, +44 01334 463996 rajr@st-andrews.ac.uk,
   University of St. Andrews – Gs
Robinson, Sarah, (719) 333-9287 sarah.robinson@usafa.edu,
   United States Air Force Academy – GgZi
Robinson, Steve, j.s.robinson@reading.ac.uk,
   University of Reading – Sf
Robinson, Stuart, +44 (1865) 272058 stuartr@earth.ox.ac.uk,
   University of Oxford – PeGr
Robinson, Walter, 919-515-7002 walter_robinson@ncsu.edu,
   North Carolina State University – As
Robinson, Walter A., (217) 333-2292 robinson@atmos.uiuc.edu,
   University of Illinois, Urbana-Champaign – As
Robinson, William, william.robinson@umb.edu,
   University of Massachusetts, Boston – Zn
Rocha, Guillermo, 7189515000 x2887 grocha@brooklyn.cuny.edu,
   Brooklyn College (CUNY) – GgeCg
Roche, Didier, +31 20 59 83077 didier.roche@vu.nl,
   Vrije Universiteit Amsterdam – AsZn
Roche, James E., (225) 388-2707 jroche@geol.lsu.edu,
   Louisiana State University – Gg
Roche, Steven L., 918-631-3307 sroche@utulsa.edu,
   The University of Tulsa – YesYg
Rocheford, MaryKathryn (Kat), 906-635-2140 mrocheford@lssu.edu,
   Lake Superior State University – GeZiGa
Rochester, Michael G., (709) 737-7565 mrochest@morgan.ucs.mun.ca,
   Memorial University of Newfoundland – Yg
Rochette, Scott M., (585) 395-2603 srochett@brockport.edu,
   SUNY, The College at Brockport – Am
Rocholl, Alexander, 089/2180 4293 rocholl@min.uni-muenchen.de,

Ludwig-Maximilians-Universitaet Muenchen – Gz
Rock, Jessie L., 701-231-7951 jessie.rock@ndsu.edu,
    North Dakota State University – ZgPcg
Rockaway, John D., (859) 572-5412 rockawayj@nku.edu,
    Northern Kentucky University – Ng
Rockwell, Thomas K., (619) 594-4441 trockwell@mail.sdsu.edu,
    San Diego State University – Gm
Rockwood, Anthony A., (303) 556-8399 Metropolitan State College of
    Denver – Am
Rodbell, Donald T., (518) 388-6034 rodbelld@union.edu,
    Union College – GlmGe
Roden, Eric E., (608) 260-0724 eroden@geology.wisc.edu,
    University of Wisconsin-Madison – PgSbCl
Roden, Gunnar I., (206) 543-5627 giroden@u.washington.edu,
    University of Washington – Op
Roden, Michael F., (706) 542-2416 mroden@uga.edu,
    University of Georgia – Gi
Rodgers, David W., (208) 282-3460 rodgdavi@isu.edu,
    Idaho State University – Gct
Rodgers, Jim, (307) 766-2286 Ext. 255 james.rodgers@wyo.gov,
    Wyoming State Geological Survey – Gg
Rodgers, John C., (662) 325-0732 jcr100@msstate.edu,
    Mississippi State University – Zy
Rodgers, N, RodgersN@cf.ac.uk,
    Cardiff University – Am
Rodgers, Nick, +44(0)29 208 79064 RodgersN@cardiff.ac.uk,
    University of Wales – Xm
Rodland, David L., (740) 826-8425 drodland@muskingum.edu,
    Muskingum University – PieGr
Rodolfo, Kelvin S., krodolfo@uic.edu,
    University of Illinois at Chicago – GueGs
Rodrigues, Cyril G., 519-253-3000 ext. 2499 cgr@uwindsor.ca,
    University of Windsor – Pm
Rodriguez, Lizzette A., (787) 265-3845 lizzette.rodriguez1@upr.edu,
    University of Puerto Rico – GviZr
Rodríguez, Marta, martarm@ucm.es,
    Univ Complutense de Madrid – Ggs
Rodriguez, Vanessa del S., (787) 722-2526 Puerto Rico Bureau of Geology
    – Zn
Rodriguez-Blanco, Juan Diego, +353 1 8961691 j.d.rodriguez-blanco@tcd.ie,
    Trinity College – GzClZm
Roe, Carol, 757-221-2440 crroex@wm.edu,
    College of William & Mary – Zn
Roe, Gerard H., (206) 543-4980 gerard@ess.washington.edu,
    University of Washington – As
Roe, Gerard H., 206-697-3298 groe@uw.edu,
    University of Washington – AsGml
Roecker, Steven W., (518) 276-6773 roecks@rpi.edu,
    Rensselaer Polytechnic Institute – Yg
Roegiers, Jean-Claude, (405) 255-5459 jroegiers@ou.edu,
    University of Oklahoma – Nr
Roelofse, Frederick, +27 (0)51 401 9001 roelofsef@ufs.ac.za,
    University of the Free State – Gip
Roemer, Elizabeth, (520) 621-2897 eroemer@pirlmail.lpl.arizona.edu,
    University of Arizona – Zn
Roemmele, Christopher, (610) 436-2108 croemmele@wcupa.edu,
    West Chester University – ZeGgZg
Roemmich, Dean H., (858) 534-2307 droemmich@ucsd.edu,
    University of California, San Diego – Op
Roering, Joshua J., (541) 346-5574 jroering@uoregon.edu,
    University of Oregon – Gm
Roeske, Sarah M., (530) 752-4933 smroeske@ucdavis.edu,
    University of California, Davis – GtcGp
Roesler, Collin, 207-725-3842 croesler@bowdoin.edu,
    Bowdoin College – OgpZr
Roethel, Frank J., (631) 632-8732 frank.roethel@stonybrook.edu,
    SUNY, Stony Brook – Oc
Rogers, Garry C., (250) 363-6450 University of Victoria – Ys
Rogers, Jefferson S., (731) 881-7442 jrogers@utm.edu,
    University of Tennessee, Martin – Zn
Rogers, Jeffrey C., (614) 292-0148 rogers.21@osu.edu,
    Ohio State University – As
Rogers, Joe D., 806-651-2570 West Texas A&M University – Ga
Rogers, Karyn L., (518) 276-2372 rogerk5@rpi.edu,
    Rensselaer Polytechnic Institute – PgCl
Rogers, Pamela Z., (505) 667-1765 Los Alamos National Laboratory – Cl
Rogers, Raymond R., (651) 696-6434 rogersk@macalester.edu,
    Macalester College – Gs
Rogers, Robert D., (209) 667-3466 rrogers1@csustan.edu,
    California State University, Stanislaus – GcmGt
Rogers, Steven L., (+44) 01782 733752 s.l.rogers@keele.ac.uk,
    Keele University – PgGs
Rogers, William J., 806-651-2581 West Texas A&M University – Zn
Rogerson, Robert J., (403) 329-5117 rogerson@uleth.ca,
    University of Lethbridge – Gm
Roggenthen, William M., (605) 394-2461 William.Roggenthen@sdsmt.edu,
    South Dakota School of Mines & Technology – NgYg
Rogova, Galina L., (716)645-3489 rogova@buffalo.edu,
    SUNY, Buffalo – Gq
Rogowski, Andrew S., (814) 863-8758 asr@psu.edu,
    Pennsylvania State University, University Park – Sp
Rohay, Alan C., (509) 376-6925 alan.rohay@pnl.gov,
    Pacific Northwest National Laboratory – Ys
Rohe, Randall, randall.rohe@uwc.edu,
    University of Wisconsin Colleges – ZyGgZn
Rohr, David M., (432) 837-8167 drohr@sulross.edu,
    Sul Ross State University – PiGd
Rohrssen, Megan, 989-774-2079 rohrs1m@cmich.edu,
    Central Michigan University – Co
Rohs, C. Renee, (660) 562-1201 rrohs@nwmissouri.edu,
    Northwest Missouri State University – Cg
Roinstad, Lori L., 605-677-6154 lori.roinstad@usd.edu,
    South Dakota Dept of Environment and Natural Resources – Zn
Rokop, Donald J., (505) 667-4299 Los Alamos National Laboratory – Cc
Rollins, Kyle, (573) 368-2171 kyle.rollins@dnr.mo.gov,
    Missouri Dept of Natural Resources – ZyGg
Rollinson, Hugh, +44 01332 591786 h.rollinson@derby.ac.uk,
    University of Derby – Cg
Rollinson, Paul A., paulrollinson@missouristate.edu,
    Missouri State University – Zn
Rolston, Dennis E., (916) 752-2113 University of California, Davis – So
Roman, Aubrecht, aubrecht@fns.uniba.sk,
    Comenius University in Bratislava – Gdr
Roman, Charles T., (401) 874-6885 croman@gso.uri.edu,
    University of Rhode Island – Ob
Roman, Diana C., 202-478-8834 droman@carnegiescience.edu,
    Carnegie Institution for Science – Gv
Roman, Eric W., (609) 984-6587 eric.roman@dep.state.nj.us,
    New Jersey Geological and Water Survey – Hw
Romanak, Katherine D., (512) 471-6136 katherine.romanak@beg.utexas.edu,
    University of Texas at Austin, Jackson School of Geosciences – CgScGe
Romanovsky, Vladimir, (907) 474-7459 ffver@uaf.edu,
    University of Alaska, Fairbanks – Yg
Romanowicz, Barbara A., 510-643-5690 barbara@seismo.berkeley.edu,
    University of California, Berkeley – Ys
Romanowicz, Edwin A., (518) 564-2152 romanoea@plattsburgh.edu,
    SUNY, Plattsburgh – HwGcYu
Romans, Brian W., 540-231-2234 romans@vt.edu,
    Virginia Polytechnic Institute & State University – GsrEo
Ronayne, Michael J., (970) 491-0666 Michael.Ronayne@colostate.edu,
    Colorado State University – Hwq
Ronck, Catherine, (254) 968-1862 ronck@tarleton.edu,
    Tarleton State University – God
Rondot, Beth, (516) 299-2318 beth.rondot@liu.edu,
    Long Island University, C.W. Post Campus – Zn
Rood, Richard B., (734) 647-3530 rbrood@umich.edu,
    University of Michigan – ZcAsm
Rooney, Neil, (519) 824-4120 Ext.52573 nrooney@uoguelph.ca,
    University of Guelph – Zn
Rooney, Tyrone, 517-432-5522 rooneyt@msu.edu,
    Michigan State University – Gi
Rooney-Varga, Juliette, (978) 934-4715 juliette_rooneyvarga@uml.edu,
    University of Massachusetts, Lowell – ZcRc
Root, Tara L., 561 297-3253 troot@fau.edu,
    Florida Atlantic University – Hg
Roozeboom, Kraig, (785) 532-3781 kraig@ksu.edu,
    Kansas State University – So
Rose, Arthur W., (814) 238-2838 awr1@psu.edu,
    Pennsylvania State University, University Park – Cge
Rose, Candace M., (630) 252-3499 cmrose@anl.gov,
    Argonne National Laboratory – Zn
Rose, Catherine V., +353 1 8961165 crose@tcd.ie,
    Trinity College – Cg
Rose, Dan, +353 21 4902189 University College Cork – Zn

Rose, Leanna S., (678) 839-4067 srose@westga.edu,
University of West Georgia – ZyAsm
Rose, Peter E., (801) 585-7785 prose@egi.utah.edu,
University of Utah – Np
Rose, Seth E., 404413-5750 geoser@langate.gsu.edu,
Georgia State University – Hw
Rose, Timothy, (202) 633-1398 roset@si.edu,
Smithsonian Inst / Natl Museum of Natural Hist– CaGv
Rose, William I., (906) 487-2367 raman@mtu.edu,
Michigan Technological University – Gv
Roselle, Gregory T., (208) 496-7683 roselleg@byui.edu,
Brigham Young University - Idaho – GpCaHg
Rosen, Carl J., (612) 625-8114 crosen@umn.edu,
University of Minnesota, Twin Cities – Sc
Rosen, Michael R., 775.887.7683 mrosen@usgs.gov,
University of Nevada, Reno – GnClHg
Rosen, Peter S., p.rosen@neu.edu,
Northeastern University – OnZu
Rosenberg, Gary, 215.299.1033 gr347@drexel.edu,
Drexel University – Pi
Rosenberg, Gary D., grosenbe@iupui.edu,
Milwaukee Public Museum – RhPg
Rosenberg, Philip E., (509) 335-4368 rosenberg@wsu.edu,
Washington State University – GzCg
Rosenfeld, Carla, rose0859@umn.edu,
University of Minnesota, Twin Cities – Pg
Rosenfeld, John L., (310) 825-1505 rosenfel@ucla.edu,
University of California, Los Angeles – Gp
Rosengaus, Rebeca, (617) 373-7032 r.rosengaus@neu.edu,
Northeastern University – Zn
Rosenthal, Yair, 848-932-6555X227 rosentha@marine.rutgers.edu,
Rutgers, The State University of New Jersey – CmlOc
Ross, Andrew, +44(0) 113 34 37590 a.n.ross@leeds.ac.uk,
University of Leeds – Am
Ross, David A., (508) 289-2578 dross@whoi.edu,
Woods Hole Oceanographic Institution – GuOg
Ross, Gerald M., (403) 292-7000 University of Calgary – Gt
Ross, Kirstin, kirstin.ross@flinders.edu.au,
Flinders University – Zn
Ross, Martin E., (617) 373-3263 m.ross@neu.edu,
Northeastern University – Gie
Ross, Nancy L., (540) 231-6356 nross@vt.edu,
Virginia Polytechnic Institute & State University – Gz
Ross, Robert M., (607) 273-6623 x18 rmr16@cornell.edu,
Paleontological Research Institution – PgZePc
Ross, Tetjana, (902) 494-1327 tetjana.ross@dal.ca,
Dalhousie University – Op
Ross, Thomas E., (910) 521-6218 tom.ross@uncp.edu,
University of North Carolina, Pembroke – Zy
Rossby, Hans T., (401) 874-6521 trossby@gso.uri.edu,
University of Rhode Island – Op
Rosscoe, Steven, (325) 670-1387 srosscoe@hsutx.edu,
Hardin-Simmons University – PmGsPs
Rossell, Irene M., (828) 232-5185 irossell@unca.edu,
University of North Carolina, Asheville – Zn
Rossetti, Piergiorgio, piergiorgio.rossetti@unito.it,
Università di Torino – Eg
Rossman, George R., (626) 395-6471 grr@gps.caltech.edu,
California Institute of Technology – GzCa
Rost, Sebastian, +44(0) 113 34 35212 s.rost@leeds.ac.uk,
University of Leeds – Ysp
Rostam-Abadi, Massoud, 217-244-4977 massoud@isgs.uiuc.edu,
Illinois State Geological Survey – Zn
Rostoker, Gordon, (780) 492-5286 rostoker@space.ualberta.ca,
University of Alberta – Xy
Rostron, Ben, (780) 492-2178 ben.rostron@ualberta.ca,
University of Alberta – Zn
Roth, Danica, 303-273-3802 droth@mines.edu,
Colorado School of Mines – GmYs
Roth, Gregory W., (814) 863-1018 gwr@psu.edu,
Pennsylvania State University, University Park – So
Roth, Leonard T., troth@hccfl.edu,
Hillsborough Community College – Zg
Rothman, Daniel H., (617) 253-7861 dhr@mit.edu,
Massachusetts Institute of Technology – Yg
Rothrock, David A., (206) 545-2262 rothrock@apl.washington.edu,
University of Washington – Op

Rothstein, Lewis, (401) 874-6517 lrothstein@gso.uri.edu,
University of Rhode Island – Op
Rouff, Ashaki, 718-997-3073 ashaki.rouff@qc.cuny.edu,
Queens College (CUNY) – Cac
Rouff, Ashaki, 973-353-2511 ashaki.rouff@rutgers.edu,
Rutgers, The State University of New Jersey, Newark – Cl
Roughgarden, Joan, (650) 723-3648 Stanford University – Ob
Rougvie, James R., (608) 363-2268 rougviej@beloit.edu,
Beloit College – GxpGz
Rouhani, Farhang, (540) 654-1895 frouhani@mwc.edu,
Mary Washington College – Zn
Rouleau, Alain, 4185455011x5213 arouleau@uqac.ca,
Universite du Quebec a Chicoutimi – Hw
Rounds, Steven W., (916) 278-7828 rounds@csus.edu,
California State University, Sacramento – Gg
Rounsevell, Mark D., +44 (0) 131 651 4468 mark.rounsevell@ed.ac.uk,
Edinburgh University – Zu
Rouse, Gregory W., (858) 534-7943 grouse@ucsd.edu,
University of California, San Diego – Ob
Rouse, Jesse, 910-521-6387 jesse.rouse@uncp.edu,
University of North Carolina, Pembroke – Zni
Rouse, Joseph D., 16717352685 jdrouse@triton.uog.edu,
University of Guam – Hg
Rouse, Roland C., (734) 763-0952 rousec@umich.edu,
University of Michigan – Gz
Rouse, Jr., Lawrence J., (225) 388-2953 Louisiana State University – Op
Rousell, Don H., 7056751151 x2265 drousell@laurentian.ca,
Laurentian University, Sudbury – Gc
Roussel-Dupre, R., (505) 667-9228 rroussel-dupre@lanl.gov,
Los Alamos National Laboratory – Zn
Roussenov, Vassil M., +44 0151 794 4099 v.roussenov@liverpool.ac.uk,
University of Liverpool – OpZo
Rovey, Charles W., (417) 836-6890 charlesrovey@missouristate.edu,
Missouri State University – HyGle
Rowan, Christopher J., 330-672-7428 crowan5@kent.edu,
Kent State University – Gt
Rowden, Robert, Robert.Rowden@dnr.iowa.gov,
Iowa Dept of Natural Resources – Gg
Rowe, Charlotte A., 505-665-6404 char@lanl.gov,
New Mexico Institute of Mining and Technology – Ys
Rowe, Christie, (514) 398-2769 christie.rowe@mcgill.ca,
McGill University – GctEm
Rowe, Clinton M., (402) 472-1946 crowe1@unl.edu,
University of Nebraska, Lincoln – As
Rowe, Gilbert T., (409) 740-4458 roweg@tamug.edu,
Texas A&M University – Ob
Rowe, Timothy B., (512) 471-1725 rowe@mail.utexas.edu,
University of Texas, Austin – Pv
Rowell, Albert J., (785) 864-2747 arowell@ku.edu,
University of Kansas – Pi
Rowland, Scott K., (808) 956-3150 University of Hawai'i, Manoa – Zn
Rowland, Stephen, (702) 895-3625 steve.rowland@unlv.edu,
University of Nevada, Las Vegas – Pi
Rowley, David B., (773) 702-8146 rowley@geosci.uchicago.edu,
University of Chicago – GtgGc
Rowley, Rex J., 309-438-7832 rjrowle@ilstu.edu,
Illinois State University – Zi
Roy, Denis W., (418) 545-5011 dwroy@uqac.ca,
Universite du Quebec a Chicoutimi – Gc
Roy, Martin, 514-987-3000 #7619 roy.matin@uqam.ca,
Universite du Quebec a Montreal – Gl
Roy, Suzanne, (418) 723-1986 x1748 suzanne_roy@uqar.ca,
Universite du Quebec a Rimouski – Ob
Roy, William R., (217) 840-9769 wroy@illinois.edu,
University of Illinois, Urbana-Champaign – ScGeCl
Roychoudhury, Alakendra N., +27 21 808 3124 roy@sun.ac.za,
Stellenbosch University – ClmOc
Royden, Leigh H., (617) 253-1292 lhroyden@mit.edu,
Massachusetts Institute of Technology – Gt
Royer, Dana, (860) 685-2836 droyer@wesleyan.edu,
Wesleyan University – PegPb
Royer, Todd, 812-855-0563 Indiana University, Bloomington – HsZg
Rozmus, Wojciech, (780) 492-8486 rozmus@phys.ualberta.ca,
University of Alberta – Zn
Ruark, Matthew D., (608) 263-2889 mdruark@wisc.edu,
University of Wisconsin, Madison – So
Rubin, Alan E., (310) 825-3202 rubin@igpp.ucla.edu,

University of California, Los Angeles – Xm
Rubin, Allan M., (609) 258-1506 arubin@princeton.edu,
  Princeton University – Yg
Rubin, Charles M., (509) 963-2827 rbeling@wvu.edu,
  Central Washington University – Gt
Rubin, Kenneth H., (808) 956-8973 krubin@hawaii.edu,
  University of Hawai'i, Manoa – CcGvCg
Rubio-Sierra, Javier, 089/2180 4317 rubio@lrz.uni-muenchen.de,
  Ludwig-Maximilians-Universitaet Muenchen – Gz
Rucklidge, John C., (416) 978-2061 jcr@quartz.geology.utoronto.ca,
  University of Toronto – Ge
Ruddick, Barry R., (902) 494-2505 barry.ruddick@dal.ca,
  Dalhousie University – Op
Ruddiman, William F., 540-348-1963 wfr5c@virginia.edu,
  University of Virginia – Gu
Rudge, John, +44 (0) 1223 765545 jfr23@cam.ac.uk,
  University of Cambridge – YgGt
Rudnick, Daniel L., (858) 534-7669 drudnick@ucsd.edu,
  University of California, San Diego – Op
Rudnick, Roberta L., rudnick@ucsb.edu,
  University of Maryland – CgsCt
Rudnicki, Ryan E., 210-486-0061 rrudnicki@alamo.edu,
  San Antonio Community College – Zyr
Rudolph, Maxwell L., (530) 752-3669 maxrudolph@ucdavis.edu,
  University of California, Davis – Yg
Rueger, Bruce F., (207) 859-5806 bfrueger@colby.edu,
  Colby College – PlGg
Ruehr, Thomas A., (805) 756-2552 truehr@calpoly.edu,
  California Polytechnic State University – Sb
Ruff, Larry J., (734) 763-9301 ruff@umich.edu,
  University of Michigan – Ys
Ruffel, Alice, (214) 507-9014 aruffel@dcccd.edu,
  El Centro College - Dallas Community College District – Gg
Ruffman, Alan, (902) 422-6482 Dalhousie University – Ys
Ruggiero, Peter, (541) 737-1239 ruggierp@science.oregonstate.edu,
  Oregon State University – On
Ruhl, Christine J., 918-631-3018 cruhl@utulsa.edu,
  The University of Tulsa – YsGtc
Ruhl, Laura S., 501-683-4197 lsruhl@ualr.edu,
  University of Arkansas at Little Rock – GeClGb
Ruina, Andy L., 607-255-7108 alr3@cornell.edu,
  Cornell University – Yx
Ruiz, Javier , jaruiz@ucm.es,
  Univ Complutense de Madrid – Xg
Ruiz, Joaquin, (520) 621-4090 jruiz@email.arizona.edu,
  University of Arizona – Cgc
Ruiz-Diaz, Dorivar, (785) 532-6183 ruizdiaz@ksu.edu,
  Kansas State University – So
Rumble, III, Douglas, 202-478-8990 drumble@carnegiescience.edu,
  Carnegie Institution for Science – Gp
Rumrill, Julie, 203-392-5842 rumrillj1@southernct.edu,
  Southern Connecticut State University – GgeGl
Rumstay, Kenneth S., (229) 333-5754 krumstay@valdosta.edu,
  Valdosta State University – Znn
Rundberg, Robert S., (505) 667-4559 Los Alamos National Laboratory – Gq
Rundle, John, jbrundle@ucdavis.edu,
  University of California, Davis – Yd
Rundquist, Bradley C., (701) 777-4246 bradley.rundquist@und.edu,
  University of North Dakota – ZriZy
Rundquist, Donald C., (402) 472-3471 drundquist@unl.edu,
  Unversity of Nebraska - Lincoln – Zri
Runkel, Anthony, 612-627-4780 (Ext. 222) runke001@umn.edu,
  University of Minnesota, Twin Cities – Gs
Runkel, Anthony, (612) 626-1822 runke001@umn.edu,
  University of Minnesota – Gs
Runnegar, Bruce, (310) 206-1738 University of California, Los Angeles – Pg
Runyon, Cassandra R., (843) 953-8279 runyonc@cofc.edu,
  College of Charleston – XgSoZe
Runyon, Simone, srunyon@uwyo.edu,
  University of Wyoming – EdGip
Rupert, Denise, (570) 484-2048 drupert@lockhaven.edu,
  Lock Haven University – Zn
Rupert, Gerald B., (573) 341-4616 Missouri University of Science and Technology – Ye
Rupp, David, 541-737-5222 drupp@coas.oregonstate.edu,
  Oregon State University – AsOp
Rupp, John A., (812) 855-1323 rupp@indiana.edu,
  Indiana University – Go
Ruppel, Stephen C., (512) 471-2965 stephen.ruppel@beg.utexas.edu,
  University of Texas at Austin, Jackson School of Geosciences – Gd
Ruppert, Kelly R., (657) 278-3561 kruppert@fullerton.edu,
  California State University, Fullerton – Gg
Ruprecht, Philipp P., 775-682-6084 pruprecht@unr.edu,
  University of Nevada, Reno – Giv
Rusmore, Margaret E., 323 259 2565 rusmore@oxy.edu,
  Occidental College – Gc
Russ, Jean M., (301) 934-7814 jruss@csmd.edu,
  College of Southern Maryland – Zey
Russell, Ann D., (530) 752-3311 adrussell@ucdavis.edu,
  University of California, Davis – Cm
Russell, Armistead G., (404) 894-3079 ted.russell@ce.gatech.edu,
  Georgia Institute of Technology – As
Russell, Christopher T., (310) 825-3188 ctrussell@igpp.ucla.edu,
  University of California, Los Angeles – Yg
Russell, Dale A., (919) 515-1339 dale_russell@ncsu.edu,
  North Carolina State University – Pv
Russell, James K., (604) 822-2703 krussell@eos.ubc.ca,
  University of British Columbia – GviCg
Russell, James M., (401) 863-6330 James_Russell@Brown.edu,
  Brown University – Gn
Russell, Joellen, 520-626-2194 jrussell@email.arizona.edu,
  University of Arizona – Oc
Russell, Lynn M., (858) 534-4852 lmrussell@ucsd.edu,
  University of California, San Diego – As
Russell, R. Doncaster, (604) 822-2551 drussell@eos.ubc.ca,
  University of British Columbia – Yg
Russell, Ron, (512) 471-8831 ron.russell@beg.utexas.edu,
  University of Texas at Austin, Jackson School of Geosciences – Zn
Russell, Sally, +44(0) 113 34 35279 S.Russell@leeds.ac.uk,
  University of Leeds – Ge
Russell, Terry P., (250) 721-6184 trussell@uvic.ca,
  University of Victoria – Zn
Russell, Theresa J., (501) 575-4403 trussell@comp.uark.edu,
  University of Arkansas, Fayetteville – Zn
Russelle, Michael P., (612) 625-8145 russelle@soils.umn.edu,
  University of Minnesota, Twin Cities – Sc
Russo, Mary Rose, 609 258-4101 mrusso@princeton.edu,
  Princeton University – Zn
Russo, Raymond, 352-392-6766 rrusso@ufl.edu,
  University of Florida – Yg
Russo, Tess A., 814-865-7389 russo@psu.edu,
  Pennsylvania State University, University Park – Hw
Rust, Derek, +44 023 92 842298 derek.rust@port.ac.uk,
  University of Portsmouth – Gt
Rustad, James R., jrrustad@ucdavis.edu,
  University of California, Davis – Cl
Rutberg, Randye L., (212) 772-5326 randye.rutberg@hunter.cuny.edu,
  Hunter College (CUNY) – CmOcZc
Rutford, Robert H., (972) 883-6470 rutford@utdallas.edu,
  University of Texas, Dallas – Gl
Rutherford, Malcolm J., (401) 863-1927 Malcolm_Rutherford@Brown.edu,
  Brown University – Cp
Rutledge, Steven A., rutledge@atmos.colostate.edu,
  Colorado State University – As
Ruttan, Lore, 404 7274217 lruttan@emory.edu,
  Emory University – Ob
Ruttenberg, Kathleen, 808-956-9371 kcr@soest.hawaii.edu,
  University of Hawai'i, Manoa – Cg
Rutter, Ernest H., +44 0161 275-3945 e.rutter@manchester.ac.uk,
  University of Manchester – NrGct
Rutter, Nathaniel W., (780) 492-3085 nat.rutter@ualberta.ca,
  University of Alberta – Ge
Ruzicka, Alexander (Alex) M., (503) 725-3372 ruzickaa@pdx.edu,
  Portland State University – Xm
Ruzicka, Jaromir, jaromirr@hawaii.edu,
  University of Hawai'i, Manoa – Zn
Ryall, Patrick J., (902) 494-3465 pryall@is.dal.ca,
  Dalhousie University – Yg
Ryan, Anne-Marie, 902 494 3184 amryan@dal.ca,
  Dalhousie University – GeZeCg
Ryan, Cathy, (403) 220-2793 ryan@geo.ucalgary.ca,
  University of Calgary – Hw

Ryan, Jeffrey G., (813) 974-1598 ryan@shell.cas.usf.edu,
University of South Florida, Tampa – Ct

Ryan, Peter C., (802) 443-2557 pryan@middlebury.edu,
Middlebury College – Cl

Ryan, Susan, (724) 938-4531 ryan@calu.edu,
California University of Pennsylvania – Zg

Ryan, William F., (814) 865-0478 Pennsylvania State University, University Park – Am

Ryan, William B. F., billr@ldeo.columbia.edu,
Columbia University – Yr

Rychert, Catherine A., +44 (0)23 80598663 C.Rychert@soton.ac.uk,
University of Southampton – Yg

Ryder, Isabelle, +44-151-794-5143 I.Ryder@liverpool.ac.uk,
University of Liverpool – Gt

Ryder, Roy, rryder@southalabama.edu,
University of South Alabama – SdZry

Rye, Danny M., (203) 432-3174 danny.rye@yale.edu,
Yale University – Cs

Rygel, Michael C., (315) 267-3401 rygelmc@potsdam.edu,
SUNY Potsdam – GsrGo

Rykaczewski, Ryan R., 803-777-8159 ryk@sc.edu,
University of South Carolina – ObZoc

Rysgaard, Soren, 204-272-1611 rysgaard@umanitoba.ca,
University of Manitoba – Gl

Rößner, Gertrud, 089/2180 6612 g.roessner@lrz.uni-muenchen.de,
Ludwig-Maximilians-Universitaet Muenchen – Pg

S

Sá, Artur A., asa@utad.pt,
Universidade de Trás-os-Montes e Alto Douro – PgiPs

Saal, Alberto E., (401) 863-7238 alberto_saal@brown.edu,
Brown University – CgGi

Saalfeld, Alan J., 614 292 6665 saalfeld.1@osu.edu,
Ohio State University – Zi

Saar, Martin O., 612-625-7332 saar@umn.edu,
University of Minnesota, Twin Cities – Hg

Sabine, Christopher L., (206) 526-4809 chris.sabine@noaa.gov,
University of Washington – Oc

Sabol, Martin, +421-2-60296229 sabol@fns.uniba.sk,
Comenius University in Bratislava – PvgPs

Sabra, Karim, (404) 385-6193 Georgia Institute of Technology – Yg

Sacchi, Mauricio D., (780) 492-1060 sacchi@phys.ualberta.ca,
University of Alberta – Ys

Saccocia, Peter J., 508-531-2124 psaccocia@bridgew.edu,
Bridgewater State University – CqsCm

Sack, Richard, 425-880-4418 rosack@uw.edu,
University of Washington – CgZn

Sacks, I. Selwyn, ssacks@carnegiescience.edu,
Carnegie Institution for Science – Ys

Sacramentogrilo, Isabelle, 619-594-5607 isacramentogrilo@mail.sdsu.edu,
San Diego State University – Gg

Sacristan, Felix, felix.sacristan@uab.cat,
Universitat Autonoma de Barcelona – Ge

Sadd, James L., 323 259 2518 jsadd@oxy.edu,
Occidental College – ZiGeZr

Sadiq, Abdulali A., sadiqa@qu.edu.qa,
University of Qatar – Zr

Sadler, Peter M., (951) 827-5616 peter.sadler@ucr.edu,
University of California, Riverside – GrPs

Sadowsky, Michael J., (612) 624-2706 sadowsky@soils.umn.edu,
University of Minnesota, Twin Cities – Sb

Saeidi, Ali, (418) 545-5011 x2561 asaeidi@uqac.ca,
Universite du Quebec a Chicoutimi – Nrm

Saffer, Demian M., 814-865-7965 dsaffer@geosc.psu.edu,
Pennsylvania State University, University Park – Hw

Sagebiel, J. C., 909 307-2669 237 jsagebiel@sbcm.sbcounty.gov,
San Bernardino County Museum – Pv

Sageman, Bradley B., (847) 467-2257 brad@earth.northwestern.edu,
Northwestern University – PsCbPc

Sager, William W., (979) 845-9828 wsager@ocean.tamu.edu,
Texas A&M University – Ou

Sager, William W., (713) 743-3108 wwsager@uh.edu,
University of Houston – GtYrm

Sagiroglu, Ahmet, 00904242370000-5990 sagiroglu@firat.edu.tr,
Firat University – Egm

Sahagian, Dork, 610-758-6379 dork.sahagian@lehigh.edu,
Lehigh University – PeGvr

Sahakian, Valerie J., 541-346-3401 vjs@uoregon.edu,
University of Oregon – Yg

Sahay, Pratap, pratap@cicese.mx,
Centro de Investigación Científica y de Educación Superior de Ensenada – Ys

Sahr, John D., (206) 616-7175 jdsahr@ee.washington.edu,
University of Washington – ZrXym

Saikia, Udoy, udoy.saikia@flinders.edu.au,
Flinders University – Zn

Saila, Saul B., (401) 874-6485 saila@gso.uri.edu,
University of Rhode Island – Ob

Saillet, Elodie, +33(0)3 44067563 elodie.saillet@lasalle-beauvais.fr,
Institut Polytechnique LaSalle Beauvais (ex-IGAL) – GctGo

Saini-Eidukat, Bernhardt, (701) 231-8785 bernhardt.saini-eidukat@ndsu.edu,
North Dakota State University – GiCqEg

Saint, Prem K., psaint@fullerton.edu,
California State University, Fullerton – Hw

Saito, Mak A., 508-289-3696 msaito@whoi.edu,
Woods Hole Oceanographic Institution – Oc

Saja, David B., 2162314600 x3229 dsaja@cmnh.org,
Cleveland Museum of Natural History – Gdc

Sak, Peter B., (717) 245-1423 sakp@dickinson.edu,
Dickinson College – Gcm

Sakhaee-Pour, Ahmad, 405-325-3306 sakhaee@ou.edu,
University of Oklahoma – Np

Saku, James C., (301) 687-4724 jsaku@frostburg.edu,
Frostburg State University – Zn

Salacup, Jeff, 413-545-1913 jsalacup@geo.umass.edu,
University of Massachusetts, Amherst – PcCb

Salami, B. M., 234-803-321-9685 salamibm@oauife.edu.ng,
Obafemi Awolowo University – Hg

Salami, Sikiru A., sikiru.salami@uniben.edu,
University of Benin – YgeGg

Salaun, Pascal, +44-151-794-4101 Pascal.Salaun@liverpool.ac.uk,
University of Liverpool – Em

Salaun, Rachel, +440151 795 4649 Rachel.Jeffreys@liverpool.ac.uk,
University of Liverpool – Oc

Saleeby, Jason B., (626) 395-6141 jason@gps.caltech.edu,
California Institute of Technology – Gc

Salisbury, Joseph, 603-862-0849 joe.salisbury@unh.edu,
University of New Hampshire – Og

Salje, Ekhard, +44 (0) 1223 768321 ekhard@esc.cam.ac.uk,
University of Cambridge – Gy

Sallan, Lauren, 215-898-5650 lsallan@sas.upenn.edu,
University of Pennsylvania – PqvPe

Salle, Bethan, (214) 578-1914 bsalle@dcccd.edu,
El Centro College - Dallas Community College District – GgeZg

Sallu, Susannah, +44(0) 113 34 31641 s.sallu@leeds.ac.uk,
University of Leeds – Ga

Salmon, Richard L., (858) 534-2090 rsalmon@ucsd.edu,
University of California, San Diego – Op

Salmun, Haydee, (212) 772-4159 hsalmun@hunter.cuny.edu,
Hunter College (CUNY) – OpZgAs

Salters, Vincent J., (850) 644-1934 salters@magnet.fsu.edu,
Florida State University – Cg

Saltzman, Eric S., 949-824-3936 esaltzma@uci.edu,
University of California, Irvine – As

Saltzman, Matthew R., 614 292-0481 saltzman.11@osu.edu,
Ohio State University – GsClg

Salviulo, Gabriella, +390498279157 gabriella.salviulo@unipd.it,
Università degli Studi di Padova – Gz

Salvucci, Guido D., (617) 353-8344 gdsalvuc@bu.edu,
Boston University – Hq

Salzer, Matthew W., (520) 621-2946 msalzer@ltrr.arizona.edu,
University of Arizona – PeGav

Samelson, Roger, 541-737-4752 rsamelson@coas.oregonstate.edu,
Oregon State University – Op

Sammarco, Paul W., (225) 851-2800 Louisiana State University – Zn

Sammis, Charles G., (213) 740-5836 sammis@usc.edu,
University of Southern California – Yg

Sammis, Theodore W., (505) 646-3405 New Mexico State University, Las Cruces – Sp

Samonds, Karen, 815-753-3201 ksamonds@niu.edu,
Northern Illinois University – Pv

Sample, James C., (928) 523-0881 james.sample@nau.edu,
Northern Arizona University – GtClOu

Samra, Charles, (307) 766-2286 Ext. 243 charles.samra@wyo.gov,

Wyoming State Geological Survey – GgcEg
Samson, Iain M., 519-253-3000 ext. 2489 ims@uwindsor.ca,
University of Windsor – CgEmGx
Samson, John C., (780) 492-3616 samson@space.ualberta.ca,
University of Alberta – Xy
Samson, Perry, (734) 276-0815 samson@umich.edu,
University of Michigan – AsZeAm
Samson, Scott D., 315-443-2672 sdsamson@syr.edu,
Syracuse University – Cc
Samsonov, Sergey V., (613) 656-1534 sergey.samsonov@canada.ca,
Natural Resources Canada – YdZrYg
Samuelson, Alan C., (765) 285-8270 Ball State University – Hw
Sanabia, Elizabeth R., 410-293-6556 sanabia@usna.edu,
United States Naval Academy – Am
Sanchez, Charles, (520) 782-3836 sanchez@ag.arizona.edu,
University of Arizona – Zn
Sanchez, Marcelo J., 979.862.6604 msanchez@civil.tamu.edu,
Texas A&M University – NtrNx
Sánchez, Nuria, nsanchez@ucm.es,
Univ Complutense de Madrid – Gz
Sanchez, Veronica I., (361) 593-3590 veronica.sanchez@tamuk.edu,
Texas A&M University, Kingsville – GctGm
Sánchez, Yolanda, yol@ucm.es,
Univ Complutense de Madrid – Gs
Sanchez-Azofeifa, G. Arturo, (780) 492-1822 arturo.sanchez@ualberta.ca,
University of Alberta – ZirZg
Sandau, Ken L., 406-496-4151 ksandau@mtech.edu,
Montana Tech of The University of Montana – Zi
Sandel, Bill R., (520) 621-4305 sandel@vega.lpl.arizona.edu,
University of Arizona – As
Sanders, Laura L., (773) 442-6051 l-sanders@neiu.edu,
Northeastern Illinois University – HwGe
Sanders, Ronald S., (818) 354-2867 sanders@jpl.nasa.gov,
Jet Propulsion Laboratory – Gr
Sandhu, Harpinder, harpinder.sandhu@flinders.edu.au,
Flinders University – Zn
Sandlin, Stephen H., 909-389-8644 ssandlin@sbccd.cc.ca.us,
San Bernardino Valley College – Zy
Sandor, Jonathan A., (515) 294-2209 jasandor@iastate.edu,
Iowa State University of Science & Technology – Sd
Sandoval, Samuel, (530) 754-9646 samsandoval@ucdavis.edu,
University of California, Davis – Hg
Sandoval Solis, Samuel, 530-750-9722 samsandoval@ucdavis.edu,
University of California, Davis – Hg
Sandvol, Eric A., (573) 884-9616 sandvole@missouri.edu,
University of Missouri – Ys
Sandwell, David T., (858) 534-7109 dsandwell@ucsd.edu,
University of California, San Diego – Yr
Sandwith, Zoe, (508) 289-3942 zsandwith@whoi.edu,
Woods Hole Oceanographic Institution – Oc
Sandy, Michael R., (937) 229-3436 University of Dayton – Pi
Sanford, Lawrence P., (410) 221-8429 lsanford@hpl.umces.edu,
University of Maryland – Op
Sanford, Robert A., 217-244-7250 rsanford@illinois.edu,
University of Illinois, Urbana-Champaign – CbSbPo
Sanford, Thomas B., (206) 543-1365 sanford@apl.washington.edu,
University of Washington – Op
Sanford, William E., (970) 491-5929 bills@cnr.colostate.edu,
Colorado State University – Hw
Sanislav, Ioan, ioan.sanislav@jcu.edu.au,
James Cook University – GctGp
Sanjurjo, Jorge, 0034981167000 jsanjurjo@udc.es,
Coruna University – GeCcGa
Sankey, Julia, (209) 667-3466 jsankey@csustan.edu,
California State University, Stanislaus – Pve
Sansone, Francis J., 808-956-2912 sansone@hawaii.edu,
University of Hawai'i, Manoa – Cm
Santander, Erma, (520) 621-7120 erma@hwr.arizona.edu,
University of Arizona – Zn
Santelli, Cara M., santelli@umn.edu,
University of Minnesota, Twin Cities – Pg
Santelmann, Mary V., (541) 737-1215 santelmm@geo.oregonstate.edu,
Oregon State University – Zy
Santi, Paul M., 303-273-3108 psanti@mines.edu,
Colorado School of Mines – Ng
Santillan, Marcelo, (509) 963-1107 marcelo@geology.cwu.edu,
Central Washington University – Yd

Santisteban, Juan Ignacio, juancho@ucm.es,
Univ Complutense de Madrid – Gr
Santos, Eduardo, (785) 532-6932 esantos@ksu.edu,
Kansas State University – So
Santos, Hernan, (787) 265-3845 hernan.santos@upr.edu,
University of Puerto Rico – PiGdr
Santschi, Peter H., (409) 740-4476 santschip@tamug.edu,
Texas A&M University – Oc
Sanudo, Sergio, 213-821-1302 sanudo@usc.edu,
University of Southern California – Pg
Sanz, Esther, mesanz@ucm.es,
Univ Complutense de Madrid – Gd
Sapigao, Gladys, 718-997-3300 gladys.sapigao@qc.cuny.edu,
Queens College (CUNY) – ZnnZn
Sar, Abdullah, 00904242370000-5961 asasr@firat.edu.tr,
Firat University – GviGp
Sarachik, Edward, 206-543-6720 sarachik@atmos.washington.edu,
University of Washington – As
Sarah, Willig B., (215) 898-5724 sawillig@verizon.net,
University of Pennsylvania – Ge
Saravanan, R., 979-845-0175 r.saravanan@tamu.edu,
Texas A&M University – As
Sargent, Kenneth A., (864) 294-3362 ken.sargent@furman.edu,
Furman University – Hy
Sarmiento, Jorge L., (609) 258-6585 jls@princeton.edu,
Princeton University – Oc
Sarrionandia, Fernando, fernando.sarrionandia@ehu.eus,
University of the Basque Country UPV/EHU – GigGv
Sartain, Jerry B., (352) 392-1803 (Ext. 330) sartain@ufl.edu,
University of Florida – So
Sarwar, A. K. Mostofa, (504) 280-6717 asarwar@uno.edu,
University of New Orleans – Ye
Sasmaz, Ahmet, +905323406642 asasmaz@firat.edu.tr,
Firat University – EgGeCe
Sasowsky, Ira D., (330) 972-5389 ids@uakron.edu,
University of Akron – HwGmo
Sasowsky, Kathryn, (216) 987-5227 Kathryn.Sasowsky@tri-c.edu,
Cuyahoga Community College - Western Campus – Zg
Sass, Henrik, +44(0)29 208 76001 SassH@cardiff.ac.uk,
University of Wales – Co
Sassenrath, Gretchen, (620) 421-4826 gsassenrath@ksu.edu,
Kansas State University – So
Sassi, Raffaele, +390498279149 raffaele.sassi@unipd.it,
Università degli Studi di Padova – Ggx
Satkoski, Alina A., 518-265-7183 alina.satkoski@austincc.edu,
Austin Community College District – GemGg
Sato, Yoko, (973) 655-4448 satoyo@mail.montclair.edu,
Montclair State University – Gg
Satterfield, Joseph I., 325-486-6766 joseph.satterfield@angelo.edu,
Angelo State University – Gc
Sauck, William A., (269) 567-8282 bill.sauck@wmich.edu,
Western Michigan University – YgeYv
Sauer, Nancy N. S., (505) 665-3759 Los Alamos National Laboratory – Zn
Sauer, Peter, 812-855-6591 pesauer@indiana.edu,
Indiana University, Bloomington – Pe
Saunders, Andy D., +440116 252 3923 ads@le.ac.uk,
Leicester University – GiCgGe
Saunders, Charles, CharlesSaunders@rmc.edu,
Randolph-Macon College – ZgGrPg
Saunders, Kate, +44 (0) 131 650 2544 kate.saunders@ed.ac.uk,
Edinburgh University – GviGz
Saunders, Ralph, (310) 243-3284 rsaunders@csudh.edu,
California State University, Dominguez Hills – Zg
Saunders, W. Bruce, (610) 526-5114 wsaunder@brynmawr.edu,
Bryn Mawr College – Pi
Saura, Eduard, eduard.saura@uab.cat,
Universitat Autonoma de Barcelona – Gc
Sautter, Leslie R., (843) 953-5586 sautterl@cofc.edu,
College of Charleston – Ob
Sava, Diana, 720-981-0522 diana.sava@beg.utexas.edu,
University of Texas at Austin, Jackson School of Geosciences – Yg
Sava, Diana C., 512-475-9507 diana.sava@beg.utexas.edu,
University of Texas, Austin – Yg
Sava, Paul, 303-384-2362 psava@mines.edu,
Colorado School of Mines – Ye
Savage, Brian, (401) 874-5392 savage@uri.edu,
University of Rhode Island – Ys

Savage, Heather M., hsavage@ldeo.columbia.edu,
   Columbia University – Ys
Savage, Norman M., (541) 346-4585 nmsavage@uoregon.edu,
   University of Oregon – Pi
Savard, Dany, 418-545-5011 ddsavard@uqac.ca,
   Universite du Quebec a Chicoutimi – Ca
Savigny, K. Wayne, (604) 684-5900 University of British Columbia – Ng
Savin, Samuel M., (216) 368-6592 sms7@po.cwru.edu,
   Case Western Reserve University – Cs
Savina, Mary E., 507-222-4404 msavina@carleton.edu,
   Carleton College – Gm
Savov, Ivan, +44(0) 113 34 35199 earis@leeds.ac.uk,
   University of Leeds – Cg
Savrda, Charles E., (334) 844-4887 savrdce@auburn.edu,
   Auburn University – GdPeGr
Sawin, Robert S., (785) 864-2099 bsawin@kgs.ku.edu,
   University of Kansas – Gg
Sawyer, Audrey , 614-292-8383 sawyer.143@osu.edu,
   Ohio State University – HwSpHs
Sawyer, Carol F., 251-460-6169 sawyer@southalabama.edu,
   University of South Alabama – GmZye
Sawyer, Dale S., (713) 348-5106 dale@rice.edu,
   Rice University – Yr
Sawyer, Derek, 614-292-7243 sawyer.144@osu.edu,
   Ohio State University – YgGoRn
Sawyer, Edward W., (418) 545-5011 x5636 edward-w_sawyer@uqac.ca,
   Universite du Quebec a Chicoutimi – GptGi
Sawyer, J. F., (605) 394-2462 Foster.Sawyer@sdsmt.edu,
   South Dakota School of Mines & Technology – GsoHw
Sawyer, Suzanne, 801-537-3333 ssawyer@utah.gov,
   Utah Geological Survey – Zn
Saxena, Surenda K., (718) 951-5416 Graduate School of the City University of New York – Cg
Saxena, Surendra K., (305) 348-3030 saxenas@fiu.edu,
   Florida International University – Gy
Saxon, Christopher, (254) 968-9739 saxon@tarleton.edu,
   Tarleton State University – Gco
Sayles, Frederick L., (508) 289-2561 fsayles@whoi.edu,
   Woods Hole Oceanographic Institution – Oc
Saylor, Beverly Z., 216 368 3763 bzs@case.edu,
   Case Western Reserve University – Gs
Sbar, Marc, msbar@email.arizona.edu,
   University of Arizona – Yes
Scanlin, Michael A., 717-361-1323 jte2@psu.edu,
   Elizabethtown College – Ye
Scanlon, Bridget R., 512-471-8241 bridget.scanlon@beg.utexas.edu,
   University of Texas, Austin – Hg
Scanlon, Todd M., (434) 924-3382 tms2v@virginia.edu,
   University of Virginia – Hg
Scannapieco, Evan, (480) 727-6788 evan.scannapieco@asu.edu,
   Arizona State University – Xa
Scapelli, Krista, (817) 257-7270 k.l.scarpelli@tcu.edu,
   Texas Christian University – Zn
Scarnato, Barbara V., bvscarna@nps.edu,
   Naval Postgraduate School – Am
Scarpino, Samuel, 617.373.2482 s.scarpino@northeastern.edu,
   Northeastern University – Zn
Scarselli, Nicola, +44 1784 443597 nicola.scarselli@rhul.ac.uk,
   Royal Holloway University of London – Pgi
Schaal, Barbara, (314) 935-6822 schaal@wustl.edu,
   Washington University in St. Louis – Zn
Schaap, Marcel, (520) 626-4532 mschapp@cals.arizona.edu,
   University of Arizona – So
Schade, Gunnar, (979) 845-7671 schade@ariel.met.tamu.edu,
   Texas A&M University – As
Schaef, Herbert T., (509) 373-9949 todd.schaef@pnl.gov,
   Pacific Northwest National Laboratory – Cg
Schaefer, Janet R. G., (907) 451-5005 janet.schaefer@alaska.gov,
   Alaska Division of Geological & Geophysical Surveys – Gv
Schaefer, Joerg, schaefer@ldeo.columbia.edu,
   Columbia University – Pe
Schaefer, Joerg, 845-365-8703 schaefer@ldeo.columbia.edu,
   Columbia University – Cg
Schafer, Carl M., (586) 286-2154 schaferc@macomb.edu,
   Macomb Community College, Center Campus – Gg
Schafer, Tom, 785-628-5969 tschafer@fhsu.edu,
   Fort Hays State University – ZyiRn

Schaff, David, dschaff@ldeo.columbia.edu,
   Columbia University – Ys
Schaffrin, Burkhard A., 614 292-0502 schaffrin.1@osu.edu,
   Ohio State University – Yd
Schagrin, Zachery, 717.702.2045 zschagrin@pa.gov,
   Pennsylvania Bureau of Topographic & Geologic Survey – Gug
Schaller, Mirjam, 734-615-4286 mirjam@umich.edu,
   University of Michigan – Gm
Schaller, Morgan F., (518) 276-3358 schall@rpi.edu,
   Rensselaer Polytechnic Institute – CsPe
Schapaugh, Bill T., (785) 532-7242 wts@ksu.edu,
   Kansas State University – So
Schardt, Christian, (218) 726-7899 cschardt@d.umn.edu,
   University of Minnesota, Duluth – EgCg
Schardt, Lawrence A., (814) 863-7655 las233@psu.edu,
   Pennsylvania State University, University Park – SoHw
Scharnberger, Charles K., Charles.scharnberger@millersville.edu,
   Millersville University – Gc
Schauble, Edwin A., (310) 825-3880 schauble@ucla.edu,
   University of California, Los Angeles – Cgs
Schauss, Kim E., (812) 464-1701 keschauss@usi.edu,
   University of Southern Indiana – Zn
Scheel, Patrick, 573-368-2243 patrick.scheel@dnr.mo.gov,
   Missouri Dept of Natural Resources – Zn
Scheer, Clemens, clemens.scheer@qut.edu.au,
   Queensland University of Technology – SbGe
Scheidemen, Kathy J., (805) 893-7615 kathys@icess.ucsb.edu,
   University of California, Santa Barbara – Zn
Scheidt, Brian, 573-518-2314 bscheidt@mineralarea.edu,
   Mineral Area College – Hw
Scheingross, Joel S., jscheingross@unr.edu,
   University of Nevada, Reno – Gsm
Schellart, Wouter , +31 20 59 8610 w.p.schellart@vu.nl,
   Vrije Universiteit Amsterdam – Gt
Schellenberg, Stephen A., 61959421039 saschellenberg@mail.sdsu.edu,
   San Diego State University – Pe
Schelske, Claire L., (352) 392-9617 University of Florida – Pe
Schenck, William S., 302-831-8262 rockman@udel.edu,
   University of Delaware – ZiGi
Schenck, William S., (302) 831-2833 rockman@udel.edu,
   University of Delaware – GgpGc
Schenk, Anton, (716)645-3489 afshenko@buffalo.edu,
   SUNY, Buffalo – Zr
Scher, Howie, 803-777-2410 hscher@geol.sc.edu,
   University of South Carolina – GuCl
Scherer, Reed P., (815) 753-7951 reed@niu.edu,
   Northern Illinois University – Pm
Schermer, Elizabeth R., (360) 650-3658 schermer@geol.wwu.edu,
   Western Washington University – Gt
Schiappa, Tamra A., (724) 738-2829 tamra.schiappa@sru.edu,
   Slippery Rock University – PiGrZe
Schieber, Juergen, (812) 856-4740 jschiebe@indiana.edu,
   Indiana University, Bloomington – Gs
Schiebout, Judith A., (225) 578-2717 schiebout@geol.lsu.edu,
   Louisiana State University – Pv
Schiefer, Erik, (928) 523-6535 erik.schiefer@nau.edu,
   Northern Arizona University – GmZyi
Schiffbauer, James, (573) 882-9501 schiffbauerj@missouri.edu,
   University of Missouri – Pgi
Schiffman, Peter, (530) 752-3669 pschiffman@ucdavis.edu,
   University of California, Davis – Gp
Schilling, Jean-Guy E., (401) 423 1417 jgchilling@yahoo.com,
   University of Rhode Island – OuCuc
Schilling, Keith, Keith.Schilling@dnr.iowa.gov,
   Iowa Dept of Natural Resources – Gg
Schilling, Keith E., (319) 335-1575 keith.schilling@dnr.iowa.gov,
   University of Iowa – Hgs
Schimmelmann, Arndt, (812) 855-7645 aschimme@indiana.edu,
   Indiana University, Bloomington – CsGsPe
Schimmrich, Steven, (845) 687-7683 schimmrs@sunyulster.edu,
   SUNY, Ulster County Community College – GgZgn
Schincariol, Robert A., (519) 661-3732 schincar@uwo.ca,
   Western University – HwsNg
Schindler, Michael, 7056751151 x2368 mschindler@laurentian.ca,
   Laurentian University, Sudbury – Gz
Schlautman, Mark, (864) 656-4059 mschlau@clemson.edu,
   Clemson University – ClHgSc

Schlegel, Alan, (620) 376-4761 schlegel@ksu.edu,
Kansas State University – So
Schleifer, Stanley, 718-262-2726 sschleifer@york.cuny.edu,
York College (CUNY) – Ge
Schlesinger, William H., (919) 613-8004 schlesin@duke.edu,
Duke University – Zn
Schlichting, Hilke, (626) 316-3629 hilke@ucla.edu,
University of California, Los Angeles – Xy
Schlische, Roy W., 848-445-3445 schlisch@eps.rutgers.edu,
Rutgers, The State University of New Jersey – GctEo
Schlogl, Jan , schlogl@fns.uniba.sk,
Comenius University in Bratislava – Pig
Schlosser, C. Adam, 617-253-3983 casch@mit.edu,
Massachusetts Institute of Technology – As
Schlosser, Peter, (845) 365-8707 schlosser@ldeo.columbia.edu,
Columbia University – Cg
Schlue, John W., jwschlue@yahoo.com,
New Mexico Institute of Mining and Technology – YsgZg
Schlumpberger, Debbie, 320-308-3260 dmschlumpberger@stcloudstate.edu,
Saint Cloud State University – Zn
Schmahl, Wolfgang, 089/2180 4311 Wolfgang.Schmahl@lrz.uni-muenchen.de,
Ludwig-Maximilians-Universitaet Muenchen – Gz
Schmandt, Brandon, 505.277.4204 bschmandt@unm.edu,
University of New Mexico – Yg
Schmerr, Nicholas C., 301-405-4385 nschmerr@umd.edu,
University of Maryland – XyYs
Schmid, Dieter, 089/2180 6635 d.schmid@lrz.uni-muenchen.de,
Ludwig-Maximilians-Universitaet Muenchen – Pg
Schmid, Katherine W., 412-442-4232 kschmid@pa.gov,
Pennsylvania Bureau of Topographic & Geologic Survey – GoCc
Schmidt, Amanda H., 440-775-8351 amanda.schmidt@oberlin.edu,
Oberlin College – Gm
Schmidt, Bennetta, bennetta.schmidt@lamar.edu,
Lamar University – Gg
Schmidt, Dale R., 217-300-1169 schmidt2@illinois.edu,
Illinois State Geological Survey – Ge
Schmidt, David , 206-685-3799 dasc@uw.edu,
University of Washington – Gt
Schmidt, David, 937 775-3539 david.schmidt@wright.edu,
Wright State University – PiGd
Schmidt, Jonathan, (519) 824-4120 Ext.53966 jonschm@uoguelph.ca,
University of Guelph – Zn
Schmidt, Keegan L., (208) 790-2283 klschmidt@lcsc.edu,
Lewis-Clark State College – Gc
Schmidt, Lisa, lschmidt@sbccd.cc.ca.us,
San Bernardino Valley College – Zy
Schmidt, Matthew, 757-683-4285 mwschmid@odu.edu,
Old Dominion University – Ou
Schmitt, Danielle M., (609) 258-7015 dschmitt@princeton.edu,
Princeton University – Ze
Schmitt, Douglas R., (780) 492-3985 schmitt@purdue.edu,
Purdue University – YxNrYe
Schmitt, Michael A., (612) 625-7017 mschmitt@soils.umn.edu,
University of Minnesota, Twin Cities – Sc
Schmittner, Andreas, 541-737-9952 aschmittner@coas.oregonstate.edu,
Oregon State University – AsYr
Schmitz, Darrel W., 662-268-1032 Ext 241 schmitz@geosci.msstate.edu,
Mississippi State University – Hw
Schmitz, Mark D., 208-426-5907 markschmitz@boisestate.edu,
Boise State University – CcaCg
Schmutz, Phillip P., 850-474-3418 pschmutz@uwf.edu,
University of West Florida – Gm
Schneider, David, (613) 562-5800 x6155 david.schneider@uottawa.ca,
University of Ottawa – GtCcGg
Schneider, Edwin K., (703) 993-5364 eschnei1@gmu.edu,
George Mason University – As
Schneider, John F., (630) 252-8923 Argonne National Laboratory – Ca
Schneider, Julius, 089/2180 4354 julius.schneider@lrz.uni-muenchen.de,
Ludwig-Maximilians-Universitaet Muenchen – Gz
Schneider, Niklas, (808) 956-8383 nschneid@hawaii.edu,
University of Hawai'i, Manoa – Og
Schneider, Robert J., 508-289-2756 rschneider@whoi.edu,
Woods Hole Oceanographic Institution – Cc
Schneider, Robert V., 361-593-3589 robert.schneider@tamuk.edu,
Texas A&M University, Kingsville – GoYes
Schneider, Tapio, (626) 395-6143 tapio@caltech.edu,
California Institute of Technology – AsmZg
Schneiderman, Jill S., 845-437-5542 schneiderman@vassar.edu,
Vassar College – Gs
Schnetzer, Astrid, (919) 515-7837 aschnet@ncsu.edu,
North Carolina State University – Ob
Schoenberger, Erica, (410) 516-6158 ericas@jhu.edu,
Johns Hopkins University – Zn
Schoene, R B., (609) 258-5747 bschoene@princeton.edu,
Princeton University – Cc
Schoenemann, Spruce W., 406 683-7624 spruce.schoenemann@umwestern.edu,
University of Montana Western – GeCgs
Scholtz, Theresa C., (630) 252-6499 Argonne National Laboratory – Ge
Scholz, Christopher A., 315-443-2672 cascholz@syr.edu,
Syracuse University – Gs
Scholz, Christopher H., scholz@ldeo.columbia.edu,
Columbia University – Ys
Scholz, Christopher H., (845) 365-8360 scholz@ldeo.columbia.edu,
Columbia University – YxNr
Schoof, Christian, (604) 822-3063 cschoof@eos.ubc.ca,
University of British Columbia – Gl
Schoonen, Martin A., (631) 632-8007 martin.schoonen@sunysb.edu,
Stony Brook University – Cl
Schoonmaker, Adam, 792-2577 adschoonmaker@utica.edu,
Utica College – GczGx
Schoonmaker, Jane E., (808) 956-9935 jane@soest.hawaii.edu,
University of Hawai'i, Manoa – Cl
Schopf, J. William, (310) 825-1170 schopf@epss.ucla.edu,
University of California, Los Angeles – Pg
Schopf, Paul S., (703) 993-5394 pschopf@gmu.edu,
George Mason University – OpAs
Schouten, Hans, (508) 289-2574 hschouten@whoi.edu,
Woods Hole Oceanographic Institution – Yr
Schrader, Christian M., 315-267-2285 schradcm@potsdam.edu,
SUNY Potsdam – GiEmGv
Schrag, Daniel P., (617) 495-7676 schrag@eps.harvard.edu,
Harvard University – Cg
Schrage, Jon M., (402) 280-5759 schragej@gmail.com,
Creighton University – Am
Schrank, Christoph, +61 7 3138 1583 christoph.schrank@qut.edu.au,
Queensland University of Technology – GcqGt
Schreiber, B. Charlotte, (718) 997-3300 Queens College (CUNY) – Gd
Schreiber, Charlotte, 206-297-1454 geologo1@uw.edu,
University of Washington – Gsd
Schreiber, Madeline E., (540) 231-3377 Virginia Polytechnic Institute & State University – Hw
Schreier, Hans D., (604) 822-4401 University of British Columbia – Zg
Schrenk, Matt, (517) 884-7966 schrenkm@msu.edu,
Michigan State University – Pg
Schrieber, B. Charlotte, (718) 997-3300 Graduate School of the City University of New York – Gd
Schriver, David, (310) 825-6663 dave@igpp.ucla.edu,
University of California, Los Angeles – Yg
Schroder-Adams, Claudia, csadams@ccs.carleton.ca,
Carleton University – Pm
Schroeder, Dustin M., (650) 725-7861 dustin.m.schroeder@stanford.edu,
Stanford University – Gl
Schroeder, John L., 806-834-5678 john.schroeder@ttu.edu,
Texas Tech University – As
Schroeder, Kathleen, (828) 262-7055 schroederk@appstate.edu,
Appalachian State University – Zn
Schroeder, Norman C., (505) 667-0967 Los Alamos National Laboratory – Cc
Schroeder, Paul A., (706) 542-2384 schroe@uga.edu,
University of Georgia – GzEnCl
Schroeder, Stefan, +44 0161 306-6870 stefan.schroeder@manchester.ac.uk,
University of Manchester – Go
Schroeder, William W., (334) 861-7528 wschroeder@disl.org,
Dauphin Island Sea Lab – Og
Schroth, Andrew W., (802)656 3481 aschroth@uvm.edu,
University of Vermont – ClqCb
Schubert, Brian, 337-482-6967 bas9777@louisiana.edu,
University of Louisiana at Lafayette – Cg
Schubert, Gerald, (310) 825-4577 University of California, Los Angeles – Yg
Schuberth, Bernhard, 089/2180 4220 bernhard@geophysik.uni-muenchen.de,
Ludwig-Maximilians-Universitaet Muenchen – Yg
Schulingkamp, Arren, 225-578-3412 warrenii@lsu.edu,

Louisiana State University – Gg
Schulmeister, Marcia K., (620) 341-5983 mschulme@emporia.edu, Emporia State University – HwClGe
Schulte, Kimberly D., kschulte1@gsu.edu, Georgia State University, Perimeter College, Online – Gg
Schultz, Adam, 541-737-9832 adam.schultz@oregonstate.edu, Oregon State University – YgmRn
Schultz, David, +44 0161 306-3909 david.schultz@manchester.ac.uk, University of Manchester – Am
Schultz, Gerald E., 806-651-2580 gschultz@wtamu.edu, West Texas A&M University – PvGzZg
Schultz, Jan, (805) 965-0581 (Ext. 2313) schultz@sbcc.edu, Santa Barbara City College – GgePg
Schultz, Peter H., (401) 863-2417 Peter_Schultz@Brown.edu, Brown University – Xg
Schultze, Steven R., schultze@southalabama.edu, University of South Alabama – At
Schulz, Layne D., 605-67-76161 layne.schulz@usd.edu, South Dakota Dept of Environment and Natural Resources – Glg
Schulz, William J., 410-293-6563 schulz@usna.edu, United States Naval Academy – OpAm
Schulze, Anja, 409 740 4540 schulzea@tamug.edu, Texas A&M University – Ob
Schulze, Daniel J., 905-8283970 dschulze@utm.utoronto.ca, University of Toronto – Gi
Schulze, Karl, (630) 466-2652 kschulze@waubonsee.edu, Waubonsee Community College – ZgAms
Schulze-Makuch, Dirk, 509-335-1180 dirksm@wsu.edu, Washington State University – Hw
Schumacher, Courtney, (979) 845-5522 cschu@tamu.edu, Texas A&M University – As
Schumacher, Matthew, mschumac@stfx.ca, Saint Francis Xavier University – Zg
Schumann, Arnold W., (863) 956-1151 schumaw@ufl.edu, University of Florida – So
Schumer, Rina, (775) 673-7414 rina.schumer@dri.edu, University of Nevada, Reno – HqwGm
Schutt, Derek L., (970) 491-5786 schutt@warnercnr.colostate.edu, Colorado State University – Ysg
Schwab, Brandon E., bes21@humboldt.edu, Humboldt State University – GizGv
Schwab, Brandon E., (828) 227-7495 beschwab@wcu.edu, Western Carolina University – GizGv
Schwab, Fred, (310) 825-3123 schwab@igpp.ucla.edu, University of California, Los Angeles – Ys
Schwab, Frederick L., (540) 458-5830 schwabf@wlu.edu, Washington & Lee University – Gdg
Schwab, James J., (518) 437-8754 jschwab@albany.edu, SUNY, Albany – Acs
Schwab, Paul, 979-845-3663 pschwab@tamu.edu, Texas A&M University – Sc
Schwankl, Larry J., (530) 752-1130 ljschwankl@ucdavis.edu, University of California, Davis – Zn
Schwarcz, Henry P., (905) 525-9140 (Ext. 24186) schwarcz@mcmaster.ca, McMaster University – Cs
Schwartz, David, (408) 479-6495 daschwar@cabrillo.edu, Cabrillo College – Gu
Schwartz, Frank W., (614) 292-6196 schwartz.11@osu.edu, Ohio State University – Hw
Schwartz, Hilde, (831) 459-5429 hschwartz@pmc.ucsc.edu, University of California, Santa Cruz – Pv
Schwartz, Joshua J., (818) 677-5813 joshua.schwartz@csun.edu, California State University, Northridge – GiCc
Schwartz, Matthew C., 850-474-3469 mschwartz@uwf.edu, University of West Florida – OcCl
Schwartz, Susan, 831-459-3133 syschwar@ucsc.edu, University of California, Santa Cruz – Ys
Schwartz, Theresa M., (814) 332-2873 tschwartz@allegheny.edu, Allegheny College – Gs
Schwarz, Karen, (610)436-2788 kschwarz@wcupa.edu, West Chester University – XaZe
Schweitzer, Carrie E., 330-672-2505 cschweit@kent.edu, Kent State University – Pi
Schweitzer, Carrie E., 330-244-3303 cschweit@kent.edu, Kent State University at Stark – PiGg
Schwerdtner, Walfried M., (416) 978-5080 fried@quartz.geology.utoronto.ca, University of Toronto – GctGq

Schwert, Donald P., (701) 231-7496 donald.schwert@ndsu.edu, North Dakota State University – PeGe
Schwimmer, David R., 706 569-3028 schwimmer_david@columbusstate.edu, Columbus State University – Pv
Schwimmer, Reed A., 609-896-5346 rschwimmer@rider.edu, Rider University – GsOgn
Schwob, Stephanie L., 214-768-2770 Southern Methodist University – Zn
Sclater, John G., (858) 534-3051 jsclater@ucsd.edu, University of California, San Diego – Yh
Scoates, James S., (604) 822-3667 jscoates@eoas.ubc.ca, University of British Columbia – GizGv
Scotese, Christopher R., (817) 272-2987 cscotese@exchange.uta.edu, University of Texas, Arlington – Gt
Scott, Andrew C., +44 1784 443608 A.Scott@rhul.ac.uk, Royal Holloway University of London – Pb
Scott, Christopher A., cascott@email.arizona.edu, University of Arizona – Hw
Scott, Craig, (403) 823-7707 Royal Tyrrell Museum of Palaeontology – Pv
Scott, Darren M., (905) 525-9140 (Ext. 24953) scottdm@mcmaster.ca, McMaster University – Zi
Scott, David B., (902) 494-3604 Dalhousie University – Pm
Scott, James M., +64 3 479-7520 james.scott@otago.ac.nz, University of Otago – Gp
Scott, Kathy, (604) 822-5606 kscott@eos.ubc.ca, University of British Columbia – Zn
Scott, Larry, 303-384-2631 lmscott@mines.edu, Colorado Geological Survey – Zy
Scott, Robert B., 512-471-0375 rscott@ig.utexas.edu, University of Texas, Austin – Og
Scott, Robert W., (918) 230-3436 rwscott@cimtel.net, The University of Tulsa – GsrPi
Scott, Steven D., (416) 978-5424 scottsd@es.utoronto.ca, University of Toronto – GuEmCm
Scott, Thomas A., (951) 827-5115 thomas.scott@ucr.edu, University of California, Riverside – Zy
Scott, Tim, (401) 254-3108 Roger Williams University – Ob
Scott, Verne H., (530) 752-0453 vhscott@ucdavis.edu, University of California, Davis – Hs
Scott, Vernon, 405-744-6358 vscott@okstate.edu, Oklahoma State University – Ze
Scott Smith, Barbara H., (604) 984-9609 bhssmith@allstream.net, University of British Columbia – Gz
Scott-Dupree, Cynthia, (519) 824-4120 Ext.52477 cscottdu@uoguelph.ca, University of Guelph – Zn
Scotton, Paolo, +390498279120 paolo.scotton@unipd.it, Università degli Studi di Padova – Hg
Scow, Kate, (530) 752-4632 kmscow@ucdavis.edu, University of California, Davis – So
Scranton, Mary I., (631) 632-8735 mary.scranton@stonybrook.edu, SUNY, Stony Brook – Oc
Screaton, Elizabeth J., (352) 3924612 screaton@ufl.edu, University of Florida – Hg
Scudder, Sylvia J., 3523921721 x246 scudder@flmnh.ufl.edu, University of Florida – Pg
Scuderi, Louis A., (505) 277-4204 tree@unm.edu, University of New Mexico – GmPcGs
Scyphers, Steven, 781.581.7370 s.scyphers@northeastern.edu, Northeastern University – Zn
Seager, Richard, seager@ldeo.columbia.edu, Columbia University – Am
Seager, Sara, 617-253-6775 seager@mit.edu, Massachusetts Institute of Technology – Xy
Seal, Thom, (775) 682-8813 tseal@unr.edu, University of Nevada, Reno – NxEmNm
Seal, Thom, 775-784-1813 tseal@unr.edu, University of Nevada, Reno – Nx
Seaman, Nelson L., (814) 863-1583 Pennsylvania State University, University Park – Am
Seaman, Sheila J., (413) 545-2822 sjs@geo.umass.edu, University of Massachusetts, Amherst – Gi
Searcy, Steven, (619) 260-2793 ssearcy@sandiego.edu, University of San Diego – Ob
Searle, Mike, +44 (1865) 272022 mike.searle@earth.ox.ac.uk, University of Oxford – Gt
Searls, Mindi L., 402-472-6934 University of Nebraska, Lincoln – YgGg
Sears, Heather, 7815817370 x316 h.sears@neu.edu, Northeastern University – Zn

Sears, James W., (406) 243-5251 james.sears@umontana.edu,
    University of Montana – GctEg
Sebol, Lesley, (303) 384-2633 lsebol@mines.edu,
    Colorado Geological Survey – HwGeZg
Secco, Luciano, I+390498279158 luciano.secco@unipd.it,
    Università degli Studi di Padova – Gz
Secco, Richard A., (519) 661-4079 secco@uwo.ca,
    Western University – Gy
Secord, Ross, (402) 472-2663 rsecord2@unl.edu,
    University of Nebraska, Lincoln – PvCs
Sediek, Kadry N., 002-03-3921595 kknsed@yahoo.com,
    Alexandria University – GgsGd
Sedivy, Robert, 402-465-9021 rasedivy@anl.gov,
    Argonne National Laboratory – Hw
Sedivy, Robert A., (630) 252-1897 rasedivy@anl.gov,
    Argonne National Laboratory – Hw
Sedlacek, Alexa, 319-273-3072 alexa.sedlacek@uni.edu,
    University of Northern Iowa – CsGr
Seeber, Leonardo, nano@ldeo.columbia.edu,
    Columbia University – Ys
Seedorff, Eric, (520) 626-3921 seedorff@email.arizona.edu,
    University of Arizona – Eg
Seeger, Cheryl M., (573) 368-2184 cheryl.seeger@dnr.mo.gov,
    Missouri Dept of Natural Resources – Gig
Seeley, Mark W., (612) 625-4724 mseeley@umn.edu,
    University of Minnesota, Twin Cities – As
Seewald, Jeffrey S., (508) 289-2966 jseewald@whoi.edu,
    Woods Hole Oceanographic Institution – Cp
Segall, Marylin, 801-585-5730 mpsegall@egi.utah.edu,
    University of Utah – GeOu
Segall, Paul, (650) 725-7241 Stanford University – Yg
Segars, William P., (706) 542-9072 University of Georgia – Sc
Seibel, Erwin, (415) 338-2061 San Francisco State University – On
Seibt, Ulrike, (310) 206-4442 useibt@ucla.edu,
    University of California, Los Angeles – PgCg
Seidemann, David E., (718) 951-5761 Graduate School of the City University of New York – Cc
Seidemann, David E., 7189515000 x2882 dseidemann@earthlink.net,
    Brooklyn College (CUNY) – Cc
Seifert, Karl E., (515) 294-5265 kseifert@iastate.edu,
    Iowa State University of Science & Technology – Ct
Seifoullaev, Roustam K., (512) 232-3223 roustam@utig.ig.utexas.edu,
    University of Texas, Austin – Ye
Seifu, Abiye, 706-568-2187 seifu_abiye@columbusstate.edu,
    Columbus State University – Zn
Seigley, Lynette S., 319-335-1598 lynette.seigley@dnr.iowa.gov,
    Iowa Dept of Natural Resources – Gg
Seirer, Jami J., jjseirer2@fhsu.edu,
    Fort Hays State University – ZriZn
Seitz, Jeffery C., (510) 885-3438 jeff.seitz@csueastbay.edu,
    California State University, East Bay – CgZeGx
Selby, Dave, +44 (0) 191 33 42294 david.selby@durham.ac.uk,
    Durham University – Cc
Selden, Paul A., selden@ku.edu,
    University of Kansas – Pig
Selim, Hussein M., (504) 388-2110 Louisiana State University – Sp
Selin, Noelle, (617) 324-2592 selin@mit.edu,
    Massachusetts Institute of Technology – As
Sellmeier, Bettina, +49 (89) 289 25822 sellmeier@tum.de,
    Technical University of Munich – Ngr
Selph, Karen E., (808) 956-7941 selph@hawaii.edu,
    University of Hawai'i, Manoa – Ob
Selverstone, Jane E., selver@unm.edu,
    University of New Mexico – Gpt
Semazzi, Fred H. M., (919) 515-1434 fred_semazzi@ncsu.edu,
    North Carolina State University – As
Semken, Steven, (480) 965-7965 semken@asu.edu,
    Arizona State University – ZeGgZg
Semken, Jr., Holmes A., (319) 335-1830 holmes-semken@uiowa.edu,
    University of Iowa – Pv
Sen, Gautam, (305) 348-2299 seng@fiu.edu,
    Florida International University – Gi
Sen, Mrinal K., (512) 471-0466 University of Texas, Austin – Ye
Sen Gupta, Barun K., (225) 388-5984 barun@geol.lsu.edu,
    Louisiana State University – Pm
Send, Uwe, (858) 822-6710 usend@ucsd.edu,
    University of California, San Diego – Op

Senko, John M., 330 972-8047 senko@uakron.edu,
    University of Akron – Cg
Sennert, Sally K., (202) 633-1805 kuhns@si.edu,
    Smithsonian Inst / Natl Museum of Natural Hist – GvZr
Seong, Jeong C., (678) 839-4069 jseong@westga.edu,
    University of West Georgia – Zir
Sephton, Mark, +44 20 759 46542 m.a.sephton@imperial.ac.uk,
    Imperial College – Xm
Sepúlveda, Julio C., jsepulveda@Colorado.EDU,
    University of Colorado – Co
Sericano, Jose L., (979) 862-2323 jose@gerg.tamu.edu,
    Texas A&M University – Cm
Serne, R. Jeffrey, (360) 539-6162 jeff.serne@pnnl.gov,
    Pacific Northwest National Laboratory – CgSoCl
Serpa, Laura F., (504) 280-6801 lserpa@uno.edu,
    University of New Orleans – Ys
Serpa, Laura F., (915) 747-6058 lfserpa@utep.edu,
    University of Texas, El Paso – Ye
Serreze, Mark, mark.serreze@colorado.edu,
    University of Colorado – As
Sertich, Joseph, 303-370-6331 Joe.Sertich@dmns.org,
    Denver Museum of Nature & Science – Pv
Sessa, Jocelyn A., (215) 299-1149 jsessa@drexel.edu,
    Drexel University – PeCsPq
Sessions, Alex L., 626.395.6445 als@gps.caltech.edu,
    California Institute of Technology – CosCb
Sethi, Parvinder S., (540) 831-5619 psethi@radford.edu,
    Radford University – Gz
Setterholm, Dale, (612) 626-5119 sette001@umn.edu,
    University of Minnesota – Gs
Severinghaus, Jeffrey P., (858) 822-2483 jseveringhaus@ucsd.edu,
    University of California, San Diego – Pe
Severs, Matthew, matthew.severs@stockton.edu,
    Stockton University – Gg
Severs, Matthew R., 609-626-6857 matthew.severs@stockton.edu,
    Stockton University – GxCgEg
Sevilla, Paloma, psevilla@ucm.es,
    Univ Complutense de Madrid – Pi
Sewall, Jacob, 484-646-5864 sewall@kutztown.edu,
    Kutztown University of Pennsylvania – GeAs
Sexton, John L., (618) 453-7374 sexton@geo.siu.edu,
    Southern Illinois University Carbondale – Ye
Sexton, Philip, philip.sexton@open.ac.uk,
    The Open University – Ge
Seyfang, Gill, +44 (0)1603 59 2956 g.seyfang@uea.ac.uk,
    University of East Anglia – Og
Seyfferth, Angelia L., 302-831-4865 angelias@udel.edu,
    University of Delaware – Sc
Seyfried, Jr., William E., 612-624-1333 wes@umn.edu,
    University of Minnesota, Twin Cities – Cm
Seyler, Beverly, 217-244-2389 seyler@isgs.uiuc.edu,
    Illinois State Geological Survey – Eo
Seyoum, Wondwosen M., (309) 438-2833 wmseyou@ilstu.edu,
    Illinois State University – HwqZr
Shaaban, Mohamad N., 002-03-3921595 Moshaaban@yahoo.com,
    Alexandria University – GgsGd
Shaak, Graig D., 3523921721 x257 gdshaak@flmnh.ufl.edu,
    University of Florida – Pe
Shackley, Simon J., +44 (0) 131 650 7862 Simon.Shackley@ed.ac.uk,
    Edinburgh University – Eg
Shade, Harry, 4087412045 x 3678 geology1@earthlink.net,
    West Valley College – Gg
Shade, Janet, 814-362-7560 jas144@pitt.edu,
    University of Pittsburgh, Bradford – Zn
Shafer, Erik, shafere@pdx.edu,
    Portland State University – Gg
Shafer, Jason, 802-626-6225 jason.shafer@northernvermont.edu,
    Northern Vermont University-Lyndon – As
Shah, Subhash N., (405) 325-6871 subhash@ou.edu,
    University of Oklahoma – Np
Shahar, Anat, (202) 478-8929 ashahar@carnegiescience.edu,
    Carnegie Institution for Science – CgYxCs
Shail, Robin, +44 01326 371826 r.k.shail@exeter.ac.uk,
    Exeter University – GtcEm
Shakoor, Abdul, (330) 672-2968 ashakoor@kent.edu,
    Kent State University – Ng
Shakun, Jeremy D., 617-552-1625 jeremy.shakun@bc.edu,

Boston College – Gg
Shalimba, Ester, +264-61-2063839 eshalimba@unam.na,
   University of Namibia – Ggp
Shamberger, Kathryn E., (979) 845-5752 katie.shamberger@tamu.edu,
   Texas A&M University – OcZc
Shams, Asghar, +44 (0)131 451 3904 a.shams@hw.ac.uk,
   Heriot-Watt University – Yg
Shanafield, Margaret, margaret.shanafield@flinders.edu.au,
   Flinders University – Hw
Shanahan, Timothy M., 512-232-7051 tshanahan@jsg.utexas.edu,
   University of Texas, Austin – PeGsCg
Shane, Tyrrell, +353 (0)91 494387 shane.tyrrell@nuigalway.ie,
   National University of Ireland Galway – Gs
Shang, Congxiao, +44 (0)1603 59 3123 c.shang@uea.ac.uk,
   University of East Anglia – Hw
Shank, Stephen G., 717-702-2021 stshank@pa.gov,
   Pennsylvania Bureau of Topographic & Geologic Survey – GziGp
Shankland, Thomas J., (505) 667-4907 shanklan@lanl.gov,
   Los Alamos National Laboratory – Yx
Shannon, Jack D., (630) 252-5807 jack_shannon@anl.gov,
   Argonne National Laboratory – Am
Shannon, Jeremy, (906) 487-3573 jmshanno@mtu.edu,
   Michigan Technological University – Gg
Shapiro, Russell S., (530) 898-4300 rsshapiro@csuchico.edu,
   California State University, Chico – PgGdXb
Shapley, Mark, shap0029@umn.edu,
   University of Minnesota, Twin Cities – Gn
Sharkey, Debra, (916) 691-7210 Cosumnes River College – Zy
Sharma, Govind, (205) 851-5462 aamaxs01@aamu.edu,
   Alabama A&M University – Zn
Sharma, Mukul, 603-646-0024 Mukul.Sharma@dartmouth.edu,
   Dartmouth College – Ge
Sharma, Shikha, 304-293-5603 Shikha.Sharma@mail.wvu.edu,
   West Virginia University – Csa
Sharma, Shiv K., (808) 956-8476 University of Hawai'i, Manoa – Gx
Sharma, Suresh, ssharma@ou.edu,
   University of Oklahoma – Np
Sharma, Vasudha, 612-626-4986 vasudha@umn.edu,
   University of Minnesota, Twin Cities – SpHg
Sharman, Glenn R., 479-575-4471 gsharman@uark.edu,
   University of Arkansas, Fayetteville – Go
Sharp, Jonathan H., (302) 645-4259 jsharp@udel.edu,
   University of Delaware – Oc
Sharp, Martin J., (780) 492-4156 martin.sharp@ualberta.ca,
   University of Alberta – Gl
Sharp, Patricia S., (936) 468-2095 psharp@sfasu.edu,
   Stephen F. Austin State University – Gg
Sharp, S. L., 540-231-4080 llyn@vt.edu,
   Virginia Polytechnic Institute & State University – Ze
Sharp, Thomas G., 480-965-3071 tom.sharp@asu.edu,
   Arizona State University – Gz
Sharp, Zachary D., (505) 277-4204 zsharp@unm.edu,
   University of New Mexico – Cs
Sharp, Jr., John M., (512) 471-3317 jmsharp@jsg.utexas.edu,
   University of Texas, Austin – HqwGe
Sharples, Jonathan, +44 0151 794 4093   Jonathan.Sharples@liverpool.ac.uk,
   University of Liverpool – Ocp
Shaulis, James R., 717.702.2037 jshaulis@pa.gov,
   Pennsylvania Bureau of Topographic & Geologic Survey – EcZe
Shaver, Stephen A., (931) 598-1116 sshaver@sewanee.edu,
   Sewanee: University of the South – Eg
Shaw, Bruce, shaw@ldeo.columbia.edu,
   Columbia University – Ys
Shaw, Christopher A., (323) 857-6317 chrissha@bcf.usc.edu,
   Los Angeles County Museum of Natural History – Pv
Shaw, Cliff S., 506-447-3195 cshaw@unb.ca,
   University of New Brunswick – Giv
Shaw, Colin, 406 994 6760 colin.shaw1@montana.edu,
   Montana State University – GcEmGp
Shaw, Fred, 573-368-2479 fred.shaw@dnr.mo.gov,
   Missouri Dept of Natural Resources – Zn
Shaw, Frederick C., (212) 642-2202 Lehman College (CUNY) – Ps
Shaw, Frederick C., (718) 960-8565 fcshaw@verizon.net,
   Graduate School of the City University of New York – Pi
Shaw, George H., (518) 388-6310 shawg@union.edu,
   Union College – Yx
Shaw, Glenn D., 406-496-4809 gshaw@mtech.edu,
   Montana Tech of the University of Montana – HwCsHs
Shaw, Joey N., (334) 844-3957 shawjo1@auburn.edu,
   Auburn University – Sd
Shaw, John, (780) 492-3573 john.shaw@ualberta.ca,
   University of Alberta – Gl
Shaw, John, (617) 495-8008 shaw@eps.harvard.edu,
   Harvard University – Gt
Shaw, John B., 479-575-7489 shaw84@uark.edu,
   University of Arkansas, Fayetteville – GsrGg
Shaw, Katy, 2538339111 x 4325 kshaw@greenriver.edu,
   Green River Community College – ZgOgGg
Shaw, Kenneth L., (212) 642-2202 kshaw@egi.utah.edu,
   University of Utah – Ye
Shaw, Liz, e.j.shaw@reading.ac.uk,
   University of Reading – Sb
Shaw, Ping-Tung, (919) 515-7276 ping-tung_shaw@ncsu.edu,
   North Carolina State University – Op
Shaw, Richard F., (225) 388-6734 Louisiana State University – Ob
Shcherbakov, Robert, 519-661-2111 ext.84212 rshcherb@uwo.ca,
   Western University – YgsYd
Shchukin, Eugene D., (410) 516-5079 shchukin@jhuvms.hcf.jhu.edu,
   Johns Hopkins University – Zn
Shea, Erin, 907-786-6846 eshea2@uaa.alaska.edu,
   University of Alaska, Anchorage – CcGcCg
Shearer, Peter M., (858) 534-2260 pshearer@ucsd.edu,
   University of California, San Diego – Ysg
Shearer, Jr., Charles K., (505) 277-9159 cshearer@unm.edu,
   University of New Mexico – Gi
Shearman, Kipp, 541-737-1866 shearman@coas.oregonstate.edu,
   Oregon State University – Opn
Shears, Andrew, ashears@mansfield.edu,
   University of Wisconsin Colleges – Zin
shedied, ahmad g., 01005856505 ags00@fayoum.edu.eg,
   Fayoum University – GgHww
Sheehan, Anne F., (303) 492-4597 anne.sheehan@colorado.edu,
   University of Colorado – Yse
Sheehan, Peter M., (414) 278-2741 sheehan@mpm.edu,
   Milwaukee Public Museum – Pi
Sheehan, Peter M., (414) 988-9128 sheehan@uwm.edu,
   University of Wisconsin, Milwaukee – Pie
Sheets, Julie, (614) 406-3298 sheets.2@osu.edu,
   Ohio State University – Gz
Sheldon, Amy L., (585) 245-5988 sheldon@geneseo.edu,
   SUNY, Geneseo – Hw
Shell, Karen M., (541) 737-0980 kshell@coas.oregonstate.edu,
   Oregon State University – As
Shellito, Lucinda, (970) 351-2491 lucinda.shellito@unco.edu,
   University of Northern Colorado – AsPeAm
Shelton, Dale W., (410) 554-5505   dale.shelton@maryland.gov,
   Maryland Department of Natural Resources – ZeGg
Shelton, Kevin L., (573) 882-1004 SheltonKL@missouri.edu,
   University of Missouri – EmCs
Shelton, Sally Y., (605) 394-2487 Sally.Shelton@sdsmt.edu,
   South Dakota School of Mines & Technology – PgvGe
Shem, Linda M., (630) 252-3857 Argonne National Laboratory – Ge
Shen, Jian, (804) 684-7359 shen@vims.edu,
   College of William & Mary – Op
Shen, Linhan, 609-258-4101 linhans@princeton.edu,
   Princeton University – Cas
Shen, Samuel S., (780) 492-0216 shen@ualberta.ca,
   University of Alberta – As
Shen, Xinhua, (319) 273-2536 xinhua.shen@uni.edu,
   University of Northern Iowa – Asm
Shen, Yang, (401) 874-6848 yshen@gso.uri.edu,
   University of Rhode Island – Ou
Sheng, Jinyu, (902) 494-2718 jinyu.sheng@dal.ca,
   Dalhousie University – Op
Shepard, Michael K., (570) 389-4568 mshepard@bloomu.edu,
   Bloomsburg University – XgYgZr
Shepardson, Daniel P., (765) 494-5284 dshep@purdue.edu,
   Purdue University – Ze
Shepherd, Gordon G., 4167362100 x33221 gordon@yorku.ca,
   York University – AsZrAp
Shepherd, Mark A., (512) 574-4227 mark.shepherd@austincc.edu,
   Austin Community College District – AsObZn
Shepherd, Stephanie L., 334-844-4926 sls0070@auburn.edu,
   Auburn University – GmZi

Sheppard, Scott S., (202) 478-8854 ssheppard@carnegiescience.edu,
Carnegie Institution for Science – Xa

Shepson, Paul B., (765) 494-7441 pshepson@purdue.edu,
Purdue University – As

Sheridan, Michael F., mfs@.buffalo.edu,
SUNY, Buffalo – Gv

Sheridan, Robert E., (848) 445-2015 rsheridan@eps.rutgers.edu,
Rutgers, The State University of New Jersey – YrGu

Sherman-Morris, Kathleen M., (662) 268-1032 x242 kms5@geosci.msstate.edu,
Mississippi State University – As

Sherrell, Robert M., (848) 932-6555 (Ext. 252) sherrell@ahab.rutgers.edu,
Rutgers, The State University of New Jersey – CmOu

Sherriff, Barbara L., (204) 474-9786 bl_sherriff@umanitoba.ca,
University of Manitoba – Gz

Sherrod, Brian L., 253-653-8358 bsherrod@uw.edu,
University of Washington – YsGt

Sherrod, Laura, (484) 646-4113 sherrod@kutztown.edu,
Kutztown University of Pennsylvania – YgHwGa

Shervais, John W., (435) 797-1274 john.shervais@usu.edu,
Utah State University – Gi

Sherwood, Owen, 902 494 3604 Owen.Sherwood@dal.ca,
Dalhousie University – CsOoGo

Sherwood, William C., (540) 568-6473 sherwowc@jmu.edu,
James Madison University – So

Sherwood Lollar, Barbara, 416-978-0770 bslollar@chem.utoronto.ca,
University of Toronto – Cs

Shevenell, Lisa, (775) 784-1779 lisaas@unr.edu,
University of Nevada, Reno – Hw

Shevenell, Lisa, (775) 784-6691 lisas@atlasgeoinc.com,
University of Nevada – Cg

Shew, Roger D., 910-962-7676 shewr@uncw.edu,
University of North Carolina, Wilmington – Eo

Shi, Wei, 919-513-4641 wei_shi@ncsu.edu,
North Carolina State University – Sb

Shieh, Sean R., 5196612111 x82467 sshieh@uwo.ca,
Western University – YxGzZm

Shieh, Yuch-Ning, (765) 494-3272 ynshieh@purdue.edu,
Purdue University – CsGpv

Shiel, Alyssa E., (541) 737-5209 ashiel@coas.oregonstate.edu,
Oregon State University – ClsCt

Shields, Robin, +44 (0) 131 451 8215 robin.shields@hw.ac.uk,
Heriot-Watt University – Cg

Shields-Zhou, Graham, +44 020 7679 7821 g.shields@ucl.ac.uk,
University College London – Cg

Shiels, Christine, 604-527-5217 shielsc@douglascollege.ca,
Douglas College – GgxZg

Shiller, Alan M., (228) 688-1178 alan.shiller@usm.edu,
University of Southern Mississippi – OcCtHs

Shiller, Thomas, (432) 837-8238 Thomas.Shiller@sulross.edu,
Sul Ross State University – PvGrd

Shillington, Donna, djs@ldeo.columbia.edu,
Columbia University – Yr

Shim, Sang-Heon, (480) 727-2876 SHDShim@asu.edu,
Arizona State University – Gz

Shim-Chim, Richard, (609) 984-6587 rich.shim-chim@dep.state.nj.us,
New Jersey Geological and Water Survey – Hw

Shimer, Grant, 435 865 8429 grantshimer@suu.edu,
Southern Utah University – Gs

Shimizu, Melinda, mel2066254@maricopa.edu,
Mesa Community College – Gg

Shimizu, Nobumichi, (508) 289-2963 Woods Hole Oceanographic Institution – Ca

Shinn, Eugene, (727) 893-3100 University of South Florida – Gg

Shinoda, Toshiaki, 361-825-3636 Toshiaki.Shinoda@tamucc.edu,
Texas A&M University, Corpus Christi – As

Shipley, Thomas H., (512) 471-0430 tom@ig.utexas.edu,
University of Texas, Austin – Yr

Shipley, Thomas H., 512-471-6156 tom@utig.ig.utexas.edu,
University of Texas, Austin – Ys

Shirer, Hampton N., (814) 863-1992 hns@psu.edu,
Pennsylvania State University, University Park – Am

Shirey, Steven B., (202) 478-8473 sshirey@carnegiescience.edu,
Carnegie Institution for Science – CcGiz

Shirgaokar, Manish, 780.492.2802 shirgaokar@ualberta.ca,
University of Alberta – Zu

Shirley, Terry, 704-687-5925 trshirle@uncc.edu,
University of North Carolina, Charlotte – Asm

Shock, Everett, 480-965-0631 everett.shock@asu.edu,
Arizona State University – Zn

Shoemaker, Kurt A., (740) 351-3395 kshoemaker@shawnee.edu,
Shawnee State University – GxzGm

Sholkovitz, Edward R., (508) 289-2346 esholkovitz@whoi.edu,
Woods Hole Oceanographic Institution – Oc

Shorey, Christian V., (303) 273-3556 cshorey@mines.edu,
Colorado School of Mines – Gg

Shortt, Niamh K., +44 (0) 131 651 7130 Niamh.Shortt@ed.ac.uk,
Edinburgh University – Eg

Shoup, Doug, (620) 421-1530 dshoup@ksu.edu,
Kansas State University – So

Showers, William J., (919) 515-7143 w_showers@ncsu.edu,
North Carolina State University – Cs

Showman, Adam, (520) 621-4021 showman@lpl.arizona.edu,
University of Arizona – Zn

Shragge, Jeffrey, jshragge@mines.edu,
Colorado School of Mines – Ys

Shreve, Ronald L., (310) 825-5273 shreve@ess.ucla.edu,
University of California, Los Angeles – Gm

Shreve, Ronald L., shreve@ess.ucla.edu,
University of Washington – Yr

Shroat-Lewis, Rene A., 501-683-7743 rashroatlew@ualr.edu,
University of Arkansas at Little Rock – PeZe

Shroba, Cynthia S., 702-651- 7427 cindy.shroba@csn.edu,
College of Southern Nevada - West Charleston Campus – Gg

Shroder, Jr., John F., (402) 554-2770 jshroder@unomaha.edu,
University of Nebraska, Omaha – GmlZy

Shropshire, K. Lee, (970) 351-2285 leeshrop@att.net,
University of Northern Colorado – Pg

Shroyer, Emily L., 541-737-1298 eshroyer@coas.oregonstate.edu,
Oregon State University – On

Shroyer, Jim, (785) 532-6101 jshroyer@ksu.edu,
Kansas State University – So

Shu, Jinfu, (202) 478-8963 j.shu@gl.ciw.edu,
Carnegie Institution for Science – Yx

Shuford, James W., (205) 851-5462 jshufard@aamu.edu,
Alabama A&M University – So

Shugar, Dan, dshugar@uw.edu,
University of Washington – GmlZr

Shugart, Jr., Herman H., (804) 924-7642 hhs@virginia.edu,
University of Virginia – Cg

Shuib, Mustaffa Kamal, 03-79674227 mustaffk@um.edu.my,
University of Malaya – Gc

Shukla, Jagadish, (703) 993-1983 jshukla@gmu.edu,
George Mason University – As

Shuller-Nickles, Lindsay C., 864656-1448 lshulle@clemson.edu,
Clemson University – Gz

Shulski, Martha D., mshulski3@unl.edu,
University of Nebraska, Lincoln – As

Shum, CK, 614 292 7118 ckshum@osu.edu,
Ohio State University – YdHg

Shumaker, Robert C., (304) 293-5603 rshumaker@wvu.edu,
West Virginia University – Go

Shuman, Bryan N., 307-766-6442 bshuman@uwyo.edu,
University of Wyoming – PeGne

Shuman, Larry M., (404) 228-7276 University of Georgia – Sc

Shuman, Randy, (206) 296-8243 rshuman531@gmail.com,
University of Washington – Ob

Shumway, Matthew J., (801) 422-2707 jms7@byu.edu,
Brigham Young University – Zn

Shumway, Sandra, (860) 405-9282 sandra.shumway@uconn.edu,
University of Connecticut – Ob

Shuster, Robert D., (402) 554-2457 rshuster@unomaha.edu,
University of Nebraska, Omaha – GiaZe

Shuttleworth, W. James, (520) 621-8787 shuttle@hwr.arizona.edu,
University of Arizona – Hs

Siahcheshm, Kamal, +98 (411) 339 2699 University of Tabriz – Egm

Sibeko, Skhumbuzo, 012 382 6271 sibekosg@tut.ac.za,
Tshwane University of Technology – GggGg

Sibert, John R., sibert@hawaii.edu,
University of Hawai'i, Manoa – Ob

Sibley, Paul, (519) 824-4120 Ext.52707 psibley@uoguelph.ca,
University of Guelph – Zn

Sibray, Steven S., (308) 632-1382 ssibray@unl.edu,
Unversity of Nebraska - Lincoln – HwGg

Sibson, Rick H., +64 3 479-7506 rick.sibson@otago.ac.nz,
University of Otago – Gc
Sicard, Karri, 907-451-5040 karri.sicard@alaska.gov,
Alaska Division of Geological & Geophysical Surveys – Gg
Siddoway, Christine S., (719) 389-6717 csiddoway@coloradocollege.edu,
Colorado College – GctGg
Sidor, Christian A., 206-221-3285 casidor@uw.edu,
University of Washington – Pv
Siebach, Kirsten L., 713.348.6751 ksiebach@rice.edu,
Rice University – Xg
Siegel, Donald I., 315-443-2672 disiegel@syr.edu,
Syracuse University – Hw
Siegler, Matthew, 214-768-2745 msiegler@smu.edu,
Southern Methodist University – YgXy
Siegrist, jr, Henry G., 410 827 8095 g.a.siegrist@gmail.com,
University of Guam – GdzHy
Siemens, Michael A., (573) 368-2134 mike.siemens@dnr.mo.gov,
Missouri Dept of Natural Resources – GgHgZi
Siesser, William G., 615-298-5659 william.g.siesser@vanderbilt.edu,
Vanderbilt University – PmGsu
Siewers, Fredrick D., (270) 745-5988 fred.siewers@wku.edu,
Western Kentucky University – GsPiZc
Sigler, William V., (419) 530-2897 University of Toledo – Zn
Sigloch, Karin, +44 (1865) 272000 karin.sigloch@earth.ox.ac.uk,
University of Oxford – YgsGt
Sigman, Daniel M., (609) 258-2194 sigman@princeton.edu,
Princeton University – ClsPc
Sigurdsson, Haraldur, (401) 874-6596 haraldur@gso.uri.edu,
University of Rhode Island – Ou
Sikora, Todd D., 717-872-3292 Todd.Sikora@millersville.edu,
Millersville University – As
Silliman, James, (361) 825-3718 James.Silliman@tamucc.edu,
Texas A&M University, Corpus Christi – Co
Silva, Lucas C., lcsilva@ucdavis.edu,
University of California, Davis – Sb
Silva, Michael A., (916) 324-0768 California Geological Survey – Ng
Silva-Castro, Jhon, 859-257-1173 john.silva@uky.edu,
University of Kentucky – Nm
Silveira, Maria, 8637351314 x209 mlas@ufl.edu,
University of Florida – So
Silver, Eli A., (831) 459-2266 esilver@pmc.ucsc.edu,
University of California, Santa Cruz – Yr
Silver, Leon T., (626) 395-6490 lsilver@gps.caltech.edu,
California Institute of Technology – Cc
Silver, Mary W., (831) 459-2908 msilver@cats.ucsc.edu,
University of California, Santa Cruz – Ob
Silvertooth, Jeffrey C., (520) 621-7145 silver@ag.arizona.edu,
University of Arizona – So
Silvestri, Alberta, +390498279142 alberta.silvestri@unipd.it,
Università degli Studi di Padova – Gz
Simandl, George J., (250) 952-0413 gsimandle@galaxy.gov.bc.ca,
University of Victoria – En
Simard, Renee-Luce ., (705) 670-5924 renee-luce.simard@ontario.ca,
Ontario Geological Survey – GgEgZe
Simard, Suzanne, 604–822–1955 suzanne.simard@ubc.ca,
University of British Columbia – Sb
Simila, Gerald W., (818) 677-3543 gerry.simila@csun.edu,
California State University, Northridge – Yg
Simjouw, Jean-Paul, 203-932-1253 JSimjouw@newhaven.edu,
University of New Haven – OcCma
Simmons, Craig, craig.simmons@flinders.edu.au,
Flinders University – Hw
Simmons, F. William, (217) 333-4424 University of Illinois, Urbana-Champaign – Sp
Simmons, Lizanne V., (949) 589-0562 Santiago Canyon College – Gr
Simmons, M. G., gene.simmons29@outlook.com,
Massachusetts Institute of Technology – Yg
Simmons, Peter, +44 (0)1603 59 3122 p.simmons@uea.ac.uk,
University of East Anglia – Eg
Simmons, Stuart, 801-581-4122 ssimmons@egi.utah.edu,
University of Utah – EmCqHg
Simmons, William B., (504) 280-6791 wsimmons@uno.edu,
University of New Orleans – Gz
Simmons, William B., 7342866325 , 5042806325 wsimmons@uno.edu,
University of Michigan – Gz
Simms, Alexander, (805) 893-7292 asimms@geol.ucsb.edu,
University of California, Santa Barbara – GsuGm

Simms, Janet E., (601) 634-3493 Mississippi State University – Yg
Simon, Kathy, kasimon@calpoly.edu,
California Polytechnic State University – Zn
Simons, Frederik J., (609) 258-2598 fjsimons@princeton.edu,
Princeton University – YsdYm
Simons, Mark, (626) 395-6984 simons@caltech.edu,
California Institute of Technology – Ys
Simonson, Bruce M., bruce.simonson@oberlin.edu,
Oberlin College – GsXmHw
Simony, Philip S., (403) 220-6679 University of Calgary – Gc
Simpkins, William W., (515) 294-7814 bsimp@iastate.edu,
Iowa State University of Science & Technology – HwGlCs
Simpson, Edward L., (610) 683-4447 simpson@kutztown.edu,
Kutztown University of Pennsylvania – Gs
Simpson, Frank -., 519-253-3000 ext. 2487 franks@uwindsor.ca,
University of Windsor – GsrGe
Simpson, Robert M., (731) 881-7439 msimpson@utm.edu,
University of Tennessee, Martin – AsmZi
Simpson, Jr., H. James, (703) 779-2043 simpsonj@ldeo.columbia.edu,
Columbia University – Cm
Sims, Albert L., (218) 281-8619 simsx008@umn.edu,
University of Minnesota, Twin Cities – Sc
Sims, David, D.W.Sims@soton.ac.uk,
University of Southampton – Ob
Sims, Douglas , 702-651-4840 douglas.sims@csn.edu,
College of Southern Nevada - West Charleston Campus – GeSp
Sims, Kenneth W., 307-766-3386 ksims7@uwyo.edu,
University of Wyoming – CgGvCa
Sinclair, Alastair J., (604) 822-3086 asinclair@eos.ubc.ca,
University of British Columbia – EmGg
Sinclair, Hugh D., +44 (0) 131 650 2518 Hugh.Sinclair@ed.ac.uk,
Edinburgh University – So
Singer, Bradley S., (608) 262-6366 bsinger@geology.wisc.edu,
University of Wisconsin-Madison – Cc
Singer, David M., 330-672-3006 dsinger4@kent.edu,
Kent State University – Gze
Singer, Jared W., (518) 276-4095 singej2@rpi.edu,
Rensselaer Polytechnic Institute – CaZm
Singer, Jill K., (716) 878-4724 singerjk@buffalostate.edu,
Buffalo State College – GsOp
Singer, Michael, +44 01334 462874 michael.singer@st-andrews.ac.uk,
University of St. Andrews – Hg
Singh, Amrita, amrita4@ualberta.ca,
University of Alberta – Zu
Singh, Hanumant, 617.373.7286 ha.singh@neu.edu,
Northeastern University – Zn
Singh, Harbans, (973) 655-7383 singhh@mail.montclair.edu,
Montclair State University – Zn
Singha, Kamini, 303-273-3822 ksingha@mines.edu,
Colorado School of Mines – Hw
Singler, Charles R., 330-941-3611 crsingler@ysu.edu,
Youngstown State University – Gs
Singleton, John, (970) 491-0740 john.singleton@colostate.edu,
Colorado State University – Gct
Sinha, A. Krishna, (540) 231-5580 pitlab@vt.edu,
Virginia Polytechnic Institute & State University – Gt
Sinha, Ashish, (310) 243-3166 asinha@csudh.edu,
California State University, Dominguez Hills – PeCs
Sinton, John M., (808) 956-7751 sinton@hawaii.edu,
University of Hawai'i, Manoa – Gi
Sintubin, Manuel, manuel.sintubin@ees.kuleuven.be,
Katholieke Universiteit Leuven – GtcGa
Sipos, György, +3662544156 gysipos@geo.u-szeged.hu,
Univesity of Szeged – HqYm
Sirbescu, Mona, 989-774-4497 sirbe1mc@cmich.edu,
Central Michigan University – GiCup
Sîrbu, Smaranda D., doina.sirbu@uaic.ro,
Alexandru Ioan Cuza – Ge
Sirk, Robert A., (931) 221-7473 sirkr@apsu.edu,
Austin Peay State University – RnZyu
Sisson, Virginia, 713-743-7634 vbsisson@uh.edu,
University of Houston – Gpt
Sisterson, Doug, 630-252-5836 dlsisterson@anl.gov,
Argonne National Laboratory – As
Sit, Stefany , 312-413-1868 ssit@uic.edu,
University of Illinois at Chicago – Ygs
Sitwell, O.F. George, sitwell@telus.net,

University of Alberta – Zn
Siuta, David, 802-626-6238 david.siuta@northernvermont.edu,
　Northern Vermont University-Lyndon – AsZn
Size, William B., (404) 727-0203 wsize@emory.edu,
　Emory University – GivGa
Sjoblom, Megan, (208) 486-7678 pickardm@byui.edu,
　Brigham Young University - Idaho – GiZeGe
Sjostrom, Derek, (406) 238-7387 derek.sjostrom@rocky.edu,
　Rocky Mountain College – ClGsHs
Skalbeck, John D., 262-595-2490 skalbeck@uwp.edu,
　University of Wisconsin, Parkside – HwYgGg
Skarke, Adam, (662) 268-1032 ext 258 adam.skarke@msstate.edu,
　Mississippi State University – GuYrOn
Skehan, James W., 617-552-8312 james.skehan@bc.edu,
　Boston College – Gc
Skelton, Lawrence H., mskelton5@cox.net,
　University of Kansas – Gge
Skemer, Philip, 314-935-3584 pskemer@levee.wustl.edu,
　Washington University in St. Louis – GcCp
Skidmore, Mark L., (406) 994-7251 skidmore@montana.edu,
　Montana State University – ClPgGl
Skillen, James, 616-526-7546 jrs39@calvin.edu,
　Calvin College – Zu
Skinner, Brian J., (203) 432-3175 brian.skinner@yale.edu,
　Yale University – CgEmGz
Skinner, Christopher , (978) 934-3901 christopher_skinner@uml.edu,
　University of Massachusetts, Lowell – At
Skinner, Lisa, 928-523-5814 Lisa.Skinner@nau.edu,
　Northern Arizona University – ZeGvZi
Skinner, Luke C., +44 (0) 1223 764912 luke00@esc.cam.ac.uk,
　University of Cambridge – GeCmOg
Skinner, Randall, 801-422-6083 randy_skinner@byu.edu,
　Brigham Young University – Gg
Skippen, George B., gskippen@ccs.carleton.ca,
　Carleton University – Cg
Sklar, Leonard, 415-338-1204 leonard@sfsu.edu,
　San Francisco State University – Gm
Skoda, Radek, +420 549 49 7392 rskoda@sci.muni.cz,
　Masaryk University – GzZmGi
Skoog, Annelie, 860-405-9220 annelie.skoog@uconn.edu,
　University of Connecticut – Oc
Skubis, Steven T., (315) 312-2799 steven.skubis@oswego.edu,
　SUNY, Oswego – Am
Skyllingstad, Eric, 541-737-5697 skylling@coas.oregonstate.edu,
　Oregon State University – Am
Slade, Jr., Raymond M., rslade@austincc.edu,
　Austin Community College District – HgsHq
Slater, Brian, 518-473-9988 bslater@mail.nysed.gov,
　New York State Geological Survey – GoEoGr
Slater, Brian K., (614) 292-7314 slater.39@osu.edu,
　Ohio State University – SdZi
Slater, David, (775) 784-4893 dslater@seismo.unr.edu,
　University of Nevada, Reno – Ys
Slater, Gregory F., (905) 525-9140 (Ext. 26388) gslater@mcmaster.ca,
　McMaster University – Cg
Slater, Lee S., 973-353-5109 lslater@newark.rutgers.edu,
　Rutgers, The State University of New Jersey, Newark – YeHw
Slater, Tom, +44 (0) 131 650 9506 tom.slater@ed.ac.uk,
　Edinburgh University – Zn
Slatkin, Montgomery, (510) 642-6300 University of California, Berkeley – Pg
Slatt, Roger M., (405) 325-3253 sgg@hoth.gcn.ou.edu,
　University of Oklahoma – Go
Slattery, William, 937 775-3455 william.slattery@wright.edu,
　Wright State University – ZeGr
Slaughter, Richard, 608/262-2399 rich@geology.wisc.edu,
　University of Wisconsin-Madison – Pg
Slavetskas, Carol, cslavets@binghamton.edu,
　Binghamton University – Zn
Slawinski, Michael A., (709) 737-7541 mslawins@mun.ca,
　Memorial University of Newfoundland – Ys
Sledzinski, Grazyna, (313) 577-2506 ab4478@wayne.edu,
　Wayne State University – Gg
Sleep, Norman H., (650) 723-0882 Stanford University – Yg
Sleezer, Richard O., rsleezer@emporia.edu,
　Emporia State University – ZiSdHs
Sletten, Ronald S., 206-543-0571 sletten@uw.edu,
　University of Washington – Sob
Slingerland, Rudy L., (814) 865-6892 sling@geosc.psu.edu,
　Pennsylvania State University, University Park – Gs
Sloan, Doris, (510) 527-5710 dsloan@berkeley.edu,
　University of California, Berkeley – Pme
Sloan, Heather, 718-960-8008 Heather.Sloan@lehman.cuny.edu,
　Lehman College (CUNY) – Yr
Sloan, Jon R., (818) 677-4880 jon.sloan@csun.edu,
　California State University, Northridge – Pm
Sloan, Lisa C., (831) 459-3693 lsloan@pmc.ucsc.edu,
　University of California, Santa Cruz – Pe
Slobodnik, Marek, +420 549 49 7055 marek@sci.muni.cz,
　Masaryk University – GeEg
Sloss, Craig, 610731382610 c.sloss@qut.edu.au,
　Queensland University of Technology – GsrGm
Slovinsky, Peter, (207) 287-7173 peter.a.slovinsky@maine.gov,
　Dept of Agriculture, Conservation, and Forestry – Gu
Slowey, Niall C., (979) 845-8478 nslowey@ocean.tamu.edu,
　Texas A&M University – Ou
Smaglik, Suzanne M., 307-855-2146 ssmaglik@cwc.edu,
　Central Wyoming College – GgZeCg
Smale, Jody L., 717-702-2020 jsmale@pa.gov,
　Pennsylvania Bureau of Topographic & Geologic Survey – Zn
Small, Christopher, small@ldeo.columbia.edu,
　Columbia University – Zr
Small, Christopher, (845) 365-8354 small@ldeo.columbia.edu,
　Columbia University – ZrYr
Smalley, Gabriela W., 609-896-5097 gsmalley@rider.edu,
　Rider University – ObcOp
Smalley, Glendon W., (615) 598-5714 Sewanee: University of the South – Sf
Smalley, Jr., Robert, (901) 678-2007 rsmalley@memphis.edu,
　University of Memphis – Yd
Smart, Christopher, +44 1752 584764 C.Smart@plymouth.ac.uk,
　University of Plymouth – Pg
Smart, Katie A., 011 717 6549 katie.smart2@wits.ac.za,
　University of the Witwatersrand – CsGx
Smay, Jessica J., 4082982181 x3933 jessica.smay@sjcc.edu,
　San Jose City College – ZeGm
Smayda, Theodore J., (401) 874-6171 tsmayda@gso.uri.edu,
　University of Rhode Island – Ob
Smedes, Harry W., (541) 552-6479 Southern Oregon University – Gi
Smedley, Andrew, +44 0161 306-8770 Andrew.Smedley@manchester.ac.uk,
　University of Manchester – As
Smilnak, Roberta A., (303) 556-3144 Metropolitan State College of Denver – Zg
Smirnov, Aleksey V., (906) 487-2365 asmirnov@mtu.edu,
　Michigan Technological University – YmgGt
Smirnov, Anna, (418) 654-3711 asmirnov@nrcan.gc.ca,
　Universite du Quebec – Cs
Smith, Alan L., (909) 537-5409 alsmith@csusb.edu,
　California State University, San Bernardino – GviGz
Smith, Alison J., (330) 672-3709 alisonjs@kent.edu,
　Kent State University – GnPce
Smith, Ben , 206-616-9176 bsmith@apl.washington.edu,
　University of Washington – GlZr
Smith, Betty E., 217-581-6340 besmith@eiu.edu,
　Eastern Illinois University – Zi
Smith, Brianne, 718-951-5000 x2689 Brianne.Smith43@brooklyn.cuny.edu,
　Brooklyn College (CUNY) – Hs
Smith, C. K., 210-486-0062 csmith55@alamo.edu,
　San Antonio Community College – Zy
Smith, C. Ken, (931) 598-3219 ksmith@sewanee.edu,
　Sewanee: University of the South – Sf
Smith, Catherine H., (505) 667-0113 chsmith@lanl.gov,
　Los Alamos National Laboratory – Ca
Smith, Charles K., csmith55@alamo.edu,
　Alamo Colleges, San Antonio College – Zy
Smith, Craig R., (808) 956-7776 csmith@soest.hawaii.edu,
　University of Hawai'i, Manoa – Ob
Smith, Dan, +440116 252 5355 djs40@le.ac.uk,
　Leicester University – Ge
Smith, David C., (401) 874-6172 dcsmith@gsosun1.gso.uri.edu,
　University of Rhode Island – Ob
Smith, David E., (804) 982-3058 des3e@virginia.edu,
　University of Virginia – Ob
Smith, David R., 410-293-6553 drsmith@usna.edu,

United States Naval Academy – Am
Smith, Dena M., 303-735-2011 dena.smith@colorado.edu,
　University of Colorado – Pg
Smith, Dena M., dena@colorado.edu,
　University of Colorado – PeiPb
Smith, Diane R., (210) 999-7656 dsmith@trinity.edu,
　Trinity University – Gi
Smith, Donald R., (831) 459-5041 dsmith@scipp.ucsc.edu,
　University of California, Santa Cruz – Zn
Smith, Douglas, (512) 471-4261 doug@jsg.utexas.edu,
　University of Texas, Austin – GipGz
Smith, Douglas L., (210) 458-5751 University of Florida – Yh
Smith, Dwight E., 410-293-6566 United States Naval Academy – AmOg
Smith, Elizabeth Y., 702-895-4065 elizabeth.smith@unlv.edu,
　University of Nevada, Las Vegas – ZnnZn
Smith, Eugene I., (702) 895-3971 gsmith@ccmail.nevada.edu,
　University of Nevada, Las Vegas – Gi
Smith, Florence P., (630) 252-7980 Argonne National Laboratory – Ca
Smith, Gerald R., (734) 764-0491 grsmith@umich.edu,
　University of Michigan – Ps
Smith, Grant, 503-838-8862 smithg@wou.edu,
　Western Oregon University – Zg
Smith, H D., SmithHD@cf.ac.uk,
　Cardiff University – Zn
Smith, J. Leslie, (604) 822-4108 lsmith@eos.ubc.ca,
　University of British Columbia – Hw
Smith, James A., (609) 258-4615 jsmith@princeton.edu,
　Princeton University – Hg
Smith, James E., (905) 525-9140 (Ext. 24534) smithja@mcmaster.ca,
　McMaster University – Hw
Smith, Janet S., (717) 477-1757 jssmit@ship.edu,
　Shippensburg University – Zi
Smith, Jason J., 607-778-5116 smithjj@sunybroome.edu,
　Broome Community College – Ggs
Smith, Jeffery L., (509) 335-7648 Washington State University – Sb
Smith, Jen R., jensmith@wustl.edu,
　Washington University in St. Louis – Ga
Smith, Jennifer E., (858) 246-0803 smithj@ucsd.edu,
　University of California, San Diego – Obn
Smith, Jennifer R., 314-935-9451 jensmith@levee.wustl.edu,
　Washington University in St. Louis – Ga
Smith, Jim, j.smith@flinders.edu.au,
　Flinders University – Zn
Smith, Jim, +353 21 4902151 j.smith@ucc.ie,
　University College Cork – Pl
Smith, Jim, +44 023 92 842416 jim.smith@port.ac.uk,
　University of Portsmouth – Ge
Smith, Jon, 785-864-2179 jjsmith@kgs.ku.edu,
　University of Kansas – GsPe
Smith, Joseph P., 410-293-6568 jpsmith@usna.edu,
　United States Naval Academy – CmOg
Smith, K. L., (702) 895-3971 Wichita State University – Zn
Smith, Karen, 630-252-0136 smithk@anl.gov,
　Argonne National Laboratory – Hw
Smith, Karl A., (805) 756-2262 ksmith@tc.umn.edu,
　University of Minnesota, Twin Cities – Nx
Smith, Kathlyn M., (912) 478-5398 KSmith@GeorgiaSouthern.edu,
　Georgia Southern University – Pv
Smith, Ken D., (775) 784-4218 ken@seismo.unr.edu,
　University of Nevada, Reno – YsGt
Smith, Langhorne, 518-473-6262 lsmith@remove@this.mail.nysed.gov,
　New York State Geological Survey – Go
Smith, Larry N., (406) 496-4859 lsmith@mtech.edu,
　Montana Tech of the University of Montana – GsmEo
Smith, Laurence C., (310) 825-3154 lsmith@geog.ucla.edu,
　University of California, Los Angeles – Hs
Smith, Martin, martin.smith@durham.ac.uk,
　Durham University – Pe
Smith, Matthew C., (352) 392-2106 mcsmith@ufl.edu,
　University of Florida – Gi
Smith, Michael, 205-247-3724 msmith@gsa.state.al.us,
　Geological Survey of Alabama – Hg
Smith, Michael E., 928-523-8965 Michael.E.Smith@nau.edu,
　Northern Arizona University – Gsr
Smith, Michael S., (910) 962-3496 smithms@uncw.edu,
　University of North Carolina, Wilmington – GzpGa
Smith, Mike, +44020 8417 2500 Michael.Smith@kingston.ac.uk,
　Kingston University – Zr
Smith, Mike, 970-204-8134 Mike.Smith@frontrange.edu,
　Front Range Community College - Larimer – GgZe
Smith, Mikel, 701-777-4246 mikel.smith@und.edu,
　University of North Dakota – Zn
Smith, Norman D., (402) 472-5362 nsmith3@unl.edu,
　University of Nebraska, Lincoln – Gsm
Smith, Paul H., (505) 667-7494 Los Alamos National Laboratory – Zn
Smith, Paul L., (604) 822-6456 psmith@eoas.ubc.ca,
　University of British Columbia – Pg
Smith, Paul L., 605-394-1990 Paul.Smith@sdsmt.edu,
　South Dakota School of Mines & Technology – As
Smith, Peter, (520) 621-2725 psmith@lpl.arizona.edu,
　University of Arizona – Zr
Smith, Peter C., (902) 426-3474 smithp@mar.dfo-mpo.gc.ca,
　Dalhousie University – Op
Smith, Peter J., peterjsmith@shaw.ca,
　University of Alberta – Zn
Smith, Phillip J., (765) 494-3286 pjsmith@purdue.edu,
　Purdue University – As
Smith, Richard, 479-575-3355 University of Arkansas, Fayetteville – Gm
Smith, Richard S., 7056751151 x2364 rsmith@laurentian.ca,
　Laurentian University, Sudbury – Ye
Smith, Robert, (208) 885-2560 smithbob@uidaho.edu,
　University of Idaho – Cb
Smith, Robert L., (541) 752-6597 rsmith@coas.oregonstate.edu,
　Oregon State University – OpgZn
Smith, Ronald B., (203) 432-3129 ronald.smith@yale.edu,
　Yale University – As
Smith, Ronald M., (509) 376-5831 rmsmith@pnl.gov,
　Pacific Northwest National Laboratory – Hg
Smith, Russell, 915-831-8875 RussellS@epcc.edu,
　El Paso Community College – Gg
Smith, Sean, 207-581-2198 sean.m.smith@maine.edu,
　University of Maine – Hs
Smith, Shane V., 330-941-1752 svsmith@ysu.edu,
　Youngstown State University – Gs
Smith, Stephen V., svsmith@hawaii.edu,
　University of Hawai'i, Manoa – Ou
Smith, Steve, +44 1784 443635 Steve.Smith-1@rhul.ac.uk,
　Royal Holloway University of London – Ge
Smith, Steven C., 916-435-9180 smithsc@arc.losrios.edu,
　American River College – ZynZn
Smith, Steven J., 319-273-6495 smithsj@uni.edu,
　University of Northern Iowa – Zg
Smith, Stewart W., stews@uw.edu,
　University of Washington – Ys
Smith, Susan M., 406-496-4173 ssmith@mtech.edu,
　Montana Tech of The University of Montana – Zn
Smith, Terence E., tsmith@uwindsor.ca,
　University of Windsor – Gi
Smith, Terry L., (805) 756-2262 tsmith@calpoly.edu,
　California Polytechnic State University – So
Smith, Thomas M., (804) 924-3107 tms9a@virginia.edu,
　University of Virginia – Zg
Smith, William H., (314) 935-5638 whsmith@dasi.wustl.edu,
　Washington University in St. Louis – Xc
Smith-Engle, Jennifer M., (361) 825-2436 jennifer.smith-engle@tamucc.edu,
　Texas A&M University, Corpus Christi – Gs
Smith-Konter, Bridget R., 808-956-3618 brkonter@hawaii.edu,
　University of Hawai'i, Manoa – YdXy
Smithson, Jayne, 5102357800 x.4284 jsmithson@contracosta.edu,
　Contra Costa College – Gg
Smithyman, Brenden, bsmithym@uwo.ca,
　Western University – Ye
Smosna, Richard A., (304) 293-9664 rsmosna@wvu.edu,
　West Virginia University – GsdGo
Smrekar, Suzanne E., (818) 354-4192 Jet Propulsion Laboratory – Xy
Smukler, Sean, 604–822–2795 sean.smukler@ubc.ca,
　University of British Columbia – So
Smylie, Douglas E., (416) 736-2100 #66438 doug@core.yorku.ca,
　York University – Yg
Smyth, Joseph R., 303-492-5521 joseph.smyth@colorado.edu,
　University of Colorado – Gz
Smyth, Rebecca C., (512) 471-0232 rebecca.smyth@beg.utexas.edu,
　University of Texas at Austin, Jackson School of Geosciences – Hg
Smyth, Thomas J., (919) 515-2838 North Carolina State University – Sc

Smyth, William D., (541) 737-3029 smyth@coas.oregonstate.edu, Oregon State University – Op
Smythe, William, (310) 825-2434 University of California, Los Angeles – Ys
Snead, John I., (225) 578-3454 snead@lsu.edu, Louisiana State University – Gm
Snider, Henry I., (860) 228-9815 snider@easternct.edu, Eastern Connecticut State University – Ge
Snieder, Roel, 303-273-3456 rsnieder@mines.edu, Colorado School of Mines – Ye
Snoeckx, Hilde, (850) 474-3377 snoeckx@uwf.edu, University of West Florida – OuZg
Snoke, J. Arthur, (540) 231-6028 snoke@vt.edu, Virginia Polytechnic Institute & State University – Ys
Snow, Eleanour, (813) 974-0319 snow@usf.edu, University of South Florida, Tampa – Gz
Snow, Jonathan, 713-743-3298 jesnow@uh.edu, University of Houston – Ca
Snow, Julie A., (724) 738-2503 julie.snow@sru.edu, Slippery Rock University – As
Snyder, Daniel, 478-274-7806 dsnyder@mgc.edu, Middle Georgia College – Gg
Snyder, Jeffrey A., (419) 372-0533 jasnyd@bgnet.bgsu.edu, Bowling Green State University – Gm
Snyder, Jennifer L., (610) 359-5291 jsnyder2@dccc.edu, Delaware County Community College – Zg
Snyder, Lori D., (715) 836-5086 snyderld@uwec.edu, University of Wisconsin, Eau Claire – Gx
Snyder, Noah, 617-552-0839 snyderno@bc.edu, Boston College – GgHg
Snyder, Peter K., 612-625-8207 pksnyder@umn.edu, University of Minnesota, Twin Cities – As
Snyder, Richard L., rlsnyder@ucdavis.edu, University of California, Davis – As
Snyder, Walter S., (208) 426-3645 wsnyder@boisestate.edu, Boise State University – GrcGt
Sobel, Adam H., (212) 854-6587 ahs129@columbia.edu, Columbia University – As
Sobotka, Jerzy, jerzy.sobotka@uwr.edu.pl, University of Wroclaw – YeuEo
Soeder, Daniel J., (605) 394-2802 Dan.Soeder@sdsmt.edu, South Dakota School of Mines & Technology – GoeEo
Sohi, Sran P., +44 (0) 131 651 4471 Saran.Sohi@ed.ac.uk, Edinburgh University – So
Sohn, Robert A., (508) 289-3616 rsohn@whoi.edu, Woods Hole Oceanographic Institution – Yr
Soja, Constance M., (315) 228-7200 csoja@colgate.edu, Colgate University – PiGs
Sokolik, Irina, (404) 894-6180 isokolik@eas.gatech.edu, Georgia Institute of Technology – AsZr
Sokolova, Elena, (204) 747-8252 elena_sokolova@umanitoba.ca, University of Manitoba – Gz
Solan, Martin, +44 (0)23 80593755 M.Solan@soton.ac.uk, University of Southampton – Ob
Solana, Camen, +44 023 92 842394 carmen.solana@port.ac.uk, University of Portsmouth – Gv
Solar, Gary S., (716) 878-4900 solargs@buffalostate.edu, Buffalo State College – GcpGt
Soldat, Douglas J., (608) 263-3631 djsoldat@wisc.edu, University of Wisconsin, Madison – So
Solecki, William, (973) 655-5129 soleckiw@mail.montclair.edu, Montclair State University – Zu
Solecki, William, 212-772-5450 wsolecki@hunter.cuny.edu, Graduate School of the City University of New York – Zy
Solecki, William, 212-7724536 wsolecki@hunter.cuny.edu, Hunter College (CUNY) – Zu
Solferino, Giulio, +44 1784 443585 giulio.solferino@rhul.ac.uk, Royal Holloway University of London – Gg
Solis, Michael P., 614 265 6597 michael.solis@dnr.state.oh.us, Ohio Dept of Natural Resources – GgcGt
Soll, Wendy E., (505) 665-6930 weasel@lanl.gov, Los Alamos National Laboratory – Hw
Solomatov, Viatcheslav S., 314-935-7882 slava@wustl.edu, Washington University in St. Louis – Yg
Solomon, Douglas K., (801) 581-7231 kip.solomon@utah.edu, University of Utah – Hw
Solomon, Keith, (519) 824-4120 Ext 58792 ksolomon@uoguelph.ca, University of Guelph – Zn
Solomon, Sean , 845-365-8714 solomon@ldeo.columbia.edu, Columbia University – Xy
Solomon, Sean C., director@ldeo.columbia.edu, Columbia University – Xgy
Somarin, Alireza, 204-727-9680 somarina@brandonu.ca, Brandon University – Gig
Somasundaran, Ponisseril, (212) 854-2926 ps24@columbia.edu, Columbia University – Nx
Somayazulu, Maddury, (202) 478-8911 zulu@gl.ciw.edu, Carnegie Institution for Science – Yx
Somenhally, Anil, ASomenahally@tamu.edu, Texas A&M University – Sb
Somers, Jr., Arnold E., (229) 333-5664 gsomers@valdosta.edu, Valdosta State University – Zn
Somerville, James M., +44 (0) 131 451 3162 j.m.somerville@hw.ac.uk, Heriot-Watt University – Np
Somerville, Richard C., (858) 534-4644 rsomerville@ucsd.edu, University of California, San Diego – As
Sommer, Ulrich , usommer@geomar.de, Dalhousie University – Ob
Sommerfield, Christopher K., 302-645-4255 cs@udel.edu, University of Delaware – On
Sonder, Ingo, (716)645-6366 ingomark@buffalo.edu, SUNY, Buffalo – Gv
Sonder, Leslie J., 603-646-2372 Leslie.Sonder@dartmouth.edu, Dartmouth College – Gq
Sondergeld, Carl H., (405) 325-6870 csondergeld@ou.edu, University of Oklahoma – NpYex
Sone, Hiroki, (608) 890-0531 hsone@wisc.edu, University of Wisconsin, Madison – NrYxGt
Sonett, Charles P., (520) 621-6935 sonett@dakotacom.net, University of Arizona – Yg
Song, Alex, +44 07472526049 alex.song@ucl.ac.uk, University College London – Ysg
Song, Xiaodong, (217) 333-1841 xsong@illinois.edu, University of Illinois, Urbana-Champaign – Ys
Sonnenberg, Stephen A., 303-384-2182 ssonnenb@mines.edu, Colorado School of Mines – Go
Sonnenwald, Margreta, ‭+49 89 28925822‬ margreta.sonnenwald@tum.de, Technical University of Munich – Ng
Soreghan, Gerilyn S., (405) 325-4482 lsoreg@ou.edu, University of Oklahoma – GsPcZg
Soreghan, Michael J., (405) 325-3393 msoreg@ou.edu, University of Oklahoma – Gs
Sorensen, Sorena S., (202) 633-1820 sorensens@si.edu, Smithsonian Inst / Natl Museum of Natural Hist– Gp
Sorenson, Mary Clare, 715-346-2629 msorenso@uwsp.edu, University of Wisconsin, Stevens Point – Zn
Sorkhabi, Rasoul, 801-581-9070 rsorkhabi@egi.utah.edu, University of Utah – GctGc
Sorlien, Christopher C., 845-359-7631 chris@crustal.ucsb.edu, University of California, Santa Barbara – Yr
Sorooshian, Soroosh, (949) 824-8825 soroosh@uci.edu, University of California, Irvine – Hg
Sorooshian, Soroosh, soroosh@uci.edu, University of Arizona – Hg
Soster, Frederick M., 765-658-4670 fsoster@depauw.edu, DePauw University – Gs
Sottile, Jr., Joseph, 859-257-4616 joe.sottile@uky.edu, University of Kentucky – Nm
Souch, Catherine J., (317) 274-1103 Indiana University, Indianapolis – Gm
Soule, Melissa, (508) 289-2879 msoule@whoi.edu, Woods Hole Oceanographic Institution – Oc
Soule, Peter T., (828) 262-7056 soulept@appstate.edu, Appalachian State University – As
Soule, S. Adam, (508) 289-3213 ssoule@whoi.edu, Woods Hole Oceanographic Institution – Gvu
Soupios, Panteleimon , +966138602689 panteleimon.soupios@kfupm.edu.sa, King Fahd University of Petroleum and Minerals – Yg
Sousa, Luís M., lsousa@utad.pt, Universidade de Trás-os-Montes e Alto Douro – EnGgNg
Sousa, Wayne P., (510) 642-2435 University of California, Berkeley – Ob
Southam, Gordon, (519) 661-3197 gsoutham@uwo.ca, Western University – Pg
Southam, John R., (305) 284-1898 jsoutham@miami.edu,

University of Miami – Op
Southard, John B., (617) 253-3397 southard@mit.edu,
  Massachusetts Institute of Technology – Gs
Southard, Randal J., (530) 752-2199 rjsouthard@ucdavis.edu,
  University of California, Davis – So
Southon, John, (949) 824-3674 jsouthon@uci.edu,
  University of California, Irvine – CcOcPe
Southwell, Benjamin, (906) 635-2076 bsouthwell@lssu.edu,
  Lake Superior State University – Zn
Sowers, Todd, (814) 863-8093 sowers@geosc.psu.edu,
  Pennsylvania State University, University Park – Cs
Spaeth, Matthew P., 217-265-6578 spaeth@illinois.edu,
  Illinois State Geological Survey – Ge
Spahr, Paul, 614 265 6577 paul.spahr@dnr.state.oh.us,
  Ohio Dept of Natural Resources – GeHwZi
Spalding, Brian P., bps@ornl.gov,
  Oak Ridge National Laboratory – Sc
Spane, Frank A., (509) 371-7081 frank.spane@pnl.gov,
  Pacific Northwest National Laboratory – Hgw
Spangler, Daniel P., (352) 392-2106 University of Florida – Hw
Spanos, T.J.T (Tim), (780) 435-5245 tim@phys.ualberta.ca,
  University of Alberta – YgsEo
Sparks, David, (979) 458-1051 david-w-sparks@geos.tamu.edu,
  Texas A&M University – YdGq
Sparks, Thomas N., 859-323-0552 sparks@uky.edu,
  University of Kentucky – Gg
Spayd, Steven E., (609) 984-6587 steve.spayd@dep.state.nj.us,
  New Jersey Geological and Water Survey – HwGb
Spear, Frank S., (518) 276-6103 spearf@rpi.edu,
  Rensselaer Polytechnic Institute – GpCpGc
Specker, Jennifer, (401) 874-6858 jspecker@gsosun1.gso.uri.edu,
  University of Rhode Island – Ob
Speece, Marvin A., (406) 496-4188 mspeece@mtech.edu,
  Montana Tech of the University of Montana – YeuYs
Speed, Don, (602) 285-7244 d.speed@phoenixcollege.edu,
  Phoenix College – Gg
Speer, James, (812) 237-2257 jim.speer@indstate.edu,
  Indiana State University – PeZyGe
Speidel, David H., (718) 261-9524 david.speidel@qc.cuny.edu,
  Queens College (CUNY) – Cgp
Speijer, Robert, robert.speijer@kuleuven.be,
  Katholieke Universiteit Leuven – PmcOo
Spell, Terry L., (702) 895-1171 terry.spell@unlv.edu,
  University of Nevada, Las Vegas – CcGvi
Spence, George D., (250) 721-6187 gspence@uvic.ca,
  University of Victoria – Ys
Spencer, Edgar W., (540) 458-8866 spencere@wlu.edu,
  Washington & Lee University – GcgZe
Spencer, Jeremy M., 330 972-2394 jspencer@uakron.edu,
  University of Akron – AtRnZe
Spencer, Joel Q., (785) 532-2249 joelspen@ksu.edu,
  Kansas State University – CcGs
Spencer, Matt, (906) 635-2085 mspencer1@lssu.edu,
  Lake Superior State University – Gl
Spencer, Matthew, +44 0151 795 4399   M.Spencer@liverpool.ac.uk,
  University of Liverpool – Og
Spencer, Patrick K., (509) 527-5222 spencerp@whitman.edu,
  Whitman College – Pg
Spencer, Ronald A., (403) 220-6447 spencer@geo.ucalgary.ca,
  University of Calgary – Cg
Spera, Frank J., (805) 893-4880 University of California, Santa Barbara – Cp
Sperazza, Michael, 631-632-1687 michael.sperazza@sunysb.edu,
  Stony Brook University – Zn
Sperone, Felice, (313) 577-2506 fgsperone@wayne.edu,
  Wayne State University – GgZr
Spetzler, Hartmut A., 303-492-6715 spetzler@colorado.edu,
  University of Colorado – Ys
Spiegelman, Marc, mspieg@ldeo.columbia.edu,
  Columbia University – Ys
Spiegelman, Marc W., (845) 365-8425 mspieg@ldeo.columbia.edu,
  Columbia University – GqxCg
Spieler, Oliver, 089/2180 4221 spieler@min.uni-muenchen.de,
  Ludwig-Maximilians-Universitaet Muenchen – Gz
Spiess, Richard, +390498279150 richard.spiess@unipd.it,
  Università degli Studi di Padova – Gp
Spigel, Lindsay, (207) 287-7177 lindsay.spigel@maine.gov,
  Dept of Agriculture, Conservation, and Forestry – Gm
Spikes, Kyle T., 512-471-7674 kyle.spikes@jsg.utexas.edu,
  University of Texas, Austin – Ye
Spilde, Michael N., (505) 277-5430 mspilde@unm.edu,
  University of New Mexico – Gz
Spilker, Linda J., (818) 354-1647 linda.j.spilker@jpl.nasa.gov,
  Jet Propulsion Laboratory – XyZr
Spindel, Robert C., (206) 543-1310 spindel@apl.washington.edu,
  University of Washington – Op
Spinelli, Glenn, 575.835.6512 spinelli@nmt.edu,
  New Mexico Institute of Mining and Technology – Hw
Spinler, Joshua C., (501) 569-3544 jxspinler@ualr.edu,
  University of Arkansas at Little Rock – YdGt
Spinosa, Claude, (208) 426-5905 cspinosa@boisestate.edu,
  Boise State University – Pi
Spinrad, Richard, 541-220-1915 rick.spinrad@oregonstate.edu,
  Oregon State University – Op
Spitz, Yvette H., 541-737-3227 yspitz@coas.oregonstate.edu,
  Oregon State University – Yr
Splinter, Dale, splinted@uww.edu,
  University of Wisconsin, Whitewater – Hs
Spokas, Kurt, 612-626-2834 kurt.spokas@ars.usda.gov,
  University of Minnesota, Twin Cities – Spo
Spongberg, Alison L., (419) 530-4091 alison.spongberg@utoledo.edu,
  University of Toledo – Co
Spooner, Alecia, 425-388-9003 aspooner@everettcc.edu,
  Everett Community College – Zg
Spooner, Edward T. C., (416) 978-3280 etcs@geology.utoronto.ca,
  University of Toronto – En
Spooner, Ian S., (902) 585-1312 ian.spooner@acadiau.ca,
  Acadia University – Ge
Spotila, James A., (540) 231-2109 spotila@vt.edu,
  Virginia Polytechnic Institute & State University – Gtm
Spracklen, Dominick, +44(0) 113 34 37488 d.v.spracklen@leeds.ac.uk,
  University of Leeds – Zn
Spratt, Deborah A., (403) 220-6446 spratt@geo.ucalgary.ca,
  University of Calgary – Gc
Spray, John G., (506) 999-3544 jgs@unb.ca,
  University of New Brunswick – GpXgGz
Spreng, Alfred C., (573) 341-4669 aspreng@umr.edu,
  Missouri University of Science and Technology – Gr
Sprenke, Kenneth F., (208) 885-5791 ksprenke@uidaho.edu,
  University of Idaho – Ye
Springer, Abraham E., (928) 523-7198 abe.springer@nau.edu,
  Northern Arizona University – Hw
Springer, Everett P., (505) 667-0569 everetts@lanl.gov,
  Los Alamos National Laboratory – Hq
Springer, Gregory S., (740) 593-9431 springeg@ohio.edu,
  Ohio University – GmPcGa
Springer, Kathleen B., (909) 307-2669 242 kspringer@sbcm.sbcounty.gov,
  San Bernardino County Museum – Zg
Springer, Robert K., springer@brandonu.ca,
  Brandon University – Gi
Springston, George E., (802) 485-2734 gsprings@norwich.edu,
  Norwich University – GmZiYg
Sprinkel, Douglas A., (801) 391-1977 douglassprinkel@utah.gov,
  Utah Geological Survey – GosGg
Sprinkle, James T., (512) 471-4264 echino@jsg.utexas.edu,
  University of Texas, Austin – PigGr
Spry, Paul G., (515) 294-9637 pgspry@iastate.edu,
  Iowa State University of Science & Technology – Em
Spurr, Aaron, (319) 273-3789 aaron.spurr@uni.edu,
  University of Northern Iowa – Zeg
Squelch, Andrew P., +61 8 9266 2324 a.squelch@curtin.edu.au,
  Curtin University – NmYeZe
Squires, Richard L., (818) 677-2514 richard.squires@csun.edu,
  California State University, Northridge – Pi
Squyres, Steven W., (607) 255-3508 sws6@cornell.edu,
  Cornell University – Xg
Sremac, Jasenka, +38514606108 jsremac@yahoo.com,
  University of Zagreb – PesPb
Srimal, Neptune, (305) 919-5969 srimal@fiu.edu,
  Florida International University – Gt
Srinivasan, Balakrishnan, +914132655008 sbala.esc@pondiuni.edu.in,
  Pondicherry University – CcGiz
Sritharan, Subramania I., (937) 376-6275 sri@centralstate.edu,
  Central State University – Hg

SRIVASTAVA, HARI B., +919415353606 hbsrivastava@gmail.com,
  Banaras Hindu University – GctGp
Srivastava, Ramesh C., (312) 502-7139 srivast@geosci.uchicago.edu,
  University of Chicago – AsmAm
Sriver, Ryan, 217-300-0364 rsriver@illinois.edu,
  University of Illinois, Urbana-Champaign – As
Srnka, Len, (858) 822-1510 lsrnka@ucsd.edu,
  University of California, San Diego – Ym
Srogi, LeeAnn, (610) 436-2721 esrogi@wcupa.edu,
  West Chester University – Gp
St. John, James C., (404) 894-1754 jim.stjohn@eas.gatech.edu,
  Georgia Institute of Technology – As
St. John, Kristen E., 540-568-6675 stjohnke@jmu.edu,
  James Madison University – Ou
Stachel, Thomas, (780) 492-0865 thomas.stachel@ualberta.ca,
  University of Alberta – GiCs
Stachnik, Joshua, 610-758-2581 jcs612@lehigh.edu,
  Lehigh University – Ys
Stack, Andrew, (404) 894-3895 stackag@ornl.gov,
  Georgia Institute of Technology – Cg
Stadnyk, Leona, (403) 440-6165 lstadnyk@mtroyal.ca,
  Mount Royal University – Zn
Stafford, C. Russell, (812) 237-3989 Russell.Stafford@indstate.edu,
  Indiana State University – Gam
Stafford, Emily, (828) 227-7367 esstafford@wcu.edu,
  Western Carolina University – PgGg
Stafford, James, (307) 766-2286 Ext. 252 james.stafford@wyo.gov,
  Wyoming State Geological Survey – HgyEg
Stafford, Kevin W., (936) 468-2429 staffordk@sfasu.edu,
  Stephen F. Austin State University – Hw
Stahle, David W., (479) 575-3703 dstahle@uark.edu,
  University of Arkansas, Fayetteville – GeZon
Stahlman, Phillip, (785) 625-3425 stahlman@ksu.edu,
  Kansas State University – So
Stahmann, Paul, (815) 479-7593 pstahmann@mchenry.edu,
  McHenry County College – Zyg
Stakes, Debra, (805) 546-3100 dstakes@cuesta.edu,
  Cuesta College – Gg
Staley, Amie, (612) 626-4819 astaley@umn.edu,
  University of Minnesota – Gl
Staley, Andrew, (410) 260-8818 andrewstaley@maryland.gov,
  Maryland Department of Natural Resources – Hw
Stamm, Alfred J., (315) 312-2806 stamm@oswego.edu,
  SUNY, Oswego – Am
Stampone, Mary D., 603-862-3136 mary.stampone@unh.edu,
  University of New Hampshire – Zy
Stamps, D. S., 540-231-3651 dstamps@vt.edu,
  Virginia Polytechnic Institute & State University – YdGtRn
Stan, Cristiana, (703) 993-5391 cstan@gmu.edu,
  George Mason University – As
Stan, Oana, 0040232201467 cristina.stan@uaic.ro,
  Alexandru Ioan Cuza – GedCb
Standridge, Debbie, Deborah.Standridge@gsw.edu,
  Georgia Southwestern State University – Zn
Stanford, Scott D., (609) 292-2576 scott.stanford@dep.nj.gov,
  New Jersey Geological and Water Survey – Gml
Stanley, Clifford R., (902) 585-1344 cliff.stanley@acadiau.ca,
  Acadia University – Cea
Stanley, Daniel J., (202) 633-1354 Smithsonian Inst / Natl Museum of
  Natural Hist– Ou
Stanley, George R., (210) 486-0045 gstanley@alamo.edu,
  San Antonio Community College – ZgRm
Stanley, Jessica, (208) 885-4704 jessicastanley@uidaho.edu,
  University of Idaho – Gt
Stanley, Thomas M., (405) 325-7281 tmstanley@ou.edu,
  University of Oklahoma – PiGsc
Stanley, Jr., George D., (406) 243-5693 george.stanley@umontana.edu,
  University of Montana – Pi
Stansell, Nathan D., 815-753-1943 nstansell@niu.edu,
  Northern Illinois University – Gl
Stanton, Kathryn, 916-558-2343 stantok@scc.losrios.edu,
  Sacramento City College – GgPg
Stanton, Kelsay, kstanton@wvc.edu,
  Wenatchee Valley College – GgZe
Stanton, Timothy P., (831) 656-3144 stanton@nps.edu,
  Naval Postgraduate School – Op
Staples, Reid, 604-527-5226 rstaple4@douglascollege.ca,
  Douglas College – GpxGg
Stapleton, Michael G., (724) 738-2495 micheal.stapleton@sru.edu,
  Slippery Rock University – So
Starek, Michael , 361-825-3978 Michael.Starek@tamucc.edu,
  Texas A&M University, Corpus Christi – Zi
Stark, Robert, 089/2180 4329 stark@lrz.uni-muenchen.de,
  Ludwig-Maximilians-Universitaet Muenchen – Gz
Stark, Robin, (847) 491-3238 robin.stark@northwestern.edu,
  Northwestern University – Zu
Starr, Richard, 831-771-4400 starr@mlml.calstate.edu,
  Moss Landing Marine Laboratories – Zn
Starrfield, Sumner, (480) 965-7569 sumner.starrfield@asu.edu,
  Arizona State University – YhsYv
Starrs, Paul F., (775) 784-6930 starrs@unr.edu,
  University of Nevada, Reno – Zy
Stasko, Stanislaw, stanislaw.stasko@uwr.edu.pl,
  University of Wroclaw – Hg
Staten, Paul W., (812) 856-5135 pwstaten@indiana.edu,
  Indiana University, Bloomington – As
Staub, James R., (406) 243-4953 james.staub@umontana.edu,
  University of Montana – GsrEo
Stauffer, David R., (814) 863-3932 stauffer@essc.psu.edu,
  Pennsylvania State University, University Park – Am
Stauffer, Mel R., (306) 966-5708 mel.stauffer@usask.ca,
  University of Saskatchewan – Gc
Stead, Douglas, 778-782-6670 Simon Fraser University – Ng
Steadman, David W., (352) 273-1969 dws@flmnh.ufl.edu,
  University of Florida – PvGaPs
Steadman, Todd A., 864-656-2536 tsteadm@clemson.edu,
  Clemson University – ZnnZn
Stearley, Ralph F., (616) 526-6370 rstearle@calvin.edu,
  Calvin College – PvOgRh
Stearman, Will, +61 7 3138 4165 w.stearman@qut.edu.au,
  Queensland University of Technology – Geg
Stearns, David W., (405) 325-3253 mtdstearns@aol.com,
  University of Oklahoma – Gc
Stearns, Leigh, (785) 864-4202 stearns@ku.edu,
  University of Kansas – GlZrf
Stearns, Michael, (801) 863-8498 mstearns@uvu.edu,
  Utah Valley University – Gip
Stearns, Richard G., 615-322-2976 stearnrg@ctrvax.vanderbilt.edu,
  Vanderbilt University – Yg
Steart, David, +61 3 9479 5641 D.Steart@latrobe.edu.au,
  La Trobe University – GgPb
Stebbins, Jonathan F., (650) 723-1140 stebbins@pangea.stanford.edu,
  Stanford University – Cg
Stechmann, Samuel, 608-263-4351 stechmann@wisc.edu,
  University of Wisconsin, Madison – ZnAs
Steck, Lee, (505) 665-3528 Los Alamos National Laboratory – Gg
Steckler, Michael, steckler@ldeo.columbia.edu,
  Columbia University – Yv
Steel, Ron J., (512) 831-8238 rsteel@jsg.utexas.edu,
  University of Texas, Austin – GsoGr
Steel, Ronald J., (512) 471-0954 rsteel@mail.utexas.edu,
  University of Texas, Austin – Gs
Steele, Amber , 573-368-7152 amber.steele@dnr.mo.gov,
  Missouri Dept of Natural Resources – Sd
Steele, Andrew, 202-478-8974 asteele@carnegiescience.edu,
  Carnegie Institution for Science – Xy
Steele, Bill, 206-685-5880 wsteele@uw.edu,
  University of Washington – Ys
Steele, Ken, 479-575-6790 ksteele@uark.edu,
  University of Arkansas, Fayetteville – HgCgGg
Steele, Kenneth F., (501) 575-4403 ksteele@comp.uark.edu,
  University of Arkansas, Fayetteville – Cg
Steele, Michael A., (570) 408-4763 michael.steele@wilkes.edu,
  Wilkes University – Pg
Steen, Andrew, 865-974-0821 asteen1@utk.edu,
  University of Tennessee, Knoxville – CoOb
Steenberg, Julia, (612) 626-1830 and01006@umn.edu,
  University of Minnesota – Gs
Steenburgh, Jim, 801-581-8727 jim.steenburgh@utah.edu,
  University of Utah – As
Steenhuis, Tammo S., (607) 255-2489 tss1@cornell.edu,
  Cornell University – HgSo
Steeples, Don W., (785) 737-3399 don@ku.edu,
  University of Kansas – YseGe

Steer, David N., 330 972-2099 steer@uakron.edu,
  University of Akron – Yes
Stefani, Cristina, +390498279195 cristina.stefani@unipd.it,
  Università degli Studi di Padova – Gd
Stefano, Christopher J., 906-487-3028 cjstefano@mtu.edu,
  Michigan Technological University – GzxCg
Stefanova, Ivanka, 651-646-0665 stefa014@umn.edu,
  University of Minnesota, Twin Cities – Pl
Steffens, Katja, 089/2180 6533 katja.steffens@lmu.de,
  Ludwig-Maximilians-Universitaet Muenchen – Gg
Steger, John M., 305-237-2609 jsteger@mdc.edu,
  Miami Dade College (Kendall Campus) – OgAmZg
Stegman, Dave, (858) 822-0767 dstegman@ucsd.edu,
  University of California, San Diego – Yd
Steidl, Gregg M., (609) 984-6587 gregg.steidl@dep.state.nj.us,
  New Jersey Geological and Water Survey – Ge
Steidl, Jamison H., (805) 893-4905 steidl@eri.ucsb.edu,
  University of California, Santa Barbara – YsNeYg
Steig, Eric J., (206) 685-3715 steig@uw.edu,
  University of Washington – CsGlAs
Steiger, Scott, (315) 312-2802 steiger@oswego.edu,
  SUNY, Oswego – Am
Stein, Carol A., (312) 996-9349 cstein@uic.edu,
  University of Illinois at Chicago – Yg
Stein, Seth, (847) 491-5265 seth@earth.northwestern.edu,
  Northwestern University – YsdYg
Steinacker, Reinhold, 0043 1 4277 53730 reinhold.steinacker@univie.ac.at,
  University of Vienna – Ams
Steinberg, Roger T., (361) 698-1665 rsteinb@delmar.edu,
  Del Mar College – GgoPg
Steinberger, Julia, +44 (0)785 607 9625 J.K.Steinberger@leeds.ac.uk,
  University of Leeds – Zu
Steinen, Randolph P., (860) 933-2590 randolph.steinen@ct.gov,
  Dept of Energy and Environmental Protection – Gsg
Steiner, Jeffrey, (212) 650-6465 City College (CUNY) – Cp
Steinker, Don C., (419) 372-7200 Bowling Green State University – Pg
Steinman, Byron A., 218-726-7435 bsteinma@d.umn.edu,
  University of Minnesota, Duluth – Gn
Steinschneider, Scott, 607-255-2155 ss3378@cornell.edu,
  Cornell University – Rw
Steller, Diana, 831-771-4400 dsteller@mlml.calstate.edu,
  Moss Landing Marine Laboratories – Zn
Stelling, Pete, 360-650-4095 pete.stelling@wwu.edu,
  Western Washington University – GxEgGv
Steltenpohl, Mark G., (334) 844-4893 steltmg@auburn.edu,
  Auburn University – Gtc
Stensrud, David J., (814) 863-7714 djs78@psu.edu,
  Pennsylvania State University, University Park – Am
Stephen, Daniel, (801) 863-8584 Daniel.Stephen@uvu.edu,
  Utah Valley University – PiGsPe
Stephen, Ralph A., (508) 289-2583 rstephen@whoi.edu,
  Woods Hole Oceanographic Institution – Ys
Stephens, Jason H., (512) 484-0874 jason.stephens@austincc.edu,
  Austin Community College District – YrGuYe
Stephenson, Gerry, gerry.stephenson@rogers.com,
  University of Guelph – Zn
Stephenson, Randell A., +44 (0)1224 274817 r.stephenson@abdn.ac.uk,
  University of Aberdeen – YgGtZg
Stepien, Carol A., 419-530-8362 carol.stepien@utoledo.edu,
  University of Toledo – Zn
Steppe, Cecily N., 410-293-6558 natunewi@usna.edu,
  United States Naval Academy – Ob
Sterling, Shannon, 902 494 7741 shannon.sterling@dal.ca,
  Dalhousie University – HgCbHs
Stermer, Ed, estermer@icc.cc.il.us,
  Illinois Central College – Zg
Stern, Charles R., 303-492-7170 charles.stern@colorado.edu,
  University of Colorado – Gi
Stern, Herschel I., hstern@miracosta.edu,
  MiraCosta College – Zy
Stern, Robert J., (972) 883-2442 rjstern@utdallas.edu,
  University of Texas, Dallas – Gt
Sternberg, Ben K., (520) 621-2439 bkslasi@u.arizona.edu,
  University of Arizona – Ye
Sternberg, Richard W., (206) 543-6487 rws@ocean.washington.edu,
  University of Washington – Ou
Sternberg, Robert S., (717) 291-4133 rob.sternberg@fandm.edu,
  Franklin and Marshall College – YmGa
Sternberg, Rolf, (973) 655-7386 Montclair State University – Zn
Stetler, Larry D., (605) 394-2461 Larry.Stetler@sdsmt.edu,
  South Dakota School of Mines & Technology – NxGsm
Stevens, Calvin H., (408) 924-5029 calvin.stevens@sjsu.edu,
  San Jose State University – PsGrs
Stevens, De Anne S., (907) 451-5014 deanne.stevens@alaska.gov,
  Alaska Division of Geological & Geophysical Surveys – Ng
Stevens, Joan M., (805) 756-2261 jstevens@calpoly.edu,
  California Polytechnic State University – Zn
Stevens, Liane M., (936) 468-2024 stevenslm@sfasu.edu,
  Stephen F. Austin State University – Gpc
Stevens, Philip S., (812) 855-0732 pstevens@indiana.edu,
  Indiana University, Bloomington – As
Stevens, Robert G., (509) 786-9231 stevensr@wsu.edu,
  Washington State University – Sb
Stevens, Stan, 413-545-0773 sstevens@geo.umass.edu,
  University of Massachusetts, Amherst – Zy
Stevens (Landon), Lora R., 562-985-4817 lsteven2@csulb.edu,
  California State University, Long Beach – Gn
Stevenson, David, +44 (0) 131 650 6750 David.S.Stevenson@ed.ac.uk,
  Edinburgh University – As
Stevenson, David J., (626) 395-6108 djs@gps.caltech.edu,
  California Institute of Technology – Xy
Stevenson, John A., +44 (0) 131 650 7526 john.stevenson@ed.ac.uk,
  Edinburgh University – Gv
Stevenson, Ross, 514-987-3000 #7835 stevenson.ross@uqam.ca,
  Universite du Quebec a Montreal – Cg
Steward, Grieg F., 808-956-6775 grieg@hawaii.edu,
  University of Hawai'i, Manoa – Ob
Stewart, Alexander K., (315) 229-5087 astewart@stlawu.edu,
  St. Lawrence University – GlmHw
Stewart, Brian W., (412) 624-8883 bstewart@pitt.edu,
  University of Pittsburgh – CclCq
Stewart, Dion C., 678-240-6227 dstewart29@gsu.edu,
  Georgia State University, Perimeter College, Alpharetta Campus – Gzi
Stewart, Esther K., (608) 263-3201 esther.stewart@wgnhs.uwex.edu,
  University of Wisconsin, Extension – Gg
Stewart, Gary, 405-372-6063 Oklahoma State University – Go
Stewart, Gillian, 718-997-3104 gillian.stewart@qc.cuny.edu,
  Graduate School of the City University of New York – Oc
Stewart, Kevin G., (919) 962-0683 kgstewar@email.unc.edu,
  University of North Carolina, Chapel Hill – Gc
Stewart, Mark T., (813) 974-8749 mark@usf.edu,
  University of South Florida, Tampa – Hw
Stewart, Michael A., (217) 244-5025 stewart1@illinois.edu,
  University of Illinois, Urbana-Champaign – Gi
Stewart, Robert, 713-743-3081 rrstewart@uh.edu,
  University of Houston – Ye
Stewart, Robert H., rstewart@ocean.tamu.edu,
  Texas A&M University – Op
Stewart, Robert R., 713-743-3399 rrstewart@uh.edu,
  University of Houston – YesYu
Stewart, Robert R., (403) 220-3265 stewart@ucalgary.ca,
  University of Calgary – Ye
Steyn, Douw G., (604) 364-1266 dsteyn@eos.ubc.ca,
  University of British Columbia – Asm
Stickel, Robert, (404) 385-4413 robert.stickel@eas.gatech.edu,
  Georgia Institute of Technology – As
Stickney, Michael C., (406) 496-4332 mstickney@mtech.edu,
  Montana Tech of The University of Montana – YsGtm
Stickney, Robert R., (979) 845-3854 stickney@tamu.edu,
  Texas A&M University – Ob
Stidham, Christiane W., (631) 632-8059 christiane.stidham@sunysb.edu,
  Stony Brook University – YgGe
Stieglitz, Marc, (404) 385-6530 marc.stieglitz@ce.gatech.edu,
  Georgia Institute of Technology – Hsg
Stieglitz, Ronald D., (920) 465-2371 stieglir@uwgb.edu,
  University of Wisconsin, Green Bay – GrlHy
Stierle, Andrea, (406) 496-4117 andrea.stierle@umontana.edu,
  Montana Tech of the University of Montana – Zn
Stierman, Donald J., (419) 530-2860 donald.stierman@utoledo.edu,
  University of Toledo – YgGeYs
Stigall, Alycia L., (740) 593-0393 stigall@ohio.edu,
  Ohio University – Pg
Stillings, Lisa, (775) 784-5803 University of Nevada, Reno – Cm
Stimer, Debra, 330-244-3511 Kent State University at Stark – Zn

Stimpson, Ian G., (+44) 01782 733182 i.g.stimpson@keele.ac.uk, Keele University – YeuRc
Stinchcomb, Gary E., (270) 809-6761 gstinchcomb@murraystate.edu, Murray State University – Ge
Stine, Alexander, 415-338-1209 stine@sfsu.edu, San Francisco State University – At
Stine, Scott W., 510-885-3159 scott.stine@csueastbay.edu, California State University, East Bay – Gm
Stinger, Lindsay C., +44(0) 113 34 37530 l.stringer@leeds.ac.uk, University of Leeds – Zn
Stinson, Amy L., 949-451-5622 astinson@ivc.edu, Irvine Valley College – Gc
Stinson, Amy L., (949) 361-1260 mesiem@aol.com, Santiago Canyon College – Gc
Stix, John, (514) 398-5391 john.stix@mcgill.ca, McGill University – Gv
Stixrude, Lars, +44 020 7679 37929 l.stixrude@ucl.ac.uk, University College London – Yg
Stock, Carl W., 205-348-1883 cstock@geo.ua.edu, University of Alabama – Pi
Stock, Joann M., (626) 395-6938 jstock@gps.caltech.edu, California Institute of Technology – GtcYr
Stocker, Michelle, 540-231-8417 stockerm@vt.edu, Virginia Polytechnic Institute & State University – Pv
Stockinger, Georg M., +49 89 289 25885 georg.stockinger@tum.de, Technical University of Munich – NrtNx
Stockli, Daniel, 512-475-6037 stockli@jsg.utexas.edu, University of Texas, Austin – Gc
Stocks, Lee, (570) 662-4612 lstocks@mansfield.edu, Mansfield University – ZyGgm
Stockton, Charles W., (520) 621-7680 stockton@ltrr.arizona.edu, University of Arizona – Hq
Stoddard, Edward F., (919) 515-7939 skip_stoddard@ncsu.edu, North Carolina State University – Gp
Stoecker, Diane, (410) 221-8407 stoecker@hpl.umces.edu, University of Maryland – Ob
Stoelinga, Mark T., (206) 708-8588 mstoelinga@3tier.com, University of Washington – As
Stoessel, Achim, (979) 862-4170 astoessel@ocean.tamu.edu, Texas A&M University – OpAmGl
Stoessel, Marion, 979 845 7662 mstoessel@ocean.tamu.edu, Texas A&M University – Op
Stoessell, Ronald K., (504) 280-6795 rstoesse@uno.edu, University of New Orleans – Cl
Stofan, Ellen R., (818) 354-2076 ellen.r.stofan@jpl.nasa.gov, Jet Propulsion Laboratory – Xg
Stoffa, Paul L., (512) 471-6405 pauls@ig.utexas.edu, University of Texas, Austin – Ye
Stokes, Martin, +44 1752 584772 M.Stokes@plymouth.ac.uk, University of Plymouth – Gm
Stolper, Edward M., (626) 395-6504 ems@gps.caltech.edu, California Institute of Technology – Cp
Stone, Alan T., (410) 516-8476 astone@jhu.edu, Johns Hopkins University – ClSc
Stone, Alexander, alexm@umn.edu, University of Minnesota, Twin Cities – Gn
Stone, Glenn D., (314) 935-5239 stone@artsci.wustl.edu, Washington University in St. Louis – Zn
Stone, Jim, 970-339-6664 jim.stone@aims.edu, Aims Community College – Zg
Stone, John O., 206-221-6332 stn@uw.edu, University of Washington – Cca
Stone, Loyd, (785) 532-6101 stoner@ksu.edu, Kansas State University – Sp
Stone, Peter H., phstone@mit.edu, Massachusetts Institute of Technology – As
Stoner, Betsy, (781) 891-7106 Bentley University – ObPe
Stoner, Joseph, 541-737-9002 jstoner@coas.oregonstate.edu, Oregon State University – Gsr
Storey, Craig D., +44 023 92 842245 craig.storey@port.ac.uk, University of Portsmouth – GptCc
Stork, Allen L., (970) 943-3044 astork@western.edu, Western Colorado University – Giv
Stormer, Jr., John C., (904) 456-7884 Rice University – Gi
Storrs, Glenn W., (513) 345-8500 University of Cincinnati – Pv
Storrs, Glenn W., (513) 455-7141 gstorrs@cincymuseum.org, Cincinnati Museum Center – PvGgr

Stotler, Randy, (785) 864-6048 rstotler@ku.edu, University of Kansas – Hw
Stott, Lowell D., (213) 740-5120 stott@usc.edu, University of Southern California – Pm
Stout, James H., (612) 624-4344 jstout@umn.edu, University of Minnesota, Twin Cities – GpzGt
Stover, Susan G., 785-864-2063 sstover@kgs.ku.edu, University of Kansas – Hw
Stowell, Harold H., (205) 348-5098 hstowell@ua.edu, University of Alabama – GpCcGt
Stowell Gale, Julia, 512-232-7957 julia.gale@beg.utexas.edu, University of Texas at Austin, Jackson School of Geosciences – Gc
Stoykova, Kristalina C., +35929792213 stoykova@geology.bas.bg, Bulgarian Academy of Sciences – PemPi
Strachan, Rob, +44 023 92 842279 rob.strachan@port.ac.uk, University of Portsmouth – Cc
Strachan, Scotty, strachan@unr.edu, University of Nevada, Reno – AtPeZy
Stracher, Glenn B., (478) 289-2073 stracher@ega.edu, East Georgia State College – GycEc
Strack, Otto D., (612) 625-3009 strac001@tc.umn.edu, University of Minnesota, Twin Cities – Hw
Straffin, Eric, (814) 732-1574 estraffin@edinboro.edu, Edinboro University of Pennsylvania – GmsSo
Strahler, Alan, (617) 353-5984 Boston University – Zr
Straight, William, (703) 948-7750 wstraight@nvcc.edu, Northern Virginia Community College - Loudoun Campus – GgPv
Strain, Priscilla L., 202-633-2481 Smithsonian Institution / National Air & Space Museum – Zr
Stramski, Dariusz, (858) 534-3353 dstramski@ucsd.edu, University of California, San Diego – Op
Strangeway, Robert J., (310) 206-6247 strange@igpp.ucla.edu, University of California, Los Angeles – Zn
Strasser, Jeffrey C., (309) 794-7218 jeffreystrasser@augustana.edu, Augustana College – GmlGe
Strasser, Stefan, 089/2180 4340 stefan.strasser@vr-web.de, Ludwig-Maximilians-Universitaet Muenchen – Gz
Stratton, James F., (217) 581-2626 jfstratton@eiu.edu, Eastern Illinois University – PgGu
Straub, David, (514) 398-8995 david.straub@mcgill.ca, McGill University – Op
Straub, Derek J., 570-372-4767 straubd@susqu.edu, Susquehanna University – As
Straub, Katherine H., (570) 372-4318 straubk@susqu.edu, Susquehanna University – Am
Straub, Kyle M., 504-862-3273 kmstraub@tulane.edu, Tulane University – GgCg
Straub, Susanne, smstraub@ldeo.columbia.edu, Columbia University – Gv
Straus, David M., (703) 993-5719 dstraus@gmu.edu, George Mason University – As
Strauss, Harald, +49-251-83-33932 hstrauss@uni-muenster.de, Universitaet Muenster – CsGg
Strauss, Justin V., (603) 646-6954 justin.v.strauss@dartmouth.edu, Dartmouth College – GsCgGt
Straw, Byron, (970) 351-2470 byron.straw@unco.edu, University of Northern Colorado – ZeGgl
Strawn, Daniel G., (208) 885-2713 dgstrawn@uidaho.edu, University of Idaho – Sc
Strayer, Luther M., (510) 885-3083 luther.strayer@csueastbay.edu, California State University, East Bay – Gc
Streck, Martin J., (503) 725-3379 streckm@pdx.edu, Portland State University – GivCt
Strecker, Manfred, strecker@geo.uni-potsdam.de, Cornell University – Gt
Streepey Smith, Meg, (765) 973-2168 streeme@earlham.edu, Earlham College – Gt
Streig, Ashley, 503-725-3371 streig@pdx.edu, Portland State University – YsGt
Strick, James E., (717) 291-3856 james.strick@fandm.edu, Franklin and Marshall College – RhGe
Strickland, Richard M., (206) 543-3131 strix@ocean.washington.edu, University of Washington – ObgZg
Stright, Lisa, (970) 491-4296 lisa.stright@colostate.edu, Colorado State University – GoNpYe
Strmiæ Palinkaš, Sabina, +38514605961 sabina.strmic@inet.hr, University of Zagreb – CgeCc

Strobel, Darrell F., (410) 516-7829 strobel@jhu.edu,
Johns Hopkins University – AsZnn
Strobel, Timothy A., 202-478-8943 tstrobel@carnegiescience.edu,
Carnegie Institution for Science – Zm
Strock, Jeffrey S., (507) 752-7372 stroc001@umn.edu,
University of Minnesota, Twin Cities – So
Stroeve, Julienne, j.stroeve@ucl.ac.uk,
University College London – GlZr
Strom, Kyle, (540) 231-0979 strom@vt.edu,
Virginia Polytechnic Institute & State University – HsGs
Strom, Robert G., 520- 621-2720 rstrom@LPL.arizona.edu,
University of Arizona – Xg
Stromberg, Caroline, 206-543-1687 caestrom@uw.edu,
University of Washington – Pbe
Strong, Courtenay, 801-585-0049 court.strong@utah.edu,
University of Utah – As
Strong, Ellen, stronge@si.edu,
Smithsonian Inst / Natl Museum of Natural Hist– ZnPi
Strother, Paul K., (617) 552-8395 strother@bc.edu,
Boston College – PlbPg
Stroup, Justin, justin.stroup@oswego.edu,
SUNY, Oswego – GmHwGe
Struzhkin, Viktor V., 202-478-8952 vstruzhkin@carnegiescience.edu,
Carnegie Institution for Science – Gy
Stuart, Graham, +44(0) 113 34 35217 g.w.stuart@leeds.ac.uk,
University of Leeds – Ys
Stuart, Neil, +44 (0) 131 650 2549 N.Stuart@ed.ac.uk,
Edinburgh University – Zi
Stubbins, Aron, 617.373.5872 a.stubbins@northeastern.edu,
Northeastern University – CgbZc
Stubler, Craig, (805) 756-2188 cstubler@calpoly.edu,
California Polytechnic State University – So
Stucker, James D., (614) 265-6601 james.stucker@dnr.state.oh.us,
Ohio Dept of Natural Resources – EgCgGl
Stucki, Joseph W., (217) 333-9636 University of Illinois, Urbana-Champaign – Sc
Student, James J., 989-774-2295 stude1jj@cmich.edu,
Central Michigan University – CaGiz
Stuiver, Minze, minze@uw.edu,
University of Washington – Cc
Stull, Robert J., (323) 343-2408 rstull@calstatela.edu,
California State University, Los Angeles – Gi
Stull, Roland B., 604-822-5901 rstull@eos.ubc.ca,
University of British Columbia – As
Stumbea, Dan, 0040232201464 dan.stumbea@uaic.ro,
Alexandru Ioan Cuza – Gez
Stump, Brian W., (214) 768-1223 Southern Methodist University – Ys
Stunz, Greg, 361-825-3254 Greg.Stunz@tamucc.edu,
Texas A&M University, Corpus Christi – Zn
Stupazzini, Marco, 089/2180 4143 stupa@geophysik.uni-muenchen.de,
Ludwig-Maximilians-Universitaet Muenchen – Yg
Sturchio, Neil C., (630) 252-3986 Argonne National Laboratory – Cg
Sturchio, Neil C., (302) 831-8706 sturchio@udel.edu,
University of Delaware – ClcCa
Sturges, Bill, +44 (0)1603 59 2018 w.sturges@uea.ac.uk,
University of East Anglia – As
Sturm, Diana, 251-460-6381 dsturm@southalabama.edu,
University of South Alabama – GeZge
Sturmer, Daniel M., (513) 556-3718 daniel.sturmer@uc.edu,
University of Cincinnati – YeGdo
Stute, Martin, (845) 365-8704 martins@ldeo.columbia.edu,
Columbia University – CsHgGe
Stutz, Jochen P., (310) 825-1217 jochen@atmos.ucla.edu,
University of California, Los Angeles – As
Su, Haibin, 361-593-3590 Haibin.Su@tamuk.edu,
Texas A&M University, Kingsville – Zir
Su, Xiaobo, (541) 346-4568 xiaobo@uoregon.edu,
University of Oregon – Zn
Suarez, Celina, (479) 575-4866 casuarez@uark.edu,
University of Arkansas, Fayetteville – ClPqGg
Suárez, Luis Eugenio, luisesua@ucm.es,
Univ Complutense de Madrid – Zn
Sublette, Kerry, 918-631-3085 kerry-sublette@utulsa.edu,
The University of Tulsa – Ge
Subramaniam, Ajit, ajit@ldeo.columbia.edu,
Columbia University – ObZr
Subulwa, Angela G., (920) 424-4105 subulwaa@uwosh.edu,
University of Wisconsin Oshkosh – Zn
Suchy, Daniel R., (785) 864-2160 datares@kgs.ku.edu,
University of Kansas – Gg
Suen, C. J., (559) 278-8656 john_suen@csufresno.edu,
California State University, Fresno – HwCsGe
Sugarman, Peter J., (609) 292-2576 pete.sugarman@dep.state.nj.us,
New Jersey Geological and Water Survey – Gr
Sugarman, Peter P., (609) 292-6842 petes@eps.rutgers.edu,
Rutgers, The State University of New Jersey – Gr
Sugihara, George, (858) 534-5582 gsugihara@ucsd.edu,
University of California, San Diego – Ob
Sui, Daniel, 479-575-2470 University of Arkansas, Fayetteville – Zf
Sui, Daniel Z., (409) 845-7154 sui@geog.tamu.edu,
Texas A&M University – Zf
Sukop, Michael C., (305) 348-3117 sukopm@fiu.edu,
Florida International University – HwqSp
Sulanowska, Margaret, (508) 289-2306 msulanowska@whoi.edu,
Woods Hole Oceanographic Institution – Cm
Sullivan, Jack B., (301) 405-0106 jsull@umd.edu,
University of Maryland – Zu
Sullivan, Raymond, (415) 338-2061 sullivan@sfsu.edu,
San Francisco State University – Go
Sullivan, Walter, (207) 859-5800 wasulliv@colby.edu,
Colby College – Gct
Sullivan-Watts, Barbara K., (401) 874-6659 bsull@gso.uri.edu,
University of Rhode Island – Og
Sultan, Mohamed, (630) 252-1929 Argonne National Laboratory – Cg
Sultan, Mohamed, (269) 387-5487 mohamed.sultan@wmich.edu,
Western Michigan University – ZriGe
Sumida, Stuart S., (909) 537-5346 California State University, San Bernardino – Pv
Summers, Robert, 780-492-0342 robert.summers@ualberta.ca,
University of Alberta – Zn
Summons, Roger, (617) 452-2791 rsummons@mit.edu,
Massachusetts Institute of Technology – CobBCs
Sumner, Dawn Y., (530) 752-5353 dysumner@ucdavis.edu,
University of California, Davis – Gs
Sumner, Esther, +44 (0)23 80592067 E.J.Sumner@soton.ac.uk,
University of Southampton – Gs
Sumrall, Colin, (865) 974-2366 csumrall@utk.edu,
University of Tennessee, Knoxville – Pi
Sumrall, Jeanne L., jlsumrall@fhsu.edu,
Fort Hays State University – Zgn
Sumrall, Jonathan B., 785-628-5348 jbsumrall@fhsu.edu,
Fort Hays State University – GgxZn
Sun, Alexander, 512-475-6190 alex.sun@beg.utexas.edu,
University of Texas, Austin – Gq
Sun, Fengpeng, 816-235-2973 sunf@umkc.edu,
University of Missouri, Kansas City – ZoiZc
Sun, Hongbing, 609-895-5185 hsun@rider.edu,
Rider University – HwCgSc
Sun, Jeff, 307 268 3560 jsun@caspercollege.edu,
Casper College – Zi
Sun, Jingru, 609-258-4101 jingrus@princeton.edu,
Princeton University – As
Sun, Wen-Yih, (765) 494-7681 wysun@purdue.edu,
Purdue University – AsZoAm
Sun, Yifei, (818) 677-3532 yifei.sun@csun.edu,
California State University, Northridge – Zi
Sun, Yuefeng, (979) 845-0635 sun@geos.tamu.edu,
Texas A&M University – GoYe
Sundareshwar, P. V., 605-394-2492 psundareshwar@usaid.gov,
South Dakota School of Mines & Technology – ZeCgSb
Sundby, Bjorn, (418) 723-1986 (Ext. 1767) b.sundby@uquebec.ca,
Universite du Quebec a Rimouski – Oc
Sundby, Bjorn, 514-398-4883 bjorn.sundby@mcgill.ca,
McGill University – Cm
Sundell, Ander, 208.562.3354 andersundell@cwidaho.cc,
College of Western Idaho – Ggc
Sundell, Kent A., (307) 268-2498 ksundell@caspercollege.edu,
Casper College – GgvPv
Sunderlin, David, 610-330-5198 sunderld@lafayette.edu,
Lafayette College – Gg
Suneson, Neil , (405) 325-1472 nsuneson@ou.edu,
University of Oklahoma – Grc
Suneson, Neil H., 405-325-7315 nsuneson@ou.edu,
University of Oklahoma – Gc

Suntharalingam, Parvadha, +44 (0)1603 59 1423 p.suntharalingam@uea.ac.uk,
    University of East Anglia – Oc
Superchi-Culver, Tonia, 303-492-5211 toni.culver@colorado.edu,
    University of Colorado – Pg
Suppe, John, jsuppe@uh.edu,
    University of Houston – Gtc
Surge, Donna M., 919-843-1994 donna64@unc.edu,
    University of North Carolina, Chapel Hill – Pe
Surian, Nicola, +390498279125 nicola.surian@unipd.it,
    Università degli Studi di Padova – Gm
Suriano, Zac, 402-554-2662 zsuriano@unomaha.edu,
    University of Nebraska, Omaha – Atm
Surpless, Benjamin E., (210) 999-7110 bsurples@trinity.edu,
    Trinity University – GctGi
Surpless, Kathleen D., (210) 999-7365 ksurples@trinity.edu,
    Trinity University – Gst
Susak, Nicholas J., (506) 453-4803 nsusak@unb.ca,
    University of New Brunswick – CglEm
Sushama, Laxmi, 514-987-3000 #2414 sushama.laxmi@uqam.ca,
    Universite du Quebec a Montreal – Hy
Sussman, Robert W., (314) 935-5264 rwsussma@artsci.wustl.edu,
    Washington University in St. Louis – Zn
Suszek, Thomas J., suszek@uwosh.edu,
    University of Wisconsin, Oshkosh – Gg
Suter, Simeon, 717-702-2047 ssuter@pa.gov,
    Pennsylvania Bureau of Topographic & Geologic Survey – HwGg
Sutherland, Bruce, 780-492-0573 bruce.sutherland@ualberta.ca,
    University of Alberta – Cm
Sutherland, Dave, 541-346-8753 dsuth@uoregon.edu,
    University of Oregon – Op
Sutherland, Mary K., (406) 496-4410 msutherland@mtech.edu,
    Montana Tech of The University of Montana – Hws
Sutherland, Stuart, (604) 822-0176 ssutherland@eos.ubc.ca,
    University of British Columbia – Pg
Sutherland, Wayne, (307) 766-2286 Ext. 247 wayne.sutherland@wyo.gov,
    Wyoming State Geological Survey – Eng
Suttle, Curtis, (604) 822-8610 csuttle@eos.ubc.ca,
    University of British Columbia – Ob
Suttner, Lee J., (812) 855-4957 suttner@indiana.edu,
    Indiana University, Bloomington – Gd
Sutton, Mark, +44 20 759 47487 m.sutton@imperial.ac.uk,
    Imperial College – Pg
Sutton, Paul C., (303) 871-2399 psutton@du.edu,
    University of Denver – Zc
Sutton, Sally J., (970) 491-5995 sallys@warnercnr.colostate.edu,
    Colorado State University – GdCl
Suyker, Andrew E., asuyker@unl.edu,
    University of Nebraska, Lincoln – As
Sverdrup, Keith A., (414) 229-4017 sverdrup@uwm.edu,
    University of Wisconsin, Milwaukee – YsGt
Sverjensky, Dimitri A., (410) 516-8568 sver@jhu.edu,
    Johns Hopkins University – Cl
Svitra, Zita V., (505) 667-7616 Los Alamos National Laboratory – Zn
Swain, Geoffrey, (321) 674-8096 swain@fit.edu,
    Florida Institute of Technology – Ob
Swanger, Kate, (978) 934-2664 kate_swanger@uml.edu,
    University of Massachusetts, Lowell – Gl
Swann, Abigail L., (206) 616-0486 aswann@atmos.washington.edu,
    University of Washington – AsZn
Swanson, Basil I., (505) 667-5814 Los Alamos National Laboratory – Zn
Swanson, Brian, 250-537-4330 brians@uw.edu,
    University of Washington – Asp
Swanson, Donald A., (808) 967-8863 donswan@usgs.gov,
    University of Hawai'i, Manoa – Gv
Swanson, Karen, (201) 595-2589 William Paterson University – Cc
Swanson, R. L., (631) 632-8704 larry.swanson@stonybrook.edu,
    SUNY, Stony Brook – Og
Swanson, Sherman, (775) 784-4057 sswanson@agnt1.ag.unr.edu,
    University of Nevada, Reno – Hs
Swanson, Susan K., (608) 363-2132 swansons@beloit.edu,
    Beloit College – HwGm
Swanson, Terry W., tswanson@uw.edu,
    University of Washington – CcGe
Swapp, Susan M., (307) 766-2513 swapp@uwyo.edu,
    University of Wyoming – Gp
Swarr, Gretchen, (508) 289-2558 gswarr@whoi.edu,
    Woods Hole Oceanographic Institution – Oc
Swart, Peter K., (305) 421-4103 pswart@rsmas.miami.edu,
    University of Miami – ClPsCs
Swaters, Gordon E., (780) 492-7159 gordon.swaters@ualberta.ca,
    University of Alberta – Op
Sweeney, Mark D., (509) 373-0703 mark.sweeney@pnl.gov,
    Pacific Northwest National Laboratory – Yg
Sweeney, Mark R., (605) 677-6142 mark.sweeney@usd.edu,
    University of South Dakota – Gms
Sweet, Alisan C., 806-834-2398 alisan.sweet@ttu.edu,
    Texas Tech University – Gs
Sweet, Dustin E., (806) 834-8390 dustin.sweet@ttu.edu,
    Texas Tech University – GsdGr
Sweetman, Steve, +44 023 9284 2257 steve.sweetman@port.ac.uk,
    University of Portsmouth – Pg
Swennen, Rudy, rudy.swennen@ees.kuleuven.be,
    Katholieke Universiteit Leuven – Gso
Swenson, John B., 218-726-6844 jswenso2@d.umn.edu,
    University of Minnesota, Duluth – Gr
Swetnam, Thomas W., tswetnam@email.arizona.edu,
    University of Arizona – Zn
Swetnam, Thomas W., (520) 621-2112 tswetnam@ltrr.arizona.edu,
    University of Arizona – Pe
Swett, Keene, (319) 351-4644 keene-swett@uiowa.edu,
    University of Iowa – Gs
Swidinsky, Andrei, aswidins@mines.edu,
    Colorado School of Mines – Ye
Swift, Elijah V., (401) 874-6146 lige@gso.uri.edu,
    University of Rhode Island – Ob
Swift, Robert P., (505) 665-7871 bswift@lanl.gov,
    Los Alamos National Laboratory – Nr
Swift, Stephen A., (508) 289-2626 sswift@whoi.edu,
    Woods Hole Oceanographic Institution – Yr
Swindle, Timothy, 520-621-4128 tswindle@lpl.arizona.edu,
    University of Arizona – Xc
Swisher III, Carl C., 848-445-5363 cswish@eps.rutgers.edu,
    Rutgers, The State University of New Jersey – Cc
Swyrtek, Sheila, (810)232-9312 sheila.swyrtek@mcc.edu,
    Charles Stewart Mott Community College – Ga
Sydora, Richard D., (780) 492-3624 rsydora@phys.ualberta.ca,
    University of Alberta – Xy
Sykes, Lynn R., (845) 359-7428 sykes@ldeo.columbia.edu,
    Columbia University – YsGtZn
Sylva, Sean, 508-289-3546 ssylva@whoi.edu,
    Woods Hole Oceanographic Institution – Oc
Sylvan, Jason B., (979) 845-5105 jasonsylvan@tamu.edu,
    Texas A&M University – ObXb
Sylvester, Paul J., (709) 737-4736 sylvester@sparky2.esd.mun.ca,
    Memorial University of Newfoundland – Ca
Sylvia, Elizabeth R., 410 554-5542 elizabeth.sylvia@maryland.gov,
    Maryland Department of Natural Resources – Ge
Symbalisty, E.M.D., (505) 667-9670 esymbalisty@lanl.gov,
    Los Alamos National Laboratory – Zn
Symons, David T., (519) 253-3000 x2493 dsymons@uwindsor.ca,
    University of Windsor – YmGtEm
Syrett, William, (814) 865-6172 wjs1@psu.edu,
    Pennsylvania State University, University Park – Am
Syrup, Krista A., (708) 974-5615 syrup@morainevalley.edu,
    Moraine Valley Community College – Cs
Syverson, Kent M., (715) 836-3676 syverskm@uwec.edu,
    University of Wisconsin, Eau Claire – GlEnGe
Szatmári, József, +3662544156 szatmari@geo.u-szeged.hu,
    Univesity of Szeged – ZriZy
Szczepanski, Jacek, jacek.szczepanski@uwr.edu.pl,
    University of Wroclaw – Ggc
Szecsody, James E., (509) 372-6080 jim.szecsody@pnl.gov,
    Pacific Northwest National Laboratory – So
Székely, Balázs, balazs.szekely@ttk.elte.hu,
    Eotvos Lorand University – GmZri
Szekielda, Karl-Heinz, 212-772-4019 szeloe.da@aol.com,
    Graduate School of the City University of New York – Zg
Szeliga, Walter, 509-963-2705 walter@geology.cwu.edu,
    Central Washington University – YgsYd
Szente, Istvan, szente@ludens.elte.hu,
    Eotvos Lorand University – Pig
Szeto, Anthony M. K., (416)736-2100 #77703 szeto@yorku.ca,
    York University – Yg

Szilagyi, Jozsef, jszilagyi@unl.edu,
  Unversity of Nebraska - Lincoln – HqAs
Szlavecz, Katalin, (410) 516-8947 szlavecz@jhu.edu,
  Johns Hopkins University – Pi
Szulczewski, Melanie, (540) 654-2345 mszulcze@umw.edu,
  University of Mary Washington – ScClZg
Szunyogh, Istvan, (979) 458-0553 szunyogh@tamu.edu,
  Texas A&M University – As
Szymanowski, Dawid, 609-258-4101 Princeton University – CcGiv
Szymanski, David, (781) 891-2901 dszymanski@bentley.edu,
  Bentley University – GifCg
Szymanski, Jason, 585-292-2423 jszymanski@monroecc.edu,
  Monroe Community College – GlPe
Szynkiewicz, Anna, 865-974-6006 aszynkie@utk.edu,
  University of Tennessee, Knoxville – CslXg
Söllner, Frank, 089/2180 6519 fank.soellner@iaag.geo.uni-muenchen.de,
  Ludwig-Maximilians-Universitaet Muenchen – Gg

T

Tabidian, M. Ali, (818) 677-2536 ali.tabidian@csun.edu,
  California State University, Northridge – Hw
Taboada Castro, María T., 00 34 981 167000 teresat@udc.es,
  Coruna University – Sd
Taboga, Karl, (307) 766-2286 Ext. 226 karl.taboga@wyo.gov,
  Wyoming State Geological Survey – HwGe
Tabrum, Alan R., (412) 622-3265 tabruma@carnegiemnh.org,
  Carnegie Museum of Natural History – Pv
Tacinelli, John C., (507) 285-7501 john.tacinelli@rctc.edu,
  Rochester Community & Technical College – Gig
Tackett, Lydia S., 701-231-6164 lydia.tackett@ndsu.edu,
  North Dakota State University – PeGsPi
Tadesse, Tsegaye, ttadesse2@unl.edu,
  University of Nebraska, Lincoln – AsZr
Taggart, Christopher T., (902) 494-7144 chris.taggart@dal.ca,
  Dalhousie University – Ob
Taggart, Ralph E., 517-353-5175 taggart@msu.edu,
  Michigan State University – Pb
Tagliabue, Alessandro, +44 0151 794 4651 A.Tagliabue@liverpool.ac.uk,
  University of Liverpool – Oc
Tagne, Gilles V., 630 752 7556 gilles.tagne@wheaton.edu,
  Wheaton College – HwZiGe
Taib, Samsudin Hj., 03-79674235 samsudin@um.edu.my,
  University of Malaya – Yg
Taillefert, Martial, (404) 894-6043 mtaillef@eas.gatech.edu,
  Georgia Institute of Technology – ClmCa
Tait, C. Drew, (505) 667-7603 Los Alamos National Laboratory – Cp
Tait, Kim, ktait@rom.on.ca,
  Western University – Xm
Tait, Kimberly T., (416) 586-5820 ktait@rom.on.ca,
  Royal Ontario Museum – GzyXm
Tajik, Atieh, 404 413 5790 geoatt@langate.gsu.edu,
  Georgia State University – Gg
Takahashi, Taro, taka@ldeo.columbia.edu,
  Columbia University – Cm
Takahashi, Taro, (845) 365-8537 taka@ldeo.columbia.edu,
  Columbia University – OcCg
Takeuchi, Akira, 81-76-445-6654 takeuchi@sci.u-toyama.ac.jp,
  University of Toyama – GtcYe
Takle, Eugene S., (515) 294-9871 gstakle@iastate.edu,
  Iowa State University of Science & Technology – Asm
Talbot, Helen, +44 (0) 191 208 6426 helen.talbot@ncl.ac.uk,
  University of Newcastle Upon Tyne – Co
Talbot, James L., (360) 733-4282 talbot@wwu.edu,
  Western Washington University – Gc
Talbot, Robert, 713-743-2725 rtalbot@uh.edu,
  University of Houston – ActAm
Talley, Drew, (619) 260-6810 dtalley@sandiego.edu,
  University of San Diego – ObZe
Talley, John H., 302-831-2833 waterman@udel.edu,
  University of Delaware – As
Talley, Lynne D., (858) 534-6310 ltalley@ucsd.edu,
  University of California, San Diego – OpZc
Talling, Peter, peter.j.talling@durham.ac.uk,
  Durham University – Rn
Talwani, Manik, (713) 348-6067 manik@rice.edu,
  Rice University – Yr
Talwani, Pradeep, talwani@geol.sc.edu,
  University of South Carolina – Ys
Tamish, Mohamed M., 002-03-3921595 mtamish@hotmail.com,
  Alexandria University – GgsCg
Tan, Chunyang, tanc@umn.edu,
  University of Minnesota, Twin Cities – Cm
Taney, R. Marieke, (928) 523-2384 Marieke.Taney@nau.edu,
  Northern Arizona University – ZnnZn
Tanhua, Toste, ttanhua@geomar.de,
  Dalhousie University – Oc
Tanimoto, Toshiro, (805) 893-8375 toshiro@geol.ucsb.edu,
  University of California, Santa Barbara – Ys
Tanner, Benjamin R., (828) 227-3915 btanner@wcu.edu,
  Western Carolina University – Cos
Tapanila, Leif, 208-282-3871 tapaleif@isu.edu,
  Idaho State University – Pi
Tapanila, Lori, 208-282-5024 tapalori@isu.edu,
  Idaho State University – Gg
Tape, Carl, 907-474-5456 carltape@gi.alaska.edu,
  University of Alaska, Fairbanks – Ys
Tapp, J. B., jbt@utulsa.edu,
  The University of Tulsa – GceZi
Tarduno, John A., (585) 275-5713 john@earth.rochester.edu,
  University of Rochester – YmGtXm
Tarka, Robert, robert.tarka@uwr.edu.pl,
  University of Wroclaw – Hg
Tarr, Alexander, 508 929-8474 alexander.tarr@worcester.edu,
  Worcester State University – Zg
Taskey, Ronald D., (805) 756-1160 rtaskey@calpoly.edu,
  California Polytechnic State University – Sf
Tassier-Surine, Stephanie, Stephanie.Surine@dnr.iowa.gov,
  Iowa Dept of Natural Resources – Gg
Tatarskii, Viatcheslav, (404) 894-9224 vvt@eas.gatech.edu,
  Georgia Institute of Technology – As
Tate, Garrett W., garrett.tate@vanderbilt.edu,
  Vanderbilt University – Gc
Tatham, Robert H., (512) 471-9129 tatham@mail.utexas.edu,
  University of Texas, Austin – Yg
Tauxe, John D., (423) 574-5348 Oak Ridge National Laboratory – Hw
Tauxe, Lisa, (858) 534-6084 ltauxe@ucsd.edu,
  University of California, San Diego – Ym
Tavener, Kristi, (807) 343-8677 ktavener@lakeheadu.ca,
  Lakehead University – Zn
Tawabini, Bassam S., +96638607643 bassamst@kfupm.edu.sa,
  King Fahd University of Petroleum and Minerals – GeHsCa
Taylor, Brian, (808) 956-6182 taylorb@hawaii.edu,
  University of Hawai'i, Manoa – Gtu
Taylor, Carolyn, carolyn.taylor@mcmail.maricopa.edu,
  Mesa Community College – Gg
Taylor, Charles J., 859-323-0523 charles.taylor@uky.edu,
  University of Kentucky – HwgCl
Taylor, Danny L., (775) 784-6922 dtaylor@mines.unr.edu,
  University of Nevada, Reno – Nm
Taylor, Edith, 785-864-3621 etaylor@ku.edu,
  University of Kansas – Pb
Taylor, Elwynn, (808) 956-3899 setaylor@iastate.edu,
  Iowa State University of Science & Technology – As
Taylor, Frank J. R., (604) 822-4587 mtaylor@eos.ubc.ca,
  University of British Columbia – Ob
Taylor, Frederick W., (512) 471-0453 fred@ig.utexas.edu,
  University of Texas, Austin – GtmGe
Taylor, G. Jeffrey, (808) 956-3899 University of Hawai'i, Manoa – Xm
Taylor, Gordon T., (631) 632-8688 gordon.taylor@stonybrook.edu,
  SUNY, Stony Brook – Obc
Taylor, Graeme, +44 1752 584770 G.Taylor@plymouth.ac.uk,
  University of Plymouth – Yg
Taylor, Hugh P., (626) 395-6116 hptaylor@gps.caltech.edu,
  California Institute of Technology – Cs
Taylor, John F., (724) 357-4469 jftaylor@iup.edu,
  Indiana University of Pennsylvania -- PsiGs
Taylor, Kenneth B., (919) 707-9211 kenneth.b.taylor@ncdenr.gov,
  North Carolina Geological Survey – YsGgEo
Taylor, Kevin, +44 0161 275-8557 kevin.taylor@manchester.ac.uk,
  University of Manchester – Go
Taylor, Lansing, 801-581-8430 ltaylor@egi.utah.edu,
  University of Utah – Gc
Taylor, Lawrence D., (517) 629-0308 ltaylor@albion.edu,
  Albion College – Gl

Taylor, Mackenzie, mtaylo10@ashland.edu,
    Ashland University – RnGg
Taylor, Matthew, 303-871-2656 mtaylor7@du.edu,
    University of Denver – Zc
Taylor, Michael, michael.taylor@flinders.edu.au,
    Flinders University – Zn
Taylor, Michael H., 785-864-5828 mht@ku.edu,
    University of Kansas – GtmZr
Taylor, Nathan H., 501 683-1085 nathan.taylor@arkansas.gov,
    Arkansas Geological Survey – ZiyGe
Taylor, Penny M., 413-538-3236 pmtaylor@mtholyoke.edu,
    Mount Holyoke College – Gg
Taylor, Peter, +44(0) 113 34 37169 P.G.Taylor@leeds.ac.uk,
    University of Leeds – Ge
Taylor, Peter A., (416)736-2100 #77707 pat@yorku.ca,
    York University – As
Taylor, Rex N., +44 (0)23 80592007 rex@noc.soton.ac.uk,
    University of Southampton – GvCa
Taylor, Richard P., richard_taylor@carleton.ca,
    Carleton University – Cg
Taylor, Robert W., (973) 655-4129 Montclair State University – Zn
Taylor, Ryan W., 914 251 6652 ryan.taylor@purchase.edu,
    SUNY, Purchase – ZiyZg
Taylor, Sid, (902) 867-2299 staylor@stfx.ca,
    Saint Francis Xavier University – Gg
Taylor, Stephen B., 503-838-8398 taylors@wou.edu,
    Western Oregon University – Gm
Taylor, Steven R., (505) 667-1007 taylor@lanl.gov,
    Los Alamos National Laboratory – Ys
Taylor, Ta-Shana A., (305) 284-4254 t.taylor2@miami.edu,
    University of Miami – ZePvGe
Taylor, Wanda J., (702) 895-4615 wanda.taylor@unlv.edu,
    University of Nevada, Las Vegas – Gc
Taylor, Wayne A., (505) 667-4253 Los Alamos National Laboratory – Cl
Taylor, Witt, (781) 891-2158 wtaylor@bentley.edu,
    Bentley University – GgPbe
Tchakerian, Vatche P., (409) 845-7997 Texas A&M University – Gm
Teagle, Damon A., +44 (0)23 80592723 damon.teagle@southampton.ac.uk,
    University of Southampton – CgEmGp
Teasdale, Rachel, 530-898-5547 rteasdale@csuchico.edu,
    California State University, Chico – Gv
Tedford, Richard H., (212) 769-5809 American Museum of Nat Hist – Pv
Teed, Rebecca, 937 775-3446 rebecca.teed@wright.edu,
    Wright State University – ZePeGn
Teeuw, Richard, +44 023 92 842267 richard.teeuw@port.ac.uk,
    University of Portsmouth – AsGmZi
Tefend, Karen S., ktefend@westga.edu,
    University of West Georgia – CgGe
Teixeira, Bernardo, 337-482-1132 bernardo.teixeira@louisiana.edu,
    University of Louisiana at Lafayette – GdrGn
Teixell, Antonio, antonio.teixell@uab.cat,
    Universitat Autonoma de Barcelona – Gtc
Teixell Cácharo, Antoni, ++935811163 antonio.teixell@uab.cat,
    Universitat Autonoma de Barcelona – GtcYx
Tellam, John H., +44 (0)121 41 46138 j.h.tellam@bham.ac.uk,
    University of Birmingham – Hw
Teller, James T., (204) 474-9270 tellerjt@ms.umanitoba.ca,
    University of Manitoba – Gs
Telmer, Kevin, ktelmer@uvic.ca,
    University of Victoria – Cl
Telyakovskiy, Aleksey S., 775.784.1364 alekseyt@unr.edu,
    University of Nevada, Reno – Hqy
Temples, Tommy, (803) 348-0472 ttemples@sc.rr.com,
    Clemson University – GoYg
Templeton, Alan R., (314) 935-6868 temple_a@wustl.edu,
    Washington University in St. Louis – ZicZr
Templeton, Alexis, 303-735-6069 alexis.templeton@colorado.edu,
    University of Colorado – ClZaCq
Templeton, Jeffrey H., (503) 838-8858 templej@wou.edu,
    Western Oregon University – GivZe
Ten Brink, Norman W., tenbrinn@gvsu.edu,
    Grand Valley State University – Gm
Tencate, James, (505) 665-6667 tencate@lanl.gov,
    Los Alamos National Laboratory – Zn
Teng, Fangzhen, 206-543-7615 fteng@uw.edu,
    University of Washington – Cg
Teng, Ta-liang, (213) 740-5838 lteng@usc.edu,
    University of Southern California – YsGtYe
Tenorio, Victor O., (520) 621-3858 vtenorio@email.arizona.edu,
    University of Arizona – Nmx
Teodoriu, Catalin, 405-325-6872 cteodoriu@ou.edu,
    University of Oklahoma – Np
Tepley, III, Frank J., 541-737-2064 ftepley@coas.oregonstate.edu,
    Oregon State University – GiCs
Teplitski, Max, 352-392-1951 maxtep@ufl.edu,
    University of Florida – Zn
Tepper, Jeffrey H., (253) 879-3820 jtepper@pugetsound.edu,
    University of Puget Sound – GiCgGv
ter Voorde, Marlies, +31 20 59 87343 m.ter.voorde@vu.nl,
    Vrije Universiteit Amsterdam – YgGt
Ter-Simonian, Vardui, (213) 740-6106 tersimon@usc.edu,
    University of Southern California – Zn
Tera, Fouad, (202) 478-8472 ftera@carnegiescience.edu,
    Carnegie Institution for Science – CcgXm
Terry, Dennis O., (215) 204-8226 doterry@temple.edu,
    Temple University – SaPcGr
Tertyshnikov, Konstantin, +61 8 9266 2297 Konstantin.Tertyshnikov@curtin.edu.au,
    Curtin University – Ye
Terwey, Wes, terwey@southalabama.edu,
    University of South Alabama – Am
Terzian, Yervant, 607-255-4935 terzian@astro.cornell.edu,
    Cornell University – Xya
Tesso, Tesfaye, (785) 532-7238 ttesso@ksu.edu,
    Kansas State University – So
Tester, Jefferson W., 607-254-7211 jwt54@cornell.edu,
    Cornell University – NgCg
Tetrault, Denis, 519-253-3000 ext. 2495 deniskt@uwindsor.ca,
    University of Windsor – Gg
Tett, Simon F., +44 (0) 131 650 5341 Simon.Tett@ed.ac.uk,
    Edinburgh University – Am
Tettenhorst, Rodney T., 614 247-4246 tettenhorst.2@osu.edu,
    Ohio State University – Gz
Tew, Berry H., 205-247-3679 ntew@gsa.state.al.us,
    Geological Survey of Alabama – GroGs
Tew, Nick, 205.348.4558 bhtew@ua.edu,
    University of Alabama – EoGro
Tewksbury, Barbara J., (315) 859-4713 btewksbu@hamilton.edu,
    Hamilton College – Gc
Textoris, Daniel A., (919) 962-0690 dtextori@email.unc.edu,
    University of North Carolina, Chapel Hill – Gd
Teyssier, Christian P., (612) 624-6801 teyssier@umn.edu,
    University of Minnesota, Twin Cities – GctGg
Tezcan, Levent, +90 (312) 2977750 tezcan@hacettepe.edu.tr,
    Hacettepe University – Hw
Thackray, Glenn D., (208) 282-3565 thacglen@isu.edu,
    Idaho State University – Gl
Thakurta, Joyashish, (269) 387-3667 joyashish.thakurta@wmich.edu,
    Western Michigan University – GiEg
Thale, Paul R., 406-496-4653 pthale@mtech.edu,
    Montana Tech of The University of Montana – Zi
Tharp, Thomas M., (765) 494-8678 ttharp1@purdue.edu,
    Purdue University – Nm
Thatje, Sven, +44 (0)23 80592009 svth@noc.soton.ac.uk,
    University of Southampton – Ob
Thayer, Paul A., (910) 520-8719 thayer@uncw.edu,
    University of North Carolina, Wilmington – GdoHg
Thayn, Jonathan B., 309-438-8112 jthayn@ilstu.edu,
    Illinois State University – Zir
Theiling, Bethany P., 918-631-2754 bethany-theiling@utulsa.edu,
    The University of Tulsa – Cs
Theis, Karen, +44 0161 275-0407 Karen.Theis@manchester.ac.uk,
    University of Manchester – Zn
Theissen, Kevin, 651-962-5243 kmtheissen@stthomas.edu,
    University of Saint Thomas – GnOg
Them, II, Theodore R., themtr@cofc.edu,
    College of Charleston – CmOcZc
Themelis, Nickolas J., (212) 854-2138 njt1@columbia.edu,
    Columbia University – Nx
Theo, Lisa L., 715-346-3737 ltheo@uwsp.edu,
    University of Wisconsin, Stevens Point – Zy
Thériault, Julie Mireille, 514 987-3000 #4276 theriault.julie@uqam.ca,
    Universite du Quebec a Montreal – Zn
Therrien, Francois, (403) 823-7707 Royal Tyrrell Museum of Palaeontol-

ogy – Pv
Therrien, Pierre, therrien@ggl.ulaval.ca,
    Universite Laval – Gq
Therrien, Rene, (418) 656-5400 rene.therrien@ggl.ulaval.ca,
    Universite Laval – HwGe
Theuerkauf, Ethan, ejtheu@illinois.edu,
    University of Illinois at Chicago – On
Thibault, Yves, 613-992-1376 yves.thibault@nrcan.gc.ca,
    Western University – Gz
Thibodeau, Alyson M., (717) 245-8337 thibodea@dickinson.edu,
    Dickinson College – CcGaCs
Thiel, Dr., Volker, +49 (0)551 3914395 vthiel@gwdg.de,
    Georg-August University of Goettingen – CooCo
Thieme, Donald, (229) 333-5752 dmthieme@valdosta.edu,
    Valdosta State University – GaSdGm
Thien, Steve J., (785) 532-7207 sjthien@ksu.edu,
    Kansas State University – Sc
Thigpen, Ryan, 859-2181532 ryan.thigpen@uky.edu,
    University of Kentucky – Gtc
Thirlwall, Matthew, +44 1784 443609 M.Thirlwall@rhul.ac.uk,
    Royal Holloway University of London – Cs
Thiruvathukal, John V., (973) 655-4417 Montclair State University – Yg
Thole, Jeffrey T., (651) 696-6426 thole@macalester.edu,
    Macalester College – GgCaHw
Thomas, Amanda, amthomas@uoregon.edu,
    University of Oregon – Ys
Thomas, Andrew D., +44 (0)1970 622 781 ant23@aber.ac.uk,
    University of Wales – SbGmSf
Thomas, Debbie, (979) 845-3651 dthomas@tamu.edu,
    Texas A&M University – Ou
Thomas, Donald M., (808) 956-6482 dthomas@soest.hawaii.edu,
    University of Hawai'i, Manoa – CaHwRn
Thomas, Elizabeth K., (716)645-4329 ekthomas@buffalo.edu,
    SUNY, Buffalo – Cos
Thomas, Ellen, (860) 685-2238 ellen.thomas@yale.edu,
    Yale University – Pm
Thomas, Florence, (808) 236-7418 fithomas@hawaii.edu,
    University of Hawai'i, Manoa – Ob
Thomas, Helmuth, (902) 494-7177 helmuth.thomas@dal.ca,
    Dalhousie University – Oc
Thomas, Jay, (315)443-7631 jthom102@syr.edu,
    Syracuse University – Gi
Thomas, Jeremy, 206-947-2678 jnt@uw.edu,
    University of Washington – ZrXa
Thomas, Jim, (775) 887-7648 tom_j_smith@usgs.gov,
    University of Nevada, Reno – Cg
Thomas, John, 352-392-1951 ext 216 thomas@ufl.edu,
    University of Florida – Sp
Thomas, Kimberly W., (505) 667-4379 Los Alamos National Laboratory
    – Ct
Thomas, Margaret A., (860) 424-3583 margaret.thomas@ct.gov,
    Dept of Energy and Environmental Protection – Gg
Thomas, Mark, +44(0) 113 34 35233 m.e.thomas@leeds.ac.uk,
    University of Leeds – Ng
Thomas, Megan D., 410-293-6574 mdthomas@usna.edu,
    United States Naval Academy – Og
Thomas, Paul, 360-650-7796 Paul.Thomas@wwu.edu,
    Western Washington University – Gm
Thomas, Peter, 361749-6768 peter.thomas@utexas.edu,
    University of Texas, Austin – Ob
Thomas, Ray G., (352) 392-7984 rgthomas@ufl.edu,
    University of Florida – Zg
Thomas, Robert C., (406) 683-7615 rob.thomas@umwestern.edu,
    University of Montana Western – GseZe
Thomas, Roger D. K., (717) 358-4135 roger.thomas@fandm.edu,
    Franklin and Marshall College – PgRh
Thomas, Sara, 609-258-4101 st18@princeton.edu,
    Princeton University – Cb
Thomas, Valerie, (404) 385-7254 valerie.thomas@isye.gatech.edu,
    Georgia Institute of Technology – Zu
Thomas, William A., (205) 247-3547 geowat@uky.edu,
    Geological Survey of Alabama – GtcGr
Thomason, Jason, 217 244-2508 jthomaso@illinois.edu,
    University of Illinois – Gl
Thomasson, Joseph R., (785) 628-5665 Fort Hays State University – Pb
Thomka, James R., 330 972-5749 jthomka@uakron.edu,
    University of Akron – PggGg

Thompson, Andrew F., 626.395.8345 andrewt@caltech.edu,
    California Institute of Technology – OpgZc
Thompson, Anu, +440151 794 4095   Anu@liverpool.ac.uk,
    University of Liverpool – Oc
Thompson, Christopher J., (509) 376-6602 chris.thompson@pnl.gov,
    Pacific Northwest National Laboratory – Ca
Thompson, Curtis, (785) 532-5776 cthompso@ksu.edu,
    Kansas State University – So
Thompson, David W. J., davet@atmos.colostate.edu,
    Colorado State University – Yr
Thompson, Geoffrey, (508) 289-2397 gthompson@whoi.edu,
    Woods Hole Oceanographic Institution – Cm
Thompson, Glennis, (510) 642-7025 University of California, Berkeley
    – Zn
Thompson, Jann W. M., (202) 633-1357 Smithsonian Inst / Natl Museum
    of Natural Hist– Pg
Thompson, Joel B., (727) 864-8991 thompsjb@eckerd.edu,
    Eckerd College – Pg
Thompson, John F., jft66@cornell.edu,
    Cornell University – EmgNx
Thompson, John F. H., (604) 687-1117 University of British Columbia –
    Em
Thompson, Joseph L., (505) 667-4559 Los Alamos National Laboratory
    – Ct
Thompson, Keith R., (902) 494-3491 keith.thompson@dal.ca,
    Dalhousie University – Op
Thompson, Kenneth W., 620-341-5985 kthompso@emporia.edu,
    Emporia State University – Ze
Thompson, Lonnie G., (614) 292-6652 thompson.3@osu.edu,
    Ohio State University – GlZgPc
Thompson, LuAnne, (206) 543-9965 luanne@ocean.washington.edu,
    University of Washington – Op
Thompson, Margaret D., (781) 283-3029 mthompson@wellesley.edu,
    Wellesley College – Gc
Thompson, Michael D., (630) 252-9269 Argonne National Laboratory – Yg
Thompson, Michael L., (515) 294-2415 mlthomps@iastate.edu,
    Iowa State University of Science & Technology – Sc
Thompson, Nils W., nthompson@gsu.edu,
    Georgia State University, Perimeter College, Clarkston – GgZgGe
Thompson, Todd A., (812) 855-7428 tthomps@indiana.edu,
    Indiana University – GsmGu
Thompson, William G., (508) 289-2630 wthompson@whoi.edu,
    Woods Hole Oceanographic Institution – PeCc
Thoms, Richard E., (503) 725-3379 dick@pdx.edu,
    Portland State University – Ps
Thomsen, Charles E., (916) 484-8184 thomsec@arc.losrios.edu,
    American River College – Zy
Thomsen, Laurenz A., l.thomsen@iu-bremen.de,
    University of Washington – Ob
Thomsen, Leon, 713-743-3386 lathomsen@uh.edu,
    University of Houston – Ye
Thomson, Cary, 604-822-0653 cthomson@eos.ubc.ca,
    University of British Columbia – Zn
Thomson, Cynthia, 212-854-9896 cthomson@iri.columbia.edu,
    Columbia University – Zn
Thomson, Dennis W., dwt2@psu.edu,
    Pennsylvania State University, University Park – Am
Thomson, Jack, +44 0151 794 3594   Jack.Thomson@liverpool.ac.uk,
    University of Liverpool – Ob
Thomson, Jennifer A., (509) 359-7478 jthomson@ewu.edu,
    Eastern Washington University – Gpz
Thomson, Richard E., (250) 363-6555 ThomsonR@pac.dfo-mpo.gc.ca,
    University of British Columbia – Op
Thomson, Vivian E., (434) 924-3964 vet4y@virginia.edu,
    University of Virginia – Zn
Thorbjarnarson, Kathryn W., (619) 594-5586 kthorbjarnarson@mail.sdsu.edu,
    San Diego State University – Hq
Thorkelson, Derek J., 778-782-5390 dthorkel@sfu.ca,
    Simon Fraser University – Gt
Thorleifson, Harvey, (612) 626-2150 thorleif@umn.edu,
    University of Minnesota – Gl
Thorleifson, Harvey, 612-627-4780 ext 224 thorleif@umn.edu,
    University of Minnesota, Twin Cities – Gg
Thornberry-Ehrlich, Trista L., (406) 420-2391 trista.thornberry@colostate.edu,
    Colorado State University – GgcGx
Thorncroft, Christopher D., (518) 442-4555 SUNY, Albany – As
Thorne, Michael S., (801) 585-9792 michael.thorne@utah.edu,

University of Utah – YsxYv
Thorne, Paul D., (509) 372-4482 paul.thorne@pnl.gov,
    Pacific Northwest National Laboratory – HgyYg
Thorne, Richard M., (310) 614-6630 rmt@atmos.ucla.edu,
    University of California, Los Angeles – As
Thornton, Daniel C., 979 845 4092 dthornton@ocean.tamu.edu,
    Texas A&M University – Ob
Thornton, Edward B., (831) 656-2847 thornton@nps.edu,
    Naval Postgraduate School – On
Thornton, Joel A., (206) 543-4010 thornton@atmos.washington.edu,
    University of Washington – As
Thorson, Robert M., (860) 486-1396 robert.thorson@uconn.edu,
    University of Connecticut – Gm
Thul, David, 801-585-7013 dthul@egi.utah.edu,
    University of Utah – Gg
Thurber, Andrew, 541-737-4500 athurber@coas.oregonstate.edu,
    Oregon State University – Cb
Thurber, Clifford, (608) 262-6027 thurber@geology.wisc.edu,
    University of Wisconsin, Madison – Ysx
Thurber, David L., (718) 997-3300 Graduate School of the City University
    of New York – Cl
Thurber, David L., (718) 997-3325 dlthumper@aol.com,
    Queens College (CUNY) – Cl
Thurnherr, Andreas M., (845) 365-8816 ant@ldeo.columbia.edu,
    Columbia University – Op
Thuro, Kurosch, +498928925850 thuro@tum.de,
    Technical University of Munich – NgrGg
Thurow, Juergen, +44 020 7679 32416 j.thurow@ucl.ac.uk,
    University College London – Gs
Thurston, Phillips C., 7056751151 x2372 pthurston@laurentian.ca,
    Laurentian University, Sudbury – ZgCgGr
Thurtell, George W., (519) 824-4120 (Ext. 52453) gthurtel@uoguelph.ca,
    University of Guelph – As
Thy, Peter, pthy@ucdavis.edu,
    University of California, Davis – Gx
Tian, Di, (334) 844-4100 dzt0025@auburn.edu,
    Auburn University – At
Tibljaš, Darko, +38514605970 dtibljas@geol.pmf.hr,
    University of Zagreb – GzScGx
Tibuleac, Ileana, (775) 784-6256 ileana@seismo.unr.edu,
    University of Nevada, Reno – Ys
Tice, Mike, 979-845-3138 tice@geo.tamu.edu,
    Texas A&M University – Pg
Tick, Geoffrey, gtick@ua.edu,
    University of Alabama – Hw
Tidwell, Allan, (850) 526-2761 tidwella@chipola.edu,
    Chipola College – Zg
Tidwell, David, 205-247-3698 dtidwell@gsa.state.al.us,
    Geological Survey of Alabama – Gm
Tidwell, Vincent C., (505) 844-6025 vctidwe@sandia.gov,
    New Mexico Institute of Mining and Technology – Hwq
Tierney, Jessica, 520-621-5377 jesst@email.arizona.edu,
    University of Arizona – Co
Tierney, Kate, 319.335.0670 kate-tierney@uiowa.edu,
    University of Iowa – GrCsGg
Tierney, Kate E., 740-587-6487 tierneyk@denison.edu,
    Denison University – CgGrZg
Tierney, Sean, 303-871-3972 sean.tierney@du.edu,
    University of Denver – Zc
Tiffney, Bruce H., (805) 893-2959 bruce.tiffney@ccs.ucsb.edu,
    University of California, Santa Barbara – Pb
Tijani, Moshood N., +2348023252339 mn.tijani@mail.ui.edu.ng,
    University of Ibadan – HwCgGe
Tikoff, Basil, (608) 262-4678 basil@geology.wisc.edu,
    University of Wisconsin-Madison – Gc
Tikoo, Sonia M., (848) 445-3444 sonia.tikoo@rutgers.edu,
    Rutgers, The State University of New Jersey – YmXy
Tillman, James E., mars@atmos.washington.edu,
    University of Washington – As
Tilt, Jenna, 541-737-1232 tiltj@oregonstate.edu,
    Oregon State University – Zu
Tilton, Eric E., 303-735-5033 eric.small@colorado.edu,
    University of Colorado – Hw
Timár, Gábor, +36 1 3722700/1762 timar@caesar.elte.hu,
    Eotvos Lorand University – YgdZi
Timmer, Jaqueline R., 406-496-4842 jtimmer@mtech.edu,
    Montana Tech of The University of Montana – Ca

Timmermann, Axel, (808) 956-2720 axel@hawaii.edu,
    University of Hawai'i, Manoa – Op
Timmermans, Mary-Louise, 203 432-3167 mary-louise.timmermans@yale.edu,
    Yale University – Oc
Timmons, J M., 575-835-5237 mtimmons@nmbg.nmt.edu,
    New Mexico Institute of Mining and Technology – GctGs
Timmons, Stacy, (575) 835-6951 stacyt@nmbg.nmt.edu,
    New Mexico Institute of Mining & Technology – Gg
Tindall, Sarah E., (610) 683-4446 tindall@kutztown.edu,
    Kutztown University of Pennsylvania – Gc
Ting, Mingfang, (845) 365-8374 ting@ldeo.columbia.edu,
    Columbia University – As
Ting, Mingfang, ting@ldeo.columbia.edu,
    Columbia University – As
Tingey, David G., 801-422-7752 david_tingey@byu.edu,
    Brigham Young University – Zn
Tingley, Joseph V., jtingley@unr.edu,
    University of Nevada – Em
Tinjum, James M., 608/262-0785 tinjum@epd.engr.wisc.edu,
    University of Wisconsin, Madison – Ng
Tinker, Scott W., (512) 471-0209 scott.tinker@beg.utexas.edu,
    University of Texas at Austin, Jackson School of Geosciences – Gor
Tinker, Scott W., (512) 471-1534 scott.tinker@beg.utexas.edu,
    University of Texas, Austin – Go
Tinkham, Doug, 7056751151 x2270 dtinkham@laurentian.ca,
    Laurentian University, Sudbury – Gp
Tinkler, Dorothy, 541-881-5967 dtinkler@tvcc.cc,
    Treasure Valley Community College – Zi
Tinsley, Mark, (434) 832-7708 TinsleyM@cvcc.vccs.edu,
    Central Virginia Community College – Zg
Tinto, Kirsteen, tinto@ldeo.columbia.edu,
    Columbia University – Gu
Tipping, Robert, (612) 626-5437 tippi001@umn.edu,
    University of Minnesota – Hw
Tissot, Philippe , 361-825-3776 Philippe.Tissot@tamucc.edu,
    Texas A&M University, Corpus Christi – Zn
Titley, Spencer R., (520) 621-6018 stitley@email.arizona.edu,
    University of Arizona – Eg
Titus, Robert C., (607) 431-4733 titusr@hartwick.edu,
    Hartwick College – Ps
Titus, Sarah J., (507) 222-4419 stitus@carleton.edu,
    Carleton College – Gc
Tivey, Margaret K., (508) 289-3362 mktivey@whoi.edu,
    Woods Hole Oceanographic Institution – Cm
Tivey, Maurice A., (508) 289-2265 mtivey@whoi.edu,
    Woods Hole Oceanographic Institution – Yrm
Tiwari, Chetan, 940-369-8103 chetan.tiwari@unt.edu,
    University of North Texas – Zi
Tlusty, Michael, mtlusty@neaq.org,
    University of Massachusetts, Boston – Ob
Tobias, Craig, 860-405-9140 craig.tobias@uconn.edu,
    University of Connecticut – Oc
Tobin, Harold, 608-265-5796 htobin@geology.wisc.edu,
    University of Wisconsin-Madison – Yg
Tobin, Tom S., (205) 348-1878 ttobin@ua.edu,
    University of Alabama – PsCsPe
Tobisch, Othmar T., otobisch@yahoo.com,
    University of California, Santa Cruz – Gct
Todd, Brenda, (402) 554-2662 btodd@unomaha.edu,
    University of Nebraska, Omaha – Zn
Todd, Claire E., (253) 535-5163 toddce@plu.edu,
    Pacific Lutheran University – Glm
Todd, Steve, stephen.todd15@pcc.edu,
    Portland Community College - Sylvania Campus – Am
Todhunter, Paul, (701) 777-4593 paul.todhunter@und.edu,
    University of North Dakota – Zy
Toke', Nathan, (801) 863-8117 nathan.toke@uvu.edu,
    Utah Valley University – GtmZi
Tokmakian, Robin T., (831) 656-3255 rtt@nps.edu,
    Naval Postgraduate School – Op
Toksoz, M N., (617) 253-7852 toksoz@mit.edu,
    Massachusetts Institute of Technology – Ys
Tolhurst, Trevor, +44 (0)1603 59 3124 t.tolhurst@uea.ac.uk,
    University of East Anglia – Sb
Tolstoy, Maria, 845-365-8791 tolstoy@ldeo.columbia.edu,
    Columbia University – Yr

Tolstoy, Maria, tolstoy@ldeo.columbia.edu,
 Columbia University – Yr
Tomascak, Paul B., (315) 312-2285 tomascak@oswego.edu,
 SUNY, Oswego – CgGzi
Tomasko, Martin G., (520) 621-6969 mtomasko@lpl.arizona.edu,
 University of Arizona – As
Tomaso, Matthew S., (973) 655-7990 thomasw@mail.montclair.edu,
 Montclair State University – Ga
Tomašiæ, Nenad, +38514605968 ntomasic@geol.pmf.hr,
 University of Zagreb – GziCg
Tomiæ, Vladimir, +38514606094 vtomic@geol.pmf.hr,
 University of Zagreb – GgPgZi
Tomkin, Jonathan H., 217-244-2928 tomkin@illinois.edu,
 University of Illinois, Urbana-Champaign – Gm
Tomlinson, Emma L., + 353 1 8963856 tomlinse@tcd.ie,
 Trinity College – Cc
Tomlinson, Jaime L., 302-831-2649 jaimet@udel.edu,
 University of Delaware – Zn
Tomlinson, Peter, (785) 532-3198 ptomlin@ksu.edu,
 Kansas State University – So
Tompson, Andrew F., (925) 422-6348 tompson1@llnl.gov,
 Lawrence Livermore National Laboratory – HwSpCc
Tomson, Mason B., (713) 527-6048 mtomson@rice.edu,
 Rice University – Cg
Toner, Brandy M., (612) 624-1362 toner@umn.edu,
 University of Minnesota, Twin Cities – ClmSc
Toner, Rachel, (307) 766-2286 Ext. 248 rachel.toner@wyo.gov,
 Wyoming State Geological Survey – GoEoZi
Tong, Daoqin, daoqin@email.arizona.edu,
 University of Arizona – Zi
Tong, Vincent C., +44020 3073 8035 vincent.tong@ucl.ac.uk,
 Birkbeck College – Yg
Tongdee, Poetchanaporn, ptongdee@hccfl.edu,
 Hillsborough Community College – Zg
Toni, Rousine T., 002-03-3921595 dr_rosine@hotmail.com,
 Alexandria University – GgPg
Tonon, Marco Davide, marco.tonon@unito.it,
 Università di Torino – Ze
Toomey, Douglas R., (541) 346-5576 drt@uoregon.edu,
 University of Oregon – YsGt
Toonen, Robert, (808) 236-7425 toonen@hawaii.edu,
 University of Hawai'i, Manoa – Ob
Toor, Gurpal, 813-6334152 gstoor@ufl.edu,
 University of Florida – Sc
Toran, Laura, (215) 204-2352 ltoran@temple.edu,
 Temple University – Hw
Torcellini, Paul, 860-465-0368 torcellinip@easternct.edu,
 Eastern Connecticut State University – Zn
Torlaschi, Enrico, 514-987-3000 #6848 torlaschi.enrico@uqam.ca,
 Universite du Quebec a Montreal – Am
Tornabene, Livio, ltornabe@uwo.ca,
 Western University – XgZr
Toro, Jaime, 3042935603 ext. 4327 jtoro@wvu.edu,
 West Virginia University – GctGo
Torreano, Scott, (615) 598-1271 storrean@sewanee.edu,
 Sewanee: University of the South – Sf
Torrents, Alba, (301) 405-1979 alba@eng.umd.edu,
 University of Maryland – Sb
Torres, Joseph J., (727) 553-1169 jtorres@marine.usf.edu,
 University of South Florida – Ob
Torres, Mark, 713.348.2371 mt61@rice.edu,
 Rice University – Cb
Torres, Marta E., 541-737-2902 mtorres@coas.oregonstate.edu,
 Oregon State University – Cm
Torres, Raymond, (803) 777-4506 torres@geol.sc.edu,
 University of South Carolina – Hy
Torres Parisian, Cathleen, ctorresp@umn.edu,
 University of Minnesota, Twin Cities – Zr
Torri , Giuseppe, 808-956-2564 gtorri@hawaii.edu,
 University of Hawai'i, Manoa – ApsAm
Tortell, Philippe, (604) 822-4728 ptortell@eos.ubc.ca,
 University of British Columbia – Ob
Torvela, Taija, +44(0) 113 34 36620 t.m.torvela@leeds.ac.uk,
 University of Leeds – GctEg
Tosdal, Richard, rtosdal@eos.ubc.ca,
 University of British Columbia – Em
Toth, Charles K., 614-292-7681 toth.2@osu.edu,
 Ohio State University – ZrYdZi
Toth, Emoke, tothemoke.pal@gmail.com,
 Eotvos Lorand University – PmsPb
Tóth, József, (780) 492-1115 joe.toth@ualberta.ca,
 University of Alberta – Hw
Toth, Natalie, Natalie.Toth@dmns.org,
 Denver Museum of Nature & Science – Pg
Totsche, Kai U., +49(0)3641 948650 kai.totsche@uni-jena.de,
 Friedrich-Schiller-University Jena – HwClSo
Totten, Matthew W., (785) 532-2227 mtotten@ksu.edu,
 Kansas State University – Gso
Totten, Stanley M., (812) 866-7245 totten@hanover.edu,
 Hanover College – Gl
Touchan, Ramzi, (520) 621-2992 rtouchan@ltrr.arizona.edu,
 University of Arizona – PcHsAs
Toullec, Renaud, +33(0)3 44069333 renaud.toullec@lasalle-beauvais.fr,
 Institut Polytechnique LaSalle Beauvais (ex-IGAL) – GsoEo
Towery, Brooke L., (850) 484-2056 btowery@pensacolastate.edu,
 Pensacola Junior College – Gg
Towner, Ronald, (520) 621-6465 rht@email.arizona.edu,
 University of Arizona – Gae
Townsend, Meredith, 541-346-4573 University of Oregon – Gv
Townsend-Small, Amy, (513) 556-3762 amy.townsend-small@uc.edu,
 University of Cincinnati – Co
Toy, Virginia G., 64 3 479 7506 virginia@geology.co.nz,
 University of Otago – Gct
Tracy, Matthew, matthew.tracy@usafa.edu,
 United States Air Force Academy – Zn
Tran, Linda C., 281-765-7865 Linda.C.Tran@nhmccd.edu,
 Lonestar College - North Harris – Zi
Tranel, Lisa M., (309) 438-7966 ltranel@ilstu.edu,
 Illinois State University – Gmt
Trapp, Robert J., 217-300-0967 jtrapp@illinois.edu,
 University of Illinois, Urbana-Champaign – As
Travers, Steven, (309) 268-8640 steve.travers@heartland.edu,
 Heartland Community College – Zg
Travis, Bryan J., (505) 667-1254 bjtravis@lanl.gov,
 Los Alamos National Laboratory – Hq
Trehu, Anne M., 541-737-2655 trehu@coas.oregonstate.edu,
 Oregon State University – YrsGt
Treloar, Peter, +44020 8417 2525 P.Treloar@kingston.ac.uk,
 Kingston University – Gc
Trembanis, Art, 302-831-2498 art@udel.edu,
 University of Delaware – Ou
Tremblay, Alain, 514-987-3000 #1397 tremblay.alain@uqam.ca,
 Universite du Quebec a Montreal – Gct
Tremblay, Bruno, (514) 398-4369 bruno.tremblay@mcgill.ca,
 McGill University – Op
Tremblay, Thomas A., (512) 475-9537 tom.tremblay@beg.utexas.edu,
 University of Texas at Austin, Jackson School of Geosciences – Zy
Trenhaile, Alan S., 519-253-3000 ext. 2184 tren@uwindsor.ca,
 University of Windsor – Gm
Trentham, Robert C., 432 552-2432 trentham_r@utpb.edu,
 University of Texas, Permian Basin – Go
Treude, Tina, (310) 267-5213 ttreude@g.ucla.edu,
 University of California, Los Angeles – ObPgCm
Trevino, Ramon H., (512) 471-3362 ramon.trevino@beg.utexas.edu,
 University of Texas at Austin, Jackson School of Geosciences – GsoGe
Trevors, Jack, (519) 824-4120 Ext.53367 jtrevors@uoguelph.ca,
 University of Guelph – Zn
Trexler, Charles C., (740) 368-3506 cctrexle@owu.edu,
 Ohio Wesleyan University – Gct
Triay, Ines, (505) 665-1755 Los Alamos National Laboratory – Hg
Tribe, Selina, 604-527-5471 tribes@douglascollege.ca,
 Douglas College – EgGge
Trigoso, Erika, 303-871-3936 erika.trigoso@du.edu,
 University of Denver – Zc
Tripati, Aradhna, ripple@epss.ucla.edu,
 University of California, Los Angeles – CmPeCs
Triplett, Laura, (507) 933-7442 ltriplet@gustavus.edu,
 Gustavus Adolphus College – GemCl
Tripoli, Gregory J., (608)262-3700 tripoli@aos.wisc.edu,
 University of Wisconsin, Madison – Am
Trixler, Frank, 089/2179 509 trixler@lrz.uni-muenchen.de,
 Ludwig-Maximilians-Universitaet Muenchen – GzZmXc
Troch, Peter A., (520) 626-1277 patroch@hwr.arizona.edu,
 University of Arizona – Hs

Trofimovs, Jessica, +61 7 3138 2766 jessica.trofimovs@qut.edu.au,
    Queensland University of Technology – GsvGu
Tromp, Jeroen, (609) 258-4128 jtromp@princeton.edu,
    Princeton University – Yss
Troost, Kathy G., 206-221-1770 ktroost@uw.edu,
    University of Washington – Zun
Trop, Jeffrey M., (570) 577-3027 jtrop@bucknell.edu,
    Bucknell University – Gs
Trost, G K., (501) 575-7317 garrett@dicksonstreet.com,
    University of Arkansas, Fayetteville – Zn
Trouet, Valerie, (520) 626-8004 trouet@email.arizona.edu,
    University of Arizona – PeZyAs
Trout, Jennifer, (269) 387-8633 jennifer.l.trout@wmich.edu,
    Western Michigan University – ZnnZn
Troy, Marleen, (570) 408-4615 marleen.troy@wilkes.edu,
    Wilkes University – Ht
Trudgill, Bruce D., (303) 273-3883 btrudgil@mines.edu,
    Colorado School of Mines – Gc
Trueman, Clive, +44 (0)23 80596571 trueman@noc.soton.ac.uk,
    University of Southampton – Cg
Trujillo, Alan P., 7607441150x2734 atrujillo@palomar.edu,
    Palomar College – OguZe
Trullenque, Ghislain, +33(0)3 44069327 ghislain.trullenque@laslle-beauvais.fr,
    Institut Polytechnique LaSalle Beauvais (ex-IGAL) – GcNrGg
Trumbore, Susan E., (949) 824-6142 setrumbo@uci.edu,
    University of California, Irvine – Sc
Trussell, Geoffrey, 7815817370 x300 g.trussell@neu.edu,
    Northeastern University – Zn
Tsai, Victor, 626.395.6993 tsai@caltech.edu,
    California Institute of Technology – Yg
Tshudy, Dale, (814) 732-2453 dtshudy@edinboro.edu,
    Edinboro University of Pennsylvania – Pi
Tsige, Meaza, meaza@ucm.es,
    Univ Complutense de Madrid – Ng
Tsikos, Hari, +27 (0)46-603-8511 h.tsikos@ru.ac.za,
    Rhodes University – CgEg
Tso, Jonathan L., (540) 831-5638 jtso@radford.edu,
    Radford University – Gcp
Tsoflias, George P., (785) 864-4584 tsoflias@ku.edu,
    University of Kansas – YgeGe
Tsujita, Cameron J., (519) 661-2111 (Ext. 86740) ctsujita@uwo.ca,
    Western University – Pi
Tsvankin, Ilya D., (303) 273-3060 ilya@dix.mines.edu,
    Colorado School of Mines – Ye
Tu, Wei, (912) 478-5233 wtu@georgiasouthern.edu,
    Georgia Southern University – Zi
Tubana, Brenda S., 225-578-9420 btubana@agcenter.lsu.edu,
    Louisiana State University – So
Tubía, Jose M., (+94) 601-5392 jm.tubia@ehu.eus,
    University of the Basque Country UPV/EHU – GctYg
Tucholke, Brian E., (508) 289-2494 btucholke@whoi.edu,
    Woods Hole Oceanographic Institution – Gut
Tucker, Carla M., (409) 880-8236 ctucker@lamar.edu,
    Lamar University – Hw
Tucker, Gregory E., (303) 492-6985 gregory.tucker@colorado.edu,
    University of Colorado – GmtGs
Tudhope, Alexander, +44 (0) 131 650 8508 sandy.tudhope@ed.ac.uk,
    University of Washington – AtPc
Tufillaro, Nicholas, 541-737-2064 nbt@coas.oregonstate.edu,
    Oregon State University – Cb
Tulaczyk, Slawek, (831) 459-5207 stulaczy@ucsc.edu,
    University of California, Santa Cruz – GlmNg
Tull, James F., (850) 644-1448 jtull@fsu.edu,
    Florida State University – Gct
Tuller, Markus, (520) 621-7225 mtuller@cals.arizona.edu,
    University of Arizona – Zn
Tullis, Jan A., (401) 863-1921 jan_tullis@brown.edu,
    Brown University – Gcy
Tullis, Jason A., (479) 575-8784 jatullis@uark.edu,
    University of Arkansas, Fayetteville – ZriZf
Tullis, Terry E., (401) 863-3829 Terry_Tullis@Brown.edu,
    Brown University – Yx
Tung, Ka-Kit, tung@amath.washington.edu,
    University of Washington – As
Tung, Wen-wen, (765) 494-0272 wwtung@purdue.edu,
    Purdue University – As

Tunnicliffe, Verena, (250) 721-7135 verenat@uvic.ca,
    University of Victoria – Ob
Tura, Ali, alitura@mines.edu,
    Colorado School of Mines – YsNr
Turbeville, John, (760) 944-4449, x6413 JTurbeville@miracosta.edu,
    MiraCosta College – GgOg
Turchyn, Alexandra, +44 (0) 1223 333479 atur07@esc.cam.ac.uk,
    University of Cambridge – Ge
Turco, Richard, (310) 825-6936 turco@atmos.ucla.edu,
    University of California, Los Angeles – As
Turek, Andrew, (519) 969-6710 turek@uwindsor.ca,
    University of Windsor – Cc
Turner, A. Keith, kturner@mines.edu,
    Colorado School of Mines – Ng
Turner, Derek, 604-777-6568 turnerd1@douglascollege.ca,
    Douglas College – GemGl
Turner, Elizabeth C., 7056751151 x2267 eturner@laurentian.ca,
    Laurentian University, Sudbury – GrdEm
Turner, Grenville, +44 0161 275-0401 grenville.turner@manchester.ac.uk,
    University of Manchester – XcCc
Turner, Jay, (314) 935-5480 jrturner@wustl.edu,
    Washington University in St. Louis – Zn
Turner, Jenni, +44 (0)1603 59 3109 jenni.turner@uea.ac.uk,
    University of East Anglia – Gt
Turner, Kerry, +44 (0)1603 59 2551 r.k.turner@uea.ac.uk,
    University of East Anglia – GeEg
Turner, Robert E., (225) 388-6454 Louisiana State University – Ob
Turner, Robert S., rtt@ornl.gov,
    Oak Ridge National Laboratory – Cl
Turner, Wesley L., (936) 468-1049 turnerwl@sfasu.edu,
    Stephen F. Austin State University – Gg
Turner, III, Henry, (479) 575-7295 hturner@uark.edu,
    University of Arkansas, Fayetteville – GgtGc
Turnock, Allan C., (204) 474-6911 ac_turnock@umanitoba.ca,
    University of Manitoba – Gp
Turpening, Roger M., (906) 487-1784 roger@mtu.edu,
    Michigan Technological University – Ye
Turrin, Brent D., 848-445-3177 bturrin@eps.rutgers.edu,
    Rutgers, The State University of New Jersey – CcGvYm
Turski, Mark P., (603) 535-2749 markt@plymouth.edu,
    Plymouth State University – Eg
Turtle, Elizabeth, (520) 621-8284 turtle@lpl.arizona.edu,
    University of Arizona – Xg
Tuttle, Samuel, stuttle@mtholyoke.edu,
    Mount Holyoke College – Gq
Tvelia, Sean, (631) 451-4303 tvelias@sunysuffolk.edu,
    Suffolk County Community College, Ammerman Campus – Ggl
Twelker, Evan, 907-451-5086 evan.twelker@alaska.gov,
    Alaska Division of Geological & Geophysical Surveys – Eg
Twine, Tracy E., 612-625-7278 twine@umn.edu,
    University of Minnesota, Twin Cities – As
Twiss, Robert J., (530) 752-1860 rjtwiss@ucdavis.edu,
    University of California, Davis – Gc
Tyburczy, James A., (480) 965-2637 jim.tyburczy@asu.edu,
    Arizona State University – GyzYx
Tyce, Robert, (401) 874-6879 tyce@oce.uri.edu,
    University of Rhode Island – Ou
Tyler, Carrie, 513-529-8311 tylercl@miamioh.edu,
    Miami University – Pb
Tyler, E. Jerry, (608) 262-0853 ejtyler@wisc.edu,
    University of Wisconsin, Madison – Sd
Tyler, Scott, (775) 784-6250 styler@unr.edu,
    University of Nevada, Reno – Hg
Tynan, Cynthia T., (206) 860-6793 ctynan@coastalstudies.org,
    University of Washington – Ob
Tyning, Thomas F., (413) 236-4502 ttyning@berkshirecc.edu,
    Berkshire Community College – Zn
Tyrrell, Toby, +44 (0)23 80596110 toby.tyrrell@soton.ac.uk,
    University of Southampton – OgPe
Tziperman, Eli, (617) 384-8381 eli@eps.harvard.edu,
    Harvard University – Op
Tzortziou, Maria, 212-650-5769 mtzortziou@ccny.cuny.edu,
    Graduate School of the City University of New York – As
Törnqvist, Torbjörn E., 504-314-2221 tor@tulane.edu,
    Tulane University – Gs

U

Uahengo, Collen, +8613021208022 cuahengo@unam.na,
   University of Namibia – GgoEg
Uchupi, Elazar, (508) 289-2830 Woods Hole Oceanographic Institution – Ou
Uddin, Ashraf, (334) 844-4885 uddinas@auburn.edu,
   Auburn University – GdoGt
Ufkes, Els , +31 20 5989953 els.ufkes@falw.vu.nl,
   Vrije Universiteit Amsterdam – PmZgGu
Ugland, Richard, 541-552.6479 lane@sou.edu,
   Southern Oregon University – Gg
Uhen, Mark, (703) 993-5264 muhen@gmu.edu,
   George Mason University – Pv
Ukstins Peate, Ingrid, 319-335-1824 ingrid-peate@uiowa.edu,
   University of Iowa – GviXg
Ulanski, Stanley L., (540) 568-6130 ulanskl@jmu.edu,
   James Madison University – Og
Ullman, David J., 715.682.1312 dullman@northland.edu,
   Northland College – GlPeGm
Ullman, William J., (302) 645-4302 ullman@udel.edu,
   University of Delaware – Cm
Ullrich, Alexander D., 770-274-5083 aullrich@gsu.edu,
   Georgia State University, Perimeter College, Dunwoody – GglGu
Ulmer-Scholle, Dana S., (575) 835-5673 dana.ulmer-scholle@nmt.edu,
   New Mexico Institute of Mining and Technology – GsdCl
Umhoefer, Paul J., (928) 523-6464 paul.umhoefer@nau.edu,
   Northern Arizona University – Gt
Underwood, Ben, bsunderw@umail.iu.edu,
   Indiana University, Bloomington – Ca
Underwood, Charlie, +44 020 3073 8036 Birkbeck College – Pg
Underwood, Michael B., (573) 882-4685 underwoodm@missouri.edu,
   University of Missouri – GsuGt
Underwood, Stephen J., (912) 478-5361 SJUnderwood@GeorgiaSouthern.edu,
   Georgia Southern University – Am
Unger, Corey, (801) 538-4810 coreyunger@utah.gov,
   Utah Geological Survey – Zi
Unkefer, Clifford J., (505) 665-2560 Los Alamos National Laboratory – Zn
Uno, Kevin, kevinuno@ldeo.columbia.edu,
   Columbia University – Pe
Unrug, Kot F., (606) 257-1883 University of Kentucky – Nr
Unsworth, Martyn, (780) 492-3041 martyn.unsworth@ualberta.ca,
   University of Alberta – GtvYe
Unsworth, Martyn, unsworth@ualberta.ca,
   Cornell University – Ye
Upchurch, Paul, +44 020 7679 37947 p.upchurch@ucl.ac.uk,
   University College London – Pg
Ural, Melek, 00904242370000-5960 melekural@firat.edu.tr,
   Firat University – GivCc
Urban-Rich, Juanita, juanita.urban-rich@umb.edu,
   University of Massachusetts, Boston – Ob
Urbanczyk, Kevin, (432) 837-8110 kevinu@sulross.edu,
   Sul Ross State University – GiZi
Uriarte, Jesus Angel, jesusangel.uriarte@ehu.eus,
   University of the Basque Country UPV/EHU – NgHg
Urosevic, Milovan, +61 8 9266-2296 M.Urosevic@curtin.edu.au,
   Curtin University – YerYx
Urquhart, Mary, 972-883-2496 urquhart@utdallas.ede,
   University of Texas, Dallas – Xy
Urzua, Alfredo, 617-552-8339 alfredo.urzua@bc.edu,
   Boston College – Ng
Ustaszewski, Kamil M., 0049(0)3641/948623 kamil.u@uni-jena.de,
   Friedrich-Schiller-University Jena – Gct
Ustin, Susan L., (530) 752-0621 slustin@ucdavis.edu,
   University of California, Davis – Zr
Ustunisik, Gokce K., (605) 394-2461 Gokce.Ustunisik@sdsmt.edu,
   South Dakota School of Mines & Technology – Cp
Utgard, Russell O., 614 247-4246 utgard.1@osu.edu,
   Ohio State University – YgEnZe
Utter, James M., (914) 251-6642 SUNY, Purchase – SfZcRc
Uzochukwu, Godfrey A., (336) 285-4866 uzo@ncat.edu,
   North Carolina Agricultural & Tech State University – Gz
Uzunlar, Nuri, (605) 394-2494 Nuri.Uzunlar@sdsmt.edu,
   South Dakota School of Mines & Technology – EmCeGc

V

Vacher, H. Leonard, (813) 974-5267 vacher@chuma.cas.usf.edu,
   University of South Florida, Tampa – Hw
Vacquier, Victor D., (858) 534-4803 vvacquier@ucsd.edu,
   University of California, San Diego – Ob
Vaezi, Reza, +98 (411) 339 2721 vaezi@tabrizu.ac.ir,
   University of Tabriz – GeHgs
Vail, Peter R., (713) 348-4888 vail@rice.edu,
   Rice University – Gr
Vaillancourt, Robert, 717-872-3294 Robert.Vaillancourt@millersville.edu,
   Millersville University – Obc
Valdes, Juan B., (520) 621-2266 jvaldes@email.arizona.edu,
   University of Arizona – Hs
Valencia, Victor A., (520) 300-1605 victor.valencia@wsu.edu,
   Washington State University – ZgCcEg
Valenti, Christine, (973) 655-4448 Montclair State University – Gg
Valenti, Christine D., cvalenti@gsu.edu,
   Georgia State University, Perimeter College, Online – Gg
Valentine, Gregory, (716) 645-4295 gav4@buffalo.edu,
   SUNY, Buffalo – Gv
Valentine, James W., (510) 643-5791 University of California, Berkeley – Pg
Valentine, Michael , mvalentine@highline.edu,
   Highline College – Gt
Valentine, Michael, 253-566-5060 mvalentine@tacomacc.edu,
   Tacoma Community College – Zg
Valentine, Michael J., (253) 879-3129 mvalentine@pugetsound.edu,
   University of Puget Sound – YmGc
Valentino, David W., (315) 312-2798 david.valentino@oswego.edu,
   SUNY, Oswego – GtcYu
Valerio, Mitch, 614 265 6616 mitch.valerio@dnr.state.oh.us,
   Ohio Dept of Natural Resources – Hw
Valiunas, Jonas K., (807) 343-8677 jvaliuna@lakeheadu.ca,
   Lakehead University – Zm
Valley, John W., (608) 263-5659 valley@geology.wisc.edu,
   University of Wisconsin-Madison – CsGxz
Valsami-Jones, Eugenia, +44 (0)121 414 5537 e.valsamijones@bham.ac.uk,
   University of Birmingham – ClGeCa
van Alstine, James, +44(0) 113 34 37531 J.VanAlstine@leeds.ac.uk,
   University of Leeds – Ge
Van Alstine, James B., (320) 589-6313 vanalstj@mrs.umn.edu,
   University of Minnesota, Morris – Pg
Van Arsdale, Roy B., (901) 678-4356 rvanrsdl@memphis.edu,
   University of Memphis – Gcm
Van Avendonk, Harm, 512-471-0429 harm@ig.utexas.edu,
   University of Texas, Austin – Yg
van Balen, Ronald R., +31 20 59 87324 r.t.van.balen@vu.nl,
   Vrije Universiteit Amsterdam – GgmGt
van Bever Donker, Jan M., +27 021 959 3263 jvanbeverdonker@uwc.ac.za,
   University of the Western Cape – GcpGt
Van Brocklin, Matthew F., (315) 229-5197 mvanbrocklin@stlawu.edu,
   St. Lawrence University – GgZgn
Van Buer, Nicholas J., 909-869-3457 njvanbuer@csupomona.edu,
   California State Polytechnic University, Pomona – GxtGz
van de Flierdt, Tina, +44 20 759 41290 tina.vandeflierdt@imperial.ac.uk,
   Imperial College – Cs
van de Gevel, Saskia, (828) 262-7028 gevelsv@appstate.edu,
   Appalachian State University – Zg
Van De Poll, Henk W., (506) 453-4804 University of New Brunswick – Gs
Van de Water, Peter, 559-278-2912 pvandewater@csufresno.edu,
   California State University, Fresno – GrPbGf
van den Akker, Ben, ben.vandenakker@flinders.edu.au,
   Flinders University – Zn
Van Den Broeke, Matthew S., (402) 472-2418 mvandenbroeke2@unl.edu,
   University of Nebraska, Lincoln – AsZr
van den Heever, Sue, sue@atmos.colostate.edu,
   Colorado State University – As
Van Den Hoek, Jamon, 541-737-1229 vandenhj@oregonstate.edu,
   Oregon State University – Zur
van der Berg, Stan, +440151 794 4096   Vandenberg@liverpool.ac.uk,
   University of Liverpool – Em
Van Der Flier-Keller, Eileen, (250) 472-4019 fkeller@uvic.ca,
   University of Victoria – Cl
van der Hilst, Robert, (617) 253-3382 hilst@mit.edu,
   Massachusetts Institute of Technology – Ys
van der Horst, Dan, +44 (0) 131 651 4467 Dan.vanderHorst@ed.ac.uk,
   Edinburgh University – Eg
van der Lee, Suzan, (847) 491-8183 suzan@earth.northwestern.edu,
   Northwestern University – YsxYg
van der Lubbe, Jeroen, +31 20 59 87366 h.j.l.vander.lubbe@vu.nl,

Vrije Universiteit Amsterdam – Gs
van der Merwe, Barend, +27 (0)12 420 3699 barend.vandermerwe@up.ac.za,
   University of Pretoria – Gm
van der Pluijm, Ben, (734) 763-0373 vdpluijm@umich.edu,
   University of Michigan – GcZce
Van der Putten, Nathalie, n.n.l.vanderputten@vu.nl,
   Vrije Universiteit Amsterdam – PecZy
van der Velde, Ype, +31 20 59 87402 y.vander.velde@vu.nl,
   Vrije Universiteit Amsterdam – Hw
Van der Voo, Rob, (734) 764-8322 voo@umich.edu,
   University of Michigan – YmGtc
van der Werf, Guido R., +31 20 59 85687 guido.vander.werf@vu.nl,
   Vrije Universiteit Amsterdam – Zr
van Dijk, Deanna, (616) 526-6510 dvandijk@calvin.edu,
   Calvin College – GmOnZy
van Dongen, Bart, +44 0161 306-7460 Bart.VanDongen@manchester.ac.uk,
   University of Manchester – Co
Van Eerd, Laura, (519) 674-1644 lvaneerd@ridgetownc.uoguelph.ca,
   University of Guelph – Zn
Van Geen, Alexander, avangeen@ldeo.columbia.edu,
   Columbia University – Cg
Van Horn, Stephen R., (740) 826-8306 svanhorn@muskingum.edu,
   Muskingum University – GeZiEo
van Hunen, Jeroen, +44 (0) 191 33 42293 jeroen.van-hunen@durham.ac.uk,
   Durham University – Gt
Van Iten, Heyo, (812) 866-7303 vaniten@hanover.edu,
   Hanover College – PgHgRn
van Keken, Peter J., (734) 764-1497 keken@umich.edu,
   University of Michigan – Yg
Van Kooten, Gerald K., (616) 526-6374 gvankooten1@gmail.com,
   Calvin College – GoCePi
Van Loon, Lisa, 306-657-3585 lisa.vanloon@lightsource.ca,
   Western University – Ca
Van Meter, Kimberly, kvanmete@uic.edu,
   University of Illinois at Chicago – Hs
Van Mooy, Benjamin, (508) 289-2740 bvanmooy@whoi.edu,
   Woods Hole Oceanographic Institution – Oc
Van Niewenhuise, Donald, 713-743-3423 donvann@uh.edu,
   University of Houston – PsGoPe
van Norden, Maxim F., (228) 688-7123 maxim.vannorden@usm.edu,
   University of Southern Mississippi – Zn
Van Oostende, Nicolas, 609-258-1052 oostende@princeton.edu,
   Princeton University – Ob
Van Orman, James A., jav12@case.edu,
   Case Western Reserve University – CgGyCp
Van Roosendaal, Susan, (801) 581-8218 University of Utah – Zn
Van Ry, Michael, mvanry@occ.cccd.edu,
   Orange Coast College – Gv
Van Ryswick, Stephen, 410-554-5544 stephen.vanryswick@maryland.gov,
   Maryland Department of Natural Resources – Ge
Van Rythoven, Adrian, (570) 389-3912 avanrythov@bloomu.edu,
   Bloomsburg University – GxzEg
Van Schmus, W. Randall, (785) 864-2727 rvschmus@ku.edu,
   University of Kansas – Cc
Van Stan, John, (912) 478-8040 jvanstan@georgiasouthern.edu,
   Georgia Southern University – HgClSb
Van Straaten, H. Peter, (519) 824-4120 x52454 pvanstra@uoguelph.ca,
   University of Guelph – EnScCb
Van Tongeren, Jill A., (848) 445-5363 jvantongeren@eps.rutgers.edu,
   Rutgers, The State University of New Jersey – GiCgGt
Van Vleet, Edward S., (727) 553-1165 vanvleet@usf.edu,
   University of South Florida – Co
van Westrenen, Wim, w.van.westrenen@vu.nl,
   Vrije Universiteit Amsterdam – CpGz
Van Wijk, Jolante, (575) 835-6661 jolante.vanwijk@nmt.edu,
   New Mexico Institute of Mining and Technology – GtoYe
Vanacore, Elizabeth A., (787) 265-3845 elizabeth.vanacore@upr.edu,
   University of Puerto Rico – Ysx
VanAller-Hernick, Linda A., (518) 486-3699 lhernick@mail.nysed.gov,
   New York State Geological Survey – Pgb
Vance, Robert K., (912) 478-5353 rkvance@georgiasouthern.edu,
   Georgia Southern University – GiEgGg
Vandeberg, Gregory S., 701-777-4588 gregory.vandeberg@und.edu,
   University of North Dakota – GmZiGl
Vandemark, Douglas C., 603-862-0195 doug.vandemark@unh.edu,
   University of New Hampshire – Zr
Vanden Bergy, Michael, 801-538-5419 michaelvandenberg@utah.gov,
   Utah Geological Survey – Eo
Vandenbroucke, Thijs, thijs.vandenbroucke@ugent.be,
   Ghent University – Ps
Vander Auwera, Jacqueline, 32 4 366 22 53 jvdauwera@ulg.ac.be,
   Universite de Liege – Giv
Vanderkluysen, Loyc, (215) 571-4673 loyc@drexel.edu,
   Drexel University – GviGz
Vanderlip, Richard, (785) 532-6101 vanderrl@ksu.edu,
   Kansas State University – So
VanDervoort, Dane S., (205) 247-3626 dvandervoort@gsa.state.al.us,
   Geological Survey of Alabama – GcpZi
VanDerwerker, Tiffany J., 410-554-5558 tiffany.vanderwerker@maryland.gov,
   Maryland Department of Natural Resources – Hw
Vandike, James E., (573) 341-4616 Missouri University of Science and
   Technology – Hg
VanDorpe, Paul E., Paul.VanDorpe@dnr.iowa.gov,
   Iowa Dept of Natural Resources – Gg
VanGundy, Robert D., (276) 376-4656 rdv4v@uvawise.edu,
   The University of Virginia's College at Wise – GeHgZg
VanHorn, Jason E., (616) 526-7623 jev35@calvin.edu,
   Calvin College – ZirRh
Vaniman, David T., (505) 667-1863 vaniman@lanl.gov,
   Los Alamos National Laboratory – Gx
vanKeken, Peter E., pvankeken@carnegiescience.edu,
   Carnegie Institution for Science – GtYsCg
Vanko, David A., (410) 704-2121 dvanko@towson.edu,
   Towson University – Gxz
Vann, David R., (215) 898-4906 drvann@sas.upenn.edu,
   University of Pennsylvania – GeSfXm
Vann, Jamie, (610) 738-0598 jvann@wcupa.edu,
   West Chester University – CgGg
Vannucchi, Paola, +44 01784 443616   Paola.Vannucchi@rhul.ac.uk,
   Royal Holloway University of London – Gu
Vanos, Jennifer K., (806) 786-2577 jkvanos@ucsd.edu,
   University of California, San Diego – As
VanRegenmorter, John M., 616-331-3164 vanregjo@gvsu.edu,
   Grand Valley State University – PvGsr
Vaqueiro Rodriguez, Marcos, 0034637756150 mvaqueiro@frioya.es,
   Coruna University – GcmYd
Varadi, Ferenc D., (310) 206-6484 University of California, Los Angeles
   – Zn
Varady, Robert G., (520) 626-4393 rvarady@email.arizona.edu,
   University of Arizona – Zn
Varas, María Josefa, mjvaras@ucm.es,
   Univ Complutense de Madrid – Gd
Varekamp, Johan C., (860) 685-2248 jvarekamp@wesleyan.edu,
   Wesleyan University – CgGzv
Varela, Diana, 250-721-6120 dvarela@uvic.ca,
   University of Victoria – Ob
Vargo, Gabriel A., (727) 864-2683 vargo@mail.usf.edu,
   University of South Florida – Ob
Varner, Ruth K., (603) 862-0853 ruth.varner@unh.edu,
   University of New Hampshire – Cl
Varricchio, David J., (406) 994-6907 djv@montana.edu,
   Montana State University – Pv
Vassiliou, Andreas H., 973-353-5100 ahvass@andromeda.rutgers.edu,
   Rutgers, The State University of New Jersey, Newark – Gz
Vaudrey, Jamie, 860-405-9149 jamie.vaudrey@uconn.edu,
   University of Connecticut – Ob
Vaughan, David, +44 0161 275-3935 david.vaughan@manchester.ac.uk,
   University of Manchester – GzClGe
Vaughan, Geraint, +44 0161 306-3931 Geraint.Vaughan@manchester.ac.uk,
   University of Manchester – As
Vaughan, Michael T., 631 632-8030 michael.vaughan@sunysb.edu,
   Stony Brook University – Yx
Vaughan, Naomi, +44 (0)1603 59 3904 n.vaughan@uea.ac.uk,
   University of East Anglia – Am
Vautier, Yannick, +33(0)3 44068991 yannick.vautier@lasalle-beauvais.fr,
   Institut Polytechnique LaSalle Beauvais (ex-IGAL) – GodCo
Vaz, Nuno M., nunovaz@utad.pt,
   Universidade de Trás-os-Montes e Alto Douro – PmlYg
Veblen, Thomas T., (303) 492-8528 veblen@colorado.edu,
   University of Colorado – Zg
Vecchi, Gabriel, 609-258-4101 gvecchi@princeton.edu,
   Princeton University – ZoOp
Veeder, Glenn J., (818) 354-7388 Jet Propulsion Laboratory – Zn
Veeger, Anne, (401) 874-4184 veeger@uri.edu,

University of Rhode Island – Hw
Veeger, Anne I., (401) 874-2187 veeger@uri.edu,
University of Rhode Island – HwCl
Vega, Anthony J., (814) 226-2023 avega@clarion.edu,
Clarion University – AtmOp
Vegas, Nestor, 0034946015374 nestor.vegas@ehu.eus,
University of the Basque Country UPV/EHU – GctYm
Veizer, Jan, 613-562-5800 6461 Veizer@uottawa.ca,
University of Ottawa – Cl
Velasco, Aaron A., (915) 747-5101 aavelasco@utep.edu,
University of Texas, El Paso – Yx
Velbel, Michael A., (517) 355-4626 velbel@msu.edu,
Michigan State University – ClGdSc
Velicogna, Isabella, 949-824-2685 isabella.velicogna@gmail.com,
University of California, Irvine – Zr
Velinsky, David, 215 299 1147 velinsky@drexel.edu,
Drexel University – OcCms
Velli, Marco, 310-206-4892 mvelli@ucla.edu,
University of California, Los Angeles – Yg
Vengosh, Avner, 919 684 5847 vengosh@duke.edu,
Duke University – Hw
Venkatesan, M I., (310) 206-2561 University of California, Los Angeles – Co
Venn, Cynthia, (570) 389-4141 cvenn@bloomu.edu,
Bloomsburg University – OgClOc
Venter, Marcie, (270) 809-3457 mventer@murraystate.edu,
Murray State University – Ga
Venterea, Rodney T., (612) 624-7842 venterea@soils.umn.edu,
University of Minnesota, Twin Cities – Sp
Venti, Nicholas, 413-687-3515 nventi@geo.umass.edu,
University of Massachusetts, Amherst – Og
Vento, Frank, 814-824-2581 fvento@mercyhurst.edu,
Mercyhurst University – GaSoGm
Ventura, Stephen J., (608) 262-6416 sventura@wisc.edu,
University of Wisconsin, Madison – ZrSo
Venzke, Edward, (202) 633-1822 venzkee@si.edu,
Smithsonian Inst / Natl Museum of Natural Hist– Gv
Vepraskas, Michael J., (919) 515-1458 North Carolina State Univ – Sd
Ver Straeten, Charles, (518) 486-2004 cverstrae@mail.nysed.gov,
New York State Geological Survey – Grs
Verardo, Stacey, (703) 993-1045 sverardo@gmu.edu,
George Mason University – Pei
Veraverbeke, Sander, +31 20 59 82975 s.n.n.veraverbeke@vu.nl,
Vrije Universiteit Amsterdam – Zr
Verburg, Paul, 775-784-4511 pverburg@cabnr.unr.edu,
University of Nevada, Reno – SdcSb
Verlinde, Johannes, (814) 863-9711 verlinde@essc.psu.edu,
Pennsylvania State University, University Park – Am
Verma, Sumit, 432-552-2242 verma_s@utpb.edu,
University of Texas, Permian Basin – YeGo
Vermeesch, Pieter, +44 02076792418 p.vermeesch@ucl.ac.uk,
University College London – Cc
Vermeij, Geerat J., (530) 752-2234 gjvermeij@ucdavis.edu,
University of California, Davis – Pe
Vermeul, Vincent R., (509) 376-8316 vince.vermeul@pnl.gov,
Pacific Northwest National Laboratory – Hg
Vernhes, Jean-David, +33(0)3 44062547 jean-david.vernhes@lasalle-beauvais.fr,
Institut Polytechnique LaSalle Beauvais (ex-IGAL) – YgNgr
Veronis, George, (203) 432-3148 george.veronis@yale.edu,
Yale University – Op
Verosub, Kenneth L., (530) 752-6911 klverosub@ucdavis.edu,
University of California, Davis – Ym
Versteeg, Roelof, versteeg@ldeo.columbia.edu,
Columbia University – Yg
Vervoort, Jeffrey D., (509) 335-5597 vervoort@wsu.edu,
Washington State University – CcaGt
Vesovic, Velisa, +44 20 759 47352 v.vesovic@imperial.ac.uk,
Imperial College – Np
Vesper, Dorothy, 304 293 5603 dorothy.vesper@mail.wvu.edu,
West Virginia University – Hw
Vetter, Scott K., (318) 869-5055 svetter@centenary.edu,
Centenary College of Louisiana – Gi
Vice, Mari A., (608) 342-1055 vice@uwplatt.edu,
University of Wisconsin, Platteville – Gd
Vicens Batet, Enric, ++935811783 enric.vicens@uab.cat,
Universitat Autonoma de Barcelona – Pg
Vickery, Nancy, nvicker2@une.edu.au,
University of New England – GgeYg
Vidal Romaní, Juan Ramon, 00 34 981 167000 xemoncho@udc.es,
Coruna University – GmlGc
Vidale, John E., (206) 543-6790 vidale@uw.edu,
University of Southern California – YsGtYg
Vidale, Jon , 310-210-2131 vidale@uw.edu,
University of Washington – RnYs
Videtich, Patricia E., 616-331-3887 videticp@gvsu.edu,
Grand Valley State University – Gd
Vidoviæ, Jelena, +38514606090 jelena.vidovic@geol.pmf.hr,
University of Zagreb – Gg
Viedma, Cristóbal, viedma@ucm.es,
Univ Complutense de Madrid – Gz
Viegas, Anthony V., +91-8669609198 anthonyviegas@yahoo.com,
Goa University – GiEmGx
Vieira, David J., (505) 667-7231 Los Alamos National Laboratory – Zn
Viens, Rob, (425) 564-3158 rob.viens@bellevuecollege.edu,
Bellevue College – GgZg
Vierrether, Chris, 573-368-2370 chris.vierrether@dnr.mo.gov,
Missouri Dept of Natural Resources – GgoGe
Viertel, Dave, 217-581-6244 dviertel@eiu.edu,
Eastern Illinois University – Zru
Vietti, Laura, (307) 314-2024 lvietti@uwyo.edu,
University of Wyoming – Pv
Vig, Pradeep K., 618.545.3373 pvig@kaskaskia.edu,
Kaskaskia College – Gg
Vigouroux-Caillibot, Nathalie, 604-527-5860 vigourouxcaillibotn@douglascollege.ca,
Douglas College – GviCg
Vilcaez, Javier, 405744-6358 vilcaez@okstate.edu,
Oklahoma State University – NpCl
Villalard-Bohnsack, Martine, (401) 254-3243 Roger Williams University – Ob
Villalobos, Joshua, (915) 831-7001 jvillal6@epcc.edu,
El Paso Community College – ZeYgu
Villareal, Tracy A., (361) 749-6732 tracyv@austin.utexas.edu,
University of Texas, Austin – Ob
Villarroy, Fermín, ferminv@ucm.es,
Univ Complutense de Madrid – Hgy
Villegas, Monica B., 543584676198 mvillegas@exa.unrc.edu.ar,
Universidad Nacional de Rio Cuarto – Gsc
Villeneuve, Marlene C., +64 3 3667001 Ext 45682 marlene.villeneuve@canterbury.ac.nz,
University of Canterbury – Ng
Vimont, Daniel J., (608)262-2828 dvimont@wisc.edu,
University of Wisconsin, Madison – As
Vincent, Dayton G., dvincent@purdue.edu,
Purdue University – As
Vincent, Paul C., 229.333.5752 pvincent@valdosta.edu,
Valdosta State University – Zig
Vincent, Robert K., (419) 372-0160 rvincen@bgnet.bgsu.edu,
Bowling Green State University – Yg
Vinciguerra, Sergio, +44 0116 252 3634 sv127@le.ac.uk,
Leicester University – Nr
Vinciguerra, Sergio Carmelo, sergiocarmelo.vinciguerra@unito.it,
Università di Torino – Nr
Vinton, Bonita L., (724) 738-2048 bonita.vinton@sru.edu,
Slippery Rock University – Gg
Visaggi, Christy, cvissagi@gsu.edu,
Georgia State University – PiZe
Viskupic, Karen, (208)426-3658 karenviskupic@boisestate.edu,
Boise State University – ZeGc
Visscher, Pieter, (860) 486-4434 pieter.visscher@uconn.edu,
University of Connecticut – Co
Visscher, Pieter T., 860-405-9159 pieter.visscher@uconn.edu,
University of Connecticut – Co
Visser, Jenneke M., (337) 482-6966 jvisser@louisiana.edu,
University of Louisiana at Lafayette – Ne
Vitale, Sarah A., 715-836-4300 vitalesa@uwec.edu,
University of Wisconsin, Eau Claire – Hq
Vitale Brovarone, Alberto, alberto.vitale@unito.it,
Università di Torino – Gx
Vitek, John D., (405) 780-0623 jvitek@neo.tamu.edu,
Texas A&M University – GmZe
Vitek, John D., 405-744-6358 jvitek@okstate.edu,
Oklahoma State University – Gm

Viti, Cecilia, cecilia.viti@unisi.it,
University of Siena – Gz
Vivoni, Enrique, (480) 727-3575 vivoni@asu.edu,
Arizona State University – Hg
Vlahos, Penny, (860) 405-9269 penny.vlahos@uconn.edu,
University of Connecticut – OcGeu
Vlahovic, Gordana, (919) 530-5172 gvlahovic@nccu.edu,
North Carolina Central University – YsZeRn
Vlcan, Jennifer, 775-682-8759 mauldin@unr.edu,
University of Nevada – Zi
Vocadlo, Lidunka, +44 020 7679 37919 l.vocadlo@ucl.ac.uk,
University College London – Gy
Voelker, Bettina, (303) 273-3152 voelker@mines.edu,
Colorado School of Mines – Ca
Vogel, Eve, 413-545-0778 evevogel@geo.umass.edu,
University of Massachusetts, Amherst – Zy
Vogl, Jim, (352) 392-6987 jvogl@ufl.edu,
University of Florida – Gct
Voglesonger, Kenneth M., 773-442-6053 K-Voglesonger@neiu.edu,
Northeastern Illinois University – ClGe
Vogt, Steven, (831) 459-2151 vogt@ucolick.org,
University of California, Santa Cruz – Xc
Voice, Peter J., (269) 387-5446 peter.voice@wmich.edu,
Western Michigan University – GsZn
Voicu, Gabriel-Constantin, 514-987-3000 #1648 gvoicu@videotron.ca,
Universite du Quebec a Montreal – GcCg
Voight, Barry, (814) 238-4431 voight@ems.psu.edu,
Pennsylvania State University, University Park – GveGg
Voigt, Vicki, 573-368-2128 vicki.voigt@dnr.mo.gov,
Missouri Dept of Natural Resources – Gg
Vojtko, Rastislav , vojtko@fns.uniba.sk,
Comenius University in Bratislava – GctGm
Volborth, Alexis, (406) 496-4134 Montana Tech of the University of Montana – Ca
Volk, Michael, mvolk@umn.edu,
University of Minnesota, Twin Cities – Ym
Voller, Vaughan R., (612) 625-0764 volle001@tc.umn.edu,
University of Minnesota, Twin Cities – Zm
Vollmer, Frederick W., 845-257-3760 vollmerf@newpaltz.edu,
SUNY, New Paltz – GctGx
Vollmer, Steve, 7815817370 x312 s.vollmer@neu.edu,
Northeastern University – Zn
von Bitter, Peter H., 416-586-5591 peterv@rom.on.ca,
University of Toronto – Pm
von der Handt, Anette, 612-624-7370 avdhandt@umn.edu,
University of Minnesota, Twin Cities – GiCt
von Frese, Ralph R., (614) 292-5635 von-frese.3@osu.edu,
Ohio State University – Ye
von Glasow, Roland, +44 (0)1603 59 3204 r.von-glasow@uea.ac.uk,
University of East Anglia – Yg
Von Reden, Karl F., (508) 289-3384 kvonreden@whoi.edu,
Woods Hole Oceanographic Institution – Ct
von Salzen, Knut, knut.vonsalzen@canada.ca,
University of Victoria – As
von Salzen, Knut, (250) 363-8287 knut.vonsalzen@canada.ca,
University of British Columbia – As
von Seggern, David H., (775) 303 8461 vonseg@seismo.unr.edu,
University of Nevada, Reno – YsgEo
Vondra, Carl F., (515) 294-5867 cfvondra@iastate.edu,
Iowa State University of Science & Technology – Gr
Vonk, Jorien, +31 20 59 87336 j.e.vonk@vu.nl,
Vrije Universiteit Amsterdam – Hg
Voorhees, David H., (630) 466-2783 dvoorhees@waubonsee.edu,
Waubonsee Community College – ZgGg
Voorhees, Kent J., kvoorhee@mines.edu,
Colorado School of Mines – Co
Vopson, Melvin M., +44 023 92 842246 melvin.vopson@port.ac.uk,
University of Portsmouth – Yg
Voroney, R. Paul, (519) 824-4120 (Ext. 53057) pvoroney@uoguelph.ca,
University of Guelph – Sb
Vorosmarty, Charles, 212-650-7042 Graduate School of the City University of New York – Ng
Vorwald, Brian, vorwalb@sunysuffolk.edu,
Suffolk County Community College, Ammerman Campus – Zg
Voss, Regis D., (515) 294-1923 Iowa State University of Science & Technology – So
Votaw, Robert, rvotaw@iun.edu,
Indiana University Northwest – PiGrs
Voulgaris, George, (803) 777-2549 gvoulgaris@geol.sc.edu,
University of South Carolina – OnpOg
Vrba, Elisabeth S., (203) 432-5008 elisabeth.vrba@yale.edu,
Yale University – Pv
Vroon, Pieter, p.z.vroon@vu.nl,
Vrije Universiteit Amsterdam – GvCcl
Vuke, Susan M., (406) 496-4326 svuke@mtech.edu,
Montana Tech of The University of Montana – GgrGs
Vukovich, George, (416)736-2100 #30090 vukovich@yorku.ca,
York University – As
Vulava, Vijay M., (843) 953-1922 vulavav@cofc.edu,
College of Charleston – CgHwSc

W

Waag, Charles J., (208) 426-3658 cwaag@boisestate.edu,
Boise State University – Gc
Wach, Grant D., (902) 494-8019 grant.wach@dal.ca,
Dalhousie University – GorGs
Wada, Ikuko, (612) 301-9535 iwada@umn.edu,
University of Minnesota, Twin Cities – ZnYgGt
Waddington, David C., (604) 527-5230 waddingtond@douglascollege.ca,
Douglas College – ZgGuEd
Waddington, Edwin D., 206-543-4585 edw@uw.edu,
University of Washington – Gl
Waddington, J. M., (905) 525-9140 (Ext. 23217) wadding@mcmaster.ca,
McMaster University – Pe
Wade, Bridget S., +44 020 3108 6359 b.wade@ucl.ac.uk,
University College London – PmeCl
Wade, Phillip, 503-838-8225 wadep@wou.edu,
Western Oregon University – Zge
Wade, Terry L., (979) 862-2325 t-wade@geos.tamu.edu,
Texas A&M University – OcCao
Wadhwa, Meenakshi, (480) 965-0796 meenakshi.wadhwa@asu.edu,
Arizona State University – Zn
Wadhwa, Meenakshi, (312) 665-7639 mwadhwa@fieldmuseum.org,
Field Museum of Natural History – Xm
Wadsworth, William B., (562) 907-4200 Whittier College – Gi
Wagger, Michael G., (919) 515-4269 North Carolina State University – Sb
Waggoner, Karen, 432-685-5540 kwaggoner@midland.edu,
Midland College – GgRhGc
Wagner, John R., (864) 656-5024 jrwgnr@clemson.edu,
Clemson University – ZeGm
Wagner, Kaleb, (612) 626-6901 kewagner@umn.edu,
University of Minnesota – Gl
Wagner, Lara S., 202-478-8838 lwagner@carnegiescience.edu,
Carnegie Institution for Science – YsgGt
Wagner, Peter J., 312-665-7634 pwagner@fieldmuseum.org,
Field Museum of Natural History – Pi
Wagner, Rick, 205-247-3622 rwagner@gsa.state.al.us,
Geological Survey of Alabama – Cg
Wagner, Timothy J., (402) 280-2239 timothywagner@creighton.edu,
Creighton University – As
Wagner Riddle, Claudia, (519) 824-4120 (Ext. 2787) cwagnerr@uoguelph.ca,
University of Guelph – Am
Wagoner, Jeffrey L., (925) 422-1374 wagoner1@llnl.gov,
Lawrence Livermore National Laboratory – Gs
Waheed, Umair B., +966138602297 Umair.waheed@kfupm.edu.sa,
King Fahd University of Petroleum and Minerals – Yg
Waid, Christopher, 614 265 6627 christopher.waid@dnr.state.oh.us,
Ohio Dept of Natural Resources – Eo
Waite, Cynthia H., (213) 740-6109 waite@usc.edu,
University of Southern California – Zn
Waite, Gregory P., (906) 487-3554 gpwaite@mtu.edu,
Michigan Technological University – YsGv
Waithaka, Hunja, hunja@eng.jkuat.ac.ke,
Jomo Kenyatta University of Agriculture & Technology – YdZgi
Wakabayashi, John, 559-278-6459 jwakabayashi@csufresno.edu,
California State University, Fresno – GmcGt
Wake, Cameron P., (603) 862-2329 cameron.wake@unh.edu,
University of New Hampshire – Gl
Wake, David B., (510) 642-3059 wakelab@uclink.berkeley.edu,
University of California, Berkeley – Pg
Wake, Marvalee H., (510) 642-4743 University of California, Berkeley – Pg
Wakefield, Kelli, kelliwakefield@hotmail.com,
Mesa Community College – Gg

Wala, Andrew M., 859-257-2959 University of Kentucky – Nm
Walcek, Christopher J., (518) 437-8720 SUNY, Albany – As
Waldbusser, George, 541-737-8964 waldbuss@coas.oregonstate.edu,
    Oregon State University – Ob
Walden, John, +44 01334 463688 jw9@st-andrews.ac.uk,
    University of St. Andrews – Ym
Waldhauser, Felix, (845) 365-8538 felixw@ldeo.columbia.edu,
    Columbia University – YsgGt
Waldhauser, Felix, felixw@ldeo.columbia.edu,
    Columbia University – Ys
Waldman, John, 718-997-3603 john.waldman@qc.cuny.edu,
    Graduate School of the City University of New York – ObHsZc
Waldron, John W., john.waldron@ualberta.ca,
    University of Alberta – Gc
Walford, Nigel, +44020 8417 2512 nwalford@kingston.ac.uk,
    Kingston University – Zi
Walker, Charles T., (562) 985-4818 California State University, Long
    Beach – Cl
Walker, David, dwalker@ldeo.columbia.edu,
    Columbia University – Gx
Walker, David, (845) 365-8658 dwalker@ldeo.columbia.edu,
    Columbia University – CpGzCg
Walker, J. Douglas, (785) 864-7711 jdwalker@ku.edu,
    University of Kansas – Gc
Walker, Jeffrey R., 845-437-5546 jewalker@vassar.edu,
    Vassar College – GzScGv
Walker, John L., (630) 252-6803 jlwalker@anl.gov,
    Argonne National Laboratory – Cg
Walker, Kenneth R., (865) 974-6017 kwalker5@utk.edu,
    University of Tennessee, Knoxville – Pe
Walker, Mark, (775) 784-1938 mwalker@cabnr.unr.edu,
    University of Nevada, Reno – Zn
Walker, Nan D., (225) 388-5331 Louisiana State University – Og
Walker, Peter, (541) 346-4541 pwalker@oregon.uoregon.edu,
    University of Oregon – Zn
Walker, Raymond J., (310) 825-7685 rwalker@igpp.ucla.edu,
    University of California, Los Angeles – Xy
Walker, Richard, +440416 252 3628 Leicester University – Gs
Walker, Richard, +44 (1865) 282115 richard.walker@earth.ox.ac.uk,
    University of Oxford – Zr
Walker, Richard J., (301) 405-4089 rjwalker@umd.edu,
    University of Maryland – CcXcCg
Walker, Sally E., (706) 542-2396 swalker@gly.uga.edu,
    University of Georgia – Pm
Wallace, Adam F., (302) 831-1950 afw@udel.edu,
    University of Delaware – ClGzZm
Wallace, Davin, (228) 688-3060 davin.wallace@usm.edu,
    University of Southern Mississippi – Ou
Wallace, Douglas, (902) 494-4132 douglas.wallace@dal.ca,
    Dalhousie University – Oc
Wallace, Douglas W. R., (631) 344-2945 wallace@notes.cc.sunysb.edu,
    SUNY, Stony Brook – Oc
Wallace, Janae, (801) 537-3387 nrugs.jwallace@state.ut.us,
    Utah Geological Survey – Hw
Wallace, John M., (206) 543-7390 wallace@atmos.washington.edu,
    University of Washington – As
Wallace, Laura, lwallace@ig.utexas.edu,
    University of Texas, Austin – GtYdZr
Wallace, Paul, (541) 346-5985 pwallace@uoregon.edu,
    University of Oregon – GviCg
Wallace, Tim, 662-268-1032 Ext 244 tjw5@msstate.edu,
    Mississippi State University – Am
Wallace, William , 718-982-3876 Graduate School of the City University
    of New York – Ge
Wallace, Jr., Terry C., (505) 667-3644 wallacet@lanl.gov,
    Los Alamos National Laboratory – Ys
Wallender, Wes W., 530.752.0688 wwwallender@ucdavis.edu,
    University of California, Davis – Hg
Waller, Thomas R., (202) 357-2127 Smithsonian Inst / Natl Museum of
    Natural Hist– Pi
Wallis, Ilka, ilka.wallis@flinders.edu.au,
    Flinders University – HqCg
Walraevens, Kristine E., (329) 264-4648 kristine.walraevens@ugent.be,
    Ghent University – HwCqYu
Walrod, Amanda G., (501) 575-7317 awalrod@comp.uark.edu,
    University of Arkansas, Fayetteville – Zn
Walsh, Christopher, (301) 405-4351 cswalsh@umd.edu,
    University of Maryland – Zn
Walsh, Daniel E., (907) 474-6746 dewalsh@alaska.edu,
    University of Alaska, Fairbanks – Zn
Walsh, Ellen c., (920) 832-6739 ellen.c.walsh@lawrence.edu,
    Lawrence University – Zn
Walsh, Emily O., (319) 895-4302 ewalsh@cornellcollege.edu,
    University of Iowa – GptCc
Walsh, John J., (727) 553-1164 jwalsh@marine.usf.edu,
    University of South Florida – Ob
Walsh, Maud M., (225) 578-1211 evwals@lsu.edu,
    Louisiana State University – GeRcZe
Walsh, Tim R., 806-291-1123 Wayland Baptist University – GsPmGo
Walter, Julien, 41854550112680 Julien_Walter@uqac.ca,
    Universite du Quebec a Chicoutimi – HwqGm
Walter, Lynn M., (734) 763-4590 lmwalter@umich.edu,
    University of Michigan – Cl
Walter , Michael , 0117 9515007 m.j.walter@bristol.ac.uk,
    University of Bristol – GiCpg
Walter, Nathan A., (678) 839-4070 awalter@westga.edu,
    University of West Georgia – Zg
Walter, Robert C., 717 358-7198 robert.walter@fandm.edu,
    Franklin and Marshall College – Cc
Walter, Thomas, (212) 772-5457 twalter@hunter.cuny.edu,
    Hunter College (CUNY) – ZyeAm
Walter-Shea, Elizabeth A., (402) 472-1553 ewalter-shea1@unl.edu,
    University of Nebraska, Lincoln – As
Walters, James C., james.walters@uni.edu,
    University of Northern Iowa – Gm
Waltham, Dave, +44 1784 443617 D.Waltham@rhul.ac.uk,
    Royal Holloway University of London – Yg
Walther, Ferdinand, 089/2180 4346 macferdi@gmx.org,
    Ludwig-Maximilians-Universitaet Muenchen – Gz
Walther, John V., (214) 768-3174 Southern Methodist University – Cg
Walther, Suzanne, (619) 260-4787 swalther@sandiego.edu,
    University of San Diego – GmZi
Walton, Anthony W., (785) 864-2726 twalton@ku.edu,
    University of Kansas – GsEoGv
Walton, Gabriel, (303) 273-2235 gwalton@mines.edu,
    Colorado School of Mines – NgrYg
Walton, Ian, 801-581-8497 iwalton@egi.utah.edu,
    University of Utah – Gq
Walton, Nick, +44 023 92 842263 nick.walton@port.ac.uk,
    University of Portsmouth – Hw
Walworth, James, (520) 626-3364 walworth@ag.arizona.edu,
    University of Arizona – Soc
Wampler, J. Marion, kayargon@earthlink.net,
    Georgia State University – Cc
Wampler, Peter J., 616-331-2834 wamplerp@gvsu.edu,
    Grand Valley State University – HsZiGm
Wanamaker, Alan D., (515) 294-5142 adw@iastate.edu,
    Iowa State University of Science & Technology – CsPeOg
Wang, Alian, (314) 935-5671 alianw@levee.wustl.edu,
    Washington University in St. Louis – Ca
Wang, Bin, wangbin@hawaii.edu,
    University of Hawai'i, Manoa – Am
Wang, Chi-Yuen, (510) 642-2288 chiyuen@seismo.berkeley.edu,
    University of California, Berkeley – Yg
Wang, Dongmei, (701) 777-6143 dongmei.wang@engr.und.edu,
    University of North Dakota – Ng
Wang, Enru, enru.wang@und.edu,
    University of North Dakota – Zgi
Wang, Harry, (804) 684-7215 wang@vims.edu,
    College of William & Mary – Op
Wang, Herbert F., (608) 262-5932 hfwang@wisc.edu,
    University of Wisconsin-Madison – YgHwYx
Wang, Hong, 217-244-7692 hongwang@illinois.edu,
    Illinois State Geological Survey – Cg
Wang, Hsiang-Jui, (404) 894-3748 raywang@eas.gatech.edu,
    Georgia Institute of Technology – As
Wang, Jeen-Hwa, 886-2-27839910-326 jhwang@earth.sinica.edu.tw,
    Academia Sinica – Ysn
Wang, Jianhua, (202) 478-8457 jwang@carnegiescience.edu,
    Carnegie Institution for Science – Cs
Wang, Jianwei, (225) 578-5532 jianwei@lsu.edu,
    Louisiana State University – ClZmGy
Wang, Jim, (225) 578-1360 jjwang@agcenter.lsu.edu,
    Louisiana State University – Sc

Wang, Kun, 314-935-3855 kunwang@levee.wustl.edu,
  Washington University in St. Louis – XcCsXm
Wang, Lei, lwang@ldeo.columbia.edu,
  Columbia University – Og
Wang, Lillian T., 302-831-1096 lillian@udel.edu,
  University of Delaware – Zi
Wang, Pao-Kuan, (608)263-6479 pao@windy.aos.wisc.edu,
  University of Wisconsin, Madison – As
Wang, Qing, 831-656-7716 qwang@nps.edu,
  Naval Postgraduate School – Am
Wang, Wei-Chyung, (518) 437-8708 wcwang@albany.edu,
  SUNY, Albany – As
Wang, Weihong, (801) 863-7607 Weihong.Wang@uvu.edu,
  Utah Valley University – CmGeZi
Wang, Yang, (850) 644-1121 ywang@magnet.fsu.edu,
  Florida State University – CgsPc
Wang, Yanghua, +44 20 759 41171 yanghua.wang@imperial.ac.uk,
  Imperial College – Yg
Wang, Yuhang, (404) 894-3995 yuhang.wang@eas.gatech.edu,
  Georgia Institute of Technology – As
Wang, Yumei, (971) 673-1555 Oregon Dept of Geology and Mineral
  Industries – Ne
Wang, Yuqing, yuqing@hawaii.edu,
  University of Hawai'i, Manoa – As
Wang, Yuxuan, 713-743-9049 ywang140@uh.edu,
  University of Houston – As
Wang, Z. Aleck, (508) 289-3676 zawang@whoi.edu,
  Woods Hole Oceanographic Institution – Cm
Wang, Zhankun, 979 458 3464 zhankunwang@tamu.edu,
  Texas A&M University – Op
Wang, Zhenming, 859-323-0564 zmwang@uky.edu,
  University of Kentucky – Ys
Wang, Zhi (Luke), (559) 278-4427 zwang@csufresno.edu,
  California State University, Fresno – SpHwZi
Wang, Zhuo, 217-244-4270 zhuowang@illinois.edu,
  University of Illinois, Urbana-Champaign – As
Wanke, Ansgar, awanke@unam.na,
  University of Namibia – GsCl
Wankel, Scott, (508) 289-3944 sdwankel@whoi.edu,
  Woods Hole Oceanographic Institution – Cm
Wanless, Dorsey, dwanless@boisestate.edu,
  Boise State University – Gi
Wanless, Harold R., (305) 284-2697 hwanless@miami.edu,
  University of Miami – GseOn
Wannamaker, Phillip E., (801) 581-3547 pewanna@egi.utah.edu,
  University of Utah – Ye
Warburg, Helena F., helena.f.warburg@williams.edu,
  Williams College – Zn
Warburton, David L., (561) 297-3312 warburto@fau.edu,
  Florida Atlantic University – CgGe
Ward, Ben A., +44 023 80596041 B.A.Ward@soton.ac.uk,
  University of Southampton – Ob
Ward, Bess B., (609) 258-5150 bbw@princeton.edu,
  Princeton University – ObCb
Ward, Brent C., (604) 291-4229 bcward@sfu.ca,
  Simon Fraser University – GlmGe
Ward, Calvin H., (713) 348-4086 wardch@rice.edu,
  Rice University – Zg
Ward, Colin R., + 61 2 9385 4718 c.ward@unsw.edu.au,
  University of New South Wales – EcGd
Ward, David M., (406) 994-3401 umbdw@montana.edu,
  Montana State University – Zn
Ward, Dylan, (513) 556-2174 dylan.ward@uc.edu,
  University of Cincinnati – GmqZi
Ward, J. Evan, 860-405-9073 evan.ward@uconn.edu,
  University of Connecticut – Ob
Ward, Kevin M., (605) 394-2461 Kevin.Ward@sdsmt.edu,
  South Dakota School of Mines & Technology – YsGtYg
Ward, Larry G., (603) 862-5132 larry.ward@unh.edu,
  University of New Hampshire – Gu
Ward, Peter D., 206-543-2962 swift@ocean.washington.edu,
  University of Washington – Pg
Wardlaw, Norman C., (403) 220-6429 nwardlaw@geo.ucalgary.ca,
  University of Calgary – Go
Waren, Kirk B., 406-496-4866 kwaren@mtech.edu,
  Montana Tech of The University of Montana – HwsZe
Warger, Jane, (909) 652-6485 jane.warger@chaffey.edu,
  Chaffey College – ZgGg
Warhaft, Zellman, 607-255-3898 zw16@cornell.edu,
  Cornell University – Zn
Warke, Patricia, p.warke@qub.ac.uk,
  Queen's University Belfast – GmZym
Warkentin, Alicia, 604-822-3146 awarkentin@eoas.ubc.ca,
  University of British Columbia – Zn
Warland, Jon, (519) 824-4120 (Ext. 6374) jwarland@uoguelph.ca,
  University of Guelph – As
Warme, John E., jwarme@mines.edu,
  Colorado School of Mines – GsPeGu
Warner, Mark J., (206) 543-0765 warner@u.washington.edu,
  University of Washington – Op
Warner, Michael, +44 20 759 46535 m.warner@imperial.ac.uk,
  Imperial College – Ys
Warner, Nicholas H., 585-245-5291 warner@geneseo.edu,
  SUNY, Geneseo – XgGrHg
Warner, Richard D., (864) 656-5023 wrichar@ces.clemson.edu,
  Clemson University – Gz
Warner, Timothy A., (304) 293-4725 tim.warner@mail.wvu.edu,
  West Virginia University – Zr
Warnock, Jonathan P., (724) 357-5627 jwarnock@iup.edu,
  Indiana University of Pennsylvania – PemPv
Warny, Sophie, (225) 578-5089 swarny@lsu.edu,
  Louisiana State University – Pl
Warren, Jessica, 302-831-2569 warrenj@udel.edu,
  University of Delaware – GxCpGt
Warren, Joseph, (631) 632-5045 Joe.Warren@stonybrook.edu,
  SUNY, Stony Brook – Ob
Warren, Lesley A., (905) 525-9140 (Ext. 27347) warrenl@mcmaster.ca,
  McMaster University – CgHg
Warren, Linda M., (314) 977-3197 linda.warren@slu.edu,
  Saint Louis University – Ygs
Warren, Paul, (310) 825-2015 University of California, Los Angeles – Xm
Warren, Rachel, +44 (0)1603 59 3912 r.warren@uea.ac.uk,
  University of East Anglia – Eg
Warren, Richard G., (505) 667-7063 rgw@lanl.gov,
  Los Alamos National Laboratory – Gi
Warren, Stephen G., (206) 543-7230 sgw@atmos.washington.edu,
  University of Washington – ApGl
Warrick, Arthur W., (520) 621-1516 aww@ag.arizona.edu,
  University of Arizona – Sp
Wartes, Marwan A., (907) 451-5056 marwan.wartes@alaska.gov,
  Alaska Division of Geological & Geophysical Surveys – GsoGt
Warwick, Phillip, +44 (0)23 80592780 phil.warwick@noc.soton.ac.uk,
  University of Southampton – CaGe
Wash, Carlyle H., 831-656-7776 wash@nps.edu,
  Naval Postgraduate School – Am
Wassenaar, Len, (306) 239-2270 University of Saskatchewan – Hw
Wassermann, Joachim, 08141/5346762 jowa@geophysik.uni-muenchen.de,
  Ludwig-Maximilians-Universitaet Muenchen – Yg
Wasson, John T., (310) 825-1986 jtwasson@ucla.edu,
  University of California, Los Angeles – Xm
Wasylenki, Laura, Laura.Wasylenki@nau.edu,
  Northern Arizona University – ClOoCs
Wasylenki, Laura E., (812) 855-7508 lauraw@indiana.edu,
  Indiana University, Bloomington – ClOoCs
Waszek, Lauren, lwaszek@nmsu.edu,
  New Mexico State University, Las Cruces – Ys
Watanabe, Tohru, 81-76-445-6650 twatnabe@sci.u-toyama.ac.jp,
  University of Toyama – YxsGv
Waters, Dave, +44 (1865) 282457 dave.waters@earth.ox.ac.uk,
  University of Oxford – GpzGc
Waters, Laura E., watersla@sonoma.edu,
  Sonoma State University – GviGz
Waters, Matthew N., (334) 844-4100 mwaters@auburn.edu,
  Auburn University – Gn
Waters, Michael R., (979) 845-5246 mwaters@tamu.edu,
  Texas A&M University – Ga
Waters-Tormey, Cheryl, (828) 227-3696 cherylwt@wcu.edu,
  Western Carolina University – Gct
Waterstone, Marvin, marvinw@email.arizona.edu,
  University of Arizona – Zn
Waterstone, Marvin, (520) 621-1478 marvinw@u.arizona.edu,
  University of Arizona – Zn
Watkins, David K., (402) 472-2177 dwatkins1@unl.edu,
  University of Nebraska, Lincoln – PmGu

Watkins, James, watkins4@uoregon.edu,
    University of Oregon – Cg
Watkinson, A. John, (509) 335-2470 watkinso@mail.wsu.edu,
    Washington State University – Gc
Watkinson, Andrew, +44 (0)1603 59 2267 a.watkinson@uea.ac.uk,
    University of East Anglia – Ge
Watkinson, David H., dwatkson@ccs.carleton.ca,
    Carleton University – Em
Watkinson, Ian M., +44 1784 414046 ian.watkinson@rhul.ac.uk,
    Royal Holloway University of London – GtcRn
Watkinson, Matthew, +44 1752 584765 M.P.Watkinson@plymouth.ac.uk,
    University of Plymouth – Gs
Watney, Lynn W., (785) 864-2184 lwatney@kgs.ku.edu,
    University of Kansas – Gr
Watson, Alan, awatson@uoguelph.ca,
    University of Guelph – Zn
Watson, David B., (865) 241-4749 v6i@ornl.gov,
    Oak Ridge National Laboratory – Hw
Watson, E. Bruce, (518) 276-8838 watsoe@rpi.edu,
    Rensselaer Polytechnic Institute – CpcCt
Watson, Elizabeth, 215.299.1109 elizabeth.b.watson@drexel.edu,
    Drexel University – Zc
Watson, James R., 541-737-2519 jrwatson@coas.oregonstate.edu,
    Oregon State University – Zy
Watson, Kelly, 859-622-1419 kelly.watson@eku.edu,
    Eastern Kentucky University – ZriZn
Watters, Robert J., (775) 784-6069 watters@mines.unr.edu,
    University of Nevada, Reno – Nrg
Watters, Thomas R., 202-633-2483 Smithsonian Institution / National Air
    & Space Museum – Xg
Wattrus, Nigel J., (218) 726-7154 nwattrus@d.umn.edu,
    University of Minnesota, Duluth – Yr
Watts, Anthony B., +44 (1865) 272032 tony@earth.ox.ac.uk,
    University of Oxford – GutYv
Watts, Chester F., (540) 831-5637 cwatts@radford.edu,
    Radford University – NgrHw
Watts, D. Randolph, (401) 874-6507 rwatts@gso.uri.edu,
    University of Rhode Island – Op
Watts, Doyle, 937 775-3455 doyle.watts@wright.edu,
    Wright State University – YeZr
Waugh, Darryn W., (410) 516-8344 waugh@jhu.edu,
    Johns Hopkins University – As
Waugh, John, 757-822-7436 tcwaugj@tcc.edu,
    Tidewater Community College – Gg
Waugh, Richard A., (608) 342-1386 waugh@uwplatt.edu,
    University of Wisconsin, Platteville – Gc
Waugh, Truman, 785-864-2119 twaugh@ku.edu,
    University of Kansas – Ca
Wauthier, Christelle, 814-865-6711 cuw25@psu.edu,
    Pennsylvania State University, University Park – Zn
Wax, Charles L., (662) 325-3915 wax@geosci.msstate.edu,
    Mississippi State University – As
Wayne, William J., 402/476-0440 wwayne3@unl.edu,
    University of Nebraska, Lincoln – GmlZu
Wdowinski, Shimon, (305) 348-6826 shimon.wdowinski@fiu.edu,
    Florida International University – YdGt
Weaver, Andrew J., (250) 472-4001 weaver@ocean.seos.uvic.ca,
    University of Victoria – Op
Weaver, Barry L., (405) 325-4492 bweaver@ou.edu,
    University of Oklahoma – Ct
Weaver, Douglas J., (702) 295-5916 douglas_weaver@lanl.gov,
    Los Alamos National Laboratory – Zn
Weaver, John T., (250) 721-6155 weaver@phys.uvic.ca,
    University of Victoria – Ym
Weaver, Justin E., 806-834-4610 justin.e.weaver@ttu.edu,
    Texas Tech University – As
Weaver, Stephen G., (719) 389-6954 sweaver@coloradocollege.edu,
    Colorado College – Gx
Weaver, Thomas A., (505) 667-8464 tweaver@lanl.gov,
    Los Alamos National Laboratory – Yg
Webb, Christine R., (410) 507-3070 webbc@si.edu,
    Smithsonian Inst / Natl Museum of Natural Hist– GzgZn
Webb, Craig, Craig.Webb@eku.edu,
    Eastern Kentucky University – Zyn
Webb, Elizabeth A., 5196612111 x80208 ewebb5@uwo.ca,
    Western University – CslSa
Webb, John A., +61 3 9479 1273 john.webb@latrobe.edu.au,
    La Trobe University – GeHwGa
Webb, Laura E., (802) 656-8136 lewebb@uvm.edu,
    University of Vermont – GtCcGc
Webb, Nathan D., (217) 244-2426 ndwebb2@illinois.edu,
    Illinois State Geological Survey – EoGo
Webb, Peter N., (614) 292-7285 webb.3@osu.edu,
    Ohio State University – PmRhGr
Webb, Robert H., (520) 626-3293 rhwebb@usgs.gov,
    University of Arizona – Gm
Webb, Spahr, scw@ldeo.columbia.edu,
    Columbia University – Yr
Webb, Spahr C., (845) 365-8439 scw@ldeo.columbia.edu,
    Columbia University – Yrg
Webb, III, Thompson, (401) 863-3128 thompson_webb_iii@brown.edu,
    Brown University – PelAs
Webber, Andrew, (513) 455-7160 Cincinnati Museum Center – Pi
Webber, Jeffrey, jeffrey.webber@stockton.edu,
    Stockton University – Gg
Webber, Jeffrey R., (609) 652-4213 jeffrey.webber@stockton.edu,
    Stockton University – GcpGt
Webber, Karen L., (504) 280-7395 kwebber@uno.edu,
    University of Michigan – Gv
Webber, Karen L., (504) 280-6791 kwebber@uno.edu,
    University of New Orleans – Gv
Weber, Bodo, bweber@cicese.mx,
    Centro de Investigación Científica y de Educación Superior de
    Ensenada – Gp
Weber, John C., 616-331-3191 weberj@gvsu.edu,
    Grand Valley State University – GctYd
Weber, Karrie A., (402) 472-2739 kweber@unl.edu,
    University of Nebraska, Lincoln – PgCl
Weber, Keith, (208) 282-2757 webekeit@isu.edu,
    Idaho State University – Zi
Weber, Rodney J., (404) 894-1750 rweber@eas.gatech.edu,
    Georgia Institute of Technology – As
Weber-Diefenbach, Klaus, 089/2180 6549 klaus.diefenbach@iaag.geo.
    uni-muenchen.de,
    Ludwig-Maximilians-Universitaet Muenchen – Gg
Weborg-Benson, Kimberly, (716) 673-3293 kim.weborg-benson@fredonia.edu,
    SUNY, Fredonia – GgAsPg
Webre, Cherri B., (225) 388-8328 cherri@lgs.bri.lsu.edu,
    Louisiana State University – Zn
Webster, Ferris, (302) 645-4266 University of Delaware – Op
Webster, Gary D., (509) 335-4369 webster@wsu.edu,
    Washington State University – Pis
Webster, James D., 212-769-5401 jdw@amnh.org,
    American Museum of Natural History – Eg
Webster, John R., (701) 858-3873 john.webster@minotstateu.edu,
    Minot State University – Giz
Webster, Peter J., (404) 894-1748 pjw@eas.gatech.edu,
    Georgia Institute of Technology – AsOp
Wedding, William C., 859-257-1883 chad.wedding@uky.edu,
    University of Kentucky – Nm
Weeden, Lori, (978) 934-3344 lori_weeden@uml.edu,
    University of Massachusetts, Lowell – GgeGs
Weeraratne, Dayanthie, (818) 677-2046 dsw@csun.edu,
    California State University, Northridge – Yg
Weglein, Arthur B., (713) 743-3848 aweglein@uh.edu,
    University of Houston – Ye
Wegmann, Karl, 919-515-0380 karl_wegmann@ncsu.edu,
    North Carolina State University – Gm
Wegner, John, 404 727 4206 jwegner@emory.edu,
    Emory University – Sf
Wehmiller, John F., (302) 831-2926 jwehm@udel.edu,
    University of Delaware – Cl
Wehmiller, John F., jwehm@udel.edu,
    University of Delaware – Gu
Wehner, Peter J., (512) 223-0201 pwehner@austincc.edu,
    Austin Community College District – Gv
Wehrmann, Laura, 089/2180 Ludwig-Maximilians-Universitaet Muenchen
    – Pg
Wei, Xiaofang, 937 376 6193 xwei@centralstate.edu,
    Central State University – ZrOgZe
Weiblen, Paul W., 612-625-3477 pweib@umn.edu,
    University of Minnesota, Twin Cities – Gi
Weidner, Donald J., (631) 632-8211 donald.weidner@sunysb.edu,
    Stony Brook University – Gy

Weil, Arlo B., (610) 526-5113 aweil@brynmawr.edu,
    Bryn Mawr College – GctYm
Weiland, Thomas J., tjw@canes.gsw.edu,
    Georgia Southwestern State University – Gi
Weimer, Paul, (303) 492-3809 paul.weimer@colorado.edu,
    University of Colorado – Gor
Weinberger, Alycia J., (202) 478-8820 aweinberger@carnegiescience.edu,
    Carnegie Institution for Science – Xa
Weinstein, Charles E., (413) 236-4556 cweinste@berkshirecc.edu,
    Berkshire Community College – Zn
Weintraub, Michael N., 419-530-2585 michael.weintraub@utoledo.edu,
    University of Toledo – Zn
Weirich, Frank H., (319) 335-0156 frank-weirich@uiowa.edu,
    University of Iowa – Gm
Weis, Dominique A., (604) 822-1697 dweis@eos.ubc.ca,
    University of British Columbia – CaGeCu
Weisberg, Michael, 718-997-3366 Graduate School of the City University of New York – Zy
Weisberg, Robert H., (727) 553-1568 rweisberg@marine.usf.edu,
    University of South Florida – Op
Weisener, Christopher, 5192533000 x3753 weisener@uwindsor.ca,
    University of Windsor – CoGePg
Weisenfluh, Gerald A., 859-323-0505 jerryw@uky.edu,
    University of Kentucky – Ec
Weislogel, Amy L., 304-293-5603 Amy.Weislogel@mail.wvu.edu,
    West Virginia University – Gs
Weisman, Robert A., 320-308-3247 raweisman@stcloudstate.edu,
    Saint Cloud State University – Am
Weiss, Alfred W., 630.466.2720 aweiss@waubonsee.edu,
    Waubonsee Community College – Ziy
Weiss, Benjamin, (617) 324-0224 bpweiss@mit.edu,
    Massachusetts Institute of Technology – Ym
Weiss, Chester J., 540-231-6521 cjweiss@unm.edu,
    University of New Mexico – Ye
Weiss, Christopher C., 806-834-4712 chris.weiss@ttu.edu,
    Texas Tech University – AssAs
Weiss, Dennis, (212) 650-6849 Graduate School of the City University of New York – Pg
Weiss, Dominik, +44 20 759 46383 d.weiss@imperial.ac.uk,
    Imperial College – ClaCq
Weiss, Ray F., (858) 534-2598 rfweiss@ucsd.edu,
    University of California, San Diego – Cm
Weiss, Robert, 540-231-2334 weiszr@vt.edu,
    Virginia Polytechnic Institute & State University – RnOn
Weissmann, Gary S., 505-277-4204 weissman@unm.edu,
    University of New Mexico – HwGsr
Welby, Charles W., (919) 515-7158 cwwelby@unity.ncsu.edu,
    North Carolina State University – Hw
Welch, Steve M., (785) 532-7236 welchsm@ksu.edu,
    Kansas State University – Sp
Welch, Susan A., 614-292-9059 welch.318@osu.edu,
    Ohio State University – ClGeCs
Weldeab, Syee, 805-893-4903 weldeab@geol.ucsb.edu,
    University of California, Santa Barbara – PeCm
Weldon, Ray J., (541) 346-4584 ray@uoregon.edu,
    University of Oregon – Gc
Welford, Mark R., (912) 478-5943 mwelfgeog@gsvms2.cc.gasou.edu,
    Georgia Southern University – Zy
Welhan, John A., (208) 282-4254 welhjohn@isu.edu,
    Idaho State University – Hw
Welling, Tim, 845-431-8536 welling@sunydutchess.edu,
    Dutchess Community College – GeSo
Wellner, Julia, 713-743-2887 jwellner@uh.edu,
    University of Houston – GslOu
Wells, David E., (228) 688-3389 dew@unb.ca,
    University of Southern Mississippi – YdZi
Wells, Michael L., (702) 895-0828 michael.wells@unlv.edu,
    University of Nevada, Las Vegas – Gc
Wells, Neil A., (330) 672-2951 nwells@kent.edu,
    Kent State University – Gs
Wells, Neil C., +44 023 80592428 N.C.Wells@soton.ac.uk,
    University of Southampton – Op
Wells, Randy, (941) 388-4441 University of South Florida – Ob
Wells, Steve G., (775) 673-7470 sgwells@dri.edu,
    University of Nevada, Reno – Zn
Welp, Lisa, (765) 496-6896 lwelp@purdue.edu,
    Purdue University – Cs

Welsh, James L., 507-933-7333 welsh@gac.edu,
    Gustavus Adolphus College – Gxc
Wen, Lianxing, (631) 632-1726 lianxing.wen@sunysb.edu,
    Stony Brook University – Ys
Wen, Xianyun, +44(0) 113 34 31425 x.wen@leeds.ac.uk,
    University of Leeds – NoAsm
Wen, Yuming, 16717352685 ywen@uguam.uog.edu,
    University of Guam – Zir
Wendlandt, Richard F., (303) 273-3809 rwendlan@mines.edu,
    Colorado School of Mines – GizCp
Weng, Kevin, 808-956-4109 kevin.weng@hawaii.edu,
    University of Hawai'i, Manoa – Ob
Weng, Qihao, 812-237-2255 qweng@indstate.edu,
    Indiana State University – Zru
Wenk, Hans-Rudolf, (510) 642-7431 wenk@berkeley.edu,
    University of California, Berkeley – Gz
Wennberg, Paul O., (626) 395-2447 wennberg@gps.caltech.edu,
    California Institute of Technology – As
Wenner, David B., (706) 542-2393 dwenner@uga.edu,
    University of Georgia – Cs
Wenner, Jennifer M., (920) 424-7003 wenner@uwosh.edu,
    University of Wisconsin, Oshkosh – Gi
Wentzcovitch, Renata, rmw2150@columbia.edu,
    Columbia University – Ys
Wenzel, Christopher, (307) 532-8293 chris.wenzel@ewc.wy.edu,
    Eastern Wyoming College – SbfHg
Werdon, Melanie B., (907) 451-5082 melanie.werdon@alaska.gov,
    Alaska Division of Geological & Geophysical Surveys – Eg
Werhner, Matthew J., (813) 253-7263 mwerhner@hccfl.edu,
    Hillsborough Community College – Gg
Werne, Josef, (412) 624-8775 jwerne@pitt.edu,
    University of Pittsburgh – CosGn
Werner, Adrian, adrian.werner@flinders.edu.au,
    Flinders University – Hw
Werner, Alan, (413) 538-2134 awerner@mtholyoke.edu,
    Mount Holyoke College – GleGm
Werner, Bradley T., (858) 534-0583 bwerner@ucsd.edu,
    University of California, San Diego – Gm
Wernicke, Brian P., (626) 395-6192 brian@gps.caltech.edu,
    California Institute of Technology – Gtc
Werts, Scott P., 803-323-4930 wertss@winthrop.edu,
    Winthrop University – ScbPg
Wesely, Marvin L., (630) 252-5827 mlwesely@anl.gov,
    Argonne National Laboratory – As
Wesley Wood, Charley, 850-9837126 woodwes@ufl.edu,
    University of Florida – So
Wesnousky, Steve, (775) 784-6067 stevew@seismo.unr.edu,
    University of Nevada, Reno – Ys
Wessel, Paul, (808) 956-4778 pwessel@hawaii.edu,
    University of Hawai'i, Manoa – Yr
West, A. Joshua (Josh), (213) 740-6736 joshwest@usc.edu,
    University of Southern California – ClGmHs
West, Alan, 276-328-0203 maw4v@uvawise.edu,
    The University of Virginia's College at Wise – Zn
West, David P., (802) 443-3476 dwest@middlebury.edu,
    Middlebury College – Gc
West, Jared, +44(0) 113 34 35253 L.J.West@leeds.ac.uk,
    University of Leeds – Hg
West, Keith, (715) 735-4352 keith.west@uwc.edu,
    University of Wisconsin Colleges – ZynZn
West, Larry T., (706) 542-0906 University of Georgia – Sd
West, Nicole, 989-774-4475 west2n@cmich.edu,
    Central Michigan University – Gm
West, Olivia M., (615) 576-0505 qm5@ornl.gov,
    Oak Ridge National Laboratory – Ng
West, Robert, 323-260-8115 westrb@elac.edu,
    East Los Angeles College – GgmSd
West, Ronald R., (785) 532-2248 rrwest@ksu.edu,
    Kansas State University – PgeGs
West, Terry R., (765) 494-3296 trwest@purdue.edu,
    Purdue University – Ng
Westerink, Joannes J., (574) 631-6475 University of Notre Dame – On
Westerman, David S., (802) 485-2337 westy@norwich.edu,
    Norwich University – Gxi
Westervelt, Daniel, danielmw@ldeo.columbia.edu,
    Columbia University – Og
Westgate, James W., (409) 880-8237 james.westgate@lamar.edu,

Lamar University – Pev
Westgate, John A., (416) 287-7235 University of Toronto – Gl
Westhoff, Martijn C., +31 20 59 85741 m.c.westhoff@vu.nl,
  Vrije Universiteit Amsterdam – Hq
Westrop, Stephen R., (405) 325-0542 swestrop@ou.edu,
  University of Oklahoma – Pi
Wetherelt, Andrew, +44 01326 371827 A.Wetherelt@exeter.ac.uk,
  Exeter University – Nm
Wettlaufer, John S., (203) 432-0892 john.wettlaufer@yale.edu,
  Yale University – Yg
Wettstein, Justin, (541) 737-5177 justinw@coas.oregonstate.edu,
  Oregon State University – As
Wetz, Michael, 361-825-2132 Michael.Wetz@tamucc.edu,
  Texas A&M University, Corpus Christi – Ob
Wetzel, Dan L., (907) 488-3746 University of Alaska, Fairbanks – Zn
Wetzel, Laura R., (727) 864-8484 wetzellr@eckerd.edu,
  Eckerd College – YrOuGg
Whalen, Michael T., 907-474-5302 ffmtw@uaf.edu,
  University of Alaska, Fairbanks – Gs
Whaler, Kathryn A., +44 (0) 131 650 4904 kathy.whaler@ed.ac.uk,
  Edinburgh University – YmRn
Whalley, John, +44 023 9284 2257 john.whalley@port.ac.uk,
  University of Portsmouth – Gc
Whatley, Robin L., 312-344-8604 rwhatley@colum.edu,
  Columbia College Chicago – PvZr
Whattam, Scot , +9661368602993 scott.whattam@kfupm.edu.sa,
  King Fahd University of Petroleum and Minerals – GxXc
Whattam, Scott A., 82-2-3290-3182 whattam@korea.ac.kr,
  Korea University – Gc
Wheatcraft, Steve, (775) 784-1973 steve@wheatcraftandassociates.com,
  University of Nevada, Reno – Hw
Wheatcroft, Rob, 541-737-3891 raw@coas.oregonstate.edu,
  Oregon State University – OgGg
Wheaton, John R., (406) 272-1603 jwheaton@mtech.edu,
  Montana Tech of The University of Montana – HwEc
Wheeler, Andy J., +353 21 4903951 a.wheeler@ucc.ie,
  University College Cork – GusGg
Wheeler, Greg, 916-278-3919 wheelergr@csus.edu,
  California State University, Sacramento – Em
Wheeler, John, +440151 794 5172 johnwh@liverpool.ac.uk,
  University of Liverpool – GpcGz
Wheeler, Paul, +44 01326 255778 P.Wheeler@exeter.ac.uk,
  Exeter University – Nx
Wheeler, Richard F., (931) 221-1004 wheelerr@apsu.edu,
  Austin Peay State University – Ggo
Whelan, Jean K., (508) 289-2819 jwhelan@whoi.edu,
  Woods Hole Oceanographic Institution – Co
Whelan, Michael, (616) 620-3637 michael.whelan@scu.edu.au,
  Southern Cross University – Zir
Whelan, Peter M., (320) 589-6315 whelan@mrs.umn.edu,
  University of Minnesota, Morris – Gx
Whiley, Harriet, harriet.whiley@flinders.edu.au,
  Flinders University – Zn
Whipkey, Charles, (540) 654-1428 cwhipkey@umw.edu,
  University of Mary Washington – Cg
Whipple, Kelin, 480-965-9508 kxw@asu.edu,
  Arizona State University – Gt
Whisner, Jennifer B., (570) 389-4139 jwhisner@bloomu.edu,
  Bloomsburg University – GmHw
Whisner, S. Christopher, (570) 389-4746 swhisner@bloomu.edu,
  Bloomsburg University – GiCtg
Whisonant, Robert C., (540) 831-5224 rwhisona@radford.edu,
  Radford University – GsaGg
Whitaker, Rodney W., (505) 667-7672 rww@lanl.gov,
  Los Alamos National Laboratory – As
White, Bekki C., 501 296-1877 bekki.white@arkansas.gov,
  Arkansas Geological Survey – GoEoGg
White, Craig M., (208) 426-3633 cwhite@boisestate.edu,
  Boise State University – Giv
White, George W., (301) 687-4264 gwhite@frostburg.edu,
  Frostburg State University – Zn
White, James D., +64 3 479-9009 james.white@otago.ac.nz,
  University of Otago – Gvs
White, James W. C., 303-492-5494 jwhite@colorado.edu,
  University of Colorado – CsAt
White, Jeffrey, (812) 855-0731 whitej@indiana.edu,
  Indiana University, Bloomington – Hs
White, Jeffrey G., (919) 515-2389 jeffrey_white@ncsu.edu,
  North Carolina State University –
White, Jeffrey R., (812) 855-0731 whitej@indiana.edu,
  Indiana University, Bloomington – Hg
White, John C., (859) 622-1276 john.white@eku.edu,
  Eastern Kentucky University – GiCgt
White, Jonathan, (303) 384-2650 jwhite@mines.edu,
  Colorado Geological Survey – NgZuGg
White, Joseph C., (506) 453-4804 clancy@unb.ca,
  University of New Brunswick – Gc
White, Joseph D., (254) 710-2141 Joseph_D_White@baylor.edu,
  Baylor University – Zr
White, Justin, 801-863-6864 justin.white@uvu.edu,
  Utah Valley University – ZyiZu
White, Lisa D., 510-664-4966 ldwhite@berkeley.edu,
  University of California, Berkeley – ZePim
White, Lisa D., (415) 338-1963 lwhite@sfsu.edu,
  San Francisco State University – Pg
White, Mark D., (509) 372-6070 mark.white@pnl.gov,
  Pacific Northwest National Laboratory – Hy
White, Martin, +353 (0)91 493 214 martin.white@nuigalway.ie,
  National University of Ireland Galway – Op
White, Nicky, +44 (0) 1223 337063 njw10@cam.ac.uk,
  University of Cambridge – YgGt
White, Robert, +44 (0) 1223 337187 rsw1@cam.ac.uk,
  University of Cambridge – YgGt
White, Scott, (970) 247-7475 white_s@fortlewis.edu,
  Fort Lewis College – ZirZy
White, Scott M., (803) 777-6304 swhite@geol.sc.edu,
  University of South Carolina – YrZr
White, Susan, just19@earthlink.net,
  Los Angeles Harbor College – Zg
White, Susan Q., 61 3 9479 5641 susanqwhite@netspace.net.au,
  La Trobe University – GmgGe
White, William B., (814) 667-2709 wbw2@psu.edu,
  Pennsylvania State University, University Park – CgGzy
White, William M., (607) 255-7466 wmw4@cornell.edu,
  Cornell University – CctCa
White, William M., wmw4@cornell.edu,
  Cornell University – Ce
Whiteaker, Timothy L., 512-471-0570 twhit@mail.utexas.edu,
  University of Texas, Austin – Zi
Whitehead, James, (506) 453-4593 University of New Brunswick – Gg
Whitehead, Peter W., +61 742321200 peter.whitehead@jcu.edu.au,
  James Cook University – Ggv
Whitehead, Robert E., rwhitehead@laurentian.ca,
  Laurentian University, Sudbury – Ce
Whitehill, Matthew, 218-733-5981 m.whitehill@lsc.edu,
  Lake Superior College – Gg
Whiteley, Martin, +44 01332 593752 m.whiteley@derby.ac.uk,
  University of Derby – Go
Whiteside, Jessica H., +44 (0)23 80593199 J.Whiteside@soton.ac.uk,
  University of Southampton – Pe
Whiteway, James, (416)736-2100 #22310 whiteway@yorku.ca,
  York University – As
Whiticar, Michael J., (250) 721-6514 whiticar@uvic.ca,
  University of Victoria – CosGo
Whiting, Peter J., (216) 368-3989 peter.whiting@case.edu,
  Case Western Reserve University – Gm
Whitlock, Cathy, (406) 994-6910 whitlock@montana.edu,
  Montana State University – Peg
Whitman, Brian E., (570) 408-4882 brian.whitman@wilkes.edu,
  Wilkes University – HqwSb
Whitman, Dean, (305) 348-3089 whitmand@fiu.edu,
  Florida International University – YgGet
Whitman, Jill M., (253) 535-8720 whitmaj@plu.edu,
  Pacific Lutheran University – OuGuZe
Whitmeyer, Steven J., whitmesj@jmu.edu,
  James Madison University – GctZi
Whitmore, John H., (937) 766-7947 johnwhitmore@cedarville.edu,
  Cedarville University – PgGsg
Whitney, D.A., (785) 532-6101 whitney@ksu.edu,
  Kansas State University – So
Whitney, Donna L., (612) 626-7582 dwhitney@umn.edu,
  University of Minnesota, Twin Cities – Gpt
Whitney, Earl M., (505) 667-3595 whitney@lanl.gov,
  Los Alamos National Laboratory – Zn

Whitney, James A., (706) 548-6894 jamesawhitney@hotmail.com,
University of Georgia – GivCg
Whitney, Michael, 860-405-9157 michael.whitney@uconn.edu,
University of Connecticut – Op
Whittaker, Alex, +44 20 759 47491 a.whittaker@imperial.ac.uk,
Imperial College – Gt
Whittaker, Amber, 207-287-2803 amber.whittaker@maine.gov,
Dept of Agriculture, Conservation, and Forestry – ZiGcCg
Whittecar, Jr., G. Richard, (757) 683-5197 rwhittec@odu.edu,
Old Dominion University – GmHwZe
Whittemore, Donald O., 785-864-3965 donwhitt@kgs.ku.edu,
University of Kansas – Cl
Whittemore, Donald O., (785) 864-2182 donwhitt@kgs.ku.edu,
University of Kansas – Hw
Whittier, Michael, MWhittier@csustan.edu,
California State University, Stanislaus – Ggz
Whittington, Alan, (573) 884-7625 whittingtona@missouri.edu,
University of Missouri – GivCg
Whittington, Carla, (206) 878-3710 cwhittin@highline.edu,
Highline College – GgvZg
Whitton, Mark, +44 023 9284 2257 mark.witton@port.ac.uk,
University of Portsmouth – Pg
Wiberg, Patricia L., (434) 924-7546 pw3c@virginia.edu,
University of Virginia – Cg
Wickert, Andrew D., (612) 625-6878 awickert@umn.edu,
University of Minnesota, Twin Cities – Gml
Wickham, John S., (817) 272-2987 wickham@uta.edu,
University of Texas, Arlington – Gco
Wickham, Thomas, (724) 938-4180 wickham@calu.edu,
California University of Pennsylvania – Zu
Wicks, Carol M., (225) 578-2692 cwicks@lsu.edu,
Louisiana State University – HwClHy
Wicks, Frederick J., fredw@rom.on.ca,
Royal Ontario Museum – Gz
Wicks, Frederick J., (416) 978-5395 fredw@rom.on.ca,
University of Toronto – Gz
Widanagamage, Inoka, 662-915-2154 ihwidana@olemiss.edu,
University of Mississippi – ClGep
Widdowson, Mark A., (540) 231-7153 mwiddows@vt.edu,
Virginia Polytechnic Institute & State University – HwSpHq
Widom, Elisabeth, (513) 529-5048 widome@miamioh.edu,
Miami University – CcGve
Widory, David, 514-987-3000 #1968 widory.david@uqam.ca,
Universite du Quebec a Montreal – Cs
Wiedenmann, Jorg, +44 (0)23 80596497 joerg.wiedenmann@noc.soton.ac.uk,
University of Southampton – Ob
Wiederspahn, Mark, (512) 471-0406 markw@ig.utexas.edu,
University of Texas, Austin – Zn
Wiederwohl, Chrissy, 979-845-7191 chrissyw@tamu.edu,
Texas A&M University – Op
Wielicki, Matthew, (205) 348-0548 mmwielicki@ua.edu,
University of Alabama – CcgGt
Wielicki, Michelle, mdwielicki@ua.edu,
University of Alabama – CcXgGi
Wiens, Douglas A., (314) 935-6517 doug@wustl.edu,
Washington University in St. Louis – YsrYg
Wierenga, Peter J., (520) 792-9591 wierenga@ag.arizona.edu,
University of Arizona – Sp
Wierenga, Peter J., (520) 621-1899 pjw@gmail.com,
University of Arizona – Sp
Wiese, Katryn, (415) 452-5061 katryn.wiese@mail.ccsf.edu,
City College of San Francisco – GiOg
Wiesenburg, Denis A., (601) 266-4937 denis.wiesenburg@usm.edu,
University of Southern Mississippi – Og
Wiesner, Mark R., (713) 348-5129 wiesner@rice.edu,
Rice University – Zg
Wiggert, Jerry D., (228) 688-3491 jerry.wiggert@usm.edu,
University of Southern Mississippi – Op
Wigley, Rochelle, 603-862-1135 rochelle@ccom.unh.edu,
University of New Hampshire – CgGu
Wignall, Paul, +44(0) 113 34 35247 P.B.Wignall@leeds.ac.uk,
University of Leeds – PgGs
Wijbrans, Jan, j.r.wijbrans@vu.nl,
Vrije Universiteit Amsterdam – CcGtv
Wijesinghe, Ananda M., (925) 423-0605 Lawrence Livermore National Laboratory – Nr
Wilbur, Bryan, 626 585-3118 bcwilbur@pasadena.edu,
Pasadena City College – Gg
Wilch, Thomas I., (517) 629-0759 twilch@albion.edu,
Albion College – GlvGm
Wilcock, Peter W., (410) 516-5421 Johns Hopkins University – Gm
Wilcock, William S., (206) 543-6043 wilcock@u.washington.edu,
University of Washington – Ou
Wilcox, Andrew, 406-243-4761 andrew.wilcox@mso.umt.edu,
University of Montana – Gm
Wilcox, Jeffrey D., (828) 232-5184 jwilcox@unca.edu,
University of North Carolina, Asheville – HwClGg
Wildeman, Thomas R., twildema@mines.edu,
Colorado School of Mines – Ct
Wilder, Lee, (603) 271-1976 geology@des.nh.gov,
New Hampshire Geological Survey – GgZee
Wilder, Margaret, mwilder@email.arizona.edu,
University of Arizona – Zn
Wilderman, Candie, (717) 245-1573 wilderma@dickinson.edu,
Dickinson College – Hs
Wiles, Gregory C., (330) 263-2298 gwiles@wooster.edu,
College of Wooster – Gl
Wilf, Peter D., (814) 865-6721 pwilf@geosc.psu.edu,
Pennsylvania State University, University Park – Pg
Wilhelmson, Robert B., (217) 333-8651 bw@ncsa.uiuc.edu,
University of Illinois, Urbana-Champaign – As
Wilhelmy, Jerry B., (505) 665-3188 Los Alamos National Laboratory – Zn
Wilhite, Donald A., (402) 472-4270 dwilhite2@unl.edu,
University of Nebraska, Lincoln – AsZyn
Wilkerson, M. S., (765) 658-4666 mswilke@depauw.edu,
University of Illinois, Urbana-Champaign – Gc
Wilkey, Patrick L., (630) 252-6258 Argonne National Laboratory – Ng
Wilkie, Ann C., (352) 392-8699 acwilkie@ifas.ufl.edu,
University of Florida – Sb
Wilkie, Richard W., (413) 253-5752 rwilkie@geo.umass.edu,
University of Massachusetts, Amherst – ZynZn
Wilkins, Colin, +44 1752 584773 C.Wilkins@plymouth.ac.uk,
University of Plymouth – Eg
Wilkins, David E., (208) 426-2390 dwilkins@boisestate.edu,
Boise State University – ZyGm
Wilkinson, Bruce, 315-443-3869 eustasy@syr.edu,
Syracuse University – Gs
Wilkinson, Bruce H., (734) 764-1435 eustasy@umich.edu,
University of Michigan – GseGt
Wilks, Daniel, (607) 255-1750 dsw5@cornell.edu,
Cornell University – As
Wilks, Maureen, (505) 835-5322 mwilks@gis.nmt.edu,
New Mexico Institute of Mining & Technology – Gp
Willahan, Duane, (408) 848-4702 Gavilan College – Gg
Willenbring, Jane, 215-746-8197 erosion@sas.upenn.edu,
University of Pennsylvania – GmCcGl
William, Nancy, (785) 532-7257 nkw@ksu.edu,
Kansas State University – Zn
Williams, Aaron, (251) 460-6915 bwilliams@southalabama.edu,
University of South Alabama – Am
Williams, Allison M., (905) 525-9140 (Ext. 24334) awill@mcmaster.ca,
McMaster University – Zn
Williams, Alton P., williams@ldeo.columbia.edu,
Columbia University – Pc
Williams, Bruce A., (509) 372-3799 bruce.williams@pnl.gov,
Pacific Northwest National Laboratory – Ng
Williams, Christopher J., (717) 291-3814 chris.williams@fandm.edu,
Franklin and Marshall College – PeSb
Williams, Curtis J., (714) 484-7000 (Ext. 48181) Cypress College – Gg
Williams, Dave E., williams.david@interchange.ubc.ca,
University of British Columbia – Oc
Williams, David A., (270) 827-3414 williams@uky.edu,
University of Kentucky – Eg
Williams, Erik, (775) 784-1396 eswilliams@unr.edu,
University of Nevada, Reno – Ys
Williams, Forrest, 831-656-3274 fwillia@comcast.net,
Naval Postgraduate School – Am
Williams, Harris, (919) 530-7127 hewhew@nccu.edu,
North Carolina Central University – OgZg
Williams, Harry F. L., (940) 565-3317 harryf.williams@unt.edu,
University of North Texas – Gm
Williams, Ian S., (709) 737-8395 ian.williams@uwrf.edu,
University of Wisconsin, River Falls – Yg
Williams, James H., (573) 341-4616 Missouri University of Science and

Technology – Gg
Williams, Jeremy C., (330) 672-1459 jwill243@kent.edu,
 Kent State University – Cg
Williams, John W., (408) 924-5050 john.williams@sjsu.edu,
 San Jose State University – Ng
Williams, John W., (608) 265-5537 jww@geography.wisc.edu,
 University of Wisconsin, Madison – PeZyPl
Williams, Kay R., (717) 477-1602 krwill@ship.edu,
 Shippensburg University – As
Williams, Kevin K., 716-878-5116 williakk@buffalostate.edu,
 Buffalo State College – GmXg
Williams, Kevin K., 716-878-4911 williakk@buffalostate.edu,
 Buffalo State College – Xg
Williams, Kevin L., (206) 543-3949 williams@apl.washington.edu,
 University of Washington – Op
Williams, Kim R., (303) 273-3245 krwillia@mines.edu,
 Colorado School of Mines – Ca
Williams, Mark, +44 0116 252 3642 mri@le.ac.uk,
 Leicester University – Pg
Williams, Mathew, +44 (0) 131 650 7776 mat.williams@ed.ac.uk,
 Edinburgh University – ZcrZu
Williams, Michael L., (413) 545-0745 mlw@geo.umass.edu,
 University of Massachusetts, Amherst – Gc
Williams, Paul, +44 0161 306-3905 paul.i.williams@manchester.ac.uk,
 University of Manchester – As
Williams, Paul F., (506) 453-5185 pfw@unb.ca,
 University of New Brunswick – Gc
Williams, Quentin, (831) 459-3132 qwilliams@pmc.ucsc.edu,
 University of California, Santa Cruz – Gy
Williams, Ric, +44-151-794-5136 Ric@liverpool.ac.uk,
 University of Liverpool – Og
Williams, Roger T., rtwillia@nps.edu,
 Naval Postgraduate School – Am
Williams, Rosa, 706-649-1474 williams_rosa@columbusstate.edu,
 Columbus State University – Xa
Williams, Stanley N., (480) 965-1438 stanley.williams@asu.edu,
 Arizona State University – Gv
Williams, Thomas, (208) 885-6656 tomw@uidaho.edu,
 University of Idaho – Gf
Williams, Wayne K., (423) 425-4427 wayne-williams@utc.edu,
 University of Tennessee, Chattanooga – GdgGo
Williams, Wyn, +44 (0) 131 650 4909 Wyn.Williams@ed.ac.uk,
 Edinburgh University – Ym
Williams, II, Richard T., (865) 974-6169 rwilliams@utk.edu,
 University of Tennessee, Knoxville – Yg
Williams-Bruinders, Leizel, +27 (0)41 504 4361 leizel.williams@nmmu.ac.za,
 Nelson Mandela Metropolitan University – Zn
Williams-Jones, Anthony E., (514) 398-1676 willyj@eps.mcgill.ca,
 McGill University – Ce
Williams-Jones, Glyn, (778) 782-3306 glynwj@sfu.ca,
 Simon Fraser University – GvCeYe
Williamson, Ben, +44 01326 371856 B.J.Williamson@exeter.ac.uk,
 Exeter University – EmGv
Williamson, Douglas, 212-772-5265 douglas.williamson@hunter.cuny.edu,
 Hunter College (CUNY) – Zi
Willis, Grant C., (801) 537-3355 grantwillis@utah.gov,
 Utah Geological Survey – Gg
Willis, Julie B., 208-496-1905 willisj@byui.edu,
 Brigham Young University - Idaho – GtcZi
Willis, Marc, 714-992-7446 mwillis@fullcoll.edu,
 Fullerton College – Gg
Willoughby, Hugh E., 305-348-0243 hugh.willoughby@fiu.edu,
 Florida International University – As
Wills, William V., 813-253-7809 wwills@hccfl.edu,
 Hillsborough Community College – Zg
Willsey, Shawn P., (208) 732-6421 swillsey@csi.edu,
 College of Southern Idaho – GgcGv
Willson, Lee, 713.348.6219 Rice University – Zn
Wilmut, Michael, (250) 472-4343 University of Victoria – Zn
Wilson, Alicia M., (803) 777-1240 awilson@geol.sc.edu,
 University of South Carolina – Hw
Wilson, Blake B., (785) 864-2118 bwilson@kgs.ku.edu,
 University of Kansas – Zi
Wilson, Carol A., carolw@lsu.edu,
 Louisiana State University – GsOnCc
Wilson, Charlie, +44 (0)1603 59 1386 charlie.wilson@uea.ac.uk,
 University of East Anglia – Ge

Wilson, Clark R., (512) 471-5008 crwilson@jsg.utexas.edu,
 University of Texas, Austin – Yg
Wilson, Emily, 717-358-3821 emily.wilson@fandm.edu,
 Franklin and Marshall College – CgGi
Wilson, Fred L., (325) 486-6984 fwilson@angelo.edu,
 Angelo State University – Gm
Wilson, Greg, 541-737-4015 wilsongr@coas.oregonstate.edu,
 Oregon State University – Onp
Wilson, Greg C., 616-331-2392 wilsong@gvsu.edu,
 Grand Valley State University – Gm
Wilson, Gregory, 206-543-8917 gpwilson@uw.edu,
 University of Washington – Pvg
Wilson, James R., jwilson@weber.edu,
 Weber State University – Gz
Wilson, Jeffey S., (317) 274-1128 jeswilso@iupui.edu,
 Indiana University, Indianapolis – Zy
Wilson, Jeffrey A., (734) 647-7461 wilsonja@umich.edu,
 University of Michigan – Gg
Wilson, John D., (780) 492-0353 University of Alberta – Am
Wilson, John L., (575) 835-5308 jwilson@nmt.edu,
 New Mexico Institute of Mining and Technology – Hw
Wilson, John R., (610) 330-5197 wilsonj@lafayette.edu,
 Lafayette College – Zi
Wilson, Laura E., (785) 639-6192 lewilson6@fhsu.edu,
 Fort Hays State University – PvePg
Wilson, Lorne G., (520) 621-9108 lorne@email.arizona.edu,
 University of Arizona – Hw
Wilson, Lucy A., (506) 648-5607 lwilson@unbsj.ca,
 University of New Brunswick Saint John – Ga
Wilson, Mark A., (330) 263-2247 mwilson@wooster.edu,
 College of Wooster – Pi
Wilson, Melissa L., 612-625-4276 mlw@umn.edu,
 University of Minnesota, Twin Cities – Sco
Wilson, Merwether, +44 (0) 131 650 8636 Meriwether.Wilson@ed.ac.uk,
 Edinburgh University – Ob
Wilson, Michael C., wilsonmi@douglascollege.ca,
 Douglas College – GaPgGs
Wilson, P. Christopher, 772-468-3922 ext 119 pcwilson@ufl.edu,
 University of Florida – So
Wilson, Paul A., +44 (0)23 80596164 paul.wilson@noc.soton.ac.uk,
 University of Southampton – GsCg
Wilson, Rick I., (709) 737-8386 California Geological Survey – Ge
Wilson, Robert, +44 01334 463914 rjsw@st-andrews.ac.uk,
 University of St. Andrews – Zn
Wilson, Robert E., (631) 632-8689 robert.wilson@stonybrook.edu,
 SUNY, Stony Brook – Op
Wilson, Roy R., (860) 465-4370 wilsonr@easternct.edu,
 Eastern Connecticut State University – Zi
Wilson, Sarah B., (540) 458-8800 wilsons@wlu.edu,
 Washington & Lee University – Zn
Wilson, Steven D., (217) 333-0956 sdwilson@illinois.edu,
 University of Illinois – Hw
Wilson, Terry J., 614-292-0723 wilson.43@osu.edu,
 Ohio State University – GctGl
Wilson, Thomas, +64 3 3667001 Ext 45511 thomas.wilson@canterbury.ac.nz,
 University of Canterbury – Gv
Wilson, Thomas B., (520) 621-9308 twilson@ag.arizona.edu,
 University of Arizona – Zn
Wilson, Thomas H., (304) 293-6431 tom.wilson@mail.wvu.edu,
 West Virginia University – YeGcYg
Wilton, Derek H., (709) 737-8389 dwilton@sparky2.esd.mun.ca,
 Memorial University of Newfoundland – EgCae
Wilton, Robert D., (905) 525-9140 (Ext. 24536) wiltonr@mcmaster.ca,
 McMaster University – Zn
Wiltshire, John C., (808) 956-6042 johnw@soest.hawaii.edu,
 University of Hawai'i, Manoa – Og
Wimbush, Mark, (401) 874-6515 m.wimbush@gso.uri.edu,
 University of Rhode Island – Op
Winant, Clinton D., (858) 534-2067 cwinant@ucsd.edu,
 University of California, San Diego – On
Winberry, Paul, winberry@geology.cwu.edu,
 Central Washington University – Yg
Winchell, Robert E., (562) 985-4920 California State University, Long Beach – Gz
Windsor, Jr., John G., 321-674-8096 jwindsor@fit.edu,
 Florida Institute of Technology – Oc
Wine, Paul H., (404) 894-3425 pw7@prism.gatech.edu,

Georgia Institute of Technology – As
Winebrenner, Dale P., 206-543-1393 dpw@apl.washington.edu,
    University of Washington – GlZr
Wing, Scott L., (202) 357-2649 Smithsonian Inst / Natl Museum of Natural
    Hist– Pb
Winglee, Robert M., 206-685-8160 winglee@uw.edu,
    University of Washington – XyYx
Winguth, Arne M., 817 272 2987 awinguth@uta.edu,
    University of Texas, Arlington – OpAs
Winkelstern, Ian Z., 616-331-9219 winkelsi@gvsu.edu,
    Grand Valley State University – GsOo
Winkler, Dale A., (214) 768-2750 Southern Methodist University – Pv
Winklhofer, Michael, 089/2180 4143 michael@geophysik.uni-muenchen.de,
    Ludwig-Maximilians-Universitaet Muenchen – Yg
Winnick, Matthew, 413-545-1715 mwinnick@geo.umass.edu,
    University of Massachusetts, Amherst – Cbq
Winsor, Roger A., (828) 262-7053 winsorra@appstate.edu,
    Appalachian State University – Zn
Winston, Barbara, (314) 935-7047 Washington University in St. Louis – Zn
Winterbottom, Wesley, winterbottomw@easternct.edu,
    Eastern Connecticut State University – Zg
Winterkamp, Judith L., (505) 667-1264 judyw@lanl.gov,
    Los Alamos National Laboratory – Zn
Winton, Alison, (806) 834-0497 alison.winton@ttu.edu,
    Texas Tech University – ZnnZn
Wintsch, Robert P., (812) 855-4018 wintsch@indiana.edu,
    Indiana University, Bloomington – GpcGt
Wirth, Erin, ewirth@uw.edu,
    University of Washington – YsRn
Wirth, Karl R., (651) 696-6449 wirth@macalester.edu,
    Macalester College – GiZn
Wisdom, Jack, (617) 253-7730 wisdom@mit.edu,
    Massachusetts Institute of Technology – Zn
Wise, Dana, (618) 453-4362 dwise@siu.edu,
    Southern Illinois University Carbondale – Zn
Wise, Michael A., (202) 633-1826 wisem@si.edu,
    Smithsonian Inst / Natl Museum of Natural Hist– Gz
Wise, Jr, Sherwood W., (850) 644-6265 swise@fsu.edu,
    Florida State University – PmGu
Wisely, Beth , 307-268 2233 bwisely@caspercollege.edu,
    Casper College – YgHwGc
Wishart, De Bonne N., dwishart@centralstate.edu,
    Central State University – YgCa
Wishner, Karen, (401) 874-6402 kwishner@gso.uri.edu,
    University of Rhode Island – Ob
Withers, Kim, 361-825-5907 Kim.Withers@tamucc.edu,
    Texas A&M University, Corpus Christi – Zn
Withers, Mitchell M., (901) 678-2007 mwithers@memphis.edu,
    University of Memphis – Ys
Withers, Tony, 519-661-2111 x.88627 tony.withers@uwo.ca,
    Western University – Cp
Withjack, Martha O., (848) 445-3142 drmeow3@eps.rutgers.edu,
    Rutgers, The State University of New Jersey – GctEo
Witkowski, Christine, (860) 343-5781 cwitkowski@mxcc.edu,
    Middlesex Community College – GeZg
Witt, Emma, emma.witt@stockton.edu,
    Stockton University – Hg
Witter, Rob, 541-574-7969 rwitter@usgs.gov,
    Oregon Dept of Geology & Mineral Industries – Zg
Wittke, James, 9285239565/9044 james.wittke@nau.edu,
    Northern Arizona University – Gi
Wittke, Seth, (307) 766-2286 Ext. 244 seth.wittke@wyo.gov,
    Wyoming State Geological Survey – Rn
Wittkop, Chad, (507) 389-6929 chad.wittkop@mnsu.edu,
    Minnesota State University – GsCgGl
Witzke, Brian J., (319) 335-1590 brian-witzke@uiowa.edu,
    University of Iowa – GrPsg
Wixman, Ronald, (541) 346-4568 rwixman@uoregon.edu,
    University of Oregon – Zn
Wizevich, Michael, wizevichmic@ccsu.edu,
    Central Connecticut State University – GsdGm
Wobus, Reinhard A., (413) 597-2470 reinhard.a.wobus@williams.edu,
    Williams College – GivGz
Woerheide, Prof. Dr. Gert, geobiologie@geo.lmu.de,
    Ludwig-Maximilians-Universitaet Muenchen – Pgi
Wofsy, Steven C., (617) 495-4566 scw@io.harvard.edu,
    Harvard University – As

Wogelius, Roy, (+44)-(0)161-275 3841 Roy.Wogelius@manchester.ac.uk,
    University of Manchester – Cg
Wohl, Ellen E., (970) 491-5298 ellen.wohl@colostate.edu,
    Colorado State University – GmHs
Wohletz, Kenneth H., (505) 667-9202 wohletz@lanl.gov,
    Los Alamos National Laboratory – YxGvRm
Wojewoda, Jurand, jurand.wojewoda@uwr.edu.pl,
    University of Wroclaw – Gs
Wojtal, Steven F., (440) 775-8352 steven.wojtal@oberlin.edu,
    Oberlin College – Gc
Wolak, Jeannette, (931) 372-3695 jwolak@tntech.edu,
    Tennessee Tech University – Gsr
Wolaver, Brad, (512) 471-1368 brad.wolaver@beg.utexas.edu,
    University of Texas, Austin – HwGe
Wolcott, Donna L., (919) 515-7866 donna_wolcott@ncsu.edu,
    North Carolina State University – Ob
Wolcott, Ray, 6196447454 x3099 rwolcott@palomar.edu,
    Cuyamaca College – Og
Wolcott, Thomas G., (919) 515-7866 tom_wolcott@ncsu.edu,
    North Carolina State University – Ob
WoldeGabriel, Giday, 505-667-8749 wgiday@lanl.gov,
    Los Alamos National Laboratory – Gx
Wolf, Aaron T., (541) 737-2722 wolfa@geo.oregonstate.edu,
    Oregon State University – Hg
Wolf, Lorraine W., (334) 844-4878 wolflor@auburn.edu,
    Auburn University – Ysg
Wolf, Michael B., (309) 794-7304 MichaelWolf@augustana.edu,
    Augustana College – Gi
Wolfe, Amy, 513-529-5009 wolfeal4@miamioh.edu,
    Miami University – Cs
Wolfe, Ben, (816) 604-6622 ben.wolfe@mcckc.edu,
    Metropolitan Community College-Kansas City – GgZyg
Wolfe, Christopher, christopher.wolfe@stonybrook.edu,
    SUNY, Stony Brook – Op
Wolfe, Karen M., (615) 898-2726 karen.wolfe@mtsu.edu,
    Middle Tennessee State University – Zn
Wolfe, Paul J., 937 775-2201 paul.wolfe@wright.edu,
    Wright State University – Ye
Wolff, Eric W., +44 (0) 1223 333486 ew428@cam.ac.uk,
    University of Cambridge – GlPe
Wolff, George, +44-151-794-4094 Wolff@liverpool.ac.uk,
    University of Liverpool – Og
Wolff, John A., (509) 335-2825 jawolff@wsu.edu,
    Washington State University – Giv
Wolfgram, Diane, (406) 496-4353 dwolfgram@mtech.edu,
    Montana Tech of the University of Montana – EgoNm
Wolfsberg, Andrew V., (505) 667-3599 awolf@lanl.gov,
    Los Alamos National Laboratory – Hw
Wolfsberg, Kurt, (505) 667-4464 Los Alamos National Laboratory – Cc
Wolken, Gabriel J., (907) 451-5018 gabriel.wolken@alaska.gov,
    Alaska Division of Geological & Geophysical Surveys – Gm
Wollan, Jacinda, mandy.looser@ndsu.edu,
    North Dakota State University – Zn
Wolny, Dave, (970) 248-1154 dwolmy@mesastate.edu,
    Colorado Mesa University – Ys
Wolosz, Thomas H., (518) 564-4031 woloszth@plattsburgh.edu,
    SUNY, Plattsburgh – Pe
Woltemade, Christopher J., (717) 477-1143 cjwolt@ship.edu,
    Shippensburg University – HgRwGm
Wolter, Calvin, Calvin.Wolter@dnr.iowa.gov,
    Iowa Dept of Natural Resources – GgZi
Wolverton, Steve, 940-565-4987 steven.wolverton@unt.edu,
    University of North Texas – Ga
Wong, Chi S., (250) 363-6407 WongCS@pac.dfo-mpo.gc.ca,
    University of British Columbia – Cm
Wong, Cindy, (716) 878-6731 solargs@buffalostate.edu,
    Buffalo State College – Zn
Wong, Martin, (315) 228-7203 mswong@colgate.edu,
    Colgate University – Gtc
Wong, Teng-fong, (631) 632-8212 teng-fong.wong@stonybrook.edu,
    Stony Brook University – Yx
Woo, David, 510-885-3160 david.woo@csueastbay.edu,
    California State University, East Bay – Zr
Woo, Ming-Ko, (905) 525-9140 woo@mcmaster.ca,
    McMaster University – HgZy
Wood, Aaron R., (515) 294-8862 awood@iastate.edu,
    Iowa State University of Science & Technology – PvGrs

Wood, Bernard, +44 (1865) 272014 Bernie.Wood@earth.ox.ac.uk, University of Oxford – Gz
Wood, David A., 210-486-0063 dwood30@alamo.edu, San Antonio Community College – Znn
Wood, Eric F., (609) 258-4675 efwood@princeton.edu, Princeton University – HgZrHq
Wood, Howard, (361) 825-3335 tony.wood@tamucc.edu, Texas A&M University, Corpus Christi – ZeHsZn
Wood, Ian, +44 020 7679 32405 ian.wood@ucl.ac.uk, University College London – Gz
Wood, Jacqueline, (504) 671-6485 jwood@lakelandcc.edu, Delgado Community College – Gg
Wood, James R., (906) 487-2894 jrw@mtu.edu, Michigan Technological University – Cl
Wood, Kim, 662-268-1032 kimberly.wood@msstate.edu, Mississippi State University – AsmZr
Wood, Lesli, 303-273-3801 lwood@mines.edu, Colorado School of Mines – GsoGm
Wood, Lesli J., 512-471-0328 lesli.wood@beg.utexas.edu, University of Texas, Austin – Eo
Wood, Neill, +44 01326 255163 n.a.wood@exeter.ac.uk, Exeter University – YuZen
Wood, Robert, (206) 543-1203 robwood@atmos.washington.edu, University of Washington – As
Wood, Spencer H., (208) 426-3629 swood@boisestate.edu, Boise State University – Gmm
Woodall, Debra W., (386) 506-3765 woodald@daytonastate.edu, Daytona State College – GgOg
Woodard, Gary C., (520) 621-5399 gwoodard@sahra.arizona.edu, University of Arizona – Zn
Woodard, Jeremy, (031) 260-2804 woodardj@ukzn.ac.za, University of KwaZulu-Natal – CcGiz
Woodburne, Michael O., (909) 787-5028 michael.woodburne@ucr.edu, University of California, Riverside – Pv
Woodcock, Curtis, curtis@bu.edu, Boston University – Zr
Woodcock, Curtis E., (617) 353-5746 Boston University – Zr
Woodcock, Nigel H., +44 (0) 1223 333430 nhw1@esc.cam.ac.uk, University of Cambridge – GcsGt
Woodgate, Rebecca A., (206) 221-3268 woodgate@apl.washington.edu, University of Washington – Op
Woodhouse, Connie, 520-626-0235 conniew1@email.arizona.edu, University of Arizona – Pe
Woodhouse, Iain H., +44 (0) 131 650 2527 i.h.woodhouse@ed.ac.uk, Edinburgh University – Zi
Woodhouse, John, +44 (1865) 272021 john.woodhouse@earth.ox.ac.uk, University of Oxford – Yg
Woodin, Sarah A., (803) 782-9727 swoodin@gmail.com, University of South Carolina – Ob
Wooding, Frank B., (508) 289-3334 Woods Hole Oceanographic Institution – Yr
Woodland, Bertram G., (312) 665-7648 Field Museum of Natural History – Gp
Woodley, Teresa, (604) 822-3146 twoodley@eos.ubc.ca, University of British Columbia – Zn
Woodruff, Jonathan D., 413-577-3831 woodruff@geo.umass.edu, University of Massachusetts, Amherst – Gs
Woodruff, William H., (505) 665-2557 Los Alamos National Laboratory – Zn
Woods, Adam D., (657) 278-2921 awoods@fullerton.edu, California State University, Fullerton – GsPe
Woods, Andy, +44 (0) 1223 765702 andy@bpi.cam.ac.uk, University of Cambridge – YgGt
Woods, Juliette, juliette.woods@flinders.edu.au, University of South Australia – Hwq
Woods, Karen M., (409) 880-2251 karen.woods@lamar.edu, Lamar University – Zg
Woods, Neal, woodsn@mailbox.sc.edu, University of South Carolina – Zn
Woods, Rachel A., +44 (0) 131 650 6014 Rachel.Wood@ed.ac.uk, Edinburgh University – Gs
Woodward, Lee A., (505) 277-5309 University of New Mexico – Gc
Woodward, Mac B., 501 683-0113 mac.woodward@arkansas.gov, Arkansas Geological Survey – Gc
Woodwell, Grant R., (540) 654-1427 gwoodwel@mwc.edu, University of Mary Washington – Gc
Wooldridge, C F., wooldridge@cf.ac.uk, Cardiff University – Zn
Woolery, Edward W., (859) 257-3016 woolery@uky.edu, University of Kentucky – Ys
Wooten, Richard M., 828-296-4632 rick.wooten@ncdenr.gov, North Carolina Geological Survey – RnGc
Worcester, Peter, Peter.Worcester@eku.ed, Eastern Kentucky University – GgOuZe
Worcester, Peter A., (812) 866-7306 worcestr@hanover.edu, Hanover College – Gi
Worden, Richard, 0151 794 5184 R.Worden@liverpool.ac.uk, University of Liverpool – Gs
Wordsworth, Robin, rwordsworth@seas.harvard.edu, Harvard University – Xg
Worrall, Fred, 0191 3742525 fred.worrall@durham.ac.uk, Durham University – Cg
Wortel, Matthew J., (319) 335-3992 matthew-wortel@uiowa.edu, University of Iowa – Gx
Worthington, Lindsay L., 505.277.4204 lworthington@unm.edu, University of New Mexico – Yg
Wortmann, Ulrich B., 416-978-2084 University of Toronto – Gg
Wraith, Jon M., (406) 994-1997 jean.dixon@montana.edu, Montana State University – Sp
Wright, Alan, 561-992-1555 alwr@ufl.edu, University of Florida – Sc
Wright, Carrie L., (812) 465-1145 clwright@usieagles.org, University of Southern Indiana – Ze
Wright, Chris, 614 265 6632 chris.wright@dnr.state.oh.us, Ohio Dept of Natural Resources – Ec
Wright, James D., (848) 445-5722 jdwright@eps.rutgers.edu, Rutgers, The State University of New Jersey – CsOuPm
Wright, James (Jim) E., (706) 542-4394 jwright@gly.uga.edu, University of Georgia – Gt
Wright, Kathyrn, (269) 387-5486 kathyrn.wright@wmich.edu, Western Michigan University – Zn
Wright, Stephen F., (802) 656-4479 swright@uvm.edu, University of Vermont – Glc
Wright, Tim, +44(0) 113 34 35258 t.j.wright@leeds.ac.uk, University of Leeds – Yd
Wright, V P., wrightVP@cf.ac.uk, Cardiff University – Gs
Wright, William C., 845-938-2063 william.wright@westpoint.edu, United States Military Academy – ZiYd
Wronkiewicz, David J., (630) 252-4385 Argonne National Laboratory – Cg
Wu, Chin, (608) 263-3078 chinwu@engr.wisc.edu, University of Wisconsin, Madison – Onp
Wu, David T., (303) 273-2066 dwu@mines.edu, Colorado School of Mines – Zn
Wu, Francis T., (607) 777-2512 Binghamton University – Ys
Wu, Jonny, (713) 743-9624 jwu40@central.uh.edu, University of Houston – GtcGo
Wu, Patrick, (403) 220-7855 ppwu@ucalgary.ca, University of Calgary – Ys
Wu, Shiliang, (906) 487-2590 slwu@mtu.edu, Michigan Technological University – AsCg
Wu, Shuang-Ye, (937) 229-1720 Shuang-Ye.Wu@notes.udayton.edu, University of Dayton – Zy
Wu, Xiaoqing, (515) 294-9872 wuxq@iastate.edu, Iowa State University of Science & Technology – As
Wu, Xingru, xingru.wu@ou.edu, University of Oklahoma – Np
Wu, Yutian, (765) 494-8677 wu640@purdue.edu, Purdue University – As
Wu, Yutian, yutianwu@ldeo.columbia.edu, Columbia University – Og
Wuebbles, Donald J., (217) 244-1568 wuebbles@illinois.edu, University of Illinois, Urbana-Champaign – As
Wuerthele, Norman, (412) 622-3265 wuerthelen@carnegiemnh.org, Carnegie Museum of Natural History – Pv
Wulamu, Wasit, (314) 977-5156 awulamu@slu.edu, Saint Louis University – Zi
Wulff, Andrew, (270) 745-5976 andrew.wulff@wku.edu, Western Kentucky University – Gv
Wulff, Andrew H., (562) 907-4220 Whittier College – Gi
Wunsch, Carl, (617) 496-2732 cwunsch@fas.harvard.edu, Harvard University – Op
Wunsch, Carl I., (617) 253-5937 cwunsch@mit.edu, Massachusetts Institute of Technology – Opo

Wunsch, David R., (302) 831-8258 dwunsch@udel.edu,
University of Delaware – ClHwNg
Wust-Bloch, Gilles H., (03-) 640-5475 Tel Aviv University – Ys
Wyckoff, John W., (303) 556-2590 john.wyckoff@ucdenver.edu,
University of Colorado, Denver – Zy
Wyckoff, William K., (406) 994-6914 bwyckoff@montana.edu,
Montana State University – Zn
Wygal, Brian, (516) 877-4170 bwygal@adelphi.edu,
Adelphi University – GaRmGe
Wykel, Andy, 803.896.7703 wykelc@dnr.sc.gov,
South Carolina Dept of Natural Resources – Gg
Wylie, Ann G., (301) 405-4079 awylie@umd.edu,
University of Maryland – Gz
Wyllie, Peter J., (626) 395-6461 wyllie@gps.caltech.edu,
California Institute of Technology – Cp
Wyman, Derek A., (618) 938-0117 University of Saskatchewan – Cg
Wymore, Adam, adam.wymore@unh.edu,
University of New Hampshire – Hg
Wynn, Elizabeth A., (205) 247-3671 awynn@gsa.state.al.us,
Geological Survey of Alabama – HsZyi
Wynn, Thomas C., (570) 484-2081 twynn@lockhaven.edu,
Lock Haven University – GsPiEo
Wypych, Alicja, 907-451-5010 alicja.wypych@alaska.gov,
Alaska Division of Geological & Geophysical Surveys – Gg
Wyse Jackson, Patrick N., + 353-1-8961477 wysjcknp@tcd.ie,
Trinity College – PiRh
Wysession, Michael E., (314) 935-5625 michael@wucore.wustl.edu,
Washington University in St. Louis – Ys
Wysocki, Mark, (607) 255-2568 mww3@cornell.edu,
Cornell University – As
Wysong, Jr., James F., (813) 253-7805 jwysong@hccfl.edu,
Hillsborough Community College – Am
Wyss, Andre R., (805) 893-8628 wyss@geol.ucsb.edu,
University of California, Santa Barbara – Pv

X

Xiao, Dana, (570) 389-3902 dxiao@bloomu.edu,
Bloomsburg University – Zi
Xiao, Shuhai, 540-231-1366 xiao@vt.edu,
Virginia Polytechnic Institute & State University – Pi
Xie, Lian, (919) 515-1435 xie@ncsu.edu,
North Carolina State University – As
Xie, Xiangyang, (817) 257-4395 x.xie@tcu.edu,
Texas Christian University – GoEog
Xie, Zhixiao, (561) 297-2852 xie@fau.edu,
Florida Atlantic University – ZirZy
Xu, Huifang, 608/265-5587 hfxu@geology.wisc.edu,
University of Wisconsin-Madison – Gz
Xu, Jie, 915-747-7556 jxu2@utep.edu,
University of Texas, El Paso – ClPoGz
Xu, Li, (508)289-3673 lxu@whoi.edu,
Woods Hole Oceanographic Institution – Yr
Xu, Shangping, 414-229-6148 xus@uwm.edu,
University of Wisconsin, Milwaukee – Hg
Xu, Wei, wei.xu@uleth.ca,
University of Lethbridge – Zi
XU, Yingfeng, 337-482-1113 yingfeng@Louisiana.edu,
University of Louisiana at Lafayette – ZgGe
Xuan, Chuang, +44 (0)23 80596401 c.xuan@soton.ac.uk,
University of Southampton – Ym

Y

Yacobucci, Peg M., (419) 372-7982 mmyacob@bgsu.edu,
Bowling Green State University – Pgi
Yacucci, Mark, (309) 268-8640 mark.yacucci@heartland.edu,
Heartland Community College – Zg
Yacucci, Mark A., 217-265-0747 yacucci@illinois.edu,
University of Illinois – Zi
Yalcin, Kaplan, (541) 737-1230 yalcink@geo.oregonstate.edu,
Oregon State University – GgeZe
Yalcin, Rebecca, yalcinr@onid.orst.edu,
Oregon State University – Gg
Yalda, Sepideh, 717-872-3293 Sepi.Yalda@millersville.edu,
Millersville University – As
Yamanaka, Tsuyoneri, +44 0151 795 5291 T.Yamanaka@liverpool.ac.uk,
University of Liverpool – Ob
Yan, Beizhan, yanbz@ldeo.columbia.edu,
Columbia University – Cg
Yan, Eugene, (630) 252-6322 eyan@anl.gov,
Argonne National Laboratory – Hw
Yan, Jun, (270) 745-8952 jun.yan@wku.edu,
Western Kentucky University – Zi
Yan, Xiao-Hai, (302) 831-3694 University of Delaware – Zr
Yan, Y E., (630) 252-6322 eyan@anl.gov,
Argonne National Laboratory – Gg
Yancey, Thomas E., (979) 845-0643 yancey@geo.tamu.edu,
Texas A&M University – Pg
Yandle, Tracy, 404 727 5652 tyandle@emory.edu,
Emory University – Ob
Yanes, Yurena, yurena.yanes@uc.edu,
University of Cincinnati – PeCsPg
Yang, Changbing, 512-471-4364 changbing.yang@beg.utexas.edu,
University of Texas, Austin – Hw
Yang, Gang, (970)-491-3789 gang.yang@colostate.edu,
Colorado State University – CcaCg
Yang, Jianwen, 519-253-3000 x2181 jianweny@uwindsor.ca,
University of Windsor – HwYg
Yang, Panseok, (204) 474-6910 panseok.yang@umanitoba.ca,
University of Manitoba – CaGp
Yang, Ping, (979) 845-7679 pyang@tamu.edu,
Texas A&M University – As
Yang, Qiang, qyang@ldeo.columbia.edu,
Columbia University – Cg
Yang, Wan, 316 -978-3140 Wichita State University – Gs
Yang, Wenchang, 609-258-4101 wenchang@princeton.edu,
Princeton University – As
Yang, Y, YangY6@cf.ac.uk,
Cardiff University – Hg
Yang, Zhiming, (919) 530-5296 zyang@nccu.edu,
North Carolina Central University – Zry
Yang, Zong-Liang, (512) 471-3824 liang@mail.utexas.edu,
University of Texas, Austin – AmsHq
Yankovsky, Alexander E., (803) 777-3550 ayankovsky@geol.sc.edu,
University of South Carolina – OpnAm
Yao, Wensheng, (727) 553-3922 University of South Florida – Oc
Yao, Yon, +27 (0)46-603-7393 y.yao@ru.ac.za,
Rhodes University – Eg
Yapp, Crayton J., (214) 768-3897 Southern Methodist University – Csl
Yarbrough, Lance D., 662-915-7499 ldyarbro@olemiss.edu,
University of Mississippi – Ng
Yarbrough, Robert A., 912-478-0846 RYarbrough@GeorgiaSouthern.edu,
Georgia Southern University – Zn
Yardley, Bruce W., b.w.d.yardley@leeds.ac.uk,
University of Leeds – CgEgGg
Yarger, Douglas N., (515) 294-9872 doug@iastate.edu,
Iowa State University of Science & Technology – As
Yarger, Douglas N., doug@iastate.edu,
Iowa State University of Science & Technology – As
Yassin, Barbara E., (601) 961-5571 barbara_yassin@mdeq.ms.gov,
Mississippi Office of Geology – ZyiGm
Yates, Martin G., (207) 581-2154 yates@maine.edu,
University of Maine – EgGzx
Yates, Mary Anne, (505) 667-7090 Los Alamos National Laboratory – Zn
Yau, Man Kong, (514) 398-3719 peter.yau@mcgill.ca,
McGill University – As
Ye, Hengchun, (323) 343-2229 hye2@calstatela.edu,
California State University, Los Angeles – ZyHs
Ye, Ming, mingye@scs.fsu.edu,
Florida State University – Hg
Yeager, Kevin M., (859) 257-5431 kevin.yeager@uky.edu,
University of Kentucky – GsCmc
Yeager, Lauren A., 361-749-6776 lyeager@utexas.edu,
University of Texas, Austin – Og
Yeats, Robert S., 541-740-3806 yeatsr@geo.oregonstate.edu,
Oregon State University – GtYsNg
Yee, Nathan, 848-932-5714 nyee@envsci.rutgers.edu,
Rutgers, The State University of New Jersey – PgCo
Yegulalp, Tuncel M., (212) 854-2984 yegulalp@columbia.edu,
Columbia University – Nm
Yeh, Jim, (520) 621-5943 yeh@hwr.arizona.edu,
University of Arizona – Sp
Yeh, Joseph S., (512) 471-3323 joseph.yeh@beg.utexas.edu,
University of Texas at Austin, Jackson School of Geosciences – Gm
Yeh, Tian-Chyi J., (520) 621-5943 ybiem@hwr.arizona.edu,

Yelderman, Jr., Joe C., (254) 710-2185 joe_yelderman@baylor.edu,
Baylor University – HwgZu

Yelisetti, Subbarao, 361-593-4894 subbarao.yelisetti@tamuk.edu,
Texas A&M University, Kingsville – Ygs

Yelle, Roger, (520) 621-6243 yelle@lpl.arizona.edu,
University of Arizona – Zn

Yellen, Brian, byellen@geo.umass.edu,
University of Massachusetts, Amherst – Gs

Yellich, John, (269) 387-8649 john.a.yellich@wmich.edu,
Western Michigan University – EgGeHw

Yeung, Laurence Y., (713) 348-6304 ly19@rice.edu,
Rice University – CsAsCl

Yi, Chuixiang, 718-997-3366 chuixiang.yi@qc.cuny.edu,
Queens College (CUNY) – Asm

Yi, Yuchan, 614 292 6005 yi.3@osu.edu,
Ohio State University – Ydv

Yiannakoulias, Niko, (905) 525-9140 (Ext. 20117) yiannan@mcmaster.ca,
McMaster University – Ge

Yildirim, Ismail, 00904242370000-5970 iyildirim@firat.edu.tr,
Firat University – ScGix

Yilmaz, Alper, 614-247-4323 yilmaz.15@osu.edu,
Ohio State University – ZfrZu

Yin, An, (310) 825-8752 yin@epss.ucla.edu,
University of California, Los Angeles – Gt

Yin, Jianjun, 520-626-7453 yin@email.arizona.edu,
University of Arizona – Og

Yin, Qing-zhu, 530-752-0934 qyin@ucdavis.edu,
University of California, Davis – Cc

Yin, Zhi-Yong, (619) 260-8864 zyin@sandiego.edu,
University of San Diego – Zir

Yoakum, Barbara, 410-293-6928 yoakum@usna.edu,
United States Naval Academy – Zn

Yogodzinskli, Gene M., 803-777-9524 gyogodzin@geol.sc.edu,
University of South Carolina – GiCuc

Yoh, Shing, (908) 737-3692 syoh@kean.edu,
Kean University – Am

Yon, Lisa, 7607441150 x2369 lyon@palomar.edu,
Palomar College – OgZg

Yong, Wenjun, 519-661-2111 ext 86628 wyong4@uwo.ca,
Western University – Yx

Yonkee, W. A., (801) 626-7419 ayonkee@weber.edu,
Weber State University – Gct

Yoo, Kyungsoo, 612-624-7784 kyoo@umn.edu,
University of Minnesota, Twin Cities – Sd

Yool, Stephen R., yools@email.arizona.edu,
University of Arizona – Zri

Yoshida, Glenn, (323) 241-5296 yoshidgy@lasc.edu,
Los Angeles Southwest College – Gg

Yoshinobu, Aaron S., (806) 834-7715 aaron.yoshinobu@ttu.edu,
Texas Tech University – GctGu

Yoskowitz, David, 361-825-2966 david.yoskowitz@tamucc.edu,
Texas A&M University, Corpus Christi – Zn

You, Jinsheng, jyou2@unl.ed,
University of Nebraska, Lincoln – Hg

Young, Donald W., (602) 542-5025 University of Arizona – Zn

Young, Edward D., (310) 267-4930 eyoung@epss.ucla.edu,
University of California, Los Angeles – CsXc

Young, George S., (814) 863-4228 young@ems.psu.edu,
Pennsylvania State University, University Park – Am

Young, Graham A., (204) 988-0648 ggyoung@cc.umanitoba.ca,
University of Manitoba – Pi

Young, Grant M., (519) 473-5692 gyoung@uwo.ca,
Western University – GrCgGt

Young, Harvey R., 204-727-9798 young@brandonu.ca,
Brandon University – Gd

Young, James E., (828) 262-8482 youngje@appstate.edu,
Appalachian State University – Ze

Young, Jeffrey, (204) 474-8863 jeff.young@umanitoba.ca,
University of Manitoba – GcZe

Young, Jeri J., 602.708.8558 jeribenhorin@email.arizona.edu,
Arizona Geological Survey – Ys

Young, John A., (608)263-2374 jayoung@wisc.edu,
University of Wisconsin, Madison – AstOp

Young, Michael H., (512) 475-8830 michael.young@beg.utexas.edu,
University of Texas, Austin – SpHwGe

Young, Mike, 902 494 2364 mdyoung@dal.ca,
Dalhousie University – GgcGp

Young, Nicolas, nicolasy@ldeo.columbia.edu,
Columbia University – Cg

Young, Patrick, (480) 727-6581 patrick.young.1@asu.edu,
Arizona State University – Zn

Young, Priscilla E., 605-677-6144 priscilla.young@usd.edu,
South Dakota Dept of Environment and Natural Resources – Zn

Young, Richard A., (585) 245-5296 young@geneseo.edu,
SUNY, Geneseo – GmXgZr

Young, Richard E., (808) 956-7024 ryoung@hawaii.edu,
University of Hawai'i, Manoa – Ob

Young, Robert S., (828) 227-3822 ryoung@wcu.edu,
Western Carolina University – GsOn

Young, William, +44(0) 113 34 31640 c.w.young@leeds.ac.uk,
University of Leeds – Ge

Young, William R., (858) 534-1380 wryoung@ucsd.edu,
University of California, San Diego – Op

Youngs, Edward, edward.youngs@open.ac.uk,
The Open University – Sp

Yow, Donald M., (859) 622-1420 don.yow@eku.edu,
Eastern Kentucky University – AtZye

Yu, Jin-Yi, (949) 824-3878 jyyu@uci.edu,
University of California, Irvine – AsOpZc

Yu, Qian, (413) 545-2095 qyu@geo.umass.edu,
University of Massachusetts, Amherst – Zri

Yu, Wen-che, fgsyw@earth.sinica.edu.tw,
Academia Sinica – Ys

Yu, Zicheng, 610-758-6751 ziy2@lehigh.edu,
Lehigh University – Pe

Yuan, Xiaojun, xyuan@ldeo.columbia.edu,
Columbia University – Op

Yue, Stephen, (514) 398-4755 McGill University – Nx

Yuen, Cheong-yip R., (630) 252-4869 yuenr@anl.gov,
Argonne National Laboratory – Ge

Yuen, David A., 612-624-1868 davey@umn.edu,
University of Minnesota, Twin Cities – Yg

Yule, J. Douglas, (818) 677-6238 j.d.yule@csun.edu,
California State University, Northridge – Gt

Yun, Misuk, (204) 474-8870 yun@cc.umanitoba.ca,
University of Manitoba – Cs

Yun, Seong-Taek, 82-2-3290-3176 styun@korea.ac.kr,
Korea University – CqHyCl

Yung, Yuk L., (626) 395-6940 yly@gps.caltech.edu,
California Institute of Technology – As

Yunwei, Sun, (925)422-1587 sun4@llnl.gov,
Lawrence Livermore National Laboratory – Ng

Yuretich, Richard F., (413) 545-0538 yuretich@geo.umass.edu,
University of Massachusetts, Amherst – Cl

Yurkovich, Steven P., (828) 227-7367 yurkovich@wcu.edu,
Western Carolina University – Gx

Yurtsever, Ayhan, 089/2180 4346 ayhan.yurtsever@ltz.uni-uemchen.de,
Ludwig-Maximilians-Universitaet Muenchen – Gz

Yusoff, Ismail, 03-79674153 ismaily70@um.edu.my,
University of Malaya – Hg

Yuter, Sandra, 919-513-7963 sandra_yuter@ncsu.edu,
North Carolina State University – As

Yutzy, Gale, (301) 687-4369 gyutzy@frostburg.edu,
Frostburg State University – Zn

Yvon-Lewis, Shari A., 979 458 1816 syvonlewis@ocean.tamu.edu,
Texas A&M University – Oc

Z

Zabielski, Victor, (703) 845-6507 vzabielski@nvcc.edu,
Northern Virginia Community College - Alexandria – GgOg

Zaccaria, Daniele, 530-219-7502 dzaccaria@ucdavis.edu,
University of California, Davis – Hg

Zachos, James, 831-459-4644 jzachos@ucsc.edu,
University of California, Santa Cruz – Pc

Zachos, Louis G., (662) 915-8827 lgzachos@olemiss.edu,
University of Mississippi – PiGeNx

Zachry, Doy L., (479) 575-2785 dzachry@uark.edu,
University of Arkansas, Fayetteville – Gr

Zaffos, Andrew A., 520.621.3635 azafazaffos@email.arizona.edu,
Arizona Geological Survey – Zf

Zafiriou, Oliver C., (508) 289-2342 ozafiriou@whoi.edu,
Woods Hole Oceanographic Institution – Oc

Zahm, Christopher K., 512-471-3159 chris.zahm@beg.utexas.edu,

University of Texas, Austin – Yg

Zaitchik, Benjamin, 410-516-7135 bzaitch1@jhu.edu,
Johns Hopkins University – As

Zaja, Annalisa, +390498279193 annalisa.zaja@unipd.it,
Università degli Studi di Padova – Yg

Zajac, Roman N., (203) 932-7114 rzajac@newhaven.edu,
University of New Haven – Zni

Zajacz, Zoltan, 416-946-0278 zajacz@es.utoronto.ca,
University of Toronto – Cu

Zakardjian, Bruno, (418) 723-1986 (Ext. 1570) bruno_zakardjian@uqar.qc.ca,
Universite du Quebec a Rimouski – Ob

Zakharova, Natalia, 989-774-4496 zakha1n@cmich.edu,
Central Michigan University – YgGe

Zakrzewski, Richard J., (785) 628-5389 rzakrzew@fhsu.edu,
Fort Hays State University – PvsPe

Zalasiewicz, Jan, +440116 252 3928 jaz1@le.ac.uk,
Leicester University – Pg

Zaleha, Michael J., (937) 327-7331 mzaleha@wittenberg.edu,
Wittenberg University – Gs

Zaleha, Robert, (216) 987-5278 Robert.Zaleha@tri-c.edu,
Cuyahoga Community College - Western Campus – Zg

Zaman, Musharraf, 405-325-2626 zaman@ou.edu,
University of Oklahoma – Np

Zamani, Behzad, +98 (411) 339 2700 zamani@tabrizu.ac.ir,
University of Tabriz – GtYs

Zambito, James J., (608) 363-2223 zambitoj@beloit.edu,
Beloit College – PcGsPg

Zampieri, Dario, +390498279179 dario.zampieri@unipd.it,
Università degli Studi di Padova – Gc

Zamstein, Lavi, 706-507-8089 zamstein_lavi@columbusstate.edu,
Columbus State University – Zn

Zamzow, Craig E., czamzow@clarion.edu,
Clarion University – Gi

Zanazzi, Alessandro, (801) 863-5395 alessandro.zanazzi@uvu.edu,
Utah Valley University – CgsPe

Zandt, George, (520) 621-2273 gzandt@email.arizona.edu,
University of Arizona – Ys

Zanella, Elena, elena.zanella@unito.it,
Università di Torino – Ym

Zanetti, Kathleen, (702) 895-4789 University of Nevada, Las Vegas – Gg

Zaniewski, Kazimierz J., (920) 424-4105 zaniewsk@uwosh.edu,
University of Wisconsin Oshkosh – Zn

Zappa, Christopher, zappa@ldeo.columbia.edu,
Columbia University – Op

Zappa, Christopher J., 845-365-8547 zappa@ldeo.columbia.edu,
Columbia University – OpAsZr

Zarillo, Gary , 321-674-7378 zarillo@fit.edu,
Florida Institute of Technology – OupOn

Zarnetske, Jay, (517) 353-3249 jpz@msu.edu,
Michigan State University – Hwg

Zarroca Hernández, Mario, ++935812033 mario.zarroca.hernandez@uab.cat,
Universitat Autonoma de Barcelona – YgNg

Zaspel, Craig E., (406) 683-7366 craig.zaspel@umwestern.edu,
University of Montana Western – Yg

Zattin, Massimiliano, +390498279186 massimiliano.zattin@unipd.it,
Università degli Studi di Padova – Gs

Zawadzki, Isztar, (514) 398-1034 isztar.zawadzki@mcgill.ca,
McGill University – AsZr

Zawiskie, John M., (313) 577-2506 jzawiskie@cranbrook.edu,
Wayne State University – Gg

Zayac, John M., (818) 710-2218 zayacjm@piercecollege.edu,
Los Angeles Pierce College – Ggv

Zebker, Howard A., (650) 723-8067 zebker@stanford.edu,
Stanford University – Yx

Zeebe, Richard E., (808) 956-6473 zeebe@soest.hawaii.edu,
University of Hawai'i, Manoa – Ou

Zehnder, Joseph A., 402-280-2448 zehnder@creighton.edu,
Creighton University – As

Zehr, Jonathan P., (831) 459-4009 zehrj@cats.ucsc.edu,
University of California, Santa Cruz – Ob

Zeidouni, Mehdi, mehdi.zeidouni@beg.utexas.edu,
University of Texas, Austin – Eo

Zeiger, Elaine, (312) 665-7627 Field Museum of Natural History – Zn

Zeigler, E. Lynn, 678-891-3767 ezeigler1@gsu.edu,
Georgia State University, Perimeter College, Clarkston Campus – Gd

Zeitler, Peter K., (610) 758-3671 pkz0@lehigh.edu,
Lehigh University – Cc

Zeitlhöfler, Matthias, 089/2180 6570 matthias@zeitlhoefler.de,
Ludwig-Maximilians-Universitaet Muenchen – Gg

Zelazny, Lucian W., (540) 231-9781 Virginia Polytechnic Institute & State University – Sc

Zelizer, Nora, 609 258-5809 nzelizer@princeton.edu,
Princeton University – Zn

Zelt, Colin A., (713) 348-4757 czelt@rice.edu,
Rice University – Ys

Zeman, Josef, +420 549 49 8295 jzeman@sci.muni.cz,
Masaryk University – CgpCe

Zender, Charles, 949-824-2987 zender@uci.edu,
University of California, Irvine – As

Zeng, Hongliu, 512-475-6382 hongliu.zeng@beg.utexas.edu,
University of Texas, Austin – YsGs

Zeng, Ning, (301) 405-5377 zeng@umd.edu,
University of Maryland – AsCgGl

Zentilli, Marcos, (902) 494-3873 Dalhousie University – Eg

Zentner, Nick, (509) 963-2828 nick@geology.cwu.edu,
Central Washington University – Gg

Zerkle, Aubrey, +44 01334 464949 az29@st-andrews.ac.uk,
University of St. Andrews – Co

Zhai, Xiaoming, 847 543-2504 College of Lake County – Gp

Zhai, Xiaoming, +44 (0)1603 59 3762 xiaoming.zhai@uea.ac.uk,
University of East Anglia – Og

Zhan, Hongbin, (979) 862-7961 zhan@geos.tamu.edu,
Texas A&M University – Hw

Zhan, Zhongwen, 626-395-6906 zwzhan@caltech.edu,
California Institute of Technology – Ys

Zhang, Bo, 205-348-4544 bzhang33@ua.edu,
University of Alabama – YeGoEo

Zhang, Caiyun, 561-297-2648 czhang3@fau.edu,
Florida Atlantic University – Zr

Zhang, Chi, 785-864-4974 chizhang@ku.edu,
University of Kansas – YgxYu

Zhang, Chidong, (206) 526-6239 chidong.zhang@noaa.gov,
University of Washington – AsmOp

Zhang, Guiming, 303-871-7908 Guiming.Zhang@du.edu,
University of Denver – Zi

Zhang, Guorong, (785) 625-3425 gzhang@ksu.edu,
Kansas State University – So

Zhang, Henian, (404) 894-1738 henian.zhang@eas.gatech.edu,
Georgia Institute of Technology – As

Zhang, Huan, 860-405-9237 huan.zhang@uconn.edu,
University of Connecticut – Ob

Zhang, Jin , 505-277-1607 University of New Mexico – Gyp

Zhang, Jin, jinzhang@unm.edu,
University of New Mexico – Gy

Zhang, Jinhong, (520) 626-9656 jhzhang@email.arizona.edu,
University of Arizona – Nm

Zhang, Lin, 361-825-2436 Lin.Zhang@tamucc.edu,
Texas A&M University, Corpus Christi – Cs

Zhang, Max, 607-254-5402 kz33@cornell.edu,
Cornell University – Zn

Zhang, Minghua , 530-752-4953 mhzhang@ucdavis.edu,
University of California, Davis – Hg

Zhang, Minghua, (631) 632-8318 minghua.zhang@stonybrook.edu,
SUNY, Stony Brook – As

Zhang, Ning, 937 376 6043 nzhang@centralstate.edu,
Central State University – RwEo

Zhang, Qiaofeng (Robin), (270) 809-6760 qzhang@murraystate.edu,
Murray State University – ZirZu

Zhang, Ren, (254)7102496 Ren_Zhang@baylor.edu,
Baylor University – Cs

Zhang, Renyi, (979) 845-7671 renyi-zhang@geos.tamu.edu,
Texas A&M University – As

Zhang, Rong, (609) 987-5061 rong.zhang@noaa.gov,
Princeton University – Op

Zhang, Rui, 337-482-6920 University of Louisiana at Lafayette – Ye

Zhang, Tongwei, 512-232-1496 tongwei.zhang@beg.utexas.edu,
University of Texas, Austin – Cgs

Zhang, Xi, (831) 502-8126 xiz@ucsc.edu,
University of California, Santa Cruz – AsXy

Zhang, Xinning, 609-258-7438 Xinningz@princeton.edu,
Princeton University – Cb

Zhang, Y. J., 804-684-7466 yjzhang@vims.edu,
College of William & Mary – Opn

Zhang, Yang, 919-515-9688 yang_zhang@ncsu.edu,

North Carolina State University – As
Zhang, Ye, 307-223-2292 yzhang9@uwyo.edu,
University of Wyoming – HyGeYu
Zhang, Yige, yigezhang@fas.harvard.edu,
Texas A&M University – Co
Zhang, Yong, (205) 348-3317 yzhang264@ua.edu,
University of Alabama – Hqw
Zhang, You-Kuan, (319) 335-1806 you-kuan-zhang@uiowa.edu,
University of Iowa – Hw
Zhang, Youxue, (734) 763-0947 youxue@umich.edu,
University of Michigan – Cu
Zhao, Bo, 541-737-3497 bzhao@coas.oregonstate.edu,
Oregon State University – Zi
Zhao, Jingxia, 808-956-3736 jingxiaz@hawaii.edu,
University of Hawai'i, Manoa – AmmAc
Zhaohui, Li, (262) 595-2487 li@uwp.edu,
University of Wisconsin, Parkside – ClGzHw
Zhdanov, Michael S., (801) 581-7750 michael.zhdanov@utah.edu,
University of Utah – Ye
Zheng, Chunmiao, 205-348-0579 czheng@ua.edu,
University of Alabama – Hw
Zheng, Yan, (718) 997-3329 Queens College (CUNY) – CgHwCm
Zhou, Hua-Wei, (713) 743-3440 hzhou@uh.edu,
University of Houston – YseYg
Zhou, Wendy W., (303) 384-2181 wzhou@mines.edu,
Colorado School of Mines – NgZir
Zhou, Xiaobing, (406) 496-4350 xzhou@mtech.edu,
Montana Tech of the University of Montana – Zr
Zhou, Xiaolu, (912) 478-5943 xzhou@georgiasouthern.edu,
Georgia Southern University – Zr
Zhou, Ying, 540-231-6521 yingz@vt.edu,
Virginia Polytechnic Institute & State University – Ys
Zhou, Yu, 845-437-5543 yuzhou@vassar.edu,
Vassar College – Zun
Zhou, Yuyu, (515) 294-2842 yuyuzhou@iastate.edu,
Iowa State University of Science & Technology – Zyi
Zhu, Chen, (812) 856-1884 chenzhu@indiana.edu,
Indiana University, Bloomington –
Zhu, Hejun, (972) 883-2406 Hejun.Zhu@utdallas.edu,
University of Texas, Dallas – Ysx
Zhu, Junfeng, 859-323-0530 junfeng.zhu@uky.edu,
University of Kentucky – Hw
Zhu, Lupei, (314) 977-3118 zhul@slu.edu,
Saint Louis University – YsGt
Zhu, Ping, 305-348-7096 zhup@fiu.edu,
Florida International University – As
Zhu, Qingzhi, (631) 632-8747 Qing.Zhu@stonybrook.edu,
SUNY, Stony Brook – Oc
Zhu, Wenlu, (301) 405-1831 wzhu@umd.edu,
University of Maryland – YrNrHg
Zhuang, Guangsheng, gzhuang@lsu.edu,
Louisiana State University – CocGs
Zhuang, Qianlai, (765) 494-9610 qzhuang@purdue.edu,
Purdue University – As
Zieg, Michael J., (724) 738-2501 michael.zieg@sru.edu,
Slippery Rock University – Gi
Ziegler, Brady A., bziegler@trinity.edu,
Trinity University – HwCqGe
Ziegler, Karen, (505) 277-1437 kziegler@unm.edu,
University of New Mexico – CsXc
Ziemba, Tim, 206-685-7437 ziemba@uw.edu,
University of Washington – Xa
Ziemer, Robert R., rrz7001@humboldt.edu,
Humboldt State University – HgGmZn
Zierenberg, Robert A., (530) 752-1863 razierenberg@ucdavis.edu,
University of California, Davis – Cs
Zimba, Paul, 361-825-2768 Paul.Zimba@tamucc.edu,
Texas A&M University, Corpus Christi – Zn
Zimbelman, James R., 202-633-2471 Smithsonian Institution / National Air & Space Museum – Xg
Zimmer, Brian W., 828-262-7517 zimmerbw@appstate.edu,
Appalachian State University – Gg
Zimmer, Margaret, (831) 502-8714 margaret.zimmer@ucsc.edu,
University of California, Santa Cruz – Hw
Zimmerman, Aaron, (970) 491-3789 aaron.zimmerman@colostate.edu,
Colorado State University – Cc
Zimmerman, Andrew, 352-392-0070 azimmer@ufl.edu,
University of Florida – ComSb
Zimmerman, Brian S., (814) 732-2529 bzimmerman@edinboro.edu,
Edinboro University of Pennsylvania – Gzx
Zimmerman, Mark, 612-626-0572 zimme030@umn.edu,
University of Minnesota, Twin Cities – Yx
Zimmerman, R. Eric, (630) 252-6816 Argonne National Laboratory – Ng
Zimmerman, Robert W., +44 20 7594 7412 r.w.zimmerman@imperial.ac.uk,
Imperial College – NrHwNp
Zimmerman, Ronald K., (225) 388-8302 ron@lgs.bri.lsu.edu,
Louisiana State University – Go
Zimmerman, Jr., Jay, (618) 453-3351 zimmerman@geo.siu.edu,
Southern Illinois University Carbondale – Gc
Zimmermann, George, (609) 652-4308 george.zimmermann@stockton.edu,
Stockton University – Zn
Zink, Albert, 089/2180 4340 Ludwig-Maximilians-Universitaet Muenchen – Gz
Zinsmeister, William J., (765) 494-0279 wjzins@purdue.edu,
Purdue University – Ps
Ziolkowski, Anton M., +44 (0) 131 650 8511 anton.ziolkowski@ed.ac.uk,
Edinburgh University – YesYg
Ziolkowski, Lori A., 803-777-0035 lziolkowski@geol.dc.edu,
University of South Carolina – CoOcCs
Zipf, Karl, 303-384-2371 rzipf@mines.edu,
Colorado School of Mines – Nm
Zipper, Carl E., (540) 231-9782 czip@vt.edu,
Virginia Polytechnic Institute & State University – Ze
Zipser, Edward, 801-585-0467 ed.zipser@utah.edu,
University of Utah – As
Ziramov, Sasha, +61 8 9266-4973 Sasha.Ziramov@curtin.edu.au,
Curtin University – Ye
Zlotkin, Howard, 12012003161 hzlotkin@njcu.edu,
New Jersey City University – Ze
Zlotnik, Vitaly A., (402) 472-2495 vzlotnik1@unl.edu,
University of Nebraska, Lincoln – HwyHq
Zobaa, Mohamed, zobaa_m@utpb.edu,
University of Texas, Permian Basin – Pl
Zoback, Mark D., (650) 725-9295 zoback@stanford.edu,
Stanford University – Yg
Zodrow, Erwin L., erwin_zodrow@uccb.ca,
Cape Breton University – Pi
Zois, Constantine S., (908) 737-3693 czois@kean.edu,
Kean University – Am
Zollweg, James A., (585) 395-2352 jzollweg@brockport.edu,
SUNY, The College at Brockport – Hs
Zolotov, Mikhail Y., (480) 965-4739 zolotov@asu.edu,
Arizona State University – Cg
Zonneveld, John-Paul, 780-492-3287 zonneveld@ualberta.ca,
University of Alberta – Go
Zotti, Maria, maria.zotti@flinders.edu.au,
Flinders University – Zn
Zou, Haibo, (334) 844-4315 haibo.zou@auburn.edu,
Auburn University – CcGiv
Zouhri, Lahcen, +33(0)3 44068976 lahcen.zouhri@lasalle-beauvais.fr,
Institut Polytechnique LaSalle Beauvais (ex-IGAL) – HwgGe
Zreda, Marek, 520-621-4072 marek@hwr.arizona.edu,
University of Arizona – Cc
Zuber, Maria T., 617-253-6397 zuber@mit.edu,
Massachusetts Institute of Technology – Xy
Zuend, Andreas, (514) 398-7420 andreas.zuend@mcgill.ca,
McGill University – Asm
Zume, Joseph T., (717) 477-1548 jtzume@ship.edu,
Shippensburg University – HgYug
Zumwalt, Gary S., gzumwalt@latech.edu,
Louisiana Tech University – Pi
Zurawski, Ronald P., (615) 532-1502 ronald.zurawski@tn.gov,
Tennessee Geological Survey – GgEg
Zurevinski, Shannon, (807) 343-8015 shannon.zurevinski@lakeheadu.ca,
Lakehead University – Gz
Zurick, David, (859) 622-1427 david.zurick@eku.edu,
Eastern Kentucky University – Zn
Zuza, Andrew , (775) 784-1446 azuza@unr.edu,
University of Nevada, Reno – Gc
Zuza, Andrew V., (775) 682-8752 azuza@unr.edu,
University of Nevada – Gct
Zwiefelhofer, Luke, 507-457-2778 lzwiefelhofer@winona.edu,
Winona State University – Zn
Zyvoloski, George A., (505) 667-1581 Los Alamos National Lab – Ng

www.ingramcontent.com/pod-product-compliance
Lightning Source LLC
Chambersburg PA
CBHW080719300426
44114CB00019B/2420